BIOLOGY
The Dynamics of Life

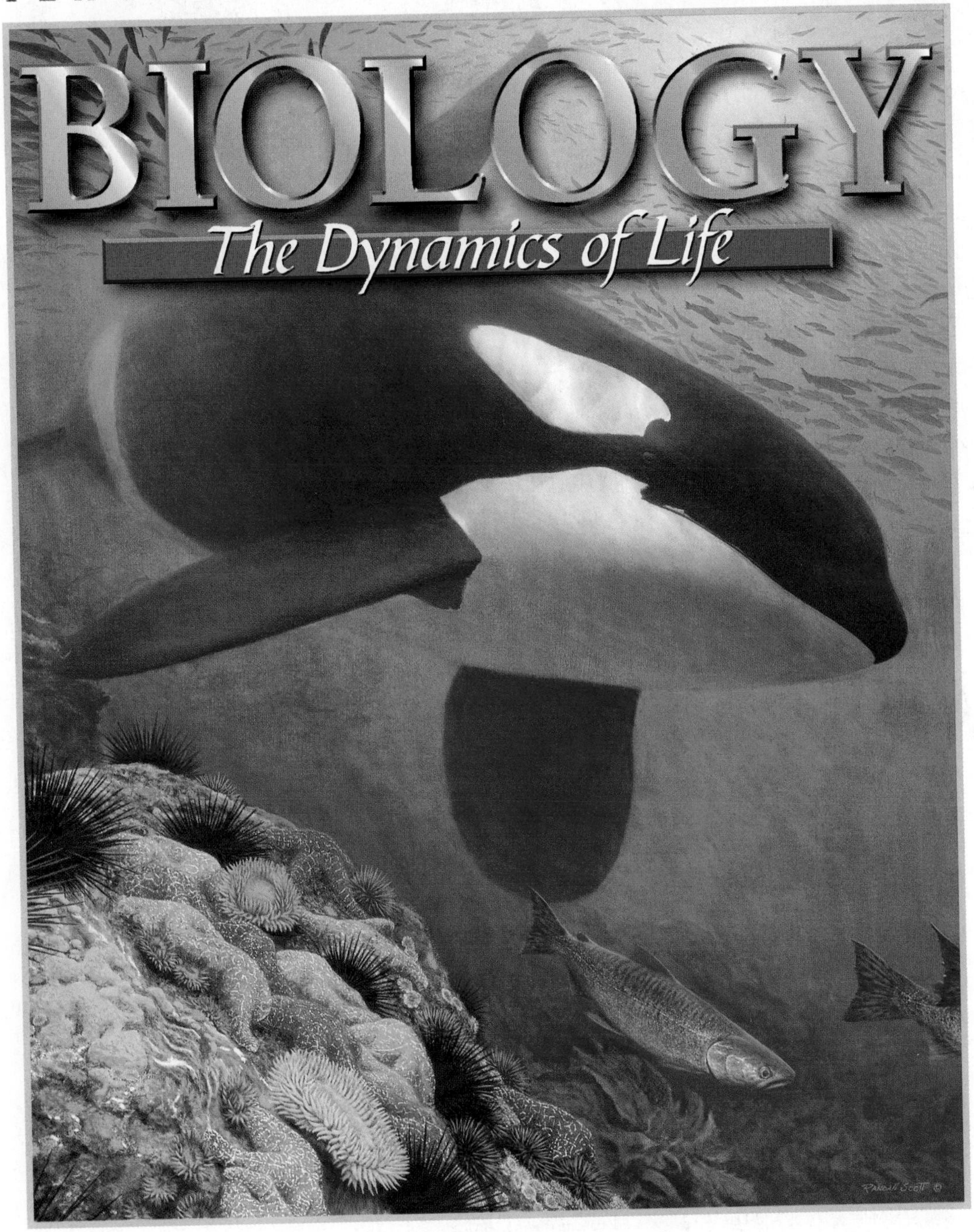

AUTHORS

Alton Biggs, M.S. • **Kathleen Gregg, Ph.D.**

Whitney Crispen Hagins, M.A., M.A.T. • **Chris Kapicka, Ph.D.**

Linda Lundgren, M.S. • **Peter Rillero, Ph.D.** • **National Geographic Society**

Glencoe
McGraw-Hill

New York, New York   Columbus, Ohio   Woodland Hills, California   Peoria, Illinois

A GLENCOE PROGRAM
BIOLOGY: THE DYNAMICS OF LIFE

Student Edition
Teacher Wraparound Edition
Laboratory Manual, SE and TE
Reinforcement and Study Guide,
    SE and TE
Content Mastery, SE and TE
Section Focus Transparencies and Masters
Reteaching Skills Transparencies and Masters
Basic Concepts Transparencies and Masters
BioLab and MiniLab Worksheets
Concept Mapping
Chapter Assessment
Critical Thinking/Problem Solving
Spanish Resources
Tech Prep Applications

Biology Projects
ExamView® Pro Software
Lesson Plans
Block Scheduling
Inside Story Poster Package
English/Spanish Audiocassettes
MindJogger Videoquizzes
Interactive CD-ROM
Videodisc Program
Glencoe Science Professional Series:
    Exploring Environmental Issues
    Performance Assessment in the Biology Classroom
    Alternate Assessment in the Science Classroom
    Cooperative Learning in the Science Classroom
    Using the Internet in the Science Classroom

Glencoe/McGraw-Hill

A Division of The **McGraw·Hill** Companies

The Test-Taking Tips in this book were written by The Princeton Review, the nation's leader in test preparation. Through its association with McGraw-Hill, The Princeton Review offers the best way to help students excel on standardized assessments.

The Princeton Review is not affiliated with Princeton University or Educational Testing Service.

Send all inquiries to:
Glencoe/McGraw-Hill
8787 Orion Place
Columbus, OH 43240

Special Education Consultant
Shay Guffey
Rutherfordton, NC

ISBN 0-07-825924X
Printed in the United States of America.
3 4 5 6 7 8 9 10 071 043 08 07 06 05 04 03 02

Contents in Brief

Teacher Wraparound Edition

Student Edition

National Science Education Standards

The *National Science Education Standards*, published by the National Research Council and representing the contribution of thousands of educators and scientists, offer a comprehensive vision of a scientifically literate society. The standards describe not only what students should know but also offer guidelines for biology teaching and assessment. If you are using, or plan to use, the standards to guide changes in your biology curriculum, you can be assured that *Biology: The Dynamics of Life* aligns with the *National Science Education Standards*.

Biology: The Dynamics of Life is an example of how Glencoe's commitment to effective science education is changing the materials used in biology classrooms today. More than just a collection of facts in a textbook, *Biology: The Dynamics of Life* is a program that provides numerous opportunities for students, teachers, and school districts to meet the *National Science Education Standards*.

National Science Content Standards

Correlations in each Chapter Organizer and BioDigest show the close alignment between *Biology: The Dynamics of Life* and the content standards. Correlations are designated according to the numbering system in the table of science content standards shown on the opposite page. The approach of *Biology: The Dynamics of Life* allows students to discover concepts within each of the content standards, giving them opportunities to make connections between biology concepts and the real world. Hands-on labs and inquiry-based lessons reinforce the science processes emphasized in the standards.

Teaching Standards

Alignment with the *National Science Education Standards* requires much more than alignment with the outcomes in the content standards. The way in which concepts are presented is critical to effective learning. The teaching standards within the *National Science Education Standards* recommend an inquiry-based program facilitated and guided by teachers. *Biology: The Dynamics of Life* provides such opportunities through activities and discussions that allow students to discover critical concepts by inquiry and apply the knowledge they've gained to their own lives. Throughout the program, students are building critical skills that will be available to them for lifelong learning. The *Teacher Wraparound Edition* helps you make the most of every instructional moment. It offers an abundance of effective strategies and suggestions for guiding students as they explore biology.

Assessment Standards

The assessment standards are supported by many of the components that make up the *Biology: The Dynamics of Life* program. The *Teacher Wraparound Edition* and *Teacher Classroom Resources* provide multiple chances to assess students' understanding of important concepts as well as their abilities to perform a wide range of skills. Ideas for portfolios, performance assessment, written reports, and other assessment activities accompany every lesson. For more suggestions about assessment ideas and resources, see pages 32T-33T of the Teacher Guide.

National Science Education Standards

Unifying Concepts and Processes

UCP.1 Systems, order, and organization
UCP.2 Evidence, models, and explanation
UCP.3 Change, constancy, and measurement
UCP.4 Evolution and equilibrium
UCP.5 Form and function

Science as Inquiry

A.1 Abilities necessary to do scientific inquiry
A.2 Understandings about scientific inquiry

Physical Science

B.1 Structure of atoms
B.2 Structure and properties of matter
B.3 Chemical reactions
B.4 Motions and forces
B.5 Conservation of energy and increase in disorder
B.6 Interactions of energy and matter

Life Science

C.1 The cell
C.2 Molecular basis of heredity
C.3 Biological evolution
C.4 Interdependence of organisms
C.5 Matter, energy, and organization in living systems
C.6 Behavior of organisms

Earth and Space Sciences

D.1 Energy in the earth system
D.2 Geochemical cycles
D.3 Origin and evolution of the earth system
D.4 Origin and evolution of the universe

Science and Technology

E.1 Abilities of technological design
E.2 Understandings about science and technology

Science in Personal and Social Perspectives

F.1 Personal and community health
F.2 Population growth
F.3 Natural resources
F.4 Environmental quality
F.5 Natural and human-induced hazards
F.6 Science and technology in local, national, and global challenges

History and Nature of Science

G.1 Science as a human endeavor
G.2 Nature of scientific knowledge
G.3 Historical perspectives

Program Overview

Philosophy and Themes

Biology: The Dynamics of Life is a course in biology that follows a phylogenetic approach in its organization. This approach allows you to explain the diversity of life-forms in depth while revealing their relationships and fundamental unity in form and function.

How do you attract and hold the attention of students unless you find topics that are of interest to students? ***Biology: The Dynamics of Life*** capitalizes on topics, such as ecology and human genetics, that interest students. Ecology, for example, is of popular and political interest and often the lead story of local and national news broadcasts. Rather than present ecology at the end of the text, the authors have chosen to move it to the front, where it can be emphasized. Heredity and human genetics, too, have emerged as biological topics with meaningful and far-ranging consequences for society. Students will find it easy to see how and why biological science is an important component of their studies and their futures.

In ***Biology: The Dynamics of Life,*** there is an emphasis on six themes, as summarized in the chart below. Emphasis on themes contributes to the big picture by focusing learning on connections among major ideas and concepts. The thematic approach contributes to students' comprehension of fundamental life processes, understanding of interactions among organisms, and appreciation of how scientists work.

Biology: The Dynamics of Life emphasizes the following themes.

Unity Within Diversity

Life on Earth is represented by millions of species. In order to survive, all species must possess the same basic functions. General patterns of carrying out life functions are easily recognized, and these patterns form a common bond that unites all forms of life. This theme is evident in discussions of topics ranging from cell structure and function, to the genetic code, to the six kingdoms of life.

Evolution

Similarities and differences among species indicate evolutionary relationships. All organisms are related in that they are descendants of the first forms of life on Earth. Evolution is the process by which organisms are adapted to changing environments and by which one species gives rise to another. The diversity of life on Earth is a product of evolutionary change. This theme is developed not only where

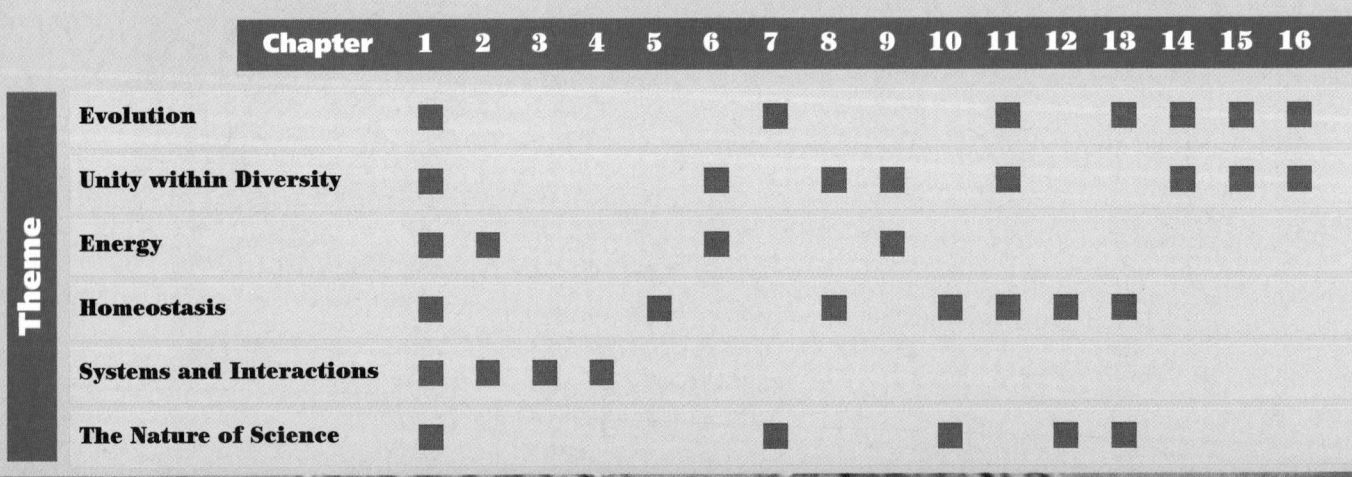

Chapter	1	2	3	4	5	6	7	8	9	10	11	12	13	14	15	16
Evolution	■						■				■		■	■	■	■
Unity within Diversity	■					■		■	■		■			■	■	■
Energy		■	■			■			■							
Homeostasis		■					■		■		■	■	■	■		
Systems and Interactions	■	■	■	■												
The Nature of Science		■					■						■		■	■

the process of evolution is discussed, but also in discussions of related groups of organisms, the six kingdoms of life, and comparisons of adaptations among species.

Energy

In order to carry out basic functions, organisms require a continual source of energy and a means of extracting and utilizing that energy. Energy flow pervades every level of biological organization from a cell to the biosphere. This theme is explored in numerous contexts: photosynthesis and respiration, ATP and its uses, and energy flow through ecosystems.

Homeostasis

In the strictest sense, homeostasis refers to the tendency to maintain stability in the internal environment of an organism. However, the theme of homeostasis is extended here to the balance of chemistry at the cellular level and maintenance of stability within an ecosystem. The theme has been applied to the balance within organisms and to that which exists among organisms and between organisms and their environments.

Systems and Interactions

No single part at any level of organization operates independently. Proper functioning depends on coordinated relationships. Parts of organisms interact with one another, organisms are interdependent, and populations interact with their environments. This theme is stressed in the study of the interaction of organelles within cells, cells within tissues, organs within systems, and systems within organisms. It is also apparent in such topics as feeding relationships, symbiotic associations, recycling of nutrients, life cycles, patterns of population growth, and ecological succession.

The Nature of Science

Students learn about the fundamental methods common to scientific inquiry and see evidence of those methods as classic experiments leading to the development of major principles are discussed and analyzed. More significantly, students are afforded the opportunity to practice the methods of science themselves throughout the course. Far from the "cookbook" labs that require students to confirm what they have already learned, many of the lab activities require students to formulate hypotheses and then design their own controlled experiments.

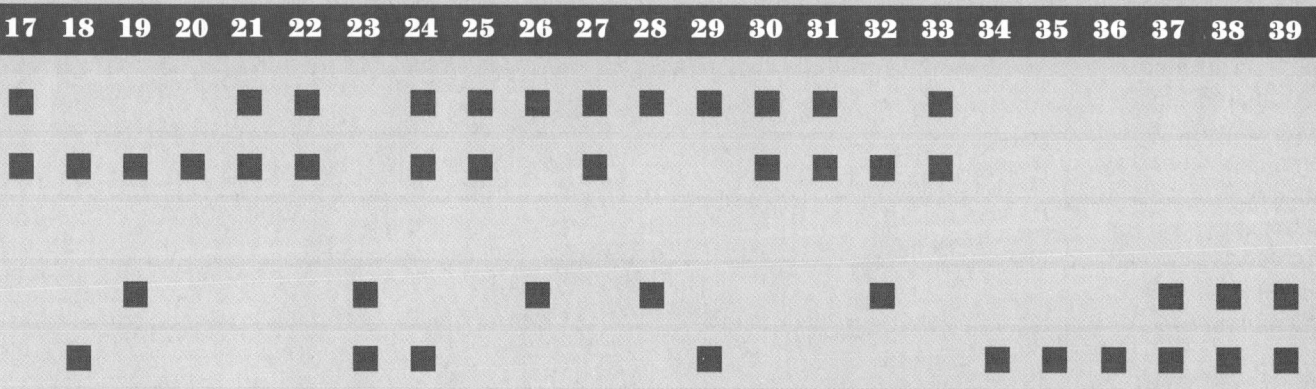

What You'll Learn

Introduces the main concepts of the chapter before students begin to study the material.

Why It's Important

Provides an answer to the "Why do we have to learn this?" question that students ask.

Getting Started

This easy-to-perform activity provides an inquiry approach to introducing the chapter and gets students involved right from the first page.

Internet Connection

Connects students to the resources available on the Glencoe Science Web Site. Encourages individual Internet research on science topics at home, or in class for those who finish the lesson early. Assign a different site to each student group for group reports, and use to enrich text information.

Chapter

7 A View of the Cell

What You'll Learn

- You will distinguish eukaryotic and prokaryotic cells.
- You will learn the structure and function of the plasma membrane.
- You will relate the structure of cell parts to their functions.

Why It's Important

A knowledge of cell structure and function is essential to a basic understanding of life.

READING BIOLOGY

Go through the chapter, and note the figures that depict different types of cell structures. For each figure, note the various components of the cell. As you read the text, write the characteristics of each cell type by its name.

BIOLOGY Online

To find out more about cells, visit the Glencoe Science Web site.
science.glencoe.com

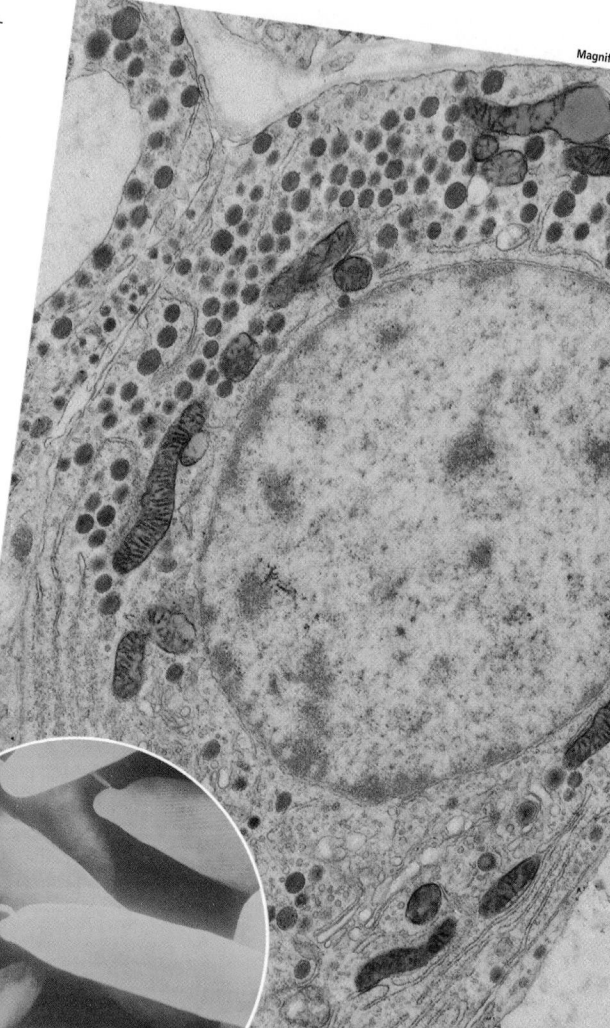

Magnification: 17 270×

Cells are amazingly diverse. Yet for all their diversity, cells such as this nerve cell and the protist *Euglena* share many common features.

Magnification: 9310×

174 A VIEW OF THE CELL

Working in the lab is often the most enjoyable part of biology. When students feel inquisitive but there's a shortage of time, assign a *MiniLab*. *BioLabs* give your students an opportunity to act like biologists, develop their own plans for studying a question or problem, and share their data with others. Whether they're designing experiments or following well-tested procedures, they'll have fun doing these lab activities.

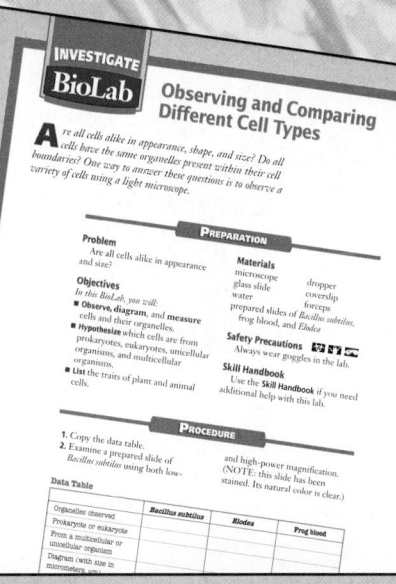

INVESTIGATE BioLab

Observing and Comparing Different Cell Types

Are all cells alike in appearance, shape, and size? Do all cells have the same organelles present within their cell boundaries? One way to answer these questions is to observe a variety of cells using a light microscope.

PREPARATION

Problem
Are all cells alike in appearance and size?

Objectives
In this BioLab, you will:
- **Observe, diagram, and measure** cells and their organelles.
- **Hypothesize** which cells are from prokaryotes, eukaryotes, unicellular organisms, and multicellular organisms.
- **List** the traits of plant and animal cells.

Materials
microscope
glass slide
water
prepared slides of *Bacillus subtilis*, frog blood, and *Elodea*
dropper
coverslip
forceps

Safety Precautions
Always wear goggles in the lab.

Skill Handbook
Use the **Skill Handbook** if you need additional help with this lab.

PROCEDURE

1. Copy the data table.
2. Examine a prepared slide of *Bacillus subtilis* using both low- and high-power magnification. (NOTE: this slide has been stained. Its natural color is clear.)

Data Table

Organelles observed	*Bacillus subtilis*	*Elodea*	Frog blood
Prokaryote or eukaryote			
From a multicellular or unicellular organism			
Diagram (with size in micrometers)			

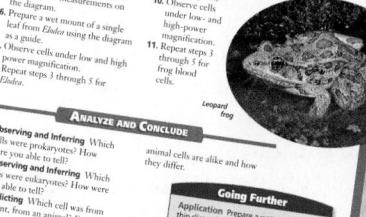

Leopard frog

ANALYZE AND CONCLUDE

1. **Observing and Inferring** Which cells were prokaryotes? How were you able to tell?
2. **Observing and Inferring** Which cells were eukaryotes? How were you able to tell?
3. **Predicting** Which cell was from a plant, from an animal? Explain your answer.
4. **Measuring** Are prokaryote or eukaryote cells larger? Give specific measurements to support your answer.
5. **Defining Operationally** Describe how plant and animal cells are alike and how they differ.

Going Further

Application Prepare a wet mount of very thin slices of bamboo (saxophone reed). Observe under low and high power. Explain what structures you are looking at. Explain the absence of all other organelles from this material.

BIOLOGY Online To find out more about microscopy and cell types, visit the Glencoe science Web site. science.glencoe.com

---

Investigate BioLab

Guide students in following the procedure steps to arrive at the answer to the lab problem. Labs can also be set up as teacher demos if time or materials are short.

DESIGN YOUR OWN BioLab

How do earthworms respond to their environment?

An earthworm spends its time eating its way through soil, digesting organic matter, and passing inorganic matter through the digestive system and out of its body. Because earthworms are dependent on soil for food and shelter, they respond to stimuli in a way that will ensure a continuous supply of food and a safe place in which to live. These responses are genetically controlled. In this BioLab, you will design an experiment to determine the responses of earthworms to various stimuli.

PREPARATION

Problem
How do earthworms respond to light, different surfaces, moist and dry environments, and warm and cold environments?

Hypotheses
Place your worm in a tray with some moist soil. Watch your worm for about 5 minutes, and record what

you observe. Make a hypothesis based on your observations about what the worm might do under conditions of light and dark, rough and smooth surfaces, moist and dry surfaces, and warm and cold conditions. Limit your investigation as time requires.

Objectives
In this BioLab, you will:
- **Measure** the sensitivity of earthworms to different stimuli, including light, water, and temperature.
- **Interpret** earthworm responses according to terms of adaptations that promote their survival.

Possible Materials
live earthworms
glass pan
culture dishes
thermometer
dropper
ice
black paper
hand lens or stereomicroscope
paper towels
sandpaper
warm tap water
water
penlight
ruler
cotton swab

Safety Precautions
Be sure to treat the earthworm in a humane manner at all times. Wet your hands before handling earthworms. Always wear goggles in the lab.

Skill Handbook
Use the **Skill Handbook** if you need additional help with this lab.

PLAN THE EXPERIMENT

1. As a group, make a list of possible ways you might test your hypothesis. Keep the available materials in mind as you plan your procedure.
2. Be sure to design an experiment that will

1. What data will you collect, and how will they be recorded?
2. Does each test have one variable and a control? What are they?

Design Your Own BioLab

Challenge your students to design an experiment that provides an answer to the lab problem.

INTERNET BioLab

Zebra Fish Development

The zebra fish (*Danio rerio*) is a common freshwater fish sold in pet shops. They are ideal animals for study because they undergo developmental changes quickly and major stages can be observed within hours after fertilization.

PREPARATION

Problem
What do the developmental stages of the zebra fish look like?

Materials
aquarium
zebra fish
turkey baster
beaker
dropper
petri dish
binocular microscope
wax pencil or labels

Safety Precautions
Always wear safety goggles in the lab. Use caution when working with a binocular microscope and glassware.

PROCEDURE

5. Your teacher will advise you as to the approximate time that fertilization took place. All ages should be reported in your data table as hpf (hours past fertilization).
6. Using a binocular microscope, observe the embryos. Make a diagram in your data table of the embryos' appearances and indicate the age of the embryos in hpf.

7. Go to the Glencoe Science Web Site as shown in the Sharing Your Data box below to post your data.
8. Continue to observe your embryos daily for a minimum of one week. Note the appearance of new organs and when movement is first seen. If you wish to continue watching developmental changes, consult with your teacher for directions. CAUTION: Wash your hands with soap and water immediately after completing observations.

Data Table

Date	hpf	Diagram	Observations

ANALYZE AND CONCLUDE

1. **Communicating** Explain why zebra fish are ideal animals for studying embryonic development.
2. **Thinking Critically** Explain why you may not have been able to see stages such as a blastula or gastrula.
3. **Thinking Critically** Suggest how you could change your experiment's design to observe these stages.
4. **Using the Internet** Use the Glencoe Science web site to Internet sites that complete sequences of changes during the development of zebra fish.
   a. between 1 labeled diagrams changes.
   b. between 10 and 28 hpf. Include labeled diagrams.
   c. between 28 and 72 hpf. Include labeled diagrams.

Sharing Your Data Find this BioLab as well

Internet BioLab

Students use the Internet to share data with and retrieve data from others around the country and the world. Teaches students the importance of having large numbers of trials in scientific research and emphasizes the importance of sharing scientific information.

MiniLab 7-1 Measuring in SI

Magnification: 260×

Measuring Objects Under a Microscope Knowing the diameter of the circle of light you see when looking through a microscope allows you to measure the size of objects that are being viewed. For most microscopes, the diameter of the circle of light is 1.5 mm, or 1500 µm (micrometers), under low power and 0.375 mm, or 375 µm, under high power.
Refer to *Practicing Scientific Methods* in the *Skill Handbook* if you need help with SI units.

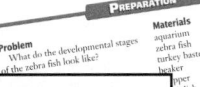

Human hair

Procedure

1. Look at diagram A that shows an object viewed under low power. Knowing the circle diameter to be 1500 µm, the estimated length of object (a) is 400 µm. What is the estimated length of object (b)?

2. Look at diagram B that shows an object viewed under high power. Knowing the circle diameter to be 375 µm, the estimated length of object (c) is 100 µm. What is the estimated length of object (d)?

3. Prepare a wet mount of a strand of your hair. Your teacher can help with this procedure. **CAUTION: Use caution when handling microscopes and glass slides.** Measure the width of your hair strand while viewing it under low and then high power.

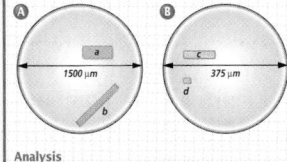

A 1500 µm a b

B 375 µm c d

Analysis

1. An object can be magnified 100, 200, or 1000 times when viewed under a microscope. Does the object's actual size change with each magnification? Explain.

2. Do your observations of the size of your hair strand under low and high power support the answer to question 1? If not, offer a possible explanation why.

MiniLabs

Answers the how, what, and why about the living world in just a few minutes. Assign to individuals as homework or in-class work or to student pairs or groups in a lab setting. Can be used as a class demonstration.

Student Edition

Visual Learning

Inside Story

Use the *Inside Story* features as a preview to a microscope activity or as a dissection guide. Struggling readers or English Language Learners can use *Inside Story* features instead of full-text information.

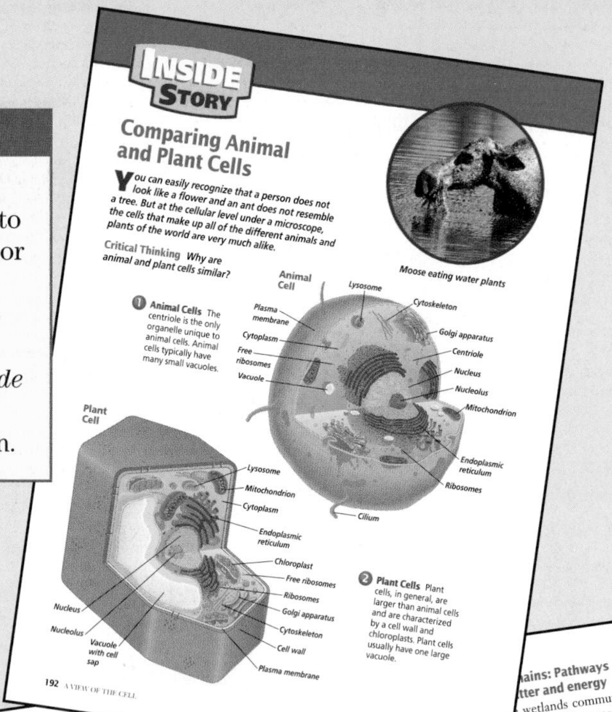

INSIDE STORY

Comparing Animal and Plant Cells

You can easily recognize that a person does not look like a flower and an ant does not resemble a tree. But at the cellular level under a microscope, the cells that make up all of the different animals and plants of the world are very much alike.

Critical Thinking Why are animal and plant cells similar?

Moose eating water plants

Animal Cell

1 **Animal Cells** The centriole is the only organelle unique to animal cells. Animal cells typically have many small vacuoles.

Plasma membrane
Cytoplasm
Free ribosomes
Vacuole
Lysosome
Cytoskeleton
Golgi apparatus
Centriole
Nucleus
Nucleolus
Mitochondrion

Plant Cell

Lysosome
Mitochondrion
Cytoplasm
Endoplasmic reticulum
Chloroplast
Free ribosomes
Ribosomes
Golgi apparatus
Cytoskeleton
Cell wall
Nucleus
Nucleolus
Vacuole with cell sap
Plasma membrane
Cilium
Endoplasmic reticulum
Ribosomes

2 **Plant Cells** Plant cells, in general, are larger than animal cells and are characterized by a cell wall and chloroplasts. Plant cells usually have one large vacuole.

192 A VIEW OF THE CELL

Figure References

Each figure and table is referenced in the text, fully integrating the text with all graphics—the ideal visual learning strategy.

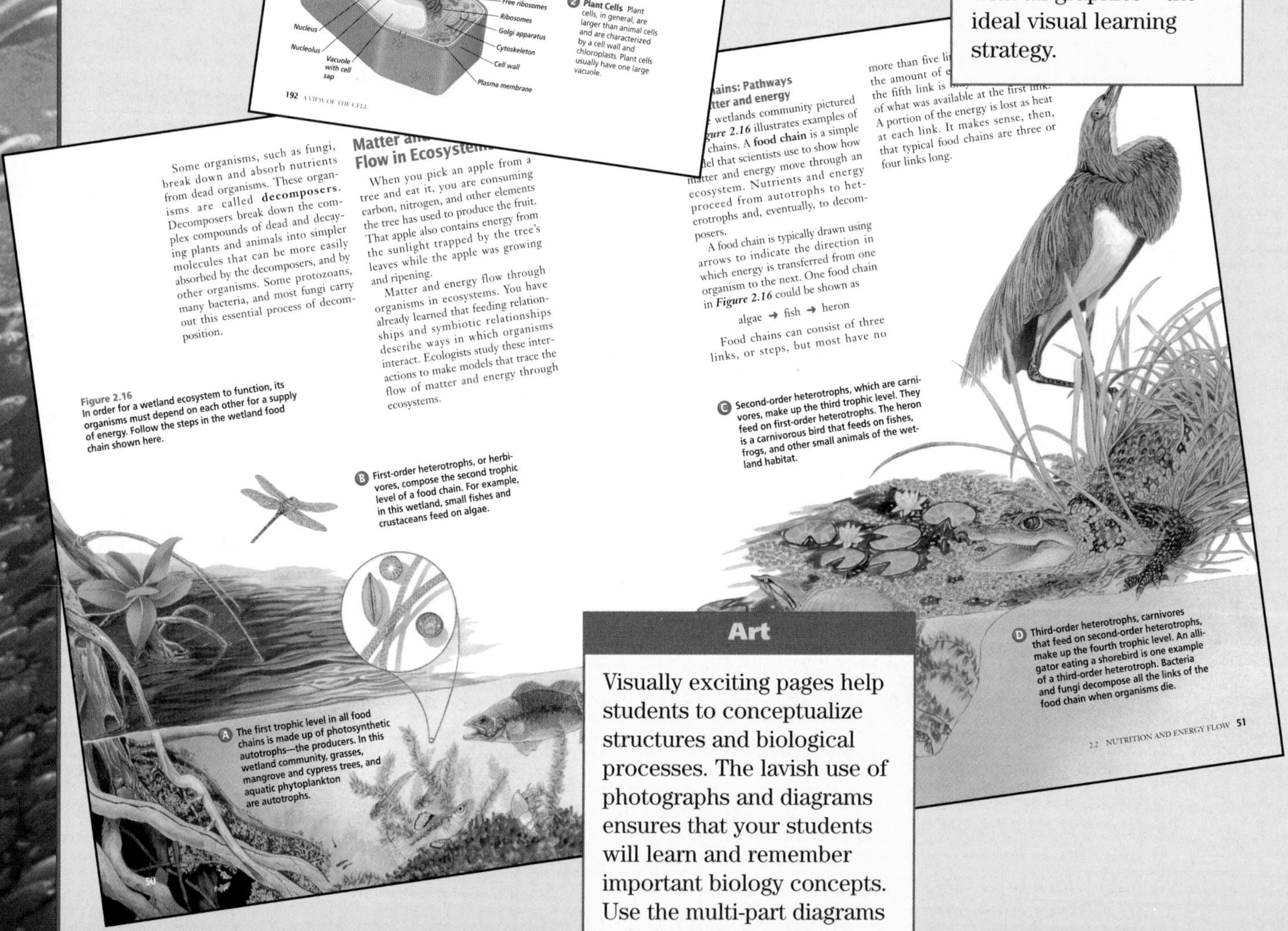

Some organisms, such as fungi, break down and absorb nutrients from dead organisms. These organisms are called **decomposers**. Decomposers break down the complex compounds of dead and decaying plants and animals into simpler molecules that can be more easily absorbed by the decomposers, and by other organisms. Some protozoans, many bacteria, and most fungi carry out this essential process of decomposition.

Matter and Flow in Ecosystems

When you pick an apple from a tree and eat it, you are consuming carbon, nitrogen, and other elements the tree has used to produce the fruit. That apple also contains energy from the sunlight trapped by the tree's leaves while the apple was growing and ripening.

Matter and energy flow through organisms in ecosystems. You have already learned that feeding relationships and symbiotic relationships describe ways in which organisms interact. Ecologists study these interactions to make models that trace the flow of matter and energy through ecosystems.

Chains: Pathways of Matter and energy

The wetlands community pictured in Figure 2.16 illustrates examples of food chains. A **food chain** is a simple model that scientists use to show how matter and energy move through an ecosystem. Nutrients and energy proceed from autotrophs to heterotrophs and, eventually, to decomposers.

A food chain is typically drawn using arrows to indicate the direction in which energy is transferred from one organism to the next. One food chain in *Figure 2.16* could be shown as

algae → fish → heron

Food chains can consist of three links, or steps, but most have no more than five links... the amount of e... the fifth link is only... of what was available at the first link. A portion of the energy is lost as heat at each link. It makes sense, then, that typical food chains are three or four links long.

Figure 2.16
In order for a wetland ecosystem to function, its organisms must depend on each other for a supply of energy. Follow the steps in the wetland food chain shown here.

B First-order heterotrophs, or herbivores, compose the second trophic level of a food chain. For example, in this wetland, small fishes and crustaceans feed on algae.

C Second-order heterotrophs, which are carnivores, make up the third trophic level. They feed on first-order heterotrophs. The heron is a carnivorous bird that feeds on fishes, frogs, and other small animals of the wetland habitat.

D Third-order heterotrophs, carnivores that feed on second-order heterotrophs, make up the fourth trophic level. An alligator eating a shorebird is one example of a third-order heterotroph. Bacteria and fungi decompose all the links of the food chain when organisms die.

2.2 NUTRITION AND ENERGY FLOW 51

A The first trophic level in all food chains is made up of photosynthetic autotrophs—the producers. In this wetland community, grasses, mangrove and cypress trees, and aquatic phytoplankton are autotrophs.

Art

Visually exciting pages help students to conceptualize structures and biological processes. The lavish use of photographs and diagrams ensures that your students will learn and remember important biology concepts. Use the multi-part diagrams to help students understand more challenging biological processes.

The invention and development of the light microscope some 300 years ago allowed scientists to see cells for the first time. Improvements have vastly increased the range of visibility of microscopes. Today researchers can use these powerful tools to study cells at the molecular level.

FOCUS ON
Microscopes

THIS EARLY COMPOUND MICROSCOPE, housed in a gold-embossed leather case, was designed by English scientist Robert Hooke about 1665. Using it, he observed and made drawings of cork cells. Although the microscope has three lenses, they are of poor quality and Hooke could see little detail. ▶

ANTON VAN LEEUWENHOEK

HOOKE'S MICROSCOPE

◀ **THIS HISTORIC MICROSCOPE—** held by a modern researcher— was designed by Anton van Leeuwenhoek (above). By 1700, Dutch scientist Leeuwenhoek had greatly improved the accuracy of microscopes. Grinding the lenses himself, Leeuwenhoek built some 240 single-lens versions. He discovered—and described for the first time—red blood cells and bacteria, taken from scrapings from his teeth. By 1900, problems with lenses that had once limited image quality had been overcome, and the compound microscope had evolved essentially into its present form.

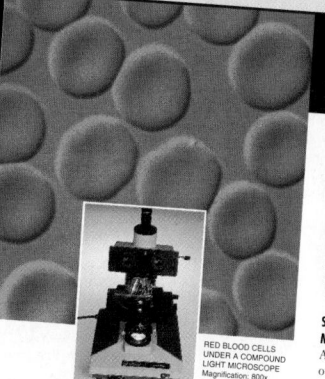

RED BLOOD CELLS UNDER A COMPOUND LIGHT MICROSCOPE Magnification: 800x

HOW IT WORKS The magnifying power of a microscope is determined by multiplying the magnification of the eyepiece and the objective lens.

A **COMPOUND LIGHT MICROSCOPE** (above) uses two or more glass lenses to magnify objects. Light microscopes are used to look at living cells, such as red blood cells (top), small organisms, and preserved cells. Compound light microscopes can magnify up to about 1500 times.

RESEARCHER USING A TEM

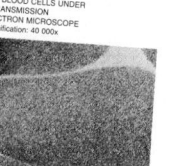

RED BLOOD CELLS UNDER A SCANNING ELECTRON MICROSCOPE Magnification: 10 000x

SCANNING ELECTRON MICROSCOPE IMAGE OF A MOSQUITO Magnification: 50x

SCANNING ELECTRON MICROSCOPE
An SEM sweeps a beam of electrons over the surface of a specimen, such as red blood cells (above), causing electrons to be emitted from the specimen. SEMs produce a realistic, three-dimensional picture—but only the surface of an object can be observed. An SEM can magnify about 60 000 times without losing clarity.

RED BLOOD CELLS UNDER A TRANSMISSION ELECTRON MICROSCOPE Magnification: 40 000x

SCANNING TUNNELING MICROSCOPE IMAGE OF A DNA FRAGMENT Magnification: 2 000 000x

SCANNING TUNNELING MICROSCOPE The STM revolutionized microscopy in the mid-1980s by allowing scientists to see atoms on an object's surface. A very fine metal probe is brought near a specimen. Electrons flow between the tip of the probe and atoms on the specimen's surface. As the probe follows surface contours, such as those on this DNA molecule (above), a computer creates a three-dimensional image. An STM can magnify up to one hundred million times.

◀ **TRANSMISSION ELECTRON MICROSCOPE** A TEM aims a beam of electrons through a specimen. Denser portions of an object allow fewer electrons to pass through. These denser areas appear darker in the image. Two-dimensional TEM images are used to study details of cells such as these red blood cells (above). A TEM can magnify hundreds of thousands of times.

EXPANDING Your View

1 **THINKING CRITICALLY** Can live specimens be examined with an electron microscope? Explain. Consider how the specimen must be prepared for viewing.

2 **COMPARING AND CONTRASTING** Compare the images seen with an SEM and with a TEM.

78

Focus on Features Authored by National Geographic Society

Some topics of biology deserve more attention than others do because they're unusual, informative, or just plain interesting. National Geographic Society has created visually exciting multi-page *Focus On* features that inform, excite, and motivate your students. *Focus On* features are relevant to the biology content of the chapter in which they are found. Assign them as a lead-in to special research projects and in-depth studies for extra credit. Use them as a basis for colorful visual displays and bulletin boards.

Student Edition

Skills

SKILL HANDBOOK

PRACTICING SCIENTIFIC METHODS

Caring for and Using the Microscope

1. Always carry the microscope by holding the arm with one hand and supporting the base with the other hand.

2. Place the microscope on a flat surface. The ___ should be toward you.

___ ok through the eyepiece. Adjust the ___ hragm so that light comes through the ___ ing in the stage.

___ a slide on the stage so that the ___

specimen is in the field of view. Hold it firmly in place by using the stage clips.

5. Always focus first with the coarse adjustment and the low-power objective lens. Once the object is in focus on low power, turn the nosepiece until the high-power objective is in place. Then use ONLY the fine adjustment to focus with this lens.

6. Store the microscope covered.

Eyepieces
Contain magnifying lenses to look through

___ jective
___ ns with ___ nification

Arm

___ cope

___ t
___ under

___ under ___ cation

Revolving nosep___
Holds and turns ___ objectives into vi___ position

High-power object___
Contain lenses with ___ powers of magnifica___

Stage
Platform used to supp___ the microscope slide

Diaphragm
Regulates the amount of ___ light that passes through ___ the specimen

Light source
Allows light to reflect upward through the diaphragm, the specimen, and the lenses.

SKILL HANDBOOK **SH1**

Skill Handbook

Biology: The Dynamics of Life encourages the interaction between content and critical-thinking processes and experiences with scientific methods. All the labs and activities require students to make observations and collect a variety of data. Skill Review questions at the end of each section and in the Chapter Assessment provide students with another opportunity to practice the thinking processes relevant to the material they are studying. The *Skill Handbook* provides students with examples of all the process skills that they need to practice during these activities.

Problem-Solving Lab

Problem-Solving Labs, located in the Student Edition at point of use, offer a unique opportunity for students to evaluate another scientist's experiments and data. Can be assigned as homework or class work. Use these labs to reinforce the steps of scientific methods.

Problem-Solving Lab

What organelle directs cell ___
of marine alga, grows as single, large cells 2 to 5 cm in height. The nuclei of these cells are in the "feet." Different species of these algae have different kinds of caps, some petal-like and others that look like umbrellas. If a cap is removed, it quickly grows back. If both cap and foot are removed from the cell of one species and a foot from another species is attached, a new cap will grow. This new cap will have a structure with characteristics of both species. If this new cap is removed, the cap that grows back will be like the cell that donated the nucleus.

The scientist who discovered these properties was Joachim Hämmerling. He wondered why the first cap that grew had characteristics of both species, yet the second cap was clearly like that of the cell that donated the nucleus.

Analysis
Look at the diagram below and interpret the data to explain the results.

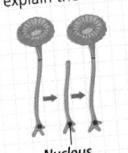  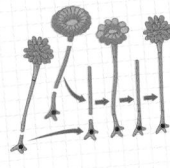

Nucleus   *Nucleus*

Thinking Critically
Why is the final cap like that of the cell from which the ___ cleus was taken? (HINT: Recall the function of the nucleus.)

WORD Origin

chloroplast
From the Greek words *chloros,* meaning "green," and *platos,* meaning "formed object." Chloroplasts capture light energy and produce food for plant cells.

Word Origin

Enhances the meaning and understanding of biology terms. Use these *Word Origin* margin features as a pre-reading activity, or have students keep a vocabulary journal. Additional word origins can be found in Appendix B.

THE PRINCETON REVIEW TEST-TAKING TIP

Maximize Your Score
Ask how your test will be scored. In order to do your best, you need to know if there is a penalty for guessing, and if so, how much of a penalty. If there is no rand___ ___ g penalty at all, you ___

Test-Taking Tips

Test-Taking Tips, written by The Princeton Review, help prepare students for taking the tests that are an integral part of their school experience. Found in every Chapter Assessment, these tips can be collected in a journal by students and reviewed before they take end-of-course, PSAT, or state exams.

CAREERS IN BIOLOGY

Science Reporter

Does science fascinate you? Can you explain complex ideas and issues in a clear and interesting way? If so, you should consider a career as a science reporter.

Skills for the Job

As a science reporter, you are a writer first and a scientist second. A degree in journalism and/or a scientific field is usually necessary, but curiosity and good writing skills are also essen... You might work for newspapers, national magazines, ... or scientific publications, television networks, or Interne... news services. You could work as a full-time employee on ... freelance writer. You must read constantly to stay up-to-d... Many science reporters attend scientific conventions and ... events to find news of interest to the public. Then they ca... fully and accurately translate what's new so nonscientists ... understand it.

For more careers in related fields, be sure to check the Glencoe Science Web site. **science.glencoe.com**

Careers in Biology

What do biologists do? Use the *Careers in Biology* features to emphasize that biological applications can be used in everyday life.

---

a large surface area that fits in a small space. Energy-storing molecules are produced on the inner folds. Mitochondria occur in varying numbers depending on the function of the cell. For example, liver cells may have up to 2500 mitochondria.

Although the process by which energy is produced and used in the cells is a technical concept to learn, the *Literature Connection* at the end of this chapter explains how cellular processes can also be inspiring. Look at the *Inside Story* on the next page to compare plant and animal cells. Notice how similar they are.

In-text references to labs and features

In-text references call out all labs and features in the chapter and show how they are integrated into the chapter.

Biology & SOCIETY

Why are the corals dying?

Coral reefs are some of Earth's most spectacularly beautiful and productive ecosystems. A reef is composed of hundreds of corals that together create a structure of brightly colored shapes and patterns. In the reef's cracks and crevices live a dazzling array of fishes and invertebrates.

Coral reefs protect nearby shore areas from erosion by breaking up the energy of incoming waves. But worldwide, coral reefs are increasingly being damaged and destroyed.

Physical Damage to Coral Reefs Hurricanes cause serious damage to coral reefs, as do large ships that run aground on reefs. Coral reefs lie close to the water's surface, and make attractive anchoring sites for boats. But when an anchor is pulled up, it may rip away a piece of the reef. In some parts of the world, explosives are used to mine coral limestone for building materials and fertilizers. Tropical aquarium fishes are sometimes collected by poisoning with cyanide, which stuns fishes and makes them easier to collect, but also kills corals. Collectors take pieces of coral for jewelry and souvenirs.

Damage from Disease In the 1970s, marine scientists began to realize that the world's coral reefs were being attacked by diseases no one had seen before.

Black-band disease is caused by several species of cyanobacteria that combine to form a band of black filaments. This invading community slowly moves across the coral. White-band disease causes the living tissue of a coral to peel away from its skeleton; this may be caused by bacteria. Rapid-wasting disease, possibly caused by a fungus, forms white patches that consume not ... the living tissue but also the top layers of the coral skeleton.

Many of the world's coral reefs are losing their beautiful colors in a process called bleaching. The corals become gray or white in color. Some scientists hypothesize that coral bleaching is the result of a loss of zooxanthellae, the ... otic protists that live in coral and give it much of its color as well as nutrients.

Different Viewpoints

It is not easy to tell whether microorganisms are causing the diseases or are attacking already damaged or ailing corals that have lost ... natural defenses. Most researchers hy... esize that coral diseases are on the ... increase because environmental ... changes, such as pollution in coastal ... runoff, higher water levels, or ... changes in ocean temperature, ... make corals more vulnerable to ... opportunistic diseases.

Healthy coral reef (below), and diseased reef (right)

ANALYZING THE ISSUE

Analyzing the Issue What ... might the death of a coral reef ... on nearby ocean and coastal l...

BIOLOGY Online To find out ... about coral ... visit the Glencoe Science Web... science.glencoe.com

---

BIO Technology

Scanning the Mind

Advancements in medical technology have led to instruments—such as X-ray and magnetic resonance imaging (MRI) machines—that can examine the human body in a noninvasive way. Recently, another technology has been added to the medical toolbox—a positron emission tomography (PET). This instrument is unique in that it allows a physician to view internal body tissues while they carry out their normal daily functions.

PET scanners are excellent tools for studying the human brain. By monitoring either the blood flow to an area or the amount of glucose being metabolized there, doctors are able to pinpoint active sections of the brain.

Here's how it works: The patient is injected with a compound containing radioactive isotopes. Because these isotopes emit detectable radiation, they can be tracked by the sensitive PET scanner. Computers create a picture of brain activity by converting the energy emitted from the radioisotopes into a colorful map. The image indicates the location of an activity, such as glucose utilization, and its relative intensity in various regions.

Applications for the Future

PET scanners are important in brain research, including the detection and diagnosis of brain tumours, the evaluation of damage due to stroke, and the mapping of brain functions.

PET scans can also be used to see how learning takes place in the brain. The images on this page show activity in the left and right brains of two different people. Each person was given a list of nouns and asked to visualize them. The ... and brain (top) had no previous experi... ... thus was forced to ...

about their physical and emotional status while the scanner records metabolic activity in the brain. Researchers hope that information gained from the study of drug addiction and treating manic-depress... help in diagnosing and treating manic-depress... psychosis and schizophrenia.

PET scans

INVESTIGATING THE TECHNOLOGY

Critically What other body ... with the aid of ...

Literature Connection

The Lives of a Cell
by Lewis Thomas

You may think of yourself as a body made up of parts. Arms, legs, skin, stomach, eyes, brain, heart, lungs. Your mind controls the whole, and you probably believe that you own all the parts that make up your body. In actual fact, you are a community of living structures that work together for growth and survival.

Your body is made up of eukaryotic cells with organelles that work together for each cell's survival. Organelles may work closely together, such as a ribosome and the endoplasmic reticulum, or they may perform a unique function within the cell, such as the mitochondrion.

An organism is similar to a cell in that several parts work together. Groups of cells work together as tissues. Several tissues form an organ and many organs form an organ system. For example, in an organ system such as the digestive system, cells and tissues form an organ such as the stomach, but several organs such as the intestines, the pancreas, and the liver, are needed to completely digest and absorb the food you eat.

In the same way, you might also consider how all the organisms in a community are interconnected and how the whole planet Earth is a collection of interdependent ecosystems. Lewis Thomas pondered this thought.

"I have been trying to think of the earth as a kind of organism, but it is no go. I cannot think of it this way. It is too big, too complex, with too many working parts lacking visible connection.... I wondered about this. If not like an organism, what is it like, what is it most like? Then, satisfactorily for the moment, it came to me: it is most like a single cell."

Words are like organelles Just as a cell is a group of organelles working together, so is a paragraph composed of words that together

convey thoughts and ideas. Despite all his technical knowledge, Dr. Thomas, a physician and medical researcher, writes simply and engagingly about everything from the tiny universe inside a single cell to the possibility of visitors from a distant planet.

Medicine, a young science Dr. Thomas grew up with the practice of medicine. As a boy, he accompanied his father, a family physician, on house calls to patients. Years later, Dr. Thomas described those days in his autobiography *The Younger Science*. The title reflects his belief that the practice of medicine is "still very early on" and that some basic problems of disease are just now yielding to exploration.

CONNECTION TO BIOLOGY

After you have studied this chapter, write a paragraph using Dr. Thomas's style to describe how the organelles of a cell work together for cell survival.

BIOLOGY Online To find out more about the works of Dr. Lewis Thomas, visit the Glencoe Science Web site. science.glencoe.com

Earth "is most like a single cell."

196 A VIEW OF THE CELL

Biology & Society, BioTechnology, Interdisciplinary Connection

Help students to realize that biology is connected to all their courses and that it impacts their lives and society as a whole. These features cover topics that students hear about from the media. *Biology & Society* gives students the opportunity to understand the many different sides of issues, and—as future decision-makers and voters—develop their own viewpoints. Use these features as topics for debate or persuasive writing. *BioTechnology* features show students how technology may affect their lives. Encourages students to take other science courses. *Interdisciplinary Connections* show students that even if they don't major in science, they will use science in other areas. Share these features with other teachers for cross-disciplinary instruction.

Student Edition

BioDigest

Plants

Seed Plants

A seed is a reproductive structure that contains a sporophyte embryo and a food supply enclosed in a protective coating. The food supply nourishes the young plant during the first stages of growth. Like spores, seeds can survive harsh conditions. The seed develops into the sporophyte generation of the plant. Seed plants include conifers and flowering plants.

Conifers

Conifers, ... produce seeds, usually ... have needle... seeds are no... are evergre... leaves all ye...

Adapted for Cold and Dry Climates

Conifers are common in cold or dry habitats throughout the world. Conifer needles have a compact shape and a thick, waxy covering that helps reduce evaporation and conserve water. Conifer stems are covered with a thick layer of bark that insulates the tissues inside. These adaptations enable conifers to carry on life processes even when tempera-...

Flowering Plants

The flowering plants, division Anthophyta, form the largest and most diverse group of plants on Earth today. They provide much of the food eaten by humans. Anthophytes produce flowers and develop seeds enclosed in a fruit.

Monocots and Dicots

The Anthophytes are classified into two classes: the monocotyledons and the dicotyledons. Cotyledons, or "seed leaves," are contained in the seed along with the plant embryo. Monocots have a single seed leaf that absorbs food for the embryo. The two seed leaves of dicots store food for the embryo.

Monocots (left) include grasses, orchids, and palms. Dicots (below) include many flower-...

Flowers

Flowers are the organs of reproduction in anthophytes. Sepals enclose the flower bud and protect it until it opens; petals, which are often brightly colored or perfumed, attract pollinators. Inside the circle of petals are the pistil and stamens.

The pistil is the female reproductive organ. Inside the ovary at the base of the pistil are the ovules. Ovules contain the female gametophyte generation of the plant. Female gametes—egg cells—form in each ovule.

Pollen — Anther — Stamen — Stigma — Style — Filament — Pistil — Ovary

Pollen

In seed plants, the sperm are enclosed in the thick-coated pollen grains, which are the male gametophyte generation of the plant. Pollen is one of the important adaptations that has enabled seed plants to live in a wide variety of land habitats.

Pollinators

Flowers can be pollinated by wind, insects, birds, and even bats. Some flowers have colorful or perfumed petals that attract pollinators. Flowers may also contain sweet nectar, as well as pollen, which provides pollinators with food.

Plants that depend on the wind to carry pollen from anther to stigma tend to have small, inconspicuous flowers. The flowers of grasses and this alder are pollinated by the wind.

Plants that depend on insects for pollination may be brightly colored and fragrant. Pollen rubs off on the bee that visits a flower to feed on nectar. When it moves to another flower, some of the pollen may rub off onto the stigma.

Many plants produce fruits that are eaten by animals.

Fruit

Following fertilization, the ovary develops into a fruit with seeds inside. Some flowering plants develop fleshy fruits, such as apples, melons, tomatoes, and squash. Other flowering plants develop dry fruits, such as peanuts, almonds, or sunflowers. Fruits help protect seeds until they are mature. Fruits also help scatter seeds into new habitats.

Maple trees produce fruits with a winglike shape that can be carried long distances by the breeze.

Pollen is carried to the stigma of a flower. The pollen grain grows a tube down the style to the ovary. Two sperm travel down the tube.

Pollen tube — Ovary — Sperm nuclei — Ovule — Egg nucleus and endosperm nucleus

... double fertilization, one of the egg and the other unit... n nucleus.

Plant Responses

Plants respond to changes in their environment such as light, temperature, and water availability. Chemicals called hormones control some of these responses by increasing cell division and growth.

The phototropic response shown here is the result of increased cell growth on the side of the stem away from the light.

Stems growing up

Roots exhibit positive geotropism. Stems show a negative geotropic response.

Roots growing down

BioDigest Assessment

Understanding Main Ideas
1. Which of the following is a Bryophyte?
... club moss

7. Mosses, ferns, and club mosses are alike because they require _____.
a. water for fertilization
b. adaptations for conserving water

BioDigest

Use the two-to-ten page *BioDigests* as an introduction or a summary of a unit.

Use the *BioDigest* in each unit to give students a quick overview of important information in the unit.

Can be used as a year-end review for exams by assigning a different *BioDigest* to each student group for a report.

Contents in Brief

iv

Teacher Wraparound Edition

Chapter Organizer

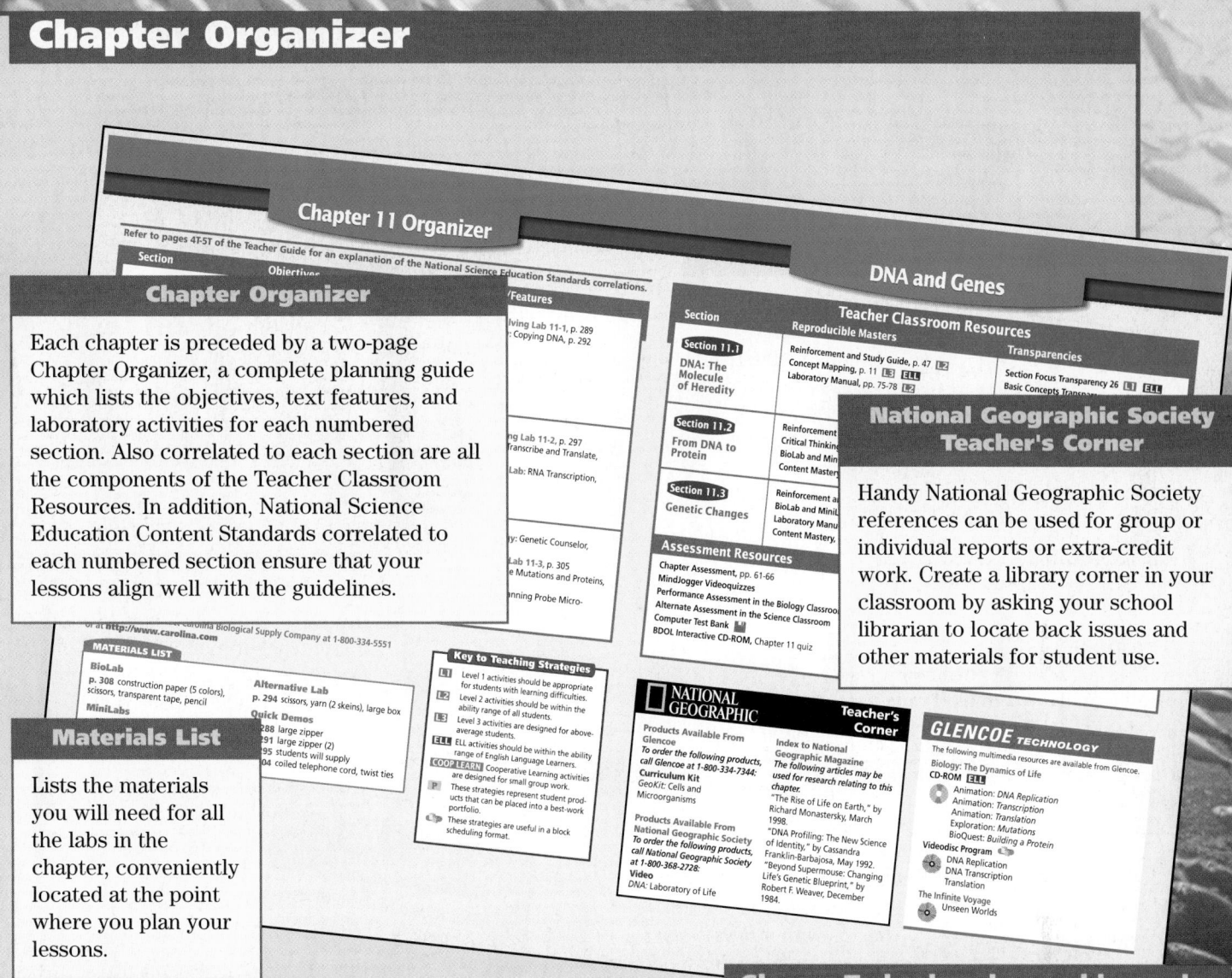

Chapter Organizer

Each chapter is preceded by a two-page Chapter Organizer, a complete planning guide which lists the objectives, text features, and laboratory activities for each numbered section. Also correlated to each section are all the components of the Teacher Classroom Resources. In addition, National Science Education Content Standards correlated to each numbered section ensure that your lessons align well with the guidelines.

Materials List

Lists the materials you will need for all the labs in the chapter, conveniently located at the point where you plan your lessons.

National Geographic Society Teacher's Corner

Handy National Geographic Society references can be used for group or individual reports or extra-credit work. Create a library corner in your classroom by asking your school librarian to locate back issues and other materials for student use.

Key to Teaching Strategies

The activities in the Teacher Wraparound Edition are broken down into three levels to accommodate all student ability levels. In addition, certain teaching strategies are designated as suitable for English Language Learners or cooperative learning groups. Finally, strategies that produce work suitable for placement in student portfolios and strategies that are useful in a block scheduling format are called out with easy-to-find icons. Use this coding to select strategies and reproducible masters or transparencies for different learners.

Glencoe Technology box and bar codes

A *Glencoe Technology* box for each chapter summarizes what CD-ROM, videodisc, and videotape materials are available for the chapter from Glencoe. Look for more detailed information about these technology materials at point of use in the margins of the *Teacher Wraparound Edition.* All references to videodiscs include bar codes for easy access to the segments you want. Make a talking poster by gluing bar codes to the bottom of poster illustrations. Place the videodisc player nearby.

Teacher Wraparound Edition

Classroom Management

OUT OF TIME?

If time does not permit teaching the entire chapter, use the BioDigest at the end of the unit as an overview.

Out of Time

The Out of Time logo on the *Chapter Opener* page reminds you to use the BioDigest if time does not permit teaching the entire chapter or unit.

1 Focus

Bellringer

Before presenting the lesson, display **Section Focus Transparency 41** on the overhead projector and have students answer the accompanying question.

L1 **ELL**

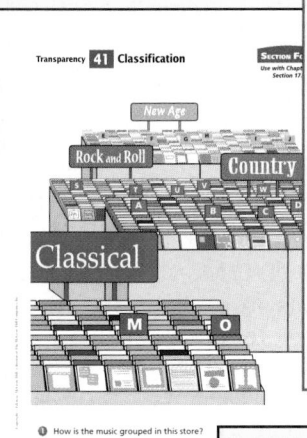

Bellringer

The first step to teaching any set of concepts or skills is to focus the students' attention. Each numbered section begins with a *Bellringer* suggestion for using a *Section Focus Transparency* to focus the class on the key topic of the text section. A copy of the transparency in the teacher margin lets you preview the transparency.

Resource Manager boxes

Resource Manager boxes located in the teacher margin at point of use show you where to integrate the many Teacher Classroom Resources materials into your lessons.

Resource Manager

BioLab and MiniLab Worksheets, p. 81 **L2**
Tech Prep Applications, pp. 25-26 **L2**
Laboratory Manual, pp. 117-120 **L2**

Internet Address Book

Do you jot down interesting web sites on little scraps of paper, only to lose them before you can use them? The *Teacher Wraparound Edition* provides several *Biology Online* boxes in the teacher margin of the chapter where you can write down those URLs you want to investigate.

Note Internet addresses that you find for quick reference.

16T

Hands-On Experiences

Quick Demos

Sometimes, the easiest way to teach students a concept is to show them. Many *Quick Demos* are provided in the teacher margin for your use in helping students visualize biology concepts.

Quick Demo

Kinesthetic Build a bacteriophage model. Screw two nuts onto a large bolt so that they touch the top of the bolt. Take three 14-cm pieces of #22 gauge wire and twist them around the bottom of the bolt. Bend the ends of the wires to resemble the tails of a bacteriophage. Ask students what structure the wires represent? *tail fibers* What structure do the nuts and top of the bolt represent? *protein head* L1

GETTING STARTED DEMO

Kinesthetic Instruct students to pick up pencils without using their thumbs, and use them to write their names. Then, ask students to describe how their thumbs differ from their other fingers. L1 ELL

Getting Started Demo

Located on every chapter opener, the *Getting Started Demo* provides an inquiry approach to starting the chapter. Like the Quick Demos, the *Getting Started Demo* may involve student participation.

Display
Add pictures of seed plants to the bulletin board.

Chalkboard Activity
Visual-Spatial Have students construct a time line that marks the era when different types of land plants evolved. Ask students to describe what the climate was like during each time period. L2

 The BioLab at the end of the chapter can be used at this point in the lesson.

Figure 22.19
Each scale of a conifer's cone is a modified branch.

A Male cones are made up of thin papery scales that open to shed clouds of yellow pollen grains into the wind.

B In many conifers, including the pine family, two seeds develop at the base of each of the woody scales that make up a female cone.

same tree. The male cones produce pollen. They are small and easy to overlook. Female cones are much larger. They stay on the tree until the seeds ha... both typ... *Figure 2...*

Conifers...
Most... that is, ... year. Alt... off as th... tree neve... Pine nee... remain... from two... the speci... Trees... photosyn... soon as... Evergree... coating... material... where th... short, so... gives the... a whole... they are... are scar... leaves of... find out...

Deciduo...
A few conifers, including larches and bald cypress trees, are deciduous, *Figure 22.20.* **Deciduous plants** lose all their leaves at the same time. Dropping all leaves is an adaptation for reducing water loss when water is unavailable as it can be the case in the tundra, in deserts, or during wintertime. Plants lose most of their water through the leaves; very little is lost through bark or roots. However, a tree with no leaves cannot photosynthesize and must remain dormant during this time.

... meaning to fall off." Deciduous trees drop all of their leaves at the same time.

612 THE DIVERSITY OF PLANTS

BioLab placement icon

The BioLab for each chapter is located at the end of the chapter so as not to disrupt the flow of text. How do you know where to teach the BioLab? Look for the *BioLab placement icon* in the teacher margin, which suggests the place to teach the lab.

DESIGN YOUR OWN BioLab

The BioLab at the end of the chapter can be used at this point in the lesson. →

Unit Projects

Each unit opens with four individual or group unit projects that address different learning styles. Use them as semester-long projects.

Alternative Lab

The Conifer Leaf

Purpose
Students will observe, diagram, and label the structures of a typical conifer leaf.

Materials
microscope, prepared slide of conifer leaf cross section.

612

Procedure
Give students the following directions.
1. Observe the slide under low power.
2. Diagram what you see. Label the following structures using these descriptions as a guide: (a) cuticle—noncellular layer, thin covering over leaf; (b) epidermis—below cuticle, one cell in thickness; (c) mesophyll (spongy layer)—large, thick layer just below epidermis; (d) endodermis—

one-cell-layer thick, surrounds large oval central are... many cells in th... endodermis; (f)... center of leaf, tw...
3. Look again at th... scope. Check fo... on the epidermis... several ringlik... ducts, in the m... parts and labels f...

Alternative Lab

One or two additional labs in each chapter are available for use as make-up labs or as

Teacher Wraparound Edition

Diversity in the Classroom

Multiple Learning Styles

Look for the following logos for strategies that emphasize different learning modalities.

Kinesthetic Project, pp. 288, 300; Reteach, p. 293; Quick Demo, p. 295; Extension, p. 301
Visual-Spatial Quick Demo, p. 291; Project, p. 292; Reinforcement, p. 298; Reteach, p. 301; Meeting Individual Needs, p. 304
Interpersonal Tech Prep, p. 303

Intrapersonal Enrichment, pp. 288, 296; Biology Journal, p. 306; Going Further, p. 310
Linguistic Portfolio, pp. 289, 300, 305; Enrichment, p. 290; Biology Journal, pp. 291, 299, 302
Logical-Mathematical Enrichment, p. 303; Reinforcement, p. 304

Multiple Learning Styles

The *Teacher Wraparound Edition* provides a number of strategies for encouraging students with diverse learning styles. Look for the various learning style icons that identify strategies suitable for each style of learning. Strategies in each chapter are summarized in the *Multiple Learning Styles* box at the bottom of the chapter opener page.

MEETING INDIVIDUAL NEEDS

Learning Disabled/Visually Impaired

Kinesthetic Make large cardboard X and Y chromosomes. Have students manipulate them to see how each parent donates one of his/her sex chromosomes to each gamete and how they come together during fertilization to produce offspring of each sex. **L1 ELL**

Meeting Individual Needs

Use these teaching strategies and student activities to structure the learning environment in your classroom to meet students' special needs.

Cultural Diversity

Har Gobind Khorana

Teach students about some of the experimental methods scientists have used to decipher [...]. In particular, [...] explain the [...] Har [...] tured in the laboratory. Khorana's artificial genes were able to code for the synthesis of proteins just as they do in living cells. This work ultimately led to the cracking of a portion of the genetic code. For this work, Khorana and colleagues received the Nobel Prize for Physiology or Medicine in 1968.

Cultural Diversity

Respond to diversity by sharing with students the *Cultural Diversity* connections found at the bottom of the page in every chapter of the *Teacher Wraparound Edition*.

Tech Prep

Tech Prep teaching activities, identified by the Tech Prep bar at the top of the box, get your students involved in everyday applications of biology. Tech Prep icons identify other teaching strategies that are particularly suited

TECHPREP

Plant Breeding

Kinesthetic Have students visit a grocery store and take notes and make diagrams [of] squash varieties. Include any [in]formation the store might [pr]ovide. Then, after the dia[gr]ams and information are [re]corded, advise students that

Assessment

GLENCOE'S ASSESSMENT ADVANTAGE

Assessment Planner

All the assessments in a chapter are summarized in the *Assessment Planner* located at the beginning of every chapter. Skill, Knowledge, Performance, and Portfolio assessment strategies provide a variety of assessments for all your students.

Assessment
Knowledge Have students explain in their own words how energy is stored within an ATP molecule. **L2**

Assessment

Every *BioLab*, *MiniLab*, *Problem-Solving Lab*, and *Alternative Lab* includes a ✔Assessment strategy in the teacher margin. In addition, you will find many additional ✔Assessment strategies within a section. Finally, each chapter contains many ✔Portfolio suggestions that can be used as assessments.

Assessment Planner

Portfolio Assessment
 Portfolio, TWE, pp. 289, 300, 306
 BioLab, TWE, pp. 308-309
Performance Assessment
 Assessment, TWE, pp. 291, 304
 Problem-Solving Lab, TWE, p. 305
 MiniLab, TWE, p. 306
 MiniLab, SE, pp. 299, 306
 BioLab, SE, pp. 308-309
 Alternative Lab, TWE, pp. 294-295

Knowledge Assessment
 Problem-Solving Lab, TWE, pp. 289, 297
 Assessment, TWE, pp. 293, 300, 301, 307
 Section Assessment, SE, pp. 293, 301, 307
 Chapter Assessment, SE, pp. 311-313
Skill Assessment
 Alternative Lab, TWE, pp. 294-295
 MiniLab, TWE, p. 299

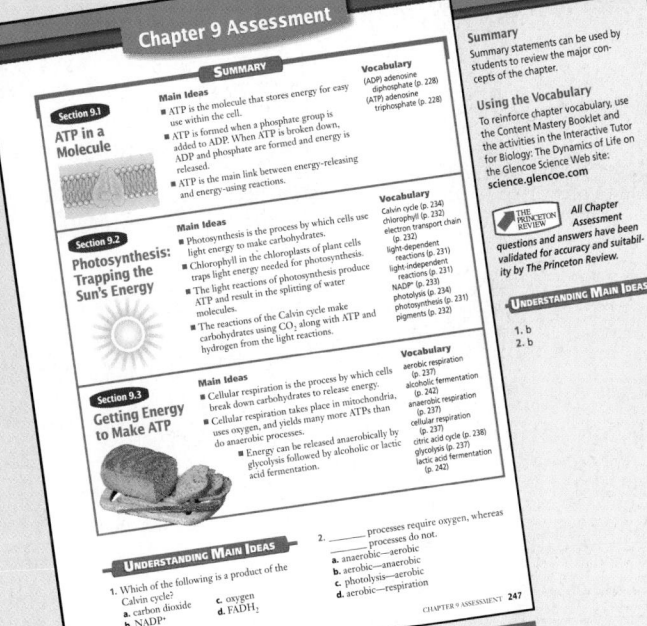

The Princeton Review

All Chapter Assessment questions and answers have been validated for accuracy by The Princeton Review. All Chapter Assessment questions have been checked against the chapter's objectives to ensure that the objectives are adequately tested. You can be sure when you assign a Chapter Assessment question that it reflects the content of the chapter in an accurate and suitable way for your class.

Teacher Classroom Resources

Hands-On Learning

Resources For All Your Needs

In addition to the wide array of instructional options provided in the student and teacher editions, *Biology: The Dynamics of Life* offers an extensive list of support materials and program resources. Some of these materials offer alternative ways of presenting your biology program, others provide tools for reinforcing core concepts and evaluating student learning, and still others will help you extend and enrich your course. You won't have time to use them all, but the ones you use will help you make the best use of the time you have.

Laboratory Manual

GLENCOE

BIOLOGY
The Dynamics of Life

Lab Manual, Student Edition and Teacher Edition

If you want more hands-on options, the *Laboratory Manual* offers you two more labs for each chapter. Choose from a selection of traditional labs, Design Your Own labs, and computer-based labs to provide students with a variety of laboratory experiences that reinforce the biology principles in the text. The Teacher Edition provides full support.

BioLab and MiniLab Worksheets

Each of the BioLabs and MiniLabs in the student text is also available in copymaster form in *BioLab and MiniLab Worksheets*. As students proceed with a lab, have them copy their observations and data on the expanded data tables provided in these worksheets.

BioLab and MiniLab Worksheets

GLENCOE

BIOLOGY
The Dynamics of Life

NATIONAL GEOGRAPHIC SOCIETY

Contents and Features
- An activity master for every BioLab and MiniLab in the student text
- Students record data directly on lab worksheets
- Answer pages

Biology Projects

GLENCOE

BIOLOGY
The Dynamics of Life

NATIONAL GEOGRAPHIC SOCIETY

Contents and Features
- Long-term biology projects for each unit of the student text
- ... for teaching science
- ...

Biology Projects

Biology Projects provide suggestions for long-term group or individual activities that expand on a major topic of each unit. These projects are designed to involve and excite students about biology in their lives. They require students to be creative, design and conduct experiments, and report on their results using a variety of methods.

Applications and Enrichments

Critical Thinking/Problem Solving

Challenge students to apply their critical thinking and problem-solving skills with the *Critical Thinking/Problem Solving* book. It is especially suited to average and above-average ability students.

Critical Thinking/Problem Solving

GLENCOE
BIOLOGY
The Dynamics of Life

NATIONAL GEOGRAPHIC SOCIETY

Contents and Features
- Critical thinking/problem solving activity for each chapter
- Answer pages

Tech Prep Applications

GLENCOE
BIOLOGY
The Dynamics of Life

Contents
- 27 Ex...
  every...
- 9 ha...
- Tea...
  ap...

Tech Prep Applications

Tech Prep Applications encourage your students to become familiar with everyday applications of biology. These worksheets show students how the main ideas of their biology course are essential to the understanding of modern developments in technology and medicine.

Exploring Environmental Issues

GLENCOE SCIENCE PROFESSIONAL SERIES

CONTENTS AND FEATURES
- More than 20 student-directed projects
- Focus on current environmental issues and problems
- Involves students in data collection and analysis
- Ide... for small or large gro... use

Exploring Environmental Issues

Exploring Environmental Issues incorporates many of the environmental issues covered daily by the media into open-ended projects. The activities will reveal how these issues can affect students' lives.

Inside Story Poster Package

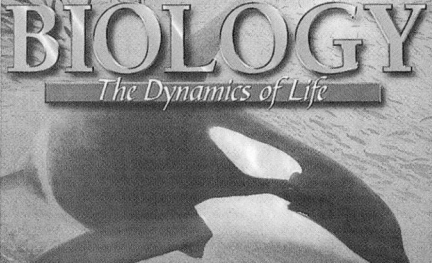

GLENCOE
BIOLOGY
The Dynamics of Life

NATIONAL GEOGRAPHIC SOCIETY

Contents and Features
- 10 full-color posters of major concepts and processes of biology
- Activity masters for each poster
- Teaching strategies

ISBN 0-

Inside Story Poster Package

The *Inside Story Poster Package* includes ten full-color posters, enlargements of ten Inside Story features from the text. A booklet contains an enrichment worksheet and teacher guide for each poster. Additional subject material and related activities make up the enrichment worksheet. The teacher guide provides suggestions for how and when to use the material.

Teacher Classroom Resources

✓ Assessment

Chapter Assessment

Chapter Assessment includes a four-part assessment tool containing questions of various levels for every chapter, which helps you assess process as well as content objectives. Unique simulations require students to analyze experimental designs and interpret data.

Multimedia Assessment Options (MindJogger, ExamView® Pro Software, CD-ROM Quizzes)

Multimedia Assessment Options offer you assessment options for every chapter in the form of MindJogger videotape quizzes, quizzes on the *Biology: The Dynamics of Life* Interactive CD-ROM, and ExamView® Pro Software. You will never have to write another test question with the options you have.

CD-ROM Quizzes

CD-ROM Quizzes are also available on Glencoe's web site. www.glencoe.com/sec/science

GLENCOE'S ASSESSMENT ADVANTAGE

Performance Assessment in the Biology Classroom

The *Performance Assessment in the Biology Classroom* book offers a unique variety of performance-based task and skill assessments especially designed for the biology classroom. Use them to measure the skills and knowledge of your students as problem solvers.

CONTENTS AND FEATURES
• 30 performance-based assessment activities for high school biology
• Complete information for setting up, conducting, and scoring performance-based tasks

Performance Assessment in the Science Classroom

Use the performance task assessment lists and rubrics in *Performance Assessment in the Science Classroom* to assess students' work. Every MiniLab, Problem-Solving Lab, BioLab, and Alternative Lab in the text contains an assessment that is correlated to an assessment list in this book.

INCLUDES:
• Rationale for Assessing Performance
• Strategies for Implementation
• Performance Task Assessment Lists
• Rubrics

CONTENTS INCLUDE SUGGESTIONS AND IDEAS FOR:
• Assessment Techniques
• Observing and Questioning
• Projects and Investigations
• Portfolios and Journals Performance Assessment
• Much More!

Alternate Assessment in the Science Classroom

Alternate Assessment in the Science Classroom provides practical strategies for using alternate forms of assessment such as performance-based methods and portfolios. In addition, there are report forms and scoring rubrics.

Teacher Classroom Resources

Review and Reinforcement

Reinforcement and Study Guide, Student and Teacher Editions

Use section-by-section masters from the *Reinforcement and Study Guide* book to reinforce the core concepts presented in the student text. These worksheets are ideal for students of average and below-average ability. A consumable student edition of the guide is also available.

Concept Mapping

The *Concept Mapping* book gives students the opportunity to incorporate major biology concepts into the concept maps provided. Concept maps make abstract information more concrete and useful by visually representing relationships among concepts.

Basic Concepts Transparencies

The *Basic Concepts Transparency Package* includes full-color transparencies designed to teach basic biology concepts. A transparency master and worksheet accompany each transparency.

Content Mastery, Student and Teacher Editions

Content Mastery aids students with reading difficulties, ELL students, and students requiring extra help. The worksheets allow students to master the basic concepts of each chapter while reinforcing their reading and study skills. One activity in each chapter is interactive on the Glencoe web site.

Reteaching Skills Transparencies

Reteaching Skills Transparencies, many with label overlays, help you present, reinforce, or review key concepts developed in the text. Many overheads are designed to reinforce scientific methods and processes. An accompanying workbook provides additional opportunities for review and reinforcement.

Teaching Aids

Block Scheduling

GLENCOE

BIOLOGY

Lesson Plans

GLENCOE

BIOLOGY
The Dynamics of Life

Lesson Plans and Block Scheduling

Lesson Plans offer you a complete planning resource by correlating objectives, activities, and program resources for every lesson. Each plan is geared to the teaching cycle employed in the *Teacher Wraparound Edition* and contains a complete list of all the resources available for that numbered section. If you use block scheduling, the *Block Scheduling* book offers the same resources for your schedule.

Cooperative Learning in the Science Classroom

Provides strategies for implementing cooperative learning techniques that put the effectiveness of group learning to work in your biology classroom.

INCLUDES:
- Strategies for Implementing and Evaluating Cooperative Learning Activities in Science
- Blackline Masters and Facsimiles

Section Focus Transparency Package

GLENCOE

BIOLOGY
The Dynamics of Life

Contents and Fea
- 97 full-color transpare one for every numbere the student text
- Blackline copy master transparency
- Teaching strategies
- 3-ring binder for conve

Section Focus Transparencies and Masters

The 97 full-color transparencies in the *Section Focus Transparency Package* provide a way for you to begin each lesson by capturing the attention of your students with simple activities. Each numbered section begins with a Bellringer strategy that shows a photo of the Section Focus Transparency for that section. The package also contains a booklet of copymasters of all transparencies.

Spanish Resources

GLENCOE

BIOLOGY
The Dynamics of Life

Contents and
- Spanish translat objectives, sum and key terms
- Complete Engli

Spanish Resources

Help your Spanish-speaking students get more out of your science lessons by reproducing pages from the *Spanish Resources* book. In addition to the complete English/Spanish glossary, the book contains translations of all objectives, vocabulary, and Main Ideas for each chapter of the student text.

Teacher Classroom Resources

Glencoe Technology

Interactive CD-ROM

A creative combination of sound, video, photographs, and animation makes Glencoe's *Biology: The Dynamics of Life CD-ROM Multimedia System* an exciting addition to your biology program. Offers an enormous array of interactive simulations that engage and help students to learn. Full-motion videos, animations, explorations, and lab simulations let students explore biology concepts and practice critical-thinking skills, all at the computer.

Videodisc Program

Provide your students with visual explanations of major biology concepts by showing them the wide variety of videos on *Biology: The Dynamics of Life Videodisc Program*. Chapter-by-chapter video clips and animations help you bring biology to life.

English/Spanish Audiocassettes

Comprehensive chapter overviews in Spanish and English are a way for auditory learners, students with reading difficulties, and ELL students to review key chapter concepts. Students can listen individually during class or at home. Chapter overviews are printed in the *Content Mastery* and *Reinforcement and Study Guide* books for students who prefer to read rather than listen. You may find the overviews useful for reviewing the chapter as you plan lessons.

ExamView® Pro Software

Design and create your own test instruments from our ExamView® Pro Software for Windows and Macintosh. Select your own test items by objective from two different levels of difficulty, or write and edit your own.

Glencoe Online Science

The *Glencoe Biology: The Dynamics of Life Web Site* provides both you and your students with a wide range of materials. Students can access previewed Internet sites (Web Links) that correlate to text chapters, tutor themselves with interactive games that review chapter content, and practice science skills. The Biology Updates feature helps you and your students keep current with new findings in biology. Internet BioLabs give students an opportunity to experiment and share data with other students.

MindJogger Videoquizzes

The interactive quiz-show format of *Biology: The Dynamics of Life MindJogger Videoquizzes* provides fun for your students while reviewing core concepts for every chapter.

Teaching Strategies

Development of Thinking Processes and Skills

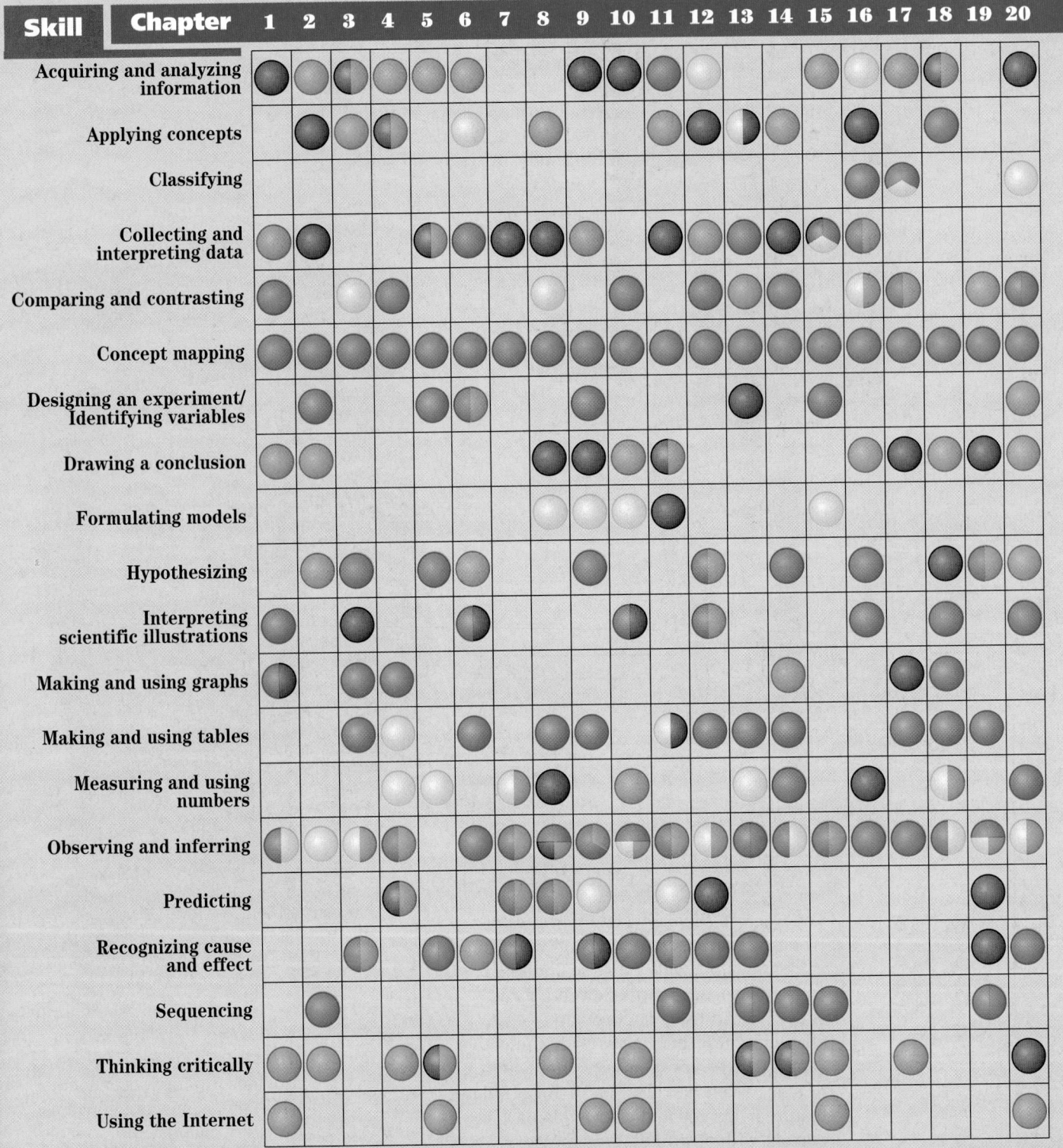

Skill \ Chapter	1	2	3	4	5	6	7	8	9	10	11	12	13	14	15	16	17	18	19	20
Acquiring and analyzing information																				
Applying concepts																				
Classifying																				
Collecting and interpreting data																				
Comparing and contrasting																				
Concept mapping																				
Designing an experiment/Identifying variables																				
Drawing a conclusion																				
Formulating models																				
Hypothesizing																				
Interpreting scientific illustrations																				
Making and using graphs																				
Making and using tables																				
Measuring and using numbers																				
Observing and inferring																				
Predicting																				
Recognizing cause and effect																				
Sequencing																				
Thinking critically																				
Using the Internet																				

Look for the skills listed above in the following elements of the text

Key: = BioLab, = MiniLab, = Problem-Solving Lab, = Chapter Assessment, = Section Assessment

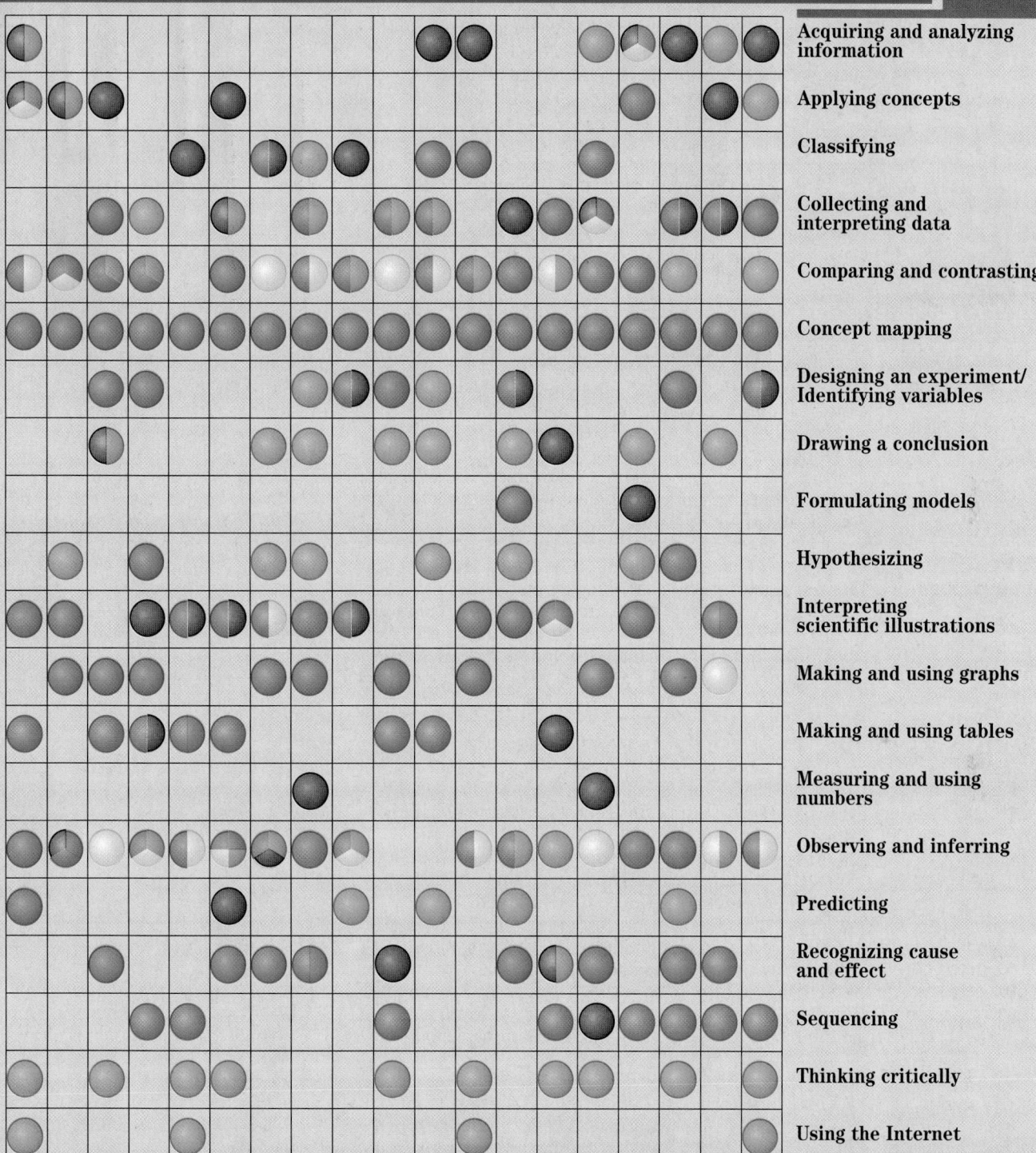

	21	22	23	24	25	26	27	28	29	30	31	32	33	34	35	36	37	38	39	Skill

Look for the skills listed above in the following elements of the text

Key: ● = BioLab, ● = MiniLab, ● = Problem-Solving Lab, ● = Chapter Assessment, ● = Section Assessment

Skills listed (rows):
- Acquiring and analyzing information
- Applying concepts
- Classifying
- Collecting and interpreting data
- Comparing and contrasting
- Concept mapping
- Designing an experiment/ Identifying variables
- Drawing a conclusion
- Formulating models
- Hypothesizing
- Interpreting scientific illustrations
- Making and using graphs
- Making and using tables
- Measuring and using numbers
- Observing and inferring
- Predicting
- Recognizing cause and effect
- Sequencing
- Thinking critically
- Using the Internet

29T

Teaching Strategies

Meeting Individual Needs

	Description	Sources of Help/Information
Learning Disabled	All learning disabled students have an academic problem in one or more areas, such as academic learning, language, perception, social-emotional adjustment, memory, or attention.	*Journal of Learning Disabilities* *Learning Disability Quarterly*
Behaviorally Disordered	Children with behavior disorders deviate from standards or expectations of behavior and impair the functioning of others and themselves. These children may also be gifted or learning disabled.	*Exceptional Children* *Journal of Special Education*
Physically Challenged	Children who are physically disabled fall into two categories —those with orthopedic impairments and those with other health impairments. Orthopedically impaired children have the use of one or more limbs severely restricted, so the use of wheelchairs, crutches, or braces may be necessary. Children with other health impairments may require the use of respirators or other medical equipment.	Batshaw, M.L. and M.Y. Perset. *Children with Handicaps: A Medical Primer.* Baltimore: Paul H. Brooks, 1981. Hale, G. (Ed.). *The Source Book for the Disabled.* New York: Holt, Rinehart & Winston, 1982. *Teaching Exceptional Children*
Visually Impaired	Children who are visually disabled have partial or total loss of sight. Individuals with visual impairments are not significantly different from their sighted peers in ability range or personality. However, blindness may affect cognitive, motor, and social development, especially if early intervention is lacking.	*Journal of Visual Impairment and Blindness* *Education of Visually Handicapped* *American Foundation for the Blind*
Hearing Impaired	Children who are hearing impaired have partial or total loss of hearing. Individuals with hearing impairments are not significantly different from their hearing peers in ability range or personality. However, the chronic condition of deafness may affect cognitive, motor, and social development if early intervention is lacking. Speech development also is often affected.	*American Annals of the Deaf* *Journal of Speech and Hearing Research* *Sign Language Studies*
English Language Learners	Multicultural and/or bilingual children often speak English as a second language or not at all. The customs and behavior of people in the majority culture may be confusing for some of these students. Cultural values may inhibit some of these students from full participation.	*Teaching English as a Second Language Reporter* R.L. Jones (Ed.). *Mainstreaming and the Minority Child.* Reston, VA: Council for Exceptional Children, 1976.
Gifted	Although no formal definition exists, these students can be described as having above-average ability, task commitment, and creativity. Gifted students rank in the top 5% of their class. They usually finish work more quickly than other students, and are capable of divergent thinking.	*Journal for the Education of the Gifted* *Gifted Child Quarterly* *Gifted Creative/Talented*

Tips for Instruction

1. Provide support and structure; clearly specify rules, assignments, and duties.
2. Practice skills frequently. Use games and drills to help maintain student interest.
3. Allow students to record answers on tape and allow extra time to complete tests and assignments.
4. Provide outlines or tape lecture material.
5. Pair students with peer helpers, and provide class time for pair interaction.

---

1. Provide a clearly structured environment with regard to scheduling, rules, room arrangement, and safety.
2. Clearly outline objectives and how you will help students obtain objectives. Seek input from them about their strengths, weaknesses, and goals.
3. Reinforce appropriate behavior and model it for students.
4. Do not expect immediate success. Instead, work for long-term improvement.
5. Balance individual needs with group requirements.

---

1. Openly discuss with the student any uncertainties you have about when to offer aid.
2. Ask parents or therapists and students what special devices or procedures are needed, and whether any special safety precautions need to be taken.
3. Allow physically disabled students to do everything their peers do, including participating in field trips, special events, and projects.
4. Help nondisabled students and adults understand physically disabled students.

---

1. As with all students, help the student become independent. Some assignments may need to be modified.
2. Teach classmates how to serve as guides.
3. Limit unnecessary noise in the classroom.
4. Encourage students to use their sense of touch. Provide tactile models whenever possible.
5. Describe people and events as they occur in the classroom.
6. Provide taped lectures and reading assignments.
7. Team the student with a sighted peer for laboratory work.

---

1. Seat students where they can see your lip movements easily, and avoid visual distractions.
2. Avoid standing with your back to the window or light source.
3. Using an overhead projector allows you to maintain eye contact while writing.
4. Seat students where they can see speakers.
5. Write all assignments on the board, or hand out written instructions.
6. If the student has a manual interpreter, allow both student and interpreter to select the most favorable seating arrangements.

---

1. Remember that students' ability to speak English does not reflect their academic ability.
2. Try to incorporate the student's cultural experience into your instruction. The help of a bilingual aide may be effective.
3. Include information about different cultures in your curriculum to help build students' self-image. Avoid cultural stereotypes.
4. Encourage students to share their cultures in the classroom.

---

1. Make arrangements for students to take selected subjects early and to work on independent projects.
2. Let students express themselves in art forms such as drawing, creative writing, or acting.
3. Make public services available through a catalog of resources, such as agencies providing free and inexpensive materials, community services and programs, and people in the community with specific expertise.
4. Ask "what if" questions to develop high-level thinking skills. Establish an environment safe for risk taking.
5. Emphasize concepts, theories, ideas, relationships, and generalizations.

The *Student Edition* and *Teacher Wraparound Edition* contain a variety of strategies for addressing the individual needs of students:

- **Meeting Individual Needs** strategies
- Both structured and open-ended **BioLabs**
- Hands-on **MiniLabs**
- **Internet Connections**
- **Visual Learning** strategies

The following Glencoe products also can help you tailor your instruction to meet the individual needs of your students:

- **Content Mastery**
- **Reinforcement and Study Guide**
- **Concept Mapping**
- **Critical Thinking/Problem Solving**
- **Tech Prep Applications**
- **Reteaching Skills Transparencies**
- **Basic Concepts Transparencies**
- **English/Spanish Audiocassettes**
- Various multimedia products include **Glencoe *Biology: The Dynamics of Life* Videodisc, Interactive CD-ROM, and MindJogger Videoquizzes; The Infinite Voyage Videodisc Series; National Geographic Society Videodisc Series; and The Secret of Life Videodisc and Videotape Series**

Assessment

Practical Strategies and Tools

ASSESSMENT ADVANTAGE
GLENCOE'S

The **Biology: The Dynamics of Life** program has been designed to provide you with a variety of assessment tools, both formal and informal, to help you develop a clearer picture of your students' progress.

Performance Assessment

Various methods of assessing individual student performance are becoming more common in today's schools. These performance assessments differ in formality and complexity, but in most cases, the teacher observes a student or group of students involved in an activity and rates the performance and/or the products that result from the activity. Background information and specific examples of performance assessment are included in Glencoe's **Alternate Assessment in the Science Classroom.**

Biology: The Dynamics of Life provides numerous opportunities to observe student behavior both in informal and formal settings. Each *BioLab*, *Problem-Solving Lab*, *MiniLab* and *Alternative Lab* contains assessment suggestions that will enable you to assess students' understanding of both concepts and process skills.

Another approach for assessing student mastery of concepts and skills in the laboratory is provided in Glencoe's **Performance Assessment in the Biology Classroom.** It features 30 laboratory exercises that enable you to evaluate students' skills in handling laboratory equipment and students' knowledge of laboratory processes.

Group Performance Assessment

Recent research has shown that cooperative learning structures produce improved student learning outcomes for students of all ability levels. **Biology: The Dynamics of Life** provides many opportunities for cooperative learning and,

as a result, many opportunities to observe group work processes and products. Glencoe's **Cooperative Learning in the Science Classroom** provides strategies and resources for implementing and evaluating group activities.

In cooperative group assessment, all members of the group contribute to the work process and its products. For example, if a mixed-ability, four-member laboratory work group conducts an activity, you can use a rating scale or checklist to assess the quality of both group interaction and work skills. All four members of the group are expected to review and agree on the data sheet produced by the group. You can require each member to certify the group's results by signing the data sheet or lab report. In this approach, all members of the group receive the same grade on the work product. Research shows that cooperative group assessment is as valid as individual assessment. Additionally, it reduces the marking and grading workload of the teacher.

Portfolios: Putting It All Together

The purpose of a student or cooperative group portfolio is to present examples of the individual's or group's work in a nontesting environment. A portfolio is simply a method for assembling and presenting selected examples of work products. The process of assembling the portfolio should be both integrative (of process and content) and reflective. The performance portfolio is not a complete collection of all worksheets and other assignments for a grading period.

At its best, the portfolio should include integrated performance products that show growth in concept attainment and skill development. You can structure the portfolio development process by establishing categories and other limiting specifications. An essential component in portfolio development is the composition of a submission letter or reflective paper that lists the contents of the portfolio and discusses growth in knowledge, attitudes, and skills.

Biology: The Dynamics of Life presents a wealth of opportunities for performance portfolio development. Each chapter contains projects; enrichment activities; laboratory investigations; skill reviews; suggestions for library research; features with critical-thinking opportunities; and connections to life, social studies, and the arts. Each of these student activities results in a product. A mixture of these products can be used to document student growth during the grading period.

In addition, *Biology: The Dynamics of Life* strongly suggests the use of student journals. Students are encouraged to write observations, descriptions, and reflections in their journals. They are also encouraged to include diagrams and drawings. Excerpts from the student journal can be included in an individual or group portfolio. Additionally, as many writers have discovered, the journal will be an excellent resource for developing the reflective submission letter or paper.

Content Assessment

The *Biology: The Dynamics of Life* program contains numerous strategies and an assortment of traditional aids for evaluating student progress toward mastery of science concepts. Concluding each numbered section in the *Student Edition, Section Assessment* questions are presented. This spaced review process helps build learning bridges that allow all students to progress confidently from one lesson to the next.

After instruction for the chapter is completed, the formal review process for the written content assessment can begin. *Biology: The Dynamics of Life* presents a three-page

Chapter Assessment at the end of each chapter. By evaluating the student responses to this extensive review, you can determine whether any substantial reteaching is needed.

For the formal content assessment, a six-page test is provided in the *Chapter Assessment* booklet for each chapter. If your individual assessment plan requires a test that differs from this test in the resource package, customized tests can be easily produced using the *ExamView® Pro Software,* available in Windows and MacIntosh formats.

Also available for formal or informal assessments are chapter quizzes on the *Interactive CD-ROM. Mindjogger Videoquizzes* provide an informal assessment in a game show format. Finally, interactive chapter quizzes are available on Glencoe's web site at www.glencoe.com/sec/science.

Skill Assessment

The *Biology: The Dynamics of Life* program contains many assessment strategies that require students to use skills such as making and using graphs, tables, and concept maps. The *Skill Handbook*, located at the back of the *Student Edition*, can help students master these skills.

The Princeton Review

All *Chapter Assessment* questions and answers have been validated for accuracy and suitability by **The Princeton Review.** In addition, each *Chapter Assessment* contains a test-taking tip that will aid students in preparing for and taking tests.

THE PRINCETON REVIEW

Course Planning Guide

Biology: The Dynamics of Life provides a complete selection of core concepts that are presented in a way to meet the needs of all your students. As the teacher, you are in the best position to design a biology course that meets the needs of individual students and classes, sets the pace at which the content is covered, and determines what material should be given the most emphasis. To assist you in planning the course, the following Course Planning Guide is provided.

Biology: The Dynamics of Life may be used in a full-year, two-semesters program that is comprised of 180 periods of approximately 45 minutes each. This type of schedule is represented in the table under the heading of Single-Class Scheduling. This table also outlines a plan under the heading Block Scheduling for schools that use a block scheduling system. With block scheduling, it is assumed that the course will be taught for 90 class periods of approximately 90 minutes each.

Please remember that the planning guide is provided as an aid in planning the best course for your students. Use this guide in relation to your curriculum and the ability levels of the classes you teach.

Course Planning Guide for *Biology: The Dynamics of Life*

Chapter/Section	Single-Class Scheduling (180 days)	Block Scheduling (90 days)
1 Biology: The Study of Life	3	1 ½
1.1 What Is Biology?	½	—
1.2 The Methods of Biology	1	½
1.3 The Nature of Biology	1	½
Chapter Assessment	½	½
BioDigest: What Is Biology?	½	½
2 Principles of Ecology	5	2 ½
2.1 Organisms and Their Environment	2	1
2.2 Nutrition and Energy Flow	2	1 ½
Chapter Assessment	1	1
3 Communities and Biomes	5	2 ½
3.1 Communities	2	1
3.2 Biomes	2	1
Chapter Assessment	2	½
4 Population Biology	5	2
4.1 Population Dynamics	3	1
4.2 Human Population Growth	1	½
Chapter Assessment	1	½
5 Biological Diversity and Conservation	5	2
5.1 Vanishing Species	2	1
5.2 Conservation of Biodiversity	2	½
Chapter Assessment	1	½
BioDigest: Ecology	½	½

Chapter/Section	Single-Class Scheduling (180 days)	Block Scheduling (90 days)
6 The Chemistry of Life	5	2
6.1 Atoms and Their Interactions	1	½
6.2 Water and Diffusion	1	½
6.3 Life Substances	2	½
Chapter Assessment	1	½
7 A View of the Cell	6	3
7.1 The Discovery of Cells	1	½
7.2 The Plasma Membrane	2	1
7.3 Eukaryotic Cell Structure	2	1
Chapter Assessment	1	½
8 Cellular Transport and the Cell Cycle	6	3 ½
8.1 Cellular Transport	2	1
8.2 Cell Growth and Reproduction	2	2
8.3 Control of the Cell Cycle	1	—
Chapter Assessment	1	½
9 Energy in a Cell	6	3
9.1 ATP in a Molecule	1	½
9.2 Photosynthesis: Trapping the Sun's Energy	2	1
9.3 Getting Energy to Make ATP	2	1
Chapter Assessment	1	½
BioDigest: The Life of a Cell	½	1
10 Mendel and Meiosis	6	4
10.1 Mendel's Laws of Heredity	3	2
10.2 Meiosis	2	1 ½
Chapter Assessment	1	½

Course Planning Guide

Chapter/Section	Single-Class Scheduling (180 days)	Block Scheduling (90 days)
25 What Is an Animal?	3	2
25.1 Typical Animal Characteristics	1	1/2
25.2 Body Plans and Adaptations	1	1
Chapter Assessment	1	1/2
26 Sponges, Cnidarians, Flatworms, and Roundworms	4	—
26.1 Sponges	1/2	—
26.2 Cnidarians	1	—
26.3 Flatworms	1	—
26.4 Roundworms	1/2	—
Chapter Assessment	1	—
27 Mollusks and Segmented Worms	3	1 1/2
27.1 Mollusks	1	1/2
27.2 Segmented Worms	1	1/2
Chapter Assessment	1	1/2
28 Arthropods	3	1 1/2
28.1 Characteristics of Arthropods	1	1/2
28.2 Diversity of Arthropods	1	1/2
Chapter Assessment	1	1/2
29 Echinoderms and Invertebrate Chordates	3	1
29.1 Echinoderms	1	1/2
29.2 Invertebrate Chordates	1	—
Chapter Assessment	1	1/2
BioDigest: Invertebrates	1/2	2
30 Fishes and Amphibians	4	1 1/2
30.1 Fishes	1 1/2	1/2
30.2 Amphibians	1 1/2	1/2
Chapter Assessment	1	1/2
31 Reptiles and Birds	4	1 1/2
31.1 Reptiles	1 1/2	1/2
31.2 Birds	1 1/2	1/2
Chapter Assessment	1	1/2
32 Mammals	4	1 1/2
32.1 Mammal Characteristics	1	1
32.2 Diversity of Mammals	2	—
Chapter Assessment	1	1/2
33 Animal Behavior	3	—
33.1 Innate Behavior	1	—
33.2 Learned Behavior	1	—
Chapter Assessment	1	—

Chapter/Section	Single-Class Scheduling (180 days)	Block Scheduling (90 days)
BioDigest: Vertebrates	1/2	2
34 Protection, Support, and Locomotion	5	3
34.1 Skin: The Body's Protection	1	1/2
34.2 Bones: The Body's Support	2	1
34.3 Muscles for Locomotion	1	1
Chapter Assessment	1	1/2
35 The Digestive and Endocrine Systems	4	2
35.1 Following Digestion of a Meal	1	1
35.2 Nutrition	1/2	—
35.3 The Endocrine System	1 1/2	1/2
Chapter Assessment	1	1/2
36 The Nervous System	6	1 1/2
36.1 The Nervous System	2	1
36.2 The Senses	1	—
36.3 The Effects of Drugs	1	—
Chapter Assessment	1	1/2
37 Respiration, Circulation, and Excretion	5	3
37.1 The Respiratory System	1 1/2	1
37.2 The Circulatory System	1 1/2	1
37.3 The Urinary System	1	1/2
Chapter Assessment	1	1/2
38 Reproduction and Development	6	3
38.1 Human Reproductive Systems	2	1 1/2
38.2 Development Before Birth	2	1
38.3 Birth, Growth, and Aging	1	—
Chapter Assessment	1	1/2
39 Immunity from Disease	4	—
39.1 The Nature of Disease	1	—
39.2 Defense Against Infectious Diseases	2	—
Chapter Assessment	1	—
BioDigest: The Human Body	1/2	3
Total Sessions	**180**	**90**

Managing Activities in the Biology Classroom

The activities in **Biology: The Dynamics of Life** are designed to minimize dangers in the laboratory. Careful planning and preparation as well as being aware of hazards can keep accidents to a minimum. Numerous books and pamphlets are available on laboratory safety with detailed instructions on preventing accidents. Know the rules and be familiar with the common violations that occur. Know the Safety Symbols used in this book (see p. 38T). Know where emergency equipment is stored and how to use it. Practice good laboratory housekeeping and management by observing these guidelines:

Classroom/Laboratory

1. Store chemicals properly. (See p. 39T.)
2. Store equipment properly.
   a. Clean and dry all equipment before storing.
   b. Protect electronic equipment and microscopes from dust, humidity, and extreme temperatures.
   c. Label and organize equipment so that it is accessible.
3. Provide adequate work space.
4. Provide adequate room ventilation.
5. Post safety and evacuation guidelines.
6. Be sure safety equipment is accessible and works.
7. Provide containers for disposing of chemicals, waste products, and biological specimens. Disposal methods must meet local guidelines.
8. Use hot plates as a heat source whenever possible. If burners are used, a central shutoff valve for the gas supply should be available to the teacher. Never use open flames when a flammable solvent is in the same room.

First Day of Class (with students)

1. Distribute and discuss safety rules, safety symbols, and first aid guidelines. Have students refer to Appendix C on pages A7-A8 to review safety symbols and guidelines.
2. Review safe use of equipment and chemicals.
3. Review use and location of safety equipment.
4. Discuss safe disposal of materials and laboratory cleanup policy.

5. Discuss proper laboratory attitude and conduct.
6. Document students' understanding of the preceding points. Have students sign a safety contract and return it.

Before Each BioLab

1. Perform each investigation yourself before assigning it.
2. Arrange the lab in such a way that equipment and supplies are clearly labeled and easily accessible.
3. Have available only equipment and supplies needed to complete the assigned investigation.
4. Review the procedure with students, emphasizing any caution statements or safety symbols that appear.
5. Be sure all students know the proper procedures to follow if an accident should occur.

During the BioLab

1. Make sure the lab is clean and free of clutter.
2. Insist that students wear goggles and aprons.
3. Never allow a student to work alone in the lab.
4. Never allow students to use a cutting device with more than one edge.
5. Students should not point the open end of a heated test tube toward anyone.
6. Remove broken glassware or frayed cords from use. Also clean up any spills immediately. Dilute solutions with water before removing.
7. Be sure all glassware that is to be heated is of a heat-treated type that will not shatter.
8. Remind students that hot glassware looks cool.
9. Prohibit eating and drinking in the lab.

After the BioLab

1. Be sure that the lab is clean.
2. Be certain that students have returned all equipment and disposed of broken glassware and chemicals properly.
3. Be sure that all hot plates and electrical connections are off.
4. Insist that each student wash his or her hands when lab work is completed.

Biology in the Laboratory

Safety Symbols

The **Biology: The Dynamics of Life** program uses safety symbols to alert you and your students to possible laboratory dangers. These symbols are provided in the student text on pages A7-A8 and are explained below. Be sure your students understand each symbol before they begin an activity that displays a symbol.

SAFETY SYMBOLS	HAZARD	EXAMPLES	PRECAUTION	REMEDY
Disposal	Special disposal considerations required	chemicals, broken glass, living organisms such as bacterial cultures, protists, etc.	Plan to dispose of wastes as directed by your teacher.	Ask your teacher how to dispose of laboratory materials
Biological	Organisms or organic materials that can harm humans	bacteria, fungus, blood, raw organs, plant material	Avoid skin contact with organisms or material. Wear dust mask or gloves. Wash hands thoroughly.	Notify your teacher if you suspect contact.
Extreme	Objects that can burn skin by being too cold or too hot	boiling liquids, hot plates, liquid nitrogen, dry ice, all burners	Use proper protection when handling. Remove flammables from area around open flames or spark sources.	Go to your teacher for first aid.
Sharp Object	Use of tools or glassware that can easily puncture or slice skin	razor blade, scalpel, awl, nails, push pins	Practice common sense behavior and follow guidelines for use of the tool.	Go to your teacher for first aid.
Fume	Potential danger to olfactory tract from fumes	ammonia, heating sulfur, moth balls, nail polish remover, acetone	Make sure there is good ventilation and never smell fumes directly.	Leave foul area and notify your teacher immediately.
Electrical	Possible danger from electrical shock or burn	improper grounding, liquid spills, short circuits	Double-check setup with instructor. Check condition of wires and apparatus.	Do not attempt to fix electrical problems. Notify your teacher immediately.
Irritant	Substances that can irritate the skin or mucus membranes	pollen, mothballs, steel wool, potassium permanganate	Dust mask or gloves are advisable. Practice extra care when handling these materials.	Go to your teacher for first aid.
Corrosive	Substances (acids and bases) that can react with and destroy tissue and other materials	acids such as vinegar, hydrochloric acid, hydrogen peroxide; bases such as bleach, soap, sodium hydroxide	Wear goggles and an apron.	Immediately begin to flush with water and notify your teacher.
Toxic	Poisonous substance that can be acquired through skin absorption, inhalation, or ingestion	mercury, many metal compounds, iodine, Poinsettia leaves	Follow your teacher's instructions. Always wash hands thoroughly after use.	Go to your teacher for first aid.
Flammable	Flammable and combustible materials that may ignite if exposed to an open flame or spark	alcohol, powders, kerosene, potassium permanganate	Avoid flames and heat sources. Be aware of locations of fire safety equipment.	Notify your teacher immediately. Use fire safety equipment if applicable.

 **Eye Safety**
This symbol appears when a danger to eyes exists.

 **Clothing Protection**
This symbol appears when substances could stain or burn clothing.

 **Animal Safety**
This symbol appears when safety of live animals and students must be ensured.

Chemical Storage and Disposal

General Guidelines

Be sure to store all chemicals properly. The following are guidelines commonly used. Your school, city, county, or state may have additional requirements for handling chemicals. It is the responsibility of each teacher to become informed of the rules or guidelines in effect in his or her area.

1. Separate chemicals by reaction type. Strong acids should be stored together. Likewise, strong bases should be stored together and should be separated from acids. Oxidants should be stored away from easily oxidized materials, and so on.
2. Be sure all chemicals are stored in labeled containers indicating contents, concentration, source, date purchased (or prepared), any precautions for handling and storage, and expiration date.
3. Dispose of any outdated or waste chemicals properly according to accepted disposal procedures.
4. Do not store chemicals above eye level.
5. Wood shelving is preferable to metal. All shelving should be firmly attached to the wall and should have anti-roll edges.
6. Store only those chemicals that you plan to use.
7. Hazardous chemicals require special storage containers and conditions. Be sure to know which chemicals those are and the accepted practices for your area. Some substances must be stored outside the building.
8. When working with chemicals or preparing solutions, observe the same general safety precautions that you would expect from students. These include wearing an apron and goggles. Wear gloves and use the fume hood when necessary. Students will want to do as you do whether they admit it or not.
9. If you are a new teacher in a particular laboratory, it is your responsibility to survey the chemicals stored there to be sure they are stored properly. If not, they should be disposed of. Consult the rules and laws in your area concerning which chemicals can be kept in your classroom. For disposal, consult up-to-date disposal information from state and federal governments.

Disposal of Chemicals

Local, state, and federal laws regulate the proper disposal of chemicals. These laws should be consulted before chemical disposal is attempted. Although most substances encountered in high school biology can be flushed down the drain with plenty of water, it is not safe to assume that this is always true. Teachers who use chemicals should consult the following book from the National Research Council:

Prudent Practices in the Laboratory. Washington, DC: National Academy Press, 1995. This book is useful and was revised in 1995. Current laws in your area would, of course, supersede the information in this book.

DISCLAIMER

Glencoe Publishing Company makes no claims to the completeness of this discussion of laboratory safety and chemical storage. The material presented is not all-inclusive, nor does it address all of the hazards associated with handling, storing, and disposing of chemicals, or with laboratory management.

Biology in the Laboratory

Preparation of Solutions

It is most important to use safe laboratory techniques when handling all chemicals. Many substances may appear harmless but are, in fact, toxic, corrosive, or very reactive. Always check with the supplier. Chemicals should never be ingested. Be sure to use proper techniques to smell solutions or other agents. Always wear safety goggles and an apron. Observe the following precautions.

1. Poisonous/corrosive liquid and/or vapor. Use in the fume hood. Examples: acetic acid, nitric acid, hydrochloric acid, ammonium hydroxide.
2. Poisonous and corrosive to eyes, lungs, and skin. Examples: acids, limewater, iron(III) chloride, bases, silver nitrate, iodine, potassium permanganate.
3. Poisonous if swallowed, inhaled, or absorbed through the skin. Examples: glacial acetic acid, copper compounds, barium chloride, lead compounds, chromium compounds, lithium compounds, cobalt(II) chloride, silver compounds.
4. Always add acids to water, never the reverse.
5. When sulfuric acid and sodium hydroxide are added to water, a large amount of thermal energy is released. Sodium metal reacts violently with water. Use extra care when handling any of these substances.

Aceto-orcein stain: Dissolve 1 g orcein in 45 mL hot glacial acetic acid in a large beaker. Foaming may result when the orcein is added. Cool the solution and add 55 mL distilled water. Store in a capped bottle. Filter before use.

Alcohol testing solution: Wear goggles, gloves, and an apron. In a fume hood, add 20 g of potassium dichromate powder to a Pyrex beaker. Pour 20 mL concentrated H_2SO_4 into the beaker and stir with a glass stirring rod to dissolve most of the powder. Slowly and carefully pour the solution into 60 mL distilled water in a Pyrex beaker and continue to stir. The solution becomes VERY HOT. Allow to cool. Powder may precipitate out after cooling. Pour only the liquid solution into dropper bottles for student use. Solution has a shelf life of one year.

Baking soda (sodium bicarbonate) solution: To prepare a 0.25% solution, dissolve 0.5 g baking soda (sodium hydrogen carbonate) in 200 mL of water.

Benedict's solution: Dissolve 173 g sodium citrate and 100 g sodium carbonate in 700 mL water over a hot plate. Filter. Dissolve 17.3 g copper sulfate in 100 mL water. Slowly add to the first solution. Add water to a total volume of 1 L.

Bromothymol blue: Add 0.5 g bromothymol blue powder to 500 mL distilled water to make a BTB stock solution. Dilute 40 mL BTB stock solution to 2 L with distilled water. Solution should be bright blue. If not, add one drop of NaOH at a time, swirling to mix. Check color.

Carnoy's fluid: Mix 6 parts absolute alcohol, 3 parts chloroform, and 1 part glacial acetic acid.

Chalkey's solution: Dissolve 1 g sodium chloride, 0.04 g potassium chloride, and 0.06 g calcium chloride in 1 L of distilled water. Dilute the prepared solution by adding 100 mL of solution to 900 mL distilled water.

Cola, dilute solution: Add 1 part cola to 1 part distilled water.

Congo red: Add 0.1 g Congo red powder to 50 mL distilled water.

Cough medicine, dilute: Add 2 mL of cough medicine (syrup) to 98 mL distilled water. Stir before use.

Ethyl alcohol, dilute: Add 2 mL ethyl alcohol to 98 mL distilled water. Stir.

Fertilizer solution: To make a 1% fertilizer solution, mix 1 g 5-10-5 fertilizer with 99 mL water. For a 0.1 % serial dilution, mix 1 mL 1% solution with 9 mL water. For a 0.01% serial dilution, mix 1 mL 0.1% solution with 9 mL water.

Frog testes solution: Slowly add 0.5 mL (10 drops) sodium acetate to 0.5 mL frog testes solution.

Gelatin solution: Soften 1 g gelatin in 20 mL water; then add 80 mL hot, not boiling, water to dissolve. Cool to room temperature before using.

Glucose solution: For 1% glucose solution, dissolve 1 g of glucose in 99 mL water.

Gum arabic solution: Dissolve 1 g gum arabic in 100 mL warm water. Cool to room temperature before use.

Hydrochloric acid (HCl) solution: To make a 10% solution, add 27 mL concentrated hydrochloric acid to 73 mL water while stirring. To make a $0.1M$ solution, add 1 mL concentrated hydrochloric acid to 100 mL water while stirring.

Hydrochloric acid-alcohol solution: Mix 1 part $0.1M$ hydrochloric acid and 1 part absolute alcohol.

Iodine solution/Iodine stain: Dilute 1 part Lugol's solution with 15 parts water.

Lugol's solution: Dissolve 10 g potassium iodide in 100 mL distilled water; then add and dissolve 5 g iodine. Store in dark bottle. Keeps indefinitely.

Methylene blue stain: Dissolve 1.5 g methylene blue in 100 mL ethyl alcohol. Dilute by adding 10 mL of solution to 90 mL water.

Methyl cellulose solution: Add 20 g methyl cellulose to 40 mL of boiling distilled water. Let stand for 30 minutes, then add 40 mL distilled water. Stir until uniform. Solution will be very thick.

Pancreatic solution: Blend a pig or sheep pancreas with 150 mL 30% ethyl alcohol. Allow the solution to stand for 24 hours, shaking occasionally. Strain the solution through cheesecloth and then filter. Neutralize with KOH until you get near the end point, then use 0.5% sodium carbonate.

Potassium chloride (KCl) solution: To make a $0.5M$ solution, dissolve 3.73 g of potassium chloride in 60 mL of distilled water, then add distilled water to make 100 mL final volume.

Potassium permanganate: For a $0.02M$ solution of potassium permanganate, dissolve 0.3 g $KMnO_4$ in 100 mL water.

Ringer's solution for frogs: To prepare Ringer's solution for frogs, add the following to 1 L of distilled water: 0.14 g KCl, 6.50 g NaCl, 0.12 g $CaCl_2$, 0.20 g $NaHCO_3$

Salmon testes solution: Add 0.5 mL sodium acetate to 0.5 mL salmon testes solution.

Salt solution: For a 3.5% salt solution that simulates the concentration of ocean water, dissolve 35 g salt in 965 mL water. For a 1% solution, dissolve 1 g of salt in 99 mL of water. For a 3% solution, dissolve 3 g of salt in 97 mL of water. For a 5% solution, dissolve 5 g of salt in 95 mL of water. For a 6% solution, dissolve 6 g of salt in 94 mL of water.

Sodium hydroxide (NaOH) solution: To make a 1% solution, dissolve 1 g NaOH in 99 mL of water. For a 0.04% serial dilution, mix 4 mL 1% solution with 96 mL water.

Starch solution: Make a 1% solution by stirring a slurry of 1 g cornstarch and 50 mL cold water into 1 L boiling water. Cool before using.

Sterile pond water: Filter pond water and place it in flat pans. Boil for 15 minutes. Allow to cool before using.

Sucrose solution: For a 1% sucrose solution, dissolve 1 g sucrose in 99 mL water. For a 2% sucrose solution, dissolve 2 g sucrose in 98 mL water. For a 5% sucrose solution, dissolve 5 g sucrose with 95 mL water. For a 10% sucrose solution, dissolve 10 g of sucrose in 90 mL water. For a 20% sucrose solution, dissolve 20 g of sucrose in 80 mL of water. For a 30% sucrose solution, dissolve 30 g of sucrose in 70 mL of water. For a 40% sucrose solution, dissolve 40 g of sucrose in 60 mL of water.

Sugar solution: Add 1 tablespoon of sugar to 1 cup of warm water in a deep jar or flask. Stir to dissolve.

Tetrazolium solution: Dissolve 1 g of 2,3,5-triphenyl tetrazolium chloride in 100 mL of water. Store in dark glass bottle.

Tobacco solution: Grind tobacco from one cigarette into a fine powder. Mix the powder with 100 mL of a 1% glucose solution.

Urine (artificial) solutions: *Normal:* Add 1 tsp. of salt and 4 drops of yellow food coloring to 500 mL of tap water. Stir to dissolve. *Abnormal:* Add 1 tsp. of salt, 2 tsp. of glucose or honey, and 4 drops of yellow food coloring to 500 mL of tap water. Stir to dissolve.

Yeast culture: Add 1/5 package dry baker's yeast to 200 mL distilled water.

Biology in the Laboratory

Equipment and Materials List

These easy-to-use tables of equipment and consumable materials can help you prepare for your biology classes for the year. Quantities listed for BioLabs, Alternative Labs, and MiniLabs are the maximum quantities you will need for one student group for the year. The pages on which each item is used are listed in parentheses after the quantities. Refer to the *Chapter Organizer* in front of each chapter for a list of equipment and materials used for each laboratory activity in the chapter.

Non-Consumables

Item	BioLab	MiniLab	Alternative Lab
aquarium	1 (p. 706)		1 (p. 822)
balance, laboratory	1 (pp. 780, 860)		1 (pp. 80, 202, 628, 714, 762)
ball, golf	1 (p. 860)		
ball, Ping-Pong	1 (p. 860)		
beaker, 100-mL		4 (pp. 38, 155, 204, 242)	2 (p. 202)
beaker, 250-mL	5 (pp. 60, 560, 706, 780, 800)	2 (pp. 23, 126, 654, 818)	4 (pp. 240, 714)
beaker, 400-mL	1 (p. 168)		
beaker, 1000-mL	1 (p. 244)		1 (p. 80)
brush, camel hair	1 (p. 734)		
Bunsen burner			1 (p. 414)
clock or watch with second hand	1 (pp. 26, 168, 244, 940, 1020)	1 (pp. 159, 900, 1013)	1 (p. 870)
coverslip	5 (pp. 60, 194, 538, 646, 678, 800)	10 (pp. 6, 75, 177, 187, 260, 379, 522, 527, 546, 606, 629, 640, 695, 703, 871)	5 (pp. 188, 390, 576)
culture dish	4 (pp. 754, 834)	1 (p. 719)	
dichotomous key for trees		1 (p. 456)	
dichotomous key transparency		1 (p. 746)	
dropper	5 (pp. 60, 88, 194, 538, 646, 678, 706, 754, 800, 904, 996)	4 (pp. 6, 14, 75, 260, 522, 527, 546, 609, 629, 640, 695, 703, 719, 871, 1019, 1038)	4 (pp. 188, 390, 950, 1014)
earthworm cross-section diagrams		1 (p. 750)	
earthworm longitudinal diagrams		1 (p. 750)	
flashlight or penlight	1 (pp. 754, 834)		
flask, 250-mL			1 (p. 1014)
flower pots	2 (p. 280)		
forceps	1 (pp. 168, 194, 336, 512)	1 (pp. 159, 546, 606, 609, 763, 774, 797, 871)	1 (pp. 188, 576, 714, 1064)
glassware, laboratory, assorted			1 (p. 468)
graduated cylinders	1 (pp. 560, 780)	1 (p. 126)	3 (pp. 12, 40, 202, 950)
hand lens	1 (pp. 280, 678, 754)	1 (pp. 333, 853, 925)	1 (p. 576)
hole punch		1 (p. 406)	
hot plate	1 (pp. 168, 560)		1 (p. 240)
incubator, 37°C	1 (p. 512)		1 (p. 1064)
inoculation loop			1 (p. 414)
jar with lid			5 (p. 894)
jar, glass	3 (p. 88)	1 (p. 96)	
knife, kitchen	1 (p. 168)		1 (p. 202)
lamp, 60-watt	9 (pp. 280, 834)		1 (pp. 504, 744, 842)
lamp, 150-watt with reflector	1 (p. 244)		

Item	BioLab	MiniLab	Alternative Lab
light source, fluorescent	1 (p. 336)		
meterstick		1 (pp. 384, 980)	
microscope	1 (pp. 60, 88, 194, 220, 538, 646, 678, 800, 996)	1 (pp. 6, 69, 75, 177, 187, 215, 260, 379, 506, 522, 527, 546, 606, 629, 640, 677, 695, 703, 719, 732, 788, 869, 871, 964, 1038, 1067)	1 (pp. 188, 332, 358, 390, 576, 612, 702, 796, 924)
microscope slides	5 (pp. 60, 88, 194, 538, 646, 678, 734, 800, 996)	10 (pp. 6, 14, 75, 177, 187, 260, 379, 522, 527, 546, 606, 629, 640, 695, 703, 797, 871, 1019, 1038)	5 (pp. 188, 358, 390, 576)
objects, miscellaneous small			1 (p. 870)
overhead projector		1 (p. 746)	
paintbrush	2 (p. 336)		
pan, glass, shallow	1 (p. 754)		
pennies	100 (p. 394)		20 (p. 762)
petri dish	3 (pp. 26, 512, 706, 734, 800)	1 (p. 654)	2 (pp. 100, 322, 414, 576, 714, 1064)
petri dish, plastic	1 (p. 904)	1 (p. 890)	1 (p. 762)
pH color chart		1 (p. 155)	
pipette, glass			1 (p. 100)
plate		1 (p. 546)	
protractor	1 (pp. 446, 860)		
razor blade, single-edged	1 (pp. 336, 646, 678, 734)	1 (pp. 159, 677)	
refrigerator			1 (p. 164)
ring stand			1 (p. 744)
ruler, metric	1 (pp. 26, 446, 474, 512, 538, 646, 754, 860, 966, 1020)	1 (pp. 159, 415, 490, 797, 937)	1 (pp. 442, 744, 842, 980)
scissors	1 (pp. 308, 362, 538, 904, 1048)		1 (pp. 272, 294, 322, 576)
step stool	1 (p. 940)		
stereomicroscope	1 (pp. 706, 734, 754, 834, 904)	1 (pp. 577, 797)	1 (pp. 534, 714)
stirring rod		2 (p. 23)	6 (pp. 202, 390)
stoppers, one-hole, with glass tube	3 (p. 560)		
stopwatch	1 (p. 940)		1 (p. 822)
syringe with needle	1 (p. 800)		
terrarium and cover			1 (pp. 870, 894)
test tube	5 (pp. 560, 800)	2 (pp. 56, 654)	4 (pp. 390, 414, 950)
test tube, large	3 (pp. 560)	2 (p. 242)	
test tube, screw-top			4 (p. 504)
test-tube rack	1 (p. 560)		
thermometer	1 (pp. 168, 280, 560, 754, 834)	1 (p. 818)	1 (p. 842)
tongs	1 (p. 168)		
towel, cloth			1 (p. 762)
tray, clear plastic	1 (p. 780)		
tray, plastic		1 (p. 126)	
trowel, garden			1 (p. 628)
tubing, rubber, 10-cm length	3 (p. 560)		
turkey baster	1 (p. 706)		
washers, metal	2 (p. 244)	3 (p. 242)	
watch glass		1 (p. 719)	
watering bottle, plant	1 (p. 280)		
weights, small	2 (p. 940)		

Biology in the Laboratory

Consumables

Item	BioLab	MiniLab	Alternative Lab
Accent seasoning			1 g (p. 504)
adding machine tape		1 (p. 384)	
aged tap water	25 mL (p. 996)		
aluminum foil, 30 cm × 30 cm	1 (p. 560)		4 (pp. 202, 546)
ammonia solution, household		5 mL (p. 155)	
antacid tablet		1 (p. 56)	
antibiotic disks (3 types)	3 (p. 512)		3 (p. 1064)
apples, fresh		4 (p. 1060)	1 (p. 894)
apples, rotting		1 (p. 1060)	
bag, paper, small	1 (pp. 108, 422)	1 (p. 554)	
bag, plastic sandwich		1 (p. 204)	6 (p. 40)
bag, plastic, zipper-lock		8 (pp. 38, 546, 677, 1060)	4 (p. 122)
balloon, round	1 (p. 1020)		
banana		1 (p. 96)	
bar codes			5 (p. 358)
bean seeds	175 (p. 108)	5 (p. 677)	
beans, kidney, canned			10 (p. 674)
beans, kidney, dried			10 (p. 674)
beans, pinto	50 (p. 422)		60 (p. 40)
beans, white navy	50 (p. 422)		
bird seed, assorted varieties		500 g (p. 856)	
bottle, plastic milk, 1-gallon		3 (p. 856)	
box, cardboard			100 cm x 50 cm (p. 294)
Brassica rapa seeds	40 (p. 336)		
bread, bakery, fresh		2 slices (p. 546)	
brine shrimp eggs	100 (p. 780)		
can, small metal			2 (p. 842)
cardboard	30 cm × 100 cm (p. 860)		10 cm × 10 cm (p. 546)
carrot		1 (p. 654)	
celery stalk		1 (p. 629)	
cellophane, assorted colors	1 (p. 244)		
clay, garden			1 kg (p. 80)
clay, modeling assorted colors	100 g (p. 860)	500 g (pp. 234, 274)	30 g (p. 842)
cloth squares, 30 cm × 30 cm			3 (p. 80)
coffee, caffeinated	1 mL (p. 996)		30 mL (p. 12)
coffee, decaffeinated			30 mL (p. 12)
cola	1 mL (p. 996)		
colored pencils (4 colors)	1 (pp. 362, 422, 678, 966)	1 (p. 333)	1 (p. 1030)
conifer branches, assorted	3 (p. 618)		
conifer cones, assorted	3 (p. 618)		
conifer twigs, assorted	3 (p. 618)		
corn seeds		5 (p. 677)	60 (p. 40)
cosmetics, assorted		1 (p. 14)	
cotton balls		1 (p. 1060)	
cotton swabs	3 (pp. 512, 754)		2 (pp. 546, 1064)
cough medicine	2 mL (p. 996)		
data sheet			1 (p. 1030)
detergent, liquid		5 mL (p. 155)	
diatomaceous earth		1 g (p. 379)	
dinosaur eggs, artificial, from cereal		2 (p. 23)	
egg, hard-boiled	1 (p. 860)		
envelopes			4 (p. 358)

Item	BioLab	MiniLab	Alternative Lab
filter paper disk, sterile	3 (p. 512)		
food coloring, yellow		1 mL (p. 1019)	
garlic clove		1 (p. 654)	
gelatin dessert mix, 6-oz. box			1 (p. 164)
glucose test paper strips		6 (p. 1019)	6 (p. 950)
glue		5 mL (p. 456)	5 mL (p. 272)
grapes			10 (p. 164)
graph paper, sheet of	4 (pp. 394, 940, 966)	1 (p. 1042)	1 (p. 1030)
grass seed		1 g (p. 609)	
gravel, aquarium, black and white			2 kg (p. 822)
hair, human		1 (p.177)	2 (p. 358)
ice	400 g (pp. 168, 754)		200 g (p. 744)
ice cubes	2 (p. 560)	2 (p. 818)	
index cards	2 (p. 538)	1 (p. 925)	
ink pad		1 (p. 925)	
ink, red			1 mL (p. 358)
juice, apple		15 mL (p. 242)	
juice, lemon		5 mL (p. 155)	
juice, red grape			1 mL (p. 358)
labels	12 (pp. 88, 336, 706, 734, 780)	2 (p. 38)	15 (pp. 12, 40, 122, 202, 240, 322, 576, 674)
Lactaid liquid			1 mL (p. 950)
lima beans		10 (p. 609)	
macaroni			100 g (p. 546)
marking pens	1 (pp. 108, 512)		1 (pp. 202, 1064)
matchbox			4 (p. 894)
milk			15 mL (p. 950)
mold spores			1 g (p. 546)
mushroom, fresh		4 (p. 554)	
mustard seeds			1 g (p. 576)
nail polish (4 colors)			1 (p. 894)
oat cereal flakes			10 g (pp. 534, 546, 894)
onion		1 (p. 187)	
orange			1 (p. 164)
owl pellet		1 (p. 871)	
paper clips			10 (pp. 240, 272, 980)
paper cup, small			15 (pp. 40, 122, 164, 546, 674)
paper napkins			5 (p. 100)
paper towels	2 (p. 754)	8 (pp. 38, 677, 871, 890)	5 (pp. 40, 122, 322)
paper, black, sheet of	1 (p. 754)	3 (pp. 406, 890)	12 (pp. 240, 576, 744, 822, 842)
paper, construction (5 colors)	1 (p. 308)		
paper, heavy, sheet of	1 (p. 1048)	1 (p. 456)	
paper, notebook, sheet of	4 (pp. 26, 108, 1074)	10 (pp. 6, 22, 23, 96, 105, 527, 1067)	1 (p. 100)
paper, tracing, sheet of	1 (p. 1048)		
paper, white, sheet of	1 (p. 362)	2 (pp. 406, 554)	1 (p. 842)
peas		10 (p. 609)	
personal hygiene products		1 (p. 14)	
pH paper strip		6 (p. 155)	10 (p. 390)
pineapple, canned chunk			1 (p. 164)
pineapple, fresh			1 (p. 164)
pipette, plastic disposable		1 (p. 242)	
plastic wrap, 30 cm × 30 cm	1 (p. 88)		2 (pp. 202, 546)

Item	BioLab	MiniLab	Alternative Lab
pond water	200 mL (p. 88)	300 mL (p. 527)	
pond water, sterile	300 mL (p. 60)		
poster board, 30-cm × 30-cm			9 (p. 272)
potato	1 (p. 168)	2 (pp. 159, 654)	1 (p. 202)
puzzle, paper		1 (p. 900)	
radish seeds			1 g (p. 100)
rice	10 g (p. 88)	1 g (p. 609)	
rubber bands	1 (p. 904)	1 (p. 96)	
rye seeds		1 g (p. 609)	
sample keys from guidebooks	1 (p. 474)		
sand			2 kg (pp. 80, 870, 894)
sandpaper, sheet of	1 (p. 754)		1 (p. 744)
sawdust			500 g (p. 870)
sea urchin eggs		1 mL (p. 1038)	
sea urchin sperm		1 mL (p. 1038)	
sea water	500 mL (p. 800)		
seedling flat	2 (p. 280)		
seeds		40 (p. 38)	40 (p. 122)
shampoo		5 mL (p. 155)	
shoebox with lid	1 (p. 394)		
skull diagrams	1 (p. 446)		
soil samples (4 different)			1 (p. 504)
soil, lawn with grass		1 kg (p. 126)	
soil, lawn without grass		1 kg (p. 126)	
soil, potting	1 kg (pp. 280, 336)		1 kg (p. 80)
sponge, cellulose			1 (p. 894)
sponge, synthetic, four types			1 (p. 714)
spoon, plastic	1 (p. 88)		
spring water	150 mL (pp. 734, 904)		
spring water, pasteurized	600 mL (p. 88)		
sterile untreated disks			2 (p. 1064)
straw, drinking	1 (p. 560)	1 (p. 56)	1 (p. 1014)
string	1 m (pp. 244, 860, 1020)		1 m (p. 272)
T-shirt			1 (p. 358)
tape, masking	10 cm (p. 904)		3 m (p. 822)
tape, transparent	10 cm (pp. 308, 362)	10 cm (p. 890)	40 cm (pp. 272, 762, 842, 1064)
tea	1 mL (p. 996)		
thread, heavy			30 cm (p. 762)
tobacco	1 g (p. 996)		
tobacco seeds	20 (p. 280)		20 (p. 322)
toothbrush bristles		10 (p. 695)	
toothpicks	4 (p. 538)	4 (p. 654)	4 (p. 272)
tray, deli, clear plastic			1 (p. 744)
tray, potting	1 (p. 336)		
twist ties		5 (pp. 204, 274)	2 (p. 80)
unshelled peanuts		30 (p. 415)	
wax marking pencil	1 (pp. 560, 706, 734)	1 (pp. 14, 1019)	1 (pp. 414, 714, 744)
wax paper, 30 cm × 30 cm	1 (p. 168)		2 (pp. 164, 674)
wheat seeds		1 g (p. 522)	
wire		2 m (p. 856)	
wire mesh, fine, 15 cm × 15 cm		1 (p. 96)	
yarn, skein of			3 (pp. 272, 294)
yeast, baker's		5 g (p. 242)	

Living Organisms

Item	BioLab	MiniLab	Alternative Lab
Armadillidium (pill bug)	1 (p. 26)	5 (p. 890)	
Bacillus subtilis culture			1 (p. 414)
bacteria cultures	1 (p. 512)		
bess beetles			2 (p. 762)
brine shrimp culture		1 (p. 719)	
Coleus plant			2 (p. 240)
crickets, male and female			5 (p. 894)
Daphnia culture	1 (p. 996)		
Didinium culture	1 (p. 60)		
E. coli culture			1 (p. 1064)
earthworms	2 (p. 754)		
Elodea plant	3 (p. 244)		1 (p. 188)
Equisetum		1 (p. 586)	
Euglena culture	1 (p. 538)		
fern frond		1 (p. 606)	
flower (complete)	1 (p. 678)	1 (p. 260)	
frog eggs, fertilized	10 (p. 834)		
goldfish		1 (p. 818)	
hydra culture		1 (p. 719)	
land snails			2 (p. 744)
lichen		4 (p. 69)	
Marchantia		1 (p. 577)	
marine plankton culture		1 (p. 75)	
mildew sample		1 (p. 6)	
onion plant	1 (p. 646)		
Paramecium culture	1 (pp. 60, 538)	1 (p. 522)	
Physarum polycephalum			1 (p. 534)
planarian culture	1 (p. 734)		
plant leaves		6 (pp. 456, 640)	
rotifer culture		1 (p. 695)	
sea urchins (male and female)	2 (p. 800)		
Snails	1 (p. 904)		
tropical fishes (gourami, catfish, zebra)			10 (p. 822)
vinegar eel culture		1 (p. 703)	
zebra fish	10 (p. 706)		

Chemicals

Item	BioLab	MiniLab	Alternative Lab
2,3,5-triphenyl tetrazolium chloride solution			1 mL (p. 674)
acetone		10 mL (p. 925)	
agar plate, sterile nutrient	1 (p. 512)		1 (p. 1064)
agar, sterile nutrient, tube			3 (p. 414)
agar, streptomycin, tube			1 (p. 414)
alcohol testing solution		1 mL (p. 14)	
alcohol, isopropyl (2-propanol)		1 mL (p. 14)	
baking soda (sodium hydrogen carbonate)			5 g (p. 504)
baking soda solution, 0.25%	50 mL (p. 244)		
bromothymol blue (BTB) solution	30 mL (p. 560)	30 mL (p. 56)	
distilled water			700 mL (pp. 504, 1014)
ethanol, 95%	2 mL (p.996)	5 mL (p. 1060)	100 mL (pp. 240, 1064)

Item	BioLab	MiniLab	Alternative Lab
gelatin solution			5 mL (p. 390)
glucose (corn syrup)	20 mL (p. 560)		10 mL (p. 950)
glycerol (glycerin)		1 mL (p. 606)	
gum arabic solution			3 mL (p. 390)
hydrochloric acid, 0.1M			1 mL (p. 390)
hydrogen peroxide, 3%	1 mL (p. 168)		
iodine solution		150 mL (p. 204)	100 mL (p. 240)
iodine stain		1 mL (p. 609)	
methyl cellulose solution	1 mL (p. 538)		
nail polish remover		10 mL (p. 925)	
NaOH solution, 0.04%			1 mL (p. 1014)
phenolphthalein indicator			1 mL (p. 1014)
potassium chloride solution	1 mL (p. 800)		
potassium permanganate solution		100 mL (p. 159)	
Ringer's solution for frogs	400 mL (p. 834)		
salt solution, 5%		1 mL (p. 640)	
salt solution, 10%		50 mL (p. 38)	
salt, table			15 g (p. 202)
salt, table, non-iodized	30 g (p. 780)		
Schultz liquid plant food			1 mL (p. 504)
starch solution		100 mL (p. 204)	
vinegar		5 mL (p. 155)	60 mL (pp. 122, 504)

Preserved Specimens

Item	BioLab	MiniLab	Alternative Lab
Bacillus subtilus slides	1 (p. 194)		
bacteria slides		1 (p. 506)	
blood cells, red and white slides		1 (p. 1067)	
Branchiostoma californiense (*Amphioxus*)		1 (p. 797)	
conifer leaf cross-section slides			1 (p. 612)
crayfish		1 (p. 763)	
earthworm cross-section slides			1 (p. 702)
Elodea slides	1 (p. 194)		
feather, contour		1 (p. 853)	
feather, down		1 (p. 853)	
fish mitosis slides		1 (p. 215)	
frog blood slides	1 (p. 194)		
frog, adult		1 (p. 830)	
grasshopper, life stage specimens		1 (p. 774)	
human cheek cells, male and female, slides			1 (pp. 188, 332, 358)
human skin slides			1 (p. 924)
human tooth cross-section slides		1 (p. 869)	
Hydra cross-section slides			1 (p. 702)
lancelet slides			1 (p. 796)
Lycopodium		1 (p. 586)	
moth, life stage specimens		1 (p. 774)	
nematode cross-section slides			1 (p. 702)
onion root tip slides	1 (p. 220)		
planarian cross-section slides			1 (p. 702)
porkworm larvae slides		1 (p. 732)	
sea star pedicellariae slides		1 (p. 788)	
sea urchin development slides			1 (p. 796)
shells, marine, assorted		1 (p. 746)	
sponges, sea, four types			1 (p. 714)
tadpole		1 (p. 830)	
thyroid and parathyroid slides		1 (p. 964)	

Suppliers

Equipment Suppliers

American Science & Surplus
3605 Howard St.
Skokie, IL 60076
(800) 934-0722

Bio-Rad Laboratories
2000 Alfred Nobel Dr.
Life Science Group
Hercules, CA 94547
(800) 876-3425
ron_mardigian@bio-rad.com

Carolina Biological Supply Co.
2700 York Road
Burlington, NC 27215
(800) 334-5551
www.carolina.com

Edmund Scientific Company
101 E. Gloucester Pike
Barrington, NJ 08007
(609) 547-3488
scientifics@edsci.com

Fisher Science Education
Educational Materials
  Division
485 S. Frontage Rd.
Burr Ridge, IL 60521
(800) 955-1177
www.fisheredu.com/

Nasco Science
901 Janesville Avenue
P.O. Box 901
Fort Atkinson, WI 53538-
  0901
(800) 558-9595
www.nascofa.com

Nebraska Scientific
3823 Leavenworth St.
Omaha, NE 68105-1180
(800) 228-7117
nescientif@aol.com

PASCO Scientific
10101 Foothills Blvd.
P.O. Box 619011
Roseville, CA 95747
(800) 772-8700
sales@pasco.com

Sargent-Welch/VWR Scientific
  Products
P.O. Box 5229
Buffalo Grove, IL 60089-5229
(800) SAR-GENT
www.SargentWelch.com

WARD's Natural Science Est.
5100 W. Henrietta Road
P.O. Box 92912
Rochester, NY 14692-9012
(800) 962-2660
www.wardsci.com

Audiovisual Distributors

Bullfrog Films
P.O. Box 149
Oley, PA 19547
(800) 543-FROG
www.bullfrogfilms.com

Coronet/MTI Film & Video
2349 Chaffee Dr.
St. Louis, MO 63146
(800) 221-1274
bfaeduc@worldnet.att.net

Discovery Channel
  Education/TLC
7700 Wisconsin Avenue
Bethesda, MD 20814
(301)986-0444
www.school.discovery.com

Films for the Humanities and
  Sciences
P.O. Box 2053
Princeton, NJ 08543
(800) 257-5126

Flinn Scientific
P.O. Box 219
770 N. Raddant Rd.
Batavia, IL 60510
(800) 452-1261
www.flinnsci.com

Frey Scientific, Div. of
  Beckley Cardy
100 Paragon Parkway
Mansfield, OH 44903
(800) 235-3739

Media Design Associates
1093 Albion Rd.
P.O. Box 3189
Boulder, CO 80307-3189
(800) 228-8854
www.indra.com/mediades

National Geographic Society
Educational Services
1145 17th Street, N.W.
Washington, DC 20036
(800) 368-2728

Optical Data School Media
512 Means St., N.W.
Atlanta, GA 30318
(800) 524-2481
www.opticaldata.com

Scholastic, Inc.
555 Broadway
New York, NY 10012-3999
(800) 325-6149
www.scholastic.com

Time-Life Education
P.O. Box 85026
Richmond, VA 23285-5026
(800) 449-2010

Videodiscovery
1700 Westlake Ave., N.
Suite 600
Seattle, WA 98109-3012
(800) 548-3472
www.videodiscovery.com

Software Distributors

Boreal Laboratories, Ltd.
399 Vansickle Rd.
St. Catharines, Ontario,
  L2S 3T4
Canada
(800) 387-9393
boreal@niagara.com

Cross Educational Software
508 E. Kentucky Ave.
P.O. Box 1536
Ruston, LA 71270
(800) 768-1969
MarkCross@aol.com

Educational Activities, Inc.
1937 Grand Ave.
Baldwin, NY 11510
(800) 645-3739
www.edact.com

IBM Global Education
4111 Northside Parkway
Atlanta, GA 30301-2150
(800) 426-4968
www.solutions.ibm.com/k12

Intellimation
130 Cremora Drive
P.O. Box 1922
Santa Barbara, CA 93116-
  1922
(800) 346-8355

J. Weston Walch, Publisher
321 Valley St.
P.O. Box 658
Portland, ME 04104-0658
(800) 341-6094
www.walch.com

TEK-GEAR, Inc.
30153 Arena Dr.
Evergreen, CO 80439
(800) 677-5609
leapsys@aol.com

Scholastic, Inc.
555 Broadway
New York, NY 10012-3999
(800) 325-6149
www.scholastic.com

Science Kit and Boreal
  Laboratories
777 East Park Dr.
Tonawanda, NY 14150
(800) 828-7777
www.sciencekit.com

Sunburst Communications
101 Castleton St.
Pleasantville, NY 10570
(800) 321-7511
www.SUNBURST.com

Bibliography

Student Readings

Attenborough, D., *The Private Life of Plants*, Boston: Little, Brown and Company, 1995.

Berrill, Michael, *The Plundered Seas: Can the World's Fish Be Saved?* Vancouver: Greystone Books, 1997.

Caldwell, Mark, "How Does a Single Cell Become a Whole Body?" *Discover*, November 1992.

Cohen, Joel E., "How Many People Can Earth Hold?" *Discover*, November 1992.

Crews, David, "Animal Sexuality," *Scientific American*, January 1994.

Crow, James F., "The Odds of Losing at Genetic Roulette," *Nature*, January 28, 1999.

Duellman, William E., "Reproductive Strategies of Frogs," *Scientific American*, July 1992.

Freedman, David H., "The Aggressive Egg," *Discover*, June 1992.

Gould, Stephen J., "The Reversal of Hallucigenia," *Natural History*, January 1992.

Hanson, Betsy, "Message in a Barrel," *Discover*, June 1992.

Howlett, Rory, "Monkey Business in the Aquarium," *Nature*, January 21, 1999.

Jaroff, Leon, "Happy Birthday Double Helix," *Time*, March 15 1993.

Kunzig, Robert, "Climbing Through the Brain," *Discover*, February 1998.

Madigan, Michael T. and Marrs, Barry L., "Extremophiles," *Scientific American*, April 1997.

Martin, Glen, "Spring Fever," *Discover*, October 1992.

Ravven, Wallace, "In the Beginning," *Discover*, October 1992.

Schiebinger, L., "The Loves of the Plants," *Scientific American*, February 1996.

Schrof, Joanne M., "Miracle Vaccines," *U.S. News & World Report*, November 23, 1998.

Shimeck, Ronald L., "Sex Among the Sessile," *Natural History*, March 1987.

"Special Issue 1995 The Year in Science," *Discover*, January 1996.

Sword, Gregory A., "Density-dependent Warning Coloration," *Nature*, January 21, 1999.

Taubes, Gary, "Malarial Dreams," *Discover*, March 1998.

Vickers-Rich, P., and Hewilt-Rich, T., "Australia's Polar Dinosaurs," *Scientific American*, July 1993.

Watson, James D., *The Double Helix*, New York: Antheneum, 1968.

Wilson, Edward O., "The Diversity of Life," *Discover*, September 1992.

Zimmer, Carl, "Masters of an Ancient Sky," *Discover*, February 1994.

Teacher Readings

Allard, David W., and Royce L. Granberry, "Osmosis Revisited," *The American Biology Teacher*, November/December 1992.

Beardsley, Tim, "Teaching Real Science," *Scientific American*, October 1992.

Calver, M.C. and Wooler, R.D., "A Non-Destructive Laboratory Exercise for Teaching Some Principles of Predation," *Journal of Biological Education*, Winter 1998.

Carson, Rachel, *Silent Spring*, New York: Houghton Mifflin Company, 1962.

Chiras, Daniel D., "Teaching Critical Skills in the Biology and Environmental Science Classrooms," *The American Biology Teacher*, November/December 1992.

DiSpezio, Michael A., "Retroviruses—Gaining an Understanding," *The Science Teacher*, October 1990.

Eigen, M., "Viral Quasispecies," *Scientific American*, July 1993.

Fredrickson, James K and Onstott, Tullis C., "Microbes Deep Inside the Earth," *Scientific American*, October 1996.

Glasgow, Neal A., *New Curriculum for New Times: A guide to student-centered, problem-based learning*, Thousand Oaks, CA: Corwin Press, Inc., 1997.

Gould, Stephen J., and Niles Eldredge, *Punctuated Equilibria: An Alternative to Phyletic Gradualism*, San Francisco: Freeman, Cooper, and Co., 1972.

Guilfoile, P., "Wrinkled Peas & White-Eyed Fruit Flies: The Molecular Basis of Two Classical Genetic Traits," *The American Biology Teacher*, February 1997.

Johanson, D., and Blake Edgar, *From Lucy to Language*, New York: Simon & Schuster Editions, 1996.

Kreuzer, H. and Massey, A., *Recombinant DNA and Biotechnology: A Guide for Teachers*, Washington: ASM Press, 1996.

Levinton, Jeffrey S., "The Big Bang of Animal Evolution," *Scientific American*, November 1992.

Levy, Stuart B., "The Challenge of Antibiotic Resistance," *Scientific American*, March 1998.

Okebukola, Peter A., "Concept Mapping with a Cooperative Learning Flavor," *The American Biology Teacher*, April 1992.

Pittman, Kim M., "Student Generated Analogies: Another Way of Knowing?" *Journal of Research in Science Teaching*, January 1999.

Roach, Mary, "Secrets of the Shamans," *Discover*, November 1993.

Stern, K., *Introductory Plant Biology*, Boston: The McGraw-Hill Companies, Inc., 1997.

Wilson, E.O., *The Diversity of Life*, Cambridge: Belknap Press, 1992.

BIOLOGY
The Dynamics of Life

AUTHORS

Alton Biggs, M.S. • **Kathleen Gregg, Ph.D.**
Whitney Crispen Hagins, M.A., M.A.T. • **Chris Kapicka, Ph.D.**
Linda Lundgren, M.S. • **Peter Rillero, Ph.D.** • **National Geographic Society**

Glencoe
McGraw-Hill

New York, New York   Columbus, Ohio   Woodland Hills, California   Peoria, Illinois

A GLENCOE PROGRAM
BIOLOGY: THE DYNAMICS OF LIFE

Student Edition
Teacher Wraparound Edition
Laboratory Manual, SE and TE
Reinforcement and Study Guide,
*   SE and TE*
Content Mastery, SE and TE
Section Focus Transparencies and Masters
Reteaching Skills Transparencies and Masters
Basic Concepts Transparencies and Masters
BioLab and MiniLab Worksheets
Concept Mapping
Chapter Assessment
Critical Thinking/Problem Solving
Spanish Resources
Tech Prep Applications

Biology Projects
ExamView® Pro Software
Lesson Plans
Block Scheduling
Inside Story Poster Package
English/Spanish Audiocassettes
MindJogger Videoquizzes
Interactive CD-ROM
Videodisc Program
Glencoe Science Professional Series:
*   Exploring Environmental Issues*
*   Performance Assessment in the Biology Classroom*
*   Alternate Assessment in the Science Classroom*
*   Cooperative Learning in the Science Classroom*
*   Using the Internet in the Science Classroom*

Glencoe/McGraw-Hill

A Division of The McGraw-Hill Companies

THE PRINCETON REVIEW

The Test-Taking Tips in this book were written by The Princeton Review, the nation's leader in test preparation. Through its association with McGraw-Hill, The Princeton Review offers the best way to help students excel on standardized assessments.

The Princeton Review is not affiliated with Princeton University or Educational Testing Service.

Send all inquiries to:
Glencoe/McGraw-Hill
8787 Orion Place
Columbus, OH 43240

ISBN 0-07-825925-8
Printed in the United States of America.
2 3 4 5 6 7 8 9 10 071 08 07 06 05 04 03 02 01

Alton Biggs teaches biology at Allen High School, Allen, Texas. He has a B.S. in Natural Sciences and an M.S. in Biology from Texas A & M University-Commerce. Mr. Biggs received NABT's Outstanding Biology Teacher Award for Texas in 1982 and 1995, he was president of the Texas Association of Biology Teachers in 1985, and in 1992 was the president of the National Association of Biology Teachers.

Kathleen Gregg is a lecturer in anatomy and neuroanatomy at The Ohio State University, College of Medicine, Columbus, Ohio. She has a B.A. in Biology and Chemistry from Point Loma Nazarene College, and a Ph.D. in Neuroscience from The Ohio State University. In 1996, she was awarded an N.I.H. predoctoral training grant in Development, Plasticity and Regeneration.

Whitney Crispen Hagins teaches biology at Medfield High School in Medfield, Massachusetts. She has a B.A. and an M.A. in Biological Sciences from Mount Holyoke College and an M.A.T. from Duke University. In 1991, Ms. Hagins was awarded a research fellowship by the American Society of Clinical Investigation. Since 1997, she has been an instructor for Project STIR Technology Institute.

Chris Kapicka is a biology professor at Northwest Nazarene College, Nampa, Idaho, and does collaborative heart research with the Veteran's Administration Hospital in Boise, Idaho. She has a B.S. in Biology from Boise State University, an M.S. in Microbiology from Washington State University, and a Ph.D. in Cell Physiology and Pharmacology from the University of Nevada-Reno.

Linda Lundgren is a research associate in the Mathematics, Science, and Technology Program at The University of Colorado at Denver. She taught biology at Bear Creek High School, Lakewood, Colorado, for 10 years. Ms. Lundgren has a B.A. in Journalism and Zoology from the University of Massachusetts and an M.S. in Zoology from The Ohio State University. In 1991, she was named Colorado Science Teacher of the Year.

Peter Rillero is a professor of science education at Arizona State University West in Phoenix. He has an M.A. in Science Education from Teachers College, Columbia University; an M.A. in Biology from City University of New York; and a Ph.D. in Science Education from The Ohio State University. In 1998, Dr. Rillero was awarded a Fulbright fellowship to teach science methods at Akureyri University in Iceland.

National Geographic Society, founded in 1888 for the increase and diffusion of geographic knowledge, is the world's largest nonprofit scientific and educational organization. The School Publishing division supports the Society's mission by developing innovative educational programs—ranging from print materials to multimedia programs, including CD-ROMS, videodiscs, and software.

Contributing Authors
Albert Kaskel, Evanston Township High School, Evanston, IL
Rebecca Johnson, Science Writer, Sioux Falls, SD

Contents in Brief

FOUNDATION

Contents in Brief

ENRICHMENT

v

Contents

Contents

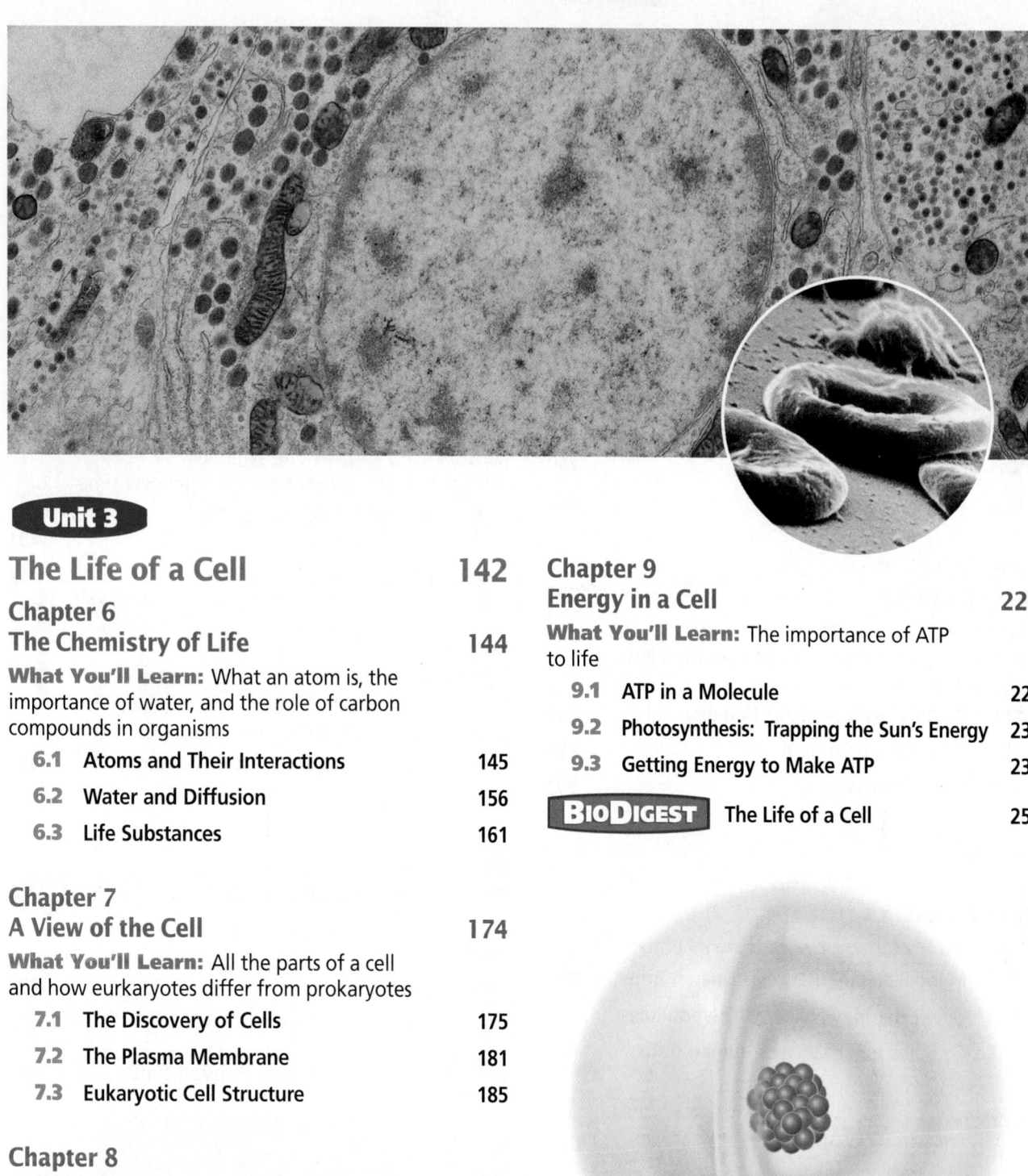

Contents

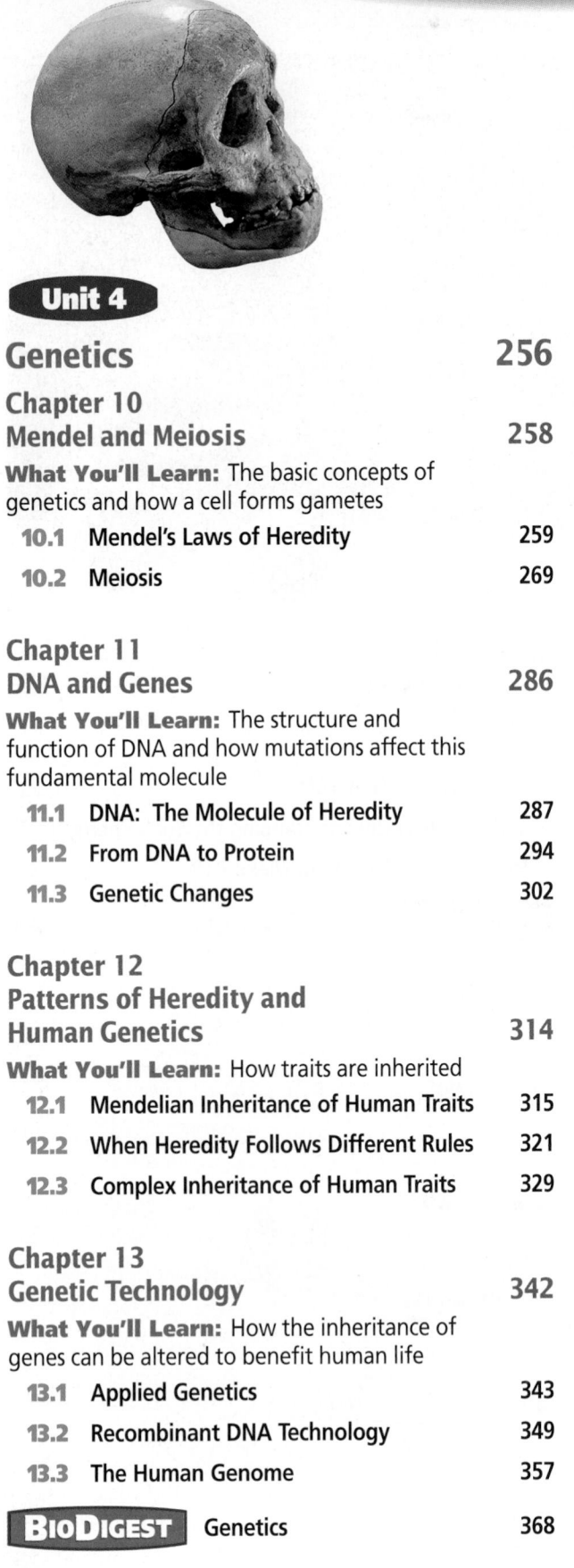

Contents

Contents

Contents

Appendices

BioLab

Working in the lab is often the most enjoyable part of biology. BioLabs give you an opportunity to act like a biologist and develop your own plans for studying a question or problem. Whether you're designing experiments or following well-tested procedures, you'll have fun doing these lab activities.

BioLab

MiniLab

Do you often ask how, what, or why about the living world around you? Sometimes it takes just a little time to find out the answers for yourself. These short activities can be tried on your own at home or with help from a teacher at school. When you're feeling inquisitive, try a MINILAB.

MiniLab

MiniLab

Problem-Solving Lab

Sharpen up your pencil and your wit because you'll need them to solve these PROBLEM-SOLVING LABS. These labs offer a unique opportunity to evaluate another scientist's experiments and data without lab bench cleanup.

Problem-Solving Lab

Problem-Solving Lab

INSIDE STORY

As you study biology, you will discover that some concepts are better explained by detailed illustrations. The INSIDE STORY features display elaborate diagrams with step-by-step explanations of complex structures or processes in biology. They are designed to help you remember the fine details of biology.

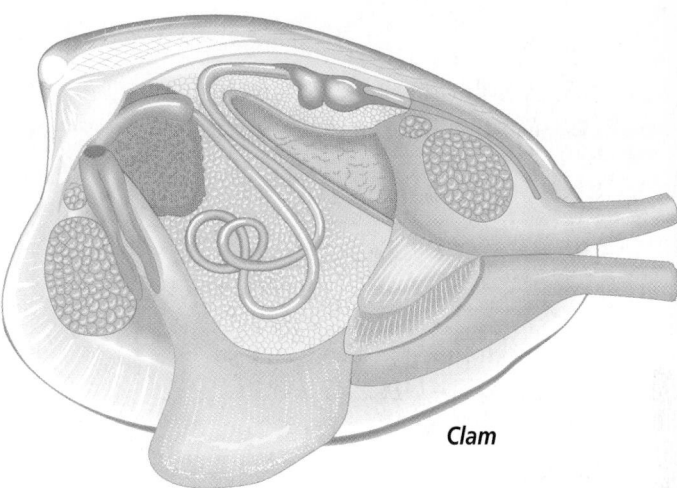

Clam

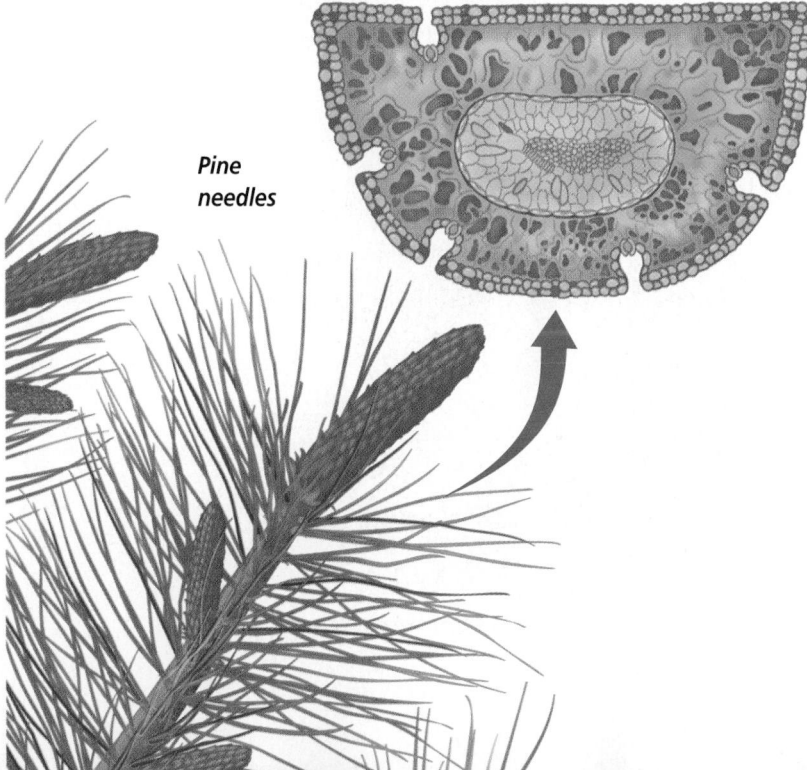

Pine needles

Connections

It may not have occurred to you that biology is connected to all your courses. Learn in these features how biology is connected to art, literature, and other subjects that interest you.

Some topics of biology deserve more attention than others because they're unusual, informative, or just plain interesting. Here are several features that FOCUS ON these fascinating topics.

CAREERS IN BIOLOGY

What can you do with a knowledge of biology? Why is biology so important to you? Knowing about biology can start you on a journey into any one of these exciting careers.

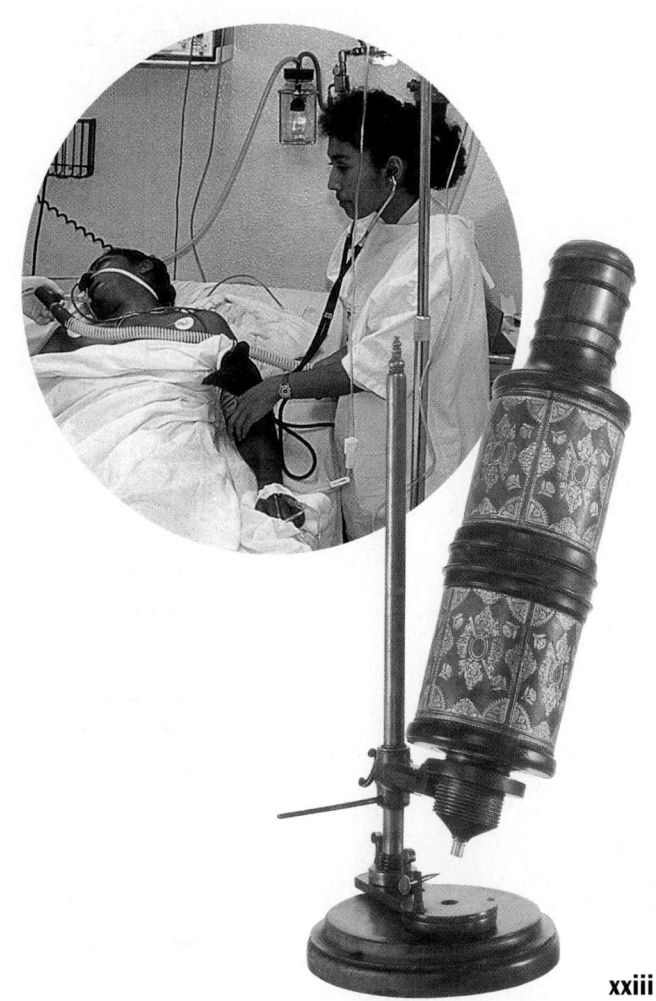

Biology & SOCIETY

How does biology impact our society? How does biology affect what you eat, the world around you, and your future? Take this opportunity to understand the many different sides of issues, and learn how biotechnology may affect your life.

BIO Technology

Teacher Reviewers and Content Specialists

Teacher Reviewers

Becky Ashe
West High School
Knoxville, TN

Dawn Day Boyd, M.Ed.
Hoover High School
Hoover, AL

Priscilla B. Cheek, M.Ed.
South Cobb High School
Austell, GA

Ron Dodson, M.A.
Hoover High School
Hoover, AL

Linda Franklin, M.Ed.
Myers Park High School
Charlotte, NC

Eve S. Jordan, M.S.
Halls High School
Knoxville, TN

John Kerr, M.S.
The Columbus Academy
Gahanna, Ohio

Mike Nelson
Lamar High School
Arlington, TX

Warren Orloff, M.A.
Upper Arlington High School
Columbus, OH

Michael H. Tally, M.Ed.
Science Supervisor
Wake County Schools
Raleigh, NC

Marc C. Taylor
Clovis High School
Clovis, CA

Mike Trimble
Corona Del Sol High School
Tempe, AZ

Johnetta D. Wiley, M.Ed.
Columbus South High School
Columbus, OH

Richard Wood, M.S.
Brookfield East High School
Brookfield, WI

Lilly Yap
Worthington Kilbourne High School
Columbus, OH

Content Specialists

Neil W. Blackstone, Ph.D
Department of Biological Sciences
Northern Illinois University
DeKalb, IL

Timothy J. Cain, Ph.D
Department of Cell Biology,
    Neurobiology, and Anatomy
The Ohio State University College
    of Medicine and Public Health
Columbus, OH

Gerald D. Carr, Ph.D
Department of Botany
University of Hawaii at Manoa
Honolulu, HI

James W. Fourqurean, Ph.D
Department of Biological Sciences
Florida International University
Miami, FL

Derek J. Girman, Ph.D
Department of Biology
Sonoma State University
Rohnert Park, CA

Daniel F. Gleason, Ph.D
Department of Biology
Georgia Southern University
Statesboro, GA

Paula E. Gregory, Ph.D
Department of Molecular Virology,
    Immunology & Medical Genetics
The Ohio State University College of
    Medicine and Public Health
Columbus, OH

Randy D. Krauss, Ph.D
Department of Biochemistry
Boston University
School of Medicine
Boston, MA

Gary L. Lindberg, Ph.D
Nutritional Physiology Group
Iowa State University of Science
    and Technology
College of Agriculture
Ames, IA

Charles F. Lytle, Ph.D
Biology Outreach Programs
North Carolina State University
Raleigh, NC

Mark D. Olson, Ph.D
Department of Anatomy and Cell Biology
University of North Dakota
School of Medicine and Health Sciences
Grand Forks, ND

Wojciech Pawlina, M.D
Department of Anatomy and Cell
    Biology
University of Florida
College of Medicine
Gainesville, FL

Seth H. Pincus, M.D
Department of Microbiology
Montana State University
Bozeman, MT

Marilyn J. Stapleton, Ph.D
Gene Tec Corporation
Durham, NC

Lawrence Williams, Ph.D
Primate Research Laboratory
University of South Alabama
Mobile, AL

Lorne Wolfe, Ph.D
Department of Biology
Georgia Southern University
Statesboro, GA

Christopher J. Woolverton, Ph.D
Department of Biological Sciences
Kent State University
Kent, OH

Safety Consultant

Lucille Daniel, Ed.D.
Rutherfordton, NC

Unit Overview

This unit includes one chapter that introduces students to the nature, excitement, and methods of biology. Students are first introduced to the characteristics of living organisms. The nature of science and the methods of science are then discussed by using examples intended to spark student interest as they attempt to answer questions and solve problems concerning the world of life.

Introducing the Unit

Ask students to look at the photo and describe some of the adaptations of sunflowers that make them successful. Tell students that they will learn in this unit how to identify the characteristics of all living organisms, including sunflowers. Explain to students that during their studies in this course they will use the methods of science described later in the chapter.

Unit 1

What Is Biology?

Biologists seek answers to questions about living things. For example, a biologist might ask how plants, such as sunflowers, convert sunlight into chemical energy that can be used by the plants to maintain life processes. Biologists use many methods to answer their questions about life. During this course, you will gain an understanding of the questions and answers of biology, and how the answers are learned.

UNIT CONTENTS

1 Biology: The Study of Life

BIODIGEST What Is Biology?

UNIT PROJECT

BIOLOGY *Online* Use the Glencoe Science Web site for more project activities that are connected to this unit.
science.glencoe.com

Unit Projects

Discover the Diversity of Living Things

Have students do one of the projects for this unit as described on the Glencoe Science Web site. As an alternative, students can do one of the projects described on these two pages.

Display

Visual-Spatial Students can use photographs or illustrations from magazines and science journals to make a collage showing as many different living things as possible in the available time. **L1** **ELL**

Pond Study

Visual-Spatial Have student groups study a local pond or stream to find out what living organisms are present. Students may wish to make drawings or take photos to illustrate their findings. **L1** **ELL** **COOP LEARN**

Advance Planning

Chapter 1

■ Purchase preserved mildew or collect mildew samples for MiniLab 1-1.

■ Purchase specially marked packages of Quaker Oatmeal for MiniLab 1-3.

■ Order or collect pill bugs and other materials for the BioLab.

■ Prepare leaning plants for the Getting Started Demo.

■ Obtain a culture of mealworms for the Assessment activity.

Unit Projects

Tape Production

Visual-Spatial Have students use a video camera to make a tape about living things in your area. Students may tape the same area through the seasons to show changes. **L1** **ELL**

Make a Model

Kinesthetic Students can make a three-dimensional model that depicts the kinds of living things found in your area. Alternatively, students can model another area, such as a rain forest. **L2** **ELL**

Final Report

Have students present their group's findings in oral reports that could be understood by students at your local middle school. **L3**

Chapter 1 Organizer

Refer to pages 4T-5T of the Teacher Guide for an explanation of the National Science Education Standards correlations.

Section	Objectives	Activities/Features
Section 1.1 **What Is Biology?** National Science Education Standards UCP.2; A.1, A.2; C.1, C.3, C.4, C.5, C.6; E.1, E.2; F.3, F.4, F.6; G.1 ($^1/_2$ session)	1. **Recognize** some possible benefits from studying biology. 2. **Summarize** the characteristics of living things.	**MiniLab 1-1:** Predicting Whether Mildew is Alive, p. 6 **Careers in Biology:** Nature Preserve Interpreter, p. 8 **Internet BioLab:** Collecting Biological Data, p. 26
Section 1.2 **The Methods of Biology** National Science Education Standards UCP.2; A.1, A.2; B.2; C.6; F.4, F.5; G.1, G.2 (1 session, $^1/_2$ block)	3. **Compare** different scientific methods. 4. **Differentiate** among hypothesis, theory, and principle.	**MiniLab 1-2:** Testing for Alcohol, p. 14 **Focus On** Scientific Theories, p. 16 **Problem-Solving Lab 1-1,** p. 18 **Inside Story:** Scientific Methods, p. 19
Section 1.3 **The Nature of Biology** National Science Education Standards UCP.2; A.1, A.2; E.1, E.2; F.3, F.4, F.5; G.1, G2 (1 session, 1 block)	5. **Compare and contrast** quantitative and descriptive research. 6. **Explain** why science and technology cannot solve all problems.	**Problem-Solving Lab 1-2,** p. 22 **MiniLab 1-3:** Hatching Dinosaurs, p. 23 **Biology & Society:** Organic Food: Is it healthier? p. 28

Need Materials? Contact Carolina Biological Supply Company at 1-800-334-5551 or at **http://www.carolina.com**

MATERIALS LIST

BioLab

p. 26 *Armadillidium* (pill bug), watch or clock, petri dish, paper, pencil, metric ruler, Internet access

MiniLabs

p. 6 mildew sample, microscope, microscope slide, coverslip, dropper, water, paper, pencil

p. 14 microscope slides, dropper, water, isopropyl alcohol, alcohol testing solution, personal hygiene products or cosmetics, wax pencil, paper, pencil

p. 23 artificial dinosaur eggs (2),

stirring rods (2), beakers (2), cold and boiling water, paper, pencil

Alternative Lab

p. 12 small jars or beakers (2), graduated cylinder, caffeinated coffee, decaffeinated coffee, labels, pencil

Quick Demos

p. 7 *Lithops* plant
p. 12 sugar cube, ashes, matches
p. 15 assorted scientific journals
p. 22 pine needles or leaves, metric ruler, paper, pencil

Key to Teaching Strategies

L1	Level 1 activities should be appropriate for students with learning difficulties.
L2	Level 2 activities should be within the ability range of all students.
L3	Level 3 activities are designed for above-average students.
ELL	ELL activities should be within the ability range of English Language Learners.
COOP LEARN	Cooperative Learning activities are designed for small group work.
P	These strategies represent student products that can be placed into a best-work portfolio.
📦	These strategies are useful in a block scheduling format.

Biology: The Study of Life

Teacher Classroom Resources

Section	Reproducible Masters	Transparencies
Section 1.1 **What Is Biology?**	Reinforcement and Study Guide, pp. 1-2 **L2** BioLab and MiniLab Worksheets, p. 1 **L2** Concept Mapping, p. 1 **L3** **ELL** Critical Thinking/Problem Solving, p. 1 **L3** Content Mastery, pp. 1-2, 4 **L1**	Section Focus Transparency 1 **L1** **ELL**
Section 1.2 **The Methods of Biology**	Reinforcement and Study Guide, p. 3 **L2** BioLab and MiniLab Worksheets, pp. 3-4 **L2** Laboratory Manual, pp. 1-4 **L2** Tech Prep Applications, pp. 1-2 **L2** Content Mastery, pp. 1, 3-4 **L1**	Section Focus Transparency 2 **L1** **ELL**
Section 1.3 **The Nature of Biology**	Reinforcement and Study Guide, p. 4 **L2** BioLab and MiniLab Worksheets, pp. 5, 7-8 **L2** Content Mastery, pp. 1, 3-4 **L1** Laboratory Manual, pp. 5-8 **L2**	Section Focus Transparency 3 **L1** **ELL**

Assessment Resources

Chapter Assessment, pp. 1-6
MindJogger Videoquizzes
Performance Assessment in the Biology Classroom
Alternate Assessment in the Science Classroom
ExamView® Pro Software 💾
BDOL Interactive CD-ROM, Chapter 1 quiz

Additional Resources

Spanish Resources **ELL**
English/Spanish Audiocassettes **ELL**
Cooperative Learning in the Science Classroom **COOP LEARN**
Lesson Plans/Block Scheduling

 **NATIONAL GEOGRAPHIC** — **Teacher's Corner**

Index to National Geographic Magazine
The following articles may be used for research relating to this chapter.
"The Rise of Life on Earth," by Richard Monastersky, March 1998.
"Life Grows Up," by Richard Monastersky, April 1998.

GLENCOE TECHNOLOGY

The following multimedia resources are available from Glencoe.

Biology: The Dynamics of Life
CD-ROM **ELL**

Video: *Biologist at Work*
Video: *How Organisms Interact*
Video: *Adapted For Survival*
Video: *Bioengineering*
Exploration: *Interpreting Data*

Videodisc Program 📼

Biologist at Work
How Organisms Interact
Bioengineering

The Infinite Voyage

The Keepers of Eden
Unseen Worlds

Theme Development

Students are introduced to six major themes of biology: **systems and interactions, homeostasis, the nature of science, unity within diversity, evolution, and energy.** These themes are presented as the characteristics of life are discussed and the nature of science and scientific methods are explained.

0:00 OUT OF TIME?

If time does not permit teaching the entire chapter, use the BioDigest at the end of the unit as an overview.

READING BIOLOGY

Glencoe's *Biology: The Dynamics of Life* contains many resources to assist a student's reading skills. Each chapter contains figures with expanded captions that expand on written material. Word Origins, located along the side of text, expand knowledge of biology vocabulary. Glencoe's Content Mastery Booklet helps develop reading skills while reinforcing content. In addition, use the Interactive Tutor for *Biology: The Dynamics of Life* on the Glencoe Web site to reinforce vocabulary. **science.glencoe.com**

Biology: The Study of Life

What You'll Learn

- You will identify the characteristics of life.
- You will recognize how scientific methods are used to study living things.

Why It's Important

Recognizing life's characteristics and the methods used to study life provide a basis for understanding the living world.

READING BIOLOGY

Scan the chapter's section headings and make a short list of examples of ways you think scientists might work. For instance, under data collection, a list might include: computer filing, taking photos, or taking notes. Think of how businesses, schools or communities might benefit from these experiments and research.

BIOLOGY Online

To find out more about the characteristics of living things, visit the Glencoe Science Web site.
science.glencoe.com

The plants and animals of this forest exhibit all the characteristics of life. The mushrooms, and the unseen bacteria living in the soil and leaf litter of the forest floor, also share the basic characteristics of life.

Multiple Learning Styles

Look for the following logos for strategies that emphasize different learning modalities.

 Kinesthetic Meeting Individual Needs, p. 5; Activity, p. 7

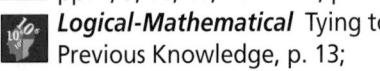 *Visual-Spatial* Getting Started Demo, p. 2; Portfolio, pp. 4, 7, 9; Concept Development, p. 4; Microscope Activity, p. 9; Project, p. 24

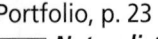 *Interpersonal* Cultural Diversity, p. 8; Meeting Individual Needs, p. 14; Project, p. 19

Linguistic Biology Journal, pp. 4, 9, 16, 18; Portfolio, p. 17

Logical-Mathematical Tying to Previous Knowledge, p. 13; Portfolio, p. 23

 *Naturalist* Quick Demo, p. 22

1.1 What Is Biology?

How far do monarch butter-flies travel during their annual migration? Why do viceroy butterflies look so much like monarchs? The natural world often poses questions like these. Usually, such questions have simple explana-tions, but sometimes the study of biology reveals startling surprises. Whether nature's puzzles are simple or complex, many can be explained using the concepts and principles of biological science.

SECTION PREVIEW

Objectives
Recognize some possible benefits from studying biology.
Summarize the characteristics of living things.

Vocabulary
biology
organism
organization
reproduction
species
growth
development
environment
stimulus
response
homeostasis
energy
adaptation
evolution

Monarch butterfly (above) and Viceroy butterfly (inset)

Prepare

Key Concepts
In this section, students learn that biology is an organized study and that many questions remain unanswered. The characteristics that living things share in common are presented in relationship to the themes of biology.

Planning
■ Purchase a *Lithops* plant from a nursery for the Quick Demo.
■ Gather materials needed for the Activity on page 7.

1 Focus

Bellringer 🖐
Before presenting the lesson, display **Section Focus Trans-parency 1** on the overhead pro-jector and have students answer the accompanying questions. **L1 ELL**

The Science of Biology

People have always been curious about living things—how many dif-ferent species there are, where they live, what they are like, how they relate to each other, and how they behave. These and many other ques-tions about life can be answered. The concepts, principles, and theories that allow people to understand the natural environment form the core of **biology,** the study of life. What will you, as a young biologist, learn about in your study of biology?

A key aspect of biology is simply learning about the different types of living things around you. With all the facts in biology textbooks, you might think that biologists have answered almost all the questions about life. Of course, this is not true. There are undoubtedly many life forms yet to be discovered; millions of life forms haven't even been named yet, let alone studied. Life on Earth includes not only the common organisms you notice every day, but also distinctive life forms that have unusual behaviors.

When studying the different types of living things, you'll ask what, why, and how questions about life. You might ask, "Why does this living thing possess these particular fea-tures? How do these features work?"

WORD Origin
biology
From the Greek words *bios*, meaning "life," and *logos*, meaning "study." Biology is the study of life.

✓ Assessment Planner

Portfolio Assessment
 Assessment, TWE, p. 18
 Portfolio, TWE, pp. 4, 7, 9, 17, 18, 23
 Alternative Lab, TWE, pp. 12-13
Performance Assessment
 Assessment, TWE, pp. 8, 13
 Alternative Lab, TWE, pp. 12-13
 BioLab, TWE, pp. 26-27
 MiniLab, TWE, pp. 6, 14, 23
 MiniLab, SE, pp. 6, 14, 23

Knowledge Assessment
 Section Assessment, SE, pp. 10, 20, 25
 Chapter Assessment, TWE, pp. 29-31
Skill Assessment
 Assessment, pp. 10, 20, 22, 25
 Problem-Solving Lab, pp. 18, 22
 BioLab, SE, p. 27

Resource Manager

Section Focus Transparency 1 and Master L1 ELL

Visual Learning

Figure 1.1 Ask students to discuss other strange creatures they have read about or seen on television. Explain that even with such diversity, all living things have certain characteristics in common.

Concept Development

Visual-Spatial Take students around the school grounds. Ask them to list in their notebooks all the different kinds of organisms they observe. They should indicate the characteristics they used to categorize each organism as a living thing. **L2**

Display

Make a bulletin board display that shows unusual structural and behavioral adaptations of plants and animals.

GLENCOE
TECHNOLOGY

CD-ROM
Biology: The Dynamics of Life
Video: *Biologist at Work*
Disc 1

VIDEODISC
Biology: The Dynamics of Life
Biologist at Work (Ch. 2)

Disc 1, Side 1
28 sec.

Figure 1.1
Few of the creatures you read about in works of fantasy and fiction are as unusual as some of the organisms that actually live on Earth.

A Orcas, also known as killer whales, are highly intelligent marine mammals that live and hunt in social units called pods.

B Leaf-cutter ants feed on fungus. They carry bits of leaves to their nest, then chew the bits and form them into moist balls on which the fungus grows.

C Leaves of the insect-eating pitcher plant form a tube lined with downward-pointing hairs that prevent insects from escaping. Trapped insects fall into a pool of water and digestive juices at the bottom of the tube.

The answers to such questions lead to the development of general biological principles and rules. As strange as some forms of life such as those shown in *Figure 1.1* may appear to be, there is order in the natural world.

Biologists study the interactions of life

One of the most general principles in biology is that living things do not exist in isolation; they are all functioning parts in the delicate balance of nature. As you can see in *Figure 1.2*,

Figure 1.2
Questions about living things can sometimes be answered only by finding out about their interactions with their surroundings.

A The seahorse is well hidden in its environment. Its body shape blends in with the shapes of the seaweeds in which it lives.

B The spadefoot toad burrows underground and encases itself in a waterproof envelope to prevent water loss during extended periods of dry weather.

4 BIOLOGY: THE STUDY OF LIFE

BIOLOGY JOURNAL

Investigating New Life Forms

Linguistic Ask students to find out about some of the most recent discoveries of new species. Ask them to report in their journals about how and where these new life forms were discovered and who discovered them. Ask them to discuss the significance of each discovery. **L3**

Portfolio

Identifying Habitats

Visual-Spatial Provide students with pictures of unusual organisms. Ask them to speculate on the type of habitat in which each might live. Have them explain their reasoning and place their pictures and habitat descriptions in their portfolios. **L2**

living things depend upon other living and nonliving things in a variety of ways and for a variety of reasons.

Biologists Study the Diversity of Life

Many people study biology simply for the pleasure of learning about the world of living things. As you've seen, the natural world is filled with examples of living things that can be amusing or amazing, and that challenge one's thinking. Through your study of biology, you will come to appreciate the great diversity of species on Earth and the way each species fits into the dynamic pattern of life on our planet.

Biologists study the interactions of the environment

Because no species, including humans, can exist in isolation, the study of biology must include the investigation of interactions among species. For example, learning about a population of wild rabbits would require finding out what plants they eat and what animals prey on them. The study of one species always involves the study of other species with which it interacts.

Human existence, too, is closely intertwined with the existence of other organisms living on Earth. Plants and animals supply us with food and with raw materials like wood, cotton, and oil. Plants also replenish the essential oxygen in the air. The students in *Figure 1.3* are studying organisms that live in a local stream. Activities like this help provide a thorough understanding of living things and the intricate web of nature. It is only through such knowledge that humans can expect to understand how to preserve the health of our planet.

Biologists study problems and propose solutions

The future of biology holds many exciting promises. Biological research can lead to advances in medical treatment and disease prevention in humans and in other species. It can reveal ways to help preserve species that are in danger of disappearing, and solve other problems, including the one described in *Figure 1.4.* The study of biology will teach you how humans function and how we fit in with the rest of the natural world. It will also equip you with the knowledge you need to help sustain this planet's web of life.

Figure 1.3
By understanding the interactions of living things, you will be better able to impact the planet in a positive way.

Figure 1.4
Honeybees and many other insects are important to farmers because they pollinate the flowers of crop plants, such as fruit trees. In the 1990s, populations of many pollinators declined, raising worries about reduced crop yields.

Purpose

Students will look at mildew under the microscope and decide if it is living.

Process Skills

predict, compare and contrast, observe and infer

Teaching Strategies

■ Prior to the activity, ask how many students believe mildew is living/not living. Post the results on the chalkboard. Poll students again after the MiniLab and compare both sets of responses.

■ Scrape mildew samples from shower grout or plastic shower curtains and bring them to class. Store samples in water to prevent them from drying out. **CAUTION**: *If live specimens are used, seal them in a plastic bag to protect students who might be allergic to mildew spores.*

■ Preserved mildew is available from biological supply houses and can be used in place of live specimens.

■ Instruct students on proper use of microscope and microscope slides.

■ Demonstrate the technique for making a wet mount. Include proper handling and cleaning of cover glasses and glass slides.

Expected Results

Students may initially predict that mildew is not alive. During microscopic examination, students will see long filaments (hyphae) and tiny circular objects (spores). Some students may see spores enclosed in a spore sac.

Characteristics of Living Things

Most people feel confident that they can tell the difference between a living thing and a nonliving thing, but sometimes it's not so easy. In identifying life, you might ask, "Does it move? Does it grow? Does it reproduce?" These are all excellent questions, but consider a flame. A flame can move, it can grow, and it can produce more flames. Are flames alive?

Biologists have formulated a list of characteristics by which we can recognize living things. Sometimes, nonliving things have one or more of life's characteristics, but only when something has all of them can it be considered living. Anything that possesses all of the characteristics of life is known as an **organism,** like the plants shown in *Figure 1.5.* All living things have an orderly structure, produce offspring, grow and develop, and adjust to changes in the environment. Practice identifying the characteristics of life by carrying out the *MiniLab* on this page.

MiniLab 1-1   Observing

Mildew

Predicting Whether Mildew is Alive What is mildew? Is it alive? We see it "growing" on plastic shower curtains or on bathroom grout. Does it show the characteristics associated with living things?

Procedure

1. Copy the data table below.

Data Table	
Prediction	**Life characteristics**
First	none
Second	
Third	

2. Predict whether or not mildew is alive. Record your prediction in the data table under "First Prediction."
3. Obtain a sample of mildew from your teacher. Examine it for life characteristics. Make a second prediction and record it in the data table along with any observed life characteristics. **CAUTION: Wash hands thoroughly after handling the mildew sample.**
4. Following your teacher's directions, prepare a wet mount of mildew for viewing under the microscope. **CAUTION: Use caution when working with a microscope, microscope slides, and cover slips.**
5. Are there any life characteristics visible through the microscope that you could not see before? Make a third prediction and include any observed life characteristics.

Analysis

1. Describe any life characteristics you observed.
2. Compare your three predictions and explain how your observations may have changed them.
3. Explain the value of using scientific tools to extend your powers of observation.

Figure 1.5
These plants are called *Lithops* from the Greek *lithos,* meaning "stone." Although they don't appear to be so, *Lithops* are just as alive as elephants. Both species possess all of the characteristics of life.

Analysis

1. organization, reproduction, growth, adjusting to environment
2. Answers will vary, depending on the observations made by individual students.
3. Certain life characteristics could not have been seen without the use of a microscope.

✓ Assessment

Performance Have students design an experiment to determine if commercial products reduce or kill mildew as claimed. Use the Performance Task Assessment List for Designing an Experiment in **PASC,** p. 23. **L2**

Living things are organized

When biologists search for signs of life, one of the first things they look for is structure. That's because they know that all living things show an orderly structure, or **organization.**

The living world is filled with organisms. All of them, including the earthworm pictured in *Figure 1.6,* are composed of one or more cells. Each cell contains the genetic material, or DNA, that provides all the information needed to control the organism's life processes.

Although living things are very diverse—there may be five to ten million species, perhaps more—they are unified in having cellular organization. Whether an organism is made up of one cell or billions of cells, all of its parts function together in an orderly, living system.

Living things make more living things

One of the most obvious of all the characteristics of life is **reproduction,** the production of offspring. The litter of mice in *Figure 1.7* is just one example. Organisms don't live forever. For life to continue, they must replace themselves.

Reproduction is not essential for the survival of an individual organism, but it is essential for the continuation of the organism's **species** (SPEE sheez). A species is a group of organisms that can interbreed and produce fertile offspring in nature. If individuals in a species never reproduced, it would mean an end to that species' existence on Earth.

Figure 1.6
Like all organisms, earthworms are made up of cells. The cells form structures that carry out essential functions, such as feeding or digestion. The interaction of these structures and their functions result in a single, orderly, living organism.

Figure 1.7
A variety of mechanisms for reproduction have evolved that ensure the continuation of each species. Some organisms, including mice, produce many offspring in one lifetime.

Discussion

Lead students in a discussion of general developmental stages of various organisms. Examples include complete and incomplete metamorphosis in insects, or birth, infancy, childhood, puberty, adulthood, old age, and death in humans.

8

CAREERS IN BIOLOGY

Nature Preserve Interpreter

I**f you like people as much as you love nature, you can combine your skills and interests in a career as a nature preserve interpreter.

Skills for the Job

Interpreters are also called naturalists, ecologists, and environmental educators. They might work for a nature preserve or a state or national park, where they give talks, conduct tours, offer video presentations, and teach special programs. Some interpreters are required to have a degree in biology, botany, zoology, forestry, environmental science, education, or a related field. They must also be skilled in communicating with others.

Many interpreters begin as volunteers who have no degrees, just a love of what they do. Over time, volunteers may become interns, and eventually be hired. Interpreters often help restore natural habitats and protect existing ones. Part of their job is to make sure visitors do not harm these habitats.

For example, many tidepool organisms find protection from too much sunlight by crawling under rocks. A naturalist can explain the importance of replacing rocks exactly as they were found.

BIOLOGY *Online*

For more careers in related fields, be sure to check the Glencoe Science Web site. **science.glencoe.com**

Figure 1.8
All life begins as a single cell. As cells multiply, each organism grows and develops and begins to take on the characteristics that identify it as a member of a particular species.

Living things change during their lives

An organism's life begins as a single cell, and over time, it grows and takes on the characteristics of its species. **Growth** results in an increase in the amount of living material and the formation of new structures.

All organisms grow, with different parts of the organism growing at different rates. Organisms made up of only one cell may change little during their lives, but they do grow. On the other hand, organisms made up of numerous cells go through many changes during their lifetimes, such as the changes that will take place in the young nestlings shown in *Figure 1.8.* Think about some of the structural changes your body has already undergone since you were born. All of the changes that take place during the life of an organism are known as its **development.**

Living things adjust to their surroundings

Organisms live in a constant interface with their surroundings, or **environment,** which includes the air, water, weather, temperature, any other organisms in the area, and

---

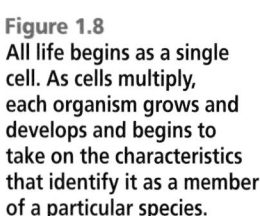

many other factors. For example, the fox in *Figure 1.9* feeds on small animals such as rabbits and mice. The fox responds to the presence of a rabbit by quietly moving toward it, then pouncing. Trees adjust to cold, dry winter weather by losing their leaves. Any condition in the environment that requires an organism to adjust is a **stimulus.** A reaction to a stimulus is a **response.**

The ability to respond to stimuli in the environment is an important characteristic of living things. It's one of the more obvious ones, as well. That's because many of the structures and behaviors that you see in organisms enable them to adjust to the environment. Try the *BioLab* at the end of this chapter to find out more about how organisms respond to environmental stimuli.

Regulation of an organism's internal environment to maintain conditions suitable for its survival is called **homeostasis** (hoh mee oh STAY sus). Homeostasis is a characteristic of life because it is a process that occurs in all living things. In addition to responding to external stimuli, living things

respond to internal changes. For example, organisms must make constant adjustments to maintain the correct amount of water and minerals in their cells and the proper internal temperature. Without this ability to adjust to internal changes, organisms die.

Living things reproduce themselves, grow and develop, respond to external stimuli, and maintain homeostasis by using **energy.** Energy is the ability to do work. Organisms get their energy from food. Plants make their own food, whereas animals, fungi, and other organisms get their food from plants or from organisms that consume plants.

Living things adapt and evolve

Any structure, behavior, or internal process that enables an organism to respond to stimuli and better survive in an environment is called an **adaptation** (ad ap TAY shun).

Adaptations are inherited from previous generations. There are always some differences in the adaptations of individuals within any population of organisms. As the environment changes, some adaptations are more

Figure 1.9
Living things respond to stimuli and make adjustments to environmental conditions.

A By dropping their leaves in the fall, these trees conserve water and avoid freezing during winter.

B Keen senses of smell and hearing enable a fox to find prey. Fur allows foxes and other mammals to regulate body temperature.

1.1   WHAT IS BIOLOGY?   **9**

Figure 1.10
Living things adapt to their environments in a variety of ways.

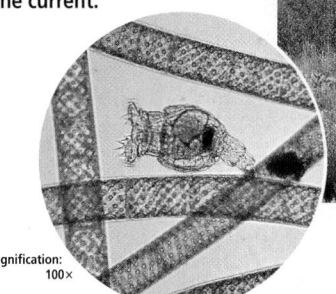

A Rotifers are microscopic organisms that create a water current with their wheels of cilia. They feed on microscopic food particles brought in with the current.

Magnification: 100×

B The desert *Octillo* has leaves for only a few days after a good rain. This adaptation helps conserve water.

C Many nocturnal animals, such as this owl, possess large eyes for efficient vision at night.

suited to the new conditions than others. Individuals with more suitable adaptations are more likely to survive and reproduce. As a result, individuals with these adaptations become more numerous in the population. *Figure 1.10* shows some examples of adaptation.

The gradual accumulation of adaptations over time is **evolution** (ev uh LEW shun). Clues to the way the present diversity of life came about may be understood through the study of evolution. In later chapters of this book, you will study how the theory of evolution can help answer many of the questions people have about living things.

As you learn more about Earth's organisms in this book, reflect on the general characteristics of life. Rather than simply memorizing facts about organisms or the vocabulary terms, try to see how these facts and vocabulary are related to the characteristics of living things.

Section Assessment

Understanding Main Ideas

1. What are some important reasons for studying biology?
2. Explain the difference between a stimulus and a response and give an example of each. How do these terms relate to an organism's internal environment?
3. Why is energy required for living things? How do living things obtain energy?
4. How are evolution and reproduction related?

Thinking Critically

5. How are energy and homeostasis related in living organisms?

SKILL REVIEW

6. **Observing and Inferring** Suppose you discover an unidentified object on your way home from school one day. What characteristics would you study to determine whether the object is a living or nonliving thing? For more help, refer to *Thinking Critically* in the **Skill Handbook**.

10 BIOLOGY: THE STUDY OF LIFE

1.2 The Methods of Biology

W hy do earthworms crawl onto sidewalks after it rains? Why do mosses grow only in wet, shady areas? Biologists ask questions like these every day. Different approaches may be used to answer different questions. Scientists who discovered that earthworms crawl out of rain-soaked soil to avoid drowning used different methods from those who learned that mosses require water for reproduction. Yet all scientific inquiries share some methods in common.

Mosses are tiny plants that grow in dense clumps.

SECTION PREVIEW

Objectives
Compare different scientific methods.
Differentiate among hypothesis, theory, and principle.

Vocabulary
scientific methods
hypothesis
experiment
control
independent variable
dependent variable
safety symbol
data
theory

Observing and Hypothesizing

Curiosity is often what motivates biologists to try to answer simple questions about everyday observations, such as why earthworms leave their burrows after it rains. Earthworms obtain oxygen through their skin, and will drown in water-logged soil. Sometimes, answers to questions like these also provide better understanding of general biological principles and may even lead to practical applications, such as the discovery that a certain plant can be used as a medicine. The knowledge obtained when scientists answer one question often generates other questions or proves useful in solving other problems.

The methods biologists use

To answer questions, different biologists may use many different approaches, yet there are some steps that are common to all approaches. The common steps that biologists and other scientists use to gather information and answer questions are **scientific methods.**

Scientists often identify problems to solve—that is, questions to ask and answer—simply by observing the world around them. For example, a laboratory scientist who is investigating questions about the reproduction of pea plants may come up with a new question about their development. Or, a scientist may ask a question about the feeding habits of prairie dogs after first observing prairie dog behavior in the field.

Prepare

Key Concepts
Students study the methods used in scientific investigations. They identify the methods as they relate to problems in various areas of biology.

Planning
■ Obtain several kinds of cough syrup and mouthwash to test in MiniLab 1-2.

1 Focus

Bellringer
Before presenting the lesson, display **Section Focus Transparency 2** on the overhead projector and have students answer the accompanying questions.
L1 **ELL**

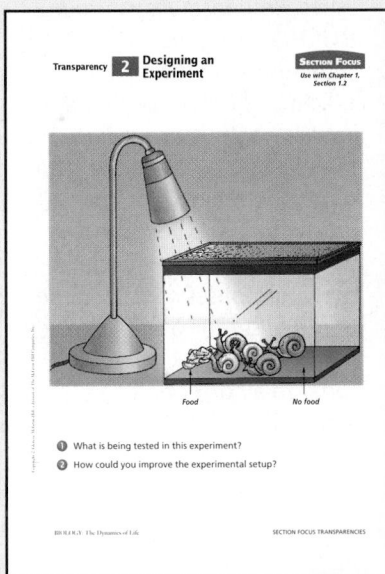

Note Internet addresses that you find useful in the space below for quick reference.

_____
_____
_____
_____
_____

Resource Manager

Section Focus Transparency 2 and Master **L1** **ELL**

2 Teach

Brainstorming

Ask students to brainstorm a list of questions they have that relate to biology. Write the questions as a list on the board. Allow students to give examples of how they might find answers to their questions if they were given access to any equipment or technology they can imagine.

Using Scientific Terms

Ask students to state a hypothesis about some question. Point out that a hypothesis is not a question, but rather a statement that answers a question.

Quick Demo

Try to set a sugar cube on fire using a lighted match. *It does not burn.* Rub the edge of the cube in ashes and attempt to ignite it. *It burns.* Ask students to hypothesize why the cube burns after it is rubbed in ashes. *Responses may include that the ashes—not the sugar cube—burn, that the ashes served as kindling, or that the ashes catalyze, or speed up, the reaction between the sugar and the oxygen in air (burning).* Ask students how they might test their explanations. *Responses might include trying to set the ashes on fire separately or exposing the sugar to a hotter flame for a longer time.*

The question of brown tree snakes

WORD *Origin*
hypothesis
From the Greek words *hypo*, meaning "under," and *thesis*, meaning a "placing." A hypothesis is a testable explanation of a natural phenomenon.

Have you ever been told that you have excellent powers of observation? This is one trait that is required of biologists. The story of the brown tree snake in *Figure 1.11* serves as an example. During the 1940s, this species of snake was accidentally introduced to the island of Guam from the Admiralty Islands in the Pacific Ocean. In 1965, it was reported in a local newspaper that the snake might be considered beneficial to the island because it is a predator that feeds on rats, mice, and other small rodents. Rodents are often considered pests because they carry disease and contaminate food supplies.

Shortly after reading the newspaper report, a young biologist walking through the forests of Guam made an important observation. She noted that there were no bird songs echoing through the forest. Looking into the trees, she saw a brown tree snake hanging from a branch. After learning that the bird population of Guam had declined rapidly since the introduction of the snake, she hypothesized that the snake might be responsible.

Figure 1.11
Brown tree snakes *(Boiga irregularis)* were introduced to Guam more than 50 years ago. Since then, their numbers have increased to a population of more than a million and they have severely reduced the native bird population of the island.

A **hypothesis** (hi PAHTH us sus) is an explanation for a question or a problem that can be formally tested. Hypothesizing is one of the methods most frequently used by scientists. A scientist who forms a hypothesis must be certain that it can be tested. Until then, he or she may propose suggestions to explain observations.

As you can see from the brown tree snake example, a hypothesis is not a random guess. Before a scientist makes a hypothesis, he or she has developed some idea of what the answer to a question might be through personal observations, extensive reading, or previous experiments.

After stating a hypothesis, a scientist may continue to make observations and form additional hypotheses to account for the collected data. Eventually, the scientist may test a hypothesis by conducting an experiment. The results of the experiment will help the scientist draw a conclusion about whether or not the hypothesis is correct.

Experimenting

People do not always use the word *experiment* in their daily lives in the same way scientists use it in their work. As an example, you may have heard someone say that he or she was going to experiment with a cookie recipe. Perhaps the person is planning to substitute raisins for chocolate chips, use margarine instead of butter, add cocoa powder, reduce the amount of sugar, and bake the cookies for a longer time. This is not an experiment in the scientific sense because there is no way to know what effect any one of the changes alone has on the resulting cookies. To a scientist, an **experiment** is a procedure that tests a hypothesis by the process of collecting information under controlled conditions.

Alternative Lab

Conducting an Experiment

Purpose
Students will follow the steps of scientific methods to solve a problem. The question to be answered is, Will the caffeine present in coffee prevent mold growth?

Preparation
Have students wear lab aprons, safety goggles, and disposable latex gloves, and wash their hands after checking for mold growth. Some students may be allergic to mold spores.

Materials
2 small jars (baby food jars), graduated cylinder, labels, caffeinated coffee, decaffeinated coffee

Procedure
Give the following directions to students.
1. Form and record a hypothesis.
2. Write your name and the date on the labels. Write #1 on one label and #2 on the other. Place the labels on the jars.
3. Add the following to each jar. Jar 1: 30 mL caffeinated coffee; Jar 2: 30 mL decaffeinated coffee.
4. Place both jars in the same location.

What is a controlled experiment?

Some experiments involve two groups: the control group and the experimental group. The **control** is the group in which all conditions are kept the same. The experimental group is the test group, in which all conditions are kept the same except for the single condition being tested.

Suppose you wanted to learn how bacteria affect the growth of different varieties of soybean plants. Your hypothesis might state that the presence of certain bacteria will increase the growth rate of each plant variety. An experimental setup designed to test this hypothesis is shown in *Figure 1.12.* Bacteria are present on the roots of the experimental plants, but not the controls. All other conditions—including soil, light, water, and fertilizer—are held constant for both groups of plants.

Designing an experiment

In a controlled experiment, only one condition is changed at a time. The condition in an experiment that is changed is the **independent variable,** because it is the only variable that affects the outcome of the experiment. In the case of the soybeans, the presence of bacteria is the independent variable. While changing the independent variable, the scientist observes or measures a second condition that results from the change. This condition is the **dependent variable,** because any changes in it depend on changes made to the independent variable. In the soybean experiment, the dependent variable is the growth rate of the plants. Controlled experiments are most often used in laboratory settings.

However, not all experiments are controlled. Suppose you were on a group of islands in the Pacific that is the only nesting area for a large sea

bird known as a waved albatross, shown in *Figure 1.13.* Watching the nesting birds, you observe that the female leaves the nest when her mate flies back from a foraging trip. The birds take turns sitting on the eggs or caring for the chicks, often for two weeks at a time. You might hypothesize that the birds fly in circles around the island, or that they fly to some distant location. To test one of these hypotheses, you might attach a satellite transmitter to some of the birds and record their travels. An experiment such as this, which

Figure 1.12
This experiment tested the effect of bacteria on the growth of several varieties of soybeans. For each experiment there are three rows of each variety. The center rows are the experimental plants. The outer rows are the controls.

Figure 1.13
The waved albatross is a large bird that nests only on Hood Island in the Galapagos Islands. By tagging the birds with satellite transmitters, scientists have learned where these birds travel.

5. Make daily observations of your jars for one week. Check for the presence of mold.
6. Record your observations in a suitable data table by making diagrams of the coffee liquid surfaces.

Analysis

1. Which type of coffee allows mold to grow? *both*
2. Was your hypothesis supported by your data? *will depend on hypothesis*
3. What was the control, independent variable, and dependent variable? *decaffeinated coffee, caffeinated coffee, mold growth*

Purpose

Students will experiment to determine which products contain alcohol.

Process Skills

draw a conclusion, experiment, interpret data, observe and infer

Safety Precautions

■ Review safety precautions regarding goggles and the use of an acid solution.

Teaching Strategies

■ For preparation instructions for the alcohol-testing reagent, see page 40T of the Teacher Guide.

■ Make circles on slides with a China marking or eyebrow pencil. Students could use small test tubes rather than glass slides. Use caution when cleaning glassware.

■ For circle C, use isopropyl (rubbing) alcohol. Products to be tested may include aftershave lotion, cough syrups, mouthwash.

Expected Results

Student data will vary with the products tested. Circle A will appear yellow-orange, circle B green to blue, and circle C yellow.

Analysis

1. to determine which colors indicate the presence or absence of alcohol
2. Answers will depend on products tested.

✔ **Assessment**

Performance Provide students with several different types of alcohol (rubbing, ethyl, methyl) and ask them to determine which types the alcohol-testing chemical can detect. Use the Performance Task Assessment List for Carrying Out a Strategy and Collecting Data in **PASC**, p. 25. L2 ELL

14

MiniLab 1-2 Experimenting

Testing for Alcohol Commercials for certain over-the-counter products may not tell you that one of the ingredients is alcohol. How can you verify whether or not a certain product contains alcohol? One way is to simply rely on the information provided during a commercial. Another way is to experiment and find out for yourself.

Procedure

1 Copy the data table.

Data Table		
	Color of liquid	Alcohol present
Circle A		
Circle B		
Circle C		
Product name		
Product name		

2 Draw three circles on a glass slide. Label them A, B, and C. **CAUTION: *Put on safety goggles.***

3 Add one drop of water to circle A, one drop of alcohol to circle B, and one drop of alcohol-testing chemical to circles A, B, and C. **CAUTION: *Rinse immediately with water if testing chemical gets on skin or clothing.***

4 Wait 2-3 minutes. Note in the data table the color of each liquid and the presence or absence of alcohol.

5 Record the name of the first product to be tested.

6 Draw a circle on a clean glass slide. Add one drop of the product to the circle.

7 Add a drop of the alcohol-testing chemical to the circle. Wait 2-3 minutes. Record the color of the liquid.

8 Repeat steps 5-7 for each product to be tested. **CAUTION: *Wash your hands with soap and water immediately after using the alcohol testing chemical.***

9 Complete the last column of the data table. If alcohol is present, the liquid turns green, deep green, or blue. A yellow or orange color means no alcohol is present.

Analysis

1. Explain the purpose of using the alcohol-testing chemical with water, with a known alcohol, and by itself.
2. Which products did contain alcohol? No alcohol?

14 BIOLOGY: THE STUDY OF LIFE

---

has no control, is the type of biological investigation most often used in field work.

The experimental design that is selected depends on what other experimenters have done and what information the biologist hopes to gain. Sometimes, a biologist will design a second experiment even while a first one is being conducted, if he or she thinks the new experiment will help answer the question. Try your hand at experimenting in the *MiniLab* on this page.

Using tools

To carry out experiments, scientists need tools that enable them to record information. The growth rate of plants and the information from satellite transmitters placed on albatrosses are examples of important information gained from experiments.

Biologists use a wide variety of tools to obtain information in an experiment. Some common tools include beakers, test tubes, hot plates, petri dishes, thermometers, dissecting instruments, balances, metric rulers, and graduated cylinders. More complex tools include microscopes, centrifuges, radiation detectors, spectrophotometers, DNA analyzers, and gas chromatographs. *Figure 1.14* shows some of these more complex tools.

Maintaining safety

Safety is another important factor that scientists consider when carrying out experiments. Biologists try to minimize hazards to themselves, the people working around them, and the organisms they are studying.

In the experiments in this textbook, you will be alerted to possible safety hazards by safety symbols like those shown in *Table 1.1*. A **safety symbol** is a symbol that warns you

---

📋 **Resource Manager**

Tech Prep Applications, pp. 1-2 L2
Critical Thinking/Problem Solving, p. 1 L3
BioLab and MiniLab Worksheets, pp. 3-4 L2

MEETING INDIVIDUAL NEEDS

Learning Disabled

Interpersonal Group students appropriately. Have them use scientific methods to alter a cookie recipe. Ask students to name the control to be used, the independent variable, and the dependent variable. L1 ELL COOP LEARN

about a danger that may exist from chemicals, electricity, heat, or procedures you will use. Refer to the safety symbols in *Appendix C* at the back of this book before beginning any lab activity in this text. It is your responsibility to maintain the highest safety standards to protect yourself as well as your classmates.

Data gathering

To answer their questions about scientific problems, scientists seek information from their experiments. Information obtained from experiments is called **data.** Sometimes, data from experiments are referred to as experimental results.

Often, data are in numerical form, such as the distance covered in an albatross's trip or the height that soybean plants grow per day. Numerical data may be measurements of time, temperature, length, mass, area, volume, or other factors. Numerical data may also be counts, such as the number of bees that visit a flower per day or the number of wheat seeds that germinate at different soil temperatures.

Sometimes data are expressed in verbal form, using words to describe observations made during an experiment. Scientists who first observed the behavior of pandas in China obtained data by recording what these animals do in their natural habitat and how they respond to their environment. Learning that pandas are solitary animals with large territories helped scientists understand how to provide better care for them in zoos and research centers.

Having the data from an experiment does not end the scientific process. Read the *Focus On* on the next page to see how data collection relates to other important aspects of research.

Figure 1.14
Biologists use many tools in their studies.

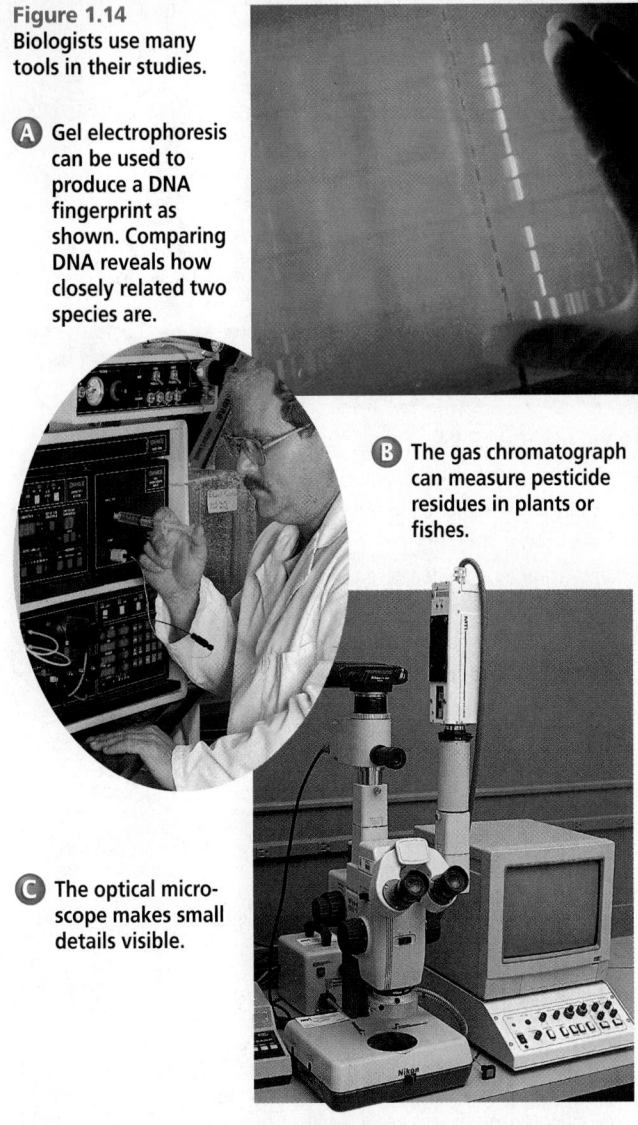

 A Gel electrophoresis can be used to produce a DNA fingerprint as shown. Comparing DNA reveals how closely related two species are.

B The gas chromatograph can measure pesticide residues in plants or fishes.

C The optical microscope makes small details visible.

Table 1.1 Safety symbols	
	Sharp Object Safety This symbol appears when a danger of cuts or punctures caused by the use of sharp objects exists.
	Clothing Protection Safety This symbol appears when substances used could stain or burn clothing.
	Eye Safety This symbol appears when a danger to the eyes exists. Safety goggles should be worn when this symbol appears.
	Chemical Safety This symbol appears when chemicals used can cause burns or are poisonous if absorbed through the skin.

Bring copies of scientific journals to class and allow students to examine them. Ask students to speculate as to which branch of science each journal addresses.

Brainstorming

A student repeats an experiment several times and each time records different data. Have students offer possible reasons why an experiment might yield different data for different trials. *Reasons include failure to keep all factors but one the same, errors in data recording, errors in mathematical treatment of the data, or naturally occurring variability in experimental outcome.* **L2**

GLENCOE TECHNOLOGY

VIDEODISC
The Secret of Life
Microscopy Segment

The Infinite Voyage
Unseen Worlds, Technology Reconstructs Egyptian Mummies (Ch. 1), 6 min. 30 sec.

MEETING INDIVIDUAL NEEDS

Hearing Impaired

Interpersonal Supply students with the following data showing how many seeds were germinated by three laboratory groups over a three-day period. Group A had two seeds germinate on day 1, four seeds on day 2, and four seeds on day 3. Group B had one seed germinate on day 1, six on day 2, and five on day 3. Group C had two seeds germinate on day 1, four on day 2, and three on day 3. Have students work in groups to prepare a class histogram of the number of seeds germinating each day for all three groups. **L2**

FOCUS ON
Scientific Theories

IGUANODON

What is a scientific theory? In casual usage, "theory" means an unproven assumption about a set of facts. A scientific theory is an explanation of a natural phenomenon supported by a large body of scientific evidence obtained from various investigations and observations. The scientific process begins with observations of the natural world. These observations lead to hypotheses, data collection, and experimentation. If weaknesses are observed, hypotheses are rejected or modified and then tested again and again. When little evidence remains to cause a hypothesis to be rejected, it may become a theory. Follow the scientific process described here that led to new theories about dinosaurs.

HADROSAUR

16

OBSERVING

People have been unearthing fossils for hundreds of years. The first person to reconstruct a dinosaur named it *Iguanodon*, meaning "iguana tooth," because its bones and teeth resembled those of an iguana. By 1842 these extinct animals were named *dinosaurs*, meaning "terrible lizards."

FIELD MUSEUM OF NATURAL HISTORY, CHICAGO

MAKING HYPOTHESES

Reptiles are ectotherms—animals with body temperatures influenced by their external environments. Early in the study of dinosaur fossils, many scientists assumed that because dinosaur skeletons resembled those of some modern reptiles, dinosaurs, too, must have been ectotherms. This assumption led scientists to conclude that many dinosaurs, being both huge and ectothermic, were slow-growing, slow-moving, and awkward on land.

Because the most complete dinosaur skeletons occurred in rocks formed at the bottom of bodies of water, scientists hypothesized that dinosaurs lived in water and that water helped to support their great weight. When skeletons of duck-billed dinosaurs, called hadrosaurs, were discovered, this hypothesis gained support. Hadrosaurs had broad, flat ducklike bills, which, scientists suggested, helped them collect and eat water plants.

BREAD PALM, CYCAD FAMILY

THINKING CRITICALLY

In the 1960s, paleontologist Robert Bakker (right) hypothesized that dinosaurs were not sluggish ectotherms but fast-moving, land-dwelling endotherms—animals like birds and mammals. Bakker observed that many dinosaurs had feet and legs built for life on land. If hadrosaurs had led a semiaquatic life, Bakker reasoned, their feet would have been webbed with long, thin, widely spaced toes. But hadrosaurs had short, stubby toes and feet, obviously suited for land. In addition to Bakker's observations, studies of fossilized stomach contents revealed that hadrosaurs dined on the cones and leaves of cycads (above) and other terrestrial plants. After considering these data carefully, Bakker proposed that many dinosaurs were quick, agile endotherms that roamed Earth's ancient landscape.

ROBERT BAKKER

COLLECTING DATA

DINOSAUR BONE SHOWING CHANNELS FOR BLOOD VESSELS
Magnification: 25 x

To test his hypotheses, Bakker intensified his research on dinosaur skeletons and bone structure. He found reports from the 1950s comparing thousands of cross sections of dinosaur bones with those of reptiles, birds, and mammals. These reports noted that many dinosaur bones were less dense than those of modern reptiles and riddled with channels for blood vessels. In short, many dinosaur bones resembled those of endotherms not ectotherms. Bakker confirmed his observations by collecting supporting evidence from other sources.

FORMING THEORIES

Bakker's hypotheses—supported by data gathered by other paleontologists and by dinosaur bones, growth patterns, and behavior—prompted scientists to reexamine theories about dinosaurs. Were some dinosaurs endotherms and others ectotherms? Did dinosaurs have their own unique physiology resembling neither reptiles nor mammals? Scientific theories about dinosaurs continue to evolve as new fossils are discovered and new tools to study those fossils are developed.

PALEONTOLOGIST WORKING ON FOSSIL

ROBERT BAKKER WITH BRONTOSAUR FEMUR

EXPANDING Your View

1 **APPLYING CONCEPTS** Robert Bakker's research led to a different theory regarding the physiology of dinosaurs. As new fossils are found and new tools developed to study them, paleontologists will continue to replace existing theories with newer ones. What are some reasons for a scientific theory to be changed?

Visual Learning

Have students study the art of *Iguanodon* and the hadrosaur. Have them list the structures that support the hypotheses that hadrosaurs lived in water and on land. **L1**

Answers to Expanding Your View

1. **Applying Concepts** Scientific theories must be changed to accommodate new or conflicting data.

✓ Portfolio

New Research on Dinosaurs

Linguistic Although Dr. Bakker's bone evidence shows a link between dinosaurs and endotherms, other researchers dispute this and point to factors, such as nasal cavities, that show dinosaurs were more like ectotherms. Have students research the newest hypotheses about dinosaurs by reading recent articles in magazines. Ask them to write a summary of the articles for their portfolios and list the factors that support both ectothermy and endothermy as the homeostatic mechanisms for dinosaur metabolism. **L3** **P**

Purpose

Students will analyze the claims made in a commercial.

Process Skills

analyze information, draw a conclusion, experiment, think critically

Teaching Strategies

■ You may wish to have the entire class analyze the same commercial. Tape a specific commercial and play it back to the class. Or provide students with the printed dialogue from a radio commercial. **L2**

Thinking Critically

1. Student answers will vary. Many commercials make claims without specifying any scientific backing.
2. Student answers will vary. Many commercials do not use experimental evidence but base their claims on inference.
3. Students should describe a hypothesis, control, dependent and independent variables, trials, data accumulation, and a conclusion based on data.

✔ Assessment

Portfolio Ask students to design a video commercial that bases its claims on experimental data. Have them include their data and analyses as part of the commercial. Use the Performance Task Assessment List for Video in **PASC**, p. 81. **L2**

Problem-Solving Lab 1-1 — Analyzing Information

Are the claims valid? "Our product is new and improved." "Use this mouth wash and your mouth will feel clean all day." Sound familiar? TV and radio commercials constantly tell us how great certain products are. Are these claims always based on facts?

Analysis

Listen to or view a commercial for a product that addresses a medical problem such as heartburn, allergies, or bad breath. If possible, tape the commercial so that you can replay it as often as needed. Record the following information:

1. What is the major claim made in the commercial?
2. Is the claim based on experimentation?
3. What data, if any, are used to support the claim?

Thinking Critically

1. In general, was the claim based on scientific methods? Explain your answer.
2. In general, are product claims made in commercials based on experimental evidence? Explain your answer.
3. Describe a scientific experiment that could be conducted to establish claims made for the product in your commercial.

Thinking about what happened

Often, the thinking that goes into analyzing experimental data takes the greatest amount of a scientist's time. After careful review of the results, the scientist must come to a conclusion: Was the hypothesis supported by the data? Was it not supported? Are more data needed? Data from an experiment may be considered confirmed only if repeating that experiment several times yields similar results. To review how scientific methods are carried out, read the *Inside Story*.

After analyzing the data, most scientists have more questions than they had before the experiment. They compare their results and conclusions with the results of other studies by researching the published literature for more information. They also begin to think of other experiments they might carry out. Are all the claims you hear on TV commercials based on data gathered by the scientific method? Find out by conducting the *Problem-Solving Lab* here.

Reporting results

Results and conclusions of experiments are reported in scientific journals, where they are open to examination by other scientists. Hundreds of scientific journals are published weekly or monthly. In fact, scientists usually spend a large part of their time reading journal articles to keep up with new information as it is reported. The amount of information published every day in scientific journals is more than any single scientist could read. Fortunately, scientists also have access to computer databases that contain summaries of scientific articles, both old and new.

Verifying results

Data and conclusions are shared with other scientists for an important reason. After results of an investigation have been published, other scientists can try to verify the results by repeating the experiment. If they obtain similar results, there is even more support for the hypothesis. When a hypothesis is supported by data from additional experiments, it is considered valid and is generally accepted by the scientific community. When a scientist publishes the results of his or her investigation, other scientists can relate their own work to the published data.

Field Research

Linguistic Have students look in *Discover, National Geographic,* or similar publications to find an article that describes field research. Ask them to write a report on the article, contrasting the scientist's work in the field with the work the scientist had to continue back in the laboratory. **L3**

Working as a Food Chemist

Linguistic Tell students to imagine they are the chief food chemist working in a candy factory. They are trying to improve on the taste of a "Smacking Good Bar." Have them describe some things they would suggest to their staff to attempt, in a controlled manner, to find a better candy bar recipe. **L1**

Scientific Methods

Scientific methods are used by scientists to answer questions and solve problems. The development of the cell theory, one of the most useful theories in biological science, illustrates how the methods of science work. In 1665, Robert Hooke first observed cells in cork.

Thinking Critically *What is the function of other scientists in the scientific process?*

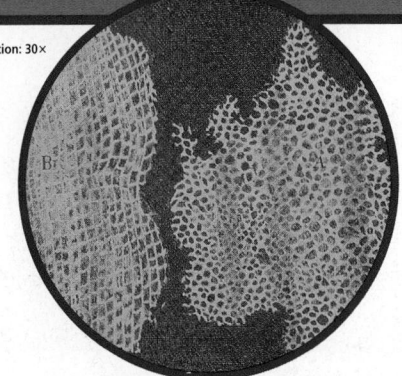
Magnification: 30×

Cork cells as drawn by Robert Hooke

1 Observing The first step toward scientific discovery takes place when a scientist observes something no one has noticed before. After Hooke's discovery, other scientists observed cells in a variety of organisms.

2 Making a hypothesis A hypothesis is a testable answer to a question. In 1824, René Dutrochet hypothesized that cells are the basic unit of life.

3 Collecting data Data can support or disprove a hypothesis. Over the years, scientists who used microscopes to examine organisms found that cells are always present.

4 Publishing results Results of an experiment are useful only if they are made available to other scientists. Many scientists published their observations of cells in the scientific literature.

5 Forming a theory A theory is a hypothesis that is supported by a large body of scientific evidence. By 1839, many scientific observations supported the hypothesis that cells are fundamental to life. The hypothesis became a theory.

6 Developing new hypotheses A new theory may prompt scientists to ask new questions or form additional hypotheses. In 1833, Robert Brown hypothesized that the nucleus is an important control center of the cell.

7 Revising the theory Theories are revised as new information is gathered. The cell theory gave biologists a start for exploring the basic structure and function of all life. Important discoveries, including the discovery of DNA, have resulted.

19

PROJECT

Using Scientific Methods

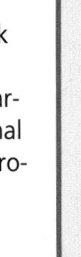 *Interpersonal* Have students work in groups on a biological problem they select, such as the behavior of a particular species of bird, reptile, or mammal in your area. At the conclusion of the project, students should be able to explain the methods of science they used. **L2**

COOP LEARN

Resource Manager

Laboratory Manual, pp. 1-4 **L2**

Purpose
Students will review methods used in scientific investigations.

Teaching Strategies

■ Point out that not every scientific investigation uses every method, nor do all investigations lead to a published theory.

■ Provide students with biological journals such as *Nature* or *Scientific American*. Ask them to read the articles and identify the methods of science used.

Visual Learning

■ Have students look at cork cells through a microscope and compare them with the photo on this page, then ask them to compare the capabilities of their microscopes with the one used by Robert Hooke.

Critical Thinking

Other scientists repeat the same experiment to validate original results. Scientists extend understanding by performing experiments to answer related questions.

---

3 Assess

Check for Understanding

Provide students with scientific methods listed in a scrambled order. Ask students to sequence steps in the correct order. **L1**

Reteach

Ask students to outline the steps used in scientific methods. For each level of the outline, have them provide an example taken from studies described in the text. **L2**

Extension

Have students look up cell theory in this text. Ask them to speculate about the hypotheses that may have been made by each of the scientists who first discovered cells. **L2**

Assessment

Skill Provide each student with a piece of laboratory equipment. Have them list five observations about the equipment and suggest how the equipment might be used. **L1**

4 Close

Demonstration

Display laboratory equipment and safety equipment and clothing students will use in their study of biology. For each item, identify its function and proper use. As a follow-up, set up lab stations at which students are required to demonstrate their knowledge of each item. **L2**

Resource Manager

Reinforcement and Study Guide, p. 3 **L2**

Figure 1.15
Experiments have shown that male elephants communicate with other males using threat postures and low-frequency vibrations that warn rival males away.

For example, biologists studying the behavior of elephants in Africa published their observations. Other scientists, who were studying elephant communication, used that data to help determine which of the elephants' behaviors are related to communication. Further experiments showed that female elephants emit certain sounds in order to attract mates, and that some of the sounds

produced by bull elephants warn other males away from receptive females, as described in *Figure 1.15*.

Theories and laws

People use the word *theory* in everyday life very differently from the way scientists use this word. You may have heard someone say that he or she has a theory that a particular football team will win the Super Bowl this year. What the person really means is that he or she believes one team will play better for some reason. Much more evidence is needed to support a scientific theory.

In science, a hypothesis that is supported by many separate observations and experiments, usually over a long period of time, becomes a theory. A **theory** is an explanation of a natural phenomenon that is supported by a large body of scientific evidence obtained from many different investigations and observations. A theory results from continual verification and refinement of a hypothesis.

In addition to theories, scientists also recognize certain facts of nature, called laws or principles, that are generally known to be true. The fact that a dropped apple falls to Earth is an illustration of the law of gravity.

Section Assessment

Understanding Main Ideas

1. Suppose you made the observation that bees seem to prefer a yellow flower that produces abundant amounts of pollen and nectar over a purple flower that produces less pollen and nectar. List two separate hypotheses that you might make about bees and flowers.
2. Describe a controlled experiment you could perform to determine whether ants are more attracted to butter or to honey.
3. What is the difference between a theory and a hypothesis?
4. Why do experiments usually require a control?

Thinking Critically

5. Describe a way that a baker might conduct a controlled experiment with a cookie recipe.

SKILL REVIEW

6. **Interpreting Scientific Illustrations** Review the *Inside Story*. What happens when a hypothesis is not confirmed? What does the position of the word *theory* indicate about the strength of a scientific theory compared to the strength of a hypothesis? For more help, refer to *Thinking Critically* in the **Skill Handbook**.

Section Assessment

1. Students might hypothesize that bees prefer yellow flowers to purple flowers or that bees prefer flowers with more abundant pollen.
2. Set up an experimental chamber. Within a specific amount of time, count and record how many ants move to butter placed a specific distance from the ants. Repeat several times. Repeat using honey in place of the butter.
3. A hypothesis is a testable explanation for a question. A theory is a refined explanation supported by many different experiments.
4. A control provides greater certainty that observed results are not due to chance or other variables.
5. Prepare one batch of cookies (the control) by following a recipe and another batch of cookies (the experimental group) by varying a single variable in the recipe, such as amount of sugar.
6. A new, revised hypothesis is tested, or the experiment may be changed. Theories are supported by the results of a variety of experiments.

1.3 The Nature of Biology

SECTION PREVIEW

Compare and contrast quantitative and descriptive research.

Explain why science and technology cannot solve all problems.

Vocabulary
ethics
technology

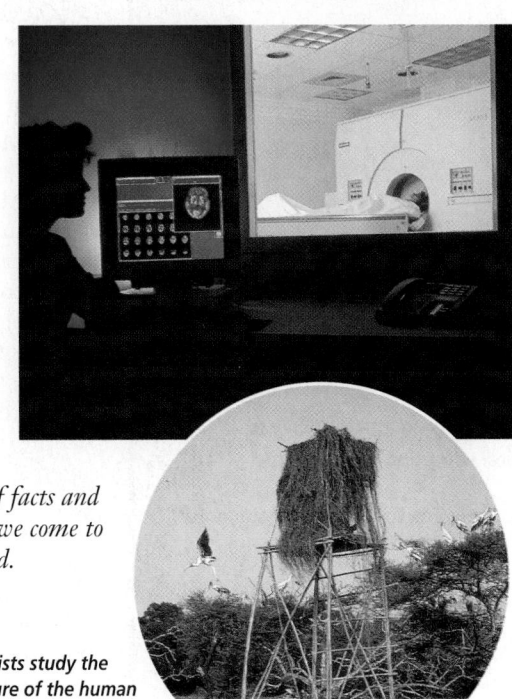

S *cientific study includes learning many known facts about the world around us. Biologists use known facts to discover new problems, make hypotheses, design experiments, interpret data, and draw conclusions. Biology is also an active process that includes making observations and conducting experiments in the laboratory and in the field. Science is both a body of facts and ideas and a process by which we come to understand the natural world.*

Biologists study the structure of the human brain (top) and observe animal behavior (inset).

Kinds of Research

You have learned that scientists use a variety of methods to test their hypotheses about the natural world. Scientific research can usually be classified into one of two main types, quantitative or descriptive.

Quantitative research

Biologists sometimes conduct controlled experiments that result in counts or measurements—that is, numerical data. These kinds of experiments occur in quantitative research. The data are analyzed by comparing numerical values.

Data obtained in quantitative research may be used to make a graph or table. Graphs and tables communicate large amounts of data in a form that is easy to understand. Suppose, for example, that a biologist is studying the effects of climate on freshwater life. He or she may count the number of microscopic organisms, called *Paramecium*, that survive at a given temperature. This study is an example of quantitative research.

Section 1.3

Key Concepts

Students differentiate between quantitative and descriptive research methods. The role of research and its application and use by society as technology is examined.

Planning

■ Obtain materials for the Mini-Lab.

---

1 Focus

Bellringer

Before presenting the lesson, display **Section Focus Transparency 3** on the overhead projector and have students answer the accompanying questions.
L1 **ELL**

BIOLOGY JOURNAL

Quantitative Data

Linguistic Have students review articles on biological research in *Scientific American, Science News, Newsweek,* or local newspapers. Ask students to write a short essay identifying the quantitative measurements taken in each study and describing how the measurements were taken. **L2**

2 Teach

Naturalist Provide students with a metric ruler and a clump of pine needles. Ask them to make two lists of observations: one that describes the needles with words and another that uses measurements. Have students compare their lists with classmates. **L1**

Problem-Solving Lab 1-2

Purpose

Students will analyze a graph and determine that the amount of information obtained from a graph is limited.

Process Skills

acquire information, interpret data, think critically

Teaching Strategies

■ Review the terminology associated with graphs or refer students to the **Skill Handbook.**

Thinking Critically

1. No. No bar extends to the 100% line.
2. The number of students enrolled in physical education declines as students progress from freshman to senior year.
3. Data needed to answer the question are not supplied.
4. No. The graph does not provide this information.

Skill Have students design a graph to illustrate why fewer seniors take physical education. Students can make up the questions to be graphed and estimate the number of 'yes' responses. Use the Performance Task Assessment List for Conducting a Survey and Graphing the Results in **PASC**, p. 35. **L2**

22

Problem-Solving Lab 1-2 — Making and Using Graphs

What can be learned from a graph? One way to express information is to present it in the form of a graph. The amount of information available from a graph depends on the nature of the graph itself.

U.S. Students Enrolled in Physical Education

■ Male
■ Female

Percent (0, 20, 40, 60, 80, 100) vs *Grade* (9, 10, 11, 12)

Source: Youth Risk Behavior Survey, 1995

Analysis

Study the graph at right. Answer the questions that follow and note the type of information that can and cannot be answered from the graph itself.

Thinking Critically

1. Is there ever a year in high school when all students are enrolled in physical education? Explain your answer.
2. Is there a relationship between the number of students enrolled in physical education and their year of high school? Explain your answer.
3. Can you tell which states in the country have the largest number of students enrolled in physical education?
4. Based on the graph, can you explain why so few students take physical education in their senior year?

The data obtained from the *Paramecium* study is presented as a graph in *Figure 1.16.* You can practice using graphs by carrying out the *Problem-Solving Lab* on this page.

Measuring in the International System

It is important that scientific research be understandable to scientists around the world. For example, what if scientists in the United States reported quantitative data in inches, feet, yards, ounces, pounds, pints, quarts, and gallons? People in many other countries would have trouble understanding these data because they are unfamiliar with the English system of measurement. Instead, scientists always report measurements in a form of the metric system called the International System of Measurement, commonly known as SI.

One advantage of SI is that there are only a few basic units, and nearly all measurements can be expressed in these units or combinations of them. The greatest advantage is that SI, like the metric system, is a decimal system. Measurements can be expressed

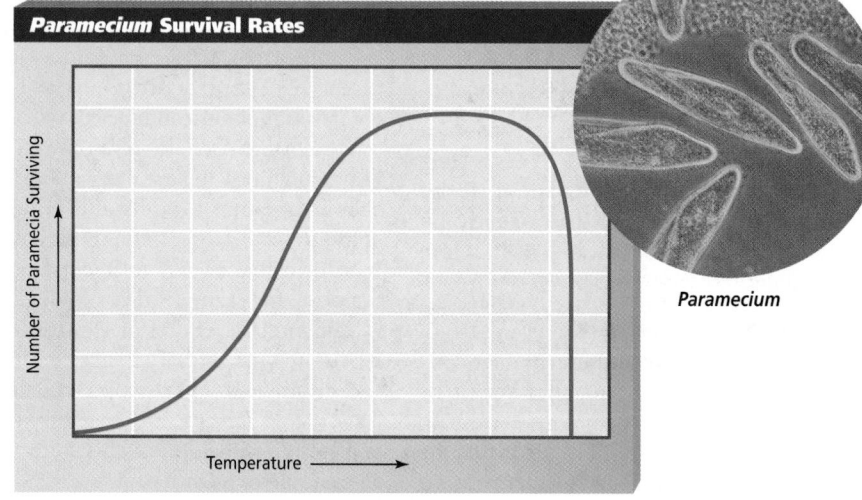

Magnification: 65×

Figure 1.16
This graph shows how many paramecia survive as the temperature increases.

Paramecium Survival Rates

Number of Paramecia Surviving vs Temperature →

Paramecium

22 BIOLOGY: THE STUDY OF LIFE

Cultural Diversity

Units and Standards

The SI system is used in 95% of the countries in the world. It provides a standardized system of measurement that makes scientific communication easier. Many early systems of measurement were not standardized. The ancient Egyptians used a unit called the cubit, which was based on the length of the arm from elbow to fingertips. Because sizes of individuals varied, the size of the unit varied. In England, the foot was equal to the length of the foot of the king. When a new king came to power, the length changed according to the size of his foot. Have students research measuring systems used around the world and create a visual display of their findings. **L3**

in multiples of tens or tenths of a basic unit by applying a standard set of prefixes to the unit. In biology, the metric units you will encounter most often are meter (length), gram (mass), liter (volume), second (time), and Celsius degree (temperature). For a thorough review of measurement in SI, see Practicing Scientific Methods in the **Skill Handbook.**

Descriptive research

Do you think the behavior of the animals shown in *Figure 1.17* would be easier to explain with numbers or with written descriptions of what the animals did? Observational data—that is, written descriptions of what scientists observe—are often just as important in the solution of a scientific problem as numerical data.

When biologists use purely observational data, they are carrying out descriptive research. Descriptive research is useful because some phenomena aren't appropriate for quantitative research. For example, how a particular wild animal reacts to events in its environment cannot easily be illustrated with numbers. Practice your descriptive research skills in the *MiniLab* on this page.

MiniLab 1-3   Observing and Inferring

Hatching Dinosaurs "Dinosaur eggs" can be found in specially marked packages of oatmeal. You will conduct an investigation to determine what causes these pretend eggs to hatch.

Procedure
1. Copy the data table below.

Data Table	Before treatment	Hot water treatment	Cold water treatment
Appearance after one minute			

2. Observe the dinosaur eggs provided and record their characteristics in your table.
3. Place an egg in each of two containers.
4. Make a hypothesis about the water temperature that will cause the eggs to hatch.
5. Pour boiling water into one container and cold water in the other. Stir for one minute. Record your observations.

Analysis
1. Infer whether heat or moisture was more important for hatching eggs.
2. Design an experiment that would test either heat or moisture as the variable. What kind of quantitative data will you gather?
3. What will be your control?
4. How many trials will you run and how many eggs will you test? If time permits, conduct your experiment.

Figure 1.17
Do you think these animals behave in the same way in zoos as they do in nature?

A Penguins cannot fly. They use their wings for swimming in the oceans of the southern hemisphere.

B Toucans live in the rain forests of South America.

✔ Portfolio

Making Predictions

Logical-Mathematical Ask students to carry out the following activity. Have them predict the chance that a coin, when flipped, should come up heads. *50%* How many heads should appear if a coin is flipped 10 times? *1/2 x 10 or 5* How many heads will appear if a coin is flipped 100 times? *1/2 x 100 or 50* Ask them to carry out the coin tosses and record their results. Have them use the activity to explain in their portfolios if scientists can predict the results of an experiment with 100% certainty. Have them explain the advantage of using a large sample or many trials in an experiment. *Large samples increase the likelihood that the sample is representative.* **L2** **P**

23

Science and Society

The road to scientific discovery includes making observations, formulating hypotheses, performing experiments, collecting and analyzing data, drawing conclusions, and reporting results in scientific journals. No matter what methods scientists choose, their research often provides society with important information that can be put to practical use.

Maybe you have heard people blame scientists for the existence of nuclear bombs or controversial drugs. To comprehend the nature of science in general, and biology in particular, people must understand that knowledge gained through scientific research is never inherently good or bad. Notions of good and bad arise out of human social, ethical, and moral concerns. **Ethics** refers to the moral principles and values held by humans. Scientists might not consider all the possible applications for the products of their research when planning their investigations. Society as a whole must take responsibility for the ethical use of scientific discoveries.

WORD *Origin*

technology
From the Greek words *techne*, meaning an "art or skill," and *logos*, meaning "study." Technology is the application of science in our daily lives.

Figure 1.18
If bad luck caused by black cats occurred as reliably and as swiftly in real life as it does in cartoons, it really would be scientifically testable.

Can science answer all questions?

Some questions are simply not in the realm of science. Such questions may involve decisions regarding good versus evil, ugly versus beautiful, or similar judgments. There are also scientific questions that cannot be tested using scientific methods. However, this does not mean that these questions are unimportant.

Consider a particular question that is not testable. Some people assert that if a black cat crosses your path, as shown in the cartoon in **Figure 1.18,** you will have bad luck. On the surface, that hypothesis appears to be one that you could test. But what is bad luck, and how long would you have to wait for the bad luck to occur? How would you distinguish between bad luck caused by the black cat and bad luck that occurs at random? Once you examine the question, you can see there is no way to test it scientifically because you cannot devise a controlled experiment that would yield valid data.

Can technology solve all problems?

Science attempts to explain how and why things happen. Scientific study that is carried out mainly for the sake of knowledge—with no immediate interest in applying the results to daily living—is called pure science.

However, much of pure science eventually does have an impact on people's lives. Have you ever thought about what it was like to live in the world before the development of water treatment plants, vaccinations, antibiotics, or high-yielding crops? These and other life-saving developments are indirect results of research done by scientists in many different fields over hundreds of years.

Other scientists work in research that has obvious and immediate applications. **Technology** (tek NAHL uh jee) is the application of scientific research to society's needs and problems. It is concerned with making improvements in human life and the world around us. Technology has helped increase the production of food, reduced the amount of manual labor needed to make products and raise crops, and aided in the reduction of wastes and environmental pollution.

The advance of technology has benefited humans in numerous ways, but it has also resulted in some serious problems. For example, irrigation technology is often used to boost the production of food crops. If irrigation is used over too many years in one area, the soil may become depleted of minerals or the evaporation of the irrigation water may leave deposits of mineral salts in the soil. Eventually the soil may become useless for growing crops, as illustrated in *Figure 1.19*.

Science and technology will never answer all of the questions we ask, nor will they solve all of our problems. However, during your study of biology you will have many of your

Figure 1.19
One example of a possible harmful side effect of technology is the deterioration of soil caused by irrigation. In the field shown here, irrigation technology initially appeared to solve the problem of low crop yield, but later caused a different problem—the buildup of excess mineral salts that prevent crop growth.

questions answered and you will explore many new concepts. As you learn more about living things, remember that you are a part of the living world and you can use the processes of science to ask and answer questions about that world.

Check for Understanding
Have students provide an example of: quantitative research, descriptive research, a contribution of technology, and an ethical issue in science. **L2**

Reteach
Ask students to prepare an outline of the major concepts of this section. **L2**

Extension
Logical-Mathematical Have students research the idea of "being able to beat cancer with a strong positive mental attitude." Have them explain why it may be difficult to evaluate scientifically how a positive mental attitude contributes to recovery from disease. **L3**

Assessment
 Skill Have students measure their arm spans and palm widths in centimeters. Convert these measurements to millimeters and meters. **L2**

4 Close

Discussion
Describe one possible benefit or spin-off that might come from the study of: how birds find their way during migration; longer lasting batteries; bat echolocation. **L3**

Section Assessment

Understanding Main Ideas
1. Why is it important that scientific experiments be repeated?
2. Compare and contrast quantitative and descriptive research.
3. Why is science considered to be a combination of fact and process?
4. Why is technology not the solution to all scientific problems?

Thinking Critically
5. Biomedical research has led to the development of technology that can keep elderly, very ill patients alive. How does the statement "The results of research aren't good or bad; they just are," apply to such research?

SKILL REVIEW
6. **Making and Using Graphs** Look at the graph in *Figure 1.16*. Why do you think the high-temperature side of the graph drops off more sharply than the low-temperature side? For more help, refer to *Organizing Information* in the **Skill Handbook**.

Section Assessment

1. to see if the results are repeatable, thus confirming their authenticity
2. Quantitative research reports data in numerical values based on measuring. Descriptive research reports data in written descriptions based on observations.
3. A scientist needs a background of knowledge in his or her field. The scientific process increases that knowledge.
4. Some problems do not have a scientific basis. Some technological solutions may pose more problems than they solve.
5. The biomedical researchers sought to increase knowledge. The application of the resulting technology is a question society must answer.
6. Paramecia die above a certain temperature. This results in a rapid drop in numbers once this temperature is reached. They are better able to survive as low temperatures rise, thus the graph reflects this increased survival.

Time Allotment
One class period

Process Skills
collect data, define operationally, experiment, observe and infer, acquire information, think critically, communicate

Safety Precautions
Remind students to treat live animals gently and follow directions carefully. Have students wear goggles, lab aprons, and disposable latex gloves.

PREPARATION

■ Pill bugs—also called sow bugs, wood lice, or isopods—may be collected locally or purchased from a biological supply house. *Armadillidium vulgare* is the only species that will roll tightly into a "pill," so it is preferred for this activity.

ANALYZE AND CONCLUDE

1. Student answers will vary and may include the following: shows organization; has specific parts for certain jobs; all pill bugs look alike. Yes, orderly structure also applies to nonliving things. Many nonliving things are organized, including buildings, books, computers.

2. The graph shows that pill bugs vary in length, but most reach a maximum size of about 10 mm.

3. Pill bugs remain curled for an average of approximately 20 seconds.

4. The pill bug's outer shell is rather tough; rolling into a ball could prevent predators from attacking its soft underside and fragile appendages.

INTERNET

BioLab

Collecting Biological Data

Seeing different life forms, and even interacting with them, is pretty much part of a typical day. Petting a dog, swatting at a fly, cutting the grass, and talking to your friends are common examples. But, have you ever asked yourself the question, "What do all of these different life forms have in common?" Let's try to find out.

PREPARATION

Problem
What life characteristics can be observed in a pill bug?

Objectives
In this BioLab, you will:
■ **Observe** whether life characteristics are present in a pill bug.
■ **Measure** the length of a pill bug.
■ **Experiment** to determine if a pill bug responds to changes in its environment.
■ **Use the Internet** to collect and compare data from other students.

Materials
pill bugs, *Armadillidium*
watch or classroom clock
container, glass or plastic
pencil with dull point
ruler
computer with Internet connection

Safety Precautions
Always wear goggles in the lab.

Skill Handbook
Use the **Skill Handbook** if you need additional help with this lab.

PROCEDURE

Data Table

Organization and growth and development	
Orderly structure?	
Pill bug length in mm	

Response to environment	
Trial	Time in seconds
1	
2	
3	
4	
5	
Total	
Average time	

1. Make copies of the data table and graph outlines shown here.
2. Obtain a pill bug from your teacher and place it in a small container.
3. Observe your pill bug to determine whether or not it has an orderly structure. Record your answer in the data table.
4. Using millimeters, measure and record the length of your pill bug in the data table.
5. Using your data and data from your classmates, complete the

PROCEDURE

Troubleshooting
Students may have difficulty assessing whether pill bugs show an orderly structure. Advise them to make a "best guess" to answer this question.

Teaching Strategies
■ Allow students to work in small groups of two or three.

■ Make sure that students have already covered the section dealing with life characteristics before attempting this laboratory activity.
■ Laboratory finger bowls or plastic dishes from supermarkets or fast food restaurants can be used as small containers.
■ Review the procedure for determining an average. Have students round off their

graph "Pill Bug Length: Classroom Data."

6. Go to the Glencoe Science Web Site at the address shown below to **post your data.**

7. *Gently* touch the underside of the pill bug with a *dull* pencil point. It may be necessary to gently flip the pill bug over with the pencil to get at its underside. **CAUTION:** *Use care to avoid injuring the pill bug.*

8. Note its response and time, in seconds, how long the animal remains curled up. Record the time in the data table as Trial 1.

9. Repeat steps 7-8 four more times, recording each trial in the data table.

10. Calculate the average length of time your pill bug remains curled up in a ball.

11. **Post your data** on the Glencoe Science Web Site.

12. Return the pill bug to your teacher. **CAUTION:** *Wash your hands with soap and water after working with pill bugs.*

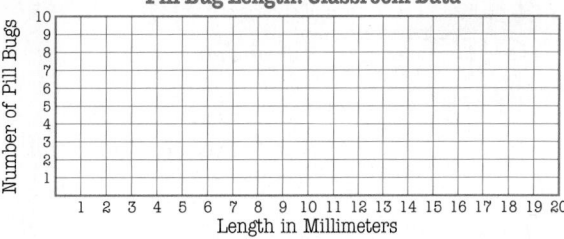

Pill Bug Length: Classroom Data

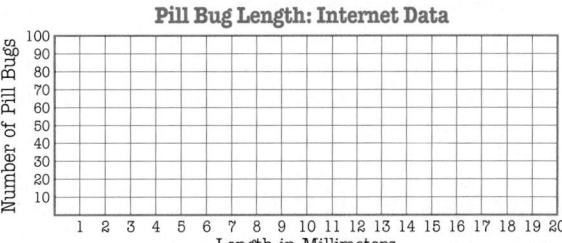

Pill Bug Length: Internet Data

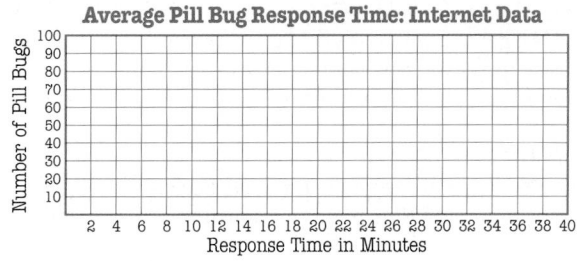

Average Pill Bug Response Time: Internet Data

ANALYZE AND CONCLUDE

1. **Thinking Critically** Explain how you would define the term "orderly structure." Explain how this trait might also pertain to nonliving things.

2. **Using the Internet** Explain how data from the classroom and Internet graphs support the idea that pill bugs grow and develop.

3. **Interpreting Data** What was the most common length of time pill bugs remained curled in response to being touched?

4. **Drawing a Conclusion** Explain how the response to being

touched is an adaptation.

5. **Experimenting** How might you design an experiment to determine whether or not pill bugs reproduce?

Sharing Your Data

BIOLOGY *Online* Find this BioLab on the Glencoe Science Web site at **science.glencoe.com** and post your data in the data table provided for this activity. Use the additional data from other students on the Internet, analyze the combined data, and complete your graphs.

5. Student answers will vary. Place several pill bugs in a sealed container and add food and moisture as needed. Observe pill bugs weekly and compare numbers present to those originally placed in container. Or, look for immature forms that may appear in the container.

✓ Assessment

Skill Ask students to prepare a graph of data that would be representative of the experimental results from "Sharing Your Data." Refer students to *Making and Using Graphs* in the **Skill Handbook.** Use the Performance Task Assessment List for Graph from Data in **PASC,** p.39. **L2**

Sharing Your Data

BIOLOGY *Online* To navigate to the Internet BioLabs, choose the *Biology: The Dynamics of Life* icon at the Glencoe Science Web site. Click on the student site icon, then the BioLabs icon. To expand this activity, have students design and conduct an experiment to determine if pill bugs have preferences for certain food types. **L2**

Resource Manager

BioLab and MiniLab Worksheets, pp. 7-8 **L2**

---

average data on the length of time in seconds the pill bug remains rolled up.

■ Review the technique for preparing a histogram.

Data and Observations

Students should conclude that pill bugs show organization, undergo growth and development, and adjust to their environment. Pill bugs will remain rolled in a ball for approximately 20 seconds. Typical length of a mature pill bug is 10 mm.

Organic Food: Is it healthier?

The produce section of the supermarket has two bins of leafy lettuce that look very much alike. One is labeled "organic" and has a higher price. More and more consumers are willing to pay extra for organically grown fruits, vegetables, meats, and dairy products. What are they paying that extra money for?

Purpose

This feature allows students to analyze claims made about organically grown produce. Students are encouraged to formulate and present their own views.

Teaching Strategies

■ Provide recent newspaper and magazine articles regarding claims about organically grown produce.

■ Ask students to discuss the advantages and disadvantages of using herbicides, pesticides, and fertilizers on crops.

■ Have students interview produce managers in several supermarkets to obtain their views concerning organically grown produce. **L3**

■ Have students investigate the methods used for displaying produce in supermarkets. Ask them to write a summary of their findings. **L2**

Investigating the Issue

Student standards will differ, and this produces fuel for the debate. Standards might include the exclusion of all agricultural chemicals, or the reduction of chemicals.

The term "organic" usually refers to foods that are produced without the use of chemical pesticides, herbicides, or fertilizers. Organic farmers use nonchemical methods to control pests and encourage crop growth. Beneficial insects, such as ladybugs and trichogramma wasps, are brought in to feed on aphids, caterpillars, and other damaging insects. Instead of applying herbicides, organic farmers pull weeds by hand or by machine. In place of fertilizers, they use composting and crop rotation to enrich the soil. Organic farming is very labor intensive, so organic foods are usually more expensive than those produced by conventional methods.

Different Viewpoints

People usually buy organic products because they want to be sure they're getting nutritious food with no chemical residues. But there are differences of opinion about how much better organic food actually is, and even which foods should be called organic.

Is organic food healthier? Agricultural chemicals can leave residues on food and contaminate drinking water supplies. Since exposure to some chemicals is known to cause health problems, including cancer, many consumers think that organic foods are healthier. Chemical pest controls kill beneficial organisms as well as unwanted pests, and can adversely affect the health of other animals, especially those that feed on insects. Organic pest control methods usually target specific pests and have little effect on beneficial organisms.

Is conventionally grown food healthier? Chemical fertilizers and pesticides make it possible to grow larger crops at lower cost, which makes more food available to more people. Making sure everyone can afford an adequate supply of fruits and vegetables may be more important than the risk of disease posed by agricultural chemicals.

Not everyone agrees about what is organic and what isn't. Should genetically engineered plant or animal foods be considered organic? What about herbs or meats preserved by irradiation, or lettuce and tomatoes fertilized with sewage sludge?

Produce from an organic farm

INVESTIGATING THE ISSUE

Comparing and Contrasting Propose your own set of standards for defining organic fruits and vegetables. Organize a debate in which you and your classmates present arguments to support your proposed standards.

 To find out more about the organic food debate, visit the Glencoe Science Web site. **science.glencoe.com**

Going Further

Bring samples of organic and conventional produce to class. Ask students to compare the appearance and taste of the samples and discuss how price, appearance, and taste might affect consumer purchase decisions. **L1**

SUMMARY

Section 1.1

What Is Biology?

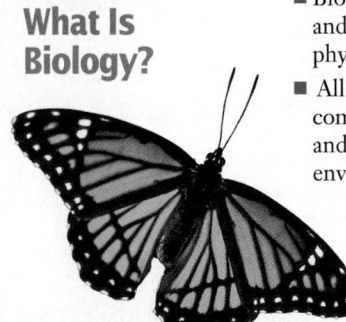

Main Ideas

■ Biology is the organized study of living things and their interactions with their natural and physical environments.

■ All living things have four characteristics in common: organization, reproduction, growth and development, and the ability to adjust to the environment.

Vocabulary

adaptation (p. 9)
biology (p.3)
development (p.8)
energy (p.9)
environment (p.8)
evolution (p.10)
growth (p.8)
homeostasis (p.9)
organism (p.6)
organization (p.7)
reproduction (p.7)
response (p.9)
species (p.7)
stimulus (p.9)

Section 1.2

The Methods of Biology

Main Ideas

■ Biologists use controlled experiments to obtain data that either do or do not support a hypothesis. By publishing the results and conclusions of an experiment, a scientist allows others to try to verify the results. Repeated verification over time leads to the development of a theory.

Vocabulary

control (p.13)
data (p.15)
dependent variable (p.13)
experiment (p.12)
hypothesis (p.12)
independent variable (p.13)
safety symbol (p.14)
scientific methods (p.11)
theory (p.20)

Section 1.3

The Nature of Biology

Main Ideas

■ Biologists do their work in laboratories and in the field. They collect both quantitative and descriptive data from their experiments and investigations.

■ Scientists conduct investigations to increase knowledge about the natural world. Scientific results may help solve some problems, but not all.

Vocabulary

ethics (p.24)
technology (p.25)

UNDERSTANDING MAIN IDEAS

1. For experiments to be considered valid, the results must be _____.
   a. verified
   b. inductive
   c. published
   d. repeatable

2. Reproduction is an important life characteristic because all living things _____.
   a. replace themselves
   b. show structure
   c. grow
   d. adjust to surroundings

Main Ideas

Summary statements can be used by students to review the major concepts of the chapter.

Using the Vocabulary

To reinforce chapter vocabulary, use the Content Mastery Booklet and the activities in the Interactive Tutor for Biology: The Dynamics of Life on the Glencoe Science Web site. **science.glencoe.com**

 *All Chapter Assessment questions and answers have been validated for accuracy and suitability by The Princeton Review.*

UNDERSTANDING MAIN IDEAS

1. d
2. a

GLENCOE TECHNOLOGY

 **VIDEOTAPE**
MindJogger Videoquizzes
Chapter 1: *Biology: The Study of Life*
Have students work in groups as they play the videoquiz game to review key chapter concepts.

Resource Manager

Chapter Assessment, pp. 1-6
MindJogger Videoquizzes
ExamView® Pro Software
BDOL Interactive CD-ROM,
    Chapter 1 quiz

3. c
4. a
5. b
6. d
7. a
8. c
9. b
10. d
11. data
12. technology
13. experiment
14. scientific methods
15. theory
16. control
17. independent
18. stimulus; response
19. theory
20. control

3. The photograph to the right is an example of which characteristic of life?
   a. evolution
   b. reproduction
   c. development
   d. response to a stimulus

4. Which of the following is an appropriate scientific question?
   a. How do paramecia behave when a pond begins to dry up?
   b. Which perfume smells the best?
   c. Which religion is most sound?
   d. Are llamas less valuable than camels?

5. If data from repeated experiments do not support the hypothesis, what is the scientist's next step?
   a. Give up.
   b. Revise the hypothesis.
   c. Repeat the experiment.
   d. Overturn the theory.

6. Similar-looking organisms, such as the dogs shown below, that can interbreed and produce fertile offspring are called _____.
   a. a living system   c. organization
   b. an adaptation   d. a species

 **THE PRINCETON REVIEW** **TEST-TAKING TIP**

Words Are Easy to Learn
Make a huge stack of vocabulary flashcards and study them. Use your new words in daily conversation. The great thing about learning new words is the ability to express yourself more specifically.

7. The environment includes _____.
   a. air, water, weather
   b. response to a stimulus
   c. adaptations
   d. evolution

8. Which of the following terms are most related to each other?
   a. adaptation—response
   b. stimulus—growth
   c. adaptation—evolution
   d. stimulus—evolution

9. Which of the following is not an appropriate question for science to consider?
   a. How many seals can a killer whale consume in a day?
   b. Which type of orchid flower is most beautiful?
   c. What birds prefer seeds as a food source?
   d. When do hoofed mammals in Africa migrate northward?

10. The single factor that is altered in an experiment is the _____.
    a. control
    b. dependent variable
    c. hypothesis
    d. independent variable

11. The information gained from an experiment is called _____.

12. The application of scientific research to society's needs is _____.

13. A procedure that tests a hypothesis is a(n) _____.

14. Processes that scientists use to solve a problem are called _____.

15. An explanation of a natural phenomenon with a high degree of confidence is a(n) _____.

16. The group that is not altered in an experiment is the _____.

17. The single change in the manipulated group in an experiment is a(n) _____ variable.

18. When a horse swats a fly with its tail, the fly is a _____ and the swat of the tail is a _____.

19. The idea that germs are the cause of disease has been continuously supported by experiments and has, therefore, been elevated to the status of a _____.

20. The standard group against which others are measured in an experiment is a _____.

APPLYING MAIN IDEAS

21. Describe how the human body shows the life characteristic of organization.

22. Explain the relationships among an organism's environment, adaptations, and evolution.

THINKING CRITICALLY

23. **Comparing and Contrasting** Consider the following items: a flame, bubbles blown from a bubble wand, and a balloon released into the air. List characteristics of each that might indicate life and those that indicate they are not alive.

24. **Concept Mapping** Complete the concept map by using the following vocabulary terms: experiment, theory, hypothesis, scientific methods

may begin with an observation, which leads to a(n)

that is tested by a(n)

and, after many repetitions, may result in a(n)

CD-ROM

For additional review, use the assessment options for this chapter found on the *Biology: The Dynamics of Life Interactive CD-ROM* and on the Glencoe Science Web site. science.glencoe.com

ASSESSING KNOWLEDGE & SKILLS

A team of students measured the number of seeds that germinated over ten days in a control group at 18°C and in an experimental group at 25°C. They graphed their data as shown below.

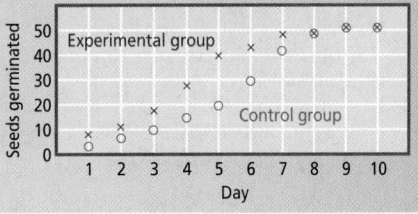

The Effect of Temperature on Germination

Interpreting Data Study the graph and answer the following questions.

1. Which of the following would represent the hypothesis tested?
   a. Black seeds are best.
   b. Seeds germinate faster at warmer temperatures.
   c. Fertilization of seeds requires heat.
   d. Seeds germinate when freezing.

2. When was the first appropriate day to end the experiment?
   a. day 3   c. day 7
   b. day 6   d. day 9

3. Which of the following was the independent variable?
   a. kind of seeds   c. temperature
   b. number germinating d. time

4. Which of the following was the dependent variable?
   a. kind of seeds   c. temperature
   b. number germinating d. time

5. **Interpreting Data** Describe the germination rate between days 3 and 5 in the control group.

APPLYING MAIN IDEAS

21. It is composed of cells, which are organized into tissues and organs, which are organized into body systems.

22. Evolution is the result of organisms adapting to environmental changes.

THINKING CRITICALLY

23. A flame has energy and may appear to grow and reproduce. Bubbles blown from a wand move and may grow. A balloon released into air moves. These objects cannot adapt to changes in the environment or maintain homeostasis.

24. 1. Scientific methods; 2. Hypothesis; 3. Experiment; 4. Theory

ASSESSING KNOWLEDGE & SKILLS

1. b
2. d
3. c
4. b
5. Between days 3 and 5, approximately 5 seeds germinated per day.

National Science Education Standards
UCP,.1, A.2, G.1, G.2

Prepare

Purpose

This BioDigest can be used as a brief overview of the nature of science and the characteristics of life. If time is limited, you may wish to use this unit summary to teach these concepts in place of Chapter 1.

Key Concepts

Students are introduced to the characteristics of life and the methods of science. They learn about the nature and limitations of science and technology.

1 Focus

Bellringer

Show students a candle flame and a caged mouse. Ask students to explain the similarities and differences of the two to elicit the characteristics of living things. **L1**

2 Teach

Quick Demo

📦 *Visual-Spatial* Display a small animal such as an earthworm. Ask students to describe its observable characteristics of life. Then ask them how they could observe life characteristics they cannot see. **L2** 📦

For a **preview** of the what is biology unit, study this BioDigest before you read the chapter. After you have studied the chapter, you can use the BioDigest to **review** the unit.

What Is Biology?

L *iving things abound almost everywhere on Earth—in deep ocean trenches, atop the highest mountains, in dry deserts, and in wet tropical forests. Biology is the study of living organisms and the interactions among them. Biologists use a variety of scientific methods to study the details of life.*

All living things share certain characteristics. What characteristics tell you these robins are living organisms?

Characteristics of Life

Biologists have formulated a list of characteristics by which we can recognize living things.

Organization

All living things are organized into cells. Organisms may be composed of one cell or many cells. Cells are like rooms in a building. You can think of a many-celled organism as a building containing many rooms. Groups of rooms in different areas of the building are used for different purposes. These areas are analogous to the tissues, organs, and body systems of plants and animals.

Homeostasis

A stable internal environment is necessary for life. Organisms maintain this stability through homeostasis, which is a process that requires the controlled use of energy in cells. Plants obtain energy by converting light, water, and carbon dioxide into food. Other organisms obtain their energy indirectly from plants.

Response to a Stimulus

Living things respond to changes in their external environment. Any change, such as a rise in temperature or the presence of food, is a stimulus.

FOCUS ON CAREERS

Biology at Work

T housands of career opportunities are available in the biological sciences. Some of these careers require only a high school education. Others require a college degree, or even an advanced degree. Many careers in biology involve work as a research biologist in the field or in a laboratory setting.

If you enjoy working outdoors, a career in field biology may be for you.

Other careers also rely on skills or knowledge about biology. Doctors and dentists, nurses and laboratory technicians, florists, foresters, and zookeepers all must have a knowledge of biology. Careers related to the biological sciences also include food processing, farming, and ranching. Can you think of some other careers in which biology plays an important role?

32

Multiple Learning Styles

Look for the following logos for strategies that emphasize different learning modalities.

📦 *Visual-Spatial* Quick Demo, p. 32; Microscope Activity, p. 33; Reteach, p. 33; Extension, p. 33

Resource Manager

Reinforcement and Study Guide, p. 5-6 **L2**

Content Mastery, p. 5-8 **L1**

Growth and Development

When living things grow, their cells enlarge and divide. As organisms age, other changes also take place. Development consists of the changes in an organism that take place over time.

Reproduction

Living things reproduce by transmitting their hereditary information from one generation to the next.

Scientific Methods

Scientists employ a variety of scientific methods to answer questions and solve problems. Not all investigations will use all methods, and the order in which they are used will vary.

Observation

Curiosity leads scientists to make observations that raise questions about natural phenomena.

Hypothesis

A statement that can be tested and presents a possible solution to a question is a hypothesis.

Experiment

After making a hypothesis, the next step is to test it. An experiment is a formal method of testing a hypothesis. In a controlled experiment, two groups are tested and all conditions except one are kept the same for both groups. The single condition that changes is the independent variable. A condition caused by the change in the independent variable is a dependent variable.

Theory

When a hypothesis has been confirmed by many experiments, it may become a theory. Theories explain natural phenomena.

Many experiments are conducted in the laboratory, where conditions can be easily controlled.

BioDigest Assessment

Understanding Main Ideas

1. The basic unit of organization of living things is _____.
   a. an atom   **c.** a cell
   b. an organism   **d.** an organ

2. Storing energy obtained from food is an example of _____.
   a. evolution   **c.** response
   b. homeostasis   **d.** growth

3. A hypothesis that is supported many times becomes _____.
   a. an experiment   **c.** a theory
   b. a conclusion   **d.** an observation

4. All of the procedures scientists use to answer questions are _____.
   a. life characteristics   **c.** research
   b. scientific methods   **d.** hypotheses

5. To test a hypothesis, a scientist may _____.
   a. do an experiment
   b. write a theory
   c. do research in a library
   d. make some observations

Thinking Critically

1. List the characteristics you would check to see if a pine tree is a living thing. Give an example that shows how the tree exhibits each characteristic.

2. Compare the characteristics of life with the flames of a fire. How are they similar and different?

3. Why do most experiments have a control? Describe an experiment that does not have a control.

33

Microscope Activity

☑ *Visual-Spatial* Have students examine slides of protists. Ask what characteristics of life the organisms share.' **L2** **ELL**

3 Assess

Check for Understanding

Have students explain why science and technology cannot answer all questions. **L2**

Reteach

☑ *Visual-Spatial* Ask students to observe a seed and make a hypothesis about whether it is alive. Have them plant the seed and record the characteristics of life they note over the period of a week or two. **L1** **ELL**

Extension

☑ *Visual-Spatial* Have students make a bulletin board collage of pictures of living organisms. **L1** **ELL**

✔ Assessment

Skill Provide students with examples of living organisms and nonliving objects. Ask them to group the living organisms together and the nonliving objects together. Include some once-living objects to make a third group. **L2**

4 Close

Discussion

Ask students to discuss specific examples of a stimulus and the resulting response they may have observed in common animals. Have students identify both the stimulus and response, and speculate about how the response might help the organism maintain homeostasis.

Understanding Main Ideas

1. c
2. b
3. c
4. b
5. a

Thinking Critically

1. Cells would show cellular structure; homeostasis is shown by making and using energy from sunlight; growing toward the light is a response to stimulus; growth and development would be shown in the changes since the plant was a seed; reproduction occurs when it produces new seeds.

2. Fires can grow, use energy, and reproduce. Fires are not composed of cells.

3. A control is a basis for comparison. A behavior experiment may not have a control.

33

Ecology

Unit Overview

This unit focuses on the relationships and interactions that exist among organisms and their environments. In Chapter 2, students are introduced to ecology and the biotic and abiotic factors that exist in an ecosystem. Chapter 3 centers on the development of communities and describes major world biomes. In Chapter 4, environmental factors that limit population growth are presented, and students study the effects of demographics. Chapter 5 brings the unit to a close with a discussion of people's impact on the environment and threats to biodiversity. Strategies of conservation biology are described.

Introducing the Unit

Ask students to look at the scarlet macaws in the photograph and describe how these birds are dependent on both living and nonliving things in their environments. Explain that ecology focuses on the interactions that take place in an environment.

Unit 2

Ecology

A tropical rain forest ecosystem consists of interactions among organisms, and between organisms and their environment. For example, rain forest plants are adapted to use the ample water and sunlight in the production of nutrients. The plants use these nutrients for their own growth and development, and, in turn, the nutrients that make up the plants may then be passed to animals that feed on them. Scarlet macaws eat seeds and fruits from rain forest trees, but they also eat clay soil that helps to detoxify many of the poisonous plants that they eat.

UNIT CONTENTS

2 Principles of Ecology

3 Communities and Biomes

4 Population Biology

5 Biological Diversity and Conservation

BioDigest Ecology

UNIT PROJECT

BIOLOGY *Online* Use the Glencoe Science Web site for more project activities that are connected to this unit. **science.glencoe.com**

34

Unit Projects

Develop a Model of an Ecosystem

Have students do one of the projects for this unit as described on the Glencoe Science Web site. As an alternative, students can do one of the projects listed on these two pages.

Interview

Linguistic Students can interview a pet shop owner to find out how to keep fish in an aquarium. Students can describe how to maintain a healthy environment for the fish. **L1**

Display

Visual-Spatial Ask students to make a bulletin board that describes the experimental Biosphere II in Arizona and explains why it did not work. **L1** **ELL**

Unit 2

Advance Planning

Chapter 2
- Purchase seeds and gather materials for MiniLab 2-1.
- Order bromothymol blue and antacid for MiniLab 2-2.
- Set up or borrow a fish tank for a Quick Demo.
- Gather or purchase lichens for a Quick Demo.
- Order cultures of *Paramecium* and *Didinium* for the BioLab.
- Gather materials for the Alternative Lab.

Chapter 3
- Gather lichen and other materials for MiniLab 3-1.
- Purchase or borrow plants for a Quick Demo.
- Purchase plankton and gather materials for Mini Lab 3-2.
- Gather pond water and other materials for the BioLab.
- Gather sod for a Quick Demo.
- Gather materials for the Alternative Lab.

Chapter 4
- Purchase bananas and gather materials for MiniLab 4-1.
- Purchase radish seeds. Gather petri dishes and napkins for the Alternative Lab.
- Purchase materials for the BioLab.

Chapter 5
- Purchase or gather seeds and other materials for the Alternative Lab.
- Gather gingko leaves for a Quick Demo.
- Gather soil, water, and containers for MinLab 5-2.

Make a Model

Kinesthetic Students can create a working aquatic ecosystem that contains both biotic and abiotic elements in a healthy balance. **L1** **ELL**

Use the Library

Intrapersonal Have students use the library to find out why space exploration may be dependent on the quest for artificial closed ecosystems. **L2**

Final Report

Have students present their group's findings in an oral report that could be understood by students at your local middle school. **L3**

Chapter 2 Organizer

Refer to pages 4T-5T of the Teacher Guide for an explanation of the National Science Education Standards correlations.

Section	Objectives	Activities/Features
Section 2.1 **Organisms and Their Environment** National Science Education Standards UCP.1-3; A.2; C.4, C.5, C.6; F.3; G.1-3 (2 sessions, 1 block)	1. **Distinguish** between the biotic and abiotic factors in the environment. 2. **Compare** the different levels of biological organization and living relationships important in ecology. 3. **Explain** the difference between a niche and a habitat.	**MiniLab 2-1:** Salt Tolerance of Seeds, p. 38 **Problem-Solving Lab 2-1:** p. 39 **Careers in Biology:** Science Reporter, p. 40
Section 2.2 **Nutrition and Energy Flow** National Science Education Standards UCP.1-3; A.1, A.2; B.6; C.4, C.5, C.6; D.2; F.3-5; G.1, G.2 (3 sessions, 2½ blocks)	4. **Compare** how organisms satisfy their nutritional needs. 5. **Trace** the path of energy and matter in an ecosystem. 6. **Analyze** how nutrients are cycled in the abiotic and biotic parts of the biosphere.	**Problem-Solving Lab 2-2.** p. 52 **MiniLab 2-2:** Detecting Carbon Dioxide, p. 56 **Inside Story:** The Carbon Cycle, p. 57 **Design Your Own BioLab:** How can one population affect another? p. 60 **Biology & Society:** The Florida Everglades—An Ecosystem at Risk, p. 62

Need Materials? Contact Carolina Biological Supply Company at 1-800-334-5551 or at **http://www.carolina.com**

MATERIALS LIST

BioLab
p. 60 microscope, microscope slides, coverslips, droppers, beakers or jars, sterile pond water, culture of *Didinium*, culture of *Paramecium*

MiniLabs
p. 38 seeds (40), small beaker (2), paper towels, zipper-lock plastic bags (2), labels, water, 10% salt water solution
p. 56 test tubes (2), bromthymol blue solution, antacid tablet, drinking straw

Alternative Lab
p. 40 corn seeds (60), pinto bean seeds (60), paper cups (6), plastic sandwich bags (6), water, paper towels, graduated cylinder, labels, pencil

Quick Demos
p. 40 aquarium setup
p. 45 lichens
p. 55 glass bowl, soil, plants, water, plastic wrap

Key to Teaching Strategies

L1 Level 1 activities should be appropriate for students with learning difficulties.

L2 Level 2 activities should be within the ability range of all students.

L3 Level 3 activities are designed for above-average students.

ELL ELL activities should be within the ability range of English Language Learners.

COOP LEARN Cooperative Learning activities are designed for small group work.

P These strategies represent student products that can be placed into a best-work portfolio.

These strategies are useful in a block scheduling format.

Teacher Classroom Resources

Section	Reproducible Masters	Transparencies
Section 2.1 **Organisms and Their Environment**	**Reinforcement and Study Guide,** pp. 7-8 `L2` **Critical Thinking/Problem Solving,** p. 2 `L3` **BioLab and MiniLab Worksheets,** p. 9 `L2` **Laboratory Manual,** pp. 9-10 `L2` **Tech Prep Applications,** pp. 3-4 `L2` **Content Mastery,** pp. 9-10, 12 `L1`	**Section Focus Transparency 4** `L1` `ELL`
Section 2.2 **Nutrition and Energy Flow**	**Reinforcement and Study Guide,** pp. 9-10 `L2` **BioLab and MiniLab Worksheets,** pp. 10-12 `L2` **Concept Mapping,** p. 2 `L3` `ELL` **Content Mastery,** pp. 9, 11-12 `L1` **Laboratory Manual,** pp. 11-14 `L2` **Inside Story Poster** `ELL`	**Section Focus Transparency 5** `L1` `ELL` **Basic Concepts Transparency 1** `L2` `ELL` **Basic Concepts Transparency 2** `L2` `ELL` **Reteaching Skills Transparencies 1, 2, 3** `L1` `ELL`

Assessment Resources

Chapter Assessment, pp. 7-12
MindJogger Videoquizzes
Performance Assessment in the Biology Classroom
Alternate Assessment in the Science Classroom
ExamView® Pro Software 🖫
BDOL Interactive CD-ROM, Chapter 2 quiz

Additional Resources

Spanish Resources `ELL`
English/Spanish Audiocassettes `ELL`
Cooperative Learning in the Science Classroom `COOP LEARN`
Lesson Plans/Block Scheduling

NATIONAL GEOGRAPHIC — **Teacher's Corner**

Index to National Geographic Magazine
The following articles may be used for research relating to this chapter.
"Rain Forest Canopy: The High Frontier,"
by Edward O. Wilson, December 1991.

GLENCOE TECHNOLOGY

The following multimedia resources are available from Glencoe.

Biology: The Dynamics of Life
CD-ROM `ELL`

Video: *How Organisms Interact*
Video: *Symbiosis*
BioQuest: *Antarctic Food Web*
Exploration: *Pyramid of Energy*

Videodisc Program

How Organisms Interact
Symbiosis
The Everglades

The Infinite Voyage

Secrets From a Fozen World

The Secret of Life Series

 Niche
 Predator–Prey
Mutualism

GETTING STARTED DEMO

Have students examine the chapter opener photographs. Discuss the feeding relationships shown. *The trout eats the mosquito. Human blood nourishes the mosquito.* Help students realize that humans, like the mosquito and trout, depend on other living things for food. Have students choose an animal product they eat and then show this feeding relationship in a food chain—for example, sun to grass to cow to human. **L1**

Theme Development

Systems and interactions is a theme of this chapter. Organisms have niches because of interactions among biotic and abiotic factors. A critical aspect of niche is how organisms obtain nutrients and energy. **Energy** flow, a major theme of Section 2, is traced through the trophic levels of a food chain.

Resource Manager

Section Focus Transparency 4 and Master **L1** **ELL**

READING BIOLOGY

Glencoe's *Biology: The Dynamics of Life* contains many resources to assist a student's reading skills. Each chapter contains figures with expanded captions that expand on written material. Word Origins, located along the side of text, expand knowledge of biology vocabulary. Glencoe's Content Mastery Booklet helps develop reading skills while reinforcing content. In addition, use the Interactive Tutor for *Biology: The Dynamics of Life* on the Glencoe Web site to reinforce vocabulary. **science.glencoe.com**

36

What You'll Learn

- You will describe ecology and the work of ecologists.
- You will identify important aspects of an organism's environment.
- You will trace the flow of energy and nutrients in the living and nonliving worlds.

Why It's Important

To understand life, you need to know how organisms meet their needs in their natural environments. To reduce the impact of an expanding human population on the natural world, it is important to understand how living things depend on their environments.

READING BIOLOGY

As you look through Chapter 2, pick out one or two new vocabulary words from the text. While reading the chapter, try to see how charts or diagrams demonstrate the vocabulary words' meanings.

To find out more about ecology, visit the Glencoe Science Web site.
science.glencoe.com

You might think mosquitoes are pests, but for trout and other animals, mosquitoes and their larvae are a major food source.

36 PRINCIPLES OF ECOLOGY

Multiple Learning Styles

Look for the following logos for strategies that emphasize different learning modalities.

Visual-Spatial Portfolio, pp. 39, 53; Quick Demo, p. 40; Enrichment, p. 41; Reinforcement, p. 55

Interpersonal Project, p. 43; Reteach, p. 47; Meeting Individual Needs, pp. 51, 54

Intrapersonal Portfolio, p. 42; Meeting Individual Needs, p. 42

Linguistic Biology Journal, pp. 44, 50, 56; Project, p. 46; Extension, p. 47; Portfolio, p. 58

Logical-Mathematical Discussion Question, p. 54

Naturalist Reinforcement, pp. 46, 49; Activity, p. 47; Biology Journal, p. 52; Going Further, p. 62

Organisms and Their Environment

A s shown in the photographs, people can impact plant and animal communities in both positive and negative ways. Learning how ecological principles explain interaction between organisms and their environment can help you understand environmental issues and form your own opinions about them. In this section, you will learn some of the history and the focus of ecology.

Animals wander into cities in search of food (above). A wildlife rehabilitator releases an owl (inset).

SECTION PREVIEW

Objectives

Distinguish between the biotic and abiotic factors in the environment.

Compare the different levels of biological organization and living relationships important in ecology.

Explain the difference between a niche and a habitat.

Vocabulary

ecology
biosphere
abiotic factor
biotic factor
population
community
ecosystem
habitat
niche
symbiosis
commensalism
mutualism
parasitism

What Is Ecology?

Do you know anyone who likes to observe nature? Perhaps it is a person who knows the names of many animals, plants, or rocks. People have enjoyed studying nature for thousands of years. Birdwatchers know the names and behaviors of the birds in their area. Some people carefully record observations of rainfall and temperature. Others make it a hobby to study plants; they keep log books with records of when plants produced leaves, flowers, and fruit, as shown in *Figure 2.1*. Some people who are interested in nature record observations, discuss their results, and note how patterns change from year to year.

Figure 2.1
An amateur nature study log book from the 17th century.

37

Section 2.1

Prepare

Key Concepts

Students are provided with an overview of the history of ecology, living and nonliving factors in an environment, and close relationships among organisms that enhance survival.

Planning

■ Gather seeds, salt, and other materials for MiniLab 2-1. Mustard seeds work very well. You can find them in the spice section of any supermarket.

■ Set up or borrow a fish tank for the Quick Demo.

■ Gather or purchase lichens for the Quick Demo.

■ Gather cups, bags, and seeds (pinto beans or corn) for the Alternative Lab.

1 Focus

Bellringer

Before presenting the lesson, display **Section Focus Transparency 4** on the overhead projector and have students answer the accompanying questions.
L1 ELL

Assessment Planner

Portfolio Assessment
    Portfolio, TWE, pp. 39, 42, 53
Performance Assessment
    MiniLab, TWE, p. 38
    Alternative Lab, TWE, pp. 40-41
    Assessment, TWE, pp. 42, 58, 59
    MiniLab, SE, pp. 38, 56
    BioLab, SE, pp. 60-61

Knowledge Assessment
    Assessment, TWE, pp. 47, 55
    Section Assessment, SE, pp. 47, 59
    Chapter Assessment, SE, pp. 63-65
Skill Assessment
    Problem-Solving Lab, TWE, pp. 39, 52
    MiniLab, TWE, p. 56
    BioLab, TWE, pp. 60-61

2 Teach

MiniLab 2-1

Purpose
Students will experiment with the abiotic factor of salinity to determine if seed germination is affected.

Process Skills
collect data, experiment, interpret data

Teaching Strategies
■ Prepare the 10% sodium chloride (table salt) solution by dissolving 100 g of table salt in 900 mL of tap water.
■ Allow students to form a hypothesis prior to the experiment. At the conclusion of the activity, ask if their hypothesis was supported.
■ Ask students to identify the control and the independent and dependent variables in the experiment.
■ Have students work in small groups.

Expected Results
Seeds soaked in water will show germination. Seeds in salt water will show little or no germination.

Analysis
1. Yes. Seeds soaked in tap water germinate. Seeds soaked in salt water do not.
2. salinity of water; germination
3. No, each seed type may respond differently to salinity. Experimentation is needed.

✓ Assessment
Performance Have students design an experiment that would determine the minimum percentage of salinity that can be tolerated by a specific seed type and still allow germination to occur. Use the Performance Task Assessment List for Assessing a Whole Experiment and Planning the Next Experiment in **PASC**, p. 33. **L2**

38

MiniLab 2-1 Experimenting

Salt Tolerance of Seeds Salinity, or the amount of salt dissolved in water, is an abiotic factor. Might salt water affect how certain seeds sprout or germinate? Experiment to find out.

Salt marsh

Freshwater pond

Procedure
1. Soak 20 seeds in freshwater and 20 seeds in salt water overnight.
2. The next day, wrap the seeds in two different moist paper towels. Slide the towels into separate self-sealing plastic bags.
3. Label the bags "fresh" and "salt."
4. Examine all seeds two days later. Count the number of seeds in each treatment that show signs of root growth or sprouting, which is called germination. Record your data. **CAUTION: Be sure to wash your hands after handling seeds.**

Analysis
1. Did the germination rates differ between treatments? If yes, how?
2. What abiotic factor was tested in this experiment? What biotic factor was influenced?
3. Might all seeds respond to salt in a similar manner? How could you find out?

Ecology defined
The branch of biology that developed from natural history is called ecology. **Ecology** is the scientific study of interactions among organisms and their environments. Ecological study reveals relationships among living and nonliving parts of

---

the world. Ecology combines information and techniques from many scientific fields, including mathematics, chemistry, physics, geology, and other branches of biology.

You have learned that scientific research includes both descriptive and quantitative methods. Most ecologists use both types of research. They obtain descriptive information by observing organisms in the field and laboratory. They obtain quantitative data by making measurements and carrying out carefully controlled experiments. Using these methods, ecologists learn a great deal about relationships, such as what organisms a coyote eats, how day length influences the behavior of migrating birds, how tiny shrimp help rid ocean fishes of parasites, or how acid rain threatens some of Earth's forests.

Aspects of Ecological Study

As far as we know, life exists only on Earth. Living things can be found in the air, on land, and in both fresh- and salt water. The **biosphere** (BI uh sfihr) is the portion of Earth that supports life. It extends from high in the atmosphere to the bottom of the oceans. This life-supporting layer may seem extensive to us, but if you could shrink Earth to the size of an apple, the biosphere would be thinner than the apple's peel.

Although it is thin, the biosphere is very diverse and supports a wide range of organisms. The climate, soils, plants, and animals in a desert are very different from those in a tropical rain forest. Living things are affected by both the physical environment and by other living things. Ecologists study these interactions among different organisms and their environments.

TechPrep

People and Habitats
People may alter habitats, which changes the abiotic factors. Have students survey the school grounds and find five examples of specific changes that people made and how each has affected the abiotic environment. For example, a parking lot covered with asphalt reduces the amount of water entering the ground. On sunny days the asphalt gets hotter than unpaved land. **L1** **ELL**

The nonliving environment: Abiotic factors

Ecology includes the study of features of the environment that are not living because these features are an important part of an organism's life. For example, a complete study of the ecology of moles would include an examination of the types of soil in which these animals dig their tunnels. Similarly, a thorough investigation of the life cycle of trout would need to include whether these fish lay their eggs on rocky or sandy stream bottoms. The nonliving parts of an organism's environment are the **abiotic factors** (ay bi AHT ihk). Examples of abiotic factors include air currents, temperature, moisture, light, and soil.

Abiotic factors have obvious effects on living things and often determine which species survive in a particular environment. For example, lack of rainfall can cause drought in a grassland, as shown in *Figure 2.2*. Can you think of changes in a grassland that might result from a drought? Grasses would grow more slowly, wildflowers would produce fewer seeds, and the animals that depend on plants for food would find it harder to survive. Examine other ways that abiotic factors affect living things in the *MiniLab* and *Problem-Solving Lab* shown on these pages.

Problem-Solving Lab 2-1   Interpreting Data

How does an abiotic factor affect food production?
Green plants carry out the process of photosynthesis. Glucose, a sugar, is one of the products produced during this process. Thus, glucose production can be used as a means for judging the rate at which the process of photosynthesis is occurring.

Analysis
Examine the following graph of a plant called salt bush *(Atriplex)*. It shows how this plant's glucose production is influenced by temperature.

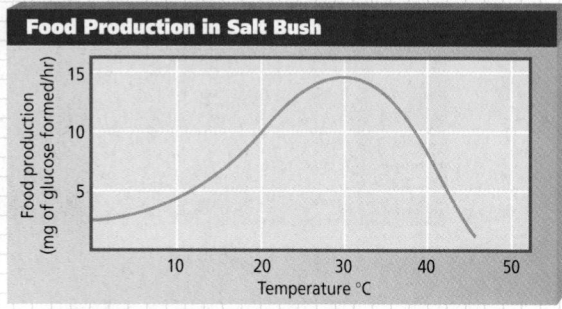

Food Production in Salt Bush

(Graph: x-axis Temperature °C from 10 to 50; y-axis Food production (mg of glucose formed/hr) from 5 to 15)

Thinking Critically
1. Name the abiotic factor influencing photosynthesis and describe the influence of this factor on photosynthesis.
2. Name the biotic factor being influenced.
3. Based on the graph, describe the type of ecosystem this plant might live in. Explain your answer.
4. Does the graph tell you how the rate of photosynthesis might vary for plants other than salt bushes? Explain your answer.
5. Hypothesize why the formation of glucose drops quickly after reaching 30°C.

Figure 2.2
Droughts are common in grasslands. As the grasses dry out, they turn yellow and appear to be dead, but new shoots grow in the low-lying areas soon after it rains. Some animal species are adapted to living in grasslands by their ability to burrow underground and sleep through the dry periods.

39

Problem-Solving Lab 2-1

Purpose
Students will determine how temperature influences the rate at which photosynthesis occurs for a specific plant.

Process Skills
analyze information, draw a conclusion, interpret data

Teaching Strategies
■ You may wish to introduce the process of photosynthesis by describing the raw materials needed and the role the process plays in providing food for all life forms.
■ You might explain the term *optimum* as it relates to the optimum temperature at which photosynthesis occurs for salt bush.

Thinking Critically
1. Temperature. As temperature increases, the photosynthesis rate also increases until a maximum of 30°C (optimum temperature) is reached. Above 30°, the photosynthesis rate decreases.
2. food production
3. Salt bush appears to benefit from warm temperatures because it produces maximum food amounts at higher temperatures. It may, therefore, be found living in the desert.
4. The graph is specific for salt bush. The responses of other plants to temperature would have to be determined experimentally.
5. High temperatures may damage or kill the cells responsible for photosynthesis.

✔ Assessment
Skill Have students write a lab report summarizing the results of the lab. Use the Performance Task Assessment List for Lab Report in **PASC**, p. 47. **L2**

MEETING INDIVIDUAL NEEDS

Learning Disabled
Visual-Spatial Provide students who may require reinforcement of the concepts of biotic and abiotic factors with photographs from old nature magazines. Ask students to identify all the biotic factors in each photograph. Ask them to explain why they identified these factors as biotic. **L1 ELL**

✔ Portfolio

Studying the Local Environment
Visual-Spatial Ask students to observe the natural environment of the area in which they live on several different days. Tell them to prepare a table in which they list the biotic and abiotic factors of the environment. Have students summarize the importance of the abiotic factors listed in their tables. **L2 ELL P**

Figure 2.3 How might other living things affect this goldfish?

CAREERS IN BIOLOGY

Science Reporter

Does science fascinate you? Can you explain complex ideas and issues in a clear and interesting way? If so, you should consider a career as a science reporter.

Skills for the Job

As a science reporter, you are a writer first and a scientist second. A degree in journalism and/or a scientific field is usually necessary, but curiosity and good writing skills are also essential. You might work for newspapers, national magazines, medical or scientific publications, television networks, or Internet news services. You could work as a full-time employee or a freelance writer. You must read constantly to stay up-to-date. Many science reporters attend scientific conventions and events to find news of interest to the public. Then they carefully and accurately translate what's new so nonscientists can understand it.

BIOLOGY *Online*

For more careers in related fields, be sure to check the Glencoe Science Web site. **science.glencoe.com**

The living environment: Biotic factors

Look at the goldfish in *Figure 2.3.* Now consider its relationships with other organisms. It may depend on other living things for food, or it may be food for other life. The goldfish needs members of the same species to reproduce. To meet its needs, the goldfish may compete with organisms of the same or different species.

A key consideration of ecology is that living organisms affect other organisms. All the living organisms that inhabit an environment are called **biotic factors** (by AHT ihk). Ecologists investigate how biotic factors affect different species. To help them understand the interactions of the biotic and abiotic parts of the world, ecologists have organized the living world into levels.

Levels of Organization in Ecology

The study of an individual organism, such as a male deer, known as a buck, might reveal what food items it prefers, how often it eats, and how far it roams to search for food or shelter. Although it spends a large part of its time alone, it does interact with other individuals of its species. For example, it periodically goes in search of a mate, which may require battling with other bucks.

All organisms depend on others for food, shelter, reproduction, or protection. So you can see that the study of an individual would provide only part of the story of its life cycle. To get a more complete picture requires studying its relationships with other organisms.

Ecologists study interactions among organisms at several different

---

Organism

Figure 2.4
Ecology deals with several levels of biological organization, including organisms, populations, communities, ecosystems, biomes, and the biosphere.

Populations

Communities

Ecosystems

Biosphere

levels, as shown in *Figure 2.4.* They study individual organisms, interactions among organisms of the same species, and interactions among organisms of different species. Ecologists also study how abiotic factors affect groups of interacting species.

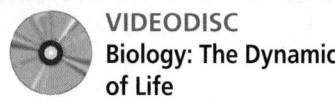

Figure 2.5
These marsh marigolds represent a population of organisms. What characteristics are shared by this group of flowers that make them a population?

Figure 2.6
Adult frogs and their young have different food requirements. This limits competition for food resources for the species.

Interactions within populations

The marsh marigolds shown in *Figure 2.5* form a population. A **population** is a group of organisms of one species that interbreed and live in the same place at the same time.

Members of the same population may compete with each other for food, water, or other resources. Competition occurs only if resources are in short supply. How organisms in a population share the resources of their environment determines how far apart organisms live and how large the populations become.

Some species have adaptations that reduce competition within a popula-

A Eggs that adult frogs lay in the water hatch into tadpoles. Tadpoles have gills, live in water, and eat algae and small aquatic creatures.

B Adult frogs live both on land and in the water. They breathe air and eat insects such as dragonflies, grasshoppers, and beetles.

42 PRINCIPLES OF ECOLOGY

tion. An example is the life cycle of a frog, shown in *Figure 2.6*. The juvenile stage of the frog, called the tadpole, not only looks very different from the adult but also has completely different food requirements. Many species of insects, including dragonflies and moths, also produce juveniles that differ from the adult in body form and food requirements.

Individuals interact within communities

No species lives independently of other species. Just as a population is made up of individuals, a community is made up of several populations. A **community** is a collection of interacting populations. An example of a community is shown in *Figure 2.7*.

A change in one population in a community will cause changes in the other populations. Some of these changes can be minor, such as when a small increase in the number of individuals of one population causes a small decrease in the size of another population.

For example, if the population of mouse-eating hawks increases slightly, the population of mice will, as a result, decrease slightly. Other changes might be more extreme, as when the size of one population

grows so large it begins affecting the food supply for another species in the community.

Interactions among living things and abiotic factors form ecosystems

In a healthy forest community, interactions among populations might include birds eating insects, squirrels eating nuts from trees, mushrooms growing from decaying leaves or bark, and raccoons fishing in a stream. In addition to population interactions, ecologists also study interactions among populations and their physical surroundings in ecosystems. An **ecosystem** is made up of the interactions among the populations in a community and the community's physical surroundings, or abiotic factors.

There are three major kinds of ecosystems. Terrestrial ecosystems are those located on land. Examples include forests, meadows, and desert scrub. Aquatic ecosystems occur in both fresh- and saltwater. Freshwater ecosystems include ponds, lakes, and streams. Saltwater ecosystems, also called marine ecosystems, make up approximately 75 percent of Earth's surface. *Figure 2.8* shows a marine and a freshwater ecosystem.

Figure 2.7
Beech and maple trees dominate this forest community; therefore, it is called a beech-maple forest. Beech-maple forests are found in the eastern United States, Europe, and northeast China.

Figure 2.8
There may be hundreds of populations interacting in a pond or tide pool. How do you think the abiotic factors in these environments affect the biotic factors?

Ⓑ Organisms living in tide pools must survive dramatic changes in abiotic factors. When the tide is high, ocean waves replenish the water in the pool. When the tide is low, water in the pool evaporates.

Ⓐ Dragonflies live near moist meadows and ponds. They feed on small insects they catch while flying. Dragonflies lay their eggs in the pond or on pond plants.

Organisms in Ecosystems

A prairie dog living in a grassland makes its home in burrows it digs underground. Some species of birds make their homes in the trees of a beech-maple forest. In these forests, they find food, avoid enemies, and reproduce. The grassland and beech-maple forests are examples of habitats. A **habitat** (HAB uh tat) is the place where an organism lives out its life. Organisms of different species use a variety of strategies to live and reproduce in their habitats. Habitats can change, and even disappear, from an area. Examples of how habitats change due to both natural and human causes are presented in *Biology and Society* at the end of this chapter.

Niche

Although several species may share a habitat, the food, shelter, and other essential resources of that habitat are often used in different ways. For example, if you turn over a log like the one shown in *Figure 2.9*, you will find millipedes, centipedes, insects, and worms living there. At first, it looks like this community of animals is competing for food because they all live in the same habitat. But close inspection reveals that each feeds in different ways, on

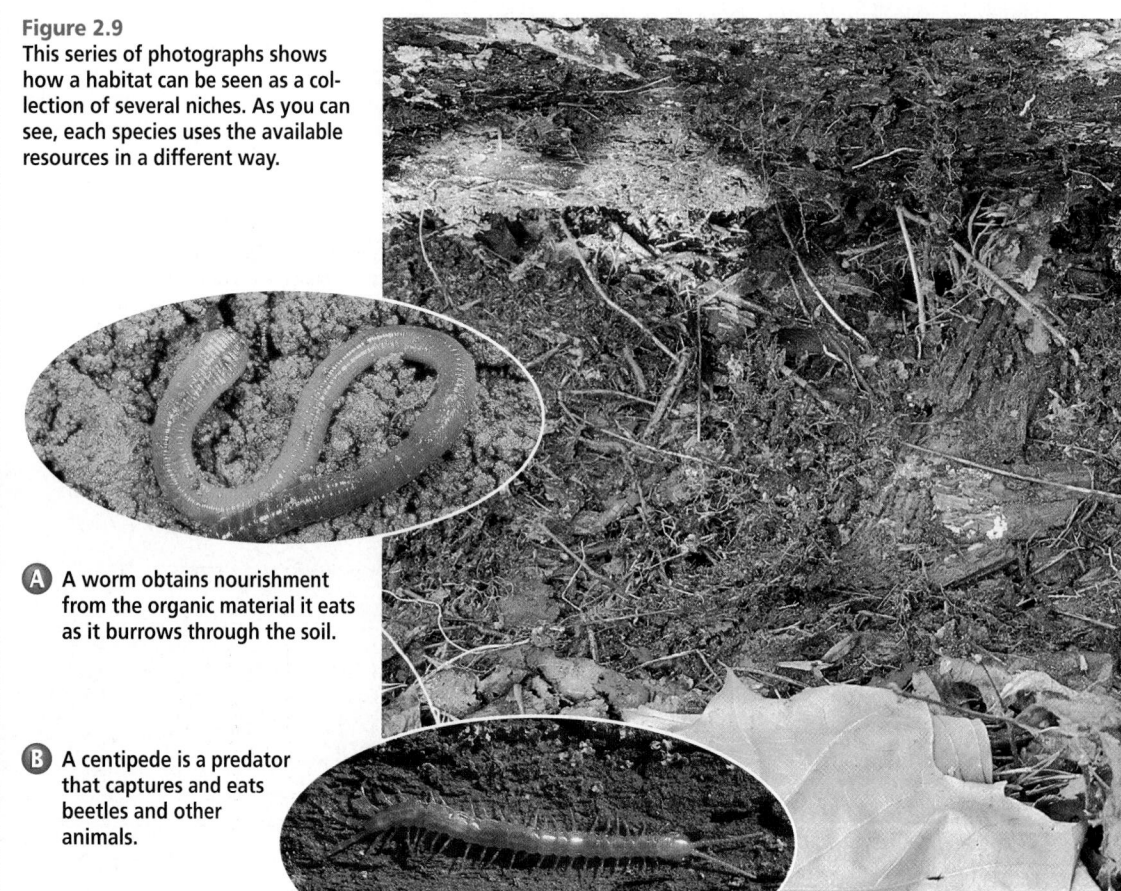

Figure 2.9
This series of photographs shows how a habitat can be seen as a collection of several niches. As you can see, each species uses the available resources in a different way.

A A worm obtains nourishment from the organic material it eats as it burrows through the soil.

B A centipede is a predator that captures and eats beetles and other animals.

different materials, and at different times. These differences lead to reduced competition.

Each species is unique in satisfying all its needs; each species occupies a niche. A **niche** (nich) is the role and position a species has in its environment—how it meets its needs for food and shelter, how it survives, and how it reproduces. A species' niche includes all its interactions with the biotic and abiotic parts of its habitat. It is an advantage for a species to occupy a niche different from those of other species. Life may be harsh in the polar regions, but the polar bear, with its thick coat, flourishes there. Nectar may be deep in the flower, inaccessible to most species, but the hummingbird, with its long beak, gets it. Unique strategies and structures are important to a species' niche and important for reducing competition with other species.

Living relationships

Some species enhance their chances of survival by forming relationships with other species. Biologists once assumed that all organisms living in the same environment are in a continuous battle for survival. Some interactions are harmful to one species, yet beneficial to another. Predators are animals such as lions and insect-eating birds that

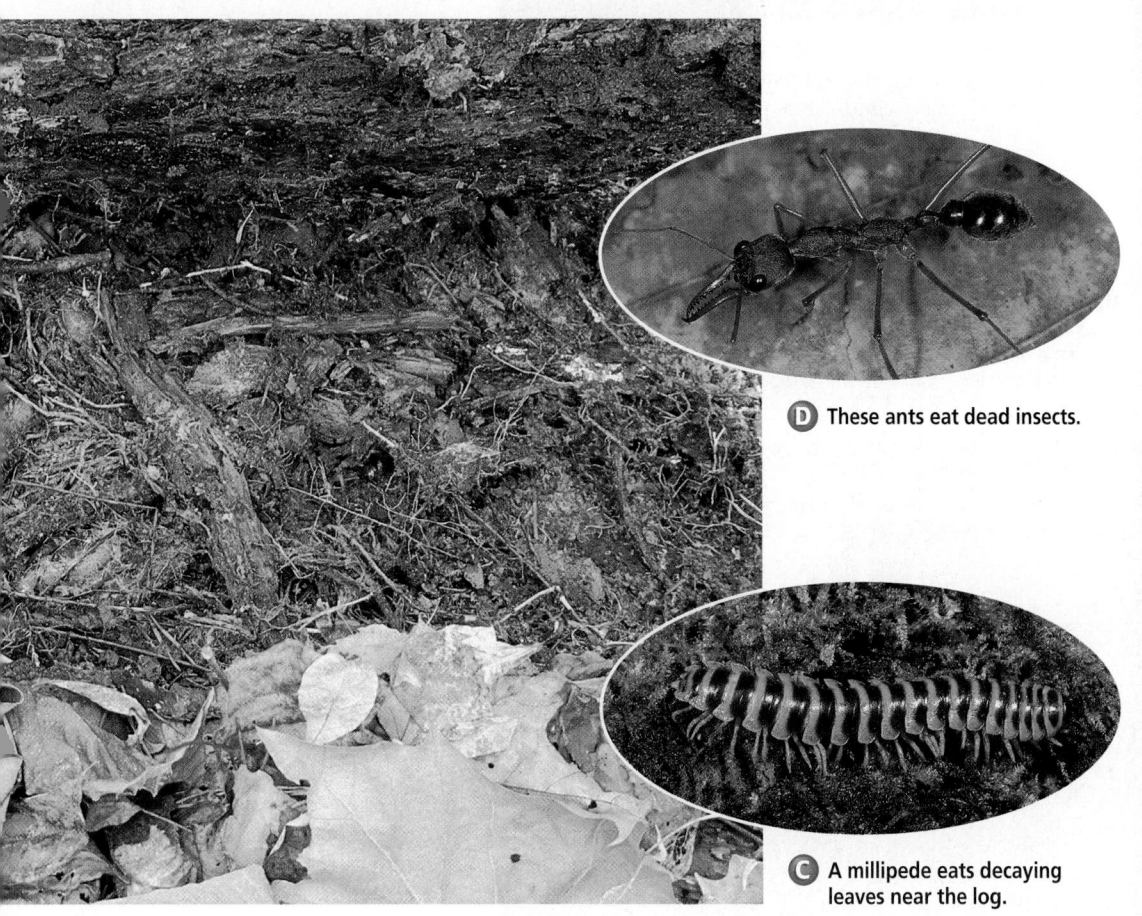

D These ants eat dead insects.

C A millipede eats decaying leaves near the log.

BIOLOGY
Online
Note Internet addresses that you find useful in the space below for quick reference.

_____

_____

_____

_____

_____

3 Assess

Figure 2.10
Red-breasted geese (a) and peregrine falcons (b) both nest in the Siberian arctic in the spring. They share a symbiotic relationship.

Figure 2.11
Spanish moss grows on and hangs from the limbs of trees but does not obtain any nutrients or cause any harm to the trees.

kill and eat other animals. The animals that predators eat are called prey. Predator-prey relationships such as the one between lions and wildebeests involve a fight for survival. Use the *BioLab* at the end of this chapter to more closely examine a predator-prey relationship. But there are other relationships among organisms that help maintain survival in many species. The relationship in which there is a close and permanent association among organisms of different species is called **symbiosis** (sihm by OH sus). Symbiosis means living together.

There are several kinds of symbiosis. A symbiotic relationship between the peregrine falcon and red-breasted goose has evolved in the cold arctic region of Siberia in Russia, as shown in *Figure 2.10.* Normally, the peregrine falcon preys upon the red-breasted goose, but the falcon hunts away from its nesting area. During the nesting season, the falcon fiercely defends its territory from predators. The geese take advantage of this, choosing nesting areas close to those of the falcons, and are thereby protected from predators. The geese benefit from the relationship, and the falcon is neither benefited nor harmed. This is called a commensal relationship. **Commensalism** (kuh MEN suh lihz um) is a symbiotic relationship in which one species benefits and the other species is neither harmed nor benefited.

Commensal relationships also occur among plant species. Spanish moss, a kind of flowering plant that grows on the branch of a tree, is shown in *Figure 2.11.* Orchids, ferns, mosses, and other plants sometimes grow on the branches of larger plants. The larger plants are not harmed, but the smaller plants benefit from the additional habitat.

Sometimes, two species of organisms benefit from living in close association. A symbiotic relationship in which both species benefit is called **mutualism** (MYEW chuh lihz um). Ants and acacia trees living in the subtropical regions of the world illustrate mutualism, as shown in *Figure 2.12.* The ants protect the tree by attacking any animal that tries to feed on it. The tree provides nectar and a home for the ants. In an experiment, ecologists removed the ants from some acacia trees. Results showed that the trees with ants grew faster and survived longer than trees without ants.

Sometimes, one organism harms another. Have you ever owned a dog or cat that was attacked by ticks or fleas? Ticks and fleas, shown in *Figure 2.13*, are examples of parasites. A symbiotic relationship in which one organism derives benefit at the expense of the other is called **parasitism** (PER uh suh tihz um). Parasites have evolved in such a way that they harm, but usually do not kill, the host. If the host dies, the parasite also will die unless it can quickly find another host. Some parasites, such as tapeworms and roundworms, live inside other organisms.

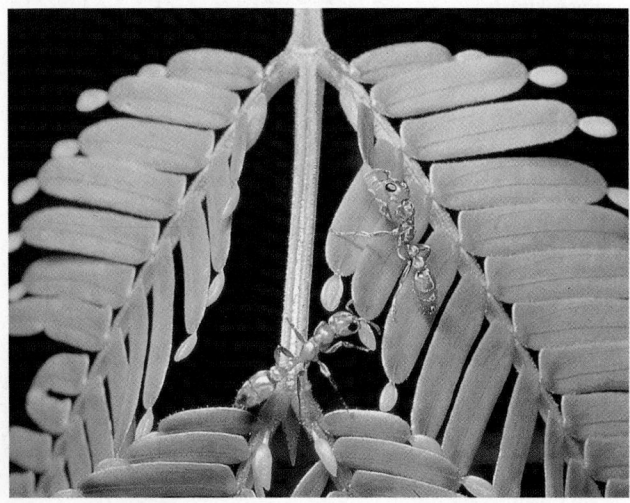

Figure 2.12
These ants and acacia trees both benefit from living in close association. This mutualistic relationship is so strong that in nature the trees and ants are never found apart.

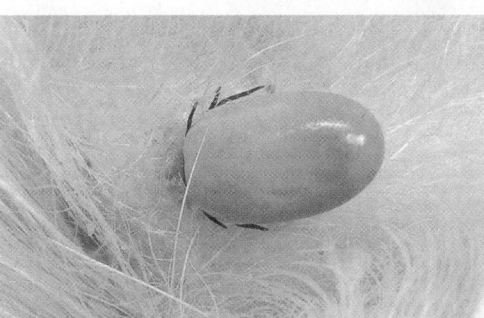

Figure 2.13
Ticks cause harm to the animals they live on when they obtain nutrients from their host animal. This relationship is called parasitism.

WORD *Origin*
ecology
From the Greek words *oikos*, meaning "homestead," and *logos*, meaning "the study of." Ecology is the study of how organisms interact with their environments.

Reteach

Interpersonal Ask students to work in groups to provide examples of biotic and abiotic factors within the classroom or school. **L2** **COOP LEARN**

Extension

Linguistic Ask students to research how abiotic factors limit life in the Arctic tundra or in a desert environment. Have them include a written summary of the information they gather in their portfolios. **L3** **P**

✔ **Assessment**
Knowledge Ask students to consider the school grounds an ecosystem. Have them explain and give examples of populations and communities that live in this "ecosystem." **L2**

4 Close

Activity

Naturalist Have students work in groups to invent four pairs of organisms that display all four symbiotic relationships. Students should name the organisms, describe the interactions, and identify the symbiotic relationships. **L1** **COOP LEARN**

Resource Manager

Reinforcement and Study Guide, pp. 7-8 **L2**
Content Mastery, p. 10 **L1**

Section Assessment

Understanding Main Ideas
1. List several different biotic and abiotic factors in an ecosystem.
2. Compare and contrast populations and communities.
3. Give examples that would demonstrate the differences between the terms niche and habitat.
4. A leaf-eating caterpillar turns into a nectar-eating butterfly. How is this feeding behavior an advantage for this species?

Thinking Critically
5. Clownfish are small, tropical marine fish usually found swimming among the stinging tentacles of sea anemones. What type of symbiotic relationship do these animals have if the clownfish are protected by the sea anemone, but the anemone does not benefit from the clownfish?

SKILL REVIEW
6. **Designing an Experiment** Design an experiment to test the hypothesis that clownfish and sea anemones have a mutualistic relationship. For more help, refer to *Practicing Scientific Methods* in the **Skill Handbook**.

Section Assessment

1. Responses may include the following: *Biotic—tree, grass, human, dog, ant Abiotic—daylight hours, amount of rainfall, humidity, air, soil.*
2. A population consists of a single species that can interbreed and is present in the same place at the same time. A community consists of several populations that interact with one another.
3. Responses may be similar to the following: *Squirrel—habitat: forest; niche: gathers, eats, and stores nuts. Mushroom—habitat: moist forest soil; niche: digests and absorbs organic matter. Bat—habitat: cave; niche: fertilizes flowers, eats insects.*
4. The caterpillar and the butterfly do not compete with each other for food.
5. Commensalism. The clownfish benefits, but the sea anemone is not helped nor hurt.
6. The experiment would compare the growth and health of a sea anemone and clownfish when they live together and when they live separately.

Prepare

Key Concepts

Energy is needed for survival. The ways that organisms obtain and pass energy are depicted with food chains and food webs. This section also addresses trophic levels and the nitrogen, carbon, and water cycles.

Planning

■ Set up a terrarium to help show the water cycle.
■ Gather bromothymol blue and antacid for MiniLab 2-2.
■ Bring in a fertilizer label for the Enrichment.
■ Purchase duckweed for the Project on plant growth.
■ Prepare sterile pond water or follow the directions in the BioLab for substitutions.

1 Focus

Bellringer

Before presenting the lesson, display **Section Focus Transparency 5** on the overhead projector and have students answer the accompanying questions.
L1 **ELL**

Transparency **5** Energy Pathways

SECTION FOCUS
Use with Chapter 2, Section 2.2

❶ What is the source of all energy in this ecosystem?
❷ What path does this energy take to get to the hawk?

BIOLOGY: The Dynamics of Life   SECTION FOCUS TRANSPARENCIES

Objectives

Compare how organisms satisfy their nutritional needs.
Trace the path of energy and matter in an ecosystem.
Analyze how nutrients are cycled in the abiotic and biotic parts of the biosphere.

Vocabulary

autotroph
heterotroph
scavenger
decomposer
food chain
trophic level
food web

Section

2.2 Nutrition and Energy Flow

What eats what? The oriole eats the grasshopper. The grasshopper eats the grass. Organisms, such as the oriole, grasshopper, and grass, need nutrition for growth, repair, and energy. How they satisfy their nutritional needs is an important part of their niche, and an important focus of ecology.

Orioles (above) and grasshoppers (inset) form part of a food chain.

WORD Origin

autotroph
From the Greek words *auto*, meaning "self," and *trophe*, meaning "nourishment." Autotrophs are self-nourishing; they make their own food.

heterotroph
From the Greek words *hetero*, meaning "other," and *trophe*, meaning "nourishment." Heterotrophs consume other organisms for their nutrition.

How Organisms Obtain Energy

A roadrunner sprints, a cactus flowers, an aphid reproduces. Energy drives all these events. One of the most important characteristics of a species' niche is how the species obtains its energy. Ecologists trace the flow of energy through communities to discover nutritional relationships. The ultimate source of the energy is the sun, which supplies the energy that fuels life.

The producers: Autotrophs

Plants use the sun's energy to manufacture food in a process called photosynthesis. Organisms that use energy from the sun or energy stored in chemical compounds to manufacture their own nutrients are called **autotrophs** (AWT uh trohfs). The grass in *Figure 2.14* is an autotroph. Although plants are the most familiar terrestrial autotrophs, some unicellular organisms also make their own nutrients. Most other organisms depend on autotrophs for nutrients and energy.

The consumers: Heterotrophs

A deer nibbles the leaves of a clover plant, a bison munches grass, an owl swallows a mouse. The deer, buffalo, and owl are incapable of producing their own food. They obtain nutrients by eating other organisms. Organisms that cannot make their own food and must feed on other

---

MEETING INDIVIDUAL NEEDS

Learning Disabled

Linguistic Have students who understand the concept of producer and consumer work with students who are having difficulty. Group the terms *producer, plant,* and *autotroph* and *consumer, heterotroph,* and *animal.* Have students analyze the meanings of the groups. **L1**
ELL **COOP LEARN**

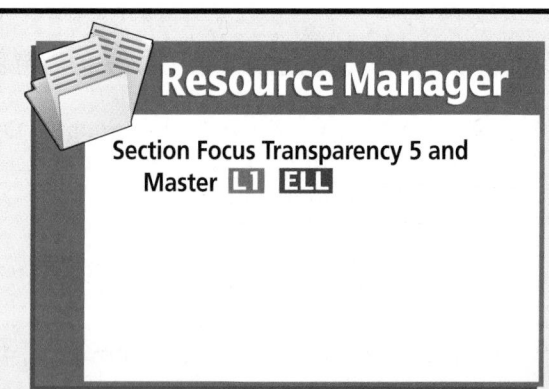

Resource Manager

Section Focus Transparency 5 and Master **L1** **ELL**

Figure 2.14
Many kinds of organisms live in the savanna of East Africa. Identify the autotrophs and the heterotrophs.

organisms are called **heterotrophs** (HET uh ruh trohfs). Heterotrophs include organisms that feed only on autotrophs, organisms that feed only on other heterotrophs, and organisms that feed on both autotrophs and heterotrophs.

Some heterotrophs, such as grazing, seed-eating, and algae-eating animals, feed directly on autotrophs. The wildebeests in *Figure 2.14* depend on plants for their food. A heterotroph that feeds only on plants is called a herbivore. Herbivores include rabbits, grasshoppers, beavers, squirrels, bees, elephants, and fruit-eating bats.

Some heterotrophs eat other heterotrophs. Animals such as lions that kill and eat only other animals are called carnivores. Some animals do not kill for food; instead, they eat animals that have already died. **Scavengers** such as black vultures feed on carrion and refuse, and play a beneficial role in the ecosystem. Imagine for a moment what the environment would be like if there were no vultures to devour animals killed on the African plains, no buzzards to clean up dead animals along roads, and no ants and beetles to remove dead insects and small animals from sidewalks and basements.

Humans are an example of a third type of heterotroph. The teenagers in *Figure 2.15* are eating a variety of foods that include both animal and plant materials. They are omnivores. Raccoons, opossums, and bears are other examples of omnivores.

Figure 2.15
People are omnivores because they eat both autotrophs and heterotrophs.

WORD Origin

herbivore
From the Latin words *herba*, meaning "grass," and *vorare*, meaning "to devour." Herbivores feed on grass and other plants.

carnivore
From the Latin words *caro*, meaning "flesh," and *vorare*, meaning "to devour." Carnivores eat animals.

omnivore
From the Latin words *omnis*, meaning "all," and *vorare*, meaning "to devour." Omnivores eat both plants and animals.

Cultural Diversity

Cultural Adaptations to the Environment

Humans occupy all types of habitats, adapting to Earth's varying environments in many ways. For example, people have designed clothing suited to a wide range of climate conditions, from heavy rainfall to sub-zero temperatures. People around the world use available materials to create shelters adapted to their environments. Ask students to describe some examples of using available materials to meet needs created by the environment. *Inuit groups in North America built homes from snow and ice to conserve heat. Groups living in the southwestern United States built homes using a mud and clay mixture called adobe.*

Ask students to explain what the arrow in all food chains represents. *The arrow shows in which direction matter and energy are moving through the food chain. Why must all second-level organisms be consumers? By definition, these organisms feed on or consume other organisms. Why must all third-level organisms be carnivores and not herbivores? By definition, these organisms feed on other animals and are therefore meat or flesh eaters.* L2

Some organisms, such as fungi, break down and absorb nutrients from dead organisms. These organisms are called **decomposers.** Decomposers break down the complex compounds of dead and decaying plants and animals into simpler molecules that can be more easily absorbed by the decomposers, and by other organisms. Some protozoans, many bacteria, and most fungi carry out this essential process of decomposition.

Figure 2.16
In order for a wetland ecosystem to function, its organisms must depend on each other for a supply of energy. Follow the steps in the wetland food chain shown here.

Matter and Energy Flow in Ecosystems

When you pick an apple from a tree and eat it, you are consuming carbon, nitrogen, and other elements the tree has used to produce the fruit. That apple also contains energy from the sunlight trapped by the tree's leaves while the apple was growing and ripening.

Matter and energy flow through organisms in ecosystems. You have already learned that feeding relationships and symbiotic relationships describe ways in which organisms interact. Ecologists study these interactions to make models that trace the flow of matter and energy through ecosystems.

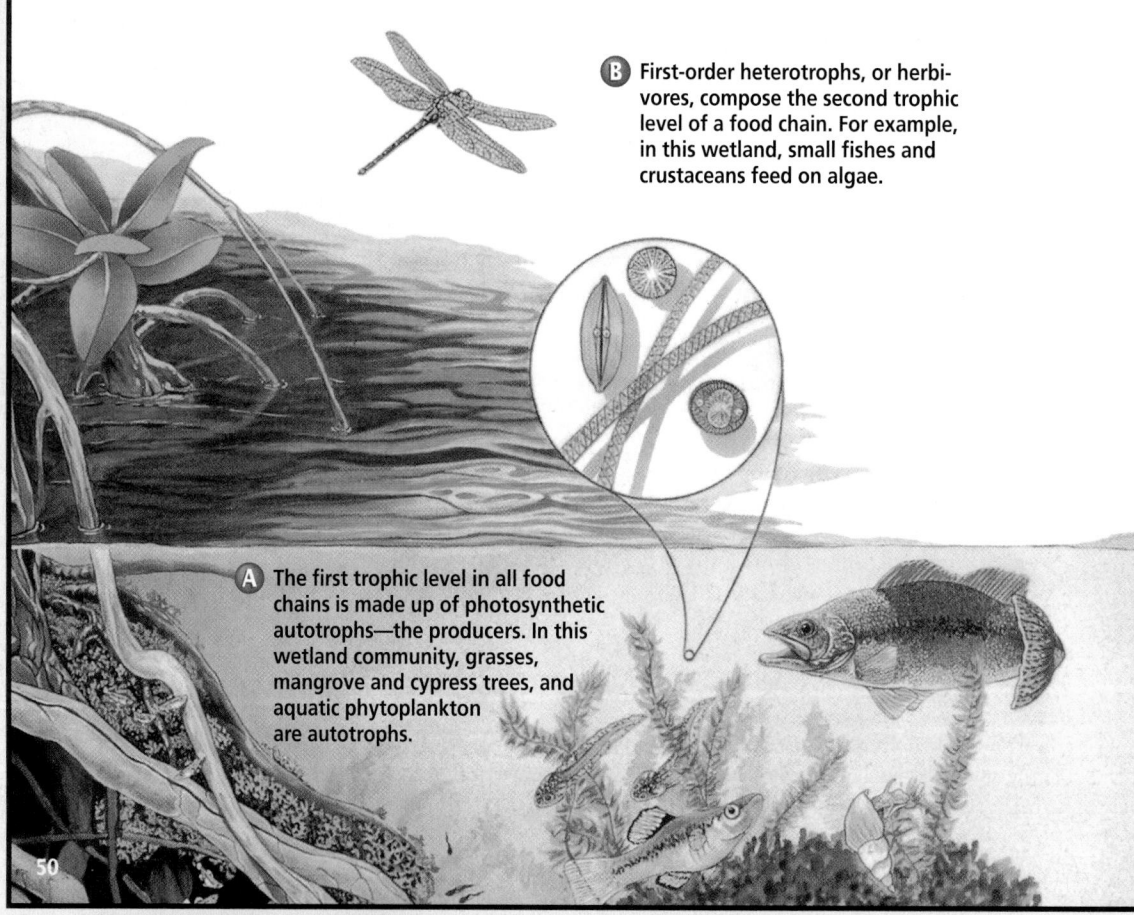

B First-order heterotrophs, or herbivores, compose the second trophic level of a food chain. For example, in this wetland, small fishes and crustaceans feed on algae.

A The first trophic level in all food chains is made up of photosynthetic autotrophs—the producers. In this wetland community, grasses, mangrove and cypress trees, and aquatic phytoplankton are autotrophs.

50

BIOLOGY JOURNAL

Living as a Decomposer

Linguistic Have students assume they are decomposers. Ask them to write a short paragraph that describes: a) what they look like, b) where they live, c) what they are going to eat. L2

Food chains: Pathways for matter and energy

The wetlands community pictured in *Figure 2.16* illustrates examples of food chains. A **food chain** is a simple model that scientists use to show how matter and energy move through an ecosystem. Nutrients and energy proceed from autotrophs to heterotrophs and, eventually, to decomposers.

A food chain is typically drawn using arrows to indicate the direction in which energy is transferred from one organism to the next. One food chain in *Figure 2.16* could be shown as

algae ➜ fish ➜ heron

Food chains can consist of three links, or steps, but most have no more than five links. This is because the amount of energy remaining in the fifth link is only a small portion of what was available at the first link. A portion of the energy is lost as heat at each link. It makes sense, then, that typical food chains are three or four links long.

C Second-order heterotrophs, which are carnivores, make up the third trophic level. They feed on first-order heterotrophs. The heron is a carnivorous bird that feeds on fishes, frogs, and other small animals of the wetland habitat.

D Third-order heterotrophs, carnivores that feed on second-order heterotrophs, make up the fourth trophic level. An alligator eating a shorebird is one example of a third-order heterotroph. Bacteria and fungi decompose all the links of the food chain when organisms die.

Ask students to describe possible food chains, other than the one shown in Figure 2.16. *They are likely to substitute different organisms in place of those shown or mentioned in the text. Accept all logical responses.*

Visual Learning

Ask students to use Figure 2.16 to identify other animals that might occupy the third trophic level. *Responses may include other birds of prey such as owls or eagles or land animals such as lions, bears, or cats. Accept all logical responses.*

GLENCOE TECHNOLOGY

VIDEODISC
The Infinite Voyage: *Secrets from a Frozen World*
Krill: The Vital Link of a Food Chain (Ch. 2)
5 min. 30 sec.

Resource Manager

Basic Concepts Transparency 1 and Master
L2 ELL

Purpose

Students use their knowledge of trophic levels to organize and summarize information.

Process Skills

think critically, classify, sequence

Teaching Strategies

■ You may wish to provide the outline diagrams to students.
■ Allow students to work in small groups to complete the lab.
■ Remind students that there is only one correct placement for each label.
■ Review and/or define any term that is still not clear to students.

Thinking Critically

1. Starting at the lowest level:
   autotroph;
   1st-order heterotroph;
   2nd-order heterotroph;
   3rd-order heterotroph.

2. Starting at the lowest level on the left: producer, herbivore, carnivore (top two levels). On the right, the lowest level is autotroph and the top three levels are heterotrophs.

3. Small arrows show direction of energy from one trophic level to next.

✔ Assessment

Skill Ask students to prepare a concept map of the ideas covered in this lab. Their maps must include all terms used on the diagram. Use the Performance Task Assessment List for Concept Map in **PASC**, p. 89.
L3 **ELL**

Problem-Solving Lab 2-2 Applying Concepts

How can you organize trophic level information?
Diagrams or charts may help to summarize information or concepts in a more logical and simpler manner. This is the case with information that shows relationships among trophic levels.

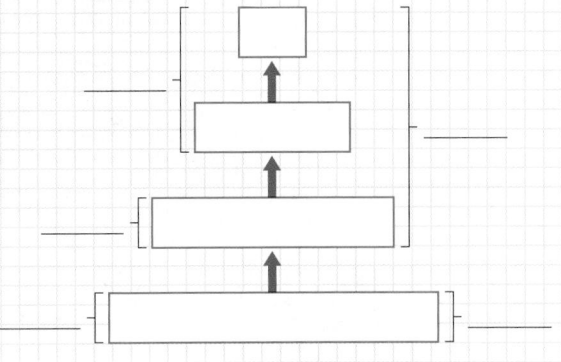

Analysis

Copy the diagram above. It will show, when completed correctly, the various relationships in a food chain.

Thinking Critically

1. Each box represents a trophic level. Write the name for each trophic level in the proper box. Use these choices: 1st order heterotroph, autotroph; 3rd order heterotroph; 2nd order heterotroph.

2. Each bracket identifies one or more traits of the trophic levels. Use the following labels to identify them in their proper order: herbivore, autotroph, carnivore, heterotroph, producer.

3. What is being represented by the small arrows connecting trophic levels?

Trophic levels represent links in the chain

Each organism in a food chain represents a feeding step, or **trophic level** (TROHF ihk), in the passage of energy and materials. Examine how energy flows through trophic levels in the *Problem-Solving Lab* shown here. A food chain represents only one possible route for the transfer of matter and energy in an ecosystem. Many other routes may exist. As

---

Figure 2.16 indicates, many different species occupy each trophic level in a wetlands ecosystem. In addition, many different kinds of organisms eat a variety of foods, so a single species may feed at several trophic levels. For example, the great blue heron eats largemouth black bass, but it also eats minnows, bluegills, and frogs. The alligator may feed on the heron, fish, or even a deer that comes too close. Can you think of other possible food chains in this ecosystem?

Food webs

Simple food chains are easy to study, but they cannot indicate the complex relationships that exist among organisms that feed on more than one species. Ecologists who are particularly interested in energy flow in an ecosystem set up experiments with as many organisms in the community as they can. The model they create, a **food web,** expresses all the possible feeding relationships at each trophic level in a community. A food web is a more realistic model than a food chain because most organisms depend on more than one other species for food. Notice how the food web of the forest ecosystem shown in *Figure 2.17* represents a network of interconnected food chains. In an actual ecosystem, many more plants and animals would be involved in the food web.

Energy and trophic levels: Ecological pyramids

How can you show how energy is used in an ecosystem? Ecologists use food chains and food webs to model the distribution of matter and energy within an ecosystem. They also use another kind of model, called an ecological pyramid. An ecological pyramid shows how energy flows through an ecosystem. The

---

GLENCOE
TECHNOLOGY

CD-ROM
Biology: The Dynamics of Life
BioQuest: *Antarctic Food Web*
Disc 1

BIOLOGY JOURNAL

It's a Jungle

Naturalist Have students compare trophic levels to the organization of a business. Ask them to diagram a specific business to show how each level of employee supports the next level. **L2**

Labels on the illustration:

Red tail hawk

Grizzly bear

Marmot

Grouse

Insects

Chipmunk

Berries

Deer

Grasses

Seeds

Chalkboard Example

Write the word *human* at the top of the chalkboard. Ask students to complete a food web that includes the two trophic levels below humans, using as many different organisms for each level as possible. *Examples of first-order consumers may include chickens, cows, sheep, and pigs; examples of producers may include grass, shrubs, lettuce, pears, and corn. Ask students to explain why this represents a food web rather than a food chain. A food chain involves only one organism for each trophic level.*

Tying to Previous Knowledge

Have students review the meanings of the terms *scavenger* and *decomposer*. Ask them to describe the role of each type of organism in a food chain or food web. **L1** **ELL**

GLENCOE
TECHNOLOGY

CD-ROM
Biology: The Dynamics of Life
Exploration: *Pyramid of Energy*
Disc 1

Resource Manager

Basic Concepts Transparency 2 and Master
L2 **ELL**

base of the ecological pyramid represents the autotrophs, or first trophic level. Higher trophic levels are layered on top of one another. Examine each type of ecological pyramid in *Figures 2.18, 2.19* and *2.20.* Each pyramid gives different information

about an ecosystem. Observe that each pyramid summarizes interactions of matter and energy at each trophic level. Notice that the initial source of energy for all three of these ecological pyramids is energy from the sun.

Figure 2.17
A forest community food web includes many organisms at each trophic level. Arrows indicate the flow of materials and energy.

Portfolio

Making a Food Web

Visual-Spatial Ask students to design a food web using the following organisms: wheat, rat, fox, human, cow, corn, rabbit, hawk, grass. Have them use a colored pencil or marker to outline one food chain in this web. Ask them to indicate tropic levels as well as omnivores, herbivores, and carnivores. **L1**

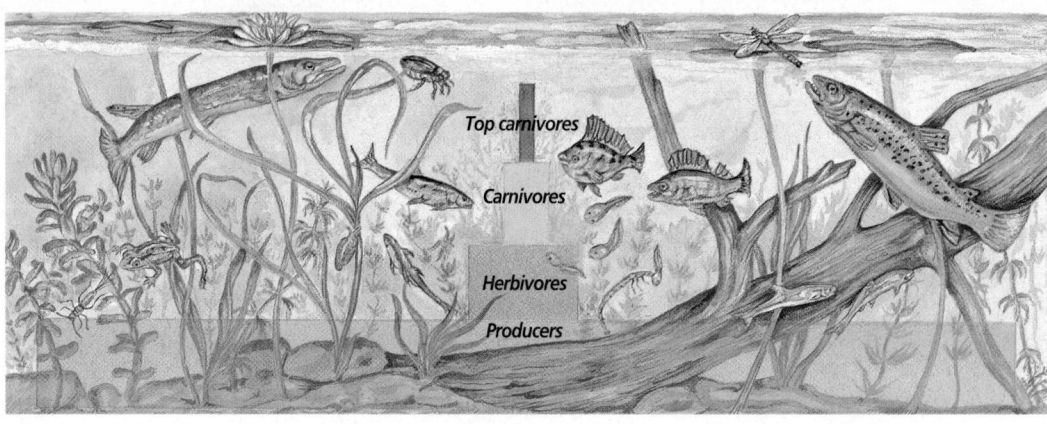

Figure 2.18
Pyramid of energy Each bar in the pyramid represents energy available within a trophic level. Notice that energy decreases as the trophic level increases.

The pyramid of energy shown in *Figure 2.18* illustrates that energy decreases at each succeeding trophic level. The total energy transfer from one trophic level to the next is only about ten percent because organisms fail to capture and eat all the food available at the trophic level below them. When an organism consumes food, it uses some of the energy in the food for metabolism, some for building body tissues, and some is given off as waste. When the organism is eaten, the energy that was used to build body tissue is available as energy to be used by the organism that consumed it. The energy lost at each successive trophic level enters the environment as heat.

Ecologists construct a pyramid of numbers based on the population sizes of organisms in each trophic level. The pyramid of numbers in *Figure 2.19* shows that population sizes decrease at each higher trophic level. This is not always true. For example, one tree can be food for 50 000 insects. In this case, the pyramid would be inverted.

A pyramid of biomass, such as the one shown in *Figure 2.20,* expresses the weight of living material at each trophic level. Ecologists calculate the biomass at each trophic level by finding the average weight of each species at that trophic level and multiplying by the estimated number of organisms in each population.

Figure 2.19
Pyramid of numbers Each bar in the pyramid represents population size within a trophic level. Notice that population size decreases as the trophic level increases.

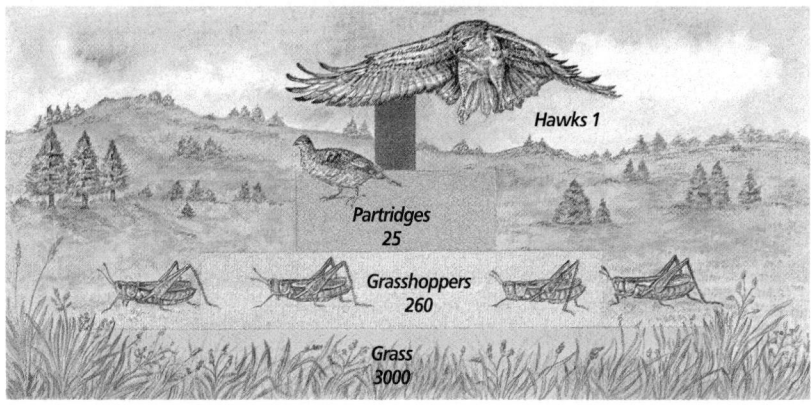

Cycles in Nature

Food chains, food webs, and ecological pyramids all show how energy moves in only one direction through the trophic levels of an ecosystem. Ecological pyramids also show how energy is lost from one trophic level to the next. This energy is lost to the environment as heat generated by the body processes of organisms. Sunlight is the primary source of all this energy, so energy is always being replenished.

Matter, in the form of nutrients, also moves through the organisms at each trophic level. But matter cannot be replenished like the energy from sunlight. The atoms of carbon, nitrogen, and other elements that make up the bodies of organisms alive today are the same atoms that have been on Earth since life began. Matter is constantly recycled.

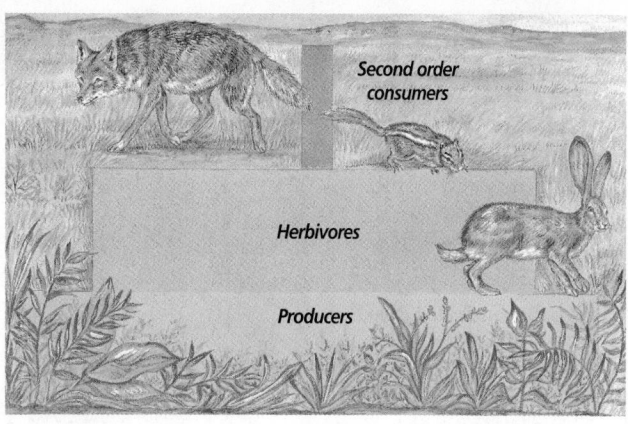

The water cycle

Life on Earth depends on water. Even before there was life on Earth, water cycled through stages, as shown in *Figure 2.21.* Have you ever left a glass of water out and a few days later observed there was less water in the glass? This is the result of evaporation. Just as the water

Figure 2.20
Pyramid of biomass Each bar in the pyramid represents the amount of biomass within a trophic level. Notice that biomass decreases as the trophic level increases.

Figure 2.21
In the water cycle, water is constantly moving between the atmosphere and Earth.

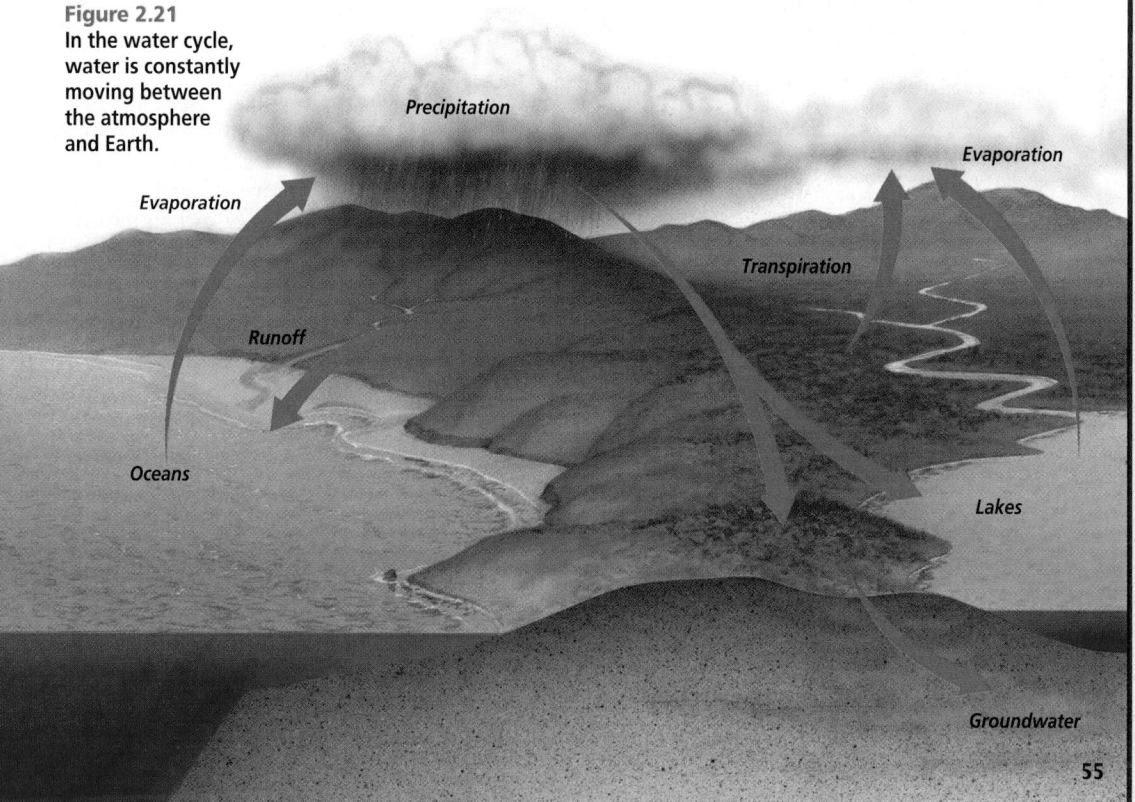

55

Purpose

Students will use bromothymol blue to test for the presence of carbon dioxide gas.

Process Skills

acquire information, draw a conclusion, observe and infer, recognize cause and effect

Safety Precautions

Caution students not to inhale or drink the bromothymol solution. Have them wear goggles and a lab apron.

Teaching Strategies

■ Prepare bromothymol blue solution as follows: Stock solution—add 0.5 g bromothymol blue powder to 500 mL distilled water. Dilute 10 mL stock solution with 500 mL of distilled water. (If solution is green, add one or more drops of NH_4OH until a blue color appears. If solution is deep blue, add one or more drops of HCl until a light blue color appears.)

■ An effervescent antacid must be used. Check the label to be sure the ingredients include a carbonate or bicarbonate and a weak acid such as citric acid.

■ Bromothymol blue does not actually indicate the presence of carbon dioxide. It changes color in the presence of any acid solution, such as the carbonic acid formed by carbon dioxide and water.

Expected Results

Bromothymol blue will change to green or yellow when an antacid tablet is added or exhaled air is bubbled through the indicator. The color change shows that carbon dioxide is present.

Analysis

1. Bromothymol blue changes to green or yellow when carbon dioxide gas is added.
2. carbon dioxide
3. Yes, the color change in the bromothymol blue indicated the presence of carbon dioxide.

56

MiniLab 2-2   Observing and Inferring

Detecting Carbon Dioxide
Carbon dioxide is given off during respiration. When carbon dioxide is dissolved in water, an acid is formed. Certain chemicals called indicators can be used to detect acids. One indicator, called bromothymol blue, will change from its normal blue color to green or yellow if an acid is present.

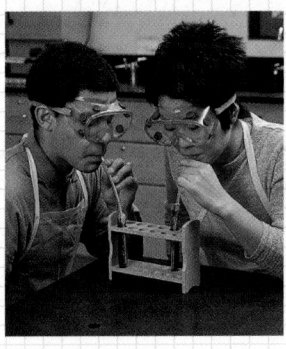

Procedure

1. Half fill a test tube with bromothymol blue solution.
2. Add a quarter of an effervescent antacid tablet to the tube and note any color change.
3. Half fill a test tube with bromothymol blue solution. Using a straw, exhale into the bromothymol blue at least 30 times. **CAUTION:** *Do not inhale the bromothymol blue.* Record any color change in the test tube.

Analysis

1. Describe the color change that occurs when carbon dioxide is added to bromothymol blue.
2. What was the chemical composition of the bubbles seen in the tube with the antacid tablet?
3. Does exhaled air contain carbon dioxide? Explain.

---

evaporated from the glass, water evaporates from lakes and oceans and becomes water vapor in the air.

Have you ever noticed the drops of water that form on a cold can of soda? The water vapor in the air condenses on the surface of the can because the can is colder than the surrounding air. Just as water vapor condenses on cans, it also condenses on dust in the air and forms clouds. Further condensation makes small drops that build in size until they fall from the clouds as precipitation in the form of rain, ice, or snow. The water falls on Earth and accumulates in oceans and lakes where evaporation continues.

Plants and animals need water to live. Plants pull water from the ground and lose water from their leaves through transpiration. This puts water vapor into the air. Animals breathe out water vapor in every breath; when they urinate, water is returned to the environment. Natural processes constantly recycle water throughout the environment.

The carbon cycle

All life on Earth is based on carbon molecules. Atoms of carbon form the framework for proteins, carbohydrates, fats, and other important molecules. More than any other element, carbon is the molecule of life.

The carbon cycle starts with the autotrophs. During photosynthesis, energy from the sun is used to convert carbon dioxide gas into energy-rich carbon molecules. Autotrophs use these molecules for growth and energy. Heterotrophs, which feed either directly or indirectly on the autotrophs, also use the carbon molecules for growth and energy. When the autotrophs and heterotrophs use the carbon molecules for energy, carbon dioxide is released and returned to the atmosphere. Learn how to detect the presence of carbon dioxide in the *MiniLab* shown here. The carbon cycle is described in the *Inside Story* on the next page.

The nitrogen cycle

If you add nitrogen fertilizer to a lawn, houseplants, or garden, you may see that it makes them greener, bushier, and taller. Even though the air is 78 percent nitrogen, plants seem to do better when they receive nitrogen fertilizer. This is because plants cannot use the nitrogen in the air. They use nitrogen in the soil that has been converted into more usable forms.

✔ Assessment

Skill Have students make a hypothesis for an experiment using bromothymol blue and a decomposer. Use the Performance Task Assessment List for Formulating a Hypothesis in **PASC**, p. 21. **L2**

BIOLOGY JOURNAL

A Wet Life Cycle

Linguistic Have students write a description of what it would be like to be a molecule of water that is cycled through an ecosystem. Where would they spend most of their time, what sites would they visit, and how many changes in phase (gas, liquid, solid) might they experience? **L2**

The Carbon Cycle

From proteins to sugars, carbon is the building block of the molecules of life. Linked carbon atoms form the frame for molecules produced by plants and other living things. Organisms use these carbon molecules for growth and energy.

Critical Thinking *How is carbon released from the bodies of organisms?*

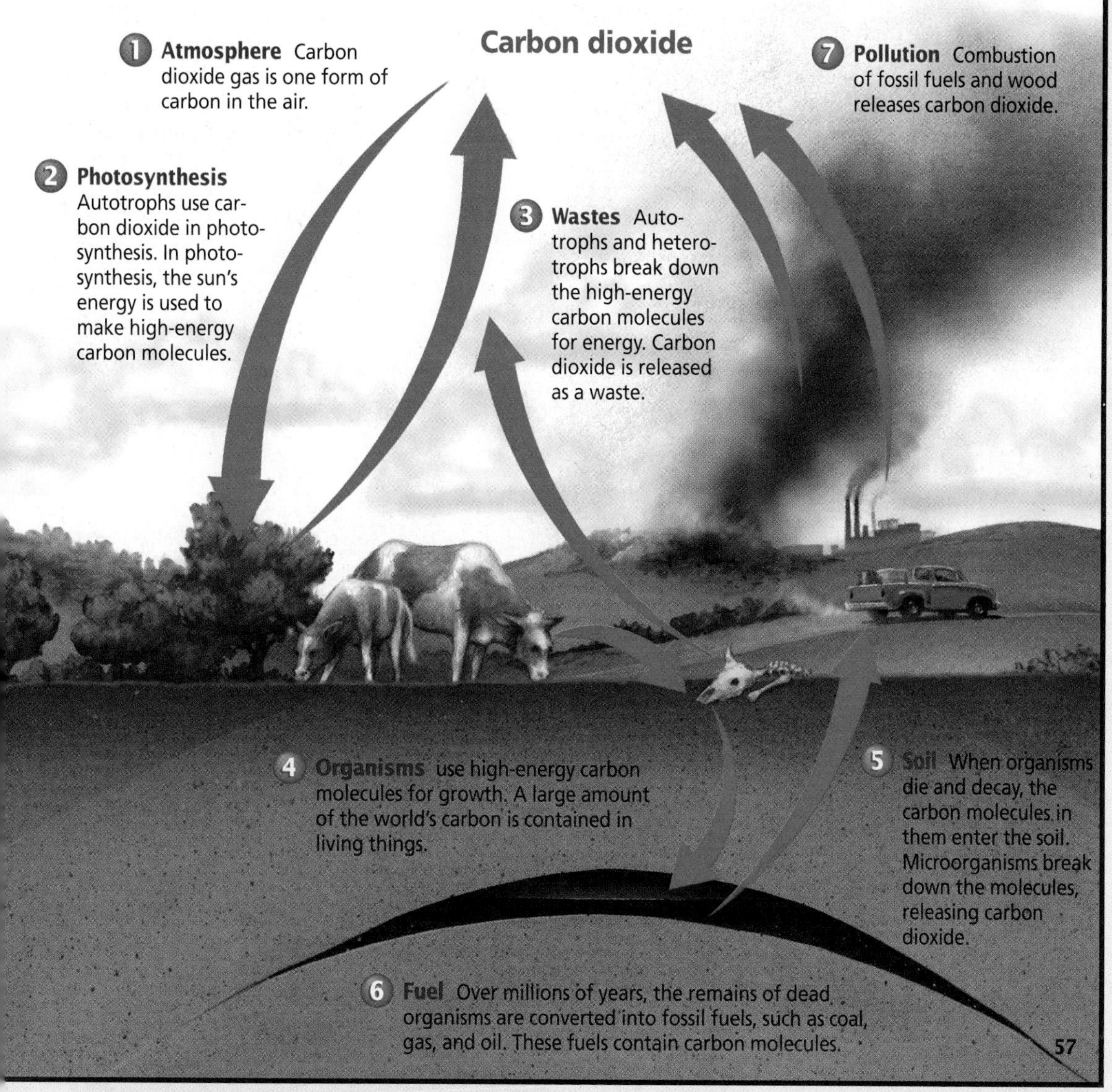

Forests use carbon dioxide.

Carbon dioxide

① Atmosphere Carbon dioxide gas is one form of carbon in the air.

② Photosynthesis Autotrophs use carbon dioxide in photosynthesis. In photosynthesis, the sun's energy is used to make high-energy carbon molecules.

③ Wastes Autotrophs and heterotrophs break down the high-energy carbon molecules for energy. Carbon dioxide is released as a waste.

⑦ Pollution Combustion of fossil fuels and wood releases carbon dioxide.

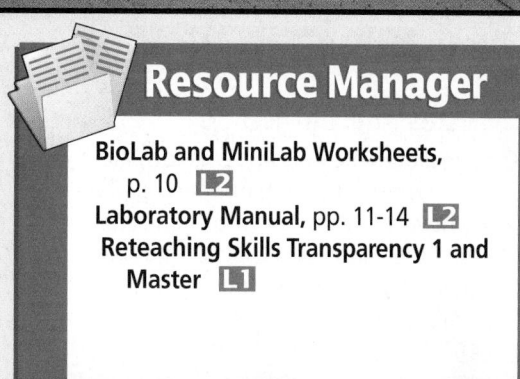

④ Organisms use high-energy carbon molecules for growth. A large amount of the world's carbon is contained in living things.

⑤ Soil When organisms die and decay, the carbon molecules in them enter the soil. Microorganisms break down the molecules, releasing carbon dioxide.

⑥ Fuel Over millions of years, the remains of dead organisms are converted into fossil fuels, such as coal, gas, and oil. These fuels contain carbon molecules.

57

Purpose 🎞

Students study the cycling of carbon in the environment.

Teaching Strategies

■ Help students understand that plants remove carbon dioxide from the atmosphere and use it to create nutrients. When plants and animals use the nutrients for energy, carbon dioxide is returned to the atmosphere.

■ Help students understand other ways carbon dioxide enters the atmosphere.

Visual Learning

■ Have students trace a carbon molecule from the atmosphere, through two trophic levels, and back into the atmosphere. **L1**

Critical Thinking

The carbon is released in the form of carbon dioxide through respiration, decay, or burning.

GLENCOE TECHNOLOGY

💿 **VIDEODISC**
The Secret of Life
Carbon Cycle

‖‖‖‖‖‖‖‖‖‖‖‖‖

✓ Assessment

Portfolio Have students summarize the carbon cycle and draw their own diagrams. Ask them to label the parts of the cycle and use arrows to show the direction of movement.

MEETING INDIVIDUAL NEEDS

English Language Learners

☑ *Visual-Spatial* Have students work in groups to prepare a concept map describing the carbon cycle. Included in this map should be terms such as *consumers, photosynthesis, respiration, decay,* and *producers.* Group students who are learning English with those who are fluent in the language. **L2**

Resource Manager

BioLab and MiniLab Worksheets, p. 10 **L2**
Laboratory Manual, pp. 11-14 **L2**
Reteaching Skills Transparency 1 and Master **L1**

Reinforcement

Ask students to trace the roles of producers, consumers, and decomposers in the cycling of nitrogen. *Producers take in nitrogen compounds in soil and pass these compounds to consumers that eat the producers. Decomposers break down the nitrogen compounds and release nitrogen gas to the air.*

Enrichment

Show students plant fertilizer labels. These fertilizers add nitrogen and phosphorous to the soil because these substances are frequently in short supply. Explain the numbers on the label. For example, a plant fertilizer that is labeled 20-20-10 contains 20 percent nitrogen, 20 percent phosphorous, and 10 percent potassium compounds.

3 Assess

Check for Understanding

Have students explain the relationship between the words in each of the following pairs.
- **a.** autotroph—producer
- **b.** heterotroph—consumer
- **c.** recycling—carbon
- **d.** lightning—nitrogen cycle

L1 **ELL** 🔲

As *Figure 2.22* shows, lightning and certain bacteria convert the nitrogen in the air into these more usable forms. Chemical fertilizers also give plants nitrogen in a form they can use.

Plants use the nitrogen to make important molecules such as proteins. Herbivores eat plants and convert nitrogen-containing plant proteins into nitrogen-containing animal proteins. After you eat your food, you convert the proteins in your food into human proteins. Urine, an animal waste, contains excess nitrogen. When an animal urinates, nitrogen returns to the water or soil. When organisms die, their nitrogen molecules return to the soil. Plants reuse these nitrogen molecules. Bacteria

Figure 2.22
In the nitrogen cycle, nitrogen is converted from a gas to compounds important for life and back to a gas.

also act on these molecules and put nitrogen back into the air.

The phosphorus cycle

Materials other than water, carbon, and nitrogen cycle through ecosystems. Substances such as sulfur, calcium, and phosphorus, as well as others, must also cycle through an ecosystem. One essential element, phosphorus, cycles in two ways.

All organisms require phosphorus for growth and development. Plants obtain phosphorus from the soil. Animals get phosphorus by eating plants, as shown in *Figure 2.23*. When these animals die, they decompose and the phosphorus is returned to the soil to be used again. This is the short-term phosphorus

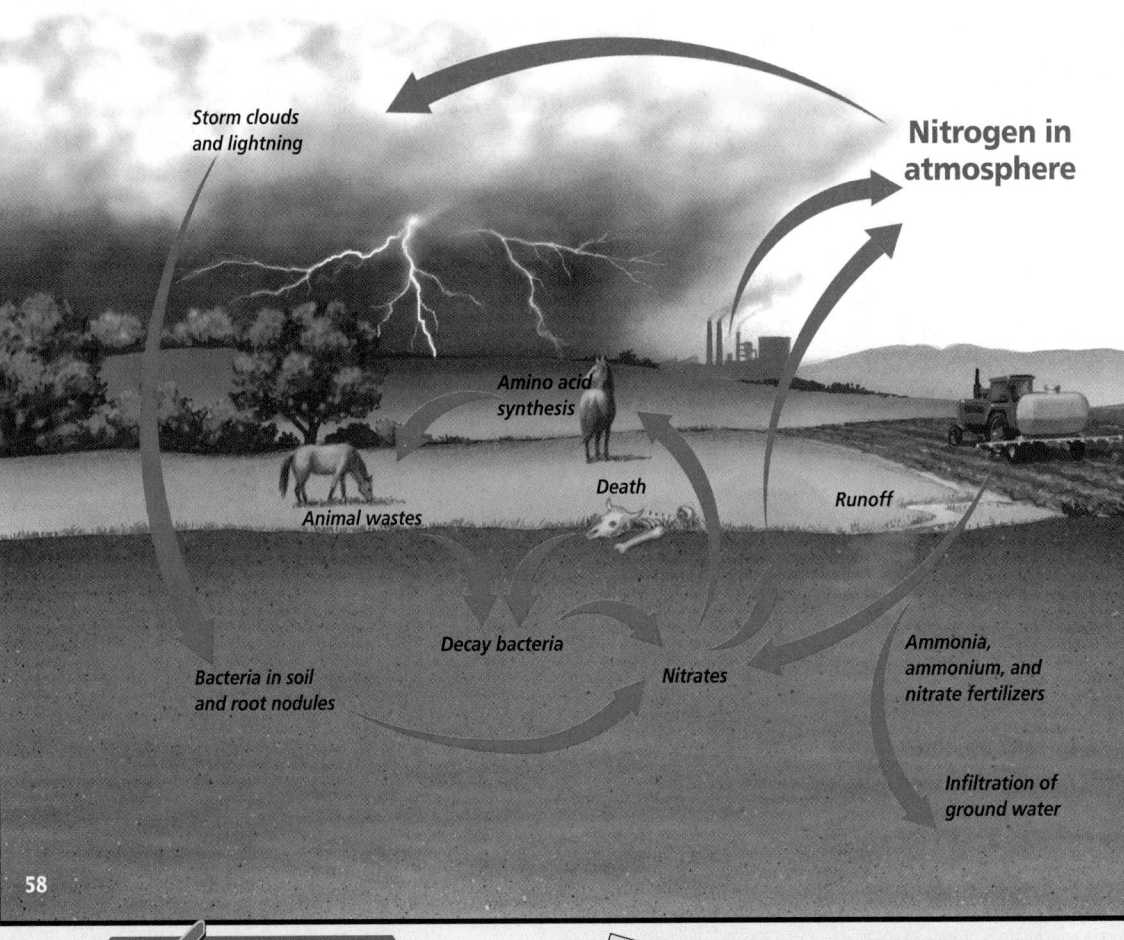

Storm clouds and lightning

Nitrogen in atmosphere

Amino acid synthesis

Death

Runoff

Animal wastes

Decay bacteria

Nitrates

Ammonia, ammonium, and nitrate fertilizers

Bacteria in soil and root nodules

Infiltration of ground water

58

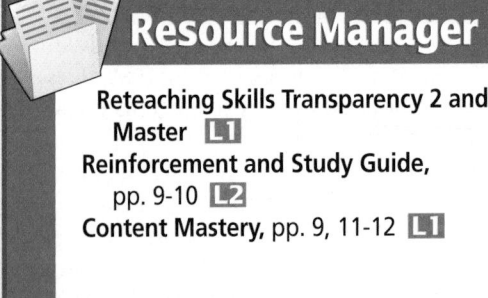

Figure 2.23
In the phosphorus cycle, phosphorus moves between the living and non-living parts of the environment.

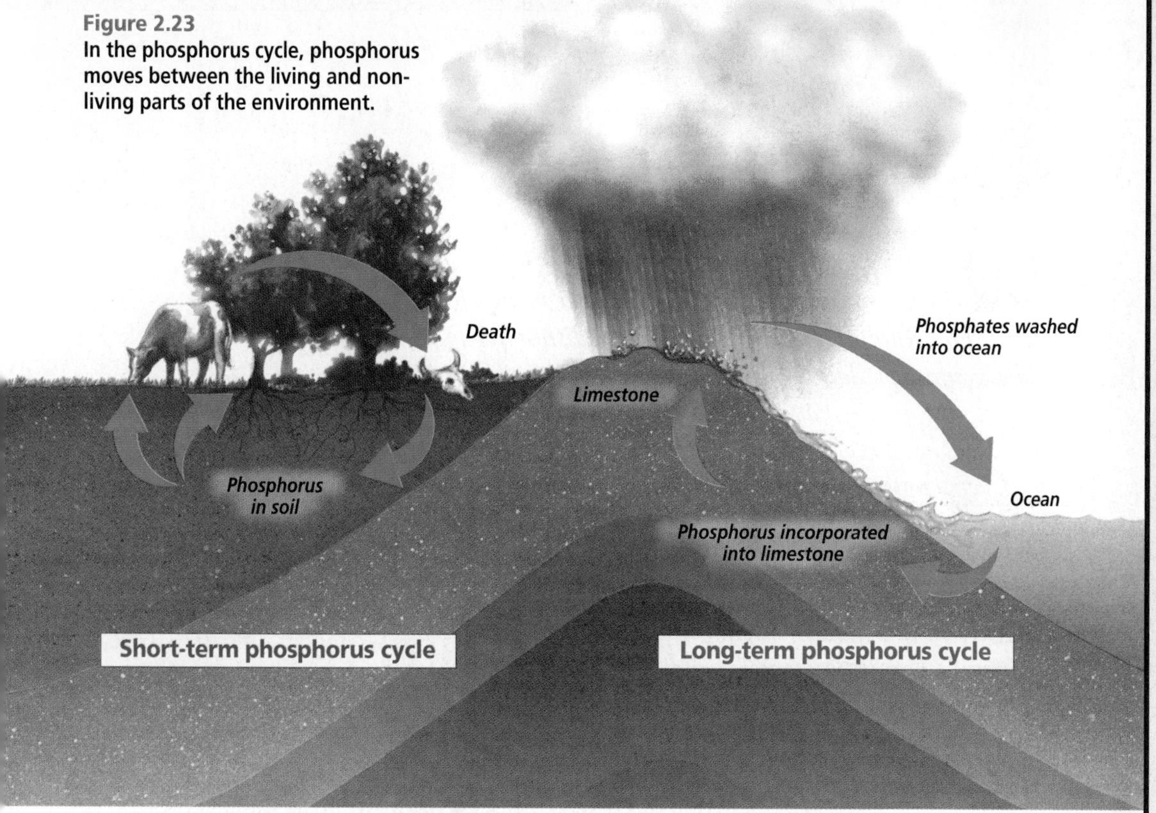

Death

Phosphates washed into ocean

Limestone

Phosphorus in soil

Ocean

Phosphorus incorporated into limestone

Short-term phosphorus cycle

Long-term phosphorus cycle

cycle. Phosphorus also has a long-term cycle, where phosphates washed into the sea are incorporated into rock as insoluble compounds. Millions of years later, as the environment changes, the rock containing phosphorus is exposed. As the rock erodes and disintegrates, the phosphorus again becomes part of the local ecological system.

Section Assessment

Understanding Main Ideas
1. What is the difference between an autotroph and a heterotroph?
2. Why do autotrophs always occupy the lowest level of ecological pyramids?
3. Give two examples of how nitrogen cycles from the abiotic portion of the environment into living things and back.
4. Describe a food chain that was not presented in this section.

Thinking Critically
5. The country of Avorare has many starving people. Should you encourage the people to grow crops such as vegetables, wheat, and corn, or is it better to encourage them to use the land to raise cattle for beef?

SKILL REVIEW

6. **Designing an Experiment** Suppose there is a fertilizer called GrowFast. It contains extra nitrogen and phosphorus. Design an experiment to see if GrowFast increases the growth rate of plants. For more help, refer to the *Practicing Scientific Methods* in the **Skill Handbook**.

Reteach
Have students identify the trophic level of each organism in a food chain and indicate the direction of energy flow. L1

Extension
Naturalist Have students speculate about the consequences to a food web if an organism at the second trophic level were to be eliminated. *This would eliminate a food source for some organisms at the third trophic level.* L2

✔ Assessment
Performance Have students diagram, label, and explain one of the following: a food web, water cycle, energy or biomass pyramid, or carbon cycle. L2

4 Close

Activity
Visual-Spatial List on the chalkboard 20 different organisms. Have students use these organisms to create a food web. L1 ELL

Section Assessment

1. An autotroph makes its own nutrients. A heterotroph must consume another organism to meet its nutritional needs.
2. Autotrophs capture light energy and create nutrients. When eaten, they provide nutrients for other organisms.
3. (1) Nitrogen in the air passes to bacteria, which form compounds used by plants. Decay of dead plants returns nitrogen to air. (2) Nitrogen in air passes to bacteria and again moves to plants. Plants are eaten by consumers, passing along nitrogen. Decay of dead consumers returns nitrogen to air.
4. Any realistic food chain is acceptable. Arrows should show the correct flow of energy.
5. Crops are better because growing them takes less land than raising cattle. Crops provide more energy for people than cattle, at a higher trophic level.
6. Designs should include control and experimental groups. Sample size should be greater than one plant. Students should define *growth*, as the increase in plant height.

DESIGN YOUR OWN
BioLab

Time Allotment

Initial preparation: one class session; ten-minute sessions for one to two weeks every other day.

Process Skills

observe, record and analyze data, design an experiment, separate and control variables

Safety Precautions

Use oven mitts when handling hot, sterile pond water. Have students wash their hands after the lab.

PREPARATION

Alternative Materials

■ Artificial pond water, called Chalkey's solution, may be prepared as follows. Dissolve 1 g sodium chloride, with 0.04 g potassium chloride, and 0.06 g calcium chloride in 1 L of distilled water. Dilute the prepared solution by adding 100 mL of solution to 900 mL of distilled water.

■ To prepare sterile pond water, filter the water and place it in flat pans. Boil for 15 minutes. Allow to cool before using.

Possible Hypotheses

In general, if a predator population is added to a prey population, the size of the predator population will increase while the prey population decreases.

However, if a small predator population is added to a large prey population, no observable difference will occur in the sizes of the populations.

DESIGN YOUR OWN
BioLab

How can one population affect another?

Why don't prey populations disappear when predators are present? Prey organisms have evolved a variety of defenses to avoid being eaten. For example, some caterpillars are distasteful to birds, and some fishes confuse predators by appearing to have eyes at both ends of their bodies. Just as prey have evolved defenses to avoid predators, predators have evolved mechanisms to overcome those defenses.

Even single-celled protists such as Paramecium have predators. Didinium is another unicellular protist that attacks and devours Paramecium larger than itself. Do populations of Paramecium change when a population of Didinium is present? In this investigation, you will use various methods to determine how both of these species interact.

Didinium

PREPARATION

Problem

How does a population of *Paramecium* react to a population of *Didinium?*

Hypotheses

Have your group agree on an hypothesis to be tested. Record your hypothesis.

Objectives

In this BioLab, you will:

■ **Design** an experiment to establish the relationships between *Paramecium* and *Didinium*.

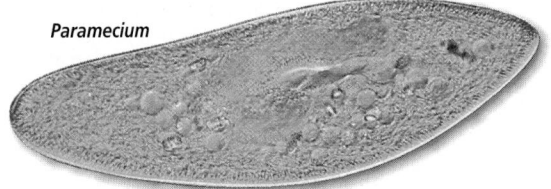
Paramecium

■ **Use** appropriate variables, constants, and controls in experimental design.

Possible Materials

microscope
microscope slides
coverslips
culture of *Didinium*
culture of *Paramecium*
beakers or jars
eyedroppers
sterile pond water

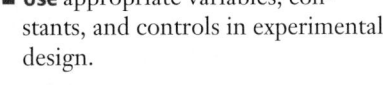

Safety Precautions

Take care when using electrical equipment. Always use goggles in the lab. Handle slides and coverslips carefully. Dispose of broken glass in a container provided by your teacher.

Skill Handbook

Use the **Skill Handbook** if you need additional help with this lab.

60 PRINCIPLES OF ECOLOGY

PLAN THE EXPERIMENT

Teaching Strategies

■ A lower magnification provides a wider field of view, making counting easier.
■ Methyl cellulose, available from supply houses, may be used to slow the protozoans.

Troubleshooting

■ Cover or stack culture dishes to prevent drying out.

■ Have students examine unmixed cultures first so they can later distinguish between *Paramecium* and *Didinium*.
■ Suggest that several low-power field counts be made and an average of these counts be used in the data tables.

Possible Procedures

Controls will consist of *Paramecium* cultures

PLAN THE EXPERIMENT

1. Review the discussion of feeding relationships in this chapter.
2. Decide which materials you will use in your investigation. Record your list.
3. Be sure that your experimental plan contains a control, tests a single variable such as population size, and allows for the collection of quantitative data.
4. Prepare a list of numbered directions. Explain how you will use each of your materials.

Check the Plan

Discuss the following points with other group members to decide final procedures. Make any needed changes to your plan.

1. What will you measure to determine the effect of the *Didinium* on *Paramecium?* If you count *Paramecia*, will you count all you can see in the field of vision of the micro-

scope at a certain power? Will you have multiple trials? If so, how many?
2. What single factor will you vary? For example, will you put no *Didinium* in one culture of *Paramecium* and 5 mL of *Didinium* culture in another culture of *Paramecium?*
3. How long will you observe the populations?
4. How will you estimate the changes in the populations of *Paramecium* and *Didinium* during the experiment?
5. ***Your teacher must approve your plan before you proceed.***
6. Carry out your experiment.
7. Make a data table that has Date, Number of *Paramecium*, and Number of *Didinium* across the top. Place the data obtained for each culture in rows. Design and complete a graph of your data.

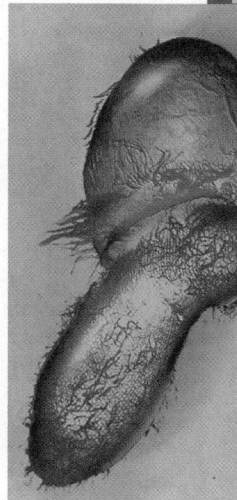

A *Didinium* captures a *Paramecium*.

ANALYZE AND CONCLUDE

1. **Analyzing Data** What differences did you observe among the experimental groups? Were these differences due to the presence of *Didinium?* Explain.
2. **Drawing Conclusions** Did the *Paramecium* die out in any culture? Why or why not?
3. **Checking Your Hypothesis** Was your hypothesis supported by your data? If not, suggest a new hypothesis.
4. **Thinking Critically** List several ways that your methods may have

affected the outcome of the experiment.

Going Further

Project Based on this lab experience, design another experiment that would help you answer any questions that arose from your work. What factors might you allow to vary if you kept the number of *Didinium* constant?

BIOLOGY *Online* To find out more about population biology, visit the Glencoe Science Web site.
science.glencoe.com

ANALYZE AND CONCLUDE

1. Only cultures containing both *Didinium* and *Paramecium* showed a decline in numbers after a period of time. This difference was due to the presence of *Didinium* because they preyed upon the *Paramecium*.
2. *Paramecium* died out in the mixed culture. They were preyed upon by *Didinium*.
3. In most cases, hypotheses will be supported.
4. The list may include: counting errors, too few samples, or cultures becoming contaminated or being affected by temperature.

Error Analysis

The amount of initial culture of the two species should be equal and can be quantified by premeasuring the volume in the pipettes.

✔ Assessment

Skill Ask students to prepare a summary of this experiment in their journals. Use the Performance Task Assessment List for Writing in Science in **PASC**, p. 87. **L2**

Going Further

Have students alter the type of protozoans used or change the initial volume of predator culture. **L2**

Resource Manager

BioLab and MiniLab Worksheets, p. 11 **L2**

with no predators added. Food will have to be added to these cultures. Use one alfalfa pill (available from pharmacies) per liter of water. Note: If one culture receives food, all cultures must receive food to maintain control conditions.

Data and Observations

Depending on the experiment, data and observations will vary. Typically, when the populations are mixed together, both will initially increase in number. A decrease in prey will then be detected as the predator population feeds upon them, with a final drop in the population of predators as their food supply runs out.

Biology & SOCIETY

The Florida Everglades— An Ecosystem at Risk

The Florida Everglades ecosystem covers the southern portion of the Florida peninsula. As with any wetlands, water is the critical factor that defines the ecology of the area.

Purpose

Students gain an understanding of the complexity of ecosystems and learn some ways in which human activities can affect the environment.

Background

Everglades water that flows all the way to the Gulf coast carries with it nutrients that support the growth of mangrove swamps and sea grass beds. These two estuarine communities feed and shelter young fish, shellfish, and other important links in the marine food chain. Changes in the water flow of the area sometimes result in the discharge of large amounts of freshwater into Florida Bay, changing the salinity of the water and endangering the health of mangroves and sea grass. Pollutants and excessive amounts of fresh water may also be endangering the coral reefs off the Florida coast.

Teaching Strategies

After students have completed the reading, invite them to discuss issues involved in balancing the needs of both human and wildlife populations. Point out that one of the main reasons why south Florida's human population has increased over the past 100 years is the strong attraction of the Everglades. Also, many Florida residents work in the tourism industry, and the Everglades brings large numbers of tourists to the state. **L2**

Investigating the Issue

The park was cut off from its water source. Even though land and wildlife inside the park were protected from development, not enough clean water was provided to keep the park healthy.

This subtropical region receives between 100 and 165 cm (40-65 inches) of rain each year, but only during the rainy season, which lasts from May to October. The heavy rainfall causes shallow Lake Okeechobee to overflow. A wide, thin sheet of water spreads out from the lake, creating an extensive marshy area.

Early in the twentieth century, the slow-moving river that flows out of Lake Okeechobee was 80 km (50 miles) wide in some places, and only 15-90 cm (six inches to three feet) deep. This wetland teemed with fishes, amphibians, and other animals that fed millions of wading birds. Healthy populations of crocodiles, alligators, and other large animals also lived here. During the dry season, from December to April, water levels in the marshes gradually dropped. Fishes and other water dwellers moved into deeper pools that held water all year long.

Changing the Everglades Water from Lake Okeechobee is no longer allowed to flood the countryside. Instead, it is diverted into channels to create dry land for agriculture and homes, and stored behind levees to supply water for cities. As a result, half the acreage of the original Everglades has been drained. Habitats have disappeared.

Different Viewpoints

Everglades National Park was established to preserve a portion of the Everglades. But the land that forms the park is an island surrounded by farms and towns and cut off from the waters of Lake Okeechobee. Human needs determine how much water comes into the park. When reserves are low, water is held back for people to use, depriving Everglades habitats of the moisture they need. If floods threaten, large amounts of water are released quickly. These sudden flows destroy the nests of wading birds and other animals.

Restoring the Everglades In 1993, Florida developed a restoration plan for rescuing the Everglades. The goals of the plan are to restore, as much as possible, the natural flow of unpolluted water through the area, recover native habitats and species, and create a sustainable ecosystem that permits both humans and native species to flourish.

INVESTIGATING THE ISSUE

Analyzing the Issue When Everglades National Park was established, scientists and government officials believed a portion of the Everglades ecosystem could be preserved by drawing boundaries around it and declaring it off limits to development. Why did this approach fail to preserve the Everglades?

BIOLOGY Online To find out more about the Everglades, visit the Glencoe Science Web site. **science.glencoe.com**

ATLANTIC OCEAN

Florida

Lake Okeechobee

Big Cypress National Parks

Everglades National Park

The map (above) shows the location of the Everglades (inset).

NATIONAL GEOGRAPHIC

VIDEODISC
STV: Water, *Water Management*
Unit 3, Side 1, 3 min. 9 sec.
Everglades

Going Further

Naturalist Everglades habitats include sea grass beds, coastal prairies, mangrove swamps, freshwater sloughs, marl prairies, hammocks, and pinelands. Invite students to find out what kinds of environmental conditions characterize one or more of these habitats and what kinds of organisms live there. **L3**

SUMMARY

Section 2.1

Organisms and Their Environment

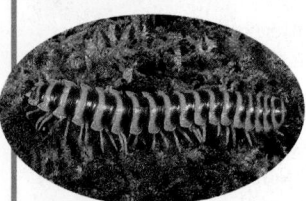

Main Ideas

- Natural history, the observation of how organisms live out their lives in nature, led to the development of the science of ecology—the study of the interactions of organisms with one another and with their environments.

- Ecologists classify and study the biological levels of organization from the individual up to the ecosystem. Ecologists study the abiotic and biotic factors that are a part of an organism's habitat. They investigate the strategies an organism uses to exist in its niche. An aspect of its niche may involve symbiosis with other organisms.

Vocabulary

abiotic factor (p. 39)
biosphere (p. 38)
biotic factor (p. 40)
commensalism (p. 46)
community (p. 42)
ecology (p. 38)
ecosystem (p. 43)
habitat (p. 44)
mutualism (p. 46)
niche (p. 45)
parasitism (p. 47)
population (p. 42)
symbiosis (p. 46)

Section 2.2

Nutrition and Energy Flow

- Autotrophs, such as plants, make nutrients that can be used by the plants and by heterotrophs. Heterotrophs include herbivores, carnivores, omnivores, and decomposers.

- Food chains are simple models that show one way that materials move from autotrophs to heterotrophs and eventually to decomposers.

- Food webs represent many interconnected food chains and illustrate possible ways materials are transferred within an ecosystem. Energy from the sun fuels life in the ecosystems. Although the sun adds new energy, the materials of life, such as carbon and nitrogen, do not increase. These materials are used and reused as they cycle through the ecosystem.

Vocabulary

autotroph (p. 48)
decomposer (p. 50)
food chain (p. 51)
food web (p. 52)
heterotroph (p. 49)
scavenger (p. 49)
trophic level (p. 52)

UNDERSTANDING MAIN IDEAS

1. Which of the following would be abiotic factors for a polar bear?
   a. extreme cold, floating ice
   b. eating only live prey
   c. large body size
   d. paws with thick hair

2. Organisms that use the sun's energy to make food are called _____.
   a. herbivores   **c.** autotrophs
   b. animals   **d.** heterotrophs

3. In the food web below, which of the organisms, X, Y, or Z, is a herbivore?
   a. Z   **c.** both X and Y
   b. Y   **d.** X

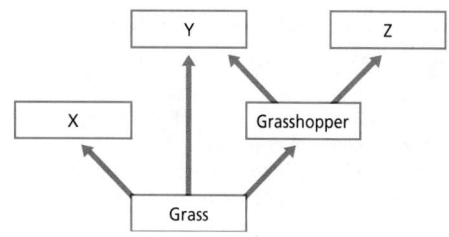

Main Ideas

Summary statements can be used by students to review the major concepts of the chapter.

Using the Vocabulary

To reinforce chapter vocabulary, use the Content Mastery Booklet and the activities in the Interactive Tutor for Biology: The Dynamics of Life on the Glencoe Science Web site.
science.glencoe.com

 *All Chapter Assessment questions and answers have been validated for accuracy and suitability by The Princeton Review.*

UNDERSTANDING MAIN IDEAS

1. a
2. c
3. d

GLENCOE TECHNOLOGY

 VIDEODISC
Biology: The Dynamics of Life
The Everglades (Ch. 6)
 Disc 1, Side 1, 29 sec.

GLENCOE TECHNOLOGY

 **VIDEOTAPE**
MindJogger Videoquizzes
Chapter 2: *Principles of Ecology*
Have students work in groups as they play the videoquiz game to review key chapter concepts.

 ## Resource Manager

Chapter Assessment, pp. 7-12
MindJogger Videoquizzes
ExamView® Pro Software 💾
BDOL Interactive CD-ROM, Chapter 2 quiz

4. c
5. b
6. a
7. d
8. d
9. b
10. c
11. commensalism, parasitic
12. biotic
13. symbiosis, mutualism
14. population, community
15. omnivore, herbivore
16. food, trophic
17. Ecology
18. pyramid of energy, pyramid of numbers
19. carbon dioxide
20. nitrogen

APPLYING MAIN IDEAS

21. The pesticide may be carried in the water to other bodies of water or may be carried from one organism to another.

22. Algae ⇒ Caterpillar/Moth ⇒ Bird. The symbiotic relationship between the algae and the sloth is commensal because the algae have a place to grow undisturbed, while the sloth is neither helped nor harmed.

---

4. Which organism is a carnivore?
   a. human   **c.** lion
   b. rabbit   **d.** opossum

5. Biotic factors in a wetland community might include _____.
   a. water   **c.** temperature
   b. crayfishes   **d.** soil type

6. Which of the following would decrease the amount of carbon dioxide in the air?
   a. a maple tree growing
   b. a dog running
   c. a person driving a car to work
   d. a forest burning

7. As energy flows through an ecosystem, energy _____ at each trophic level.
   a. remains the same
   b. increases
   c. decreases then increases
   d. decreases

8. An elk eats grass. A grizzly bear eats the elk. This is an example of a _____.
   a. pyramid of numbers
   b. commensal relationship
   c. food web
   d. food chain

9. Which of the following is true concerning the flow of energy and matter in an ecosystem?
   a. Both energy and matter are recycled and used again.
   b. Matter is recycled and used again, energy is lost.
   c. Energy is recycled and used again, matter is lost.
   d. Neither energy nor matter are recycled and used again.

THE PRINCETON REVIEW   **TEST-TAKING TIP**

Quiet Zone
It's best to study in an environment similar to the one in which you'll be tested. Blaring stereos, video game machines, chatty friends, and beepers are NOT allowed in the classroom during test time. So why get used to them?

---

10. Cowbirds get their name because they follow cows and eat the insects disturbed by the walking cows. Cowbirds have an unusual method for reproducing. The brown-headed cowbird goes to the nest of a different bird species, such as the red-wing blackbird. The cowbird rolls one of the blackbird's eggs out of the nest and lays its own egg in place. The blackbird protects the cowbird egg and feeds the chick when it hatches. This description best describes part of the cowbird's _____.
    a. community   **c.** niche
    b. habitat   **d.** tropic level

11. For the cowbird description in question 10, the symbiotic relationship between the cow and the cowbird is _____. The association between the cowbird and the blackbird is a(n) _____ relationship.

12. The presence of predators, prey, and parasites are examples of _____ factors in an organism's habitat.

13. A close and permanent relationship between two organisms is called _____. If both organisms benefit it is _____.

14. A group of organisms of the same species living in the same area is called a(n) _____. When the group includes different species, it is called a _____.

15. A(n) _____ eats both plants and animals. A(n) _____ eats only plants.

16. A(n) _____ chain is a model of how matter and energy pass through organisms. Each organism is at a different _____ level.

17. _____ is a branch of biology that studies the interactions of organisms and their environment.

18. An ecological pyramid that shows the amount of energy for different trophic levels is called a(n) _____. If it shows how many organisms are at each tropic level, it is called a(n) _____.

19. Plants absorb _____ from the air, and with the sun's light energy they make high-energy carbon molecules.

20. Lightning and bacteria act to move and convert _____ from the air into compounds in the soil that can be readily used by plants.

APPLYING MAIN IDEAS

21. Explain how pesticides sprayed on the water in a wetland ecosystem could affect a different ecosystem.

22. Sloths are slow-moving herbivores that have algae growing in their fur. Caterpillars of certain kinds of moths eat the algae. Birds eat the moths. Using this example, draw a food chain and describe one symbiotic relationship.

THINKING CRITICALLY

23. Sequencing Place the following organisms in correct order in a food chain: mouse, hawk, wheat, snake.

24. Sequencing Describe one example of feeding relationships that cycle matter through an ecosystem.

25. Concept Mapping Complete the concept map by using the following vocabulary terms: autotrophs, decomposers, food webs, heterotrophs.

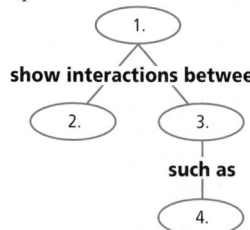

1.

show interactions between

2.   3.

such as

4.

CD-ROM

For additional review, use the assessment options for this chapter found on the *Biology: The Dynamics of Life Interactive CD-ROM* and on the Glencoe Science Web site. **science.glencoe.com**

ASSESSING KNOWLEDGE & SKILLS

The graph below compares the growth rates of two organisms when grown together and when grown separately.

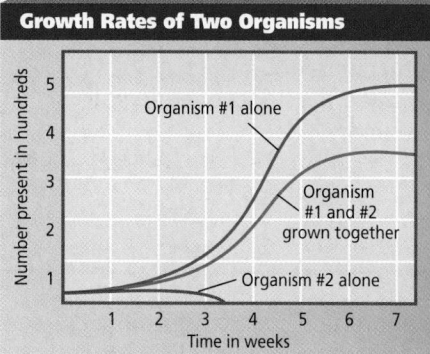

Growth Rates of Two Organisms

Organism #1 alone

Organism #1 and #2 grown together

Organism #2 alone

Number present in hundreds

Time in weeks

Interpreting Data Use the graph and information to answer the following questions.

1. When grown separately, approximately how long after the extinction of Organism 2 did it take the population of Organism 1 to reach its highest point?
a. 3 days   **c.** 3 weeks
b. 1 week   **d.** 5 weeks

2. When the organisms were grown together, what was the approximate rate of growth between weeks 2 and 6?
a. 75 per week   **c.** 50 per week
b. 100 per month   **d.** 25 per day

3. Observing and Inferring From the data, it is clear that the association between the organisms is _____.
a. commensalism   **c.** mutualism
b. parasitism   **d.** socialism

4. Hypothesizing Describe one possible benefit that Organism 2 gets from its association with Organism 1. Explain a possible reason why the association lowers the population of Organism 1.

THINKING CRITICALLY

23. wheat ⇒ mouse ⇒ snake ⇒ hawk

24. There are a variety of acceptable answers such as the following. Plants provide nitrogen to animals that eat the plants. Bacteria and other decomposers release nitrogen and carbon dioxide when they feed upon decaying organisms. Animals obtain carbon when eating other organisms.

25. 1. food webs 2. autotrophs 3. heterotrophs 4. decomposers

ASSESSING KNOWLEDGE & SKILLS

1. c
2. a
3. b
4. Organism #2 most likely benefits by obtaining nutrients from Organism #1. Other possible benefits include anything that helps Organism #2 satisfy its needs. The population size of Organism #1 is probably lowered because Organism #1 is weakened and cannot grow or reproduce as quickly. The interactions of the species may also cause more deaths in the population of Organism #1.

Chapter 3 Organizer

Refer to pages 4T-5T of the Teacher Guide for an explanation of the National Science Education Standards correlations.

Section	Objectives	Activities/Features
Section 3.1 **Communities** National Science Education Standards UCP.1, UCP.3, UCP.4; A.1, A.2; C.4, C.5; D.3; F.5; G.3 (2 sessions, 1 block)	1. **Explain** how limiting factors and ranges of tolerance affect distribution of organisms. 2. **Sequence** the stages of ecological succession.	**MiniLab 3-1:** Looking at Lichens, p. 69 **Problem-Solving Lab 3-1,** p. 70 **Investigate BioLab:** Succession in a Jar, p. 88
Section 3.2 **Biomes** National Science Education Standards UCP.1-3; A.1, A.2; C.4, C.5, C.6; F.3-5; G.1, G.3 (3 sessions, 1½ blocks)	3. **Compare and contrast** the photic and aphotic zones of marine biomes. 4. **Identify** the major limiting factors affecting distribution of terrestrial biomes. 5. **Distinguish** among biomes.	**Problem-Solving Lab 3-2,** p. 74 **MiniLab 3-2:** Looking at Marine Plankton, p. 75 **Focus On** Biomes, p. 78 **Inside Story:** A Tropical Rain Forest, p. 86 **Literature Connection:** *The Yellowstone National Park* by John Muir, p. 90

Need Materials? Contact Carolina Biological Supply Company at 1-800-334-5551 or at **http://www.carolina.com**

MATERIALS LIST

BioLab

p. 88 glass jar (3), pasteurized spring water, pond water with plant material, labels, microscope slides, coverslips, droppers, plastic wrap, cooked rice, plastic teaspoon, microscope

MiniLabs

p. 69 microscope, lichen samples
p. 75 microscope, microscope slide, coverslip, dropper, marine plankton culture

Alternative Lab

p. 80 cloth squares (3), large beaker, water, sand, clay, potting soil, balance, twist ties

Quick Demos

p. 68 cactus, broad-leafed houseplant
p. 82 paper towels, wax paper, water
p. 83 grass sod
p. 85 paper, scissors

Key to Teaching Strategies

L1 Level 1 activities should be appropriate for students with learning difficulties.

L2 Level 2 activities should be within the ability range of all students.

L3 Level 3 activities are designed for above-average students.

ELL ELL activities should be within the ability range of English Language Learners.

COOP LEARN Cooperative Learning activities are designed for small group work.

P These strategies represent student products that can be placed into a best-work portfolio.

These strategies are useful in a block scheduling format.

Communities and Biomes

Teacher Classroom Resources

Section	Reproducible Masters	Transparencies
Section 3.1 Communities	Reinforcement and Study Guide, pp. 11-12 **L2** Concept Mapping, p. 3 **L3** **ELL** BioLab and MiniLab Worksheets, p. 13 **L2** Laboratory Manual, pp. 15-22 **L2** Content Mastery, pp. 13-14, 16 **L1**	Section Focus Transparency 6 **L1** **ELL** Basic Concepts Transparency 3 **L2** **ELL** Reteaching Skills Transparency 4 **L1** **ELL**
Section 3.2 Biomes	Reinforcement and Study Guide, pp. 13-14 **L2** Critical Thinking/Problem Solving, p. 3 **L3** BioLab and MiniLab Worksheets, pp. 14-16 **L2** Content Mastery, pp. 13, 15-16 **L1** Inside Story Poster **ELL**	Section Focus Transparency 7 **L1** **ELL** Reteaching Skills Transparency 5 **L1** **ELL**

Assessment Resources

Chapter Assessment, pp. 13-18
MindJogger Videoquizzes
Performance Assessment in the Biology Classroom
Alternate Assessment in the Science Classroom
ExamView® Pro Software
BDOL Interactive CD-ROM, Chapter 3 quiz

Additional Resources

Spanish Resources **ELL**
English/Spanish Audiocassettes **ELL**
Cooperative Learning in the Science Classroom **COOP LEARN**
Lesson Plans/Block Scheduling

NATIONAL GEOGRAPHIC
Teacher's Corner

Products Available From Glencoe
To order the following products, call Glencoe at 1-800-334-7344:
CD-ROM
NGS PictureShow: Looking at Ecosystems
Transparency Set
NGS PicturePack: Looking at Ecosystems

Index to National Geographic Magazine
The following articles may be used for research relating to this chapter.
"Chesapeake Bay—Hanging in the Balance," by Tom Horton, June 1993.

GLENCOE TECHNOLOGY

The following multimedia resources are available from Glencoe.

Biology: The Dynamics of Life
CD-ROM **ELL**

Exploration: *World Biomes*
Video: *Tundra*
Video: *Tiaga*
Video: *Desert*
Video: *Temperate Grassland*
Video: *Temperate Forest*
Video: *Tropical Rain Forest*

Videodisc Program

Tundra
Tiaga
Desert
Temperate Grassland
Temperate Forest
Tropical Rain Forest

The Infinite Voyage

Secrets From a Frozen World
The Living Clock

3 Communities and Biomes

Theme Development

The theme of **systems and interactions** is illustrated as students learn about the changes involved in primary and secondary succession. Succession results from changes in interactions between biotic and abiotic factors. The system reaches relative stability when a climax community is formed. Biomes and their interactions are also discussed.

READING BIOLOGY

Glencoe's *Biology: The Dynamics of Life* contains many resources to assist a student's reading skills. Each chapter contains figures with expanded captions that expand on written material. Word Origins, located along the side of text, expand knowledge of biology vocabulary. Glencoe's Content Mastery Booklet helps develop reading skills while reinforcing content. In addition, use the Interactive Tutor for *Biology: The Dynamics of Life* on the Glencoe Web site to reinforce vocabulary.
science.glencoe.com

What You'll Learn

- You will identify factors that limit the existence of species to certain areas.
- You will describe how and why different communities form.
- You will compare and contrast biomes of planet Earth.

Why It's Important

Life is found in communities made of different species. To understand life on Earth, it is important to understand the interactions and growth of communities.

READING BIOLOGY

Make a list of new vocabulary words from the chapter. Read the list, looking for words that are familiar. Then go back to the text to see how the words are used.

BIOLOGY Online

To find out more about communities and biomes, visit the Glencoe Science Web site.
science.glencoe.com

This forest is a community of life. The inset photo shows the same area 50 years ago. Plants and animals return to an area in stages. Because communities depend on the climate and other abiotic factors, different regions of the world have different biomes.

66 COMMUNITIES AND BIOMES

66

3.1 Communities

Most organisms are adapted to maintain homeostasis in their native environments. A cactus can live in the desert, but it still needs water to survive. Its cells and tissues can absorb and store large amounts of water. Chipmunks can survive cold winters in the forest by going into hibernation.

But what if the ecosystem changes? What happens when a flash flood sends torrents of water through the desert? What happens when a forest fire destroys hundreds of acres of trees?

Mount St. Helens before the eruption in 1980 (inset) and Mount St. Helens after the eruption (above).

SECTION PREVIEW

Objectives

Explain how limiting factors and ranges of tolerance affect distribution of organisms.

Sequence the stages of ecological succession.

Vocabulary

limiting factor
succession
primary succession
climax community
secondary succession

Living in the Community

Look closely at a green lawn. At first glance, you might think there is only one species of plant, a grass. However, with closer examination you will find other organisms, such as insects, worms, weeds, and other species of grasses. Recall that communities are interacting populations of different species. How do species interact in your lawn?

Have you ever wondered why plants, animals, and other organisms live where they do? Why do lichens grow on bare rock but not on rich soil? Why do polar bears, such as

shown in *Figure 3.1*, live only in cold, snowy polar regions? How do catfish manage to live in waters that are too warm for trout to survive? Abiotic and biotic factors interact and result in conditions that are suitable for life for some organisms and unsuitable for other organisms.

Figure 3.1
Polar bears live near the north pole. Their white fur makes them hard to distinguish from the surrounding ice and snow, enabling them to stalk the seals and walruses that serve as their primary food.

3.1 COMMUNITIES **67**

Section 3.1

Prepare

Key Concepts

The concept of limiting factors, the biotic and abiotic factors that restrict life activities, is introduced. The orderly successions in ecosystems are discussed.

Planning

■ Gather lichen, microscopes, and other materials for Mini-Lab 3-1.

■ Purchase or borrow a cactus and broad-leafed houseplant for the Quick Demo.

---

1 Focus

Bellringer 📖

Before presenting the lesson, display **Section Focus Transparency 6** on the overhead projector and have the students answer the accompanying questions. **L1** **ELL**

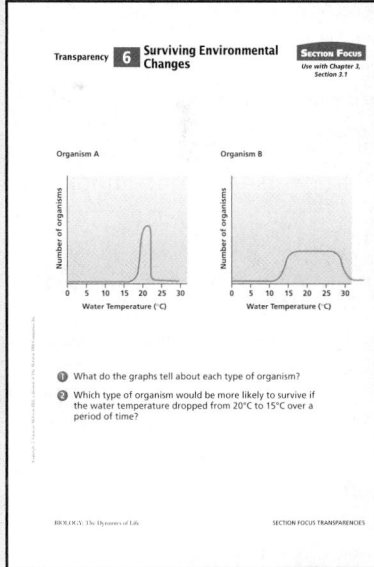

2 Teach

68

Limiting factors

Environmental factors that affect an organism's ability to survive in its environment, such as food availability, predators, and temperature, are **limiting factors.** A limiting factor is any biotic or abiotic factor that restricts the existence, numbers, reproduction, or distribution of organisms. The timberline in *Figure 3.2* illustrates how limiting factors affect the plant life of an ecosystem. At high elevations, temperatures are too low, winds too strong, and the soil too thin to support the growth of large trees. Vegetation is limited to small, shallow-rooted plants, mosses, ferns, and lichens.

Factors that limit one population in a community may also have an indirect effect on another population. For example, a lack of water could limit the growth of grass in a grassland, reducing the number of seeds produced. The population of mice dependent on those seeds for food

Figure 3.2
The timberline is the upper limit of tree growth on this mountainside.

68 COMMUNITIES AND BIOMES

---

will also be reduced. What about hawks that feed on mice? Their numbers may be reduced, too, as a result of a decrease in their food supply.

Ranges of tolerance

Farmers will tell you that corn plants need two to three months of sunny weather and a steady supply of water to produce a good yield. Corn grown in the shade or during a long dry period may survive, but probably won't produce much of a crop. The ability of an organism to withstand fluctuations in biotic and abiotic environmental factors is known as tolerance. *Figure 3.3* illustrates how the size of a population varies according to its tolerance for environmental change.

Some species can tolerate conditions that another species cannot. For example, catfish can live in warm water with low amounts of dissolved oxygen, which other fish species, such as bass or trout, could not tolerate. The bass or trout would have to swim to cooler water with more dissolved oxygen to avoid exceeding their range of tolerance.

Succession: Changes over Time

If grass were no longer cut on a lawn, what would it look like in one year, five years, and 20 years? Ecologists can accurately predict the changes that take place. The grass gets taller; weeds start to grow. The area resembles a meadow. Later, bushes grow, trees appear and different animals enter the area to live. The bushes and trees change the environment; less light reaches the ground. The grass slowly disappears. Thirty years later, the area is a forest. Ecologists refer to the orderly, natural changes and species replacements

---

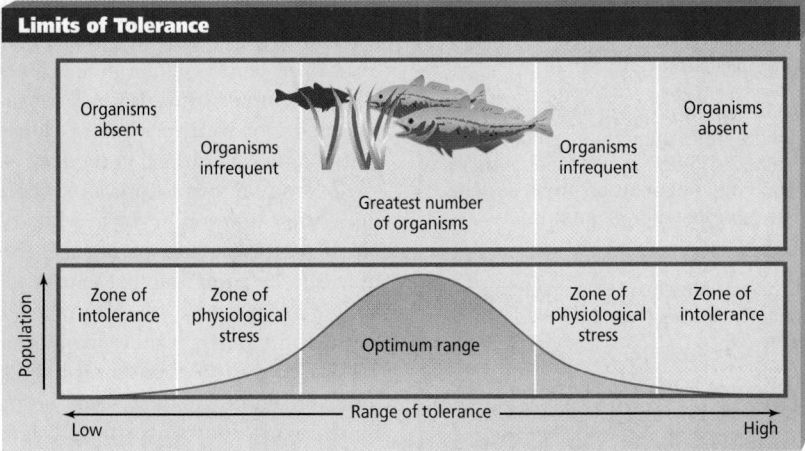

Limits of Tolerance

Organisms absent

Organisms infrequent

Greatest number of organisms

Organisms infrequent

Organisms absent

Population

Zone of intolerance

Zone of physiological stress

Optimum range

Zone of physiological stress

Zone of intolerance

Low ← Range of tolerance → High

Figure 3.3
The limits of an organism's tolerance are reached when the organism receives too much or too little of some environmental factor. Organisms become fewer as conditions move toward either extreme of the range of tolerance.

that take place in the communities of an ecosystem as **succession** (suk SESH un).

Succession occurs in stages; different species at different stages create conditions that are suitable for some organisms and unsuitable for others. Succession is often difficult to observe. It can take decades, or even centuries, for one type of community to completely succeed another. Observe the effects of succession in the *BioLab* at the end of this chapter.

Primary succession

Lava flowing from the mouth of a volcano is so hot it destroys everything in its path, but when it cools it forms new land. An avalanche exposes rock and creates ledges and gullies even as it buries the areas below. The colonization of new sites like these by communities of organisms is called **primary succession.** The first species in an area are called pioneer species. An example of a pioneer species is a lichen. Examine lichens more closely in the *MiniLab* on this page.

After some time, primary succession slows down, and, after many changes in species composition, the

MiniLab 3-1   Observing

Looking at Lichens
Lichens have the reputation for being a pioneer species when it comes to succession. They often inhabit rocky areas and start the process of soil formation. How is it possible for lichens to grow on a rock?

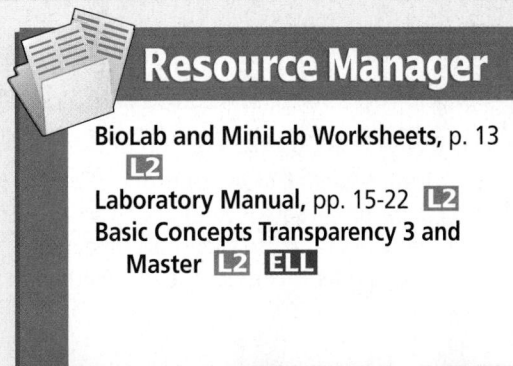

A lichen, note the alga and fungus in the close-up below.

Magnification: 700×

Procedure
1. Examine the lichen samples provided by your teacher. Note their color, shape, and texture.
2. Use a microscope to examine a prepared slide of a stained section of a lichen. Use low-power magnification and then change to high power as needed.
3. Observe the dark bodies that are cells containing chloroplasts. Notice that lichens are composed of an alga and a fungus. Diagram what you see.

Analysis
1. Describe the general appearance of a whole lichen and of the lichen under a microscope.
2. How does a lichen illustrate mutualism?
3. Explain how mutualism explains why lichens are able to survive on rocks.

3.1 COMMUNITIES **69**

Resource Manager

BioLab and MiniLab Worksheets, p. 13 **L2**

Laboratory Manual, pp. 15-22 **L2**
Basic Concepts Transparency 3 and Master L2 ELL

MiniLab 3-1

Purpose
Students will observe the gross and microscopic appearance of a lichen.

Process Skills
observe, apply concepts, define operationally, draw a conclusion

Safety Precautions
Remind students to wear goggles and aprons and to wash their hands after handling lichens.

Teaching Strategies
■ Prepared slides and whole lichens are available from biological supply houses. If available, gather whole lichens locally.
■ Teasing apart small sections of a whole lichen for microscopic observation is an alternative.
■ Review symbiosis with students prior to starting this lab.

Expected Results
Students will be able to observe the algae and the fungi in a lichen.

Analysis
1. Color may be dull green, red, orange, or yellow; shape may be crusty or flat. Fungus portion may be long, clear strands; alga portion, small green cells
2. The fungus receives food from the alga, and the alga receives moisture from the fungus.
3. Rocks offer harsh living conditions. The algae and fungi in lichens overcome this by making their own food and retaining moisture.

✔ Assessment

Portfolio Have students design an experiment to determine if each component of a lichen could survive without the other. Use the Performance Task Assessment List for Designing an Experiment in **PASC,** p. 23. **L2**

Purpose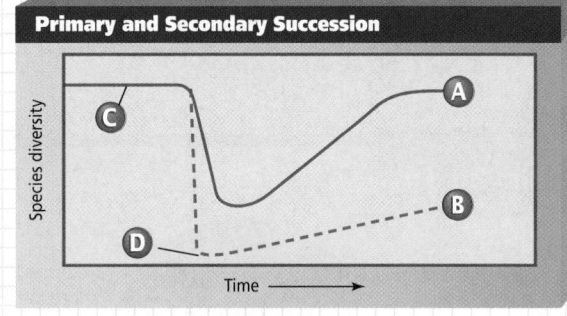

Students will use a graphic representation to determine differences and similarities between primary and secondary succession.

Process Skills

think critically, apply concepts, compare and contrast, draw a conclusion, interpret scientific illustrations

Teaching Strategies

■ If necessary, review the meanings of *primary* and *secondary* before students do this activity.

■ Have students work in small groups to complete this activity.

Thinking Critically

1. B; primary succession takes longer to reach stability.

2. A; secondary succession takes less time to reach stability than primary succession.

3. C; climax communities are stable and exist prior to a disturbance. D; pioneer organisms are not stable.

4. a sudden, disruptive event, such as a fire

✔ Assessment

Skill Have students draw a graph with proper time units along the *x*-axis to depict the expected appearance of a vacant lot as it undergoes succession. Use the Performance Task Assessment List for Graph from Data in **PASC**, p. 39. **L2**

Resource Manager

Problem-Solving Lab 3-1 — Interpreting Scientific Illustrations

How do you distinguish between primary and secondary succession? Succession is the series of gradual changes that occur in an ecosystem. Ecologists recognize two types of succession—primary and secondary. The events occurring during these two processes can be represented by a graph.

Analysis

Examine the graph. The two lines marked A and B represent primary and secondary succession. Note, however, that neither line is identified for you.

Primary and Secondary Succession

(graph: Species diversity vs. Time, with points C, A, D, B)

Thinking Critically

1. Which line best represents primary succession? Explain.
2. Which line best represents secondary succession? Explain.
3. Which label, C or D, might best represent a climax community? Pioneer organisms? Explain.
4. What does the sudden drop of line C represent?

---

community becomes fairly stable. A stable, mature community that undergoes little or no change in species is called a **climax community.** Primary succession of bare rock into a climax community is illustrated in *Figure 3.4.*

As pioneer organisms die, their decaying bodies cling to the bits of rock accumulating in cracks and crevices, initiating the first patches of soil. The presence of soil makes it possible for weedy plants, small ferns, and insects to become established. The soil builds up, and seeds borne by the wind blow into these larger patches of soil and begin to grow.

Over time, as the community of organisms changes and develops, additional habitats emerge, new species move in, and old species disappear. Eventually, the area becomes a forest of vines, trees, and shrubs inhabited by birds and other forest-dwelling animals.

Secondary succession

What happens when a natural disaster such as a forest fire or hurricane destroys a community? What happens when farmers abandon a field or when a building is demolished in a

Figure 3.4
The first organisms to colonize a new, rocky site are hardy pioneer species such as lichens. Larger plants eventually replace the pioneer species.

Beeches and maples

Pines

Ferns, shrubs, and grasses

Mosses

Lichens

70

⚡ TECHPREP

Highway Succession

Visual-Spatial Many transportation departments are letting grassy areas near highways revert to their natural condition. Have students contact the local highway department and find such areas. They should then make a photo journal showing the different areas and the dates of the last mow. **L3**

city and nothing is built on the site? **Secondary succession** refers to the sequence of community changes that takes place after a community is disrupted by natural disasters or human actions.

During secondary succession, as in primary succession, the community of organisms inhabiting an area gradually changes. Secondary succession, however, occurs in areas that previously contained life, and on land that contains soil. Therefore, the pioneer species involved in secondary succession are different from those in primary succession, but the same climax community will be reached in areas with a similar climate. Because soil already exists, secondary succession usually takes less time than primary succession to reach a climax community. Learn more about the differences between primary and secondary succession in the *Problem-Solving Lab.*

In 1988, a forest fire burned out of control in Yellowstone National Park. Thousands of acres of trees, shrubs, and grasses were burned. As you can see in *Figure 3.5,* the fire has given biologists an excellent opportunity to study secondary succession in a community. They have

been able to observe and compare secondary succession in areas that suffered damage of different levels of severity. Annual wildflowers were the first plants to grow back. Previously, the shade of the trees inhibited wildflower growth. Within three years, perennial wildflowers, grasses, ferns, and pine seedlings began to replace the annuals. Once the pine seedlings grow above the shade cast by the grasses and perennials, the trees will grow more quickly, and eventually a mature forest of lodge pole pines, the same community that was destroyed, will once again develop.

Figure 3.5
After Yellowstone National Park's forest fire of 1988, the pioneer species were wildflowers.

Section Assessment

Understanding Main Ideas
1. Give an example of a limiting factor for a pine tree.
2. Some species of fishes can survive in both fresh- and salt water. What does this say about their range of tolerance?
3. Give an example of secondary succession. Include plants and animals in your example.
4. Give an abiotic factor, and explain how it could be a limiting factor for a coyote population.

Thinking Critically
5. Explain how the growth of one population

can bring about the disappearance of another population during succession.

SKILL REVIEW

6. **Making and Using Graphs** Using the following data, graph the limits of tolerance for temperature for carp. The first number in each pair is temperature in degrees Celsius; the second number is the number of carp surviving at that temperature: 0, 0; 10, 5; 20, 25; 30, 34; 40, 27; 50, 2; 60, 0. For more help, refer to *Organizing Information* in the **Skill Handbook.**

Section Assessment

1. Answers will vary, but pine trees require water, deep soil rich in nutrients, and proper temperatures.
2. These fish have either a wide range of tolerance or during their life history their range of tolerance shifts.
3. Examples should describe the return of life to a damaged ecosystem. Examples should include both plants and animals

returning to the area.
4. Examples may include absence of water or extreme temperatures that limit the places where coyotes can survive.
5. One species can crowd, block the sun, eat the available food, or absorb the nutrients and water needed by the other species.
6. Check student graphs for logic and accuracy.

3 Assess

Check for Understanding

Have students explain how the terms in the following pairs are related. **L1**
  a. limiting factors—range of tolerance
  b. primary succession—secondary succession
  c. pioneer community—climax community

Reteach

Visual-Spatial Have students prepare a chart showing similarities and differences between primary and secondary succession. **L2**

Extension

Visual-Spatial Have students use library references to prepare a flow chart showing the sequence of changes that occur during succession of a pond into a hardwood forest. **L2**

✔ Assessment

Knowledge Ask students to recall the opening discussion regarding what a football field lot might look like if it were not used for 30 years. Ask them to rethink the changes they described and state whether they would predict the same changes now. Ask them to name this process of change. *succession* **L2**

4 Close

Discussion

Have students explain how human activities may disrupt or contribute to succession. Then have them list examples of natural events that bring about or hasten succession. **L2**

Prepare

Key Concepts

Students are introduced to world biomes—both aquatic and terrestrial. Limiting factors such as annual rainfall, temperature range, and sunlight availability are discussed in terms of how they result in the establishment of life zones throughout the world.

Planning

■ Gather grass sod for the Quick Demo.

■ Gather pond water and sediment, jars, and other materials for the BioLab.

■ Purchase or gather cloth squares, sand, clay, and potting soil for the Alternative Lab.

■ Purchase plankton and gather other materials for Mini-Lab 3-2.

1 Focus

Bellringer 🖑

Before presenting the lesson, display **Section Focus Transparency 7** on the overhead projector and have the students answer the accompanying questions. **L1** **ELL**

Objectives

Compare and contrast the photic and aphotic zones of marine biomes.

Identify the major limiting factors affecting distribution of terrestrial biomes.

Distinguish among biomes.

Vocabulary

biome
photic zone
aphotic zone
estuary
intertidal zone
plankton
tundra
permafrost
taiga
desert
grassland
temperate forest
tropical rain forest

Section

3.2 Biomes

Climate, *a combination of temperature, sunlight, prevailing winds, and precipitation, is an important factor in determining which climax community will develop at any spot on Earth. Soil type is also important. Many regions of the world share similar soil and climate characteristics and, as a result, also share similar types of climax communities. Although the species of organisms living in each desert ecosystem may vary, all are adapted for life in an environment with dry weather and poor soil.*

Cardon cactus (above) and kangaroo rat (inset)

Aquatic Biomes: Life in the Water

Ecosystems that have similar kinds of climax communities can be grouped into a broader category of organization called a biome. A **biome** is a large group of ecosystems that share the same type of climax community. Biomes located on land are called terrestrial biomes; those located in oceans, lakes, streams, ponds, or other bodies of water are called aquatic biomes.

As a human who lives on land, you may tend to think of Earth as a primarily terrestrial planet. But one look at a globe, a world map, or a photograph of Earth taken from space tells you there is an aquatic world, too; approximately 75 percent

of Earth's surface is covered with water. Most of that water is salty. Oceans, seas, and even some inland lakes contain salt water. Freshwater is confined to rivers, streams, ponds, and most lakes. Saltwater and freshwater environments have similarities, but they also have important differences. As a result, aquatic biomes are separated into marine biomes and freshwater biomes.

Marine biomes

If you've watched TV programs about ocean life, you may have gotten the impression that the oceans are mostly full of great white sharks, whales, and other large animals. However, different parts of the ocean differ in physical factors and in the organisms found there. The oceans

Resource Manager

Section Focus Transparency 7 and
Master **L1** **ELL**

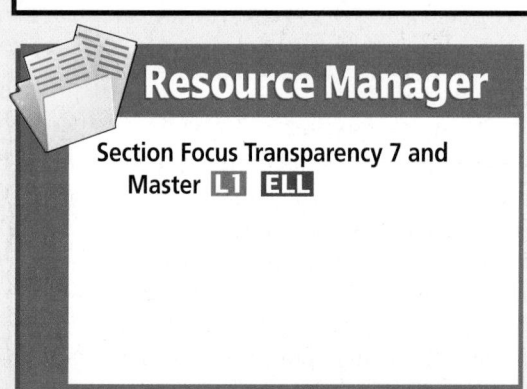

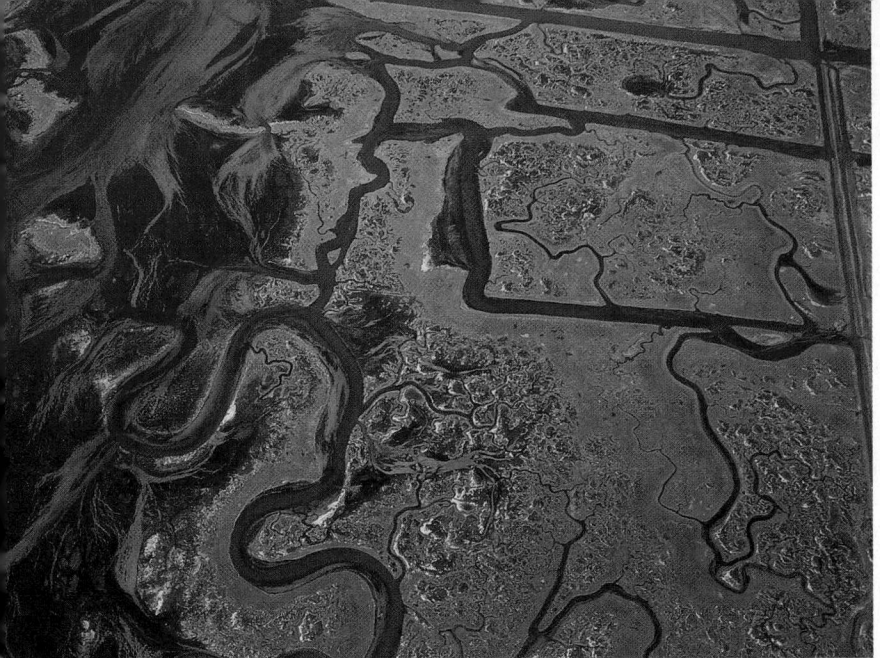

contain the largest amount of biomass, or living material, of any biome on Earth, but most of this biomass is made up of extremely small, often microscopic, organisms that humans usually don't see.

One of the ways ecologists study marine biomes is to separate them into shallow, sunlit zones and deeper, unlighted zones. The portion of the marine biome that is shallow enough for sunlight to penetrate is called the **photic zone.** Shallow marine environments exist along the coastlines of all landmasses on Earth. These coastal ecosystems include rocky shores, sandy beaches, and mudflats, and all are part of the photic zone. Deeper water that never receives sunlight makes up the **aphotic zone.** The aphotic zone includes the deepest, least explored areas of the ocean.

A mixing of waters

If you were to follow the course of any river, you would eventually reach a sea or ocean. Wherever rivers join oceans, freshwater mixes with salt water. In many such places, an estuary is formed. An **estuary** (ES chuh wer ee) is a coastal body of water, partially surrounded by land, in which freshwater and saltwater mix. It may extend many miles inland. The salinity in an estuary ranges between that of seawater and that of freshwater, and depends on how much freshwater the river brings into the estuary. Salinity in the estuary also changes with the tide. Because of these changes in salinity, a wide range of organisms can live in estuaries. Estuaries may contain salt marsh ecosystems, which are dominated by salt-tolerant grasses, as illustrated in *Figure 3.6.* These plants often grow so thick that their stems and roots form a tangled mat that traps food material and provides additional habitat for small organisms. These small organisms attract a wide range of predators, including cranes and other birds. The decay of dead organisms proceeds quickly, recycling nutrients through the food web.

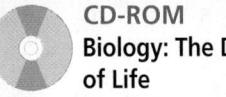 **Origin**

photic
From the Greek word *phos*, meaning "light." The marine photic zone receives light from the sun.

aphotic
From the Greek words *a*, meaning "without," and *phos*, meaning "light." The aphotic zone doesn't receive light.

2 Teach

Visual Learning

Ask students to explain which organisms in Figure 3.6 are likely to benefit most from the availability of light and nutrients in estuary waters. *Producers are most likely to benefit because they require light and nutrients for growth and development.*

✔ **Assessment**

Performance Assessment in the Biology Classroom, p. 57, *Investigating Salinity and Marine Algae.* Have students carry out this activity to determine the effect of salinity on marine algae. **L2**

GLENCOE TECHNOLOGY

 CD-ROM
Biology: The Dynamics of Life
Exploration: *World Biomes*
Disc 1

 **VIDEODISC**
The Secret of Life
Ocean Zones

 **Resource Manager**

Critical Thinking/Problem Solving 3, p. 3 **L3**

✔ **Portfolio**

Graphing Earth's Surface

Logical-Mathematical Ask students to prepare a circle graph that depicts the composition of Earth's surface. Provide these data for their graphs: salt water 73.5%, freshwater 1.5%, land 25.0 %. **L2**

P

BIOLOGY JOURNAL

Comparing Estuaries and Oceans

Naturalist Have students compare estuaries and oceans. As part of this comparison, ask students to focus on both biotic and abiotic factors associated with each ecosystem. **L3**

Purpose

Students will read a graph and analyze the extent and limitations of the information in it.

Process Skills

analyze information, draw a conclusion, hypothesize, interpret data, make and use graphs, predict, think critically

Teaching Strategies

■ To do this activity with the entire class, ask for their input on each question.

■ This activity is suitable for cooperative group analysis.

■ Explain to students that the final analysis of results is least important.

■ Review the concept of tide pools if necessary.

Thinking Critically

1. Answers may include that oxygen levels were measured in ppm, the experiment was conducted for 18 hours, the dependent variable (oxygen levels) is plotted on the *y*-axis, and the independent variable (time of day) is plotted on the *x*-axis.

2. Answers may include that both tide pool and ocean were sampled, samples were taken about every two hours for ocean oxygen, samples were taken almost every hour for tide pool oxygen, and 17 samples were taken for tide pool analysis, while 10 samples were taken for ocean analysis.

3. Answers may include how the samples were tested for oxygen, where the experiment was conducted, when during the year the experiment was conducted, when the tide pool was cut off from the ocean, and from what depth the ocean samples were taken.

Problem-Solving Lab 3-2 — Analyzing Information

What information can be learned from studying a graph?
Tide pools are depressions along rocky coasts that are covered by ocean waters during high tide. However, when oceans retreat during low tide, these tide pools are stranded and become temporarily cut off from ocean waters.

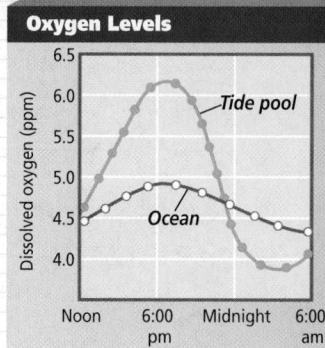

Oxygen Levels

Analysis

The graph shows results from tests of water samples taken in a tide pool and in the surrounding ocean. A scientist measured oxygen levels in ppm (parts per million). Both the ocean and tide pool have the same producer present, a green algae called Cladophora.

Thinking Critically

1. What can you tell about how the experiment was done using only the *x*- and *y*-axis information?
2. What can you tell about how the experiment was done from studying the graph?
3. What can't you tell about the experiment from the data provided?
4. What specific information was learned as a result of the experiment?

The effects of the tides

Twice a day, the gravitational pull of the sun and moon causes the rise and fall of ocean tides. The portion of the shoreline that lies between the high and low tide lines is called the **intertidal zone.** The size of this zone depends upon the slope of the land and the height of the tide. Intertidal ecosystems have high levels of sunlight, nutrients, and oxygen, but productivity may be limited by waves crashing against the shore. Intertidal zones differ in rockiness and wave action. *Figure 3.*7 shows examples of different types of intertidal zones. If the shore is rocky, waves constantly threaten to wash organisms into deeper water. Snails, sea stars, and other intertidal animals of the rocky shore have body parts that act as suction cups for holding onto the wave-beaten rocks. Other animals, such as mussels and barnacles, secrete a strong glue that helps them remain anchored. If the shore is sandy, wave action keeps the bottom in constant motion. Most of the clams, worms, snails, crabs, and other organisms that live along sandy shores survive by burrowing into the sand.

Figure 3.7
Waves crashing against a rocky shore are a constant threat to life in the intertidal zone **(a)**. Wave action churns the bottom of a sandy shore **(b)**.

---

4. The amount of dissolved oxygen in the tide pool rose sharply during the early afternoon and then fell sharply. The amount in the ocean rose and fell only slightly during the same periods.

✔ Assessment

Knowledge Ask students to explain the scientists' findings. Have them correlate increases or decreases in oxygen levels with the fact that there is a photosynthetic organism present. Use the Performance Task Assessment List for Analyzing the Data in **PASC,** p. 27. **L3**

Tide pools, pools of water left when the water recedes at low tide, can landlock the organisms that live in the intertidal zone until the next high tide. These areas vary greatly in nutrient and oxygen levels from the nearby ocean. Compare and contrast oxygen content between tide pools and the ocean in the *Problem-Solving Lab*.

In the light

As you move away from the intertidal zone and into deeper water, the ocean bottom is less and less affected by waves or tides. Many organisms live in this shallow-water region that surrounds most continents and islands. Nutrients washed from the land by rainfall contribute to the abundant life and high productivity of this region of the photic zone.

The photic zone of the marine biome also includes the vast expanse of open ocean that covers most of Earth's surface. Most of the organisms that live in the marine biome are plankton. **Plankton,** shown in *Figure 3.8,* are small organisms that live in the waters of the photic zone. Examine plankton more closely in the *MiniLab* shown here. Plankton include autotrophs, such as diatoms, and heterotrophs, such as juvenile stages of many marine animals.

In the dark

Imagine a darkness blacker than night and pressure so intense it exerts hundreds of pounds of weight on every square centimeter of your body's surface. Does this sound like a hospitable place to live? Almost 90 percent of the ocean is more than a kilometer deep. In some places, it may extend kilometers below the sunlit surface. Even though the animals living there are very far below the photic zone where plankton abound, many of them still depend

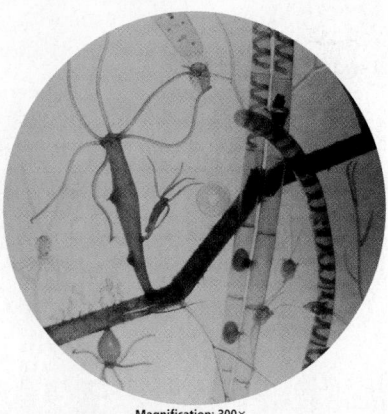

Figure 3.8
Plankton forms the base of all aquatic food chains, but not all organisms that eat plankton are small. Baleen whales and whale sharks, some of the largest organisms that have ever lived, consume vast amounts of plankton.

Magnification: 300×

MiniLab 3-2 — Comparing and Contrasting

Looking at Marine Plankton Plankton is the term used to define the microscopic life forms present in an aquatic environment. Plankton consists mainly of protists and animal larvae.

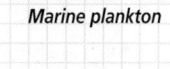

Magnification: 15×

Marine plankton

Procedure

1. Use a medicine dropper to obtain a small sample of marine plankton.
2. Prepare a wet mount of the material. **CAUTION:** *Handle microscope slides and coverslips carefully.*
3. Observe under low-power magnification of the microscope.
4. Look for a variety of organisms and diagram several different types.

Analysis

1. Describe the appearance of specific planktonic organisms. Draw what you see.
2. Are both autotrophs and heterotrophs present? How can you distinguish them?
3. Why are plankton important in food chains?

---

on plankton for food, either directly or indirectly by eating organisms that feed directly on plankton. Fishes living in the deep areas of the ocean are adapted to a life of darkness and a scarcity of food. What adaptations might help these organisms survive in this environment?

Purpose
Students will study the diverse nature of ocean plankton.

Process Skills
observe and infer, classify, compare and contrast

Safety Precautions
Have students wash their hands after the MiniLab.

Teaching Strategies
■ Marine plankton samples are available from biological supply houses. Freeze-dried plankton can be purchased at pet shops.
■ Because of the preservative used, many producer organisms may have lost their green color in the plankton samples.
■ You may wish to collect your own plankton samples. Pond plankton would be suitable. Small clusters of plants or algae from a pond system will provide good examples of both producer and consumer organisms.

Expected Results
Students will observe a variety of young or larval stages of consumer organisms in plankton.

Analysis
1. Plankton consists of many small shrimplike, wormlike, or insectlike organisms.
2. Answers may vary. Producers are green, but consumers are not.
3. Small fish eat plankton, and larger fish eat the small fish. Plankton are the basis of many marine food chains.

✓ Assessment
Performance Have students use references to identify some of the organisms observed in the plankton samples. Use the Performance Task Assessment List for Making Observations and Inferences in **PASC**, p. 17. **L2**

PROJECT

Measuring Oxygen

Kinesthetic Purchase kits from a biological supply house that measure the dissolved oxygen in water. Have students use these kits to test water samples from different local bodies of water. Ask students to record the source of each sample. **L2** **ELL**

Resource Manager

BioLab and MiniLab Worksheets, p. 14 **L2**

Figure 3.9
The shallow waters in which these plants grow are highly productive and include fishes, algae, protists, insect larvae, tadpoles, and crayfishes. As you move from the margins of a lake or pond toward the center, you find concentric bands of different species of plants.

Freshwater biomes

Have you ever gone swimming or boating in a lake or pond? If so, you may have noticed different kinds of plants, such as cattails and sedges, growing around the shoreline and even into the water, as shown in *Figure 3.9*. The shallow water in which these plants grow serves as home for tadpoles, aquatic insects, turtles that bask on fallen tree trunks, and worms and crayfishes that burrow into the muddy bottom. Insect larvae, whirligig beetles, dragonflies, and fishes such as minnows, bluegill, and carp also live here.

If you have ever jumped into a deep lake on a warm summer day, you probably got a cold surprise the instant you entered the water. Although the summer sun heats the lake's surface, the water a few feet below the surface remains cold because cold water is denser than warm water. If you were to dive all the way to the bottom of the lake, you would discover more layers of increasingly cold water as you descended. These temperature variations within a lake are an abiotic factor that limits the kinds of organisms that can survive in deep lakes.

Another abiotic factor that limits life in deep lakes is light. Not enough sunlight penetrates to the bottom to support photosynthesis, so few aquatic plants or algae grow. As a result, population density is lower in deeper waters. Decay takes place at the bottom of a lake. As dead organisms drift to the bottom, bacteria break them down and recycle the nutrients they contain.

Terrestrial Biomes

Imagine that you are setting off on an expedition beginning at the north pole and traveling south to the equator, as *Figure 3.10* shows. What kinds of environmental changes do you notice? The weather gets warmer, of course. You also see a gradual change in the kinds of plants that cover the ground. At the snow- and ice-covered polar cap, temperatures are always freezing and no plants exist. A little farther south, where temperatures sometimes rise above freezing but the soil never thaws, you might see soggy ground with just a few small cushions of low-growing lichens and plants.

As you continue on your journey, temperatures rise a little and you enter forests of coniferous trees. Farther south are grasslands and

---

Cultural Diversity

Saving the World's Biomes

The treaties negotiated at the Earth Summit held in Rio de Janeiro in 1992 represented a major step in protecting the world's biomes. A global warming treaty included firm guidelines to reduce gas emissions that contribute to the greenhouse effect. A biodiversity treaty called for protection of the millions of species that inhabit Earth, half of which live in rain forest regions. Many developing nations resent having to suppress their economies to protect the environment when many environmental problems were created by industrialized nations. Initiate a debate on this subject and discuss the importance of international cooperation in the effort to preserve Earth and its resources.

deserts, with higher summertime temperatures and little rain, and deciduous forests, with more rain and lower temperatures than the grasslands and deserts. Finally, as you approach the equator, you may find yourself surrounded by the lush growth of a tropical forest, where it rains almost every day.

As you move south from the north pole, you find yourself traveling through one biome after another. The graph in *Figure 3.11* shows how two abiotic factors—temperature and precipitation—influence the kind of climax community that develops in a particular part of the world. Small differences in temperature or precipitation can create many different climax communities, ecosystems, and biomes. Look at the distribution of the six most common terrestrial biomes in *Focus On Biomes*.

Figure 3.11
If you know the average annual temperature and rate of precipitation of a particular area, you should be able to determine the climax community that will develop.

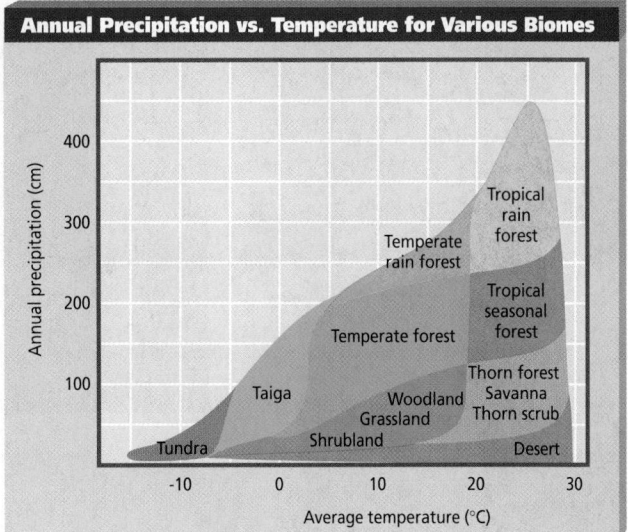

Figure 3.10
Many different terrestial biomes are encountered as you travel southward from the north pole to the equator.

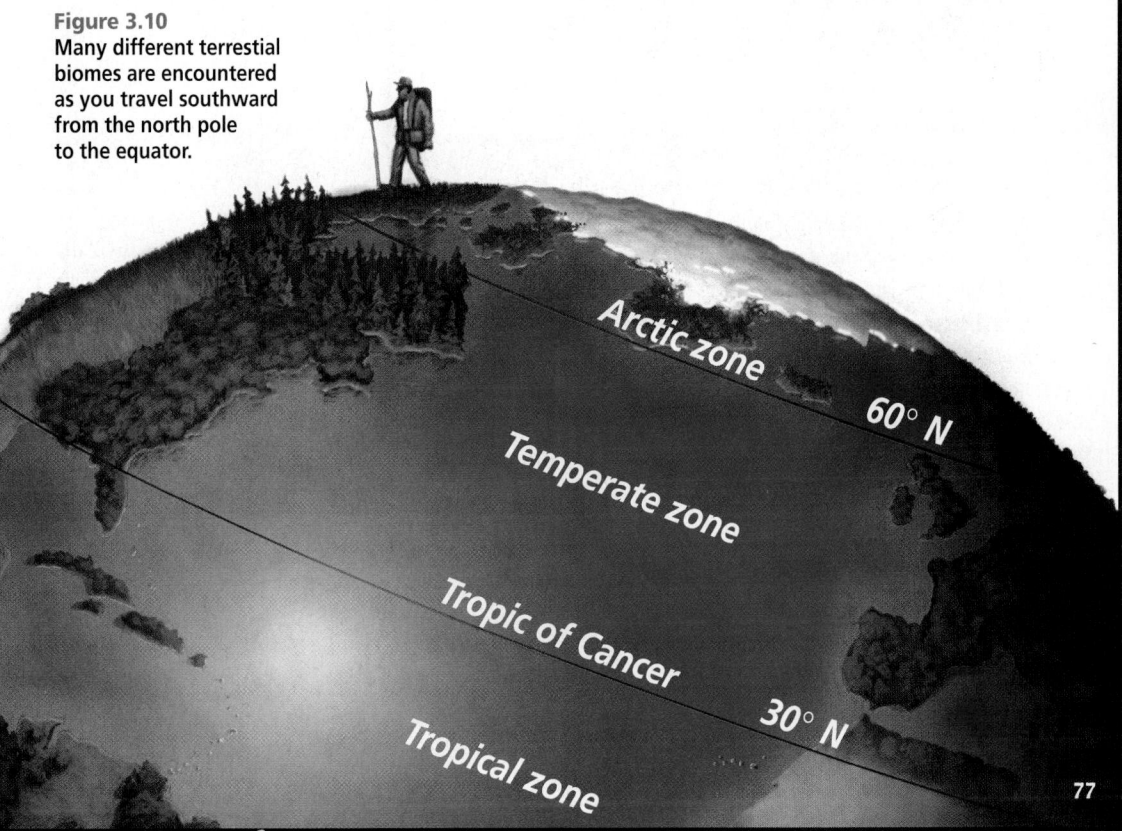

Focus On

Biomes

Purpose

Students will associate terrestrial biomes with their geographic locations.

Background

The major biomes of the world are shown on this map. Latitude influences biomes because generally the closer to the equator, the warmer the area. Precipitation is also a major influence on the formation of biomes.

Teaching Strategies

■ Ask students to name organisms that live in the different biomes.

■ Remove a globe from its stand. Have one student gently toss it to another student. Have them look at their right thumbs and record which biomes they are touching. Repeat 10 times. Most of the time, the thumbs will be on the ocean, helping students realize that oceans cover the greatest area.

■ Show a video or slides for each biome.

■ Discuss how air currents carrying moisture drop this moisture as precipitation as they approach a mountain. Therefore, one side of a mountain range has a lot of precipitation, and the other side is a desert.

WATER LILY

SALTWATER

DESERT

Focus On Biomes

A biome is a large group of ecosystems that share the same type of mature climax community undergoing little or no succession. When you think of a biome, you may imagine lions on an African grassland or monkeys in the rain forest. But ecologists look at climax communities of plants rather than animals. Because plants don't migrate, they are a better indicator of the long-term characteristics of a biome.

GRASSLAND

TAIGA

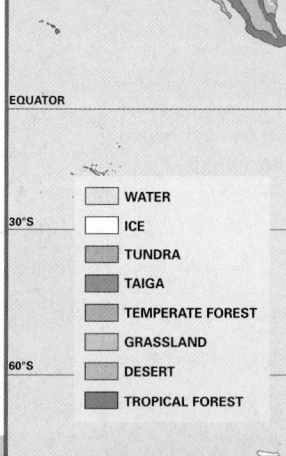

	WATER
	ICE
	TUNDRA
	TAIGA
	TEMPERATE FOREST
	GRASSLAND
	DESERT
	TROPICAL FOREST

60°N
30°N
EQUATOR
30°S
60°S

78

TEMPERATE FOREST

BIOMES

Earth's surface is marvelously diverse. Millions of species find a home here. But their distribution is not random. As the world map shows, Earth's biomes exhibit a definite pattern.

In general, three factors—latitude, altitude, precipitation—determine which biome dominates a terrestrial location. A rainy, low-lying area near the Equator will have a tropical rain forest as climax vegetation. A few kilometers away on a mountainside, ecologists may find plants typical of a biome thousands of kilometers to the north or south.

Look at the world map. Notice that Earth is more than two-thirds water. This water is mostly oceans, which make up the saltwater biome. Freshwater from precipitation makes up the other major water biome on land.

60°N

30°N

EQUATOR

30°S

60°S

TUNDRA

TROPICAL RAIN FOREST

EXPANDING Your View

1 **THINKING CRITICALLY** Which biome do you think would recover most slowly from destruction arising from natural events or human causes? Explain.

2 **COMPARING AND CONTRASTING** Think about the general pattern of biome types that exists from the equator to the poles. Do you think you would find a similar pattern if you climbed from the foot of a very high mountain to its peak? Explain.

Tying to Previous Knowledge

Have students name examples of survival adaptations used by organisms living in the tundra. *Responses may include color camouflage, migration, heavy fur coats, hibernation, and flat leaves on plants to reduce water loss from wind.* **L1**

Revealing Misconceptions

A popular belief is that lemmings, small mammals common in the tundra, periodically march into the ocean in mass suicides to reduce their large population. Actually, lemmings are migratory. After severely depleting an area of food, they move in large numbers to other areas. Sometimes, the animals stumble into the ocean during these mass migrations. However, this event is not a programmed effort to reduce their population.

GLENCOE TECHNOLOGY

VIDEODISC
The Secret of Life
Biome Distribution

Figure 3.12
Grasses, grasslike sedges, small annuals, and reindeer moss, a type of lichen on which reindeer feed, are the most numerous producers of the tundra. The short growing season may last fewer than 60 days.

Figure 3.13
Snowy owls (a) are lemming (b) predators. Lemmings are the most numerous mammals living in tundra communities. Populations of these small, furry animals rise to exceedingly high numbers. As the lemming population increases, so does the population of snowy owls.

a

b

Life on the tundra

As you begin traveling south from the north pole, you reach the first of two biomes that circle the pole. This first area is the **tundra** (TUN druh), a treeless land with long summer days and short periods of winter sunlight.

Because temperatures in the tundra never rise above freezing for long, only the topmost layer of soil thaws during the summer. Underneath this topsoil is a layer of permanently frozen ground called **permafrost.** Some areas of permafrost have remained frozen for so long that the frozen bodies of animals that have been extinct for thousands of years, such as the elephantlike mammoth, are sometimes found there.

In most areas of the tundra, the topsoil is so thin it can support only shallow-rooted grasses and other small plants. The soil is also lacking in nutrients. The process of decay is so slow due to the cold temperatures that nutrients are not recycled quickly.

Summer days on the tundra may be long, but the growing season is short. Because all food chains depend on the producers of the community, the short growing season is a limiting factor for life in this biome. For example, typical flowering tundra plants, *Figure 3.12,* are grasses, dwarf shrubs, and cushion plants. These organisms live a long time and are resistant to drought and cold.

Mosquitoes and other biting insects are some of the most common tundra animals, at least during the short summer. The tundra is also home to a variety of small animals, including ratlike lemmings, weasels, arctic foxes, snowshoe hares, snowy owls, and hawks. Musk oxen, caribou, and reindeer are among the few large animals that inhabit this biome during the summer months. *Figure 3.13* shows two common tundra animals.

Alternative Lab

Water-Holding Capacity of Soils

Purpose
Students are introduced to an abiotic factor that affects plant life in different biomes.

Safety Precautions
Remind students to wash their hands after the lab.

Materials
3 cloth squares (30 cm per side), large beaker, water, sand, clay, potting soil, balance, twist ties

Procedure
Give students the following directions.
1. Wrap a sample of sand into a cloth square. Fold the ends to form a bag and secure it with a twist tie.
2. Determine and record the mass of the sand bag.
3. Place the sand bag into a large beaker filled with water and allow it to soak for 5 minutes.
4. Remove the bag and allow it to drain for 1 minute. Determine and record the mass of the sand bag. Calculate the gain in mass of the wet sand.
5. Calculate and record the water-holding capacity (WHC) of sand using this formula:
   WHC = mass gain x 100/dry mass

Life on the taiga

Just south of the tundra lies another biome that circles the north pole. The **taiga** (TI guh), also called the northern coniferous forest, is a land of larch, fir, hemlock, and spruce trees, as shown in *Figure 3.14*.

How can you tell when you leave the tundra and enter the taiga? The line between any two biomes is indistinct, and patches of one blend almost imperceptibly into the other. For example, if the soil in the taiga is waterlogged, a peat swamp habitat develops that looks much like tundra. Taiga communities are usually somewhat warmer and wetter than tundra, but the prevailing climatic conditions are still harsh, with long, severe winters and short, mild summers.

In the taiga, which stretches across much of Canada, Northern Europe, and Asia, permafrost is usually absent. The topsoil, which develops from decaying coniferous needles, is acidic and poor in minerals. When fire or logging disrupt the taiga community, the first trees to recolonize the land may be birch, aspen, or

other deciduous species because the new soil conditions are within their ranges of tolerance. The abundance of trees in the taiga provides more food and shelter for animals than the tundra. More large species of animals are found in the taiga as compared with the tundra. *Figure 3.15* shows some animals of the taiga.

Figure 3.14
The dominant climax plants of the taiga in North America are primarily fir and spruce trees.

Figure 3.15
The taiga stretches across most of Canada, Northern Europe, and Asia.

A The lynx is a predator that depends on the snowshoe hare as its primary source of food.

B During the winter, the snowshoe hare grows a thick, white winter coat that includes extra hair on its feet for warmth.

C Caribou are large, herbivorous mammals of the taiga, where they may be found during most of the year.

81

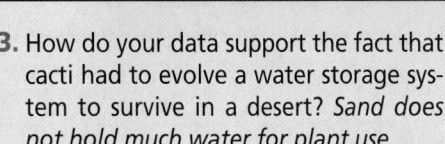
81

Figure 3.16
Creosote bushes cover many square kilometers of desert in the southwestern United States. These plants have yellow flowers and small leaves coated with a waxy resin that helps reduce water loss.

Life in the desert

The driest biome is the desert biome. A **desert** is an arid region with sparse to almost nonexistent plant life. Deserts usually get less than 25 cm of precipitation annually. One desert, the Atacama Desert in Chile, is the world's driest place. This desert receives an annual rainfall of zero.

Vegetation in deserts varies greatly, depending on precipitation levels. Areas that receive more rainfall produce a shrub community that may include drought-resistant trees such as mesquite. Less rainfall supports scattered plant life and produces an environment with large areas of bare ground. The driest deserts are drifting sand dunes with virtually no life at all. Plants have developed various adaptations for living in arid areas, as shown in *Figure 3.16.* Many desert plants are annuals that germinate from seed and grow to maturity quickly after sporadic rainfall. Cacti have leaves reduced to spines, photosynthetic stems, and thick waxy coatings that reduce water loss. The leaves of some desert plants curl up, or even drop off altogether, to reduce water loss during extremely dry spells. Desert plants sometimes have spines, thorns, or poisons that act to discourage herbivores.

Most desert mammals are small herbivores that remain under cover during the heat of the day, emerging at night to forage on plants. The kangaroo rat is a desert herbivore that does not have to drink water. These rodents obtain all the water they need to live from the water content in their food. Coyotes, hawks, owls, and roadrunners are carnivores that feed on the snakes, lizards, and small mammals of the desert. Scorpions are an example of a desert carnivore that uses venom to capture prey. Two of the many reptiles that make the desert their home are shown in *Figure 3.17.*

Figure 3.17
Lizards, tortoises, and snakes are numerous in desert communities. Desert tortoises feed on insects and plants (**a**). Venomous snakes such as the diamondback rattlesnake are major predators of small rodents (**b**).

a

b

Note Internet addresses that you find useful in the space below for quick reference.

_____

_____

_____

_____

_____

Life in the grassland

If an area receives between 25 and 75 cm of precipitation annually, a grassland usually forms. **Grasslands** are large communities covered with grasses and similar small plants. Grasslands, such as the ones shown in *Figure 3.18,* occur principally in climates that experience a dry season, where insufficient water exists to support forests. Called prairies in Australia, Canada, and the United States, these communities are called steppes in Russia, savannas in Africa, and pampas in Argentina. A grassland in the United States can be found in Yellowstone National Park. Find out how Yellowstone became our first national park in the *Literature Connection* at the end of this chapter.

Grasslands contain fewer than ten to 15 trees per hectare, though larger numbers of trees are found near streams and other water sources. This biome occupies more area than any other terrestrial biome, and it has a higher biological diversity than deserts, often with more than 100 species per acre.

The soils of grasslands have considerable humus content because many grasses die off each winter, leaving decay byproducts to build up in the soil. Grass roots survive through the winter, enlarging every year to form a continuous underground mat called sod.

Because they are ideal for growing cereal grains such as oats, rye, and wheat, which are different species of grasses, grasslands have become known as the breadbaskets of the world. Many other plant species live in this environment, including drought-resistant and late-summer-flowering species of wildflowers, such as blazing stars and sunflowers.

Most grasslands are populated by large herds of grazing animals, such as bison, a species of mammal native to the American prairies, shown in *Figure 3.18.* Millions of bison, commonly known as buffalo, once ranged over the American prairie, where they were preyed upon by wolves, coyotes, and humans. Other important prairie animals include prairie dogs, which are seed-eating rodents that build underground "towns" known to stretch across mile after mile of grassland, and the foxes and ferrets that prey on them. Many species of insects, birds, and reptiles, also make their homes in grasslands.

Figure 3.18
Summers are hot, winters are cold, and rainfall is often uncertain in a grassland (**a**). The prairies of America support bison as well as many species of birds and insects (**b**).

Quick Demo

Hold up a square of grass sod taken from a lawn. Explain that the roots hold the soil together and survive harsh winters to sprout new grass in spring.

GLENCOE
TECHNOLOGY

 **VIDEODISC**
Biology: The Dynamics of Life
Temperate Grassland (Ch. 10) Disc 1, Side 1, 28 sec.

 **CD-ROM**
Biology: The Dynamics of Life
Video: *Temperate Grassland* Disc 1

MEETING INDIVIDUAL NEEDS

Visually Impaired

Kinesthetic Bring samples of sandy, clay, and loam soils to class. Allow visually impaired students to feel the texture of the soils. Have them work with peers to rank the samples in terms of decayed material present. Ask them to explain how such material contributes to the fertility of each soil. **L1**

Figure 3.19
There are many types of temperate forests, each described by the two or three dominant species of trees. Typical trees of the temperate forest include birch, hickory, oak, beech, and maple.

Figure 3.20
Black bears and deer have always been residents of temperate forests in the United States. Other abundant animals in temperate forests are squirrels and salamanders.

Life in the temperate forest

When precipitation ranges from about 70 to 150 cm annually in the temperate zone, temperate deciduous forests develop. **Temperate forests** are dominated by broad-leaved hardwood trees that lose their foliage annually, *Figure 3.19.*

When European settlers first arrived on the east coast of North America, they cleared away vast tracts of temperate forest for farmland. The thin soil of the mountainous regions was soon depleted by crops, and farmers abandoned their land. Since then, secondary succession has restored much of the original forest.

The soil of temperate forests usually consists of a top layer that is rich in humus and a deeper layer of clay. If mineral nutrients released by the decay of the humus are not immediately absorbed by the roots of the living trees, they may be washed into the clay and lost from the food web for many years.

The animals that live in the temperate deciduous forest, as shown in *Figure 3.20,* include familiar squirrels, mice, rabbits, deer, and bears. Many birds, such as bluejays, live in the forest all year long, whereas others, such as the great crested flycatcher, migrate south to tropical regions during the winter.

84

Life in tropical rain forests

Tropical rain forests, such as the one shown in *Figure 3.21,* are home to more species of organisms than any other place on Earth. For example, one small national park in Costa Rica has more species of butterflies than all of North America. One tree in a South American rain forest was found to contain more species of ants than exist in all of the British Isles. The huge number of species in tropical rain forests has made their protection from human destruction an important mission of many people.

As their name implies, **tropical rain forests** have warm temperatures, wet weather, and lush plant growth. These forests are warm because they are near the equator. The average temperature is about 25°C. They are moist because wind patterns drop a lot of precipitation on them. Rain forests receive at least 200 cm of rain annually; some rain forests receive 600 cm.

Why do tropical rain forests contain so many species? The following hypotheses have been proposed by ecologists:

1. Due to their location near the equator, tropical rain forests were not covered with ice during the last ice age. Thus, the communities of species had more time to evolve.
2. Unlike the temperate forests—where deciduous trees drop their leaves in autumn—the warm weather near the equator gives tropical rain forest plants year-round growing conditions. This creates a greater food supply in tropical rain forests, which can support larger numbers of organisms.
3. Tropical rain forests provide a multitude of possible habitats for diverse organisms.

One reason for the large number of niches is the vertical layering of the tropical rain forest. Just as a library has shelves to hold more books, a tropical rain forest has layers that allow more species to exist. How are these layers arranged? Find out by reading the *Inside Story*. From bottom to top, these layers are the ground, understory, and canopy layers. The layers often blend together, but their differences allow many organisms to find a niche.

Most of the nutrients in a tropical rain forest are tied up in the living material. There are very few nutrients held in the soil. The hot humid climate enables ants, termites, fungi, and other decomposers to break down dead plants and animals rapidly. Plants must quickly absorb these

Figure 3.21
Warm temperatures, high humidity, and abundant rainfall allow the lush growth and great species diversity found in the tropical rain forest.

85

INSIDE STORY

Purpose

Students study the layers of rain forests and how they act to increase biodiversity.

Teaching Strategies

■ Have students create their own mural of the rain forest layers. **L1** **ELL**

■ Have students make a table that for each layer shows the abiotic factors and organisms in that layer.

Visual Learning

■ Have students describe how conditions above the canopy differ from those below it. **L2**

■ Refer students to Figure 3.18. Have them explain why rain forests support greater biodiversity than grasslands.

Critical Thinking

Life in the canopy is exposed to the sun, wind, and temperature extremes. Below the canopy, there are fewer sun, wind, and temperature extremes. Organisms with different niches live in different layers.

3 Assess

Check for Understanding

Have students explain the following word relationships. **L1** **ELL**

a. photic zone—aphotic zone
b. intertidal zone—estuary
c. river—estuary—ocean
d. food chain—plankton
e. biome—limiting factor

INSIDE STORY

A Tropical Rain Forest

The layers of a tropical rain forest provide niches for thousands of species. Ecologists generally consider rain forests to have three layers. The illustration shows organisms in a Central American tropical rain forest.

Critical Thinking How are the niches in the canopy different from those in the understory?

1 Canopy The canopy layer, 25-45 meters high, is a living roof. The tree tops are exposed to rain, sunlight, and strong winds. A few giant trees called emergents poke through the canopy. Monkeys frequently pass through. Birds, such as the beautiful scarlet macaw, live on the fruits and nuts of the trees.

2 Understory In the understory, the air is still, humid, and dark. Vines and trees grow from the soil to the canopy. Leaf cutter ants harvest leaves and bring them to the ground. Plants include ferns, broad-leaved shrubs, and dwarf palms. Insects are common in the understory. The limbs of the trees are coated with a thick layer of mosses and other epiphytes. Birds and bats prey upon the insects. Tree frogs are common understory amphibians. Reptiles include chameleons and snakes.

3 Ground The ground layer is a moist forest floor. Leaves and other organic materials decay quickly. Roots spread throughout the top 18 inches of soil, competing for nutrients. Mammals living on the ground include rodents and cats, such as the jaguar. Ants, termites, earthworms, bacteria, and fungi live in the soil and decompose organic materials.

MEETING INDIVIDUAL NEEDS

Gifted

Visual-Spatial Have students prepare an illustrated chart that describes the major features of the three zones of the tropical rain forest: the canopy, understory, and forest floor. Have students include typical plant and animal life as well as the limiting factors that produce the zonation effect. **L3**

NATIONAL GEOGRAPHIC

VIDEODISC

STV: Rain Forest, *Forest Floor*
Unit 2, Side 1, 8 min. 13 sec.

Forest Floor
(in its entirety)

Forest Canopy Unit 3, Side 1, (in its entirety)

a

b

c

nutrients before they are leached out of the soil by the rain. Rain forest trees have roots and mycorrhizae near the surface that enable them to absorb the nutrients. The shallow tree roots form a thick mat on the surface of the soil. Roots that support tall trees are sometimes greatly enlarged or may form buttresses, which resemble the fins of a rocket.

Tropical rain forest habitats, such as those of the species shown in *Figure 3.22,* are being destroyed by human activities. Rain forests are cut for their hardwoods, such as mahogany. Farmers clear the rain forest land to grow crops. After a few years, the crops deplete the soil of nutrients, and the farmers must then

move on and clear a different rain forest area. Rain forests are also cleared to produce grasslands for cattle.

Fortunately, people are realizing the importance of tropical rain forests and efforts are aimed at protecting these species-rich environments. In some areas, logging is now prohibited. But in areas where people need to use the land, they are being taught how to preserve the land they currently have cleared through crop rotation and fertilization.

Section Assessment

Understanding Main Ideas
1. Explain why the photic and aphotic zones of marine biomes are interdependent.
2. What is the most important abiotic factor that limits distribution of the tundra biome?
3. Describe some common plants and animals from a tropical rain forest and a grassland biome.
4. Describe three changes you would observe as you travel south from a taiga into a temperate forest.

Thinking Critically
5. Shaneka and her family were planning a trip to a foreign country. In reading before

the trip, Shaneka found that the winter was cold, the summer was hot, and most of the land was planted in fields of wheat. Infer which biome Shaneka's family would visit. Explain your answer.

SKILL REVIEW

6. **Making and Using Tables** Make a table to show the climate, plant types, plant adaptations, animal types, and animal adaptations for the terrestrial biomes. For more help, refer to *Organizing Information* in the **Skill Handbook**.

Reteach
*Naturalist* Ask students to pick a biome and describe its location, climatic characteristics, typical animals and plants, and special or unusual traits. **L2**

Extension
Naturalist Have students explain the relationship among the following: greenhouse effect, rain forest destruction, and the carbon cycle. **L2**

✔ **Assessment**
Portfolio Have students describe how the destruction of a biome such as the ocean would affect them directly. Have them suggest one or two ways that they personally can make a difference in the saving of a specific biome. **L2** **P**

4 Close

Activity
Advise students that they are to lead a group of tourists to a biome of their choice. Have them: (a) pick a specific biome as a destination; (b) suggest clothing the tourists should pack; (c) explain what the tourists should be prepared to see and experience. **L1**

Resource Manager

Content Mastery, pp. 13, 15-16 **L1**
Reinforcement and Study Guide, pp. 13-14 **L2**

Section Assessment

1. The photic zone provides food for the scavengers and decomposers in the aphotic zone. The decomposers of the aphotic zone return nutrients to water that can be used by plants in the photic zone.
2. temperature
3. tropical rain forest: trees, vines, ferns, insects, monkeys; grasslands: grass,

wildflowers, bison, prairie dogs, foxes
4. It gets warmer, forests turn from coniferous to deciduous, soil becomes less acidic, summers are longer, and winters are less harsh.
5. Grasslands; most of the grassland biome has been replaced with grass-like crop plants such as wheat. The conditions of this biome support the

cultivation of commercial crops.
6. Students' tables should resemble the following example. Taiga biome: long, harsh winters and short, cool summers; coniferous trees with needlelike leaves that resist drought; snowshoe hares grow white fur in winter and dark fur in summer; moose have heavy fur for warmth.

Time Allotment

First day—one class period. Every third day for the next three to four weeks—15 minutes.

Process Skills

apply concepts, classify, experiment, compare and contrast, draw a conclusion, interpret data, observe and infer, predict

Safety Precautions

Have students wash their hands well each time they examine the jars.

PREPARATION

Alternative Materials

- Jar C may be "seeded" with protist cultures from a biological supply house.
- Beakers or culture dishes may be used in place of glass jars.
- Cooked wheat or pea seeds may be used in place of rice. Do not use instant rice.
- Boiled and cooled tap water may be used in place of pasteurized water (available from supply houses).
- Plastic culture tubes, from supply houses, will enable students to grow the organisms in a microenvironment.

PROCEDURE

Teaching Strategies

- A good reference for identifying organisms is *Guide to Microlife*, available through Connecticut Valley Biological.
- Students may work in groups of four.
- Examine the jars once a week if time is short.
- Continue the experiment for at least three weeks.
- You may wish to present overhead diagrams of the most common organisms to help

students identify them.
- You may ask students to make and display large labeled diagrams of their organisms to help classmates identify similar organisms.
- Hold each jar against newspaper print. Inability to read the print through the jar indicates turbidity.
- Have students add new pasteurized water to replace any lost through evaporation.

INVESTIGATE
BioLab
Succession in a Jar

*S*uccession is usually described as changes seen in ecosystems following forests destroyed by fire or in farmlands left to lie fallow. Succession, however, can also be observed in a micro ecosystem, such as in a jar of pond water. The type and number of organisms will change over time. The advantage of studying this type of succession is that it will take only weeks, not years.

PREPARATION

Problem

Can you observe succession in a pond water ecosystem?

Objectives

In this BioLab, you will:

- **Observe** changes in three pond water environments.
- **Count** the number of each type of organism seen.
- **Determine** if the changes observed illustrate succession.

Materials

glass jars, 3
pasteurized spring water
pond water containing plant material
labels
glass slides and cover glasses
droppers
plastic wrap
cooked white rice
teaspoon, plastic
microscope

Safety Precautions

Always wear goggles in the lab.

Skill Handbook

Use the **Skill Handbook** if you need additional help with this lab.

PROCEDURE

1. Examine the pond water sample provided by your teacher.
2. Fill three glass jars with equal amounts of pasteurized (sterilized) spring water.
3. Label them A, B, and C. Add your name and today's date to each label.
4. Add the following to each of your three jars:
   Jar A: Nothing else
   Jar B: 3 grains of cooked white rice
   Jar C: 3 grains of cooked white rice, one teaspoon of pond sediment, and a small amount of any plant material present in the pond water.
5. Record the turbidity of each jar in the data table below. Turbidity means cloudiness and can best be judged by comparing jar A to B to C. Score turbidity on a scale of 1 to 10, with 1 meaning very clear water and 10 meaning very cloudy water.

Troubleshooting

- If students have trouble counting organisms, drops of *Protoslo* (available from supply houses) may be used in preparing wet mounts. This chemical slows the organisms down.

6. Gently swirl the contents of each jar.
7. Using a different clean dropper for each jar, remove a sample and prepare a wet mount of the liquid from each jar. Label each glass slide A, B, or C to avoid any mix-up. **CAUTION:** *Handle glass slides, coverslips, and glassware carefully.*
8. Observe each slide under low power. Look for autotrophic and heterotrophic organisms. Identify these organisms by name, describe their appearance, or make a sketch of what they look like.

9. Report the number of each type of organism as viewed in a low-power field of view (the circle of light seen when looking through the microscope under low power is your field of view).
10. Complete the data table for your first observations.
11. Cover each jar with either a lid or plastic wrap and place them in a lighted area.
12. Observe the jars every three days for several weeks, repeating steps 5-11 each time an observation is made. Use your data table to record your observations.

Data Table

Date	Jar	Turbidity	Name, description, or diagram of organism seen	Autotroph or heterotroph?	Number seen per low-power field
	A				
	B				
	C				
	A				
	B				

ANALYZE AND CONCLUDE

1. **Applying Concepts** Which of the three jars was a control? Explain why.
2. **Observing and Inferring** What might have been the role of the cooked rice?
3. **Recognizing Cause and Effect** Turbidity was a means of indirectly measuring the amount of bacterial growth in the jars. Why was there little, if any, turbidity in jar A?
4. **Analyzing Information** Describe the changes over time in the number and type of heterotrophs. Could these changes be described as succession? Explain.
5. **Observing and Inferring** Were you able to observe a climax ecosystem during this experiment? Explain your answer.

Going Further

Experimenting Design and carry out an experiment that tests the effect of temperature on the rate at which succession occurs in pond water.

 **BIOLOGY** *Online* To find out more about succession, visit the Glencoe Science Web site. **science.glencoe.com**

ANALYZE AND CONCLUDE

1. A; only pasteurized water was added.
2. food for heterotrophs
3. It had no living organisms.
4. See Data and Observations. Yes; any change in population is a key element of succession.
5. Answers will depend on how long the experiment continues.

✓ Assessment

Portfolio Have students pick three organisms (preferably some auto- and heterotrophs) and graph the changes in their population during the experiment. Use the Performance Task Assessment List for Graph from Data in **PASC**, p. 39. **L2**

Going Further

Have students determine the effect on succession when the jars are placed in the dark rather than in light.

Resource Manager

BioLab and MiniLab Worksheets, p. 15 **L2**

Data and Observations

■ Jar A should remain clear, showing no turbidity or life forms. Jar B may show turbidity caused by bacteria but should show no auto- or heterotrophs.

■ Jar C will show increasing and then decreasing turbidity. Autotrophic organisms will appear and increase in number. Heterotrophs will increase and decrease as follows: steady decrease in bacteria; sharp rise and then decline in *Colpidium* within first 20 days; sharp rise and then decline in *Euplotes* from days 40-60; slow and steady rise in *Paramecium* from days 10-50.

Literature Connection

Purpose

Students learn how a writer can influence public opinion and government policy towards the conservation of natural resources.

Teaching Strategies

■ Ask students what they can do to protect the environment. *Responses may include recycling, using public transportation, and eating organically grown foods.*

■ Have students debate whether public use of national parks should be restricted in order to protect the land from damage caused by overuse.

Connection to Biology

Authors might include Rachel Carson, Ralph Waldo Emerson, Henry David Thoreau, Charles Darwin, E.O. Wilson, Annie Dillard.

Literature Connection

The Yellowstone National Park by John Muir

The first, and largest, national park in the world was commissioned by an act of the United States Congress in 1872 as Yellowstone National Park. Because of the writing and influence of a man named John Muir, Congress also created the National Parks System, which includes Yellowstone, to preserve the lands that we still enjoy today. In recognition of his contributions, Muir is often called "The Father of our National Park System."

Although it includes waterfalls, a high-elevation lake with one hundred and ten miles of shoreline, and one of the world's largest volcanic explosion craters, Yellowstone National Park is probably most famous for its hot springs and geysers. In fact, more boiling caldrons and spouting plumes of hot water and mud are found in Yellowstone than in all of the rest of the world. In his book, *The Yellowstone National Park*, Muir provides his readers with the following description of the boiling basins and geysers:

Many of these pots and caldrons have been boiling thousands of years. Pots of sulphurous mush, stringy and lumpy, and pots of broth as black as ink, are tossed and stirred with constant care, and thin transparent essences, too pure and fine to be called water, are kept simmering gently in beautiful sinter cups and bowls that grow ever more beautiful the longer they are used.

Muir's Dream As a young man Muir had a vision of a "wildlands set aside by the government." The purpose of these lands would be simply to preserve the scenery and to educate people about the natural wonders of the land. As an adult, he was an avid explorer and prolific writer whose goals were to educate the public about the value of nature and the destructive effects man had on the natural environment. Muir felt that the beauty of nature was as essential to the well-being of man as was food.

"Old Faithful" geyser (above) and boiling mud pots (inset)

Over his lifetime, Muir wrote ten books and three hundred articles popularizing the idea of wilderness conservation. His writings so affected the attitude of the public, and even presidents of the United States, towards preservation of our natural resources that Muir has been called the United States' most famous and influential naturalist and conservationist.

CONNECTION TO BIOLOGY

What other authors have written material that influenced public opinion about the value of nature and preserving our natural environment?

BIOLOGY Online To find out more about Yellowstone National Park, visit the Glencoe Science Web site. **science.glencoe.com**

BIOLOGY Online Note Internet addresses that you find useful in the space below for quick reference.

_____

_____

_____

_____

_____

SUMMARY

Section 3.1
Communities

Main Ideas
- Communities, populations, and individual organisms occur in areas where biotic or abiotic factors fall within their range of tolerance. Abiotic or biotic factors that define whether or not an organism can survive are limiting factors.
- The sequential development of living communities from bare rock is an example of primary succession. Secondary succession occurs when communities are disrupted. Left undisturbed, both primary succession and secondary succession will eventually result in a climax community.

Vocabulary
climax community (p. 70)
limiting factor (p. 68)
primary succession (p. 69)
secondary succession (p. 71)
succession (p. 69)

Section 3.2
Biomes

Main Ideas
- Biomes are large areas that have characteristic climax communities. Aquatic biomes may be marine or freshwater. Estuaries occur at the boundaries of marine and freshwater biomes. Approximately three-quarters of Earth's surface is covered by aquatic biomes, and the vast majority of these are marine communities.
- Terrestrial biomes include tundra, taiga, desert, grassland, deciduous forest, and tropical rain forest. Two climatic factors, temperature and precipitation, are the major limiting factors for the formation of terrestrial biomes.

Vocabulary
aphotic zone (p. 73)
biome (p. 72)
desert (p. 82)
estuary (p. 73)
grassland (p. 83)
intertidal zone (p. 74)
permafrost (p. 80)
photic zone (p. 73)
plankton (p. 75)
taiga (p. 81)
temperate forest (p. 84)
tropical rain forest (p. 85)
tundra (p. 80)

Main Ideas
Summary statements can be used by students to review the major concepts of the chapter.

Using the Vocabulary
To reinforce chapter vocabulary, use the Content Mastery Booklet and the activities in the Interactive Tutor for Biology: The Dynamics of Life on the Glencoe Science Web site.
science.glencoe.com

 **THE PRINCETON REVIEW** *All Chapter Assessment questions and answers have been validated for accuracy and suitability by The Princeton Review.*

UNDERSTANDING MAIN IDEAS

1. d
2. c
3. a
4. b

UNDERSTANDING MAIN IDEAS

1. The removal of which of the following organisms would have the biggest impact on a marine ecosystem?
   a. fishes
   b. whales
   c. shrimp
   d. plankton

2. An undersea volcano erupts creating a new island off the coast of South Carolina. Life slowly starts appearing on the island. What would probably be the first species to grow and survive?
   a. maple trees
   b. finches
   c. lichens
   d. grasses

3. The changes in communities that take place on a new island would best be described as _____.
   a. primary succession
   b. secondary succession
   c. tertiary succession
   d. none of the above

4. Which of the following is true?
   a. Temperate forests have more rainfall than tropical rain forests.
   b. Tropical rain forests have more species of trees than temperate forests.
   c. Temperate forests are closer to the equator than tropical rain forests.
   d. Tropical rain forests are younger than temperate forests.

CHAPTER 3 ASSESSMENT **91**

GLENCOE TECHNOLOGY

 **VIDEOTAPE**
MindJogger Videoquizzes
Chapter 3: *Communities and Biomes*
Have students work in groups as they play the videoquiz game to review key chapter concepts.

Resource Manager

Chapter Assessment, pp. 13-18
MindJogger Videoquizzes
ExamView® Pro Software
BDOL Interactive CD-ROM, Chapter 3 quiz

5. c
6. b
7. c
8. b
9. b
10. b
11. permafrost
12. pioneer
13. estuaries, grasses
14. grassland, prairies, savannas
15. canopy
16. tundra, lemmings
17. climax
18. tropical rain forest
19. taiga
20. desert

5. The photograph shows a forest in the United States. The annual rainfall is 300 cm and the average temperature is 15°C. What type of forest is shown in the photograph?
a. tropical rain forest
b. coniferous forest
c. temperate rain forest
d. temperate forest

6. A lack of food prevents further growth in a deer population. This is an example of a
_____.
a. range of tolerance
c. photic zone
b. limiting factor
d. biome

7. Most species of ocean fishes spend their entire life in salty water. However, some fishes, such as the flounder and salmon, travel up freshwater rivers to reproduce. Which statement is most likely true about the salmon and flounder?
a. They aren't affected by physiological stress.
b. They have no zones of intolerance.
c. They have wider optimum ranges.
d. They are pioneer species.

8. A deep sea fisherman catches an ocean fish. This fish, like others of the same species, has no developed eyes. Its habitat is most likely the _____.
a. intertidal zone
c. photic zone
b. aphotic zone
d. zone of intolerance

THE PRINCETON REVIEW **TEST-TAKING TIP**

Beat the Clock and Then Go Back
As you take a practice test, pace yourself to finish a few minutes early so you can go back and check over your work. You'll usually find a mistake or two. Don't worry. It is better to make corrections than to hand in a test with wrong answers.

9. Locations of biomes are usually determined by _____.
a. temperature and altitude
b. temperature and precipitation
c. altitude and precipitation
d. soil type and temperature

10. The kangaroo rat conserves its water and obtains all its water from the food it eats. In what ecosystem is the kangaroo rat most likely to live?
a. tropical rain forest   **c.** taiga
b. desert   **d.** savanna

11. The layer of frozen soil found in the tundra is called the _____.

12. A species that first occupies and lives in an area is called a _____ species.

13. Aquatic areas with a mix of salt water and freshwater are called _____. They contain salt marsh ecosystems where the dominant plant life is _____.

14. Cereal grains grow best in the biome called the _____. These areas are called _____ in the United States and _____ in Africa.

15. An ecologist climbs the tallest tree in a patch of rain forest. She is probably at the _____ layer.

16. Between the north pole and the taiga lies the _____. In this area there are more _____, a furry, ratlike animal, than any other mammals.

17. Beech trees and maple trees dominate a forest that has stayed the same for 100 years. The ecological term for this stable community is a _____ community.

18. The terrestrial biome with lush plant growth but poor soil is the _____.

19. The biome shown here is most likely _____.

20. A diamondback rattlesnake rattles his rattle and you back into a cactus. You are most likely stuck in a _____ biome.

APPLYING MAIN IDEAS

21. How may agriculture lead to soil erosion in a grassland biome?

22. How might annual fires affect the succession of a temperate deciduous forest?

23. Compare the biodiversity of the temperate forest biome with that of a tropical rain forest biome.

THINKING CRITICALLY

24. Forming a Hypothesis What would be the characteristics of a successful pioneer species in a secondary succession?

25. Recognizing Cause and Effect Some plant seeds need fire to germinate. Describe the niche of these organisms in secondary succession.

26. Concept Mapping Complete the concept map by using the following vocabulary terms: biomes, desert, grassland, temperate forest, tropical rain forest, tundra.

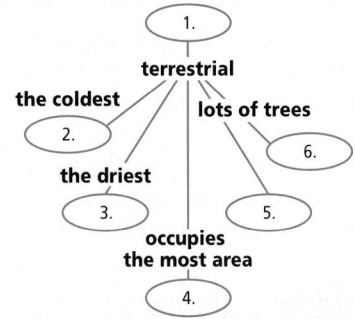

CD-ROM

For additional review, use the assessment options for this chapter found on the *Biology: The Dynamics of Life Interactive CD-ROM* and on the Glencoe Science Web site.
science.glencoe.com

ASSESSING KNOWLEDGE & SKILLS

The table below shows rates of decomposition for terrestrial ecosystems.

Table 3.1 Rates of Decomposition	
Ecosystem	**Rate**
Tundra	– –
Taiga	–
Desert	–
Temperate Forest	+
Tropical Rain Forest	+++
Grassland	++

Interpreting Data Study the table and answer the following questions.

1. Which ecosystem has the slowest rate of decomposition?
a. tundra
b. taiga
c. tropical rain forest
d. desert

2. Where would a twig decay fastest?
a. tundra
b. taiga
c. tropical rain forest
d. grassland

3. Sequencing Put the ecosystems in a list from fastest to slowest rates of decomposition.

4. Hypothesizing Suggest two possible reasons why there are differences in rates of decomposition between the biomes.

5. Predicting What rating do you think the following aquatic ecosystems would have? Why?
a. estuaries
b. interdidal zone
c. aphotic zone
d. pond

APPLYING MAIN IDEAS

21. After crops are harvested, the bare soil may be subject to erosion by wind and water.
22. Annual fires would prevent the growth of the climax species of deciduous trees.
23. Biodiversity is much higher in a tropical forest biome than in a temperate forest biome.

THINKING CRITICALLY

24. Pioneers are fast-growing plants that produce many seeds. They are sun-tolerant and hardy.
25. They are among the first to germinate and grow after a fire, making them a pioneer species.
26. 1. biomes; 2. tundra; 3. desert; 4. grassland; 5. and 6. tropical rain forest and temperate forest

ASSESSING KNOWLEDGE & SKILLS

1. a
2. c
3. (1) tropical rain forest, (2) grassland, (3) temperate forest, (4) desert and (4) taiga, (5) tundra
4. The different rates could be caused by different temperatures, amounts of moisture, and biomasses of decomposers.
5. Answers may vary, but students should support their responses with logical reasons.

Chapter 4 Organizer

Refer to pages 4T-5T of the Teacher Guide for an explanation of the National Science Education Standards correlations.

Section	Objectives	Activities/Features
Section 4.1 **Population Dynamics** National Science Education Standards UCP.1-3; A.1, A.2; C.4, C.5; F.2, F.4, F.5, F.6; G.1, G.2 (3 sessions, 1 block)	1. **Compare and contrast** exponential and linear population growth. 2. **Relate** the reproductive patterns of different populations of organisms to models of population growth. 3. **Predict** effects of environmental factors on population growth.	**MiniLab 4-1:** Fruit Fly Population Growth, p. 96 **Inside Story:** Population Growth, p. 98 **Problem-Solving Lab 4-1,** p. 99 **Investigate BioLab:** How can you determine the size of an animal population? p. 108 **Chemistry Connection:** Polystyrene: Friend or Foe? p. 110
Section 4.2 **Human Population Growth** National Science Education Standards UCP.1-3; A.1, A.2; C.4, C.5, C.6; F.4, F.5; G.1, G.3 (2 sessions, 1 block)	4. **Relate** population characteristics to population growth rates. 5. **Compare** the age structure of rapidly growing, slow-growing, and no-growth countries. 6. **Hypothesize** about problems that can be caused by immigration and emigration.	**MiniLab 4-2:** Doubling Time, p. 105 **Problem-Solving Lab 4-2,** p. 106

Need Materials? Contact Carolina Biological Supply Company at 1-800-334-5551 or at **http://www.carolina.com**

MATERIALS LIST

BioLab

p. 108 paper bag, beans (175), magic marker, paper, pencil, calculator (optional)

MiniLabs

p. 96 banana, glass jar, rubber band, fine cloth or wire mesh, paper, pencil

p. 105 paper, pencil, calculator (optional)

Alternative Lab

p. 100 petri dish (2), radish seeds, paper napkins, wax paper, pipette, water, paper, pencil

Quick Demos

p. 101 overhead projector, cardboard, checkers

p. 106 none

Key to Teaching Strategies

L1 Level 1 activities should be appropriate for students with learning difficulties.

L2 Level 2 activities should be within the ability range of all students.

L3 Level 3 activities are designed for above-average students.

ELL ELL activities should be within the ability range of English Language Learners.

COOP LEARN Cooperative Learning activities are designed for small group work.

P These strategies represent student products that can be placed into a best-work portfolio.

These strategies are useful in a block scheduling format.

Population Biology

Teacher Classroom Resources

Section	Reproducible Masters	Transparencies
Section 4.1 **Population Dynamics**	Reinforcement and Study Guide, pp. 15-16 **L2** Concept Mapping, p. 4 **L3** **ELL** Critical Thinking/Problem Solving, p. 4 **L3** BioLab and MiniLab Worksheets, p. 17 **L2** Laboratory Manual, pp. 23-30 **L2** Content Mastery, pp. 17-18, 20 **L1**	Section Focus Transparency 8 **L1** **ELL** Reteaching Skills Transparency 6 **L1** **ELL**
Section 4.2 **Human Population Growth**	Reinforcement and Study Guide, pp. 17-18 **L2** BioLab and MiniLab Worksheets, pp. 18-20 **L1** Content Mastery, pp. 17, 19-20 **L1**	Section Focus Transparency 9 **L1** **ELL**

Assessment Resources

Chapter Assessment, pp. 19-24
MindJogger Videoquizzes
Performance Assessment in the Biology Classroom
Alternate Assessment in the Science Classroom
ExamView® Pro Software 💾
BDOL Interactive CD-ROM, Chapter 4 quiz

Additional Resources

Spanish Resources **ELL**
English/Spanish Audiocassettes **ELL**
Cooperative Learning in the Science Classroom **COOP LEARN**
Lesson Plans/Block Scheduling

NATIONAL GEOGRAPHIC · Teacher's Corner

Products Available From Glencoe
To order the following products, call Glencoe at 1-800-334-7344:
Videodisc
GTV: Planetary Manager

Index to National Geographic Magazine
The following articles may be used for research relating to this chapter.
"Feeding the Planet," by T. R. Reid, October 1998.
"Human Migration," by Michael Parfit, October 1998.
"Population," by Joel L. Swerdlow, October 1998.
"Making Sense of the Millennium," by Joel L. Swerdlow, January 1998.
"The World's Food Supply at Risk," by Robert E. Rhoades, April 1991.
"Beyond Supermouse: Changing Life's Genetic Blueprint," by Robert F. Weaver, December 1984.
"World's Urban Explosion," by Robert W. Fox, August 1984.

GLENCOE TECHNOLOGY

The following multimedia resources are available from Glencoe.

Biology: The Dynamics of Life
CD-ROM **ELL**
 BioQuest: *Antarctic Food Web*
Animation: *Carrying Capacity*

Videodisc Program
 Carrying Capacity

The Infinite Voyage
 The Keepers of Eden

The Secret of Life Series
Competition
 Predator–Prey
Gone Before You Know It: *The Biodiversity Crisis*

Chapter

4 Population Biology

What You'll Learn

- You will explain how populations grow.
- You will identify factors that inhibit the growth of populations.
- You will summarize forces behind and issues in human population growth.

Why It's Important

How a population grows is critical to its niche. A population that becomes too large too quickly may run out of food and space, and diseases spread more easily through large populations; a population that grows too slowly may become extinct.

READING BIOLOGY

Carefully read the "Inside Story: Population Growth" on page 98. Observe the five numbered stages of population growth. Choose a specific animal and its habitat. Describe how each stage of population growth would unfold for the animal and impact the surrounding environment.

To find out more about population biology, visit the Glencoe Science Web site. **science.glencoe.com**

The walruses (*Odobenus rosmarus*) in this large group make up one population.

4.1 Population Dynamics

SECTION PREVIEW

Objectives
Compare and contrast exponential and linear population growth.
Relate the reproductive patterns of different populations of organisms to models of population growth.
Predict effects of environmental factors on population growth.

Vocabulary
exponential growth
carrying capacity
density-dependent factor
density-independent factor

Weeds! You've probably observed a scene like this before. What was recently a clean, grass-filled lawn is now crowded with hundreds, perhaps thousands, of bright yellow dandelions. Why do these plants appear so quickly and in such large numbers? Each season a dandelion plant produces hundreds of seeds. In contrast, the lion produces only two to four cubs when it successfully mates. Despite the different reproductive modes, dandelion populations and lion populations have lived in their habitats for thousands of years.

Dandelions (above) and a lioness with cub (inset)

Principles of Population Growth

How and why do populations grow? Population growth is defined as an increase in the size of a population over time. Scientists use a variety of methods to investigate population growth in organisms, as shown in *Figure 4.1.* One method involves placing microorganisms, such as bacteria or yeast cells, into a tube or bottle of nutrient solution and observing how rapidly the population grows. Another method involves introducing

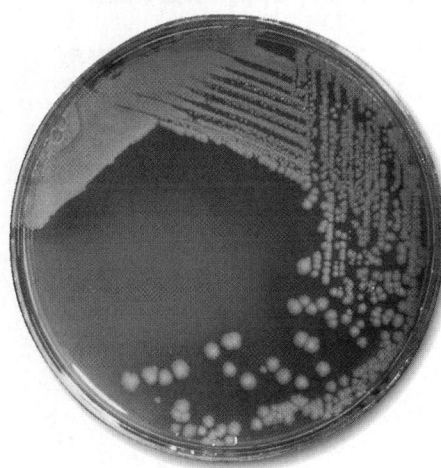

Figure 4.1
Ecologists can study population growth by inoculating a petri dish containing a nutrient medium with a few organisms and watching their growth.

Prepare

Key Concepts
Population growth is the increase in population size over time. Students learn that population growth, while exponential at times, is controlled by limiting factors that determine the carrying capacity of the environment. Such limits to population growth may result from predator-prey interactions or overcrowding.

Planning
■ Purchase bananas and gather jars and mesh for MiniLab 4-1.
■ Purchase radish seeds. Gather petri dishes and napkins for the Alternative Lab.

1 Focus

Bellringer ✍
Before presenting the lesson, display **Section Focus Transparency 8** on the overhead projector and have the students answer the accompanying questions. **L1** **ELL**

Transparency **8** Predator-Prey Relationships — SECTION FOCUS Use with Chapter 4, Section 4.1

❶ How is the number of sea urchins in this community affected by the number of sea otters?
❷ How is the number of sea otters affected by the number of sea urchins?

BIOLOGY: The Dynamics of Life — SECTION FOCUS TRANSPARENCIES

✔ Assessment Planner

Portfolio Assessment
 Portfolio, TWE, pp. 98, 104
 Problem-Solving Lab, TWE, p. 99
 Assessment, TWE, p. 103
 MiniLab, TWE, p. 105
Performance Assessment
 MiniLab, SE, pp. 96, 105
 Assessment, TWE, pp. 100, 107
 Alternative Lab, TWE, pp. 100-101
 BioLab, SE, p. 109

Knowledge Assessment
 Section Assessment, SE, pp. 103, 107
 Problem-Solving Lab, TWE, p. 106
 Chapter Assessment, SE, pp. 111-113
Skill Assessment
 MiniLab, TWE, p. 96
 Assessment, TWE, pp. 102,105
 BioLab, TWE, p. 109

2 Teach

Purpose
Students will learn that food is a factor in population growth.

Process Skills
observe and infer, interpret data

Safety Precautions
Have students wash their hands after each observation.

Teaching Strategies
■ If temperatures are warm enough, fruit flies should be attracted quickly to the bananas.
■ You can order wingless fruit flies from a biological supply company.
■ Have students record their data and observations in a table.

Expected results
Fruit flies will arrive and quickly reproduce. Larvae will be evident on the jar and the banana. In a few weeks, the jar will be full of flies. Soon the population will decline because the food supply is limited.

Analysis
1. Answers will vary.
2. lack of food due to over-crowding
3. The population quickly increased and then quickly decreased.

✔ Assessment
Skill Have students describe how they could determine whether food type affects the type of animal attracted. Use the Performance Task Assessment List for Designing an Experiment in **PASC**, p. 23. **L2**

MiniLab 4-1 — Making and Using Tables

Fruit Fly Population Growth Fruit flies (*Drosophila melanogaster*) and similar insects have rapid rates of reproduction. Fruit flies are frequently used in biological research because they reproduce quickly and are easy to keep and count. In this activity you will observe the growth of a fruit fly population as it exploits a food supply.

Procedure
1. Place half of a banana in a jar and allow it to sit outside in a warm shaded area, or put it in a warm area in your classroom.
2. Leave the jar for one day or until you have at least three fruit flies in it. Put the mesh on top of the jar and fasten with the rubber band.
3. Each day record how many adult fruit flies are alive in the jar. Record for at least three weeks. Put your data into table form. **CAUTION:** *Return the fruit flies to your teacher for proper disposal.*

Analysis
1. How many fruit flies did you start with? On what day were there the most fruit flies? How many were there?
2. Why did the number of fruit flies decrease?
3. Based on this investigation, why are insects considered to display a rapid reproduction pattern?

Figure 4.2
The way you earn money at an hourly rate is a straight line graph. Other examples might include the growth of your weekly allowance or the number of cars produced by an assembly line each month.

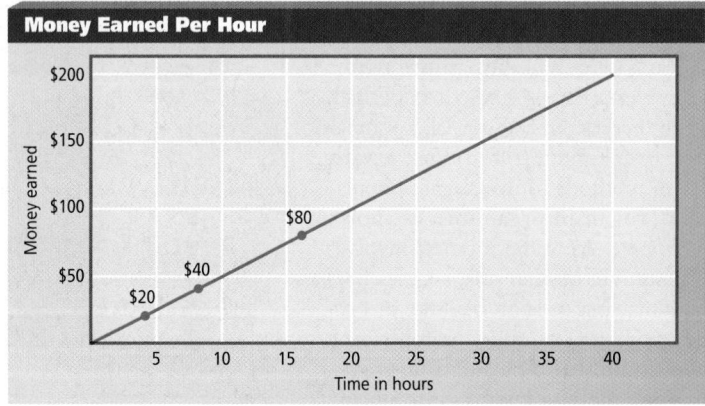

Money Earned Per Hour

---

a plant or animal species into a new environment that contains abundant resources and then observing the population growth of that species.

Use the *MiniLab* on this page to demonstrate this method of measuring population growth. Through studies such as these, scientists have identified clear patterns showing how and why populations grow.

How fast do populations grow?
What's interesting about the growth of populations is that it is unlike the growth of some other familiar things. Consider, for example, the growth of a weekly paycheck for an after-school job. Suppose you are working for a company that pays you $5 per hour. You know if you work for two hours, you will be paid $10; if you work for four hours, you will be paid $20; if you work for eight hours, you will be paid $40; and so on. When you plot this rate of increase on a graph, as shown in *Figure 4.2*, you can see that the result is a steady, linear increase; that is, growth occurs in a straight line when graphed.

Populations of organisms do not experience this linear growth. Rather, the resulting graph of a growing population first resembles a J-shaped curve. Whether the population is one

---

GLENCOE TECHNOLOGY

CD-ROM
Biology: The Dynamics of Life
BioQuest: *Antarctic Food Web,* Disc 1

96

MEETING INDIVIDUAL NEEDS

Learning Disabled
Review the proper construction of a graph with students and/or refer them to the **Skill Handbook**. Assign students to cooperative groups of mixed ability levels. Have them prepare graphs from data that you supply to illustrate linear and exponential changes. **L1**

PROJECT

Fruit Fly Demographics
Kinesthetic Fruit flies are easy to maintain in captivity. Allow students to carry out an experiment that will compare a population of fruit flies kept in balance with their environment with one not kept in balance. Students will first have to determine how this balance will be maintained. **L3**

of weeds in a field, frogs in a pond, or humans in a city, the initial increase in the number of organisms is slow because the number of reproducing organisms is small. Soon, however, the rate of population growth increases rapidly because the total number of potentially reproducing organisms increases. This pattern illustrates the exponential nature of population growth. **Exponential growth** means that as a population gets larger, it also grows faster. Exponential growth, as illustrated in *Figure 4.3*, results in a population explosion.

Limits of the environment

Can a population of organisms grow indefinitely? What prevents the world from being overrun with all kinds of living things? Through population experiments, scientists have found that, fortunately, population growth does have limits. Eventually, limiting factors, such as availability of food and space, will cause a population to stop increasing. In time, this leveling-off of population size results in an S-shaped growth curve.

The number of organisms of one species that an environment can support is its **carrying capacity.** When populations are under the carrying capacity of a particular environment, births exceed deaths until the carrying capacity is reached. If the population temporarily overshoots the carrying capacity, deaths exceed births until population levels are once again below carrying capacity. Thus, the number of organisms in a population is sometimes more than the carrying capacity and sometimes less. Learn how to determine population size by completing the *BioLab* at the end of this chapter. When the population falls below the carrying capacity, the population tends to increase; when it is above the carrying capacity, the

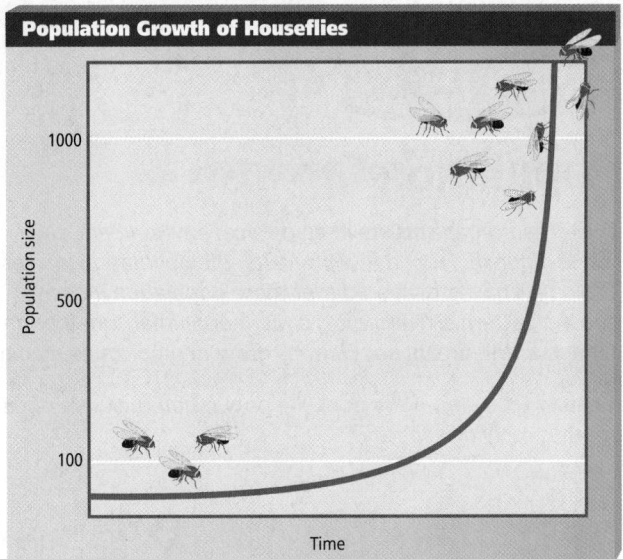

Population Growth of Houseflies

Population size: 100, 500, 1000

Time

population tends to decrease. Do all populations follow the same growth pattern? Find out in the *Inside Story* on the next page.

Patterns of population growth

In nature, many animal and plant populations change in size. Why, for example, does it seem like mosquitoes are more numerous at certain times of the year? Why don't populations reach carrying capacity and become stable? To answer these questions, population biologists study the most important factor that determines population growth—an organism's reproductive pattern.

A range of population growth patterns are possible in nature. The two extremes of this range are demonstrated by the population growth rates of both mosquitoes and elephants. Mosquitoes reproduce very rapidly and produce many offspring in a short period of time. Elephants have a slow rate of reproduction and produce relatively few young over their lifetimes. What causes species to have different life-history patterns?

Figure 4.3
Because they grow exponentially, populations such as houseflies have the potential for explosive growth.

Reinforcement
Explain that while linear growth may be shown by height increases in children, it does not reflect the growth pattern of most populations.

Chalkboard Example
Graph two lines onto the same axis using different colors of chalk. Have one line show linear growth, and the other line show exponential growth. Ask students to describe how the patterns differ. Point out that when the curves begin, differences are smaller than at the end points.

Enrichment
Logical-Mathematical Have students make the calculations needed to decide if they would rather be paid a linear salary of $5.00 per hour for a 40-hour week or an exponential salary that starts at 1 cent the first hour and doubles each hour up to 40 hours. *The exponential salary will exceed the linear salary many times over.* **L2**

The BioLab at the end of the chapter can be used at this point in the lesson.

INVESTIGATE BioLab

GLENCOE
TECHNOLOGY

CD-ROM
Biology: The Dynamics of Life
Animation: *Carrying Capacity*
Disc 1

VIDEODISC
Biology: The Dynamics of Life
Carrying Capacity (Ch. 13)
Disc 1, Side 1
17 sec.

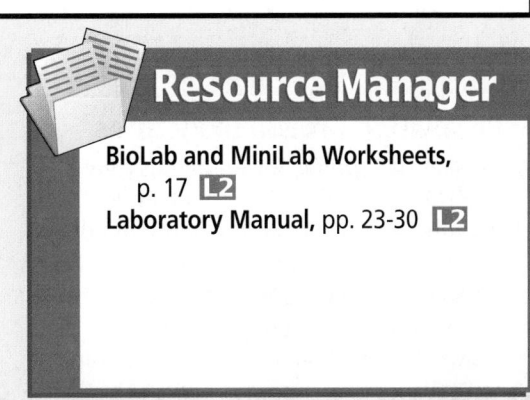

BIOLOGY JOURNAL
Evaluating Population Growth
Linguistic Have students write a scenario that depicts what life in the United States might be like if the population doubled. Have them consider available recreational space, demands made on natural resources, and food and housing needs. **L2**

Resource Manager
BioLab and MiniLab Worksheets, p. 17 **L2**
Laboratory Manual, pp. 23-30 **L2**

Purpose

Students study the general pattern of population growth in stable environments.

Teaching Strategies

■ Have students write a one-paragraph summary of how a population grows, including why the population levels off. **L1**

■ Have students become one of the original animals in a population that grows quickly. Ask them to write a fictional account of what happens from the animal's perspective.

■ The Project Wild simulation "Oh, Deer!" is an excellent method of simulating that fluctuations above and below the carrying capacity are normal.

■ Give each group of students a die. For time period 1, they roll the die and record the number that shows up as a population size. Do the same for time period 2, 3, all the way to 10. On the x-axis, plot the time periods 1 to 10. On the y-axis, plot the population size. Have them find the average population size for the 10 time periods and draw a straight line to reflect this average. Relate this line to carrying capacity.

Critical Thinking

The population begins exponential growth. This resembles the letter **J**. Eventually growth slows down and the population levels off. After it levels off, it fluctuates above and below the carrying capacity.

INSIDE STORY

Population Growth

When organisms are in an optimal environment, they flourish. From a few pioneers, the population increases. Ecologists have discovered that these population increases show a pattern. Whether it is a plant or animal, whether on land or in the ocean, populations grow in predictable manners.

Critical Thinking Why does the population fluctuate once it reaches carrying capacity?

Humpback whales have a long life-history, living up to thirty years.

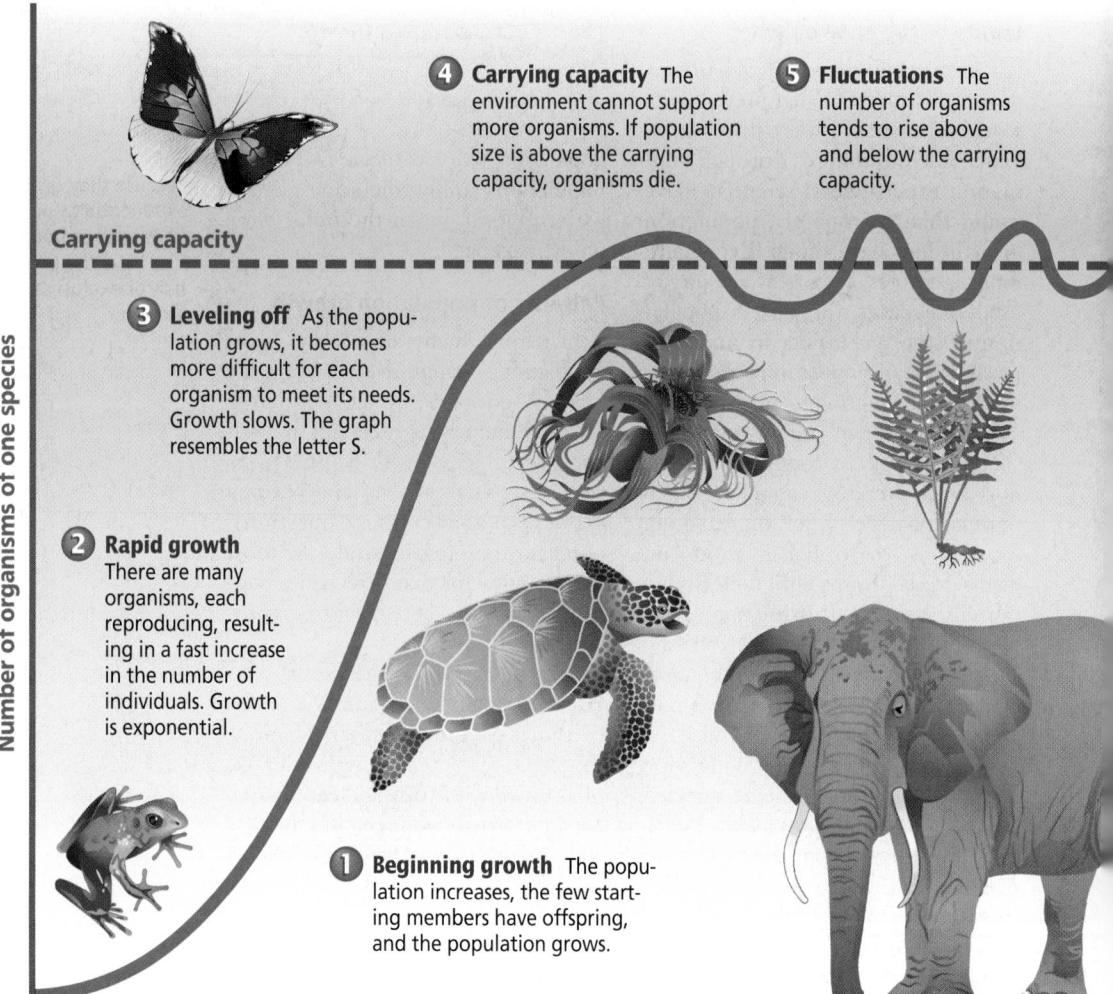

4 Carrying capacity The environment cannot support more organisms. If population size is above the carrying capacity, organisms die.

5 Fluctuations The number of organisms tends to rise above and below the carrying capacity.

Carrying capacity

3 Leveling off As the population grows, it becomes more difficult for each organism to meet its needs. Growth slows. The graph resembles the letter S.

2 Rapid growth There are many organisms, each reproducing, resulting in a fast increase in the number of individuals. Growth is exponential.

1 Beginning growth The population increases, the few starting members have offspring, and the population grows.

Number of organisms of one species

Time

98 POPULATION BIOLOGY

PROJECT

Population Growth in Pictures

 Visual-Spatial Have students create flipbooks that show population growth. Draw scenes on the right sides of index cards. On the first cards, show a few organisms, then exponential growth. Show competition for resources. Then show a constant population size. A line on a graph can advance in each scene. **L1** **ELL**

Portfolio

Interpreting Graphs

 Visual-Spatial Ask students to draw an S-shaped curve. Have them identify the following: slow growth phase, exponential growth phase, plateau, point where carrying capacity (K) is reached. Ask how this graph might change with a sudden increase or decrease in food supply. **L2** **P**

The kind of reproductive pattern a species has depends mainly on environmental conditions. For example, species such as mosquitoes are successful in environments that are unpredictable and change rapidly. Rapid life-history patterns are found in organisms from unpredictable environments. Typically, these organisms have a small body size, mature rapidly, reproduce early, and have a short life span. Populations of these organisms increase rapidly, then decline rapidly as environmental conditions suddenly change and become unsuitable. The small surviving population will begin reproducing exponentially when conditions are again favorable. The *Problem-Solving Lab* on this page will allow you to observe an organism with this type of a life-history pattern, bacteria.

Species that live in more stable environments, such as elephants, often have a different life-history pattern. Elephants, humans, bears, whales, and long-lived plants, such as cacti and bristlecone pine shown in *Figure 4.4*, are large, reproduce and mature slowly, and are long-lived. These organisms maintain population sizes near the carrying capacities of their environments.

Although populations could display a variety of life-histories, under

Problem-Solving Lab 4-1   Predicting

How rapidly can bacteria reproduce? The faster an organism reproduces, the quicker you can see population increases. Bacteria are examples of rapidly reproducing organisms. Thus, they are often used in experiments that deal with population studies or trends.

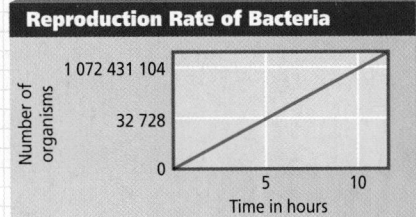

Reproduction Rate of Bacteria

Analysis
Here are some facts regarding unchecked bacterial reproduction:
1. A single bacterium can reproduce to yield two bacteria under ideal conditions every 20 minutes.
2. Ideal conditions for bacterial reproduction include proper temperature, unlimited food, space to grow, and dispersion of waste materials.

Thinking Critically
1. Suppose you start with one bacterium under ideal conditions. If no bacteria die, compute the number of bacteria present after 1 hour, 5 hours, and 10 hours.
2. What environmental factors might affect a bacterial population's reproduction?
3. The above graph is an example of one group's data:
   a. What error did they make in the y-axis of the graph?
   b. Redraw the graph correctly.
4. An elephant reproduces once every four to six years. Why are elephants not likely to be used in laboratory population studies?

Figure 4.4
This bristlecone pine is an example of a long-lived species with a slow life-history pattern.

Purpose
Students will calculate the change in number of individuals present in a bacterial population.

Process Skills
calculate, think critically, predict, apply concepts, make and use graphs, organize data

Teaching Strategies
■ Allow students to use calculators.
■ Suggest that they prepare their own data table to record the number of bacteria present during each time period.
Example:

Time (in minutes)	Number Present
0	1
20	2
40	4
60 (1 hour)	8
80	16

Thinking Critically
1. 8; 32 728; 1 072 431 104
2. Answers may include temperature, light, and pollution.
3. a. The y-axis is not properly scaled.
   b. Check students' graphs for accuracy.
4. Answers may include the long time period between generations and the size of the animals.

✔ Assessment
Portfolio Ask students to write a paragraph for their portfolios, evaluating the use of yeast, mice, and primates in population studies. Use the Performance Task Assessment List for Writing in Science in **PASC**, p. 87. **L3** **P**

BIOLOGY JOURNAL

Life as a Mosquito
Linguistic Have students imagine they are breeding mosquitoes. Tell them to describe the conditions that make their environment unpredictable and subject to rapid change. Students will have to research life stages and breeding habits of mosquitoes to complete this task. **L2**

GLENCOE TECHNOLOGY

VIDEODISC
The Infinite Voyage: *The Keepers of Eden, Extinction and the National Zoo's Tamarin Monkey Project* (Ch. 4) 13 min. 30 sec.

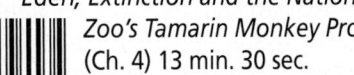

The Cheetah: Using DNA Profiles to Research Reproduction (Ch. 5) 9 min.

Resource Manager

Reteaching Skills Transparency 6 and Master
L1 **ELL**

Revealing Misconceptions

Exponential growth may be misunderstood. Some students may think that (1) the growth rate must be very fast and (2) the growth rate increases. Use this analogy to help. You put $100 in the bank and get 10% interest a year. How much will you have after the first five years? *$110, $121, $133, $146, $161*

The big idea is that although the account grows only at a slow and constant 10% a year, growth builds upon growth. This is what exponential growth is all about.

Revealing Misconceptions

Help students realize that most organisms fall in a continuum between rapid life-histories and slow life-histories. These are the extremes of the scale.

✓ **Assessment**

Performance Assessment in the Biology Classroom, p. 53, *Estimating Populations.* Have students carry out the activity to show their knowledge of how population size is determined. **L2**

NATIONAL GEOGRAPHIC

VIDEODISC
GTV: Planetary Manager
Agriculture, Side 2

Figure 4.5
Wild mustard plants taking over an abandoned field represent a species with a rapid life-history pattern (**a**). Organisms that have a slow life-history pattern, such as these Canada geese, provide much parental care for their young in order to ensure species survival (**b**).

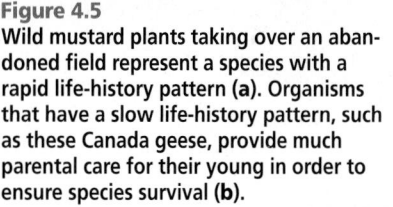

uncrowded conditions, such as the pioneer stage in succession, rapid population growth seems to be most common. *Figure 4.5* shows organisms that represent both extremes in life-history patterns. Which of these organisms would be most successful in a rapidly changing environment?

Figure 4.6
Corn smut is a fungus that produces large, deformed growths on the ears of corn. To prevent it from spreading through a cornfield, affected plants must be burned or buried before the fungus reproduces.

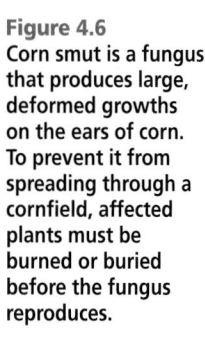

Environmental limits to population growth

Limiting factors, you may remember, include biotic or abiotic factors that determine whether or not an organism can live in a particular environment. Limiting factors also regulate the size of a population. Limited food supply, extreme temperatures, and even storms can affect population size. Ecologists have identified two kinds of limiting factors: density-dependent and density-independent factors.

Density-dependent factors include disease, competition, parasites, and food. These have an increasing effect as the population increases. Disease, for example, spreads more quickly in a population with members that live close together, as indicated in *Figure 4.6,* than in smaller populations with members that live farther apart. In very dense populations, disease may quickly wipe out an entire population. In crops such as corn or soybeans in which large numbers of the same plant are grown together, a disease can spread rapidly throughout the

---

Alternative Lab

Germinating Radishes

Purpose
Students will study the percent of germination of radish seeds.

Safety Precautions
Have students wash their hands after handling the seeds.

Materials per group
two petri dishes, radish seeds, napkins, pipette

Procedure
Model the method of germination: Cut a napkin into two circles that will fit into a petri dish. Place one circle into the dish, put some seeds on top, and place the other circle on top of the seeds. Add a pipette or two of water. Cover the dish. Then give students these instructions

1. With your group, choose one variable to

change, such as the amount or type of solution used for watering. You could also investigate the effects of heat, light, or music.

2. Complete this problem statement: What is the effect of _____ on the germination of a radish seed? Write a hypothesis to answer this question.

3. Write out your procedures before starting. Plan for a control and an experimental group.

whole crop. In less dense populations, fewer individuals may be affected.

Density-independent factors affect all populations, regardless of their density. Most density-independent factors are abiotic factors, such as temperature, storms, floods, drought, and habitat disruption, shown in *Figure 4.*7. No matter how many earthworms live in a field, they will drown in a flood. It doesn't matter if there are many or few mosquitoes; a cold winter will kill them. Another example of a density-independent factor is pollution. How does pollution affect a habitat? Find out in the *Chemistry Connection* at the end of this chapter.

Organism Interactions Limit Population Size

Population sizes are limited not only by abiotic factors, but also are controlled by various interactions among organisms that share a community.

Predation affects population size

A barn owl kills and eats a mouse. A swarm of locusts eats and destroys acres of lettuce on a farm. When the brown tree snake was introduced in Guam, an island in the South Pacific, it wiped out most of the birds on the island. These examples demonstrate how predation can affect population sizes in both minor and major ways. When a predator consumes prey, it can affect the population size of the prey population. For this reason, predation may be a limiting factor of population size.

Populations of predators and prey experience changes in their numbers over a period of years. Predator-prey relationships often show a cycle of population increases and decreases over time. One classic example of this has been demonstrated by graphing 90 years of data concerning the populations of the Canadian lynx and the snowshoe hare. A member of the cat family, the lynx stalks, attacks,

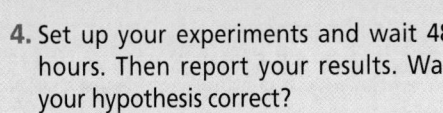

Figure 4.7
Populations are affected by both density-dependent and density-independent factors.

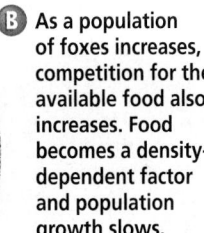

A Hurricane Mitch, which hit south Florida in 1998, did extensive damage to both heavily populated areas and less populated ones. Catastrophic weather patterns are density-independent factors.

B As a population of foxes increases, competition for the available food also increases. Food becomes a density-dependent factor and population growth slows.

4.1 POPULATION DYNAMICS **101**

4. Set up your experiments and wait 48 hours. Then report your results. Was your hypothesis correct?

Analysis

1. Why do plants produce so many seeds? *Not all seeds germinate.*
2. Was your variable a biotic or abiotic factor? *Most will be abiotic factors.*
3. What was your conclusion for your experiment? *Answers will vary.*

Explain how the decrease in available energy at the top trophic level of a food chain serves as a limiting factor on predator population size.

Visual Learning

Ask students to study the graph in Figure 4.8 and identify the point at which the numbers of snowshoe hare and lynx began to increase. Have them also indicate where decreasing numbers occur. Use the graph to point out the cyclic nature of these events. Elicit from students why these events are cyclic.

✓ Assessment

Skill Have students examine Figure 4.8 and answer these questions: In what approximate year were there the most hares? *1865* In what year were there the most lynxes? *1935* After the hare population spikes, how many years pass until it crashes? *approximately 5 years*

3 Assess

Check for Understanding

Ask students to explain the difference between the words in each of the following pairs. **L1** **ELL**
 a. linear growth—exponential growth
 b. carrying capacity—population size
 c. density-dependent factors—density-independent factors
 d. predation—competition

NATIONAL GEOGRAPHIC

VIDEODISC
GTV: Planetary Manager
Animal, Side 2

---

and eats the snowshoe hare as a primary source of food.

The data in *Figure 4.8* show the lynx and hare populations rise and fall almost together. When the hare population increases, there is more food for the lynx population, and the lynx population increases. When the lynx population rises, predation on the hares increases, and the hare population decreases. With fewer hares available for food, the lynx population also declines. With fewer predators, the hare population increases, and the cycle continues. This example shows how predator populations affect the size of the prey populations. At the same time, prey populations affect the size of the predator populations. As the snowshoe hare's food supply of grasses and herbs dwindles during the fall and winter months, the hare population decreases. Because there are now fewer hares to hunt, the lynx population also decreases. With the return of spring, the hare's food supply and

its population recover. This leads to more hares, allowing the lynx population to increase as well.

Predator-prey relationships are important for the health of natural populations. Usually, in prey populations, the young, old, or injured members are caught. Predation helps improve the odds that there will be sufficient resources for the healthiest individuals in a population.

The effects of competition

Organisms within a population constantly compete for resources. When population numbers are low, resources are plentiful. However, as population size increases, competition for resources such as food, water, and territory can become fierce. Competition is a density-dependent factor. When only a few individuals compete for resources, no problem arises. When a population increases to the point at which demand for

Figure 4.8
The data in this graph reflect the number of hare and lynx pelts sold to the Hudson Bay Company in northern Canada. Notice that as the number of hares increased, so did the number of lynx.

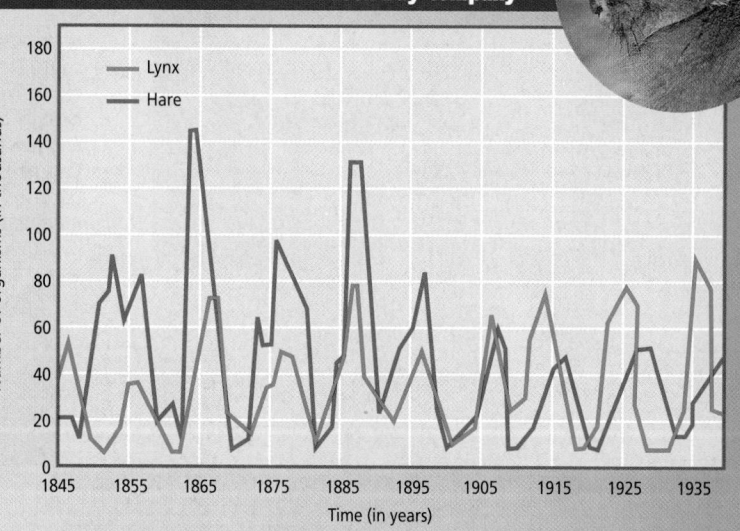

Lynx and Hare Pelts Sold to the Hudson Bay Company

MEETING INDIVIDUAL NEEDS

Learning Disabled

Visual-Spatial Have students tape the outlines of two squares, measuring 10 cm per side. Ask them to use pennies to illustrate a dense population in one square and a less dense population in the other square. Ask them to predict which population is more vulnerable to disease. *the dense population* **L1**

GLENCOE TECHNOLOGY

VIDEOTAPE
The Secret of Life
Gone Before You Know It: The Biodiversity Crisis

Figure 4.9
Stress caused by overcrowding in a rat population can limit population size. When overcrowded, animals fight and kill each other, they reproduce less, and they stop caring for offspring.

resources exceeds the supply, the population size decreases.

The effects of crowding and stress

When populations of organisms become crowded, individuals may exhibit stress. The factors that create stress are not well understood, but the effects have been documented experimentally in populations of rats and mice, as shown in *Figure 4.9*. As populations increase in size, individual animals begin to exhibit a variety of symptoms, including aggression, decrease in parental care, decreased fertility, and decreased resistance to disease. All of these symptoms can lead to a decrease in population size.

Section Assessment

Understanding Main Ideas
1. How are graphs of exponential growth and linear growth different?
2. Explain how short and long life-history patterns differ.
3. Describe how density-dependent and density-independent factors regulate population growth.
4. How can a density-independent factor, such as a flood, influence carrying capacity?

Thinking Critically
5. An organic farmer does not use pesticides on her farm. Instead of growing one crop on her farm, as many farmers do, she grows ten different crops. Explain how this may decrease insect damage to her plants.

SKILL REVIEW
6. **Making and Using Graphs** Graph the following population growth for the unknown organism shown in *Table 4.1* and state whether the organism has a population growth pattern closer to a rapid or slow life-history pattern. For more help, refer to *Organizing Information* in the **Skill Handbook.**

Table 4.1 Population of unknown organisms

Year	Spring	Summer	Autumn	Winter
1995	564	14 598	25 762	127
1996	750	16 422	42 511	102
1997	365	14 106	36 562	136

Section Assessment

1. Linear growth graphs form a straight line, and exponential graphs form a curved line described as J-shaped.
2. Short life-history organisms show a rapid increase and rapid decline caused by their unpredictable environments. Long life-history organisms show a slow population change within stable environments and usually maintain sizes near the carrying capacity of the environment.
3. Density-dependent factors have an increasing effect as population size increases. Density-independent factors affect a population no matter what its size.
4. The flood could damage a habitat, thus lowering its carrying capacity.
5. Each kind of insect usually damages specific crops. When a farmer grows ten crops, an infestation of one kind of insect is unlikely to affect all of her crops.
6. The graphs should show a rapid life-history pattern. The organism reproduces rapidly and declines rapidly.

Prepare

Key Concepts

Population growth is the increase in population size over time. Students learn that population growth, while exponential at times, is controlled by limiting factors that determine the carrying capacity of the environment. Such limits to population growth may result from predator-prey interactions or overcrowding.

Planning

■ Purchase beans, markers, and bags for the BioLab.

1 Focus

Bellringer 🖑

Before presenting the lesson, display **Section Focus Transparency 9** on the overhead projector and have the students answer the accompanying questions. **L1** **ELL**

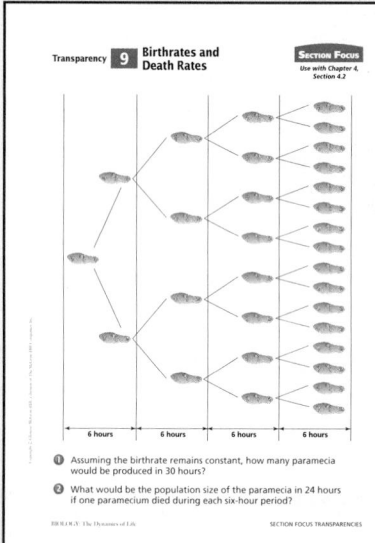

SECTION PREVIEW

Objectives

Relate population characteristics to population growth rates.

Compare the age structure of rapidly growing, slow-growing, and no-growth countries.

Hypothesize about problems that can be caused by immigration and emigration.

Vocabulary

demography
age structure
immigration
emigration

Section

4.2 Human Population Growth

Does Earth have a carrying capacity for the human population? How many people can live on Earth? No one knows how many people Earth can support, and it is presently impossible to tell when the human population will stop growing. However, demographers suggest that food production will not always keep pace with the population increase.

Earth from space

Figure 4.10
Ten thousand years ago, approximately 10 million people inhabited Earth. Today, there are more than 6 billion, and scientists estimate that by the year 2050, there will be more than 10 billion people on Earth.

Demographic Trends

A good way to predict the future of the human population is to look at past population trends. For example, are there observable patterns in the growth of populations? That is, are there any similarities among the population growths of different countries—similarities that might help scientists predict, and therefore control, future population catastrophes? As you have seen, some populations tend to increase until the environment cannot support any additional growth. **Demography** (dem AH graf ee) is the study of human population growth characteristics. Demographers study such population characteristics as growth rate, age structure, and geographic distribution.

What is the history of population growth for humans? Although local human populations often show fluctuations, the worldwide human population has increased exponentially over the past several hundred years, as shown in *Figure 4.10.* Unlike other organisms, humans are able to reduce environmental effects by eliminating competing organisms, increasing food production, and controlling disease organisms.

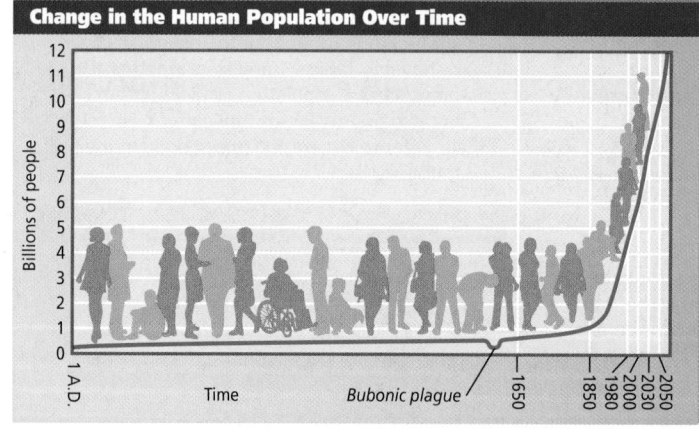

Change in the Human Population Over Time

Billions of people (y-axis: 0–12)

Time — Bubonic plague — 1650 1850 1980 2000 2030 2050 — 1 A.D.

104 POPULATION BIOLOGY

TECHPREP

Graphing Population Growth

Logical-Mathematical Have students make a bar graph that shows the population growth of each country in Table 4.2, from highest to lowest percentage. **L2**

✓ Portfolio

Interpreting Graphs

Linguistic Ask students to work with partners to write a true and a false statement that are both based on Table 4.2. Have partners exchange statements and identify which is true and which is false. **L2**

Effects of birthrates and death rates

How can you tell if a population is growing? A population's growth rate is the difference between the birthrate and the death rate. One way of calculating a population's growth rate is by calculating its doubling rate. Learn how to calculate doubling rates in the *MiniLab* on this page. In many industrialized countries, such as the United States, declining death rates have a greater effect on total population growth than increasing birthrates. For example, in the United States, life expectancy increases almost every year. This means that you are more likely to live slightly longer than students who are presently in college.

Although people in the United States are living longer, the fertility rate is decreasing. This is because more people are waiting until their thirties to have children. Today's families also have fewer children than they did in previous decades. Fertility rate is the number of offspring a female produces during her reproductive years. When fertility rates are high, populations grow more rapidly unless the death rate is also high. *Table 4.2* shows the birthrate, death rate, and fertility rate of some rapidly

MiniLab 4-2   Using Numbers

Doubling Time The time needed for any population to double its size is known as its "doubling time." For example, if a population grows slowly, its doubling time will be long. If it is growing rapidly, its doubling time will be short.

Procedure

1. The following formula is used to calculate a population's doubling time:

$$\text{Doubling time (in years)} = \frac{70}{\text{annual percent growth rate}}$$

2. Copy the data table below.
3. Complete the table by calculating the doubling time of human populations for the listed geographic regions.

Data Table

Geographic region	Annual percent growth rate	Doubling time
Africa	2.8	
Latin America	2.2	
Asia	1.9	
North America	0.7	
Europe	0.3	

Analysis

1. Which region has the fastest doubling time? Slowest doubling time?
2. How might this type of information be useful to governments of these regions?
3. What are some of the ecological implications for an area with a fast doubling time?

Table 4.2 Birthrates and death rates around the world

	Birthrate (per 1000)	Death Rate (per 1000)	Fertility (per woman)	Population increase (percent)
Rapid Growth Countries				
Jordan	38.8	5.5	3.3	3.3
Uganda	50.8	21.8	7.1	2.9
Zimbabwe	34.3	9.4	5.2	2.5
Slow Growth Countries				
Germany	9.4	10.8	1.2	−1.5
Italy	9.4	9.7	1.2	−0.5
Sweden	10.8	10.6	1.5	0.1
North American Countries				
Mexico	27.0	5.2	3.1	2.2
United States	14.8	8.8	2.0	0.6

105

Cultural Diversity

Fertile Grounds

Have the class discuss reasons why developing countries have high fertility rates. Cultural reasons could include (a) people prefer large families; (b) large families may have more prestige; and (c) religions may ban the use of birth control. There could be socioeconomic reasons as well. Statistics suggest that there tend to be more babies per woman in economically depressed countries. Reasons might include (a) higher infant mortality may encourage people to have more children; (b) families may have more children to help earn money; and (c) families don't have access to medical services or birth control.

If the information is available, help students construct an age structure graph for an aquarium of fish or cage of mice that you have in your classroom. Discuss what the graph may reveal about expected population growth. **L2**

Problem-Solving Lab 4-2

Purpose

Students will analyze trends in human population changes.

Process Skills

think critically, apply concepts, analyze information, use numbers, draw a conclusion, make and use graphs

Teaching Strategies

■ Review the meanings of density-dependent factors, density-independent factors, and carrying capacity.
■ Explain doubling time and annual percent growth if students did not complete or discuss MiniLab 4-2.
■ Allow students to use calculators.

Thinking Critically

1. Density-dependent: disease, war, immigration, emigration, birth rate, death rate, availability of food or water. Density-independent: pollutants in water, soil, or air.
2. No. Humans are able to provide food and water, remove wastes, and prevent disease.
3. over 7 billion
4. 200 years; 70 years; 55 years
5. It is taking less and less time for the human population to double in size.
6. The values show an annual percent growth rate ranging from 0.35% (1650–1850) to 1.27% (1920–1975). The trend is one of higher growth

---

Problem-Solving Lab 4-2 — Applying Concepts

What trends are seen in the human population? Human population trends present some interesting ideas and concepts. Because of our intelligence, we are better able to control our population size, regulate our food supply, and remove waste products from our environment. Thus, human population trends may differ from those of other organisms.

A human population

Analysis

Figure 4.10 shows human population changes over time. Study this graph and answer the questions below.

Thinking Critically

1. What density-dependent factors can influence human population growth? What density-independent factors can influence human population growth?
2. Has the human population reached its carrying capacity? Explain why.
3. Based on the graph in Figure 4.10, what will be Earth's population in the year 2010?
4. Determine the human population's doubling time as it increases from half a billion to one billion, from one to two billion, and from two to four billion.
5. Explain the significance of the trend shown in question 4.
6. Using the values for doubling time provided in question 4, calculate the human population's annual percent growth rate using the method described in the *MiniLab* on the previous page. Use the formula below:

$$\text{Doubling time (in years)} = \frac{70}{\text{annual percent growth rate}}$$

7. Explain the significance of the trend shown by your answer to question 6.

---

growing and slower growing countries. Some countries, such as Uganda, have high death rates among children because of disease and malnutrition. However, these countries have high birthrates, and their populations are growing rapidly. Some other countries, such as Sweden and Italy, have low death rates, but their birthrates are also low, so these coun-

rate for each period of time. (Note: The annual percent growth rate is actually for a span of many years rather than only one year.)

7. The human population continues to show an increase in its annual growth rate and has not yet reached its carrying capacity.

---

tries' populations are growing slowly, if at all.

As you can see, different combinations of birthrates and death rates have different effects on populations. The birthrate, death rate, and fertility rate of a country provide clues to that country's rate of population growth. Learn more about human population trends in the *Problem-Solving Lab* on this page.

Does age affect population growth?

Imagine a country filled mostly with teenagers. Does it make a difference to population growth if the largest proportion of the population is in one age group? In order to make predictions about populations of the future, demographers must know the age structure of a population. Age structure refers to the proportions of a population that are at different age levels. **Age structure** can be visualized by the use of graphs, as shown in *Figure 4.11,* and can help predict if a population is growing rapidly, growing slowly, or not growing at all. Rapid growth countries have an age structure with a wide base because a large percentage of the population is teenagers and children. These young people are likely to mature into adults and have their own children. If the percentage of people in each category is fairly equal, this indicates the population is stable, neither increasing nor decreasing.

Mobility has an effect on population size

Humans can move in and out of different communities. The effects of human migrations can make it difficult for a demographer to make predictions, but patterns do exist. Movement of individuals into a population is **immigration.**

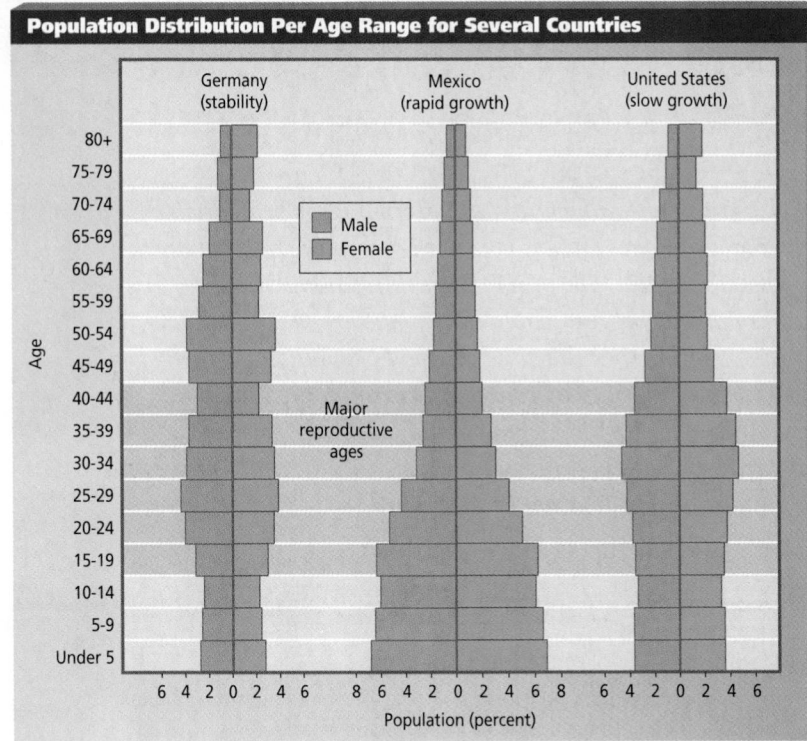

Population Distribution Per Age Range for Several Countries

Germany (stability)

Mexico (rapid growth)

United States (slow growth)

■ Male
■ Female

Age: 80+, 75-79, 70-74, 65-69, 60-64, 55-59, 50-54, 45-49, 40-44, 35-39, 30-34, 25-29, 20-24, 15-19, 10-14, 5-9, Under 5

Major reproductive ages

Population (percent)

Figure 4.11
Notice that in a rapidly developing country such as Mexico, the large number of individuals in their pre-reproductive years will add significantly to the population when they reach reproductive age. Populations that are not growing, such as Germany's, have an almost even distribution of ages among the population.

Movement from a population is **emigration.** Obviously, movement of people between countries has no effect on total world population, but it does affect national population growth rates. Local populations can also feel the effects of a moving population. Many suburbs of large cities are expanding rapidly. This places stress on schools, roads, and police and fire services. What other problems result from suburban growth?

 **WORD** *Origin*
emigration
From the Latin words *e*, meaning "out," and *migrate*, meaning "to migrate." People emigrate when they move out of a country.

Check for Understanding

Have students explain the relationship between the words in each of the following pairs. **L1**
ELL
a. population growth—birth and death rate
b. age of population—population growth

Reteach

Ask students to explain how a high fertility rate with low death rate will influence a population's size. *Population will increase.* **L2**

Extension

Have students research the pros and cons of allowing large numbers of immigrants to enter the United States. **L3**

✔ Assessment

Performance Have students write a paragraph that expresses their opinion as to whether the future size of human population should be controlled.

4 Close

Activity

Linguistic Have students list changes that can increase and decrease a population's size. **L1**

Resource Manager

Section Focus Transparency 9 and Master
L1 ELL
BioLab and MiniLab Worksheets, p. 18 **L2**
Content Mastery, p. 17, 19-20 **L1**
Reinforcement and Study Guide, pp. 17-18 **L2**

Section Assessment

Understanding Main Ideas
1. What characteristics of populations do demographers study?
2. How does life expectancy affect death rate?
3. What clues can an age structure graph provide about a country's population growth?
4. Discuss some possible problems for local populations caused by immigration and emigration of people.

Thinking Critically
5. Using the age structure graph for the United States in *Figure 4.11*, explain which gender has a higher life expectancy and then suggest a hypothesis for why this difference exists.

SKILL REVIEW

6. **Making and Using Graphs** Construct a bar graph showing the age structure of Kenya using the following data: pre-reproductive years (0-14)—42 percent; reproductive years (15-44)—39 percent; post-reproductive years (45-85+)—19 percent. For more help, refer to *Organizing Information* in the **Skill Handbook.**

Section Assessment

1. Demographers study growth rate, age structure, and geographic distribution.
2. If life expectancies rise, death rates will decrease.
3. It can determine if a population is growing rapidly, slowly, or not at all.
4. Answers will vary. Immigration can increase the need for schools, medical facilities, and public transportation. Emigration may result in unused facilities and vacant homes.
5. Female; hypotheses will vary, but may include a healthier life style or different hormone concentrations.
6. Evaluate students' graphs for logic and accuracy.

How can you determine the size of an animal population?

Scientists can determine the number of animals in a large population by using a sampling technique. Here is how it works. They trap and mark a few animals in a specified area. The animals are released and the traps are reset. Among the animals caught this second time, some are marked and some are unmarked. Scientists then calculate the total population for their specified area based on the ratio of marked animals to unmarked animals.

In this activity, a bag represents the area of land where the population study is being conducted. Beans will represent animals. All the beans in the bag represent the total animal population being studied.

Time Allotment

One class period

Process Skills

apply concepts, collect data, experiment, formulate models, interpret data, predict, use numbers, write about biology

Safety Precautions

Have students wash their hands after handling the bean seeds.

PREPARATION

■ Prepare bags in advance. Place 150-175 beans in each bag.
■ Shortcut to counting beans—count out 165 beans. Then determine the volume of this number of beans by placing them into a graduated cylinder. Read the volume and use this as a guide in preparing the other bags. All bags DO NOT have to have the same number of beans.
■ Remove all marked beans from the bag at the end of each class and replace with unmarked beans.

Resource Manager

BioLab and MiniLab Worksheets, p. 19 **L2**

PREPARATION

Problem

How can you model a measuring technique to determine the size of an animal population?

Objectives

In this BioLab, you will:
■ **Model** the procedure used to measure an animal population.
■ **Collect** data on a modeled animal population.
■ **Calculate** the size of a modeled animal population.

Materials

paper bag containing beans
magic marker
calculator (optional)

Safety Precautions

Always wear goggles in the lab. Wash hands after working with plant material.

Skill Handbook

Use the **Skill Handbook** if you need additional help with this lab.

PROCEDURE

1. Copy the data table.
2. Reach into your bag and remove 20 beans.
3. Use a dark magic marker to color these beans. These will represent your caught and marked animals.
4. When the ink has dried, return the beans to the bag.

PROCEDURE

Teaching Strategies

■ Do not tell students the total number of beans in each bag.
■ Any type of marker is suitable. Make sure the color contrasts with the bean color. Make sure all sides of the bean are colored.
■ The number of beans removed in step 5 does not have to be exactly 30 but should be close.

■ Review the technique for calculating an average.

Data and Observations

Sample data may appear as shown on page 109.

5. Shake the bag. Without looking into the bag, reach in and remove 30 beans.
6. Record the number of marked beans (recaught and marked) and the number of unmarked beans (caught and unmarked) in your data table as trial 1.
7. Return all the beans to the bag.
8. Repeat steps 5 to 7 four more times for trials 2 to 5.
9. Calculate averages for each of the columns.
10. Using average values, calculate the original size of the bean population in the bag by using the following formula:
    M = number initially marked
    CwM = average number caught during the trials with marks
    Cw/oM = average number caught during the trials without marks

$$\text{Calculated Population Size} = \frac{M \times (CwM + Cw/oM)}{CwM}$$

Data Table

Trial	Total caught	Number caught with marks	Number caught without marks
1	30		
2	30		
3	30		
4	30		
5	30		
Averages	**30**		

Calculated population size = _____

Actual population size = _____

11. Record the calculated population size in the data table.
12. To verify the actual population size, count the total number of beans in the bag and record this value in the data table.

ANALYZE AND CONCLUDE

1. **Thinking Critically** This experiment is a simulation. Explain why this type of activity is best done as a simulation.
2. **Applying Concepts** Give an example of how this technique could actually be used by a scientist.
3. **Analyzing Data** Compare the calculated to the actual population size. Explain why they may not agree exactly. What changes to the procedure would improve the accuracy of the activity?
4. **Making Inferences** Explain why this technique is used more often with animals than with plants when calculating population size.
5. **Making Predictions** Assume you were doing this experiment with living animals. What would you be doing in step 2? Step 3? Step 5?

Going Further

Writing about Biology Assume that you are a field biologist on Mackinaw Island, Michigan. Explain in detail how you would go about determining the deer population on the island. Include a data table that could be used in your procedure.

BIOLOGY Online To find out more about populations of organisms, visit the Glencoe Science Web site.
science.glencoe.com

ANALYZE AND CONCLUDE

1. Students may note that it's difficult to obtain enough traps, find a suitable site for study, and work with trapped animals.
2. Possible answers: to determine the size of a population in a specific area, such as a national park, or to measure a change in a population size over time
3. The calculated value is based on counting a representation of the population, not every individual in the population. To increase the accuracy of the activity, one could increase the number of trials, the number of animals recaught, or the number of animals caught and marked.
4. Plants do not move about and so are easier to count.
5. Step 2, capture animals in cages; step 3, tag or mark the animals in some manner; step 5, reset traps and capture more animals

✓ Assessment

Skill Provide students with raw data and have them calculate the size of a population. Use the Performance Task Assessment List for Using Math in Science in **PASC**, p. 29. **L3**

Going Further

Have students design a long-term experiment that would determine the effectiveness of a "roach motel" on reducing the size of a cockroach population.

Data Table

Trial	Total caught	Number caught with marks	Number caught without marks
1	30	3	27
2	30	2	28
3	30	5	25
4	30	4	26
5	30	6	24
Totals	**150**	**20**	**130**
Average	**30**	**4**	**26**

Calculated population size = _150_
Actual population size = _160_
(This number will vary per bag.)

Chemistry
Connection

Polystyrene: Friend or Foe?

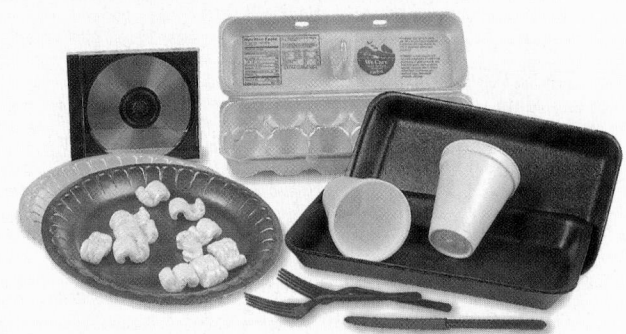

Polymerization is a process in which single molecules known as monomers are joined to form long chains called polymers. Polystyrene is a petrochemical polymer. Manufacturers of this plastic argue that it has revolutionized many industries, especially those related to food service. Environmentalists contend that this plastic, like many others, is a dangerous substance that harms Earth and its inhabitants.

The basic chemical unit, or monomer, of polystyrene is styrene. Pure polystyrene is brittle, but by combining styrene with other monomers, an impact-resistant plastic is formed. Polystyrene is commonly used as building insulation and flotation devices, as packaging materials, and to make a wide assortment of disposable cups, plates, bowls, containers, and cutlery.

Foamed plastics Foamed plastics, often used in food packaging, are made by blowing tiny holes into a polymer. Polystyrene was the first foamed plastic developed. Initially, chlorofluorocarbons, or CFCs, were used as blowing agents to produce this closed-cell foam in which gas cells are completely isolated from each other by thin walls of plastic. Today, most foamed polystyrene is made using pentane or carbon dioxide as the blowing agent. Unlike CFCs, these gases do not destroy Earth's ozone layer.

Polystyrene products Foam beverage cups and plates, plastic forks, some packaging "peanuts," and the jewel cases that house compact discs are a few of the many polystyrene products on the market today. In the food service industry, polystyrene products are touted for their safety because they are used only once and thrown away. Because of its insulating properties, foamed polystyrene keeps hot foods hot and cold foods cold, thus reducing the risk of contamination

by bacteria. Foam packaging peanuts and polystyrene blocks protect electronics and other fragile items during shipping.

The perils of polystyrene Despite its convenience and popularity in various industries, polystyrene poses its share of environmental concerns. The main monomer in this plastic—styrene—is classified as a neurotoxin, as it impairs the central and peripheral nervous systems. Long-term exposure to even small amounts of styrene can also cause abnormal hematological, lymphatic, and cytogenetic effects. People do not eat plastic containers, but studies have shown that styrene is common in human fat tissue because it is often leached from foods and beverages stored, eaten, and drunk from such containers.

CONNECTION TO BIOLOGY

Polystyrene makes up less than 1 percent of our waste by weight, yet it constitutes 25-30 percent by volume. What does this say about the average consumer in this country? Also, polystyrene does not degrade in landfills. Suggest two reasons why this is both a good thing and a bad thing.

BIOLOGY Online To find out more about plastics, visit the Glencoe Science Web site. **science.glencoe.com**

Note Internet addresses that you find useful in the space below for quick reference.

_____
_____
_____
_____
_____

SUMMARY

Section 4.1

Population Dynamics

Main Ideas

- Some populations grow exponentially until they reach the carrying capacity of the environment. Populations may exhibit slow growth that tends to approach the carrying capacity with minor fluctuations, rapid growth that tends to expand exponentially and then experiences massive diebacks, or fall somewhere in between.

- Density-dependent factors such as disease and food supply, and density-independent factors such as weather, have effects on population size. Interactions among organisms such as predation, competition, stress, and crowding also limit population size.

Vocabulary

carrying capacity (p. 97)
density-dependent factor (p. 100)
density-independent factor (p. 101)
exponential growth (p. 97)

Section 4.2

Human Population Growth

Main Ideas

- Demography is the study of population characteristics, such as growth rate, age structure, and movement of individuals. Birthrate, death rate, and fertility differ considerably among different countries, resulting in uneven population growth patterns across the world.

Vocabulary

age structure (p. 106)
demography (p. 104)
emigration (p. 107)
immigration (p. 106)

Main Ideas

Summary statements can be used by students to review the major concepts of the chapter.

Using the Vocabulary

To reinforce chapter vocabulary, use the Content Mastery Booklet and the activities in the Interactive Tutor for Biology: The Dynamics of Life on the Glencoe Science Web site.
science.glencoe.com

THE PRINCETON REVIEW

All Chapter Assessment questions and answers have been validated for accuracy and suitability by The Princeton Review.

UNDERSTANDING MAIN IDEAS

1. c
2. c
3. d
4. d
5. c
6. d

UNDERSTANDING MAIN IDEAS

1. Which of the following factors is density-dependent?
   a. drought   **c.** food
   b. flood   **d.** wind speed

2. When populations increase, resource depletion may result in _____.
   a. exponential growth
   b. straight-line growth
   c. competition
   d. increase in predators

3. Storms, cold temperatures, and drought are all _____.
   a. density dependent
   b. biotic factors
   c. exponential
   d. density independent

4. Between A.D. 1 and A.D. 1650, the world's population had a major dip because of _____.
   a. fertility
   b. decreased death rate
   c. density-independent factors
   d. bubonic plague

5. Which of the following environments would be more likely to have organisms that exhibit fast growth?
   a. hot deserts
   b. large, deep lakes
   c. prairies that often flood
   d. tropical rain forests

6. A female's fertility rate is the number of offspring she produces _____.
   a. in a year   **c.** by age 50
   b. in a decade   **d.** over her lifetime

CHAPTER 4 ASSESSMENT   **111**

GLENCOE TECHNOLOGY

VIDEOTAPE
MindJogger Videoquizzes
Chapter 4: *Population Biology*
Have students work in groups as they play the videoquiz game to review key chapter concepts.

Resource Manager

Chapter Assessment, pp. 19-24
MindJogger Videoquizzes
ExamView® Pro Software
BDOL Interactive CD-ROM, Chapter 4 quiz

7. c
8. a
9. b
10. c
11. exponential growth
12. carrying capacity
13. birth, death, emigration
14. limiting factor
15. Jordan, Uganda
16. carrying capacity
17. demographer
18. 4, 8
19. life expectancy, increases
20. exponential growth

7. According to the graph, the growth rate of a house fly population _____.
   a. increases, then drops suddenly
   b. increases, at a steady rate
   c. increases rapidly
   d. levels off after a certain amount of time

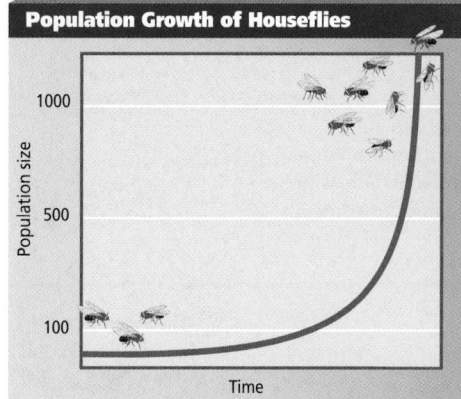

Population Growth of Houseflies

Population size — 1000, 500, 100

Time

8. A person breeds rabbits in a cage. After a few generations, she observes that the rabbits are more aggressive towards each other, the young are less healthy, and more young rabbits die. This population is:
   a. under stress
   b. under the carrying capacity
   c. density independent
   d. both choices a and c

9. A life-history pattern that tends to approach carrying capacity with minor fluctuations is _____.
   a. population dependent
   b. slow growth
   c. fast growth
   d. exhibited by most insects

THE PRINCETON REVIEW   **TEST-TAKING TIP**

Pace Yourself
Many test questions look more complicated than they really are. If you find yourself having to do a great deal of work to answer a question, take a second look and consider whether there might be a simpler way to find the answer.

10. What can be said about the growth of a country with an age structure graph that approximates a rectangle?
    a. It is decreasing.
    b. It is increasing slowly.
    c. It is stable.
    d. It is increasing rapidly.

11. A J-shaped growth curve indicates a population is experiencing _____ _____.

12. The highest level at which a population can be sustained is its _____ _____.

13. Population growth rates are affected by _____ and _____ rates. It is also influenced by immigration and _____.

14. Predators that help control the size of a population represent a _____.

15. From *Table 4.2,* the fastest growing country is _____. The women of _____ give birth to the most children.

16. The solid-line graph below shows the changes in deer population over time. The dotted line represents the _____ _____ for the deer population.

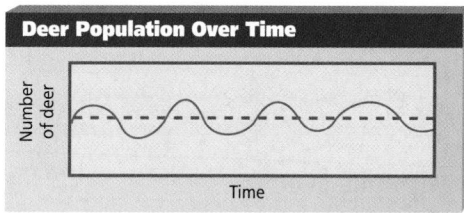

Deer Population Over Time

Number of deer

Time

17. A person who studies human population growth characteristics is a _____.

18. A bacterial cell divides every 30 minutes and produces two cells. Starting with one cell, after 30 minutes there would be two cells, after 60 minutes _____ cells, and after 90 minutes, _____ cells.

19. A demographic estimate of how long a person of a particular age will live is called his/her _____ _____. In the United States, this estimate _____ every year.

20. Populations that grow without restriction are experiencing _____ _____.

APPLYING MAIN IDEAS

21. Which environmental factors would most affect the populations of developing countries?

22. As human populations grow, what might happen to the populations of other species? Discuss the causes for your hypotheses.

23. Assume that a female black rat gives birth every month and produces eight young. In each litter, four are female and four are male. Starting in month zero with one newborn litter of four males and four females for a total population of eight, calculate the total population size at months 1, 2, and 3.

THINKING CRITICALLY

24. Observing and Inferring Why are short life-history species, such as mosquitoes and some weeds, successful even though they often experience massive population declines?

25. Comparing and Contrasting Compare and contrast the characteristics of species having long life-history patterns with those that have short life-history patterns.

26. Concept Mapping Complete the concept map by using the following vocabulary terms: density-dependent factors, density-independent factors, exponential growth, immigration.

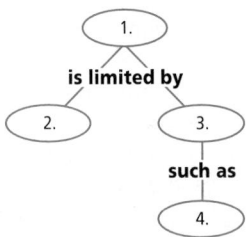

CD-ROM

For additional review, use the assessment options for this chapter found on the *Biology: The Dynamics of Life Interactive CD-ROM* and on the Glencoe Science Web site.
science.glencoe.com

ASSESSING KNOWLEDGE & SKILLS

The following bar graph indicates the number of dandelions counted by five groups of students (groups A-E) on school grounds. Assume that the school grounds have 21 950 square meters planted in grass.

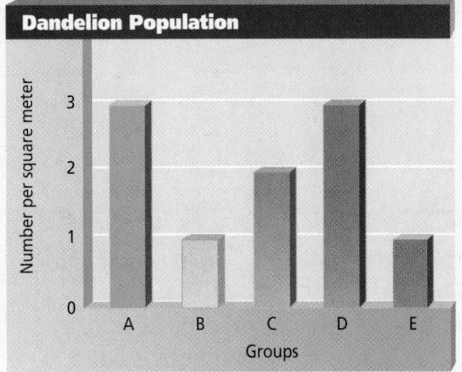

Interpreting Data Use the graph to answer the following questions.

1. How many dandelions per square meter did group A count?
  a. one   **c.** three
  b. two   **d.** four

2. On average, how many dandelions did the students count per square meter?
  a. one   **c.** three
  b. two   **d.** four

3. Using the average number of dandelions per square meter, what is the estimated number of dandelions on the school grounds?
  a. 21 950   **c.** 50 485
  b. 43 900   **d.** 65 850

4. Interpreting Data The study is repeated two months later and now the average is five dandelions per square meter. Now, what is the estimated size of the population?

APPLYING MAIN IDEAS

21. density-dependent factors such as limited food or water and increased chances for disease to spread

22. The populations of other species may decrease. Humans may disrupt the habitats of other species, so that they do not have enough space for mating, rearing offspring, or finding shelter and food.

23. Month one = 8 parents + (4 females times 8 offspring) = 40. Month two = 40 parents + (20 females times 8 offspring) = 200. Month three = 200 parents + (100 females times 8 offspring) = 1000.

THINKING CRITICALLY

24. Short life-history species provide enough potential offspring in the form of seeds, eggs, or larvae to produce the next generation, even when there are massive die-offs.

25. Species with long life-histories usually have few young that take a long time to mature, while those species with short life-histories have many offspring that mature quickly.

26. 1. Exponential growth; 2. Density-independent factors; 3. Density-dependent factors; 4. Immigration

ASSESSING KNOWLEDGE & SKILLS

1. c
2. b
3. b
4. 109 750

Chapter 5 Organizer

Refer to pages 4T-5T of the Teacher Guide for an explanation of the National Science Education Standards correlations.

Section	Objectives	Activities/Features
Section 5.1 **Vanishing Species** National Science Education Standards UCP.1-3; A.1, A.2; C.4, C.5, C.6; E.1, E.2; F.1-6; G.1-3 (2 sessions, 1 block)	1. **Explain** biodiversity and its importance. 2. **Relate** various threats to the loss of biodiversity.	**MiniLab 5-1:** Measuring Species Diversity, p. 116 **Problem-Solving Lab 5-1,** p. 119 **Internet BioLab:** Researching Information on Exotic Pets, p. 130
Section 5.2 **Conservation of Biodiversity** National Science Education Standards UCP.1-3; A.1, A.2; C.4, C.5, C.6; E.1, E.2; F.2-6; G.1-3 (3 sessions, 1 block)	3. **Describe** strategies used in conservation biology. 4. **Relate** success in protecting an endangered species to the methods used to protect it.	**MiniLab 5-2:** Conservation of Soil, p. 126 **Art Connection:** Wildlife Photography of Art Wolfe, p. 132

Need Materials? Contact Carolina Biological Supply Company at 1-800-334-5551 or at **http://www.carolina.com**

MATERIALS LIST

BioLab
p. 130 Internet access, paper, pencil

MiniLabs
p. 116 paper, pencil, calculator (optional)
p. 126 beaker, plastic tray, graduated cylinder, water, lawn soil, lawn soil with grass

Alternative Lab
p. 122 self-sealing plastic bags (2), paper towels, small cups (2), labels, water, vinegar, small seeds (40)

Quick Demos
p. 121 cardboard, scissors, masking tape
p. 122 test tube (2), one-hole stopper, glass tube, distilled water, Alka-Seltzer tablet, pH paper
p. 129 *Ginko biloba* leaves

Key to Teaching Strategies

L1 Level 1 activities should be appropriate for students with learning difficulties.
L2 Level 2 activities should be within the ability range of all students.
L3 Level 3 activities are designed for above-average students.
ELL ELL activities should be within the ability range of English Language Learners.
COOP LEARN Cooperative Learning activities are designed for small group work.
P These strategies represent student products that can be placed into a best-work portfolio.
These strategies are useful in a block scheduling format.

Biological Diversity and Conservation

Teacher Classroom Resources

Section	Reproducible Masters	Transparencies
Section 5.1 **Vanishing Species**	Reinforcement and Study Guide, pp. 19-21 **L2** Concept Mapping, p. 5 **L3** **ELL** Critical Thinking/Problem Solving, p. 5 **L3** BioLab and MiniLab Worksheets, pp. 21-22 **L2** Laboratory Manual, pp. 31-38 **L2** Tech Prep Applications, pp. 5-6 **L2** Content Mastery, pp. 21-22, 24 **L1**	Section Focus Transparency 10 **L1** **ELL** Reteaching Skills Transparencies 7a, 7b, 7c **L1** **ELL**
Section 5.2 **Conservation of Biodiversity**	Reinforcement and Study Guide, p. 28 **L2** BioLab and MiniLab Worksheets, p. 29 **L2** Laboratory Manual, pp. 23, 25-26 **L2** Content Mastery, pp. 21, 23-24 **L1** Tech Prep Applications, pp. 7-8 **L2**	Section Focus Transparency 11 **L1** **ELL**

Assessment Resources

Chapter Assessment, pp. 25-30
MindJogger Videoquizzes
Performance Assessment in the Biology Classroom
Alternate Assessment in the Science Classroom
ExamView® Pro Software 💾
BDOL Interactive CD-ROM, Chapter 5 quiz

Additional Resources

Spanish Resources **ELL**
English/Spanish Audiocassettes **ELL**
Cooperative Learning in the Science Classroom **COOP LEARN**
Lesson Plans/Block Scheduling

NATIONAL GEOGRAPHIC — Teacher's Corner

Products Available From National Geographic Society
To order the following products, call National Geographic Society at 1-800-368-2728:
CD-ROM
NGS PictureShow: Earth's Endangered Environments
Videodiscs
GTV: Planetary Manager
GTV: Biodiversity

Index to National Geographic Magazine
The following articles may be used for research relating to this chapter:
"Making Sense of the Millennium," by Joel L. Swerdlow, January 1998.
"Sanctuary: U. S. National Wildlife Refuges," by Douglas H. Chadwick, October 1996.

GLENCOE TECHNOLOGY

The following multimedia resources are available from Glencoe.
Biology: The Dynamics of Life
CD-ROM **ELL**
 BioQuest: *Biodiversity Park*

The Infinite Voyage
Life in the Balance
Crisis in the Atmosphere
Secrets From a Frozen World

The Secret of Life Series
Gone Before You Know It: *The Biodiversity Crisis*

Chapter 5
Biological Diversity and Conservation

What You'll Learn

- You will explain the importance of biological diversity.
- You will distinguish environmental changes that may result in the loss of species.
- You will describe the work of conservation biologists.

Why It's Important

When all the members of a species die, that species' place in the ecosystem is gone forever. Knowledge of biological diversity leads to strategies to protect the permanent loss of species from Earth.

READING BIOLOGY

Compile a list of different types of problems conservationists face, such as water pollution or habitat loss. For each problem on your list, think of ways to combat the problem. Take examples from your community or examples you may have heard about in the news.

To find out more about biological diversity and conservation biology, visit the Glencoe Science Web site.
science.glencoe.com

Earth may lose all of its giant pandas (above). Their loss of habitat, due to the encroachment of humans, has put them in peril. The passenger pigeon (inset), once so common it filled the skies of North America, was hunted to extinction.

114

5.1 Vanishing Species

Imagine yourself standing in a cornfield and then standing in a rain forest. On this farmland in Iowa, one species dominates—corn. However, in this temperate rain forest in Washington State, you can see and hear hundreds of different species. The rain forest is a richer ecosystem; it is home to more species of organisms. The rain forest is more likely to survive disease, insects, and drought than the corn on the farmland.

A temperate rain forest in Washington (above) and a cornfield in Iowa (inset)

SECTION PREVIEW

Objectives
Explain biodiversity and its importance.
Relate various threats to the loss of biodiversity

Vocabulary
biodiversity
extinction
threatened species
endangered species
habitat fragmentation
edge effect
habitat degradation
acid precipitation
ozone layer
exotic species

Biological Diversity

A rain forest has a greater amount of biological diversity, or biodiversity, than a cornfield. **Biodiversity** refers to the variety of life in an area. This area could be Mississippi, Mexico, the Sonoran Desert, or the entire planet Earth. The simplest and most common measure of biodiversity is the number of species that live in a certain area. For example, one acre of farmland may be dominated by only one species of plant; one acre of rain forest may contain 400 species of plants. The cornfield may contain two species of beetle, and the rain forest may have 5000 species of beetles.

Where is biodiversity found?

Areas around the world differ in biodiversity. A hectare of tropical rain forest in Amazonian Peru may have 300 tree species. Yet, one hectare of forest in the United States is more likely to have 30 tree species or less. Consider the number of species of mammals; Canada has 163 species, the United States has 367, and Mexico has 439. These examples illustrate that terrestrial biodiversity tends to increase as you move towards the equator. In fact, tropical regions contain two-thirds of all land species on Earth. The richest environments for biodiversity are all warm places: tropical rain forests,

5.1 VANISHING SPECIES **115**

Prepare

Key Concepts

Students will explore the concept of biodiversity and factors that affect it. They will learn about human-created threats to biodiversity.

Planning

- Gather cardboard and tape for the first Quick Demo.
- Gather materials for the second Quick Demo.
- Purchase or gather plastic bags, paper towels, small cups, vinegar, and seeds for the Alternative Lab.

1 Focus

Bellringer

Before presenting the lesson, display **Section Focus Transparency 10** on the overhead projector and have the students answer the accompanying questions. **L1** **ELL**

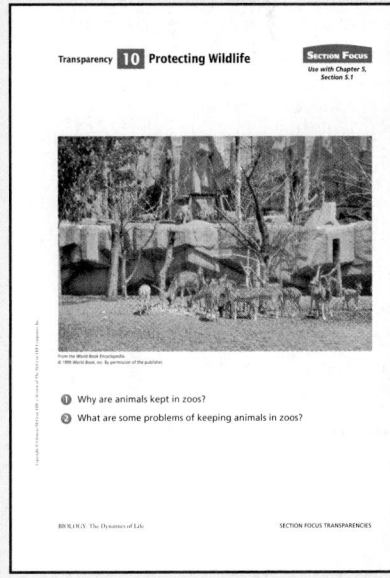

Assessment Planner

Portfolio Assessment
   Portfolio, TWE, pp. 118, 120, 121, 126
   Problem-Solving Lab, TWE, p. 119
   Assessment, TWE, pp. 124, 127, 129
Performance Assessment
   MiniLab, SE, pp. 116, 126
   Assessment, TWE, p. 123
   Alternative Lab, TWE, pp. 122-123
   Problem-Solving Lab, TWE, p. 128

BioLab, SE, pp. 130-131
BioLab, TWE, pp. 130-131
Knowledge Assessment
   MiniLab, TWE, pp. 116, 126
   Assessment, TWE, p. 121
   Section Assessment, SE, pp. 124, 129
   Chapter Assessment, SE, pp. 133-135
Skill Assessment
   Alternative Lab, TWE, pp. 122-123

2 Teach

Purpose

Students will survey an area near school and calculate an index of diversity (I.D.).

Process Skills

apply concepts, collect data, compare and contrast, interpret data, observe and infer, organize data, predict, use numbers

Teaching Strategies

■ Any plant type can be used. For example, the study could be conducted with flowering plants or cacti. Simply have students follow a marked path through any community and record their observations.

■ If possible, locate and mark off an area that contains about 10 trees of different species. The trees do not have to be in a straight line but will have to be marked so they can be numbered in order of observation.

■ If you wish, draw a map of the area to be visited and give copies to students. Include a birds-eye diagram of the trees and pre-number them.

■ If a field trip is not possible, provide students with a map of trees labeled by species.

Expected Results

Numbers below one (decimals) indicate a low index of diversity. Numbers over 1 show greater diversity. A typical city street may yield an I.D. of slightly over 1.

Analysis

1. A vacant lot might have a higher I.D. because many different species might be present in a small area. A grass lawn would have a much lower I.D., perhaps with only one species present.
2. A higher I.D. would indicate greater species diversity. Communities with a low I.D. may be prone to species extinction if environmental conditions change.

116

MiniLab 5-1   Using Numbers

Measuring Species Diversity
Index of diversity (I.D.) is a mathematical way of expressing the amount of biodiversity and species distribution in a community. Communities with many different species (a high index of diversity) will be able to withstand environmental changes better than communities with only a few species (a low index of diversity).

A tree-lined street

Procedure

1. Copy the data table below.
2. Walk a city block or an area designated by your teacher and record the number of *different species* of trees present (you don't have to know their names, just that they differ by species). Record this number in your data table.
3. Walk the block or area again. This time, make a list of the trees by assigning each a number as you walk by it. Place an X under Tree 1 on your list. If Tree 2 is the same species as Tree 1, mark an X below it. Continue to mark an X under the trees as long as the species is the same as the previous one. When a different species is observed, mark an O under that tree on your list. Continue to mark an O if the next tree is the same species as the previous. If the next tree is different, mark an X.
4. Record in your data table:
   a. the number of "runs." Runs are represented by a group of similar symbols in a row. Example—XXXXOOXO would be 4 runs (XXXX = 1 run, OO = 2 runs, X = 3, O = 4).
   b. the total number of trees counted.
5. Calculate the Index of Diversity (I.D.) using the formula in the data table.

Data Table
Number of species =
Number of runs =
Number of trees =
Index of diversity = $\dfrac{\text{Number of species} \times \text{number of runs}}{\text{Number of trees}}$ =

Analysis

1. Compare how your tree I.D. might compare with that of a vacant lot and with that of a grass lawn. Explain your answer.
2. If humans were concerned about biological diversity, would it be best to have a low or high I.D. for a particular environment? Explain your answer.

116 BIOLOGICAL DIVERSITY AND CONSERVATION

---

coral reefs, and large tropical lakes. Learn one way to measure species distribution in the *MiniLab* on this page.

The study of islands has led to additional understanding of factors that influence biodiversity. Large islands tend to have a higher biodiversity than smaller islands, as shown in *Figure 5.1.* These Caribbean islands are near each other; however, they differ in the number of species they each contain. For example, Redonda—a small island—has fewer species than Saba—a large island. Why do larger islands tend to have a greater biodiversity than smaller islands? The larger islands have more space and are more likely to have a greater variety of environments and ecosystems. In some cases, however, smaller islands can have a larger biodiversity than larger islands. Compare the huge island of Iceland to the much smaller island of Maui in *Figure 5.2.* Maui has more biodiversity because it is warmer and has

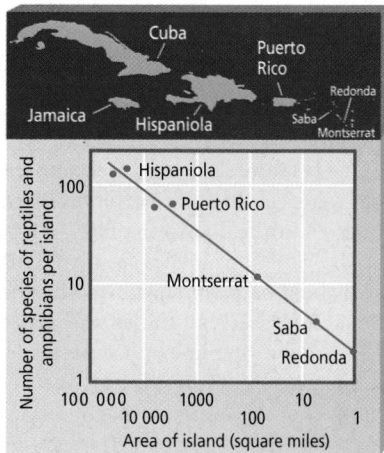

Figure 5.1
The relative number of species on an island can be predicted from the size of the island. For example, Puerto Rico is larger than Montserrat, and it has more species.

✔ Assessment

Knowledge Ask students why a city with one major industry may suffer if there is a downturn in that industry, whereas a city with a diverse industrial base will be better able to withstand an economic downturn. Use the Performance Task Assessment List for *Making Observations and Inferences* in **PASC,** p. 17. **L2**

MEETING INDIVIDUAL NEEDS

Learning Disabled

 *Logical-Mathematical* The simplest way of calculating biodiversity is to count the number of species in an area. Have students count the number of species in a fish tank to understand this concept. Caution them to count the number of species, not the number of fish. **L1 ELL**

more nutrient-rich soil. However, if everything else is the same, the larger the island the greater the biodiversity it contains. These findings from island research have become important for managing and designing national parks and protected areas. Such areas have become "islands," not surrounded by oceans, but surrounded by human populations with buildings and roads.

Importance of Biodiversity

Compare a parking lot having nothing but asphalt to your favorite place in nature, perhaps a meadow, a forest, or a thriving lake. Which environment do you think is more pleasant? The presence of different forms of life makes our planet beautiful. You may go to the natural area to relax or to think. Artists get inspiration from these areas for songs, paintings, and literature. Looking at the beauty of one of Art Wolfe's photograph in the *Art Connection* can help you appreciate the beauty biodiversity gives our world. Beyond beauty, why is biodiversity important?

Importance to nature

Organisms are adapted to live together in communities. Although ecologists know of many complex relationships among organisms, many relationships are yet to be discovered. Scientists do know that if a species is lost from an ecosystem, the loss may have consequences for other living things. Other organisms suffer when an organism they feed upon is removed permanently from a food chain or food web. A population may soon exceed the area's carrying capacity if its predators are removed. If the symbiotic relationships among organisms are broken due to the loss of one of the dependent species, then the other species will soon be affected.

Life depends on life. Animals could not exist without green plants, many flowering plants could not exist without animals to pollinate them, and plants need decomposers to break dead or decaying material into nutrients they can use. A rain forest tree grows from nutrients in the soil released by decomposers. A sloth eats the leaves of this tree. Moss grows on the back of the sloth. Living things create niches for other living things.

Figure 5.2
Although Iceland (left) is a bigger island than Maui (right), Maui has a greater biodiversity due to its warm climate and good soils. Nevertheless, all other things being equal, the larger the island the greater its biodiversity.

118

Figure 5.3
Diverse living things provide many important medical drugs.

B Rosy periwinkle is the source of drugs for Hodgkin's disease and leukemia.

C Willow bark was the original source of aspirin.

A Taxol, a strong anti-cancer drug, was first discovered in the Pacific yew.

WORD *Origin*

Extinction
From the Latin words *ex*, meaning "out," and *stinguere*, meaning "to quench." A species becomes extinct when its last organisms die.

Biodiversity also brings stability to an ecosystem. A pest could destroy all the corn in a farmer's field, but it would be far more difficult for an insect or disease to destroy all individuals of a plant species in a rain forest. These plants would exist in many parts of the rain forest, making it more difficult for the insect or disease to spread. Although ecosystems are stronger due to their biodiversity, losing species may weaken them. One hypothesis suggests species in an ecosystem are similar to rivets holding an airplane together. A few rivets might break, and nothing will happen, but at some point enough rivets break and the airplane falls apart.

Importance to people

Humans depend on other organisms for their needs. Oxygen is supplied and carbon dioxide is removed from the air by diverse species of plants and algae living in a variety of ecosystems. Biodiversity gives people a diverse diet. Beef, chicken, tuna, shrimp, and pork are only a few of the meats and fish we eat. Think of all the plant products that people eat, from almonds to zucchini. When you consider all the foods eaten by all the people in the world, you realize that hundreds of species help nourish the human population. Biodiversity can help breeders make food crops grow better. For example, through cross-breeding with a wild plant, a food crop can be made pest-resistant or drought-tolerant. People also rely on the living world for materials used in clothes, furniture, and buildings.

Biodiversity can also be used to improve people's health. Living things supply the world pharmacy. Although drug companies may manufacture synthetic drugs, active compounds are usually first isolated from living things, such as shown in *Figure 5.3*. The antibiotic penicillin came from the mold, *Penicillium;* the antimalarial drug quinine came from the bark of the cinchona tree. Preserving biodiversity ensures there will be a large supply of living things, some of which may provide future drugs. Maybe a cure for cancer or HIV is contained in the leaves of a small rain forest plant.

118 BIOLOGICAL DIVERSITY AND CONSERVATION

Loss of Biodiversity

Have you ever seen a flock of passenger pigeons? How about a woodland caribou, relic leopard frog, or Louisiana prairie vole? Unless you have seen a photograph or a stuffed museum specimen, your answer will be no. These animals are extinct. **Extinction** (ek STINGK shun) is the disappearance of a species when the last of its members dies. Since 1980, almost 40 species of plants and animals living in the United States have become extinct. Although extinction can occur as a result of natural processes, humans have been responsible for the extinction of many species. Is there a link between the land area a species can inhabit and extinction? Find out in the *Problem-Solving Lab*.

When the population of a species begins declining rapidly, the species is said to be a **threatened species.** African elephants, for example, are listed as a threatened species. In the early 1970s, the wild elephant population numbered about three million. Twenty years later, it numbered only about 700 000. Elephants have traditionally been hunted for their ivory tusks, which are used to make jewelry and ornamental carvings. Many countries have banned the importation and sale of ivory, and this has helped slow

the decline of the elephant population. *Figure 5.4* shows some threatened animal species under protection in the United States.

Problem-Solving Lab 5-1   Interpreting Data

Does species extinction correlate with land area?
Species are at risk of extinction when their habitats are destroyed through human action. Is there a better chance for survival when the land area is large?

Analysis

A study of land mammals was conducted by a scientist to determine the effect of land area on species extinction. His research was confined to a group of South Pacific islands of Indonesia. The scientist's basis for determining the initial number of species present was based on research conducted by earlier scientists and from fossil evidence.

Table 5.1 Relationship of land area to extinctions

Island	Area in km²	Initial number of species	Extinctions	Percent of loss
Borneo	751 709	153	30	20
Java	126 806	113	39	35
Bali	5 443	66	47	71

Thinking Critically

1. In general, how does land area correlate with loss of species?
2. What is the relationship between island size and the initial number of species?
3. Hypothesize why the study was conducted on only land mammals.

Problem-Solving Lab 5-1

Purpose
Students will determine that reduced land area contributes to the extinction of species.

Process Skills
think critically, analyze information, compare and contrast, draw a conclusion, interpret data, predict, recognize cause and effect, use numbers

Teaching Strategies
■ You may wish to ask students to locate these island areas on a map.
■ Review the procedure for calculating percent. Allow students to use calculators.

Thinking Critically
1. There is a higher percent of species loss for smaller areas.
2. The larger the island, the more initial species it has.
3. Answers will vary, but students may mention that it's easier to trap and observe land mammals, and the initial data on species numbers may be more reliable.

✓ Assessment
Portfolio Have students write a newspaper article describing some possible ways to reduce the number of extinctions of land mammal species on a small island such as Bali. Use the Performance Task Assessment List for Newspaper Article in **PASC**, p. 69.
L2 P

Figure 5.4
Several animal species in the United States are threatened.

A Sea otters *(Enhydra lutris* subspecies *nereis)* have been hunted for centuries for their fur.

B Wildlife experts classify loggerhead turtles *(Caretta caretta),* as well as five other sea turtle species, as threatened.

119

PROJECT

Finding Answers

Visual-Spatial Have students choose an extinct organism and research possible reasons for its extinction. Create a classroom bulletin board on extinction. L3

Revealing Misconceptions

While biologists suspect that humans have increased the rate of natural extinctions, help students realize that extinction is a natural process. Dinosaurs were extinct before people evolved. Extinction and evolution often go hand-in-hand.

Concept Development

Ask students to explain ways that people alter habitats. Try to identify whether the changes affect the biotic or abiotic factors of that habitat. For example, cutting trees can have biotic affects such as reducing the food supply for certain animals. It can also have abiotic effects, such as increasing the amount of sunlight that reaches an area.

VIDEODISC
GTV: Planetary Manager
Side 1: *Vanishing Act*

STV: Biodiversity
Loss of Diversity
Unit 1, Side 1, 6 min. 20 sec.

Figure 5.5
In the United States, scientists have developed programs designed to save endangered species.

A Urban growth has destroyed much of the California condor's habitat. To protect the species, the few remaining California condors (*Gymnogyps californianus*) were captured and placed in reserves.

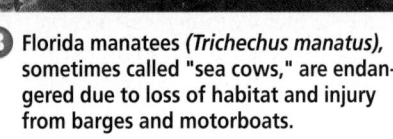

B Florida manatees (*Trichechus manatus*), sometimes called "sea cows," are endangered due to loss of habitat and injury from barges and motorboats.

A species is considered to be an **endangered species** when its numbers become so low that extinction is possible. In Africa, the black rhinoceros has become an endangered species. Poachers hunt and kill these animals for their horns. Rhinoceros horns are composed of fused hair rather than bone or ivory. In the Middle East, the horns are carved into handles for ceremonial daggers. In parts of Asia, some people believe powdered horn is an herbal medicine. *Figure 5.5* shows just two endangered species in the United States. Unfortunately, there are many more.

Threats to Biodiversity

Complex interactions among species make ecosystems unique and species are usually well adapted to their habitats. Changes to habitats can threaten organisms with extinction. As populations of people increase, the impact from their growth and development is altering the face of Earth and pushing many other species to the brink of extinction.

Habitat loss

The biggest threat to biodiversity is habitat loss. When a rain forest is made into a cattle pasture, a meadow made into a mall parking lot, or a swamp drained for housing, habitats are lost. With their habitats gone, the essentials of life are lost for species dependent upon these habitats, and species disappear. In tropical rain forests, the Earth's richest source of biodiversity, an area the size of Florida is cut or burned every year. Coral reefs, such as shown in *Figure 5.6*, are also very rich in biodiversity. Water and temperature changes can damage coral reefs. People remove large sections of coral reef for a variety of reasons. In some areas, coral is mined for building materials. Coral reef is often collected for souvenirs and aquarium decorations. Through habitat loss, as well as pollution and disease, many species that live in coral reefs are in danger of extinction.

Habitat fragmentation

As roads cut across wilderness areas, and as building projects expand into new areas, many habitats are

---

Portfolio

Observing Habitat Destruction

 Visual-Spatial Have students record any direct evidence they observe of habitat destruction brought on by humans. Ask them to describe how the conditions they observe might be contributing to the reduction in numbers of some native plant or animal species. **L2**

Figure 5.6
Coral reefs are rich in biodiversity. Removal of coral results in a loss of habitat for reef organisms.

becoming virtual islands. **Habitat fragmentation** is the separation of wilderness areas from other wilderness areas. Habitat fragmentation presents problems for many organisms. Recall how the study of islands revealed that the smaller the island, the fewer species it supports. Fragmented areas are like islands, and the smaller the fragment, the less biodiversity it will support.

Biotic issues

Habitat fragmentation, as shown in *Figure 5.7*, presents problems for organisms that need large areas to gather food. Large predators cannot obtain enough food if they are restricted to a small area. Some organisms, such as zebra and wildebeest, migrate with the seasons to ensure a constant grass supply. If their range is restricted, they will starve. Habitat fragmentation also makes it difficult for species to reestablish themselves in an area. Imagine a small fragment of forest where a species of salamander lives. A fast burning fire destroys the trees and all the salamanders. In nonfragmented land, when new trees grow, new salamanders would eventually move back into the land. However, in the fragmented land, there is no migratory route for the salamanders to reestablish their population.

Abiotic issues

Another problem with habitat fragmentation is that it can change the climate of the area. Consider a tropical rain forest. Recall that the area under the canopy is moist and shady. Now suppose loggers come in and cut everything down, except for a plot of land equal in size to a football field. The once shady, moist area is now exposed from the sides to sunlight and winds. The area dries out, and organisms that evolved in rain

Figure 5.7
Wildlife areas that become surrounded by human development result in habitat fragmentation.

121

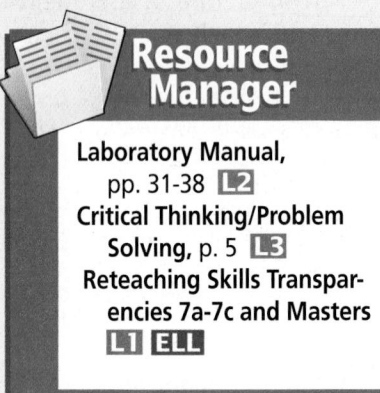

Figure 5.8
Acid precipitation and acid fog are believed to be major contributors of this damage to trees in Mount Mitchell, North Carolina.

forest conditions can no longer survive at the edges. The different conditions along the boundaries of an ecosystem are called an **edge effect.**

Habitat degradation

Another threat to biodiversity is **habitat degradation,** the damage to a habitat by pollution. Three types of pollution are air, water, and land pollution. Air pollution can have negative effects on living organisms, such as causing breathing problems and irritating membranes in the eyes and nose. Although pollutants can enter the atmosphere in many ways—including volcanic eruptions and forest fires—the burning of fossil fuels is the greatest source of air pollution.

Acid precipitation—rain, snow, sleet, and fog with low pH values—is responsible for the deterioration of forests and lakes. Sulfur dioxide from coal-burning factories and nitrogen oxides from automobile exhausts combine with water vapor in the air to form acidic droplets of water vapor. When the acidic droplets of

water fall from the sky as precipitation, the moisture leaches calcium, potassium, and other valuable nutrients from the soil. This loss of nutrients can lead to the death of trees. Acid precipitation also damages plant tissues and interferes with plant growth. Many trees in the forests of the United States are dying because of acid rain and fog, as shown in *Figure 5.8.* Acid precipitation also has severe effects on lake ecosystems. Acid precipitation falling into a lake, or entering it as runoff from streams, causes the pH of the lake water to fall below normal. The excess acidity can cause the death of plants, animals, and other organisms.

The atmosphere contains a sort of sunscreen—known as the **ozone layer**—that helps to protect living organisms on Earth's surface from receiving damaging or lethal doses of ultraviolet radiation. The chemical formula for ozone is O_3, meaning it contains three oxygen atoms. Chlorofluorocarbons, or CFCs, are synthetic chemicals that break down the ozone layer. CFCs are used as coolants in refrigerators and air conditioners, and in the production of polystyrene. Some scientists have hypothesized that the loss of the ozone layer, and the resulting increased ultraviolet radiation, may partially explain why more frogs and toads are born with deformities, and why many amphibian populations are declining.

Water pollution

Water pollution degrades aquatic habitats in streams, rivers, lakes, and oceans. A variety of pollutants can affect aquatic life. Excess fertilizers and animal wastes from farms are often carried by rain into streams and lakes. The sudden availability of nutrients causes algal blooms, the excessive growth of algae, such as the

122 BIOLOGICAL DIVERSITY AND CONSERVATION

one shown in *Figure 5.9*. As the algae die, they sink and decay, removing needed oxygen from the water. Silt from eroded soils can also enter water and clog the gills of fishes. In coral reefs, silt can cover the coral and prevent sufficient light from reaching the photosynthetic organisms within the coral polyps. Detergents, heavy metals, and industrial chemicals in runoff can cause sickness and death in aquatic organisms. Abandoned drift nets in oceans can entangle and kill dolphins, whales, and other sea life.

Land pollution

How much garbage does your family produce every day? Trash, or solid waste, is made up of the cans, bottles, paper, plastic, metals, dirt, and spoiled food that people throw away every day. The average American produces about 1.8 kg of solid waste daily. That's a total of about 657 kg of waste per person per year. Does it ever decompose? Although some of it might, most of our trash becomes part of the billions of tons of solid waste that are buried in landfills all over the world. As you might expect, these landfills destroy wildlife habitats by taking up space and polluting the immediate area.

The use of pesticides and other chemicals can also lead to habitat degradation. For many years, DDT was commonly sprayed on crops to control insects and sprayed on water to kill mosquito larvae. Birds that fed on insects, fish, and other small animals exposed to DDT were observed to have high levels of DDT in their bodies. The DDT was passed on to the predators that ate these birds. Because of the DDT in their bodies, some species of predators, such as the American bald eagle and the peregrine falcon, tended to lay eggs with very thin shells that cracked easily,

Figure 5.9
Cattle manure contains nitrogen and other nutrients that make it valuable as a plant fertilizer. But too much of a good thing can cause pollution problems.

Ⓐ The cattle on this feed lot produce more waste than the decomposers in the soil can handle. If the waste is not contained, it can be washed into nearby streams by rainfall.

Ⓑ The large amounts of nitrogen in runoff from a feed lot stimulate the rapid growth of algae in waterways downstream. This lush growth of algae consumes all or most of the oxygen in the water, making it impossible for insects, fishes, and other animals to survive.

123

3 Assess

Check for Understanding
Ask students to explain how the following could lead to a loss of species.
   a. habitat loss
   b. habitat fragmentation
   c. acid rain
   d. introduction of exotics

and the word *vinegar*.
5. Prepare a data table to record your observations. Record the number of seeds that germinate during the next four days. Germinating seeds show a root growing from the seed.

Expected Results
Seeds soaked in water will germinate, but most of those soaked in vinegar will not.

Analysis
1. Compare the number of germinated seeds after four days. How do the water-soaked and vinegar-soaked seeds compare? *There is less germination of the vinegar-soaked seeds.*
2. Vinegar is an acid. Explain how acid precipitation might affect seed germination. *Acid precipitation could delay or prevent seed germination.*

Figure 5.10
Exotic species can cause many problems when introduced into new ecosystems.

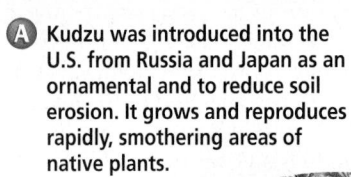

A Kudzu was introduced into the U.S. from Russia and Japan as an ornamental and to reduce soil erosion. It grows and reproduces rapidly, smothering areas of native plants.

B Zebra mussels were introduced into the Great Lakes from the ballast of ships. These fast-growing mussels filter food from the water, blocking many food chains.

killing the chicks inside. The populations of these species showed sharp declines. These observations led to the banning of DDT, and helped more people become aware of conservation biology issues.

Introduction of exotics

People, either on purpose or by accident, sometimes introduce a new species into an ecosystem. This can cause problems for the native species. When people brought goats to Santa Catalina Island, located off the coast of California, 48 native species of plants soon disappeared from the local environment. Building the Erie canal in the nineteenth century made it possible for the sea lamprey to swim into the Great Lakes. The sea lamprey resembles an eel, with a round clamp for a mouth. It swims up to a fish, clamps onto its body, and sucks the fluids out of the fish using its sharp teeth and tongue. The lamprey has totally eliminated certain fish species from some lakes. **Exotic species,** such as the goat and the lamprey, are organisms that are not native to a particular area. Other examples of exotic species are shown in *Figure 5.10*. When exotic species are introduced into an area, these species can grow at an exponential rate due to a lack of competitors and a lack of predators. They may take over niches of native species, and can eventually replace the native species completely.

Section Assessment

Understanding Main Ideas
1. What are two causes for a species to become threatened or endangered?
2. How does acid precipitation kill trees?
3. What is an edge effect?
4. How do exotic species affect populations of native species?

Thinking Critically
5. Suggest reasons why warm tropical areas have more biodiversity than cooler areas.

SKILL REVIEW
6. **Designing an Experiment** Thinning of the ozone layer allows increased ultraviolet radiation from the sun to penetrate Earth's atmosphere. Your hypothesis is that increased ultraviolet radiation results in higher mortality rates for frogs. Design an experiment to test your hypothesis. Remember to set up a control group. For more help, refer to *Practicing Scientific Methods* in the **Skill Handbook.**

124 BIOLOGICAL DIVERSITY AND CONSERVATION

Section Assessment

1. Answers may include habitat loss, degradation, or fragmentation; introduction of exotic species; or excessive hunting or collecting.
2. by destroying their leaves or needles through acid fog and precipitation or by changing the soil pH, which causes leaching of nutrients

3. different conditions along the boundaries of a natural area
4. The exotics can out-compete the native species for food or shelter. Or the exotic species may eat a native species.
5. Possible reasons: They have longer growing seasons, greater biomass, and a longer time since their last ice age.

6. The control and experimental groups of frogs should both include adequate numbers. The experimental group will be exposed to a greater amount of ultraviolet radiation. Mortality rates of the groups will be compared.

5.2 Conservation of Biodiversity

SECTION PREVIEW

Objectives

Describe strategies used in conservation biology.

Relate success in protecting an endangered species to the methods used to protect it.

Vocabulary

conservation biology
sustainable use
habitat corridors
reintroduction
    programs
captivity

The loss of species from Earth is a permanent tragedy. Fortunately, people are working to protect species from extinction. One strategy is to bring organisms back into areas where they once lived, as is being done in Yellowstone National Park with gray wolves. Organisms are endangered for a variety of reasons, and perhaps through a variety of strategies, these organisms can be protected.

The gray wolf

Strategies of Conservation Biology

Conservation biology is a new field that studies methods and implements plans to protect biodiversity. Effective conservation strategies are based on principles of ecology. These strategies include natural resource conservation and species conservation. Learn about soil conservation in the *MiniLab* on the next page.

Many species are in danger due to the actions of humans, so working with people is an important part of conservation biology. Conservation biologists not only focus on ecology, but also seek to understand law, politics, sociology, and economics to find effective strategies for conserving Earth's biodiversity.

Legal protections of species

In response to concern about extinction, President Nixon signed the U.S. Endangered Species Act into law in 1973. This law made it illegal to harm any species on the endangered or threatened species lists. Further, the law made it illegal for federal agencies to fund any project that would harm organisms on these lists. Harm includes changing an ecosystem where endangered or threatened species live. This law has been partially responsible for recoveries in the populations of the American bald eagle, the American alligator, and the brown pelican. Other countries have enacted similar laws. International agreements also protect endangered or threatened species. For example, the Convention on

5.2 CONSERVATION OF BIODIVERSITY **125**

VIDEODISC
GTV: Planetary Manager
Side 2: *Tidy World*

VIDEODISC
STV: Biodiversity
Preserving Diversity (in its entirety)
Unit 2, Side 1, 12 min. 10 sec.

Section 5.2

Prepare

Key Concepts

Students will learn how conservation biology strategies are used to prevent the loss of species. Throughout the section, students will learn success stories in conservation biology.

Planning

■ Gather soil, trays, beakers, and water for the MiniLab.

■ Gather ginkgo leaves for the Quick Demo.

■ Arrange access to the Internet for the BioLab.

1 Focus

Bellringer

Before presenting the lesson, display **Section Focus Transparency 11** on the overhead projector and have the students answer the accompanying questions. **L1** **ELL**

Resource Manager

Section Focus Transparency 11 and Master **L1** **ELL**

2 Teach

MiniLab 5-2

Purpose
Students will determine that soil erosion is reduced on land areas with vegetation compared with land areas with no vegetation.

Process Skills
compare and contrast, draw a conclusion, interpret data, measure in SI, recognize cause and effect, formulate a model

Safety Precautions
Have students wear a lab apron and safety goggles.

Teaching Strategies
■ Obtain soil with grass growing in it from a lawn area (with permission) or purchase sod from a garden store.
■ Try to keep the volume of soil equal in both trays. Water should be poured slowly.
■ Review the use of a graduated cylinder for reading volumes. Remind students to allow the soil to settle after they pour it into the cylinder.
■ You may wish to conduct this activity as a demonstration.

Expected Results
There will be less measured soil erosion when vegetation is present than when soil is bare.

Analysis
1. water and soil draining into the dish
2. to reduce soil erosion

✓ Assessment
Knowledge Pull a small plant from the ground and have students examine the root system and the soil clinging to it. Have them explain how roots reduce soil erosion. Use the Performance Task Assessment List for Making Observations and Inferences in PASC, p. 17. **L1**

126

MiniLab 5-2 Experimenting

Conservation of Soil Soil is an important natural resource and should be conserved. How does one conserve soil? As a start, we should be aware of factors that promote or speed up its unnecessary loss or erosion.

Procedure
1 Copy the data table below.

Data Table			
Source of sample	Volume of original water	Volume of collected water	Volume of eroded soil
Bare soil			
Soil with grass			

2 Measure 200 mL of water in a beaker.
3 Fill a plastic tray with soil as shown in the diagram.
4 Pour the water onto the soil, tilting the tray and placing a dish underneath the end of it as indicated in the diagram.
5 Wait several minutes for all water and soil to drain into the dish.
6 Pour the soil and water from the dish into a graduated cylinder. Wait several minutes for the soil and water to settle. Measure the volume of soil and water that washed or eroded into the dish. Record these values in your data table.
7 Repeat steps 2-6, only this time place a section of soil in which grass is growing onto the tray. **CAUTION: Always wash your hands with soap and water after working with soil.**

Analysis
1. What part of the experiment simulated soil erosion?
2. Based on this experiment, explain why farmers usually plant unused fields with some type of crop cover.

International Trade in Endangered Species (CITES) has established lists of species for which international trade is prohibited or controlled. This agreement has been endorsed by more than 120 countries.

Preserving habitats
Another conservation biology approach focuses on protecting whole communities and whole ecosystems as the best way to conserve species. An effective way to do this is by creating nature preserves. The United States established its first national park—Yellowstone National Park—in 1872. You might look at Yellowstone, *Figure 5.11*, and think of bear, elk, bison, and moose; however, the park was created to preserve the unique geology of the area, not its biodiversity. Back in 1872, bear, bison, moose, and elk were widely distributed across the United States as far east as Pennsylvania. As the human population in America expanded, large parks such as Yellowstone preserved habitats and prevented the extinction of many species.

Establishing parks and other protected areas can be an effective way to preserve an ecosystem and the communities of species that live there. Yet, only 6 percent of Earth's land surface is protected in this way. Although this is only a small fraction of our planet's land, these areas contain a large amount of biodiversity. For example, in Zaire, Africa, only 3.9 percent of the land is protected, but 89 percent of Zaire's bird species are found there. Because the biggest threat to biodiversity is habitat loss, preserving land is an effective means of preserving species.

Saying an area is protected does not automatically make all species safe. Parks and protected areas often hire people, such as rangers, to manage the parks and ensure the protection of organisms. In some areas, access by people is restricted. In other lands, people can harvest food or obtain materials but this use is managed. The philosophy of **sustainable use** strives

---

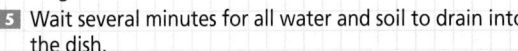

MEETING INDIVIDUAL NEEDS

Hearing Impaired
Visual-Spatial Photocopy a map of a country and a map of the same region that shows protected areas. Give both maps to students. Have them color in the protected areas on the unmarked map. Ask them to estimate what percent of the country's land is protected. **L2**
ELL

✓ Portfolio

Preserving Natural Habitats
Linguistic Have students research a national park. Ask them to record in their portfolios the date it was established, its location, and common plants and animals in the park. **L3** **P**

to let people use the resources of wilderness areas in ways that will not damage the ecosystem. For example, in some tropical rain forests in the Amazon, people harvest Brazil nuts, as shown in *Figure 5.12,* to eat and to sell. This practice gives people the opportunity to earn a living, and it helps them appreciate the value of preserving the area. When local people benefit from a natural area, through jobs or resources, they are more likely to cooperate with the area's preservation. For this reason, the preferred choice of people hired to protect natural areas are local people.

If a conservation group had enough money to buy 1000 acres of forest to set aside for protection, what would be their best strategy? Should they buy one big 1000-acre plot or ten 100-acre plots? Recall the research concerning number of species on islands of different sizes. In general, larger islands have more species than smaller islands. Therefore, to best protect the biodiversity of an area, the conservation group should generally buy one large plot of land. However, if the ten plots protected a variety of ecosystems,

they may be a better choice for conserving biodiversity. Each decision needs to be made on a case-by-case basis. There are general guidelines, however. It is usually better to buy land that is connected to other protected areas so bigger "islands" are created. It is also a good idea to connect protected areas with corridors. **Habitat corridors** are natural strips that allow the migration of organisms from one area to another. Corridors can help overcome some of the problems of habitat fragmentation.

Figure 5.11
Yellowstone National Park was the first national park in the United States. It is home to a wide variety of life.

Figure 5.12
Allowing sustainable use of wildlife areas provides additional motivation for local people to protect these areas. The sale of Brazil nuts, sustainably harvested from rain forests in the Amazon, provides income for people living in and near rain forests.

Purpose

Students will explore opinions regarding the preservation of certain endangered species.

Process Skills

think critically, analyze information, communicate, draw a conclusion, recognize cause and effect

Teaching Strategies

■ You may wish to have teams of volunteers debate the issues explained in this lab.

■ Remind students to show respect for others' opinions.

Thinking Critically

1. No. Ranchers may think that any wolf they see is threatening their livestock and kill it.

2. Yes. The native wolves probably evolved from the reintroduced species and might look similar to them.

3. They are killing ranchers' livestock.

✓ Assessment

Performance Have volunteers give persuasive speeches to convince "decision-makers" to continue reintroducing wolves or to end the program. Use the Performance Assessment List for Investigating an Issue Controversy in **PASC,** p. 65. **L2**

Revealing Misconceptions

Students may not realize that reintroduction is successful only when the problems affecting the reintroduced species have been corrected. For example, if urban sprawl led to the disappearance of jaguars, reintroduced jaguars would probably still not have enough land to meet their needs.

Problem-Solving Lab 5-2 — Thinking Critically

Is everyone pleased with conservation efforts? There have been many attempts to breed wild animals or move wildlife from one area to another. The goal is to preserve wildlife species if they appear to be in danger of extinction. These programs may result from the Endangered Species Act. How do groups with opposing interests or opinions view these programs?

A gray wolf

Analysis

Case 1: In March 1998, the U.S. Fish and Wildlife Service reintroduced 11 captive-bred Mexican gray wolves *(Canus lupus baileyi)* into parts of Arizona. They had been extinct from the area for 20 years. By law, ranchers are not allowed to kill native wolves. However, reintroduced wolves received special legal status under the Endangered Species Act and can be killed by ranchers only if they threaten livestock.

Case 2: Gray wolves *(Canus lupus)* had been extirpated from Yellowstone National Park since 1926. In 1995, the first group of gray wolves was captured in Canada and introduced to Yellowstone National Park, nearly 30 years after being listed as endangered in the lower 48 states. Despite a program to compensate ranchers for livestock losses to wolves, legal pressure was mounted to have the wolves removed as a threat to livestock. In December 1997, a United States district court ruled that the wolves should be removed from the park. However, a nonprofit group called Defenders of Wildlife has appealed the decision and the fate of the wolves remains with the courts.

Thinking Critically

1. Should supporters of the Endangered Species Act be pleased with the special legal status in Case 1? Explain.

2. In Case 1, ranchers pointed out that they cannot tell reintroduced species from native wolves. Is their concern justified? Explain your answer.

3. In Case 2, what is the main reason for the opposition against wolves in the park? Explain your answer.

Reintroduction programs

The year is 1992 in Wyoming. A wildlife manager carries a cage with a black-footed ferret, as shown in *Figure 5.13.* She opens the cage door, and the ferret cautiously steps

---

Figure 5.13
A black-footed ferret is reintroduced into its native habitat in Wyoming. The caregiver wears a mask to reduce the chances she will transmit a human disease to the ferret.

out onto the land. Black-footed ferrets used to live here, but the population declined steadily. Farmers had poisoned prairie dog colonies, thinking that the prairie dogs were competing with their livestock for food. Prairie dogs are one of the ferret's main foods. Black-footed ferret survivors were captured and bred. Some of the ferrets that were released back to the land survived to reproduce. **Reintroduction programs,** such as this one, release organisms into an area where their species once lived. Form your own opinions about these programs by reading the *Problem-Solving Lab.* The factors that lead to the decline of the organisms must be removed if the reintroduction is to be a success. For example, the prairie dog population must be healthy for the ferrets to survive.

The most successful reintroductions occur when organisms are taken from an area and transported to a new suitable habitat. The brown pelican was once common along the shores of the Gulf of Mexico. DDT caused this bird's eggs to break, and the brown pelican completely disappeared from these areas. After DDT

MEETING INDIVIDUAL NEEDS

Learning Disabled

Interpersonal Organize students into groups and have each group draw a large protected area and a system of several small protected areas. Then have the groups explain the advantages and disadvantages of each. **L1**

Resource Manager

Content Mastery, pp. 21, 23-24 **L1**
Reinforcement and Study Guide, p. 22 **L2**
Tech Prep Applications, p. 7 **L2**

was banned, 50 birds were taken from Florida and put on Grand Terre Island in Louisiana. The population grew and spread, and today over 7000 brown pelicans live in the area.

Sometimes all the organisms in wild areas are gone, and the only members of the species are in zoos. An organism that is held by people is said to be in **captivity**. The ginkgo tree, *Figure 5.14,* is an example of a species surviving extinction because it was kept by people. The ginkgo is an ancient tree; all similar species became extinct long ago. However, Chinese monks planted the ginkgo tree around their temples. By keeping these trees in "captivity," they prevented the tree from becoming extinct. Ginkgos now add beauty to modern cities. The ginkgo shows that not only can a species be protected, but that a protected species may be useful to people.

The best place to keep plants is in their natural ecosystems in the wild. If they must be kept in captivity, an economical way to do this is to store their seeds. Seeds can be cooled and stored for long periods of time. By establishing seed banks for threatened and endangered plants, these species can be reintroduced if they become extinct.

Reintroductions of captive animals is more difficult than for plants. Keeping animals in captivity, with enough space, adequate care, and proper food, is expensive. Animals kept in captivity may lose the necessary behaviors to survive and reproduce in the wild. Despite the difficulties involved, some species held in captivity, such as the Arabian oryx and the California condor, have been successfully reintroduced to their native habitats after becoming extinct in the wild.

Figure 5.14
The ginkgo tree would probably be extinct if not for the care given to it by Chinese monks. It is a beautiful tree that survives pollution well, making it a good tree for urban landscapes.

Section Assessment

Understanding Main Ideas

1. Contrast the fields of conservation biology and ecology.
2. Describe the U.S. Endangered Species Act. When did it become law, and how does it help protect or preserve endangered species?
3. What are the difficulties with reintroduction programs using captive-born animals?
4. Choose one species that you have read about, either here or in your library, and explain how conservation strategies lead to its recovery.

Thinking Critically

5. Why is it necessary for a conservation biologist to understand economics?

SKILL REVIEW

6. **Recognizing Cause and Effect** For a wildlife area near you, what would be one use of the area that fits into the philosophy of sustainable use? What is one use that would damage the ecosystem? For more help, refer to *Thinking Critically* in the **Skill Handbook**.

5.2 CONSERVATION OF BIODIVERSITY **129**

Quick Demo

Ginkgo leaves are distinctive. Bring in a few leaves so students can identify the tree.

3 Assess

Check for Understanding

Have students explain which conservation biology strategy they feel is the most important and why.

Reteach

Have students list the three conservation biology strategies in this section. For each strategy, they should describe one organism that was helped. **L2**

Extension

Visual-Spatial Have students create endangered species posters. Each poster should have a picture, a description of the organism's niche, and what could be done to help the organism. **L2** **ELL**

Assessment

Portfolio Have students write a letter to a local newspaper expressing their views about preserving endangered species. **L2**

4 Close

Discussion

Hold a class discussion about ways that individuals can help conserve biodiversity. Discuss programs and projects that can be undertaken by groups and communities to meet the same goal.

Section Assessment

1. Conservation biology focuses on protecting biodiversity. Ecology is the study of interactions in nature.
2. This 1973 law made it illegal to harm organisms on the threatened or endangered species lists.
3. Captive-born animals are expensive to raise and may lose their wild behaviors.
4. Students might describe the California condor being helped by reintroduction programs or elephants being helped by bans on selling ivory.
5. Wildlife areas are often destroyed for economic reasons, so providing economic reasons for their preservation is an important conservation strategy.
6. Hunting and nut collecting are examples of sustainable use. Damage could include building a mall or cutting trees for lumber.

INTERNET
BioLab

Time Allotment 🗂

If conducted at home, the activity will require a 15-minute orientation in class. If computers with Internet access are used in class, the activity will take one to two periods.

Process Skills

acquire information, collect data, compare and contrast, draw a conclusion, think critically, define operationally

PREPARATION

Alternative Materials

- Use current magazines if Internet access is not available.
- Arrange for a guest lecturer, such as a veterinarian. Ask the speaker to discuss the issue and provide specific information for students to record in their data tables.

ANALYZE AND CONCLUDE

1. *Domesticated* means bred for an extended period of time in captivity.
2. Answers will depend on the information students located.

Resource Manager

BioLab and MiniLab Worksheets, p. 25 L2

INTERNET
BioLab

Researching Information on Exotic Pets

What would it be like to own a pet like a snake or a ferret? Sound glamorous? Maybe it's a lot of hard work for the owner and maybe it's not so fair to the pet. Use the Internet as a research tool to locate information on exotic pets. Consider any animal as exotic if it is not commonly domesticated and is not native to your area.

A rhesus monkey

PREPARATION

Problem
How can you use the Internet to gather information on keeping an exotic animal as a pet?

Objectives
In this BioLab, you will:
- **Select** one animal that is considered an exotic pet.

- **Use the Internet** to collect and compare information from other students.
- **Conclude** if the animal you have chosen would or would not make a good pet.

Materials
access to the Internet

Scarlet macaw

PROCEDURE

1. Copy the data table and use it as a guide for the information to be researched.

2. Pick an exotic pet from the following list of choices: hedgehog, snake, ferret, large cat such as tiger or panther, monkey, ape, iguana, tropical bird.
3. Go to the Glencoe Science Web Site **to find links** that will provide you with information for this BioLab. Note: You are not limited to the pet suggestions provided. If a different organism appeals to you, research it instead.
4. Record your findings in the data table.

A hedgehog

PROCEDURE

Teaching Strategies
- Students with home computers and access to the Internet should be encouraged to complete this activity at home. Students lacking access to computers can gain the information needed through articles in current periodicals.
- If necessary, review how to research a topic.

Troubleshooting
Students may need to be directed to narrow their searches or ignore certain web sites.

Data and Conclusions
Student data and conclusions will differ, depending on the animals they researched, the web sites they visited, and other references they used.

Data Table

Category	Response
Exotic pet choice	
Scientific name	
Natural habitat (where found in nature)	
Adult size	
Dietary needs	
Special health problems	
Source of medical care, if needed	
Safety issues for humans	
Size of cage area needed	
Special environmental needs	
Social needs	
Cost of purchase	
Cost of maintaining (monthly estimate)	
Care issues (high/low maintenance)	
Additional information	
Additional sources	

A ferret

Iguanas

ANALYZE AND CONCLUDE

1. **Defining Operationally** What is meant by the term *domesticated?*
2. **Using the Internet** Look at the findings posted by other students. Which of the animals researched would make the best pet? Which would not be a wise choice? Why?
3. **Interpreting Data** What do you consider the most important information gained from your research that:
   a. supports keeping your exotic pet choice?
   b. does not support keeping your exotic pet choice?
4. **Thinking Critically** What positive contribution might be made to the cause of conservation when keeping an exotic pet? Explain.

5. **Thinking Critically** How can keeping exotic pets be a negative influence on conservation biology efforts?
6. **Analyzing Information** What are some reasons why zoos rather than individuals are better able to handle exotic animals?

Sharing Your Data

BIOLOGY *Online* Find this BioLab on the Glencoe Science Web site at **science.glencoe.com.** Post your findings in the data table provided for this activity. Use additional data from other students on the Internet to answer the questions for this BioLab.

5.2 CONSERVATION OF BIODIVERSITY **131**

INTERNET **BioLab**

3. a. Answers may include that exotic pets are interesting, have a good market for resale, attract attention, and are challenging to raise.
   b. Answers may include that these animals are expensive and difficult to feed and care for, including finding veterinarian services.
4. Answers may include that keeping these pets allows for possible breeding, thus helping to save an endangered species.
5. Students may note that buying these pets endangers species even more by encouraging black-market trade.
6. Zoos can better meet the animals' needs for food, shelter, and medical care.

✔ Assessment

Performance Ask volunteers to debate this issue. One group should support the keeping of exotic pets, and the other group should be opposed to the idea. Use the Performance Task Assessment List for Investigating an Issue Controversy in **PASC**, p. 65. **L2**

Sharing Your Data

BIOLOGY *Online* To navigate to the Internet BioLabs, choose the *Biology: The Dynamics of Life* icon at the Glencoe Science Web site. Click on the student icon, then the BioLabs icon. Have students determine whether capture, breeding, and release programs have been successful for certain species of endangered animals. **L2**

Art
Connection

Art
Connection

Wildlife Photography of Art Wolfe

Purpose

Students will learn about a renowned photographer and his unique way of "preserving" nature —in photographs.

Teaching Strategies

■ Obtain some of Wolfe's books from a library and encourage the class to study the photographs. Have students each choose a favorite photograph and explain their choice. Challenge students to classify several photographs with respect to Wolfe's goals while he is on location.

■ Have students search the web to find a current list of endangered species. Have them identify which of Wolfe's subjects are endangered.

Connection to Biology

Wildlife photographs capture interesting phenomena of nature as well as document what might not exist in the future.

Art Wolfe received a degree in fine arts from the University of Washington where he was trained as a painter. He applied his knowledge of art and painting to become one of the world's best wildlife photographers. With more than 1 million images to his credit, Wolfe's photographs of Earth and its inhabitants capture the best nature has to offer.

Although he has captured many an awesome landscape on film, photographer Art Wolfe is probably most well-known for his images of wildlife. These include just about every organism from A to Z. Wolfe, for example, has traveled to Antarctica to photograph playful Adélie penguins as they waddle across the ice sheet. Journeys to the northern hemisphere brought Wolfe into contact with brown bears, mule deer, and wolves. Treks to Africa have allowed the photographer to capture the intricate patterns and symmetry of zebras, both alone and in herds.

Capturing the moment Wolfe generally has four goals in mind when out on a shoot. One of the goals is to get as close as possible to his subject. This, according to the artist, allows him to freeze the instant of contact between himself and his subject. Another of Wolfe's goals is to try to capture an animal in its natural surroundings. Says Wolfe of one of his images of a black bear, ". . . habitat is as important as the animal." Wolfe's third goal—to capture patterns in nature—is perhaps fueled by his knowledge of art and his admiration of the German artist, Martin Escher. A photograph of a herd of zebras taken from above the herd exemplifies this particular goal. His last goal while out on a shoot is to try to capture an animal's behavior. Among Wolfe's photographs illustrating this goal are brown bears sparring in a cold Alaskan river, a Northern gannet meticulously constructing its nest, a snowy owl chick practicing a fierce stare, and mule deer foraging in a grassy field somewhere in Montana.

Snowy owl chick

Preserving nature for posterity Wolfe's work has been seen by wide audiences. His photographs help people appreciate the natural world and want to protect it. In 1998, the North American Nature Photography Association awarded Wolfe its prestigious Outstanding Photographer of the Year Award. That year, Wolfe also received the first Rachel Carson Award presented by the National Audubon Society for his work in calling attention to animals and habitats that are in danger of disappearing forever.

CONNECTION TO BIOLOGY

Why is it important for wildlife to be photographed by Art Wolfe and others?

BIOLOGY Online To find out more about endangered species, visit the Glencoe Science Web site. **science.glencoe.com**

BIOLOGY Online Note Internet addresses that you find useful in the space below for quick reference.

_____
_____
_____
_____
_____

SUMMARY

Section 5.1

Vanishing Species

Main Ideas

- Biodiversity refers to the variety of life in an area.
- The most common measure of biodiversity is the number of species in an area.
- Extinctions occur when the last members of species die.
- Human actions have resulted in habitat loss, fragmentation, and degradation that has accelerated the rate of extinctions.

Vocabulary

acid precipitation (p. 122)
biodiversity (p. 115)
edge effect (p. 122)
endangered species (p. 120)
exotic species (p. 124)
extinction (p. 119)
habitat degradation (p. 122)
habitat fragmentation (p. 121)
ozone layer (p. 122)
threatened species (p. 119)

Section 5.2

Conservation of Biodiversity

Main Ideas

- Conservation biology is the study and implementation of methods to preserve Earth's biodiversity.
- Legal protection of species and habitat preservation have provided effective strategies in conservation biology.
- Larger protected areas generally have greater biodiversity than smaller protected areas.
- Animal reintroduction programs have been more successful when the reintroduced organisms come from the wild rather than from captivity.

Vocabulary

captivity (p. 129)
conservation biology (p. 125)
habitat corridors (p. 127)
reintroduction programs (p. 128)
sustainable use (p. 126)

Main Ideas

Summary statements can be used by students to review the major concepts of the chapter.

Using the Vocabulary

To reinforce chapter vocabulary, use the Content Mastery Booklet and the activities in the Interactive Tutor for Biology: The Dynamics of Life on the Glencoe Science Web site.
science.glencoe.com

 **THE PRINCETON REVIEW** *All Chapter Assessment questions and answers have been validated for accuracy and suitability by The Princeton Review.*

UNDERSTANDING MAIN IDEAS

1. c
2. c
3. d

UNDERSTANDING MAIN IDEAS

1. In a study, aquatic ecologists counted the number of species in equal volumes of water. Where would you expect them to find the smallest number of species?
   a. Lake Victoria, a very large tropical lake in East Africa
   b. the Great Barrier Reef, a coral reef off the coast of Australia
   c. Lake Champlain, a large lake between New York and Vermont
   d. coral reefs in the Red Sea, between Israel and Egypt

2. The water hyacinth is from South America. People brought it to the waterways of Florida where it is growing out of control, killing native water plants, and blocking the waterways. The water hyacinth is a(n) _____ species.
   a. threatened   c. exotic
   b. extinct   d. endangered

3. A protected wildlife area allows local hunters to shoot deer when their population rises above a certain level. This is an example of _____.
   a. habitat degradation   c. habitat loss
   b. habitat fragmentation   d. sustainable use

GLENCOE TECHNOLOGY

 **VIDEOTAPE**
MindJogger Videoquizzes
Chapter 5: *Biological Diversity and Conservation*
Have students work in groups as they play the videoquiz game to review key chapter concepts.

Resource Manager

Chapter Assessment, pp. 25-30
MindJogger Videoquizzes
ExamView® Pro Software
BDOL Interactive CD-ROM, Chapter 5 quiz

4. b
5. d
6. d
7. c
8. b
9. b
10. c
11. conservation biology
12. sustainable use
13. biodiversity, species
14. degradation
15. reintroduction
16. exotic species
17. habitat corridor
18. crack
19. threatened, endangered
20. Reintroduction programs

4. If a researcher found that a higher incidence of eye damage in kangaroos was due to increased ultraviolet radiation hitting the Earth, which pollutant would most likely be suspected as the cause?
   a. sulfur dioxide   **c.** ozone
   b. chloroflourocarbons   **d.** nitrogen dioxide

5. The California condor was extinct in the wild. Later they were bred in captivity and released in their old habitat. This is an example of:
   a. corridors
   b. sustainable use
   c. an exotic species
   d. a reintroduction program

6. The federally funded project of building a telescope in Arizona was stopped because an endangered species was found on the land. This probably occurred due to:
   a. the Convention on International Trade in Endangered Species
   b. reintroduction programs
   c. habitat fragmentation
   d. the U.S. Endangered Species Act

7. Two species of animals, Prezwalski's horse and Pere David's deer, no longer exist in the wild; they exist only in zoos. These species:
   a. would be classified as a threatened species
   b. are making a comeback
   c. exist only in captivity
   d. are extinct

THE PRINCETON REVIEW   **TEST-TAKING TIP**

Stumbling Is Not Falling
Every once in a while you'll hit a question that will completely throw you. It happens. You read the question eight times over, and you still can't understand it. Eliminate, guess, and then move on.

8. A national park has four ways to have roads cross the park. Which method would produce the least habitat fragmentation?
   a.   **b.**   **c.**   **d.**
   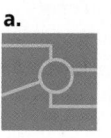 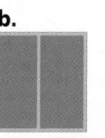 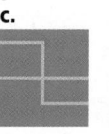 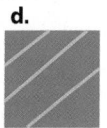

9. DDT was effective in killing insects. However, which organism was endangered most by its use?
   a. passenger pigeon   **c.** elephant
   b. American bald eagle   **d.** sea lamprey

10. The biggest threat to global biodiversity is:
    a. habitat degradation   **c.** habitat loss
    b. habitat fragmentation   **d.** exotic species

11. A branch of biology that seeks to preserve the world's biodiversity is _____ _____.

12. Allowing people to remove resources from a protected area as long as damage is not done to the ecosystem is an example of the philosophy of _____ _____.

13. The variety of life in an area is its _____. The number of _____ in the area usually provides a measure of this.

14. Water pollution is an example of habitat _____.

15. If Venus fly trap plants were planted in their native habitat, this could be part of a(n) _____ program.

16. Rats came to Hawaii on European boats; their populations grew very quickly. The rats are an example of a(n) _____ _____.

17. In Costa Rica, a thin strip of land links two protected wildlife areas. This is an example of a(n) _____ _____.

18. The insecticide DDT caused the American bald eagle and peregrine falcon eggs to _____, killing the chicks.

19. When a population of organisms falls drastically, that species may become a(n) _____ species. If its population falls almost to the point of extinction it is said to be a(n) _____ species.

20. _____ _____ release organisms into an area where the species was once found.

APPLYING MAIN IDEAS

21. Why is habitat loss such a big threat to biodiversity?
22. How do habitat corridors help overcome some problems with habitat fragmentation?
23. Would building a road across a wilderness area, without building on the land, help preserve wild habitats? Explain your answer.

THINKING CRITICALLY

24. **Recognizing Cause and Effect** Using the idea of carrying capacity, design a plan for sustainable use of trout in a midwestern lake by fishers.
25. **Forming a Hypothesis** Suggest reasons why large carnivores have a greater chance of becoming extinct than other organisms. Provide examples.
26. **Concept Mapping** Complete the concept map by using the following vocabulary terms: biodiversity, conservation biology, habitat degradation, extinctions.

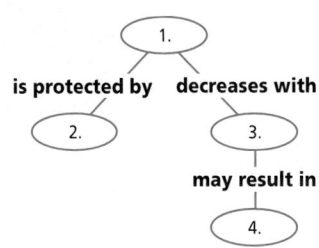

1.

is protected by   **decreases with**

2.   3.

may result in

4.

CD-ROM

For additional review, use the assessment options for this chapter found on the *Biology: The Dynamics of Life Interactive CD-ROM* and on the Glencoe Science Web site.
science.glencoe.com

ASSESSING KNOWLEDGE & SKILLS

Making and Using Graphs The graph below shows extinction rates for birds and mammals since 1600.

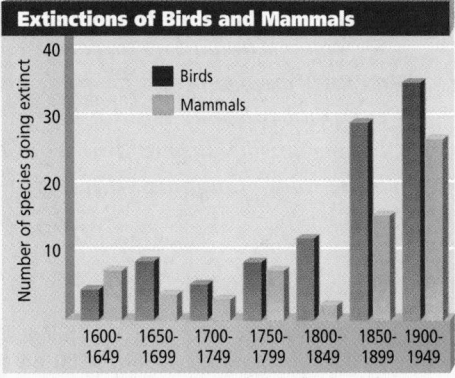

Extinctions of Birds and Mammals

Use the graph to answer the following questions.

1. In what interval were there more extinctions for mammals than for birds?
   a. 1600–1649   c. 1750–1799
   b. 1650–1699   d. 1850–1859
2. In what interval were there the most mammal extinctions?
   a. 1600–1649   c. 1850–1999
   b. 1650–1699   d. 1900–1949
3. Approximately how many birds became extinct in the interval 1650–1699?
   a. 7   c. 15
   b. 10   d. 20
4. Approximately how many birds became extinct during the interval from 1600–1949?
   a. 37   c. 115
   b. 70   d. 300
5. **Predicting** How many species of birds and how many species of mammals do you predict became extinct in the years 1959–1999?

APPLYING MAIN IDEAS

21. Loss of habitat results in less area in which organisms can try to meet their needs. As habitats shrink, they support fewer species.
22. They allow the movement of organisms between areas so they can better meet their needs.
23. No, because the road would cause an edge effect and divide the area into two smaller areas.

THINKING CRITICALLY

24. Any plan that limits the catching of trout is acceptable. Plans include allowing harvesting of a limited number of trout, of certain size trout, or only in certain seasons.
25. Reasonable hypotheses include that carnivores need a greater range to obtain their prey since they are at a higher trophic level, that they depend on the healthy existence of a larger number of organisms, and that their large size makes them more likely to be killed by people.
26. 1. Biodiversity; 2. Conservation biology; 3. Habitat degradation; 4. Extinctions

ASSESSING KNOWLEDGE & SKILLS

1. a
2. d
3. b
4. c
5. Students might predict an increase over the 1900-1949 numbers, perhaps 45-50 species of birds and 40-45 species of mammals.

BIODIGEST

National Science Education Standards
UCP.I, UCP.3, UCP.4, C.3, C.4, C.5, C.6, F.2, F.4, F.5

Prepare

Purpose

This BioDigest can be used as an overview for ecology. If time is limited, you may wish to use this unit summary in place of the chapters in the Ecology unit.

Key Concepts

Students learn about abiotic and biotic factors in ecosystems. They are introduced to trophic levels, population growth, succession, and biodiversity.

1 Focus

Bellringer

Before presenting the lesson, write the term *food chain* on the chalkboard. Then write these terms with connecting arrows: *rose* ⇒ *aphid* ⇒ *ladybug*. Ask students what they think the arrows show. *how one organism consumes or depends on another; the movement of matter and energy*

GLENCOE
TECHNOLOGY

CD-ROM
Biology: The Dynamics of Life

Exploration: *World Biomes*
Disc 1

Video: *Symbiosis*
Disc 1

For a **preview** of the ecology unit, study this BioDigest before you read the chapters. After you have studied the ecology chapters, you can use the BioDigest to **review** the unit.

Ecology

An organism's environment is the source for all its needs and all its threats. Living things depend on their environments for food, water, and shelter. Yet, environments may contain things that can injure or kill organisms, such as diseases or predators. Ecology is the study of organisms in their environments.

This coral reef's survival strategies include methods of obtaining needs and avoiding dangers.

Ecosystems

The relationships among living things and how the nonliving environment affects life are the key aspects of ecology. An ecosystem is composed of all the interactions among organisms and their environment that occur within a defined space.

Abiotic Influences

Around the world there are many types of biomes, such as rain forests and grasslands. The nonliving parts of the environment, called abiotic factors, influence life. For example, temperature and precipitation influence the type of life in a terrestrial biome.

Hot temperatures and little precipitation are abiotic factors in this desert biome.

FOCUS ON ADAPTATIONS

Symbiosis

Relationships formed between organisms are important biotic factors in an environment. Adaptations, which can be physical structures or behaviors, enable organisms to profit from relationships. In symbiosis, the close relationship between two species, at least one species profits. There are three categories of symbiosis that depend on whether the other species profits, suffers, or is unaffected by the relationship.

Bees pollinate flowers, and in return bees obtain nectar.

Cling fish hide in the spines of a sea urchin.

Mutualism In mutualism, both species benefit from their relationship. For example, bees have a mutualistic relationship with flowers. As the bee eats nectar from the flower, pollen becomes attached to the bee. The bee moves to another flower, and some of the pollen from the first flower pollinates the second flower. The bee gets food, and the flower is able to reproduce.

136

Multiple Learning Styles

Look for the following logos for strategies that emphasize different learning modalities.

Kinesthetic Meeting Individual Needs, p. 138; Biology Journal, p. 140

Visual-Spatial Quick Demo, p. 137; Tech Prep, p. 139;

Interpersonal Reinforcement, p. 137

Linguistic Portfolio, p. 138; Concept Development, p. 139; Project, p. 140

In photosynthesis, plants use nonliving materials, including water, carbon dioxide, and light energy, to produce nutrients.

The water cycle, featuring evaporation from lakes and oceans, condensation to produce clouds, and precipitation, provides an understanding of how water flows through an ecosystem. Nitrogen and carbon also cycle through ecosystems. In the carbon cycle, plants produce nutrients from carbon dioxide in the atmosphere. When these nutrients are broken down for energy, carbon dioxide returns to the atmosphere.

Biotic Factors

The ways organisms affect each other are called biotic factors. Within a population of one species, parents nurture their young, and organisms compete for food or mates. Within a community of species, organisms compete for needs, predators kill prey, and diseases spread.

Food for Life

One of the most important ecological characteristics of organisms is how they meet their nutritional needs. Unlike animals that must eat other organisms to obtain their energy, plants usually make their own food using the energy from the sun.

Autotrophs and Heterotrophs

Organisms that make their own food, such as plants and algae, are called autotrophs. Organisms that cannot make their food must consume other organisms. These organisms are called heterotrophs. Heterotrophs that consume only plants are called herbivores. Heterotrophs that consume only animals are called carnivores. Omnivores consume both plants and animals.

Energy Flow

Life on Earth depends on energy from the sun. Plants use this light energy to make food. Animals eat plants or other animals for food.

The path the nutrients and energy take can be shown in a food chain such as:

rose → aphid → lady bug

This shows that the aphid eats the rose and obtains its nutrients and energy. The ladybug eats the aphid and obtains its nutrients and energy. This food chain is a simple representation. Feeding relationships are often more complex and are represented with a food web.

Commensalism In commensalism, one species benefits, while the other species is neither helped nor hurt. A cling fish hiding in the spines of a sea urchin is commensalism. The cling fish hides in the spines of the sea urchin to hide from predators because the sea urchin's sting deters many organisms. The cling fish benefits from the relationship, but the sea urchin is not harmed.

Parasitism Another form of symbiosis is parasitism, which exists when a smaller parasite gets its nutrition from a larger host. The relationship benefits the parasite, and is harmful to the host. Examples of parasitism include mistletoe on a tree and a tapeworm in the intestines of a human.

A tree is parasitized by mistletoe.

Trophic Levels

Nutrients and energy move from autotroph to herbivore to carnivore. Each of these layers is called a trophic level. If a forest area were roped off and three piles created—autotrophs, herbivores, and carnivores, the autotroph pile would be larger than the herbivore pile, which would be larger than the carnivore pile. The mass of the piles indicates the biological mass, or biomass, of the three trophic levels. The mass of autotrophs is usually about ten times the mass of the herbivores, and the mass of the herbivores is about ten times the mass of the carnivores.

In a stable ecosystem, the biomass of plants must be greater than the biomass of herbivores. If the herbivores were more abundant then the plants, all the plants would be consumed and the herbivores would die. Likewise, if carnivores outnumbered herbivores, there would not be enough food for the carnivores.

Populations may be kept below their carrying capacity due to predation.

Carnivores

Herbivores

Producers

VITAL STATISTICS

Vital Statistics
Energy in a 100 x 100 m section of forest:
Producers—24,055,000 kilocalories
Herbivores—2,515,000 kilocalories
Carnivores—235,000 kilocalories

Biomass of a 100m x 100m square section of deciduous forest.

FOCUS ON ADAPTATIONS

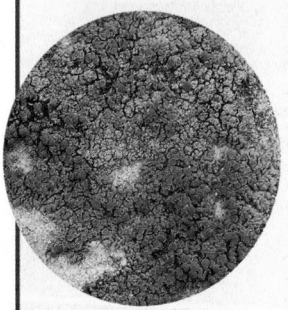

Pioneer Species

The first organisms to colonize new areas are pioneer species. Rocky areas, such as land recently covered by a lava flow, have different pioneer species than areas that already have soil. Rocky areas are usually first colonized by lichens.

On Rock Lichens are made of two organisms, a species of fungus and a species of photosynthetic algae or bacteria. The fungus holds its photosynthetic partner between thick fiber layers, allowing just enough light to penetrate to allow photosynthesis without drying out the lichen. The fungus provides a tough case and the photosynthetic partner supplies nutrition. Through this mutualistic relationship, lichens are able to survive in the harshest of climates such as high on mountains, in cold arctic regions, and in hot deserts.

As lichens grow, they break down rocks and produce soil.

138

Population Size

A population is defined as the number of organisms of one species living in an area. The size of a population is influenced by the environment. For example, a lack of food could limit the number of organisms. Other limiting factors in population growth are water, shelter, and space. As population size increases, competition for some needed items intensifies.

Trees provide shade, which favors the growth of ferns.

Carrying Capacity

The maximum population size an environment can support is called its carrying capacity. If the population size rises above the carrying capacity, organisms die because they cannot meet all their needs.

Exponential Growth

If a population had no predators, and the organisms were able to satisfy their needs, population size would grow quickly. This fast growth is called exponential growth. This exponential growth cannot continue forever; at some point, some need will be scarce and become a limiting factor in the population's growth.

Animals often die due to disease or lack of food if population size exceeds carrying capacity.

In Soil After land is disturbed, such as after a fierce forest fire, pioneer species appear. Most pioneer plants produce many small seeds that are dispersed over wide areas, so that when land is disturbed, the seeds will be there, ready to grow. Another characteristic of pioneer species is they tend to grow and reproduce quickly. When a fresh patch of soil is disturbed, pioneer species sprout, grow quickly, and produce many new seeds to colonize other areas. Finally, pioneer species tend to like much sun and they tend to be robust.

Dandelions are effective pioneer species because they grow fast and disperse many seeds.

139

Revealing Misconceptions

Help students realize that a population refers to the number of organisms of one species in a given area.

Concept Development

 Linguistic Ask students to describe some needs that an organism might not be able to meet if its population rises too high. **L2**

Enrichment

Explain that exponential growth results in a J-shaped curve. Exponential growth that levels off as a population rises to its carrying capacity results in an S-shaped curve.

Revealing Misconceptions

Make sure that students realize that, although humans have increased the rates of extinction for other species, some extinctions are a natural process. For example, dinosaurs were extinct before humans walked on Earth.

GLENCOE TECHNOLOGY

 CD-ROM
Biology: The Dynamics of Life
Animation: *Carrying Capacity*
Disc 1

 **VIDEODISC**
Biology: The Dynamics of Life
Carrying Capacity (Ch. 13)
Disc 1, Side 1, 17 sec.

TECHPREP

Encouraging Pioneers

Visual-Spatial In the schoolyard or a nearby vacant lot, remove all plants from a patch of soil that is about 1 yard (1 m) square. Have students draw pictures of the first plants that begin to grow in the empty space. **L2** **ELL**

BioDigest

3 Assess

Check for Understanding

Ask students to compare and contrast these terms. **L1**
- a. autotrophs—heterotrophs
- b. population size—carrying capacity
- c. pioneer species—succession

Reteach

Remind students that population is the number of organisms of a species in a given area. Have them contrast population with biodiversity—the number of species in a given area.

Extension

Ask students to describe natural conditions that could lead to succession. *volcanoes, floods, fires* **L2**

✓ Assessment

Performance Ask students to research in groups and find nominees for the "toughest" pioneer species. Have them present arguments, in pro-wrestling style posters, about why their nominee is the toughest. **L2**

NATIONAL GEOGRAPHIC

VIDEODISC
STV: Biodiversity
Destroying Diversity
Unit 1, Side 1, 12 min. 15 sec.
Destroying Diversity
(in its entirety)

BioDigest

Succession: Changes in a Community

What would happen if a farmer decided to quit farming a plot of land? No doubt, one of the first things to grow in the land would be those pesky weeds that the farmer had been fighting for years. These weeds are pioneer species, the first organisms to thrive in the new environment.

Changing Species

After the weeds take hold, other plants appear, including annual flowers, grasses, and then bushes. These provide shade, so now when a tree seed germinates, it receives protection from the hot, drying sun. As the tree saplings grow, they block the sun from the plants underneath. The pioneer plants and the annual flowers, without the strong light they need, begin to die back. The

changes that take place are called ecological succession. As species appear, they change the ecosystem and help create conditions suitable for other organisms.

VITAL STATISTICS

Extinction Estimates
Shallow water mussels of the Tennessee River Shoal—44 of original 68 species are extinct.
Plant species extinct in the United States—200
Plant species in the United States in danger of extinction—680
Current loss from tropical rain forests—27,000 species a year

The California condor is one of many species threatened with extinction.

After the extinction of the dinosaurs, mammals, such as this wooly mammoth, thrived.

The most dramatic examples of extinctions are the dinosaurs. When they lived on Earth, they were the dominant animals, and now they are all extinct.

Polacanthus

140

PROJECT

Checking Theories

Linguistic Have students hypothesize why a specific organism became extinct and research whether most scientists agree with their hypotheses. **L3**

BIOLOGY JOURNAL

Preserving Pioneers

Kinesthetic Encourage students to create a leaf press of one or two common weeds in your area. Have them put these into their portfolios and explain why the weeds are good pioneer soil species. **L1** **P**

BioDigest

Biodiversity

In succession, the species living in an ecosystem change over time. Earth's biosphere, the part of our planet that supports life, has experienced changes over time. As new forms of life evolved, other species became extinct.

Some ecosystems took millions of years to form. Millions of different organisms evolved to live in these areas. They are rich in biological diversity, or biodiversity. The most common measure of biodiversity is the number of species that live in an area.

Earth's biodiversity is facing a new crisis due to a new dominant animal—people. As land is converted for human use, organisms lose their habitat and the likelihood of extinctions increases.

Biodiversity adds beauty to our planet. The living world also provides oxygen and food and materials for clothes, drugs, and buildings.

BioDigest

VIDEODISC
STV: Biodiversity
Preserving Diversity
Unit 2, Side 1, 12 min. 10 sec.
Preserving Diversity
(in its entirety)

GLENCOE TECHNOLOGY

VIDEOTAPE
The Secret of Life
Gone Before You Know It: The Biodiversity Crisis

4 Close

Activity

Have students make concept maps of these groups of terms.
a. parasitism, symbiosis, mutualism, commensalism
b. carrying capacity, population size, exponential growth

Resource Manager

Content Mastery, pp. 25-28 **L1**
Reinforcement and Study Guide, pp. 23-24 **L2**

BioDigest Assessment

Understanding Main Ideas

1. A species of fluke called Schistoma lives and obtains nutrients from cells in the human bladder. It produces about 300 to 3 000 eggs per day. This is an example of:
   a. mutualism **c.** commensalism
   b. parasitism **d.** extinction

2. Evaporation, condensation, and precipitation are all part of the _____ cycle.
   a. carbon **c.** nitrogen
   b. water **d.** biomass

3. Which of the following is an abiotic factor in an ecosystem?
   a. amount of light received
   b. number of predators
   c. the biomass of the area
   d. the number of endangered species

4. You observe an animal. It "oinks," eats plants, and eats insects. It is probably:
   a. a producer **c.** an autotroph
   b. a herbivore **d.** an omnivore

5. You are on land that has very warm temperatures and little rain. This biome is most likely a:
   a. deciduous forest **c.** desert
   b. tropical rain forest **d.** grassland

6. You sample an area and find the biomass of the producers is 3 143 kilograms and the biomass of carnivores is 37 kilograms. Which number is the best estimate for the biomass of herbivores in the same area?
   a. 4.1 kilograms **c.** 39 kilograms
   b. 299 kilograms **d.** 32 128 kilograms

7. You played in a meadow as a child. When you visit the area again as an adult, it is a forest. Which of the following accounts for the change?
   a. extinction of species
   b. ecological succession
   c. pyramid of biomass
   d. carrying capacity

Thinking Critically

1. Describe how a specific abiotic factor and a specific biotic factor could affect the life of a deer.

2. Explain the difference between commensalism and mutualism.

3. Is a lichen more of a heterotroph or more of an autotroph? Explain your answer.

4. What eventually happens to a population experiencing exponential growth? Why?

141

The Life of a Cell

Unit Overview

Unit 3 introduces students to basic chemistry, the structure and function of cells, and cell energetics. In Chapter 6, students learn the basic concepts of chemistry that are important in biology. Chapter 7 introduces cell structure and function of organelles. This discussion is expanded upon in Chapter 8 through an in-depth view of cellular transport and the cell cycle. Finally, Chapter 9 acquaints students with the details of energy flow that result from photosynthesis and respiration.

Introducing the Unit

Ask students to describe some of the things they see in the photograph of cells of the retina in terms of color, shape, and other visual characteristics. We appreciate the amazing variety and details of the visual world around us through the work of specialized cells in the retina. Each retinal cell receives a small piece of an image from another cell, such as the pyramidal neuron from another part of the brain, and only by working together can cells piece together all of the information to give an accurate view of the world around us.

Unit 3

The Life of a Cell

A cell is the most basic unit of living organisms. No matter how complex an organism becomes, at its core is a collection of cells. As a cell grows in size, it eventually divides to form two identical cells. In many organisms, cells work together, forming more complex structures.

UNIT CONTENTS

6 **The Chemistry of Life**

7 **A View of the Cell**

8 **Cellular Transport and the Cell Cycle**

9 **Energy in a Cell**

BIODIGEST   The Life of a Cell

UNIT PROJECT

BIOLOGY Online   Use the Glencoe Science Web site for more project activities that are connected to this unit.
science.glencoe.com

142

Develop a Model of a Cell

Have students do one of the projects for this unit as described on the Glencoe Science Web site. As an alternative, students can do one of the projects described on these two pages.

Display

Visual-Spatial Instruct students to collect examples of cells from magazines and science journals. They should make a display that describes the different parts of a cell. **L1 ELL**

Building a Model

Kinesthetic Have student groups design and make a model of a cell that they might find in any part of a plant. They may use any materials they wish. Small motors can be used to show the motion of cell products. **L1 ELL**

Advance Planning

Chapter 6
- Obtain potassium peranganate for the BioLab, mint oil for the Activity on p. 160, and ball-and-stick models for the Quick Demo and Revealing Misconceptions.

Chapter 7
- Purchase prepared slides for Microscopy project, Portfolio activity, and BioLab.
- Purchase animal and plant cell models and an *Elodea* plant.

Chapter 8
- Schedule guest speakers (cell biologist and school nurse).
- Order prepared frog blastula slides for MiniLab and onion root slides for BioLab.

Chapter 9
- Obtain filter paper for the chromatography project.
- Purchase a molecular model of a lipid and dialysis bags for the BioDigest activities.

Unit Projects

Using the Library
Intrapersonal Encourage students to find out how a cell uses energy and make a poster showing the flow of energy through a cell. **L2**

Microscopy
Visual-Spatial Have students examine slides of different types of human cells (muscle, nerve, skin, etc.). They can draw diagrams of each and postulate why each has its unique shape. **L1 ELL**

Final Report
Have student groups compile their findings about cells in reports that could be presented to students at your local middle school. **L3**

Chapter 6 Organizer

Refer to pages 4T-5T of the Teacher Guide for an explanation of the National Science Education Standards correlations.

Section	Objectives	Activities/Features
Section 6.1 **Atoms and Their Interactions** National Science Education Standards UCP.1, UCP.2, UCP.3; A.1, A.2; B.1-3; C.5; D.2; E.1, E.2; F.1; G.1, G.2 (1 session, $\frac{1}{2}$ block)	1. **Relate** the particle structure of an atom to the identity of elements. 2. **Relate** the formation of covalent and ionic chemical bonds to the stability of atoms. 3. **Distinguish** mixtures and solutions. 4. **Define** acids and bases and relate their importance to biological systems.	**Problem-Solving Lab 6-1,** p. 149 **Careers in Biology:** Weed/Pest Control Technician, p. 154 **MiniLab 6-1:** Determine pH, p. 155 **Design Your Own BioLab:** Does temperature affect an enzyme reaction? p. 168
Section 6.2 **Water and Diffusion** National Science Education Standards UCP.2, UCP.3, UCP.4; A.1, A.2; C.5; G.1, G.3 (1 session, $\frac{1}{2}$ block)	5. **Relate** water's unique features to polarity. 6. **Explain** how the process of diffusion occurs and why it is important to cells.	**Problem-Solving Lab 6-2,** p. 158 **MiniLab 6-2:** Examine the Rate of Diffusion, p. 159
Section 6.3 **Life Substances** National Science Education Standards UCP.1, UCP.2; A.1, A.2; B.2, B.3; C.5; E.1, E.2; G.1-3 (3 sessions, 1 block)	7. **Classify** the variety of organic compounds. 8. **Describe** how polymers are formed and broken down in organisms. 9. **Compare** the chemical structures of carbohydrates, lipids, proteins, and nucleic acids, and relate their importance to living things.	**Inside Story:** Action of Enzymes, p. 166 **BioTechnology:** Are fake fats for real? p. 170

Need Materials? Contact Carolina Biological Supply Company at 1-800-334-5551 or at **http://www.carolina.com**

MATERIALS LIST

BioLab
p. 168 timer or clock, 400-mL beaker, kitchen knife, tongs or forceps, potato, 3% hydrogen peroxide, ice, hot plate, waxed paper, thermometer

MiniLabs
p. 155 small beakers (4), lemon juice, household ammonia solution, liquid detergent, shampoo, vinegar, pH paper, pH color chart
p. 159 raw potato, single-edge razor blade, beaker, forceps, metric ruler, timer or clock, potassium permanganate solution

Alternative Lab
p. 164 paper cups (4), gelatin dessert mix, fresh pineapple, canned chunk pineapple, grapes or orange sections, refrigerator, waxed paper, knife, water

Quick Demos
p. 146 iron nails, copper pipe, aluminum foil, coal or charcoal, mercury thermometer
p. 150 beaker, water, table salt, 9-volt battery, electrical wire
p. 157 paper towel, bowl; cooking pot, Bunsen burner, ring stand
p. 163 model of methane molecule

Key to Teaching Strategies

L1 Level 1 activities should be appropriate for students with learning difficulties.

L2 Level 2 activities should be within the ability range of all students.

L3 Level 3 activities are designed for above-average students.

ELL ELL activities should be within the ability range of English Language Learners.

COOP LEARN Cooperative Learning activities are designed for small group work.

P These strategies represent student products that can be placed into a best-work portfolio.

These strategies are useful in a block scheduling format.

The Chemistry of Life

Teacher Classroom Resources

Section	Reproducible Masters	Transparencies
Section 6.1 **Atoms and Their Interactions**	Reinforcement and Study Guide, pp. 25-26 L2 BioLab and MiniLab Worksheets, p. 27 L2 Content Mastery, pp. 29-30, 32 L1	Section Focus Transparency 12 L1 ELL Basic Concepts Transparency 4 L2 ELL Basic Concepts Transparency 5a, 5b L2 ELL
Section 6.2 **Water and Diffusion**	Reinforcement and Study Guide, p. 27 L2 Concept Mapping, p. 6 L3 ELL BioLab and MiniLab Worksheets, p. 28 L2 Content Mastery, pp. 29-32 L1	Section Focus Transparency 13 L1 ELL
Section 6.3 **Life Substances**	Reinforcement and Study Guide, p. 28 L2 Critical Thinking/Problem Solving, p. 6 L3 BioLab and MiniLab Worksheets, pp. 29-30 L2 Laboratory Manual, pp. 39-46 L2 Content Mastery, pp. 29, 31-32 L1 Tech Prep Applications, pp. 9-10 L2	Section Focus Transparency 14 L1 ELL Reteaching Skills Transparency 8 L1 ELL

Assessment Resources

Chapter Assessment, pp. 31-36
MindJogger Videoquizzes
Performance Assessment in the Biology Classroom
Alternate Assessment in the Science Classroom
ExamView® Pro Software 💾
BDOL Interactive CD-ROM, Chapter 6 quiz

Additional Resources

Spanish Resources ELL
English/Spanish Audiocassettes ELL
Cooperative Learning in the Science Classroom COOP LEARN
Lesson Plans/Block Scheduling

NATIONAL GEOGRAPHIC
Teacher's Corner

Index to National Geographic Magazine
The following articles may be used for research relating to this chapter:
"Worlds Within the Atom," by John Boslough, May 1985.

GLENCOE TECHNOLOGY

The following multimedia resources are available from Glencoe.

Biology: The Dynamics of Life
CD-ROM ELL

Animation: *The Covalent Bond*
Animation: *The Ionic Bond*
Animation: *Enzyme Action*
Exploration: *Acid Base Test*
Video: *Properties of Water*

Videodisc Program

Covalent Bonding
Ionic Bonding
Properties of Water

The Infinite Voyage

Unseen Worlds
The Future of the Past

6 The Chemistry of Life

GETTING STARTED DEMO

Direct students' attention to the photographs shown here. Ask students to list the features of living things in their journals. Also ask students to speculate and write in their journals about why early scientists thought a mysterious force controlled chemical changes in organisms. **L2**

Theme Development

Unity within diversity is stressed through the discussion that although living things and nonliving things differ, they are alike in that all are made up of elements. **Energy** should be discussed along with chemical reactions. Be sure to stress that although elements unite with other matter or break apart during chemical reactions, matter is neither created nor destroyed.

0:00 OUT OF TIME?

If time does not permit teaching the entire chapter, use the BioDigest at the end of the unit as an overview.

READING BIOLOGY

Glencoe's *Biology: The Dynamics of Life* contains many resources to assist a student's reading skills. Each chapter contains figures with expanded captions that expand on written material. Word Origins, located along the side of text, expand knowledge of biology vocabulary. Glencoe's Content Mastery Booklet helps develop reading skills while reinforcing content. In addition, use the Interactive Tutor for *Biology: The Dynamics of Life* on the Glencoe Web site to reinforce vocabulary. **science.glencoe.com**

144

What You'll Learn

- You will relate the structure of an atom to how it interacts with other atoms.
- You will explain how water is important to life.
- You will compare the role of carbon compounds in organisms.

Why It's Important

Living organisms are made of simple elements as well as complex carbon compounds. With an understanding of these elements and compounds, you will be able to relate them to how living organisms function.

READING BIOLOGY

Go through the chapter and observe the structure diagrams. Write down any questions you have. As you read the chapter, try to answer the questions from the reading material.

BIOLOGY Online

To find out more about cell chemistry, visit the Glencoe Science Web site. **science.glencoe.com**

Magnification: 32 900×

The gold shown in this scanning-tunneling microscope picture is made of atoms. Atoms make up the colorful frogs and the element gold.

144 THE CHEMISTRY OF LIFE

Multiple Learning Styles

Look for the following logos for strategies that emphasize different learning modalities.

Kinesthetic Meeting Individual Needs, p. 147; Project, p. 151; Reteach, p. 153; Activity, p. 154

Visual-Spatial Portfolio, p. 146; Extension, p. 153; Biology Journal, pp. 157, 158; Reteach, p. 150; Tech Prep, p. 162; Check for Understanding, p. 166

Interpersonal Meeting Individual Needs, p. 161;

Portfolio, p. 163

Intrapersonal Enrichment, pp. 148, 165; Portfolio, p. 153; Extension, p. 160;

Linguistic Biology Journal, pp. 149, 161; Portfolio, p. 159

Logical-Mathematical Quick Demo, p. 157; Activity, p. 160; Portfolio, p. 162; Going Further, p. 169

6.1 Atoms and Their Interactions

What makes a living thing different from a nonliving thing? Are the particles that make up a rock different from those of a frog or a clam? The difference between living and nonliving things may be readily apparent to you. For example, you may have played a game of basketball yesterday, something you would not expect a rock to do. We know, however, that living things have a great deal in common with rocks, CDs, computer chips, and other nonliving objects. Both living and nonliving things are composed of the same basic building blocks called atoms.

All things are similar at the atomic level.

SECTION PREVIEW

Objectives

Relate the particle structure of an atom to the identity of elements.

Relate the formation of covalent and ionic chemical bonds to the stability of atoms.

Distinguish mixtures and solutions.

Define acids and bases and relate their importance to biological systems.

Vocabulary

element
atom
nucleus
isotope
compound
covalent bond
molecule
ion
ionic bond
metabolism
mixture
solution
pH
acid
base

Elements

Everything—whether it is a rock, frog, or flower—is made of substances called elements. Suppose you find a nugget of pure gold. You could grind it into a billion bits of powder and every particle would still be gold. You could treat the gold with every known chemical, but you could never break it down into simpler substances. That's because gold is an element. An **element** is a substance that can't be broken down into simpler chemical substances. On Earth, 90 elements occur naturally.

Natural elements in living things

Of the 90 naturally occurring elements, only about 25 are essential to living organisms. *Table 6.1* lists some elements found in the human body. Notice that only four of the 90 elements—carbon, hydrogen, oxygen, and nitrogen—make up more than 96 percent of the mass of a human. Each element is identified by a one- or two-letter abbreviation called a symbol. For example, the symbol C represents the element carbon, Ca represents calcium, and Cl represents chlorine.

6.1 ATOMS AND THEIR INTERACTIONS **145**

Section 6.1

Prepare

Key Concepts

Students are introduced to the subatomic particles that make up the atoms of elements. In addition, students become acquainted with isotopes in biology. Students also study compounds and bonding. Students compare the properties of acids and bases.

Planning

- Gather items for the Quick Demo and demonstrations.
- Purchase sand, salt, and gelatin for the Enrichment.
- Obtain common household solutions for MiniLab 6-1.
- Purchase beans and toothpicks and gumdrops or marshmallows for the activities.

1 Focus

Bellringer

Before presenting the lesson, display **Section Focus Transparency 12** on the overhead projector and have students answer the accompanying questions. **L1 ELL**

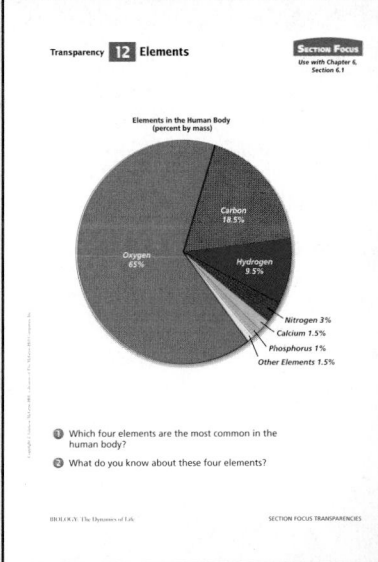

Assessment Planner

Portfolio Assessment
 MiniLab, TWE, pp. 155, 159
 BioLab, TWE, p. 169
Performance Assessment
 MiniLabs, SE, pp. 155, 159
 BioLab, SE, p. 169
 Alternative Lab, TWE, p. 165
Knowledge Assessment
 Assessment, TWE, pp. 147, 160, 162, 167

Problem-Solving Labs, TWE, pp. 149, 158
Alternative Lab, TWE, p. 165
Section Assessments, TWE, pp. 155, 160, 167
Chapter Assessment, TWE, p. 171-173
Skill Assessment
 Assessment, TWE, p. 153

Display various elements using common objects such as iron nails, a piece of copper pipe, a piece of aluminum foil, a piece of coal or charcoal, and a mercury thermometer. Tell students that each material is composed of a different kind of element. Direct students' attention to the appropriate object when discussing each element.

Visual Learning

Visual-Spatial Discuss possible dietary sources of each element listed in Table 6.1. Have students redesign the table to include pictures that show two sources of each element listed. Have students conduct additional research if necessary. **L2** **ELL**

Concept Development

Stress that the elements carbon, hydrogen, nitrogen, oxygen, phosphorus, and sulfur are the main components of living matter. All these elements form molecules through covalent bonding.

GLENCOE TECHNOLOGY

VIDEODISC
The Secret of Life
Composition of Living Organisms

Table 6.1 Elements that make up the human body

Element	Symbol	Percent by mass in human body	Element	Symbol	Percent by mass in human body
Oxygen	O	65.0	Iron	Fe	trace
Carbon	C	18.5	Zinc	Zn	trace
Hydrogen	H	9.5	Copper	Cu	trace
Nitrogen	N	3.3	Iodine	I	trace
Calcium	Ca	1.5	Manganese	Mn	trace
Phosphorus	P	1.0	Boron	B	trace
Potassium	K	0.4	Chromium	Cr	trace
Sulfur	S	0.3	Molybdenum	Mo	trace
Sodium	Na	0.2	Cobalt	Co	trace
Chlorine	Cl	0.2	Selenium	Se	trace
Magnesium	Mg	0.1	Fluorine	F	trace

Trace elements

Notice that some of the elements listed in *Table 6.1*, such as iron and magnesium, are present in living things in very small amounts. Such elements are known as trace elements. They play a vital role in maintaining healthy cells in all organisms, as shown by the examples in *Figure 6.1*. Plants obtain trace elements by absorbing them through their roots. Animals get these important elements from the foods they eat.

Figure 6.1
Trace elements are needed in small amounts to control cell metabolism.

Atoms: The Building Blocks of Elements

Elements, whether they are found in living things or not, are made up of atoms. An **atom** is the smallest particle of an element that has the characteristics of that element.

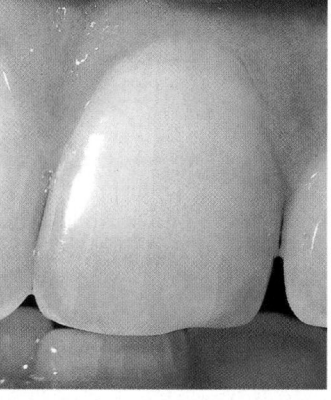

A Mammals use iodine (I), an essential element, to produce substances that affect the rates of growth, development, and chemical activities in the body.

B Plants must have magnesium (Mg) in order to form the green pigment chlorophyll that captures light energy for the production of sugars.

C A trace of fluorine, another essential element, binds with the surface structure of teeth, making them resistant to decay.

Portfolio

Diagramming Atomic Structure

Visual-Spatial Have students make a detailed drawing of an aluminum atom, showing the contents of the nucleus of the atom and the energy levels of the electrons. **L3** **ELL** **P**

Resource Manager

Section Focus Transparency 12 and Master **L1**
Basic Concepts Transparency 4 and Master **L2** **ELL**

Figure 6.2
Electrons move rapidly around atoms, forming electron clouds that have several energy levels.

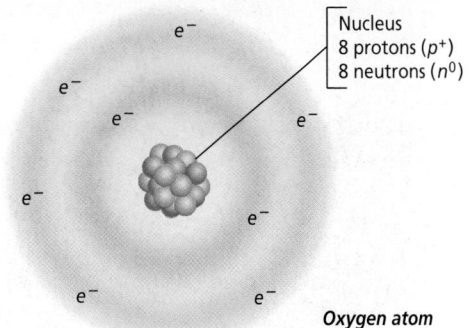

Nucleus
8 protons (p^+)
8 neutrons (n^0)

Oxygen atom

B Oxygen has two electrons in its first energy level and six electrons in the second level.

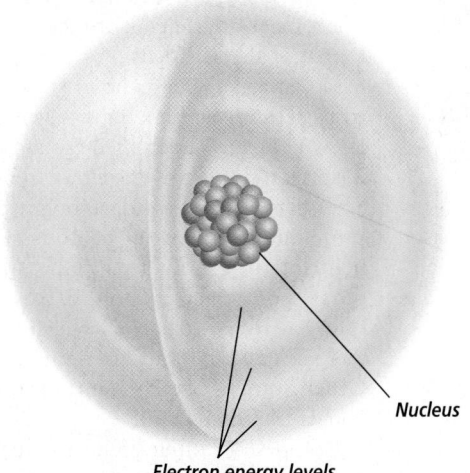

Nucleus

Electron energy levels

A An atom has a nucleus and electrons in cloudlike energy levels.

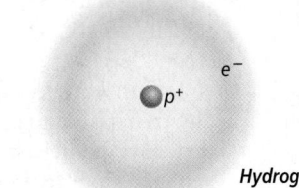

p^+

Hydrogen atom

C Hydrogen, the simplest atom, has just one electron in its first energy level and one proton in its nucleus.

The structure of an atom

Each element has distinct characteristics that result from the structure of the atoms that compose the element. For example, iron differs from aluminum because the structure of iron atoms differs from that of aluminum atoms. Still, all atoms have the same general arrangement. The center of an atom is called the **nucleus** (NEW klee us) (plural, nuclei). It is made of positively charged particles called protons (p^+) and particles that have no charge, called neutrons (n^0). All nuclei are positively charged because of the presence of protons.

Forming a cloud around the nucleus are even smaller, negatively charged particles called electrons (e^-). If you've ever looked at a spinning fan, you've probably noticed that as the fan blades turn, they appear to form a blurry disk that occupies a space around the center of the fan. Similarly, an electron cloud is the space around the atom's nucleus that is occupied by these fast-moving electrons. Although it is impossible to pinpoint the exact location of an electron because it is moving so quickly, the electron cloud is an area where it is most likely to be found.

Electron energy levels

Electrons travel around the nucleus in certain regions known as energy levels, as indicated by *Figure 6.2*. Each energy level has a limited capacity for electrons. Because the first energy level is the smallest, it can hold a maximum of only two electrons. The second level is larger and can hold a maximum of eight electrons. The third level is larger yet and can hold 18 electrons. For example,

Tying to Previous Knowledge

Relate the discussion of the chemistry of life to the characteristics of living things discussed in Chapter 1. Emphasize that the presence of carbon is a characteristic shared by all organisms.

Display

Make a bulletin board display that models the structure of an atom. Label the nucleus, protons, neutrons, and electrons of the model. Refer to the display when discussing atomic structure.

✓ **Assessment**

Knowledge On the chalkboard, prepare a table with the heads: Particle, Location, Charge, and Symbol. Beneath the Particle head, list: Electron, Neutron, and Proton. Have volunteers come to the chalkboard to complete the table. **L2**

GLENCOE
TECHNOLOGY

VIDEODISC
The Secret of Life
Hydrogen/Carbon Atom

Oxygen Atom

Figure 6.3
The properties of an element are determined by the structure of its atoms. As you can see, copper, gold, and carbon have very different properties.

the oxygen atom in *Figure 6.2* has a total of eight electrons. Two electrons fill the first energy level. The remaining six electrons occupy the second level.

Atoms contain equal numbers of electrons and protons; therefore, they have no net charge. The hydrogen (H) atom in *Figure 6.2* has just one electron and one proton. Oxygen (O) has eight electrons and eight protons. *Figure 6.3* shows three other elements whose properties differ because of their atomic structure.

Isotopes of an Element

Atoms of an element sometimes contain different numbers of neutrons in the nucleus. Atoms of the same element that have different numbers of neutrons are called **isotopes** (I suh tohps) of that element. For example, most carbon nuclei contain six neutrons. However, some have seven or eight neutrons. Each of these atoms is an isotope of the element carbon. Scientists refer to isotopes by giving the combined total of protons and neutrons in the nucleus. Thus, the most common carbon atom is referred to as carbon-12 because it has six protons and six neutrons. Other isotopes of carbon include carbon-13 and carbon-14.

Isotopes are often useful to scientists. The nuclei of some isotopes, such as carbon-14, are unstable and tend to break apart. As nuclei break, they give off radiation. These isotopes are said to be radioactive. Because radiation is detectable and

Figure 6.4
Radioactive isotopes are used in medicine to diagnose and/or treat some diseases.

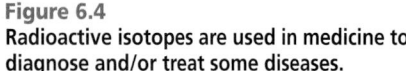

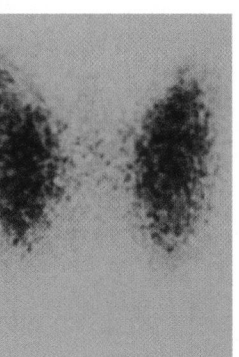
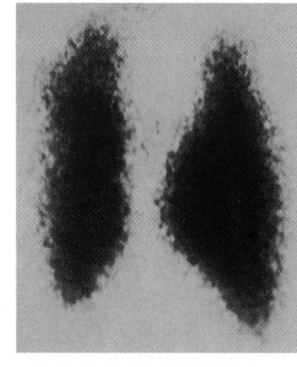

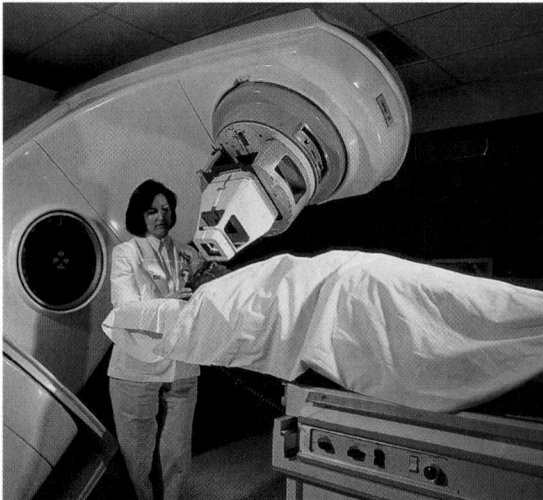

Ⓐ Radioactive iodine (I) introduced into the body is absorbed by the thyroid gland. By detecting the radioactive iodine taken up, the function of the thyroid gland can be measured. The thyroid scan on the left is normal and the scan on the right is overactive.

Ⓑ Radiation given off when radioactive isotopes break apart is deadly to rapidly growing cancer cells. The patient is being treated with radiation from a radioactive isotope of cobalt (Co).

BIOLOGY Online Note Internet addresses that you find useful in the space below for quick reference.

_____

_____

_____

_____

_____

can damage or kill cells, scientists have developed some useful applications for radioactive isotopes, as described in *Figure 6.4*.

Atomic models like those discussed in the *Problem-Solving Lab* on this page help scientists and students visualize the structure of atoms and understand complex intermolecular interactions.

Compounds and Bonding

Water is a substance that everyone is familiar with; however, water is not an element. Rather, water is a type of substance called a compound. A **compound** is a substance that is composed of atoms of two or more different elements that are chemically combined. Water (H_2O) is a compound composed of the elements hydrogen and oxygen. If you pass an electric current through water, it breaks

down into these elements. Just as the combined ingredients in a pizza result in a tasty meal, you can see in *Figure 6.5* that the properties of a compound are different from those of its individual elements.

How covalent bonds form

Most matter is in the form of compounds. But how and why do atoms combine, and what is it that holds the atoms of unlike components together in a compound? Atoms combine with other atoms only when conditions are right, and they do so to become more stable.

Figure 6.5
Table salt is made from the elements sodium (Na) and chlorine (Cl). The flask contains the poisonous, yellow-green chlorine gas. The lump of silver-white metal is the element sodium. The white crystals of table salt no longer resemble either sodium or chlorine.

Problem-Solving Lab 6-1 — Interpreting Scientific Illustrations

What information can be gained from seeing the nucleus of an atom? Looking at a model of an atom's nucleus can reveal certain information about that particular atom. Models may help predict electron number, position of electrons in energy levels, and how isotopes of an element differ from each other.

Analysis

Examine diagrams A and B. Both are models of an atom of beryllium. Only the nucleus of each atom is shown.

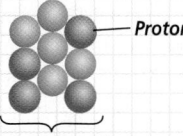 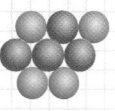

Beryllium nucleus — Proton — Most common form — Beryllium nucleus

Thinking Critically

1. What is the neutron number for A? For B?
2. Which diagram represents an isotope of beryllium? Explain how you were able to tell.
3. How many electrons are present in atoms A and B? Explain how you were able to tell.
4. How many energy levels would be present for A and B? How might the electrons in A and B be distributed in these levels?

Purpose
Students will use diagrams of a nucleus of beryllium to predict electron number, position of electrons in energy shells, and how differences in the number of neutrons affects the element.

Process Skills
think critically, apply concepts, compare and contrast, define operationally, interpret data

Teaching Strategies
■ Review the concept of isotopes with students. Provide other examples for them to analyze.
■ Explain or emphasize the concept that all atoms of the same element have the same electron and proton number. They may differ only in the number of neutrons.
■ Review the charges associated with each atomic particle.

Thinking Critically
1. 5, 4
2. Both A and B have four protons, so they are both isotopes of beryllium.
3. 4. Electron and proton numbers are always the same.
4. 2. Two electrons in each level.

✓ Assessment
Knowledge Provide students with two diagrams of the nucleus for the element fluorine. Have one nucleus with 9 neutrons and one with 10 neutrons. Advise them of the most common form (with 9 neutrons). Ask them to provide the same information for beryllium. Use the Performance Task Assessment List for Making Observations and Inferences in **PASC**, p. 17. L2

BIOLOGY JOURNAL

Life as an Atom
Linguistic Ask students to pretend that they are a particular type of atom. Ask them to write a paragraph or two describing their basic structure. Encourage the students to use their imaginations, while retaining scientific accuracy. L2

Resource Manager

Basic Concepts Transparency 5a and Master L2 ELL

Figure 6.6
Sometimes atoms combine by sharing electrons to form covalent bonds.

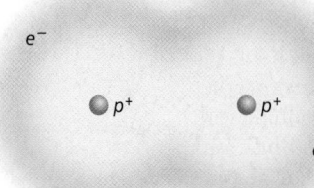 Hydrogen gas (H_2) exists as two hydrogen atoms sharing electrons with each other. The electrons move around the nuclei of both atoms.

Hydrogen molecule

B When two hydrogens share electrons with oxygen, they form covalent bonds to produce a molecule of water.

Water molecule

CD-ROM
View an animation of covalent bonding in the Presentation Builder of the Interactive CD-ROM.

Remember electron energy levels? For most elements, an atom becomes stable when its outermost energy level is full, such as having eight electrons in the second level. An exception is hydrogen, which becomes stable when its first energy level is full (two electrons). How do elements fill the energy levels and become stable? One way is to share electrons with other atoms.

For example, two hydrogen atoms can combine with each other by sharing their electrons, as shown in *Figure 6.6.* As you know, individual atoms of hydrogen contain only one electron. Each atom becomes stable by sharing its electron with the other atom. The two shared electrons move about the first energy level of both atoms. The attraction of the positively charged nuclei for the shared, negatively charged electrons holds the atoms together. When two atoms share electrons, such as hydrogen sharing with oxygen in water, the force that holds them together is called a **covalent bond** (koh VAY lunt). Most compounds in organisms have covalent bonds. Examples include sugars, fats, proteins, and water.

A **molecule** is a group of atoms held together by covalent bonds and having no overall charge. A molecule of water is represented by the chemical formula H_2O. The subscript 2 represents two atoms of hydrogen (H) combined with one atom of oxygen (O). As you will see, many compounds in living things have more complex formulas.

Cultural Diversity

Kenichi Fukui and Chemical Reactions

In the 1950s, Japanese chemist Kenichi Fukui (1918–1998) developed the idea that chemical reactions occur as a result of interactions of the outer-level electrons of one atom or molecule with the outer-level electrons of another atom or molecule. In 1981, Fukui received the Nobel Prize for Chemistry for his investigations of the mechanisms of chemical reactions. Discuss with students the work of Kenichi Fukui toward understanding chemical interactions.

How ionic bonds form

Not all atoms bond with each other by sharing electrons. Sometimes atoms combine with each other by gaining or losing electrons in their outer energy levels. An atom (or group of atoms) that gains or loses electrons has an electrical charge and is called an ion. An **ion** is a charged particle.

A different type of chemical bond holds ions together. The bond formed between a sodium atom (Na) and chlorine atom (Cl) is a good example of this. A sodium atom contains 11 electrons, including one in the third energy level. A chlorine atom has 17 electrons, with the outer level holding seven electrons. When sodium and chlorine combine, the sodium atom loses one electron, and the chlorine atom gains it. Thus, with eight electrons in its outer level, the chlorine ion formed is stable and has a negative charge. Sodium has lost the one electron that was in its third energy level. Thus, the sodium ion is stable and has a positive charge. The attractive force between two ions of opposite charge is known as an **ionic bond.** The bond between sodium and chlorine is an ionic bond, as shown in *Figure 6*.7.

Ionic compounds are less abundant in living things than are covalent molecules, but ions are important in biological processes. For example, sodium and potassium ions are required for transmission of nerve impulses. Calcium ions are necessary for muscles to contract. Plant roots absorb essential minerals in the form of ions.

Chemical Reactions

When chemical reactions occur, bonds between atoms are formed or broken, causing substances to combine and recombine as different molecules. In organisms, chemical reactions occur over and over inside cells. All of the chemical reactions that occur within an organism are referred to as that organism's **metabolism.** These reactions break down and build molecules that are important

Figure 6.7
The positive charge of a sodium ion attracts the negative charge of a chlorine ion, and the elements combine with an ionic bond that forms explosively, as shown in the photograph.

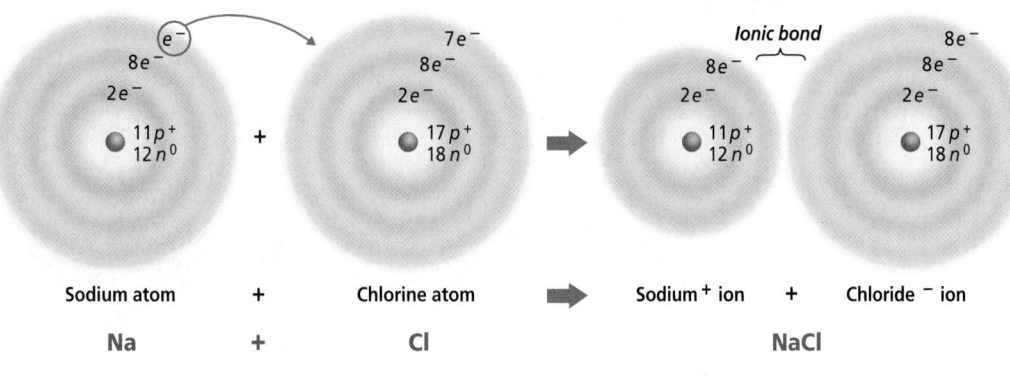

Ionic bond

$8e^-$	$7e^-$	$8e^-$	$8e^-$
$2e^-$	$8e^-$	$8e^-$	$8e^-$
$11p^+$ $12n^0$	$2e^-$	$2e^-$	$2e^-$
	$17p^+$ $18n^0$	$11p^+$ $12n^0$	$17p^+$ $18n^0$

Sodium atom + Chlorine atom → Sodium$^+$ ion + Chloride$^-$ ion

Na + Cl → NaCl

for the functioning of organisms. Scientists represent chemical reactions by writing chemical equations. Chemical equations use symbols and formulas to represent each element or substance.

Writing chemical equations

The events that take place when hydrogen gas combines with oxygen gas are shown in *Figure 6.8.* Substances that undergo chemical reactions, such as hydrogen and oxygen, are called reactants. Substances formed by chemical reactions, such as water, are called products.

It's easy to tell how many molecules are involved in a reaction because the number before each chemical formula indicates the number of molecules of each substance. The subscript numbers in a formula indicate the number of atoms of each element in a molecule of the substance. A molecule of table sugar can be represented by the formula $C_{12}H_{22}O_{11}$. The lack of a number before a formula or under a symbol indicates that only one atom or molecule is present.

Looking at the equation in *Figure 6.8,* you can see that each molecule of hydrogen gas is composed of two atoms of hydrogen.

Likewise, a molecule of oxygen gas is made of two atoms. Perhaps the easiest way to understand chemical equations is to know that atoms are neither created nor destroyed in chemical reactions. They are simply rearranged. Therefore, an equation is written so that the same numbers of atoms of each element appear on both sides of the arrow. In other words, equations must always be written so that they are balanced.

Mixtures and Solutions

When elements combine to form a compound, the elements no longer have their original properties. What happens if substances are just mixed together and do not combine chemically? A **mixture** is a combination of substances in which the individual components retain their own properties. *Figure 6.9* shows a mixture of sand and sugar. If you stirred sand and sugar together, you could still tell the sand from the sugar. Neither component of the mixture would change, nor would they combine chemically. You could easily separate them by adding water to dissolve the sugar and then filtering the mixture to collect the sand.

Figure 6.8
This balanced equation shows two molecules of hydrogen gas reacting with one molecule of oxygen gas to produce two molecules of water.

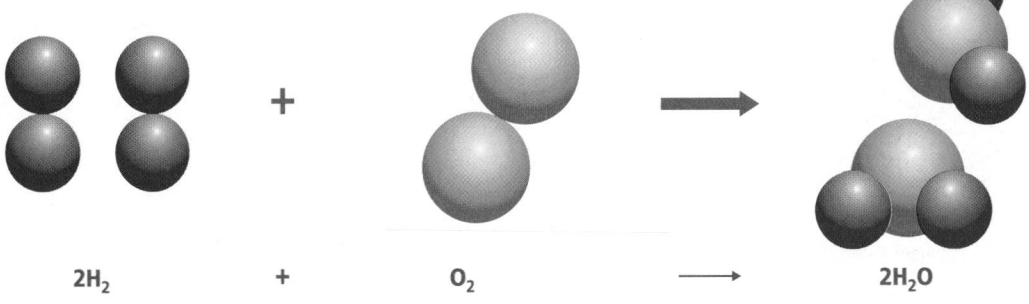

$2H_2$ + O_2 ⟶ $2H_2O$

A **solution** is a mixture in which one or more substances (solutes) are distributed evenly in another substance (solvent). In other words, one substance is dissolved in another and will not settle out of solution. You may remember making Kool-Aid when you were younger. The sugar molecules in Kool-Aid dissolve easily in water to form a solution, as shown in *Figure 6.10.*

Solutions are important in living things. In organisms, many vital substances, such as sugars and mineral ions, are dissolved in water. The more solute that is dissolved in a given amount of solvent, the greater is the solution's concentration (strength). The concentration of a solute is important to organisms. Organisms can't live unless the concentration of dissolved substances stays within a specific, narrow range. Organisms have many mechanisms to keep the

Figure 6.9
This combination of sand and sugar illustrates a mixture. Both components retain their original properties.

concentrations of molecules and ions within this range. For example, the pancreas and other organs in your body produce substances that keep the amount of sugar dissolved in your bloodstream within a critical range.

Figure 6.10
The sugar molecules in the Kool-Aid dissolve in the water, making a solution. Here, sugar is the solute and water is the solvent.

Water molecules

Sugar molecules

Sugar crystal

Using Science Terms
Tell students to imagine they are making a glass of lemonade from a mix. Ask: "Which part of the resulting solution is the solute? Which is the solvent?" *The lemonade mix is the solute because it is the material being dissolved. The water in which the mix is dissolved is the solvent.* **L1**

3 Assess

Check for Understanding
Visual-Spatial Refer students to the periodic table. Provide students with a list of several common elements. For each element, ask students to give the correct symbol and diagram the atom's structure to show its protons, neutrons, and electron energy levels. **L2**

Reteach
Kinesthetic Using gumdrop and toothpick molecules, demonstrate a chemical reaction such as $CH_4 + 2O_2 \Rightarrow CO_2 + 2H_2O$. Stress the conservation of matter as students tear the original molecules apart to build new molecules. **L1** **ELL**

Extension
Visual-Spatial Ask students to make a display to show the chemical formulas of ten common substances. **L3**

✔ Assessment
Skill Give students the following equations to balance. **L2**
$$N_2O_4 \Rightarrow NO_2$$
$$C_3H_8 + O_2 \Rightarrow CO_2 + H_2O$$

✔ Portfolio

Mixtures and Compounds
Intrapersonal Have students construct a table to compare and contrast mixtures and compounds. Encourage students to reread the section entitled "Mixtures and Compounds" to find the information needed to complete their tables. **L2** **P** ⬡

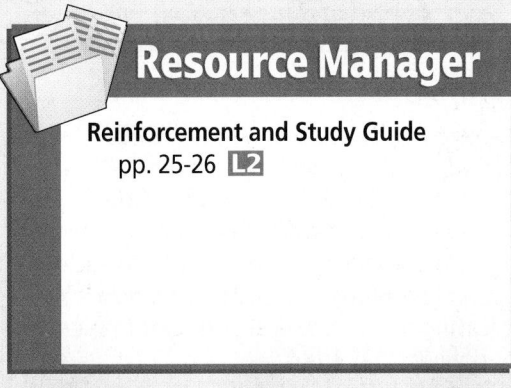

Resource Manager

Reinforcement and Study Guide
pp. 25-26 **L2**

4 Close

Activity

Kinesthetic Have students build a model of an element, such as oxygen, using navy beans for electrons, pinto beans for neutrons, and kidney beans for protons. **L1** **ELL**

Figure 6.11
The pH scale ranges from 0 to 14. Lemon and tomato are acidic (pH < 7). Pure water is neutral (pH 7). Ammonia (pH 11) is basic.

Lemon pH 2
Tomato pH 4
Milk pH 6
Water pH 7
Egg pH 8
Antacid pH 10

Neutral

More acidic

CAREERS IN BIOLOGY

Weed/Pest Control Technician

A career working with chemicals does not always require a Ph.D. Weed and pest control technicians use chemicals to get rid of unwanted weeds, insects, and other pests.

Skills for the Job
After high school, most technicians receive on-the-job training in pest control or take correspondence courses to earn a degree in this field. In many states, you must pass a test to become licensed.

As a technician, you may visit homes, office buildings, restaurants, hotels, and other places where insects, animals, or weeds have become a problem. You will choose the correct chemical and form, such as a spray or gas, to get rid of or prevent infestations of flies, roaches, termites, or other creatures. You will select different chemicals to deal with weeds. You might also set traps to catch rats, mice, moles, or other animals.

To find out more about careers in related fields, visit the Glencoe Science Web site.
science.glencoe.com

Acids and bases

Chemical reactions can occur only when conditions are right. For example, a reaction might depend on temperature, the availability of energy, or a certain concentration of a substance dissolved in solution. Chemical reactions in organisms also depend on the pH of the environment. The **pH** is a measure of how acidic or basic a solution is. A scale with values ranging from 0 to 14 is used to measure pH. Indicators like pH paper describe the pH of substances like those shown in *Figure 6.11*.

Substances with a pH below 7 are acidic. An **acid** is any substance that forms hydrogen ions (H^+) in water. When the compound hydrogen chloride (HCl) is added to water, hydrogen ions (H^+) and chloride ions (Cl^-) are formed. Thus, hydrogen chloride in solution is called hydrochloric acid. This acidic solution contains an abundance of H^+ ions and has a pH below 7.

Substances with a pH above 7 are basic. A **base** is any substance that forms hydroxide ions (OH^-) in water. For example, if sodium hydroxide (NaOH) is dissolved in water, it forms

154 THE CHEMISTRY OF LIFE

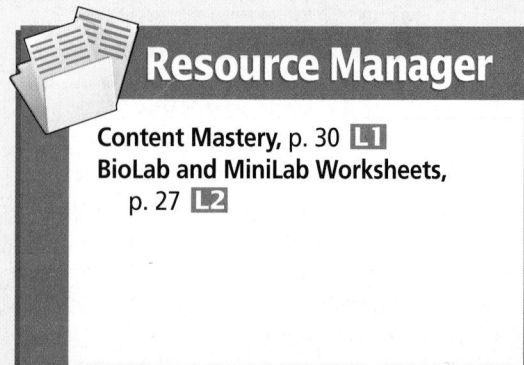

Household
ammonia
pH 11

Drain
cleaner
pH 13

| 11 | 12 | 13 | 14 |

More basic

sodium ions (Na⁺) and hydroxide ions (OH⁻). This basic solution contains an abundance of OH⁻ ions and has a pH above 7.

Acids and bases are important to living systems, but strong acids and bases can also be dangerous. For example, some plants grow well only in acidic soil, whereas others require soil that is basic. Another example, orange juice, is a common acid that can corrode immature teeth if the acid is not later rinsed away. The

---

MiniLab 6-1 — Experimenting

Determine pH The pH of a solution is a measurement of how acidic or basic that solution is. An easy way to measure the pH of a solution is to use pH paper.

Procedure

1. Pour a small amount (about 5 mL) of each of the following into separate clean, small beakers or other small glass containers: lemon juice, household ammonia solution, liquid detergent, shampoo, and vinegar.
2. Dip a fresh strip of pH paper briefly into each solution and remove.
3. Compare the color of the wet paper with the pH color chart; record the pH of each material. **CAUTION:** *Wash your hands after handling lab materials.*

Household Solutions

Analysis

1. Which solutions are acids?
2. Which solutions are bases?
3. What ions in the solution caused the pH paper to change? Which solution contained the highest concentration of hydroxide ions? How do you know?

MiniLab shown here describes how you can investigate several household solutions to determine if they are acids or bases.

Section Assessment

Understanding Main Ideas

1. Describe where the electrons are located in an atom.
2. A nitrogen atom contains seven protons, seven neutrons, and seven electrons. Describe the structure of a nitrogen atom. Use a labeled drawing to help you explain this structure.
3. How does the formation of an ionic bond differ from the formation of a covalent bond?
4. What can you say about the proportion of hydrogen ions and hydroxide ions in a solution that has a pH of 2?

Thinking Critically

5. A fluorine atom has nine electrons. Make an energy level diagram of fluorine. How many electrons would be needed to fill its outer level?

SKILL REVIEW

6. **Interpreting Scientific Illustrations** Study the diagram in *Figure 6.10,* which shows the process of a polar compound dissolving in water. Describe the process step-by-step. Tell what the water molecules are doing and why. Describe what is happening to the sugar molecules and why. Describe the nature of the mixture after the compound dissolves. For more help, refer to *Thinking Critically* in the **Skill Handbook.**

MiniLab 6-1

Purpose
Students will determine the pH of common solutions.

Process Skills
observe and infer, interpret data

Safety Precautions
Have students wear aprons and safety goggles. **CAUTION:** *Both high and low pH solutions can injure the skin and eyes.* Have students wash hands after this activity.

Teaching Strategies
■ Use a pH paper that measures from pH 0 to 14.

Expected Results
The approximate pH of the solutions are: lemon juice, pH 3; household ammonia, pH 11; liquid detergent, pH 10; shampoo, pH 7; and vinegar, pH 3.

Analysis
1. lemon juice and vinegar
2. household ammonia and liquid detergent
3. H⁺ ions and OH⁻ ions. Household ammonia contains the most OH⁻ ions; it had the highest pH.

✓ Assessment

Portfolio Have students make a pH scale in their portfolios and show where each solution falls on the scale. Use the Performance Task Assessment List for Scientific Drawing in **PASC,** p. 55. **L1** **ELL**

Section Assessment

1. Electrons move around the nucleus in regions known as energy levels.
2. The nucleus contains seven protons and seven neutrons. The first energy level contains two electrons. The next energy level contains five electrons.
3. Ionic bonds form as one atom gains electrons from another or gives up electrons to another. Covalent bonds involve sharing of electrons.
4. There are more hydrogen than hydroxide ions in an acidic solution of pH 2.
5. The diagram should show two energy levels containing two and seven electrons, respectively. One electron is needed to fill the outer level.
6. Polar water molecules attract and surround the polar sugar molecules. Eventually, this attraction pulls the molecules of the sugar crystal apart. A solution of a molecular substance consists of water molecules and solute molecules.

Prepare

Key Concepts

Students will develop an understanding of the properties of water that make it an excellent solvent and a necessary biological component. Students will also study the process of diffusion.

Planning

■ Buy a roll of paper towels for the Quick Demo.
■ Obtain marbles and plastic container for Reteach.
■ Buy potatoes for MiniLab 6-2.

1 Focus

Bellringer

Before presenting the lesson, display **Section Focus Transparency 13** on the overhead projector and have students answer the accompanying questions. **L1** **ELL**

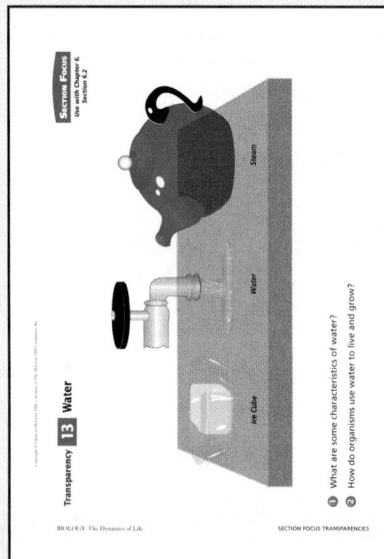

SECTION PREVIEW

Objectives
Relate water's unique features to polarity.
Explain how the process of diffusion occurs and why it is important to cells.

Vocabulary
polar molecule
hydrogen bond
diffusion
dynamic equilibrium

Section
6.2 Water and Diffusion

Most of us take water for granted. We turn on the kitchen faucet at home to get a drink and expect water to come out of the faucet. When we hike in a forest, we expect water to be flowing in the streams and filling the lakes. Most of the time, we don't think about how important water is to our life and the life of other organisms on Earth.

Water is vital to the living world.

Water and Its Importance

Water is perhaps the most important compound in living organisms. Most life processes can occur only when molecules and ions are free to move and collide with one another. This condition exists when they are dissolved in water. Water also serves as a means of material transportation in organisms. For example, blood and plant sap, which are mostly water, transport materials in animals and plants. In fact, water makes up 70 to 95 percent of most organisms.

Water is polar

Sometimes, when atoms form covalent bonds, they do not share the electrons equally. The water molecule pictured in *Figure 6.12* shows that the shared electrons are attracted by the oxygen atom more strongly than by the hydrogen atoms. As a result, the electrons spend more time near the oxygen atom than they do near the hydrogen atoms.

When atoms in a covalent bond do not share the electrons equally, they form a polar molecule. A **polar molecule** is a molecule with an unequal distribution of charge; that is, each molecule has a positive end and a negative end. Water is an example of a polar molecule. Polar water molecules attract each other as well as ions and other polar molecules. Because of this attraction, water can dissolve many ionic compounds, such as salt, and many other polar molecules, such as sugar.

Water molecules also attract other water molecules. The positively

---

Resource Manager

Section Focus Transparency 13 and Master **L1** **ELL**
Concept Mapping, p. 6 **L3** **ELL**

MEETING INDIVIDUAL NEEDS

Gifted

Intrapersonal Some bacteria can live on the underside of snow. Ask gifted students to research how these bacteria keep from freezing. **L3**

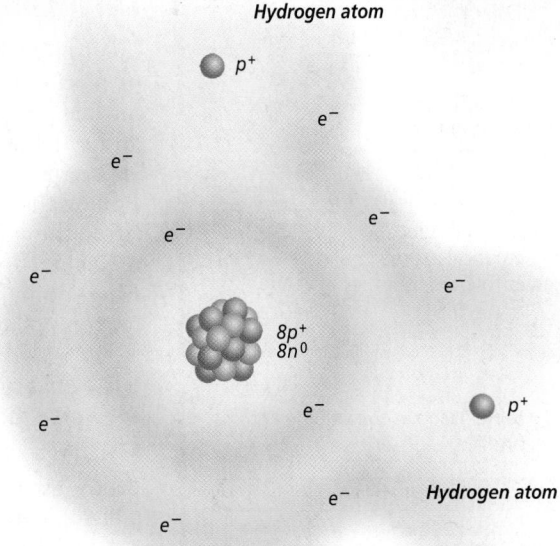

Figure 6.12
Because of water's bent shape, the protruding oxygen end of the molecule has a slight negative charge, and the ends with protruding hydrogen atoms have a slight positive charge.

Hydrogen atom

p^+

e^-

e^-

e^-

e^-

$8p^+$
$8n^0$

e^-

e^-

e^-

p^+

e^-

Hydrogen atom

e^-

Oxygen atom

Positively-
charged end

+

+

–

Negatively-
charged end

charged hydrogen atoms of one water molecule attract the negatively charged oxygen atoms of another water molecule. This attraction of opposite charges between hydrogen and oxygen forms a weak bond called a **hydrogen bond.** Hydrogen bonds are important to organisms because they help hold many large molecules, such as proteins, together.

Also because of its polarity, water has the unique property of being able to creep up thin tubes. Plants in particular take advantage of this property, called capillary action, to get water from the ground. Capillary action and the tension on the water's surface, which is also a result of polarity, play major roles in getting water from the soil to the tops of even the tallest trees.

Water resists temperature changes

Water resists changes in temperature. Therefore, water requires more heat to increase its temperature than do most other common substances. Likewise, water loses a lot of heat when it cools. In fact, water is like an insulator that helps maintain a steady environment when conditions may fluctuate. Because cells exist in an aqueous environment, this property of water is extremely important to cellular functions as it helps cells maintain an optimum environment.

Water expands when it freezes

Water is one of the few substances that expands when it freezes. Because of this property, ice is less dense than liquid water and floats as it forms in a pond. Use the *Problem-Solving Lab* on the next page to investigate this property. Water expands as it freezes inside the cracks of rocks. As it expands, it often breaks apart the rocks. Over long time periods, this process forms soil.

The properties of water make it an excellent vehicle for carrying substances in living systems. Another way to move substances is by diffusion.

Quick Demo

Logical-Mathematical Ask students to explain the water movement as you dip the edge of a paper towel in a bowl of water. **L2** **ELL**

Concept Development
The solubility of oxygen in water increases as water temperature decreases. Ask students to explain why having water at a temperature of 2°C under a sheet of ice would be important for living organisms in terms of oxygen solubility. *The water at 2°C will dissolve more oxygen than the warmer layers of water beneath this layer. As the surface of the lake is sealed by ice, the temperature of the water becomes important to the ability of the water to provide enough oxygen to the aquatic organisms during winter.*

GLENCOE
TECHNOLOGY

VIDEODISC
Biology: The Dynamics of Life
Properties of Water (Ch. 19)
Disc 1, Side 1,
45 sec.

BIOLOGY JOURNAL

"Water, water everywhere..."
 *Visual-Spatial* Have students cut out pictures from magazines that illustrate uses of water. Ask them to prepare a display of their pictures. Students may want to include industrial uses of water, uses of water by organisms, or environmental uses of water. **L1** **ELL**

GLENCOE TECHNOLOGY

CD-ROM
Biology: The Dynamics of Life
Video: *Properties of Water*
Disc 1

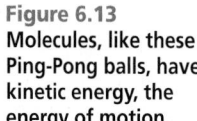

Purpose

Students will calculate the densities for water and ice and will correlate this information with the fact that water expands as it freezes.

Process Skills

compare and contrast, draw a conclusion, measure in SI, think critically, use numbers, recognize cause and effect

Teaching Strategies

■ Review the equation for calculating density. Provide some examples for students to use as practice.
■ Review the use of units such as cm³ and mL if necessary.
■ Remind students that cm³ is the same as cubic centimeters or cc.
■ Review the procedure for arriving at the proper units to express density.

Thinking Critically

1. The density of water is 1 g/cm³; density of ice is 0.9 g/cm³. Ice is less compact; mass in example is the same, but volume for ice is greater. A lower density indicates a greater volume with the same mass as water.

2. farther apart. This pattern accounts for increased volume.

3. Water expands as it freezes and eventually breaks glass.

4. Formation of ice crystals and the expansion within delicate cells and tissues could damage a living organism.

Problem-Solving Lab 6-2 — Using Numbers

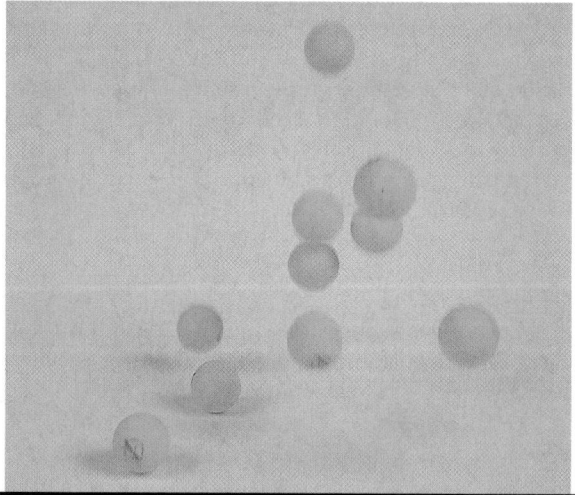

Why does ice float? Most liquids contract when frozen. Water is different; it expands. Freezing changes the density of water to that of ice, which allows ice to float. Density refers to compactness and is often described as the mass of a substance per unit of volume. A mathematical expression of density would read as follows:

$$\text{Density} = \frac{\text{Mass}}{\text{Volume}}$$

Analysis

Examine the following table. It shows the volume and mass for a sample of water and ice.

Data Table		
Source of sample	Volume (cm³)	Mass (g)
Water	126	126
Ice	140	126

Thinking Critically

1. How does the density of ice compare with the density of water? Use specific values and proper units expressing density in your answer. Which of the two, ice or water, is less compact? Explain your answer.

2. Are the molecules of water moving closer together toward one another or farther apart as water freezes? Explain your answer.

3. Explain why a glass bottle filled with water will shatter if placed in a freezer.

4. Explain why ice forming within a living organism may result in its death.

Diffusion

All objects in motion have energy of motion called kinetic energy. A moving particle of matter moves in a straight line until it collides with another particle, much like the Ping-Pong balls shown in *Figure 6.13*. After the collision, both particles rebound. Particles of matter, like the Ping-Pong balls, are in constant motion, colliding with each other.

Early observations: Brownian motion

In 1827, Scottish scientist Robert Brown used a microscope to observe pollen grains suspended in water. He noticed that the grains moved constantly in little jerks, as if being struck by invisible objects. This motion, he thought, was the result of a life force hidden within the pollen grains. However, when he repeated his experiment using dye particles, which are nonliving, he saw the same erratic motion. This motion is now called Brownian motion. Brown had no explanation for the motion he saw, but today we know that Brown was observing evidence of the random motion of molecules. This was the invisible "life" that was moving the tiny visible particles. The random

Figure 6.13
Molecules, like these Ping-Pong balls, have kinetic energy, the energy of motion.

movement that Brown observed is characteristic of gas, liquid, and some solid molecules.

The process of diffusion

Molecules of different substances that are in constant motion have an effect on each other. For example, if you layer pure corn syrup on top of corn syrup colored with food coloring in a beaker as illustrated in *Figure 6.14,* over time you will observe the colored corn syrup mixing with the pure corn syrup. This mixture is the result of the random movement of corn syrup molecules. **Diffusion** is the net movement of particles from an area of higher concentration to an area of lower concentration. Diffusion results because of the random movement of particles (Brownian motion). The corn syrup in *Figure 6.14* will begin to diffuse in hours but will take months to mix completely.

Diffusion is a slow process because it relies on the random molecular motion of atoms. Three key factors, concentration, temperature, and pressure, affect the rate of diffusion. The concentration of the substances involved is the primary controlling factor. The more concentrated the substances, the more rapidly diffusion occurs. For example, loose sugar placed in water will diffuse more rapidly than will a more concentrated cube of sugar. Two external factors, temperature and pressure, can speed the process of diffusion. An increase in temperature or agitation will cause more rapid molecular movement and will speed diffusion. Similarly, increasing pressure will accelerate molecular movement and, therefore, diffusion. With common materials, you can use the *MiniLab* shown here to learn more about diffusion in a cell.

Figure 6.14
The random movement of particles of corn syrup will cause the colored sample to diffuse into the uncolored sample.

MiniLab 6-2   Applying Concepts

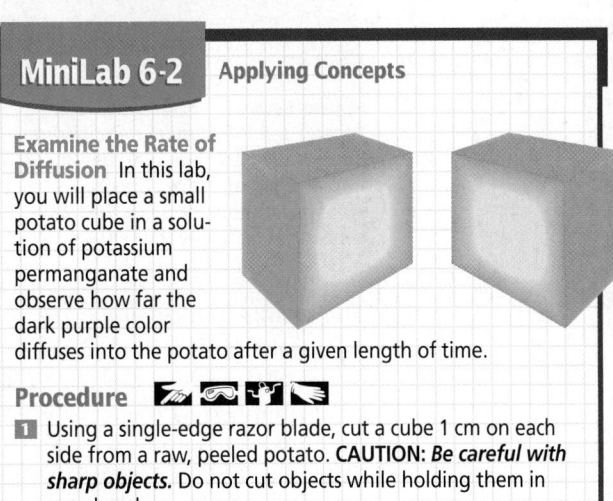

Examine the Rate of Diffusion In this lab, you will place a small potato cube in a solution of potassium permanganate and observe how far the dark purple color diffuses into the potato after a given length of time.

Procedure

1. Using a single-edge razor blade, cut a cube 1 cm on each side from a raw, peeled potato. **CAUTION:** *Be careful with sharp objects.* Do not cut objects while holding them in your hand.
2. Place the cube in a cup or beaker containing the purple solution. The solution should cover the cube. Note and record the time. Let the cube stand in the solution for between 10 and 30 minutes.
3. Using forceps, remove the cube from the solution and note the time. Cut the cube in half.
4. Measure, in millimeters, how far the purple solution has diffused, and divide this number by the time you allowed your potato to remain in the solution. This is the diffusion rate.

Analysis

1. How far did the purple solution diffuse?
2. What was the rate of diffusion per minute?

MiniLab 6-2

Purpose
Students will determine the rate of diffusion of a solution.

Process Skills
measure in SI, collect and organize data, interpret data, experiment, and analyze

Teaching Strategies
■ Regarding safety, caution students that the solution can be caustic. If they get some on their hands, they should wash immediately. To keep the solution off their hands, students should set the cube on waxed paper or foil when they remove it from the solution and hold the cube with the forceps as it is cut. Also remind students to cut away from their body.

Expected Results
The color will diffuse only a few millimeters into the cube, the exact distance depending upon the amount of time it is in the solution.

Analysis
1. Answers will depend on the amount of time the cube is in the solution.
2. The rate will be in tenths to hundredths of millimeters per minute.

✓ Assessment

Portfolio Have students write a report of the MiniLab for their portfolios. Use the Performance Task Assessment List for Lab Report in **PASC**, p. 47. **L2**

✓ Portfolio

The Importance of Water

Linguistic Have students write an essay that explains why water is important to them. Ask them to think about their families' daily use of water. Challenge them to think about what uses are especially important in view of the fact that they might someday be asked to cut their water use in half. **L2 P**

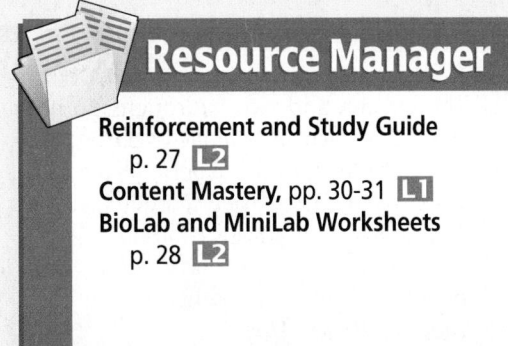

Resource Manager

Reinforcement and Study Guide
    p. 27 **L2**
Content Mastery, pp. 30-31 **L1**
BioLab and MiniLab Worksheets
    p. 28 **L2**

Check for Understanding

Quiz students orally about the importance of water to living organisms and test their understanding of each of the properties of water. **L2**

Reteach

Visual-Spatial In a clear container with a lid, place 10–15 marbles of one color. On top of those marbles, place an equal number of different colored marbles. Ask students to predict what will occur if you shake the container continuously. *Like diffusion, the marbles will eventually disperse among each other, reaching an equilibrium of mixed color.* **L1 ELL**

Extension

Intrapersonal Have students research how rapidly molecules can actually move over a particular distance. **L3**

✔ Assessment

Knowledge Have students list the properties of water and give an example of how each property is useful in a living organism. **L2**

4 Close

Activity

Logical-Mathematical Place a watch glass containing mint oil on the front desk. Have students discuss how the molecules of mint are reaching them. *Besides diffusion, which is a slow process, air currents are also carrying the molecules to their noses.* **L2**

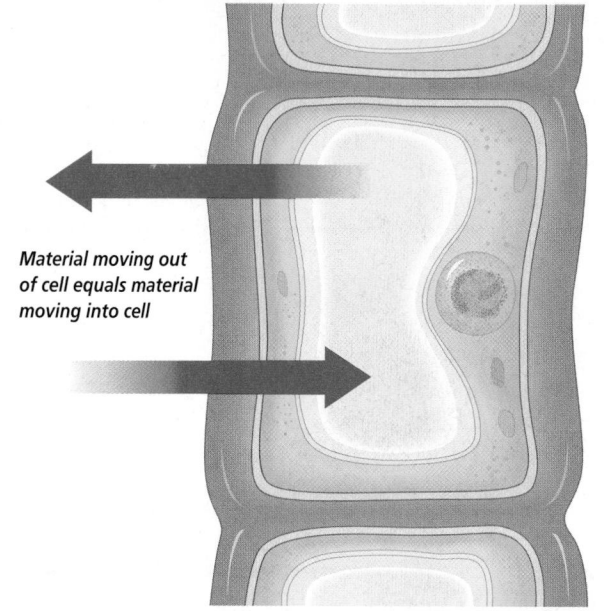

Material moving out of cell equals material moving into cell

Figure 6.15
When a cell is in dynamic equilibrium with its environment, materials move into and out of the cell at equal rates. As a result, there is no net change in concentration inside the cell.

The results of diffusion

As the colored corn syrup continues to diffuse into the pure corn syrup, the two will become evenly distributed eventually. After this point, the atoms continue to move randomly and collide with one another; however, no further change in concentration will occur. This condition, in which there is continuous movement but no overall concentration change, is called **dynamic equilibrium.** *Figure 6.15* illustrates dynamic equilibrium in a cell.

Diffusion in living systems

Most substances in and around a cell are in water solutions where the ions and molecules of the solute are distributed evenly among water molecules, like the Kool-Aid and water example. The difference in concentration of a substance across space is called a concentration gradient. Because ions and molecules diffuse from an area of higher concentration to an area of lower concentration, they are said to move with the gradient. If no other processes interfere, diffusion will continue until there is no concentration gradient. At this point, dynamic equilibrium occurs. Diffusion is one of the methods by which cells move substances in and out of the cell.

Diffusion in biological systems is also evident outside of the cell and can involve substances other than molecules in an aqueous environment. For example, oxygen (a gas) diffuses into the capillaries of the lungs because there is a greater concentration of oxygen in the air sacs of the lungs than in the capillaries.

Section Assessment

Understanding Main Ideas
1. Explain why water is a polar molecule.
2. How does a hydrogen bond compare to a covalent bond?
3. What property of water explains why it can travel to the tops of trees?
4. What is the eventual result of diffusion? Describe concentration prior to and at this point.

Thinking Critically
5. Explain why water dissolves so many different substances.

SKILL REVIEW
6. **Inferring** If a substance is known to enter a cell by diffusion, what effect would raising the temperature have on the cell? For more help, refer to *Thinking Critically* in the **Skill Handbook.**

Section Assessment

1. The oxygen and two hydrogens do not share the electrons equally (electrons are more often near the oxygen). As a result, the oxygen is negatively charged and the hydrogens are positively charged.
2. Hydrogen bonds are very weak compared with covalent bonds.
3. capillary action
4. The particles will reach dynamic equilibrium.
5. Because water molecules are polar and attract other charged particles, water easily dissolves many substances.
6. An increase in temperature causes an increase in kinetic energy and the rate of diffusion of the substance into the cell.

6.3 Life Substances

Did you ever hear the saying, *"You are what you eat"?* It's at least partially true because the compounds that form the cells and tissues of your body are produced from similar compounds in the foods you eat. Common to most of these foods and to most substances in organisms is the element carbon. The first carbon compounds that scientists studied came from organisms and were called organic compounds.

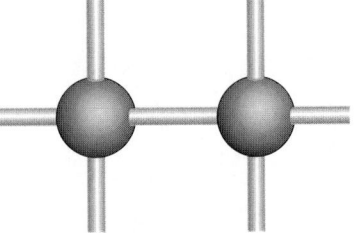

Carbon defines living organisms.

SECTION PREVIEW

Objectives

Classify the variety of organic compounds.

Describe how polymers are formed and broken down in organisms.

Compare the chemical structures of carbohydrates, lipids, proteins, and nucleic acids, and relate their importance to living things.

Vocabulary

isomer
polymer
carbohydrate
lipid
protein
amino acid
peptide bond
enzyme
nucleic acid
nucleotide

Role of Carbon in Organisms

A carbon atom has four electrons available for bonding in its outer energy level. In order to become stable, a carbon atom forms four covalent bonds that fill its outer energy level. Look at the illustration showing carbon atoms and bond types in *Figure 6.16.* Carbon can bond with other carbon atoms, as well as with many other elements. When each atom shares two electrons, a double bond is formed. A double bond is represented by two bars between carbon atoms. When each shares three electrons, a triple bond is formed. Triple bonds are represented by three bars drawn between carbon atoms.

Single bond

Double bond

Triple bond

Figure 6.16
When two carbon atoms bond, they share one, two, or three electrons each and form a covalent bond.

Section 6.3

Prepare

Key Concepts

Students will examine the classes of carbon compounds present in organisms. The structural and functional aspects of carbohydrates, lipids, proteins, and nucleic acids will be studied.

Planning

- Collect potatoes, knives, hydrogen peroxide and waxed paper for the BioLab.
- Purchase grocery and paper items for the Alternative Lab.
- Make flash cards for the Check for Understanding.

---

1 Focus

Bellringer

Before presenting the lesson, display **Section Focus Transparency 14** on the overhead projector and have students answer the accompanying questions. **L1** **ELL**

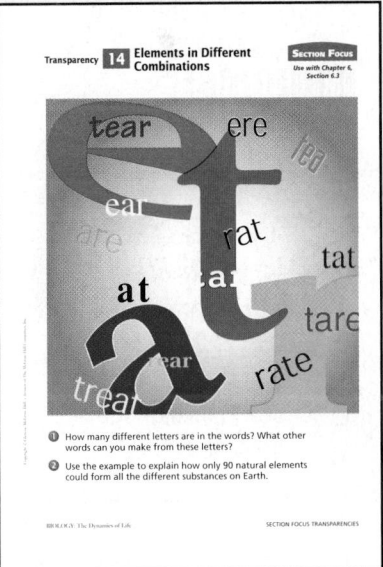

---

2 Teach

Visual Learning

Figure 6.17 Direct students' attention to the illustration. Ask them what compound in addition to sucrose is formed when glucose and fructose combine. *water*

VIDEODISC
The Secret of Life
Carbon Bonds

Resource Manager

Section Focus Transparency 14 and Master **L1** **ELL**
Critical Thinking/Problem Solving, p. 6 **L3**

When carbon atoms bond to each other, they can form straight chains, branched chains, or rings. In addition, these chains and rings can have almost any number of carbon atoms and can include atoms of other elements as well. This property makes a huge number of carbon structures possible. In addition, compounds with the same simple formula often differ in structure. Compounds that have the same simple formula but different three-dimensional structures are called **isomers** (I suh murz). The glucose and fructose molecules shown in *Figure 6.17* have the same simple formula, $C_6H_{12}O_6$, but different structures.

Molecular chains

Carbon compounds also vary greatly in size. Some compounds contain just one or two carbon atoms, whereas others contain tens, hundreds, or even thousands of carbon atoms. These large molecules are called macromolecules. Proteins are examples of macromolecules in organisms. Cells build macromolecules by bonding small molecules together to form chains called polymers. A **polymer** is a large molecule formed when many smaller molecules bond together.

Condensation is the chemical reaction by which polymers are formed. In condensation, the small molecules that are bonded together to make a polymer have an −H and an −OH group that can be removed to form H−O−H, a water molecule. The subunits become bonded by a covalent bond as shown in *Figure 6.18*.

Hydrolysis is a method by which polymers can be broken apart. Hydrogen ions and hydroxide ions from water attach to the bonds between the subunits that make up the polymer, thus breaking the polymer as shown in *Figure 6.18*.

The structure of carbohydrates

You may have heard of runners eating large quantities of spaghetti or other starchy foods the day before a race. This practice is called "carbohydrate loading." It works because carbohydrates are used by cells to store and release energy. A **carbohydrate** is an organic compound composed of

Figure 6.17
The different arrangement of hydrogen and oxygen atoms around each carbon atom gives glucose and fructose molecules different chemical properties. When glucose and fructose combine, they form the disaccharide sucrose, also known as table sugar.

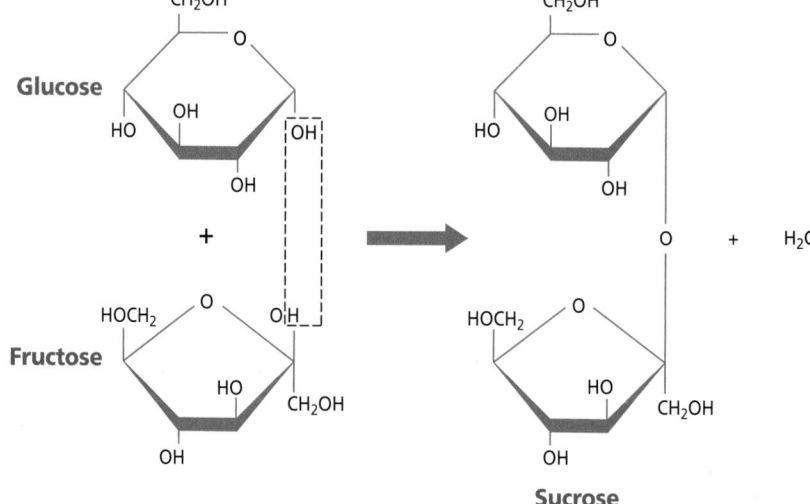

Figure 6.18
Condensation and Hydrolysis

A Polymers are formed by condensation and can be broken by hydrolysis, the reaction that occurs when water is introduced to a polymer.

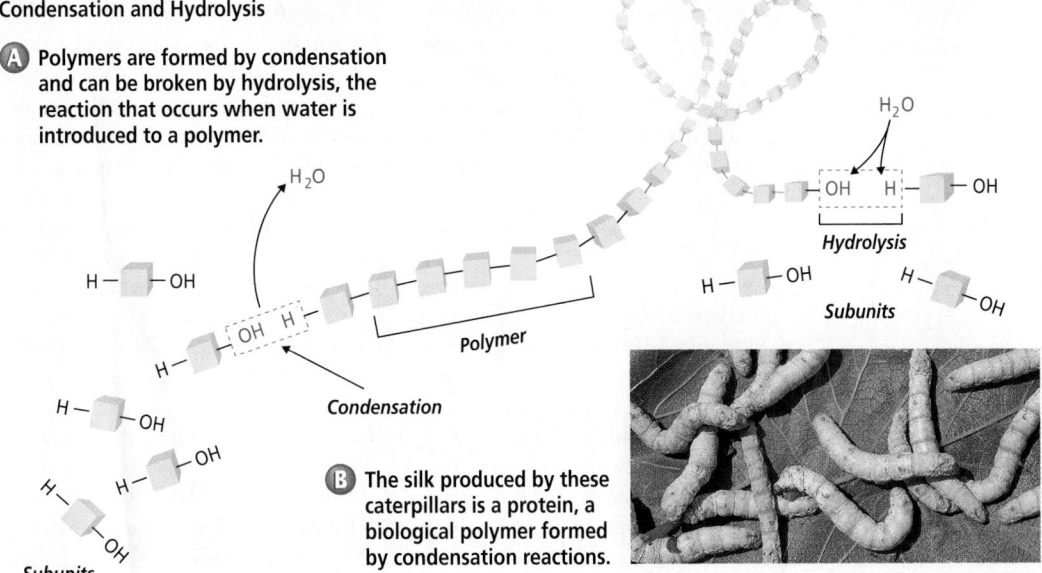

H_2O

Hydrolysis

Subunits

Polymer

Condensation

Subunits

B The silk produced by these caterpillars is a protein, a biological polymer formed by condensation reactions.

carbon, hydrogen, and oxygen with a ratio of about two hydrogen atoms and one oxygen atom for every carbon atom.

The simplest type of carbohydrate is a simple sugar called a monosaccharide (mahn uh SAK uh ride). Common examples are the isomers glucose and fructose. Two monosaccharide molecules can link together to form a disaccharide, a two-sugar carbohydrate. When glucose and fructose combine by a condensation reaction, a molecule of sucrose, known as table sugar, is formed.

The largest carbohydrate molecules are polysaccharides, polymers composed of many monosaccharide subunits. The starch, glycogen, and cellulose pictured in *Figure 6.19* are examples of polysaccharides. Starch consists of highly branched chains of glucose units and is used as food storage by plants in food reservoirs such as seeds and bulbs. Mammals store food in the liver in the form of glycogen, a glucose polymer similar to

Figure 6.19
Look at the structural differences among the polysaccharides starch, glycogen, and cellulose. Notice that all three are polymers of glucose.

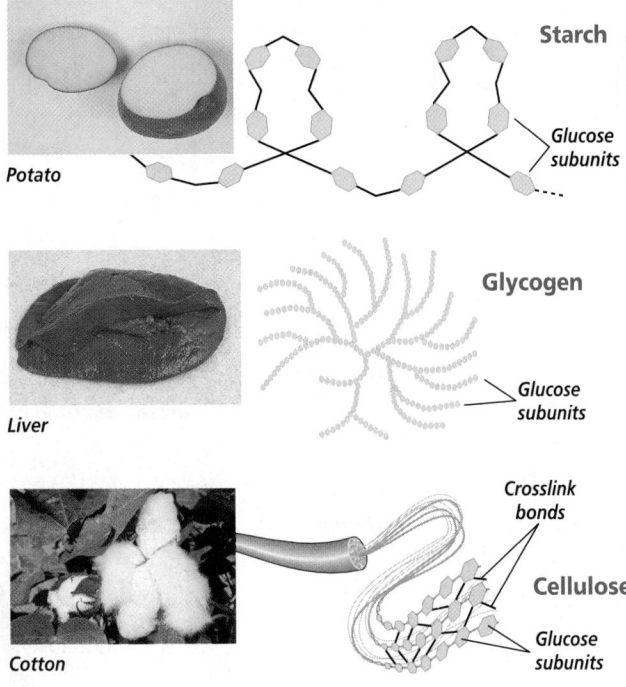

Potato

Starch

Glucose subunits

Liver

Glycogen

Glucose subunits

Cotton

Crosslink bonds

Cellulose

Glucose subunits

6.3 LIFE SUBSTANCES **163**

Students often think molecules are flat, two-dimensional structures because the formulas in books are flat. Show students structural models of various molecules so they can observe the three-dimensional appearance of molecules.

Visual Learning

Figure 6.20 Ask students to use Figure 6.20 to answer the following questions. (a) What compound serves as the backbone for lipid molecules? *glycerol* (b) How do the bonds in saturated fats differ from those in unsaturated fats? *Saturated fats have single bonds; unsaturated fats have double bonds.*

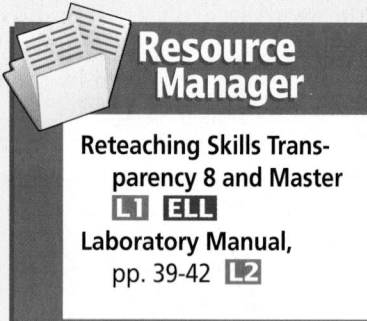

Resource Manager

Reteaching Skills Transparency 8 and Master
L1 **ELL**
Laboratory Manual, pp. 39–42 **L2**

---

starch but more highly branched. Cellulose is another glucose polymer that forms the cell walls of plants and gives plants structural support. Cellulose is also made of long chains of glucose units hooked together in arrangements somewhat like a chain-link fence.

The structure of lipids

If you've ever tried to lose weight, you may have wished that lipids (fats) never existed. Lipids, however, are extremely important for the proper functioning of organisms. **Lipids** are organic compounds that have a large proportion (much greater than 2 to 1) of C–H bonds and less oxygen than carbohydrates. For example, a lipid found in beef fat has the formula $C_{57}H_{110}O_6$.

Lipids are commonly called fats and oils. They are insoluble in water because their molecules are nonpolar. Recall that a nonpolar molecule is one in which there is no net electrical charge and, therefore, lipids are not attracted by water molecules. Cells use lipids for energy storage, insulation, and protective coatings. In fact, lipids are the major components of the membranes that surround all living cells. Lipids are also used in food preparation as discussed in the *BioTechnology* feature. The most common type of lipid, shown in *Figure 6.20*, consists of three fatty acids bound to a molecule of glycerol.

The structure of proteins

Proteins are essential to all life. They provide structure for tissues and organs and carry out cell metabolism. A **protein** is a large, complex polymer

Figure 6.20
Glycerol is a three-carbon molecule that serves as a backbone for the lipid molecule. Attached to the glycerol are three fatty acids.

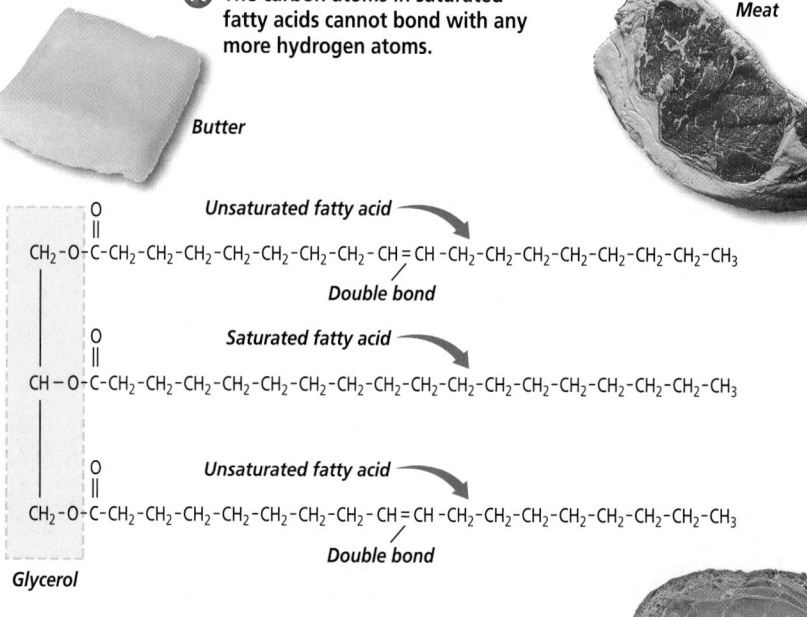

Ⓐ The carbon atoms in saturated fatty acids cannot bond with any more hydrogen atoms.

Meat

Butter

Unsaturated fatty acid
Double bond

Saturated fatty acid

Unsaturated fatty acid
Double bond

Glycerol

Ⓑ The carbon atoms in unsaturated fatty acids can bond with more hydrogen atoms.

Peanut butter

Alternative Lab

What fruits contain enzymes that act on protein?

Purpose
Students will investigate the effect of the enzyme bromelin on gelatin.

Safety Precautions
Remind students to use caution with the knife.

Preparation
One pineapple should be enough for 20 students working in groups of four. A 6-ounce box of gelatin makes enough for 2 groups. Make gelatin and pour 100 mL into each paper cup. Use hot/cold cups large enough to hold 200 mL.

Materials
paper cups (4) with gelatin, fresh pineapple, knife, waxed paper, canned chunk pineapple, refrigerator, grapes or orange sections

Procedure
Give students the following directions.
1. Number the cups 1 to 4.
2. Select 3 chunks of canned pineapple. Cut 3 chunks of fresh pineapple the same size as the canned chunks. Cut 3

H Hydrogen atom
|
Amino group NH₂ — C — COOH Carboxyl group
|
R Variable group

a

$$NH_2 - \underset{\underset{R}{|}}{C} - \underset{\underset{O}{\|}}{C} \xrightarrow{\text{Peptide bond}} \underset{\underset{H}{|}}{N} - \underset{\underset{R}{|}}{C} - COOH$$

b

composed of carbon, hydrogen, oxygen, nitrogen, and usually sulfur. The basic building blocks of proteins are called **amino acids,** shown in *Figure 6.21a.* There are 20 common amino acids. These 20 building blocks, in various combinations, make literally thousands of proteins. Therefore, proteins come in a large variety of shapes and sizes. In fact, proteins vary more in structure than any other class of organic molecules.

Amino acids are linked together when an –H from one amino acid and an –OH group from another amino acid are removed to form a water molecule. The covalent bond formed between the amino acids, like the bond labeled in *Figure 6.21b,* is called a **peptide bond.** The number and order of amino acids in protein chains determine the kind of protein.

Figure 6.21
Each amino acid contains a central carbon atom to which are attached a carboxyl group, a hydrogen atom, an amino group (–NH₂), and a group (–R) that makes each amino acid different (a). Amino acids are linked together by peptide bands (b).

Many proteins consist of two or more amino acid chains that are held together by hydrogen bonds.

Proteins are the building blocks of many structural components of organisms, as illustrated in *Figure 6.22.* Proteins are also important in the contracting of muscle tissue, transporting oxygen in the bloodstream, providing immunity, regulating other proteins, and carrying out chemical reactions.

Enzymes are important proteins found in living things. An **enzyme** is a protein that changes the rate of a chemical reaction. In some cases,

Figure 6.22
Proteins make up much of the structure of organisms, such as hair, fingernails, horns, and hoofs.

Enrichment

🔲 *Intrapersonal* Have students research the differences in the structures of starch, glycogen, and cellulose to determine how there can be more than one polymer of glucose. **L3**

Using an Analogy

To increase understanding of how only 20 amino acids can create such a variety of proteins, relate the 20 naturally occurring amino acids to the letters of the alphabet. Elicit from students why the letters of the alphabet can be used to create such a large number of words. Relate this phenomenon to how a similar number of amino acids can create so many different proteins.

The BioLab at the end of the chapter can be used at this point in the lesson.

GLENCOE TECHNOLOGY

CD-ROM
Biology: The Dynamics of Life
Animation: *Enzyme Action*
Disc 1

---

grapes in half, or choose three orange slices. Set these aside on waxed paper.
3. Add the following to each cup. Make sure the fruits are submerged.
Cup 1—nothing
Cup 2—canned pineapple chunks
Cup 3—fresh pineapple chunks
Cup 4—grape halves or orange slices
4. Set cups in the refrigerator. Check the cups at the end of the period.

Expected Results
Students should observe that the gelatin to which the fresh pineapple was added remained liquefied.

Analysis
1. What was the purpose of cup 1? *Control*
2. Gelatin is a protein. Bromelin is a protein-digesting enzyme. What happened to the bromelin in the canned pineapple? *It was destroyed by heat during the canning process.*

✓ Assessment
Knowledge Ask students this question: Based on this activity, what can you conclude about which fruits have enzymes that act on protein? Use the Performance Task Assessment List for Analyzing the Data in **PASC**, p. 27. **L2**

INSIDE STORY

Purpose 🔲

Students will study the way enzymes function in a reaction.

Teaching Strategy

■ Ask students to make an analogy of a lock and key to how enzymes function. Students should also describe where the analogy fails. Even though the lock and key fit together specifically, the lock does not change shape when the key fits in the keyhole. **L2**

Visual Learning

Visual-Spatial Show a computerized video sequence of lysozyme activity in a cell.

Critical Thinking

The enzyme is not changed by the reaction. Once it releases the substrate, the enzyme can bind to another substrate. This process can be repeated over and over.

---

3 Assess

Check for Understanding

Visual-Spatial Use flash cards containing the names of monomers on one side and the corresponding polymer on the other side. First, show students the polymer name (e.g., protein) and then have them respond with the appropriate monomer name (amino acid), and vice versa. **L1 ELL**

Resource Manager

Reinforcement and Study Guide, p. 28 **L2**
Content Mastery, pp. 31, 29-32 **L1**
Laboratory Manual, pp. 43-46 **L2**

INSIDE STORY

Action of Enzymes

An enzyme enables molecules, called substrates, to undergo a chemical change to form new substances, called products. The enzyme works due to an area on its surface that fits the shape of the substrate, called an active site. When the substrate fits the active site, it causes the enzyme to alter its shape slightly as shown below.

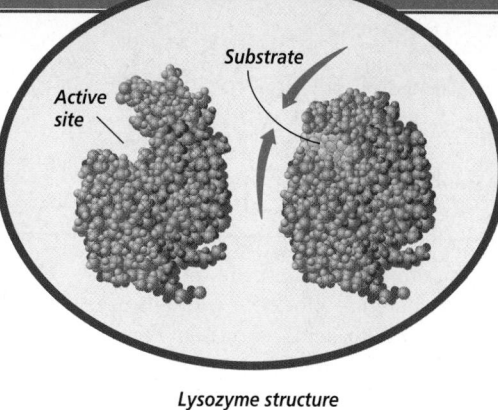

Lysozyme structure

Critical Thinking *How can an enzyme participate over and over in chemical reactions?*

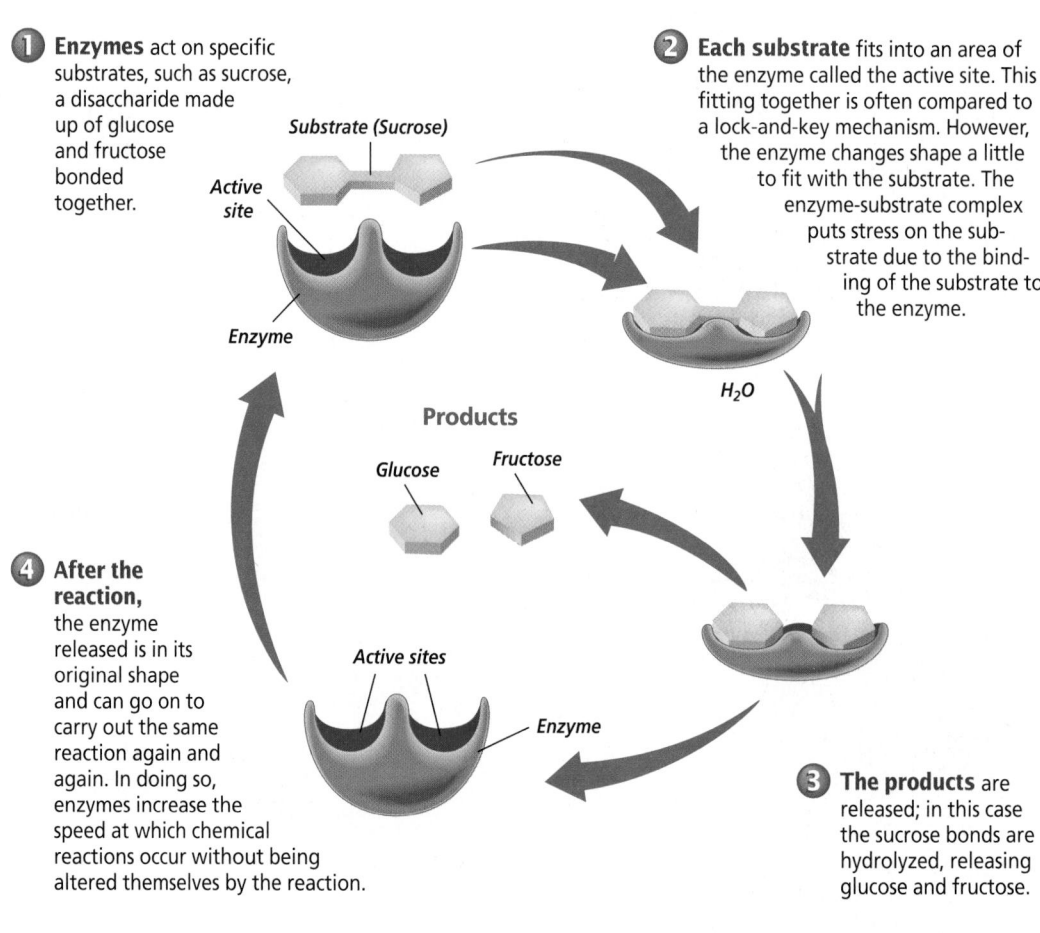

① **Enzymes** act on specific substrates, such as sucrose, a disaccharide made up of glucose and fructose bonded together.

Substrate (Sucrose)

Active site

Enzyme

② **Each substrate** fits into an area of the enzyme called the active site. This fitting together is often compared to a lock-and-key mechanism. However, the enzyme changes shape a little to fit with the substrate. The enzyme-substrate complex puts stress on the substrate due to the binding of the substrate to the enzyme.

H_2O

Products

Glucose Fructose

④ **After the reaction**, the enzyme released is in its original shape and can go on to carry out the same reaction again and again. In doing so, enzymes increase the speed at which chemical reactions occur without being altered themselves by the reaction.

Active sites

Enzyme

③ **The products** are released; in this case the sucrose bonds are hydrolyzed, releasing glucose and fructose.

166 THE CHEMISTRY OF LIFE

GLENCOE TECHNOLOGY

VIDEODISC
The Infinite Voyage
The Future of the Past
Winterthur Museum: New Cleaning Techniques of Old Paintings (Ch. 5)

6 min.

VIDEODISC
The Secret of Life
Structure of DNA

enzymes increase the speed of reactions that would otherwise occur so slowly you might think they wouldn't occur at all.

Enzymes are involved in nearly all metabolic processes. They speed the reactions in digestion of food. Enzymes also affect synthesis of molecules, and storage and release of energy. How do enzymes act like a lock and key to facilitate chemical reactions within a cell? Read the *Inside Story* to find out. The *BioLab* at the end of this chapter also experiments with enzymes.

The structure of nucleic acids

Nucleic acids are another important type of organic compound that is necessary for life. A **nucleic acid** (noo KLAY ihk) is a complex macromolecule that stores cellular information in the form of a code. Nucleic acids are polymers made of smaller subunits called **nucleotides.**

Nucleotides consist of carbon, hydrogen, oxygen, nitrogen, and phosphorus atoms arranged in three groups—a base, a simple sugar, and a phosphate group—as shown in *Figure 6.23*. You have probably heard of the nucleic acid DNA, which stands for deoxyribonucleic

Phosphate **Ribose sugar** **Nitrogen base**

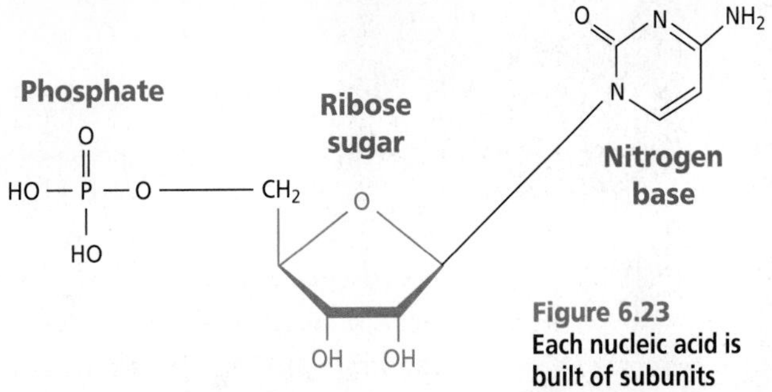

Figure 6.23
Each nucleic acid is built of subunits called nucleotides that are formed from a sugar molecule bonded to a phosphate group and a nitrogen base.

acid. DNA is the master copy of an organism's information code. The information coded in DNA contains the instructions used to form all of an organism's enzymes and structural proteins. Thus, DNA forms the genetic code that determines how an organism looks and acts. DNA's instructions are passed on every time a cell divides and from one generation of an organism to the next.

Another important nucleic acid is RNA, which stands for ribonucleic acid. RNA is a nucleic acid that forms a copy of DNA for use in making proteins. The chemical differences between RNA and DNA are minor but important. A later chapter discusses how DNA and RNA work together to produce proteins.

CD-ROM
View an animation of enzyme action in the Presentation Builder of the Interactive CD-ROM.

Reteach
The concept of large polymers being composed of repeating units of monomers can be reinforced by having students list items that are composed of smaller units, such as beads making up a necklace, chain links, jigsaw puzzle pieces, or letters making up words. **L1 ELL**

Extension
Linguistic Encourage above-level students to read *The Double Helix* by James Watson (Atheneum, 1968), which tells the story of the discovery of the DNA structure. **L3**

✓ Assessment
Knowledge Prepare a handout showing structural formulas for lipids, proteins, carbohydrates, and nucleic acids. Ask students to identify the type of organic compound shown in each diagram. **L2**

4 Close

Discussion
Ask students to explain the differences between saturated and unsaturated fats. *Saturated fats are composed of lipids containing fatty acids with only single bonds. Unsaturated fats are composed of fatty acid chains of carbon with double bonds.* **L2**

Section Assessment

Understanding Main Ideas
1. List three important functions of lipids in living organisms.
2. Describe the process by which polymers in living things are formed from smaller molecules.
3. How does a monosaccharide differ from a disaccharide?

Thinking Critically
4. Enzymes are proteins that facilitate chemical reactions. Based on your knowledge of enzymes, what might the result be if

one particular enzyme malfunctioned or was not present?

SKILL REVIEW

5. **Making and Using Tables** Make a table comparing polysaccharides, lipids, proteins, and nucleic acids. List these four types of biological substances in the first column. In the next two columns, list the subunits that make up each substance and the functions of each in organisms. In the last column, provide some examples of each from the chapter. For more help, refer to *Organizing Information* in the **Skill Handbook.**

Section Assessment

1. long-term energy storage, insulation, protective coatings
2. Polymers form when one monomer loses an H+ ion and another loses an OH− to form water. A covalent bond forms between the monomers.
3. A disaccharide is made of two simple sugars called monosaccharides.
4. The chemical reaction would proceed extremely slowly.
5. Subunits and functions: polysaccharides, monosaccharides—for energy storage and structural components; lipids, glycerol, and fatty acids—for long-term energy storage; proteins, amino acids—structure and enzymes;

nucleic acids, nucleotides—store information in cells. Examples: polysaccharides—starch, glycogen, and cellulose; lipids—animal fats and vegetable oils; proteins—muscle proteins, immunity proteins, enzymes; nucleic acids—DNA and RNA.

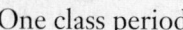

DESIGN YOUR OWN BioLab

Time Allotment

One class period

Process Skills

form a hypothesis, design an experiment, interpret data, recognize cause and effect

Safety Precautions

Students should wear aprons and safety goggles. Remind students to use caution with heat sources and handle glassware with tongs.

PREPARATION

- Obtain potatoes, knives, hydrogen peroxide, and waxed paper.

Alternative Materials

Pieces of raw liver can be used instead of potato.

Possible Hypotheses

- If temperatures are very high or very low, the enzymes will be deactivated.
- If the temperature is raised, the speed at which the enzyme will work will increase.

Resource Manager

BioLab and MiniLab Worksheets, p. 29 **L2**

168

DESIGN YOUR OWN BioLab

Does temperature affect an enzyme reaction?

The compound hydrogen peroxide, H_2O_2, is a by-product of metabolic reactions in most living things. However, hydrogen peroxide is damaging to delicate molecules inside cells. As a result, nearly all organisms contain the enzyme peroxidase, which breaks down H_2O_2 as it is formed. Potatoes are one source of peroxidase. Peroxidase speeds up the breakdown of hydrogen peroxide into water and gaseous oxygen. This reaction can be detected by observing the oxygen bubbles generated.

PREPARATION

Problem

Does the enzyme peroxidase work in cold temperatures? Does peroxidase work better at higher temperatures? Does peroxidase work after being frozen or boiled?

Hypotheses

Make a hypothesis regarding how you think temperature will affect the rate at which the enzyme peroxidase breaks down hydrogen peroxide. Consider both low and high temperatures.

Objectives

In this BioLab, you will:
- **Observe** the activity of an enzyme.
- **Compare** the activity of the enzyme at various temperatures.

Possible Materials

clock or timer
400-mL beaker
kitchen knife
tongs or large forceps
5-mm thick potato slices
3% hydrogen peroxide

ice
hot plate
waxed paper
thermometer

Safety Precautions

Be sure to wash your hands before and after handling the lab materials. Always wear goggles in the lab.

Skill Handbook

Use the **Skill Handbook** if you need additional help with this lab.

PLAN THE EXPERIMENT

Teaching Strategies

- Discuss the factors that might affect the rate of a reaction controlled by an enzyme.
- Students who cool the potato to low temperatures should be sure to run the test while the potato is still cool.
- Allow groups to discuss how their results differed when different experimental procedures were used.

Possible Procedures

- Students may place a piece of potato on ice for 5 minutes, boil a second piece for 5 minutes, and allow a third piece of potato to sit at room temperature for 5 minutes. Each potato will then be tested for enzyme activity.

PLAN THE EXPERIMENT

1. Decide on a way to test your group's hypothesis. Keep the available materials in mind.
2. When testing the activity of the enzyme at a certain temperature, consider the length of time it will take for the potato to reach that temperature, and how the temperature will be measured.
3. To test for peroxidase activity, add 1 drop of hydrogen peroxide to the potato slice and observe what happens.
4. When heating a thin potato slice, first place it in a small amount of water in a beaker. Then heat the beaker slowly so that the temperature of the water and the temperature of the slice are always the same. Try to make observations at several temperatures between 10°C and 100°C.

Check the Plan

Discuss the following points with other groups to decide on the final procedure for your experiment.

1. What data will you collect? How will you record them?
2. What factors should be controlled?
3. What temperatures will you test?
4. How will you achieve those temperatures?
5. *Make sure your teacher has approved your experimental plan before you proceed further.*
6. Carry out your experiment. **CAUTION:** *Be careful with chemicals and heat. Wash hands after lab.*

ANALYZE AND CONCLUDE

1. **Checking Your Hypothesis** Do your data support or reject your hypothesis? Explain your results.
2. **Analyzing Data** At what temperature did peroxidase work best?
3. **Identifying Variables** What factors did you need to control in your tests?
4. **Recognizing Cause and Effect** If you've ever used hydrogen peroxide as an antiseptic to treat a cut or scrape, you know that it foams as soon as it touches an open

wound. How can you account for this observation?

Going Further

Changing Variables To carry this experiment further, you may wish to use hydrogen peroxide to test for the presence of peroxidase in other materials, such as cut pieces of different vegetables. Also, test raw beef and diced bits of raw liver.

BIOLOGY *Online* To find out more about enzymes, visit the Glencoe Science Web site.
science.glencoe.com

Data and Observations

Cooling will not deactivate the enzymes but can slow the overall reaction. Potato slices heated over 70°C will not generate oxygen bubbles.

ANALYZE AND CONCLUDE

1. Students should explain whether their data support or reject their hypotheses.
2. Between 20°C–50°C
3. Answers may include the amount of time each potato was exposed to the temperature, the sizes of the potato slices, and the amount of peroxide added.
4. Human tissue contains peroxidase, so the hydrogen peroxide is broken down and releases oxygen.

Error Analysis

Advise students that potato slices must reach the desired temperature they are testing, which will take time. Samples need to be removed and observed carefully for bubbles.

✓ Assessment

Portfolio In their portfolios, have students summarize the results, especially the cold treatment. Discuss how results differed between cool pieces tested immediately and those allowed to warm to room temperature. Use the Performance Task Assessment List for Evaluating a Hypothesis in **PASC,** p. 31.

Going Further

Logical-Mathematical Ask students why vegetables are boiled for a short time before freezing. *One reason is that boiling inactivates enzymes that begin to break down the other molecules in the vegetables.*

Purpose

Students learn about the molecular structure of dietary fats and fat substitutes. They also learn about some of the health effects that can result from the consumption of processed foods.

Background

Olestra is made by Procter & Gamble and has the trade name Olean™. Olestra prevents the absorption of fat-soluble nutrients because of its large number of fatty acid molecules. The nutrients are absorbed by the Olestra molecules, and because these are not digested, neither are the nutrients.

Teaching Strategies

■ On the chalkboard, draw a diagram of a sucrose molecule, a glycerol molecule, and a fatty acid molecule. Point out that all are formed from carbon, hydrogen, and oxygen atoms. Simple diagrams that use letter symbols for each atom and straight lines to signify chemical bonds can be found in an introductory chemistry textbook.

Investigating the Technology

1. Olestra contributes fewer calories and fats, which is helpful to people who are watching their weight or need to limit fats because of cholesterol problems or heart disease. But Olestra also prevents the body from obtaining certain nutrients from food, which could have negative health consequences. Fats are an essential part of a healthy diet, so replacing all fats with fat substitutes could present a health risk.

2. To conduct a "blind" tasting, have someone who is not participating in the test put a

170

Are fake fats for real?

Most of us love snacks like chips, cookies, candy, fries, and ice cream. But these foods are typically high in fat, and most of us also realize that limiting our consumption of fat is one of the keys to a healthy lifestyle.

In 1996, the Federal Food and Drug Administration (FDA) approved the use of a new fat substitute called Olestra. Other fat substitutes may contain fewer calories than fat, but they break down when exposed to high temperatures. Olestra can withstand the high temperatures needed to produce fried foods like potato chips.

If it isn't fat, what is it? Most fat substitutes are molecules that are similar to fat but do not have fat's high calorie content. The oldest fat substitutes are based on carbohydrates and are used in salad dressings, dips, spreads, candy, and other foods.

Protein-based fat substitutes are also common. The proteins go through a process called microparticulation, in which they are formed into microscopic round particles. This round shape gives the substances a pleasing smoothness. Protein-based fat substitutes can be used in some cooked foods, but not fried foods.

A fat substitute made from fat Unlike other fat substitutes, Olestra is based on actual fat molecules. It is made by surrounding a sugar molecule, sucrose, with six to eight fatty acid molecules. Naturally occurring dietary fats are made of a glycerol molecule with three fatty acids attached. Because Olestra contains many fatty acid molecules, the digestive system cannot break it down and it passes through the body undigested.

Olestra does have drawbacks. As Olestra passes through the digestive system, it can absorb and carry some fat-soluble vitamins and nutrients. These include vitamins A and E, plus beta carotene, which has been shown to help prevent some forms of cancer.

Fat substitutes will not replace natural fats entirely, but products like Olestra give food

Snack foods made with fat substitutes

scientists some options when they are developing reduced-fat and reduced-calorie foods.

INVESTIGATING THE TECHNOLOGY

1. What are some of the pros and cons of including foods made with Olestra in your diet? Do you think it's a good idea to eat foods made only with fat substitutes rather than true fats? Why or why not?

2. Set up a blind taste test to compare chips or other snack foods made with naturally occurring fats to snacks made with different fat substitutes. Record class preferences. How many tasters could tell the difference between "fake" fats and "real" fats?

For more information on food technology, visit the Glencoe Science Web site.
science.glencoe.com

sample of each product in a separate container and give it an identifying letter or number. Tasters are not allowed to see the product packages or learn whether each sample contains true fats or fat substitutes until after they have recorded their taste preferences. **CAUTION:** *Advise students who have had adverse reactions to Olestra not to participate in the blind taste test.*

Going Further

Have students conduct research to find out which nutrients Olestra prevents the body from absorbing and why they are essential. Encourage students to find out about the long-term health effects of Olestra and other fat substitutes. **L3**

SUMMARY

Section 6.1

Atoms and Their Interactions

Main Ideas

- Atoms are the basic building blocks of all matter.
- Atoms consist of a nucleus containing protons and neutrons. The positively charged nucleus is surrounded by a cloud of rapidly moving, negatively charged electrons.
  - Atoms become stable by bonding to other atoms through covalent or ionic bonds.
  - Components of mixtures retain their properties—components of solutions do not.

Vocabulary

acid (p. 154)
atom (p. 146)
base (p. 154)
compound (p. 149)
covalent bond (p. 150)
element (p. 145)
ion (p. 151)
ionic bond (p. 151)
isotope (p. 148)
metabolism (p. 151)
mixture (p. 152)
molecule (p. 150)
nucleus (p. 147)
pH (p. 154)
solution (p. 153)

Section 6.2

Water and Diffusion

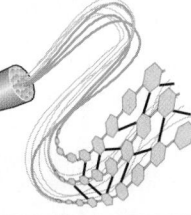

Main Ideas

- Water is the most abundant compound in living things.
- Water is an excellent solvent due to the polar property of its molecules.
- Particles of matter are in constant motion.
- Diffusion occurs from areas of higher concentration to areas of lower concentration.

Vocabulary

diffusion (p. 159)
dynamic equilibrium (p. 160)
hydrogen bond (p. 157)
polar molecule (p. 156)

Section 6.3

Life Substances

Main Ideas

- All organic compounds contain carbon atoms.
- There are four principal types of organic compounds that make up living things: carbohydrates, lipids, proteins, and nucleic acids.

Vocabulary

amino acid (p. 165)
carbohydrate (p. 162)
enzyme (p. 165)
isomer (p. 162)
lipid (p. 164)
nucleic acid (p. 167)
nucleotide (p. 167)
peptide bond (p. 165)
polymer (p. 162)
protein (p. 164)

Main Ideas

Summary statements can be used by students to review the major concepts of the chapter.

Using the Vocabulary

To reinforce chapter vocabulary, use the Content Mastery Booklet and the activities in the Interactive Tutor for Biology: The Dynamics of Life on the Glencoe Science Web site.
science.glencoe.com

 *All Chapter Assessment questions and answers have been validated for accuracy and suitability by The Princeton Review.*

UNDERSTANDING MAIN IDEAS

1. c
2. c

UNDERSTANDING MAIN IDEAS

1. What are the basic building blocks of all matter?
   a. electrons
   b. protons
   c. atoms
   d. molecules

2. Which feature of water explains why water has high surface tension?
   a. water diffuses into cells
   b. water's resistance to temperature changes
   c. water is a polar molecule
   d. water expands when it freezes

GLENCOE TECHNOLOGY

 **VIDEOTAPE**
MindJogger Videoquizzes
Chapter 6: *The Chemistry of Life*
Have students work in groups as they play the videoquiz game to review key chapter concepts.

 **Resource Manager**

Chapter Assessment, pp. 31-36
MindJogger Videoquizzes
ExamView® Pro Software
BDOL Interactive CD-ROM, Chapter 6 quiz

3. a
4. c
5. d
6. d
7. d
8. c
9. d
10. d
11. speeds
12. 20
13. covalent
14. 2
15. hydrogen; oxygen
16. active site
17. elements
18. peptide; protein
19. high; low
20. hydrogen; oxygen

APPLYING MAIN IDEAS

21. The underlying energy level is a filled level.
22. Water is a polar molecule; it will not attract the nonpolar grease.
23. Carbon is the building block element of the four basic substances (carbohydrates, lipids, proteins, and nucleic acids) found in all known living organisms.
24. The substance was a compound because two new substances were formed by the chemical reaction.

---

3. Which of the following describes an isotope of the commonly occurring oxygen atom which has 8 electrons, 8 protons, and 8 neutrons?
    a. 8 electrons, 8 protons, and 9 neutrons
    b. 7 electrons, 8 protons, and 8 neutrons
    c. 8 electrons, 7 protons, and 8 neutrons
    d. 7 electrons, 7 protons, and 8 neutrons

4. Which of the following will form a solution?
    a. sand and water   **c.** salt and water
    b. oil and water   **d.** salt and sand

5. Which of the following applies to a water molecule?
    a. Water is a nonpolar molecule.
    b. The atoms in water are bonded by ionic bonds.
    c. The weak bond between two water molecules is a covalent bond.
    d. Water molecules have a negatively charged end and a positively charged end.

6. Which of the following carbohydrates is a polysaccharide?
    a. glucose   **c.** sucrose
    b. fructose   **d.** starch

7. Which of the following pairs is unrelated?
    a. sugar—carbohydrate
    b. fat—lipid
    c. amino acid—protein
    d. starch—nucleic acid

8. An acid is any substance that forms _____ in water.
    a. hydroxide ions   **c.** hydrogen ions
    b. oxygen ions   **d.** sodium ions

9. Which of these is NOT made up of proteins?
    a. hair   **c.** fingernails
    b. enzymes   **d.** cellulose

 THE PRINCETON REVIEW   **TEST-TAKING TIP**

When Eliminating, Cross It Out
Cross out choices you've eliminated with your pencil. List the answer choice letters on the scratch paper and cross them out there. You'll stop yourself from choosing an answer you've mentally eliminated.

---

10. Which of the following is NOT a smaller subunit of a nucleotide?
    a. phosphate   **c.** ribose sugar
    b. nitrogen base   **d.** glycerol

11. An enzyme _____ chemical reactions.

12. A calcium atom has 20 protons and _____ electrons.

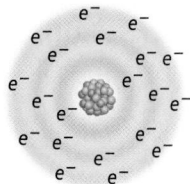

13. A _____ bond involves sharing of electrons.

14. The first energy level of an atom holds _____ electrons; the second energy level holds 8 electrons.

15. In a water molecule, each _____ atom shares one electron with the single _____ atom.

16. A substrate fits into an area of an enzyme called the _____.

17. Hydrogen, chlorine, and sodium are examples of _____.

18. Long chains of amino acids connected to each other by a _____ bond form a _____.

19. Diffusion is the process in which molecules move from a _____ concentration to a _____ concentration.

20. The positively charged _____ atoms of one water molecule attract the negatively charged _____ atom of another water molecule to form a hydrogen bond.

APPLYING MAIN IDEAS

21. A magnesium atom has 12 electrons. When it reacts, it usually loses two electrons. How does this loss make magnesium more stable?

22. Explain why water and a sponge would not be effective in cleaning up a grease spill.

23. Explain why carbon is the most critical element to living things.

24. If heating a white substance produces a vapor

and black material, how do you know the substance was not an element?

THINKING CRITICALLY

25. Interpreting Data The following graph compares the abundance of four elements in living things to their abundance in Earth's crust, oceans, and atmosphere. Which element is the most abundant in organisms? What can you say about the general composition of living things compared to nonliving matter near Earth's surface?

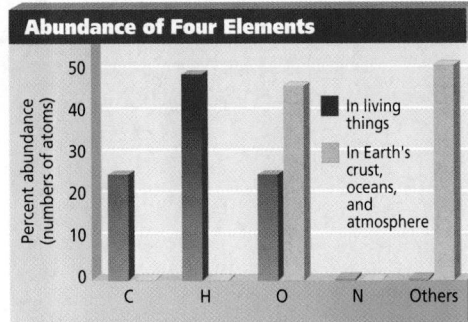

Abundance of Four Elements

26. Designing an Experiment The enzyme peroxidase triggers the breakdown of hydrogen peroxide to form water and oxygen gas. Design an experiment that measures the rate of the reaction.

27. Concept Mapping Complete the concept map by using the following vocabulary terms: protein, amino acid, peptide bond.

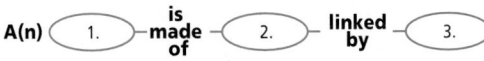

CD-ROM

For additional review, use the assessment options for this chapter found on the *Biology: The Dynamics of Life Interactive CD-ROM* and on the Glencoe Science Web site.
science.glencoe.com

ASSESSING KNOWLEDGE & SKILLS

Two students were studying the effect of temperature on two naturally occurring enzymes. The graph below summarizes their data.

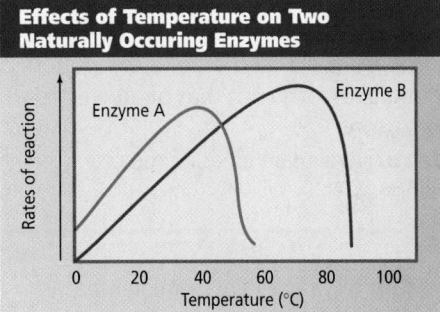

Effects of Temperature on Two Naturally Occuring Enzymes

Using a Graph Study the graph and answer the following questions.

1. At what temperature does the maximum activity of enzyme B occur?
   a. 0°   **c.** 60°
   b. 35°   **d.** 70°

2. At what temperature do both enzymes have an equal rate of reaction?
   a. 10°   **c.** 45°
   b. 20°   **d.** 60°

3. Which of the following descriptions best explains the patterns of temperature effects shown on this graph?
   a. Each enzyme has its own optimal temperature range.
   b. Both enzymes have the same optimal temperature ranges.
   c. Each enzyme will function at room temperature.
   d. Both enzymes are inactivated by freezing temperatures.

4. Designing an Experiment Design an experiment to test the optimal pH of enzyme B.

THINKING CRITICALLY

25. Hydrogen is the most abundant element in organisms. Living things contain much more hydrogen and carbon, about half the oxygen, and similar amounts of nitrogen when compared with nonliving substances.
26. Possible answers might include counting the bubbles given off or collecting and measuring the volume of oxygen given off.
27. 1. Protein; 2. Amino acids; 3. Peptide bonds

ASSESSING KNOWLEDGE & SKILLS

1. d
2. c
3. a
4. Place an equal amount of enzyme and substrate at different pH levels and assess the rate of the reaction at each pH.

Chapter 7 Organizer

Refer to pages 4T-5T of the Teacher Guide for an explanation of the National Science Education Standards correlations.

Section	Objectives	Activities/Features
Section 7.1 **The Discovery of Cells** National Science Education Standards UCP.1, UCP.5; B.2; C.1, C5; E.1, E.2; G.1-3 (1 session, ½ block)	1. **Relate** advances in microscope technology to discoveries about cells and cell structure. 2. **Compare** the operation of a compound light microscope with that of an electron microscope. 3. **Identify** the main ideas of the cell theory.	**MiniLab 7-1:** Measuring Objects Under a Microscope, p. 177 **Focus On** Microscopes, p. 178 **Investigate BioLab:** Observing and Comparing Different Cell Types, p. 194
Section 7.2 **The Plasma Membrane** National Science Education Standards UCP.1-3, UCP.5; A.1, A.2; B.2; C.1, C.5; G.1 (2 sessions, 1 block)	4. **Explain** how a cell's plasma membrane functions. 5. **Relate** the function of the plasma membrane to the fluid mosaic model.	**Problem-Solving Lab 7-1,** p. 182
Section 7.3 **Eukaryotic Cell Structure** National Science Education Standards UCP.1-3, UCP.5; A.1, A.2; C.1, C.5; E.1, E.2; G.1-3 (3 sessions, 1½ blocks)	6. **Understand** the structure and function of the parts of a typical eukaryotic cell. 7. **Explain** the advantages of highly folded membranes in cells. 8. **Compare and contrast** the structures of plant and animal cells.	**Problem-Solving Lab 7-2,** p. 186 **MiniLab 7-2:** Cell Organelles, p. 187 **Inside Story:** Comparing Animal and Plant Cells, p. 192 **Literature Connection:** *The Lives of a Cell* by Lewis Thomas, p. 196

Need Materials? Contact Carolina Biological Supply Company at 1-800-334-5551 or at **http://www.carolina.com**

MATERIALS LIST

BioLab
p. 194 microscope, microscope slide, coverslip, water, dropper, forceps, prepared slides of *Bacillus subtilis,* frog blood, and *Elodea*

MiniLabs
p. 177 microscope, microscope slide, coverslip, human hair, water
p. 187 microscope, microscope slide, coverslip, onion, water

Alternative Lab
p. 188 microscope, microscope slide, coverslip, dropper, forceps, *Elodea* plant, prepared slide of human cheek cells

Quick Demos
p. 176 cork
p. 183 water, alcohol, salad oil, beaker, stirring rod
p. 191 gelatin, grapes, assorted fruit, water

Key to Teaching Strategies

L1	Level 1 activities should be appropriate for students with learning difficulties.
L2	Level 2 activities should be within the ability range of all students.
L3	Level 3 activities are designed for above-average students.
ELL	ELL activities should be within the ability range of English Language Learners.
COOP LEARN	Cooperative Learning activities are designed for small group work.
P	These strategies represent student products that can be placed into a best-work portfolio.
	These strategies are useful in a block scheduling format.

A View of the Cell

Teacher Classroom Resources

Section	Reproducible Masters	Transparencies
Section 7.1 — The Discovery of Cells	Reinforcement and Study Guide, p. 29 L2 BioLab and MiniLab Worksheets, p. 31 L2 Laboratory Manual, pp. 47-54 L2 Content Mastery, pp. 33-34, 36 L1	Section Focus Transparency 15 L1 ELL Reteaching Skills Transparency 9 L1 ELL
Section 7.2 — The Plasma Membrane	Reinforcement and Study Guide, p. 30 L2 Tech Prep Applications, pp. 11-14 L2 Content Mastery, pp. 33, 35-36 L1	Section Focus Transparency 16 L1 ELL Basic Concepts Transparency 6 L2 ELL
Section 7.3 — Eukaryotic Cell Structure	Reinforcement and Study Guide, pp. 31-32 L2 Concept Mapping, p. 7 L3 ELL Critical Thinking/Problem Solving, p. 7 L3 BioLab and MiniLab Worksheets, pp. 32-34 L2 Content Mastery, pp. 33, 35-36 L1 Inside Story Poster ELL	Section Focus Transparency 17 L1 ELL Basic Concepts Transparency 7 L2 ELL Reteaching Skills Transparency 10 L1 ELL

Assessment Resources

Chapter Assessment, pp. 37-42
MindJogger Videoquizzes
Performance Assessment in the Biology Classroom
Alternate Assessment in the Science Classroom
ExamView® Pro Software 💾
BDOL Interactive CD-ROM, Chapter 7 quiz

Additional Resources

Spanish Resources ELL
English/Spanish Audiocassettes ELL
Cooperative Learning in the Science Classroom COOP LEARN
Lesson Plans/Block Scheduling

NATIONAL GEOGRAPHIC — Teacher's Corner

Products Available From Glencoe
To order the following products, call Glencoe at 1-800-334-7344:
CD-ROM
NGS PictureShow: The Cell
Curriculum Kit
GeoKit: Cells and Microorganisms
Transparency Set
NGS PicturePack: The Cell

Products Available From National Geographic Society
To order the following products,
call National Geographic Society at 1-800-368-2728:
Video
Discovering the Cell

Index to National Geographic Magazine
The following articles may be used for research relating to this chapter:
"Life Grows Up," by Richard Monastersky, April 1998.
"The Rise of Life on Earth," by Richard Monastersky, March 1998.

GLENCOE TECHNOLOGY

The following multimedia resources are available from Glencoe.

Biology: The Dynamics of Life
CD-ROM ELL

Video: *Light Microscope*
Video: *The SEM*
Exploration: *Parts of a Cell*
BioQuest: *Cellular Pursuit*

The Infinite Voyage

Unseen Worlds

The Secret of Life Series

Cells: *Microscopy*
Prokaryotic Cell
Eukaryotic Cell
Cell Membranes: *Membranes and Transport*

7   A View of the Cell

Theme Development

The first section stresses the **nature of science** through a discussion of the development of the microscope and the cell theory. **Evolution** is apparent in the discussion of the number, variety, and complexity of cellular organelles.

Resource Manager

Section Focus Transparency 15 and Master **L1** **ELL**
BioLab and MiniLab Worksheets, p. 31 **L2**
Laboratory Manual, pp. 47-50 **L2**

READING BIOLOGY

Glencoe's *Biology: The Dynamics of Life* contains many resources to assist a student's reading skills. Each chapter contains figures with expanded captions that expand on written material. Word Origins, located along the side of text, expand knowledge of biology vocabulary. Glencoe's Content Mastery Booklet helps develop reading skills while reinforcing content. In addition, use the Interactive Tutor for *Biology: The Dynamics of Life* on the Glencoe Web site to reinforce vocabulary.
science.glencoe.com

What You'll Learn

- You will distinguish eukaryotic and prokaryotic cells.
- You will learn the structure and function of the plasma membrane.
- You will relate the structure of cell parts to their functions.

Why It's Important

A knowledge of cell structure and function is essential to a basic understanding of life.

READING BIOLOGY

Go through the chapter, and note the figures that depict different types of cell structures. For each figure, note the various components of the cell. As you read the text, write the characteristics of each cell type by its name.

BIOLOGY *Online*

To find out more about cells, visit the Glencoe Science Web site.
science.glencoe.com

Cells are amazingly diverse. Yet for all their diversity, cells such as this nerve cell and the protist *Euglena* share many common features.

Magnification: 9310×

174   A VIEW OF THE CELL

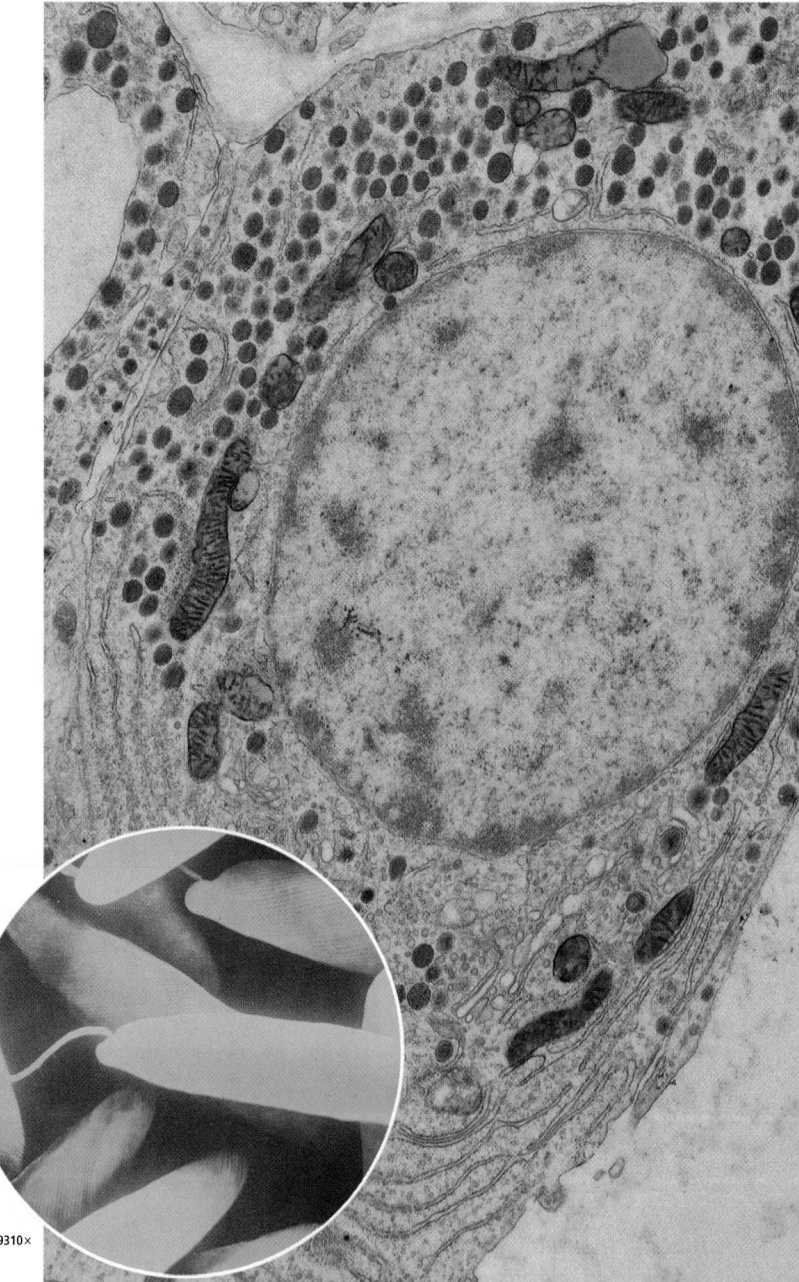

Magnification: 17 270×

7.1 The Discovery of Cells

SECTION PREVIEW

Objectives

Relate advances in microscope technology to discoveries about cells and cell structure.

Compare the operation of a compound light microscope with that of an electron microscope.

Identify the main ideas of the cell theory.

Vocabulary

cell
compound light microscope
cell theory
electron microscope
prokaryote
eukaryote
organelle
nucleus

Have you ever used a magnifying glass to examine something you'd found? Hundreds of years ago, scientists, too, were fascinated and motivated to study their environment. The first microscopes were not much different from hand-held magnifying glasses, consisting of only one lens. Through a single-lens microscope, the Dutch scientist Anton van Leeuwenhoek in the mid-1600s was the first person to record looking at water under a microscope. He was amazed to find it full of living things.

Magnification reveals minute details.

The History of the Cell Theory

Before microscopes were invented, people believed that diseases were caused by curses and supernatural spirits. They had no idea that organisms such as bacteria existed. As scientists began using microscopes, they quickly realized they were entering a new world—one of microorganisms (my kroh OR guh nihz umz). Microscopes enabled scientists to view and study **cells,** the basic units of living organisms.

Development of light microscopes

The microscope van Leeuwenhoek (LAY vun hook) used is considered a simple light microscope because it contained one lens and used natural light to view objects. Over the next 200 years, scientists greatly improved microscopes by grinding higher quality lenses and creating the compound light microscope. **Compound light microscopes** use a series of lenses to magnify objects in steps. These microscopes can magnify objects up to 1500 times. As the observations of plants and animals viewed under a microscope expanded, scientists began to draw conclusions about the organization of living matter. With the microscope established as a valid scientific tool, scientists had to learn the size relationship of magnified objects to their true size. Look at *Focus on Microscopes* to see what specimens look like at different magnifications.

WORD *Origin*

microscope
From the Greek words *mikros*, meaning "small", and *skopein*, meaning to "look." A microscope is used to examine small objects.

7.1 THE DISCOVERY OF CELLS **175**

Section 7.1

Prepare

Key Concepts

Students are provided with an overview of the historical development of the microscope and the events that led to the formation of the cell theory.

Planning

■ Collect pictures of microscopes for the microscope collage.

■ Gather cork for the Quick Demo.

■ Collect materials needed for the microscopy project. Some cells should be purchased in advance.

1 Focus

Bellringer 🖋

Before presenting the lesson, display **Section Focus Transparency 15** on the overhead projector and have students answer the accompanying questions. **L1** **ELL**

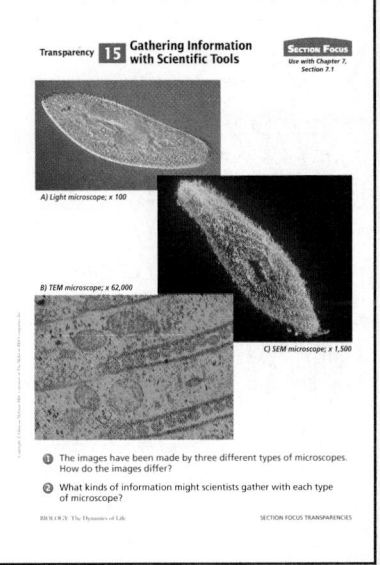

Assessment Planner

Portfolio Assessment
Problem-Solving Lab, TWE, p. 186
BioLab, TWE, p. 195

Performance Assessment
Assessment, TWE, p. 184
Alternative Lab, TWE, p. 189
MiniLabs, TWE, pp. 177, 187
BioLab, SE, p. 195

Knowledge Assessment
Assessment, TWE, pp. 180, 183
Problem-Solving Lab, TWE, p. 182
Section Assessments, SE, pp. 180, 184, 193
Chapter Assessment, SE, pp. 197-199

Skill Assessment
MiniLab, TWE, p. 187
Assessment, TWE, p. 193

Enrichment

 *Linguistic* Have students do library research on how various electron microscopes work. Students can write a report or make an oral presentation to the class. **L3**

Quick Demo

 *Kinesthetic* Show students a piece of cork and have them examine the cellular structure. Emphasize that cork was a living part of an organism. It is the dead cells of oak bark.

Visual Learning

Figure 7.1 Have students examine this figure and determine whether organelles are present in the cork cells.

✔ Assessment

Performance Have students study a cell under the microscope at increasing magnifications (or use appropriate photographs). Have them describe as much detail as possible at each magnification and draw diagrams of the cells at different magnifications. **L2** **ELL**

GLENCOE TECHNOLOGY

VIDEODISC
The Secret of Life
Cells: Microscopy

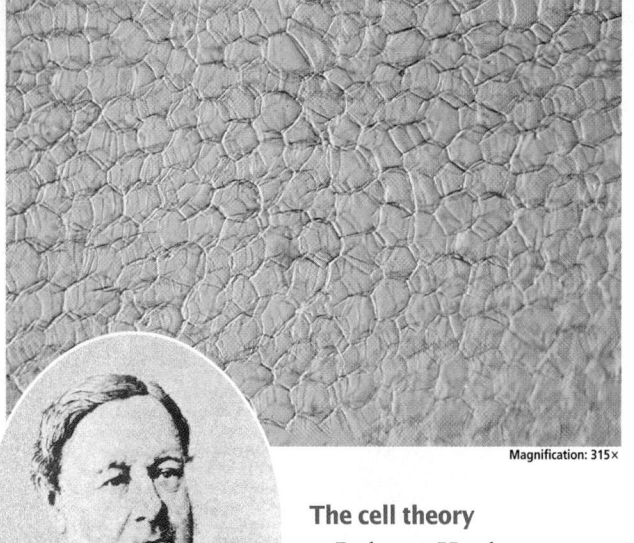

Magnification: 315×

Figure 7.1
Cork cells (top) were observed by Robert Hooke using a crude compound light microscope that magnified structures only 30 times. Theodore Schwann (inset) made similar observations in animals.

The cell theory

Robert Hooke was an English scientist who lived at the same time as van Leeuwenhoek. Hooke used a compound light microscope to study cork, the dead cells of oak bark. In cork, Hooke observed small geometric shapes, like those shown in *Figure 7.1.* Hooke gave these box-shaped structures the name cells because they reminded him of the small rooms monks lived in at a monastery. Cells are the basic building blocks of all living things. Hooke published his drawings and descriptions, which encouraged other scientists to search for cells in the materials they were studying.

Several scientists extended Hooke's observations and drew some important conclusions. In the 1830s, the German scientist Matthias Schleiden observed a variety of plants and concluded that all plants are composed of cells. Another German scientist, Theodore Schwann, *Figure 7.1,* made similar observations on animals. The observations and conclusions of these scientists are summarized as the cell theory, one of the fundamental ideas of modern biology.

The **cell theory** is made up of three main ideas:

1. *All organisms are composed of one or more cells.* An organism may be a single cell, such as the organisms van Leeuwenhoek saw in water. Others, like most of the plants and animals with which you are most familiar, are multicellular, or made up of many cells.
2. *The cell is the basic unit of organization of organisms.* Although organisms can become very large and complex, such as humans, dogs, and trees, the cell remains the simplest, most basic component of any organism.
3. *All cells come from preexisting cells.* Before the cell theory, no one knew how cells were formed, where they came from, or what determined the type of cell they became. The cell theory states that a cell divides to form two identical cells.

Development of electron microscopes

The microscopes we have discussed so far use a light source and can magnify an object up to about 1500 times its actual size. Although light microscopes continue to be valuable tools, scientists knew another world existed within a cell that they could not yet see. In the 1940s a new type of microscope, the **electron microscope,** was invented. This microscope uses a beam of electrons instead of natural light to magnify structures up to 500 000 times their actual size, allowing scientists to see structures within a cell.

There are two basic types of electron microscopes. Scientists commonly use the scanning electron microscope (SEM) to scan the surfaces of cells to learn their three-dimensional shape. The transmission

---

BIOLOGY JOURNAL

Building Vocabulary

 *Linguistic* Have students define the following Greek and Latin words: *micro, scop, eu, karyon, pro, plasma, chrom, endo, lysis, mitos, chondros, chloros,* and *plastos.* Ask students to identify words from the chapter that use these and explain how the Greek or Latin meaning relates to the meaning of the term. **L2** **ELL**

GLENCOE TECHNOLOGY

VIDEODISC
The Infinite Voyage
Unseen Worlds
Technology Reconstructs Egyptian Mummies
(Ch. 1) 6 min. 30 sec.

electron microscope (TEM) allows scientists to study the structures contained within a cell.

New types of microscopes and new techniques are continually being designed. The scanning tunneling microscope (STM) uses the flow of electrons to investigate atoms on the surface of a molecule. New techniques using the light microscope have increased the information scientists can learn with this basic tool. Most of these new techniques seek to add contrast to structures within the cells, such as adding dyes that stain some parts of a cell, but not others. Try the *Minilab* on this page to practice measuring objects under a microscope.

Two Basic Cell Types

With the invention of light microscopes, scientists noticed that cells could be divided into two broad groups: those with internal, membrane-bound structures and those without. Cells lacking internal membrane-bound structures are called prokaryotic (proh KER ee oh tik) cells. The cells of most unicellular organisms such as bacteria do not have membrane-bound structures and are therefore called **prokaryotes.**

Cells of the basic second type, those containing membrane-bound structures, are called eukaryotic (yew KER ee oh tik) cells. Most of the multicellular plants and animals we know have cells containing membrane-bound structures and are therefore called **eukaryotes.** It is important to note, however, that some eukaryotes, such as some algae and yeast, are unicellular organisms.

The membrane-bound structures within eukaryotic cells are called **organelles.** Each organelle has a specific function for cell survival.

MiniLab 7-1   Measuring in SI

Magnification: 260✓

Measuring Objects Under a Microscope Knowing the diameter of the circle of light you see when looking through a microscope allows you to measure the size of objects that are being viewed. For most microscopes, the diameter of the circle of light is 1.5 mm, or 1500 µm (micrometers), under low power and 0.375 mm, or 375 µm, under high power.
   Refer to *Practicing Scientific Methods* in the **Skill Handbook** if you need help with SI units.

Human hair

Procedure

1. Look at diagram A that shows an object viewed under low power. Knowing the circle diameter to be 1500 µm, the estimated length of object (a) is 400 µm. What is the estimated length of object (b)?

2. Look at diagram B that shows an object viewed under high power. Knowing the circle diameter to be 375 µm, the estimated length of object (c) is 100 µm. What is the estimated length of object (d)?

3. Prepare a wet mount of a strand of your hair. Your teacher can help with this procedure. **CAUTION:** *Use caution when handling microscopes and glass slides.* Measure the width of your hair strand while viewing it under low and then high power.

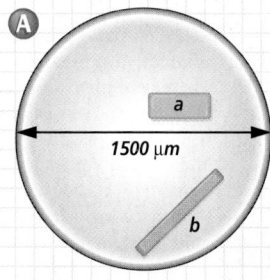
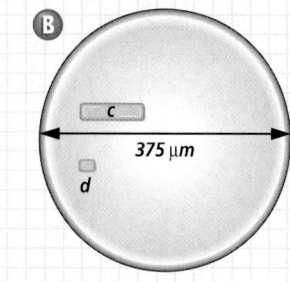

A   1500 µm   a   b

B   375 µm   c   d

Analysis

1. An object can be magnified 100, 200, or 1000 times when viewed under a microscope. Does the object's actual size change with each magnification? Explain.

2. Do your observations of the size of your hair strand under low and high power support the answer to question 1? If not, offer a possible explanation why.

7.1   THE DISCOVERY OF CELLS   **177**

MiniLab 7-1

Purpose
Students will learn how to estimate an object's size under the microscope.

Process Skills
measure in SI, observe and infer, use numbers, estimate

Safety Precautions
Remind students to handle slides carefully and dispose of broken glass properly.

Teaching Strategies
■ Review the procedure for making a wet mount.
■ Give students some sample problems to practice estimating size and converting from mm to µm
■ Provide other objects, such as diatoms and yeast cells, for viewing and measuring.

Expected Results
The size of object (b) is 700 µm and object (d) is 25 µm. Hair width may be within the range of 60–100µm (depending on race and/or hair color) and will be the same (or very close) under low and high power magnification.

Analysis
1. No. An object's size does not change, only its magnification changes.
2. Answers will vary, but observations under low and high power should be close. Error in estimating may be the cause of differing measurements.

✓ Assessment
   Performance Have students compare the width of cranial and arm hair; blonde and black hair; Asian and Caucasian hair; and Caucasian and African American hair. Ask them to prepare a data table with their results. Use the Performance Task Assessment List for Data Table in **PASC**, p. 37. **L2 ELL**

✓ Portfolio

Microscopy and Staining

Kinesthetic Have students examine various types of cells to develop their microscope skills. Set out iodine solution and methylene blue and encourage students to experiment with the stains. For iodine solution preparation instructions, see page 40T of the Teacher Guide. Possible cells to examine might include: onion, *Elodea* leaf, cheek, potato, *Lactobacillus* in cultured yogurt, teeth bacteria, *Paramecium*, amoeba, *Volvox*, tomato skin, tomato pulp, and pear pulp. You may wish to have students work in their cooperative groups to prepare their stained slides. Have students review the proper procedures for using the microscope in the **Skill Handbook**. **L2 ELL**
COOP LEARN

Focus On

Microscopes

Purpose

Students will be provided with an historical perspective of the events that led to the development of the compound microscope, transmission electron microscope (TEM), and scanning electron microscope (SEM) and the uses of each microscope type.

Background

Magnifying lenses were developed hundreds of years ago but required live specimens. Further progress was delayed due to a lack of knowledge about how to prepare specimens.

Teaching Strategies

■ Ask students whether they think the development of a successful microscope was delayed because of problems with creating lenses or a lack of knowledge about how to prepare specimens for study. *Problems resulted more from a lack of knowledge about how to prepare specimens since lens construction dates back to the time of Galileo.*

■ Discuss the techniques involved in preparing specimens for electron microscopy studies. The use of metals for shadowing, freeze-fracture techniques, and ultramicrotomes are all good topics for discussion.

■ Have students research the backgrounds and training of the individuals involved in the development of the microscope *or* the scientists who contributed to the development of the cell theory. **L3**

178

The invention and development of the light microscope some 300 years ago allowed scientists to see cells for the first time. Improvements have vastly increased the range of visibility of microscopes. Today researchers can use these powerful tools to study cells at the molecular level.

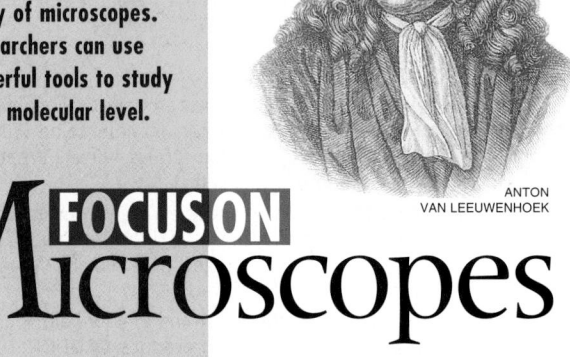
ANTON VAN LEEUWENHOEK

FOCUS ON Microscopes

THIS EARLY COMPOUND MICROSCOPE, housed in a gold-embossed leather case, was designed by English scientist Robert Hooke about 1665. Using it, he observed and made drawings of cork cells. Although the microscope has three lenses, they are of poor quality and Hooke could see little detail. ▶

◀ **THIS HISTORIC MICROSCOPE**— held by a modern researcher— was designed by Anton van Leeuwenhoek (above). By 1700, Dutch scientist Leeuwenhoek had greatly improved the accuracy of microscopes. Grinding the lenses himself, Leeuwenhoek built some 240 single-lens versions. He discovered—and described for the first time—red blood cells and bacteria, taken from scrapings from his teeth. By 1900, problems with lenses that had once limited image quality had been overcome, and the compound microscope had evolved essentially into its present form.

HOOKE'S MICROSCOPE

178

✓ Portfolio

 The Compound Microscope

Visual-Spatial Have students draw and label the parts of their microscope. Under each label have them describe the function of the part. For the objective lenses, have students calculate and identify the power of magnification for an object viewed with each lens. **L1** **ELL** **P**

GLENCOE TECHNOLOGY

 CD-ROM
Biology: The Dynamics of Life
Video: *The SEM*
Disc 1
Video: *The Light Microscope*
Disc 1

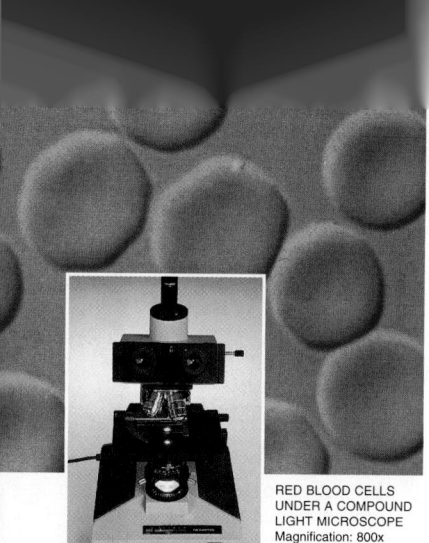

RED BLOOD CELLS UNDER A COMPOUND LIGHT MICROSCOPE
Magnification: 800x

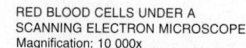

RED BLOOD CELLS UNDER A SCANNING ELECTRON MICROSCOPE
Magnification: 10 000x

SCANNING ELECTRON MICROSCOPE IMAGE OF A MOSQUITO
Magnification: 50x

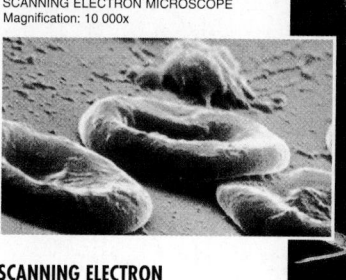

SCANNING ELECTRON MICROSCOPE

An SEM sweeps a beam of electrons over the surface of a specimen, such as red blood cells (above), causing electrons to be emitted from the specimen. SEMs produce a realistic, three-dimensional picture—but only the surface of an object can be observed. An SEM can magnify about 60 000 times without losing clarity.

HOW IT WORKS
The magnifying power of a microscope is determined by multiplying the magnification of the eyepiece and the objective lens.

A COMPOUND LIGHT MICROSCOPE

(above) uses two or more glass lenses to magnify objects. Light microscopes are used to look at living cells, such as red blood cells (top), small organisms, and preserved cells. Compound light microscopes can magnify up to about 1500 times.

RED BLOOD CELLS UNDER A TRANSMISSION ELECTRON MICROSCOPE
Magnification: 40 000x

SCANNING TUNNELING MICROSCOPE IMAGE OF A DNA FRAGMENT
Magnification: 2 000 000x

SCANNING TUNNELING MICROSCOPE
The STM revolutionized microscopy in the mid-1980s by allowing scientists to see atoms on an object's surface. A very fine metal probe is brought near a specimen. Electrons flow between the tip of the probe and atoms on the specimen's surface. As the probe follows surface contours, such as those on this DNA molecule (above), a computer creates a three-dimensional image. An STM can magnify up to one hundred million times.

◀ **TRANSMISSION ELECTRON MICROSCOPE** A TEM aims a beam of electrons through a specimen. Denser portions of an object allow fewer electrons to pass through. These denser areas appear darker in the image. Two-dimensional TEM images are used to study details of cells such as these red blood cells (above). A TEM can magnify hundreds of thousands of times.

EXPANDING Your View

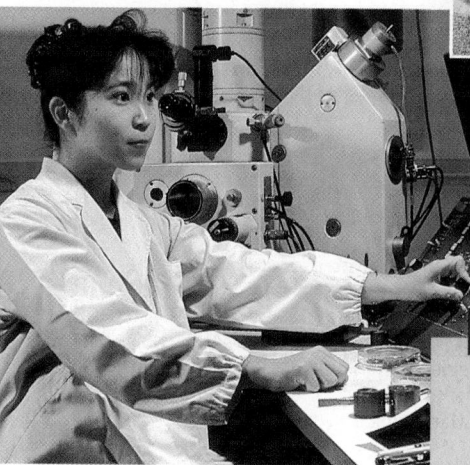

RESEARCHER USING A TEM

1 **THINKING CRITICALLY** Can live specimens be examined with an electron microscope? Explain. Consider how the specimen must be prepared for viewing.

2 **COMPARING AND CONTRASTING** Compare the images seen with an SEM and with a TEM.

3 Assess

Check for Understanding

Quiz students orally about the importance of the cell theory and its acceptance by the scientific community.

Reteach

Review the parts of the cell theory and the significance of each statement making up the theory as a reinforcement exercise.

Extension

 **Intrapersonal** Have interested students research new types of microscopes, such as the atomic force microscope. **L3**

✔ **Assessment**

Knowledge Quiz students on the names and functions of the parts of the compound microscope. **L2**

---

4 Close

Activity

Intrapersonal In a class discussion, summarize the contributions of each scientist mentioned in this section. Have students make a table that lists the name of each scientist and his contributions. Encourage students to retain their tables in their journals and to add information about other scientists throughout the remainder of the course. **L2**

📑 **Resource Manager**

Reinforcement and Study Guide, p. 29 **L2**
Content Mastery, p. 34 **L1**
Reteaching Skills Transparency 9 and Master **L1** **ELL**
Laboratory Manual, pp. 51-54 **L2**

180

Figure 7.2
Bacteria and archaebacteria are prokaryotes. All other organisms are eukaryotes.

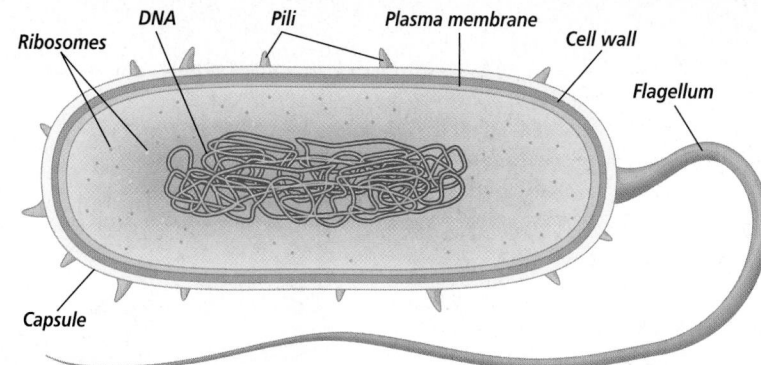

Ribosomes · DNA · Pili · Plasma membrane · Cell wall · Flagellum · Capsule

A A **Prokaryotic cell** does not have internal organelles surrounded by a membrane. Most of a prokaryote's metabolic functions take place in the cytoplasm.

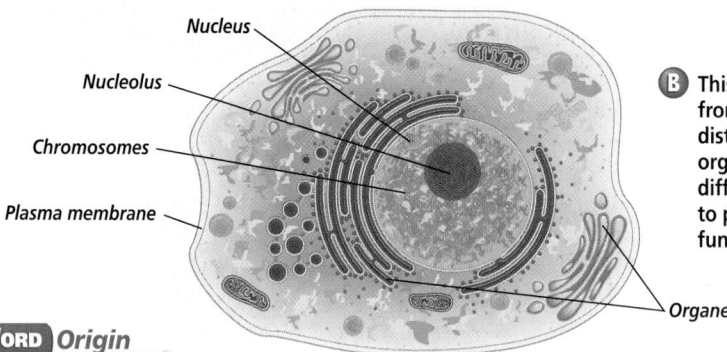

Nucleus · Nucleolus · Chromosomes · Plasma membrane · Organelles

B This **eukaryotic cell** from an animal has distinct membrane-bound organelles that allow different parts of the cell to perform different functions.

WORD Origin

organelle
From the Greek words *organon*, meaning "tool" or "implement" and *ella*, meaning "small." Organelles are small, membrane-bound structures in cells.

Compare the prokaryotic and eukaryotic cells in *Figure* 7.2. Separation of organelles into distinct compartments benefits the eukaryotic cell. One benefit is that chemical reactions that would normally not occur in the same area of the cell can now be carried out at the same time.

Robert Brown, a Scottish scientist, first observed a prominent structure in cells that Rudolf Virchow later concluded was the structure responsible for cell division. We now know this structure as the **nucleus,** the central membrane-bound organelle that manages cellular functions.

Section Assessment

Understanding Main Ideas
1. Why was the development of microscopes necessary for the study of cells?
2. How does the cell theory describe the organization of living organisms?
3. Compare the light sources of light microscopes and electron microscopes.
4. How are prokaryotic and eukaryotic cells different?

Thinking Critically
5. Suppose you discovered a new type of fern.

Applying the cell theory, what can you say for certain about this organism?

SKILL REVIEW

6. **Care and Use of a Microscope** Most compound light microscopes have four objective lenses with magnifications of 4×, 10×, 40×, and 100×. What magnifications are available if the eyepiece magnifies 15 times? For more help, refer to *Practicing Scientific Methods* in the **Skill Handbook**.

180 A VIEW OF THE CELL

Section Assessment

1. Microscopes were needed because most cells are not visible with the unaided eye.
2. Cells are the basic unit of organization of living organisms. All organisms are composed of cells. Cells are derived from pre-existing cells.
3. Light microscopes use natural light to penetrate objects. Electron microscopes use a beam of electrons.
4. Eukaryotic cells have membrane-bound organelles and a nucleus surrounded by a double bilayer. Prokaryotic cells have neither.
5. The fern is made of one or more cells, which are the basic units of its organization. The fern's cells came from preexisting cells.
6. 60×, 150×, 600×, 1500×

7.2 The Plasma Membrane

Cells, like your entire body, require a constant environment.

SECTION PREVIEW

Objectives

Explain how a cell's plasma membrane functions.

Relate the function of the plasma membrane to the fluid mosaic model.

Vocabulary

plasma membrane
homeostasis
selective permeability
phospholipid
fluid mosaic model
transport protein

*T*hink of trudging home from school on a cold wintry day. When you finally arrive, you enter a room where you are sheltered from the wind and surrounded by a warm, comfortable environment. Your house is a controlled environment just like the cells in your body. Similar to the way the walls of your home act as a barrier against the elements, the plasma membrane provides a barrier between the internal components of a cell and its external environment.

Maintaining a Balance

You are comfortable in your house largely because the thermostat maintains the temperature within a limited range regardless of what's happening outside. Similarly, all living cells must maintain a balance regardless of internal and external conditions. Survival depends on the cell's ability to maintain the proper conditions within itself.

Why cells must control materials

Your cells need nutrients such as glucose, amino acids, and lipids to function. It is the job of the **plasma membrane,** the boundary between the cell and its environment, to allow a steady supply of these nutrients to come into the cell no matter what the

external conditions are. However, too much of any of these nutrients or other substances, especially ions, can be harmful to the cell. If levels become too high, the plasma membrane removes the excess. The plasma membrane also allows waste and other products to leave the cell. This process of maintaining the cell's environment is called **homeostasis**.

How does the plasma membrane maintain homeostasis? One mechanism is **selective permeability,** a process in which the plasma membrane of a cell allows some molecules into the cell while keeping others out. Thinking back to your home, a screen in a window can perform selective permeability in a similar way. When you open the window, the screen lets fresh air inside and keeps

WORD Origin

permeable
From the Latin words *per,* meaning "through," and *meare,* meaning "to glide." Materials move easily (glide) through permeable membranes.

7.2 THE PLASMA MEMBRANE **181**

TECHPREP

Kinesthetic Group students and have them build models of a plasma membrane using materials such as Styrofoam "peanuts," yarn, pipe cleaners, and popsicle sticks. Encourage students to be creative. **L2** **ELL**

Resource Manager

Section Focus Transparency 16 and Master **L1** **ELL**
Tech Prep Applications, pp. 11-12 **L2**

Section 7.2

Prepare

Key Concepts

Students will examine the fluid mosaic model of the plasma membrane and discover how this model explains the functions of the membrane.

Planning

■ Collect the materials for the Tech Prep modeling project.

■ Obtain glass jars, water, alcohol, food coloring, and salad oil for the Quick Demo.

■ Purchase Ping-Pong balls or apples for the Reteach. Also obtain a suitable tub.

---

1 Focus

Bellringer

Before presenting the lesson, display **Section Focus Transparency 16** on the overhead projector and have students answer the accompanying questions. **L1** **ELL**

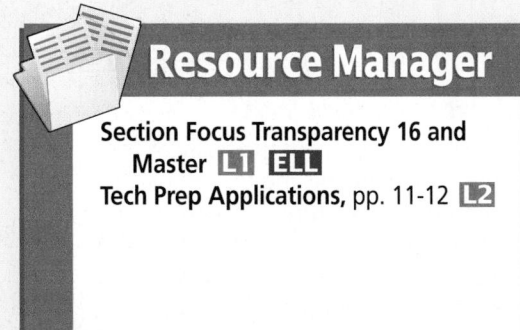

181

2 Teach

Purpose 🔲

Students will determine how the plasma membrane of living cells is a selective barrier to certain molecules.

Process Skills

compare and contrast, interpret scientific illustrations, recognize cause and effect, think critically

Teaching Strategies

■ Remind students that yeast cells are living organisms. Elicit the composition of a blue stain (molecules of a chemical compound).

■ Have students describe some of the organelles that should be present in yeast and verify that the plasma membrane is included in the discussion. **L2**

Thinking Critically

1. Boiling killed the cells; it disrupted the intact plasma membrane.

2. Answers will vary. For example, if yeast cells are boiled and killed, then their membrane can no longer serve as a barrier to the blue stain.

3. Yes, while alive, yeast cell membranes prevent the blue stain from entering the interior. Once cells are killed, the membranes can no longer keep the blue stain out.

✔ Assessment

Knowledge Have students make diagrams depicting the before and after appearance of the yeast cell membranes. Have them show the blue stain in both diagrams. Students are to assume that no carrier molecules were involved. Use the Performance Task Assessment List for Scientific Drawing in **PASC**, p. 55. **L2 ELL**

182

Problem-Solving Lab 7-1   Recognizing Cause and Effect

Is the cell membrane a selective barrier? Yeast cells are living and contain a plasma membrane. How is it possible to show that living yeast plasma membranes are capable of limiting what enters the cell? Conduct an experiment to find an answer.

Analysis

Diagram A shows the appearance of yeast cells in a solution of blue stain. Note their color as well as the color of the surrounding stain.

Diagram B also shows yeast cells in a solution of blue stain. These cells, however, were boiled for 10 minutes before being placed in the stain. Again, note the color of the yeast cells as well as the color of the surrounding stain.

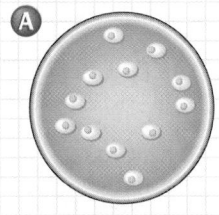 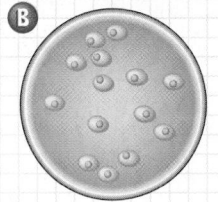

Thinking Critically

1. Explain how boiling affected the yeast cells.
2. Hypothesize why the color of the cells differs under different conditions. Be sure that your hypothesis takes the role of the plasma membrane into consideration.
3. Are plasma membranes selective barriers? Explain.

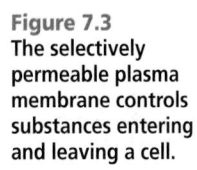

Figure 7.3
The selectively permeable plasma membrane controls substances entering and leaving a cell.

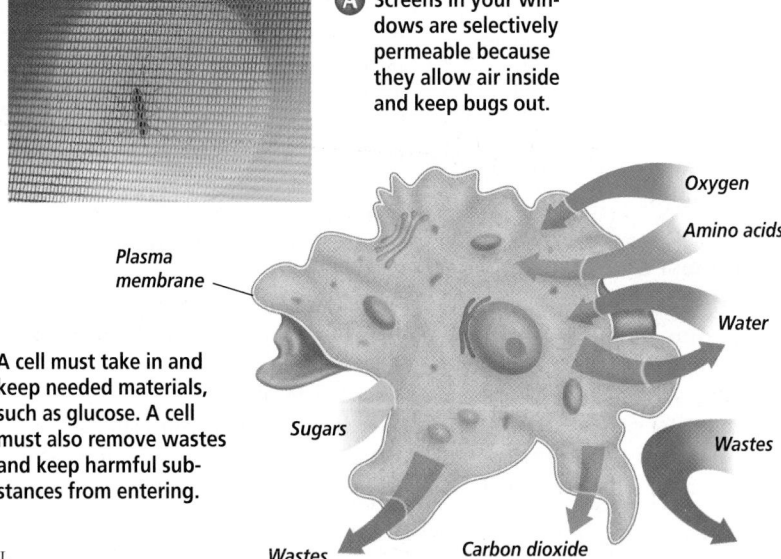

A Screens in your windows are selectively permeable because they allow air inside and keep bugs out.

Plasma membrane

B A cell must take in and keep needed materials, such as glucose. A cell must also remove wastes and keep harmful substances from entering.

Oxygen

Amino acids

Water

Wastes

Sugars

Wastes   Carbon dioxide

182 A VIEW OF THE CELL

---

most insects out. Some molecules, such as water, freely enter the cell through the plasma membrane, as shown in *Figure 7.3*. Other particles, such as sodium and calcium ions, must be allowed into the cell only at certain times, in certain amounts, and through certain channels. The plasma membrane must be selective in allowing these ions to enter. Use the *Problem-Solving Lab* here to evaluate the plasma membrane of a yeast cell.

Structure of the Plasma Membrane

Now that you understand the basic function of the plasma membrane, you can study its structure. Recall that lipids are insoluble molecules that are the primary components of cellular membranes. The plasma membrane is composed of a phospholipid bilayer, which is two layers of phospholipid back-to-back. **Phospholipids** are lipids with a phosphate group attached to them. The lipids in a plasma membrane have a glycerol backbone, two fatty acid chains, and a phosphate group.

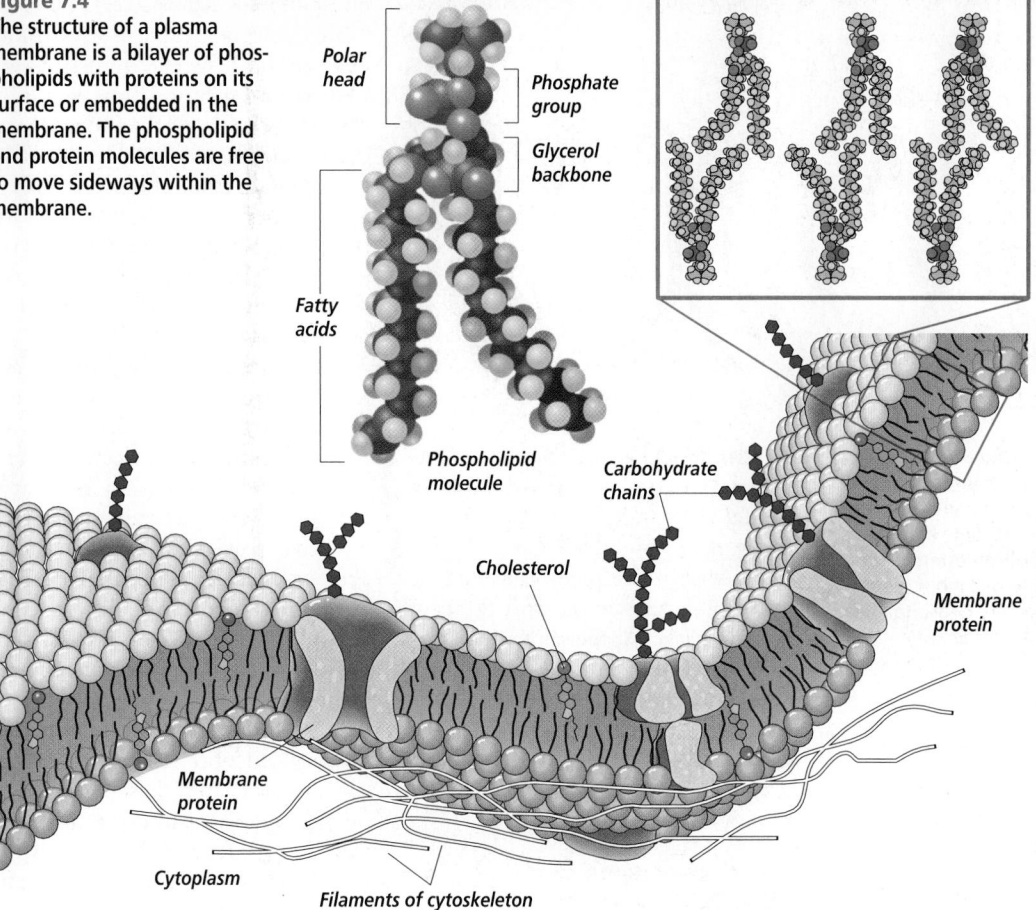

Figure 7.4
The structure of a plasma membrane is a bilayer of phospholipids with proteins on its surface or embedded in the membrane. The phospholipid and protein molecules are free to move sideways within the membrane.

Polar head

Phosphate group

Glycerol backbone

Fatty acids

Phospholipid molecule

Carbohydrate chains

Cholesterol

Membrane protein

Membrane protein

Cytoplasm

Filaments of cytoskeleton

Makeup of the phospholipid bilayer

The addition of the phosphate group does more than change the name of the lipid. The phosphate group is critical for the formation and function of the plasma membrane. *Figure 7.4* illustrates phospholipids and their place within the structure of the plasma membrane. The two fatty acid tails of the phospholipids are nonpolar, whereas the head of the phosphate molecule is polar.

Water is a key component of living organisms, both inside and outside the cell. The polar phosphate group allows the cell membrane to interact

with its watery environment because, as you recall, water is also polar. The fatty acid tails, on the other hand, avoid water. The two layers of phospholipid molecules make a sandwich with the fatty acid tails forming the interior of the membrane and the phospholipid heads facing the watery environment outside the cell. When many phospholipid molecules come together in this manner, a barrier is created that is water-soluble at its outer surfaces and water-insoluble in the middle. Water-soluble molecules will not easily move through the membrane because they are stopped by this water-insoluble layer.

183

✓ Assessment

Knowledge Quiz students on how the plasma membrane controls the passage of materials into and out of the cell. **L1**

Quick Demo

Mix polar substances such as colored water and alcohol to show that the substances dissolve in each other. Then add a small amount of salad oil to the colored water. Point out that the oil is nonpolar. Stir the mixture vigorously. Explain that the oil forms spheres because the oil molecules have an affinity for each other but not for the water molecules.

3 Assess

Check for Understanding

Ask students to write a description of the fluid mosaic model including in their descriptions the following terms: plasma membrane, phospholipid, bilayer, polar, nonpolar, and proteins. Ask volunteers to present their summaries to the class. **L2**

GLENCOE TECHNOLOGY

CD-ROM
Biology: The Dynamics of Life
Exploration: *Parts of a Cell* Disc 1

BIOLOGY JOURNAL

Biology and Art

Linguistic Have students use a dictionary to define *mosaic*. Ask them to describe how this term applies to a piece of art and to the plasma membrane. Have students write about the fluid nature of the membrane and explain why its fluidity is important to living cells. **L2** **ELL**

✓ Portfolio

Plasma Membrane Structures

Visual-Spatial Have students draw and label the parts of the plasma membrane. For the lipids, have the students identify whether the parts are polar or nonpolar. For the proteins, have the students indicate some possible functions.
L2 **ELL** **P**

Reteach

Model the action of a fluid mosaic by filling a plastic tub half full with water. Add just enough Ping-Pong balls to the tub to completely cover the water's surface. Move one ball across the surface and have students note how the other balls jostle each other to make way for the moving ball. A similar demonstration could be done with a tub of red apples and one yellow apple.

Extension

 *Linguistic* Have students research and report on the development of the fluid mosaic model. Ask them to include the work of Gorter and Grendel, Danielli and Davison, and Singer and Nicholson in their reports. **L3**

✔ Assessment

Performance Ask a volunteer to use colored chalk to draw the fluid mosaic model on the chalkboard and describe it to the class. **L2** **ELL**

4 Close

Discussion

Ask students to evaluate the strengths and weaknesses of the analogy that a plasma membrane is similar to the walls of a house. **L1**

Resource Manager

Basic Concepts Transparency 6 and Master **L2** **ELL**
Reinforcement and Study Guide, p. 30 **L2**

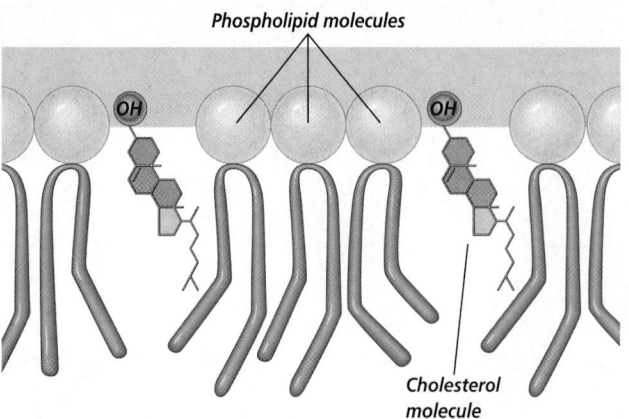

Figure 7.5
Eukaryotic plasma membranes can contain large amounts of cholesterol—up to one molecule for every phospholipid molecule.

This model of the plasma membrane is called the **fluid mosaic model.** It is fluid because the membrane is flexible. The phospholipids move within the membrane just as water molecules move with the currents in a lake. At the same time, proteins embedded in the membrane also move among the phospholipids like boats with their decks above water and hulls below water. These proteins create a "mosaic," or pattern, on the membrane surface.

Other components of the plasma membrane

Cholesterol, shown in *Figure 7.5,* is also found in the plasma membrane where it helps stabilize the phospholipids. Cholesterol is a common topic in health issues today because high levels are associated with reduced blood flow in blood vessels. Yet, for all the emphasis on cholesterol-free foods, it is important to recognize that cholesterol plays a critical role in the stability of the plasma membrane. Cholesterol prevents the fatty acid chains of the phospholipids from sticking together.

You've learned that proteins are found within the lipid membrane. Some proteins span the entire membrane, creating the selectively permeable membrane that regulates which molecules enter and which molecules leave a cell. These proteins are called transport proteins. **Transport proteins** allow needed substances or waste materials to move through the plasma membrane. Other proteins and carbohydrates that stick out from the cell surface help cells identify each other. As you will discover later, these characteristics are important in protecting your cells from infection. Proteins at the inner surface of a plasma membrane play an important role in attaching the plasma membrane to the cell's internal support structure, giving the cell its flexibility.

Section Assessment

Understanding Main Ideas
1. How is the plasma membrane a bilayer structure?
2. Explain how selective permeability maintains homeostasis within the cell.
3. What are the components of the phospholipid bilayer, and how are they organized to form the plasma membrane?
4. Why is the plasma membrane referred to as a fluid mosaic?

Thinking Critically
5. Suggest what might happen if cells grow and reproduce in an environment where no cholesterol is available.

SKILL REVIEW
6. **Recognizing Cause and Effect** Consider that plasma membranes allow materials to pass through them. Explain how this property contributes to homeostasis. For more help, refer to *Thinking Critically* in the **Skill Handbook.**

Section Assessment

1. Phospholipids that form the membrane consist of a double layer.
2. Selective permeability lets cells get rid of wastes, lets critical molecules in, and keeps harmful molecules out.
3. Two lipid layers consisting of a glycerol backbone, two fatty acid tails, and a phosphate group are positioned with the heads facing out and the tails facing in.
4. The lipids and proteins in the membrane are free to move, making a pattern like a mosaic.
5. The membranes would be very fragile and would not hold together.
6. Selective permeability allows materials that are needed for metabolism and survival to enter the cell.

7.3 Eukaryotic Cell Structure

When you work on a group project, each person has his or her own skills and talents that add a particular value to the group's work. In the same way, each component of a eukaryotic cell has a specific job, and all of the parts of the cell work together to help the cell survive.

Cell structures, like this team of students, work together for a common purpose.

SECTION PREVIEW

Objectives

Understand the structure and function of the parts of a typical eukaryotic cell.

Explain the advantages of highly folded membranes in cells.

Compare and contrast the structures of plant and animal cells.

Vocabulary

cell wall
chromatin
nucleolus
ribosome
cytoplasm
endoplasmic reticulum
Golgi apparatus
vacuole
lysosome
chloroplast
plastid
chlorophyll
mitochondria
cytoskeleton
microtubule
microfilament
cilia
flagella

Cellular Boundaries

When a group works together, someone on the team decides what resources are necessary for the project and provides these resources. In the cell, the plasma membrane, shown in *Figure 7.6*, performs this task by acting as a selectively perme- able membrane. The fluid mosaic model describes the plasma membrane as a flexible boundary of a cell. However, plant cells, fungi, most bacteria, and some protists have an additional boundary. The **cell wall** is a fairly rigid structure located outside the plasma membrane that provides additional support and protection.

Figure 7.6
The plasma membrane is made up of two layers, which you can distinguish in this photomicrograph.

Magnification: 375 000×

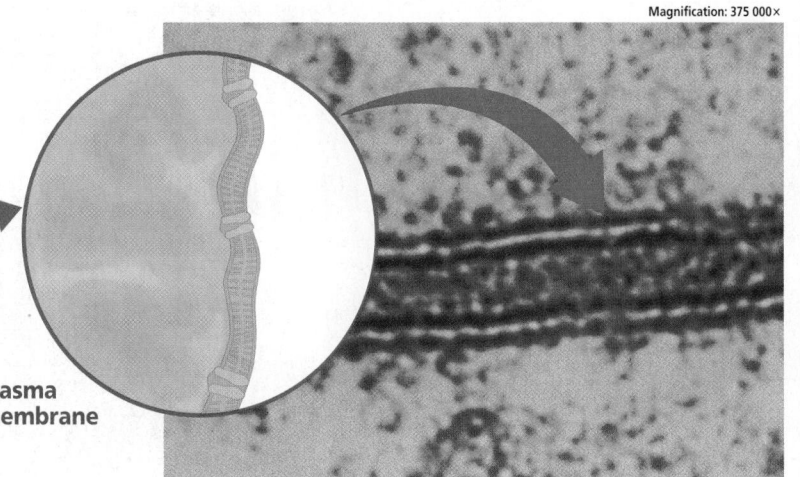

Plasma membrane

7.3 EUKARYOTIC CELL STRUCTURE **185**

BIOLOGY *Online* Note Internet addresses that you find useful in the space below for quick reference.

_____

_____

_____

_____

Resource Manager

Section Focus Transparency 17 and Master **L1** **ELL**

Prepare

Key Concepts

The section describes the structure and function of eukaryotic cells. Detailed art and photos illustrate organelles and show the complexity of eukaryotic cells.

Planning

■ Obtain an onion and iodine solution for the MiniLab.
■ Purchase *Elodea* for the Alternative Lab.
■ Gather materials for the classroom "cell" as suggested in the Portfolio strategy.
■ Purchase gelatin and fruit for the Quick Demo.
■ Collect materials for the BioLab.

---

1 Focus

Bellringer

Before presenting the lesson, display **Section Focus Transparency 17** on the overhead projector and have students answer the accompanying questions. **L1** **ELL**

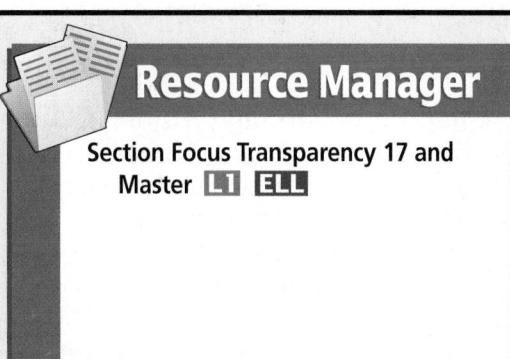

2 Teach

Purpose

Students will become familiar with the role of the nucleus in directing cell activities.

Process Skills

relate cause and effect, interpret data

Background

Two species of *Acetabularia* both have nuclei in their feet but produce different caps. Cutting and replacing the caps allows one to determine the role the nucleus plays in directing production of cellular proteins.

Teaching Strategies

■ You may wish to draw diagrams of the *Acetabularia* species used in this study on the chalkboard or show a transparency and review the results as a class. Lead a discussion about why the final cap ended as it did.

Thinking Critically

The nucleus produces substances that control the type of cap the cell has. The first cap was of intermediate form because the cytoplasm contained substances from both the previous nucleus and the present nucleus. The cytoplasm of the second cap had only substances from the present nucleus.

✓ Assessment

Portfolio Ask students to write a summary of this lab. Encourage them to include diagrams with their summaries. Use the Performance Task Assessment List for Lab Report in **PASC**, p. 47. **L2** **P**

Problem-Solving Lab 7-2   Interpret the Data

What organelle directs cell activity? *Acetabularia,* a type of marine alga, grows as single, large cells 2 to 5 cm in height. The nuclei of these cells are in the "feet." Different species of these algae have different kinds of caps, some petal-like and others that look like umbrellas. If a cap is removed, it quickly grows back. If both cap and foot are removed from the cell of one species and a foot from another species is attached, a new cap will grow. This new cap will have a structure with characteristics of both species. If this new cap is removed, the cap that grows back will be like the cell that donated the nucleus.

The scientist who discovered these properties was Joachim Hämmerling. He wondered why the first cap that grew had characteristics of both species, yet the second cap was clearly like that of the cell that donated the nucleus.

Analysis
Look at the diagram below and interpret the data to explain the results.

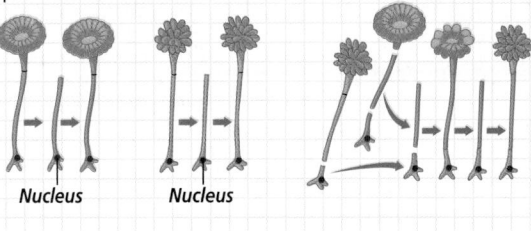

Nucleus   *Nucleus*

Thinking Critically
Why is the final cap like that of the cell from which the nucleus was taken? (HINT: Recall the function of the nucleus.)

The cell wall

The cell wall forms an inflexible barrier that protects the cell and gives it support. *Figure* 7.7 shows a plant cell wall that is made up of a carbohydrate called cellulose. The fibers of cellulose form a thick mesh of fibers. This fibrous cell wall is very porous and allows molecules to pass through, but unlike the plasma membrane, it does not select which molecules can enter into the cell.

Nucleus and cell control

Just as every team needs a leader to direct activity, so the cell needs a leader to give directions. The nucleus is the leader of the eukaryotic cell because it contains the directions to make proteins. Every part of the cell depends on proteins to do its job, so by containing the blueprint to make proteins, the nucleus controls the activity of the organelles. Read the *Problem-Solving Lab* on this page and consider how the *Acetabularia* nucleus controls the cell.

The master set of directions for making proteins is contained in **chromatin,** which are strands of the genetic material, DNA. When the

Magnification: 4500×

Figure 7.7
The cell wall is a firm structure that protects the cell and gives the cell its shape. Plant cell walls are made mainly of cellulose.

Plasma membrane

Cell wall

186 A VIEW OF THE CELL

✓ Portfolio

Observing Plant and Animal Cells

Visual-Spatial Have students observe several kinds of plant and animal cells with a microscope. Ask them to make labeled diagrams of each cell they observe. Beside each label, have them describe the function of each organelle. Have students use their observations to create a table in which they compare the organelles of plant and animal cells. Students' tables should include all organelles shown in their diagrams and identify whether the organelle is common to both plant and animal cells or unique to only one kind of cell. **L2** **P**

cell divides, the chromatin condenses to form chromosomes. Within the nucleus is another organelle called the **nucleolus** that makes ribosomes. **Ribosomes** are the sites where the cell assembles enzymes and other proteins according to the directions of DNA. Unlike other organelles, ribosomes are not bound by a membrane within the cell. Look at some cells as described in the *MiniLab* shown here and try to identify the nucleus in cells of an onion.

For proteins to be made, ribosomes must move out of the nucleus and into the cytoplasm, and the blueprints contained in DNA must be copied and sent to the cytoplasm. **Cytoplasm** is the clear, gelatinous fluid inside a cell. As the ribosomes and the copied DNA are transported to the cytoplasm, they pass through the nuclear envelope—a structure that separates the nucleus from the cytoplasm as shown in *Figure 7.8*. The nuclear envelope is a double membrane made up of two phospholipid bilayers containing small nuclear pores for substances to pass through. Ribosomes and the DNA copy pass into the cytoplasm through the nuclear envelope.

MiniLab 7-2   Experimenting

Cell Organelles Adding stains to cellular material helps you distinquish cell organelles.

Procedure

CAUTION: *Be sure to wash hands before and after this experiment.*

1. Prepare a water wet mount of onion skin. Do this by using your finger nail to peel off the inside of a layer of onion bulb. The layer must be almost transparent. Use the following diagram as a guide.

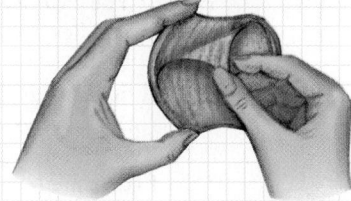

2. Make sure that the onion layer is lying flat on the glass slide and not folded.
3. Observe the onion cells under low- and high-power magnification. Identify as many organelles as possible.
4. Repeat steps 1 through 3, only this time use an iodine stain instead of water.

Analysis

1. What organelles were easily seen in the unstained onion cells? Cells stained with iodine?
2. How are stains useful for viewing cells?

Figure 7.8
The transmission electron photomicrograph shows the nucleus of a eukaryotic cell. The large holes in the nuclear envelope are pores.

Magnification: 17 130×

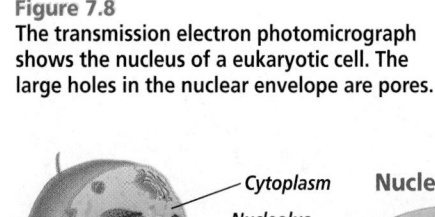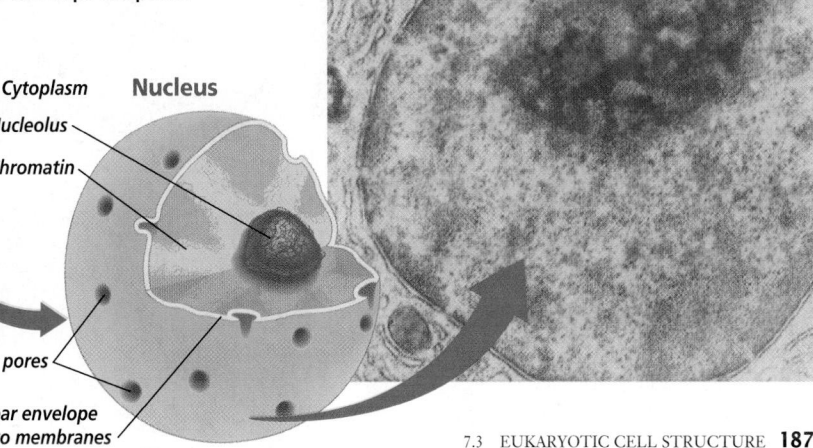

Cytoplasm   **Nucleus**
Nucleolus
Chromatin
Nuclear pores
Nuclear envelope of two membranes

7.3 EUKARYOTIC CELL STRUCTURE **187**

MiniLab 7-2

Purpose
Students will learn the technique and value of using stains for doing microscope observations.

Process Skills
compare and contrast, experiment, observe and infer

Teaching Strategies
■ Review the procedure for making a wet mount.
■ Review proper microscope procedures. Emphasize that reducing light may be critical for viewing onion tissue.
■ Demonstrate how to peel an onion epidermis from the bulb.
■ Iodine stain may be placed in dark dropper bottles—add 1.5 g potassium iodide and 0.3 g iodine to 1 L water.

Expected Results
Students will observe more organelles when cells are stained than when cells are unstained.

Analysis
1. cell wall, cytoplasm; cell wall, cytoplasm, nucleus
2. Staining allows certain organelles to be more easily observed.

✔ Assessment

Skill Ask students to diagram two onion cells, one with and one without stain. Have them label the organelles that were visible under each condition. Use the Performance Task Assessment List for Scientific Drawing in **PASC**, p. 55. **L1** **ELL**

 **Resource Manager**

BioLab and MiniLab Worksheets, p. 32 **L2**
Basic Concepts Transparency 7 and Master **L2** **ELL**

Cultural Diversity

Santiago Ramón y Cajal

Introduce Spanish cell biologist Santiago Ramón y Cajal and his work on nerve cells in the early 1900s. Cajal, together with Italian biologist Camillo Golgi, received the Nobel Prize for Physiology in 1906 for establishing that neurons are the basic units of the nervous system. This research was important to understanding the transmission of nerve impulses. Ramón y Cajal was also responsible for developing cell staining techniques that are still used in today's laboratories.

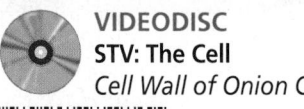
Assembly, Transport, and Storage

You have begun to follow the trail of protein production as directed by the cell manager—the nucleus. But what happens to the blueprints for proteins once they pass into the cytoplasm?

Structures for assembly and transport of proteins

The cytoplasm suspends the cell's organelles. One particular organelle in a eukaryotic cell, the **endoplasmic reticulum** (ER), is the site of cellular chemical reactions. *Figure 7.9* shows how the ER is a series of highly folded membranes suspended in the cytoplasm. The ER is basically a large workspace within the cell. Its folds are similar to the folds of an accordion in that if you spread the folds out it would take up tremendous space. But by pleating and folding it up, the accordion fits its surface area into a compact unit. So by folding the membrane over and over again, a large amount of membrane is available to do work.

Ribosomes in the cytoplasm attach to areas on the endoplasmic reticulum, called rough endoplasmic reticulum, where they carry out the function of protein synthesis. The ribosome's only job is to make proteins. Each protein made in the rough ER has a particular function; it may become the protein that forms a part of the plasma membrane, the protein released from the cell, or the protein transported to other organelles. Ribosomes can also be found floating freely in the cytoplasm where they make proteins that perform tasks within the cytoplasm itself.

Areas of the ER that are not studded with ribosomes are known as smooth endoplasmic reticulum. The smooth ER is involved in numerous biochemical activities, including the production and storage of lipids.

After proteins are produced, they are transferred to another organelle called the **Golgi apparatus** (GAWL jee). The Golgi apparatus as shown in *Figure 7.10* is a flattened system of tubular membranes that modifies the proteins. The Golgi apparatus and membrane-bound structures called vesicles sort the proteins into packages to be sent to the appropriate destination, like mail being sorted at the post office.

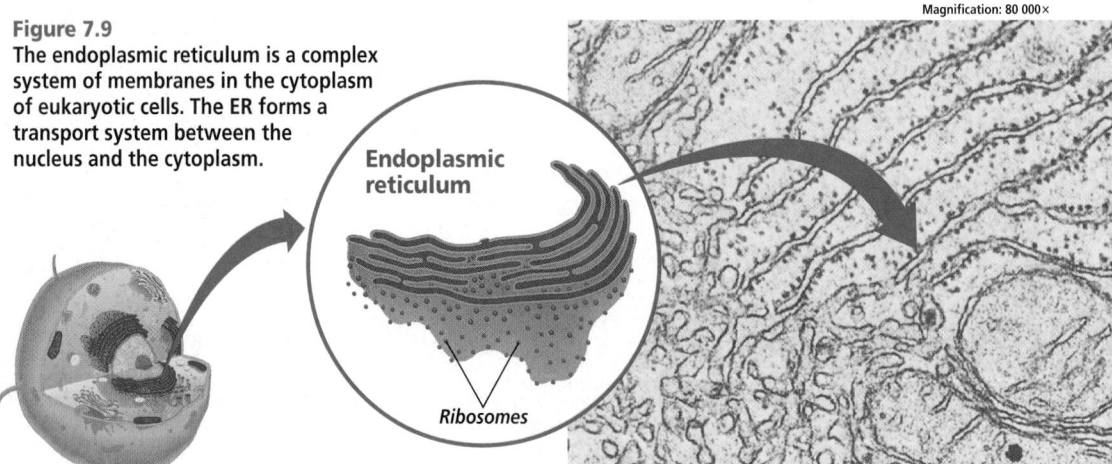

Magnification: 80 000×

Figure 7.9
The endoplasmic reticulum is a complex system of membranes in the cytoplasm of eukaryotic cells. The ER forms a transport system between the nucleus and the cytoplasm.

Endoplasmic reticulum

Ribosomes

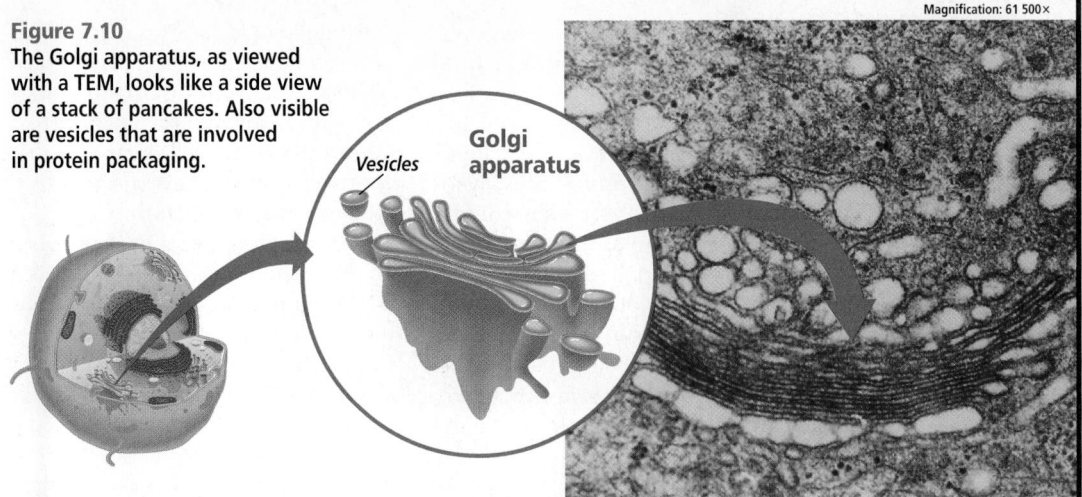

Figure 7.10
The Golgi apparatus, as viewed with a TEM, looks like a side view of a stack of pancakes. Also visible are vesicles that are involved in protein packaging.

Magnification: 61 500×

Vesicles

Golgi apparatus

Vacuoles and storage

Now let's look at some of the other members of the cell team important to the cell's functioning. Cells have membrane-bound spaces, called **vacuoles,** for temporary storage of materials. A vacuole, like those in *Figure 7.11,* is a sac surrounded by a membrane. Vacuoles often store food, enzymes, and other materials needed by a cell, and some vacuoles store waste products. Notice the difference between vacuoles in plant and animal cells.

Lysosomes and recycling

Did anyone ever ask you to take out the trash? You probably didn't consider that action as part of a team effort, but in a cell, it is. **Lysosomes** are organelles that contain digestive enzymes. They digest excess or worn out organelles, food particles, and engulfed viruses or bacteria. The membrane surrounding a lysosome prevents the digestive enzymes inside from destroying the cell. Lysosomes can fuse with vacuoles and dispense their enzymes into the vacuole,

Figure 7.11
Plant cells usually have one large vacuole (a); animal cells contain many smaller vacuoles (b).

Magnification: 1810×

a

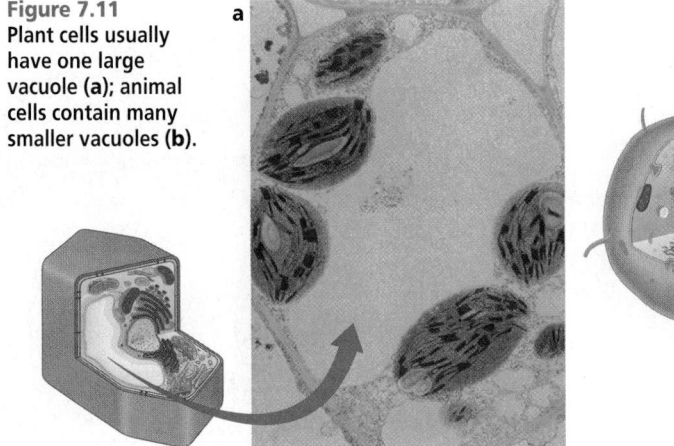

Magnification: 21 840×

b

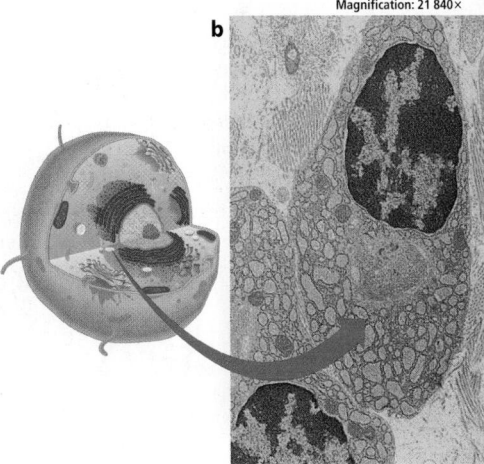

---

4. Place the prepared slide of human cheek cells on the microscope stage and view under low power, moving the slide to center a single cell. Change to high power and observe the cell. Draw the cell and label the structures.

Analysis

1. What structures did you see in each cell? What are their functions? *Likely responses may include cell walls, cell membranes, vacuoles, and nuclei.*

Accept all logical descriptions of the functions of each cell part named.

2. Methylene blue is one of many stains used when observing cells. What is the function of this stain? *Stain makes it possible to see some parts of the cell that might not be visible.*

Revealing Misconceptions

Students often believe that because plants carry on photosynthesis, they do not respire. Point out that, like animals, the cells of plants contain mitochondria. Mitochondria present in all eukaryotic cells produce the ATP that provides the energy for the cells. Therefore, plant cells carry on respiration as well as photosynthesis.

VIDEODISC
STV: The Cell
Parts of the Cell
Unit 2, 7 min. 7 sec.
Parts of the Cell (in its entirety)

Chloroplasts

---

digesting its contents. For example, when an amoeba engulfs a food morsel and encloses it in a vacuole, a lysosome fuses to the vacuole and releases its enzymes, which helps digest the food. Sometimes, lysosomes digest the cells that contain them. For example, when a tadpole develops into a frog, lysosomes within the cells of the tadpole's tail cause its digestion. The molecules thus released are used to build different cells, perhaps in the newly formed legs of the adult frog.

Energy Transformers

Now that you know about a number of the cell parts and have learned what they do, it's not difficult to imagine that each of these cell team members requires a lot of energy. Protein production, modification, transportation, digestion—all of these require energy. Two other organelles, chloroplasts and mitochondria, provide that energy.

Chloroplasts and energy

When you walk through a field or pick a vegetable from the garden, you may not think of the plants as energy

WORD Origin

chloroplast
From the Greek words *chloros*, meaning "green," and *platos*, meaning "formed object." Chloroplasts capture light energy and produce food for plant cells.

generators. In fact, that is exactly what you see. Located in the cells of green plants and some protists, chloroplasts are the heart of the generator. **Chloroplasts** are cell organelles that capture light energy and produce food to store for a later time.

A chloroplast, like a nucleus, has a double membrane. A diagram and a TEM photomicrograph of a chloroplast with an outer membrane and a folded inner membrane system are shown in *Figure 7.12*. It is within these thylakoid membranes that the energy from sunlight is trapped. These inner membranes are arranged in stacks of membranous sacs called grana, which resemble stacks of coins. The fluid that surrounds the grana membranes is called stroma.

The chloroplast belongs to a group of plant organelles called **plastids**, which are used for storage. Some plastids store starches or lipids, whereas others contain pigments, molecules that give color. Plastids are named according to their color or the pigment they contain. Chloroplasts contain the green pigment chlorophyll. **Chlorophyll** traps light energy and gives leaves and stems their green color.

Figure 7.12
Chloroplasts are usually disc shaped but have the ability to change shape and position in the cell as light intensity changes. The pigment chlorophyll is embedded in the inner series of thylakoid membranes.

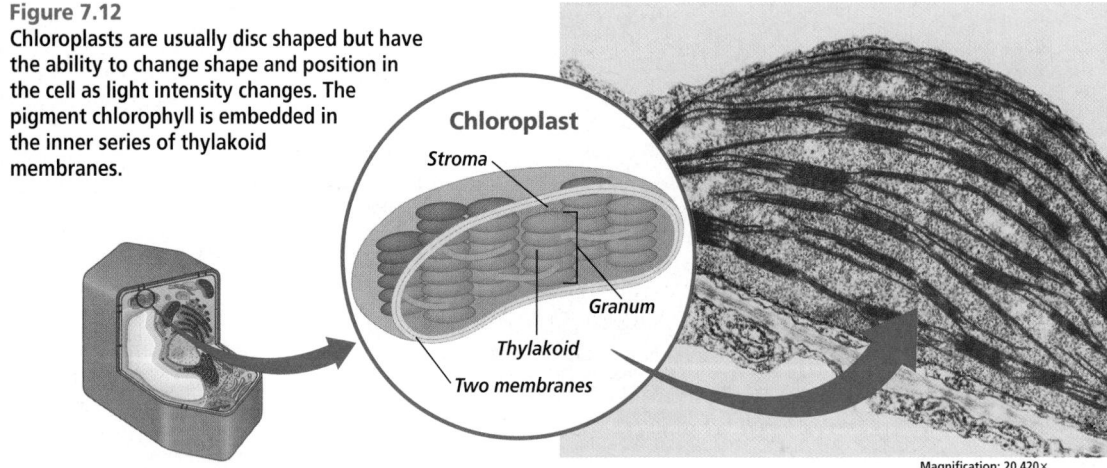

Chloroplast
Stroma
Granum
Thylakoid
Two membranes

Magnification: 20 420×

Portfolio

Formulating Models

Kinesthetic Turn part of your classroom into a giant animal cell. From the ceiling, hang four strings to define the size of your cell. For ease in calculation, make the cell a cube. Have students calculate the magnification factor by dividing the length of a side (in μm) by 20 μm (the length of a liver cell). Have students research the sizes of organelles and calculate how large they need to make each organelle using the magnification factor (cell organelle size × magnification factor). Have students use paper and other materials to build cell organelles to scale and then hang them inside the classroom cell. **L2** **ELL** **P**

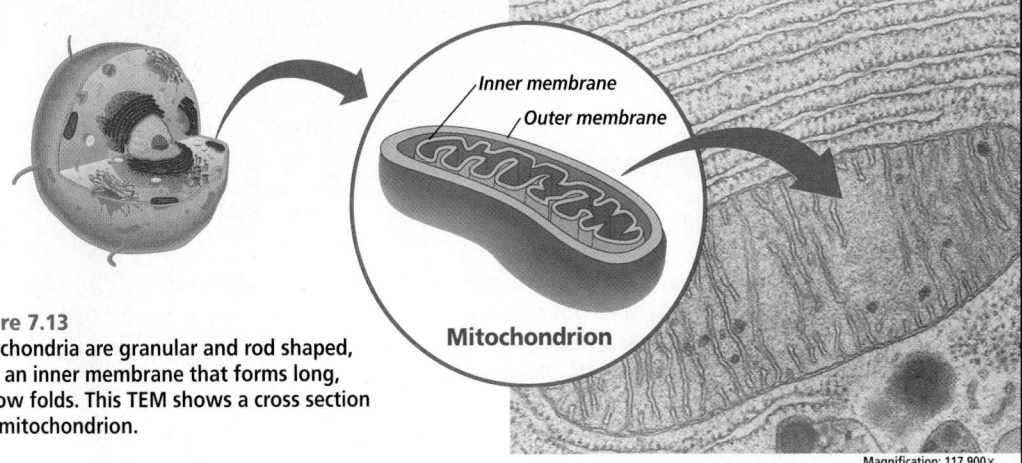

Figure 7.13
Mitochondria are granular and rod shaped, with an inner membrane that forms long, narrow folds. This TEM shows a cross section of a mitochondrion.

Inner membrane
Outer membrane
Mitochondrion

Magnification: 117 900×

Mitochondria and energy

The food energy generated by choloroplasts is stored until it is broken down and released by mitochondria, shown in *Figure 7.13*. **Mitochondria** are membrane-bound organelles in plant and animal cells that transform energy for the cell. This energy is then stored in other molecules that allow the cell organelles to use the energy easily and quickly when it is needed.

A mitochondrion has an outer membrane and a highly folded inner membrane. As with chloroplasts, the folds of the inner membrane provide a large surface area that fits in a small space. Energy-storing molecules are produced on the inner folds. Mitochondria occur in varying numbers depending on the function of the cell. For example, liver cells may have up to 2500 mitochondria.

Although the process by which energy is produced and used in the cells is a technical concept to learn, the *Literature Connection* at the end of this chapter explains how cellular processes can also be inspiring. Look at the *Inside Story* on the next page to compare plant and animal cells. Notice how similar they are.

Structures for Support and Locomotion

Scientists once thought that cell organelles just floated in a sea of cytoplasm. More recently, cell biologists have discovered that cells have a support structure called the **cytoskeleton** within the cytoplasm. The cytoskeleton is composed of a variety of tiny rods and filaments that form a framework for the cell, like the skeleton that forms the framework for your body. However, unlike your bones, the cytoskeleton is a constantly changing structure.

Cellular support

The cytoskeleton is a network of thin, fibrous elements that acts as a sort of scaffold to provide support for organelles. It maintains cell shape similar to the way that poles maintain the shape of a tent. The cytoskeleton is composed of microtubules and microfilaments that are associated with cell shape and assist organelles in moving from place to place within the cell. **Microtubules** are thin, hollow cylinders made of protein. **Microfilaments** are thin, solid protein fibers.

WORD *Origin*
cytoskeleton
From the Latin word *cyte*, meaning "cell." The cytoskeleton provides support and structure for the cell.

The BioLab at the end of the chapter can be used at this point in the lesson. →

Purpose

Students will compare plant and animal cells and discover the similarities.

Teaching Strategies

■ Review cell organelles with your class, then use this Inside Story to compare plant and animal cells with regard to organelles. Be sure to point out the number and type of organelles these cells have in common. Also emphasize the organelles and structures that are unique to plant and animal cells.

Visual Learning

■ Have students draw an animal cell and a plant cell on the board. Label the cell wall, vacuole, and centrioles.

■ Make a table with the headings: nucleus, ribosomes, centrioles, endoplasmic reticulum, vacuoles, mitochondria, plasma membrane, and cell wall. Underneath the headings list animal cell and plant cell. Have the class complete the table using this Inside Story.

Critical Thinking

Most cells, regardless of whether they belong to a plant or an animal, must perform the same critical functions to survive, grow, and reproduce. Therefore, plant *and* animal cells have many of the same organelles to meet these needs.

192

Moose eating water plants

Comparing Animal and Plant Cells

You can easily recognize that a person does not look like a flower and an ant does not resemble a tree. But at the cellular level under a microscope, the cells that make up all of the different animals and plants of the world are very much alike.

Critical Thinking *Why are animal and plant cells similar?*

① **Animal Cells** The centriole is the only organelle unique to animal cells. Animal cells typically have many small vacuoles.

② **Plant Cells** Plant cells, in general, are larger than animal cells and are characterized by a cell wall and chloroplasts. Plant cells usually have one large vacuole.

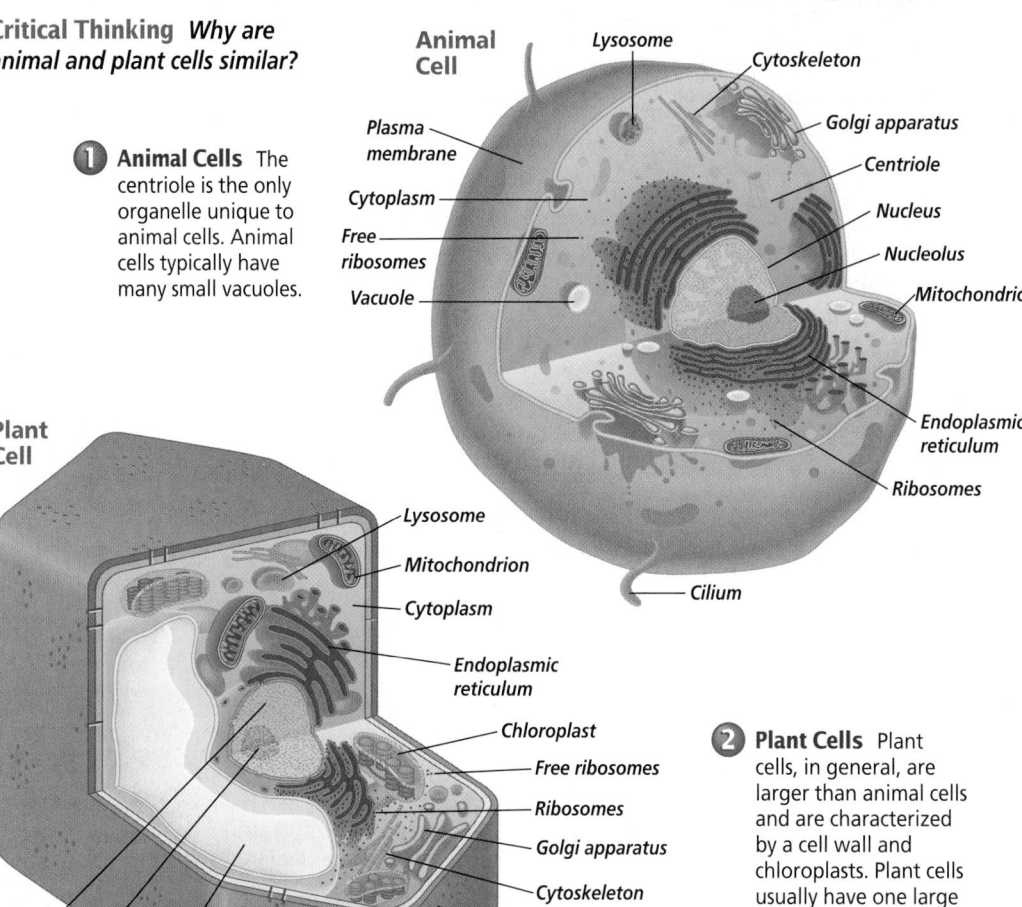

Animal Cell — Lysosome, Cytoskeleton, Plasma membrane, Golgi apparatus, Cytoplasm, Centriole, Free ribosomes, Nucleus, Vacuole, Nucleolus, Mitochondrion, Endoplasmic reticulum, Ribosomes, Cilium

Plant Cell — Lysosome, Mitochondrion, Cytoplasm, Endoplasmic reticulum, Chloroplast, Free ribosomes, Ribosomes, Golgi apparatus, Cytoskeleton, Cell wall, Plasma membrane, Nucleus, Nucleolus, Vacuole with cell sap

192 A VIEW OF THE CELL

GLENCOE TECHNOLOGY

VIDEODISC
The Secret of Life

Animal Cell

Plant Cell

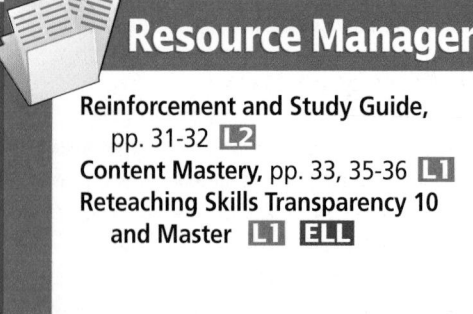

Resource Manager

Reinforcement and Study Guide, pp. 31-32 **L2**
Content Mastery, pp. 33, 35-36 **L1**
Reteaching Skills Transparency 10 and Master **L1** **ELL**

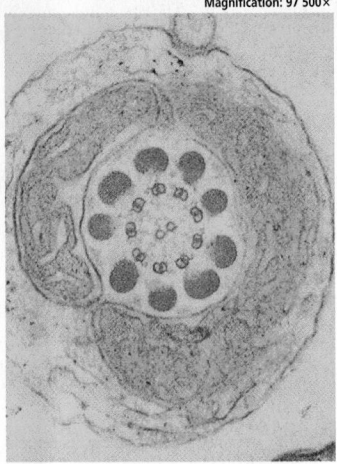
Magnification: 97 500×

Magnification: 4610×

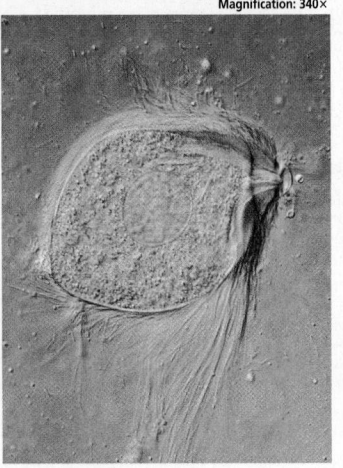
Magnification: 340×

A In eukaryotic cells, both cilia and flagella are composed of microtubules arranged in a ring.

B Cilia in the windpipe beat and propel particles of dirt and mucus toward the mouth and nose where they are expelled.

C The flagella of this *Trichonympha* move the parasitic flagellate forward with their whiplike action.

Cilia and flagella

Some cell surfaces have cilia and flagella, which are structures that aid in locomotion or feeding. Cilia and flagella are composed of pairs of microtubules, with a central pair surrounded by nine additional pairs, as shown in *Figure 7.14*. The entire structure is enclosed by the plasma membrane. The outer microtubules have a protein that allows a pair of microtubules to slide along an adjacent pair. This causes the cilium or flagellum to bend.

Cilia and flagella can be distinguished by their structure and by the nature of their action. **Cilia** are short, numerous, hairlike projections that move in a wavelike motion. **Flagella** are longer projections that move with a whiplike motion. In unicellular organisms, cilia and flagella are the major means of locomotion.

Figure 7.14
Many cells of animals and protists are covered with cilia or flagella.

Section Assessment

Understanding Main Ideas
1. What is the advantage of highly folded membranes in a cell? Name an organelle that uses this strategy.
2. What organelles would be especially numerous in a cell that produces large amounts of a protein product?
3. Why are digestive enzymes in a cell enclosed in a membrane-bound organelle?
4. Why might a cell need a cell wall in addition to a plasma membrane?

Thinking Critically
5. How do your cells and the cells of other organisms that are not green plants obtain food energy from the chloroplasts of green plants?

SKILL REVIEW
6. **Observing and Inferring** Some cells have large numbers of mitochondria with many internal folds. Other cells have few mitochondria and, therefore, fewer internal folds. What can you conclude about the functions of these two types of cells? For more help, refer to *Observing and Inferring* in the **Skill Handbook**.

Section Assessment

1. The folding increases the surface area where chemical reactions occur. Examples: endoplasmic reticulum and mitochondria.
2. rough endoplasmic reticulum and Golgi apparatus
3. If these enzymes were free in the cytoplasm, they could damage the cell.
4. Cell walls provide additional support and protection for plant cells.

5. Organisms that are not green plants obtain the food energy of green plants by eating them. For example, humans eat salads or other animals that have ingested green plants.
6. The cells with many mitochondria produce more energy and perform more work than the cells with fewer mitochondria.

---

3 Assess

Check for Understanding
Ask students what would happen if a cell had a decreased number of mitochondria. *The cell would probably not have enough energy to carry out all life functions.* **L2**

Reteach
Interpersonal Ask students working in groups to role-play the organelles of a typical animal cell. Students should show how organelles work together to keep the entire cell functioning. **L1** **ELL** **COOP LEARN**

Extension
Interpersonal Have students play vocabulary football with cell terms. Divide the class into two teams. Ask a question of one team. If they answer correctly, advance the ball 10 yards on a football field drawn on the chalkboard. If they miss the question, the question should be given to the other team. The team that scores the most touchdowns wins. **L2**

✓ Assessment
Skill Have students construct a table to summarize cell structures and their functions. Working in pairs, students can quiz each other on information in their tables. **L2**

---

4 Close

Discussion
Discuss the analogy of comparing a cell to a team. Have students explain where it is accurate and where it fails. **L1**

GLENCOE TECHNOLOGY

CD-ROM
Biology: The Dynamics of Life
Bioquest: *Cellular Pursuit* Disc 1

Observing and Comparing Different Cell Types

Are all cells alike in appearance, shape, and size? Do all cells have the same organelles present within their cell boundaries? One way to answer these questions is to observe a variety of cells using a light microscope.

Time Allotment
One class period

Process Skills
observe and infer, measure in SI, hypothesize, compare and contrast, use numbers

Safety Precautions
Students must handle prepared slides, glass slides, and especially cover slides with care.

PREPARATION

Gather the materials for the lab and review microscopy procedures prior to performing the lab.

Alternative Materials
■ *Elodea* is available in pet shops. Prepared slides may be substituted.
■ Epithelium slides are available as prepared frog skin or human cheek cells in place of frog blood. Do not substitute human blood.
■ Any prepared slide of stained bacteria may be used.

PREPARATION

Problem
Are all cells alike in appearance and size?

Objectives
In this BioLab, you will:
■ **Observe, diagram**, and **measure** cells and their organelles.
■ **Hypothesize** which cells are from prokaryotes, eukaryotes, unicellular organisms, and multicellular organisms.
■ **List** the traits of plant and animal cells.

Materials
microscope dropper
glass slide coverslip
water forceps
prepared slides of *Bacillus subtilus*, frog blood, and *Elodea*

Safety Precautions
Always wear goggles in the lab.

Skill Handbook
Use the **Skill Handbook** if you need additional help with this lab.

PROCEDURE

1. Copy the data table.
2. Examine a prepared slide of *Bacillus subtilus* using both low- and high-power magnification. (NOTE: this slide has been stained. Its natural color is clear.)

Data Table

	Bacillus subtilus	*Elodea*	Frog blood
Organelles observed			
Prokaryote or eukaryote			
From a multicellular or unicellular organism			
Diagram (with size in micrometers, μm)			

PROCEDURE

Teaching Strategies
■ Review the technique used for preparing a wet mount.
■ Review the technique used for measuring objects under the microscope (see MiniLab 7-1). Review the conversion of millimeters to micrometers.

194

■ Do not tell students in advance the nature of each cell type being observed. That is, do not tell them that *Bacillus subtilus* is a bacterium.

Troubleshooting
■ Individual *Elodea* cells are best observed along the leaf edge where the thickness of tissue is usually one cell.

■ If students cannot see organelles, have them practice changing their depth of view by rapidly moving the fine adjustment back and forth while viewing a cell under high-power magnification.
■ Mark the diameter of this circle as 1500 μm. Draw several objects in the circle and have students estimate the size of each.

CAUTION: *Use care when handling slides. Dispose of any broken glass in a container provided by your teacher.*

3. Look for and record the names of any observed organelles. Hypothesize if these cells are prokaryotes or eukaryotes. Hypothesize if these cells are from a unicellular or multicellular organism. Record your findings in the table.
4. Diagram one cell as seen under high-power magnification.
5. While using high power, determine the length and width in micrometers of this cell. Refer to *Practicing Scientific Methods* in the **Skill Handbook** for help with determining magnification. Record your measurements on the diagram.
6. Prepare a wet mount of a single leaf from *Elodea* using the diagram as a guide.
7. Observe cells under low and high power magnification.
8. Repeat steps 3 through 5 for *Elodea*.

9. Examine a prepared slide of frog blood. (NOTE: This slide has been stained. Its natural color is pink.)
10. Observe cells under low- and high-power magnification.
11. Repeat steps 3 through 5 for frog blood cells.

Leopard frog

ANALYZE AND CONCLUDE

1. **Observing and Inferring** Which cells were prokaryotes? How were you able to tell?
2. **Observing and Inferring** Which cells were eukaryotes? How were you able to tell?
3. **Predicting** Which cell was from a plant, from an animal? Explain your answer.
4. **Measuring** Are prokaryote or eukaryote cells larger? Give specific measurements to support your answer.
5. **Defining Operationally** Describe how plant and animal cells are alike and how they differ.

Going Further

Application Prepare a wet mount of very thin slices of bamboo (saxophone reed). Observe under low and high power. Explain what structures you are looking at. Explain the absence of all other organelles from this material.

BIOLOGY Online To find out more about microscopy and cell types, visit the Glencoe Science Web site.
science.glencoe.com

Data and Observations

Bacterial cells are prokaryotic with no visible internal organelles; size ranges between 10 and 20 μm. *Elodea* cells are eukaryotic with visible organelles such as chloroplasts and a nucleus; size ranges between 40 and 80 μm. Blood cells are from a multicellular, eukaryotic organism–and are approximately 40–50 μm.

Resource Manager

Literature
Connection

The Lives of a Cell
by Lewis Thomas

Purpose

Students will be introduced to the thoughts and scientific writings of Lewis Thomas.

Teaching Strategies

■ Lewis Thomas said that his writings are "Notes of a Biology Watcher." Have students discuss what this phrase means and explain how all people are "Biology Watchers."

■ Have students read all of Lewis Thomas's *Lives of a Cell* or his book *Medusa and the Snail*. Ask students to prepare a written report on the book they read.

Connection to Biology

Writing and expressing scientific ideas are important to every branch of science, including biology. Although much of science reporting is done in a technical language, creative and expressive language also has a place in science communication. **L2**

You may think of yourself as a body made up of parts. Arms, legs, skin, stomach, eyes, brain, heart, lungs. Your mind controls the whole, and you probably believe that you own all the parts that make up your body. In actual fact, you are a community of living structures that work together for growth and survival.

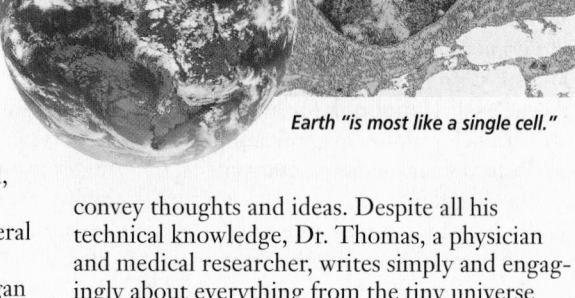

Magnification: 15 600×

Earth "is most like a single cell."

Your body is made up of eukaryotic cells with organelles that work together for each cell's survival. Organelles may work closely together, such as a ribosome and the endoplasmic reticulum, or they may perform a unique function within the cell, such as the mitochondrion.

An organism is similar to a cell in that several parts work together. Groups of cells work together as tissues. Several tissues form an organ and many organs form an organ system. For example, in an organ system such as the digestive system, cells and tissues form an organ such as the stomach, but several organs such as the intestines, the pancreas, and the liver, are needed to completely digest and absorb the food you eat.

In the same way, you might also consider how all the organisms in a community are interconnected and how the whole planet Earth is a collection of interdependent ecosystems. Lewis Thomas pondered this thought.

"I have been trying to think of the earth as a kind of organism, but it is no go. I cannot think of it this way. It is too big, too complex, with too many working parts lacking visible connections.... I wondered about this. If not like an organism, what is it like, what is it most like? Then, satisfactorily for that moment, it came to me: it is most like a single cell."

Words are like organelles Just as a cell is a group of organelles working together, so is a paragraph composed of words that together

convey thoughts and ideas. Despite all his technical knowledge, Dr. Thomas, a physician and medical researcher, writes simply and engagingly about everything from the tiny universe inside a single cell to the possibility of visitors from a distant planet.

Medicine, a young science Dr. Thomas grew up with the practice of medicine. As a boy, he accompanied his father, a family physician, on house calls to patients. Years later, Dr. Thomas described those days in his autobiography, *The Youngest Science*. The title reflects his belief that the practice of medicine is "still very early on" and that some basic problems of disease are just now yielding to exploration.

CONNECTION TO BIOLOGY

After you have studied this chapter, write a paragraph using Dr. Thomas's style to describe how the organelles of a cell work together for cell survival.

BIOLOGY *Online* To find out more about the works of Dr. Lewis Thomas, visit the Glencoe Science Web site. **science.glencoe.com**

BIOLOGY *Online*  Note Internet addresses that you find useful in the space below for quick reference.

_____

_____

_____

_____

SUMMARY

Section 7.1

The Discovery of Cells

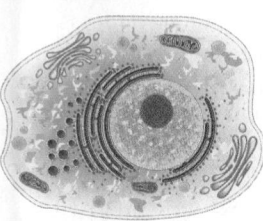

Main Ideas

- Microscopes enabled biologists to see cells and develop the cell theory.
- The cell theory states that the cell is the basic unit of organization, all organisms are made up of one or more cells, and all cells come from preexisting cells.
- Using electron microscopes, scientists can study cell structure in detail.
- Cells are classified as prokaryotic or eukaryotic based on whether or not they have membrane-bound organelles.

Vocabulary

cell (p. 175)
cell theory (p. 176)
compound light microscope (p. 175)
electron microscope (p. 176)
eukaryote (p. 177)
nucleus (p. 180)
organelle (p. 177)
prokaryote (p. 177)

Section 7.2

The Plasma Membrane

Main Ideas

- Through selective permeability, the plasma membrane controls what enters and leaves a cell.
- The fluid mosaic model describes the plasma membrane as a phospholipid bilayer with embedded proteins.

Vocabulary

fluid mosaic model (p. 184)
homeostasis (p. 181)
phospholipid (p. 182)
plasma membrane (p. 181)
selective permeability (p. 181)
transport proteins (p. 184)

Section 7.3

Eukaryotic Cell Structure

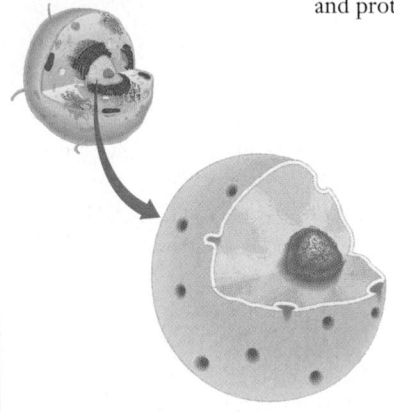

Main Ideas

- Eukaryotic cells have a nucleus and organelles, are enclosed by a plasma membrane, and some have a cell wall that provides support and protection.
- Cells make proteins on ribosomes that are often attached to the highly folded endoplasmic reticulum. Cells store materials in the Golgi apparatus and vacuoles.
- Mitochondria break down food molecules to release energy. Chloroplasts convert light energy into chemical energy.
- The cytoskeleton helps maintain cell shape, is involved in the movement of organelles and cells, and resists stress placed on cells.

Vocabulary

cell wall (p. 185)
chlorophyll (p. 190)
chloroplast (p. 190)
chromatin (p. 186)
cilia (p. 193)
cytoplasm (p. 187)
cytoskeleton (p. 191)
endoplasmic reticulum (p. 188)
flagella (p. 193)
Golgi apparatus (p. 188)
lysosome (p. 189)
microfilament (p. 191)
microtubule (p. 191)
mitochondria (p. 191)
nucleolus (p. 187)
plastid (p. 190)
ribosome (p. 187)
vacuole (p. 189)

Main Ideas

Summary statements can be used by students to review the major concepts of the chapter.

Using the Vocabulary

To reinforce chapter vocabulary, use the Content Mastery Booklet and the activities in the Interactive Tutor for Biology: The Dynamics of Life on the Glencoe Science Web site: **science.glencoe.com**

 *All Chapter Assessment questions and answers have been validated for accuracy and suitability by The Princeton Review.*

CHAPTER 7 ASSESSMENT **197**

GLENCOE TECHNOLOGY

 **VIDEOTAPE**
MindJogger Videoquizzes
Chapter 7: *A View of the Cell*
Have students work in groups as they play the videoquiz game to review key chapter concepts.

 ## Resource Manager

Chapter Assessment, pp. 37-42
MindJogger Videoquizzes
ExamView® Pro Software 💾
BDOL Interactive CD-ROM, Chapter 7 quiz

UNDERSTANDING MAIN IDEAS

1. c
2. b
3. c
4. d
5. b
6. a
7. c
8. b
9. d
10. d
11. vacuoles
12. protein synthesis
13. transmission electron microscope (TEM)
14. plasma membrane
15. cell walls, cellulose
16. ribosomes

UNDERSTANDING MAIN IDEAS

1. What type of cell would you examine to find a chloroplast?
   a. prokaryote   **c.** plant
   b. animal   **d.** fungus

2. Which of the following structures utilizes the sun's energy to make carbohydrates?

   a. 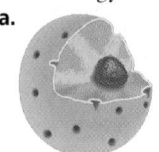   **c.**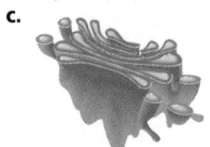

   b.    **d.**

3. Which of the following pairs of terms is NOT related?
   a. nucleus—DNA
   b. chloroplasts—chlorophyll
   c. flagella—chromatin
   d. cell wall—cellulose

4. Magnifications greater than 10 000 times can be obtained when using _____.
   a. light microscopes
   b. metric rulers
   c. hand lenses
   d. electron microscopes

5. A bacterium is classified as a prokaryote because it _____.
   a. has cilia
   b. has no membrane-bound nucleus
   c. is a single cell
   d. has no DNA

 THE PRINCETON REVIEW   **TEST-TAKING TIP**

Maximize Your Score
Ask how your test will be scored. In order to do your best, you need to know if there is a penalty for guessing, and if so, how much of a penalty. If there is no random-guessing penalty at all, you should always fill in an answer.

6. Which of the following structures is NOT found in both plant and animal cells?
   a. chloroplast   **c.** ribosomes
   b. cytoskeleton   **d.** mitochondria

7. Which component is NOT stored in plastids?
   a. lipids   **c.** amino acids
   b. pigments   **d.** starches

8. Which is a main idea of the cell theory?
   a. All cells have a plasma membrane.
   b. All cells come from preexisting cells.
   c. All cells are microscopic.
   d. All cells are made of atoms.

9. Electron microscopes can view only dead cells because _____.
   a. only dead cells are dense enough to be seen
   b. a magnetic field is needed to focus the electrons
   c. the fluorescent screen in the microscope kills the cells
   d. the specimen must be in a vacuum

10. Ribosomes _____.
   a. do not have a cell wall
   b. are not surrounded by a membrane
   c. do not contain cytoplasm
   d. all of the above

11. _____ are membrane-bound spaces that serve as temporary storage areas.

12. The small bumps shown in this photomicrograph are the site of _____.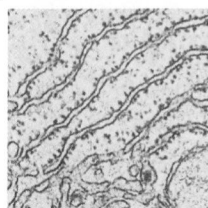

13. The photomicrograph in question 12 was probably taken using a _____ microscope.

14. The _____ maintains a chemical balance within a cell by regulating the materials that enter and leave the cell.

15. Plants are able to grow tall because their cells have rigid _____ that contain a strong network of _____.

16. Smooth ER is different from rough ER in that smooth ER has no _____.

17. Although prokaryotes lack _____, they still contain DNA.

18. Microtubules and microfilaments, which make up the cell cytoskeleton, are composed of _____.

19. A plant cell has a green color due to the presence of _____, a pigment that is embedded in the _____ membranes of the _____.

20. Cilia and flagella are an arrangement of _____ and allow the cell to _____.

APPLYING MAIN IDEAS

21. Explain why packets of proteins collected by the Golgi apparatus merge with lysosomes.

22. How does the structure of the plasma membrane allow materials to move across it in both directions?

THINKING CRITICALLY

23. Making Predictions Predict whether you would expect muscle or fat cells to contain more mitochondria and explain why.

24. Concept Mapping Complete the concept map using the following vocabulary terms: cytoplasm, mitochondria, Golgi apparatus, ribosomes, plasma membrane, nucleus.

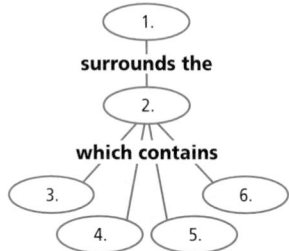

CD-ROM

For additional review, use the assessment options for this chapter found on the *Biology: The Dynamics of Life Interactive CD-ROM* and on the Glencoe Science Web site.
science.glencoe.com

ASSESSING KNOWLEDGE & SKILLS

The diagram below shows the parts of a cell.

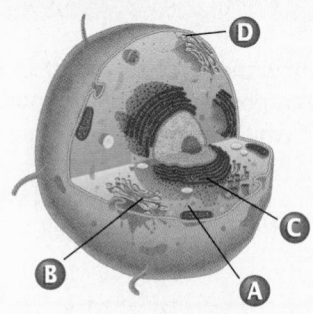

Interpreting Scientific Illustrations Use the diagram to answer the following questions.

1. The structure labeled C represents the _____.
 a. plasma membrane
 b. nuclear membrane
 c. endoplasmic reticulum
 d. nucleolus

2. The function of the circular structures on membrane C is to _____.
 a. synthesize cellulose
 b. transform energy
 c. synthesize proteins
 d. capture the sun's energy

3. The structure labeled B represents the _____.
 a. lysosome   **c.** nucleus
 b. Golgi apparatus   **d.** vacuole

4. The type of cell shown is a _____ cell.
 a. plant   **c.** animal
 b. fungal   **d.** prokaryotic

5. Sequencing Structures A, B, C, and D are involved in making a product to be released to the outside of the cell. What is the sequence of the production of this product?

---

17. organelles/a nucleus
18. proteins
19. chlorophyll, thylakoid membranes, chloroplasts
20. microtubules, move

APPLYING MAIN IDEAS

21. If a particular protein is not needed immediately, its building blocks can be recycled into other proteins.

22. Some protein transporters carry materials out of the cell and some carry materials into the cell. The lipid bilayer itself allows diffusion of small molecules in either direction.

THINKING CRITICALLY

23. Muscle cells are very active cells that use a lot of energy, whereas fat cells are used mainly for storage of fat. Mitochondria are the cell organelles that transform energy for the cell. Therefore, muscle cells will contain more mitochondria.

24. 1. Plasma membrane; 2. Cytoplasm; 3. Mitochondria; 4. Golgi apparatus; 5. Ribosomes; 6. Nucleus

---

ASSESSING KNOWLEDGE & SKILLS

Structure A: ribosomes
Structure B: Golgi apparatus
Structure C: endoplasmic reticulum
Structure D: vesicles

Interpreting Scientific Illustrations

1. c
2. c
3. b
4. c
5. The ribosomes (A) synthesize proteins,

which are transported by the endoplasmic reticulum (C) to the Golgi apparatus (B), which packages the proteins into vesicles (D), which are then transported to cell membranes for release to the outside of the cell.

Chapter 8 Organizer

Refer to pages 4T-5T of the Teacher Guide for an explanation of the National Science Education Standards correlations.

Section	Objectives	Activities/Features
Section 8.1 **Cellular Transport** National Science Education Standards UCP.1-3, UCP.5; A.1, A.2; B.6; C.1, C.5 (2 session, 1 block)	1. **Explain** how the processes of diffusion, passive transport, and active transport occur and why they are important to cells. 2. **Predict** the effect of a hypotonic, hypertonic, or isotonic solution on a cell.	**MiniLab 8-1:** Cell Membrane Simulation, p. 204
Section 8.2 **Cell Growth and Reproduction** National Science Education Standards UCP.1-3, UCP.5; A.1, A.2; C.1, C.5; G.1-3 (2 sessions, 2 blocks)	3. **Sequence** the events of the cell cycle. 4. **Relate** the function of a cell to its organization as a tissue, organ, and an organ system.	**Problem-Solving Lab 8-1,** p. 209 **Problem-Solving Lab 8-2,** p. 210 **Inside Story:** The Cell Cycle, p. 211 **MiniLab 8-2:** Seeing Asters, p. 215 **Investigate BioLab:** Where is mitosis most common? p. 220
Section 8.3 **Control of the Cell Cycle** National Science Education Standards UCP.1, UCP.2; A.1, A.2; C.1, C.6; E.1, E.2; F.1, F.4, F.5, F.6; G.1, G.2 (2 sessions, $\frac{1}{2}$ block)	5. **Describe** the role of enzymes in the regulation of the cell cycle. 6. **Distinguish** between the events of a normal cell cycle and the abnormal events that result in cancer. 7. **Identify** ways to potentially reduce the risk of cancer.	**Problem-Solving Lab 8-3,** p. 218 **Health Connection:** Skin Cancer, p. 222

Need Materials? Contact Carolina Biological Supply Company at 1-800-334-5551 or at **http://www.carolina.com**

MATERIALS LIST

BioLab
p. 220 microscope, prepared slide of onion root tip

MiniLabs
p. 204 small beaker, starch solution, iodine solution, small plastic bag, twist tie
p. 215 microscope, prepared slide of "fish mitosis"

Alternative Lab
p. 202 potato, 100-mL beakers (2), table salt, water, graduated cylinder, labels, pen, stirring rod, balance, plastic wrap or aluminum foil, knife

Quick Demos
p. 202 microprojector, microscope slide, coverslip, India ink, water
p. 208 lamp cord, string, rubber band

Key to Teaching Strategies

L1 Level 1 activities should be appropriate for students with learning difficulties.

L2 Level 2 activities should be within the ability range of all students.

L3 Level 3 activities are designed for above-average students.

ELL ELL activities should be within the ability range of English Language Learners.

COOP LEARN Cooperative Learning activities are designed for small group work.

P These strategies represent student products that can be placed into a best-work portfolio.

These strategies are useful in a block scheduling format.

Cellular Transport and the Cell Cycle

Teacher Classroom Resources

Section	Reproducible Masters	Transparencies
Section 8.1 **Cellular Transport**	Reinforcement and Study Guide, p. 33 **L2** Concept Mapping, p. 8 **L3** **ELL** BioLab and MiniLab Worksheets, p. 35 **L2** Laboratory Manual, pp. 55-56 **L2** Content Mastery, pp. 37-38, 40 **L1**	Section Focus Transparency 18 **L1** **ELL** Basic Concepts Transparency 8 **L2** **ELL** Basic Concepts Transparency 9 **L2** **ELL** Reteaching Skills Transparency 11 **L1** **ELL** Reteaching Skills Transparency 12 **L1** **ELL**
Section 8.2 **Cell Growth and Reproduction**	Reinforcement and Study Guide, pp. 34-35 **L2** BioLab and MiniLab Worksheets, p. 36 **L2** Laboratory Manual, pp. 57-60 **L2** Content Mastery, pp. 37, 39-40 **L1**	Section Focus Transparency 19 **L1** **ELL** Basic Concepts Transparency 10 **L2** **ELL** Reteaching Skills Transparency 13 **L1** **ELL**
Section 8.3 **Control of the Cell Cycle**	Reinforcement and Study Guide, p. 36 **L2** Critical Thinking/Problem Solving, p. 8 **L3** BioLab and MiniLab Worksheets, pp. 37-38 **L2** Content Mastery, pp. 37, 39-40 **L1**	Section Focus Transparency 20 **L1** **ELL**

Assessment Resources

Chapter Assessment, pp. 43-48
MindJogger Videoquizzes
Performance Assessment in the Biology Classroom
Alternate Assessment in the Science Classroom
ExamView® Pro Software 💾
BDOL Interactive CD-ROM, Chapter 8 quiz

Additional Resources

Spanish Resources **ELL**
English/Spanish Audiocassettes **ELL**
Cooperative Learning in the Science Classroom **COOP LEARN**
Lesson Plans/Block Scheduling

⬛ NATIONAL GEOGRAPHIC — Teacher's Corner

Products Available From Glencoe
To order the following products, call Glencoe at 1-800-334-7344:
CD-ROM
NGS PictureShow: The Cell
Curriculum Kit
GeoKit: Cells and Microorganisms
Transparency Set
NGS PicturePack: The Cell

Products Available From National Geographic Society
To order the following products, call National Geographic Society at 1-800-368-2728:
Video
Discovering the Cell

GLENCOE TECHNOLOGY

The following multimedia resources are available from Glencoe.

Biology: The Dynamics of Life
CD-ROM **ELL**

   Animation: *The Cell Cycle*
   Exploration: *Phases of Mitosis*

Videodisc Program

   Passive Transport
   Active Transport
   The Cell Cycle

The Infinite Voyage

   The Living Clock

The Secret of Life Series

   Osmosis Demonstration

Theme Development

A major theme of the chapter is **homeostasis** as it relates to the function of the plasma membrane in regulating cellular transport and as a critical factor in successful cellular reproduction.

Another theme of the chapter is **unity within diversity.** This theme is evident as the striking similarities of the process of mitosis in cells of both plants and animals are presented.

0:00 OUT OF TIME?

If time does not permit teaching the entire chapter, use the BioDigest at the end of the unit as an overview.

READING BIOLOGY

Glencoe's *Biology: The Dynamics of Life* contains many resources to assist a student's reading skills. Each chapter contains figures with expanded captions that expand on written material. Word Origins, located along the side of text, expand knowledge of biology vocabulary. Glencoe's Content Mastery Booklet helps develop reading skills while reinforcing content. In addition, use the Interactive Tutor for *Biology: The Dynamics of Life* on the Glencoe Web site to reinforce vocabulary. **science.glencoe.com**

---

8 Cellular Transport and the Cell Cycle

Magnification: 14 000×

What You'll Learn

- You will discover how molecules move across the plasma membrane.
- You will sequence the stages of cell division.

Why It's Important

Transportation of substances through the plasma membrane and cell reproduction are two important functions that help cells maintain homeostasis and keep you healthy.

READING BIOLOGY

Look through the list of new vocabulary words for the chapter. Break down the words and try to determine their origins. As you read the chapter more thoroughly, write down definitions for any unfamiliar words.

BIOLOGY Online

To find out more about cellular transport and the cell cycle, visit the Glencoe Science Web site. **science.glencoe.com**

Magnification: 36 000×

Cells divide in plant root tips (above) in normal growth. Some cells are cancerous (inset) and divide indefinitely.

Multiple Learning Styles

Look for the following logos for strategies that emphasize different learning modalities.

Kinesthetic Enrichment, pp. 205, 212; Meeting Individual Needs, p. 208; Portfolio, p. 209; Reinforcement, p. 214; Going Further, p. 221

Visual-Spatial Meeting Individual Needs, p. 204; Biology Journal, p. 205; Visual Learning, p. 214; Reteach, pp. 206, 216; Activity, p. 216

Interpersonal Tech Prep, p. 213

Intrapersonal Portfolio, p. 211; Meeting Individual Needs, p. 211; Extension, p. 219; Activity, p. 219

Linguistic Tech Prep, p. 203; Biology Journal, pp. 207, 217; Discussion, p. 213

Logical-Mathematical Discussion, p. 206; Portfolio, p. 212

8.1 Cellular Transport

SECTION PREVIEW

Objectives

Explain how the processes of diffusion, passive transport, and active transport occur and why they are important to cells.

Predict the effect of a hypotonic, hypertonic, or isotonic solution on a cell.

Vocabulary
osmosis
isotonic solution
hypotonic solution
hypertonic solution
passive transport
facilitated diffusion
active transport
endocytosis
exocytosis

This dam is a barrier that, when opened, allows water to pass to the other side of the floodgates. In contrast, to move water from the well and out through the pump, someone must physically move the handle that draws the water up against gravity. The plasma membrane of a cell can act as both a dam and a pump as it regulates the traffic of ions and molecules into and out of the cell.

A dam and a pump regulate the flow of water.

Osmosis: Diffusion of Water

Although the plasma membrane of a cell can act as a dam or pump for water-soluble molecules that cannot pass freely through the membrane, it does not limit the diffusion of water. Recall that diffusion is the movement of particles from an area of higher concentration to an area of lower concentration. In a cell, water always tries to reach an equal concentration on both sides of the membrane. The diffusion of water across a selectively permeable membrane is called **osmosis** (ahs MOH sus). Regulating the water flow through the plasma membrane

is an important factor in maintaining homeostasis within the cell.

What controls osmosis

If you add sugar to water, the water becomes sweeter as you add more sugar. As the number of sugar molecules increases, the number of water molecules decreases. If a strong sugar solution and a weak sugar solution are placed in direct contact, water molecules diffuse in one direction and sugar molecules diffuse in the other direction until all molecules are evenly distributed throughout.

If the two solutions are separated by a selectively permeable membrane that allows only water to diffuse across

WORD *Origin*

osmosis
From the Greek word *osmos*, meaning "pushing." Osmosis can push out a cell's plasma membrane.

Assessment Planner

Portfolio Assessment
  Alternative Lab, TWE, p. 203
  Assessment, TWE, p. 208
  MiniLab, TWE, p. 215
  Problem-Solving Lab, TWE, p. 218
Performance Assessment
  Alternative Lab, TWE, p. 203
  MiniLab, SE, pp. 204, 215
  Problem-Solving Lab, TWE, pp. 209, 210
  Assessment, TWE, p. 216

BioLab, SE, p. 221
Knowledge Assessment
  Assessment, pp. 202, 206, 217
  BioLab, TWE, p. 221
  Section Assessment, SE, pp. 206, 216, 219
  Chapter Assessment, SE, pp. 223-225
Skill Assessment
  MiniLab, TWE, p. 204
  Assessment, TWE, p. 219

Prepare

Key Concepts

Students will recognize how the structure of the plasma membrane permits diffusion, passive transport, and active transport. They will develop an understanding of the importance of these processes in maintaining homeostasis and proper cell function.

Planning

- Obtain India ink for the Quick Demo.
- Obtain celery sticks for the Reteach.
- Purchase raisins for the Getting Started Demo.
- Collect potato and measuring spoons for the Alternative Lab.

---

1 Focus

Bellringer

Before presenting the lesson, display **Section Focus Transparency 18** on the overhead projector and have students answer the accompanying questions.
L1 **ELL**

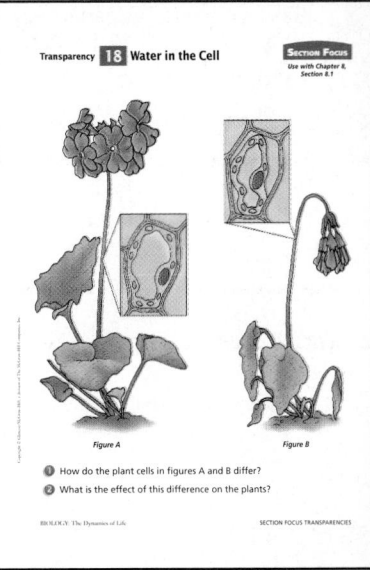

Transparency **18** Water in the Cell

Figure A   *Figure B*

1 How do the plant cells in figures A and B differ?
2 What is the effect of this difference on the plants?

2 Teach

Figure 8.1
During osmosis, water diffuses across a selectively permeable membrane. Notice that the number of sugar molecules did not change on each side of the membrane, but the number of water molecules did change.

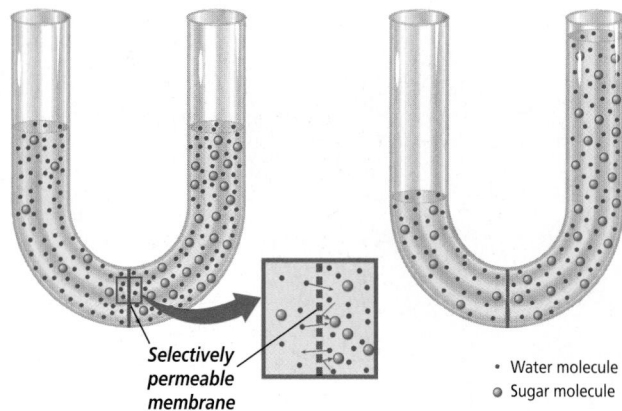

Before osmosis   **After osmosis**

Selectively permeable membrane

• Water molecule
• Sugar molecule

it, as shown in *Figure 8.1*, water flows to the side of the membrane where the water concentration is lower. The water continues to diffuse until it is in equal concentration on both sides of the membrane. Therefore, we know that unequal distribution of particles, called a concentration gradient, is one factor that controls osmosis.

With your knowledge of osmosis, it is important to understand how osmosis affects cells.

Figure 8.2
In an isotonic solution, water molecules move into and out of the cell at the same rate, and cells retain their normal shape (**a**). Notice the concave disc shape of a red blood cell (**b**). A plant cell has its normal shape and pressure in an isotonic solution (**c**).

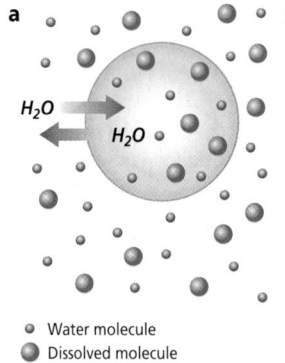

H_2O
H_2O

• Water molecule
• Dissolved molecule

a

Magnification: 2000×

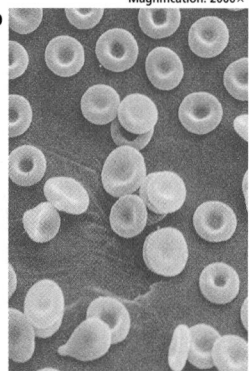 **b**

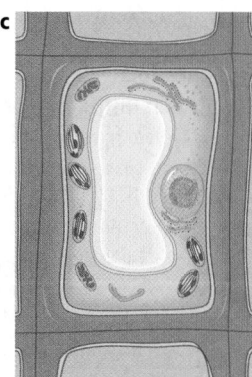 **c**

Cells in an isotonic solution

Most cells, whether in multicellular or unicellular organisms, are subject to osmosis because they are surrounded by water solutions. In an **isotonic solution,** the concentration of dissolved substances in the solution is the same as the concentration of dissolved substances inside the cell. Likewise, the concentration of water in the solution is the same as the concentration of water inside the cell.

Cells in an isotonic solution do not experience osmosis and they retain their normal shape, as shown in *Figure 8.2*. Most solutions, including the immunizations your doctor gives, are isotonic so that cells are not damaged by the loss or gain of water.

Cells in a hypotonic solution

In a **hypotonic solution,** the concentration of dissolved substances is lower in the solution outside the cell than the concentration inside the cell. Therefore, there is more water outside the cell than inside. Cells in a hypotonic solution experience osmosis that causes water to move through the plasma membrane into the cell. With osmosis, the cell swells and its internal pressure increases.

As the pressure increases inside animal cells, the plasma membrane

202 CELLULAR TRANSPORT AND THE CELL CYCLE

Figure 8.3
In a hypotonic solution, water enters a cell by osmosis, causing the cell to swell (a). Animal cells, like these red blood cells, may continue to swell until they burst (b). Plant cells swell beyond their normal size as pressure increases (c).

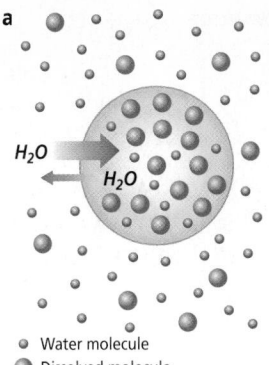
H₂O
H₂O
● Water molecule
● Dissolved molecule

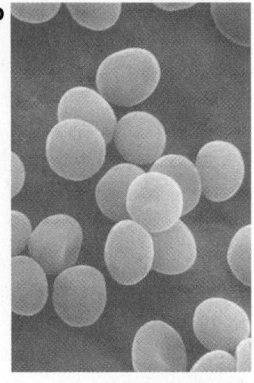
Magnification: 1800×

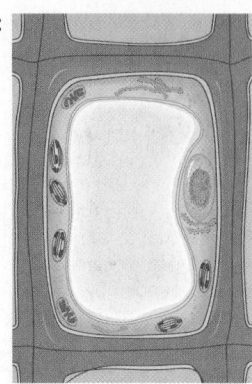

swells, like the red blood cells shown in *Figure 8.3*. If the solution is extremely hypotonic, such as distilled water, the plasma membrane may be unable to withstand this pressure and may burst.

Because plant cells contain a rigid cell wall that supports the cell, they do not burst when in a hypotonic solution. As the pressure increases inside the cell, the plasma membrane is pressed against the cell wall. Instead of bursting, the plant cell becomes more firm. Grocers use this reaction to keep produce looking fresh by misting the fruits and vegetables with water.

Cells in a hypertonic solution

In a **hypertonic solution,** the concentration of dissolved substances outside the cell is higher than the concentration inside the cell. Cells in a hypertonic solution experience osmosis that causes water to flow out.

Animal cells in a hypertonic solution shrivel because of decreased pressure in the cells, as indicated in *Figure 8.4*. This explains why you should not salt meat before cooking. The salt forms a hypertonic solution on the meat's surface. Water inside the meat cells diffuses out, leaving the cooked meat dry and tough.

Plant cells in a hypertonic environment lose water, mainly from the central vacuole. The plasma membrane and cytoplasm shrink away from the cell wall, as shown in *Figure 8.4.* Loss of water in a plant cell results in a drop in pressure and explains why plants wilt.

WORD *Origin*
iso-, hypo-, hyper-
From the Greek words *isos*, meaning "equal," *hypo*, meaning "under," and *hyper*, meaning "over," respectively.

Concept Development

Explain to students that *net movement* is the overall movement of a material and not the specific movement of each particle.

Figure 8.4
In a hypertonic solution, water leaves a cell by osmosis, causing the cell to shrink (a). Animal cells like these red blood cells shrivel up as they lose water (b). Plant cells lose pressure as the plasma membrane shrinks away from the cell wall (c).

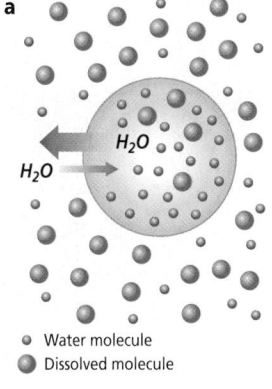
H₂O
H₂O
● Water molecule
● Dissolved molecule

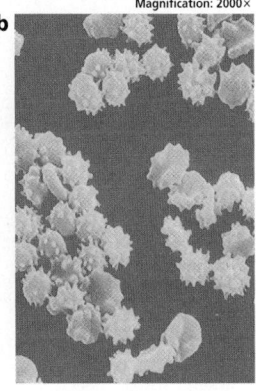
Magnification: 2000×

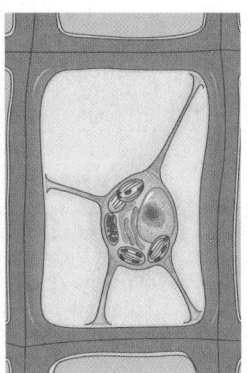

aluminum foil and allow them to sit undisturbed for two days.
6. On the second day, carefully remove the potato cubes one at a time and blot them dry on the outside. Weigh the pieces and record their masses. Observe any changes in the texture of each cube.

Analysis
1. Describe what happened to the mass of each cube after soaking. *The mass of the potato placed in salt water decreased, while the one in plain water increased.*
2. Describe what happened to the texture of each cube after soaking. *The potato in the salt water became softer than that in the plain water.*
3. Explain the changes you observed in terms of osmosis. *Water in the potato placed in salt water left the potato by osmosis because of the high salt content of the water outside the potato.*

 **Assessment**
Portfolio Have students prepare a written laboratory report containing a data table and the answers to the analysis questions. Use the Performance Task Assessment List for Lab Report in **PASC,** p. 47. L2 P

Purpose

Students will determine if a plastic membrane is selectively permeable.

Process Skills

formulate models, draw conclusions, observe and infer, recognize cause and effect

Safety Precautions

Have students wear gloves, aprons, and goggles. Remind students to rinse immediately if iodine gets on skin or clothing. If iodine gets in eyes, rinse thoroughly at the eyewash station.

Teaching Strategies

■ Allow students to work in teams of two or three.

■ See p. 40T of the Teacher Guide for preparation of starch and iodine solutions. Lightweight, inexpensive bags work best for this lab.

Expected Results

The inside of the bag will be purple indicating passage of iodine into the bag. The outside of the bag will be rust color, indicating starch did not pass out of the bag.

Analysis

1. Start—starch was clear, iodine was rust; end—starch was purple, iodine was rust.

2. **a.** Iodine moved into the bag as shown by the purple color. **b.** Starch did not move out of bag as shown by no color change outside of the bag.

3. The plastic membrane let iodine pass into the bag but blocked starch from passing out.

✔ Assessment

Skill Ask students to make a diagram depicting pore size of the plastic membrane in relation to molecule size of iodine and starch. Use the Performance Task Assessment List for Scientific Drawing in **PASC**, p. 55. **L2**

MiniLab 8-1 Formulating Models

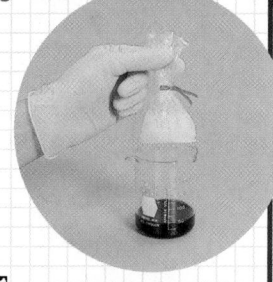

Cell Membrane Simulation If membranes show selective permeability, what might happen if a plastic bag (representing a cell's membrane) were filled with starch molecules on the inside and surrounded by iodine molecules on the outside?

Procedure  *Selective permeability*

1 Fill a plastic bag with 50 mL of starch. Seal the bag with a twist tie.

2 Fill a beaker with 50 mL of iodine solution. **CAUTION:** *Rinse with water if iodine gets on skin. Iodine is toxic.*

3 Note and record the color of the starch and iodine.

4 Place the bag into the beaker. **CAUTION:** *Wash your hands after handling lab materials.*

5 Note and record the color of the starch and iodine 24 hours later.

Analysis

1. Describe and compare the color of the iodine and starch at the start and at the conclusion of the experiment.

2. Fact: Starch mixed with iodine forms a purple color.
   a. In which direction did the iodine move? What is your evidence?
   b. In which direction did the starch move? What is your evidence?

3. Explain how this experiment illustrates selective permeability.

Figure 8.5
Channel proteins provide the openings through which small, dissolved particles, especially ions, diffuse by passive transport.

💿 CD-ROM

View an animation of passive transport in the Presentation Builder of the Interactive CD-ROM.

Passive Transport

Water, lipids, and lipid-soluble substances are some of the few compounds that can pass through the plasma membrane by diffusion. The cell uses no energy to move these particles; therefore, this movement of particles across membranes by diffusion is called **passive transport.**

Passive transport of other substances that are not attracted to the phospholipid bilayer or are too large to pass through can still occur by other mechanisms as long as the substance is moving with the concentration gradient.

You can investigate passive transport by performing the *MiniLab* shown here.

Passive transport by proteins

Recall that transport proteins help substances move through the plasma membrane. These proteins function in a variety of ways to transport molecules and ions across the membrane.

The passive transport of materials across the plasma membrane with the aid of transport proteins is called **facilitated diffusion.** As illustrated in *Figure 8.5*, the transport proteins

Channel protein

Carrier proteins

Passive diffusion

Facilitated diffusion

● Water molecule
○ Dissolved molecule

MEETING INDIVIDUAL NEEDS

Learning Disabled

Visual-Spatial Using colored chalk, draw a U-tube on the chalkboard similar to the "before osmosis" diagram in Figure 8.1. Draw molecules in at least two colors, showing one that cannot cross the selectively permeable membrane. Have students make a similar diagram using colored pencils. Challenge students to diagram the "after osmosis" stage. Walk around the room, checking to see if students understand the concept of osmosis as they draw. Help students as needed. **L1** **ELL**

Figure 8.6
Carrier proteins are used in active transport to pick up ions or molecules from near the cell membrane, carry them across the membrane, and release them on the other side. Active transport requires energy.

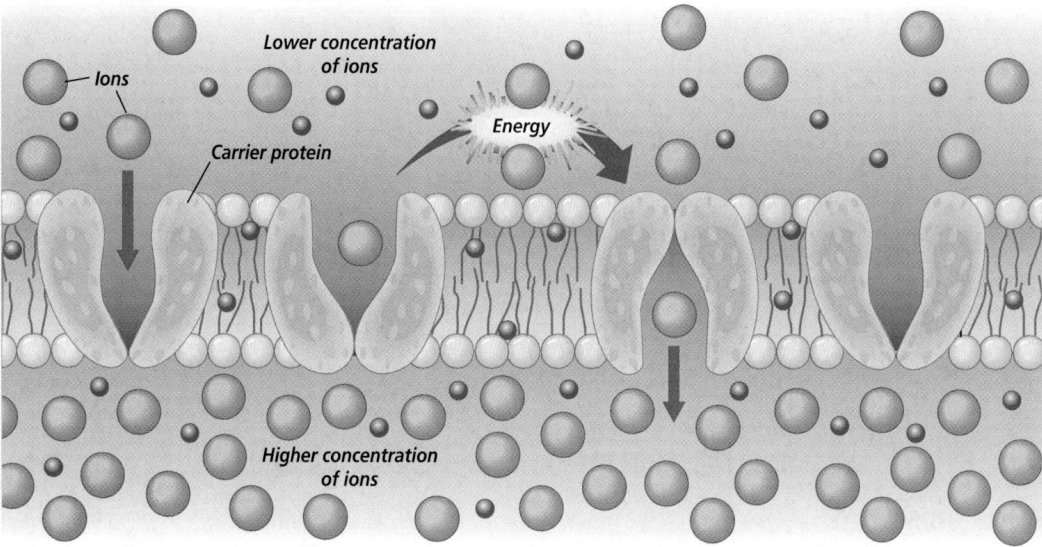

Lower concentration of ions

Ions

Energy

Carrier protein

Higher concentration of ions

provide convenient openings for particles to pass through. Facilitated diffusion is a common method of moving of sugars and amino acids across membranes. Facilitated diffusion, like simple diffusion, is driven by a concentration gradient; substances on both sides of the membrane are trying to reach equal concentrations.

Active Transport

A cell can move particles from a region of lower concentration to a region of higher concentration, but it must expend energy to counteract the force of diffusion that is moving the particles in the opposite direction. Movement of materials through a membrane against a concentration gradient is called **active transport** and requires energy from the cell.

How active transport occurs

In active transport, a transport protein called a carrier protein first binds with a particle of the substance to be transported. In general, each type of carrier protein has a shape that fits a specific molecule or ion. When the proper molecule binds with the protein, chemical energy allows the cell to change the shape of the carrier protein so that the particle to be moved is released on the other side of the membrane, something like the opening of a door. Once the particle is released, the protein's original shape is restored, as illustrated in *Figure 8.6.* Active transport allows particle movement into or out of a cell against a concentration gradient.

Transport of large particles

Some cells can take in large molecules, groups of molecules, or even

CD-ROM
View an animation of active transport in the Presentation Builder of the Interactive CD-ROM.

8.1 CELLULAR TRANSPORT **205**

---

3 Assess

Check for Understanding
Evaluate students' understanding of the following terms: passive transport, active transport, diffusion, facilitated diffusion, isotonic, hypotonic, and hypertonic. Have students predict the direction of movement between cells and solutions. **L2**

BIOLOGY JOURNAL

Comparing Modes of Transport

 *Visual-Spatial* Have students create a table that lists and compares the modes of passive transport with those of active transport. Students should identify the kinds of materials transported by each mode. *For passive transport, tables should* *include diffusion through a bilayer (osmosis and diffusion of small molecules) and facilitated diffusion (channel transport proteins and carrier proteins). Active transport modes should include transport proteins, endocytosis, and exocytosis.* **L2**

4 Close

Discussion

Figure 8.7
Some unicellular organisms ingest food by endocytosis and release wastes or cell products from a vacuole by exocytosis.

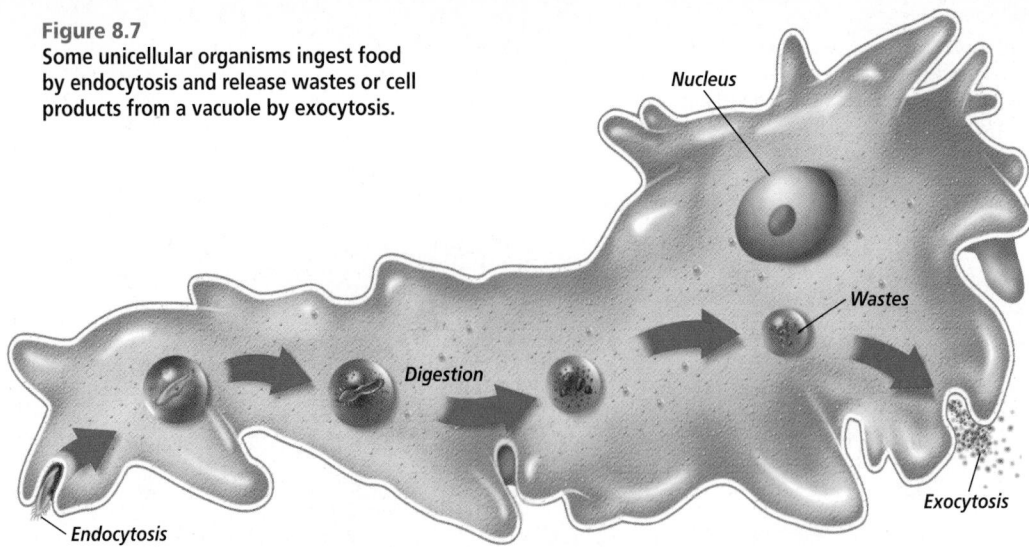

Nucleus

Wastes

Digestion

Exocytosis

Endocytosis

whole cells. **Endocytosis** is a process by which a cell surrounds and takes in material from its environment. This material does not pass directly through the membrane. Instead, it is engulfed and enclosed by a portion of the cell's plasma membrane. That portion of the membrane then breaks away, and the resulting vacuole with its contents moves to the inside of the cell.

Figure 8.7 shows the reverse process of endocytosis, called exocytosis. **Exocytosis** is the expulsion or secretion of materials from a cell.

Cells use exocytosis to expel wastes, such as indigestible particles, from the interior to the exterior environment. They also use this method to secrete substances, such as hormones produced by the cell. Because endocytosis and exocytosis both move masses of material, they both require energy and are, therefore, both forms of active transport.

With the various mechanisms the cell uses to transport materials in and out, cells must also have mechanisms to regulate size and growth.

WORD *Origin*
endo-, exo-
From the Greek words *endon*, meaning "within," and *exo*, meaning "out." Endocytosis moves materials into the cell; exocytosis moves materials out of the cell.

Section Assessment

Understanding Main Ideas
1. What factors affect the diffusion of water through a membrane by osmosis?
2. How do animal cells and plant cells react differently to osmosis in a hypotonic solution?
3. Compare and contrast active transport and facilitated diffusion.
4. How do carrier proteins facilitate passive transport of molecules across a membrane?

Thinking Critically
5. A paramecium expels water when the organism is surrounded by freshwater. What can you deduce about the concentration gradient in the organism's environment?

SKILL REVIEW
6. **Observing and Inferring** Osmosis is a form of diffusion. What effect do you think an increase in temperature has on osmosis? For more help, refer to *Thinking Critically* in the **Skill Handbook**.

206 CELLULAR TRANSPORT AND THE CELL CYCLE

Section Assessment

1. The concentration of water on either side of the membrane and the permeability of the membrane.
2. In a hypotonic solution, water moves into the cell. In an animal cell, the extra water may cause the plasma membrane to burst. In a plant cell, the

206

plasma membrane pushes against the cell wall, providing added support.
3. Facilitated diffusion and active transport use carrier proteins. Facilitated diffusion does not require energy; active transport does.
4. Carrier proteins move substances that cannot diffuse through the plasma membrane from an area of higher to

lower concentration.
5. The organism is in a hypotonic environment and the concentration gradient is from outside to inside.
6. Increasing temperature will increase the rate of osmosis, but it will not change the final outcome because it cannot change the membrane permeability to other solutes.

8.2 Cell Growth and Reproduction

Picture this unlikely scene. As the movie begins, people run screaming madly in the streets. In the background, a huge cell towers above the skyscrapers, its cilia-covered surface slowly waving to propel it through the city. Flagella flail along its side, smashing the buildings. Proteins on the plasma membrane form a crude face with a sneer. Although this scene is ridiculous, how do you know that giant cells are not possible? What limits the size of a cell?

An impossibly large cell

SECTION PREVIEW

Objectives

Sequence the events of the cell cycle.

Relate the function of a cell to its organization as a tissue, organ, and an organ system.

Vocabulary

chromosome
chromatin
cell cycle
interphase
mitosis
prophase
sister chromatid
centromere
centriole
spindle
metaphase
anaphase
telophase
cytokinesis
tissue
organ
organ system

Prepare

Key Concepts

Students will learn that there are limits to cell size. Limiting factors include, among others, diffusion, DNA content, and surface area-to-volume ratio. Students learn that cells react by dividing when they reach maximum size. The events of the cell cycle are considered, including the stages of mitosis.

Planning

- Obtain a telephone cord for Meeting Individual Needs.
- Purchase pipe cleaners or yarn for the Reinforcement.
- Collect old insulated electrical cords for the Quick Demo.

1 Focus

Bellringer

Before presenting the lesson, display **Section Focus Transparency 19** on the overhead projector and have students answer the accompanying questions. **L1** **ELL**

Cell Size Limitations

Although a giant cell will never threaten a city, cells do come in a wide variety of sizes. Some cells, such as red blood cells, measure only 8 micrometers (µm) in diameter. Other cells, such as nerve cells in large animals, can reach lengths of up to 1 m but with small diameters. The cell with the largest diameter is the yolk of an ostrich egg measuring 8 cm! Most living cells, however, are between 2 and 200 µm in diameter. Considering this wide range of cell sizes, why then can't most organisms be just one giant cell?

Diffusion limits cell size

You know that the plasma membrane allows a steady supply of nutrients such as glucose and oxygen to enter the cell and allows wastes to leave. Within the bounds of the plasma membrane, these nutrients and wastes move by diffusion.

Although diffusion is a fast and efficient process over short distances, it becomes slow and inefficient as the distances become larger. For example, a mitochondrion at the center of a hypothetical cell with a diameter of 20 cm would have to wait months before receiving molecules entering the cell. Because of the slow rate of

8.2 CELL GROWTH AND REPRODUCTION **207**

BIOLOGY JOURNAL

Limits to Cell Size

Linguistic Using what they know about the relationship of surface area to the volume of an object, ask students to write a paragraph explaining why the existence of a single-celled giant creature such as the one in the movie *The Blob* would be impossible. **L2**

Resource Manager

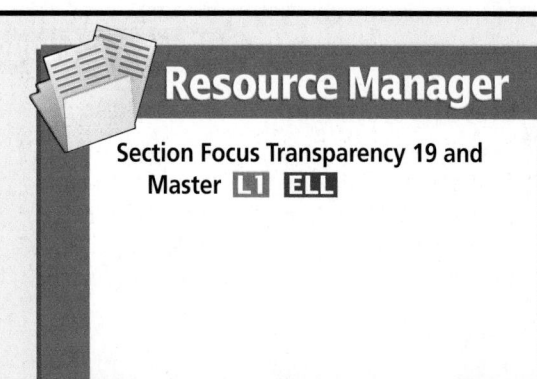

Section Focus Transparency 19 and Master **L1** **ELL**

Portfolio Have students draw interphase and the four stages of mitosis. Make sure they include the chromosomes with sister chromatids, mitotic spindle fibers, and centrioles. **L2** **ELL**

Quick Demo

Show how chromatids separate during anaphase. Slightly separate the middle portion between the two covered wires of an old piece of insulated electrical cord. Tie a piece of string to each piece of the separated cord. Slip a rubber band around the cord and through the strings. Pull the strings slowly apart until the wire splits in two, similar to the way chromatids are pulled along the spindle in a dividing cell. 🔲

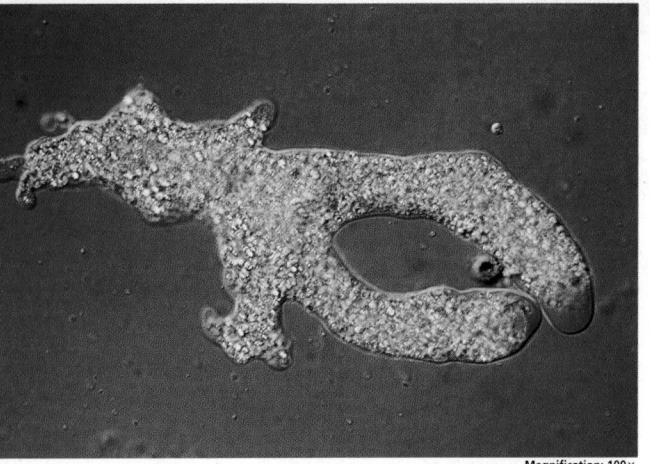

Magnification: 100×

Figure 8.8
This giant amoeba, *Pelomyxa*, is several millimeters in diameter. It can have up to 1000 nuclei.

diffusion, organisms can't be just one giant-sized cell. They would die long before nutrients could reach the organelles that needed them.

DNA limits cell size

You have learned that the nucleus contains blueprints for the cell's proteins. Proteins are used throughout the cell by almost all organelles to perform critical cell functions. But there is a limit as to how quickly the blueprints for these proteins can be copied in the nucleus and made into proteins in the cytoplasm. The cell cannot survive unless there is enough DNA to support the protein needs of the cell.

What happens in larger cells where an increased amount of cytoplasm requires increased supplies of enzymes? In many large cells, such as the giant amoeba *Pelomyxa* shown in *Figure 8.8*, more than one nucleus has evolved. Large amounts of DNA in many nuclei ensure that cell activities are carried out quickly and efficiently.

Surface area-to-volume ratio

Another size-limiting factor is the cell's surface area-to-volume ratio. As a cell's size increases, its volume increases much faster than its surface area. Picture a cube-shaped cell like those shown in *Figure 8.9*. The smallest cell has 1 mm sides, a surface area of 6 mm², and a volume of 1 mm³. If the side of the cell is doubled to 2 mm, the surface area will increase fourfold to $6 \times 2 \times 2 = 24$ mm². Observe what happens to the volume; it increases eightfold to 8 mm³.

Figure 8.9
Surface area-to-volume ratio is one of the factors that limits cell size. Note how the surface area and the volume change as the sides of a cell double in length from 1 mm to 2 mm. Calculate the change in surface area and volume as the cell doubles in size again to 4 mm on a side.

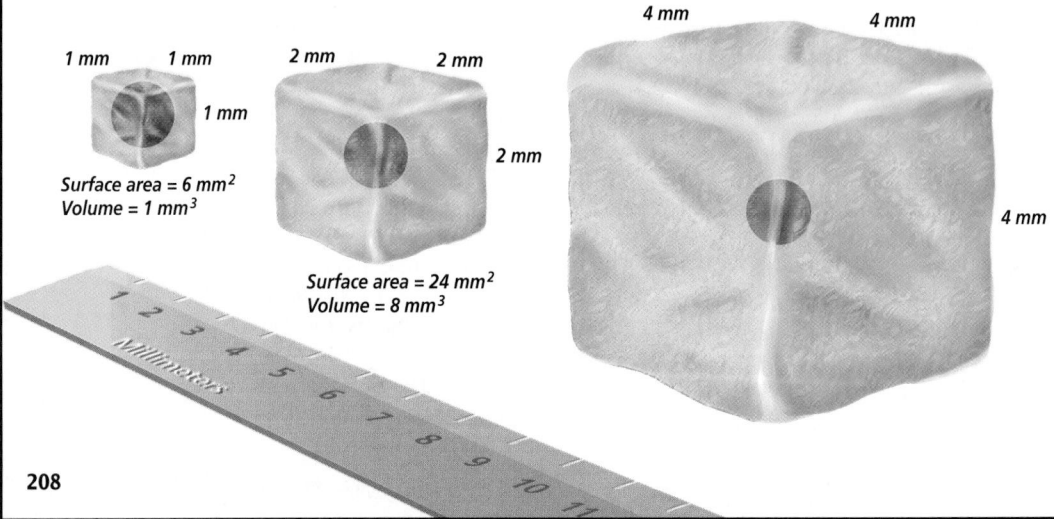

1 mm   1 mm
1 mm
Surface area = 6 mm²
Volume = 1 mm³

2 mm   2 mm
2 mm
Surface area = 24 mm²
Volume = 8 mm³

4 mm   4 mm
2 mm
4 mm
4 mm

208

Visually Impaired

🔲 *Kinesthetic* To demonstrate chromosome coiling and thickening to students who are visually impaired, remove a short, coiled telephone cord from a telephone. Have students stretch out the cord. Explain that the stretched cord represents an interphase chromosome. Have students allow the cord to return to its normal shape. Explain that this coiling is what happens to chromosomes during prophase, when the chromatin condenses. **L1** **ELL** 🔲

What does this mean for cells? How does the surface area-to-volume ratio affect cell function? If cell size doubled, the cell would require eight times more nutrients and would have eight times more waste to excrete. The surface area, however, would increase by a factor of only four. Thus, the plasma membrane would not have enough surface area through which oxygen, nutrients, and wastes could diffuse. The cell would either starve to death or be poisoned from the buildup of waste products. You can investigate surface area-to-volume ratios yourself in the *Problem-Solving Lab* shown here.

Because cell size can have dramatic and negative effects on a cell, cells must have some method of maintaining optimum size. In fact, cells divide before they become too large to function properly. Cell division accomplishes other purposes, too, as you will read next.

Cell Reproduction

Recall that the cell theory states that all cells come from preexisting cells. Cell division is the process by which new cells are produced from one cell. Cell division results in two cells that are identical to the original, parent cell. Right now, as you are reading this page, many of the cells in your body are growing, dividing, and dying. Old cells on the soles of your feet and on the palms of your hands are being shed and replaced, cuts and bruises are healing, and your intestines are producing millions of new cells each second. New cells are produced as tadpoles become frogs, and as an ivy vine grows and wraps around a garden trellis. All organisms grow and change; worn-out tissues are repaired or are replaced by newly produced cells.

The discovery of chromosomes

Most interesting to the early biologists was their observation that just before cell division, several short, stringy structures suddenly appeared in the nucleus. Scientists also noticed that these structures seemed to vanish as mysteriously as they appeared soon after division of a cell. These structures, which contain DNA and become darkly colored when stained, are called **chromosomes** (KROH muh sohmz).

Eventually, scientists learned that chromosomes are the carriers of the genetic material that is copied and passed from generation to generation of cells. This genetic material is crucial to the identity of the cell.

Problem-Solving Lab 8-1 — Drawing Conclusions

What happens to the surface area of a cell as its volume increases? One reason cells are small is that, as they grow, their volume increases faster than their surface area.

Analysis

Look at the cubes shown below. Note the size and magnitude of difference in surface area and volume among the cubes.

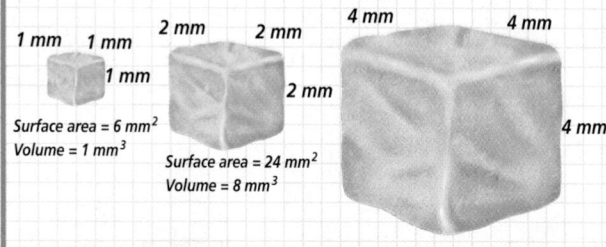

1 mm   1 mm
1 mm
Surface area = 6 mm²
Volume = 1 mm³

2 mm   2 mm
2 mm
Surface area = 24 mm²
Volume = 8 mm³

4 mm   4 mm
4 mm
4 mm

Thinking Critically

1. How many small cubes (1 mm) do you think it would take to fill the largest cube (4 mm)?
2. Relating this example to cells, describe how a cell is affected by its size.
3. Explain how a small change in cell size can have a huge impact on the cell and its normal functions.

Portfolio

Surface Area Demonstration

Kinesthetic Using a few small boxes and one large box that is approximately the same size as the small boxes combined and wrapping paper, ask the students to figure out which set will need more paper, the large box or the small boxes each wrapped separately.

The students should wrap all of the boxes, then unwrap them to demonstrate the difference in the amounts of paper needed. Students should see how the volumes of the large box and the set of small boxes are approximately equal, but the total surface area is much larger for the set of small boxes than for the large box. Have them include illustrations and a summary of the demonstration in their portfolios. **L3** **P**

Problem-Solving Lab 8-1

Purpose
Students will compare the increase in volume of an object with the increase in its surface area.

Background
Cell volume increases much faster than cell surface area. In cells, this fact contributes to cell size limitation because cells do not have sufficient surface area to accommodate the influx of nutrients to support the volume of a large size.

Process Skills
measure in SI, use numbers, recognize cause and effect, interpret data, analyze

Teaching Strategies
■ Ask students if they have ever wondered why cells can't continue to grow larger and larger to become giant cells. Then ask them to consider the fact that most cells, whether from an elephant or an earthworm, are microscopic in size.

Thinking Critically

1. 64
2. As the cell surface area grows, its volume increases dramatically. More resources are needed by organelles and more waste is produced.
3. As the cell grows, it reaches a point where the surface area is not large enough to transport resources and wastes to allow the cell to survive.

Assessment

Performance Students should write a summary of the MiniLab, including the Analysis questions, for their journals. Use the Performance Task Assessment List for Lab Report in **PASC**, p. 47. **L2**

Purpose 🖥

Students will compare cell cycles of two different types of cells.

Background

Mitosis requires the same relative amount of time no matter what type of cell is dividing. The amount of time spent in interphase determines the length of the cell cycle.

Process Skills

compare and contrast, interpret data, analyze

Teaching Strategies

■ Relate this lab to the uncontrolled growth of cancer cells, which spend a very short time in interphase.

Thinking Critically

1. The first part of interphase, in which the cell is growing, is the most variable in length.
2. The cell with the longer period of growth would carry on more metabolic activities than the more rapidly dividing cell.
3. Students should justify their answers with statements such as that certain types of cells are always being damaged and need to be replaced.

✔ Assessment

Performance Have students write three questions related to this lab in their journals. Use the Performance Task Assessment List for Asking Questions in **PASC,** p. 19. **L2**

How does the length of the cell cycle vary? The cell cycle varies greatly in length from one kind of cell to another. Some kinds of cells divide rapidly, while others divide more slowly.

Analysis

Examine the cell cycle diagrams of two different types of cells. Observe the total length of each cell cycle and the length of time each cell spends in each phase of the cell cycle.

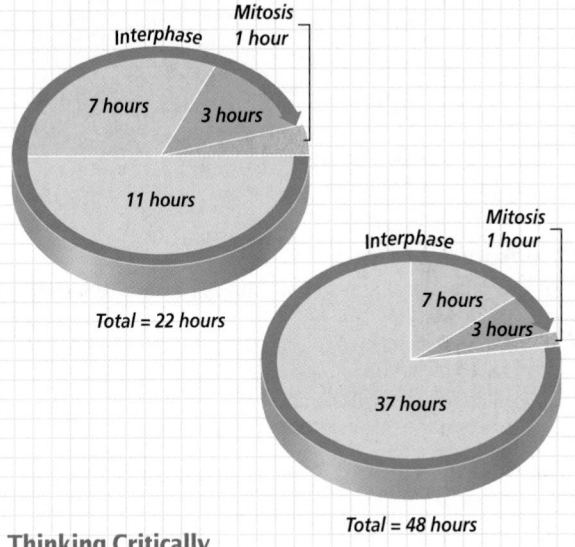

Thinking Critically

1. Which part of the cell cycle is most variable in length?
2. What can you infer about the functions of these two types of cells?
3. Why do you think the cycle of some types of cells is faster than in others? Explain your answer.

Accurate transmission of chromosomes during cell division is critical.

The structure of eukaryotic chromosomes

For most of a cell's lifetime, chromosomes exist as **chromatin,** long strands of DNA wrapped around proteins. Under an electron microscope, chromatin looks somewhat chaotic, resembling a plate of tangled-up spaghetti. This loose, seemingly unorganized arrangement is necessary for the protein blueprints to be copied. However, before a cell can divide, the long strands of chromatin must be reorganized, just as you would coil a long strand of rope before storing it. As the nucleus begins to divide, chromosomes take on a different structure in which the chromatin becomes tightly packed.

The Cell Cycle

Fall follows summer, night follows day, and low tide follows high tide. Many events in nature follow a recurring, cyclical pattern. Living organisms are no exception. One cycle common to most living things is the cycle of the cell. The **cell cycle** is the sequence of growth and division of a cell.

As a cell proceeds through its cycle, it goes through two general periods: a period of growth and a period of division. The majority of a cell's life is spent in the growth period known as **interphase.** During interphase, a cell grows in size and carries on metabolism. Also during this period, chromosomes are duplicated in preparation for the period of division.

Following interphase, a cell enters its period of nuclear division called **mitosis** (mi TOH sus). Mitosis is the process by which two daughter cells are formed, each containing a complete set of chromosomes. Interphase and mitosis make up the bulk of the cell cycle. One final process, division of the cytoplasm, takes place after mitosis. Look at the *Inside Story* to find out how many stages of growth are involved in interphase. You can use the *Problem-Solving Lab* on this page and the *BioLab* at the end of this chapter to investigate the rate of mitosis.

Jane Cooke Wright

Discuss with students the role of African-American scientist Jane Cooke Wright in the development of chemotherapy techniques to treat cancer. Wright's work in the 1950s and 1960s involved testing various anti-cancer drugs on people with different kinds of cancer. In the 1970s, Wright found that examinations of cancer cells grown in tissue culture could help predict which drugs would be most effective against that type of cancer. Since that time, Wright has been an active publisher of work in the field, and in 1975 she was honored by the American Association for Cancer Research for her contributions.

The Cell Cycle

The cell cycle is divided into interphase, when most of the cell's metabolic functions are carried out and the chromosomes are replicated, and mitosis, when nuclear division occurs, leading to the formation of two daughter cells. The division of cytoplasm, called cytokinesis, follows mitosis.

Critical Thinking *During which stage of the interphase does a cell spend most of its time? Why?*

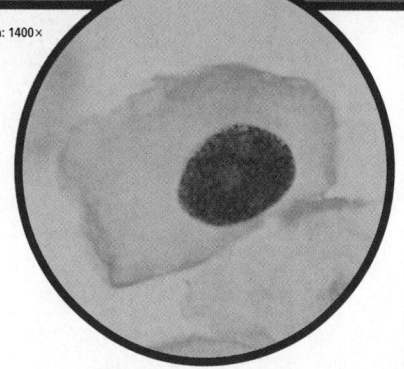

Magnification: 1400×

Interphase

Magnification: 625×

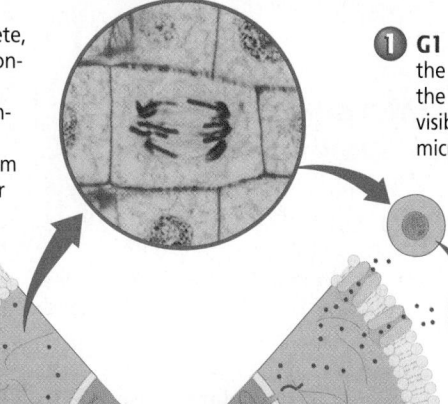

④ Mitosis When interphase is complete, the cell undergoes mitosis. Mitosis consists of four stages (*Figure 8.12*) that result in the formation of two daughter cells with identical copies of the DNA. Following mitosis, the cytoplasm divides, separating the two daughter cells.

③ G2 The chromosomes begin to shorten and coil, and protein synthesis is in high gear. In this stage of interphase, most of the proteins being synthesized are needed for mitosis and the cell organizes and prepares for mitosis. In animals, the centriole pair replicates and prepares to form the mitotic spindle.

① G1 Interphase begins with the G1 stage. At this point the chromosomes are not visible under the light microscope because they are uncoiled. Protein synthesis is rapidly occurring as the cell grows and develops.

② S Stage During this stage of interphase, the chromosomes are replicated in the nucleus. Chromosomes divide to form identical sister chromatids connected by a centromere.

Purpose

Students will see the relationship between interphase and mitosis and learn the stages of interphase.

Background

Most of a cell's life is spent in interphase. Interphase can be divided into discrete phases, and each phase carries out specific cellular activities.

Teaching Strategy

■ Ask students to identify and describe the stages of the cell cycle. **L1**

Visual Learning

■ Have students draw a cell in each of the G1, S, and G2 stages of interphase. Students should include the structure of chromosomes and indicate whether they are coiled or uncoiled and where they are replicated. **L2 ELL**

■ Have students create a table with columns labeled for the three stages of interphase, the four stages of mitosis, and cytokinesis. Have them fill the columns with descriptions of the activities occurring during that stage. **L2**

Critical Thinking

A cell spends more time in the G1 stage than in S1 or G2. It is in G1 where intense cellular activity, including rapid cell growth and protein synthesis, takes place prior to the onset of mitosis.

Nuclear Envelope During Mitosis

🔲 *Intrapersonal* Have the students hypothesize as to what might be happening to the nuclear envelope during mitosis. Have students design an experiment that could test their hypotheses. **L2** 🔳

MEETING INDIVIDUAL NEEDS

Gifted

🔲 *Intrapersonal* Have students research the newest findings on what triggers the onset of mitosis and how microtubules are assembled and disassembled during spindle formation. Students could also research the formation of cell walls after cell division in plants. **L3** 🔳

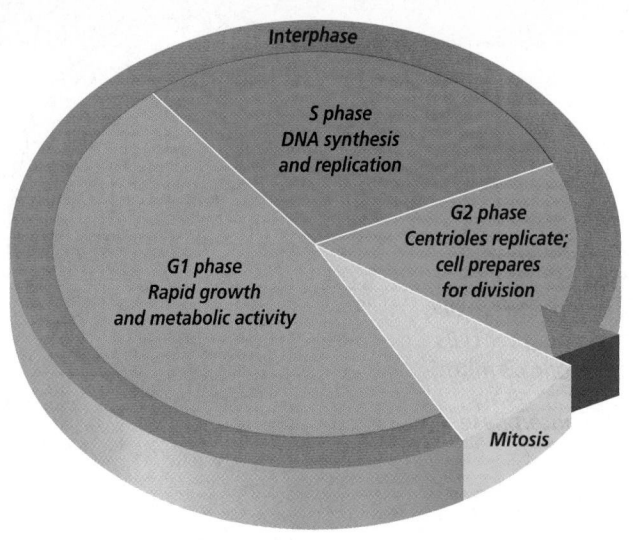

Figure 8.10
In preparation for mitosis, most of the time spent in the cell cycle is in interphase. The process of mitosis, represented here by the yellow wedge, is shown in detail in *Figure 8.12*

Figure 8.11
This photomicrograph shows a fully coiled chromosome. The two sister chromatids are held together by a centromere.

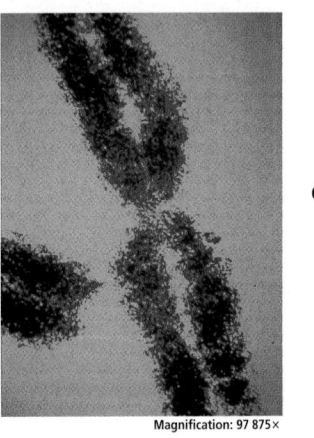

Magnification: 97 875×

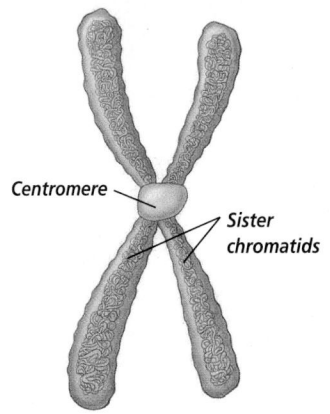

Centromere

Sister chromatids

Interphase: A Busy Time

Interphase, the busiest phase of the cell cycle, is divided into three parts as shown in *Figure 8.10*. During the first part, the cell grows and protein production is high. In the next part of interphase, the cell copies its chromosomes. DNA synthesis does not occur all through interphase but is confined to this specific time. After the chromosomes have been duplicated, the cell enters another shorter growth period in which mitochondria and other organelles are manufactured and cell parts needed for cell division are assembled. Following this activity, interphase ends and mitosis begins.

The Phases of Mitosis

Cells undergo mitosis as they approach the maximum cell size at which the nucleus can provide blueprints for proteins and the plasma membrane can efficiently transport nutrients and wastes into and out of the cell.

Although cell division is a continuous process, biologists recognize four distinct phases of mitosis—each phase merging into the next. The four phases of mitosis are prophase, metaphase, anaphase, and telophase. Refer to *Figure 8.12* to help you understand the process as you read about mitosis.

Prophase: the first phase of mitosis

During **prophase**, the first and longest phase of mitosis, the long, stringy chromatin coils up into visible chromosomes. At this point the chromosomes look hairy. As you can see in *Figure 8.11*, each duplicated chromosome is made up of two halves. The two halves of the doubled structure are called **sister chromatids.** Sister chromatids and the DNA they contain are exact copies of each other and are formed when DNA is copied during interphase. Sister chromatids are held together by a structure called a **centromere,** which plays a role in chromosome movement during mitosis. By their characteristic location, centromeres also help scientists identify and study chromosomes.

BIOLOGY Online Note Internet addresses that you find useful in the space below for quick reference.

_____

_____

_____

_____

_____

Figure 8.12
Mitosis begins after interphase. Follow the stages of mitosis as you read the text. The diagrams describe mitosis in animal cells and the photos show mitosis in plant cells.

A **Interphase** precedes mitosis. Refer to the *Inside Story.*

Magnification: 1065×

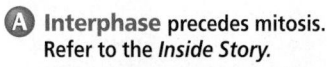
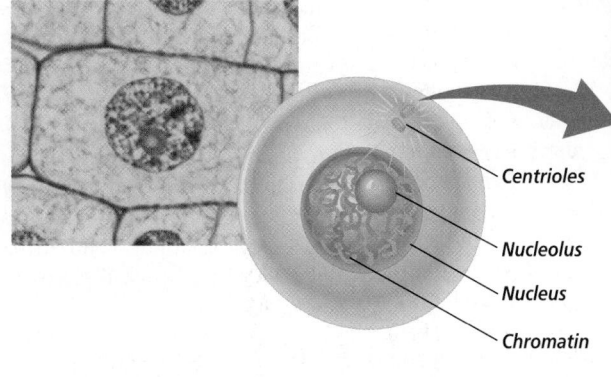

— Centrioles

— Nucleolus

— Nucleus

— Chromatin

B **Prophase** The chromatin coils to form visible chromosomes.

Magnification: 1065×

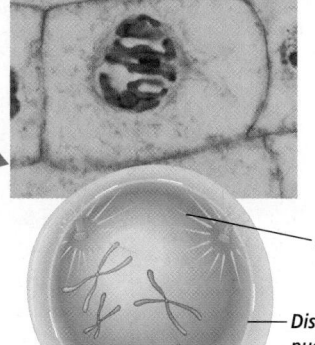

— Spindle fibers

— Disappearing nuclear envelope

Doubled chromosome

Nuclear envelope reappears

Two daughter cells are formed

Pole

Centromere

Sister chromatids

Magnification: 1250×

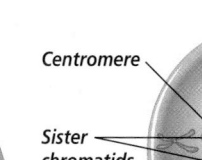

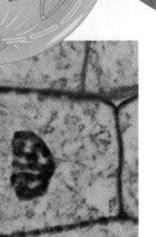

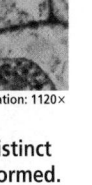

Magnification: 1120×

E **Telophase** Two distinct daughter cells are formed. The cells separate as the cell cycle proceeds into the next interphase.

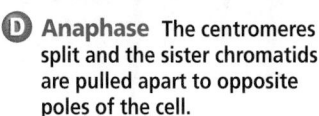

Magnification: 1120×

D **Anaphase** The centromeres split and the sister chromatids are pulled apart to opposite poles of the cell.

C **Metaphase** The chromosomes move to the equator of the spindle.

8.2   CELL GROWTH AND REPRODUCTION   **213**

Discussion

Linguistic Discuss the details of the timing mechanism of the cell cycle with the class.

The S phase of interphase and mitosis may be triggered by an enzyme called "cdc2 kinase." Another protein called "cyclin" is also involved in cell division control. In a new cell, cyclin is synthesized continuously and once it builds up to a critical level, it links up with cdc2 kinase protein to form a complex.

Several other biochemical steps then serve to activate the cdc2–cyclin complex, transforming it into maturation-promoting factor (MPF). Mitosis then takes place.

Now ask the students to write about another process (biological or otherwise) and the timing mechanism that governs the steps or phases of the process. Students should include a discussion of what happens if one portion of the process breaks down. **L3**

NATIONAL GEOGRAPHIC

VIDEODISC
STV: The Cell
How Cells Reproduce
Unit 3, 10 min. 13 sec.
How Cells Reproduce
(in its entirety)

Resource Manager

Basic Concepts Transparency 10 and Master **L2** **ELL**
Laboratory Manual,
pp. 57-60 **L2**

TECHPREP

Biology in Practice

Interpersonal Ask a cell biologist to talk to the class about how basic cell biology is critical to advanced research. Ask the speaker to describe how his or her investigations revolve intricately around what the class is learning in this unit. Have the class prepare some questions in advance to ask the guest. **L2**

CD-ROM

View an animation of the cell cycle in the Presentation Builder of the Interactive CD-ROM.

As prophase continues, the nucleus begins to disappear as the nuclear envelope and the nucleolus disintegrate. By late prophase, these structures are completely absent. In animal cells, two important pairs of structures, the centrioles, begin to migrate to opposite ends of the cell. **Centrioles** are small, dark, cylindrical structures that are made of microtubules and are located just outside the nucleus, *Figure 8.13*. Centrioles play a role in chromatid separation.

As the pairs of centrioles move to opposite ends of the cell, another important structure, called the spindle, begins to form between them. The **spindle** is a football-shaped, cagelike structure consisting of thin fibers made of microtubules. In plant cells, the spindle forms without centrioles. The spindle fibers play a vital role in the separation of sister chromatids during mitosis.

Figure 8.13
Centrioles duplicate during interphase. In the photomicrograph, one centriole is cut crosswise and the other longitudinally.

Microtubule

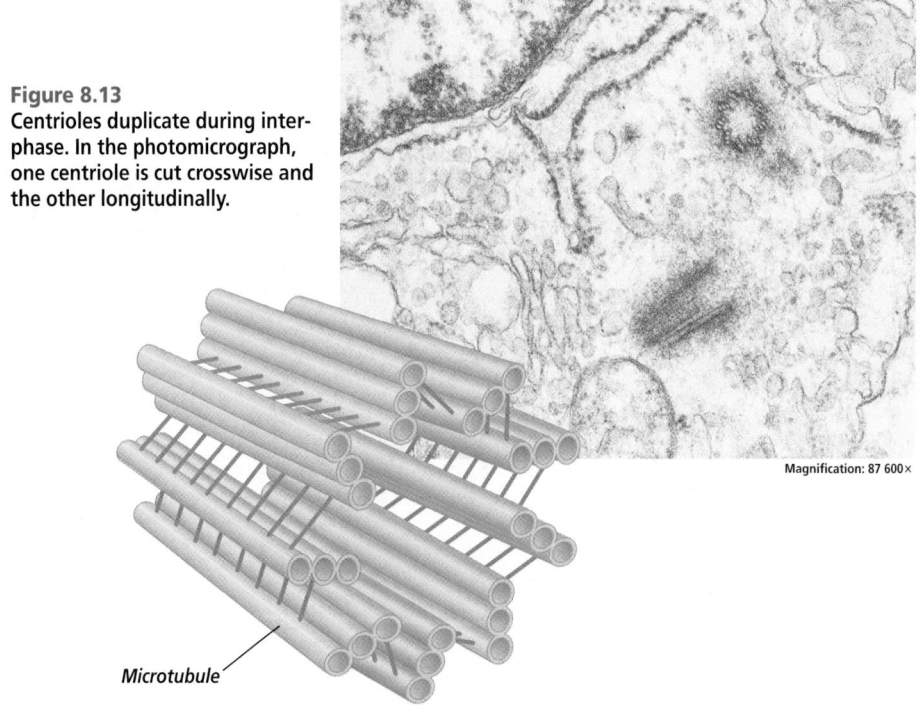

Magnification: 87 600×

Metaphase: the second stage of mitosis

During **metaphase,** the short second phase of mitosis, the doubled chromosomes become attached to the spindle fibers by their centromeres. The chromosomes are pulled by the spindle fibers and begin to line up on the midline, or equator, of the spindle. Each sister chromatid is attached to its own spindle fiber. One sister chromatid's spindle fiber extends to one pole, and the other extends to the opposite pole. This arrangement is important because it ensures that each new cell receives an identical and complete set of chromosomes.

Anaphase: the third phase of mitosis

The separation of sister chromatids marks the beginning of **anaphase,** the third phase of mitosis.

During anaphase, the centromeres split apart and chromatid pairs from each chromosome separate from each other. The chromatids are pulled apart by the shortening of the microtubules in the spindle fibers.

Telophase: the fourth phase of mitosis

The final phase of mitosis is **telophase.** Telophase begins as the chromatids reach the opposite poles of the cell. During telophase, many of the changes that occurred during prophase are reversed as the new cells prepare for their own independent existence. The chromosomes, which had been tightly coiled since the end of prophase, now unwind so they can begin to direct the metabolic activities of the new cells. The spindle begins to break down, the nucleolus reappears, and a new nuclear envelope forms around each set of chromosomes. Finally, a new double membrane begins to form between the two new nuclei.

Division of the cytoplasm

Following telophase, the cell's cytoplasm divides in a process called **cytokinesis** (site uh kih NEE sus). Cytokinesis differs between plants and animals. Toward the end of telophase in animal cells, the plasma membrane pinches in along the equator as shown in *Figure 8.14.* As the cell cycle proceeds, the two new cells are separated. Find out more about mitosis in animal cells in the *MiniLab.*

Plant cells have a rigid cell wall, so the plasma membrane does not pinch in. Rather, a structure known as the cell plate is laid down across the cell's equator. A cell membrane forms around each cell, and new cell walls form on each side of the cell plate until separation is complete.

MiniLab 8-2 — Comparing and Contrasting

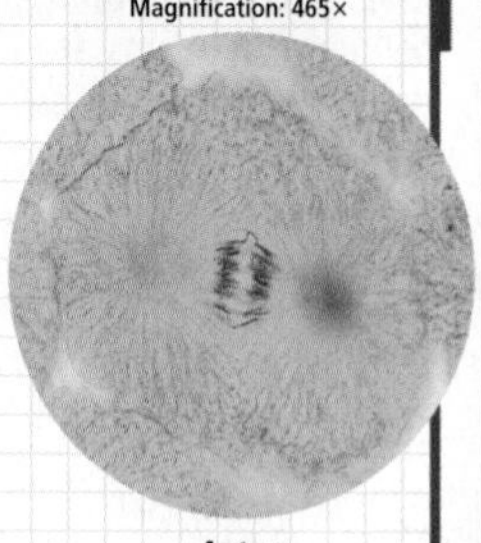

Magnification: 465×

Asters

Seeing Asters The result of the process of mitosis is similar in plant and animal cells. However, animal cells have asters whereas plant cells do not. Animal cells undergoing mitosis clearly show these structures.

Procedure

1. Examine a slide marked "fish mitosis" under low- and high-power magnification. **CAUTION:** *Use care when handling prepared slides.*
2. Find cells that are undergoing mitosis. You will be able to see dark-stained rodlike structures within certain cells. These structures are chromosomes.
3. Note the appearance and location of asters. They will appear as ray or starlike structures at opposite ends of cells that are in metaphase.
4. Asters may also be observed in cells that are in other phases of mitosis.

Analysis

1. Describe the appearance and location of asters in cells that are in prophase.
2. Explain how you know that asters are not critical to mitosis.
3. Design an experiment that tests the hypothesis that asters are not essential for mitosis in animal cells.

Magnification: 1100×

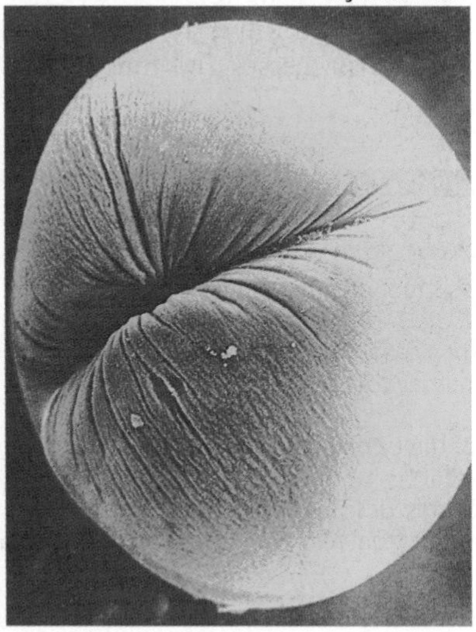

Figure 8.14
At the end of telophase in animal cells, such as this frog egg, proteins positioned just under the plasma membrane at the equator of the cell contract and slide past each other to cause a deep furrow. The furrow deepens until the cell is pinched in two.

MiniLab 8-2

Purpose
Students will observe animal cells that are undergoing mitosis and will note the location and appearance of the aster.

Process Skills
compare and contrast, observe and infer, critical thinking

Teaching Strategies
■ Advise students that the slide material is taken from fish blastulas. Ask students why this material is ideal for the study of cells undergoing mitosis.
■ To reduce cost of purchasing a class set of prepared slides, purchase one slide and use a microvideo camera or a microprojector if available. Alternatively, purchase 35 mm slides and project them onto the screen with a slide projector.
■ If necessary, review the stages of mitosis with students.

Analysis
1. Asters are starlike projections of microtubules associated with centrioles. Asters are found at the cell poles in prophase.
2. Asters are not critical because plant cells undergo mitosis without the structures.
3. Laser microbeams can be aimed to destroy specific organelles. Destroy the asters in a dividing cell and note if the cell completes mitosis.

✔ Assessment
Portfolio Have students research the role and function of the aster and summarize the scientific opinion in their portfolios. Use the Performance Task Assessment List for Writing in Science in **PASC**, p. 87. **L3**

Resource Manager

Reinforcement and Study Guide,
  pp. 34-35 **L2**
BioLab and MiniLab Worksheets,
  p. 36 **L2**

Check for Understanding

Test the students' ability to recognize the various phases of mitosis. Place photomicrographs on an overhead projector and ask the class to identify each stage. **L2**

Reteach

Visual-Spatial Review the phases of mitosis, emphasizing that the process is continuous and that one phase blends into the next. Use photomicrographs and diagrams to help students identify the phases and learn the terms associated with structures in mitosis. **L2** **ELL**

Extension

Encourage students to research the stages of the cell cycle. They may find information on how long each stage lasts for various species and what events occur at each stage. **L3**

✔ Assessment

Performance Call out various stages of mitosis and have students find and show that stage to their lab partners using onion root slides under the microscope. Walk around the room to check their results. **L1**

4 Close

Activity

Visual-Spatial Prepare a worksheet with drawings of various stages of mitosis. Ask students to draw the next stage for each. **L1** **ELL**

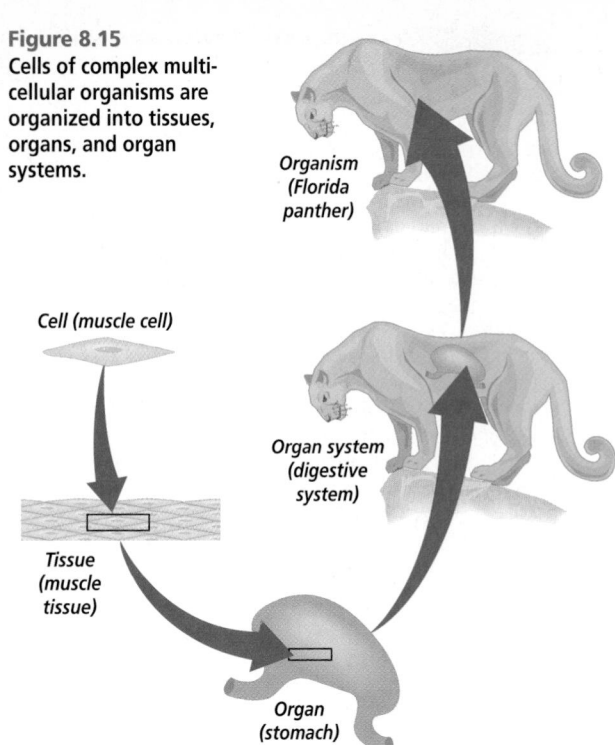

Figure 8.15
Cells of complex multicellular organisms are organized into tissues, organs, and organ systems.

Organism (Florida panther)

Cell (muscle cell)

Organ system (digestive system)

Tissue (muscle tissue)

Organ (stomach)

Results of mitosis

Mitosis is a process that guarantees genetic continuity, resulting in the production of two new cells with chromosome sets that are identical to those of the parent cell. These new daughter cells will carry out the same cellular processes and functions as

those of the parent cell and will grow and divide just as the parent cell did.

When mitosis is complete, unicellular organisms remain as single cells—the organism simply multiplied. In multicellular organisms, cell growth and reproduction result in groups of cells that work together as **tissue** to perform a specific function. Tissues organize in various combinations to form **organs** that perform more complex roles within the organism. For example, cells make up muscle tissue, then muscle tissue works with other tissues in the organ called the stomach to mix up food. Multiple organs that work together form an **organ system.** The stomach is one organ in the digestive system, which functions to break up and digest food.

All organ systems work together for the survival of the organism, whether the organism is a fly or a human. *Figure 8.15* shows an example of cell specialization and organization for a complex organism. In addition to its digestive system, the panther has a number of other organ systems that have developed through cell specialization. It is important to remember that no matter how complex the organ system or organism becomes, the cell is still the most basic unit of that organization.

Section Assessment

Understanding Main Ideas
1. Describe how a cell's surface area-to-volume ratio limits its size.
2. Why is it necessary for a cell's chromosomes to be distributed to its daughter cells in such a precise manner?
3. How is the division of the cytoplasm different in plants and in animals?
4. In multicellular organisms, describe two cellular specializations that result from mitosis.

Thinking Critically
5. At one time, interphase was referred to as the resting phase of the cell cycle. Why do you think this description is no longer used?

SKILL REVIEW
6. **Making and Using Tables** Make a table showing the phases of the cell cycle. Mention one important event that occurs at each phase. For more help, refer to *Organizing Information* in the **Skill Handbook.**

216 CELLULAR TRANSPORT AND THE CELL CYCLE

Section Assessment

1. As volume increases, surface area does not increase sufficiently to support large cells.
2. Each daughter must get an identical copy of the set of chromosomes.
3. A cell plate forms when a plant cell divides; the plasma membrane of an animal cell pinches in to divide the cytoplasm.
4. Possible answers include tissues, organs, organ systems.

5. The cell is not resting, but growing, producing proteins, and replicating its chromosomes in preparation for division.
6. Interphase, chromosomes are copied; prophase, nuclear envelope disappears; metaphase, chromosomes line up; anaphase, chromatids separate; telophase, chromosomes move to the poles.

8.3 Control of the Cell Cycle

Accurate cell division and regulation of the cell cycle is critical to the health of an organism. Some cells, such as the cells lining the intestine, complete the cell cycle in 24 to 48 hours. Other cells, such as the cells in a frog embryo, complete the cell cycle in less than an hour. Some cells, such as nerve cells, never divide once they mature. Despite this diversity, the factors that control the cell cycle are generally similar. A mistake in the cell cycle can lead to cancer.

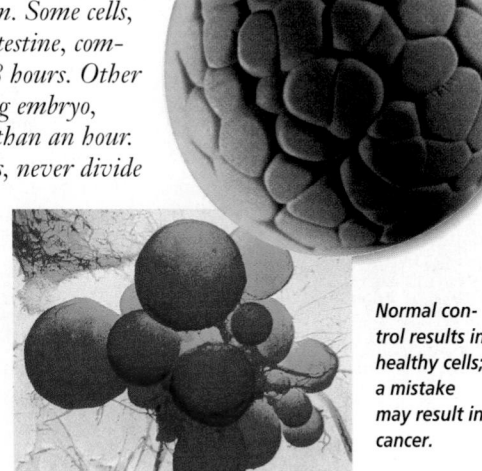

Normal control results in healthy cells; a mistake may result in cancer.

Magnification: 16 500×

SECTION PREVIEW

Objectives

Describe the role of enzymes in the regulation of the cell cycle.

Distinguish between the events of a normal cell cycle and the abnormal events that result in cancer.

Identify ways to potentially reduce the risk of cancer.

Vocabulary
cancer
gene

Normal Control of the Cell Cycle

For more than a quarter of a century, scientists have worked long and hard to discover the factors that initiate and control cell division. A clear understanding of these control factors can, among others, benefit medical research. Today, the full story is still not known; however, scientists do have some clues.

Enzymes control the cell cycle

Most biologists agree that a series of enzymes monitors a cell's progress from phase to phase during the cell cycle. Certain enzymes are necessary to begin and drive the cell cycle, whereas other enzymes control the cycle through its phases. Occasionally, cells lose control of the cell cycle. This uncontrolled dividing of cells can result from the failure to produce certain enzymes, the overproduction of enzymes, or the production of other enzymes at the wrong time. **Cancer** is one result of uncontrolled cell division. This loss of control may be caused by environmental factors or by changes in enzyme production.

Enzyme production is directed by genes located on the chromosomes. A **gene** is a segment of DNA that controls the production of a protein.

8.3 CONTROL OF THE CELL CYCLE **217**

Section 8.3

Prepare

Key Concepts
Students will learn about the events that regulate the cell cycle and compare these normal events with the abnormal events that result in cancer.

Planning
- Obtain blocks for Meeting Individual Needs.

1 Focus

Bellringer
Before presenting the lesson, display **Section Focus Transparency 20** on the overhead projector and have students answer the accompanying questions. **L1 ELL**

2 Teach

Assessment
Knowledge Determine the students' awareness of the causes of cancer and discuss with the class what each student can do to lead a healthy lifestyle.

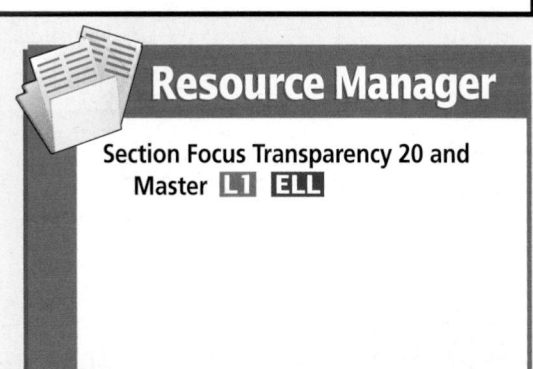

BIOLOGY JOURNAL

Cancer
Linguistic Have students research a specific type of cancer and write a summary in their journals. Have them include the role of genes in producing the cancer, if this is known. **L2**

Resource Manager
Section Focus Transparency 20 and Master **L1 ELL**

217

Purpose

Students will analyze a graph showing incidence of certain body organ cancers.

Process Skills

interpret data, think critically, analyze information, sequence, make and use graphs

Teaching Strategies

■ Allow students to work in small groups.
■ Explain the difference between skin melanoma and basal cell or squamous skin cancers.
■ Help student to determine % survival if they are having difficulty.

Thinking Critically

1. basal cell and squamous, skin melanoma

2. lung, basal cell and squamous

3. Student answers may vary: use of suntan parlors and overexposure to sun are common answers.

4. Calculation: deaths – new cases = survivors. So (survivors ÷ new cases) × 100 = % survival. 180 000 – 40 000 = 140 000. Then (140 000 ÷ 180 000) × 100 = 78%.

✔ Assessment

Portfolio Have students gather information on a specific type of cancer and prepare a brief oral report. Students should be encouraged to find information on the Internet. Use the Performance Task Assessment List for Oral Presentation in PASC, p. 71. L2

Problem-Solving Lab 8-3   Interpreting Data

How does the incidence of cancer vary? Cancer affects many different body organs. In addition, the same body organ, such as our skin, can be affected by several different types of cancer. Some types of cancer are more treatable than others. Use the following graph to analyze the incidence of cancer.

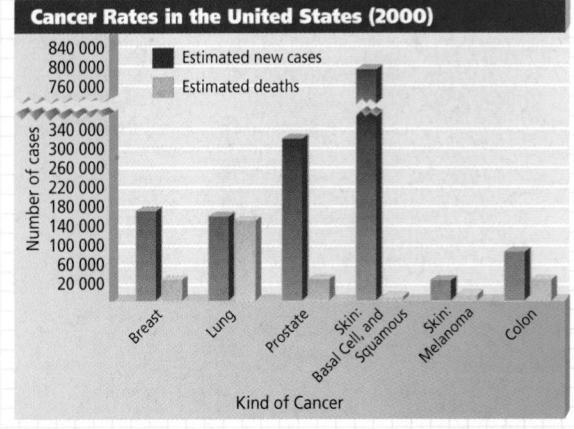

Cancer Rates in the United States (2000)

Thinking Critically

1. Which cancer type is most common? Least common?
2. Which cancer type seems to be least treatable? Most treatable?
3. Offer a possible explanation for why the incidence of basal and squamous skin cancer is so high.
4. Using breast cancer as an example, calculate the percent of survival for this cancer type.

Many studies point to the portion of interphase just before DNA replication as being a key control period in the cell cycle. Scientists have identified several enzymes that trigger DNA replication.

Cancer: A Mistake in the Cell Cycle

Currently, scientists consider cancer to be a result of changes in one or more of the genes that produce enzymes that are involved in controlling the cell cycle. These changes are

---

expressed as cancer when something prompts the damaged genes into action. Cancerous cells form masses of tissue called tumors that deprive normal cells of nutrients. In later stages, cancer cells enter the circulatory system and spread throughout the body, a process called metastasis, forming new tumors that disrupt the function of organs, organ systems, and ultimately, the organism.

Cancer is the second leading cause of death in the United States, exceeded only by heart disease. Cancer can affect any tissue in the body. In the United States, lung, colon, breast, and prostate cancers are the most prevalent types. Use the *Problem-Solving Lab* on this page to estimate the number of people in the United States who will develop these kinds of cancers in this decade, and how many people are expected to die from cancers. The *Health Connection* feature at the end of this chapter further discusses skin cancer.

The causes of cancer

The causes of cancer are difficult to pinpoint because both genetic and environmental factors are involved. The environmental influences of cancer become obvious when you consider that people in different countries develop different types of cancers at different rates. For example, the rate of breast cancer is relatively high in the United States, but relatively low in Japan. Similarly, stomach cancer is common in China, but rare in the United States.

In addition, when people move from one country to another, cancer rates appear to follow the pattern of the country in which they are currently living, not their country of origin. Other environmental factors, such as cigarette smoke, air and water pollution, and exposure to ultraviolet

---

GLENCOE TECHNOLOGY

VIDEODISC
The Infinite Voyage: *The Living Clock, Time Therapy: Curing Cancer Through Chronobiology* (Ch. 9)
7 min. 30 sec.

✔ Portfolio

Interview

Have students interview the school nurse to find out the warning signs of cancer. Have the students compile the class results in their portfolios and then discuss how these signs relate to the rapid cell division that occurs in cancer. L2 P

radiation from the sun, are all known to damage the genes that control the cell cycle. Cancer may also be caused by viral infections that damage genes.

Cancer prevention

From recent and ongoing investigations, scientists have established a clear link between a healthy lifestyle and the incidence of cancer.

Physicians and dietary experts agree that diets low in fat and high in fiber content can reduce the risk of many kinds of cancer. For example, diets high in fat have been linked to increased risk for colon, breast, and prostate cancers, among others. People who consume only a minimal amount of fat reduce the potential risk for these and other cancers and may also maintain a healthy body weight more easily. In addition, recent studies suggest that diets high in fiber are associated with reduced risk for cancer, especially colon cancer. Fruits, vegetables, and grain products are excellent dietary options because of their fiber content and because they are naturally low in fat. The foods displayed in *Figure 8.16* illustrate some of the choices that are associated with cancer prevention.

Vitamins and minerals may also help prevent cancer. Key in this category are carotenoids, vitamins A, C,

Figure 8.16
A healthy diet may reduce your risk of cancer.

and E, and calcium. Carotenoids are found in foods such as yellow and orange vegetables and green leafy vegetables. Citrus fruits are a great source of vitamin C, and many dairy products are rich in calcium.

In addition to diet, other healthy choices such as daily exercise and not using tobacco also are known to reduce the risk of cancer.

Section Assessment

Understanding Main Ideas
1. Do all cells complete the cell cycle in the same amount of time?
2. Describe how genes control the cell cycle.
3. How can disruption of the cell cycle result in cancer?
4. How does cancer affect normal cell functioning?

Thinking Critically
5. What evidence shows that the environment influences the occurrence of cancer?

SKILL REVIEW
6. **Observing and Inferring** Although breast cancer is more prevalent than lung cancer, more deaths are caused by lung cancer than breast cancer. Using your knowledge of how cancer spreads and factors that influence cancer, provide an explanation for this difference. For more help, refer to *Thinking Critically* in the **Skill Handbook**.

Section Assessment

1. No, the rate of completion varies widely depending on the cell and its function.
2. Genes code for proteins and enzymes that monitor and control the cell cycle.
3. Uncontrolled cell division could cause tumor formation and cancer.
4. Cancer cells form masses of tissue that deprive normal cells of nutrients.
5. People in different countries get cancer at different rates. Immigrants are influenced by the cancer pattern of the new environment.
6. Lung tissue has a much larger blood supply than does breast tissue; therefore, lung cancer spreads more rapidly than breast cancer.

---

3 Assess

Check for Understanding
Have students compare a normal and a cancer cell cycle. **L2**

Reteach
Write the following cell reproduction times (in minutes) on the board. Normal chicken stomach cells: interphase 120, prophase 60, metaphase 10, anaphase 3, telophase 12. Cancerous chicken stomach cells: interphase 16, prophase 15, metaphase 2, anaphase 1, telophase 3. Ask students to suggest possible reasons why they are different.

Extension
Intrapersonal Ask students to find out how chemotherapy drugs work. Ask them to find out why hair follicles and the lining of the digestive system are affected by these drugs, resulting in hair loss and nausea. **L3**

✔ Assessment
Skill Ask students to sequence the events that regulate the cell cycle and describe how these events change in the growth of cancer cells. **L2**

4 Close

Activity
Intrapersonal Ask students to find out what types of cancer can affect a particular organ and what treatments are available. **L2**

Resource Manager

Reinforcement and Study Guide, p. 36 **L2**
Content Mastery, pp. 37, 39-40 **L1**
Critical Thinking/Problem Solving, p. 8 **L3**

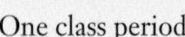

Time Allotment
One class period

Process Skills
collect data, compare and contrast, apply concepts, think critically, interpret data, observe and infer

PREPARATION

■ Obtain prepared slides of Allium (onion) root tip from a biological supply house.

PROCEDURE

Teaching Strategies
■ Review the phases of mitosis with students.

■ Allow students to look at the onion root tip slide macroscopically or through a dissecting microscope first to locate areas X and Y. Area X is a site of high mitotic activity. Area Y is a site of almost no mitotic activity.

■ Remind students that when slides are placed on the microscope stage, everything is reversed. The tip end of the root should be facing away from the student, and when moving from area X to Y, the slide must be moved away from the student.

■ You may wish to verify that students are looking at the correct areas before they begin their counts.

■ If materials are in short supply, consider doing the lab using a microprojector or with a video camera setup.

■ Advise students that the cells on the prepared slide are not alive. The cells have been caught in the various stages of mitosis.

220

INVESTIGATE
BioLab

Where is mitosis most common?

Mitosis and the resulting multiplication of cells are responsible for the growth of an organism. Does mitosis occur in all areas of an organism at the same rate, or are there certain areas within an organism where mitosis occurs more often? You will answer this question in this BioLab. *Your organism will be an onion, and the areas you are going to investigate will be different locations in its root.*

PREPARATION

Problem
Does mitosis occur at the same rate in all parts of an onion root?

Objectives
In this BioLab, you will:
■ **Observe** cells in two different root areas.
■ **Identify** the stages of mitosis in each area.

Materials
prepared slide of onion root tip microscope

Skill Handbook
Use the **Skill Handbook** if you need additional help with this lab.

PROCEDURE

1. Copy the data table.
2. Using Diagram **A** as a guide, locate area X on a prepared slide of onion root tip.
3. Place the prepared slide under your microscope and use low power to locate area X.
   CAUTION: *Use care when handling prepared slides.*

4. Switch to high power.
5. Using Diagram **B** as a guide:
   a. Identify those cells that are in mitosis and in interphase.
   b. Record in the data table the number of cells observed in each phase of mitosis and interphase for area X.

Data Table

Phase	Area X	Area Y
Interphase		
Prophase		
Metaphase		
Anaphase		
Telophase		

Data and Observations
Student data will vary but the following sample data can be used as a guide: Mitosis should be observed in cells in area X but not in area Y.

Data Table

Phase	Area X	Area Y
Interphase	95	36
Prophase	15	0
Metaphase	5	0
Anaphase	2	0
Telophase	4	0

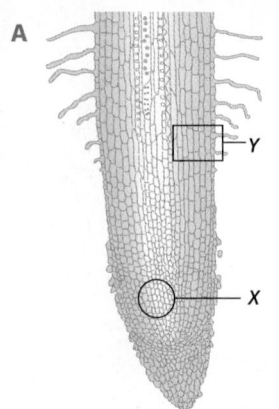

A

Y

X

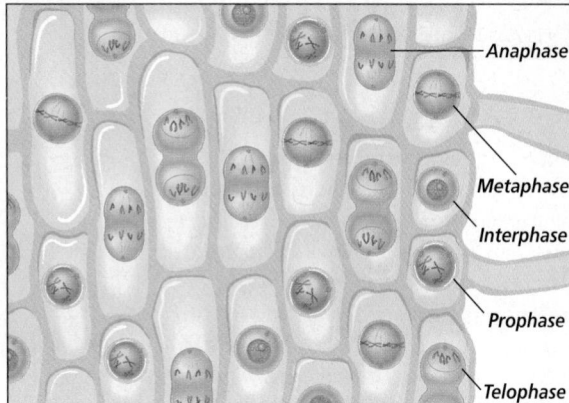

B

Anaphase

Metaphase

Interphase

Prophase

Telophase

Note: It will be easier to count and keep track of cells by following rows. See Diagram **C** as a guide to counting.

6. Using Diagram **A** again, locate area Y on the same prepared slide.

7. Place the prepared slide under your microscope and use low power to locate area Y.

8. Switch to high power.

9. Using Diagram **B** as a guide:
   a. Identify those cells that are in mitosis and in interphase.
   b. Record in the data table the number of cells observed in each phase of mitosis and interphase for area Y.

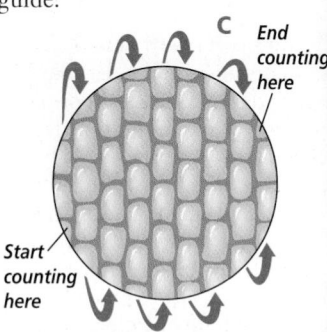

C

End counting here

Start counting here

High power view

ANALYZE AND CONCLUDE

1. **Observing** Which area of the onion root tip (X or Y) had the greatest number of cells undergoing mitosis? The fewest? Use specific totals from your data table to support your answer.

2. **Predicting** If mitosis is associated with rapid growth, where do you believe is the location of most rapid root growth, area X or Y? Explain your answer.

3. **Applying** Where might you look for cells in the human body that are undergoing mitosis?

4. **Calculating** According to your data, which phase of mitosis is most common? Least common?

5. **Thinking Critically** Assume that you were not able to observe cells in every phase of mitosis? Explain why this might be.

Going Further

Application Prepare a circle graph that shows the total number of cells counted in area X and the percentage of cells in each phase of mitosis.

BIOLOGY *Online* To find out more about mitosis, visit the Glencoe Science Web site.
science.glencoe.com

Resource Manager

BioLab and MiniLab Worksheets, p. 37 **L2**

ANALYZE AND CONCLUDE

1. X, Y; student totals will vary to support the conclusions.

2. X; this was the area showing cells undergoing mitosis at the highest rate.

3. An area of rapid growth such as skin, hair follicles, intestine lining.

4. prophase (150); anaphase (2). The appearance of few cells in metaphase, anaphase and telophase may be the result of the speed at which these phases occur.

5. Answers may vary—the phase has already occurred, the phase has not yet occurred, area of view is not rapidly growing, incorrect observation or recording stage.

✓ Assessment

Knowledge Ask students to explain the steps that could be taken to make the gathering of data more accurate. Use the Performance Task Assessment List for Designing an Experiment in **PASC**, p. 23. **L2**

Going Further

Kinesthetic Have students prepare their own slides of onion mitosis. Kits that provide the onion tips, stain, and directions are available through biological supply houses. (See Carolina Biological Supply catalog #D8-17-1130.) **L2**

Health Connection

Skin Cancer

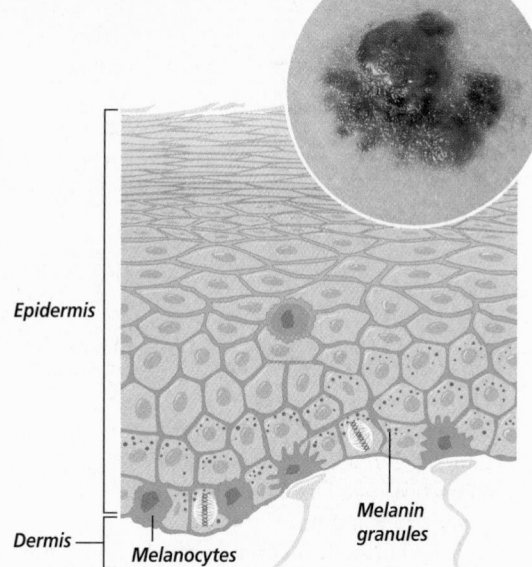

Skin cancer accounts for one-third of all malignancies diagnosed in the United States, and the incidence of skin cancer is increasing. Most cases are caused by exposure to harmful ultraviolet rays emitted by the sun, so skin cancer most often develops on the exposed face or neck. The people most likely at risk are those whose fair skin contains smaller amounts of a protective pigment called melanin.

Epidermis

Dermis

Melanocytes

Melanin granules

Structure of the skin

Skin is composed of two layers of tissue, the epidermis and the dermis. The epidermis is the part that we see on the surface of our bodies and is composed of multiple layers of closely packed cells. As the cells reach the surface, they die and become flattened. Eventually they flake away. To replace the loss, cells on the innermost layer of the epidermis are constantly dividing.

Your body has a natural protection system to shield skin cells from potentially harmful rays of the sun. A pigment called melanin is produced by cells called melanocytes and absorbs the UV rays before they reach basal cells.

Types of skin cancers

Uncontrolled division of epidermal cells leads to skin cancer. Squamous cell carcinoma is a common type of skin cancer that affects cells throughout the epidermis. Squamous cell cancer takes the form of red or pink tumors that can grow rapidly and spread. Precancerous growths produced by sun-damaged basal cells can become basal cell carcinoma, another common type of skin cancer. In basal cell carcinoma, the cancerous cells are from the layer of the epidermis that replenishes the shed epithelial cells. Both squamous cell carcinoma and basal cell carcinoma are usually discovered when they are small and can be easily removed in a doctor's office. Both types also respond to treatment such as surgery, chemotherapy, and radiation therapy.

The most lethal skin cancer is malignant melanoma. Melanomas are cancerous growths of the melanocytes that normally protect other cells in the epithelium from the harmful rays of the sun. An important indication of a melanoma can be a change in color of an area of skin to a variety of colors including black, brown, red, dark blue, or gray. A single melanoma can have several colors within the tumor. Melanomas can also form at the site of moles. Melanomas can be dangerous because cancerous cells from the tumor can travel to other areas of the body before the melanoma is detected. Early detection is essential, and melanomas can be surgically removed.

CONNECTION TO BIOLOGY

Scientists know that the UV rays of sunlight can contribute to skin cancer. How can you minimize the risk?

 To find out more about skin cancer, visit the Glencoe Science Web site. **science.glencoe.com**

BIOLOGY Online Note Internet addresses that you find useful in the space below for quick reference.

_____

_____

_____

_____

_____

SUMMARY

Section 8.1

Cellular Transport

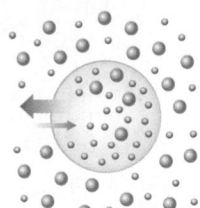

Main Ideas
- Osmosis is the diffusion of water through a selectively permeable membrane.
- Passive transport moves a substance with the concentration gradient and requires no energy from the cell.
- Active transport moves materials against the concentration gradient and requires energy to overcome the opposite flow of materials with the concentration gradient.
- Large particles may enter a cell by endocytosis and leave by exocytosis.

Vocabulary

active transport (p. 205)
endocytosis (p. 206)
exocytosis (p. 206)
facilitated diffusion (p. 204)
hypertonic solution (p. 203)
hypotonic solution (p. 202)
isotonic solution (p. 202)
osmosis (p. 201)
passive transport (p. 204)

Section 8.2

Cell Growth and Reproduction

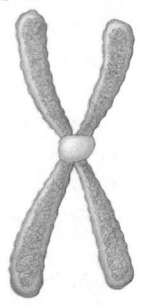

Main Ideas
- Cell size is limited largely by the diffusion rate of materials into and out of the cell, the amount of DNA available to program the cell's metabolism, and the cell's surface area-to-volume ratio.
- The life cycle of a cell is divided into two general periods: a period of active growth and metabolism known as interphase, and a period of cell division known as mitosis.
- Mitosis is divided into four phases: prophase, metaphase, anaphase, and telophase.
- The cells of most multicellular organisms are organized into tissues, organs, and organ systems.

Vocabulary

anaphase (p. 214)
cell cycle (p. 210)
centriole (p. 214)
centromere (p. 212)
chromatin (p. 210)
chromosome (p. 209)
cytokinesis (p. 215)
interphase (p. 210)
metaphase (p. 214)
mitosis (p. 210)
organ (p. 216)
organ system (p. 216)
prophase (p. 212)
sister chromatid (p. 212)
spindle (p. 214)
telophase (p. 215)
tissue (p. 216)

Section 8.3

Control of the Cell Cycle

Main Ideas
- The cell cycle is controlled by key enzymes that are produced at specific points in the cell cycle.
- Cancer is caused by genetic and environmental factors that change the genes that control the cell cycle.
- For some types of cancer, research has shown that lifestyle choices like eating a healthy diet and exercising regularly can reduce the incidence of cancer.

Vocabulary

cancer (p. 217)
gene (p. 217)

Main Ideas

Summary statements can be used by students to review the major concepts of the chapter.

Using the Vocabulary

To reinforce chapter vocabulary, use the Content Mastery Booklet and the activities in the Interactive Tutor for Biology: The Dynamics of Life on the Glencoe Science Web site. **science.glencoe.com**

All Chapter Assessment questions and answers have been validated for accuracy and suitability by The Princeton Review.

GLENCOE TECHNOLOGY

VIDEOTAPE
MindJogger Videoquizzes
Chapter 8: *Cellular Transport and the Cell Cycle*
Have students work in groups as they play the videoquiz game to review key chapter concepts.

Resource Manager

Chapter Assessment, pp. 43-48
MindJogger Videoquizzes
ExamView® Pro Software
BDOL Interactive CD-ROM, Chapter 8 quiz

UNDERSTANDING MAIN IDEAS

1. b
2. b
3. c
4. d
5. d
6. d
7. b
8. d
9. b
10. d
11. Active
12. shrink
13. hypertonic
14. pressure
15. interphase
16. chromosomes
17. interphase
18. centriole
19. tissues
20. gene

UNDERSTANDING MAIN IDEAS

1. What kind of environment is described when the concentration of dissolved substances is greater outside the cell than inside?
   a. hypotonic   **c.** isotonic
   b. hypertonic   **d.** saline

2. Osmosis is defined how?
   a. as an active process
   b. as diffusion of water through a selectively permeable membrane
   c. as an example of facilitated diffusion
   d. as requiring a transport protein

3. An amoeba ingests large food particles by what process?
   a. osmosis   **c.** endocytosis
   b. diffusion   **d.** exocytosis

4. Considering the surface area-to-volume ratio, what structure does surface area represent?
   a. cytoplasm   **c.** ER
   b. mitochondria **d.** plasma membrane

5. Chromosomes are made of what?
   a. cytoplasm   **c.** RNA
   b. centrioles   **d.** DNA

6. Which of the following does NOT occur during interphase?
   a. excretion of wastes   **c.** protein synthesis
   b. cell repair   **d.** nuclear division

7. If a cell that has eight chromosomes goes through mitosis, how many chromosomes will the daughter cells have?
   a. 4   **c.** 16
   b. 8   **d.** 32

THE PRINCETON REVIEW   **TEST-TAKING TIP**

Become an Expert on What You Fear the Most
If you just can't remember all those different parts of those important processes, don't run away. Instead, consider it a challenge, meet the problem head on, and you'll probably be surprised at how easy it is to conquer the toughest concepts.

8. During metaphase, the chromosomes move to the equator of what structure (shown here)?
   a. poles
   b. cell plate
   c. centriole
   d. spindle

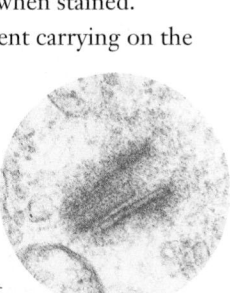

9. All but which of the following factors limit cell size?
   a. time required for diffusion
   b. elasticity of the plasma membrane
   c. presence of only one nucleus
   d. surface area-to-volume ratio

10. Which of the following is NOT a known cause of cancer?
    a. environmental influences
    b. certain viruses
    c. cigarette smoke
    d. bacterial infections

11. _____ transport requires energy.

12. A red blood cell placed in a 3% salt solution will _____.

13. Sprinkling sugar on a bowl of strawberries creates a _____ solution surrounding the strawberries.

14. Grocers spray water on produce to increase the _____ inside the cells.

15. Chromosomes are replicated during the _____ stage of the cell cycle.

16. The _____ inside cells contain DNA and become darkly colored when stained.

17. Most of a cell's life is spent carrying on the activities of _____.

18. The _____ (shown here) is present only in an animal cell.

19. An organ consists of several kinds of _____.

20. A _____ is a segment of DNA that controls the production of a protein.

APPLYING MAIN IDEAS

21. How would you expect the number of mitochondria in a cell to be related to the amount of active transport it carries out?

22. Explain why drinking quantities of ocean water is dangerous to humans. (Hint: The body excretes salt as a water solution.)

23. Suppose that all of the enzymes that control the normal cell cycle were identified. Suggest some ways that this information might be used to fight cancer.

THINKING CRITICALLY

24. Making Predictions What do you think will happen when a freshwater paramecium is placed in salt water?

25. Observing and Inferring How does cell division in adult animals help maintain homeostasis?

26. Concept Mapping Complete the concept map using the following vocabulary terms: mitosis, cell cycle, metaphase, prophase, telophase, interphase, anaphase.

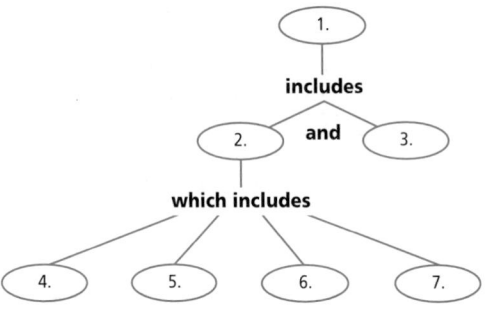

CD-ROM

For additional review, use the assessment options for this chapter found on the *Biology: The Dynamics of Life Interactive CD-ROM* and on the Glencoe Science Web site.
science.glencoe.com

ASSESSING KNOWLEDGE & SKILLS

Different species of organisms vary in the number of chromosomes found in body cells.

Table 8.1 Chromosome comparison of four organisms				
Organism	Human	Rye	Potato	Guinea Pig
Number of chromosomes in body cells	46	14	48	64
Number of chromatids during metaphase of mitosis	92	A	96	128
Number of chromosomes in daughter cells	46	14	48	64

Interpreting Data Examine the table then answer the following questions.

1. During late interphase, the chromosomes double to form chromatids that are attached to each other. During which phase do the chromatids separate?
  a. prophase   **c.** anaphase
  b. metaphase   **d.** telophase

2. What number belongs in the space labeled A under Rye in the table?
  a. 14   **c.** 7
  b. 28   **d.** 21

3. If one pair of chromatids failed to separate during mitosis in rye cells, how many chromosomes would end up in the daughter cells?
  a. 28 and 28   **c.** 7 and 8
  b. 14 and 14   **d.** 15 and 13

4. Thinking Critically Using the information presented in the table, explain how the number of chromosomes in body cells is related to the complexity of an organism.

APPLYING MAIN IDEAS

21. Cells that carry on a great deal of active transport would have more mitochondria to supply the necessary amounts of energy.

22. In order to excrete the excess salt, the body excretes more water than it takes in.

23. By being able to control the enzymes, scientists may be able to modify the rapid cell division in cancer cells.

THINKING CRITICALLY

24. The amoeba, being in a hypertonic solution, would probably die because water diffuses out.

25. Cells divide to maintain optimum size and surface-area-to-volume ratios so that the cell can receive all of the nutrients it needs and excrete wastes sufficiently.

26. 1. Cell cycle; 2. Mitosis; 3. Interphase; 4. Prophase; 5. Metaphase; 6. Anaphase; 7. Telophase

ASSESSING KNOWLEDGE & SKILLS

1. c
2. b
3. d
4. There is no relationship between the number of chromosomes in body cells and the complexity of an organism.

Chapter 9 Organizer

Refer to pages 4T-5T of the Teacher Guide for an explanation of the National Science Education Standards correlations.

Section	Objectives	Activities/Features
Section 9.1 **ATP in a Molecule** National Science Education Standards UCP.2, UCP.3; A.1, A.2; B.3, B.6; C.1, C.5 (1 session, $\frac{1}{2}$ block)	1. **Explain** why organisms need a supply of energy. 2. **Describe** how energy is stored and released by ATP.	Problem-Solving Lab 9-1, p. 228
Section 9.2 **Photosynthesis: Trapping the Sun's Energy** National Science Education Standards UCP.2, UCP.3; A.1, A.2; B.3, B.6; C.1, C.5; G.1 (2 sessions, 1 block)	3. **Relate** the structure of chloroplasts to the events in photosynthesis. 4. **Describe** light-dependent reactions. 5. **Explain** the reactions and products of the light-independent Calvin cycle.	Problem-Solving Lab 9-2, p. 232 MiniLab 9-1: Use Isotopes to Understand Photosynthesis, p. 234 Inside Story: The Calvin Cycle, p. 235 Careers in Biology: Biochemist, p. 236 Internet BioLab: What factors influence photosynthesis? p. 244 Chemistry Connection: Plant Pigments, p. 246
Section 9.3 **Getting Energy to Make ATP** National Science Education Standards UCP.1-3; A.1, A.2; B.3, B.6; C.1, C.5; F.6 (3 sessions, $\frac{1}{2}$ block)	6. **Compare and contrast** cellular respiration and fermentation. 7. **Explain** how cells obtain energy from cellular respiration.	Inside Story: The Citric Acid Cycle, p. 239 Problem-Solving Lab 9-3, p. 241 MiniLab 9-2: Determine if Apple Juice Ferments, p. 242

Need Materials? Contact Carolina Biological Supply Company at 1-800-334-5551 or at **http://www.carolina.com**

MATERIALS LIST

BioLab

p. 244 1000 mL beaker, *Elodea* plants (3), string, washers, colored cellophane, 150-watt light with reflector, 0.25% baking soda solution, watch with second hand

MiniLabs

p. 234 clay (various colors)
p. 242 small beaker, plastic pipette, large test tube, water, metal washers, baker's yeast, apple juice

Alternative Lab

p. 240 black paper, labels, iodine solution, paper clips, hot plate, beaker, 95% ethanol, small bowl, *Coleus* plants

Quick Demos

p. 229 potato, water, beaker, Bunsen burner
p. 233 prism
p. 238 sugar (sucrose), water, flask, baker's yeast
p. 251 potato, meat, vegetable oil, molecular model of lipid

Key to Teaching Strategies

L1 Level 1 activities should be appropriate for students with learning difficulties.

L2 Level 2 activities should be within the ability range of all students.

L3 Level 3 activities are designed for above-average students.

ELL ELL activities should be within the ability range of English Language Learners.

COOP LEARN Cooperative Learning activities are designed for small group work.

P These strategies represent student products that can be placed into a best-work portfolio.

These strategies are useful in a block scheduling format.

226A

Energy in a Cell

Teacher Classroom Resources

Section	Reproducible Masters	Transparencies
Section 9.1 **ATP in a Molecule**	Reinforcement and Study Guide, p. 37 **L2** Tech Prep Applications, pp. 15-16 **L2** Content Mastery, pp. 41-42, 44 **L1**	Section Focus Transparency 21 **L1** **ELL** Basic Concepts Transparency 11 **L2** **ELL**
Section 9.2 **Photosynthesis: Trapping the Sun's Energy**	Reinforcement and Study Guide, p. 38-39 **L2** Concept Mapping, p. 9 **L3** **ELL** Critical Thinking/Problem Solving, p. 9 **L3** BioLab and MiniLab Worksheets, p. 39 **L2** Content Mastery, pp. 41, 43-44 **L1**	Section Focus Transparency 22 **L1** **ELL** Basic Concepts Transparency 12 **L2** **ELL**
Section 9.3 **Getting Energy to Make ATP**	Reinforcement and Study Guide, p. 40 **L2** BioLab and MiniLab Worksheets, pp. 40-42 **L2** Laboratory Manual, pp. 61-68 **L2** Content Mastery, pp. 41, 43-44 **L1**	Section Focus Transparency 23 **L1** **ELL** Basic Concepts Transparency 13 **L2** **ELL** Reteaching Skills Transparency 14 **L1** **ELL** Reteaching Skills Transparency 15 **L1** **ELL**

Assessment Resources

Chapter Assessment, pp. 49-54
MindJogger Videoquizzes
Performance Assessment in the Biology Classroom
Alternate Assessment in the Science Classroom
ExamView® Pro Software 💾
BDOL Interactive CD-ROM, Chapter 9 quiz

Additional Resources

Spanish Resources **ELL**
English/Spanish Audiocassettes **ELL**
Cooperative Learning in the Science Classroom **COOP LEARN**
Lesson Plans/Block Scheduling

■ NATIONAL GEOGRAPHIC
Teacher's Corner

Products Available From Glencoe
To order the following products, call Glencoe at 1-800-334-7344:
CD-ROMs
NGS PictureShow: The Cell
NGS PictureShow: Plants: What It Means to Be Green
Curriculum Kit
GeoKit: Cells and Microorganisms
Transparency Sets
NGS PicturePack: The Cell
NGS PicturePack: Plants: What It Means to Be Green

Products Available From National Geographic Society
To order the following products, call National Geographic Society at 1-800-368-2728:
Video
Discovering the Cell
Photosynthesis: Life Energy

Index to National Geographic Magazine
The following articles may be used for research relating to this chapter:
"How the Sun Gives Life to the Sea," by Paul A. Zahl, February 1961.

GLENCOE TECHNOLOGY

The following multimedia resources are available from Glencoe.

Biology: The Dynamics of Life
CD-ROM **ELL**
 Animation: *The Light Reactions*
Exploration: *Parts of the Cell*
Exploration: *Phases of Mitosis*
BioQuest: *Cellular Pursuit*

Videodisc Program 📼
 The Light Reactions

The Infinite Voyage
 Unseen World
The Champion Within

The Secret of Life Series
 ATP Structure
ATP Function
ATP Serves as an Energy Carrier

Theme Development

The theme of this chapter is **energy**. ATP is presented as the energy storage molecule and the most prevalent source of energy for cells. The synthesis and breakdown of ATP are covered. An underlying theme is that there is cellular **unity within diversity**.

Resource Manager

Section Focus Transparency 21 and Master **L1** **ELL**
Basic Concepts Transparency 11 and Master **L2** **ELL**

READING BIOLOGY

Glencoe's *Biology: The Dynamics of Life* contains many resources to assist a student's reading skills. Each chapter contains figures with expanded captions that expand on written material. Word Origins, located along the side of text, expand knowledge of biology vocabulary. Glencoe's Content Mastery Booklet helps develop reading skills while reinforcing content. In addition, use the Interactive Tutor for *Biology: The Dynamics of Life* on the Glencoe Web site to reinforce vocabulary.
science.glencoe.com

---

9 Energy in a Cell

What You'll Learn

- You will learn what ATP is.
- You will explain how ATP provides energy for the cell.
- You will describe how chloroplasts trap the sun's energy to make ATP and complex carbohydrates.
- You will compare ATP production in mitochondria and chloroplasts.

Why It's Important

Every cell in your body needs energy in order to function. The energy your cells produce and store is the fuel for basic body functions such as eating and breathing.

READING BIOLOGY

Scan the chapter, and write down the titles of the section headings. Compose a few questions about each heading. Look over the headings, and try to determine general ways in which energy is stored. For headings with unfamiliar terms, scan the text and try to find the definitions.

BIOLOGY Online

To find out more about how cells use and produce energy, visit the Glencoe Science Web site.
science.glencoe.com

Energy can be captured through photosynthesis in these cabbage leaves, or burned into electricity at this power plant.

Multiple Learning Styles

Look for the following logos for strategies that emphasize different learning modalities.

Kinesthetic Portfolio, p. 228; Reteach, p. 230; Project, p. 233

Visual-Spatial Portfolio, p. 239; Check for Understanding, p. 243

Interpersonal Portfolio, p. 232;

Intrapersonal Meeting Individual Needs, pp. 229, 238; Extension, p. 236

Linguistic Biology Journal, pp. 229, 234; Extension, p. 230; Meeting Individual Needs, p. 231

9.1 ATP in a Molecule

SECTION PREVIEW

Objectives

Explain why organisms need a supply of energy.

Describe how energy is stored and released by ATP.

Vocabulary

ATP (adenosine triphosphate)
ADP (adenosine diphosphate)

A *spring stores energy when it is compressed. When the compressed spring is released, energy is also released, energy that sends this smiley-face toy flying into the air. Like this coiled spring, chemical bonds store energy that can be released when the bond is broken. Just as some springs are tighter than others, some chemical bonds store more energy than others.*

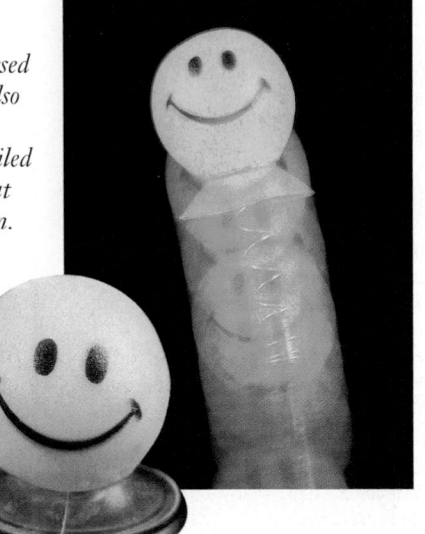

Stored energy

Figure 9.1
Active transport requires energy to bind and pump this molecule across the plasma membrane.

Cell Energy

Energy is essential to life. All living organisms must be able to produce energy from the environment in which they live, store energy for future use, and use energy in a controlled manner.

Work and the need for energy

You've learned about several cell processes that require energy. Active transport, cell division, movement of flagella or cilia, and the production and storage of proteins are some examples. The transport of proteins is shown in *Figure 9.1.* You can probably come up with other examples of biological work, such as muscles contracting during exercise, your heart pumping, and your brain

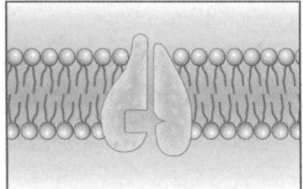

A Carrier protein shown with two binding sites.

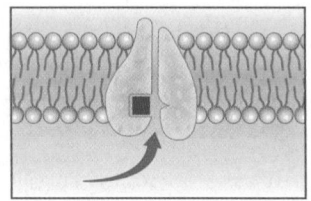

B A molecule binds to the carrier protein.

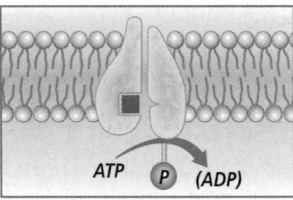

ATP **P** *(ADP)*

C Energy released as a phosphate group from ATP is transferred to the protein.

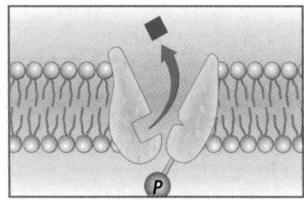

P

D The phosphate group and energy trigger the protein to pump the molecule through the membrane.

Section 9.1

Prepare

Key Concepts

Students will examine the source of cellular energy—the ATP molecule. They will also learn about the processes in which cells use the energy stored in ATP.

Planning

- Obtain a potato, peanuts, a dissection needle, a cork, and a Bunsen burner for the Quick Demo and closing Demonstration.
- Gather toothpicks, gumdrops, and construction paper for the modeling activities.

1 Focus

Bellringer

Before presenting the lesson, display **Section Focus Transparency 21** on the overhead projector and have students answer the accompanying questions.
L1 **ELL**

Transparency **21** Using Energy

SECTION FOCUS
Use with Chapter 9, Section 9.1

❶ How is each of these organisms using energy?

❷ In what other ways do organisms use energy?

BIOLOGY: The Dynamics of Life SECTION FOCUS TRANSPARENCIES

✓ Assessment Planner

Portfolio Assessment
Assessment, TWE, p. 243
BioLab, TWE, pp. 244-245
Performance Assessment
Problem-Solving Lab, TWE, pp. 228, 232
MiniLab, TWE, p. 242
Assessment, TWE, p. 236
Alternative Lab, TWE, pp. 240-241
MiniLab, SE, pp. 234, 242
BioLab, SE, pp. 244-245

Knowledge Assessment
Assessment, TWE, pp. 229, 230
MiniLab, TWE, p. 234
Section Assessment, SE, pp. 230, 236, 243
Chapter Assessment, SE, pp. 247-249
Skill Assessment
Assessment, TWE, pp. 233, 240
Problem-Solving Lab, TWE, p. 241

2 Teach

Purpose

Students will determine that fat, rather than carbohydrates, is the preferred compound for energy storage in humans.

Process Skills

acquire information, compare and contrast, think critically, draw a conclusion

Background

Fats yield more ATPs than carbohydrates because they have more C–H bonds. A six-carbon fat fragment has a molecular weight of about 100, while the same carbohydrate size has a molecular weight of 180. Fats are hydrophobic while carbohydrates are hydrophilic.

Teaching Strategies

■ Emphasize that this lab discusses the storage of excess energy and reiterate that carbohydrates are also important compounds.

■ Elicit from students if they believe early humans were typically very agile. *Heavy body weight would not have been an asset to the survival of hunters. Therefore, if early humans were storing energy as fat, they were not likely very agile.*

Thinking Critically

1. Because the metabolism of water yields zero ATP, water is "excess baggage." Fat, with no water, carries less "excess baggage," helping to make it more efficient.

2. Because we store excess energy as fat, we do not carry the weight of water that is associated with carbohydrates.

228

Problem-Solving Lab 9-1

Recognizing Cause and Effect

Why is fat the choice?
Humans store their excess energy as fat rather than as carbohydrates. Why is this? From an evolutionary and efficiency point of view, fats are better for storage than carbohydrates. Find out why.

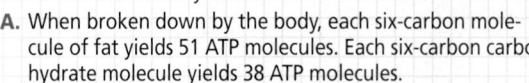

Analysis
The following facts compare certain characteristics of fats and carbohydrates:

A. When broken down by the body, each six-carbon molecule of fat yields 51 ATP molecules. Each six-carbon carbohydrate molecule yields 38 ATP molecules.

B. Carbohydrates bind and store water. The metabolism of water yields zero ATP. Fat has zero grams of water bound to it.

C. An adult who weighs 70 kg can survive on the energy derived from stored fat for 30 days without eating. The same person would have to weigh nearly 140 kg to survive 30 days on stored carbohydrates.

Thinking Critically

1. From an ATP production viewpoint, use fact B to make a statement regarding the efficiency of fats vs. carbohydrates.

2. Explain why the average weight for humans is close to 70 kg and not 140 kg.

WORD *Origin*

mono-, di-, tri
From the Latin words *mono*, *di*, and *tri*, meaning "one," "two," and "three," respectively. Adenosine triphosphate contains three phosphate groups.

controlling your entire body. This work cannot be done without energy. Read the *Problem-Solving Lab* on this page and think about how the human body stores energy.

When you finish strenuous physical exercise, such as running cross country, your body wants a quick source of energy, so you may eat a candy bar. Similarly, there is a molecule in your cells that is a quick source of energy for any organelle in the cell that needs it. This energy is stored in the chemical bonds of the molecule and can be used quickly and easily by the cell.

The name of this energy molecule is **adenosine triphosphate** (uh DEN uh seen • tri FAHS fayt), or **ATP** for short. ATP is composed of an adenosine molecule with three phosphate groups attached. Recall that phosphate groups are charged molecules, and remember that molecules with the same charge do not like being too close to each other.

Forming and Breaking Down ATP

The charged phosphate groups act like the positive poles of two magnets. If like poles of a magnet are placed next to each other, it is difficult to force the magnets together. Likewise, bonding phosphate groups to adenosine requires considerable energy. When only one phosphate group is attached, a small amount of energy is required and the chemical bond does not store much energy. A molecule of this sort is called adenosine monophosphate (AMP). When a second phosphate is added, a more substantial amount of energy is required to force the two phosphate groups together. A molecule of this sort is called **adenosine diphosphate**, or **ADP**. When the third phosphate group is added, a tremendous amount of energy is required to force the third charged phosphate close to the two other phosphate groups. The third phosphate group is so eager to get away from the other two that, when that bond is broken, a great amount of energy is released.

The energy of ATP becomes available when the molecule is broken down. In other words, when the chemical bond between phosphate groups in ATP is broken, energy is released and the resulting molecule is ADP. At this point, ADP can reform

---

ATP by bonding with another phosphate group. This creates a renewable cycle of ATP formation and breakdown. *Figure 9.2* illustrates the chemical reactions that are involved in the cycle.

The formation/breakdown recycling activity is important because it relieves the cell of having to store all of the ATP it needs. As long as phosphate molecules are available, the cell has an unlimited supply of energy. Another benefit of the formation/breakdown cycle is that ADP can also be used as an energy source. Although most cell functions require the amount of energy in ATP, some cell functions do not require as much energy and can use the energy stored in ADP.

How cells tap into the energy stored in ATP

When ATP is broken down and the energy is released, cells must have a way to capture that energy and use it efficiently. Otherwise, it is wasted. ATP is a small compound. Cellular proteins have a specific site where ATP can bind. Then, when the phosphate bond is broken and the energy released, the cell can use the energy for activities such as making a protein or transporting molecules through the plasma membrane. This cellular process is similar to the way energy in batteries is used by a radio. Batteries sitting on a table are of little use if the energy stored within the batteries cannot be accessed. When the batteries are snapped into the holder on the radio, the radio then has access to the stored energy and can use it. Likewise, when the energy in the batteries has been used, the batteries can be taken out, recharged, and replaced in the holder. In a similar fashion in a cell, when ATP has been broken down to ADP, ADP is

Figure 9.2
The addition and release of a phosphate group on adenosine diphosphate creates a cycle of ATP formation and breakdown.

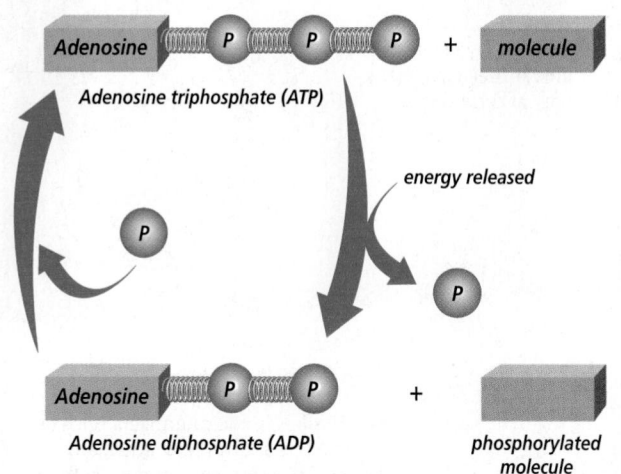

released from the binding site in the protein and the binding site may then be filled by another ATP molecule. ATP binding and energy release in a protein is shown in *Figure 9.3*.

Figure 9.3
To access the energy stored in ATP, proteins bind ATP and allow the phosphate group to be released. The ADP that is formed is released, and the protein binding site can once again bind ATP.

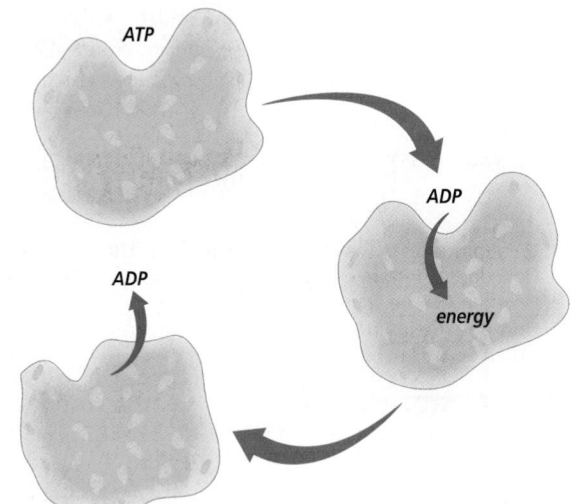

Quick Demo

To demonstrate the transfer of energy in small amounts, heat a potato in boiling water or in a microwave oven in the school cafeteria. When it is cool enough to handle safely, have students pass the heated potato to one another. Ask students who first handle the potato to describe its temperature as hot, warm, cool, or cold. Have a student near the end make the same observation.

GLENCOE TECHNOLOGY

VIDEODISC
The Secret of Life
ATP Structure

ATP Function

ATP Serves as an Energy Carrier

Resource Manager

Reinforcement and Study Guide, p. 37 **L2**
Content Mastery, p. 42 **L1**
Tech Prep Applications, p. 15 **L2**

BIOLOGY JOURNAL

Surviving Without Bread or Water

 *Linguistic* Have students find out how long a person can survive without food and without water. Have students report their findings and evaluate why people perish without water before they perish without food. **L2**

MEETING INDIVIDUAL NEEDS

Hearing Impaired

Intrapersonal Give hearing impaired students a photograph of an organism from a magazine or similar source and ask them to list the activities of the organism that require energy from ATP. **L1**

230

3 Assess

Check for Understanding

Ask students why energy must be stored in small amounts and relate the answer to paying for a small purchase with a $1 bill rather than with a $50 bill.

Reteach

Kinesthetic Have students join like poles of four small magnets to simulate the storage of energy in ATP. Explain how energy is released when the magnets are pulled apart. **L1** **ELL**

Extension

Linguistic Have interested students write a report about a bioluminescent organism and include facts about that organism's energy requirements. **L2**

✔ Assessment

Knowledge Ask students to summarize how each organism or structure in Figure 9.4 uses energy. **L2**

4 Close

Demonstration 👓

Burn a peanut to demonstrate that energy is stored in food. Impale a peanut on the end of a dissection needle. Stick the other end of the needle into a cork. Ignite the peanut with a Bunsen burner. Ask students to describe the form of energy released during burning. *Energy is released as light and heat.* 📦

Figure 9.4
ATP fuels the cellular activity that drives the organism.

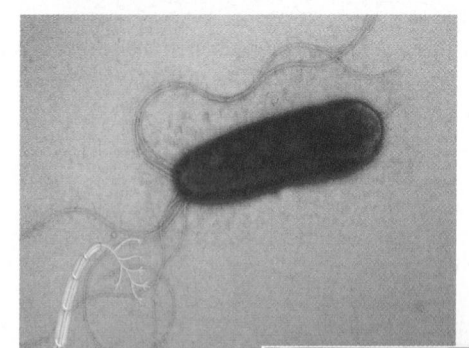

Ⓐ Nerve cells transmit impulses by using ATP to power the active transport of certain ions.

Ⓑ Some organisms with cilia or flagella (left) use energy from ATP to move.

Ⓒ Fireflies, some caterpillars, such as the one shown here, and many marine organisms produce light by a process called bioluminescence. The light results from a chemical reaction that is powered by the breakdown of ATP.

Uses of Cell Energy

You can probably think of hundreds of physical activities that require energy, but energy is equally important at the cellular level for nearly all of the cell's activities.

Making new molecules is one way that cells use energy. Some of these molecules are enzymes, which carry out chemical reactions. Other molecules build membranes and cell organelles. Cells use energy to maintain homeostasis. Kidneys use energy to move molecules and ions in order to eliminate waste substances while keeping needed substances in the bloodstream. *Figure 9.4* shows several organisms and activities that illustrate ways that cells use energy.

Section Assessment

Understanding Main Ideas
1. What processes in the cell need energy from ATP?
2. How does ATP store energy?
3. How can ADP be "recycled" to form ATP again?
4. How do proteins in your cells access the energy stored in ATP?

Thinking Critically
5. Phosphate groups in ATP repel each other because they have negative charges.

What charge might be present in the ATP binding site of proteins to attract another ATP molecule?

SKILL REVIEW
6. **Observing and Inferring** When animals shiver in the cold, muscles move almost uncontrollably. Suggest how shivering helps an animal survive in the cold. For more help, refer to *Thinking Critically* in the **Skill Handbook**.

Section Assessment

1. Active transport, movement and protein synthesis are examples.
2. ATP stores energy in its phosphate–phosphate bonds.
3. A phosphate group can be added to ADP, reforming ATP.
4. They may have a pocket that ATP will fit

into so that when ATP releases energy, the protein can use it.
5. Opposite charges attract, so an ATP binding site might have a positive charge.
6. When muscles move during shivering, heat is generated. This heat helps to warm the animal.

9.2 Photosynthesis: Trapping the Sun's Energy

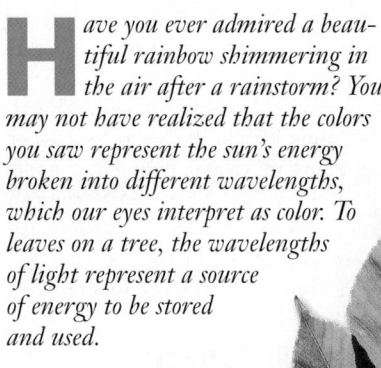

Have you ever admired a beautiful rainbow shimmering in the air after a rainstorm? You may not have realized that the colors you saw represent the sun's energy broken into different wavelengths, which our eyes interpret as color. To leaves on a tree, the wavelengths of light represent a source of energy to be stored and used.

Photosynthesis requires sunlight.

SECTION PREVIEW

Objectives
Relate the structure of chloroplasts to the events in photosynthesis.
Describe the light-dependent reactions.
Explain the reactions and products of the light-independent Calvin cycle.

Vocabulary
photosynthesis
light-dependent reactions
light-independent reactions
pigments
chlorophyll
electron transport chain
NADP⁺
photolysis
Calvin cycle

Trapping Energy from Sunlight

To use the energy of the sun's light, plant cells must trap light energy and store it in a form that is readily usable by cell organelles—that form is ATP. However, light energy is not available 24 hours a day, so the plant cell must also store some of the energy for the dark hours. **Photosynthesis** is the process plants use to trap the sun's energy and build carbohydrates, called glucose, that store energy. To accomplish this, photosynthesis happens in two phases. The **light-dependent reactions** convert light energy into chemical energy. The molecules of ATP

produced in the light-dependent reactions are then used to fuel the **light-independent reactions** that produce glucose. The general equation for photosynthesis is written as follows:

$$6CO_2 + 6H_2O \rightarrow C_6H_{12}O_6 + 6O_2$$

The *BioLab* at the end of this chapter describes an experiment you can perform to investigate the rate of photosynthesis.

The chloroplast and pigments

Recall that the chloroplast is the cell organelle where photosynthesis occurs. It is in the membranes of the thylakoid discs in chloroplasts that the light-dependent reactions take place.

WORD Origin
photosynthesis
From the Greek words *photo*, meaning "light," and *syn-tithenai*, meaning "to put together." Photosynthesis puts together sugar molecules using energy from light, water, and carbon dioxide.

MEETING INDIVIDUAL NEEDS

Gifted

 *Linguistic* The citric acid cycle is sometimes called the Krebs cycle. Have students research the work of Melvin Calvin or Hans Krebs and write a report.
L3

Resource Manager

Section Focus Transparency 22 and Master **L1** **ELL**
Concept Mapping, p. 9 **L3** **ELL**
Basic Concepts Transparency 12 and Master **L2** **ELL**

Section 9.2

Prepare

Key Concepts

Students will relate the structure of the chloroplast to the process of photosynthesis. They will also explore the overall reactions that occur in the light reactions and the Calvin cycle.

Planning

■ Acquire a prism to use in the Quick Demo.
■ Acquire colored lights for the plant growth project.
■ Purchase *Elodea* for the BioLab.
■ Collect objects to model photosynthesis for MiniLab 9-1.
■ Obtain leaves and acetone for the leaf pigment project.

---

1 Focus

Bellringer

Before presenting the lesson, display **Section Focus Transparency 22** on the overhead projector and have students answer the accompanying questions.
L1 **ELL**

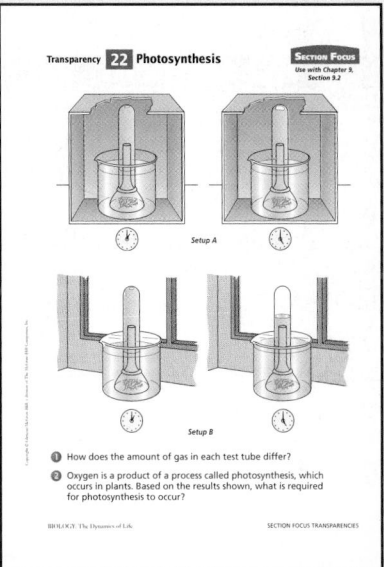

2 Teach

Purpose

Students will examine the effect of increasing light intensity on the rate of photosynthesis.

Process Skills

recognize cause and effect, interpret data, analyze data

Background

Plants depend on the energy derived from light to synthesize organic compounds from carbon dioxide and water.

Teaching Strategies

■ Ask students how horticulturists increase light intensity, especially during winter. *They use grow lights and sun reflectors.*
■ Ask students why keeping temperature constant during the experiment is important. *This avoids introducing another variable.*

Thinking Critically

The rate of photosynthesis increases with increasing light intensity until another factor limits the rate. Limiting factors include the availability of water, carbon dioxide, phosphate, ADP, and enzymes.

✓ Assessment

Performance Have students work in groups to design and carry out an experiment that uses grow lights to test whether increasing light intensity increases the rate of photosynthesis. Use the Performance Task Assessment List for Designing an Experiment in **PASC**, p. 23. **L2** **COOP LEARN**

232

Problem-Solving Lab 9-2 Drawing Conclusions

How does photosynthesis vary with light intensity? Photosynthesis is the process by which green plants synthesize organic compounds from water and carbon dioxide using energy absorbed by chlorophyll from sunlight.

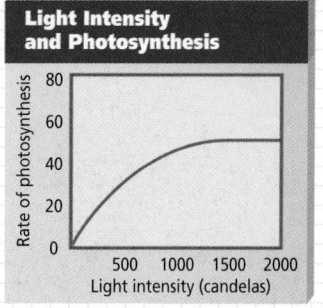

Light Intensity and Photosynthesis

Rate of photosynthesis (y-axis: 0, 20, 40, 60, 80)
Light intensity (candelas) (x-axis: 500, 1000, 1500, 2000)

Analysis

Green plants were exposed to increasing light intensity, as measured in candelas, and the rate of photosynthesis was measured. The temperature of the plants was kept constant during the experiment. The graph shown here depicts the data obtained.

Thinking Critically

Considering the overall equation for photosynthesis, make a statement summarizing what the graph shows. Under normal field conditions, what factors may limit the relative rate of photosynthesis when the light intensity is increased and temperature remains constant?

Figure 9.5
The red, yellow, and purple pigments are visible in the autumn when trees reabsorb chlorophyll.

WORD Origin

chlorophyll
From the Greek words *chloros,* meaning "pale yellowish-green," and *phyllon,* meaning "leaf." Chlorophyll is a green pigment found in leaves.

232 ENERGY IN A CELL

To trap the energy in the sun's light, the thylakoid membranes contain **pigments,** molecules that absorb specific wavelengths of sunlight. The most common pigment in chloroplasts is **chlorophyll.** Chlorophyll in forms *a* and *b* absorbs most wavelengths of light except for green. Because chloroplasts have no means to absorb this wavelength, it is reflected, giving leaves a green appearance. In the fall, trees reabsorb chlorophyll from the leaves and other pigments are visible, giving leaves like those in *Figure 9.5* a wide variety of colors. Read the *Chemistry Connection* at the end of this chapter to find out more about biological pigments.

Light-Dependent Reactions

The first phase of photosynthesis requires sunlight. As sunlight strikes the chlorophyll molecules in the thylakoid membrane, the energy in the light is transferred to electrons. These highly energized, or excited, electrons are passed from chlorophyll to an **electron transport chain,** a series of proteins embedded in the thylakoid membrane. Use the *Problem-Solving Lab* shown here to consider how light intensity affects photosynthesis.

Each protein in the chain passes energized electrons along from protein to protein, similar to a bucket brigade in which a line of people pass a bucket of water from person to person to fight a fire. At each step along the transport chain, the electron loses energy, just as some of the water might be spilled from buckets in the fire-fighting chain. The electron transport chain allows small amounts of the electron's energy to be released at a time. This energy can be used to form ATP

✓ Portfolio

Communicating about Science

 *Interpersonal* Divide the class into three groups. Ask each group to discuss chloroplasts, the light reactions, or the Calvin cycle and present a short report. Individual students should place copies of their group report in their portfolios. **L2**
P **COOP LEARN**

GLENCOE TECHNOLOGY

VIDEODISC
The Secret of Life
Chloroplast Membrane Structure

from ADP, or to pump hydrogen ions into the center of the thylakoid disc.

After the electron has traveled down the first electron transport chain, it is passed down a second electron transport chain. Following the second electron transport chain, the electron is still very energized. So that this energy is not wasted, the electron is transferred to the stroma of the chloroplast. To do this, an electron carrier molecule called **NADP+** (nicotinamide adenine dinucleotide phosphate) is used. When carrying the excited electron, NADP+ combines with a hydrogen ion and becomes NADPH.

Just as proteins contain a binding site where they can bind ATP, so the NADP+ molecule and other electron carrier molecules like it have a binding site for energized electrons. However, in this case, NADPH does not use the energy present in the energized electron; it simply stores the energy until it can transfer it to another series of reactions that will take place in the stroma. There, NADPH will play an important role in the formation of carbohydrates. The light-dependent reactions are summarized in *Figure 9.6*.

Figure 9.6
The Light Reactions

A Chlorophyll molecules absorb light energy and energize electrons for producing ATP or NADPH.

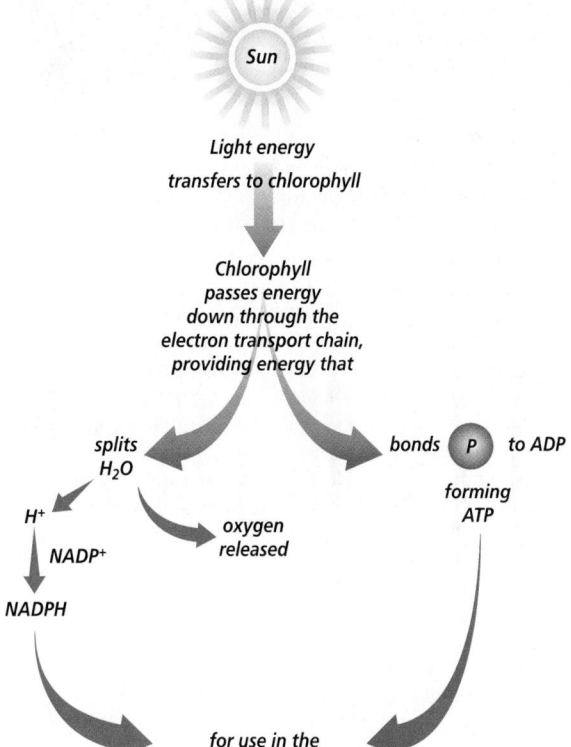

B This expanded view shows energized electrons lose some energy as they are passed from protein to protein through the electron transport chain. This energy can be used to form ATP or NADPH.

CD-ROM
View an animation of the light-dependent reactions in the Presentation Builder of the Interactive CD-ROM.

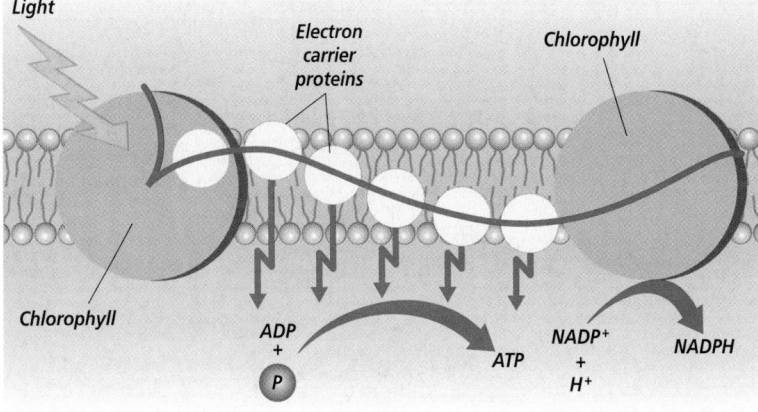

Electron Transport Chains

Assessment

Skill Have students write the equation that summarizes photosynthesis. Ask them to identify the raw materials (reactants) in the process and the products. *Carbon dioxide and water are the raw materials; simple sugars and oxygen are the products.* Guide students to an understanding of what the equation means in terms of energy capture and conversion. **L2**

Quick Demo

Use a prism to show how visible light can be separated into a spectrum. Students may recall from their study of physical science that the different colors of the visible spectrum represent different wavelengths of light.

The BioLab at the end of the chapter can be used at this point in the lesson.

INTERNET BioLab

GLENCOE TECHNOLOGY

 CD-ROM
Biology: The Dynamics of Life
Animation: *The Light Reactions*
Disc 1

 VIDEODISC
Biology: The Dynamics of Life
Light Reactions (Ch. 26)
Disc 1, Side 1, 51 sec.

PROJECT

Growing Plants in Colored Light

Kinesthetic Have students place a plant in each of several closed boxes containing a colored light bulb. Using caution, tape the opening through which the electric cord passes so no outside light can enter. Provide the plants in each box with equal amounts of water. Have students hypothesize how the plants in each box will grow compared with a control plant.

Ask students to write a summary of the experiment. Students should include results and observations and an evaluation of the hypothesis. **L2**

233

Purpose

Students will build a model to trace the fate of molecules involved in photosynthesis.

Process Skills

analyze information, define operationally, formulate models, think critically, predict

Teaching Strategies

■ Models may be constructed from colored beads, Legos, Tinkertoys, or colored clay. Make sure that students provide a key (probably by color) to show the molecules being referenced. **L1** **ELL**

■ If materials for modeling are not available, students can trace molecules from the left to right side of the equation by using lines and arrows.

Expected Results

Student models will show that all carbon from carbon dioxide ends up in glucose (CH_2O); all oxygen from carbon dioxide ends up in glucose and water; all hydrogen from water (on the left side) ends up in glucose and water.

Analysis

1. Student answers may vary. Isotopes are radioactive and thus can be traced by their radioactivity.
2. **a.** incorporated into glucose or water
   b. incorporated into glucose
   c. incorporated into glucose and water

✔ Assessment

Knowledge Have students model the equation for photosynthesis starting with an isotope of hydrogen in the water molecule to the left of the arrow. Use the Performance Task Assessment List for Model in **PASC,** p. 51. **L2**

MiniLab 9-1 — Formulating Models

van Niel

Use Isotopes to Understand Photosynthesis C. B. van Niel discovered the steps of photosynthesis when he used radioactive isotopes of oxygen as tracers. Radioactive isotopes are used to follow a particular molecule through a chemical reaction.

Procedure

1. Study the following equation for photosynthesis that resulted from the van Niel experiment:

$$6CO_2 + 6H_2O^* \rightarrow C_6H_{12}O_6 + 6O_2^*$$

2. Radioactive water, water tagged with an isotope of oxygen as a tracer (shown with the *), was used. Note where the tagged oxygen in water ends up on the right side of the chemical reaction.
3. Assume that van Niel repeated his experiment, but this time he put a radioactive tag on the oxygen in CO_2.
4. Using materials provided by your teacher, model what you would predict the appearance of his results would be. Your model must include a "tag" to indicate the oxygen isotope on the left side of the arrow as well as where it ends up on the right side of the arrow.
5. You must use labels or different colors in your model to indicate also the fate of carbon and hydrogen.

Analysis

1. Explain how an isotope can be used as a tag.
2. Using your model, predict:
   a. the fate of all oxygen molecules that originated from carbon dioxide.
   b. the fate of all carbon molecules that originated from carbon dioxide.
   c. the fate of all hydrogen molecules that originated from water.

WORD Origin

photolysis
From the Greek words *photos*, meaning "light," and *lyein*, meaning "to split." Light energy splits water molecules in photolysis.

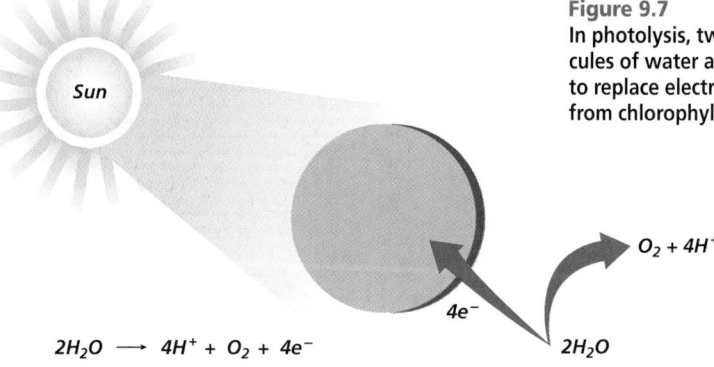

Sun

$$2H_2O \longrightarrow 4H^+ + O_2 + 4e^-$$

$O_2 + 4H^+$

$4e^-$

$2H_2O$

Restoring electrons to chlorophyll

Although some of the light-energized electrons may be returned to chlorophyll after they've moved through the electron transport chain, many leave with NADPH for the light-independent reactions. If these electrons are not replaced, the chlorophyll will be unable to absorb light and the light-dependent reactions will stop, as will the production of ATP.

To replace the lost electrons, molecules of water are split. Each water molecule produces one-half molecule of oxygen gas, two hydrogen ions, and two electrons. This reaction is shown in *Figure 9.*7 and is called **photolysis** (FO tohl ih sis). The oxygen of photolysis supplies the oxygen we breath. The *MiniLab* on this page describes how scientists traced oxygen through photosynthesis.

Light-Independent Reactions

The second phase of photosynthesis does not require light. It is called the **Calvin cycle**, which is a series of reactions that use carbon dioxide to form carbohydrates. The Calvin cycle takes place in the stroma of the chloroplast. What are the stages of the Calvin cycle? To find out, read the *Inside Story.*

Figure 9.7
In photolysis, two molecules of water are split to replace electrons lost from chlorophyll.

BIOLOGY JOURNAL

Oxygen Journey

Linguistic After the class has performed MiniLab 9-1, have the students write a story describing the pathway taken by the radioactively labeled oxygen in the van Niel experiment. Encourage creativity while at the same time conveying biological accuracy. **L2**

Resource Manager

Reinforcement and Study Guide pp. 38-39 **L2**

BioLab and MiniLab Worksheets, p. 39 **L2**

Critical Thinking/Problem Solving, p. 9 **L3**

The Calvin Cycle

The Calvin cycle takes the carbon in CO_2 and forms carbohydrates through a series of reactions in the stroma of the chloroplast. NADPH and the ATP produced during the earlier light-dependent reactions are important molecules for this series of reactions.

Critical Thinking *Environmentalists are concerned about the increasing loss of Earth's forests. How is the Calvin cycle connected to this concern?*

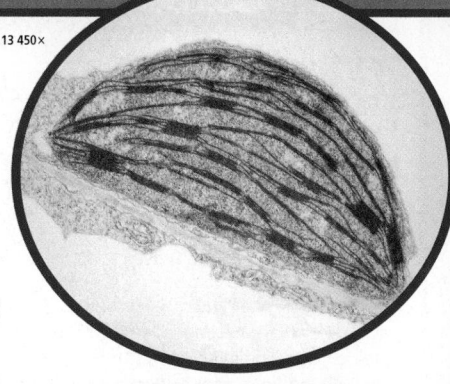

Magnification: 13 450×

The stroma in chloroplasts host the Calvin cycle

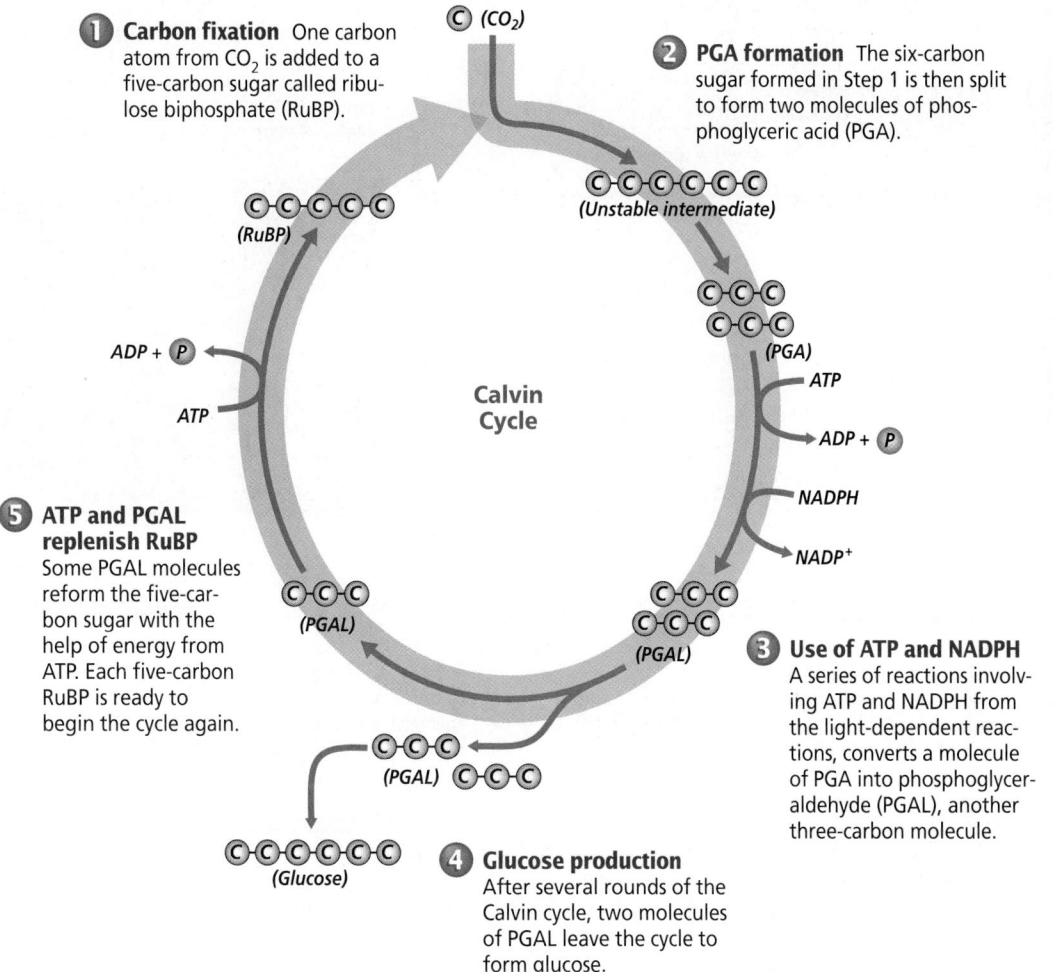

1 Carbon fixation One carbon atom from CO_2 is added to a five-carbon sugar called ribulose biphosphate (RuBP).

2 PGA formation The six-carbon sugar formed in Step 1 is then split to form two molecules of phosphoglyceric acid (PGA).

(CO₂)

(Unstable intermediate)

(RuBP)

ADP + P

ATP

Calvin Cycle

(PGA)

ATP

ADP + P

NADPH

NADP⁺

5 ATP and PGAL replenish RuBP Some PGAL molecules reform the five-carbon sugar with the help of energy from ATP. Each five-carbon RuBP is ready to begin the cycle again.

(PGAL)

(PGAL)

(PGAL)

(Glucose)

3 Use of ATP and NADPH A series of reactions involving ATP and NADPH from the light-dependent reactions, converts a molecule of PGA into phosphoglyceraldehyde (PGAL), another three-carbon molecule.

4 Glucose production After several rounds of the Calvin cycle, two molecules of PGAL leave the cycle to form glucose.

Purpose

Students will learn the steps of the Calvin cycle and how CO_2 is used to make glucose.

Background

Energy from the light reactions is used to make sugar molecules that will serve as a long-term storage for energy.

Teaching Strategies

■ Have students identify the main steps of the Calvin cycle.

Visual Learning

■ Have students make a chart of everything required for one round of the Calvin cycle, including the number of molecules. **L2**

Critical Thinking

The Calvin cycle is critical to photosynthesis and carbohydrate production. If less photosynthesis occurs, fewer food molecules are produced.

3 Assess

Check for Understanding

Students should know the general equation for photosynthesis and understand that the light reactions feed the Calvin cycle. **L1**

Reteach

Explain that the Calvin cycle uses CO_2 and hydrogen to form the simple sugars that make more complex carbohydrates.

PROJECT

Extracting and Testing Pigments

Have students soak ground, fresh leaves in warm alcohol and apply the extract to a 12-cm-long strip of filter paper in a narrow band about 2 cm from the bottom of the strip. Place the filter paper into a test tube containing a small amount of acetone (clear fingernail polish remover). Tightly seal the test tube. Just the bottom edge of the paper should dip into the solvent; the leaf extract should not be immersed.

As the solvent is drawn up the paper, less soluble compounds remain near the bottom and more soluble compounds are carried higher. Remove the paper strip before the solvent reaches the top. **L3**

GLENCOE
TECHNOLOGY

 VIDEODISC
The Secret of Life
The Calvin Cycle

235

CAREERS IN BIOLOGY

Biochemist

If you are curious about what makes plants and animals grow and develop, consider a career as a biochemist. The basic research of biochemists is to understand how processes in an organism work to ensure the organism's survival.

Skills for the Job

A bachelor's degree in chemistry or biochemistry will qualify you to be a lab assistant. For a more involved position, you will need a master's degree; advanced research requires a Ph.D. Some biochemists work with genes to create new plants and new chemicals from plants. Others research the causes and cures of diseases or the effects of poor nutrition. Still others investigate solutions for urgent problems, such as finding better ways of growing, storing, and caring for crops.

 To find out more about careers in biology and related fields, visit the Glencoe Science Web site. science.glencoe.com

The Calvin cycle

The Calvin cycle, named after Melvin Calvin shown in *Figure 9.8,* is called a cycle because one of the last molecules formed in the series of chemical reactions is also one of the molecules needed for the first reaction of the cycle. Therefore, one of the products can be used again to initiate the cycle.

In the electron transport chain, you learned that an energized electron is passed from protein to protein and the energy is slowly released. You can imagine that making a complex carbohydrate from a molecule of CO_2 would be a large task for a cell, so the light-independent reactions in the stroma of the chloroplast break down the complicated process into small steps. Each sugar molecule made by the Calvin cycle contains six carbon atoms, and because only one molecule of CO_2 is added to the cycle each time, it takes a total of six rounds of the cycle to form one sugar.

Figure 9.8
Melvin Calvin

Section Assessment

Understanding Main Ideas
1. Why do you see green when you look at a leaf on a tree? Why do you see other colors in the fall?
2. How do the light-dependent reactions of photosynthesis relate to the Calvin cycle?
3. What is the function of water in photosynthesis? Explain the reaction that achieves this function.
4. How does the electron transport chain transfer light energy in photosynthesis?

Thinking Critically
5. In photosynthesis, is chlorophyll considered a reactant, a product, or neither? How does the role of chlorophyll compare with those of CO_2 and H_2O?

SKILL REVIEW
6. **Designing an Experiment** Design an experiment that would simulate photosynthesis. For more help, refer to *Practicing Scientific Methods* in the **Skill Handbook**.

Section Assessment

1. The chlorophyll in the leaf reflects green and yellow while absorbing other colors. In the fall, chlorophyll is absorbed, revealing other leaf pigments.
2. ATP and hydrogen ions from the light reactions are used in the Calvin cycle.
3. Photolysis splits water to provide hydrogen ions for the Calvin cycle and restore electrons to chlorophyll.
4. Proteins convey energized electrons through the chloroplast.
5. Chlorophyll is neither a reactant nor a product. It contributes electrons to photosynthesis but is not changed during the reaction. CO_2 and H_2O are reactants.
6. Students could demonstrate how solar panels convert sunlight to electricity.

Section
9.3 Getting Energy to Make ATP

You know that the chlorophyll in green plants is the key to photosynthesis. You also know that your body needs energy to survive. What would it be like if you had chlorophyll in your skin and could convert light energy to carbohydrates to produce ATP? People would look quite different. Fortunately, the mitochondria in your cells can convert the carbohydrates that green plants formed by the Calvin cycle into ATP. This allows you to access the sun's energy through the work done by plant cells.

Magnification:
221 875×

Human cells produce energy in mitochondria (inset).

Cellular Respiration

The process by which mitochondria break down food molecules to produce ATP is called **cellular respiration.** There are three stages of cellular respiration: glycolysis, the citric acid cycle, and the electron transport chain. The first stage, glycolysis, is **anaerobic**—no oxygen is required. The last two stages are **aerobic** and require oxygen to be completed.

Glycolysis

Glycolysis (gli KOL ih sis) is a series of chemical reactions in the cytoplasm of a cell that break down glucose, a six-carbon compound, into two molecules of pyruvic (pie RUE vik) acid, a three-carbon compound. Because two molecules of ATP are used to start glycolysis, and only four ATP molecules are produced, glycolysis is not very efficient, giving a net profit of only two ATP molecules for each glucose molecule broken down.

In the electron transport chain of photosynthesis, an electron carrier called NADP⁺ was described as carrying energized electrons to another location in the cell for further chemical reactions. Glycolysis also uses an

WORD Origin
anaerobic
From the Greek words *an*, meaning "without," and *aeros*, meaning "air." Anaerobic organisms can live without oxygen.

9.3 GETTING ENERGY TO MAKE ATP **237**

Cultural Diversity

Severo Ochoa

Have students research the efforts of Spanish-American biochemist Severo Ochoa (1905–1993) toward the modern understanding of the citric acid cycle and photosynthesis. Ochoa showed how the oxidation of one glucose molecule could yield 38 ATP molecules. He also elucidated the mechanisms of the citric acid cycle and photosynthesis by identifying the function of key enzymes.

Ochoa's research in cellular respiration in the 1930s and 1940s resulted ultimately in the discovery of the mechanisms of RNA and DNA synthesis, for which Ochoa and colleague Arthur Kornberg received a Nobel prize in 1959. **L3**

SECTION PREVIEW

Objectives
Compare and contrast cellular respiration and fermentation.
Explain how cells obtain energy from cellular respiration.

Vocabulary
cellular respiration
anaerobic
aerobic
glycolysis
citric acid cycle
lactic acid fermentation
alcoholic fermentation

Section 9.3

Prepare

Key Concepts

Students will learn the similarities and differences between cellular respiration and anaerobic processes that release energy. They will also learn how these reactions are related to photosynthesis.

Planning

- Purchase baker's yeast for the Quick Demo and MiniLab 9-2.
- Obtain *Coleus* plants and iodine solution and other materials for the Alternative Lab.

1 Focus

Bellringer

Before presenting the lesson, display **Section Focus Transparency 23** on the overhead projector and have students answer the accompanying questions. **L1 ELL**

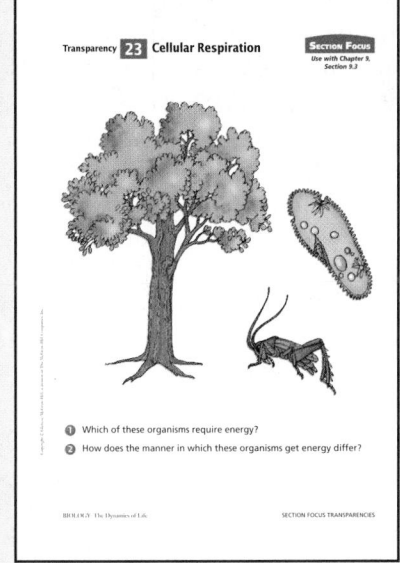

Transparency **23** Cellular Respiration   SECTION FOCUS
Use with Chapter 9, Section 9.3

1. Which of these organisms require energy?
2. How does the manner in which these organisms get energy differ?

BIOLOGY: The Dynamics of Life   SECTION FOCUS TRANSPARENCIES

237

2 Teach

Concept Development

Read a recipe for making bread. Explain to students that the yeast used in the recipe are microorganisms classified as fungi. Discuss how, unlike most organisms, yeast carry out processes that release energy in the absence of oxygen. Explain that bakers use yeast in their recipes for almost all breads, because as yeast function anaerobically, they release CO_2, which causes the bread to rise.

Quick Demo

Prepare a sugar solution by mixing a tablespoon of sugar with a cup of warm water in a deep jar or flask. Add some baker's yeast a few hours before class. Have students note the odor of alcohol and the bubbles of carbon dioxide. Point out that these products result as the yeast carry out alcoholic fermentation.

Reinforcement

Make sure students understand that pyruvic acid is the intermediate product for both types of fermentation. Challenge students to explain how pyruvic acid changes in each type of fermentation.

Concept Development

For students having difficulty with the concepts of this chapter, review the roles of a producer and consumer. Relate the processes in this chapter to the concepts of Unit 2, Ecology.

Resource Manager

Section Focus Transparency 23 and Master **L1** **ELL**
Basic Concepts Transparency 13 and Master **L2** **ELL**

Figure 9.9

Glycolysis breaks down a molecule of glucose into two molecules of pyruvic acid. In the process, it forms a net profit of two molecules of ATP, two molecules of NADH , and two hydrogen ions.

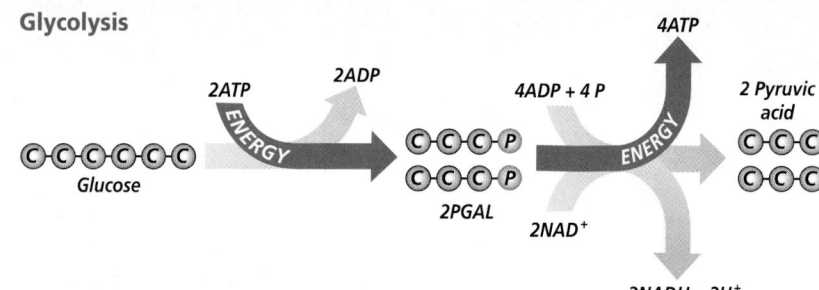
Glycolysis

electron carrier called NAD^+ (nicotinamide adenine dinucleotide). NAD forms NADH when it is carrying an electron. Glycolysis is an anaerobic process as you can see by the absence of oxygen in the equation shown in *Figure 9.9*.

Following glycolysis, the pyruvic acid molecules move to the mitochondria, the organelles that produce ATP for the cell. In the presence of oxygen, two more stages complete cellular respiration: the citric acid cycle (also known as the Krebs cycle) and the electron transport chain of the mitochondrion. Before these two stages can begin, however, pyruvic acid undergoes a series of reactions in which it loses a molecule of CO_2 and combines with coenzyme A to form acetyl-CoA. The reaction with coenzyme A produces a molecule of NADH and H^+. These reactions are shown in *Figure 9.10*.

The citric acid cycle

The **citric acid cycle** is a series of chemical reactions similar to the Calvin cycle in that one of the molecules needed for the first reaction is also one of the end products. However, in the Calvin cycle, glucose molecules are formed; in the citric acid cycle, glucose is broken down. What is the first compound that acetyl-CoA combines with in the citric acid cycle? To learn the answer, look at the *Inside Story*.

In the process of breaking down glucose, one molecule of ATP is produced for every turn of the cycle. Two electron carriers are used, NAD^+ and FAD (flavin adenine dinucleotide). A total of three NADH + H^+ molecules and one $FADH_2$ molecule are formed. These electron carriers pass the energized electrons along to the electron transport chain in the inner membrane of the mitochondrion.

Figure 9.10
As a result of the reactions that convert pyruvic acid to acetyl-CoA within the mitochondrion, a molecule of NADH and H^+ are formed.

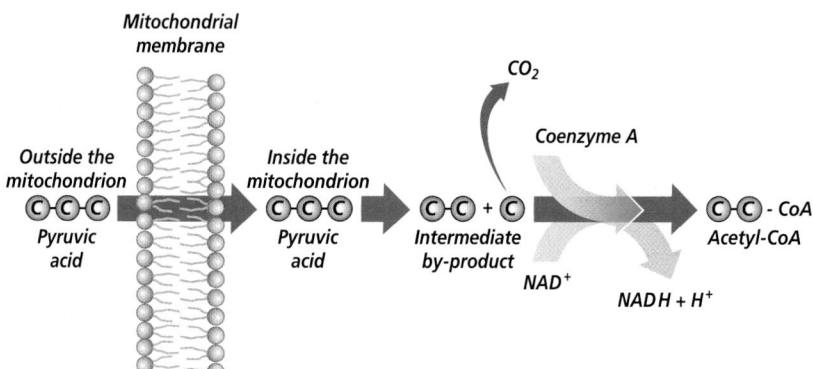

MEETING INDIVIDUAL NEEDS

Gifted

Intrapersonal Have students research the role of the inner mitochondrial membrane in cellular respiration. **L3**

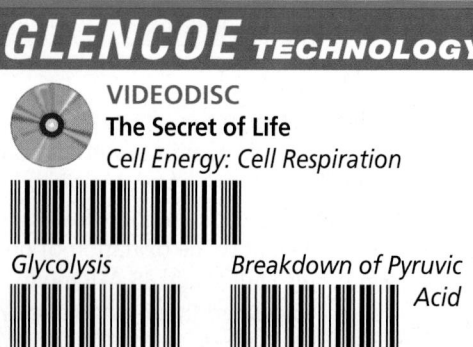

The Citric Acid Cycle

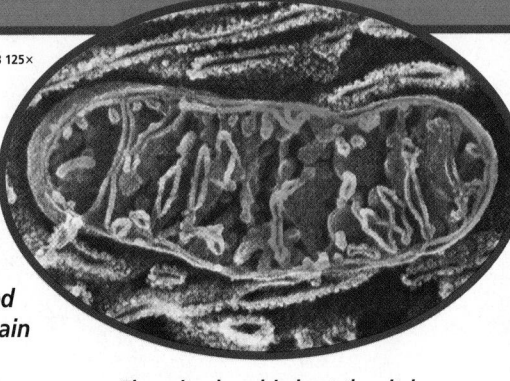

Magnification: 33 125×

The citric acid cycle takes a molecule of acetyl-CoA and breaks it down, forming ATP and CO_2. The electron carriers NAD^+ and FAD pick up energized electrons and pass them to the electron transport chain in the inner mitochondrial membrane.

The mitochondria host the citric acid cycle.

Critical Thinking How many ATP molecules are produced by a single turn of the citric acid cycle?

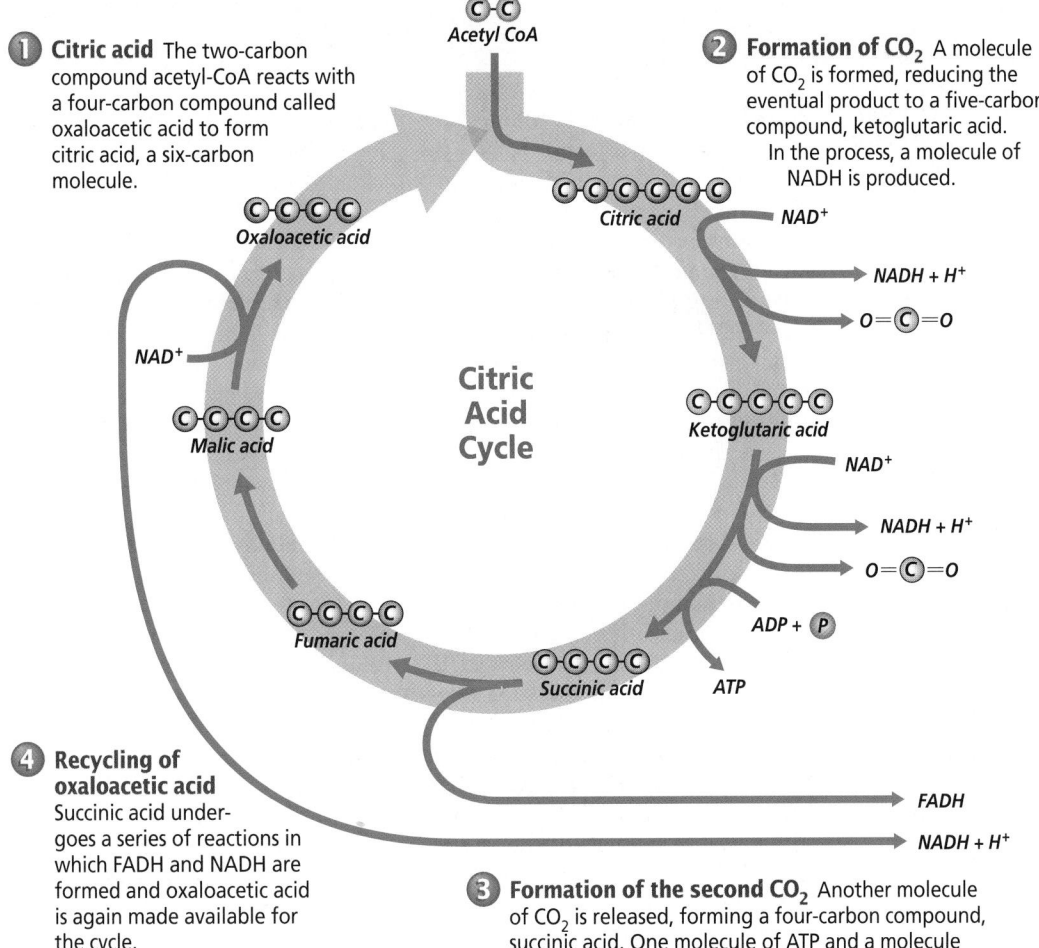

1 **Citric acid** The two-carbon compound acetyl-CoA reacts with a four-carbon compound called oxaloacetic acid to form citric acid, a six-carbon molecule.

2 **Formation of CO_2** A molecule of CO_2 is formed, reducing the eventual product to a five-carbon compound, ketoglutaric acid. In the process, a molecule of NADH is produced.

3 **Formation of the second CO_2** Another molecule of CO_2 is released, forming a four-carbon compound, succinic acid. One molecule of ATP and a molecule of NADH are also produced.

4 **Recycling of oxaloacetic acid** Succinic acid undergoes a series of reactions in which FADH and NADH are formed and oxaloacetic acid is again made available for the cycle.

9.3 GETTING ENERGY TO MAKE ATP **239**

Purpose

To teach students the main events of the citric acid cycle that break down acetyl CoA to form ATP, NADH, $FADH_2$, and carbon dioxide.

Background

Glycolysis initiates the breakdown of glucose. Pyruvate, the end product of glycolysis, undergoes a series of reactions to form acetyl CoA, which enters the citric acid cycle. The citric acid cycle is important for its large energy yield.

Teaching Strategies

■ Ask students to describe the four stages of the citric acid cycle. Include the products for each stage. **L2**

Visual Learning

Make a table with the following headings: ATP, NADH, and $FADH_2$. Under the headings list one round of the citric acid cycle. Have the students fill in the number of each molecule produced during one round.

Critical Thinking

Each round produces:
1 ATP = 1 ATP
3 NADH × 3 ATP/NADH = 9 ATP
1 $FADH_2$ × 2 ATP/$FADH_2$ = 2 ATP
TOTAL = 12 ATP

GLENCOE **TECHNOLOGY**

VIDEODISC
The Secret of Life
Citric Acid Cycle

239

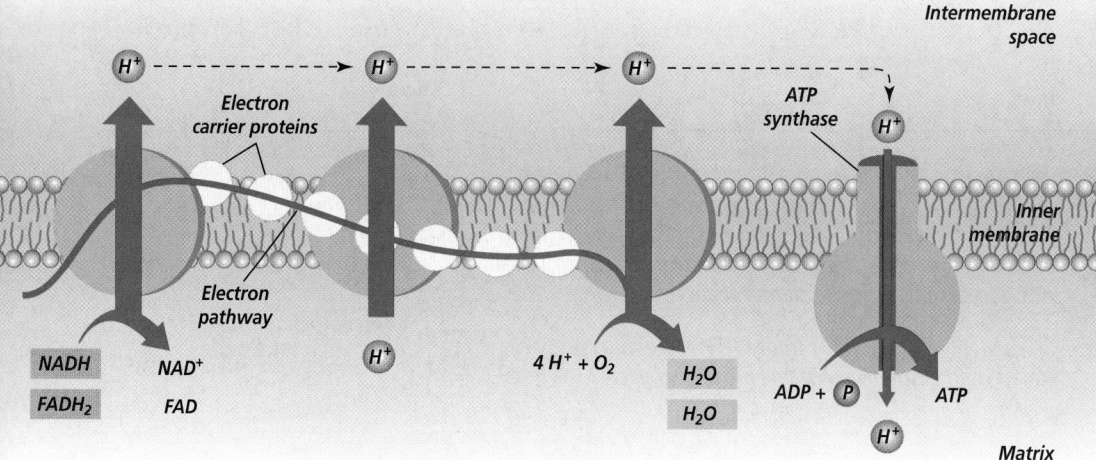

Figure 9.11
In the electron transport chain, the carrier molecules NADH and FADH$_2$ give up electrons that pass through a series of reactions. Oxygen is the final electron acceptor.

The electron transport chain

The electron transport chain in the inner membrane of the mitochondrion is very similar to the electron transport chain of the thylakoid membrane in the chloroplast of plant cells during photosynthesis. NADH and FADH$_2$ pass energized electrons from protein to protein within the membrane, slowly releasing small amounts of the energy contained within the electron. Some of that energy is used directly to form ATP; some is used to pump H$^+$ ions into the center of the mitochondrion. Consequently, the mitochondrion inner membrane becomes positively charged because of the high concentration of positively charged hydrogen ions. At the same time, the exterior of the membrane is negatively charged, which further attracts hydrogen ions. The gradient that results is both electrical and chemical: electrical because of the charge difference; chemical because of the concentration difference.

The inner membrane of the mitochondrion forms ATP from this electrochemical gradient of H$^+$ ions across the membrane, just as the thylakoid membranes did in the chloroplasts. The electron transport chain and the formation of ATP are shown in *Figure 9.11.*

The final electron acceptor in the chain is oxygen, which reacts with four hydrogen ions (4H$^+$) to form two molecules of water (H$_2$O). This is why oxygen is so important to our bodies. Without oxygen, the proteins in the electron transport chain cannot pass along the electrons. If a protein can't pass an electron along, it cannot accept another electron either. Very quickly, the entire chain becomes blocked and the aerobic processes of cellular respiration cannot occur.

Overall, the electron transport chain produces 32 ATP molecules. Obviously, the aerobic method of ATP production is very effective. However, an anaerobic process can produce ATP for short periods of time in the absence of oxygen, but it is not efficient enough to generate sufficient quantities of ATP for all of the cell's needs.

Alternative Lab

Carbon Fixation

Purpose
Students will demonstrate that without light carbon fixation slows or stops.

Materials
black paper, labels, iodine solution, *Coleus* plants, paper clip, hot plate, beaker, 95%

alcohol, small bowl

Procedure
1. Cut out two identical pieces of black paper in the shape of a small square.
2. Stick a label on one piece and write your initials and the date on it.
3. Use a paper clip to fasten the black shapes to the top and bottom surfaces of a *Coleus* leaf so that they are matched up exactly.
4. Leave the plant in sunlight for 7 days.

Then remove the leaf that was partially covered and take off the paper clip and papers.

5. Place the leaf into a beaker of boiling 95% alcohol enclosed in a fume hood and boil it until it turns white. **CAUTION: *Do not use a heat source that has an open flame or a hot plate with an unsealed element.***
6. Remove the leaf and place it in a small bowl. Pour iodine solution on the leaf

Fermentation

There are times when your cells are without oxygen for a short period of time. When this happens, an anaerobic process called fermentation follows glycolysis and provides a means to continue producing ATP until oxygen is available again. Some organisms exist in anaerobic environments and use fermentation to produce energy. There are two major types of fermentation: lactic acid fermentation and alcoholic fermentation. *Figure 9.12* and *Table 9.1* compare the two processes. Perform the *Problem-Solving Lab* shown here to further compare and contrast cellular respiration and fermentation.

Problem-Solving Lab 9-3 — Acquiring Information

Is cellular respiration better than fermentation? The methods by which organisms derive ATP from their food may differ; however, the result, the production of ATP molecules, is similar.

Analysis

Study *Table 9.1* and evaluate cellular respiration, lactic acid fermentation, and alcoholic fermentation.

Thinking Critically

1. Describe some of the reasons why cellular respiration produces so much more ATP than does fermentation.
2. Describe a situation when a human would use more than one of the above processes.
3. Think of an organism that might generate ATP only by fermentation and consider why fermentation is the best process for the organism.

Figure 9.12
Fermentation produces ATP when oxygen for respiration is scarce.

A Lactic acid and alcoholic fermentation are comparable in the production of ATP, but compared to cellular respiration, it is obvious that fermentation is far less efficient in ATP production.

Table 9.1 Comparison of lactic acid and alcoholic fermentation		
Lactic acid	**Alcoholic**	**Cellular respiration**
glucose	glucose	glucose
↓	↓	↓
glycolysis (pyruvic acid)	glycolysis (pyruvic acid)	glycolysis (pyruvic acid)
↓	↓	↓
	carbon dioxide +	carbon dioxide +
lactic acid + 2 ATP	alcohol + 2 ATP	water + 38 ATP

B Lactic acid fermentation occurs in some bacteria, in plants, and in most animals, including humans.

Purpose

Students will compare and contrast the process of aerobic respiration, lactic acid fermentation, and alcoholic fermentation.

Process Skills

think critically, compare and contrast, acquire information, analyze information, sequence, define operationally

Teaching Strategies

■ It will be necessary for students to review the sections dealing with aerobic respiration, lactic acid fermentation, and alcoholic fermentation before they attempt to complete this lab.

Thinking Critically

1. The chemical reactions of the citric acid cycle provide more ATP than does fermentation. Cellular respiration better meets the energy requirements of complex organisms.
2. Humans normally carry out cellular respiration. During a time of intense exercise, we revert to lactic acid fermentation.
3. For an organism that lives in anaerobic conditions and uses small amounts of energy, cellular respiration may not be economical if fermentation consistently provides sufficient energy.

✔ Assessment

Skill Provide an incomplete table similar to Table 9.1 and have the students complete it. Use the Performance Task Assessment List for Data Table in **PASC,** p. 37. **L1**

---

and let it absorb the iodine for a few minutes. **CAUTION:** *Iodine is an irritant. Rinse thoroughly if iodine gets on skin or clothing.*

7. Rinse the leaf with tap water and observe.

Analysis

1. What happens when the leaf is boiled in alcohol? *Its chlorophyll dissolves.*
2. In what part of the leaf did carbon fixation slow or stop? *The covered part*

received no light and therefore could not carry on photosynthesis, which is a carbon fixation process.

✔ Assessment

Performance Have students write a summary of the lab. Use the Performance Task Assessment List for Lab Report in **PASC,** p. 47. **L2**

MiniLab 9-2 — Predicting

Determine if Apple Juice Ferments Organisms such as yeast have the ability to break down food molecules and synthesize ATP when no oxygen is available. When the appropriate food is available, yeast can carry out alcoholic fermentation, producing CO_2. Thus, the production of CO_2 can be used to judge whether alcoholic fermentation is taking place.

Test tube
Water
Plastic pipette
Metal washers
Yeast and apple juice

Procedure
1. Carefully study the diagram and set up the experiment as shown.
2. Hold the test tube in a beaker of warm (not hot) water and observe.

Analysis
1. What were the gas bubbles that came from the plastic pipette?
2. Predict what would happen to the rate of bubbles given off if more yeast were present in the mixture.
3. Why was the test tube placed in warm water?
4. On the basis of your observations, was this process aerobic or anaerobic?

Figure 9.13
Alcoholic fermentation by the yeast in a bread recipe produces CO_2 bubbles that raise the bread dough.

Lactic acid fermentation

Lactic acid fermentation is one of the processes that supplies energy when oxygen is scarce. You know that under anaerobic conditions, the electron transport chain backs up because oxygen is not present as the final electron acceptor. As NADH and $FADH_2$ arrive with energized electrons from the citric acid cycle and glycolysis, they cannot release their energized electrons to the electron transport chain. The citric acid cycle and glycolysis cannot continue without a steady supply of NAD^+ and FAD.

The cell does not have a method to replace FAD during anaerobic conditions; however, NAD^+ can be replaced through lactic acid fermentation. In lactic acid fermentation, two molecules of pyruvic acid use NADH to form two molecules of lactic acid. This releases NAD^+ to be used in glycolysis, allowing two ATP molecules to be formed for each glucose molecule. The lactic acid is transferred from muscle cells, where it is produced during strenuous exercise, to the liver that converts it back to pyruvic acid. The lactic acid that builds up in muscle cells results in muscle fatigue.

Alcoholic fermentation

Another type of fermentation, **alcoholic fermentation,** is used by, among others, yeast cells to produce CO_2 and ethyl alcohol. When making bread, like that shown in *Figure 9.13,* yeast cells produce CO_2 that forms bubbles in the dough. Eventually the heat of baking the bread kills the yeast and the bubble pockets are left to lighten the bread. You can do the activity in the *MiniLab* on this page to examine fermentation in apple juice.

Comparing Photosynthesis and Cellular Respiration

The production and breakdown of food molecules are accomplished by distinct processes that bear certain similarities. Both photosynthesis and cellular respiration use electron carriers and a cycle of chemical reactions to form ATP. Both use an electron transport chain to form ATP and to create a chemical and a concentration gradient of H^+ within a cell. This hydrogen gradient can be used to form ATP by an alternative process.

However, despite using such similar tools, the two cellular processes accomplish quite different tasks. Photosynthesis produces high-energy carbohydrates and oxygen from the sun's energy, whereas cellular respiration uses oxygen to break down carbohydrates to form ATP and compounds with a much lower level of energy. Also, one of the end products of cellular respiration is CO_2, which is one of the beginning products for photosynthesis. The oxygen produced during photosynthesis is a critical molecule necessary for cellular respiration. *Table 9.2* in *Figure 9.14* compares these complementary processes.

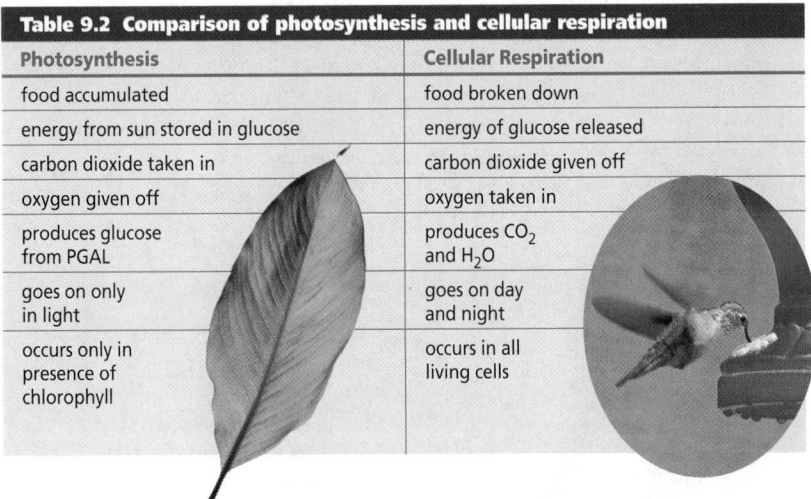

Table 9.2 Comparison of photosynthesis and cellular respiration	
Photosynthesis	**Cellular Respiration**
food accumulated	food broken down
energy from sun stored in glucose	energy of glucose released
carbon dioxide taken in	carbon dioxide given off
oxygen given off	oxygen taken in
produces glucose from PGAL	produces CO_2 and H_2O
goes on only in light	goes on day and night
occurs only in presence of chlorophyll	occurs in all living cells

Figure 9.14
Photosynthesis and cellular respiration are complementary processes. The requirements of one process are the products of the other.

Section Assessment

Understanding Main Ideas

1. Compare the ATP yields of glycolysis and aerobic respiration.
2. How do alcoholic and lactic acid fermentation differ?
3. How is most of the ATP from aerobic respiration produced?
4. When is lactic acid fermentation important to the cell?

Thinking Critically

5. Compare the energy-producing processes in a jogger's leg muscles with those of a sprinter's leg muscles. Which is likely to build up more lactic acid? Which runner is more likely to be out of breath after running? Explain.

SKILL REVIEW

6. **Making and Using Tables** Use the section called Comparing Photosynthesis and Cellular Respiration to make a set diagram summarizing the similarities and differences between the two processes. For more help, refer to *Organizing Information* in the **Skill Handbook**.

Section Assessment

1. Glycolysis produces 2 ATP molecules; aerobic respiration, as many as 38.
2. Alcoholic fermentation produces alcohol and carbon dioxide. Lactic acid fermentation produces lactic acid.
3. Most of the ATP is produced by the reactions of the electron transport chain.
4. When oxygen is unavailable.
5. Aerobic respiration occurs in the leg muscles of both runners. The sprinter may build up more lactic acid and be out of breath because of an oxygen debt associated with the quick burst of energy.
6. *Differences:* Photosynthesis forms sugars and oxygen, uses water and CO_2. Respiration uses oxygen to break down sugar, forms water and CO_2. Photosynthesis uses light to form chemical bond energy. Respiration uses chemical bond energy to form ATP. *Similarities:* Both are complex reactions, require enzymes, occur in specific organelles, and involve electron transport chains.

243

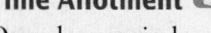

Process Skills

use variables and controls, think critically, observe and infer, collect data, interpret data

Safety Precautions

Students should always wear goggles in the lab. Be sure hands and work areas are dry when handling electrical equipment and have students wash hands at the end of the lab.

PREPARATION

Prepare sodium hydrogen carbonate solution by mixing 2.5 g sodium hydrogen carbonate with 1000 mL of water.

Alternative Materials

■ If *Elodea* is unavailable, you can use any other green aquatic plant.
■ Students can use the classroom clock if the second hand is visible.

Resource Manager

BioLab and MiniLab Worksheets, p. 41 L2

INTERNET
BioLab

What factors influence photosynthesis?

Oxygen is one of the products of photosynthesis. Because oxygen is only slightly soluble in water, aquatic plants such as Elodea give off visible bubbles of oxygen as they carry out photosynthesis. By measuring the rate at which bubbles form, you can measure the rate of photosynthesis.

PREPARATION

Problem

How do different wavelengths of light a plant receives affect its rate of photosynthesis?

Objectives
In this BioLab, you will:
■ **Observe** photosynthesis in an aquatic organism.
■ **Measure** the rate of photosynthesis.
■ **Observe** how various wavelengths of light influence the rate of photosynthesis.
■ **Use the Internet** to collect and compare data from other students.

Materials
1000-mL beaker
three *Elodea* plants
string
washers
colored cellophane, assorted colors
lamp with reflector and 150-watt bulb
0.25% sodium hydrogen carbonate (baking soda) solution
watch with second hand

Safety Precautions
Always wear goggles in the lab.

Skill Handbook
Use the **Skill Handbook** if you need additional help with this lab.

PROCEDURE

Teaching Strategies

■ Place the *Elodea* sprigs in a large bowl and place them under a lamp for about 10 minutes before students begin the lab. This reduces the time the students will wait to begin seeing evidence of photosynthesis.
■ You may wish to circulate through the room during this activity to ensure that the setups are constructed properly and that the students are seeing evidence of photosynthesis.

Data and Observations

Students should record data under control conditions and then under experimental conditions for the same amount of time—at least 5 minutes.

PROCEDURE

1. Construct a basic setup like the one shown opposite.
2. Create a data table to record your measurements. Be sure to include a column for each color of light you will investigate and a column for the control experiment.
3. Place the *Elodea* plants in the beaker, then completely cover the plants with water. Add some of the baking soda solution. The solution provides CO_2 for the aquarium plants. **Be sure to use the same amount of water and solution for each trial.**
4. Conduct a control experiment by directing the lamp (without colored cellophane) on the plant and notice when you see the bubbles.
5. Observe and record the number of oxygen bubbles that *Elodea* generates in five minutes.
6. Repeat steps 4 and 5 with a piece of colored cellophane. Record your observations.
7. Repeat steps 4 and 5 with a different color of cellophane and record your observations.
8. Go to the Glencoe Science Web Site at the address shown below to **post your data.**

Data Table

	Control	Color 1	Color 2
Bubbles observed in five minutes			

ANALYZE AND CONCLUDE

1. **Interpreting Observations** From where did the bubbles of oxygen emerge? Why?
2. **Making Inferences** Explain how counting bubbles measures the rate of photosynthesis.
3. **Using the Internet** Make a graph of your data and data posted by other students with the rate of photosynthesis per minute plotted against the wavelength of light you tested for both the control and experimental setups.

Write a sentence or two explaining the graph.

Sharing Your Data

BIOLOGY Online Find this BioLab on the Glencoe Science Web site at **science.glencoe.com**. Post your data in the data table provided for this activity. Use the additional data from other students who tested wavelengths other than those you tested to expand your graph.

ANALYZE AND CONCLUDE

1. The bubbles emerged from the end of the stem of the *Elodea* plant.
2. Oxygen is an end product of photosynthesis. As the rate of photosynthesis changes, so will the rate at which oxygen is produced.
3. The rate for the control setup should be the greatest and therefore the highest line. The rates corresponding to colored, filtered light will be lower than the control. Results will vary depending on the color of cellophane the students use.

✓ Assessment

Portfolio Have students write an evaluation of the lab. Their evaluations should include an overview of the experiment, data analysis, and conclusion statements. Have them also write hypotheses about how photosynthesis might be affected if another condition, such as temperature, were changed. Use the Performance Task Assessment List for Formulating a Hypothesis in **PASC**, p. 21. **L2**

Sharing Your Data

BIOLOGY Online To navigate the Internet BioLabs, choose the Biology: The Dynamics of Life icon at the site. Click on the student site icon, then the Biolabs icon. The advantage of using the Internet for an experiment of this nature is that students can collect considerably more and varied data than they could collect on their own in the allotted time. They also gain the experience of communicating scientific information using a technology that was originally intended for the dissemination of scientific knowledge.

Purpose

Students should recognize the role that pigments such as chlorophyll play in the process of photosynthesis.

Background

There are far more chlorophyll molecules in a green leaf than carotenoids, and for most of the growing season, chlorophyll masks the presence of those accessory pigments. In the autumn, however, chlorophyll breaks down and the other pigments in leaves are visible as "fall colors." The cause of this chlorophyll breakdown is not completely understood, but it seems to be tied to the gradual reduction in daylight that takes place as summer ends.

Teaching Strategies

■ Have students examine live (or preserved) specimens of green, brown, and red algae to observe the difference in pigments. **L2** **ELL**

■ Melanin is a pigment in animals that absorbs ultraviolet radiation and protects skin and other tissues from sun damage. Have students research melanin along with a plant pigment and create posters that compare the two. **L3**

Connection to Biology

Accessory pigments have made it possible for photosynthetic organisms to use a broader range of the visible light spectrum, and in so doing, survive in places where the amount or quality of visible light is minimal.

Plant Pigments

In photosynthesis, light energy is converted into chemical energy. To begin the process, light is absorbed by colorful pigment molecules contained in chloroplasts.

A pigment is a substance that can absorb the various wavelengths of visible light. You can observe the colors of these wavelengths by letting sunlight pass through a prism to create a "rainbow," or spectrum, that has red light on one end, violet on the other, and orange, yellow, green, and blue light in between.

Every photosynthetic pigment is distinctive in that it absorbs certain wavelengths in the visible light spectrum.

Chlorophylls *a* and *b* The principal pigment of photosynthesis is chlorophyll. Chlorophyll exists in two forms designated as *a* and *b*. Chlorophyll *a* and *b* both absorb light in the violet to blue and red to red-orange parts of the spectrum, although at somewhat different wavelengths. These pigments also reflect green light, which is why plant leaves appear green.

When chlorophyll *b* absorbs light, it transfers the energy it acquires to chlorophyll *a*, which then feeds that energy into the chemical reactions that lead to the production of ATP and NADP. In this way, chlorophyll *b* acts as an "accessory" pigment by making it possible for photosynthesis to occur over a broader spectrum of light than would be possible with chlorophyll *a* alone.

Carotenoids and phycobilins Carotenoids and phycobilins are other kinds of accessory pigments that absorb wavelengths of light different from those absorbed by chlorophyll *a* and *b*, and so extend the range of light that can be used for photosynthesis.

Carotenoids are yellow-orange pigments. They are found in all green plants, but their color is usually masked by chlorophyll.

Magnification: 1630×

Pigments color cyanobacteria (above) and red algae (inset).

Carotenoids are also found in cyanobacteria and in brown algae. A particular carotenoid called fucoxanthin gives brown algae their characteristic dark brown or olive green color.

Phycobilins are blue and red. Red algae get their distinctive blood-red coloration from phycobilins. Some phycobilins can absorb wavelengths of green, violet, and blue light that penetrate into deep water. One species of red algae that contains these pigments is able to live at ocean depths of 269 meters (884 feet). The algae's pigments absorb enough of the incredibly faint light that penetrates to this depth—only 0.0005 percent of what is available at the water's surface—to power photosynthesis.

CONNECTION TO BIOLOGY

How do you think accessory pigments may have influenced the spread of photosynthetic organisms into diverse habitats such as the deep sea?

 To find out more about pigments, visit the Glencoe Science Web site. **science.glencoe.com**

BIOLOGY Online Note Internet addresses that you find useful in the space below for quick reference.

_____

_____

_____

_____

_____

SUMMARY

Section 9.1

ATP in a Molecule

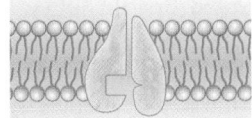

Main Ideas
- ATP is the molecule that stores energy for easy use within the cell.
- ATP is formed when a phosphate group is added to ADP. When ATP is broken down, ADP and phosphate are formed and energy is released.
- ATP is the main link between energy-releasing and energy-using reactions.

Vocabulary
(ADP) adenosine diphosphate (p. 228)
(ATP) adenosine triphosphate (p. 228)

Section 9.2

Photosynthesis: Trapping the Sun's Energy

Main Ideas
- Photosynthesis is the process by which cells use light energy to make carbohydrates.
- Chlorophyll in the chloroplasts of plant cells traps light energy needed for photosynthesis.
- The light reactions of photosynthesis produce ATP and result in the splitting of water molecules.
- The reactions of the Calvin cycle make carbohydrates using CO_2 along with ATP and hydrogen from the light reactions.

Vocabulary
Calvin cycle (p. 234)
chlorophyll (p. 232)
electron transport chain (p. 232)
light-dependent reactions (p. 231)
light-independent reactions (p. 231)
NADP+ (p. 233)
photolysis (p. 234)
photosynthesis (p. 231)
pigments (p. 232)

Section 9.3

Getting Energy to Make ATP

Main Ideas
- Cellular respiration is the process by which cells break down carbohydrates to release energy.
- Cellular respiration takes place in mitochondria, uses oxygen, and yields many more ATPs than do anaerobic processes.
- Energy can be released anaerobically by glycolysis followed by alcoholic or lactic acid fermentation.

Vocabulary
aerobic respiration (p. 237)
alcoholic fermentation (p. 242)
anaerobic respiration (p. 237)
cellular respiration (p. 237)
citric acid cycle (p. 238)
glycolysis (p. 237)
lactic acid fermentation (p. 242)

UNDERSTANDING MAIN IDEAS

1. Which of the following is a product of the Calvin cycle?
   a. carbon dioxide   **c.** oxygen
   b. NADP+   **d.** FADH$_2$

2. _____ processes require oxygen, whereas _____ processes do not.
   a. anaerobic—aerobic
   b. aerobic—anaerobic
   c. photolysis—aerobic
   d. aerobic—respiration

Summary
Summary statements can be used by students to review the major concepts of the chapter.

Using the Vocabulary
To reinforce chapter vocabulary, use the Content Mastery Booklet and the activities in the Interactive Tutor for Biology: The Dynamics of Life on the Glencoe Science Web site:
science.glencoe.com

All Chapter Assessment questions and answers have been validated for accuracy and suitability by The Princeton Review.

UNDERSTANDING MAIN IDEAS

1. b
2. b

GLENCOE **TECHNOLOGY**

VIDEOTAPE
MindJogger Videoquizzes
Chapter 9: *Energy in a Cell*
Have students work in groups as they play the videoquiz game to review key chapter concepts.

Resource Manager

Chapter Assessment, pp. 49-54
MindJogger Videoquizzes
ExamView® Pro Software
BDOL Interactive CD-ROM, Chapter 9 quiz

3. c
4. a
5. d
6. d
7. b
8. b
9. a
10. a
11. carbon dioxide; oxygen
12. chlorophyll
13. thylakoid; mitochondrion
14. Photolysis
15. Citric acid cycle
16. Electron transport chain
17. lactic acid; ethanol
18. phosphates; repel
19. Lactic acid fermentation; alcoholic fermentation
20. Pigments

APPLYING MAIN IDEAS

21. During physical activity, muscle cells must release energy at a higher rate than skin cells.
22. Photosynthesis stores energy, whereas respiration releases energy, and the products of one process are the reactants of the other process.
23. Other pigments in the plant absorb some of the light of other wavelengths and pass the energy to chlorophyll for use in photosynthesis. The remaining light that is not trapped is reflected or absorbed as heat.
24. Possible answer: The amount of oxygen in the atmosphere would start to decrease because it would be used up in respiration and not replaced.

3. During all energy conversions, some of the energy is converted to _____.
  a. carbon dioxide   **c.** heat
  b. water   **d.** sunlight

4. Four molecules of glucose would give a net yield of _____ ATP following glycolysis.
  a. 8   **c.** 4
  b. 16   **d.** 12

5. In which of the following structures do the light-independent reactions of photosynthesis take place?

a. 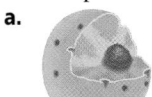   **c.**

b.    **d.**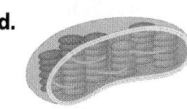

6. What is the first process in the cell to be affected by anaerobic conditions?
  a. citric acid cycle
  b. fermentation
  c. glycolysis
  d. electron transport chain

7. Which molecule provides the most accessible source of energy for cell organelles?
  a. glucose   **c.** starch
  b. ATP   **d.** carbon dioxide

8. Which of the following uses no energy?
  a. glycolysis   **c.** light reactions
  b. Calvin cycle   **d.** muscle contraction

9. Which of the following transports high-energy electrons in photosynthesis?

 **TEST-TAKING TIP**

Take 5 and Stay Sharp
Wanting to perform well on your exam is praiseworthy. But if you study for long periods of time, you could actually end up hurting your chances for a good score. Remember to take frequent short breaks in your studies to keep your mind fresh.

  a. NADP⁺   **c.** FAD
  b. NAD⁺   **d.** ATP

10. When yeast ferments the sugar in a bread mixture, what is produced that causes the bread dough to rise?
  a. carbon dioxide   **c.** ethyl alcohol
  b. water   **d.** oxygen

11. Plants must have a constant supply of _____ for photosynthesis, but they provide _____ for cellular respiration.

12. _____ is the first molecule to provide electrons for photosynthesis.

13. The _____ membrane is the site of photosynthesis, whereas the _____ is the site for cellular respiration.

14. _____ is the process by which electrons are restored to chlorophyll after photosynthesis.

15. _____ acts as a source of CO_2 in cellular respiration.

16. _____ occurs in the inner membrane of the mitochondrion.

17. In fermentation, pyruvic acid can form _____ or _____ in addition to NAD⁺.

18. ATP stores energy because the three _____ have a negative charge and they _____ each other.

19. _____ produces muscle soreness, whereas _____, another type of fermentation, produces ethyl alcohol.

20. _____ such as chlorophyll are the chemical substances that give color to plants.

APPLYING MAIN IDEAS

21. Why would human muscle cells contain many more mitochondria than skin cells?

22. How are cellular respiration and photosynthesis complementary processes?

23. What happens to sunlight that strikes a leaf but is not trapped by photosynthesis?

24. What might happen to Earth's atmosphere if photosynthesis suddenly stopped?

25. If you were planning on studying the compounds that could possibly be the source of the oxygen released during photosynthesis, which compounds would you need to consider?

THINKING CRITICALLY

26. Formulating Hypotheses *Elodea* sprigs were placed under a white light, and the rate of photosynthesis was measured by counting the number of oxygen bubbles per minute for ten minutes. Predict the rate of photosynthesis if a piece of red cellophane were placed over the white light.

27. Observing and Inferring Yeast cells must be forced to ferment by placing them in an environment without any oxygen. Why would the yeast cells carry out aerobic respiration rather than fermentation when oxygen is present?

28. Recognizing Cause and Effect A window plant native to the desert of South Africa has leaves that grow almost entirely underground with only the transparent tip of the leaf protruding above the soil surface. Suggest how this adaptation aids the survival of this plant.

29. Concept Mapping Make a concept map using the following vocabulary terms: cellular respiration, glycolysis, citric acid cycle, electron transport chain.

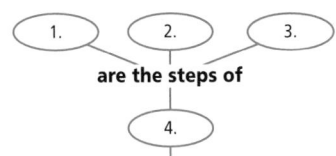

are the steps of

which takes place in mitochondria

CD-ROM

For additional review, use the assessment options for this chapter found on the ***Biology: The Dynamics of Life Interactive CD-ROM*** and on the Glencoe Science Web site. **science.glencoe.com**

ASSESSING KNOWLEDGE & SKILLS

Yeast cells and sucrose were placed in a test tube, and the tube was then plugged. The yeast–sucrose mixture incubated for 24 hours. Gas bubbles began to rise to the top of the tube. After 24 hours, no sucrose was left in the solution.

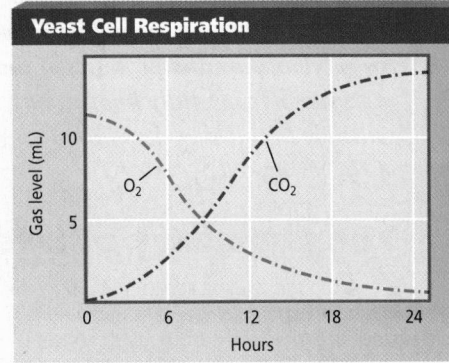

Interpreting Data Use the graph to answer the questions below.

1. What process was the yeast using to digest the sucrose at the beginning of the experiment?
   a. photosynthesis
   b. anaerobic respiration
   c. aerobic respiration
   d. light-dependent reactions

2. Which of the following would be left in the solution after 24 hours?
   a. sucrose   **c.** oxygen
   b. lactic acid   **d.** ethyl alcohol

3. What gas would be found in the top of the tube after the incubation period?
   a. carbon dioxide   **c.** hydrogen
   b. oxygen   **d.** nitrogen

4. Making a Table Construct a table from the graph showing the oxygen levels and carbon dioxide levels.

THINKING CRITICALLY

25. carbon dioxide and water
26. Photosynthetic rate would decrease. Although the cellophane transmits red light, it would cut down the light in the blue end of the spectrum that is also used in photosynthesis.
27. Aerobic respiration is a much more efficient process and produces many more ATPs per sugar molecule.
28. This adaptation conserves water while allowing the plant to get light for photosynthesis, thereby solving the main problem of desert plants.
29. 1. Glycolysis; 2. Citric acid cycle; 3. Electron transport chain; 4. Cellular respiration

ASSESSING KNOWLEDGE & SKILLS

1. c
2. d
3. a
4. Student tables should be consistent with the information in the graph.

BioDigest

National Science Education Standards:
UCP.1, UCP.2, UCP.3, UCP.5,
B.3, B.6, C.1, C.5, F.1, G.1, G.3

Prepare

Purpose

This BioDigest can be used as an introduction to or as an overview of the structure and function of the cell. If time is limited, you may wish to use this unit summary to teach about the cell in place of the chapters in the Cell unit.

Key Concepts

Students will learn that chemistry is an integral component of living organisms. According to the cell theory, the cell is the basic unit of organization of all living matter. This means that even a complex organism, such as a tree or an elephant, is made up of cells. Students will learn that the structure and function of a cell and its parts are similar, even for cells that serve very different functions. The concepts of how a cell transforms energy from the sun and food sources and how a cell reproduces will also be described.

1 Focus

Bellringer

Display an overhead transparency of a eukaryotic cell. Point out each of the organelles and ask students to identify the functions of each organelle. **L1** **ELL**

For a **preview** of the cell unit, study this BioDigest before you read the chapters. After you have studied the cell chapters, you can use the BioDigest to **review** the unit.

The Life of a Cell

All organisms are made up of cells, and each cell is like a complex, self-contained machine that can perform all of the life functions of the cell. Yet as small as they are, all of the mechanisms and processes of these little machines are not fully known, and scientists continue to unravel the marvelous mysteries of the living cell.

Cells are microscopic machines

The Chemistry of Life

Although you are studying biology, chemistry is fundamental to all biological functions. Understanding some of the basic concepts of chemistry will enhance your understanding of the biological world.

Elements and Atoms

Every substance in and on Earth is a combination of elements. An atom, the smallest component of an element, is formed by layers of electrons around a nucleus made of protons and neutrons. Atoms come together to form molecules.

VITAL STATISTICS

Carbon Isotopes
Isotopes of carbon contain different numbers of neutrons.
Carbon 12: six protons and six neutrons
Carbon 13: six protons and seven neutrons
Carbon 14: six protons and eight neutrons

FOCUS ON HISTORY

The Cell Theory

Magnification: 200×

van Leeuwenhoek might have viewed microorganisms like these found in a droplet of pond water.

In the 1600s, Anton van Leeuwenhoek was the first person to view living organisms through a microscope. Another scientist, Robert Hooke, named the structures cells. Two hundred years later, several scientists, including Matthias Schleiden, Theodor Schwann, and Rudolph Virchow continued to study animal and plant tissues under the microscope. Conclusions from all scientists were combined to form the cell theory:

1. All organisms are composed of one or more cells.
2. The cell is the basic unit of organization of organisms.
3. All cells come from preexisting cells.

250

Multiple Learning Styles

Look for the following logos for strategies that emphasize different learning modalities.

Kinesthetic Meeting Individual Needs, p. 253

Visual-Spatial Portfolio, p. 252; Microscope Activity, p. 252, 253

Intrapersonal Project, p. 254

Linguistic Biology Journal, p. 252; Check for Understanding, p. 255; Extension, p. 255

A polysaccharide is a type of carbohydrate made up of a chain of monosaccharides.

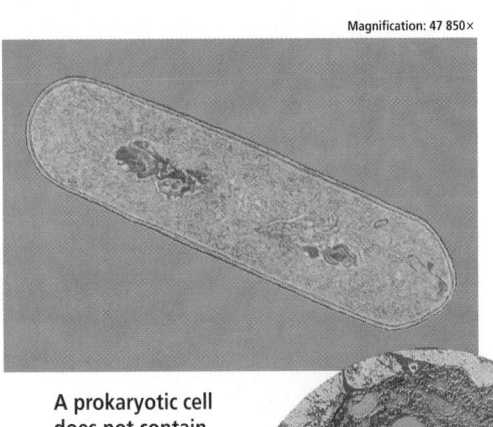

Polysaccharides

Organic Compounds

Carbohydrates are chemical compounds made up of carbon, hydrogen, and oxygen molecules. Common carbohydrates include sugars, starches, and cellulose. Lipids, known as fats and oils, contain a glycerol backbone and three fatty acid chains. Proteins are a combination of amino acids connected by peptide bonds.

Lipids are made up of a glycerol backbone and three fatty acid chains.

Glycerol

Fatty acid chains

Amino acids can be joined with peptide bonds to form proteins.

Eukaryotes and Prokaryotes

All cells are surrounded by a plasma membrane. Eukaryotic cells contain membrane-bound structures within the cell called organelles. Cells without internal membrane-bound structures are called prokaryotic cells.

Magnification: 47 850×

A prokaryotic cell does not contain membrane-bound organelles.

A eukaryotic cell contains membrane-bound organelles.

Organelles

Magnification: 8050×

251

2 Teach

✔ Assessment

Portfolio Have students cut out pictures of trees from magazines and place them in their portfolios. Then have them write about the tree from a cellular point of view. They should include details such as whether the cells of the trees are eukaryotic or prokaryotic, a list of organelles that may be present, information about cells and photsynthesis, ideas on cellular growth and reproduction, and cellular differentiation within the tree. Then have the students write a comparison with an animal cell using the same information. **L2** **P**

Quick Demo

Bring in food items to demonstrate organic compounds. A potato or bread can demonstrate carbohydrates, meat can demonstrate proteins, and vegetable oil or shortening can demonstrate lipids. Combine this experience with a molecular model of a lipid and relate the model to the example lipid and to the overall structure of the plasma membrane.

NATIONAL GEOGRAPHIC

VIDEODISC
STV: The Cell
Viewing the Cell
Early Pioneers - Unit 1

38 sec.

Cell Theorists - Unit 1

1 min. 15 sec.

✔ Assessment Planner

Portfolio Assessment
Assessment, TWE, p. 251
Performance Assessment
Assessment, TWE, p. 252
Knowledge Assessment
BioDigest Assessment, SE, p. 255
Skill Assessment
Assessment, TWE, p. 255

GLENCOE TECHNOLOGY

VIDEODISC
The Infinite Voyage *Unseen Worlds, Studying the Basic Building Blocks: The Atom* (Ch. 9) 2 min.

The Scanning Tunneling Microscope: Observing Atomic Particles (Ch. 10) 2 min.

BioDigest

Microscope Activity

 Visual-Spatial Set up microscopes with slides demonstrating animal and plant cells. Have students look at the cells and identify differences between the two cells.

Help the students by explaining that animal cells have centrioles and plant cells do not. Centrioles play a role during cell division and in the formation of microtubules. Also explain that plant cells contain a cell wall outside the plasma membrane and only one or two large vacuoles that store water. Make sure the students understand that animal cells also have vacuoles, but there are more of them and they are smaller than those found in plants. **L2** **ELL**

✓ Assessment

Performance Have students look at slides of animal and plant cells and identify the structures.

GLENCOE TECHNOLOGY

CD-ROM
Biology: The Dynamics of Life
Exploration: *Parts of the Cell*
Disc 1
BioQuest: *Cellular Pursuit*
Disc 1

BioDigest

Cell Organelles

The organelles of a cell work together to carry out the functions necessary for cell survival.

Plasma Membrane

According to the fluid mosaic model, the plasma membrane is formed by two layers of phospholipids with the fatty acid chains back-to-back; the phosphate groups face the external environment, and proteins are embedded in the membrane.

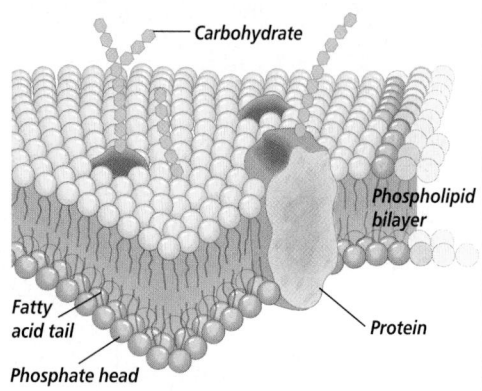

Carbohydrate

Phospholipid bilayer

Fatty acid tail

Protein

Phosphate head

The plasma membrane is composed of a lipid bilayer with embedded proteins.

Control of Cell Functions

The nucleus contains the master plans for proteins, which are then produced by organelles called ribosomes. The nucleus also controls cellular functions.

Assembly, Transport, and Storage

The cytoplasm suspends the cell's organelles, including endoplasmic reticulum, Golgi apparatus, vacuoles, and lysozomes. The endoplasmic reticulum and Golgi apparatus transport and modify proteins.

Energy Transformers

Chloroplasts are found in plant cells and capture the sun's light energy so it can be transformed into useable chemical energy. Mitochondria are found in both animal and plant cells and transform the food you eat into a useable energy form.

Support and Locomotion

A network of microfilaments and microtubules attach to the plasma membrane to give the cell structure. Cilia are short, numerous projections that move like a wave. Flagella are longer projections that move in a whiplike fashion to propel a cell.

Plant cell

Animal cell

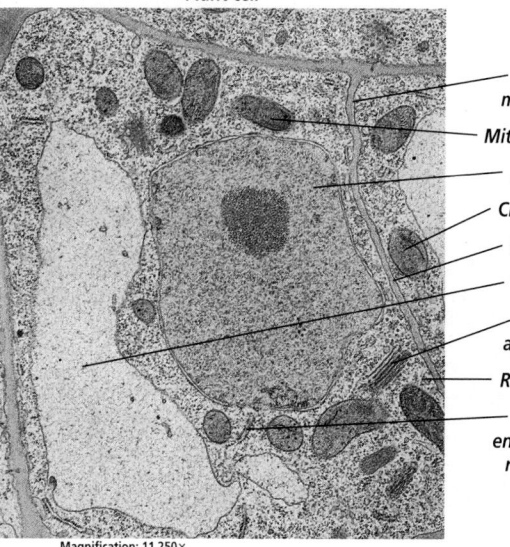
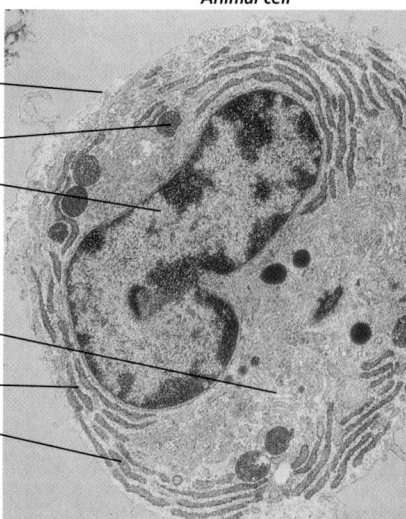

Plasma membrane

Mitochondrion

Nucleus

Chloroplast

Cell wall

Vacuole

Golgi apparatus

Ribosomes

Rough endoplasmic reticulum

Magnification: 11 250×

Magnification: 46 000×

252

✓ Portfolio

The Plasma Membrane

 *Visual-Spatial* In their portfolios, have students draw a plasma membrane. They should include phospholipids showing the polar head and fatty acid tails forming a bilayer with the heads and tails properly oriented. Have them include several transmembrane proteins. **L1** **ELL** **P**

BIOLOGY JOURNAL

Organelles

Linguistic Have students write a paragraph about how the cellular organelles act like a factory to produce proteins and energy. They should include how organelles regulate the intake of necessary materials and the export of products and wastes. **L2**

Diffusion and Osmosis

The selectively permeable plasma membrane allows only certain substances to cross. Diffusion is the movement of a substance from an area of higher concentration to an area of lower concentration. Diffusion of water across a selectively permeable membrane is called osmosis.

Simple diffusion across a membrane occurs by random movement. Facilitated diffusion requires proteins to bind and help move molecules across the membrane. Active transport requires energy to move molecules across a concentration gradient.

VITAL STATISTICS

Cellular Environments
Isotonic solution: same number of molecules inside and outside the cell
Hypotonic solution: more molecules inside the cell; water enters the cell
Hypertonic solution: fewer molecules inside the cell; water leaves the cell

Membrane proteins can transport substances against the concentration gradient in active transport.

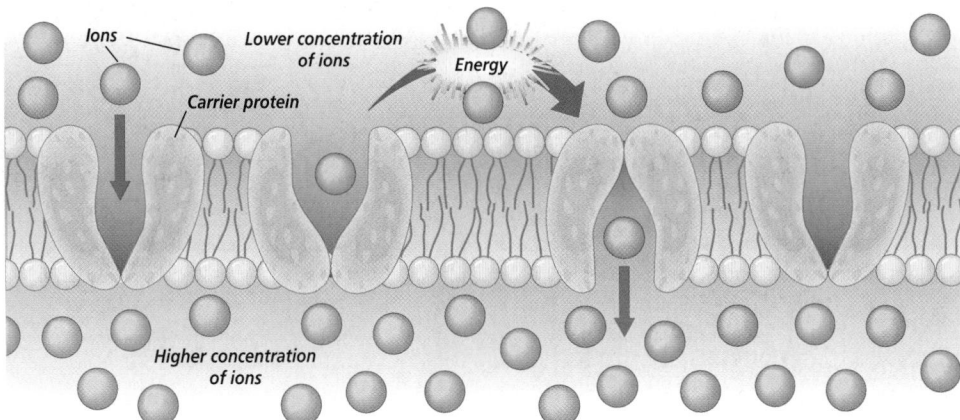

Mitosis

As cells grow, they reach a size where the plasma membrane cannot transport enough nutrients and wastes to maintain cell growth. At this point, the cell undergoes mitosis and divides.

Prior to mitosis is a period of intense metabolic activity called interphase. The first stage of mitosis is prophase, when the duplicated chromosomes condense and the mitotic spindle forms on the two opposite ends of the cell. The chromosomes line up in the center of the cell during metaphase and slowly separate during anaphase. In telophase the nucleus divides followed by cytokinesis, which separates the daughter cells.

During the stages of mitosis, chromosomes are replicated and separated into two daughter cells.

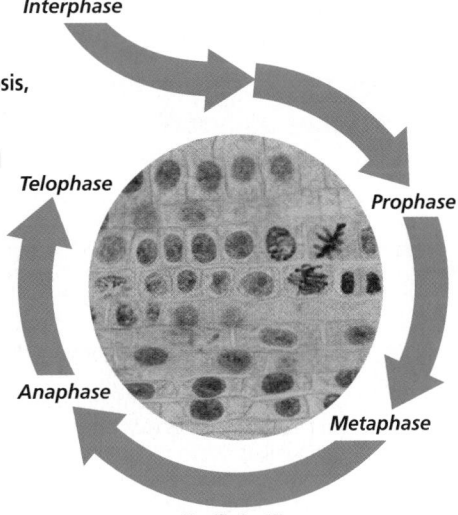

Interphase · Prophase · Metaphase · Anaphase · Telophase

Magnification: 160×

253

Demonstration

Linguistic Use this demonstration to simulate osmosis in hypotonic, isotonic, and hypertonic solutions. Fill three pieces of dialysis tubing with red 15% sucrose solution. Tie the top of each dialysis bag around a hollow piece of glass tubing. Clamp each bag/tubing apparatus to a ring stand and place in one of three solutions: distilled water (hypotonic), 15% clear sucrose solution (isotonic), and a 30% sucrose solution (hypertonic). At the beginning of the demonstration, mark the initial level of the red sucrose in the glass tubing. At regular intervals, such as the end of each class period, again mark the level of the red sucrose solution. Use the changes in volume of the three dialysis bags to quantify osmosis in hypotonic, isotonic, and hypertonic solutions.

Microscope Activity

Visual-Spatial Set up a series of microscopes with prepared slides of onion root tip demonstrating the stages of mitosis. Have students look at the slides before discussing the stages of mitosis. After naming and discussing the stages, have students look at the slides again and ask them to explain what they see. **L2**

GLENCOE TECHNOLOGY

CD-ROM
Biology: The Dynamics of Life
Exploration: *Phases of Mitosis* Disc 1

MEETING INDIVIDUAL NEEDS

Visually Impaired

 **Kinesthetic** Tie pairs of socks together in the middle to simulate chromosomes undergoing mitosis. Create a series of cells and place the pairs of socks in the appropriate places to mimic the stages of mitosis. Have visually impaired students manipulate the sock chromosomes. **L1 ELL**

GLENCOE TECHNOLOGY

VIDEODISC
The Secret of Life
Cell Membranes: Membranes and Transport

BIODIGEST

BIODIGEST

BIODIGEST

Classroom Activity

Use play money to demonstrate the analogy that ATP is the currency for energy. Assign a cost to several objects in the classroom and point out that some items "cost" more than others. Have students purchase these objects using large bills such as $100s or $500s to demonstrate that large denominations must be broken down to smaller amounts. Then relate the cost to the energy expense to perform cellular activity. An example of an inexpensive transaction in a cell is transporting two Na+ ions outside of the cell. An example of an expensive activity is making a protein. Relate that glucose is a large denomination of energy currency that cells break down into smaller amounts to fuel cellular activity.

Visual Learning

Display overheads of the Calvin cycle and the citric acid cycle. Explain that the cycles can be repeated again and again to produce different products.

GLENCOE TECHNOLOGY

VIDEODISC
The Secret of Life
Cell Energy: Cell Respiration

254

Energy in a Cell

Adenosine triphosphate (ATP) is the most commonly used source of energy in a cell. Two organelles are involved in forming ATP from other sources of energy.

Chloroplasts in plant cells harvest energy from the sun's rays and convert it to ATP using light-dependent reactions. Light-independent reactions convert energy into carbohydrates such as starch through the Calvin cycle.

Mitochondria, found in both plant and animal cells, convert food energy into ATP through a series of chemical reactions that include glycolysis, the citric acid cycle, and the electron transport chain.

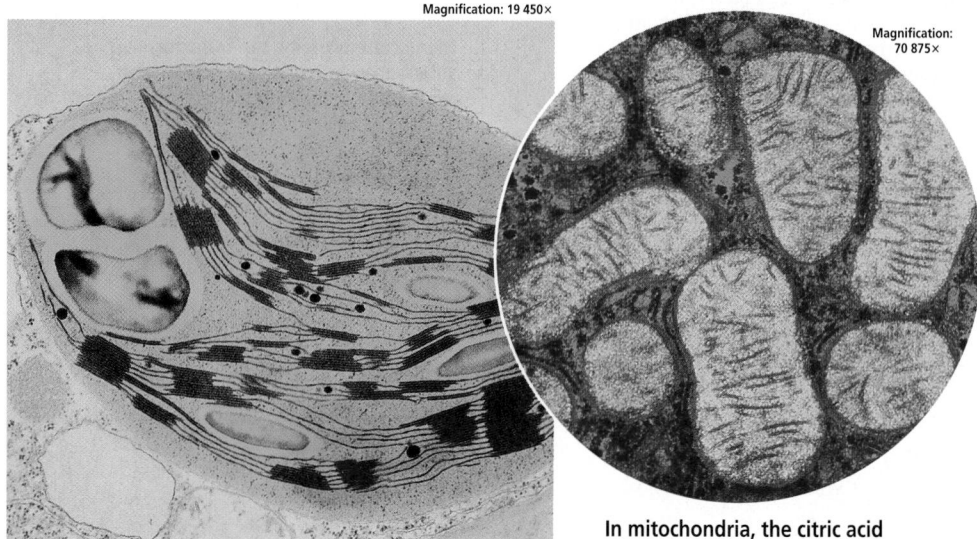

Magnification: 19 450×

Magnification: 70 875×

In chloroplasts, the Calvin cycle allows the sun's energy to be stored as carbohydrates for later use.

In mitochondria, the citric acid cycle allows the breakdown of food sources to form ATP.

FOCUS ON HEALTH

Cancer

Cancer is one condition that can result when a cell can no longer control the rate of mitosis. Cancer may be due to factors inside the cell, such as enzyme overproduction or production at the wrong time. Cancer can also be a result of environmental factors.

In recent years, much emphasis has been placed on healthy lifestyles that can help prevent cancer. Eating diets rich in fruits and vegetables, exercising regularly, and quitting or avoiding smoking can all help reduce the incidence of cancer.

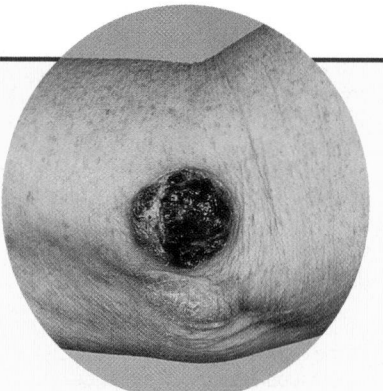

This melanoma is an example of skin cancer.

254

PROJECT

Cancer Facts Research

Intrapersonal Have students research one type of cancer that can be prevented by a lifestyle such as not smoking, sun avoidance, exercise, or diet. Have students present their report to the class. L3

BIODIGEST ASSESSMENT

Understanding Main Ideas

1. b	4. c	7. a	10. a
2. c	5. d	8. c	
3. a	6. d	9. c	

Thinking Critically

1. Chloroplasts trap light energy and store it by producing complex sugars. Mitochondria break down sugars to release ATP. Both are involved in energy production for the cell.

BIODIGEST

VITAL STATISTICS

ATP production for each molecule of glucose

Glycolysis: produces a net gain of two ATP, two NADH, and two H⁺

Acetyl-CoA formation: produces two NADH

Citric acid cycle (Krebs cycle): produces two ATP, six NADH, and two FADH

Electron transport chain: produces 32 ATP from NADH and FADH

Lactic acid fermentation: produces two ATP

Energy from ATP

BIODIGEST ASSESSMENT

Understanding Main Ideas

1. What is a starch?
   - **a.** a lipid
   - **b.** a carbohydrate
   - **c.** a protein
   - **d.** a peptide

2. The building blocks of proteins are _____.
   - **a.** fatty acids
   - **b.** monosaccharides
   - **c.** amino acids
   - **d.** glycerol

3. A cell that does not contain internal, membrane-bound structures is a _____.
   - **a.** prokaryote
   - **b.** eukaryote
   - **c.** yeast cell
   - **d.** a cell with organelles

4. The organelle that produces proteins is the _____.
   - **a.** nucleus
   - **b.** lysozome
   - **c.** ribosomes
   - **d.** vacuole

5. What structure is part of the cell's skeleton?
   - **a.** Golgi apparatus
   - **b.** nucleus
   - **c.** endoplasmic reticulum
   - **d.** microfilaments

6. The movement of a substance across the plasma membrane against the concentration gradient by binding to a protein is _____.
   - **a.** facilitated transport
   - **b.** osmosis
   - **c.** simple diffusion
   - **d.** active transport

7. What structure is involved in movement?
   - **a.** cilia
   - **b.** lysozomes
   - **c.** vacuole
   - **d.** chloroplasts

8. The stage of mitosis when the chromatids separate and move to opposite sides of the cell is _____.
   - **a.** prophase
   - **b.** metaphase
   - **c.** anaphase
   - **d.** telophase

9. The organelle that transforms light energy to ATP is _____.
   - **a.** mitochondrion
   - **b.** endoplasmic reticulum
   - **c.** chloroplast
   - **d.** nucleus

10. Which is involved in capturing light energy from the sun?
    - **a.** light-dependent reactions
    - **b.** Citric acid cycle
    - **c.** Calvin cycle
    - **d.** light-independent reactions

Thinking Critically

1. **Compare and contrast** the functions of chloroplasts and mitochondria in energy transformation.

2. **Describe** the organelles of a eukaryotic cell.

3. **Relate** transport across the plasma membrane in isotonic, hypertonic, and hypotonic solutions to changes in cell shape due to water movement across the membrane.

255

BIODIGEST

3 Assess

Check for Understanding

Linguistic Ask students to name as many organelles as they can and identify the function of each. **L1**

Reteach 🖎

Put up an overhead of the stages of mitosis and have students identify the stages and describe the events. **L2**

Extension

Linguistic Have students research melanoma and prepare a presentation illustrating its occurrence and treatment. Have them also develop a prevention plan. **L3**

 **Assessment**

Skill Ask students to describe what will happen to a cell's volume if it is placed in a hypotonic, isotonic, or hypertonic environment. **L2**

4 Close

Discussion

Discuss how a plant cell obtains energy and stores it and how an animal cell obtains energy and transforms it to a usable form for cell organelles.

 Resource Manager

Reinforcement and Study Guide, pp. 41-42 **L2**
Content Mastery, pp. 45-48 **L1**

2. Most organelles of a eukaryotic cell are surrounded by a membrane. This membrane allows very different chemical reactions to be carried out in a cell at the same time.

3. In an isotonic solution, the concentration of solutes on both sides of the membrane is the same, so there will be no net movement across the membrane and the cell shape will not change.

In a hypertonic solution, there is a greater concentration of solutes outside the cell, there is a net movement of water from the cell into the surrounding fluid, and the cell size will shrink. In a hypotonic solution, there is a greater concentration of solutes inside the cell; therefore, water will flow into the cell and the cell size will increase.

Unit 4

Genetics

Unit Overview

Unit 4 presents an overview of genetics and its role in determining the traits of organisms. Chapter 10 introduces genetics through a short historical presentation of the work of Gregor Mendel. Meiosis is then introduced and discussed. In Chapter 11, students learn about the structure of DNA and how it is replicated. The processes of transcription and translation are explained. Various kinds of mutations are described. In Chapter 12, students examine non-Mendelian patterns of heredity and the principles of genetics as applied to humans. Finally, selective breeding, DNA technology, and the Human Genome Project are discussed in Chapter 13.

Introducing the Unit

Have students look at the picture of the snapdragons and describe the traits they see. Then have them look around the room and describe the traits they see in each other. Just as the snapdragons are all the same kind of plant with different colored flowers, so are all the students alike, yet with different traits. Unit 4 will discuss how genes determine an organism's physical traits, as well as how those traits are inherited.

Unit 4

Genetics

Physical traits, such as the colors of these snapdragons, are encoded in small segments of a chromosome called genes, which are passed from one generation to the next. By studying the inheritance pattern of a trait through several generations, the probability that future offspring will express that trait can be predicted.

UNIT CONTENTS

UNIT PROJECT

BIOLOGY *Online* Use the Glencoe Science Web site for more project activities that are connected to this unit. **science.glencoe.com**

256

Unit Projects

Human Genetics

Have students do one of the projects for this unit as described on the Glencoe Science Web site. As an alternative, students can do one of the projects described on these two pages.

Make a Graph

Logical-Mathematical Measure the height of every student in the class and then graph the heights. How is height inherited? **L1**

Modeling

Kinesthetic Create several chromatids by tying two socks together. Now model what happens to them during meiosis. **L1 ELL**

Advance Planning

Chapter 10
- Purchase the following seed types from a biological supply house for the BioLab: pure breeding green tobacco seeds and green-albino tobacco seed mixed in 3:1 ratio (heterozygous parents).
- Purchase Wisconsin Fast Plants for the Project and the Tech Prep activity.

Chapter 11
none

Chapter 12
- Order *Brassica rapa* seeds for the BioLab.
- Obtain photos of a red shorthorn bull, a white shorthorn cow, and a roan shorthorn cow for the Discussion.
- Collect articles and pamphlets on genetic disorders for the Display.
- Order tobacco seeds for Alternative Lab 12-1.
- Obtain slides of normal and sickled blood cells for the Microscope Activity.
- Obtain slides of male and female body cells for Alternative Lab 12-2.

Chapter 13
- Order the Chromosome Simulation Biokit for the Quick Demo in Section 13.2.
- Purchase cloning kits for the Extension in Section 13.3.
- Obtain a slide of female cheek cells for the Alternative Lab.

Unit Projects

Using the Library
Intrapersonal Research one disease currently being treated by gene therapy. **L2**

Interview
Linguistic Interview someone with a family pedigree that demonstrates an inherited disease. **L1**

Final Report
Have student groups present their findings about genetics in reports that could be presented to students at your local middle school. **L3**

Chapter 10 Organizer

Refer to pages 4T-5T of the Teacher Guide for an explanation of the National Science Education Standards correlations.

Section	Objectives	Activities/Features
Section 10.1 **Mendel's Laws of Heredity** National Science Education Standards UCP.1-3, UCP.5; A.1, A.2; G.1-3 (3 sessions, 2 blocks)	1. **Analyze** the results obtained by Gregor Mendel in his experiments with garden peas. 2. **Predict** the possible offspring of a genetic cross by using a Punnett square.	**MiniLab 10-1:** Looking at Pollen, p. 260 **Problem-Solving Lab 10-1,** p. 268 **Internet BioLab:** How can phenotypes and genotypes of plants be determined? p. 280 **Math Connection:** A Solution from Ratios, p. 282
Section 10.2 **Meiosis** National Science Education Standards UCP.1-3; C.1, C.2; E.1, E.2; F.6; G.1-3 (3 sessions, 2 blocks)	3. **Analyze** how meiosis maintains a constant number of chromosomes within a species. 4. **Infer** how meiosis leads to variation in a species. 5. **Relate** Mendel's laws of heredity to the events of meiosis.	**Problem-Solving Lab 10-2,** p. 270 **MiniLab 10-2:** Modeling Crossing Over, p. 274 **Inside Story:** Genetic Recombination, p. 277

Need Materials? Contact Carolina Biological Supply Company at 1-800-334-5551 or at **http://www.carolina.com**

MATERIALS LIST

BioLab
p. 280 potting soil, small flowerpots or seedling flats, 2 groups of tobacco seeds, hand lens, light source, thermometer, plant watering bottle

MiniLabs
p. 260 flower, microscope, microscope slide, coverslip, dropper, water
p. 274 clay, twist ties (2), pencil

Alternative Lab
p. 272 9 sheets of unlined paper or poster board (30-cm square), long length of yarn, paper clips, string, toothpicks, tape or glue, scissors

Quick Demos
p. 265 black beans (300), white beans (100), paper cups
p. 272 none

Key to Teaching Strategies

L1 Level 1 activities should be appropriate for students with learning difficulties.

L2 Level 2 activities should be within the ability range of all students.

L3 Level 3 activities are designed for above-average students.

ELL ELL activities should be within the ability range of English Language Learners.

COOP LEARN Cooperative Learning activities are designed for small group work.

P These strategies represent student products that can be placed into a best-work portfolio.

These strategies are useful in a block scheduling format.

Mendel and Meiosis

Teacher Classroom Resources

Section	Reproducible Masters	Transparencies
Section 10.1 **Mendel's Laws of Heredity**	Reinforcement and Study Guide, pp. 43-44 **L2** Critical Thinking/Problem Solving, p. 10 **L3** BioLab and MiniLab Worksheets, p. 43 **L2** Tech Prep Applications, pp. 17-18 **L2** Content Mastery, pp. 49-50, 52 **L1**	Section Focus Transparency 24 **L1** **ELL** Basic Concepts Transparency 14 **L2** **ELL** Reteaching Skills Transparency 16 **L1** **ELL**
Section 10.2 **Meiosis**	Reinforcement and Study Guide, pp. 45-46 **L2** Concept Mapping, p. 10 **L3** **ELL** BioLab and MiniLab Worksheets, pp. 45-48 **L2** Laboratory Manual, pp. 69-74 **L2** Content Mastery, pp. 49, 51-52 **L1**	Section Focus Transparency 25 **L1** **ELL** Basic Concepts Transparency 15 **L2** **ELL** Reteaching Skills Transparency 17 **L1** **ELL**

Assessment Resources

Chapter Assessment, pp. 55-60
MindJogger Videoquizzes
Performance Assessment in the Biology Classroom
Alternate Assessment in the Science Classroom
ExamView® Pro Software 💾
BDOL Interactive CD-ROM, Chapter 10 quiz

Additional Resources

Spanish Resources **ELL**
English/Spanish Audiocassettes **ELL**
Cooperative Learning in the Science Classroom **COOP LEARN**
Lesson Plans/Block Scheduling

GLENCOE TECHNOLOGY

The following multimedia resources are available from Glencoe.

Biology: The Dynamics of Life
CD-ROM **ELL**

  Exploration: *Trait Inheritance*
  Animation: *Meiosis*
Videodisc Program
  Meiosis

The Infinite Voyage
  The Geometry of Life

The Secret of Life Series

  Heredity in Mendel's Peas
  Dominant vs. Recessive
  Segregation
  Sex and the Single Gene: *Cell Development*
  Independent Assortment
  Meiosis Ia, Ib, IIa, IIb

Theme Development

The theme of **nature of science** is developed within the chapter as Mendel's laws of segregation and independent assortment are explained. Mendel's laws are shown to be supported by current knowledge of meiosis. The theme of **homeostasis** is illustrated by the knowledge that diploid chromosome numbers are maintained when gametes join at fertilization. The gametes are formed as the result of meiosis, and their chromosome numbers are half the diploid number.

READING BIOLOGY

Glencoe's *Biology: The Dynamics of Life* contains many resources to assist a student's reading skills. Each chapter contains figures with expanded captions that expand on written material. Word Origins, located along the side of text, expand knowledge of biology vocabulary. Glencoe's Content Mastery Booklet helps develop reading skills while reinforcing content. In addition, use the Interactive Tutor for *Biology: The Dynamics of Life* on the Glencoe Web site to reinforce vocabulary. **science.glencoe.com**

---

10 Mendel and Meiosis

What You'll Learn
- You will identify the basic concepts of genetics.
- You will examine the process of meiosis.

Why It's Important
Genetics explains why you have inherited certain traits from your parents. If you understand how meiosis occurs, you can see how these traits were passed on to you.

READING BIOLOGY

Look over the chapter, and make a list of several vocabulary words that are familiar. Review the list. For each word, think of where and how you may have heard the word used before. As you read the text, note if the words are used differently in a science context than elsewhere.

BIOLOGY Online

To find out more about genetics, visit the Glencoe Science Web site. **science.glencoe.com**

Organisms usually resemble their parents because they inherit certain characteristics from them. These characteristics, also called traits, are determined by genetic information on chromosomes such as those shown in the inset photo.

258 MENDEL AND MEIOSIS

Multiple Learning Styles

Look for the following logos for strategies that emphasize different learning modalities.

Kinesthetic Meeting Individual Needs, pp. 260, 271; Tech Prep, p. 265; Reinforcement, p. 266; Quick Demo, p. 272

Visual-Spatial Getting Started Demo, p. 258; Quick Demo, p. 265; Meeting Individual Needs, pp. 265, 267; Biology Journal, p. 266;

Activity, p. 268; Portfolio, p. 274; Visual Learning, p. 277

Intrapersonal Meeting Individual Needs, p. 260

Linguistic Enrichment, p. 261; Portfolio, pp. 264, 275; Biology Journal, p. 276; Tech Prep, p. 278

Logical-Mathematical Concept Development, p. 266

10.1 Mendel's Laws of Heredity

An Austrian monastery in the mid-nineteenth century might seem an unusual place to begin your search for the answer to why offspring resemble their parents. Yet, it was in this community of scholars that young Gregor Mendel began to breed garden pea plants so that he could study the inheritance of their characteristics.

Gregor Mendel and pea plants

SECTION PREVIEW

Objectives

Analyze the results obtained by Gregor Mendel in his experiments with garden peas.

Predict the possible offspring of a genetic cross by using a Punnett square.

Vocabulary
heredity
genetics
trait
gamete
pollination
fertilization
hybrid
allele
dominant
recessive
law of segregation
phenotype
genotype
homozygous
heterozygous
law of independent assortment

Why Mendel Succeeded

Gregor Mendel carried out the first important studies of **heredity,** the passing on of characteristics from parents to offspring. Although people had noticed for thousands of years that family resemblances were inherited from generation to generation, a complete explanation required the careful study of **genetics**—the branch of biology that studies heredity. Characteristics that are inherited are called **traits.** Mendel was the first person to succeed in predicting how traits would be transferred from one generation to the next. How was he able to solve this problem of heredity?

Mendel chose his subject carefully

Mendel studied many plants before deciding to use the garden pea in his experiments. Garden pea plants reproduce sexually, which means they have two distinct sex cells—male and female. Sex cells are called **gametes.**

In peas, both male and female gametes are in the same flower. The male gamete is in the pollen grain, which is produced by the anther. The female gamete is in the ovule, which is located in the pistil. The transfer of the male pollen grains to the pistil of a flower is called **pollination.** The uniting of male and female gametes, in a process called **fertilization,** occurs

WORD *Origin*

heredity
From the Latin word *hered-,* meaning "heir." Heredity describes the genetic qualities you receive from your ancestors.

Section 10.1

Prepare

Key Concepts
Students are led through mono- and dihybrid crosses, applying Mendel's laws of segregation and independent assortment.

Planning
■ Purchase smooth and wrinkled peas for the Getting Started Demo.
■ Purchase fresh or preserved flowers for MiniLab 10-1.
■ Gather soil and pots for the BioLab, the Project, and the Tech Prep activity. Locate thermometers, hand lenses, and possible light banks.
■ Purchase black and white beans for the Quick Demo.
■ Gather candy and buttons for the Reinforcement.

1 Focus

Bellringer
Before presenting the lesson, display **Section Focus Transparency 24** on the overhead projector and have students answer the accompanying questions.
L1 **ELL**

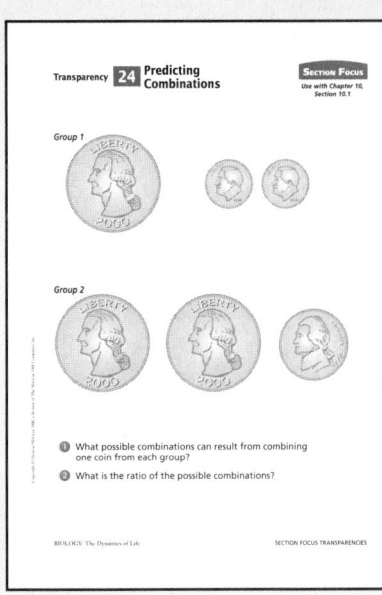

Transparency **24** Predicting Combinations

SECTION FOCUS
Use with Chapter 10, Section 10.1

Group 1

Group 2

❶ What possible combinations can result from combining one coin from each group?
❷ What is the ratio of the possible combinations?

BIOLOGY: The Dynamics of Life   SECTION FOCUS TRANSPARENCIES

Assessment Planner

2 Teach

Purpose
Students will observe pollen grains in plants.

Process Skills
observe and infer

Teaching Strategies
■ Preserved or fresh flower material may be used. Preserved material is available from biological supply houses.

■ Have students wash their hands after handling flowers and flower parts.

■ Forceps may be used to remove stamens from the flower.

■ If coverslips or slides break during the squashing process, have students place them in a container for broken glass and start over with new material.

Expected Results
Students will observe that each anther contains thousands of small pollen grains.

Analysis
1. Numbers should be in the several thousands.
2. Student answers will vary depending on species used. Pollen grains are microscopic cells, often round in shape.
3. Pollen grains contain the male reproductive cells needed for fertilization. They provide the chromosomes and genes of the male parent.

✔ Assessment
Performance Have students make a diagram of several pollen cells under high power. Ask them to determine the size of pollen cells in micrometers. Pollen size should be included on their diagrams. If students have not done any previous measuring, refer them to MiniLab 7-1. Use the Performance Task Assessment List for Using Math in Science in **PASC**, p. 29. **L2**

260

MiniLab 10-1   Observing and Inferring

Looking at Pollen Pollen grains are formed within the male anthers of flowers. What is their role? Pollen contains the male gametes or sperm cells needed for fertilization. This means that pollen grains carry the hereditary units from male parent plants to female parent plants. The pollen grains that Mendel transferred from the anther of one pea plant to the pistil of another plant carried the hereditary traits that he so carefully observed in the next generation.

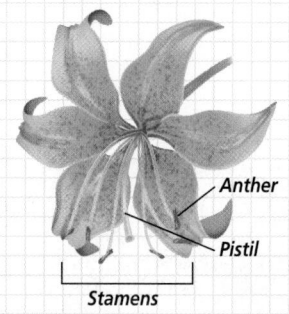

Anther

Pistil

Stamens

Procedure
1. Examine a flower. Using the diagram as a guide, locate the stamens of your flower. There are usually several stamens in each flower.
2. Remove one stamen and locate the enlarged end—the anther.
3. Add a drop of water to a microscope glass slide. Place the anther in the water. Add a coverslip. Using the eraser end of a pencil, tap the coverslip several times to squash the anther.
4. Observe under low power. Look for numerous small round structures. These are pollen grains.

Analysis
1. Provide an estimate of the number of pollen grains present in an anther.
2. Describe the appearance of a single pollen grain.
3. Explain the role of pollen grains in plant reproduction.

Figure 10.1
In his experiments, Mendel often had to transfer pollen from one plant to another plant with different traits. This is called making a cross.

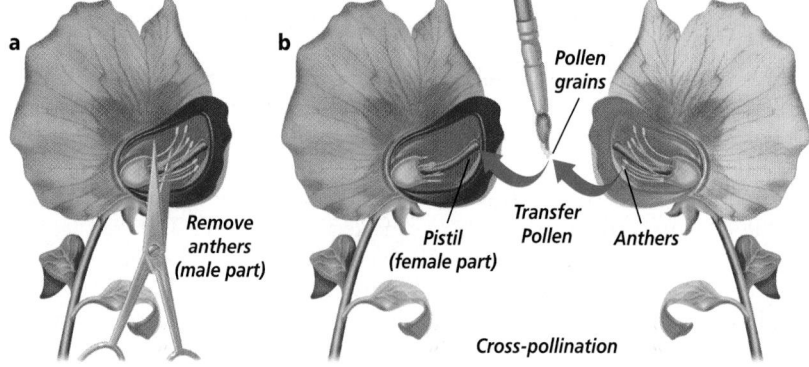

a

Remove anthers (male part)

b

Pistil (female part)

Pollen grains

Transfer Pollen   **Anthers**

Cross-pollination

260 MENDEL AND MEIOSIS

---

when the male gamete in the pollen grain meets and fuses with the female gamete in the ovule. After the ovule is fertilized, it matures into a seed.

The reproductive parts of the pea flower are tightly enclosed in petals, preventing the pollen of other flowers from entering. As a result, peas normally reproduce by self-pollination; that is, the male and female gametes come from the same plant. In many of Mendel's experiments, this is exactly what he wanted. When he needed to breed—or cross—one plant with another, Mendel opened the petals and removed the anthers from a flower, *Figure 10.1a*. He then dusted the pistil with pollen from the plant he wished to cross it with, *Figure 10.1b*, and covered the flower with a small bag to prevent pollen in the air from landing on the pistil. This process is called cross-pollination. By using this technique, Mendel could be sure of the parents in his cross. You can observe anthers and their pollen grains in the *MiniLab* on this page.

Mendel was a careful researcher
Mendel carefully controlled his experiments and the peas he used. He studied only one trait at a time to control variables, and he analyzed his data mathematically. The tall pea plants

---

he worked with were from populations of plants that had been tall for many generations and had always produced tall offspring. Such plants are said to be true breeding for tallness. Likewise, the short plants he worked with were true breeding for shortness.

Mendel's Monohybrid Crosses

What did Mendel do with the tall and short pea plants he so carefully selected? He crossed them to produce new plants. Mendel referred to the offspring of this cross as hybrids. A **hybrid** is the offspring of parents that have different forms of a trait, such as tall and short height. Mendel's first experiments are called monohybrid crosses because *mono* means "one" and the two parent plants differed by a single trait—height.

The first generation

Mendel selected a six-foot-tall pea plant that came from a population of pea plants, all of which were over six feet tall. He cross-pollinated this tall pea plant with a short pea plant that was less than two feet tall and which came from a population of pea plants that were all short. When he planted the seeds from this cross, he found that all of the offspring grew to be as tall as the taller parent. In this first generation, it was as if the shorter parent had never existed!

The second generation

Next, Mendel allowed the tall plants in this first generation to self-pollinate. After the seeds formed, he planted them and counted more than 1000 plants in this second generation. Mendel found that three-fourths of the plants were as tall as the tall plants in the parent and first generations. He also found that one-

fourth of the offspring were as short as the short plants in the parent generation. In other words, in the second generation, tall and short plants occurred in a ratio of approximately three tall plants to one short plant, *Figure 10.2*. The short trait had reappeared as if from nowhere!

The original parents, the true-breeding tall and short plants, are known as the P_1 generation. The P stands for "parent." The offspring of the parent plants are known as the F_1

Figure 10.2
When Mendel crossed true-breeding tall pea plants with true-breeding short pea plants, all the offspring were tall. When he allowed first-generation tall plants to self-pollinate, three-fourths of the offspring were tall and one-fourth were short.

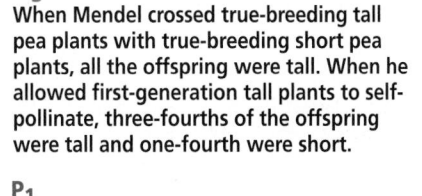
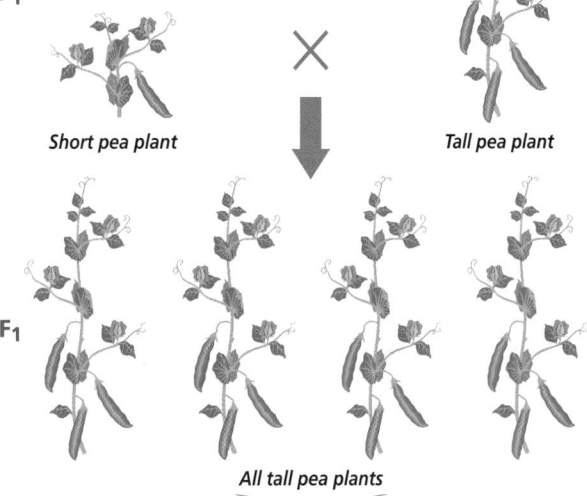

P_1

Short pea plant ✕ *Tall pea plant*

F_1

All tall pea plants

F_2

3 tall : 1 short

WORD *Origin*

allele
From the Greek word *allelon*, meaning "of each other." Genes exist in alternative forms called alleles.

Figure 10.3
Mendel chose seven traits of peas for his experiments. Each trait had two clearly different forms; no intermediate forms were observed.

generation. The *F* stands for "filial"—son or daughter. When you cross two F_1 plants with each other, their offspring are called the F_2 generation—the second filial generation. You might find it easier to understand these terms if you look at your own family. Your parents are the P_1 generation. You are the F_1 generation, and any children you might have in the future would be the F_2 generation.

Mendel did similar monohybrid crosses with a total of seven pairs of traits, studying one pair of traits at a time. These pairs of traits are shown in *Figure 10.3*. In every case, he found that one trait of a pair seemed to disappear in the F_1 generation, only to reappear unchanged in one-fourth of the F_2 plants.

The rule of unit factors

Mendel concluded that each organism has two factors that control each of its traits. We now know that

these factors are genes and that they are located on chromosomes. Genes exist in alternative forms. We call these different gene forms **alleles** (uh LEELZ). For example, each of Mendel's pea plants had two alleles of the gene that determined its height. A plant could have two alleles for tallness, two alleles for shortness, or one allele for tallness and one for shortness. An organism's two alleles are located on different copies of a chromosome—one inherited from the female parent and one from the male parent.

The rule of dominance

Remember what happened when Mendel crossed a tall P_1 plant with a short P_1 plant? The F_1 offspring were all tall. In other words, only one trait was observed. In such crosses, Mendel called the observed trait **dominant** and the trait that disappeared **recessive**. We now know that

	Seed shape	Seed color	Flower color	Flower position	Pod color	Pod shape	Plant height
Dominant trait	round	yellow	purple	axial (side)	green	inflated	tall
Recessive trait	wrinkled	green	white	terminal (tips)	yellow	constricted	short

BIOLOGY *Online* Note Internet addresses that you find useful in the space below for quick reference.

_____

_____

_____

_____

_____

in Mendel's pea plants, the allele for tall plants is dominant to the allele for short plants. Plants that had one allele for tallness and one for shortness were tall because the allele for tallness is dominant to the allele for shortness. Expressed another way, the allele for short plants is recessive to the allele for tall plants. Pea plants that had two alleles for tallness were tall, and those that had two alleles for shortness were short. You can see in *Figure 10.4* how the rule of dominance explained the resulting F_1 generation.

When recording the results of crosses, it is customary to use the same letter for different alleles of the same gene. An uppercase letter is used for the dominant allele, and a lowercase letter for the recessive allele. The dominant allele is always written first. So the allele for tallness is written as *T*, and the allele for shortness as *t*, as it is in *Figure 10.4*.

The law of segregation

Now recall the results of Mendel's cross between F_1 tall plants, when the trait of shortness reappeared. To explain this result, Mendel formulated the first of his two laws of heredity. He concluded that each tall plant in the F_1 generation carried one dominant allele for tallness and one unexpressed recessive allele for shortness. It received the allele for tallness from its tall parent and the allele for shortness from its short parent in the P_1 generation. Because each F_1 plant has two different alleles, it can produce two different types of gametes— "tall" gametes and "short" gametes. During fertilization, these gametes randomly pair to produce four combinations of alleles. This conclusion, illustrated in *Figure 10.5* on the next page, is called the **law of segregation.**

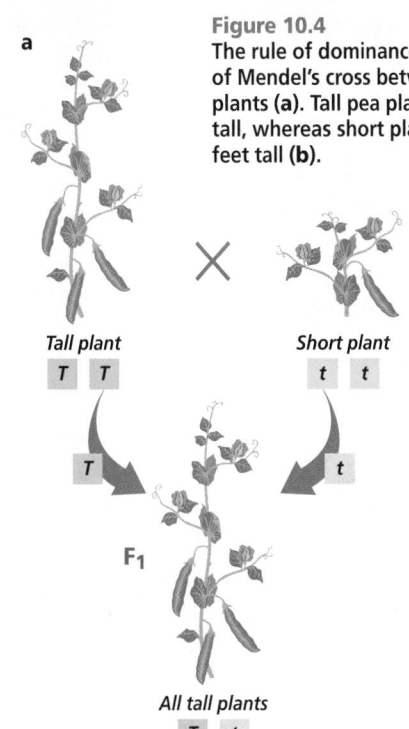

Figure 10.4
The rule of dominance explains the results of Mendel's cross between P_1 tall and short plants (**a**). Tall pea plants are about six feet tall, whereas short plants are less than two feet tall (**b**).

a

Tall plant
T T

Short plant
t t

T

t

F_1

All tall plants
T t

b

Revealing Misconceptions
Point out to students that the dominant trait is not necessarily the more common or desirable trait. For example, in humans, Huntington's disease and hypercholesterolemia (having dangerously high levels of blood cholesterol) are both dominant traits and both are rare.

Visual Learning
Figure 10.4 Have students write out the three important written conventions that are described with this diagram. *Use the same letter for different alleles of the same gene; use uppercase letters for dominant alleles and lowercase letters for recessive alleles; and always write the dominant allele first.* **L1**

PROJECT

Experimental Crosses
Seeds of a plant called *Brassica rapa* are available from Carolina Biological Supply Company under the name of Wisconsin Fast Plants. These plants, grown from seed, complete their life cycle in 30 to 35 days. They are ideal for use in the classroom because within this short time, the plants flower and form seeds for the next generation. Genetic studies can be carried out using different traits, such as petal-less flowers or hairy stems. Students will find it easy to carry out the pollinating between plants. These plants are an ideal experimental organism for genetic studies. **L2**

It is sometimes easier for students to remember the meaning of the term *genotype* by reversing the word so it becomes "type of gene." When stated this way, students can associate the term *genotype* with its definition. *Pheno* means "to show." Thus, the phenotype shows the type of trait or how it appears.

Concept Development

■ Ask students to supply the correct term—genotype or phenotype—to the following examples: (a) *LL*, (b) blond hair, (c) dimpled chin, (d) blue eyes, (e) *Dd*, (f) *ss*, (g) white and green leaves. *a, e, and f are genotypes; b, c, d, and g are phenotypes.*

■ Have students provide the following information for this example: *G* = green pea pod, *g* = yellow pea pod. (a) Give the phenotypes of plants with these genotypes: *Gg*, *GG*, and *gg*. (b) Use the terms *homozygous* or *heterozygous* to describe each of the three examples above. *(a) green, green, yellow; (b) heterozygous, homozygous, homozygous.*

GLENCOE TECHNOLOGY

VIDEODISC
VIDEOTAPE
The Secret of Life
Sex and the Single Gene: Cell Development

Figure 10.5
Mendel's law of segregation explains the results of his cross between F₁ tall plants. He concluded that the two alleles for each trait must separate when gametes are formed. A parent, therefore, passes on at random only one allele for each trait to each offspring.

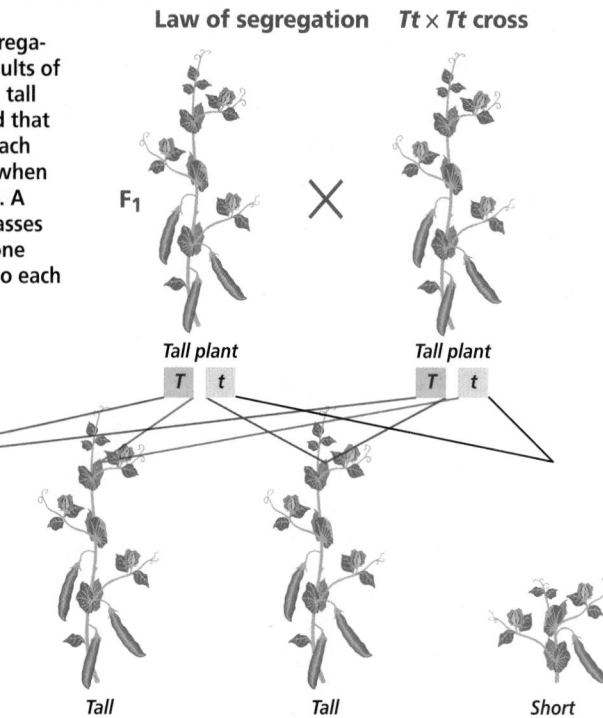

Law of segregation   *Tt × Tt* **cross**

Phenotypes and Genotypes

Mendel showed that tall plants are not all the same. Some tall plants, when crossed with each other, yielded only tall offspring. These were Mendel's original P₁ true-breeding tall plants. Other tall plants, when crossed with each other, yielded both tall and short offspring. These were the F₁ tall plants in *Figure 10.5* that came from a cross between a tall plant and a short plant.

Two organisms, therefore, can look alike but have different underlying gene combinations. The way an organism looks and behaves is called its **phenotype** (FEE nuh tipe). The phenotype of a tall plant is tall, whether it is *TT* or *Tt*. The gene combination an organism contains is known as its **genotype** (JEE nuh tipe). The genotype of a tall plant that has two alleles for tallness is *TT*. The genotype of a tall plant that has one allele for tallness and one allele for shortness is *Tt*.

You can see that you can't always know an organism's genotype simply by looking at its phenotype. An organism is **homozygous** (hoh muh ZI gus) for a trait if its two alleles for the trait are the same. The true-breeding tall plant that had two alleles for tallness (*TT*) would be homozygous for the trait of height. Because tallness is dominant, a *TT* individual is homozygous dominant for that trait. A short plant would always have two alleles for shortness (*tt*). It would, therefore, always be

264 MENDEL AND MEIOSIS

✓ Portfolio

Martian Traits

Linguistic Have students imagine that they have encountered their first Martian. Assume that Martian traits are inherited exactly the same way as Earthling traits. Have students provide examples of five Martian traits through the use of a diagram and a written statement describing the possible genotypes and phenotypes for these five traits. **L2** **P**

homozygous recessive for the trait of height.

An organism is **heterozygous** (het uh roh ZI gus) for a trait if its two alleles for the trait differ from each other. Therefore, the tall plant that had one allele for tallness and one allele for shortness (*Tt*) is heterozygous for the trait of height.

Now look at *Figure 10.5* again. Can you identify the phenotype and genotype of each plant? Is each homozygous or heterozygous? You can practice determining genotypes and phenotypes in the *BioLab* at the end of this chapter.

Mendel's Dihybrid Crosses

Mendel performed another set of crosses in which he used peas that differed from each other in two traits rather than only one. Such a cross involving two different traits is called a dihybrid cross because *di* means "two." In a dihybrid cross, will the two traits stay together in the next generation or will they be inherited independently of each other?

The first generation

Mendel took true-breeding pea plants that had round yellow seeds (*RRYY*) and crossed them with true-breeding pea plants that had wrinkled green seeds (*rryy*). He already knew that when he crossed plants that produced round seeds with plants that produced wrinkled seeds, all the plants in the F₁ generation produced seeds that were round. In other words, just as tall plants were dominant to short plants, the round-seeded trait was dominant to the wrinkled-seeded trait. Similarly, when he crossed plants that produced yellow seeds with plants that produced green seeds, all the plants

Figure 10.6
When Mendel crossed true-breeding plants with round yellow seeds and true-breeding plants with wrinkled green seeds, the seeds of all the offspring were round and yellow. When the F₁ plants were allowed to self-pollinate, they produced four different kinds of plants in the F₂ generation.

Dihybrid cross   round yellow × wrinkled green

P₁   Round yellow   ×   Wrinkled green

F₁   All round yellow

F₂
9 Round yellow   3 Round green   3 Wrinkled yellow   1 Wrinkled green

in the F₁ generation produced yellow seeds—yellow was dominant. Therefore, Mendel was not surprised when he found that the F₁ plants of his dihybrid cross all had the two dominant traits of round and yellow seeds, as *Figure 10.6* shows.

The second generation

Mendel then let the F₁ plants pollinate themselves. As you might expect, he found some plants that produced round yellow seeds and others that produced wrinkled green seeds. But that's not all. He also found some plants with round green seeds and others with wrinkled yellow seeds. When Mendel sorted and counted the plants of the F₂ generation, he found they appeared in a definite ratio of phenotypes—9 round yellow: 3 round green: 3 wrinkled yellow: 1 wrinkled green. To explain the results of this dihybrid cross, Mendel formulated his second law.

WORD Origin
heterozygous
From the Greek words *heteros*, meaning "other," and *zygotos*, meaning "joined together." A trait is heterozygous when an individual has two different alleles for that trait.

10.1 MENDEL'S LAWS OF HEREDITY **265**

Concept Development

Logical-Mathematical Have students construct Punnett squares and solve these problems, giving the genotypic and phenotypic ratios expected. (a) Homozygous tall plant bred to a homozygous short plant. *All offspring will be tall and heterozygous.* (b) Homozygous tall plant bred to a heterozygous tall plant. *All offspring will be tall, half being* TT *and half being* Tt. (c) Heterozygous tall plant bred to a homozygous short plant. *Half will be tall and half will be short; all tall offspring will be* Tt *and all short offspring will be* tt. **L2 ELL**

Reinforcement

Kinesthetic Provide students with two large and two small round pieces of candy plus two large and two small buttons. Advise them that the buttons and candy represent two different genes and the large-sized objects represent dominant alleles. Ask them to prepare two sets of objects (alleles) for two traits in two parents, making both parents heterozygous for both traits. Then have students arrange their parental gene sets into independently assorted gametes of two alleles each. Advise them that they cannot have two candies or two buttons in their final groups. Have them explain how this illustrates the law of independent assortment. **L2 ELL**

The law of independent assortment

Mendel's second law states that genes for different traits—for example, seed shape and seed color—are inherited independently of each other. This conclusion is known as the **law of independent assortment.** When a pea plant with the genotype *RrYy* produces gametes, the alleles *R* and *r* will separate from each other (the law of segregation) as well as from the alleles *Y* and *y* (the law of independent assortment), and vice versa. These alleles can then recombine in four different ways. If the alleles for seed shape and color were inherited together, only two kinds of pea seeds would have been produced: round yellow and wrinkled green.

Punnett Squares

In 1905, Reginald Punnett, an English biologist, devised a shorthand way of finding the expected proportions of possible genotypes in the offspring of a cross. This method is called a Punnett square. It takes account of the fact that fertilization occurs at random, as Mendel's law of segregation states. If you know the genotypes of the parents, you can use a Punnett square to predict the possible genotypes of their offspring.

Monohybrid crosses

Consider the cross between two F_1 tall pea plants, each of which has the genotype *Tt*. Half the gametes of each parent would contain the *T* allele, and the other half would contain the *t* allele. A Punnett square for this cross is two boxes tall and two boxes wide because each parent can produce two kinds of gametes for this trait. The two kinds of gametes from one parent are listed on top of the square, and the two kinds of gametes from the other parent are listed on the left side, *Figure 10.7A*. It doesn't matter which set of gametes is on top and which is on the side, that is, which parent contributes the *T* and which contributes the *t*. Refer to the Punnett square in *Figure 10.7B* to determine the possible genotypes of the offspring. Each box is filled in with the gametes above and to the left side of that box. You can see that each box then contains two alleles—one possible genotype.

After the genotypes have been determined, you can determine the

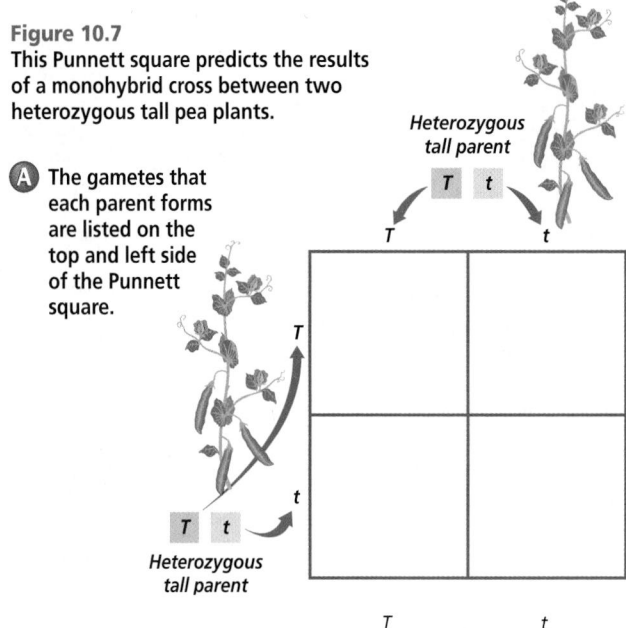

Figure 10.7
This Punnett square predicts the results of a monohybrid cross between two heterozygous tall pea plants.

A The gametes that each parent forms are listed on the top and left side of the Punnett square.

B You can see that there are three different possible genotypes—*TT, Tt,* and *tt*—and that *Tt* can result from two different combinations.

BIOLOGY JOURNAL

Punnett Square

Visual-Spatial Have students construct a Punnett square to illustrate the probable bean parents from the Quick Demo on the previous page. Ask students to write a prediction of what the outcome would be if two white beans were crossed. **L2**

phenotypes. Looking at the Punnett square for this cross in *Figure 10.7B,* you can see that three-fourths of the offspring are expected to be tall because they have at least one dominant allele. One-fourth are expected to be short because they lack a dominant allele. Of the tall offspring, one-third will be homozygous dominant *(TT)* and two-thirds will be heterozygous *(Tt)*. Note that whereas the genotype ratio is 1*TT*: 2*Tt*: 1*tt,* the phenotype ratio is 3 tall: 1 short. You can practice doing calculations such as Mendel did in the *Math Connection* at the end of this chapter.

Dihybrid crosses

What happens in a Punnett square when two traits are considered? Think again about Mendel's cross between pea plants with round yellow seeds and plants with wrinkled green seeds. All the F$_1$ plants produced seeds that were round and yellow and were heterozygous for each trait *(RrYy)*. What kind of gametes will these F$_1$ plants form?

Mendel explained that the traits for seed shape and seed color would be inherited independently of each other. This means that each F$_1$ plant will produce gametes containing the following combinations of genes with equal frequency: round yellow *(RY)*, round green *(Ry)*, wrinkled yellow *(rY)*, and wrinkled green *(ry)*. A Punnett square for a dihybrid cross will then need to be four boxes on each side for a total of 16 boxes, as *Figure 10.8* shows.

Probability

Punnett squares are good for showing all the possible combinations of gametes and the likelihood that each will occur. In reality, however, you don't get the exact ratio of

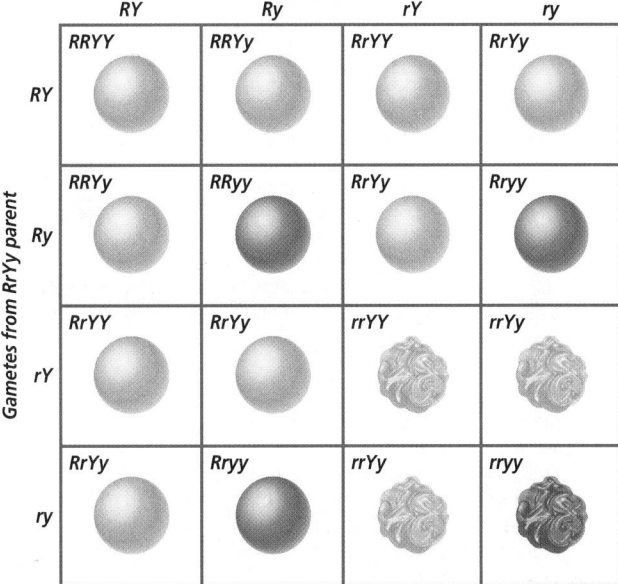

Punnett Square of Dihybrid Cross
Gametes from RrYy parent

Figure 10.8
A Punnett square for a dihybrid cross between heterozygous pea plants with round yellow seeds shows clearly that the offspring fulfill Mendel's observed ratio of 9 round yellow: 3 round green: 3 wrinkled yellow: 1 wrinkled green.

F$_1$ cross: *RrYy × RrYy*
○ round yellow
● round green
○ wrinkled yellow
● wrinkled green

results shown in the square. That's because, in some ways, genetics is like flipping a coin—it follows the rules of chance.

When you toss a coin, it lands either heads up or tails up. The probability or chance that an event will occur can be determined by dividing the number of desired outcomes by the total number of possible outcomes. So the probability of getting heads when you toss a coin would be one in two chances, written as 1: 2 or 1/2. A Punnett square can be used to determine the probability of getting a pea plant that produces round seeds when two plants that are heterozygous *(Rr)* are crossed. Because this Punnett

Figure 10.8 Ask students how many different genotypes and phenotypes resulted from this cross? *9 genotypes and 4 phenotypes*

3 Assess

Check for Understanding
Have students describe the relationship between or among the following terms. **L1** **ELL**
   a. pollination—fertilization
   b. allele—dominant—recessive
   c. genotype—phenotype
   d. homozygous—heterozygous
   e. monohybrid—dihybrid

Reteach
Have students provide an example of each relationship provided in Check for Understanding. **L1**

Extension
Have students list the genotypes and phenotypes resulting from (a) an *RrYy* plant cross-pollinated by an *RRyy* plant; (b) an *rrYy* plant cross-pollinated by a *RrYy* plant. *(a) genotypes:* RrYy, RRyy, RRYy, Rryy; *phenotypes: 1/2 round yellow, 1/2 round green; (b) genotypes:* RrYY; RrYy, Rryy, rrYY, rrYy, rryy; *phenotypes: 3 round yellow; 3 wrinkled yellow; 1 round green; 1 wrinkled green* **L2**

✓ Assessment
Knowledge Have students explain what each of the following represents in a Punnett square: (a) the letters written at the top and side of the square; (b) the letters written within each box; (c) the boxes. *(a) gametes; (b) genotype of an offspring; (c) possible offspring* **L1** ▣

MEETING INDIVIDUAL NEEDS

Learning Disabled/English Language Learners ✋

▣ *Visual-Spatial* Provide students with blank Punnett square outlines. Project a similar copy onto a screen with an overhead projector. Lead students through the steps showing a cross between two *Tt* parents. Reinforce how the letters placed to the

side and across the top represent all the possible gametes for each parent and how this illustrates the law of segregation. Reinforce the significance of the four squares and what the letter combinations within them represent. Provide students with a variety of problems *(TT × tt, Tt × tt, Tt × TT, tt × tt)*. **L1** ▣

Resource Manager

Critical Thinking/Problem Solving, p. 10 **L3**
Reinforcement and Study Guide, pp. 43-44 **L2**
Content Mastery, p. 50 **L1**

Problem Solving Lab 10-1

Purpose
Students will convert and express two sets of related numbers as a ratio.

Process Skills
use and interpret data, calculate

Background
The round seeds contain an abundance of well-formed starch grains and have higher water retention than the wrinkled seeds. A lack of well-formed starch grains and lower water retention account for the wrinkled appearance.

Teaching Strategies
■ Have students work in groups. Place students with poor math skills with students working at grade level.
■ Provide students with an example if they are having difficulty in determining a ratio.

Thinking Critically
1. The ratio was 2.96 : 1.
2. The observed ratio is slightly lower than expected.
3. The observed and expected ratios may differ slightly due to chance.

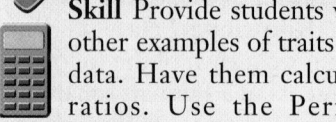
Assessment

Skill Provide students with other examples of traits and data. Have them calculate ratios. Use the Performance Task Assessment List for Using Math in Science in **PASC**, p. 29. **L2**

4 Close

Activity

Visual-Spatial Have students illustrate Mendel's law of segregation using as much of a Punnett square as needed. Do the same with the law of independent assortment. **L2**

268

Problem-Solving Lab 10-1 — Analyzing Information

How did Mendel analyze his data?
In addition to crossing tall and short pea plants, Mendel crossed plants that formed round seeds with plants that formed wrinkled seeds. He found a 3 : 1 ratio of round-seeded plants to wrinkled-seeded plants in the F_2 generation.

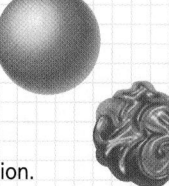

Analysis
Mendel's actual results in the F_2 generation are shown to the right.

Mendel's results	
Kind of Plants	**Number of Plants**
Round-seeded	5474
Wrinkled-seeded	1850

1. Calculate the actual ratio of round-seeded plants to wrinkled-seeded plants. To do this, divide the number of round-seeded plants by the number of wrinkled-seeded plants (round to the nearest hundredth). Your answer tells you how many more times round-seeded plants resulted than wrinkled-seeded plants.

2. To express your answer as a ratio, write the number from step 1 followed by a colon and the numeral 1.

Thinking Critically
1. What was the actual ratio Mendel observed for this cross?
2. How does Mendel's observed ratio compare with the expected 3 : 1 ratio?
3. Why was the actual ratio different from the expected ratio?

square shows three plants with round seeds out of four total plants, the probability is 3/4, as **Figure 10.9** shows. Yet, if you calculate the fraction of round-seeded plants from Mendel's actual data in the *Problem-Solving Lab* on this page, you will see that slightly less than three-fourths of the plants were round-seeded. It is important to remember that the results predicted by probability are more likely to be seen when there is a large number of offspring.

Figure 10.9
The probability that the offspring from a mating of two heterozygotes will show a dominant phenotype is 3 out of 4, or 3/4.

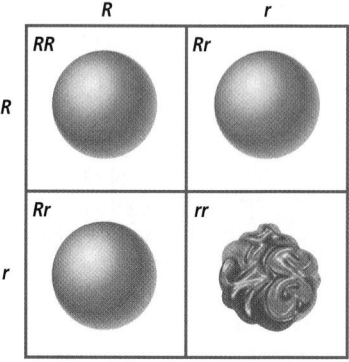

Section Assessment

Understanding Main Ideas
1. What structural features of pea plant flowers made them suitable for Mendel's genetic studies?
2. What are the genotypes of a homozygous and a heterozygous tall pea plant?
3. One parent is homozygous tall and the other parent is heterozygous tall. Make a Punnett square to determine what fraction of their offspring is expected to be heterozygous.
4. How many different gametes can an *RRYy* parent form? What are they?

Thinking Critically
5. In garden peas, the allele for yellow peas is dominant to the allele for green peas. Suppose you have a plant that produces yellow peas, but you don't know whether it is homozygous dominant or heterozygous. What experiment could you do to find out? Draw a Punnett square to help you.

SKILL REVIEW
6. **Observing and Inferring** The offspring of a cross between a purple-flowered plant and a white-flowered plant are 23 plants with purple flowers and 26 plants with white flowers. Use the letter *P* for purple and *p* for white. What are the genotypes of the parent plants? Explain your reasoning. For more help, refer to *Thinking Critically* in the **Skill Handbook**.

268 MENDEL AND MEIOSIS

Section Assessment

1. They are self-pollinating, and male flower parts can be easily removed to allow for cross-pollination.
2. homozygous tall = *TT*, heterozygous = *Tt*
3. One-half of all offspring will be heterozygous. (Note: It will not matter if the homozygous parent is homozygous dominant or recessive.)
4. two; *RY* and *Ry*

5. Cross the unknown yellow plant with a recessive green parent. If the offspring are all yellow, the unknown genotype is homozygous yellow. If half the offspring are yellow and half green, the unknown genotype is heterozygous.
6. Genotypes of parents are *Pp* for the purple-flowered plant and *pp* for the white-flowered plant. Purple is dominant.

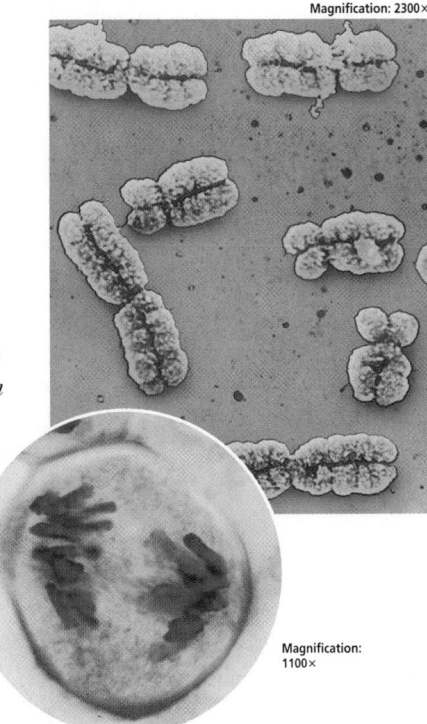

Magnification: 2300×

Metaphase chromosomes (top) and a plant cell in early anaphase of meiosis

Magnification: 1100×

Mendel's study of inheritance was based on careful observations of pea plants, but pieces of the hereditary puzzle were still missing. Modern technologies such as high-power microscopes allow us a glimpse of things that Mendel could only imagine. You can now look inside a cell to see the chromosomes on which the traits described by Mendel are carried. You can also examine the process by which these traits are transmitted to the next generation.

SECTION PREVIEW

Objectives

Analyze how meiosis maintains a constant number of chromosomes within a species.

Infer how meiosis leads to variation in a species.

Relate Mendel's laws of heredity to the events of meiosis.

Vocabulary
diploid
haploid
homologous
   chromosome
meiosis
sperm
egg
zygote
sexual reproduction
crossing over
genetic recombination
nondisjunction

Genes, Chromosomes, and Numbers

Organisms have tens of thousands of genes that determine individual traits. Genes do not exist free in the nucleus of a cell; they are lined up on chromosomes. Typically, a chromosome can contain a thousand or more genes.

Diploid and haploid cells

If you examined the nucleus in a cell of one of Mendel's pea plants, you would find it had 14 chromosomes—seven pairs. In the body cells of animals and most plants, chromo-

somes occur in pairs. One chromosome in each pair came from the male parent, and the other came from the female parent. A cell with two of each kind of chromosome is called a **diploid** cell and is said to contain a diploid, or $2n$, number of chromosomes. This pairing supports Mendel's conclusion that organisms have two factors—alleles—for each trait. One allele is located on each of the paired chromosomes.

Organisms produce gametes that contain one of each kind of chromosome. A cell with one of each kind of chromosome is called a **haploid** cell and is said to contain a haploid, or n,

10.2 MEIOSIS **269**

Section 10.2

Prepare

Key Concepts

This section develops the concepts of diploid and haploid chromosome numbers and homologous chromosomes. Events that occur during meiosis are illustrated, and their role in genetic recombination is explained.

Planning

- Gather pipe cleaners or jellybeans for the Meeting Individual Needs.
- Gather materials for the Alternative Lab.
- Purchase modeling clay or Plasticene for MiniLab 10-2.
- Collect pictures of karotypes for the Display.
- Prepare line drawings for the Portfolio.

1 Focus

Bellringer

Before presenting the lesson, display **Section Focus Transparency 25** on the overhead projector and have students answer the accompanying questions.
L1 **ELL**

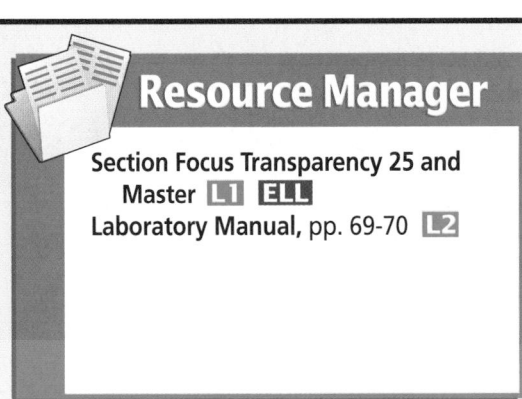

Resource Manager

Section Focus Transparency 25 and
   Master **L1** **ELL**
Laboratory Manual, pp. 69-70 **L2**

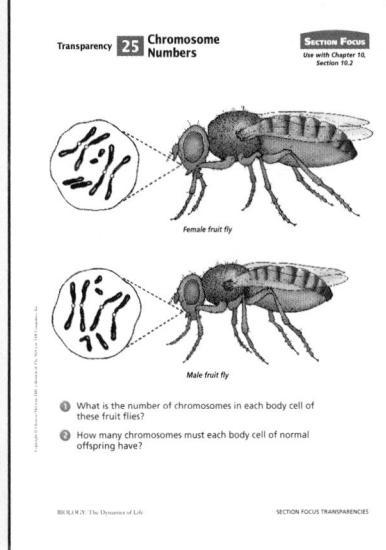

2 Teach

Problem-Solving Lab 10-2

Purpose
Students will compare alleles on homologous chromosomes.

Process Skills
interpret scientific diagrams, apply concepts, compare and contrast, think critically

Teaching Strategies
■ Review the meaning of alleles and homologous chromosomes.

Thinking Critically
1. 2 is homologous with 1. Alleles may or may not be identical, as long as they are positioned at the same location on matching chromosomes.
2. 3 is not homologous with 1. Genes must be identical. Gene K is not the same as gene J.
3. 4 is not homologous with 1. Chromosomes must match in physical size and location of genes in order to be homologous. (Sex chromosomes are an exception to the rule of matching physical size.)
4. 5 is homologous with 1 for the same reasons as in question 1.

✓ Assessment
Performance Ask students to make a diagram of a chromosome with three marked gene locations. Have them diagram all the homologous chromosomes that are possible for this chromosome. Use the Performance Task Assessment List for Scientific Drawing in **PASC**, p. 55. **L2**

270

Problem-Solving Lab 10-2 — Interpreting Scientific Illustrations

Can you identify homologous chromosomes?
Homologous chromosomes are paired chromosomes having genes for the same trait located at the same place on the chromosome. The gene itself, however, may have different alleles, producing different forms of the trait.

Analysis
The diagram below shows chromosome 1 with four different genes present. These genes are represented by the letters *F, g, h,* and *J*. Possible homologous chromosomes of chromosome 1 are labeled 2-5. Examine the five chromosomes and the genes they contain to determine which of chromosomes 2-5 are homologous with chromosome 1.

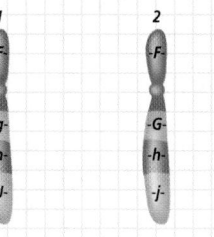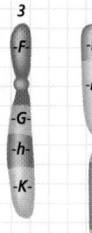

Thinking Critically
1. Could chromosome 2 be homologous with chromosome 1? Explain why.
2. Could chromosome 3 be homologous with chromosome 1? Explain why.
3. Could chromosome 4 be homologous with chromosome 1? Explain why.
4. Could chromosome 5 be homologous with chromosome 1? Explain why.

270 MENDEL AND MEIOSIS

---

number of chromosomes. This fact supports Mendel's conclusion that parent organisms give one factor, or allele, for each trait to each of their offspring.

Each species of organism contains a characteristic number of chromosomes. *Table 10.1* shows the diploid and haploid numbers of chromosomes of some species. Note the large range of chromosome numbers. Note also that the chromosome number of a species is not related to the complexity of the organism.

Homologous chromosomes

The two chromosomes of each pair in a diploid cell help determine what the individual organism looks like. These paired chromosomes are called **homologous chromosomes** (huh MAHL uh gus). Each of a pair of homologous chromosomes has genes for the same traits, such as pod shape. On homologous chromosomes, these genes are arranged in the same order, but because there are different possible alleles for the same gene, the two chromosomes in a homologous pair are not always identical to each other. Identify the homologous chromosomes in the *Problem-Solving Lab*.

Table 10.1 Chromosome Numbers of Some Common Organisms

Organism	Body Cell (2n)	Gamete (n)
Fruit fly	8	4
Garden pea	14	7
Corn	20	10
Tomato	24	12
Leopard frog	26	13
Apple	34	17
Human	46	23
Chimpanzee	48	24
Dog	78	39
Adder's tongue fern	1260	630

Adder's tongue fern

Corn

Leopard frog

Let's look at the seven pairs of homologous chromosomes in Mendel's peas. These chromosome pairs are numbered 1 through 7. Each pair contains certain genes located at specific places on the chromosome. Chromosome 4 contains the genes for three of the traits that Mendel studied. Many other genes can be found on this chromosome as well.

Every pea plant has two copies of chromosome 4. It received one from each of its parents and will give one at random to each of its offspring. Remember, however, that the two copies of chromosome 4 in a pea plant may not necessarily have identical alleles. Each can have one of the different alleles possible for each gene. The homologous chromosomes diagrammed in *Figure 10.10* show both alleles for each of three traits. Thus, the plant represented by these chromosomes is heterozygous for each of the traits.

Why meiosis?

When cells divide by mitosis, the new cells have exactly the same number and kind of chromosomes as the original cells. Imagine if mitosis were the only means of cell division. Each pea plant parent, which has 14 chromosomes, would produce gametes that contained a complete set of 14 chromosomes. That means that each offspring formed by fertilization of gametes would have twice the number of chromosomes as each of its parents. The F_1 pea plants would have cell nuclei with 28 chromosomes, and the F_2 plants would have cell nuclei with 56 chromosomes. The nuclei would certainly be crowded! What do you think these plants might look like?

Clearly, there must be another form of cell division that allows offspring to have the same number of

Homologous chromosome 4

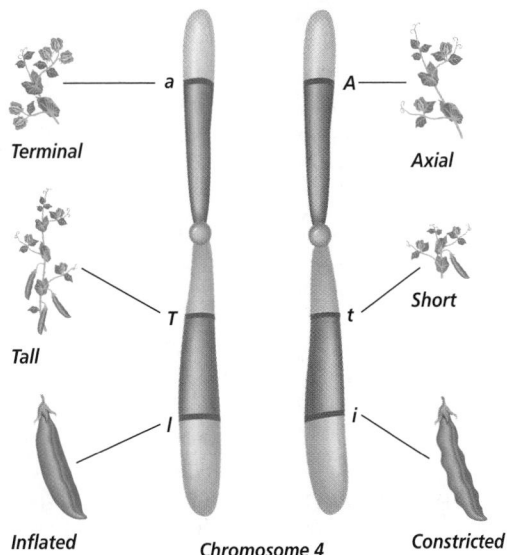

Figure 10.10
Each chromosome 4 in garden peas contains genes for flower position, height, and pod shape. Flower position can be either axial (flowers located along the stems) or terminal (flowers clustered at the top of the plant). Plant height can be either tall or short. Pod shape can be either inflated or constricted.

chromosomes as their parents. This kind of cell division, which produces gametes containing half the number of chromosomes as a parent's body cell, is called **meiosis** (mi OH sus). Meiosis occurs in the specialized body cells of each parent that produce gametes.

Meiosis consists of two separate divisions, known as meiosis I and meiosis II. Meiosis I begins with one diploid ($2n$) cell. By the end of meiosis II, there are four haploid (n) cells. These haploid cells are called sex cells—gametes. Male gametes are called **sperm.** Female gametes are called **eggs.** When a sperm fertilizes an egg, the resulting cell, called a **zygote** (ZI goht), once again has the diploid number of chromosomes.

WORD *Origin*
meiosis
From the Greek word *meioun*, meaning "to diminish." Meiosis is cell division that results in a gamete containing half the number of chromosomes of its parent.

Visual Learning

Table 10.1 Ask students whether there is any evidence to support the idea that plants have fewer chromosomes than animals. Tell students to use examples from the table to support their answers. *No; apples have 34 chromosomes, which is more than fruit flies or frogs have.* Ask for the chromosome numbers in skin cells of a leopard frog and a dog, *26 and 78,* and in root cells of tomatoes and garden peas, *24 and 14.*

Discussion

Show students an egg and explain that it is a gamete. Based on previous information, students should be able to tell how many alleles are present for each trait. *only one allele for each trait* Ask why an organism cannot produce gametes by mitosis. *Mitosis produces a cell with both members of each pair of chromosomes.*

GLENCOE TECHNOLOGY

CD-ROM
Biology: The Dynamics of Life
Animation: *Meiosis* Disc 2

VIDEODISC
Biology: The Dynamics of Life
Disc 1, Side 1, 1 min. 47 sec. *Meiosis* (Ch. 30)

MEETING INDIVIDUAL NEEDS

Learning Disabled

Kinesthetic Have students use pipe cleaners or jelly beans to show why meiosis is necessary to prevent the number of chromosomes from doubling in each generation. Students should start with a body cell containing two pipe cleaners or jelly beans, 2*n*. After meiosis, the gametes should each contain one, *n*. After fertilization, the zygote again has two. Students can then repeat the process for several generations, assuming that gametes are formed by mitosis. In the first generation, the gametes will be 2*n*. In the next generation, they will be 4*n*. Students will see that the number of chromosomes doubles in each generation.

L1 **ELL** **COOP LEARN**

The zygote then can develop by mitosis into a multicellular organism. The pattern of reproduction that involves the production and subsequent fusion of haploid sex cells is called **sexual reproduction.** This reproductive pattern is illustrated in *Figure 10.11.*

The Phases of Meiosis

During meiosis, a spindle forms and the cytoplasm divides in the same ways they do during mitosis. However, what happens to the chromosomes in meiosis is very different. *Figure 10.12* illustrates interphase and the phases of meiosis. Examine the diagram and photo of each phase as you read about it.

Interphase

Recall from Chapter 8 that, during interphase, the cell replicates its chromosomes. During interphase that precedes meiosis I, the cell also replicates its chromosomes. After replication, each chromosome consists of two identical sister chromatids, held together by a centromere.

Prophase I

A cell entering prophase I behaves in a similar way to one entering prophase of mitosis. The chromosomes coil up and a spindle forms. Then, in a step unique to meiosis, each pair of homologous chromosomes comes together, matched gene by gene, to form a four-part structure called a tetrad. A tetrad consists of two homologous chromosomes, each made up of two sister chromatids. The chromatids in a tetrad pair tightly. In fact, they pair so tightly that nonsister chromatids from homologous chromosomes sometimes actually exchange genetic material in a process known as **crossing over.** Crossing over can occur at any location on a chromosome, and it can occur at several locations at the same time. It is estimated that during prophase I of meiosis in humans, there is an average of two to three crossovers for each pair of homologous chromosomes.

Figure 10.11
In sexual reproduction, the doubling of the chromosome number that results from fertilization is balanced by the halving of the chromosome number that results from meiosis.

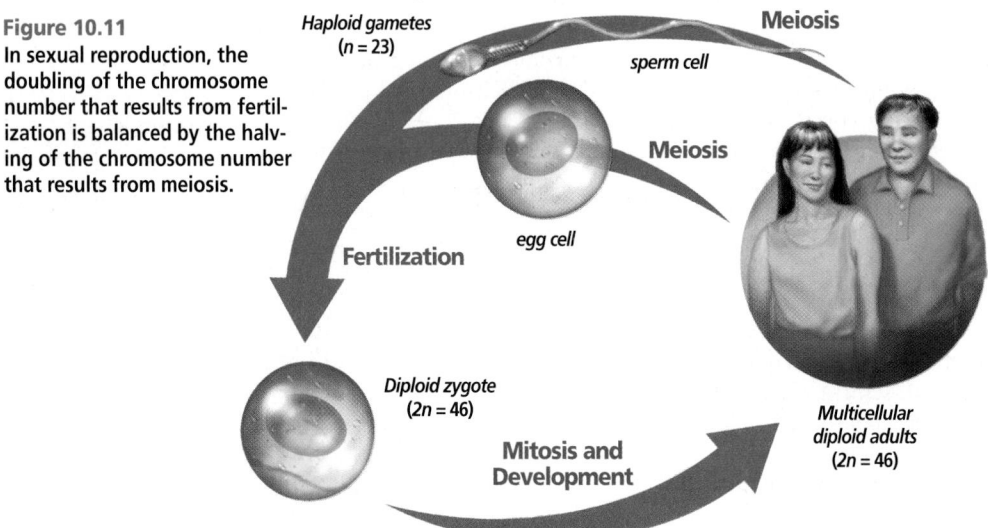

Haploid gametes
(*n* = 23)

sperm cell

Meiosis

Meiosis

egg cell

Fertilization

Diploid zygote
(2*n* = 46)

Mitosis and
Development

Multicellular
diploid adults
(2*n* = 46)

Figure 10.12
Follow the diagrams showing interphase and meiosis as you read about each phase. Compare these diagrams with those of mitosis in Chapter 8. Note that after telophase II, meiosis is finished and gametes form. In what other ways are mitosis and meiosis different?

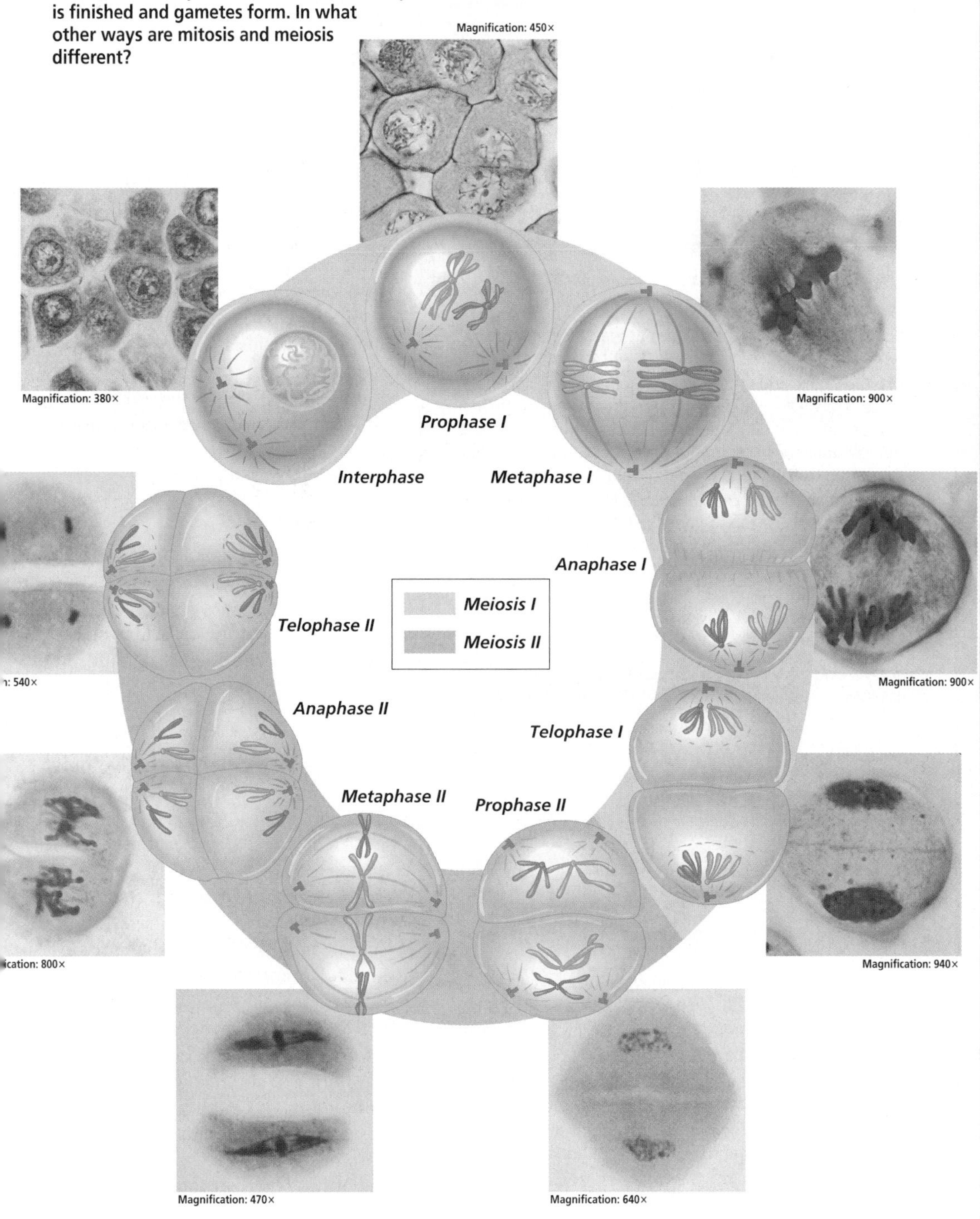

Magnification: 450×

Magnification: 380×

Magnification: 900×

Prophase I

Interphase   *Metaphase I*

Anaphase I

Meiosis I
Meiosis II

Telophase II

Magnification: 900×

Anaphase II

Telophase I

Metaphase II   *Prophase II*

Magnification: 940×

Magnification: 800×

Magnification: 470×

Magnification: 640×

10.2 MEIOSIS **273**

Using an Analogy

To reinforce the concept of homologous chromosomes, sister chromatids, tetrad formation, crossing over, and anaphase I, try the following analogy: A magic pair of shoes (left and right) is found on a shelf (homologous chromosomes in a cell). These shoes, being magic, can and do replicate (interphase replication). Each copy is tied to its original with its shoelaces (centromere; both lefts are tied together and both rights are tied together). Both rights are now called right sister shoes (sister chromatids). Both lefts are now called left sister shoes (sister chromatids). All four shoes line up next to one another (tetrad formation). While next to one another, part of one nonsister shoe (a left shoe) exchanges its innersole with another nonsister shoe (a right shoe) (crossing over). Right shoes move away from left shoes to different shelves (anaphase I, with homologous chromosomes separating and going to two different cells. Both rights are still tied together and both lefts are still tied together). To improve the analogy, locate two identical pairs of shoes to demonstrate the events as they are described.

Resource Manager

BioLab and MiniLab Worksheets, pp. 45-46 **L2**
Basic Concepts Transparency 15 and Master **L2** **ELL**
Reteaching Skills Transparency 17 and Master **L1** **ELL**

5. Model interphase, prophase I, metaphase I, anaphase I, and telophase I, prophase II, metaphase II, anaphase II, and telophase II.

Expected Results
Students will gain an understanding of meiosis when visualizing each phase.

Analysis
1. What happens to the chromosome number during meiosis? *reduced by 1/2*

2. How many cells are formed during meiosis? *4*
3. What is the fate of the cells formed during meiosis? *They form either egg or sperm cells.*

Assessment
Knowledge Ask students to write a paragraph explaining the value of making models such as the ones in this lab. Use the Performance Task Assessment List for Writing in Science in **PASC**, p. 87. **L2**

Purpose

Students will model the process of crossing over.

Process Skills

apply concepts, compare and contrast, formulate models, recognize cause and effect, think critically, define operationally

Teaching Strategies

■ Plasticene or clay is available from biological supply houses or craft stores.

■ String or twine may be used as a substitute for twist ties.

■ Review the concept that gametes contain only one chromosome from each pair.

■ Make sure students wash their hands after the lab.

Expected Results

All gametes will show the same combination of genes if no crossing over occurs and different combinations of genes if crossing over does occur.

Analysis

1. Only two chromosomes should be drawn. The sequence of genes would be *C B A* on one chromosome and *c B a* on the other.

2. Crossing over is the exchange of genetic material between nonsister chromatids during prophase I of meiosis.

3. All gamete cells show the same pattern of genes on chromosomes as in the original diagram if no crossing over occurs. If crossing over occurs between *B* and *C*, gamete cells with gene arrangements *C B a* and *c B A* will be formed as well as cells with *C B A* and *c B a*.

4. Student answers will vary. There is a mixing of the gene traits from their original order.

5. There would be no mixing of gene traits when compared with the original chromosomes because sister chromatids are identical.

274

MiniLab 10-2 — Formulating Models

Modeling Crossing Over Crossing over occurs during meiosis and involves only the nonsister chromatids that are present during tetrad formation. The process is responsible for the appearance of new combinations of alleles in gamete cells.

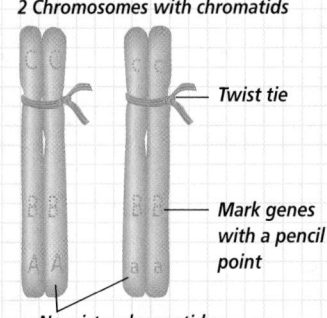

2 Chromosomes with chromatids

Twist tie

Mark genes with a pencil point

Nonsister chromatids

Procedure

1. Copy the data table.
2. Roll out four long strands of clay at least 10 cm long to represent two chromosomes, each with two chromatids.
3. Use the figure above as a guide in joining and labeling these model chromatids. Although there are four chromatids, assume that they started out as a single pair of homologous chromosomes prior to replication. The figure shows tetrad formation during prophase I of meiosis.
4. First, assume that no crossing over takes place. Model the appearance of the four gamete cells that will result at the end of meiosis. Record your model's appearance by drawing the gametes' chromosomes and their genes in your data table.
5. Next, repeat steps 2-4. This time, however, assume that crossing over occurs between genes B and C.

Data Table	
No crossing over	**Crossing over**
Appearance of gamete cells	Appearance of gamete cells

Analysis

1. Predict and diagram the appearance of the chromosomes prior to replication.
2. Define crossing over and explain when it occurs.
3. Compare any differences in the appearance of genes on chromosomes in gamete cells when crossing over occurs and when it does not occur.
4. Crossing over has been compared to "shuffling the deck" in cards. Explain what this means.
5. What would be accomplished if crossing over occurred between sister chromatids? Explain your answer.

274 MENDEL AND MEIOSIS

This exchange of genetic material is diagrammed as an X-shaped configuration in *Figure 10.13b.* Crossing over results in new combinations of alleles on a chromosome, as shown in *Figure 10.13c.* You can practice modeling crossing over in the *MiniLab* at the left.

Metaphase I

As prophase I ends, the centromere of each chromosome becomes attached to a spindle fiber. The spindle fibers pull the tetrads into the middle, or equator, of the spindle. This is an important step unique to meiosis. Note that homologous chromosomes are lined up side by side as tetrads. In mitosis, on the other hand, homologous chromosomes line up on the equator independently of each other.

Anaphase I

Anaphase I begins as homologous chromosomes, each with its two chromatids, separate and move to opposite ends of the cell. This occurs because the centromeres holding the sister chromatids together do not split as they do during anaphase in mitosis. This critical step ensures that each new cell will receive only one chromosome from each homologous pair.

Telophase I

Events occur in the reverse order from the events of prophase I. The spindle is broken down, the chromosomes uncoil, and the cytoplasm divides to yield two new cells. Each cell has only half the genetic information of the original cell because it has only one chromosome from each homologous pair. However, another cell division is needed because each chromosome is still doubled, consisting of two sister chromatids.

✓ Assessment

Performance Ask students to model the appearance of gamete cells if crossing over occurred between sister chromatids. This should confirm their answer to question 5. Use the Performance Task Assessment List for Model in **PASC,** p. 51. **L2** **ELL**

✓ Portfolio

Modeling the Membrane

Visual-Spatial Have student groups invent a mechanism for demonstrating how the plasma membrane of an animal cell pinches together during telophase I and II of meiosis. Students can use a variety of common objects as models, such as balloons, string, and wire. **L2** **ELL**
COOP LEARN

The phases of meiosis II

The newly formed cells in some organisms undergo a short interphase in which the chromosomes do not replicate. In other organisms, however, the cells go from late anaphase of meiosis I directly to metaphase of meiosis II, skipping telophase I, interphase, and prophase II.

The second division in meiosis consists of prophase II, metaphase II, anaphase II, and telophase II. During prophase II, a spindle forms in each of the two new cells and the spindle fibers attach to the chromosomes. The chromosomes, still made up of sister chromatids, are pulled to the center of the cell and line up randomly at the equator during metaphase II, just as they do in mitosis. Anaphase II begins as the centromere of each chromosome splits, allowing the sister chromatids to separate and move to opposite poles. Finally, nuclei re-form, the spindles break down, and the cytoplasm divides during telophase II. The events of meiosis II are identical to those you studied for mitosis.

At the end of meiosis II, four haploid sex cells have been formed from one original diploid cell. Each haploid cell contains one chromosome from each homologous pair. These haploid cells will become gametes, transmitting the genes they contain to offspring.

Meiosis Provides for Genetic Variation

Cells that are formed by mitosis are identical to each other and to the parent cell. Meiosis, however, provides a mechanism for shuffling the chromosomes and the genetic information they carry. By shuffling the chromosomes, genetic variation is produced.

Genetic recombination

How many different kinds of sperm can a pea plant produce? Each cell undergoing meiosis has seven pairs of chromosomes. Because each of the seven pairs of chromosomes can line up at the cell's equator in two different ways, 128 different kinds of sperm are possible ($2^7 = 128$).

WORD *Origin*

pro-
From the Greek word *pro*, meaning "before."

meta-
From the Greek word *meta*, meaning "after."

ana-
From the Greek word *ana*, meaning "away, onward."

telo-
From the Greek *telos*, meaning "end."

The four phases of cell division are prophase, metaphase, anaphase, and telophase.

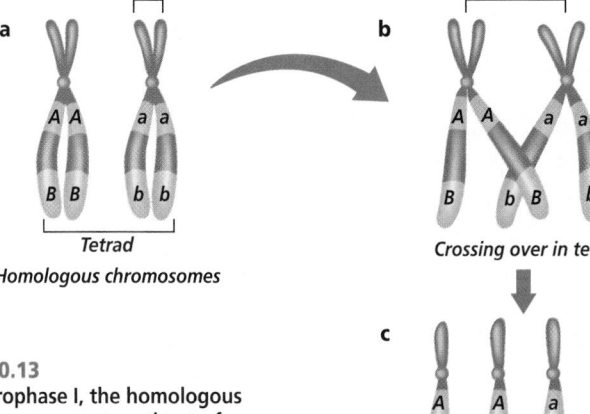

Figure 10.13
Late in prophase I, the homologous chromosomes come together to form tetrads (**a**). Arms of nonsister chromatids wind around each other (**b**), and genetic material may be exchanged (**c**).

Using an Analogy

Continue the shoe analogy or demonstration of meiosis to illustrate telophase I, metaphase II, and telophase II events. Start with two right shoes tied together. Both sets are separate from each other on different closet shelves (two new cells formed after telophase I). The shoes untie (centromere splits after metaphase II). They move to different shelves in the closet (two new cells formed in telophase II). Ask students: (a) How many shoes are now on separate shelves (separate cells)? *4* (b) How many shoes were present at the start of this analogy? *2* (c) Is the chromosome number in a cell diploid or haploid before a cell undergoes meiosis? *diploid* (d) How many shoes are present on each shelf at the end of the process? *1* (e) How many shoes were on the original shelf? *2* (f) Is the chromosome number in each of the four new cells formed in meiosis reduced by half? *yes*

Concept Development

The chances of two humans being born exactly alike is almost an impossibility (except for identical twins). Explain why children in the same family can nevertheless resemble one another rather closely.

GLENCOE TECHNOLOGY

VIDEODISC
The Secret of Life
Meiosis Ia

Meiosis Ib

Comparing Mitosis and Meiosis

Linguistic Provide students with two sets of simple line drawings, one showing interphase and the phases of mitosis and the other showing interphase and meiosis. Place comparable phases next to each other when possible (prophase of mitosis next to prophase I of meiosis, and so on). Ask students to describe in their portfolios the similarities and differences between processes. Also have them indicate the type of cell formed at the end of the process—body or gamete—and whether its chromosome number would be diploid or haploid. **L2**

Performance Assessment in the Biology Classroom, p. 13. *Investigating Mitosis and Meiosis,* Have students carry out this activity to determine which slides in an unlabeled set show mitosis and which show meiosis, and which were made from animal cells and which from plant cells. **L2**

VIDEODISC
The Secret of Life
Meiosis IIa

Meiosis IIb

Resource Manager

Concept Mapping, p. 10
L3 **ELL**
Laboratory Manual,
pp. 71-74 **L2**

Figure 10.14
If a cell has two pairs of chromosomes ($n = 2$), four kinds of gametes (2^2) are possible, depending on how the homologous chromosomes line up at the equator during meiosis I **(a)**. This event is a matter of chance. When zygotes are formed by the union of these gametes, 4×4 or 16 possible combinations may occur **(b)**.

In the same way, any pea plant can form 128 different eggs. Because any egg can be fertilized by any sperm, the number of different possible offspring is 16 384 (128×128). *Figure 10.14a* shows a simple example of how genetic recombination occurs. You can see that the gene combinations in the gametes vary depending on how each pair of homologous chromosomes lines up during metaphase I, a random process.

These numbers increase greatly as the number of chromosomes in the species increases. In humans, $n = 23$, so the number of different kinds of eggs or sperm a person can produce is more than 8 million (2^{23}). When fertilization occurs, $2^{23} \times 2^{23}$, or 70 trillion, different zygotes are possible! It's no wonder that each individual is unique.

In addition, crossing over can occur anywhere at random on a chromosome. Typically, two or three crossovers per chromosome occur during meiosis. This means that an almost endless number of different possible chromosomes can be produced by crossing over, providing additional variation to that already produced by

the random assortment of chromosomes. This reassortment of chromosomes and the genetic information they carry, either by crossing over or by independent segregation of homologous chromosomes, is called **genetic recombination.** It is a major source of variation among organisms. Variation is important to a species because it is the raw material that forms the basis for evolution. How does crossing over increase genetic variability? To answer this question, read the *Inside Story* on the next page.

Meiosis explains Mendel's results

Meiosis provides the physical basis for explaining Mendel's results. The segregation of chromosomes in anaphase I of meiosis explains Mendel's observation that each parent gives one allele for each trait at random to each offspring, regardless of whether the allele is expressed. The segregation of chromosomes at random during anaphase I explains Mendel's observation that factors, or genes, for different traits are inherited independently of each other. Today, Mendel's laws of heredity form the foundation of modern genetics.

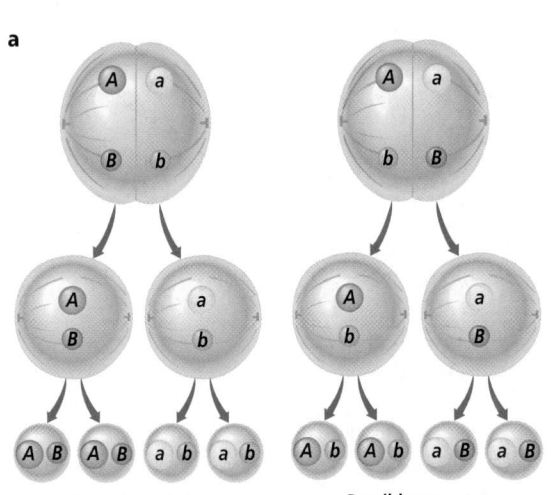

a

Possible gametes

Possible gametes

b

Possible arrangement of chromosomes in sperm

AB Ab aB ab

Possible arrangement of chromosomes in eggs

	AB	Ab	aB	ab
AB	AABB	AABb	AaBB	AaBb
Ab	AABb	AAbb	AaBb	Aabb
aB	AaBB	AaBb	aaBB	aaBb
ab	AaBb	Aabb	aaBb	aabb

Possible combinations of chromosomes in zygotes (in boxes)

BIOLOGY JOURNAL

Gamete Formation

Linguistic Ask students to imagine that they are pea gametes who have just met their cousin leaf cells from the same plant. You want to explain to your cousins how you are different from them. Write a story that describes how you became a gamete and how you differ from your cousins. **L1**

Genetic Recombination

One source of genetic recombination is a process known as crossing over, the exchange of genetic material by nonsister chromatids during meiosis.

Critical Thinking *How can the frequency of crossing over be used to map the location of genes on chromosomes?*

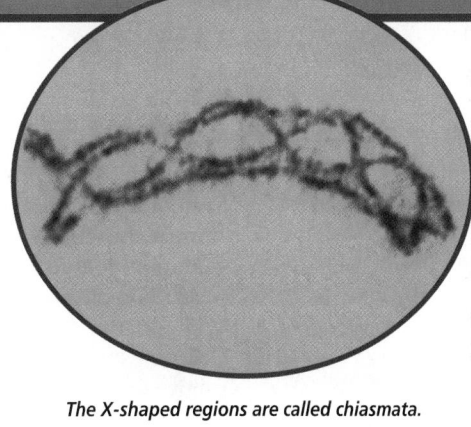

The X-shaped regions are called chiasmata.

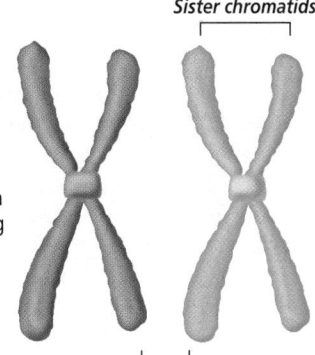

1 Tetrads During late prophase I, homologous chromosomes come together to form a tetrad. This pairing of homologous chromosomes is called synapsis.

Sister chromatids

Nonsister chromatids

4 Mapping Geneticists have used the frequency of crossing over to map the relative location of alleles on chromosomes. Alleles that are further apart on the chromosomes are more likely to have chiasmata between them than alleles that are close together.

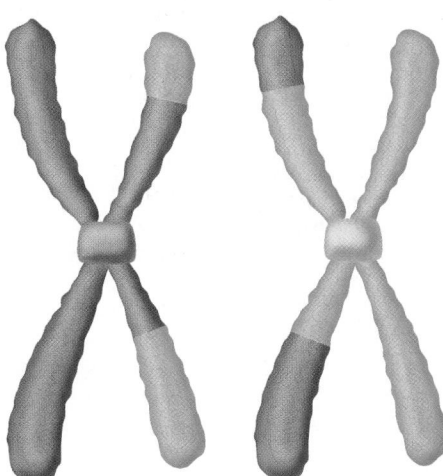

2 Chiasmata The pairing of homologous chromosomes during synapsis is precise— they line up with each other allele by allele. The nonsister chromatids twist around each other, forming X-shaped regions called chiasmata (ki az MAH tuh). The two nonsister chromatids break and exchange genetic material.

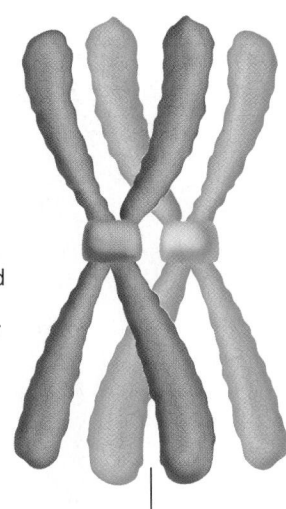

Chiasmata

3 New allele combinations Crossing over causes a shuffling of allele combinations, just as shuffling a deck of cards produces new combinations of cards dealt in a hand. Rather than the alleles from each parent staying together on their homologous chromosome, new combinations of alleles can form. Thus, variability is increased.

10.2 MEIOSIS **277**

Purpose
Students study the process of crossing over and its importance in genetic recombination.

Teaching Strategies
■ Ask students to describe the process of crossing over. **L2**

Visual Learning
■ Have students examine the photograph and count the chiasmata.
■ *Visual-Spatial* Have students draw chromosomes before and after crossing over. The use of colored pencils will help students visualize how this process leads to genetic recombination. **L2 ELL**

Critical Thinking
The relative position of genes on a chromosome can be determined by the frequency of crossing over. The greater the frequency of crossing over, the greater the distance between the two genes.

BIOLOGY Online
Note Internet addresses that you find useful in the space below for quick reference.

_____

_____

_____

_____

_____

Make a bulletin board display of pictures of karotypes showing abnormal numbers of chromosomes.

3 Assess

Check for Understanding

Ask students to explain how the words in the following combinations are related. **L1** **ELL**
   a. diploid—haploid
   b. homologous chromosomes—allele
   c. sperm—egg—zygote
   d. meiosis—gamete
   e. crossing over—genetic recombination

Reteach

Ask students to prepare a list of the important characteristics of each step in the process of meiosis. Then have them prepare a second list of all the reasons why meiosis is important to organisms that reproduce sexually. **L2**

Mistakes in Meiosis

Although the events of meiosis usually proceed accurately, sometimes an accident occurs and chromosomes fail to separate correctly. The failure of homologous chromosomes to separate properly during meiosis is called **nondisjunction**. Recall that during meiosis I, one chromosome from each homologous pair moves to each pole of the cell. Occasionally, both chromosomes of a homologous pair move to the same pole of the cell.

Trisomy, monosomy, and triploidy

In one form of nondisjunction, two kinds of gametes result. One has an extra chromosome, and the other is missing a chromosome. The effects of nondisjunction are often seen after gametes fuse. For example, when a gamete with an extra chromosome is fertilized by a normal gamete, the zygote will have an extra chromosome. This condition is called trisomy (TRI soh mee). In humans, if a gamete with an extra chromosome number 21 is fertilized by a normal gamete, the resulting zygote has 47 chromosomes instead of 46. This

Figure 10.15
Follow the steps to see how a tetraploid plant such as this chrysanthemum is produced.

zygote will develop into a baby with Down syndrome.

Although organisms with extra chromosomes often survive, organisms lacking one or more chromosomes usually do not. When a gamete with a missing chromosome fuses with a normal gamete during fertilization, the resulting zygote lacks a chromosome. This condition is called monosomy. In humans, most zygotes with monosomy do not survive. If a zygote with monosomy does survive, the resulting organism usually does not. An example of monosomy that is not lethal is Turner syndrome, in which human females have only a single X chromosome instead of two.

Another form of nondisjunction involves a total lack of separation of homologous chromosomes. When this happens, a gamete inherits a complete diploid set of chromosomes, as shown in *Figure 10.15*. When a gamete with an extra set of chromosomes is fertilized by a normal haploid gamete, the offspring has three sets of chromosomes and is triploid. The fusion of two gametes, each with an extra set of chromosomes, produces offspring with four

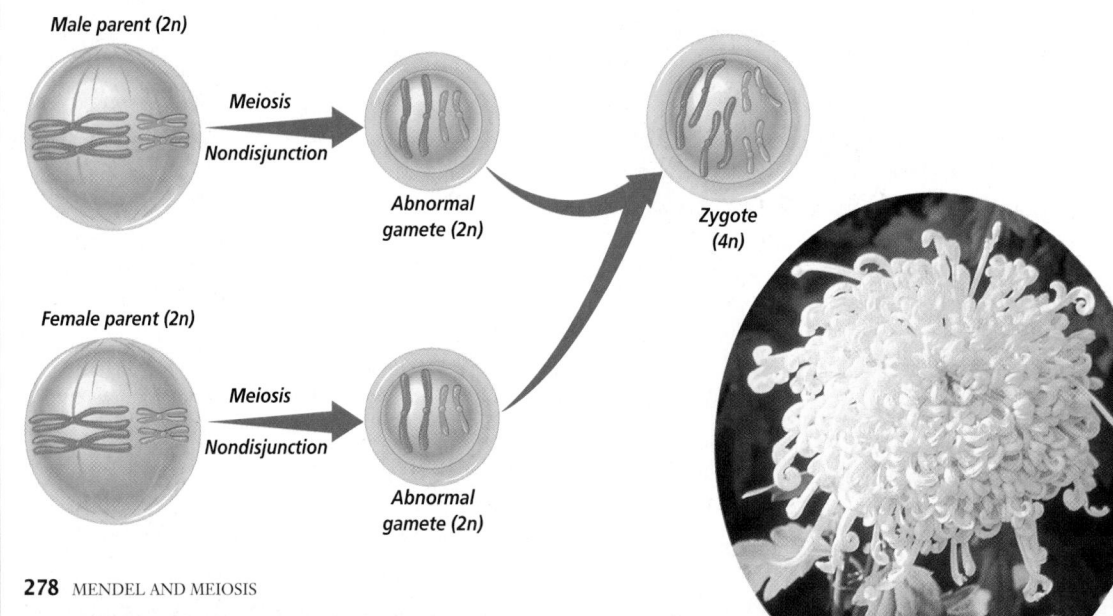

Male parent (2n)

Meiosis
Nondisjunction

Abnormal gamete (2n)

Female parent (2n)

Meiosis
Nondisjunction

Abnormal gamete (2n)

Zygote (4n)

TECHPREP

Polyploidy in Plants

Linguistic Have students research and report on how plant breeders create polyploid flowers. Students should name some polyploid cultivars and indicate how many sets of chromosomes each has. **L3**

Figure 10.16
The banana plant is an example of a triploid plant (**a**). This day lily is a tetraploid plant (**b**).

sets of chromosomes—a tetraploid. This can be seen in *Figure 10.15*

Organisms with more than the usual number of chromosome sets are called polyploids. Polyploidy is rare in animals and almost always causes death of the zygote. However, polyploidy frequently occurs in plants. Often, the flowers and fruits of these plants are larger than normal, and the plants are healthier. Many polyploid plants, such as the sterile banana plant and the day lily shown in *Figure 10.16,* are of great commercial value.

Meiosis is a complex process, and the results of an error occurring are sometimes unfortunate. However, mistakes in meiosis can be beneficial, such as those that have occurred in agriculture. Tetraploid (4*n*) wheat, triploid (3*n*) apples, and polyploid chrysanthemums all are available commercially. You can see that a thorough understanding of meiosis and genetics would be very helpful to plant breeders. In fact, plant breeders have learned to artificially produce polyploid plants using chemicals that cause nondisjunction.

Section Assessment

Understanding Main Ideas

1. How are the cells at the end of meiosis different from the cells at the beginning of meiosis? Use the terms *chromosome number, haploid,* and *diploid* in your answer.
2. What is the role of meiosis in maintaining a constant number of chromosomes in a species?
3. Why are there so many varied phenotypes within a species such as humans?
4. If the diploid number of a plant is 10, how many chromosomes would you expect to find in its triploid offspring?

Thinking Critically

5. How do the events of meiosis explain Mendel's law of independent assortment?

SKILL REVIEW

6. **Interpreting Scientific Illustrations** Compare *Figures 10.12* and *8.12.* Explain why crossing over between nonsister chromatids of homologous chromosomes cannot occur during mitosis. For more help, refer to *Thinking Critically* in the **Skill Handbook.**

10.2 MEIOSIS **279**

Section Assessment

1. The chromosome number in a cell at the end is half the chromosome number in a parent cell. The original cell has a diploid number of chromosomes and each of the new cells has a haploid number.
2. The reduction of chromosome numbers by half allows for the return to the constant chromosome number

when a zygote is formed at fertilization.
3. Crossing over as well as the reassortment of the 46 chromosomes both contribute to the large number of phenotypes that are possible.
4. 15
5. After meiosis, only one member of each homologous chromosome pair

can be found in a gamete. Thus, no gamete will end up with two homologues. Alleles on different chromosomes will sort independently from one another.
6. Tetrad formation does not occur during mitosis. This prevents crossing over from taking place.

INTERNET BioLab

How can phenotypes and genotypes of plants be determined?

Time Allotment

Initial session: one class period; follow-up session: 5 minutes each day for watering, 20 minutes on last day for counting

Process Skills

hypothesize, observe and infer, collect data

Safety Precautions

Some seed materials are poisonous. Do not allow students to eat the seeds. Have students wash their hands after the lab.

It's difficult to predict the traits of plants if all that you see is their seeds. But if these seeds are planted and allowed to grow, certain traits will appear. By observing these traits, you might be able to determine the possible phenotypes and genotypes of the parent plants that produced these seeds. In this lab, you will determine the genotypes of plants that grow from two groups of tobacco seeds. Each group of seeds came from different parents. Plants will be either green or albino (white) in color. Use the following genotypes for this cross. CC = green, Cc = green, and cc = albino

PREPARATION

Alternative Materials

Seeds can be germinated and observed in petri dishes. This will eliminate the need for soil, flats, or pots. Place seeds on moistened paper towels in the bottom of the dish and keep the dish covered.

Possible Hypotheses

- If the parent plants were true breeding for green color, then all offspring will be green.
- If the parent plants were heterozygous for green color, then offspring will show an approximate ratio of 3 green to 1 white.

ANALYZE AND CONCLUDE

1. Leaf color cannot be observed in the seed but appears only after the plant has emerged from the seed.
2. The gene for green color is a dominant trait.
3. No, one parent may have been true breeding for green (*CC*), the other may have been heterozygous (*Cc*). This would have yielded all green offspring. Offspring may be *CC* or *Cc* but still appear green.

PREPARATION

Problem

Can the phenotypes and genotypes of the parent plants that produced two groups of seeds be determined from the phenotypes of the plants grown from the seeds?

Hypotheses

Have your group agree on a hypothesis to be tested that will answer the problem question. Record your hypothesis.

Objectives

In this BioLab, you will:

- **Analyze** the results of growing two groups of seeds.
- **Draw conclusions** about phenotypes and genotypes based on those results.
- **Use the Internet** to collect and compare data from other students.

Possible Materials

potting soil
small flowerpots or seedling flats
two groups of tobacco seeds
hand lens
light source
thermometer
plant-watering bottle

Safety Precautions

Always wash your hands after handling plant materials. Always wear goggles in the lab.

Skill Handbook

Use the **Skill Handbook** if you need additional help with this lab.

PLAN THE EXPERIMENT

Teaching Strategies

- When supplying seeds to students, make sure that the two types are kept separate from each other. Stick seeds onto a piece of tape for dispensing.
- An ideal quantity of seeds to use is 20-30 per type.
- Cotyledons (seed leaves) will appear after about 10 days.

Possible Procedures

- Students should keep growing conditions for the two seed groups as constant as possible. Soil should be kept moist at all times. Natural window light should be sufficient.
- Seeds should be planted about 1 cm below the soil. Planting is easier if the soil is moist.

PLAN THE EXPERIMENT

1. Examine the materials provided by your teacher. As a group, make a list of the possible ways you might test your hypothesis.
2. Agree on one way that your group could investigate your hypothesis.
3. Design an experiment that will allow you to collect quantitative data. For example, how many plants do you think you will need to examine?
4. Prepare a numbered list of directions. Include a list of materials and the quantities you will need.
5. Make a data table for recording your observations.

Check the Plan
1. Carefully determine what data you are going to collect. How many seeds do you think you will need? How long will you carry out the experiment?
2. What variables, if any, will have to be controlled? (Hint: Think about the growing conditions for the plants.)
3. *Make sure your teacher has approved your experimental plan before you proceed.*
4. Carry out your experiment. Make any needed observations, such as the numbers of green and albino plants in each group, and complete your data table.
5. Go to the Glencoe Science Web Site at the address shown below to **post your data.**

4. Yes, both parents must be heterozygous to yield a ratio of 3 green to 1 albino. For the offspring genotypes, you can conclude only that the albino offspring are *cc*. Green are either *CC* or *Cc*.
5. Answers may vary. Genetic ratios are governed by the laws of probability. The larger the population size, the closer the calculated value will be to the theoretical.

✓ Assessment

Portfolio Ask students to make diagrams that show the parental and offspring genotypes and phenotypes for both groups of seeds used in this experiment. Use the Performance Task Assessment List for Scientific Drawing in **PASC**, p. 55. **L2** **P**

ANALYZE AND CONCLUDE

1. **Thinking Critically** Why was it necessary to grow plants from the seeds in order to determine the phenotypes of the plants that formed the seeds?
2. **Drawing Conclusions** Using the information in the introduction, describe how the gene for green color (*C*) is inherited.
3. **Making Inferences** For the group of seeds that yielded all green plants, are you able to determine exactly the genotypes of the parents that formed these seeds? Can you determine the genotype of each plant observed? Explain.
4. **Making Inferences** For the group of seeds that yielded some green and some albino plants, are you able to determine exactly the genotypes of the plants that formed these seeds? Can you determine the genotype of each plant observed? Explain.
5. **Using the Internet** Compare your experimental design with that of other students. Were your results similar? What might account for the differences?

Sharing Your Data

BIOLOGY Online Find this BioLab on the Glencoe Science Web site at **science.glencoe.com.** Briefly describe your experimental design. Post your results in the table provided.

Sharing Your Data

BIOLOGY Online To navigate to the Internet BioLabs choose the *Biology: The Dynamics of Life* icon at Glencoe's Web site. Click on the student site icon, then the BioLabs icon. Students should go to the Glencoe Science Web site only after they have done the experiment and collected their data to post their experimental design and compare it with that of other students. Students should be careful that only data from identical crosses are pooled.

■ Be sure students mark the seed type planted in the flat or pot. Popsicle sticks can serve as markers.

Data and Observations

Seeds that came from true breeding plants will produce plants that are all green. Seeds from heterozygous parents will produce both green and albino seedlings in the ratio of about 3 green to 1 albino. Have students review Problem-Solving Lab 10-1 for help in calculating the ratio of green to albino plants.

Purpose 🔖

Students will gain insight into the importance of mathematics to the study of biology. They will learn how mathematics aided Mendel in understanding the laws of heredity.

Teaching Strategies

■ Students may be surprised to learn that a biologist of Charles Darwin's stature missed a great opportunity in his brilliant career. Had he interpreted his data about snapdragon flower shapes correctly, the whole world might have understood the laws of heredity 40 years earlier.

■ Discuss how ratios helped Mendel see that definite factors were being passed on from parents to offspring. He didn't know what these factors were, nor how they operated. He knew nothing about chromosomes and meiosis, yet was able to show how traits were transmitted because of his mathematical analysis.

Connection to Biology

Students may say that the ratios revealed that a dominant trait showed up three times more often because it was always able to overcome the effect of a recessive trait that accompanied it.

A Solution from Ratios

Gregor Mendel was an Austrian monk who experimented with garden peas. In 1866, he published the results of eight years of experiments. His work was ignored until 1900, when it was rediscovered.

Mendel had three qualities that led to his discovery of the laws of heredity. First, he was curious, impelled to find out why things happened. Second, he was a keen observer. Third, he was a skilled mathematician. Mendel was the first biologist who relied heavily on statistics for solutions to how traits are inherited.

Darwin missed his chance

About the same time that Mendel was carrying out his experiments with pea plants, Charles Darwin was gathering data on snapdragon flowers. When Darwin crossed plants that had normal-shaped flowers with plants that had odd-shaped flowers, all the offspring had normal-shaped flowers. He thought the two traits had blended. When he allowed the F_1 plants to self-pollinate, his results were 88 plants with normal-shaped flowers and 37 plants with odd-shaped flowers. Darwin was puzzled by the results and did not continue his studies with these plants. Lacking Mendel's statistical skills, Darwin failed to see the significance of the ratio of normal-shaped flowers to odd-shaped flowers in the F_2 generation. What was this ratio? Was this ratio similar to Mendel's ratio of dominant to recessive traits in pea plants?

Finding the ratios for four other traits

Figure 10.3 on page 262 shows seven traits that Mendel studied in pea plants. You have already looked at Mendel's data for plant height and seed shape. Now use the data for seed color, flower position, pod color, and pod shape to find the ratios of dominant to recessive for these traits in the F_2 generation.

Draw *Table B* in your notebook or journal. Calculate the ratios for the data in Table A and complete Table B by following these steps:

- Step 1 Divide the larger number by the smaller number.
- Step 2 Round to the nearest hundredth.
- Step 3 To express your answer as a ratio, write the number from step 2 followed by a colon and the number *1*.

Table A Mendel's Results

Seed Color	Flower Position	Pod Color	Pod Shape
Yellow 6022	Lateral 651	Green 428	Inflated 882
Green 2001	Terminal 207	Yellow 152	Constricted 299

Table B Calculating Ratios for Mendel's Results

	Seed Color	Flower Position	Pod Color	Pod Shape
Calculation	$\frac{6022}{2001} = 3.00$			
Ratio	3:1 yellow:green			

CONNECTION TO BIOLOGY

Why were ratios so important in understanding how dominant and recessive traits are inherited?

 **BIOLOGY** *Online* — To find out more about Mendel's work, visit the Glencoe Science Web site. **science.glencoe.com**

Main Ideas

Summary statements can be used by students to review the major concepts of the chapter.

SUMMARY

Section 10.1

Mendel's Laws of Heredity

Main Ideas

- Genes are located on chromosomes and exist in alternative forms called alleles. A dominant allele can mask the expression of a recessive allele.
- When Mendel crossed pea plants differing in one trait, one form of the trait disappeared until the second generation of offspring. To explain his results, Mendel formulated the law of segregation.
  - Mendel formulated the law of independent assortment to explain that two traits are inherited independently.
  - Events in genetics are governed by the laws of probability.

Vocabulary

allele (p. 262)
dominant (p. 262)
fertilization (p. 259)
gamete (p. 259)
genetics (p. 259)
genotype (p. 264)
heredity (p. 259)
heterozygous (p. 265)
homozygous (p. 264)
hybrid (p. 261)
law of independent assortment (p. 266)
law of segregation (p. 263)
phenotype (p. 264)
pollination (p. 259)
recessive (p. 262)
trait (p. 259)

Section 10.2

Meiosis

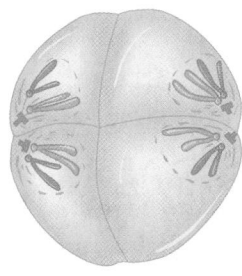

Main Ideas

- In meiosis, one diploid (2n) cell produces four haploid (n) cells, providing a way for offspring to have the same number of chromosomes as their parents.
- Mendel's results can be explained by the distribution of chromosomes during meiosis.
- Random assortment and crossing over during meiosis provide for genetic variation among the members of a species.
- Mistakes in meiosis may result from nondisjunction, the failure of chromosomes to separate properly during cell division.

Vocabulary

crossing over (p. 272)
diploid (p. 269)
egg (p. 271)
genetic recombination (p. 276)
haploid (p. 269)
homologous chromosome (p. 270)
meiosis (p. 271)
nondisjunction (p. 278)
sexual reproduction (p. 272)
sperm (p. 271)
zygote (p. 271)

Main Ideas

Summary statements can be used by students to review the major concepts of the chapter.

Using the Vocabulary

To reinforce chapter vocabulary, use the Content Mastery Booklet and the activities in the Interactive Tutor for Biology: The Dynamics of Life on the Glencoe Science Web site.
science.glencoe.com

 THE PRINCETON REVIEW *All Chapter Assessment questions and answers have been validated for accuracy and suitability by The Princeton Review.*

UNDERSTANDING MAIN IDEAS

1. a
2. d
3. b
4. a

UNDERSTANDING MAIN IDEAS

1. An organism that is true breeding for a trait is said to be _____.
   a. homozygous   **c.** a monohybrid
   b. heterozygous   **d.** a dihybrid

2. At the end of meiosis, how many haploid cells have been formed from the original cell?
   a. one   **c.** three
   b. two   **d.** four

3. When Mendel transferred pollen from one pea plant to another, he was _____ the plants.
   a. self-pollinating   **c.** self-fertilizing
   b. cross-pollinating   **d.** cross-fertilizing

4. A short pea plant is _____.
   a. homozygous recessive
   b. homozygous dominant
   c. heterozygous
   d. a dihybrid

CHAPTER 10 ASSESSMENT **283**

 *GLENCOE* **TECHNOLOGY**

VIDEOTAPE
MindJogger Videoquizzes
Chapter 10: *Mendel and Meiosis*
Have students work in groups as they play the videoquiz game to review key chapter concepts.

 **Resource Manager**

Chapter Assessment, pp. 55-60
MindJogger Videoquizzes
ExamView® Pro Software 💾
BDOL Interactive CD-ROM, Chapter 10 quiz

5. d
6. c
7. a
8. d
9. d
10. c
11. 3:1
12. recessive
13. heterozygous
14. nondisjunction
15. 23
16. homologous chromosomes
17. telophase I
18. recessive
19. haploid
20. gametes

APPLYING MAIN IDEAS

21. Like plants, humans reproduce sexually, have chromosomes and genes, and have traits controlled by genes.
22. The likelihood that close relatives share the same recessive genes is greater than in the general population, thus raising the risk of a child being homozygous for those traits.
23. 50%; the probability of any one child being a certain sex is unaffected by the birth of previous children.
24. The order of lining up at the equator during metaphase I of meiosis will vary, thus providing additional variation when the chromosomes separate during anaphase I.

5. Which of these shows a dominant trait in garden peas?

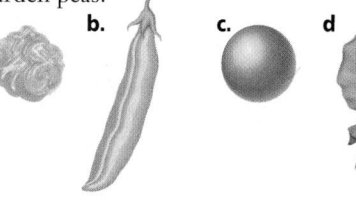

a.   b.   c.   d

6. During what phase of meiosis do sister chromatids separate?
   a. prophase I   c. anaphase II
   b. telophase I   d. telophase II
7. During what phase of meiosis do homologous chromosomes cross over?
   a. prophase I   c. telophase I
   b. anaphase I   d. telophase II
8. Recessive traits appear only when an organism is _____.
   a. mature
   b. different from its parents
   c. heterozygous
   d. homozygous
9. Mendel's use of peas was a good choice for genetic study because _____.
   a. they produce many offspring
   b. they are easy to grow
   c. they can be self-pollinated
   d. all of the above
10. A dihybrid cross between two heterozygous parents produces a phenotypic ratio of _____.
   a. 3 : 1   c. 9 : 3 : 3 : 1
   b. 1 : 2 : 1   d. 1 : 6 : 9
11. If two heterozygous organisms for a single dominant trait mate, the ratio of their offspring should be about _____.

THE PRINCETON REVIEW   **TEST-TAKING TIP**

Use the Buddy System
Study in a group. A small gathering of people works well because it allows you to draw from a broader base of skills and expertise. Keep it small and keep on target.

12. A trait that is hidden in the heterozygous condition is said to be a _____ trait.
13. An organism that has two different alleles for a trait is called _____.
14. The process that results in Down syndrome is called _____.
15. If a species normally has 46 chromosomes, the cells it produces by meiosis will each have _____ chromosomes.
16. Metaphase I of meiosis occurs when _____ line up next to each other at the cell's equator.
17. The stage of meiosis shown here is _____.

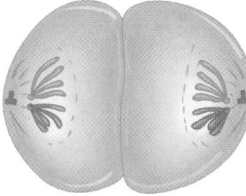

18. In the first generation of Mendel's experiments with a single trait, the _____ trait disappeared, only to reappear in the next generation.
19. A cell that has successfully completed meiosis has a chromosome number called _____.
20. Meiosis results in the direct production of _____.

APPLYING MAIN IDEAS

21. Why do you think Mendel's results are also valid for humans?
22. On the average, each human has about six recessive alleles that would be lethal if expressed. Why do you think that human cultures have laws against marriage between close relatives?
23. Assume that a couple has four children who are all boys. What are the chances that their next child will also be a boy? Explain your answer.
24. How does separation of homologous chromosomes during anaphase I of meiosis increase variation among offspring?

25. Relating to the methods of science, why do you think it was important for Mendel to study only one trait at a time during his experiments?

26. Observing and Inferring Why is it possible to have a family of six girls and no boys, but extremely unlikely that there will be a public school with 500 girls and no boys?

27. Comparing and Contrasting Compare metaphase of mitosis with metaphase I of meiosis.

28. Recognizing Cause and Effect Why is it sometimes impossible to determine the genotype of an organism that has a dominant phenotype?

29. Observing and Inferring While examining a cell in prophase I of meiosis, you observe a pair of homologous chromosomes pairing tightly. What is the significance of the places at which the chromosomes are joined?

30. Concept Mapping Complete the concept map by using the following vocabulary terms: recessive, zygote, homozygous, fertilization, heterozygous.

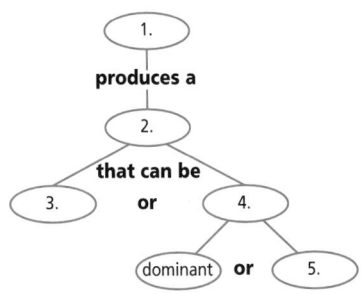

CD-ROM

For additional review, use the assessment options for this chapter found on the *Biology: The Dynamics of Life Interactive CD-ROM* and on the Glencoe Science Web site.
science.glencoe.com

ASSESSING KNOWLEDGE & SKILLS

In fruit flies, the allele for long wings is dominant to the allele for short wings.

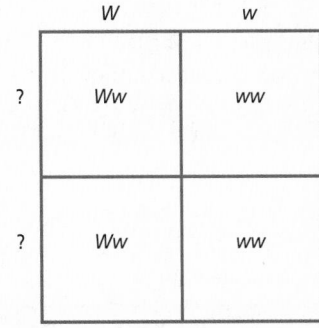

Interpreting Data Study the Punnett square and answer the following questions.

1. What term is given to the parent fly whose genotype is shown?
  a. heterozygous   **c.** recessive
  b. homozygous   **d.** haploid

2. What is the phenotype of each parent?
  a. both dominant
  b. both recessive
  c. one dominant and one recessive
  d. unable to tell

3. What is the genotype of each parent?
  a. *WW—Ww*   **c.** *Ww—Ww*
  b. *Ww—ww*   **d.** *WW—WW*

4. What are the phenotypes of the offspring?
  a. all long wings
  b. all short wings
  c. mostly long wings
  d. half short and half long

5. Interpreting Data Suppose the fruit fly parents in the Punnett square above were both heterozygous for an eye color trait in which *R* is red and *r* is white. What genotypes appear in the offspring? What fraction of the offspring will have short wings and white eyes?

25. Controlled experiments, such as Mendel's, require that no more than one variable be manipulated at a time. By doing many experiments, Mendel was able to determine the principles that govern genetics. These principles would not have been as easily observable without controlled experiments.

THINKING CRITICALLY

26. The probability of a family with six girls is $(1/2)^6$, but the probability of an entire school of girls would be $(1/2)^{500}$.

27. In metaphase of mitosis, all of the chromosomes align randomly at the equator of the cell. In metaphase I of meiosis, pairs of homologous chromosomes align at the equator of the cell.

28. If the dominant allele completely masks the recessive allele, you cannot tell if an organism with the dominant trait is homozygous or heterozygous for the dominant allele.

29. These are the places where crossing over takes place and chromosomal material is exchanged.

30. 1. Fertilization; 2. Zygote; 3. Heterozygous; 4. Homozygous; 5. Recessive

ASSESSING KNOWLEDGE & SKILLS

1. a
2. c
3. b
4. d
5. Genotypes include *WwRr*, *WwRR*, *Wwrr*, *wwRr*, *wwRR*, *wwRr*. The genotype of a fruit fly with short wings and white eyes is *wwrr*. Two of the sixteen possible outcomes, 1/8, in this Punnett square are *wwrr*.

Chapter 11 Organizer

Refer to pages 4T-5T of the Teacher Guide for an explanation of the National Science Education Standards correlations.

Section	Objectives	Activities/Features
Section 11.1 **DNA: The Molecule of Heredity** National Science Education Standards UCP.1-3, UCP.5; A.1, A.2; B.2, B.3; C.2, C.5; G.1-3 (2 sessions, 1 block)	1. **Analyze** the structure of DNA. 2. **Determine** how the structure of DNA enables it to reproduce itself accurately.	**Problem-Solving Lab 11-1,** p. 289 **Inside Story:** Copying DNA, p. 292
Section 11.2 **From DNA to Protein** National Science Education Standards UCP.1-3, UCP.5; A.1, A.2; B.2, B.3; C.1, C.2 (2 sessions, 2 blocks)	3. **Relate** the concept of the gene to the sequences of nucleotides in DNA. 4. **Sequence** the steps involved in protein synthesis.	**Problem-Solving Lab 11-2,** p. 297 **MiniLab 11-1:** Transcribe and Translate, p. 299 **Investigate BioLab:** RNA Transcription, p. 308
Section 11.3 **Genetic Changes** National Science Education Standards UCP.1-3; A.1, A.2; B.3; C.1, C.2; E.1, E.2; F.1, F.4, F.5; G.1, G.2 (2 sessions, 1 block)	5. **Categorize** the different kinds of mutations that can occur in DNA. 6. **Compare** the effects of different kinds of mutations on cells and organisms.	**Careers in Biology:** Genetic Counselor, p. 303 **Problem-Solving Lab 11-3,** p. 305 **MiniLab 11-2:** Gene Mutations and Proteins, p. 306 **BioTechnology:** Scanning Probe Microscopes, p.310

Need Materials? Contact Carolina Biological Supply Company at 1-800-334-5551 or at **http://www.carolina.com**

MATERIALS LIST

BioLab
p. 308 construction paper (5 colors), scissors, transparent tape, pencil

MiniLabs
p. 299 pencil, paper, Table 11.2
p. 306 pencil, paper, Table 11.2

Alternative Lab
p. 294 scissors, yarn (2 skeins), large box

Quick Demos
p. 288 large zipper
p. 291 large zipper (2)
p. 295 students will supply
p. 304 coiled telephone cord, twist ties

Key to Teaching Strategies

L1 Level 1 activities should be appropriate for students with learning difficulties.

L2 Level 2 activities should be within the ability range of all students.

L3 Level 3 activities are designed for above-average students.

ELL ELL activities should be within the ability range of English Language Learners.

COOP LEARN Cooperative Learning activities are designed for small group work.

P These strategies represent student products that can be placed into a best-work portfolio.

These strategies are useful in a block scheduling format.

Teacher Classroom Resources

Section	Reproducible Masters	Transparencies
Section 11.1 **DNA: The Molecule of Heredity**	Reinforcement and Study Guide, p. 47 L2 Concept Mapping, p. 11 L3 ELL Laboratory Manual, pp. 75-78 L2	Section Focus Transparency 26 L1 ELL Basic Concepts Transparency 16 L2 ELL
Section 11.2 **From DNA to Protein**	Reinforcement and Study Guide, pp. 48-49 L2 Critical Thinking/Problem Solving, p. 11 L3 BioLab and MiniLab Worksheets, pp. 49-50 L2 Content Mastery, pp. 53, 55-56 L1	Section Focus Transparency 27 L1 ELL Basic Concepts Transparency 17 L2 ELL Basic Concepts Transparency 18 L2 ELL Reteaching Skills Transparency 18 L1 ELL
Section 11.3 **Genetic Changes**	Reinforcement and Study Guide, p. 50 L2 BioLab and MiniLab Worksheets, pp. 51-56 L2 Laboratory Manual, pp. 79-82 L2 Content Mastery, pp. 53-55, 56 L1	Section Focus Transparency 28 L1 ELL Reteaching Skills Transparency 19a, 19b L1 ELL

Assessment Resources

Chapter Assessment, pp. 61-66
MindJogger Videoquizzes
Performance Assessment in the Biology Classroom
Alternate Assessment in the Science Classroom
ExamView® Pro Software
BDOL Interactive CD-ROM, Chapter 11 quiz

Additional Resources

Spanish Resources ELL
English/Spanish Audiocassettes ELL
Cooperative Learning in the Science Classroom COOP LEARN
Lesson Plans/Block Scheduling

NATIONAL GEOGRAPHIC — Teacher's Corner

Products Available From Glencoe
To order the following products, call Glencoe at 1-800-334-7344:
Curriculum Kit
GeoKit: Cells and Microorganisms

Products Available From National Geographic Society
To order the following products, call National Geographic Society at 1-800-368-2728:
Video
DNA: Laboratory of Life

Index to National Geographic Magazine
The following articles may be used for research relating to this chapter.
"The Rise of Life on Earth," by Richard Monastersky, March 1998.
"DNA Profiling: The New Science of Identity," by Cassandra Franklin-Barbajosa, May 1992.
"Beyond Supermouse: Changing Life's Genetic Blueprint," by Robert F. Weaver, December 1984.

GLENCOE TECHNOLOGY

The following multimedia resources are available from Glencoe.

Biology: The Dynamics of Life
CD-ROM ELL

Animation: *DNA Replication*
Animation: *Transcription*
Animation: *Translation*
Exploration: *Mutations*
BioQuest: *Building a Protein*

Videodisc Program

DNA Replication
DNA Transcription
Translation

The Infinite Voyage

Unseen Worlds

11 DNA and Genes

Show students photographs of other fruit fly mutations such as various eye colors and wing shapes. Ask students to share ways that the flies could have received the mutations. Explain that various environmental factors, such as radiation and chemicals, can cause mutations. Guide the discussion to the concept of the DNA molecule, which serves as the blueprint for life.

Theme Development

The first section of this chapter stresses **homeostasis**, or stability. DNA contains the blueprints for life, and replication processes make exact copies of these blueprints. The second section illustrates **unity within diversity;** the process and the code by which a cell makes proteins are the same in all species. The third section illustrates how **homeostasis** can be disrupted when mutations occur. Change in the DNA can be harmful, but changes are responsible for the **evolution** of a species.

Glencoe's *Biology: The Dynamics of Life* contains many resources to assist a student's reading skills. Each chapter contains figures with expanded captions that expand on written material. Word Origins, located along the side of text, expand knowledge of biology vocabulary. Glencoe's Content Mastery Booklet helps develop reading skills while reinforcing content. In addition, use the Interactive Tutor for *Biology: The Dynamics of Life* on the Glencoe Web site to reinforce vocabulary.
science.glencoe.com

What You'll Learn
- You will relate the structure of DNA to its function.
- You will explain the role of DNA in protein production.
- You will distinguish among different types of mutations.

Why It's Important
An understanding of birth defects, viral diseases, cancer, aging, genetic engineering, and even criminal investigations depends upon knowing about DNA, how it holds information, and how it plays a role in protein production.

Scan the vocabulary words listed in the Section Preview at the start of each section. Note familiar words. Make a list of current events issues you may have heard about that involve genetics, DNA, or cloning. As you read the chapter, refer back to your list to add notes from the material.

BIOLOGY
Online

To find out more about DNA and genes, visit the Glencoe Science Web site.
science.glencoe.com

The appearance of these two flies depends on the type of genes they contain. Chromosomes, made of genes, which are made of DNA, determine how an organism looks and how it functions.

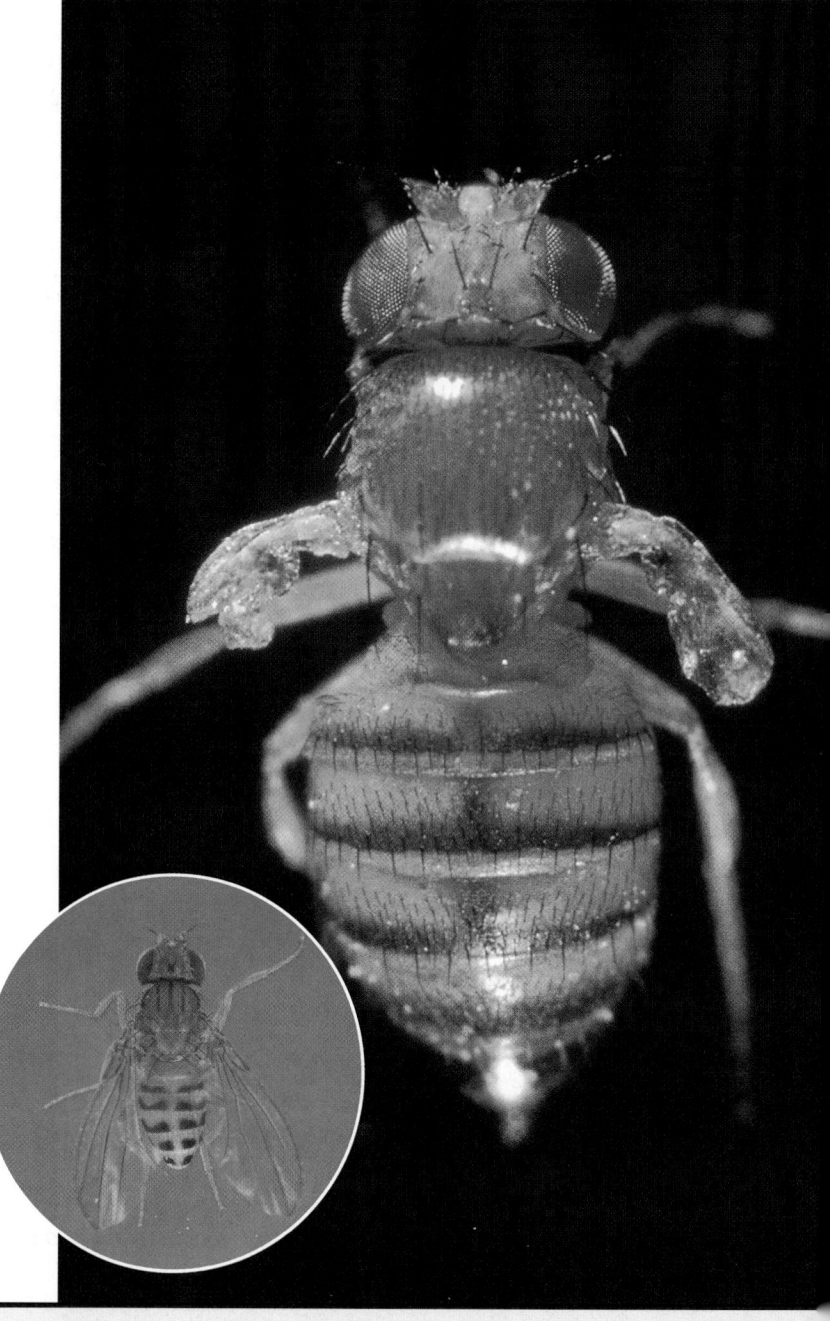

Multiple Learning Styles

Look for the following logos for strategies that emphasize different learning modalities.

Kinesthetic Project, pp. 288, 300; Reteach, p. 293; Quick Demo, p. 295; Extension, p. 301

Visual-Spatial Quick Demo, p. 291; Project, p. 292; Reinforcement, p. 298; Reteach, p. 301; Meeting Individual Needs, p. 304

Interpersonal Tech Prep, p. 303

Intrapersonal Enrichment, pp. 288, 296; Biology Journal, p. 306; Going Further, p. 310

Linguistic Portfolio, pp. 289, 300, 305; Enrichment, p. 290; Biology Journal, pp. 291, 299, 302

Logical-Mathematical Enrichment, p. 303; Reinforcement, p. 304

11.1 DNA: The Molecule of Heredity

SECTION PREVIEW

Objectives

Analyze the structure of DNA.

Determine how the structure of DNA enables it to reproduce itself accurately.

Vocabulary

nitrogen base
double helix
DNA replication

Section 11.1

Can you imagine all of the information that could be contained in 1000 textbooks? Remarkably, that much information—and more—is carried by the genes of a single organism. Scientists have found that the substance DNA, contained in genes, holds this information. Because of the unique structure of DNA, new copies of the information can be easily reproduced.

Model of a DNA molecule

What is DNA?

Although the environment influences how an organism develops, the genetic information that is held in the molecules of DNA ultimately determines an organism's traits. DNA achieves its control by producing proteins. Living things contain proteins. Your skin contains protein, your muscles contain protein, and your bones contain protein mixed with minerals. All actions, such as eating, running, and even thinking, depend on proteins called enzymes. Enzymes are critical for an organism's function because they control the chemical reactions needed for life. Within the structure of DNA is the information for life—the complete instructions for manufacturing all the proteins for an organism.

The structure of DNA

DNA is capable of holding all this information because it is a very long molecule. Recall that DNA is a polymer made of repeating subunits called nucleotides. Nucleotides have three parts: a simple sugar, a phosphate group, and a nitrogen base. The simple sugar in DNA, called

11.1   DNA: THE MOLECULE OF HEREDITY   **287**

Prepare

Key Concepts

The structure and composition of DNA are presented. The process of replication of DNA and its importance to organisms are emphasized.

Planning

■ Collect photos of *Drosophila* mutations for the Getting Started Demo.

■ Purchase two large zippers for the Quick Demos.

■ Collect modeling clay, colored paperclips, and twist ties or pipe cleaners for the DNA Model Project.

■ Gather index cards and rubber bands for the Flip Books Project.

■ Make a set of cards for Reteach.

1 Focus

Bellringer

Before presenting the lesson, display **Section Focus Transparency 26** on the overhead projector and have students answer the accompanying questions.
L1 **ELL**

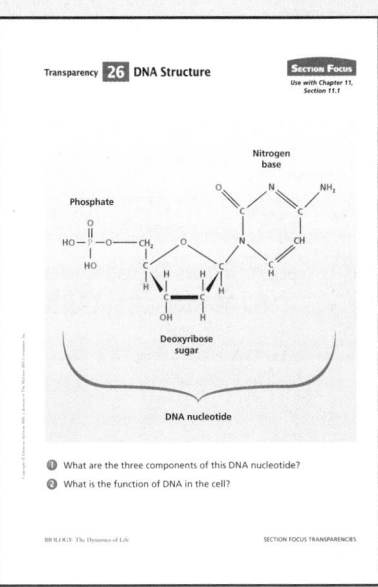

✔ **Assessment Planner**

Portfolio Assessment
   Portfolio, TWE, pp. 289, 300, 306
   BioLab, TWE, pp. 308-309
Performance Assessment
   Assessment, TWE, pp. 291, 304
   Problem-Solving Lab, TWE, p. 305
   MiniLab, TWE, p. 306
   MiniLab, SE, pp. 299, 306
   BioLab, SE, pp. 308-309
   Alternative Lab, TWE, pp. 294-295

Knowledge Assessment
   Problem-Solving Lab, TWE, pp. 289, 297
   Assessment, TWE, pp. 293, 300, 301, 307
   Section Assessment, SE, pp. 293, 301, 307
   Chapter Assessment, SE, pp. 311-313
Skill Assessment
   Alternative Lab, TWE, pp. 294-295
   MiniLab, TWE, p. 299

Discussion

Ask students whether they know why each of them is a unique individual. *Students may suggest that hereditary factors determine their uniqueness.* Why has there been no other human on Earth exactly like any of them? *Each individual has different DNA and thus different traits.*

Quick Demo

Hold up a large unzipped zipper and relate the zipper teeth to the nucleotides and the cloth band to the deoxyribose-phosphate backbone. Explain that in DNA, hydrogen bonds between the nucleotides hold the two strands together. Zip the zipper to show how the DNA looks when the nucleotides are hydrogen-bonded. Then twist the ends of the zipper to show how DNA is twisted into a helix.

Enrichment

 **Intrapersonal** Have students research one of the people involved in the discovery of the structure and function of DNA: Fred Griffith, O.T. Avery, Alfred Hershey, Linus Pauling, Martha Chase, Erwin Chargaff, Rosalind Franklin, or Maurice Wilkins. Have students prepare short oral reports on what these scientists contributed. **L3**

Resource Manager

Section Focus Transparency 26 and Master **L1** **ELL**

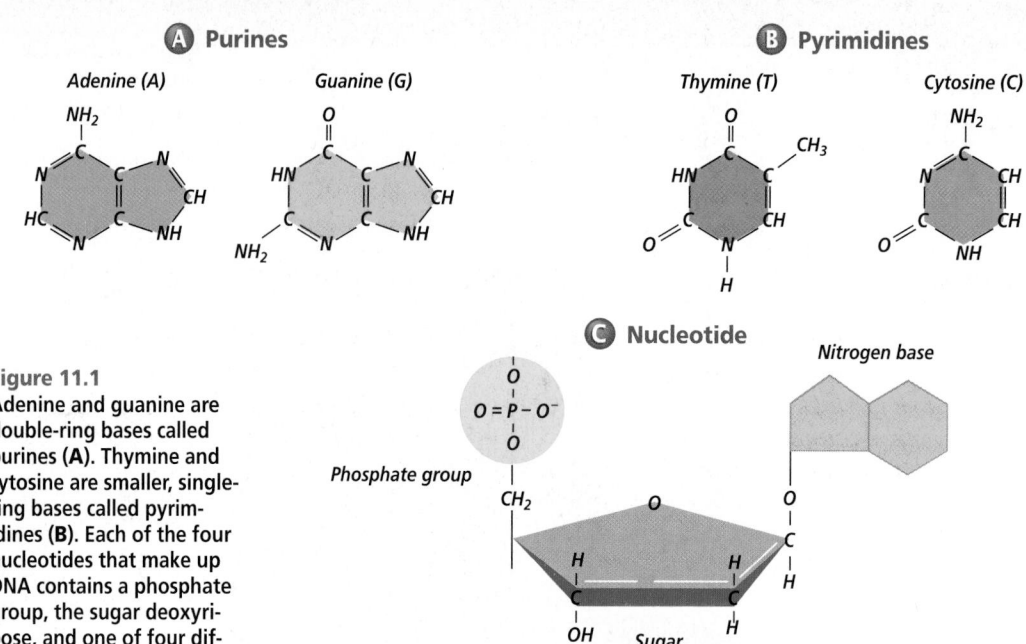

A Purines

Adenine (A)   Guanine (G)

B Pyrimidines

Thymine (T)   Cytosine (C)

C Nucleotide

Nitrogen base

Phosphate group

CH_2

Sugar (deoxyribose)

Figure 11.1
Adenine and guanine are double-ring bases called purines (**A**). Thymine and cytosine are smaller, single-ring bases called pyrimidines (**B**). Each of the four nucleotides that make up DNA contains a phosphate group, the sugar deoxyribose, and one of four different nitrogen bases (**C**).

deoxyribose (dee ahk sih RI bos), gives DNA its name—deoxyribonucleic acid. The phosphate group is composed of one atom of phosphorus surrounded by four oxygen atoms. A **nitrogen base** is a carbon ring structure that contains one or more atoms of nitrogen. In DNA, there are four possible nitrogen bases: adenine (A), guanine (G), cytosine (C), and thymine (T). Thus, in DNA there are four possible nucleotides, each containing one of these four bases, as shown in *Figure 11.1*.

Nucleotides join together to form long chains, with the phosphate group of one nucleotide bonding to the deoxyribose sugar of an adjacent nucleotide. The phosphate groups and deoxyribose molecules form the backbone of the chain, and the nitrogen bases stick out like teeth on a zipper. In DNA, the amount of adenine is always equal to the amount of thymine, and the amount of guanine is always equal to the amount of

cytosine. You can see this in the *Problem-Solving Lab* on the next page.

In 1953, James Watson and Francis Crick published a journal article that was only one page in length, yet monumental in importance. Watson and Crick proposed that DNA is made of two chains of nucleotides joined together by the nitrogen bases. Just as the teeth of a zipper hold the two sides of the zipper together, the nitrogen bases of the nucleotides hold the two strands of DNA together with weak hydrogen bonds. The two strands can be held together in this way because they are complementary to each other; that is, the bases on one strand determine the bases on the other strand. Specifically, adenine on one strand bonds with a thymine on the other strand, and guanine on one strand bonds with a cytosine on the other strand. These two bonded bases, called a complementary base pair, explain why adenine and

---

PROJECT

DNA Model

Kinesthetic Have student groups make two strands of modeling clay, each about 8 cm long and 1 cm thick. Each group will need five each of four different colors of paper clips to represent the DNA bases. Assign letters to the colors so that each group has 5 A-clips, 5 T-clips, 5 C-clips, and 5 G-clips. Students should poke half of the clips in a row into one of the clay strands. On the second clay strand, they should line up the complementary clips in the proper order (A = T, C = G). Have them connect the bases with pipe cleaners or twist ties. Once the model is bonded together, it can be twisted to suggest the double helix structure of DNA. **L1** **ELL** **COOP LEARN**

thymine are always present in equal amounts. Likewise, the guanine and cytosine base pairs result in equal amounts of these nucleotides in DNA. Watson and Crick also proposed that DNA is shaped like a long zipper that is twisted. When something is twisted like a coiled spring, the shape is called a helix. Because DNA is composed of two strands twisted together, its shape is called a **double helix.** This shape is shown in *Figure 11.2*.

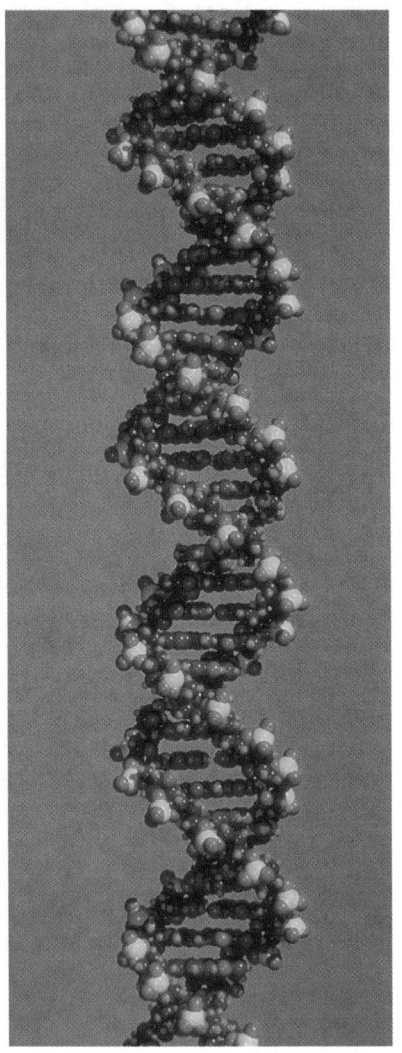

Figure 11.2
DNA normally exists in the shape of a double helix. This shape is similar to that of a twisted zipper.

Problem-Solving Lab 11-1 Interpreting the Data

What does chemical analysis reveal about DNA? Much of the early research on the structure and composition of DNA was done by carrying out chemical analyses. The data from these experiments provide evidence of a relationship among the nitrogen bases of DNA.

Analysis

Examine *Table 11.1*. Compare the amounts of adenine, guanine, cytosine, and thymine found in the DNA of each of the cells studied.

Table 11.1 Percent of each base in DNA samples				
Source of sample	A	G	C	T
Human liver	30.3	19.5	19.9	30.3
Human thymus	30.9	19.9	19.8	29.4
Herring sperm	27.8	22.2	22.6	27.5
Yeast	31.7	18.2	17.4	32.6

Thinking Critically

1. Compare the amounts of A, T, G, and C in each kind of DNA. Why do you think the relative amounts are so similar in human liver and thymus cells?
2. How do the relative amounts of each base in herring sperm compare with the relative amounts of each base in yeast?
3. What fact can you state about the overall composition of DNA, regardless of its source?

The importance of nucleotide sequences

An elm, an elk, and an eel are all different organisms composed of different proteins. If you compare the chromosomes of these organisms, you will find that they all contain DNA made up of nucleotides with adenine, thymine, guanine, and cytosine bases. How can organisms be so different from each other if their genetic material is made of the same four nucleotides? Their differences result

WORD *Origin*

helix
From the Latin word *helix*, meaning "spiral." A double helix has two twisted strands that form a spiral.

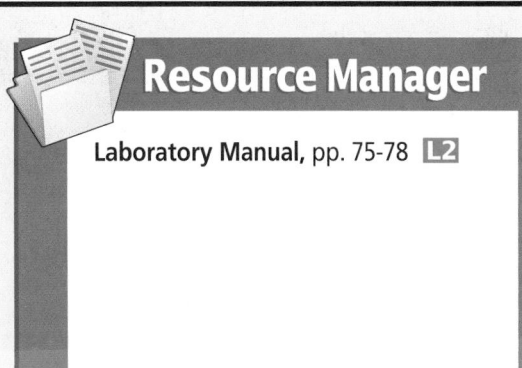
Concept Development

Point out to students that the words *elm, elk,* and *eel* differ from each other by only one letter but that the meaning of the words is drastically different. This observation can introduce the importance of base sequences.

Problem-Solving Lab 11-1

Purpose

Students will analyze a table showing the percentages of four bases in DNA samples.

Process Skills

compare and contrast, observe and infer, interpret data

Background

In 1950, American biochemist Erwin Chargaff first showed that there is a 1:1 ratio between adenine and thymine and between guanine and cytosine in all DNA.

Teaching Strategies

■ Guide students to compare the numbers within each sample before comparing different samples.

Thinking Critically

1. The ratio of A:T and of C:G is approximately 1:1. The relative amounts of each base are so similar in human liver and thymus because they come from the same species.
2. The relative amounts of each base are different in herring and in yeast because they are different species.
3. The ratio of A:T and G:C is 1:1 for all sources.

✓ Assessment

Knowledge Have students summarize their analysis of the data and write a paragraph explaining why the ratio of bases was important in determining the structure of DNA. Use the Performance Task Assessment List for Writing in Science in **PASC,** p. 87. **L2**

289

290

---

from the sequence of the four different nucleotides along the DNA strands, as you can see in *Figure 11.3*.

The sequence of nucleotides forms the unique genetic information of an organism. For example, a nucleotide sequence of A-T-T-G-A-C carries different information from a sequence of T-C-C-A-A-A. In a similar way, two six-letter words made of the same letters but arranged in different order have different meanings. The closer the relationship between two organisms, the greater the similarity in their order of DNA nucleotides. The DNA sequences of a chimpanzee are similar to those of a gorilla, but different from those of a rose bush. Scientists use nucleotide sequences to determine evolutionary relationships among organisms. Nucleotide sequences can also be used to determine whether two people are related, or whether the DNA from a crime scene matches the DNA of a suspected criminal.

Replication of DNA

A sperm cell and an egg cell of a fruit fly, both produced through meiosis, unite to form a fertilized egg. From one fertilized egg, a fruit fly with billions of cells is produced by the process of mitosis. Each cell has a copy of the DNA that was in the original fertilized egg. As you have learned, before a cell can divide by mitosis or meiosis, it must first make a copy of its chromosomes. The DNA in the chromosomes is copied in a process called **DNA replication.** Without DNA replication, new cells

Figure 11.3
The structure of DNA is shown here.

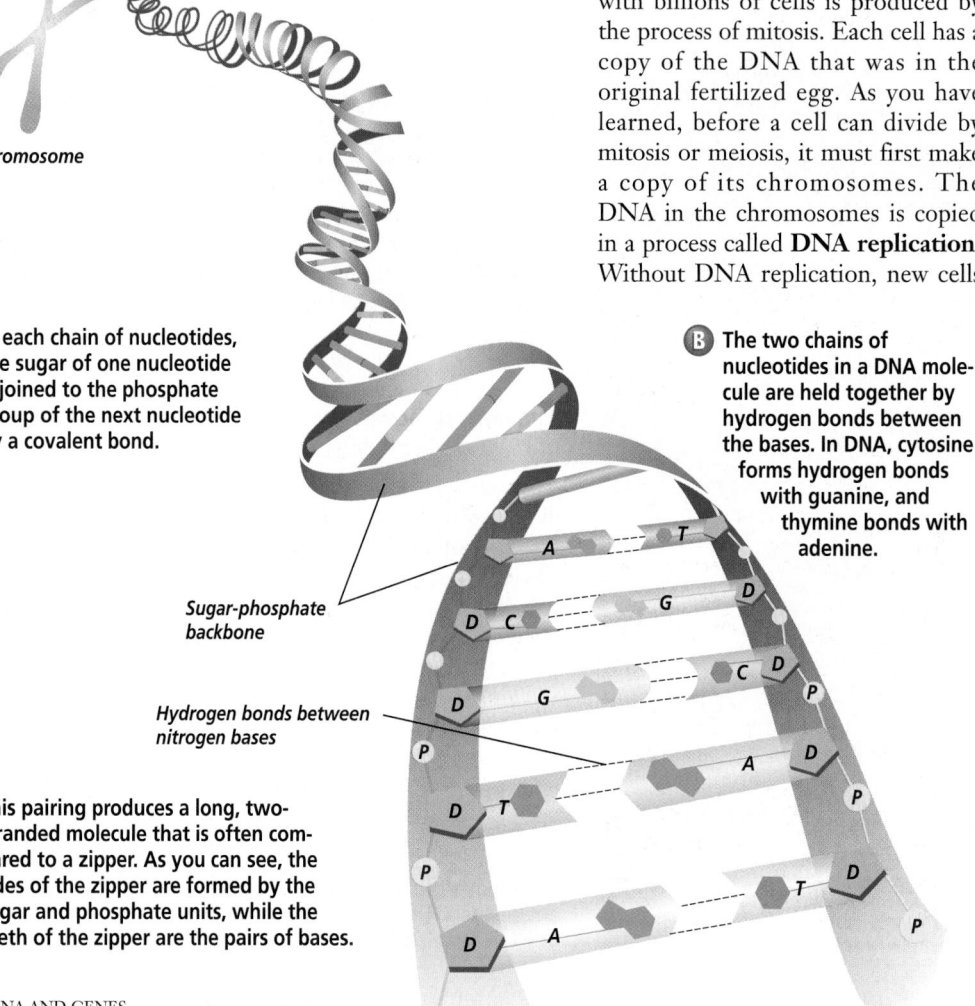

Chromosome

A In each chain of nucleotides, the sugar of one nucleotide is joined to the phosphate group of the next nucleotide by a covalent bond.

B The two chains of nucleotides in a DNA molecule are held together by hydrogen bonds between the bases. In DNA, cytosine forms hydrogen bonds with guanine, and thymine bonds with adenine.

Sugar-phosphate backbone

Hydrogen bonds between nitrogen bases

C This pairing produces a long, two-stranded molecule that is often compared to a zipper. As you can see, the sides of the zipper are formed by the sugar and phosphate units, while the teeth of the zipper are the pairs of bases.

290 DNA AND GENES

Figure 11.4
DNA replication produces two molecules from one.

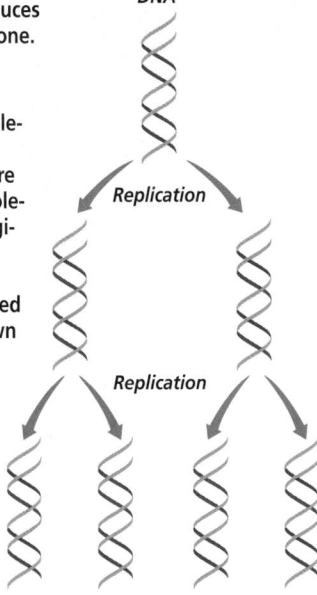

DNA

A When a DNA molecule replicates, two molecules are formed. Each molecule has one original strand and one new strand. Newly-synthesized strands are shown in red.

Replication

Replication

B This circular bacterial DNA is replicating. The photo shows two loops. The bottom loop is twisted into a figure-8 shape. Replication is taking place at the intersections of the two loops, as indicated by the arrows.

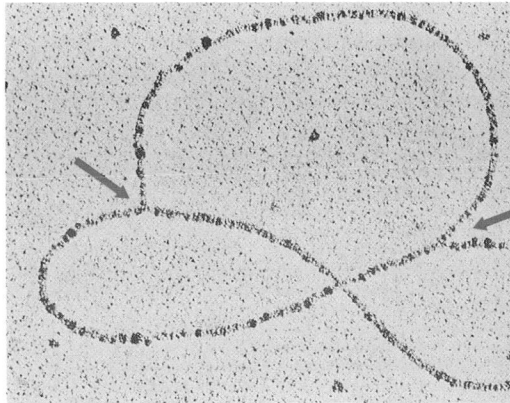

Magnification: 188 000×

would have only half the DNA of their parents. Species could not survive, and individuals could not grow or reproduce successfully. All organisms undergo DNA replication. *Figure 11.4B* shows bacterial DNA replicating.

How DNA replicates

You have learned that a DNA molecule is composed of two strands, each containing a sequence of nucleotides. As you know, an adenine on one strand pairs with a thymine on the other strand. Similarly, guanine pairs with cytosine. Therefore, if you knew the order of bases on one strand, you could predict the sequence of bases on the other, complementary strand. In fact, part of the process of DNA replication is done in just the same way. During replication, each strand serves as a pattern to make a new DNA molecule. How can a molecule serve as a pattern? Read the *Inside Story* on the next page to find out.

DNA replication begins as an enzyme breaks the hydrogen bonds between nitrogen bases that hold the two strands together, thus unzipping the DNA molecule. As the DNA continues to unzip, nucleotides that are floating free in the surrounding medium bond to the single strands by base pairing. Another enzyme bonds these new nucleotides into a chain.

This process continues until the entire molecule has been unzipped and replicated. Each new strand formed is a complement of one of the original, or parent, strands. The result is the formation of two DNA molecules, each of which is identical to the original DNA molecule.

When all the DNA in all the chromosomes of the cell has been copied by replication, there are two copies of the organism's genetic information. In this way, the genetic makeup of an organism can be passed on to new cells during mitosis or to new generations through meiosis followed by sexual reproduction.

Purpose

Students study the steps involved in the replication of DNA.

Teaching Strategies

■ Have students write a one-paragraph summary of the events in replication. **L2**

■ Have students write a single strand sequence and exchange it with a partner. The partner should write the complementary strand and return it to the first student to be checked. **L1 COOP LEARN**

Visual Learning

■ Have students draw the steps of replication as separate diagrams. **L1 ELL**

Critical Thinking

The most likely occurrence is that both daughter cells would be missing some of their chromosomes and would not be able to live. Another possibility is that one cell would get the complete set of chromosomes and could live and function. The other cell would get no chromosomes and would die.

Resource Manager

Basic Concepts Transparency 16 and Master **L2 ELL**
Reinforcement and Study Guide, p. 47 **L2**
Content Mastery, p. 54 **L1**

Copying DNA

DNA is copied during interphase prior to mitosis and meiosis. It is important that the new copies are exactly like the original molecules. The structure of DNA provides a mechanism for accurate copying of the molecule. The process of making copies of DNA is called DNA replication.

Critical Thinking *What would be the outcome if mitosis occurred before replication took place?*

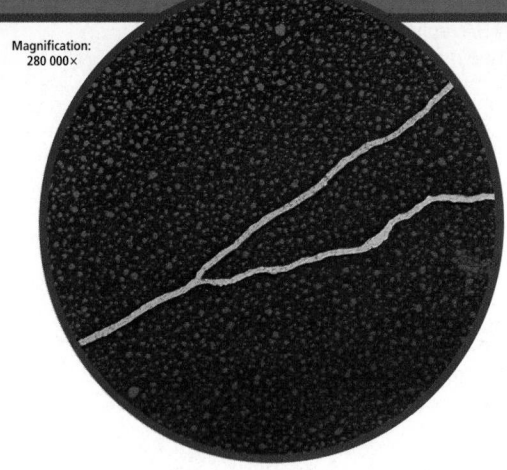

Magnification: 280 000×

Two molecules of DNA from one

1 **Separation of strands** When a cell begins to copy its DNA, the two nucleotide strands of a DNA molecule first separate at their base pairs when the hydrogen bonds connecting the base pairs are broken. As the DNA molecule unzips, the nucleotides are exposed.

Original DNA strand

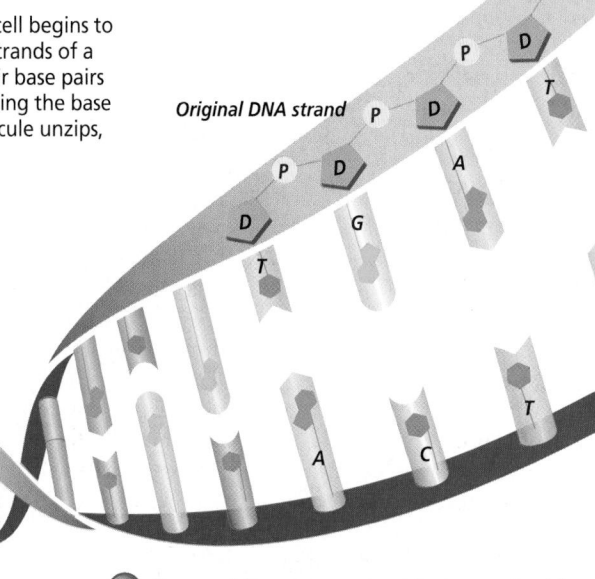

Original DNA

2 **Base pairing** Free nucleotides base pair with exposed nucleotides. If one nucleotide on a strand has thymine as a base, the free nucleotide that pairs with it would be adenine. If the strand contains cytosine, a free guanine nucleotide will pair with it. Thus, each strand builds its complement by base pairing with free nucleotides.

292 DNA AND GENES

PROJECT

Flip Books

 *Visual-Spatial* Have students create animated flip books showing replication by drawing scenes near the edges of a stack of index cards. They should show DNA unzipping, individual nucleotides bonding to the exposed strands, and the formation of two DNA molecules. The stack can be held together with a rubber band. Have students share their "films" with classmates. **L1 ELL**

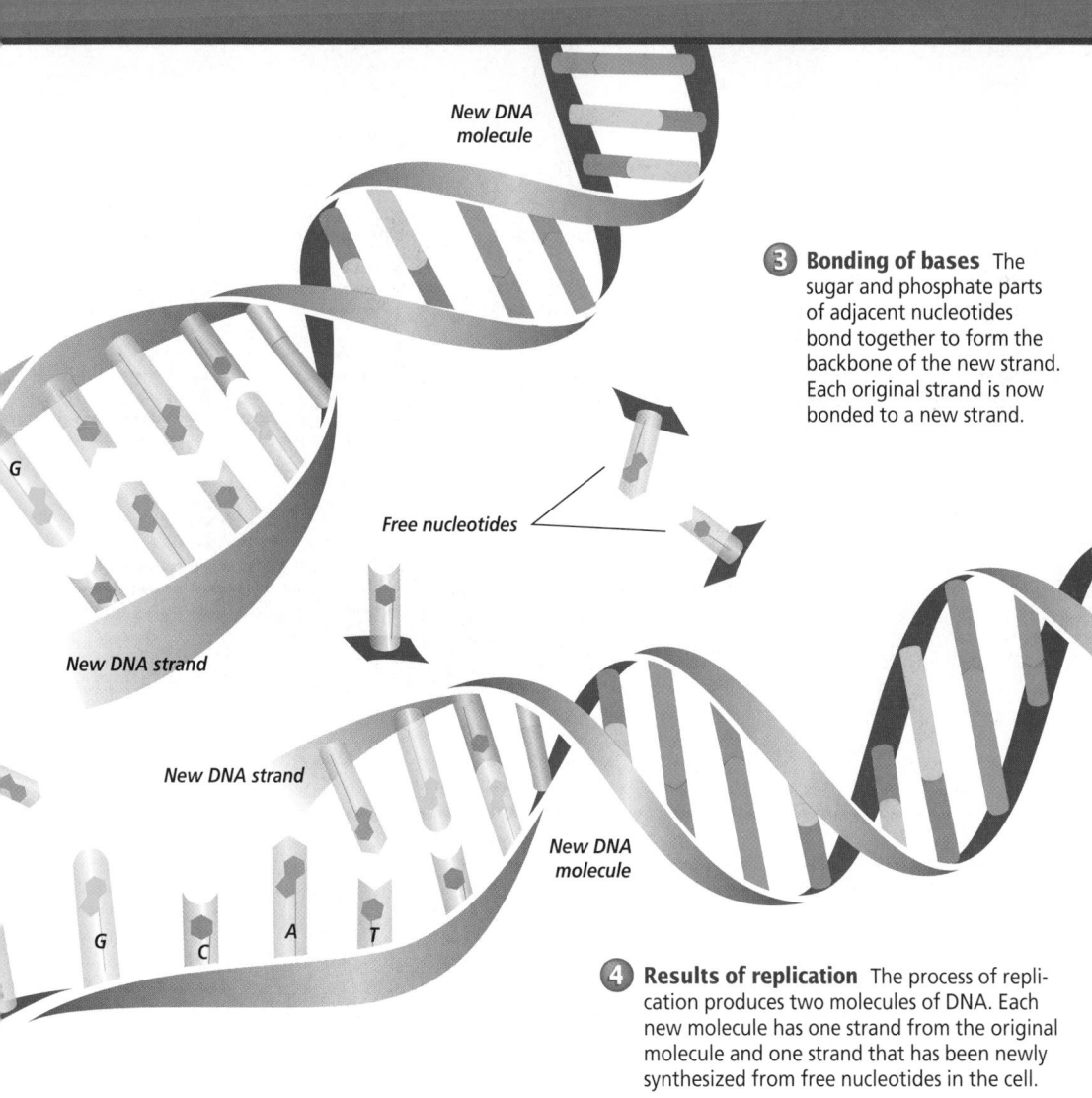

New DNA molecule

G

Free nucleotides

New DNA strand

New DNA strand

New DNA molecule

G   C   A   T

3 Bonding of bases The sugar and phosphate parts of adjacent nucleotides bond together to form the backbone of the new strand. Each original strand is now bonded to a new strand.

4 Results of replication The process of replication produces two molecules of DNA. Each new molecule has one strand from the original molecule and one strand that has been newly synthesized from free nucleotides in the cell.

Section Assessment

Understanding Main Ideas
1. Describe the structure of a nucleotide.
2. How do the nucleotides in DNA pair?
3. Explain why the structure of a DNA molecule is often described as a zipper.
4. How does DNA hold information?

Thinking Critically
5. The sequence of nitrogen bases on one strand of a DNA molecule is GGCAGTTCATGC. What would be the sequence of bases on the complementary strand?

SKILL REVIEW
6. **Sequencing** Sequence the steps that occur during DNA replication. For more help, refer to *Organizing Information* in the **Skill Handbook**.

Section Assessment

1. A nucleotide consists of a sugar, a phosphate group, and a nitrogen base.
2. Cytosine forms hydrogen bonds with guanine; thymine bonds with adenine.
3. The molecule is shaped like a twisted zipper with the sides formed by the sugar and phosphate molecules. The teeth of the zipper are the pairs of bases.
4. The information is held in the sequence of nucleotides.
5. CCGTCAAGTACG
6. The two strands separate at the base pairs. Free complementary nucleotides are attracted to those on the strands. Enzymes join the new nucleotides to form strands complementary to the original strands. Two new double-stranded molecules separate.

3 Assess

Check for Understanding
Have students each write two questions about the process of DNA replication. Collect the questions and use them to quiz the class. **L1**

Reteach
Kinesthetic Divide the class into four equal groups. Give each group a letter name—A, T, G, or C. Give each student a card with his or her letter. Write a base sequence for one strand of a DNA molecule on the chalkboard. Be sure to write as many bases as there are students. Give students two minutes to line up next to the appropriate complementary letter on the chalkboard. Tell them they have just replicated a DNA molecule. **L1** **ELL**

Extension
Have students research the work of Frederick Griffith in 1928 and Oswald Avery and his colleagues in 1944. These scientists demonstrated that DNA is the genetic material. **L3**

✓ Assessment
Knowledge Prepare a sheet showing a short section of two-stranded DNA. Ask students to diagram the steps this short section would go through in order to replicate. **L2**

4 Close

Discussion
Ask students how the DNA structure lends itself to replication. Why is accuracy so important in replication? **L2**

293

Prepare

Key Concepts

Students will learn how DNA, genes, and proteins are related. The relationship between genes and the nucleotide sequences in DNA is discussed. Finally, the steps involved in the formation of mRNA and the role of tRNA in translation are explained.

Planning

- Buy yarn and bring in a box for the Alternative Lab.
- Purchase or photocopy blank bingo cards for the Reinforcement.
- Purchase or gather five colors of construction paper for the BioLab.
- Locate a combination lock for the Meeting Individual Needs.

1 Focus

Bellringer 📖

Before presenting the lesson, display **Section Focus Transparency 27** on the overhead projector and have the students answer the accompanying questions. **L1** **ELL**

Transparency **27** Using Codes

SECTION FOCUS
Use with Chapter 11, Section 11.2

① How are these pieces of music similar?
② How do they differ? What is the result of this difference?

BIOLOGY: The Dynamics of Life   SECTION FOCUS TRANSPARENCIES

294

SECTION PREVIEW

Objectives
Relate the concept of the gene to the sequences of nucleotides in DNA.
Sequence the steps involved in protein synthesis.

Vocabulary
messenger RNA
ribosomal RNA
transfer RNA
transcription
codon
translation

Morse code was a method of communicating that was developed in the nineteenth century. This code used a pattern of dots and dashes to represent letters of the alphabet. In this way, long sequences of dots and dashes could produce an infinite number of different messages. Living organisms have their own code, called the genetic code, in which the sequence of nucleotides in DNA can be converted to the sequence of amino acids in proteins.

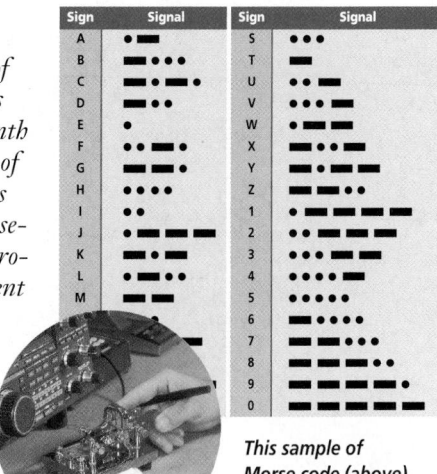

This sample of Morse code (above) is being sent (inset).

Genes and Proteins

The sequences of nucleotides in DNA contain information. This information is put to work through the production of proteins. Proteins form into complex three-dimensional shapes to become key cell structures and regulators of cell functions. Some proteins become important structures, such as the filaments in muscle tissue, walls of blood vessels, and transport proteins in membranes. Other proteins, such as enzymes, control chemical reactions that perform key life functions—breaking down glucose molecules in cellular respiration, digesting food, or making spindle fibers during mitosis. In fact, enzymes control all the chemical reactions of an organism. Thus, by encoding the instructions for making proteins, DNA controls cells.

You learned earlier that proteins are polymers of amino acids. The sequence of nucleotides in each gene contains information for assembling the string of amino acids that make up a single protein. It is estimated that each human cell contains about 80 000 genes.

RNA

RNA, like DNA, is a nucleic acid. However, RNA structure differs from DNA structure in three ways, shown in *Figure 11.5*. First, RNA is single stranded—it looks like only one-half a zipper—whereas DNA is double stranded. The sugar in RNA is ribose; DNA has deoxyribose.

294 DNA AND GENES

Alternative Lab

Gene and Chromosome Size

Purpose 🖐️
Students will conceptualize the size of a gene and a bacterial chromosome by constructing yarn models that are scaled up a million times in size.

Materials 🔪✂️
2 skeins of 3-ply yarn for a class of 20-30,

scissors, box about 100 cm × 50 cm

Procedure
Give students the following directions.

1. One nucleotide in a molecule of DNA occupies a length of 3.4×10^{-10} m. If an average gene coding for an average-sized protein contains 1200 base pairs, how long is an average gene?
3.4×10^{-10} m × 1200 = 4.08×10^{-7} m
How many amino acids would these nucleotides code for? *1200 bases ÷ 3*

Figure 11.5
The three chemical differences between DNA and RNA are shown here.

A An RNA molecule usually consists of a single strand of nucleotides, not a double strand. This single-stranded structure is closely related to its function.

Phosphate

$O = P - O$

Ribose

B The sugar in RNA is ribose, rather than the deoxyribose sugar of DNA.

Uracil

Hydrogen bonds

Adenine

C RNA contains the nitrogen base uracil (U) in place of thymine (T). Uracil base pairs with adenine just as thymine does in DNA.

Finally, both DNA and RNA contain four nitrogen bases, but rather than thymine, RNA contains a similar base called uracil (U). The uracil forms a base pair with adenine, just as thymine does in DNA.

What is the role of RNA in the cell? Let's look at an analogy. Perhaps you have seen a car being built on an automobile assembly line. Complex automobiles are built in many simple steps. Engineers tell workers how to make the cars, and the workers follow directions to build the cars on the assembly line. Suppliers bring parts to the assembly line so they can be installed in the car. Protein production is similar to car production. DNA provides workers with the

instructions for making the proteins, and the workers build the proteins. Other workers bring parts, the amino acids, over to the assembly line. The workers for protein synthesis are RNA molecules. They take from DNA the instructions on how the protein should be assembled, then— amino acid by amino acid—they assemble the protein.

There are three types of RNA that help to build proteins. Extending the car-making analogy, you can consider these RNA molecules to be the workers in the protein assembly line. One type of RNA, **messenger RNA** (mRNA) brings information from the DNA in the nucleus to the cell's factory floor, the cytoplasm. On the

---

Quick Demo

Kinesthetic Have a few students simulate the assembly line manufacture of "widgets." Some students bring design plans to the assembly line, others bring supplies to the line, and still others do the assembly on the line. Relate this to the manufacture of proteins. **L2**

Tying to Previous Knowledge

Have students recall the structure of proteins, amino acids, polypeptides, and peptide bond formation from Chapter 6. Make sure students understand that most proteins require the synthesis of two or more polypeptide chains.

Resource Manager

Critical Thinking/Problem Solving, p. 11 **L3**
Concept Mapping, p. 11 **L3** **ELL**
Section Focus Transparency 27 and Master **L1** **ELL**

---

bases per codon = 400 codons = 400 amino acids

2. Cut a piece of yarn 40 cm long to represent the average gene. This length is 1 000 000 times that of a gene.

3. Cut a piece of yarn that would represent 150 genes. *150 × 40 cm = 60 m* Tie the end of your 60-m piece of yarn to another until all yarn strands in the class are connected.

Expected Results
The total length of all the yarn pieces, 1500 m, represents the length of DNA in a bacterium if the cell were scaled up to the size of the box.

Analysis
1. Compare the length of DNA with the size of the "cell" (box). How does all the DNA fit inside the cell? *The DNA is coiled and twisted.*

2. A human cell contains about 80 000

genes. How long would the yarn be that represents all the DNA in a human cell? *80 000 × 40 cm = 32 000 m*

✔ Assessment

Skill Ask students to write a lab report and calculate the answers to the analysis questions. Use the Performance Task Assessment List for Using Math in Science in **PASC,** p. 29.

Intrapersonal Scientists have developed a technique that uses fluorescence microscopes to film DNA. The technique involves attaching a fluorescent dye to DNA to form a complex that glows. Then, just as you can watch a firefly travel across the yard, researchers can follow the DNA and its movements. Scientists are using this method to investigate how chromosomal DNA is tightly bound to histones and other proteins and how they fold into their functional forms. Have interested students investigate this technique and propose other possible uses for it. **L3**

INVESTIGATE BioLab The BioLab at the end of the chapter can be used at this point in the lesson. →

GLENCOE
TECHNOLOGY

CD-ROM
Biology: The Dynamics of Life
Animation: *Transcription*
Disc 2

VIDEODISC
Biology: The Dynamics of Life
DNA Transcription (Ch.32)
Disc 1, Side 1, 55 sec.

Resource Manager

Basic Concepts Transparency 17 and Master
L2 **ELL**

---

factory floor, the mRNA becomes part of the assembly line. Ribosomes, made of **ribosomal RNA** (rRNA), clamp onto the mRNA and use its information to assemble the amino acids in the correct order. The third type of RNA, **transfer RNA** (tRNA) is the supplier. Transfer RNA transports amino acids to the ribosome to be assembled into a protein.

Transcription

How does the information in DNA, which is found in the nucleus, move to the ribosomes in the cytoplasm? Messenger RNA carries this information from the DNA to the cell's ribosomes for manufacturing proteins, just as a worker brings information from the engineers to the factory floor for manufacturing a car. In the nucleus, enzymes make an RNA copy of a portion of a DNA strand in a process called **transcription** (trans KRIHP shun). Follow the steps in *Figure 11.6* as you read about transcription.

The process of transcription is similar to that of DNA replication.

Figure 11.6
Messenger RNA is made during the process of transcription.

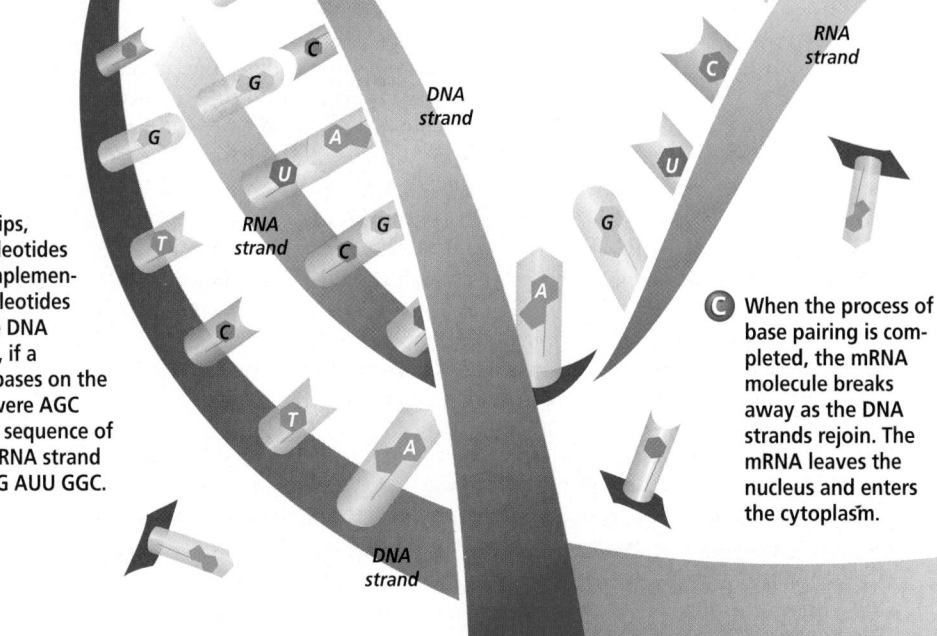

A The process of transcription begins as enzymes unzip the molecule of DNA, just as they do during DNA replication.

B As the DNA molecule unzips, free RNA nucleotides pair with complementary DNA nucleotides on one of the DNA strands. Thus, if a sequence of bases on the DNA strand were AGC TAA CCG, the sequence of bases on the RNA strand would be UCG AUU GGC.

C When the process of base pairing is completed, the mRNA molecule breaks away as the DNA strands rejoin. The mRNA leaves the nucleus and enters the cytoplasm.

DNA strand

RNA strand

DNA strand

RNA strand

DNA strand

BIOLOGY Online Note Internet addresses that you find useful in the space below for quick reference.

The main difference is that transcription results in the formation of one single-stranded RNA molecule rather than a double-stranded DNA molecule. You can find out how scientists use new microscopes to "watch" transcription take place by reading the *BioTechnology* at the end of the chapter. Modeling the process of transcription in the *BioLab* will help you to understand this process.

The Genetic Code

The nucleotide sequence transcribed from DNA to a strand of messenger RNA acts as a genetic message, the complete information for the building of a protein. Think of this message as being written in a language that uses nitrogen bases as its alphabet. As you know, proteins are built from chains of smaller molecules called amino acids. You could say that the language of proteins uses an alphabet of amino acids. A code is needed to convert the language of mRNA into the language of proteins. There are 20 different amino acids, but mRNA contains only four types of bases. How can these bases form a code for proteins? Biochemists began to crack the code when they discovered that a group of three nucleotides codes for one amino acid. For example, a sequence of three uracil nucleotides in mRNA (U-U-U) results in the amino acid phenylalanine being placed in a protein. Each set of three nitrogen bases in mRNA coding for an amino acid is known as a **codon.** You can follow the biochemists' reasoning for why three bases are needed by doing the *Problem-Solving Lab* on this page.

The order of nitrogen bases in the mRNA will determine the type and order of amino acids in a protein. Sixty-four combinations are possible

---

Problem-Solving Lab 11-2 — Formulating Models

How many nitrogen bases determine an amino acid?
After the structure of DNA had been discovered, scientists tried to predict the number of nucleotides that code for a single amino acid. It was already known that there were 20 amino acids, so at least 20 codons were needed. If one nucleotide coded for an amino acid, then only four amino acids could be represented. How many nucleotides are needed?

Analysis
Examine the three safes. Letters representing nitrogen bases have replaced numbers on the dials. Copy the data table. Calculate the possible

 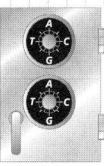 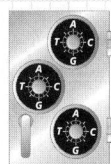

Safe 1 Safe 2 Safe 3

number of combinations that will open the safe in each diagram using the formula provided in the table. The 4 corresponds to the number of letters on each dial; the superscript refers to the number of available dials.

Data Table				
	Number of dials	Number of letters per dial	Total possible combinations	Formula
Safe 1				4^1
Safe 2				4^2
Safe 3				4^3

Thinking Critically

1. Using safe 1, write down several examples of dial settings that might open the safe. Do the total possible combinations seen in safe 1 equal or surpass the total number of amino acids known?
2. Could a nitrogen base (A, T, C, or G) taken one at a time code for 20 different amino acids? Explain.
3. Using safe 2, write down several examples of dial combinations that might open the safe. Do the total possible combinations seen in safe 2 equal or surpass the total number of amino acids known?
4. Could nitrogen bases taken two at a time code for 20 different amino acids? Explain.
5. Using the same procedure for safe 3, see whether the total possible combinations equal or surpass the total number of amino acids known.
6. Could nitrogen bases taken three at a time code for 20 different amino acids? Explain.
7. Does the analogy prove that three bases code for an amino acid? Explain.

Learning Disabled

A memory device, "You are single," can help students differentiate RNA from DNA. "You" stands for "U"; RNA has uracil. "Are" is for "R"; RNA has ribose sugar. Single refers to RNA being single stranded.

✔ Assessment

Knowledge Ask students to determine which letter in the sequence of three is most important in coding for a specific amino acid. Have them give specific examples while referring to Table 11.2. Use the Performance Task Assessment List for Analyzing the Data in **PASC,** p. 27. **L2**

Problem-Solving Lab 11-2

Purpose
Students will use an analogy to see that at least three nitrogen bases are required to code for a single amino acid.

Process Skills
analyze information, apply concepts, collect data, compare and contrast, draw a conclusion, interpret data, predict, think critically, use numbers

Background
The coding for amino acids using three nitrogen bases is said to be redundant. That is, there can be several different codons that code for the same amino acid. Thus, the number of combinations is as high as 61 (64 if stops are considered). The number 64 agrees perfectly when three of any four nitrogen bases combine in any order to code for an amino acid.

Teaching Strategies
■ Remind students that there are 20 amino acids.
■ You may wish to have students determine the number of two-letter combinations without the use of the formula. It may not be practical, however, to have them actually determine the number of combinations for a three-dial safe.

Thinking Critically
1. A or T or C or G; no
2. No, there are 20 amino acids and only 4 different codes using 1 letter.
3. AA, AT, CT, CG, etc; no
4. No, there are 20 amino acids and only 16 different codes using 2 letters.
5. ATT, ATC, ATG, ATA, etc; yes
6. Yes, there are 20 amino acids and 64 possible combinations using 3 letters.
7. No, it shows only that at least 3 bases are needed.

Reinforcement

Visual-Spatial Give students a blank four square by four square bingo card. Students write one amino acid into each square. An amino acid should be used only once. Play bingo, but don't call out amino acids; instead, pick and call mRNA codons. For example, if you call UCU, students use Table 11.2 and cross out serine on their cards. **L1**

Discussion

Have students compare the number of letters in the alphabet with the number of amino acids available for protein formation. How many different words can be made using 26 letters? How many different proteins can be made with the amino acids available? Ask students to consider that proteins contain many more amino acids than words do letters. **L2**

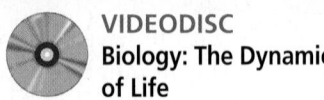

GLENCOE
TECHNOLOGY

VIDEODISC
Biology: The Dynamics of Life
Translation (Ch. 33)
Disc 1, Side 1
1 min. 29 sec.

Resource Manager

BioLab and MiniLab Worksheets, pp. 49-50 **L2**
Basic Concepts Transparency 18 and Master **L2** **ELL**
Reteaching Skills Transparency 18 and Master **L1** **ELL**

when a sequence of three bases is used; thus, 64 different mRNA codons are in the genetic code, shown in *Table 11.2*. Some codons do not code for amino acids; they provide instructions for assembling the protein. For example, UAA is a *stop* codon indicating that protein production ends at that point. AUG is a *start* codon as well as the codon for the amino acid methionine. As you can see, more than one codon can code for the same amino acid. However, for any one codon, there can be only one amino acid.

All organisms use the same genetic code for amino acids and assembling proteins; UAC codes for tyrosine in the messenger RNA of bacteria, birch trees, and bison. For this reason, the genetic code is said to be universal, and this provides evidence that all life on Earth evolved from a common origin. From the chlorophyll of a

WORD *Origin*

codon
From the Latin word *codex*, meaning "a tablet for writing." A codon is the three-nucleotide sequence that codes for an amino acid.

birch tree to the digestive enzymes of a bison, a large number of proteins are produced from DNA. It may be hard to imagine that only four nucleotides can produce so many diverse proteins; yet, think about computer programming. You may have seen computer code, such as 00010101110000110. Through a binary language with only two options—zeros and ones—many types of software are created. From computer games to World Wide Web browsers, complex software is built by stringing together the zeros and ones of computer code into long chains. Likewise, complex proteins are built from the long chains of DNA carrying the genetic code. If the DNA in all the human cells of an adult were lined up end-to-end, it would stretch to about 60 billion miles—about 16 times the distance from the sun to Pluto, the outermost

Table 11.2 The messenger RNA genetic code					
First letter	Second letter				Third letter
	U	C	A	G	
U	Phenylalanine (UUU)	Serine (UCU)	Tyrosine (UAU)	Cysteine (UGU)	U
	Phenylalanine (UUC)	Serine (UCC)	Tyrosine (UAC)	Cysteine (UGC)	C
	Leucine (UUA)	Serine (UCA)	Stop (UAA)	Stop (UGA)	A
	Leucine (UUG)	Serine (UCG)	Stop (UAG)	Tryptophan (UGG)	G
C	Leucine (CUU)	Proline (CCU)	Histadine (CAU)	Arginine (CGU)	U
	Leucine (CUC)	Proline (CCC)	Histadine (CAC)	Arginine (CGC)	C
	Leucine (CUA)	Proline (CCA)	Glutamine (CAA)	Arginine (CGA)	A
	Leucine (CUG)	Proline (CCG)	Glutamine (CAG)	Arginine (CGG)	G
A	Isoleucine (AUU)	Threonine (ACU)	Asparagine (AAU)	Serine (AGU)	U
	Isoleucine (AUC)	Threonine (ACC)	Asparagine (AAC)	Serine (AGC)	C
	Isoleucine (AUA)	Threonine (ACA)	Lysine (AAA)	Arginine (AGA)	A
	Methionine; Start (AUG)	Threonine (ACG)	Lysine (AAG)	Arginine (AGG)	G
G	Valine (GUU)	Alanine (GCU)	Aspartate (GAU)	Glycine (GGU)	U
	Valine (GUC)	Alanine (GCC)	Aspartate (GAC)	Glycine (GGC)	C
	Valine (GUA)	Alanine (GCA)	Glutamate (GAA)	Glycine (GGA)	A
	Valine (GUG)	Alanine (GCG)	Glutamate (GAG)	Glycine (GGG)	G

298 DNA AND GENES

MEETING INDIVIDUAL NEEDS

Visually Impaired

Kinesthetic Help visually impaired students understand Problem-Solving Lab 11-2 by having them manipulate combination locks. Then make large copies of the safe diagrams. **L1**

Gifted

The direction of replication, transcription, and translation is a college topic. For advanced students, you may wish to explain that these processes always proceed in the 5' ⇒ 3' direction. **L3**

planet in our solar system. With proteins, as in software, elaborate things are constructed from a simple code.

Translation: From mRNA to Protein

How is the language of mRNA translated into the language of proteins? The process of converting the information in a sequence of nitrogen bases in mRNA into a sequence of amino acids that make up a protein is known as **translation.** You can summarize transcription and translation by completing the *MiniLab.*

Translation takes place at the ribosomes in the cytoplasm. In prokaryotic cells that have no nucleus, the mRNA is made in the cytoplasm. In eukaryotic cells, mRNA leaves the nucleus through an opening in the nuclear membrane and travels to the cytoplasm. When the strands of mRNA arrive in the cytoplasm, ribosomes attach to them like clothespins clamped onto a clothesline.

The role of transfer RNA

For proteins to be built, the 20 different amino acids dissolved in the cytoplasm must be brought to the ribosomes. This is the role of transfer RNA (tRNA), *Figure 11.7.* Each

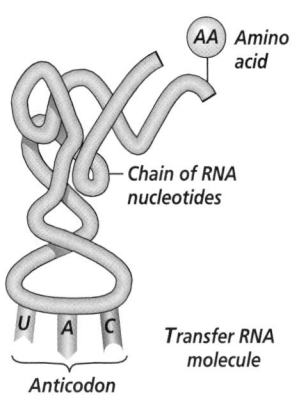

AA *Amino acid*

Chain of RNA nucleotides

U A C

Transfer RNA molecule

Anticodon

Figure 11.7
A tRNA molecule is composed of about 80 nucleotides. Each tRNA recognizes only one amino acid. The amino acid becomes bonded to one side of the tRNA molecule. Located on the other side of the tRNA molecule are three nitrogen bases, called an anticodon, that pair up with an mRNA codon during translation.

MiniLab 11-1 Predicting

Transcribe and Translate Molecules of DNA carry the genetic instructions for protein formation. Converting these DNA instructions into proteins requires a series of coordinated steps in transcription and translation.

Procedure

1. Copy the data table.
2. Complete column B by writing the correct mRNA codon for each sequence of DNA bases listed in the column marked *DNA Base Sequence.* Use the letters A, U, C, or G.
3. Identify the process responsible by writing its name on the arrow in column A.
4. Complete column D by writing the correct anticodon that bonds to each codon from column B.
5. Identify the process responsible by writing its name on the arrow in column C.
6. Complete column E by writing the name of the correct amino acid that is coded by each base sequence. Use *Table 11.2* on page 298 to translate the mRNA base sequences to amino acids.

Data Table

	A	B	C	D	E
DNA base sequence	Process	mRNA codon	Process	tRNA anticodon	Amino acid
AAT	→		→		
GGG	→		→		
ATA	→		→		
AAA	→		→		
GTT	→		→		

Analysis

1. Where within the cell:
   a. are the DNA instructions located?
   b. does transcription occur?
   c. does translation occur?
2. Describe the structure of a tRNA molecule.
3. Explain why specific base pairing is essential to the processes of transcription and translation.

MiniLab 11-1

Purpose
Students will follow a series of DNA base codes through transcription and translation.

Process Skills
apply concepts, analyze information, compare and contrast, draw a conclusion

Teaching Strategies
■ Make sure that students have read and reviewed the section in the text dealing with transcription and translation before attempting this activity.
■ If you feel it is necessary, you may want to walk students through the first example.

Expected Results
See the table below.

Analysis
1. a. on chromosomes in the nucleus
   b. in the nucleus
   c. in the ribosomes
2. tRNA is a small molecule that has a three-base anticodon at one end and an amino-acid attachment site at the opposite end.
3. Precise base pairing is essential to transcription and translation so that the correct genetic information in DNA is transferred to the forming protein.

✓ Assessment

Skill Provide students with a series of amino acids and have them make a poster, working backwards from these amino acids to the tRNA anticodon to the mRNA codon to DNA. Use the Performance Task Assessment List for Poster in **PASC**, p. 73.

BIOLOGY JOURNAL

Converting Languages

Linguistic Translation is a term that is used for converting words in one language to words in a different language. Have students write hypotheses for why the process of converting a base sequence in mRNA to an amino acid sequence in a protein is also called translation. **L2**

Data Table

	A	B	C	D	E
DNA base sequence	Process	mRNA codon	Process	tRNA anticodon	Amino acid
AAT	transcription	UUA	translation	AAU	leucine
GGG	transcription	CCC	translation	GGG	proline
ATA	transcription	UAU	translation	AUA	tyrosine
AAA	transcription	UUU	translation	AAA	phenylalanine
GTT	transcription	CAA	translation	GUU	glutamine

Assessment

Knowledge On the chalkboard, write the sequence for one strand of DNA. Have students copy the sequence and write the corresponding sequences for mRNA, the tRNA anticodons, and the coded protein. Students should follow the changes in logical steps from the original DNA through transcription and translation. Afterward, go through the correct answer on the board. **L2**

3 Assess

Check for Understanding

Review the processes of transcription and translation orally. Ask students to supply missing words, descriptions, and process words, including DNA, complementary codons, mRNA, tRNA, ribosomal RNA, transcription, translation, and anticodons. Discuss the definitions. **L1**

GLENCOE TECHNOLOGY

CD-ROM
Biology: The Dynamics of Life
BioQuest: *Building a Protein*
Disc 2
Animation: *Translation*
Disc 2

Resource Manager

Reinforcement and Study Guide, pp. 48-49 **L2**
Content Mastery, p. 55 **L1**

Figure 11.8
A protein is formed by the process of translation.

A As translation begins, the starting end of the mRNA strand attaches to a ribosome. Then, tRNA molecules, each carrying a specific amino acid, approach the ribosome. When a tRNA anticodon pairs with the first mRNA codon, the two molecules temporarily join together.

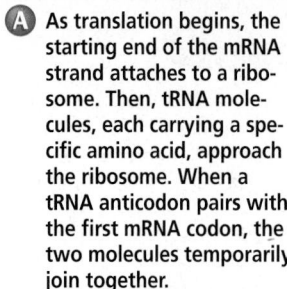
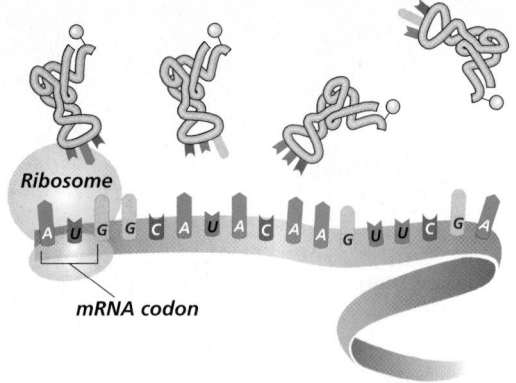

Ribosome

A U G G C A U A C A A G U U C G A

mRNA codon

tRNA molecule attaches to only one type of amino acid.

Correct translation of the mRNA message depends upon the joining of each mRNA codon with the correct tRNA molecule. How does a tRNA molecule carrying its amino acid recognize which codon to attach to? The answer again involves base pairing. On the opposite side of the transfer-RNA molecule from the amino-acid attachment site, there is a sequence of three nucleotides that are the complement of the nucleotides in the codon. These three nucleotides are called an anticodon because they bond to the codon of the messenger RNA. The tRNA carries only the amino acid that the anticodon specifies. For example, one tRNA molecule for the amino acid cysteine has an anticodon of A-C-A. This anticodon bonds with the mRNA codon U-G-U. Now, use *Table 11.2* to find the mRNA codon for tryptophan, then determine its tRNA anticodon.

Translating the mRNA code

Follow the steps in *Figure 11.8* as you read how translation occurs. As translation begins, a tRNA molecule brings the first amino acid to the mRNA strand that is attached to the ribosome, *Figure 11.8A*. The anticodon forms a temporary bond with the codon of the mRNA strand, *Figure 11.8B*. This places the amino acid in the correct position for forming a bond with the next amino acid. The ribosome slides down the mRNA chain to the next codon, and a new tRNA molecule brings another amino acid, *Figure 11.8C*. The amino acids bond, the first tRNA releases its amino acid and detaches from the mRNA, *Figure 11.8D*. The tRNA molecule is now free to pick up and deliver another molecule of its specific amino acid to a ribosome. Again, the ribosome slides down to the next codon; a new tRNA molecule arrives, and its amino acid bonds to the previous one. A chain of amino acids begins to form. When a *stop* codon is reached, translation ends, and the amino acid strand is released from the ribosome, *Figure 11.8E*.

Like Silly String sprayed into a friend's hair, amino acid chains become proteins when they are freed from the ribosome and twist and curl into complex three-dimensional shapes. Unlike Silly String, however, each protein chain forms the same shape every time it is produced. These proteins become enzymes and cell and tissue structures. The formation of protein, originating from the DNA code, produces the diverse and magnificent living world.

Portfolio

Translation

Linguistic Have students write a summary of the events that are being shown in Figure 11.8. Have them include the events that led up to and that will follow translation. **L2** **P**

PROJECT

Building a Model

Kinesthetic Have students build a model demonstrating protein synthesis. They may wish to use various types of macaroni on poster board, colored pipe cleaners, beads, colored building blocks, or yarn. **L1** **ELL**

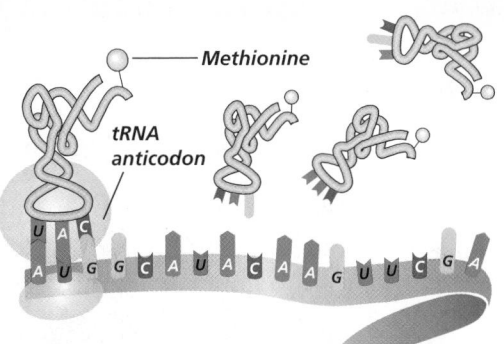

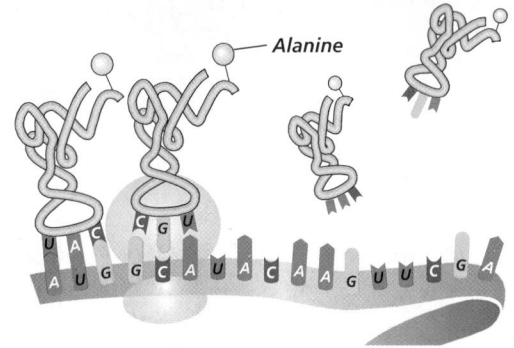

B Usually, the first codon on mRNA is AUG, which codes for the amino acid methionine. AUG signals the start of protein synthesis. When this signal is given, the ribosome slides along the mRNA to the next codon.

C A new tRNA molecule carrying an amino acid pairs with the second mRNA codon.

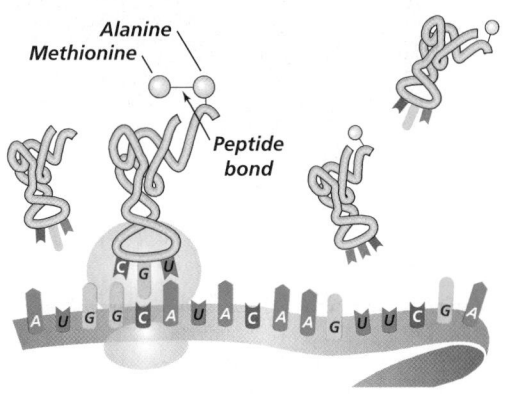

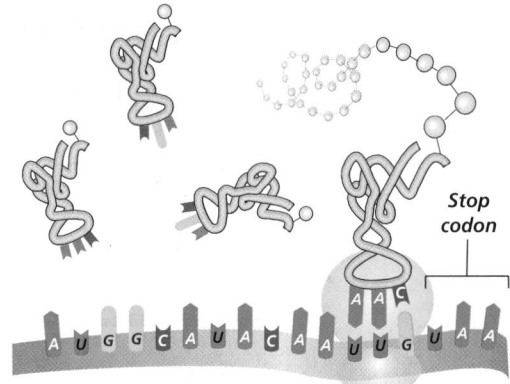

D When the first and second amino acids are in place, an enzyme joins them by forming a peptide bond between them.

E As the process continues, a chain of amino acids is formed until the ribosome reaches a *stop* codon on the mRNA strand.

Section Assessment

Understanding Main Ideas

1. In what ways do the chemical structures of DNA and RNA differ?
2. What is a codon, and what does it represent?
3. What is the role of tRNA in protein synthesis?
4. Compare and contrast the final products of DNA replication and transcription.

Thinking Critically

5. You have learned that there is a *stop* codon that signals the end of an amino acid chain. Why is it important that a signal to stop translation be part of protein synthesis?

SKILL REVIEW

6. **Sequencing** Sequence the steps involved in protein synthesis from the production of mRNA to the final translation of the DNA code. For more help, refer to *Organizing Information* in the **Skill Handbook**.

Section Assessment

1. RNA contains ribose and uracil and DNA contains deoxyribose and thymine. RNA is usually single stranded.
2. A codon is a 3-base sequence of mRNA that codes for a single amino acid.
3. Transfer RNA brings a specific amino acid to a ribosome by matching a codon on the messenger RNA strand.
4. Replication produces 2 molecules of double-stranded DNA. Transcription results in the production of 1 molecule of single-stranded RNA.
5. Because a protein's 3-dimensional structure depends on its length as well as its amino acid sequence, translation must start and stop at precise positions.
6. The DNA strands separate and free RNA nucleotides pair with complementary DNA bases on one of the DNA strands, forming mRNA. mRNA leaves the nucleus and attaches to a ribosome. tRNA molecules, carrying specific amino acids, pair with the appropriate mRNA codons. When 2 amino acids are in place, an enzyme joins them together.

Prepare

Key Concepts

Mutations are changes in the sequence of DNA. Their effect on body cells is different than on reproductive cells. The causes and results of gene and chromosome mutations are discussed. The section ends with a discussion of DNA repair mechanisms.

Planning

- Make copies of Figures 11.10 and 11.11 without captions for the Reteach.
- Gather a coiled telephone cord and twist ties for the Quick Demo.

1 Focus

Bellringer

Before presenting the lesson, display **Section Focus Transparency 28** on the overhead projector and have the students answer the accompanying questions. **L1** **ELL**

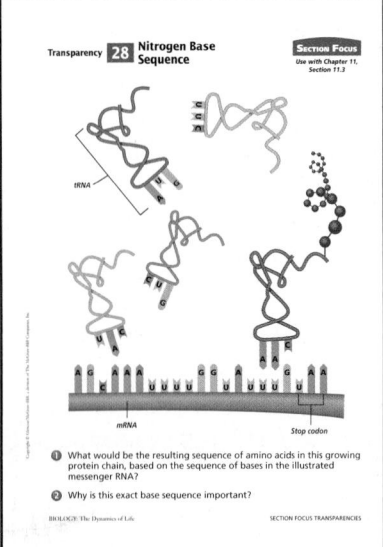

SECTION PREVIEW

Objectives

Categorize the different kinds of mutations that can occur in DNA.

Compare the effects of different kinds of mutations on cells and organisms.

Vocabulary

mutation
point mutation
frameshift mutation
chromosomal mutation
mutagen

Section

11.3 Genetic Changes

DNA controls the structures and functions of a cell. What happens when the sequence of DNA nucleotides is changed in a gene? Sometimes it may have little or no harmful effect, as in this tailless Manx cat, and the DNA changes are passed on to offspring of the organism. At other times, however, the change can cause the cell to behave differently. For example, in the type of skin cancer shown here, UV rays from the sun change the DNA and cause the cells to grow and divide rapidly.

Manx cat (above) and melanoma (inset)

Mutation: A Change in DNA

Radiation may be given off in the reactor areas of nuclear power plants. If a person is exposed to this radiation, serious problems may result. The gamma radiation found in nuclear reactors can break apart a molecule of DNA, causing the nucleotide sequence to be changed. For this reason, nuclear power plant workers wear radiation-detecting devices such as the ones shown in *Figure 11.9*. As you know, the sequences of nucleotides in DNA molecules control the structure and function of cells. Any change in the DNA sequence that also changes

WORD Origin

mutation
From the Latin word *mutare*, meaning "to change." Mutations are changes in DNA.

the protein it codes for is called a **mutation**.

Mutations in reproductive cells

Mutations can affect the reproductive cells of an organism by changing the sequence of nucleotides within a gene in a sperm or an egg cell. If these cells take part in fertilization, the altered gene would become part of the genetic makeup of the offspring. The mutation may produce a new trait or it may result in a protein that does not work correctly, resulting in structural or functional problems in cells and in the organism. Sometimes, the mutation is so severe that the

302 DNA AND GENES

BIOLOGY JOURNAL

Mutations

Linguistic Ask students to describe in their journals the images that the word *mutation* conjures up in their minds. They may describe fantastic beings they have encountered in movies and stories. Point out that real mutations are usually much less spectacular. **L1**

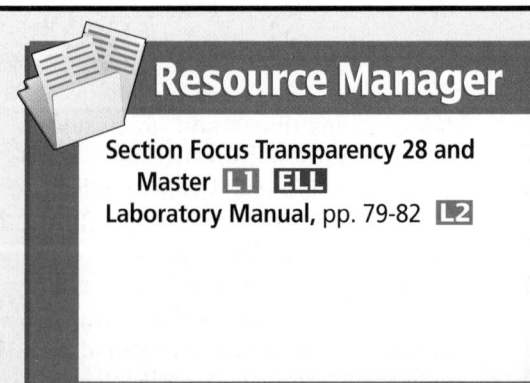

Resource Manager

Section Focus Transparency 28 and Master **L1** **ELL**
Laboratory Manual, pp. 79-82 **L2**

Figure 11.9
Nuclear power plant workers wear radiation badges (a) and pocket dosimeters (b) to monitor their exposure to radiation.

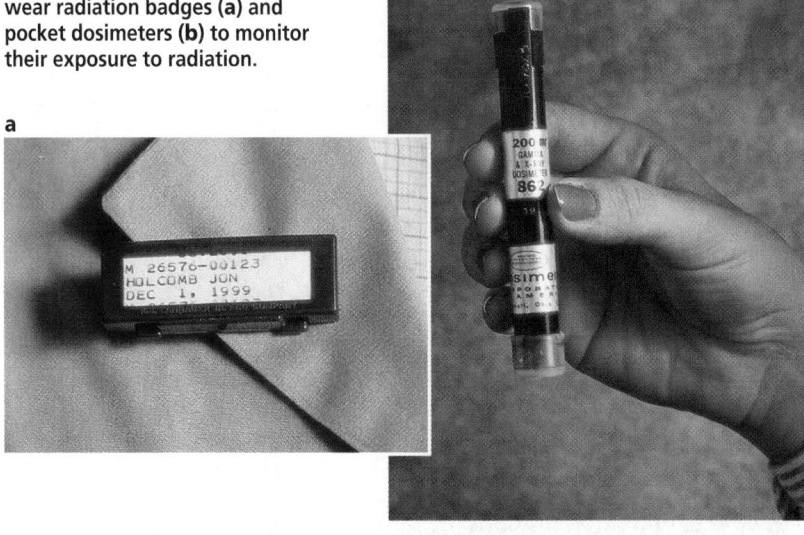

a

b

resulting protein is nonfunctional, and the embryo does not survive.

In some rare cases, a gene mutation may have positive effects. An organism may receive a mutation that makes it faster or stronger; such a mutation may help an organism—and its offspring—better survive in its environment. You will learn later that mutations that benefit a species play an important role in the evolution of that species.

Mutations in body cells

What happens if powerful radiation, such as gamma radiation, hits the DNA of a nonreproductive cell, a cell of the body such as in skin, muscle, or bone? If the cell's DNA is changed, this mutation would not be passed on to offspring. However, the mutation may cause problems for the individual. Damage to a gene may impair the function of the cell; for example, it may make a muscle cell lose its ability to make a protein that contracts, or a skin cell may lose its elasticity. When that cell divides, the

---

CAREERS IN BIOLOGY

Genetic Counselor

Are you fascinated by how you inherit traits from your parents? If so, you could become a genetic counselor and help people assess their risk of inheriting or passing on genetic disorders.

Skills for the Job

Genetic counselors are medical professionals who work on a health care team. They analyze families' medical histories to determine their risk of having children with genetic disorders, such as hemophilia or cystic fibrosis. Counselors also educate the public and help families with genetic disorders find support and treatment. These counselors may work in a medical center, a private practice, research, or a commercial laboratory. To become a counselor, you must earn a two-year master's degree in medical genetics and pass a test to become certified. The most important requirement is the ability to listen and to help families make difficult decisions.

For more careers in related fields, be sure to check the Glencoe Science Web site.
science.glencoe.com

11.3 GENETIC CHANGES **303**

new cells also will have the same mutation. Many scientists suggest that the buildup of cells with less than optimal functioning is an important cause of aging.

Some mutations of DNA in body cells affect genes that control cell division. This can result in the cells growing and dividing rapidly, producing the disease called cancer. As you learned earlier, cancer is the uncontrolled dividing of cells. Cancer may result from gene mutations. For example, ultraviolet radiation in sunlight can change the DNA in skin cells, altering their behavior. The cells reproduce rapidly, causing skin cancer.

The effects of point mutations

Consider what might happen if an incorrect amino acid were inserted into a growing protein chain during the process of translation. The mistake might affect the structure of the entire molecule. Such a problem can occur if a point mutation arises. A **point mutation** is a change in a single base pair in DNA.

A simple analogy can illustrate point mutations. Read the two sentences below to see what happens when a single letter in the first sentence is changed.

THE DOG BIT THE CAT.
THE DOG BIT THE CAR.

As you can see, changing a single letter changes the meaning of the above sentence. Similarly, a change in a single nitrogen base can change the entire structure of a protein because a change in a single amino acid can affect the shape of the protein. *Figure 11.10A* shows what can happen with a point mutation.

Figure 11.10
The results of a point mutation and a frameshift mutation are different. The diagrams show the mRNA that would be formed from each corresponding DNA.

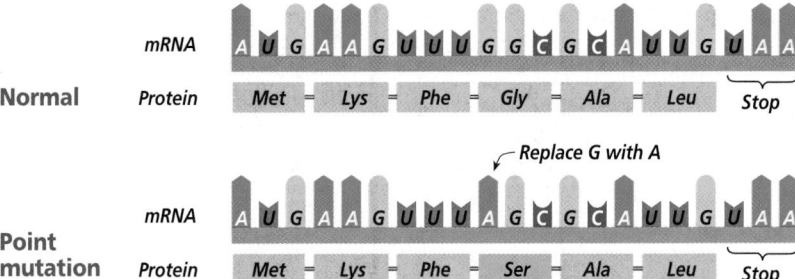

A In this point mutation, the mRNA produced by the mutated DNA had the base guanine changed to adenine. This change in the codon caused the insertion of serine rather than glycine into the growing amino acid chain. The error may or may not interfere with protein function.

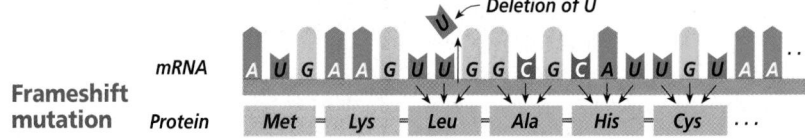

B Proteins that are produced as a result of frameshift mutations seldom function properly because such mutations usually change many amino acids. Adding or deleting one base of a DNA molecule will change nearly every amino acid in the protein after the addition or deletion.

304 DNA AND GENES

Frameshift mutations

When the mRNA strand moves across the ribosome, a new amino acid is added to the protein for every codon on the mRNA strand. What would happen if a single base were lost from a DNA strand? This new sequence with the deleted base would be transcribed into mRNA. But then, the mRNA would be out of position by one base. As a result, every codon after the deleted base would be different, as shown in *Figure 11.10B.* This mutation would cause nearly every amino acid in the protein after the deletion to be changed. In the sentence THE DOG BIT THE CAT, deleting a G would produce the sentence THE DOB ITT HEC AT. The same effect would also result from the addition of a single base. A mutation in which a single base is added or deleted from DNA is called a **frameshift mutation** because it shifts the reading of codons by one base. In general, point mutations are less harmful to an organism because they disrupt only a single codon. The *MiniLab* on the next page will help you distinguish point mutations from frameshift mutations, and the *Problem-Solving Lab* on this page will show you an example of an actual human mutation.

Chromosomal Mutations

Changes may occur at the level of chromosomes as well as in genes. Mutations to chromosomes may occur in a variety of ways. For example, sometimes parts of chromosomes are broken off and lost during mitosis or meiosis. Often, chromosomes break and then rejoin incorrectly. Sometimes, the parts join backwards or even join to the wrong chromosome. These changes in chromosomes are called **chromosomal mutations.**

Problem-Solving Lab 11-3 — Making and Using Tables

What type of mutation results in sickle-cell anemia? A condition called sickle-cell anemia results from a genetic change in the base sequence of DNA. Red blood cells in patients with sickle-cell anemia have molecules of hemoglobin that are misshapen. As a result of this change in protein shape, sickled blood cells clog capillaries and prevent normal flow of blood to body tissues, causing severe pain.

Analysis

Table 11.3 shows the sequence of bases in a short segment of the DNA that controls the order of amino acids in the protein, hemoglobin.

Table 11.3 DNA base sequences	
Normal hemoglobin	GGG CTT CTT TTT
Sickled hemoglobin	GGG CAT CTT TTT

Thinking Critically

1. Use *Table 11.2* on page 298 to transcribe and translate the DNA base sequence for normal hemoglobin and for sickled hemoglobin into amino acids. Remember that the table lists mRNA codons, not DNA base sequences.
2. Does this genetic change illustrate a point mutation or frameshift mutation? Explain your answer.
3. Explain why the correct sequence of DNA bases is important to normal development of proteins.
4. Assume that the base sequence reads GGG CTT CTT AAA instead of the normal sequence for hemoglobin. Would this result in sickled hemoglobin? Explain your answer.

Effects of chromosomal mutations

Chromosomal mutations occur in all living organisms, but they are especially common in plants. Such mutations affect the distribution of genes to gametes during meiosis because they cause nondisjunction, the failure of chromosomes to separate. Nondisjunction occurs because homologous chromosomes cannot pair correctly when one has extra or missing parts. Gametes that should have a complete set of genes may end up with extra copies or a complete lack of some genes.

Problem-Solving Lab 11-3

Purpose

Students will compare the base sequence of a segment of normal DNA with the base sequence of DNA that codes for sickled human hemoglobin.

Process Skills

analyze information, apply concepts, compare and contrast, recognize cause and effect

Teaching Strategies

■ Make sure that students are familiar with Table 11.2 and understand how it is to be used.
■ Discuss sickle-cell anemia briefly and point out its greater occurrence among African Americans.

Thinking Critically

1. normal: proline, glutamate, glutamate, lysine; sickled: proline, valine, glutamate, lysine
2. Point mutation; one codon has been altered and one amino acid is different.
3. In this example, the correct sequence of amino acids is needed to form a molecule of normal hemoglobin. With a change of only one amino acid, the protein molecule no longer functions in a normal manner.
4. Student answers may vary. Probably not; protein formation is very specific and the substitution of a different amino acid could result in a totally different form of hemoglobin.

BIOLOGY *Online* Note Internet addresses that you find useful in the space below for quick reference.

_____
_____
_____
_____

Portfolio

Protein Building

Linguistic Have students write an essay comparing the building of proteins to the building of a house and what will occur if there are problems with the blueprint guiding the builders. **L3**

 **P**

✔ Assessment

Performance Ask students to write an informative newspaper article in which they describe sickle-cell anemia and explain its cause to the public. Use the Performance Task Assessment List for Newspaper Article in **PASC**, p. 69. **L2**

Purpose

Students will examine the effect of gene mutations on proteins.

Process Skills

recognize cause and effect, analyze

Teaching Strategies

■ Individual students can do this MiniLab at their desks as you walk around the room and check their work.

Expected Results

Base sequence of mRNA: UUACGGUCACCAAGCGUG; amino acids: leucine-arginine-serine-proline-serine-valine; Changing the fourth base in the DNA from G to C would change the second amino acid from arginine to glycine. Adding G at the fourth position of the DNA would result in the mRNA base sequence UUACCGGUCAC-CAAGCGUG and the amino acid sequence leucine-proline-valine-threonine-lysine-arginine.

Analysis

1. Changing G to C was a point mutation; adding G to the chain was a frameshift mutation.
2. The point mutation changed only one amino acid.
3. The frameshift mutation changed every amino acid following the addition of G.

✔ Assessment

Performance Ask students to give an oral presentation explaining the effect on the amino acid sequence if a C is substituted for a T in the third position of the DNA strand above. *The amino acid sequence will be the same because both sequences code for leucine.* Use the Performance Task Assessment List for Oral Presentation in **PASC**, p. 71 L1

MiniLab 11-2 — Making and Using Tables

Gene Mutations and Proteins Gene mutations often have serious effects on proteins. In this activity, you will demonstrate how such mutations affect protein synthesis.

DNA segment

Procedure

1. Copy the following base sequence of one strand of an imaginary DNA molecule: AATGCCAGTGGTTCGCAC.
2. Then, write the base sequence for an mRNA strand that would be transcribed from the given DNA sequence.
3. Use *Table 11.2* to determine the sequence of amino acids in the resulting protein fragment.
4. If the fourth base in the original DNA strand were changed from G to C, how would this affect the resulting protein fragment?
5. If a G were added to the original DNA strand after the third base, what would the resulting mRNA look like? How would this addition affect the protein?

Analysis

1. Which change in DNA was a point mutation? Which was a frameshift mutation?
2. In what way did the point mutation affect the protein?
3. How did the frameshift mutation affect the protein?

Figure 11.11
Study the four kinds of chromosomal mutations.

Ⓐ Deletions occur when part of a chromosome is left out.

Ⓑ Insertions occur when a part of a chromatid breaks off and attaches to its sister chromatid. The result is a duplication of genes on the same chromosome.

Ⓒ Inversions occur when part of a chromosome breaks off and is reinserted backwards.

Ⓓ Translocations occur when part of one chromosome breaks off and is added to a different chromosome.

306 DNA AND GENES

Few chromosome mutations are passed on to the next generation because the zygote usually dies. In cases where the zygote lives and develops, the mature organism is often sterile and thus incapable of producing offspring. The most important of these mutations—deletions, insertions, inversions, and translocations—are illustrated in *Figure 11.11.*

Causes of Mutations

Some mutations seem to just happen, perhaps as a mistake in base pairing during DNA replication. These mutations are said to be spontaneous. However, many mutations are caused by factors in the environment. As you learned earlier, gamma radiation is capable of causing mutations. Any agent that can cause a change in DNA is called a **mutagen** (MYEWT uh jun). Mutagens include high energy radiation, chemicals, and even high temperatures.

BIOLOGY JOURNAL

Mutagens

Intrapersonal Have students do research on various mutagenic agents such as X rays, ultraviolet light, radioactive substances, and other chemicals. Students should report on the mechanism by which each agent causes mutations. L3

Resource Manager

BioLab and MiniLab Worksheets, p. 51 L2
Reinforcement and Study Guide, p. 50 L2
Content Mastery, pp. 53, 56 L1

Forms of radiation, such as X rays, cosmic rays, ultraviolet light, and nuclear radiation, are dangerous mutagens because they contain a large amount of energy that can blast DNA apart. The breaking and reforming of a double-stranded DNA molecule can result in deletions. Radiation can also cause substitutions of incorrect nucleotides in the DNA.

Chemical mutagens include dioxins, asbestos, benzene, cyanide, and formaldehyde, compounds that are commonly found in buildings and in the environment, *Figure 11.12*. These mutagens are highly reactive compounds that interact with the DNA molecule and cause changes. Chemical mutagens usually result in a substitution mutation.

Repairing DNA

The cell processes that copy genetic material and pass it from one generation to the next are usually accurate. This accuracy is important to ensure the genetic continuity of both new cells and offspring. Yet, mistakes sometimes do occur. There are many sources of mutagens in an organism's environment. Although many of these are due to human activities, others—such as cosmic

rays from outer space—have affected living things since the beginning of life. Repair mechanisms that fix mutations in cells have evolved. Much like a book editor, enzymes proofread the DNA and replace incorrect nucleotides with correct nucleotides. These repair mechanisms work extremely well, but they are not perfect. The greater the exposure to a mutagen such as UV light, the more likely is the chance that a mistake will not be corrected. Thus, it is wise for people to limit their exposure to mutagens.

Figure 11.12
Asbestos was formerly used to insulate buildings. It is now known to cause lung cancer and other lung diseases and must be removed from these buildings, as shown here.

Section Assessment

Understanding Main Ideas
1. What is a mutation?
2. Describe how point mutations and frameshift mutations affect the synthesis of proteins.
3. Describe why a mutation of a sperm or egg cell has different consequences than a mutation of a heart cell.
4. What is the relationship between mutations and cancer?

Thinking Critically
5. Why do you think low levels of mutation might

be an adaptive advantage to a species, whereas high levels of mutation might be a disadvantage?

SKILL REVIEW

6. **Recognizing Cause and Effect** In an experiment with rats, the treatment group receives radiation while the control group does not. Months later, the treatment group has a greater percentage of rats with cancer and newborn rats with birth defects than the control group. Explain these observations. For more help, refer to *Thinking Critically* in the **Skill Handbook**.

---

3 Assess

Check for Understanding
Ask students to name the type of mutation involved in each of the following cases. (a) One kind of base in DNA takes the place of another. (b) Some genes are duplicated on the same chromosome. (c) Part of one chromosome breaks off and is attached to a different one. *(a) point mutation, (b) insertion, (c) translocation* **L1**

Reteach
Use an overhead projector to show Figures 11.10 and 11.11 without captions. Have students describe in their own words what is taking place in each type of mutation. **L1**

Extension
Ask students to research polyploidy in plants and find out the effects of extra sets of chromosomes on plants. **L3**

✓ Assessment
Knowledge Ask students to explain the effects of point, frameshift, and chromosomal mutations. **L1**

4 Close

Discussion
Have each student form a hypothesis about how mutations may be involved in evolution of a species. Discuss the hypotheses as a class. Correct any misconceptions students may have. **L2**

Section Assessment

1. A mutation is any mistake or change in the DNA sequence.
2. A point mutation may change a single amino acid in a protein. A frameshift mutation may alter every amino acid after the mutation because the shift in bases occurs all along the strand.
3. A mutation of a reproductive cell affects the individual's offspring.

A mutation of a heart cell affects only the individual.
4. Mutations of a body cell may cause cancer by altering the cell processes that control cell division. This makes the cells divide rapidly.
5. A change in the DNA might result in a better adaptation. A high rate of mutation could cause rapid speciation

or lead to extinction of a species. A low level of mutations provides stability.
6. Greater incidence of cancer can be explained by the radiation causing mutations in the processes that control cell division in body cells. More birth defects can be explained by mutations in the reproductive cells.

RNA Transcription

Time Allotment
One class period

Process Skills
sequence, observe and infer, recognize cause and effect

Safety Precautions
Remind students to use care when cutting with scissors and to cut away from their bodies.

PREPARATION

■ Instead of having students copy models onto construction paper, you may wish to use the masters provided in the BioLab and MiniLab Worksheets booklet. Copy these onto white paper and have students color the models with pencils or crayons.

Resource Manager

BioLab and MiniLab Worksheets, pp. 53-56 L2

*A*lthough DNA remains in the nucleus of a cell, it passes its information into the cytoplasm by way of another nucleic acid, messenger RNA. The base sequence of this mRNA is complementary to the sequence in the strand of DNA, and is produced by base pairing during transcription. In this activity, you will demonstrate the process of transcription through the use of paper DNA and mRNA models.

PREPARATION

Problem
How does the order of bases in DNA determine the order of bases in mRNA?

Objectives
In this BioLab, you will:
■ **Formulate a model** to show how the order of bases in DNA determines the order of bases in mRNA.
■ **Infer** why the structure of DNA enables it to be easily copied.

Materials
construction paper, 5 colors
scissors
clear tape

Safety Precautions
Be careful when using scissors. Always use goggles in the lab.

Skill Handbook
Use the **Skill Handbook** if you need additional help with this lab.

Parts for DNA Nucleotides

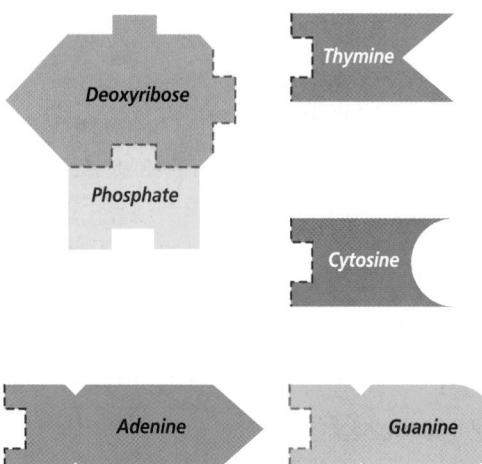

Extra Parts for RNA Nucleotides

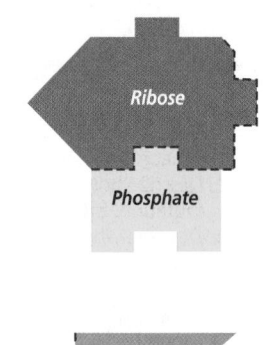

PROCEDURE

Teaching Strategies
■ You may wish to give each student two envelopes to hold the nucleotide pieces. One can be used for the DNA nucleotides and the other for the RNA nucleotides.
■ To make larger, more varied models, the lab groups could pool their models.

Data and Observations
Students should construct a DNA molecule and a complementary mRNA molecule.

PROCEDURE

1. Copy the illustrations of the four different DNA nucleotides onto your construction paper, making sure that each different nucleotide is on a different color paper. You should make ten copies of each nucleotide.

2. Using scissors, carefully cut out the shapes of each nucleotide.

3. Using any order of nucleotides that you wish, construct a double-stranded DNA molecule. If you need more nucleotides, copy them as before.

4. Fasten your molecule together using clear tape. Do not tape across base pairs.

5. As in step 1, copy the illustrations of A, G, and C nucleotides. Use the same colors of construction paper as in step 1. Use the fifth color of construction paper to make copies of uracil nucleotides.

6. With scissors, carefully cut out the nucleotide shapes.

7. With your DNA molecule in front of you, demonstrate the process of transcription by first pulling the DNA molecule apart between the base pairs.

8. Using only one of the strands of DNA, begin matching complementary RNA nucleotides with the exposed bases on the DNA model to make mRNA.

9. When you are finished, tape your new mRNA molecule together.

ANALYZE AND CONCLUDE

1. **Observing and Inferring** Does the mRNA model more closely resemble the DNA strand from which it was transcribed or the complementary strand that wasn't used? Explain your answer.

2. **Recognizing Cause and Effect** Explain how the structure of DNA enables the molecule to be easily transcribed. Why is this important for genetic information?

3. **Relating Concepts** Why is RNA important to the cell? How does an mRNA molecule carry information from DNA?

Going Further

Biology Journal Do library research to find out more about how the bases in DNA were identified and how the base pairing pattern was determined.

BIOLOGY *Online* To find out more about DNA, visit the Glencoe Science Web site.

science.glencoe.com

ANALYZE AND CONCLUDE

1. mRNA more closely resembles the complementary DNA. They have the same base sequence except that mRNA has uracil in place of thymine.

2. Because DNA is double-stranded, sections can be unzipped to allow complementary bases to hydrogen bond while the remaining DNA stays zipped. Thus, only the information needed at one time is being transcribed.

3. The mRNA is formed as a complementary copy of the genetic information. The RNA copy can leave the nucleus while the "master copy" stays protected within the nucleus.

✓ Assessment

Portfolio Have students write a lab report summarizing the lab and including answers to Analyze and Conclude. Use the Performance Task Assessment List for Lab Report in **PASC**, p. 47. **L2** **P**

Going Further

Have students use Table 11.2 to determine which amino acids they "produced" in the BioLab. **L1**

MEETING INDIVIDUAL NEEDS

Visually Impaired

Cut out very large copies of the five different nucleotides (A, T, C, G, and U) so visually impaired students can participate in the BioLab.

Scanning Probe Microscopes

Purpose 🗂️

Students learn about the operation and the capabilities of scanning probe microscopy.

Background

If the current flowing through the probe is increased slightly above the levels required to produce an image, a scanning probe microscope can pick up atoms and move them around. Before the 1990s, molecules had to be removed from the cell for observation.

Teaching Strategies

■ Explain to students that this is the first type of microscopy capable of producing images at the scale of the nanometer, the scale at which molecular bonding occurs. A nanometer is 10^{-9} meters, or 0.001 micrometers. These microscopes can also produce images at somewhat larger scales, including 10^{-8}, the scale of the DNA molecule, 10^{-7}, the scale of the nucleus and other organelles, and 10^{-6}, the scale of intact cells.

■ Point out that biology is not the only field that has benefited from the invention of scanning probe microscopes. For example, physicists use them to look at the crystal lattice structure of metals and other materials. Chemists use them to observe how molecules behave during chemical reactions. Materials scientists use them to explore methods of fabricating ever-tinier computer chips.

Investigating the Technology

Scanning probe microscopes can show the arrangement of atoms on the surface of a molecule. Because specimens can be examined in liquid or air, biological molecules can be observed functioning as they would inside the cell.

Have you ever heard of someone dissecting a chromosome to get a closer look at DNA? Imagine an instrument small enough to allow you to grab hold of one of the nucleotides in a strand of DNA, yet powerful enough to provide a detailed image.

Scanning probe microscopes can show the arrangement of atoms on the surface of a molecule. They make it possible for scientists to pick up molecules, and even atoms, and move them around. They can also be used to observe how biological molecules interact. There are many types of scanning probe microscopes. All of them use a very sharp probe that may be only a single atom wide at its tip. The probe sits very close to the specimen, but does not actually touch it. As the probe moves across, or scans, the specimen, it measures some property of the specimen.

The scanning tunneling microscope (STM)

The STM uses a probe through which a tiny amount of electric current flows. As the probe scans a molecule, it encounters ridges and valleys formed by the different kinds of atoms on the molecule's surface. The probe moves up and down as needed to keep the current constant. The movements of the probe are recorded by a computer, which produces an image of the molecule.

The atomic force microscope (AFM)

The AFM can measure many different properties, including electricity, magnetism, and heat. As the probe moves across the specimen, changes in the property being measured move the probe. These changes are used to create the image.

What can they do? One of the primary advantages of scanning probe microscopy, besides its atomic-level resolution, is the ability to observe molecules in air or liquid. This means that biologists can "watch" molecules interact as they would inside a cell. In 1998, for example, biologists used the AFM to observe the behavior of an enzyme called RNA polymerase. This enzyme is involved in transcription. The AFM images show how the polymerase molecule binds to a strand

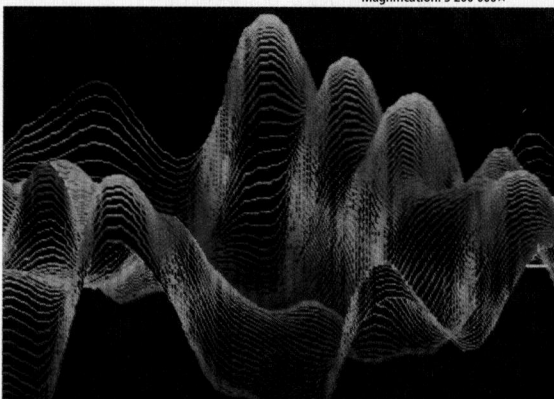

Magnification: 3 200 000×

STM image of DNA

of DNA and creates a strand of mRNA by gathering nucleotides from the surrounding liquid.

Applications for the Future

Biologists have used a combination of lasers and an AFM to study the physical properties of DNA molecules. Laser "tweezers" hold down one end of a coiled DNA helix and pull on the other end. The AFM measures the forces that hold the strand together and the forces that cause it to coil and uncoil as it performs its functions in the cell.

INVESTIGATING THE TECHNOLOGY

Thinking Critically What advantages do scanning probe microscopes have over other types of microscopes?

 To find out more about microscopes, visit the Glencoe Science Web site. **science.glencoe.com**

Going Further

 **Intrapersonal** Improvements to this microscope technology continue to be made. Invite students to use the library or Internet to research recent images produced by scanning probe microscopes. **L2**

GLENCOE TECHNOLOGY

VIDEODISC
The Infinite Voyage
Unseen Worlds
The Scanning Tunneling Microscope: Observing Atomic Particles (Ch. 10)

 2 min.

Chapter 11 Assessment

SUMMARY

Section 11.1

DNA: The Molecule of Heredity

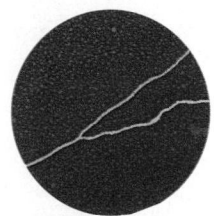

Main Ideas

■ DNA, the genetic material of organisms, is composed of four kinds of nucleotides. A DNA molecule consists of two strands of nucleotides with sugars and phosphates on the outside and bases paired by hydrogen bonding on the inside. The paired strands form a twisted-zipper shape called a double helix.

■ Because adenine can pair only with thymine, and guanine can pair only with cytosine, DNA can replicate itself with great accuracy. This process keeps the genetic information constant through cell division and during reproduction.

Vocabulary

DNA replication (p. 290)
double helix (p. 289)
nitrogen base (p. 288)

Section 11.2

From DNA to Protein

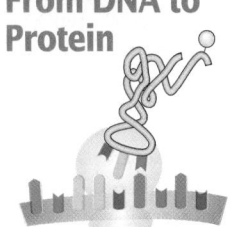

Main Ideas

■ Genes are small sections of DNA. Most sequences of three bases in the DNA of a gene code for a single amino acid in a protein.

■ The order of nucleotides in DNA determines the order of nucleotides in messenger RNA in a process called transcription.

■ Translation is a process through which the order of bases in messenger RNA codes for the order of amino acids in a protein.

Vocabulary

codon (p. 297)
messenger RNA (p. 295)
ribosomal RNA (p. 296)
transcription (p. 296)
transfer RNA (p. 296)
translation (p. 299)

Section 11.3

Genetic Changes

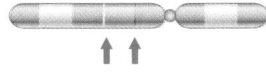

Main Ideas

■ A mutation is a change in the base sequence of DNA. Mutations may affect only one gene, or they may affect whole chromosomes.

■ Mutations in eggs or sperm affect future generations by producing offspring with new characteristics. Mutations in body cells affect only the individual and may result in cancer.

Vocabulary

chromosomal mutation (p. 305)
frameshift mutation (p. 305)
mutagen (p. 306)
mutation (p. 302)
point mutation (p. 304)

Main Ideas

Summary statements can be used by students to review the major concepts of the chapter.

Using the Vocabulary

To reinforce chapter vocabulary, use the Content Mastery Booklet and the activities in the Interactive Tutor for Biology: The Dynamics of Life on the Glencoe Science Web site. **science.glencoe.com**

All Chapter Assessment questions and answers have been validated for accuracy and suitability by The Princeton Review.

UNDERSTANDING MAIN IDEAS

1. c
2. b

UNDERSTANDING MAIN IDEAS

1. Which of the following processes requires DNA replication?
   a. transcription   **c.** mitosis
   b. translation   **d.** protein synthesis

2. In which of the following processes does the DNA unzip?
   a. transcription and translation
   b. transcription and replication
   c. replication and translation
   d. all of these

GLENCOE TECHNOLOGY

 **VIDEOTAPE**
MindJogger Videoquizzes
Chapter 11: *DNA and Genes*
Have students work in groups as they play the videoquiz game to review key chapter concepts.

Resource Manager

Chapter Assessment, pp. 61-66
MindJogger Videoquizzes
ExamView® Pro Software
BDOL Interactive CD-ROM, Chapter 11 quiz

3. a

4. d

5. c

6. a

7. d (Remember the genetic code is the same for all living things.)

8. c

9. b

10. a

11. transcription

12. messenger

13. replication

14. transfer RNA, ribosomes or mRNA

15. double helix

16. deoxyribose

17. hydrogen

18. reproductive

19. point mutation, mutagen

20. replication

◄ APPLYING MAIN IDEAS ►

21. The lung cell mutation does not affect the reproductive cells. Thus, this mutation would not be passed on to offspring.

---

3. Which DNA strand can base pair to the following DNA strand?

a. T-A-C-G-A-T
b. A-T-G-C-T-A
c. U-A-C-G-A-U
d. A-U-G-C-U-A

4. Which of the following nucleotide chains could be part of a molecule of RNA?
a. A-T-G-C-C-A
b. A-A-T-A-A-A
c. G-C-C-T-T-G
d. A-U-G-C-C-A

5. Which of the following mRNA codons would cause synthesis of a protein to terminate? Refer to *Table 11.2*.
a. G-G-G
b. U-A-C
c. U-A-G
d. A-A-G

6. A DNA sequence of A-C-C would create an mRNA codon for which amino acid? Refer to *Table 11.2*.
a. tryptophan
b. serine
c. leucine
d. phenylalanine

7. The genetic code for an oak tree is _____.
a. more similar to an ash tree than to a squirrel
b. more similar to a chipmunk than to a maple tree
c. more similar to a mosquito than to an elm tree
d. exactly the same as for an octopus

8. Which of the following base pairs would not be found in a cell?
a. adenine—thymine
b. cytosine—guanine
c. thymine—uracil
d. adenine—uracil

9. A protein is assembled amino acid-by-amino acid during the process of _____.
a. replication
b. translation
c. transcription
d. mutation

THE PRINCETON REVIEW **TEST-TAKING TIP**

Use as Much Time as You Can
You will not get extra points for finishing early. Work slowly and carefully on any test and make sure you don't make careless errors because you are hurrying to finish.

---

10. A deer is born normal, but UV rays cause a mutation in its retina. Which of the following statement is *least* likely to be true?
a. The mutation may be passed on to the offspring of the deer.
b. The mutation may cause retinal cancer.
c. The mutation may interfere with the function of the retinal cell.
d. The mutation may interfere with the structure of the retinal cell.

11. In the process of _____, enzymes make an RNA copy of a DNA strand.

12. The RNA copy that carries information from DNA in the nucleus into the cytoplasm is _____ RNA.

13. DNA is copied before a cell divides in the process called _____.

14. Molecules of _____ bring amino acids to the _____ for assembly into proteins.

15. The shape of a molecule of DNA is called a _____.

16. RNA has a ribose sugar, whereas DNA has a _____ sugar.

17. Nucleotides form base pairs through a weak bond called a _____ bond.

18. A female lab rat is exposed to X rays. Its future offspring will be affected only if a mutation occurs in the rat's _____ cells.

19. Chemical Q causes the following change in the sequences of nucleotides. This change is an example of a _____. Chemical Q is a _____.

A T G A A A ➔ A T T A A A

20. DNA _____ is necessary before a cell divides so each cell has a complete copy of the chromosomes.

◄ APPLYING MAIN IDEAS ►

21. Explain why a mutation in a lung cell would not be passed on to offspring.

22. Explain why codons can't consist of two bases instead of three for each amino acid.

23. A bricklayer has an assistant who brings bricks to the bricklayer so she can build a wall. What part of translation most closely resembles the assistant's job? What do the bricks represent?

THINKING CRITICALLY

24. **Making Inferences** Explain how the universality of the genetic code is evidence that all organisms alive today evolved from a common ancestor in the past.

25. **Analyzing** Identify the type of chromosomal mutation illustrated in each diagram below.

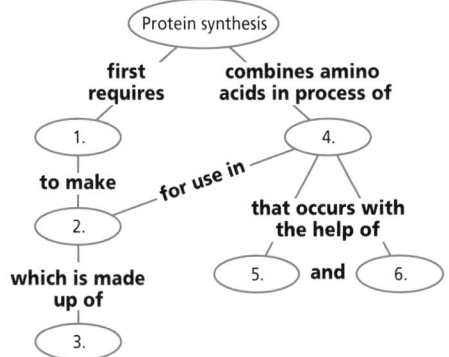

a K L M N O   P R Q S

b F G H I J   K M N

c K L M N O   O P Q X Y R

26. **Concept Mapping** Complete the concept map by using the following vocabulary terms: transfer RNA, codons, messenger RNA, transcription, translation, ribosomal RNA.

Protein synthesis

first requires — combines amino acids in process of

1. — 4.

to make — for use in — that occurs with the help of

2. — 5. and 6.

which is made up of

3.

CD-ROM

For additional review, use the assessment options for this chapter found on the *Biology: The Dynamics of Life Interactive CD-ROM* and on the Glencoe Science Web site.
science.glencoe.com

ASSESSING KNOWLEDGE & SKILLS

The following graph records the amount of DNA in liver cells that have been grown in a culture so that all the cells are at the same phase in the cell cycle.

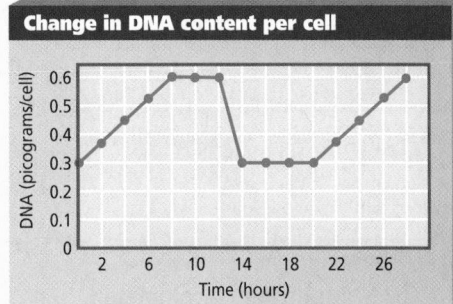

Change in DNA content per cell

DNA (picograms/cell) vs. Time (hours)

Interpreting Data Use the data in the graph to answer the following questions.

1. During the course of the experiment, these cells went through cell division. What is this type of division called?
 a. transcription   **c.** mitosis
 b. translation   **d.** meiosis

2. During which hours were the cells carrying out cell division?
 a. 0-8 hours   **c.** 12-14 hours
 b. 8-10 hours   **d.** 14-20 hours

3. Which phase of the cell cycle were the cells in during hours 2-6?
 a. interphase   **c.** prophase
 b. telophase   **d.** anaphase

4. If you added radioactive thymine to the culture at 0 hour, what would happen to the amount incorporated into the DNA between hours 20 and 28 relative to the amount at 0 hour?
 a. stay the same   **c.** double
 b. divide in half   **d.** triple

5. **Predicting** Predict what will be the DNA content of the cells, in picograms, at 33 hours.

22. A sequence of two bases of four different kinds will produce 16 different combinations. The nucleotides must code for at least 20 amino acids.

23. The bricklayer's assistant is analogous to transfer RNA. The bricks represent amino acids.

THINKING CRITICALLY

24. The genetic code of all organisms is based on the same four nucleotides and the same sequences, suggesting a link to a common ancestor.

25. (a) inversion; (b) deletion; (c) insertion

26. 1. Transcription; 2. Messenger RNA; 3. Codons; 4. Translation; 5. and 6. Transfer RNA, Ribosomal RNA

ASSESSING KNOWLEDGE & SKILLS

1. c
2. c
3. a
4. c
5. The DNA content will drop to 0.3 picograms/cell.

Chapter 12 Organizer

Refer to pages 4T-5T of the Teacher Guide for an explanation of the National Science Education Standards correlations.

Section	Objectives	Activities/Features
Section 12.1 **Mendelian Inheritance of Human Traits** National Science Education Standards UCP.2, UCP.3; A.1, A.2; C.2; F.1; G.1, G.2 (1 session, $\frac{1}{2}$ block)	1. **Interpret** a pedigree. 2. **Determine** human genetic disorders that are caused by inheritance of recessive alleles. 3. **Predict** how a human trait can be determined by a simple dominant allele.	**MiniLab 12-1:** Illustrating a Pedigree, p. 316 **Problem-Solving Lab 12-1,** p. 317
Section 12.2 **When Heredity Follows Different Rules** National Science Education Standards UCP.2, UCP.3; A.1, A.2; C.2; F.4; G.1-3 (3 sessions, $2\frac{1}{2}$ blocks)	4. **Distinguish** between incompletely dominant and codominant alleles. 5. **Compare** multiple allelic and polygenic inheritance. 6. **Analyze** the pattern of sex-linked inheritance. 7. **Summarize** how internal and external environments affect gene expression.	**Problem-Solving Lab 12-2,** p. 324 **Design Your Own BioLab:** What is the pattern of cytoplasmic inheritance? p. 336
Section 12.3 **Complex Inheritance of Human Traits** National Science Education Standards UCP.2, UCP.3, UCP.5; A.1, A.2; C.2; F.1; G.1-3 (2 sessions, $1\frac{1}{2}$ blocks)	8. **Compare** codominance, multiple allelic, sex-linked, and polygenic patterns of inheritance in humans. 9. **Distinguish** among conditions in which extra autosomal or sex chromosomes exist.	**Inside Story:** The ABO Blood Group, p. 331 **Problem-Solving Lab 12-3,** p. 332 **MiniLab 12-2:** Detecting Colors and Patterns in Eyes, p. 333 **Social Studies Connection:** Queen Victoria and Royal Hemophilia, p. 338

Need Materials? Contact Carolina Biological Supply Company at 1-800-334-5551 or at **http://www.carolina.com**

MATERIALS LIST

BioLab
p. 336 *Brassica* rapa seeds (normal and variegated), potting soil, potting trays, paintbrushes, forceps, single-edge razor blade, light source, labels

MiniLabs
p. 316 pencil, paper
p. 333 hand lens, colored pencils, paper

Alternative Labs
p. 322 petri dish, label, paper towels, scissors, tobacco seeds
p. 332 microscope, prepared slides of male and female human cheek cells

Quick Demos
p. 318 none
p. 323 none
p. 334 none

Key to Teaching Strategies

L1 Level 1 activities should be appropriate for students with learning difficulties.

L2 Level 2 activities should be within the ability range of all students.

L3 Level 3 activities are designed for above-average students.

ELL ELL activities should be within the ability range of English Language Learners.

COOP LEARN Cooperative Learning activities are designed for small group work.

P These strategies represent student products that can be placed into a best-work portfolio.

These strategies are useful in a block scheduling format.

Patterns of Heredity and Human Genetics

Teacher Classroom Resources

Section	Reproducible Masters	Transparencies
Section 12.1 **Mendelian Inheritance of Human Traits**	Reinforcement and Study Guide, p. 51 `L2` Critical Thinking/Problem Solving, p. 12 `L3` BioLab and MiniLab Worksheets, p. 57 `L2` Tech Prep Applications, pp. 19-20 `L2` Content Mastery, pp. 57-58, 60 `L1`	Section Focus Transparency 29 `L1` `ELL` Reteaching Skills Transparency 20 `L1` `ELL`
Section 12.2 **When Heredity Follows Different Rules**	Reinforcement and Study Guide, pp. 52-53 `L2` Concept Mapping, p. 12 `L3` `ELL` Laboratory Manual, pp. 83-86 `L2` Content Mastery, pp. 57, 59-60 `L1`	Section Focus Transparency 30 `L1` `ELL` Reteaching Skills Transparency 21 `L1` `ELL`
Section 12.3 **Complex Inheritance of Human Traits**	Reinforcement and Study Guide, p. 54 `L2` BioLab and MiniLab Worksheets, pp. 58-60 `L2` Laboratory Manual, pp. 87-90 `L2` Content Mastery, pp. 57, 60 `L1`	Section Focus Transparency 31 `L1` `ELL`

Assessment Resources

Chapter Assessment, pp. 67-72

MindJogger Videoquizzes

Performance Assessment in the Biology Classroom

Alternate Assessment in the Science Classroom

ExamView® Pro Software 💾

BDOL Interactive CD-ROM, Chapter 12 quiz

Additional Resources

Spanish Resources `ELL`

English/Spanish Audiocassettes `ELL`

Cooperative Learning in the Science Classroom `COOP LEARN`

Lesson Plans/Block Scheduling

NATIONAL GEOGRAPHIC — Teacher's Corner

Index to National Geographic Magazine
The following articles may be used for research relating to this chapter:
"The Family Line: The Human-Cat Connection," by Stephen J. O'Brien, June 1997.

GLENCOE TECHNOLOGY

The following multimedia resources are available from Glencoe.

Biology: The Dynamics of Life

CD-ROM `ELL`

  Exploration: *Trait Inheritance*
  Video: *Fruit Fly Genetics*
  Animation: *Sex-linked Traits*
  Exploration: *Blood Types*

Videodisc Program

  Sex-linked Traits

The Infinite Voyage

  A Taste of Health
  The Geometry of Life

Many genes are needed to code for all the traits shown by the family in the photo. Demonstrate how this large amount of DNA can fit into the small volume of a chromosome by coiling a long piece of wire. First, coil the wire so that it looks like a telephone cord, then coil the coiled wire to make it even more compact.

Theme Development

The main theme of the chapter is **homeostasis**, which is normally maintained during the transmission of genetic material but is disrupted by the inheritance of particular alleles that result in genetic disorders. The **nature of science** is illustrated by the concepts developed by Morgan as he worked with and interpreted data from sex-linked traits.

0:00 OUT OF TIME?

If time does not permit teaching the entire chapter, use the BioDigest at the end of the unit as an overview.

READING BIOLOGY

Glencoe's *Biology: The Dynamics of Life* contains many resources to assist a student's reading skills. Each chapter contains figures with expanded captions that expand on written material. Word Origins, located along the side of text, expand knowledge of biology vocabulary. Glencoe's Content Mastery Booklet helps develop reading skills while reinforcing content. In addition, use the Interactive Tutor for *Biology: The Dynamics of Life* on the Glencoe Web site to reinforce vocabulary. **science.glencoe.com**

Chapter

12 Patterns of Heredity and Human Genetics

What You'll Learn

- You will compare the inheritance of recessive and dominant traits in humans.
- You will analyze the inheritance of incompletely dominant and codominant traits.
- You will determine the inheritance of sex-linked traits.

Why It's Important

The transmission of traits from generation to generation affects your appearance, your behavior, and your health. Understanding how these traits are inherited is important in understanding the traits you may pass on to a future generation.

READING BIOLOGY

Before starting the chapter, make a list of some physical characteristics that may appear in your family members. As you read the chapter, note whether your family members have dominant or recessive traits discussed in the text.

BIOLOGY Online

To find out more about human genetics, visit the Glencoe Science Web site. **science.glencoe.com**

Magnification: 4500×

It is difficult to imagine how the information for such varied traits as eye or hair color and athletic talent could be contained in the nucleic acids composing this chromosome.

314

12.1 Mendelian Inheritance of Human Traits

As you learn about traits, you will see that some, such as tongue rolling or a widow's peak hairline, are relatively harmless. Other traits produce devastating disorders and even death. All of these traits demonstrate how genes are inherited, and this is what you need to learn. The disorders caused by genetic transmission of traits are the motivation that drives scientists to do research to discover treatments and cures.

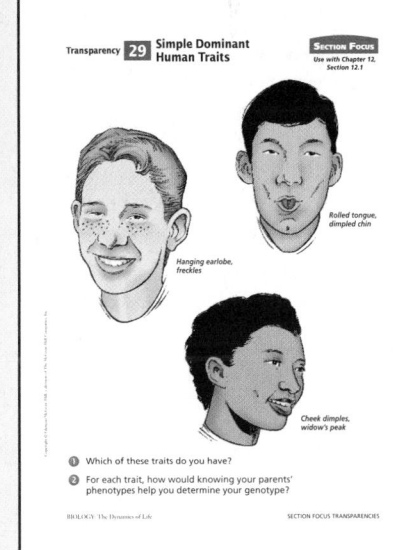

Tongue rolling (above) and widow's peak hairline (inset)

SECTION PREVIEW

Objectives

Interpret a pedigree.

Determine human genetic disorders that are caused by inheritance of recessive alleles.

Predict how a human trait can be determined by a simple dominant allele.

Vocabulary

pedigree
carrier
fetus

Making a Pedigree

At some point, you have probably seen a family tree, either for your family or for someone else's. A family tree traces a family name and various family members through successive generations. Through a family tree, you can trace your cousins, aunts, uncles, grandparents, and great-grandparents.

Pedigrees illustrate inheritance

Geneticists often need to map the inheritance of genetic traits from generation to generation. A **pedigree** is a graphic representation of genetic inheritance. At a glance, it looks very similar to any family tree.

A pedigree is made up of a set of symbols that identify males and females, individuals affected by the trait being studied, and family relationships. Some commonly used symbols are shown in *Figure 12.1*. A circle represents a female; a square

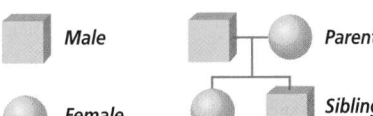

Male — Parents

Female — Siblings

Affected male — Known heterozygotes for recessive allele

Affected female

Mating — Death

Figure 12.1
Geneticists use these symbols to make and analyze a pedigree.

Prepare

Key Concepts

The section begins with a discussion of pedigrees and their interpretation. Then the inheritance of autosomal recessive disorders such as cystic fibrosis, phenylketonuria, and Tay-Sachs disease is described. The section closes with a discussion of autosomal traits such as tongue rolling, widow's peak, and Huntington's disease.

Planning

■ Obtain a long piece of wire for the Getting Started Demo.

---

1 Focus

Bellringer

Before presenting the lesson, display **Section Focus Transparency 29** on the overhead projector and have students answer the accompanying questions.
L1 **ELL**

Transparency **29** Simple Dominant Human Traits

Hanging earlobe, freckles

Rolled tongue, dimpled chin

Cheek dimples, widow's peak

❶ Which of these traits do you have?

❷ For each trait, how would knowing your parents' phenotypes help you determine your genotype?

Assessment Planner

Portfolio Assessment
> **Portfolio, TWE,** pp. 316, 321, 325, 330
> **Problem-Solving Lab, TWE,** p. 324
> **Assessment, TWE,** p. 330

Performance Assessment
> **Problem-Solving Lab, TWE,** p. 317
> **Assessment, TWE,** pp. 318, 323, 328, 334
> **Alternative Lab, TWE,** p. 332-333
> **MiniLab, SE,** pp. 316, 333
> **BioLab, SE,** pp. 336-337

Alternative Lab, pp. 322-323, 332-333

Knowledge Assessment
> **MiniLab, TWE,** pp. 316, 333
> **Problem-Solving Lab, TWE,** p. 332
> **Section Assessment, SE,** pp. 320, 328, 335
> **Chapter Assessment, SE,** pp. 339-341

Skill Assessment
> **Assessment, TWE,** pp. 320, 335
> **Alternative Lab, TWE,** pp. 322-323
> **BioLab, TWE,** pp. 336-337

2 Teach

Purpose
Students will observe a specific human trait and prepare a pedigree.

Process Skills
observe and infer, interpret scientific illustrations

Teaching Strategies
■ If any of your students are adopted, be sure he or she is paired with a student who is not adopted and can contribute family information to the pair's pedigree.

■ Provide students with information about various human traits that are inherited in a simple Mendelian pattern.

Expected Results
Students will construct pedigrees of a human trait.

Analysis
1. Answers may include earlobe shape, widow's peak, tongue rolling, or ability to taste PTC paper.
2. The number of individuals may be too small to determine the inheritance pattern.

Assessment
Knowledge Provide a pedigree that consists of only a few individuals and has enough information so that students can determine genotypes for the given phenotypes. Use the Performance Task Assessment List for Analyzing the Data in **PASC**, p. 27. **L1**

BioLab and MiniLab Worksheets, p. 57 **L2**
Section Focus Transparency 29 and Master **L1** **ELL**

316

MiniLab 12-1 Analyzing Information

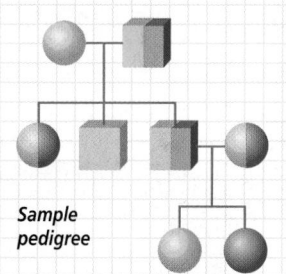

Illustrating a Pedigree The pedigree method of studying a trait in a family uses records of phenotypes extending for two or more generations. Studies of pedigrees can be used to yield a great deal of genetic information about a related group.

Sample pedigree

Procedure
1. Working with a partner, choose one human trait, such as attached and free-hanging earlobes or tongue rolling, that interests both of you.
2. Using either your or your partner's family, collect information about your chosen trait. Include whether each individual is male or female, does or does not have the trait, and the relationship of the individual to others in the family.
3. Use your information to draw a pedigree for the trait.
4. Try to determine how your trait is inherited.

Analysis
1. What trait did you study? Can you determine from your pedigree what the apparent inheritance pattern of the trait is?
2. How is the study of inheritance patterns limited by pedigree analysis?

Figure 12.2
This pedigree shows how a rare, recessive allele is transmitted from generation to generation.

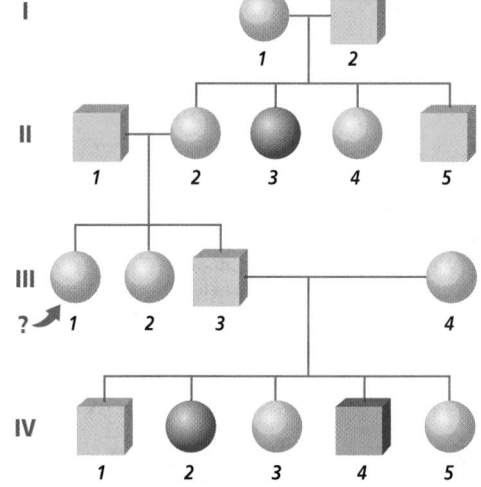

---

represents a male. Shaded circles and squares represent individuals showing the trait being studied. Unshaded circles and squares designate individuals that do not show the trait. A half-shaded circle or square represents a **carrier,** a heterozygous individual. A horizontal line connecting a circle and a square indicates that the individuals are parents, and a vertical line connects a set of parents with their offspring. Each horizontal row of circles and squares in a pedigree designates a generation, with the most recent generation shown at the bottom. The generations are identified in sequence by Roman numerals, and each individual is given an Arabic number. You can practice using these symbols to make a pedigree in the *MiniLab* on this page.

Analyzing a pedigree

An example of a pedigree for a fictitious rare, recessive disorder in humans is shown in *Figure 12.2*. This disorder could be any of several recessive disorders in which the disorder shows up only if the affected person carries two recessive alleles for the trait. Follow the pedigree as you read how to analyze it.

Suppose individual III-1 in the pedigree wants to know the likelihood of passing on this allele to her children. By studying the pedigree, the individual will be able to determine the likelihood that she carries the allele. Notice that information can also be gained about other members of the family by studying the pedigree. For example, you know that I-1 and I-2 are both carriers of the recessive allele for the trait because they have produced II-3, who shows the recessive phenotype. If you drew a Punnett square for the mating of individuals I-1 and I-2, you

---

Portfolio

Blue People

Linguistic Have students read "The Blue People of Troublesome Creek," by Cathy Trost, *Science 82,* Nov. 1982, pp. 34-39, and construct a pedigree from the article. These people have an autosomal recessive gene that causes their skin to appear dark blue. Have students explain this disorder. **L2**

GLENCOE TECHNOLOGY

CD-ROM
Biology: The Dynamics of Life
Exploration: *Trait Inheritance*

Disc 2

would find, according to Mendelian segregation, that the ratio of homozygous dominant to heterozygous to homozygous recessive genotypes among their children would be 1: 2: 1. Of those genotypes possible for the members of generation II, only the homozygous recessive genotype will express the trait, which is the case for II-3.

You can't tell the genotypes of II-4 and II-5, but they have a normal phenotype. If you look at the Punnett square you made, you can see that the probability of II-4 and of II-5 being a carrier is each two out of three because they can have only two possible genotypes—homozygous normal and heterozygous. The homozygous recessive genotype is not a possibility in these individuals because neither of them shows the affected phenotype.

Because none of the children in generation III are affected and because the recessive allele is rare, it is reasonably safe to assume that II-1 is not a carrier. You know that individual II-2 must be a carrier like her parents because she has passed on the recessive allele to subsequent generation IV. Because individual III-1 has one parent who is heterozygous and the other parent who is assumed to be homozygous normal, III-1 most likely has a one-in-two chance of being a carrier. If her parent II-1 had been heterozygous instead of homozygous normal, III-1's chances of being a carrier are increased to two in three.

Simple Recessive Heredity

Most genetic disorders are caused by recessive alleles. Many of these alleles are relatively rare, but a few are common in certain ethnic groups. You can practice calculating the chance that offspring will be born with some of these genetic traits in the *Problem-Solving Lab* above.

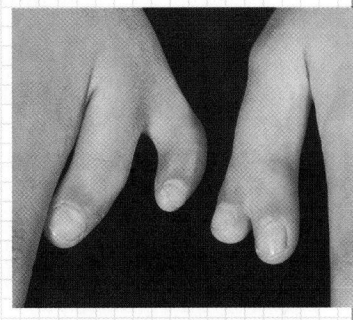

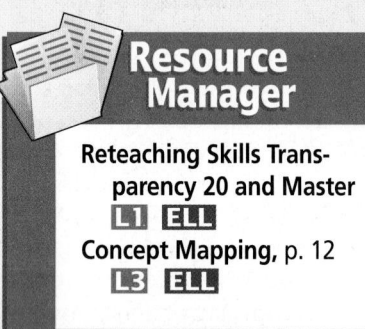
Cystic fibrosis

Cystic fibrosis (CF) is a fairly common genetic disorder among white Americans. Approximately one in 20 white Americans carries the recessive allele, and one in 2000 children born to white Americans inherits the disorder. Due to a defective protein in the plasma membrane, cystic fibrosis results in the formation and accumulation of thick mucus in the lungs and digestive tract. Physical therapy, special diets, and new drug therapies have continued to raise the average life expectancy of CF patients.

Tay-Sachs disease

Tay-Sachs (tay saks) disease is a recessive disorder of the central nervous system. In this disorder, a recessive allele results in the absence of an enzyme that normally breaks down a lipid produced and stored in tissues of the central nervous system. Therefore, this lipid fails to break

Figure 12.3
A study of families who have children with Tay-Sachs disease shows typical pedigrees for traits inherited as simple recessives. Note that the trait appears to skip generations, a characteristic of a recessive trait.

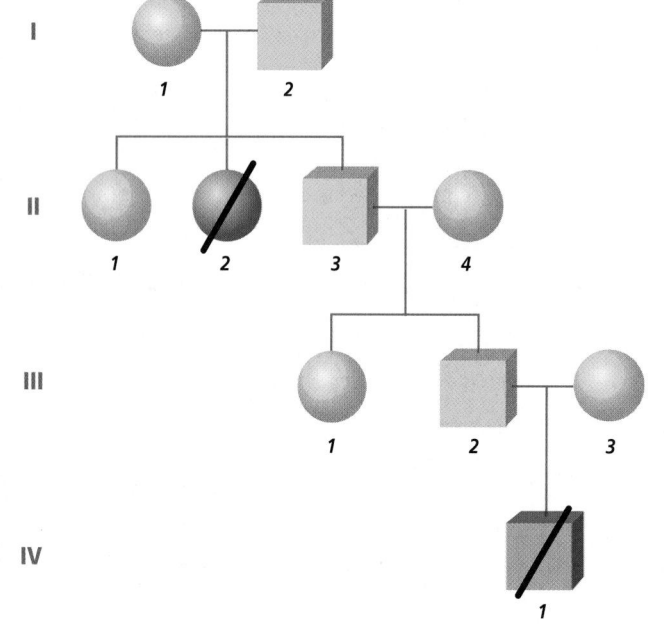

down properly and accumulates in the cells. The allele for Tay-Sachs is especially common in the United States among Ashkenazic Jews, whose ancestors came from eastern Europe. *Figure 12.3* shows a typical pedigree for Tay-Sachs disease.

Phenylketonuria

Phenylketonuria (fen ul keet un YOOR ee uh), also called PKU, is a recessive disorder that results from the absence of an enzyme that converts one amino acid, phenylalanine, to a different amino acid, tyrosine. Because phenylalanine cannot be broken down, it and its by-products accumulate in the body and result in severe damage to the central nervous system. The PKU allele is most common among people whose ancestors came from Norway or Sweden.

A homozygous PKU newborn appears healthy at first because its mother's normal enzyme level prevented phenylalanine accumulation during development. However, once the infant begins drinking milk, which is rich in phenylalanine, the amino acid accumulates and mental retardation occurs. Today, a PKU test is normally performed on all infants a few days after birth. Infants affected by PKU are given a diet that is low in phenylalanine until their brains are fully developed. With this special diet, the toxic effects of the disorder can be avoided.

Ironically, the success of treating phenylketonuria infants has resulted in a new problem. If a female who is homozygous recessive for PKU becomes pregnant, the high phenylalanine levels in her blood can damage her **fetus**—the developing baby. This problem occurs even if the fetus is heterozygous and would be phenotypically normal. You may have

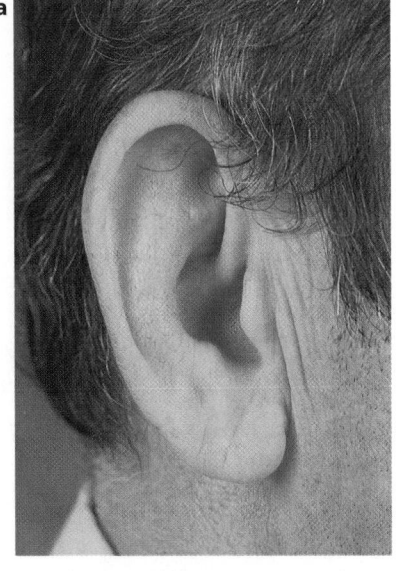
a

b

Figure 12.4
The allele *F* for freely hanging earlobes (**a**) is dominant to the allele *f* for attached earlobes (**b**). The Hapsburg lip, a protruding lower lip that results in a half-open mouth, has been traced back to the fourteenth century through portraits of members of the Hapsburg Dynasty of Europe (**c**).

c

noticed PKU warnings on cans of diet soft drinks. Because most diet drinks are sweetened with an artificial sweetener that contains phenylalanine, a pregnant woman who is homozygous recessive must limit her intake of diet foods.

Simple Dominant Heredity

Unlike the inheritance of recessive traits in which a recessive allele must be inherited from both parents for a person to show the recessive phenotype, many traits are inherited just as the rule of dominance predicts. Remember that in Mendelian inheritance, a single dominant allele inherited from one parent is all that is needed for a person to show the dominant trait.

Simple dominant traits

Tongue rolling is one example of a simple dominant trait. If you can roll your tongue, you've inherited the dominant allele from at least one of your parents. A Hapsburg lip is shown in *Figure 12.4* along with earlobe types, another dominant trait that is determined by simple Mendelian inheritance. Having earlobes that are attached to the head is a recessive trait (*ff*), whereas heterozygous (*Ff*) and homozygous dominant (*FF*) individuals have earlobes that hang freely.

There are many other human traits that are inherited by simple dominant inheritance. *Figure 12.5* shows one of these traits—hitchhiker's thumb, the ability to bend your thumb tip

Figure 12.5
Hitchhiker's thumb is a dominant trait.

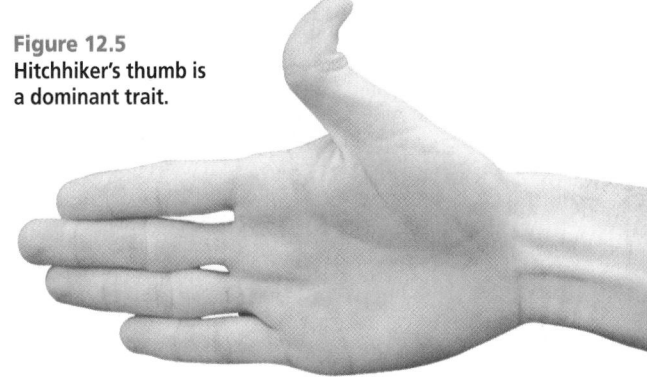

3 Assess

Check for Understanding

Ask students to summarize why the study of genetics is important to couples considering having children.

Reteach

Visual-Spatial Have students make a table of the genetic disorders described in this section, including the type of inheritance and the effects of the disorder on an affected individual. **L1**

Extension

Ask student groups to contact the local March of Dimes organization to gather information on the help it gives individuals with genetic disorders. Students can give a report on their findings. **L1** **COOP LEARN**

✔ Assessment

Skill Ask students to design a handout that tells about a genetic disorder. Students could draw Punnett squares and illustrate the chances of offspring inheriting the disorder from parents who are carriers. **L1**

4 Close

Discussion

Discuss with students whether genetic testing should be required before obtaining a marriage license.

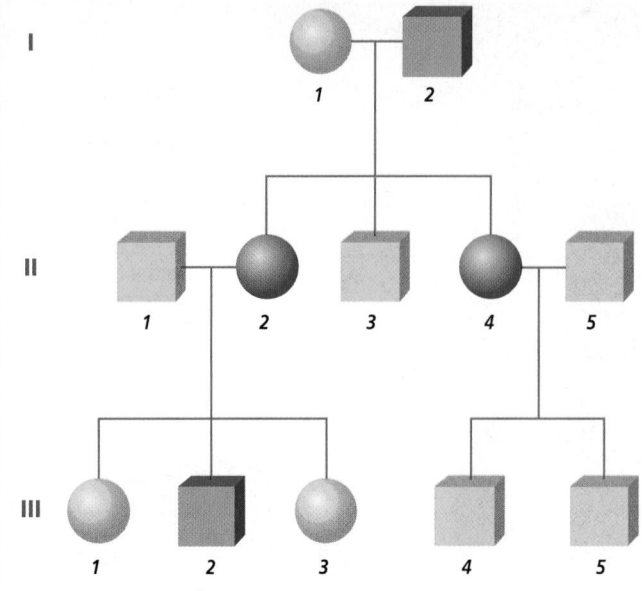

Figure 12.6
This is a typical pedigree for a simple, dominant inheritance of Huntington's disease. This particular chart shows the disorder in each generation and equally distributed among males and females.

backward more than 30 degrees. A straight thumb is recessive. Other dominant traits in humans include almond-shaped eyes (round eyes are recessive), thick lips (thin lips are recessive), and the presence of hair on the middle section of your fingers.

Huntington's disease

Huntington's disease is a lethal genetic disorder caused by a rare dominant allele. It results in a breakdown of certain areas of the brain. No effective treatment exists.

Ordinarily, a dominant allele with such severe effects would result in death before the affected individual could have children and pass the allele on to the next generation. But because the onset of Huntington's disease usually occurs between the ages of 30 and 50, an individual may have children before knowing whether he or she carries the allele. A genetic test has been developed that allows individuals to check their DNA. Although this test allows carriers to decide whether they want to have children and risk passing the trait on to future generations, it also places a tremendous burden on them in knowing they will develop the disease. For this reason, some people may choose not to be tested. The pedigree in *Figure 12.6* shows a typical pattern of occurrence of Huntington's disease in a family.

Notice that every child of an affected individual has a 50 percent chance of being affected and then a 50 percent chance of passing the defective allele to his or her own child.

Section Assessment

Understanding Main Ideas

1. In your own words, define the following symbols used in a pedigree: a square, a circle, an unshaded circle, a shaded square, a horizontal line, and a vertical line.
2. Describe one genetic disorder that is inherited as a recessive trait.
3. How are the cause and onset of symptoms of Huntington's disease different from those of PKU and Tay-Sachs disease?
4. Describe one trait that is inherited as a dominant allele. If you carried that trait, would you necessarily pass it on to your children?

Thinking Critically

5. Suppose that a child with free-hanging earlobes has a mother with attached earlobes. Can a man with attached earlobes be the child's father?

SKILL REVIEW

6. **Interpreting Scientific Illustrations** Make a pedigree for three generations of a family that shows at least one member of each generation who demonstrates a particular trait. Would this trait be dominant or recessive? For more help, refer to *Thinking Critically* in the **Skill Handbook.**

Section Assessment

1. A square represents a male, a circle a female. An unshaded circle is an unaffected female. A shaded square is an affected male. A horizontal line indicates two parents. A vertical line indicates offspring (children).
2. Students could describe cystic fibrosis, Tay-Sachs, or phenylketonuria.

3. Huntington's disease is an autosomal dominant disorder with onset between the ages of 30 and 50, whereas PKU and Tay-Sachs disease are autosomal recessive disorders with onset at birth.
4. Huntington's disease, tongue rolling, widow's peak, a Hapsburg lip are all examples of dominant traits. If your children inherit even one dominant

allele from you, they will express a dominant trait.
5. The man cannot be the father because the child had to receive an allele for free-hanging earlobes from one parent; the father would have to have at least one dominant allele for this trait.
6. This trait would be dominant. See Figure 12.6 for a sample pedigree.

12.2 When Heredity Follows Different Rules

V *ariations in the pattern of inheritance explained by Mendel became known soon after his work was discovered. What do geneticists do when observed patterns of inheritance, such as kernel color in this ear of corn, do not appear to follow Mendel's laws? They often use a strategy of piecing together bits of a puzzle until the basis for the unfamiliar inheritance pattern is understood.*

The genetics of Indian corn (above) is often like a puzzle (inset).

SECTION PREVIEW

Objectives

Distinguish between incompletely dominant and codominant alleles.

Compare multiple allelic and polygenic inheritance.

Analyze the pattern of sex-linked inheritance.

Summarize how internal and external environments affect gene expression.

Vocabulary

incomplete dominance
codominant alleles
multiple alleles
autosome
sex chromosome
sex-linked trait
polygenic inheritance

Complex Patterns of Inheritance

Patterns of inheritance that are explained by Mendel's experiments are often referred to as simple Mendelian inheritance—the inheritance controlled by dominant and recessive paired alleles. However, many inheritance patterns are more complex than those studied by Mendel. As you will learn, most traits are not simply dominant or recessive. The *BioLab* at the end of this chapter investigates a type of inheritance that doesn't even involve chromosomes.

Incomplete dominance: Appearance of a third phenotype

When inheritance follows a pattern of dominance, heterozygous and homozygous dominant individuals both have the same phenotype. When traits are inherited in an **incomplete dominance** pattern, however, the phenotype of the heterozygote is intermediate between those of the two homozygotes. For example, if a homozygous red-flowered snapdragon plant *(RR)* is crossed with a homozygous white-flowered snapdragon plant *(R'R')*, all of the F_1 offspring will have pink

---

BIOLOGY JOURNAL

Codominance

Provide students with practice working with codominance by having them determine the phenotypes of offspring resulting from the following crosses. (a) a checkered rooster mated to a checkered hen; (b) a checkered rooster mated to a white hen; (c) a checkered rooster mated to a black hen. **L2**

✓ **Portfolio**

Comparing Inheritance Patterns

Have students show through a series of Punnett squares how the genotypic and phenotypic ratios of the offspring would differ if the trait for chicken feather color were inherited through Mendelian dominance and incomplete dominance compared with the actual pattern of codominance. **L3** **P**

Prepare

Key Concepts

Students are shown the difference between codominance and incomplete dominance and are given examples of multiple-allelic traits, sex-linked traits, and polygenic inheritance. The section ends with a brief description of how internal and external environmental factors can affect the appearance of certain traits.

Planning

■ Gather small pots, soil, and lights for the BioLab.

■ Make cardboard X and Y chromosomes for Meeting Individual Needs.

■ Buy mustard seeds for the Project.

---

1 Focus

Bellringer

Before presenting the lesson, display **Section Focus Transparency 30** on the overhead projector and have students answer the accompanying questions. **L1** **ELL**

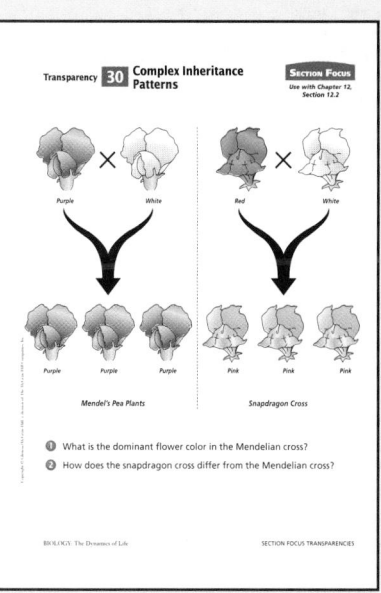

2 Teach

Discussion

Exhibit photos of a red shorthorn bull, a white shorthorn cow, and a roan shorthorn cow. Tell students that the roan cow, offspring of the red bull and the white cow, has both red and white hairs. Ask students whether Mendel's law of dominance applies to the inheritance of coat color in cattle. Why or why not? *No, because a third phenotype appears when the red and white cattle are crossed, and traits that Mendel studied had only two phenotypes.* Ask students what type of inheritance pattern explains how roan cattle express their coat color. *codominance*

GLENCOE TECHNOLOGY

VIDEODISC
The Secret of Life
Incomplete Dominance

DESIGN YOUR OWN BioLab
The BioLab at the end of the chapter can be used at this point in the lesson.

Resource Manager
Section Focus Transparency 30 and Master **L1** **ELL**

Figure 12.7
A Punnett square of snapdragon color shows that the red snapdragon is homozygous for the allele *R*, and the white snapdragon is homozygous for the allele *R′*. All of the pink snapdragons are heterozygous, or *RR′*.

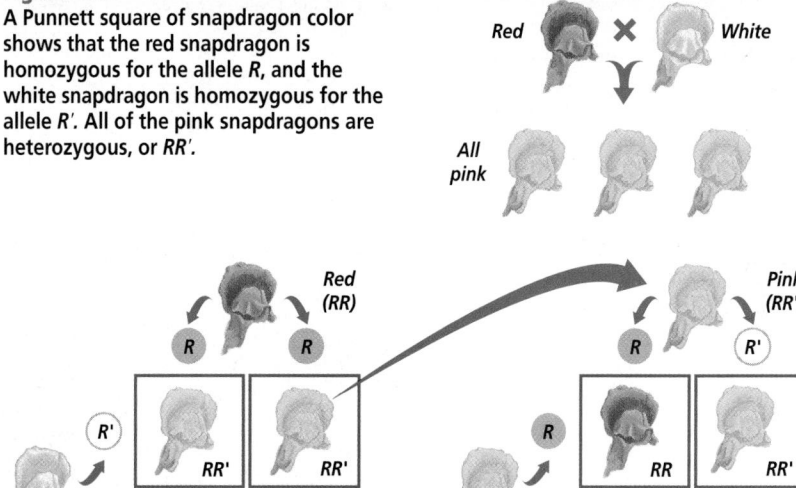

Red × White

All pink

Red (RR) Pink (RR′)

White (R′R′) Pink (RR′)

All pink flowers   1 red: 2 pink: 1 white

flowers, as shown in *Figure 12.7.* The intermediate pink form of the trait occurs because neither allele of the pair is completely dominant. Note that the letters *R* and *R′*, rather than *R* and *r*, are used to show that these alleles are incompletely dominant.

The new phenotype occurs because the flowers contain a colored pigment. The *R* allele codes for an enzyme that produces a red pigment. The *R′* allele codes for a defective enzyme that makes no pigment. Because the heterozygote has only one copy of the *R* allele, its flowers produce only half the amount of red pigment that the flowers of the red homozygote produce, and they appear pink. The *R′R′* homozygote has no normal enzyme, produces no red pigment, and appears white.

Note that the segregation of alleles is the same as in simple Mendelian inheritance. However, because neither allele is dominant, the plants of

the F$_1$ generation all have pink flowers. When pink-flowered F$_1$ plants are crossed with each other, the offspring in the F$_2$ generation appear in a 1: 2: 1 phenotypic ratio of red to pink to white flowers. This result supports Mendel's law of independent assortment, which states that the alleles are inherited independently.

Codominance: Expression of both alleles

In chickens, black-feathered and white-feathered birds are homozygotes for the *B* and *W* alleles, respectively. Two different uppercase letters are used to represent the alleles in codominant inheritance.

One of the resulting heterozygous offspring in a breeding experiment between a black rooster and a white hen is shown in *Figure 12.8.* You might expect that heterozygous chickens, *BW,* would be black if the pattern of inheritance followed Mendel's law of dominance, or gray if the trait were

---

Alternative Lab 12.1

Incomplete Dominance

Purpose
Students will observe the phenotypic ratio that appears with incomplete dominance.

Materials
petri dish, paper toweling, tobacco seeds

Procedure
Give the following directions to students.
1. Label the top of a petri dish with your name and the date. Place several layers of paper toweling inside the dish.
2. Moisten the toweling with water and place 20 tobacco seeds on it.
3. Cover the dish and place it where it will receive light.
4. Check the seeds for the next 10 days. Keep the toweling moist but not

soaked. After 8-10 days, count the number of plants with green, yellow-green, and yellow leaves.
5. Design a data table to record your results and class totals.
6. Wash your hands after handling seeds.

Expected Results
Data will approach a 1: 2: 1 ratio of phenotypes.

Analysis
1. What ratio of phenotypes does your

incompletely dominant. Notice, however, that the heterozygote is neither black nor gray. Instead, all of the offspring are checkered; some feathers are black and other feathers are white. In such situations, the inheritance pattern is said to be codominant. **Codominant alleles** cause the phenotypes of both homozygotes to be produced in heterozygous individuals. In codominance, both alleles are expressed equally.

Multiple phenotypes from multiple alleles

Although each trait has only two alleles in the patterns of heredity you have studied thus far, it is common for more than two alleles to control a trait in a population. This is understandable when you recall that a new allele can be formed any time a mutation occurs in a nitrogen base somewhere within a gene. Although only two alleles of a gene can exist within a diploid cell, multiple alleles for a single gene can be studied in a population of organisms.

Traits controlled by more than two alleles have **multiple alleles.** The pigeons pictured in *Figure 12.9* show the effects of multiple alleles for feather color. Three alleles of a single gene govern their feather color, although each pigeon can have only two of these alleles. The number of alleles for any particular trait is not limited to three, and there are instances in which more than 100

Figure 12.8
When a certain variety of black chicken is crossed with a white chicken, all of the offspring are checkered. Both feather colors are produced by codominant alleles.

Figure 12.9
In pigeons, a single gene that controls feather color has three alleles. An enzyme that activates the production of a pigment is controlled by the *B* allele. This enzyme is lacking in *bb* pigeons.

A The dominant B^A allele produces ash-red colored feathers.

B The *B* allele produces wild-type blue feathers. *B* is dominant to *b* but recessive to B^A.

C The allele *b* produces a chocolate-colored feather and is recessive to both other alleles.

data show? Is this ratio close to the expected ratio of 1: 2: 1 with incomplete dominance? *ratio of phenotypes will vary; not very close to the expected ratio*

2. Why might your data not be close to a 1: 2: 1 ratio? *small sample size*

3. What ratio of phenotypes do you get when using class totals? Why is this ratio closer to what was expected? *close to 1: 2: 1; large sample size*

Purpose

Students will work with problems that deal with multiple alleles.

Process Skills

acquire information, apply concepts, predict, think critically

Teaching Strategies

■ Have students sequence the genotypes in order from dominant to recessive. **L2**

■ Emphasize again that only two alleles at a time govern coat color. However, there are 10 different combinations that are possible when dealing with four different alleles in a multiple allele situation.

■ Have students define in their own words the meaning of multiple allele.

Thinking Critically

1. a. CC, Cc^{ch}, Cc^h, Cc
   b. $c^{ch}c^{ch}$, $c^{ch}c^h$, $c^{ch}c$
   c. c^hc^h, c^hc
   d. cc

2. Chinchilla; the c^{ch} allele for chinchilla is dominant to the c^h allele for Himalayan. Dark gray; the C allele is dominant to all other alleles.

3. Yes, if the chinchilla rabbit is $c^{ch}c$.

4. No; Himalayan and white rabbits have no alleles for chinchilla.

5. The chinchilla parent is $c^{ch}c$ and the Himalayan parent is c^hc.

✓ Assessment

Portfolio Ask students to plan a breeding program that would produce all Himalayan rabbit offspring when starting with one white parent and one dark gray parent known to be Cc^{ch}. Punnett squares should be included where appropriate. All work should be placed in student portfolios. Use the Performance Task Assessment List for Designing an Experiment in **PASC**, p. 23. **L3** **P**

324

Problem-Solving Lab 12-2 Predicting

How is coat color in rabbits inherited? Coat color in rabbits is inherited as a series of multiple alleles. This means that there can be more than just two alleles for a single gene. In the case of coat color in rabbits, there are four alleles, and each one is expressed with a different phenotype.

Analysis

Examine **Table 12.2**. Use this information to answer the questions. Remember, each rabbit can have only two alleles for coat color.

Table 12.2 Coat color in rabbits

Phenotype	Allele	Pattern of inheritance
Dark gray coat	C	dominant to all other alleles
Chinchilla	c^{ch}	dominant to Himilayan and to white
Himalayan	c^h	dominant to white
White	c	recessive

Thinking Critically

1. List all possible genotypes for a
   a. dark gray-coated rabbit (there are 4).
   b. chinchilla rabbit (there are 3).
   c. Himalayan rabbit (there are 2).
   d. white rabbit (there is 1).
2. Predict the phenotype for a rabbit with a c^hc^{ch} and with a Cc^h genotype. Explain your answer.
3. Would it be possible to obtain white rabbits if one parent is white and the other is chinchilla? Explain your answer.
4. Would it be possible to obtain chinchilla rabbits if one parent is Himalayan and the other is white? Explain.
5. A chinchilla rabbit is mated with a Himalayan. Some offspring are white. What are the parents' genotypes?

Figure 12.10
The sex chromosomes in humans are called X and Y.

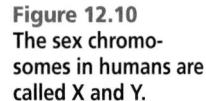

Ⓐ The sex chromosomes are named for the letters they resemble.

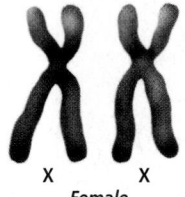
X X
Female

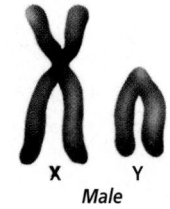
X Y
Male

Ⓑ Half the offspring of any mating between humans will have two X chromosomes, XX, which is female. The other half of the offspring will have one X and one Y chromosome, XY, which is male.

324 PATTERNS OF HEREDITY AND HUMAN GENETICS

---

alleles are known to exist for a single trait! You can learn about another example of multiple alleles in the *Problem-Solving Lab* shown here.

Sex determination

Recall that in humans the diploid number of chromosomes is 46, or 23 pairs. There are 22 pairs of matching homologous chromosomes called **autosomes.** Homologous autosomes look exactly alike. The 23rd pair of chromosomes differs in males and females. These two chromosomes, which determine the sex of an individual, are called **sex chromosomes.** In humans, the chromosomes that control the inheritance of sex characteristics are indicated by the letters X and Y. If you are a human female, XX, your 23rd pair of chromosomes are homologous and look alike, as shown in *Figure 12.10A*. However, if you are a male, XY, your 23rd pair of chromosomes look different. Males, which have one X and one Y chromosome, produce two kinds of gametes, X and Y, by meiosis. Females have two X chromosomes, so they produce only X gametes. *Figure 12.10B* shows that after fertilization, a 1:1 ratio of males to females is expected. Because fertilization is governed by the laws of

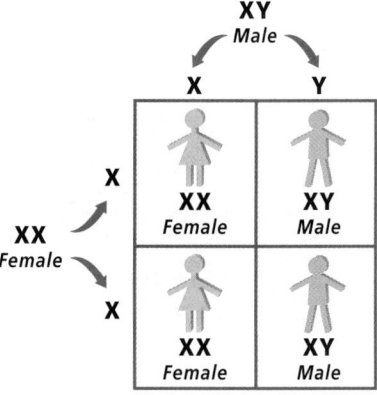

probability, the ratio usually is not exactly 1: 1 in a small population.

Sex-linked inheritance

Drosophila (droh SAHF uh luh), commonly known as fruit flies, inherit sex chromosomes in the same way as humans do. Traits controlled by genes located on sex chromosomes are called **sex-linked traits.** The alleles for sex-linked traits are written as superscripts of the X or Y chromosome. Because the X and Y chromosomes are not homologous, the Y chromosome has no corresponding allele to one on the X chromosome and no superscript is used. Also remember that any allele on the X chromosome of a male will not be masked by a corresponding allele on the Y chromosome.

In 1910, Thomas Hunt Morgan discovered traits linked to sex chromosomes. Morgan noticed one day that one male fly had white eyes rather than the usual red eyes. He crossed the white-eyed male with a homozygous red-eyed female. All of the F$_1$ offspring had red eyes, indicating that the white-eyed trait is recessive. Then Morgan allowed the F$_1$ flies to mate among themselves. According to simple Mendelian inheritance, if the trait were recessive, the offspring in the F$_2$ generation would show a 3: 1 ratio of red-eyed to white-eyed flies. As you can see in *Figure 12.11,* this is what Morgan observed. However, he also noticed that the trait of white eyes appeared only in male flies.

Morgan hypothesized that the red-eye allele was dominant and the white-eye allele was recessive. He also reasoned that the gene for eye color was located on the X chromosome and was not present on the Y chromosome. In heterozygous females, the dominant allele for red eyes masks the recessive allele for white eyes. In males, however, a single recessive allele is expressed as a white-eyed phenotype. When Morgan crossed a heterozygous red-eyed

Visual Learning
Figure 12.10 Ask students why the X and Y chromosomes are not homologous? *They do not have the same size, shape, or similar genes. However, they are able to pair during meiosis because they have a small homologous region at one end of each of the chromosomes.*

Make it clear to students that there is a 50% chance that each offspring will receive XX chromosomes, which is female, and a 50% chance that each offspring will receive XY chromosomes, which is male.

Reinforcement
Draw a Punnett square. Place sex chromosomes X and X along the left side, and X and Y along the top to represent gametes. Ask students why each gamete has only one sex chromosome. *The sex chromosomes are separated during meiosis.*

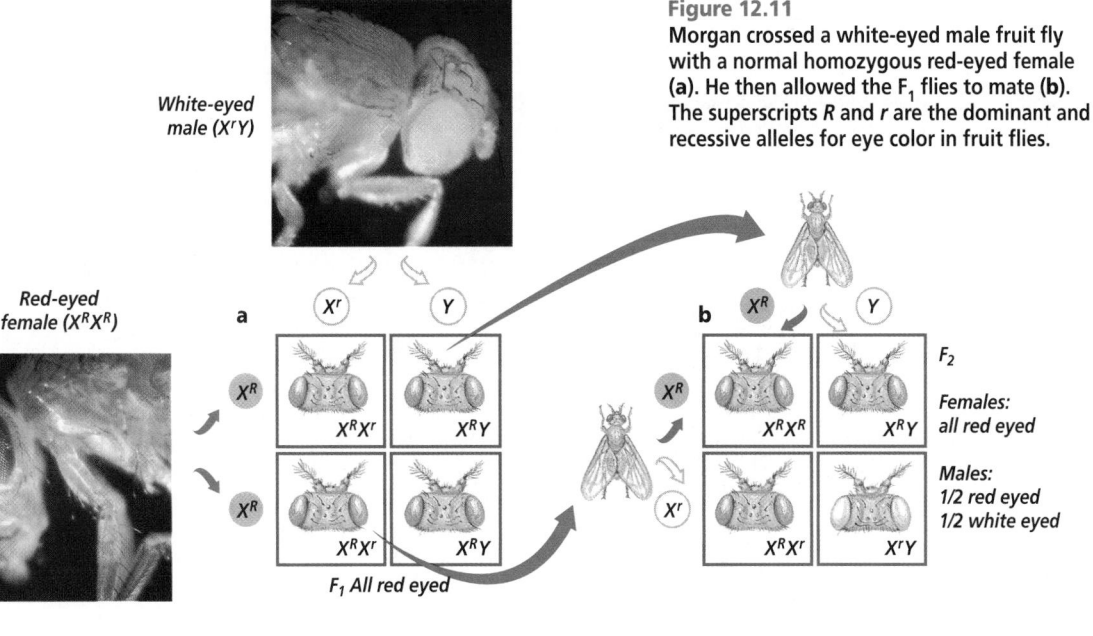

Figure 12.11
Morgan crossed a white-eyed male fruit fly with a normal homozygous red-eyed female (**a**). He then allowed the F$_1$ flies to mate (**b**). The superscripts *R* and *r* are the dominant and recessive alleles for eye color in fruit flies.

White-eyed male (X^rY)

Red-eyed female (X^RX^R)

a

X^r Y

X^R X^RX^r X^RY

X^R X^RX^r X^RY

F$_1$ All red eyed

b

X^R Y

X^R X^RX^R X^RY

X^r X^RX^r X^rY

F$_2$

Females: all red eyed

Males: 1/2 red eyed 1/2 white eyed

✓ Portfolio

Eye Color
Visual-Spatial Provide students with a pedigree outline that involves a sex-linked trait such as white eye color in *Drosophila*. Have students record above each symbol the sex chromosomes and alleles that each fly possesses for the trait. **L1** **P** 🗂

female with a white-eyed male, half of all the males and half of all the females inherited white eyes. The only explanation of these results is Morgan's hypothesis: The allele for eye color is carried on the X chromosome and the Y chromosome has no allele for eye color.

Traits dependent on genes that follow the inheritance pattern of a sex chromosome are called sex-linked traits. Eye color in fruit flies is an example of an X-linked trait. Y-linked traits are passed only from male to male.

Polygenic inheritance

Some traits, such as skin color and height in humans, and cob length in corn, vary over a wide range. Such ranges occur because these traits are governed by many different genes. **Polygenic inheritance** is the inheritance pattern of a trait that is controlled by two or more genes. The genes may be on the same chromosome or on different chromosomes, and each gene may have two or more alleles. For simplicity, uppercase and lowercase letters are

used to represent the alleles, as they are in Mendelian inheritance. Keep in mind, however, that the allele represented by an uppercase letter is not dominant. All heterozygotes are intermediate in phenotype.

In polygenic inheritance, each allele represented by an uppercase letter contributes a small, but equal, portion to the trait being expressed. The result is that the phenotypes usually show a continuous range of variability from the minimum value of the trait to the maximum value.

Suppose, for example, that stem length in a plant is controlled by three different genes: *A*, *B*, and *C*. Each gene is on a different chromosome and has two alleles, which can be represented by uppercase or lowercase letters. Thus, each diploid plant has a total of six alleles for stem length. A plant that is homozygous for short alleles at all three gene locations (*aabbcc*) might grow to be only 4 cm tall, the base height. A plant that is homozygous for tall alleles at all three gene locations (*AABBCC*) might be 16 cm tall. The difference between the tallest possible plant and the shortest possible plant is 12 cm, or 2 cm per each of the six tall alleles. You could say that each allele represented by an uppercase letter contributes 2 cm to the total height of the plant.

Suppose a 16-cm-tall plant were crossed with a 4-cm-tall plant. In the F_1 generation, all the offspring would be *AaBbCc*. If each of the three tall genes *A*, *B*, and *C* contributed 2 cm of height to the base height of 4 cm, the plants would be 10 cm tall (4 cm + 6 cm)—intermediate in height. If they are allowed to interbreed, the F_2 offspring

WORD Origin

polygenic
From the Greek words *polys*, meaning "many," and *genos*, meaning "kind." Polygenic inheritance involves many genes.

Figure 12.12
In this example of polygenic inheritance, three genes each have two alleles that contribute to the trait. When the distribution of plant heights is graphed, a bell-shaped curve is formed. Intermediate heights occur most often.

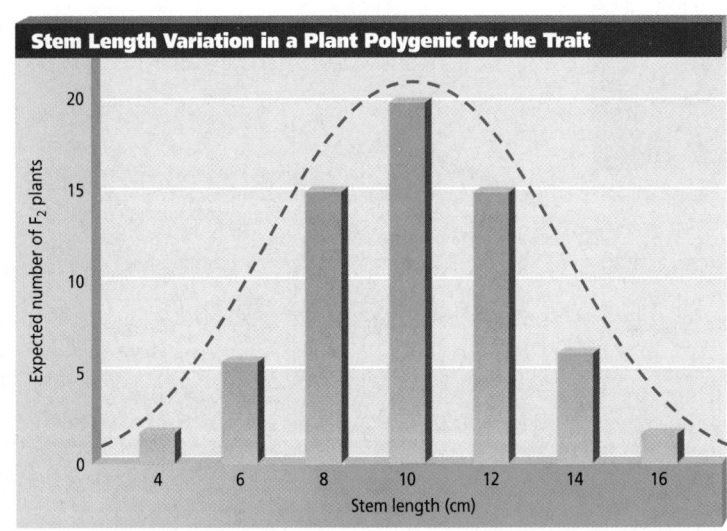

Stem Length Variation in a Plant Polygenic for the Trait

Expected number of F_2 plants

Stem length (cm)

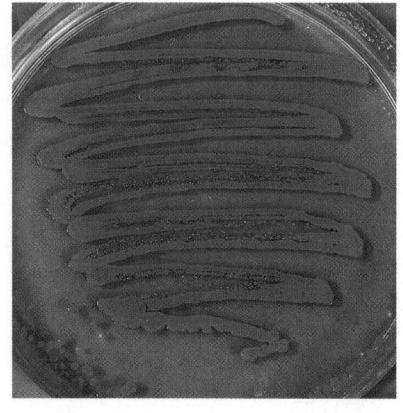

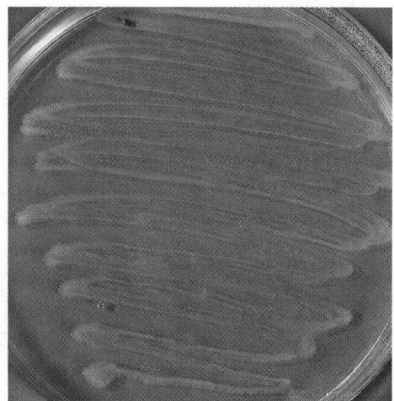

will show a broad range of heights. A Punnett square of this trihybrid cross would show that 10-cm-tall plants are most often expected, and the tallest and shortest plants are seldom expected. Notice in *Figure 12.12* that when these results are graphed, the shape of the graph confirms the prediction of the Punnett square.

Environmental Influences

Even when you understand dominance and recessiveness and you have solved the puzzles of the other patterns of heredity, the inheritance picture is not complete. The genetic makeup of an organism at fertilization determines only the organism's potential to develop and function. As the organism develops, many factors can influence how the gene is expressed, or even whether the gene is expressed at all. Two such influences are the organism's external and internal environments.

Influence of external environment

Sometimes, individuals known to have a particular gene fail to express the phenotype specified by that gene. Temperature, nutrition, light, chemicals, and infectious agents all can

influence gene expression. In certain bacteria, temperature has an effect on the expression of color, as shown in *Figure 12.13*. External influences can also be seen in leaves. Leaves on a tree can have different sizes and shapes depending on the amount of light they receive.

Influence of internal environment

The internal environments of males and females are different because of hormones and structural differences, *Figure 12.14*. For example, traits such as horn size in mountain sheep, male-pattern baldness in humans, and feather color in peacocks are expressed differently in the

Figure 12.14
The horns of a ram (male) are much heavier and more coiled than those of a ewe (female) although their genotypes are identical.

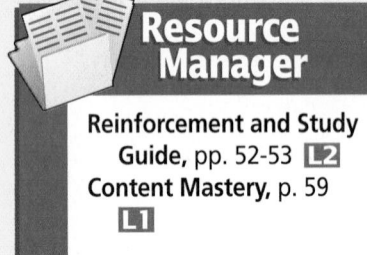
Figure 12.15
Some traits are expressed differently in the sexes.

Ⓐ The plumage of the male peacock is highly decorated and colored.

Ⓑ The plumage of the female peahen is dull by comparison.

Ⓒ Human male-pattern baldness, premature balding that occurs in a characteristic pattern, affects males but not females.

sexes, as you can see in *Figure 12.15.* These differences are controlled by different hormones, which are determined by different sets of genes.

An organism's age also affects gene function. The nature of such a pattern is not well understood, but it is known that the internal environment of an organism changes with age.

You can now see that you must learn how genes interact with each other and with the environment to form a more complete picture of inheritance. Mendel's idea that heredity is a composite of many individual traits still holds. Later researchers have filled in more details of Mendel's great contributions.

Section Assessment

Understanding Main Ideas
1. A cross between a purebred animal with red hairs and a purebred animal with white hairs produces an animal that has both red hairs and white hairs. What type of inheritance pattern is involved?
2. In a cross between individuals of a species of tropical fish, all of the male offspring have long tail fins, and none of the females possess the trait. Mating of the F_1 fish fails to produce females with the trait. Explain a possible inheritance pattern of the trait.
3. A red-flowered sweet pea plant is crossed with a white-flowered sweet pea plant. All of the offspring are pink. What is the inheritance pattern being expressed?
4. The color of wheat grains shows a wide variability between red and white with multiple phenotypes. What type of inheritance pattern is being expressed?

Thinking Critically
5. Armadillos always have four offspring that have identical genetic makeup. Suppose that, within a litter, each young armadillo is found to have a different phenotype for a particular trait. How could you explain this phenomenon?

SKILL REVIEW
6. **Forming a Hypothesis** An ecologist observes that a population of plants in a meadow has flowers that may be red, yellow, white, pink, or purple. Hypothesize what the inheritance pattern might be. For more help, refer to *Practicing Scientific Methods* in the **Skill Handbook**.

Section Assessment

1. Alleles for both red and white hairs are expressed, which is typical of a pattern of codominance.
2. The inheritance pattern of this trait is that of a sex-linked gene located on the Y chromosome, which is possessed by males only.
3. Incomplete dominance is being expressed because the heterozygous plant is an intermediate form between red and white.
4. polygenic inheritance
5. The environment in the mother's uterus before birth may have been different for each of the offspring. The external environment can affect gene expression.
6. This is probably a case of multiple allelic or polygenic inheritance.

Section

12.3 Complex Inheritance of Human Traits

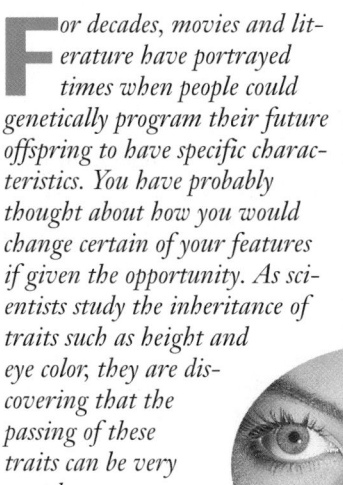

For decades, movies and literature have portrayed times when people could genetically program their future offspring to have specific characteristics. You have probably thought about how you would change certain of your features if given the opportunity. As scientists study the inheritance of traits such as height and eye color, they are discovering that the passing of these traits can be very complex.

Eye color and height are complex traits.

SECTION PREVIEW

Objectives

Compare codominance, multiple allelic, sex-linked, and polygenic patterns of inheritance in humans.

Distinguish among conditions in which extra autosomal or sex chromosomes exist.

Vocabulary
karyotype

Codominance in Humans

Remember that in codominance, the phenotypes of both homozygotes are produced in the heterozygote. One example of this type of inheritance in humans is the disorder sickle-cell anemia.

Sickle-cell anemia

Sickle-cell anemia is a major health problem in the United States and in Africa. In the United States, it is most common in black Americans whose families originated in Africa and in white Americans whose families originated in the countries surrounding the Mediterranean Sea. About one in 12 African Americans, a much larger proportion than in most populations, is heterozygous for the disorder.

In an individual who is homozygous for the sickle-cell allele, the oxygen-carrying protein hemoglobin differs by one amino acid from normal hemoglobin. This defective hemoglobin forms crystal-like structures that change the shape of the red blood cells. The abnormal red blood cells are shaped like a sickle, or half-moon. The change in shape occurs in the body's narrow capillaries after the hemoglobin releases oxygen to the cells. Abnormally shaped blood cells,

12.3 COMPLEX INHERITANCE OF HUMAN TRAITS **329**

Cultural Diversity

Mary Styles Harris

In your discussion of sickle-cell anemia, point out contributions of contemporary African American scientists in the treatment and etiology of this disorder. One of the more prominent workers in this field has been geneticist Mary Styles Harris (born 1949). From 1977 to 1979, Harris was the executive director of the Sickle Cell Foundation of Georgia, and she has published many papers on this subject. In 1980, Harris was honored as one of *Glamour* magazine's Outstanding Women Scientists. Around this time, she also wrote, narrated, and produced an educational science series for Georgia TV through a grant from the National Science Foundation.

Section 12.3

Prepare

Key Concepts

Complex inheritance patterns in humans are presented. Sickle-cell anemia is used as an example of codominance, blood types as an example of multiple allelic inheritance, color-blindness and hemophilia as examples of sex-linked patterns, and skin color as an example of polygenic inheritance. The section concludes with a discussion of changes in chromosome numbers.

Planning

- Obtain magnifying glasses and colored pencils for MiniLab 12-2.
- Obtain color blindness testing charts for Meeting Individual Needs.

1 Focus

Bellringer

Before presenting the lesson, display **Section Focus Transparency 31** on the overhead projector and have students answer the accompanying questions.
L1 **ELL**

Microscope Activity

Visual-Spatial Have students view a prepared slide of sickled blood cells and compare it with a slide of normal blood cells. **L1** **ELL**

✓ Assessment

Portfolio Have students write a short paragraph explaining why it is so important to type blood before it is used for a transfusion. **L1** **P**

GLENCOE
TECHNOLOGY

VIDEOTAPE
The Secret of Life
It's in the Genes: Evolution

CD-ROM
Biology: The Dynamics of Life
Exploration: *Blood Types*
Disc 5

Resource Manager

Critical Thinking/Problem Solving, p. 12 **L3**
Section Focus Transparency 31 and Master **L1** **ELL**
Laboratory Manual, pp. 83-86 **L2**

Figure 12.16
In sickle-cell anemia, the gene for hemoglobin produces a protein that is different by one amino acid. This hemoglobin crystallizes when oxygen levels are low, forcing the red blood cells into a sickle shape (**a**). A normal red blood cell is disc shaped (**b**).

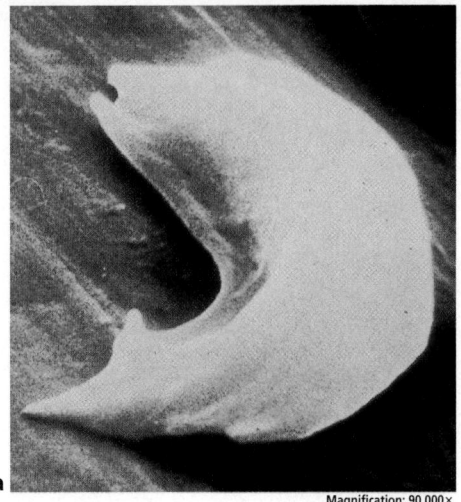

a   Magnification: 90 000×

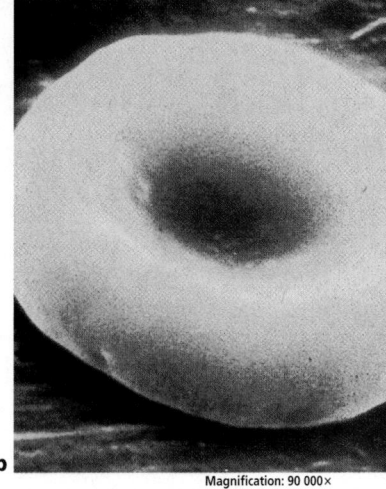

b   Magnification: 90 000×

Figure 12.16, slow blood flow, block small vessels, and result in tissue damage and pain. Because sickled cells have a shorter life span than normal red blood cells, the person suffers from anemia, a condition in which there is a low number of red blood cells.

Individuals who are heterozygous for the allele produce both normal and sickled hemoglobin, an example of codominance. They produce enough normal hemoglobin that they do not have the serious health problems of those homozygous for the allele and can lead relatively normal lives. Individuals who are heterozygous are said to have the sickle-cell trait because they can show some signs of sickle-cell anemia if the availability of oxygen is reduced.

Multiple Alleles in Humans

Traits that are governed by simple Mendelian heredity have only two alleles. However, you have learned that more than two alleles of a gene are possible for certain traits. The ABO blood group is a classic example of a single gene that has multiple alleles in humans. How many alleles does this gene have? Read the *Inside Story* to answer this question.

Multiple alleles govern blood type

Human blood types, listed in *Table 12.3,* are determined by the presence or absence of certain molecules on the surfaces of red blood cells. As the determinant of blood types A, B, AB, and O, the gene *I* has three alleles: I^A, I^B, and *i.*

The importance of blood typing

Determining blood type is necessary before a person can receive a blood transfusion because the red blood cells of incompatible blood types could clump together, causing death. Blood typing can also be helpful in solving cases of disputed parentage. For example, if a child has type AB blood and its mother has type A, a man with type O blood could not possibly be the father. But blood tests cannot prove that a certain man definitely is the father; they indicate only that he could be. DNA tests are necessary to determine actual parenthood.

✓ Portfolio

Alzheimer's Disease

Linguistic Have students research the connection between chromosome 21 and Alzheimer's disease. Ask them to include a copy of their findings in their portfolios. **L3** **P**

The ABO Blood Group

The gene for blood type, gene I, codes for a membrane protein found on the surface of red blood cells. Each of the three alleles codes for a different protein. Your immune system recognizes the red blood cells as belonging to you. If cells with a different protein enter your body, your immune system will attack them.

Critical Thinking *If you inherit ii from your parents, what is your blood type?*

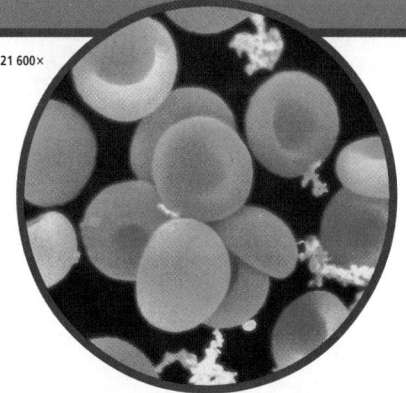

Magnification: 21 600×

Red blood cells

① **Phenotype A** The I^A allele is dominant to *i*, so inheriting either the $I^A i$ alleles or the $I^A I^A$ alleles from your two parents will give you type A blood. Surface protein A is produced.

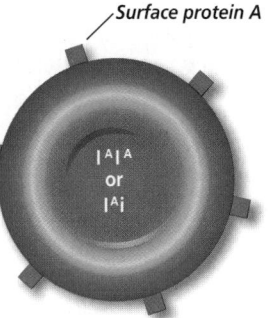

Surface protein A

$I^A I^A$ or $I^A i$

④ **Phenotype O** The *i* allele is recessive and produces no surface molecule. Therefore, if you are homozygous *ii*, your blood cells have no surface proteins and you have blood type O.

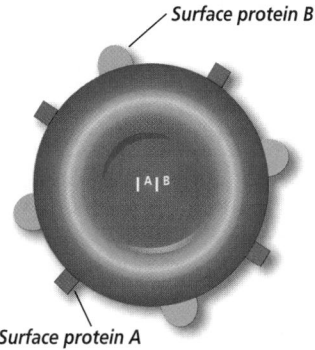

ii

② **Phenotype B** The I^B allele is also dominant to *i*. To have type B blood, you must inherit the I^B allele from one parent and either another I^B allele or the *i* allele from the other. Surface protein B is produced.

Surface protein B

$I^B I^B$ or $I^B i$

Surface protein B

$I^A I^B$

Surface protein A

③ **Phenotype AB** The I^A and I^B alleles are codominant to each other. This means that if you inherit the I^A allele from one parent and the I^B allele from the other, your red blood cells will produce both surface proteins and you will have type AB blood.

Table 12.3 Human blood types

Genotypes	Surface Proteins	Phenotypes
$I^A I^A$ or $I^A i$	A	A
$I^B I^B$ or $I^B i$	B	B
$I^A I^B$	A and B	AB
ii	none	O

12.3 COMPLEX INHERITANCE OF HUMAN TRAITS **331**

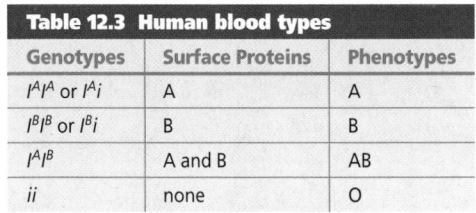
BIOLOGY JOURNAL

"Improved" Children

Linguistic Some day, scientists may have the capability to change genes, making offspring smarter, physically stronger, and better looking. Ask students to write a short essay on whether they would have their genes changed to "improve" their children if they were given the opportunity. **L2**

INSIDE STORY

Purpose
Students will learn the genetic basis of ABO blood types

Teaching Strategies
■ Give students a genotype and ask them to identify the phenotype on red blood cells.

Visual Learning
■ Have students create a table with genotypes as the headings. Under each genotype, have them correctly place the resulting phenotype. **L1**
■ Ask students to draw red blood cells with cell surface proteins in their journals. Have them draw red blood cells for all four blood types and write possible genotypes underneath the cells. **L1**

Critical Thinking
You will be blood type O.

GLENCOE TECHNOLOGY

VIDEODISC
The Secret of Life
Blood Types

Biology: The Dynamics of Life
Sex-Linked Traits (Ch. 35)
Disc 1, Side 1,
1 min. 49 sec.

CD-ROM
Biology: The Dynamics of Life
Animation: *Sex-Linked Traits*
Disc 2

331

Purpose

Students will determine the pattern of inheritance for Duchenne's muscular dystrophy.

Process Skills

observe and infer, recognize cause and effect

Background

Because various forms of muscular dystrophy can be inherited as an autosomal dominant, an autosomal recessive, or a sex-linked disorder, a family pedigree must be analyzed to determine the exact pattern of inheritance. From the pedigree, the mode of inheritance of Duchenne's muscular dystrophy can be inferred to be X-linked.

Teaching Strategies

■ Ask students which individuals in the pedigree were keys to determining the type of inheritance involved.

Thinking Critically

There is a 100% probability that the daughter will be a carrier because the only X chromosome she can inherit from her father would have the defective gene. The son would not inherit the disorder because his X chromosome would come from his mother, who presumably is not a carrier.

✔ Assessment

Knowledge Ask students what mating would have to occur to produce a female child with Duchenne's muscular dystrophy. *mating a female carrier or an affected female with an affected male* Use the Performance Task Assessment List for Formulating a Hypothesis in **PASC**, p. 21. **L2**

Problem-Solving Lab 12-3 — Drawing a Conclusion

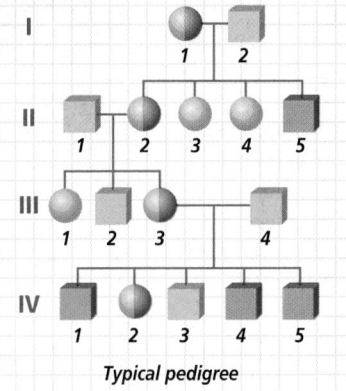

How is Duchenne's muscular dystrophy inherited? Muscular dystrophy is often thought of as a single disorder, but it is actually a group of genetic disorders that produce muscular weakness, progressive deterioration of muscular tissue, and loss of coordination. Different forms of muscular dystrophy can be inherited as a dominant autosomal, a recessive autosomal, or a sex-linked disorder. These three patterns of inheritance appear different from one another when a pedigree is made. One rare form of muscular dystrophy, called Duchenne's muscular dystrophy, affects three in 10 000 American males.

Typical pedigree

Analysis

The pedigree shown here represents the typical inheritance pattern for Duchenne's muscular dystrophy. Refer to *Figure 12.1* if you need help interpreting the symbols. Analyze the pedigree to determine the pattern of inheritance. Is this an autosomal or a sex-linked disorder?

Thinking Critically

If individual IV-1 had a daughter and a son, what would be the probability that the daughter is a carrier? That the son inherited the disorder?

Sex-Linked Traits in Humans

Several human traits are determined by genes that are carried on the sex chromosomes; most of these genes are located on the X chromosome. The pattern of sex-linked inheritance is explained by the fact that males, who are XY, pass an X chromosome to each of their daughters and a Y chromosome to each son. Females, who are XX, pass one of their X chromosomes to each child, *Figure 12.17*. If a son receives an X chromosome with a recessive allele from his mother, he will express the recessive phenotype because he has no chance of inheriting from his father a dominant allele that would mask the expression of the recessive allele.

Two traits that are governed by X-linked inheritance in humans are certain forms of color blindness and hemophilia. Determine whether Duchenne's muscular dystrophy is sex-linked by reading the *Problem-Solving Lab* on this page.

Red-green color blindness

People who have red-green color blindness can't differentiate these two

---

Figure 12.17
If a trait is X-linked, males pass the X-linked allele to all of their daughters but to none of their sons **(a)**. Heterozygous females have a 50 percent chance of passing a recessive X-linked allele to each child **(b)**.

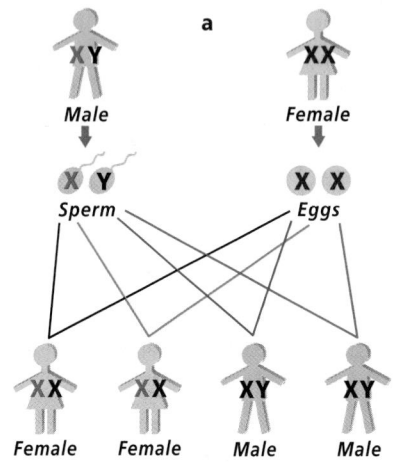

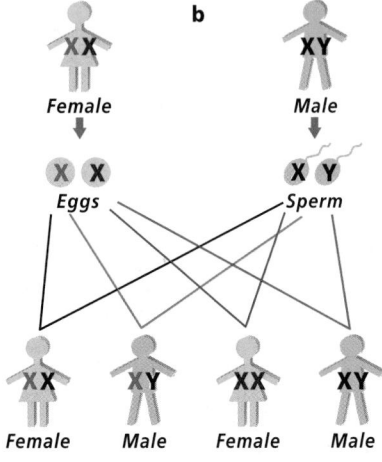

Alternative Lab 12.2

Barr Bodies

Purpose
Students will locate Barr bodies in cells.

Materials
prepared slides of male and female cheek cells, microscope

Background
Females have two X chromosomes, but early in development of a female embryo, one of the X chromosomes in each cell becomes inactive. The inactive chromosome becomes condensed and can be seen as a Barr body. Cells of males do not contain Barr bodies.

Procedure
Give students the following directions.

1. Place the slide of female cheek cell on

colors. Red-green color blindness was first described in a boy who could not be trained to harvest only the ripe, red apples from his father's orchard. Instead, he chose green apples as often as he chose red.

Other more serious problems can result from this disorder, such as the inability of color-blind people to identify red and green traffic lights by color. Color blindness is caused by the inheritance of either of two recessive alleles at two gene sites on the X chromosome that affect red and green receptors in the cells of the eyes.

Hemophilia: An X-linked disorder

Did you ever wonder about why a cut stops bleeding so quickly? This human adaptation is essential. If your blood didn't have the ability to clot at all, any cut could take a long time to stop bleeding. Of greater concern would be internal bleeding resulting in a bruise, which a person may not immediately notice.

Hemophilia A is an X-linked disorder that causes just such a problem with blood clotting. About one male in every 10 000 has hemophilia, but only about one in 100 million females inherits the same disorder. Why? Males inherit the allele for hemophilia on the X chromosome from their carrier mothers. A single recessive allele for hemophilia will cause the disorder in males. Females would need two recessive alleles to inherit hemophilia. The family of Queen Victoria, pictured in the *Social Studies Connection* at the end of this chapter, is the best-known study of hemophilia A, also called royal hemophilia.

Hemophilia A can be treated with blood transfusions and injections of Factor VIII, the blood-clotting enzyme that is absent in people affected by the condition. However, both treatments are expensive. New methods of DNA technology are being used to develop a cheaper source of the clotting factor.

Polygenic Inheritance in Humans

Think of all the traits you inherited from your parents. Although many of your traits were inherited through simple Mendelian patterns or through multiple alleles, many other human traits are determined by polygenic inheritance. These kinds of traits usually represent a range of variation that is measurable. The *MiniLab* shown here examines one of these traits—the color variations in human eyes.

12.3 COMPLEX INHERITANCE OF HUMAN TRAITS **333**

MiniLab 12-2 Observing and Inferring

Detecting Colors and Patterns in Eyes Human eye color, like skin color, is determined by polygenic inheritance. You can detect several shades of eye color, especially if you look closely at the iris with a magnifying glass. Often, the pigment is deposited so that light reflects from the eye, causing the iris to appear blue, green, gray, or hazel (brown-green). In actuality, the pigment may be yellowish or brown, but not blue.

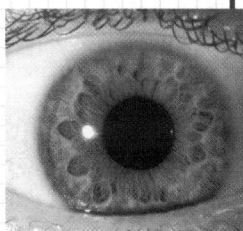
Hazel eye color

Procedure

1. Use a magnifying glass to observe the patterns and colors of pigment in the eyes of five classmates.
2. Use colored pencils to make drawings of the five irises.
3. Describe your observations in your journal.

Analysis

1. How many different pigments were you able to detect in each eye?
2. From your data, do you suspect that eye color might not be inherited by simple Mendelian rules? Explain.
3. Suppose that two people have brown eyes. They have two children with brown eyes, one with blue eyes, and one with green eyes. What pattern might this suggest?

MiniLab 12-2

Purpose
Students will observe the colors and patterns in the eyes of several classmates in order to hypothesize how eye color is inherited.

Process Skills
observe and infer, recognize cause and effect

Teaching Strategies
■ Provide students with five cards, each with a circle and a line under the circle. Students should write the name of the person on the line and then draw the iris within the circle. Provide colored pencils. **L1** **ELL**

Expected Results
Students will detect many different pigments, suggesting polygenic inheritance.

Analysis

1. Students may detect brown, blue, yellow, gray, green, and black pigments in each eye.
2. Eye color is probably not inherited by simple Mendelian rules because there are so many phenotypes.
3. The pattern suggests polygenic inheritance.

✔ Assessment

Knowledge Ask students to write a paragraph telling how they could determine the number of genes involved in the inheritance of eye color. Use the Performance Task Assessment List for Writing in Science in **PASC**, p. 87. **L2**

---

the microscope.

2. Examine cells under high power. A Barr body will be seen in each cell as a darkly stained mass just inside the nuclear membrane.
3. Examine the slide of male cheek cells for the presence of Barr bodies.
4. Draw a cell with a Barr body and one without a Barr body.
5. Wash your hands when you have finished your observations.

Expected Results
Students will see Barr bodies in cells from females but not in those from males.

Analysis

1. Do all your body cells have Barr bodies? *yes for females; no for males*
2. What percentage of a female's cells have visible Barr bodies? *100%*
3. How many Barr bodies would you find in a cell of an XXX female? *2*

✔ Assessment

Performance Students should include a summary of the lab, their drawings, and the answers to the Analysis questions in their journals. Use the Performance Task Assessment List for Lab Report in **PASC**, p. 47. **L3**

Concept Development

Many different types of color-blindness are possible. Make color-blindness testing charts available to students. Color-blindness testing kits are also available. Have pairs of students use the kits.

Visual Learning

Figure 12.18 The graph of what number of genes most closely matches the observed distribution of skin color? *The actual data match the 4-gene graph best.*

Figure 12.18
This graph (**b**) shows the expected distribution of human skin color if controlled by one, three, or four genes. The observed distribution of skin color (**a**) closely matches the distribution shown by four genes.

a

b

Genes Involved in Skin Color

Expected distribution

Number of individuals

Observed distribution of skin color

4 genes
3 genes
1 gene

I   II   III   IV   V   VI   VII   VIII   IX

Classes of skin color

WORD Origin

autosome
From the Greek words *autos,* meaning "self," and *soma,* meaning "body." An autosome is a chromosome that is not a sex chromosome.

Skin color: A polygenic trait

In the early 1900s, the idea that polygenic inheritance occurs in humans was first tested using data collected on skin color. Scientists found that when light-skinned people mate with dark-skinned people, their offspring have intermediate skin colors. When these children produce the F_2 generation, the resulting skin colors range from the light-skin color to the dark-skin color of the grandparents (the P_1 generation), with most children having an intermediate skin color. As *Figure 12.18* shows, the variation in skin color indicates that between three and four genes are involved.

Changes in Chromosome Numbers

You have been reading about traits that are caused by one or several genes on chromosomes. What would happen if an entire chromosome or part of a chromosome were missing from the complete set? What if cells had an extra chromosome? As you have learned, abnormal numbers of chromosomes usually, but not always, result from accidents of meiosis. Many abnormal phenotypic effects result from such mistakes.

Unusual numbers of autosomes

You know that a human usually has 23 pairs of chromosomes, or 46 chromosomes altogether. Of these 23 pairs of chromosomes, 22 pairs are autosomes. Humans who have an unusual number of autosomes all are trisomic—that is, they have three of a particular autosome instead of just two. In other words, they have 47 chromosomes. Recall that trisomy usually results from nondisjunction, which occurs when paired homologous chromosomes fail to separate properly during meiosis.

To identify an abnormal number of chromosomes, a sample of cells is obtained from an individual or from a fetus. Metaphase chromosomes are photographed, and the chromosome pictures are then enlarged, cut apart, and arranged in pairs on a chart according to length and location of the centromere, as *Figure 12.19* shows. This chart of chromosome pairs is called a **karyotype,** and it is valuable in pinpointing unusual chromosome numbers in cells.

Down syndrome: Trisomy 21

Most disorders of chromosome number that occur in humans cause symptoms so severe that the developing fetus dies, often before the woman even realizes she is pregnant. Fortunately, these disorders occur only rarely. Down syndrome is the only autosomal trisomy in which affected individuals survive to adulthood. It occurs in about one in 700 live births.

Down syndrome is a group of symptoms that results from trisomy of chromosome 21. Individuals who have Down syndrome have at least some degree of mental retardation. The incidence of Down syndrome births is higher in older mothers, especially those over 40.

Unusual numbers of sex chromosomes

Many abnormalities in the number of sex chromosomes are known to

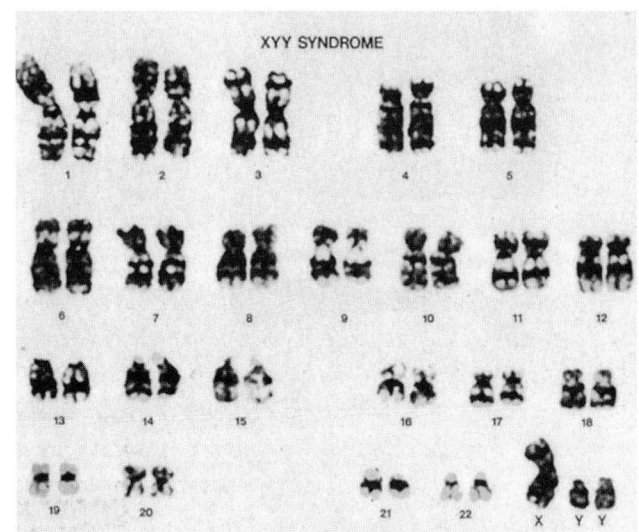

XYY SYNDROME

exist. An X chromosome may be missing (designated as XO) or there may be an extra one (XXX or XXY). There may also be an extra Y chromosome (XYY), as you can see by examining *Figure 12.19*. Any individual with at least one Y chromosome is a male, and any individual without a Y chromosome is a female. Most of these individuals lead normal lives, but they cannot have children and some have varying degrees of mental retardation.

Figure 12.19
This karyotype demonstrates XYY syndrome, where two Y chromosomes are inherited in addition to an X chromosome.

Section Assessment

Understanding Main Ideas
1. Why are sex-linked traits such as red-green color blindness and hemophilia more commonly found in males than in females? Explain your answer in terms of the X chromosome.
2. In addition to revealing chromosome abnormalities, what other information would a karyotype show?
3. What would the genotypes of parents have to be for them to have a color-blind daughter? Explain.
4. Describe a genetic trait in humans that is inherited as codominance. Describe the phenotypes of the two homozygotes and that of the heterozygote. Why is this trait an example of codominance?

Thinking Critically
5. A man is accused of fathering two children, one with type O blood and another with type A blood. The mother of the children has type B blood. The man has type AB blood. Could he be the father of both children? Explain your answer

SKILL REVIEW
6. **Making and Using Tables** Construct a table of the traits discussed in this section. For column heads, use Trait, Pattern of inheritance, and Characteristics. For more help, refer to *Organizing Information* in the **Skill Handbook.**

3 Assess

Check for Understanding

Linguistic Ask students to summarize the patterns shown by multiple allelic, polygenic, and sex-linked inheritance. **L1**

Reteach

Visual-Spatial Have students draw the phases of meiosis to demonstrate how various changes in chromosome numbers occur. **L1 ELL**

Extension

Ask students to research the rare sex-linked disorder severe combined immune deficiency (SCID). The "boy in the bubble" had this disorder. What experimental treatments are currently being used to treat or cure this disorder? **L3**

Assessment

Skill Provide students with filled-in Punnett squares showing genotypes and phenotypes of offspring. Have students determine the pattern of inheritance from the information given. **L1**

4 Close

Discussion

Ask students to explain why hemophilia is extremely rare in females. *Because the allele is rare in the general population, there is only a small likelihood that a male with hemophilia would marry a carrier female or a female with hemophilia.*

Section Assessment

1. Males inherit only one X chromosome. If it carries an allele for a disorder, the male will show the trait.
2. the sex of the child
3. The father must be color-blind (have recessive allele on his X chromosome). The mother must have at least one X chromosome with the color-blind allele since a female receives one X from each parent.
4. In sickle-cell anemia, an individual homozygous for normal hemoglobin will produce only normal hemoglobin. An individual homozygous for sickle-cell anemia will produce only abnormal hemoglobin. A heterozygote will produce both types of hemoglobin.
5. The mother could be $I^B i$ or $I^B I^B$ and the father is $I^A I^B$. If a child received I^A from the father and i from the mother, it would be type A so the man could have fathered the type A child. This man could not father a type O child because he has no i allele.
6. Material for the table can be found on pages 329-335 of the text.

Time Allotment

Initial session: one class period for planting of P_1 generation, then 5 minutes daily for watering; one class period for cross-pollination; 10 days later, one class period for collecting and planting of F_1 seeds; 10 days later: one class period for examination of F_2 plants.

Process Skills

form a hypothesis, observe and infer, collect data

Safety Precautions

Have students use caution when handling, using, and plugging in light fixtures or light banks. Students should use caution when using a razor blade.

PREPARATION

■ Seeds of normal and variegated *Brassica rapa* (Wisconsin Fast Plants) can be ordered from biological supply houses. The variegated gene is carried on chloroplast DNA.
■ It is critical to provide light in fluorescent banks (cool-white, 40 watts/bulb) in order to achieve the complete life cycle in such a short time. Light banks should be adjustable so that they remain about 5-8 cm above the plants' growing tips at all times. Lights remain on continuously for 24 hours each day.
■ Soil must be kept constantly moist, especially during the germination period.

Possible Hypotheses

If the trait is inherited through the cytoplasm, then the female parent contributes the trait.

If the trait is inherited through the cytoplasm, then the male parent contributes the trait.

What is the pattern of cytoplasmic inheritance?

The mitochondria of all eukaryotes and the chloroplasts of plants and algae contain DNA. This DNA is not coiled into chromosomes, but it still carries genes that control genetic traits. Many of the mitochondrial genes control steps in the respiration process.

The DNA in chloroplasts controls traits such as chlorophyll production. Lack of chlorophyll in some cells causes the appearance of white patches in a leaf. This trait is known as variegated leaf. In this BioLab, you will carry out an experiment to determine the pattern of cytoplasmic inheritance of the variegated leaf trait in Brassica rapa.

PREPARATION

Problem

What inheritance pattern does the variegated leaf trait in *Brassica* show?

Hypotheses

Consider the possible evidence you could collect that would answer the problem question. Among the people in your group, form a hypothesis that you can test to answer the question, and write the hypothesis in your journal.

Objectives

In this BioLab, you will:
■ **Determine** which crosses of *Brassica* plants will reveal the pattern of cytoplasmic inheritance.
■ **Analyze** data from *Brassica* crosses.

Possible Materials

Brassica rapa seeds, normal and variegated
potting soil and trays
paintbrushes
forceps
single-edge razor blade
light source
labels

Safety Precautions

Always wear goggles in the lab. Handle the razor blade with extreme caution. Always cut away from you. Wash your hands with soap and water after working with plant material.

Skill Handbook

Use the **Skill Handbook** if you need additional help with this lab.

PLAN THE EXPERIMENT

Teaching Strategies

■ Variegated plants tend to grow a little slower than nonvariegated plants. Therefore, these seeds should be started about 4 days earlier.
■ Review the process of fertilization and make sure students realize that the egg contributes most of the cytoplasm and organelles to the zygote. The sperm is almost all nucleus.
■ Students should work in groups.
■ *Brassica* will not self-pollinate. Therefore, keep the two plant types separate from each other to avoid random cross-pollination, or have students remove flowers that were not used in cross-pollination.

Possible Procedures

■ Both variegated and normal seed types

PLAN THE EXPERIMENT

1. Decide which crosses will be needed to test your hypothesis.
2. Keep the available materials in mind as you plan your procedure. How many seeds will you need?
3. Record your procedure, and list the materials and quantities you will need.
4. Assign a task to each member of the group. One person should write data in a journal, another can pollinate the flowers, while a third can set up the plant trays. Determine who will set up and clean up materials.
5. Design and construct a data table for recording your observations.

Check the Plan

Discuss the following points with other group members to decide the final procedure for your experiment.

1. What data will you collect, and how will data be recorded?
2. When will you pollinate the flowers? How many flowers will you pollinate?
3. How will you transfer pollen from one flower to another?
4. How and when will you collect the seeds that result from your crosses?
5. What variables will have to be controlled? What controls will be used?
6. When will you end the experiment?
7. *Make sure your teacher has approved your experimental plan before you proceed further.*
8. Carry out your experiment.

ANALYZE AND CONCLUDE

1. **Checking Your Hypothesis** Did your data support your hypothesis? Why or why not?
2. **Interpreting Observations** What is the inheritance pattern of variegated leaves in *Brassica?*
3. **Making Inferences** Explain why genes in the chloroplast are inherited in this pattern.
4. **Drawing Conclusions** Which parent is responsible for passing the variegated trait to its offspring?
5. **Making Scientific Illustrations** Draw a diagram tracing the

inheritance of this trait through cell division.

Going Further

Project Make crosses between normal *Brassica* plants and genetically dwarfed, mutant *Brassica* plants to determine the inheritance pattern of the dwarf mutation.

BIOLOGY *Online* To find out more about inheritance of traits, visit the Glencoe Science Web site.

science.glencoe.com

12.3 COMPLEX INHERITANCE OF HUMAN TRAITS **337**

ANALYZE AND CONCLUDE

1. Student answers will vary depending on their hypotheses.
2. Variegation is inherited as a cytoplasmic trait from the female parent.
3. Sperm cells contain little to no cytoplasm. Cytoplasm containing chloroplasts is contributed only by the egg.
4. female
5. Diagrams should show the trait being transmitted in an egg cell of the female but not in pollen of the male.

Error Analysis

Sufficient light must be supplied to the plants to maximize the contrast between green and white leaves.

✓ Assessment

Skill Have students draw Punnett squares to explain the inheritance pattern in offspring if this trait were inherited as a simple dominant allele. Have them show how cytoplasmic inheritance deviates from this pattern. Use the Performance Task Assessment List for Scientific Drawing in **PASC**, p. 55. **L2**

Going Further

Have students carry out crosses between nonvariegated male and female parents as well as variegated male and female parents. **L3 ELL**

should be planted and grown. After 13-15 days, the plants will flower. Students will then perform the following cross-pollinations depending on their hypotheses: variegated female with nonvariegated male (transfer of pollen from nonvariegated male anther via brush to pistil of variegated female) and nonvariegated female with variegated male. Pods of seeds will mature between days 28 and 30. These

seeds will then be planted, and new offspring will be observed for the presence of the variegated trait.

Data and Observations

Variegated F_1 plants will appear in crosses where the female parent was variegated.

Purpose

Students will learn why hemophilia has been called the royal disease.

Teaching Strategies

■ Inform students that about 1 in 10 000 males has hemophilia, whereas about 1 in 100 000 000 females inherits this disorder. Make sure students realize that the reason for this is the fact that hemophilia is an X-linked disorder.

■ Have students hypothesize how Queen Victoria became a carrier of the disease when neither of her parents had the disorder. Students should be able to deduce that one of Victoria's parents developed a spontaneous mutation in his or her X chromosome, which was then passed to her. The mutation could not have occurred during the production of one of Victoria's egg cells because so many of her children were affected. **L2**

Connection to Biology

Four genotypes are possible in the offspring. Half the girls will be carriers and half will be normal. Among the boys, half will have hemophilia and half will be normal. Thus, one-fourth of all children of a female carrier and a normal male will be carriers of a sex-linked trait.

Social Studies
Connection

Queen Victoria and Royal Hemophilia

One of the most famous examples of a pedigree demonstrating inheritance of a sex-linked trait is the family of Queen Victoria of England and hemophilia.

Queen Victoria and her family

Queen Victoria had four sons and five daughters. Her son Leopold had hemophilia and died as a result of a minor fall. Two of her daughters, Alice and Beatrice, were carriers for the trait and passed the disorder to royal families in Spain, Prussia, and Russia over four generations.

The Spanish royal family Victoria's daughter Beatrice, a carrier for the trait, married Prince Henry of Battenberg, a descendent of Prussian royalty. Two of their sons inherited the trait, both dying before the age of 35. Her daughter, Victoria, was a carrier and married King Alfonso XIII of Spain, thus transmitting the allele to the Spanish royal family. Two of their sons died of hemophilia, also by their early thirties.

The Prussian royal family Alice, another of Victoria's daughters, married Louis IV of Hesse, part of the Prussian royal family and related to Prince Henry of Battenberg. One of Alice's sons, Frederick, died at the age of three from hemophilia. One of her daughters, Irene, continued to pass the trait to the next generation of Prussian royalty by giving it to two of her sons.

The Russian royal family Irene's sister and Queen Victoria's granddaughter, Alix (Alexandra), married Czar Nicholas II of Russia. Four healthy daughters were born, but the first male heir, Alexis, showed signs of bleeding and bruising at only six weeks of age. Having a brother, an uncle, and two cousins who had suffered from the disorder and died at early ages, you can imagine the despair Alix felt for her son and the future heir. In desperation, the family turned to Rasputin, a man who claimed to have healing abilities and used Alexis' illness for his

own political power. The series of events surrounding Alexis and his hemophilia played a role in the downfall of the Russian monarchy.

The British throne today Queen Elizabeth II, the current English monarch, is descended from Queen Victoria's eldest son, Edward VII. Because he did not inherit the trait, he could not pass it on to his children. Therefore, the British monarchy today does not carry the recessive allele for hemophilia, at least not inherited from Queen Victoria.

CONNECTION TO BIOLOGY

If you were the child of a female carrier for a sex-linked trait such as hemophilia, what would be your chances of carrying the trait?

 To find out more about hemophilia, visit the Glencoe Science Web site. **science.glencoe.com**

BIOLOGY *Online* Note Internet addresses that you find useful in the space below for quick reference.

_____

_____

_____

_____

SUMMARY

SUMMARY

Section 12.1

Mendelian Inheritance of Human Traits

Main Ideas
- A pedigree is a family tree of inheritance.
- Most human genetic disorders are inherited as rare recessive alleles, but a few are inherited as dominant alleles.

Vocabulary
carrier (p. 316)
fetus (p. 318)
pedigree (p. 315)

Section 12.2

When Heredity Follows Different Rules

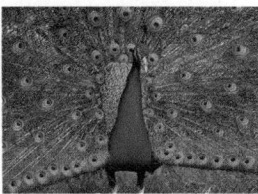

Main Ideas
- Alleles can be incompletely dominant or codominant.
- There may be many alleles for one trait or many genes that interact to produce a trait.
- Inheritance patterns of genes located on sex chromosomes are due to differences in the number and kind of sex chromosomes in males and in females.
- The expression of some traits is affected by the internal and external environments of the organism.

Vocabulary
autosome (p. 324)
codominant alleles (p. 323)
incomplete dominance (p. 321)
multiple alleles (p. 323)
polygenic inheritance (p. 326)
sex chromosome (p. 324)
sex-linked trait (p. 325)

Section 12.3

Complex Inheritance of Human Traits

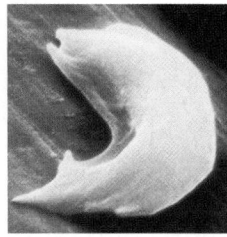

Main Ideas
- The majority of human traits are controlled by multiple alleles or by polygenic inheritance. The inheritance patterns of these traits are highly variable.
- Sex-linked traits are determined by inheritance of sex chromosomes. X-linked traits are usually passed from carrier females to their male offspring. Y-linked traits are passed only from male to male.
- Mistakes in meiosis, usually due to nondisjunction, may result in an abnormal number of chromosomes. Autosomes or sex chromosomes can be affected.

Vocabulary
karyotype (p. 335)

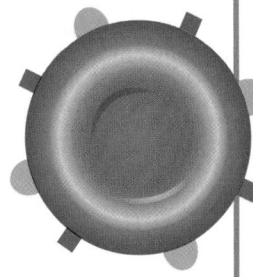

Main Ideas
Summary statements can be used by students to review the major concepts of the chapter.

Using the Vocabulary
To reinforce chapter vocabulary, use the Content Mastery Booklet and the activities in the Interactive Tutor for Biology: The Dynamics of Life on the Glencoe Science Web site.
science.glencoe.com

THE PRINCETON REVIEW

All Chapter Assessment questions and answers have been validated for accuracy and suitability by The Princeton Review.

UNDERSTANDING MAIN IDEAS

1. a
2. d

UNDERSTANDING MAIN IDEAS

1. If a trait is X-linked, males pass the X-linked allele to _____ of their daughters.
   a. all
   b. half
   c. none
   d. 1/4

2. Stem length demonstrates a range of phenotypes. This is an example of _____.
   a. autosomal dominant
   b. autosomal recessive
   c. sex-linkage
   d. polygenic inheritance

GLENCOE TECHNOLOGY

 **VIDEOTAPE**
MindJogger Videoquizzes
Chapter 12: *Patterns of Heredity and Human Genetics*
Have students work in groups as they play the videoquiz game to review key chapter concepts.

 ## Resource Manager

Chapter Assessment, pp. 67-72
MindJogger Videoquizzes
ExamView® Pro Software 💾
BDOL Interactive CD-ROM, Chapter 12 quiz

3. b
4. d
5. c
6. d
7. b
8. d
9. b
10. b
11. two
12. Down syndrome
13. son
14. XY, XX
15. numbers of chromosomes, karyotype
16. Nondisjunction
17. autosomal recessive
18. Yy
19. multiple allelic
20. incomplete dominance, codominance

APPLYING MAIN IDEAS

21. If the woman's father had the allele on his X chromosome, he would have had hemophilia, but he did not. Thus, the X chromosome she received from him is free of the allele. It is possible, but unlikely, that she received an X chromosome that carries the allele from her mother. In that case, she could pass it to her son.

3. Two parents with normal phenotypes have a daughter with a genetically inherited disorder. This is an example of a(n) _____ trait.
  a. autosomal dominant **c.** sex-linked
  b. autosomal recessive **d.** polygenic

4. Which of the following disorders would be inherited according to the pedigree shown here?
  a. Tay-Sachs disease
  b. sickle-cell anemia
  c. cystic fibrosis
  d. Huntington's disease

5. Which of the following disorders is likely to be inherited by more males than females?
  a. Huntington's disease
  b. Down syndrome
  c. hemophilia
  d. cystic fibrosis

6. Infants with PKU cannot break down the amino acid _____.
  a. tyrosine   **c.** methionine
  b. lysine   **d.** phenylalanine

7. A karyotype reveals _____.
  a. an abnormal number of genes
  b. an abnormal number of chromosomes
  c. polygenic traits
  d. multiple alleles for a trait

8. A mother with blood type $I^B i$ and a father with blood type $I^A I^B$ have children. Which of the following genotypes would be possible for their children?
  a. AB   **c.** B
  b. O   **d.** a and c are correct

THE PRINCETON REVIEW **TEST-TAKING TIP**

Get to the Root of Things
If you don't know a word's meaning, you can still get an idea of its meaning if you focus on its roots, prefixes, and suffixes. For instance, words that start with non-, un-, a-, dis-, and in- generally reverse what the rest of the word means.

9. Normally, lethal autosomal dominant traits are eliminated from a population because they _____.
  a. have a late onset
  b. have an early onset
  c. don't produce phenotypes that affect a carrier's health
  d. aren't dominant

10. Whose chromosomes determine the sex of offspring in humans?
  a. mother's   **c.** both parents'
  b. father's   **d.** neither parents'

11. A single individual carries _____ alleles for a trait.

12. _____ is a disorder that results from trisomy of chromosome 21.

13. Most sex-linked traits are passed from mother to _____.

14. The normal sex chromosomes of human males are _____, and the normal sex chromosomes of females are _____.

15. To analyze _____, geneticists make a chart of chromosomes called a(n) _____.

16. _____ during meiosis might result in monosomy or trisomy.

17. The inheritance pattern that occurs equally in both sexes and skips generations is _____.

18. The genotype of the individual represented by this pedigree symbol is _____. Use the letters Y and y to represent alleles.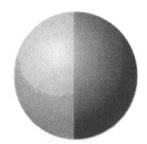

19. Feather colors in pigeons are produced by _____ inheritance.

20. If a trait has three different phenotypes, the trait is inherited by _____ or _____.

APPLYING MAIN IDEAS

21. The brother of a woman's father has hemophilia. Her father was unaffected, but she worries that she may have an affected son. Should she worry? Explain.

22. If a child has type O blood and its mother has type A, could a man with type B be the father? Why couldn't a blood test be used to prove that he is the father?

23. Why do certain human genetic disorders, such as sickle-cell anemia and Tay-Sachs disease, occur more frequently among one ethnic group than another?

24. How can a single gene mutation in a protein such as hemoglobin affect several body systems?

THINKING CRITICALLY

25. Recognizing Cause and Effect Explain why a male with a recessive X-linked trait usually produces no female offspring with the trait.

26. Comparing and Contrasting Compare multiple allelic with polygenic inheritance.

27. Concept Mapping Complete the concept map by using the following vocabulary terms: sex-linked trait, autosomes, karyotype, sex chromosomes, polygenic inheritance, codominant alleles.

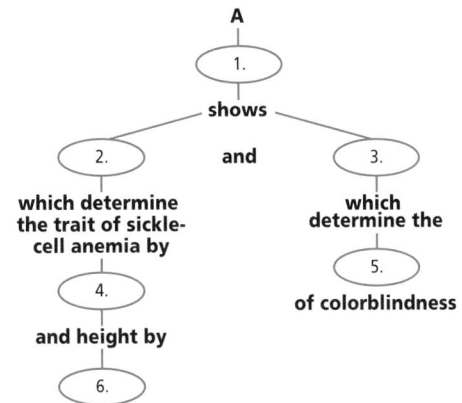

CD-ROM

For additional review, use the assessment options for this chapter found on the *Biology: The Dynamics of Life Interactive CD-ROM* and on the Glencoe Science Web site.
science.glencoe.com

ASSESSING KNOWLEDGE & SKILLS

The following graph illustrates the number of flowers produced per plant by a certain plant population.

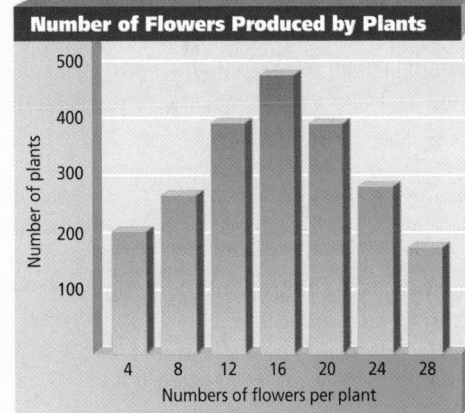

Interpreting Data Use the graph to answer the questions that follow.

1. How many flowers are produced by plants that have only dominant genes for flower production?
  a. 4   **c.** 16
  b. 12   **d.** 28

2. How many flowers are produced by plants that have half the possible number of dominant genes for flower production?
  a. 4   **c.** 16
  b. 12   **d.** 28

3. What pattern of inheritance is suggested by the graph?
  a. multiple alleles
  b. incomplete dominance
  c. polygenic inheritance
  d. sex-linkage

4. Observing and Inferring From the above graph, estimate the number of gene pairs that control the number of flowers in these plants.

22. The man could be $I^B i$ and be the father. A blood test can show only that he could possibly be the father, but not that he is the father.

23. People tend to marry within their own ethnic groups, allowing recessive alleles to show up more frequently in the populations.

24. The protein may be one that is transported or used in many systems. Malfunctions of one body system usually affect others.

THINKING CRITICALLY

25. All X chromosomes from the male parent are contributed to female offspring. If the female parent is homozygous dominant for the trait, all female offspring will be heterozygous and will not show the trait. The female parent would have to be heterozygous for that trait in order for half her female offspring to show the trait.

26. In multiple allelic inheritance, many forms of a trait may be found in a population, but they are based on only two alleles for the trait in each individual. In polygenic inheritance, the forms of the trait appear continuous, and individuals have many genes that add to the inheritance of the trait.

27. 1. Karyotype; 2. Autosomes; 3. Sex chromosomes; 4. Codominant alleles; 5. Sex-linked trait; 6. Polygenic inheritance

ASSESSING KNOWLEDGE & SKILLS

1. d
2. c
3. c
4. There are three gene pairs governing the number of flowers produced per plant.

Chapter 13 Organizer

Refer to pages 4T-5T of the Teacher Guide for an explanation of the National Science Education Standards correlations.

Section	Objectives	Activities/Features
Section 13.1 **Applied Genetics** National Science Education Standards UCP.2, UCP.3; A.1, A.2; C.2; E.1, E.2; F.1; G.1-3 (1 session)	1. **Predict** the outcome of a test cross. 2. **Evaluate** the importance of plant and animal breeding to humans.	**Focus On** Selective Breeding of Cats, p. 344 **Problem-Solving Lab 13-1**, p. 347
Section 13.2 **Recombinant DNA Technology** National Science Education Standards UCP.2, UCP.3, UCP.5; A.1, A.2; C.2; E.1, E.2; F.1, F.4, F.5, F.6; G.1, G.2 (2 sessions)	3. **Summarize** the steps used to engineer transgenic organisms. 4. **Give examples** of applications and benefits of genetic engineering.	**MiniLab 13-1:** Matching Restriction Enzymes to Cleavage Sites, p. 351 **Inside Story:** Gel Electrophoresis, p. 354 **Problem-Solving Lab 13-2**, p. 355 **Investigate BioLab:** Modeling Recombinant DNA, p. 362 **BioTechnology:** How to Clone a Mammal, p.364
Section 13.3 **The Human Genome** National Science Education Standards UCP.2, UCP.3; A.1, A.2; C.2; E.1, E.2; F.1, F.6; G.1-3 (2 sessions)	5. **Analyze** how the effort to completely map and sequence the human genome will advance human knowledge. 6. **Predict** future applications of the Human Genome Project.	**MiniLab 13-2:** Storing the Human Genome, p. 358 **Careers in Biology:** Forensic Analyst, p. 359 **Problem-Solving Lab 13-3**, p. 361

Need Materials? Contact Carolina Biological Supply Company at 1-800-334-5551 or at **http://www.carolina.com**

MATERIALS LIST

BioLab
p. 362 paper, transparent tape, scissors, red and green pencils

MiniLabs
p. 351 paper, pencil
p. 358 paper, pencil, book, calculator (optional)

Alternative Lab
p. 358 microscope, microscope slide, prepared slide of female cheek cells, T-shirt, red ink, red grape juice, human hair (blond and dark), envelopes (4), bar codes (5)

Quick Demos
p. 346 photographs of pets
p. 350 Chromosome Simulation Biokit
p. 359 human chromosome map

Key to Teaching Strategies

L1	Level 1 activities should be appropriate for students with learning difficulties.
L2	Level 2 activities should be within the ability range of all students.
L3	Level 3 activities are designed for above-average students.
ELL	ELL activities should be within the ability range of English Language Learners.
COOP LEARN	Cooperative Learning activities are designed for small group work.
P	These strategies represent student products that can be placed into a best-work portfolio.
	These strategies are useful in a block scheduling format.

Genetic Technology

Teacher Classroom Resources

Section	Reproducible Masters	Transparencies
Section 13.1 **Applied Genetics**	Reinforcement and Study Guide, p. 55 **L2** Laboratory Manual, pp. 91-94 **L2** Content Mastery, pp. 61, 64 **L1**	Section Focus Transparency 32 **L1** **ELL** Basic Concepts Transparency 19 **L2** **ELL**
Section 13.2 **Recombinant DNA Technology**	Reinforcement and Study Guide, pp. 56-57 **L2** BioLab and MiniLab Worksheets, pp. 61-62 **L2** Laboratory Manual, pp. 95-98 **L2** Tech Prep Applications, pp. 21-22 **L2** Content Mastery, pp. 61-62, 64 **L1**	Section Focus Transparency 33 **L1** **ELL** Reteaching Skills Transparency 22 **L1** **ELL**
Section 13.3 **The Human Genome**	Reinforcement and Study Guide, p. 58 **L2** Concept Mapping, p. 13 **L3** **ELL** Critical Thinking/Problem Solving, p. 13 **L3** BioLab and MiniLab Worksheets, pp. 63-66 **L2** Content Mastery, pp. 61, 63-64 **L1** Tech Prep Applications, pp. 21-22 **L2**	Section Focus Transparency 34 **L1** **ELL**

Assessment Resources

Chapter Assessment, pp. 73-78
MindJogger Videoquizzes
Performance Assessment in the Biology Classroom
Alternate Assessment in the Science Classroom
ExamView® Pro Software 💾
BDOL Interactive CD-ROM, Chapter 13 quiz

Additional Resources

Spanish Resources **ELL**
English/Spanish Audiocassettes **ELL**
Cooperative Learning in the Science Classroom **COOP LEARN**
Lesson Plans/Block Scheduling

NATIONAL GEOGRAPHIC — Teacher's Corner

Products Available From National Geographic Society

To order the following products, call National Geographic Society at 1-800-368-2728:

Video
DNA: Laboratory of Life

GLENCOE TECHNOLOGY

The following multimedia resources are available from Glencoe.

Biology: The Dynamics of Life
CD-ROM **ELL**

Animation: *Gene Cloning*
Video: *Bioengineering*
Animation: *Recombinant DNA*

Videodisc Program 📼

Gene Cloning
Bioengineering
Recombinant DNA

The Infinite Voyage

The Geometry of Life
Testcross–Homozygous
Testcross–Heterozygous
Miracles by Design

Bring in several vegetables and fruits such as corn or strawberries and ask how genetic technology could improve them. Answers may include making produce larger, making them bruise less easily, or helping them survive adverse weather conditions such as frost.

Theme Development

The theme of **evolution** is alluded to as students are introduced to selective breeding techniques that achieve new and different traits in offspring. The theme **nature of science** is developed in this chapter as the techniques for changing the genetic makeup of organisms are discussed. Some of the techniques may be used to restore **homeostasis** to organisms afflicted with genetic disorders.

0:00 OUT OF TIME?

If time does not permit teaching the entire chapter, use the BioDigest at the end of the unit as an overview.

READING BIOLOGY

Glencoe's *Biology: The Dynamics of Life* contains many resources to assist a student's reading skills. Each chapter contains figures with expanded captions that expand on written material. Word Origins, located along the side of text, expand knowledge of biology vocabulary. Glencoe's Content Mastery Booklet helps develop reading skills while reinforcing content. In addition, use the Interactive Tutor for *Biology: The Dynamics of Life* on the Glencoe Web site to reinforce vocabulary. **science.glencoe.com**

What You'll Learn
- You will evaluate the importance of plant and animal breeding to humans.
- You will summarize the steps used to engineer transgenic organisms.
- You will analyze how mapping the human genome will benefit human life.

Why It's Important
Genetic technology will continue to impact every aspect of your life, from growing the food you eat to treating a disease you might inherit.

READING BIOLOGY

Go through the chapter, and make a list of several new vocabulary words. For each word, find the definition in the text. Compare them with dictionary definitions, then write a new sentence using each word.

BIOLOGY *Online*

To find out more about genetic technology, visit the Glencoe Science Web site. **science.glencoe.com**

Genetic technology provides opportunities for changing plants and animals. Dolly the sheep was the first cloned mammal. Strawberry plants have been made frost-resistant by genetic engineering.

342 GENETIC TECHNOLOGY

13.1 Applied Genetics

SECTION PREVIEW

Objectives
Predict the outcome of a test cross.
Evaluate the importance of plant and animal breeding to humans.

Vocabulary
inbreeding
test cross

For thousands of years, humans have admired the strength, power, and grace of animals like this mountain lion. Through years of selective breeding, people have been able to select certain qualities in plants and animals and breed them so that these qualities are common and more useful to humans. The traits of the mountain lion would not be desirable in a domesticated cat. Instead, gentleness and the ability to provide companionship are traits that have been selectively bred into these pets.

Mountain lion (above) and domesticated cat (inset)

Selective Breeding

The same principle of selective breeding that applies to cats also applies to much of the food we eat and to the animals such as horses that help us with hard labor. You can read about the selective breeding of domesticated cats in the *Focus On* feature on the next page. The process of selective breeding requires time, patience, and several generations of offspring before the desired trait becomes common in a population. Although our ancestors did not realize it, their efforts at selective breeding increased the frequency of a desired allele within a population. Increasing the frequency of desired alleles in a population is the essence of genetic technology.

Selective breeding produces organisms with desired traits

From ancient times, breeders have chosen the plants and animals with the most desired traits to serve as parents of the next generation. Farmers use for seed the largest heads of grain, the juiciest berries, and the most disease-resistant clover. They raise the calves of the best milk producer and save the eggs of the best egg-laying hen for hatching. Breeders of plants and animals want to be sure that their populations breed consistently so that each member shows the desired trait.

13.1 APPLIED GENETICS **343**

Section 13.1

Prepare

Key Concepts
Students will study the role of the test cross as a tool in determining genotypes. They investigate means of achieving desirable traits in plants and animals through the practices of selective breeding and hybridization.

Planning
■ Purchase fruits and vegetables for the Getting Started Demo.

1 Focus

Bellringer
Before presenting the lesson, display **Section Focus Transparency 32** on the overhead projector and have students answer the accompanying questions. **L1** **ELL**

Resource Manager

Section Focus Transparency 32 and Master **L1** **ELL**
Tech Prep Applications, pp. 19-20 **L2**

Assessment Planner

Portfolio Assessment
Portfolio, TWE, pp. 345, 353, 360
MiniLab, TWE, pp. 351, 358
Performance Assessment
Assessment, TWE, p. 353
Problem-Solving Lab, TWE, p. 355
Alternative Lab, TWE, pp. 358-359
BioLab, TWE, pp. 362-363
BioLab, SE, pp. 362-363
MiniLab, SE, pp. 351, 358

Knowledge Assessment
Assessment, TWE, pp. 346, 348, 356
Problem-Solving Lab, TWE, p. 360
Section Assessment, SE, pp. 348, 356, 361
Chapter Assessment, SE, pp. 365-367
Skill Assessment
Problem-Solving Lab, TWE, p. 347
Assessment, TWE, p. 361

2 Teach

Focus On

Selective Breeding of Cats

Purpose
Students will gain some insight into the breeding of cats and the influence of cats on human life throughout history.

Background
Cats can be bred for a number of characteristics, including physical traits such as hair length or color. Temperament can also be selected by breeding. Certain breeds have a reputation for being quiet and loving, others for being noisy.

Teaching Strategies
■ Make sure that students are familiar with the meaning of the terms *breed* and *selective breeding*.
■ Ask students to bring in photographs of their pet cats. If any student's cat is a pure breed, have the student describe distinctive traits or qualities of the breed.

GLENCOE TECHNOLOGY

VIDEODISC
The Infinite Voyage
The Geometry of Life
Selective Breeding (Ch. 7)
2 min. 30 sec.

VIDEOTAPE
The Secret of Life
Gone Before You Know It: The Biodiversity Crisis

FOCUS ON

Selective Breeding of Cats

KITTENS (ABOVE AND BELOW) WITH DOMINANT WHITE MARKINGS ON FACE, PAWS, AND THROAT

Graceful, agile, and independent, cats are popular pets. In the United States alone, more than 55 million cats are kept as pets. Although the origin of the domestic cat is lost in antiquity, archeological evidence indicates that an association between cats and people existed as much as 3500 years ago in ancient Egypt. Unlike dogs, cattle, and many other domesticated animals, however, cats have only recently been bred selectively to exhibit specific traits. Currently about 40 recognized breeds exist—developed by selectively mating cats having especially desirable or distinctive characteristics. Different breeds vary primarily in color, in length and texture of fur, and in temperament.

COLORFUL COATS
Cats come in many colors, but the most common coats are tabby (a striped or blotchy pattern), black, and orange. Cats with "orange" coats range in color from creamy yellow to dark ginger red. The genetic control of cat fur color is complex and only partially understood. Solid white fur is dominant to all other fur colors. Spots of white—especially on the face, throat, and paws—are also dominant to solid color coats. Some breeds such as the Siamese (below) have been bred for a light-colored body with dark legs, tail, ears, and face—the perfect frame for bright blue eyes.

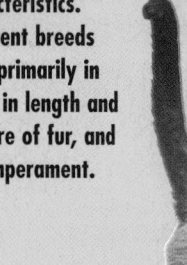

SIAMESE

SOMALI—A LONGHAIRED ABYSSINIAN BREED—AT CAT SHOW

344

Cultural Diversity

Cultural Taboos Against Inbreeding

Although inbreeding can result in high rates of stillbirths and children with congenital disorders, scientists are not sure whether taboos against incest are culturally or naturally selected. Explain to students that taboos against inbreeding are not universal, indicating that their origins are cultural rather than genetic. However, individuals brought up in close proximity to each other as family members usually are less likely to develop a sexual interest in each other. This suggests that incest taboos may have a biological component.

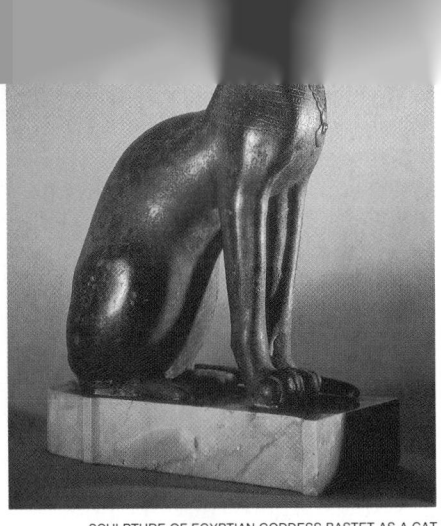

SCULPTURE OF EGYPTIAN GODDESS BASTET AS A CAT

SHORT VERSUS LONG

Cat breeds can be divided into two major groups: those with short hair and those with long. The Abyssinian—slender and regal-looking with large ears and almond-shaped eyes—is a popular short-haired breed. The ancestry of Abyssinians is unclear, but they may be descended from the sacred cats of ancient Egypt. Certainly, their similarity to Egyptian cat sculptures, such as the one at left, is striking. The American shorthair, on the other hand, is a sturdy muscular breed developed from cats that accompanied European settlers to the American colonies.

There are about a dozen breeds of longhaired cats, ranging from the large (up to 13.5 kg, or 30 pounds), shaggy Maine coon cat to the ever-popular Persian. Persian cats (below) are prized for their extremely long fur that stands out from their bodies, especially on the neck, face, and tail. Hundreds of years of careful breeding have refined the distinct "powder puff" appearance of the modern Persian.

TRAITS AND TEMPERAMENTS

Some cats have been bred for special traits. The Manx (right), for example, is tailless. Manx cats trace their roots to the Isle of Man off the coast of England. With hind legs longer than front legs, Manx cats run with a rabbitlike, hopping gait. The breed known as Ragdoll gets its name from the fact that it relaxes its muscles and goes completely limp when picked up. Fearless and calm, the Ragdoll is a fairly new breed of cat, which originated in the United States in the 1960s. Different breeds of cats have different temperaments, or personalities. Siamese tend to be vocal and demanding. The Japanese bobtail—thought to bring good luck—is playful and adaptable. The elegant Abyssinian is known for being quiet and very affectionate.

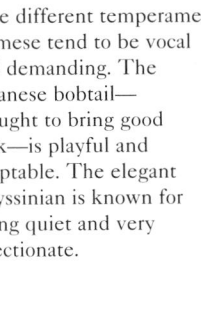

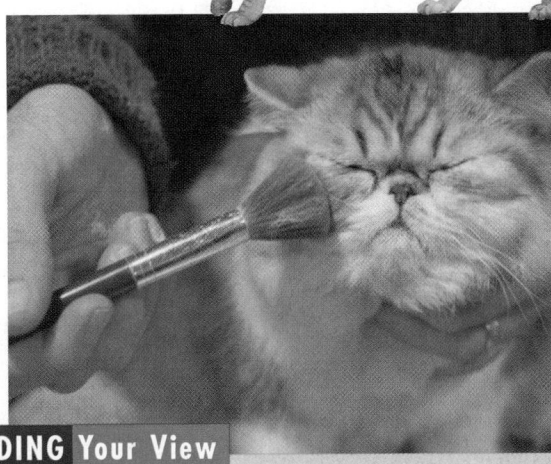

SILVER MANX TABBY (ABOVE), AND WHITE PERSIAN GETTING GROOMED ▶

EXPANDING Your View

1 THINKING CRITICALLY The ancient Egyptians stored large amounts of grain near their cities to ensure there would be enough to eat when crops failed. Speculate on why the ancient Egyptians may have been motivated to domesticate cats.

2 JOURNAL WRITING Research a breed of domestic cat. In your journal, write about the breed's history and specific traits for which it was bred.

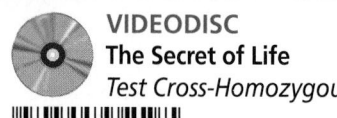

Figure 13.1
A pure breed, such as this German shepherd dog, is homozygous for the particular characteristics for which it has been bred.

One example of the effectiveness of selective breeding is seen in a comparison of milk production in cattle in 1947 and 1997. In 1947, an average milk cow produced 4997 pounds of milk per year. In 1997, 50 years later, an average milk cow produced 16 915 pounds of milk in a year, more than three times more milk per cow. Fewer than half the number of cows are now needed to produce the same amount of milk, resulting in savings for dairy farmers.

Figure 13.2
These roses, all different cultivars, have been hybridized to combine traits such as color, aroma, and flower shape.

Inbreeding develops pure lines

To make sure that breeds consistently exhibit a trait and to eliminate any undesired traits from their breeding lines, breeders often use the method of inbreeding. **Inbreeding** is mating between closely related individuals. It ensures that the offspring are homozygous for most traits. However, inbreeding also brings out harmful, recessive traits because there is a greater chance that two closely related individuals may both carry a harmful recessive allele for the trait.

Horses and dogs are two examples of animals that breeders have developed as pure breeds. A breed (called a cultivar in plants) is a selected group of organisms within a species that has been bred for particular characteristics. For example, the pure breed German shepherd dog has long hair, is black with a buff-colored base, has a black muzzle, and resembles a wolf, *Figure 13.1*.

Hybrids are usually bigger and better

Selective breeding of plants can increase productivity of food for humans. For example, plants that are disease resistant can be crossed with others that produce larger and more numerous fruit. The result is a plant that will produce a lot of fruit and be more disease resistant. Recall that a hybrid is the offspring of parents that have different forms of a trait. When two cultivars or closely related species are crossed, their offspring will be hybrids. Hybrids produced by crossing two purebred plants are often larger and stronger than their parents. Many crop plants such as wheat, corn, and rice, and garden flowers such as roses and dahlias have been developed by hybridization. *Figure 13.2* shows some examples.

Determining Genotypes

A good breeder must be careful to determine which plants or animals will have the greatest chances of transmitting a desired trait to the next generation. Choosing the best parents may be difficult. The genotype of an organism that is homozygous recessive for a trait is obvious to an observer because the recessive trait is expressed. However, organisms that are either homozygous dominant or heterozygous for a trait controlled by Mendelian inheritance have the same phenotype. How can a breeder learn which genotype should be used for breeding?

Test crosses can determine genotypes

One way to determine the genotype of an organism is to perform a test cross. A **test cross** is a cross of an individual of unknown genotype with an individual of known genotype. The pattern of observed phenotypes in the offspring can help determine the unknown genotype of the parent. Usually, the parent with the known genotype is homozygous recessive for the trait in question.

Many traits, such as disease vulnerability in rose plants and progressive blindness in German shepherd dogs, are inherited as recessive alleles. These undesired traits are maintained in the population by carriers of the trait. A carrier, or heterozygous individual, appears to have the same phenotype as an individual that is homozygous dominant.

What are the possible results of a test cross? If the known parent is homozygous recessive and the unknown parent is homozygous dominant, all of the offspring will be heterozygous for the trait and will show the dominant trait (be phenotypically

dominant), as shown in *Figure 13.3* on the next page. However, if the organism being tested is heterozygous, the predicted 1: 1 phenotypic ratio will be observed. If any of the offspring have the undesired trait, the parent in question must be heterozygous. Doing the *Problem-Solving Lab* will show you how to set up and analyze a test cross.

13.1 APPLIED GENETICS **347**

Problem-Solving Lab 13-1 — Designing an Experiment

When is a test cross practical? How can you tell the genotype of an organism that has a dominant phenotype? There are two ways. The first is through the use of pedigree studies. This technique works well as long as a family is fairly large and records are accurate. The second technique is a test cross. Test crosses help determine whether an organism is homozygous or heterozygous for a dominant trait.

Analysis

Your pet guinea pig has black hair. This trait is dominant and can be represented by a *B* allele. Your neighbor has a white guinea pig. This trait is recessive and can be represented by a *b* allele. You want to breed the two guinea pigs but want all offspring from the mating to be black. You are not sure, however, of the genotype of your black guinea pig and want to find out before starting the breeding program.

Thinking Critically

1. What may be the possible genotypes of your black guinea pig? Explain why you are unable to tell even though the animal has a black phenotype.
2. What is the genotype of the white guinea pig? Explain how you are able to tell.
3. Outline a procedure that will determine the coat color genotype for your black guinea pig. Include Punnett squares to illustrate the conclusions that you will reach. (Hint: You will be doing a test cross.)
4. What options do you have for breeding all black offspring if you determine that your guinea pig is heterozygous for black color?
5. Explain why a test cross is not practical when trying to determine human genotypes.

Problem-Solving Lab 13-1

Purpose
Students will describe a procedure for determining a genotype.

Process Skills
think critically, apply concepts, draw a conclusion

Teaching Strategies

■ Pedigree studies will work in determining genotypes only when there are many offspring and when certain individuals are homozygous recessive. Illustrate this point with a pedigree that shows a small family of two parents (both with a dominant phenotype) and one offspring (with a dominant phenotype). There is no way to predict whether any individuals are heterozygous.

Thinking Critically

1. *BB* or *Bb*; both genotypes show a black phenotype.
2. *bb*; the trait is recessive and the white phenotype can show only when both alleles are recessive.
3. Mate an unknown black guinea pig with a white guinea pig. If all offspring are black, you may conclude that the black guinea pig is *BB*. If any white offspring are born, however, the black guinea pig must be heterozygous.
4. Continue to breed your guinea pig with a white guinea pig but understand that half of all offspring will be white, or do not breed your guinea pig, or breed your black guinea pig with a homozygous black guinea pig. In the last case, all offspring will be black but half will be heterozygous.
5. Because humans usually have only one offspring per birth, it would require too much time for determining genotype. The person would have to mate with a homozygous recessive individual for the trait in question.

347

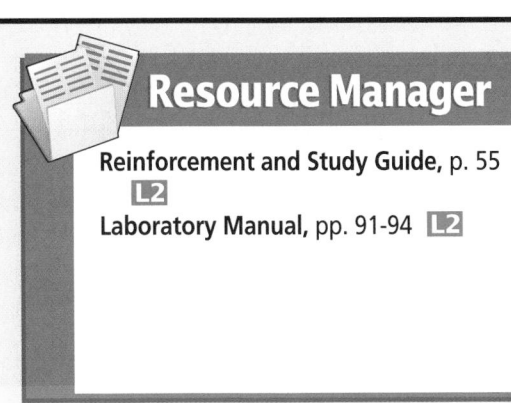

Resource Manager

Reinforcement and Study Guide, p. 55
L2

Laboratory Manual, pp. 91-94 **L2**

✓ Assessment

Skill Ask students to prepare Punnett squares that would illustrate the results of a cross between two heterozygous black guinea pigs and between one heterozygous and one homozygous black. Which mating would be preferred if only black offspring were desired? Use the Performance Task Assessment List for Scientific Drawing in **PASC**, p. 55. **L1**

3 Assess

Check for Understanding

Have students explain the role of a test cross and what breeding programs attempt to accomplish. **L1**

Reteach

Albinism, or white fur, in rabbits is due to a recessive allele. Let *A* = normal fur pigment and *a* = albinism (no fur pigment). A breeder wishes to know if a male rabbit is homozygous or heterozygous for pigmented fur. Have students diagram the possible crosses in Punnett squares and interpret the results. **L1**

Extension

Have students explore the problems that occur when horses are bred to donkeys to produce mules and hinnies. Have them speculate as to how Mendel's laws might have differed if he had worked with these animals. **L3**

✔ **Assessment**

Knowledge Provide students with a test cross for a trait. Ask students to complete the genotypes for the test cross. **L1**

---

4 Close

Discussion

Discuss with students that there is currently an interest in eggs with low cholesterol content. Then ask students how they might proceed to breed chickens that produce such low-cholesterol eggs.

Figure 13.3
In this test cross of Alaskan malamutes, the known test dog is homozygous recessive for a dwarf allele *(dd)*, and the other dog's genotype is unknown.

A The unknown dog can be either homozygous dominant *(DD)* or heterozygous *(Dd)* for the trait.

B If the unknown dog's genotype is homozygous dominant, all of the offspring will be phenotypically dominant.

Homozygous × Homozygous

	DD	dd
	d	d
D	Dd	Dd
D	Dd	Dd

Offspring: all dominant

C If the unknown dog's genotype is heterozygous, half the offspring will express the recessive trait and appear dwarf. The other half will express the dominant trait and be of normal size.

Heterozygous × Homozygous

	Dd	dd
	d	d
D	Dd	Dd
d	dd	dd

Offspring: 1/2 dominant
1/2 recessive

Section Assessment

Understanding Main Ideas

1. A test cross made on a cat that may be heterozygous for a recessive trait produces ten kittens, none of which has the trait. What is the presumed genotype of the cat? Explain.
2. Suppose you want to produce a plant cultivar that has red flowers and speckled leaves. You have two cultivars, each having one of the desired traits. How would you proceed?
3. Why is inbreeding rarely a problem among animals in the wild?
4. Hybrid corn is produced that is resistant to bacterial infection and is highly productive. What might have been the phenotypes of the two parents ?

Thinking Critically

5. What effect might selective breeding of plants and animals have on the size of Earth's human population? Why?

SKILL REVIEW

6. **Making and Using Tables** A bull is suspected of carrying a rare, recessive allele. Following a test cross with a homozygous recessive cow, four calves are born, two that express the recessive trait and two that do not. Draw a Punnett square that shows the test cross, and determine the genotype of the bull. For more help, refer to *Organizing Information* in the **Skill Handbook.**

348 GENETIC TECHNOLOGY

Section Assessment

1. The cat is probably homozygous dominant. If it were heterozygous, there would probably have been some offspring with the recessive trait.
2. First breed the red flower plants among themselves to ensure that they are breeding true. Do the same for the plants with speckled leaves. Then, hybridize the two cultivars by breeding them together.
3. In nature, mate selection is random and the chance that a mate will be closely related is greatly reduced.
4. One parent was highly productive. The other was resistant to bacterial infection.
5. Selective breeding can increase crop size and provide for more nutritious and more disease-resistant crops, and this might increase the human population.
6. Students' Punnett squares should show a cross between a homozygous recessive individual and a heterozygous individual. Half the offspring will be normal and half will express the recessive trait. This is what was observed, so the bull was heterozygous.

13.2 Recombinant DNA Technology

Y ou have learned that DNA can function like a zipper, opening up to allow replication and transcription. Scientists have found a series of enzymes that can cut DNA at specific locations, sometimes unzipping the strands as they cut. These enzymes allow scientists to insert genes from other sources into DNA. The glowing plant shown here was created by inserting a firefly gene into the DNA of a tobacco plant.

Transgenic tobacco plant (above) and firefly (inset)

SECTION PREVIEW

Objectives
Summarize the steps used to engineer transgenic organisms.
Give examples of applications and benefits of genetic engineering.

Vocabulary
genetic engineering
recombinant DNA
transgenic organism
restriction enzyme
vector
plasmid
gene splicing
clone

Prepare

Key Concepts

Students will learn that genetic engineering involves a three-step process that produces a transgenic organism. Students will explore techniques used to sequence DNA and applications of DNA technology in agriculture, industry, and medicine.

Planning

■ Make cutouts of bases for the Meeting Individual Needs.

1 Focus

Bellringer

Before presenting the lesson, display **Section Focus Transparency 33** on the overhead projector and have students answer the accompanying questions. **L1 ELL**

Genetic Engineering

You learned that selective breeding is a form of genetic technology because it increases the frequency of an allele in a population. You also learned that it may take many generations of breeding for a trait to become homozygous and consistently expressed in the population. **Genetic engineering** is a much faster and more reliable method for increasing the frequency of a specific allele in a population. This method involves cutting—or cleaving—DNA from one organism into small fragments and inserting the fragments into a host organism of the same or a different species. You may also hear genetic engineering referred to as recombinant (ree KAHM buh nunt) DNA technology. **Recombinant DNA** is made by connecting, or recombining, fragments of DNA from different sources.

Transgenic organisms contain recombinant DNA

Recombinant DNA can be inserted into a host organism's chromosomes and that organism will use the foreign DNA as if it were its own. Plants and animals that contain functional recombinant DNA, such as the glowing tobacco plant, are known as **transgenic organisms** because they contain foreign DNA.

The glowing tobacco plant is the result of a three-step process that is used to produce a transgenic organism.

WORD Origin

transgenic
From the Latin word *trans*, meaning "across," and the Greek word *genos*, meaning "race." A transgenic organism contains genes from another species.

13.2 RECOMBINANT DNA TECHNOLOGY **349**

BIOLOGY Online Note Internet addresses that you find useful in the space below for quick reference.

_____

_____

_____

_____

_____

Resource Manager

Section Focus Transparency 33 and Master **L1 ELL**
Basic Concepts Transparency 19 and Master **L2 ELL**

2 Teach

Discussion

Ask students what genetic changes people might want to engineer into their pets. *Students might suggest stamina, intelligence, hair retention, and disease resistance.*

Concept Development

To help students remember EcoR1, explain that this enzyme was discovered in the bacterium *Escherichia coli.* E comes from *Escherichia*, co from *coli*, R from Restriction enzyme, and 1 indicates that this enzyme was the first restriction enzyme found in this bacterium.

Quick Demo

The use of restriction enzymes and sticky ends can be demonstrated by using the materials in the Chromosome Simulation Biokit available from Carolina Biological Supply Company.

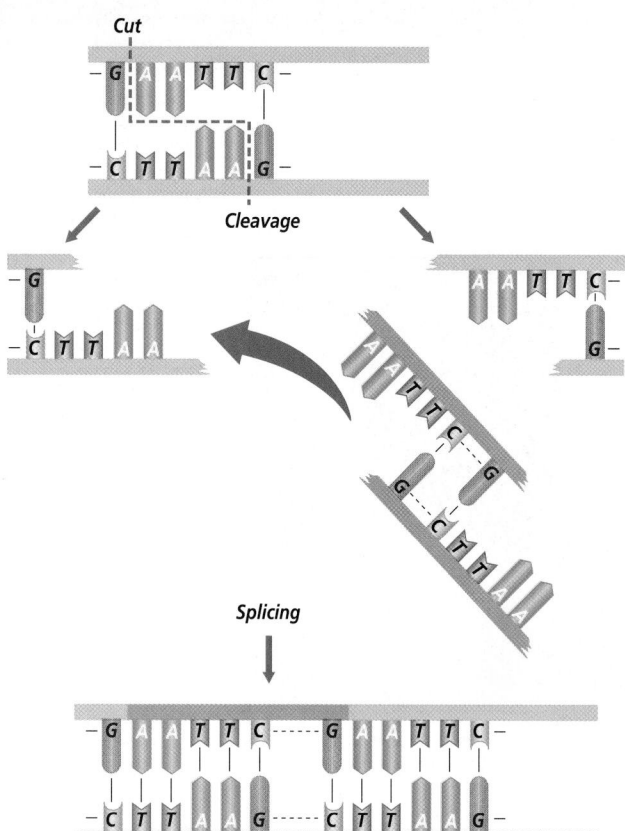

Figure 13.4
In the presence of the restriction enzyme *Eco*R1, a double strand of DNA containing the sequence—GAATTC— is cleaved between the G and the A.

The first step is to isolate the foreign DNA fragment that will be inserted. The second step is to attach the DNA fragment to a "vehicle." The third step is the transfer of the vehicle into the host organism. Each of these three steps now will be discussed in greater detail.

Restriction enzymes cleave DNA

To isolate a DNA fragment, small pieces of DNA must be cut from a larger chromosome. In the example of the glowing tobacco plant, this fragment is a section of firefly DNA that codes for the light-producing enzyme. The discovery in the early 1970s of DNA-cleaving enzymes made it possible to cut DNA. These

restriction enzymes are bacterial proteins that have the ability to cut both strands of the DNA molecule at a specific nucleotide sequence. There are hundreds of restriction enzymes, each capable of cutting DNA at a specific point in a specific nucleotide sequence. The resulting DNA fragments are of different lengths. When restriction enzymes cut DNA, it is similar to cutting a zipper into pieces by cutting only between certain teeth of the zipper. Note in *Figure 13.4* that the same sequence of bases is found on both DNA strands, but running in opposite directions. This arrangement is called a palindrome (pal uhn drohm). Palindromes are words or sentences that read the same forwards or backwards. The words *mom* and *dad* are two very simple examples of palindromes.

Some enzymes produce fragments in which the DNA is cut straight across both strands. These are called blunt ends. Other enzymes, such as the enzyme called *Eco*R1, cut palendromic sequences of DNA by unzipping them for a few nucleotides, as shown in *Figure 13.4.* When this DNA is cut, double-stranded fragments with single-stranded ends are formed. The single-stranded ends have a tendency to join with other single-stranded ends to become double stranded, so they attract DNA they can join with. For this reason, these ends are called sticky ends. This is the key to recombinant DNA because if the same enzyme is used to cleave DNA from two organisms, such as firefly DNA and bacterial DNA, the two pieces of DNA will have matching sticky ends and will join together at these ends. When the firefly DNA joins with bacterial DNA, recombinant DNA is formed. The *MiniLab* on the opposite page models the way restriction enzymes work.

350 GENETIC TECHNOLOGY

350

MEETING INDIVIDUAL NEEDS

Visually Impaired

Kinesthetic Have students who are visually impaired use large pieces of cardboard cut into four shapes (several pieces of each shape) to represent each of the four bases: A, T, C, and G. Use them to demonstrate a palindrome and to model the action of restriction enzymes. **L1**
ELL

Vectors transfer DNA

Loose fragments of DNA do not readily become part of a host organism's chromosomes. To make this process easier, the fragments are first attached to a vehicle that will carry them into the host organism's cells. In the case of the transgenic tobacco plant, the light-producing firefly DNA has to be inserted into bacterial DNA before it can be placed inside the plant. The bacterial DNA is a **vector**. A vector is a means by which DNA from another species can be carried into the host cell. Vectors may be biological or mechanical.

Biological vectors include viruses and plasmids. A **plasmid,** shown in *Figure 13.5,* is a small ring of DNA found in a bacterial cell. The genes it carries are different from those on the larger bacterial chromosome.

Two mechanical vectors carry foreign DNA into the nucleus of a cell. One, a micropipette, is inserted into a cell; the other is a tiny metal bullet coated with DNA that is shot into the cell using a device called a gene gun.

Figure 13.5
Plasmids are small rings of DNA. The large ring is the bacterium's chromosome.

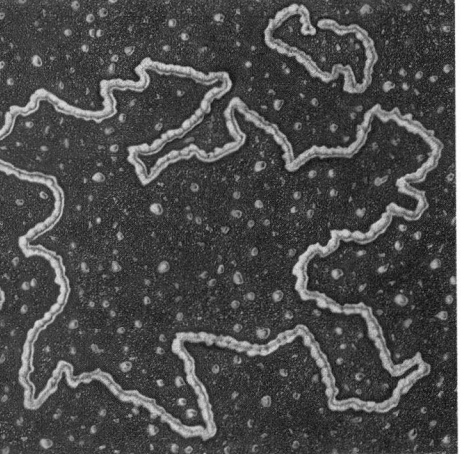

Magnification: 157 500×

MiniLab 13-1 — Applying Concepts

Matching Restriction Enzymes to Cleavage Sites Many restriction enzymes cut sequences of DNA that are palindromes. As a result of cuts to the DNA, single-stranded sequences of DNA are left dangling at the ends of a fragment. These ends are available for pairing with their complementary bases in a plasmid or piece of viral DNA.

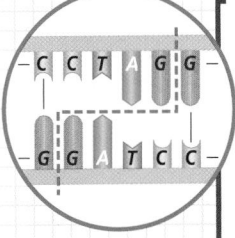

DNA fragment

Procedure
1. Copy the data table below.

Data Table

DNA fragment	Enzyme letter (D-F)	Action of restriction enzyme	Cleaved fragment of DNA
—GGTACC— —CCATGG—	E	—G¦GTACC— —CCATG¦G—	—G GTACC— —CCATG G—
—CCATGG— —GGTACC—			
—CAATTG— —GTTAAC—			
—GATATC— —CTATAG—			

2. Figure out which restriction enzyme will cleave each DNA fragment. Use the following guides.

   Enzyme D cleaves at an A-A site and leaves 3 single-stranded bases on each end.

   Enzyme E cleaves at a G-G site and leaves 4 single-stranded bases on each end.

   Enzyme F cleaves at a G-A site and leaves 4 single-stranded bases on each end.

3. Draw in the action of each enzyme. Record its letter.
4. Diagram each fragment of DNA as it would appear if cleaved by the proper restriction enzyme.
5. Use the top row in the table as an example and guide.

Analysis
1. Use the example provided in the data table to illustrate a single-stranded dangling end of DNA.
2. Record the DNA base sequence that must be present on a piece of viral DNA if these ends could "stick to" the dangling bases in the example shown in the data table.
3. Are restriction enzymes very specific as to where they cleave DNA? Explain your answer and give an example.

Resource Manager

BioLab and MiniLab Worksheets, pp. 61-62 **L2**
Reteaching Skills Transparency 22 and Master **L1** **ELL**

Data Table

DNA fragment	Enzyme letter (D-F)	Action of restriction enzyme	Cleaved fragment of DNA
—GGTACC— —CCATGG—	E	—G¦GTACC— —CCATG¦G—	—G GTACC— —CCATG G—
—CCATGG— —GGTACC—	E	—CCATG¦G— —G¦GTACC—	—CCATG G— —G GTACC—
—CAATTG— —GTTAAC—	D	—CA¦ATTG— —GTTAA¦C—	—CA ATTG— —GTTAA C—
—GATATC— —CTATAG—	F	—G¦ATATC— —CTATA¦G—	—G ATATC— —CTATA G—

MiniLab 13-1

Purpose
Students will match restriction enzymes with their proper base sequences in a DNA strand.

Process Skills
think critically, analyze information, apply concepts, define operationally, draw a conclusion

Teaching Strategies
■ Be sure that students have read over the applicable material in the text before starting this MiniLab.
■ Allow students to work in small groups to complete this activity.
■ Remind students that the DNA segments shown are only a small portion of the DNA molecule.
■ Point out to students that in the example provided, the complementary bases are joined together to form a double strand but are not joined together after being cleaved by the enzyme. Thus, single-stranded sticky ends are created.

Expected Results
See data table below.

Analysis
1. GTAC– is a dangling end.
2. –CATG
3. Yes; each enzyme can cleave only a specific site and can leave only a specific number of single-stranded DNA bases. Enzyme F cleaves between G and A but must leave 4 bases in the dangling end.

Assessment
Portfolio Ask students to invent a plasmid strand of DNA that would fit the cleaved ends of the second example in the data table. Use the Performance Task Assessment List for Invention in **PASC,** p. 45. **L2**

Enrichment

Intrapersonal Mice with mutant genes are particularly valuable to scientists studying human diseases. Using a grant from the National Institute of General Medical Services, scientists have developed what they call knockout mice. Have capable students find out how these mice are produced and how scientists study the effect of missing genes in the mice on conditions such as aging or cancer. **L3**

INVESTIGATE BioLab The BioLab at the end of the chapter can be used at this point in the lesson.

GLENCOE TECHNOLOGY

CD-ROM
Biology: The Dynamics of Life
Animation: *Gene Cloning*
Disc 2

VIDEODISC
Biology: The Dynamics of Life
Gene Cloning (Ch. 38)
Disc 1, Side 1, 23 sec.

The Infinite Voyage *Miracles by Design, Biodegradable Plastic: The Miracle Material?* (Ch. 4)
7 min.

The Infinite Voyage *The Geometry of Life, Manipulating Genetic Engineering* (Ch. 10)
3 min.

Figure 13.6
Foreign DNA is spliced into a plasmid vector. The recombined plasmid then carries the foreign DNA into the bacterial cell, where it replicates independently of the bacterial chromosome. If the foreign DNA contained a gene for human growth hormone, each cell will make the hormone.

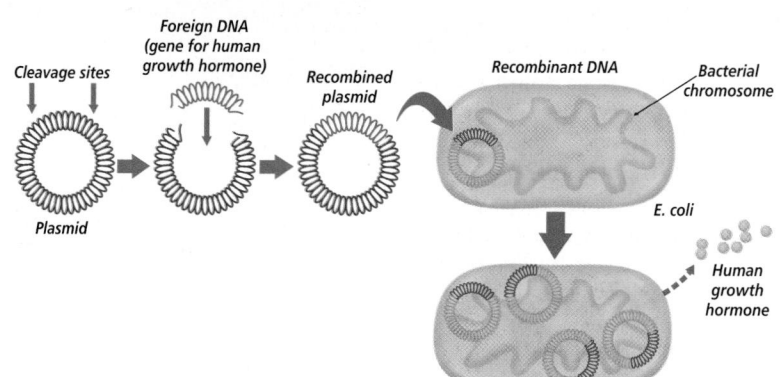

Gene splicing

As you have learned, if a plasmid and foreign DNA have been cleaved with the same restriction enzyme, the ends of each will match and they will join together, reconnecting the plasmid ring. Rejoining of DNA fragments is called **gene splicing.** The foreign DNA is recombined into a plasmid or viral DNA with the help of a second enzyme. You can practice modeling gene splicing in the *BioLab* at the end of this chapter.

Gene cloning

After the foreign DNA has been spliced into the plasmid, the recombined DNA is transferred into a bacterial cell. The plasmid is capable of replicating separately from the bacterial host and can produce up to 500 copies per bacterial cell. An advantage to using bacterial cells to clone DNA is that they reproduce quickly; therefore, millions of bacteria are produced and each bacterium contains hundreds of recombinant DNA molecules. **Clones** are genetically identical copies. Each identical recombinant DNA molecule is called a gene clone.

Plasmids can also be used to deliver genes to animal or plant cells, which incorporate the recombinant DNA. Each time the host cell divides it copies the recombinant DNA along with its own. The host cell can also produce the protein coded for by the recombinant DNA, and scientists can study the function of this protein in cells that don't normally produce such proteins. They can also produce mutant forms of a protein and determine how that mutation alters the protein's function within the cell.

Why clones are possible

Bacteria can express genes that are introduced using plasmids. Each cell makes the desired product. For example, if the plasmid contains the gene for insulin (something bacteria do not produce) taken from human DNA, each bacterium containing that DNA will make human insulin. The billions of cloned cells could generate the large quantities of hormone needed for diabetic patients who require it. Furthermore, the insulin produced would be less likely to cause a reaction in patients since it is human insulin (pig insulin was used in the past). It would be impossible to obtain such large quantities of the product directly from humans because hormones are normally produced in the body in very small amounts. *Figure 13.6* summarizes the formation and cloning of recombinant DNA in a bacterial host cell.

PROJECT

Enzyme Model

Kinesthetic Have students build a model demonstrating the use of restriction enzymes on DNA. They could use various materials to represent the DNA nucleotides, such as colored push pins, colored paper clips, or M & Ms. **L1**
ELL **COOP LEARN**

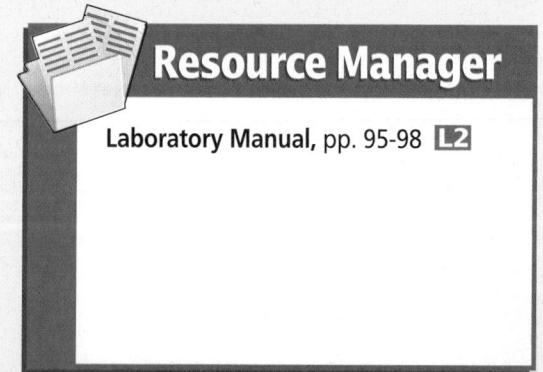

Resource Manager

Laboratory Manual, pp. 95-98 **L2**

Cloning of animals

So far, you have read about cloning a single gene. For decades, scientists attempted to expand the technique from a single gene to an entire organism. The most famous cloned animal is Dolly, the sheep. You can read more about Dolly in the *BioTechnology* feature at the end of this chapter. Although their techniques differ, scientists are coming closer to perfecting the technique of cloning animals. One of the benefits for humans in cloning animals is that ranchers and dairy farmers can clone particularly productive, healthy animals to increase yields.

Sequencing DNA

Genetic engineering techniques can also provide pure DNA for use in determining the sequence or correct order of the DNA bases. This information can allow scientists to identify mutations.

In DNA sequencing, millions of copies of a double-stranded DNA fragment are cloned using plasmids. Then, the strands are separated from each other. The single-stranded fragments are placed in four different (one for each DNA base) test tubes. Each tube contains four normal nucleotides (A,C,G,T) and an enzyme that can catalyze the synthesis of a complementary strand. One nucleotide in each tube is tagged with a different fluorescent color. The reactions produce complementary strands of varying lengths. These strands are separated according to size by gel electrophoresis (ih lek troh fuh REE sus), producing a pattern of fluorescent bands in the gel. The bands are visualized using a laser scanner. How do the DNA fragments separate from each other in the gel? Read the *Inside Story* on the next page to find out.

Applications of DNA Technology

Once it became possible to transfer genes from one organism to another, large quantities of hormones and other products could be produced. How is this technology of use to humans? Many species of bacteria have been engineered to produce chemical compounds that are of use to humans. The three main areas proposed for recombinant bacteria are in industry, medicine, and agriculture.

Recombinant bacteria in industry

In industry, recombinant bacteria have been engineered to break down pollutants into harmless products. Laboratory experiments with recombinant bacteria first showed that these engineered bacteria could degrade oil more rapidly than the same species of naturally occurring bacteria, as seen in *Figure 13*.7. These recombinant bacteria were used with some success in the Gulf of Mexico to clean up an oil spill off the coast of Texas. Mining companies also are interested in bioengineering bacteria that will extract valuable minerals from ores.

Figure 13.7
The first patented organism, a bacterium that breaks down oil, was engineered by Dr. Ananda Chakrabarty. The flask in the rear contains oil and natural bacteria. The flask at the front contains the engineered bacteria and is almost free of oil.

353

Purpose

Students will learn how gel electrophoresis separates DNA fragments of different sizes.

Background

Gel electrophoresis is a basic technique used in the preparation of DNA samples for other techniques such as DNA sequencing.

Teaching Strategies

■ Give students several paper "DNA fragments" of different sizes. Have them draw what a gel electrophoresis of these fragments would look like. Make sure they indicate at which end of the gel the fragments were added. **L1**

■ Have students describe orally the steps necessary to run a gel.

Visual Learning

■ Use colored chalk to demonstrate different sized DNA fragments that have been run in a gel.

Critical Thinking

DNA sequencing can be done only on small pieces of DNA. Gel electrophoresis separates the small pieces of DNA from large pieces before sequencing takes place.

Resource Manager

Reinforcement and Study Guide, pp. 56-57 **L2**
Content Mastery, p. 62 **L1**

Gel Electrophoresis

Restriction enzymes are the perfect tools for cutting DNA. However, once the DNA is cut, a scientist needs to determine exactly what fragments have been formed. Once DNA fragments have been separated on a gel, many other techniques, such as DNA sequencing, can be used to specifically identify a DNA fragment.

Critical Thinking *Why might gel electrophoresis be an important step before DNA sequencing can be done?*

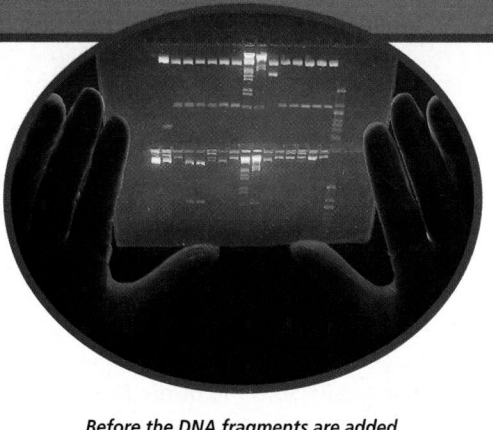

Before the DNA fragments are added to the wells, they are treated with a dye that glows under ultraviolet light, allowing the bands to be studied.

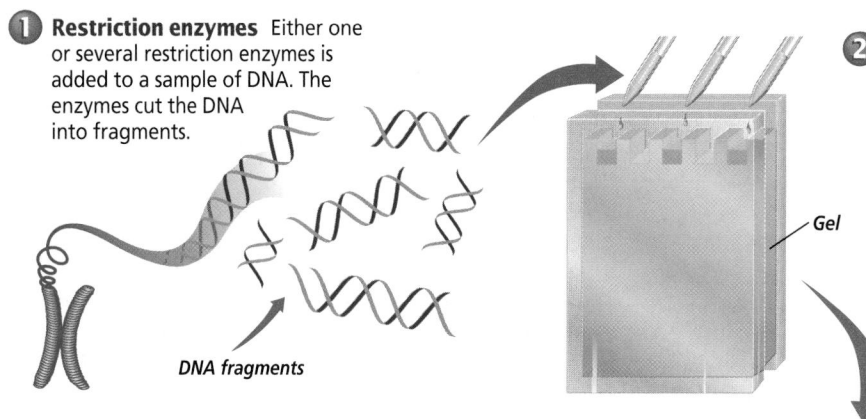

1 Restriction enzymes Either one or several restriction enzymes is added to a sample of DNA. The enzymes cut the DNA into fragments.

DNA fragments

2 The gel A gel, with a consistency similar to gelatin, is formed so that small wells are left at one end. Into these wells, small amounts of the DNA sample are placed.

Gel

4 The fragments move The negatively charged DNA fragments travel toward the positive end. The smaller the fragment, the faster it moves through the gel. Fragments that are the farthest from the well are the smallest.

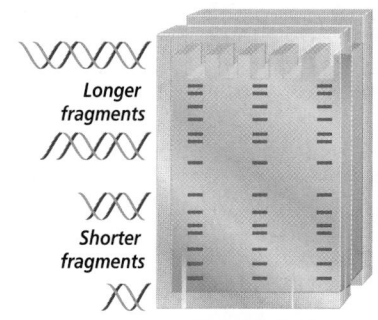

Longer fragments

Shorter fragments

Completed gel

(−) Negative end

Power source

(+) Positive end

3 The electrical field The gel is placed in a solution, and an electrical field is set up so that one end of the gel is positive and the other end is negative.

354 GENETIC TECHNOLOGY

GLENCOE TECHNOLOGY

VIDEODISC
The Secret of Life
Electrophoresis and PCR Segment

Recombinant bacteria in medicine

In medicine, pharmaceutical companies already are producing molecules made by recombinant bacteria to treat human diseases. Recombinant bacteria are employed in the production of human growth hormone to treat dwarfism, and of insulin to treat diabetes. The human gene for insulin is inserted into a bacterial plasmid by genetic engineering techniques. Recombinant bacteria produce such large quantities of insulin that they bulge. Another strain of recombinant bacteria is being used to produce phenylalanine, an amino acid that is needed to make aspartame, an artificial sweetener found in many diet products.

Transgenic animals

Transgenic animals are opening up new avenues through which scientists can study diseases and the role specific genes play in an organism. A favorite animal for transgenic studies is the mouse because mice reproduce quickly, within three weeks, and mouse chromosomes are similar to human chromosomes. In addition, scientists know a lot about where genes are located on mouse chromosomes. The roundworm *Caenorhabditis elegans* is another organism with simple genetics that is well understood and is used for transgenic studies. A third animal commonly used for transgenic studies is the fruit fly, *Drosophila melanogaster.*

Although animals such as a worm and a fruit fly may seem very different from humans, it is surprising how many genes are common to all. One way in which scientists use genetic engineering is to create animals with human diseases. By studying these animals, scientists hope to treat and cure these diseases in humans.

Problem-Solving Lab 13-2 Thinking Critically

How might gene transfer be verified? When you spray weeds with a chemical herbicide, they die. Glyphosate is the active ingredient in some herbicides. The problem with herbicides, however, is that glyphosate often gets sprayed accidentally onto crops and they also die. A certain gene will confer resistance to glyphosate. If this gene can be genetically engineered into crop plants, they will survive when sprayed with this herbicide. The following experiment is a test to determine to which plants the gene has been successfully transferred.

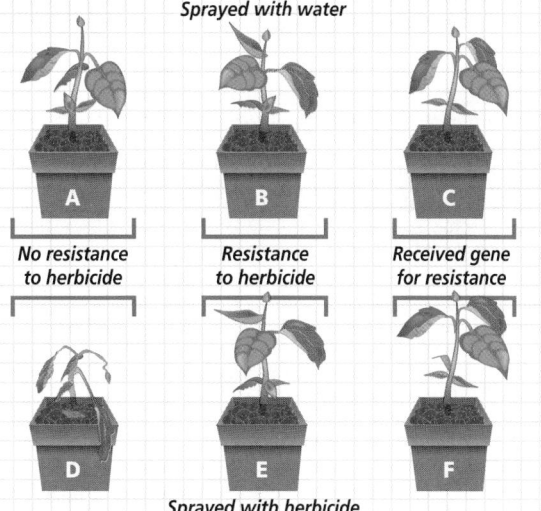

Sprayed with water

A — No resistance to herbicide
B — Resistance to herbicide
C — Received gene for resistance

D
E
F

Sprayed with herbicide

Analysis

Two groups of pots are each planted with three plants, according to the diagram above. Plants A, B, and C are sprayed with water. Plants D, E, and F are sprayed with a herbicide containing glyphosate.

Thinking Critically

1. Assume that the transfer of glyphosate resistance was successful in plant F. Predict whether each of plants D, E, and F will remain healthy after being sprayed with glyphosate. Explain your prediction.
2. Predict whether plant F will remain healthy if the transfer of glyphosate resistance was not successful.
3. Which plants are transgenic organisms? Explain your answer.
4. Explain why a glyphosate-resistant plant will produce seeds that show resistance to the herbicide.
5. Why were plants A, B, and C sprayed with water? What was the purpose of these plants in the experiment?

Problem-Solving Lab 13-2

Purpose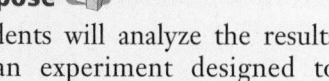

Students will analyze the results of an experiment designed to illustrate genetic transfer.

Process Skills

think critically, analyze information, apply concepts, draw a conclusion, interpret data, predict

Background

Glyphosate blocks the synthesis of amino acids phenylalanine, tyrosine, and tryptophan. Cells unable to form these amino acids die. Glyphosate inhibits the enzyme EPSP synthase from synthesizing these amino acids.

Teaching Strategies

■ You may wish to discuss the technique that is used to transfer the resistant gene to its host plant. The organism or vector is *Agrobacterium tumefaciens*, a bacterium that infects plants.

■ Make sure that students understand that plants sprayed with glyphosate will die if they do not contain genes making them resistant to the chemical.

■ Students may work in small cooperative groups to complete this activity.

Thinking Critically

1. Plants E and F will remain healthy because they have the gene that imparts resistance. Plant D will die because it is not resistant to glyphosate.

2. If resistance is not transferred, plant F will die because it is not resistant to glyphosate.

3. Plants C and F are transgenic organisms. They contain foreign DNA.

4. All the cells of a resistant plant, including cells that produce gametes, contain the gene for resistance.

5. Plants A, B, and C were controls. They were sprayed with water to show that spraying does not transfer resistance to the chemical.

BIOLOGY JOURNAL

Ethics

Linguistic Have students write an essay about the ethics of using transgenic animals to study human diseases.
L2

✔ Assessment

Performance Have students design and carry out an experiment to test whether glyphosate (Roundup) affects seeds of plants. Use the Performance Task Assessment List for Designing an Experiment in **PASC**, p. 23.
L3

Check for Understanding

Write a sequence of DNA that includes two of the sequences GAATTC and ask students to demonstrate where EcoR1 will cut the strand of DNA. **L1**

Reteach

Kinesthetic Have students model how restriction enzymes create sticky ends, using other restriction enzymes so they can see that choice of the particular enzyme is important. For example, BamH1 recognizes GGATCC and cuts between the two Gs. Hind III recognizes AAGCTT and cuts between the two As. **L1**

Extension

Interested students can look into the applications of protein engineering and computer modeling to produce modified natural proteins or to create entirely new proteins. **L3**

✔ Assessment

Knowledge Have students summarize the formation of recombinant DNA. **L2**

4 Close

Discussion

Lead into the next section by discussing how DNA technology may provide therapy for many genetic disorders.

Figure 13.8
The bacterium that causes these tumor-like plant galls is the only known biological plant vector. Unfortunately, it does not work on many kinds of plants.

Recombinant bacteria in agriculture

One species of bacteria has already been engineered successfully for use on agricultural crops. These particular bacteria normally occur on strawberry plants and cause frost damage to the leaves and fruits because ice crystals form around a protein on the surface of each bacterium. After engineering the bacteria to remove the gene for this protein, the recombinant bacteria are applied to the leaves of the plants. They replace the natural bacteria, and frost damage is prevented.

Farmers hope that another species of bacteria that lives in soil and in the roots of some plants can be engineered to increase the rate of conversion of atmospheric nitrogen to nitrates, a natural fertilizer used by plants. If this can be accomplished, farmers will be able to save money by cutting back on fertilizer.

Transgenic plants

The *Problem-Solving Lab* on the previous page shows one way plants could be improved by genetic bioengineering. Plants are more difficult to genetically engineer than bacteria because plant cells do not have the plasmids or kinds of viruses needed for taking up foreign pieces of DNA that bacterial cells have. Also, because plant cells are surrounded by thick cell walls, it is difficult to insert the foreign DNA.

A bacterium that normally causes tumorlike growths in the tissues of certain plants, *Figure 13.8*, has been used to carry foreign genes into plant cells. Most engineering of plants uses mechanical vectors such as DNA-coated "bullets." Brief jolts of high-voltage electricity also can be used. The jolts cause temporary pores to form in the plasma membrane, through which the DNA can enter.

Plants have been genetically engineered to resist herbicides, produce internal pesticides, or increase their protein production. These plants have been used in field trials, and are now used extensively.

Section Assessment

Understanding Main Ideas
1. How are transgenic organisms different from natural organisms of the same species?
2. How are sticky ends important in making recombinant DNA?
3. Why is it presently more difficult to engineer transgenic plants and animals than it is to engineer bacteria?
4. Explain two ways in which recombinant bacteria are used for human applications.

Thinking Critically
5. Many scientists consider genetic engineering to be simply an efficient method of selective breeding. Explain.

SKILL REVIEW
6. **Sequencing** Order the steps in producing recombinant DNA in a bacterial plasmid. For more help, refer to *Organizing Information* in the **Skill Handbook**.

Section Assessment

1. Transgenic organisms contain DNA from other sources.
2. Sticky ends allow the cleaved DNA to join to complementary single strands on DNA molecules from another organism.
3. Plants and animals do not readily accept plasmids into their DNA, and plants have tough cell walls that make it difficult to insert DNA.
4. Recombinant bacteria produce human growth hormone, insulin, and phenyl-alanine. They are also used for better crop production such as frost-resistant strawberries.
5. Answers may include that both genetic engineering and selective breeding involve increasing the frequency of a desired gene in a population by human intervention
6. Cleave the plasmid and foreign DNA with the same restriction enzyme, and the sticky ends will join, reconnecting the plasmid ring.

13.3 The Human Genome

Magnification: 1750×

Y̲ou have already learned about several genetic disorders that affect humans—Huntington's disease, cystic fibrosis, Tay Sach's disease, and sickle-cell anemia. In the last section, you learned that transgenic bacteria are being designed to treat disorders such as diabetes and dwarfism. Scientists hope someday to be able to treat more disorders. To accomplish this, they are sequencing the entire human genome. How are they doing this and what do they hope to accomplish?

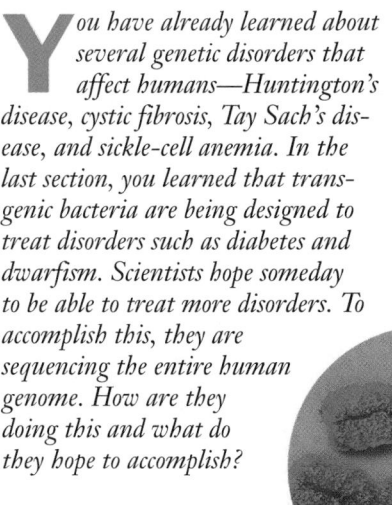
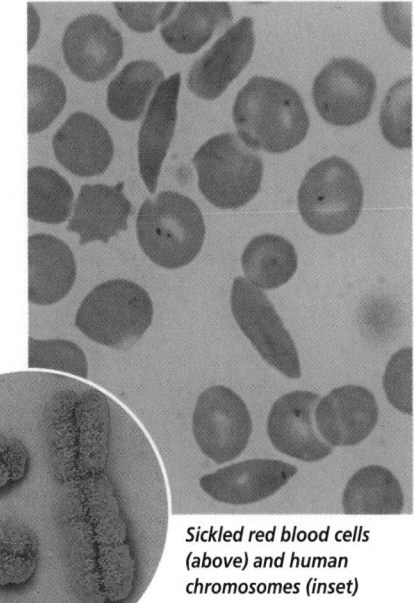

Magnification: 16 800×

Sickled red blood cells (above) and human chromosomes (inset)

Mapping and Sequencing the Human Genome

In 1990, scientists in the United States organized the Human Genome Project (HGP). It is an international effort to completely map and sequence the **human genome,** the approximately 60-100 000 genes on the 46 human chromosomes. In the summer of 2000, the HGP completed the first draft of the 3 billion base pairs of DNA in every human cell. The *MiniLab* on the next page gives you an idea of the size of the human genome.

Linkage maps

The locations of only a few thousand of the total number of known human genes have been mapped on particular chromosomes. This means that for most human genes, scientists don't know the exact or even the approximate locations on chromosomes. The genetic map that shows the location of genes on a chromosome is called a **linkage map.**

The historical method used to assign genes to a particular human chromosome was to study linkage data from human pedigrees. Recall from your study of meiosis that crossing over occurs during prophase I.

WORD Origin

genome
From the Greek word *genos*, meaning "race." A genome is the total number of genes in an individual.

13.3 THE HUMAN GENOME **357**

SECTION PREVIEW

Objectives
Analyze how the effort to completely map and sequence the human genome will advance human knowledge.
Predict future applications of the Human Genome Project.

Vocabulary
human genome
linkage map
gene therapy

Section 13.3

Prepare

Key Concepts
The organized effort to map and sequence the human genome using DNA technology is presented. Students will also explore applications of this project.

Planning
■ Obtain a chromosome linkage map for the Quick Demo.
■ Gather the materials for the Alternative Lab.

1 Focus

Bellringer
Before presenting the lesson, display **Section Focus Transparency 34** on the overhead projector and have students answer the accompanying questions. **L1** **ELL**

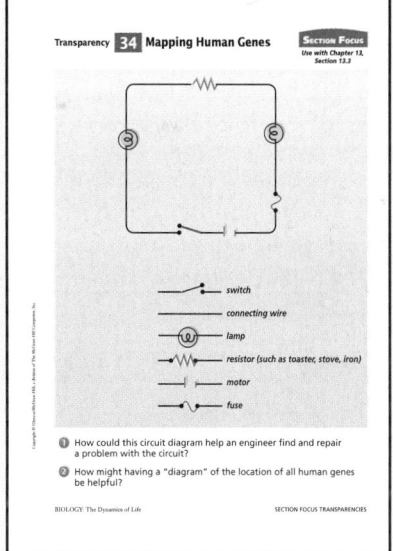

Transparency **34** Mapping Human Genes

GLENCOE TECHNOLOGY

VIDEODISC
The Infinite Voyage
The Geometry of Life
The Human Genome Project
(Ch. 11), 4 min.

Resource Manager

Section Focus Transparency 34 and Master **L1** **ELL**

2 Teach

MiniLab 13-2

Purpose 🖩
Students will use an analogy to calculate the size of the human genome.

Process Skills
acquire information, interpret data, think critically, use numbers

Teaching Strategies
■ 👆 Go through calculations with students. Place the data table on an overhead projector as you work through each step.
■ To provide consistency of data, photocopy a page from a novel and allow all students to use it to gather their initial data. Advise them of the total number of pages in the book.

Expected Results
Sample data: A = 80; B = 45; C = 3600; D = 1800; E = 400; F = 720 000; G = 4167

Analysis
1. Take an average of several pages to arrive at total number of characters per page.
2. The human genome is made up of 3 billion bases.
3. DNA bases are paired (A-T, C-G). Thus, the total genome is expressed in pairs so the number of characters must be expressed in pairs.
4. **a.** 4167 books **b.** 4.17 books

✔ Assessment
Portfolio Have students calculate the number of base pairs on each chromosome, assuming that all chromosomes are the same length. Use the Performance Task Assessment List for Using Math in Science in **PASC**, p. 29. [L2]

358

MiniLab 13-2 — Using Numbers

Storing the Human Genome It has been estimated that the human genome consists of three billion nitrogen base pairs. How much room would all of the genetic information in a single cell take up if it were printed in a book the size of a typical novel?

Procedure
1. Copy the right column of the data table marked "Letters and numbers."
2. Select a random page from a novel.
3. Follow the directions in the table. Record your calculations in your data table.

Data Table

Directions	Letters and numbers
A. Count the number of characters (letters, punctuation marks, and spaces) across one entire line of your selected page.	
B. Count the number of lines on the page.	
C. Calculate the number of characters on the page. (Multiply A × B.)	
D. Let one nitrogen base equal one character. Knowing that DNA is made of nitrogen base pairs, divide C by 2.	
E. Record the number of pages in your novel.	
F. Calculate the number of base pairs in your novel. (Multiply E × D.)	
G. Calculate the number of books the size of your novel needed to hold the human genome. (Divide 3 billion by F.)	

Analysis
1. What changes could be taken to improve the accuracy of this activity at steps A-C?
2. What assumption is being made at step G?
3. Explain the logic for step D.
4. **a.** How many books the size of your novel would be needed to store the human genome?
   b. How many books the size of your novel would be needed to store a typical bacterial genome? Assume there are three million base pairs in the genome of a bacterium.

As a result of crossing over, the offspring have a combination of alleles not found in either parent. The frequency with which these alleles occur together is a measure of the distance between the genes. Genes that cross over frequently must be farther apart than genes that rarely cross over. The percentage of these crossed-over traits appearing in offspring is then used to determine the relative position of genes on the chromosome, and thus to create a linkage map.

Because humans have only a few offspring compared with the larger numbers of offspring in other species and because the generation time is so long, mapping by linkage data is extremely inefficient. Biotechnology has now provided scientists with new methods of mapping genes. Using a technique called polymerase chain reaction (PCR), millions of copies of tiny DNA fragments are cloned in a matter of a few hours. Scientists use PCR to amplify genetic markers that are spread throughout the genome. A genetic marker is a segment of DNA with an identifiable physical location on a chromosome and whose inheritance can be followed. A marker can be a gene, or it can be some section of DNA with no known function. Because DNA segments that lie near each other on a chromosome tend to be inherited together, markers are often used as indirect ways of tracking the inheritance pattern of a gene that has not yet been identified, but whose approximate location is known. This technique is called genotyping.

Sequencing the human genome
The difficult job of sequencing the human genome is accomplished by cleaving samples of DNA into fragments using restriction enzymes, as described earlier in this chapter. Each

---

Alternative Lab

Crime Analysis

Purpose 🖩
Students will solve a crime using human genetics and DNA technology.

Materials 📞 👕
glass slide, red juice, 5 bar codes (2 that match), 2 strands of human hair (blond and dark), T-shirt with spot of red ink, stained

slide of female cheek cells, microscope

Preparation
Table 1: hair from the scene. Table 2: shirt with a blood stain (red ink) and a slide with two drops of juice to represent blood typing (no clotting with the addition of anti-A and anti-B sera). Table 3: slide of stained cheek cells with microscope pointer on a Barr body. Table 4: bar code representing the DNA fingerprint of blood from the crime scene (one of two that match). Place

fragment is then individually cloned and sequenced. The cloned fragments are aligned in the proper order by overlapping matching sequences, thus determining the sequence of a longer fragment. Automated machines can perform this work, greatly increasing the speed of map development.

Applications of the Human Genome Project

As chromosome maps are made, how can they be used? Improved techniques for prenatal diagnosis of human disorders, use of gene therapy, and development of new methods of crime detection are current areas of research.

Diagnosis of genetic disorders

Once it is clearly understood where a gene is located and the gene's DNA sequence is known, a diagnosis of a genetic disorder may be made before birth. The DNA of people with and without the disorder is analyzed for common patterns that may be associated with the disorder. Cells are obtained from the fluid surrounding the fetus. DNA is isolated and PCR is used to analyze the area where the mutation is found. If the gene is normal, the PCR product will be a standard size—a deviation means a mutation is present. For some diagnostic tests, the DNA must be analyzed using gel electrophoresis only. This is usually when the disease-causing mutation alters a restriction enzyme-cutting site, producing DNA fragments of different sizes than normal. Thus, when DNA from fetal cells is examined and found to have the mutation associated with the disorder, the fetus will develop the disorder.

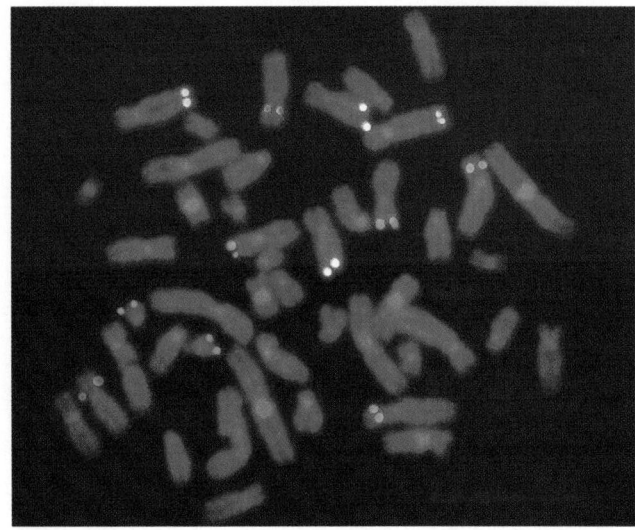

Figure 13.9
Fluorescently labeled complementary DNA for the gene to be mapped is made and added to metaphase chromosomes. The labeled DNA binds to the gene and indicates its location as a spot. In this photo, six genes are mapped simultaneously.

a bar code in each of four envelopes marked: blond female, type O blood (matching bar code); blond male, type B blood; dark-haired male, type O blood; and dark-haired male, type B blood.

Procedure
1. Roger Trueblood, dark hair, type B blood, was shot and killed. You must find which suspect is guilty.
2. Rotate to the tables and collect information. At the last table, decide who is

guilty and open the envelope to see if the DNA fingerprint matches that at the crime scene.

Analysis
1. What information did you learn from the table displays? *The clues included blond hair, type O blood, and female cheek cells (presence of Barr bodies).*
2. How are the bar codes like DNA fingerprinting? *They are a series of lines of varying thickness.*

Purpose

Students will analyze DNA fingerprints to determine identity of an unidentified soldier.

Process Skills

think critically, analyze information, apply concepts, compare and contrast, interpret data, interpret scientific illustrations

Teaching Strategies

■ Advise students that photos of DNA fingerprints are X rays of electrophoresis results. The DNA is treated with radioactive dyes that expose X-ray film.

■ Remind students that each person has unique DNA (except for identical twins). Each contains DNA from both parental DNA. Ask them to recall the number of chromosomes contributed at fertilization by each parent to future offspring.

■ Photocopy the diagram for student use so that they do not write in the book.

■ Suggest that a ruler be placed along the bottom edge of each band for ease in matching.

Thinking Critically

1. Parents C and D; half of the soldier's DNA matched parent C and the other half matched parent D.

2. 50% from each parent; each parent contributes half of their DNA to an offspring.

3. Yes, the soldier's DNA would still match the living parent's half.

✓ Assessment

Knowledge Ask students: Could the unidentified soldier be identified if he were adopted? *not unless his birth parents are known* Can the unknown soldier from WW I be identified through DNA technology? *Not likely; his parents must be deceased.* Use the Performance Task Assessment List for Writing in Science in **PASC**, p. 87. **L1**

Gene therapy

Individuals who inherit a serious genetic disorder now have a reason to hope for a bright future—gene therapy. **Gene therapy** is the insertion of normal genes into human cells to correct genetic disorders. This technology has already entered trial stages in a number of attempts to treat genetic and acquired diseases. Some early, less-successful work with treating cystic fibrosis, shown in *Figure 13.10*, laid the groundwork for other experiments. The first successful trials were on patients suffering from SCID (severe combined immunodeficiency disease). One particular type of SCID is caused by a defect in an enzyme within a specific cell in the immune system. Without the normal enzyme, a substrate of the enzyme builds up in the cells of the immune system and eventually kills the cells. The immune system is shut down with the result that in these patients, even slight colds can be deadly. In gene therapy for this disorder, the cells of the immune system are separated from blood samples and the functional gene is added to them. The cells are then injected back into the patient. It is not clear whether the gene will stay active for very long periods of time, or if the treated cells themselves have long life spans. So the treatments may or may not have to be repeated often. Still, it seems likely that the next decade will see the use of DNA technology to treat many different disorders using gene therapy.

DNA fingerprinting

Law-enforcement workers use unique fingerprint patterns to determine whether suspects have been at a crime scene. In the past ten years, biotechnologists have developed a method that determines DNA fingerprints. DNA fingerprinting can be used to convict or acquit individuals of criminal offenses because every person is genetically unique.

Small DNA samples can be obtained from blood, hair, skin, or semen and copied millions of times using polymerase chain reaction techniques. When an individual's DNA is cleaved with a restriction enzyme, the DNA is cut into fragments of different lengths. DNA fragments can then be separated by electrophoresis and compared with those obtained from a crime scene.

Chromosomes consist of genes that are separated by segments of noncoding DNA, DNA that doesn't code for proteins. The genes follow fairly standard patterns from person

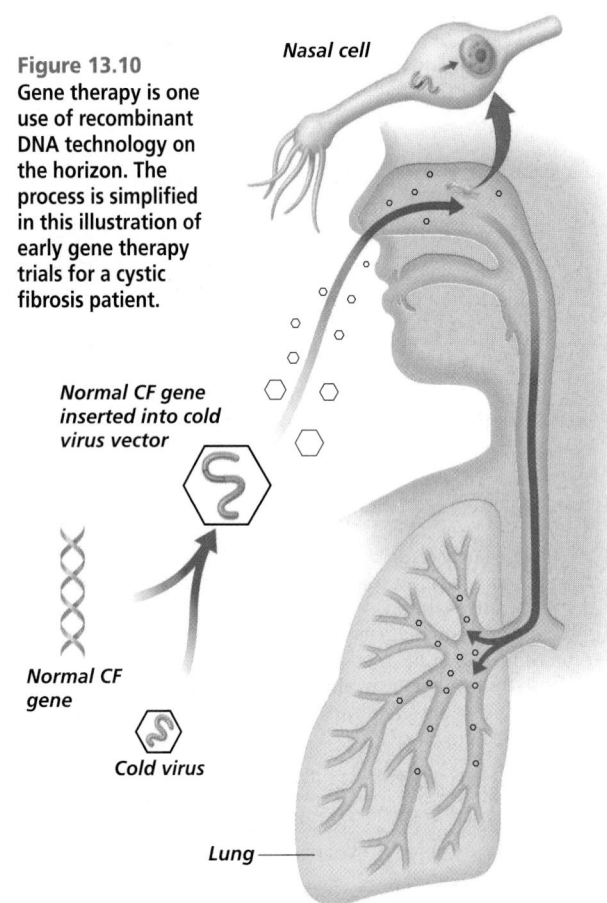

Figure 13.10
Gene therapy is one use of recombinant DNA technology on the horizon. The process is simplified in this illustration of early gene therapy trials for a cystic fibrosis patient.

Nasal cell

Normal CF gene inserted into cold virus vector

Normal CF gene

Cold virus

Lung

BIOLOGY JOURNAL

Gene Therapy

Linguistic Have students research the current progress being made on gene therapy in the treatment of cystic fibrosis or other genetic disorders. They can then write reports to include in their journals. **L2**

✓ Portfolio

Bioethics

Ask students to write an essay giving their opinions on whether parents should change a child's genes before birth for each of the following: to correct a child's genetic disorder, to make a child more attractive, to change the child's gender. **L1 P**

to person, but the non-coding segments produce distinct combinations of patterns unique to each individual—so distinct, in fact, that DNA patterns can be used like fingerprints to identify the person (or other organism) from whom they came. DNA fingerprinting works because no two individuals (except identical twins) have the same DNA sequences, and because all cells (except gametes) of an individual have the same DNA. You can read about a real example of DNA fingerprinting in the *Problem-Solving Lab*.

PCR techniques have been used to clone DNA from many sources. Geneticists are cloning DNA from mummies and analyzing it in order to better understand ancient life. Abraham Lincoln's DNA has been taken from the tips of a lock of his hair and studied for evidence of a possible genetic disorder. The DNA from fossils has been analyzed and used to compare extinct species with living species, or even two extinct species with each other. The uses of DNA technology are unlimited.

Problem-Solving Lab 13-3 Applying Concepts

How is identification made from a DNA fingerprint?
DNA fingerprint analysis requires a sample of DNA from a person, either living or dead. The DNA is first cut up into smaller segments with enzymes. The segments are then separated according to size using electrophoresis. When stained, the DNA segments appear as colored bands in a specific order. These bands form a person's DNA fingerprint.

Analysis

An unidentified soldier from the Vietnam War who had been placed in the Tomb of the Unknowns at Arlington National Cemetery was identified through DNA fingerprinting. The soldier could have been one of four possible individuals. A DNA sample from his body was analyzed. The DNA from the parents of the four suspected soldiers was analyzed. The diagram shows a DNA fingerprint pattern analysis similar to the one that was actually done. Find the correct match between the soldier's DNA fingerprint pattern and those of his parents. Remember, a child's DNA is a combination of parental DNA.

Soldier	Parents A	B	Parents C	D	Parents E	F	Parents G	H

Thinking Critically

1. Which parental DNA matched the unknown soldier? How did the DNA pattern allow you to decide?
2. What percent of the soldier's DNA matched his father's DNA? His mother's? Explain why.
3. Could an exact identification have been made if only one parent were alive? Explain why.

Check for Understanding

Ask students to explain what the Human Genome Project proposes to do. *map and sequence all the genes on human chromosomes*

Reteach

Ask students to list applications of the Human Genome Project. *diagnosis of genetic disorders, gene therapy, and DNA fingerprinting*

Extension

Kinesthetic Cloning kits can be purchased from biological supply companies. **L3**

✔ Assessment

Skill Ask students to predict future applications of the Human Genome Project. **L1**

4 Close

Discussion

Ask students whether couples should be able to change features of their unborn children if the features are not disorders, such as eye color, hair color, or gender.

Resource Manager

Concept Mapping, p. 13 **L3** **ELL**

Critical Thinking/Problem Solving, p. 13 **L3**

BioLab and MiniLab Worksheets, pp. 63-64 **L2**

Reinforcement and Study Guide, p. 58 **L2**

Content Mastery, pp. 63-64 **L1**

Tech Prep Applications, pp. 21-22 **L2**

Section Assessment

Understanding Main Ideas
1. What is the Human Genome Project?
2. Compare a linkage map and a sequencing map.
3. What is gene therapy?
4. Explain why DNA fingerprinting can be used as evidence in law enforcement.

Thinking Critically
5. Describe some possible benefits of the Human Genome Project.

SKILL REVIEW

6. **Observing and Inferring** Suppose a cystic fibrosis patient has been treated with gene therapy, in which a normal allele is inserted into lung cells of the patient using a virus vector. Does this person still run the risk of passing the disorder to his or her offspring? Explain. For more help, refer to *Thinking Critically* in the **Skill Handbook**.

13.3 THE HUMAN GENOME **361**

Section Assessment

1. an international effort to completely map and sequence the human genome
2. A sequencing map shows the sequence of DNA bases; a linkage map shows the position of genes on a chromosome.
3. the insertion of normal alleles into human cells to correct genetic disorders
4. The DNA of every individual, except identical twins, is unique. It can be used to identify individuals with great accuracy.
5. If the location of genes is known, then gene therapy can be used to replace defective genes and cure a disorder.
6. Because only the lung cells had the allele replaced, the sex cells would still contain a cystic fibrosis allele and the disorder would be passed on to the offspring.

Modeling Recombinant DNA

Experimental procedures have been developed that allow recombinant DNA molecules to be engineered in a test tube. From a wide variety of restriction enzymes available, scientists choose one or two that recognize particular sequences of DNA within a longer DNA sequence of a chromosome. The enzymes are added to the DNA, which is cleaved at the recognition sites. Because the cleaved fragments have ends that are available for attachment to complementary strands, the fragments can be added to plasmids or to viral DNA that has been similarly cut. When the fragment has been incorporated into the DNA of the plasmid or virus, it is called recombinant DNA.

Time Allotment
One class period

Process Skills
compare and contrast, observe and infer, recognize cause and effect

Safety Precautions
Be careful with sharp edges of scissors. Always cut away from the body.

PREPARATION

- Review the chemical structure of DNA from Chapter 11 before students begin the BioLab.
- Additional background material on modeling recombinant DNA can be found in the article, "Recombinant Paper Plasmids" by Christie L. Jenkins, *The Science Teacher*, April 1987. You may wish to read this article before you teach the BioLab.

Resource Manager

BioLab and MiniLab Worksheets, pp. 65-66 **L2**

PREPARATION

Problem
How can you model recombinant DNA technology?

Objectives
In this BioLab, you will:
- **Model** the process of preparing recombinant DNA.
- **Analyze** a model for preparation of recombinant DNA.

Materials
white paper
colored pencils, red and green
tape
scissors

Safety Precautions
Always wear goggles in the lab. Be careful with sharp objects.

Skill Handbook
Use the **Skill Handbook** if you need additional help with this lab.

PROCEDURE
---

1. Cut a lengthwise strip of paper from a sheet of white paper into a rectangle about 3 cm by 28 cm. This strip of paper represents a long sequence of DNA containing a particular gene that you wish to combine with a plasmid.
2. Cut another lengthwise strip of paper into a rectangle about 3 cm by 10 cm. When taped into a ring, this piece of paper will represent a bacterial plasmid.
3. Use your colored pencils to color the longer strip red and the shorter strip green.
4. Write the following DNA sequence once on the shorter strip of paper and two times

---
PROCEDURE

Teaching Strategies
- Students can do this lab alone or with partners.
- If red and green colored paper is available, then colored pencils will not be necessary.
- If you make cardboard models of DNA bases for the Meeting Individual Needs on page 350, use them in this BioLab.

Data and Observations
Data for completing the table should be as follows: Gene splicing—process of taping green and red paper together; Plasmid—green strip; Restriction enzyme—scissors; Sticky ends—cut ends on paper; Recombinant DNA—red and green strips taped together.

```
-G-G-A-T-C-C-              -G-G-A-T-C-C-
 | | | | | |                | | | | | |
-C-C-T-A-G-G-              -C-C-T-A-G-G-
```

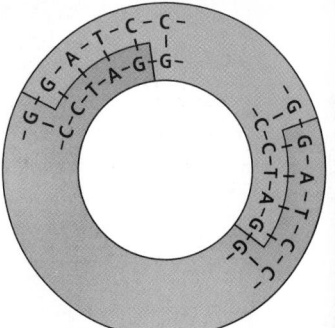

about 5 cm apart on the longer strip of paper.

-G-G-A-T-C-C-
-C-C-T-A-G-G-

5. After coloring the shorter strip of paper and writing the sequence on it, tape the ends together.

6. Assume that a particular restriction enzyme is able to cleave DNA in a staggered way as illustrated here.

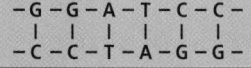

```
 -G              G-A-T-C-C-
-C-C-T-A-G              G-
```

Cut the longer strand of DNA in both places as shown. You now have a cleaved foreign DNA fragment containing a gene that can be inserted into the plasmid.

7. Once the sequence containing the foreign gene has been cleaved, cut the plasmid in the same way.

8. Splice the foreign gene into the plasmid by taping the paper together where the sticky ends pair properly. The new plasmid represents recombinant DNA.

Data Table

Term	BioLab model
Gene splicing	
Plasmid	
Restriction enzyme	
Sticky ends	
Recombinant DNA	

9. Copy the data table. Relate the steps of producing recombinant DNA to the activities of the modeling procedure by explaining how the terms relate to the model.

ANALYZE AND CONCLUDE

1. **Comparing and Contrasting** How does the paper model of a plasmid resemble a bacterial plasmid?

2. **Comparing and Contrasting** How is cutting with the scissors different from cleaving with a restriction enzyme?

3. **Thinking Critically** Enzymes that modify DNA, such as restriction enzymes, have been discovered and isolated from living cells. What functions do you think they have in living cells?

Going Further

Project Design and construct a three-dimensional model that illustrates the process of preparing recombinant DNA. Consider using clay, plaster of Paris, or other materials in your model. Label the model and explain it to your classmates.

BIOLOGY Online To find out more about recombinant DNA, visit the Glencoe Science Web site.
science.glencoe.com

INVESTIGATE BioLab

ANALYZE AND CONCLUDE

1. It is a small circular piece of DNA.

2. The scissors can cut the DNA anywhere, but the restriction enzyme recognizes a particular sequence.

3. These enzymes may function in a cell by cutting up and destroying invading viral DNA.

✓ Assessment

Performance Have students write a paragraph summarizing what they have learned about the use of restriction enzymes. Have students explain how the BioLab model is like the terms listed in the data table. Use the Performance Task Assessment List for Writing in Science in **PASC**, p. 87.

Going Further

 Kinesthetic Students could use different colors of clay, or use food coloring to dye plaster of Paris, to show recombinant DNA in their models. **L1** **ELL**

GLENCOE TECHNOLOGY

CD-ROM
Biology: The Dynamics of Life
Animation: *Recombinant DNA*
Disc 2

Purpose 📖

Students learn about the technique used to produce the first successful clone of an adult mammal and are introduced to some of the potential benefits of cloning technology.

Background

Frogs were the first vertebrate animals to be cloned. During the 1950s and 1960s, scientists even managed to clone adult frogs, but none of the clones survived past the tadpole stage. Clones of mammals may have the same genes, but they are not necessarily identical. The uterine environment in which an individual develops, and the conditions under which it is raised, have a profound effect on its appearance and behavior. Identical twins may actually be more alike than clones because they share the same uterine environment and are usually raised under the same conditions.

Teaching Strategies

■ Review with students the steps in the reproductive process. An egg cell and sperm cell each contain half the normal complement of chromosomes. When fertilization occurs, the two cells fuse to form a zygote with a full complement of chromosomes. The zygote divides by mitosis to form a multicellular embryo. As the embryo grows, cell division continues and the cells begin to differentiate as the body systems are formed. When all body systems are present, the embryo becomes a fetus. The fetus continues to grow and develop until it is mature enough to survive outside the mother's uterus.

Investigating the Technology

The genes had to be turned on in order for all organs and systems to develop.

How to Clone a Mammal

A cloned organism has exactly the same genes as its parent. A single cell from the parent is used to produce the clone. During the 1980s and early 1990s, the first successful clones of mammals were produced from embryo cells of mice, sheep, cattle, goats, rabbits, and monkeys. Creating a clone from an adult mammal cell proved to be more difficult, but was achieved in 1997.

When an egg is fertilized by a sperm, a zygote is formed that rapidly divides into a multicellular embryo. During the very early stages of development, all the genes present in each embryo cell are functional, or turned on. At this point, each embryo cell has the capacity to develop into a separate and complete individual. But during later stages of development, as the embryo grows into a fetus, its cells begin to differentiate. They become specialized for different tasks. Liver cells become distinct from muscle cells. Skin cells become different from blood cells. Once a cell has become specialized, most of its genes turn off and stop functioning. Only the genes needed to operate that particular kind of cell remain turned on, and the cell can no longer perform any other tasks. Because all the cells of an adult animal are specialized, cloning an animal from an adult cell requires finding a way to turn on all the genes in the cell.

A sheep called Dolly Dolly, a lamb born in Scotland in February 1997, was the first mammal to be cloned from an adult cell. To produce Dolly, researchers took an udder cell from a full-grown ewe, a female sheep. They used an electric charge to force the udder cell to fuse with an unfertilized sheep egg cell from which the nucleus had been removed. The electric charge created openings in the membrane of the egg cell large enough to admit the udder cell. The fused cell began to divide and develop into an embryo. The embryo was implanted into the uterus of another sheep, which gave birth to Dolly several months later. The method worked, but it took 277 cloning attempts to produce a single lamb. Since Dolly was cloned, mice and other animals have been cloned by slightly different techniques. These new techniques are much more efficient.

Cloned mice

Applications for the Future

Cloning experiments help to increase our knowledge of how genes and cells operate. Learning how specific genes are turned on and off can provide clues about how a fetus grows and develops. It may also contribute to our understanding of genetic diseases. For example, if the activity of a particular gene is found to be responsible for a disease, scientists could treat the disease by turning off the gene. Cloning techniques could also be used to change the function of a cell, such as altering skin cells to create blood cells for a transfusion or to grow bone or cartilage to replace damaged tissue.

INVESTIGATING THE TECHNOLOGY

Analyzing Concepts Do you think the genes in the fused cell that produced Dolly were turned on or turned off? Explain.

 To find out more about genetic technology, visit the Glencoe Science Web site.
science.glencoe.com

Going Further

Encourage students to research the latest developments in cloning technology and its uses in medicine and other industries. Invite them to discuss or write about ethical issues arising from the development of cloning.

SUMMARY

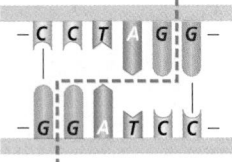

Section 13.1
Applied Genetics

Main Ideas

■ Geneticists use test crosses to determine the genotypes of individuals and the probability that offspring will have a particular allele.

■ Plant and animal breeders use genetics to selectively breed organisms with desirable traits.

Vocabulary

inbreeding (p. 346)
test cross (p. 347)

Section 13.2
Recombinant DNA

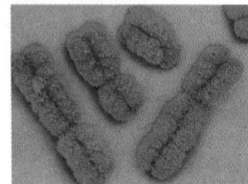

Main Ideas

■ Scientists have developed methods to move genes from one species into another. This process uses restriction enzymes to cleave one organism's DNA into fragments and other enzymes to splice the DNA fragment into a plasmid or viral DNA. Transgenic organisms are able to manufacture genetic products foreign to themselves using recombinant DNA.

■ Genetic engineering has already been applied to bacteria, plants, and animals. These organisms are engineered to be of use to humans.

Vocabulary

clone (p. 352)
gene splicing (p. 352)
genetic engineering (p. 349)
plasmid (p. 351)
recombinant DNA (p. 349)
restriction enzyme (p. 350)
transgenic organism (p. 349)
vector (p. 351)

Section 13.3
The Human Genome

Main Ideas

■ International efforts are presently underway to sequence the DNA of the entire human genome and to determine the chromosome location for every gene.

■ Applications of the Human Genome Project include the goals of detecting, treating, and curing genetic disorders. DNA fingerprinting can be used to identify persons responsible for crimes and to provide evidence that certain persons are not responsible for crimes.

Vocabulary

gene therapy (p. 360)
human genome (p. 357)
linkage map (p. 357)

Main Ideas

Summary statements can be used by students to review the major concepts of the chapter.

Using the Vocabulary

To reinforce chapter vocabulary, use the Content Mastery Booklet and the activities in the Interactive Tutor for Biology: The Dynamics of Life on the Glencoe Science Web site.
science.glencoe.com

 *All Chapter Assessment questions and answers have been validated for accuracy and suitability by The Princeton Review.*

UNDERSTANDING MAIN IDEAS

1. b
2. d
3. c

UNDERSTANDING MAIN IDEAS

1. What is the purpose of a test cross?
 a. produce offspring that consistently exhibit a specific trait
 b. check for carriers of a trait
 c. explain recessiveness
 d. show polygenic inheritance

2. Two closely related cattle are mated. Their offspring are _____.
 a. vectors   **c.** mutants
 b. carriers   **d.** inbred

3. A section of mouse DNA is joined to a bacterial plasmid. The DNA formed is _____.
 a. translated DNA   **c.** recombinant DNA
 b. restricted DNA   **d.** transcribed DNA

CHAPTER 13 ASSESSMENT **365**

 GLENCOE *TECHNOLOGY*

VIDEOTAPE
MindJogger Videoquizzes
Chapter 13: *Genetic Technology*
Have students work in groups as they play the videoquiz game to review key chapter concepts.

 **Resource Manager**

Chapter Assessment, pp. 73-78
MindJogger Videoquizzes
ExamView® Pro Software
BDOL Interactive CD-ROM, Chapter 13 quiz

4. c
5. d
6. c
7. b
8. a
9. c
10. a
11. Inbreeding
12. DNA sequencing
13. palindrome
14. Bacterial
15. Three
16. cloning or PCR
17. vector
18. linkage
19. clones
20. test cross

APPLYING MAIN IDEAS

21. treating and curing human genetic disorders
22. Bacteria are capable of making large quantities of human proteins when they carry human genes. These proteins have potential medicinal and industrial usages.
23. This may upset the balance between the natural nitrogen-converting bacteria and the plants, possibly disrupting the soil ecosystem. Plant growth would increase because nitrogen is usually a limiting factor.

4. When a bacterial plasmid transfers a piece of foreign DNA to a host cell, the plasmid serves as a(n) _____.
   a. transgenic cell   c. vector
   b. hybrid   d. enzyme

5. The goal of gene therapy is to insert a _____ into cells to correct a genetic disorder.
   a. recessive allele   c. dominant allele
   b. growth hormone   d. normal allele

6. _____ are genetically identical copies of DNA.
   a. Vectors   c. Clones
   b. Plasmids   d. Spliced genes

7. Plant cells are difficult to genetically engineer because they do not have the _____ needed for taking up foreign DNA.
   a. amino acids   c. clones
   b. plasmids   d. enzymes

8. The process of gel electrophoresis separates _____ fragments by using an electric field.
   a. DNA   c. gel
   b. cell   d. enzyme

9. An organism that contains functional recombinant DNA is called a _____ organism.
   a. cleaved   c. transgenic
   b. cloned   d. spliced

10. Restriction enzyme *Eco*R1 cuts DNA strands, leaving _____ ends.
    a. sticky   c. blunt
    b. smooth   d. spliced

11. _____ usually increases the appearance of genetic disorders.

12. _____ is a method used to determine the order of the DNA nucleotides.

THE PRINCETON REVIEW **TEST-TAKING TIP**

Let Bygones Be Bygones
Once you have read a question, considered the answers, and chosen one, put that question behind you. Don't try to keep the question in the back of your mind, thinking that maybe a better answer will come to you as the test continues.

13. A _____ occurs when the same sequence of bases, running in opposite directions, is found on both DNA strands.

14. _____ cells are the primary source of the plasmids used in DNA technology today.

15. *Eco*R1 restriction enzyme recognizes the sequence GAATTC in double-stranded DNA. _____ pieces of DNA would result if *Eco*R1 were added to the following DNA.

16. The technique _____ is used to make millions of copies of DNA fragments.

17. A gene gun is used to deliver a mechanical type of _____.

18. The genetic map that shows the location of genes on a chromosome is a _____ map.

19. Cells in a cell culture all have the same genetic material because they are _____.

20. The Punnett square to the right illustrates a _____.

	a	a
A	Aa	Aa
a	aa	aa

APPLYING MAIN IDEAS

21. What is the potential use of a map showing the sequence of DNA bases in a human chromosome?

22. Why might it be important to be able to have a human gene expressed in a bacterium?

23. Assume that transgenic organisms can be developed to speed the conversion of nitrogen from the air into nitrates that plants can use as fertilizer. How might use of this organism affect an ecosystem?

Chapter 13 Assessment

THINKING CRITICALLY

24. Observing and Inferring Explain why the use of bacterial plasmids for gene splicing does not interfere with normal cell functions such as growth and reproduction.

25. Recognizing Cause and Effect How may using biotechnology to engineer many different transgenic organisms of a given species alter the course of evolution for that species?

26. Sequencing Once a foreign gene has been inserted into a plasmid to form a recombinant plasmid, what would be the next step if you were to continue the model of recombinant DNA technology?

27. Interpreting Data If all human genes have similar patterns, how can DNA fragments from hair or skin be used to identify distinct individuals by DNA fingerprinting?

28. Concept Mapping Complete the concept map by using the following vocabulary words: genetic engineering, transgenic organisms, restriction enzymes, vector, plasmid, gene splicing.

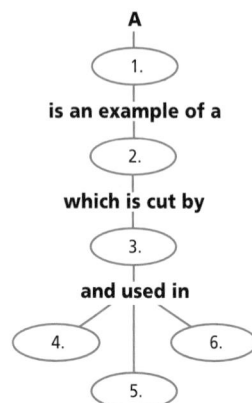

A

1.

is an example of a

2.

which is cut by

3.

and used in

4.   6.

5.

CD-ROM

For additional review, use the assessment options for this chapter found on the *Biology: The Dynamics of Life* Interactive CD-ROM and on the Glencoe Science Web site.
science.glencoe.com

ASSESSING KNOWLEDGE & SKILLS

The following graph shows the results of an experiment using natural and bioengineered bacteria of the same species that can break down oil. Each culture had 40 mL of oil added on Day 1.

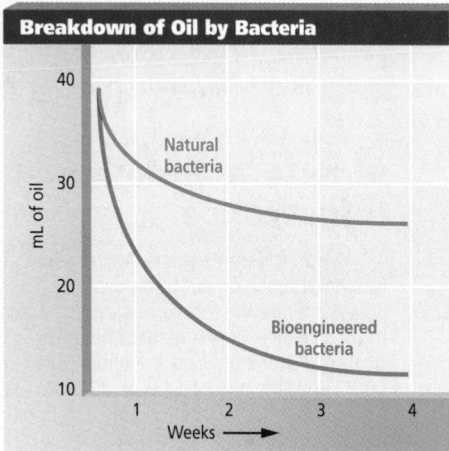

Breakdown of Oil by Bacteria

Natural bacteria

Bioengineered bacteria

mL of oil

Weeks

Interpreting Data Use the graph to answer the following questions.

1. Approximately how much oil had been converted into harmless products by the natural bacteria after four weeks?
a. 4 mL   **c.** 24 mL
b. 14 mL   **d.** 40 mL

2. How much oil had been converted by bioengineered bacteria after four weeks?
a. 4 mL   **c.** 28 mL
b. 14 mL   **d.** 40 mL

3. How much more efficient are the bioengineered bacteria than the naturally occurring species?
a. 1×   **c.** 2×
b. 1.5×   **d.** 3×

4. Interpreting Data How can this technology be applied to an oil spill?

THINKING CRITICALLY

24. Plasmids used for gene splicing are not part of the cell's chromosome. Thus, they can be spliced and manipulated without affecting normal cell functions.

25. If many alleles were to be transferred into the normal population, their frequency would change. This would affect the genetic makeup of the population.

26. The recombinant DNA is then transferred into a host cell.

27. DNA fingerprinting uses the segments of noncoding DNA, which are in distinct, individual patterns.

28. 1. Plasmid; 2. Vector; 3. Restriction enzymes; 4. Genetic engineering; 5. Transgenic organisms; 6. Gene splicing

ASSESSING KNOWLEDGE & SKILLS

1. b
2. c
3. c
4. Because these bacteria have the ability to digest oil into a harmless product, the bacteria could be added to oil spills to clean them up quickly.

CHAPTER 13 ASSESSMENT **367**

National Science Education Standards
UCP.1, UCP.2, UCP.3, UCP.5, C.1, C.5, E.2, F.1, F.5, F.6, G.1, G.3

Prepare

Purpose

This BioDigest can be used as an introduction to or as an overview of genetics. If time is limited, you may wish to use this unit summary to teach genetics in place of the chapters in the Genetics unit.

Key Concepts

Students will learn about Mendel and his insights into heredity. The process by which genetic information is copied and proteins are formed by the genetic code is described. Finally, the patterns of human heredity and DNA technology are discussed.

1 Focus

Bellringer

 Visual-Spatial Have students closely examine the flower of a pea plant so they can become familiar with the materials with which Mendel worked. **L1** **ELL**

GLENCOE TECHNOLOGY

CD-ROM
Biology: The Dynamics of Life
Exploration: *Trait Inheritance*
Exploration: *Punnett Square*
Animation: *Meiosis*
Disc 2

BioDigest

For a **preview** of the genetics unit, study this BioDigest before you read the chapters. After you have studied the genetics chapters, you can use the BioDigest to **review** the unit.

Genetics

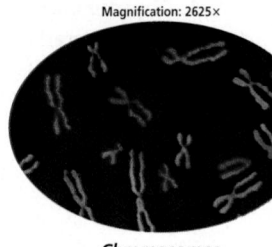

Magnification: 2625×

Chromosomes

Genetics is the study of inheritance. The physical traits, or phenotype, of an individual are encoded in small segments of chromosomes called genes. Not all genes are expressed as a phenotype. Therefore, the genotype, the traits encoded in the genes, may be different from the expressed phenotype.

Simple Mendelian Inheritance

A trait is dominant if only one allele of a gene is needed for that trait to be expressed. If two alleles are needed for expression, the trait is said to be recessive. In pea plants, the allele for purple flowers is dominant and the allele for white flowers is recessive. Any plant with *PP* or *Pp* alleles will have purple flowers. Any plant with *pp* alleles will have white flowers.

When a *PP* purple pea plant is crossed with a *pp* white plant, all the offspring are purple, *Pp*. When two *Pp* plants are crossed, three-fourths of the plants in the next generation will be purple and one-fourth will be white.

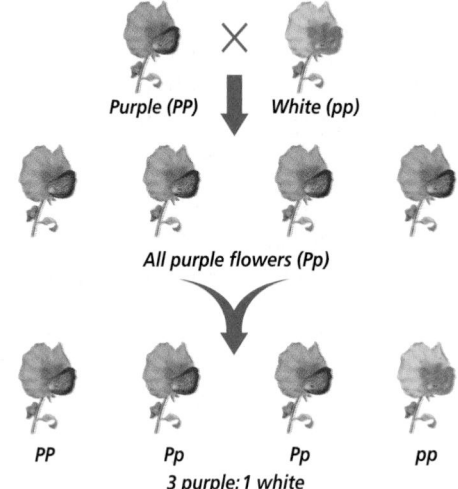

Purple (PP) × White (pp)

All purple flowers (Pp)

PP   Pp   Pp   pp

3 purple: 1 white

FOCUS ON HISTORY

Mendel

Gregor Mendel

To investigate the genetic inheritance of pea plant traits, the Austrian monk Gregor Mendel used critical thinking skills to design his experiments. When he collected data, he considered not only the qualitative characteristics, such as whether the plants were tall or short, but also the quantitative data by analyzing the ratios of tall to short plants in each generation. Mendel observed that there were two variations for each trait, such as tall and short plants. He formed the hypothesis that alleles transmitted these traits from one generation to the next. After studying several traits for many generations, Mendel formed two laws. The law of segregation states that the two alleles for each trait separate when gametes are formed. The law of independent assortment states that genes for different traits are inherited independently of each other.

368

Multiple Learning Styles

Look for the following logos for strategies that emphasize different learning modalities.

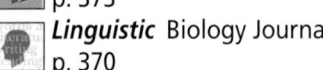

Kinesthetic Quick Demo, p. 372; Meeting Individual Needs, p. 372

Visual-Spatial Bellringer, p. 368; Quick Demo, p. 369; Activity, p. 369; Demonstration, p. 371; Portfolio, p. 371

Intrapersonal Extension, p. 373

Linguistic Biology Journal, p. 370

BIODIGEST

Meiosis

Meiosis produces gametes that contain only one copy of each chromosome instead of two. Some stages of meiosis are similar to those of mitosis, but in meiosis, homologous chromosomes come together as tetrads to exchange genes during a process called crossing over. Meiosis also provides a mechanism for rearranging the genetic information carried by cells. Both crossing over and the rearrangement of genes during meiosis produce genetic variability, which may give offspring a survival advantage if the environment changes.

Meiosis consists of two divisions, meiosis I and meiosis II. During meiosis I, the homologous chromosomes separate from each other. In meiosis II, the sister chromatids of each chromosome separate from each other.

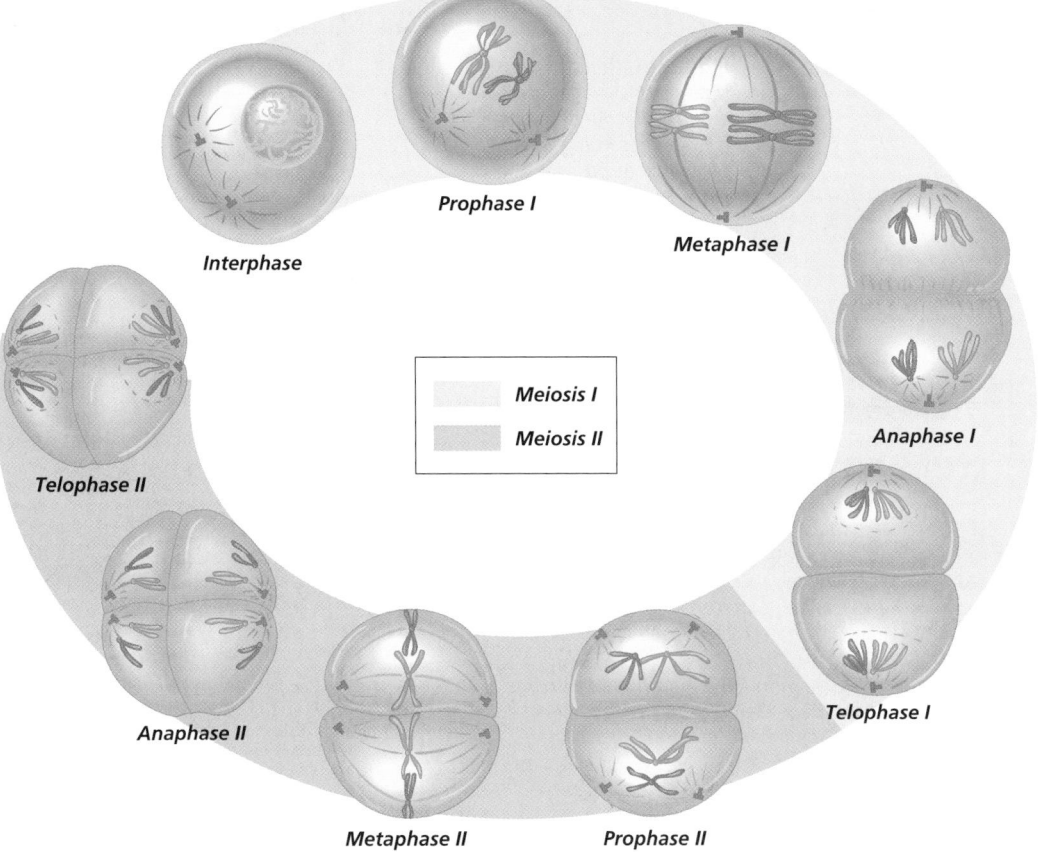

Interphase

Prophase I

Metaphase I

Anaphase I

Telophase I

Prophase II

Metaphase II

Anaphase II

Telophase II

Meiosis I

Meiosis II

BIODIGEST

2 Teach

✓ Assessment

Skill Have students calculate the ratios from Mendel's data as listed in the Vital Statistics box. Ask students why the ratios are not exactly 3:1. **L2**

Discussion

Have students discuss Mendelian inheritance. Begin by comparing and contrasting genotype and phenotype. What are the possible genotype(s) of a dominant trait? A recessive trait? *A genotype is the alleles for a trait that an individual possesses. A phenotype is the physical appearance of that trait. The possible genotypes of a dominant trait are homozygous, PP, or heterozygous, Pp for the purple flower color. A recessive trait must be homozygous, pp, for the recessive phenotype.*

Quick Demo

Visual-Spatial Draw a Punnett square for another cross of Mendel's experiments mentioned in the Vital Statistics box. Correlate this to the pea crosses for petal color. Ask students whether the ratios of offspring are the same.

Activity

Visual-Spatial Have students draw a cell undergoing meiosis. Make sure they label homologous chromosomes and sister chromatids. **L1** **ELL**

Genetics

Visual Learning

Put up an overhead of transcription and ask students questions about the process. Ask them to define a gene, transcription, and mRNA. Have them identify the bases of DNA. *A gene is a segment of DNA containing the information for making a protein. Transcription is the process whereby a DNA sequence of bases is copied into a sequence of bases in mRNA. mRNA will carry the information for proteins into the cytoplasm. The bases of DNA are adenine, thymine, cytosine, and guanine.*

Discussion

Discuss the idea that some traits such as sickle-cell anemia may confer an advantage for survival to a heterozygote. How might this idea have been used throughout evolution to define characteristics of living species? *Traits that confer an advantage to individuals in a species increase the chances of that individual surviving to reproduce and pass on the traits.*

Producing Physical Traits

Deoxyribonucleic acid (DNA) is a double-stranded molecule made up of a sequence of nucleotide base pairs that encode each gene on a chromosome. There are four bases in DNA: A, T, C, and G. Because of their molecular shape, A can pair only with T, and C can pair only with G. This precise pairing allows the DNA molecule to copy itself in a process called DNA replication.

Transcription

To make a protein, the segment of DNA containing the gene for that protein must be transcribed. First, the sequence of bases in the DNA segment is copied into a molecule of messenger ribonucleic acid (mRNA), which then moves from the nucleus to the cytoplasm. RNA is similar to DNA except that RNA is single-stranded and contains the base U in place of T.

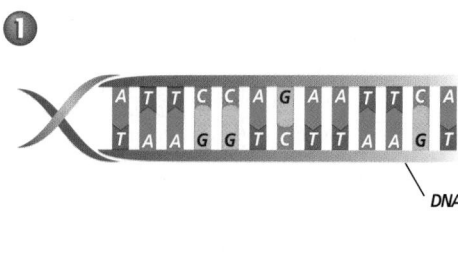

DNA

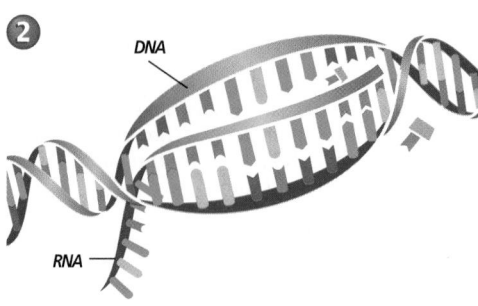

DNA

RNA

In transcription, the two strands of DNA separate and a molecule of mRNA is made according to the sequence of bases in the DNA. After transcription, the mRNA strand leaves the nucleus to travel to the ribosomes in the cytoplasm.

RNA

FOCUS ON ADAPTATIONS

Heterozygote Advantage

Some recessive alleles, such as the sickle-cell allele, remain in a population because they give the heterozygote a genetic advantage in certain environments. Sickle-cell anemia affects red blood cells so that they change shape when oxygen levels are low, blocking blood vessels. The disorder is prevalent in Africa, where malaria is common. When individuals who are heterozygous for the sickle-cell trait contract malaria, they suffer fewer and milder symptoms of malaria, whereas normal individuals who contract malaria may die. Therefore, sickle-cell alleles in heterozygotes remain in the population.

Magnification: 6950×

In sickle-cell anemia, red blood cells form a sickle shape compared to the round shape of normal red blood cells.

Magnification: 6950×

370

BIOLOGY JOURNAL

Mendel's Experiments

Linguistic Have students write about Mendel's experiments and his conclusions. How did he follow the scientific method in testing his hypotheses? What two laws did he derive from his studies? L2

Translation

A codon, a sequence of three bases in the mRNA, serves as a code for an amino acid. In a process called translation, a ribosome "reads" the codons on the molecule of mRNA. Transfer RNA (tRNA) molecules bring the appropriate amino acids to the mRNA at the ribosome. The amino acids are bonded together to form a protein.

In translation, the sequence of bases in the mRNA is translated into a sequence of amino acids in a protein chain. Every three nucleotides code for a specific amino acid.

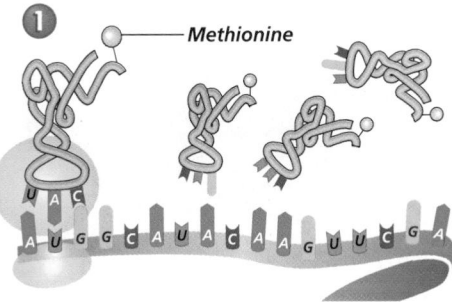

① Methionine

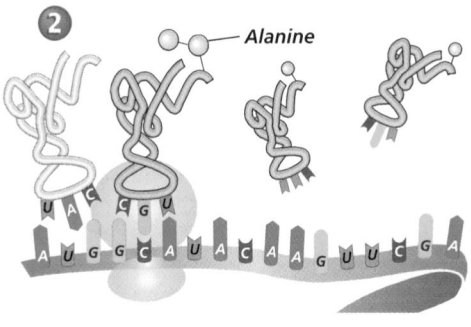

② Alanine

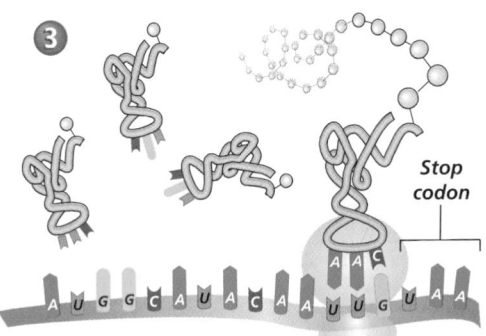

③ Stop codon

Complex Inheritance Patterns

An incomplete dominance pattern of inheritance produces an intermediate phenotype in the heterozygote. In codominant inheritance, the heterozygote expresses both alleles. Some traits, such as human blood types, are governed by multiple alleles, although any individual can carry only two of those alleles.

Flower color in snapdragons is an example of incomplete dominance. The flower color of the heterozygote is intermediate to that of the two homozygotes.

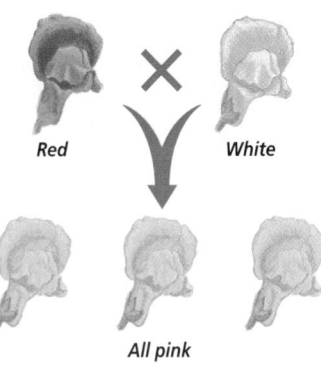

Red   White

All pink

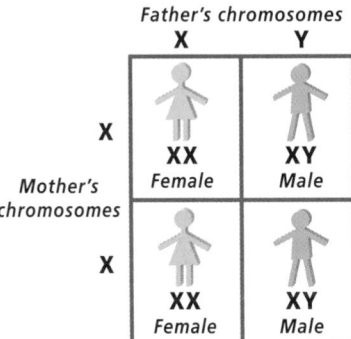

Father's chromosomes

	X	Y
X	XX *Female*	XY *Male*
X	XX *Female*	XY *Male*

Mother's chromosomes

In any mating between humans, half the offspring will have the XX genotype, which are females, and half the offspring will have the genotype XY, which are males.

Visual Learning

Put up an overhead of translation and ask students questions. Have them differentiate between mRNA and tRNA. Give students a sequence of several mRNA codons and ask them to "translate" the RNA bases into an amino acid sequence. *mRNA contains a copy of a gene from DNA. tRNA carries amino acids to ribosomes.*

Demonstration

Visual-Spatial Grow red, white, and pink snapdragons. Ask students to draw Punnett squares for various matings of the snapdragons. Have them identify this trait as incomplete dominance. **L1** **ELL**

GLENCOE TECHNOLOGY

 **VIDEODISC**
Biology: The Dynamics of Life

Sex-Linked Traits (Ch. 35)
Disc 1, Side 1, 1 min. 49 sec.

Fruit Fly Genetics (Ch. 34)
Disc 1, Side 1, 50 sec.

Portfolio

Punnett Squares

Visual-Spatial Make Punnett squares demonstrating various crosses between individuals with a dominant trait, a recessive trait, incomplete dominance, codominance, and a sex-linked trait. **L1** **ELL** **P**

GLENCOE TECHNOLOGY

 CD-ROM
Biology: The Dynamics of Life
Animation: *DNA Replication*
Animation: *DNA Transcription*
Animation: *Translation*
BioQuest: *Building a Protein*
Disc 2

BioDigest

Reinforcement

Have students make a table with the following headings: Dominant, Recessive, Incomplete dominance, Codominance, Multiple alleles, Polygenic inheritance, and X-linked traits. Under each heading, they should list the following: genotype(s), whether the trait is more common in males or females, one specific example from the text, and how many phenotypes are possible.

Quick Demo

 Kinesthetic Take two pieces of electrical wire of different colors. Use clippers as restriction enzymes to cut the wires. Peel some of the colored coating from the cut edges of the wires to demonstrate sticky ends. Then splice the two wires together to simulate recombinant DNA.

GLENCOE TECHNOLOGY

 **VIDEODISC VIDEOTAPE** The Secret of Life *In the Land of Milk and Money: Biotechnology*

Tinkering With Our Genes: Genetic Medicine

Resource Manager

Reinforcement and Study Guide, pp. 59-60 **L2**
Content Mastery, pp. 65-68 **L1**

The X chromosome, one of two sex chromosomes, carries many genes, including the genes for hemophilia and for color blindness. Most X-linked disorders appear in males because they inherit only one X chromosome. In females, a normal allele on one X chromosome can mask the expression of a recessive allele on the other X chromosome. Finally, some traits, such as skin color, are polygenic—governed by several genes.

VITAL STATISTICS

Frequency of Genetic Disorders in the Population
Cystic Fibrosis: 1 in 2000 white Americans is homozygous recessive.
Sickle-Cell Anemia: 1 in 500 African Americans is homozygous recessive.
Tay-Sachs Disease: 1 in 3600 Jews of eastern European ancestry is homozygous recessive.

Recombinant DNA can be cloned to produce many copies of a specific segment of DNA. The inset photo shows crystals of insulin that were purified from a culture of recombinant bacteria.

BioDigest

Recombinant DNA Technology

To make recombinant DNA, a small segment of DNA containing a desired gene is inserted into a bacterial plasmid, a small ring of DNA. The plasmid acts as a vector to carry the DNA segment into a host bacterial cell. Every time the bacterium reproduces, the plasmid containing the inserted DNA is duplicated, producing copies of the recombinant DNA along with the host chromosome. Because these new DNA segments are identical to the original, they are called clones. The host cell produces large quantities of the protein encoded by the recombinant DNA it contains.

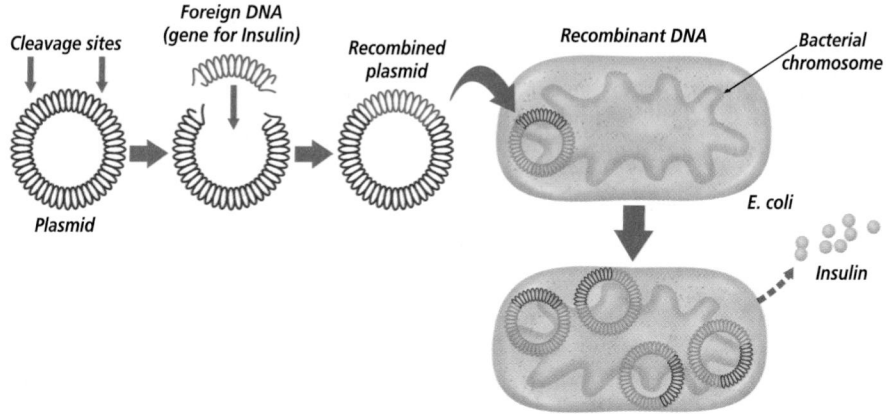

Cleavage sites
Foreign DNA (gene for Insulin)
Recombined plasmid
Recombinant DNA
Bacterial chromosome
Plasmid
E. coli
Insulin

MEETING INDIVIDUAL NEEDS

Learning Disabled

 Kinesthetic Have students use a model of a human respiratory system to simulate gene therapy for cystic fibrosis. Students should show the pathway taken by viral DNA containing a normal allele.
L1 ELL

GLENCOE TECHNOLOGY

VIDEODISC The Secret of Life *Recombinant DNA Segment*

BioDigest

BioDigest

Gene Therapy

Gene therapy involves the insertion of normal alleles into human cells to correct genetic disorders. Gene therapy is used to treat cystic fibrosis, a serious genetic disorder caused by a missing protein in the cell membrane. In gene therapy for this disorder, the normal allele is inserted into a virus that causes the common cold. The virus carries the gene into the patient's cells, which begin to produce the missing protein.

In gene therapy for cystic fibrosis, the normal allele is inserted into a virus that can be inhaled through a nasal spray.

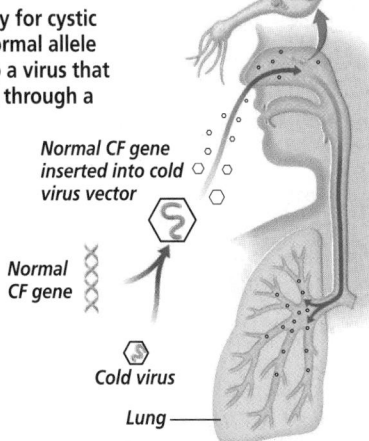

Nasal cell

Normal CF gene inserted into cold virus vector

Normal CF gene

Cold virus

Lung

BioDigest Assessment

Understanding Main Ideas

1. An example of a dominant trait is _____.
   a. cystic fibrosis
   b. white pea flowers
   c. purple pea flowers
   d. sickle-cell anemia

2. The three-nucleotide base sequence that codes for an amino acid is called _____.
   a. a codon   c. tRNA
   b. a base pair   d. mRNA

3. Something that carries a piece of DNA into a host cell is called a _____.
   a. clone   c. vector
   b. bacterial cell   d. recombinant DNA

4. A base sequence that would be found in RNA but not in DNA is _____.
   a. ATTCGA   c. AUUCCG
   b. TTTGGC   d. TUCCGT

5. A broad range of phenotypes for a given trait is usually the result of _____.
   a. multiple alleleic inheritance
   b. incomplete dominance
   c. codominance
   d. polygenic inheritance

6. A base pair that can be found in a DNA molecule is _____.
   a. A—T   c. A—G
   b. A—C   d. C—C

7. The process of copying DNA into mRNA is called _____.
   a. translation   c. replication
   b. transcription   d. cloning

8. A trait that can be expressed only as a homozygous genotype is _____.
   a. dominant   c. codominant
   b. recessive   d. X-linked

9. Meiosis differs from mitosis in that the cells produced by meiosis have _____ the number of chromosomes as the parent cell.
   a. half   c. twice
   b. three times   d. the same as

10. In crossing over, genetic material is exchanged between _____, resulting in genetic variability of offspring.
    a. cells   c. genes
    b. chromatids   d. alleles

Thinking Critically

1. State Mendel's law of independent assortment and explain what it means.

2. Compare the phenotypes of the heterozygotes in simple Mendelian inheritance, incomplete dominance, and codominance.

3. Describe the process for producing a bacterium that contains recombinant DNA.

373

3 Assess

Check for Understanding

Ask students to define transcription and translation and to explain each process.

Reteach

Put up an overhead of the stages of meiosis and ask students to identify the stages.

Extension

Intrapersonal Have students research some of the uses of genetic technology, such as the production of human insulin and human growth hormone. **L3**

✓ Assessment

Knowledge Put an example of incomplete dominance or codominance on the board and ask students to identify the pattern of inheritance. **L1**

4 Close

Discussion

Discuss the implications of gene therapy using cystic fibrosis as an example. Will effective treatment alter the possibility of passing this trait to future generations?

BioDigest Assessment

Understanding Main Ideas

1. c	4. c	7. b	9. a
2. a	5. d	8. b	10. b
3. c	6. a		

Thinking Critically

1. Each gene on a chromosome will separate independently of any other gene during meiosis.

2. In simple Mendelian inheritance, a heterozygote will carry the phenotype of the dominant allele. In incomplete dominance, the phenotype will be a blend of that of the two homozygotes. In codominance, the phenotype shows the products of both alleles.

3. Restriction enzymes must be used to cut the DNA fragment and the bacterial plasmid. Then the two may be spliced together and inserted into a bacterium.

Change Through Time

Unit Overview

In this unit, students will study the concepts and principles of evolution and classification. Chapter 14 deals with the history of life on Earth, and some hypotheses about how life began. Students will learn about fossils—what they are, how they are formed, and how they can be used to reconstruct the history of life on Earth. In Chapter 15, Darwin's theory of evolution by natural selection is discussed. The role of natural selection in the evolution of new species is presented. In Chapter 16, evidence of the ancestry of humans is explored. Chapter 17 introduces taxonomy and the diversity of organisms.

Introducing the Unit

Have students look at the birds in the photo and describe their successful adaptations. Tell students that they will learn in this unit why the ancestors of modern birds may be a group of extinct dinosaurs. Explain to students that as the environment changes populations may adapt, migrate, or become extinct.

Unit 5

Change Through Time

Life on Earth has a history of change that is called evolution. An enormous variety of fossils, such as those of early birds, provide evidence of evolution. Genetic studies of populations of bacteria, protists, plants, insects, and even humans provide further evidence of the history of change among organisms that live or have lived on Earth.

UNIT CONTENTS

- **14** The History of Life
- **15** The Theory of Evolution
- **16** Primate Evolution
- **17** Organizing Life's Diversity
- **BioDigest** Change Through Time

UNIT PROJECT

BIOLOGY *Online* Use the Glencoe Science Web site for more project activities that are connected to this unit.
science.glencoe.com

374

Unit Projects

Diversity of Organisms

Have students do one of the projects for this unit as described on the Glencoe Science Web site. As an alternative, students can do one of the projects described on these two pages.

Display

Visual-Spatial Students can use photographs or illustrations from magazines and science journals to make a collage showing different living things. **L1** **ELL**

Making a Collection

Naturalist Have student groups study leaves or insects in your area and add them to the school collection. Students should identify differences in habitats, behaviors, and needs that contribute to the diversity among the organisms. **L2** **ELL**

Advance Planning

Chapter 14
- Order diatomaceous earth for MiniLab 14-1.
- Order the chemicals for the Alternative Lab.
- Order a live culture or preserved slides of *Oscillatoria* for the Quick Demo.
- Order live cultures or preserved slides of bacteria and cyanobacteria for the Biology Journal.

Chapter 15
- Order bacterial cultures, nutrient agar, and other materials for the Alternative Lab.

Chapter 16
- Obtain casts of various fossil hominids and ape skulls for the BioLab.

Chapter 17
- Obtain guidebooks that have dichotomous keys for local trees and shrubs, perhaps from your state's Bureau of Forestry, for MiniLab 17-1.
- Obtain identification guides to insects and other organisms for the Biolab.

Unit Projects

Interview

Linguistic Students can interview professionals at nature preserves, parks, or local environmental departments to find out how they care for the diverse life forms under their protection. Have students prepare their interview questions in advance. **L1**

Using the Library

Intrapersonal Students can read Chapter 17, which is "Galapagos Archipelago," of *Voyage of the Beagle* by Charles Darwin to find out about the diversity of birds and reptiles in the Galapagos. **L3**

Final Report

Have students present oral reports of their findings about the diversity of organisms. **P** **COOP LEARN**

Chapter 14 Organizer

Refer to pages 4T-5T of the Teacher Guide for an explanation of the National Science Education Standards correlations.

Section	Objectives	Activities/Features
Section 14.1 **The Record of Life** National Science Education Standards UCP.2-4; A.1, A.2; C.3, C.6; D.3; G.1-3 (2 sessions, 1 block)	1. **Identify** the different types of fossils and how they are formed. 2. **Summarize** the major events of the Geologic Time Scale.	**MiniLab 14-1:** Marine Fossils, p. 379 **Problem-Solving Lab 14-1:** p. 380 **Inside Story:** The Fossilization Process, p. 381 **Careers in Biology:** Animal Keeper, p. 382 **MiniLab 14-2:** A Time Line, p. 384 **Investigate BioLab:** Determining a Fossil's Age, p. 394
Section 14.2 **The Origin of Life** National Science Education Standards UCP.2-5; A.1, A.2; B.2, B.3; C.1, C.3, C.6; D.2; F.3, F.4; G.1-3 (2 sessions, 1 block)	3. **Analyze** early experiments that support the concept of biogenesis. 4. **Compare and contrast** modern theories of the origin of life. 5. **Relate** hypotheses about the origin of cells to the environmental conditions of early Earth.	**Problem-Solving Lab 14-2:** p. 392 **Biology & Society:** How Did Life Begin: Different Viewpoints, p. 396

Need Materials? Contact Carolina Biological Supply Company at 1-800-334-5551 or at http://www.carolina.com

MATERIALS LIST

BioLab

p. 394 shoebox with lid, pennies, graph paper

MiniLabs

p. 379 microscope, microscope slide, coverslip, diatomaceous earth, water
p. 384 meterstick, adding machine tape

Alternative Lab

p. 390 gelatin solution, droppers (3), gum arabic solution, pH paper, microscope, microscope slides, coverslips, hydrochloric acid, test tube, stirring rod

Quick Demos

p. 381 igneous rock, sedimentary rock, metamorphic rock
p. 385 microscope, prepared slide of *Oscillatoria*
p. 389 beef bouillon cube (2), flasks (2), water, rubber stopper

Key to Teaching Strategies

L1 Level 1 activities should be appropriate for students with learning difficulties.

L2 Level 2 activities should be within the ability range of all students.

L3 Level 3 activities are designed for above-average students.

ELL ELL activities should be within the ability range of English Language Learners.

COOP LEARN Cooperative Learning activities are designed for small group work.

P These strategies represent student products that can be placed into a best-work portfolio.

These strategies are useful in a block scheduling format.

Teacher Classroom Resources

Section	Reproducible Masters	Transparencies
Section 14.1 **The Record of Life**	Reinforcement and Study Guide, pp. 61-62 **L2** Concept Mapping, p. 14 **L3** **ELL** Critical Thinking/Problem Solving, p. 14 **L3** BioLab and MiniLab Worksheets, pp. 67-68 **L2** Laboratory Manual, pp. 99-102 **L2** Content Mastery, pp. 69-70, 72 **L1**	Section Focus Transparency 35 **L1** **ELL** Reteaching Skills Transparency 23 **L1** **ELL**
Section 14.2 **The Origin of Life**	Reinforcement and Study Guide, pp. 63-64 **L2** BioLab and MiniLab Worksheets, pp. 69-70 **L2** Content Mastery, pp. 69, 71-72 **L1**	Section Focus Transparency 36 **L1** **ELL** Basic Concepts Transparency 20 **L2** **ELL**

Assessment Resources

Chapter Assessment, pp. 79-84
MindJogger Videoquizzes
Performance Assessment in the Biology Classroom
Alternate Assessment in the Science Classroom
ExamView® Pro Software 💾
BDOL Interactive CD-ROM, Chapter 14 quiz

Additional Resources

Spanish Resources **ELL**
English/Spanish Audiocassettes **ELL**
Cooperative Learning in the Science Classroom **COOP LEARN**
Lesson Plans/Block Scheduling

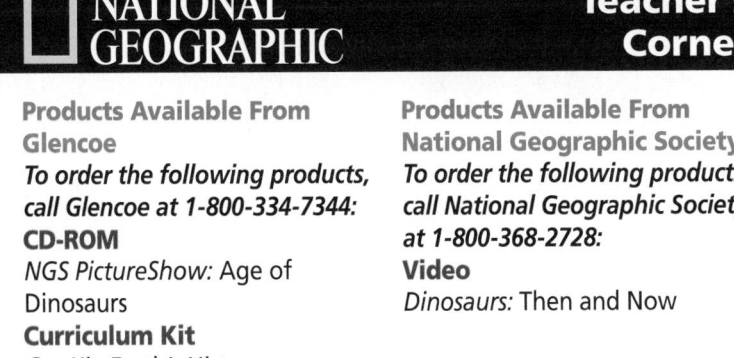

NATIONAL GEOGRAPHIC

Teacher's Corner

Products Available From Glencoe
To order the following products, call Glencoe at 1-800-334-7344:
CD-ROM
NGS PictureShow: Age of Dinosaurs
Curriculum Kit
GeoKit: Earth's History

Products Available From National Geographic Society
To order the following products, call National Geographic Society at 1-800-368-2728:
Video
Dinosaurs: Then and Now

GLENCOE TECHNOLOGY

The following multimedia resources are available from Glencoe.

Biology: The Dynamics of Life
CD-ROM **ELL**
Video: *Discovering Dinosaurs*
Exploration: *The Record of Life*
Videodisc Program
Discovering Dinosaurs
Plate Tectonics
The Infinite Voyage
The Dawn of Humankind
The Great Dinosaur Hunt
The Secret of Life Series
Diatom
Layers in Time
Plate Tectonics
Archaebacteria
What's in Stetter's Pond: *The Basics of Life*

14 The History of Life

Theme Development

In this chapter, the themes of **unity within diversity** and **evolution** are interwoven in a discussion of today's diversity among organisms and how diversity results from the evolution of unicellular organisms that lived billions of years ago.

READING BIOLOGY

Glencoe's *Biology: The Dynamics of Life* contains many resources to assist a student's reading skills. Each chapter contains figures with expanded captions that expand on written material. Word Origins, located along the side of text, expand knowledge of biology vocabulary. Glencoe's Content Mastery Booklet helps develop reading skills while reinforcing content. In addition, use the Interactive Tutor for *Biology: The Dynamics of Life* on the Glencoe Web site to reinforce vocabulary. **science.glencoe.com**

What You'll Learn

- You will examine how rocks and fossils provide evidence of changes in Earth's organisms.
- You will correlate the Geologic Time Scale with biological events.
- You will sequence the steps by which small molecules may have produced living cells.

Why It's Important

Knowing the geological history of Earth and understanding ideas about how life began provide background for an understanding of the theory of evolution.

READING BIOLOGY

Before reading the chapter, write down several section headings. Make a list of any new vocabulary words you see in the sections. As you read, write down the words' meanings, as well as a few key points in each section.

BIOLOGY *Online*

To find out more about fossils and early Earth's history, visit the Glencoe Science Web site. science.glencoe.com

Erupting lava fountains such as this one in Hawaii indicate what Earth may have been like soon after it formed. Scientists discover that living things change over time.

14.1 The Record of Life

Y ou may have seen thrilling movies and read books that described travel in time machines. The characters in such stories speed forward or backward through time, often encountering unusual organisms and strange environments. The differences you saw and read about probably didn't surprise you. After all, everything changes over time. Even Earth has changed considerably during the estimated 4.6 billion years it has existed. It's difficult to imagine what Earth might have been like that long ago.

A model of a velociraptor

SECTION PREVIEW

Objectives
Identify the different types of fossils and how they are formed.
Summarize the major events of the Geologic Time Scale.

Vocabulary
fossil
plate tectonics

Early History of Earth

To learn more about ancient Earth, step into an imaginary time machine and punch a few buttons. Get ready to visit a place to which you might never want to return—primitive Earth.

Early Earth was inhospitable

What was early Earth like? Some scientists suggest that it was probably very hot. The friction of colliding meteorites could have heated its surface, while both the compression of minerals and the decay of radioactive materials heated its interior. Volcanoes might have frequently spewed lava and gases, relieving some of the pressure in Earth's hot interior. These gases helped form Earth's ancient atmosphere, which probably contained little free oxygen, but a lot of water vapor and other gases, such as carbon dioxide, and nitrogen. If ancient Earth's atmosphere was like this, you would not have survived in it.

By about 3.9 billion years ago, Earth might have cooled enough for the water in its atmosphere to condense. This might have led to millions of years of rainstorms with lightning—enough rain to fill Earth's oceans. It is in the oceans, probably between 3.9 and 3.5 billion years ago, some scientists propose, that the first organisms appeared.

14.1 THE RECORD OF LIFE **377**

Prepare

Key Concepts

Students will explore different types of fossils and their scientific value. They will learn how scientists use the fossil record to reconstruct the history of life on Earth.

Planning

- Gather meter sticks and rolls of adding machine tape for MiniLab 14-2.
- Gather shoe boxes, pennies, and graph paper for the BioLab.

1 Focus

Bellringer

Before presenting the lesson, display **Section Focus Transparency 35** on the overhead projector and have students answer the accompanying questions. **L1** **ELL**

Transparency **35** Inferring from Fossils

SECTION FOCUS
Use with Chapter 14,
Section 14.1

❶ The diagram shows a set of fossilized footprints. What observations can you make about the footprints?

❷ What can you infer from the observations?

BIOLOGY: The Dynamics of Life

SECTION FOCUS TRANSPARENCIES

Resource Manager

Section Focus Transparency 35
and Master **L1** **ELL**

✓ Assessment Planner

Portfolio Assessment
BioLab, TWE, p. 395
Performance Assessment
MiniLabs, TWE, p. 379, 384
MiniLabs, SE, pp. 379, 384
Performance Assessment, TWE, p. 387
Alternative Lab, TWE, p. 390-391
BioLab, SE, pp. 394-395

Knowledge Assessment
Section Assessments, SE, pp. 387, 393
Knowledge Assessments, TWE, p. 390, 393
Alternative Lab, TWE, p. 390-391
Problem-Solving Lab 14-2, TWE, p. 392
Chapter Assessment, SE, p. 397-399

Skill Assessment
Problem-Solving Lab 14-1, TWE, p. 380
Skill Assessment, TWE, p. 385

2 Teach

Brainstorming

Ask students to brainstorm a list of questions they must answer in order to gain a more accurate picture of ancient life.

Visual Learning

Figure 14.1 Ask students to list the information scientists can obtain about each type of fossil. Responses may include how the fossil compares with a modern-day organism.

Display

Naturalist Obtain samples of the types of fossils in Figure 14.1. Have students try to identify the fossils on display and explain how each may have formed. **L2** **ELL**

History in Rocks

Can scientists be sure the Earth formed in this way? No, they cannot. There is no direct evidence of the earliest years of Earth's history. The physical processes of Earth constantly destroy and reform rock. The oldest rocks that have been found on Earth formed only about 3.9 billion years ago. Although rocks cannot provide information about Earth's infancy, they are an important source of information about the diversity of life that has existed on the planet.

Figure 14.1
There are many types of fossils that provide clues about ancient organisms.

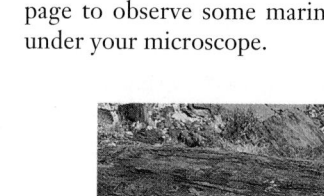

Ⓐ **Trace fossils** A trace fossil is the marking left by an animal and may include a footprint, a trail, and a burrow.

Ⓑ **Casts** When minerals in rocks fill a space left by a decayed organism, they make a replica, or cast, of the organism.

Ⓒ **Petrified fossils** Minerals sometimes penetrate and replace the hard parts of an organism, producing copies of them.

Ⓓ **Imprints** A thin object, such as a leaf, that falls into sediment can leave an imprint when the sediment hardens into rock.

Ⓔ **Amber-preserved and frozen fossils** At times, an entire organism was quickly trapped in ice or tree sap that hardened into amber.

Ⓕ **Molds** A mold forms when an organism is buried in sediment and then decays, leaving an empty space.

Fossils—Clues to the past

If you've ever visited a zoo or botanical garden, you've seen evidence of the diversity of life. But the millions of species living today are probably only a small fraction of all the species that ever existed. About 99 percent of species are extinct—they no longer live on Earth. Among other techniques, scientists study fossils to learn about the ancient species. A **fossil** is evidence of an organism that lived long ago.

Because fossils can form in many different ways, there are many types of fossils, as you can see in *Figure 14.1*. Use the *MiniLab* on the next page to observe some marine fossils under your microscope.

Cultural Diversity

Motonori Matuyama and Paleomagnetism

For unexplained reasons, Earth's polarity has reversed at times so that the magnetic north pole became the south pole and vice versa. Scientists assess volcanic rocks to study polarity changes and use the data they gather to date rocks. The Japanese geologist, Motonori Matuyama (1884-1956) discovered the magnetic reversals. Discuss some examples of how paleomagnetic studies have been important in understanding the evolution of some organisms.

Paleontologists—Detectives to the past

The study of fossils is a lot like solving a mystery. Paleontologists (pay lee ahn TAHL uh justs), scientists who study ancient life, are like detectives who use fossils to understand events that happened long ago. They use fossils to determine the kinds of organisms that lived in the past and sometimes to learn about their behavior. For example, fossil bones and teeth can indicate the size of animals, how they moved, and what they ate.

Paleontologists also study fossils to gain knowledge about ancient climate and geography. For example, when scientists find a fossil like the one in *Figure 14.2*, which resembles a present-day plant that lives in a mild climate, they may reason that the ancient environment was also mild.

By studying the condition, position, and location of rocks and fossils, geologists and paleontologists can make deductions about the geography of past environments. For example, if only the heaviest bones of an animal's skeleton are found in an area, it might mean that a river once ran through the area and carried

MiniLab 14-1 — Observing and Inferring

Present-day diatoms

Marine Fossils Certain sedimentary rocks are formed totally from the fossils of once-living ocean organisms called diatoms. The diatom fossils are often 1000 meters thick. These sedimentary rocks were at one time in the past under ocean water and were then lifted above sea level during periods of geological change.

Procedure

1. Prepare a wet mount of a small amount of diatomaceous earth. **CAUTION:** *Use care in handling microscope slides and coverslips.*
2. Examine the material under low-power magnification.
3. Draw several of the different shapes you see.
4. Compare the shapes of the fossils you observe to present-day diatoms shown in the photograph. Remember, however, that the fossils you observe are probably only pieces of the whole organism.

Analysis

1. Describe the appearance of fossil diatoms.
2. How are fossil diatoms similar to and different from the diatoms in the photo? Can you use these similarities and differences to predict how diatoms have changed over time? Explain your answer.
3. What part of the original diatom did you observe under the microscope? How did this part survive millions of years? Why were the fossils you observed in pieces?

Figure 14.2
These fossil leaves are from rocks about 200 million years old. They are remarkably similar to the leaves of *Ginkgo biloba*, trees that are planted as ornamentals throughout the United States.

14.1 THE RECORD OF LIFE **379**

Purpose
Students will observe fossil diatoms and compare them with modern species of diatoms.

Process Skills
observe and infer, compare and contrast, apply concepts

Teaching Strategies
■ Remind students to handle glass microscope slides carefully and place broken slides in the container for broken glass.
■ Remind students that they are looking at fragments of the silica-containing cell walls of diatoms.
■ Have students use a very small amount of the diatomaceous earth for better viewing.

Expected Results
Students should see broken cell walls of many different shapes.

Analysis
1. Answers will vary—rod-shaped, glasslike, circular, boatlike, needle-shaped, ridged or scored surface
2. Although broken, the fossil diatoms look similar and therefore have probably changed little over time.
3. The cell walls were visible because they did not decompose. The weight of water and sediments crushed them.

✔ **Assessment**
Performance Have students determine the shapes and sizes of the diatoms in the sample they observed. Use the Performance Task Assessment List for Making Observations and Inferences in **PASC**, p. 17. **L1** **ELL**

BIOLOGY *Online* Note Internet addresses that you find useful in the space below for quick reference.

_____

_____

_____

_____

_____

GLENCOE TECHNOLOGY

VIDEODISC
The Secret of Life
Diatom

Resource Manager

BioLab and MiniLab Worksheets, p. 67 **L2**

Purpose

Students will analyze a situation and evaluate its explanations.

Process Skills

analyze information, draw a conclusion, judge, think critically

Teaching Strategies

■ Review the appearance of Gondwanaland.

Thinking Critically

1. It's unlikely that ferns could grow in cold temperatures.
2. Not reasonable; it's unlikely that mutations could result in ferns adapted to extreme cold.
3. Reasonable; areas near the poles would have been warmer.

Assessment

Skill Ask students to suggest other reasons why fern fossils are in Antarctica. Use the Performance Task Assessment List for Designing an Experiment in **PASC**, p. 33. **L2**

Problem-Solving Lab 14-1 · Thinking Critically

Could ferns have lived in Antarctica?
Scientists have discovered fossil remains of ferns in the rocks of Antarctica. These fern fossils are related to ferns that grow in temperate climates on Earth today.

Fern Fossil from Antarctica

Analysis

Read each statement below and judge whether or not the statement is reasonable. Explain the reason for each of your judgments.

Thinking Critically

1. Fern fossils in Antarctica are of plants that could withstand freezing temperatures.
2. The ferns in Antarctica may have been mutated forms of ferns that grew in warm climates.
3. The temperature of Earth may have been much warmer millions of years ago than it is today.

Figure 14.3
Most sedimentary rocks form in horizontal layers with the younger layers closer to the surface.

away the lighter bones. You can use the *Problem-Solving Lab* on this page to try to solve a fossil mystery.

Fossils occur in sedimentary rocks

For fossils to form, organisms usually have to be buried in small particles of mud, sand, or clay soon after they die. These particles are com-

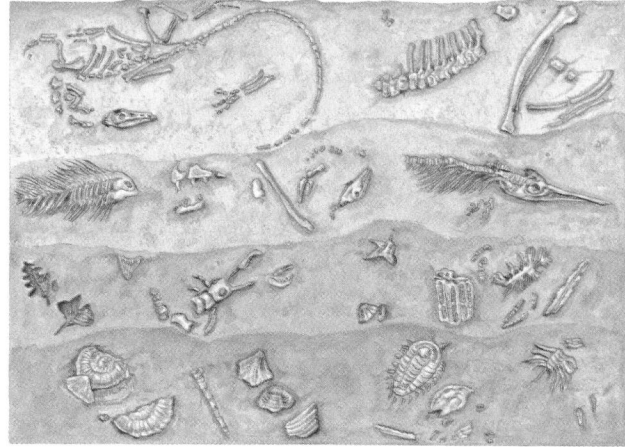

---

pressed over time and harden into a type of rock called sedimentary rock. Today, fossils still form at the bottoms of lakes, streams, and oceans.

Most fossils are found in sedimentary rocks. Layers of these rocks form by processes that prevent damage to the organism. How do these fossils become visible millions of years later? To answer the question, look at the *Inside Story.* Fossils are not usually found in other types of rock because of the ways those rocks form. For example, metamorphic rocks form when heat, pressure, and chemical reactions change other rocks. The conditions under which metamorphic rocks form would destroy any fossils that were in the sedimentary rock.

The Age of a Fossil

The fossils in different layers of sedimentary rock vary in age. Scientists use a variety of methods to determine the age of fossils.

Relative dating

One method is a technique called relative dating. To understand relative dating, imagine yourself stacking newspapers at home. As each day's newspaper is added to the stack, the stack becomes taller. If the stack is left undisturbed, the newspapers at the bottom are older than ones at the top.

The relative dating of rock layers uses the same principle. In *Figure 14.3,* you see fossils in different layers of rock. If the rock layers have not been disturbed, the layers at the surface must be younger than the deeper layers. The fossils in the top layer must also be younger than those in deeper layers. This principle is a geological law. Using this law, scientists can determine the order of appearance and extinction of the species that formed fossils in the layers.

INSIDE STORY

The Fossilization Process

Few organisms become fossilized because, without burial, bacteria and fungi immediately decompose their dead bodies. Occasionally, however, organisms do become fossils in a process that usually takes many years. Most fossils are found in sedimentary rocks.

Critical Thinking Describe how the movements of Earth might expose a fossil.

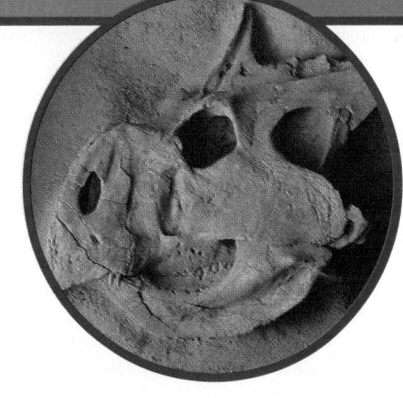

A Protoceratops

1 A **Protoceratops** drinking at a river falls into the water and drowns.

5 **After discovery,** scientists carefully extract the fossil from the surrounding rock.

2 **Sediments from upstream** rapidly cover the body, preventing its decomposition. Minerals from the sediments seep into the body.

3 **Over time,** additional layers of sediment compress the sediments around the body, forming rock. Minerals eventually replace all the body's bone material.

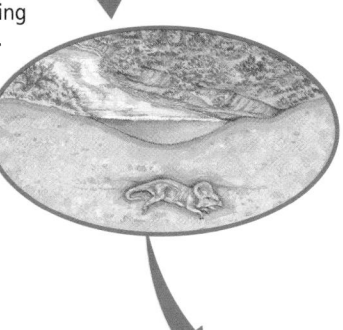

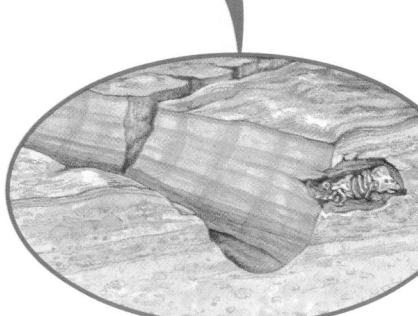

4 **Earth movements** or erosion may expose the fossil millions of years after it formed.

14.1 THE RECORD OF LIFE **381**

INSIDE STORY

Purpose
Students will understand the geological and biological processes involved in how fossils form.

Teaching Strategies

■ Point out that an organism's remains are subjected to destructive environmental factors, such as heat, cold, and pressure. Therefore, the number and quality of fossils is limited.

■ Have students make a flow-chart of the events that lead to the formation of fossils. **L2**

Visual Learning

Visual-Spatial Sedimentation allows fossils to form. Model the process using objects and sediments of various sizes and weights placed in a 2L plastic container three-quarters full of water. Shake the container. Have students observe what occurs. Explain that under pressure the objects may be fossilized. **L2** **ELL**

Critical Thinking

The Earth's surface can rupture during earthquakes and some rock layers can fold over others. The slow erosion of rock may uncover fossils.

Quick Demo

Logical-Mathematical Show students pieces of igneous, metamorphic, and sedimentary rocks. Explain how each is formed. Ask students to explain why fossils in metamorphic rocks are rare. **L2** **ELL**

PROJECT

Fossil Preservation

Kinesthetic Student groups can carry out the following activity to learn more about the preservation of fossils.

1. Place fresh fruit, such as strawberries or orange slices, in an open plastic container. Cover the fruit with water and place the container in a freezer.

2. Put the same amount and types of fruit in a similar container. Leave the container undisturbed at room temperature.

3. After three days, observe all the fruit. Summarize your observations and draw a conclusion about any differences you observe between the containers. **L1** **ELL** **COOP LEARN**

Concept Development

Explain to students that radioactive dating is an accurate technique for geological timekeeping because the decay rate of a radioactive element is constant.

Enrichment

 Visual-Spatial If possible, schedule a visit to a local museum of natural history. Most have displays of the Geological Time Scale. **L2** **ELL**

CAREERS IN BIOLOGY

Career Path

Courses in high school: biology, mathematics, and English
College: degree in biology or zoology for zoo/aquarium work
Other educational sources: on-the-job-training; two-year program in animal health

Career Issue

Ask students whether they think a person can love animals so much that he or she is unable to care for them effectively. Have them explain their answers.

For More Information

For more information, write to:
American Association of Zoo Keepers
635 Gage Boulevard
Topeka, Kansas 66606-2066
or
Animal Caretakers Information
Humane Society of the United States
Companion Animals Division
2100 L Street NW
Washington, DC 20037

382

Radiometric dating

Just as you cannot tell the actual age of a specific newspaper in the stack by just glancing at the stack, you cannot determine the actual age of a fossil by using relative dating techniques. To find the specific ages of fossils, scientists use radiometric dating techniques utilizing the radioactive isotopes in rocks or fossils. Recall that radioactive isotopes are atoms with unstable nuclei that break down, or decay, over time, giving off radiation. A radioactive isotope forms a new element after it decays. Because every radioactive isotope has a characteristic decay rate, scientists use the rate of decay as a type of clock. The decay rate of a radioactive isotope is called its half-life.

Just as you can judge how long ago an hourglass was turned by comparing the amounts of sand in the top and bottom containers, scientists try to determine the approximate ages of rocks by comparing the amount of a radioactive isotope and the new element into which it decays. For example, suppose that when a rock forms it contains a radioactive isotope that decays to half its original amount in one million years. Today, if the rock contains equal amounts of the radioactive isotope and the element into which it decays, then the rock must be about 1 million years old.

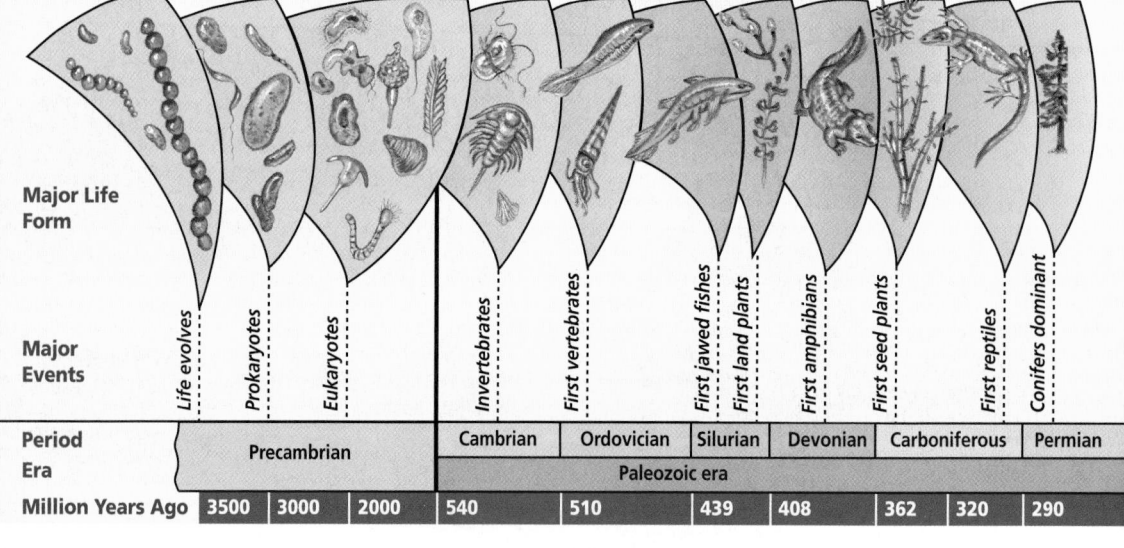

Major Life Form										
Major Events	Life evolves	Prokaryotes	Eukaryotes	Invertebrates	First vertebrates	First jawed fishes / First land plants	First amphibians	First seed plants	First reptiles	Conifers dominant
Period	Precambrian			Cambrian	Ordovician	Silurian	Devonian	Carboniferous		Permian
Era				Paleozoic era						
Million Years Ago	3500	3000	2000	540	510	439	408	362	320	290

Portfolio

Calculating Age Using Half-Lives

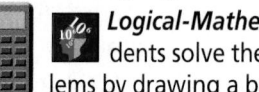 **Logical-Mathematical** Have students solve the following problems by drawing a bar graph. Have them place their graphs in their portfolios.

- A radioactive element has a known half-life of 20 days. How much of a 16-g sample of the radioactive element will remain unchanged after 80 days? *1 g*

- A fossil contains a radioactive element with a half-life of 10 000 years. If the ratio of radioactive element to decay element is 1:3, how old is the fossil? *20 000 years or two half-lives* **L3** **P**

Scientists use potassium-40, a radioactive isotope that decays to argon-40, to date rocks containing very old fossils. Based on chemical analysis, chemists have determined that potassium-40 decays to half its original amount in 1.3 billion years. Scientists use carbon-14 to date fossils less than 50 000 years old. Again, based on chemical analysis, they know that carbon-14 decays to half its original amount in 5730 years.

In the *BioLab* at the end of this chapter, you can simulate this dating technique. Keep in mind, however, that this dating technique frequently produces inconsistent dates because the initial amount of radioisotope in the rock can never be known with certainty. For this reason, scientists always analyze many samples of a rock using as many methods as possible to obtain reasonably consistent results about the rock's age.

A Trip Through Geologic Time

By examining layers of sedimentary rock and dating the fossils that are found in the layers, scientists have been able to put together a chronology, or calendar, of Earth's history. This chronology, called the Geologic Time Scale, is based on evidence from Earth's rocks.

The Geologic Time Scale

Rather than being based on months or even years, the Geologic Time Scale is divided into the four eras that you see in *Figure 14.4:* the Precambrian (pree KAM bree un) era, the Paleozoic (pay lee uh ZOH ihk) era, the Mesozoic (mez uh ZOH ihk) era, and the Cenozoic (sen uh ZOH ihk) era. An era is the largest division in the scale, and each represents a very long period of time. Each era is subdivided into periods.

The divisions in the Geologic Time Scale are distinguished by the organisms that lived during the time period. The fossil record indicates that there were several occurrences of mass extinction that fall between time divisions. A mass extinction is an event that occurs when entire groups of organisms disappear from the fossil record almost at once.

Figure 14.4
The Geologic Time Scale is a calendar of Earth's history based on evidence found in rocks. Life probably first appeared on Earth between 3.9 and 3.5 billion years ago.

The BioLab at the end of the chapter can be used at this point in the lesson.

INVESTIGATE BioLab

Chalkboard Example

To illustrate the concept of geological time, reproduce on the chalkboard or an overhead transparency a page from a monthly calendar. Remind students that the total area of the page represents a unit of time equal to one month. Divide the calendar into four (or five) horizontal strips. Ask students what amount of time each strip represents. *one week* Cut one of the strips into seven pieces and ask students what each square represents. *one day* Then, ask students how minutes and seconds could be shown. Elicit from students how geologic time, like calendars, is also divided into units.

GLENCOE TECHNOLOGY

CD-ROM
Biology: The Dynamics of Life
Exploration: *The Record of Life*
Disc 2

Resource Manager

Concept Mapping, p. 14
L3 **ELL**
Laboratory Manual, pp. 101-102 **L2**

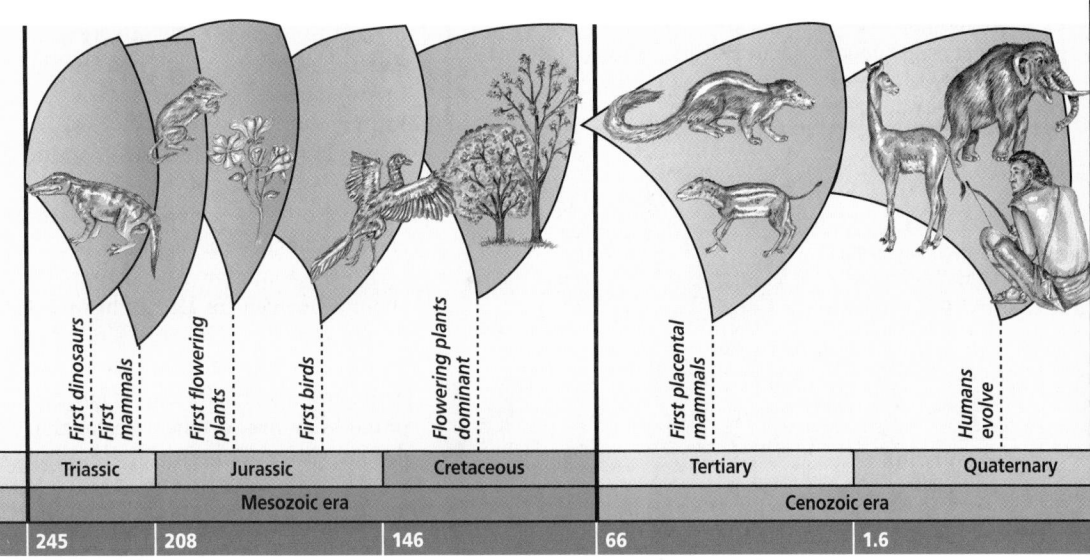

First dinosaurs / First mammals	First flowering plants	First birds	Flowering plants dominant	First placental mammals	Humans evolve
Triassic	Jurassic		Cretaceous	Tertiary	Quaternary
Mesozoic era				Cenozoic era	
245	208		146	66	1.6

BIOLOGY JOURNAL

Summarizing Geologic Time

Visual-Spatial Have students draw a large box on a sheet of paper in their journals. Then, have them divide the box into four smaller boxes with the first box occupying about one-third of the total area and the last box occupying about one-eighth of the total. Have students label each box with the name of one of the eras of geologic time, placing the Precambrian in the largest box and the Cenozoic in the smallest. As students read about these eras, have them write terms or phrases inside the box that describe that era. Students can use their sheets as study tools. **L1** **ELL**

Purpose

Students will model the Geologic Time Scale.

Process Skills

use a table, sequence, measure in SI

Teaching Strategies

■ Review how to relate distance to a time scale.
■ Review the Geologic Time Scale's two major divisions.

Expected Results

Students perform the calculations that establish their scales. They plot major events on the scale.

Analysis

1. the longest era—Paleozoic the shortest era—Cenozoic
2. Mesozoic era
3. primates

✓ Assessment

Performance Have student groups collect pictures of 15-20 organisms. Ask them to illustrate the Geologic Time Scale on poster board, and glue each picture where the organism first appeared in the fossil record. Use the Performance Task Assessment List for Poster in **PASC,** p. 73.
L1 ELL COOP LEARN

Resource Manager

BioLab and MiniLab Worksheets, p. 68 L2

MiniLab 14-2 Organizing Data

A Time Line In this activity, you will construct a time line that is a scale model of the Geologic Time Scale. Use a scale in which 1 meter equals 1 billion years. Each millimeter then represents 1 million years.

Procedure

1. Use a meterstick to draw a continuous line down the middle of a 5 m strip of adding-machine tape.
2. At one end of the tape, draw a vertical line and label it "The Present."
3. Measure off the distance that represents 4.6 billion years ago. Draw a vertical line at that point and label it "Earth's Beginning."
4. Using the table at right, plot the location of each event on your time line. Label the event and the number of years ago it occurred.

Geologic Time Scale	
Event	**Estimated years ago**
Earliest evidence of life	3.5 billion
Paleozoic era begins	540 million
First land plants	430 million
Mesozoic era begins	245 million
Triassic period begins	245 million
Jurassic period begins	208 million
First dinosaurs	225 million
First birds	150 million
Cretaceous period begins	146 million
Dinosaurs become extinct	66 million
Cenozoic era begins	66 million
Primates appear	60 million
Humans appear	200 000

Analysis

1. Which era is the longest? The shortest?
2. In which eras did dinosaurs and birds appear on Earth?
3. What major group first appeared after dinosaurs became extinct?

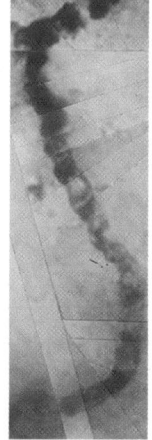
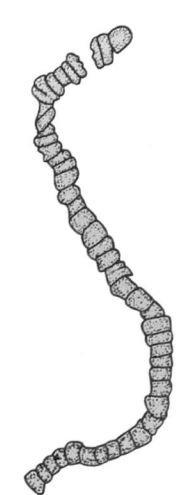

Figure 14.5
The filamentous fossils of these ancient organisms resemble some modern organisms.

The Geologic Time Scale begins with the formation of Earth about 4.6 billion years ago. To understand the large size of this number, try the *MiniLab* and also try scaling down the history of Earth into a familiar, but hypothetical, calendar year.

Life in the Precambrian

In your hypothetical calendar year, the first day of January becomes the date on which Earth formed. The oldest fossils are found in Precambrian rocks that are about 3.5 billion years old—near the end of March on the hypothetical calendar. Scientists found these fossils, which are shown in *Figure 14.5,* in rocks found in the deserts of western Australia. They have found more examples of similar types of fossils on other continents. The fossils resemble the forms of modern species of photosynthetic bacteria called cyanobacteria (si a noh bak TIHR ee uh). You will read more about cyanobacteria in a later chapter.

Scientists have also found fossils of dome-shaped structures called stromatolites (stroh MAT ul ites) in Australia and on other continents. Stromatolites still form today in Australia from mats of cyanobacteria, *Figure 14.6.* Thus, the stromatolite fossils are evidence of the existence of photosynthetic bacteria on Earth about 3.5 billion years ago.

The Precambrian era accounts for about 87 percent of Earth's history—until about the middle of October in the hypothetical calendar year. At the beginning of the Precambrian era, unicellular prokaryotes—cells that do not have a membrane-bound nucleus—appear to have been the only life forms on Earth. Then, about 1.8 billion years ago, the fossil record

---

BIOLOGY JOURNAL

Extinct Animals

Linguistic Ask students to research an extinct animal, such as a woolly mammoth, brachiosaur, pterodactyl, or saber-toothed cat. Have them describe in their journals what foods, environments, enclosures, or other factors might be required to keep the animal alive in a zoo.
L2

shows that more complex eukaryotic organisms, living things with membrane-bound nuclei in their cells, appeared. The eukaryotes flourished. By the end of the Precambrian, about 544 million years ago, multicellular eukaryotes, such as sponges and jellyfishes, filled the oceans.

Diversity in the Paleozoic

In the Paleozoic era, which lasted until 245 million years ago, many more types of animals and plants appeared on Earth, and some were embedded in the fossil record. The earliest part of the Paleozoic era is called the Cambrian period. Paleontologists often refer to a Cambrian explosion of life because the fossil record shows an enormous increase in the diversity of life forms at this time. During the Cambrian period, the oceans teemed with many types of animals, including worms, sea stars, and unusual arthropods, such as the one shown in *Figure 14*.7.

During the first half of the Paleozoic, fishes, the oldest animals with backbones, appeared in Earth's waters. There is also fossil evidence of ferns and early seed plants existing on land about 400 million years ago. Around the middle of the Paleozoic, four-legged animals such as amphibians appeared on Earth. During the last half of the era, the fossil record shows that reptiles appeared and began to flourish on land.

The largest mass extinction recorded by the fossil record marked the end of the Paleozoic. About 90 percent of Earth's marine species and 70 percent of the land species disappeared at this time.

Life in the Mesozoic

The Mesozoic era began about 245 million years ago, which would be about December 10 on the hypo-

thetical one-year calendar. Many changes, in both Earth's organisms and its geology, occurred over the span of this era.

The Mesozoic era is divided into three periods. Fossils from the Triassic period, the oldest period, show that mammals appeared on Earth at this time. These fossils of mammals indicate that early mammals were small and mouselike. They probably scurried around in the shadows of huge fern forests, trying to avoid dinosaurs, reptiles that also appeared during this time.

Figure 14.6
Fossils of stromatolites, like the modern Australian specimens shown here, provide evidence that photosynthetic bacteria lived on Earth 3.5 billion years ago.

Figure 14.7
Arthropods, such as this trilobite, were among the many groups of animals that appeared during the Cambrian explosion.

14.1   THE RECORD OF LIFE   **385**

Quick Demo

Set up a microscope with a living culture or prepared slide of *Oscillatoria*. Point out that early cyanobacteria are hypothesized to have produced much of the oxygen that changed the initial composition of Earth's atmosphere.

Enrichment

Some students may comment on the smooth gliding movement of *Oscillatoria*. Explain that these organisms expel jets of slime through holes in their cell walls. This pushes the cells through their environment.

Visual Learning

Ask students what living group of animals the trilobite in Figure 14.7 most resembles. *arthropods or insects*

✓ Assessment

Skill Assess the students' understanding of geologic time by having them sequence some of the major events of the Geologic Time Scale with the era in which they occurred. **L2**

GLENCOE TECHNOLOGY

 VIDEOTAPE
The Secret of Life
Gone Before You Know It: The Biodiversity Crisis

 VIDEODISC
The Infinite Voyage
The Great Dinosaur Hunt, The Evolution of Extinction (Ch. 2) 3 min.

The Great American Bone Rush (Ch. 3) 2 min.

3 Assess

Figure 14.8
Both fossil evidence like this *Archaeopteryx* **(a)** and some characteristics of present-day birds like this hoatzin **(b)** suggest that dinosaurs might have been the ancestors of today's birds.

WORD Origin

tectonics
From the Greek word *tecton*, meaning "builder." Plate tectonics can build mountains.

The middle of the Mesozoic is called the Jurassic period, which began about 208 million years ago, or mid-December on the hypothetical calendar. The Jurassic period is often referred to as the Age of the Dinosaurs.

Recent fossil discoveries support the idea that modern birds evolved from some of these dinosaurs toward the end of this period. For example in **Figure 14.8,** you see the fossil of *Archaeopteryx*, a small dinosaur discovered in Germany. The fossil reveals that *Archaeopteryx* had feathers, a birdlike feature. You also see a present-day bird, the hoatzin, in **Figure 14.8.** This bird has a reptilian feature, claws on its wings, for its first few weeks of life. It also flies poorly, as the earliest birds probably did. Paleontologists suggest that such evidence supports the idea that modern birds evolved from dinosaurs.

A mass extinction

The last period in the Mesozoic, the Cretaceous, began about 144 million years ago. During this period, many new types of mammals and the first flowering plants appeared on Earth. The mass extinction of the dinosaurs marked the end of the Cretaceous period about 66 million years ago. Scientists estimate that not only dinosaurs but almost two-thirds of all living species at the time became extinct. Based on geological evidence of a large crater in the waters off eastern Mexico, some scientists propose that a large meteorite collision caused this mass extinction. Such a collision could have filled the atmosphere with thick, toxic dust that, in turn, changed the climate to one in which many species could no longer survive.

Changes during the Mesozoic

Geological events during the Mesozoic changed the places in which species lived and affected their distribution on Earth. The theory of continental drift, which is illustrated in **Figure 14.9,** suggests that Earth's continents have moved during Earth's history and are still moving today at a rate of about six centimeters per year. This is about the same rate at which your hair grows. Early in the Mesozoic, the continents were merged into one large land mass. During the era the continent broke up and the pieces drifted.

The geological explanation for how the continents move is called **plate tectonics** (tek TAHN ihks). According to this idea, Earth's crust consists of several rigid plates that drift on top of molten rock. These plates are continually moving—spreading apart, sliding by, or pushing against each another. The movements affect organisms. For example, after a long time, the descendants of organisms living on plates that are moving apart may be living in areas with very different climates.

The Cenozoic era

The Cenozoic began about 66 million years ago—around December 26 on the hypothetical calendar of Earth's history. It is the era in which you now live. Mammals began to flourish during the early part of this era. Among the mammals that appeared was a group of animals to which you belong, the primates. Primates first appeared approximately 30 million years ago and have diversified greatly. The modern human species appeared perhaps as recently as 200 000 years ago. On the hypothetical calendar of Earth's history, 200 000 years ago is late in the evening of December 31.

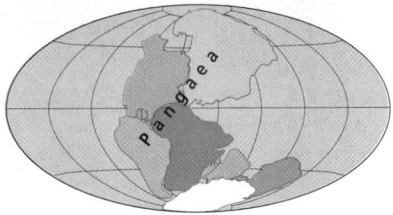

A About 245 million years ago, the continents were joined in a landmass known as Pangaea.

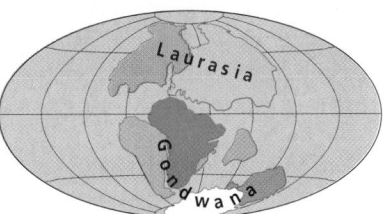

B By 135 million years ago, Pangaea broke apart resulting in two large landmasses.

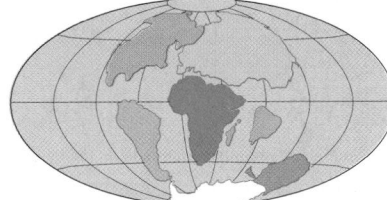

C By 66 million years ago, the end of the Mesozoic, most of the continents had taken on their modern shapes.

Figure 14.9
The theory of continental drift describes the movement of the land masses over geological time. The modern continents are shown in different colors.

CD-ROM
View an animation of Plate Tectonics located in the Presentation Builder of the Interactive CD-ROM.

Section Assessment

Understanding Main Ideas
1. Describe what some scientists propose Earth was like before life arose.
2. Why are most fossils found in sedimentary rocks?
3. What do paleontologists learn from fossils?
4. Explain the difference between relative dating and radiometric dating.

Thinking Critically
5. Suppose you are examining layers of sedimentary rock. In one layer, you discover the remains of an extinct relative of the polar bear. In a deeper layer, you discover the fossil of an extinct alligator. What can you hypothesize about changes over time in this area's environment?

SKILL REVIEW

6. **Making and Using Tables** Construct a table listing the four geologic eras, their time spans, and the major forms of life that appeared during each era. Use the information in the table to construct a time line based on the face of a clock. For more help, refer to *Organizing Information* in the **Skill Handbook**.

Section Assessment

1. Earth was hot; meteors bombarded its surface, and volcanoes erupted. As Earth cooled, it rained constantly.
2. Sedimentary rocks form slowly, preventing damage to a fossilizing organism.
3. Paleontologists can infer diet, size, habitat, and other information from fossils.
4. Relative dating relies on the position of rock layers and results in relative date.

Radiometric dating relies on half-lives of radioactive elements.

5. The region may have changed from a warm to a cold climate.
6. Student tables should be constructed from the information in Figure 14.4 and under the heading A Trip Through Geologic Time in the text. Check tables and student time lines for accuracy.

387

Prepare

Key Concepts

Students will learn about scientific hypotheses of the origin of life. They will read about some classic experiments designed to prove biogenesis. Then, they will study modern hypotheses about the origin of cells.

Planning

- Obtain beef bouillon cubes for the Quick Demo.
- Collect the materials needed for the Alternative Lab.

1 Focus

Bellringer

Before presenting the lesson, display **Section Focus Transparency 36** on the overhead projector and have students answer the accompanying questions.
L1 ELL

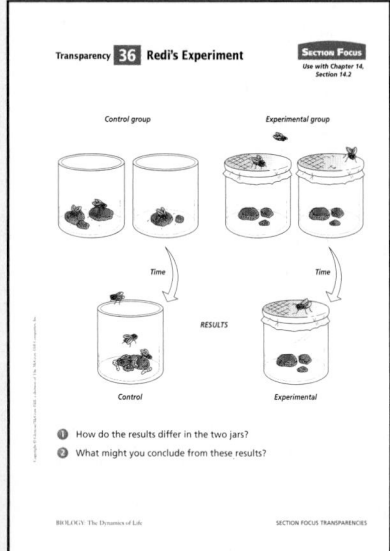

SECTION PREVIEW

Objectives
Analyze early experiments that support the concept of biogenesis.
Compare and contrast modern theories of the origin of life.
Relate hypotheses about the origin of cells to the environmental conditions of early Earth.

Vocabulary
spontaneous generation
biogenesis
protocell
archaebacteria

Section
14.2 The Origin of Life

Will rotting meat give rise to maggots? Can mud produce live fish like the ones shown here? Will a bag of wheat give birth to mice? You'd probably answer all these questions with a no. But not too long ago, people did not know how life begins. However, evidence provided by scientists such as Louis Pasteur finally solved the mystery. To obtain the evidence, scientists relied on the scientific methods of observation, hypothesis, and controlled experimentation.

Mudskippers

Origins: The Early Ideas

In the past, the ideas that decaying meat produced maggots, mud produced fish, and grain produced mice were reasonable explanations for what people observed occuring in their environment. After all, they saw maggots appear on meat and young mice just appear in sacks of grain. Such observations led people to believe in **spontaneous generation**—the idea that nonliving material can produce life.

Figure 14.10
Francesco Redi's controlled experiment tested the spontaneous generation of maggots from decaying meat.

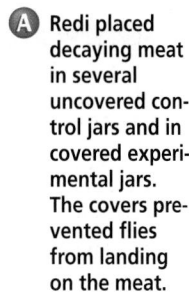 **A** Redi placed decaying meat in several uncovered control jars and in covered experimental jars. The covers prevented flies from landing on the meat.

Control group *Time*

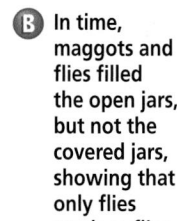 **B** In time, maggots and flies filled the open jars, but not the covered jars, showing that only flies produce flies.

Experimental group *Time*

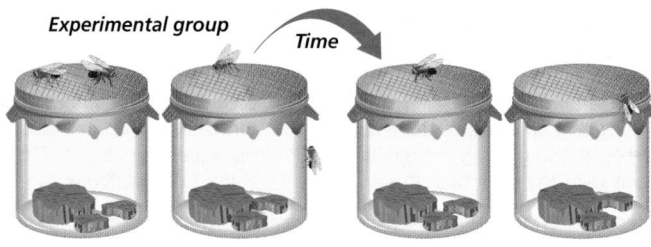

388 THE HISTORY OF LIFE

MEETING INDIVIDUAL NEEDS

Gifted

Intrapersonal Have interested students prepare a report or poster presentation about the work of Italian scientist Lazzaro Spallanzani in the debate over spontaneous generation. **L3**

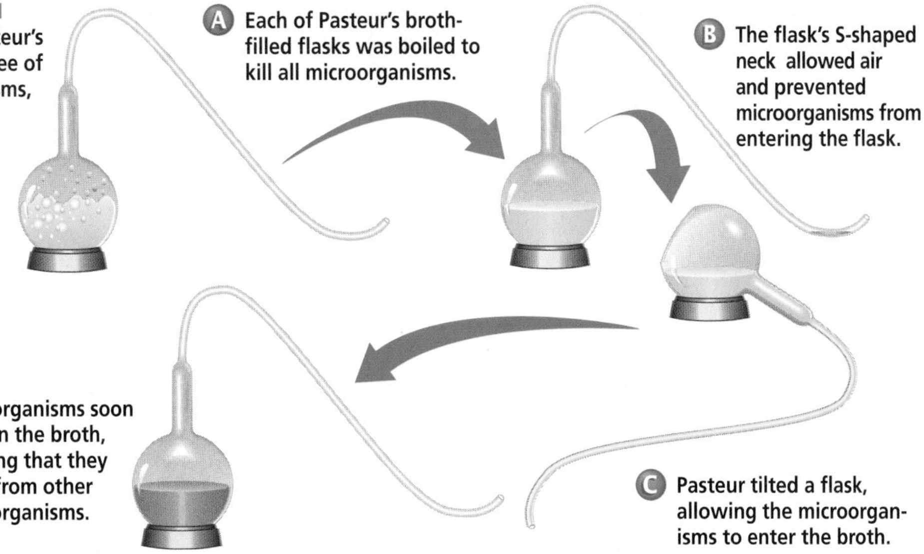
A Each of Pasteur's broth-filled flasks was boiled to kill all microorganisms.
B The flask's S-shaped neck allowed air and prevented microorganisms from entering the flask.
C Pasteur tilted a flask, allowing the microorganisms to enter the broth.
D Microorganisms soon grew in the broth, showing that they come from other microorganisms.

Figure 14.11
Some of Pasteur's flasks, still free of microorganisms, are at the Pasteur Institute in Paris.

Spontaneous generation is disproved

In 1668, an Italian physician, Francesco Redi, disproved a commonly held belief at the time—the idea that decaying meat produced maggots, which are immature flies. You can follow the steps of Redi's experiment in ***Figure 14.10.*** Redi's well-designed, controlled experiment successfully convinced many scientists that maggots, and probably most large organisms, did not arise by spontaneous generation.

However, during Redi's time, scientists began to use the latest tool in biology—the microscope. With the microscope, they saw that microorganisms live everywhere. Although Redi had disproved the spontaneous generation of large organisms, many scientists thought that microorganisms were so numerous and widespread that they must arise spontaneously—probably from a vital force in the air.

Pasteur's experiments

Disproving the existence of a vital force in air proved difficult. Finally, in the mid-1800s, Louis Pasteur designed an experiment that disproved the spontaneous generation of microorganisms. Pasteur set up an experiment in which air, but no microorganisms, was allowed to contact a broth that contained nutrients. You can see how Pasteur carried out his experiment in ***Figure 14.11.***

Pasteur's experiment showed that microorganisms do not simply arise in broth, even in the presence of air. From that time on, **biogenesis** (bi oh JEN uh sus), the idea that living organisms come only from other living organisms, became a cornerstone of biology.

Origins: The Modern Ideas

Biologists have accepted the concept of biogenesis for more than 100 years. However, biogenesis does not answer the question: How did life begin on Earth? No one will ever know for certain how life began on Earth. However, many scientists have developed theories about the origin

WORD Origin
biogenesis
From the Greek word *bios*, meaning "life," and the Latin word *genesis*, meaning "birth." Biogenesis proposes that living organisms come only from other living organisms.

2 Teach

Quick Demo

Kinesthetic Explain that people once believed that life spontaneously arose from nonliving matter, such as broth. Dissolve two bouillon cubes in separate flasks of hot water. Boil the bouillon for several minutes. After the broth cools, seal one flask with a rubber stopper and label it "boiled." Leave the second flask open. Set the flasks on a shelf in the classroom. Have students observe the flasks every day and record their observations in their portfolios. L2 ELL P

Visual Learning

Figures 14.10 and 14.11 The experiments by Redi, Pasteur, and others to disprove spontaneous generation are examples of the application of scientific methods. Use the illustrations to reinforce students' understanding of scientific methods. Discuss the purpose of each step of the procedures with students.

Concept Development

Point out that the spontaneous generation of cells was harder to disprove than the spontaneous generation of whole organisms because microscopes were not powerful enough to show cells dividing until 1875.

Resource Manager

Section Focus Transparency 36 and Master L1 ELL
Basic Concepts Transparency 20 and Master L2 ELL

MEETING INDIVIDUAL NEEDS

English Language Learners

Visual-Spatial Have students use block diagrams on paper to model how the experiments of Redi or Pasteur demonstrate the scientific method. Have students record their procedures and observations. L1 ELL

Portfolio

Evaluating Pasteur's Experiment

Linguistic After setting up the Quick Demo using broth, have students predict what will happen in each flask, observe the flasks for a week, and then evaluate their prediction. Have them write their conclusions and put them in their portfolios. L2 P

Review the chemistry concepts that students studied previously, particularly the role of carbon in organic molecules. Review the general structure of proteins and nucleic acids. Relate this information to the hypotheses about the origin of life.

Using Science Terms

Have students list synonyms and antonyms for *primordial*. *first, original, early; last, final, ultimate* Ask them to explain why Oparin suggested that molecules in the oceans formed "primordial soup." *They were in a liquid medium.*

TECHPREP

Primordial Soup

Visual-Spatial Have students design a can and label for "Primordial Soup." The student design should have a creative graphic and a section of contents. **L2** **ELL**

✓ Assessment

Knowledge Have students compare the early ideas of biogenesis and Oparin's hypothesis in an essay. **L2**

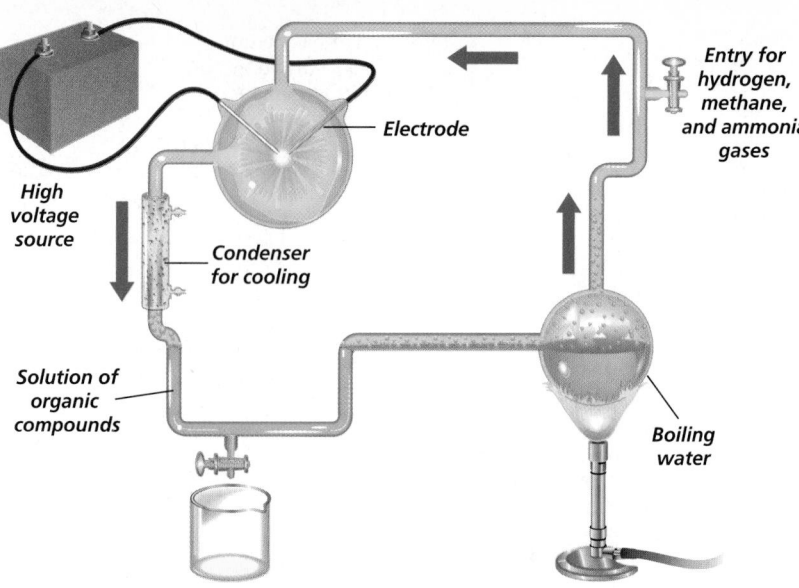

Figure 14.12
Miller and Urey's experiments showed that under the proposed conditions on early Earth, small organic molecules, such as amino acids, could form.

Labels in figure: High voltage source; Electrode; Condenser for cooling; Solution of organic compounds; Entry for hydrogen, methane, and ammonia gases; Boiling water

WORD Origin

primordial
From the Latin word *primordium*, meaning "origin." The origin of life may have been in the primordial soup.

of life on Earth from testing scientific hypotheses about conditions on early Earth. The *Biology & Society* at the end of this chapter summarizes some important viewpoints about the origin of life on Earth.

Simple organic molecules formed

Scientists hypothesize that two developments must have preceded the appearance of life on Earth. First, simple organic molecules, or molecules that contain carbon, must have formed. Then these molecules must have become organized into complex organic molecules such as proteins, carbohydrates, and nucleic acids that are essential to life.

Remember that Earth's early atmosphere probably contained little free oxygen. Instead, the atmosphere was probably composed of water vapor, carbon dioxide, nitrogen, and perhaps methane and ammonia. Many scientists have tried to explain how these substances could have joined together and formed the simple organic molecules that are found in all organisms today.

In the 1930s, a Russian scientist, Alexander Oparin, hypothesized that life began in the oceans that formed on early Earth. He suggested that energy from the sun, lightning, and Earth's heat triggered chemical reactions to produce small organic molecules from the substances present in the atmosphere. Then, rain probably washed the molecules into the oceans to form what is often called a primordial soup.

In 1953, two American scientists, Stanley Miller and Harold Urey, tested Oparin's hypothesis by simulating the conditions of early Earth in the laboratory. In an experiment similar to the one shown in *Figure 14.12,* Miller and Urey mixed water vapor (steam) with ammonia, methane, and hydrogen gases. They then sent an electric current that simulated lightning through the mixture. Then, they cooled the mixture of gases, produced a liquid that simulated rain, and collected the liquid in a flask. After a week, they analyzed the chemicals in the flask and found several kinds of amino acids, sugars, and other small

Alternative Lab

Making Coacervates

Purpose

Students will investigate the conditions under which the first cells may have evolved.

Materials
1% gelatin solution, droppers (3), 1% gum arabic solution, pH papers, microscopes, microscope slides, coverslips, 0.1*M* hydrochloric acid (HCl), test tubes, stirring rods. For preparation instructions, see page 40T of the Teacher Guide.

Procedure
Give the following directions.
1. Remind students to wear safety goggles, a lab apron, and disposable gloves. Remind them to use care when working with a microscope and glass slides.
2. Measure 5 mL of 1% gelatin solution. Pour it into the test tube. Add 3 mL of 1% gum arabic solution. Mix gently.
3. Record the pH of the mixture.
4. Make a wet mount of the mixture. Observe under high power and record your observations.

organic molecules, providing evidence that supported Oparin's hypothesis.

The formation of protocells

The next step in the origin of life, as proposed by some scientists, was the formation of complex organic compounds. In the 1950s, various experiments were performed and showed that, if the amino acids are heated without oxygen, they link and form complex molecules called proteins. A similar process produces ATP and nucleic acids from small molecules. These experiments convinced many scientists that complex organic molecules might have originated in pools of water where small molecules had concentrated and been warmed.

How did these complex chemicals combine to form the first cells? The work of American biochemist Sidney Fox showed how the first cells may have occurred. As you can see in *Figure 14.13*, Fox produced protocells by heating solutions of amino acids. A **protocell** is a large, ordered structure, enclosed by a membrane, that carries out some life activities, such as growth and division.

The Evolution of Cells

Although the origin of life may always be a mystery, fossils indicate that by about 3.5 billion years ago, photosynthetic prokaryotic cells existed on Earth. But these were probably not the earliest cells. What were the earliest cells like, and how did they evolve?

The first true cells

The first forms of life may have been prokaryotes that evolved from a protocell. Because Earth's atmosphere lacked oxygen, scientists have proposed that these organisms were most likely anaerobic. For food, the first prokaryotes probably used some of the organic molecules that were abundant in Earth's early oceans. Because they obtained food rather than making it themselves, they would have been heterotrophs.

Over time, these heterotrophs would have used up the food supply. However, organisms that could make food had probably evolved by the

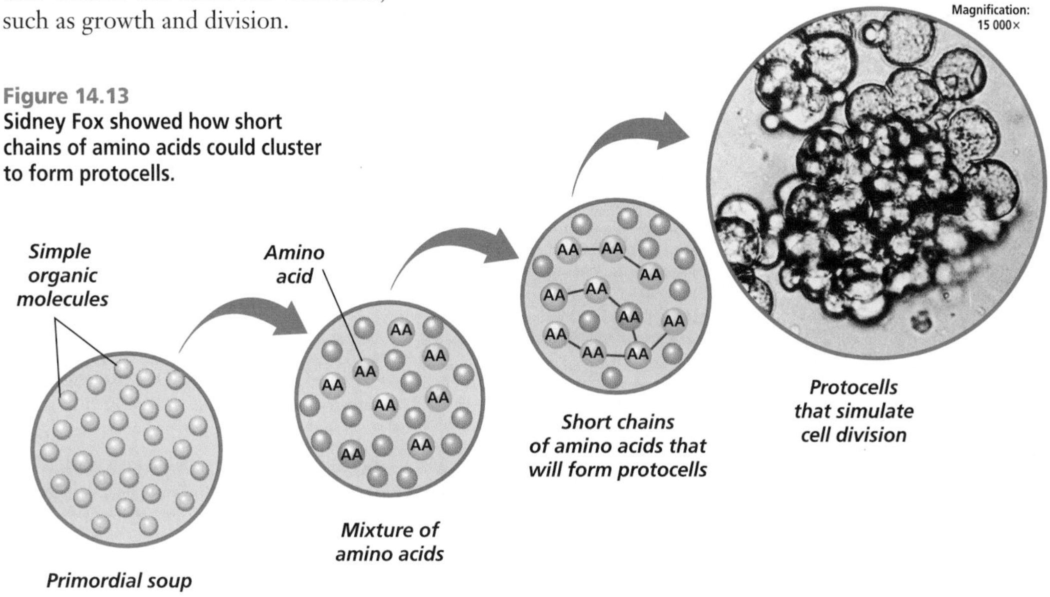

Figure 14.13
Sidney Fox showed how short chains of amino acids could cluster to form protocells.

Magnification: 15 000×

Simple organic molecules

Amino acid

AA

Primordial soup

Mixture of amino acids

Short chains of amino acids that will form protocells

Protocells that simulate cell division

Tying to Previous Knowledge

Review the basic structure and function of cells. Emphasize the substances cells need to live and the different ways cells release energy. Relate this information to the evolution of cells.

Discussion

Ask students why studying modern archaebacteria and cyanobacteria is important for determining the origin of life. *Such bacteria resemble the earliest known living things and indicate the conditions required to support life then.*

GLENCOE
TECHNOLOGY

VIDEODISC
The Secret of Life
Archaebacteria

---

5. Stir one drop of 0.1*M* HCl solution into the mixture. Record the mixture's pH.
6. Repeat step 4 for the HCl mixture.
7. Repeat steps 5 and 6 until you see coacervates, clumps that look like cells. Record the mixture's pH and describe the appearance of the coacervates.
8. Repeat steps 5 and 6 until the coacervates disappear.

Analysis

1. What living things do coacervates resemble? *cells*
2. Around what pH did you observe coacervates? *pH 5*
3. What conditions of early Earth does the mixture of gelatin (a protein) and gum arabic (a carbohydrate) simulate? *the amino acids and simple sugars in the "primordial soup"*

✓ Assessment

Knowledge Have students present an oral report of the investigation. Ask: What was the role of hydrochloric acid? How does this investigation relate to hypotheses about the origin of life? Use the Performance Task Assessment List for Oral Presentation in **PASC**, p. 71. **L1**

Purpose 🎲

Students will use a model to show events on the Geologic Time Scale.

Process Skills

observe and infer, sequence, use an illustration

Teaching Strategies

■ Question students about the history of life to assess their knowledge before they start.

Thinking Critically

Prokaryotes appeared around 4:00 to 4:30 a.m. and eukaryotes around 10:00 a.m.

Knowledge Have students approximate the time of evolution of three other groups of organisms by finding out when they appear in the fossil record. **L1**

3 Assess

Check for Understanding

Have students list the three main conclusions of this section in their journals. *Life comes from existing life; life probably originated on Earth through the reaction of chemicals in Earth's atmosphere and their further reaction on Earth's surface; and cells probably evolved as the chemicals on early Earth became more organized.* Ask students to list the scientific evidence that supports each conclusion. **L2**

Problem-Solving Lab 14-2   Interpreting Data

Can a clock model Earth's history? As a result of studying fossils and analyzing geological events, scientists have been able to construct the Geologic Time Scale, a timetable that shows the appearance of organisms during the history of Earth.

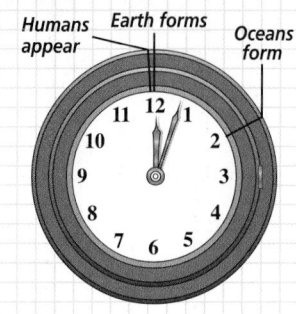

Analysis

The diagram shown here compresses the history of Earth into a 12-hour clock face. On the clock, assume that the formation of Earth occurred at midnight. The oceans formed at 2:00 A.M. Use this information to help you answer the following questions.

Thinking Critically

Based on fossil evidence, at what time on the face of the clock did prokaryotes evolve? At what time did the first eukaryotes appear?

Figure 14.14
Present-day archaebacteria live in places like this hot spring in Yellowstone National Park.

time the food was gone. These first autotrophs were probably similar to present-day archaebacteria. **Archaebacteria** (ar kee bac TIHR ee uh) are prokaryotes that live in harsh environments, such as deep-sea vents and hot springs like the one shown in *Figure 14.14.* The earliest autotrophs probably made glucose by chemosynthesis rather than by photosynthesis, which requires light-trapping pigments. In chemosynthesis, autotrophs release the energy of inorganic compounds in their environment to make their food.

Photosynthesizing prokaryotes

Photosynthesizing prokaryotes might have been the next type of organism to evolve. Recall that the process of photosynthesis produces oxygen. As the first photosynthetic organisms increased in number, the concentration of oxygen in Earth's atmosphere began to increase. Organisms that could respire aerobically would have evolved and thrived. In fact, the fossil record indicates that there was a very large increase in the diversity of prokaryotes about 2.8 billion years ago.

The presence of oxygen in Earth's atmosphere probably affected life on Earth in another important way. Lightning would have converted much of the oxygen into ozone molecules that would then have formed the thick layer of gas that now exists 10 to 15 miles above Earth's surface. The ozone layer probably shielded organisms from the harmful effects of ultraviolet radiation and enabled the evolution of more complex organisms, the eukaryotes.

The endosymbiont theory

The complex cells of eukaryotes probably evolved from prokaryote cells. Use the *Problem-Solving Lab* to determine how long the event might have taken. The endosymbiont theory, proposed by American biologist Lynn Margulis in the early 1960s, explains how eukaryote cells may have arisen.

BIOLOGY JOURNAL

Photosynthetic Prokaryotes

Visual-Spatial Have students compare slides or living cultures of both bacteria and cyanobacteria to photographs of stromatolite sections and early cells, such as those discovered in western Australia. Have students record their observations in their journals. **L3** **ELL**

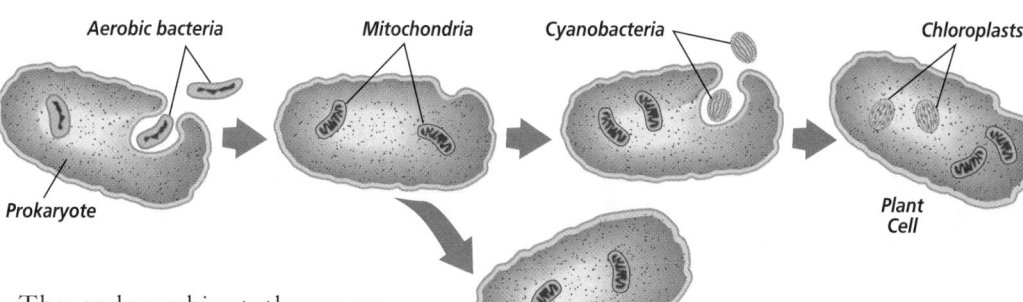

A A prokaryote ingested some aerobic bacteria. The aerobes were protected and produced energy for the prokaryote.

B Over a long time, the aerobes become mitochondria, no longer able to live on their own.

C Some primitive prokaryotes also ingested cyanobacteria, which contain photosynthetic pigments.

D The cyanobacteria become chloroplasts, no longer able to live on their own.

Aerobic bacteria

Mitochondria

Cyanobacteria

Chloroplasts

Prokaryote

Plant Cell

Animal Cell

Figure 14.15
The eukaryotic cells of plants and animals probably evolved by endosymbiosis.

The endosymbiont theory as shown in *Figure 14.15,* proposes that eukaryotes evolved through a symbiotic relationship between ancient prokaryotes. Margulis based her hypothesis on observations and experimental evidence of present-day unicellular organisms. For example, some bacteria that are similar to cyanobacteria and chloroplasts resemble each other in size and in the ability to photosynthesize. Likewise, mitochondria and some other bacteria look similar. Experimental evidence revealed that both chloroplasts and mitochondria contain DNA that is very similar to the DNA in prokaryotes and very unlike the DNA in eukaryotic nuclei.

Since Margulis first proposed the endosymbiont theory, new evidence has been found to support it. For example, scientific research has shown that chloroplasts and mitochondria have their own ribosomes that are similar to the ribosomes in prokaryotes. In addition, both chloroplasts and mitochondria reproduce independently of the cells that contain them. The fact that some modern prokaryotes live in close association with eukaryotes, a relationship from which the endosymbiont theory stems, also supports the theory.

Reteach

Ask students to list the scientists who experimented about the origin of life and their conclusions. Have them use their lists to quiz each other. **L1** **COOP LEARN**

Extension

Intrapersonal Ask students who have mastered this section to find out how scientists think that nucleic acids may have developed from elements already present on Earth. **L3**

✓ Assessment

Knowledge Ask each student to give an oral report about how life originated. **L2**

4 Close

Discussion

Have students discuss whether new life could originate on Earth today. They should consider how modern Earth differs from early Earth, including having an atmosphere rich in oxygen. **L2**

GLENCOE TECHNOLOGY

VIDEOTAPE
The Secret of Life
What's in Stetter's Pond: The Basics of Life

Resource Manager

Reinforcement and Study Guide, pp. 63-64 **L2**
Content Mastery, pp. 69, 71-72 **L1**

Section Assessment

Understanding Main Ideas

1. How did Pasteur's experiment finally disprove spontaneous generation?
2. What was Oparin's hypothesis, and how was it tested experimentally?
3. Why do scientists think the first living cells to appear on Earth were probably anaerobic heterotrophs?
4. How would the increasing number of photosynthesizing organisms on Earth have affected both Earth and its other organisms?

Thinking Critically

5. Some scientists speculate that lightning was not present on early Earth. How could you modify the Miller-Urey experiment to reflect this new idea? What energy source would you use to replace lightning?

SKILL REVIEW

6. **Sequencing** Make a flow chart sequencing the evolution of life from protocells to eukaryotes. For more help, refer to *Organizing Information* in the **Skill Handbook.**

Section Assessment

1. Life would not appear in sterile broth unless previously existing organisms entered the broth.
2. Life began in Earth's oceans after organic molecules formed from ocean chemicals. Miller and Urey tested the hypothesis.
3. There was no oxygen in the atmosphere and there were food molecules in the "organic soup" of the ocean.
4. They eventually created an oxygen-rich atmosphere. Organisms evolved that could use the oxygen.
5. Possible answers include sunlight, radioactivity, or heat from volcanoes.
6. protocells ⇒ anaerobic prokaryotes ⇒ chemosynthetic prokaryotes ⇒ photosynthetic prokaryotes ⇒ eukaryotes.

Determining a Fossil's Age

Time Allotment
One class period

Process Skills

make and use graphs, collect data, define operationally, formulate models, interpret data, use numbers

PREPARATION

- Any box large enough to hold the 100 pennies may be used.
- Provide graph paper.

Resource Manager

BioLab and MiniLab
Worksheets, p. 69 **L2**

Radioactive isotopes are used to date fossils and rocks. The dating is based on knowing four things: the amount of the radioactive isotope in the rock when it formed; the element into which the isotope decays; the rate of decay; and the amounts of isotope and new element in a rock or fossil. It takes 1.3 billion years for half of the radioactive isotope K 40 in a sample to decay—change into Ar 40. This time is K 40's half-life. When a rock such as volcanic lava forms, it is assumed that the amount of K 40 in the rock is 100 percent and the amount of Ar 40 is 0 percent.

PREPARATION

Problem
How can you simulate radioactive half-life?

Objectives
In this BioLab you will:
- **Formulate models** Simulate the radioactive decay of K 40 into Ar 40 with pennies.
- **Collect data** Collect data to determine the amount of K 40 present after several half-lives.

- **Make and use a graph** Graph your data and use its values to determine the age of rocks.

Materials
shoe box
with lid
100 pennies
graph paper

Skill Handbook
Use the **Skill Handbook** if you need additional help with this lab.

PROCEDURE

1. Copy the data table.
2. Place 100 pennies in a shoe box.

3. Arrange the pennies so that their "head" sides are facing up. Each "head" represent an atom of K 40, and each "tail" an atom of Ar 40.
4. Record the number of "heads" and "tails" present at the start of the experiment. Use the row marked "0" in the data table.
5. Cover the box. Then shake the box well. Let the shake represent one half-life of K 40, which is 1.3 billion years.

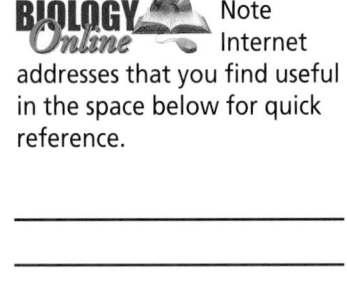

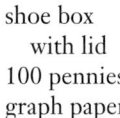

394

PROCEDURE

Teaching Strategies

- Have students work in groups to reduce the quantity of materials needed.
- Be sure that students understand the value of a simulation experiment such as this one. You may want to discuss with students why this particular lab would have to be run as a simulation.

- Remind students not to return the "tail" pennies to the box.
- Review the experimental need for several trials to collect data.
- Review how to determine an average.
- Review the meaning of a symbol such as K40 and how to determine the number of protons in an element.

Data Table

Number of Shakes (half lives)	Number of Heads (K 40 atoms left)				
	Trial 1	Trial 2	Trial 3	Totals	Average
0					
1					
2					
3					
4					
5					

6. Remove the lid and record the number of "heads" you see facing up. Remove all the "tail" pennies.
7. To complete the first trial, repeat steps 5 and 6 four more times.
8. Run two more trials and determine an average for the number of "heads" present at each half-life.

9. Draw a full-page graph. Plot your average values on the graph. Plot the number of half-lives for K 40 on the x axis and the number of "heads" on the y axis. Connect the points with a line. Remember, each half-life mark on the graph axis for K 40 represents 1.3 billion years.

ANALYZE AND CONCLUDE

1. **Applying Concepts** What symbol represented an atom of K 40 in this experiment? What symbol represented an atom of Ar 40?
2. **Thinking Critically** Compare the numbers of protons and neutrons of K 40 and Ar 40. (Consult the Periodic Chart in the Appendix for help.) Can Ar 40 change back to K 40? Explain your answer, pointing out what procedural part of the experiment supports your answer.
3. **Defining Operationally** Define the term half-life. What procedural part of the simulation represented a half-life period of time in the experiment?
4. **Communicating** Explain how scientists use radioactive dating to approximate a fossil's age.

5. **Making and Using Graphs** You are attempting to determine the age of a rock sample. Use your graph to read the rock's age if it has:
   a. 70% of its original K 40 amount.
   b. 35% of its original K 40 amount.
   c. 10% of its original K 40 amount.

Going Further

Application Suppose you had calculated the same data for an element with a half-life of 5000 years rather than 1.3 billion years. Plot a graph for the hypothetical isotope. How do the graphs compare?

BIOLOGY *Online* To find out more about radioactive dating, visit the Glencoe Science Web site.
science.glencoe.com

INVESTIGATE
BioLab

ANALYZE AND CONCLUDE

1. K-40—penny with "head" side up; Ar-40—penny with "tail" side up
2. Potassium has 19 protons and 21 neutrons, and Argon has 18 protons and 22 neutrons; No. K40 decays or changes into Ar-40. Remove "tail" pennies.
3. A half-life is the time needed for half the number of atoms of a radioactive element to change into atoms of a different element. Shaking the box represented a half-life.
4. They compare the amount of a radioactive element present now in a fossil or rock to the amount originally present, and use the element's half-life to calculate the sample's age.
5. a. 650 000 000
   b. 1 950 000 000
   c. 4 550 000 000

✓ Assessment

Portfolio Have students write a summary of this experiment in their portfolios that emphasizes the value of using the simulation to illustrate the concept of radioactive decay. Use the Performance Task Assessment List for Lab report in **PASC**, p. 47. **L1** **P**

Going Further

Intrapersonal The graphs will appear the same. Ask students to research how a radioactive element decays to form a different element. **L2**

Data and Observations

Half-life	Average number of K40
0	100
1	50
2	25
3	12
4	6
5	3

Biology & SOCIETY

How Did Life Begin: Different Viewpoints

How life originated on Earth is a fascinating and challenging question. Many have proposed answers, but the mystery remains unsolved.

Purpose

Students will explore a variety of ideas about the origin of life.

Teaching Strategies

■ Organize students into teams for a debate on the origins of life. Have each team defend one point of view. Ask students to research the strengths of their viewpoint and the weaknesses of the opposing viewpoints. **L2**

COOP LEARN

■ Review the chemical composition of nucleic acids, amino acids, lipids, carbohydrates, and other organic molecules in cells.

Investigating the Issue

The views assume the conditions and substances found on early Earth are the same. It could more directly lead to an explanation of a DNA world.

Going Further

Linguistic Have students research and report to the class about organic molecules on planets, meteors, and comets. Students may also research current technology in SETI (search for extraterrestrial intelligence) projects. Have students prepare models, videos, or posters for a class presentation. **L3** **ELL**

Because it is impossible to travel in time, the question of how life originated on Earth may never be answered. However, many ideas and beliefs have been proposed.

Divine origins Common to human cultures throughout history is the belief that life on Earth did not arise spontaneously but was placed here by a creator. Many major religions teach that life was created by a supreme being. Many people believe that life could only have arisen with the intervention of a divine force.

Meteorites One scientific idea about life's origin on Earth proposes that life arrived here on meteorites, space-borne rocks that may collide with Earth's surface. Many meteorites contain some organic matter, which could help explain how organic molecules considered necessary for the formation of cells might have arrived on Earth and entered its oceans.

Primordial soup A. I. Oparin proposed that Earth's ancient atmosphere probably contained

the gases nitrogen, methane, and ammonia, but little free oxygen. Energy from the sun, volcanoes, and lightning fueled chemical reactions among these gases, which eventually combined into small organic molecules such as amino acids. Rain trapped and then washed these molecules into the oceans, making a primordial soup of organic molecules. In this soup, proteins, lipids, and other complex organic molecules found in present-day cells formed. Harold Urey and Stanley Miller provided the first experimental evidence to support this idea.

Bubbles In 1986, Louis Lerman proposed that the chemical reactions of the primordial soup took place inside tiny bubbles of lipid molecules, where they occurred more quickly. Wind, waves, and rainfall created the bubbles.

An RNA world Some scientists propose that the formation of self-replicating molecules preceded the formation of cells. Today's self-replicating molecules, DNA and RNA, provide clues about the earliest self-replicating molecules. Scientists propose that RNA, which is central to the functioning of a cell, probably predated DNA on Earth. However, because RNA is a more complex molecule than protein, it is not easy to obtain data that supports the idea that RNA could have been formed on early Earth.

INVESTIGATING THE ISSUE

Thinking Critically How do the primordial soup and the bubble views of life's origin on Earth support each other? What is the strength of the RNA world hypothesis?

BIOLOGY Online To find out more about the origin of life, visit the Glencoe Science Web site. **science.glencoe.com**

Stanley Miller

BIOLOGY JOURNAL

Exploring Life's Origins

Linguistic Have students record their opinions on research about the origin of life. Ask if the research of Oparin, Miller and Urey, Fox, and others is conclusive, or should more work be done in this area? Students can use evidence discussed in the text or in class to support their arguments. **L2**

MEETING INDIVIDUAL NEEDS

Gifted

Linguistic Have students research other experiments about the origin of life. Have them summarize in their journals the experiments they research. Ask them to evaluate the evidence from each experiment in light of other research findings. **L3**

SUMMARY

Section 14.1

The Record of Life

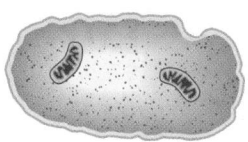

Main Ideas

- Fossils provide a record of life on Earth. Fossils come in many forms, such as the imprint of a leaf, the burrow of a worm, or an actual bone.
- By studying fossils, scientists learn about the diversity of life and about the behavior of ancient organisms and their environments.
- Earth's history is divided into the four eras of the Geologic Time Scale, a calendar of Earth's history based on evidence in rocks and fossils.

Vocabulary

fossil (p. 378)
plate tectonics (p. 387)

Section 14.2

The Origin of Life

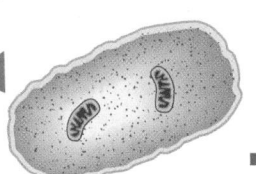

Main Ideas

- Francesco Redi and Louis Pasteur designed controlled experiments to disprove spontaneous generation. Their experiments and others like them convinced scientists to accept biogenesis.
- Small organic molecules might have formed from substances present in Earth's early atmosphere and oceans. Small organic molecules can form complex organic molecules.
- The earliest organisms were probably anaerobic, heterotrophic prokaryotes. Over time, chemosynthetic prokaryotes evolved and then photosynthetic prokaryotes that produced oxygen, changing the atmosphere and triggering the evolution of aerobic cells and eukaryotes.
- The endosymbiont hypothesis proposes that eukaryotes evolved through a symbiotic relationship between prokaryotes.

Vocabulary

archaebacteria (p. 392)
biogenesis (p. 389)
protocell (p. 391)
spontaneous generation (p. 388)

Main Ideas

Summary statements can be used by students to review the major concepts of the chapter.

Using the Vocabulary

To reinforce chapter vocabulary, use the Content Mastery Booklet and the activities in the Interactive Tutor for Biology: The Dynamics of Life on the Glencoe Science Web site.
science.glencoe.com

THE PRINCETON REVIEW

All Chapter Assessment questions and answers have been validated for accuracy and suitability by The Princeton Review.

UNDERSTANDING MAIN IDEAS

1. c
2. a
3. b
4. c

UNDERSTANDING MAIN IDEAS

1. About how many years ago do scientists suggest that Earth cooled enough for water vapor to condense?
   a. 20 million years   **c.** 3.9 billion years
   b. 4.6 billion years   **d.** 5.5 billion years

2. Most fossils occur in layers of _____ rocks.
   a. sedimentary   **c.** igneous
   b. metamorphic   **d.** volcanic

3. The endosymbiont theory suggests how _____ evolved from _____.
   a. protocells—organelles
   b. eukaryotes—prokaryotes
   c. archaebacteria—eubacteria
   d. prokaryotes—protocells

4. Who was the scientist who showed that microscopic life is not produced by spontaneous generation?
   a. Francesco Redi   **c.** Louis Pasteur
   b. Stanley Miller   **d.** Harold Urey

GLENCOE TECHNOLOGY

VIDEOTAPE
MindJogger Videoquizzes
Chapter 14: *The History of Life*
Have students work in groups as they play the videoquiz game to review key chapter concepts.

Resource Manager

Chapter Assessment, pp. 79-84
MindJogger Videoquizzes
ExamView® Pro Software
BDOL Interactive CD-ROM,
   Chapter 14 quiz

5. a
6. b
7. c
8. d
9. b
10. b
11. trace fossils
12. archaebacteria
13. unicellular prokaryotes
14. Mesozoic
15. protocell
16. spontaneous generation
17. fossil
18. Precambrian
19. plate tectonics
20. biogenesis

APPLYING MAIN IDEAS

21. Because fossils are found in sedimentary rock, paleontologists need to know the rock's history.
22. From fossils, scientists may be able to determine whether animals lived in large groups, what they ate, how they obtained their food, and what their home ranges were.

5. The Geologic Time Scale is based on
   _____.
   a. different organisms that appeared during Earth's history
   b. various rock layers that occur in Europe and Canada
   c. landforms such as mountains and faults in California
   d. when oceans and seas formed

6. Shallow seas that teemed with life could first be found _____.
   a. in the Precambrian era
   b. in the Paleozoic era
   c. in the Mesozoic era
   d. 6.5 billion years ago

7. The mass of continents in the diagram shown here is called _____.

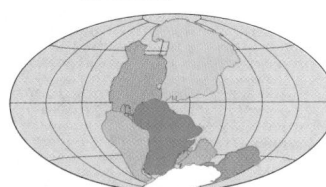

   a. Laurasia   c. Pangaea
   b. Gondwana   d. North America

8. An entire, intact organism may be preserved in _____ and _____.
   a. casts—trace fossils
   b. molds—casts
   c. imprints—petrified fossils
   d. amber—ice

9. Scientists theorize that oxygen buildup in the atmosphere resulted from _____.
   a. respiration   c. chemosynthesis
   b. photosynthesis   d. rock weathering

THE PRINCETON REVIEW — **TEST-TAKING TIP**

Do Some Reconnaissance
Find out what the conditions will be for taking the test. Is it timed? Will you be allowed a break? Know these things in advance so that you can practice taking tests under the same conditions.

10. Which of the following cell organelles may have evolved from a plant cell's symbiotic association with a photosynthetic bacterium?

   a. 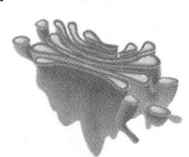   c.

   b.   d.

11. Fossil markings such as footprints or animal burrows are called _____.
12. The group of prokaryotes that live in harsh conditions, such as the near-boiling water in hot springs, is _____.
13. Fossils of organisms found in the early Precambriam are all of _____.
14. Dinosaurs, mammals, and flowering plants appeared during the _____ era.
15. Sidney Fox showed that heating solutions of amino acids can produce a large, associated structure called a _____.
16. The idea of _____ suggests that nonliving matter can produce living things.
17. Any evidence of an organism that lived long ago is a(n) _____.
18. The earliest evidence of life so far recorded is found in rocks dated to the _____ era.
19. The geological explanation for how continents move is _____.
20. The idea that life comes only from preexisting life is _____.

APPLYING MAIN IDEAS

21. Why is knowledge of geology important to paleontologists?
22. Explain how fossils might help paleontologists to learn about the important behaviors of different types of animals. Which social behaviors might they provide information about?

23. How might the way organisms obtain energy have evolved over time?

24. Explain why there might be similar fossils on the east coast of South America and the west coast of Africa.

THINKING CRITICALLY

25. Observing and Inferring Why are amber-preserved and frozen fossils of great value to scientists?

26. Comparing and Contrasting What details about an organism might casts show that are not displayed in molds?

27. Formulating Hypotheses Why do scientists propose that the 3.5 billion-year-old fossils of cyanobacteria-like prokaryotic cells found in Australia, were not the first species to have evolved on Earth?

28. Measuring in SI Assume that a particular radioactive element has a half-life of 1 year. If 1000.0 g of the radioactive element are present in a sample today, how much will be in the sample after 10 years?

29. Concept Mapping Complete the concept map by using the following vocabulary terms: archaebacteria, biogenesis, spontaneous generation, protocells.

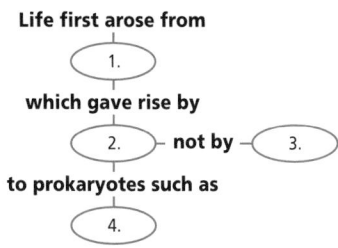

Life first arose from

1.

which gave rise by

2. **— not by —** 3.

to prokaryotes such as

4.

CD-ROM

For additional review, use the assessment options for this chapter found on the **Biology: The Dynamics of Life Interactive CD-ROM** and on the Glencoe Science Web site.
science.glencoe.com

ASSESSING KNOWLEDGE & SKILLS

The following graph illustrates the decay rate of a particular radioactive element.

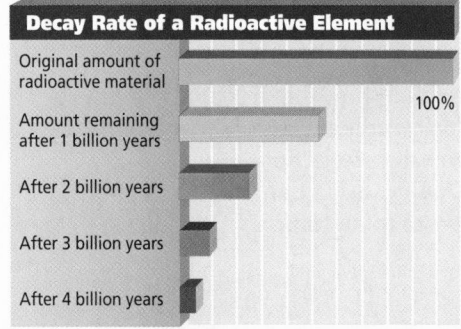

Decay Rate of a Radioactive Element

Original amount of radioactive material — 100%
Amount remaining after 1 billion years
After 2 billion years
After 3 billion years
After 4 billion years

Interpreting Data Study the graph and answer the following questions.

1. How long does it take for half of the element to decay?
a. 1 billion years **c.** 3 billion years
b. 2 billion years **d.** 4 billion years

2. How much of the original material is left after 2 billion years?
a. 100% **c.** 25%
b. 50% **d.** 12.5%

3. How much of the original material is left after 4 billion years?
a. 50% **c.** 12.5%
b. 25% **d.** less than 10%

4. This element would best be used to date fossils that are _____ years old.
a. a few thousand **c.** a few million
b. less than a million **d.** a billion

5. Interpreting Scientific Illustrations Use both **Figure 14.4** and the above graph to identify the era in which this radioactive element would be the most useful in dating fossils.

23. The earliest organisms probably were heterotrophic, obtaining energy through anaerobic processes. Some later anaerobic organisms may have been autotrophic, increasing the amount of oxygen in the atmosphere. The presence of oxygen in the atmosphere and the evolution of cells with mitochondria allowed organisms to use oxygen in aerobic respiration.

24. The theory of plate tectonics explains that the two continents were once joined.

THINKING CRITICALLY

25. The entire organism is preserved, including its soft tissues.

26. Casts are three-dimensional representations of molds and show more details, such as surface texture, than do molds.

27. The earliest organisms were probably anaerobic heterotrophs, not autotrophs.

28. 0.9766 grams

29. 1. Protocells; 2. Spontaneous generation; 3. Biogenesis; 4. Archaebacteria

ASSESSING KNOWLEDGE & SKILLS

1. a
2. c
3. d
4. d
5. the Precambrian era

Refer to pages 4T-5T of the Teacher Guide for an explanation of the National Science Education Standards correlations.

Section	Objectives	Activities/Features
Section 15.1 **Natural Selection and the Evidence for Evolution** National Science Education Standards UCP.1-5; A.1, A.2; C.3, C.4, C.6; F.4; G.1, G.3 (2 sessions, 1 block)	1. **Summarize** Darwin's theory of natural selection. 2. **Explain** how the structural and physiological adaptations of organisms relate to natural selection. 3. **Distinguish** among the types of evidence for evolution.	**MiniLab 15-1:** Camouflage Provides an Adaptive Advantage, p. 406 **Problem-Solving Lab 15-1,** p. 407
Section 15.2 **Mechanisms of Evolution** National Science Education Standards UCP.1-5; A.1, A.2; C.1-4, C.6; F.4; G.1-3 (3 sessions, 2 blocks)	4. **Summarize** the effects of the different types of natural selection on gene pools. 5. **Relate** changes in genetic equilibrium to mechanisms of speciation. 6. **Explain** the role of natural selection in convergent and divergent evolution.	**MiniLab 15-2:** Detecting a Variation, p. 415 **Internet BioLab:** Natural Selection and Allelic Frequency, p. 422 **Math Connection:** Mathematics and Evolution, p. 424

Need Materials? Contact Carolina Biological Supply Company at 1-800-334-5551 or at **http://www.carolina.com**

MATERIALS LIST

BioLab

p. 422 colored pencils (2), paper bag, graph paper, pinto beans, white navy beans

MiniLabs

p. 406 hole punch, paper, white and black
p. 415 ruler, unshelled peanuts (30)

Alternative Lab

p. 414 culture of *Bacillus subtilis*, 3 tubes of nutrient agar, tube of streptomycin agar, inoculation loop, petri dishes (2), Bunsen burner, wax pencil, test tube

Quick Demos

p. 402 photographs of automobile model
p. 418 overhead projector

Key to Teaching Strategies

L1 Level 1 activities should be appropriate for students with learning difficulties.

L2 Level 2 activities should be within the ability range of all students.

L3 Level 3 activities are designed for above-average students.

ELL ELL activities should be within the ability range of English Language Learners.

COOP LEARN Cooperative Learning activities are designed for small group work.

P These strategies represent student products that can be placed into a best-work portfolio.

These strategies are useful in a block scheduling format.

Teacher Classroom Resources

Section	Reproducible Masters	Transparencies
Section 15.1 **Natural Selection and the Evidence for Evolution**	Reinforcement and Study Guide, pp. 65-66 **L2** Concept Mapping, p. 15 **L3** **ELL** Critical Thinking/Problem Solving, p. 15 **L3** BioLab and MiniLab Worksheets, p. 71 **L2** Laboratory Manual, pp. 103-108 **L2** Content Mastery, pp. 73-74, 76 **L1**	Section Focus Transparency 37 **L1** **ELL**
Section 15.2 **Mechanisms of Evolution**	Reinforcement and Study Guide, pp. 67-68 **L2** Critical Thinking/Problem Solving, p. 15 **L3** BioLab and MiniLab Worksheets, pp. 72-74 **L2** Content Mastery, pp. 73, 75-76 **L1**	Section Focus Transparency 38 **L1** **ELL** Basic Concepts Transparency 21 **L2** **ELL** Basic Concepts Transparency 22 **L2** **ELL** Reteaching Skills Transparency 24 **L1** **ELL**

Assessment Resources

Chapter Assessment, pp. 85-90
MindJogger Videoquizzes
Performance Assessment in the Biology Classroom
Alternate Assessment in the Science Classroom
ExamView® Pro Software
BDOL Interactive CD-ROM, Chapter 15 quiz

Additional Resources

Spanish Resources **ELL**
English/Spanish Audiocassettes **ELL**
Cooperative Learning in the Science Classroom **COOP LEARN**
Lesson Plans/Block Scheduling

NATIONAL GEOGRAPHIC

Teacher's Corner

Products Available From National Geographic Society
To order the following products, call National Geographic Society at 1-800-368-2728:
Book
National Geographic Atlas of World History

Index to National Geographic Magazine
The following articles may be used for research relating to this chapter:
"The Dawn of Humans: Redrawing Our Family Tree?" by Lee Berger, August 1998.
"Dinosaurs Take Wing," by Jennifer Ackerman, July 1998.
"A Curious Kinship: Apes and Humans," by Eugene Linden, May 1992.

GLENCOE TECHNOLOGY

The following multimedia resources are available from Glencoe.

Biology: The Dynamics of Life
CD-ROM **ELL**
 Video: *Galapagos*
Video: *Adapted for Survival*
Exploration: *The Record of Life*
Exploration: *Selection Pressure*

Videodisc Program
 Geographic Isolation
Adapted for Survival

The Infinite Voyage
The Great Dinosaur Hunt

The Secret of Life Series
 It's in the Genes: *Evolution*
Camouflage: *Caterpillars*
Camouflage: *Spider*
Horse Evolution
Patterns of Descent
Gone Before You Know It: *The Biodiversity Crisis*

15 The Theory of Evolution

Chapter 15

GETTING STARTED DEMO

Kinesthetic Blindfold a student volunteer. Then place a couple of small, familiar objects, such as a stapler and a pen, on a desk and ask the volunteer to identify the objects. Point out to the class that some organisms that live in dark caves successfully use senses other than sight to monitor their environments. **L1** **ELL**

Theme Development

The **unity within diversity** theme is apparent in this chapter. The theme of **evolution** is also evident. The theory of evolution can explain the diversity of organisms.

0:00 OUT OF TIME?

If time does not permit teaching the entire chapter, use the BioDigest at the end of the unit as an overview.

READING BIOLOGY

Glencoe's *Biology: The Dynamics of Life* contains many resources to assist a student's reading skills. Each chapter contains figures with expanded captions that expand on written material. Word Origins, located along the side of text, expand knowledge of biology vocabulary. Glencoe's Content Mastery Booklet helps develop reading skills while reinforcing content. In addition, use the Interactive Tutor for *Biology: The Dynamics of Life* on the Glencoe Web site to reinforce vocabulary. **science.glencoe.com**

What You'll Learn

- You will analyze the theory of evolution.
- You will compare and contrast the processes of evolution.

Why It's Important

Evolution is a key concept for understanding biology. Evolution explains the diversity of species and predicts changes.

READING BIOLOGY

Look closely at the figures that appear in the chapter. For several of the drawings, make notes on how the figure shows an aspect of evolution. As you read the chapter, review the figures and add new information to your notes.

To find out more about evolution, visit the Glencoe Science Web site.
science.glencoe.com

This crayfish (above) and cricket (inset) live in dark caves and are blind. They have sighted relatives that live where there is light. Both the cave-dwelling species and their relatives are adapted to different environments. As populations adapt to new or changing environments, individuals in the population that are adapted successfully survive.

400 THE THEORY OF EVOLUTION

Multiple Learning Styles

Look for the following logos for strategies that emphasize different learning modalities.

Kinesthetic Getting Started Demo, p. 400; Visual Learning, p. 413

Visual-Spatial Reteach, p. 421

Interpersonal Project, pp. 408, 417; Activity, p. 411

Intrapersonal Reteach, p. 410

Linguistic Portfolio, pp. 402, 406, 407; Meeting Individual Needs, pp. 403, 419; Biology Journal, pp. 405, 413, 416; Extension, p. 421

Logical-Mathematical Project, p. 404; Portfolio, p. 410; Tech Prep, p. 412; Reinforcement, p. 416

Naturalist Meeting Individual Needs, p. 409

15.1 Natural Selection and the Evidence for Evolution

Y*ou need only to look around you to see the diversity of organisms on Earth. About 150 years ago, Charles Darwin, who had studied an enormous variety of life forms, proposed an idea to explain how organisms probably change over time. Biologists still base their work on this idea because it explains the living world they study.*

An Asian leopard and a cheetah (inset)

SECTION PREVIEW

Objectives

Summarize Darwin's theory of natural selection.

Explain how the structural and physiological adaptations of organisms relate to natural selection.

Distinguish among the types of evidence for evolution.

Vocabulary

artificial selection
natural selection
mimicry
camouflage
homologous structure
analogous structure
vestigial structure
embryo

Charles Darwin and Natural Selection

The modern theory of evolution is a fundamental concept in biology. Recall that evolution is the change in populations over time. Learning the principles of evolution makes it easier to understand modern biology. One place to start is by learning about the ideas of English scientist Charles Darwin (1809–1882)—ideas supported by fossil evidence.

Fossils shape ideas about evolution

Biologists have used fossils in their work since the eighteenth century. In fact, fossil evidence formed the basis of the early evolutionary concepts.

Scientists wondered how fossils formed, why many fossil species were extinct, and what kinds of relationships might exist between the extinct and the modern species.

When geologists provided evidence indicating that Earth was much older than many people had originally thought, biologists began to suspect that life slowly changes over time, or evolves. Many explanations about how species evolve have been proposed, but the ideas first published by Charles Darwin are the basis of modern evolutionary theory.

Darwin on HMS *Beagle*

It took Darwin years to develop his theory of evolution. He began in 1831 at age 21 when he took a job as

---

Key Concepts

Students will study Charles Darwin's concept of natural selection. They will also learn about scientific evidence that supports the theory of evolution.

Planning

- Collect photos of automobiles for the Quick Demo.
- Purchase pinto beans for the Project.
- Obtain black and white construction paper and paper punches for MiniLab 15-1.
- Obtain bird bones (chicken, turkey, quail) for the Display.
- Gather photos of a variety of organisms for the Activity.

---

1 Focus

Bellringer

Before presenting the lesson, display **Section Focus Transparency 37** on the overhead projector and have students answer the accompanying questions.
L1 **ELL**

Transparency **37** Camouflage

SECTION FOCUS
Use with Chapter 15, Section 15.1

Snowshoe hare in summer

Snowshoe hare in winter

❶ What is the advantage of this snowshoe hare's seasonal color change?

❷ The adaptation that allows an animal to blend in with its environment is called camouflage. What examples of camouflage are you familiar with?

BIOLOGY: The Dynamics of Life   SECTION FOCUS TRANSPARENCIES

✓ Assessment Planner

Portfolio Assessment

Portfolio, TWE, pp. 402, 406, 407, 410, 420
Assessment, TWE, pp. 411, 419
MiniLab, TWE, p. 415

Performance Assessment

Assessment, TWE, pp. 403, 408
MiniLabs, SE, pp. 406, 415
Alternative Lab, TWE, pp. 414-415
BioLab, SE, pp. 422–423

Knowledge Assessment

MiniLab, TWE, p. 406
Problem-Solving Lab, TWE, p. 407
Alternative Lab, TWE, p. 415
BioLab, TWE, p. 423
Section Assessments, SE, pp. 411, 421
Chapter Assessment, SE, pp. 425-427

Skill Assessment

Assessment, TWE, p. 421

2 Teach

Concept Development

A significant influence on Darwin's thinking was the book *The Principles of Geology* by Charles Lyell. This book proposed that Earth is very old and that the forces that have produced changes on Earth's surface in the past are the same ones that continue to operate today. Discuss how Darwin was influenced by other ideas of his day.

Visual Learning

Figure 15.1 Have the students examine the photos of the finch, tortoise, and iguana. Discuss each organism, asking students to identify its adaptations.

a naturalist on the English ship HMS *Beagle*, which sailed to South America and the South Pacific on a five-year scientific journey.

As the ship's naturalist, Darwin studied and collected biological specimens at every port along the route. As you might imagine, these specimens were quite diverse. Studying the specimens made Darwin curious about possible relationships among species. His studies provided the foundation for his theory of evolution by natural selection.

Darwin in the Galapagos

The Galapagos (guh LAHP uh gus) Islands are a group of small islands near the equator, about 1000 km off the west coast of South America. The observations that Darwin made and the specimens that he collected there were especially important to him.

On the Galapagos Islands, Darwin studied many species of animals and plants, *Figure 15.1*, that are unique to the islands, but similar to species elsewhere. These observations led Darwin to consider the possibility that species can change over time. However, after returning to England, he could not at first explain how such changes occur.

Darwin continues his studies

For the next 22 years, Darwin worked to find an explanation for how species change over time. He read, studied, collected specimens, and conducted experiments.

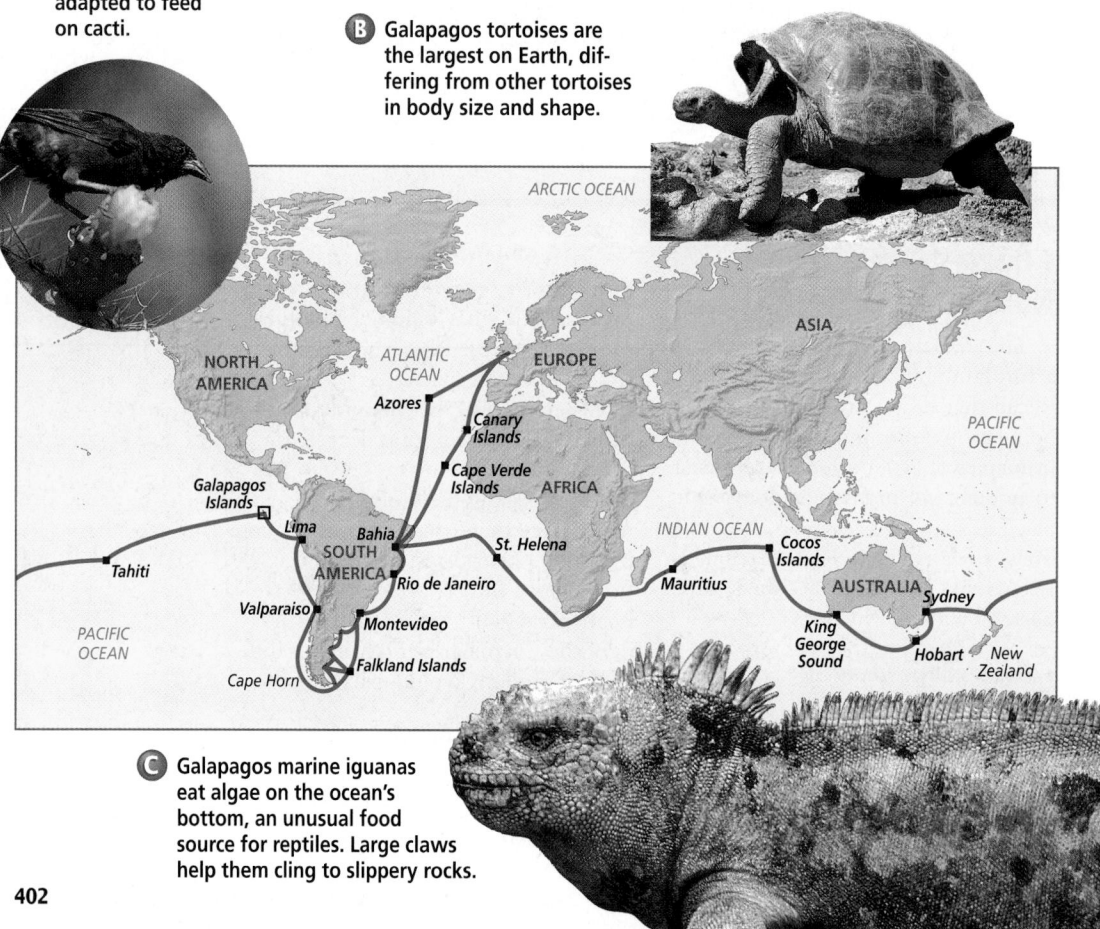

Figure 15.1
The five-year voyage of HMS *Beagle* took Darwin around the world. Animal species in the Galapagos Islands have unique adaptations.

A The beak of this Galapagos finch is adapted to feed on cacti.

B Galapagos tortoises are the largest on Earth, differing from other tortoises in body size and shape.

C Galapagos marine iguanas eat algae on the ocean's bottom, an unusual food source for reptiles. Large claws help them cling to slippery rocks.

402

Finally, English economist Thomas Malthus proposed an idea that Darwin modified and used in his explanation. Malthus's idea was that the human population grows faster than Earth's food supply. How did this help Darwin? He knew that many species produce large numbers of offspring. He also knew that such species had not overrun Earth. He realized that individuals struggle to survive. There are many kinds of struggles, such as competing for food and space, escaping from predators, finding mates, and locating shelter. Only some individuals survive the struggle and produce offspring. Which individuals survive?

Darwin gained insight into the mechanism that determined which organisms survive in nature from his pigeon-breeding experiments. Darwin observed that the traits of individuals vary in populations—even in a population of pigeons. Sometimes variations are inherited. By breeding pigeons with desirable variations, Darwin produced offspring with these variations. Breeding organisms with specific traits in order to produce offspring with identical traits is called **artificial selection.** Darwin hypothesized that there was a force in nature that worked like artificial selection.

Darwin explains natural selection

Using his collections and observations, Darwin identified the process of natural selection, the steps of which you can see summarized in *Figure 15.2.* **Natural selection** is a mechanism for change in populations. It occurs when organisms with certain variations survive, reproduce, and pass their variations to the next generation. Organisms without these variations are less likely to survive and reproduce. As a result, each generation consists largely of offspring from parents with these variations that aid survival.

Figure 15.2
Darwin proposed the idea of natural selection to explain how species change over time.

A In nature, organisms produce more offspring than can survive. Fishes, for example, can sometimes lay millions of eggs.

B In any population, individuals have variations. Fishes, for example, may differ in color, size, and speed.

C Individuals with certain useful variations, such as speed, survive in their environment, passing those variations to the next generation.

D Over time, offspring with certain variations make up most of the population and may look entirely different from their ancestors.

Concept Development

Before Darwin developed his theory of evolution by natural selection, French biologist Jean-Baptiste de Lamarck (1744-1829) proposed a different mechanism for evolutionary change. Lamarck's idea rested on two assumptions: (1) the more an organism uses a part of its body, the more that part develops, and (2) the physical characteristics that an organism develops in this way can be passed to offspring. Discuss Lamarck's hypothesis with students, asking them to list its weaknesses.

Visual Learning

Figure 15.2 shows the four principal ideas of natural selection. Discuss each principle to reinforce the ideas. Provide other examples of natural selection, using alternative organisms and habitats, to review the concept.

✔ Assessment

Performance Assessment in the Biology Classroom, p. 23, *Investigating Variations in Populations.* Have students carry out this activity to explore what variations occur in a population. **L2**

Different Viewpoints in Biology

Provide students with a set of class data, such as the data they gathered in this chapter's Getting Started. Ask different students to interpret the data to show how the same information can be interpreted differently. **L2**

Visual Learning

Figure 15.3 illustrates the probable evolution of the common mole-rat from a member of the rodent family Bathyergidae. After students have studied each step of the illustration, ask them to list the steps that may have occurred during the evolution of the sightless, cave-dwelling fish genus *Amblyopsis*, and the blind, burrowing snake genus *Typhlops*.

Darwin was not the only one to recognize the significance of natural selection for populations. As a result of his studies on islands near Indonesia in the Pacific Ocean, Alfred Russell Wallace, another British naturalist, had reached a similar conclusion. After Wallace wrote Darwin to share his ideas about natural selection, Darwin and Wallace had their similar ideas jointly presented to the scientific community. However, it was Darwin who published the first book about evolution called *On the Origin of Species by Natural Selection* in 1859. The ideas detailed in Darwin's book are today a basic unifying theme of biology.

Interpreting evidence after Darwin

Volumes of scientific data have been gathered as evidence for evolution since Darwin's time. Much of this evidence is subject to interpretation by different scientists. One of the problems is that evolutionary processes are difficult for humans to observe directly. The short scale of human life spans makes it difficult to comprehend evolutionary processes that occur over millions of years. For some people the theory of evolution is contradictory to their faith, and they offer other interpretations of the data. Many biologists, however, have suggested that the amount of scientific evidence supporting the theory of evolution is overwhelming. Almost all of today's biologists accept the theory of evolution by natural selection. However, biologists are also now more aware of genetics. Evolution is more commonly defined by modern biologists as any change in the gene pool of a population.

Adaptations: Evidence for Evolution

Have you noticed that some plants have thorns and some plants don't?

A The ancestors of today's common mole-rats probably resembled African rock rats.

B Some ancestral rats may have avoided predators better than others because of variations such as the size of teeth and claws.

PROJECT

Variation in Beans

 Logical-Mathematical Students can study the effects of individual variations by planting a pinto bean garden. Have them wash their hands after handling bean seeds. Obtain some pinto bean seeds and ask the students to measure and observe them, placing the seeds into categories, such as short, long, wide, thin, etc. Have them write hypotheses that predict how each category of bean seed will grow. Then plant 3 or 4 beans from each category. Students should observe the plants each day, recording their observations. Have them write a brief summary after 4-5 weeks of plant growth. **L1**
ELL

Have you noticed that some animals have distinctive coloring but others don't? Have you ever wondered how such variations arose? Recall that an adaptation is any variation that aids an organism's chances of survival in its environment. Thorns are an adaptation of some plants and distinctive colorings are an adaptation of some animals. Darwin's theory of evolution explains how adaptations may develop in species.

Structural adaptations arise over time

According to Darwin's theory, adaptations in species develop over many generations. Learning about adaptations in mole-rats can help you understand how natural selection has affected them. Mole-rats that live underground in darkness are blind. These blind mole-rats have many adaptations that enable them to live successfully underground. Look at *Figure 15.3* to see how

these modern mole-rat adaptations might have evolved over millions of years from characteristics of their ancestors.

The structural adaptations of common mole-rats include large teeth and claws. These are body parts that help mole-rats survive in their environment by, for example, enabling them to dig better tunnels. Structural adaptations such as the teeth and claws of mole-rats are often used to defend against predators. Some adaptations of other organisms that keep predators from approaching include a rose's thorns or a porcupine's quills.

Some other structural adaptations are subtle. **Mimicry** is a structural adaptation that enables one species to resemble another species. In one form of mimicry, a harmless species has adaptations that result in a physical resemblance to a harmful species. Predators that avoid the harmful species also avoid the similar-looking, harmless species. See if you can tell

Figure 15.3
Darwin's ideas about natural selection can explain some adaptations of mole-rats.

C Ancestral rats that survived passed their variations to offspring. After many generations, most of the population's individuals would have these adaptations.

D Over time, natural selection produced modern mole-rats. Their blindness may have evolved because vision had no survival advantage for them.

Enrichment

Using Figure 15.3 as a model, have the students illustrate or describe possible evolutionary sequences of one of the following: (1) the evolution of long necks in giraffes from short-necked ancestors, (2) the evolution of whales from terrestrial carnivores, (3) the evolution of flight in birds from bipedal dinosaurs, (4) the evolution of high-speed running in cheetahs from slower movements of their ancestors.

GLENCOE
TECHNOLOGY

VIDEOTAPE
The Secret of Life
It's in the Genes: Evolution

Resource Manager

Concept Mapping, p. 15 **L3**
ELL
Laboratory Manual, pp. 105-108 **L2**

BIOLOGY JOURNAL

Evidence for Natural Selection

Linguistic Have students describe the main evidence Darwin used in formulating his concept of natural selection. Next, have them select an organism and, in their own words, use the main ideas of the concept of natural selection to explain the evolution of the organism.
L3

BIOLOGY *Online* Note Internet addresses that you find useful in the space below for quick reference.

_____

_____

_____

_____

_____

405

Purpose
Students will model how a camouflage adaptation can aid an organism's survival.

Process Skills
observe and infer, form a hypothesis

Teaching Strategies
■ Have students do this activity after studying camouflage.

■ Explain that students will simulate how natural selection might operate on a population of insects that vary in color.

Expected Results
Most groups will have picked up more white dots than black dots.

Analysis
1. white dots
2. Light-colored insects may be seen and preyed on more easily than dark-colored insects. Therefore, dark-colored insects have a higher survival rate.
3. Over time, an insect population might become dark-colored because light-colored insects were eliminated from the population.

MiniLab 15-1   Formulating Models

Camouflage Provides an Adaptive Advantage
Camouflage is a structural adaptation that allows organisms to blend with their surroundings. In this activity, you'll discover how natural selection can result in camouflage adaptations in organisms.

Procedure
1. Working with a partner, punch 100 dots from a sheet of white paper with a paper hole punch. Repeat with a sheet of black paper. These dots will represent black and white insects.
2. Scatter both white and black dots on a sheet of black paper.
3. Decide whether you or your partner will role-play a bird.
4. The "bird" looks away from the paper, then turns back, and immediately picks up the first dot he or she sees.
5. Repeat step 4 for one minute.

Analysis
1. What color dots were most often collected?
2. How does color affect the survival rate of insects?
3. What might happen over many generations to a similar population in nature?

Figure 15.4
Mimicry and camouflage are protective adaptations of organisms. The colors and body shape of a yellow jacket wasp (a) and a harmless syrphid fly (b) are similar. Predators avoid both insects. Camouflage enables organisms, such as this leaf frog (c), to blend with their surroundings.

406

---

the difference between a harmless fly and the wasp it mimics when you look at *Figure 15.4*.

In another form of mimicry, two or more harmful species resemble each other. For example, yellow jacket hornets, honeybees, and many other species of wasps all have harmful stings and similar coloration and behavior. Predators may learn quickly to avoid any organism with their general appearance.

Another subtle adaptation is **camouflage** (KAM uh flahj), an adaptation that enables species to blend with their surroundings, as shown in *Figure 15.4*. Because well-camouflaged organisms are not easily found by predators, they survive to reproduce. Try the *MiniLab* to experience how camouflage can help an organism survive. Then use the *Problem-Solving Lab* on the next page to analyze data from an English study of camouflaged peppered moths.

✔ Portfolio

Camouflage and Mimicry

Linguistic Have the students write about an organism that has camouflage or mimicry adaptations. The report should include the organism's name, details about its environment and predators, and a description of its camouflage or mimicry adaptations. **L3** **P**

GLENCOE TECHNOLOGY

VIDEODISC
The Secret of Life
Camouflage: Caterpillars

Camouflage: Spider

Physiological adaptations can develop rapidly

In general, most structural adaptations develop over millions of years. However, there are some adaptations that evolve much more rapidly. For example, do you know that some of the medicines developed during the twentieth century to fight bacterial diseases are no longer effective? When the antibiotic drug penicillin was discovered about 50 years ago, it was called a wonder drug because it killed many types of disease-causing bacteria and saved many lives. Today, penicillin no longer affects as many species of bacteria because some species have evolved physiological (fihz ee uh LAHJ ih kul) adaptations to prevent being killed by penicillin. Look at *Figure 15.5* to see how resistance develops in bacteria.

Physiological adaptations are changes in an organism's metabolic processes. In addition to species of bacteria, scientists have observed these adaptations in species of insects and weeds that are pests. After years of exposure to specific pesticides, many species of insects and weeds have become resistant to these chemicals that used to kill them.

Figure 15.5
The development of bacterial resistance to antibiotics is direct evidence for evolution.

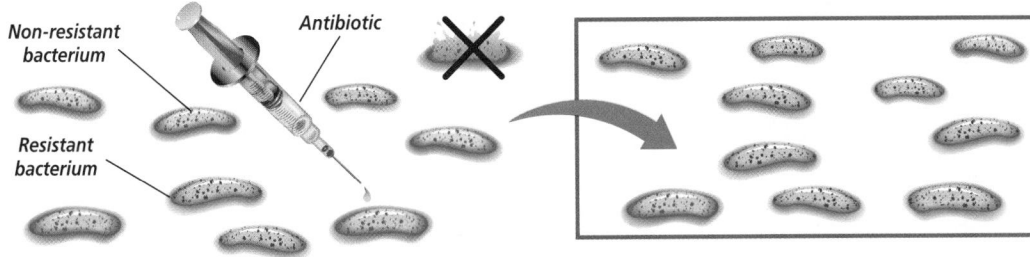

A The bacteria in a population vary in their ability to resist antibiotics.

B When the population is exposed to an antibiotic, only the resistant bacteria survive.

C The resistant bacteria live and produce more resistant bacteria.

Problem-Solving Lab 15-1 — Interpreting Data

How can natural selection be observed? In some organisms that have a short life cycle, biologists have observed the evolution of adaptations to rapid environmental changes. In the early 1950s, English biologist H. B. Kettlewell studied camouflage adaptations in a population of light- and dark-colored peppered moths, *Biston betularia.* The moths rested on the trunks of trees that grew in both the country and the city. Moths are usually speckled gray-brown, and dark moths, which occur occasionally, are black. Birds pluck the moths from the trees for food. Urban industrial pollution had blackened the bark of city trees with soot. In the photo, you see a city tree with dark bark similar to the color of one of the moths.

Biston betularia

Analysis

Kettlewell raised more than 3000 caterpillars to provide adult moths. He marked the wings of the moths these caterpillars produced so he would recapture only his moths. In a series of trials in the country and the city, he released and recaptured the moths. The number of moths recaptured in a trial indicates how well the moths survived in the environment. Examine the table below.

Table 15.1 Comparison of country and city moths

Location		Numbers of light moths	Numbers of dark moths
Country	Released	496	488
	Recaptured	62	34
City	Released	137	493
	Recaptured	18	136

Thinking Critically

Calculate the percentage of moths recaptured in each experiment and explain any differences in survival rates in the country and the city moths in terms of natural selection.

Other Evidence for Evolution

The development of physiological resistance in species of bacteria, insects, and plants is direct evidence of evolution. However, most of the evidence for evolution is indirect, coming from sources such as fossils and studies of anatomy, embryology, and biochemistry.

Fossils

Fossils are an important source of evolutionary evidence because they provide a record of early life and evolutionary history. For example, paleontologists conclude from fossils that the ancestors of whales were probably land-dwelling, doglike animals.

Although the fossil record provides evidence that evolution occurred, the record is incomplete. Working with an incomplete fossil record is something like trying to put together a jigsaw puzzle with missing pieces. But, after the puzzle is together, even with missing pieces, you will probably still understand the overall picture. It's

the same with fossils. Although paleontologists do not have intermediate forms of most species, they can often still understand the overall picture of how a species evolved.

Fossils are found throughout the world. As the fossil record becomes more complete, the sequences of evolution become more clear. For example, in *Figure 15.6* you can see how paleontologists sequenced the possible forms that led to today's camel after piecing together fossil skulls, teeth, and limb bones.

Anatomy

Structural features with a common evolutionary origin are called **homologous structures.** Homologous structures can be similar in arrangement, in function, or in both. For example, look at the forelimb bones of the animals shown in *Figure 15.*7. Although the bones of each forelimb are modified for their function, the basic arrangement of the bones in each limb is similar. Evolutionary biologists view homologous structures as evidence that organisms evolved

Figure 15.6
Paleontologists have used fossils to trace the evolution of the modern camel. About 66 million years ago the ancestors of camels were as small as rabbits.

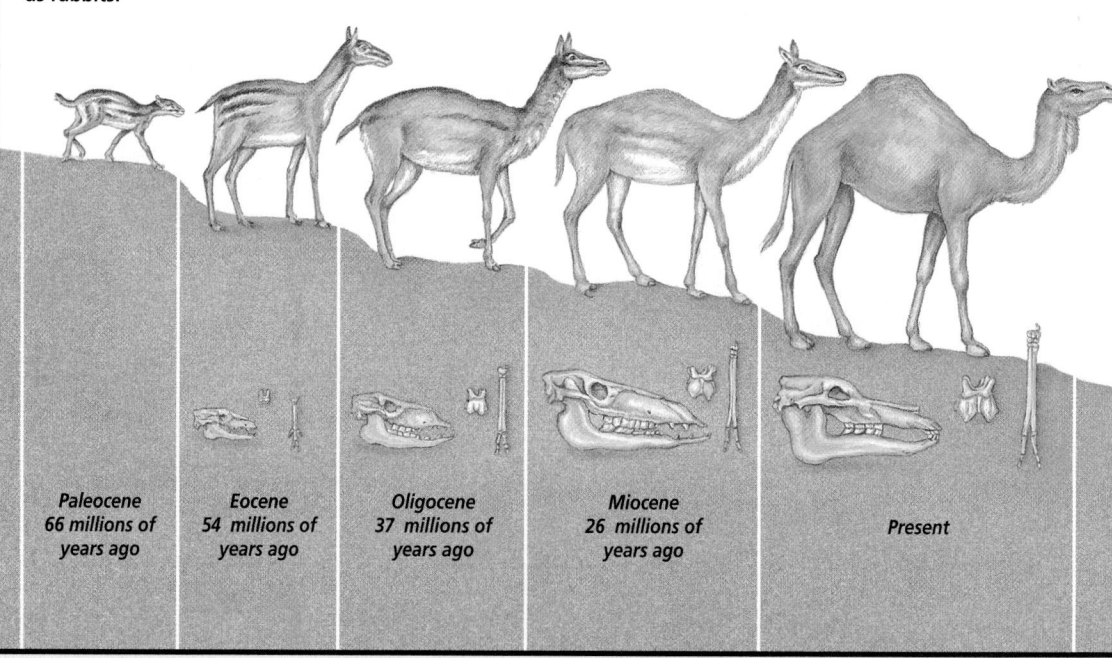

Paleocene 66 millions of years ago	*Eocene* 54 millions of years ago	*Oligocene* 37 millions of years ago	*Miocene* 26 millions of years ago	*Present*

PROJECT

Evolving Bacteria

 *Interpersonal* Have student groups research for a class presentation a bacterium that has evolved quickly to develop resistance to antibiotics. Possible bacteria include those that cause staph and strep infections, TB, and childhood ear infections.

Students should identify the bacterium, the disease it causes, how it is transmitted, and the data that suggest the bacterium is resistant to antibiotic treatment. Students can make visuals—graphs, data tables, and time lines—to illustrate their presentations. **L2**
COOP LEARN

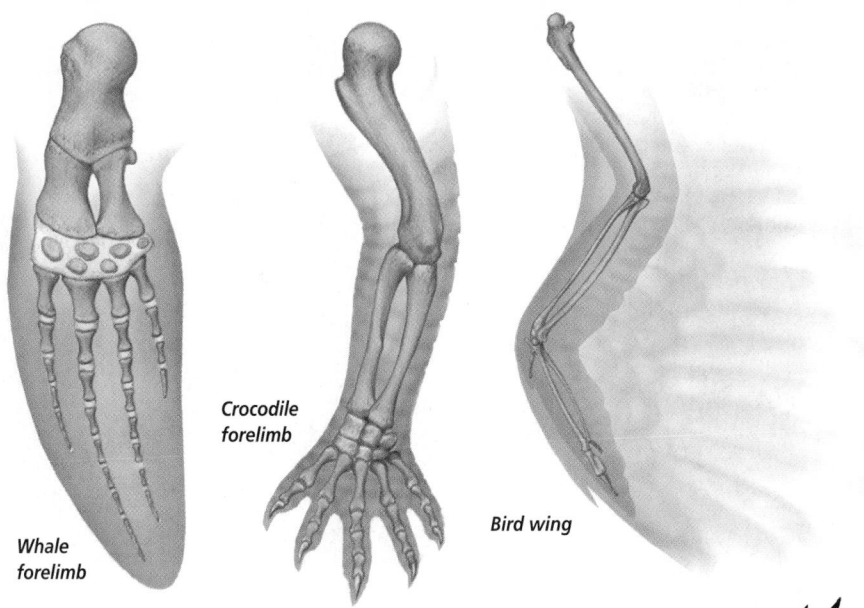

Figure 15.7
The forelimbs of crocodiles, whales, and birds are homologous structures. The bones of each are modified for their function.

Whale forelimb

Crocodile forelimb

Bird wing

from a common ancestor. It would be unlikely for so many animals to have similar structures if each species arose separately.

The structural or functional similarity of a body feature doesn't always mean that two species are closely related. In *Figure 15.8,* you can compare the wing of a butterfly with the wing of a bird. Bird and butterfly wings are not similar in structure, but they are similar in function. The wings of birds and insects evolved independently of each other in two distantly related groups of ancestors. The body parts of organisms that do not have a common evolutionary origin but are similar in function are called **analogous structures.**

Although analogous structures don't shed light on evolutionary relationships, they do provide evidence of evolution. For example, insect and bird wings probably evolved separately when their different ancestors adapted independently to similar ways of life.

Figure 15.8
Insect and bird wings are similar in function but not in structure. Bones are the framework of bird wings, whereas a tough material called chitin composes insect wings.

Display

Obtain samples of different bird wings (turkey, chicken, duck, or guinea hen) from a supermarket. Display the wings and have students identify the homologous structures. Discuss the structure and function of a bird's wing. Ask students if their observations support the idea that the organisms are closely related.

Tying to Previous Knowledge

Point out that the theory of evolution predicts that organisms with similar physical characteristics will also have similar DNA. Briefly review the structure and function of DNA. Remind students that DNA makes up the genes of an organism that are part of the organism's chromosomes.

Figure 15.9
Vestigial structures, such as the forelimbs of the extinct elephant bird (**a**) and those of the present-day ostrich (**b**), are evidence of evolution because they show structural change over time.

WORD *Origin*

vestigial
From the Latin word *vestigium*, meaning "sign." The forelimbs of ostriches are vestigial structures.

Figure 15.10
Comparing embryos can reveal their evolutionary relationships. The presence of gill slits and tails in early vertebrate embryos shows that they may share a common ancestor.

Another type of body feature that suggests evolutionary relationship is a **vestigial structure** (veh STIHJ ee ul)—a body structure that has no function in a present-day organism but was probably useful to an ancestor. A structure becomes vestigial when the species no longer needs the feature. Although the structure has no function, it is still inherited as part of the body plan for the species.

Many organisms have structures with no apparent function. The eyes of blind mole-rats and cave fish are vestigial structures because they have no function. In *Figure 15.9*, you see two flightless birds—an extinct elephant bird and an African ostrich—with extremely reduced forelimbs. Their ancestors probably foraged on land for food and nested on the ground. As a result, over time, the ancestral birds probably became quite large and unable to fly, features evident in fossils of the elephant bird and present in the African ostrich.

Embryology

It's very easy to see the difference between an adult bird and an adult mammal, but can you distinguish between them by looking at their embryos? An **embryo** is the earliest stage of growth and development of both plants and animals. The embryos of a fish, a reptile, a bird, and a mammal are shown in *Figure 15.10.* You can see a tail and gill slits in all the embryos. You know that reptiles, birds, and mammals do not have gills when they are mature. As development continues, the differences in the embryos will increase until you can distinguish among them. However, the similarities among the young embryos suggest evolution from a distant, common ancestor.

Biochemistry

Biochemistry also provides evidence for evolution. It reveals information about relationships between

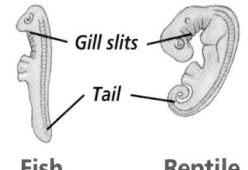

Fish **Reptile**   **Bird** **Mammal**

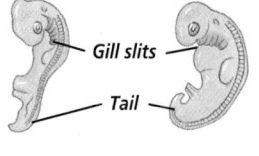

410 THE THEORY OF EVOLUTION

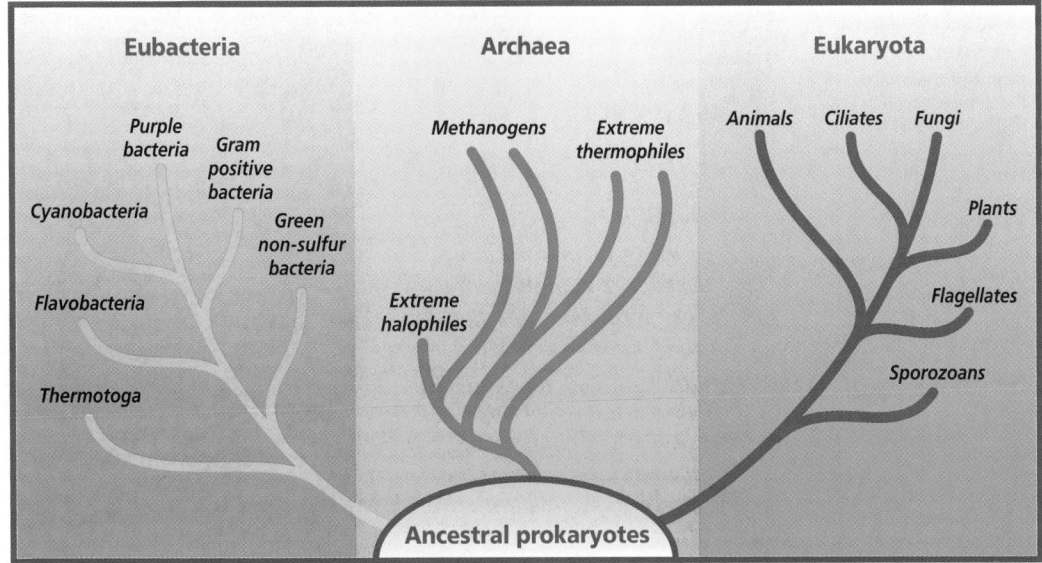

| Eubacteria | Archaea | Eukaryota |

Eubacteria:
- Purple bacteria
- Gram positive bacteria
- Cyanobacteria
- Green non-sulfur bacteria
- Flavobacteria
- Thermotoga

Archaea:
- Methanogens
- Extreme thermophiles
- Extreme halophiles

Eukaryota:
- Animals
- Ciliates
- Fungi
- Plants
- Flagellates
- Sporozoans

Ancestral prokaryotes

Extension

Have students answer the problem: Anteaters, toothless mammals that live in South American rain forests, feed on termites. If the termites they normally feed on are replaced by termites that are too large to swallow whole, how might the anteaters change over time?

individuals and species. Comparisons of the DNA or RNA of different species produce biochemical evidence for evolution. Today, many scientists use the results of biochemical studies to help determine the evolutionary relationships of species.

Since Darwin's time, scientists have constructed evolutionary diagrams that show levels of relationships among species. In the 1970s, some biologists began to use RNA

and DNA nucleotide sequences to construct evolutionary diagrams. The evolutionary diagram you see in *Figure 15.11* was constructed using the results of biochemical analysis and other data, including anatomical data. Notice that it divides all species into three domains: the Archaea, the Eubacteria, and the Eukaryota—an idea that underlies one of the most recently developed classification systems for organisms.

Figure 15.11
This evolutionary diagram is based on comparisons of organisms' RNA and supported by other data.

✔ Assessment

Portfolio Ask the students to research an organism and describe five of its adaptations. Then, have them select a new environment for the organism and predict how natural selection would affect the organism there. **L2** **P**

4 Close

Activity

Interpersonal Divide the class into groups and show photos of different organisms. Have groups brainstorm a list of the organisms' adaptations and three explanations for each adaptation. **L1** **COOP LEARN**

Resource Manager

Reinforcement and Study Guide, pp. 65-66 **L2**
Content Mastery, p. 74 **L1**

Section Assessment

Understanding Main Ideas
1. Briefly explain Darwin's ideas about natural selection.
2. Some snakes have vestigial legs. Why is this considered evidence for evolution?
3. Explain how mimicry and camouflage help species survive.
4. How do homologous structures provide evidence for evolution?

Thinking Critically
5. A parasite that lives in red blood cells causes the disease called malaria. In recent years, new strains of the parasite have appeared that are resistant to the drugs used to treat the disease. Explain how this could be an example of natural selection occurring.

SKILL REVIEW
6. **Sequencing** Fossils indicate that whales evolved from ancestors that had legs. Using your knowledge of natural selection, sequence the steps that may have occurred during the evolution of whales from their terrestrial, doglike ancestors. For more help, refer to *Organizing Information* in the **Skill Handbook.**

Section Assessment

1. Organisms produce many offspring with variations, some of which enable longer survival than others. Variations with a survival advantage are widespread among descendants.
2. They suggest that snake ancestors had functional legs and today's snakes may have evolved from them.

3. They reduce a species' visibility to predators or mimic the appearance of an organism that predators avoid.
4. They suggest common ancestry.
5. Some parasites had a variation that made them resistant to drugs. They survived and passed this variation to their offspring.

6. Ancestral whales were forced to live in water. Individuals with variations that had survival advantages in water reproduced, passing on these variations. Over time the variations became common in the population.

Prepare

Key Concepts

Students will learn about gene pools and how natural selection affects them. Then they will study factors that may contribute to speciation. Finally, specific examples of different patterns of evolution will enhance the students' understanding of the theory of evolution.

Planning

■ Purchase beans for the Visual Learning and BioLab, unshelled peanuts for MiniLab 15-2, and carnations for the Tech Prep.
■ Gather equipment for the Alternative Lab.

1 Focus

Bellringer 🔥

Before presenting the lesson, display **Section Focus Transparency 38** on the overhead projector and have students answer the accompanying questions.
L1 **ELL**

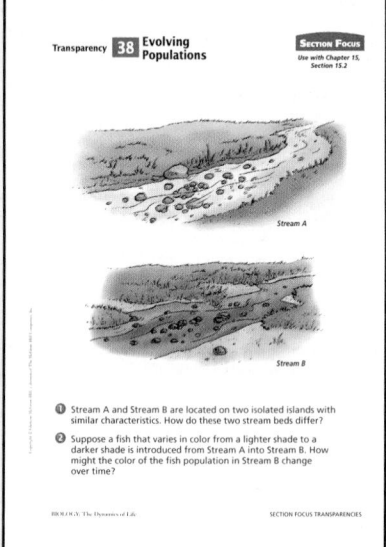

Section

15.2 Mechanisms of Evolution

Y*ou may recognize the birds shown here as meadowlarks. These birds range throughout much of the United States. Meadowlarks look so similar that it's often difficult to tell them apart. But if you listen, you'll hear a melodious bubbling sound from the Western meadowlark, whereas the Eastern meadowlark produces a clear whistle. Although they are closely related and occupy the same ranges in parts of the central United States, these different meadowlarks do not normally interbreed and are classified as distinct species.*

Western meadowlark—
Sturnella neglecta

Eastern meadowlark—
Sturnella magna

Population Genetics and Evolution

When Charles Darwin developed his theory of natural selection in the 1800s, he did so without knowing about genes. Since Darwin's time, scientists have learned a great deal about genes and modified Darwin's ideas accordingly. At first, genetic information was used to explain the variation among individuals of a population. Then, studies of the complex behavior of genes in populations of plants and animals developed into the field of study called population genetics. The principles of today's modern theory of evolution are rooted in population genetics and other related fields of study and are expressed in genetic terms.

Populations, not individuals, evolve

Can individuals evolve? That is, can an organism respond to natural selection by acquiring or losing characteristics? Recall that genes determine most of an individual's features, such as tooth shape or flower color. If an organism has a feature—a variation called a phenotype in genetic terms—that is poorly adapted to its environment, the organism may be unable to survive and reproduce. However, within its lifetime, it cannot evolve a new phenotype in response to its environment.

Rather, natural selection acts on the range of phenotypes in a population. Recall that a population consists of all the members of a species that live in an area. Each member has the

412 THE THEORY OF EVOLUTION

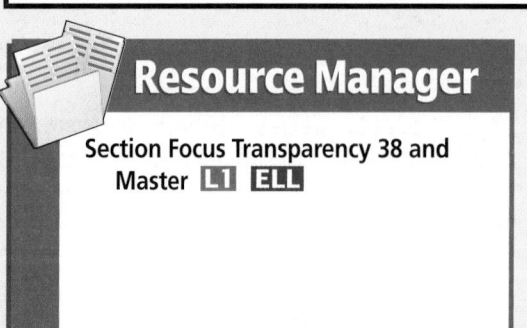

Resource Manager

Section Focus Transparency 38 and Master **L1** **ELL**

TECHPREP

Logical-Mathematical Have students use a bouquet of carnations (red, pink, and white) to calculate the frequency of alleles that determine flower color. Students should prepare a chart similar to the one in Figure 15.12. **L2** 🖐

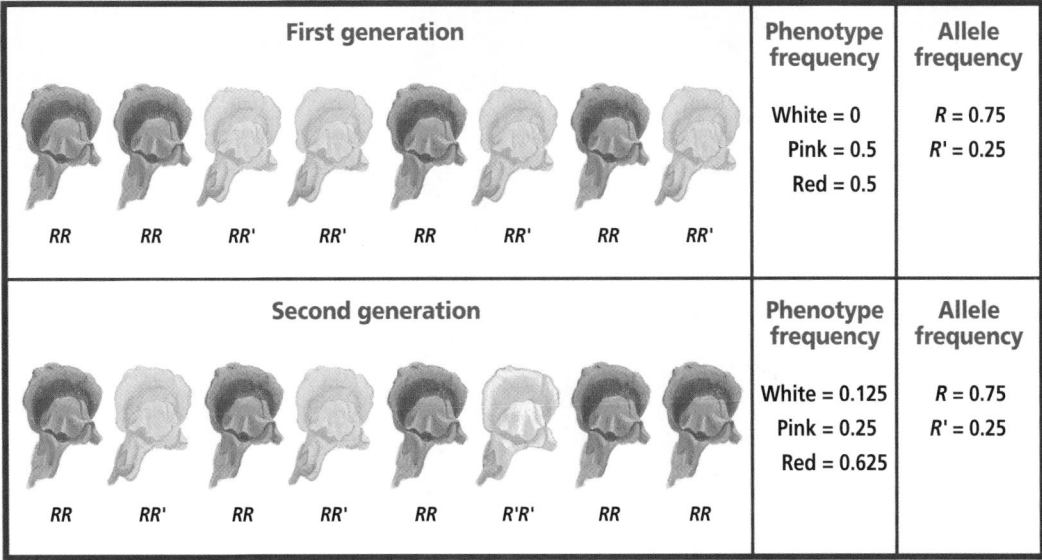

First generation								Phenotype frequency	Allele frequency
RR	RR	RR'	RR'	RR	RR'	RR	RR'	White = 0 Pink = 0.5 Red = 0.5	R = 0.75 R' = 0.25

Second generation								Phenotype frequency	Allele frequency
RR	RR'	RR	RR'	RR	R'R'	RR	RR	White = 0.125 Pink = 0.25 Red = 0.625	R = 0.75 R' = 0.25

genes that characterize the traits of the species, and these genes exist as pairs of alleles. Just as all of the individuals make up the population, all of the genes of the population's individuals make up the population's genes. Evolution occurs as a population's genes and their frequencies change over time.

How can a population's genes change over time? Picture all of the alleles of the population's genes as being together in a large pool called a **gene pool.** The percentage of any specific allele in the gene pool is called the **allelic frequency.** Scientists calculate the allelic frequency of an allele in the same way that a baseball player calculates a batting average. They refer to a population in which the frequency of alleles remains the same over generations as being in **genetic equilibrium.** In the *Math Connection* at the end of the chapter, you can read about the mathematical description of genetic equilibrium. You can study the effect of natural selection on allelic frequencies in the *BioLab* at the end of the chapter.

Look at the population of snapdragons shown in *Figure 15.12*. A pattern of heredity called incomplete dominance, which you learned about earlier, governs flower color in snapdragons. If you know the flower-color genotypes of the snapdragons in a population, you can calculate the allelic frequency for the flower-color alleles. The population of snapdragons is in genetic equilibrium when the frequency of its alleles for flower color is the same in all its generations.

Changes in genetic equilibrium

A population that is in genetic equilibrium is not evolving. Because allelic frequencies remain the same, phenotypes remain the same, too. Any factor that affects the genes in the gene pool can change allelic frequencies, disrupting a population's genetic equilibrium, which results in the process of evolution.

You have learned that one mechanism for genetic change is mutation. Environmental factors, such as radiation or chemicals, cause many mutations, but other mutations occur by

Figure 15.12
Incomplete dominance produces three phenotypes: red flowers (*RR*), white flowers (*R'R'*), and pink flowers (*RR'*). Although the phenotype frequencies of the generations vary, the allelic frequencies for the *R* and *R'* alleles do not vary.

2 Teach

Visual Learning

Kinesthetic **Figure 15.12** Students can use beans to model allelic frequency. Mix red pinto beans, black beans, and white navy beans in a large container. Have students withdraw 20 random beans to represent the gene pool of a population with the genotypes *BB* (black bean), *BB** (white bean), and *B*B** (red bean). Have them calculate the phenotype frequencies by dividing the number of each phenotype by 20, and the allelic frequencies by counting the numbers of each allele and dividing by 40. **L1** **ELL**

The BioLab at the end of the chapter can be used at this point in the lesson.

Quick Demo

Use a world map and explain that many human populations are isolated for geographical, political, or other reasons. Ask students how this might affect these nations' gene pools.

GLENCOE TECHNOLOGY

 **VIDEOTAPE** **The Secret of Life**
Gone Before You Know It: The Biodiversity Crisis

 CD-ROM **Biology: The Dynamics of Life**
Exploration: *Selection Pressure* Disc 2

 BIOLOGY JOURNAL

Desert Adaptions

Linguistic Ask students to imagine changes in food, clothing, shelter, and other factors that would be useful if their environment suddenly became desertlike. Have them write a short story about life in the new environment. **L1**

Genetic Drift

Linguistic Have students write a short summary of how genetic drift may affect small populations. **L3**

Concept Development

Point out that in small populations that interbreed, such as certain religious groups and royal families, gene pools change quickly because the number of potential mates is limited.

Tying to Previous Knowledge

Review meiosis. Explain how random factors involved in some of the steps of meiosis can contribute to genetic drift.

Using An Analogy

Flip a coin to show how small populations can be affected by genetic drift. If you flip a coin 100 times, the chances of getting 100 heads and 0 tails—or even 80 heads and 20 tails—is unlikely. The result will probably be close to 50-50. But if you flip the coin 10 times, the chance of getting 8 heads and 2 tails—or even 10 heads and 0 tails—is likely to occur. Similarly, the loss of alleles by chance is lower in large populations than in small ones.

Figure 15.13 Genetic drift can result in an increase of rare alleles in a small population such as in the Amish community of Lancaster County, Pennsylvania.

chance. Of the mutations that affect organisms, many are lethal, and the organisms do not survive. Thus, lethal mutations are quickly eliminated. However, occasionally, a mutation results in a useful variation, and the new gene becomes part of the population's gene pool by the process of natural selection.

Another mechanism that disrupts a population's genetic equilibrium is **genetic drift**—the alteration of allelic frequencies by chance events. Genetic drift can greatly affect small populations that include the descendants of a small number of organisms. This is because the genes of the

original ancestors represent only a small fraction of the gene pool of the entire species and are the only genes available to pass on to offspring. The distinctive forms of life that Darwin found in the Galapagos Islands may have resulted from genetic drift.

Genetic drift has been observed in some small human populations that have become isolated due to reasons such as religious practices and belief systems. For example, in Lancaster County, Pennsylvania, there is an Amish population of about 12 000 people who have a unique lifestyle and marry other members of their community. By chance, at least one of the original 30 Amish settlers in this community carried a recessive allele that results in short arms and legs and extra fingers and toes in offspring, *Figure 15.13*. Because of the small gene pool, many individuals inherited the recessive allele over time. Today, the frequency of this allele among the Amish is high—1 in 14 rather than 1 in 1000 in the larger population of the United States.

Genetic equilibrium is also disrupted by the movement of individuals in and out of a population. The transport of genes by migrating individuals is called gene flow. When an individual leaves a population, its genes are lost from the gene pool.

Papilio ajax ajax

Papilio ajax ampliata

414 THE THEORY OF EVOLUTION

Alternative Lab

Bacterial Resistance

Purpose

Students will study variation in bacterial resistance to antibiotics. Dispose of used dishes after autoclaving or incinerating.

Materials 🔌 🧤 ☣️ 🚫
culture of *Bacillus subtilis,* 3 tubes of nutrient agar, tube of streptomycin agar, inoculation loop, 2 petri dishes, Bunsen burner, wax pencil, test tube

Procedure

Give the following directions.

1. Wear goggles, aprons, and disposable gloves when working with bacteria.
2. Write *A* and *B* on the halves of one petri dish and *C* and *D* on another.

3. Pour streptomycin agar into dish *A-B*. Place the dish on a pencil, so the liquid flows to one side to solidify. **CAUTION: Liquid agar is hot.**
4. After the agar solidifies, pour a tube of hot nutrient agar into the dish and cover the dish after it solidifies.
5. Pour the other tubes of agar into the *C-D* dish. Cover the dish after it cools.
6. Sterilize the inoculation loop in the Bunsen burner's flame. Remove the

When individuals enter a population, their genes are added to the pool.

Mutation, genetic drift, and gene flow may significantly affect the evolution of small and isolated gene pools, such as those on islands. However, their effect is often insignificant in larger, less isolated gene pools. Natural selection is usually the most significant factor that causes changes in established gene pools—small or large.

Natural selection acts on variations

As you've learned, traits have variation, as shown in the butterflies pictured in *Figure 15.14*. If you measured the thumb lengths of everyone in your class, you'd find average, long, and short thumbs. Try measuring the variations in peanut shells in the *MiniLab* on this page.

Recall that some variations increase or decrease an organism's chance of survival in an environment. These variations can be inherited and are controlled by alleles. Thus, the allelic frequencies in a population's gene pool will change over generations due to the natural selection of variations. There are three different types of natural selection that act on variation: stabilizing, directional, and disruptive.

MiniLab 15-2   Collecting Data

Detecting a Variation Pick almost any trait—height, eye color, leaf width, or seed size—and you can observe how the trait varies in a population. Some variations are an advantage to an organism and some are not.

Procedure
1. Copy the data table shown here, but include the lengths in millimeters (numbers 25 through 45) that are missing from this table.

Data Table											
Length in mm	20	21	22	23	24	—	46	47	48	49	50
Checks											
My Data—Number of shells											
Class Data—Number of shells											

2. Use a millimeter ruler to measure a peanut shell's length. In the Checks row, check the length you measured.
3. Repeat step 2 for 29 more shells.
4. Count the checks under each length and enter the total in the row marked My Data.
5. Use class totals to complete the row marked Class Data.

Analysis
1. Was there variation among the lengths of peanut shells? Use specific class totals to support your answer.
2. If larger peanut shells were a selective advantage, would this be stabilizing, directional, or disruptive selection? Explain your answer.

Papilio ajax curvifascia

Papilio ajax ehrmanni

Figure 15.14
These swallowtail butterflies live in different areas of North America. Despite their slight variations, they can interbreed to produce fertile offspring.

stopper and flame the lip of the bacterial culture.
7. Dip the loop into the culture and remove it. Flame the lip of the container again and replace the stopper.
8. Streak the agar in dish *A-B* with the loop. Do not break the agar surface.
9. Repeat steps 6-8 on the *C-D* dish.
10. Recover the dishes, invert them, and place them in a dark drawer or closet. Do not open the dishes again after you

have inoculated and recovered them. Observe them after 24 hours. Dispose of used petri dishes as your teacher instructs.

Expected Results
There should be more bacterial growth in the C-D dish than in the A-B dish.

Analysis
Have students sketch and describe the appearance of both plates.

Visual Learning

Figure 15.15 illustrates the three main types of natural selection. Refer to each type and offer students several other examples of each.

Reinforcement

Logical-Mathematical Have students describe the type of natural selection in each of the following examples.

- Members of a population of Amazon tree frogs hop from tree to tree searching for food in the rain forest. They vary in leg length. Events result in massive destruction of the forest's trees. After several generations, only long-legged tree frogs remain alive. *directional selection*
- Different grass plants in a population range in length from 8 cm to 28 cm. The 8-10 cm grass blades receive little sunlight, and the 25-28 cm grass blades are eaten quickly by grazing animals. *stabilizing selection*
- The spines of a sea urchin population's members vary in length. The short-spined sea urchins are camouflaged easily on the seafloor. However, long-spined sea urchins are well defended against predators. *disruptive selection*

Have students illustrate each situation and predict what will happen to the members of each population if natural selection continues to operate. **L1**

Resource Manager

Basic Concepts Transparency 22 and Master **L2** **ELL**
Critical Thinking/Problem Solving **L3**

Figure 15.15

Different types of natural selection act over the range of a trait's variation. The red, bell-shaped curve indicates a trait's variation in a population. The blue, bell-shaped curve indicates the effect of a natural selection.

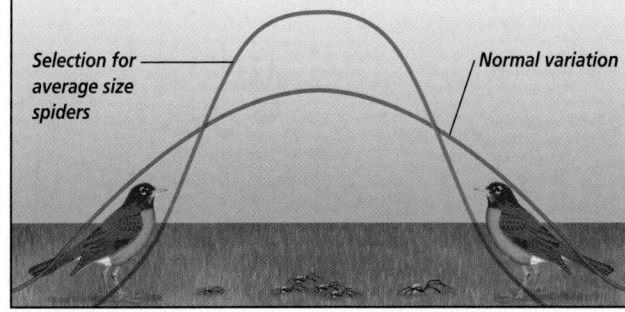

Selection for average size spiders

Normal variation

A **Stabilizing selection** favors average individuals. This type of selection reduces variation in a population.

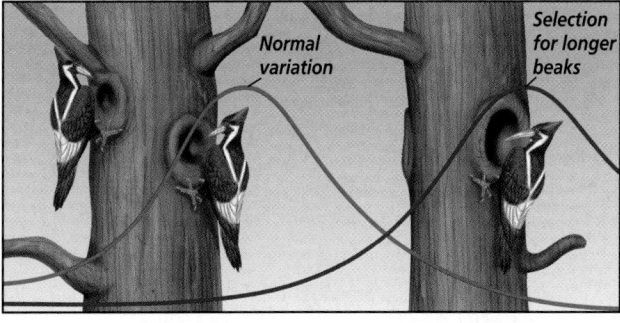

Normal variation

Selection for longer beaks

B **Directional selection** favors one of the extreme variations of a trait and can lead to the rapid evolution of a population.

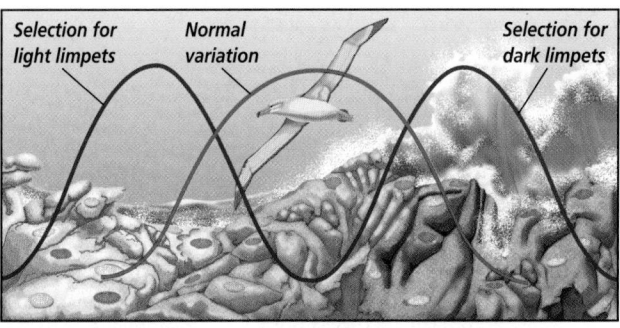

Selection for light limpets

Normal variation

Selection for dark limpets

C **Disruptive selection** favors both extreme variations of a trait, resulting eventually in no intermediate forms of the trait and leading to the evolution of two new species.

416 THE THEORY OF EVOLUTION

Stabilizing selection is natural selection that favors average individuals in a population as shown in *Figure 15.15*. Consider a population of spiders in which average size is a survival advantage. Predators in the area might easily see and capture spiders that are larger than average. However, small spiders may find it difficult to find food. Therefore, in this environment, average-sized spiders are more likely to survive—they have a selective advantage, or are "selected for."

Directional selection occurs when natural selection favors one of the extreme variations of a trait. For example, imagine a population of woodpeckers pecking holes in trees to feed on the insects living under the bark. Suppose that a species of insect that lives deep in tree tissues invades the trees in a woodpecker population's territory. Only woodpeckers with long beaks could feed on that insect. Therefore, the long-beaked woodpeckers in the population would have a selective advantage over woodpeckers with very short or average-sized beaks.

Finally, in **disruptive selection,** individuals with either extreme of a trait's variation are selected for. Consider, for example, a population of marine organisms called limpets. The shell color of limpets ranges from white, to tan, to dark brown. As adults, limpets live attached to rocks. On light-colored rocks, white-shelled limpets have an advantage because their bird predators cannot easily see them. On dark-colored rocks, dark-colored limpets have the advantage because they are camouflaged. On the other hand, birds easily see tan-colored limpets on either the light or dark backgrounds. Disruptive selection tends to eliminate the intermediate phenotypes.

BIOLOGY JOURNAL

Populations and Natural Selection

Linguistic Have students write and illustrate a short story about the effects of natural selection on a specific population. Have them predict what happens to the population if all three types of natural selection occur in it. **L2**

BIOLOGY Online Note Internet addresses that you find useful in the space below for quick reference.

_____

_____

_____

_____

_____

Natural selection can significantly alter the genetic equilibrium of a population's gene pool over time. Significant changes in the gene pool can lead to the evolution of a new species over time.

The Evolution of Species

You've just read about how natural processes such as mutation, genetic drift, gene flow, and natural selection can change a population's gene pool over time. But how do the changes in the makeup of a gene pool result in the evolution of new species? Recall that a species is defined as a group of organisms that look alike and can interbreed to produce fertile offspring in nature. The evolution of new species, a process called **speciation** (spee shee AY shun), occurs when members of similar populations no longer interbreed to produce fertile offspring within their natural environment.

Physical barriers can prevent interbreeding

In nature, physical barriers can break large populations into smaller ones. Lava from volcanic eruptions can isolate populations. Sea-level changes along continental shelves can create islands. The water that surrounds an island isolates its populations. **Geographic isolation** occurs whenever a physical barrier divides a population.

A new species can evolve when a population has been geographically isolated. For example, imagine a population of tree frogs living in a rain forest, *Figure 15.16*. If small populations of tree frogs were geographically isolated, they would no longer be able to interbreed and exchange genes. Over time, each small population might adapt to its environment through natural selection and develop its own gene pool. Eventually, the gene pools of each population might become so different that new species

WORD *Origin*
speciation
From the Latin word *species*, meaning "kind." Speciation is a process that produces two species from one.

Figure 15.16
When geographic isolation divides a population of tree frogs, the individuals no longer mate across populations.

CD-ROM
View an animation of Figure 15.16 in the Presentation Builder of the Interactive CD-ROM.

Discussion
Remind students that scientists used to classify organisms only on the basis of morphological comparisons. This type of classification is useful but limited. For example, using morphological classification, North American yellow flickers, red-shafted flickers, and their hybrid offspring could be considered three different species.

According to the biological species concept, organisms are classified by whether or not they can naturally interbreed with one another to produce fertile offspring, as the yellow flickers and red-shafted flickers can do. Elicit from students how many species of North American flickers exist based on this biological species definition. *one*

GLENCOE
TECHNOLOGY

VIDEODISC
Biology: The Dynamics of Life
Geographic Isolation (Ch. 6)
Disc 1, Side 2, 17 sec.

PROJECT

Mammalian Evolution

Interpersonal Groups of students can prepare a written report, oral report, or poster project about the geographic and reproductive isolation effect of plate tectonics on a mammalian family. Some families to research are: Bradypodidae, Myrmecophagi-dae, Camelidae, Mustelidae, Felidae, and Ursidae. Projects should contain details about the mammals, such as their structure and behavior, a brief summary of their fossil record, and explanations for how they evolved. **L2** **ELL** **COOP LEARN**

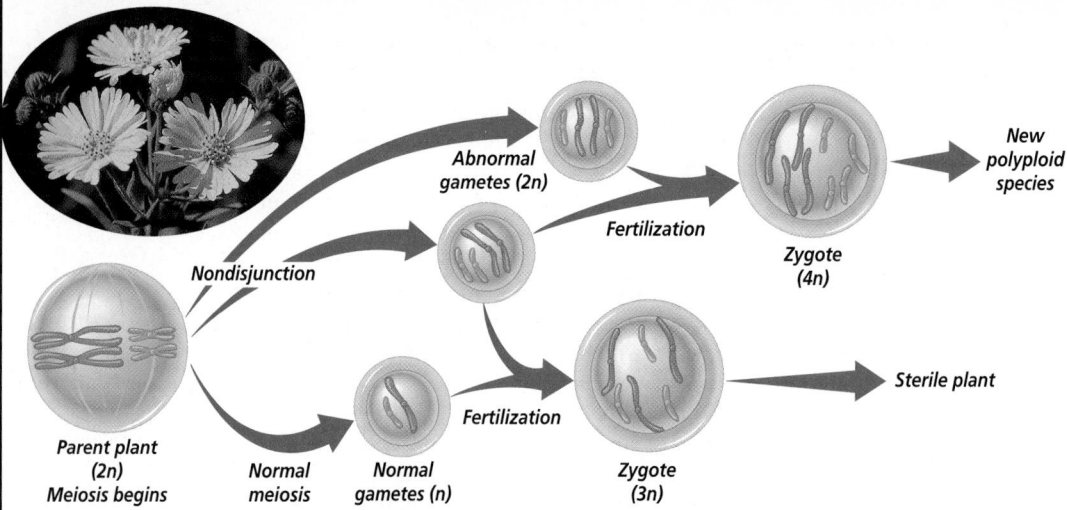

Figure 15.17
Many flowering plants, such as this California tarweed, are polyploids—individuals that result from mistakes made during meiosis.

WORD *Origin*

polyploidy
From the Greek word *polys*, meaning "many." Polyploid plants contain multiple sets of chromosomes.

of tree frogs would evolve in the different forest patches or the populations might become extinct.

Reproductive isolation can result in speciation

As populations become increasingly distinct, reproductive isolation can arise. **Reproductive isolation** occurs when formerly interbreeding organisms can no longer mate and produce fertile offspring.

There are different types of reproductive isolation. One type occurs when the genetic material of the populations becomes so different that fertilization cannot occur. Some geographically separated populations of salamanders in California have this type of reproductive isolation. Another type of reproductive isolation is behavioral. For example, if one population of tree frogs mates in the fall, and another mates in the summer, these two populations will not mate with each other and are reproductively isolated.

A change in chromosome numbers and speciation

Chromosomes can also play a role in speciation. Many new species of plants and some species of animals have evolved in the same geographic area as a result of polyploidy (PAHL ih ployd ee), which is illustrated in *Figure 15.17*. Any species with a multiple of the normal set of chromosomes is known as a **polyploid.**

Mistakes during mitosis or meiosis can result in polyploid individuals. For example, if chromosomes do not separate properly during the first meiotic division, diploid ($2n$) gametes can be produced instead of the normal haploid (n) gametes. Polyploidy results in immediate reproductive isolation. When a polyploid mates with an individual of the normal species, the resulting zygotes may not develop normally because of the difference in chromosome numbers. In other cases, the zygotes develop into adults that probably cannot reproduce. However, polyploids within a population may interbreed and form a separate species.

Polyploids can arise from within a species or from hybridization between species. Many flowering plant species, as well as many important crop plants, such as wheat, cotton, apples, and bananas, originated by polyploidy.

418 THE THEORY OF EVOLUTION

Speciation can occur quickly or slowly

Although polyploid speciation takes only one generation, most other mechanisms of speciation do not occur as quickly. What is the usual rate of speciation?

Scientists once argued that evolution occurs at a slow, steady rate, with small, adaptive changes gradually accumulating over time in populations. **Gradualism** is the idea that species originate through a gradual change of adaptations. Some evidence from the fossil record supports gradualism. For example, fossil evidence shows that camels evolved slowly and steadily over time.

In 1971, Stephen J. Gould and Niles Eldridge proposed another hypothesis known as **punctuated equilibrium.** This hypothesis argues that speciation occurs relatively quickly, in rapid bursts, with long periods of genetic equilibrium in between. According to this hypothesis, environmental changes, such as higher temperatures or the introduction of a competitive species, lead to rapid changes in a population's gene pool. Speciation happens quickly—in about 10 000 years or less. Like gradualism, punctuated equilibrium is supported by fossil evidence as shown in *Figure 15.18.*

Biologists generally agree that both gradualism and punctuated equilibrium can result in speciation, depending on the circumstances. It shouldn't surprise you to see scientists offer alternative hypotheses to explain observations. The nature of science is such that new evidence or new ideas can modify theories.

Figure 15.18
The fossil record of elephant evolution supports the view of punctuated equilibrium. Three elephant species may have evolved from an ancestral population in a short time.

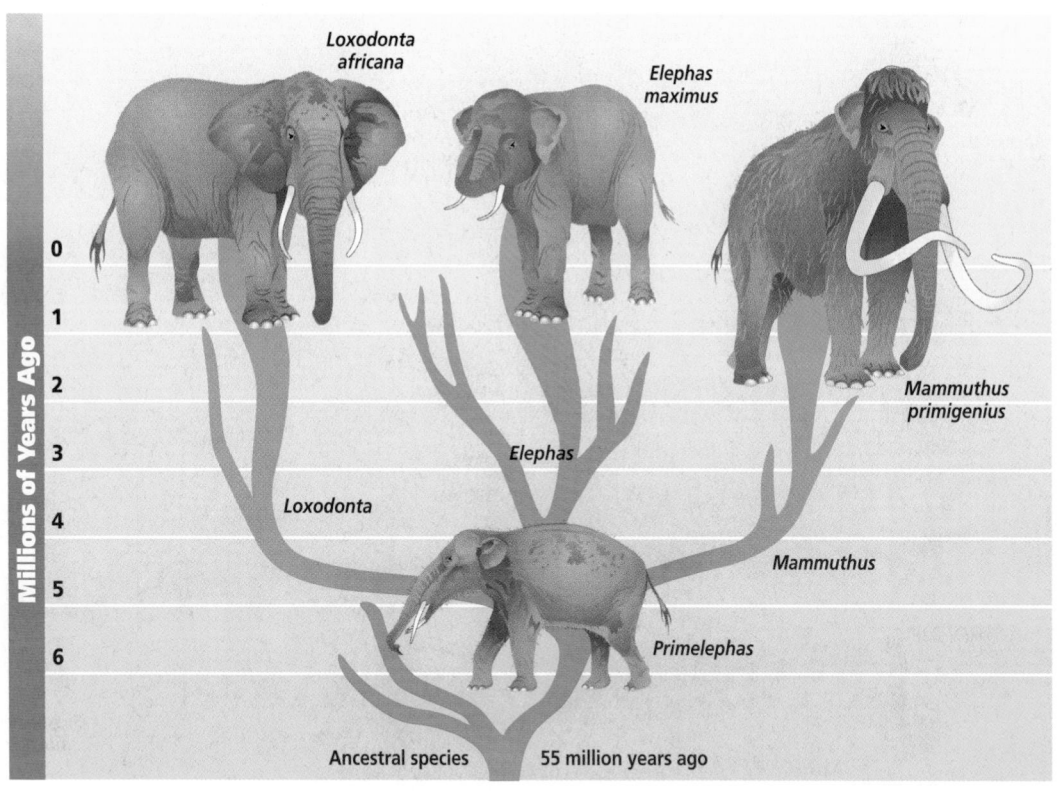

Loxodonta africana

Elephas maximus

Mammuthus primigenius

Elephas

Loxodonta

Mammuthus

Primelephas

Ancestral species   55 million years ago

Millions of Years Ago
0
1
2
3
4
5
6

Concept Development

Darwin believed that species evolve slowly over long periods of time. For example, fossils show that today's horseshoe crabs, genus *Limulus*, are nearly identical to ancestors that lived hundreds of millions of years ago. If possible, show a modern specimen of the horseshoe crab and a fossil counterpart to demonstrate their similarities. Remind students that scientists have found limited support for the idea of gradualism in the fossil record. However, point out some examples that do support it.

✔ Assessment

Portfolio Discuss the Abert and Kaibab squirrels of the Grand Canyon area. Have students prepare short summaries of the discussion to put in their portfolios. Summaries should describe the environment that each squirrel lives in, the characteristics of each species, and possible hypotheses for how the differences evolved. **L2**

GLENCOE
TECHNOLOGY

VIDEODISC
The Secret of Life
Punctuated Equilibrium

MEETING INDIVIDUAL NEEDS

Gifted

Linguistic Have the students research speciation rate. Students should first read Gould and Eldredge's 1972 article concerning punctuated equilibrium entitled "Punctuated Equilibria: An alternative to phyletic gradualism," found in *Models in Paleobiology,* T. J. M. Schopf (ed.), Freeman, Cooper, and Co. Have students compare the evidence for both punctuated equilibrium and gradualism. Their report should draw a conclusion about which hypothesis best supports the available evidence. **L3**

Reinforcement

 Linguistic Reinforce the concept of the niche by asking students to write brief autobiographies in which they describe where they live and something about their school lives, activities, and hobbies. **L1**

Concept Development

Use the student autobiographies to develop the niche concept. Discuss some autobiographies and point out that, just as no two autobiographies are alike, no two niches on Earth are alike. Tie the concepts of niche and struggle for existence to adaptive radiation.

Display

Display examples of convergent structures in organisms, such as models or actual specimens of bird, bat, and insect wings.

GLENCOE
TECHNOLOGY

VIDEODISC
Biology: The Dynamics of Life
Adapted for Survival (Ch. 5)
Disc 1, Side 2, 33 sec.

3 Assess

Check for Understanding

Develop five questions about types of natural selection and patterns of evolution. Have students work in groups to answer the questions. **L1** **COOP LEARN**

Patterns of Evolution

Biologists have observed different patterns of evolution that occur throughout the world in different natural environments. These patterns support the idea that natural selection is an important agent for evolution.

Diversity in new environments

An extraordinary diversity of unique plants and animals live or have lived on the Hawaiian Islands, among them a group of birds called Hawaiian honeycreepers. This group of birds is interesting because, although similar in body size and shape, they differ sharply in color and beak shape. Different species of honeycreepers evolved to occupy their own niches.

Despite their differences, scientists hypothesize that honeycreepers, as shown in *Figure 15.19*, evolved from a single ancestral species that lived on the Hawaiian Islands long ago. When an ancestral species evolves into an array of species to fit a number of diverse habitats, the result is called **adaptive radiation.**

Adaptive radiation in both plants and animals has occurred and continues to occur throughout the world and is common on islands. For example, the many species of finches that Darwin observed on the Galapagos Islands are a typical example of adaptive radiation.

Adaptive radiation is a type of **divergent evolution,** the pattern of evolution in which species that once were similar to an ancestral species diverge, or become increasingly distinct. Divergent evolution occurs when populations adapting to different environmental conditions change,

Figure 15.19
Evolutionary biologists have suggested that the ancestors of all Hawaiian Island honeycreepers migrated from North America about 5 million years ago. As this ancestral bird population settled in the diverse Hawaiian niches, adaptive radiation occurred.

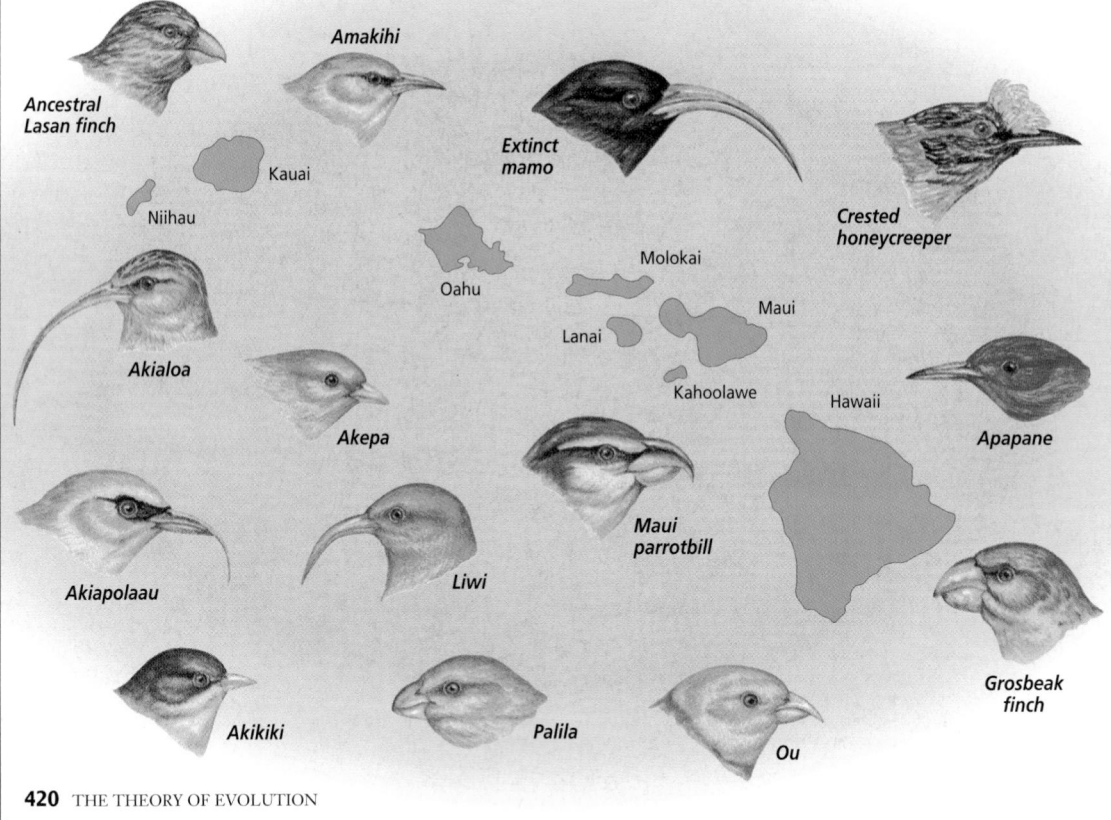

420 THE THEORY OF EVOLUTION

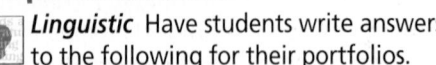

Adaptive Radiation

 *Linguistic* Have students write answers to the following for their portfolios.
- Describe some adaptive changes that might occur in a population of gray squirrels over time if the population became deposited suddenly on an island that contained a swamp, a desert, a tropical rain forest, and a snow-covered mountain.
- How do examples of adaptive radiation support the concept of evolution by natural selection? **L2** **P**

420

Figure 15.20
Unrelated species of plants such as the organ pipe cactus (a) and this *Euphorbia* (b) share a similar fleshy body type and no leaves.

a

b

Reteach

Visual-Spatial Have students make a concept map to demonstrate how natural selection acts on the variation of a trait. **L1** **ELL**

Extension

Linguistic Have the students write a summary about how Darwin's finches illustrate adaptive radiation. **L2**

Assessment

Skill Have students analyze the following data on rabbit population allele frequencies.

1st generation: $A = 0.5$, $a = 0.5$
2nd generation: $A = 0.6$, $a = 0.4$
3rd generation: $A = 0.7$, $a = 0.3$
4th generation: $A = 0.8$, $a = 0.2$
5th generation: $A = 0.9$, $a = 0.1$

Have students describe what is happening in the population if A represents an allele for white fur, and a represents an allele for brown fur. **L2**

4 Close

Discussion

Discuss some of the human cultural adaptations that "shield" us from the effects of natural selection, such as clothing, medicine, automobiles, etc.

Resource Manager

Reinforcement and Study Guide, pp. 68-69 **L2**
Content Mastery, pp. 73, 75-76 **L1**

becoming less alike as they adapt, eventually resulting in new species.

Different species can look alike

A pattern of evolution in which distantly related organisms evolve similar traits is called **convergent evolution.** Convergent evolution occurs when unrelated species occupy similar environments in different parts of the world. Because they share similar environmental pressures, they share similar pressures of natural selection.

For example, in *Figure 15.20,* you see a cactus of the family Cactaceae, an organ pipe cactus, that grows in the deserts of North and South America and a plant of the family Euphorbiaceae that looks like the cactus and lives in African deserts. Although these plants are unrelated species, their environments are similar. You can see that they both have fleshy bodies and no leaves. That convergent evolution has apparently occurred in unrelated species, is further evidence for natural selection.

Section Assessment

Understanding Main Ideas
1. Explain why the evolution of resistance to antibiotics in bacteria is an example of directional selection.
2. How can geographic isolation change a population's gene pool?
3. Why is rapid evolutionary change more likely to occur in small populations?
4. How do gradualism and punctuated equilibrium differ? How are they similar?

Thinking Critically
5. Hummingbird moths are night-flying insects whose behavior and appearance are

similar to those of hummingbirds. Explain how these two organisms demonstrate the concept of convergent evolution.

SKILL REVIEW

6. **Designing an Experiment** Biologists have discovered two species of squirrels living on opposite sides of the Grand Canyon. They hypothesize that they both evolved from a recent, common ancestor that lived in the area before the Grand Canyon formed. What observations or experiments could provide evidence for this hypothesis? For more help, refer to *Practicing Scientific Methods* in the **Skill Handbook.**

Section Assessment

1. Only bacteria that are totally resistant to antibiotics survive.
2. It may result in different local environments for a separated population. Different adaptations are useful in different environments, and the gene pool will in time reflect the differences.
3. It occurs because of genetic drift and the limited number of mates.
4. Gradualism takes a much longer time than punctuated equilibrium, but they both result in evolution.
5. Although not closely related, they share similar environments and have evolved similar behaviors and appearances.
6. Analyze the DNA, structure, and behavior of each. Examine fossils.

Time Allotment
One class period

Process Skills
make and use tables, observe and infer, make and use graphs

PREPARATION

Alternative Materials
- Beads or other small objects may be substituted for beans.

Resource Manager

BioLab and MiniLab Worksheets, pp. 73-74 **L2**

INTERNET
BioLab

Natural Selection and Allelic Frequency

*E*volution can be described as the change in allelic frequencies of a gene pool over time. Natural selection can place pressure on specific phenotypes and cause a change in the frequency of the alleles that produce the phenotypes. In this activity, you will simulate the effects of eagle predation on a population of rabbits, where GG represents the homozygous condition for gray fur; Gg is the heterozygous condition for gray fur; and gg represents the homozygous condition for white fur.

PREPARATION

Problem
How does natural selection affect allelic frequency?

Objectives
In this BioLab, you will:
- **Simulate** natural selection by using beans of two different colors.
- **Calculate** allelic frequencies over five generations.
- **Demonstrate** how natural selection can affect allelic frequencies over time.
- **Use the Internet** to collect and compare data from other students.

Materials
colored pencils (2)
paper bag
graph paper
pinto beans
white navy beans

Skill Handbook
Use the **Skill Handbook** if you need additional help with this lab.

PROCEDURE

1. Copy the data table shown on the next page.
2. Place 50 pinto beans and 50 white navy beans into the paper bag.
3. Shake the bag. Remove two beans. These represent one rabbit's genotype. Set the pair aside, and continue to remove 49 more pairs.
4. Arrange the beans on a flat surface in two columns representing the two possible rabbit phenotypes, gray (genotypes *GG* or *Gg*) and white (genotype *gg*).
5. Examine your columns. Remove 25 percent of the gray rabbits and 100 percent of the white rabbits. These numbers represent a random selection pressure on your rabbit population. If the number you calculate is a fraction, remove a whole rabbit to make whole numbers.

PROCEDURE

Teaching Strategies
- Tell students that they will simulate natural selection on a population to see how allelic frequency changes.
- You may wish to circulate during this activity to ensure that students are following the procedure correctly.
- Have students wash their hands after handling the beans.

Data and Observations
Make sure students are correctly calculating allelic frequency after each "generation" and recording these data in their data tables. Students should observe changes in the allelic frequencies of the rabbit population. Student graphs should show an increase in the frequency of the *G* allele and a decrease in the *g* allele.

6. Count the number of pinto and navy beans remaining. Record this number in your data table.

7. Calculate the allelic frequencies by dividing the number of beans of one type by 100. Record these numbers in your data table.

8. Begin the next generation by placing 100 beans into the bag. The proportions of pinto and navy beans should be the same as the percentages you calculated in step 7.

9. Repeat steps 3 through 8, collecting data for five generations.

10. Go to the Glencoe Science Web Site at the address shown below to **post your data.**

11. Graph the frequencies of each allele over five generations. Plot the frequency of the allele on the vertical axis and the number of the generation on the horizontal axis. Use a different colored pencil for each allele.

Data Table

| Generation | Allele *G* | | | Allele *g* | | |
	Number	Percentage	Frequency	Number	Percentage	Frequency
Start	50	50	0.50	50	50	0.50
1						
2						
3						
4						
5						

ANALYZE AND CONCLUDE

1. **Analyzing Data** Did either allele disappear? Why or why not?
2. **Thinking Critically** What does your graph show about allelic frequencies and natural selection?
3. **Making Inferences** What would happen to the allelic frequencies if the number of eagles declined?
4. **Using the Internet** Explain any differences in allelic frequencies you observed between your data and the data from the Internet. What advantage is there to having a large amount of data? What problems might there be in using data from the internet?

Sharing Your Data

BIOLOGY Online Find this *BioLab* on the Glencoe Science Web site at **science.glencoe.com.** Post your data in the data table provided for this BioLab. Use the additional data from other students on the Internet, and graph and analyze the combined data.

ANALYZE AND CONCLUDE

1. Neither allele disappeared from the population because the *g* allele is also in the heterozygous (*Gg*) rabbits.
2. The graph shows an increase in the frequency of the *G* allele and a decrease in the frequency of the *g* allele due to natural selection against white rabbits.
3. There would be less selective pressure on white rabbits and, therefore, less decline in the frequency of the *g* allele.
4. Students should notice little difference in the allelic frequencies posted on the Internet and the frequencies they calculated. By combining data students may get more accurate results.

Sharing Your Data

BIOLOGY Online To navigate to the Internet BioLabs choose the *Biology: The Dynamics of Life* icon at Glencoe's Web site. Click on the student site icon, then the BioLabs icon. The data from many trials supports a student's data and the conclusions the student may draw from the data.

✓ Assessment

Knowledge Ask students whether allele frequencies would change as fast if only 60% of the white rabbits were removed from the population each generation. Why or why not? *No, because the gene pool would contain more g alleles that could produce more white rabbits.* Use the Performance Task Assessment List for Analyzing the Data in **PASC,** p. 27. **L2**

Math Connection

Mathematics and Evolution

In the early 1900s, G. H. Hardy, a British mathematician, and W. Weinberg, a German doctor, independently discovered how the frequency of a trait's alleles in a population could be described mathematically.

Suppose that in a population of pea plants, 36 plants are homozygous dominant for the tall trait (*TT*), 48 plants are heterozygous tall (*Tt*), and 16 plants are short plants (*tt*). In the homozygous tall plants, there are (36) (2), or 72, *T* alleles and in the heterozygous plants there are 48 *T* alleles, for a total of 120 *T* alleles in the population. There are 48 *t* alleles in the heterozygous plants plus (16) (2), or 32, *t* alleles in the short plants, for a total of 80 *t* alleles in the population. The number of *T* and *t* alleles in the population is 200. The frequency of *T* alleles is 120/200 or 0.6, and the frequency of *t* alleles in 80/200, or 0.4.

The Hardy-Weinberg principle The Hardy-Weinberg principle states that the frequency of the alleles for a trait in a stable population will not vary. This statement is expressed as the equation $p + q = 1$, where p is the frequency of one allele for the trait and q is the frequency of the other allele. The sum of the frequencies of the alleles always includes 100 percent of the alleles, and is therefore stated as 1.

Squaring both sides of the equation produces the equation $p^2 + 2pq + q^2 = 1$. You can use this equation to determine the frequency of genotypes in a population: homozygous dominant individuals (p^2), heterozygous individuals ($2pq$), and recessive individuals (q^2). For example, in the pea plant population described above, the frequency of the genotypes would be

$$(0.6) (0.6) + 2(0.6) (0.4) + (0.4) (0.4) = 1$$

The frequency of the homozygous tall genotype is 0.36, the heterozygous genotype is 0.48, and the short genotype is 0.16.

424 THE THEORY OF EVOLUTION

In any sexually reproducing, large population, genotype frequencies will remain constant if no mutations occur, random mating occurs, no natural selection occurs, and no genes enter or leave the population.

Implications of the principle The Hardy-Weinberg principle is useful for several reasons. First, it explains that the genotypes in populations tend to remain the same. Second, because a recessive allele may be masked by its dominant allele, the equation is useful for determining the recessive allele's frequency in the population. Finally, the Hardy-Weinberg principle is useful in studying natural populations to determine how much natural selection may be occurring in the population.

CONNECTION TO BIOLOGY

The general population of the United States is getting taller. Assuming that height is a genetic trait, does this observation violate the Hardy-Weinberg principle? Explain your answer.

BIOLOGY *Online* To find out more about the Hardy-Weinberg principle, visit the Glencoe Science Web site. **science.glencoe.com**

A population of penguins

SUMMARY

Section 15.1

Natural Selection and the Evidence for Evolution

Main Ideas

- After many years of experimentation and observation, Charles Darwin proposed the idea that species originated through the process of natural selection.
- Natural selection is a mechanism of change in populations. In a specific environment, individuals with certain variations are likely to survive, reproduce, and pass these variations to future generations.
- Evolution has been observed in the lab and field, but much of the evidence for evolution has come from studies of fossils, anatomy, and biochemistry.

Vocabulary

analogous structure (p. 409)
artificial selection (p. 403)
camouflage (p. 406)
embryo (p. 410)
homologous structure (p. 408)
mimicry (p. 405)
natural selection (p. 403)
vestigial structure (p. 410)

Section 15.2

Mechanisms of Evolution

Main Ideas

- Evolution can occur only when a population's genetic equilibrium changes. Mutation, genetic drift, and gene flow can change a population's genetic equilibrium, especially in a small, isolated population. Natural selection is usually a factor that causes change in established gene pools—both large and small.
- The separation of populations by physical barriers can lead to speciation.
- There are many patterns of evolution in nature. These patterns support the idea that natural selection is an important mechanism of evolution.

Vocabulary

adaptive radiation (p. 420)
allelic frequency (p. 413)
convergent evolution (p. 421)
directional selection (p. 416)
disruptive selection (p. 416)
divergent evolution (p. 420)
gene pool (p. 413)
genetic drift (p. 414)
genetic equilibrium (p. 413)
geographic isolation (p. 417)
gradualism (p. 419)
polyploid (p. 418)
punctuated equilibrium (p. 419)
reproductive isolation (p. 418)
speciation (p. 417)
stabilizing selection (p. 416)

Main Ideas

Summary statements can be used by students to review the major concepts of the chapter.

Using the Vocabulary

To reinforce chapter vocabulary, use the Content Mastery Booklet and the activities in the Interactive Tutor for Biology: The Dynamics of Life or the Glencoe Science Web site: science.glencoe.com

THE PRINCETON REVIEW

All Chapter Assessment questions and answers have been validated for accuracy and suitability by The Princeton Review.

UNDERSTANDING MAIN IDEAS

1. d
2. a
3. d

UNDERSTANDING MAIN IDEAS

1. Two closely related species of squirrels live on opposite sides of the Grand Canyon. The ancestral species probably evolved into two species because of _____.
   a. structural isolation
   b. punctuated isolation
   c. behavioral isolation
   d. geographic isolation

2. What type of evolutionary evidence do fossils provide?
   a. structural
   b. functional
   c. physiological
   d. critical

3. Which of the following is an example of direct evidence for evolution?
   a. fossils
   b. embryology
   c. vestigial structures
   d. bacterial resistance to penicillin

GLENCOE TECHNOLOGY

VIDEOTAPE
MindJogger Videoquizzes
Chapter 15: *The Theory of Evolution*
Have students work in groups as they play the videoquiz game to review key chapter concepts.

Resource Manager

Chapter Assessment, pp. 86-90
MindJogger Videoquizzes
ExamView® Pro Software
BDOL Interactive CD-ROM, Chapter 15 quiz

4. c
5. c
6. d
7. b
8. d
9. c
10. b
11. punctuated equilibrium
12. geographical isolation
13. gene pool
14. natural selection
15. variations
16. Mimicry
17. artificial
18. camouflage
19. homologous
20. adaptive radiation

APPLYING MAIN IDEAS

21. Marine biomes are very stable and slow to change such environmental factors as salinity, temperature, light penetration, etc. In stable environments, natural selection pressures tend to remain stable as well. Shark populations may be close to genetic equilibrium. They have stable relationships with both the environment and other organisms in the environment.

22. Many adaptations are related to escaping from predators. Poisons are a natural defense. If a predator eats a brightly colored insect and becomes ill, it will avoid such an organism the next time. Bright colors indicate that the organism may be poisonous and deter predators.

4. Which of the structures shown below is not homologous with the others?

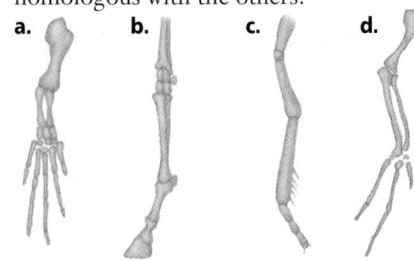
a.   **b.**   **c.**   **d.**

5. Which type of natural selection favors the average individuals in a population?
   a. directional   **c.** stabilizing
   b. disruptive   **d.** divergent

6. Which of the following pairs of terms is not related?
   a. analogous structures—butterfly wings
   b. evolution—natural selection
   c. vestigial structure—appendix
   d. adaptive radiation—convergent evolution

7. Unlike any other birds, hummingbirds have wings that allow them to hover and to fly backwards. This is an example of a _____ adaptation.
   a. physiological   **c.** reproductive
   b. structural   **d.** embryological

8. Which of the following is a true statement about evolution?
   a. Individuals evolve more slowly than populations.
   b. Individuals evolve; populations don't.
   c. Individuals evolve by changing the gene pool.
   d. Populations evolve; individuals don't.

THE PRINCETON REVIEW   **TEST-TAKING TIP**

Wear a Watch
If you are taking a timed test, you should make sure that you pace yourself and do not spend too much time on any one question, but don't spend time staring at the clock. When the test begins, place your watch on the desk and check it after each section of the test.

9. An example of a vestigial human structure is the _____.
   a. eye   **c.** appendix
   b. big toe   **d.** ribs

10. The fish and whale shown here are not closely related. Their structural similarities appear to be the result of _____.

   a. adaptive radiation
   b. convergent evolution
   c. divergent evolution
   d. punctuated equilibrium

11. The scientific hypothesis that explains how an ancestral population of elephants speciated quite rapidly after a long period of stability is _____.

12. Speciation due to physical barriers occurs as a result of _____.

13. An understanding of population genetics depends on an understanding of the _____, which is a collection of all the alleles in a population.

14. The mechanism Darwin proposed to explain how species adapt to their environment over many generations is _____.

15. The differences in the size of the peanuts in a bag are called _____.

16. _____ is the structural adaptation of an organism that enables it to resemble another harmful or distasteful species.

17. The existence of desirable characteristics in both crops and domestic animals results from the process called _____ selection.

18. A subtle adaptation that allows an organism to blend in with its surroundings is known as _____.

19. The wings of bats and the forelimbs of crocodiles are examples of _____ structures.

20. A species may find its way to an island and then evolve into many species in a process called _____.

APPLYING MAIN IDEAS

21. The structural characteristics of many species, such as sharks, have changed little over time. What evolutionary factors might be affecting their stability?

22. How might the bright colors of poisonous species aid in their survival?

23. Why is DNA a useful tool for determining possible relationships among the species of organisms?

THINKING CRITICALLY

24. **Observing and Inferring** Describe adaptive radiation as a form of divergent evolution.

25. **Interpreting Data** In a population of clams, let two alleles, *T* and *t*, represent shell color. The population consists of ten *TT* clams and ten *tt* clams. What are the allelic frequencies of the *T* and *t* in the population?

26. **Concept Mapping** Complete the concept map by using the following vocabulary terms: allelic frequency, geographic isolation, gradualism, natural selection, punctuated equilibrium, reproductive isolation, speciation.

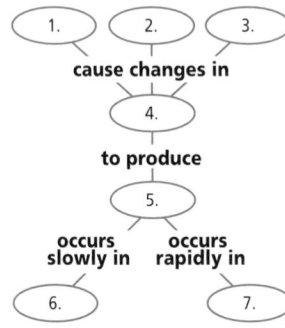

CD-ROM

For additional review, use the assessment options for this chapter found on the *Biology: The Dynamics of Life Interactive CD-ROM* and on the Glencoe Science Web site. **science.glencoe.com**

ASSESSING KNOWLEDGE & SKILLS

The following graph shows leaf length in a population of maple trees.

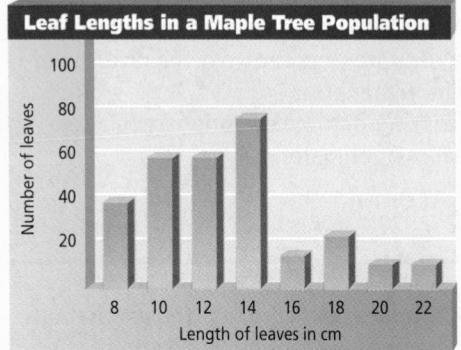

Interpreting Data Study the graph and answer the following questions.

1. What was the range of leaf lengths?
   - **a.** 14 cm
   - **c.** 20-100 cm
   - **b.** 8-22 cm
   - **d.** 10-14 cm

2. What was the average leaf length?
   - **a.** 8 cm
   - **c.** 14 cm
   - **b.** 12 cm
   - **d.** 6 cm

3. What type of evolutionary pattern does the graph most closely match?
   - **a.** artificial selection
   - **b.** stabilizing selection
   - **c.** disruptive evolution
   - **d.** directional evolution

4. **Interpreting Data** Use the graph below to explain what might be occurring in this shark population.

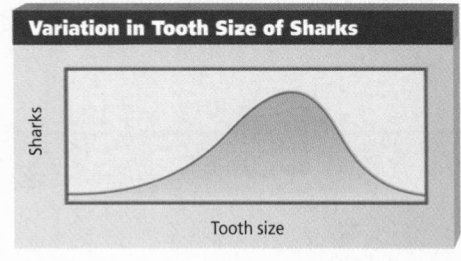

23. It is easier to quantify differences in DNA than differences in behavior or morphology.

THINKING CRITICALLY

24. In adaptive radiation, a generalized ancestor encounters an area of many available niches and eventually diverges into many species. This is an example of divergent evolution, in which species similar to ancestral species adapt to different environmental conditions.

25. *T* = 0.5; *t* = 0.5

26. 1. Natural selection; 2. Reproductive isolation; 3. Geographic isolation; 4. Allelic frequency; 5. Speciation; 6. Gradualism; 7. Punctuated equilibrium

ASSESSING KNOWLEDGE & SKILLS

1. b
2. b
3. d
4. Directional selection is occurring in favor of larger teeth in sharks.

CHAPTER 15 ASSESSMENT **427**

Chapter 16 Organizer

Refer to pages 4T-5T of the Teacher Guide for an explanation of the National Science Education Standards correlations.

Section	Objectives	Activities/Features
Section 16.1 **Primate Adaptation and Evolution** National Science Education Standards UCP.1-5; C.3, C.4, C.6; G.2, G.3 (1 session)	1. **Recognize** the adaptations of primates. 2. **Compare and contrast** the diversity of living primates. 3. **Distinguish** the evolutionary relationships of primates.	**Inside Story:** A Primate, p. 430 **MiniLab 16-1:** Comparing Old and New World Monkeys, p. 433 **Problem-Solving Lab 16-1,** p. 434 **Focus On** Primates, p. 436 **Earth Science Connection:** The Land Bridge to the New World, p. 448
Section 16.2 **Human Ancestry** National Science Education Standards UCP.2-5; A.1, A.2; C.3, C.6; E.1, E.2; G.1-3 (2 sessions)	4. **Compare and contrast** the adaptations of australopithecines with those of apes and humans. 5. **Summarize** the major anatomical changes in hominids during human evolution.	**MiniLab 16-2:** Compare Human Proteins with Those of Other Primates, p. 439 **Problem-Solving Lab 16-2,** p. 443 **Investigate BioLab:** Comparing Skulls of Three Primates, p. 446

Need Materials? Contact Carolina Biological Supply Company at 1-800-334-5551 or at **http://www.carolina.com**

MATERIALS LIST

BioLab

p. 446 metric ruler, protractor, copy of skull diagrams of *Australopithecus africanus*, *Gorilla gorilla*, and *Homo sapiens*

MiniLabs

p. 433 none
p. 439 calculator (optional)

Alternative Lab

p. 442 metric ruler or tape measure

Quick Demos

p. 432 skull models or photos
p. 441 video tape of sculptor doing paleoanthropological reconstruction

Key to Teaching Strategies

L1 Level 1 activities should be appropriate for students with learning difficulties.

L2 Level 2 activities should be within the ability range of all students.

L3 Level 3 activities are designed for above-average students.

ELL ELL activities should be within the ability range of English Language Learners.

COOP LEARN Cooperative Learning activities are designed for small group work.

P These strategies represent student products that can be placed into a best-work portfolio.

These strategies are useful in a block scheduling format.

Primate Evolution

Teacher Classroom Resources

Section	Reproducible Masters	Transparencies
Section 16.1 **Primate Adaptation and Evolution**	Reinforcement and Study Guide, pp. 69-70 L2 Concept Mapping, p. 16 L3 ELL BioLab and MiniLab Worksheets, p. 75 L2 Laboratory Manual, pp. 109-112 L2 Inside Story Poster ELL Content Mastery, pp. 77-80 L1	Section Focus Transparency 39 L1 ELL
Section 16.2 **Human Ancestry**	Reinforcement and Study Guide, p. 71-72 L2 Critical Thinking/Problem Solving, p. 16 L3 BioLab and MiniLab Worksheets, pp. 76-79 L2 Laboratory Manual, pp. 113-116 L2 Content Mastery, p. 77-80 L1 Tech Prep Applications, pp. 23-24 L2	Section Focus Transparency 40 L1 ELL Basic Concepts Transparency 23 L2 ELL Reteaching Skills Transparency 25 L1 ELL

Assessment Resources

Chapter Assessment, pp. 91-96
MindJogger Videoquizzes
Performance Assessment in the Biology Classroom
Alternate Assessment in the Science Classroom
ExamView® Pro Software 💾
BDOL Interactive CD-ROM, Chapter 16 quiz

Additional Resources

Spanish Resources ELL
English/Spanish Audiocassettes ELL
Cooperative Learning in the Science Classroom COOP LEARN
Lesson Plans/Block Scheduling

NATIONAL GEOGRAPHIC — Teacher's Corner

Products Available From Glencoe
To order the following products, call Glencoe at 1-800-334-7344:
Curriculum Kit
GeoKit: Earth's History
Videodiscs
STV: Animals
STV: Biodiversity

Products Available From National Geographic Society
To order the following products, call National Geographic Society at 1-800-368-2728:
Videos
The Diversity of Life
Fossils: Clues to the Past

Index to National Geographic Magazine
The following articles may be used for research relating to this chapter:
"The Dawn of Humans: Redrawing Our Family Tree?" by Lee Berger, August 1998.
"Expanding Worlds: The Dawn of Humans," by Rick Gore, May 1997.
"The First Steps: The Dawn of Humans," by Rick Gore, February 1997.

GLENCOE TECHNOLOGY

The following multimedia resources are available from Glencoe.
Biology: The Dynamics of Life
CD-ROM ELL

Video: *Primate Characteristics*
BioQuest: *Biodiveristy Park*
Video: *Gorilla*

Videodisc Program

Primate Charactersitics

The Infinite Voyage
The Keepers of Eden
The Dawn of Humankind

The Secret of Life Series

Gone Before You Know It: *The Biodiversity Crisis*
What's in Stetter's Pond: *The Basics of Life*
Homosapiens–Origin (a), (b)

Chapter

16 Primate Evolution

Theme Development

The main theme of this chapter is **evolution**. Students learn how the availability of human fossils and archaeological evidence affect the development of hypotheses about human evolution. Another theme, **unity within diversity**, is developed through descriptions of shared primate characteristics.

0:00 OUT OF TIME?

If time does not permit teaching the entire chapter, use the BioDigest at the end of the unit as an overview.

READING BIOLOGY

Glencoe's *Biology: The Dynamics of Life* contains many resources to assist a student's reading skills. Each chapter contains figures with expanded captions that expand on written material. Word Origins, located along the side of text, expand knowledge of biology vocabulary. Glencoe's Content Mastery Booklet helps develop reading skills while reinforcing content. In addition, use the Interactive Tutor for *Biology: The Dynamics of Life* on the Glencoe Web site to reinforce vocabulary.
science.glencoe.com

What You'll Learn

- You will compare and contrast primates and their adaptations.
- You will analyze the evidence for the ancestry of humans.

Why It's Important

Humans are primates. A knowledge of primates and their evolution can provide an understanding of human origins.

READING BIOLOGY

Scan the chapter, noting the new vocabulary words that appear. Try to determine the origin of the new words by breaking them down. As you read, make illustrations or diagrams to help clarify their meanings.

BIOLOGY Online

To find out more about primate evolution, visit the Glencoe Science Web site.
science.glencoe.com

Humans are not the only animals that construct and use tools. This chimpanzee has broken off a twig to use like a fishing pole to catch termites, a favorite food.

428 PRIMATE EVOLUTION

16.1 Primate Adaptation and Evolution

Section 16.1

Monkeys have always fascinated humans, perhaps because of the many structural and behavioral similarities we share with them. In 1871, in his book The Descent of Man, Charles Darwin proposed that there might be an evolutionary link among monkeys, apes, and humans. Today, scientists examine both living primates and primate fossils in the search for information about how primates may have evolved.

Baboons (top) and a *Homo habilis* skull

SECTION PREVIEW

Objectives
Recognize the adaptations of primates.
Compare and contrast the diversity of living primates.
Distinguish the evolutionary relationships of primates.

Vocabulary
primate
opposable thumb
anthropoid
prehensile tail

Section 16.1

Prepare

Key Concepts

Students discover the shared characteristics of primates. The characteristics of each primate group are explored and the evolution of primates is discussed.

Planning

- Obtain a human skeleton and collect pictures or specimens of primate and nonprimate mammalian skeletons for the Visual Learning.
- Obtain pictures or models of prosimian and anthropoid skulls for the Quick Demo.
- Obtain slides or videos that show examples of each primate group for the Visual Learning.

1 Focus

Bellringer

Before presenting the lesson, display **Section Focus Transparency 39** on the overhead projector and have students answer the accompanying questions.

What Is a Primate?

Have you ever gone to a zoo and seen monkeys, chimpanzees, gorillas, or baboons? If you have, then you've observed some different types of primates. A **primate** is a group of mammals that includes lemurs, monkeys, apes, and humans. Primates come in a variety of shapes and sizes, but, despite their diversity, they share common traits.

What characteristic accounts for the complex behaviors of primates? Find out by reading the *Inside Story* on the next page. Perhaps the most distinctive trait of all primates is the rounded shape of their heads. They also have a flattened face when compared with

faces of other mammals. Fitting snugly inside the rounded head is a brain that, relative to body size, is the largest brain of any terrestrial mammal. Primate brains are also more complex than those of other animals. The diverse behaviors and social interactions of primates reflect the complexity of their brains.

The majority of primates are arboreal, meaning they live in trees, and have several adaptations that help them survive there. For example, the primate skeleton is well adapted for movement among trees. All primates have relatively flexible shoulder and hip joints. These flexible joints are important for climbing and swinging among branches.

16.1 PRIMATE ADAPTION AND EVOLUTION **429**

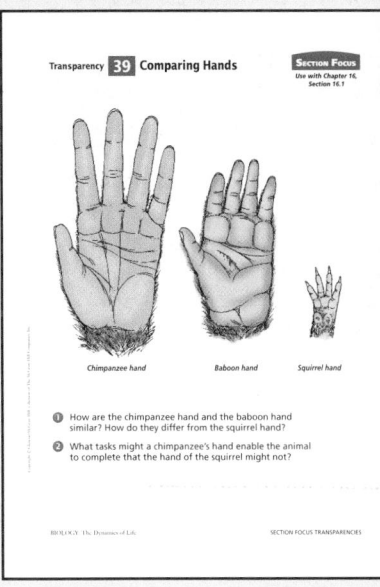

Transparency **39** Comparing Hands — **SECTION FOCUS** Use with Chapter 16, Section 16.1

Chimpanzee hand — Baboon hand — Squirrel hand

❶ How are the chimpanzee hand and the baboon hand similar? How do they differ from the squirrel hand?

❷ What tasks might a chimpanzee's hand enable the animal to complete that the hand of the squirrel might not?

BIOLOGY: The Dynamics of Life — SECTION FOCUS TRANSPARENCIES

Assessment Planner

Portfolio Assessment
Problem-Solving Lab, TWE, p. 434
Portfolio, TWE, pp. 434, 436
MiniLab, TWE, p. 439
BioLab, TWE, p. 447

Performance Assessment
Problem-Solving Lab, TWE, p. 443
Assessment, TWE, p. 445
MiniLabs, SE, pp. 433, 439

Alternative Lab, TWE, pp. 442-443
BioLab, SE, pp. 446-447

Knowledge Assessment
Assessment, TWE, p. 432
Section Assessments, SE, pp. 435, 445
Chapter Assessment, SE, pp. 449-451

Skill Assessment
MiniLab, TWE, p. 433
Assessments, TWE, pp. 435, 442
Alternative Lab, TWE, pp. 442-443

Purpose 🔲

Students examine traits common to primates.

Teaching Strategies

■ Ask students why primates are popular zoo attractions and what makes primates unique. List responses on the chalkboard.

■ Remind students that humans are primates and, although not arboreal, share these adaptations.

■ Discuss each major primate characteristic, having students explain how each is an important human adaptation.

Visual Learning

■ Display pictures or specimens of primate (including human) and other mammalian skeletons and teeth to compare and contrast them.

■ Show how a human skeleton's hands, feet, and joints are similar to those of other primates.

Critical Thinking

Binocular vision allows primates to judge depth, a useful adaptation for arboreal species.

Resource Manager

Section Focus Transparency 39 and Master **L1** **ELL**
Laboratory Manual, pp. 109-112 **L2**

A Primate

Primates are a diverse group of mammals, but they share some common features. For example, you can see in the drawing of an orangutan that primates have rounded heads and flattened faces, unlike most other groups of mammals.

Critical Thinking *Why would binocular vision be an adaptive advantage for primates?*

Orangutans

1 Opposable thumbs The primate's opposable thumbs enable it to grasp and manipulate objects. The thumb is also flexible, which increases the primate's ability to manipulate objects.

2 Vision Vision is the dominant sense in a primate. In addition to good visual perception, a primate has binocular vision, which provides it with a stereoscopic view of its surroundings.

3 Brain volume A primate's brain volume is large relative to its body size. The complex behaviors of a primate reflect its large brain.

4 Arm movement The shoulders of a primate are adapted for arm movement in different directions. Flexible arm movement is an important advantage for arboreal primates.

5 Flexible joints The flexible joint in a primate's elbow allows the primate to turn its hand in many directions.

6 Feet A primate's feet can grasp objects. However, modern primates have different degrees of efficiency for grasping objects with their feet.

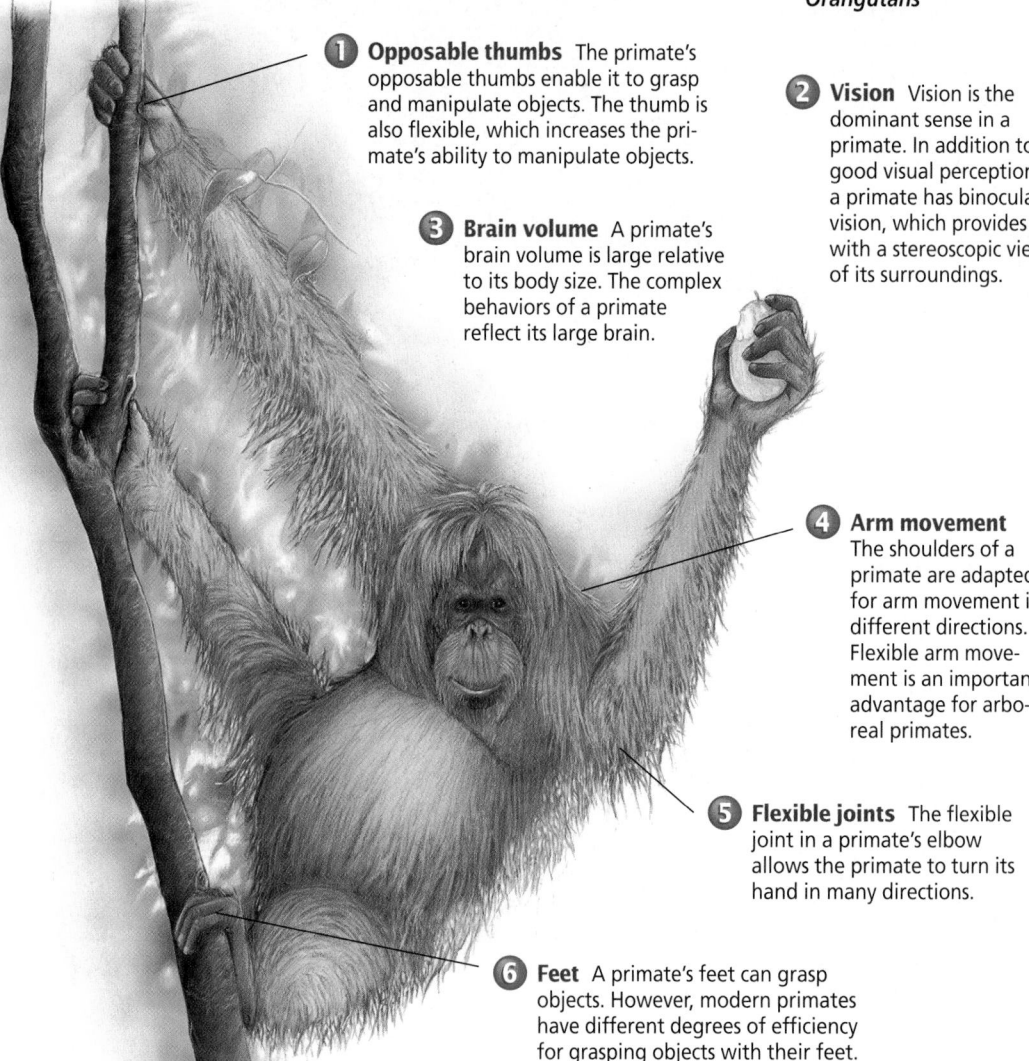

430

BIOLOGY JOURNAL

Observing a Primate

Intrapersonal Have students find the scientific name of a primate they have seen and then write answers to the following questions: Where did you first see the primate? What was it doing? What about the primate interested you? What more about the primate would you like to learn? **L1** 🔲

BIOLOGY *Online* Note Internet addresses that you find useful in the space below for quick reference.

_____

_____

_____

_____

_____

Primate hands and feet are unique among mammals. Their digits, fingers and toes, have nails rather than claws and their joints are flexible. In addition, primates have an adaptive **opposable thumb**—a thumb that can cross the palm to meet the other fingertips. Opposable thumbs enable primates to grasp and cling to objects, such as the branches of trees, *Figure 16.1.* They also enable primates to manipulate tools.

Primates have a highly developed sense of vision, called binocular vision. Primate eyes face forward so that they see an object simultaneously from two viewpoints. This eye positioning enables primates to perceive depth and thus gauge distances. As you might imagine, this type of vision is helpful for an animal jumping from tree to tree. Primates also have color vision that aids depth perception, enhances their ability to detect predators, and helps them find ripe fruits.

Primate Origins

The similarities among the many primates is evidence that primates share an evolutionary history. Scientists use fossil evidence and comparative anatomical, genetic, and

Figure 16.1
Notice the thumb of the chimpanzee. An opposable thumb helps a primate grasp and cling to objects and manipulate them.

biochemical studies of modern primates to propose ideas about how primates are related and how they evolved. Biologists classify primates into two major groups: prosimians and anthropoids, as shown in *Figure 16.2.*

Prosimianlike primates evolved first

Prosimians are small, present-day primates that include, among others, the lemurs, aye-ayes, and tarsiers. Most prosimians have large eyes and are nocturnal. They live in the tropical forests of Africa and Southeast Asia, where they prowl through the leafy canopy in search of insects, seeds, and small fruits. The earliest

WORD Origin

anthropoid
From the Greek words *anthropos*, meaning "man," and *eidos*, meaning "shape." The anthropoid apes resemble humans in their general appearance.

Figure 16.2
Primates are divided into two groups: the prosimians and the anthropoids, which are subdivided into monkeys and hominoids.

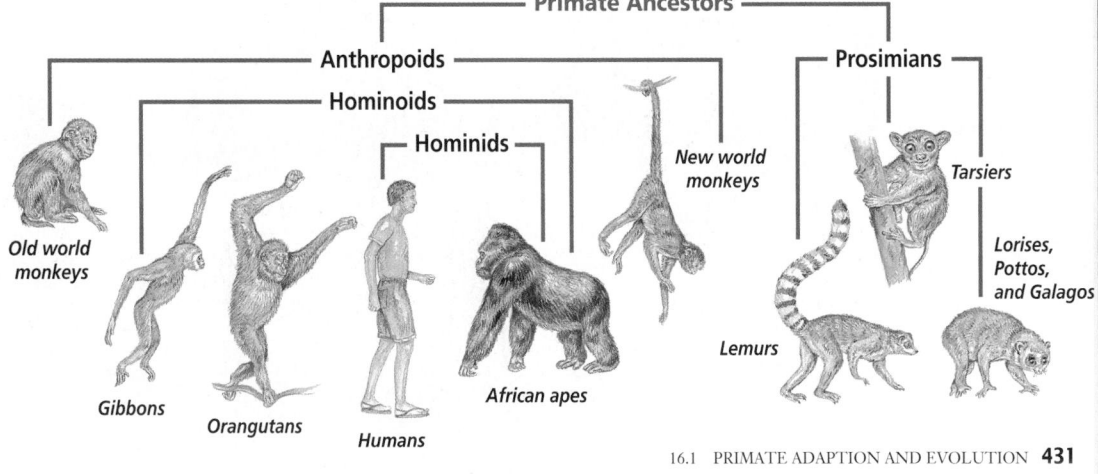

Primate Ancestors

Anthropoids
Hominoids
Hominids

Old world monkeys
Gibbons
Orangutans
Humans
African apes
New world monkeys

Prosimians
Tarsiers
Lorises, Pottos, and Galagos
Lemurs

Concept Development

Primate hands are divided into three regions: the carpus, the metacarpus, and the phalanges. Point out each region and describe its anatomy. Explain that the wrist consists of eight or nine bones aligned in two rows. Between the two rows is the midcarpal joint that provides flexibility. The joints at the juncture of the metacarpals and most phalanges lack mobility. The joint at the thumb's base is extremely mobile.

Then, remind students that, although most primate hands have the same numbers of bones, the relative bone sizes vary with the species' needs for locomotion or manipulation. For example, the slow-climbing loris has a strong thumb and long, lateral digits for grasping branches. In contrast, gibbons and spider monkeys have long, slender digits that function almost like hooks as they hang under branches.

GLENCOE TECHNOLOGY

CD-ROM
Biology: The Dynamics of Life
Video: *Primate Characteristics*
Disc 2

VIDEODISC
Biology: The Dynamics of Life
Primate Characteristics (Ch. 7)
Disc 1, Side 2, 47 sec.

Resource Manager

Concept Mapping, p. 16
L3 ELL

PROJECT

Interpreting Behavior

Paleoanthropologists use a variety of methods to interpret primate behavior from fossil evidence. Have student groups report on one of the following methods:

■ Analyzing the wear on teeth to learn what extinct primates ate **L2 ELL COOP LEARN**

■ Using cladistics to determine phylogeny **L2 COOP LEARN**

■ Analyzing muscle function in living primates **L3 COOP LEARN**

Student reports should include one or two examples of how the method answered a question about primate evolution.

432

✓ Assessment

Knowledge Have students write an essay about how their lives would be affected if their shoulder and hip joints moved only back and forth, like those of dogs and horses. **L2** 📖

Tying to Previous Knowledge

Review the environmental conditions and the types of organisms living in the early Cenozoic Era, when primates evolved. Remind students that many mammalian groups diversified at this time. Discuss how these factors may have influenced primate evolution.

Quick Demo

Use photos, illustrations, models, or actual skulls to show the differences between prosimians and anthropoids. For example, contrast the fused frontal bone of anthropoids to the unfused one of prosimians. 📖

Figure 16.3
Most prosimians are small, nocturnal animals that live in tropical environments.

A The aye-aye, a prosimian found in Madagascar, uses its long middle finger to dig for grubs.

B Tarsiers are prosimians that live in the Philippines, Borneo, and Sumatra.

fossils of prosimians are about 50 to 55 million years old.

Some scientists consider fossils of an organism called *Purgatorius* to be the earliest of primate fossils. *Purgatorius*, which probably resembled a squirrel, was a prosimianlike animal that lived about 66 million years ago. Although there are no living species of *Purgatorius*, present-day prosimians, ***Figure 16.3***, are probably quite similar.

Humanlike primates evolve

Humanlike primates are called **anthropoids** (AN thruh poydz). Anthropoids, some of which are shown in ***Figure 16.4***, include monkeys and hominoids. In turn, hominoids include apes and humans. Many features distinguish anthropoids from prosimians. In particular, anthropoids have more complex brains than prosimians. Anthropoids are also larger and have different

Figure 16.4
Monkeys and hominoids are classified as anthropoids.

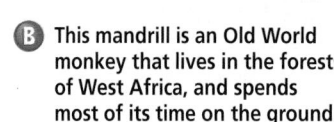

A Golden lion tamarins are arboreal New World monkeys that live in South America.

B This mandrill is an Old World monkey that lives in the forests of West Africa, and spends most of its time on the ground.

BIOLOGY JOURNAL

Cartoon Evolution

Visual-Spatial Ask students to collect some cartoons that relate to human evolution. Have them work in groups to develop their own cartoons.
L1 **ELL** 📖 **COOP LEARN**

TECHPREP

Training Animals

Ask students to suppose that they train animals for people who have visual, auditory, and physical challenges. Have them list the tasks that they could train monkeys, but not dogs, to perform. **L1** 📖

skeletal features, such as a more or less upright posture, than prosimians.

Monkeys are classified as New World monkeys or Old World monkeys. Try the *MiniLab* to compare some characteristics of these two groups of monkeys.

New World monkeys, which live in the rain forests of South America and Central America, are all arboreal. A long, muscular **prehensile tail** (pree HEN sul), characterizes many of these primates. They use the tail as a fifth limb, grasping and wrapping it around branches as they move from tree to tree. Among the New World monkeys are tiny marmosets and larger spider monkeys.

Old World monkeys are generally larger than New World monkeys. They include the arboreal monkeys, such as the colobus monkeys and guenons, the terrestrial monkeys, such as baboons, and monkeys, such as macaques, which are equally at home in trees or on the ground. Old World monkeys do not have prehensile tails. They are adapted to many environments that range from the hot, dry savannas of Africa to the cold mountain forests of Japan.

Hominoids are classified as apes or humans. Apes include gibbons, orangutans, chimpanzees, bonobos, and gorillas. Apes lack tails and have different adaptations for arboreal life from those of the prosimians and monkeys. For example, apes have long, muscled forelimbs for climbing in trees and swinging from branches.

C Gibbons are small apes that live in Southeast Asia. They have long arms and long, curved fingers.

MiniLab 16-1 — Comparing and Contrasting

Comparing Old and New World Monkeys In this activity, you will gather and then compare data about Old World monkeys and New World monkeys.

Procedure
1. Copy the data table.
2. Examine the diagrams.
3. Complete the data table.

Analysis
1. Why would a low body weight be helpful for an animal with an arboreal life style?
2. Which group appears to be more closely related to humans? Explain your answer. (Note: Humans have eight premolars, twelve molars, and a total of 32 teeth.)

New World Monkey / Old World Monkey
— Premolars
— Molars

Nostrils far apart and open up / Nostrils close and open down

Tail can grasp / Tail cannot grasp

Data Table

Characteristic	New World monkey	Old World monkey
Number of premolars ($\frac{1}{4}$ jaw)		
Number of molars ($\frac{1}{4}$ jaw)		
Total teeth in mouth		
Nostril position		
Tail		
Body weight	0.14 to 11 kg	1.2 to 30 kg

Purpose
Students will compare and contrast traits of Old World and New World monkeys.

Process Skills
compare and contrast, analyze information, draw a conclusion, think critically

Teaching Strategies
- Explain the meaning of the term "prehensile."
- Review the number and type of human teeth. Ask students to count their teeth. Discuss any differences, such as unerupted molars.
- Use photos of Old World and New World monkeys to illustrate their nostrils.

Expected Results
Number of premolars (1/4 jaw) =3, 2; number of molars (1/4 jaw)=3, 3; total teeth in mouth=36, 32; nostril position=up, down; tail=prehensile, not prehensile.

Analysis
1. It would allow for efficient movement in trees.
2. Old World monkeys share more traits with humans.

✔ Assessment
Skill Ask students to visit a zoo and determine if the monkeys are Old World or New World by their features. Use the Performance Task Assessment List for Making Observations and Inferences in **PASC**, p. 17. **L2**

Resource Manager
BioLab and MiniLab Worksheets, p. 75 **L2**

MEETING INDIVIDUAL NEEDS

Gifted
Intrapersonal Many unanswered questions about primate evolution remain. Have students report on one of the following topics. **L3**
- What group gave rise to the primates?
- What is the phylogeny of tarsiers?
- How did New World monkeys reach South America?

NATIONAL GEOGRAPHIC

VIDEODISC
GTV: Planetary Manager
Animal, Side 2

Purpose

Students will observe that the weights of several body regions of an infant and adult primate correlate with different percentages of their body weights.

Process Skills

analyze information, compare and contrast, draw a conclusion, make and use graphs, think critically, use numbers

Teaching Strategies

■ Explain to students that non-human infant primates typically go everywhere with their mother, clinging at first to her belly, and later to her back.

■ Remind students that, unlike a primate's other organs, its brain does not grow much after birth.

Thinking Critically

1. Major changes in brain size do not occur during growth.

2. The percentage decreases because the need for strong muscles to continually grasp their mothers diminishes. No. Human infants do not cling to their mothers.

✓ Assessment

Portfolio Ask students to prepare a graph similar to the one they just used. The graph should compare the percent body weights for a human infant and adult. Use the Performance Task Assessment List for Graph from Data in **PASC,** p. 39. **L2 P**

434

Problem-Solving Lab 16-1   Using Numbers

How do primate infants and adults compare? Some infant primates, such as macaques, cling to their mothers for their first few months of life. Therefore, muscles associated with clinging may represent a higher percentage of total body weight in infant macaques than in adult macaques.

Analysis

The graph shows the percentages of body weight for specific body parts of adult and infant macaques.

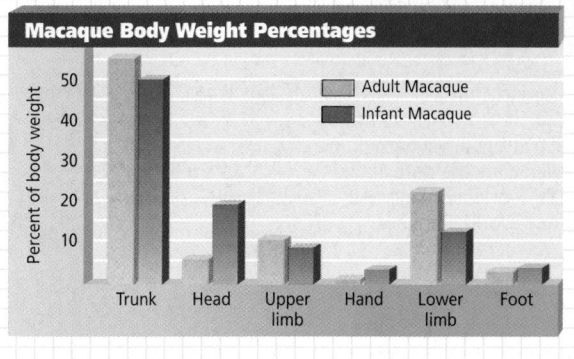

Macaque Body Weight Percentages

Percent of body weight — Adult Macaque, Infant Macaque

Trunk, Head, Upper limb, Hand, Lower limb, Foot

Thinking Critically

1. Explain the difference between the percentage of body weight of infant heads and adult heads.

2. Explain why the percentage of body weight for hands and feet change as macaques mature. Would you expect the same pattern in humans? Explain your answer.

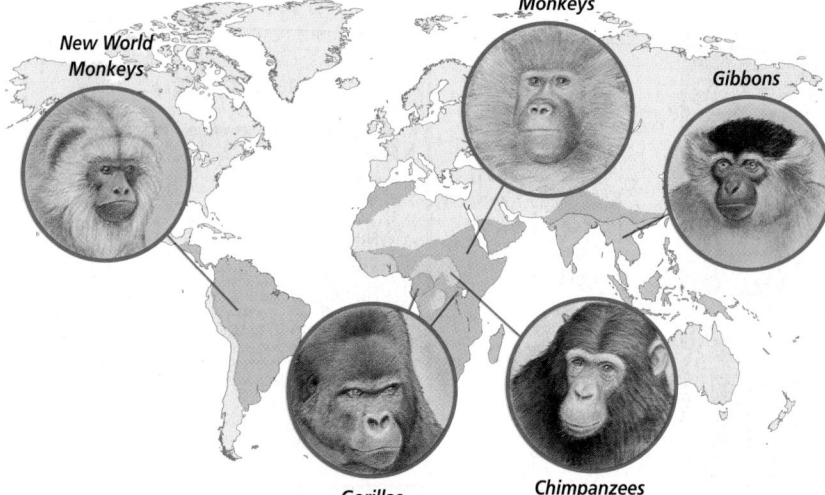

Figure 16.5
The present-day, worldwide distribution of monkeys and apes suggests that they probably evolved a long time ago from a common ancestor.

New World Monkeys

Old World Monkeys

Gibbons

Gorillas

Chimpanzees

Although many apes are arboreal, most also spend time on the ground. Gorillas, the largest of the apes, live in social groups on the ground. Among the apes, social interactions and long-term parental care indicate a large brain capacity.

Humans have an even larger brain capacity and walk upright. You will read more about human primates in the next section. Anthropologists have suggested that monkeys, apes, and humans share a common anthropoid ancestor based on their structural and social similarities. Use the *Problem-Solving Lab* to explore this idea. The oldest anthropoid fossils are from Africa and Asia and date to about 37 to 40 million years ago.

Anthropoids evolved worldwide

The oldest monkey fossils are of New World monkeys and are 30 to 35 million years old. Although New World monkeys probably share a common anthropoidlike ancestor with the Old World monkeys, they evolved independently of the Old World monkeys because of geographic isolation. In *Figure 16.5*, you

---

✓ Portfolio

Ape and Baboon Evolution

 *Visual-Spatial* Have students draw the probable events in baboon and ape evolution. Their illustrations should include the time period and the environmental factors that may have influenced ape and baboon evolution. **L3 ELL P**

MEETING INDIVIDUAL NEEDS

Learning Disabled

Visual-Spatial Give students photographs, illustrations, or videos of each group of primates. Have students write the names of each group and at least five of its characteristics in their journals. Then, have students review their lists to compare and contrast the characteristics. **L1 ELL**

Figure 16.6
Modern apes are diverse, and fossils indicate that ancient apes were even more diverse. Orangutans are arboreal apes that live in the forests of Borneo and Sumatra **(a)**. Gorillas are ground-dwelling African apes that live in small social groups **(b)**.

can see the worldwide geographic distribution of monkeys and apes.

Old World monkeys probably evolved more recently than New World monkeys. Scientists suspect this is true because the oldest fossils of Old World monkeys are only about 20 to 22 million years old. The fossils indicate that the earliest Old World monkeys were arboreal like today's New World monkeys.

Hominoids evolved in Asia and Africa

According to the fossil record, there was a global cooling when the hominoids evolved in Asia and Africa. At about the same time, the Old World monkeys evolved and became adapted to this climatic cooling. Fossils indicate how the apes adapted and diversified. You can see the modern-day diversity of apes in *Figure 16.6.*

Remember that hominoids include the apes and humans. By examining the DNA of each of the modern hominoids, scientists have evaluated the probable order in which the different apes and humans evolved. From this type of evaluation, it appears that gibbons were probably the first apes to evolve, followed by the orangutans that are found in southeast Asia. Finally, the African apes, chimpanzees, and gorillas evolved. Some anthropologists suggest that one of the groups of African apes was the ancestor of modern humans.

Section Assessment

Understanding Main Ideas
1. What adaptations help primates live in the trees?
2. Describe how hominoids are classified.
3. What features distinguish anthropoids from prosimians?
4. What is the major physical difference between Old World monkeys and New World monkeys?

Thinking Critically
5. Imagine you are a world famous primatologist, a scientist who studies primates. An unidentified,

complete fossil skeleton arrives at your lab. You suspect that it's a primate fossil. What observations would you make to determine if your suspicions are accurate?

SKILL REVIEW
6. **Classifying** Make a table listing the different types of primates, key facts about each group, and how the groups might be related. For more help, refer to *Organizing Information* in the **Skill Handbook.**

Check for Understanding
Have students write the shared characteristics of primates and describe beside each its adaptive significance. **L1**

Reteach
Visual-Spatial Give students outline maps of the world. Have them develop a key to show where the groups of primates live. **L1** **ELL**

Extension
Naturalist Have students prepare a primate phylogenetic tree that contains the name of each primate group, when the group evolved, and the group's unique characteristics. **L2** **ELL**

✔ Assessment
Skill Have students create in their journals a diagram that shows the possible evolutionary relationships of the primate groups. **L2** **ELL**

---

4 Close

Discussion
Review the highlights of primate evolution. Have students identify the major evolutionary developments. **L1**

Resource Manager

Reinforcement and Study Guide, pp. 69-70 **L2**
Content Mastery, p. 78 **L1**

Section Assessment

1. Opposable thumbs, digits with nails, and flexible feet help primates grip branches. Flexible skeletons and binocular vision enable them to gauge depth and distance.
2. Hominoids are classified as apes or humans.
3. Anthropoids have larger brains, different skull and skeletal structures, and larger sizes than prosimians.
4. New World monkeys have prehensile tails. Old World monkeys are generally larger and lack prehensile tails.
5. Observations could include the presence of an opposable thumb, a large brain size, nails rather than claws, binocular vision, and flexible joints.
6. Check student lists for accuracy and understanding of the main concepts.

Focus On

Primates

Purpose

Students will learn about the characteristics of the major primate groups.

Background

By investigating the functional relationships between primate features, such as size, tooth structure, bone shape, and behavioral habits, the probable evolution of many anatomical differences can be better understood.

Teaching Strategies

■ Discuss the characteristics of each group, emphasizing how each adaptation is important for its environment.

■ Guide students to hypothesize how natural selection may have affected the groups' evolution.

■ Discuss some of the methods used to analyze primate fossils, such as biomechanical studies to infer locomotion and scanning-electron microscopy to infer diet.

436

SQUIRREL MONKEY

FOCUS ON
Primates

Catch the gaze of an orangutan and you'll be staring into a face very much like your own. Similarities between apes and humans are striking—expressive eyes, fingers that can grasp, keen intelligence, and complex social systems. The resemblance is no coincidence. Apes and other primates are humans' closest relatives. The Primate order is made up of 13 families, including *Hominidae*, to which *Homo sapiens*, our species, belongs.

436

PROSIMIANS ▶

These small, tree-dwelling prosimians look least like other primates. Their triangular faces, set off by large round eyes, lack muscles needed to make facial expressions that other primates use for communication. They can be as small as a mouse or as big as a large house cat. Perhaps the best known prosimians are the lemurs (right), which live only in Madagascar and neighboring islands off the coast of eastern Africa.

AYE-AYE

RING-TAILED LEMUR

OLD WORLD MONKEYS

MANDRILL

Monkeys found in Europe, Asia, and Africa are called Old World monkeys. They grow larger than their New World relatives and have no prehensile tail for grasping. Pads of tough skin on their rumps cushion them while they are seated. Among Old World monkeys, the mandrill (above) wins the prize for the most colorful face. The Japanese macaque (right), or snow monkey, lives farther north than any other species of monkey.

JAPANESE MACAQUES

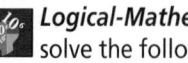
Portfolio

Fossil Problems

Logical-Mathematical Have students solve the following problems.

■ A fossil skeleton of a primate known as *Apidium* is discovered in Egypt. It is 31 million years old and shows similarities to New World monkeys. Could this animal be an ancestor of New World monkeys? *It is unlikely because the fossil is from the same period when the New World monkeys evolved.* **L2** **P**

■ A jaw fragment discovered in eastern Africa is 8 million years old. Scientists suggest that it might be from the common ancestor of African apes and humans. What tests might you perform to verify this idea? *analysis of skeletal structures and brain size* **L2** **P**

CAPUCHINS

NEW WORLD MONKEYS ▲

Unlike their Old World counterparts, New World monkeys have prehensile tails. Sometimes compared to an extra hand, the tail can wrap around a tree limb and support the monkey's weight. Thus, the animal can dangle upside down to eat. The capuchins (above) of Central and South America have thumbs that can move to touch other fingers and help them pick up food.

ORANGUTAN

CHIMPANZEE

APES

Unlike monkeys, apes have no tails, and they are usually larger than monkeys. While monkeys run on all fours, apes walk on two legs with support from their hands. Chimpanzees (left), gibbons, gorillas, and orangutans (above) are all apes. Living in Africa and Asia, these primates have large brains and are considered to be more like humans than any other animal. They are subject to many of the same diseases as humans, can use simple tools, and some have been taught to communicate with humans using sign language.

EXPANDING Your View

1 THINKING CRITICALLY Examine the photo of New World monkeys. How is having a tail an adaptive advantage for these primates?

2 COMPARING AND CONTRASTING Which of the species in this feature probably live in trees? Explain your answer.

Visual Learning

■ Prepare a video or slide presentation of one or more species of each primate group to provide detail about the groups' behaviors and ecologies.

■ Show pictures of primate fossils and discuss how each fossil provides evidence of primate evolution.

Answers to Expanding Your View

1. The tail aids in balance and allows them to grasp a branch and keep their hands free.
2. Long-tailed primates probably live in trees: prosimians, New World monkeys, and some Old World monkeys.

GLENCOE TECHNOLOGY

CD-ROM
Biology: The Dynamics of Life
Video: *Gorilla*
Disc 4

PROJECT

All About a Primate

Interpersonal Have student groups make presentations about particular species of primates. Presentations can be models, posters, videos, or skits and must include behavior, anatomy, ecology, taxonomy, evolution, and importance to humans. **L3** **ELL** 🔲 **COOP LEARN**

MEETING INDIVIDUAL NEEDS

English Language Learners

Linguistic Have students define the following terms: opposable thumb, binocular vision, prosimian, anthropoid, and ape. Have students circle the word in the following group that does not belong: New World monkey, *opposable thumb*, Old World monkey, ape, prosimian **L1** **ELL** 🔲

437

Prepare

Key Concepts

Students will study the archaeological evidence of human evolution. They will investigate ideas about the behavior of different groups of probable human ancestors and how cultural adaptations, such as the use of tools, fire, and language, may have originated.

Planning

Gather several round stones to use as tools for the Project.

1 Focus

Bellringer

Before presenting the lesson, display **Section Focus Transparency 40** on the overhead projector and have students answer the accompanying questions.

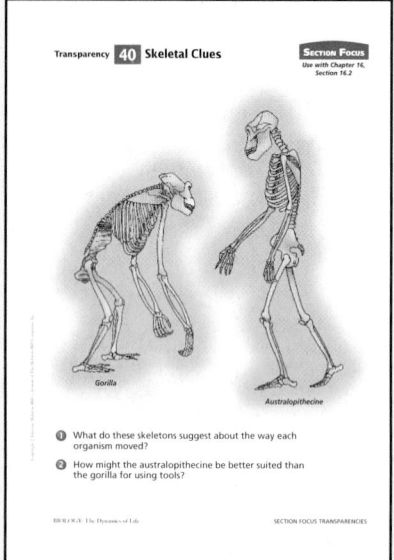

Transparency **40** Skeletal Clues

SECTION FOCUS
Use with Chapter 16,
Section 16.2

Gorilla

Australopithecine

① What do these skeletons suggest about the way each organism moved?

② How might the australopithecine be better suited than the gorilla for using tools?

SECTION FOCUS TRANSPARENCIES

SECTION PREVIEW

Objectives

Compare and contrast the adaptations of australopithecines with those of apes and humans.

Summarize the major anatomical changes in hominids during human evolution.

Vocabulary

hominid
bipedal
australopithecine
Neanderthal
Cro-Magnon

Section

16.2 Human Ancestry

What would it be like to discover an ancient primate skull? Well, you'd probably have to spend weeks at the excavation site brushing small sections of the area each day. Some days you'd find a piece of bone; other days you wouldn't. After weeks of tedious labor, you might be ready to assemble the pieces of bone in your laboratory. After more weeks of work, you might finally be looking into the empty orbits of a very human-looking skull that generates more questions than answers. What organism was this? How did it live?

An australopithecine reconstruction (above) and an australopithecine skull

Hominids

Some scientists propose that between 5 and 8 million years ago in Africa, a population that was ancestral to the African apes and humans diverged into two lines. According to this hypothesis, one line evolved into the African apes—gorillas and chimpanzees. The other line evolved into modern humans. These two lines are collectively called the **hominids** (hoh MIHN udz)—primates that can walk upright on two legs and include gorillas, chimpanzees, bonobos, and humans. Hominids do not include the other types of apes—the gibbons and orangutans.

There are relatively few fossils to support this hypothesis, but DNA studies of the modern hominids provide data that support the idea. You can work with some of these data in the *MiniLab* on the next page.

Some anthropologists suggest that the divergence of the African population of ancestral hominids might have occurred in response to environmental changes that forced some ancestral apes to leave their treetop environments and move onto the ground to find food. In order to move efficiently on the ground while avoiding predators, it was helpful for the apes to be **bipedal,** meaning able to walk on two legs. In addition to speed, walking on two legs leaves the arms and hands free for other activities, such as hunting, feeding, protecting young, and using tools.

Resource Manager

Section Focus Transparency 40 and
Master **L1** **ELL**
Laboratory Manual, pp. 113-116 **L2**

GLENCOE TECHNOLOGY

VIDEODISC The Infinite Voyage: *The Dawn of Humankind, Dating Fossils: Effects of Dating Methods and Interbreeding Theories* (Ch. 5) 4 min.

DNA Studies Create Controversy (Ch. 7) 3 min. 30 sec.

Therefore, apes with the ability to walk upright better and more often than others probably survived more successfully on the ground and lived to reproduce and pass the characteristic to offspring. According to this reasoning, the bipedal organisms that evolved might have been the earliest forms of a hominid.

Although the fossil record is incomplete, more hominid fossils are found every year. The many fossils that scientists have found reveal much about the anatomy and behavior of early hominids. Fossils of skulls provide scientists with information about the appearance and brain capacity of the early hominid types. Complete the *BioLab* at the end of the chapter to learn more about the kinds of information scientists gather from skulls of hominids.

Early hominids walked upright

In *Figure 16.7*, you see a South African anatomist, Raymond Dart, who, in 1924, discovered a skull of a

Figure 16.7
Raymond Dart discovered the first australopithecine fossil, the Taung child, *Australophithecus africanus.* The skull has features of both apes and humans.

MiniLab 16-2 Analyzing Information

Compare Human Proteins with Those of Other Primates
Scientists use differences in amino acid sequences in proteins to determine the evolutionary relationships of living species. In this activity, you'll compare representative short sequences of amino acids of a protein among groups of primates to determine their evolutionary history.

Table 16.1 Aminio Acid Sequences in Primates

Baboon	Chimp	Lemur	Gorilla	Human
ASN	SER	ALA	SER	SER
THR	THR	THR	THR	THR
THR	ALA	SER	ALA	ALA
GLY	GLY	GLY	GLY	GLY
ASP	ASP	GLU	ASP	ASP
GLU	GLU	LYS	GLU	GLU
VAL	VAL	VAL	VAL	VAL
ASP	GLU	GLU	GLU	GLU
ASP	ASP	ASP	ASP	ASP
SER	THR	SER	THR	THR
PRO	PRO	PRO	PRO	PRO
GLY	GLY	GLY	GLY	GLY
GLY	GLY	SER	GLY	GLY
ASN	ALA	HIS	ALA	ALA
ASN	ASN	ASN	ASN	ASN

Procedure
1. Copy the data table.
2. For each primate listed in the table above, determine how many amino acids differ from the human sequence. Record these numbers in the data table.
3. Calculate the percentage differences by dividing the numbers by 15 and multiplying by 100. Record the numbers in your data table.

Analysis
1. Which primate is most closely related to humans? Least closely related?
2. Construct a diagram of primate evolutionary relationships that most closely fits your results.

Data Table		
Primate	Amino acids different from humans	Percent difference
Baboon		
Chimpanzee		
Gorilla		
Lemur		

---

2 Teach

The BioLab at the end of the chapter can be used at this point in the lesson.  **INVESTIGATE BioLab**

MiniLab 16-2

Purpose
Students will learn how comparing amino acid sequences indicates phylogenetic relationships.

Process Skills
compare and contrast, make and use tables, interpret data

Teaching Strategies
■ Explain the molecular and biochemical methods used to determine phylogeny.

Expected Results
The human, chimpanzee, and gorilla sequences are identical. Baboons differ by 33 percent and lemurs by 47 percent.

Analysis
1. gorilla and chimpanzee; lemur
2. Baboons should branch off lemurs. Gorillas, chimpanzees, and humans should be close together.

✓ Assessment
Portfolio Ask students to summarize this activity. Have them predict the results of analyzing another protein. Use the Performance Task Assessment List for Lab Report in **PASC**, p. 47 **L1** **P**

Resource Manager

BioLab and MiniLab Worksheets, p. 76 **L2**
Tech Prep Applications, p. 23 **L2**

Cultural Diversity

The Hominid Gang

Linguistic Finding human fossils is difficult and requires considerable expertise. Introduce the contributions of Kenyan fossil expert Kamoya Kimeu, the leader of a team of fossil hunters known as the Hominid Gang. Since the 1960s, Kimeu's work has led to the discovery of many important hominid fossils, including "Lucy" and the 12-year-old *Homo erectus* male, the "Strapping Youth."

Have students read sections from *Origins Reconsidered* by Richard Leakey and Roger Lewin or other similar books to learn more about Kimeu and the techniques involved in finding human fossils. **L3**

young hominid with an apelike braincase and facial structure. However, the skull also had an unusual feature for an ape skull—the position of the *foramen magnum*, the opening in the skull through which the spinal cord passes as it leaves the brain.

In the fossil, the opening was located on the bottom of the skull, as it is in humans but not in apes. Because of this feature, Dart proposed that the organism had walked upright. He classified the organism as a new primate species, *Australopithecus africanus* (aw stray loh PIHTH uh kus • af ruh KAHN us), which means "southern ape from Africa." The skull that Dart found has been dated at about 1 to 2 million years old.

Since Dart's discovery, paleoanthropologists, scientists who study human fossils, have recovered many more australopithecine specimens. They describe an **australopithecine** as an early hominid that lived in Africa and possessed both apelike and humanlike characteristics.

WORD Origin

paleoanthropology
From the Greek words *paleo*, meaning "ancient," *anthropo*, meaning "human," and *logos*, meaning "study." Paleoanthropology is the study of human fossils.

Early hominids:
Apelike and humanlike

Later, in East Africa in 1974, an American paleoanthropologist, Donald Johanson, discovered a nearly complete australopithecine skeleton that he called "Lucy" after a popular song of the time. Radiometric dating shows that Lucy probably lived about 3.5 million years ago. Johanson proposed that the Lucy skeleton was a new species, *Australopithecus afarensis*. Other fossils of *A. Afarensis* indicate that this species probably existed between 3 and 5 million years ago, making it the earliest known hominid species.

Although the fossils show that *A. afarensis* individuals had apelike shoulders and forelimbs, the structure of the pelvis, as shown in *Figure 16.8*, indicates that these individuals were bipedal, like humans. On the other hand, the size of the braincase suggests that their brains had a small, apelike volume and not a larger human volume.

Figure 16.8
Some skeletal features of an australopithecine are intermediate between those of modern apes and humans. Compare the skull and pelvic bone of *Australopithecus afarensis* (**b**) with those of the chimpanzee (**a**) and the human (**c**).

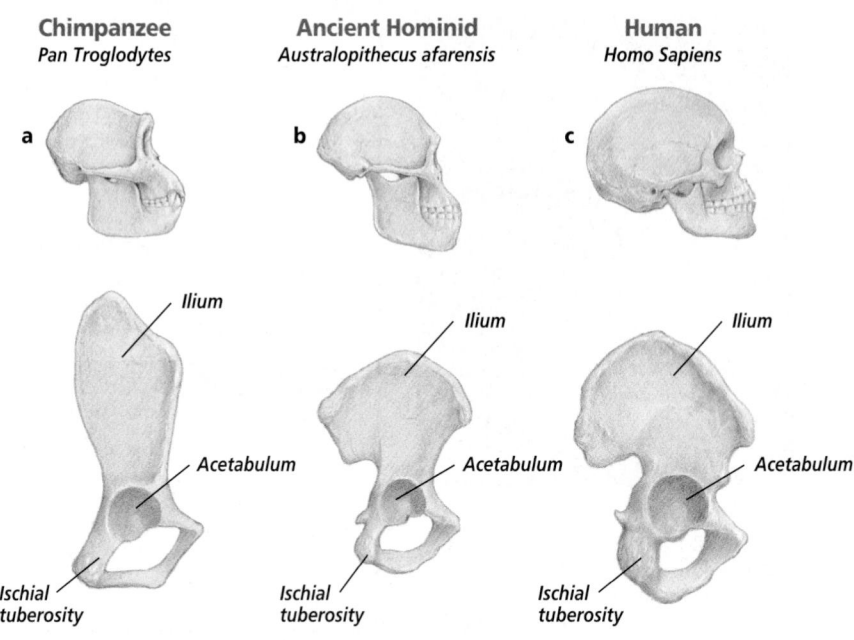

Chimpanzee
Pan Troglodytes

Ancient Hominid
Australopithecus afarensis

Human
Homo Sapiens

440 PRIMATE EVOLUTION

440

You may be wondering what life was like for hominids like Lucy. Because of the combination of ape-like and humanlike features, one idea is that *A. afarensis* and other species of australopithecines might have lived in small family groups, sleeping and eating in trees. But, to travel, they walked upright on the ground. The fossil record indicates that an *A. afarensis* individual rarely survived longer than 25 years.

In addition to fossils of *A. afarensis* and *A. africanus*, fossils of two, or perhaps three, other species of australopithecines have been found. These other species, discovered at sites in East Africa and South Africa, are dated about 1 to 2.5 million years old. Overall, the later species are similar to the earlier ones. However, these later hominids are grouped into the genus *Paranthropus* because their fossils suggest that the individuals were more robust and had larger teeth and jaws than earlier species.

The evolutionary relationships among australopithecines are not clear from the fossil record. However, the genus disappears from the record at about 1 million years ago. Although australopithecines became extinct, some paleoanthropologists propose that an early population of these hominids might have been ancestral to modern hominids.

The Emergence of Modern Humans

Any ideas about the evolution of modern hominids must include how bipedalism and a large brain evolved. Australopithecine fossils provide support for the idea that bipedalism evolved first. But when did a large brain evolve in a hominid species? When did hominids begin to use tools and develop culture?

Figure 16.9
Louis and Mary Leakey discovered many fossils in the Olduvai Gorge area of Tanzania, Africa.

Early members of the genus *Homo* made stone tools

In 1964, anthropologists Louis and Mary Leakey, *Figure 16.9*, described skull portions belonging to another type of hominid in Tanzania, Africa. This skull was more humanlike than those of australopithecines. In particular, the braincase was larger and the teeth and jaws were smaller, more like those of modern humans. Because of the skull's human similarities, the Leakeys classified the hominid with modern humans in the genus *Homo*. Because stone tools were found near the fossil skull, they named the species *Homo habilis*, which means "handy human."

Radiometric dating indicates that *Homo habilis* lived between about 1.5 and 2 million years ago. It is the earliest known hominid to make and use stone tools. These tools suggest that *Homo habilis* might have been a scavenger who used the stone tools to cut meat from carcasses of animals that had been killed by other animals.

Figure 16.10
The average brain volume of *Homo habilis* was 600 to 700 cm³, smaller than the average 1350 cm³ volume of modern humans, but larger than the 400 to 500 cm³ volume of australopithecines.

You can see a *Homo habilis* skull in *Figure 16.10*.

Hunting and using fire

Some anthropologists propose that a *H. habilis* population gave rise to another species about 1.6 million years ago. This new hominid species was called *Homo erectus*, which means "upright human." *H. erectus* had a larger brain and a more humanlike face than *H. habilis*. However, it had prominent browridges and a lower jaw without a chin, as shown in *Figure 16.11*, which are apelike characteristics.

Some scientists interpret the stone tools called hand axes that they find at some *H. erectus* excavation sites as an indication that *H. erectus* hunted. In caves at these sites, they have also found hearths with charred bones. This evidence suggests that these hominids used fire and lived in caves.

The distribution of fossils indicates that *H. erectus* migrated from Africa about 1 million years ago. Then this hominid spread through Africa and Asia, and possibly migrated into Europe, before becoming extinct between 300 000 and 500 000 years

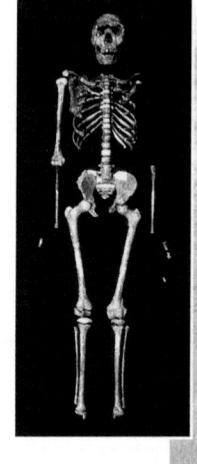

Figure 16.11
An almost complete *Homo erectus* skeleton of a 12-year-old male was discovered in East Africa in 1985. *H. erectus* had a brain volume of about 900 cm³ and long legs like modern humans.

Alternative Lab

Measuring Human Variation

Purpose
Students will analyze variations in three human traits.

Materials
metric ruler or tape measure

Procedure
Give students the following directions.
1. Prepare a data table with the following headings: Height (cm), Length of left index finger (cm), and Length of left forearm (cm).
2. Working with a partner, measure in centimeters your height, the length of your left index finger, and length of your left forearm (elbow to wrist). Record the results in your data table and on a class data table. Calculate a class average for each measurement.
3. Divide the class measurements for each trait into five equal intervals. Count the number of students within each interval. Make bar graphs for each measurement.

Expected Results
The data will vary. Students will note a wide range of variation in each trait.

ago. However, some scientists propose that more human-looking hominids might have arisen from *H. erectus* before it disappeared.

Culture developed in modern humans

The incomplete fossil record and the inaccuracies of dating fossils produce controversy about dating the emergence of modern humans. However, paleoanthropologists have enough evidence to propose possible paths to the origin of the modern human species, *Homo sapiens*. A description of one such path follows.

The fossil record indicates that the species *H. sapiens* appeared in Europe, Africa, the Middle East, and Asia about 100 000 to 400 000 years ago. The early forms of the species are called archaic *H. sapiens* because their skulls resemble those of *H. erectus* but have less prominent browridges, more bulging foreheads, and smaller teeth. Also, the braincases of archaic *H. sapiens* are larger than those of *H. erectus*, and can contain a brain volume of 1000 to 1400 cm³, which is within the modern human range.

Best known among these archaic *H. sapiens* were Neanderthals (nee AN dur tawlz), illustrated in *Figure 16.12*.

You can compare the sizes of two skulls of the genus *Homo* in the *Problem-Solving Lab* on this page.

The **Neanderthals** lived from about 35 000 to 100 000 years ago in Europe, Asia, and the Middle East.

Figure 16.12
Neanderthals were skilled hunters. They had many tools, including spears, scrapers, and knives.

443

Analysis

1. What is the average height of your class members? The average forearm length? The average finger length? *Answers depend on measurements.*
2. Explain the variations. *Traits and growth rates vary among humans.*
3. How might the data change if you measured the same traits of a group of adults? *fewer extreme measurements*

Visual Learning

Figure 16.14 illustrates two possible phylogenetic trees of human evolution. Point out that some paleoanthropologists have suggested alternative phylogenies.

Enrichment

 Visual-Spatial Have students draw probable evolutionary pathways of humans that include time periods and three characteristics of each species. **L3**

GLENCOE
TECHNOLOGY

VIDEODISC
The Secret of Life
Homo sapiens—
Origin (a)

Homo sapiens—Origin (b)

The Infinite Voyage: *The Dawn of Humankind, Bridging of Fossils and Genetic Research* (Ch. 9), 2 min. 30 sec.

Evolution of the Mind (Ch. 10) 4 min.

Resource Manager

Basic Concepts Transparency 23 and Master **L2** **ELL**
Reinforcement and Study Guide, pp. 71-72 **L2**
Content Mastery, pp. 77, 79-80 **L1**

Figure 16.13
The dwelling sites of Cro-Magnons, full of cave paintings, detailed stone and bone artifacts, and tools, have been excavated in Europe.

Fossils reveal that Neanderthals had thick bones and large faces with prominent noses. The brains of Neanderthals were at least as large as those of modern humans.

The fossil records also indicate that Neanderthals lived in caves during the ice ages of their time. In addition, the tools, figurines, flowers, and other evidence from excavation sites, such as burial grounds, suggest that Neanderthals may have had religious views and communicated through spoken language.

What happened to Neanderthals?

Could Neanderthals have evolved into modern humans? The fossil record shows that a more modern type of *H. sapiens* spread throughout Europe between 35 000 to 40 000 years ago. This type of *H. sapiens* is called Cro-Magnon (kroh MAG nun). **Cro-Magnons** were identical to modern humans in height, skull structure, tooth structure, and brain size. Paleoanthropologists suggest that Cro-Magnons were

toolmakers and artists, as shown in *Figure 16.13*. Cro-Magnons probably also used language, as their skulls contain a bulge that corresponds to the area of the brain that is involved in speech in modern humans.

Did Neanderthals evolve into Cro-Magnons? Current genetic and archaeological evidence indicates that this is unlikely. Current dates for hominid fossils suggest that modern *H. sapiens* appeared in both South Africa and the Middle East about 100 000 years ago, which was about the same time the Neanderthals appeared. In addition, genetic evidence supports the idea of an African origin of modern *H. sapiens*, perhaps as early as 200 000 years ago. This idea suggests that the African *H. sapiens* migrated to Europe and Asia.

Most fossil evidence supports the idea that Neanderthals were most likely a side branch of *H. sapiens*, and not an ancestral branch of modern humans. Look at *Figure 16.14* to see two proposed evolutionary paths to modern humans.

444 PRIMATE EVOLUTION

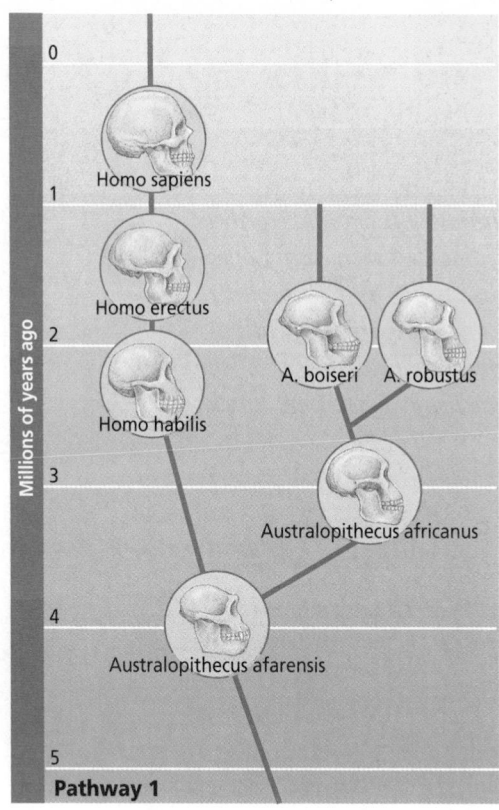

Pathway 1

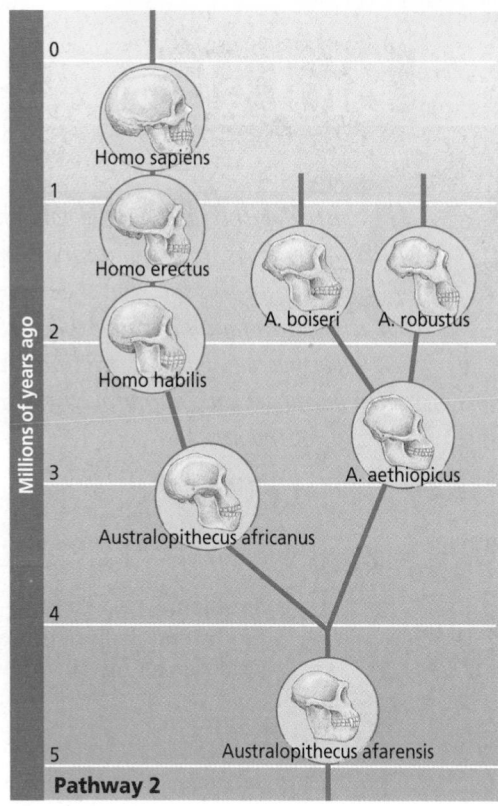

Pathway 2

Fossil evidence shows that humans have not changed much anatomically over the last 200 000 years. Humans probably first established themselves in Africa, Europe, and Asia. Then, about 12 000 years ago, evidence shows that they crossed a land bridge into North America. You can read more abouth this event in the *Earth Science Connection* at the end of the chapter. By 8000 to 10 000 years ago, Native Americans had built permanent settlements and were domesticating animals and farming.

Figure 16.14
These diagrams represent two possible pathways for the evolution of *Homo sapiens*.

Section Assessment

Understanding Main Ideas

1. What evidence supports the idea that species of australopithecines were intermediate forms between apes and humans?
2. Why was the development of bipedalism a very important event in the evolution of hominids?
3. What evidence supports the idea that *H. habilis* was an ancestor of *H. erectus*?
4. Describe the evidence that supports the idea that Neanderthals were not the ancestors of Cro-Magnon people?

Thinking Critically

5. What kind of animal bones might you expect to find at the site of *Homo habilis* remains if *H. habilis* was a scavenger? A hunter?

SKILL REVIEW

6. **Interpreting Scientific Illustrations** Draw a time line to show the evolution of hominids. Indicate each species of hominid that evolved and where it evolved. For more help, refer to *Thinking Critically* in the **Skill Handbook**.

Section Assessment

1. They have a brain capacity larger than apes and smaller than humans. They climbed trees like apes, but walked like humans.
2. Bipedalism allowed hominids to better scan the horizon for predators and freed their forelimbs for carrying objects.
3. *Homo habilis* was the first hominid to purposefully construct stone tools.
4. Genetic and archaeological evidence suggests that Neanderthals were not the ancestors of Cro-Magnon people.
5. If *H. habilis* were scavengers, bones that contain little meat, such as ribs, would be at their sites. If they were hunters, there would be meat-containing bones at their sites.
6. Evaluate the time lines for accuracy.

3 Assess

Check for Understanding

Visual-Spatial Have students construct a time line of the following events; use of fire, bipedal movement, tool use, language, art, burial of the dead. **L1** **ELL**

Reteach

Visual-Spatial Have students make a bar graph showing the evolution of human brain size. **L2**

Extension

Have students illustrate a human phylogenetic tree. **L3**

✔ Assessment

Performance Have students give an oral report about a day in the life of a hominid species. **L1** **ELL**

4 Close

Discussion

Discuss what future paleoanthropologists might determine about today's humans. Have students identify probable artifacts and their information. **L1**

GLENCOE TECHNOLOGY

VIDEODISC
The Infinite Voyage:
The Dawn of Humankind, "Out-of-Africa" vs. Multiregional Debate on Origination of Modern Man (Ch. 8)
6 min.

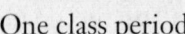

INVESTIGATE BioLab

ANALYZE AND CONCLUDE

1. Humans have a small facial area compared with brain area. Apes have a large facial area compared with brain area. Australopithecines were intermediate between apes and humans but closer to apes.

Resource Manager

BioLab and MiniLab Worksheets, pp. 77-78 **L2**

INVESTIGATE BioLab — Comparing Skulls of Three Primates

Australopithecines are the earliest hominids in the fossil record. In many ways, their anatomy is intermediate between living apes and humans. In this lab, you'll determine the apelike and humanlike characteristics of an australopithecine skull, and compare the skulls of australopithecines, gorillas, and modern humans. The diagrams of skulls shown below are one-fourth natural size. The heavy black lines indicate the angle of the jaw.

PREPARATION

Problem

How do skulls of primates provide evidence for human evolution?

Objectives

In this BioLab, you will:
■ **Determine** how paleoanthropologists study early human ancestors.
■ **Compare and contrast** the skulls of australopithecines, gorillas, and modern humans.

Materials

metric ruler
protractor
copy of skull diagrams

Skill Handbook

Use the **Skill Handbook** if you need additional help with this lab.

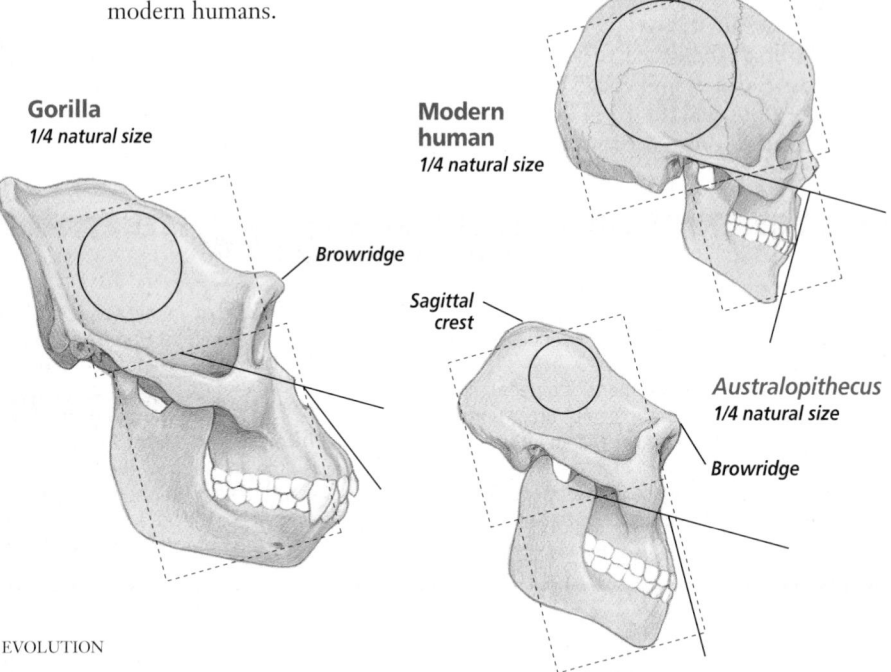

Gorilla
1/4 natural size

Modern human
1/4 natural size

Browridge

Sagittal crest

Australopithecus
1/4 natural size

Browridge

PROCEDURE

Teaching Strategies

■ Remind students that australopithecines existed between 1 and 5 million years ago, around the time that African apes and humans probably diverged. Because of their age, australopithecines were probably primitive anatomically.

■ To estimate cranial capacities, students should draw circles just inside each skull on the worksheets provided or on the copied and enlarged skulls from the student edition page.

■ Estimating cranial capacity by multiplying by the factor of 200 is suitable only for these drawings because the factor is based on the scale of the drawings at one-half natural size.

PROCEDURE

1. Your teacher will provide copies of the skulls (1/2 natural size) of *Australopithecus africanus*, *Gorilla gorilla*, and *Homo sapiens*.
2. The rectangles drawn over the skulls represent the areas of the brain (upper rectangle) and face (lower rectangle). On each skull, determine and record the area of each rectangle (length × width).
3. Measure the diameters of the circles in each skull. Multiply these numbers by 200 cm². The result is the cranial capacity (brain volume) in cubic centimeters.
4. The two heavy lines projected on the skulls are used to measure how far forward the jaw protrudes. Use your protractor to measure the outside angle (toward the right) formed by the two lines.
5. Complete the data table.

Data Table

	Gorilla	Australopithecus	Modern human
1. Face area in cm²			
2. Brain area in cm²			
3. Is brain area smaller or larger than face area?			
4. Is brain area 3 times larger than face area?			
5. Cranial capacity in cm³			
6. Jaw angle			
7. Does lower jaw stick out in front of nose?			
8. Is sagittal crest present?			
9. Is browridge present?			

ANALYZE AND CONCLUDE

1. **Comparing and Contrasting** How would you describe the similarities and differences in face-to-brain area in the three primates?
2. **Interpreting Observations** How do the cranial capacities compare among the three skulls? How do the jaw angles compare?
3. **Drawing Conclusions** Based on your findings, what statements can you make about the placement of australopithecines in human evolution?

Going Further

Application Different parts of the australopithecine skeleton are also intermediate between apes and humans. Obtain diagrams of primate skeletons to determine the similarities and differences.

BIOLOGY *Online* To find out more about primate evolution, visit the Glencoe Science Web site.
science.glencoe.com

2. Apes have a small cranial capacity, whereas humans have a large cranial capacity, and that of australopithecines was intermediate but closer to the apes. An ape has a small jaw angle, and a human has a large jaw angle. The jaw angle of australopithecines was intermediate but closer to that of the ape.
3. Many australopithecine skull traits were intermediate between those of apes and humans, and some were more similar to those of apes. Australopithecines represent very early human ancestors.

✔ Assessment

Portfolio Have students formulate three hypotheses about the natural selection pressures that may have taken place during australopithecine evolution. Have them place their hypotheses in their portfolios. Use the Performance Task Assessment List for Formulating a Hypothesis in **PASC**, p. 21. **L2** **P**

Going Further

Logical-Mathematical Obtain illustrations or actual fossil casts of australopithecine limb bones, hands, or feet. Have students compare measurements and other observations of the postcranial skeletons of apes, humans, and australopithecines. **L1**

Data Table

	Gorilla	Australopithecus	Modern human
1. Face area in cm²	32 cm²	19 cm²	12 cm²
2. Brain area in cm²	23 cm²	23 cm²	40 cm²
3. Is brain area smaller or larger than face area?	Smaller	Larger	Larger
4. Is brain area 3 times larger than face area?	No	No	Yes
5. Cranial capacity in cm³	600 cm³	600 cm³	1060 cm³
6. Jaw angle	35°	55°	90°
7. Does lower jaw stick out in front of nose?	Yes	Yes	No
8. Is sagittal crest present?	Yes	Yes	No
9. Is browridge present?	Yes	Yes	No

Earth Science
Connection

Earth Science
Connection

The Land Bridge to the New World

The Bering Land Bridge, or Beringia, is a strip of land that connects Asia and North America. During the last Ice Age, Beringia was dry land above sea level. Human ancestors may have walked across this land to reach North America.

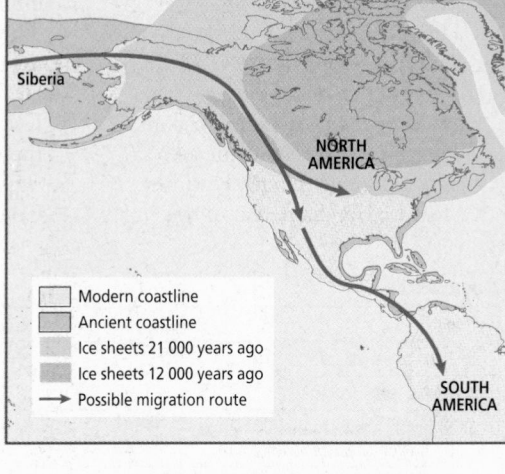

Modern coastline
Ancient coastline
Ice sheets 21 000 years ago
Ice sheets 12 000 years ago
→ Possible migration route

The 1500 km-wide piece of land known as the Bering Land Bridge is located between the Bering and Chukchi Seas and links northeastern Siberia and northwestern North America. Today, the land bridge is about 267 meters below the ocean's surface. However, during the last ice age, sea level was much lower than it is today. At that time, this land bridge was above the water's surface. Humans could have migrated from Asia to North America across this land bridge. Recent evidence indicates that such a human migration probably occurred 11 330 to 11 000 years ago.

Dating the land bridge Anthropologists compared two kinds of data to determine the 11 000-year date for human migration across Beringia. They used radiometric dating methods on fossils and data tables that indicate the sea levels at different times in geological history. Both data reveal that Beringia was last above sea level about 11 000 years ago, which is about 4000 years earlier than previous calculations had determined.

Pollen reveals plant life Pollen found in sediments dredged from the bottoms of the Bering and Chukchi Seas indicates that the land bridge and the surrounding areas were tundra ecosystems. Willows, birch, sedge tussocks, and spring flowers were the dominant plants of the area, and caribou probably roamed over the frozen soil.

The pollen studies also showed that the temperature at the time was warmer than it is in present-day Alaska. Scientists have used this finding to propose that perhaps the Ice Age was

ending. The glaciers would have melted in a warming climate and the sea level would have risen, covering the land bridge with water.

An alternate route In addition to these findings, recent archeological studies in South America have uncovered new information about human migration to the New World. Researchers from the University of Illinois have discovered evidence that prehistoric humans lived in Brazil about 12 800 years ago. The question paleoanthropologists now ask is did humans enter the New World only via the Bering Land Bridge, or was there another route to the Americas?

CONNECTION TO BIOLOGY

Study the map shown above. Suggest another way that prehistoric humans might have entered the New World.

 To find out more about human origins, visit the Glencoe Science Web site. **science.glencoe.com**

BIOLOGY Online Note Internet addresses that you find useful in the space below for quick reference.

_____

_____

_____

_____

Chapter 16 Assessment

SUMMARY

Section 16.1

Primate Adaptation and Evolution

Main Ideas

- Primates are primarily an arboreal group of mammals. They have adaptations, such as binocular vision, opposable thumbs, and flexible joints, that help them survive in trees.

- There are two groups of primates: prosimians, such as lemurs and tarsiers; and anthropoids, which include monkeys and hominoids.

- Fossils indicate that primates appeared on Earth about 65 to 70 million years ago. Major trends in primate evolution include an increasing brain size and walking upright.

Vocabulary

anthropoid (p. 432)
opposable thumb (p. 431)
prehensile tail (p. 433)
primate (p. 429)

Section 16.2

Human Ancestry

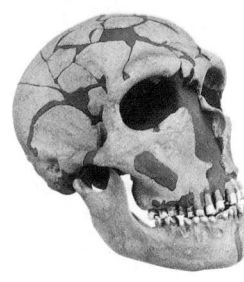

Main Ideas

- One idea about the evolution of humans is that the earliest hominids arose in Africa approximately 5 to 8 million years ago. Australopithecine fossils indicate that these individuals were bipedal, but also climbed trees.

- The fossil record indicates that humans developed over time. Their brain and body size gradually increased, bipedalism became more efficient, and their jaws and teeth decreased in size.

- The appearance of stone tools in the fossil record coincided with the appearance of the genus *Homo* about 2 million years ago. The use of fire and language, as well as the development of culture, probably developed in more recent *Homo* species.

Vocabulary

australopithecine (p. 440)
bipedal (p. 438)
Cro-Magnon (p. 444)
hominid (p. 438)
Neanderthal (p. 443)

Main Ideas

Summary statements can be used by students to review the major concepts of the chapter.

Using the Vocabulary

To reinforce chapter vocabulary, use the Content Mastery Booklet and the activities in the Interactive Tutor for Biology: The Dynamics of Life on the Glencoe Science Web site: **science.glencoe.com**

 **THE PRINCETON REVIEW** *All Chapter Assessment questions and answers have been validated for accuracy and suitability by The Princeton Review.*

UNDERSTANDING MAIN IDEAS

1. c
2. d
3. d
4. b

UNDERSTANDING MAIN IDEAS

1. Which living primate group is probably the most similar to the earliest primates?
   a. apes   **c.** prosimians
   b. monkeys   **d.** hominids

2. The first *Homo sapiens* were _____.
   a. Cro-Magnon people
   b. *Homo erectus*
   c. *Australopithecus afarensis*
   d. Neanderthals

3. Which of the following pairs of terms is most closely related?
   a. primate—squirrel
   b. arboreal—gorilla
   c. prosimian—hominid
   d. Cro-Magnon—*Homo sapiens*

4. Because the eyes of primates face forward, they _____.
   a. have color vision
   b. have binocular vision
   c. can climb trees well
   d. see well to the sides

GLENCOE TECHNOLOGY

 **VIDEOTAPE**
MindJogger Videoquizzes
Chapter 16: *Primate Evolution*
Have students work in groups as they play the videoquiz game to review key chapter concepts.

Resource Manager

Chapter Assessment, pp. 91-96
MindJogger Videoquizzes
ExamView® Pro Software
BDOL Interactive CD-ROM, Chapter 16 quiz

5. b
6. a
7. a
8. b
9. b
10. d
11. bipedal
12. prehensile
13. Neanderthals
14. australopithecine
15. primates
16. *Homo*
17. Cro-Magnon humans
18. binocular vision
19. opposable thumb
20. australopithecines

APPLYING MAIN IDEAS

21. Yes. The first evidence of humans in the Americas dates from about 12 000 years ago. A 25 000-year-old arrowhead would more than double the time that humans are known to be in North America.
22. The adaptations that evolved in nonhuman primates laid the groundwork for adaptations that arose during hominid evolution.
23. Answers could include that scientists might compare the DNA of modern humans and that of Neanderthals, if such DNA exists. Or scientists could compare the skulls or skeletons of Neanderthals and modern humans.
24. Apes evolved their unique traits as a result of adapting to conditions in the environments they inhabited.

5. Primates native to the area indicated by the map below are _____.
   a. Old World monkeys   c. apes
   b. New World monkeys   d. prosimians

PACIFIC OCEAN   ATLANTIC OCEAN

6. The science of studying the fossils of humans is _____.
   a. paleoanthropology   c. paleontology
   b. geology   d. anthropology

7. The dominant sense in primates is _____.
   a. vision   c. smell
   b. taste   d. hearing

8. Which of these is NOT a primate?
   a. human   c. lemur
   b. squirrel   d. orangutan

9. The earliest primates were most like _____.

a.

c.

b.

d.

THE PRINCETON REVIEW   **TEST-TAKING TIP**

Dress Comfortably
Loose, layered clothing is best. Whatever the temperature, you're prepared. Important test scores do not take climate into consideration.

10. The study of the fossil Lucy helped scientists determine that _____.
    a. both primates and hominids have color vision
    b. hominids are primates with opposable thumbs
    c. hominids had large brains before they walked upright
    d. hominids walked upright before they had large brains

11. Organisms that walk upright on two legs are _____.

12. The term used to describe the tails of New World monkeys is _____.

13. An early group of humans who may have been the first to develop religious views and a spoken language were _____.

14. The fossil of the Taung child represents the first _____ skull ever discovered.

15. A group of mammals that includes lemurs, monkeys, apes, and humans is _____.

16. Humanlike hominids are grouped in the genus _____.

17. Toolmakers and artists who lived in Europe about 30 000 years ago were the _____.

18. _____ enables primates to perceive depth and thereby gauge distances.

19. The primate structure that allows for the grasping of objects is _____.

20. African hominids who possessed apelike and humanlike qualities are classified as _____.

APPLYING MAIN IDEAS

21. Suppose that you were told that a scientist found a 25 000-year-old arrowhead in Arizona. Would you be surprised? Why or why not?
22. Why is it important for a paleoanthropologist to know about all primates?
23. Some scientists suggest that Neanderthals evolved into modern humans. What information should they gather to support their idea?
24. How is the evolution of apes an example of adaptive radiation?

Chapter 16 Assessment

THINKING CRITICALLY

25. Observing and Inferring How could you tell from the position of the foramen magnum that an animal walked upright? Explain.

26. Formulating Hypotheses How would you test the idea that opposable thumbs are beneficial adaptations for arboreal mammals?

27. Interpreting Data The data in *Table 16.2* are from an experiment comparing amino acid sequences in apes. What conclusions can you draw from such data?

Table 16.2 Comparisons of Amino Acid Sequences	
Primate	**Percentage amino acid sequence difference from humans**
Gibbon	5.2
Orangutan	3.7
Chimpanzee	1.8
Gorilla	2.1

28. Concept Mapping Complete the concept map by using the following vocabulary terms: australopithecine, hominids, Cro-Magnon, bipedal, Neanderthal, opposable thumbs

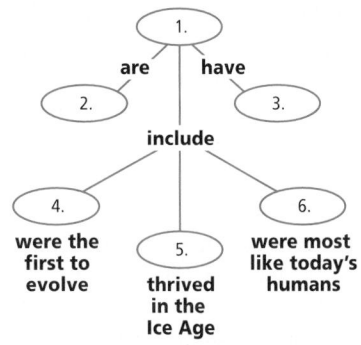

CD-ROM

For additional review, use the assessment options for this chapter found on the *Biology: The Dynamics of Life Interactive CD-ROM* and on the Glencoe Science Web site.
science.glencoe.com

ASSESSING KNOWLEDGE & SKILLS

The intermembral index is the ratio of forelimb length to hindlimb length. Primates with a high index are good branch swingers. Primates with a low index tend to walk on all four legs.

Table 16.3 Intermembral Index of Some Primates	
Species	**Intermembral Index** $\frac{\text{FORELIMB LENGTH}}{\text{HINDLIMB LENGTH}} \times 100$
Prosimians	
Indri	64
Slow loris	88
New World monkeys	
Squirrel monkey	80
Black spider monkey	105
Old World monkey	
Pig-tailed macaque	92
Hanuman langur	83
Hominoids	
Orangutan	139
Common chimpanzee	103

Interpreting Data Study *Table 16.3* and answer the following questions.

1. Which group appears to have the greatest range of intermembral distance?
   a. prosimians
   b. New World monkeys
   c. Old World monkeys
   d. hominoids
2. Which group appears to have the narrowest range of intermembral distance?
   a. prosimians
   b. New World monkeys
   c. Old World monkeys
   d. hominoids
3. **Interpreting Data** To which group might a primate with an intermembral index of 92 belong? Explain your answer.

THINKING CRITICALLY

25. For an animal to walk erect, its spinal cord must enter the skull at its base. The foramen magnum would be located more toward the back of the skull than at its base in an animal that used four appendages for locomotion.
26. Students can design an experiment to test and compare the success of two arboreal animals, a primate and a nonprimate, in a situation unique to arboreal life, such as obtaining fruit at the ends of branches.
27. Chimpanzees are the most closely related to humans, then gorillas, then orangutans, and finally gibbons.
28. 1. Hominids; 2. Bipedal; 3. Opposable thumbs; 4. Australopithecine; 5. Neanderthal; 6. Cro-Magnon

ASSESSING KNOWLEDGE & SKILLS

1. d
2. c
3. An intermembral index of 92 could place this new primate with Old World monkeys. It could also place the primate with New World monkeys as 92 is in the range shown. More information would be needed to correctly place this new primate.

Chapter 17 Organizer

Refer to pages 4T-5T of the Teacher Guide for an explanation of the National Science Education Standards correlations.

Section	Objectives	Activities/Features
Section 17.1 **Classification** National Science Education Standards UCP.1, UCP.2, UCP.4; A.2; C.3, C.5; G.1-3 (1 session, $1/2$ block)	1. **Evaluate** the history, purpose, and methods of taxonomy. 2. **Explain** the meaning of a scientific name. 3. **Describe** the organization of taxa in a biological classification system.	**MiniLab 17-1:** Using a Dichotomous Key, p. 456 **Problem-Solving Lab 17-1,** p. 457 **Focus On** Kingdoms of Life, p. 460 **Investigate BioLab:** Making a Dichotomous Key, p. 474
Section 17.2 **The Six Kingdoms** National Science Education Standards UCP.1, UCP.2, UCP.4, UCP.5; A.1, A.2; C.1, C.3, C.5, C.6; E.1, E.2; G.1-3 (3 sessions, 2 blocks)	4. **Describe** how evolutionary relationships are determined. 5. **Explain** how cladistics reveals phylogenetic relationships. 6. **Compare** the six kingdoms of organisms.	**MiniLab 17-2:** Using a Cladogram to Show Relationships, p. 467 **Problem-Solving Lab 17-2,** p. 470 **BioTechnology:** Molecular Clocks, p. 476

Need Materials? Contact Carolina Biological Supply Company at 1-800-334-5551 or at http://www.carolina.com

MATERIALS LIST

BioLab
p. 474 sample keys from guidebooks, metric ruler

MiniLabs
p. 456 miscellaneous leaves, dichotomous key for trees, paper, glue
p. 467 pencil, paper

Alternative Lab
p. 468 flasks, test tubes, beakers, stirring rods, thermometers, bottle stoppers, pipettes, droppers, funnels, petri dishes, graduated cylinders, bottles, jars

Quick Demos
p. 454 sponges, slime molds, mosses, mildew, lichens
p. 461 leaf, feather, moss, mushroom, seeds, seaweed, sponge, poultry wishbone
p. 468 organisms from the six kingdoms
p. 481 microscope, prepared slide of *Euglena*
p. 483 small mammal, cage

Key to Teaching Strategies

L1 Level 1 activities should be appropriate for students with learning difficulties.

L2 Level 2 activities should be within the ability range of all students.

L3 Level 3 activities are designed for above-average students.

ELL ELL activities should be within the ability range of English Language Learners.

COOP LEARN Cooperative Learning activities are designed for small group work.

P These strategies represent student products that can be placed into a best-work portfolio.

These strategies are useful in a block scheduling format.

Organizing Life's Diversity

Teacher Classroom Resources

Section	Reproducible Masters	Transparencies
Section 17.1 **Classification**	Reinforcement and Study Guide, p. 73-74 **L2** Concept Mapping, p. 17 **L3** **ELL** Critical Thinking/Problem Solving, p. 17 **L3** BioLab and MiniLab Worksheets, p. 81 **L2** Laboratory Manual, pp. 117-120 **L2** Tech Prep Applications, pp. 25-26 **L2** Content Mastery, pp. 81-82, 84 **L1**	Section Focus Transparency 41 **L1** **ELL**
Section 17.2 **The Six Kingdoms**	Reinforcement and Study Guide, pp. 75-76 **L2** Concept Mapping, p. 17 **L3** **ELL** BioLab and MiniLab Worksheets, pp. 82-84 **L2** Laboratory Manual, pp. 121-124 **L2** Content Mastery, pp. 81, 83-84 **L1**	Section Focus Transparency 42 **L1** **ELL** Basic Concepts Transparency 24 **L2** **ELL** Reteaching Skills Transparency 26 **L1** **ELL**

Assessment Resources

Chapter Assessment, pp. 97-102

MindJogger Videoquizzes

Performance Assessment in the Biology Classroom

Alternate Assessment in the Science Classroom

ExamView® Pro Software 💾

BDOL Interactive CD-ROM, Chapter 17 quiz

Additional Resources

Spanish Resources **ELL**

English/Spanish Audiocassettes **ELL**

Cooperative Learning in the Science Classroom **COOP LEARN**

Lesson Plans/Block Scheduling

◼ NATIONAL GEOGRAPHIC
Teacher's Corner

Products Available From Glencoe
To order the following products, call Glencoe at 1-800-334-7344:
CD-ROMs
Mammals: A Multimedia Encyclopedia
NGS PictureShow: Classifying Plants and Animals
Curriculum Kits
GeoKit: Cells and Organisms
GeoKit: Fish, Reptiles, and Amphibians
GeoKit: Plants

Transparency Set
NGS PicturePack: Classifying Plants and Animals

Products Available From National Geographic Society
To order the following products, call National Geographic Society at 1-800-368-2728:
Books
National Geographic Book of Mammals
Field Guide to the Birds of North America
Video
Plant Classification

GLENCOE TECHNOLOGY

The following multimedia resources are available from Glencoe.

Biology: The Dynamics of Life
CD-ROM **ELL**

BioQuest: *Biodiversity Park*
Video: *Museum Collections*
Exploration: *The Six Kingdoms*

Videodisc Program

Museum Collections

The Infinite Voyage

The Great Dinosaur Hunt
Insects: *The Ruling Class*
The Geometry of Life
The Dawn of Humankind

The Secret of Life Series

Gone Before You Know It: *The Biodiversity Crisis*
What's in Stetter's Pond: *The Basics of Life*
Using Cladistics

Theme Development

The theme of **evolution** underlies the phylogenetic basis of classification. The theme of **unity within diversity** emerges when the shared features of the species in each kingdom are described.

0:00 OUT OF TIME?

If time does not permit teaching the entire chapter, use the BioDigest at the end of the unit as an overview.

READING BIOLOGY

Glencoe's *Biology: The Dynamics of Life* contains many resources to assist a student's reading skills. Each chapter contains figures with expanded captions that expand on written material. Word Origins, located along the side of text, expand knowledge of biology vocabulary. Glencoe's Content Mastery Booklet helps develop reading skills while reinforcing content. In addition, use the Interactive Tutor for *Biology: The Dynamics of Life* on the Glencoe Web site to reinforce vocabulary.
science.glencoe.com

Chapter

17 Organizing Life's Diversity

What You'll Learn

- You will both identify and compare various methods of classification.
- You will distinguish among six kingdoms of organisms.

Why It's Important

Biologists use a system of classification to organize living things. Understanding classification helps you study organisms and their evolutionary relationships.

READING BIOLOGY

Look through some of the photos and illustrations in the chapter. Make a list of some of the obvious differences used to classify organisms. As you read the text, compare this list to the classification systems described in the reading.

BIOLOGY Online

To find out more about classification and taxonomy, visit the Glencoe Science Web site.
science.glencoe.com

Biologists have classified all the organisms you see in these photos as well as millions of others.

Multiple Learning Styles

Look for the following logos for strategies that emphasize different learning modalities.

Kinesthetic Building a Model, p. 458; Meeting Individual Needs, pp. 462, 467; Portfolio, p. 471

Visual-Spatial Portfolio, pp. 456, 458; Meeting Individual Needs, p. 465; Time Line, p. 469; Project, p. 471; Activity, p. 473

Interpersonal Activity, p. 465

Intrapersonal Tech Prep, pp. 455, 456; Reteach, p. 473

Linguistic Biology Journal, pp. 454, 458

Logical-Mathematical Meeting Individual Needs, p. 454

Naturalist Activity, pp. 455, 472; Biology Journal, pp. 457, 466; Reteach, p. 459; Extension, p. 459; Going Further, p. 475

17.1 Classification

Every day you see items that are grouped, and you group items yourself. In a supermarket, you find all the fresh produce in one area, baked goods in another, and dairy products in still another. In a music store, the type of music is the basis for shelving a CD. When you put away the dishes, you probably place the dinner plates on one shelf and the glasses on another. You group similar articles so often that you probably never think about why you do it. However, grouping things creates order, and order saves time and energy when you look for an item.

A display of fruit for sale

SECTION PREVIEW

Objectives

Evaluate the history, purpose, and methods of taxonomy.

Explain the meaning of a scientific name.

Describe the organization of taxa in a biological classification system.

Vocabulary

classification
taxonomy
binomial nomenclature
genus
family
order
class
phylum
kingdom
division

How Classification Began

Organizing items can help you understand them better and find them more easily. For example, you probably order your clothes drawers and your CD collection. Biologists want to better understand organisms so they organize them into groups. One tool that they use to do this is **classification**—the grouping of objects or information based on similarities. **Taxonomy** (tak SAHN uh mee) is the branch of biology that groups and names organisms based on studies of their different characteristics. Biologists who study taxonomy are called taxonomists.

Aristotle's system

The Greek philosopher Aristotle (384–322 B.C.) developed the first widely accepted system of biological classification. He classified all the organisms he knew into two groups: plants and animals. He subdivided plants into the three groups, herbs, shrubs, and trees, depending on the size and structure of a plant. He grouped animals according to where they lived or spent a great deal of time: on land, in the air, or in water.

The basis for Aristotle's groups was useful but did not group organisms according to their evolutionary history. According to his system, birds, bats, and flying insects are classified together even though they have

WORD *Origin*

taxonomy
From the Greek words *taxo*, meaning to "arrange," and *nomy*, meaning "ordered knowledge." Taxonomy is the science of classification.

17.1 CLASSIFICATION **453**

Section 17.1

Prepare

Key Concepts

Students will examine the history, purpose, and methods of classification and taxonomy. They will compare the contributions of Aristotle and Linnaeus and learn about taxonomic categories.

Planning

■ Obtain samples of unusual organisms, such as some slime molds, moss, and lichens for the Quick Demo.

■ Collect supermarket advertisements of produce sales for the Tech Prep.

■ Collect leaves on lab day for MiniLab 17-1.

1 Focus

Bellringer

Before presenting the lesson, display **Section Focus Transparency 41** on the overhead projector and have students answer the accompanying questions. **L1** **ELL**

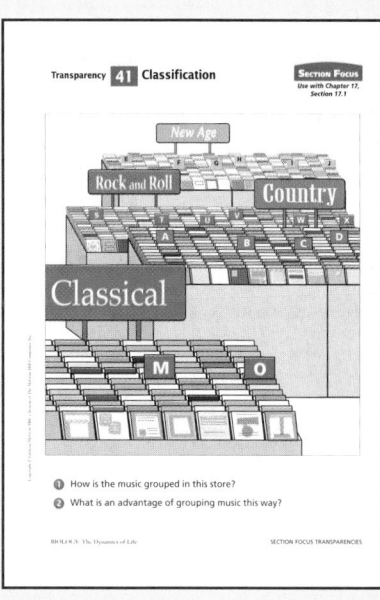

Transparency **41** Classification

SECTION FOCUS
Use with Chapter 17, Section 17.1

New Age
Rock and Roll
Country
Classical

❶ How is the music grouped in this store?
❷ What is an advantage of grouping music this way?

BIOLOGY: The Dynamics of Life   SECTION FOCUS TRANSPARENCIES

✓ Assessment Planner

Portfolio Assessment
  Portfolio, TWE, pp. 456, 458, 471
  Alternative Lab, TWE, pp. 468-469
Performance Assessment
  MiniLab, SE, pp. 456, 467
  MiniLab, TWE, pp. 456, 467
  Assessment, TWE, pp. 458, 459, 473
  Alternative Lab, TWE, pp. 468-469
  BioLab, SE, pp. 474-475

BioLab, TWE, pp. 474-475
Knowledge Assessment
  Problem-Solving Lab, TWE, p. 457
  Section Assessment, SE, pp. 459, 473
  Assessment, TWE, p. 466
  Chapter Assessment, SE, pp. 477-479
Skill Assessment
  Problem-Solving Lab, TWE, p. 470

Visual Learning

Figure 17.1 Ask students to identify features they could use to classify these flowers. *Possible answers include the numbers and arrangements of petals or male and female reproductive organs.*

little in common besides the ability to fly. As time passed, more organisms were discovered and some did not fit easily into Aristotle's groups, but many centuries passed before Aristotle's system was replaced.

Linnaeus's system

In the late eighteenth century, a Swedish botanist, Carolus Linnaeus (1707–1778), developed a method of grouping organisms that was more useful than Aristotle's. Linnaeus's system was based on physical and structural similarities of organisms. For example, he might use the similarities in flower parts as a basis for classifying flowering plants, *Figure 17.1*. As a result, the groupings revealed the relationships of the organisms.

Eventually, some biologists proposed that structural similarities reflect the evolutionary relationships of species. For example, although bats fly like birds, they also have hair and produce milk for their young. Therefore, bats are classified as mammals rather than as birds, reflecting the evolutionary history that bats share with other mammals. This way of organizing organisms is the basis of modern classification systems.

Two names for a species

Modern classification systems use a two-word naming system called **binomial nomenclature** that Linnaeus developed to identify species. In this system, the first word identifies the genus of the organism. A **genus** (JEE nus) (plural, genera) consists of a group of similar species. The second word, which often describes a characteristic of the organism, immediately follows the genus name. Thus, the scientific name for each species is a combination of the genus and descriptive names. For example, the scientific name of modern humans is *Homo sapiens*. Modern humans are in the genus *Homo*, and one of their characteristics is intelligence. The Latin word *sapiens* means "wise."

> **WORD** *Origin*
> **binomial nomenclature**
> From the Latin words *bi*, meaning "two," *nomen*, meaning "name," and *calatus*, meaning "list." The system of binomial nomenclature assigns two words to the name of each species.

Figure 17.1 Linnaeus classified flowering plants according to their flower structures.

MEETING INDIVIDUAL NEEDS

Learning Disabled

Logical-Mathematical Ask students how they would find the phone number of a music store whose name they'd forgotten in the yellow pages. Then ask how they would use the white pages to do the same thing. Have them name other situations in which classification is useful. **L1**

BIOLOGY JOURNAL

Using Binomials

Linguistic Have students reread the paragraphs titled "Two names for a species" and use the information to write the names of their family members as binomials. **L1** **ELL**

Latin is the language of scientific names. Taxonomists are required to give each newly discovered species a Latin scientific name. They use Latin because the language is no longer used in conversation and, therefore, does not change. Scientific names should be italicized in print and underlined when handwritten. The first letter of the genus name is uppercase, but the first letter of the descriptive name is lowercase.

Although a scientific name gives information about the relationships of an organism and how it is classified, many organisms have common names just like you and your friends might have nicknames. However, a common name can be misleading. For example, a sea horse is a fish, not a horse. In addition, it is confusing when a species has more than one common name. The bird in *Figure 17.2* lives not only in the United States but also in several countries in Europe. In each country it has a different common name. Therefore, if an English scientist publishes an article about the bird's behavior and uses the bird's English common name, a Spanish scientist looking for information might not recognize the bird as the same species also living in Spain.

Biological Classification

Expanding on Linnaeus's work, today's taxonomists try to identify the underlying natural relationships of organisms and use the information as a basis for classification. They compare the external and internal structures of organisms, as well as their geographical distribution and chemical makeup to reveal their probable evolutionary relationships. Grouping organisms on the basis of their evolutionary relationships makes it easier to understand biological diversity.

Taxonomy: A framework

Just as similar food items in a supermarket are stacked together, taxonomists group similar organisms, both living and extinct. Classification provides a framework in which to study the relationships among living and extinct species.

For example, biologists study the relationship between birds and dinosaurs within the framework of classification. Are dinosaurs more closely related to birds or reptiles? The bones of some dinosaurs have large internal spaces like those in birds. Some paleontologists who study dinosaur fossils propose that some dinosaurs may have been endothermic—able to maintain a constant body temperature—which is a characteristic of all birds. Because of such evidence, they suggest that dinosaurs are more closely related to ostriches, which are birds, than to lizards, which are reptiles.

Taxonomy: A useful tool

Classifying organisms is a useful tool for scientists who work in agriculture, forestry, and medicine. For

Figure 17.2
In the United States and England, this bird is called the house sparrow, in Spain the gorrion, in Holland the musch, and in Sweden the hussparf. However, the bird has only one scientific name, *Passer domesticus.*

MiniLab 17-1

Purpose
Students will use a dichotomous key to identify organisms.

Process Skills
classify, observe and infer, compare and contrast

Teaching Strategies
■ Have students wash their hands after touching the leaves.
■ Use keys that identify trees and shrubs by leaf structure.
■ Obtain leaves on the day students will do the activity.
■ Make a transparency of a key to demonstrate the key's use.

Expected Results
Students will identify trees and shrubs based on leaf structure.

Analysis
1. identification of organisms
2. vein and margin structure, leaf shape and size, number of lobes
3. more specific

✓ Assessment
Performance Provide students with several algae and a dichotomous key. Ask them to examine, diagram, and then identify the algae. Use the Performance Task Assessment List for Making and Using a Classification System in **PASC**, p. 49. **L3**

Resource Manager

BioLab and MiniLab Worksheets, p. 81 **L2**
Tech Prep Applications, pp. 25-26 **L2**
Laboratory Manual, pp. 117-120 **L2**

MiniLab 17-1 — Classifying

Using a Dichotomous Key How could you identify a tree growing in front of your school? You might ask a local expert, or you could use a manual or field guide that contains descriptive information and keys about trees. A key is a set of descriptive sentences that is subdivided into steps. A dichotomous key has two descriptions at each step. You follow the steps until the key reveals the name of the tree.

Procedure
1. Using a few leaves from local trees and a dichotomous key for trees of your area, identify the tree from which each leaf came. To use the key, study one leaf. Then choose the one statement from the first pair that most accurately describes the leaf. Continue following the key until you identify the leaf's tree. Repeat the process for each leaf.
2. Glue each leaf on a separate sheet of paper. For each leaf, record the tree's name.

Analysis
1. What is the function of a dichotomous key?
2. List three different characteristics used in your key.
3. As you used the key, did the characteristics become more general or more specific?

Figure 17.3
Taxonomists can easily distinguish among this poison ivy (**a**) and other plants, such as Virginia creeper, with which it is often confused. The red berries produced by a holly bush are poisonous to humans (**b**).

---

example, suppose a child eats berries from the holly plant that you see in *Figure 17.3*. The child's parents would probably rush the child and some of the plant and its berries to the nearest hospital. Someone working at a poison control center could identify the plant, and the physicians would then know how to treat the child.

Anyone can learn to identify many organisms. The *MiniLab* on this page will guide you through a way of identifying some organisms in your own neighborhood. Then try the *BioLab* at the end of this chapter.

Taxonomy and the economy

It often happens that the discovery of new sources of lumber, medicines, and energy results from the work of taxonomists. The characteristics of a familiar species are frequently similar to those found in a new, related species. For example, if a taxonomist knows that a certain species of pine tree contains chemicals that make good disinfectants, it's likely that another pine species will also contain these useful substances.

TECHPREP

Classification in Daily Life
Intrapersonal Provide students with newspaper classified ads. Ask them to describe their dream car and explain where they would find it listed in the ads. **L1**

✓ Portfolio

Zoo and Garden Classification
Visual-Spatial Send students to a nearby zoo or botanical garden to find an example of classification in use. Have them make a photo or video essay of the example for their portfolios. **L2** **ELL** **P**

How Living Things Are Classified

In any classification system, items are categorized, making them easier to find and discuss. For example, in a newspaper's classified advertisements, you'll find a section listing autos for sale. This section frequently subdivides the many ads into two smaller groups—domestic autos and imported autos. In turn, these two groups are subdivided by more specific criteria, such as different car manufacturers and the year and model of the auto. Although biologists group organisms, not cars, they subdivide the groups on the basis of more specific criteria. Any group of organisms is called a taxon (plural, taxa).

Taxonomic rankings

Organisms are ranked in arbitrary taxa that range from having very broad characteristics to very specific ones. The broader a taxon, the more general its characteristics, the more species it contains. You can think of the taxa as fitting together like nested boxes of increasing sizes. You already know about two taxa. The smallest taxon is that of species. Organisms that look alike and successfully interbreed belong to the same species. The next largest taxon is a genus—a group of similar species that have similar features and are closely related.

It is not always easy to determine the species of an organism. For example, over many years, taxonomists have debated how to classify the red wolf, the coyote, and the gray wolf. Some biologists wanted to classify them as separate species, and others wanted to classify them as a single species. Use the *Problem-Solving Lab* on this page to explore the evidence for and against classifying these three organisms as separate species.

Problem-Solving Lab 17-1 — Drawing a Conclusion

Is the red wolf a separate species? The work of taxonomists results in changing views of species. This is due to both the discovery of new species and the development of new techniques for studying classification.

Red wolf   Coyote   Gray wolf

Analysis

A. The red wolf (*Canis rufus*) can breed and produce offspring with both the coyote (*Canus latrans*) and the gray wolf (*Canis lupus*). Despite this fact, the three animal types have been classified as separate species.

B. A biologist measured their skulls and concluded that in size and structure the red wolf's measurements fell midway between gray wolves and coyotes.

C. Based on these data, the biologist concluded that they are separate species.

D. Geneticists, attempting to determine if the three animal types were separate species, found that the nucleotide sequences from the red wolf's DNA were not distinctively different from those of gray wolves or coyotes.

E. The geneticists concluded that the red wolf is a hybrid of the gray wolf and coyote.

Thinking Critically

1. A species can be defined as a group of animals that can mate with one another to produce fertile offspring but cannot mate successfully with members of a different group. Does statement (A) support or reject this definition? Explain your answer.

2. What type of evidence was the biologist using (B)? The geneticists (D)? Explain your answer.

3. A hybrid is the offspring from two species. Which sentence, beside (D) and (E), supports hybrid evidence? Explain.

4. If you supported the biologist's work, would you use the three different scientific names for coyotes, gray wolves, and red wolves? Explain your answer.

5. If you supported the geneticists' conclusions, would you use the three different scientific names? Explain your answer.

6. How does this example support the idea that the work of scientists results in changing views of species?

Problem-Solving Lab 17-1

Purpose
Students will analyze why taxonomy can change.

Process Skills
analyze information, apply concepts, define operationally, draw a conclusion, think critically

Background
The red wolf is an endangered species. Programs to reintroduce the red wolf into states such as North Carolina have received federal funding. But now that the red wolf is considered a hybrid, some question spending endangered species funds on it.

Teaching Strategies
■ Point out that each animal has its own scientific name.
■ Use large photos to show the animals' similarities.

Thinking Critically

1. Rejects it. The red wolf mates with two other species.
2. biologist, structural; geneticists, biochemical
3. B; intermediate skull size implies hybridism.
4. Yes; each species must have its own scientific name.
5. No; the red wolf is a hybrid.
6. Since 1995, the red wolf is considered to be a hybrid, not a distinct species.

Assessment

Knowledge The red fox's scientific name is *Vulpes vulpes*. Have students classify the red fox and red wolf. Use the Performance Task Assessment List for Making and Using a Classification System in **PASC**, p. 49. L2

BIOLOGY JOURNAL

Organizing Information

Naturalist Provide students with a list of three to five familiar organisms, such as a common fish, amphibian, reptile, bird, and mammal. Have them research how the organisms are classified and use the information to make a table in their journals. L1

GLENCOE TECHNOLOGY

VIDEODISC
The Infinite Voyage
Insects: The Ruling Class The Rothschild Legacy: Study of Fleas and the Bubonic Plague (Ch. 2), 6 min. 30 sec.

CAREERS IN BIOLOGY

Biology Teacher

Are you intrigued by the actions and interactions of plants, animals, and other organisms? Would you like to share this interest with others? Maybe you should become a biology teacher.

Skills for the Job

Biology teachers help students learn about organisms through discussions and activities both inside and outside the classroom. As a biology teacher, you might also teach general science and health. To become a biology teacher, you must earn a bachelor's degree in science, biology, or a closely related field. You sometimes have to spend several months student teaching. Many positions require a master's degree. In addition, you have to pass a national test for teachers in many states. This national test includes a test in biology or in a combination of biology and general science. After all this education, testing, and work, you will be ready to teach others!

BIOLOGY *Online*

To find out more about careers in related fields, be sure to check the Glencoe Science Web site.
science.glencoe.com

In *Figure 17.4,* you can compare the appearance of a lynx, *Lynx rufus,* a bobcat, *Lynx canadensis,* and a mountain lion, *Panthera concolor.* The scientific names of the lynx and bobcat tell you that they belong to the same genus, *Lynx.* All species in the genus *Lynx* share the characteristic of having a jaw that contains 28 teeth. Mountain lions and other lions, which are similar to bobcats and lynxes, are not classified in the *Lynx* genus because their jaws contain 30 teeth.

Bobcats, lynxes, lions, and mountain lions belong to the same family called Felidae. **Family,** the next larger taxon in the biological classification system, consists of a group of similar genera. In addition to domesticated cats, bobcats, lynxes, and lions belong to the family Felidae. All members of the cat family share certain characteristics. They have short faces, small ears, forelimbs with five toes, and hindlimbs with four toes. Most can retract their claws.

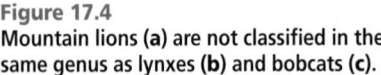

Figure 17.4
Mountain lions (**a**) are not classified in the same genus as lynxes (**b**) and bobcats (**c**).

458 ORGANIZING LIFE'S DIVERSITY

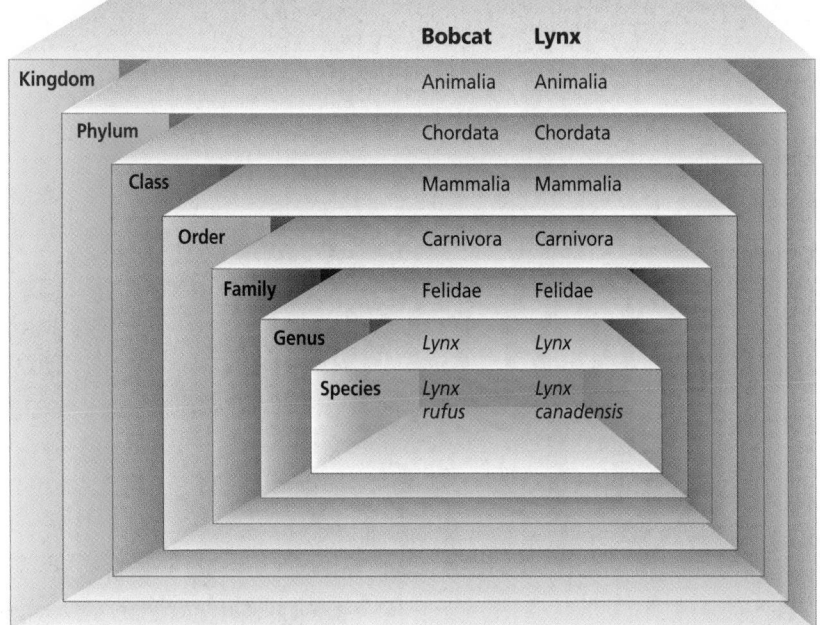

		Bobcat	Lynx
Kingdom		Animalia	Animalia
Phylum		Chordata	Chordata
Class		Mammalia	Mammalia
Order		Carnivora	Carnivora
Family		Felidae	Felidae
Genus		*Lynx*	*Lynx*
Species		*Lynx rufus*	*Lynx canadensis*

Figure 17.5
A lynx, called *Lynx canadensis*, has both a short tail with black fur circling its tip and highly visible tufts of hair on its ears. A bobcat, *Lynx rufus*, has a short tail with black fur only on the top of its tail's tip and inconspicuous tufts of ear hair. However, these species share many characteristics.

The larger taxa

There are four larger taxa. An **order** is a taxon of similar families. A **class** is a taxon of similar orders. A **phylum** (FI lum) (plural, phyla) is a taxon of similar classes. Plant taxonomists use the taxon **division** instead of phylum. A **kingdom** is a taxon of similar phyla or divisions. The six kingdoms are described in the *Focus on* beginning on the next page.

As shown in *Figure 17.5*, bobcats and lynxes belong to the order, Carnivora. Carnivores have similar arrangements of teeth and belong to the class, Mammalia. Mammals have hair or fur covering their bodies and produce milk for their young. The phylum Chordata, to which mammals belong, includes mostly animals with backbones. Kingdom Animalia includes all phyla of animals.

Section Assessment

Understanding Main Ideas

1. For what reasons are biological classification systems needed?
2. Give two reasons why binomial nomenclature is useful.
3. What did Linnaeus contribute to the field of taxonomy?
4. What are the taxa used in biological classification? Which taxon contains the largest number of species? Which taxon contains the fewest number of species?

Thinking Critically

5. Use categories that parallel the taxa of a biological classification system to organize the items you can borrow from a library.

SKILL REVIEW

6. **Classifying** Make a list of all the furniture in either your classroom or your room at home. Classify it into groups based on function. For more help, refer to *Organizing Information* in the **Skill Handbook**.

Section Assessment

1. With them, it is easier to study organisms and their relationships.
2. Common names do not indicate relationships among organisms. The genus name is the same for closely related species; the species name describes the organism.
3. He developed the binomial system for naming organisms and the basis of today's biological classification system.
4. Kingdom, phylum, class, order, family, genus, species; kingdom taxa contain the largest numbers of species, and species taxa contain the fewest.
5. Answers will vary, but students may use categories like reference books, novels, biographies, audio tapes, etc.
6. Functions may include resting, sleeping, writing, working, eating, and storing.

Section Assessment

1. With them, it is easier to study organisms and their relationships.
2. Common names do not indicate relationships among organisms. The genus name is the same for closely related species; the species name describes the organism.
3. He developed the binomial system for naming organisms and the basis of today's biological classification system.
4. Kingdom, phylum, class, order, family, genus, species; kingdom taxa contain the largest numbers of species, and species taxa contain the fewest.
5. Answers will vary, but students may use categories like reference books, novels, biographies, audio tapes, etc.
6. Functions may include resting, sleeping, writing, working, eating, and storing.

Reinforcement

Have students use the mnemonic "King Philip came over from Geneva Switzerland" to remember the order of the taxa.

3 Assess

Check for Understanding

Have students compare and contrast the classification systems of species and library books. *Both systems provide a way of organizing information. They differ in the criteria used to create categories and in the number of categories.* **L1**

Reteach

Naturalist Have students list the taxa biologists use to classify organisms and describe each taxon they list. **L1**

Extension

Naturalist Have students find the scientific names of the wolf, fox, domestic dog, and coyote and explain how the animals are related. **L2**

✔ Assessment

Performance Ask students to collect five weeds and use field guides to identify them. **L3**

4 Close

Activity

Give pairs of students some fruits or vegetables. Ask each team to make up binomials for these "species." **L1**

Resource Manager

Reinforcement and Study Guide, pp. 73-74 **L2**
Content Mastery, p. 82 **L1**

Focus On

Kingdoms of Life

Purpose

Students will learn the major differences among organisms in the six kingdoms. They will study the history of classification and different classification systems.

Background

Taxonomists work continually to improve the organizational system of about 1.5 million known species of organisms. Recent attempts at improvement include the division of prokaryotes into Kingdoms Archaebacteria and Eubacteria to better reflect the genetic differences of these two types of prokaryotes. Incorporating a taxon called domain, which is more inclusive than the kingdom taxon, into the modern classification system has been supported by some taxonomists. Such an organizational system would divide all known organisms into three domains that reflect today's knowledge about the evolutionary relationships of living things.

FOCUS ON
Kingdoms of Life

PUFFIN

DROMEDARY

The great diversity of life on Earth—estimated at 3 to 10 million species and counting—can be overwhelming. To make sense of this bewildering array of living things, biologists use classification systems to group organisms in ways that highlight their similarities, differences, and relationships. The systematic grouping of living things originated in the 4th century B.C. But biological classification has changed a great deal over the years, as new tools and technologies have made it possible to examine organisms in increasing detail and trace their complex evolutionary pathways through time.

MOTH COLLECTION

ARISTOTLE RECOGNIZES PLANTS AND ANIMALS

Taxonomy, as the science of biological classification is called, began with the Greek philosopher Aristotle (384–322 B.C.). A keen observer of nature, Aristotle separated all living things into two major groups: plants and animals. He grouped plants into herbs, shrubs, and trees, and classified animals on the basis of size, where they lived—on the land or in the water, and how they moved. Although Aristotle's system of classification did little to reveal natural relationships among living things, it was widely accepted and used, with few modifications, into the Middle Ages.

CONEFLOWER

RHODODENDRON

OAK TREE

BIOLOGY JOURNAL

Looking at Classification From a Historical Perspective

Have students research the work of Aristotle, Linnaeus, Haeckel, Whittaker, and Woese and then write an essay that compares their work. **L3**

LINNAEUS IDENTIFIES TWO KINGDOMS

Modern classification began with the work of John Ray (1627–1705), an English naturalist who outlined the idea of species. In the mid-1700s, Swedish botanist Carolus Linnaeus (1707–1778) picked up on this idea and developed a classification scheme that formed the basis of the system we use today. Linnaeus divided all living things between two kingdoms—plants and animals. But he subdivided these kingdoms into a hierarchy of smaller and more specific groups: classes, orders, genera, and species. Linnaeus placed organisms in these groups primarily on the basis of their physical similarities and differences.

QUEEN ANGELFISH

BARREL SPONGE WITH CRINOIDS

PROTISTS: THE THIRD KINGDOM

Linnaeus' classification system revolutionized taxonomy, but from the start there were problems. Organisms such as mushrooms and sponges resemble plants but do not make their own food. To which kingdom did they belong? As light microscopes improved, the situation became much more complex as biologists discovered a vast assortment of minute, primarily one-celled organisms. In 1866, German zoologist Ernst Haeckel (1834–1919) proposed giving these unicellular organisms—named protists—a kingdom of their own.

FLY AGARIC MUSHROOMS

PROKARYOTES AND FOUR KINGDOMS

The three-kingdom classification system persisted, however, until the middle of the 20th century when the electron microscope and advances in biochemistry made it possible to study living things at the subcellular level. These new tools revealed that there are two fundamentally different kinds of cells in the living world—prokaryotes and eukaryotes. Prokaryotes, such as the bacterium *Salmonella* (below left), lack the membrane-bound nuclei and most of the organelles characteristic of eukaryotic cells. All prokaryotes were then recognized as a separate kingdom that contained all the bacteria.

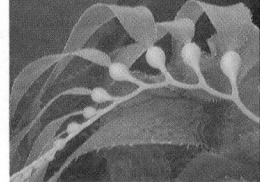
KELP (EUKARYOTE)

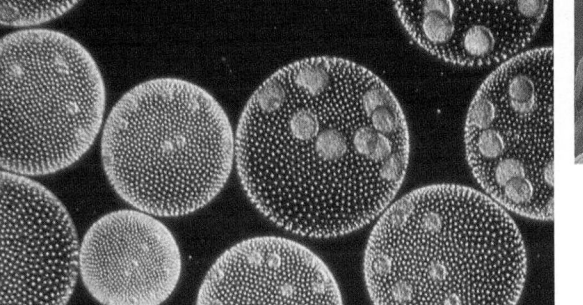

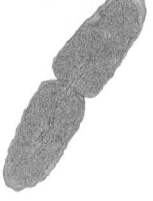
SALMONELLA (PROKARYOTE)
Magnification: 34 300 X

VOLVOX (EUKARYOTE) Magnification: 15 x

461

Teaching Strategies

■ As you review the kingdoms with your students, ask them to explain the meaning of the word *diversity* in order to identify any preconceptions.

■ Ask students to identify both the similarities and differences among the organisms in the photos on these pages.

■ Have students locate some books, journals, magazines, and newspaper articles that estimate the numbers of species in each kingdom. Ask them to research the numbers and discuss how the numbers were derived.

Quick Demo

Place the parts of different organisms, such as a leaf, a mushroom, moss, a feather, seeds, a poultry wishbone, seaweed, and a piece of sponge, in a box. As you remove each item and display it, ask students to identify the organism that produced the item and classify it in one of the six kingdoms.

Using Science Terms

On the chalkboard, write the names of the six kingdoms and the two words *eukaryote* and *prokaryote*. Have students use dictionaries to find the meanings of the Latin or Greek prefixes and suffixes in each word. Then ask them to explain how each name describes its group. **L1** **ELL**

Display

Make a bulletin board display from photos and illustrations of organisms from each kingdom. Group the visuals according to kingdom. Invite students to add photographs they find in magazines and newspapers.

Activity

Have students use a microscope to observe a variety of prokaryotic and eukaryotic cells. Ask students to explain differences they observe. **L2** **ELL**

GLENCOE
TECHNOLOGY

CD-ROM
BioQuest: *Biodiversity Park*
Disc 3, 4

THE FIVE-KINGDOM SYSTEM

A flurry of ideas for new classification systems followed close on the heels of the discovery of prokaryotes. In 1959, American biologist R.H. Whittaker (1924–1980) proposed a five-kingdom system (right) that soon became universally accepted. The five kingdoms were Monera (bacteria), Protista (algae and other protists), Fungi (mushrooms, molds, and lichens), Plantae (mosses, ferns, and cone-bearing and flowering plants), and Animalia (invertebrate and vertebrate animals). The kingdom Monera included all the prokaryotes; the other four kingdoms consisted of eukaryotes. Fungi, plants, and animals were easily distinguished by their modes of nutrition. But the kingdom Protista was a grab bag, a diverse assortment of living things—some plantlike, some animal-like, some funguslike—that did not fit clearly into any of the other eukaryotic kingdoms.

EVOLUTIONARY RELATIONSHIPS

With the five-kingdom system in place, many taxonomists focused their research on reclassifying living things in terms of their evolutionary relationships rather than on their structural similarities. Present-day organisms, such as the millipede (below), were compared with extinct forms preserved in the fossil record, such as the trilobite (below right). New biochemical

MILLIPEDE

techniques made it possible to compare nucleotide sequences in genes and amino-acid sequences in proteins from different organisms to determine how closely those organisms were related.

462

TRILOBITE

WHITTAKER'S SYSTEM

PLANTS FUNGI ANIMALS

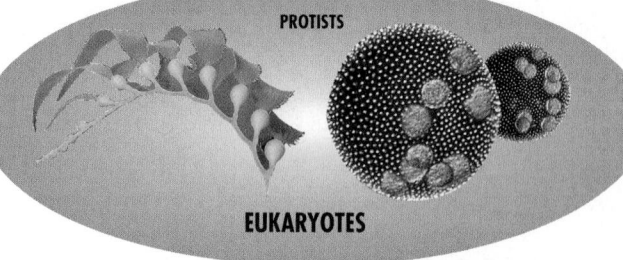

PROTISTS

EUKARYOTES

PROKARYOTES

MONERANS

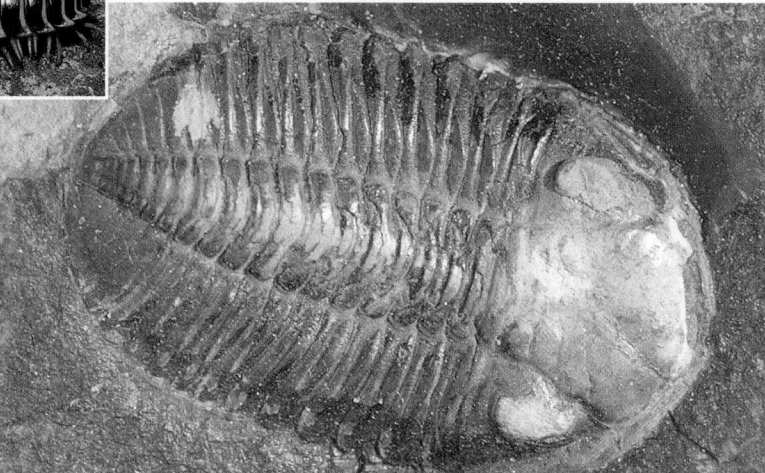

MEETING INDIVIDUAL NEEDS

Visually Impaired

Kinesthetic When discussing eukaryotic cells and prokaryotic cells, provide visually impaired students with cell models. Provide them with a scale that compares the size difference between prokaryotic and eukaryotic cells.

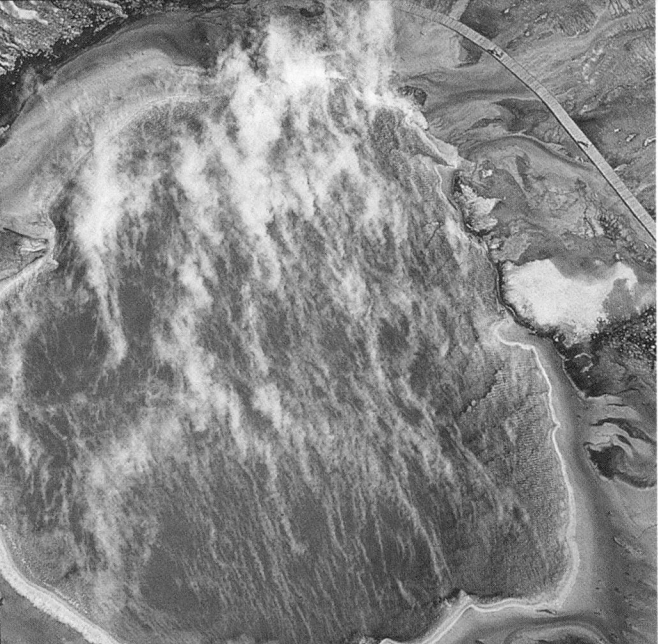

GRAND PRISMATIC SPRING, YELLOWSTONE NATIONAL PARK

THE SIXTH KINGDOM

In the 1970s, genetic tests showed that members of the kingdom Monera were far more diverse than anyone had suspected. One group of bacteria, originally called archaebacteria (ancient bacteria), seemed especially unusual. Archaebacteria, or archaeans, as most biologists now refer to them, often live in extreme environments—very hot or salty places—such as the Grand Prismatic Spring (left) in Yellowstone National Park. In 1996, researchers sequenced the archaean genome and discovered that these tiny cells are as different from bacteria as you are. A sixth kingdom was formed.

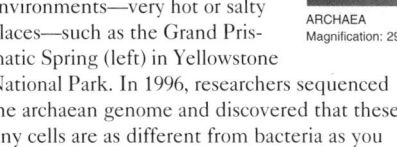

ARCHAEA
Magnification: 29 000×

DOMAINS AND SUPERKINGDOMS

The discovery of the nature of Archaea led C. R. Woese and his colleagues at the University of Illinois to propose a new classification scheme (left) made up of three domains. The domain Bacteria has one kingdom, Eubacteria (true bacteria). The domain Archaea contains two kingdoms. The domain Eukarya consists of the kingdoms Fungi, Plantae, and Animalia. Woese recognized that there were unresolved questions regarding where to place protists in Eukarya.

Recent efforts to establish the most natural groupings of organisms that show evolutionary relationships use molecular genetics. Genbank is a federal agency that gathers genetic data for all of Earth's organisms. Genbank's system (below left) recognizes three superkingdoms: Archaea, Eubacteria, and Eukaryota. The eukaryotes are not grouped into kingdoms because the six kingdoms we are now familiar with show many different origins.

WOESE'S SYSTEM

DOMAIN	DOMAIN	DOMAIN
Bacteria	**Archaea**	**Eukarya**
KINGDOM **Eubacteria**	KINGDOMS **Euryarchaeota** (Methanogens)	KINGDOMS **Animalia**
	Crenarchaeota (Thermophiles)	**Plantae**
		Fungi
		Protista?

GENBANK'S SYSTEM

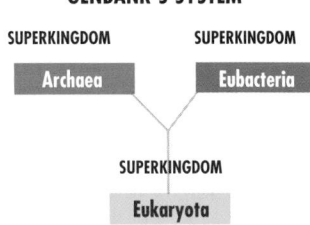

SUPERKINGDOM **Archaea**   SUPERKINGDOM **Eubacteria**

SUPERKINGDOM **Eukaryota**

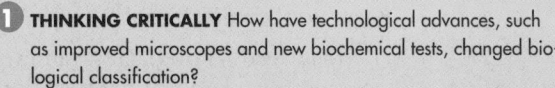
EXPANDING Your View

1 THINKING CRITICALLY How have technological advances, such as improved microscopes and new biochemical tests, changed biological classification?

2 JOURNAL WRITING The kingdom Protista contains very diverse organisms—from unicellular "animallike" amebas to multicellular "plantlike" giant kelp. In your journal, predict what might happen to the protist kingdom in the next few years as biologists study its members in more detail at biochemical and genetic levels.

Enrichment

Have students identify which species of archaebacteria was sequenced. Current information suggests that archaebacteria may live in subsurface pores of rocks, as well as in other less severe environments. Ask how such data might affect the view of archaebacteria as "extremophiles."

Answers to Expanding Your View

1. Improved microscopes allow scientists to differentiate between prokaryotic and eukaryotic cells and among the cell organelles. Biochemical tests determine the presence of specific molecules in cells.
2. Answers will vary. Possible ideas are that organisms currently classified in Kingdom Protista may be reclassified into other kingdoms or that the subgroupings in Kingdom Protista may become additional kingdoms.

Going Further

Have students choose a kingdom to research. They should explore the kingdom's diversity by determining its number of phyla and listing representative organisms. Urge the students to include other information they discover. **L2**

Resource Manager

Critical Thinking/Problem Solving, p. 17 **L3**

MEETING INDIVIDUAL NEEDS

English Language Learners

Have the students complete a time line about the historical evolution of biological classification. Provide students with a baseline of major events to include and a word bank of names and key terms. **L1**
ELL

Prepare

Key Concepts

Students will learn how to interpret phylogenetic classification models. They will compare the characteristics of organisms in the six kingdoms.

Planning

- Obtain pictures of seven dinosaurs for MiniLab 17-2.
- Collect laboratory glassware for the Alternative Lab.
- Gather some protractors for Problem-Solving Lab 17-2.
- Obtain metric rulers and guidebooks with dichotomous keys of insects for the Biolab.

1 Focus

Bellringer

Before presenting the lesson, display **Section Focus Transparency 42** on the overhead projector and have students answer the accompanying questions. **L1** **ELL**

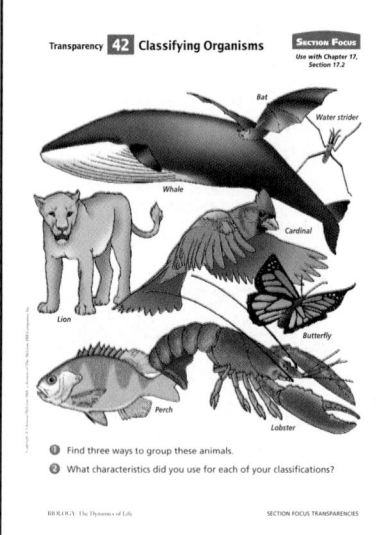

Transparency **42** Classifying Organisms — **Section Focus** Use with Chapter 17, Section 17.2

❶ Find three ways to group these animals.
❷ What characteristics did you use for each of your classifications?

SECTION PREVIEW

Objectives

Describe how evolutionary relationships are determined.
Explain how cladistics reveals phylogenetic relationships.
Compare the six kingdoms of organisms.

Vocabulary

phylogeny
cladistics
cladogram
eubacteria
protist
fungus

Section

17.2 The Six Kingdoms

Suppose you entered a room full of strangers and were asked to identify two related people. What clues would you look for? You might listen for similar-sounding voices. You might look for similar hair, eye, and skin coloration. You might watch for shared behaviors and mannerisms among individuals. When taxonomists want to identify evolutionary relationships among species, they examine certain characteristics of species.

Western sword fern *(Polystichum munitum)* (above) and Northern holly fern *(Polystichum lonchitis)* (inset)

How Are Evolutionary Relationships Determined?

Evolutionary relationships are determined on the basis of similarities in structure, breeding behavior, geographical distribution, chromosomes, and biochemistry. Because these characteristics provide the clues about how species evolved, they also reveal the probable evolutionary relationships of species.

Structural similarities

Structural similarities among species reveal relationships. For example, the presence of many shared physical structures implies that species are closely related and may have evolved from a common ancestor. For example, because lynxes and bobcats have structures more similar to each other than to members of any other groups, taxonomists suggest that they share a common ancestor. Likewise, plant taxonomists use structural evidence to classify dandelions and sunflowers in the same family, Asteraceae, because they have similar flower and fruit structures.

If you observe an unidentified animal that can retract its claws, you can infer that it belongs to the cat family. You can then assume that the animal has other characteristics in common with cats. Taxonomists observe and compare features among members of different taxa and use this information to infer their evolutionary history.

464 ORGANIZING LIFE'S DIVERSITY

GLENCOE TECHNOLOGY

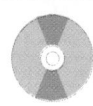

CD-ROM
Biology: The Dynamics of Life
Video: *Museum Collections*
Disc 3

VIDEODISC
Biology: The Dynamics of Life
Museum Collections (Ch. 8)
Disc 1, Side 2, 20 sec.

a

b

Breeding behavior

Sometimes, breeding behavior provides important clues to relationships among species. For example, two species of frogs, *Hyla versicolor* and *H. chrysoscelis*, live in the same area and look similar. During the breeding season, however, there is an obvious difference in their mating behavior. The males of each species make different sounds to attract females, and therefore attract and mate only with members of their own group. Scientists concluded that the frogs were two separate species.

Geographical distribution

The location of species on Earth helps biologists determine their relationships with other species. For example, many different species of finches live on the Galapagos Islands off the coast of South America. Biologists propose that in the past some members of a finchlike bird species that lived in South America reached the Galapagos Islands, where they became isolated. These finches probably spread into different niches on the volcanic islands and evolved over time into many distinct species. The fact that they share a common ancestry is supported by their geographical distribution in addition to their genetic similarities.

Chromosome comparisons

Both the number and structure of chromosomes, as seen during mitosis and meiosis, provide evidence about relationships among species. For example, cauliflower, cabbage, kale, and broccoli look different but have chromosomes that are almost identical in structure. Therefore, biologists propose that these plants are related. Likewise, the similar appearance of chromosomes among chimpanzees, gorillas, and humans suggest a common ancestry.

Biochemistry

Powerful evidence about relationships among species comes from biochemical analyses of organisms. Closely related species have similar DNA sequences and, therefore, similar proteins. In general, the more inherited nucleotide sequences that two species share, the more closely related they are. For example, the DNA sequences in giant pandas and red pandas differ. They differ so much that many scientists suggest that giant pandas are more closely related to bears than to red pandas such as the one shown in *Figure 17.6*. Read the *BioTechnology* feature at the end of this chapter to learn more about how chemical similarities can reveal evolutionary relationships.

Figure 17.6
DNA sequences in red pandas (a) and giant pandas (b) suggest that red pandas are related to raccoons and giant pandas are related to bears.

2 Teach

Activity

Interpersonal Make a set of cards containing the features biologists use to classify organisms. Have a student select a card, describe the feature, and explain why the feature is useful in classification. Return the card to the set and continue until each student has chosen a card. **L2**

Visual Learning

Figure 17.6 Ask students in what ways the giant panda and red panda are similar and different. *They have similar names and body shape and features, but have different DNA sequences.*

BIOLOGY Online Note Internet addresses that you find useful in the space below for quick reference.

_____

_____

_____

_____

_____

Resource Manager

Section Focus Transparency 42 and Master **L1** **ELL**

Phylogenetic Classification: Models

Species that share a common ancestor also share an evolutionary history. The evolutionary history of a species is called its **phylogeny** (fy LOH juh nee). A classification system that shows the evolutionary history of species is a phylogenetic classification and reveals the evolutionary relationships of species.

Early classification systems did not reflect the phylogenetic relationships among organisms. As scientists learned more about geologic time, they modified the early classification schemes to reflect the phylogeny of species.

Cladistics

One biological system of classification that is based on phylogeny is **cladistics** (kla DIHS tiks). Scientists who use cladistics assume that as groups of organisms diverge and evolve from a common ancestral group, they retain some unique

WORD *Origin*

phylogeny
From the Greek words *phylon*, meaning "related group," and *geny*, meaning "origin." Organisms are classified based on their phylogeny.

cladistics
From the Greek word *klados*, meaning "sprout" or "branch." Cladistics is based on phylogeny.

inherited characteristics that taxonomists call derived traits. Biologists identify a group's derived traits and use them to make a branching diagram called a **cladogram** (KLAD eh gram). A cladogram is a model of the phylogeny of a species, and models are important tools for understanding scientific concepts.

Cladograms are similar to the pedigrees, or family trees, you studied in an earlier chapter. Branches on both pedigrees and cladograms show proposed ancestry. In a cladogram, two groups on diverging branches probably share a more recent ancestor than those groups farther away. If two organisms are near each other on a pedigree's branch, they also share an ancestor. However, an important difference between cladograms and pedigrees is that, whereas pedigrees show the direct ancestry of an organism from two parents, cladograms show a probable phylogeny of a group of organisms from ancestral groups.

In *Figure 17.7*, you see the cladogram for modern birds, such as robins. How was the cladogram developed? First, taxonomists identified the derived traits of modern birds—flight feathers, light bones, a wishbone, down feathers, and feathers with

Figure 17.7
This cladogram uses the derived traits of a modern bird, such as the robin, to model its phylogeny. Groups that are closer together on the cladogram probably share a more recent common ancestor.

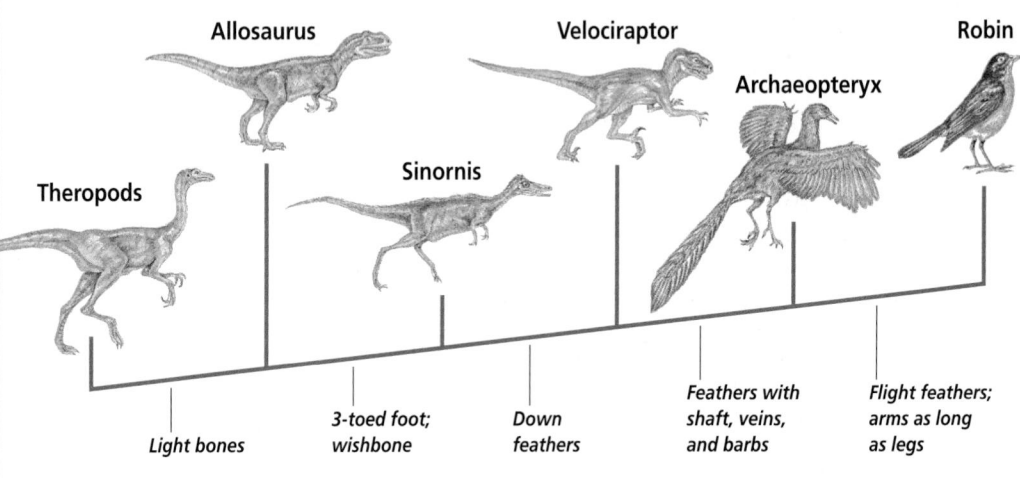

BIOLOGY JOURNAL

Vertebrate Phylogeny

Naturalist Provide students with the following list of vertebrate classes: birds, reptiles, mammals, amphibians, and fish. In addition, give them a listing of derived traits, such as legs, feathers, and hair. Have them construct a vertebrate cladogram from the lists. L3

shafts. Next, they identified ancestral species that have at least some of these traits. Most biologists agree that the ancestors of birds are a group of dinosaurs called theropods. Some of these theropods are *Allosaurus*, *Archaeopteryx*, *Velociraptor*, *Sinornis*, and *Protoarchaeopteryx*. Each of these ancestors has a different number of derived traits. Some groups share more derived traits than others.

Finally, taxonomists constructed the robin's cladogram from this information. They assume that if groups share many derived traits, they share common ancestry. Thus, *Archaeopteryx* and the robin, which share four derived traits, are on adjacent branches, indicating a recent common ancestor. Use the *MiniLab* on this page to construct a cladogram for another species.

Another type of model

In this book, you will see cladograms and other types of models that provide information about the phylogenetic relationships among species. One type of model resembles a fan. Unlike a cladogram, a fanlike model may communicate the time organisms became extinct or the relative number of species in a group. A fanlike diagram incorporates fossil information and the knowledge gained from anatomical, embryological, genetic, and cladistic studies.

In *Figure 17.8* on the next page, you can see a fanlike model of the six-kingdom classification system. For easy reference, the same diagram is located inside the back cover of this textbook. This model includes both Earth's geologic time and the probable evolution of organisms during that timespan. In addition, this fanlike diagram helps you to find relationships between modern and extinct species.

MiniLab 17-2 Classifying

Using a Cladogram to Show Relationships Cladograms were developed by Willie Hennig. They use derived characteristics to illustrate evolutionary relationships.

Procedure

1. The following table shows the presence or absence of six derived traits in the seven dinosaurs that are labeled A-G.
2. Use the information listed in *Table 17.1* to answer the questions below.

Table 17.1 Derived traits of dinosaurs

Dinosaur trait	A	B	C	D	E	F	G
Hole in hip socket	yes	yes	yes	yes	yes	yes	yes
Extension of pubis bone	no	no	no	yes	yes	yes	yes
Unequal enamel on teeth	no	no	no	no	yes	yes	yes
Skull has "shelf" in back	no	no	no	no	no	yes	yes
Grasping hand	yes	yes	yes	no	no	no	no
Three-toed hind foot	yes	yes	no	no	no	no	no

Analysis

1. Copy the partially completed cladogram. Complete the missing information on the right side.

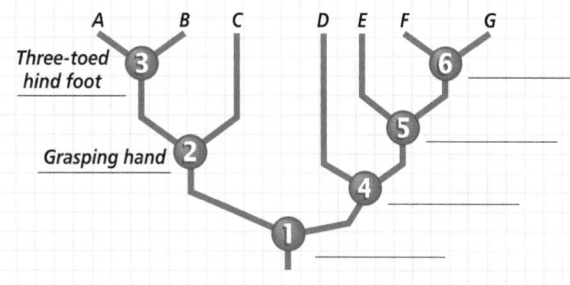

2. How many traits does dinosaur F share with dinosaur C, with dinosaur D, and with dinosaur E?
3. Dinosaurs A and B form a grouping called a clade. The dinosaurs A, B, and C form another clade. What derived trait is shared only by the A and B clade? By the A, B, and C clade? By the D, E, F, and G clade?
4. Traits that evolved very early, such as the hole in the hip socket, are called primitive traits. The traits that evolved later, such as a grasping hand, are called derived traits. Are primitive traits typical of broader or smaller clades? Are derived traits typical of broader or smaller clades? Give an example in each case.

Purpose
Students will make a cladogram.

Process Skills
acquire information, analyze information, apply concepts, classify, compare and contrast, interpret data

Teaching Strategies
■ Check student answers to question 1 before they complete the remaining analysis questions.
■ Show students pictures of dinosaurs A-G.

Archaeopteryx	*Allosaurus*
Plateosaurus	*Stegosarus*
Parasaurolophus	*Triceratops*
Pachycephalosaurus	

Expected Results
Students will complete the cladogram as follows: node 1 = hole in hip socket, node 4 = extension of pubis bone, node 5 = unequal enamel, node 6 = skull shelf.

Analysis
1. See "Expected Results."
2. 1 trait with C, 2 traits with D, and 3 traits with E.
3. three-toed hind foot; grasping hand; extension of pubis bone
4. broader—hole in hip socket extends to all 7 animals; smaller—three-toed hind foot extends to only 1 clad

✔ Assessment
Performance Ask students to classify the seven dinosaurs according to their external similarities and record their scheme on the chalkboard. Then, show them how a cladogram classifies these dinosaurs. Use the Performance Task Assessment List for Making and Using a Classification System in **PASC**, p. 49. **L2**

MEETING INDIVIDUAL NEEDS

Visually Impaired

Kinesthetic Fold paper into a fan shape. Open the fan and allow visually impaired students to feel its shape. Explain that in the fanlike phylogenetic model, prokaryotes are located where the fan forms a point (its base). Just above this point are eukaryotes, which evolved from prokaryotes. **L1**

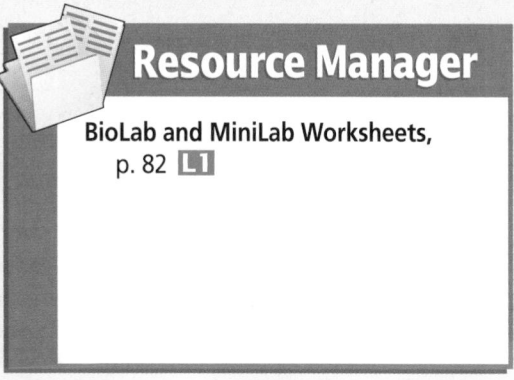

Resource Manager

BioLab and MiniLab Worksheets, p. 82 **L1**

Show students actual examples of organisms in each kingdom. Briefly describe the characteristics of the organisms. You may have to use photos or drawings of different archaebacteria and eubacteria

Visual Learning

Figure 17.8 Ask the students to explain how time is indicated on the model. *Earth's earliest organisms are near the center of the model, and those that appeared later are near the rim.*

Resource Manager

Concept Mapping, p. 17 **L3**
ELL

LIFE'S SIX KINGDOMS

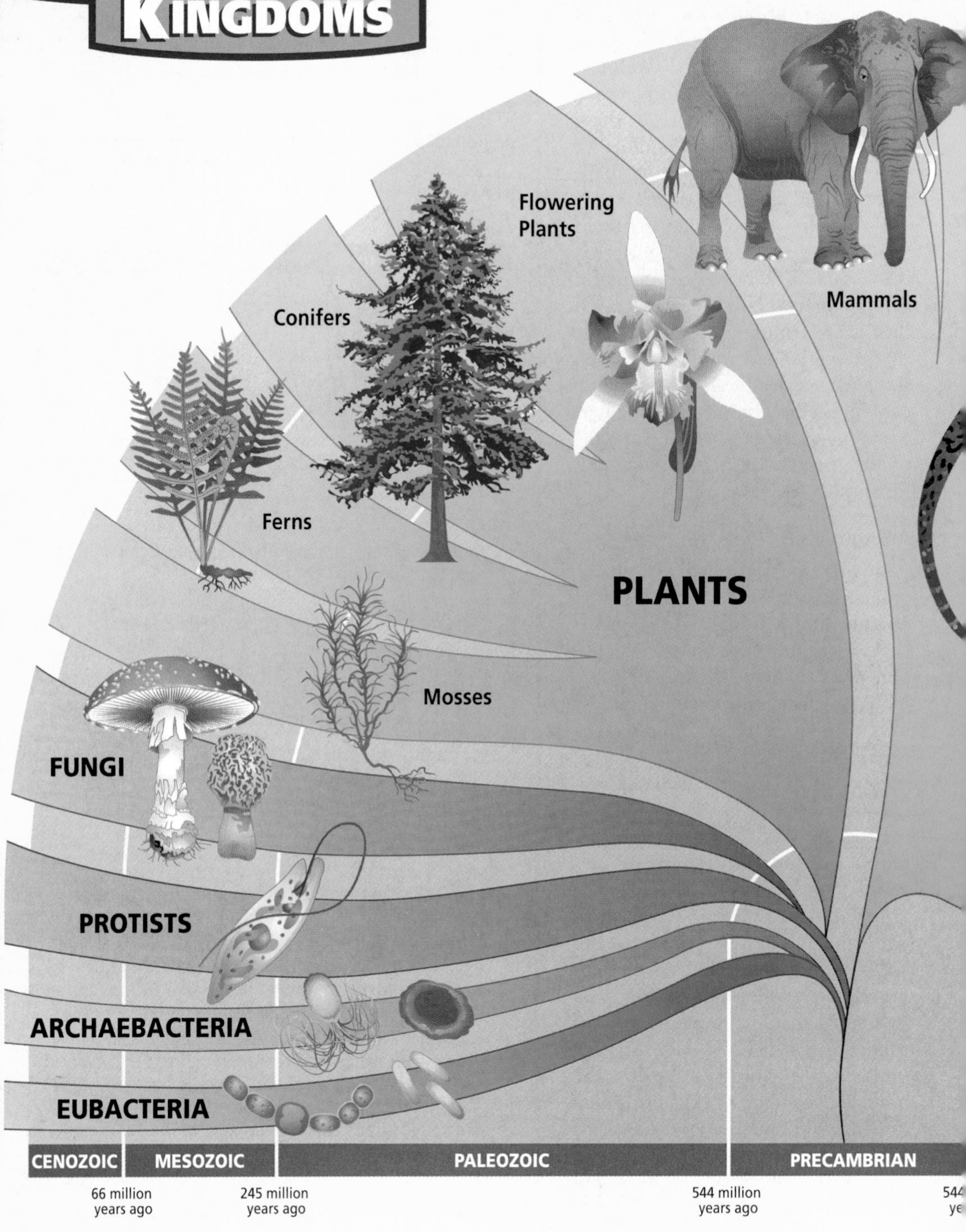

Flowering Plants

Conifers

Mammals

Ferns

PLANTS

Mosses

FUNGI

PROTISTS

ARCHAEBACTERIA

EUBACTERIA

| CENOZOIC | MESOZOIC | PALEOZOIC | PRECAMBRIAN |

66 million years ago   245 million years ago   544 million years ago   544 ye...

Alternative Lab

Classification of Glassware

Purpose
Students will classify laboratory glassware.

Materials
laboratory glassware such as flasks, test tubes, beakers, stirring rods, thermometers, bottle stoppers, pipets, droppers, funnels, petri dishes, graduated cylinders, and jars

Procedure
Give students the following directions.

1. Obtain a set of glassware to classify according to its structure and function. **CAUTION:** *Handle glassware carefully because it can break easily. Dispose of broken glass as your teacher directs.*

2. Classify the glassware into two groups. Then subdivide the groups. Note each group's characteristics and the names of its glassware.

3. Place the glassware groups in a phylogenetic model such as in Figure 17.8. Consider how the glassware may have "evolved," and why some groups may be more closely "related" than others. Closely related groups should be closer together on your model.

4. Make a large diagram of your phylogenetic model.

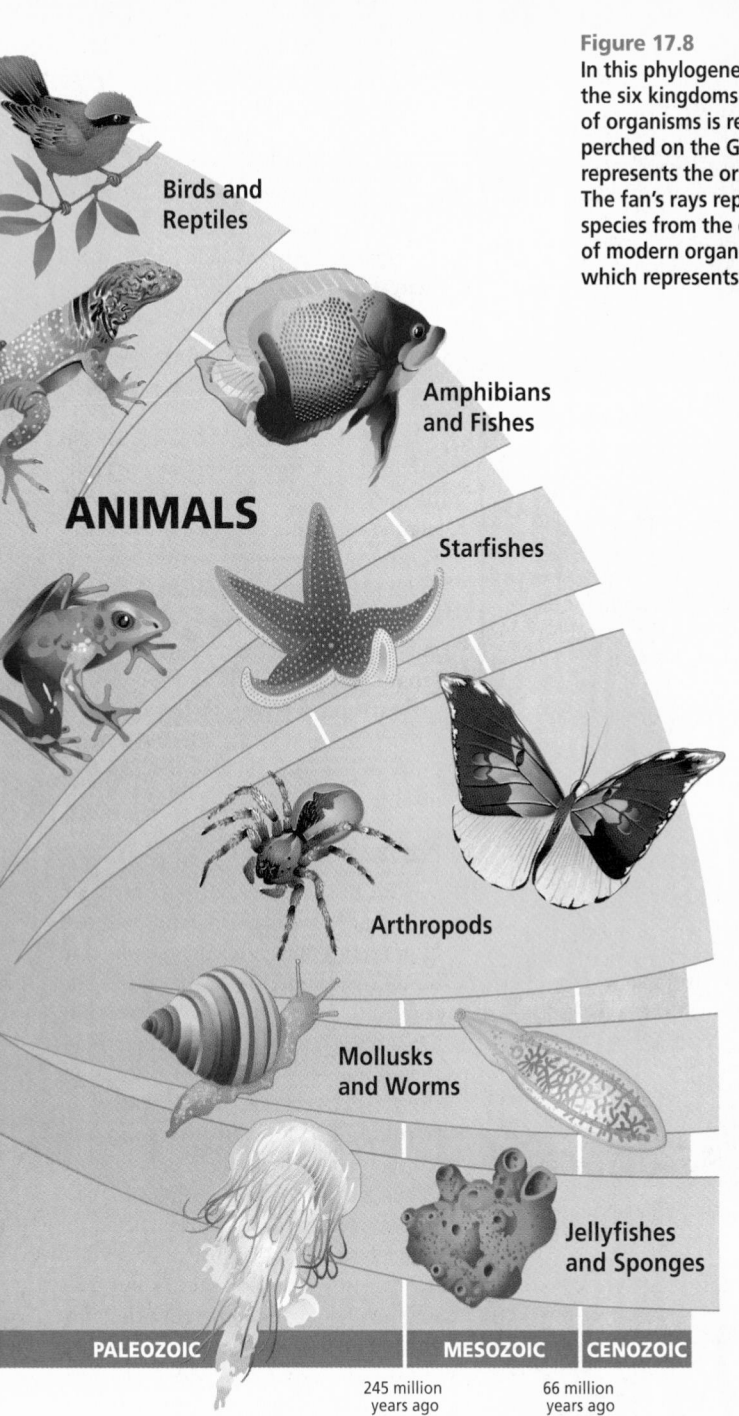

Figure 17.8
In this phylogenetic diagram, six colors represent the six kingdoms of living things. The phylogeny of organisms is represented by a fanlike structure perched on the Geologic Time Scale. The fan's base represents the origin of life during the Precambrian. The fan's rays represent the probable evolution of species from the common origin. The major groups of modern organisms occupy the fan's outer edge, which represents present time.

Birds and Reptiles

Amphibians and Fishes

ANIMALS

Starfishes

Arthropods

Mollusks and Worms

Jellyfishes and Sponges

PALEOZOIC | MESOZOIC | CENOZOIC

245 million years ago | 66 million years ago

Analysis

1. What glassware features were most useful in your model? *Responses are likely to indicate functional traits.*
2. How did your phylogenetic model differ from the model in Figure 17.8? *The glassware model lacks dates, which would be purely speculative.*

✓ **Assessment**

Portfolio Have students write an essay about the "evolution" of their glassware. They should use the glassware's features to support their statements. Use the Performance Task Assessment List for Writing in Science in **PASC,** p. 87. **L2** **P**

Brainstorming

Have students use Figure 17.8 to identify other organisms that belong in each group. Ask them to explain their choices.

Time Line

 Visual-Spatial Have students make a time line of the evolutionary history of life on Earth using the information in Figure 17.8. Remind students that their time lines represent millions of years. **L1**

GLENCOE
TECHNOLOGY

 **VIDEODISC**
The Secret of Life
Domains/Six Kingdoms

 **CD-ROM**
Biology: The Dynamics of Life
Explorations: *The Six Kingdoms* Disc 3

 **Resource Manager**

Reteaching Skills Transparency 26 and Master **L1** **ELL**

Purpose

Students will determine the approximate number of species in each kingdom.

Process Skills

classify, make and use graphs, use numbers, think critically

Teaching Strategies

■ Have a supply of protractors.
■ Review how to use a protractor and calculate percent.

Background

The number of named species is around 1.4 million, but estimates of as many as 30 million living species have been proposed. Of the species named, about 75% are animals. This value probably does not reflect their relative abundance, but rather their size and accessibility for study.

Thinking Critically

1. Answers should be close to the following. Archaebacteria = 500, Eubacteria = 10 000, Protista = 70 000, Fungi = 110 000, Plants = 510 000, Animals = 2 900 000.
2. 3 580 500
3. Archaebacteria = 0.028%, Eubacteria = 0.3%, Protista = 2.8%, Fungi = 5.6%, Plants = 15%, Animals = 76%.
4. Answers may vary—the number of species may vary due to the extinction and discovery of species.

✔ Assessment

Skill Have students find the approximate number of species that become extinct each day, week, month, and year and summarize their findings in a data table. Then ask them to list the factors that contribute to the loss of species. Use the Performance Task Assessment List for Data Table in **PASC,** p. 37. **L3**

Problem-Solving Lab 17-2 Using Graphs

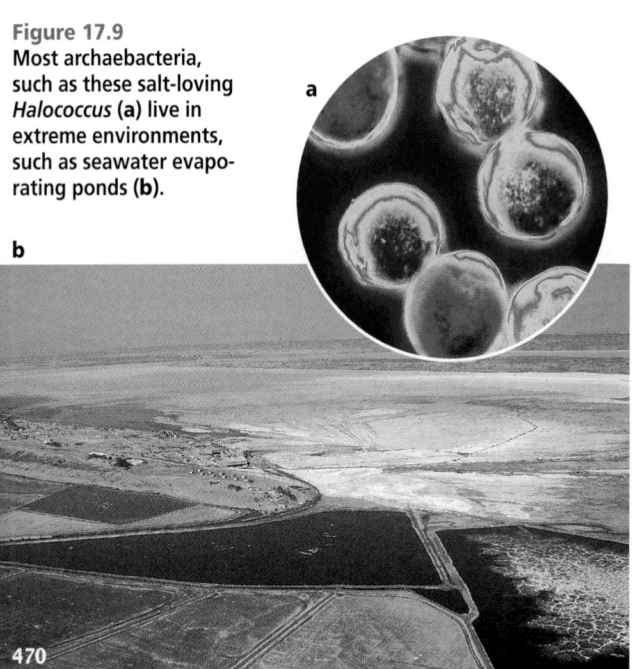

How many species are there in each kingdom? You may not realize it, but you probably already have seen more than 1000 different species since you were born. How close might you be to having seen all the different species that exist?

Analysis

The circle graph above shows Earth's six kingdoms.

Thinking Critically

1. List the approximate number of species for each of the six kingdoms. Each degree of the circle graph is equal to 10 000 species. Archaebacteria represent 1/20 of a degree, and Eubacteria represent 1 degree.
2. What is the approximate total number of species for all life forms on Earth?
3. What approximate percent of the total life forms on Earth are in each kingdom?
4. Why do all questions refer to the number of species as "approximate?"

Figure 17.9
Most archaebacteria, such as these salt-loving *Halococcus* (a) live in extreme environments, such as seawater evaporating ponds (b).

470

Groups of organisms that are closer in the same colored ray share more inherited characteristics and are probably more closely related than groups that are farther apart. For example, find the jellyfishes, the fishes, and the reptiles on the model. Notice that fishes and reptiles are closer to each other than they are to the jellyfishes, indicating that they are more closely related to each other than they are to jellyfishes.

The Six Kingdoms of Organisms

As you saw in *Figure 17.8,* the six kingdoms of organisms are archaebacteria, eubacteria, protists, fungi, plants, and animals. In general, differences in cellular structures and methods of obtaining energy are the two main characteristics that distinguish among the members of the six kingdoms. The six kingdoms reflect evolutionary history. Learn more about the number of species in each kingdom in the *Problem-Solving Lab* on this page.

Prokaryotes

The prokaryotes, organisms with cells that lack distinct nuclei bounded by a membrane, are microscopic and unicellular. Some are heterotrophs and some are autotrophs. In turn, some prokaryotic autotrophs are chemosynthetic, whereas others are photosynthetic. Prokaryotes are classified in two kingdoms: Archaebacteria and Eubacteria. The oldest prokaryotic fossils are about 3.5 billion years old.

There are a few hundred species of archaebacteria and most of them live in extreme environments such as swamps, deep-ocean hydrothermal vents, and seawater evaporating ponds, *Figure 17.9.* Most of these environments are oxygen-free. The

---

BIOLOGY Online

Note Internet addresses that you find useful in the space below for quick reference.

_____

_____

_____

_____

_____

Figure 17.10
Eubacteria are a diverse kingdom of prokaryotes. Both their cellular structure and the way they obtain food vary widely.

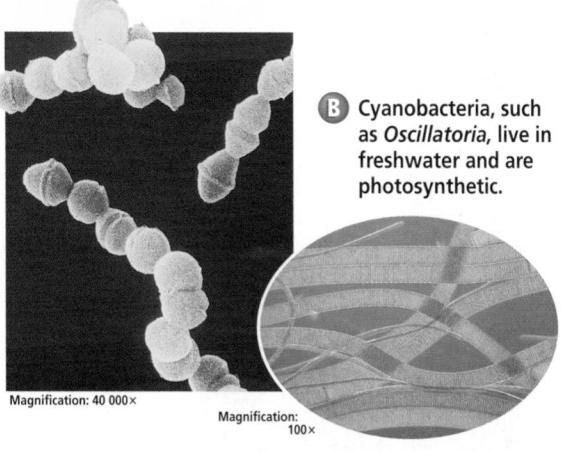

A The bacteria that cause strep throat are heterotrophs called *Streptococcus*.

B Cyanobacteria, such as *Oscillatoria*, live in freshwater and are photosynthetic.

Magnification: 40 000×

Magnification: 100×

WORD *Origin*

eubacteria
From the Greek words *eu*, meaning "true," and *bakterion*, meaning "small rod." Eubacteria are prokaryotes.

protist
From the Greek word *protistos*, meaning "very first" or "superlative." The first eukaryote cells were protists.

lipids in the cell membranes of archaebacteria, the composition of their cell walls, and the sequence of nucleic acids in their ribosomal RNA differ considerably from those of other prokaryotes. However, their genes have a similar structure to those in eukaryotes.

All of the other prokaryotes, more than 10 000 species of bacteria, are classified in Kingdom Eubacteria. **Eubacteria,** such as the ones you see in *Figure 17.10,* have very strong cell walls and a less complex genetic makeup than found in archaebacteria or eukaryotes. They live in most habitats except the extreme ones where the archaebacteria live. Although some eubacteria cause diseases, such as strep throat and pneumonia, most bacteria are harmless and many are actually helpful.

Protists: A diverse group

Kingdom Protista contains diverse species as you can see in *Figure 17.11,* but protists do share some characteristics. A **protist** is a eukaryote that

Figure 17.11
Although these three protists look different, they are all eukaryotes and live in moist environments.

Magnification: 150×

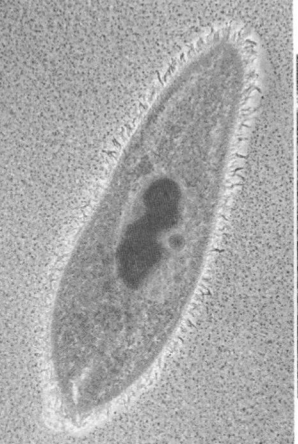

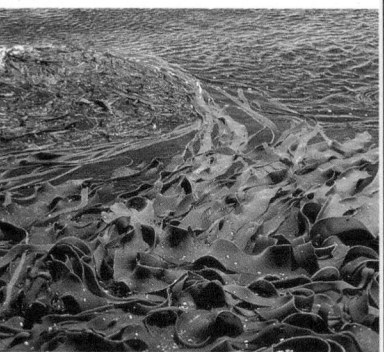

A Funguslike slime molds often creep through damp forests, feeding on microorganisms.

B The paramecium is an animal-like protist that moves through water.

C These kelps are multicellular plantlike protists. Although they look like plants, they do not have organs or organ systems.

17.2 THE SIX KINGDOMS **471**

Quick Demo
Display pictures or actual organisms from each kingdom. Have students observe the organisms and identify common traits of each kingdom's organisms.

Visual Learning
Use the organisms in Figures 17.10 and 17.11 to point out the differences between bacteria and protists. Be certain that students understand that, unlike protists, bacteria lack a nucleus and organelles bound by membranes.

Resource Manager
Basic Concepts Transparency 24 and Master L2 ELL
Laboratory Manual, pp. 121-124 L2

Portfolio

Gathering Protists
Kinesthetic Have students obtain water samples from local ponds. Tell them to scrape the underneath surfaces of rocks into their collecting jars. Have them use microscopes to observe their samples and sketch any protists they see, explaining why each is a protist. L1
ELL P

PROJECT

Classifying Organisms
Visual-Spatial Ask students to make a photo essay of the six kingdoms. They should label each organism with its scientific and common name and explain the essay to their group members. L1
ELL

Visual Learning

Use Figures 17.12 and 17.13 to compare the features of fungi and plants. Have students explain why fungi used to be classified in the plant kingdom. *The fungi are plantlike and have cell walls, but, unlike plants, the fungi are not autotrophic.*

Activity

Naturalist Have students collect fungi from their local environment. Caution them not to ingest any of the fungi as some are poisonous and to wash their hands after collecting. Ask them to describe the similarities and differences among the fungi. *The differences include the shapes and colors of fruiting bodies, the gills, and other features. Similarities include cell walls made of chitin and heterotrophy.* **L3**

Brainstorming

Ask students to brainstorm a list of familiar plants. Be sure they include trees, shrubs, and grasses, as well as annual and perennial garden plants.

Visual Learning

Figure 17.14 Ask students what senses the cheetah uses while running. *sight, touch, and smell*

3 Assess

Check for Understanding

Visual-Spatial Show students pictures of organisms. Ask them to identify the kingdom to which each belongs. **L1**

Figure 17.12
Morels are edible fungi that grow for only a few days in only a few places.

lacks complex organ systems and lives in moist environments. Fossils of plantlike protists show that protists existed on Earth up to two billion years ago. Although some protists are unicellular, others are multicellular. Some are plantlike autotrophs, some are animal-like heterotrophs, and others are funguslike heterotrophs that produce reproductive structures like those of fungi.

Figure 17.13
A *Hibiscus* is just one kind of flowering plant (**a**). Tropical tree ferns do not produce flowers (**b**).

Fungi: Earth's decomposers

Organisms in Kingdom Fungi are heterotrophs that do not move from place to place. A **fungus** is either a unicellular or multicellular eukaryote that absorbs nutrients from organic materials in the environment. Fungi first appeared in the fossil record about 400 million years ago. There are more than 100 000 known species of fungi, including the one you see in *Figure 17.12*.

Plants: Multicellular oxygen producers

All of the organisms in Kingdom Plantae are multicellular, photosynthetic eukaryotes. None moves from place to place. A plant's cells usually contain chloroplasts and have cell walls composed of cellulose. Plant cells are organized into tissues that, in turn, are organized into organs and organ systems. You can see two of the many diverse types of plants in *Figure 17.13*.

The oldest plant fossils are a little more than 400 million years old.

Cultural Diversity

Classifying Lion Tamarins

Introduce students to the work of Adelmar Coimbra-Filho, Brazil's leading primatologist. Since the 1960s, Coimbra-Filho has published many articles on the systematics of the lion tamarin, *Leontopithecus*, a newly discovered primate species. Coimbra-Filho directs Rio de Janeiro's Primate Center. He has coordinated efforts with the World Wildlife Fund to save lion tamarins from extinction. Elicit from students the importance of the work done by taxonomists and systematicists in discovering new species.

Figure 17.14
Many animals have well-developed nervous and muscular systems.

 **A** The luna moth's antennae are sense organs, which are part of its nervous system. The moth's antennae detect tiny quantities of chemicals in the air.

B The cheetah uses many of its organ systems, especially its nervous and muscular systems, to speed through the grasslands, perhaps chasing prey.

However, some scientists propose that plants existed on Earth's land masses much earlier than these fossils indicate. Plants do not fossilize as often as organisms that contain hard structures, such as bones, which more readily fossilize than soft tissues.

There are more than 500 000 known species of plants. Although you may be most familiar with flowering plants, there are many other types of plants, including mosses, ferns, and evergreens.

Animals: Multicellular consumers

Animals are multicellular heterotrophs. Nearly all are able to move from place to place. Animal cells do not have cell walls. Their cells are organized into tissues that, in turn, are organized into organs and complex organ systems. Some organ systems in animals are the nervous, circulatory, and muscular systems, *Figure 17.14*. Animals first appeared in the fossil record 600 million years ago.

Section Assessment

Understanding Main Ideas

1. How do members of the different kingdoms obtain nutrients?
2. Make a list of the characteristics that archaebacteria and eubacteria share. Then make a list of their differences.
3. For what reasons is a fanlike diagram a useful model for phylogenetic classification? Explain each of your reasons.
4. How do cladograms and fanlike diagrams differ?

Thinking Critically

5. Why is phylogenetic classification more natural than a system based on characteristics such as medical usefulness, or the shapes, sizes, and colors of body structures?

SKILL REVIEW

6. **Making and Using Tables** Make a table that compares the characteristics of members of each of the six kingdoms. For more help, refer to *Organizing Information* in the **Skill Handbook**.

Section Assessment

1. Prokaryotes and protists use photosynthesis or chemosynthesis. Fungi absorb nutrients. The plants photosynthesize. Animals eat other organisms.
2. They are unicellular prokaryotes with cell walls. They differ in their RNA and cell wall composition.
3. The location of species shows their relationships and when they evolved.
4. Cladograms show phylogeny; fanlike diagrams show evolutionary history in geologic time.
5. Phylogenetic classification relies on inherited features and can provide more reliable information than other systems that do not.
6. Information for tables can be found on pages 470-473.

INVESTIGATE
BioLab

Time Allotment

One class period

Process Skills

classify, observe and infer, compare and contrast, interpret data, sequence

PREPARATION

Alternative Materials

■ Students can make a key from Styrofoam packing pieces or kitchen utensils with a variety of sizes, shapes, and colors.

ANALYZE AND CONCLUDE

1. The keys may or may not have been alike. The groups may have first divided the beetles into groups based on different features, such as size rather than color.

2. Useful: size, color, and shape of various body parts, number of body sections, and antennae features; not useful: number of legs, number of antennae, habitat.

3. Having only two choices makes it easy to analyze organisms. In many keys, the choice is that the organism either has or does not have a particular characteristic.

Resource Manager

BioLab and MiniLab Worksheets, pp. 83-84 L2

INVESTIGATE
BioLab

Making a Dichotomous Key

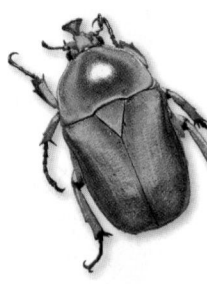

Do you remember the first time you saw a beetle? You may have asked someone nearby, "What is it?" You may still be naturally curious and want to know the names of insects you find. To help identify organisms, taxonomists have developed dichotomous keys. A dichotomous key is a set of paired statements that can be used to identify organisms. When you use a dichotomous key, you choose one statement from each pair that best describes the organism. At the end of each statement you chose, you are directed to the next set of statements to use. Finally, you will read a statement that contains at the end the name of the organism or the group to which it belongs.

Scarab beetle

Longhorned woodboring beetle

PREPARATION

Problem

How is a dichotomous key made?

Objectives

In this BioLab, you will:
■ **Classify** organisms on the basis of structural characteristics.
■ **Develop** a dichotomous key.

Materials

sample keys from guidebooks
metric ruler

Skill Handbook

Use the **Skill Handbook** if you need additional help with this lab.

PROCEDURE

1. Study the drawings of beetles.
2. Choose one characteristic of the beetles and classify the beetles into two groups based on that characteristic. Take measurements if you wish.
3. Record the chosen characteristic in a diagram like the one shown. Write the numbers of the beetles in each group on your diagram.
4. Continue to form subgroups within your two groups based on different characteristics. Record the characteristics and numbers

of the beetles in your diagram until you have only one beetle in each group.
5. Using the diagram you have just made, make a dichotomous key for the beetles. Remember that each numbered step should contain two choices for classification. Begin with 1A and 1B. For help, examine sample keys provided by your teacher.
6. Exchange dichotomous keys with another team. Use their key to identify the beetles.

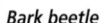
Bark beetle

PROCEDURE

Teaching Strategies

■ Have the students examine samples of dichotomous keys.
■ Explain that a well-designed key has contrasting traits from which to chose.
■ Point out that after a beetle is identified, its characteristics are not used to develop additional choices.

■ 👆 Have groups make transparencies of their keys and explain them. Ask students to list the keys' strengths and weaknesses.

1 Variegated mud-loving beetle

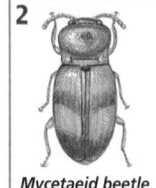

2 Mycetaeid beetle

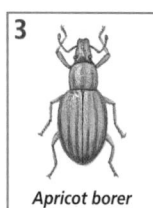

3 Apricot borer

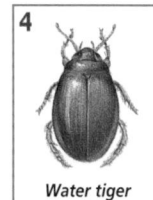

4 Water tiger

5 Predaceous diving beetle

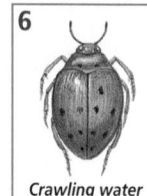

6 Crawling water beetle

7 Flathead apple beetle

8 Red-necked cane beetle

9 Cucumber snout beetle

10 Whirligig beetle

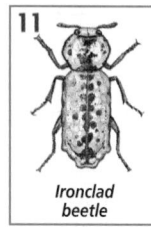

11 Ironclad beetle

12 Broad-horned flour beetle

13 Red flour beetle

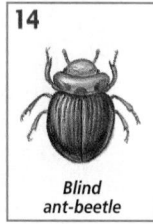

14 Blind ant-beetle

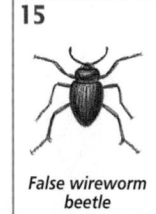

15 False wireworm beetle

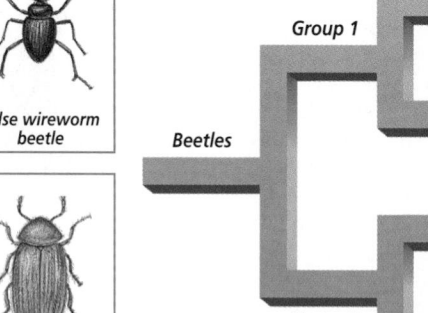

Group 1

Beetles

Group 2

16 White-marked spider beetle

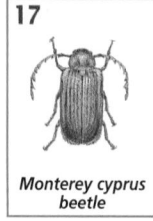

17 Monterey cyprus beetle

18 Drug store beetle

ANALYZE AND CONCLUDE

1. **Comparing and Contrasting** Was the dichotomous key you constructed exactly like those of other students? Explain your answer.
2. **Analyzing Data** What characteristics were most useful for making a classification key for beetles? What characteristics were not useful?
3. **Thinking Critically** Why do keys typically offer only two choices and not more?

Going Further

Application Using the same procedure that you just used to make a beetle key, make a dichotomous key to identify the students in your class.

BIOLOGY Online To find out more about identifying organisms, visit the Glencoe Science Web site. **science.glencoe.com**

✓ Assessment

Performance Give each group a bag of mixed beans. Ask them to make a dichotomous key for the beans. Have the groups exchange their keys to see if they work. Students can place the keys in their portfolios. Use the Performance Task Assessment List for Making and Using a Classification system in **PASC**, p. 49. **L2** **P**

Going Further

Naturalist Have students use 10-20 features of their classmates, such as hair color or some personal belongings, such as notebooks, shoes, and backpacks, to classify them. You may want to select neutral features in advance to avoid sensitive issues. **L1**

BIOLOGY Online Note Internet addresses that you find useful in the space below for quick reference.

_____

_____

_____

_____

_____

Data and Observations

Have students exchange their keys. If another group can use them, their accuracy is confirmed.

BIO
Technology

Molecular Clocks

How long ago did animals first appear on Earth? Did the giant panda evolve along the same family line as bears or raccoons? To help answer questions like these, biologists have learned how to use DNA, proteins, and other biological molecules as "clocks" that reveal details about evolutionary relationships.

Accumulated molecular differences in the DNA of two species can indicate how long they have been separate species. Comparing both the DNA base sequences and the amino acid sequence of a specific protein of two species can indicate the closeness of their relationship.

Comparing DNA One way to compare DNA is to measure how strongly the single strands of DNA from two species will bond. This method is known as DNA-DNA hybridization. Double-stranded DNA from each species is heated to separate the complementary strands. Then the single strands of DNA from each species are mixed and allowed to cool. As the DNA cools, the single strands from the two species bond, or hybridize. If the species are closely related, more of their DNA base pairs will match, and their DNA strands will bond strongly.

Another method of comparing the DNA of species is called DNA sequencing. Biologists select a gene that species have in common and compare the genes' bases. Counting how many base pairs differ can indicate approximately how long ago each species became distinct. Estimates obtained by DNA sequencing show that many of the animal phyla began to appear on Earth about 1.2 billion years ago.

Protein Clocks A specific protein is assumed to evolve at about the same rate in all species that contain the protein. Comparing the amino acid sequences of the protein in several species can show about how long ago the species diverged. For example, cytochrome *c* is a protein in the cells of aerobic organisms. Both human and chimpanzee cytochrome *c* has the same amino acid sequence. The cytochrome *c* of other primates has a different amino acid sequence.

Flamingoes and storks are closely related.

Application for the Future

DNA-DNA hybridization has shown that flamingoes are more closely related to storks than they are to geese. Protein clock data suggest that humans and chimpanzees became distinct species recently in the history of Earth. These biotechnological methods are useful in determining phylogenetic relationships.

INVESTIGATING THE TECHNOLOGY

Analyzing Information The cytochrome *c* found in humans and chimpanzees differs from that found in dogs by 13 amino acids, in tuna by 31 amino acids, and in rattlesnakes by 20 amino acids. What assumptions can you make based on this information?

BIOLOGY *Online* To find out more about molecular clocks, visit the Glencoe Science Web site. **science.glencoe.com**

GLENCOE TECHNOLOGY

VIDEODISC
The Infinite Voyage: *The Geometry of Life, Evolution, Molecular Genetics, and DNA Sequencing* (Ch. 5)

 8 min.

The Dawn of Humankind, DNA Studies Create Controversy (Ch. 7)

 3 min. 30 sec.

Bridging of Fossils and Genetic Research (Ch. 9)

 2 min. 30 sec.

Chapter 17 Assessment

SUMMARY

Section 17.1
Classification

Main Ideas

■ Although Aristotle developed the first classification system, Linnaeus laid the foundation for modern classification systems by using structural similarities to organize species and by developing a binomial naming system for species.

■ Scientists use a two-word system called binomial nomenclature to give species scientific names.

■ Classification provides an orderly framework in which to study the relationships among living and extinct species.

■ Organisms are classified in a hierarchy of taxa: kingdom, phylum or division, class, order, family, genus, and species.

Vocabulary

binomial nomenclature (p. 454)
class (p. 459)
classification (p. 453)
division (p. 459)
family (p. 458)
genus (p. 454)
kingdom (p. 459)
order (p. 459)
phylum (p. 459)
taxonomy (p. 453)

Section 17.2
The Six Kingdoms

Main Ideas

■ Biologists use similarities in body structures, breeding behavior, geographic distribution, chromosomes, and biochemistry to determine evolutionary relationships.

■ Modern classification systems are based on phylogeny. Both cladograms and the fanlike models include information about phylogeny.

■ Taxonomists organize organisms into six kingdoms. Kingdoms Archaebacteria and Eubacteria contain only unicellular prokaryotes that differ chemically from each other. Kingdom Protista contains eukaryotes that lack complex organ systems. Kingdom Fungi includes heterotrophic eukaryotes that absorb their nutrients. Kingdom Plantae includes multicellular eukaryotes that are photosynthetic. Kingdom Animalia includes multicellular, eukaryotic heterotrophs with cells that lack cell walls.

Vocabulary

cladistics (p. 466)
cladogram (p. 466)
eubacteria (p. 471)
fungus (p. 472)
phylogeny (p. 466)
protist (p. 471)

Main Ideas
Summary statements can be used by students to review the major concepts of the chapter.

Using the Vocabulary
To reinforce chapter vocabulary, use the Content Mastery booklet and the activities in the Interactive Tutor for Biology: The Dynamics of Life on the Glencoe Science Web site.
science.glencoe.com

THE PRINCETON REVIEW *All Chapter Assessment questions and answers have been validated for accuracy and suitability by The Princeton Review.*

UNDERSTANDING MAIN IDEAS

1. d
2. a

UNDERSTANDING MAIN IDEAS

1. Which of the following is a scientific name of a species?
a. *bison bison*
b. Mimus Polyglottis
c. *homo Sapiens*
d. *Quercus alba*

2. Which of the following would be a useful characteristic to use in cladistics?
a. derived characteristic
b. similar habitat
c. mutations
d. random differences

CHAPTER 17 ASSESSMENT **477**

GLENCOE TECHNOLOGY

VIDEOTAPE
MindJogger Videoquizzes
Chapter 17: *Organizing Life's Diversity*
Have students work in groups as they play the videoquiz game to review key chapter concepts.

Resource Manager

Chapter Assessment, pp. 97-102
MindJogger Videoquizzes
ExamView® Pro Software 💾
BDOL Interactive CD-ROM, Chapter 17 quiz

3. d
4. b
5. d
6. c
7. a
8. b
9. b
10. c
11. phylogeny; taxonomy
12. protist
13. division; phylum
14. genus
15. order
16. phylogeny
17. Fungi
18. binomial nomenclature
19. classification
20. family

APPLYING MAIN IDEAS

21. Aristotle based his system on a small number of organisms and used criteria, such as habitat, that do not reveal phylogenetic relationships. Linnaeus classified large numbers of organisms based on similarities in physical characteristics that often reflect relationships.

22. It belongs in Kingdom Fungi because it is eukaryotic, multicellular, not autotrophic, and has cell walls.

3. Which taxon contains the others?
   a. family
   b. species
   c. order
   d. phylum

4. Unlike a pedigree, a cladogram _____.
   a. shows ancestry
   b. shows hypothesized phylogeny
   c. indicates ancestry from two parents
   d. explains relationships

5. Which of the following pairs of terms are most closely related?
   a. Linnaeus—DNA analysis
   b. Aristotle—binomial nomenclature
   c. protist—prokaryote
   d. taxonomy—classification

6. Linnaeus based most of his classification system on _____.
   a. cell organelles
   b. biochemical comparisons
   c. structural comparisons
   d. embryology

7. A group of prokaryotes that often live in extreme environments is the _____.
   a. archaebacteria
   b. protists
   c. eubacteria
   d. fungi

8. Based on the structures of the organisms you see, which of the following is most closely related to a domestic cat?

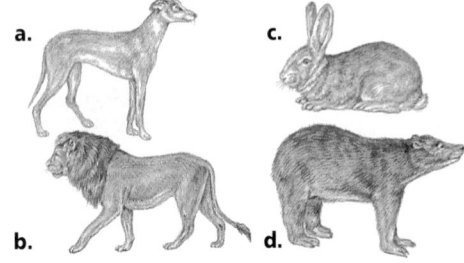

a.   c.

b.   d.

TEST-TAKING TIP

You Are Smarter Than You Think
Nothing on these tests is rocket science. You can learn to master any of it. When you admit that, you're 90 percent of the way home. Just keep practicing.

9. A flaw in Aristotle's classification system was that _____.
   a. it included too many organisms
   b. it did not show natural relationships
   c. large organisms were not included
   d. it was based on Greek instead of Latin

10. Which of the following describes the organism shown to the right?
   a. unicellular consumer
   b. unicellular producer
   c. multicellular consumer
   d. multicellular producer

11. The evolutionary history of a species is its _____, but _____ is the science that groups and names species.

12. A diverse group of eukaryotes that lack complex organ systems and live in moist places is the _____.

13. A group of related classes in the plant kingdom is a _____, but in the animal kingdom it is called a _____.

14. From their scientific names, *Quercus alba* and *Quercus rubrum*, you know that these species of oak trees are in the same _____.

15. _____ is the name given to the taxon that contains all the families of carnivores, such as bears and cats.

16. Modern classification systems are based on _____.

17. A multicellular heterotrophic eukaryote that absorbs nutrients would be classified in Kingdom _____.

18. Linnaeus devised a two-word naming system for organisms that is called _____.

19. The process of grouping similar objects or information is called _____.

20. All genera of dogs are classified in the taxon called a _____.

APPLYING MAIN IDEAS

21. Explain why Linnaeus's system of classification is more useful than Aristotle's.

22. You find an unusual organism growing on the bark of a dying tree. Under a microscope, you observe that its cells are eukaryotic, have cell walls, and do not contain chloroplasts. Into what kingdom would you classify this organism? Explain your decision.

THINKING CRITICALLY

23. **Observing and Inferring** In what way does the work of Linnaeus illustrate the nature of science?

24. **Classifying** Make a list of a minimum of five physical features you could use to classify trees.

25. **Comparing and Contrasting** Compare the classification system of your school library with that of organisms.

26. **Concept Mapping** Complete the concept map by using the following vocabulary terms: divisions, taxonomy, kingdoms, binomial nomenclature, phyla.

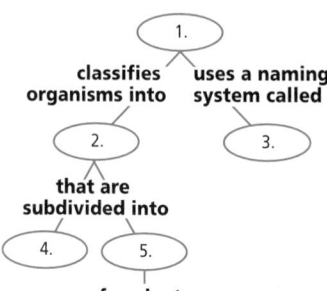

CD-ROM

For additional review, use the assessment options for this chapter found on the *Biology: The Dynamics of Life Interactive CD-ROM* and on the Glencoe Science Web site.
science.glencoe.com

ASSESSING KNOWLEDGE & SKILLS

Identify the organisms in these photographs.

Key

1A Front and hind wings similar in size and shape, and folded parallel to the body when at rest...damselflies

1B Hind wings wider than front wings near base, and extended on either side of the body when at rest.................................dragonflies

Classifying Study the dichotomous key and answer the following questions.

1. **Interpreting Data** The insect in the photo on the right is a damselfly because it has _____.
   a. wings that are opaque
   b. wings folded at rest
   c. smaller eyes
   d. wings not similar in size

2. The insect in the photo on the left is a dragonfly because it has _____.
   a. wings that are opaque
   b. wings folded at rest
   c. larger eyes
   d. wings not similar in size

3. **Classifying** From the key and the photographs above, identify traits that indicate dragonflies and damselflies may have evolved from a common ancestor.

THINKING CRITICALLY

23. Linnaeus developed his classification system based on observation and experimentation.

24. The features might include bark types, buds, leaves, root systems, branch patterns, and the average height and girth of each species.

25. Librarians group books by type and content similarities, but not by their chronology. Classification of organisms is based on both similarities and evolutionary history, a factor that implies chronology.

26. 1. Taxonomy; 2. Kingdoms; 3. Binomial nomenclature; 4. Phyla; 5. Divisions

ASSESSING KNOWLEDGE & SKILLS

1. b
2. d
3. two pairs of wings located on same body segment; wings transparent with clearly defined veins; long, thin body. Accept any reasonable answers.

National Science Education Standards:
UCP.1, UCP.2, UCP.3, UCP.4,
UCP.5, A.2, C.3, C.4, C.6, D.3,
E.2, G.1, G.2, G.3

Prepare

Purpose

This BioDigest can be used as an overview of the concepts of evolution and classification. If time is limited, you may wish to use this unit summary to teach these concepts in place of the chapters in the Change Through Time unit.

Key Concepts

Students are introduced to the formation of Earth and the origins of life. They learn about the theory of evolution by natural selection and the classification of diverse organisms.

1 Focus

Bellringer

Before beginning the lesson, pass around iron, nickel, silica sand, and charcoal samples. Ask students to infer why Earth's core contains iron and nickel, and the crust contains silicon and carbon. *Dense elements fell into Earth's core, and lighter materials rose to the crust.* **L1**

GLENCOE
TECHNOLOGY

CD-ROM
Biology: The Dynamics of Life
Video: *Discovering Dinosaurs*
Disc 2

BioDigest

For a **preview** of the change through time unit, study this BioDigest before you read the chapters. After you have studied the change through time chapters, you can use the BioDigest to **review** the unit.

Change Through Time

Scientists propose that about five billion years ago Earth was extremely hot. As Earth slowly cooled, water vapor in its atmosphere fell as rain, forming today's oceans. Life had appeared in these oceans by 3.6 billion years ago. Since then, millions of species have evolved and then become extinct.

Geologic Time Scale

The four eras of the Geologic Time Scale span about 4.6 billion years of Earth's history.

The Precambrian Era

The Precambrian era encompasses approximately the first four billion years of the scale. Prokaryotic cells appear in rocks dated 3.6 billion years old. By the end of the Precambrian, the first eukaryotic cells had evolved.

The Geologic Time Scale illustrates major events that have occurred during Earth's 4.6-billion-year history. Each era is subdivided into smaller time spans called periods.

The Paleozoic Era

The following 300 million years make up the Paleozoic era. Many plant groups such as ferns and conifers appeared. Animal groups such as worms, insects, fishes, and reptiles evolved.

The Mesozoic Era

From 245 million years ago to 66 million years ago, the Mesozoic era, reptiles diversified, and mammals and flowering plants evolved. The Mesozoic, the Age of Dinosaurs, ended with a rapid extinction of the dinosaurs.

The Cenozoic Era

The current Cenozoic era, which has encompassed the previous 66 million years, is often referred to as the Age of Mammals. Primates, including humans, evolved during this era.

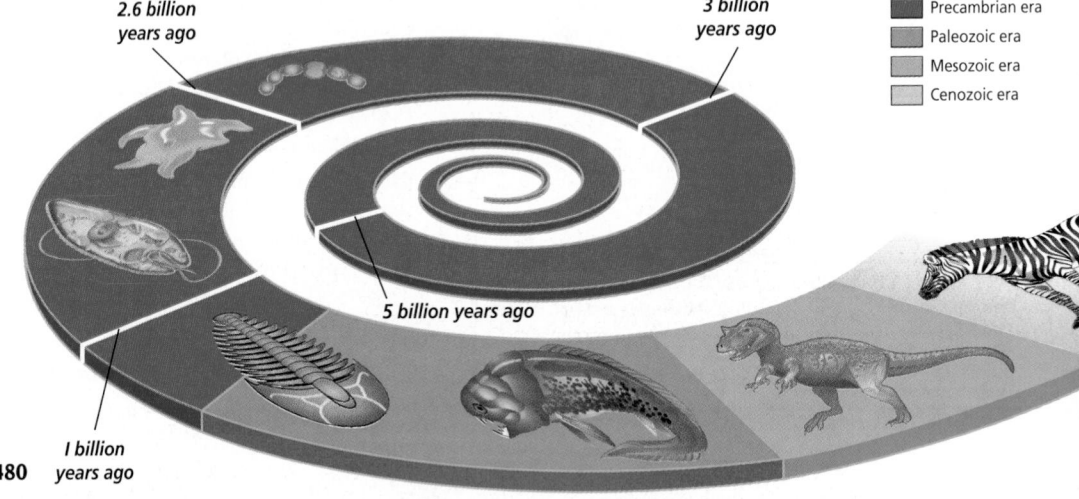

2.6 billion years ago

3 billion years ago

■	Precambrian era
■	Paleozoic era
■	Mesozoic era
■	Cenozoic era

5 billion years ago

1 billion years ago

Multiple Learning Styles

Look for the following logos for strategies that emphasize different learning modalities.

 Kinesthetic Quick Demo, p. 481

 *Visual-Spatial* Activity, pp. 482, 483; Biology Journal, p.482; Reteach, p. 484

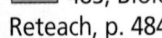 *Intrapersonal* Project, p. 483; Biology Journal, p. 483; Meeting Individual Needs, p. 484

 Linguistic Meeting Individual Needs, p. 482; Extension, p. 484

 Logical-Mathematical Visual Learning, p. 482

 Naturalist Activity, p. 485

Origin of Life Theories

People once thought that life was able to arise spontaneously from nonliving material. Two scientists, Francesco Redi and Louis Pasteur, designed controlled experiments to try to disprove spontaneous generation. Their experiments convinced scientists to accept the theory of biogenesis—that life comes only from preexisting life.

Modern Ideas About the Origin of Life

Most scientists agree that small organic molecules formed from substances present in Earth's early atmosphere and oceans. At some point, nucleic acids must have formed. Then, clusters of organic molecules might have formed protocells that may have evolved into the first true cells.

Louis Pasteur disproved the idea of spontaneous generation by conducting experiments using broth in swan-necked flasks like this one.

Heterotrophic, anaerobic prokaryotes were probably the earliest organisms to live on Earth. Chemosynthetic prokaryotes evolved over time, followed by oxygen-producing photosynthetic prokaryotes. As the amount of oxygen in the atmosphere increased, aerobically respiring eukaryotes probably evolved.

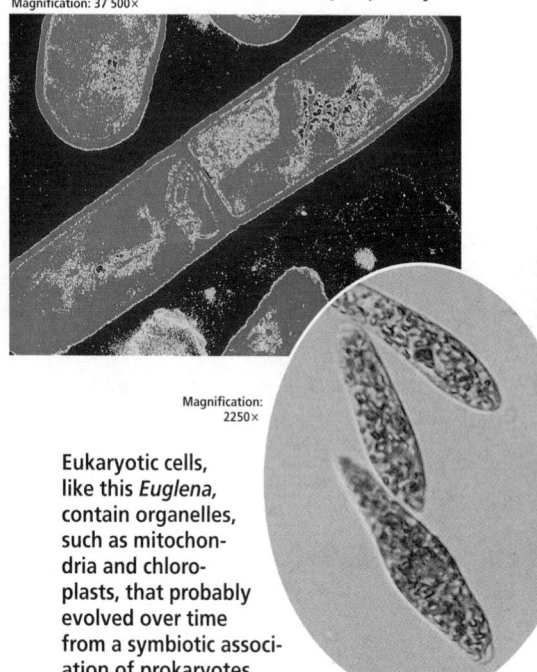

A heterotrophic prokaryote

Magnification: 37 500×

Magnification: 2250×

Eukaryotic cells, like this *Euglena*, contain organelles, such as mitochondria and chloroplasts, that probably evolved over time from a symbiotic association of prokaryotes.

FOCUS ON HISTORY

Pioneers

Two scientists, Stanley Miller and Harold Urey, pioneered work about the origin of Earth's life. Their experiments showed that small molecules can form complex organic materials under conditions that may have existed on early Earth. Other scientists demonstrated how these complex chemicals could form protocells, which are large, organized structures that carry out some activities associated with life, such as growth and division.

The American biologist Lynn Margulis proposed the endosymbiont theory. This theory suggests that cell organelles, such as mitochondria and chloroplasts, may have evolved when small prokaryotes entered larger prokaryotes and began to live symbiotically inside these larger cells.

Lynn Margulis

481

BioDigest

2 Teach

Quick Demo

Kinesthetic Set up microscopes with slides of *Euglena* showing chloroplasts. Then, use the slides and photos on this page to explain Lynn Margulis's endosymbiont theory.

Field Trip

Arrange a class trip to a local natural history museum where the students can see much of the material discussed in this unit. Ask students to describe how the fossils they observe relate to today's species **L1**

✔ Assessment

Skill Direct the students' attention to the Geologic Time Scale at the bottom of page 480. Ask them to order the eras from longest to shortest. **L1**

GLENCOE TECHNOLOGY

 **VIDEOTAPE**
The Secret of Life
Gone Before You Know It: The Biodiversity Crisis
What's in Stetter's Pond: The Basics of Life
It's in the Genes: Evolution

✔ Assessment Planner

Portfolio Assessment
  Assessment, TWE, p. 482
Performance Assessment
  Assessment, TWE, p. 484
Knowledge Assessment
  BioDigest Assessment, SE, p. 485
Skill Assessment
  Assessment, TWE, p. 481

BIODIGEST

Visual Learning

 Logical-Mathematical Point out the photos and the Vital Statistics on this page. Ask students which isotopes they would use to date the shells that are several thousand years old and the dinosaur that lived more than a million years ago. *carbon-14 for the shells, and uranium-235 for the dinosaur* L2

Activity

☑ *Visual-Spatial* Have students look at slides of diatomaceous earth. Point out that the shells they observe are fossils of protists that lived thousands of years ago. L1 ELL

✓ Assessment

Portfolio Have students write an essay in which they distinguish between the two types of fossil dating methods and explain the value of each method. L2 P

GLENCOE TECHNOLOGY

💿 CD-ROM
Biology: The Dynamics of Life
Exploration: *The Record of Life,* Disc 2

BIODIGEST

Evidence of Evolution

Charles Darwin and Alfred Wallace proposed the idea of natural selection as a mechanism of evolution. Natural selection occurs because all organisms, which produce many more young than can survive, compete for mates, food, space, and other resources. Such competition favors the survival of individuals with variations that help them compete successfully in a specific environment. Individuals that survive to reproduce can pass their traits to the next generation.

Fossils of dinosaurs similar to this *Tyrannosaurus rex* have been found in North and South America, and also in China.

Fossils

The fossil record contains evidence for evolution and provides a record of life on Earth. Fossils come in many forms, such as imprints, the burrow of a worm, or an actual bone. By studying fossils, scientists learn how organisms changed over time.

Scientists use relative and radiometric dating methods to determine the age of fossils. Relative dating assumes that in undisturbed layers of rock, the deepest rock layers contain the oldest fossils. Radiometric dating analysis compares the known half-lives of radioactive isotopes to a ratio of the amount of radioactive isotope originally in a rock or fossil with the amount of isotope in the rock or fossil today.

The members of related species have variations such as those you see in the shells of these snails.

VITAL STATISTICS

Half-Lives of Radioactive Isotopes
Radium226—1620 years
Carbon14—5710 (± 30) years
Potassium40—1.3 billion years
Rubidium87—4.7 billion years
Uranium235—710 million years
Uranium238—4.5 billion years

BIOLOGY JOURNAL

The Endosymbiont Theory

☑ *Visual-Spatial* Have students read more about the endosymbiont theory and make a labeled drawing in their journals that illustrates Margulis's ideas about how mitochondria, chloroplasts, and flagella evolved. L3

MEETING INDIVIDUAL NEEDS

English Language Learners

🧠 *Linguistic* To enhance students' understanding of the information in this BioDigest, have them write a summary in their native language and in English. L2 ELL

Additional Evidence

Similar anatomical structures, called homologous structures, in different organisms might indicate possible shared ancestry. For example, both vertebrate limbs and developmental stages show how vertebrates might be related. In addition, similarities among the nucleic acid sequences of species provide evidence for evolution. Direct evidence for evolution has been observed in the laboratory among species of bacteria that have developed resistance to antibiotics.

The bones that make up a penguin's wings are homologous to those that form the wings of an albatross. The forelimb bones of four-legged vertebrates are also homologous.

Mechanics of Evolution

Evolution occurs when a population's genetic equilibrium changes. Mutations, genetic drift, and migration may slightly disrupt the genetic equilibrium of large populations, but they will greatly alter that of small populations. Natural selection affects the genetic equilibrium of all populations.

Three Patterns of Evolution

Three patterns of natural selection lead to speciation. Stabilizing selection favors the survival of a population's average individuals for a feature. Directional selection naturally selects for an extreme feature. Disruptive selection, which usually occurs when a physical barrier divides one population into two, eventually produces two populations, each with one of a feature's extreme characteristics.

In California, there are seven subspecies of reproductively isolated salamanders, *Ensatina eschscholtzi*.

Activity

Visual-Spatial Have students make a bulletin board collage of organisms from magazine photos. Ask each student to point out an adaptation that helps one of the organisms survive. **L1**

Visual Learning

Ask students how albatross, penguin, and other vertebrate forelimbs are homologous. *They all possess bones from the same embryonic structures that have evolved for different purposes.*

Quick Demo

Display a small, caged mammal, such as a hamster or a mouse. Have students describe the animal's variations and its adaptations for survival in its natural environment.

GLENCOE TECHNOLOGY

CD-ROM
Biology: The Dynamics of Life, Exploration: *Selection Pressure,* Disc 2 Animation: *Geographic Isolation and Speciation,* Disc 2

VIDEODISC
Biology: The Dynamics of Life
Geographic Isolation (Ch.6) Disc 1, Side 2, 17 sec.

FOCUS ON ADAPTATIONS

Adaptive Radiation in Galapagos Finches

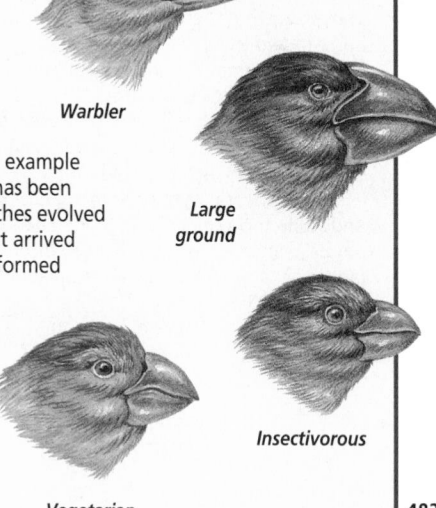

Tool-using

Warbler

Large ground

Cactus ground

Insectivorous

Vegetarian

The finches in the Galapagos Islands are an example of the rapid development of a species. It has been proposed that the 13 species of Galapagos finches evolved from just a few ancestral species of finches that arrived from South America and colonized the newly formed habitats of these volcanic islands.

The adaptive radiation of finch species occurred as the original finch population adapted to the different niches found in the islands. The pressures of natural selection produced different species, each with their own feeding and habitat adaptations.

483

PROJECT

Collecting Fossils

Intrapersonal Have students make a collection of fossils from the local area. Ask them to identify each type of fossil and the organism that was fossilized and research additional information. **L2**

BIOLOGY JOURNAL

Patterns of Evolution

Intrapersonal Have the students describe the three patterns of evolution using examples other than those presented in the text. **L2**

BioDigest

3 Assess

Check for Understanding

Have students explain adaptations that are advantageous for a primate in its environment. **L1**

Reteach

 Visual-Spatial Have students make a time line of the Geologic Time eras and their life forms. **L1**

Extension

Linguistic Have students read a chapter from one of Darwin's books and then discuss what they learned. **L3**

✔ Assessment

Performance Give students examples of classification levels and have them identify the examples' taxonomic levels. For example, *Homo sapiens* is the species, cats is the family, and mollusks is the phylum. **L2**

MEETING INDIVIDUAL NEEDS

Gifted

Intrapersonal Have students research the Galapagos finches and tortoises and compare these two examples of divergent evolution. **L3**

BioDigest

Primate Evolution

Primates are a grouping of mammals with adaptations such as binocular vision, opposable thumbs, and mobile skeletal joints. These adaptations help arboreal animals survive in forest trees, where all primates may have originally lived and where most primates still live.

There are two categories of primates: the prosimians, including lemurs and tarsiers, and the anthropoids, including humans, apes, and monkeys. Monkeys are subdivided further into two groups that are called Old World monkeys and New World monkeys.

Primates first appear in the fossil record in the Cenozoic era. Fossils indicate that increasing brain size and bipedal locomotion are the two major trends in primate evolution.

Unlike most Old World monkeys, New World monkeys, such as this howler monkey, have prehensile tails that are used as a fifth limb.

Human Ancestry

Fossils of the possible human ancestors, called *Australopithecines,* were discovered in Africa and date from 5 to 8 million years ago. They show that these ancestors were bipedal and climbed trees.

After examining more recently discovered hominid fossils, paleoanthropologists suggest that the increasing efficiency of bipedal locomotion and the decreasing size of jaws and teeth were two directions of human evolution.

The appearance of both the genus *Homo* and stone tools coincides in the fossil record about 2 million years ago. The use of fire, tools, language, and ceremonies developed in later species of *Homo*.

Taxonomists classify the bobcat within a hierarchy of taxa.

Organizing Life's Diversity

Biologists use a classification system to study and communicate about both the three to ten million species living on Earth today and the many extinct species represented by fossils. Although Aristotle produced the first system of classification, Linnaeus developed the basic structure of the modern-day classification system. Linnaeus also developed a naming system, termed binomial nomenclature, that is still used today.

Today's phylogenetic classification uses a hierarchy of taxa to classify organisms. From largest to smallest, this hierarchy is kingdom, phylum or division, class, order, family, genus, and species. The most useful systems of classification show evolutionary relationships among species.

VITAL STATISTICS

Hominids
Cranial Capacity of Hominids
Australopithecus—range of 375–550 cc
Homo habilis—range of 500–800 cc
Homo erectus—range of 750–1225 cc
Homo sapiens (archaic)—average of 1200 cc
Homo neanderthalensis—average of 1450 cc
Homo sapiens (modern)—average of 1350 cc

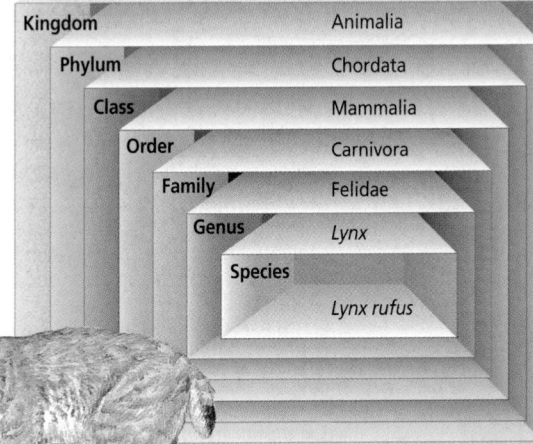

	Bobcat
Kingdom	Animalia
Phylum	Chordata
Class	Mammalia
Order	Carnivora
Family	Felidae
Genus	*Lynx*
Species	*Lynx rufus*

GLENCOE TECHNOLOGY

VIDEODISC
Biology: The Dynamics of Life,
Primate Characteristics (Ch. 7)
Disc 1, Side 2, 47 sec.

CD-ROM
Biology: The Dynamics of Life
Exploration: *The Six Kingdoms*
Disc 3
Video: *Primate Characteristics*
Disc 2

Six Kingdoms of Classification

Species are classified into one of six kingdoms. Prokaryotes belong to Kingdom Archaebacteria or Kingdom Eubacteria depending on their RNA sequences. Kingdom Protista contains the eukaryotes that lack complex organ systems and live in moist environments. The kingdom Fungi includes heterotrophic eukaryotes that absorb nutrients. Multicellular autotrophs with complex organ systems are placed in Kingdom Plantae. Kingdom Animalia includes multicellular heterotrophs.

Criteria for Classification

Biologists use criteria such as body structure, breeding behavior, and geographic distribution to classify organisms. Biochemistry and chromosome analysis are also important for explaining the relationships among organisms.

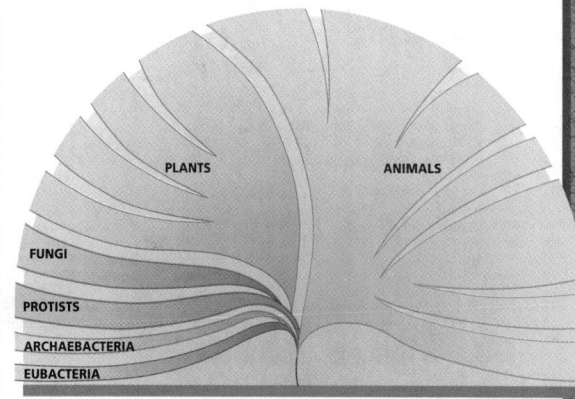

The phylogenetic relationship of the six kingdoms can be represented by a fanlike diagram.

BioDigest Assessment

Understanding Main Ideas

1. The Geologic Time Scale's longest era was the _____.
   a. Precambrian   c. Mesozoic
   b. Paleozoic   d. Cenozoic

2. When did humans and primates evolve?
   a. the Precambrian   c. the Mesozoic
   b. the Paleozoic   d. the Cenozoic

3. Who disproved spontaneous generation?
   a. Louis Pasteur   c. Charles Darwin
   b. Lynn Margulis   d. Miller and Urey

4. Which scientist proposed the endosymbiont theory?
   a. Francesco Redi   c. Harold Urey
   b. Lynn Margulis   d. Stanley Miller

5. Direct evidence of evolution includes _____.
   a. fossils
   b. anatomical similarities
   c. bacterial antibiotic resistance
   d. embryo

6. Which dating method relies on the position of rock layers?
   a. radiometric   c. absolute
   b. relative   d. morphology

7. Which of the following was proposed by Charles Darwin?
   a. endosymbiont hypothesis
   b. biogenesis
   c. natural selection
   d. experimentation

8. _____ is one of the two major groups of primates.
   a. Old World monkey c. apes
   b. New World monkey d. prosimians

9. Binomial nomenclature is a biological system of _____.
   a. naming species   c. evolution
   b. phylogeny   d. bipedalism

10. Multicellular heterotrophs are placed in Kingdom _____.
    a. Protista   c. Plantae
    b. Fungi   d. Animalia

Thinking Critically

1. Describe the types of organisms that existed in the Mesozoic. In the Cenozoic.

2. Explain how natural selection might be a mechanism of evolution.

3. Why do biologists classify organisms?

485

BioDigest

4 Close

Activity

 **Naturalist** Provide photos and specimens of organisms and have students classify each organism in a kingdom. Or, have students bring in photos of members of each kingdom. **L1**

BioDigest Assessment

Understanding Main Ideas

1. a
2. d
3. a
4. b
5. c
6. b
7. c
8. d
9. a
10. d

Thinking Critically

1. Reptiles, mammals, and flowering plants flourished in the Mesozoic. Primates evolved in the Cenozoic and mammals and flowering plants have diversified.

2. When natural selection occurs, organisms pass environmentally benificial traits to offspring. After a long time, a population consists of very different organisms.

3. to study and communicate about species.

GLENCOE TECHNOLOGY

 **CD-ROM**
Biology: The Dynamics of Life
BioQuest: *Biodiversity Park,* Disc 3, 4

 VIDEODISC
Biology: The Dynamics of Life
Museum Collections (Ch. 8)
Disc 1, Side 2
20 sec.

Resource Manager

Reinforcement and Study Guide, pp. 77-78 **L2**
Content Mastery, pp. 85-88 **L1**

Viruses, Bacteria, Protists, and Fungi

Unit Overview

Chapter 18 introduces students to the characteristics of viruses and to the structure, ecology, and importance of bacteria. In Chapter 19, students study the diversity and classification of protists. Finally, in Chapter 20, students learn about the characteristics and diversity of fungi.

Introducing the Unit

Naturalist Organize students into groups. Ask each group to estimate how many species live on Earth and to explain their reasoning. To emphasize the diversity of life on Earth, ask students to list all the living things that they see in the photo. **L1** **COOP LEARN**

Unit 6

Viruses, Bacteria, Protists, and Fungi

Only about 1.8 million of an estimated 10 million species have been identified. Most of the unidentified species probably belong to kingdoms that you will study in this unit. However, some members of these kingdoms, such as those shown in the photograph, are well known.

UNIT CONTENTS

- **18** Viruses and Bacteria
- **19** Protists
- **20** Fungi

BIODIGEST Viruses, Bacteria, Protists, and Fungi

UNIT PROJECT

BIOLOGY Online Use the Glencoe Science Web site for more project activities that are connected to this unit. science.glencoe.com

486

Unit Projects

Microbes and Food

Have students do one of the projects for this unit as described on the Glencoe Science Web site. As an alternative, students can do one of the projects described on these two pages.

Make a Poster

Visual-Spatial Design and produce a poster about foods that commonly carry disease-causing organisms. **L1** **ELL**

Interview a Specialist

Linguistic Interview a restaurant owner about how his or her practices ensure food safety. **L2**

Advance Planning

Chapter 18
- Order live *Oscillatoria* for the Project.
- Order prepared slides of both heterotrophic and autotrophic bacteria for the Activity.
- Order sterile agar plates for the Quick Demo and BioLab, and bacterial cultures and antibiotic disks for the BioLab.
- Order slides of cocci, bacilli, and spirilla for MiniLab 18-2.

Chapter 19
- Order live protozoans for the Getting Started.
- Order *Paramecium* for Mini-Lab 19-1 and for the BioLab.
- Order *Euglena* for the Project and BioLab and methyl cellulose for the BioLab.
- Purchase slides of some protozoans for the Quick Demo.
- Order termites for the Meeting Individual Needs.
- Order diatomaceous earth for the Activity.
- Order brown and red algae for the Quick Demo.
- Order *Physarum polycephalum* for the Alternative Lab.

Chapter 20
- Grow mold on fruit for the Getting Started.
- Purchase preserved specimens of *Peziza* for the Quick Demo.
- Order a mushroom farming kit for the Project.

Unit Projects

Display

Visual-Spatial Make a bulletin board from news articles about both harmful and beneficial microbes.
L1 **ELL**

Using the Library

Intrapersonal Find out about state and federal laws that regulate food handlers. Report on these laws and how they help to prevent food contamination.
L3

Final Report

Have each group present its findings to the class in the form of an oral report, demonstration, or poster.

487

Chapter 18 Organizer

Refer to pages 4T-5T of the Teacher Guide for an explanation of the National Science Education Standards correlations.

Section	Objectives	Activities/Features
Section 18.1 **Viruses** National Science Education Standards UCP.1, UCP.2, UCP.5; A.1, A.2; C.5; F.1, F.5; G.1-3 (2 sessions, $^1/_2$ block)	1. **Identify** the different kinds of viruses. 2. **Compare and contrast** the replication cycles of viruses.	**MiniLab 18-1:** Measuring a Virus, p. 490 **Problem-Solving Lab 18-1,** p. 494 **Careers in Biology:** Dairy Farmer, p. 495 **Focus On** Viruses, p. 498
Section 18.2 **Archaebacteria and Eubacteria** National Science Education Standards UCP.1, UCP.2, UCP.5; A.1, A.2; C.1, C.4, C.5, C.6; E.1, E.2; F.1, F.4-6; G.1-3 (3 sessions, $1^1/_2$ blocks)	3. **Compare** the types of prokaryotes. 4. **Explain** the characteristics and adaptations of bacteria. 5. **Evaluate** the economic importance of bacteria.	**Inside Story:** A Typical Bacterial Cell, p. 503 **MiniLab 18-2:** Bacteria Have Different Shapes, p. 506 **Problem-Solving Lab 18-2,** p. 508 **Design Your Own BioLab:** How sensitive are bacteria to antibiotics? p. 512 **Biology & Society:** Super Bugs Defy Drugs, p. 514

Need Materials? Contact Carolina Biological Supply Company at 1-800-334-5551 or at **http://www.carolina.com**

MATERIALS LIST

BioLab

p. 512 bacteria cultures, sterile nutrient agar, petri dishes, antibiotic disks, sterile filter paper disks, marking pen, long-handled cotton swabs, forceps, incubator, metric ruler

MiniLabs

p. 490 metric ruler, pencil, paper
p. 506 microscope, paper, prepared slides of bacteria

Alternative Lab

p. 504 screw-top test tubes, distilled water, vinegar, Schultz liquid plant food, Accent seasoning, baking soda, 60-watt light bulb, soil samples

Quick Demos

p. 491 bolt, nut (2), #22 gauge wire
p. 504 petri dishes (2), sterile nutrient agar, soap, labels
p. 509 Swiss cheese, pickles, vinegar, sauerkraut, yogurt, peas, beans, soybeans, peanuts, milk, sour cream

Key to Teaching Strategies

L1 Level 1 activities should be appropriate for students with learning difficulties.

L2 Level 2 activities should be within the ability range of all students.

L3 Level 3 activities are designed for above-average students.

ELL ELL activities should be within the ability range of English Language Learners.

COOP LEARN Cooperative Learning activities are designed for small group work.

P These strategies represent student products that can be placed into a best-work portfolio.

These strategies are useful in a block scheduling format.

Teacher Classroom Resources

Section	Reproducible Masters	Transparencies
Section 18.1 **Viruses**	Reinforcement and Study Guide, p. 79-80 L2 Concept Mapping, p. 18 L3 ELL BioLab and MiniLab Worksheets, p. 85 L2 Laboratory Manual, pp. 125-128 L2 Content Mastery, pp. 89-90, 92 L1	Section Focus Transparency 43 L1 ELL Basic Concepts Transparency 25 L2 ELL Basic Concepts Transparency 26 L2 ELL Reteaching Skills Transparency 27 L1 ELL
Section 18.2 **Archaebacteria and Eubacteria**	Reinforcement and Study Guide, pp. 81-82 L2 Critical Thinking/Problem Solving, p. 18 L3 BioLab and MiniLab Worksheets, pp. 86-88 L2 Laboratory Manual, pp. 129-132 L2 Content Mastery, pp. 89, 91-92 L1 Inside Story Poster ELL	Section Focus Transparency 44 L1 ELL Basic Concepts Transparency 27 L2 ELL Reteaching Skills Transparency 28 L1 ELL

Assessment Resources

Chapter Assessment, pp. 103-108
MindJogger Videoquizzes
Performance Assessment in the Biology Classroom
Alternate Assessment in the Science Classroom
ExamView® Pro Software
BDOL Interactive CD-ROM, Chapter 18 quiz

Additional Resources

Spanish Resources ELL
English/Spanish Audiocassettes ELL
Cooperative Learning in the Science Classroom COOP LEARN
Lesson Plans/Block Scheduling

NATIONAL GEOGRAPHIC — Teacher's Corner

Products Available From Glencoe
To order the following products, call Glencoe at 1-800-334-7344:
CD-ROM
NGS PictureShow: The Cell
Curriculum Kit
NGS PicturePack: Cells and Microorganisms

Products Available From National Geographic Society
To order the following products, call National Geographic Society at 1-800-368-2728:
Videos
Bacteria
Virus!

Index to National Geographic Magazine
The following articles may be used for research relating to this chapter:
"Body Beasts," by Richard Conniff, December 1998.
"The Rise of Life on Earth," by Richard Monastersky, March 1998.
"Viruses: On the Edge of Life, On the Edge of Death," Peter Jaret, July 1994.
"Bacteria: Teaching Old Bugs New Tricks," by Thomas Y. Canby, August 1993.
"The Disease Detectives," by Peter Jaret, January 1991.

GLENCOE TECHNOLOGY

The following multimedia resources are available from Glencoe.

Biology: The Dynamics of Life
CD-ROM ELL

  Animation: *The Lytic Cycle*
  Animation: *The Lysogenic Cycle*
  BioQuest: *Biodiversity Park*
  Video: *Binary Fission*

Videodisc Program

  Lytic Cycle
  Lysogenic Cycle
  Binary Fission

The Secret of Life Series

  Flu Virus
  Bacteria Virus–Phage

Chapter

18 Viruses and Bacteria

What You'll Learn

- You will categorize viruses and bacteria.
- You will explain how viruses and bacteria reproduce.
- You will recognize the medical and economic importance of viruses and bacteria.

Why It's Important

Viruses and bacteria are important because some cause diseases. Bacteria are also important in nature's nutrient cycles and in the food and pharmaceutical industries.

READING BIOLOGY

Draw a line down the middle of a sheet of paper. Label one side "Virus" and the other "Bacteria." As you read, write down different characteristics of the two. Also, write down any key terms or questions that arise.

BIOLOGY Online

To find out more about viruses and bacteria, visit the Glencoe Science Web site.
science.glencoe.com

Magnification: 4600×

Magnification: 30 000×

You may not recognize the landscape shown here, but it is human skin. Many bacteria, such as this *Staphylococcus epidermidis* (inset), live on your skin.

18.1 Viruses

SECTION PREVIEW

Objectives

Identify the different kinds of viruses.

Compare and contrast the replication cycles of viruses.

Vocabulary

virus
host cell
bacteriophage
capsid
lytic cycle
lysogenic cycle
provirus
retrovirus
reverse transcriptase

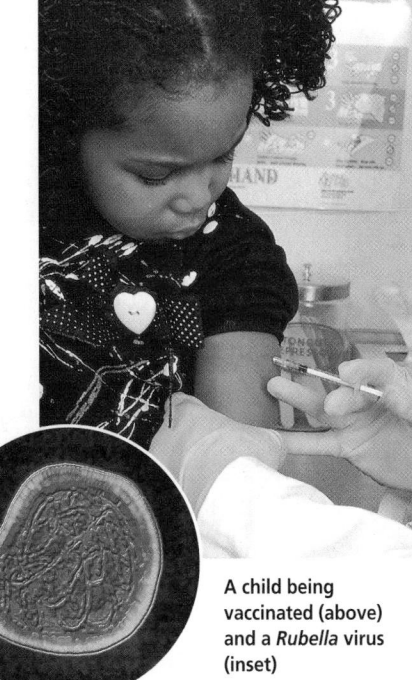

Magnification: 200 000×

A child being vaccinated (above) and a *Rubella* virus (inset)

How many childhood diseases have you had—chicken pox, mumps, German measles, whooping cough? These diseases occur mostly in children and therefore are called childhood diseases. When your grandparents were young, these childhood diseases were so common that most children got them. Today, the availability of vaccinations makes these diseases rare. However, the causes of childhood diseases still exist, and these causes will break out where people are not vaccinated against them.

What Is a Virus?

You've probably had influenza—the flu—at some time during your life. Nonliving particles called **viruses** cause influenza. Viruses are about one-half to one-hundredth the size of the smallest bacterium. To appreciate how very tiny viruses are, try the *MiniLab* on the next page.

Most biologists consider viruses to be nonliving because viruses are not cells and don't exhibit all the criteria for life. For example, they don't carry out respiration, grow, or develop. All viruses can do is replicate—make copies of themselves—and they can't even do that without the help of living cells. A cell in which a virus replicates is called the **host cell.**

Because they are nonliving, viruses were not named in the same way as organisms. Viruses, such as rabies viruses and polioviruses, were named after the diseases they cause. Other viruses were named for the organ or tissue they infect. For example, scientists first found the adenovirus (uh DEN uh vyruhs), which is one cause of the common cold, in adenoid tissue between the back of the throat and the nasal cavity.

18.1 VIRUSES **489**

Section 18.1

Prepare

Key Concepts

The structure and replication cycles of viruses are described. The origin of viruses is also discussed and their relationship to living cells.

Planning

■ Collect metric rulers and electron micrographs of different viruses for MiniLab 18-1.

■ Gather wire (#22 gauge), bolts, and nuts for the Quick Demo.

■ Prepare photocopies of viral and bacterial growth curves for the Extension.

1 Focus

Bellringer

Before presenting the lesson, display **Section Focus Transparency 43** on the overhead projector and have students answer the accompanying questions.
L1 **ELL**

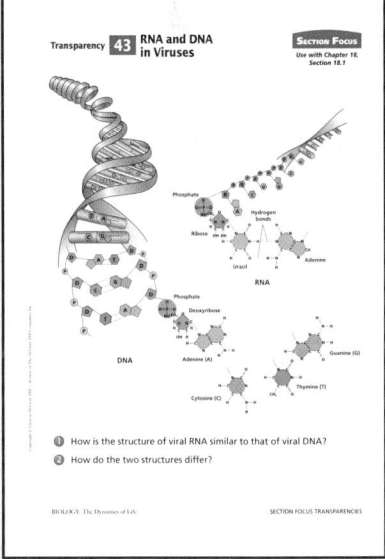

✓ Assessment Planner

Portfolio Assessment
    Portfolio, TWE, pp. 492, 496, 503
    MiniLab, TWE, p. 490
    Assessment, TWE, p. 511
    BioLab, TWE, pp. 512-513
Performance Assessment
    Assessment, TWE, pp. 493, 505
    MiniLab, SE, pp. 490, 506
    Problem-Solving Lab, TWE, p. 494

Alternative Lab, TWE, pp. 504-505
MiniLab, TWE, p. 506
BioLab, SE, pp. 512-513
Knowledge Assessment
    Section Assessment, SE, pp. 497, 511
    Problem-Solving Lab, TWE, p. 508
    Chapter Assessment, SE, pp. 515-517
Skill Assessment
    Assessment, TWE, p. 497

2 Teach

MiniLab 18-1

Purpose
Students will determine the size of a virus in a TEM photo.

Process Skills
measure in SI, use numbers, collect data

Teaching Strategies
- Remind students that viruses vary in size, and they will determine the size of only one virus.
- Use the overhead projector to preview this lab. Using a clear ruler helps to demonstrate what to measure.
- Use microbiology textbooks for other viruses to measure or order viral electron micrographs from biological supply houses.
- Have students use calculators.

Expected Results
Students should calculate the size of the poliovirus as follows.

$$\frac{69 \text{ mm}}{5 \text{ mm}} = \frac{0.4 \text{ μm}}{(x)} = 0.029 \text{ μm}$$

Analysis
1. The virus's diameter measures 0.029 μm, less than the 0.2 μm light microscope limit.
2. 345 viruses (10 μm divided by 0.029 μm)

✔ Assessment

Portfolio Have students look up the size of six viruses and make a scale drawing that shows their relative sizes. Use the Performance Task Assessment List for Scientific Drawing in **PASC**, p. 55. **L2**

Resource Manager

Section Focus Transparency 43 and Master **L1** **ELL**
BioLab and MiniLab Worksheets, p. 85 **L2**

490

MiniLab 18-1 Measuring in SI

Measuring a Virus Can you use a light microscope to view a virus? Find out by measuring the size of a poliovirus in the photo below and then comparing it to 0.2 μm, the size limit for viewing objects with a light microscope.

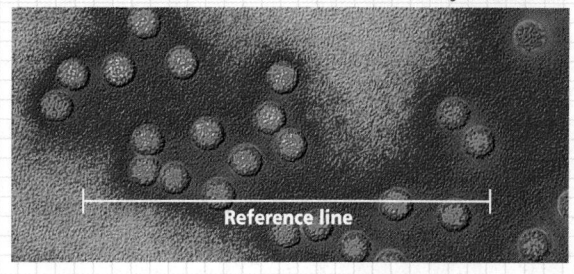

Magnification: 172 500×

Reference line

Procedure
1 Copy the data table below.

Data Table

Values to measure and calculate	Measurement
Length of photo line in mm	
Diameter of poliovirus in mm	
Diameter of poliovirus in μm	

2 Examine the photo. The horizontal line you see would measure only 0.4 micrometer (μm) in length if the photo was not magnified 172 500×. Use this line for reference.

3 Calculate the diameter of one poliovirus. First, measure the length of the reference line in millimeters. Record the value in the table. Then, measure the diameter of a poliovirus in millimeters. Record the value in the table.

4 Use the following equation to calculate the actual diameter of the poliovirus (X). Record your answer in the table.

$$\frac{\text{photo line length in mm } (A)}{\text{diameter of virus in mm } (B)} = \frac{0.4 \text{ μm}}{\text{diameter of virus in μm } (X)}$$

Analysis
1. Explain why you cannot see viruses with a light microscope. Use specific numbers in your answer.
2. A bacterial cell may be 10 μm in size. How many polioviruses could fit across the top of such a bacterium?

Today, most viruses are given a genus name ending in the word "virus" and a species name. However, sometimes scientists use code numbers to distinguish among similar

BIOLOGY JOURNAL

Viral Time Line

Visual-Spatial Have students make a time line showing the major developments in virology—instrumentation, the discovery of immune system components, the discovery of viral diseases, and vaccine availablility. Students could make a separate time line for HIV. **L2**

---

viruses that infect the same host. For example, seven similar-looking viruses that infect the common intestinal bacteria *Escherichia coli* have the code numbers T1 through T7 (*T* stands for "Type"). A virus that infects a bacterium is called a **bacteriophage** (bak TIHR ee uh fayj), or phage for short.

Viral structure

A virus has an inner core of nucleic acid, either RNA or DNA, and an outer coat of protein called a **capsid.** Some relatively large viruses, such as human flu viruses, may have an additional layer, called an envelope, surrounding their capsids. Envelopes are composed primarily of the same materials that are found in the plasma membranes of all cells. You can learn more about both viral capsids and envelopes in the *Focus On* at the end of this section.

The core of nucleic acid contains a virus's genetic material. Viral nucleic acid is either DNA or RNA and contains instructions for making copies of the virus. Some viruses have only four genes. The arrangement of proteins in the capsid of a virus determines the virus's shape. Four different viral shapes are shown in *Figure 18.1*. The protein arrangement also plays a role in determining what cell can be infected and how the virus infects the cell.

Attachment to a host cell

Before a virus can replicate, it must enter a host cell. Before it can enter, it must first recognize and attach to a receptor site on the plasma membrane of the host cell.

A virus recognizes and attaches to a host cell when one of its proteins interlocks with a molecular shape that is the receptor site on the host cell's plasma membrane. A protein in the tail fibers of the bacteriophage T4, shown in *Figure 18.1*, recog-

---

GLENCOE TECHNOLOGY

VIDEODISC

The Secret of Life
Flu Virus

Bacterial Virus—Phage

nizes and attaches the T4 to its bacterial host cell. In other viruses, the attachment protein is in the capsid or in the envelope. The recognition and attachment process is like two pieces of a jigsaw puzzle fitting together. The process might also remind you of two spaceships docking.

Figure 18.1
The different proteins in viral capsids produce a wide variety of viral shapes.

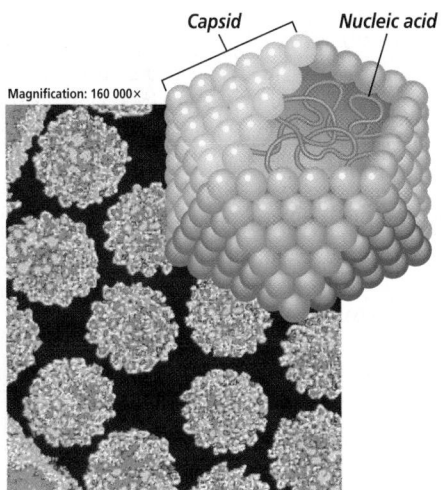

Magnification: 160 000×

Capsid *Nucleic acid*

A Polyhedral viruses, such as the papilloma virus that causes warts, resemble small crystals.

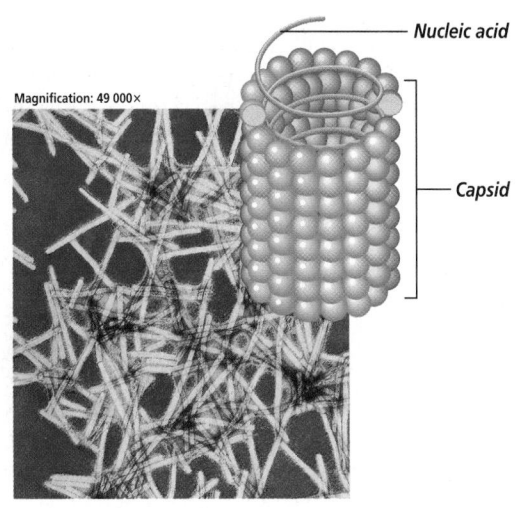

Magnification: 49 000×

Nucleic acid *Capsid*

B The tobacco mosaic virus has a long, narrow helical shape.

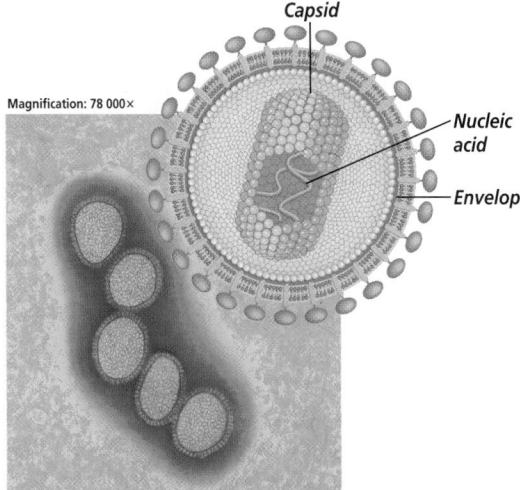

Magnification: 78 000×

Capsid *Nucleic acid* *Envelope*

C An envelope studded with projections covers some viruses, including the influenza virus (photo) and the AIDS-causing virus (inset).

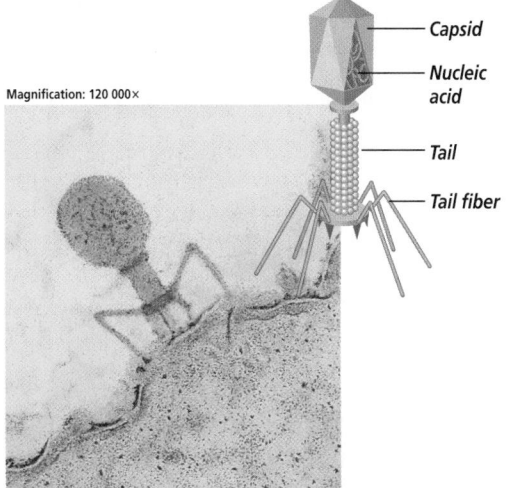

Magnification: 120 000×

Capsid *Nucleic acid* *Tail* *Tail fiber*

D This T4 virus, which infects *E. coli*, consists of a polyhedral-shaped head attached to a cylindrical tail with leglike fibers.

18.1 VIRUSES **491**

491

Using an Analogy

Use the following analogy to describe the lytic cycle. During wartime, a tank (virus) filled with enemy troops (nucleic acid) crashes through the wall (membrane) of an automobile factory (cell). The troops take over the factory's machinery (nucleus and organelles), and adapt them to produce new tanks (viruses) instead of cars (cell parts).

Revealing Misconceptions

Students may think that antibiotics will cure viral infections, such as a cold and the flu. Explain that although antibiotics cure many bacterial infections, they do not affect viral diseases.

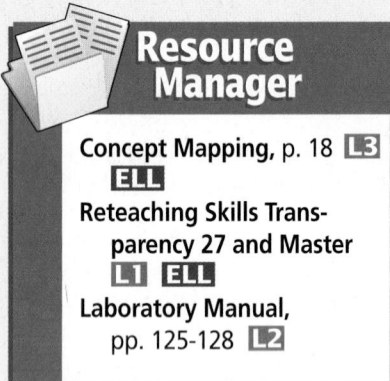

Resource Manager

Concept Mapping, p. 18 **L3**
ELL

Reteaching Skills Transparency 27 and Master
L1 **ELL**

Laboratory Manual,
pp. 125-128 **L2**

Attachment is a specific process

Each virus has a specifically shaped attachment protein. Therefore, each virus can usually attach to only a few kinds of cells. For example, the T4 phage can infect only certain types of *E. coli* because the T4's attachment protein matches a surface molecule of only these *E. coli*. A T4 cannot infect a human, animal, or plant cell, or even another bacterium. Similarly, a tobacco mosaic virus infects only a cell of a tobacco plant. In general, viruses are species specific, and some also are cell-type specific. For example, polio viruses normally infect only human intestinal and nerve cells.

The species specific characteristic of viruses is significant for controlling the spread of viral diseases. For example, by 1980, the World Health Organization had announced that smallpox, which is a deadly human viral disease, had been eradicated. The eradication was possible partly because the smallpox virus infects only humans. It is more difficult to eradicate a virus that is not species specific, such as the flu virus, which infects humans and other animals.

Viral Replication Cycles

Once attached to the plasma membrane of the host cell, the virus enters the cell and takes over its metabolism. Only then can the virus replicate. Viruses have two ways of getting into host cells. The virus may inject its nucleic acid into the host cell like a syringe injects a vaccine into your arm, as shown in *Figure 18.2*. The capsid of the virus stays attached to the outside of the host cell. An enveloped virus enters a host cell in a different way. After attachment, the plasma membrane of the host cell

WORD *Origin*

lytic
From the Greek word *lyein*, meaning to "break down." The host cell is destroyed during a lytic cycle.

Figure 18.2
In a lytic cycle, a virus uses the host cell's energy and raw materials to make new viruses. A typical lytic cycle takes about 30 minutes and produces about 200 new viruses.

CD-ROM
View an animation of *Figure 18.2* in the Presentation Builder of the Interactive CD-ROM.

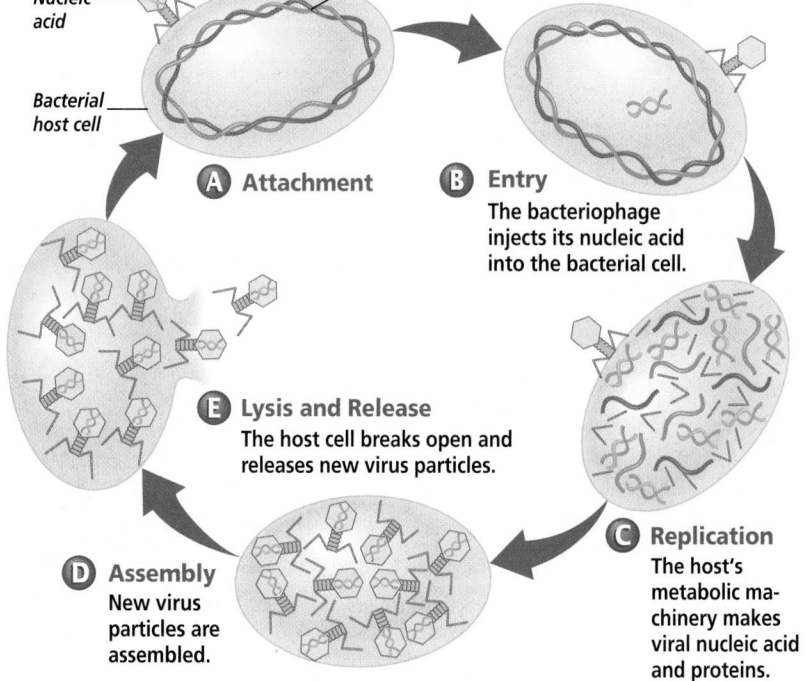

A Attachment

B Entry
The bacteriophage injects its nucleic acid into the bacterial cell.

C Replication
The host's metabolic machinery makes viral nucleic acid and proteins.

D Assembly
New virus particles are assembled.

E Lysis and Release
The host cell breaks open and releases new virus particles.

Bacteriophage — *Bacterial DNA*
Nucleic acid
Bacterial host cell

492 VIRUSES AND BACTERIA

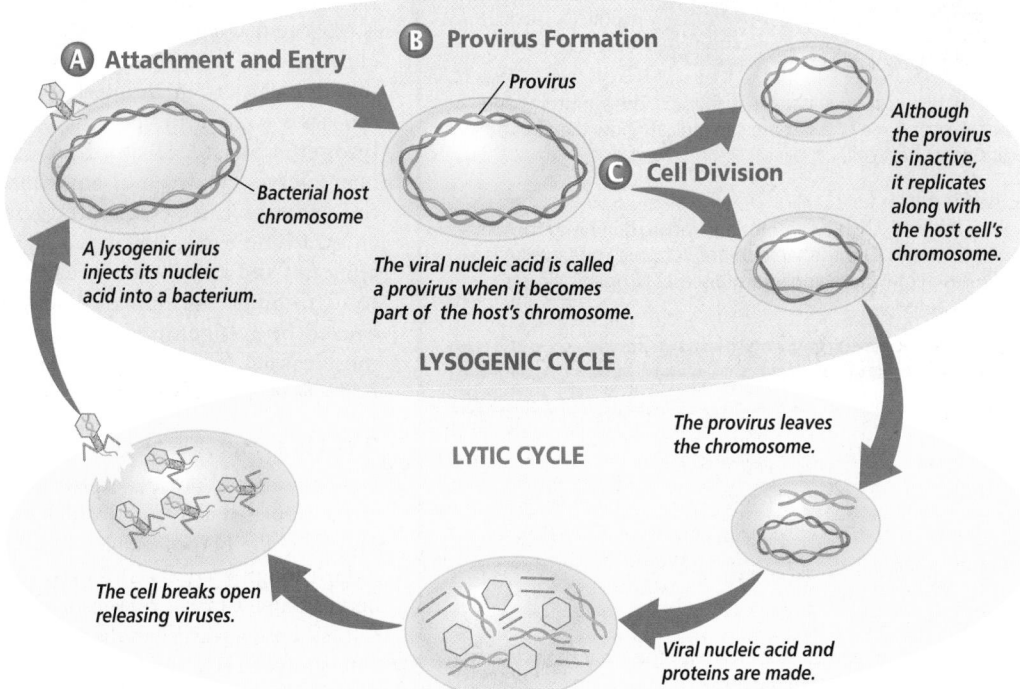

A Attachment and Entry

A lysogenic virus injects its nucleic acid into a bacterium.

Bacterial host chromosome

B Provirus Formation

Provirus

The viral nucleic acid is called a provirus when it becomes part of the host's chromosome.

C Cell Division

Although the provirus is inactive, it replicates along with the host cell's chromosome.

LYSOGENIC CYCLE

LYTIC CYCLE

The provirus leaves the chromosome.

The cell breaks open releasing viruses.

Viral nucleic acid and proteins are made.

surrounds the virus and produces a virus-filled vacuole inside the host cell's cytoplasm. Then, the virus bursts out of the vacuole and releases its nucleic acid into the cell.

Lytic cycle

Once inside the host cell, a virus's genes are expressed and the substances that are produced take over the host cell's genetic material. The viral genes alter the host cell to make new viruses. The host cell uses its own enzymes, raw materials, and energy to make copies of viral genes that along with viral proteins are assembled into new viruses, which burst from the host cell, killing it. The new viruses can then infect and kill other host cells. This process is called a **lytic cycle** (LIH tik). Follow the typical lytic cycle for a bacteriophage shown in *Figure 18.2.*

Lysogenic cycle

Not all viruses kill the cells they infect. Some viruses go through a **lysogenic cycle,** a replication cycle in which the virus's nucleic acid is integrated into the host cell's chromosome. A typical lysogenic cycle for a virus that contains DNA is shown in *Figure 18.3.*

A lysogenic cycle begins in the same way as a lytic cycle. The virus attaches to the host cell's plasma membrane and its nucleic acid enters the cell. However, in a lysogenic cycle, instead of immediately taking over the host's genetic material, the viral DNA is integrated into the host cell's chromosome.

Viral DNA that is integrated into the host cell's chromosome is called a **provirus.** A provirus may not affect the functioning of its host cell, which continues to carry out its own metabolic activity. However, every time

Figure 18.3
In a lysogenic cycle, a virus does not destroy the host cell at once. Rather, the viral nucleic acid is integrated into the genetic material of the host cell and replicates with it for a while before entering a lytic cycle.

CD-ROM

View an animation of *Figure 18.3* in the Presentation Builder of the Interactive CD-ROM.

Visual Learning

Figures 18.2 and 18.3 On the chalkboard, draw flowcharts that sequence the stages in the lytic and lysogenic cycles. Encourage students to copy the flowcharts into their journals and use them to study.

Concept Development

Ask students to list a viral disease they have had. Provide them with the resources to find out how each disease spreads and what methods of prevention are available. Students can use the lists to generate a classroom table of viral diseases. **L1**

✓ Assessment

Performance Have students create their own mnemonic devices of the steps of the lytic and lysogenic cycles and share them with others. **L2**

GLENCOE TECHNOLOGY

CD-ROM
Biology: The Dynamics of Life
Animation: *The Lytic Cycle*
Disc 3
Animation: *The Lysogenic Cycle*
Disc 3

VIDEODISC
Biology: The Dynamics of Life
The Lytic Cycle (Ch. 9)
Disc 1, Side 2, 1 min. 14 sec.

The Lysogenic Cycle (Ch. 10)
Disc 1, Side 2, 1 min. 25 sec.

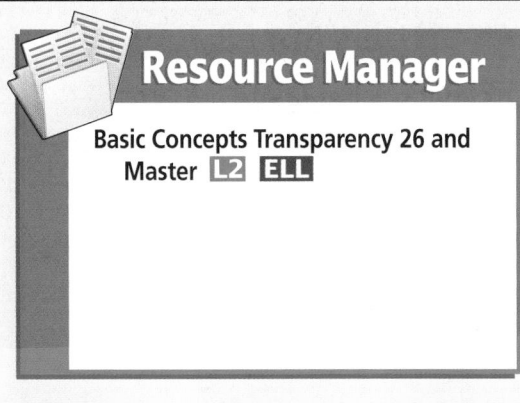

PROJECT

Vaccine Development

Interpersonal Ask student groups to prepare skits about the discovery of the following vaccines: polio, smallpox, measles, and German measles. Each skit should include information about the disease, the scientists involved in developing the vaccine, and the vaccine's impact on society. **L2** **COOP LEARN**

Resource Manager

Basic Concepts Transparency 26 and Master **L2** **ELL**

Purpose

Students will analyze data about viruses and decide if the viruses have lytic or lysogenic cycles.

Process Skills

think critically, analyze information, classify, define operationally, interpret data

Background

Measles is caused by the rubeola virus. Shingles is caused by a herpes virus (varicella-zoster virus). Human papillomavirus causes warts. Colds are caused by several viral types—the rhinovirus is the most common. AIDS is caused by the human immunodeficiency virus.

Teaching Strategies

■ If necessary, review the two viral replication cycles or ask students to reread these sections.

Thinking Critically

1. short—usually days; long—could be months or years
2. Measles, cold; a short incubation implies a short replication cycle.
3. Shingles, warts, AIDS; a long incubation implies a long replication cycle.
4. They may be unaware that they're infected and then transmit the disease to others.

✔ Assessment

Performance Ask students to research one of these diseases and write a brief report in their journal. Use the Performance Task Assessment List for Writing in Science in **PASC**, p. 87. **L2**

Problem-Solving Lab 18-1   Analyzing Information

What type of virus causes disease? The symptoms and incubation time of a disease can indicate how the virus acts inside its host cell.

Analysis

Table 18.1 lists symptoms and incubation times for some viral diseases. Use the table to predict which diseases lytic viruses might cause and which diseases lysogenic viruses might cause.

Table 18.1 Characteristics of some viral diseases

Disease	Symptom	Incubation
Measles	rash, fever	9-11 days
Shingles	pain, itching on skin	years
Warts	bumpy areas on skin	months
Coryza (cold)	sneezing, runny nose, fever	2-4 days
HIV	fatigue, weight loss, fever	2-5 years

Thinking Critically

1. How much time is associated with the replication cycle of a lytic virus? A lysogenic virus?
2. What diseases may lytic viruses cause? Explain your answer.
3. What diseases may lysogenic viruses cause? Explain your answer.
4. What is a possible consequence of the fact that a person infected with HIV may have no symptoms for years?

Figure 18.4
Before the influenza virus leaves a host cell, it is wrapped in a piece of the host's plasma membrane, making an envelope with the same structure as the host's plasma membrane

Magnification: 18 000×

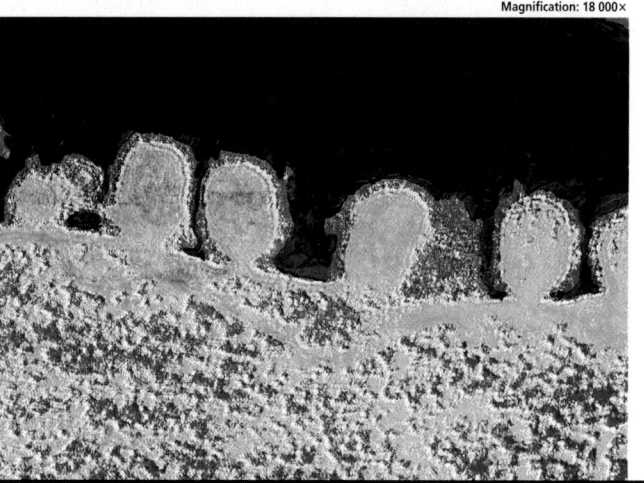

BIOLOGY JOURNAL

Smallpox Epidemic

Linguistic Ask students to write an article about an epidemic of smallpox in Boston, Massachusetts, during colonial times as it might have appeared then in a newspaper. **L2**

the host cell reproduces, the provirus is replicated along with the host cell's chromosome. Therefore, every cell that originates from an infected host cell has a copy of the provirus. The lysogenic phase can continue for many years. However, at any time, the provirus can be activated and enter a lytic cycle. Then the virus replicates and kills the host cell. Try to distinguish the human diseases caused by lysogenic viruses from those caused by lytic viruses in the *Problem-Solving Lab* on this page.

Disease symptoms of proviruses

The lysogenic process explains the reoccurrence of cold sores, which are caused by the herpes simplex I virus. Even though a cold sore heals, the herpes simplex I virus remains in your cells as a provirus. When the provirus enters a lytic cycle, another cold sore erupts. No one knows what causes a provirus to be activated, but some scientists suspect that physical stress, such as sunburn, and emotional stress, such as anxiety, play a role.

Many disease-causing viruses have lysogenic cycles. Three examples of these viruses are herpes simplex I, herpes simplex II that causes genital herpes, and the hepatitis B virus that causes hepatitis B. Another lysogenic virus is the one that causes chicken pox. Having chicken pox, which usually occurs before age ten, gives lifelong protection from another infection by the virus. However, some chicken pox viruses may remain as proviruses in some of your body's nerve cells. Later in your life, these proviruses may enter a lytic cycle and cause a disease called shingles—a painful infection of some nerve cells.

Release of viruses

Either lysis, the bursting of a cell, or exocytosis, *Figure 18.4*, the active

Figure 18.5
Retroviruses have an enzyme that transcribes their RNA into DNA. The viral DNA becomes a provirus that steadily produces small numbers of new viruses without immediately destroying the cell.

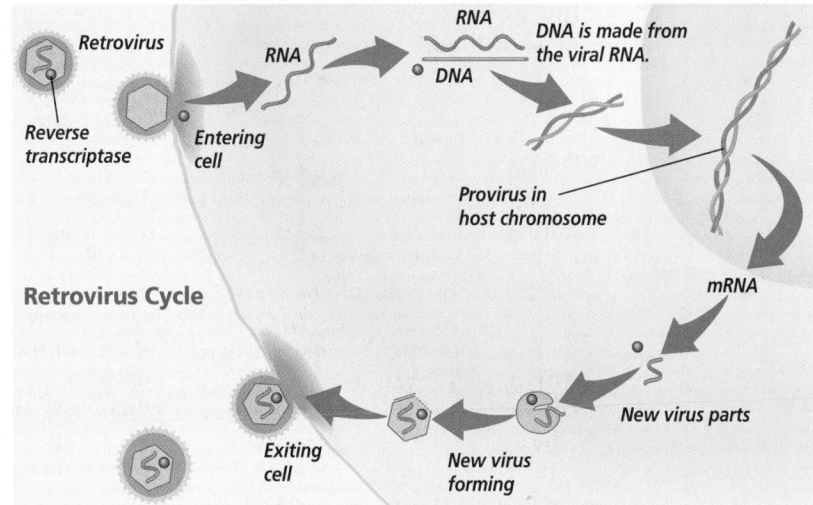

Retrovirus Cycle

Retrovirus
Reverse transcriptase
Entering cell
RNA
RNA
DNA
DNA is made from the viral RNA.
Provirus in host chromosome
mRNA
New virus parts
New virus forming
Exiting cell

transport process by which materials are expelled or secreted from a cell, release new viruses from the host cell. In exocytosis, a newly produced virus approaches the inner surface of the host cell's plasma membrane. The plasma membrane surrounds the virus, enclosing it in a vacuole that then fuses with the host cell's plasma membrane. Then, the viruses are released to the outside.

Retroviruses

Many viruses, such as the human immunodeficiency virus (HIV) that causes the disease AIDS, are RNA viruses—RNA being their only nucleic acid. The RNA virus with the most complex replication cycle is the **retrovirus** (reh tro VY rus). How can RNA be integrated into a host cell's chromosome, which contains DNA?

Once inside a host cell, the retrovirus makes DNA from its RNA. To do this, it uses **reverse transcriptase** (trans KRIHP taz), an enzyme it carries inside its capsid. This enzyme helps produce double-stranded DNA from the viral RNA. Then the double-stranded viral DNA is integrated into the host cell's chromosome and

becomes a provirus. If reverse transcriptase is found in a person, it is evidence for infection by a retrovirus. You can see how a retrovirus replicates in its host cell in *Figure 18.5.*

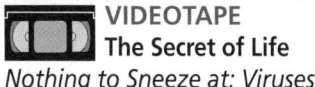

Based on viral structure and replication, have students identify approaches to the production of antiviral vaccines. **L2**

3 Assess

Check for Understanding

 *Visual-Spatial* Give students out-of-sequence sketches of the steps in the lytic and lysogenic cycles. Have them rearrange the sketches into the correct sequences. **L1**

Reteach

Ask students to make a table that lists each virus discussed in this section, its type of nucleic acid, structure, method of replication, and host organism. **L1**

NATIONAL GEOGRAPHIC

VIDEODISC
STV: Human Body Vol. 3

AIDS Virus (tinted blue)1

AIDS Virus (tinted blue) *2*

HIV: An infection of white blood cells

Once inside a human host, HIV infects white blood cells. Newly made viruses are released into the blood stream by exocytosis. Then these viruses infect other white blood cells. Infected host cells still function normally because the viral genetic material is a provirus that produces only a small number of new viruses at a time. Because the infected cells are still able to function normally, an infected person may not appear sick. However, people who are infected with HIV but have no symptoms can transmit the virus in their body fluids.

An HIV-infected person can experience no AIDS symptoms for a long time. However, most people with an HIV infection eventually get AIDS because, over time, more and more white blood cells are infected and produce new viruses, *Figure 18.6*. People will gradually lose white blood cells because proviruses enter a lytic cycle and kill their host cells. Because white blood cells are part of a body's disease-fighting system, their destruction interferes with the body's ability to protect itself from organisms that cause disease, a symptom of AIDS.

Viruses and Cancer

Retroviruses are one kind of virus that may cause some cancers. The retroviruses that convert, or transform, normal cells to tumor cells are known as tumor viruses. The first tumor virus was discovered in chickens. In addition to retroviruses, the papilloma virus, which is a DNA virus that causes warts, and the hepatitis B virus, a DNA virus thought to cause liver cancer in humans, are also tumor viruses.

Plant Viruses

The first virus to be identified was a plant virus, called tobacco mosaic virus, that causes disease in tobacco plants. Biologists know of more than 400 viruses that infect a variety of plants. These viruses cause as many as 1000 plant diseases.

Not all viral plant diseases are fatal or even harmful. For example, there are some mosaic viruses that cause striking patterns of color in the flowers

Figure 18.6
Normal white blood cells are an essential part of a human's immune system (**a**). In an HIV infected person, white blood cells are eventually destroyed by HIV proviruses that enter lytic cycles (**b**).

Magnification: 5800×

Magnification: 7800×

✔ **Portfolio**

Preventing the Spread of AIDS

Linguistic Ask student groups to prepare interview questions for a dentist or dental assistant to find out how dental professionals protect both themselves and patients from HIV. Once prepared, have one or two students from each group interview a dental professional and report back to the group. Encourage students to find out about the American Dental Association's guidelines for AIDS prevention. A group report can be prepared. **L2** **P** **COOP LEARN**

a

b

of plants such as some tulips, gladioli, and pansies. The infected flowers, like the one shown in *Figure 18.*7, have streaks of vibrant, contrasting colors in their petals. These viruses are easily spread among plants when you cut an infected stem and then cut healthy stems with the same tool.

Origin of Viruses

You might assume that viruses represent an ancestral form of life because of their relatively uncomplicated structure. This is probably not so. For replication, viruses need host cells; therefore, scientists suggest that viruses might have originated from their host cells.

Some scientists suggest that viruses are nucleic acids that break free from their host cells while maintaining an ability to replicate parasitically within the host cells. The fact that tumor viruses contain genes that are identical to ones found in normal cells is evidence for this hypothesis. According to this hypothesis, viruses are more closely related to their host cells than they are to each other.

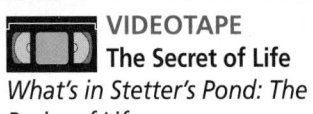

Section Assessment

Understanding Main Ideas
1. Why is a virus considered to be nonliving?
2. What is the difference between a lytic cycle and a lysogenic cycle?
3. What is a provirus?
4. How do retroviruses convert their RNA to DNA?

Thinking Critically
5. Describe the state of a herpes virus in a person who had cold sores several years ago but who does not have them now.

SKILL REVIEW

6. **Making and Using Graphs** A microbiologist added some viruses to a bacterial culture. Every hour from noon to 4:00 p.m., she determined the number of viruses present in a sample of the culture. Her data were 3, 3, 126, 585, and 602. Graph these results. How would the graph look if the culture had initially contained dead bacteria? For more help, refer to *Organizing Information* in the **Skill Handbook.**

Section Assessment

1. It does not respire, grow, move, or carry on any life function on its own.
2. In the lytic cycle, viral nucleic acid immediately takes over the cell's machinery and quickly destroys the cell. In the lysogenic cycle, it becomes part of the host cell's DNA and does not quickly kill the cell.
3. viral DNA in a host cell's chromosome
4. using the enzyme reverse transcriptase

5. a provirus in some cells of the person
6. The horizontal axis should be labeled "Time" and the vertical axis "Number of Virus Particles." Student graphs should show a stepwise increase in the number of viruses. Without living bacteria for viral replication, there would be no data and thus no graph.

Focus On

Viruses

Purpose

Students will compare and contrast the structures and shapes of viruses.

Background

The classification of viruses is based on the criteria of size, morphology, type of genetic material (DNA or RNA and double-stranded or single-stranded), and means of replication.

Teaching Strategies

■ Discuss with students the different ways viruses enter cells as illustrated by a bacteriophage and an enveloped virus.

■ Students interested in Ebola may wish to read *Ebola: A Documentary Novel of Its First Explosion* by William T. Close or *Hot Zone* by Richard Preston. **L2**

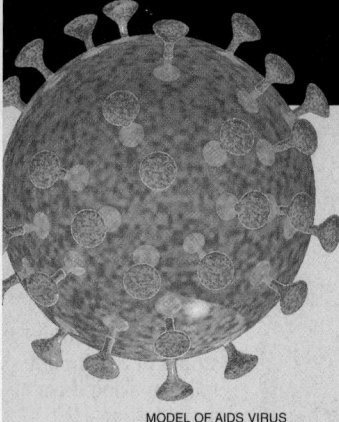

MODEL OF AIDS VIRUS

FOCUS ON Viruses

Viruses lurk everywhere—on computer keyboards, in bird droppings, under your fingernails—just waiting to get inside your body or some other living thing. Smaller than the smallest bacteria, viruses are not alive. By themselves, they cannot move, grow, or reproduce. But give viruses the chance to invade a living cell, and they will take over its metabolic machinery, reprogramming it to churn out more viruses to attack other cells.

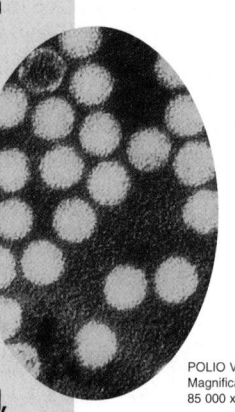

POLIO VIRUS
Magnification:
85 000 x

INVISIBLE INVADERS

Scientists have identified thousands of viruses. Some invade plants, others attack animals, and still others target bacteria.

In humans, viruses are responsible for chicken pox, warts, cold sores, and the common cold, as well as dreaded diseases such as rabies, influenza, hepatitis, and AIDS.

WOMAN EXPERIENCING SYMPTOMS OF THE COMMON COLD

STRUCTURE

A single drop of blood can contain six billion viruses. Despite their incredibly small size, many viruses, such as this tobacco mosaic virus (below), have complex structures. All viruses consist of a core of nucleic acid—either DNA or RNA—enclosed in a protein coat called a capsid. Both the type and arrangement of proteins in the capsid give different viruses characteristic shapes.

CAPSID   DNA

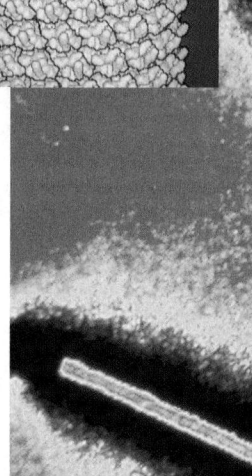

MODEL OF TOBACCO MOSAIC VIRUS

ICOSAHEDRAL VIRUSES

Many animal viruses—such as polio (above) and adenovirus—have 20-sided, or icosahedral, capsids. Viewed under an electron microscope, icosahedral viruses look like perfectly symmetrical crystals.

498

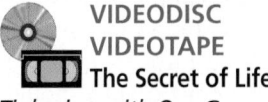

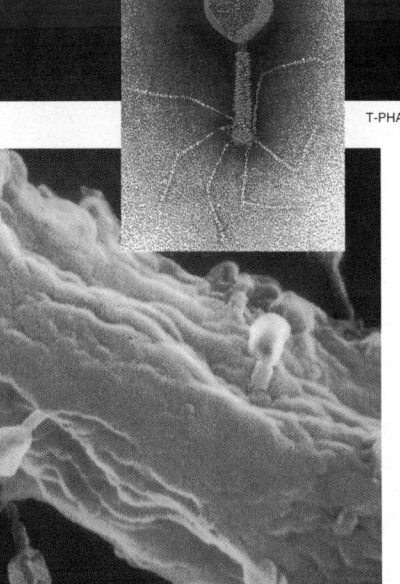
T-PHAGE

PHAGES

Bacteriophages, or phages for short, are viruses that infect bacteria. This T4-phage (top left), looks like a miniature lunar-landing module. It has a DNA-containing head, a protein tail, and protein tail fibers that attach to the surface of a bacterium. Once viruses are attached (left), the tail section contracts and pierces the cell wall, and viral DNA is injected into the host cell.

ENVELOPED VIRUSES

Some viruses, such as influenza and HIV (the virus that causes AIDS), are enclosed in an envelope composed of lipids, carbohydrates, and proteins. Envelope proteins (right) form spiky projections that help the virus gain entry to a host cell, much like keys fitting into a lock.

T-PHAGES, IN BLUE, INFECTING *E. COLI* BACTERIUM
Magnification: 90 000 x

INFLUENZA VIRUS
Magnification: 17 150 x

HELICAL VIRUSES

Helical viruses are shaped like tiny cylinders, with the viral genetic material spiraling down the center of a hollow protein tube. Tobacco mosaic virus (below), which infects plants (right), is a long helical virus.

TOBACCO MOSAIC VIRUS
Magnification: 30 000 x

PLANT INFECTED BY TOBACCO MOSAIC VIRUS

DEADLY BEAUTY

Some viruses have irregular shapes. The Ebola virus (below), which causes massive internal bleeding in humans, has a twisted, worm-like form. A strain of Ebola virus from Zaire, Africa, is one of the most deadly viruses researchers have ever studied.

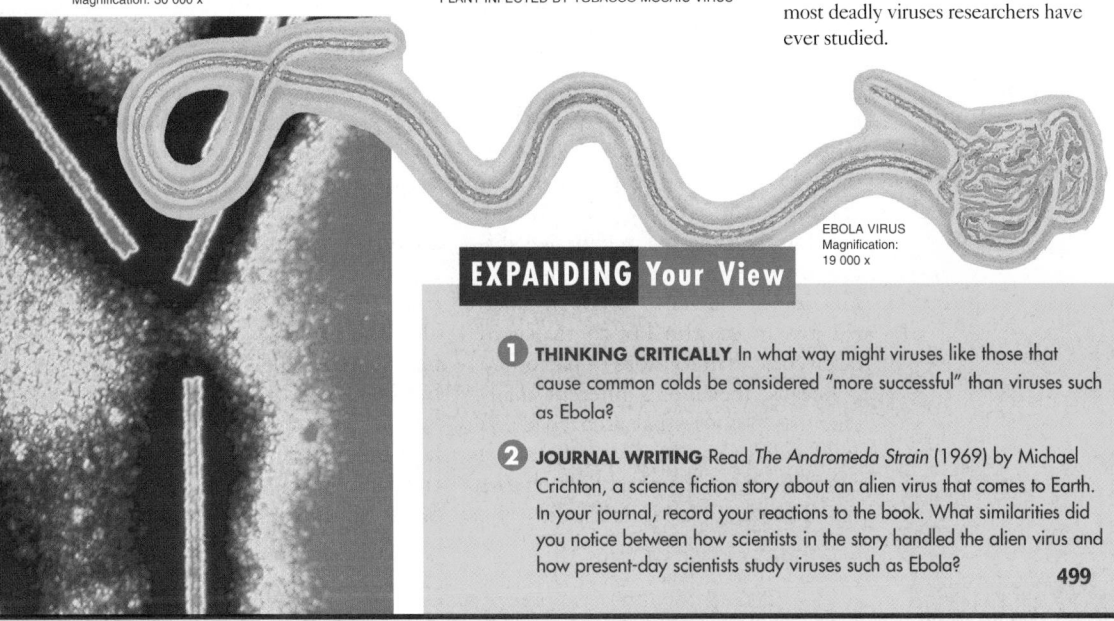
EBOLA VIRUS
Magnification: 19 000 x

EXPANDING Your View

1 **THINKING CRITICALLY** In what way might viruses like those that cause common colds be considered "more successful" than viruses such as Ebola?

2 **JOURNAL WRITING** Read *The Andromeda Strain* (1969) by Michael Crichton, a science fiction story about an alien virus that comes to Earth. In your journal, record your reactions to the book. What similarities did you notice between how scientists in the story handled the alien virus and how present-day scientists study viruses such as Ebola?

499

Visual Learning

■ Ask students to calculate the actual size of one of the tobacco mosaic viruses and one of the polio viruses by using a metric ruler to measure the size of the virus in the photo. Then, they should divide the measurement by the photo's magnification. **L2**

■ Have students make a video or computer presentation about a viral disease, such as mumps, Ebola, or rabies. Their presentations should include information about both the virus and the disease it causes. **L3**

Answers to Expanding Your View

1. They don't kill their hosts.
2. They used the highest level of containment when working with the virus, as do virologists who work with deadly and newly discovered viruses today.

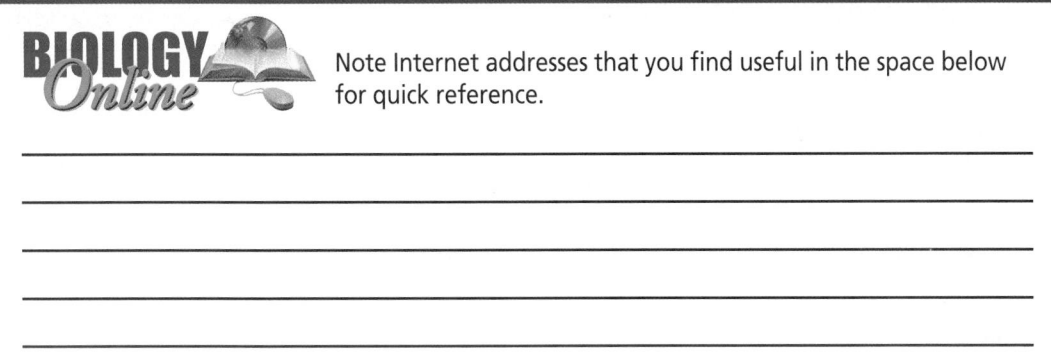
BIOLOGY Online

Note Internet addresses that you find useful in the space below for quick reference.

_____

_____

_____

_____

_____

Prepare

Key Concepts

The classification, structure, survival adaptations, and diversity of bacteria are explained. Bacterial reproduction and the importance of bacteria to people, other organisms, and the environment are discussed.

Planning

■ Purchase distilled water and yogurt containing live bacterial cultures for the Microscope Activity.

■ Purchase pickles, yogurt, Swiss cheese, soybeans, peanuts, milk, sour cream, and sauerkraut or bring in their pictures for the Quick Demo.

1 Focus

Bellringer 📖

Before presenting the lesson, display **Section Focus Transparency 44** on the overhead projector and have students answer the accompanying questions. **L1** **ELL**

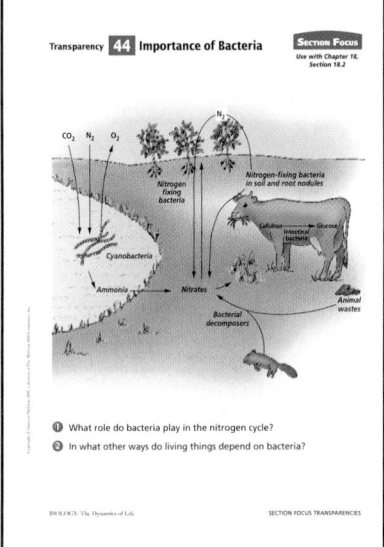

SECTION PREVIEW

Objectives

Compare the types of prokaryotes.

Explain the characteristics and adaptations of bacteria.

Evaluate the economic importance of bacteria.

Vocabulary

chemosynthesis
binary fission
conjugation
obligate aerobe
obligate anaerobe
endospore
toxin
nitrogen fixation

Section

18.2 Archaebacteria and Eubacteria

I magine yourself going back three-and-a-half billion years. You wander around the young Earth and find yourself alone with the first life on this planet. Dinosaurs? Saber-toothed tigers? No. You would be alone with what have become some of the most diverse forms of life on Earth—prokaryotes.

Magnification: 17 280×

A hot spring (above) and the archaebacterium *Thermoproteus tenax* (inset)

Diversity of Prokaryotes

Recall that prokaryotes are unicellular organisms that do not have membrane-bound organelles. They are classified in two kingdoms—archaebacteria and eubacteria. Many biochemical differences exist between these two types of prokaryotes. For example, their cell walls and the lipids in their plasma membranes differ. In addition, the structure and function of the genes of archaebacteria are more similar to those of eukaryotes than to those of bacteria.

Because they are so different, many scientists propose that archaebacteria and eubacteria arose separately from a common ancestor several billion years ago. The exact time is unknown.

Archaebacteria: Often the extremists

There are three types of archaebacteria that live mainly in extreme habitats where there is usually no free oxygen available. You can see some of these inhospitable places in **Figure 18.8.** One type of archaebacterium lives in oxygen-free environments and produces methane gas. These methane-producing archaebacteria live in marshes, lake sediments, and the digestive tracts of some mammals, such as cows. They also are found at sewage disposal plants, where they play a role in the breakdown of sewage.

A second type of archaebacterium lives only in water with high concentrations of salt, such as in Utah's Great Salt Lake and the Middle East's

500 VIRUSES AND BACTERIA

PROJECT

Bacterial Microscopy

Visual-Spatial Show students how to prepare depression slides and use the hanging drop (oil suspension) method of microscopy. Ask students to use these methods to make slides of living *Oscillatoria* and observe the cells. Have students write and sketch their observations. **L1** **ELL** 📖

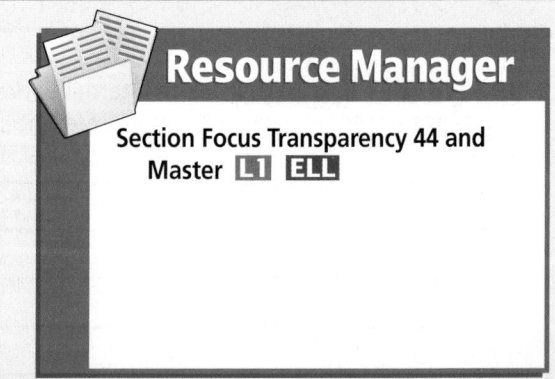

Resource Manager

Section Focus Transparency 44 and Master **L1** **ELL**

Figure 18.8
Archaebacteria live in extreme environments.

Ⓐ Methane-producing archaebacteria flourish in this swamp and also live in the stomachs of cows.

Ⓑ Salt-loving archaebacteria live in these salt pools left after this lake in British Columbia, Canada evaporated. These pools have high levels of magnesium and potassium salts.

Ⓒ Heat- and acid-loving archaebacteria live around deep ocean vents where water temperatures are often above 100°C.

Dead Sea. A third type lives in the hot, acidic waters of sulfur springs. This type of anaerobic archaebacterium also thrives near cracks deep in the Pacific Ocean's floor, where it is the autotrophic producer for a unique animal community's food chain.

Eubacteria: The heterotrophs

Eubacteria, the other kingdom of prokaryotes, includes those prokaryotes that live in more hospitable places than archaebacteria inhabit and that vary in nutritional needs. The heterotrophic eubacteria live almost everywhere and use organic molecules as a food source. Some bacterial heterotrophs are parasites, obtaining their nutrients from living organisms. They are not adapted for trapping food that contains organic molecules or for making organic molecules themselves. Others are saprophytes, which are organisms that feed on dead organisms or organic wastes. Recall that saprophytes break down and thereby recycle the nutrients locked in the body tissues of dead organisms.

18.2 ARCHAEBACTERIA AND EUBACTERIA **501**

BIOLOGY *Online*

Note Internet addresses that you find useful in the space below for quick reference.

_____

_____

_____

_____

_____

Activity

Kinesthetic Thin yogurt containing live bacterial cultures with distilled water. Have students make a wet mount of the yogurt, adding a drop of methylene blue, and look for bacteria on their slides. The bacteria should be visible under the high-power. Remind students to use care when handling a microscope and glass slides and to use special care not to break the slide when viewing it under the high-power objective. Caution students that stains may be permanent on their clothing. Have them rinse their skin immediately if they spill stain on it. **L1** **ELL** 🔲

GLENCOE
TECHNOLOGY

CD-ROM
Biology: The Dynamics of Life
BioQuest: *Biodiversity Park*
Disc 3, 4

Resource Manager

Basic Concepts Transparency 27 and Master **L2** **ELL**

Eubacteria:
Photosynthetic autotrophs

A second type of eubacterium is the photosynthetic autotroph. These eubacteria live in places with sunlight because they need light to make the organic molecules that are their food. Cyanobacteria are photosynthetic autotrophs. They contain the pigment chlorophyll that traps the sun's energy, which they then use in photosynthesis. Most cyanobacteria, like the *Anabaena* shown in *Figure 18.9*, are blue-green and some are red or yellow in color. Cyanobacteria commonly live in ponds, streams, and moist areas of land. They are composed of chains of independent cells—an exception to the unicellular form of most other bacteria.

WORD Origin
cyanobacterium
From the Greek words *kyanos*, meaning "blue," and *bakterion*, meaning "small rod." The cyanobacteria are blue-green bacteria.

Eubacteria:
Chemosynthetic autotrophs

A third type of eubacterium is the chemosynthetic autotroph. Like photosynthetic bacteria, these bacteria make organic molecules that are their food. However, unlike the photosynthetic bacteria, the chemosynthetic

Figure 18.9
Cyanobacteria, such as *Anabaena*, have a blue-green color.

Magnification: 1100×

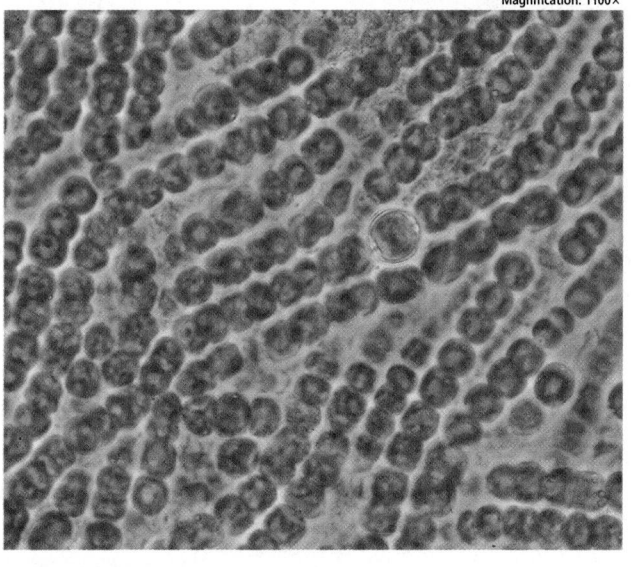

502 VIRUSES AND BACTERIA

---

bacteria do not obtain the energy they need to make food from sunlight. Instead, they break down and release the energy of inorganic compounds containing sulfur and nitrogen in the process called **chemosynthesis** (kee moh SIHN thuh sus). Some chemosynthetic bacteria are very important to other organisms because they are able to convert atmospheric nitrogen into the nitrogen-containing compounds that plants need.

What Is a Bacterium?

A bacterium consists of a very small cell. Although tiny, a bacterial cell has all the structures necessary to carry out its life functions.

The structure of bacteria

Prokaryotic cells have ribosomes, but their ribosomes are smaller than those of eukaryotes. They also have genes that are located for the most part in a single circular chromosome, rather than in paired chromosomes. What structures can protect a bacterium? Look at the *Inside Story* on the next page to learn about other structures located in bacterial cells.

One structure that supports and protects a bacterium is the cell wall. The cell wall protects the bacterium by preventing it from bursting. Because most bacteria live in a hypotonic environment, one in which there is a higher concentration of water molecules outside than inside the cell, water is always trying to enter a bacterial cell. A bacterial cell remains intact, however, and does not burst open as long as its cell wall is intact. If the cell wall is damaged, water will enter the cell by osmosis, causing the cell to burst. Scientists used a bacterium's need for an intact cell wall to develop a weapon against bacteria that cause disease.

🟦 MEETING INDIVIDUAL NEEDS

Learning Disabled

Interpersonal Have pairs of students design flash cards that show a bacterial structure on one side and its function on the other side. Later, students can make cards that show bacterial shape on one side and the name of a bacterium with that shape on the other. They can also make cards that show bacterial arrangements—clusters, chains, and pairs—to practice using the correct prefixes. Students can use the cards to learn the names of the bacteria studied in this section. **L1** **ELL** 🔲 **COOP LEARN**

INSIDE STORY

A Typical Bacterial Cell

Bacteria are microscopic, prokaryotic cells. The great majority of bacteria are unicellular. A typical bacterium would have some or all of the structures shown in this diagram of a bacterial cell.

Critical Thinking *Which structures of bacteria are involved in reproduction?*

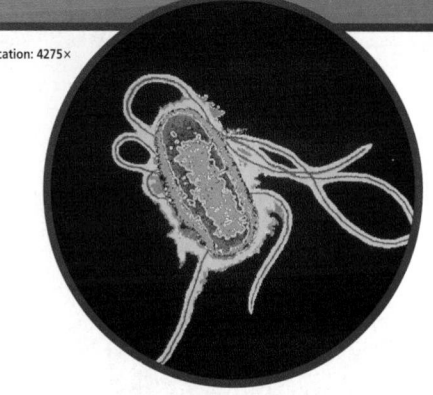

Magnification: 4275×

Escherichia coli

1 Chromosome A single DNA molecule, arranged as a circular chromosome and not enclosed in a nucleus, contains most of the bacterium's genes.

2 Cell wall A cell wall surrounds the plasma membrane. It gives the cell its shape, and prevents osmosis from bursting the cell.

3 Capsule Some bacteria have a sticky gelatinous capsule around the cell wall. A bacterium with a capsule is more likely to cause disease than a bacterium without a capsule.

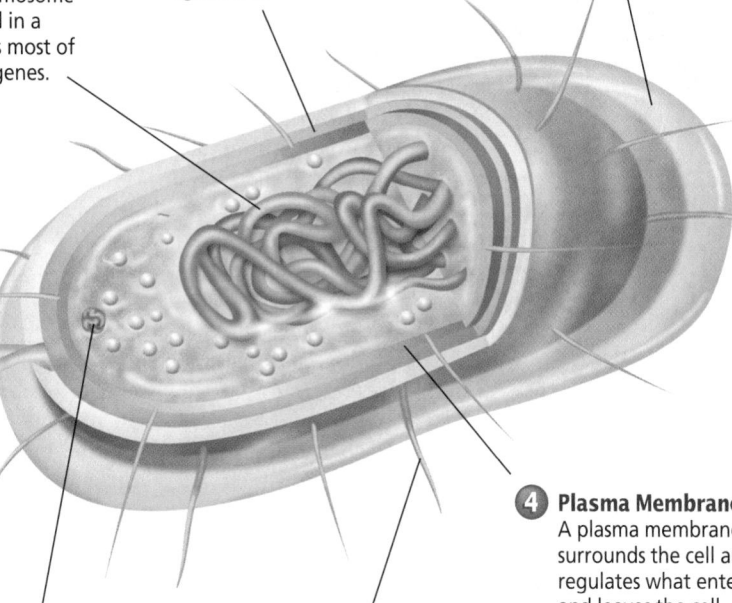

7 Flagellum Some bacteria have long, whiplike protrusions called flagella that enable them to move.

6 Plasmid A few genes are located in a small circular chromosome piece called a plasmid. A bacterium may have one or more plasmids.

5 Pilus Some bacteria have pili—extensions of their plasma membranes. A hairlike pilus helps a bacterium stick to a surface. It is also like a bridge through or on which two bacteria can exchange DNA.

4 Plasma Membrane A plasma membrane surrounds the cell and regulates what enters and leaves the cell.

18.2 ARCHAEBACTERIA AND EUBACTERIA **503**

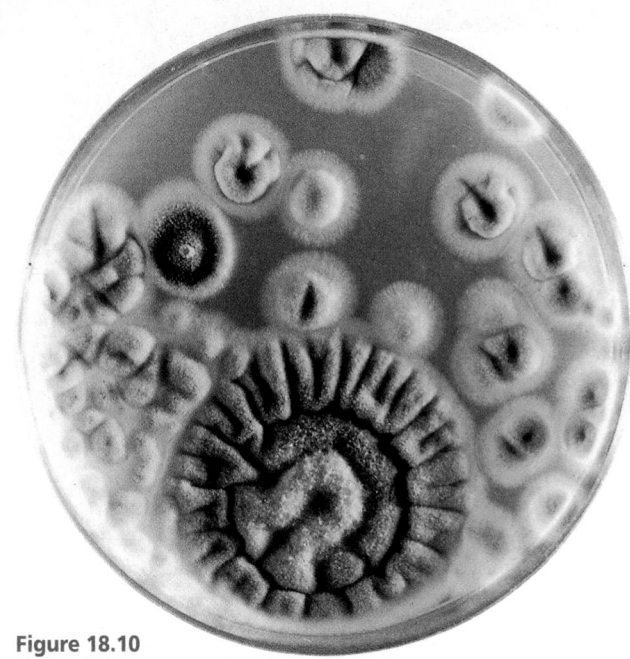

Figure 18.10
The mold known as *Penicillium notatum* produces the antibiotic penicillin.

In 1928, Sir Alexander Fleming accidentally discovered penicillin, the first antibiotic used in humans. He was growing bacteria when an airborne mold, *Penicillium notatum*, contaminated his culture plates. He noticed that the mold, shown in *Figure 18.10*, secreted a substance—now known as the antibiotic penicillin—that killed the bacteria he was growing. Later, biologists discovered that penicillin interferes with the ability of some bacteria to make cell walls. When such bacteria grow in penicillin, holes develop in their cell walls, water enters their cells, and they rupture and die.

Identifying bacteria

You may think of bacteria as all the same, but scientists have developed ways to distinguish among them. For example, one trait that helps categorize bacteria is how they react to Gram stain. Gram staining is a technique that distinguishes two groups of bacteria because the stain reflects a basic difference in the composition of bacterial cell walls. The cell walls of all bacteria are made of interlinked sugar and amino acid molecules that differ in arrangement and therefore react differently to Gram stain. After staining, gram-positive bacteria are purple and Gram-negative bacteria are pink. Gram-positive bacteria are affected by different antibiotics—substances that can destroy bacterial cells—than the antibiotics that affect Gram-negative bacteria.

Not only do bacterial cell walls react differently to Gram stain, but they also give bacteria different shapes. Shape is another way to categorize bacteria. The three most common shapes are spheres, rods, and spirals, as shown in *Figure 18.11*. In addition to having one of these shapes, bacterial cells often grow in characteristic patterns that provide another way of categorizing them. Diplo- is a prefix that refers to a paired arrangement of cell growth. The prefix staphylo- describes an arrangement of cells that resemble grapes. Strepto- is a prefix that refers to an arrangement of chains of cells.

Reproduction by binary fission

Bacteria cannot reproduce by mitosis or meiosis because they have no nucleus, and instead of pairs of chromosomes, they have one circular chromosome and varying numbers of smaller circular pieces of DNA called plasmids. Therefore, they have other ways to reproduce.

Bacteria reproduce asexually by a process known as **binary fission.** To reproduce in this way, a bacterium first copies its chromosome. Then the original chromosome and the copy become attached to the cell's plasma membrane for a while. The cell grows larger, and eventually the

504 VIRUSES AND BACTERIA

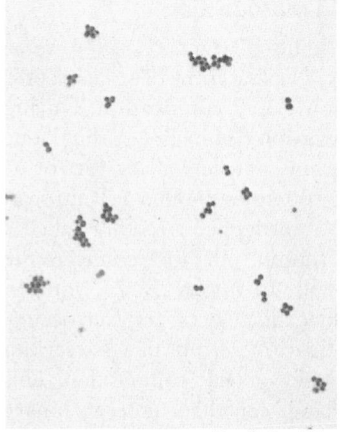
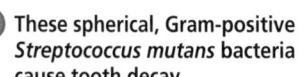

Magnification: 1100×

A These spherical, Gram-positive *Streptococcus mutans* bacteria cause tooth decay.

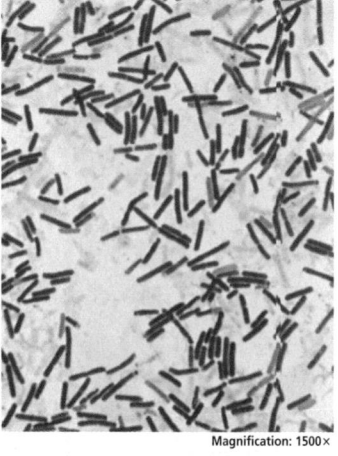

Magnification: 1500×

B This rodlike, Gram-positive bacterium, *Clostridium botulinum*, produces a poison that can result in food poisoning.

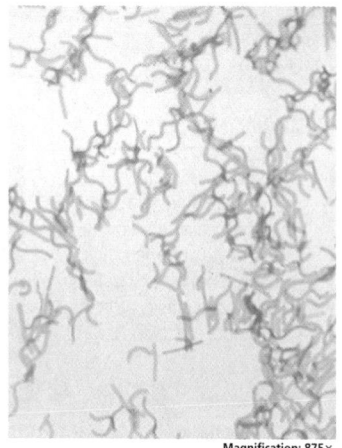

Magnification: 875×

C This spiral-shaped, Gram-negative *Spirillum volutans* bacterium has flagella.

two chromosomes separate and move to opposite ends of the cell. Then, a partition forms between the chromosomes, as shown in *Figure 18.12*. This partition separates the cell into two similar cells. Because each new cell has either the original or the copy of the chromosome, the resulting cells are genetically identical.

Bacterial reproduction can be rapid. In fact, under ideal conditions, some bacteria can reproduce every 20 minutes, producing enormous numbers of bacteria quickly. If bacteria always reproduced this fast, they would cover the surface of Earth within a few weeks. But, this doesn't happen, because bacteria don't always have ideal growing conditions. They run out of nutrients and water, they poison themselves with their own wastes, and predators eat them.

Figure 18.11
Scrub, shampoo, and gargle as you will, you'll remove only a small fraction of the bacteria that live on and in you.

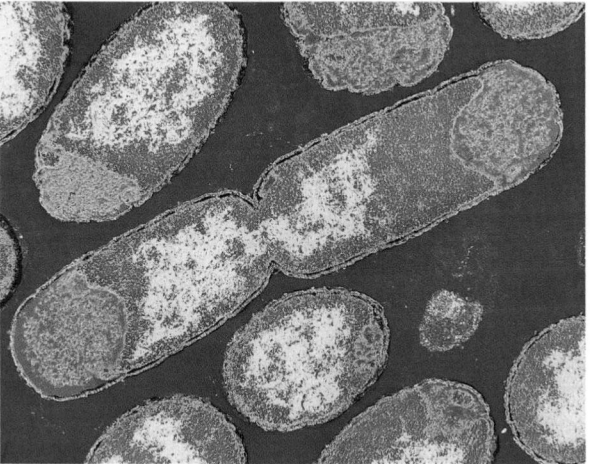

Magnification: 18 300×

Figure 18.12
This *Escherichia coli* cell is starting to divide. The newly forming partition is visible in the center of the cell.

TECHPREP

Bacterial Growth

 Linguistic Have students research how antiseptics, sterilizing, canning, freezing, salting, smoking, and cooling can reduce the growth of bacteria. Assign students individual topics and have them demonstrate their findings to the class. **L2**

Visual Learning
Figure 18.12 Ask students why cell growth is needed for binary fission. *for chromosome duplication and separation.*

✔ **Assessment**
Performance Assessment in the Biology Classroom, p. 29, *Inhibiting Bacterial Growth.* Have students carry out the activity to determine how to inhibit bacterial growth. **L2**

GLENCOE TECHNOLOGY

CD-ROM
Biology: The Dynamics of Life
Video: *Binary Fission*
Disc 3

tube. Add growth medium, screw on the test tube's top, and shake well.
2. Place the tubes about 60 cm away from a light bulb and do not disturb.
3. Hypothesize where bacteria will grow.
4. Examine the cultures daily for two weeks. Red, rust, pink, and orange colors indicate photosynthetic bacteria.

Analysis
1. In which tubes did you find photosynthetic bacteria? *from all to none*

2. Was your hypothesis supported by your data? Speculate why or why not. *Data will vary. There are hundreds of photosynthetic bacteria and only some may have been able to grow.*
3. What caused the colors you saw in the test tubes? *photosynthetic pigments*

✔ **Assessment**
Performance Have students experiment to determine the best temperature for the growth of their isolated bacteria. They must subculture every two weeks to maintain the culture. Use the Performance Task Assessment List for Assessing a Whole Experiment and Planning the Next Experiment in **PASC,** p. 33. **L3**

505

Purpose

Students will examine three typical shapes of bacteria.

Process Skills

compare and contrast, classify, observe and infer

Teaching Strategies

■ Provide students with their microscopes' high-power field of view diameter.

■ Encourage students to identify bacterial arrangements.

■ Caution students to work carefully with microscopes, slides, and coverslips especially when viewing slides under a high-power objective.

Expected Results

Students will observe spherical, rod, and spiral-shaped bacteria.

Analysis

1. The spherical bacteria were smallest. The spirilla were most likely largest.
2. the pink bacteria
3. exchange information and nuclear material.

Assessment

Performance Give students a prepared slide of unfamiliar bacteria. Ask them to identify the bacteria's shape and arrangement. Use the Performance Task Assessment List for Making Observations and Inferences in **PASC,** p. 17. **L1**

Visual Learning

Figure 18.13 Ask students why it is advantageous for bacteria to transfer genetic material. *The transfer of genetic material allows for genetic recombination, leading to variation in the species that permits some members to survive conditions that others cannot.*

MiniLab 18-2 Observing

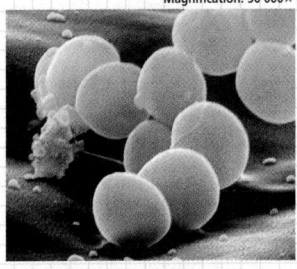

Magnification: 50 000×

***Staphylococcus* bacteria**

Bacteria Have Different Shapes Bacteria come in three shapes: spherical (coccus), rodlike (bacillus), and spiral shaped (spirillum). They may appear singly or in pairs, chains, or clusters. Each species has a typical shape and reaction to Gram stain.

Procedure

1. Obtain slides of bacteria from your teacher.
2. Using low power, locate bacteria of one shape. Switch to high power. Look for individual cells and observe their shape. Observe also the size of the cells and their color. Then look for groups of bacterial cells to determine their arrangement. **CAUTION:** *Use caution when working with a microscope and microscope slides.*
3. Repeat step 2 for bacteria with the other shapes. Then, compare the sizes of the bacteria.
4. Draw a diagram of each type of bacteria.

Analysis

1. How do the sizes of the three bacteria compare?
2. Which of the bacteria were Gram negative?
3. What adaptive advantage might there be for bacteria to form groups of cells?

Figure 18.13
The *E. coli* at the bottom is attached to the other bacteria by pili, through or on which genetic material is being transferred.

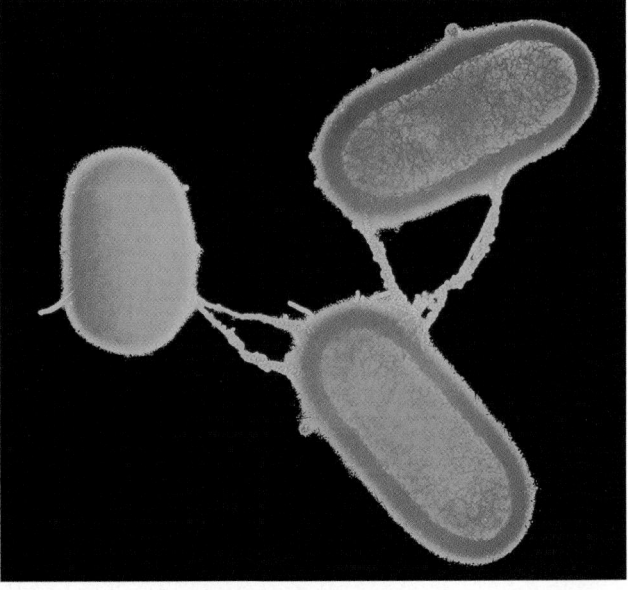

Magnification: 19 800×

Sexual reproduction

In addition to binary fission, some bacteria have a form of sexual reproduction called conjugation. During **conjugation** (kahn juh GAY shun), one bacterium transfers all or part of its chromosome to another cell through or on a bridgelike structure called a pilus (plural, pili) that connects the two cells. In *Figure 18.13,* you can see how this genetic transfer occurs. Conjugation results in a bacterium with a new genetic composition. This bacterium can then undergo binary fission, producing more cells with the same genetic makeup.

Try the *MiniLab* on this page to see some bacterial staining reactions, cell shapes, and patterns of growth.

Adaptations in Bacteria

Based on fossil evidence, some scientists propose that anaerobic bacteria were probably among the first photosynthetic organisms, producing not only their own food but also oxygen. As the concentration of oxygen increased in Earth's atmosphere,

---

MEETING INDIVIDUAL NEEDS

Gifted

Intrapersonal Ask students to research Gram positive and Gram negative bacteria. Ask them to examine prepared slides of Gram-stained bacteria and present their findings. **L3**

Resource Manager

BioLab and MiniLab Worksheets, p. 86 **L2**

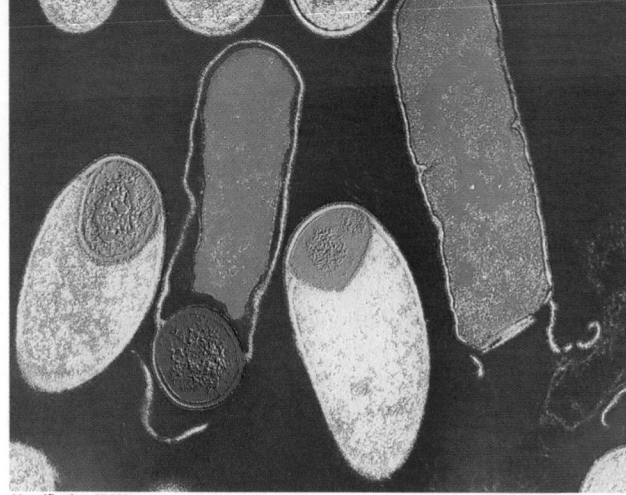

Figure 18.14
This TEM shows bacteria in three different stages of endospore production

Magnification: 25 500×

Figure 18.14 Point out the yellow bacterium in the middle of the photo that is in the early stage of spore formation and still lacks an endospore wall. Have students find the yellow bacterium nearer the edge of the photo that contains a mature endospore with a wall. Ask them to find the blue bacterium between the two yellow ones that is releasing its endospore. Ask students why a thick wall is valuable to the bacterium. *It helps the bacterium resist dehydration, boiling, and many chemicals.*

Revealing Misconceptions

Students may think that stepping on a rusty nail causes tetanus because of the rust on the nail. But, the soil that the nail touched is the source of the bacterium that causes tetanus.

GLENCOE
TECHNOLOGY

VIDEODISC
The Secret of Life
Bacterial Endospore

Resource Manager

Laboratory Manual,
pp. 129-132 **L2**

Diversity of metabolism

Recall that the breaking down food to release its energy is called cellular respiration. Modern bacteria have diverse types of respiration.

Many bacteria require oxygen for respiration. These bacteria are called **obligate aerobes.** *Mycobacterium tuberculosis*, the organism that causes the lung disease called tuberculosis, is an obligate aerobe. There are other bacteria, called **obligate anaerobes,** that are killed by oxygen. Among bacteria that are obligate anaerobes is the bacterium *Treponema pallidum* that causes syphilis, a sexually transmitted disease, and the bacterium that causes botulism, a type of food poisoning that you will learn more about soon. There are still other bacteria that can live either with or without oxygen, releasing the energy in food aerobically by cellular respiration or anaerobically by fermentation.

A survival mechanism

Some bacteria, when faced with unfavorable environmental conditions, produce endospores, shown in *Figure 18.14*. An **endospore** is a tiny structure that contains a bacterium's DNA and a small amount of its cytoplasm, encased by a tough outer covering that resists drying out, temperature extremes, and harsh chemicals. As an endospore, the bacterium rests and does not reproduce. When environmental conditions improve, the endospore germinates, or produces a cell that begins to grow and reproduce. Some endospores have germinated after thousands of years in the resting state.

Although endospores are useful to bacteria, they can cause problems for people. Endospores can survive a temperature of 100°C, which is the boiling point of water. To kill endospores, items must be sterilized—heated under high pressure in either a pressure cooker or an autoclave. Under pressure, water will boil at a higher temperature than its usual 100°C, and this higher temperature kills endospores.

Canned foods must be sterilized and acidified. This is because the endospores of the bacterium called *Clostridium botulinum* easily get into foods being canned. These bacteria belong to the group clostridia—all obligate anaerobic bacteria that form endospores. If the endospores

18.2 ARCHAEBACTERIA AND EUBACTERIA **507**

Immunization History

Linguistic Have students obtain their immunization records and write in their journals the dates of their most recent vaccinations, indicating if each vaccination was against a disease caused by bacteria or viruses. Encourage students to find out for how long the immunization is good. Provide research resources. Students who have religious beliefs that prohibit immunization can write about the medically suggested immunization schedule. **L2**

Problem-Solving Lab 18-2

Purpose 🖼️

Students are asked to apply information about one type of food poisoning.

Process Skills

think critically, analyze information, hypothesize

Background

Endospores of *Clostridium botulinum* that are left undestroyed on a vegetable being canned can germinate after the jar is sealed. The bacteria grow in the jar's anaerobic conditions, producing a powerful endotoxin that can result in respiratory failure if ingested by humans.

Teaching Strategies

■ Have students review the meaning of the terms obligate anaerobe, endospore, anaerobic conditions, germinate, toxin. **L1** **ELL**

Thinking Critically

1. Answers may vary. Endospores won't germinate in oxygen.
2. Answers may vary. Spores in soil cling to the plant.
3. Answers may vary. A high enough temperature may not have been reached during canning.
4. The jars contain no oxygen.

508

Problem-Solving Lab 18-2 · Hypothesizing

Can you get food poisoning from eating home-canned foods? *Clostridium botulinum* is a bacterial species that causes food poisoning.

Analysis

C. botulinum is an obligate anaerobic soil bacterium, and it easily spreads onto plants. It forms endospores that are highly heat-resistant and germinate only in anaerobic conditions. The bacterium produces a heat-resistant toxin that can kill humans. Commercially canned foods are heated to 121°C for a minimum of 20 minutes to ensure that all spores are killed.

Thinking Critically

1. Hypothesize why you don't get food poisoning if you eat fresh vegetables that are contaminated with the endospores of *C. botulinum*.
2. Hypothesize how the endospores of *C. botulinum* get into home-canned vegetables.
3. Hypothesize how *C. botulinum* endospores can survive inadequate home-canning procedures.
4. Explain why endospores of *C. botulinum* germinate inside canning jars.

Figure 18.15
CAUTION: When a foil-wrapped potato is baked, any *Clostridium botulinum* spores on its skin can survive. If the potato is eaten immediately, the spores cannot germinate. However, if the still-wrapped potato cools at room temperature, the spores can germinate in the anaerobic environment of the foil, and the bacteria will produce their deadly toxin.

---

of *C. botulinum* get into improperly sterilized canned food, they germinate. Bacteria grow in the anaerobic environment of the can and produce a powerful and deadly poison, called a **toxin**, as they grow. This deadly toxin saturates the food and, if eaten, causes the disease called botulism. Although rare, botulism is often fatal, and it can be transmitted in many ways other than poorly canned food, as shown in *Figure 18.15*. Try the *Problem-Solving Lab* on this page to learn more about *C. botulinum*.

Another clostridia, *Clostridium tetani*, produces a powerful nerve toxin that causes the disease called tetanus, which is often fatal. Because endospores of *C. tetani* exist almost everywhere, they will often enter a wound. Deep wounds and puncture wounds are hard to clean and provide the conditions needed for the growth of anaerobes. The endospores germinate in the wound, and the bacteria grow and produce a toxin that the blood carries to nerve cells in the spinal cord. Fortunately, there is an immunization for tetanus. You probably received this shot as a child. You

---

need a booster shot periodically, or immediately if you receive a puncture wound and are seriously injured.

The Importance of Bacteria

When you think about bacteria, your first thought may be disease. But disease-causing bacteria are few compared with the number of harmless and beneficial bacteria on Earth. Bacteria help to fertilize fields, to recycle nutrients on Earth, and to produce foods and medicines.

Nitrogen fixation

Most of the nitrogen on Earth exists in the form of nitrogen gas, N_2, which makes up 80 percent of the atmosphere. All organisms need nitrogen because the element is a component of their proteins, DNA, RNA, and ATP. Yet few organisms, including most plants, can directly use nitrogen from the air.

Several species of bacteria have enzymes that convert N_2 into ammonia (NH_3) in a process known as **nitrogen fixation.** Other bacteria then convert the ammonia into nitrite (NO_2^-) and nitrate (NO_3^-), which plants can use. Bacteria are the only organisms that can perform these chemical changes.

Some nitrogen-fixing bacteria live symbiotically within the roots of some trees and legumes—plants such as peas, peanuts, and soybeans—in swollen areas called nodules. You can see some nodules in *Figure 18.16.* Farmers grow legume crops after the harvesting of crops such as corn, which depletes the soil of nitrogen. Not only do legumes replenish the soil's nitrogen supply, they are an economically useful crop.

Recycling of nutrients

You learned that life could not exist if decomposing bacteria did not break down the organic materials in

a

b

c

Magnification: 21 500×

Figure 18.16
Soybean plants have swellings, called nodules, on their roots (a). The nodules (b) contain bacteria called *Rhizobium* (c) that convert nitrogen gas into ammonia. In this symbiotic association, the plant gains usable nitrogen, and the bacteria gain food.

509

Quick Demo

Bring to class some foods that require bacteria for processing. Include Swiss cheese, pickles, vinegar, sauerkraut, yogurt, and sour cream. Explain the importance of bacteria to each food.

Enrichment

Explain that some bacteria produce methane gas, which is the major component of natural gas. On the chalkboard, sketch the structural formula of methane and write its molecular formula, CH_4. Light a Bunsen burner to illustrate methane's properties. Tell students that natural gas companies add chemicals with odors that warn their customers of a fuel leak. Have students brainstorm advantages resulting from utilizing bacteria that produce methane. **L3**

PROJECT

Bacteria of Decay

Kinesthetic Have students model a landfill in a jar containing moist soil. Instruct them to bury carrot shavings, paper, aluminum foil, and a few other items that are in landfills. Make sure they do not bury any animal products.

After two weeks, have them unearth the items they buried and compare the way they look with their pre-burial appearances. Ask students to record their observations and use them to discuss the importance of the recycling that bacteria do.

Have students wear a lab apron, disposable latex gloves, and safety goggles when working on this project. Remind them to wash their hands thoroughly each time they work with the model landfill. **L2**

TechPrep

Food Preparation

**Kinesthetic** Use bacteria to make yogurt, sauerkraut, or cheese as a class project. Recipes can be found in cookbooks or on the Internet.
L1 **ELL**

DESIGN YOUR OWN BioLab

The BioLab at the end of the chapter can be used at this point in the lesson. ➡

3 Assess

Check for Understanding

Have students list the common characteristics of the major groups of bacteria. **L1**

Reteach

Linguistic Ask students to outline this section of the chapter using standard outline format. Beneath each of their outline heads, have students write a short summary of the section. **L2**

Resource Manager

Critical Thinking/Problem Solving, p. 18 **L3**

---

dead organisms and wastes, returning nutrients, both organic materials and inorganic materials, to the environment. Autotrophic bacteria and also plants and algae, which are at the bottom of the food chains, use the nutrients in the food they make.

This food is passed from one heterotroph to the next in food chains and webs. In the process of making food, many autotrophs replenish the supply of oxygen in the atmosphere. You can see from all this that other life depends on bacteria.

Food and medicines

Some foods that you eat—mellow Swiss cheese, shown in *Figure 18.17,* crispy pickles, tangy yogurt—would not exist without bacteria. During respiration, different bacteria produce diverse products, many of which have distinctive flavors and aromas. As a result, specific bacteria are used to make different foods, such as vinegar, cheeses, and sauerkraut.

In addition to food, some bacteria produce important antibiotics that destroy other types of bacteria. Streptomycin, erythromycin, bacitracin, and neomycin are some of these antibiotics. How do you know which antibiotic you need when you are sick? The *BioLab* at the end of this chapter will help you learn how scientists have obtained such information.

Figure 18.17
Bacteria not only give Swiss cheese (**a**) its flavor but also its holes as they produce carbon dioxide that bubbles through the cheese (**b**). Useful bacteria are grown in large industrial fermenting vats (**c**).

Cultural Diversity

Medicine in Latin America

Bacteriology is an important field of study in Latin America thanks in part to work conducted at Brazil's Institute of Experimental Pathology founded in 1902. Tell students about some twentieth century findings from these laboratories, including, in 1909, both Carlos Chagas's identification of the pathogen called trypanosomiasis that causes Chagas's disease (American sleeping sickness) and Alberto Barton's solution to Carrion's disease. Also discuss the work of Cuban scientist Carlos Finlay, a major contributor to the etiology and pathology of yellow fever.

Table 18.2 Diseases caused by bacteria

Disease	Transmission	Symptoms	Treatment
Strep throat	Inhale or ingest through mouth	Fever, sore throat, swollen neck glands	Antibiotic
Tuberculosis	Inhale	Fatigue, fever, night sweats, cough, weight loss, chest pain	Antibiotic
Tetanus	Puncture wound	Stiff jaw, muscle spasms, paralysis	Open and clean wound, antibiotic; give antitoxin
Lyme disease	Bite of infected tick	Rash at site of bite, chills, body aches, joint swelling	Antibiotic
Dental cavities (caries)	Bacteria in mouth	Destruction of tooth enamel, tooth ache	Remove and fill the destroyed area of tooth
Cholera	Drinking contaminated water	Diarrhea, vomiting, dehydration	Replace body fluids, antibiotics

Bacteria cause disease

Although only a few kinds of bacteria cause diseases, those that do greatly affect human lives. Bacteria cause about half of all human diseases, some of which you can see listed in *Table 18.2.* Disease-causing bacteria can enter human bodies through openings, such as the mouth. They are carried in air, food, and water and sometimes invade humans through skin wounds.

In the past, bacterial illnesses had a greater effect on human populations than they do now. As recently as 1900, life expectancy in the United States was only 47 years. The most dangerous diseases at that time were the bacterial illnesses tuberculosis and pneumonia. In the last 100 years, human life expectancy has increased to about 75 years. This increase is due to many factors, including better public health systems, improved water and sewage treatment, better nutrition, and better medical care. These improvements, along with antibiotics, have reduced the death rates from bacterial diseases to low levels. However, this is starting to change as you can read in the *Biology & Society* at the end of this chapter.

Extension

Linguistic Ask a group of students to write to the Centers for Disease Control and Prevention in Atlanta, Georgia, to get recent copies of the "Morbidity and Mortality Weekly Report" (MMWR). Have them summarize the information about diseases that are currently monitored in the United States and around the world. **L3**

✔ Assessment

Portfolio Have students make labeled diagrams of the bacteria they have studied in this chapter. Ask them to include captions that describe their adaptations. **L1** **P**

4 Close

Brainstorming

Ask students to summarize the importance of bacteria, including their diverse roles in the environment, in causing diseases, and in industry. **L2**

Resource Manager

Reinforcement and Study Guide, pp. 81-82 **L2**
Content Mastery, pp. 89, 91-92 **L1**

Section Assessment

Understanding Main Ideas
1. Describe six parts of a typical bacterial cell. State the function of each.
2. What are endospores? How do they help bacteria survive?
3. Explain how penicillin affects a bacterial cell.
4. Explain how bacteria avoid osmotic rupture.

Thinking Critically
5. Some scientists have proposed that bacterialike cells were probably among the earliest organisms to live on Earth. Draw up a list of reasons why such a suggestion is feasible. Then explain each reason on your list.

SKILL REVIEW
6. **Making and Using Tables** Construct a table comparing and contrasting archaebacteria and eubacteria. Include at least three ways they are alike and three ways they are different. For more help, refer to *Organizing Information* in the **Skill Handbook.**

Section Assessment

1. Chromosome has genetic information; plasma membrane regulates entry and exit; cell wall prevents osmotic rupture and gives shape; capsule protects; flagella move; pili help it adhere and conjugate.
2. Survival structure; endospores protect bacteria from dryness, chemicals, etc.
3. It interferes with the construction and repair of bacterial cell walls.
4. Cell walls protect them.
5. The oldest fossils are bacterialike. The prokaryotic cell is not complex. Today, some bacteria live in extreme conditions.
6. Differences: plasma membrane and cell wall structure, tRNA and rRNA base sequences, reactions to antibiotics. Similarities: prokaryotes, unicellular, one circular chromosome.

Time Allotment

Initial session: one class period; followup session: 15 minutes, 48 hours after lab is begun.

Process Skills

compare and contrast, observe and infer, recognize cause and effect, form a hypothesis, interpret data

Safety Precautions

Have students wear a lab apron, safety goggles, and disposable latex gloves during this lab. Students should wash their hands with soap after handling bacterial cultures. When students complete the lab, they should clean their work areas with disinfectant. Used swabs and petri dishes should be disposed of after autoclaving or by incineration.

If students are transferring bacteria from an agar culture, they should use sterile technique. Remind them that even though *E. Coli* is not pathogenic under normal conditions, it is wise to treat cultures with caution.

PREPARATION

■ Arrange for delivery of the culture of *E. coli* and the petri dishes shortly before the students will do the lab.
■ Purchase discs that contain antibiotics with which students are likely to be familiar.

Possible Hypotheses

If antibiotic disks are in a bacterial culture, no bacteria will grow. If no antibiotic disks are present, bacterial growth will be great.

How sensitive are bacteria to antibiotics?

*D*octors *must know which antibiotic kills each disease-causing bacterium. You can use a test similar to the one in this* BioLab *to provide this information. You will use sterile, agar-containing Petri dishes and sterile, antibiotic disks. When you place a disk on the agar, the antibiotic diffuses into the agar. A clear ring that develops around a disk—a zone of inhibition—is where the antibiotic killed sensitive bacteria.*

PREPARATION

Problem
How can you determine which antibiotic most effectively kills specific bacteria?

Hypotheses
Decide on one hypothesis that you will test. Your hypothesis might be that the antibiotic with the widest zone of inhibition most effectively kills bacteria.

Objectives
In this BioLab, you will:
■ **Compare** how effectively different antibiotics kill specific bacteria.
■ **Determine** the most effective antibiotic to treat an infection that these bacteria might cause.

Possible Materials
cultures of bacteria
sterile nutrient agar Petri dishes
antibiotic disks
sterile disks of blank filter paper
marking pen
long-handled cotton swabs
forceps
37°C incubator
metric ruler

Safety Precautions
Always wear goggles in the lab. Although the bacteria you will work with are not disease-causing, do not spill them. Wash your hands with soap immediately after handling any bacterial culture. Carefully clean your work area after you finish. Follow your teacher's instructions about disposal of your swabs, cultures, and Petri dishes.

Skill Handbook
Use the **Skill Handbook** if you need additional help with this lab.

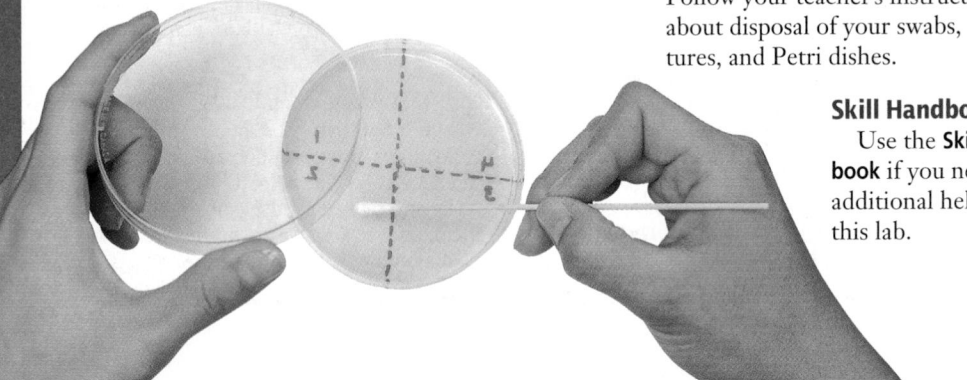

PLAN THE EXPERIMENT

Teaching Strategies
■ Show students a petri dish that you inoculated with *E. coli* and then incubated two days before they begin this lab.
■ Teach students sterile technique. Have them practice with a sterile loop on plain agar before working with live cultures. A loop and solid cultures are harder to use than a cotton swab and a liquid culture.

■ Have students hold the dish to the light to examine the clear zones of inhibition.
■ Explain that the *E. coli* that is a common, harmless intestinal dweller can cause disease in other body areas.

Possible Procedures
■ Students will likely divide the dish into four quadrants and place different

PLAN THE EXPERIMENT

1. Examine the materials provided by your teacher, and study the photos in this lab. As a group, make a list of ways you might investigate your hypothesis.
2. Agree on one way that your group could investigate your hypothesis. Design an experiment in which you can collect quantitative data.
3. Make a list of numbered directions. In your list, include the amounts of each material you will need. If possible, use no more than one Petri dish for each person.
4. Design and construct a table for recording data. To do this, carefully consider what data you need to record and how you will measure the data. For example, how will you measure what happens around the antibiotic disks as the antibiotic diffuses into the agar?

Check the Plan

Discuss the following points with other group members to decide on your final procedure.

1. How will you set up your Petri dishes? How many antibiotics can you test on one Petri dish? How will you measure the effectiveness of each antibiotic? What will be your control?
2. Will you add the bacteria or the antibiotic disks first?
3. What will you do to prevent other bacteria from contaminating the Petri dishes?
4. How often will you observe the Petri dishes?
5. *Make sure your teacher has approved your experimental plan before you proceed.*
6. Carry out your experiment. **CAUTION:** *Wash your hands with soap and water after handling dishes of bacteria.*

ANALYZE AND CONCLUDE

1. **Measuring in SI** How did you measure the zones of inhibition? Why did you do it this way?
2. **Drawing Conclusions** Suppose you were a physician treating a patient infected with these bacteria. Which antibiotic would you use? Why?
3. **Analyzing the Procedure** What limitations does this technique have? If these bacteria were infecting a person, what other tests might increase your confidence about treating the person with the antibiotic that appears most effective against these bacteria?

Going Further

Application Use a similar procedure to test the effectiveness of four commercial antibacterial soaps. Prepare your disks by soaking them in the different soap solutions.

BIOLOGY *Online* To find out more about antibiotics, visit the Glencoe Science Web site.
science.glencoe.com

antibiotic disks in each of three quadrants and a plain sterile disk in the fourth.
■ After incubating the dish for two days, students can measure the diameter of the clear areas that surround the disks.

Data and Observations

All antibiotic disks should show some inhibition. Results will vary depending on the antibiotic disks used. The untreated disks should show no zone of inhibition.

ANALYZE AND CONCLUDE

1. Most students will measure the diameter of the zone of inhibition and use the value as a basis of comparison.
2. Use the one with the largest zone of inhibition, which best inhibits bacterial growth.
3. The antibiotic may work in an agar culture, but not in human tissue. A doctor might also do blood tests to examine the kinds of white blood cells present, or look for the bacteria themselves.

✔ Assessment

Portfolio Ask students to write a paragraph about the following event: A man arrives at his doctor's office with a bacterial infection in a cut. The doctor can prescribe one of ten antibiotics. How can she learn which antibiotic will work best for the patient? Use the Performance Task Assessment List for Writing in Science in **PASC,** p. 87. **L2**

Going Further

Students can use liquid soaps or a solution of distilled water and shavings of bar soap. They may need to vary the concentrations of the soap solutions.

Resource Manager

BioLab and MiniLab Worksheets, pp. 87-88 **L2**

Purpose

Students learn why drug-resistant bacteria develop and how this threatens public health.

Background

Low levels of antibiotics in the environment promote natural selection for resistant bacteria. Without antibiotics, larger populations of nonresistant bacteria thrive because they usually outcompete resistant strains. Therefore, the presence of nonresistant bacteria might help slow the development of drug resistance.

Teaching Strategies

■ Remind students that common antibiotics include penicillin, ampicillin, and tetracycline.

■ Explain that patients who take fewer than the prescribed number of daily pills maintain low drug levels in their bodies that kill nonresistant cells.

■ Tell students that some bacteria produce substances called bacteriocins that are toxic to closely related bacteria. Sometimes bacteriocins can replace antibiotics to control bacteria that can cause food poisoning.

■ Explain that responsible use can prolong the effectiveness of current antibiotics while new ones are being tested.

Investigating the Issue

Their use maintains a low level of antibacterial agent and may promote the natural selection of resistant bacteria.

Super Bugs Defy Drugs

You have strep throat—a bacterial infection. After taking an antibiotic for six days, you feel fine, so you stop taking it. A few weeks later, you have strep throat again. Your doctor prescribes the same antibiotic. But this time the sore throat doesn't go away. You have to take a different antibiotic to get rid of the infection.

Antibiotics have prevented millions of deaths from bacterial diseases in the past. Today, however, antibiotics do not always cure disease because many disease-causing bacteria have developed resistance to many antibiotics.

Different Viewpoints

During the past 50 years, antibiotics have been used for preventive medical reasons and in agriculture. With the development of resistant bacteria, these uses are being reassessed.

How Much Is Too Much? Because antibiotics have worked well and had few side effects, some physicians prescribe them for preventive reasons. For example, physicians may prescribe antibiotics before surgery to prevent the chance of infection from bacteria during the surgery. In addition, some physicians prescribe antibiotics for patients with viral infections because a viral infection makes a body vulnerable to a bacterial infection.

Because antibiotics hasten the growth of healthy cattle, chickens, and other domestic animals, many animal feeds contain small amounts of antibiotics. Similarly, antibiotics are used to coat fruit and other agricultural products. These antibiotics may produce resistant bacteria, which pass to people when they eat the food.

A Public Health Crisis? More than 100 antibiotics are available, and many bacteria that they once killed are now resistant to one or more of them. Tuberculosis, for example, is a deadly, highly contagious disease that a combination of antibiotics usually treats effectively. But strains of resistant tuberculosis bacteria have appeared, and the number of deaths from the disease are beginning to increase.

Staphylococcus aureus and *Enterococcus faecalis* are common bacteria. *S. aureus* causes serious

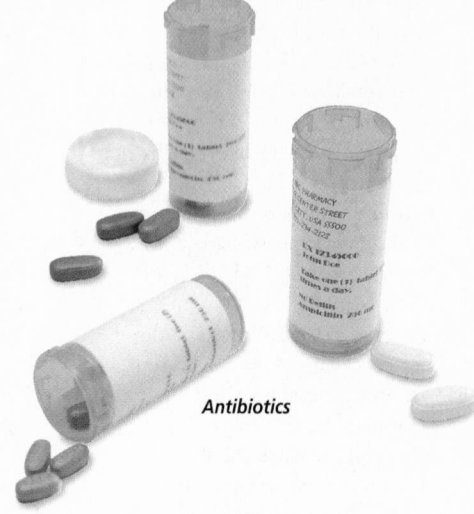

Antibiotics

infections in hospital patients and is beginning to show resistance to the last antibiotic effective against it—vancomycin. The body's immune system usually controls *E. faecalis.* But in patients with weakened immune systems, this bacterium causes life-threatening illness. Like *S. aureus*, it is resistant to most antibiotics. Bacteria often pass traits among species, and health workers fear that *S. aureus* could pass its vancomycin resistance to *E. faecalis.*

INVESTIGATING THE ISSUE

Thinking Critically Antibacterial products—from soaps and lotions to kitchen cutting boards and sponges—cram market shelves. Is it a good idea to rely on them to keep kitchens and bathrooms free of harmful bacteria? Why or why not?

BIOLOGY *Online* To find out more about bacteria that are antibiotic-resistant, visit the Glencoe Science Web site. **science.glencoe.com**

Going Further

Encourage students to explore recent developments in the use of bacteriocins to battle food-borne bacteria, the search for new antibiotics, and the occurrence of drug-resistant strains of disease.

SUMMARY

Section 18.1

Viruses

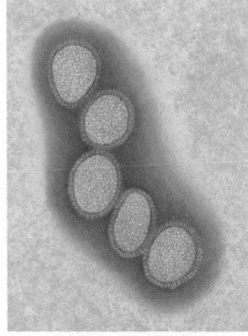

Main Ideas

- Viruses are nonliving particles that have a nucleic acid core and a protein-containing capsid.

- To replicate, a virus must first recognize a host cell, then attach to it, and finally enter the host cell and take over its metabolism.

- During a lytic cycle, a virus replicates and kills the host cell. In a lysogenic cycle, a virus's DNA is integrated into a chromosome of the host cell, but the host cell does not die.

- Retroviruses contain RNA. Reverse transcriptase is an enzyme that helps convert viral RNA to DNA, which is then integrated into the host cell's chromosome.

- Viruses probably originated from their host cells.

Vocabulary

bacteriophage (p. 490)
capsid (p. 490)
host cell (p. 489)
lysogenic cycle (p. 493)
lytic cycle (p. 493)
provirus (p. 493)
retrovirus (p. 495)
reverse transcriptase (p. 495)
virus (p. 489)

Section 18.2

Archaebacteria and Eubacteria

Main Ideas

- There are two kingdoms of prokaryotes: archaebacteria and eubacteria. Archaebacteria inhabit extreme environments. Eubacteria live almost everywhere else. They probably arose separately from a common ancestor billions of years ago.

- Some bacteria are heterotrophs, some are photosynthetic autotrophs, and others are chemosynthetic autotrophs. Some bacteria are obligate aerobes, some obligate anaerobes, and some are both aerobes and anaerobes.

- Bacteria usually reproduce by binary fission. Some have a type of sexual reproduction called conjugation. Some bacteria form endospores that enable them to survive when conditions are unfavorable.

- Bacteria fix nitrogen, recycle nutrients, and help make food products and medicines. Some bacteria cause diseases.

Vocabulary

binary fission (p. 504)
chemosynthesis (p. 502)
conjugation (p. 506)
endospore (p. 507)
nitrogen fixation (p. 509)
obligate aerobe (p. 507)
obligate anaerobe (p. 507)
toxin (p. 508)

Main Ideas

Summary statements can be used by students to review the major concepts of the chapter.

Using the Vocabulary

To reinforce chapter vocabulary, use the Content Mastery Booklet and the activities in the Interactive Tutor for Biology: The Dynamics of Life on the Glencoe Science Web site.
science.glencoe.com

THE PRINCETON REVIEW — *All Chapter Assessment questions and answers have been validated for accuracy and suitability by The Princeton Review.*

UNDERSTANDING MAIN IDEAS

1. d
2. b

UNDERSTANDING MAIN IDEAS

1. A _____ is never a part of a virus.
  a. nucleic acid   **c.** viral envelope
  b. protein coat   **d.** cell wall

2. The cell walls of bacteria _____.
  a. control what enters and leaves the cell
  b. prevent osmotic rupture
  c. are involved in penicillin synthesis
  d. are involved in protein synthesis

CHAPTER 18 ASSESSMENT **515**

GLENCOE TECHNOLOGY

VIDEOTAPE
MindJogger Videoquizzes
Chapter 18: *Viruses and Bacteria*
Have students work in groups as they play the videoquiz game to review key chapter concepts.

Resource Manager

Chapter Assessment, pp. 103-108
MindJogger Videoquizzes
ExamView® Pro Software 💾
BDOL Interactive CD-ROM, Chapter 18 quiz

3. c
4. c
5. b
6. d
7. a
8. d
9. d
10. a
11. RNA, DNA
12. *Escherichia coli*
13. lytic
14. Archaebacteria
15. staphylo-
16. binary fission
17. smallpox
18. cell wall
19. protein, nucleic acid
20. conjugation

APPLYING MAIN IDEAS

21. Bacteria recycle nutrients for living things.
22. Prokaryotic cells do not have membrane-bound organelles and have one circular chromosome. Eukaryotic cells have membrane-bound organelles and linear chromosomes.
23. Bacteria may run out of nutrients, be eaten by predators, or dehydrate.
24. Scientists cannot study them or conduct experiments to discover their characteristics and requirements.

3. Which of the following is NOT a common bacterial shape?

a.    b.   c. 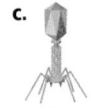   d.

4. What characteristic do viruses share with all living organisms?
   a. respiration   c. replication
   b. metabolism   d. movement

5. During a lytic cycle, after a virus enters the cell, the virus _____.
   a. forms a provirus   c. dies
   b. replicates   d. becomes inactive

6. Prokaryotic cells have _____.
   a. organelles   c. mitochondria
   b. a nucleus   d. a cell wall

7. In _____, bacteria convert gaseous nitrogen into ammonia, nitrates, and nitrites.
   a. nitrogen fixation   c. conjugation
   b. binary fission   d. attachment

8. Bacteria that require _____ for respiration are called _____.
   a. food—obligate saprophytes
   b. hydrogen—archaebacteria
   c. oxygen—obligate anaerobes
   d. oxygen—obligate aerobes

9. Some bacteria, when faced with unfavorable environmental conditions, produce structures called _____.
   a. pili   c. toxins
   b. capsules   d. endospores

10. Which of the following would be most likely to live in Utah's Great Salt Lake?
    a. Archaebacteria   c. Eubacteria
    b. staphylococci   d. viruses

 **THE PRINCETON REVIEW**   **TEST-TAKING TIP**

Investigate
Ask what kinds of questions to expect on the test. Ask for practice tests so that you can become familiar with the test-taking materials.

11. The nucleic acid core of a virus contains _____or _____.
12. Viruses are species specific so that the T4 bacteriophage can infect only organisms known as _____.
13. In the _____ cycle, viruses use the cell's energy and raw materials to copy themselves, then burst from the cell.
14. _____ are the prokaryotes that have genes most similar to those of eukaryotes.
15. _____ is a prefix that describes the pattern of growth of the bacteria shown here.

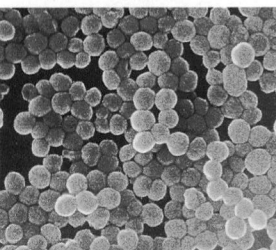

16. _____ results in two bacteria, each genetically identical to the original.
17. The World Health Organization has successfully eradicated the disease _____ from the world's population.
18. Penicillin kills bacteria by interfering with the enzymes that link the sugar chains in the _____.
19. All viruses contain a coat of _____ and core of _____.
20. Some bacteria have a form of sexual reproduction called _____.

APPLYING MAIN IDEAS

21. Why are bacteria essential to life?
22. Discuss two ways that prokaryotic cells differ from eukaryotic cells.
23. Discuss three factors that limit bacterial growth.
24. Scientists cannot grow about 99 percent of all bacteria in the laboratory. How might this inability interfere with understanding bacteria?

THINKING CRITICALLY

25. Applying Concepts Although bacteria grow on nutrient agar in a laboratory, viruses will not grow on agar. What kind of substance would you need to grow viruses in the laboratory? Explain your answer.

26. Observing and Inferring If you were offered the choice of either a million dollars or a sum of money equal to a penny that doubles every day for 64 days in a row, which would you choose? Relate your choice to the growth rate of bacteria cells.

27. Interpreting Scientific Illustrations Bacteria from an infected person were tested for their sensitivity to three antibiotics. The results of the test are shown in the petri dish at right. If you were the patient's physician, which antibiotic would you prescribe and why?

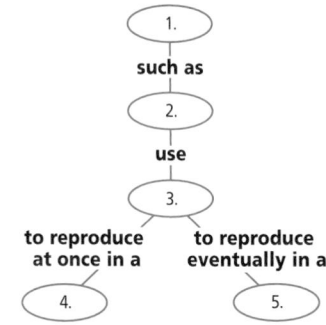

28. Concept Mapping Complete the concept map by using the following vocabulary terms: host cells, viruses, lysogenic cycle, bacteriophages, lytic cycle.

```
        ┌──────┐
        │  1.  │
        └──────┘
       such as
        ┌──────┐
        │  2.  │
        └──────┘
         use
        ┌──────┐
        │  3.  │
        └──────┘
  to reproduce      to reproduce
   at once in a     eventually in a
   ┌──────┐          ┌──────┐
   │  4.  │          │  5.  │
   └──────┘          └──────┘
```

💿 CD-ROM

For additional review, use the assessment options for this chapter found on the *Biology: The Dynamics of Life Interactive CD-ROM* and on the Glencoe Science Web site.
science.glencoe.com

ASSESSING KNOWLEDGE & SKILLS

One milliliter of *E. coli* culture was added to each of three Petri dishes (A, B, and C). The dishes were incubated for 36 hours, and then the number of bacterial colonies on each were counted.

Table 18.3 Growth of *E. coli* under various conditions		
Petri dish number	Medium	Colonies per dish
I	Agar and carbohydrates	35
II	Agar, carbohydrates, and vitamins	250
III	Agar and vitamins	0

Interpreting data Study the table and answer the following questions.

1. Which of the above dishes demonstrate that carbohydrates are necessary for the growth of *E. coli*?
a. Dish I alone
b. Dishes I and II
c. Dishes I and III
d. Dish III

2. Which of the above dishes demonstrate that vitamins enhance the growth of *E. coli*?
a. Dishes I and II
b. Dishes II and III
c. Dishes I and III
d. None of the dishes

3. Which of the following is a variable in this experiment?
a. *E. coli*
b. agar
c. carbohydrates
d. number of colonies

4. Making a Graph Construct a bar graph from the data in the above table.

THINKING CRITICALLY

25. Cell cultures because viruses need to grow and reproduce inside living cells.
26. The penny would yield more than $1 million in 64 days. Bacteria increase at an exponential rate.
27. Antibiotic 3; it shows the largest zone of inhibition.
28. 1. Viruses; 2. Bacteriophages; 3. Host cells; 4. Lytic cycle; 5. Lysogenic cycle

ASSESSING KNOWLEDGE & SKILLS

1. c
2. a
3. c
4. Student bar graphs should match the data in the table.

Chapter 19 Organizer

Refer to pages 4T-5T of the Teacher Guide for an explanation of the National Science Education Standards correlations.

Section	Objectives	Activities/Features
Section 19.1 **The World of Protists** National Science Education Standards UCP.1, UCP.2, UCP.5; A.1, A.2; C.1, C.4, C.5, C.6; F.1, F.4, F.5 (1 session, 1 block)	1. **Identify** the characteristics of Kingdom Protista. 2. **Compare and contrast** the four groups of protozoans.	**MiniLab 19-1:** Observing Ciliate Motion, p. 522 **Inside Story:** A Paramecium, p. 523 **Problem-Solving Lab 19-1,** p. 524
Section 19.2 **Algae: Plantlike Protists** National Science Education Standards UCP.1, UCP.2, UCP.5; A.1, A.2; C.1, C.4, C.5, C.6; F.1, F.4, F.5 (2 sessions, 1 block)	3. **Compare and contrast** the variety of plantlike protists. 4. **Explain** the process of alternation of generations in algae.	**MiniLab 19-2:** Going on an Algae Hunt, p. 527 **Problem-Solving Lab 19-2,** p. 530 **Design Your Own BioLab:** How do *Paramecium* and *Euglena* respond to light? p. 538
Section 19.3 **Slime Molds, Water Molds, and Downy Mildews** National Science Education Standards UCP.1, UCP.2, UCP.4, UCP.5; A.1, A.2; C.1, C.3-5, C.6; F.1, F.4, F.5; G.1-3 (2 sessions, 1/2 block)	5. **Contrast** the cellular differences and life cycles of the two types of slime molds. 6. **Discuss** the economic importance of the downy mildews and water molds.	**Problem-Solving Lab 19-3,** p. 534 **Social Studies Connection:** The Irish Potato Famine, p. 540

Need Materials? Contact Carolina Biological Supply Company at 1-800-334-5551 or at **http://www.carolina.com**

MATERIALS LIST

BioLab

p. 538 microscope, microscope slides, coverslips, dropper, metric ruler, index cards, scissors, toothpicks, methyl cellulose solution, *Euglena* culture, *Paramecium* culture

MiniLabs

p. 522 *Paramecium* culture, wheat seeds, microscope, microscope slide, coverslip, dropper

p. 527 microscope, microscope slide, coverslip, dropper, pond water, paper, pencil

Alternative Lab

p. 534 microscope, sterile agar plate, oat cereal flakes, plasmodium stage of *Physarum polycephalum*

Key to Teaching Strategies

L1 Level 1 activities should be appropriate for students with learning difficulties.

L2 Level 2 activities should be within the ability range of all students.

L3 Level 3 activities are designed for above-average students.

ELL ELL activities should be within the ability range of English Language Learners.

COOP LEARN Cooperative Learning activities are designed for small group work.

P These strategies represent student products that can be placed into a best-work portfolio.

These strategies are useful in a block scheduling format.

Protists

Teacher Classroom Resources

Section	Reproducible Masters	Transparencies
Section 19.1 **The World of Protists**	Reinforcement and Study Guide, p. 83 `L2` Critical Thinking/Problem Solving, p. 19 `L3` BioLab and MiniLab Worksheets, p. 89 `L2` Laboratory Manual, pp. 133-136 `L2` Content Mastery, pp. 93-94, 96 `L1`	Section Focus Transparency 45 `L1` `ELL` Basic Concepts Transparency 30 `L2` `ELL` Reteaching Skills Transparency 29 `L1` `ELL` Reteaching Skills Transparency 30 `L1` `ELL`
Section 19.2 **Algae: Plantlike Protists**	Reinforcement and Study Guide, pp. 84-85 `L2` BioLab and MiniLab Worksheets, p. 90 `L2` Laboratory Manual, pp. 137-140 `L2` Tech Prep Applications, pp. 27-28 `L2` Content Mastery, pp. 93, 95-96 `L1`	Section Focus Transparency 46 `L1` `ELL` Basic Concepts Transparency 28 `L2` `ELL` Basic Concepts Transparency 30 `L2` `ELL`
Section 19.3 **Slime Molds, Water Molds, and Downy Mildews**	Reinforcement and Study Guide, p. 86 `L2` Concept Mapping, p. 19 `L3` `ELL` BioLab and MiniLab Worksheets, pp. 91-92 `L2` Content Mastery, pp. 93, 95-96 `L1`	Section Focus Transparency 47 `L1` `ELL` Basic Concepts Transparency 29 `L2` `ELL` Basic Concepts Transparency 30 `L2` `ELL`

Assessment Resources

Chapter Assessment, pp. 109-114
MindJogger Videoquizzes
Performance Assessment in the Biology Classroom
Alternate Assessment in the Science Classroom
ExamView® Pro Software
BDOL Interactive CD-ROM, Chapter 19 quiz

Additional Resources

Spanish Resources `ELL`
English/Spanish Audiocassettes `ELL`
Cooperative Learning in the Science Classroom `COOP LEARN`
Lesson Plans/Block Scheduling

NATIONAL GEOGRAPHIC
Teacher's Corner

Products Available From Glencoe
To order the following products, call Glencoe at 1-800-334-7344:
CD-ROM
NGS PictureShow: The Cell

Products Available From National Geographic Society
To order the following products, call National Geographic Society at 1-800-368-2728:
Video
Protists: Threshold of Life

Index to National Geographic Magazine
The following articles may be used for research relating to this chapter:
"Slime Mold: The Fungus That Walks," by Lee Douglas, July 1981.

GLENCOE TECHNOLOGY

The following multimedia resources are available from Glencoe.

Biology: The Dynamics of Life
CD-ROM `ELL`

Exploration: *The World of Protists*
BioQuest: *Biodiversity Park*
Video: *Protists*
Video: *Kelp Forests*
Video: *Slime Mold*

Videodisc Program

Protists
Kelp Forests
Slime Mold

19 Protists

Theme Development

The theme of **unity within diversity** is evident throughout the chapter in discussions of the characteristics of protists. The theme of **homeostasis** is prominent in discussions of how the different protists carry out their life functions.

0:00 OUT OF TIME?

If time does not permit teaching the entire chapter, use the BioDigest at the end of the unit as an overview.

READING BIOLOGY

Glencoe's *Biology: The Dynamics of Life* contains many resources to assist a student's reading skills. Each chapter contains figures with expanded captions that expand on written material. Word Origins, located along the side of text, expand knowledge of biology vocabulary. Glencoe's Content Mastery Booklet helps develop reading skills while reinforcing content. In addition, use the Interactive Tutor for *Biology: The Dynamics of Life* on the Glencoe Web site to reinforce vocabulary.
science.glencoe.com

What You'll Learn

- You will differentiate among the major groups of protists.
- You will recognize the ecological niches of protists.
- You will identify some human diseases and the protists responsible for them.

Why It's Important

Because protists are responsible for much of the oxygen in the atmosphere, and are the base for most food chains in aquatic environments, most other organisms depend on protists to exist.

READING BIOLOGY

Scan the chapter, noting the illustrations of several protists. For each figure, write the name of the protist featured and a summary of the illustration. As you read through the text, add notes to your summaries.

BIOLOGY Online

To find out more about the protists, visit the Glencoe Science Web site.
science.glencoe.com

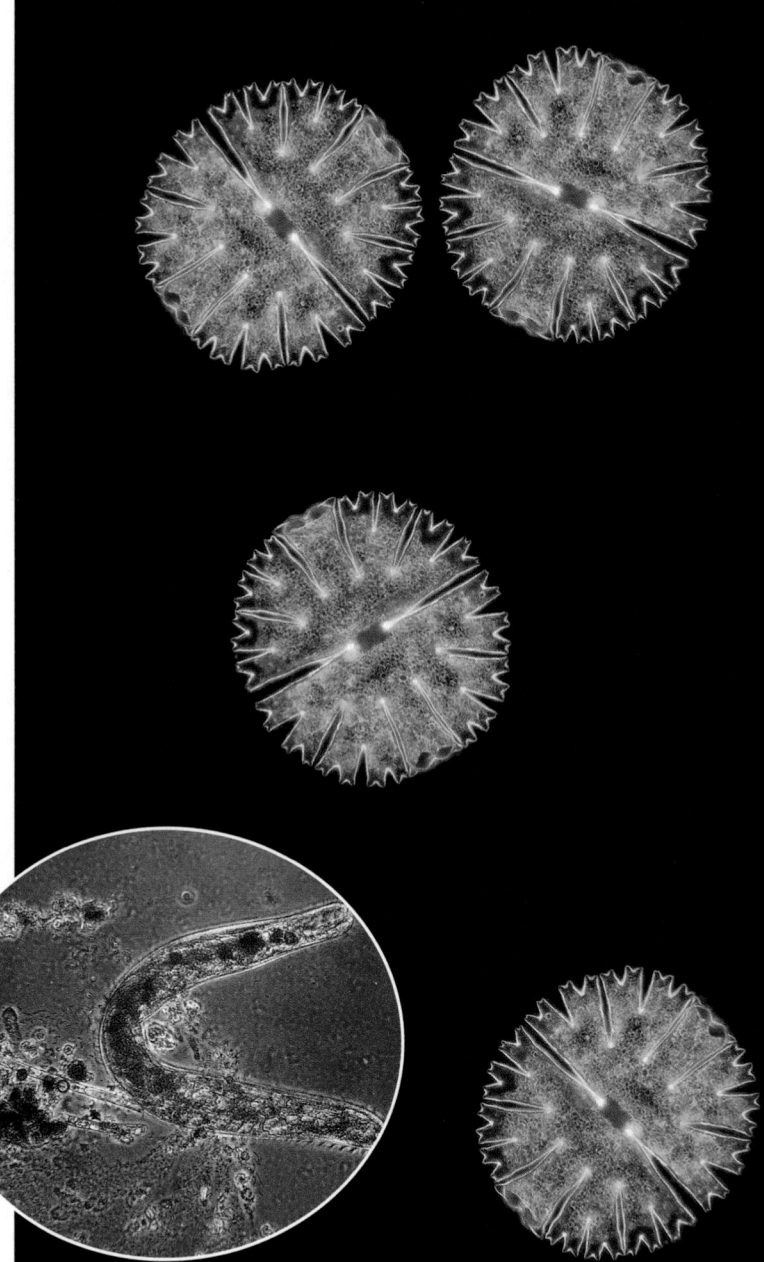

Magnification: 300×

Magnification: 475×

Desmids are plantlike protists that produce much of the oxygen you breathe. The *Spirostomum ambiguum* (inset) is an animal-like protist.

518 PROTISTS

19.1 The World of Protists

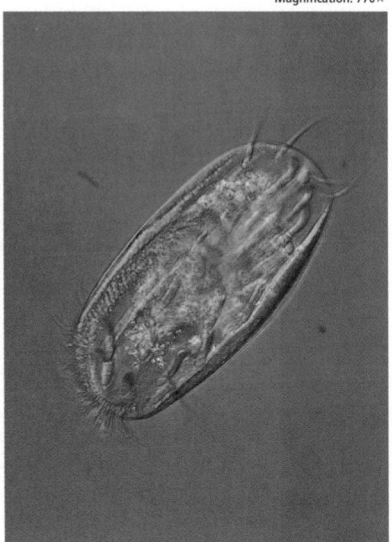

Magnification: 770×

A *Stylonchia* searching for food

I*n just a few drops of pond water, you can find an amazing collection of protists. Some protists will be moving, perhaps searching for food. Others will be using photosynthesis to make food. Still others will be decomposing organic matter in the pond water. In this section, you will read about the great diversity of protists, and why, in spite of this diversity, biologists group them together in Kingdom Protista.*

What Is a Protist?

Kingdom Protista contains the most diverse organisms of all the kingdoms. Protists may be unicellular or multicellular, microscopic or very large, and heterotrophic or autotrophic. In fact, there is no such organism as a typical protist. When you look at different protists, you may wonder how they could be grouped together. The characteristic that all protists share is that, unlike bacteria, they are all eukaryotes, which means that most of their metabolic processes occur inside their membrane-bound organelles.

Although there are no typical protists, some resemble animals in their method of nutrition. The animal-like protists are called **protozoa** (proht uh ZOH uh) (singular, protozoan). Unlike animals, though, all protozoans are unicellular. Other protists are plantlike autotrophs, using photosynthesis to make their food. Plantlike protists are called **algae** (AL jee) (singular, alga). Unlike plants, algae do not have organs such as roots, stems, and leaves. Still other protists are more like fungi because they decompose dead organisms. However, unlike fungi, funguslike protists are able to move at some point in their life and do not have chitin in their cell walls.

It might surprise you to learn how much protists affect other organisms. Some protists cause diseases, such as malaria and sleeping sickness, that

WORD *Origin*

protozoa
From the Greek words *protos*, meaning "first," and *zoa*, meaning "animals." Protozoa are animal-like protists.

2 Teach

Activity

 **Visual-Spatial** After students have read about the characteristics of protists, have them diagram a generalized protist cell. They should draw their diagrams to scale and label the cell organelles. Then have students draw and label a bacterial cell for comparison. Although this activity will be difficult, it should initiate a useful discussion about the characteristics of protists. **L3** **ELL**

Tying to Previous Knowledge

Ask students to compare and contrast the terms in the word pairs that follow: eukaryotes and prokaryotes, heterotrophs and autotrophs, and motile and non-motile. *Eukaryotes have membrane-bound organelles; the prokaryotes lack membrane-bound organelles. Heterotrophs take in food from the environment; autotrophs produce their own food. A motile organism can move; a nonmotile organism cannot move.* **L2** **ELL**

NATIONAL GEOGRAPHIC

VIDEODISC
STV: The Cell
Amoeba Changing Shape

Amoeba With Pseudopodia

result in millions of human deaths throughout the world every year. Unicellular algae produce much of the oxygen in Earth's atmosphere and are the basis of aquatic food chains. Slime molds and water molds decompose a significant amount of organic material, making the nutrients available to living organisms. Protozoans, algae, and funguslike protists play important roles on Earth. Look at *Figure 19.1* to see some protists.

WORD Origin
pseudopodia
From the Greek words *pseudo*, meaning "false," and *podos*, meaning "foot." An amoeba uses psuedopodia to obtain food.

What Is a Protozoan?

If you sat by a pond, you might notice clumps of dead leaves at the water's edge. Under a microscope, a piece of those wet decaying leaves reveals a small world, probably inhabited by animal-like protists. Although a diverse group, all protozoans are unicellular heterotrophs that feed on other organisms or dead organic matter. They usually reproduce asexually, and some also reproduce sexually.

Diversity of Protozoans

Many protozoans are grouped according to the way they move. Some protozoans use cilia or flagella to move. Others move and feed by sending out cytoplasm-containing extensions of their plasma membranes. These extensions are called **pseudopodia** (sewd uh POHD ee uh). Other protozoans are grouped together because they are parasites. There are four main groups of protozoans: the amoebas (uh MEE buz), the flagellates, the ciliates, and the sporozoans (spor uh ZOH unz).

Amoebas: Shapeless protists

The phylum Rhizopoda includes hundreds of species of amoebas and amoebalike organisms. Amoebas have no cell wall and form pseudopodia to move and feed. As a pseudopod forms, the shape of the cell changes and the amoeba moves. Amoebas form pseudopodia around their food, as you can see in *Figure 19.2*.

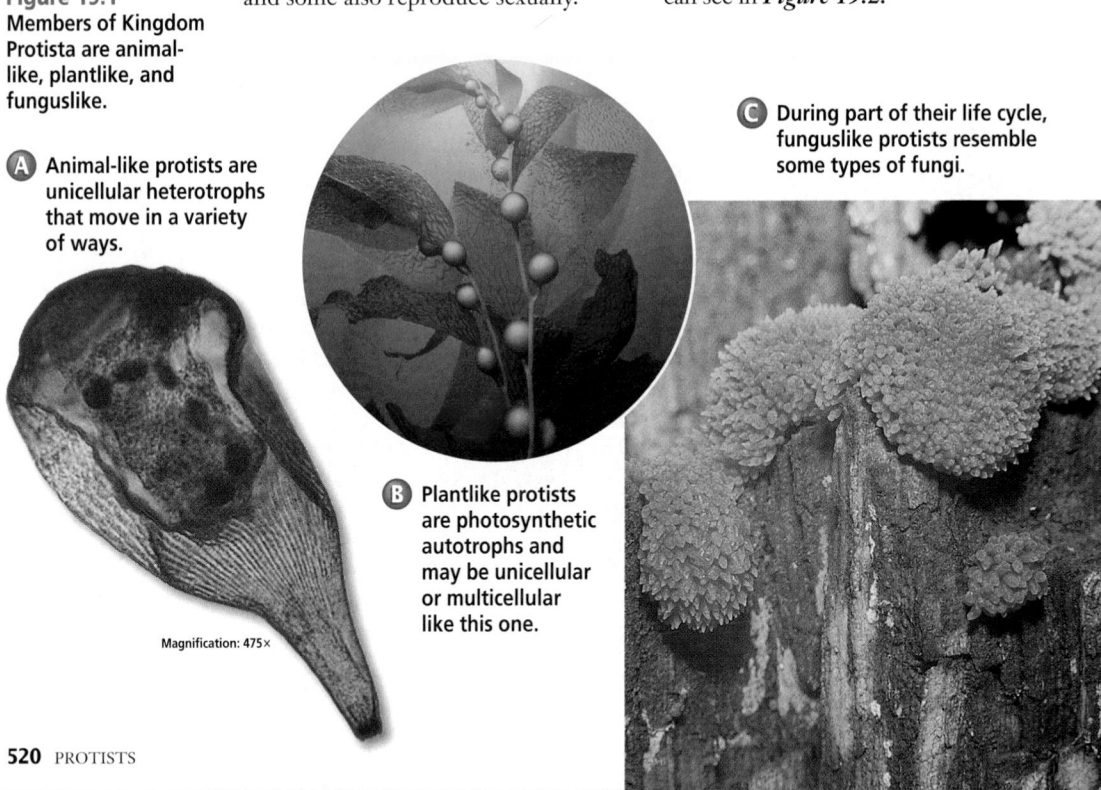

Figure 19.1
Members of Kingdom Protista are animal-like, plantlike, and funguslike.

A Animal-like protists are unicellular heterotrophs that move in a variety of ways.

Magnification: 475×

B Plantlike protists are photosynthetic autotrophs and may be unicellular or multicellular like this one.

C During part of their life cycle, funguslike protists resemble some types of fungi.

MEETING INDIVIDUAL NEEDS

Visually Impaired

Kinesthetic Ask students to build tactile models to demonstrate the difference between cilia and flagella. Velcro can model cilia, and pieces of string taped to a ball can simulate flagella. **L1** **ELL**

Portfolio

A Moving Analogy

Intrapersonal Have the students describe or draw some everyday items that are analogous in structure, appearance, or function to protozoan structures, such as cilia, flagella, and pseudopodia. **L1** **P**

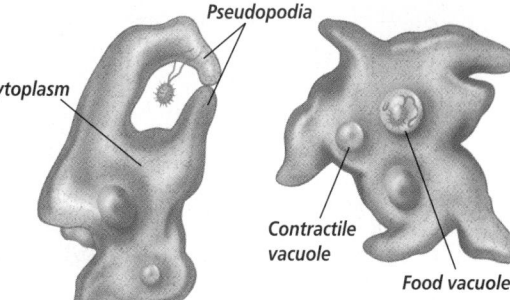

Figure 19.2
An amoeba feeds on small organisms such as bacteria.

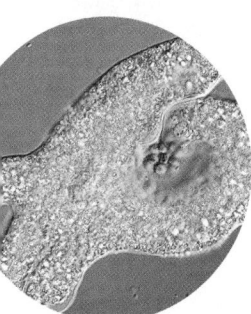

Magnification: 315×

Nucleus

Cytoplasm

Pseudopodia

Contractile vacuole

Food vacuole

(A) As an amoeba approaches food, pseudopodia form and eventually surround the food.

(B) The food becomes enclosed in a food vacuole.

(C) Digestive enzymes break down the food, and the nutrients diffuse into the cytoplasm.

Although most amoebas live in saltwater, there are freshwater ones that live in the ooze of ponds, in wet patches of moss, and even in moist soil. Because amoebas live in moist places, nutrients dissolved in the water around them can diffuse directly through their cell membranes. However, because freshwater amoebas live in hypotonic environments, they constantly take in water. Their contractile vacuoles collect and pump out excess water.

Two groupings of mostly marine amoebas, the foraminiferan and radiolarian shown in *Figure 19.3*, have shells. Foraminiferans, which are abundant on the sea floor, have hard shells made of calcium carbonate. Fossil forms of these protists help geologists determine the ages of some rocks and sediments. Unlike foraminiferans, radiolarians have shells made of silica. Under a microscope, you can see the complexity of these shells. In addition, radiolarians are an important part of marine plankton—an assortment of microscopic organisms that float in the ocean's photic zone and form the base of marine food chains.

Most amoebas commonly reproduce by **asexual reproduction**, in which a single parent produces one or

Figure 19.3
Foraminiferans **(a)** and radiolarians **(b)** are amoebas that extend pseudopodia through tiny holes in their shells. Pseudopodia act like sticky nets that trap food.

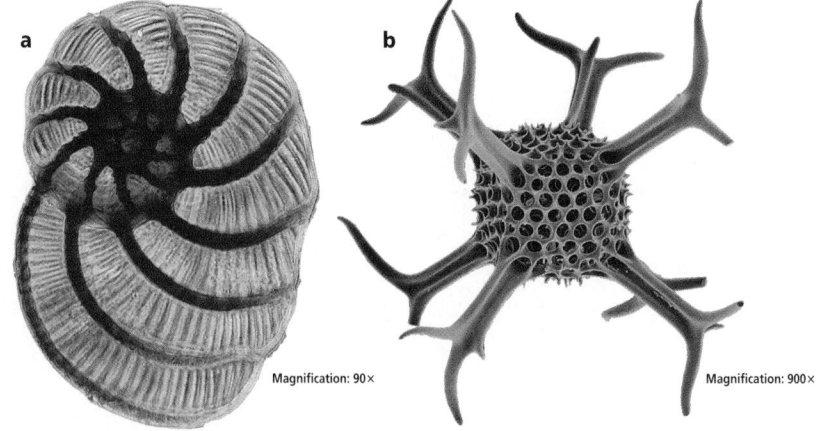

a

b

Magnification: 90×

Magnification: 900×

Purpose

Students will discover how *Paramecium* responds when it contacts a solid object.

Process Skills

observe and infer, draw a conclusion

Teaching Strategies

■ Using methyl cellulose to slow the protists interferes with *Paramecium's* normal response.

■ Toothbrush bristles or small pieces of cotton may be used to block the path of *Paramecium*.

■ Remind students to use caution when working with microscopes, slides, and coverslips.

■ Have students wash their hands after handling cultures.

Expected Results

Paramecium typically reverses the direction in which its cilia are beating and backs away from a solid object that it contacts.

Analysis

1. It backs up, then proceeds forward in a new direction.
2. briefly
3. The cell turned on its long axis, and it folded.

✔ Assessment

Portfolio Have students draw what occurs when *Paramecium* encounters solid objects and write captions for the events. Use the Performance Task Assessment List for Scientific Drawing in **PASC**, p. 55. **L2**
P

Resource Manager

BioLab and MiniLab Worksheets, p. 89 **L2**
Reteaching Skills Transparency 29 and Master **L1** **ELL**

522

MiniLab 19-1 — Observing and Inferring

Observing Ciliate Motion The cilia on the surface of a paramecium move so that the cell normally swims through the water with one end directed forward. But when this end bumps into an obstacle, the paramecium responds by changing direction.

Observing Paramecium

Procedure 🗣️ 🔬 🖐️

1. Observe a *Paramecium* culture that has had boiled, crushed wheat seeds in it for several days.
2. Carefully place a drop of water containing wheat seed particles on a microscope slide. Gently add a coverslip.
3. Using low power, locate a paramecium near some wheat seed particles. **CAUTION: Use caution when working with a microscope, glass slides, and coverslips.**
4. Watch the paramecium as it swims around among the particles. Record your observations of the organism's responses each time it contacts a particle.

Analysis

1. Describe what a paramecium does when it encounters an obstacle.
2. How long does the paramecium's response last?
3. Describe any changes in the shape of the paramecium as it moved among the particles.

---

more identical offspring by dividing into two cells. When environmental conditions become unfavorable, some types of amoebas form cysts that can survive extreme conditions.

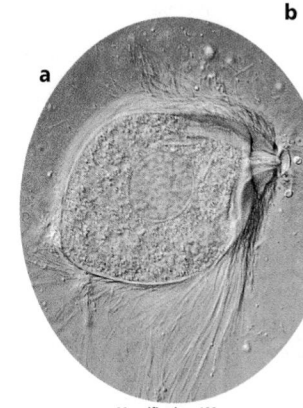

Figure 19.4
The flagellated protozoans (a) that live in the guts of termites (b) produce enzymes that digest wood, making nutrients available to their hosts.

522 PROTISTS

Magnification: 100×

Flagellates: Protozoans with flagella

The phylum Zoomastigina consists of protists called **flagellates,** which have one or more flagella. Flagellated protists move by whipping their flagella from side to side.

Some flagellates are parasites that cause diseases in animals, such as African sleeping sickness in humans. Other flagellates are helpful. For example, termites like those you see in *Figure 19.4* survive on a diet of wood. Without the help of a certain species of flagellate that lives in the guts of termites, some termites could not survive on such a diet. In a mutualistic relationship, flagellates convert cellulose from wood into a carbohydrate that both they and their termite hosts can use.

Ciliates: Protozoans with cilia

The roughly 8000 members of the protist phylum Ciliophora, known as **ciliates,** use the cilia that cover their bodies to move. Use the *MiniLab* on this page to observe a typical ciliate's motion. Ciliates live in every kind of aquatic habitat—from ponds and streams to oceans and sulfur springs. What does a typical ciliate look like? To find out, look at the *Inside Story* on the next page.

MEETING INDIVIDUAL NEEDS

Gifted

Kinesthetic Ask students to place a termite on a glass slide, grasp its head with a forceps, and gently pull to separate the head and intestines from the body. Have students add a few drops of Ringer's solution or water to the termite intestines, add a coverslip, and press gently on the coverslip to squash the intestines. Remind them to handle glass microscope slides and coverslips carefully and to use special care when viewing slides under high power. Then, ask students to observe the slide under low- and high-power magnification, looking for the termite's intestinal flagellates. Ask them to diagram and label the flagellates they see and place their drawings in their portfolios. **L3** **ELL**

A Paramecium

Paramecia are unicellular organisms, but their cells are quite complex. Within a paramecium are many organelles and structures that are each adapted to carry out a distinct function.

Critical Thinking *How might the contractile vacuoles of a paramecium respond if the organism were placed in a dilute salt solution?*

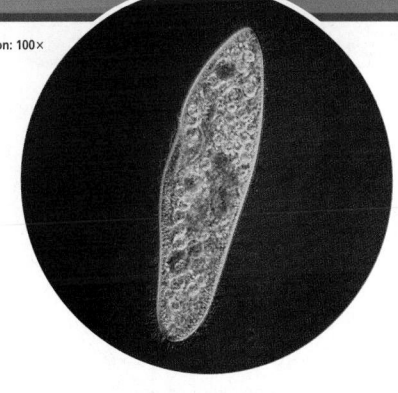

Magnification: 100×

Paramecium caudatum

1 **Cilia** The cell is encased by an outer covering called a pellicle through which thousands of tiny, hairlike cilia emerge. The paramecium can move by beating its cilia.

2 **Oral groove** Paramecia feed primarily on bacteria that are swept into the gullet by cilia that line the oral groove.

3 **Gullet** Food moves into the gullet, becoming enclosed at the end in a food vacuole. Enzymes break down the food, and the nutrients diffuse into the cytoplasm.

4 **Micronucleus and macronucleus** The small micronucleus plays a major role in sexual reproduction. The large macronucleus controls the everyday functions of the cell.

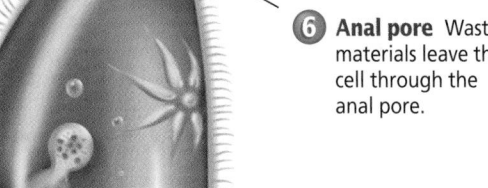

6 **Anal pore** Waste materials leave the cell through the anal pore.

5 **Contractile vacuole** Because a paramecium lives in a fresh-water, hypotonic environment, water constantly enters its cell by osmosis. A pair of contractile vacuoles pump out the excess water.

Pore

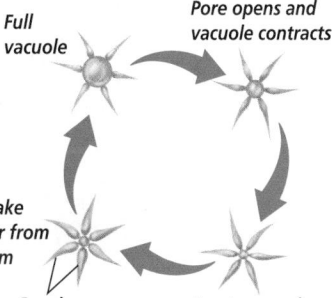

Full vacuole

Pore opens and vacuole contracts

Canals take up water from cytoplasm

Canals

Empty vacuole

TECHPREP

Charting Protist Diseases

☑ ***Visual-Spatial*** Have students use an encyclopedia to make a table about the causal agents and symptoms of the following diseases: malaria, sleeping sickness, Chagas disease, giardiasis, and amoebic dysentery. Ask them to use the column headings: Kingdom, Phylum, Means of locomotion, Method of transmission, Parasitic or free-living, and Disease symptoms. **L2**

Purpose

Analyze the role of digestive enzymes in *Paramecium*.

Process Skills

observe and infer, interpret scientific illustrations, recognize cause and effect

Teaching Strategies

■ Review pH. Inform students that there are both liquid and paper pH indicators.

Thinking Critically

It varies. The pH in a food vacuole forming at the end of the gullet is above 5. As digestion begins, the pH drops to below 3 and changes again as digestion progresses. As digestion nears completion, the pH increases until it is above 5 again. There are many digestive enzymes involved in *Paramecium* digestion, and they work at acidic and basic pHs.

✓ Assessment

Performance Give students liquids with different pHs. Have them use pH paper to find which liquids match the pH in *Paramecium's* food vacuole at the start and end of digestion. Use the Performance Task Assessment List for Carrying Out a Strategy and Collecting Data in **PASC**, p. 25. **L2** **ELL**

3 Assess

Check for Understanding

Visual-Spatial Have students make a concept map showing Kingdom Protista's three subgroups. Ask them to expand their maps to include the four protozoan phyla. **L2**

Problem-Solving Lab 19-1   **Drawing a Conclusion**

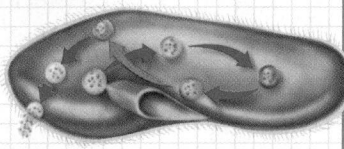

How do digestive enzymes function in paramecia?
Paramecia ingest food particles and enclose them in food vacuoles. Each food vacuole circulates in the cell as the food is digested by enzymes that enter the vacuole. Digested nutrients are absorbed into the cytoplasm.

Analysis

1. Some digestive enzymes function best at high pH levels, while others function best at low (more acidic) pH levels.
2. Congo red is a pH indicator dye; it is red when the pH is above 5 and blue when the pH is below 3 (very acidic).
3. Yeast cells that contain Congo red can be produced by adding dye to the solution in which the cells are growing.
4. When paramecia feed on dyed yeast cells, the yeast is visible inside food vacuoles.
5. Examine the drawing above. The appearance of a yeast-filled food vacuole over time is indicated by the colored circles inside the paramecium. Each arrow indicates movement and the passing of time.

Thinking Critically

What happens to the pH in the food vacuole over time? Explain what sequence of digestive enzymes might function in a paramecium.

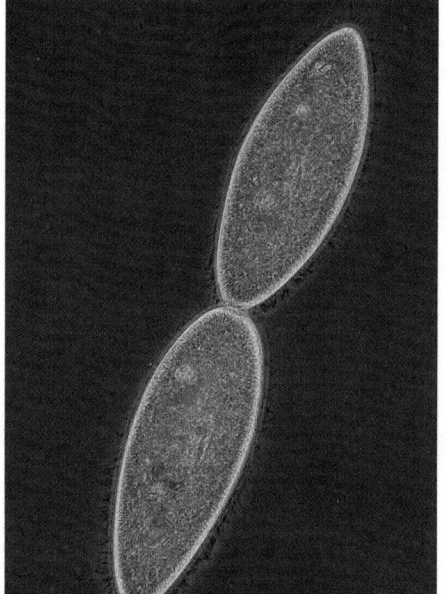

Magnification: 300×

Figure 19.5
A paramecium is dividing into two identical daughter cells.

Many structures found in ciliates' cells may work together to perform just one important life function. For example, *Paramecium* uses its cilia, oral groove, gullet, and food vacuoles in the process of digestion. Use the *Problem-Solving Lab* on this page to explore how a paramecium digests the food in a vacuole.

A paramecium usually reproduces asexually by dividing crosswise and separating into two daughter cells, as you can see in *Figure 19.5*. Whenever their food supplies dwindle or their environmental conditions change, paramecia usually undergo a form of conjugation. In this complex process, two paramecia join and exchange genetic material. Then they separate, and each divides asexually, passing on its new genetic composition.

Sporozoans: Parasitic protozoans

Protists in the phylum Sporozoa are often called **sporozoans** because most produce spores. A **spore** is a reproductive cell that forms without fertilization and produces a new organism.

All sporozoans are parasites. They live as internal parasites in one or more hosts and have complex life cycles. Sporozoans are usually found in a part of a host that has a ready food supply, such as an animal's blood or intestines. *Plasmodium*, members of the sporozoan genus, are organisms that cause the disease malaria in humans and other mammals and in birds.

Sporozoans and malaria

Throughout the world today, more than 300 million people have malaria, a serious disease that usually occurs in places that have tropical climates. The *Plasmodium* that mosquitoes transmit to people cause human malaria. As you can see in *Figure 19.6,* malaria-causing *Plasmodium* live in both humans and mosquitoes.

Resource Manager

Critical Thinking/Problem Solving, p. 19 **L3**
Laboratory Manual, pp. 133-136 **L2**

BIOLOGY JOURNAL

Malaria and Sickle-Cell Anemia

Linguistic Have students use references to report about the correlation between the genetic disease sickle-cell anemia and the protist disease malaria. They should include the evolutionary significance of such a relationship. **L3**

Until World War II, the drug quinine was used to treat malaria. Today, a combination of the drugs chloroquine and primaquine are most often used to treat this disease because they cause few serious side effects in humans. But, some species of *Plasmodium* have begun to resist these drugs. Therefore, quinine is once again being used to treat the resistant strains.

Figure 19.6
The life cycle of *Plasmodium* involves two hosts—mosquitoes and humans.

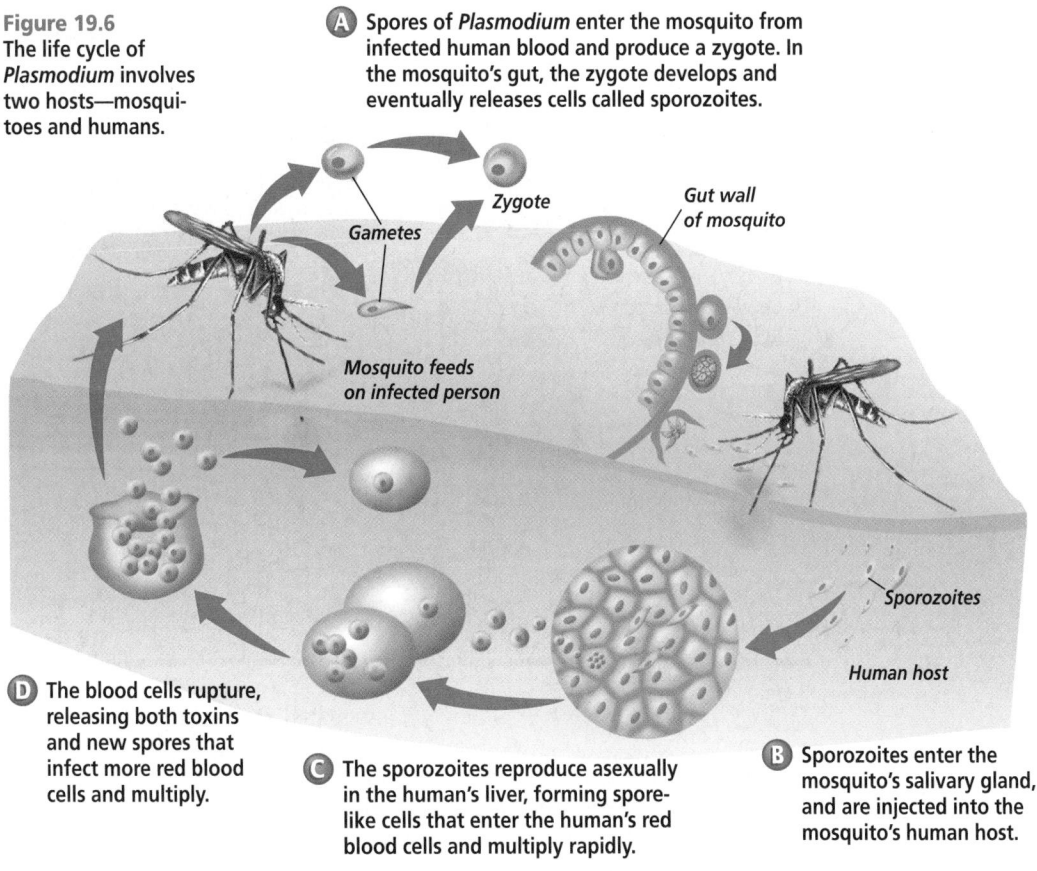

Ⓐ Spores of *Plasmodium* enter the mosquito from infected human blood and produce a zygote. In the mosquito's gut, the zygote develops and eventually releases cells called sporozoites.

Zygote

Gametes

Gut wall of mosquito

Mosquito feeds on infected person

Sporozoites

Human host

Ⓓ The blood cells rupture, releasing both toxins and new spores that infect more red blood cells and multiply.

Ⓒ The sporozoites reproduce asexually in the human's liver, forming spore-like cells that enter the human's red blood cells and multiply rapidly.

Ⓑ Sporozoites enter the mosquito's salivary gland, and are injected into the mosquito's human host.

Section Assessment

Understanding Main Ideas

1. Describe the characteristics of the organisms called protists. Then compare the characteristics of the four major groups of protozoans. How is each group of protozoans animal-like?
2. How do amoebas obtain food?
3. Explain any differences that exist between ciliates and flagellates.
4. What makes a sporozoan different from other protozoan groups?

Thinking Critically

5. What role do contractile vacuoles play in helping freshwater protozoans maintain homeostasis?

SKILL REVIEW

6. **Sequencing** Trace the life cycle of a *Plasmodium* that causes human malaria. Identify all forms of the sporozoan and the role each plays in the disease. For more help, refer to *Organizing Information* in the **Skill Handbook**.

Reteach
Have students list five general protist traits and then make a second list of protozoan-specific traits. L1

Extension
Have students research the phylum of the organism *Trypanosoma cruzi*. Ask them to draw and label this protist. L2

✔ Assessment

Knowledge Have students identify the unrelated word in each of the following groups: amoeba, foraminiferan, *Paramecium*, radiolarian; *pseudopodia*, cilia, oral groove, pellicle; dysentery, *AIDS*, sleeping sickness, Chagas disease. L2

---

4 Close

Activity

Logical-Mathematical Make five sets of ten index cards with one of the following words on each card: locomotion, cilia, pseudopodia, flagella, sarcodines, asexual reproduction, cyst, micronucleus, spore, sexual reproduction. Give a set of cards to each of five student groups, telling them to logically sequence the cards and then explain their sequences. L2

Resource Manager

Reinforcement and Study Guide, p. 83 L2
Content Mastery, p. 94 L1

Section Assessment

1. Protists are eukaryotes without tissues. Protozoans are unicellular, heterotrophic protists. Rhizopods have pseudopodia, Ciliophorans have cilia, and flagella propel Mastigophorans. Sporozoans are parasites.
2. Pseudopodia entrap prey.
3. Flagellates have long, whiplike flagella. Ciliates have many short, hairlike cilia.
4. All are parasitic, many produce spores, and some cannot move.
5. Contractile vacuoles pump out excess water that enters the cell by osmosis.
6. Student answers should include the information in Figure 19.6. Both the human and mosquito hosts and the forms that *Plasmodium* take in each host should be distinguished.

Prepare

Key Concepts

In this section, the diversity of algae is explored by focusing on the characteristics and adaptations of members of the six algae phyla. Finally, alternation of generations in algae is discussed.

Planning

- Collect pond water or fish tank scum for MiniLab 19-2.
- Buy algae for students to taste.

1 Focus

Bellringer

Before presenting the lesson, display **Section Focus Transparency 46** on the overhead projector and have students answer the accompanying questions. **L1** **ELL**

Transparency **46** Giant Kelp

SECTION FOCUS
Use with Chapter 19,
Section 19.2

❶ The large plantlike organism in the picture is giant kelp, a type of protist called a brown alga. What role does the kelp play in this ecosystem?

❷ How might the loss of the kelp affect this ecosystem?

Section
19.2 Algae: Plantlike Protists

Each time you inhale, you breathe in oxygen, much of which is being produced by plantlike protists. The algae are important in the world of living things. Just about every living thing depends either directly or indirectly on these protists for oxygen and food.

Kelp forest (above) and algae growing in a freshwater pond (inset)

What Are Algae?

Photosynthesizing protists are called algae. All algae contain up to four kinds of chlorophyll as well as other photosynthetic pigments. These pigments produce a variety of colors in algae, including purple, rusty-red, olive-brown, yellow, and golden-brown, and are a way of classifying algae into groups.

Algae include both unicellular and multicellular organisms. The photosynthesizing unicellular protists, known as phytoplankton (fite uh PLANK tun), are so numerous that they are one of the major producers of nutrients and oxygen in aquatic ecosystems in the world. It's been

estimated that algae produce more than half of the oxygen generated by Earth's photosynthesizing organisms. Although multicellular algae may look like plants because they are large and sometimes green, they have no roots, stems, or leaves. Use the *MiniLab* on the next page to observe some algae.

Diversity of Algae

Algae are classified into six phyla. Three of these phyla—the euglenoids, diatoms, and dinoflagellates—include only unicellular species. However, in the other three phyla, which are the green, red, and brown algae, most species are multicellular.

526 PROTISTS

Resource Manager

BioLab and MiniLab Worksheets, p. 90 **L2**

Section Focus Transparency 46 and Master **L1** **ELL**

BIOLOGY JOURNAL

Protozoans and Algae

Naturalist Have students prepare a table that compares and contrasts protozoans and algae. Suggest that they use the headings Similarities and Differences. The rows can include topics such as Cell organization, Nutrition, and Cell structures. **L2**

Euglenoids: Autotrophs and heterotrophs

Hundreds of species of euglenoids (yoo GLEE noydz) make up the phylum Euglenophyta. Euglenoids are unicellular, aquatic protists that have both plant and animal characteristics. Unlike plant cells, they lack a cell wall made of cellulose. However, they do have a flexible pellicle made of protein that surrounds the cell membrane. Euglenoids are plantlike in that most have chlorophyll and photosynthesize. However, they are also animal-like because, when light is not available, they can ingest food in ways that might remind you of some protozoans. In other words, euglenoids can be heterotrophs. In *Figure 19.*7, you can see a typical euglenoid.

Euglenoids might also remind you of protozoans because they have one or more flagella to move. They use their flagella to move toward light or food. In the *BioLab* at the end of this chapter you can learn more about how a euglenoid responds to light.

Diatoms: The golden algae

Diatoms (DI uh tahmz), members of the phylum Bacillariophyta, are unicellular photosynthetic organisms with

---

MiniLab 19-2 Observing

Going on an Algae Hunt Pond water may be teeming with organisms. Some are macroscopic organisms, but the majority are microscopic. Some may be heterotrophs, and others autotrophs. How can you tell them apart?

Procedure
1. Copy the data table.

Data Table		
Diagram	**Motile/Nonmotile**	**Unicellular/Multicellular**

2. Place a drop of pond water onto a glass slide and add a coverslip. **CAUTION:** *Use caution when working with a microscope, glass slides, and coverslips.*
3. Observe the pond water under low magnification of your microscope, and look for algae that may be present. Algae from a pond will usually be green or yellow-green in color.
4. Diagram several different species of algae in your data table and indicate if each is motile or nonmotile. Indicate if the algae are unicellular or multicellular.

Analysis
1. What characteristic distinguished algae from any protozoans that may have been present?
2. Explain how the characteristic in question 1 categorizes algae as autotrophs.
3. Did you observe any relationship between movement and size? Explain your answer.

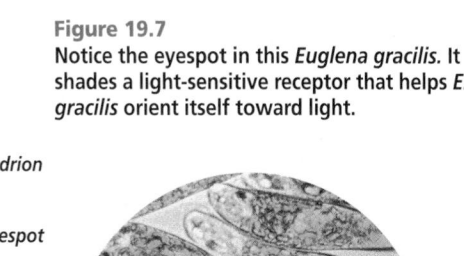
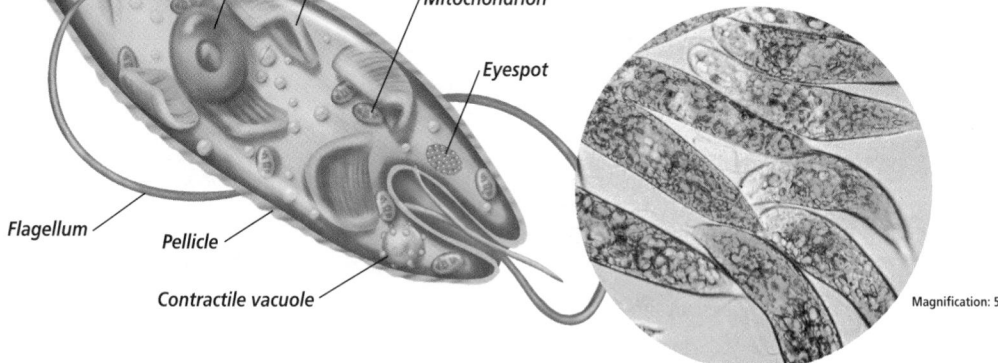

Figure 19.7
Notice the eyespot in this *Euglena gracilis*. It shades a light-sensitive receptor that helps *E. gracilis* orient itself toward light.

Nucleus
Chloroplast
Mitochondrion
Eyespot
Flagellum
Pellicle
Contractile vacuole

Magnification: 525×

527

Figure 19.8
Diatom shells have many shapes.

Magnification: 1500×

shells composed of silica. They are abundant in both marine and freshwater ecosystems, where they are a large component of the phytoplankton.

The delicate shells of diatoms, like those you see in *Figure 19.8,* might remind you of boxes with lids. Each species has its own unique shape, decorated with grooves and pores.

Diatoms contain chlorophyll as well as other pigments called carotenoids

(KER uh teen oydz) that usually give them a golden-yellow color. The food that diatoms make is stored as oils rather than starch. These oils give fishes that feed on diatoms an oily taste. They also give diatoms buoyancy so that they float near the surface where light is available.

When diatoms reproduce asexually, the two halves of the box separate; each half then produces a new half to fit inside itself. This means that half of each generation's offspring are smaller than the parent cells. When diatoms are about one-quarter of their original size, they reproduce sexually by producing gametes that fuse to form zygotes. The zygote develops into a full-sized diatom, which will divide asexually for a while. You can see both the asexual and sexual reproductive processes of diatoms in *Figure 19.9.*

Figure 19.9
Diatoms reproduce asexually for several generations before reproducing sexually.

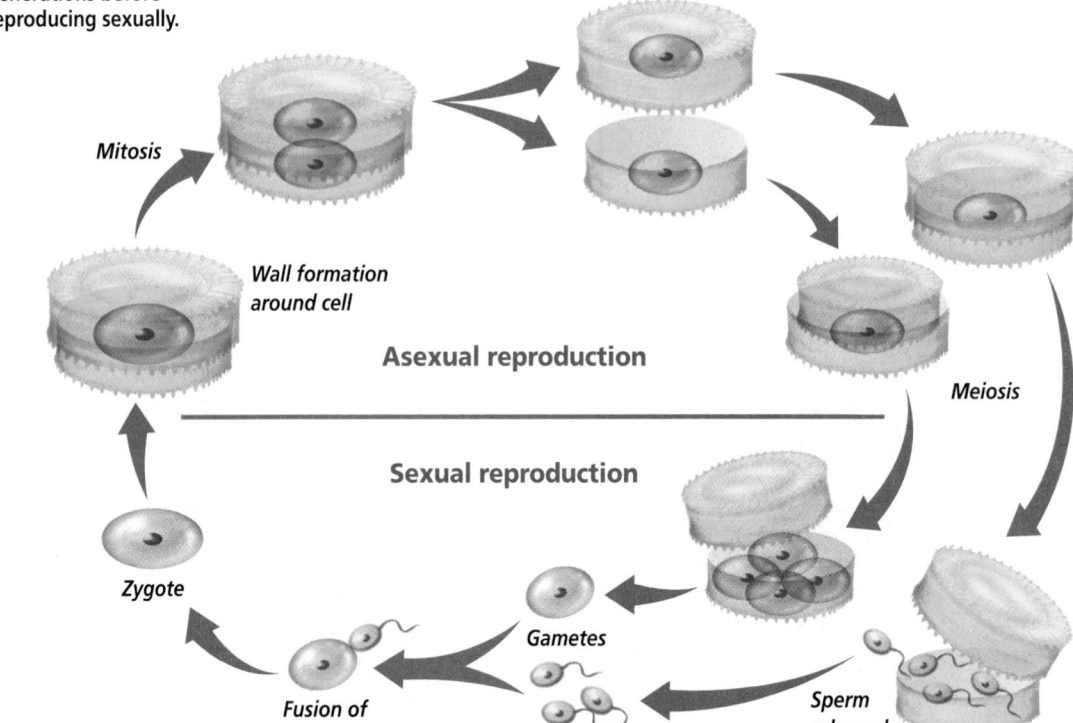

Mitosis

Wall formation around cell

Asexual reproduction

Meiosis

Sexual reproduction

Zygote

Fusion of gametes

Gametes

Sperm released

a

b

Magnification: 75×

When diatoms die, their shells sink to the ocean floor. The deposits of diatom shells—some of which are millions of years old—are dredged or mined, processed, and used as abrasives in tooth and metal polishes, or added to paint to give the sparkle that makes pavement lines more visible at night.

Dinoflagellates: The spinning algae

Dinoflagellates (di nuh FLAJ uh layts), members of the phylum Dinoflagellata, have cell walls that are composed of thick cellulose plates. They come in a great variety of shapes and styles—some even resemble helmets, and others look like suits of armor.

Dinoflagellates contain chlorophyll, carotenoids, and red pigments. They have two flagella located in grooves at right angles to each other. The cell spins slowly as the flagella beat. A few species of dinoflagellates live in freshwater, but most are marine and, like diatoms, are a major component of phytoplankton. Many species live symbiotically with jellyfishes, mollusks, and corals. Some free-living species are bioluminescent, which means that they emit light.

Several species of dinoflagellates produce toxins. One toxin-producing dinoflagellate, *Pfiesteria piscicida*, that some North Carolina researchers discovered in 1988 has caused a number of fish kills in the coastal areas of North Carolina.

Another toxic species, *Gonyaulax catanella*, produces an extremely strong nerve toxin that can be lethal. In the summer, these organisms may become so numerous that the ocean takes on a reddish color as you can see in ***Figure 19.10***. This population explosion is called a red tide. In some red tides, there can be as many as 40 to 60 million dinoflagellates per liter of seawater.

The toxins produced during a red tide may make humans ill. During red tides, the harvesting of shellfish is usually banned because shellfish feed on the toxic algae and the toxins concentrate in their tissues. People who eat such shellfish risk being poisoned. You can learn more about the causes and effects of red tides in the *Problem-Solving Lab* on the next page.

Using Science Terms

Tell students that the prefix *dino* means "whirling" and the suffix *flagellate* means "whip." Ask them why the name dinoflagellates is appropriate for these algae. *They spin as their flagellas propel them through water.* **L1**

Concept Development

Describe coral as a heterotroph and ask students to explain the role of dinoflagellates and coral when they live symbiotically. *Dinoflagellates make food for the corals, and the coral reef helps safeguard dinoflagellates from predators.*

Reinforcement

Check that students know the meaning of shellfish, bioluminescent, and toxins. Discuss why the term shellfish is a misnomer. *Shellfish are not classified as fishes, but as arthropods (shrimp and crabs) or mollusks (clams and squid).*

GLENCOE TECHNOLOGY

VIDEODISC
The Infinite Voyage
Secrets from a Frozen World
The Antarctic Peninsula: Pack Ice and Life Cycles (Ch. 6)
10 min. 30 sec.

Effects of UV Radiation on Phytoplankton (Ch. 8)
4 min. 30 sec.

Portfolio

Algal Names

Linguistic Although algae are often called "seaweed, scum, and sea moss," none of these terms are scientifically correct. Have students find the meaning of each term and then explain why it is scientifically incorrect. **L2**

NATIONAL GEOGRAPHIC

VIDEODISC
STV: The Cell

Dinoflagellates   *Red Tide*

529

Quick Demo

Purchase preserved brown and red algae for class display. If possible, buy some edible algae for students to taste.

Problem-Solving Lab 19-2

Purpose

Students will analyze real-life events associated with red tides.

Process Skills

think critically, acquire information, draw a conclusion, recognize cause and effect

Teaching Strategies

■ Have students work in small groups, discussing each event thoroughly.
■ Have groups share answers.

Thinking Critically

1. **a.** A or B—Toxins accumulate in shellfish that humans eat; mackerel eat dinoflagellates and whales eat the mackerel.
   b. E—Autotrophs use sunlight and nutrients to grow.
   c. C—Human population growth correlates with the frequency of red tides.
2. A or B—Shellfish are not affected by dinoflagellate toxins, but humans are; mackerel do not die from dinoflagellate toxins, but whales do.

✔ Assessment

Portfolio Have students diagram dinoflagellates they observe on prepared slides. Use the Performance Task Assessment List for Scientific Drawing in PASC, p. 55. **L1**

Problem-Solving Lab 19-2   Recognizing Cause and Effect

Why is the number of red tides increasing? Scientists have been aware of red tide poisoning of birds, fishes, and mammals such as whales and humans for years. Could the rise in red tide poisoning be related to human activities?

A sperm whale's carcass

Analysis

The following events are associated with the appearance of red tides.

A The dinoflagellate toxin that causes illness and sometimes death in humans accumulates in the body tissues of shellfish, such as clams and oysters.

B Within five weeks, 14 humpback whales died on beaches in Massachusetts. The whales' stomachs contained mackerel with high levels of dinoflagellate toxin.

C Between 1976 and 1986, the human population of Hong Kong increased sixfold, and its harbor had an eightfold increase in red tides. Human waste water was commonly emptied into the harbor.

D Studies show that red tides are increasing world wide.

E An algal bloom occurs when algae, using sunlight and abundant nutrients, increase rapidly in number to hundreds of thousands of cells per milliliter of water.

Thinking Critically

1. Which statement above provides evidence that supports each of the following ideas. Explain each answer.
   a. Dinoflagellate poisons flow through the food chain.
   b. Dinoflagellates are autotrophs.
   c. There is a correlation between human activities and algae growth.
2. All animals equally tolerate dinoflagellate toxins. Which statement contradicts this idea? Explain your answer.

Red algae

Red algae, members of the phylum Rhodophyta, are multicellular marine seaweeds. The body of a seaweed, as well as that of some plants and other organisms, is called a **thallus** and lacks roots, stems, or leaves. Red algae use structures called holdfasts to attach to rocks. They grow in tropical waters or along rocky coasts in cold water. You can see a red alga in *Figure 19.11.*

In addition to chlorophyll, red algae also contain photosynthetic pigments called phycobilins. These pigments absorb green, violet, and blue light—the only part of the light spectrum that penetrates water below depths of 100 m. Therefore, the red algae can live in deep water where most other seaweeds cannot thrive.

Brown algae

About 1500 species of multicellular brown algae make up the phylum Phaeophyta. Almost all of these species live in salt water along rocky coasts in cool areas of the world. Brown algae contain chlorophyll as well as a yellowish-brown carotenoid called fucoxanthin, which gives them their brown color. Many species of brown algae have air bladders that keep their bodies floating near the surface, where light is available.

Figure 19.11
This Coralline alga is only one of about 4000 species of red algae. Some species are popular foods in Japan and other countries.

Cultural Diversity

Algae Harvesting in Japan

Inform students that people in many areas of the world, particularly Asia, eat some algae. Point out that algae contain nutrients, such as protein, and many vitamins and minerals. Algae are eaten fresh and boiled or fried in many Asian recipes. One edible algae used in Japanese cooking is *Porphyra*, a red alga commonly called nori. Since the seventeenth century, the Japanese have harvested nori from Tokyo Bay. Have students research how the Japanese harvest algae. Bring in some Japanese foods that contain algae for the class to sample. **L3**

The largest and most complex of brown algae are kelp. In kelp, the thallus is divided into the holdfast, stipe, and blade. The holdfasts anchor kelp to rocks or the sea bottom. Some giant kelps may grow up to 60 meters long. In some parts of the world, such as off the California coast, giant kelps form dense, underwater forests. These kelp forests are rich ecosystems and provide a wide variety of marine organisms with their habitats.

Green algae

Green algae make up the phylum Chlorophyta. The green algae are the most diverse algae, with more than 7000 species. The major pigment in green algae is chlorophyll, but some species also have yellow pigments that give them a yellow-green color. Most species of green algae live in freshwater, but some live in the oceans, in moist soil, on tree trunks, in snow, and even in the fur of sloths—large, slow-moving mammals that live in the tropical rain forest canopy.

Green algae can be unicellular, colonial, or multicellular in organization. As you can see in *Figure 19.12*, *Chlamydomonas* is a unicellular and flagellated green alga. *Spirogyra* is a multicellular species that forms slender filaments. *Volvox* is a green alga that can form a **colony,** a group of cells that lives together in close association.

A *Volvox* colony is composed of hundreds, or thousands, of flagellated cells arranged in a single layer forming a hollow, ball-shaped structure. The cells are connected by strands of cytoplasm, and the flagella of individual cells face outward. The flagella can beat in a coordinated fashion, spinning the colony through the water. Small balls of daughter colonies form inside the large sphere. The wall of the large colony will eventually break open and release the daughter colonies.

Green algae reproduce both asexually and sexually. For example, *Spirogyra* reproduces asexually by fragmentation. During **fragmentation,** an individual breaks up into pieces and each piece grows into a new individual.

WORD *Origin*

thallus
From the Greek word *thallos*, meaning "green shoot." A thallus is a plant without stems, roots, or leaves.

Figure 19.12
Chlamydomonas is a unicellular species of green algae (a), while *Spirogyra* is a multicellular form (b). The wall of a *Volvox* colony contains hundreds of cells (c). The smaller balls inside the sphere are daughter colonies.

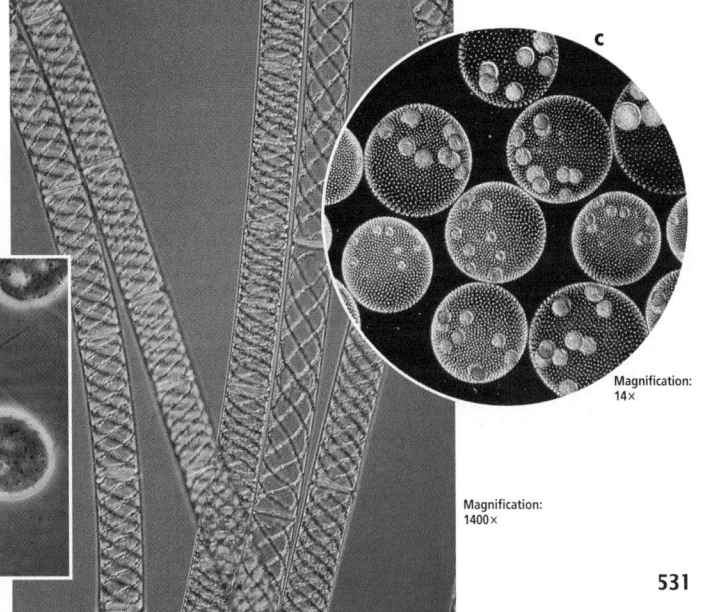

Magnification: 14×

Magnification: 1400×

Magnification: 1250×

531

3 Assess

Check for Understanding

Have students describe briefly three unicellular and three multi-cellular algae phyla. **L2**

Reteach

Have the class develop a concept map of the material in this chapter's first two sections. **L2**

Extension

Ask students to compile a list of foods that contain algae. **L2**

✓ Assessment

Knowledge Make six line drawings showing an alga from each of the six phyla. Give students sets of the drawings and ask them to list three facts or ideas about each alga. **L2**

4 Close

Discussion Questions

Have each student prepare three questions about algae. Collect the questions and have the students answer them. **L1**

GLENCOE
TECHNOLOGY

VIDEODISC
The Infinite Voyage
*The Living Clock,
Circadian Rhythm and the
Biological Clock* (Ch. 4)
5 min.

Resource Manager

Reinforcement and Study Guide, pp. 84-85 **L2**

532

Figure 19.13
In the life cycle of the sea lettuce, the generations alternate between haploid (gametophyte) and diploid (sporophyte). Both fungi and plants also alternate generations.

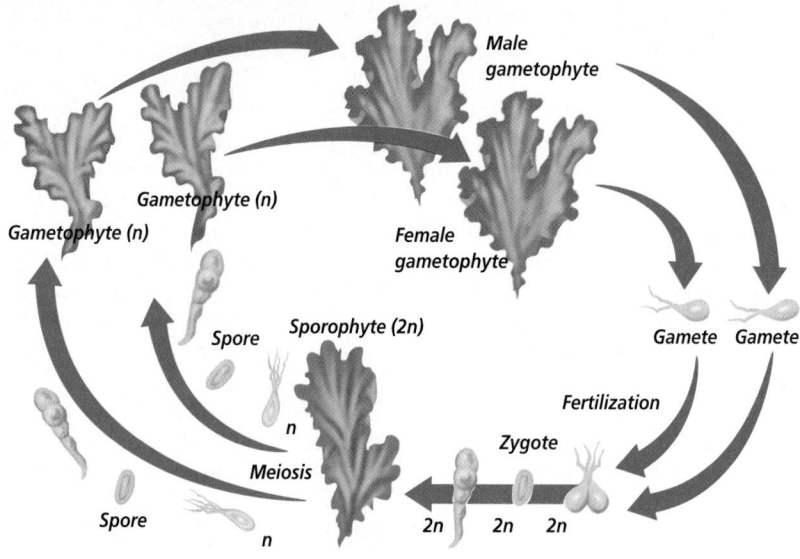

Green algae, and some other types of algae, have a complex life cycle. This life cycle consists of individuals that alternate between producing reproductive cells called spores and producing gametes.

Alternation of Generations

The life cycles of some algae and all plants have a pattern called **alternation of generations.** An organism that has this pattern alternates between a haploid and a diploid generation.

The haploid form of the organism is called the **gametophyte** because it produces gametes. The gametes fuse to form a zygote from which the diploid form of the organism, which is called the **sporophyte,** develops. Certain cells in the sporophyte undergo meiosis. Eventually, these cells become haploid spores that can develop into a new gametophyte. Look at *Figure 19.13* to see the life cycle of *Ulva*, a multicellular green alga.

Section Assessment

Understanding Main Ideas

1. In what ways are algae important to all living things on Earth?
2. Give examples that show why the green algae are considered to be the most diverse of the six phyla of algae.
3. In what ways do the sporophyte and gametophyte generations of an alga such as *Ulva* differ from each other?
4. Why are phycobilins an important pigment in red algae?

Thinking Critically

5. Use a table to list the reasons why euglenoids should be classified as protozoans and also as algae.

SKILL REVIEW

6. **Making and Using Tables** Construct a table listing the different phyla of algae. Indicate whether they have one or more cells, their color, and give an example of each. For more help, refer to *Organizing Information* in the **Skill Handbook**.

532 PROTISTS

Section Assessment

1. They are producers of oxygen and the primary autotrophs of many food chains.
2. *Chlamydomonas* is a unicellular example. *Volvox* is a colonial form. *Ulva* is a multi-cellular alga.
3. Haploid gametophytes form gametes and diploid sporophytes form spores.
4. They can absorb the green, violet, and blue light that penetrates deep water.
5. Their chloroplasts are algaelike and their locomotion is protozoanlike.
6. Euglenophyta—one, green, euglena; Bacillariophyta—one, golden, diatom; Dinoflagellata—one, green, yellow, and red, dinoflagellate; Rhodophyta—many, red, red seaweed; Phaeophyta—many, brown, brown seaweed; Chlorophyta—one or many, green, green algae.

19.3 Slime Molds, Water Molds, and Downy Mildews

SECTION PREVIEW

Objectives
Contrast the cellular differences and life cycles of the two types of slime molds.
Discuss the economic importance of the downy mildews and water molds.

Vocabulary
plasmodium

When you walk through the woods, you might notice a spot of color on a fallen log. Turning the log over, you uncover a glistening mass of yellow-orange slime that fans out over the log. What you've found is a slime mold, one of a variety of funguslike protists. Slime molds, along with water molds and downy mildews, obtain energy by decomposing organic materials, and play an important role in recycling nutrients in many ecosystems.

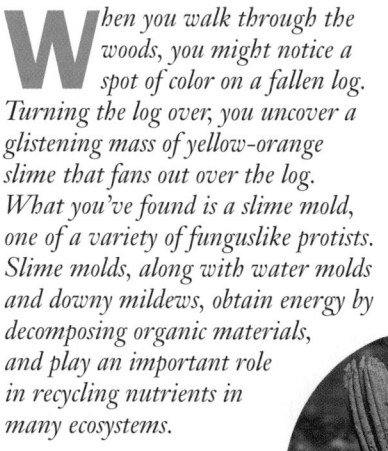

A plasmodial slime mold that is feeding (above) and reproducing (inset)

What are Funguslike Protists?

Certain groups of protists, the slime molds, the water molds, and the downy mildews, consist of organisms with some funguslike features. Recall that fungi are heterotrophic organisms that decompose organic materials to obtain energy. Like fungi, the funguslike protists decompose organic materials.

There are three phyla of funguslike protists. Two of these phyla consist of slime molds. Slime molds have characteristics of both protozoans

and fungi and are classified by the way they reproduce. Water molds and downy mildews make up the third phylum of funguslike protists. Although funguslike protists are not an everyday part of human lives, some disease-causing species damage vital crops.

Slime Molds

Many slime molds are beautifully colored, ranging from brilliant yellow or orange to rich blue, violet, and jet black. They live in cool, moist, shady places where they grow on damp,

19.3 SLIME MOLDS, WATER MOLDS, AND DOWNY MILDEWS **533**

Section 19.3

Prepare

Key Concepts

This section presents the characteristics of funguslike protists. The phyla of plasmodial slime molds, cellular slime molds, and water molds and downy mildews are described. Finally, the origin of protists is discussed.

Planning

■ Obtain meat, string, and an aquarium for the Quick Demo.

---

1 Focus

Bellringer

Before presenting the lesson, display **Section Focus Transparency 47** on the overhead projector and have students answer the accompanying questions. **L1** **ELL**

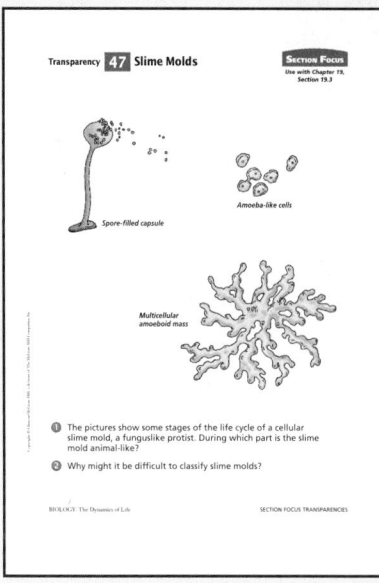

Transparency **47** Slime Molds   SECTION FOCUS
Use with Chapter 19, Section 19.3

Spore-filled capsule

Amoeba-like cells

Multicellular amoeboid mass

❶ The pictures show some stages of the life cycle of a cellular slime mold, a funguslike protist. During which part is the slime mold animal-like?

❷ Why might it be difficult to classify slime molds?

BIOLOGY: The Dynamics of Life   SECTION FOCUS TRANSPARENCIES

2 Teach

Purpose

Students will relate cell division to a slime mold's life cycle.

Process Skills

apply concepts, interpret scientific illustrations, think critically

Teaching Strategies

■ Grow slime molds and have students observe them with a stereo microscope.

■ Review mitosis, meiosis, fertilization, and spore formation.

Thinking Critically

1. mitosis—cell growth; meiosis—haploid gamete formation (gametes form at C)
2. D—two gamete cells join; C—cells with flagella; E—it results from fertilization.
3. growth stage—organisms use food to grow

✔ Assessment

Performance Have students grow slime molds from purchased spores. Use the Performance Task Assessment List for Carrying Out a Strategy and Collecting Data in **PASC,** p. 25.
L3 **ELL**

Problem-Solving Lab 19-3   Predicting

What changes occur during a slime mold's life cycle?
Plasmodial slime molds undergo a number of different stages during their life cycle. The most visible stage is the plasmodial stage, where the organism looks like a slimy mass of material. The plasmodium changes into a reproductive stage that is microscopic and, therefore, less visible.

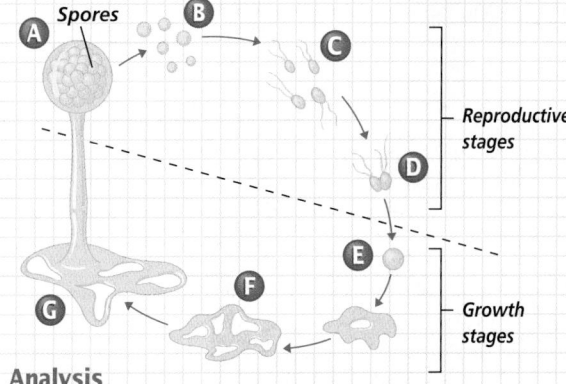

Analysis

Examine the life cycle of a plasmodial slime mold. The structures below the dashed line are diploid in chromosome number. Based on the diagram and your understanding of mitosis and meiosis, answer the questions below.

Thinking Critically

1. What cell process, mitosis or meiosis, takes place between F and G? Explain why. Between A and B? Explain why.
2. What letter best shows fertilization occurring? Motile spores? An embryo? Explain why in each case.
3. During which stage does the slime mold feed? Explain.

Figure 19.14
The moving, feeding form of a plasmodial slime mold is a multinucleate blob of cytoplasm.

---

organic matter such as rotting leaves or decaying tree stumps and logs.

There are two major types of slime molds—plasmodial slime molds and cellular slime molds. The plasmodial slime molds belong to the phylum Myxomycota, and the cellular slime molds make up another grouping, the phylum Acrasiomycota.

Slime molds are animal-like during much of their life cycle, moving about and engulfing food in a way similar to that of amoebas. However, like fungi, slime molds make spores to reproduce. Use the *Problem-Solving Lab* on this page to learn more about the life cycle of a slime mold.

Plasmodial slime molds

Plasmodial slime molds get their name from the fact that they form a **plasmodium** (plaz MOHD ee um), a mass of cytoplasm that contains many diploid nuclei but no cell walls or membranes. This slimy, multinucleate mass, like the one you see in *Figure 19.14,* is the feeding stage of the organism. The plasmodium creeps like an amoeba over the surfaces of decaying logs or leaves. Some plasmodiums move at the rate of about 2.5 centimeters per hour, engulfing microscopic organisms and

534 PROTISTS

Alternative Lab

Observing Slime Mold

Purpose

Allow students to observe a plasmodium-type slime mold.

Materials

petri dish with plasmodium stage of

Physarum polycephalum (Prepare subcultures two days in advance by cutting sections from the culture, putting the pieces on plastic agar plates that contain a few flakes of oat cereal moistened with distilled water.); stereo microscope

Procedure

Give students the following directions.
1. Wear a lab apron, gloves, and safety goggles during this lab.
2. Observe the slime mold.

3. Design a way to determine if the organism moves or not.
4. Record your macroscopic observations.
5. Use a stereo microscope to observe the slime mold. Record your observations about its microscopic appearance and behavior.
6. Wash your hands thoroughly when you finish. Dispose of the slime mold as your teacher directs.

Figure 19.15
The reproductive cycle of a cellular slime mold is complex (**a**). Single cells clump and form a structure that resembles a small garden slug (**b**). Eventually, the clump forms a stalked reproductive structure that produces spores (**c**).

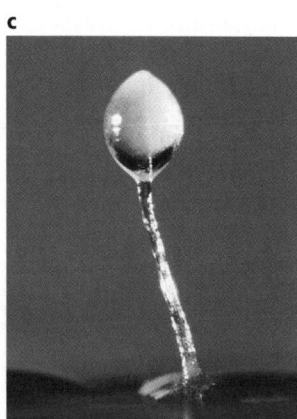

a

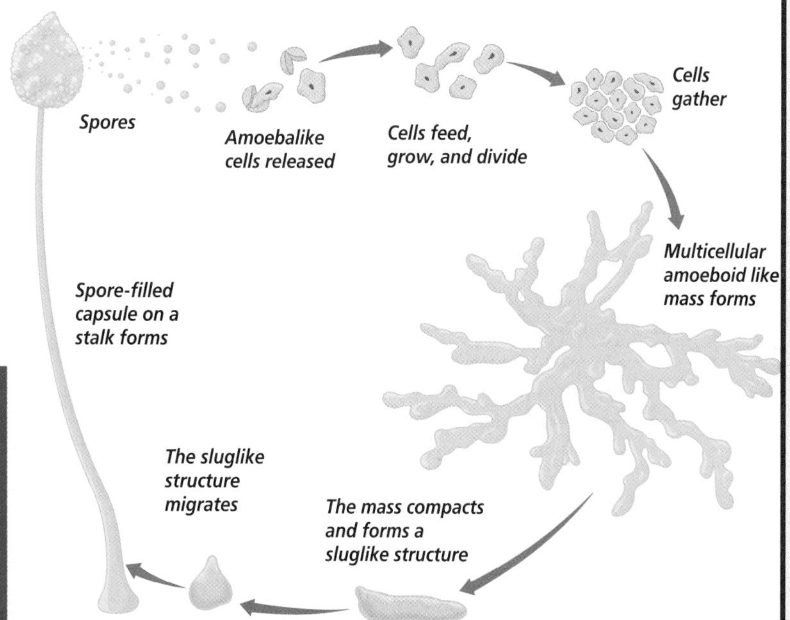

Spores

Amoebalike cells released

Cells feed, grow, and divide

Cells gather

Multicellular amoeboid like mass forms

Spore-filled capsule on a stalk forms

The sluglike structure migrates

The mass compacts and forms a sluglike structure

c

b

Magnification: 140×

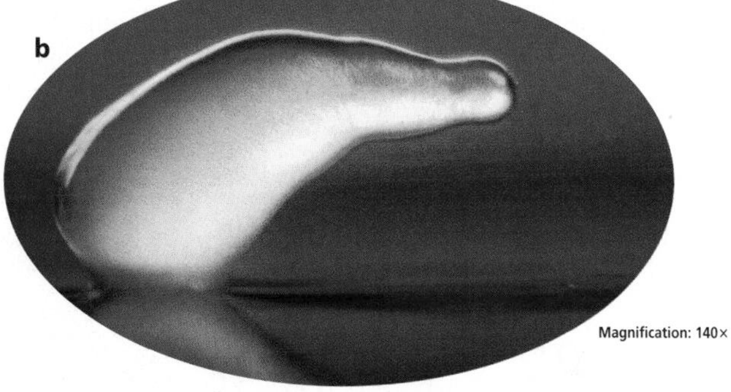

digesting them in food vacuoles. At that rate, a plasmodium would cross your textbook page in eight hours.

A plasmodium may reach more than a meter in diameter and contain thousands of nuclei. However, when moisture and food become scarce in its surroundings, a plasmodium transforms itself into many separate, stalked, spore-producing structures. Meiosis takes place within these structures and produces haploid spores, which the wind disperses. A spore germinates into either a flagellated or an amoeboid cell, or a gamete, that can fuse with another cell to form a zygote. The diploid zygote grows into a new plasmodium.

Cellular slime molds

Unlike plasmodial slime molds, cellular slime molds spend part of their life cycle as an independent amoeboid cell that feeds, grows, and divides by cell division, as shown in *Figure 19.15.* When food becomes scarce, these independent cells join with hundreds or thousands of others to reproduce. Such an aggregation of amoeboid cells resembles a plasmodium. However, this mass of cells is multicellular—made up of many individual amoeboid cells, each with a distinct cell membrane. Cellular slime molds are haploid during their entire life cycle.

WORD *Origin*

plasmodium
From the Greek word *plassein*, meaning "mold," and the Latin word *odium*, meaning "hateful." One form of a slime mold is a plasmodium.

Analysis
1. Describe the slime mold's appearance and behavior. Is it a plasmodial or cellular slime mold? *yellow, bloblike, stringy, moves very slowly—plasmodium*
2. Describe the slime mold's microscopic appearance and behavior. *Material flows within the slime mold, stops, then flows in the opposite direction.*

Assessment

Knowledge Write a paragraph explaining if you observed the feeding or the reproductive stage of the slime mold. Use the Performance Task Assessment List for Writing in Science in **PASC**, p. 87. **L2**

Reinforcement
Prepare an overview that illustrates the relationship between the phyla Myxomycota and Acrasiomycota and Kingdom Protista.

Assessment

Performance Have each student write two questions about slime molds. Divide the class into pairs and have students quiz each other. **L1**

GLENCOE
TECHNOLOGY

VIDEODISC
VIDEOTAPE
The Secret of Life
On the Brink: Portraits of Modern Science

CD-ROM
Biology: The Dynamics of Life
Video: *Slime Mold*
Disc 3

VIDEODISC
Biology: The Dynamics of Life
Slime Mold (Ch. 14)
Disc 1, Side 2, 14 sec.

Resource Manager

Concept Mapping, p. 19 **L3**
ELL
Basic Concepts Transparency 29 and Master **L2** **ELL**
Reteaching Skills Transparency 30 and Master **L1** **ELL**

Water Molds and Downy Mildews

Water molds and downy mildews are both members of the phylum Oomycota. Most members of this large and diverse group of funguslike protists live in water or moist places. As shown in *Figure 19.16,* some feed on dead organisms and others are plant parasites.

Most water molds appear as fuzzy, white growths on decaying matter. They resemble some fungi because they grow as a mass of threads over a food source, digest it, and then absorb the nutrients. But at some point in their life cycle, water molds produce flagellated reproductive cells—something that fungi never do. This is why water molds are classified as protists rather than fungi.

One economically important member of the phylum Oomycota is a downy mildew that causes disease in many plants. In the *Social Studies* *Connection* at the end of the chapter, you can read about a downy mildew called *Phytophthora infestans* that has affected the lives of many people by destroying their major food crop.

Origin of Protists

How are the many different kinds of protists related to each other and to fungi, plants, and animals? You can see the relationships of protists to each other in *Figure 19.17.*

Although taxonomists are now comparing the RNA and DNA of these groups, there is little conclusive evidence to indicate whether ancient protists were the evolutionary ancestors of fungi, plants, and animals or whether protists emerged as evolutionary lines that were separate. Because of evidence from comparative RNA sequences in modern green algae and plants, many biologists agree that ancient green algae were probably ancestral to modern plants.

Figure 19.16
Water molds and downy mildews live in moist places and cause both plant and animal diseases.

A The downy mildew *Phytophthora infestans* is killing this potato plant.

B The water mold growing on this insect is decomposing the insect's tissues and absorbing the nutrients.

Magnification: 40×

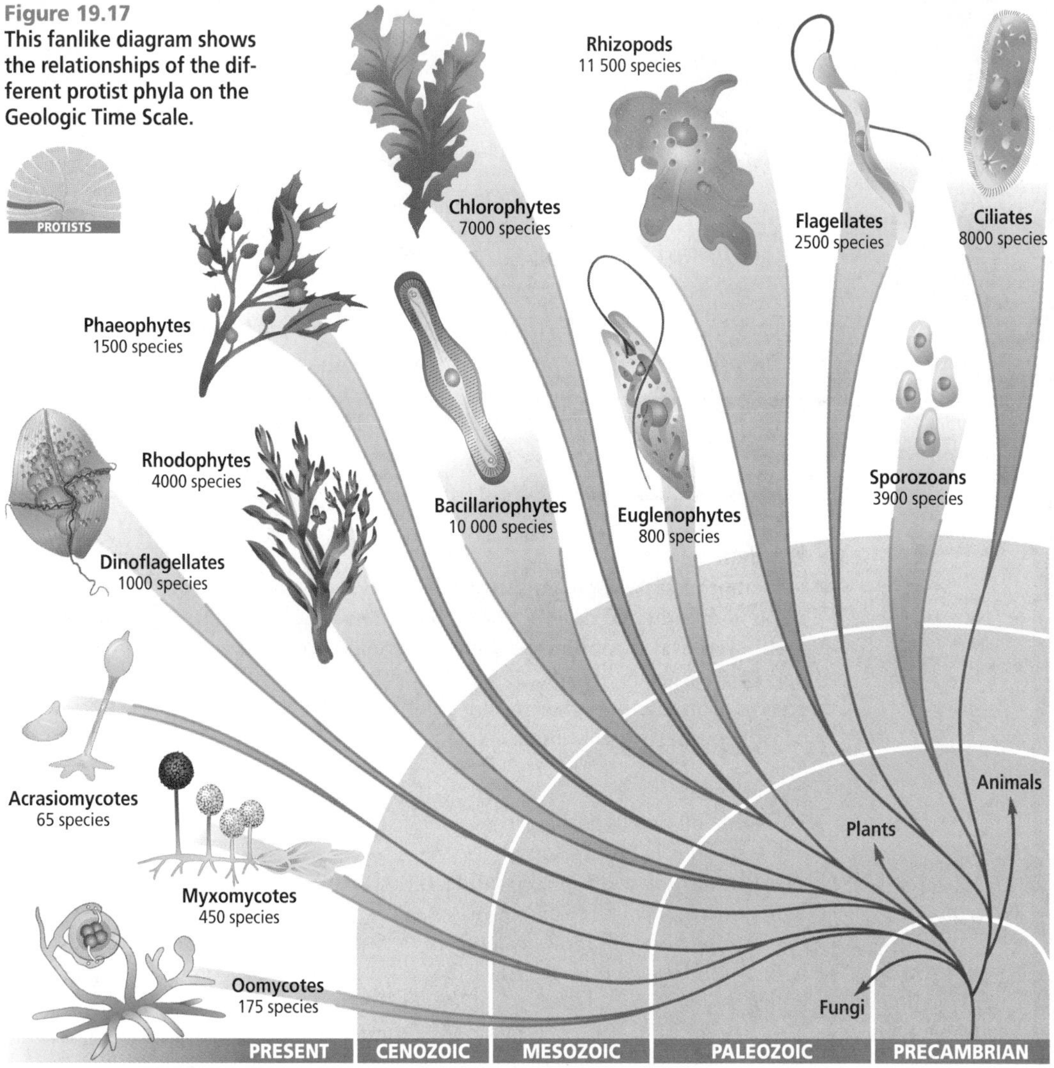
PROTISTS
Rhizopods
11 500 species
Chlorophytes
7000 species
Flagellates
2500 species
Ciliates
8000 species
Phaeophytes
1500 species
Rhodophytes
4000 species
Bacillariophytes
10 000 species
Euglenophytes
800 species
Sporozoans
3900 species
Dinoflagellates
1000 species
Acrasiomycotes
65 species
Myxomycotes
450 species
Oomycotes
175 species
Animals
Plants
Fungi
PRESENT
CENOZOIC
MESOZOIC
PALEOZOIC
PRECAMBRIAN

Figure 19.17
This fanlike diagram shows the relationships of the different protist phyla on the Geologic Time Scale.

Section Assessment

Understanding Main Ideas

1. Describe the protozoan and funguslike characteristics of slime molds.
2. Why might some biologists refer to plasmodial slime molds as acellular slime molds. (Hint: Look in Appendix B for the origins of scientific terms.)
3. How could a water mold eventually kill a fish?
4. How does a plasmodial slime mold differ from a cellular slime mold?

Thinking Critically

5. In what kinds of environments would you expect to find slime molds? Explain your answer.

Skill Review

6. **Observing and Inferring** If you know that a plasmodium consists of many nuclei within a single cell, what can you infer about the process that formed the plasmodium? For more help, refer to *Thinking Critically* in the **Skill Handbook.**

Assessment

Knowledge Ask students to list several protozoanlike traits and several funguslike traits of the funguslike protists.

4 Close

Activity

Have students list five words related to funguslike protists. Tell them that four words must be related to each other, and the fifth word must be unrelated. Collect the lists. Copy some lists onto the chalkboard and, in each case, have the class find the unrelated word and describe how the remaining words are related. L2

Resource Manager

Reinforcement and Study Guide, p. 86 L2
Content Mastery, pp. 93, 95-96 L1

Section Assessment

1. They are protozoanlike in that at different stages they have flagella and the ability to move like amoebas. They are funguslike in that they produce spores, and many are saprophytic decomposers.
2. The plasmodium is a mass of cytoplasm containing many nuclei but no cell walls or membranes that separate cells.
3. The mold digests its tissues.
4. Plasmodial slime molds feed as a multinucleated plasmodium, but cellular slime molds feed as amoeboid cells.
5. Slime molds should live in moist environments where the moisture would prevent dehydration and provide the conditions their food supply needs to thrive.
6. The process suggests mitosis without cell division.

DESIGN YOUR OWN BioLab

Time Allotment
One class period

Process Skills
observe and infer, experiment, form a hypothesis, identify and control variables

Safety Precautions
Remind students to be careful when working with microscopes, slides, and coverslips and to use special care when viewing a slide under high power. Have students wash their hands after finishing the lab.

PREPARATION

■ To demonstrate how *Euglena* and *Paramecium* respond to light, cover a test tube containing *Euglena* with black paper containing a narrow, slitlike opening. Leave the tube in bright light for 12 hours and then remove the paper. Where the slit was located, students should observe a green band of *Euglena*. Cover half of a test tube containing *Paramecium* with black paper, place the tube on its side in bright light for 12 hours. *Paramecium* should congregate in the covered side.

Possible Hypotheses
If *Paramecium* are attracted to light, then they will move to the light zone on a glass slide containing both light and dark zones.
If *Euglena* are attracted to light, then they will move to the light zone on a glass slide containing both light and dark zones.

538

DESIGN YOUR OWN BioLab

How do *Paramecium* and *Euglena* respond to light?

Members of the genus Paramecium *are ciliated protozoans—unicellular, heterotrophic protists that move around in search of small food particles.* Euglena *are unicellular algae—autotrophic protists that usually contain numerous chloroplasts. In this* BioLab, *you'll investigate how these two protists respond to light in their environment.*

PREPARATION

Problem
Do both *Paramecium* and *Euglena* respond to light and do they respond in different ways? Among the members of your group, decide on one type of protist activity that would constitute a response to light.

Hypotheses
Decide on one hypothesis that you will test. Your hypothesis might be that *Paramecium* will not respond to light and *Euglena* will respond, or that *Paramecium* will move away from light and *Euglena* will move toward light.

Objectives
In this BioLab, you will:
■ **Prepare** slides of *Paramecium* and *Euglena* cultures and observe swimming patterns in the two organisms.
■ **Compare** how these two different protists respond to light.

Possible Materials
Euglena culture
Paramecium culture
microscope
microscope slides
dropper
methyl cellulose
coverslips
metric ruler
index cards
scissors
toothpicks

Safety Precautions
Always wear goggles in the lab. Use caution when working with a microscope, glass slides, and coverslips. Wash your hands with soap and water immediately after working with protists and chemicals.

Skill Handbook
Use the **Skill Handbook** if you need additional help with this lab.

538 PROTISTS

PLAN THE EXPERIMENT

Teaching Strategies
■ Provide time for students to observe the organisms' size, speed, and mobility under a microscope before they begin.
■ Have students use low power to observe *Paramecium* (a 5× objective works best) and high power to observe *Euglena*.
■ Advise students that they must count organisms quickly.

Possible Procedures
■ Index cards can provide a light and dark microscope zone. Cut the cards to fit over part of the coverslip. Students should focus on the edge of the card to observe each organism's direction of movement. If possible, use coverslips (22 × 30 mm or 22 × 40 mm) that cover a large area.
■ Prepare or purchase methyl cellulose.

PLAN THE EXPERIMENT

1. Decide on an experimental procedure that you can use to test your hypothesis.
2. Record your procedure, step-by-step, and list the materials you will be using.
3. Design a data table in which to record your observations and results.

Check the Plan

Discuss all the following points with other group members to determine your final procedure.

1. What variables will you have to measure?
2. What will be your control?
3. What will be the shape of the light-controlled area(s) on your microscope slide?
4. Decide who will prepare materials, make observations, and record data.
5. *Make sure your teacher has approved your experimental plan before you proceed further.*

6. To mount drops of *Paramecium* culture and *Euglena* culture on microscope slides, use a toothpick to place a small ring of methyl cellulose on a clean microscope slide. Place a drop of *Paramecium* or *Euglena* culture within this ring. Place a coverslip over the ring and culture. The thick consistency of methyl cellulose should slow down the organisms for easy observation.
7. Make preliminary observations of swimming *Paramecia* and *Euglena*. Then think again about the observation times that you have planned. Maybe you will decide to allow more or less time between your observations.
8. Carry out your experiment.

ANALYZE AND CONCLUDE

1. **Checking Your Hypothesis** Did your data support your hypothesis? Why or why not?
2. **Comparing and Contrasting** Compare and contrast all the responses of the *Paramecium* and *Euglena* to both light and darkness. What explanations can you suggest for their behavior?
3. **Making Inferences** Can you use your results to suggest what sort of responses to light and darkness you might observe using other heterotrophic or autotrophic protists?

Going Further

Project You may want to extend this experiment by varying the shapes or relative sizes of light and dark areas or by varying the brightness or color of the light. In each case, make hypotheses before you begin. Keep your data in a notebook, and draw up a table of your results at the end of your investigations.

BIOLOGY Online To find out more about protists, visit the Glencoe Science Web site.
science.glencoe.com

ANALYZE AND CONCLUDE

1. Answers may vary. Data must be used to either support or reject student hypotheses.
2. *Euglena* are attracted to light. *Paramecium* avoid bright light. *Euglena* need light to make food. Heterotrophic, *Paramecium* can find food in dim and dark places.
3. Most autotrophs show a positive response to light whereas most heterotrophs show a negative response.

Error Analysis

Advise students to do several trials for each organism. Ask them why this is important.

✔ Assessment

Portfolio Have students write an evaluation of what they learned in this investigation. Use the Performance Task Assessment List for Writing in Science in **PASC,** p. 87. **L2**

Going Further

Students may wish to test the response of these protists to other factors, such as different concentrations of salt solutions. **L3** **ELL**

19.3 SLIME MOLDS, WATER MOLDS, AND DOWNY MILDEWS **539**

Data and Observations

Data should indicate that *Euglena* are attracted to light; *Paramecium* are not.

Resource Manager

BioLab and MiniLab Worksheets, pp. 91-92 **L2**

Social Studies
Connection

The Irish Potato Famine

Purpose

Students will learn why a balance between humans and the organisms in their environment is important.

Teaching Strategies

■ Ask what influences other than famine might lead to mass emigrations? *Climatic changes, particularly droughts, political upheavals, wars, economic devastation.*

Connection to Biology

With fewer infestations of *Phytophthora infestans*, it is more difficult to determine its evolutionary relationships. Outbreaks of *P. infestans* are minimized by the use of fungicides, but new resistant strains have arisen that are difficult to control and that destroy many crops. In addition, the overuse of pesticides has increased the number of crop-damaging insects, making some crops more susceptible to insect destruction.

A funguslike protist known as Phytophthora infestans *causes a disease called potato blight. Between the years 1845–1847, the disease damaged or totally destroyed the Irish potato crop—a primary food source for about one-third of the Irish population at the time. A severe seven-year famine resulted.*

The potato was first grown in Ireland in the late 1500s and within 200 years was a widespread food crop. Potatoes grew well in the temperate, rainy Irish climate. They were not vulnerable to many diseases, needed no processing, and were nutritious. In addition, most Irish people were tenant farmers with farms averaging fewer than 15 acres. Enough potatoes could be grown on a few acres to support an entire family for a year. By 1845, the potato crop fed a large percentage of the Irish population.

Invasion of a downy mildew Historians suggest that *Phytophthora infestans* arrived in Ireland in 1845 on a ship that arrived from North America. The wet conditions in Ireland during July and August that year were ideally suited for the spread of the protist, a downy mildew, which is classified in the phylum

Oomycota. In addition, the wind probably widely dispersed the *P. infestans* spores, infecting the leaves and stems of mature potato plants. The spores also washed into the soil and infected the underground stems, or tubers, of the potatoes. The potato blight damaged the Irish potato crop in 1845, and it destroyed nearly the entire potato crop in one week's time in the summer of 1846.

Mass emigration of the Irish Many Irish people starved in the years that followed. Many others emigrated from Ireland. By 1855, the population of Ireland had fallen from about eight million to four million. The Irish immigrated primarily to four countries, the United States, England, Canada, or Australia. The large numbers of immigrants greatly affected the social structure of these four countries as well as that of Ireland.

Most of the Irish quickly adapted to their new homes. For example, in the United States, some Irish became politically active. In fact, John F. Kennedy, who was president of the United States from 1960–1963, was the great-great-grandson of an Irish tenant farmer who immigrated to the United States in 1848.

Many Irish people who immigrated to the United States at the time of this famine settled in the large East Coast cities, such as New York, Philadelphia, and Boston. Their descendants are now an integral part of American society and live throughout the United States.

CONNECTION TO BIOLOGY

Today, fungicides, chemicals that prevent fungal growth, control outbreaks of *Phytophthora infestans*. How might this information affect research about the evolutionary relationships of funguslike protists?

BIOLOGY Online To find out more about the Irish potato famine and *Phytophthora*, visit the Glencoe Science Web site. **science.glencoe.com**

BIOLOGY Online Note Internet addresses that you find useful in the space below for quick reference.

_____

_____

_____

_____

SUMMARY

Section 19.1

The World of Protists

Main Ideas

- Kingdom Protista is a diverse group of living things that contains animal-like, plantlike, and funguslike organisms.
- Some protists are heterotrophs, some are autotrophs, and some get their nutrients by decomposing organic matter.
- Amoebas move by extending pseudopodia. The flagellates use one or more flagella to move. The beating of cilia produces cilliate movement. Sporozoans live as parasites and produce spores.

Vocabulary

algae (p. 519)
asexual reproduction (p. 521)
ciliate (p. 522)
flagellate (p. 522)
protozoan (p. 519)
pseudopodia (p. 520)
spore (p. 524)
sporozoan (p. 524)

Section 19.2

Algae: Plantlike Protists

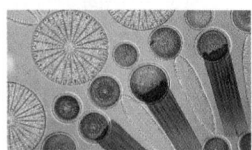

Main Ideas

- Algae are unicellular and multicellular photosynthetic autotrophs. Unicellular species include the euglenoids, diatoms, dinoflagellates, and some green algae. Multicellular species include red, brown, and green algae.
- Green, red, and brown algae, often called seaweeds, have complex life cycles that alternate between haploid and diploid generations.

Vocabulary

alternation of generations (p. 532)
colony (p. 531)
fragmentation (p. 531)
gametophyte (p. 532)
sporophyte (p. 532)
thallus (p. 530)

Section 19.3

Slime Molds, Water Molds, and Downy Mildews

Main Ideas

- Slime molds, water molds, and downy mildews are funguslike protists that decompose organic material to obtain nutrients.
- Plasmodial and cellular slime molds change in appearance and behavior before producing reproductive structures.

Vocabulary

plasmodium (p. 534)

Main Ideas

Summary statements can be used by students to review the major concepts of the chapter.

Using the Vocabulary

To reinforce chapter vocabulary, use the Content Mastery Booklet and the activities in the Interactive Tutor for Biology: The Dynamics of Life on the Glencoe Science Web site. science.glencoe.com

 **All Chapter Assessment** questions and answers have been validated for accuracy and suitability by The Princeton Review.

UNDERSTANDING MAIN IDEAS

1. a
2. c
3. d
4. a

UNDERSTANDING MAIN IDEAS

1. Which organisms cause red tides?
   a. dinoflagellates   c. green algae
   b. euglenoids   d. red algae

2. Which organelle in protists is able to eliminate excess water?
   a. anal pore   c. contractile vacuole
   b. mouth   d. gullet

3. Which of the following pairs contains terms most related to each other?
   a. paramecium—alternation of generations
   b. asexual reproduction—gametophyte
   c. sporozoan—cilia
   d. amoeba—pseudopodia

4. Producers in aquatic food chains include _____.
   a. algae   c. slime molds
   b. protozoans   d. amoebas

CHAPTER 19 ASSESSMENT **541**

GLENCOE TECHNOLOGY

 **VIDEOTAPE**
MindJogger Videoquizzes
Chapter 19: *Protists*
Have students work in groups as they play the videoquiz game to review key chapter concepts.

 ## Resource Manager

Chapter Assessment, pp. 109-114
MindJogger Videoquizzes
ExamView® Pro Software
BDOL Interactive CD-ROM, Chapter 19 quiz

5. a
6. d
7. b
8. b
9. a
10. c
11. eukaryotes
12. sporozoan
13. colony
14. flagellum
15. fragmentation
16. algae
17. gametophyte; sporophyte
18. protozoans
19. plasmodium
20. cilia; flagellum

APPLYING MAIN IDEAS

21. Finding a suitable host; producing many spores will improve the possibility of survival because some of the spores may find a host.
22. oak forest—dry conditions stimulate spore-producing structures.
23. Answers will vary, but may include cilia, pseudopodia, flagella, eyespots, contractile vacuoles, or others.

---

5. Protists are classified on the basis of their _____.
  a. nutrition
  b. method of locomotion
  c. reproductive abilities
  d. size

6. Euglenoids are unique algae because of their _____.
  a. flagella
  b. cilia
  c. silica walls
  d. heterotrophic nature

7. Which of the following is not a protist?

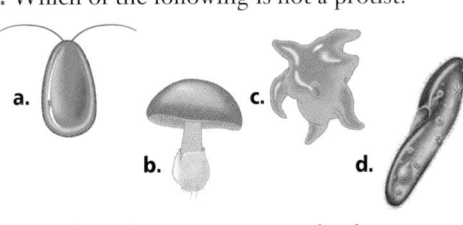

8. The algae that can survive in the deepest water are the _____?
  a. brown algae   **c.** diatoms
  b. red algae   **d.** green algae

9. The largest and most complex of brown algae are the _____.
  a. kelp   **c.** sea lettuce
  b. *Chlamydomonas*   **d.** *Spirogyra*

10. Which of the following are protected by armored plates?
  a. kelp   **c.** dinoflagellates
  b. fire algae   **d.** diatoms

11. Unlike bacteria, all protists are _____.

THE PRINCETON REVIEW | **TEST-TAKING TIP**

Work Weak Areas, Maintain Strong Ones
It's sometimes difficult to focus on all the concepts needed for a test. So ask yourself "What's my strongest area?" and "What's my weakest area?" Focus most of your energy on your weak areas. But also put in some "upkeep" time in your strongest areas.

---

12. The protozoan that causes malaria is classified as a _____ because it is a spore-producing parasite.

13. *Volvox* is an alga that lives in a _____, a group of cells in close association.

14. What type of structure does the protist shown to the right use to move?

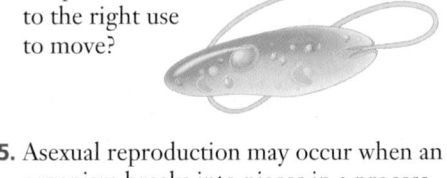

15. Asexual reproduction may occur when an organism breaks into pieces in a process known as _____.

16. Diatoms and dinoflagellates are two phyla of protists in the group collectively called _____.

17. The haploid form of an alga that has alternation of generations is known as the _____, and the diploid form is the _____.

18. Amoebas, cilliates, flagellates, and sporozoans are collectively called _____.

19. Individual cells of a cellular slime mold may fuse to form a structure that resembles a _____ and that reproduces.

20. Paramecia move about using short projections called _____, but euglenas move using a long projection, the _____.

APPLYING MAIN IDEAS

21. Why is it a disadvantage for a sporozoan to be a parasite? How might a sporozoan's method of asexual reproduction offset this disadvantage?

22. In which ecosystem would a plasmodial slime mold transform itself into spore-producing structures more frequently: a rainy forest in the Pacific Northwest or a dry, oak forest in the Midwest? Explain your answer.

23. Give three examples of organelles that help protists maintain homeostasis in their environments.

---

24. Up to the late 1800s, malaria was common in the extreme southeastern part of the United States. In an attempt to fight the disease, ponds and wetlands were often filled in or drained. How do you suppose this action helped cut down on the number of malaria cases?

THINKING CRITICALLY

25. Observing and Inferring Why do you suppose many people who own aquariums add snails to them?

26. Formulating Hypotheses In agricultural regions where farmers use large amounts of nitrogen fertilizers in their fields, local ponds and lakes often develop a thick, green scum containing algae in late summer. Hypothesize why this happens.

27. Sequencing Sequence the stages of both sexual and asexual reproduction in diatoms.

28. Concept Mapping Complete the concept map by using the following vocabulary terms: amoebas, sporozoans, flagellates, protozoans, ciliates

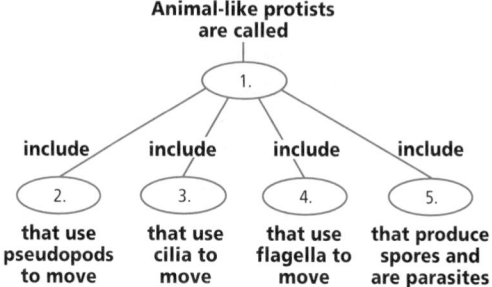

Animal-like protists are called

1.

include — include — include — include

2. — 3. — 4. — 5.

that use pseudopods to move | that use cilia to move | that use flagella to move | that produce spores and are parasites

CD-ROM

For additional review, use the assessment options for this chapter found on the *Biology: The Dynamics of Life Interactive CD-ROM* and on the Glencoe Science Web site. **science.glencoe.com**

ASSESSING KNOWLEDGE & SKILLS

During a summer ecology class, a group of high school students studied unicellular algae at a site in the middle of a pond. For three days and nights, they measured the number of cells in the water at various depths. They produced the following graph based on their data.

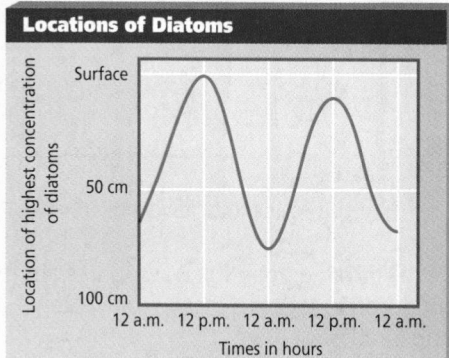

Locations of Diatoms

Location of highest concentration of diatoms (Surface, 50 cm, 100 cm) vs. Times in hours (12 a.m., 12 p.m., 12 a.m., 12 p.m., 12 a.m.)

Interpreting Data Use the graph to answer the following questions.

1. At what time were the highest concentrations of diatoms at the surface?
  a. midnight **c.** 3 a.m.
  b. noon **d.** 6 p.m.

2. At what time were the highest concentrations of diatoms about a meter below the surface?
  a. midnight **c.** 3 a.m.
  b. noon **d.** 6 p.m.

3. Which of the following is the best description of the movement of diatoms in the water column?
  a. 6-hour cycle
  b. 12-hour cycle
  c. 24-hour cycle
  d. irregular cycling

4. Interpreting Data Why might the diatoms show the pattern found by the group of high school students?

24. Mosquitoes breed in water. The elimination of breeding grounds reduces the mosquito population and reduces the risk of contracting malaria.

THINKING CRITICALLY

25. Snails help keep the growth of unwanted algae in check.

26. The nitrogen fertilizers run off into water sources, where they stimulate the growth and reproduction of algae.

27. A diatom reproduces asexually for several generations, with its cell walls getting progressively smaller and smaller. Eventually, the cell undergoes meiosis and releases gametes that grow a new cell with a large cell wall. Then, the new diatoms begin reproducing asexually.

28. 1. Protozoans; 2. Amoebas; 3. Ciliates; 4. Flagellates; 5. Sporozoans

ASSESSING KNOWLEDGE & SKILLS

1. b
2. a
3. c
4. They are photosynthetic and move to the surface to utilize sunlight during daylight hours.

Chapter 20 Organizer

Refer to pages 4T-5T of the Teacher Guide for an explanation of the National Science Education Standards correlations.

Section	Objectives	Activities/Features
Section 20.1 **What Is a Fungus?** National Science Education Standards UCP.1, UCP.2, UCP.5; A.1, A.2; C.1, C.4, C.5, C.6; F.5 (1 session, 1 block)	1. **Identify** the basic characteristics of fungi. 2. **Explain** the role of fungi as decomposers and how this role affects the flow of both energy and nutrients through food chains.	**MiniLab 20-1:** Growing Mold Spores, p. 546 **Problem-Solving Lab 20-1,** p. 550
Section 20.2 **The Diversity of Fungi** National Science Education Standards UCP.1-5; A.1, A.2; C.1, C.3, C.4, C.5, C.6; F.1, F.4, F.5; G.1-3 (3 sessions, ¹/₂ block)	3. **Identify** the four major divisions of fungi. 4. **Distinguish** among the ways spores are produced in zygomycotes, ascomycotes, and basidiomycotes. 5. **Summarize** the ecological roles of lichens and mycorrhizae.	**MiniLab 20-2:** Examining Mushroom Gills, p. 554 **Inside Story:** The Life of a Mushroom, p. 555 **Problem-Solving Lab 20-2,** p. 558 **Internet BioLab:** Does temperature affect the metabolic activity of yeast? p. 560 **Social Studies Connection:** The Dangers of Fungi, p. 562

Need Materials? Contact Carolina Biological Supply Company at 1-800-334-5551 or at **http://www.carolina.com**

MATERIALS LIST

BioLab

p. 560 bromthymol blue solution, drinking straw, small test tubes (4), large test tubes, (3), one-hole stoppers with glass tube inserts (3), yeast/white corn syrup mixture, water/white corn syrup mixture, water/yeast mixture, test-tube rack, 250-mL beakers (3), ice cubes, thermometer, hot plate, 50-mL graduated cylinder, glass-marking pencil, 10 cm length rubber tubing (3), aluminum foil

MiniLabs

p. 546 bakery bread, water, plate, self-seal plastic bags (2), forceps, micro-scope, microscope slide, coverslip, dropper, water

p. 554 fresh mushrooms, paper bag, white paper

Alternative Lab

p. 546 small paper cups, macaroni, aluminum foil, water, cardboard, oatmeal flakes, cotton swab, mold spores, plastic wrap

Quick Demos

p. 547 fresh mushrooms, paper, pencil
p. 553 microscope, microscope slide, coverslip, dropper, water, forceps, preserved specimen of *Peziza*

Key to Teaching Strategies

L1 Level 1 activities should be appropriate for students with learning difficulties.

L2 Level 2 activities should be within the ability range of all students.

L3 Level 3 activities are designed for above-average students.

ELL ELL activities should be within the ability range of English Language Learners.

COOP LEARN Cooperative Learning activities are designed for small group work.

P These strategies represent student products that can be placed into a best-work portfolio.

These strategies are useful in a block scheduling format.

Teacher Classroom Resources

Section	Reproducible Masters	Transparencies
Section 20.1 **What Is a Fungus?**	Reinforcement and Study Guide, pp. 87-88 **L2** Concept Mapping, p. 20 **L3** **ELL** Critical Thinking/Problem Solving, p. 20 **L3** BioLab and MiniLab Worksheets, p. 93 **L2** Tech Prep Applications, pp. 29-30 **L2** Content Mastery, pp. 97-98, 100 **L1**	Section Focus Transparency 48 **L1** **ELL**
Section 20.2 **The Diversity of Fungi**	Reinforcement and Study Guide, pp. 89-90 **L2** BioLab and MiniLab Worksheets, pp. 94-96 **L2** Laboratory Manual, pp. 141-144 **L2** Content Mastery, pp. 97, 99-100 **L1**	Section Focus Transparency 49 **L1** **ELL** Basic Concepts Transparency 31 **L2** **ELL** Basic Concepts Transparency 32 **L2** **ELL** Reteaching Skills Transparency 31 **L1** **ELL**

Assessment Resources

Chapter Assessment, pp. 115-120
MindJogger Videoquizzes
Performance Assessment in the Biology Classroom
Alternate Assessment in the Science Classroom
ExamView® Pro Software 💾
BDOL Interactive CD-ROM, Chapter 20 quiz

Additional Resources

Spanish Resources **ELL**
English/Spanish Audiocassettes **ELL**
Cooperative Learning in the Science Classroom **COOP LEARN**
Lesson Plans/Block Scheduling

 **NATIONAL GEOGRAPHIC** — **Teacher's Corner**

Index to National Geographic Magazine
The following articles may be used for research relating to this chapter:
"Leafcutters: Gardeners of the Ant World," by Mark W. Moffett, July 1995.
"The Wild World of Compost," by Cecil E. Johnson, August 1980.
"Bizarre World of Fungi," by Paul A. Zahl, October 1965.

GLENCOE TECHNOLOGY

The following multimedia resources are available from Glencoe.

Biology: The Dynamics of Life
CD-ROM **ELL**
 Video: *Fungal Decay*
Exploration: *The Five Kingdoms*
BioQuest: *Biodiveristy Park*
Animation: *Life Cycle of a Mushroom*

Videodisc Program
 Fungal Decay
Life Cycle of a Mushroom

The Secret of Life Series
 Six Kingdoms
Fungi
Mycorrhizae

Theme Development

The theme of **unity within diversity** is prominent in discussions of the diversity of fungi. The theme of **energy** is evident when the role of fungi as decomposers is discussed.

0:00 OUT OF TIME?

If time does not permit teaching the entire chapter, use the BioDigest at the end of the unit as an overview.

READING BIOLOGY

Glencoe's *Biology: The Dynamics of Life* contains many resources to assist a student's reading skills. Each chapter contains figures with expanded captions that expand on written material. Word Origins, located along the side of text, expand knowledge of biology vocabulary. Glencoe's Content Mastery Booklet helps develop reading skills while reinforcing content. In addition, use the Interactive Tutor for *Biology: The Dynamics of Life* on the Glencoe Web site to reinforce vocabulary.
science.glencoe.com

Chapter

20 Fungi

What You'll Learn

- You will identify the characteristics of fungi.
- You will differentiate among the divisions of fungi.

Why It's Important

By decomposing organic matter, fungi clean your environment and recycle nutrients.

READING BIOLOGY

Before starting the chapter, think of places you have seen fungi before. Write down as many characteristics about these fungi that come to mind. As you read about the types of fungi described in the text, make corrections or additions to your list.

To find out more about fungi, visit the Glencoe Science Web site.
science.glencoe.com

Chanterelles, mushrooms eaten as delicacies, grow under oak and pine trees from early spring through late fall. Other fungi such as the deadly *Amanita muscaria* mushroom (inset) should never be eaten.

544 FUNGI

Multiple Learning Styles

Look for the following logos for strategies that emphasize different learning modalities.

Kinesthetic Activity, p. 548; Quick Demo, p. 553; Project, p. 555; Portfolio, p. 558

Visual-Spatial Quick Demo, p. 547; Meeting Indivdual Needs, p. 548; Portfolio, pp. 548, 552; Biology Journal, p. 552; Enrichment, p. 557

Intrapersonal Enrichment, p. 548

Linguistic Biology Journal, p. 549; Enrichment, pp. 552, 556; Discussion, p. 559

Logical-Mathematical Portfolio, p. 557

Naturalist Reteach, pp. 549, 559; Tech Prep, p. 554; Meeting Individual Needs, p. 557

20.1 What Is a Fungus?

SECTION PREVIEW

Objectives

Identify the basic characteristics of fungi.

Explain the role of fungi as decomposers and how this role affects the flow of both energy and nutrients through food chains.

Vocabulary

hypha
mycelium
chitin
haustoria
budding
sporangium

Have you ever seen mushrooms that grow in a ring like the one shown here? The visible mushrooms are only one part of the fungus. Beneath the soil's surface are threadlike filaments that may grow a long distance away from the above-ground ring of mushrooms. These filaments can grow for a long time before they produce the surface mushrooms. Mushrooms that grow in rings are only one of many types of fungi, all of which share certain characteristics.

A ring of mushrooms

The Characteristics of Fungi

Fungi are everywhere—in the air and water, on damp basement walls, in gardens, on foods, and sometimes even between people's toes. Some fungi are large, bright, and colorful, whereas others are easily overlooked, as shown in *Figure 20.1*. Many have descriptive names such as stinkhorn, puffball, rust, or ringworm. Many

Figure 20.1
Fungi vary in form, size, and color.

A Bird's nest fungi look like nests, complete with eggs.

B Brightly colored coral fungi resemble ocean corals.

C A fungus killed this insect by feeding on its tissues.

20.1 WHAT IS A FUNGUS? **545**

![Assessment Planner]

Assessment Planner

Portfolio Assessment
  Portfolio, TWE, pp. 548, 552, 557, 558
  Problem-Solving Lab, TWE, p. 550
  Assessment, TWE, p. 556
Performance Assessment
  MiniLab, SE, pp. 546, 554
  MiniLab, TWE, p. 546
  Alternative Lab, TWE, pp. 546-547
  Assessment, TWE, p. 548
  BioLab, SE, p. 561

  BioLab, TWE, p. 561
Knowledge Assessment
  Alternative Lab, TWE, pp. 546-547
  Section Assessment, SE, pp. 550, 559
  Problem-Solving Lab, TWE, p. 558
  Assessment, TWE, p. 559
  Chapter Assessment, SE, pp. 563-565
Skill Assessment
  Assessment, TWE, p. 549
  MiniLab, TWE, p. 554

Prepare

Key Concepts

This section describes the structure of fungi and how fungi obtain nutrients and reproduce.

Planning

■ Obtain bread and plastic bags for MiniLab 20-1 and mushrooms for the Quick Demo.
■ Begin to grow molds for the Alternative Lab and the Activity one week ahead of time.

1 Focus

Bellringer

Before presenting the lesson, display **Section Focus Transparency 48** on the overhead projector and have students answer the accompanying questions.
L1 **ELL**

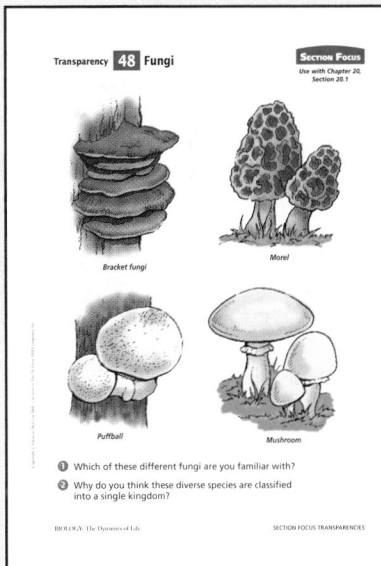

![Resource Manager]

Resource Manager

Section Focus Transparency 48 and Master **L1** **ELL**

545

2 Teach

MiniLab 20-1

Purpose

Students will identify a requirement for bread mold to grow.

Process Skills

observe and infer, experiment

Teaching Strategies

■ Mold grows quickly on bakery bread that lacks preservatives.

■ Identify calcium propionate on bread ingredient labels as a mold inhibitor. Have students design an experiment to test its effectiveness.

■ Have students wear aprons, gloves, and goggles and wash their hands after they finish. Remind them to use caution when working with glass slides and coverslips and when opening the bags containing mold.

■ Autoclave or incinerate all fungal samples to destroy them.

Expected Results

After 4-6 days, mold will grow only on moist bread slices.

Analysis

1. Only moist bread had mold. It suggests that there might be spores in the classroom air.

2. Molds require water and food for growth.

✔ Assessment

Performance Have students design an experiment to show that air contains mold spores that will grow on bread. Use the Performance Task Assessment List for Designing an Experiment in **PASC**, p. 23. **L2**

546

MiniLab 20-1 — Observing and Inferring

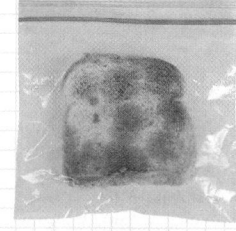

Growing Mold Spores Any mold spore that arrives in a favorable place can germinate and produce hyphae. Can you identify a condition necessary for the growth of bread mold spores?

Procedure

1. Place two slices of freshly baked bakery bread on a plate. Sprinkle some water on one slice to moisten its surface. Leave both slices uncovered for several hours.

2. Sprinkle a little more water on the moistened slice, and place both slices in their own plastic, self-seal bags. Trap air in each bag so that the plastic does not touch the bread's surface. Then seal the bags and place them in a darkened area at room temperature.

3. After five days, remove the bags and look for mold.

4. Remove a small piece of mold with a forceps, place it on a slide in a drop of water, and add a coverslip. Observe the mold under a microscope's low power and high power. **CAUTION: Use caution when working with a microscope, glass slides, and coverslips. Wash your hands with soap and water after working with mold. Dispose of the mold as your teacher directs.**

Analysis

1. Did you observe mold growth on the moistened bread? On the dry bread? How does this experiment demonstrate that there are mold spores in your classroom?

2. What conclusions can you draw about the conditions necessary for the growth of a bread mold?

---

species grow best in moist environments at warm temperatures between 20°C and 30°C. You are, however, probably familiar with molds that grow at much lower temperatures on left-over foods in your refrigerator.

Fungi used to be classified in the plant kingdom because, like plants, many fungi grow anchored in soil and have cell walls. However, as biologists learned more about fungi, they realized that fungi belong in their own kingdom.

The structure of fungi

Although there are a few unicellular types of fungi, such as yeasts, most fungi are multicellular. The basic structural units of multicellular fungi are their threadlike filaments called **hyphae** (HI fee) (singular, hypha), which develop from fungal spores, as shown in *Figure 20.2*. Hyphae elongate at their tips and branch extensively to form a network of filaments called a **mycelium** (mi SEE lee um). There are different types of hyphae in a mycelium. Some anchor the fungus, some invade the food source, and others form fungal reproductive structures. Use the *MiniLab* on this page to observe the hyphae of some bread mold you grow.

Figure 20.2
A germinating fungal spore produces hyphae that branch to form a mycelium.

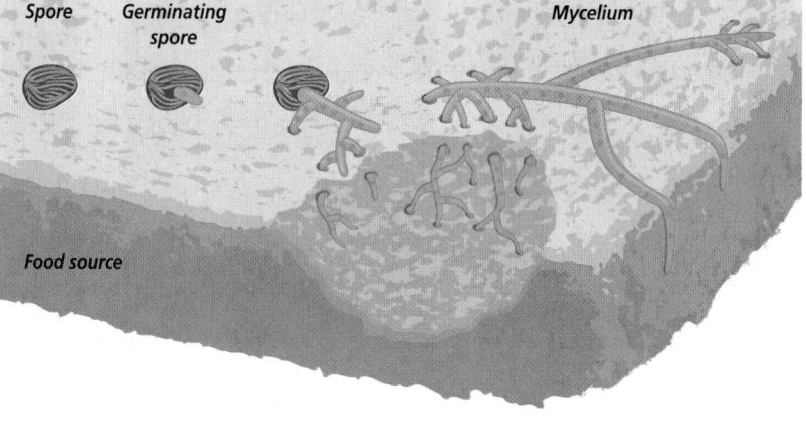

Alternative Lab

Mold Growth

Purpose

Students will show that mold grows on any moist, organic material.

Materials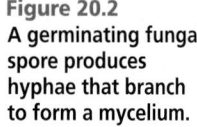

small paper cups, macaroni, aluminum foil, water, cardboard pieces, oatmeal flakes, swab with mold spores, plastic wrap

Procedure

Give students the following directions.

1. Number eight small paper cups 1-8 and label each with your name and date. Add the following to each cup (1/8 full): cup #1 cardboard, #2 moist cardboard, #3 oatmeal, #4 moist oatmeal, #5 macaroni, #6 moist macaroni, #7 aluminum foil ball, #8 wet aluminum foil ball.

2. Rub a cotton swab with mold spores on the surface of each cup's contents.

You can use a magnifying glass to see individual hyphae in molds that grow on bread. However, the hyphae of mushrooms are much more difficult to see because they are tightly packed, forming a dense mass.

Unlike plants, which have cell walls made of cellulose, the cell walls of most fungi contain a complex carbohydrate called **chitin** (KITE un). Chitin gives the fungal cell walls both strength and flexibility.

Inside hyphae

In many types of fungi, cross walls called septa (singular, septum) divide hyphae into individual cells that contain one or more nuclei, *Figure 20.3*. Septa are usually porous, allowing cytoplasm and organelles to flow freely and nutrients to move rapidly from one part of a fungus to another.

Some fungi consist of hyphae with no septa. When you look at these hyphae under a microscope, you see hundreds of nuclei streaming along in a continuous flow of cytoplasm. As in hyphae with septa, the flow of cytoplasm quickly and efficiently disperses nutrients and other materials throughout the fungus.

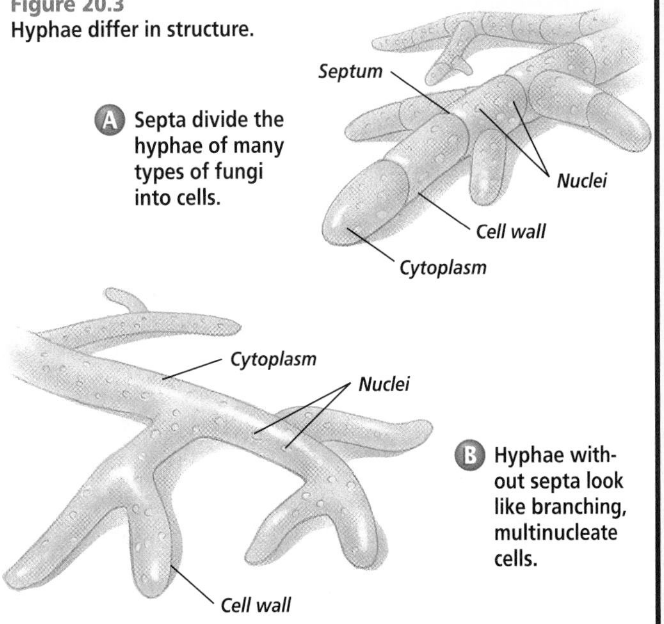

Adaptations in Fungi

Fungi can have some negative aspects—spoiled food, diseases, and poisonous mushrooms. However, they play an important role in the interactions of organisms because they decompose large quantities of Earth's organic wastes. In a world without fungi, huge amounts of wastes, dead organisms, and debris, which consist of complex organic substances, would litter Earth. Many fungi, along with several species of bacteria and protists, are decomposers. They break down complex organic substances into raw materials

Figure 20.3
Hyphae differ in structure.

A Septa divide the hyphae of many types of fungi into cells.

Septum
Nuclei
Cell wall
Cytoplasm

Cytoplasm
Nuclei

B Hyphae without septa look like branching, multinucleate cells.

Cell wall

that living organisms need. Thanks to these organic decomposers, fallen leaves, animal carcasses, and other organic materials that become waste are eventually decomposed.

How fungi obtain food

Unlike plants and some protists, fungi cannot produce their own food. Fungi are heterotrophs, and they must use a process called extracellular digestion to obtain nutrients. In this process, food is digested outside a fungus's cells, and the digested products are then absorbed. For example, as some hyphae grow into the cells of an orange, they release digestive enzymes that break down the large organic molecules of the orange into smaller molecules. These small molecules diffuse into the fungal hyphae and move in the free-flowing cytoplasm to where they are needed for growth, repair, and reproduction. The more a mycelium grows, the more surface area becomes available for nutrient absorption.

20.1   WHAT IS A FUNGUS?   **547**

3. Cover each cup with plastic wrap and place the cups in a designated area.
4. Check the cups each day for a week.
5. Design a data table to record your observations, including the amount of mold growth.
6. Dispose of any mold that grows as your teacher directs.

Analysis

1. What evidence suggests that mold feeds on organic matter? *All organic materials showed mold growth. The only inorganic material, aluminum foil, showed no growth.*
2. Explain the role of the cups without water. *They were controls, showing that molds require moisture to grow.*

Activity

 **Kinesthetic** Use any mold that grows on food to illustrate the appearance of hyphae. Have students prepare some wet mounts of the mold and view it under low- and then high-power magnification. Have students wear aprons, gloves, and goggles and wash their hands after handling the fungal samples. Caution them to use care when working with glass slides and coverslips and when viewing a slide under high power. **L2** **ELL** 📦

Enrichment

Intrapersonal A fungus called *Aspergillus flavus* can invade stores of grains and nuts. It produces a chemical called aflatoxin, a carcinogen that can destroy the liver if ingested. Have students research the ways to identify tainted grain and nuts. **L3**

Using Science Terms

Have students compare the terms *extracellular digestion* and *intracellular digestion* and provide an example of each. **L2**

✔ Assessment

Performance Have students observe a slide of budding yeast cells. Ask them to sketch what they see, label their drawings, and write a paragraph about the process. **L2**

Resource Manager

Concept Mapping, p. 20
L3 **ELL**

Different feeding relationships

Fungi have different types of food sources. A fungus may be a saprophyte, a mutualist, or a parasite depending on its food source.

WORD *Origin*

haustoria
From the Latin word *haurire*, meaning "to drink." The hyphae that invade the cells of a host to absorb nutrients are called haustoria.

Saprophytes are decomposers and feed on waste or dead organic material. Mutualists live in a symbiotic relationship with another organism, such as an alga. Parasites absorb nutrients from the living cells of their hosts. Parasitic fungi may produce specialized hyphae called **haustoria,** (huh STOR ee uh), which penetrate and grow into host cells where they directly absorb the host cells' nutrients. You can see a diagram of haustoria invading host cells in *Figure 20.4.*

Figure 20.4
Fungi may be parasites, mutualists, or saprophytes.

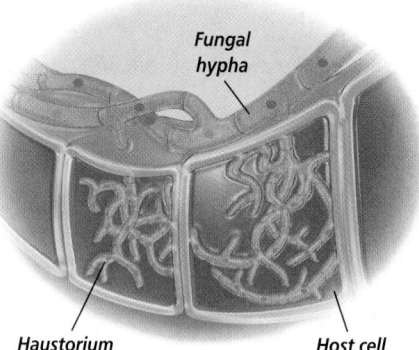

Ⓐ A parasitic fungus is killing this American elm tree.

Fungal hypha

Haustorium

Host cell

Ⓑ Fungi can produce haustoria that grow into host cells and absorb their nutrients.

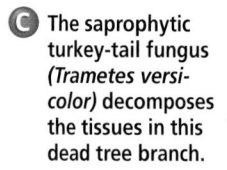

Ⓒ The saprophytic turkey-tail fungus (*Trametes versicolor*) decomposes the tissues in this dead tree branch.

Reproduction in Fungi

Depending on the species and on environmental conditions, a fungus may reproduce asexually or sexually. Fungi reproduce asexually by fragmentation, budding, or producing spores.

Fragmentation and budding

In fragmentation, pieces of hyphae that are broken off a mycelium grow into new mycelia. For example, when you prepare your garden for planting, you help fungi in the soil reproduce by fragmentation. This is because, every time you dig into the soil, your shovel slices through mycelia, fragmenting them. Most of the fragments will grow into new mycelia.

▸ MEETING INDIVIDUAL NEEDS

Learning Disabled

Visual-Spatial Give students a diagram of mold growing on bread. The diagram should also show sporangia, some of which are releasing spores. Have students label where reproduction, digestion, and feeding occur. **L1** **ELL** 📦

✔ Portfolio

Fungi Adaptations

Visual-Spatial Ask students to diagram the fungi they observed in the Getting Started and the Activity and compare them in writing, explaining their adaptations and feeding relationships. **L2** **P** 📦

Figure 20.5
Fungi reproduce asexually by budding, fragmentation, or spore production.

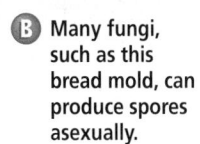 **A** Most yeasts reproduce asexually by budding.

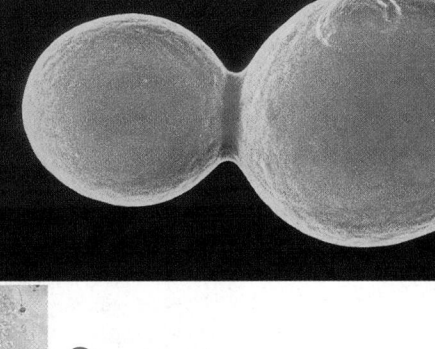

 B Many fungi, such as this bread mold, can produce spores asexually.

The unicellular fungi called yeasts often reproduce by a process called **budding**—a form of asexual reproduction in which mitosis occurs and a new individual pinches off from the parent, matures, and eventually separates from the parent. You can see a yeast cell and its bud in *Figure 20.5.*

Reproducing by spores

Recall that a spore is a reproductive cell that can develop into a new organism. Most fungi produce spores. When a fungal spore is transported to a place with favorable growing conditions, a threadlike hypha emerges and begins to grow, eventually forming a new mycelium. The mycelium becomes established in the food source.

In some fungi, after a while, specialized hyphae grow away from the rest of a mycelium and produce a spore-containing structure called a **sporangium** (spuh RAN jee uhm) (plural, sporangia)—a sac or case in which spores are produced. The tiny black spots you see in a bread mold's mycelium are a type of sporangium.

In fact, for most fungi, the specialized hyphal structures where the fungal spores are produced are usually the only part of a fungus you can see, and the sporangia often represent only a small fraction of the total organism.

Many fungi can produce two types of spores—one type by mitosis and the other by meiosis—at different times during their life cycles. One important criterion for classifying fungi into divisions is their patterns of reproduction, especially sexual reproduction, during the life cycle.

The adaptive advantages of spores

Many adaptive advantages of fungi involve spores and their production. First, the sporangia protect spores and, in some cases, prevent them from drying out until they are ready to be released. Second, most fungi produce a large number of spores at one time. For example, a puffball that measures only 25 cm in circumference produces about 1 trillion spores. Producing so many spores increases

WORD *Origin*

sporangium
From the Greek words *sporos*, meaning "seed," and *angeion*, meaning "vessel." Spores are produced in a sporangium.

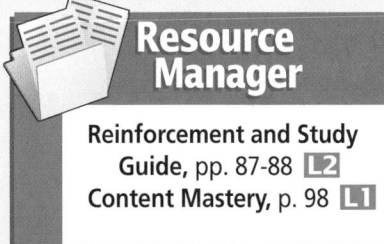

Problem-Solving Lab 20-1 — Analyzing Information

Why are chestnut trees so rare? The American chestnut tree (*Castanea dentata*) has almost disappeared from the United States, Italy, and France because of a disease known as chestnut blight, which is caused by the fungus *Cryphonectria parasitica*. Since 1900, three to four billion trees have been lost to chestnut blight.

Analysis

Fact: Spores of *C. parasitica* land on the bark of American chestnut trees and germinate. Hyphae grow below the bark and form a canker (diseased tissue) that spreads, producing large areas of dead tissue. Eventually, the nutrient and water supplies of the tree are cut off, and the tree dies.

Fact: *C. parasitica* reproduces by forming spores that are carried by wind, insects, birds, and rain to other trees that then become infected.

Fact: The Japanese chestnut tree *Castanea crenata* is resistant to the *C. parasitica* fungus. This resistance is partially due to the existence of weak fungal strains that cannot kill their host.

Thinking Critically

1. Why would it be difficult to control the disease by preventing spores from landing on healthy trees?
2. Based on how this fungus grows, why can't fungicides applied to the bark of an infected tree kill the fungus?
3. Suggest a solution to the problem in the United States knowing about the resistance of the Japanese chestnut species and the existence of weak disease-causing fungal strains. (Hint: Think about DNA technology.)

Figure 20.6
A passing animal or the pressure of raindrops may have caused these puffballs to discharge the cloud of spores that will be dispersed by the wind.

the germination rate and improves the species survival chances.

Finally, fungal spores are small and lightweight and can be dispersed by wind, water, and animals such as birds and insects. The wind will disperse the spores that the puffballs you see in *Figure 20.6* are releasing. Spores dispersed by wind can travel hundreds of kilometers. In the *Problem-Solving Lab* on this page, you can learn about the dispersal methods of a plant fungus that causes the disease called chestnut blight in chestnut trees.

Section Assessment

Understanding Main Ideas
1. What is the function of pores in hyphal septa?
2. Describe how a fungus obtains nutrients.
3. What role do fungi play in food chains?
4. How are the terms hypha and mycelium related?

Thinking Critically
5. Imagine you are a mycologist who finds an inhabited bird's nest. Explain why you would

expect to find several different types of fungi growing in the nest.

SKILL REVIEW
6. **Measuring in SI** Outline the steps you would take to calculate the approximate number of spores in a puffball fungus with a circumference of 10 cm. For more help, refer to *Practicing Scientific Methods* in the **Skill Handbook**.

Section
20.2 The Diversity of Fungi

SECTION PREVIEW

Objectives

Identify the four major divisions of fungi.

Distinguish among the ways spores are produced in zygomycotes, ascomycotes, and basidiomycotes.

Summarize the ecological roles of lichens and mycorrhizae.

Vocabulary

stolon
rhizoid
zygospore
gametangium
ascus
ascospore
conidiophore
conidium
basidium
basidiospore
mycorrhiza
lichen

Scientists who study fungi are called mycologists. A mycologist would look for certain features to help identify a mushroom like the one shown here. In particular, the mycologist would look at the spores the mushroom produced. This is because one basis for classifying fungi includes the reproductive characteristics of a fungus. As you'll see, the names of fungal divisions reflect the type of spores produced by fungi.

A *Mycena* mushroom (above) and the mushroom's gills (inset)

Zygomycotes

Have you ever taken a slice of bread from a bag and seen some black spots and a bit of fuzz on the bread's surface? If so, then you have probably seen *Rhizopus stolonifer*, a common bread mold. *Rhizopus* is probably the most familiar member of the division Zygomycota (zy goh mi KOH tuh). Many other members of about 1500 species of zygomycotes are also decomposers. Zygomycotes reproduce asexually by producing spores. They produce a different type of spore when they reproduce sexually. The hyphae of zygomycotes do not have septa that divide them into individual cells.

Growth and asexual reproduction

When a *Rhizopus* spore settles on a moist piece of bread, it germinates and hyphae begin to grow. Some hyphae called **stolons** (STOH lunz) grow horizontally along the surface of the bread, rapidly producing a mycelium. Some other hyphae form **rhizoids** (RI zoydz) that penetrate the food and anchor the mycelium in the bread. Rhizoids secrete enzymes needed for extracellular digestion and absorb the digested nutrients.

Asexual reproduction begins when some hyphae grow upward and develop sporangia at their tips. Asexual spores develop in the sporangia. When a sporangium splits open,

---

BIOLOGY Online Note Internet addresses that you find useful in the space below for quick reference.

_____

_____

_____

_____

Resource Manager

Section Focus Transparency 49 and Master L1 ELL

Section 20.2

Prepare

Key Concepts

Fungal diversity is described. Then, the characteristics of lichens and mycorrhizae are explained. Finally, possible origins of fungi are presented.

Planning

■ Buy supermarket mushrooms for MiniLab 20-2 and the Tech Prep. Obtain mushroom guides for the Tech Prep.

■ Collect lichens for the Enrichment.

■ Purchase some blue cheese for the Portfolio.

■ Purchase yeast cakes and white corn syrup and prepare a BTB solution for the Biolab. Prepare the short glass tubes.

1 Focus

Bellringer

Before presenting the lesson, display **Section Focus Transparency 49** on the overhead projector and have students answer the accompanying questions. L1 ELL

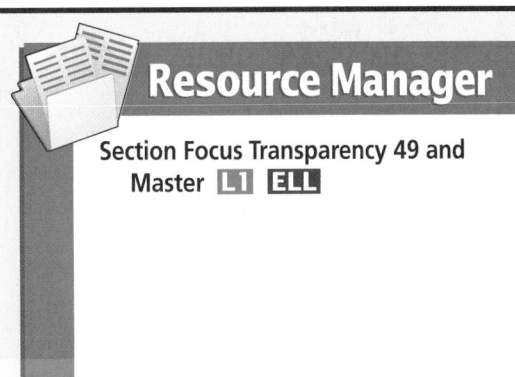

Reinforcement

Explain that hyphae have different names, depending on their function. Examples are rhizoids, stolons, and sporangia. Have students describe the role of each of these hyphae and then the role of the hyphae called a *mycelium*.

Using Science Terms

Have students find the meanings of the four terms *sporangium*, *gametangium*, *zygospore*, and *ascospore* and explain how the name of each structure suits its function.

Enrichment

Linguistic Ask students to research and report on the medical use of cyclosporine, a chemical produced from the soil fungus *Tolypocladium inflatum*. This chemical suppresses the immunity that causes the rejection of transplanted organs. More than 90% of transplant patients survive rejection today because of cyclosporine. **L3**

---

hundreds of spores are released. Those that land on a moist food supply germinate, form new hyphae, and reproduce asexually again.

Producing zygospores

Suppose that the bread on which *Rhizopus* was growing began to dry out. This unfavorable environmental condition could trigger the fungus to reproduce sexually. When zygomycotes reproduce sexually, they produce **zygospores** (ZI guh sporz), which are thick-walled spores that can withstand unfavorable conditions.

Sexual reproduction in *Rhizopus* occurs when haploid hyphae from two compatible mycelia, called plus and minus mating strains, grow together and then fuse. Where the haploid hyphae fuse, they each form a **gametangium** (gam ee TAN ghee uhm), a structure containing a haploid nucleus. When the haploid nuclei of the two gametangia fuse, a diploid zygote forms. The zygote develops a thick wall, becoming a dormant zygospore.

A zygospore may remain dormant for many months, surviving periods of drought, cold, and heat. When environmental conditions are favorable, the zygospore absorbs water, undergoes meiosis, and germinates to produce a hypha with a sporangium. Each haploid spore formed in the sporangium can grow into a new mycelium. Look at *Figure 20.7* to see how *Rhizopus* reproduces both sexually and asexually.

Figure 20.7
During its life cycle, the black bread mold, *Rhizopus stolonifera*, reproduces both asexually and sexually.

A Zygospores form where gametangia have fused.

B These *Rhizopus* sporangia are filled with thousands of haploid spores.

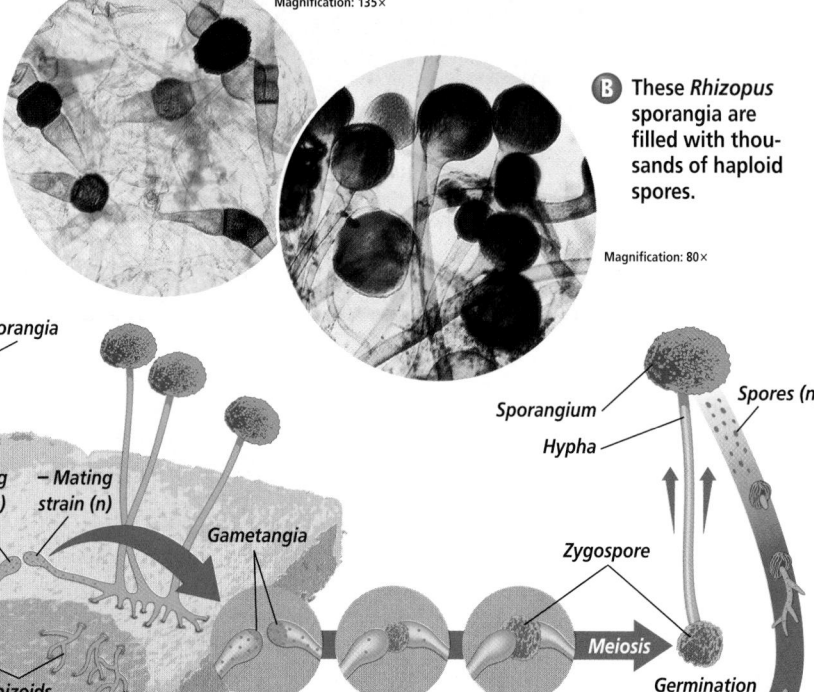

Magnification: 135×

Magnification: 80×

Asexual Reproduction · Sexual Reproduction

Fungus Reproduction

Visual-Spatial Have students make a table about the three ways that fungi reproduce asexually. Ask them to include an example of each and briefly describe and diagram each reproductive process. **L2**

Concept Map

Visual-Spatial Have students construct a concept map that uses the following structures and functions: stolons, sporangia, rhizoids, absorb nutrients, form spores, produce mycelia. **L1** **P**

Ascomycotes

The Ascomycota is the largest division of fungi, containing about 30 000 species. The ascomycotes are also called sac fungi. Both names refer to tiny saclike structures, each called an **ascus,** in which the sexual spores of the fungi develop. Because they are produced inside an ascus, the sexual spores are called **ascospores.**

During asexual reproduction, ascomycotes produce a different kind of spore. Hyphae grow up from the mycelium and elongate to form **conidiophores** (kuh NIHD ee uh forz). Chains or clusters of asexual spores called **conidia** develop from the tips of conidiophores. Wind, water, and animals disperse these haploid spores. Some conidia are shown in *Figure 20.8.*

Important ascomycotes

You've probably encountered a few types of sac fungi in your refrigerator in the form of blue-green, red, and brown molds on decaying foods. Other sac fungi are familiar to farmers and gardeners because they cause plant diseases such as apple scab and ergot of rye. Learn more about the dangers of fungi in the *Social Studies Connection* at the end of this chapter.

Not all sac fungi have a bad reputation. Ascomycotes can have many different forms, as you can see in *Figure 20.9.* Morels and truffles are two edible members of this division. Perhaps the most economically important ascomycotes are the yeasts.

Yeasts are unicellular sac fungi that rarely produce hyphae and usually reproduce asexually by budding. Yeasts are anaerobes and ferment sugars to produce carbon dioxide and ethyl alcohol. Because yeasts produce alcohol, they are used to make wine and beer. Other yeasts are used in baking because they produce carbon

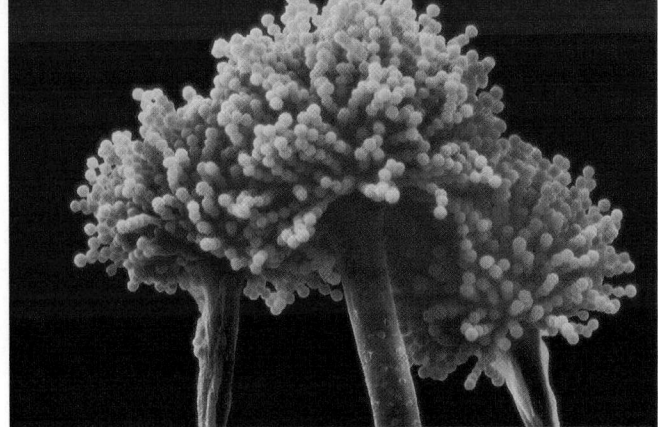

Magnification: 1200×

dioxide, the gas that causes bread dough to rise and take on a light, airy texture. Use the *BioLab* at the end of this chapter to experimentally determine the temperature at which yeasts function most efficiently.

Yeasts are also important tools for research in genetics because they have large chromosomes. A vaccine for the disease hepatitis B is produced by splicing human genes with those of yeast cells. Because yeasts multiply rapidly, they are an important source of the vaccine.

Figure 20.8
Most ascomycotes reproduce asexually by producing conidia in structures called conidiophores.

Figure 20.9
Many ascomycotes are cup shaped or have cup-shaped indentations that are lined with asci.

 **A** Morels are prized for their flavor.

B The scarlet cuplike structures of this ascomycote are visible on the dead bark.

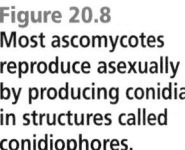

MiniLab 20-2

Purpose
Students will observe the arrangement of mushroom gills.

Process Skills
observe and infer, experiment

Teaching Strategies
- Have students wear aprons, gloves, and goggles and wash their hands after they finish.
- Find an area where nothing will disturb the caps overnight.
- Have students make wet mounts of the spores, examine them under low- and high-power magnification, and sketch their observations. L2 ELL

Expected Results
Spores will form a pattern on the paper corresponding to the gills' locations.

Analysis
1. brown
2. Spores are formed toward the edges of the gills. The patterns are the same.

Assessment

Skill Give students spore patterns from different mushrooms and use them to reconstruct the gill arrangements. Use the Performance Task Assessment List for Scientific Drawing in **PASC**, p. 55. L2 ELL

GLENCOE TECHNOLOGY

CD-ROM
Biology: The Dynamics of Life
Animation: *Life Cycle of a Mushroom,* Disc 3

VIDEODISC
Biology: The Dynamics of Life
Life Cycle of a Mushroom (Ch. 16)
Disc 1, Side 2, 1 min. 30 sec.

MiniLab 20-2 Classifying

Examining Mushroom Gills
Spore prints can often help in mushroom identification by revealing the pattern of a mushroom's gills and the color of its spores. Use this technique to see how a mushroom's gills are arranged.

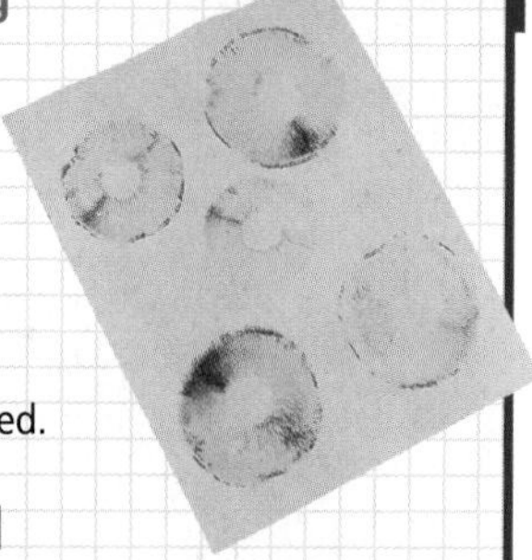

Procedure
1. Break off the stalks from some grocery-store mushrooms. Place the caps in a paper bag for a few days.
2. When the undersides of the caps are very dark brown, set the caps, gill side down, on a white sheet of paper. Be sure that the gills are touching the surface of the paper.
3. After leaving the caps undisturbed overnight, carefully lift the caps from the paper and observe the results.
4. Wash your hands with soap and water. Dispose of fungi as your teacher directs.

Analysis
1. What color are the spores on the paper?
2. How does the pattern of spores on the paper compare with the arrangement of gills on the underside of the mushroom cap that produced it?

Basidiomycotes

Of all the diverse kinds of fungi, you are probably most familiar with some of the 25 000 species in the division Basidiomycota. Mushrooms, puffballs, stinkhorns, bird's nest fungi, and bracket fungi are all basidiomycotes. So are the rust and smut fungi. Use the *MiniLab* to distinguish some mushroom species.

Basidia and basidiospores

Basidiomycotes have club-shaped hyphae called **basidia** (buh SIHD ee uh) that produce spores and give them their common name—club fungi. Basidia usually develop on short-lived, visible reproductive structures that have varied shapes and sizes, as you can see in ***Figure 20.10.*** Spores called **basidiospores** are produced in basidia during reproduction.

A basidiomycote, such as a mushroom, has a complex reproductive cycle. How does a mushroom reproduce? Study the *Inside Story* to find out.

Figure 20.10
Basidiomycotes have many different forms, and what you see are their reproductive structures.

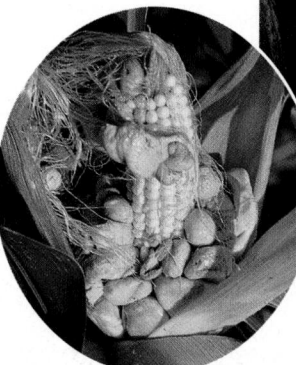

A Smuts are parasites that attack plants such as corn.

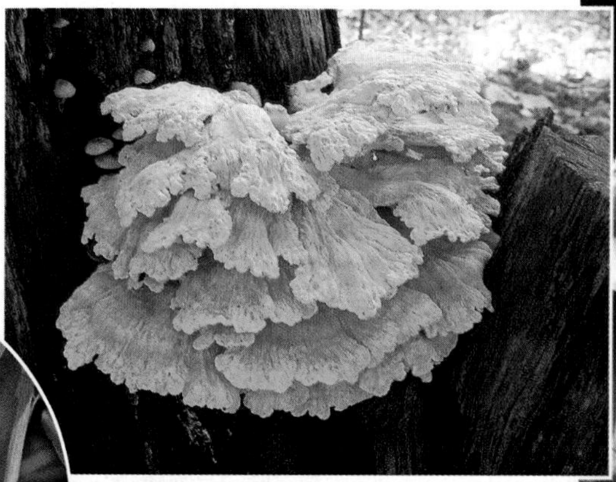

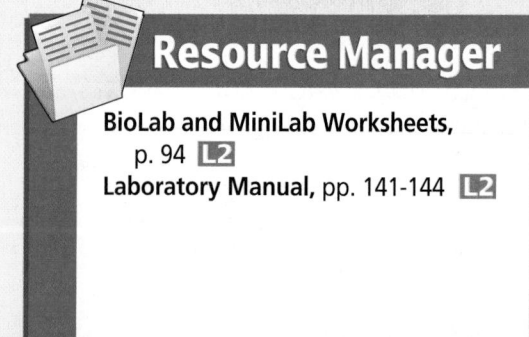

B Shelf fungi, such as this sulfur shelf, often grow on tree branches and fallen logs.

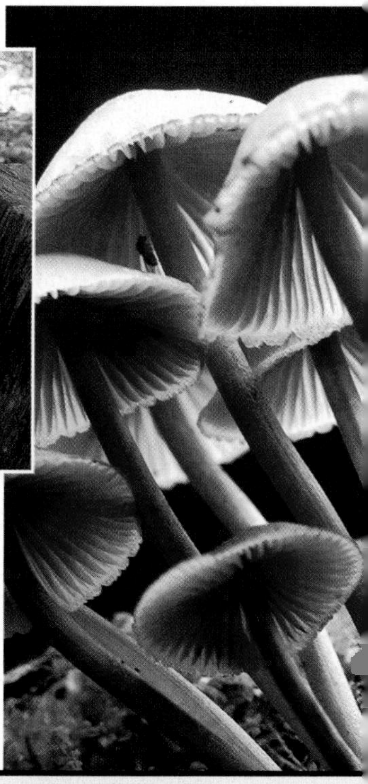

C A typical mushroom, such as this *Nycena,* has a cap that sits on top of a stalk.

TECHPREP

Mushrooms

Naturalist Provide students with an assortment of store-bought mushrooms and a mushroom field guide. Ask them to identify each mushroom. L3 ELL

Resource Manager

BioLab and MiniLab Worksheets, p. 94 L2
Laboratory Manual, pp. 141-144 L2

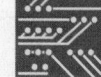

The Life of a Mushroom

What you call a mushroom is a reproductive structure of the fungus. Most of the fungus is underground and not visible. A single mushroom can produce hundreds of thousands of spores as a result of sexual reproduction. Most types of mushrooms have no asexual reproductive stages in their life cycle.

Critical Thinking Why are spores of mushrooms produced above ground?

Mycena pura

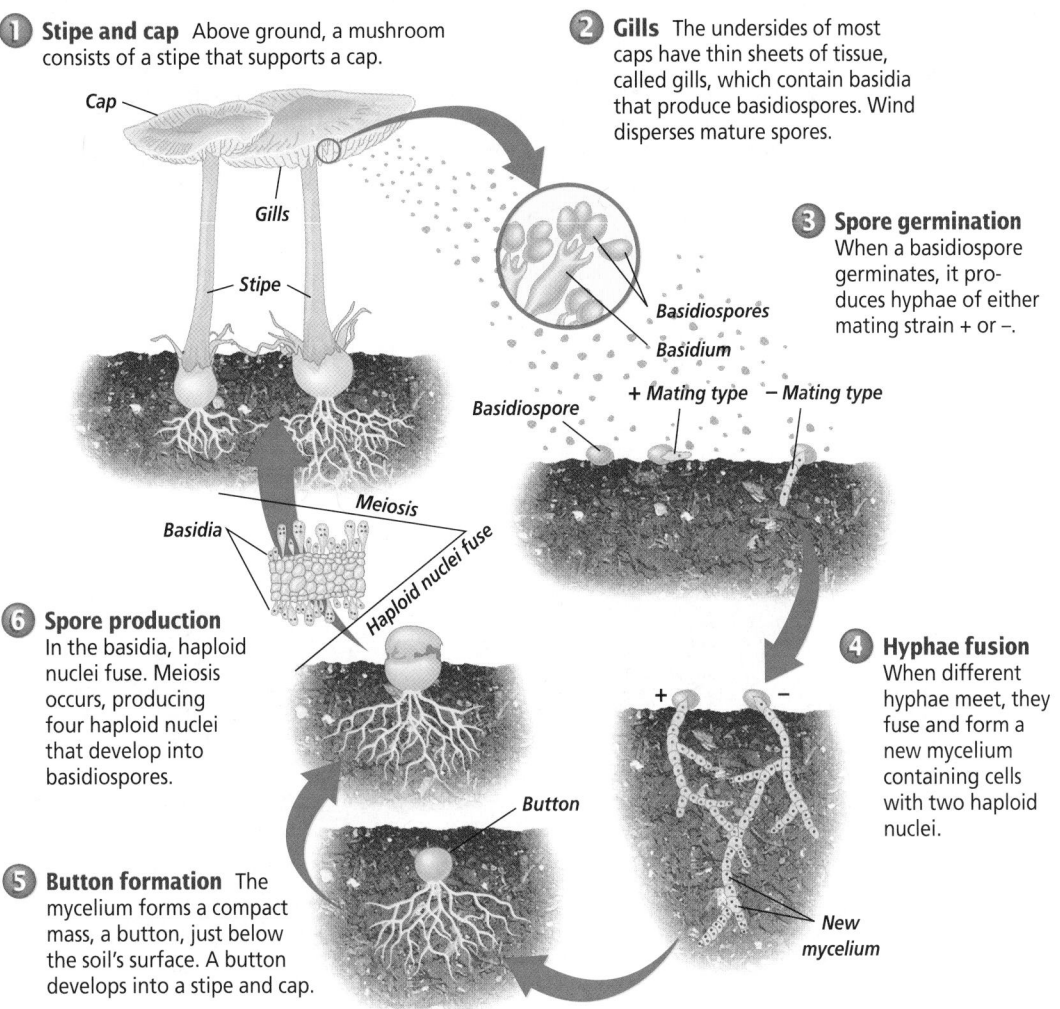

1 Stipe and cap Above ground, a mushroom consists of a stipe that supports a cap.

Cap

Gills

Stipe

Basidia

Meiosis

Haploid nuclei fuse

6 Spore production In the basidia, haploid nuclei fuse. Meiosis occurs, producing four haploid nuclei that develop into basidiospores.

5 Button formation The mycelium forms a compact mass, a button, just below the soil's surface. A button develops into a stipe and cap.

Button

2 Gills The undersides of most caps have thin sheets of tissue, called gills, which contain basidia that produce basidiospores. Wind disperses mature spores.

Basidiospores

Basidium

3 Spore germination When a basidiospore germinates, it produces hyphae of either mating strain + or –.

Basidiospore   + Mating type   – Mating type

4 Hyphae fusion When different hyphae meet, they fuse and form a new mycelium containing cells with two haploid nuclei.

+   –

New mycelium

Purpose Students will follow the steps in the life cycle of a mushroom.

Teaching Strategies

■ Remind students that Basidiomycetes spores result from sexual reproduction.

Visual Learning

■ Point out that in step 4, the nuclei from each original + and – type are in the same cell but not fused. Ask if the cells are haploid, diploid, or haploid + haploid in chromosome number. *haploid + haploid*

■ Ask students to use the diagram to explain when fertilization occurs. *just prior to spore formation in the basidium*

■ Have students explain the adaptive value of producing many basidiospores. *Many spores ensure the continued life cycle of the fungus.*

Critical Thinking

The spores are produced aboveground so that they can be easily dispersed by air currents.

Resource Manager

Basic Concepts Transparency 31 and Master L2 ELL
Reteaching Skills Transparency 31 and Master L1 ELL

Learning Disabled

Have students use the Inside Story and teacher input to guide their construction of a flowchart that shows a mushroom's life cycle. L1 ELL

PROJECT

Mushroom Farming

Kinesthetic Have students follow the directions for growing the organisms in a mushroom farm kit. Ask them to record the changes they observe as the mushrooms grow. L2 ELL COOP LEARN

Portfolio Have students make a table of the fungal divisions based on their sexual reproduction. The table should include each division's name, characteristic structures, other general information, and an illustrated representative organism. **L2**

Discussion

Tell students that, to avoid poisonous mushrooms, they should never eat ones collected from fields. Explain that even mycologists may be unsure of the identity of some mushrooms.

Enrichment

Linguistic Have interested students research mushroom folklore. One good reference is: "Who put the toad in toadstools?" A. Morgan, *New Scientist*, vol. 112, December 1986/January 1987. **L3**

Reinforcement

Review the methods by which zygomycotes, ascomycotes, and basidiomycotes reproduce sexually. Emphasize that deuteromycotes form a separate group mainly because nobody has observed them reproduce sexually.

Enrichment

Have students research the discovery of penicillin and explain how it affected the medical field. **L2**

Deuteromycotes

There are about 25 000 species of fungi, classified as the deuteromycotes, that have no known sexual stage in their life cycle, unlike the zygomycotes, ascomycotes, and basidiomycotes. Although the deuteromycotes may only be able to reproduce asexually, another possibility is that their sexual phase has not yet been observed by mycologists, biologists who study fungi.

Diverse deuteromycotes

If you've ever had strep throat, pneumonia, or another kind of bacterial infection, your doctor may have prescribed penicillin—an antibiotic produced from a deuteromycote that is commonly seen growing on fruit, as shown in *Figure 20.11*. Other deuteromycotes are used in the making of foods, such as soy sauce and some kinds of blue-veined cheese. Still other deuteromycotes are used commercially to produce substances such as citric acid, which gives jams, jellies, soft drinks, and fruit-flavored candies a tart taste.

> **WORD** *Origin*
>
> **mycorrhiza**
> From the Greek words *mykes*, meaning "fungus," and *rhiza*, meaning "root." Mycorrhizae are mutualistic relationships between fungi and plants.

Mutualism: Mycorrhizae and Lichens

Certain fungi live in a mutualistic association with other organisms. Two of these mutualistic associations that are also symbiotic are called mycorrhizae and lichens.

Mycorrhizae

A **mycorrhiza** (my kuh RHY zuh) is a mutualistic relationship in which a fungus lives symbiotically with a plant. Most of the fungi that form mycorrhizae are basidiomycotes, but some zygomycotes also form these important relationships.

How does a plant benefit from a mycorrhizal relationship? Fine, threadlike hyphae surround and often grow harmlessly into the plant's roots, as shown in *Figure 20.12*. The hyphae increase the absorptive surface of the plant's roots, resulting in more nutrients entering the plant. Phosphorus, copper, and other minerals in the soil are absorbed by the hyphae and then released into the roots. In addition, the fungus also may help to maintain water in the soil

Figure 20.11 Many deuteromycotes are useful.

Ⓐ Bleu cheese has a distinctive flavor. The blue splotches are patches of fungal spores.

Ⓑ The antibiotic penicillin is derived from *Penicillium* mold, shown here growing on an orange.

Cultural Diversity

A Fungus Called Ergot

The fungus ergot produces many chemicals, including ergotamine. In ancient Greece and Peru and also during the Middle Ages in Europe, ergot was used in small quantities for spiritual and medicinal purposes.

Ergot grows on rye and causes severe symptoms if it is milled in flour that people

ingest. Eating ergot causes Saint Anthony's fire, a disease with symptoms that include uncontrolled behavior, hallucinations, hysteria, and facial redness. Some historians suggest that ergot may have caused the bizarre behavior of the people tried for witchcraft in Salem, Massachusetts, during colonial times.

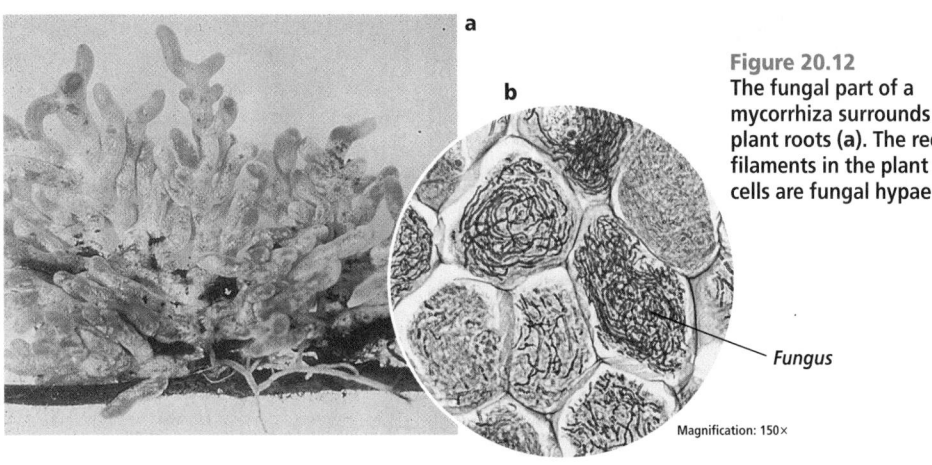

Figure 20.12
The fungal part of a mycorrhiza surrounds plant roots **(a)**. The red filaments in the plant cells are fungal hypae **(b)**.

Fungus

Magnification: 150×

around the plant. In turn, the mycorrhizal fungus benefits by receiving organic nutrients, such as sugars and amino acids, from the plant.

In addition to trees, 80 to 90 percent of all plant species have mycorrhizae associated with their roots. Plants of a species that have mycorrhizae grow larger and are more productive than those that don't. In the extreme, some species cannot survive without mycorrhizae. Orchid seeds, for example, usually do not germinate without a symbiotic fungus to provide water and nutrients.

Lichens

It's sometimes hard to believe that the orange, green, and black blotches that you see on rocks, trees, and stone walls are alive, *Figure 20.13*. They may look like flakes of old

Figure 20.13
Lichens have a variety of forms.

A Some lichens form crustlike growths on bare rocks and stone walls.

B Each stalk of a British soldier lichen is about 3 cm tall.

C Some lichens resemble leaves, like these lichens growing on a dead twig.

557

Purpose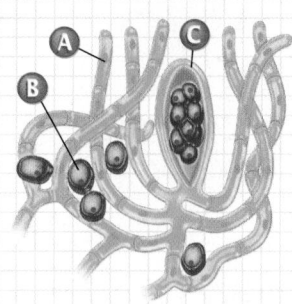

Students will differentiate between a lichen's fungal and algal components.

Process Skills

compare and contrast, think critically, design an experiment, interpret scientific diagrams

Teaching Strategies

■ Make sure that students are familiar with the composition of lichens before doing this activity.
■ Review symbiosis and discuss the life styles, including mutualism, that the term includes.

Thinking Critically

1. Structure B is the algal part because it is green, the color of chlorophyll. Structure A is the fungal part because it is not green and probably cannot make its own food.
2. Structure C shows a sac or ascus that contains spores.
3. Answers will vary. Separate the algae and fungi and try to maintain each part separately by providing nutrients and water to the fungus and light, carbon dioxide, and water to the alga.

✓ Assessment

Knowledge Have students research and write a report about the consequences of Norwegian lichens absorbing radioactive elements after the Chernobyl nuclear accident in Russia. Use the Performance Task Assessment List for Writing in Science in **PASC,** p. 87. **L3**

Problem-Solving Lab 20-2 Thinking Critically

What's inside a lichen?
A lichen consists of a fungus and an alga or cyanobacterium that live symbiotically. The prefix *sym* means "together," and *biotic* means "life." The word *symbiosis* describes the fact that there are two different life forms living together.

Analysis

You find a lichen and make a thin slice through it. You magnify the slice under the microscope and draw what you observe—the diagram above.

Thinking Critically

1. Using color as a clue, list the letters that identify the algal and fungal parts of the lichen. Explain your choices.
2. Structure C is a reproductive part. After examining it, you conclude that this is a reproductive structure of an ascomycote. Explain how you knew this information.
3. Scientists have wondered if the parts of a lichen can survive by themselves. Describe an experiment that might answer this question.

Figure 20.14
Cladina stellaris is a common lichen on the tundra and a favorite food of caribou and reindeer.

---

paint or dried moss, but they are lichens. A **lichen** (LI kun) is a symbiotic association between a fungus, usually an ascomycote, and a photosynthetic green alga or a cyanobacterium, which is an autotroph.

The fungus portion of the lichen forms a dense web of hyphae in which the algae or cyanobacteria grow. Together, the fungus and its photosynthetic partner form a structure that looks like a single organism. Use the *Problem-Solving Lab* to find out more about a lichen's structure.

Lichens need only light, air, and minerals to grow. The photosynthetic partner provides the food for both organisms. The fungus, in turn, provides its partner with water and minerals that it absorbs from rain and the air, and protects it from changes in environmental conditions.

There are about 20 000 species of lichens. They range in size from less than 1 mm to several meters in diameter. Lichens grow slowly, increasing in diameter only 0.1 to 10 mm per year. Very large lichens may be thousands of years old.

Found worldwide, lichens are pioneers, being among the first to colonize a barren area. Lichens live in arid deserts, on bare rocks exposed to bitter-cold winds, and just below the timberline on mountain peaks. On the arctic tundra, lichens, such as the one shown in *Figure 20.14,* are the dominant form of vegetation. Both caribou and musk oxen graze on lichens there, much like cattle graze on grass elsewhere.

Not only are lichens pioneers, but they are also indicators of pollution levels in the air. This is because the fungus readily absorbs materials from the air. If air pollutants are present, they kill the fungus. Without the fungal part of a lichen, the photosynthetic partner also dies.

✓ Portfolio

Checking Out Moldy Cheese

Kinesthetic Have students prepare wet mount slides of the fungus in blue cheese and observe it under a microscope. Remind them to be careful when working with microscopes and slides. Ask them to diagram and describe their observations. **L2** **P**

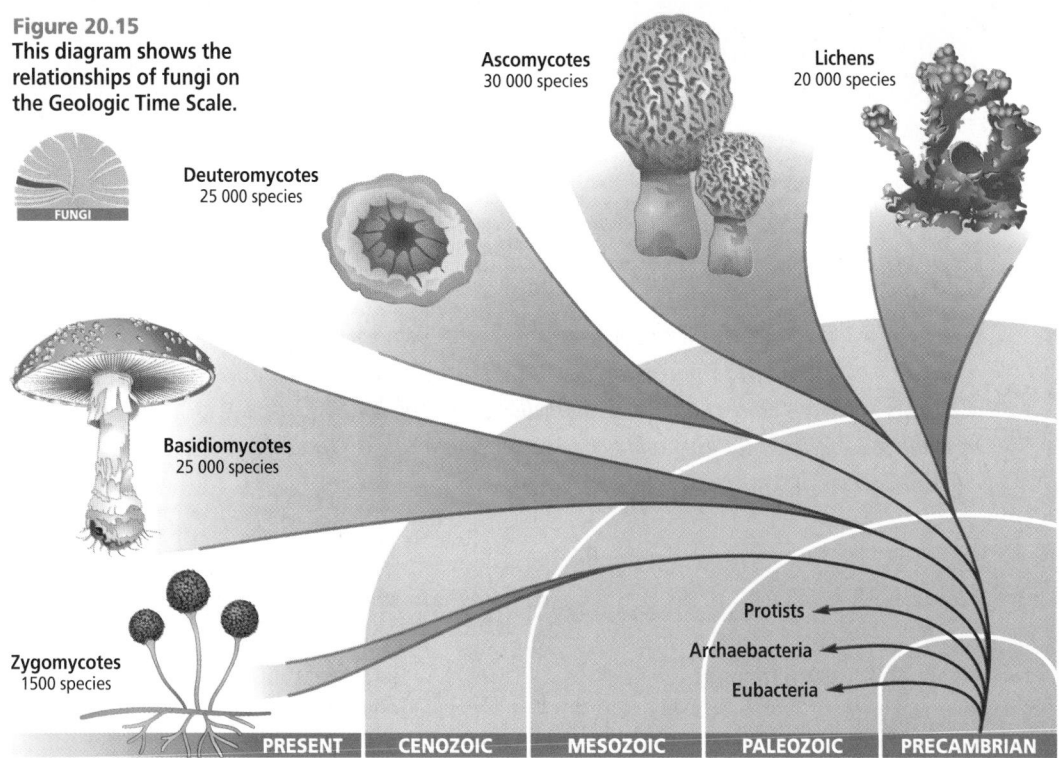

Figure 20.15
This diagram shows the relationships of fungi on the Geologic Time Scale.

FUNGI

Ascomycotes
30 000 species

Lichens
20 000 species

Deuteromycotes
25 000 species

Basidiomycotes
25 000 species

Zygomycotes
1500 species

Protists
Archaebacteria
Eubacteria

PRESENT | CENOZOIC | MESOZOIC | PALEOZOIC | PRECAMBRIAN

Origins of Fungi

Mycologists hypothesize that the ascomycotes and the basidiomycotes evolved from a common ancestor and that the zygomycotes evolved earlier, as you can see in *Figure 20.15*.

Although fossils can provide clues as to how organisms evolved, fossils of fungi are rare because fungi are composed of soft materials. The oldest fossils that have been identified as fungi are between 450 and 500 million years old.

Section Assessment

Understanding Main Ideas

1. What occurs underground between the time a basidiospore germinates and a mushroom button forms?
2. Explain how the deuteromycotes differ from members of the other divisions of fungi. Explain how they are all similar.
3. Who are the partners in a mycorrhizae? Describe how each partner benefits in a mycorrhizal relationship.
4. How does a hyphae called a stolon differ from a rhizoid?

Thinking Critically

5. You are working with a team of archaeologists on Easter Island in the Pacific Ocean. Huge stone statues were carved and erected on the island by an extinct civilization. How might you use lichens to help determine when the statues were carved?

SKILL REVIEW

6. **Comparing and Contrasting** What are the similarities and differences between ascospores and conidiophores? For more help, refer to *Thinking Critically* in the **Skill Handbook**.

Section Assessment

1. The hyphae of opposite mating types fuse and form a mycelium. Eventually, the mycelium forms a compact mass called a button.
2. Deuteromycotes have no known means of reproducing sexually. They are fungi.
3. The plant receives additional minerals and water absorbed by the fungus. The fungus receives the plant's excess food.
4. A stolon grows on a food's surface; a rhizoid invades a food source.
5. Determine the annual rate of growth for the lichens on the stones and divide it into a lichen's size to determine its approximate age.
6. Ascomycetes sexually produce ascospores inside an ascus. They produce condiophores asexually.

---

3 Assess

Check for Understanding

Have students explain the following relationships. **L1**
- stolon, rhizoid, hyphae
- mating types, zygospores, zygomycota;
- ascomycota, asci, spores
- basidiomycota, basidia spores
- penicillin, deuteromycota
- lichen, mycorrhiza, mutualism

Reteach

Naturalist Have student groups explain how each division's fungi obtain food, reproduce, and enter into symbiotic relationships. **L1**
COOP LEARN

Extension

Have students compile a list of foods derived from fungi. **L2**

✓ Assessment

Knowledge Ask students for an example, a description of the spore-forming structures and the economic importance of each fungal division. **L2**

4 Close

Discussion

TECH PREP
Linguistic Have students describe the type of buildings, conditions, and special equipment needed to grow mushrooms. **L2**

Resource Manager

Basic Concepts Transparency 32 and Master **L2** **ELL**
Reinforcement and Study Guide, pp. 89-90 **L2**
Content Mastery, pp. 97, 99-100 **L1**

Time Allotment

Initial session: 20 minutes to re-view procedure and do the setup. Second session: class period.

Process Skills

design an experiment, observe and infer, record data, relate cause and effect, communicate

PREPARATION

■ Prepare BTB solution as fol-lows: add 0.5 g BTB powder to 500 mL distilled water to make stock. Dilute 10 mL of stock in 500 mL distilled water for students to use.

■ Pretest diluted BTB solution by exhaling into it through a straw. If its color fails to change from blue to dark or light green within 60 seconds, adjust the pH by adding a drop of concentrated hydrochloric acid to the stock, diluting again, and retesting. Repeat until the desired change occurs within 60 seconds. If the stock turns green, mix in one or two drops of concentrated ammo-nium hydroxide.

■ Add a packet of yeast and 20 mL white corn syrup to 250 mL water to make a yeast/ corn syrup mixture.

Alternative Materials

Use a solution of 20% sucrose in-stead of corn syrup.Use yeast cake (found in supermarkets), which begins to ferment faster than packet yeast.

Possible Hypotheses

If yeast metabolism is not affected by temperature, then BTB solution will turn green in the same time period regardless of temperature.

Does temperature affect the metabolic activity of yeast?

Does temperature affect the rate of carbon dioxide production by yeast? Look at the experimental setup pictured at the right. As yeast metabolizes in the stoppered container, the carbon dioxide that is produced is forced out through the bent tube into the open tube, which contains a solution of bromothymol blue (BTB). Carbon dioxide causes chemical reactions that result in a color change in the BTB. Differences in the time required for this color change to occur indicate the relative rates of carbon dioxide production by yeasts.

PREPARATION

Magnification: 9100×

Problem

How can you determine the affect of temperature on the metabolism of yeast? Brainstorm ideas among the members of your group.

Hypotheses

Decide on one hypothesis that you will test. Your hypothesis might be that low temperature slows down the metabolic activity of yeast, or that a high temperature speeds up the meta-bolic activity of yeast.

Objectives

In this BioLab, you will:

■ **Measure** the rate of yeast metabo-lism using a BTB color change as a rate indicator.

■ **Compare** the rates of yeast metabo-lism at several temperatures.

■ **Use the Internet** to collect and compare data from other students.

Possible Materials

bromothymol blue solution (BTB)
straw
small test tubes (4)

large test tubes (3)
one-hole stoppers with glass tube inserts for large test tubes (3)
yeast/white corn syrup mixture
water/white corn syrup mixture
water/yeast mixture
test-tube rack
250 mL beakers (3)
ice cubes
Celsius thermometer
hot plate
50 mL graduated cylinder
glass-marking pencil
10 cm rubber tubing (3)
aluminum foil

Safety Precautions

Always wear goggles in the lab. Be careful in attaching rubber tubing to the glass tube inserts in the stoppers. Avoid touching the top of the hot plate. Wash your hands thoroughly after cleaning out test tubes at the end of your experiments.

Skill Handbook

Use the **Skill Handbook** if you need additional help with this lab.

PROCEDURE

Teaching Strategies

■ Have students wear aprons and goggles and wash their hands at the the end of the lab. Caution them to use care when working with chemicals.

■ BTB is an acid indicator. Carbon dioxide reacts with water to produce carbonic acid.

■ Students can measure the amount of gas collected in each tube rather than rely on

BTB color change. Carbon dioxide gas will collect through water displacement in an inverted tube filled with water and placed in the beaker of water with a rubber tube inserted into its mouth.

Possible Procedures

■ Control tubes may consist of stoppered tubes with no yeast/corn syrup mixture.

1. Answers will vary. The warmer the water, the more rapid the color change of BTB.
2. The Internet data should support student hypotheses.
3. It is the food for yeast.
4. The volume of water/corn syrup and the amount of yeast used were variables and had to be kept constant.
5. Because controls will vary, such as a tube with no yeast/corn syrup, some data will show more difference than other data.

✔ Assessment

Performance Have students design an experiment to measure the actual volume of gas given off by a yeast/corn syrup solution at different temperatures or to test the effect of different foods on the fermentation rate of yeast. Use the Performance Task Assessment List for Designing an Experiment in **PASC**, p. 23. **L2**

Sharing Your Data

*inter*NET
CONNECTION To navigate to the Internet BioLabs, choose the Biology: The Dynamics of Life icon at Glencoe's Web Site. Click on the student site icon, then the Bio-Labs icon. Students should pool data only from identical temperatures. If data from different temperatures are pooled inadvertently, the results will be inaccurate.

PLAN THE EXPERIMENT

1. Decide on ways to test your group's hypothesis.
2. Record your procedure, and list the materials and amounts of solutions that you will use.
3. Design a data table for recording your observations.
4. Pour 5 mL of BTB solution into a test tube. Use a straw to blow gently into the tube until you observe a series of color changes. Cover this tube with aluminum foil, and set it aside in a test-tube rack. Record your observations of the color changes caused by carbon dioxide in your breath.

Check the Plan

Discuss the following to decide on your procedure.

1. What data on color change and time will you collect? How will you record your data?
2. What variables will you control?
3. What control will you use?
4. Assign tasks for each member of your group.
5. *Make sure your teacher has approved your experimental plan before you proceed further.*
6. Carry out your experiment.
7. Visit the Glencoe Science Web Site to **post your data.**

5 mL bromothymol blue solution

30 mL yeast/corn syrup solution

ANALYZE AND CONCLUDE

1. **Checking Your Hypotheses** Explain whether your data support your hypothesis. Use your experimental data to support or reject your hypothesis concerning temperature effects on the rate of yeast metabolism.
2. **Using the Internet** How did the data you collected compare with that of other students? Compare experimental designs. Did differences in experimental design account for any differences in data collected?
3. **Making Inferences** What must be the role of white corn syrup in this experiment?
4. **Identifying Variables** Describe some variables that your group had to control in this experiment. Explain how you controlled each variable.
5. **Drawing Conclusions** Did your experiment clearly show that differences in rates of yeast metabolism were due to temperature differences?

Sharing Your Data

 **BIOLOGY** *Online* Find this BioLab on the Glencoe Science Web site at **science.glencoe.com.** Post your data in the data table provided for this activity. Briefly describe your experimental design. Compare it to that of other students.

Data and Observations

More carbon dioxide gas is produced by yeast immersed in warm water temperatures. These tubes change color more rapidly than those at cooler temperatures.

Resource Manager

BioLab and MiniLab Worksheets, pp. 95-96 **L2**

Social Studies
Connection

Social Studies
Connection
The Dangers of Fungi

Purpose

Students will learn how fungi have influenced some historical events.

Teaching Strategies

■ Ask students to discuss criteria for deciding whether an organism is beneficial or detrimental to humans. *For example, fungi that are considered pests in one culture may be used as a food in another.*

■ Discuss how events that at first seem negative can lead to positive outcomes. *The discovery of penicillin and the develpment of iron ships are examples.*

Connection to Biology

Sexual reproduction leads to genetic diversity. Asexual spores can be resistant to inhospitable conditions.

Fungi are both friend and foe. Some such as mushrooms provide food. Other fungi produce antibiotics such as penicillin. Many others break down dead tissue and recycle organic molecules, thereby keeping Earth from being buried under tons of unusable organic debris. Yet, fungi also damage crops, buildings, and animals.

Fungi cause many plant diseases that can kill plants and cause sickness and death in animals that feed on infected plants. Fungi also directly cause some human diseases.

Plant pathogens Fungi that cause the plant diseases called rusts are difficult to control. Rusts are successful because they are pleomorphic—each species produces many kinds of spores that can infect different hosts. The wind can spread their spores over hundreds of miles. For example, *Puccinia grammis* is a fungus that causes black stem rust in cereal grains, such as rice and wheat. *P. grammis* produces five kinds of spores, some of which also infect barberry plants.

Rye, another cereal plant, can host the fungus *Claviceps purpurea*, which causes the disease called ergot. Animals will contract ergot after eating infected rye. Human epidemics of ergot poisoning have occurred throughout history after people ate food made from grain infected by *C. purpurea*.

Fungi can also cause major losses of timber. For example, near the end of the nineteenth century, chestnut seedings infected with the fungus *Endothia parasitica* were brought into the United States. By 1940–1950, *E. parasitica* had destroyed most of the country's chestnut trees. Other fungi have devastated the North American populations of elm trees and eastern and western white pines.

In addition to infecting live trees, fungi damage structures built of wood. When ships were primarily wooden, dry rot always threatened their loss. Fungi cause dry rot when they grow in moist wooden structures.

Wheat attacked by the fungus *Puccinia grammis*

Human pathogens Although bacteria and viruses cause most human diseases, fungi cause their share. Most fungi are dermatophytes, that is, they invade skin, nails, and hair. Among the more common human fungal infections are ringworm and athlete's foot. Some fungal spores can be inhaled into the lungs where they can establish an infection that can spread throughout the body.

Fungi can cause substantial economic loss, disease, and even death. But their critical role in recycling organic matter and their benefit as a source of food and medicinal drugs are essential to human survival on Earth.

CONNECTION TO BIOLOGY

Many fungi reproduce both asexually and sexually. What is the advantage of having two kinds of reproduction?

 To find out more about fungi, visit the Glencoe Science Web site. **science.glencoe.com**

Internet Address Book

*inter*NET
CONNECTION Note Internet addresses that you find useful in the space below for quick reference.

_____

_____

_____

_____

_____

SUMMARY

Section 20.1

What Is a Fungus?

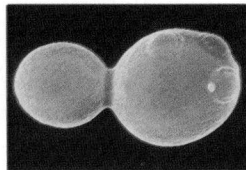

Main Ideas

■ The structural units of a fungus are hyphae, which grow and form a mycelium.

■ Fungi are heterotrophs that have extracellular digestion. A fungus may be a saprophyte, a parasite, or a mutualist in a symbiotic relationship with another organism.

■ Many fungi produce both asexual and sexual spores. One criterion for classifying fungi is their patterns of reproduction, especially sexual reproduction, during the life cycle.

Vocabulary

budding (p. 549)
chitin (p. 547)
haustoria (p. 548)
hypha (p. 546)
mycelium (p. 546)
sporangium (p. 549)

Section 20.2

The Diversity of Fungi

Main Ideas

■ Zygomycotes form asexual spores in a sporangium. They reproduce sexually by producing zygospores.

■ Ascomycotes reproduce asexually by producing spores called conidia and sexually by forming ascospores.

■ In basidiomycotes, sexual spores are produced on club-shaped structures called basidia.

■ Deuteromycotes may reproduce only asexually.

■ Fungi play an important role in decomposing organic material and recycling the nutrients on Earth.

■ Certain fungi associate with plant roots to form mycorrhizae, a symbiotic relationship between a fungus and a plant.

■ A lichen, a symbiotic association of a fungus and an alga or cyanobacterium, survives in many inhospitable habitats.

Vocabulary

ascospore (p. 553)
ascus (p. 553)
basidiospore (p. 554)
basidium (p. 554)
conidiophore (p. 553)
conidium (p. 553)
gametangium (p. 552)
lichen (p. 558)
mycorrhiza (p. 556)
rhizoid (p. 551)
stolon (p. 551)
zygospore (p. 552)

Main Ideas
Summary statements can be used by students to review the major concepts of the chapter.

Using the Vocabulary
To reinforce chapter vocabulary, use the Content Mastery Booklet and the activities in the Interactive Tutor for Biology: The Dynamics of Life on the Glencoe Science Web site.
science.glencoe.com

THE PRINCETON REVIEW

All Chapter Assessment questions and answers have been validated for accuracy and suitability by The Princeton Review.

UNDERSTANDING MAIN IDEAS

1. d
2. c
3. b

UNDERSTANDING MAIN IDEAS

1. Which of the following two terms are least related to each other?
   a. Zygomycota—gametangium
   b. Ascomycota—morel
   c. Basidiomycota—mushroom
   d. Deuteromycota—sexual spore

2. Most fungi function as _____ in their environments.
   a. consumers   **c.** decomposers
   b. producers   **d.** autotrophs

3. Which of the following is a type of asexual reproduction in fungi?
   a. sporangium   **c.** mycelium
   b. budding   **d.** haustoria

GLENCOE TECHNOLOGY

VIDEOTAPE
MindJogger Videoquizzes
Chapter 20: *Fungi*
Have students work in groups as they play the videoquiz game to review key chapter concepts.

Resource Manager

Chapter Assessment, pp. 115-120
MindJogger Videoquizzes
ExamView® Pro Software 💾
BDOL Interactive CD-ROM, Chapter 20 quiz

4. a
5. b
6. b
7. c
8. d
9. b
10. b
11. hypha
12. haustoria
13. mycelium
14. chitin
15. zygospores
16. rhizoid
17. deuteromycotes
18. ascospores; ascus
19. deuteromycote
20. sporangia

APPLYING MAIN IDEAS

21. The bulk of the mushroom lies below ground and therefore cannot be destroyed this way.
22. Fewer lichens may be an indicator of an increase in the amount of pollutants.
23. This may provide spores with an opportunity to land on new or different food sources.
24. This could prevent destruction and disruption of any mycorrhiza that may be present.

4. Mushrooms, puffballs, and bracket fungi belong to the group called _____.
   a. club fungi
   b. sac fungi
   c. Zygomycota
   d. Deuteromycota
5. Club fungi get their name from the club-shaped _____ they form.
   a. spores
   b. basidia
   c. haustoria
   d. conidia
6. Which of the following drawings represents a zygospore?

   a.   b.   c.   d.

7. Fungi sometimes live in a mutualistic relationship with a plant. They might help their host by _____.
   a. using the food supplied by the host
   b. using energy made by the host
   c. supplying water to the host
   d. providing the host with spores
8. Soy sauce, citric acid, and penicillin all come from _____.
   a. club fungi
   b. sac fungi
   c. zygomycotes
   d. deuteromycotes
9. Which of the following organisms is NOT a part of the symbiotic association called a lichen?
   a. fungus
   b. plant
   c. alga
   d. cyanobacterium
10. Fungi such as yeasts and blue molds that occur on decaying food belong to the group called _____.
    a. club fungi
    b. sac fungi
    c. zygomycotes
    d. deuteromycotes
11. The basic structural unit of a multicellular fungus is a _____.

THE PRINCETON REVIEW **TEST-TAKING TIP**

Practice, Practice, Practice
Practice to improve your performance. Don't compare yourself with anyone else.

12. Some fungi use specialized hyphae called _____, which grow into host cells.
13. A network of filaments made up of threadlike hyphae is a _____.
14. The complex carbohydrate found in the cell walls of most fungi is _____.
15. A bread mold might produce _____ if its food source dried out.
16. Suppose you study a fungus that has hyphae penetrating a food source and holding the mycelium in place. You would identify this type of hyphae as a _____.
17. Unlike other fungi, _____ are known to reproduce only asexually.
18. Sac fungi produce sexual spores called _____ in a structure known as a _____.
19. Penicillin is produced by a type of fungus called a _____.
20. The photo at right shows bread mold structures that are called _____.

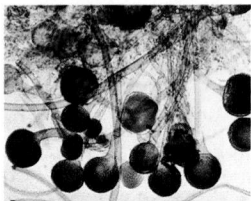

APPLYING MAIN IDEAS

21. Your neighbor is pulling up mushrooms that are growing in his lawn. He tells you that he heard mushrooms won't come back again if they are quickly removed. What would you tell him?
22. While hiking along a trail through a woods near your rapidly growing city, you notice that there are fewer lichens on the rocks and trees than there used to be. How might you interpret this change in the forest ecosystem?
23. Why is being able to produce spores that can be widely dispersed such an important adaptation for fungi?
24. When you transplant flowers, shrubs, or trees, why is it a good idea to leave the soil intact around a plant's roots?

25. Recognizing Cause and Effect When making bread, yeast is usually activated by combining it with sugar and warm water. Then, it is added to the rest of the ingredients. The resulting dough rises due to carbon dioxide released by the yeast cells. How would mixing yeast with sugar and ice water affect the way the dough rises?

26. Comparing and Contrasting Both fungi and animals are heterotrophs. Contrast the interactions of fungi and plants with the interactions of animals and plants.

27. Interpreting Scientific Illustrations To what division could the fungus in the photomicrograph at right belong? What additional information would you need before being able to place this fungus in its proper division?

28. Concept Mapping Complete the concept map by using the following vocabulary: sporangia, rhizoids, hyphae, stolons, mycellium.

Bread mold consists of

1. _____

called
2. _____

that grow horizontally and produce a
5. _____

called
3. _____

that absorb food

that grow upward and end in
4. _____

CD-ROM

For additional review, use the assessment options for this chapter found on the *Biology: The Dynamics of Life Interactive CD-ROM* and on the Glencoe Science Web site.
science.glencoe.com

The metabolic activity of yeasts at various temperatures is shown in the table below. A chemical indicator added to the yeast solution changed color when yeast cells were metabolizing.

Table 20.1 Metabolic activities of yeasts

Test tube number	Temperature of yeast solution	Time elapsed until color change
1	2°C	no color change
2	25°C	44 minutes
3	37°C	22 minutes

Interpreting Data Study *Table 20.1* and answer the following questions.

1. At what temperature was the yeast most active?
- **a.** 2°C
- **b.** 25°C
- **c.** 37°C
- **d.** 22°C

2. No color change indicated that yeasts were _____.
- **a.** metabolizing
- **b.** not metabolizing
- **c.** too hot
- **d.** dead

3. In which test tubes were yeast cells metabolizing?
- **a.** Numbers 1, 2, and 3
- **b.** Numbers 1 and 3
- **c.** Number 1
- **d.** Numbers 2 and 3

4. Observing and Inferring How was the temperature related to the rate of yeast metabolism?

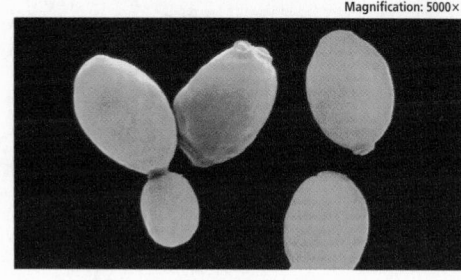

Magnification: 5000×

25. The bread would be flat. Ice water slows yeast respiration and therefore reduces the amount of carbon dioxide gas produced.

26. A few fungal species are mutualists, but fungi are mainly decomposers or parasites. They live off dead or living organic material. Plants are autotrophs and supply the food used by most fungi. Animals are consumers, depending on plants directly or indirectly for their food.

27. The fungus could be either a Basidiomycote or an Ascomycote. You would have to know if the spores are in asci or basidia to classify it correctly.

28. 1. Hyphae; 2. Stolons; 3. Rhizoids; 4. Sporangia; 5. Mycellium

ASSESSING KNOWLEDGE & SKILLS

1. c
2. b
3. d
4. Yeast metabolizes more slowly at low temperatures than at high temperatures.

National Science Education Standards
**UCP.1, UCP.2, UCP.3, UCP.5,
C.1, C.4, C.5, F.1, F.4**

Prepare

Purpose

This BioDigest can be used for an overview of viruses and bacteria, protists, and fungi. You may wish to use this summary to teach about viruses and the three types of organisms in place of the chapters in the Viruses, Bacteria, Protists, and Fungi unit.

Key Concepts

Students study the diversity of life in four kingdoms. They learn about viruses and the characteristics, ecology, and classification of bacteria, protists, and fungi.

1 Focus

Bellringer

To begin, have students compare and contrast a bacterial culture in a petri dish, *Spirogyra* in a test tube, and a bracket fungus. *Each is alive and made of cells. They do not look like each other.* **L2** **ELL**

GLENCOE
TECHNOLOGY

**VIDEOTAPE
The Secret of Life**
Nothing to Sneeze at: Viruses

**CD-ROM
Biology: The Dynamics of Life**
Animation: *The Lytic Cycle, The Lysogenic Cycle*

For a **preview** of the viruses, bacteria, protists, and fungi unit, study this BioDigest before you read the chapters. After you have studied the viruses, bacteria, protitsts, and fungi chapters, you can use the BioDigest to **review** the unit.

Viruses, Bacteria, Protists, and Fungi

Archaebacteria and eubacteria occupy most habitats on Earth, and protists and fungi are almost as diverse. But, viruses enter and take over their cells and the cells of all other organisms.

Viruses

There are many kinds of viruses, nonliving particles, most of which can cause diseases in the organisms they infect. Most viruses are much smaller than the smallest bacterium, and none respire or grow.

Structure

Viruses consist of a core of DNA or RNA surrounded by a protein coat, called a capsid. The capsid may be enclosed by a layer called an envelope that is made of phospholipids and proteins. Depending on their nucleic acid content, viruses are classified as either DNA or RNA viruses.

The lytic and lysogenic cycles

Replication

Viruses replicate only inside cells. First, a virus attaches to a specific molecule on a cell's membrane. Then, it enters the cell where it begins either a lytic or a lysogenic cycle. In the lytic cycle, the viral nucleic acid causes the host to produce new virus particles that are then released, killing the host. In the lysogenic cycle, the viral DNA becomes part of the host's chromosome for a while, and later may enter a lytic cycle.

VITAL STATISTICS

Viruses
Reports of some viral cases:
HIV—3 000 000 cases worldwide; 18 000 new cases per day
Polio—more than 200 000 cases worldwide
Mumps—fewer than 2000 cases in the U. S.
Smallpox—0 cases worldwide; eliminated by vaccine

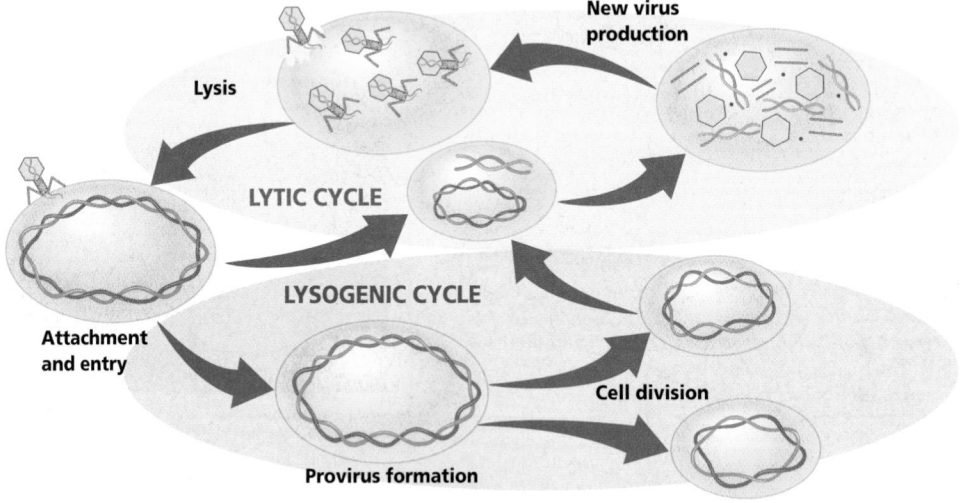

New virus production

Lysis

LYTIC CYCLE

Attachment and entry

LYSOGENIC CYCLE

Cell division

Provirus formation

Multiple Learning Styles

Look for the following logos for strategies that emphasize different learning modalities.

Kinesthetic Building a Model, p 568

Visual-Spatial Activity, pp. 567, 568, 571; Bulletin Board Display, p. 568; Quick Demo, p. 570

Interpersonal Extension, p. 571

Linguistic Meeting Individual Needs, p. 568; Biology Journal, p. 568

Naturalist Portfolio, p. 569; Using a Model, p. 570; Check for Understanding, p. 570; Reteach, p. 571

Bacteria

A bacterium is a unicellular prokaryote. Most of its genes are contained in a circular chromosome in the cytoplasm. A cell wall surrounds its plasma membrane. Bacteria may be heterotrophs, photosynthetic autotrophs, or chemosynthetic autotrophs. They reproduce asexually by binary fission and sexually by conjugation.

Adaptations

Many bacteria are obligate aerobes, needing oxygen to respire. Some bacteria called obligate anaerobes are killed by oxygen. Still other bacteria can live either with or without oxygen. Some bacteria can produce endospores to help them survive unfavorable environmental conditions.

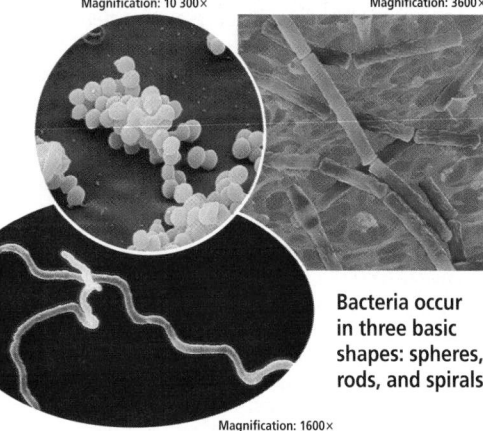

Magnification: 10 300×

Magnification: 3600×

Bacteria occur in three basic shapes: spheres, rods, and spirals.

Magnification: 1600×

Importance

Some bacteria cause diseases. Other bacteria fix nitrogen, recycle nutrients, and help make food products and medicines.

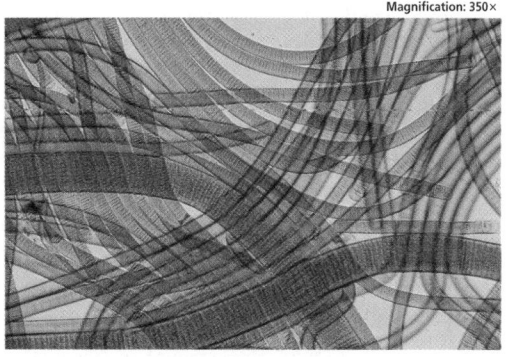

Magnification: 350×

Oscillatoria—a photosynthetic bacterium

VITAL STATISTICS

Archaebacteria and Bacteria
Numbers of Species:
Archaebacteria—approximately 600 named species
Eubacteria—more than 4000 named species
Reproduction Rates:
Slowest—*Mycobacterium tuberculosis* reproduces every 13 to 16 hours in broth.
Fastest—*Escherichia coli* reproduces every 12.5 minutes in broth.

FOCUS ON ADAPTATIONS

Archaebacteria: The Extremists

Archaebacteria are unicellular prokaryotes, most of which survive in extremely harsh environments. A group of archaebacteria that produce methane live in the intestinal tracts of animals and in sewage treatment plants. A second group thrives in hot, acidic environments, such as in the thermal springs of Yellowstone National Park or around the hot vents on ocean floors. A third group survive in extremely salty water such as that found in Utah's Great Salt Lake.

Utah's Great Salt Lake

567

Assessment Planner

Portfolio Assessment
Portfolio, TWE, p. 569
Performance Assessment
Assessment, TWE, p. 568
Knowledge Assessment
BioDigest Assessment, SE, p. 571

BIODIGEST

2 Teach

Activity

Visual-Spatial Have students view prepared slides of bacteria that show the three most common bacterial shapes: spherical, rodlike, and spiral-shaped. **L2** **ELL**

Visual Learning

- Have students compare and contrast the viral lytic and lysogenic cycles on page 566.
- Have them look at the Vital Statistics on page 566 and hypothesize reasons why the diseases vary in occurrence.
- Ask students to look at the Focus on Adaptations on page 567 and explain why archaebacteria are often considered extremists.

Reinforcement

Review the biogeochemical cycles discussed in Chapter 2. Then discuss how bacteria are important in nitrogen fixation and nutrient cycling.

GLENCOE
TECHNOLOGY

CD-ROM
Biology: The Dynamics of Life
Video: *Binary Fission*
Disc 3

BioDigest

568

BioDigest

Protists

Kingdom Protista is a diverse group of heterotrophic, autotrophic, parasitic, and saprophytic eukaryotes. Although many protists are unicellular, some are multicellular. They all live in aquatic or very moist places.

Protozoans: Animal-like protists

Animal-like protists known as protozoans are unicellular, heterotrophic organisms. Many protozoans are classified based on their adaptations for locomotion in the environment.

Phylum Rhizopoda is composed of the protozoans called amoebas that use pseudopodia, extensions of their plasma membrane, to move and engulf prey. Phylum Mastigophora is composed of protozoans that use flagella to move around. Some parasitic protozoan species that have flagella cause disease, but other flagellated species are helpful. Members of the phylum Ciliophora move by beating hairlike projections called cilia. *Paramecium* is a widely studied ciliate.

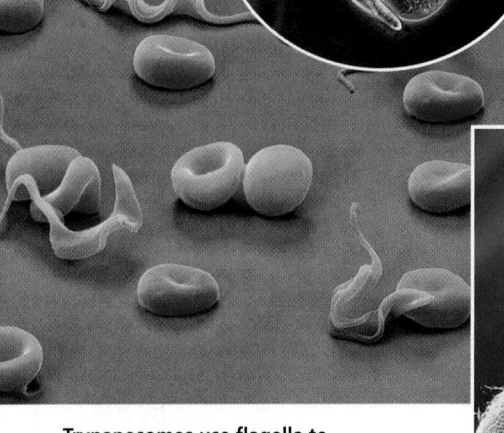

An amoeba extends pseudopodia, forming a cup-shaped trap for prey.

Magnification: 310×

Magnification: 5800×

Trypanosomes use flagella to move and cause the disease called sleeping sickness.

Plasmodium may infect more than one hundred million people every year in African and South American countries.

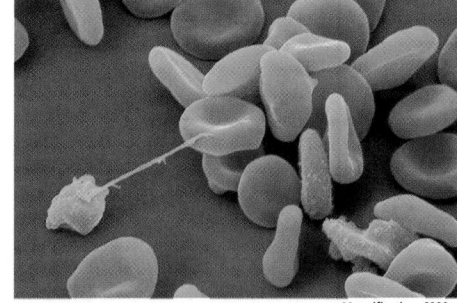

Magnification: 6000×

Sporozoans are grouped together because they are all parasites and many produce spores. Most have very complex life cycles with different stages. *Plasmodium*, the protozoans that cause the disease malaria, have a sexual stage in mosquitoes and an asexual stage in humans.

VITAL STATISTICS

Protists
Distribution: worldwide in aquatic and moist habitats
Niches: producers, herbivores, predators, parasites, and decomposers
Number of species: more than 60 000
Size range: less than 2 micrometers in length to greater than 100 meters (328 feet) in length

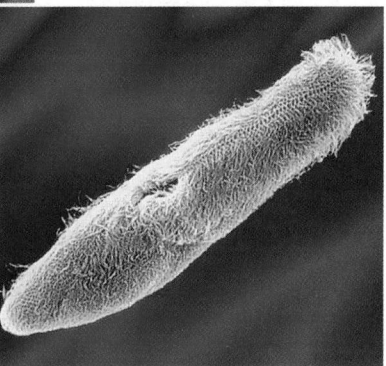

The beating cilia of a paramecium produce water currents for collecting food.

Magnification: 380×

Algae: Plantlike protists

Autotrophic protists are called algae. They are grouped on the basis of body structure and the pigments they contain. Photosynthetic algae produce a great deal of Earth's atmospheric oxygen. They are unicellular and multicellular.

Euglenas

Unicellular algae that can be both autotrophs and heterotrophs are classified in the phylum Euglenophyta. Most species have chlorophyll for photosynthesis. When there is no light, some euglenas can ingest food.

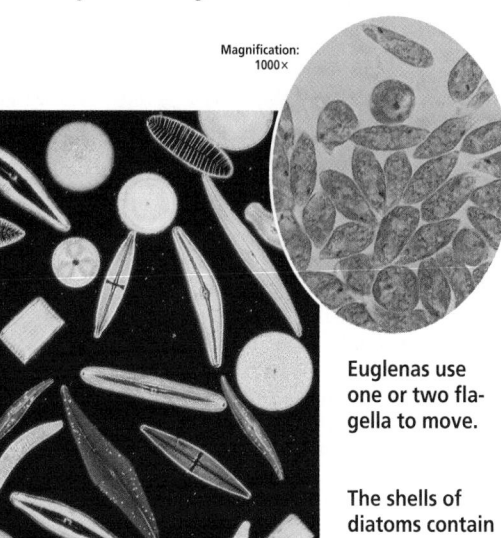

Magnification: 1000×

Euglenas use one or two flagella to move.

The shells of diatoms contain silica.

Magnification: 490×

Diatoms

Unicellular algae called diatoms are classified in phylum Bacillariophyta. In addition to chlorophyll, diatoms contain carotenoids, pigments with a golden-yellow color. Diatoms live in both saltwater and freshwater environments.

Dinoflagellates

Dinoflagellates, members of the phylum Dinoflagellata, are unicellular algae surrounded by hard, armorlike plates and propelled by flagella. They may contain a variety of pigments, including chlorophyll, caretenoids, and red pigments. Marine blooms of dinoflagellates can cause toxic red tides.

Red Algae

Members of the phylum Rhodophyta are multicellular marine algae. Because of their red and blue pigments, some species can grow at depths of 100 meters.

Brown Algae

About 1500 species of algae are classified in phylum Phaeophyta and all contain a brown pigment. The largest brown algae are the giant kelps that can grow to about 60 meters in length.

Green Algae

The green algae in the phylum Chlorophyta, may be unicellular, colonial, or multicellular. The major pigment in their cells is chlorophyll, and some also have yellow pigments.

Funguslike Protists

Funguslike protists include the slime molds, water molds, and downy mildews. They are saprophytes, decomposing organic material to obtain its nutrients.

Magnification: 350×

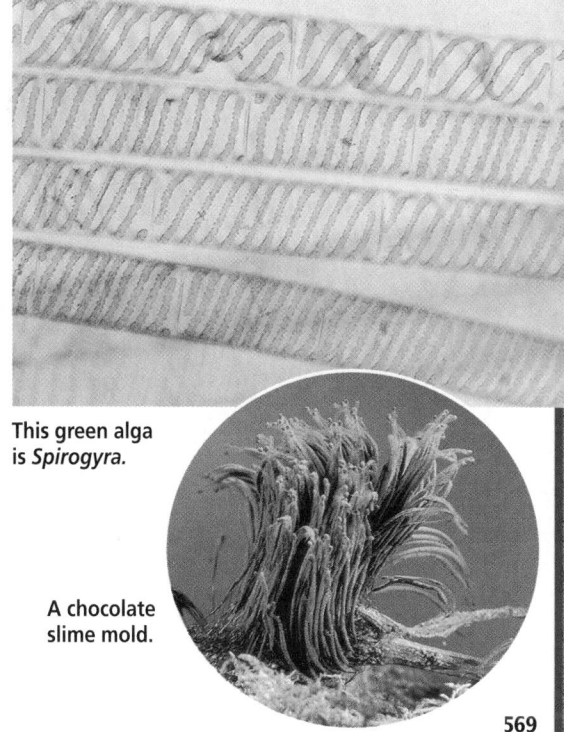

This green alga is *Spirogyra.*

A chocolate slime mold.

569

BIODIGEST

Visual Learning

- Ask students to compare the algae shown in the photographs and describe the differences they observe. *The diatoms are surrounded by silica shells, euglenas are unicellular and have at least one flagellum, and Spirogyra are filamentous and have spiral chloroplasts.*
- Ask students to describe the appearance of the slime mold in the photo on this page. *It resembles a giant amoeba.*

Quick Demo

Show students samples of both slime mold plasmodiums and fruiting bodies. Discuss why slime molds were once classified as fungi.

GLENCOE TECHNOLOGY

CD-ROM
Biology: The Dynamics of Life
Exploration: *The World of Protists; Protists; Kelp Forests*
Disc 3

Portfolio

Types of Algae

 Naturalist Have students make a table of algae phyla. The table should include information such as the phylum's name, its characteristics, general information, and an example. **L2** **P**

Visual Learning

Have students use the photos on this page to compare the reproductive structures of mushrooms and bread mold.

Using a Model

 *Naturalist* Display a model of a eukaryotic cell. Use the model to explain similarities and differences between protists and fungi.

Quick Demo

Visual-Spatial Collect some lichens and discuss how they differ with students.

GLENCOE
TECHNOLOGY

CD-ROM
Biology: The Dynamics of Life
Animation: *Life Cycle of a Mushroom; Fungal Decay*
Disc 3

---

3 Assess

Check for Understanding

Naturalist Have students differentiate among viruses, bacteria, protists, and fungi using examples and the unique characteristics of each.

Fungi

Members of Kingdom Fungi are mostly multi-cellular, eukaryotic organisms that have cell walls made of chitin. The structural units of a fungus are hyphae. Fungi secrete enzymes into a food source to digest the food and then absorb the digested nutrients.

Fungi may be saprophytes, parasites, or mutualists. They play a major role in decomposing organic material and recycling Earth's nutrients.

Club Fungi

Club fungi include mushrooms, puffballs, and bracket fungi, and all are members of phylum Basidiomycota. Club fungi have club-shaped structures called basidia in which their sexual spores are produced.

These mushrooms have gills containing basidia on the underside of their caps.

Zygospore-Forming Fungi

Members of phylum Zygomycota produce thick-walled, sexual spores called zygospores. Zygomycotes also form many asexual spores in sporangia.

Some of this bread mold's hyphae form a mat called a mycelium that is anchored in the food source by other hyphae called rhizoids.

Magnification: 63×

FOCUS ON ADAPTATIONS

Mycorrhizae

Some plants live in association with mutualistic fungi. These relationships, called mycorrhizae, benefit both the fungi and the plants. The hyphae of the fungus are entwined with the roots of the plant, and absorb sugars and other nutrients from the plant's root cells. In turn, the fungus increases the surface area of the plant's roots, allowing the roots to absorb more water and minerals.

The relationship enables the fungus to obtain food and the plant to grow larger. Some plants have grown so dependent on their mychorrhizal relationships that they cannot grow without them.

Hyphae of fungus

Plant cell

Fungal hyphae can grow among a plant's root cells.

570

BIOLOGY Online Note Internet addresses that you find useful in the space below for quick reference.

_____

_____

_____

_____

_____

Sac Fungi

Fungi that produce sexual spores called ascospores in saclike structures called asci are classified in phylum Ascomycota. Sac fungi produce asexual spores, called conidiospores, which develop in chains or clusters from the tips of elongated hyphae called conidiophores.

Lichens

A lichen is a symbiotic association of a mutualistic fungus and a photosynthetic alga or cyanobacterium. Lichens live in many inhospitable areas, such as cold climates and high altitudes, but they are sensitive to pollution and do not grow well in polluted areas.

Ascomycota, such as this scarlet cup, produce spores inside cup-shaped sacs.

Imperfect fungi, such as this *Penicillium* mold, are classified in phylum Deuteromycota. Sexual reproduction has never been observed in imperfect fungi.

BIODIGEST ASSESSMENT

Understanding Main Ideas

1. The core of a virus contains _____.
   a. phospholipids   c. amino acids
   b. nucleic acids   d. proteins

2. Photosynthetic bacteria include _____.
   a. cyanobacteria   c. methanogens
   b. anaerobes   d. chemoautotrophs

3. The most likely place to find archaebacteria would be in _____.
   a. food   c. a hot sulfur spring
   b. a DNA lab   d. a fast flowing stream

4. Mastigophorans use _____ to move.
   a. pseudopods   c. flagella
   b. cilia   d. None of these.

5. The major pigment of green algae is _____.
   a. red   c. chlorophyll
   b. carotene   d. a chloroplast

6. _____ have silica cell walls.
   a. Protozoans   c. Euglenas
   b. Diatoms   d. Funguslike protists

7. Mushrooms are classified in phylum _____.
   a. Basidiomycota   c. Ascomycota
   b. Zygomycota   d. Deuteromycota

8. Cell walls of fungi contain _____.
   a. cellulose   c. hyphae
   b. spores   d. chitin

9. A puffball is a type of _____.
   a. sac fungus   c. lichen
   b. club fungus   d. imperfect fungus

10. Lichens are sensitive to _____.
    a. pollution   c. drought
    b. cold   d. predators

Thinking Critically

1. Why are many archaebacteria called extremists?
2. Distinguish between protozoans and algae.
3. Why are mycorrhizae important to plants?
4. Compare a bacterium and a protozoan.

571

BIODIGEST

Reteach

**Naturalist** Have students prepare a table that includes examples, unique characteristics, and general information about viruses and the archaebacteria, eubacteria, protists, and fungi. **L2**

Extension

Interpersonal Have students work in groups, using information on pages 570 and 571, to explain the symbiotic relationships of lichens and mycorrhizae. **L2** **COOP LEARN**

✔ Assessment

Knowledge Provide students with pictures of viruses, bacteria, protists, and fungi from magazines, journals, or drawings. Have them identify what each picture depicts. **L1** **ELL**

4 Close

Activity

Visual-Spatial Have students identify organisms studied in this BioDigest from microscope slides. **L2**

Resource Manager

Reinforcement and Study Guide, pp. 91-92 **L2**
Content Mastery, pp. 101-104 **L1**

BIODIGEST ASSESSMENT

Understanding Main Ideas

1. b	**4.** c	**7.** a	**9.** b
2. a	**5.** c	**8.** d	**10.** a
3. c	**6.** b		

Thinking Critically

1. They survive in environments that most organisms cannot tolerate.

2. Protozoans are unicellular, animal-like protists, and algae are either unicellular or multicellular plantlike protists.

3. They provide additional surface area for water and mineral absorption by plant roots.

4. A bacterium is a small prokaryote and a protozoan is a unicellular eukaryote.

Plants

Unit Overview

Unit 7 introduces the plant kingdom. The unit begins in Chapter 21 with a brief discussion of plant origins and the adaptations of plants to life on land. This chapter also introduces the 12 plant divisions.

In Chapter 22, students learn the major characteristics of each plant division. Students learn about plant cells and tissues in Chapter 23. This chapter also describes root, stem, and leaf anatomy and plants' responses to the environment.

The unit concludes in Chapter 24 where reproduction in plants is examined.

Introducing the Unit

The majestic *Sequoia* suggests the importance of plants to life on Earth. These trees are the largest living plants on Earth. Use this photograph to initiate a class discussion on plants as living organisms and the vital role plants play in the ecosystem.

Plants

Plants, large and small, are found all over Earth—from the frozen arctic circle to harsh desert environments and lush tropical rain forests. There are characteristics shared by all plants, as well as unique features of individual species. The great beauty and diversity of plants is a source of endless wonder. In addition to their great beauty and diversity, plants provide us with food and oxygen. Life on Earth would not be possible without these amazing organisms. In this unit you will explore the fascinating world of plants.

UNIT CONTENTS

- **21** What is a Plant?
- **22** Diversity of Plants
- **23** Plant Structure and Function
- **24** Reproduction in Plants
- **BioDigest** Plants

UNIT PROJECT

BIOLOGY Online Use the Glencoe Science Web site for more project activities that are connected to this unit. **science.glencoe.com**

Unit Projects

Exploring the World of Plants

Have students do one of the projects for this unit as described on the Glencoe Science Web site. As an alternative, students can do one of the projects described on these two pages.

Display

Visual-Spatial Students can collect examples of different plants from magazines. They should make a display that illustrates the wide variety of plant species. **L1** **ELL**

Modeling

Kinesthetic Have groups of students design and create a model of a typical flowering plant. They may use any available materials. **L1** **ELL** **COOP LEARN**

Advance Planning

Chapter 21
- Purchase mustard seeds for the Alternative Lab.
- Order *Marchantia* for MiniLab 21-1 and *Lycopodium* for Mini-Lab 21-2.

Chapter 22
- Collect seeds and fruits for the Getting Started Demo.
- Order or collect fern sporangia for MiniLab 22-1 and seeds for MiniLab 22-2.

Chapter 23
- Leave some broadleaf house-plants near a sunny window to elicit a phototropic response for the Quick Demo.

Chapter 24
None

Unit Projects

Experimenting
Logical-Mathematical Have students design and perform an experiment to test one aspect of plant growth. **L2**

Interviewing
Linguistic Have students interview a florist or nursery worker to learn how they keep flowers and plants fresh and healthy. **L1**

Final Report
Have student groups present their findings about plants in reports that could be presented to students at your local middle schools. **L3** **COOP LEARN**

Refer to pages 4T-5T of the Teacher Guide for an explanation of the National Science Education Standards correlations.

Section	Objectives	Activities/Features
Section 21.1 **Adapting to Life on Land** National Science Education Standards UCP.1-5; B.2; C.5, C.6; E.1, E.2; F.3, F.4, F.6; G.1-3 (1 session, 1 block)	1. **Compare and contrast** characteristics of algae and plants. 2. **List** the characteristics of plants that adapt them for life on land. 3. **Describe** the alternation of generation in land plants.	**MiniLab 21-1:** Examining Land Plants p. 577 **Problem-Solving Lab 21-1,** p. 579 **Focus On** Plants for People, p. 580
Section 21.2 **Survey of the Plant Kingdom** National Science Education Standards UCP.1, UCP.3, UCP.4, UCP.5; A.1, A.2; C.4, C.5, C.6; E.2; F.1, F.3, F.6; G.1-3 (2 sessions, 1 block)	4. **Describe** the phylogenic relationships among divisions of plants. 5. **Identify** the twelve plant kingdom divisions.	**MiniLab 21-2:** Looking at Modern and Fossil Plants, p. 586 **Problem-Solving Lab 21-2,** p. 588 **Internet BioLab:** Researching Trees on the Internet, p. 590 **Health Connection:** Medicine from Plants, p. 592

Need Materials? Contact Carolina Biological Supply Company at 1-800-334-5551 or at **http://www.carolina.com**

MATERIALS LIST

BioLab

p. 590 Internet access

MiniLabs

p. 577 live or preserved sample of *Marchanita,* stereomicroscope

p. 586 live or preserved samples of *Lycopodium* and *Equisetum*

Alternative Lab

p. 576 petri dish, black paper, scissors, microscope, hand lens, microscope slide, coverslip, water, forceps, mustard seeds, label

Quick Demos

p. 576 pictures of land plants and green algae

p. 585 pine needles, broad-leaves

Key to Teaching Strategies

L1	Level 1 activities should be appropriate for students with learning difficulties.
L2	Level 2 activities should be within the ability range of all students.
L3	Level 3 activities are designed for above-average students.
ELL	ELL activities should be within the ability range of English Language Learners.
COOP LEARN	Cooperative Learning activities are designed for small group work.
P	These strategies represent student products that can be placed into a best-work portfolio.
	These strategies are useful in a block scheduling format.

What Is a Plant?

Teacher Classroom Resources

Section	Reproducible Masters	Transparencies
Section 21.1 **Adapting to Life on Land**	Reinforcement and Study Guide, pp. 93-94 **L2** Concept Mapping, p. 21 **L3** **ELL** Critical Thinking/Problem Solving, p. 21 **L3** BioLab and MiniLab Worksheets, p. 97 **L2** Laboratory Manual, pp. 145-152 **L2** Content Mastery, p. 105-106, 108 **L1**	Section Focus Transparency 50 **L1** **ELL**
Section 21.2 **Survey of the Plant Kingdom**	Reinforcement and Study Guide, pp. 95-96 **L2** Critical Thinking/Problem Solving, p. 21 **L3** BioLab and MiniLab Worksheets, pp. 98-100 **L2** Content Mastery, p. 105, 107-108 **L1**	Section Focus Transparency 51 **L1** **ELL** Basic Concepts Transparency 33 **L2** **ELL**

Assessment Resources

Chapter Assessment, pp. 121-126
MindJogger Videoquizzes
Performance Assessment in the Biology Classroom
Alternate Assessment in the Science Classroom
ExamView® Pro Software 💾
BDOL Interactive CD-ROM, Chapter 21 quiz

Additional Resources

Spanish Resources **ELL**
English/Spanish Audiocassettes **ELL**
Cooperative Learning in the Science Classroom **COOP LEARN**
Lesson Plans/Block Scheduling

◻ NATIONAL GEOGRAPHIC

Teacher's Corner

Products Available From Glencoe
To order the following products, call Glencoe at 1-800-334-7344:
CD-ROM
NGS PictureShow: What It Means to Be Green
Curriculum Kit
GeoKit: Plants
Transparency Set
NGS PicturePack: What It Means to Be Green
Videodisc
STV: Plants

Index to National Geographic Magazine
The following articles may be used for research relating to this chapter:
"Saving the World's Largest Power," by Willem Meijer, July 1985.
"The Poppy," by Peter T. White, February 1985.
"Stalking the West's Wild Foods," by Euell Gibbons, August 1973.
"Stalking Wild Foods on a Desert Isle," by Euell Gibbons, July 1972.
"Plants That Eat Insects," by Paul A. Zahl, May 1961.

GLENCOE TECHNOLOGY

The following multimedia resources are available from Glencoe.

Biology: The Dynamics of Life
CD-ROM **ELL**
💿 BioQuest: *Biodiversity Park*
Video: *Giant Redwoods*

The Infinite Voyage
💿 A Taste of Health
To the Edge of the Earth

Theme Development

The theme of **unity within diversity** is well illustrated in this chapter. Differences among plants in various divisions are discussed, and their similarity in reproduction via alternation of generations is stressed. The **Evolution** of plants is also a major theme in the discussion of the evolution of land plants from green algae.

0:00 OUT OF TIME?

If time does not permit teaching the entire chapter, use the BioDigest at the end of the unit as an overview.

READING BIOLOGY

Glencoe's *Biology: The Dynamics of Life* contains many resources to assist a student's reading skills. Each chapter contains figures with expanded captions that expand on written material. Word Origins, located along the side of text, expand knowledge of biology vocabulary. Glencoe's Content Mastery Booklet helps develop reading skills while reinforcing content. In addition, use the Interactive Tutor for *Biology: The Dynamics of Life* on the Glencoe Web site to reinforce vocabulary.
science.glencoe.com

Chapter

21 What Is a Plant?

What You'll Learn

- You will identify the adaptations that enable plants to live on land.
- You will distinguish among the major divisions of plants.

Why It's Important

Plants were the first multicellular organisms to colonize land almost 500 million years ago. Since then, plants have developed into an incredibly diverse group of organisms that help provide us with food, oxygen, and shelter.

READING BIOLOGY

Go through the chapter, seeking out unfamiliar words. Make flashcards of these terms, adding information as you read the chapter.

To find out more about plants, visit the Glencoe Science Web site.
science.glencoe.com

Plants provide examples of diversity within a kingdom. The banks of a pond may be covered by beautiful flowers, which you can easily see, and small mosses and liverworts, which are less prominent. The plant in the inset photo is a geranium, a common flowering plant found in houses and hanging baskets.

Multiple Learning Styles

Look for the following logos for strategies that emphasize different learning modalities.

 Kinesthetic Meeting Individual Needs, p. 578; Activity, p. 583

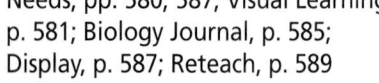 *Linguistic* Portfolio, p. 580

Visual-Spatial Getting Started Demo, p. 574; Quick Demos, pp. 576, 585; Portfolios, pp. 578, 585; Extension, p. 579; Meeting Individual

Needs, pp. 580, 587; Visual Learning, p. 581; Biology Journal, p. 585; Display, p. 587; Reteach, p. 589

 Logical-Mathematical Discussion, p. 579

 Intrapersonal Enrichment, p. 581

Section

21.1 Adapting to Life on Land

When you studied ecology, you learned that plants are important in defining biomes, ecosystems, and communities. A desert biome may seem too hot and dry for living plants, but the Saguaro cactus and many other plants flourish there. Mosses are plants that would never survive the desert. Instead, they are found in cool, damp areas. Although cacti and mosses are very different, they share characteristics common to all plants.

A *Saguaro* cactus and mosses (inset)

SECTION PREVIEW

Objectives

Compare and contrast characteristics of algae and plants.

List the characteristics of plants that adapt them for life on land.

Describe the alternation of generation in land plants.

Vocabulary

cuticle
leaf
root
stem
vascular tissue
vascular plant
nonvascular plant
seed

Origins of Plants

What is a plant? A plant is a multicellular eukaryote that can produce its own food in the form of glucose through the process of photosynthesis. In addition, plant cells are characterized by having thick cell walls made of cellulose. The stems and leaves of plants have a waxy waterproof coating called a **cuticle** (KYEWT ih kul).

Fossils and other geological evidence suggest that a billion years ago, plants had not yet begun to appear on land. No ferns, mosses, trees, grasses, or wildflowers existed. The land was barren except for some algae at the edges of inland seas and oceans. However, the shallow waters that covered much of Earth's surface at that time were teeming with bacteria, algae, and other protists, as well as simple animals such as corals, sponges, jellyfishes, and worms. Among these organisms were green algae that would eventually become adapted to life on land.

The first evidence of plants in the fossil record began to appear around 500 million years ago. These early plants were very simple in structure and did not have any leaves. They were probably instrumental in turning bare rock into rich soil. The earliest known plant fossils are those of plants called psilophytes (SI luh fites),

WORD *Origin*

cuticle
From the Greek word *cutis*, meaning "skin." The cuticle is the outermost covering of a plant.

Section 21.1

Prepare

Key Concepts

Evidence for land plants evolving from algae and the adaptations needed for land survival are covered. An overview of leaf, stem, and root anatomy is presented in conjunction with the adaptations of plants that allow them to live on land.

Planning

- Collect plants for the Getting Started Demo.
- Collect pictures of plants and algae for the Quick Demo.
- Purchase mustard seeds for the Alternative Lab.
- Collect leaf specimens for the Meeting Individual Needs.
- Purchase bean seeds for the Project.

---

1 Focus

Bellringer 🖐

Before presenting the lesson, display **Section Focus Transparency 50** on the overhead projector and have students answer the accompanying questions.
L1 **ELL**

2 Teach

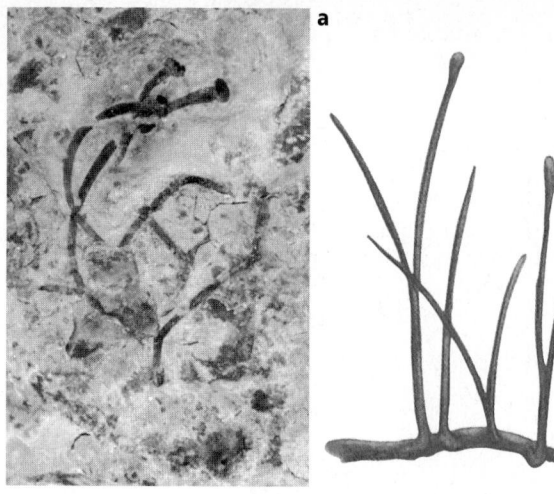

Figure 21.1
This fossil of *Cooksonia* is more than 400 million years old **(a)**. *Cooksonia* was probably one of the first vascular plants. The plant had leafless stems **(b)**.

some of which still exist today, such as those shown in *Figure 21.1*.

Scientists think that all plants probably evolved from filamentous green algae that lived in the ancient oceans. Some of the evidence for their relationship can be found in modern members of both groups. Both green algae and plants have cell walls that contain cellulose. Both groups have the same types of chlorophyll used in photosynthesis. Both algae and plants store food in the form of starch. All other major groups of organisms store food in the form of glycogen and other complex sugars.

Figure 21.2
Leaves take advantage of the plentiful supply of sunlight and carbon dioxide available to land plants.

576 WHAT IS A PLANT?

Adaptations in Plants

Life on land has advantages as well as challenges. All organisms need water to survive. A filamentous green alga floating in a pond does not need to conserve water. The alga is completely immersed in a bath of water and dissolved nutrients, which it can absorb directly into its cells. For most land plants, the only available supply of water and minerals is in the soil, and only the portion of the plant that penetrates the soil can absorb these nutrients.

When you studied protists, you learned that algae reproduce by releasing unprotected unicellular gametes into the water, where fertilization and development take place. Land plants have evolved several adaptations that help protect the gametes from drying out. In some plants, the sperm are released near the egg so they only have to travel a short distance. Other plants have protective structures to ensure the survival of the gametes. Land plants must also withstand the forces of wind and weather and be able to grow upright against the force of gravity. Over the past 500 million years or so, plants have developed a huge variety of adaptations that reflect both the challenges and advantages of living on land.

Preventing water loss

If you run your fingers over the surface of an apple, a maple leaf, or the stem of a houseplant, you'll probably find that it feels smooth and slightly slippery. Most fruits, leaves, and stems are covered with a protective, waxy layer called the cuticle. Waxes and oils are lipids, and you have previously read that lipids do not dissolve in water. The waxy cuticle creates a barrier that helps prevent the water in the plant's tissues from evaporating into the atmosphere.

Leaves carry out photosynthesis

All the cells of a filamentous green alga carry out photosynthesis. But in most land plants, the leaves are the organs usually responsible for photosynthesis. A **leaf** is a broad, flat structure of a plant that traps light energy for photosynthesis, *Figure 21.2*. They are supported by the stem and grow upward toward sunlight.

Putting down roots

Most plants depend on the soil as their primary source for water and other nutrients. Plants are able to acquire water and nutrients from the soil with their roots. In most plants, a **root** is a plant organ that absorbs water and minerals from the soil, transports those nutrients to the stem, and anchors the plant in the ground. Some roots, such as those of radishes or sweet potatoes, also accumulate starch reserves and function as organs of storage. Many people use these storage roots as a food source. Find out more about the many uses of plants in *Focus On* at the end of this section.

Practice your lab skills by using a dissecting microscope in the *MiniLab* on this page to explore some of the adaptations of plants that allow them to survive on land.

Transporting materials

Water moves from the roots of a tree to its leaves, and the sugars produced in the leaves move to the roots through the stem. A **stem** of a plant provides structural support for upright growth and contains tissues for transporting food, water, and other materials from one part of the plant to another, as shown in *Figure 21.3*. Stems may also serve as organs for food storage. Green stems contain chlorophyll and take part in photosynthesis.

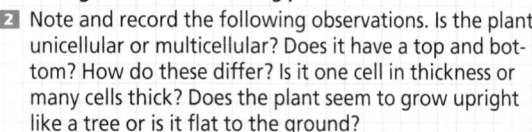
Figure 21.3
Stems can be soft and flexible like the daisy stem shown here (a). Other plants, such as this sugar maple tree, have strong, thick stems that provide support and allow the tree to grow to great heights (b).

b

a

Purpose

Students will analyze diagrams and detect errors in labeling.

Process Skills

analyze information, concept mapping, interpret scientific illustrations, sequence, think critically

Teaching Strategies

■ Make sure that students have reviewed the concept of alternation of generations prior to this lab.

■ Allow students to work in small groups.

■ Review the terms haploid and diploid with students.

■ Point out that *gamet*ophytes always form *gamet*es while *sporo*phytes always form *spores*.

Thinking Critically

1. Gametophytes always form gametes by the process of mitosis; gametophytes are already haploid in chromosome number; gametophytes do not form spores, gametes are always haploid, not diploid.

2. Sporophytes always form spores by meiosis; chromosome number of spores will be haploid or 6.

3. Diagram B follows A with an arrow connecting spores to gametophyte stage and an arrow connecting gametes to sporophyte stage. Students may also indicate fertilization of gametes prior to forming sporophyte.

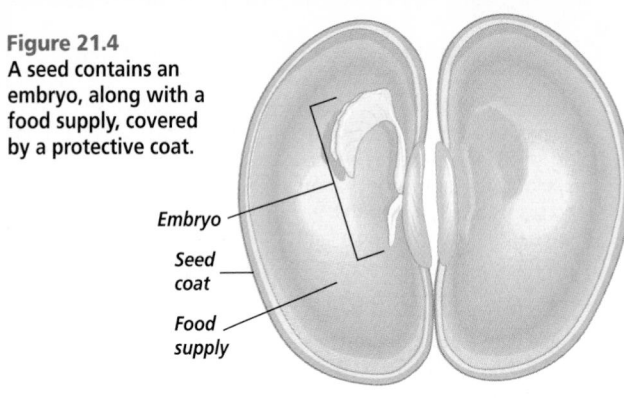

Figure 21.4
A seed contains an embryo, along with a food supply, covered by a protective coat.

Embryo

Seed coat

Food supply

The stems of most plants contain vascular tissues. **Vascular tissues** (VAS kyuh lur) are made up of tubelike, elongated cells through which water, food, and other materials are transported. Plants that possess vascular tissues are known as **vascular plants.** Most of the plants you are familiar with, including pine and maple trees, ferns, rhododendrons, rye grasses, English ivy, and sunflowers, are vascular plants.

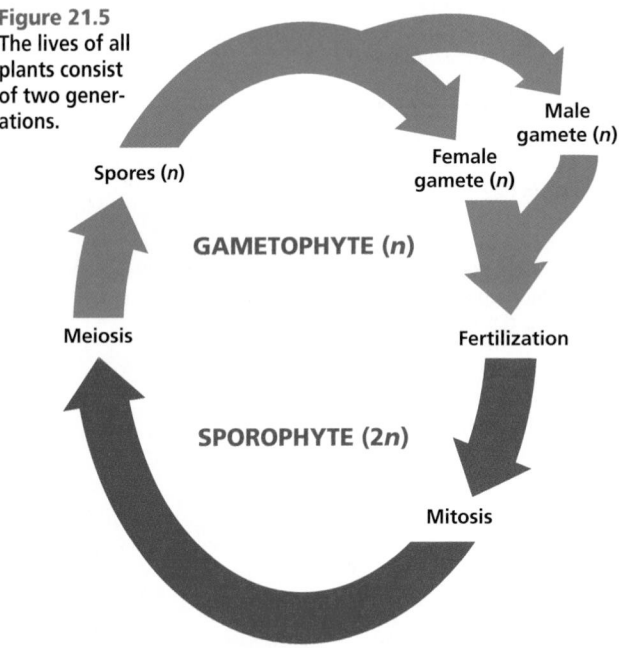

Figure 21.5
The lives of all plants consist of two generations.

Spores (*n*)

Male gamete (*n*)

Female gamete (*n*)

GAMETOPHYTE (*n*)

Meiosis

Fertilization

SPOROPHYTE (2*n*)

Mitosis

Mosses and several other small, less-familiar plants called hornworts and liverworts are usually classified as nonvascular plants. **Nonvascular plants** do not have vascular tissues. The tissues of nonvascular plants are usually no more than a few cells thick, and water and nutrients travel from one cell to another by the relatively slow processes of osmosis and diffusion.

The evolution of vascular tissues was of major importance in enabling plants to survive in the many habitats they now occupy on land. Vascular plants can live farther away from water than nonvascular plants. Also, because vascular tissues include thickened cells called fibers that help support upright growth, vascular plants can grow much larger than nonvascular plants.

Reproductive strategies

Adaptations in some land plants include the evolution of seeds. A **seed** contains an embryo, along with a food supply, covered by a protective coat, *Figure 21.4*. Seeds protect the zygote or embryo from drying out and may also aid in dispersal. Recall that a spore consists only of a single haploid cell with a hard, outer wall. Land plants reproduce by either spores or seeds.

In non-seed plants, which include mosses and ferns, the sperm swim through a film of water to reach the egg. In seed plants, which include all conifers and flowering plants, sperm are able to reach the egg without swimming through a continuous film of water. This difference is one reason why non-seed plants require moister habitats than most seed plants.

Alternation of generations

As in algae, the lives of all plants consist of two alternating stages, or generations, as shown in *Figure 21.5*. The gametophyte generation of a plant is responsible for the development of gametes. All cells of the gametophyte, including the gametes, are haploid (*n*). The sporophyte generation is responsible for the production of spores. All cells of the sporophyte are diploid (*2n*) and are produced by mitosis. The spores are produced in the sporophyte plant body by meiosis, and are therefore haploid (*n*).

The life cycles of all plants include the production of haploid spores. In non-seed vascular plants such as ferns, the spores have hard outer coverings and are released directly into the environment, where they grow into haploid gametophytes. In seed plants, such as conifers and wildflowers, the spores are retained by the parent plant and develop into gametophytes of only a few cells in size that are retained within the diploid sporophyte. Seed plants release the new sporophytes into the environment in the form of seeds. The gametophytes of non-seed plants are usually much larger than the gametophytes retained in the seeds of seed plants. Use the *Problem-Solving Lab* on this page to further explore the differences between the gametophyte and sporophyte generations of plants.

Problem-Solving Lab 21-1   Analyzing Information

How do gametophytes and sporophytes compare?
Plants have two different phases in their life cycle. One phase is called the gametophyte generation, and the other phase is called the sporophyte generation.

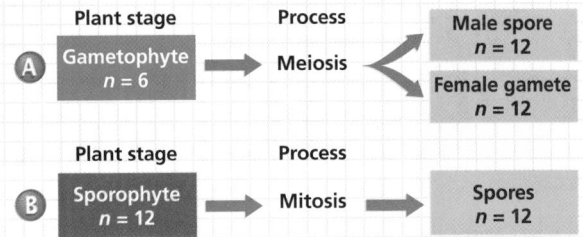

Analysis
Diagram A shows the gametophyte generation of a plant. This particular plant has a haploid chromosome number of 6. Examine the diagram carefully and look for errors that may be present.

Diagram B shows the sporophyte generation of a plant. The particular plant has a diploid chromosome number of 12. Examine the diagram carefully and look for errors that may be present.

Thinking Critically
1. List the errors observed in Diagram A and explain why they are incorrect.
2. List the errors observed in Diagram B and explain why they are incorrect.
3. Copy corrected Diagrams A and B so that they connect to one another and form a complete life cycle diagram of a plant.

Section Assessment

Understanding Main Ideas
1. What are three characteristics that plants share with algae?
2. Explain how the development of the cuticle and the vascular system influenced the evolution of plants on land.
3. How do seeds and spores differ? What are the benefits of producing seeds?
4. List the sequence of events involved in the alternation of generations in land plants. Do all plants have alternation of generations?

Thinking Critically
5. Explain why vascular plants are more likely to survive in a dry environment than nonvascular plants.

SKILL REVIEW
6. **Making and Using Tables** Make a table of the different adaptations plants have evolved that allow them to live on land. Include how each specific adaptation helps a plant survive on land. For more help, refer to *Organizing Information* in the **Skill Handbook**.

Check for Understanding
Have students explain the differences between the following word groups. **L2** **ELL**
  (a) seed—spore
  (b) vascular—nonvascular
  (c) sporophyte—gametophyte

Reteach
Have students prepare an outline that addresses the following: (a) evidence that land plants evolved from green algae, (b) adaptations of land plants that enable them to survive on land, and (c) the major functions of leaves, roots, and stems. **L2**

Extension
Visual-Spatial Have students observe a plant growing outside the school, and list traits that allow it to live on land. **L2**

✔ Assessment
Knowledge Have students explain how each of the following help plants survive on land: (a) thin leaves, (b) waxy cuticle, (c) roots, (d) supportive tissue in stems, (e) thick covering around gametes. **L2**

4 Close

Discussion
Logical-Mathematical Have students correlate the following items with a plant structure. Ask them to give reasons for their choices. Items: drinking straw, waxed paper, skin pores, Velcro, steel beams, bag of sugar, roof shingles. **L1**

Section Assessment

1. Cell walls composed of cellulose, the same types of chlorophyll for photosynthesis, and storage of food as starch.
2. A cuticle reduces water loss and a vascular system transports food, water, and minerals to all parts of the plant.
3. Seeds have a protective coat and contain a diploid embryo. Spores are a single haploid cell with a hard, outer wall. Seeds protect an embryo from drying out and aid in dispersal.
4. Sporophytes form spores via meiosis, spores form gametophytes that form gametes via mitosis, gametes fuse to form a new sporophyte. Yes, all land plants have alternation of generations.
5. Vascular plants have tubelike tissues that transport water and nutrients, whereas nonvascular plants use osmosis and diffusion, which requires a close association with water.
6. Table should include the following: preventing water loss, leaves carry out photosynthesis, roots, transporting materials, and reproductive strategies.

Focus On

Plants for People

Purpose 📦

Students will discover the role of agriculture, silviculture, and horticulture in producing plants for food, shelter, medicine, and pleasure.

Background

Scientific names for species described in this feature are as follows: corn, *Zea mays*; rice, *Oryza sativa*; wheat, *Triticum darum var. vulgare*: potato, *Solanum tuberosum*; oats, *Avena sativa*; barley, *Hordeum vulgare*; sorghum, *Sorghum bicolor*.

Teaching Strategies

■ Have students read this feature. Then have them work in cooperative groups to prepare a table of all the plants described. Along the left side, have them list the crop species. Across the top, have them use the headings: Kingdom, Division, Class, Family. Advise students to refer to the Classification Appendix A for help in the completion of their table. **Note:** Not all plants are listed in the appendix. **L1**
COOP LEARN

■ Have students write a short paragraph that describes the relative dependence of humans on plants as food sources. **L2**

FOCUS ON

Plants for People

Agriculture was perhaps the single most important development in human history. It is no accident that the beginnings of civilization occurred in productive farming areas. Today, many farmers grow just one food crop. Monocultures enable farmers to use machinery for planting, cultivating, and harvesting. However, monocultures are very vulnerable to disease. Pest infestations can spread rapidly, wiping out an entire season's crop. To combat this problem, farmers have begun to use native species that are less susceptible to disease. Scientists are also using genetic engineering techniques to make cotton, corn, and other crop plants more insect resistant.

AGRICULTURE

CORN — Farmers in the United States use more acreage to grow corn than any other crop. Livestock consume most of the corn crop, but a significant portion also goes to manufacture starch, oil, sugar, meal, breakfast cereals, and alcohol.

WHEAT, UKRAINE

WHEAT — Nearly one-third of all land in the world used for crop production is planted in wheat. Wheat probably originated in the Middle East and was an important food for the ancient Mesopotamian, Egyptian, and Indus civilizations.

RICE — Most humans equate rice with survival. In Asia, rice is the basis of almost all diets. More than 95 percent of the world's rice crop is used to feed humans. Rice is the only grain that can grow submerged in water.

RICE, BALI

580

✓ Portfolio

Commercial Agriculture

🔲 *Linguistic* Have students research the origin of a particular commercial crop and write a brief report on their findings. The report can be placed in their portfolios. **L3** **P**

MEETING INDIVIDUAL NEEDS

English Language Learners

🔲 *Visual-Spatial* Ask students to collect photos of plants from magazines and ask if any of them illustrate any of the main products obtained through agriculture. **L2** **ELL**

POTATOES, PERU

POTATOES — A South American native, the potato arrived in the United States via Europe. This nutritious root vegetable contains many essential amino acids. Potatoes also contain vitamins B and C and the minerals calcium and iron.

OATS — Oats make excellent food for both animals and humans. Containing from 10 to 16 percent protein, oats are low in fat and high in carbohydrates, proteins, B vitamins, fiber, and minerals. Native to northern Europe, oats grow well in poor soils as well as in cool, wet climates.

BARLEY — Barley was probably the first grain crop grown by humans. The Egyptians grew barley in 6000 B.C. The world's fourth largest cereal crop, barley grows fast and is able to withstand harsh growing conditions in rugged climates such as Lapland and the Himalayas.

SORGHUM — Since prehistoric times, sorghum has been a major food crop in Africa. Because of its extensive root system, sorghum is especially drought resistant, providing food for people and hay for cattle.

SORGHUM, SOMALIA

581

Teaching Strategies

■ Supply students with blank maps of the United States. Ask them to color in those areas that are primarily corn- and wheat-growing zones. Have them add labels to identify each crop. **L2**

■ Supply students with a global outline map. Have them label on the map where early agriculture began—the Tigris and the Euphrates, Nile Valley, and Indus River region. **L2**

Tying to Previous Knowledge

■ Have students review the process of fermentation and allow them to infer how corn can be used in the process to produce alcohol. *Corn serves as the food source for yeast.*

Enrichment

Intrapersonal Ask students to determine the differences between the wheat used for pasta *(T. durum)* and that used for bread *(T. vulgare)*. Have them find out what some of the different physical properties are between these starches and why one is more suitable than the other for bread or pasta products. **L2**

Visual Learning

Visual-Spatial Obtain a copy of the USDA food pyramid. Ask students to examine the text and photos of the agricultural products shown here and determine where they fit in a daily balanced diet as described by the USDA food pyramid. How many parts of the food pyramid come from plants? How important are plants for human nutrition?

TECHPREP

Nutritional Values of Foods

Have students prepare a chart comparing the fat and protein content of four of the foods mentioned in "Plants for People." Students should explain which foods would provide more energy and why. **L2**

GLENCOE TECHNOLOGY

VIDEODISC
The Infinite Voyage: *A Taste of Health, Pima Indians: Old Traditions in Nutrition* (Ch. 1)

5 min.

Pima Indians: Coping With Disease (Ch. 2)

7 min.

Background

Quinine, an antimalarial chemical derived from the bark of various species of *Cinchona* trees, was used by the Andean Indians before the European discovery of the New World.

Cinnamon is derived from the bark of *Cinnamommum zeylancium*.

Teaching Strategies

■ Have students work in cooperative groups to compile a list of different nuts that are from trees, and draw a map that shows where these are grown around the world. Responses may include almond, Brazil nut, cashew, hazelnut, pecan, pistachio, walnut, and chestnut.
L2 **COOP LEARN**

BARK FOR MEDICINAL USE, KENYA

SILVICULTURE

Often confused with forestry, silviculture includes the growing of trees for lumber and paper and for food crops, such as apples and pecans. Oranges, cloves, nuts, and olives are some of the foods and spices grown on trees. Trees are also sources of medicines, such as aspirin, originally derived from the bark of the willow tree, and quinine, from the *Cinchona* tree, used to treat malaria. Trees help reduce air pollution and replenish the oxygen we breathe.

CHEMICALS AND SPICES — Wood is the source of many chemicals including wood alcohol, latex for rubber, and cellulose used to make paper. Charcoal and rayon as well as spices, such as cinnamon and cloves, also are tree products. Taxol, an extract from the bark of a Pacific Coast yew tree, is currently being used to fight cancer.

RESEARCH ON FRUIT TREES, MARYLAND

LUMBER AND FUEL — When people think of forest products, they often think of lumber. A common building material, wood is easy to work with, durable, relatively abundant, and lightweight—making it ideal for home construction. Easy to transport and to store, wood is also a source of fuel throughout much of the world.

LUMBER INDUSTRY,
WASHINGTON STATE

582

VIDEODISC
STV: Plants, *How Plants Are Used*
Unit 3, Side 2, 13 min. 15 sec.
How Plants Are Used (in its entirety)

HORTICULTURE

Most of the plants you are familiar with were originally brought from distant lands by naturalists and explorers. By preserving and cultivating these species, horticulturalists gave us a legacy of beautiful and useful flowers and foods. Selective breeding, grafting, and more recently, genetic engineering are some methods used by plant scientists.

VEGETABLES — Vegetables were among the first plants cultivated by humans. Green cabbage, water-cress, and radishes—members of the mustard family—were known to Egyptians and Romans in the Bronze Age. Root crops such as beets, carrots, potatoes, and turnips are highly prized for fiber and nutrients and because they can be stored easily over winter.

FLOWERING BULBS, HOLLAND

FRUIT MARKET, AFGHANISTAN

FLOWERS AND MEDICINAL PLANTS— Flowering bulbs, roses, scented herbs, and ornamentals of all kinds have been a source of pleasure for generations. Plants have been used for medicinal purposes for thousands of years. Today plants continue to be a major source of pharmaceuticals and herbal remedies.

FRUITS — Trees are a source for one of the world's favorite foods—fruit. Acorns, walnuts, and pecans are among the dry fruits used as food by wildlife and humans. Fleshy fruits include oranges, apples, pears, apricots, and cherries.

EXPANDING Your View

① **APPLYING CONCEPTS** Take a piece of paper and make two columns. In the left column, list all of the products essential to your life that come from plants and/or agriculture. In the right column, list all of the essential products that *do not* come from plants and/or agriculture. Now write a paragraph summarizing the role of plants in your life.

583

Activity

Kinesthetic Have students test a variety of plant roots for the presence of starch. Have them use thin slices of radish, turnip, and potato and add a few drops of iodine/potassium iodide solution to the root. A blue-black color indicates the presence of starch. As a control, have them add a drop of iodine solution to a small clump of corn starch. Ask them to write a report of their observations for their portfolios.
L2 **P**

Answers to Expanding Your View

Except for transportation, communication, certain services, and products composed of metals, almost all human basic needs are tied to plants.

BIOLOGY *Online* Note Internet addresses that you find useful in the space below for quick reference.

_____

_____

_____

_____

_____

Prepare

Key Concepts

Students survey the 12 major plant divisions.

Planning

- Collect pictures of plants for the bulletin board.
- Collect pine needles and leaves for the Quick Demo

1 Focus

Bellringer

Before presenting the lesson, display **Section Focus Transparency 51** on the overhead projector and have students answer the accompanying questions.
L1 **ELL**

Transparency **51** Bryophytes and Hepatophytes

SECTION FOCUS
Use with Chapter 21, Section 21.2

Moss   Liverwort

❶ This moss and liverwort have no vascular tissues. How might this limit the types of habitat these plants can survive in?

❷ How else might the lack of vascular tissue limit these plants?

BIOLOGY: The Dynamics of Life   SECTION FOCUS TRANSPARENCIES

Resource Manager

Section Focus Transparency 51 and Master **L1** **ELL**

21.2 Survey of the Plant Kingdom

SECTION PREVIEW

Objectives
Describe the phylogenic relationships among divisions of plants.
Identify the twelve plant kingdom divisions.

Vocabulary
frond
cone

Members of the plant kingdom are found all over the world. Many species are widespread, but some grow in only one area. For example, *Sarracenia oreophila, commonly known as the green pitcher plant, is found only in the Southeastern United States. Others, such as the cinnamon fern, have a much wider distribution. These two plants also illustrate two different reproductive strategies. While both the fern and pitcher plant produce spores, the pitcher plant also produces seeds.*

Cinnamon ferns in forest and green pitcher plant, *Sarracenia oreophila*, with flowers (inset)

Phylogeny of Plants

Many geological and climate changes have taken place since the first plants became adapted to life on land. Landmasses have moved from place to place over Earth's surface, climates have changed, and bodies of water have formed and disappeared. Hundreds of thousands of plant species evolved, and countless numbers of these became extinct as conditions continually changed. These processes of evolution and extinction continue to be affected by local and global changes. As plant species evolved in this changing landscape, they retained many of their old characteristics and also developed new ones. These processes of evolution and extinction continue today.

Botanists use plant characteristics to classify plants into twelve divisions. Recall that a plant division is similar to a phylum in other kingdoms. The highlights of plant evolution include origins of plants from green algae, the development of vascular tissue, the production of seeds, and the formation of flowers. The production of seeds can be used as a basis to separate the divisions into two groups—non-seed plants and seed plants.

Non-seed Plants

There are seven divisions of non-seed plants, which are shown in *Figure 21.6.* These plants produce hard-walled reproductive cells called spores. Non-seed plants may be either vascular or nonvascular.

584 WHAT IS A PLANT?

Cultural Diversity

Ethnobotany

TECH PREP

Interpersonal Introduce students to the growing field of ethnobotany, the study of how native cultures use plants. Ethnobotany has become increasingly important to the development of new drugs by pharmaceutical companies. One firm, Shaman Pharmaceuticals in

California, sends researchers into South American, Asian, and African tropical forests to ask healers about the medications they derive from plants. Have students research and write reports about the medicinal uses of plants in different cultures in both contemporary and historical perspectives. **L2**

Figure 21.6
The plant kingdom includes seven divisions of non-seed plants.

A *Selaginella*, a spike moss, is a lycophyte. Lycophytes are vascular plants adapted to moist environments.

B Tree ferns are pterophytes. This *Cyathea arborea* can be found growing in damp, tropical forests.

C *Equisetum* is a sphenophyte. It has roots, stems, and leaves, but the stems are hollow and appear jointed.

D *Marchantia* is a hepatophyte. It is found on damp rocks.

E *Sphagnum* is a bryophyte. It grows in peat bogs.

F *Anthoceros*, a hornwort, is an Anthocerophyte. It is found in moist, shady habitats.

G *Psilotum* sporophytes have simple stems but no leaves.

Hepatophyta

Hepatophytes (heh PAH toh fites) are small plants commonly called liverworts because the flattened body of the plant resembles the lobes of an animal's liver. There are two kinds of liverworts: thallose liverworts and leafy liverworts. Thallose liverworts have a broad body that looks like a lobed leaf. Leafy liverworts are creeping plants with three rows of thin leaves attached to a stem. Liverworts are nonvascular plants. They grow only in moist environments because they use osmosis and diffusion to transport water and nutrients. Studies comparing the biochemistry of different plant divisions suggest that liverworts may be the ancestors of all plants.

Anthocerophyta

Anthocerophytes (an THOH ser oh fites) are also small plants. The sporophytes of these plants, which resemble the horns of an animal, give the plants their common name—hornworts. These nonvascular plants grow in damp, shady habitats and rely on osmosis and diffusion to transport nutrients.

Word Origin
hepato-
From the Greek word *hepar*, meaning "liver." Hepatophytes have liver-shaped gametophytes.

2 Teach

Display
Make a bulletin board display of a variety of plants from different divisions.

Quick Demo

Visual-Spatial Show students pine needles and broad-leaves and ask them to compare and contrast their features. **L1** **ELL**

Assessment
Knowledge Show students pictures of plants from several divisions. Have them identify the division to which each belongs. **L2**

GLENCOE TECHNOLOGY

CD-ROM
Biology: The Dynamics of Life
Exploration: *The Five Kingdoms* Disc 3

NATIONAL GEOGRAPHIC

VIDEODISC
STV: Plants
What is a Plant?
Unit 1, Side 1, 3 min. 40 sec.
Types of Plants

BIOLOGY JOURNAL

Identifying Seed Plants

Visual-Spatial Ask students to prepare a concept map that will help other students to identify seed plants. Suggest they include pictures from magazines or drawings as part of their concept map. **L2** **ELL**

Portfolio

Hepatophytes and Bryophytes

Visual-Spatial Have students draw pictures or find magazine photos of hepatophytes and bryophytes. Students should include labels that indicate the similarities and differences between the two plants. **L2** **ELL**

Purpose 📦

Students will compare living plants from two different divisions with their fossil relatives.

Process Skills

classify, collect data, compare and contrast, draw a conclusion, measure in SI, observe and infer

Teaching Strategies

■ Living and preserved samples of club moss and horsetail are available from biological supply houses.

■ Caution students about rubbing their eyes when working with preserved materials.

■ Student observations can be both qualitative and quantitative. Review the meaning of these terms before students start this activity.

■ Remind students that measurements in diagrams are in meters and the comparison asks for the same measurements in centimeters. Review conversion from meters to centimeters if necessary.

Expected Results

Students will find a number of similarities and some differences between the living plants and their fossil relatives. Numbers of similarities and differences will depend on the thoroughness of student observations.

Analysis

1. Similarities—green, grow upright, have scalelike leaves. Differences—tall and thick fossil stem, short and thin living stem, leaves on top of fossil stem, ridged surface of fossil stem, leaves on living stem. Student answers on the closeness of the relationship may vary. If going by general appearance, students may agree that they are closely related. If using the total number of similarities and differences, students may disagree.

2. Similarities—green, grow upright, leaves are in whorls,

MiniLab 21-2 Comparing and Contrasting

Looking at Modern and Fossil Plants Many modern-day plants have relatives that are known only from the fossil record. Are modern-day plants similar to their fossil relatives? Are there any differences?

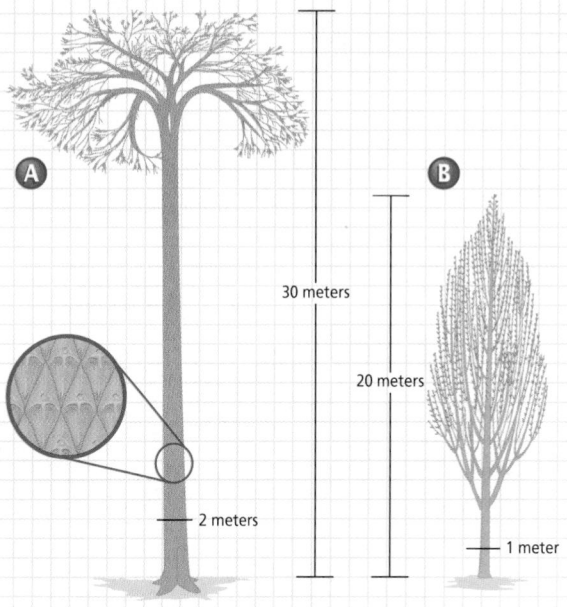

Procedure 📞 🧤 ☠

1. Examine a preserved or living sample of *Lycopodium,* a club moss. **CAUTION:** *Wear disposable latex gloves when handling preserved material.*

2. Note and record the following observations:
   a. Does the plant grow flat or upright like a tree?
   b. Describe the appearance of its leaves and its stem.
   c. Measure the plant's height and diameter in centimeters.

3. Diagram A is a representation of a fossil relative called *Lepidodendron.* Record the same observations (a–c).

4. Repeat steps a-c only this time use a preserved or living sample of *Equisetum,* a horsetail. Compare it to Diagram B, a representation of a fossil relative called *Calamites.*

Analysis

1. Describe the similarities and differences between *Lycopodium* and *Lepidodendron.* Do your observations justify their closeness as relatives? Explain.

2. Describe the similarities and differences between *Equisetum* and *Calamites.* Do your observations justify their closeness as relatives? Explain.

Bryophyta

Bryophytes (BRI uh fites), mosses, are nonvascular plants. Like liverworts and hornworts, mosses rely on osmosis and diffusion to transport water and nutrients. However, some mosses have elongated cells that conduct water and sugars. Moss plants are usually less than 5 cm tall and have leaves that are usually only one to two cells thick. Their spores are formed in capsules.

Psilophyta

Psilophytes, also known as whisk ferns, consist of thin, green, leafless stems. The psilophytes are unique among vascular plants because they have neither roots nor leaves. Small scales cover each stem. Scales are flat, rigid, overlapping structures. Most of the 30 species of psilophytes are tropical or subtropical, although one genus is found in the southern United States.

Lycophyta

Lycophytes (li KOH fites), the club mosses, are vascular plants adapted primarily to moist environments. Lycophytes have stems, roots, and leaves. Their leaves, although very small, contain vascular tissue. Species existing today are usually less than 25 centimeters high, but their ancestors grew as tall as 30 meters and formed a large part of the vegetation of Paleozoic forests. The plants of these ancient forests have become coal and are now used by people for fuel. Try the *MiniLab* on this page to explore the similarities and differences between modern and fossil lycophytes.

Sphenophyta

Sphenophytes (sfe NOH fites), the horsetails, are vascular plants. They have hollow, jointed stems surrounded by whorls of scalelike leaves. The cells covering the stems of some

---

leaves are small, rings on stem. Differences—fossil has branches while living is one single stem, fossil is taller and thicker, living plant leaves appear to be much smaller than fossil. Student answers on the closeness of the relationship may vary. If going by general appearance, students may agree that they are closely related. If using the total number of similarities and differences, students may disagree.

✔ Assessment

Knowledge Ask students to find out how long ago the fossil plant examples used in this MiniLab lived on Earth. Use the Performance Task Assessment List for Using Math in Science in **PASC**, p. 29. **L2**

sphenophytes contain large deposits of silica. Although primarily a fossil group, about 15 species of spheno-phytes exist today. All present-day horsetails are small, but their fossil relatives were the size of trees.

Pterophyta

Pterophytes (ter OH fites), ferns, are the most well-known and diverse group of non-seed vascular plants. Ferns were abundant in Paleozoic and Mesozoic forests. They have leaves that vary in length from 1 cm to 500 cm. The large size and com-plexity of these leaves, also called megaphylls, is one difference between pterophytes and other groups of seedless vascular plants. These leaves are called **fronds**. Although ferns are found nearly everywhere, most grow in the tropics.

Seed Plants

Seed plants produce seeds, which in a dry environment are a more effective means of reproduction than spores. Seeds consist of an embryonic plant and a food supply covered by a hard protective seed coat. All seed plants have vascular tissues. Examples from the seed plant divisions are shown in *Figure 21.7*.

Figure 21.7
Plants classified into the five divisions of seed plants produce seeds covered by tough, protective seed coats.

A Cycads are often mistaken for ferns or small palm trees.

D *Ginkgo biloba*, the maidenhair tree, is no longer found in the wild, although it continues to be cultivated in many countries, including the United States.

B *Welwitchsia mirabilis* is found in harsh desert environments in Africa. The leaves of this gnetophyte can grow to 2 m long.

E Wildflowers can be found in nearly every environment on Earth. This wildflower, called chicory, is an anthophyte.

C *Pinus banksiana*, the jack pine, keeps its cones closed until a fire passes over them, an example of an adaptation to extreme conditions.

587

Purpose

Students will analyze a graph and detect trends that occur in the size of the gametophyte and sporophyte generations as they move through the various plant divisions.

Process Skills

apply concepts, think critically, compare and contrast, draw a conclusion

Teaching Strategies

■ Ask students to sequence the plant divisions from Bryophyta to Anthophyta prior to doing this lab. This will allow them to refer to their lists as they complete the analysis section.

Thinking Critically

1. Gametophyte generation becomes smaller.
2. Sporophyte generation becomes larger.
3. a. large sporophyte, small gametophyte; division would be located directly to the left of Anthophytes.
   b. large gametophyte, small sporophyte; division would be located close to the right of Bryophyta
4. Redwoods are grouped in the division Coniferophyta. Therefore, the large generation is the sporophyte.

✔ Assessment

Knowledge Ask students to predict the laboratory tools needed to study the gametophyte and sporophyte generations of a moss plant and a rose. Use the Performance Task List for Designing an Experiment in PASC, p. 23. **L2**

588

Problem-Solving Lab 21-2
Applying Concepts

What trend in size is seen with gametophyte and sporophyte generations? All plants undergo alternation of generations. There is a specific trend, however, that occurs in size as one goes from one plant division to the next.

Analysis

The following graph shows the trend that occurs within the plant kingdom as one compares the size of sporophyte and gametophyte generations in three major divisions.

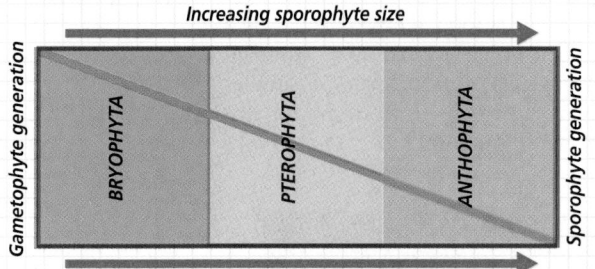

Increasing sporophyte size

Gametophyte generation — BRYOPHYTA — PTEROPHYTA — ANTHOPHYTA — Sporophyte generation

Decreasing gametophyte size

Thinking Critically

1. Describe the trend that occurs to the size of the gametophyte generation as one moves from Bryophytes to Anthophytes.
2. Describe the trend that occurs to the size of the sporophyte generation as one moves from Bryophytes to Anthophytes.
3. Estimate the size of the gametophyte generation compared with the sporophyte generation in:
   a. Coniferophyta. Explain.
   b. Lycophyta. Explain.
4. You are looking at a giant redwood tree. Which generation is it and how do you know?

Cycadophyta

WORD *Origin*

ginkgo-
From the Chinese word *yin-hing,* meaning "silver apricot." Ginkgophyta species produce apricot-colored seeds.

Cycads (SI kuds) were abundant during the Mesozoic era. Today, there are about 100 species of cycads, palmlike trees with scaly trunks. Cycads may be very short in stature or they may be 20 meters or more in height. Cycads produce male and female cones on separate trees. **Cones** are scaly structures that support male or female reproductive structures. Seeds are produced in

---

female cones. Male cones produce large clouds of pollen. The cones of cycads may be as long as 1 m.

Gnetophyta

There are three genera of gnetophytes (nee TOH fites), each of which is quite distinct. *Gnetum* includes species of trees and climbing vines; *Ephedra* includes shrubby species; and *Welwitschia,* found only in South Africa, has a short stem that grows as a large, shallow cap. The leaves of this plant grow from the base of the stem. These seed plants grow in the desert and can live to be 100 years old.

Ginkgophyta

This division has only one living species, *Ginkgo biloba,* a distinctive tree with small, fan-shaped leaves. Like cycads, ginkgos (GING kohs) have separate male and female trees. The seeds produced by the female trees have an unpleasant smell, so ginkgos planted in city parks are usually male trees. Ginkgos are hardy and resistant to insects and to air pollution. You can learn more about the suitability of different plants for various environments in the *BioLab* at the end of the chapter.

Coniferophyta

These are the conifers (KAHN uh furz), cone-bearing trees such as pine, fir, cypress, and redwood. Conifers are vascular seed plants that produce seeds in cones. Unlike anthophytes, the seeds of conifers are not protected by a fruit. Species of conifers can be identified by the characteristics of their needle-like or scaly leaves. Bristlecone pines, the oldest known living trees in the world, are members of this plant division. Another type of conifer, the Pacific yew, is a source of

---

GLENCOE TECHNOLOGY

CD-ROM
Biology: The Dynamics of Life
Video: *Giant Redwoods*
Disc 3

Resource Manager

Reinforcement and Study Guide, pp. 95-96 **L2**
Content Mastery pp. 105, 107-108 **L1**

Table 21.1 The twelve divisions of the Plant Kingdom

Division	Common Name	Approximate Number of Species	Examples
Anthocerophyta	Hornworts	100	*Anthoceros*
Hepatophyta	Liverworts	6500	*Marchantia*
Bryophyta	Mosses	10 000	*Sphagnum*
Psilophyta	Whisk Ferns	10-13	*Psilotum*
Lycophyta	Club Mosses	1000	*Lycopodium*
Sphenophyta	Horsetails	15	*Equisetum*
Pterophyta	Ferns	12 000	*Nephrolepis*
Cycadophyta	Cycads	100	*Cycas*
Ginkgophyta	Ginkgoes	1	*Ginkgo*
Coniferophyta	Conifers	550	*Pinus*
Gnetophyta	Gnetophytes	70	*Welwitschia*
Anthophyta	Flowering Plants	240 000	*Rosa*

cancer-fighting drugs. Read more about medicinal plants in the *Health Connection* at the end of this chapter.

Anthophyta

Anthophytes (an THOH fites), commonly called the flowering plants, are the largest, most diverse group of seed plants living on Earth. There are approximately 240 000 species of anthophytes. Fossils of the Anthophyta date to only the Cretaceous period, 130 million years ago. Anthophytes produce flowers and seeds enclosed in a fruit. This division has two classes: the mono-cotyledons (mahn uh kaht ul EED unz) and dicotyledons (di kaht ul EED unz). You will learn more about the distinctions between monocots and dicots when you read about anthophyte tissues.

Table 21.1 lists some information about the twelve divisions of the plant kingdom. See if you can recognize any of the common names of the plants. You can further explore the similarities and differences among the twelve divisions of plants in the *Problem-Solving Lab,* shown on the previous page.

WORD *Origin*

conifero-
From the Latin word *conifer,* meaning "cone bearing." Many plants in the division Coniferophyta produce their seeds on cones.

Section Assessment

Understanding Main Ideas

1. What is the primary difference between the seeds of conifers and anthophytes?
2. Why are seeds an important adaptation? What plant divisions produce seeds? Which plant divisions do not produce seeds?
3. Explain why pterophytes tend to be larger than bryophytes.
4. How are anthophytes and anthocerophytes similar? How are they different?

Thinking Critically

5. In which division would you expect to find apple trees? Why?

SKILL REVIEW

6. **Making a Table** Make a table of the twelve plant divisions that includes seed plants, non-seed plants, vascular plants, nonvascular plants, and seeds in fruits. For more help, refer to *Organizing Information* in the **Skill Handbook.**

21.2 SURVEY OF THE PLANT KINGDOM **589**

3 Assess

Check for Understanding
Quiz students orally about the major characteristics of the 12 divisions.

Reteach
Visual-Spatial Have students construct a table organizing the 12 divisions under two headings: non-seed and seed plants. **L2**

Extension
Have students prepare a key that will enable them to distinguish among the 12 plant divisions. Encourage students to photocopy their keys and share them with their classmates. **L3**

✔ **Assessment**
Knowledge Using the photos in Figures 21.6 and 21.7, have students explain why each plant belongs to its particular division. **L2**

The BioLab at the end of the chapter can be used at this point in the lesson.

INTERNET BioLab

4 Close

Activity
Divide the class into small groups, assigning each a different biome. Ask them to predict what plants might inhabit their biome, identify the plant divisions, and explain their reasoning. **L2**
COOP LEARN

Section Assessment

1. Anthophyte seeds are encased in fruits; conifer seeds are produced on cones.
2. Seeds provide protection and a food supply for the embryo and aid in dispersal. Seed: Cycadophyta, Gnetophyta, Ginkgophyta, Coniferophyta, and Anthophyta. Non-seed: Anthocerophyta, Hepatophyta, Bryophyta, Psilophyta, Lycophyta, Sphenophyta, and Pterophyta.
3. Ferns have vascular tissues that transport water and provide support.
4. Both show alternation of generations and are land plants. Anthocerophytes are non-seed, nonvascular plants; anthophytes are vascular seed plants.
5. Apple trees anthophytes because they produce flowers and fruit.
6. Non-seed, nonvascular: Anthocerophytes, Hepatophytes, and Bryophytes; Non-seed, vascular: Psilophytes, Lycophytes, Sphenophytes, and Pterophytes; Seed, vascular: Cycadophytes, Gnetophytes, Ginkgophyte, Coniferophytes; Anthophytes; Seeds in fruits, vascular: Anthophytes.

589

INTERNET

BioLab

Researching Trees on the Internet

Time Allotment 🖥

Two class periods (if enough computers are available in class) Two hours if done at home

Process Skills

acquire information, analyze information, collect data, define operationally, think critically, draw a conclusion, interpret data, apply concepts

Prepare different tree lists in advance for your students. This may help to speed up the initial student selection process.

Alternative Materials

If Internet resources are not available, students can still complete this BioLab through conventional library research using books and periodicals.

Resource Manager

BioLab and MiniLab Worksheets, pp. 99-100 L2

*I*magine that you are employed by a city in its department of Urban Planning. You have just been handed an assignment by the city manager. The assignment? Research the type of tree that would be most suitable for planting along the streets of your community.

Problem

Use the Internet to find different trees that would be suitable for planting in your community.

Objectives

In this BioLab, you will:
- **Research** the characteristics of five different trees.
- **Use the Internet** to collect and compare data from other students.

- **Conclude** which trees would be most suitable for planting in your community.

Materials

Internet access

Skill Handbook

Use the **Skill Handbook** if you need additional help with this lab.

1. Make a copy of the data table.
2. Pick five trees that you wish to research. Note: Your teacher may provide you with suggestions if necessary.

3. Go to the Glencoe Science Web Site at the address shown in the Sharing Your Data box to find links to information needed for this BioLab.
4. Record the information in your data table.

Northern Catalpa, *Catalpa speciosa*

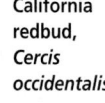

California redbud, *Cercis occidentalis*

Teaching Strategies

■ Suggest to students that they modify the space allowed in the data table for room to accommodate information or diagrams. The category marked "Additional Information" could be used to describe the trees' tolerance to city pollutants.

■ Encourage students to complete this BioLab at home as a homework assignment

if they have computers with Internet access available there.

■ Limit the selection of conifers to one or two trees out of the five selected for study.

■ To navigate to the Internet BioLabs, choose the Biology: The Dynamics of Life icon at the Glencoe Science Web Site. Click on the student site icon, then the BioLabs icon.

Data Table

	1	2	3	4	5
Tree Name (common name)					
Scientific Name					
Kingdom					
Division					
Soil/Water Preference					
Temperature Tolerance					
Height at Maturity					
Speed of Growth					
General Shape					
Disease Resistance/Problems					
Special Care					
Leaf Shape					
Shade Provider					
Additional Information					

Flowering dogwood, *Cornus florida*

Southern Magnolia, *Magnolia grandiflora*

ANALYZE AND CONCLUDE

1. **Defining Operationally** Explain the difference between trees classified as either Coniferophyta or Anthophyta.
2. **Analyzing** Was the information provided on the Internet helpful in completing your data table? Explain your answer.
3. **Thinking Critically** What do you consider to be the most important characteristic when deciding on the most suitable tree for your community? Explain your answer.
4. **Using the Internet** Using the information you gathered from the Internet, which tree species would most likely be the:
   a. most suitable for your community? Explain your answer.
   b. least suitable for your community? Explain your answer.
5. **Applying** Explain why tree selections would differ if your community were located in:
   a. Tucson, Arizona
   b. Los Angeles, California
   c. Fairbanks, Alaska

Sharing Your Data

BIOLOGY Online Find this BioLab on the Glencoe Science Web site at **science.glencoe.com** and post your findings in the table provided. Using your findings and information posted by other students, prepare a dichotomous key that allows you to identify your five trees.

INTERNET BioLab

ANALYZE AND CONCLUDE

1. Coniferophytes are cone-bearing trees, seeds formed on cones, evergreens; anthophytes are flowering plants, seeds formed within a fruit, usually deciduous.
2. Student answers will vary. The Internet (if used) will provide information for the data table.
3. Student answers will vary. Students may select speed of growth, temperature tolerance, or disease resistance as the most important qualities.
4. a. Student answers will vary.
   b. Student answers will vary.
5. a. High summer temperature and drought tolerance would be important factors.
   b. Long periods of drought during the summer would be an important factor.
   c. Temperature tolerance during the winter would be an important factor.

Sharing Your Data

BIOLOGY Online To navigate to the Internet BioLabs choose the Biology: The Dynamics of Life icon at Glencoe's Web site. Click on the student site icon, then the BioLabs icon. Make sure student's keys are truly dichotomous and are logical.

Data and Observations

Student data tables will vary depending on the initial trees selected for study.

✔ Assessment

Skill Provide students with data on two or three trees. Ask them to determine the best national site for trees with these characteristics. Use the Performance Task Assessment List for Analyzing the Data in **PASC**, p. 27.

Health
Connection

Medicines from Plants

Madagascar rosy periwinkle

What comes to mind when you hear the word plant? A vase full of flowers? A fruit and vegetable garden? An evergreen forest? Although these examples are what most people think of when they hear the word plant, plants provide us with much more than bouquets, food, and lumber. Nearly 80 percent of the world's population relies on medications derived from plants. In fact, just fewer than 100 plants provide the active ingredients used in the ten dozen or so plant-derived medicines currently on the market.

For thousands of years, the words *plants* and *medicines* were used synonymously. In the fifth century A.D., for example, doctors of the Byzantine Empire used the autumn crocus to effectively treat rheumatism and arthritis. Hundreds of years ago, certain groups of Native North Americans used the rhizomes of the mayapple as a laxative, a remedy for intestinal worms, and as a topical treatment for warts and other skin growths. The oils from peppermint leaves have long been used to settle an upset stomach. Lotions containing the thick, syrupy liquid from the succulent plant aloe vera are often used to relieve the pain associated with minor burns, including sunburn. "Herbal" medicines have again begun to play an important role in so-called modern medicine. One of the "herbal" medicines is aspirin.

Aspirin—The wonder drug Evidence suggests that almost 2500 years ago, a Greek physician named Hippocrates used a substance from the bark of a white willow tree to treat minor pains and fever. This substance, which is called salicin, unfortunately upset the stomach. Research in the late 1800s by a chemist at the company known in the United States simply as Bayer led to the discovery of acetylsalicylic acid, or aspirin. Aspirin originally was developed by chemist Felix Hoffmann to relieve the joint discomfort

associated with the condition rheumatism. Its use became widespread in the early 1900s, however, when it became possible to synthesize salicylic acid in the laboratory.

New Drugs for Cancer Several drugs proven to fight two types of cancer—Hodgkin's disease and leukemia—have been derived from the Madagascar rosy periwinkle. Drugs produced from the needles and bark of the Pacific yew have been used successfully to treat breast, ovarian, lung, and other cancers. Although the interest in medicinal plants by consumers, medical experts, and pharmaceutical companies is growing, it is estimated that less than 10 percent of the 250 000 different flowering plant species have been studied for their potential use in the field of medicine.

CONNECTION TO BIOLOGY

How do you think scientists can best decide which plants to examine for potential medicinal properties?

BIOLOGY Online To find out more about medicines derived from plants, visit the Glencoe Science Web site. **science.glencoe.com**

592 WHAT IS A PLANT?

NATIONAL GEOGRAPHIC

VIDEODISC
STV: Plants, *How Plants Are Used*
Unit 3, Side 2, 2 min.
Medicine

SUMMARY

Section 21.1
Adapting to Life on Land

Main Ideas

- Plants are multicellular eukaryotes with cuticles and cells with thick walls containing cellulose. Plants have chlorophyll for photosynthesis and store food in the form of starch.

- All plants on Earth probably evolved from filamentous green algae that lived in ancient oceans. The first plants to make the move from water to land may have been leafless.

- Adaptations for life on land include a waxy cuticle; development of leaves, roots, and stems; development of seeds; and the reduction and protection of the gametophyte generation.

Vocabulary

cuticle (p. 575)
leaf (p. 577)
nonvascular plant (p. 578)
root (p. 577)
seed (p. 578)
stem (p. 577)
vascular plant (p. 578)
vascular tissue (p. 578)

Section 21.2
Survey of the Plant Kingdom

Main Ideas

- The plant kingdom includes twelve divisions.

- Plants in the divisions Anthocerophyta, Hepatophyta, and Bryophyta are non-seed nonvascular plants.

- Non-seed vascular plant divisions are Psilophyta, Lycophyta, Sphenophyta, and Pterophyta.

- Cycadophyta, Ginkophyta, Coniferophyta, Gnetophyta, and Anthophyta are vascular seed plant divisions. Anthophytes produce flowers and protect their seeds in fruits.

Vocabulary

cone (p. 588)
frond (p. 587)

Main Ideas

Summary statements can be used by students to review the major concepts of the chapter.

Using the Vocabulary

To reinforce chapter vocabulary, use the Content Mastery Booklet and the activities in the Interactive Tutor for Biology: The Dynamics of Life on the Glencoe Science Web site.
science.glencoe.com

 *All Chapter Assessment questions and answers have been validated for accuracy and suitability by The Princeton Review.*

UNDERSTANDING MAIN IDEAS

1. c
2. a
3. c
4. a

UNDERSTANDING MAIN IDEAS

1. Which of the following organisms has vascular tissues?
   a. bacteria   **c.** ferns
   b. algae   **d.** hornworts

2. The plant organ that absorbs water and minerals from the soil is the _____.
   a. root   **c.** stem
   b. leaf   **d.** flower

3. Which of the following characteristics is NOT found in plants?
   a. eukaryotic cells
   b. cellulose cell walls
   c. prokaryotic cells
   d. waxy cuticle

4. Which of the following is NOT part of a seed?
   a. haploid cell   **c.** food supply
   b. protective coat   **d.** embryo

CHAPTER 21 ASSESSMENT **593**

GLENCOE TECHNOLOGY

 **VIDEOTAPE**
MindJogger Videoquizzes
Chapter 21: *What is a Plant?*
Have students work in groups as they play the videoquiz game to review key chapter concepts.

 ## Resource Manager

Chapter Assessment, pp. 125-130
MindJogger Videoquizzes
ExamView® Pro Software
BDOL Interactive CD-ROM, Chapter 21 quiz

5. d

6. a

7. b

8. d

9. c

10. a

11. cuticle

12. spores/meiosis

13. osmosis/diffusion

14. cones/fruit

15. vascular tissues

16. root/stem

17. moist

18. diploid or 2*n*

19. haploid or *n*

20. a frond

APPLYING MAIN IDEAS

21. Studies comparing the biochemistry of different plants suggests that the earliest plants were biochemically most similar to liverworts. Also, many of the first plants to appear during succession are often liverworts. The adaptations that allow for survival on barren land environment may be the same adaptations that allowed for survival from water to a land environment.

5. The plant in the photo is a member of the division _____.

a. Anthocerophyta   **c.** Lycophyta

b. Coniferophyta   **d.** Anthophyta

6. Plants and green algae share all of these traits EXCEPT _____.

a. reproduce by fission

b. cellulose cell walls

c. store food as starch

d. same kind of chlorophyll

7. All plant life cycles include the production of _____.

a. seeds   **c.** roots

b. spores   **d.** flowers

8. Which nonplant group is most likely to be ancestral to land plants?

a. cyanobacteria   **c.** archaebacteria

b. bryophytes   **d.** green algae

9. Which of the following adaptations was critical for plants to adapt to life in drier areas?

a. production of spores

b. loss of cuticle

c. vascular tissue

 **THE PRINCETON REVIEW**   **TEST-TAKING TIP**

Use Process of Elimination
On any multiple-choice test, you can use a process of elimination to eliminate any answers that you know are wrong. There are generally more wrong answers than right answers for any given question. Find the ones you know are wrong, eliminate them, and you'll have fewer choices from which to pick.

d. alternation of generations

10. Seeds enclosed in a fruit are adaptations of _____.

a. Anthophytes   **c.** Coniferophytes

b. Bryophytes   **d.** Pterophytes

11. The waxy covering of a leaf is called a(n) _____.

12. The sporophyte generation produces _____ by the process of _____.

13. Nonvascular plants use _____ and _____ to transport water and nutrients.

14. Conifers produce seeds in a(n) _____, whereas flowering plants produce seeds encased in a(n) _____.

15. Tissues composed of tubelike cells that transport water, food, and other materials through a plant are called _____.

16. The _____ of a plant anchors the plant, whereas the _____ provides support for upright growth.

17. Living in a(n) _____ environment is necessary for reproduction of non-seed plants.

18. The cells of the sporophyte generation are all _____.

19. The cells of the gametophyte generation are all _____.

20. What is the leafy structure of a fern shown below?

APPLYING MAIN IDEAS

21. Explain why biologists think the first plants to adapt to life on land may have been similar to liverworts?

22. Explain why nonvascular plants cannot grow as tall as vascular plants.

23. What adaptations made it possible for plants to live on land?

24. The largest division is the anthophytes. What characteristic(s) has made them so successful?

25. What advantage(s) might a seed plants have over a non-seed plant during a drought?

THINKING CRITICALLY

26. **Comparing and Contrasting** Compare and contrast conifers and flowering plants.

27. **Observing and Inferring** What role might the jack pine play in a forest area destroyed by fire?

28. **Comparing and Contrasting** Both green algae and aquatic flowering plants live in water. Are these organisms similar in any way? How are they different?

29. **Concept Mapping** Complete the concept map by using the following vocabulary terms: leaves, stems, roots, vascular tissue, vascular plant

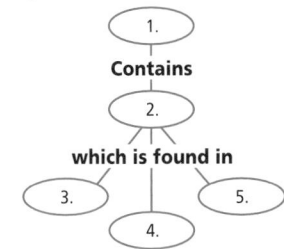

CD-ROM

For additional review, use the assessment options for this chapter found on the *Biology: The Dynamics of Life Interactive CD-ROM* and on the Glencoe Science Web site. **science.glencoe.com**

ASSESSING KNOWLEDGE & SKILLS

The diagram below illustrates the alternation of generations that is found in all plants.

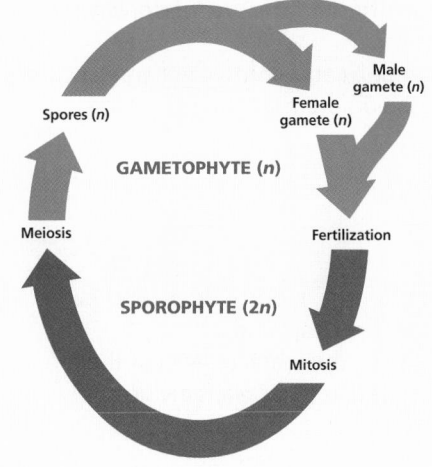

Interpreting Scientific Illustrations
Use the diagram to answer the following questions.

1. The sporophyte produces spores by the process of _____.
   a. mitosis   c. fission
   b. meiosis   d. fertilization

2. The gametes are produced by the _____.
   a. sporophytes   c. spores
   b. gametophytes d. embryo

3. Which of the following would NOT have a haploid or *n* number of chromosomes?
   a. gametophyte. c. sporophyte.
   b. spores.   d. gametes.

4. **Predicting** The sporophyte of corn has 20 chromosomes. How many chromosomes would you expect to find in the corn gametes? Explain your answer.

22. Nonvascular plants lack the vascular tissue that provides support.

23. Adaptations include cuticle, vascular tissue, root and stem structure.

24. They produce flowers and a seed protected by a fruit.

25. Seeds are more resistant to drying out, and seed-producing plants do not need water for reproduction.

THINKING CRITICALLY

26. Conifers and flowering plants are vascular, seed-producing plants. Flowering plants, unlike conifers, produce flowers and have seeds enclosed in a fruit.

27. Because it is possible for the cones to survive a fire, the jack pine could be the first plant to start growing after the fire.

28. Both the algae and aquatic flowering plant live in water, use the same type of chlorophyll, and have cellulose cell walls. However, the aquatic plant has a cuticle and produces seeds and fruits.

29. 1. Vascular plant; 2. Vascular tissue; 3. Roots; 4. Stems; 5. Leaves

ASSESSING KNOWLEDGE & SKILLS

1. b
2. b
3. c
4. 10. Gametes are formed by meiosis that reduces the number of chromosomes by one-half.

Chapter 22 Organizer

Refer to pages 4T-5T of the Teacher Guide for an explanation of the National Science Education Standards correlations.

Section	Objectives	Activities/Features
Section 22.1 **Nonvascular Plants** National Science Education Standards UCP.1, UCP.5; A.1, A.2; C.1, C.5; G.1-3 (1 session)	1. **Identify** the structures of nonvascular plants. 2. **Compare and contrast** characteristics of the different groups of nonvascular plants.	**Problem-Solving Lab 22-1,** p. 598
Section 22.2 **Non-Seed Vascular Plants** National Science Education Standards UCP.1-5; A.1, A.2; C.1, C.3, C.5, C.6; G.3 (1 session)	3. **Explain** the importance of vascular tissue to life on land. 4. **Identify** the characteristics of the non-seed vascular plant divisions.	**Problem-Solving Lab 22-2,** p. 604 **MiniLab 22-1:** Identifying Fern Sporangia, p. 606
Section 22.3 **Seed Plants** National Science Education Standards UCP.1-5; A.1, A.2; C.1, C.3, C.5, C.6; F.3-6; G.1-3 (2 sessions, $1\frac{1}{2}$ blocks)	5. **Identify** the characteristics of seed plants. 6. **Analyze** the advantages of seed and fruit production.	**MiniLab 22-2:** Comparing Seed Types, p. 609 **Inside Story:** Pine Needles, p. 613 **Careers in Biology:** Lumberjack, p. 616 **Design Your Own BioLab:** How can you make a key for identifying conifers? p. 618 **Biology & Society:** Forestry: Keeping a Balance, p. 620

Need Materials? Contact Carolina Biological Supply Company at 1-800-334-5551 or at http://www.carolina.com

MATERIALS LIST

BioLab
p. 618 conifer twigs, conifer branches, conifer cones

MiniLabs
p. 606 microscope, microscope slide, coverslip, forceps, water, glycerin, live fern frond
p. 609 lima beans, grass seeds, rice, peas, rye seeds, forceps, iodine stain, dropper, pencil, paper

Alternative Lab
p. 612 microscope, prepared slide of conifer leaf cross section

Quick Demos
p. 599 flowerpots (2), sand, peat moss, water
p. 602 capillary tube, petri dish, water, food coloring
p. 610 assorted conifer cones

Key to Teaching Strategies

L1 Level 1 activities should be appropriate for students with learning difficulties.

L2 Level 2 activities should be within the ability range of all students.

L3 Level 3 activities are designed for above-average students.

ELL ELL activities should be within the ability range of English Language Learners.

COOP LEARN Cooperative Learning activities are designed for small group work.

P These strategies represent student products that can be placed into a best-work portfolio.

These strategies are useful in a block scheduling format.

The Diversity of Plants

Teacher Classroom Resources

Section	Reproducible Masters	Transparencies
Section 22.1 Nonvascular Plants	Reinforcement and Study Guide, p. 97 **L2** Content Mastery, pp. 109-110, 112 **L1**	Section Focus Transparency 52 **L1** **ELL**
Section 22.2 Non-Seed Vascular Plants	Reinforcement and Study Guide, pp. 98-99 **L2** BioLab and MiniLab Worksheets, p. 101 **L2** Laboratory Manual, pp. 153-156 **L2** Content Mastery, pp. 109-110, 112 **L1**	Section Focus Transparency 53 **L1** **ELL**
Section 22.3 Seed Plants	Reinforcement and Study Guide, p. 100 **L2** Concept Mapping, p. 22 **L3** **ELL** Critical Thinking/Problem Solving, p. 22 **L3** BioLab and MiniLab Worksheets, pp. 102-104 **L2** Laboratory Manual, pp. 157-160 **L2** Content Mastery, pp. 109, 111-112 **L1**	Section Focus Transparency 54 **L1** **ELL** Basic Concepts Transparency 34 **L2** **ELL** Basic Concepts Transparency 35 **L2** **ELL**

Assessment Resources

Chapter Assessment, pp. 127-132

MindJogger Videoquizzes

Performance Assessment in the Biology Classroom

Alternate Assessment in the Science Classroom

ExamView® Pro Software 💾

BDOL Interactive CD-ROM, Chapter 22 quiz

Additional Resources

Spanish Resources **L1** **ELL**

English/Spanish Audiocassettes **L1** **ELL**

Cooperative Learning in the Science Classroom **COOP LEARN**

Lesson Plans/Block Scheduling

NATIONAL GEOGRAPHIC — Teacher's Corner

Products Available From Glencoe

To order the following products, call Glencoe at 1-800-334-7344:

CD-ROM
NGS PictureShow: What It Means to Be Green

Curriculum Kit
GeoKit: Plants

Transparency Set
NGS PicturePack: What It Means to Be Green

Videodisc
STV: Plants

Index to National Geographic Magazine

The following articles may be used for research relating to this chapter:

"The Gift of Gardening," by William S. Ellis, May 1992.

GLENCOE TECHNOLOGY

The following multimedia resources are available from Glencoe.

Biology: The Dynamics of Life
CD-ROM **ELL**

Animation: *Life Cycle of a Moss*
Exploration: *The Six Kingdoms*
BioQuest: *Biodiversity Park*
Video: *Fern Development*
Exploration: *Classifying Pines*
Video: *Giant Redwoods*
Exploration: *Angiosperm*
Video: *Blooming Flowers*

The Infinite Voyage

Life in the Balance

Theme Development

The theme **unity within diversity** is carried out within this chapter as students learn about features shared by nonvascular plants and vascular plants such as the plant life cycle, which includes alternation of generations. Diversity is illustrated through the unique adaptations found in members of the different divisions. **Evolution** is apparent through the discussions of the many adaptations plants exhibit that aid in their survival.

Resource Manager

Section Focus Transparency 52 and Master **L1** **ELL**

READING BIOLOGY

Glencoe's *Biology: The Dynamics of Life* contains many resources to assist a student's reading skills. Each chapter contains figures with expanded captions that expand on written material. Word Origins, located along the side of text, expand knowledge of biology vocabulary. Glencoe's Content Mastery Booklet helps develop reading skills while reinforcing content. In addition, use the Interactive Tutor for *Biology: The Dynamics of Life* on the Glencoe Web site to reinforce vocabulary.
science.glencoe.com

Chapter

22 The Diversity of Plants

What You'll Learn

- You will identify the characteristics of the major plant groups.
- You will compare the distinguishing features of vascular and nonvascular plants.
- You will analyze the advantages of seed production.

Why It's Important

We classify plants according to their characteristics. Knowing about the major characteristics of plants will help you appreciate the beauty and diversity of the plants around you.

READING BIOLOGY

Look at the photos of different plants throughout the chapter. Write down plants that look familiar, and where you may have seen them before. As you read, for each plant, note what type of plant it is, and some of its biological characteristics.

 **BIOLOGY Online**

To find out more about the diversity of plants, visit the Glencoe Science Web site. science.glencoe.com

Members of the plant kingdom exhibit a wide variety of characteristics. Some plants produce large, colorful fruits. Others, including mosses, produce tiny spores.

Multiple Learning Styles

Look for the following logos for strategies that emphasize different learning modalities.

Kinesthetic Enrichment, p. 603; Meeting Individual Needs, p. 604; Project, p. 605

Visual-Spatial Portfolio, pp. 598, 603, 614; Meeting Individual Needs, p. 602; Extension, pp. 607, 617; Quick Demo, p. 610; Chalkboard Activity, p. 612

Interpersonal Activity, p. 600

Intrapersonal Enrichment, p. 611

Linguistic Biology Journal, pp. 599, 603; Cultural Diversity, p. 609; Meeting Individual Needs, p. 611

Logical-Mathematical Chalkboard Activity, p. 605; Biology Journal, p. 610

Naturalist Meeting Individual Needs, p. 598; Discussion, p. 607

22.1 Nonvascular Plants

SECTION PREVIEW

Objectives
Identify the structures of nonvascular plants.
Compare and contrast characteristics of the different groups of nonvascular plants.

Vocabulary
antheridium
archegonium

A s you hike in a shady forest, you are sure to come across patches of soft, feathery mosses covering soil, rocks, rotting wood, or tree bark with a velvety layer of green. On closer examination, you might also notice shiny liverworts or odd-shaped hornworts along the stony bank of a stream. Mosses, liverworts, and hornworts are nonvascular plants. These small plants usually live in moist, cool environments.

Moss-covered forest floor and liverwort (inset)

What Is a Nonvascular Plant?

Nonvascular plants are not as common or as widespread in their distribution as vascular plants because life functions, including photosynthesis and reproduction, require a close association with water. Because a steady supply of water is not available everywhere, nonvascular plants are limited to moist habitats by streams and rivers or in temperate and tropical rain forests. Recall that a lack of vascular tissue also limits the size of a plant. In drier soils, nonvascular plants cannot compete with neighboring vascular plants, which can easily overgrow them and cut them off from sunlight and atmospheric gases. But even with these limitations, nonvascular plants are successful in habitats with adequate water.

Alternation of generations

As in all plants, the life cycle of nonvascular plants includes an alternation of generations between a diploid sporophyte and a haploid gametophyte. However, nonvascular plants are the only plant divisions in which the gametophyte generation is dominant. The gametophytes are dominant and the sporophytes are physically attached to the gametophytes, as shown in *Figure 22.1*, and dependent on them for most of their nutrition.

Figure 22.1
Brown stalks and spore capsules of the sporophyte generation can be seen growing from the green, leafy gametophyte of this moss.

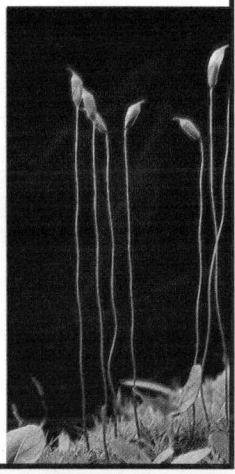

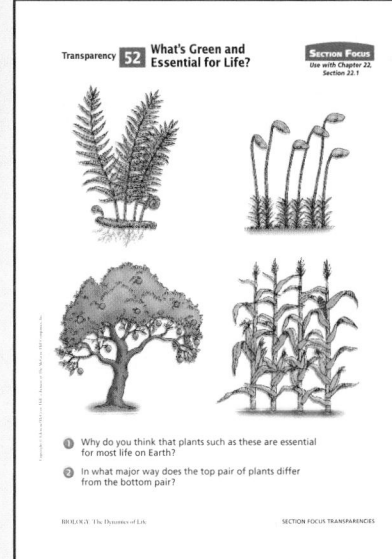

2 Teach

Purpose

Students will determine from various clues that they are observing plant gametophytes.

Process Skills

observe and infer, think critically, analyze information, interpret scientific illustrations

Background

It is not unusual to find male and female bryophyte gametophytes. The plant being used in this lab is a liverwort.

Teaching Strategies

■ Advise students to pay close attention to the scale of magnification included on the diagrams.

■ Students will have to have already studied alternation of generations and be familiar with the contributions of the gametophyte generation in order to understand this lab.

Thinking Critically

1. Sperm cell; there are numerous cells with flagella.

2. Egg cell; it is a single cell much larger than part a.

3. Gametophytes; the gametophyte generation forms gametes.

4. Mitosis; haploid—gametophytes form gametes by mitosis.

5. Haploid; all parts of the gametophyte are haploid.

✔ Assessment

Knowledge Ask students to copy the diagrams and add labels to the antheridium and archegonium. Have students diagram the rest of this plant's life cycle. Use the Performance Task Assessment List for Scientific Drawing in **PASC**, p. 55. **L2**

598

Problem-Solving Lab 22-1
Observing and Inferring

Is it a sporophyte or a gametophyte? You have just discovered a new plant species and have classified it as a nonvascular plant. You have two almost identical specimens, and microscopic examination reveals the internal structures of plants 1 and 2 below.

Analysis

Formulate a hypothesis as to the nature of plants 1 and 2.

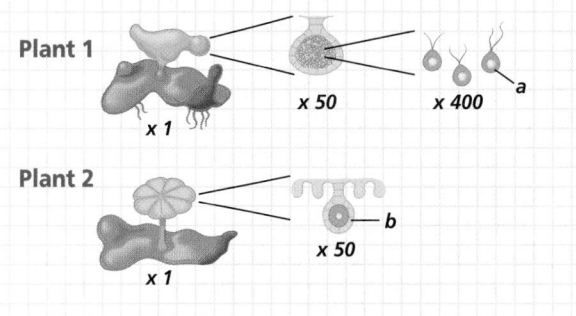

Plant 1   x 1   x 50   x 400   a

Plant 2   x 1   x 50   b

Thinking Critically

1. Compare part a to b. What is part a? How do you know?
2. What is part b? How do you know?
3. Knowing that this is a nonvascular plant, are plants 1 and 2 sporophytes or gametophytes? How do you know?
4. Are parts a and b formed by mitosis or meiosis? Are a and b haploid or diploid? Explain.
5. Are plants 1 and 2 haploid or diploid? Explain.

Figure 22.2
Mosses have a central stem surrounded by small, thin leaves. They also have rhizoids, one-cell-thick rootlike structures, which absorb water and nutrients.

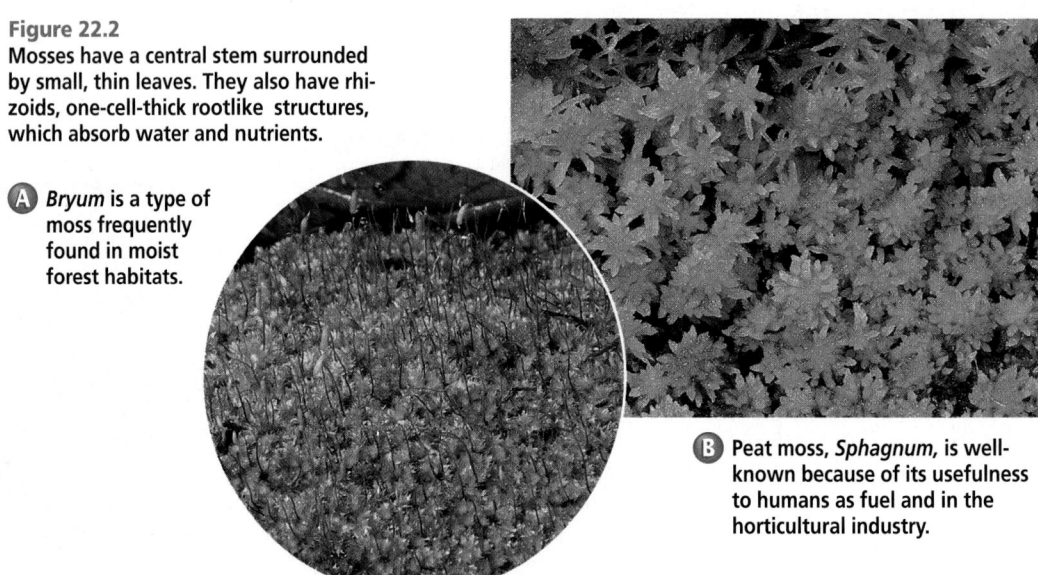

A *Bryum* is a type of moss frequently found in moist forest habitats.

B Peat moss, *Sphagnum,* is well-known because of its usefulness to humans as fuel and in the horticultural industry.

Gametophytes produce two kinds of sexual reproductive structures. The **antheridium** (an thuh RIHD ee um) is the male reproductive structure in which sperm are produced. The **archegonium** (ar kih GOH nee um) is the female reproductive structure in which eggs are produced. Think of ways you can identify the phases of a nonvascular plant's life cycle in the *Problem-Solving Lab* shown here.

Adaptations in Bryophyta

There are several divisions of nonvascular plants. The first division you'll study are the mosses, or bryophytes. Bryophytes are the most familiar of the nonvascular plant divisions. Mosses are small plants with leafy stems. The leaves of mosses are usually one cell thick. Mosses have rhizoids, colorless multicellular structures, which help anchor the stem to the soil. Although mosses do not contain true vascular tissue, some species do have a few, long water-conducting cells in their stems.

MEETING INDIVIDUAL NEEDS

Learning Disabled

Naturalist Have students list as many characteristics as possible that distinguish the nonvascular plant divisions from other plant groups. Organizing these characteristics into lists will help students with learning disabilities to distinguish these plant divisions. **L1**

✔ Portfolio

Alternation of Generations

Visual-Spatial Have students prepare a simple flowchart diagram that depicts alternation of generations. The diagram must include these terms: gametophyte generation, sporophyte generation, spore, gamete, diploid, and haploid. Have students place the flowchart in their portfolio. **L2 P**

Mosses usually grow in dense carpets of hundreds of plants. Some have upright stems; others have creeping stems that hang from steep banks or tree branches. Some mosses form extensive mats that retard erosion on exposed rocky slopes.

Mosses grow in a wide variety of habitats. They grow even in the arctic in places where there is sufficient moisture.

One of the most well-known mosses is *Sphagnum* moss, also known to gardeners as peat moss. This plant thrives in acidic bogs in northern regions of the world. It has been harvested for use as fuel and is a commonly used soil additive. Dried peat moss absorbs large amounts of water, so florists and gardeners use it to increase the water-holding ability of soil. See *Figure 22.2* to examine the characteristics of *Bryum* and *Sphagnum* mosses.

Adaptations in Hepatophyta

Another division of nonvascular plants is the liverworts, or hepatophytes. Like mosses, liverworts are small plants that usually grow in clumps or masses in moist habitats. The name of the division is derived from the word *hepatic*, which refers to the liver. The flattened body of a liverwort gametophyte is thought to resemble the shape of the lobes of an animal's liver. Liverworts occur in many environments, from the Arctic to the Antarctic. They include two groups: the thallose liverworts and the leafy liverworts, *Figure 22.3*. The body of a thallose liverwort is a thallus. It is a broad, ribbonlike body that resembles a fleshy, lobed leaf.

a
b

Figure 22.3
Liverworts may have a flattened thallus (a) or flattened leaves in three ranks borne on a stem (b).

Leafy liverworts are creeping plants with three rows of flat, thin leaves attached to a stem. Like mosses, liverworts have rhizoids; however, the rhizoids of liverworts are each composed of only one elongated cell. Most liverworts have an oily or shiny surface that helps reduce evaporation of water from the plant's tissues.

Adaptations in Anthocerophyta

Anthocerophytes are the smallest division of nonvascular plants, currently consisting of only about 100 species. Also known as hornworts, these nonvascular plants are similar

WORD *Origin*

antheridium
From the Greek word *anthera*, meaning "flowery." Sperm are produced in the antheridium.

archegonium
From the Greek word *archegonos*, meaning "originator." Eggs are produced in the archegonium.

Reteach

Linguistic Have students outline the section and write a definition for each vocabulary term. **L2**

Extension

Naturalist Have students prepare a key that will enable them to distinguish among the three divisions of nonvascular plants. **L3**

✓ Assessment

Portfolio Have students draw plants from each division. Students should label the distinguishing features on their drawings and place the drawings in their portfolios. **L1** **P**

4 Close

Activity

Interpersonal Have students write three questions based on the material covered in this section. Have students read their questions while their classmates provide the answers. **L2**

Resource Manager

Reinforcement and Study Guide, p. 97 **L2**
Content Mastery, p. 110 **L1**

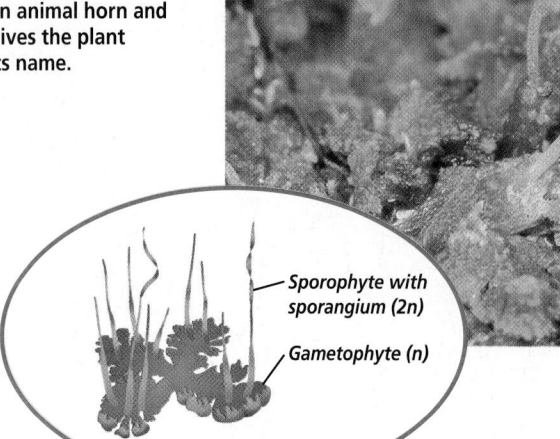

Figure 22.4
The upright sporophyte of the hornwort resembles an animal horn and gives the plant its name.

Sporophyte with sporangium (2n)

Gametophyte (n)

to liverworts in several respects. Like some liverworts, hornworts have a thallose body. As you can see in *Figure 22.4*, the sporophyte of a hornwort resembles the horn of an animal, which is why members of this division are commonly called "hornworts." Another feature unique to hornworts is the presence of a single large chloroplast in each cell. This feature suggests that hornworts may be closely related to algae, which also have only one large chloroplast in each cell.

Origins of Nonvascular Plants

Fossil and genetic evidence suggests that liverworts were the first land plants. Fossils that have been positively identified as nonvascular plants first appear in rocks from the early Paleozoic period, about 430 million years ago. However, paleobotanists think that nonvascular plants were present much earlier than current fossil evidence suggests. Both nonvascular and vascular plants probably share a common ancestor that had alternating sporophyte and gametophyte generations, cellulose in their cell walls, and chlorophyll for photosynthesis.

Section Assessment

Understanding Main Ideas

1. How can you tell a leafy liverwort from a thallose liverwort?
2. In what way is the sporophyte generation of a moss dependent on the gametophyte generation?
3. What are some characteristics shared by all nonvascular plants?
4. Explain why nonvascular plants are usually found in moist shady areas.

Thinking Critically

5. Explain why it is an advantage for mosses to grow in mats or mounds composed of many individual plants.

SKILL REVIEW

6. **Compare and Contrast** the gametophyte and sporophyte generations of nonvascular plants. For more help, refer to *Thinking Critically* in the **Skill Handbook**.

600 THE DIVERSITY OF PLANTS

Section Assessment

1. Thallose liverworts have a broad, ribbon-like body that resembles a fleshy, lobed leaf. Leafy liverworts have three rows of flat, thin leaves attached to a stem.
2. The sporophyte generation is small and obtains food, water, and minerals from the larger gametophyte generation.
3. Nonvascular plants lack vascular tissue,

600

are typically small in size, have a dominant gametophyte generation, and grow in moist, shady areas.

4. Since these plants rely on diffusion and osmosis for the transport of water and minerals, they tend to be found in areas where water is plentiful.
5. The dense mats retain water and help reduce evaporation.

6. The gametophyte and sporophyte generations are both stages of the nonvascular plant life cycle. The sporophyte generation is smaller and dependent upon the gametophyte. The gametophyte generation is haploid and produces gametes, whereas the sporophyte generation is diploid and produces spores.

SECTION PREVIEW

Objectives

Explain the importance of vascular tissue to life on land.

Identify the characteristics of the non-seed vascular plant divisions.

Vocabulary
strobilus
prothallus
rhizome
sorus

I *magine traveling back in time nearly 300 million years—50 million years before dinosaurs evolved. As you look around Earth's forests, you see a bewildering array of leafy vascular plants, some oddly familiar. Towering above the forest floor are incredibly tall, unusual-looking trees. Paleobotanists know what these ancient plants looked like because many were preserved as fossils. Living on Earth today are plants that are reminiscent of these ancient vascular plants, including the club mosses, horsetails, and ferns.*

Tree fern and horsetails (inset)

What Is a Non-Seed Vascular Plant?

The obvious difference between a vascular and a nonvascular plant is the presence of vascular tissue. As you may remember, vascular tissue is made up of tubelike, elongated cells through which water and sugars are transported. Vascular plants are able to adapt to changes in the availability of water, and thus are found in a variety of habitats. You will learn about three divisions of non-seed vascular plants: Lycophyta, Sphenophyta, and Pterophyta.

Alternation of generations

Vascular plants, like all plants, exhibit an alternation of generations. Unlike nonvascular plants, though, the spore-producing sporophyte is dominant, *Figure 22.5*. The sporophyte is much larger in size than the gametophyte. The mature sporophyte

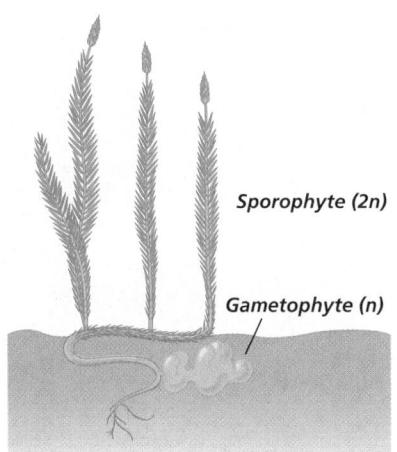

Sporophyte (2n)

Gametophyte (n)

Figure 22.5
In non-seed vascular plants, the sporophyte generation is dominant.

BIOLOGY Online

Note Internet addresses that you find useful in the space below for quick reference.

_____

_____

_____

_____

Section 22.2

Prepare

Key Concepts

The non-seed vascular plant divisions Lycophyta, Sphenophyta, and Pterophyta are presented along with the traits that distinguish these divisions from each other.

Planning

- Obtain a glass tube and colored water for the Quick Demo.
- Purchase a resurrection plant for the Enrichment.
- Collect or purchase fern spores for the MiniLab.
- Collect pictures of non-seed vascular plants for the Display.

1 Focus

Bellringer

Before presenting the lesson, display **Section Focus Transparency 53** on the overhead projector and have students answer the accompanying questions.
L1 **ELL**

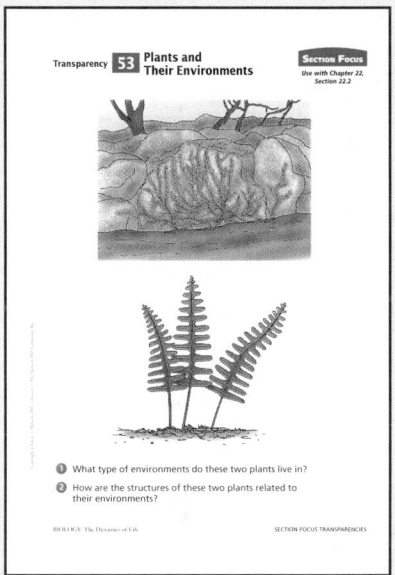

Stand a capillary tube in a petri dish filled with colored water. Have students note the rise of water within the tube. Explain that this movement occurs due to capillary action. Ask students to correlate the movement of water in the glass tube to the movement of water in vascular plants. Use this demonstration to explain the relationship between the evolution of vascular tissue and the increased size of plants, and the efficiency of water and mineral movement throughout the plants.

Assessment

Skill Have students make a concept map that illustrates the major characteristics of Lycophytes, Sphenophytes, and Pterophytes.

GLENCOE TECHNOLOGY

CD-ROM
Biology: The Dynamics of Life
Exploration: *The Six Kingdoms*
Disc 3

Resource Manager

Section Focus Transparency 53 and Master **L1** **ELL**

Figure 22.6
Spores are released from a strobilus and grow into a prothallus. The prothallus forms antheridia and archegonia. Sperm from the antheridia swim through a continuous film of water to the egg in the archegonium, where fertilization may then occur.

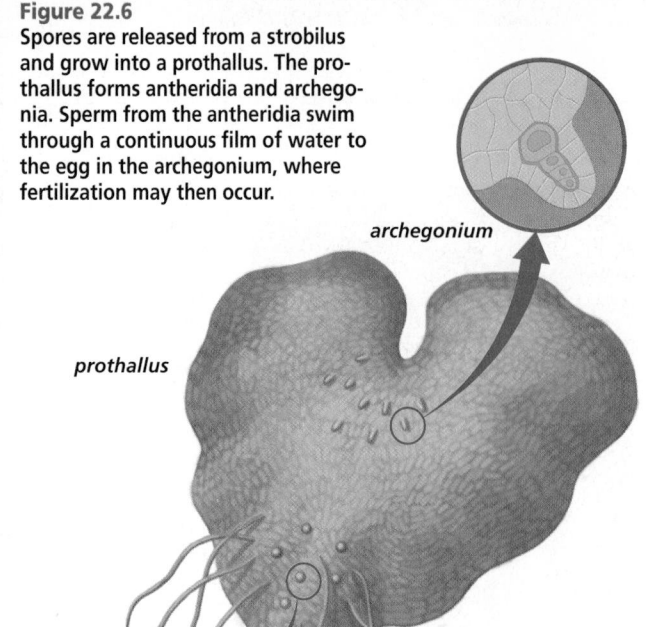

archegonium

prothallus

antheridium

Figure 22.7
Lycophytes are non-seed vascular plants with roots, stems, and leaves, and spores produced in a strobilus.

---

does not depend on the gametophyte for water or nutrients.

A major advance in this group of vascular plants was the adaptation of leaves to form structures that protect the developing reproductive cells. In some non-seed vascular plants, spore-bearing leaves form a compact cluster called a **strobilus** (STROH bih lus). The spores are released from the strobilus and then grow to form the gametophyte, called a **prothallus** (proh THAL us). The prothallus is relatively small and lives in or on the soil. The prothallus then forms antheridia and archegonia, *Figure 22.6.* Sperm are released from the antheridium and swim through a continuous film of water to the egg in the archegonium. Fertilization occurs and a large, dominant sporophyte plant grows from the fertilized zygote.

Adaptations in Lycophyta

From fossil evidence it is known that tree-sized lycophytes were once members of the early forest community. Modern lycophytes are much smaller than their early ancestors. Lycophytes are commonly called club mosses and spike mosses because their leafy stems resemble moss gametophytes, and their reproductive structures are club shaped, as shown in *Figure 22.*7. However, unlike mosses, the sporophyte generation of the lycophytes is dominant. It has roots, stems, and small leaves. The leaves occur as pairs, whorls, or spirals along the stem. A

Figure 22.8
This is the sporophyte generation of a horsetail, *Equisetum*. It has thin, narrow leaves that circle each joint of the slender, hollow stem. Plants with sporangia form a strobilus at the tips of some stems.

single vein of vascular tissue runs through each leaf. The stems of lycophytes may be upright or creeping and have roots growing from the base of the stem.

The club moss, *Lycopodium*, is commonly called ground pine because it is evergreen and resembles a miniature pine tree. Some species of ground pine have been collected for decorative uses in such numbers that the plants have become endangered.

Adaptations in Sphenophyta

Sphenophytes, or horsetails, represent a second group of ancient vascular plants. Like the lycophytes, early horsetails were tree-sized members of the forest community. There are only about 15 species in existence today, all of the genus *Equisetum*. The name horsetail refers to the bushy appearance of some species. These plants are also called scouring rushes because they contain silica, an abrasive substance, and were once used to scour cooking utensils. If you run your finger along a horsetail stem, you can feel how rough it is.

Today's sphenophytes are much smaller than their ancestors, usually growing to about 1 m tall. Most horsetails, like the one shown in *Figure 22.8,* are found in marshes, shallow ponds, stream banks, and other areas with damp soil. Some species are common in the drier soil of fields and roadsides. The stem structure of horsetails is unlike most other vascular plants; it is ribbed and hollow, and appears jointed. At each joint, there is a whorl of tiny, scale-like leaves.

Like lycophytes, sphenophyte spores are produced in a strobilus that is formed at the tip of some stems. After the spores are released, they may grow into a prothallus with antheridia and archgonia.

WORD Origin
strobilus
From the Greek word *strobos*, meaning "whirling." Spore-bearing leaves form a compact cluster called a strobilus.

Visual Learning

Figure 22.8 Point out that the strobili at the tips of the horsetail stems are formed by leaves with spores. Have students use the appearance of the plant to speculate as to why it was named a horsetail. *The plant looks roughly like the tail of a horse.*

Enrichment

Kinesthetic Purchase a resurrection plant (*Selaginella lepidophylla*). Have students examine the plant and decide if it is still alive. Soak the plant in water overnight and place it beneath a plant light. Have students observe the plant again and consider if they should revise their original conclusion about whether or not the plant is alive. **L1**

GLENCOE TECHNOLOGY

CD-ROM
Biology: The Dynamics of Life
BioQuest: *BioDiversity Park*
Disc 3, 4

Portfolio

Horsetails Cross Section

Visual-Spatial Have students use information they have read about horsetails to prepare what they believe would be cross-section diagrams of the stem of this plant. Have them add appropriate labels to structures where needed. **L2** **P**

BIOLOGY JOURNAL

Spike and Club Moss Question

Linguistic Have students write a brief essay to explain why club mosses are not common in desert habitats and why most people never see the gametophyte generation of these plants. Have students explain why these plants are not related to the mosses that were discussed in Section 22.1. **L3**

Purpose

Students will review the need for water in the process of fertilization among certain plant divisions.

Process Skills

think critically, acquire information, apply concepts, predict

Teaching Strategies

■ Students will have to consult their text to determine the answers to the column marked "Sperm must swim to egg?" You may wish to advise them of the page references to be used in researching the answer.
■ A "yes" or "no" answer is required for the completion of both columns.
■ All correct answers should be "yes."

Thinking Critically

1. yes to all
2. Sperm can swim only in a film of water. Therefore, this column must also be marked with a yes.
3. Water is needed for fertilization. Therefore, these plants must grow in areas with sufficient moisture.
4. wind, insects, birds, mammals

Problem-Solving Lab 22-2   Applying Concepts

Is water needed for fertilization? Non-seed vascular plants have a number of shared characteristics. One of these characteristics is related to certain requirements needed for reproduction.

Analysis

Examine the following data table. Notice that some of the information is incomplete.

Data Table			
Division	Example	Sperm must swim to egg?	Water needed for fertilization?
Lycophyta	club moss		
Sphenophyta	horsetail		
Pterophyta	ferns		

Thinking Critically

1. How would you complete the column marked "Sperm must swim to egg?"
2. Explain how the column marked "Water needed for fertilization" is related to answers in the previous column.
3. Explain why the three plant divisions must all grow in the same type of environment and what this environment is.
4. What other means might be possible for plant sperm delivery to eggs without the use of water?

Adaptations in Pterophyta

Ferns first appeared nearly 400 million years ago according to the fossil record, at about the same time that club mosses and horsetails were the prominent members of Earth's plant population. Ferns, division Pterophyta, grew tall and treelike, forming vast fern forests. You are probably more familiar with ferns than with club mosses and horsetails, primarily because ferns evolved into many more species and are more abundant. Ferns can be found in many types of environments. You may have seen shrub-sized ferns such as those pictured in *Figure 22.9,* on the damp forest floor or along stream banks. Some ferns inhabit dry areas, becoming dormant when moisture is scarce and resuming growth and reproduction when water becomes available again. Explore the relationship between water and non-seed vascular plants in the *Problem-Solving Lab* on this page.

Figure 22.9
There are about 12 000 species of living ferns. Ferns occupy widely diverse habitats and have a variety of different forms and sizes.

A Most modern ferns are fairly small and leafy, but many species of tall tree ferns still exist, primarily in the tropics.

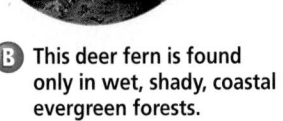

B This deer fern is found only in wet, shady, coastal evergreen forests.

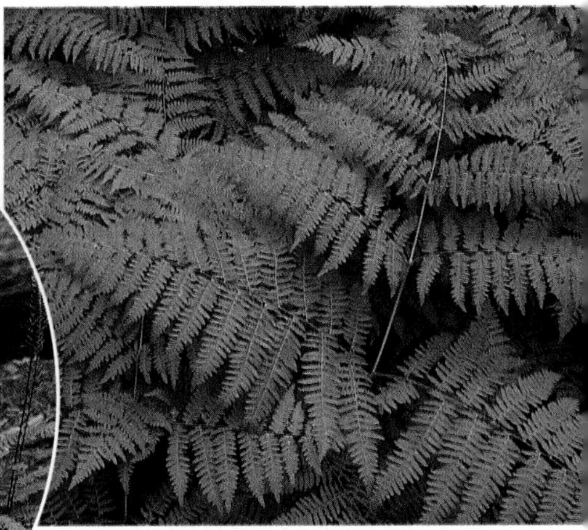

C The bracken fern thrives in the partial sun of open forests. One of the most common ferns in the world, it often takes over large areas of abandoned pasture or agricultural land.

Figure 22.10

Most ferns in warm climates are perennial plants that live from year to year, gradually enlarging in size. The fronds of ferns that live in temperate climates die back during the winter.

A The creeping, underground rhizome of a fern is a modified, condensed stem, with roots growing downward from each joint of stem and frond.

B A fern frond has a stem-like stipe and green, often finely divided leaflets called pinnae.

C Young fern fronds unfurl as they grow and are called fiddleheads because their shape is similar to the neck of a violin.

Fern structures

As with most vascular plants, it is the sporophyte generation of the fern that has roots, stems, and leaves. The part of the fern plant that we most commonly recognize is the sporophyte generation. The gametophyte in most ferns is a thin, flat structure that is independent of the sporophyte. In most ferns, the main stem is underground, *Figure 22.10.* This thick, underground stem is called a **rhizome.** It contains many starch-filled cells for storage. The leaves of a fern are called fronds and grow upward from the rhizome, as shown in *Figure 22.11.* The fronds are often divided into leaflets called pinnae, which are attached to a central stipe. Ferns are the first of the vascular plants to have evolved leaves with branching veins of vascular tissue.

The branched veins in ferns transport water and food to and from all the cells.

Figure 22.11

Fern leaves are called fronds. The fronds of the fern sporophyte grow from an underground rhizome.

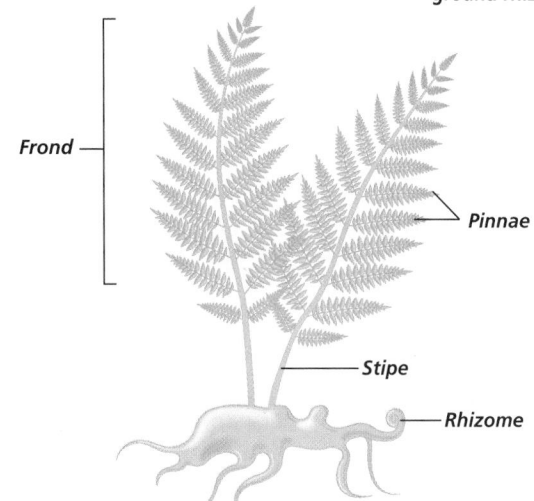

Frond

Pinnae

Stipe

Rhizome

Display

Add pictures of non-seed vascular plants to the bulletin board display.

Chalkboard Activity

Logical-Mathematical Ask students to use information about the number of plant species to prepare a bar graph. On the chalkboard, write the following information.

Bryophyta = 10 000
Hepatophyta = 6500
Anthocerophyta = 100
Lycophyta = 1000
Sphenophyta = 15
Pterophyta = 12 000

Tell students that these numbers represent the approximate number of living species for each non-seed plant division studied so far. **L2**

GLENCOE TECHNOLOGY

CD-ROM
Biology: The Dynamics of Life
Video: *Fern Development*
Disc 3

Resource Manager

Laboratory Manual, pp. 153-156 **L2**

PROJECT

Growing Fern Prothallia from Spores

Kinesthetic Have students grow their own fern prothallia by following the procedure given. C-Fern spores and growing media are available from Carolina Biological (catalog #15-6700). Spores will germinate within 5 days and prothallia within 8 to 12 days. Students can design an experiment to determine the effect of environment on spore germination and prothallia development. These cultures can be maintained and used to study fern reproduction in Chapter 24. **L3**

MiniLab 22-1   Experimenting

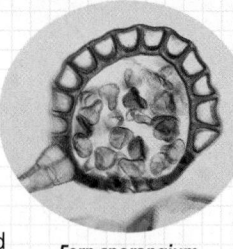
Fern sporangium

Identifying Fern Sporangia When you admire a fern growing in a garden or forest, you are admiring the plant's sporophyte generation. Upon further examination, you should be able to see evidence of spores being formed. Typically, the evidence you are looking for can be found on the underside of the fern's fronds.

Procedure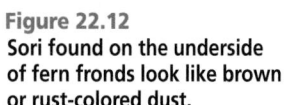

1. Place a drop of water and a drop of glycerin at opposite ends of a glass slide.
2. Use forceps to gently pick off one sorus from a frond. Place it in the drop of water and add a cover slip.
3. Add a second sorus to the glycerin and add a cover slip.
4. Observe both preparations under low-power magnification and note any similarities and differences. Look for large sporangia (resembling heads on a stalk) and spores (tiny round bodies released from a sporangium). **CAUTION:** *Use caution when working with a microscope, microscope slides, and cover slips.*

Analysis
1. Were spores more visible in water or in glycerin?
2. What did the glycerin do to the sporangium?
3. Formulate a hypothesis that may explain how sporangia naturally burst.
4. Formulate a hypothesis that may explain how sporangia were affected by glycerin.

Figure 22.12
Sori found on the underside of fern fronds look like brown or rust-colored dust.

A Most sori are found as round clusters on the pinnae. The shape of the clusters and arrangement on the pinnae vary with species.

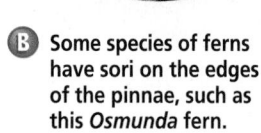

B Some species of ferns have sori on the edges of the pinnae, such as this *Osmunda* fern.

606 THE DIVERSITY OF PLANTS

The fern life cycle is representative of other non-seed vascular plants. Fern spores are produced in structures called sporangia.

Clusters of sporangia form a structure called a **sorus** (plural, sori). The sori are usually found on the undersides of the pinnae, *Figure 22.12*, but in some ferns, spores are borne on whole, modified fronds. Practice your lab skills and learn more about fern spores and sporangia in the *MiniLab* on this page.

Origins of Non-Seed Vascular Plants

The earliest evidence of non-seed vascular plants is found in fossils from the early Paleozoic period, around 390 million years ago. Large tree-sized lycophytes, sphenophytes, and pterophytes were extremely abundant in the warm, moist forests that dominated Earth during the Carboniferous period. Ancient lycophyte species grew as tall as 30 m. Many of these species of non-seed vascular plants died out about 280

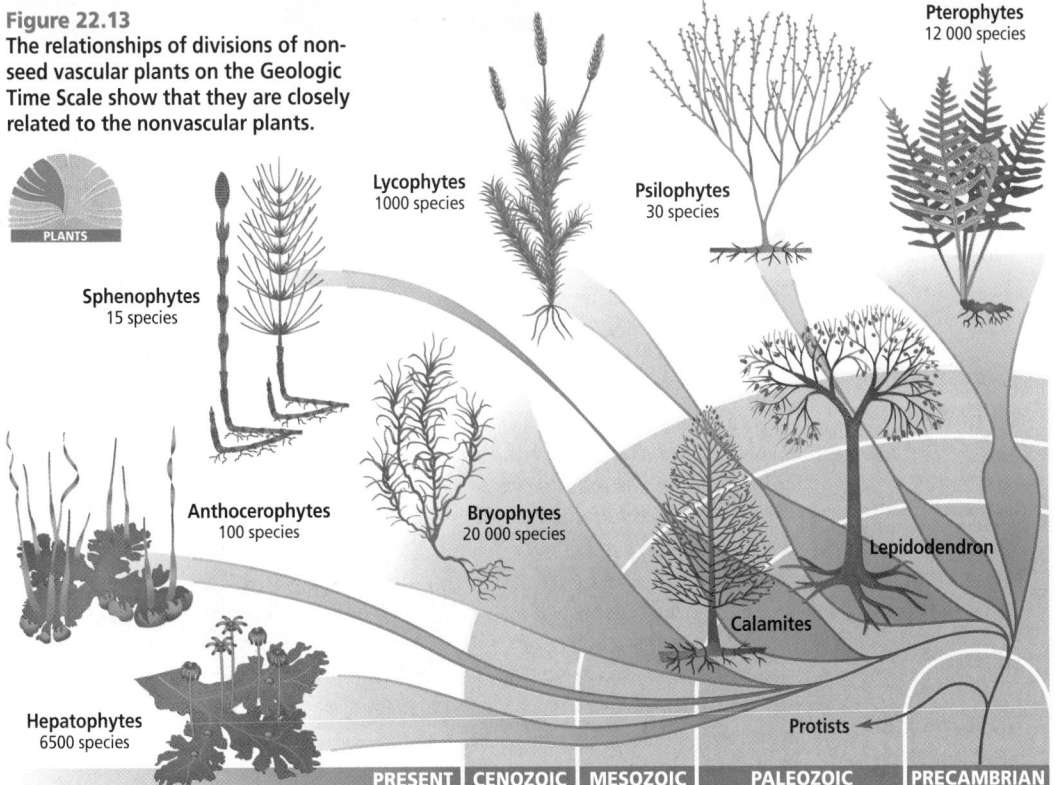

Figure 22.13
The relationships of divisions of non-seed vascular plants on the Geologic Time Scale show that they are closely related to the nonvascular plants.

PLANTS

Lycophytes
1000 species

Psilophytes
30 species

Pterophytes
12 000 species

Sphenophytes
15 species

Anthocerophytes
100 species

Bryophytes
20 000 species

Lepidodendron

Calamites

Hepatophytes
6500 species

Protists

| PRESENT | CENOZOIC | MESOZOIC | PALEOZOIC | PRECAMBRIAN |

million years ago, when it has been determined that Earth's climate became cooler and drier. Today's non-seed nonvascular plants are much smaller and less widespread in their distribution than their prehistoric ancestors.

The evolution of vascular tissue enabled these plants to live on land and to maintain larger body sizes in comparison with nonvascular plants. As you can tell from *Figure 22.13,* non-seed vascular plants are closely related to nonvascular plants.

Prepare

Key Concepts

The major characteristics of seed plants are examined. The five divisions described are Cycadophyta, Ginkgophyta, Gnetophyta, Coniferophyta, and Anthophyta.

Planning

- Collect seeds for the MiniLab.
- Collect pictures of seed plants for the Display.
- Collect examples of monocot and dicot leaves for the Visual Learning.
- Collect gardening books and seed/flower catalogs for the Tech Prep Activity.

1 Focus

Bellringer

Before presenting the lesson, display **Section Focus Transparency 54** on the overhead projector and have students answer the accompanying questions.
L1 **ELL**

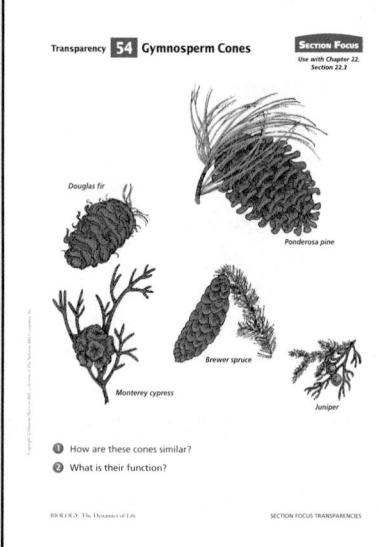

Transparency **54** Gymnosperm Cones

SECTION FOCUS
Use with Chapter 22, Section 22.3

Douglas fir

Ponderosa pine

Brewer spruce

Monterey cypress

Juniper

❶ How are these cones similar?
❷ What is their function?

BIOLOGY: The Dynamics of Life · SECTION FOCUS TRANSPARENCIES

Objectives
Identify the characteristics of seed plants.
Analyze the advantages of seed and fruit production.

Vocabulary
pollen grain
ovule
embryo
cotyledon
fruit
deciduous plant
monocotyledons
dicotyledons
annuals
biennials
perennials

Section

22.3 Seed Plants

About 280 million years ago, when club mosses, ferns, and other non-seed plants had reached their greatest numbers and diversity, it has been shown in rocks that Earth's climate changed. Freezing weather and drought may have caused many non-seed plants to become extinct, but a few of the seed plants were adapted to these extreme conditions and they were able to survive. These ancient survivors are the ancestors of today's vascular seed plants.

Anthophyte fruits and seeds and coniferophyte cones and seeds (inset)

What Is a Seed Plant?

Some vascular plants produce seeds, which are reduced sporophyte plants enclosed within a protective coat. The seeds may be surrounded by a fruit or carried naked on the scales of a cone.

Seed plants produce spores

In seed plants, as in all other plants, spores are produced by the sporophyte generation. These spores develop into the male and female gametophytes. The male gametophyte develops inside a structure called a **pollen grain** that includes sperm cells, nutrients, and a protective outer covering. The female gametophyte, which produces the egg cell, is contained within a sporophyte structure called an **ovule**.

Fertilization and reproduction

The union of the sperm and egg forms the sporophyte. In most seed plants, this process, called fertilization, does not require a continuous layer of water for fertilization, as do nonvascular and non-seed vascular plants. Remember that in non-seed plants, the sperm must swim through a continuous film of water in order to reach the egg in the archegonia of a gametophyte. Because they do not require a continuous film of water for fertilization, seed plants are able to grow and reproduce in a wide variety of habitats that have limited water availability.

After fertilization, the zygote develops into an embryo. An **embryo** is an organism at an early stage of development. In plants, an embryo is

608 THE DIVERSITY OF PLANTS

BIOLOGY Online

Note Internet addresses that you find useful in the space below for quick reference.

_____

_____

_____

_____

the young diploid sporophyte of the plant. Embryos of seed plants include one or more cotyledons. **Cotyledons** are tiny seed leaves that store or absorb food for the developing embryo. In conifers and many flowering plants, cotyledons are the plant's first leaves when it emerges from the soil.

Advantages of seeds

A seed consists of an embryo and its food supply enclosed in a tough, protective coat, *Figure 22.14.* Seed plants have several important advantages over non-seed plants. The seed contains a supply of food to nourish the young plant during the early stages of growth. This food is used by the plant until its leaves are developed enough to carry out photosynthesis. In conifers and some flowering plants, the food supply is stored in the cotyledons. The embryo is protected during harsh conditions by a tough seed coat. The seeds of many species are also adapted for easy dispersal to new areas, so the young plant does not have to compete with its parent for sunlight, water, soil nutrients, and living space. You can learn more about seed structure in the *MiniLab* shown here.

MiniLab 22-2 — Comparing and Contrasting

Comparing Seed Types Anthophytes are classified into two classes, the monocotyledons (monocots) and dicotyledons (dicots) based on the number of seed leaves.

Procedure

1. Copy the data table shown below.
2. Examine the variety of seeds given to you. Use forceps to gently remove the seed coat or covering from each seed if one is present.
3. Determine the number of cotyledons present. If two cotyledons are present, the seed will easily separate into two equal halves. If one cotyledon is present, it will not separate into halves. Record your observations in the Data Table.
4. Add a drop of iodine stain to rice and a lima bean seed. Note the color change. **CAUTION: *Wash your hands with soap and water after handling chemicals.*** Record your observations in the Data Table.

Data Table

Seed name	Number of cotyledons	Monocot or dicot	Color with iodine
Lima bean			
Rice			
Pea			—
Rye			—

Analysis

1. Starch turns purple when iodine is added to it. Describe the color change when iodine was added to rice and lima bean seeds.
2. Hypothesize why seeds contain stored starch.

Figure 22.14
Seeds exhibit a variety of structural adaptations.

A The pine embryo includes seven to nine cotyledons, which serve as an initial food supply when growth begins.

B The tough protective seed coat breaks down when growth begins.

C The feathery tuft attached to milkweed seeds aids in their dispersal.

609

Cultural Diversity

Importance of Non-flowering Seed Plants

Linguistic Many species of non-flowering seed plants play significant roles as timber trees, food plants, and sources of essential oils and soaps in the economies of countries. Initiate a discussion about the cultural importance of the ginkgo tree in China and Japan. *Ginkgo biloba* is the only living representative of the Ginkgophyta division and no longer exists in the wild. It has been cultivated for thousands of years for use as an ornamental plant in Chinese and Japanese temple gardens. Have students choose different types of non-flowering seed plants and present reports on their economic value or cultural significance in different countries.

L3

MiniLab 22-2

Purpose
Students will determine how to tell if a seed is from a monocot or dicot plant and test for the presence of starch in seeds.

Process Skills
acquire information, apply concepts, classify, compare and contrast, predict

Teaching Strategies
- Presoak all seeds overnight so that seed coats come off easily.
- Supply single-edged razor blades to remove seed coats.
- Seeds are available from grocery stores, hardware stores (rye), or pet shops (bird feed).
- Use a dilute iodine solution when testing for starch (3 g potassium iodide and 1 g iodine to 300 mL of water). Place in dropper bottles for ease in student dispensing. Caution students that iodine is toxic and an eye irritant. If it is spilled on skin or clothing it should be washed immediately.

Expected Results
The lima bean, pea, and sunflower are dicots. All others are monocots. Rice and lima beans will turn blue.

Analysis
1. The seeds turn purple.
2. Starch supplies a growing embryo with food.

✔ Assessment

Portfolio Have students diagram the lima bean seeds they dissected and color in the area that stained purple. Ask students to write a paragraph about the advantages of seed production beneath their diagram and place their work in their portfolios. Use the Performance Task Assessment List for Scientific Drawing in **PASC,** p. 55. L2 P

a

b

Figure 22.15
Cycads have a terminal rosette of leaves. The male plant has cones that produce pollen grains that are released in great masses into the air (**a**). The female plant produces cones that contain ovules with eggs (**b**).

Diversity of seed plants

Some plants produce seeds on the scales of woody strobili called cones. This group of plants is sometimes referred to as gymnosperms. The term gymnosperm means "naked seed" and is used with these plants because they are not protected by a fruit. The gymnosperm plant divisions you will learn about are Cycadophyta, Ginkgophyta, Gnetophyta, and Coniferophyta.

Flowering plants, also called angiosperms, produce seeds enclosed within a fruit. A **fruit** is the ripened ovary of a flower. The fruit provides protection and aids in seed dispersal. The Anthophyta division contains all species of flowering plants.

a

Figure 22.16
The seeds of the female ginkgo develop a fleshy outer covering (**a**). The ginkgo is sometimes called the maidenhair tree because its lobed leaves resemble the fronds of a maidenhair fern (**b**).

b

Adaptations in Cycadophyta

About 100 species of cycads exist today, exclusively in the tropics and subtropics. The only present-day species that grows wild in the United States is found in Florida, although you may see cycads cultivated in greenhouses or botanical gardens, as shown in *Figure 22.15.*

All cycads have separate male and female plants. The cones of male plants produce pollen. The cones of female plants produce seeds. Cycads are one of the few seed plants that produce motile sperm. The trunks and leaves of many species resemble palm trees, but cycads are not closely related to palms, which are anthophytes.

Adaptations in Ginkgophyta

Like cycads, ginkgos produce male and female reproductive structures on separate plants. The male ginkgo produces pollen in strobiluslike cones that grow from the bases of leaf clusters. The female ginkgo produces the seeds, which develop a fleshy, apricot-colored seed coat that covers the seeds as they ripen, *Figure 22.16.* Ginkgos are often planted in urban areas because they are able to tolerate smog and pollution. Gardeners and landscapers usually prefer the male trees because the fleshy seed coat on seeds produced by the female trees has an unpleasant smell.

Today, the division is represented by only one living species, *Ginkgo biloba.* The ginkgo tree is considered sacred in China and Japan, and has been cultivated in temple gardens for thousands of years. These Asian temple gardens probably prevented the tree from becoming extinct.

Adaptations in Gnetophyta

Most living gnetophytes can be found in the deserts or mountains of Asia, Africa, North America, and Central or South America. The division Gnetophyta contains only three genera, which are all different in structure and adaptations. The genus *Gnetum* is composed of tropical climbing plants. The genus *Ephedra* contains shrublike plants and is the only gnetophyte genus found in the United States. The third genus, *Welwitschia*, is a bizarre-looking plant found only in South Africa. It grows close to the ground, has a large tuberous root, and may live 100 years. *Ephedra* and *Welwitschia* are pictured in ***Figure 22.17.***

Figure 22.17
Most gnetophytes have separate male and female plants.

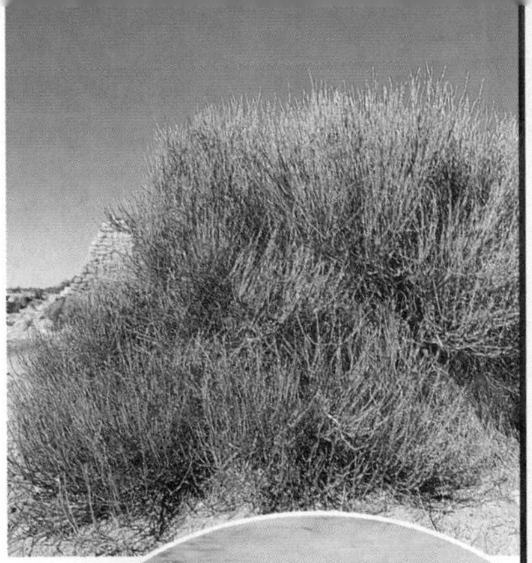

A **Members of the genus *Ephedra* are a source of ephedrine, a medicine used to treat asthma, emphysema, and hay fever.**

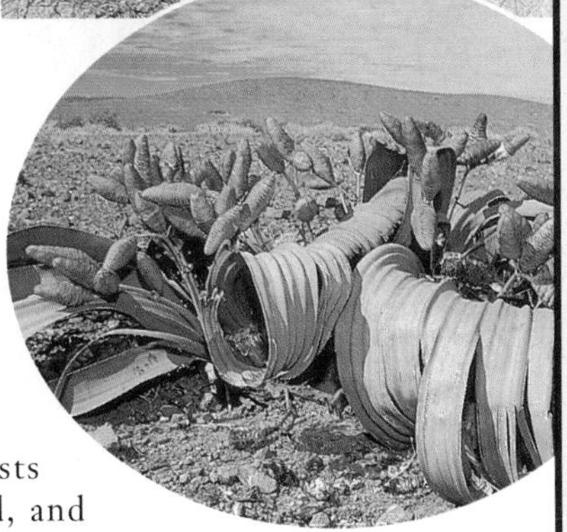

B ***Welwitschia* may live 100 years. The plant has only two leaves, which continue to lengthen as the plant grows older.**

Adaptations in Coniferophyta

The sugar pine is one of many familiar-looking forest trees that belong to the division Coniferophyta. The conifers are trees and shrubs with needle- or scalelike leaves. They are abundant in forests throughout the world, and include pine, fir, spruce, juniper, cedar, redwood, yew, and larch. A few representative conifers are shown in ***Figure 22.18.*** Think of ways that you could distinguish among the different species of conifers in the *BioLab* at the end of this chapter.

Figure 22.18
Conifers are named for the woody cones in which the seeds of most species develop.

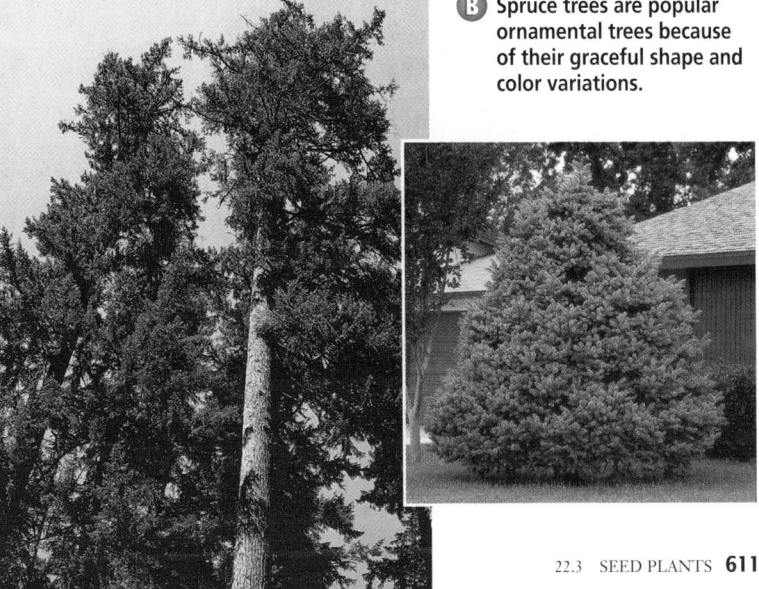

A **The Douglas fir is one of the most important lumber trees in North America. It grows straight and tall, to a height of 100 m.**

B **Spruce trees are popular ornamental trees because of their graceful shape and color variations.**

Concept Development

Ask students to compare and contrast Gnetophyta with Ginkgophyta. As a means for doing this, have students prepare a chart with the headings "Gnetophyta and Ginkgophyta Similarities" and "Gnetophyta and Ginkgophyta Differences." Have students include comparisons of reproductive structures as well as general traits. Ask them also to include where each plant is found in the world, the number of species alive today, the number of genera present today, and economic uses of these plants. L2

Enrichment

Intrapersonal The drug ephedrine has been used by the Chinese for more than five thousand years. It is called *ma huang*, which means "astringent and yellow." The words refer to the drug's taste and color. Ephedrine decreases smooth and cardiac muscle action, raises blood pressure, dilates the pupils, and reduces congestion. Have interested students research this drug and write a report on its uses and its source. L3

GLENCOE TECHNOLOGY

CD-ROM
Biology: The Dynamics of Life
Exploration: *Classifying Pines*
Disc 3
Video: *Giant Redwoods*
Disc 3

MEETING INDIVIDUAL NEEDS

Gifted

Linguistic Have students use references to determine and record the scientific names for all conifers listed in this section. Have them describe the traits that are represented in the scientific names. L3

Display

Add pictures of seed plants to the bulletin board.

Chalkboard Activity

 *Visual-Spatial* Have students construct a time line that marks the era when different types of land plants evolved. Ask students to describe what the climate was like during each time period. **L2**

DESIGN YOUR OWN BioLab The BioLab at the end of the chapter can be used at this point in the lesson. →

Figure 22.19
Each scale of a conifer's cone is a modified branch.

A Male cones are made up of thin papery scales that open to shed clouds of yellow pollen grains into the wind.

B In many conifers, including the pine family, two seeds develop at the base of each of the woody scales that make up a female cone.

The reproductive structures of most conifers are produced in cones. Most conifers have male and female cones on different branches of the same tree. The male cones produce pollen. They are small and easy to overlook. Female cones are much larger. They stay on the tree until the seeds have matured. Examples of both types of cones are shown in *Figure 22.19*.

Conifers are evergreens

Most conifers are evergreen plants, that is, they retain their leaves all year. Although individual leaves drop off as they age or are damaged, the tree never loses all its leaves at once. Pine needles, for example, may remain on the tree for anywhere from two to 40 years, depending on the species.

Trees that retain their leaves begin photosynthesis in the early spring as soon as the temperature warms. Evergreen leaves usually have a heavy coating of cutin, an insoluble waxy material. Evergreens are often found where the warm growing season is short, so keeping leaves year-round gives them a head start on growth. Because they do not need to produce a whole new set of leaves each year, they are able to grow where nutrients are scarce. How do the needlelike leaves of pines prevent water loss? To find out read the *Inside Story*.

Deciduous trees lose their leaves

A few conifers, including larches and bald cypress trees, are deciduous, *Figure 22.20*. **Deciduous plants** lose all their leaves at the same time. Dropping all leaves is an adaptation for reducing water loss when water is unavailable as it can be the case in the tundra, in deserts, or during wintertime. Plants lose most of their water through the leaves; very little is lost through bark or roots. However, a tree with no leaves cannot photosynthesize and must remain dormant during this time.

Figure 22.20
Some trees, including these bald cypress trees, lose their leaves in the fall as an adaptation for reducing water loss.

WORD *Origin*
deciduous
From the Latin word *deciduus*, meaning to "fall off." Deciduous trees drop all of their leaves at the same time.

612 THE DIVERSITY OF PLANTS

Alternative Lab

The Conifer Leaf

Purpose
Students will observe, diagram, and label the structures of a typical conifer leaf.

Materials
microscope, prepared slide of conifer leaf cross section.

612

Procedure
Give students the following directions.
1. Observe the slide under low power.
2. Diagram what you see. Label the following structures using these descriptions as a guide: (a) cuticle—noncellular layer, thin covering over leaf; (b) epidermis—below cuticle, one cell in thickness; (c) mesophyll (spongy layer)—large cells, thick layer just below epidermis; (d) endodermis—one-cell-layer thick, surrounds large oval central area; (e) palisade layer—many cells in thickness, surrounded by endodermis; (f) vascular tissue—very center of leaf, two types are present.
3. Look again at the leaf using the microscope. Check for evidence of stomata on the epidermis and the presence of several ringlike parts, called resin ducts, in the mesophyll. Add these parts and labels to your diagram.

INSIDE STORY

Pine Needles

When you look at a snow-covered pine forest, you may be surprised to learn that winter can be considered a dry time for plants. The cold temperature means that the soil moisture is unavailable because it is frozen. The needles of pines have several adaptations that enable the plants to conserve water during the cold dry winter and the dry heat of the summer.

Snow-covered pines

Critical Thinking *How do the shape and structure of pine needles enable conifers to survive hot and dry summers as well as cold and snowy winters?*

1 **Shape reduces water loss** Pine needles are usually semicircular or round. This shape reduces the surface area. Less water is lost to evaporation because of the smaller surface area from which it can evaporate compared with the large surface area of broad-leaf tree leaves, such as a maple.

2 **Modified epidermis** The epidermis has two modifications that are important in reducing water loss. The walls of the epidermal cells are very thick and the epidermis is covered with a waterproof waxy cuticle.

3 **Resin ducts** Resin is secreted from resin ducts in the leaves and bark. It protects the tree from insects.

Pine needle bundles

Resin duct — Endodermis
— Epidermis

Stoma —

4 **Recessed stomata** The stomata are recessed into the epidermal layer, the outermost layer of plant tissue. This position helps retain water in the leaf tissues.

5 **Bundles of needles** The thick, leathery needles of many species of pine grow in bundles of two, three, or five. The cylindrical nature of these bundles, as well as their flexibility, causes snow and ice to slide off the needles more easily.

Analysis

1. How does the cross-section view of a conifer leaf differ from what you would see in a deciduous tree leaf? *round rather than flat and long*

2. What may be the role of the normally green mesophyll and palisade layers? *responsible for food production*

3. What is the function of the cuticle and vascular tissue? *retain water, transport materials*

✓ Assessment

Skill Diagram or describe how a cross-section slice of a conifer leaf would be made. Use the Performance Task Assessment List for Model in **PASC,** p. 51. **L2**

INSIDE STORY

Purpose 📦
Students examine the adaptations that enable conifers to survive in harsh environments.

Teaching Strategies
■ Have students describe common everyday structures or objects that are analogous to structures found in pine needles. *Students may correlate cuticle to wax paper, or stomata to skin pores.* **L2**

Visual Learning
Have students examine a prepared slide of the cross section of a pine needle. **L1** **ELL**

Critical Thinking
Pine needles have several structural adaptations that reduce evaporation. These features help conifers survive times when water is scarce.

Figure 22.21
A florist's display is a good place to see an assortment of flowering plants. How many can you recognize?

Adaptations in Anthophyta

Flowering plants are classified in the division Anthophyta and are the most well-known plants on Earth with more than 240 000 identified species. See if you are familiar with some of the plants in *Figure 22.21.* Like other seed plants, anthophytes have roots, stems, and leaves. But unlike the other seed plants, anthophytes produce flowers and form seeds enclosed in a fruit. Many different species of flowering plants inhabit tropical forests. As you will discover in *Biology & Society* at the end of this chapter, different groups of people have different viewpoints on preserving this rich habitat.

Fruit production

Anthophytes are unique in that they are the only division of plants that produce fruits, the ripened ovary of a flower. A fruit may contain one or more seeds. One of the advantages of fruit-enclosed seeds is the added protection the fruit provides for the young embryo. Fruits often aid in the dispersal of seeds. Animals may eat them or carry them off to store for food. Seeds of some species that are eaten may pass through the animal's digestive tract unharmed and are distributed as the animal wanders. In fact, some seeds must pass through a digestive tract before they can begin to grow a new plant. Some fruits have structural adaptations that help disperse the seed by wind or water. Some examples of fruits are illustrated in *Figure 22.22.*

Monocots and dicots

The division Anthophyta is divided into two classes: monocotyledons and

Figure 22.22
Fruits exhibit a wide variety of structural adaptations that aid in seed protection and dispersal.

B The fruit of a magnolia contains many seeds surrounded with a bright red covering that attracts birds and small animals.

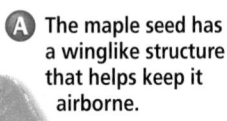

A The maple seed has a winglike structure that helps keep it airborne.

C The tough fibrous fruit of a coconut provides protection as well as a flotation device.

dicotyledons. The two classes are named for the number of seed leaves, or cotyledons, contained within the seed. **Monocotyledons** have one seed leaf; **dicotyledons** have two seed leaves. The names of these two classes are often abbreviated to monocots and dicots. Monocots are the smaller group, with about 60 000 identified species that include families such as grasses, orchids, lilies, and palms, ***Figure 22.23.*** Dicots make up the majority of flowering plants with about 170 000 identified species. They include nearly all the familiar shrubs, trees (except conifers), wildflowers, garden flowers, and herbs. Familiar dicots are shown in ***Figure 22.24.***

Figure 22.23
Monocots usually have leaves with parallel veins and flower parts in multiples of three.

A The grass family includes important cereal grains, such as rice and wheat, as well as sugarcane (left) and bamboo.

B The lily family includes asparagus and onions as well as the ornamental lilies (left) grown by many home gardeners.

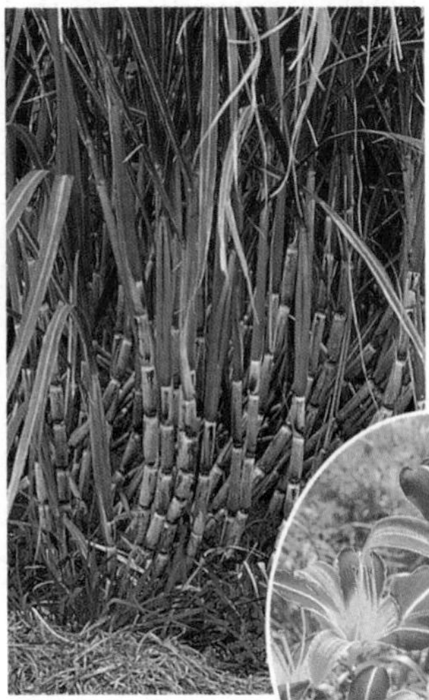

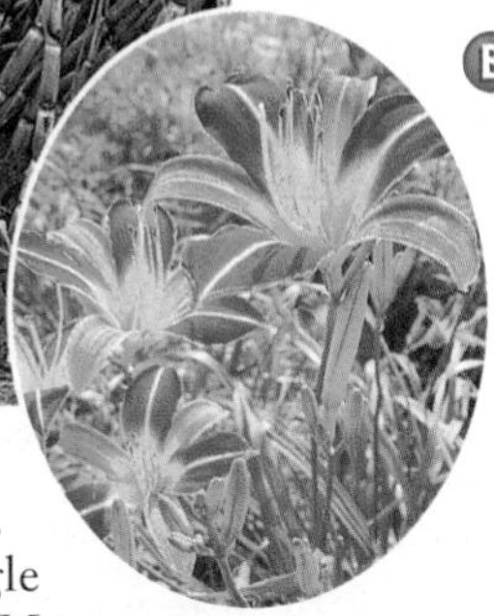

Life spans of anthophytes

Why do some plants live longer than people, and others live only a few weeks? The life span of a plant reflects its strategies for surviving periods of cold, drought, or other harsh conditions.

Annual plants live for only a year or less. They sprout from seeds, grow, reproduce, and die in a single growing season. Most annuals are herbaceous, which means their stems are green and do not contain woody tissue. Many food plants such as corn, wheat, peas, beans, and squash are annuals, as are many weeds of the temperate garden. Annuals form drought-resistant seeds that survive the winter.

Figure 22.24
Dicots usually have leaves with branched veins and flower parts in multiples of four or five.

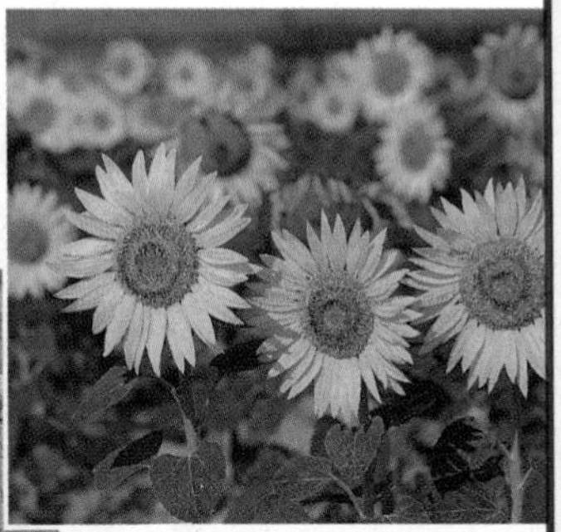

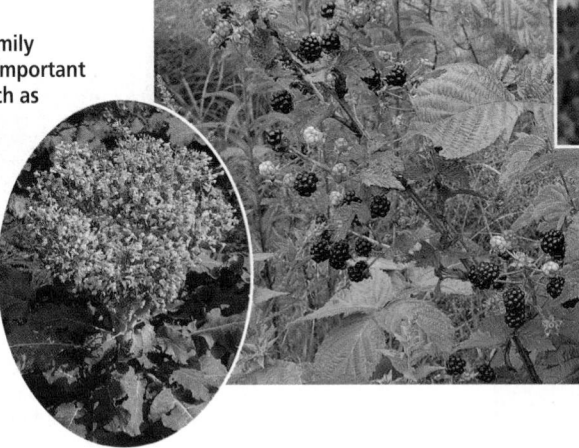

A The mustard family includes many important food plants, such as cabbage, mustard, broccoli (shown here in flower), radish, turnip, collards, and kale.

B The rose family includes blackberries (below), raspberries, apples, plums, peaches, pears, and hundreds of cultivars of garden roses.

C The daisy family includes sunflowers (above), lettuce, dandelions, chrysanthemums, and goldenrod.

Visual Learning

Show students examples of common monocot and dicot leaves. Ask students how the vein patterns in dicot leaves differ from those of monocots.

Assessment

Knowledge Have students discuss the adaptive value of annual, biennial and perennial life spans. L2

GLENCOE TECHNOLOGY

CD-ROM
Biology: The Dynamics of Life
Exploration: *Angiosperm*
Disc 3
Video: *Blooming Flowers*
Disc 3

Career Path

Courses in high school: mathematics, electronics, shop courses involving machinery **College:** courses in forestry or logging; two-year degree to become a forest technician **Other education sources:** on-the-job training

Career Issue

Some groups are trying to stop logging, especially in old-growth forests. Discuss whether students think most lumberjacks and other logging industry workers are concerned about the future of the forests.

For More Information

For more information about working in the logging industry, students can write to:

Society of American Foresters
5400 Grosvenor Lane
Bethesda, MD 20814

3 Assess

Check for Understanding

Have students explain the differences between: fruits and cones; deciduous and evergreen; monocot and dicot; annual and perennial; and gymnosperm and angiosperm. **L2**

CAREERS IN BIOLOGY

Lumberjack

If you like to spend time in the forest, consider a career as a lumberjack or a logging industry worker. Besides being outdoors, you will get lots of exercise.

Skills for the Job

The logging industry now includes many different workers. Cruisers choose which trees to cut. Fallers use chainsaws and axes to cut or "fell" the chosen trees. Buckers saw off the limbs and cut the trunk into pieces. Logging supervisors oversee these tasks. Other workers turn the tree into logs or wood chips that are used to make paper. After finishing high school, most loggers learn on the job. However, with a two-year degree, you can become a forest technician. A four-year degree qualifies you as a professional forester who manages the forest resources. Most logging jobs are in the Northwest, Northeast, South, and Great Lakes regions.

For more careers in related fields, be sure to check the Glencoe Science Web site. **science.glencoe.com.**

Figure 22.25
Anthophytes may be annuals, biennials, or perennials.

Ⓐ Vegetable gardeners grow biennial Swiss chard for its leaves.

Biennials have a life span that lasts two years. Many biennials are plants that develop large storage roots, such as carrots, beets, and turnips. During the first year, biennials grow many leaves and develop a strong root system. Over the winter, the aboveground portion of the plant dies back, but the roots remain alive. Underground roots are able to survive conditions that leaves and stems cannot endure. During the second spring, food stored in the root is used to produce new shoots that produce flowers and seeds.

Perennials live for several years, producing flowers and seeds periodically—usually once each year. They survive harsh conditions by dropping their leaves or dying back to soil level, while their woody stems or underground storage organs remain intact and dormant. Examples of plants with different lifespans are shown in *Figure 22.25*.

Ⓑ Woody perennials, like this maple, drop their leaves and become dormant during the winter.

Ⓒ These blue lupines and orange poppies are annual plants.

Ⓓ Herbaceous perennials often have underground storage organs used for overwintering.

Tree Farms

Tree farms allow foresters to meet increased demands for wood and wood products and reduce the demand for trees from natural forests. Some people oppose this monoculture technique. Monoculture refers to the practice of growing only one species of tree.

Monocultures are susceptible to insect pests and diseases. Have interested students research the types of trees harvested by foresters in their region of the United States. Ask students to find out if any of these trees are grown in monoculture stands. Have students give an oral report based on their findings.

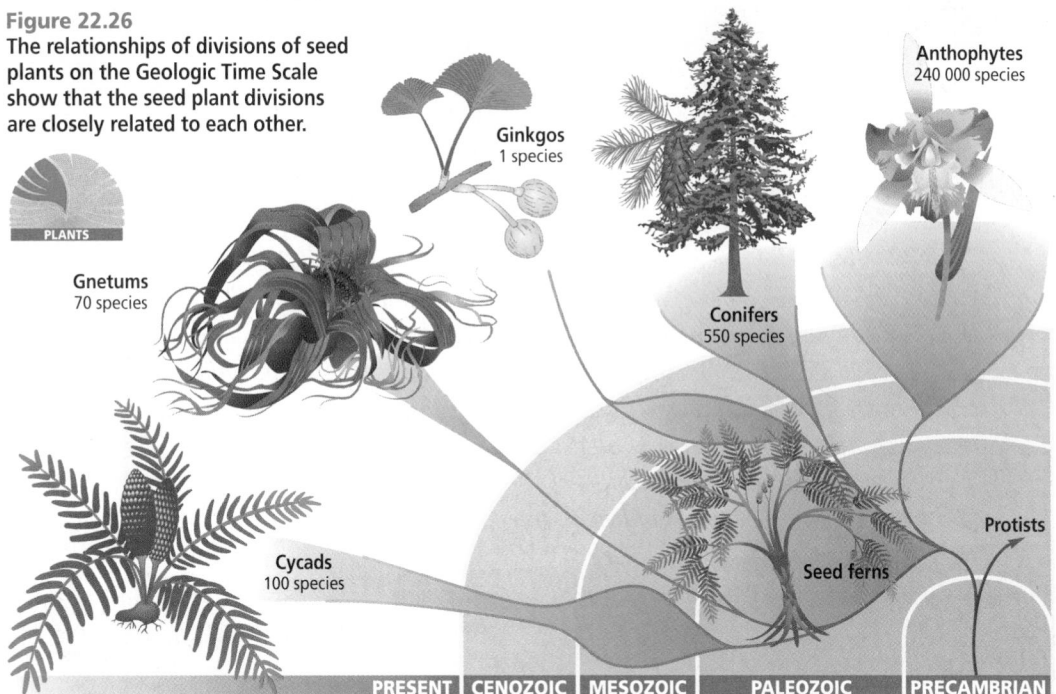

Figure 22.26
The relationships of divisions of seed plants on the Geologic Time Scale show that the seed plant divisions are closely related to each other.

PLANTS

Ginkgos
1 species

Anthophytes
240 000 species

Gnetums
70 species

Conifers
550 species

Cycads
100 species

Seed ferns

Protists

PRESENT | CENOZOIC | MESOZOIC | PALEOZOIC | PRECAMBRIAN

Origins of Seed Plants

Seed plants first appear as fossils 360 million years ago. Some seed plants, such as cycads and ginkgos, shared Earth's forests with the dinosaurs during the Triassic and Cretaceous periods. However, like the dinosaurs, most members of the Ginkgophyta died out about 66 million years ago.

According to fossil evidence, the first conifers emerged around 280 million years ago. During the Jurassic period, conifers became prominent forest inhabitants and remain so today.

Anthophytes first appear in the early Cretaceous period.

Seed plants are closely related to each other and to other groups of non-seed vascular plants as shown in *Figure 22.26.*

Section Assessment

Understanding Main Ideas
1. Name two ways that seeds help plants reproduce on land.
2. How are needlelike leaves an adaptation to life in climates where water may be limited?
3. Why are flowering plants so successful?
4. What are the major characteristics of anthophytes that distinguish them from coniferophytes?

Thinking Critically
5. How do you think the development of the seed might have affected the lives of herbivorous animals living in Earth's ancient forests?

SKILL REVIEW

6. **Compare and Contrast** Compare the formation of a spore in ferns and a seed in conifers. For more help, refer to *Thinking Critically* in the **Skill Handbook.**

22.3 SEED PLANTS **617**

Section Assessment

1. Seeds contain a food supply for the developing embryo and protect the embryo from harsh conditions.
2. The thick cuticle, reduced surface area, and sunken stomata limit evaporation.
3. Flowering plants produce fruits that protect the seeds and aid in seed dispersal.
4. Anthophytes produce flowers and fruits. Coniferophytes do not.
5. By eating seeds, herbivores were able to take advantage of the food supply intended for the young plant embryo.
6. Fern spores are produced in sporangia. Clusters of sporangia form sori that are found on the pinnae. Conifer seeds are produced on the scales of the cone.

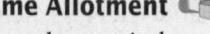

DESIGN YOUR OWN BioLab

Time Allotment
One class period

Process Skills
observe and infer, classify, compare and contrast, organize data

Safety Precautions
Some of the needle and cone specimens are very sharp. Warn students of the possibility of sticking themselves with the pointed ends.

PREPARATION

Alternative Materials
■ Specimens should contain both leaves and cones. If unavailable, substitute scale diagrams of a variety of conifer leaf and cone specimens.

Possible Hypotheses
■ The number of leaves per bundle, length of leaves, shape of leaves, color of leaves, and appearance of the sheath at the base of the leaves are useful in classifying conifers.
■ Cone shape, length, and diameter may be used to classify conifers.

Resource Manager
BioLab and MiniLab Worksheets, pp. 103-104 **L2**

How can you make a key for identifying conifers?

*M*ost conifers have cones and needle- or scalelike leaves. Different species have cones of different sizes, shapes, and thickness. The leaves of different species also have different characteristics. How would you go about identifying a conifer you are unfamiliar with? You would probably use a biological identification key. Biological keys list features of related organisms in a way that allows you to determine each organism's scientific name. Below is an example of the selections that could be found in a key that might be used to identify trees.

Needles grouped in bundles	*Needlelike leaves*
Needles not grouped in bundles	*Flat, thin leaves*
Leaves composed of three or more leaflets	*Leaves not made up of leaflets*

Pine nee[dles]

Hemlock

PREPARATION

Problem
What kinds of characteristics can be used to create a key for identifying different kinds of conifers?

Hypotheses
State your hypothesis according to the kinds of characteristics you think will best serve to distinguish among several conifer groups. Explain your reasoning.

Objectives
In this BioLab, you will:
■ **Compare** structures of several different conifer specimens.
■ **Identify** which characteristics can be used to distinguish one conifer from another.
■ **Communicate** to others the distinguishing features of different conifers.

Possible Materials
twigs, branches, and cones from several different conifers that have been identified for you

Safety Precautions
Always wash your hands after handling biological materials. Always wear goggles in the lab.

Skill Handbook
Use the **Skill Handbook** if you need additional help with this lab.

Arborvitae

618 THE DIVERSITY OF PLANTS

PLAN THE EXPERIMENT

Teaching Strategies
■ Label the samples with either their scientific names or generic labels such as Conifer A, Conifer B, etc. This will allow students to establish whether or not a key actually works when identification is attempted by another group.
■ Illustrate on the chalkboard or an overhead what an ideal key may look like. Point out the nature of the opposite trait being used at the fork.

Possible Procedures
■ Student procedures will vary. However, the general organization of the key itself will be similar from group to group. Making branches with opposite characteristics will help with the key design.

PLAN THE EXPERIMENT

1. Make a list of characteristics that could be included in your key. You might consider using shape, color, size, habitat, or other factors.
2. Determine which of those characteristics would be most helpful in classifying your conifers.
3. Determine in what order the characteristics should appear in your key.
4. Decide how to describe each characteristic.

Check the Plan

1. The traits described at each fork in a key are often pairs of contrasting characteristics. For example, the first fork in a key to conifers might compare "needles grouped in bundles" with "needles attached singly."
2. Someone who is not familiar with conifer identification

should be able to use your key to correctly identify any conifer it includes.
3. *Make sure your teacher has approved your experimental plan before you proceed further.*
4. Carry out your plan by creating your key.

Spruce

ANALYZE AND CONCLUDE

1. **Checking Your Hypothesis** Have someone outside your lab group try using your key to identify your conifer specimens. If they are unable to make it work, try to determine where the problem is and make improvements.
2. **Making Inferences** Is there only one correct way to design a key for your specimens? Explain why or why not.
3. **Relating Concepts** Give one or more examples of situations

in which a key would be a useful tool.

Going Further

Project Design a different key that would also work to identify your specimens. You may expand your key to include additional conifers.

BIOLOGY Online To find out more about conifers, visit the Glencoe Science Web site.
science.glencoe.com

ANALYZE AND CONCLUDE

1. By using student key design on the overhead, the entire class can determine whether or not the key works.
2. Student key designs placed on the overhead will illustrate the diversity of student trait choices used to organize and design each key.
3. It can distinguish poisonous from nonpoisonous plants, and harmful from nonharmful insects.

Error Analysis

Make sure students understand that their keys must distinguish among several similar conifers, each of which must be uniquely identified. Each fork in their key must be specific and identifiable.

✓ Assessment

Performance Provide students with a bag of common laboratory items such as: glass slide, coverslip, paper clip, thumbtack, rubber band, staple, and dissecting needle. Ask them to prepare a key that identifies each item. Use the Performance Task Assessment List for Making and Using a Classification System in **PASC**, p. 49. **L1**

Going Further

Ask students to explain why a key design based on leaf odor or habitat may not be workable or practical.

Data and Observations

Have students record their keys and turn them in at the end of class. Make transparencies of sample keys and use them in class the following day as a means for illustrating correct and incorrect key design.

Biology & SOCIETY

Purpose

Students are exposed to the pros and cons of traditional rain forest logging practices and sustainable harvesting alternatives.

Background

The Amazon rain forest is shrinking by nearly 5 million acres per year, losing an area the size of seven football fields every minute.

Traditional logging practices damage the forest and threaten biodiversity. In addition, logging companies build roads through forested areas. Farmers and ranchers then use these roads to gain access to new areas of forest. Trees and vegetation are burned and the land cleared alongside the roads to make room for roads and pastures.

Teaching Strategies

■ Encourage students to do an Internet search for information about rain forest conservation programs. Ask students to evaluate the feasibility of these programs in protecting rain forests while satisfying the economic needs of people living in these regions.

Investigating the Issue

If consumers can be persuaded to use nontropical woods and building materials, the demand for tropical timber would be reduced. This would reduce the need for logging and decrease the amount of land lost to farming and ranching.

Biology & SOCIETY

Forestry: Keeping a Balance

Tropical rain forests are Earth's most biologically diverse ecosystems. Large areas of these forests are lost each year due to logging and the clearing of land for crops and cattle.

Hundreds of different kinds of trees grow in tropical forests. But only a few are valuable to the logging industry. Traditionally, loggers go after only the most economically valuable timber species, but in doing so they invade and clearcut millions of acres of forest each year. When loggers remove the economically valuable trees, neighboring trees are often destroyed, and the delicate ecology of the area is upset.

A "Sustainable" Harvest Many ecologists and conservation groups have worked to promote "sustainable" forestry as an alternative to traditional logging. Sustainable forestry involves harvesting a limited number of trees from specific areas of rain forest, and then nurturing young trees to replace those that are cut. The philosophy behind sustainable forestry is that it will create a constantly regenerating supply of valuable timber and restrict logging to certain areas that are carefully managed.

620

Different Viewpoints

Although hundreds of millions of dollars have been spent on sustainable forestry, only a tiny fraction (less than 0.02 percent) of the world's tropical forests are managed sustainably. Logging companies make more money by logging forests the traditional way. Until logging companies can make profits using sustainable forestry methods, there is little economic incentive for them to change their harvesting techniques.

More harm than good? Evidence suggests that sustainable forestry may threaten biodiversity more than traditional logging. Researchers have discovered that the longer an area of forest is disturbed, the greater the threat to the plants and animals that live there. With sustainable forestry, an area is disturbed for a long period of time. This may damage the biodiversity of a forest.

Still the best hope Sustainable forestry advocates maintain that although their alternative isn't perfect, it may still be the best way to prevent widespread damage to tropical forests by restricting the damage that logging can do to smaller, controlled areas.

INVESTIGATING THE ISSUE

Analyzing the Issues Brainstorm in groups as to why some rain forest trees are more "valuable" than others. Discuss the link between the demand for certain kinds of wood and rain forest destruction. How might the demand be lessened?

BIOLOGY *Online* To find out more about rain forest destruction and sustainable forestry, visit the Glencoe Science Web site. **science.glencoe.com.**

Logged rain forest

Going Further

Linguistic Have students write an essay comparing the methods and results of sustainable harvest with traditional logging. Have students propose alternative forestry methods that might be better than either alternative. **L3**

GLENCOE TECHNOLOGY

VIDEODISC

The Infinite Voyage:
Life in the Balance
Rondonia: Home of a Dying Rain Forest
(Ch. 3) 4 min.

Chapter 22 Assessment

SUMMARY

Section 22.1

Nonvascular Plants

Main Ideas

- Nonvascular plants lack vascular tissue and reproduce by producing spores. The gametophyte generation is dominant.
- There are three divisions of nonvascular plants: Bryophyta, Hepatophyta, and Anthocerophyta.

Vocabulary

archegonium (p. 598)
antheridium (p. 598)

Section 22.2

Non-Seed Vascular Plants

Main Ideas

- The non-seed vascular plants were prominent members of Earth's ancient forests. They are represented by modern species.
- Vascular tissues provide the structural support that enables vascular plants to grow taller than nonvascular plants.
- There are three divisions of non-seed vascular plants: Lycophyta, Sphenophyta, and Pterophyta.

Vocabulary

prothallus (p. 602)
rhizome (p. 605)
sorus (p. 606)
strobilus (p. 602)

Section 22.3

Seed Plants

Main Ideas

- There are four divisions of vascular plants that produce naked seeds: Cycadophyta, Gnetophyta, Ginkgophyta, and Coniferophyta.
- Seeds contain a supply of food to nourish the young plant, protect the embryo during harsh conditions, and provide methods of dispersal.
- Anthophytes produce flowers and have seeds enclosed in a fruit.
- Anthophytes are either monocots or dicots based on the number of cotyledons present in the seed.
- Fruits provide protection for the seeds and aid in their dispersal.
- Anthophytes may be annuals, biennials, or perennials.

Vocabulary

annuals (p. 615)
biennials (p. 616)
cotyledon (p. 609)
deciduous plant (p. 612)
dicotyledons (p. 615)
embryo (p. 608)
fruit (p. 610)
monocotyledons (p. 615)
ovule (p. 608)
perennials (p. 616)
pollen grain (p. 608)

Main Ideas

Summary statements can be used by students to review the major concepts of the chapter.

Using the Vocabulary

To reinforce chapter vocabulary, use the Content Mastery Booklet and the activities in the Interactive Tutor for Biology: The Dynamics of Life on the Glencoe Science Web site. science.glencoe.com

THE PRINCETON REVIEW *All Chapter Assessment questions and answers have been validated for accuracy and suitability by The Princeton Review.*

GLENCOE TECHNOLOGY

VIDEOTAPE
MindJogger Videoquizzes
Chapter 22: *The Diversity of Plants*
Have students work in groups as they play the videoquiz game to review key chapter concepts.

Resource Manager

Chapter Assessment, pp. 127-132
MindJogger Videoquizzes
ExamView® Pro Software 💾
BDOL Interactive CD-ROM, Chapter 22 quiz

Chapter 22 Assessment

UNDERSTANDING MAIN IDEAS

1. c
2. d
3. d
4. a
5. c
6. b
7. a
8. d
9. a
10. d
11. ovule
12. rhizome
13. water
14. gametophyte
15. sorus
16. silica
17. cotyledon
18. chloroplast

UNDERSTANDING MAIN IDEAS

1. Bryophytes, hepatophytes, and anthocerophytes are the three divisions of _____ plants.
   - **a.** vascular
   - **b.** seed
   - **c.** nonvascular
   - **d.** evergreen

2. The _____ is the gametophyte of a lycophyte.
   - **a.** sorus
   - **b.** frond
   - **c.** strobilus
   - **d.** prothallus

3. Lycophytes include _____.
   - **a.** ferns
   - **b.** conifers
   - **c.** mosses
   - **d.** club mosses

4. Anthophytes and coniferophytes are divisions that are BOTH _____.
   - **a.** vascular and seed-producing
   - **b.** vascular and non-seed
   - **c.** nonvascular and non-seed
   - **d.** nonvascular and seed-producing

5. Vascular tissue is important because it helps the plant to _____.
   - **a.** anchor into the soil
   - **b.** reproduce
   - **c.** transport water and nutrients
   - **d.** photosynthesize

6. The plant in the photograph is a(n) _____.
   - **a.** Anthophyte
   - **c.** Sphenophyte
   - **b.** Pterophyte
   - **d.** Gnetophyte

THE PRINCETON REVIEW — TEST-TAKING TIP

Cramming Is Not a Good Strategy
If you don't know the material by the week before the test, you're less likely to do well. Set up a time line for your practice and preparation so that you're not rushed; then you will have time to deal with those problem areas.

7. About 280 million years ago, many of the non-seed plants became extinct because of _____.
   - **a.** long periods of freezing and drought
   - **b.** a change to a warm, wet climate
   - **c.** environmental pollution by humans
   - **d.** being eaten by dinosaurs

8. The gametophyte generation is dominant in which of the following plants?
   - **a.** pine trees
   - **b.** ferns
   - **c.** apple trees
   - **d.** mosses

9. Which of the following is NOT a part of a seed?
   - **a.** gametophyte
   - **b.** protective coat
   - **c.** food supply
   - **d.** embryo

10. An orange tree would be classified in the same division as which of the following?
    - **a.** pine tree
    - **b.** moss
    - **c.** cycad
    - **d.** sunflower

11. The sporophyte structure shown here that protects the developing female gametophyte is the _____.

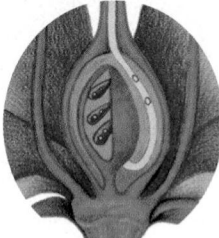

12. The thick, underground stem of a fern is the _____.

13. Seed plants do not require a continuous film of _____ to transport sperm to egg in fertilization

14. The moss sporophyte is dependent upon the _____ for nutrition.

15. A _____ is a group of sporangia found on a fern leaf.

16. The stems of sphenophytes contain _____ that gives them a rough texture.

17. Monocotyledons have only one seed leaf or _____.

18. The presence of a single large _____ in the cells of hornworts suggests they may be closely related to algae.

19. Because mosses lack vascular tissue, the amount of _____ in their environment is often the limiting factor.

20. Plants that have a life span that lasts two years are called _____.

APPLYING MAIN IDEAS

21. What is the adaptive advantage of fertilization that no longer requires a continuous film of water for the sperm to reach the eggs?

22. Why are conifers more abundant than flowering trees in Canada, Alaska, and Siberia?

23. Explain how evergreen and deciduous plants differ, and describe the adaptive value of each.

24. What might be the evolutionary advantages of having a gametophyte dependent on a sporophyte?

THINKING CRITICALLY

25. **Comparing and Contrasting** Compare and contrast cones and fruits.

26. **Observing and Inferring** What traits could you use to identify anthophytes?

27. **Concept Mapping** Complete the concept map by using the following vocabulary terms: prothallus, archegonium, antheridium.

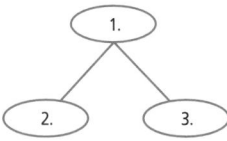

CD-ROM

For additional review, use the assessment options for this chapter found on the *Biology: The Dynamics of Life Interactive CD-ROM* and on the Glencoe Science Web site. **science.glencoe.com**

ASSESSING KNOWLEDGE & SKILLS

The germination rate of seeds is the percentage of planted seeds that eventually sprout to produce new plants. The seeds of the bristlecone pine must be exposed to cold temperatures before they will sprout.

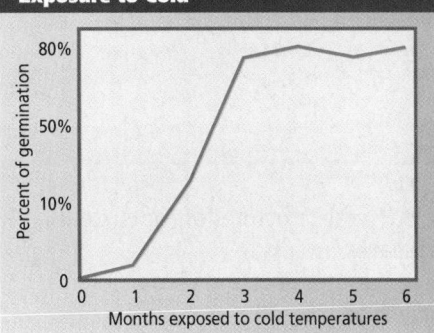

Seed Germination Rate After Exposure to Cold

Percent of germination vs *Months exposed to cold temperatures*

Using a Graph Answer the following questions based on the graph.

1. What would be the minimum time you would keep bristlecone pine seeds under refrigeration before planting?
   - **a.** 1 month
   - **c.** 6 months
   - **b.** 3 months
   - **d.** 80 months

2. How long does it take to get 50 percent germination?
   - **a.** 2½ months
   - **c.** 1 month
   - **b.** 6 months
   - **d.** 80 months

3. **Interpreting Scientific Illustrations** Examine the cross section of a conifer needle as diagrammed in the *Inside Story* and answer the following question. How do the positions of stomata and vascular tissues in the leaf help to prevent water loss?

19. moisture/water
20. biennials

APPLYING MAIN IDEAS

21. Fertilization may occur at any time because it is no longer dependent on water.

22. Conifer leaf adaptations allow them to survive in cold, harsh climates.

23. Evergreens retain leaves all year long and can start to photosynthesize as soon as the growing season starts. Deciduous plants lose their leaves, reducing water loss during the winter.

24. The sporophyte protects the gametophyte and provides nourishment.

THINKING CRITICALLY

25. Cones and fruits both contain seeds. However, the seeds in cones are borne on scales while the seeds contained within fruits are protected by the fruit.

26. Anthophytes have flowers and seeds contained within fruits.

27. 1. Prothallus; 2. Archegonium; 3. Antheridium

ASSESSING KNOWLEDGE & SKILLS

1. b
2. a
3. Sunken stomata and internal vascular tissue help prevent water loss.

Chapter 23 Organizer

Refer to pages 4T-5T of the Teacher Guide for an explanation of the National Science Education Standards correlations.

Section	Objectives	Activities/Features
Section 23.1 **Plant Cells and Tissues** National Science Education Standards UCP.1, UCP.2, UCP.5; A.1, A.2; C.1, C.5 (1 session, $1/2$ block)	1. **Identify** the major types of plant cells. 2. **Distinguish** among the functions of the different types of plant tissues.	**Inside Story:** A Plant, p. 628 **MiniLab 23-1:** Examining Plant Tissues, p. 629 **Problem-Solving Lab 23-1,** p. 630 **Investigate BioLab:** Determining the Number of Stomata on a Leaf, p. 646 **Art Connection:** *Red Poppy* by Georgia O'Keeffe (1887–1986), p. 648
Section 23.2 **Roots, Stems, and Leaves** National Science Education Standards UCP.1, UCP.2, UCP.3, UCP.5; A.1, A.2; C.1, C.5 (1 session, 1 block)	3. **Identify** the structures of roots, stems, and leaves. 4. **Describe** the functions of roots, stems, and leaves.	**Problem-Solving Lab 23-2,** p. 639 **MiniLab 23-2:** Looking at Stomata, p. 640
Section 23.3 **Plant Responses** National Science Education Standards UCP.1, UCP.2, UCP.5; A.1, A.2; C.1, C.5, C.6; G.1 (2 sessions, 1 block)	5. **Identify** the major types of plant hormones. 6. **Analyze** the different types of plant responses.	**Problem-Solving Lab 23-3,** p. 644

Need Materials? Contact Carolina Biological Supply Company at 1-800-334-5551 or at http://www.carolina.com

MATERIALS LIST

BioLab

p. 646 microscope, microscope slide, coverslip, water, dropper, metric ruler, single-edged razor blade, leaf from onion plant

MiniLabs

p. 629 microscope, microscope slides, coverslips, water, dropper, celery stalk

p. 640 microscope, microscope slides, coverslips, water, dropper, 5% salt solution, live plant leaf

Alternative Lab

p. 628 balance, garden trowel, water

Quick Demos

p. 626 slices of apple, celery and pear

p. 633 potato, iodine stain

p. 643 potted houseplants

Key to Teaching Strategies

L1 Level 1 activities should be appropriate for students with learning difficulties.

L2 Level 2 activities should be within the ability range of all students.

L3 Level 3 activities are designed for above-average students.

ELL ELL activities should be within the ability range of English Language Learners.

COOP LEARN Cooperative Learning activities are designed for small group work.

P These strategies represent student products that can be placed into a best-work portfolio.

These strategies are useful in a block scheduling format.

Plant Structure and Function

Teacher Classroom Resources

Section	Reproducible Masters	Transparencies
Section 23.1 **Plant Cells and Tissues**	Reinforcement and Study Guide, p. 101 **L2** Concept Mapping, p. 23 **L3** **ELL** BioLab and MiniLab Worksheets, p. 105 **L2** Content Mastery, pp. 113-114, 116 **L1**	Section Focus Transparency 55 **L1** **ELL** Reteaching Skills Transparency 32 **L1** **ELL**
Section 23.2 **Roots, Stems, and Leaves**	Reinforcement and Study Guide, pp. 102-103 **L2** Critical Thinking/Problem Solving, p. 23 **L3** BioLab and MiniLab Worksheets, p. 106 **L2** Laboratory Manual, pp. 161-170 **L2** Content Mastery, pp. 113-116 **L1**	Section Focus Transparency 56 **L1** **ELL** Basic Concepts Transparencies 36, 37 **L2** **ELL** Reteaching Skills Transparencies 33, 34 **L1** **ELL**
Section 23.3 **Plant Responses**	Reinforcement and Study Guide, p. 104 **L2** BioLab and MiniLab Worksheets, pp. 107-108 **L2** Content Mastery, p. 113, 115-116 **L1**	Section Focus Transparency 57 **L1** **ELL**

Assessment Resources

Chapter Assessment, pp. 133-138
MindJogger Videoquizzes
Performance Assessment in the Biology Classroom
Alternate Assessment in the Science Classroom
ExamView® Pro Software 💾
BDOL Interactive CD-ROM, Chapter 23 quiz

Additional Resources

Spanish Resources **ELL**
English/Spanish Audiocassettes **ELL**
Cooperative Learning in the Science Classroom **COOP LEARN**
Lesson Plans/Block Scheduling

NATIONAL GEOGRAPHIC — Teacher's Corner

Products Available From Glencoe
To order the following products, call Glencoe at 1-800-334-7344:
CD-ROM
NGS PictureShow: What It Means to Be Green
Curriculum Kit
GeoKit: Plants
Transparency Set
NGS PicturePack: What It Means to Be Green
Videodisc
STV: Plants

Index to National Geographic Magazine
The following articles may be used for research relating to this chapter:
"Beyond Supermouse: Changing Life's Genetic Blueprint," by Robert F. Weaver, December 1984.

GLENCOE TECHNOLOGY

The following multimedia resources are available from Glencoe.
Biology: The Dynamics of Life
CD-ROM **ELL**
 Animation: *Water Uptake in Roots*

Videodisc Program 📷
Water Uptake in Roots

23 Plant Structure and Function

Visual-Spatial Have students examine a potted plant. Look at the plant's roots, stems, and leaves. Ask students to suggest possible functions for each. **L1**

Theme Development

The theme of **systems and interaction** is apparent as the anatomy of roots, stems, and leaves is examined and the interdependence among these structures is discussed. **Homeostasis** is stressed through the discussion of how plant hormones help maintain balance in plants.

0:00 OUT OF TIME?

If time does not permit teaching the entire chapter, use the BioDigest at the end of the unit as an overview.

READING BIOLOGY

Glencoe's *Biology: The Dynamics of Life* contains many resources to assist a student's reading skills. Each chapter contains figures with expanded captions that expand on written material. Word Origins, located along the side of text, expand knowledge of biology vocabulary. Glencoe's Content Mastery Booklet helps develop reading skills while reinforcing content. In addition, use the Interactive Tutor for *Biology: The Dynamics of Life* on the Glencoe Web site to reinforce vocabulary.
science.glencoe.com

What You'll Learn

- You will describe the major types of plant cells and tissues.
- You will analyze the structure and functions of roots, stems, and leaves.
- You will identify plant hormones and determine the nature of plant responses.

Why It's Important

Plants are composed of cells, tissues, and organs. You need to be familiar with the structure of plants so you can understand how they function and how they respond to their environment.

READING BIOLOGY

Quickly scan the three sections of this chapter, and write down any questions that form. As you read, make a brief outline of the chapter. From the outline, try to find the answers to your questions.

BIOLOGY *Online*

To find out more about plants, visit the Glencoe Science Web site.
science.glencoe.com

As you look at this African tulip tree, it is important to remember that plants are composed of individual cells such as the one in the inset photograph.

624

Multiple Learning Styles

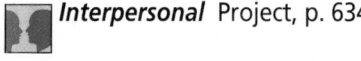

Look for the following logos for strategies that emphasize different learning modalities.

Kinesthetic Reinforcement, p. 627; Tech Prep, p. 635; Activity, p. 636; Meeting Individual Needs, pp. 637, 638

Visual-Spatial Getting Started Demo, p. 624; Microscope Activity, pp. 627, 636, 638, 643; Portfolio, pp. 627, 636, 644; Cultural Diversity, p. 630; Meeting Individual Needs, p. 640; Extension, p. 641; Quick Demo, p. 643

Interpersonal Project, p. 634

Linguistic Biology Journal, pp. 627, 639, 642; Reteach, p. 645

Logical-Mathematical Enrichment, p. 635

23.1 Plant Cells and Tissues

Objectives
Identify the major types of plant cells.
Distinguish among the functions of the different types of plant tissues.

Vocabulary
parenchyma
collenchyma
sclerenchyma
epidermis
stomata
guard cells
trichome
xylem
tracheid
vessel element
phloem
sieve tube member
companion cell
meristem
apical meristem
vascular cambium
cork cambium

Prepare

Key Concepts
This section focuses on the structure and function of plant cells and tissues. The different types of cells and their location in a plant are described. The section concludes with a discussion of plant tissues.

Planning
- Purchase apple, celery, and pear for the Quick Demo.
- Assemble materials for the Reinforcement.
- Purchase celery for MiniLab 23-1.
- Purchase or locate prepared onion root tip slides for the Inside Story.
- Purchase onion leaves for the BioLab.

T he surface of a leaf or flower reveals amazing complexity when viewed under a high-power microscope. When you pick a flower or eat a piece of fruit, you may not give its microscopic structure a minute's thought. The structure of each plant tissue is related to the function it performs. For example, the epidermis of a tomato leaf with its covering of hairs helps protect the delicate tissues of the leaf from possible insect predators.

Magnification: 350×

Tomato plant leaves and fruit (top); glandular hair on leaf (inset)

1 Focus

Bellringer
Before presenting the lesson, display **Section Focus Transparency 55** on the overhead projector and have students answer the accompanying questions.
L1 ELL

Types of Plant Cells

Like all organisms, plants are composed of cells. Plant cells can be distinguished from animal cells because they have a cell wall, a central vacuole, and can contain chloroplasts. *Figure 23.1* shows a typical plant cell. Plants, just like other organisms, are composed of many different types of cells.

Parenchyma

Parenchyma (puh RENG kuh muh) cells are the most abundant kind of plant cell. They are found throughout the tissues of a plant. These spherical cells have thin, flexible cell walls. Most parenchyma cells usually have a large central vacuole, which

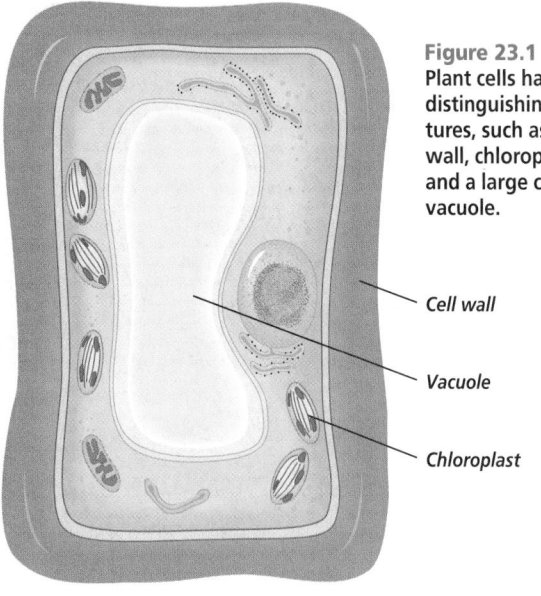

Figure 23.1
Plant cells have many distinguishing features, such as a cell wall, chloroplasts, and a large central vacuole.

Cell wall

Vacuole

Chloroplast

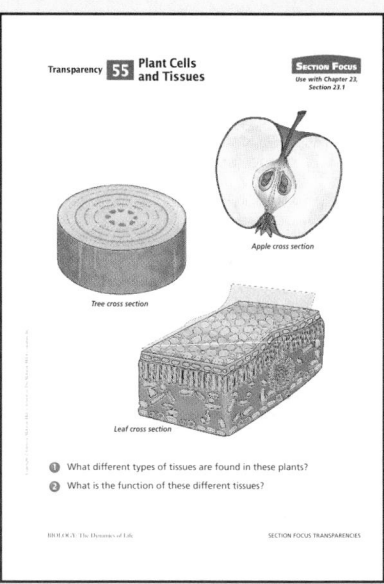

Transparency **55** Plant Cells and Tissues

SECTION FOCUS
Use with Chapter 23, Section 23.1

Apple cross section

Tree cross section

Leaf cross section

1 What different types of tissues are found in these plants?
2 What is the function of these different tissues?

BIOLOGY: The Dynamics of Life   SECTION FOCUS TRANSPARENCIES

Assessment Planner

Portfolio Assessment
Assessment, TWE, pp. 627, 633
Problem-Solving Lab, TWE, p. 644

Performance Assessment
Alternative Lab, TWE, p. 628-629
MiniLab, SE, pp. 629, 640
MiniLab, TWE, p. 629
Problem-Solving Lab, TWE, p. 639
Assessment, TWE, p. 644
BioLab, SE, p. 646-647

Knowledge Assessment
Section Assessment, SE, p. 631, 641, 645
Assessment, TWE, p. 645
Chapter Assessment, SE, p. 649-651

Skill Assessment
Problem-Solving Lab, TWE, p. 630
MiniLab, TWE, p. 640
Assessment, TWE, pp. 641, 643
BioLab, TWE, pp. 646-647

2 Teach

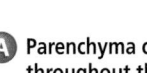

WORD *Origin*

par-
From the Greek word *para*, meaning "beside."

coll-
From the Greek word *kolla*, meaning "glue."

scler-
From the Greek words *skleros*, meaning "hard."

With *en* meaning "in," and *chymein* meaning "to pour." Parenchyma, collenchyma, and sclerenchyma are all types of plant tissues.

Figure 23.2
Plants are composed of three basic types of cells, which are shown stained with dyes here.

sometimes contains a fluid called sap, *Figure 23.2A.*

Parenchyma cells have two main functions: storage and food production. The large vacuole found in these cells can be filled with water, starch grains, or oils. The edible portions of many fruits and vegetables you eat are composed mostly of parenchyma cells. Parenchyma cells are also important in producing food for the plant. These cells contain numerous chloroplasts that produce glucose during photosynthesis.

Collenchyma

Collenchyma (coh LENG kuh muh) cells are long cells with unevenly thickened cell walls, as illustrated in *Figure 23.2B.* The structure of the cell wall is important because it allows the cells to grow. The thin portions of the wall can stretch as the cell grows while the thicker parts of the wall provide strength and support. These cells are arranged in tubelike strands or cylinders. The strands of collenchyma provide support for

surrounding tissue. The long tough strands you may have noticed in celery are composed of collenchyma.

Sclerenchyma

The walls of **sclerenchyma** (skler ENG kuh muh) cells are very thick and rigid. At maturity, these cells often die. Although their cytoplasm disintegrates, their strong, thick cell walls remain and provide support for the plant. Sclerenchyma cells can be seen in *Figure 23.2C.* There are two types of sclerenchyma cells commonly found in plants: fibers and stone cells. Fibers are long, thin cells that form strands. These strands provide support and strength for the plant and are the source of fibers used in the manufacture of linen and rope. Stone cells are circular and usually found in clusters. They are responsible for the gritty texture of pears and are a major component of the pits found in peaches and other fruits. Sclerenchyma cells are also a major part of vascular tissue, which you will learn about later in this section.

A Parenchyma cells are found throughout the plant. They have thin walls and a large vacuole.

B Collenchyma cells are often found in parts of the plant that are still growing. Notice the unevenly thickened cell walls.

C The walls of sclerenchyma cells are very thick. These dead cells are able to provide support for the plant.

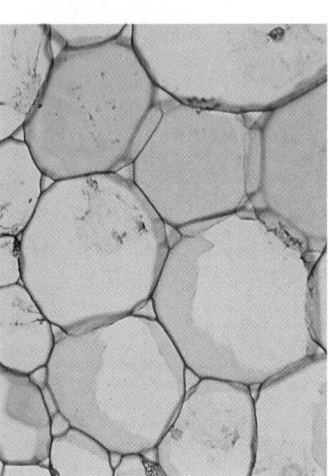

Magnification: 88×

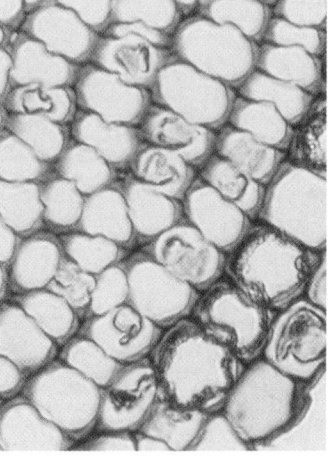

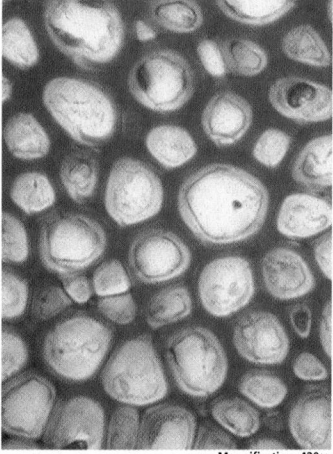

Magnification: 420×

MEETING INDIVIDUAL NEEDS

Learning Disabled

Have students prepare a table to summarize the characteristics of plant cells. Organizing these characteristics into a table will help students with learning disabilities to distinguish these cell types. Headings across the top could include: Shape of cells, Nature of cell wall, and Function. Headings down the side of the table should be: Parenchyma, Collenchyma, and Sclerenchyma. L1

Plant Tissues

Recall that a tissue is a group of cells that functions together to perform an activity. There are several different tissue types in plants. Each one is composed of cells working together.

Dermal tissues

The dermal tissue, or **epidermis,** is composed of flattened parenchyma cells that cover all parts of the plant. It functions much like the skin of an animal, covering and protecting the body of a plant. As shown in *Figure 23.3,* the cells that make up the epidermis are tightly packed and often fit together like a jigsaw puzzle. The epidermal cells produce the waxy cuticle that helps prevent water loss.

Another structure that helps control water loss from the plant, the stomata, are part of the epidermal layer. The **stomata** (STOH mut uh) are openings in the cuticle of the leaf that control the exchange of gases. Stomata are found on green stems and on the upper and lower surfaces of leaves. In many plants, fewer stomata are located on the upper surface of the leaf as a means of conserving water. Cells called **guard cells** control the opening and closing of the stomata. The opening and closing of the guard cells regulates the flow of water vapor from the leaf tissues. You can learn more about stomata in the *BioLab* at the end of this chapter.

The dermal tissue of roots may have root hairs. Root hairs are extensions of individual cells that help the root absorb water and minerals. On the stems and leaves of some plants, there are structures called trichomes. **Trichomes** are hairlike projections that extend from the epidermis and give the epidermis a "fuzzy" appear-

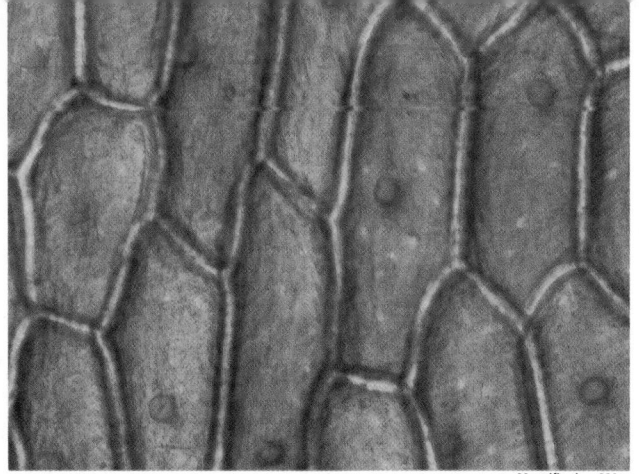

Magnification: 220×

ance. They help reduce the evaporation of water from the surface of leaves. In some cases, trichomes are glandular and secrete toxic substances that help protect the plant from predators. Stomata, root hairs, and trichomes are shown in *Figure 23.4.*

Figure 23.3
The cells of the epidermis fit together tightly to help protect the plant and prevent water loss.

Figure 23.4
Root hairs are extensions of individual cells on the root (a). Trichomes are hairlike projections from the epidermis (b). Stomata are openings in the leaf tissue surrounded by guard cells (c).

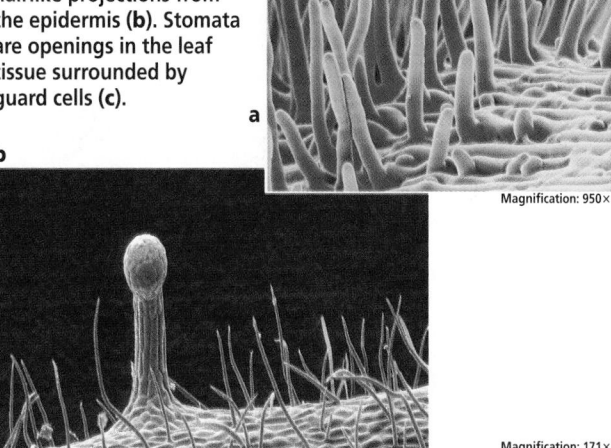

a

b

Magnification: 950×

Magnification: 263×

Magnification: 171×

c

Purpose ◆

Students will study the basic body plan of a plant.

Teaching Strategies

■ Ask students to describe the functions of the four tissue types. **L2**

Visual Learning

■ Ask students to bring in samples or pictures of leaves, stems, and roots. Have them describe the similarities and differences among the different organs. **L2**

■ Have students examine a prepared slide of an onion root tip. Point out that apical meristems produce new cells through the process of mitosis. **L1**

Critical Thinking

Apical meristems are located at the tips of stems and roots. They increase the height of the plant by producing new cells. The vascular cambium produces new phloem and xylem cells. The cork cambium produces the outer layers of bark.

A Plant

There seems to be an almost endless variety of plants. Regardless of their diversity and numerous modifications, all vascular plants have the same basic body plan. They are composed of cells, tissues, and organs.

Critical Thinking *What are the different types of meristems and how do they help the plant to grow new tissues and organs?*

House plants

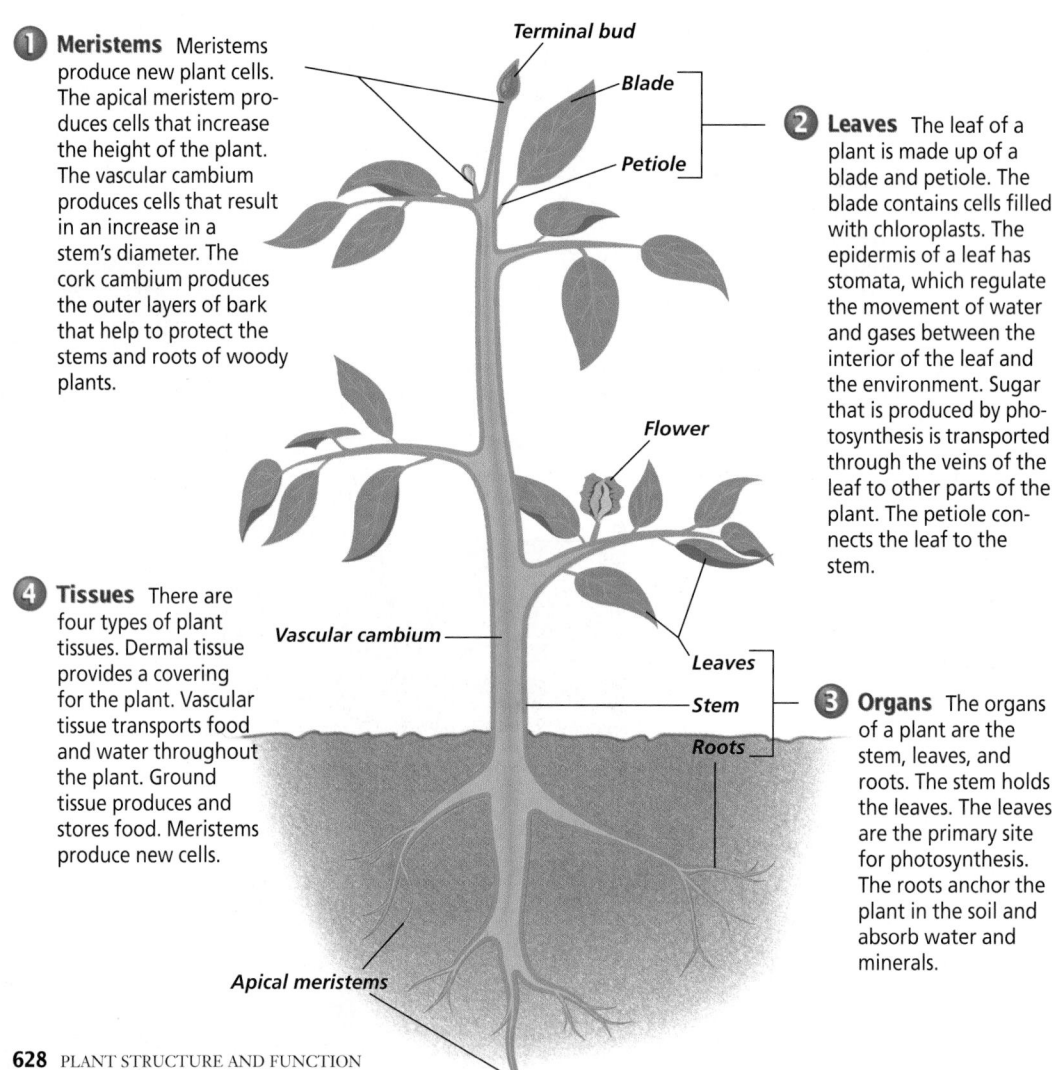

1 Meristems Meristems produce new plant cells. The apical meristem produces cells that increase the height of the plant. The vascular cambium produces cells that result in an increase in a stem's diameter. The cork cambium produces the outer layers of bark that help to protect the stems and roots of woody plants.

4 Tissues There are four types of plant tissues. Dermal tissue provides a covering for the plant. Vascular tissue transports food and water throughout the plant. Ground tissue produces and stores food. Meristems produce new cells.

2 Leaves The leaf of a plant is made up of a blade and petiole. The blade contains cells filled with chloroplasts. The epidermis of a leaf has stomata, which regulate the movement of water and gases between the interior of the leaf and the environment. Sugar that is produced by photosynthesis is transported through the veins of the leaf to other parts of the plant. The petiole connects the leaf to the stem.

3 Organs The organs of a plant are the stem, leaves, and roots. The stem holds the leaves. The leaves are the primary site for photosynthesis. The roots anchor the plant in the soil and absorb water and minerals.

Terminal bud
Blade
Petiole
Flower
Vascular cambium
Leaves
Stem
Roots
Apical meristems

628 PLANT STRUCTURE AND FUNCTION

Alternative Lab

Differences Between Taproots and Fibrous Roots

Purpose ◆

This lab will allow students to compare taproots and fibrous roots.

628

Preparation

Locate areas where grass and dandelions are growing.

Materials

garden trowel, metric balance

Procedure

Give students the following directions.

1. Examine pictures of taproots and fibrous roots in your textbook. Which type of root do you think contains more root hairs?

2. Using a trowel, loosen the soil around a dandelion plant. Gently pull the plant out of the soil. Repeat this process with a clump of grass.

3. Record the mass of each plant with its clinging ball of soil. Carefully, rinse away the soil under running water. Weigh and record the mass of the plants again.

4. Determine the ratio between plant mass and plant mass plus soil by

Vascular tissues

Food, minerals, and water are transported throughout the plant by vascular tissue. Xylem and phloem are the two types of vascular tissues. **Xylem** is plant tissue composed of tubular cells that transports water and minerals from the roots to the rest of the plant. In seed plants, xylem is composed of three types of sclerenchyma cells—tracheids, vessel elements, and fibers. Parenchyma cells are also present.

Tracheids are tubular cells tapered at each end that transport water throughout a plant. As you can see in *Figure 23.5*, both tracheids and vessel elements are cylindrical and dead at maturity. The cell walls between adjoining tracheids have pits through which water and minerals flow. Conifers have tracheids but no vessel elements in their vascular tissues.

Vessel elements are tubular cells that transport water throughout the plant. They are wider and shorter than tracheids. Vessel elements also have openings in their end walls. In some plants, mature vessel elements lose their end walls and water and minerals flow freely from one cell to another. Although almost all vascular plants have tracheids, vessel elements are most commonly found in anthophytes. This difference could be one reason why anthophytes are the most successful plants on Earth. Vessel elements are thought to transport water more efficiently than tracheids because water can flow freely from vessel element to vessel element through the openings in their end walls.

You can learn more about how vascular tissues transport water in the *MiniLab* on this page. What other types of tissues are found in vascular plants? To answer this question, look at the *Inside Story*.

MiniLab 23-1 — Observing

Examining Plant Tissues Pipes are hollow. Their shape or structure allows them to be used efficiently in transporting water. Plant vascular tissues have this same efficiency in structure.

Procedure

1. Snap a celery stalk in half and remove a small section of "stringy tissue" from its inside.
2. Place the material on a glass slide. Add several drops of water. Place a second glass slide on top. **CAUTION: Use caution when working with a microscope and slides.**
3. Press down evenly on the top glass slide with your thumb directly over the plant material.
4. Remove the top glass slide. Add more water if needed. Add a cover slip.
5. Examine the celery material under low-and high-power magnification. Diagram what you see.
6. Repeat steps 2-5 using some of the soft tissue inside the celery stalk.

Analysis

1. Describe the appearance of the stringy tissue inside the celery stalk. What may be the function of this tissue?
2. Describe the appearance of the soft tissue inside the celery stalk. What may be the function of this tissue?
3. Does the structure of these tissues suggest their functions?

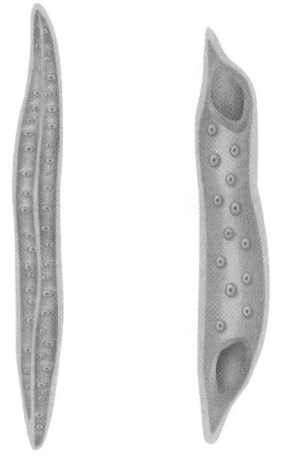

Figure 23.5
Tracheids and vessel elements make up the xylem. The xylem transports water and minerals from the roots, up the stem, and to the leaves.

Tracheid   Vessel element

MiniLab 23-1

Purpose
Students will use a squash technique to prepare, observe, and determine the function of two plant tissue types.

Process Skills
draw a conclusion, interpret data, observe and infer, predict

Safety Precautions
Caution students to be careful while using a razor blade and to cut away from their bodies. To reduce the possibility of injury resulting from glass slide breakage during the squash technique, have students place several layers of paper toweling between their thumb and the top glass slide before pressing down on the slide.

Teaching Strategies
■ One grocery store celery bunch will be sufficient for all classes.
■ Provide students with single-edged razor blades or scalpels for cutting the celery.

Expected Results
The stringy tissue will appear as nongreen tissue that resembles a track, or roadway. The soft tissue will appear green and cubelike in shape.

Analysis
1. tubelike, pipelike, stringlike, nongreen or colorless; transport of materials
2. cubelike, square cells, green in color; storage
3. Yes. Long, narrow cells are ideal for transporting water and nutrients. Cubelike cells would be most suitable for food storage.

---

dividing the plant mass by the plant mass plus soil.

Expected Results
The fibrous roots should hold more soil in close contact to its roots.

Analysis
1. Based on your ratio, which root type holds more soil in close contact to its roots?
2. Which root type would have a greater surface area for water absorption?

✔ Assessment

Performance Have students write a lab report summarizing this lab. Use the Performance Task Assessment List for Lab Report in **PASC,** p. 47. **L3**

✔ Assessment

Performance Have students observe other plant tissues. One possible activity is to prepare slide squashes of banana "strings" that remain on the fruit after peeling. Use the Performance Task Assessment List for Making Observations and Inferences in **PASC,** p. 17. **L1 ELL**

Purpose 📖

Students will determine that water and minerals move upward through stems while sugar moves downward through stems.

Process Skills

analyze information, apply concepts, interpret data, predict, recognize cause and effect

Teaching Strategies

■ Review the nature of vascular tissue within the stem of anthophytes but do not specifically mention the role of xylem and phloem.

■ Point out to students that many trees are anthophytes.

■ Diagrams may be consulted to show that vascular tissue is continuous through a plant's stem.

Thinking Critically

1. Transports food downward through plant; flow of sugar was blocked above metal sheet, causing sugar to accumulate above the metal sheet.

2. Transports water and minerals upward through plant; flow of water and minerals was blocked by metal sheet, causing water to accumulate below the metal sheet.

3. Water and minerals would not be blocked from flowing up the stem, but sugars would be blocked above the metal sheet as phloem is located just inside the bark.

✔ Assessment

Skill Ask students to write a hypothesis describing the experimental concentrations of water, minerals, and sugars 24 hours after inserting a metal sheet all the way through the stem. Have them predict the final outcome if the metal sheet were to remain in place across the tree's stem. Use the Performance Task Assessment List for Formulating a Hypothesis in **PASC**, p. 21. **L3**

Problem-Solving Lab 23-1   Applying Concepts

What happens if vascular tissue is interrupted? Anthophytes have tissues within their organs that transport materials from roots to leaves and from leaves to roots. What happens if this pathway is experimentally interrupted?

Analysis

A thin sheet of metal was inserted into the stem of a living tree as shown in the diagram. One day later, the following analysis of chemicals was made:

■ Concentration of water and minerals directly below the metal sheet was higher than water and minerals directly above the metal sheet.

■ Concentration of sugar directly above the metal sheet was higher than sugar concentration directly below the metal sheet.

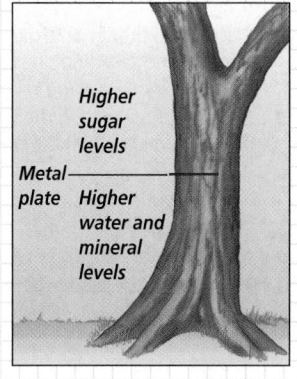

Higher sugar levels

Metal plate

Higher water and mineral levels

Thinking Critically

1. What is the function of phloem? Why was the concentration of sugars different on either side of the metal sheet?
2. What is the function of xylem? Why was the concentration of minerals and water different on either side of the metal sheet?
3. How would the experimental findings have differed if the metal sheet were inserted only into the bark of the tree?

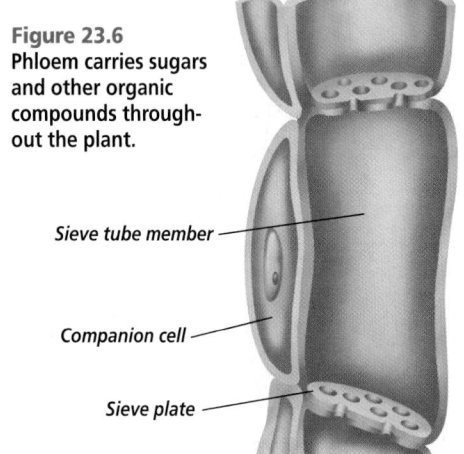

Figure 23.6
Phloem carries sugars and other organic compounds throughout the plant.

Sieve tube member

Companion cell

Sieve plate

Sugars and other organic compounds are transported throughout the plant by the phloem. **Phloem** is made up of a series of tubular cells that are still living and transports sugars from the leaves to all parts of the plant. The structure of phloem, *Figure 23.6*, is similar to xylem because it is also composed of long cylindrical cells. However the phloem cells, called **sieve tube members,** are alive at maturity. Sieve tube members are unusual because, although they contain cytoplasm, they do not have a nucleus or ribosomes. Next to each sieve tube member is a companion cell. **Companion cells** are nucleated cells that help manage the transport of sugars and other organic compounds through the sieve cells of the phloem. In anthophytes, the end walls between two sieve tube members are called sieve plates. The sieve plates have large pores. The sugar and organic compounds move from cell to cell through these pores. Phloem transports materials from the roots to the leaves as well as from the leaves to the roots.

The vascular tissue of many plants contains fibers. Although the fibers are not used for transporting materials, they are important because they provide support for the plant. You can learn more about vascular tissues in the *Problem-Solving Lab* shown here.

Ground tissue

Ground tissue includes all tissues other than the dermal tissues and vascular tissues. Ground tissue is mostly composed of parenchyma cells but may also include collenchyma and sclerenchyma cells. The functions of ground tissue include photosynthesis, storage, and support. The cells of ground tissue in leaves and green stems contain numerous chloroplasts that carry on

---

Cultural Diversity

Origin and Cultural Significance of Corn

📘 *Visual-Spatial* Corn was one of the earliest crops to be domesticated, and has been a major food source for Native Americans of both North and South America for about 7000 years. In addition to its

importance as a food source, corn also occupies a symbolic place in the culture of many Native American tribes. Ask students working in groups to research uses for corn other than as a food source. Ask each group to prepare an illustrated essay of its findings.
L2 **COOP LEARN**

photosynthesis. Cells in the stem and root contain large vacuoles that store starch grains and water. Cells, such as those seen in *Figure 23.7,* are often seen in ground tissue.

Meristematic tissues

A growing plant needs to produce new cells. These new cells are produced in areas called meristems. **Meristems** are regions of actively dividing cells. Meristematic cells are small, spherical parenchyma cells with large nuclei. There are several types of meristems.

Apical meristems are found at or near the tips of roots and stems. They produce cells that allow the roots and stems to increase in length. Lateral meristems are cylinders of dividing cells located in roots and stems. *Figure 23.8* will help you visualize the location of these meristems. The production of cells by the lateral meristems results in an increase in diameter. Most woody plants have two kinds of lateral meristems: a vascular cambium and cork cambium. The lateral meristem called the **vascular cambium** produces new xylem and phloem cells in the stems and roots. The other lateral meristem, the **cork cambium,** produces a tough covering for the surface of stems and roots. The outer bark of a tree is produced by the cork cambium.

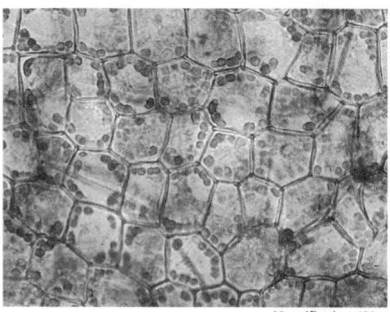

Figure 23.7
The numerous chloroplasts in this ground tissue produce food for the plant.

Magnification: 120×

Figure 23.8
The apical meristems are found in the tips of the stem and roots. The vascular cambium extends the length of the stem and roots.

Terminal meristem

Shoot apical meristem

Lateral meristems

Root apical meristems

Section Assessment

Understanding Main Ideas
1. What are the distinguishing traits of the three types of plant cells?
2. What is the function of vascular tissue? What are the two different types of vascular tissue?
3. Explain the function of stomata.
4. Draw a plant and indicate where on the plant the apical meristems would be located. How do they differ from lateral meristems?

Thinking Critically
5. What type of plant cell would you expect to find in the photosynthetic tissue of a leaf?

SKILL REVIEW

6. **Compare and Contrast** Compare and contrast the cells that make up the xylem and the phloem. For more help, refer to *Thinking Critically* in the **Skill Handbook.**

Section Assessment

1. Parenchyma are spherical, thin-walled cells. Collenchyma are cells with unevenly thickened cell walls. Sclerenchyma cells have very thick cell walls.
2. Vascular tissue transports water, food, and minerals throughout the plant. Xylem and phloem are the two types of vascular tissues.
3. Stomata control the flow of water vapor from the leaf tissue.
4. Apical meristems are found at the tips of stems and roots. Lateral meristems increase stem and root diameter; apical meristems increase root and stem length.
5. parenchyma
6. Xylem and phloem are both composed of cylindrical cells. Xylem cells are dead at maturity, whereas phloem cells are alive.

---

3 Assess

Check for Understanding
Quiz students orally about the basic characteristics of plant cells and tissues.

Reteach
Write the following terms on the board: storage, food production, support. Ask students to explain what types of cells and tissues would carry out each function. **L1**

Extension
Ask students what tissues or organs in humans are analogous to plant tissues. **L3**

✔ Assessment
Skill Have students draw a simple diagram of a plant with roots, stems, and leaves. Have them label the location of the four types of plant tissues. **L1** **ELL**

4 Close

Discussion
Have students explain what types of plant cells they would find in each of the following tissues: dermal, ground, vascular, and meristematic. **L2**

Resource Manager
Reinforcement and Study Guide, p. 101 **L2**
Content Mastery, p. 114 **L1**

Prepare

Key Concepts

Roots, stems, and leaves are the focus of this section. The function of roots as water and mineral absorbers is discussed. Next, the function of stems as conduits for water and minerals is presented. Leaf structures are described as they relate to photosynthesis.

Planning

- Purchase potato and iodine for the Quick Demo.
- Purchase bromothymol blue (BTB) solution and bean seeds for the Project.
- Find leaves for MiniLab 23-2.
- Purchase fertilizer and locate film canisters for Tech Prep.
- Locate thin slices of tree trunk sections for the Activity.

1 Focus

Bellringer

Before presenting the lesson, display **Section Focus Transparency 56** on the overhead projector and have students answer the accompanying questions.
L1 **ELL**

Transparency **56** Angiosperm Structures

Section Focus
Use with Chapter 23, Section 23.2

Tulip tree

❶ What are the main parts of this tree?
❷ What is the function of each part?

BIOLOGY: The Dynamics of Life   SECTION FOCUS TRANSPARENCIES

632

SECTION PREVIEW

Objectives

Identify the structures of roots, stems, and leaves.

Describe the functions of roots, stems, and leaves.

Vocabulary

cortex
endodermis
pericycle
root cap
sink
translocation
petiole
mesophyll
transpiration

Section

23.2 Roots, Stems, and Leaves

The next time you eat salad, look closely at your plate. The carrot is a root, the celery is a leaf stalk, lettuce is a leaf, and a bean sprout includes stems, leaves, and roots. Roots, stems, and leaves are organs of plants and your salad contains several. There are more than one-quarter million kinds of plants on Earth, and their organs exhibit an amazing variety.

Salad of roots, stems, and leaves

Figure 23.9
The taproot of the carrot can store large quantities of food and water for the plant (a). The fibrous roots of grasses absorb water and anchor the plant (b).

Roots

Roots are the underground parts of a plant. They anchor the plant in the ground, absorb water and minerals from the soil, and transport these materials up to the stem. Some plants, such as carrots, also accumulate and store food in their fleshy roots. The total surface area of a plant's roots may be as much as 50 times greater than the surface area of its leaves. As **Figure 23.9** illustrates, roots may be short or long, thick or thin, massive or threadlike. Some roots even extend above the ground.

Root systems vary according to the needs of the plant and the texture and moisture content of the soil. The two main types of root systems are taproots and fibrous roots. A taproot is a central fleshy root with smaller branch roots. For example, carrots and beets are taproots. Fibrous root systems have numerous roots branching from a central point. **Figure 23.9** shows examples of these root systems. Some plants, such as the corn in **Figure 23.10,** have adventitious roots called prop roots, which are aboveground roots that help support

a

b

BIOLOGY Online

Note Internet addresses that you find useful in the space below for quick reference.

_____

_____

_____

_____

tall plants. Many climbing plants have aerial roots that cling to objects such as walls and provide support for climbing stems. Bald cypress trees produce modified roots called pneumatophores, which are often referred to as "knees." The knees grow above the water upward from the mud and help supply oxygen to roots in waterlogged areas.

The structure of roots

If you look at the cross section of a typical root in *Figure 23.11*, you can see that the epidermis forms the outermost cell layer. A root hair is a tiny extension of a single epidermal cell that increases the surface area of the root and its contact with the soil. Root hairs absorb water, oxygen, and dissolved minerals. The next layer is a part of the ground tissue called the **cortex,** which is involved in the transport of water and ions into the vascular core at the center of the root. The cortex is made up of parenchyma cells that sometimes act as a storage area for food and water.

At the inner limit of the cortex lies the **endodermis,** a single layer of

cells that forms a waterproof seal that surrounds the root's vascular tissue. The waterproof seal between each cell of the endodermis forces all water and minerals to pass through the cells of the endodermis. Thus, the endodermis controls the flow of water and dissolved ions into the root. Just within the endodermis is the **pericycle.** The pericycle is a tissue that gives rise to lateral roots. Lateral roots are roots that are produced as offshoots of

Figure 23.10
Corn is a monocot with shallow, fibrous roots. As the plants grow to maturity, roots called prop roots grow from the stem to help keep the tall and heavy plants upright.

Figure 23.11
The root structures of dicots and monocots differ in the arrangement of xylem and phloem.

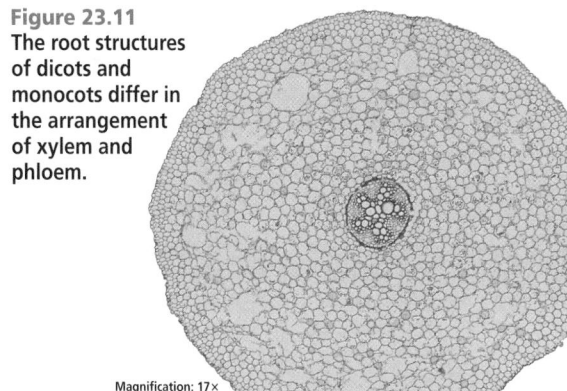

Magnification: 17×

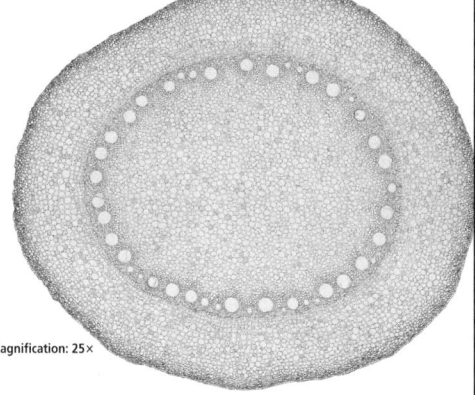

Magnification: 25×

A The xylem in a dicot root is arranged in a central star-shaped fashion. The phloem is found between the points of the star.

B In monocots, there are alternating strands of xylem and phloem that surround a pith of parenchyma cells.

23.2 ROOTS, STEMS, AND LEAVES **633**

Quick Demo

Demonstrate starch storage in root tubers by adding a few drops of an iodine stain to several slices of potato tuber. Explain that the blue-black color indicates the presence of starch.

✓ Assessment

Portfolio Tell the students to imagine themselves as a drop of water trying to get from the soil to the central vascular tissue of a root. Have them write a paragraph describing their journey. **L2** **P**

Resource Manager

Section Focus Transparency 56 and Master **L1** **ELL**
Laboratory Manual, pp. 161-166 **L2**

NATIONAL GEOGRAPHIC

VIDEODISC
STV: Plants, *What Is a Plant?*
Unit 1, Side 1, 1 min. 40 sec.
Roots and Stems

Figure 23.12
Water and mineral ions move into the root along two pathways.

A Mineral ions and water molecules enter root hairs and travel through the cells of the cortex by osmosis. Water also flows between the cells of the cortex.

Cortex

Endodermis · Pericycle · Phloem

Xylem

Endodermal cells · Waterproof seal

Root hair

B Nutrients dissolved in water can flow between the parenchyma cells, directly into the root cortex, then through the cells of the endodermis.

---

older roots. *Figure 23.12* traces the two pathways by which water and mineral ions move into the root.

Xylem and phloem are located in the center of the root. The arrangement of this xylem and phloem tissue accounts for one of the major differences between monocots and dicots. In dicot roots, the xylem forms a central star-shaped mass with phloem cells between the rays of the star. Monocot roots have strands of xylem that alternate with strands of phloem. There is usually a central core of parenchyma cells in the monocot root that is called a pith. The differences between monocot and dicot roots are illustrated in *Figure 23.11.*

Root growth

There are two meristematic regions in roots where growth is initiated by the production of new cells. Recall that meristems are areas of rapidly dividing cells. The apical meristem

produces cells that cause the root to increase in length. As cells produced by the apical meristem begin to mature, they differentiate into different types of cells. The vascular cambium, which is located between the xylem and phloem, soon begins contributing to the root's growth by adding cells that increase its diameter.

Each layer of new cells produced by the apical meristem is left farther and farther behind as more new cells are added and the root pushes forward through the soil. The tip of each root is covered by a tough, protective layer of parenchyma cells called the **root cap.** As the root pushes through the soil, the cells of the root cap wear away. Replacement cells are produced by the apical meristem so the root tip is never without its protective covering. Examine *Figure 23.13* on the following page to see if you can locate all the structures of a root.

WORD *Origin*

pericycle
From the Greek words *peri,* meaning "around," and *kykos,* meaning "circle." In vascular plants, the pericycle can produce lateral roots.

endodermis
From the Greek words *endon,* meaning "within," and *dermis,* meaning "skin." In vascular plants, the endodermis is the layer of cells forming the innermost layer of the cortex in roots.

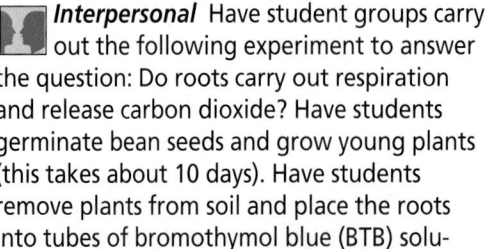

PROJECT

Do Roots Respire?

Interpersonal Have student groups carry out the following experiment to answer the question: Do roots carry out respiration and release carbon dioxide? Have students germinate bean seeds and grow young plants (this takes about 10 days). Have students remove plants from soil and place the roots into tubes of bromothymol blue (BTB) solu-

tion. Have them seal the roots from the air using a cotton or clay plug. Suggest that they prepare a control tube . Advise students that BTB turns from blue to light blue or green/yellow in the presence of carbon dioxide. Have students observe the liquid 24 hours later. Ask them to write a report summarizing what conclusions can be made regarding root tissue respiration. **L2** **ELL** **COOP LEARN**

Stems

Stems are the aboveground parts of plants that support leaves and flowers. Their form ranges from the thin, herbaceous stems of daisies, which die back every year, to the massive woody trunks of trees that may live for centuries. Green, herbaceous stems are soft and flexible and usually carry out some photosynthesis. Petunias, impatiens, and carnations are examples of plants with herbaceous stems. Trees, shrubs, and some other perennials have woody stems. Woody stems are hard and rigid and contain strands of schlerenchyma fibers and xylem.

Stems have several important functions. They provide support for all the aboveground parts of the plant. The vascular tissues that run the length of the stem transport water, mineral ions, and sugars to and from roots and leaves.

Like roots, stems are adapted to storing food. This enables the plant to survive drought, cold, or seasons with shorter days. Stems that act as food-storage organs include corms, tubers, and rhizomes. A corm is a short, thickened, underground stem surrounded by leaf scales. The term "tuber" may refer to a swollen leaf or stem. When used to refer to a stem, a tuber is a swollen, underground stem that has buds that will sprout new plants. Rhizomes are also underground stems that store food. Some examples of these food-storing stems appear in ***Figure 23.14.***

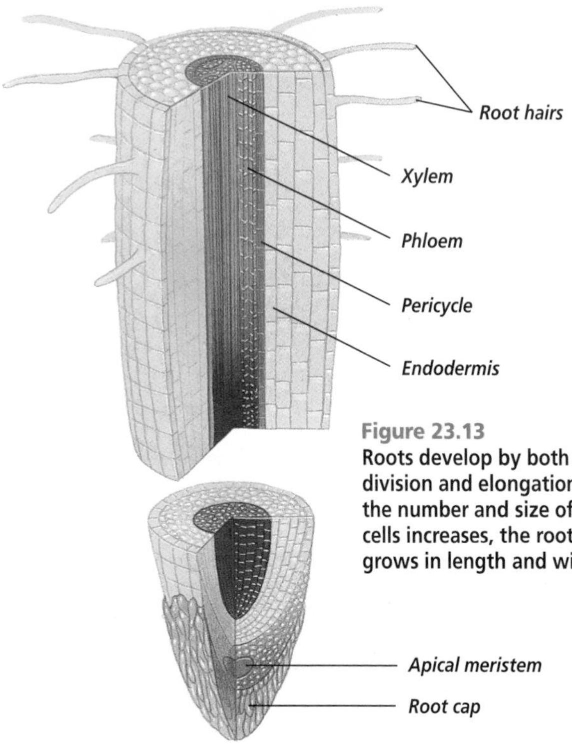
Root hairs
Xylem
Phloem
Pericycle
Endodermis
Apical meristem
Root cap

Figure 23.13
Roots develop by both cell division and elongation. As the number and size of cells increases, the root grows in length and width.

Figure 23.14
Plants can use food stored in stems to survive when conditions are less than ideal.

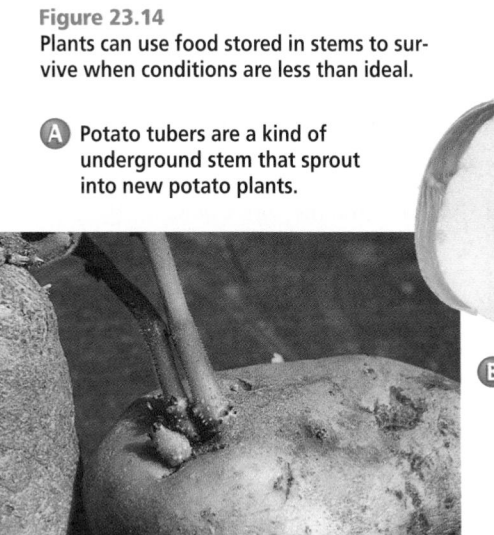

A Potato tubers are a kind of underground stem that sprout into new potato plants.

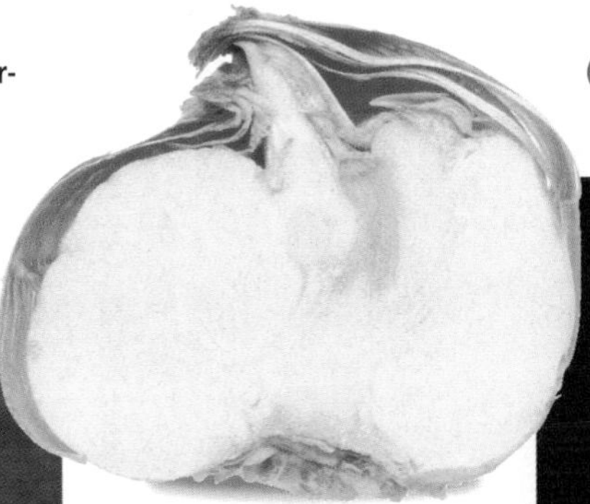

B A corm of a gladiolus is a thickened, underground stem from which roots, leaves, and flower buds arise.

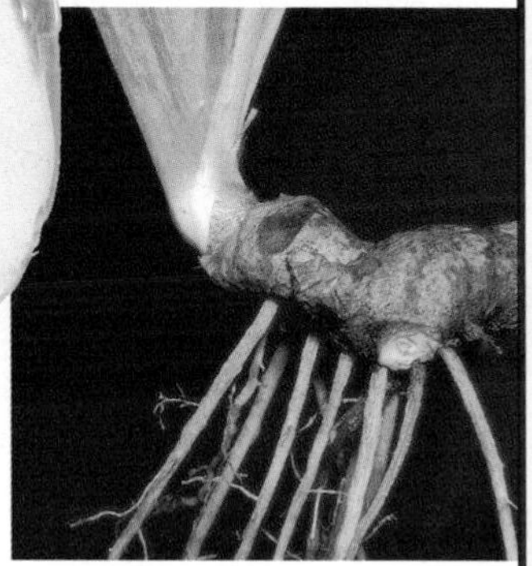

C The rhizome of an iris is an underground stem.

Enrichment

Logical-Mathematical Hydroponics is an agricultural process in which plants are grown in a nutrient solution rather than soil. Commercial hydroponics was begun in an attempt to reduce famine and eliminate the need for farmers to contend with poor soil conditions and constant weather changes. Unfortunately, hydroponics has not worked well on a large scale.

Ask students to research hydroponics, make a list of all the necessary nutrients and supplies required for a hydroponics garden, and to evaluate the pros and cons of this growth method. L3

TECHPREP

Which Fertilizers Work Best?

Kinesthetic Use liquid fertilizer to prepare solutions of the following percentages: 100, 80, 60, 40, 20, and 0. For preparation instructions, see page 32T of the Teacher Guide. Assign each student a different concentration. Have students plant seeds (turnip or mustard seeds work well) in film canisters filled with moist potting soil. Place film canisters in a warm sunny location or under fluorescent lights. Once a week, students should add 5 mL of their assigned fertilizer concentration. Students should water their plants as needed, make observations of plant height and appearance, and record observations in individual or class journals. After several weeks, students should compile their observations and draw conclusions about the effects of varying amounts of fertilizer on plant growth. L2 ELL

Activity

 **Kinesthetic** Provide thin slices of tree trunk sections to student groups. Have students count the rings to determine the age of their sample. Ask them to mark with a labeled pin the ring that represents the year when most group members were born, assuming that the tree was cut this year. **L2 ELL**

Microscope Activity

 **Visual-Spatial** Prepared slides of root cross sections are available from biological supply houses. Allow students to view such slides to help them become familiar with the various tissue names and their locations. Encourage students to sketch their observations. **L1 ELL**

Resource Manager

Critical Thinking/Problem Solving, p. 23 **L3**

Figure 23.15
One of the primary differences between roots and stems is that stems have the vascular bundles arranged in a circular pattern.

A The vascular bundles in a monocot are scattered throughout the stem.

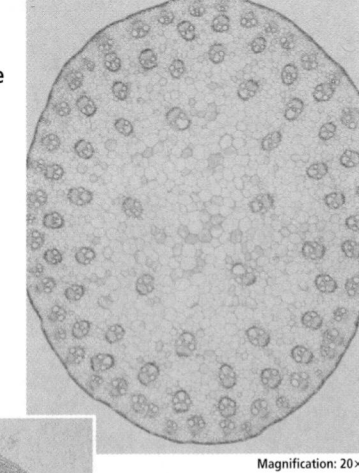

Magnification: 20×

Magnification: 20×

B In young herbaceous dicot stems, discrete bundles of xylem and phloem form a ring. In older stems, the vascular tissues form a continuous cylinder.

Figure 23.16
The inner portion of the trunk of a tree is composed primarily of dead xylem cells from the growth of the previous year.

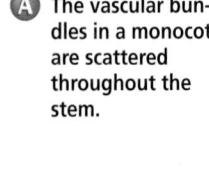

Annual Growth Rings

Cork

Phloem

Vascular Cambium

Xylem

Internal structure

Both stems and roots have vascular tissues. However, the vascular tissues in stems are arranged differently from that of roots. Stems have a bundled arrangement or circular arrangement of vascular tissues within a surrounding mass of parenchyma tissue. As you can see in *Figure 23.15*, monocots and dicots differ in the arrangement of vascular tissues in their stems. In dicots, xylem and phloem are in a circle of vascular bundles that form a ring in the cortex. The vascular bundles of monocots are scattered throughout the stem.

Woody stems

Many conifers and perennial dicots produce thick, sturdy stems, as shown in *Figure 23.16*, that may last several years, or even decades. As the stems of these plants grow in height, they also grow in thickness. This added thickness, called secondary growth, results from cell division in the vascular cambium of the stem. The xylem tissue produced by secondary growth is also called wood. In temperate regions, a tree's annual growth rings are the layers of vascular tissue produced each year by secondary growth. These annual growth rings can be used to determine the age of the plant. The vascular tissues often contain sclerenchyma fibers that provide support for the growing plant.

As secondary growth continues, the outer portion of a woody stem develops bark. Bark is composed of phloem cells and the cork cambium. Bark is a tough, corky tissue that protects the stem from damage by burrowing insects and browsing herbivores.

Stems transport materials

Water, sugars, and other organic compounds are transported through the stem. Xylem transports water from

---

✓ Portfolio

A Tree Has Good Years and Bad Years

Visual-Spatial Annual ring thickness is an indication of annual growing conditions. For example, good growing conditions yield a wider ring than years when conditions are poor. Have students use this information to make simple stylized diagrams of a woody stem cross section that illustrates years of both good and poor growth. Have them label their diagrams to show which ring corresponds to which type of growing condition. Encourage students to include additional labels that identify the general location of bark, phloem, cambium, and xylem. **L2 P**

the roots to the leaves. The water that is lost through the leaves is continually replaced by water moving up the xylem. The water molecules form an unbroken column within the xylem. As this water moves up through the xylem, it also carries minerals that are needed by all living plant cells.

Phloem transports sugars, minerals, and hormones throughout the plant. The source of these sugars is photosynthetic tissue, which is mostly in the leaves in most kinds of plants. Any portion of the plant that stores these sugars is called a **sink,** such as the parenchyma cells that make up the cortex in the root. The movement of sugars from the leaves through the phloem is called **translocation** (trans loh KAY shun). *Figure 23.1*7 shows the movement of materials through the vascular tissues of a plant.

Growth of the stem

Primary growth in a stem is similar to primary growth in a root. This increase in length is due to the production of cells by the apical meristem, which lies at the top of a stem. As mentioned earlier, secondary growth or an increase in diameter is caused by production of cells by the vascular cambium or lateral meristem. Additional meristems located at intervals along the stem give rise to leaves and branches.

Leaves

The primary function of the leaves is to trap light energy for photosynthesis. Most leaves have a relatively large surface area so they can receive plenty of sunlight. They are also often flattened, so sunlight can penetrate to the photosynthetic tissues just beneath the leaf surface.

Leaf variation

When you think of a leaf, you probably think only of a flat, broad, green structure known as the leaf blade. In fact, sizes, shapes, and types of leaves vary enormously. The giant Victoria water lily that inhabits the rivers of Guyana has leaves that may grow more than two meters in diameter.

Figure 23.17
Xylem carries water up from roots to leaves. Phloem transports sugars from the source in the leaves to sinks located throughout the plant.

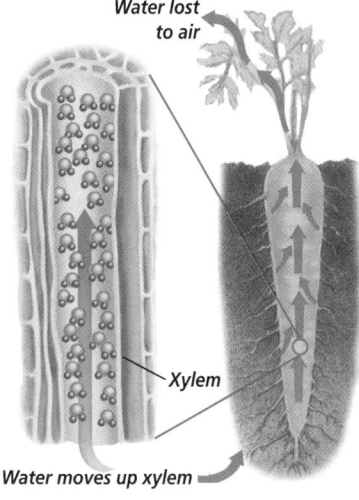

Water lost to air

Xylem

Water moves up xylem

A The open ends of xylem vessel cells form complete pipelike tubes.

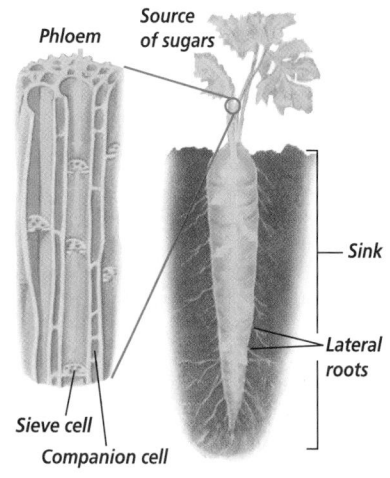

Phloem

Source of sugars

Sink

Lateral roots

Sieve cell

Companion cell

B Sugars in the phloem of this carrot plant are moving to sinks.

Figure 23.18
Leaf shapes vary, but most are adapted to receive sunlight.

Ⓐ The leaves of the walnut are compound with many leaflets.

Ⓑ The arrangement of the needlelike leaves of the evergreen yew allows them to receive the maximum amount of sunlight.

Ⓒ The tulip poplar is a deciduous tree with broad, distinctive, simple leaves.

Figure 23.19
The tissues of a leaf are adapted for photosynthesis, gas exchange, limiting water loss, and transporting water and sugars.

Cuticle
Upper epidermis
Palisade mesophyll
Vascular bundle
Xylem
Phloem
Lower epidermis
Spongy mesophyll
Stomata
Guard cells

The leaves of the tiny duckweed, a common plant of ponds and lakes, are measured in millimeters. Ferns, pines, and flowering plants commonly produce different forms of leaves on the same plant.

Some leaves, such as grass blades, are joined directly to the stem. In other leaves, a stalk joins the leaf blade to the stem. This stalk, which is part of the leaf, is called the **petiole** (PET ee ohl). The petiole contains vascular tissues that extend from the stem into the leaf to form veins. If you look closely, you will notice these veins as lines or ridges running along the leaf blade.

Leaves vary in their shape and arrangement on the stem. A simple leaf is a single leaf with a blade that is not divided. When the blade is divided into leaflets, it is called a compound leaf. Leaves also vary in their arrangement on a stem. When only one leaf is present at each point of the stem, the leaves are arranged in an alternate pattern. When two leaves are arranged in pairs along the length of the stem, the leaves form an opposite pattern. The pairs of leaves may also alternate in position along the stem forming opposite and alternate arrangements. Three or more leaves occurring at the same place on the stem are said to be whorled. *Figure 23.18* gives some examples of the variety of leaf shapes and arrangements.

Leaf structure

The internal structure of a typical leaf is shown in *Figure 23.19.* The vascular tissue of the leaf is located in the veins that run through the midrib and veins of the leaf. Just beneath the epidermal layer are two layers of mesophyll. **Mesophyll** (MEZ uh fihl) is the photosynthetic tissue of a leaf. It is usually made up of two types of parenchyma cells—palisade mesophyll and spongy mesophyll. The palisade mesophyll is made up of column-shaped cells containing many chloroplasts. These cells are found just under the upper epidermis, allowing for maximum exposure to the sun. Most photosynthesis takes place in the palisade mesophyll. Below the palisade mesophyll is the spongy mesophyll, which is composed of loosely packed, irregularly shaped cells. These cells are surrounded by many air spaces. These air spaces allow carbon dioxide, oxygen, and water vapor to freely flow around the cells. These gases can also move in and out of the stomata, which are located in the dermal layer.

Transpiration

You read previously that leaves have an epidermis with a waxy cuticle and stomata that help prevent water loss. Guard cells are cells that surround and control the size of the opening in stomata, *Figure 23.20.* The loss of water through the stomata

Figure 23.20
Guard cells regulate the size of the opening of the stomata according to the amount of water in the plant.

Problem-Solving Lab 23-2   Drawing Conclusions

What factors influence the rate of transpiration? Plants lose large amounts of water during transpiration. This process aids in pulling water up from roots to stem to leaves where it is needed for photosynthesis.

Analysis

A student was interested in seeing if a plant's surroundings might affect its rate of water loss. A geranium plant was set up as a control. A second geranium was sealed within a plastic bag and a third geranium was placed in front of a fan. All three plants were placed under lights. The student's experimental data are shown in the graph.

Rate of Water Loss

Water loss (vertical axis), Time (horizontal axis)
Lines A, B, C

Thinking Critically

1. Which line, A, B, or C, might best represent the student's control data? Explain.
2. a. Which line might best represent the data with the plant sealed within a bag? Explain.
   b. What abiotic environmental factor was being tested?
3. Which line might best represent the data with the plant in front of a fan? Explain.
4. Write a conclusion for the student's experiment.

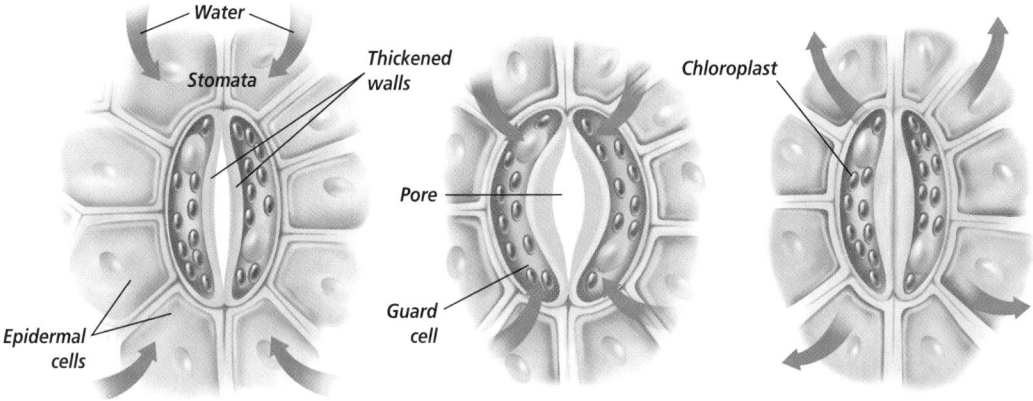

A The guard cells have a thickened inner wall.

B When water enters the guard cells, the pressure causes them to bow out, opening the pore.

C As water leaves the guard cells, the pressure is released and the cells sag together, closing the pore.

Labels: Water, Stomata, Thickened walls, Chloroplast, Pore, Guard cell, Epidermal cells

Purpose

Students will observe stomata and determine how a salt solution affects stomata.

Process Skills

observe and infer, compare and contrast, use the microscope

Safety Precautions

Remind students to be careful when working with microscopes, glass slides, and coverslips.

Teaching Strategies

■ Geranium, *Coleus*, or *Tradescantia* leaves work well. If these are unavailable, purchase fresh spinach.

■ The epidermis will appear as a clear strip of tissue at the jagged torn edge. Have students use the lower epidermis of the leaf.

Expected Results

Chloroplasts will be seen only in guard cells. The saltwater mount will show closed guard cells. Thus, stomata will appear closed in comparison to the plain water wet mount.

Analysis

1. Cells look like interlocked puzzle pieces. The cells are protective.
2. Guard cells are sausage-shaped and have chloroplasts. Epidermal cells do not have chloroplasts and are irregular in shape.
3. Stomata are closed in the salt solution. The higher water concentration inside the cells compared with that outside of the cells caused water to move out of the guards cells. This causes the guard cells to collapse and close the stomata.

✔ Assessment

Skill Provide students with a diagram of leaf epidermis cells, guard cells, and stomata. Ask them to make a display in which they label each cell type and indicate which cells are capable of carrying out photosynthesis. Use the Performance Task Assessment List for Display in **PASC**, p. 63. L2

640

MiniLab 23-2 Observing

Magnification: 3240×

Looking at Stomata The lens-shaped openings in the epidermis of a leaf allow gas exchange and help control water loss.

Procedure

1. Make a wet mount by tearing a leaf at an angle to expose a thin section of epidermis. Use tap water to make wet mounts of both the upper and lower epidermis.
2. Examine each of your slide preparations under the microscope. Draw or take down a written description of what you see. **CAUTION: *Use caution when working with a microscope, microscope slides, and coverslips.***
3. Make another wet mount using a 5 percent salt solution instead of tap water. Examine the slide under the microscope and record your observations.

Stomata

Analysis

1. What do the cells of the leaf epidermis look like? What is their function?
2. How do the epidermal cells differ from guard cells? Which cells, if any, contain chloroplasts?
3. What differences in the stomata did you notice when you used a salt solution to prepare your wet mount? Can you explain what happened in terms of osmosis?

Figure 23.21
Leaf venation patterns help distinguish between monocots and dicots. Leaves of corn plants have parallel veins, a characteristic of many monocots (**a**). Leaves of lettuce plants are net veined, a characteristic of many dicots (**b**).

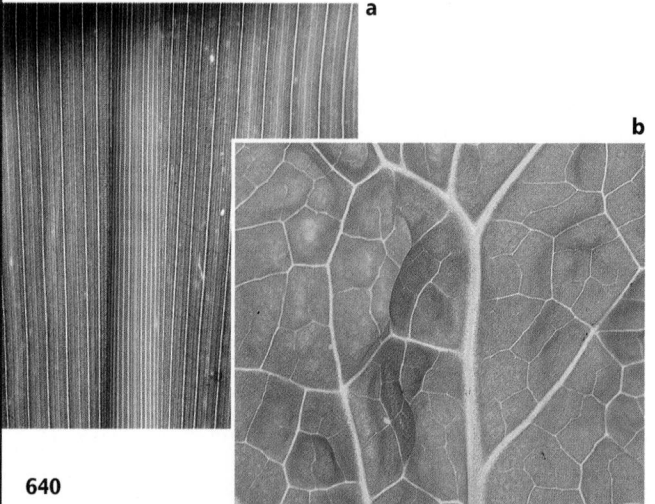

a

b

640

---

is called **transpiration.** You can learn more about how a plant's surroundings may influence transpiration in the *Problem-Solving Lab* on the previous page.

The opening and closing of guard cells regulate transpiration. As you read about how guard cells work, be sure to look carefully at the diagrams in *Figure 23.20.* Guard cells are cells scattered among the cells of the epidermis. The inner walls of these cells are thickened more than the outer walls. When there is plenty of water available in surrounding cells, guard cells take in water by osmosis. When water enters, the thicker inner walls prevent the guard cells from expanding in width, so they expand in length. Because the two guard cells are attached at either end, this expansion in length forces them to bow out and the pore opens. When water is not readily available, there is less water in tissues surrounding the guard cells. Water leaves the guard cells, thus lowering their cell pressure. The cells return to their previous shape, reducing the size of the stomatal pore. The proper functioning of guard cells is important because plants lose up to 90 percent of all the water they transport from the roots through transpiration. You can learn more about the structure and function of stomata in the *MiniLab* shown here.

Venation patterns

One way to distinguish among different groups of plants is to examine the pattern of veins in their leaves. The veins of vascular tissue run through the mesophyll of the leaf. As shown in *Figure 23.21,* leaf venation patterns may be parallel, as in many monocots, or netlike, as in most dicots. *Table 23.1* summarizes the many differences among the tissues of monocots and dicots.

MEETING INDIVIDUAL NEEDS

Learning Disabled

 ***Visual-Spatial*** Have students sequence the steps involved in the opening and closing of stomata. Sequencing each step will help learning disabled students to understand the overall functioning of the stomata. L1

Resource Manager

BioLab and MiniLab Worksheets, p. 106 L2
Reinforcement and Study Guide, pp. 102-103 L2
Content Mastery, pp. 114-115 L1
Laboratory Manual, pp. 167-170 L2

Table 23.1 Distinguishing characteristics of monocots and dicots

	Seed leaves	Veins in leaves	Vascular bundles in stems	Flower parts
Monocots	one cotyledon	usually parallel	scattered	multiples of threes
Dicots	two cotyledons	usually netlike	arranged in ring	multiples of fours and fives

Leaf modifications

Many plants have leaves that are modified for functions other than photosynthesis, such as protection and food storage. Cactus spines are modified leaves that reduce water loss and protect the plant from herbivores. A bulb is a short stem covered by enlarged, fleshy leaf bases. The first flowers of spring usually bloom from bulbs. The fleshy leaf bases of onion and daffodil bulbs are modified for food storage. The leaves of some plants are even modified to catch insects. Examples of modified leaves are shown in *Figure 23.22*.

Figure 23.22
Modified leaves serve many functions in addition to photosynthesis.

Ⓐ The leaves of the pitcher plant are modified for trapping insects.

Ⓑ The leaves of this *Aloe vera* plant are adapted to store water in a dry desert environment.

Section Assessment

Understanding Main Ideas
1. Compare and contrast the arrangement of xylem and phloem in dicot roots and stems.
2. In a plant with leaves that float on water, such as a water lily, where would you expect to find stomata? Explain.
3. What is the primary function of most leaves? What are some other functions of leaves?
4. Explain how guard cells regulate the size of the stomatal pore.

Thinking Critically
5. Compare and contrast the function and structure of the epidermis and the endodermis in a vascular plant.

SKILL REVIEW
6. **Making and Using a Table** Construct a table that summarizes the structure and functions of roots, stems, and leaves. For more help, refer to *Organizing Information* in the **Skill Handbook.**

Section Assessment

1. In dicot roots, xylem forms a star-shaped figure in the center with phloem between the star rays. In stems, xylem and phloem are arranged in bundles that form a circle with the phloem outside of the xylem.
2. Stomata would be on the upper surface that is exposed to the air. This will allow for gas exchange with the air.
3. The primary function is to trap light energy for photosynthesis. Some are modified for storage and protection.
4. When there is plenty of water, the guard cells swell and bow out, opening the stomata.
5. The endodermis and epidermis both control the movement of water. The endodermis is located within the root, whereas the epidermis is the external covering of a plant.
6. Roots anchor the plant in the ground, absorb water and minerals and transport material up to the stem. Stems provide support and transport water, mineral ions and sugars to and from roots and leaves. Leaves trap light energy for photosynthesis.

641

Prepare

Key Concepts

In this section the major types of plant hormones are described and their effects on plant growth are discussed. Tropic and nastic plant responses are explained.

Planning

- Locate onion root tip slides for Microscope Activity.
- Obtain plants and gibberellic acid for the Meeting Individual Needs - Gifted.

1 Focus

Bellringer

Before presenting the lesson, display **Section Focus Transparency 57** on the overhead projector and have students answer the accompanying questions.
L1 ELL

SECTION PREVIEW

Objectives
Identify the major types of plant hormones.
Analyze the different types of plant responses.

Vocabulary
hormone
auxin
gibberellin
cytokinin
ethylene
tropism
nastic movement

Section

23.3 Plant Responses

Plants cannot laugh or cry. They do not exhibit behaviors you commonly see in animals, but they do react to their environment. They move in response to light and gravity. The flowering heads of sunflowers can be seen to turn in response to the movement of the sun as it moves across the sky. If a plant is growing in a dark forest and a tree falls allowing in more light, the plant will grow towards the light.

Sunflowers

Plant Hormones

Plants, like animals, have hormones that regulate growth and development. A **hormone** is a chemical that is produced in one part of an organism and transported to another part, where it causes a physiological change. Only a small amount of the hormone is needed to make this change.

Figure 23.23
Auxin from the tip of the main shoot inhibits the growth of side branches (a). Once the main tip is removed, the side branches start to grow (b).

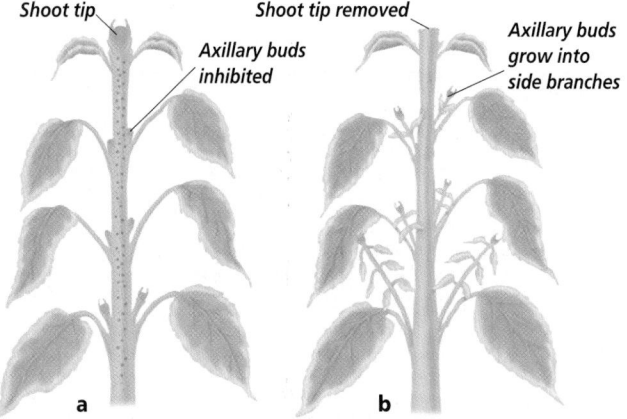

Shoot tip

Axillary buds inhibited

Shoot tip removed

Axillary buds grow into side branches

a   b

Auxins cause stem elongation

The group of plant hormones called **auxins** (AWK sunz) promote cell elongation. Indoleacetic acid (IAA) is a naturally occurring auxin that is produced in the apical meristems of a growing plant stem. It causes an increase in stem length by increasing the rate of cell division and promoting cell elongation. IAA weakens the connections between the cellulose fibers in the cell wall. This allows the cells to stretch and grow longer. The combination of new cells and increasing cell lengths leads to stem growth. Auxin is not transported in the vascular system, but rather it moves from one parenchyma cell to the next by active transport.

Auxins have a number of other effects on plant growth and development. Auxin produced in the apical meristem inhibits the growth of side branches.

Removing the stem tip reduces the amount of auxin present in the stem

642 PLANT STRUCTURE AND FUNCTION

BIOLOGY JOURNAL

Auxins and Stem Elongation

Linguistic Ask students to write a paragraph explaining how auxin causes stem elongation at the cellular level. Encourage students to draw pictures to illustrate their paragraph. **L2**

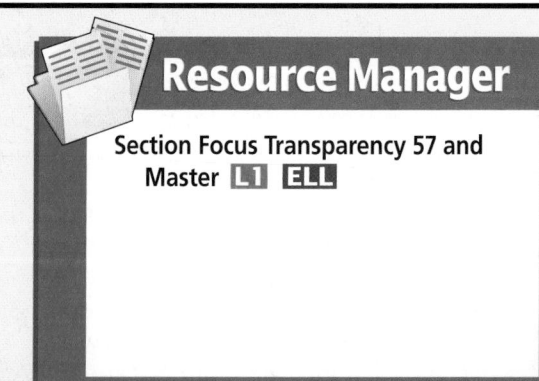

Resource Manager

Section Focus Transparency 57 and Master **L1 ELL**

and allows the formation of branches as they are no longer inhibited by auxins at the tip of the main stem, *Figure 23.23*. High concentrations of auxin also promote fruit formation and inhibit the dropping of fruit from the plant. When auxin concentrations decrease, the ripened fruits of some trees fall to the ground and deciduous trees begin to shed their leaves.

Gibberellins promote growth

The group of plant hormones called **gibberellins** (jihb uh REL uns) are growth hormones that cause plants to grow taller because, like auxins, they stimulate cell elongation. Many dwarf plants, such as the dwarf bean plants in *Figure 23.24,* are short because the plant either cannot produce gibberellins or its cells are not receptive to the hormone. If gibberellins are applied to the tip of the dwarf plant, it will grow taller. Gibberellins also increase the rate of seed germination and bud development. Farmers have learned to use gibberellins to enhance fruit formation. Florists often use gibberellins to induce flower buds to open.

Cytokinins stimulate cell division

The hormones called **cytokinins** are so named because they stimulate cell division or cytokinesis. Cytokinins increase cell division by stimulating the production of proteins needed for mitosis. Most cytokinins are produced in the meristems in the root. This hormone travels up the xylem to other parts of the plant. The effect of cytokinins is often enhanced by the presence of other hormones.

Ethylene gas promotes ripening

The plant hormone **ethylene** (ETH uh leen) is a simple, gaseous compound composed of carbon and hydrogen that speeds the ripening of fruits.

Figure 23.24
The bean plants in this picture are genetic dwarfs. However, the two plants on the right were treated with gibberellin and have grown to a normal height.

It is produced primarily by fruits, but also by leaves and stems. Ethylene is released during a specific stage of fruit ripening. It causes the cell walls to weaken and become soft. Ethylene also promotes the breakdown of complex carbohydrates to simple sugars. If you have ever enjoyed a ripe red apple you know that it tastes sweeter than an immature fruit.

Many farmers use ethylene to ripen fruits or vegetables that have been picked when they are immature as shown in *Figure 23.25*.

WORD *Origin*

auxin
From the Greek word *auxein*, meaning "to increase." Auxin causes stem elongation by increasing cell growth

Figure 23.25
Tomatoes are usually picked when they are green. Once they have reached their destination, they are treated with ethylene. Most of the ripe red tomatoes you see in the grocery stores have been ripened in this manner.

2 Teach

Quick Demo

Visual-Spatial Show students broadleaf houseplants that have been on a sunny windowsill. Ask them if they notice anything special about the leaves. Direct their attention to the orientation of the leaves. Ask students to propose ideas as to why the leaves are all facing the same direction. **L1**

✔ Assessment

Skill Have students construct a table of the different types of plant hormones and their effects on plants. **L2**

Microscope Activity

Visual-Spatial Students can view the numerous cells in various stages of mitosis in onion root tip slides. Remind students that cytokinins produced in the root tip stimulate this cell division. **L1**

✔ Assessment

Performance Assessment in the Biology Classroom, p. 47, *Controlling the Rate at Which Fruit Ripens.* Have students carry out this activity to observe plant hormones at work. **L2**

MEETING INDIVIDUAL NEEDS

Gifted
Have students research and design an experiment to test the effect of different concentrations of gibberellins on dwarf plants. Dwarf peas and the rosette variety of Wisconsin Fast Plants work well for this type of experiment. **L3**

Purpose

Students will conclude from experimental evidence that light and the tip end of a young stem are needed for phototropism.

Process Skills

predict, think critically, analyze information, compare and contrast, draw a conclusion, hypothesize, interpret data, recognize cause and effect

Teaching Strategies

■ Explain to students that the seed in each diagram is below the soil surface.
■ Define the term "opaque" if necessary.
■ Use the term "coleoptile" to refer to the young oat stem.

Thinking Critically

1. A and C; if there is no light source or the stem tip is not able to detect light, there is no phototropic response.
2. Student answers may vary. When the tip is intact (as in A), there is a phototropic response as the stem bends towards light. However, when the tip end is missing, there is no phototropic response.
3. Stem tip end; when the tip end is removed, phototropism does not occur.

✔ Assessment

Portfolio Ask students to design an experiment that would show that the tip end of the young stem does indeed contain a chemical that influences the bending of the stem toward light. Students should place their written experiment in their portfolio. Use the Performance Task Assessment List for Assessing a Whole Experiment and Planning the Next Experiment in **PASC**, p. 33. **L3**

644

Problem-Solving Lab 23-3   Drawing a Conclusion

How do plant stems respond to light? While working with young oat plants, Charles Darwin made a number of discoveries about the response of young plant stems to light. His discoveries helped to explain why plants undergo phototropism. Scientists now know that this response is the result of an auxin that causes rapid cell elongation to occur along one side of a young plant stem. However, the discovery of auxins was not yet known during Darwin's time.

Analysis

Study the before and after diagrams marked A-C. The three plants are young oat stems that are already above ground. Note that the light source is directed at the plants from one side.

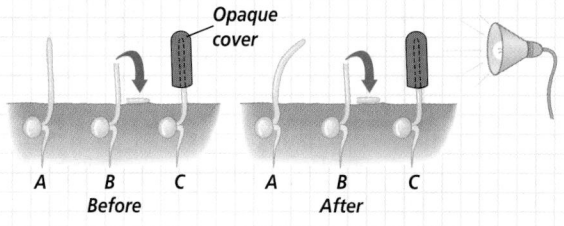

Opaque cover

A   B   C   A   B   C
Before   After

Thinking Critically

1. Which diagram (or diagrams) supports the conclusion that light is a needed factor for phototropism? Explain.
2. Which diagram (or diagrams) supports the conclusion that the stem tip is a needed factor for phototropism? Explain.
3. Where might the auxin responsible for phototropism be produced? Explain.

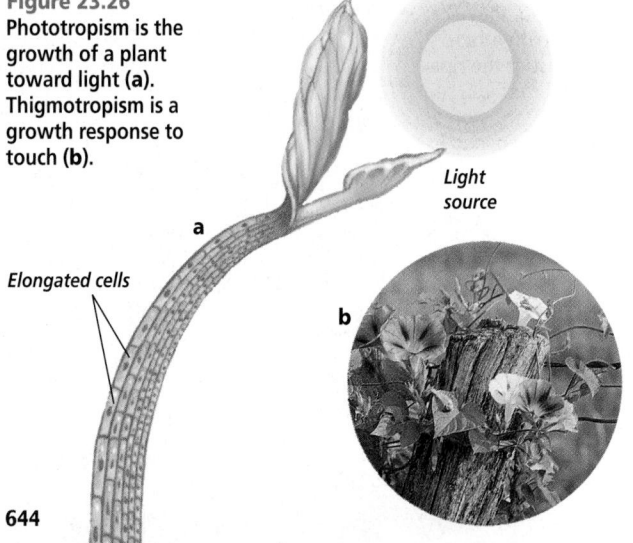

Figure 23.26
Phototropism is the growth of a plant toward light (a). Thigmotropism is a growth response to touch (b).

Light source

a

Elongated cells

b

644

Plant Responses

 **Visual-Spatial** Have students collect pictures or drawings that show different plant responses. Have them glue or tape the pictures onto construction paper and write captions for each one. Students should place their completed pictures or drawings in their portfolios. **L1 ELL**
P

Plant Responses

Why do roots grow down into the soil and stems grow up into the air? Although plants lack a nervous system and usually cannot make quick responses to stimuli, they do have mechanisms that enable them to respond to their environment. Plants grow, reproduce, and shift the position of their roots, stems, and leaves in response to environmental conditions such as gravity, sunlight, temperature, and day length.

Tropic responses in plants

If you look at the photograph of sunflowers at the beginning of this section, it is obvious that they are all responding to the same stimulus—the sun. **Tropism** is a plant's response to an external stimulus that comes from a particular direction. If the tropism is positive, the plant grows toward the stimulus. If the tropism is negative, the plant grows away from the stimulus.

The growth of a plant towards light is caused by an unequal distribution of auxin in the plant's stem. There is more auxin on the side of the stem away from the light. This results in cell elongation, but only on that side. As the cells grow, the stem bends toward the light, as shown in *Figure 23.26a*. The growth of a plant toward light is called phototropism. You can learn more about phototropism in the *Problem-Solving Lab* shown here.

There is another tropism associated with the upward growth of stems and the downward growth of roots. Gravitropism is the direction of plant growth in response to gravity. Gravitropic responses are beneficial to plants because the leaves receive more light if they grow upward. By growing down into the soil, roots are able to anchor the plant and can take in water and minerals.

📰 Resource Manager

Reinforcement and Study Guide,
p. 104 **L2**
Content Mastery, pp. 113, 115-116 **L1**

a

b

Some plants exhibit another tropism called thigmotropism, which is a growth response to touch. The tendrils of the vine in *Figure 23.26b* have coiled around a trellis after making contact during early growth.

Because tropisms involve growth, they are not reversible. The position of a stem that has grown several inches in a particular direction cannot be changed. But, if the direction of the stimulus is changed, the stem will begin growing in another direction.

Nastic responses in plants

A responsive movement of a plant that is not dependent on the direction of the stimulus is called a **nastic movement.** An example of a nastic movement is the folding up of the leaflets of a *Mimosa* leaf when the plant is touched, as shown in *Figure 23.27a.* The folding is caused by a change in turgor pressure in the cells at the base of each leaflet. A dramatic drop in pressure causes the cells to become limp. This causes the leaflets to change orientation.

Another example of a nastic response is the sudden closing of the hinged leaf of a Venus flytrap, *Figure 23.27b.* The movement of an insect on the leaf triggers the sensitive hairs on the inside of the leaf, causing the trap to snap shut. Plant responses that are due to changes in cell pressure are reversible because they do not involve growth. The *Mimosa* and Venus flytrap's leaves open once the stimulus ends.

Figure 23.27
Mimosa pudica is also known as the sensitive plant (**a**). When touched, it folds its leaves in less than one-tenth of a second. The Venus flytrap's hinged leaf snaps shut when triggered by an insect crawling on its leaf (**b**).

Section Assessment

Understanding Main Ideas
1. What is a hormone?
2. What are two differences between tropic responses and nastic movements?
3. Explain how a plant can bend towards the sun. What term describes this response?
4. Name one hormone and describe how it influences growth and development.

Thinking Critically
5. One technique that has been used for years to ripen fruit has been to put a ripened banana in a paper bag with the unripe fruit. Why does this help the unripe fruit to ripen?

SKILL REVIEW
6. **Designing an Experiment** Explain how you would design an experiment to test the effect of different colors of light on phototropism in one plant. For more help, refer to *Practicing Scientific Methods* in the **Skill Handbook.**

3 Assess

Check for Understanding
Have students explain how the words in each of the following pairs are related. **L2**
  a. auxin—hormone
  b. cytokinin—cell division
  c. ethylene—ripe fruit
  d. tropic response—nastic response

Reteach
Linguistic Have students write sentences correctly using each of the groups of words in Check for Understanding. **L2**

Extension
Have students do research on the use of plant hormones in agriculture. **L3**

✔ Assessment
Knowledge Show students pictures of different plant responses and ask them to identify the type of response. **L1**

4 Close

Biology Journal
Ask students to write a paragraph describing the relationship between plant hormones and plant responses. **L2**

Section Assessment

1. A hormone is a chemical produced in one part of an organism and transported to another part, where it then causes a physiological change.
2. Nastic movements are reversible, whereas tropic responses are not. Nastic movements are not dependent upon the direction of the stimulus, whereas tropic responses are.
3. A plant can bend towards the light as a result of cell elongation on the side of the stem away from the light. The term phototropism describes this response.
4. Possible answers: auxins cause stem elongation; gibberellins promote growth; cytokinins stimulate cell division; and ethylene gas promotes fruit ripening.
5. Ripe fruits release ethylene gas that stimulates the ripening of other fruits.
6. Expose the plant to red light for 24 hours and measure the response. Next expose the plant to blue light for 24 hours. Continue testing until all available light colors have been tested and compare results.

Determining the Number of Stomata on a Leaf

Time Allotment
One class period

Process Skills
collect data, communicate, interpret data, recognize cause and effect, draw a conclusion, compare and contrast, measure in SI, observe and infer, use numbers

Safety Precautions
Review the need for care when working with single-edged razor blades. Remind students to cut away from their bodies. Remind students that special care should be taken when viewing slides under high-power so the objective does not break the slide. Have students wash their hands after handling plant materials.

If asked to count the total number of stomata on a single leaf, you might answer by saying "that's an impossible task." It may not be necessary for you to count each and every stomate. Sampling is a technique that is used to arrive at a close answer to the actual number. You will use a sampling technique in this BioLab.

PREPARATION

Problem
How can you count the total number of stomata on a leaf?

Objectives
In this BioLab you will:
- **Measure** the area of a leaf.
- **Observe** the number of stomata seen under a high-power field of view.
- **Calculate** the total number of stomata on a leaf.

Materials

microscope	ruler
glass slide	glass cover
water and dropper	
green leaf from an onion plant	
single-edged razor blade	

Safety Precautions
Wear latex gloves when handling an onion.

Skill Handbook
Use the **Skill Handbook** if you need additional help with this lab.

PREPARATION

Alternative Materials

- Any type of leaf may be substituted. Green onion is ideal because it is inexpensive and easily available in supermarkets. It also is easily scraped with the razor blade to yield a clear epidermis.

- Other leaves may be used. Students will have to calculate the area of differently shaped leaves. Step 3 will no longer be valid if other leaves are used. Geranium and Ficus leaves also work well.

PROCEDURE

1. Copy Data Tables 1 and 2.
2. Obtain an onion leaf and carefully cut it open lengthwise using a single-edged razor blade. **CAUTION:** *Be careful when cutting with a razor blade.*

Data Table 1

Trial	Number of stomata
1	
2	
3	
4	
5	
Total	
Average	

3. Measure the length and width of your onion leaf in millimeters. Record these values in Data Table 2.
4. Remove a small section of leaf and place it on a glass slide with the dark green side facing DOWN.
5. Add several drops of water and gently scrape away all green leaf tissue using a back and forth motion with the razor blade. An almost transparent layer of leaf epidermis will be left on the slide.
6. Add water and a cover glass to the epidermis and observe under low-power magnification.

PROCEDURE

Teaching Strategies
- You may find that students will need to use a calculator.
- Review the procedure for determining an average.
- Students may want to know where the value of 0.07 mm^2 came from. This value is close to the area of the circle of light seen through high-power magnification.

- You may wish to go through all of the math steps by using sample data on the overhead projector.

Troubleshooting
- Students may have difficulty observing the stomata. Most problems are the result of not having scraped away enough of the spongy layer of the leaf.

Locate an area where guard cells and stomata can be clearly seen. **CAUTION:** *Use caution when working with a microscope, microscope slides, and coverslips.*

7. Switch to high-power magnification.

8. Count and record the number of stomata in your field of view. Consider this trial 1. Record your count in Data Table 1.

9. Move the slide to a different area. Count and record the number of stomata in this field of view. Consider this trial 2.

10. Repeat step 9 three more times. Calculate the average number of stomata observed in a high-power field of view.

11. Calculate the total number of stomata on the entire onion leaf by following the directions in Data Table 2.

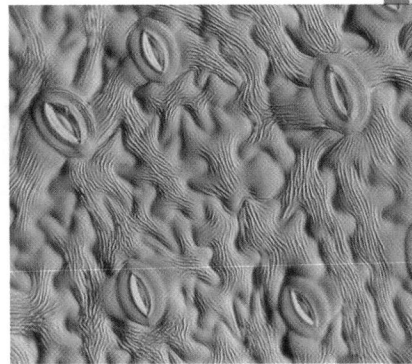

Magnification: 742×

Data Table 2

Length of leaf portion in mm	= ____ mm
Width of leaf portion in mm	= ____ mm
Calculate area of leaf (length × width)	= ____ mm²
Calculate number of high-power fields of view on leaf (area of leaf ÷ 0.07 mm², the area of one high-power field of view)	= ____
Calculate total number of stomata (number of high-power fields of view × average number of stomata per high-power field of view from Data Table 1)	= ____

ANALYZE AND CONCLUDE

1. **Communicating** Compare your data with those of your other classmates. Offer several reasons why your total number of stomata for the leaf may not be identical to your classmates.

2. **Thinking Critically** Analyze the following steps of this experiment and explain how you can change the procedure to improve the accuracy of your data.
   a. five trials in Data Table 1
   b. using 0.07 mm² as the area of your high-power field of view

3. **Concluding** Would you expect all plants to have the same number of stomata per high-power field of view? Explain your answer.

4. **Comparing and Contrasting** What are the advantages to using sampling techniques? What are some limitations?

Going Further

Application Carry out the same sampling technique to determine the total number of stomata present on leaves of a variety of different plant species in your neighborhood.

BIOLOGY Online To find out more about plant anatomy, visit the Glencoe Science Web site.

science.glencoe.com

INVESTIGATE BioLab

ANALYZE AND CONCLUDE

1. mathematical errors, different average numbers of stomata, different leaf sizes

2. a. increase the number of samples
   b. calculate microscope's high-power area

3. Different plants will have different numbers of stomata per high-power field; stomate numbers are characteristic for specific plant species.

4. Saves time and energy; you do not get an actual or true count, only an approximation.

✓ Assessment

Skill Provide students with sample data for leaf area and number of stomata per high-power field. Ask students to calculate the total number of stomata. Use the Performance Task Assessment List for Using Math in Science in **PASC,** p. 29. **L3**

Going Further

Determine if there is a difference between the number of stomata on the upper and lower epidermis of selected species of trees or shrubs. **L2**

Resource Manager

BioLab and MiniLab Worksheets, pp. 107-108 **L2**

■ On occasion, the epidermis will tear when students scrape too vigorously. Have students discard their wet mounts and try a new sample of leaf material.

■ Failure to see or find any stomata may be due to the fact that students did not place their leaf sample with the dark side down. Green onions do not have stomata on their inside surface.

Data and Observations

Student answers will vary. The number of stomata will be several thousand.

Art Connection

Purpose
To illustrate the relationship between nature and art.

Teaching Strategies
■ Posters of O'Keeffe's works are available in art shops and museum gift shops. Bring in several examples to illustrate more of her work. Have students visit an art gallery to view paintings by artists who use flowers as their subject matter. As an alternative, have students examine art books to find the names of artists who use flowers. Ask students to present oral reports on their findings. **L1**

Visual Learning
Ask students what plant hormone may have influenced the flowering of the red poppy.

Connection to Biology
One possible answer is that color attracts pollinators to the flowers.

Art Connection

Red Poppy by Georgia O'Keeffe (1887–1986)

"When you take a flower in your hand and really look at it," she said, cupping her hand and holding it close to her face, "it's your world for the moment. I want to give that world to someone else. Most people in the city rush around so, they have no time to look at a flower. I want them to see it whether they want to or not."

American artist Georgia O'Keeffe attracted much attention when the first of her many floral scenes was exhibited in New York in 1924. Everything about these paintings—their color, size, point of view, and style—overwhelmed the viewer's senses, just as their creator had intended.

In describing her huge paintings of solitary flowers, Georgia O'Keeffe said: "I decided that I wasn't going to spend my life doing what had already been done." Indeed, she did do what had not been done by painting enormous poppies, lilies, and irises on giant canvases. Her use of colors and emphasis on shapes suggests nature rather than copying it with photographic realism. Her work can be described as abstract. "I found that I could say things with color and shapes that I couldn't say in any other way—things that I had no words for," she said.

The viewer's eye is drawn into the flower's heart In this early representation of one of her familiar poppies, O'Keeffe directed the viewer's eye down into the poppy's center, much as the flower naturally attracts an insect for reproduction purposes. By contrasting the light tints of the outer ring of petals with the darkness of the poppy's center, the viewer's eye is pulled, beelike, into the heart of the flower. The overwhelming size and detailed interiors of O'Keeffe's flowers give an effect similar to a photographer's close-up camera angle.

During her long life of 98 years, Georgia O'Keeffe created hundreds of paintings. Her subjects included the flowers for which she is perhaps most famous, as well as other botanical themes. She spent many years in New Mexico. Her paintings of the New Mexico deserts are characterized by sweeping forms, portraying sunsets, rocks, and cliffs.

At the age of 90, she said of her success, "It takes more than talent. It takes a kind of nerve and a lot of hard, hard work." She died in New Mexico in 1986. Georgia O'Keeffe will be remembered for her bold, vivid paintings that are, indeed, larger than life.

CONNECTION TO BIOLOGY

Color plays a prominent role in the artistic effect of O'Keeffe's flowers. What role does color play in the life of a real flower?

BIOLOGY Online To find out more about Georgia O'Keeffe, visit the Glencoe Science Web site. **science.glencoe.com**

Chapter 23 Assessment

SUMMARY

Section 23.1

Plant Cells and Tissues

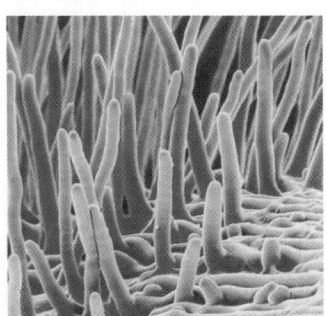

Main Ideas

- Most plant tissues are composed of parenchyma cells, collenchyma cells, and sclerenchyma cells.
  - Dermal tissue is a plant's protective covering.
  - Xylem moves water and minerals up the stem. Phloem transports sugars and organic compounds throughout the plant.
  - Ground tissue often functions in food production and storage.
  - Meristematic tissues produce new cells.

Vocabulary

apical meristem (p. 631)
collenchyma (p. 626)
companion cell (p. 630)
cork cambium (p. 631)
epidermis (p. 627)
guard cells (p. 627)
meristem (p. 631)
phloem (p. 630)
parenchyma (p. 625)
sclerenchyma (p. 626)
sieve tube member (p. 630)
stomata (p. 627)
tracheid (p. 629)
trichome (p. 627)
vascular cambium (p. 631)
vessel element (p. 629)
xylem (p. 629)

Section 23.2

Roots, Stems, and Leaves

Main Ideas

- Roots grow downward as cells elongate and transport water and minerals from the root to the rest of the plant.
- The stem supports leaves and transports food and water.
- Leaves contain chloroplasts and perform photosynthesis.

Vocabulary

cortex (p. 633)
endodermis (p. 633)
mesophyll (p. 638)
pericycle (p. 633)
petiole (p. 638)
root cap (p. 634)
sink (p. 637)
translocation (p. 637)
transpiration (p. 640)

Section 23.3

Plant Responses

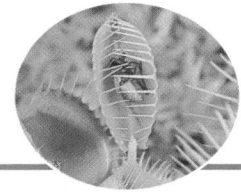

Main Ideas

- Three major plant hormones are auxins, gibberellins, and cytokinins. They promote cell division and cell elongation.
- Phototropism, gravitropism, and thigmotropism are all growth responses to external stimuli.
- Nastic responses are caused by changes in cell pressure.

Vocabulary

auxin (p. 642)
cytokinin (p. 643)
ethylene (p. 643)
gibberellin (p. 643)
hormone (p. 642)
nastic movement (p. 645)
tropism (p. 644)

Main Ideas

Summary statements can be used by students to review the major concepts of the chapter.

Using the Vocabulary

To reinforce chapter vocabulary, use the Content Mastery Booklet and activities in the Interactive Tutor for Biology: The Dynamics of Life on the Glencoe Science Web site. science.glencoe.com

All Chapter Assessment questions and answers have been validated for accuracy and suitability by The Princeton Review.

UNDERSTANDING MAIN IDEAS

1. d
2. c

UNDERSTANDING MAIN IDEAS

1. The tissue that makes up the protective covering of a plant is _____ tissue.
   a. vascular
   b. meristematic
   c. ground
   d. dermal

2. Cambium and apical meristem are examples of _____.
   a. photosynthetic tissues
   b. protective tissues
   c. growth tissues
   d. transport tissues

CHAPTER 23 ASSESSMENT **649**

GLENCOE TECHNOLOGY

VIDEOTAPE
MindJogger Videoquizzes
Chapter 23: *Plant Structure and Function*
Have students work in groups as they play the videoquiz game to review key chapter concepts.

Resource Manager

Chapter Assessment, pp. 133-142
MindJogger Videoquizzes
ExamView® Pro Software 💾
BDOL Interactive CD-ROM, Chapter 23 quiz

3. b
4. a
5. a
6. c
7. b
8. d
9. c
10. b
11. translocation
12. dicot
13. ethylene
14. water
15. gibberellin
16. parenchyma
17. auxins
18. sink
19. stem; root
20. stem

APPLYING MAIN IDEAS

21. Transpiration occurs most rapidly during the day and aids in the movement of water through the xylem. Thus, water movement is greater during the day than at night.

22. Since it is late afternoon, the sun will be in the west; the flowers will be facing west towards the light source.

3. Which of the following plant responses could be demonstrated by placing a potted plant next to a sunny window?
   a. thigmotropism   c. nastic movement
   b. phototropism   d. gravitropism

4. A cross section of a root shows a central star-shaped mass of xylem when the plant is a(n) _____.
   a. dicot   c. annual
   b. monocot   d. perennial

5. One of the primary structural differences between roots and stems is the _____.
   a. arrangement of vascular tissue within the stem and root
   b. arrangement of pith within the stem and root
   c. differences in xylem and phloem function
   d. differences in the numbers of stomata

6. What is the primary function of leaves?
   a. to provide protection for the plant
   b. to provide water for the plant
   c. to trap light energy for photosynthesis
   d. to enable the plant to grow taller

7. Which diagram correctly shows the functioning of guard cells?

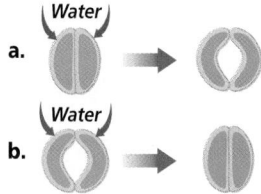

a.

b.

8. Water and mineral ions enter the root by absorption into the _____.
   a. phloem   c. stomata
   b. cuticle   d. root hairs

9. Which of the following cells is a sclerenchyma cell?
   a. meristematic cell c. vessel element
   b. companion cell   d. guard cell

10. The tissue that contains stomata is _____.
    a. vascular   c. ground
    b. dermal   d. meristematic

11. The movement of sugar through phloem is called _____.

12. The picture shown to the right is a _____ root.

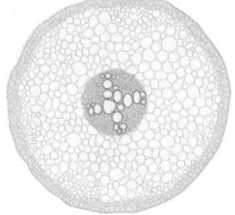

13. The plant hormone that speeds up the ripening of fruit is _____.

14. An analysis of the composition of the fluid transported by the xylem tissue would reveal that it is mostly _____.

15. A dwarf plant can be induced to reach normal height by the application of the hormone _____.

16. The thin-walled cells that make up most of the ground tissue are _____.

17. Because the tip of a stem produces _____, removal of the tip stimulates lateral growth.

18. The portion of a plant that uses sugars is referred to as a _____.

19. Apical meristems are found in the _____ and _____.

20. The potato tuber is a type of _____.

APPLYING MAIN IDEAS

21. Compare the expected rates of water movement in xylem tissues during the day and at night.

22. In late afternoon you are standing near a field of sunflowers. The flowers are all facing away from you. Explain how you know which direction they are facing.

23. Explain how the endodermis controls the flow of water and ions into the vascular tissue of the root.

24. Every spring, sap, a sugary fluid that travels through phloem, is collected from sugar maple trees to make maple syrup. Which type of vascular tissue do the farmers tap into to remove the sap? Explain.

THINKING CRITICALLY

25. **Comparing and Contrasting** Compare and contrast the flow of material through the xylem and phloem.

26. **Recognizing Cause and Effect** To allow some trees to grow straighter, foresters may remove surrounding trees by a process called girdling. To girdle a tree the forester removes a circle of bark from around the trunk of the tree. How does girdling eventually kill a tree?

27. **Comparing and Contrasting** Compare and contrast apical and lateral meristems.

28. **Concept Mapping** Complete the concept map by using these vocabulary terms: xylem, tracheid, phloem, companion cell, sieve tube member, and vessel element.

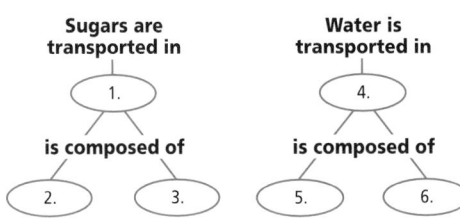

CD-ROM

For additional review, use the assessment options for this chapter found on the *Biology: The Dynamics of Life Interactive CD-ROM* and on the Glencoe Science Web site.
science.glencoe.com

ASSESSING KNOWLEDGE & SKILLS

The graph below illustrates data on daisies collected by scientists in four different states. Remember that stomata are the openings in the leaves through which water vapor escapes.

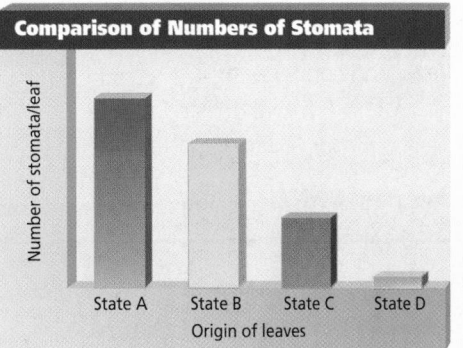

Comparison of Numbers of Stomata

Interpreting Data Examine the graph and answer the following questions.

1. In which state might there be the most rainfall?
   a. state A   **c.** state C
   b. state B   **d.** state D

2. In which state might there be the least rainfall?
   a. state A   **c.** state C
   b. state B   **d.** state D

3. How is rainfall correlated with numbers of stomata on leaves?
   a. the more stomata, the less rain
   b. the fewer the stomata, the less rain
   c. the more stomata, the more rain
   d. the fewer the stomata, the more rain

4. **Making a Graph** Make a bar graph similar to the one above that shows how the thickness of the leaf cuticle of a particular plant varies from one state to another. In state A, cuticles are thickest; state B, thinner than state A; state C, thicker than state B but not as thick as state A; state D, thinner than state B.

---

23. The endodermis creates a barrier to the flow of water and ions, forcing them to flow through the endodermal cells. These cells can then regulate the movement of these substances.

24. Phloem; Since sap contains sugar, it is transported by the phloem.

THINKING CRITICALLY

25. Xylem and phloem both transport materials throughout the plant. Water and ions are transported up the plant through the xylem. Water, sugars and other nutrients are transported up and down the plant through the phloem.

26. Girdling interrupts the passage of nutrients in phloem cells from leaves to roots. Roots are unable to receive food and eventually will die, resulting in death of the tree.

27. Both types of meristems produce new cells for growth. The apical meristem is responsible for vertical growth. Production of cells by the lateral meristem results in an increase in girth.

28. 1. Phloem; 2. Companion cell; 3. Sieve tube member; 4. Xylem; 5. Vessel element; 6. Tracheid

ASSESSING KNOWLEDGE & SKILLS

1. a
2. d
3. c
4.
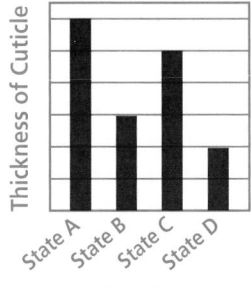

Comparison of Thickness of Cuticle

Chapter 24 Organizer

Refer to pages 4T-5T of the Teacher Guide for an explanation of the National Science Education Standards correlations.

Section	Objectives	Activities/Features
Section 24.1 **Life Cycles of Mosses, Ferns, and Conifers** National Science Education Standards UCP.1, UCP.2, UCP.3, UCP.5; A.1, A.2; C.1, C.5 (1 session)	1. **Review** the steps of alternation of generation. 2. **Describe** the life cycles of mosses, ferns, and conifers.	**MiniLab 24-1:** Growing Plants Asexually, p. 654 **Problem-Solving Lab 24-1,** p. 659 **BioTechnology:** Hybrid Plants, p. 680
Section 24.2 **Flowers and Flowering** National Science Education Standards UCP.1, UCP.2, UCP.5; A.1, C.4-6 (1 session, ¹⁄₂ block)	3. **Identify** the structures of a flower. 4. **Examine** the influence of photoperiodism on flowering.	**Problem-Solving Lab 24-2,** p. 663 **Inside Story:** Parts of a Flower, p. 665 **Investigate BioLab:** Examining the Structure of a Flower, p. 678
Section 24.3 **The Life Cycle of a Flowering Plant** National Science Education Standards UCP.1, UCP.2, UCP.3, A.1, A.2; C.1, C.4-6; E.1, E.2; F.3, F.6; G1-3 (2 sessions, 1 block)	5. **Describe** the life cycle of a flowering plant. 6. **Outline** the processes of seed and fruit formation and seed germination.	**Careers in Biology:** Greens Keeper, p. 673 **MiniLab 24-2:** Looking at Germinating Seeds, p. 677

Need Materials? Contact Carolina Biological Supply Company at 1-800-334-5551 or at **http://www.carolina.com**

MATERIALS LIST

BioLab

p. 678 microscope, microscope slide (2), coverslips (2), dropper, water, single-edged razor blade, colored pencils (red, green, and blue), hand lens, flower (complete)

MiniLabs

p. 654 potato, garlic clove, carrot, test tube, petri dish, beaker, toothpicks, water, paper, pencil
p. 677 microscope, corn kernels and bean seeds (germinating and ungerminated), paper towels, plastic zipper bags, single-edged razor blade

Alternative Lab

p. 674 canned kidney beans, paper cup, water, dried kidney beans, wax paper, labels, tetrazolium solution, dropper bottle

Quick Demos

p. 657 photomicrographs of gametophytes, metric ruler
p. 662 flower (rose or daffodil)
p. 668 peanut
p. 672 tomato, peach

Key to Teaching Strategies

L1	Level 1 activities should be appropriate for students with learning difficulties.
L2	Level 2 activities should be within the ability range of all students.
L3	Level 3 activities are designed for above-average students.
ELL	ELL activities should be within the ability range of English Language Learners.
COOP LEARN	Cooperative Learning activities are designed for small group work.
P	These strategies represent student products that can be placed into a best-work portfolio.
	These strategies are useful in a block scheduling format.

Reproduction in Plants

Teacher Classroom Resources

Section	Reproducible Masters	Transparencies
Section 24.1 **Life Cycles of Mosses, Ferns, and Conifers**	Reinforcement and Study Guide, pp. 105-106 L2 Concept Mapping, p. 24 L3 ELL BioLab and MiniLab Worksheets, p. 109 L2 Content Mastery, pp. 117-118, 120 L1	Section Focus Transparency 58 L1 ELL Basic Concepts Transparencies 38, 39, 40 L2 ELL Reteaching Skills Transparencies 35, 36 L1 ELL
Section 24.2 **Flowers and Flowering**	Reinforcement and Study Guide, p. 107 L2 Content Mastery, pp. 117, 119-120 L1 Inside Story Poster ELL	Section Focus Transparency 59 L1 ELL Basic Concepts Transparencies 41, 42, 43 L2 ELL Reteaching Skills Transparency 37 L1 ELL
Section 24.3 **The Life Cycle of a Flowering Plant**	Reinforcement and Study Guide, p. 108 L2 Critical Thinking/Problem Solving, p. 24 L3 BioLab and MiniLab Worksheets, pp. 110-112 L2 Laboratory Manual, pp. 171-178 L2 Content Mastery, pp. 117, 119-120 L1	Section Focus Transparency 60 L1 ELL

Assessment Resources

Chapter Assessment, pp. 139-144
MindJogger Videoquizzes
Performance Assessment in the Biology Classroom
Alternate Assessment in the Science Classroom
ExamView® Pro Software 💾
BDOL Interactive CD-ROM, Chapter 24 quiz

Additional Resources

Spanish Resources ELL
English/Spanish Audiocassettes ELL
Cooperative Learning in the Science Classroom COOP LEARN
Lesson Plans/Block Scheduling

NATIONAL GEOGRAPHIC — Teacher's Corner

Products Available From Glencoe
To order the following products, call Glencoe at 1-800-334-7344:
CD-ROM
NGS PictureShow: What It Means to Be Green
Curriculum Kit
GeoKit: Plants
Transparency Set
NGS PicturePack: What It Means to Be Green
Videodisc
STV: Plants

GLENCOE TECHNOLOGY

The following multimedia resources are available from Glencoe.

Biology: The Dynamics of Life
CD-ROM ELL

Animation: *Life Cycle of a Moss*
Video: *Fern Development*
Animation: *Life Cycle of a Pine*
Exploration: *Angiosperm*
Video: *Blooming Flowers*

Videodisc Program 📦

Double Fertilization
Fruit Formation
Seed Dispersal
Germination

24 Reproduction in Plants

Theme Development

Unity within diversity is explored in this chapter as the different reproductive strategies of major plant divisions are presented. The theme of **systems and interactions** is stressed in the study of the life cycles of various plants. **Evolution** is a theme that occurs throughout the chapter, especially as it relates to the coevolution of pollinators and flowers.

0:00 OUT OF TIME?

If time does not permit teaching the entire chapter, use the BioDigest at the end of the unit as an overview.

READING BIOLOGY

Glencoe's *Biology: The Dynamics of Life* contains many resources to assist a student's reading skills. Each chapter contains figures with expanded captions that expand on written material. Word Origins, located along the side of text, expand knowledge of biology vocabulary. Glencoe's Content Mastery Booklet helps develop reading skills while reinforcing content. In addition, use the Interactive Tutor for *Biology: The Dynamics of Life* on the Glencoe Web site to reinforce vocabulary. **science.glencoe.com**

What You'll Learn

- You will compare and contrast the life cycles of mosses, ferns, and conifers.
- You will sequence the life cycle of a flowering plant.
- You will describe the characteristics of flowers, seeds, and fruits.

Why It's Important

Plants are essential to Earth's biosphere. The fruits and seeds produced by flowering plants are a major food source for humans and animals, and critical for the survival of many species.

READING BIOLOGY

Carefully look at the different life cycle figures that appear in the chapter. Pick one and sketch it in your notebook. Add definitions of any new vocabulary words that are used in the diagrams.

BIOLOGY *Online*

To find out more about plants, visit the Glencoe Science Web site. **science.glencoe.com**

Animals often play an important role in pollinating flowering plants. Insects, including bees, transport pollen from flower to flower. Most nonflowering plants, such as mosses, rely on wind or water for the dispersal of spores.

652 REPRODUCTION IN PLANTS

Multiple Learning Styles

Look for the following logos for strategies that emphasize different learning modalities.

Kinesthetic Meeting Individual Needs, p. 662; Building a Model, p. 669; Tech Prep, p. 675; Portfolio, p. 676

Visual-Spatial Biology Journal, p. 657; Meeting Individual Needs, pp. 658, 665, 672; Reteach, p. 660; Display, p. 664; Discussion, p. 666; Quick Demo, p. 668; Reinforcement, p. 668; Microscope Activity, p. 668;

Portfolio, p. 670; Tech Prep, p. 672; Enrichment, p. 673

Intrapersonal Enrichment, p. 664; Project, p. 664

Linguistic Meeting Individual Needs, p. 655; Portfolio, pp. 656, 662; Biology Journal, pp. 663, 668; Extension, p. 666

Logical-Mathematical Quick Demo, p. 657

24.1 Life Cycles of Mosses, Ferns, and Conifers

SECTION PREVIEW

Objectives

Review the steps of alternation of genera-tion.

Describe the life cycles of mosses, ferns, and conifers.

Vocabulary
vegetative reproduction
protonema
megaspore
microspore
micropyle

Y ou may have seen the fine yellow dust that covers everything when pine trees release their pollen. As annoying as this pollen may seem, it has a valuable func-tion. It is an important stage in the life cycle of pine trees. Other plants have even more dramatic stages of their life cycles, such as exploding moss capsules and fern sporangia.

Male pine cone releasing pollen

Alternation of Generations

As you learned earlier, plants go through an alternation of generations during their life cycles. Remember that the two phases of the plant life cycle are the gametophyte stage and the sporophyte stage.

The cells of the sporophyte are all diploid. Certain cells of the sporo-phyte undergo meiosis and produce haploid spores. These spores grow, by mitotic division, into the gameto-phyte. The multicellular gameto-phyte that is formed is composed of haploid cells. Some cells of the game-tophyte will differentiate and form haploid gametes. The female gamete is the egg, and the male gamete is the sperm. When a sperm fertilizes an egg, a diploid zygote is formed. This zygote divides by mitosis, producing a tiny sporophyte or embryo. The development of the embryo into a

mature sporophyte allows the life cycle to begin again. *Figure 24.1* illustrates alternation of generations.

Figure 24.1

All plants exhibit an alternation of generations. The gametophyte (*n*) stage produces gametes. The sporophyte (*2n*) produces spores.

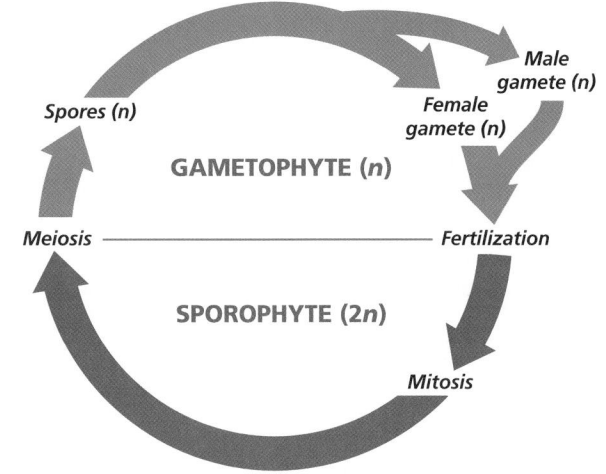

653

Prepare

Key Concepts

Alternation of generations is re-viewed and the life cycles of mosses, ferns, and conifers are presented.

Planning

- Purchase garlic, carrots, and potatoes for MiniLab 24-1.
- Locate pictures of moss, fern, and conifer reproductive struc-tures for Getting Started Demo.
- Locate pictures of moss, fern, and conifer gametophytes for the Quick Demo.

1 Focus

Bellringer

Before presenting the lesson, display **Section Focus Trans-parency 58** on the overhead pro-jector and have students answer the accompanying questions. **L1 ELL**

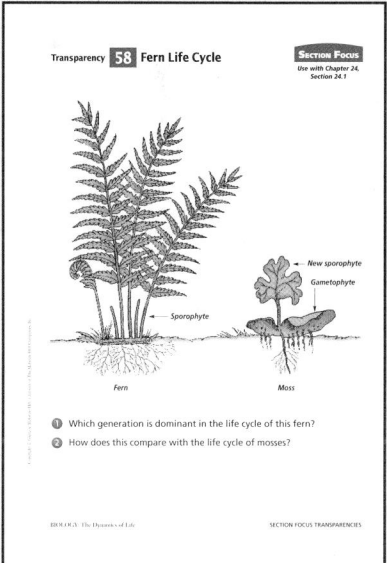

Resource Manager

Section Focus Transparency 58 and Master L1 ELL

✓ Assessment Planner

Portfolio Assessment
Assessment, TWE, p. 657
Portfolio, TWE, pp. 656, 662, 670, 676
Alternative Lab, TWE, p. 675
Performance Assessment
MiniLab, SE, pp. 654, 676
MiniLab, TWE, pp. 654, 676
Problem-Solving Lab, TWE, pp. 659, 663
Assessment, TWE, pp. 666, 676, 677

Alternative Lab, TWE, pp. 674-675
BioLab, SE, pp. 678-679
Knowledge Assessment
Section Assessment, SE, pp. 660, 666, 677
Assessment, TWE, p. 669
BioLab, TWE, p. 679
Chapter Assessment, SE, pp. 681-683
Skill Assessment
Assessment, TWE, p. 662

2 Teach

MiniLab 24-1

Purpose

Students will use several different plant parts to demonstrate the ability of plants to form new plants via asexual reproduction.

Process Skills

experiment, analyze information, collect data, compare and contrast, draw a conclusion

Teaching Strategies

■ Have students work in small groups to conserve materials.

■ Make sure that the original root end (blunt end) of the garlic clove is immersed in water. One garlic head should supply enough cloves for an entire class.

■ Both garlic and potato may be suspended over water using toothpicks as bracing. Small jars (baby food jars, plastic bathroom cups) may be used in place of beakers or test tubes.

■ Make sure that students identify their plants by placing labels on the jars and/or dishes.

■ Tell students to replace any lost water from the containers.

■ Some potatoes are chemically treated to inhibit growth of eyes. This may account for either slow or no new growth appearing within the two-week period.

Expected Results

All plant tissues will produce new growth. The garlic clove will show new root and stem/leaf growth. The carrot will show new leaves. The potato will show new stems and leaves.

Analysis

1. **a.** The experimental procedure demonstrates that different plant parts can generate new growth.
   b. The appearance of new growth occurred within several days.
   c. Only one plant was used for each experimental setup.

The basic pattern of this life cycle is the same for all plants. However, there are many variations on this pattern within the plant kingdom. For instance, recall that in mosses the gametophyte is bigger than the sporophyte. In others, such as flowering plants, the gametophyte is tiny, even microscopic. Most people have never

MiniLab 24-1 Experimenting

Growing Plants Asexually Plants are capable of reproducing asexually. Reproductive cells such as egg or sperm are not needed in asexual reproduction. Plants are able to use structures such as roots, stems, and even leaves to produce new offspring.

- Garlic clove (storage leaves)
- Water
- Test tube
- Potato with eye (stem & buds)
- Toothpick
- Water
- Beaker
- Carrot (root)
- Water

(A) (B) (C)

Procedure

1. Prepare three different plant parts for study using diagrams A, B, and C as a guide.
2. Observe any changes that occur to your plants over the next two weeks.
3. Design a data table that will provide enough room for diagrams of your observations, the number of days since the start of the experiment should be included.
4. Make your initial diagrams of the plant parts today and label these diagrams as "Day 1."
5. Observations should be made every three days. Replace any lost water as needed.

Analysis

1. What experimental evidence do you have that:
   a. plants use a variety of structures for asexual reproduction?
   b. asexual reproduction is a rapid process?
   c. asexual reproduction requires only one parent?
2. Describe several advantages of asexual reproduction in plants.

---

even seen a female gametophyte of a flowering plant. Botanists usually refer to the bigger, more obvious plant as the dominant generation. The dominant generation lives longer and can survive independently of the other generation. In most plant species the sporophyte is the dominant plant.

Asexual reproduction

Most plants can also reproduce by a process called vegetative reproduction. **Vegetative reproduction** is asexual reproduction in plants where a new plant is produced from an existing vegetative structure. For instance, liverworts produce asexual structures called gemmae that fall off and develop into new plants, *Figure 24.2*. The new plants have the same genetic make-up as the original plant, as if they were cloned. You can learn more about asexual reproduction in the *MiniLab* shown here.

Life Cycle of Mosses

Mosses belong to one of the few plant divisions in which the gametophyte plant is the dominant generation. A haploid spore germinates to

Figure 24.2
Small cups filled with tiny gemmae have formed on the thallus of this liverwort.

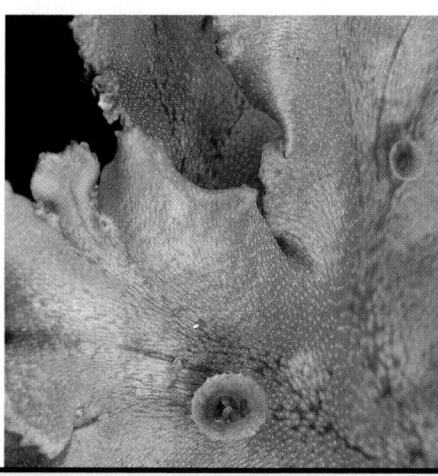

---

2. Student answers may vary; they may include faster growth and that all offspring are identical to parent.

✔ Assessment

Performance Provide students with a plant called Kalanchoe (available in most garden shops). Have them use leaves from the plant to grow new plants asexually. Have students conduct an experiment to determine what conditions stimulate the appearance of new plants along the leaf's edge. Use the Performance Task Assessment List for Carrying Out a Strategy and Collecting Data in **PASC**, p. 25. **L3**

form a structure called a protonema. The **protonema** (proht uh NEE muh) is a small green filament of cells that develops into either a male or a female gametophyte. In some mosses, the gametophyte can produce both kinds of reproductive structures. Remember that the archegonium is the female reproductive structure in which eggs are produced and that sperm are produced in the antheridium.

The motile sperm are released from the antheridium and swim through a continuous film of rainwater or dew to the archegonium. The sperm fertilizes the egg inside the archegonium, forming a diploid zygote. The zygote divides by mitosis to form a new sporophyte. The sporophyte is a stalk with a capsule at the top. It grows out of the archegonium and remains attached to the gametophyte. The sporophyte receives much of its nutrition from the gametophyte. Meiotic division within the capsule produces haploid spores.

The capsule ripens, bursts, and releases the spores, which can be carried great distances by air currents. If the spore lands in a favorable environment, it germinates, completing the life cycle. Review the moss life cycle as you examine *Figure 24.3.*

Revealing Misconceptions
Students learned in previous chapters that the process of meiosis forms gamete cells. This is true for animals. However, in plants, spores are produced directly though meiosis and then gametes are formed through mitosis of these haploid cells.

Why the difference between plants and animals? Actually, there is no difference. Plants just have an added stage or step that results from alternation of generations. In this process, meiosis still forms haploid spore cells—the gametophyte generation—that remain haploid. The gametophyte is equivalent to one large male or female animal gamete.

Visual Learning
Figure 24.3 To reinforce the concept of diploid versus haploid, ask students to assign a specific chromosome number to each stage in Figure 24.3. Explain that the diploid number for this species is 18. **L2**

Figure 24.3
The leafy green gametophyte of a moss produces gametes that fuse to form a zygote. The zygote develops into the sporophyte. The sporophyte produces spores. The spores germinate and grow into a gametophyte, completing the life cycle of a moss.

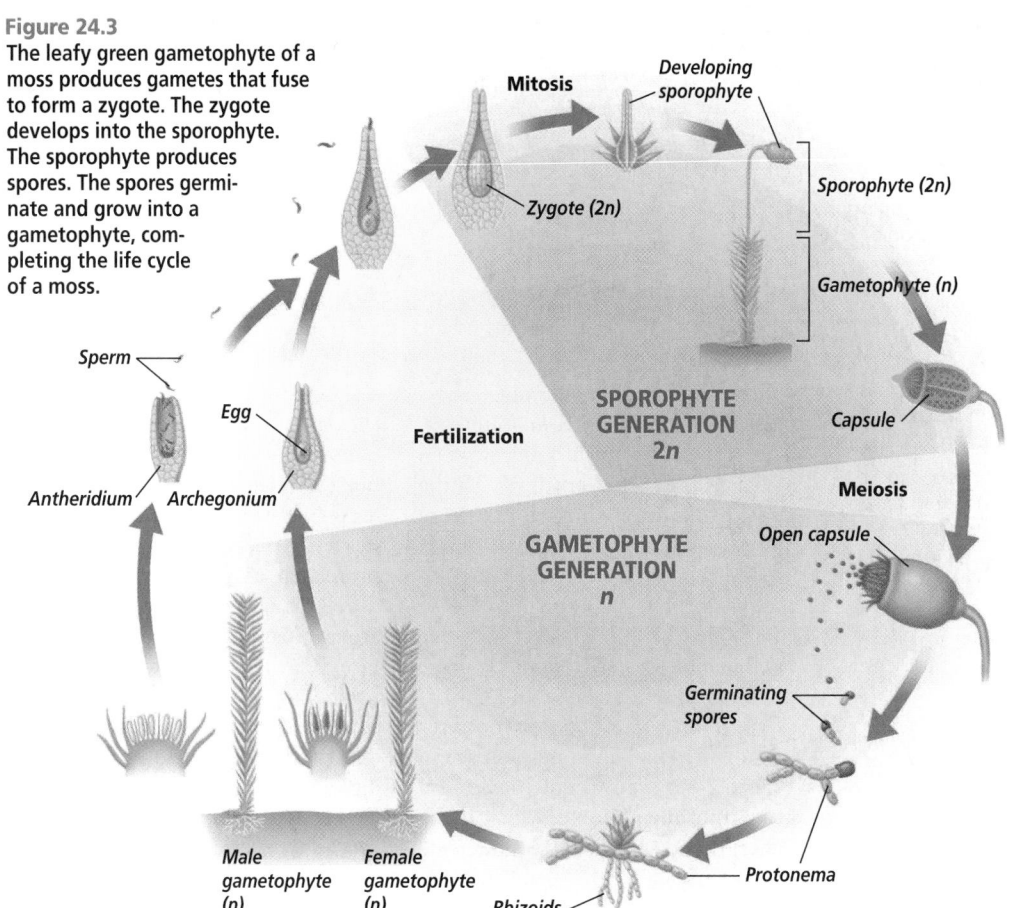

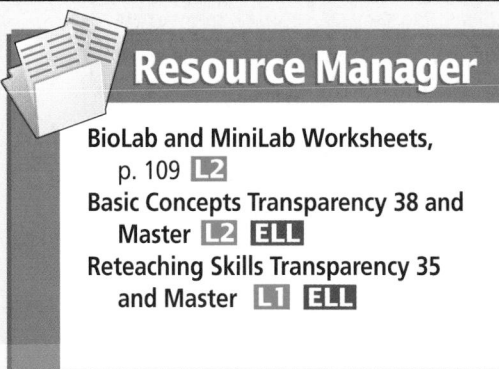

MEETING INDIVIDUAL NEEDS

English Language Learners/ Learning Disabled

Linguistic Have students review terms associated with alternation of generations by reinforcing the correlation between the name of each generation and its function. For example, *sporo*phytes form *spores,* while *gameto*phytes form *gametes.* **L1** **ELL**

Resource Manager

BioLab and MiniLab Worksheets, p. 109 **L2**
Basic Concepts Transparency 38 and Master **L2** **ELL**
Reteaching Skills Transparency 35 and Master **L1** **ELL**

Revealing Misconceptions

Students frequently think the sori present on the underside of fern fronds are some kind of insect infestation. Explain that these structures produce spores.

Chalkboard Activity

Have students write a simple moss life cycle sequence on the board. Help them get started by showing them what the cycle may look like by providing them with this sample: sporophyte ⇒ spores by meiosis ⇒ protonema ⇒ etc.

656

Figure 24.4
Fern sporophytes are easily seen by a hiker walking through a forest. However, only the very observant person would be able to find a fern gametophyte.

B The clusters of sporangia on the underside of a fern frond are called sori. Each sporangium contains spores that are released, sometimes in dramatic fashion.

A Most fern sporophytes grow 25 cm or taller.

C The heart-shaped fern gametophyte is usually less than a centimeter across.

Some mosses also reproduce asexually by vegetative reproduction. They can break up into pieces when the plant is dry and brittle. With the arrival of wetter conditions, these pieces each become a whole plant.

Life Cycle of Ferns

Unlike mosses, the dominant stage of the fern life cycle is the sporophyte plant. The fern sporophytes include the familiar fronds you see in *Figure 24.4*. The fronds of the fern grow from the rhizome, which is the underground stem. On the underside of some fronds are the sori, which are clusters of sporangia. Meiotic division within the sporangia produces the spores. When environmental conditions are right, the sporangia burst to release haploid spores.

A spore germinates to form a heart-shaped gametophyte called a prothallus, as shown in *Figure 24.5*. The prothallus produces both archegonia and antheridia on its surface. The flagellated sperm released by the antheridium swim through a film of water to the archegonium where the egg is fertilized. The diploid zygote that is the product of this fertilization develops into the sporophyte. Initially, this developing sporophyte depends upon the gametophyte for its nutrition. However,

656 REPRODUCTION IN PLANTS

Portfolio

Chemistry in Biology

Linguistic Have students do research and prepare a report about the role of chemotaxis in moss and fern reproduction. **L3** **P**

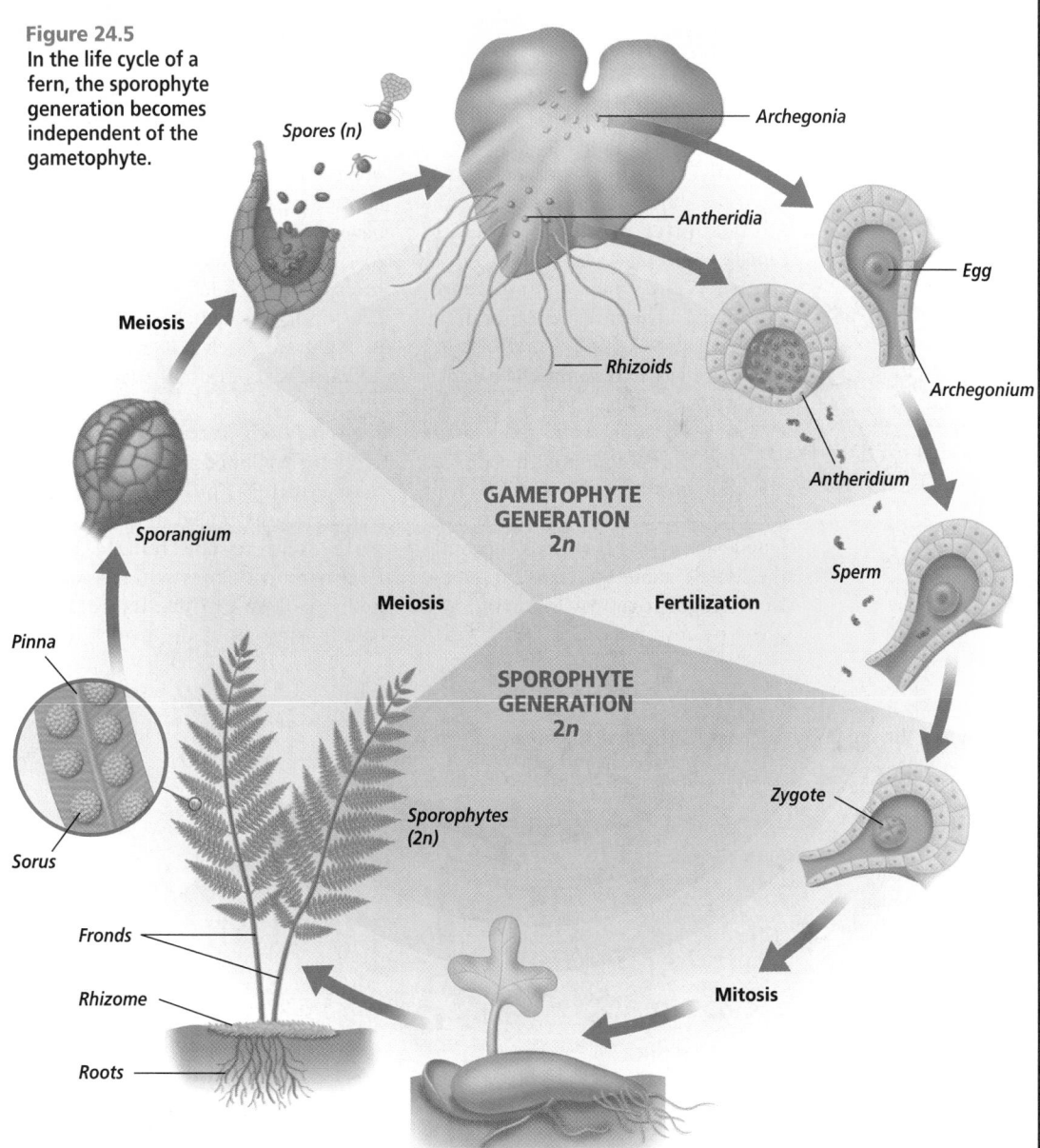

Figure 24.5
In the life cycle of a fern, the sporophyte generation becomes independent of the gametophyte.

Spores (n)

Archegonia

Antheridia

Rhizoids

Egg

Archegonium

Meiosis

GAMETOPHYTE GENERATION
2n

Antheridium

Sporangium

Meiosis

Sperm

Fertilization

Pinna

SPOROPHYTE GENERATION
2n

Sorus

Zygote

Sporophytes (2n)

Fronds

Rhizome

Roots

Mitosis

Quick Demo

Logical-Mathematical Give students pictures of moss, fern, and conifer gametophytes that have the magnification noted on the photo. Have them calculate the size of each gametophyte. The size of the drawing (in millimeters) divided by the magnification equals the size of the original gametophyte (in millimeters). Students will then be able to compare the size of the gametophytes. **L3**

Chalkboard Activity

Write the following terms on the chalkboard: spore, zygote, egg, sporophyte, and sperm. Have students identify whether each structure is haploid or diploid. **L2**

✔ Assessment

Portfolio Have students prepare a biological key that will enable others to identify mosses, ferns, and conifers based on features of the plant life cycle, including reproductive parts. **L2**

once the sporophyte produces its green fronds, it can carry on photosynthesis and survive on its own. The prothallus disintegrates as the sporophyte matures, producing a strong rhizome that can support the fronds. If pieces of rhizome break away, new

fern plants will develop from them by vegetative reproduction. New sporangia develop on the pinnae of the fronds, spores will be released, and the cycle will begin again. The life cycle of the fern is summarized in *Figure 24.5.*

BIOLOGY JOURNAL

Fern Life Cycle

Visual-Spatial Have students diagram the stages of the fern life cycle in their journals for review. Encourage students to divide the diagram so that all sporophytes are on one side and all gametophytes are on the other. Have them label the diagram Sporophyte stages, Gametophyte stages. **L1** **ELL**

The Life Cycle of Conifers

The dominant stage in conifers is the sporophyte generation. One of the more familiar conifer sporophytes is shown in **Figure 24.6.** The adult conifer produces male and female cones on separate branches of the tree. The cones contain spore-producing structures, or sporangia, on their scales. The female cones, which are larger than the male cones, develop two ovules on the upper surface of each cone scale. Each ovule contains a sporangium with a diploid cell that produces, by meiosis, four megaspores. A **megaspore** is a female spore that eventually becomes the female gametophyte. One of the four megaspores will survive and grow by mitotic cell divisions into the female gametophyte. The female gametophyte consists of hundreds of cells but is still dependent on the sporophyte for protection and nutrition. Within the female gametophyte are two or more archegonia, each containing an egg. The male cones have sporangia that undergo meiosis to produce male spores called **microspores.** Each microspore will develop into a male gametophyte, or pollen grain. Each pollen grain, with its hard, water-resistant outer covering, is a male gametophyte. Look at **Figure 24.6** to see examples of male and female conifer gametophytes.

In conifers, pollination is the transfer of the pollen grain from the male cone to the female cone. Pollination occurs when a wind-borne pollen grain falls near the opening in one of the ovules of the

Figure 24.6
In conifers, the sporophyte is immense compared with the microscopic gametophytes.

A This pine sporophyte can grow more than 25 meters tall.

B The female gametophyte in this pine ovule is less than 0.01 mm long.

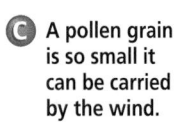

C A pollen grain is so small it can be carried by the wind.

MEETING INDIVIDUAL NEEDS

Learning Disabled

 ***Visual-Spatial*** Prepare individual cards with a drawing on each showing one single stage of the life cycle of the pine. Provide a packet of cards to students who are learning disabled. Have students arrange the cards in proper sequence to illustrate the complete life cycle for a typical pine. **L1** **ELL**

Resource Manager

Reinforcement and Study Guide,
pp. 105-106 **L2**
Content Mastery, p. 118 **L1**
Basic Concepts Transparency 40 and Master **L2** **ELL**
Reteaching Skills Transparency 36 and Master **L1** **ELL**

female cone. The pollen grain adheres to a sticky drop of fluid that covers the opening of the ovule. As the fluid evaporates, the pollen grain is drawn closer to an opening of the ovule called the **micropyle** (Mi kruh pile). Although pollination has occurred, fertilization will not take place for at least a year. The pollen grain and the female gametophyte will mature during this time.

As the pollen grain matures, it produces a pollen tube that grows through the micropyle and into the ovule. A sperm cell from the male gametophyte is transported by the pollen tube to the egg, where fertilization occurs. The zygote, which is nourished by the female gametophyte, develops inside the ovule into an embryo with several cotyledons. The cotyledons nourish the developing sporophyte. The ovule provides the seed coat as the mature seed is produced.

The seed is released when the female cone opens. When conditions are favorable, the seed germinates into a new, young sporophyte—a pine tree seedling, *Figure 24.7.* See if you can identify the stages of the life cycle in *Figure 24.8.* Use the *Problem-Solving Lab* on this page to further explore the life cycles of mosses, ferns, and conifers.

Problem-Solving Lab 24-1
Making and Using Tables

What traits do mosses, ferns, and conifers share?
Sometimes it helps to organize information in a table. The advantage of a table is that it summarizes traits and shows similarities and differences in a simple format.

Analysis
Copy the following data table. Complete the table using "yes" and "no" answers.

Data Table			
Trait	Moss	Fern	Conifer
Has alteration of generations			
Film of water needed for fertilization			
Dominant gametophyte			
Dominant sporophyte			
Sporophyte is photosynthetic			
Produces seeds			
Produces sperm			
Produces pollen grains			
Produces eggs			

Thinking Critically
1. Which two plant groups share the most characteristics? Which two share the fewest?
2. While on a woodland trail, would you easily observe:
   a. A pine gametophyte? Sporophyte? Explain.
   b. A fern gametophyte? Sporophyte? Explain.
3. Using information from your table, summarize the reproductive similarities and differences among mosses, ferns, and conifers.

Figure 24.7
Conifer seeds germinate into new, young sporophytes such as the pine tree seedling shown here.

Assessment
Performance Have students attempt to grow fern gametophytes from spores. Instructions and spores are available through biological supply houses. Use the Performance Task Assessment List for Carrying Out a Strategy and Collecting Data in **PASC**, p. 25. **L3** **ELL**

Problem-Solving Lab 24-1

Purpose
Students will identify and categorize the similarities and differences among mosses, ferns, and pines by completing a chart.

Process Skills
compare and contrast, think critically, analyze information, apply concepts, predict

Teaching Strategies
■ You may wish to photocopy the chart and pass out the copies to your students.
■ Review the meaning of any terms used in the chart that may be unfamiliar to students.
■ Allow students to work in small groups to complete the chart. Make sure they use their text as a reference to aid in verification of answers.
■ Review student answers on the chart prior to them answering the questions.
Answers going down each column of the table.
Moss: *yes, yes, yes, no, yes, no, yes, no, yes*
Fern: *yes, yes, no, yes, yes, no, yes, no, yes*
Conifer: *yes, no, no, yes, yes, yes, yes, yes, yes*

Thinking Critically
1. Mosses and ferns share the most characteristics. Mosses and conifers share the least.
2. Both ferns and mosses require a film of water for fertilization. Therefore, both would most likely be found growing in damp or moist environments.
3. All three have alternation of generations and produce sperm and eggs. Both mosses and ferns need water for fertilization; conifers do not. Mosses have a dominant gametophyte; ferns and conifers have a dominant sporophyte. Conifers produce pollen grains and seeds; Mosses and ferns do not.

3 Assess

Check for Understanding

Ask students to explain the relationships of the following word pairs. **L2** **ELL**

   a. gametophyte—sporophyte
   b. antheridium—sperm
   c. archegonium—egg
   d. megaspore—microspore

Reteach

Visual-Spatial Divide the class into small groups and assign different groups the task of making flowcharts of the life cycles of mosses, ferns, and conifers. Have them put their flowcharts on the chalkboard or poster board for class discussion. **L2** **COOP LEARN**

Extension

Have students research which animals use the cones of conifers as a food source. **L3**

✓ Assessment

Knowledge Have students explain the role of water in moss, fern, and conifer reproduction. **L2**

4 Close

Biology Journal

Have students compare the gametophyte and sporophyte generations of mosses, ferns, and conifers. **L2**

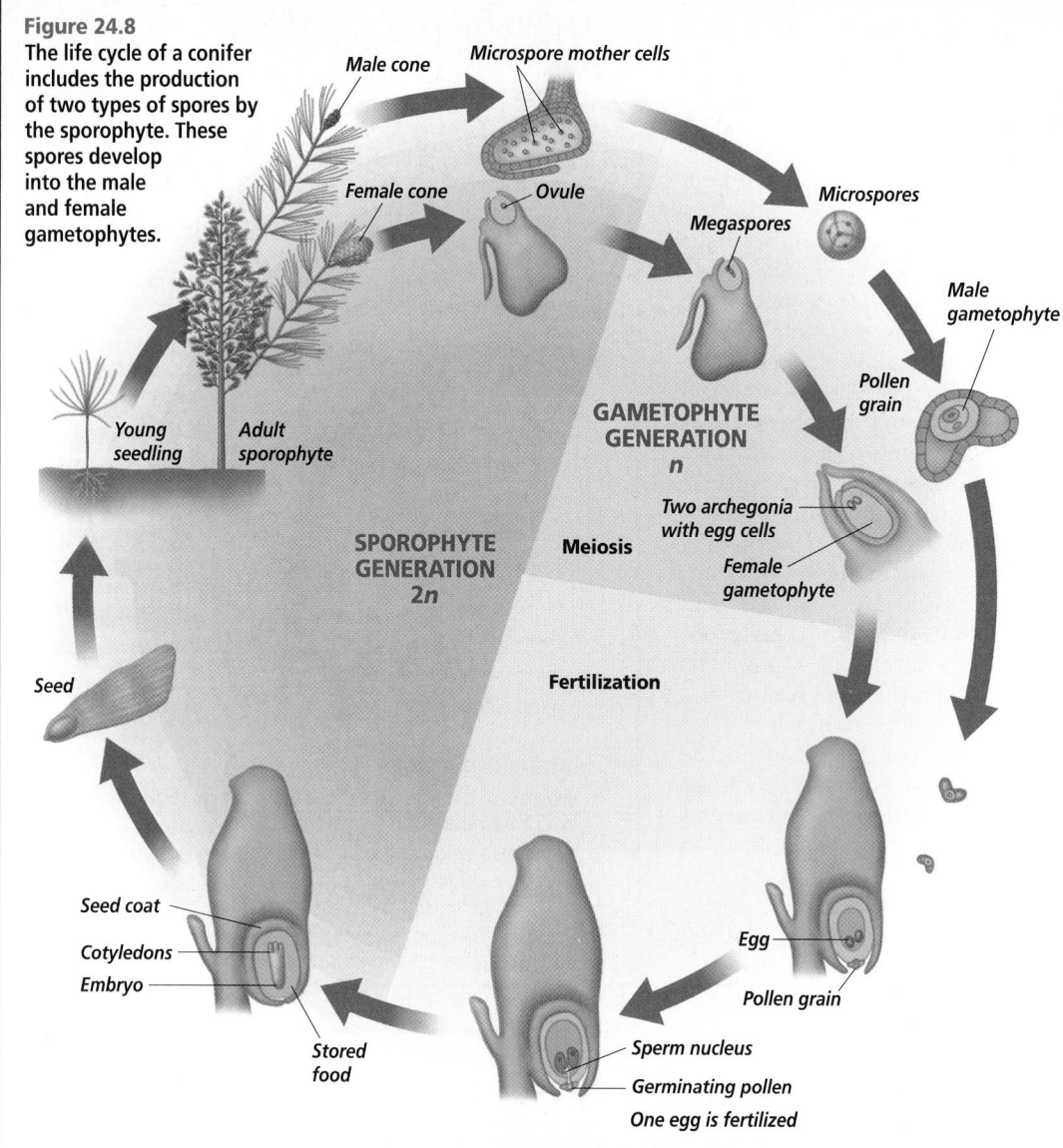

Figure 24.8
The life cycle of a conifer includes the production of two types of spores by the sporophyte. These spores develop into the male and female gametophytes.

Section Assessment

Understanding Main Ideas

1. Explain how vegetative reproduction can produce a new plant. Provide an example.
2. In what way is the sporophyte generation of a moss dependent on the gametophyte generation?
3. Describe the formation of the male gametophyte in a conifer.
4. What are two differences between the life cycle of a fern and that of a conifer?

Thinking Critically

5. Why is the term alternation of generations appropriate to describe the life cycle of a plant?

SKILL REVIEW

6. **Sequencing** Sequence the events in the life of a fern, beginning with the prothallus. For more help, refer to *Organizing Information* in the **Skill Handbook.**

660 REPRODUCTION IN PLANTS

Section Assessment

1. Vegetative reproduction is when a new plant is produced from an existing vegetative structure. An example would be the growth of liverworts from the gemmae that fall off a parent plant.
2. The sporophyte depends upon the gametophyte for support and most of its nutrition.

3. The sporangia within the male pinecone produce microspores. The microspores develop into the male gametophyte or pollen grain.
4. The gametophyte is dominant in mosses, whereas the sporophyte is dominant in ferns. In ferns, the sporophyte can exist independently of the gametophyte but in mosses the sporophyte remains

dependent upon the gametophyte.
5. The sporophyte generation alternates with the gametophyte generation in the life cycle.
6. Prothallus ⇒ forms egg and sperm ⇒ fertilization of egg by sperm forms a young sporophyte ⇒ sporophyte produces spores ⇒ spores germinate to each form a prothallus.

SECTION PREVIEW

Objectives

Identify the structures of a flower.

Examine the influence of photoperiodism on flowering

Vocabulary

petals
sepals
stamen
anther
pistil
ovary
photoperiodism
short-day plant
long-day plant
day-neutral plant

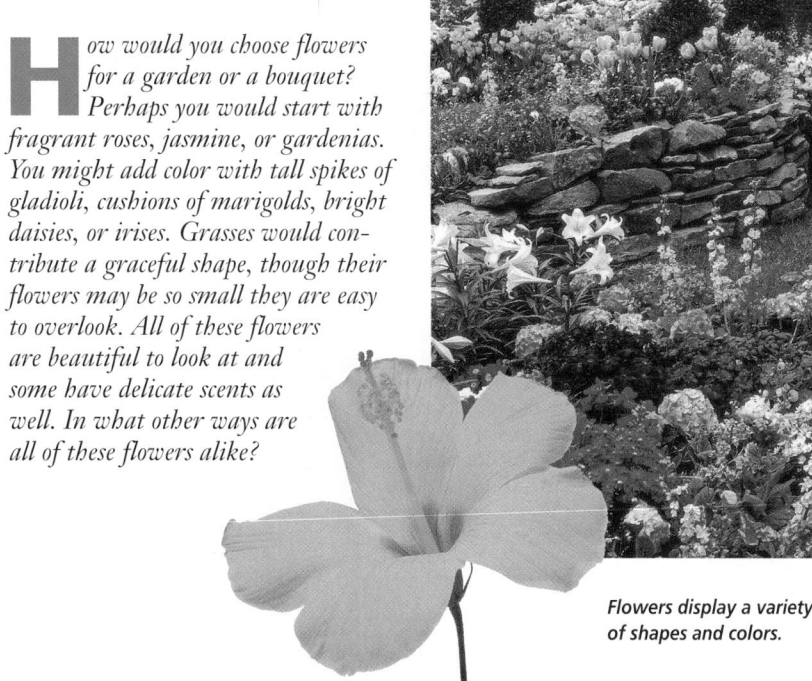

How would you choose flowers for a garden or a bouquet? Perhaps you would start with fragrant roses, jasmine, or gardenias. You might add color with tall spikes of gladioli, cushions of marigolds, bright daisies, or irises. Grasses would contribute a graceful shape, though their flowers may be so small they are easy to overlook. All of these flowers are beautiful to look at and some have delicate scents as well. In what other ways are all of these flowers alike?

Flowers display a variety of shapes and colors.

What Is a Flower?

The process of sexual reproduction in flowering plants takes place in the flower, which is a complex structure made up of several parts. Some parts of the flower are directly involved in fertilization and seed production. Other floral parts have functions in pollination. There are probably as many different shapes, sizes, colors, and configurations of flower parts as there are species of flowering plants. In fact, features of the flower are often used in plant identification.

The structure of a flower

Even though there is an almost limitless variation in flower shapes and colors, all flowers share a simple, basic structure. A flower is usually made up of four kinds of organs: sepals, petals, stamens, and pistils. The flower parts you are probably most familiar with are the petals. **Petals** are leaflike, usually colorful structures arranged in a circle around the top of a flower stem. **Sepals** are also leaflike, usually green, and encircle the flower stem beneath the petals.

24.2 FLOWERS AND FLOWERING **661**

Section 24.2

Prepare

Key Concepts

Flower anatomy and modifications are discussed in this section. The role of photoperiodism in anthophyte reproduction is explained.

Planning

■ Locate flowers for the Quick Demo.

■ Purchase flowers for the Bio-Lab.

■ Locate flower model for Meeting Individual Needs.

■ Collect pictures of different flowers for the CLOSE activity.

---

1 Focus

Bellringer

Before presenting the lesson, display **Section Focus Transparency 59** on the overhead projector and have students answer the accompanying questions. **L1** **ELL**

Transparency **59** Seed Dispersal

SECTION FOCUS
Use with Chapter 24, Section 24.2

❶ How does the structure of each seed make it suitable for its method of dispersal?

❷ Why is seed dispersal important?

BIOLOGY *Online*

Note Internet addresses that you find useful in the space below for quick reference.

_____

_____

_____

_____

Inside the circle of petals are the stamens. A **stamen** is the male reproductive structure of a flower. At the tip of the stamen is the **anther.** The anther produces pollen that contains sperm.

At the center of the flower, attached to the top of the flower stem, lie one or more pistils. The **pistil** is the female structure of the flower. The bottom portion of the pistil enlarges to form the **ovary,** a structure with one or more ovules, each containing one egg. As you read in the previous section, the female gametophyte develops inside the ovule. You can learn more about

floral structure and practice your lab skills in the *BioLab* at the end of this chapter.

Modifications in flower structure

A flower that has all four organs—sepals, petals, stamens, and pistils—is called a complete flower. The morning glory and tiger lily shown in *Figure 24.9* are examples of complete flowers. A flower that lacks one or more organs is called an incomplete flower. For example, squash plants have separate male and female flowers. The male flowers have stamens but no pistils; the female flowers bear pistils but no stamens.

Figure 24.9
The diversity of flower forms is evidence of the success of flowering plants.

B The male flowers of the walnut tree form long catkins.

C The petals of the morning glory are fused together to form a bell shape.

A The spotted petals of the tiger lily curl away from the reproductive structures at the center of the flower.

D Thistles bear clusters of tiny, tubular flowers within a mass of spiny bracts.

E The location of corn tassels at the top of the plant aids in wind pollination.

Plants such as sweet corn that are adapted for pollination by wind rather than animal pollinators have no petals. *Figure 24.9* shows some examples of the variety in flower forms. Study the structure of a typical flower in the *Inside Story*. You can explore flower adaptations further in the *Problem-Solving Lab* shown here.

There is an amazing amount of diversity in the structures of anthophyte flowers, seeds, fruits, and vegetative structures. Anthophytes are divided into different divisions and classes based on these differences. The relationships among the different classes and divisions of anthophytes are shown in *Figure 24.10*. See if you can recognize any of the different divisions of plants.

Photoperiodism

The relative length of day and night has a significant effect on the rate of growth and the timing of flower production in many species of flowering plants. For example, chrysanthemums produce flowers only during the fall, when the days are getting shorter and the nights longer. A grower who wants to produce chrysanthemum flowers in the middle of summer drapes black cloth over the plants to artificially increase the length of night. The response of flowering plants to the difference in the duration of light and dark periods in a day is called **photoperiodism.**

Plant biologists originally thought that day length controlled flowering. However, they now know that it is the length of the night, or dark period, that controls flowering. Plants can be placed in three categories depending on the conditions they require for flower production. Plants are either short-day plants, long-day plants, or day-neutral plants.

Problem-Solving Lab 24-2 — Interpreting Scientific Illustrations

How do flowers differ? There is considerable variation in flower shape. This variation occurs when certain flower parts are fused together, parts are rearranged, or when parts may be totally missing. However, with all the variation seen in flower shape, there are certain general patterns. Almost all dicot plants will have flower parts that are in fours or fives or multiples of these numbers. For example, a plant having eight or ten petals, sepals, and stamens would be a dicot. Almost all monocot plants have flower parts in threes or multiples of three.

Analysis

Diagram A shows a flower with all of its parts labeled. Imagine that the flower has been cut along the dashed line. Diagram B is a diagrammatic cross-section view seen when looking down onto the cut edge of the bottom half. Diagram B is shown a little larger than A so that any details can be more clearly seen. The flower is from a dicot plant because there are five sepals, petals, and stamens.

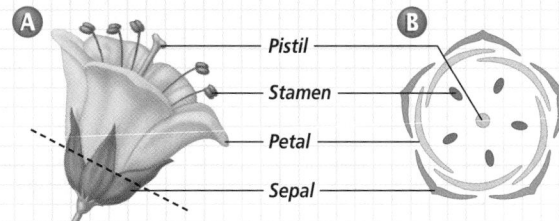

Pistil
Stamen
Petal
Sepal

Thinking Critically

1. Diagrams C, D, and E are diagrammatic cross-section views of flowers. Determine if diagram:
   a. C is a monocot or dicot. Explain.
   b. D is a monocot or dicot. Explain.
   c. E is a monocot or dicot. Explain.
2. Do diagrams A, C, D, and E show flowers that are complete or incomplete? Explain. Note: Complete flowers have sepals, petals, stamens, and pistils whereas incomplete flowers lack one or more of these parts.
3. Which flowers are capable of self-pollination? Explain.
4. Which flowers require cross-pollination? Explain.

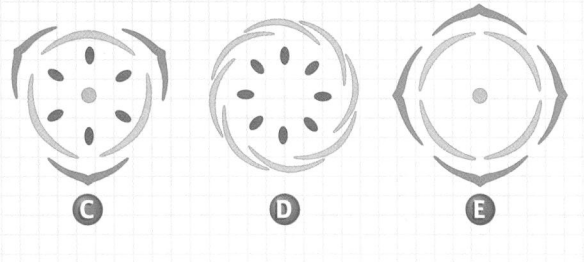

Problem-Solving Lab 24-2

Purpose

Students will study and interpret diagrams of flower cross sections.

Process Skills

think critically, compare and contrast, interpret scientific illustrations

Teaching Strategies

■ Review the basic anatomy of a flower before this activity.
■ Make sure that students understand the differences between complete and incomplete flowers.
■ Review the characteristics of monocots and dicots.
■ Provide examples of flowers having floral structures in multiples of three, four, and five.
■ Illustrate a "cross-section" cut using a cucumber as an example.

Thinking Critically

1. a. monocot; flower parts in threes
   b. dicot; flower parts in four or multiples of four
   c. dicot; flower parts in fours or multiples of four
2. A and C are complete flowers as all flower parts are present. D and E are incomplete flowers as D is missing sepals and E is missing a pistil.
3. A, C, and D as they have both stamens and pistils. Both male and female flower parts are needed for self-fertilization
4. E requires cross-pollination as it does not have a pistil.

✔ Assessment

Performance Have students make a sketch of diagrams C through D and label the flower parts. Have them make a cross-section diagram of the female flower of E. Use the Performance Task Assessment List for Scientific Drawing in **PASC**, p. 55. **L2** **ELL**

BIOLOGY JOURNAL

Keeping a Plant Diary

Linguistic Have students imagine that they are a chrysanthemum. It is springtime and they have just started growing. Ask them to describe what they look like, explain why they are not forming flowers, and indicate the relative lengths of day and night. Have students repeat the process for each of the other seasons. **L1**

Resource Manager

Reteaching Skills Transparency 37 and Master **L1** **ELL**
Basic Concepts Transparency 41 and Master **L2** **ELL**
Section Focus Transparency 59 and Master **L1** **ELL**

Enrichment

Intrapersonal Advise students that the light that triggers flower production does not have to reach the plant's flower buds, but must reach and be detected by the leaves. Have students design an experiment to prove that it is the leaf that must be stimulated by light for the plant to achieve flowering. Suggest that it is possible to place parts of the plant behind light barriers while other parts of the same plant are exposed to light. **L3**

Reinforcement

Explain to students that lily growers get their greenhouse plants to flower early in spring by subjecting the plant to artificial lighting that simulates longer days and shorter nights. Ask students if lilies are short-day plants or long-day plants. *long-day*

Display

Visual-Spatial Using Figure 24.10 as a guide, have students create a giant fan diagram on the wall or on a bulletin board. Ask them to bring in pictures of monocots and dicots to add to the display. **L2 ELL**

664

Figure 24.10
There are two classes of anthophytes—monocots and dicots. Within each class there are many different families, which show a great amount of variation in their vegetative and reproductive structures.

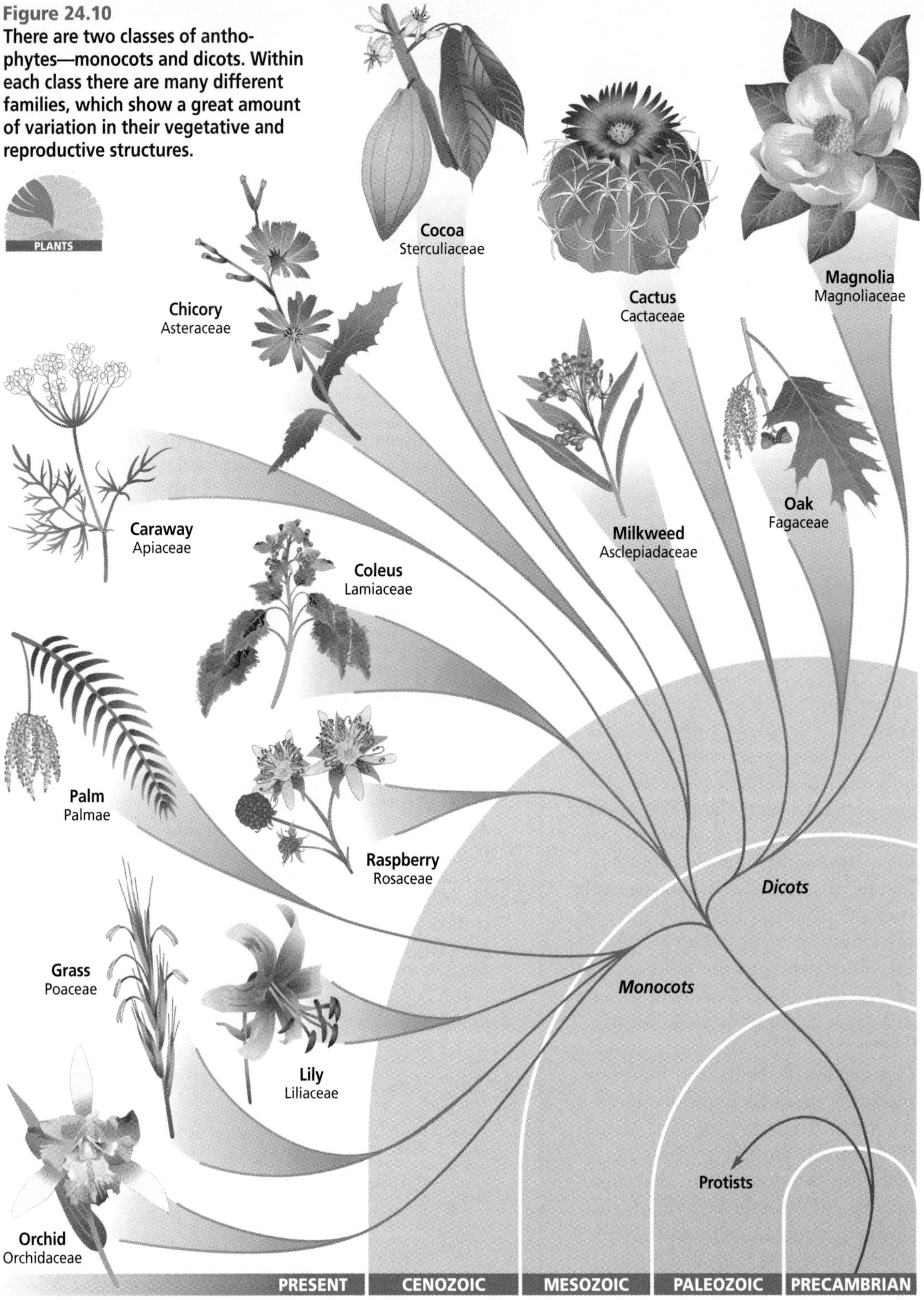

Chicory
Asteraceae

Cocoa
Sterculiaceae

Cactus
Cactaceae

Magnolia
Magnoliaceae

Caraway
Apiaceae

Coleus
Lamiaceae

Milkweed
Asclepiadaceae

Oak
Fagaceae

Palm
Palmae

Raspberry
Rosaceae

Grass
Poaceae

Lily
Liliaceae

Orchid
Orchidaceae

Dicots

Monocots

Protists

PRESENT   CENOZOIC   MESOZOIC   PALEOZOIC   PRECAMBRIAN

664 REPRODUCTION IN PLANTS

PROJECT

Are Bees the Best?

Intrapersonal Members of the *Brassica* genus can be pollinated with cotton swabs, paint brushes, and dried bees. Have students design an experiment to determine whether or not bees are the most efficient pollinators. **L3**

Parts of a Flower

Of the four major organs of a flower, only two—the stamens and pistils—are fertile structures directly involved in seed development. Sepals and petals support and protect the fertile structures and help attract pollinators. The structure of a typical flower is illustrated here by a phlox flower.

Blue phlox

Critical Thinking *How are different flower shapes important to a plant's survival?*

① **Petals** These are usually brightly colored and often have perfume or nectar at their bases to attract pollinators. In many flowers, the petal also provides a surface for insect pollinators to rest on while feeding. Petals may be fused to form a tube, or shaped in ways that make the flower more attractive to pollinators.

② **Stigma** At the top of the pistil is this sticky or feathery surface on which pollen grains land and grow. The style is the slender stalk of the pistil that connects the stigma to the ovary. The pollen tube grows down the length of the style to reach the ovary. The ovary, which will eventually become the fruit, contains the ovules. Each ovule, if fertilized, will become a seed.

③ **Sepals** A ring of sepals makes up the outermost portion of the flower. The sepals serve as a protective covering for the flower bud, helping to protect it from insect damage and prevent it from drying out. Sepals sometimes are colored and resemble petals.

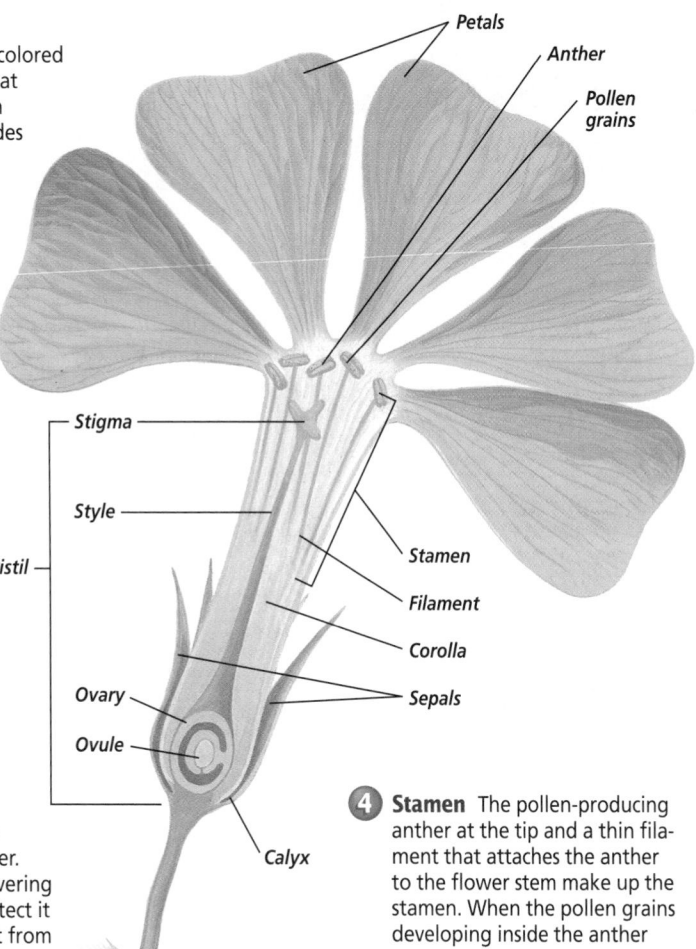

Petals
Anther
Pollen grains
Stigma
Style
Pistil
Stamen
Filament
Corolla
Ovary
Sepals
Ovule
Calyx

④ **Stamen** The pollen-producing anther at the tip and a thin filament that attaches the anther to the flower stem make up the stamen. When the pollen grains developing inside the anther reach maturity, the anther splits open to release them.

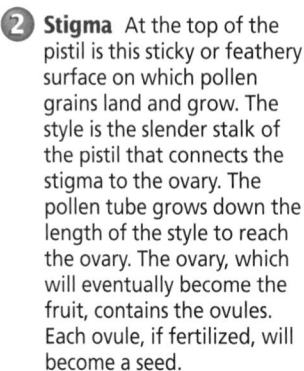
Purpose
To describe flower anatomy and identify the role flower parts play in reproduction.

Teaching Strategies
■ Ask students to explain why stamens and pistils are described as the fertile structures in a flower. *These structures are involved in the production of egg and sperm.* Elicit what role is played by flower organs that are not important in fertilization. *These organs attract pollinators or protect young or immature fertile flower parts.* **L2**

Visual Learning
■ Ask students to use the captions to explain how the following pairs of terms are related: (a) sepals and calyx; (b) petals and corolla; (c) stamen and anther; (d) stigma and style; (e) ovary and ovule. **L1**

Critical Thinking
The shape of a flower may promote different pollination methods. For example, tubular flowers are probably pollinated by animals.

Chalkboard Activity
Write the following phrases on the chalkboard: contains chlorophyll, contains anthocyanin, where meiosis takes place, contains gametophyte, is part of the sporophyte, contains diploid cells, forms spores, forms microspores, forms megaspores. Ask students to identify the flower organ or organs described by each phrase. **L2**

MEETING INDIVIDUAL NEEDS

English Language Learners/ Learning Disabled

Visual-Spatial Have students create a table with the heads Male, Female, Neither male nor female. Ask students to list the following structures beneath the appropriate head: stamen, pistil, anther, calyx, corolla, ovary, stigma, petal, sepal, pollen, egg, ovule, style. **L1 ELL**

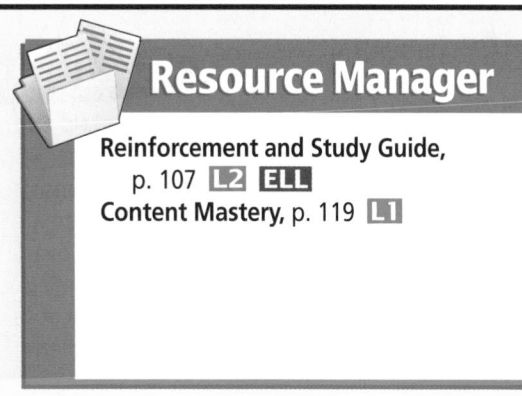

Resource Manager

Reinforcement and Study Guide, p. 107 **L2 ELL**
Content Mastery, p. 119 **L1**

3 Assess

Check for Understanding

Ask students to explain the relationships of the following word groups. **L2** **ELL**
   a. stamen—anther
   b. pistil—stigma—ovary
   c. ovary—ovule

Reteach

Obtain a flower model. As you point out structures on the model, have students name each structure and explain its function. **L1**

Extension

Linguistic Have students determine the meaning of the terms androecium, gynoecium, and perianth. Have them assign the term sterile or nonsterile to each flower part. **L3**

✓ Assessment

Performance Provide students with a diagram of a typical flower. Have them label the parts of the flower and identify reproductive parts. **L1**

4 Close

Discussion

Visual-Spatial Show students pictures of flowers with different structural modifications. Have them point out the different flower parts and speculate as to the advantage of the modification. **L2**

666

Figure 24.11
Photoperiodism refers to a plant's sensitivity to the changing length of night.

B Spinach and lettuce (above) are long-day plants that flower in midsummer.

A Short-day plants, such as pansies (above) and goldenrod, flower in late summer and fall or early spring.

C Most plants are day-neutral. Flowering in cucumbers (above), tomatoes, and corn is not influenced by a dark period.

WORD Origin

photoperiodism
From the Greek words *photos*, meaning "light," and *periodos*, meaning "a period." The flowering response of a plant to periods of dark and light is photoperiodism.

Short-day plants are induced to flower by exposure to a long night. These are plants that usually form flower buds in the fall when the days are getting shorter and the nights are long, as shown in *Figure 24.11A*. Flowering occurs in the spring, as in crocuses, or in the fall as in strawberry plants. **Long-day plants** flower when days are longer than the nights, as shown in *Figure 24.11B*. Examples of these are carnations, petunias, potatoes,

and garden peas. Most plant species are **day-neutral plants,** which means temperature, moisture or environmental factors other than day length control their flowering times, as shown in *Figure 24.11C*. The photoperiodism of flowers may ensure that a plant produces its flowers at a time when there is an abundant population of pollinators. This is important because pollination is a critical event in the life cycle of a flowering plant.

Section Assessment

Understanding Main Ideas
1. Compare and contrast sepals and petals.
2. Describe the male and female parts of a flower.
3. Explain why squash flowers are considered incomplete flowers.
4. How does photoperiodism influence flowering?

Thinking Critically
5. In the middle of the summer a florist receives a

large shipment of short-day plants. What must the florist do to induce flowering?

SKILL REVIEW

6. **Comparing and Contrasting** Explain why the structure of wind-pollinated flowers is often different from that of insect pollinated flowers. For more help, refer to *Organizing Information* in the **Skill Handbook**.

666 REPRODUCTION IN PLANTS

Section Assessment

1. Both are leaflike structures. Petals are bright colors. Sepals are usually green.
2. The male part is the stamen, which consists of the anther and filament. The female part is the pistil, which contains the ovary and ovules.
3. Squash flowers are incomplete because they do not have both pistils and stamens.
4. Flowering is controlled by photoperiodism. For example, short-day plants flower when the days are short and nights are long.
5. The florist should cover the greenhouse with tarps each afternoon to shorten the day and lengthen the night.
6. Wind-pollinated flowers are small and lack petals. Insect-pollinated flowers have brightly colored petals and nectar.

24.3 The Life Cycle of a Flowering Plant

Transferring pollen from anther to stigma is just one step in the life cycle of a flowering plant. How does pollination lead to the development of seeds encased in fruit? How do sperm cells in the pollen grain reach the egg cells in the ovary? These steps in the reproductive cycle of anthophytes take place without water—an evolutionary step that enabled flowering plants to occupy nearly every environment on Earth.

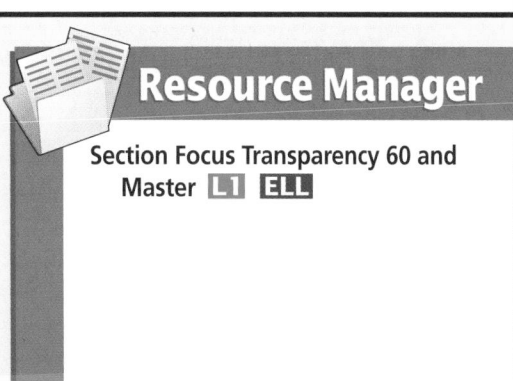

Bee orchid *Ophrys speculum* and ovary of a flower (inset)

SECTION PREVIEW

Objectives
Describe the life cycle of a flowering plant.
Outline the processes of seed and fruit formation and seed germination.

Vocabulary
polar nuclei
double fertilization
endosperm
dormancy
germination
radicle
hypocotyl

The Life Cycle of an Anthophyte

The life cycle of flowering plants is similar to that of conifers in many ways. In both coniferophytes and anthophytes, the gametophyte generation is contained within the sporophyte. Many of the reproductive structures are also similar. However, anthophytes are the only plants that produce flowers and fruits. *Figure 24.12* summarizes the life cycle of flowering plants.

Development of the female gametophyte

In anthophytes, the female gametophyte is formed in the ovule within the ovary. In the ovule, a cell undergoes meiosis, producing haploid megaspores. One of these megaspores will produce the female gametophyte. The other three spores die. In most flowering plants, the megaspore divides by mitosis three times, producing eight nuclei. These eight nuclei are the embryo sac or female gametophyte. Six of the nuclei are contained within six haploid cells, one of which is the egg cell. The two remaining nuclei, which are called **polar nuclei,** are both in one cell. This cell, the central cell, is located at the center of the embryo sac. The egg cell is near the micropyle. The other five cells are arranged as shown in *Figure 24.13*.

Prepare

Key Concepts
The formation of anthophyte gametophytes, pollination, and double fertilization are presented. Seed and fruit development are then discussed. The section ends with an explanation of seed dispersal and seed germination.

Planning
- Purchase peanuts for the Quick Demo.
- Find flowers for Assessment.
- Purchase bean seeds for the Tech Prep.
- Purchase beans and tetrazolium chloride for the Alternative Lab.
- Purchase tomato and peach for the Quick Demo.
- Purchase corn and bean seeds for MiniLab 24-2.

1 Focus

Bellringer
Before presenting the lesson, display **Section Focus Transparency 60** on the overhead projector and have students answer the accompanying questions. **L1 ELL**

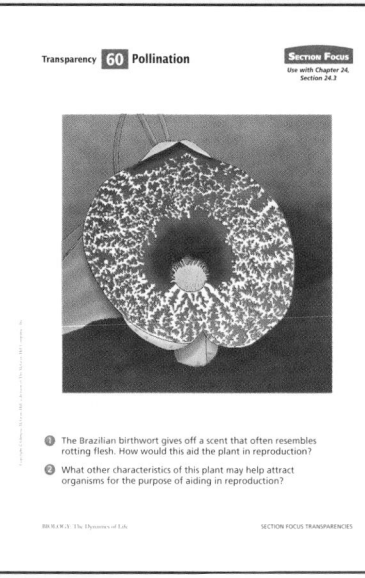

Transparency **60** Pollination

SECTION FOCUS
Use with Chapter 24, Section 24.3

1. The Brazilian birthwort gives off a scent that often resembles rotting flesh. How would this aid the plant in reproduction?

2. What other characteristics of this plant may help attract organisms for the purpose of aiding in reproduction?

Resource Manager

Section Focus Transparency 60 and Master **L1 ELL**

Quick Demo

Visual-Spatial Show students the tiny sporophyte inside a peanut. It is the structure with two tiny leaves. **CAUTION:** ***Do not allow students to eat the peanuts as some students may be allergic to them.***

Reinforcement

Visual-Spatial Provide students with a black line drawing of the life cycle of a flowering plant. Using colored pencils, have them shade in the gametophyte stage and the sporophyte stage in two different colors. L1 ELL

Microscope Activity

Visual-Spatial Have students view prepared slides of *Lilium* ovaries. Ask students to draw and label an ovule. L1

Resource Manager

Basic Concepts Transparency 42 and Master L2 ELL

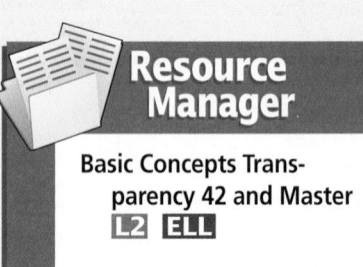

Figure 24.12
In the life cycle of a flowering plant, the sporophyte generation nourishes and protects the developing gametophyte. After fertilization, the new sporophyte, which is contained in a seed or fruit, is released from the parent plant.

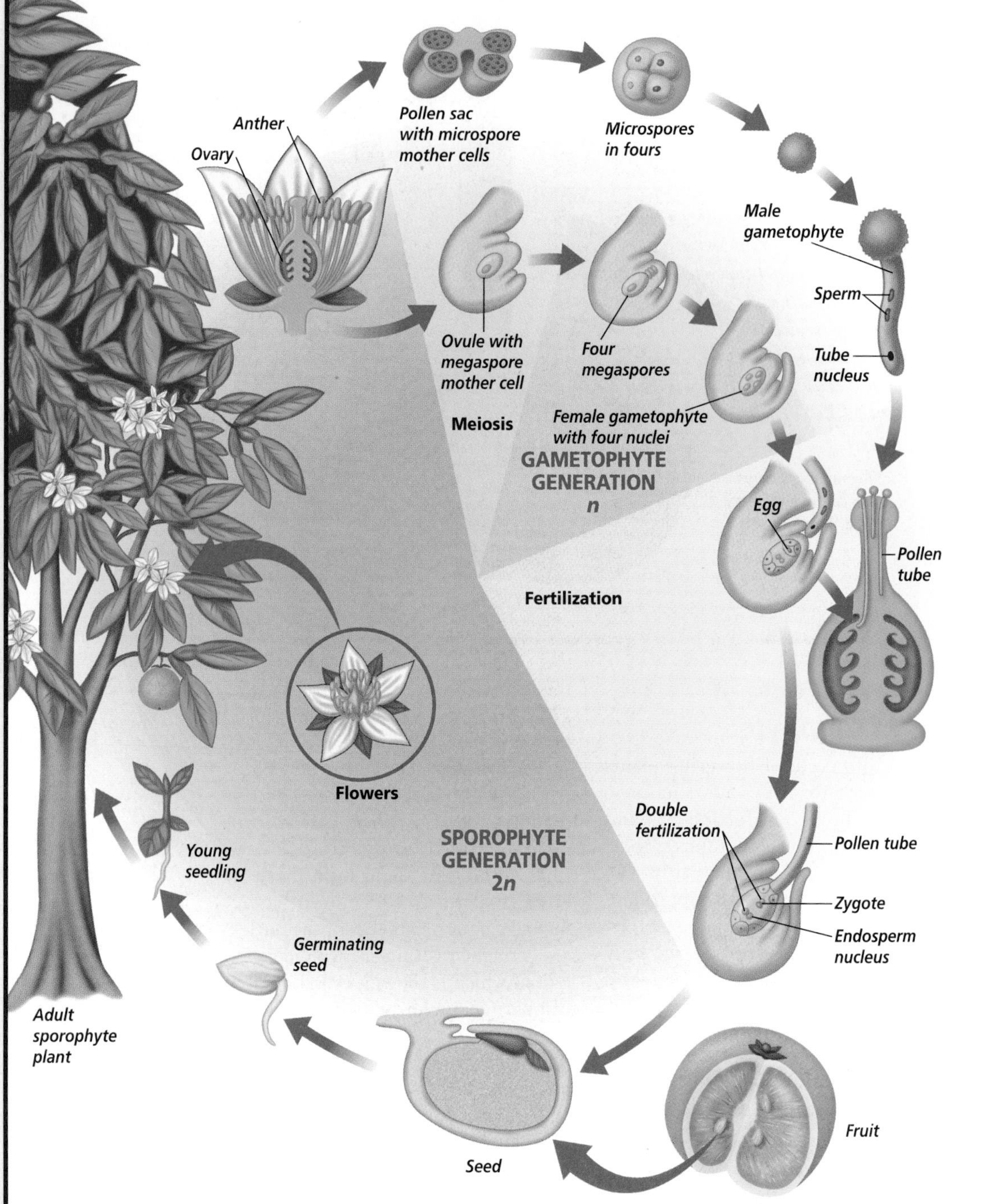

Anther
Ovary
Pollen sac
with microspore
mother cells
Microspores
in fours
Male
gametophyte
Sperm
Tube
nucleus
Ovule with
megaspore
mother cell
Four
megaspores
Meiosis
Female gametophyte
with four nuclei
GAMETOPHYTE
GENERATION
n
Egg
Pollen
tube
Fertilization
Flowers
Double
fertilization
Pollen tube
Zygote
Endosperm
nucleus
SPOROPHYTE
GENERATION
2n
Young
seedling
Germinating
seed
Adult
sporophyte
plant
Seed
Fruit

Biology Journal

Flowering Plant Life Cycle

Linguistic Have students write a paragraph in their journal outlining the steps of a flowering plant's life cycle. Encourage them to illustrate the concepts with drawings. L2

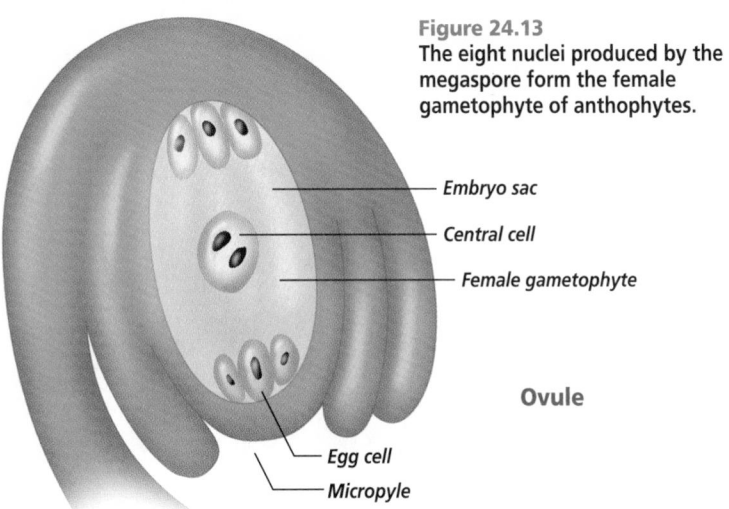

Figure 24.13
The eight nuclei produced by the megaspore form the female gametophyte of anthophytes.

- Embryo sac
- Central cell
- Female gametophyte

Ovule

- Egg cell
- Micropyle

Development of the male gametophyte

The formation of the male gametophyte begins in the anther, as seen in *Figure 24.14*. Haploid microspores are produced by meiosis within the anther. The microspores each divide into two cells. A thick, protective wall surrounds these two cells. This two-celled structure is the immature male gametophyte, or pollen grain. The cells within the pollen grain are the tube cell and the generative cell. When the pollen grains are mature the anther splits open. Depending on the type of flower, the pollen may be carried to the pistil by wind, water, or animals.

Pollination

In anthophytes, pollination is the transfer of the pollen grain from the

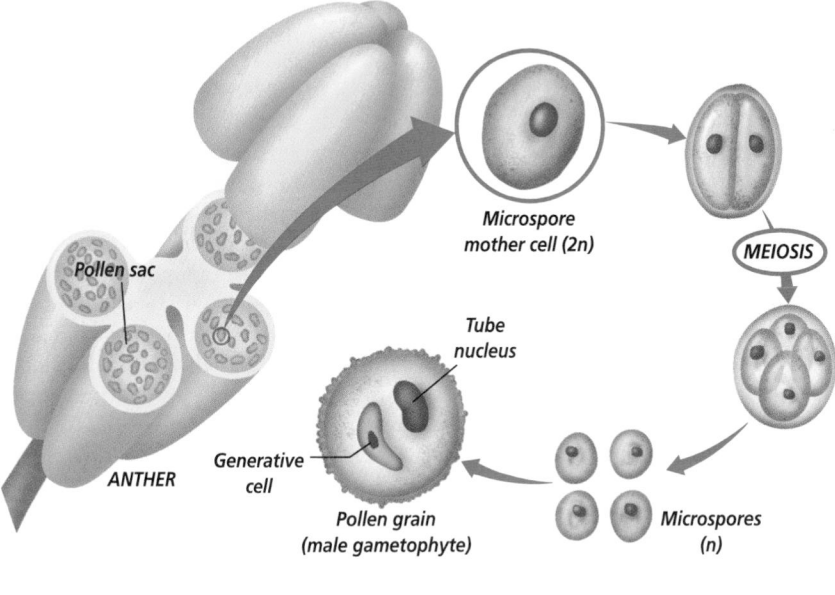

Figure 24.14
Meiotic division of each of many cells within the anther produces four microspores. These microspores develop into the male gametophyte or pollen grain

Pollen sac

Microspore mother cell (2n)

MEIOSIS

Tube nucleus

Generative cell

ANTHER

Pollen grain (male gametophyte)

Microspores (n)

Building a Model

Kinesthetic Use a plastic sandwich bag to represent a female gametophyte. Place three small balls of clay of the same color within the bag and explain that each ball represents a haploid nucleus. (One is the egg and the other two are the central cell.) Introduce two small balls of clay of a different color to the bag. Explain that each ball represents the two sperm nuclei that enter the female gametophyte. Fuse one sperm with the egg and explain that this represents the zygote, which is now diploid. Fuse the other three nuclei. Explain that this represents the triploid nucleus that forms endosperm. Point out that together the fusing of the clay illustrates double fertilization. **L1**

✔ Assessment

Knowledge Show students a flower pistil that has been cut open to reveal the ovary. Ask them to point to where the following occur: growth of pollen tube, development of female gametophyte, and double fertilization. **L2**

BIOLOGY Online
Note Internet addresses that you find useful in the space below for quick reference.

_____
_____
_____
_____
_____

Figure 24.15
The shape, color, and size of a flower reflect its relationship with a pollinator.

A The butterfly uses its long proboscis to sip nectar that bees and flies cannot reach.

B The wind-pollinated flowers of this ragweed plant are small and green and lack structures that would block wind currents.

C Bats sip nectar from night-blooming flowers with a strong, musty odor, such as bananas and some cacti.

D Flowers pollinated by hummingbirds are often tubular and colored bright red or yellow but may have little scent.

E The ultraviolet markings of some flowers guide insects to a flower's nectar.

anther to the pistil. Plant reproduction is most successful when the pollination rate is high, which means that the pistil of a flower receives enough pollen of its own species to fertilize the egg in each ovule. Many anthophytes have elaborate mechanisms that help ensure that pollen grains are deposited in the right place at the right time. Some of these are shown in *Figure 24.15.* Although it may seem wasteful for wind-pollinated plants to produce such large amounts of pollen, it does help ensure pollination. Most anthophytes pollinated by animals produce nectar, which serves as a valuable, highly concentrated food for visitors to the flowers. Nectar is a liquid made up of proteins and sugars. It usually collects in the cuplike area at the base of the petals. Animals such as insects and birds brush up against the anthers while trying to get to the nectar. The pollen that attaches to them can be carried to another flower, resulting in pollination. Some insects also gather pollen to use as food. By producing nectar and attracting animal pollinators, animal-pollinated plants are able to promote pollination without producing large amounts of pollen.

Some nectar-feeding pollinators are attracted to a flower by its color or scent or both. Some of the bright, vivid flowers attract pollinators such as butterflies and bees. Some of these flowers have markings that are invisible to the human eye but are easily seen by insects. Flowers that are pollinated by beetles and flies have a strong scent but are often dull in color.

Many flowers have structural adaptations that favor cross-pollination. This results in greater genetic variation because the sperm from one plant fertilizes the egg from another. For example, the flowers of certain species of orchids resemble female wasps. The male wasps visit the flower and attempt to mate with it and become covered with pollen, which is deposited on orchids it may visit in the future.

Fertilization

Once a pollen grain has reached the stigma of the pistil, several events take place before fertilization occurs. Inside each pollen grain are two haploid cells, the tube cell and the generative cell. The tube cell nucleus directs the growth of the pollen tube down through the pistil to the ovary, as shown in *Figure 24.16*. The generative cell divides by mitosis, producing two haploid sperm cells. The sperm cells are transported by the pollen tube through a tiny opening in the ovule called the micropyle.

Within the ovule is the female gametophyte, composed of eight haploid cells. One of the sperm unites with the egg cell forming a diploid zygote, which begins the new sporophyte generation. The other sperm cell fuses with the central cell, which contains the polar nuclei, to form a cell with a triploid ($3n$) nucleus. This process, in which one sperm fertilizes the egg and the other sperm joins with the central cell, is called **double fertilization.**

Figure 24.16
In flowering plants, the male gametophyte grows through the pistil to reach the female gametophyte. Double fertilization involves two sperm cell nuclei. One sperm cell unites with the egg nucleus and the other sperm cell unites with the diploid central cell of the female gametophyte.

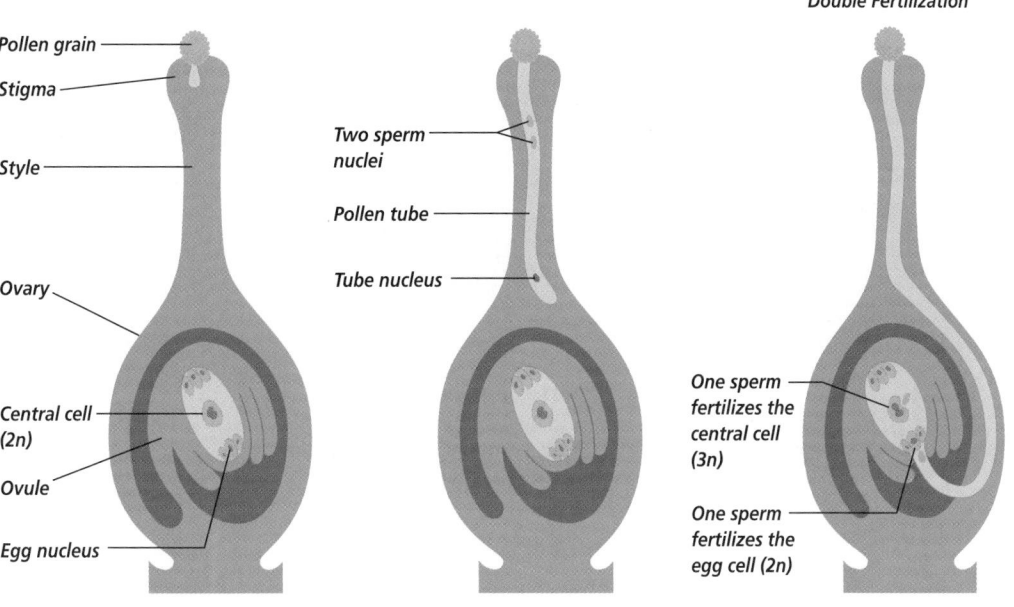

Double Fertilization

Pollen grain
Stigma
Style
Ovary
Central cell (2n)
Ovule
Egg nucleus

Two sperm nuclei
Pollen tube
Tube nucleus

One sperm fertilizes the central cell (3n)
One sperm fertilizes the egg cell (2n)

GLENCOE TECHNOLOGY

VIDEODISC
Biology: The Dynamics of Life
Double Fertilization (Ch. 24)
Disc 1, Side 2, 49 sec.

CD-ROM
Biology: The Dynamics of Life
Animation: *Double Fertilization*
Disc 3

Double fertilization is unique to anthophytes and is illustrated in *Figure 24.16*. The triploid nucleus will divide many times, eventually forming the endosperm of the seed. The **endosperm** is food storage tissue that supports development of the embryo in anthophyte seeds.

Many flowers contain more than one ovule. Pollination of these flowers requires that at least one pollen grain land on the stigma for each ovule contained in the ovary. In a watermelon plant, for example, hundreds of pollen grains are required to pollinate a single flower if each ovule is to be fertilized. You are probably familiar with the hundreds of watermelon seeds that are the result of this process.

Seeds and Fruits

The embryo contained within a seed is the next sporophyte generation. The formation of seeds and the fruits that enclose them help ensure the survival of the next generation.

Seed formation

After fertilization takes place, most of the flower parts die and the seeds begin to develop. The wall of the ovule becomes the hard seed coat, which may aid in dispersal and helps protect the embryo until it begins growing into a new plant. Inside the ovule, the zygote divides and grows into the plant embryo. The triploid central cell develops into the endosperm.

Figure 24.17
A fruit consists of the seeds and the surrounding mature ovary of a flowering plant.

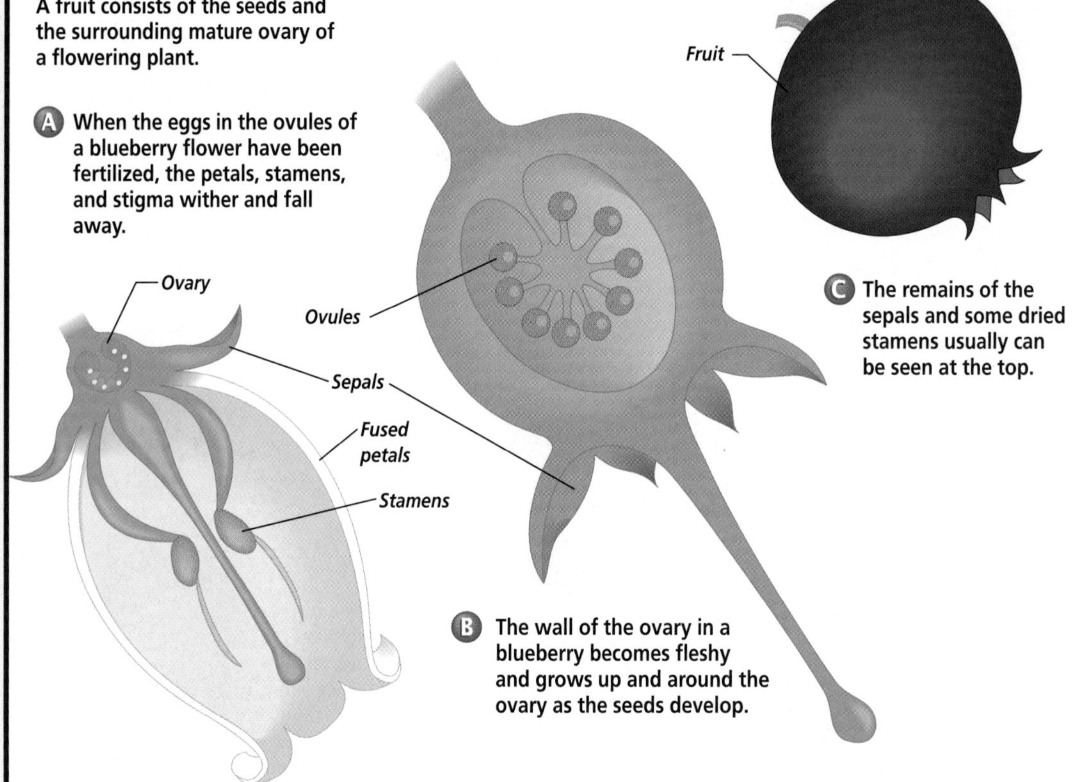

Ⓐ When the eggs in the ovules of a blueberry flower have been fertilized, the petals, stamens, and stigma wither and fall away.

Ovary
Ovules
Sepals
Fused petals
Stamens

Fruit

Ⓒ The remains of the sepals and some dried stamens usually can be seen at the top.

Ⓑ The wall of the ovary in a blueberry becomes fleshy and grows up and around the ovary as the seeds develop.

672 REPRODUCTION IN PLANTS

Fruit formation

As the seeds develop, the surrounding ovary enlarges and becomes the fruit. A fruit is the structure that contains the seeds of an anthophyte. *Figure 24.1*7 shows how the fruit of a blueberry develops from the ovary inside the flower.

A fruit is as unique to a plant as its flower, and many plants can be identified by examining the structure of their fruit. You are familiar with plants that develop fleshy fruits, such as apples, grapes, melons, tomatoes, and cucumbers. Other plants develop dry fruits such as peanuts, sunflower "seeds," and walnuts. In dry fruits, the ovary around the seeds hardens as the fruit matures. Some plant foods that we call vegetables or grains are actually fruits, as shown in *Figure 24.18*. Can you think of any vegetables that are actually fruits? For example, tomatoes are fleshy fruits that are often referred to as vegetables.

Figure 24.18
A fruit is the ripened ovary of a flower that contains the seeds or seed of the plant. The most familiar fruits are those we consume as food.

B Fleshy fruits develop a juicy fruit wall full of water and sugars.

A Dry fruits have dry fruit walls. The ovary wall may start out with a fleshy appearance, as in hickory nuts or bean pods, but when the fruit is fully matured, the ovary wall is dry.

Cultural Diversity

The History of Chocolate

Theobroma cacao is a native tree from Central America. After flowering, it produces large pods that contain about 40 cacao beans. Because of its seeds, cacao was domesticated by the Mayas and Aztecs who turned the seeds into a rich brown drink called chocolate. Researchers are suspicious that the chemicals in chocolate, theobromine and methylxanthin, may have an addictive quality. This may explain why some individuals crave chocolate and chocolate products.

Tying to Previous Knowledge

Ask students to define the term polyploidy. Explain that it is estimated that between 35 and 50% of all flowering plants are polyploid. The most common polyploid condition is tetraploid or 4*n*. Tetraploids are often bigger or more vigorous that the diploid plant and thus are more desirable for agriculture and/or horticulture.

Figure 24.19
A wide variety of seed-dispersal mechanisms have evolved among flowering plants.

B Clinging fruits, like those of the cocklebur and burdock, are covered by hooks that stick to the fur or feathers of passing animals or the clothes of passing humans.

C Wind-dispersed seeds have adaptations that enable them to be held aloft while they drift away from their parent.

A The ripe pods of violets snap open with a pop, which sends a shower of small seeds in all directions.

Seed dispersal

A fruit not only protects the seeds inside it, but also may aid in dispersing those seeds away from the parent plant and into new habitats. The dispersal of seeds, *Figure 24.19,* is important because it reduces competition for sunlight, soil, and water between the parent plant and its offspring. Animals such as raccoons, deer, bears, and birds help distribute many seeds by eating dry or fleshy fruits. They may carry the fruit some distance away from the parent plant before consuming it and spitting out the seeds. Or they may eat the fruit, seeds and all. Seeds that are eaten may pass through the digestive system unharmed and are deposited in the animal's wastes. Squirrels, birds, and other nut gatherers may drop and lose some of the seeds they collect, or even bury them only to forget where. These seeds can then germinate far from the parent plant.

Plants, such as water lilies and coconut palms that live in or near water produce fruits or seeds with air pockets in the walls, which enable them to float and drift away from the parent plant. The ripened fruits of many plants split open to release seeds designed for dispersal by the wind or by clinging to animal fur. Orchid seeds are so tiny that they resemble dust grains or feathers and are easily blown about by the wind. The fruit of the poppy flower forms a seed-filled capsule that releases sprinkles of tiny seeds like a salt shaker as it bobs about in the wind. Tumbleweed seeds are scattered by the wind as the whole plant rolls along the ground.

Seed germination

At maturity, seeds are fully formed. The seed coat dries and hardens, enabling the seed to survive conditions that are unfavorable to the parent plant. The seeds of some plant species must germinate immediately or die. However, the seeds of some plant species can remain in the soil until conditions are favorable for

---

Alternative Lab

Respiring Seeds

Purpose
To compare the rate of respiration in germinating and nongerminating seeds.

Materials
canned kidney beans, paper cup, water, dried kidney beans, wax paper, labels, tetrazolium solution in dark dropper bottles (add 1 g of 2,3,5-triphenyl tetrazolium chloride to 100 mL distilled water)

Procedure
Give the following directions to students.
1. Soak ten kidney bean seeds overnight in water. The next day, split each seed in half. Split ten canned kidney bean seeds in half.
2. Place all seeds split-side up onto two sheets of wax paper. Mark each paper with a label indicating seed treatments.
3. Add a drop of tetrazolium to each cut surface. Wait 20-30 minutes. Record in which seeds and where in the seeds a pink color appears. Note: Tetrazolium can be diluted and discarded down the sink, but the bean seeds cannot be discarded in the sink. Caution students to

growth and development of the new plant. This period of inactivity in a mature seed is called **dormancy.** The length of time a seed remains dormant can vary from one species to another. Some seeds, such as willow, magnolia, and maple remain dormant for only a few weeks after they mature. These seeds cannot survive harsh conditions for long periods of time. Other plants produce seeds that can remain dormant for remarkably long periods of time, *Figure 24.20*. Even under harsh conditions, the seeds of desert wildflowers and some conifers can survive dormant periods of 15 to 20 years. Scientists discovered ancient seeds of the East Indian Lotus, *Nelumbo nucifera*, in China, which they have radiocarbon dated to be more than a thousand years old. Imagine their amazement when these seeds germinated!

Requirements for germination

Dormancy ends when the seed is ready to germinate. **Germination** is the beginning of the development of the embryo into a new plant. The absorption of enough water and the presence of oxygen and favorable temperatures usually end dormancy, but there may be other requirements. Water is important because it activates the embryo's metabolic system. Once metabolism has begun, the seed must continue to receive water or it will die. Just before the seed coat breaks open, the plant embryo begins to respire rapidly. Many seeds germinate best at temperatures between 25°C and 30°C. Arctic species germinate at lower temperatures than do tropical species. At temperatures below 0°C or above 45°C, most seeds won't germinate at all.

Some seeds have special requirements for germination, *Figure 24.21*.

Figure 24.20
Seeds can remain dormant for long periods of time. Lupine seeds like these may germinate after remaining dormant for decades.

For example, some germinate more readily after they have passed through the acid environment of an animal's digestive system. Others require a period of freezing temperatures, as do apple seeds, extensive soaking in saltwater, as do coconut seeds, or certain day lengths. Recall that the seeds of some conifers will not germinate unless they have been exposed to fire. The same is true of

Figure 24.21
The seeds of the desert tree *Cercidium floridum* have hard seed coats that must be cracked open in order to germinate. This occurs when the seeds tumble down rocky gullies in sudden rainstorms.

use care when working with chemicals.

Analysis

1. Tetrazolium indicates cell respiration when it turns pink. Which seed treatment carried on cell respiration? Why? *Uncanned seeds are germinating and require energy; canned seeds are dead.*

2. The darker the pink color, the greater the rate of respiration. Which seed part carries on the greatest rate of respiration? Why? *Embryo; rapid growth requires more energy.*

✓ Assessment

Portfolio Design an experiment to determine how many seeds in a seed packet are alive. Use the Performance Task Assessment List for Designing an Experiment in **PASC**, p. 23. **L3**

Purpose

To compare dormant and germinating monocot and dicot seed embryos.

Process Skills

observe and infer, compare and contrast

Safety Precautions

Advise students to use caution with razor blades and always to cut away from the body.

Teaching Strategies

■ Soak some seeds 24 hours and others 72 hours prior to use. After the 72-hour seeds have been soaking for 24 hours, wrap them in moist paper towels and place them in sealed plastic bags.

■ Advise students to lay the corn on its flat side while cutting.

Expected Results

Germinating seeds have larger embryos.

Analysis

1. Labels should include cotyledon(s), embryo.
2. Diagrams of dicot should include cotyledons, embryo, epicotyl, radicle, plumule, and hypocotyl. Diagrams of monocot should include only cotyledon and embryo.
3. Monocot has one cotyledon, small embryo, is slow to germinate, and its embryo parts are hard to differentiate. Dicot seed has two cotyledons, larger embryo, and easily seen embryo parts.

✔ Assessment

Performance Provide students with seeds. Have students classify each seed after removing the seed coat. Use the Performance Task Assessment List for Making and Using a Classification System in **PASC,** p. 49. **L2**

676

Figure 24.22
Many wildflowers require fire for their seeds to germinate. This is especially true in prairie environments where fires are periodically set to induce the germination of prairie wildflower seeds.

Figure 24.23
Germination of a bean seed is stimulated by warm temperatures and water, which softens the seed coat.

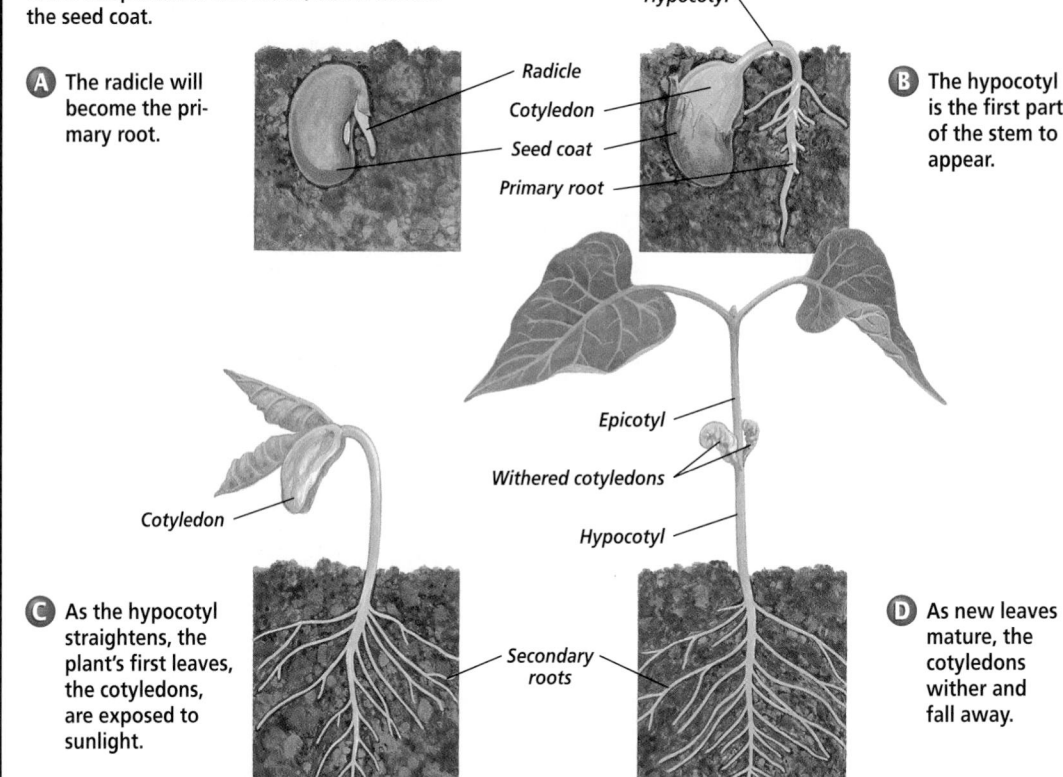

A The radicle will become the primary root.

Radicle
Cotyledon
Seed coat
Primary root

B The hypocotyl is the first part of the stem to appear.

Hypocotyl

Epicotyl
Withered cotyledons
Hypocotyl

Cotyledon

C As the hypocotyl straightens, the plant's first leaves, the cotyledons, are exposed to sunlight.

Secondary roots

D As new leaves mature, the cotyledons wither and fall away.

certain wildflower species, including lupines and gentians, *Figure 24.22.*

The germination of a typical dicot embryo is shown in *Figure 24.23.* Once the seed coat has been softened by water, the embryo starts to emerge from the seed. The first part of the embryo to appear is the embryonic root called the **radicle** (RAD ih kul). The radicle grows down into the soil and develops into a root. The portion of the stem nearest the seed is called the **hypocotyl** (HI poh kaht ul). In some plants, the first part of the stem to push above ground is an arched portion of the hypocotyl. As the hypocotyl continues growing, it straightens, bringing with it the

Resource Manager

Reinforcement and Study Guide, p. 108 **L2** **ELL**
BioLab and MiniLab Worksheets, p. 110 **L2**
Content Mastery, pp. 117, 119-120 **L1**

✔ Portfolio

Modeling Seed Dispersal

Kinesthetic Provide students with a strip of Velcro, a piece of fabric, cotton batting, tape, string, clay, stiff paper, and scissors. Have them use the materials to prepare models that show different means of seed dispersal. Models may imitate attaching to animal fur or floating in air. **L1** **ELL**

cotyledons and the plant's first leaves. As the stem grows larger and the leaves turn green, the plant can produce its own food through photosynthesis. To learn more about germinating seeds, try the *MiniLab* shown here.

Vegetative reproduction

The roots, stems, and leaves of plants are called vegetative structures. When these structures produce a new plant, it is called vegetative reproduction. Vegetative reproduction is common among anthophytes. Some modified stems of anthophytes, such as potato tubers, can produce a new plant. Potatoes will grow new stems from their "eyes" or buds. Farmers make use of this feature when they cut potato tubers into pieces and plant them. The buds on these sections grow new shoots that produce entire new plants.

Although cloning animals is a relatively new phenomenon, gardeners have relied for years on cloning to reproduce plants. Using vegetative reproduction to grow numerous plants from one plant it is frequently referred to as vegetative propagation. Some plants, such as geraniums, can be propagated by planting cuttings, which are pieces of the stem or a leaf that has been cut off another plant. Even smaller pieces of plants can be used to grow plants by tissue culture. Tiny pieces of plants are placed on nutrient agar in test tubes or petri dishes. The plants produced by cuttings and tissue cultures contain the same genetic makeup as the original plants. Therefore, they are botanical clones.

MiniLab 24-2   Observing

Looking at Germinating Seeds Seeds are made up of a plant embryo, a seed coat, and in some plants, a food-storage tissue. Monocot and dicot seeds differ in their internal structures.

Corn seed germination

Procedure
1. Obtain from your teacher a soaked, ungerminated corn kernel (monocot), a bean seed (dicot), and corn and bean seeds that have begun to germinate.
2. Remove the seed coats from each of the ungerminated seeds, and examine the structures inside. Use low-power magnification. Locate and identify each structure of the embryo and any other structures you observe.
3. Examine the germinating seeds. Locate and identify the structures you observed in the dormant seeds.

Analysis
1. Diagram the dormant embryos in the soaked seeds, and label their structures.
2. Diagram the germinating seeds, and label their structures.
3. List at least three major differences you observed in the internal structures of the corn and bean seeds.

Section Assessment

Understanding Main Ideas
1. What is the relationship between the pollination of a flower and the production of one or more seeds?
2. What part of an anthophyte flower becomes the fruit?
3. Describe the process of double fertilization in anthophytes.
4. Explain how the production of nectar could enhance the pollination of a flowering plant.

Thinking Critically
5. Describe the formation of the female gametophyte in a flowering plant.

SKILL REVIEW
6. **Making and Using Tables** Make a table that indicates whether each structure of a flower is involved in pollination, fruit formation, seed production, or seed dispersal. For more help, refer to *Organizing Information* in the **Skill Handbook**.

24.3   THE LIFE CYCLE OF A FLOWERING PLANT   **677**

Section Assessment

1. Pollination transports sperm to the female gametophyte. Inside the ovule, the sperm cells are involved in double fertilization. One pollen grain is required for the production of each seed.
2. the ovary wall
3. Double fertilization occurs when one sperm unites with the egg cell, forming the 2n zygote and the other sperm cell fuses with the central cell to form a 3n cell.
4. Nectar is a source of nutrition for animal pollinators such as insects and bats; they get pollen on them as they try to reach the nectar. As these animals travel from flower to flower, they cross-pollinate the flowers.
5. In the ovule, meiosis produces four haploid megaspores. One of these megaspores will undergo three mitotic divisions, producing eight nuclei. These eight nuclei make up the embryo sac or female gametophyte.
6. pollination: anther, stamen, stigma; fruit formation: ovary, and ovule; seed production: ovule, egg; seed dispersal: ovary

677

---

3 Assess

Check for Understanding

Have students explain how the words in each of the following pairs are related. **L2**
   a. pollen tube—double fertilization
   b. epicotyl—hypocotyl
   c. dormancy—germination

Reteach

Have students indicate when or where in the life cycle each term in the Check for Understanding applies. **L2**

Extension

Have students research the mechanism responsible for roots growing down and stems growing up. **L3**

✓ Assessment

Performance Provide students with a sliced peach or cherry half and a sliced open flower. Have them locate and name the following structures: ovule before and after fertilization, ovary before and after fertilization, embryo. **L1**

---

4 Close

Discussion

Watermelons have numerous black seeds and some small white seeds. Ask students to speculate what the small white seeds found in a watermelon might be. *They are unfertilized ovules.* **L2**

**Examining the Structure
of a Flower**

Time Allotment

One class period

Process Skills

observe and infer, use the microscope, compare and contrast, interpret data

Safety Precautions

■ Some students may be allergic to the chemicals used in preserved specimens.

■ Caution students against rubbing their eyes with their hands because of the preservative. Always have students wash their hands thoroughly after handling preserved specimens.

■ Handle the razor blade with extreme caution. Always cut away from the body.

PREPARATION

Alternative Materials

■ Do not use composite species of flowers such as sunflower, daisy, or dandelion.

■ Preserved slides that show eggs within the ovule are available from supply houses.

■ Preserved flowers are available at any time of the year from biological supply houses. A disadvantage to such flowers is that the preservative tends to discolor flower parts, especially petals and sepals, which will appear dull green.

F lowers are the reproductive structures of anthophytes. Seeds that develop within the flower are carried inside a fruit. Seeds provide an extremely important form of reproduction in flowering plants. Flowers come in many colors and shapes. Often their colors or shapes are related to the manner in which pollination takes place. The major organs of a flower include the petals, sepals, stamens, and pistils. Some flowers are incomplete, which means they do not have all four kinds of organs. You will study a complete flower.

PREPARATION

Problem

What do the parts of a flower look like? How are they arranged?

Objectives

In this BioLab, you will:
■ **Observe** the structures of a flower.
■ **Identify** the functions of flower parts.

Materials

flower—any complete flower that is available locally, such as phlox, lily, or tobacco flower
hand lens (or stereomicroscope)
colored pencils (red, green, blue)
2 microscope slides

microscope   2 coverslips
single-edged   dropper
    razor blade   water

Safety Precautions

Always wear goggles in the lab. Handle the razor blade with extreme caution. Always cut away from you. Use caution when working with a microscope and slides. Wash your hands with soap and water after handling plant material.

Skill Handbook

Use the **Skill Handbook** if you need additional help with this lab.

PROCEDURE

1. Examine your flower. Locate the sepals and petals. Note their numbers, size, color, and arrangement on the flower stem.
2. Remove the sepals and petals from your flower by gently pulling them off the stem. Locate the stamens, each of which consists of a thin filament with a

678

PROCEDURE

Teaching Strategies

■ When preparing the anther wet mount, it may be helpful for students to squash the preparation using their thumb and a piece of lens paper over the coverslip. Advise students to press down firmly to avoid cracking the coverslip, or use plastic coverslips for this wet mount.

■ Eggs will not be visible; students should

not attempt to observe ovules under the microscope in order to observe egg cells.

■ Students can work in groups of two to three for this BioLab.

■ Show students how to remove the petals and sepals.

■ You may wish to appoint certain students as lab helpers. These students can be called upon by their classmates for help and guidance.

3. Locate the pistil. The stigma at the top of the pistil is often sticky. The style is a long, narrow structure that leads from the stigma to the ovary.

4. Place an anther from one of the stamens onto a microscope slide and add a drop of water. Cut the anther into several pieces with the razor blade. **CAUTION: *Always take care when using a razor blade.***

pollen-filled anther on the tip. Note the number of stamens.

5. Examine the anther under low and high power of your microscope. The small, dotlike structures are pollen grains.

6. Slice the ovary in half lengthwise with the razor blade. Mount one half, cut side facing up, on a microscope slide.

7. Examine the ovary section with a hand lens or stereomicroscope. The many, small, dotlike structures that fill the two ovary halves are ovules. Each ovule contains an egg cell that is not visible under low power. A tiny stalk connects each ovule to the ovary wall.

8. Identify the ovary and ovules.

9. Make a diagram of the flower, labeling all its parts. Color the female reproductive parts red. Color the male reproductive parts green. Color the remaining parts blue.

ANALYZE AND CONCLUDE

1. **Observing** How many stamens are present in your flower? How many pistils, ovaries, sepals, and petals?

2. **Comparing and Contrasting** Make a reasonable estimate of the number of pollen grains in the anther and the number of ovules in an ovary of your flower.

3. **Interpreting Data** Which produces more? Pollen grains by one anther? Ovules produced by one ovary? Give a possible explanation for your answer.

Going Further

Project Use a field guide to identify common wildflowers in your area. Most field identifications are made on the basis of color, shape, numbers, and arrangement of flower parts. If collecting is permitted, pick a few common flowers to press and make into a display of local flora.

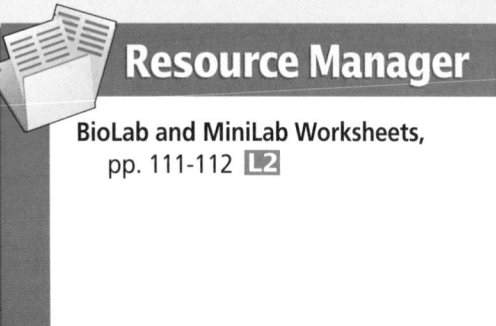

To find out more about flowers, visit the Glencoe Science Web site.

science.glencoe.com

INVESTIGATE BioLab

ANALYZE AND CONCLUDE

1. Stamens, petals, and sepals will be in multiples of 3 if a monocot, multiples of 4 or 5 if a dicot. One pistil and ovary will be present.

2. thousands of pollen grains, 10 to 100 ovules

3. Yes; the flower increases the probability of a pollen cell landing on the correct stigma by producing and releasing a large number of pollen grains.

✓ Assessment

Knowledge Provide students with diagrams of a flower other than that used in this BioLab. Have them label and indicate the general function of all important structures. Use the Performance Task Assessment List for Making Observations and Inferences in **PASC**, p. 17. **L1**

Going Further

Provide students with a composite flower such as a sunflower or daisy. Have them determine the locations of the individual florets and their various flower parts. **L2**

■ If necessary, review proper use of the stereomicroscope.

Data and Observations

Student diagrams, when colored properly, should show the petals and sepals as blue, pistil as red, and stamens as green.

Resource Manager

BioLab and MiniLab Worksheets, pp. 111-112 **L2**

Purpose

Students explore old and new methods of plant hybridization.

Background

Seed saver organizations encourage home gardeners and farmers to grow open-pollinated, or nonhybrid, varieties of vegetables and flowers. These natural varieties can reliably be reproduced from seed pollinated in the wild by wind, insects, or birds. Many growers like to raise heirloom varieties of nonhybrids that are becoming rare because an increasing number of hybrid varieties are offered every year. Open-pollinated varieties help maintain genetic diversity in plant populations. Seeds produced by hybrids are either sterile or tend to revert to unwanted characteristics of the parent generation.

Teaching Strategies

■ Review with students Mendel's experiments with garden peas and the genetic crosses discussed in Chapter 10.

■ Point out that plant breeders don't necessarily know whether a desired characteristic is regulated by one gene or many genes. This is one reason why field trials of many different crosses are often required to obtain hybrid varieties with specific characteristics.

Investigating the Technology

To ensure that a new crop will have the characteristics of the hybrid, a farmer must purchase seeds from the plant breeder each time a new crop is planted. Seed for the desired hybrid can be obtained only by cross-pollinating the same line of parent plants. F1 hybrids produce offspring with variable characteristics, not necessarily those of the hybrid. For examples, see the Punnett squares and discussion of F2 hybrids in Chapter 10.

680

BIO Technology   Hybrid Plants

If you've looked through any seed catalogs lately, you may have noticed phrases like "new this season!" or "improved yield," or "sweeter-tasting." Sometimes the designation "F_1" is given beside the names of some plant varieties. All of these plants are hybrids that have been produced in experiments conducted by plant breeders.

For thousands of years, humans have influenced the breeding of plants, especially food crops and flowers. Today's plant breeders create hybrid strains with a variety of desired characteristics, such as more colorful or fragrant flowers, tastier fruit, higher yields, or increased resistance to disease.

The perfect ear of corn The first step in creating a hybrid is the selection of parent plants with desirable characteristics. A breeder might select a corn plant that ripens earlier in the season or one that can be sown earlier in the spring because its seeds germinate well in cool, moist soil.

The next step is to grow several self-pollinated generations of each plant to form a true-breeding line that always shows the desired characteristic. To do this, each plant must be prevented from cross-pollinating with other corn. The female flowers, called silks, grow near the middle of the corn stalk. The breeder covers each flower to prevent wind-borne pollen from fertilizing it. The pollen-producing male tassels are removed and the breeder uses their pollen to hand-pollinate each flower.

Once each true-breeding line has been established, the real experimentation begins. Breeders cross different combinations of true-breeding lines to see what characteristics the resulting F_1 hybrids will have. These trials show which of the true-breeding lines reliably pass their desired characteristic to hybrid offspring, and which crosses produce seeds that the breeder can market as a new, improved variety of corn.

Applications for the future

Cell culture and genetic engineering technologies are new plant breeding techniques. Protoplast fusion removes the cell walls from the

Technician performing hybridization studies

cells of leaves or seedlings, then uses electricity or chemicals to fuse cells of two different species. Some of these fused cells have been successfully cultured in the lab and grown into adult plants, though none have produced seeds.

Recombinant DNA technology has been used to insert specific genes into the chromosomes of a plant. This technique helps produce plants that are resistant to frost, drought, or disease.

INVESTIGATING THE TECHNOLOGY

Analyzing Concepts Why do seed companies recommend not saving seeds from hybrid varieties to plant the following year? (Hint: The offspring of self-pollinated hybrids constitute the F_2 generation.)

 To find out more about hybrid seeds, visit the Glencoe Science Web site. **science.glencoe.com**

Going Further

Ask students to compare and contrast modern methods of creating hybrids with the techniques used by Gregor Mendel in his experiments with garden peas during the 1800s. **L2**

BIOLOGY Online Note Internet addresses that you find useful in the space below for quick reference.

_____

_____

_____

_____

_____

Chapter 24 Assessment

SUMMARY

Section 24.1

Life Cycles of Mosses, Ferns, and Conifers

Main Ideas

- In mosses, a gametophyte forms archegonia and antheridia. The gametophyte is dominant.
- In ferns, the prothallus forms archegonia and antheridia. The sporophyte is dominant.
- In conifers, cones produce spores that form male or female gametophytes. The pollen grain produces sperm, which fertilizes the egg. The embryo is protected by a seed.

Vocabulary

megaspore (p. 658)
micropyle (p. 659)
microspore (p. 658)
protonema (p. 655)
vegetative reproduction (p. 654)

Section 24.2

Flowers and Flowering

Main Ideas

- Flowers are made up of four organs: sepals, petals, stamens, and pistils.
- Photoperiodism affects the timing of flower production.

Vocabulary

anther (p. 662)
day-neutral plant (p. 666)
long-day plant (p. 666)
ovary (p. 662)
petals (p. 661)
photoperiodism (p. 663)
pistil (p. 662)
sepals (p. 661)
short-day plant (p. 666)
stamens (p. 662)

Section 24.3

The Life Cycle of a Flowering Plant

Main Ideas

- The male gametophyte is produced by a microspore in the anther. The female gametophyte is produced by a megaspore in the ovule.
- Sperm are transported by a pollen tube to the ovule, where fertilization takes place.
- In double fertilization, one sperm joins with the egg to form a zygote. The second sperm joins the central cell to form endosperm.
- The ovary wall becomes the fruit.
- Fruits and seeds are modified for dispersal.
- Seeds can stay dormant for long periods of time.

Vocabulary

dormancy (p. 675)
double fertilization (p. 671)
endosperm (p. 672)
germination (p. 675)
hypocotyl (p. 676)
polar nuclei (p. 667)
radicle (p. 676)

Main Ideas

Summary statements can be used by students to review the major concepts of the chapter.

Using the Vocabulary

To reinforce chapter vocabulary, use the Content Mastery Booklet and the activities in the Interactive Tutor for Biology: The Dynamics of Life on the Glencoe Science Web site.
science.glencoe.com

 *All Chapter Assessment questions and answers have been validated for accuracy and suitability by The Princeton Review.*

UNDERSTANDING MAIN IDEAS

1. b
2. c

UNDERSTANDING MAIN IDEAS

1. Pollen and nectar produced by flowers provide _____ for butterflies and bees.
   a. protection
   b. food
   c. shelter
   d. fruit

2. Flowers that are dull in color and have no nectar yet have a strong scent might be pollinated by _____.
   a. bees
   b. butterflies
   c. beetles
   d. hummingbirds

GLENCOE TECHNOLOGY

 VIDEOTAPE
MindJogger Videoquizzes
Chapter 24: *Reproduction in Plants*
Have students work in groups as they play the videoquiz game to review key chapter concepts.

Resource Manager

Chapter Assessment, pp. 139-144
MindJogger Videoquizzes
ExamView® Pro Software
BDOL Interactive CD-ROM, Chapter 24 quiz

3. b
4. d
5. a
6. c
7. b
8. d
9. d
10. a
11. endosperm
12. pistil
13. gametes
14. antheridia, archegonia
15. microspores
16. dormancy
17. ovary
18. short, long
19. micropyle
20. vegetative propagation

APPLYING MAIN IDEAS

21. Seeds store food intended for use by the embryo plant.
22. Seeds will remain alive until water becomes available for germination.

---

3. Moss gametophytes are _____ and form gametes by _____.
 a. diploid; meiosis   **c.** diploid; mitosis
 b. haploid; mitosis   **d.** haploid; meiosis

4. By eating fruit, mammals help _____.
 a. fertilize flowers   **c.** photoperiodism
 b. nastic movement   **d.** disperse seeds

5. While feeding, butterflies and bees carry pollen from flower to flower, causing

 _____.

 a. pollination   **c.** germination
 b. dormancy   **d.** photoperiodism

6. The heart-shaped structure formed by a developing fern spore is called a _____.
 a. protonema   **c.** prothallus
 b. sporophyte   **d.** frond

7. Which of the following plants do NOT produce pollen?
 a. pine tree   **c.** apple tree
 b. mosses   **d.** corn

8. Which of the following is a fern gametophyte?

 a.   **c.**

 b.   **d.**

THE PRINCETON REVIEW   **TEST-TAKING TIP**

Plan Your Work and Work Your Plan
Set up a study schedule for yourself well in advance of your test. Plan your workload so that you do a little each day rather than a lot all at once. The key to retaining information is to repeatedly review and practice it.

---

9. A(n) _____ is one that has all four organs: sepals, petals, stamens, and pistils.
 a. short-day plant   **c.** incomplete flower
 b. long-day plant   **d.** complete flower

10. The response of flowering plants to the difference in the duration of light and dark periods in a day is called _____.
 a. photoperiodism   **c.** pollination
 b. nastic movement   **d.** dormancy

11. A triploid cell resulting from double fertilization becomes the _____.

12. The structure marked A in the photograph is the _____.

13. The gametophyte produces _____ that are haploid.

14. In ferns, sperm released by the _____ fertilize eggs in the _____.

15. Male pinecones produce _____ that will develop into the male gametophyte, or pollen grain.

16. The period of inactivity in a mature seed is called _____.

17. After pollination, a pollen tube grows downward through the pistil to the _____.

18. A short-day plant is more likely to flower when the days are _____ and the nights are _____.

19. During fertilization in anthophytes, the two sperm cells enter the ovule through an opening called the _____.

20. Growing new plants from cuttings or tissue culture is called _____.

APPLYING MAIN IDEAS

21. You eat peas, beans, corn, peanuts, and cereals. Why are seeds a good source of food?

22. How does dormancy contribute to the survival of a plant species in a desert ecosystem?

23. In what ways can the relationship between the gametophyte and sporophyte in seed plants be regarded as good for the gametophyte?

24. Explain why a scientist might hypothesize that the eating habits of herbivorous mammals affected the evolution of fruits in flowering plants.

THINKING CRITICALLY

25. Observing and Inferring A plant species produces heavy, spiked pollen grains. What conclusion can you draw about the plant's pollination method?

26. Comparing and Contrasting Compare and contrast the formation of a moss sporophyte and a fern sporophyte.

27. Formulating Hypotheses Form a hypothesis that explains why the primary root is the first part of the plant to emerge from a germinating seed.

28. Concept Mapping Complete the concept map by using the following vocabulary terms: pistil, microspore, anther, stamen, ovary, megaspores.

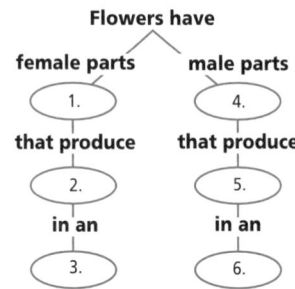

Flowers have

female parts   male parts

1.   4.

that produce   that produce

2.   5.

in an   in an

3.   6.

CD-ROM

For additional review, use the assessment options for this chapter found on the *Biology: The Dynamics of Life Interactive CD-ROM* and on the Glencoe Science Web site.
science.glencoe.com

ASSESSING KNOWLEDGE & SKILLS

The graph below provides data from an experiment that tests the effects of ionizing radiation on the germination of seeds.

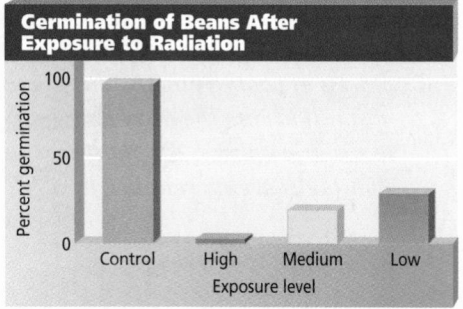

Germination of Beans After Exposure to Radiation

Percent germination — 100, 50, 0

Control   High   Medium   Low

Exposure level

Using a Graph Use the graph to answer the following questions.

1. Which group of beans had the highest percentage of germination?
   a. control
   b. high-exposure level
   c. medium-exposure level
   d. low-exposure level

2. As the radiation dose increases, germination _____.
   a. increases   **c.** stops
   b. decreases   **d.** is not affected

3. When beans are given a low dose of radiation, _____ germinate.
   a. 25 percent   **c.** none
   b. 50 percent   **d.** 100 percent

4. When beans are given a medium dose of radiation, _____ germinate.
   a. 12.5 percent   **c.** 37.5 percent
   b. 25 percent   **d.** 50 percent

5. Designing an Experiment Design an experiment on bean plants in which the following hypothesis is tested: Bean plants exposed to ionizing radiation will not grow as tall as those that are not exposed.

23. The gametophyte is protected and nourished by the sporophyte.

24. Mammals may have selected only certain fruits for their diet and thus aided in dispersal and survival of these species.

THINKING CRITICALLY

25. The plant is probably pollinated by animals.

26. The moss and fern sporophytes both develop from a diploid zygote in the archegonium of the gametophyte. The moss sporophyte will remain dependent upon the gametophyte, whereas the fern sporophyte will eventually survive on its own.

27. The primary root anchors the seed and obtains needed water from the soil for further growth and development of the plant.

28. 1. Pistil; 2. Megaspores; 3. Ovary; 4. Stamen; 5. Microspores; 6. Anther

ASSESSING KNOWLEDGE & SKILLS

1. a
2. b
3. a
4. a
5. Bean plants that have just emerged from the soil are divided into four groups: one the control that is not irradiated, another given a high dose, another given a medium dose, and another a low dose of radiation. The heights of the plants are measured and recorded for the duration of the experiment.

BioDigest

National Science Education Standards
UCP,.1, UCP.2, UCP.5, C.4, C.5,
C.6, F.4

Prepare

Purpose

This BioDigest can be used as an introduction to or an overview of the structures and functions of plants. If time is limited, you may wish to use this unit summary to teach about plants in place of the chapters in the Plants unit.

Key Concepts

Students learn about the characteristics of the major plant divisions. They are introduced to the alternation of generations; the distinctions between non-seed and seed plants; pollination and fertilization; and the adaptive value of flowers, seeds, and fruits.

1 Focus

Bellringer

 Visual-Spatial Bring an assortment of live plants into the classroom, including mosses, horsetails, and club mosses, if available. Include a small potted pine or other conifer, and several potted flowering plants. Ask students to observe each plant and make a list of its characteristics in their journals.

GLENCOE
TECHNOLOGY

CD-ROM
Biology: The Dynamics of Life
Animation: *Life Cycle of a Moss*
Disc 3

For a **preview** of the plant unit, study this BioDigest before you read the chapters. After you have studied the plant chapters, you can use the BioDigest to **review** the unit.

Plants

Earth is virtually covered with plants. Plants provide food and shelter for multitudes of organisms. Through the process of photosynthesis, they transform the radiant energy of sunlight into chemical energy in food and release oxygen to the atmosphere. All plants are multicellular eukaryotes. Plant cells are surrounded by a cell wall made of cellulose.

Mosses often grow in masses that form thick carpets

Non-Seed Plants

Non-seed plants reproduce by forming spores. A spore is a haploid (*n*) reproductive cell, produced by meiosis, which can withstand harsh environmental conditions. When conditions become favorable, a spore can develop into the haploid, gametophyte generation of a plant. A spore will become either a female or male gametophyte.

Mosses, Liverworts, and Hornworts

Mosses, liverworts, and hornworts are three divisions of non-seed, nonvascular plants that live in cool, moist habitats. Because they have no vascular tissues to move water and nutrients from one part of the plant to another, they cannot grow more than a few inches tall.

VITAL STATISTICS

Non-Seed Plants
Numbers of species:
Bryophyta—mosses, 20 000 species
Lycophyta—club mosses, 1000 species
Sphenophyta—horsetails, 15 species
Pterophyta—ferns, 12 000 species
Anthocerophyta—hornworts, 100 species
Hepatophyta—liverworts, 6500 species

FOCUS ON ADAPTATIONS

Alternation of Generations

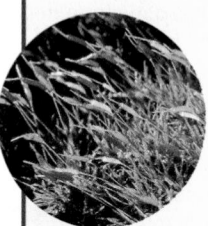

The leafy gametophyte is haploid. The spore stalks and capsules are diploid.

The life cycle of most plant species alternates between two stages, or generations. The sporophyte generation produces spores, which develop into the gametophyte generation. The gametophyte produces gametes. In nonvascular plants, the gametophyte is larger and more conspicuous than the sporophyte. In vascular plants, the sporophyte dominates. The gametophyte of a vascular plant is extremely small and may remain buried in the soil or inside the body of the sporophyte.

Gametophyte Generation

A gametophyte is haploid (*n*) and produces eggs and sperm. In mosses, the gametophyte is the familiar soft, green growth that covers rotting logs or moist soil. The tiny moss gametophytes produce male and female branches. Sperm cells produced by the male branches must swim through rain or dew to reach the egg cells produced by the female branches. Fertilization takes place inside the female reproductive organ and a diploid (*2n*) zygote is produced.

684

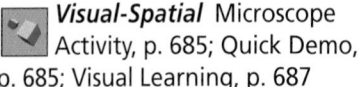

Club Mosses

Club mosses are non-seed plants in the division Lycophyta. They possess vascular tissue and are found primarily in moist environments. Species that exist today are only a few centimeters high, but they are otherwise similar to fossil Lycophytes that grew as high as 30 m and formed a large part of the vegetation of Paleozoic forests.

Ferns

Ferns, division Pterophyta, are the most well-known and diverse of the non-seed vascular plants. They have leaves called fronds that grow up from an underground stem called the rhizome. Ferns are found in many different habitats, including shady forests, stream banks, roadsides, and abandoned pastures.

Fern spores develop in clustered structures called sori, usually found on the undersides of fronds.

Horsetails

Horsetails are non-seed vascular plants in the division Sphenophyta. They are commonly found growing in areas with damp soil, such as stream banks and sometimes along roadsides. Present-day horsetails are small, but their ancestors were treelike.

The hollow, jointed stems of horsetails are surrounded by whorls of scalelike leaves.

As in all vascular plants, the sporophyte of this club moss is the dominant generation. Spores develop at the base of special leaves that form cone-shaped structures called strobili.

Sporophyte Generation The zygote develops into an embryo, which grows into the sporophyte generation of the moss. The sporophytes grow out of the tip of the female branches of the gametophyte and consist of capsule-topped stalks. Cells inside each capsule undergo meiosis to form haploid (n) spores.

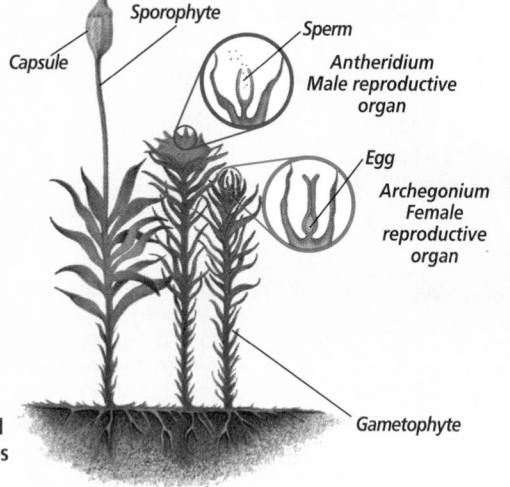

Capsule

Sporophyte

Sperm

Antheridium
Male reproductive organ

Egg

Archegonium
Female reproductive organ

Gametophyte

The green, leafy growth of this moss is the gametophyte. The brown stalks topped with spore-filled capsules are the sporophyte.

685

2 Teach

Microscope Activity

Visual-Spatial Have students view mosses, liverworts, and hornworts with a stereomicroscope. Ask students to describe their observations orally or in writing in their journals. If possible, supply sporophytes and gametophytes with male and female reproductive structures and ask students to compare the characteristics of these structures and generations. **L2**

Quick Demo

Visual-Spatial If horsetails grow in your area, bring several stems into the classroom. Show students that the stems come apart fairly easily at the nodes, and that the stems are hollow. Have them examine the strobili, if these are present. Point out that the silica in the stems gives horsetails a scouring-pad quality and is the source of the common name "scouring rush."

Quick Demo

Visual-Spatial Show students young fern fiddleheads and sori on the under-side of a fern frond. In the spring, fiddleheads are sometimes available in specialty markets as a food item. If a live fern is available, show students the underground stem (rhizome) from which roots grow down and fronds grow up.

BioDigest

Quick Demo

 Kinesthetic Bring to class and allow students to handle several varieties of conifer cones. Include male cones if available. Tell students the common name and/or Latin name of the tree each cone is from. Ask students to record their observations of each species in their journals. **L1**

Guest Speaker

TECH PREP Have a pharmacist or physician visit the class to discuss plants that are important in the production of medicines.

CD-ROM
Biology: The Dynamics of Life
Animation: *Life Cycle of a Pine*
Exploration: *Classifying Pines*
Disc 3

Resource Manager

Reinforcement and Study Guide, pp. 109-110 **L2**
Content Mastery, pp. 121-124 **L1**

Seed Plants

A seed is a reproductive structure that contains a sporophyte embryo and a food supply enclosed in a protective coating. The food supply nourishes the young plant during the first stages of growth. Like spores, seeds can survive harsh conditions. The seed develops into the sporophyte generation of the plant. Seed plants include conifers and flowering plants.

Conifers

Conifers, division Coniferophyta, produce seeds, usually in woody strobili called cones, and have needle-shaped or scale-like leaves. Conifer seeds are not enclosed in a fruit. Most conifers are evergreen plants, which means they bear leaves all year round.

Seeds of conifers develop at the base of each woody scale of female cones.

Adapted for Cold and Dry Climates

Conifers are common in cold or dry habitats throughout the world. Conifer needles have a compact shape and a thick, waxy covering that helps reduce evaporation and conserve water. Conifer stems are covered with a thick layer of bark that insulates the tissues inside. These adaptations enable conifers to carry on life processes even when temperatures are below freezing.

The leaves and branches of conifers are flexible. They bend under the weight of snow and ice, allowing any buildup to slide off before it becomes heavy enough to break the branch.

Vital Statistics

Conifers
Examples: Pine, spruce, fir, larch, yew, redwood, juniper.
Numbers: 400 species.
Size range: Giant sequoias of central California, to 99 m tall, the most massive organisms in the world; coast redwoods of California, to 117 m, the tallest trees in the world.

Focus on Adaptations

Moving from Water to Land

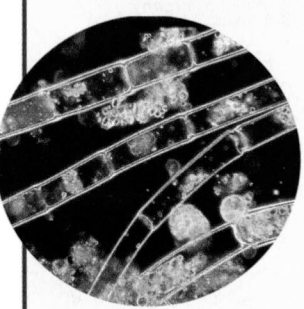

Plants probably evolved from filamentous green algae.

All plants probably evolved from filamentous green algae that lived in the nutrient-rich waters of Earth's ancient oceans. An ocean-dwelling alga can absorb water and dissolved minerals directly into its cells. As land plants evolved, new structures developed for absorbing and transporting water and minerals from the soil to all the aerial parts of the plant.

Nonvascular Plants

In nonvascular plants, water and nutrients must travel from one cell to another by the relatively slow processes of osmosis and diffusion. As a result, nonvascular plants are limited to environments where plenty of water is available.

Meeting Individual Needs

Visually Impaired

 Kinesthetic As part of the Bellringer demonstration, have students with visual impairments handle the plants. Break off small parts for them to handle. Ask them to describe what they feel as partners write down their observations. Make sure students don't injure themselves by handling the plants. **L1**

National Geographic

VIDEODISC
STV: Plants, *What Is a Seed?*
Unit 4, Side 3, 13 min. 45 sec.
What Is a Seed? (In its entirety)

Flowering Plants

The flowering plants, division Anthophyta, form the largest and most diverse group of plants on Earth today. They provide much of the food eaten by humans. Anthophytes produce flowers and develop seeds enclosed in a fruit.

Monocots and Dicots

The Anthophytes are classified into two classes: the monocotyledons and the dicotyledons. Cotyledons, or "seed leaves," are contained in the seed along with the plant embryo. Monocots have a single seed leaf that absorbs food for the embryo. The two seed leaves of dicots store food for the embryo.

Monocots (left) include grasses, orchids, and palms. Dicots (below) include many flowering trees and wildflowers.

Flowers

Flowers are the organs of reproduction in anthophytes. Sepals enclose the flower bud and protect it until it opens; petals, which are often brightly colored or perfumed, attract pollinators. Inside the circle of petals are the pistil and stamens.

The pistil is the female reproductive organ. Inside the ovary at the base of the pistil are the ovules. Ovules contain the female gametophyte generation of the plant. Female gametes—egg cells—form in each ovule.

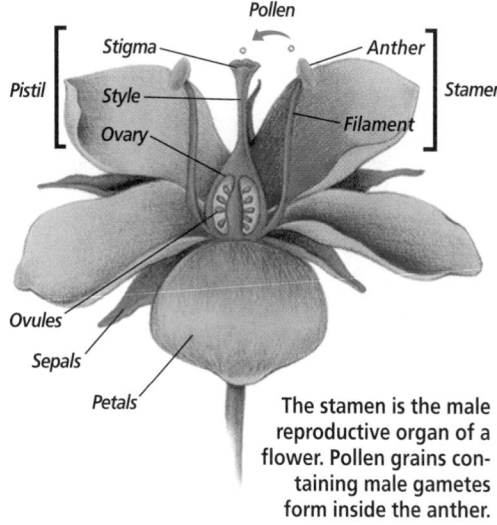

Pollen

Stigma — Anther

Pistil | Style — Stamen

Ovary — Filament

Ovules

Sepals

Petals

The stamen is the male reproductive organ of a flower. Pollen grains containing male gametes form inside the anther.

Vascular Plants The stems of most plants contain vascular tissues made up of tubelike, elongated cells through which water, food, and other materials move from one part of the plant to another. One reason vascular plants can grow larger than nonvascular plants is because vascular tissue is a much more efficient method of internal transport than osmosis and diffusion. In addition, vascular tissues include thickened fibers that can support taller upright growth.

An unbroken column of water travels from the roots in xylem tissues. Sugars formed by photosynthesis travel around the plant in phloem tissues.

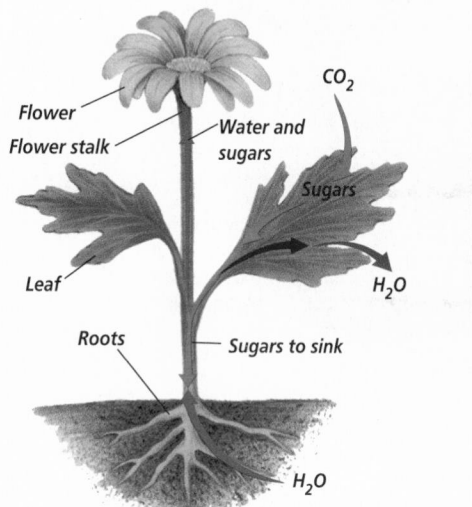

Flower

Flower stalk

Leaf

Roots

CO_2

Water and sugars

Sugars

H_2O

Sugars to sink

H_2O

687

BioDigest

BioDigest

Assessment

Knowledge Ask students to compare and contrast pollination and fertilization in conifers and flowering plants. *In conifers, pollen is carried by wind to the female cone. In flowers, pollen may also be transported by animals. In flowers, the pollen is carried to the stigma. In both conifers and flowers, the pollen grows a tube through which the sperm travels to reach the ovary. In both, sperm and egg unite to form a zygote that develops into a seed.*

Knowledge

After students have observed flowers and fruits from the same plants, ask them to discuss which parts of the flower become which parts of the fruit, and to identify the parts of the flower that wither away and do not form part of the fruit. *The ovary swells to become the fruit. The ovule may be visible around the seeds, as in apples. The sepals may be visible on the blossom end of the fruit. Petals, pistil, and stamen wither away.*

Visual Learning

Linguistic Bring into the classroom several different fruits, including both dry and fleshy varieties. If possible, bring fruits that develop from some of the flowers used in the above activity. Have students examine the exterior of each fruit, then cut it open to observe the arrangement of seeds. Ask students to describe their observations in their journals. **L1**

Pollen

In seed plants, the sperm are enclosed in the thick-coated pollen grains, which are the male gametophyte generation of the plant. Pollen is one of the important adaptations that has enabled seed plants to live in a wide variety of land habitats.

Pollinators

Flowers can be pollinated by wind, insects, birds, and even bats. Some flowers have colorful or perfumed petals that attract pollinators. Flowers may also contain sweet nectar, as well as pollen, which provides pollinators with food.

Plants that depend on the wind to carry pollen from anther to stigma tend to have small, inconspicuous flowers. The flowers of grasses and this alder are pollinated by the wind.

Plants that depend on insects for pollination may be brightly colored and fragrant. Pollen rubs off on the bee that visits a flower to feed on nectar. When it moves to another flower, some of the pollen may rub off onto the stigma.

Many plants produce fruits that are eaten by animals.

Fruit

Following fertilization, the ovary develops into a fruit with seeds inside. Some flowering plants develop fleshy fruits, such as apples, melons, tomatoes, and squash. Other flowering plants develop dry fruits, such as peanuts, almonds, or sunflowers. Fruits help protect seeds until they are mature. Fruits also help scatter seeds into new habitats.

Maple tress produce fruits with a winglike shape that can be carried long distances by the breeze.

VITAL STATISTICS

Flowering Plants
Examples: Grasses, oaks, maples, palms, irises, orchids, roses, beans.
Numbers: 230 000 species (60 000 monocots; 170 000 dicots).
Size range: A few millimeters to 75 m.

Pollen is carried to the stigma of a flower. The pollen grain grows a tube down the style to the ovary. Two sperm travel down the tube.

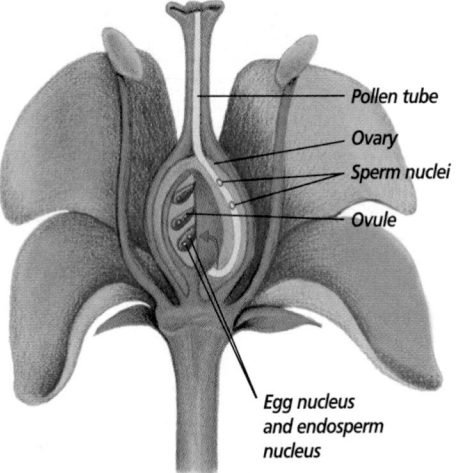

Pollen tube

Ovary

Sperm nuclei

Ovule

Egg nucleus and endosperm nucleus

In a process called double fertilization, one of the sperm fertilizes the egg and the other unites with the endosperm nucleus.

PROJECT

Street-Tree Census

Interpersonal Have students conduct a street-tree census in their community. They can analyze the results of the census and submit a class report to the community government with suggestions on locations where additional trees could be planted. **L3**
COOP LEARN

GLENCOE TECHNOLOGY

CD-ROM
Biology: The Dynamics of Life
Animation: *Double Fertilization*
Animation: *Fruit Formation*
Exploration: *Pollination*
Disc 3

Plant Responses

Plants respond to changes in their environment such as light, temperature, and water availability. Chemicals called hormones control some of these responses by increasing cell division and growth.

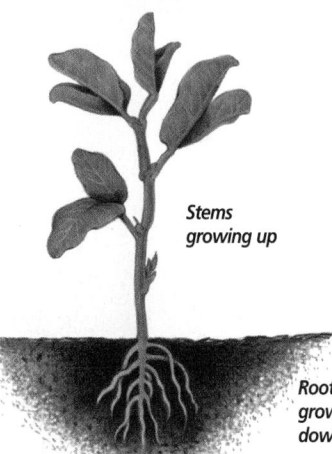

Stems growing up

The phototropic response shown here is the result of increased cell growth on the side of the stem away from the light.

Roots exhibit positive geotropism. Stems show a negative geotropic response.

Roots growing down

3 Assess

Check for Understanding

Have students explain the difference between the terms in each of the following word pairs. **L2**
  a. seed—spores
  b. vascular—nonvascular
  c. cone—fruit
  d. monocot—dicot

Reteach

Have students name plants that are or produce structures, listed in the word pairs of the Check for Understanding. Ask them to explain why each plant is representative of the term. **L2**

Extension

Linguistic Have students research one plant division of their choice and present a brief oral report to the class. **L3**

✔ Assessment

Skill Have students prepare a chart that shows the different types of plants and their major characteristics. **L2**

4 Close

Activity

Have students prepare questions about the major characteristics of plants and write them on the chalkboard. Have the class discuss each question.

BioDigest Assessment

Understanding Main Ideas

1. Which of the following is a Bryophyte?
   a. moss   c. club moss
   b. horsetail   d. conifer

2. The term for a mature fern leaf is _____.
   a. leaf   c. frond
   b. scale   d. needle

3. Nonvascular plants would most likely be found growing _____.
   a. in sandy desert soil
   b. on an ocean beach
   c. on a snowy mountain slope
   d. in a shady, moist environment

4. Which plant group has leaves adapted for life in cold environments?
   a. Anthophyta   c. Pterophyta
   b. Sphenophyta   d. Coniferophyta

5. Dicots have _____.
   a. one cotyledon
   b. two cotyledons
   c. needlelike leaves
   d. spores borne in cone-shaped strobili

6. Reproductive structures of conifers are _____.
   a. flowers   c. fruits
   b. cones   d. sori

7. Mosses, ferns, and club mosses are alike because they require _____.
   a. water for fertilization
   b. adaptations for conserving water
   c. insects for pollination
   d. warm, sunny habitats

8. Lycophytes, sphenophytes, and coniferophytes have specialized leaves that form reproductive structures known as _____.
   a. sori   c. cones
   b. flowers   d. strobili

9. Vascular plants do not include the _____.
   a. Lycophytes   c. Sphenophytes
   b. Bryophytes   d. Pterophytes

10. Which plant group produces flowers and seeds enclosed in a fruit?
   a. Anthophyta   c. Pterophyta
   b. Coniferophyta d. Lycophyta

Thinking Critically

1. Compare the spore-bearing structures of ferns with the seed-bearing structures of conifers.

2. Why do vascular plants have an adaptive advantage over nonvascular plants?

3. Describe three ways in which seeds may be dispersed.

BioDigest Assessment

Understanding Concepts

1. a	4. d	7. a	9. b
2. c	5. b	8. d	10. a
3. d	6. b		

Thinking Critically

1. Ferns produce spores in sori on fronds. Conifers develop seeds at the base of scalelike leaves that form cones.

2. Vascular plants are not limited to moist environments because they do not have to rely on osmosis and diffusion for transport of water and nutrients. Also, vascular tissue provides support so the plants can grow larger.

3. Seeds may be blown away by the wind. They may be carried by water when they fall in the ocean or in streams. They may also be transported by animals who eat the fruit and discard the seeds, or by animals who pass seeds through their digestive systems and deposit them with their droppings.

Invertebrates

Unit Overview

In this unit, students become familiar with invertebrates. Chapter 25 introduces the general characteristics of animals as well as their body plans.

In Chapter 26, students begin their examination of specific invertebrate groups through the study of the structure, adaptations, ecology, and phylogeny of sponges, cnidarians, flatworms, and roundworms. In Chapter 27, students examine the characteristics, ecology, and phylogeny of mollusks and segmented worms.

Chapter 28 allows students to explore the largest group of animals—the arthropods. In this chapter, the diversity of arthropods is examined, while features that enabled many arthropods to become land dwellers are explained.

The unit concludes in Chapter 29 with a presentation of echinoderms and invertebrate chordates.

Introducing the Unit

Ask students to examine the photo of the tidal pool. Ask them to identify several of the organisms shown in the photo. Have the class brainstorm to create lists of how these organisms are alike and how they are different. Try to lead students to include characteristics of each organism that show its adaptation to its environment. **L2**

Unit 8

Invertebrates

Almost all animals on Earth—95 percent—are invertebrates, animals without backbones. Tidal pools and the oceans that sustain them are home to many of the world's invertebrates. The enormous diversity in form and function of invertebrates is the result of a long evolutionary history. As you study invertebrates, you will discover adaptations that are evolutionary milestones.

UNIT CONTENTS

UNIT PROJECT

BIOLOGY *Online* Use the Glencoe Science Web site for more project activities that are connected to this unit.
science.glencoe.com

Unit Projects

The Life Cycle of a Butterfly

Have students do one of the projects for this unit as described on the Glencoe Science Web site. As an alternative, students can do one of the projects described on these two pages.

Use the Library

 Intrapersonal Ask students to research information about butterfly metamorphosis. **L2**

Design and Build

Kinesthetic Have students obtain the materials and develop an enclosure to observe butterfly metamorphosis. **L1** **ELL**

Advance Planning

Chapter 25
- Order slides for the Quick Demo and Alternative Lab.
- Purchase rotifers and vinegar eels for the two MiniLabs.
- Order *Daphnia* for the Project.
- Collect earthworms and planarians for the Reinforcement.

Chapter 26
- Order slides for the Quick Demo and Biology Journal.
- Order live hydra for the Mini-Lab and Close.
- Order sponges and cnidarians for the Reinforcement.
- Order a live vinegar eel culture for the Quick Demo.

Chapter 27
None

Chapter 28
- Obtain arthropods for the Quick Demo.
- Obtain bess beetles for the Alternative Lab.
- Obtain preserved crayfish for the MiniLab.
- Obtain live crayfish and a butterfly chrysalis for the Quick Demos and MiniLab.

Chapter 29
- Purchase a slide of sea star pedicellariae for the MiniLab.
- Obtain sea stars for Quick Demos.
- Purchase echinoderm fossils for the Portfolio.
- Purchase slides of a lancelet and sea urchin development for the Alternative Lab.
- Purchase *Branchiostoma californiense* for the MiniLab.
- Obtain live sea urchins for the BioLab.

Unit Projects

Display
Visual-Spatial Have students make a poster showing a plan for development of a butterfly garden that would attract local butterflies. **L2** **ELL**

Make a Map
Visual-Spatial Have students map the migration of monarch butterflies. **L3**

Final Report
Have student groups make a poster that illustrates their observations of the metamorphosis of a butterfly.

Refer to pages 4T-5T of the Teacher Guide for an explanation of the National Science Education Standards correlations.

Section	Objectives	Activities/Features
Section 25.1 **Typical Animal Characteristics** National Science Education Standards UCP.1, UCP.2, UCP.5; A.1, A.2; C.1, C.5, C.6; G.1-3 (1 session, $\frac{1}{2}$ block)	1. **Describe** the characteristics of animals. 2. **Sequence** the development of a typical animal.	**Careers in Biology:** Marine Biologist, p. 694 **MiniLab 25-1:** Observing Animal Characteristics, p. 695 **Problem-Solving Lab 25-1:** p. 696 **Inside Story:** Early Animal Development, p. 698 **Internet BioLab:** Zebra Fish Development, p. 706 **Biology & Society:** Protecting Endangered Species, p. 708
Section 25.2 **Body Plans and Adaptations** National Science Education Standards UCP.1, UCP.2, UCP.4, UCP.5; A.1, A.2; C.3, C.5; E.1, E.2; F.4, F.5, F.6; G.1-3 (2 sessions, $1\frac{1}{2}$ blocks)	3. **Compare and contrast** radial and bilateral symmetry with asymmetry. 4. **Trace** the phylogeny of animal body plans. 5. **Compare** body plans of acoelomate, pseudocoelomate, and coelomate animals.	**Problem-Solving Lab 25-2:** p. 702 **MiniLab 25-2:** Check Out a Vinegar Eel, p. 703

Need Materials? Contact Carolina Biological Supply Company at 1-800-334-5551 or at **http://www.carolina.com**

MATERIALS LIST

BioLab
p. 706 aquarium, zebra fish, turkey baster, beaker, dropper, petri dish, stereomicroscope, wax pencil or labels

MiniLabs
p. 695 microscope, microscope slide, coverslip, toothbrush bristles, rotifer culture, dropper, water
p. 703 microscope, microscope slide, coverslip, dropper, vinegar eel culture

Alternative Lab
p. 702 microscope, prepared slides of cross sections of hydra, planarian, nematode, and earthworm

Quick Demos
p. 694 microscope, prepared slides of nerve, muscle, blood, and stomach cells
p. 701 kitchen bowls, vases, and spoons; dried marine sponge, preserved jellyfish, live goldfish

Key to Teaching Strategies

L1 Level 1 activities should be appropriate for students with learning difficulties.
L2 Level 2 activities should be within the ability range of all students.
L3 Level 3 activities are designed for above-average students.
ELL ELL activities should be within the ability range of English Language Learners.
COOP LEARN Cooperative Learning activities are designed for small group work.
P These strategies represent student products that can be placed into a best-work portfolio.
These strategies are useful in a block scheduling format.

What Is an Animal?

Teacher Classroom Resources

Section	Reproducible Masters	Transparencies
Section 25.1 **Typical Animal Characteristics**	Reinforcement and Study Guide, pp. 111-112 **L2** Critical Thinking/Problem Solving, p. 25 **L3** BioLab and MiniLab Worksheets, p. 113 **L2** Laboratory Manual, pp. 179-182 **L2** Content Mastery, pp. 125-126, 128 **L1**	Section Focus Transparency 61 **L1** **ELL** Basic Concepts Transparency 44 **L2** **ELL** Reteaching Skills Transparency 38 **L1** **ELL**
Section 25.2 **Body Plans and Adaptations**	Reinforcement and Study Guide, pp. 113-114 **L2** Concept Mapping, p. 25 **L3** **ELL** BioLab and MiniLab Worksheets, pp. 114-116 **L2** Laboratory Manual, pp. 183-186 **L2** Content Mastery, pp. 125, 127-128 **L1**	Section Focus Transparency 62 **L1** **ELL** Basic Concepts Transparency 44 **L2** **ELL**

Assessment Resources

Chapter Assessment, pp. 145-150
MindJogger Videoquizzes
Performance Assessment in the Biology Classroom
Alternate Assessment in the Science Classroom
ExamView® Pro Software
BDOL Interactive CD-ROM, Chapter 25 quiz

Additional Resources

Spanish Resources **ELL**
English/Spanish Audiocassettes **ELL**
Cooperative Learning in the Science Classroom **COOP LEARN**
Lesson Plans/Block Scheduling

NATIONAL GEOGRAPHIC — Teacher's Corner

Products Available From Glencoe
To order the following products, call Glencoe at 1-800-334-7344:
CD-ROM
Mammals: A Multimedia Encyclopedia
Videodisc
STV: Animals

Products Available From National Geographic Society
To order the following products, call National Geographic Society at 1-800-368-2728:
Book
National Geographic Book of Mammals
Videos
Predators of North America
Strange Creatures of the Night

Tigers of the Snow
White Wolf
Wild Survivors: Camouflage and Mimicry

Index to National Geographic Magazine
The following articles may be used for research relating to this chapter:
"Poison-Dart Frogs: Lurid and Lethal," by Mark W. Moffet, May 1995.
"Animals at Play," by Stuart L. Brown, December 1994.
"The Amazing Frog-Eating Bat," by Merlin D. Tuttle, January 1982.

GLENCOE TECHNOLOGY

The following multimedia resources are available from Glencoe.

Biology: The Dynamics of Life
CD-ROM **ELL**
Animation: *Embryo Development*
Video: *Fetal Development*
Exploration: *Symmetry*

Videodisc Program
Embryo Development

The Infinite Voyage
The Geometry of Life

The Secret of Life Series
Dividing Cells, Early Embryo
Sex and the Single Gene: *Cell Development*
Evolution of Symmetry
Flatworm Cross Section
Developing Zebra Fish

Theme Development

This chapter stresses the themes of **unity within diversity** and **evolution**. Students will gain the understanding that within the tremendous variety of animals, there are common threads or shared features. They will also learn how animal characteristics and adaptations have evolved.

0:00 OUT OF TIME?

If time does not permit teaching the entire chapter, use the BioDigest at the end of the unit as an overview.

READING BIOLOGY

Glencoe's *Biology: The Dynamics of Life* contains many resources to assist a student's reading skills. Each chapter contains figures with expanded captions that expand on written material. Word Origins, located along the side of text, expand knowledge of biology vocabulary. Glencoe's Content Mastery Booklet helps develop reading skills while reinforcing content. In addition, use the Interactive Tutor for *Biology: The Dynamics of Life* on the Glencoe Web site to reinforce vocabulary. **science.glencoe.com**

Chapter

25 What Is an Animal?

What You'll Learn

- You will distinguish animal characteristics from those of other life forms.
- You will identify the stages of early animal development.
- You will interpret body plans of animals.

Why It's Important

The animal kingdom includes organisms as diverse as sponges, earthworms, clams, crickets, and birds. Humans are also animals. An understanding of animals will provide a better understanding of ourselves.

READING BIOLOGY

Write down a brief answer to the chapter's title, "What is an animal?" Compare your definition to that in the reading. Scan the chapter, making a list of some of the main characteristics of animals that the text highlights.

BIOLOGY
Online

To find out more about animals, visit the Glencoe Science Web site. **science.glencoe.com**

You know that the bat and jellyfish are animals, but can you name the characteristics they share? There are millions of species of animals, but they all have several features in common.

Section
25.1 Typical Animal Characteristics

What do you think of when you hear the word animal? Many people envision an organism with hair or fur and a bony skeleton. Yet more than 95 percent of the 1.5 million species of animals that have been described have neither bones nor hair.

If you saw the organism in the inset photograph for the first time, would you classify it as an animal? This organism is a sponge, an animal that remains attached to rocks or coral reefs in the ocean for all of its adult life. It doesn't have a bony skeleton or hair, yet it is still an animal.

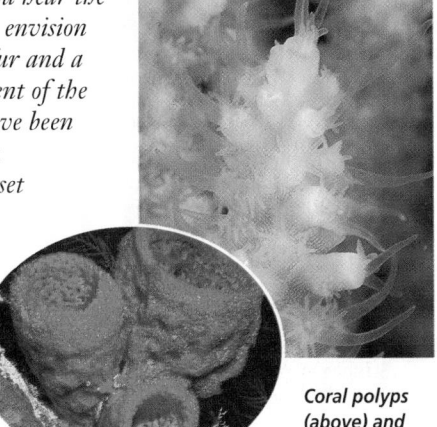

Coral polyps (above) and red sponges (left)

SECTION PREVIEW

Objectives
Describe the characteristics of animals.
Sequence the development of a typical animal.

Vocabulary
sessile
blastula
gastrula
ectoderm
endoderm
mesoderm
protostome
deuterostome

Characteristics of Animals

All animals have several characteristics in common. Animals are eukaryotic, multicellular organisms that have ways of moving that help them reproduce, obtain food, and protect themselves. Animals also have specialized cells that form tissues and organs—such as nerves and muscles. Unlike plants, animals are composed of cells that do not have cell walls. Most adult animals are fixed in size and shape; they do not continue to grow throughout their lives.

Methods for obtaining food vary

Examine the animals shown in *Figure 25.1*. One characteristic common to all animals is that they are heterotrophic, meaning they must consume food to obtain energy and nutrients. All animals depend either directly or indirectly on autotrophs for food.

Figure 25.1
The barnacle and lizard each get their food in different ways.

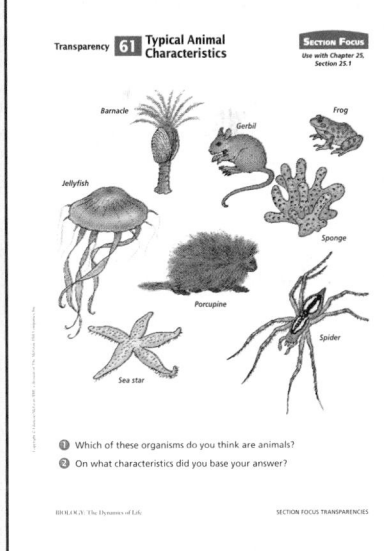

A A barnacle extends bristles from its shell to catch small organisms as they drift by in the water.

B A lizard captures flies with its long, sticky tongue.

![Assessment Planner]

Portfolio Assessment
 Portfolio, TWE, pp. 697, 705
 MiniLab, TWE, p. 703
 Alternative Lab, TWE, p. 703
 BioLab, TWE, p. 707
Performance Assessment
 Alternative Lab, TWE, p. 703
 BioLab, SE, p. 707
 MiniLabs, SE, pp. 695, 703
 Problem-Solving Lab, TWE, p. 696

Knowledge Assessment
 MiniLab, TWE, p. 695
 Section Assessment, SE, pp. 699, 705
 Chapter Assessment, SE, pp. 709-711
Skill Assessment
 Problem-Solving Lab, TWE, p. 702

Section 25.1

Prepare

Key Concepts
Students will study typical animal characteristics and the development of animals from eggs. They will compare and contrast animal features and the differences in types of development.

Planning
■ Gather live animals for the Getting Started Demo.
■ Buy zebra fish for the BioLab.

---

1 Focus

Bellringer 🖐

Before presenting the lesson, display **Section Focus Transparency 61** on the overhead projector and have students answer the accompanying questions.
L1 **ELL**

Transparency **61** Typical Animal Characteristics

SECTION FOCUS
Use with Chapter 25, Section 25.1

Barnacle, Gerbil, Frog, Jellyfish, Sponge, Porcupine, Sea star, Spider

❶ Which of these organisms do you think are animals?
❷ On what characteristics did you base your answer?

BIOLOGY: The Dynamics of Life   SECTION FOCUS TRANSPARENCIES

![Resource Manager]

Resource Manager

Tech Prep Applications,
 pp. 49-50 **L2**
Section Focus Transparency 61
and Master **L1** **ELL**

693

2 Teach

Display

 **Visual-Spatial** Make a bulletin board display of a variety of vertebrate and invertebrate animals. Calendars and catalogs from biological supply companies may have animals such as the hydra and planaria mentioned in this chapter. Ask students to bring in photographs to display that they may have taken of animals other than pets. **L2** **ELL**

Visual-Spatial Ask students to examine a prepared slide of animal cells such as nerve, muscle, blood, and cells lining the stomach. Ask them to make a labeled sketch of these cells in their journals. **L2** **ELL**

CAREERS IN BIOLOGY

Career Path
TECH PREP **Courses in high school:** biology, botany, chemistry, math, computers
College: at least bachelor's and master's degrees in marine biology or other specialized areas

Career Issue
The citizens of the United States live on land, not in the ocean. Ask students why our federal government should spend our tax money on marine biologists.

For More Information
For more information on marine biology, students can write to:
American Society of Limnology and Oceanography
Virginia Institute of Marine Science
College of William and Mary
Route 1208
Gloucester Point, VA 23062

Animals such as lizards and birds can move from place to place in an active search for food. Other animals, such as barnacles, remain stationary and are adapted to capture food from the water in which they live.

Whether an animal moves quickly or slowly depends partly on its environment. Scientists hypothesize that

Figure 25.2
Animals move in a remarkable variety of ways.

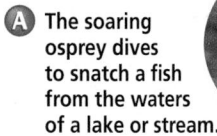

A The soaring osprey dives to snatch a fish from the waters of a lake or stream.

B A sidewinder rattlesnake barely touches the ground as it follows the trail of a mouse.

694 WHAT IS AN ANIMAL?

---

animals first evolved in water. Water is more dense and contains less oxygen than air, but water contains more suspended food. In the water, stationary animals don't expend much energy to obtain food. But there is little suspended food in the air; thus, land animals use more oxygen and expend more energy to find food.

The osprey and sidewinder snake in *Figure 25.2* are examples of vertebrates, animals with backbones, whereas the sea star is an invertebrate. Yet all these animals move about in their environments. Find out how an animal's environment affects its survival in the *Biology & Society* at the end of this chapter.

Some aquatic animals, such as sponges and corals, move about only during the early stages of their lives. They hatch from eggs into free-swimming larval forms; as adults, most attach themselves to rocks or other objects. Organisms that don't move from place to place are known as **sessile** (SES ul) organisms. They rely on water currents to carry food to them.

Animals must digest food
Animals are heterotrophs that ingest their food; after ingestion, they must digest it. In some animals,

C A sea star moves using a unique system of tube feet.

Figure 25.3
In animals such as planarians and earthworms, food is digested in a digestive tract.

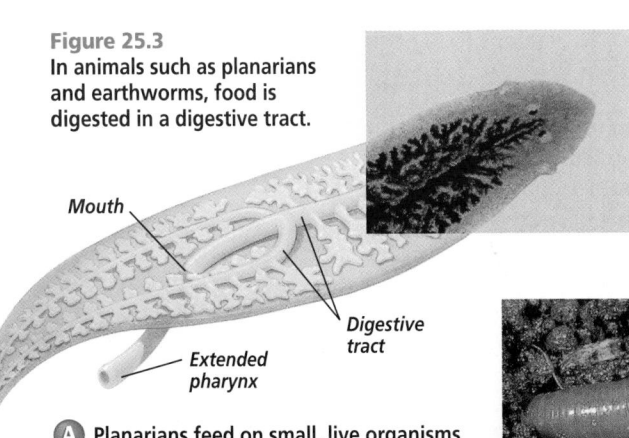

Mouth

Extended pharynx

Digestive tract

A Planarians feed on small, live organisms or on the dead bodies of larger animals. The planarian's digestive tract has only one opening through which food enters and wastes exit.

B Earthworms ingest soil and digest the organic matter contained in it. Food enters the mouth and travels along the digestive tract in one direction. Indigestible waste is eliminated at the anus.

Digestive tract

Anus

digestion is carried out within individual cells; in other animals, digestion takes place in an internal cavity. Some of the food an animal consumes and digests is stored as fat or glycogen and used when food is not available. Recall that glycogen is a polysaccharide, like starch, that is used for food storage.

Examine digestive tracts of a flatworm and earthworm in *Figure 25.3.* Notice that there is only one opening to the flatworm's digestive tract, a mouth. An earthworm has a digestive tract with two openings, with a mouth at one end and an anus at the other.

Animal cell adaptations

Most animal cells are adapted to carry out different functions. Animals have specialized cells that enable them to sense and seek out food and mates, and allow them to identify and protect themselves from predators. In the human body, nerve cells conduct information and red blood cells transport oxygen. Observe the tiny animals in the *MiniLab.* Can you identify any specialized cells in these animals?

MiniLab 25-1 Observing and Inferring

Observing Animal Characteristics Animals come in a variety of sizes and shapes, and can be found living in a number of different habitats.

Procedure
1. Copy the data table.
2. Add a few bristles from an old toothbrush to a glass slide. Add a drop of water containing rotifers to your slide. The drop should cover the bristles. Add a coverslip. **CAUTION: *Use caution when working with a microscope, slides, and coverslips.***
3. Observe your rotifers under low-power magnification.
4. Use the data table to record the characteristics that you were able to see. Describe the evidence for each trait.

Data Table		
Animal Characteristic	Observed? (Yes or No)	Evidence
Multicellular		
Feeding		
Movement		
Size in mm		

Analysis
1. Are these organisms autotrophs or heterotrophs?
2. Were you able to see evidence of feeding? Explain.
3. Are rotifers multicellular? Explain.

Cultural Diversity

Animals in a Cross-Cultural Perspective

Discuss the diverse views different societies and cultures have about the place of animals in nature. Begin your discussion by asking students how animals are viewed and used in the United States. Next, point out that not all cultures view animals in this way. For example, Native American traditions often portray animals as important as or even more important than humans. Obtain books about the legends and myths of different cultures, and discuss with students the role of animals in these stories.

Purpose

Students will determine that cytoplasm differences can influence development in frogs.

Process Skills

interpret scientific diagrams, compare and contrast, draw a conclusion, think critically, predict

Teaching Strategies

■ Explain to students that the normal pattern for cell division occurs as shown on the left in the diagram. The last phase shown in both diagrams shows the larval stage of frog development.

Thinking Critically

1. When the gray crescent is present, normal development occurs. If the gray crescent is not present, development does not occur.

2. The first cell division in development is critical for normal embryo development.

3. Neither cell would continue to develop into an embryo.

✔ Assessment

Performance Have students use a stereomicroscope or hand lens to examine preserved frog eggs. Have them diagram what they observe and label the gray crescent. (Note: Preserved frog eggs are available from biological supply houses.) Use the Performance Task Assessment List for Scientific Drawing in **PASC**, p. 55. **L1 ELL**

Resource Manager

BioLab and MiniLab Worksheets, p. 113 **L2**
Basic Concepts Transparency 44 and Master **L2 ELL**
Laboratory Manual, pp. 179–182 **L2**

Development of Animals

Most animals develop from a single, fertilized egg cell called a zygote. But how does a zygote develop into the many different kinds of cells that make up a snail, a fish, or a human? After fertilization, the zygotes of different species of animals all have similar stages of development.

Problem-Solving Lab 25-1 — Interpreting Scientific Diagrams

How important is the first cell division in frog development? During development, a fertilized egg cell divides into two cells by the process of mitosis. The first division of a cell sometimes results in two cells with an unequal amount of cytoplasm. Does the amount of cytoplasm in each cell after the first division have any impact on the development of an organism? It does in frogs.

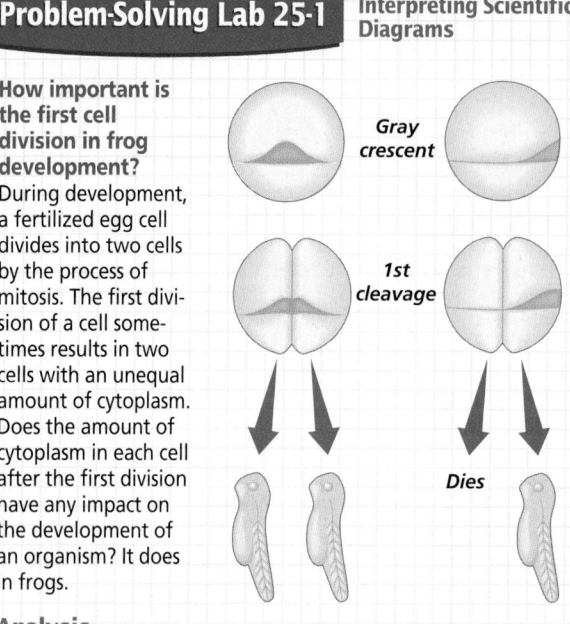

Gray crescent

1st cleavage

Dies

Analysis

In a frog cell, a small section of colored cytoplasm forms just after fertilization. This area is called the gray crescent. Note its appearance in the diagram. Follow the changes in development as the first division of cytoplasm occurs equally through the gray crescent and unequally through the gray crescent.

Thinking Critically

1. Explain how each set of diagrams illustrates the role of the gray crescent in early frog development.
2. Answer the question posed at the beginning of this lab.
3. Predict the effect on frog development if the first division occurred on the horizontal plane rather than on the vertical plane.

Fertilization

Most animals reproduce sexually. Male animals produce sperm and female animals produce eggs. Fertilization occurs when a sperm penetrates the egg, forming a unicellular zygote. In animals, fertilization may be internal or external.

Cell division

The unicellular zygote divides by mitosis to form two new cells in a process called cleavage. Find out how important this first cell division is in frog development by studying the *Problem-Solving Lab*. Once cell division has begun, the organism is known as an embryo. Recall that an embryo is an organism at an early stage of growth and development. The two cells that result from cleavage then divide to form four cells and so on, until a hollow ball of cells called a **blastula** (BLAS chuh luh) is formed. In some animals, such as a lancelet, the blastula is a single layer of cells surrounding a fluid-filled space. In other animals, such as frogs, there may be several layers of cells surrounding the space. The blastula is formed early in the development of an animal embryo. In sea urchin development, for example, the formation of a blastula is complete about ten hours after fertilization. In humans, the blastula forms about five days after fertilization.

Gastrulation

After blastula formation, cell division continues. The cells on one side of the blastula then fold inward to form a gastrula. The **gastrula** (GAS truh luh) is a structure made up of two layers of cells with an opening at one end. Gastrula formation can be compared to the way a potter creates a cup or bowl from a lump of clay,

---

BIOLOGY JOURNAL

Demonstrating Development

Naturalist Provide students with diagrams of the human body showing bone, muscle, spinal cord, brain, kidneys, liver, lungs, and pancreas. Give them colored pencils and have them shade parts of the body that develop from ectoderm yellow, parts that develop from endoderm green, and parts that develop from mesoderm red. Provide them the following information: ectoderm produces brain, spinal cord, nerves, outer skin, eye lens, nose, and ears. Endoderm produces pancreas, liver, lungs, and lining of the digestive system. Mesoderm produces skeleton, muscles, excretory system, inner skin. **L1 ELL P**

illustrated in *Figure 25.4*. First, the clay is formed into a solid ball. Then, the potter presses in on the top of the ball to form a cavity that becomes the interior of the bowl. In the same way, the cells at one end of the blastula fold inward, forming a cavity lined with a second layer of cells. The layer of cells on the outer surface of the gastrula is called the **ectoderm.** The layer of cells lining the inner surface is called the **endoderm.** The ectoderm cells of the gastrula continue to grow and divide, and eventually they develop into the skin and nervous tissue of the animal. The endoderm cells develop into the lining of the animal's digestive tract and into organs associated with digestion.

Formation of mesoderm

In some animals, the development of the gastrula progresses until a layer of cells called the mesoderm is formed. Mesoderm is found in the middle of the embryo; the term meso means "middle." The **mesoderm** (MEZ uh durm) is the third cell layer found in the developing embryo between the ectoderm and the endoderm. The mesoderm cells develop into the muscles, circulatory system, excretory system, and, in some animals, the respiratory system. How does the mesoderm form? You can find out by reading the *Inside Story* on the next page.

In some classes of animals, the opening of the indented space in the gastrula becomes the mouth. These animals, which include earthworms and insects, are called protostomes. A **protostome** (PROHT uh stohm) is an animal with a mouth that develops from the opening in the gastrula.

In other animals, such as fishes, birds, and humans, the opening of the gastrula does not develop into a mouth. A **deuterostome** (DEW tihr

Figure 25.4
You can think of a blastula as a hollow ball of cells. By pushing in on one side, a gastrula is formed.

uh stohm) is an animal in which the mouth develops from cells elsewhere on the blastula.

Scientists hypothesize that protostome animals were the first to appear in evolutionary history, and that deuterostomes followed at a later time. Biologists today often classify an unknown organism by identifying its phylogeny. Recall that phylogeny is the evolutionary history of an organism. Determining whether an animal is a protostome or deuterostome can help biologists identify its group. Even though sea urchins, for example, are invertebrates and fishes are vertebrates, both are deuterostomes and are, therefore, more closely related than you might conclude from comparing their adult body structures.

WORD Origin

protostome
From the Greek words *proto*, meaning "before," and *stoma*, meaning "mouth."

deuterostome
From the Greek words *deutero*, meaning "secondary," and *stoma*, meaning "mouth."

A protostome and a deuterostome differ in the location of the cells that become the organism's mouth.

Building a Model

Kinesthetic Give students salt dough that you have prepared (one part salt, one part flour, one part water). Have them cover a space on their desktops with a few paper towels and dip their fingers into flour so that the dough will not stick to them. Ask students to make models of a fertilized egg, a blastula, and a gastrula. They should make cross sections of the stages in sequence, with one stage turning into the next stage, rather than lining up all three stages at once. They will have to add more dough to show that the fertilized egg becomes larger; but after that time, dough should not be added because the total amount of cytoplasm does not increase as the early embryo develops. **L1** **ELL**

3 Assess

Check for Understanding

Intrapersonal Give students a live insect. Ask them to list the characteristics that show that the insect is an animal. Remind them to treat live animals gently. **L2**

Reteach

Kinesthetic Make clay models of stages of development. Ask students working in groups to place the models in sequence and describe what happens in each stage. **L1** **ELL**

Extension

Encourage students to visit a science museum that has displays about development. Have them make a photo essay of the exhibit that can be displayed in class. Have students include captions for the photos. **L2** **ELL**

Assessment

Portfolio Have students summarize the development of a sea urchin egg from fertilization to the planula stage. Have them include this summary in their portfolios. **L2** **P**

TECHPREP

Spina Bifida

Linguistic Ask students to interview a doctor about spina bifida. Ask students to write a report of their findings and to prepare diagrams that show normal and abnormal stages of development. **L3**
P **COOP LEARN**

Portfolio

Linguistic Have students work in pairs to look up words in the dictionary that begin with the prefixes ecto- and endo-. Ask students to make a list of the words. Have students exchange lists and write their own definitions of the words. Then ask each student to compare actual definitions with their own definitions. **L2** **ELL** **P** **COOP LEARN**

697

Purpose

Students study the stages of early embryonic development.

Teaching Strategies

■ Ask students to study the diagrams, then write a numbered list describing the developmental stages shown. They can use the list as a study guide.

Visual Learning

Have students describe the changes in appearance at each stage of development. **L1**

Critical Thinking

Cells continue to divide, becoming smaller and smaller with each cell division.

GLENCOE TECHNOLOGY

VIDEODISC
Biology: The Dynamics of Life
Disc 1, Side 2, 1 min. 10 sec.
Embryo Development (Ch. 30)

CD-ROM
Biology: The Dynamics of Life
Animation: *Embryo Development* Disc 4

VIDEODISC
The Secret of Life
Dividing cells, early embryo

Early Animal Development

The fertilized eggs of most animals follow a similar pattern of early development. From a single cell, many divisions occur until a hollow ball of cells forms. The hollow ball folds inward and continues to develop.

Critical Thinking *How do cells change as an embryo develops?*

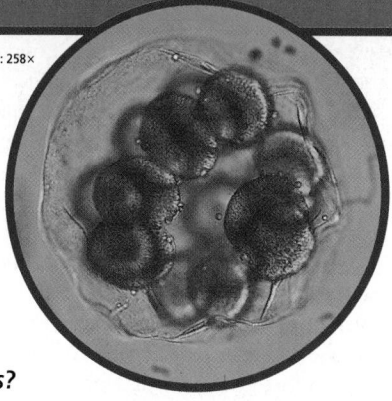

Magnification: 258×

Sea urchin blastula

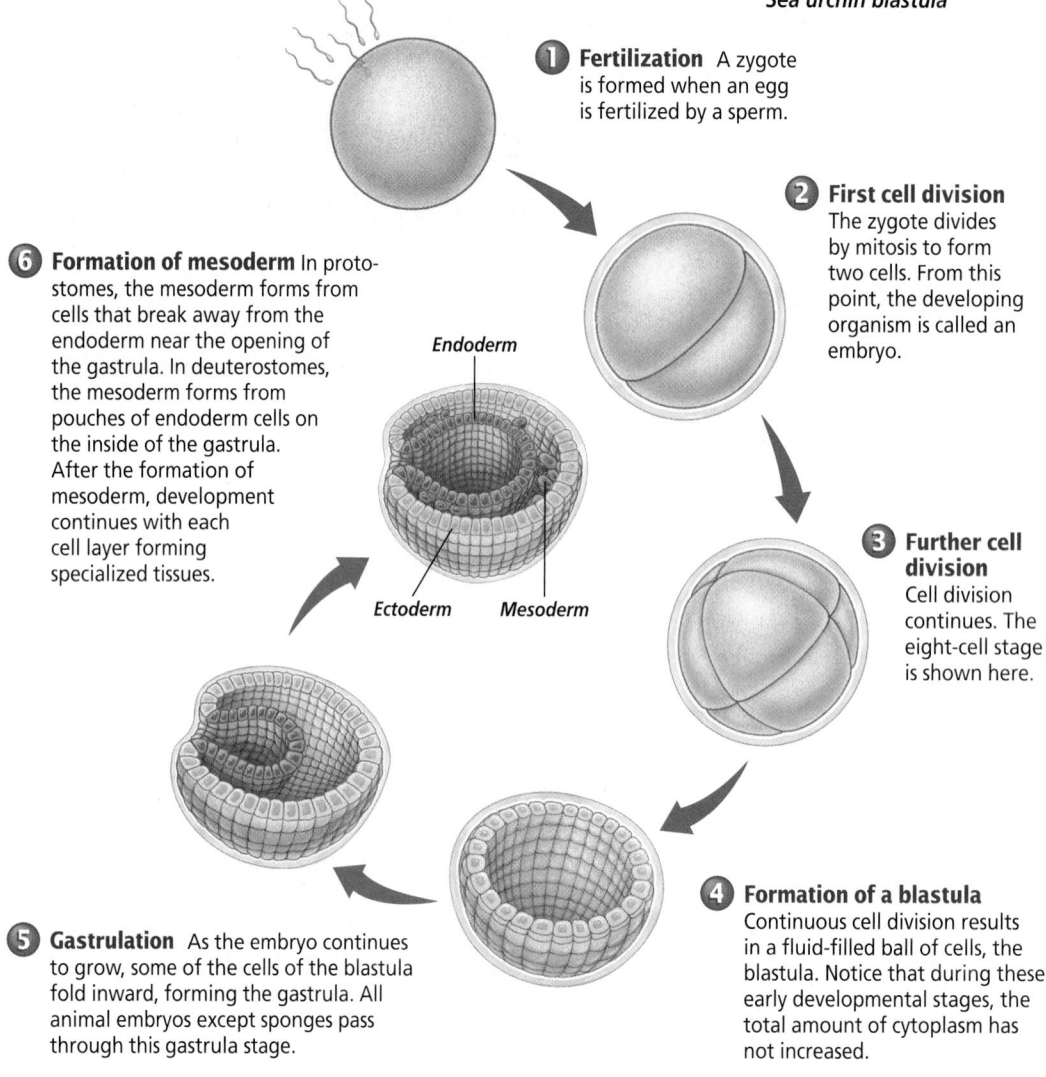

1 Fertilization A zygote is formed when an egg is fertilized by a sperm.

2 First cell division The zygote divides by mitosis to form two cells. From this point, the developing organism is called an embryo.

3 Further cell division Cell division continues. The eight-cell stage is shown here.

4 Formation of a blastula Continuous cell division results in a fluid-filled ball of cells, the blastula. Notice that during these early developmental stages, the total amount of cytoplasm has not increased.

5 Gastrulation As the embryo continues to grow, some of the cells of the blastula fold inward, forming the gastrula. All animal embryos except sponges pass through this gastrula stage.

6 Formation of mesoderm In protostomes, the mesoderm forms from cells that break away from the endoderm near the opening of the gastrula. In deuterostomes, the mesoderm forms from pouches of endoderm cells on the inside of the gastrula. After the formation of mesoderm, development continues with each cell layer forming specialized tissues.

Endoderm

Ectoderm *Mesoderm*

698 WHAT IS AN ANIMAL?

MEETING INDIVIDUAL NEEDS

Visually Impaired

Kinesthetic Obtain several old tennis balls or other type of hollow rubber balls. Have visually impaired students handle the balls and explain that the blastula appears just like the hollow balls. Then have these students hold the ball in one hand while pushing in on one side of the ball with their thumbs. Explain that now the balls represent the gastrula stage. Ask students to use their free hand to point to the parts of the ball that represent ectoderm and endoderm. **L1 ELL**

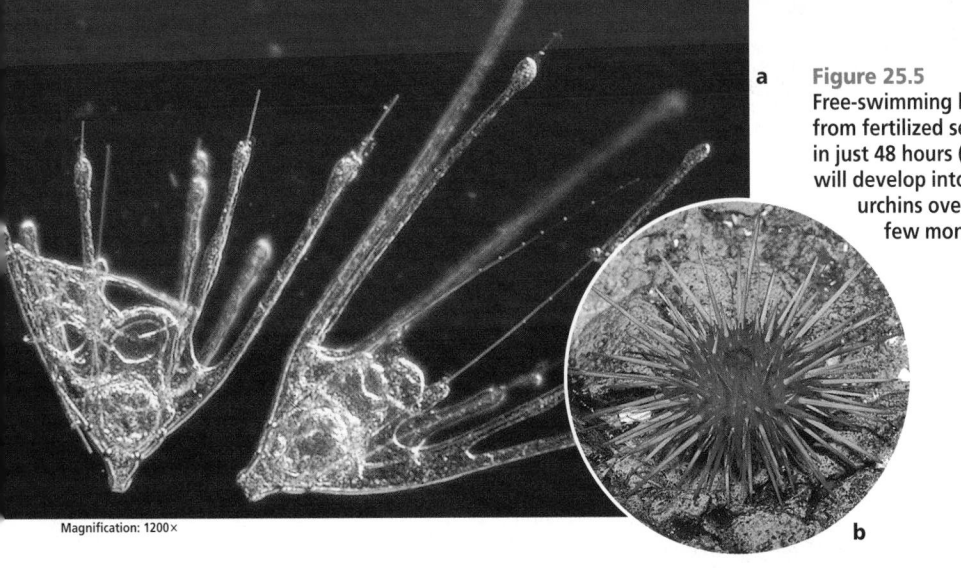

Magnification: 1200×

Continued growth and development

Cells in developing embryos continue to change shape and become specialized to perform different functions. Most animal embryos continue to develop over time, becoming juveniles that look like smaller versions of the adult animal. In some animals, such as insects and echinoderms, the embryo develops into an intermediate stage called a larva (pl. larvae) that often bears little resemblance to the adult animal. In these animals the larva is still surrounded by a membrane formed right after fertilization. When the larva hatches, it breaks through this fertilization membrane. Animals that are generally sessile as adults, such as sea urchins, often have a free-swimming larval stage, shown in **Figure 25.5**. You can observe development in fishes in the *BioLab* at the end of this chapter.

Forming an adult animal

Once the juvenile or larval stage has passed, most animals continue to grow and develop into adults. This growth and development may take just a few days in some insects, or up to fourteen years in some mammals. Eventually the adult animals reach sexual maturity, mate, and the cycle begins again.

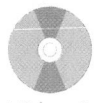
CD-ROM
View an animation of embryo development in the Presentation Builder of the Interactive CD-ROM.

Section Assessment

Understanding Main Ideas
1. Describe the characteristics that make a mouse an animal.
2. Explain why movement is an important characteristic of animals.
3. Explain the difference between a protostome and a deuterostome.
4. Compare and contrast planarian and earthworm digestive tracts.

Thinking Critically
5. Name a land animal that is sessile. Why would being sessile be a disadvantage to an animal that lives on land?

SKILL REVIEW
6. **Sequencing** Place the following words in sequence, beginning with the earliest stage: gastrula, larva, adult, fertilized egg, blastula. For more help, refer to *Organizing Information* in the **Skill Handbook.**

Section Assessment

1. It is a multicellular heterotroph that uses food for energy. It moves to get food and its cells do not have cell walls.
2. Animals do not make their own food, but depend upon other organisms for food. They must search for food, so movement is necessary for survival.
3. In protostomes the mouth develops from the opening in the gastrula. In deuterostomes, the mouth develops elsewhere on the gastrula.
4. In planarians, food is sucked in through the pharynx and ground up by movement; digestion occurs by phagocytosis in individual cells. In an earthworm, food is taken in by the mouth, ground up in the gizzard, and moves through the rest of the digestive system. Food is carried to all body cells by the circulatory system, and wastes are expelled through the anus.
5. Students will probably not be able to name a land animal that is sessile. All animals on land must move to obtain food.
6. fertilized egg, blastula, gastrula, larva, adult

Prepare

Key Concepts

Students compare and contrast types of symmetry and study basic body plans of animals.

Planning

- Collect a variety of kitchen items that display different types of symmetry for the Quick Demo.
- Gather dried marine sponges, a preserved jellyfish, and live goldfish in clear plastic containers for the Quick Demo.

1 Focus

Bellringer

Before presenting this lesson, display **Section Focus Transparency 62** on the overhead projector and have students answer the accompanying questions. **L1** **ELL**

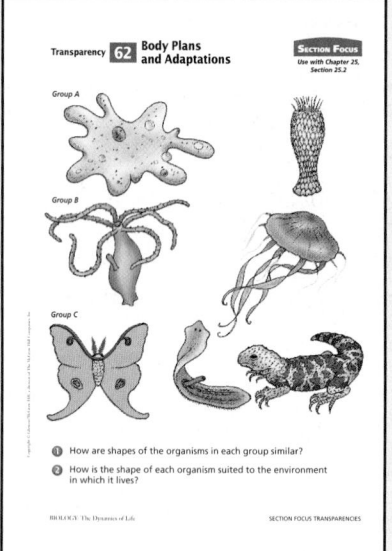

Transparency **62** Body Plans and Adaptations | **SECTION FOCUS** Use with Chapter 25, Section 25.2

Group A

Group B

Group C

① How are shapes of the organisms in each group similar?

② How is the shape of each organism suited to the environment in which it lives?

BIOLOGY: The Dynamics of Life | SECTION FOCUS TRANSPARENCIES

SECTION PREVIEW

Objectives

Compare and contrast radial and bilateral symmetry with asymmetry.

Trace the phylogeny of animal body plans.

Compare body plans of acoelomate, pseudocoelomate, and coelomate animals.

Vocabulary

symmetry
radial symmetry
bilateral symmetry
anterior
posterior
dorsal
ventral
acoelomate
pseudocoelom
coelom
exoskeleton
endoskeleton
invertebrate
vertebrate

Section

25.2 Body Plans and Adaptations

Objects made by a potter can be many different shapes and sizes. There is a plan for making each type of pottery. One plan results in a bowl, another in a vase, and still another in a flat plate. Each piece of pottery is suited for a particular function. Animals' bodies also have plans—body shapes that are suited to a particular way of life. In this section, you will study a variety of animal body plans and see how a specific body structure is adapted to life in a particular environment.

What Is Symmetry?

Look at the animals shown in *Figure 25.6.* You know that all animals share certain characteristics, but these animals don't look like they have much in common. The sponge seems to have no particular shape, whereas the fish has a head, body, and several pairs of fins. The jellyfish doesn't have a head or tail, and is circular in form. Each animal can be described in terms of symmetry according to its shape. **Symmetry** (SIH muh tree) refers to a balance in proportions of an object or organism. All animals have some kind of symmetry. Different kinds of symmetry enable animals to move about in different ways.

Asymmetry in a sponge

Many sponges have an irregularly shaped body, as seen in *Figure 25.7A.* An animal that is irregular in shape has an asymmetrical body plan. Animals with no symmetry often are sessile organisms that do not move from place to place. Most sponges do not move

Figure 25.6
A sponge **(a)**, a fish **(b)**, and a jellyfish **(c)** all exhibit different kinds of symmetry.

a

700

b

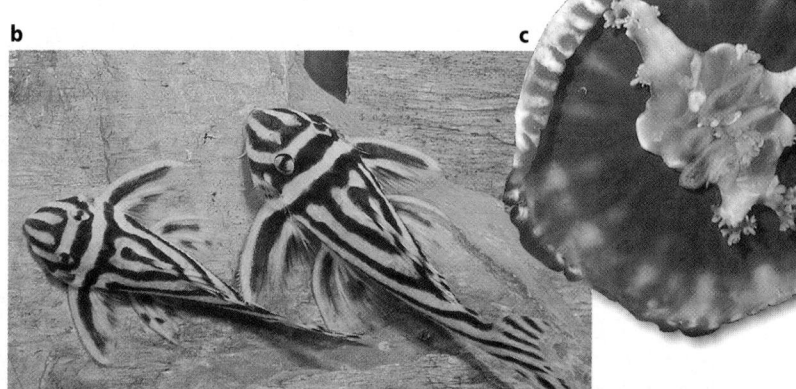

c

Resource Manager

Concept Mapping, p. 25 **L3** **ELL**
Section Focus Transparency 62 and Master **L1** **ELL**

BIOLOGY JOURNAL

Symmetry in Objects

Visual-Spatial Divide the class into groups. Have each group list in their journals as many items with radial symmetry as they can find in the classroom within a given time period. Repeat the activity with asymmetrical and bilaterally symmetrical objects. **L1** **COOP LEARN**

Dorsal

Anterior

Ventral

Posterior

A This irregularly shaped sponge is an example of an animal with an asymmetrical body plan.

B An example of an animal with radial symmetry, a hydra feeds on tiny animals it immobilizes with venomous stinging cells found along its tentacles.

C In bilaterally symmetrical animals, such as butterflies, sensory tissue is commonly concentrated in the head, or anterior, end.

about once they have reached the adult stage.

The bodies of most sponges consist of two layers of cells. Unlike all other animals, a sponge's embryonic development does not include the formation of an endoderm and mesoderm, or a gastrula stage. Fossil sponges first appeared in rocks dating back to more than 700 million years ago. They represent one of the oldest groups of animals on Earth—evidence that their two-layer body plan makes them well adapted for life in aquatic environments.

Radial symmetry in a hydra

A hydra, a tiny predator pictured in *Figure 25.7B,* feeds on small animals it snares with its tentacles. A hydra has radial symmetry. Its tentacles radiate out from around its

mouth. As you can see, animals with **radial symmetry** (RAYD ee uhl) can be divided along any plane, through a central axis, into roughly equal halves. Radial symmetry is an adaptation in hydra that enables the animal to detect and capture prey coming toward it from any direction.

Have you ever had your groceries double bagged at the store? The body plan of a hydra can be compared to a sack within a sack. These sacks are cell layers organized into tissues with distinct functions. A hydra develops from just two embryonic cell layers—ectoderm and endoderm.

Bilateral symmetry

The butterfly in *Figure 25.7C* has bilateral symmetry. An organism with **bilateral symmetry** (bi LAT uh rul) can be divided down its length into

Figure 25.7
All animals have a kind of symmetry that enables them to survive in their surroundings.

Note Internet addresses that you find useful in the space below for quick reference.

Purpose 🔲

Students identify the type of symmetry in three animals.

Process Skills

classify, observe and infer, compare and contrast, think critically

Teaching Strategies

■ Review the three types of symmetry present in animals.

Thinking Critically

1. bilateral—body can be divided along its length into two equal halves

2. head and tail regions present, body cavity present, good muscular control

3. pencil, pen, doll

4. asymmetry—body is irregular

5. does not move about, body consists of two cell layers

6. blob of clay, cottage cheese

7. radial—body can be divided along any plane through a central axis into equal halves

8. tentacles radiate from mouth, tissues present

9. wheel with spokes, open umbrella, apple corer

✔ Assessment

Skill Have students prepare a concept map that uses the following terms: sponge, hydra, flatworm, roundworm, earthworm, radial symmetry, bilateral symmetry, asymmetry, pseudocoelomate, coelomate, acoelomate. Use the Performance Task Assessment List for Concept Map in **PASC**, p. 89. **L2**

Problem-Solving Lab 25-2 Classifying

Is symmetry associated with other animal traits?
Animals show different patterns in their symmetry. Symmetry patterns are often associated with certain other characteristics or traits found in the animal.

Analysis

Study these three animal diagrams. Determine the type of symmetry being shown.

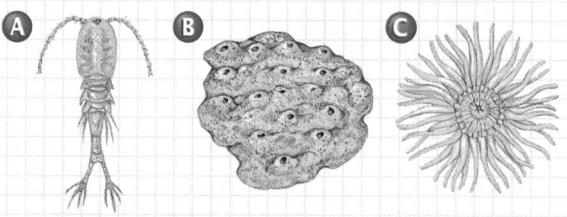

Thinking Critically

1. Animal A shows what type of symmetry? Explain your answer.
2. Describe other traits associated with animal A.
3. Name some objects other than animals that show the A pattern of symmetry.
4. Animal B shows what type of symmetry? Explain your answer.
5. Describe other traits associated with animal B.
6. Name some objects other than animals that show the B pattern of symmetry.
7. Animal C shows what type of symmetry? Explain your answer.
8. Describe other traits associated with animal C.
9. Name some objects other than animals that show the C pattern of symmetry.

---

similar right and left halves that form mirror images of one another. Bilaterally symmetrical animals can be divided only along one plane to form equal halves, in contrast to radially symmetrical animals that can be divided along any plane. In bilateral animals, the **anterior,** or head end, often has sensory organs. The **posterior** of these animals is the tail end. The **dorsal** (DOR sul), or back surface, also looks different from the **ventral** (VEN trul), or belly surface.

Animals with bilateral symmetry can find food and mates and avoid predators because they have sensory organs and good muscular control. Test your ability to identify animal symmetry in the *Problem-Solving Lab.*

Bilateral Symmetry and Body Plans

Animals that are bilaterally symmetrical also share other important characteristics. All bilaterally symmetrical animals developed from three embryonic cell layers—ectoderm, endoderm, and mesoderm. Some bilaterally symmetrical animals also have fluid-filled spaces called body cavities inside their bodies in which internal organs are found. The development of body cavities made it possible for animals to grow larger and to develop organs and organ systems, such as digestive systems.

Acoelomate flatworms have no body cavities

Flatworms are bilaterally symmetrical animals with solid, compact bodies, as shown in *Figure 25.8.* Animals that have three cell layers—ectoderm, endoderm, and mesoderm—but no body cavities are called **acoelomate** (ay SEE lum ate) animals. Acoelomate animals may have been the first group of animals in which organs evolved from cells of the mesoderm. Like other acoelomate animals, the organs of flatworms are embedded in the solid tissues of their bodies. Although acoelomate animals have no body cavities, they do have a digestive tract that extends throughout the body. A flattened body and a digestive tract allow for the diffusion of nutrients, water, and oxygen to supply all body cells and to eliminate wastes.

Alternative Lab

Body Plans

Purpose 🔲

Students examine prepared slides of cross sections of a hydra, flatworm, roundworm, and earthworm to determine if the animals are acoelomate, pseudocoelomate, or coelomate.

Materials

microscope, prepared slides of cross sections of hydra, planarian, nematode, earthworm

Procedure

Give students the following directions.

1. Examine the cross-section slides.
2. Sketch and label each cross section with the following labels: animal with two cell layers, pseudocoelomate animal, acoelomate animal, coelomate animal. Also label the drawings with the names of organisms.

Analysis

1. Order the animals' body plans from least to most complex. *hydra, planarian, nematode, earthworm*
2. In what way is each animal's body plan an adaptation to its environment? *Hydra and planarians are adapted to movement in water. Nematodes and earthworms are adapted to life in soil.*

Pseudocoelomates have a body cavity

A roundworm is another animal that has bilateral symmetry. However, unlike the solid body of a flatworm, the body of a roundworm has a body cavity that develops between the endoderm and mesoderm. This body cavity is called a pseudocoelom. A **pseudocoelom** (SEWD uh see lum) is a fluid-filled body cavity partly lined with mesoderm.

The pseudocoelom enables animals, such as roundworms, to move quickly. How does this work? Think about the way your muscles work. The muscles in your arm lift your hand by pulling against your arm bones. If there were no rigid bones in your arms, your muscles would not be able to do any work. Although the roundworm has no bones, it does have a rigid, fluid-filled space, the pseudocoelom, which is partly surrounded by mesoderm. Its muscles attach to the mesoderm and brace against the pseudocoelom. You can observe movement in a pseudocoelomate animal in the *MiniLab* on this page.

MiniLab 25-2 — Observing and Inferring

Magnification: 292×

Check Out a Vinegar Eel Vinegar eels are roundworms with pseudocoelms. They exhibit an interesting pattern of locomotion because they have only longitudinal (lengthwise) muscles.

Vinegar eel

Procedure

1. Prepare a wet mount of vinegar eels. **CAUTION:** *Use caution when working with a microscope and slides.*
2. Observe them under low-power magnification.
3. Note their pattern of locomotion. Prepare a series of diagrams that illustrate their pattern of movement.
4. Time how fast they move by timing in seconds how long it takes for one roundworm to move across the center of your field of view. Find out the diameter of your low-power field in mm. Calculate vinegar eel speed in mm/sec. You may want to time several animals and average their speed.

Analysis

1. Name the type of symmetry present in vinegar eels.
2. Describe the pattern of locomotion for vinegar eels.
3. How does the pseudocoelom aid vinegar eels in locomotion?
4. What is the speed of locomotion for a vinegar eel? Based on the speed of your vinegar eel, predict the speed in mm/sec for a flatworm. Explain your answer.

Figure 25.8 Animals with acoelomate bodies usually have a thin, somewhat flattened shape **(a)**. Pseudocoelomate animals are larger and thicker in body shape than their acoelomate ancestors **(b)**. The coelom provides a space for complex internal organs **(c)**.

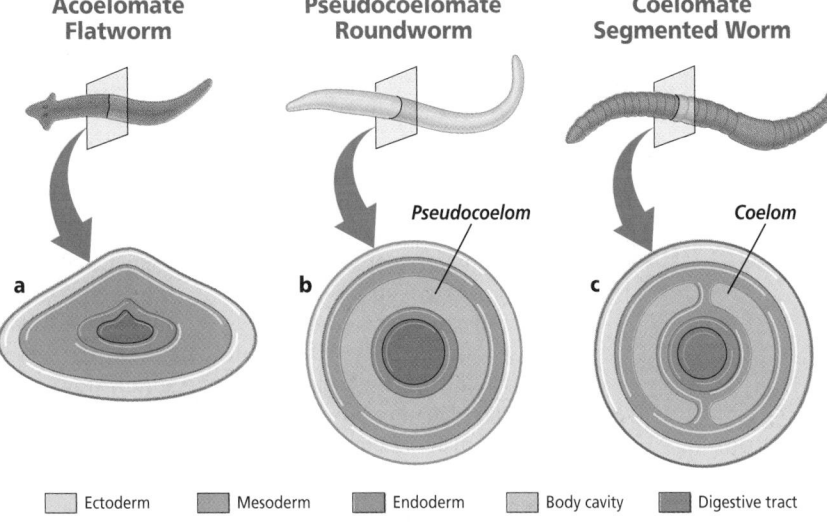

Acoelomate Flatworm	Pseudocoelomate Roundworm	Coelomate Segmented Worm
a	b *Pseudocoelom*	c *Coelom*

■ Ectoderm ■ Mesoderm ■ Endoderm ■ Body cavity ■ Digestive tract

25.2 BODY PLANS AND ADAPTATIONS **703**

3. If you were to observe these animals' movements, predict what you might see. Explain in terms of their body plans. *Students' descriptions should match those given in the chapter.*

✔ Assessment

Portfolio Ask students to summarize what they have learned about body plans in this lab. Have them consider the survival value of each type of body plan. Instruct them to put their summaries in their portfolios. Use the Performance Task Assessment List for Writing in Science in **PASC**, p. 87. **L2**

Figure 25.9
The exoskeleton of this crab has been shed in order for the crab to grow. A new exoskeleton forms in a few hours to provide protection again.

WORD *Origin*

coelom
From the Greek word *koiloma*, meaning "cavity." A coelom is a body cavity completely surrounded by mesoderm.

Pseudocoelomate animals have a complete, one-way digestive tract with organs that have specific functions. The mouth takes in food, the middle section breaks it down and absorbs nutrients, and the anus expels wastes.

The coelom provides space for internal organs

The body cavity of an earthworm develops from a **coelom** (SEE lum), a fluid-filled space that is completely surrounded by mesoderm.

Humans, insects, fishes, and many other animals have a coelomate body plan. The greatest diversity of animals is found among the coelomates.

The coelom provides space for the development of specialized organs and organ systems. In coelomate animals, the digestive tract and other internal organs are attached by double layers of mesoderm and are suspended within the coelom. Like the pseudocoelom, the coelom cushions and protects the internal organs and provides room for them to grow and move independently within an animal's body.

704 WHAT IS AN ANIMAL?

Animal Protection and Support

During the course of evolution, as development of body cavities resulted in a greater diversity of animal species, many animals became adapted to life in different environments. Some animals, such as mollusks, evolved hard shells that protected their soft bodies. Other animals, such as sponges, evolved hardened spicules between their cells that provided support. Some phyla of animals developed exoskeletons. An **exoskeleton** is a hard, waxy covering on the outside of the body that provides a framework for support, shown in *Figure 25.9*. Exoskeletons also protect soft body tissues, prevent water loss, and provide protection. Exoskeletons are secreted by the epidermis and extend into the body, where they provide a place for muscle attachment.

Other phyla of animals have evolved different structures for support and protection. Sea urchins and sea stars, for example, have an internal skeleton that is covered by layers of cells. An **endoskeleton** is an internal skeleton that provides support inside an animal's body. An endoskeleton may be made of calcium carbonate, as in echinoderms; cartilage, as in sharks; or bone. Bony fishes, amphibians, reptiles, birds, and mammals all have endoskeletons made of bone. The endoskeleton protects internal organs and provides an internal brace for muscles to pull against.

Exoskeletons are often found in invertebrates. An **invertebrate** is an animal that does not have a backbone. Many invertebrates, such as crabs, spiders, grasshoppers, dragonflies, and beetles, have exoskeletons. Echinoderms are examples of invertebrates that have endoskeletons.

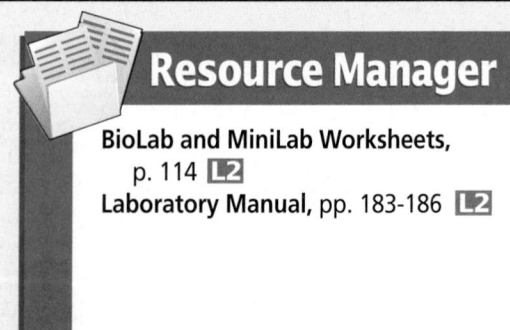

a

Figure 25.10
Invertebrate animals such as an octopus
(**a**) and a sea slug (**b**) have no backbones.
Vertebrates with backbones include a
monkey (**c**) and a flamingo (**d**).

c

b

d

A **vertebrate**
(VURT uh brayt) is
an animal with a
backbone. All verte-
brates are bilaterally
symmetrical animals that
have endoskeletons. Examples
of vertebrates include fishes, birds,
reptiles, amphibians, and mammals,
including humans. *Figure 25.10*
shows examples of invertebrate and
vertebrate animals.

Origins of Animals

Where did animals come from?
Most biologists agree that the animal
kingdom probably evolved from colonial protists. Scientists trace this evolution back in time to the beginning
of the Cambrian period.
Although it may seem that
bilaterally symmetrical
animals appeared much
later, all the major animal
body plans that exist today
were already in existence
545 million years ago.
Since then, many new
species have evolved. All
new species appear to be variations
on the animal body plans that developed during the Cambrian period.

4 Close

Discussion
Have students examine the phylogenetic diagram in Chapter 14.
Ask them to discuss the evolutionary trends of symmetry, cell
layers, and patterns of development of the animal groups shown
on the diagram.

Section Assessment

Understanding Main Ideas
1. Explain the difference between radial and
   bilateral symmetry in animals, and give an
   example of each type.
2. Compare the body plans of acoelomate and
   coelomate animals. Give an example of an
   animal with each type of body plan.
3. Explain how an adaptation such as an exoskeleton could be an advantage to land animals.
4. Compare movement in acoelomate and
   coelomate animals.

Thinking Critically
5. Explain how having a coelom enables an animal
   to have complex organ systems.

SKILL REVIEW
6. **Making and Using Tables** Construct a table
   that compares the body plans of the sponge,
   hydra, flatworm, roundworm, and earthworm.
   For more help, refer to *Organizing Information*
   in the **Skill Handbook.**

Section Assessment

1. Animals with radial symmetry, such as
   the hydra, can be divided along any
   plane into equal halves. Animals with
   bilateral symmetry, such as the flatworm, can be divided only along one
   plane to form two equal halves.
2. Animals with acoelomate bodies, such
   as a flatworm, have three cell layers
   with a digestive tract but no body
   cavity. Animals with coelomate bodies, such as an earthworm, have a
   coelom in which internal organs are
   suspended within the coelom.
3. An exoskeleton prevents water loss
   from body organs and supports an
   animal's body on land.
4. Acoelomate animals have no body
   cavity, whereas coelomate animals
   can brace their muscles against the
   coelom, thereby giving them more
   powerful movements.
5. The coelom provides space for specialized organs and organ systems.
6. Make sure students' tables indicate
   cell layers, and presence of a coelom,
   a pseudocoelom, or no body cavity.

Time Allotment

30 minutes on day 1, 15 minutes on days 2 through 5

Process Skills

interpret scientific illustrations, observe and infer, organize data, sequence, experiment

PREPARATION

- Zebra fish are available from pet shops or biological supply houses. Purchase six of each sex. Fish must be young (between 6 and 24 months old). Remind students to treat live animals gently.
- Prepare an aquarium with filter, lights, timer, and heater. Marbles should be placed along the bottom. Embryos will fall between the marbles.
- Set timer and lights as follows: 14 hours light to 10 hours dark. Set the heater at 28.5°C.
- Feed fish flakes in the morning and brine shrimp or live blood worms in the afternoon.

GLENCOE
TECHNOLOGY

VIDEODISC
The Secret of Life
Developing zebra fish

Developing zebra fish, close-up

Resource Manager

BioLab and MiniLab Worksheets, pp. 115-116 L2

706

Zebra Fish Development

The zebra fish (Danio rerio) *is a common fresh-water fish sold in pet shops. They are ideal animals for study because they undergo developmental changes quickly and major stages can be observed within hours after fertilization.*

PREPARATION

Problem

What do the developmental stages of the zebra fish look like?

Objectives

In this BioLab, you will:
- **Observe** stages of zebra fish development.
- **Record** all observations in a data table.
- **Use the Internet** to collect and compare data from other students.

Skill Handbook

Use the **Skill Handbook** if you need additional help with this lab.

Materials

aquarium
zebra fish
turkey baster
beaker
dropper
petri dish
binocular
   microscope
wax pencil or labels

Safety Precautions

Always wear safety goggles in the lab. Use caution when working with a binocular microscope and glassware.

PROCEDURE

1. Copy the data table.
2. Using the turkey baster, draw up water containing zebra fish embryos from the bottom of the aquarium.
3. Release the water into a beaker and allow the embryos to settle to the bottom.
4. Label a petri dish with your name and class period. Fill the bottom of the dish with aquarium water. Using a dropper, place several embryos in your dish.
5. Your teacher will advise you as to the approximate time that fertilization took place. All ages should be reported in your data table as hpf (hours past fertilization).
6. Using a binocular microscope, observe the embryos. Make a diagram in your data table of the embryos' appearances and indicate the age of the embryos in hpf.

PROCEDURE

Teaching Strategies

- Use the baster to remove excess debris from the bottom of the tank on the evening prior to spawning.
- Females will release eggs about 30 minutes after the light comes on. Time this light pattern to correspond with your first morning class. Students may be able to see males following females along the bottom of the tank as eggs are released. Fertilization is external.
- Classes that occur later in the day may not be able to find early stages of embryonic development because fertilization has taken place in the morning. Embryonic development is rapid.
- Feeding is not necessary for the first week. Later, feed the young live paramecia.

7. Go to the Glencoe Science Web site as shown in the Sharing Your Data box below to post your data.

8. Continue to observe your embryos daily for a minimum of one week. Note the appearance of new organs and when movement is first seen. If you wish to continue watching developmental changes, consult with your teacher for directions. **CAUTION:** *Wash your hands with soap and water immediately after completing observations.*

Data Table

Date	hpf	Diagram	Observations

ANALYZE AND CONCLUDE

1. **Communicating** Explain why zebra fish are ideal animals for studying embryonic development.

2. **Thinking Critically** Explain why you may not have been able to see stages such as a blastula or gastrula.

3. **Thinking Critically** Suggest how you could change the experiment's design to allow for observing these stages.

4. **Using the Internet** Visit the Glencoe Science Web site for links to Internet sites that will help you complete sequences of the major changes during development of zebra fish:
   a. between 1 and 10 hpf. Include labeled diagrams of these changes.
   b. between 10 and 28 hpf. Include labeled diagrams.
   c. between 28 and 72 hpf. Include labeled diagrams.

Sharing Your Data

BIOLOGY Online Find this BioLab as well as links to sites with diagrams of zebra fish development on the Glencoe Science Web site at **science.glencoe.com.** Post your data in the data table provided for this activity. Use the additional data from other students to answer the questions for this lab. Were there large variations in data posted by other students? What might have caused these differences?

INTERNET BioLab

ANALYZE AND CONCLUDE

1. They are inexpensive, easy to care for, form many embryos, and show rapid development.

2. The early stages occur rapidly after fertilization.

3. Students could reset the timer, or use embryos that were collected and preserved by an earlier class.

4. For questions a–c, student answers should agree with the information in Data and Observations. Encourage students to print out Internet diagrams and include them in their answers to this question.

✓ Assessment

Portfolio Ask students to prepare a report of their findings using their own data and diagrams as well as the diagrams available from their Internet sources. Use the Performance Task Assessment List for Writing in Science in **PASC,** p. 87. L2 P

Sharing Your Data

BIOLOGY Online To navigate to the Internet BioLabs, choose the Biology: The Dynamics of Life icon at the Glencoe Science Web site. Click on the student icon, then the BioLabs icon. The advantage to having large numbers of trials in an experiment is that the results will be more accurate.

■ Use binocular microscopes to view the development of embryos.

■ Use this Internet address if you have trouble with breeding these fish: http://weber.u.washington.edu/~fishscop/

Data and Observations

Between 1-2 hpf: dividing cells will appear above yolk sac.

Between 2-5 hpf: blastula formation.

Between 5-10 hpf: gastrulation formation is complete with appearance of somites.

Between 10-28 hpf: tissues and organs form.

Between 28-72 hpf: fins, gills, mouth form.

Biology & SOCIETY

Purpose

Students explore ways in which human actions can affect the survival of other species. They are exposed to differing opinions about the level of effort expended to rescue endangered and threatened species.

Background

The Endangered Species Act requires the preservation of habitat, often the primary source of disagreements about the law. Land use issues may be especially difficult to resolve when the species in question is small and unfamiliar, such as a wildflower, butterfly, or small fish.

Teaching Strategies

■ Encourage students to do Internet or library research to find out whether the Endangered Species Act has been altered by legislation. **L2**

■ Have students obtain current information about local or state laws that deal with protection of species that are endangered or threatened. **L2**

Investigating the Issue

Students should discover that captive breeding and reintroduction efforts, plus the banning of the pesticide DDT, are having a positive impact on the status of peregrine falcon populations in the United States.

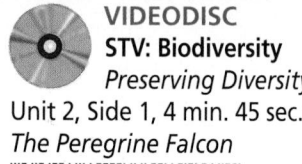

Biology & SOCIETY

Protecting Endangered Species

Peregrine falcon

The American peregrine falcon is a fast-flying, streamlined hawk that feeds primarily on birds it catches in the air. During the mid-1900s, the falcon's ability to reproduce was seriously weakened by the insecticide DDT. In 1972, DDT was banned in the United States. Banning the use of DDT gave peregrine falcons a chance to recover.

In 1973, the United States government passed a law, the Endangered Species Act, to help prevent the extinction of threatened or endangered species, and to preserve the habitats needed for their survival. An endangered species is one whose numbers have fallen so low that it could become extinct in the near future. A threatened species is one that would soon become endangered if steps are not taken to protect it. In 1998, about 965 species in the United States were listed as endangered or threatened.

Preserving or restoring habitat Under the terms of the Endangered Species Act, it is illegal to kill or harm protected species. Protection under the act also means that efforts must be made to preserve or restore the species' habitat. A recovery plan written for each species outlines steps to be taken to help reestablish it. Periodically, the status of each species is reviewed, and it may be reclassified.

Different Viewpoints

How far should we go to protect endangered species? Some people argue that the Endangered Species Act places too much importance on plants and animals at the expense of human needs. The recovery plan for the peregrine falcon included protecting any known nesting sites from disturbance by humans. For example, if falcons were found nesting in a forest that was to be logged, logging had to be delayed until after the young falcons left the nest. Loggers sometimes disagree with regulations like this because they fear losing their ability to make a living. But actions called for by the Endangered Species Act, including nest protection, elimination of DDT use, and introduction of captive-bred birds into the wild, have given the falcon a second chance.

Successful reintroduction An introduction program that began in the 1980s in New York City has been a particular success. By 1998, there were 12 breeding pairs of falcons nesting on city buildings and bridges. By the end of the twentieth century, falcons were on their way off the endangered species list.

INVESTIGATING THE ISSUE

Analyzing the Issue Use your research skills to find out the current status of peregrine falcon populations in the United States. Prepare a brief report that describes the results of reintroduction efforts to date.

BIOLOGY Online To find out more about endangered species, visit the Glencoe Science Web site. **science.glencoe.com**

Going Further

Several species listed as threatened or endangered at one time have begun to make a comeback. Invite students to find out about the current status of the bald eagle, red wolf, California sea otter, Louisiana black bear, or the American alligator. What controversies surrounded the recovery efforts for these species? **L2**

SUMMARY

Section 25.1

Typical Animal Characteristics

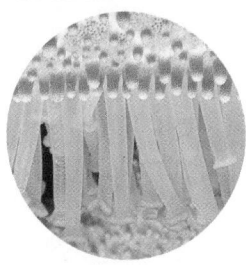

Main Ideas

- Animals are heterotrophs, digest their food inside the body, typically have a type of locomotion, and are multicellular. Animal cells have no cell walls.

- Embryonic development from a fertilized egg is similar in many animal phyla. The sequence after division of the fertilized egg is: the formation of a blastula with one layer of cells; a gastrula with two layers of cells, ectoderm and endoderm; and finally a gastrula with mesoderm, a layer of cells between the ectoderm and endoderm.

Vocabulary

blastula (p. 696)
deuterostome (p. 697)
ectoderm (p. 697)
endoderm (p. 697)
gastrula (p. 696)
mesoderm (p. 697)
protostome (p. 697)
sessile (p. 694)

Section 25.2

Body Plans and Adaptations

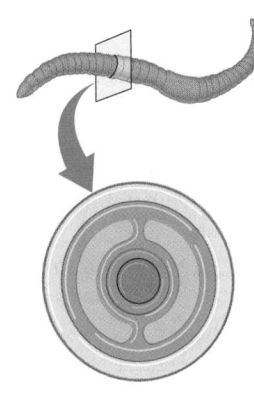

Main Ideas

- Animals have a variety of body plans and types of symmetry that are adaptations.

- Animals may be asymmetrical, radially symmetrical, or bilaterally symmetrical.

- A coelom is a fluid-filled body cavity that supports internal organs.

- Flatworms and other acoelomate animals have flattened, solid bodies with no body cavities.

- Animals such as roundworms have a pseudocoelom, a body cavity that develops between the endoderm and mesoderm.

- Coelomate animals such as humans and insects have internal organs suspended in a body cavity that is completely surrounded by mesoderm.

- Exoskeletons provide a framework of support on the outside of the body, whereas endoskeletons provide internal support.

- Animals probably evolved from colonial protists in the Cambrian period.

Vocabulary

acoelomate (p. 702)
anterior (p. 702)
bilateral symmetry (p. 701)
coelom (p. 704)
dorsal (p. 702)
endoskeleton (p. 704)
exoskeleton (p. 704)
invertebrate (p. 704)
posterior (p. 702)
pseudocoelom (p. 703)
radial symmetry (p. 701)
symmetry (p. 700)
ventral (p. 702)
vertebrate (p. 705)

Main Ideas
Summary statements can be used by students to review the major concepts of the chapter.

Using the Vocabulary
To reinforce chapter vocabulary, use the Content Mastery Booklet and the activities in the Interactive Tutor for Biology: The Dynamics of Life on the Glencoe Science Web site: **science.glencoe.com**

THE PRINCETON REVIEW
All Chapter Assessment questions and answers have been validated for accuracy and suitability by The Princeton Review.

UNDERSTANDING MAIN IDEAS

1. b
2. c

UNDERSTANDING MAIN IDEAS

1. Which of these organs forms from the ectoderm?
   a. stomach
   b. skin
   c. intestines
   d. liver

2. Which animal pair shares the most characteristics?
   a. earthworm—sea star
   b. earthworm—insect
   c. earthworm—leech
   d. earthworm—clam

GLENCOE TECHNOLOGY

VIDEOTAPE
MindJogger Videoquizzes
Chapter 25: *What is an animal?*
Have students work in groups as they play the videoquiz game to review key chapter concepts.

Resource Manager

Chapter Assessment, pp. 145-150
MindJogger Videoquizzes
ExamView® Pro Software
BDOL Interactive CD-ROM, Chapter 25 quiz

3. b

4. c

5. a

6. b

7. b

8. d

9. a

10. d

11. planarian

12. coelomate

13. exoskeleton

14. coelomate

15. gastrula

16. mesoderm

17. deuterostome

18. dorsal

19. sessile

20. radial

APPLYING MAIN IDEAS

21. No, because this pattern is that of a protostome. A bird is a deuterostome.

---

3. All animals must search for food because they are _____.
   a. autotrophic
   b. heterotrophic
   c. sessile
   d. photosynthetic

4. A(n) _____ is an example of an animal that is sessile as an adult.
   a. osprey
   b. flatworm
   c. coral
   d. sea star

5. Which of the following is NOT a characteristic of animals?
   a. cells with cell walls
   b. multicellular organisms
   c. are consumers
   d. break down food

6. When do mesoderm cells begin to form from the endoderm?
   a. during the blastula stage
   b. during the gastrula stage
   c. before the blastula stage
   d. after the gastrula stage

7. Which animal shown in the diagram below is an example of an animal with radial symmetry?

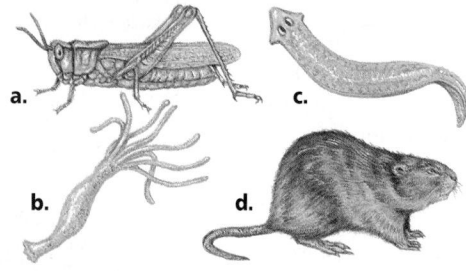

   a.
   b.
   c.
   d.

8. Fishes have one fin along their backs. Because fishes are _____ symmetrical, this fin is called the _____ fin.
   a. radially, anterior
   b. radially, ventral
   c. bilaterally, posterior
   d. bilaterally, dorsal

THE PRINCETON REVIEW **TEST-TAKING TIP**

Make Yourself Comfortable
When you take a test, try to make yourself as comfortable as possible. You will then be able to focus all your attention on the test.

---

9. Which animal obtains food by filtering it out of water?
   a. clam
   b. sea star
   c. sidewinder
   d. flatworm

10. Of these, which is NOT an example of a vetebrate animal?
   a. nurse shark
   b. black bear
   c. coral snake
   d. garden spider

11. The _____ shown in the diagram below has only one opening in its digestive system.

12. A(n) _____ body plan enables muscles to brace against a rigid structure, thereby enabling faster movement.

13. A cat has an endoskeleton, whereas a dragonfly has a(n) _____ for protection and support.

14. Your pet hamster has a(n) _____ body plan.

15. A butterfly's mouth develops from the opening of the indented space in the _____ of the developing embryo.

16. The muscles in the legs of a horse develop from the embryonic tissue called _____.

17. A swordfish is a _____ because its mouth did not develop from the opening in the gastrula.

18. A saddle is placed on the _____ surface of a horse.

19. Sponge larvae are free-swimming, whereas most adult sponges are _____.

20. An adult sea star has _____ symmetry.

APPLYING MAIN IDEAS

21. During the development of an embryo, a blastula forms first, followed by formation of a gastrula. The mouth of the embryo develops from the opening in the gastrula. Could the embryo develop into a bird? Explain.

---

22. How are body plans of animals related to the environment in which they live? Give two examples.

23. Why are asymmetrical organisms often sessile? Explain.

24. How is obtaining food different in animals and fungi?

25. Under a microscope, you observe an unfamiliar organism from a tide pool. When you offer live invertebrate food, the organism seems to pursue and consume this prey. How could you tell whether the organism is a protist or an animal?

THINKING CRITICALLY

26. **Observing and Inferring** You are looking at a preserved specimen of an unidentified animal. It is easy to see that it has radial symmetry. What would you predict about its direction of movement when alive? Explain.

27. **Observing and Inferring** Explain why the development of a body cavity enabled animals to move and feed more efficiently.

28. **Concept Mapping** Complete the concept map by using the following vocabulary terms: blastula, ectoderm, gastrula, endoderm, mesoderm.

Animals develop from a

zygote

to a

1.

stage in which these tissues form

2.   3.

which forms the

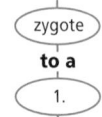 4. —**during the**— 5. —**stage**

CD-ROM

For additional review, use the assessment options for this chapter found on the *Biology: The Dynamics of Life Interactive CD-ROM* and on the Glencoe Science Web site.
science.glencoe.com

ASSESSING KNOWLEDGE & SKILLS

The following diagrams represent three different animal body plans.

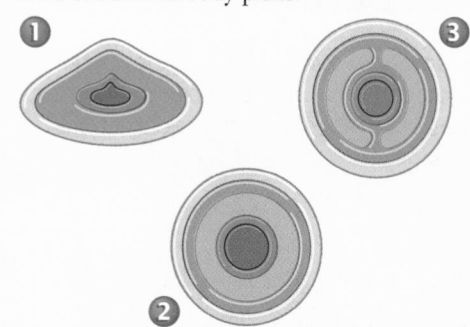

Interpreting Scientific Illustrations
Use the diagram to answer the following questions.

1. Which body plan would be capable of more complex and powerful movement?
   a. 1   c. 3
   b. 2   d. all of these

2. Which type of body plan belongs to acoelomate animals such as flatworms?
   a. 1   c. 3
   b. 2   d. none of these

3. Which type of body plan belongs to pseudocoelomate animals such as roundworms?
   a. 1   c. 3
   b. 2   d. none of these

4. Which type of body plan is more likely to be seen in animals that inhabit land environments?
   a. 1   c. 3
   b. 2   d. all of these

5. **Making a Table** Make a table that compares the three body plans. For each body plan, indicate an example of an animal with that type; whether or not a body cavity is present, and if so, which type; and how the animal's body plan affects its locomotion.

22. Fishes have coeloms that their muscles can push against as leverage in swimming motions. Acoelomate flatworms have thin bodies that aid diffusion through the cell layers.

23. Asymmetrical organisms have no front or back end; thus, they have little direction to their movement and cannot pursue food. They often wait for food to come to them—which works for a sessile animal.

24. Most animals move to get food or cause food to be drawn toward them. Most animals digest their food inside their bodies. Fungi usually live on or in their food source and do not move to find food. Fungi secrete chemicals that dissolve food outside their bodies before they take it in.

25. If it is multicellular, it is an animal.

THINKING CRITICALLY

26. Radially symmetrical animals don't have an anterior or posterior end; they don't move forward or backward as easily as bilaterally symmetrical animals.

27. Animals with no body cavities depend upon diffusion to move food into and wastes out of cells; animals with body cavities have space to develop organ systems that deliver food and take away wastes. Movement in acoelomate animals depends upon structures such as cilia, whereas coelomate animals have structures that muscles can work against for locomotion.

28. 1. Blastula; 2. Endoderm; 3. Ectoderm; 4. Mesoderm; 5. Gastrula

ASSESSING KNOWLEDGE & SKILLS

1. c
2. a
3. b
4. c

5.

Comparison of Three Body Plans			
Body Plan	**Example**	**Coelom**	**Relative Speed**
Acoelomate	Flatworm	none	slow
Pseudocoelomate	Roundworm	fluid-filled, partly lined with mesoderm	faster
Coelomate	Fish	fluid-filled, lined with mesoderm	faster

Chapter 26 Organizer

Refer to pages 4T-5T of the Teacher Guide for an explanation of the National Science Education Standards correlations.

Section	Objectives	Activities/Features
Section 26.1 **Sponges** NSES UCP.1, UCP.5; C.1, C.5-6 ($^1/_2$ session)	1. **Relate** the sessile life of sponges to their food-gathering adaptations. 2. **Describe** the reproductive adaptations of sponges.	**Inside Story:** A Sponge, p. 714 **Problem-Solving Lab 26-1:** p. 715
Section 26.2 **Cnidarians** NSES UCP.1, UCP.3, UCP.5; A.1, A.2; C.1, C.3, C.5-6; F.3, F.4 (1 session)	3. **Distinguish** the different classes of cnidarians. 4. **Sequence** the stages in the life cycle of cnidarians. 5. **Evaluate** the adaptations of cnidarians for obtaining food.	**MiniLab 26-1:** Watching Hydra Feed, p. 719 **Inside Story:** A Cnidarian, p. 720 **Problem-Solving Lab 26-2:** p. 724 **Biology & Society:** Why are the corals dying? p. 736
Section 26.3 **Flatworms** NSES UCP.1, UCP.5; A.1; C.3, C.5-6; F.1, F.5; G.1 (1 session)	6. **Distinguish** the adaptive structures of parasitic flatworms and free-living planarians. 7. **Explain** how parasitic flatworms are adapted to their way of life.	**Problem-Solving Lab 26-3:** p. 727 **Inside Story:** A Planarian, p. 728 **Investigate BioLab:** Observing Planarian Regeneration, p.734
Section 26.4 **Roundworms** NSES UCP.1, UCP.5; A.1, A.2; C.5-6; E.1, E.2; F.1, F.5; G.1, G.2 ($1^1/_2$ sessions)	8. **Compare** the structural adaptations of roundworms and flatworms. 9. **Identify** the characteristics of four roundworm parasites.	**MiniLab 26-2:** Observing the Larval Stage of a Pork Worm, p. 732 **Problem-Solving Lab 26-4:** p. 733

Need Materials? Contact Carolina Biological Supply Company at 1-800-334-5551 or at **http://www.carolina.com**

MATERIALS LIST

BioLab
p. 734 planarian culture, petri dish, springwater, camel hair brush, microscope slide, stereomicroscope, single-edged razor blade, marking pencil

MiniLabs
p. 719 microscope, watch glass, dropper, culture dish, hydra culture, brine shrimp culture, water
p. 732 microscope, prepared slide of pork worm larvae

Alternative Lab
p. 714 stereomicroscope, forceps, balance, wax pencil, beakers, petri dish, sea sponges, unused synthetic sponges

Quick Demos
p. 716 dried marine sponges
p. 720 microprojector, prepared slides of nematocysts
p. 728 slide projector, planarian culture, water, 35-mm deep-well slide

Key to Teaching Strategies

L1 Level 1 activities should be appropriate for students with learning difficulties.

L2 Level 2 activities should be within the ability range of all students.

L3 Level 3 activities are designed for above-average students.

ELL ELL activities should be within the ability range of English Language Learners.

COOP LEARN Cooperative Learning activities are designed for small group work.

P These strategies represent student products that can be placed into a best-work portfolio.

These strategies are useful in a block scheduling format.

Teacher Classroom Resources

Section	Reproducible Masters	Transparencies
Section 26.1 **Sponges**	Reinforcement and Study Guide, p. 115 **L2** Concept Mapping, p. 26 **L3** **ELL** Critical Thinking/Problem Solving, p. 26 **L3** Content Mastery, pp. 129-132 **L1**	Section Focus Transparency 63 **L1** **ELL** Basic Concepts Transparency 45 **L2** **ELL** Basic Concepts Transparency 46 **L2** **ELL**
Section 26.2 **Cnidarians**	Reinforcement and Study Guide, p. 116 **L2** BioLab and MiniLab Worksheets, p. 117 **L2** Laboratory Manual, pp. 187-188 **L2**	Section Focus Transparency 64 **L1** **ELL** Basic Concepts Transparency 45 **L2** **ELL** Basic Concepts Transparency 46 **L2** **ELL** Basic Concepts Transparency 47 **L2** **ELL**
Section 26.3 **Flatworms**	Reinforcement and Study Guide, p. 117 **L2** Laboratory Manual, pp. 189-190 **L2** Content Mastery, pp. 129-132 **L1**	Section Focus Transparency 65 **L1** **ELL** Reteaching Skills Transparency 39 **L1** **ELL**
Section 26.4 **Roundworms**	Reinforcement and Study Guide, p. 118 **L2** BioLab and MiniLab Worksheets, pp. 118-120 **L2** Content Mastery, pp. 129-132 **L1**	Section Focus Transparency 66 **L1** **ELL**

Assessment Resources

Chapter Assessment, pp. 151-156
MindJogger Videoquizzes
Performance Assessment in the Biology Classroom
Alternate Assessment in the Science Classroom
ExamView® Pro Software 🖫
BDOL Interactive CD-ROM, Chapter 26 quiz

Additional Resources

Spanish Resources **L1** **ELL**
English/Spanish Audiocassettes **L1** **ELL**
Cooperative Learning in the Science Classroom **COOP LEARN**
Lesson Plans/Block Scheduling

NATIONAL GEOGRAPHIC | **Teacher's Corner**

Index to National Geographic Magazine
The following articles may be used for research relating to this chapter:
"Consider the Sponge," by Michael E. Long, March 1977.

GLENCOE TECHNOLOGY

The following multimedia resources are available from Glencoe.

Biology: The Dynamics of Life
CD-ROM **ELL**

Video: *Ocean Cnidarians*
Video: *Coral Reefs*
BioQuest: *Biodiversity Park*

Videodisc Program 📼

Ocean Cnidarians

The Infinite Voyage
To the Edge of the Earth

26 Sponges, Cnidarians, Flatworms, and Roundworms

GETTING STARTED DEMO

 *Visual-Spatial* Have students observe the live freshwater sponge, *Spongilla*, with hand lenses. Ask students to note the asymmetrical shape of the sponge and its many pores. Explain that all sponges have a large number of pores.
L2 **ELL**

Theme Development

The themes of **evolution** and **homeostasis** are emphasized in this chapter. Evolutionary relationships among the animal phyla are stressed, as are adaptations to the environment and the homeostatic mechanisms at work in the different animals.

Resource Manager

Section Focus Transparency 63 and Master **L1** **ELL**

READING BIOLOGY

Glencoe's *Biology: The Dynamics of Life* contains many resources to assist a student's reading skills. Each chapter contains figures with expanded captions that expand on written material. Word Origins, located along the side of text, expand knowledge of biology vocabulary. Glencoe's Content Mastery Booklet helps develop reading skills while reinforcing content. In addition, use the Interactive Tutor for *Biology: The Dynamics of Life* on the Glencoe Web site to reinforce vocabulary.
science.glencoe.com

What You'll Learn

- You will compare and contrast the characteristics of sponges, cnidarians, flatworms, and roundworms.
- You will describe how sponges, cnidarians, flatworms, and roundworms are adapted to their habitats.

Why It's Important

Sponges and cnidarians are two major groups of animals that are important to aquatic biomes. Flatworms and roundworms include many species that cause diseases that affect both plants and animals.

READING BIOLOGY

Make a list of new vocabulary words that appear in the section previews. For any unfamiliar words, try to determine their origin. As you read, note how the words are used in the context of biology.

To find out more about sponges, cnidarians, flatworms, and roundworms, visit the Glencoe Science Web site.
science.glencoe.com

The sea anemone and the sponge (inset) share common traits. Both are invertebrates with simple body plans.

712

Multiple Learning Styles

Look for the following logos for strategies that emphasize different learning modalities.

Kinesthetic Quick Demo p. 716; Building a Model, p. 722

Visual-Spatial Biology Journal, p. 716; Meeting Individual Needs, pp. 716, 723; Visual Learning, p. 720, Biology Journal, pp. 721, 723, 729; Display, p. 721; Tech Prep, p. 729

Interpersonal Portfolio, p. 722; Reteach, p. 725

Intrapersonal Cultural Diversity, p. 718; Going Further, p. 736

Linguistic Enrichment, p. 716; Portfolio, p. 727

Naturalist Meeting Individual Needs, p. 728; Check for Understanding, p. 729; Portfolio, p. 731

I
s this red organism a plant or an
animal? At first glance, it may look
like a plant because it is colorful and
doesn't move from place to place, but it is
an animal. How do you know this organism is an animal? Like snakes, spiders,
and you, this organism is eukaryotic,
multicellular, and heterotrophic—
characteristics that place it in the
animal kingdom. This sessile
animal is a sponge that filters
water through many small
pores on the outside of its body.

Red sponge,
Haliclona sp.

SECTION PREVIEW

Objectives
Relate the sessile life of
sponges to their food-
gathering adaptations.
Describe the reproductive adaptations of
sponges.

Vocabulary
filter feeding
hermaphrodite
external fertilization
internal fertilization

What Is a Sponge?

Sponges are asymmetrical aquatic
animals that have a variety of colors,
shapes, and sizes. Many are bright
shades of red, orange, yellow, and
green. Some sponges are ball shaped;
others have many branches. Sponges
can be as small as a quarter or as
large as a door. Although sponges do
not resemble more familiar animals,
they carry on the same life processes
as all animals. *Figure 26.1* shows a
natural sponge harvested from the
ocean.

Sponges are pore-bearers

Sponges are classified in the invertebrate phylum Porifera, which means
pore-bearer. Of the 5000 described
species of sponges, most live in the
ocean; only 100 species can be found
in freshwater environments.

No matter where sponges live,
they are mainly sessile organisms.
Because most adult sponges are
sessile, they can't travel in search of
food. Sponges get their food by
filter feeding, a method in which
an organism feeds by filtering small
particles of food from water as it
passes by or through some part of the
organism. How does a sponge get rid
of its wastes? Find out by reading the
Inside Story.

WORD *Origin*
porifera
From the Latin
words *porus*, mean-
ing "pore," and *fera*,
meaning "bearer."
Phylum Porifera
includes animals
with pores that
allow water to flow
through their
bodies.

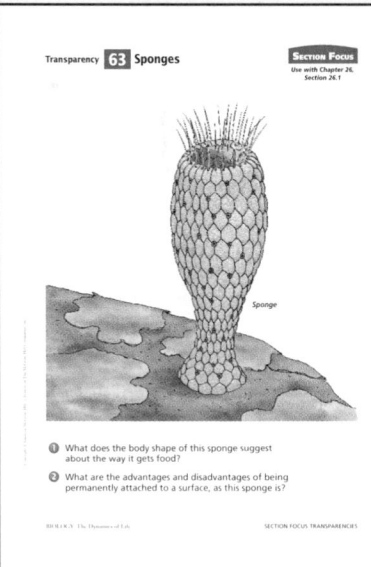

Figure 26.1
This heavy bath sponge
is dark brown or black in
its natural habitat. After
harvest, it is washed and dried
in the sun. When the process
is complete, only a pale, lightweight skeleton remains.

713

Section 26.1

Prepare

Key Concepts
Students will learn the main features of sponges and discuss their
adaptations, origins, and ecology.

Planning
■ Collect dried natural and synthetic sponges for the Alternative Lab and Quick Demo.
■ Gather butcher paper and colored markers for the Check
for Understanding.

1 Focus

Bellringer 🖐
Before presenting the lesson, display **Section Focus Transparency 63** on the overhead
projector and have students
answer the accompanying questions. **L1** **ELL**

Transparency **63** Sponges

SECTION FOCUS
Use with Chapter 26,
Section 26.1

Sponge

❶ What does the body shape of this sponge suggest
about the way it gets food?

❷ What are the advantages and disadvantages of being
permanently attached to a surface, as this sponge is?

SECTION FOCUS TRANSPARENCIES

✓ Assessment Planner

Portfolio Assessment
Problem-Solving Labs, TWE, pp. 715, 732
Portfolio, TWE, pp. 722, 727, 733
BioLab, TWE, p. 735
Performance Assessment
Performance Assessments, TWE, pp. 716,
717, 723, 725
Problem-Solving Lab, TWE, p. 724
BioLab, SE, p. 734-735
MiniLabs, SE, pp. 719, 732

Knowledge Assessment
Section Assessments, SE, pp. 717, 725,
730, 733
Problem-Solving Lab, TWE, p. 727
Chapter Assessment, SE, pp. 737-739
Knowledge Assessment, TWE, pp. 729,
730

Skill Assessment
Alternative Lab, TWE, p. 715
Problem-Solving Lab, TWE, p. 732

Purpose

Students will observe the basic features of a sponge and examine how it accomplishes filter feeding.

Teaching Strategies

■ Explain that sponges obtain food in a process called filter feeding.

Visual Learning

■ Point out that food particles in water are pulled into collar cells and digested. Nutrients from food are then distributed by amoebocytes to other body cells.

Critical Thinking

Sponges carry on the same life processes as all animals. They are multicellular eukaryotes that do not have cell walls around their cells.

Resource Manager

Basic Concepts Transparencies 45 and 46 and Masters **L2** **ELL**
Concept Mapping, p. 26 **L3**
Critical Thinking/Problem Solving, p. 26 **L3**

A Sponge

Sponges have no tissues, organs, or organ systems. The body plan of a sponge is simple, being made up of only two layers of cells with no body cavity. Between these two layers is a jellylike substance that contains other cells as well as the components of the sponge's internal support system. Sponges have four types of cells that perform all the functions necessary to keep them alive.

Orange tube sponges

Critical Thinking *Why are sponges classified as animals?*

① **Osculum** Water and wastes are expelled through the osculum, the large opening at the top of the sponge. A sponge no bigger than a pen can move more than 20 L of water through its body per day.

② **Pore cell** Surrounding each pore is a single pore cell. Pore cells bring water carrying food and oxygen into the sponge's body.

③ **Epithelial cell** Epithelial cells are thin and flat. They contract in response to touch or to irritating chemicals, and in so doing, close up pores in the sponge.

Pore cell

Direction of water flow

④ **Collar cell** Lining the interior of sponges are collar cells. Each collar cell has a flagellum that whips back and forth, drawing water through the pores of the sponge.

⑤ **Amoebocytes** Amoebocytes, located between the two cell layers of a sponge, carry nutrients to other cells, aid in reproduction, and produce chemicals that help make up the spicules of sponges.

⑥ **Spicules** Spicules are structures produced by other cells that form the hard support systems of sponges. Spicules are small, needlelike structures located between the cell layers of a sponge.

Alternative Lab

Natural and Synthetic Sponges

Purpose

Students will compare the water-holding capacity and microscopic appearance of natural and synthetic sponges. Remind students to wash their hands after handling sponges.

Materials

stereomicroscope, forceps, balance, wax marking pencil, 150 mL beakers, Petri dish bottoms, sea sponges (one piece each of four different types: grass, yellow, sheep's wool, hard head—each piece 3 cm × 3 cm × 2 cm), unused synthetic sponges (one piece each of four different types—each piece 3 cm × 3 cm × 2 cm)

Procedure

Give students the following directions.

1. Examine each piece of sponge at its thinnest point under the microscope.
2. Draw the skeletal framework of each sponge. Label each sponge piece.
3. Predict which type of sponge will hold more water. Base your prediction on microscopic examination of the sponge structures.
4. Place each sponge piece in a Petri dish and obtain its mass. Soak each sponge piece in a beaker of water for 10

Cell organization in sponges

Like all animals, sponges are multi-cellular. Sponges have different types of cells that perform functions that help the animal survive. Read the *Problem-Solving Lab* on this page to find out how sponges survive in different environments. The activities of the different types of cells are coordinated in a sponge, but sponges do not have tissues like those found in other animals. Tissues are groups of cells that are derived from the ectoderm, endoderm, and mesoderm in the embryo. Sponge embryos do not develop endoderm or mesoderm, so they do not have cells organized into tissues.

However, the cells of a sponge are organized. If you took a living sponge and put it through a sieve, you would witness a rather remarkable event. Not only would you see the sponge's many cells alive and separated out, but you also would be able to see these same cells coming together to form a whole sponge once again. It may take several weeks for the sponge's cells to reorganize themselves.

Many biologists hypothesize that sponges evolved directly from colonial, flagellated protists, such as *Volvox*. More importantly, however, sponges demonstrate what appears to have been a major step in the evolution of animals—the change from a unicellular life to a division of labor among groups of organized cells.

Reproduction in sponges

Sponges reproduce both sexually and asexually. They reproduce asexually when fragments break off from the parent animal and form new sponges, or by forming external buds. Buds may break off, float away, and eventually settle and become separate animals. Sometimes the buds remain

attached to the parents, forming a colony of sponges. You can see a colony in *Figure 26.2*.

Figure 26.2
Sponge colonies are the result of asexual reproduction. How would these sponges compare genetically? Could they be considered clones?

715

Problem-Solving Lab 26-1 — Applying Concepts

Why are there more species of marine sponges than freshwater sponges? Most sponges are marine. They live in a saltwater environment. Is there an advantage for sponges to live in a marine environment rather than in a freshwater environment? A series of statements is provided below. Read them over and then answer the questions that follow.

Analysis
A. The internal tissues of marine organisms are isotonic with their surroundings.
B. Oceans do not have problems with rapid changes in velocity (rate of flow) of water.
C. Young marine animals often spend the early part of their life cycles as free-floating organisms.

Thinking Critically
1. Using statement A, how might a freshwater environment vary? How might this be a disadvantage for freshwater sponges?
2. Using statement B, how might a freshwater environment vary? How might this be a disadvantage for freshwater sponges?
3. Using your collective answers, explain why few sponge species are found in freshwater environments.

Problem-Solving Lab 26-1

Purpose
Students will determine that freshwater environments are more difficult for sponge survival than saltwater environments.

Process Skills
think critically, analyze data, compare and contrast, draw a conclusion

Teaching Strategies
■ You may want to review osmosis with students.
■ Review the concepts of freshwater versus saltwater environments. Make sure students understand that lakes, streams, and rivers are freshwater environments.

Thinking Critically
1. The internal environment of freshwater sponges is not isotonic with its surroundings, thus energy must be used to eliminate excess water from cells of the organism.
2. The flow of water may be rapid in rivers or streams, thus washing adult or embryonic sponges away.
3. Many environmental factors in freshwater environments limit or make it impossible for sponges to survive.

✔ Assessment
Portfolio Ask students to research the number of sponge species that live in freshwater environments. **L2** **P**

---

minutes and again obtain the mass of each sponge.
5. Calculate the mass of the water absorbed by each sponge. Compare the data for all sponges.

Expected Results
Natural sponges have greater water holding capacity than synthetic sponges.

Analysis
1. Which sponges, natural or synthetic, have greater water-holding capacity?

natural sponges
2. Of what adaptive value is it for a sponge to be able to take in large amounts of water? *They are filter feeders. Taking in more water increases the chances of taking in more food.*

✔ Assessment
Skill Design and conduct an experiment to determine whether the temperature of water affects the water-holding capacity of a sponge. Use the Performance Task Assessment List for Designing an Experiment in **PASC**, p. 23. **L3** **ELL**

Linguistic Some sponges give off toxic chemicals that deter predators and the buildup of other sessile animals on their exterior surfaces. One of these chemicals, from a Caribbean sponge, is being used currently to treat cancer. Other chemicals produced by sponges are being used to fight fungal infections. Have interested students research this subject and write a report. **L3**

✓ Assessment

Performance Have students assume that a particular marine sponge harbors single-celled algae as symbionts in its cells. Ask students to make a sketch and describe in a paragraph what the best shape would be for this sponge to enable the algae to get maximum sunlight. **L2** **ELL** **P**

Quick Demo

Kinesthetic From a scientific supply house, obtain the dried skeletons of a variety of marine sponges. Ask students to examine them and describe how these animals would be different if they were alive. **L2** **ELL**

3 Assess

Check for Understanding

Visual-Spatial Give students a large piece of butcher paper and colored markers. Ask them to draw a large sponge indicating epithelial cells, pore cells, collar cells, amoebocytes, spicules, and osculum. **L2** **ELL**

Most sponges reproduce sexually. Some sponges have separate sexes, but most sponges are hermaphrodites. A **hermaphrodite** (hur MAF ruh dite) is an individual animal that can produce both eggs and sperm. Hermaphrodism increases the likelihood of fertilization in sessile or slow-moving animals. Eggs and sperm are formed from amoebocytes. During reproduction, sperm released from one sponge can be carried by water currents to another sponge, where fertilization may occur.

Fertilization in sponges may be either external or internal. A few sponges have **external fertilization,** in which the eggs and sperm are both released into the water; fertilization occurs outside the animal's body. Most sponges have **internal fertilization,** in which eggs remain inside the animal's body, and sperm are carried to the eggs in the flow of water. In sponges, the collar cells collect the sperm and transfer them to amoebocytes. The amoebocytes then transport the sperm to ripe eggs. Most sponges reproduce sexually through internal fertilization. The result is the development of free-swimming, flagellated larvae, shown in *Figure 26.3.*

Some freshwater sponges that live in temperate waters produce seedlike particles, called gemmules, in the fall. The adult sponges die over the winter, but the gemmules survive and grow into new sponges in the spring.

Figure 26.3
Sponges reproduce sexually when sperm from one sponge fertilize the eggs of another sponge.

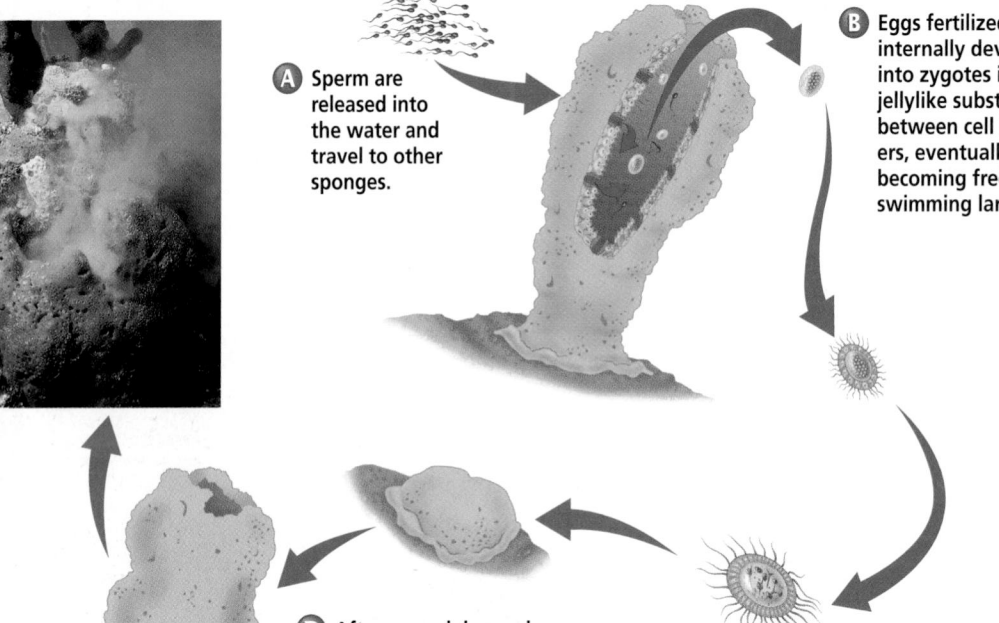

A Sperm are released into the water and travel to other sponges.

B Eggs fertilized internally develop into zygotes in the jellylike substance between cell layers, eventually becoming free-swimming larvae.

C The larvae swim from the body of the sponge out into the water.

D After several days, a larva attaches itself to a surface and develops into an adult. Most sponges can move from place to place only in their larval stages.

BIOLOGY JOURNAL

Sponge Gemmules

Visual-Spatial After observing prepared slides of sponge gemmules with a microscope, have students make drawings of gemmules in their journals. **L1** **ELL**

MEETING INDIVIDUAL NEEDS

Gifted

Visual-Spatial Have your advanced students reread the description of fragmentation. Challenge these students to diagram this process and then work with other students to review fragmentation. **L3**

a

Figure 26.4
The spicules of fresh-water sponges, such as this lake sponge, protect it from predators (a). Spicules of the deep-water glass sponges form a rigid skeleton (b).

b

Support and defense systems in sponges

Sponges are soft-bodied invertebrates, yet they can be found in waters as deep as 8500 m. You might think that the water pressure at such depths would flatten sponges, yet they all have an internal support system that enables them to withstand such pressure. Some sponges have sharp, hard spicules located between the cell layers. Spicules may be made of glasslike material or of calcium carbonate. Some species, such as the river sponge shown in *Figure 26.4*, have thousands of tiny, sharp, needle-like spicules that make them hard for animals to eat. Other sponges have an internal skeleton made of silica or of spongin, a fibrous proteinlike material. Sponges can be classified according to their spicules and/or skeletons.

Besides sharp spicules, some sponges may have other methods of defense. Some tropical sponges contain chemicals that are toxic to fishes and to other predators. Many sponges produce toxins that are poisonous to sharks. Scientists are studying sponge toxins to identify those that possibly could be used as medicines.

Reteach

Have students construct an outline to summarize this section. Have them include the phylum, symmetry, habitat, food-getting process, reproductive process, and means of protection of sponges. **L1**

Extension

Kinesthetic Ask student groups to construct a three-dimensional, cutaway model of a sponge. Have students label the parts of their model. **L2** **ELL** **COOP LEARN**

✓ **Assessment**

Performance Have students make a flowchart that shows the events involved in sexual reproduction of sponges. **L2**

4 Close

Discussion

Discuss with students situations in which natural sponges are more desirable than synthetic sponges, such as for bathing. Explain that natural sponges are not used for many cleaning purposes because of their expense.

Resource Manager

Reinforcement and Study Guide, p. 115 **L2**

Section Assessment

Understanding Main Ideas
1. How does a sponge obtain food?
2. Explain how epithelial cells control filter feeding in sponges.
3. Describe the steps involved in the sexual reproduction of sponges.
4. What are the functions of amoebocytes in sponges?

Thinking Critically
5. What advantages for obtaining food do multicellular organisms such as sponges have over unicellular organisms? Explain.

SKILL REVIEW
6. **Making and Using Tables** Make a table listing the cell types and other structures of sponges along with their functions. For more help, refer to *Organizing Information* in the **Skill Handbook**.

Section Assessment

1. Sponges take water into their bodies and filter food out of the water. They are filter feeders.
2. Epithelial cells can contract and relax, thus opening and closing pore cells.
3. Sperm are released in water and travel to other sponges. Eggs fertilized internally develop into zygotes in the jelly-like substance between cell layers, eventually becoming free-swimming larvae that later attach to a surface and develop into adult sponges.
4. Amoebocytes carry food from the collar cells to all other body cells, carry sperm to eggs for fertilization, and produce chemicals that help make up the spicules of sponges.
5. Division of labor among cells enables the organism to carry out life functions more efficiently than a single cell can.
6. Students' tables should include the following features and their functions: epithelial cells, pore cells, collar cells, amoebocytes, spicules, and osculum.

Prepare

Key Concepts

Students will learn about the important characteristics of cnidarians.

Planning

■ Prepare salt dough for Building a Model.

---

1 Focus

Bellringer 🖌

Before presenting the lesson, display **Section Focus Transparency 64** on the overhead projector and have students answer the accompanying questions.
L1 **ELL**

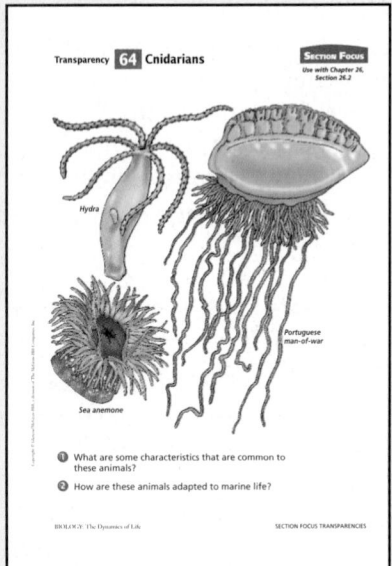

Transparency **64** Cnidarians

SECTION FOCUS
Use with Chapter 26, Section 26.2

Hydra

Portuguese man-of-war

Sea anemone

❶ What are some characteristics that are common to these animals?

❷ How are these animals adapted to marine life?

BIOLOGY: The Dynamics of Life
SECTION FOCUS TRANSPARENCIES

SECTION PREVIEW

Objectives

Distinguish the different classes of cnidarians.

Sequence the stages in the life cycle of cnidarians.

Evaluate the adaptations of cnidarians for obtaining food.

Vocabulary

polyp
medusa
nematocyst
gastrovascular cavity
nerve net

Section

26.2 Cnidarians

What's the largest structure ever built by living organisms? Is it the Sears Tower in Chicago? How about the Great Pyramid in Egypt? Actually, the largest structure ever built is the Great Barrier Reef, which extends for more than 2000 km along the northeastern coast of Australia, and it wasn't built by humans. This structure was built over many centuries by colonies of small marine invertebrate animals called corals. Corals and their relatives, jellyfishes and sea anemones, all belong to the phylum Cnidaria.

The Great Barrier Reef (above) and orange clump coral, *Tubastrea aurea* (left)

What Is a Cnidarian?

Cnidarians (ni DARE ee uns) are a group of marine invertebrates made up of more than 9000 species of jellyfishes, corals, sea anemones, and hydras. Cnidarians can be found worldwide, but coral species generally prefer the warmer oceans of the South Pacific and the Caribbean.

Cnidarians have radial symmetry

Though cnidarians are a diverse group, all possess the same basic body structure, which supports the theory that they had a single origin. All cnidarians have radial symmetry.

A cnidarian's body is made up of two cell layers with one body opening. The cell layers of cnidarians are organized into separate tissues with specific functions. Cnidarians have a simple nervous system, and both cell layers have cells that can contract as though they were muscles. The two cell layers of a cnidarian are derived from the ectoderm and endoderm of the embryo. The ectoderm of the cnidarian embryo develops into a protective outer layer of cells, and the endoderm is internal and adapted mainly to assist in digestion.

Cnidarians display only two basic body forms, which occur at different

WORD *Origin*

cnidarian
From the Greek word *knide*, meaning "nettle," a plant with stinging hairs. Cnidarians have stinging tentacles.

718 SPONGES, CNIDARIANS, FLATWORMS, AND ROUNDWORMS

Cultural Diversity

Hydra Myth

Intrapersonal Ask students to report about the Greek myth dealing with the water monster, Hydra. Students should indicate that Hydra had nine heads. Each time Hercules cut off one head, it grew back. Have students identify parts of the myth that are scientifically accurate and those that are not. **L2**

GLENCOE TECHNOLOGY

VIDEODISC
Biology: The Dynamics of Life
Ocean Cnidarians (Ch 31)
Disc 1, Side 2, 31 sec.

stages of their life cycles. These two forms are the polyp and medusa, *Figure 26.5*. A **polyp** (PAHL up) is the stage with a tube-shaped body and a mouth surrounded by tentacles. A **medusa** (mih DEW suh) is the stage with a body shaped like an umbrella with tentacles hanging downward. The hydra has a typical polyp body form. How do hydras capture their food? You can find out by reading the *Inside Story* on the next page.

In cnidarians, one body form may be more conspicuous than the other. In jellyfishes, for example, the medusa stage is the dominant body form. The polyp stage of a jellyfish is small and not very noticeable. In hydras, the polyp stage is dominant, with a small and delicate medusa stage. The corals and sea anemones have only the polyp stage.

Digestion in cnidarians

Cnidarians are predators that capture or poison their prey with nematocysts. A **nematocyst** (nih MAT uh sihst) is a capsule that contains a coiled, threadlike tube. The tube may be sticky or barbed, and it may contain toxic substances. Nematocysts, located in stinging cells, are discharged like toy popguns, but much faster, in response to touch or chemicals in the environment. Prey organisms are then taken in for digestion.

In cnidarians, you can see the origins of a digestive process similar to that of animals that evolved later. Once captured by nematocysts on the ends of tentacles, prey is brought to the mouth by contraction of the tentacles. The inner cell layer of cnidarians surrounds a space called a **gastrovascular cavity** (gas troh VAS kyuh lur) in which digestion takes place. Cells adapted for digestion line the gastrovascular cavity and release enzymes over the newly captured

prey. Any undigested materials are ejected back out through the mouth. You can observe a cnidarian feeding in the *MiniLab* on this page.

MiniLab 26-1   Observing

Watching Hydra Feed Hydras are freshwater cnidarians. They show the typical polyp body plan and symmetry associated with all members of this phylum. Observe how they capture their food.

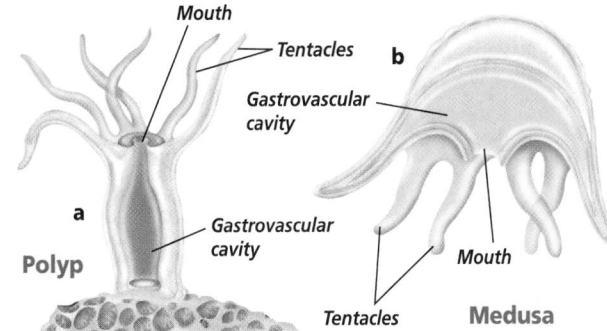

Hydra eating copepod

Procedure 🧤 🔬 ☣️

1. Use a dropper to place a hydra into a watch glass filled with water. Wait several minutes for the animal to adapt to its new surroundings. **CAUTION:** *Use caution when working with a microscope and glassware.*
2. Observe the hydra under low-power magnification.
3. Formulate a hypothesis as to how this animal obtains its food and/or catches its prey.
4. Place brine shrimp in a culture dish of freshwater to avoid introducing salt into the watch glass.
5. Add a drop of brine shrimp to the watch glass while continuing to observe the hydra through the microscope.
6. Note which structures the hydra uses to capture food.

Analysis

1. Describe how the hydra captures food.
2. Was your hypothesis supported or rejected?
3. Sequence the events that take place when a hydra captures and feeds upon its prey.
4. Explain how your observations support the fact that hydras have both nervous and muscular systems.

Figure 26.5
The two basic forms of cnidarians are the polyp form (a), and the medusa form (b).

Mouth

Tentacles   **b**

Gastrovascular cavity

a

Gastrovascular cavity

Polyp

Gastrovascular cavity

Mouth

Tentacles   **Medusa**

26.2 CNIDARIANS **719**

Purpose

Students will gain an understanding of the polyp and medusa forms of cnidarians, and learn about the tentacles and nematocysts common to these animals.

Teaching Strategies

■ Emphasize that polyps are sessile forms of cnidarians that live attached to a surface.

■ Be sure that students recognize that the mouth of a polyp points upward whereas the mouth of a medusa points downward.

Visual Learning

Visual-Spatial Provide students with prepared slides of small cnidarian polyps and medusae. Have students observe the slides and make labeled drawings of their observations. **L2** **ELL**

Critical Thinking

A sessile organism cannot hunt for prey or move away from predators. Stinging cells help sessile organisms get food and escape predation.

Quick Demo

Use a microprojector to show prepared slides of discharged nematocysts.

Resource Manager

Section Focus Transparency 64 and Master **L1** **ELL**
BioLab and MiniLab Worksheets, p. 117 **L2**

720

A Cnidarian

Cnidarians display a remarkable variety of colors, shapes and sizes. Some can be as small as the tip of a pencil. The flowerlike forms of sea anemones are often brilliant shades of red, purple, and blue. Most cnidarians go through both the polyp and medusa stages at some point in their life cycles.

Critical Thinking *How is having poisonous stinging cells an advantage for a sessile organism?*

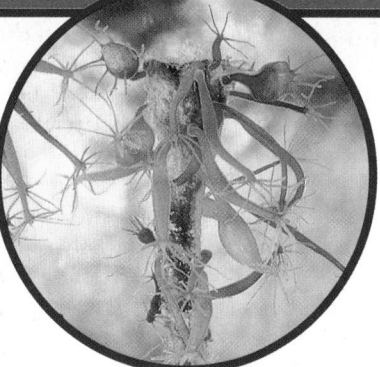

A colony of hydras

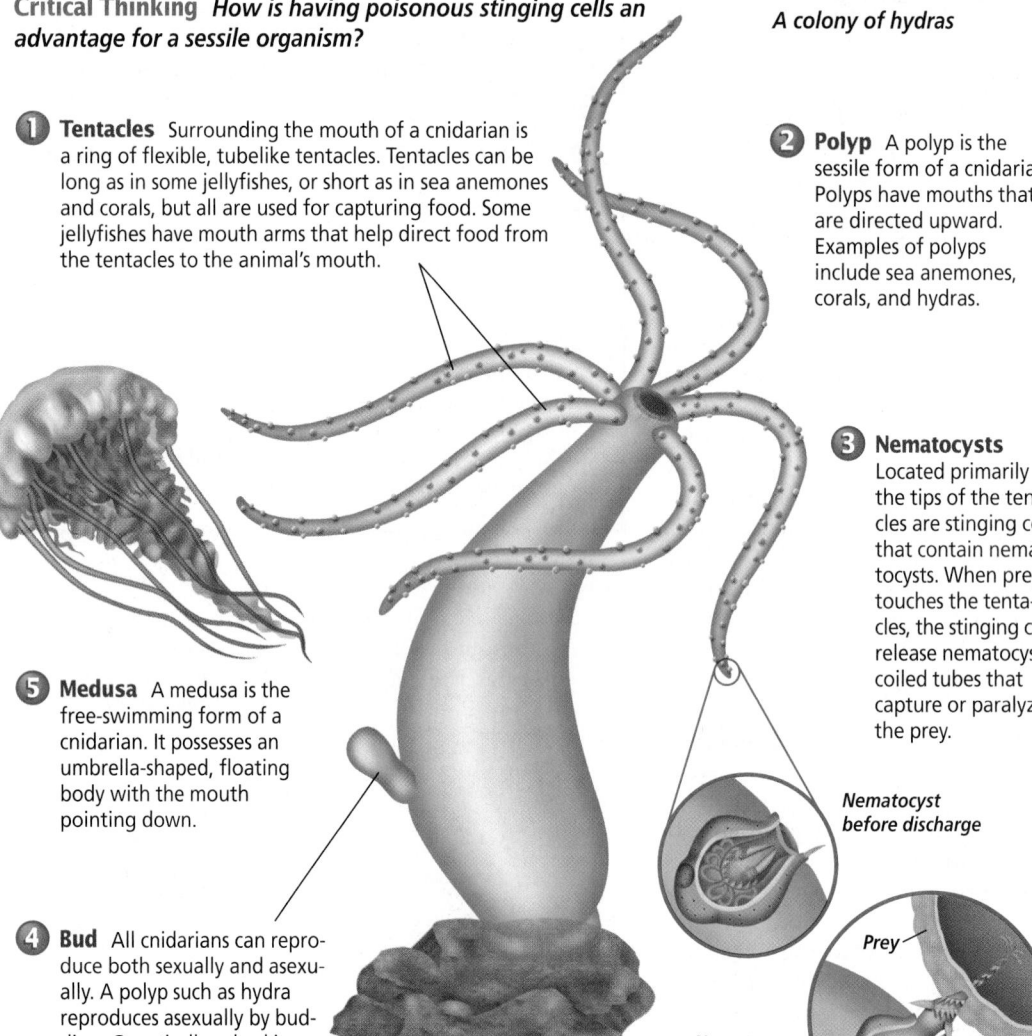

1 Tentacles Surrounding the mouth of a cnidarian is a ring of flexible, tubelike tentacles. Tentacles can be long as in some jellyfishes, or short as in sea anemones and corals, but all are used for capturing food. Some jellyfishes have mouth arms that help direct food from the tentacles to the animal's mouth.

2 Polyp A polyp is the sessile form of a cnidarian. Polyps have mouths that are directed upward. Examples of polyps include sea anemones, corals, and hydras.

3 Nematocysts Located primarily at the tips of the tentacles are stinging cells that contain nematocysts. When prey touches the tentacles, the stinging cells release nematocysts, coiled tubes that capture or paralyze the prey.

Nematocyst before discharge

5 Medusa A medusa is the free-swimming form of a cnidarian. It possesses an umbrella-shaped, floating body with the mouth pointing down.

4 Bud All cnidarians can reproduce both sexually and asexually. A polyp such as hydra reproduces asexually by budding. Genetically, a bud is a clone of its parent.

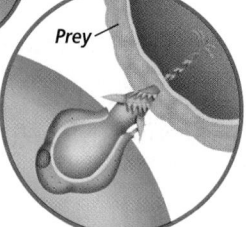

Prey

Nematocyst after discharge

720 SPONGES, CNIDARIANS, FLATWORMS, AND ROUNDWORMS

BIOLOGY Online

Note Internet addresses that you find useful in the space below for quick reference.

_____

_____

_____

_____

_____

Cnidarians are classified into groups partly based on whether or not there are divisions within the gastrovascular cavity, and if there are, how many divisions are present.

Oxygen enters cells directly

Because of a cnidarian's simple, two-cell-layer body plan, as shown in the *Inside Story*, no cell in its body is ever far from water. Oxygen dissolved in water diffuses directly into the body cells, and carbon dioxide and other wastes diffuse out of the cells directly into the surrounding water.

Nervous regulation in cnidarians

Cnidarians have a simple nervous system called a nerve net. A **nerve net** conducts nerve impulses from all parts of the cnidarian's body, but there is no control center such as the brain found in other animals. The impulses from the nerve net bring about contractions of musclelike cells in the tentacles and body of a cnidarian. For example, when touched, a hydra reacts by contracting its musclelike cells.

Reproduction in cnidarians

All cnidarians have the ability to reproduce sexually and asexually. Sexual reproduction occurs in only one phase of the life cycle, usually the medusa stage, unless there is no medusa stage. Asexual reproduction may occur in either the polyp or medusa stage. Cnidarians that remain in the polyp stage, such as hydras, corals, and sea anemones, can reproduce sexually as polyps. Polyps reproduce asexually by a process known as budding, as shown in *Figure 26.6*.

The most common form of reproduction in cnidarians can be illustrated by the life cycle of a jellyfish, shown in *Figure 26.7* on the next page. As you can see, the sexual medusa stage alternates with the asexual polyp stage, from generation to generation. Male medusae release sperm and female medusae release eggs into the water, where fertilization occurs. The resulting zygote develops into an embryo, and then into a larva. Recall that a larva is an intermediate stage in animal development. The free-swimming larva eventually settles and grows into a polyp, which, in turn, reproduces asexually to form new medusae. Even though these two stages alternate in a cnidarian life cycle, this form of reproduction is not alternation of generations as seen in plants because cnidarians are diploid animals in both medusa and polyp stages.

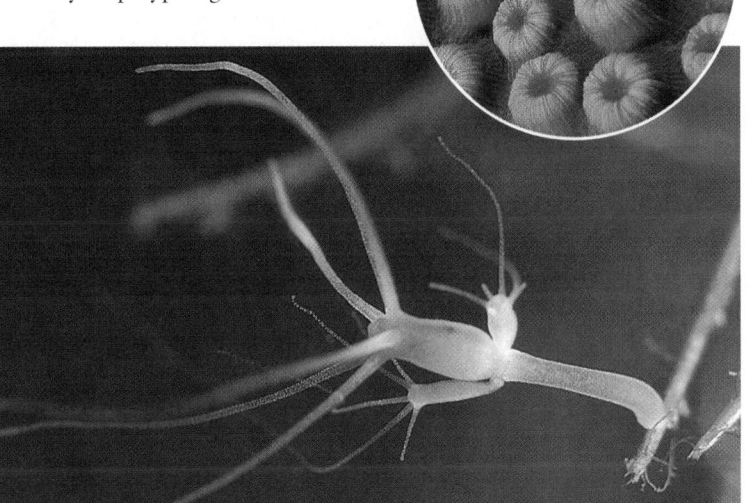

Figure 26.6
The main form of reproduction in polyps is budding. During this process, small buds grow as extensions of the body wall (**a**). In some species, such as corals, a colony develops as the buds break away and settle nearby (**b**).

Building a Model

 **Kinesthetic** Have student groups make clay or salt dough (one part salt, one part flour, one part water) models of the life cycle of a jellyfish. Tell them that when they are finished you will ask the group questions about the jellyfish life cycle. They will need to use their models to demonstrate their answers. Remind students to wear safety goggles and an apron when handling modeling materials. **L1** **ELL** **COOP LEARN**

Reinforcement

Make a transparency of Figure 26.7 with just the figures. White out the captions before you make your transparency and number each figure. Ask students to make a list of corresponding numbers on a sheet of paper and describe each stage. **L2**

GLENCOE TECHNOLOGY

VIDEODISC
The Secret of Life
Life Cycle of a
Jellyfish

Resource Manager

Basic Concepts Transparency
47 and Master **L2** **ELL**
Reinforcement and Study
Guide, p. 116 **L2**
Content Mastery, p. 130 **L1**
Laboratory Manual,
pp. 187-188 **L2**

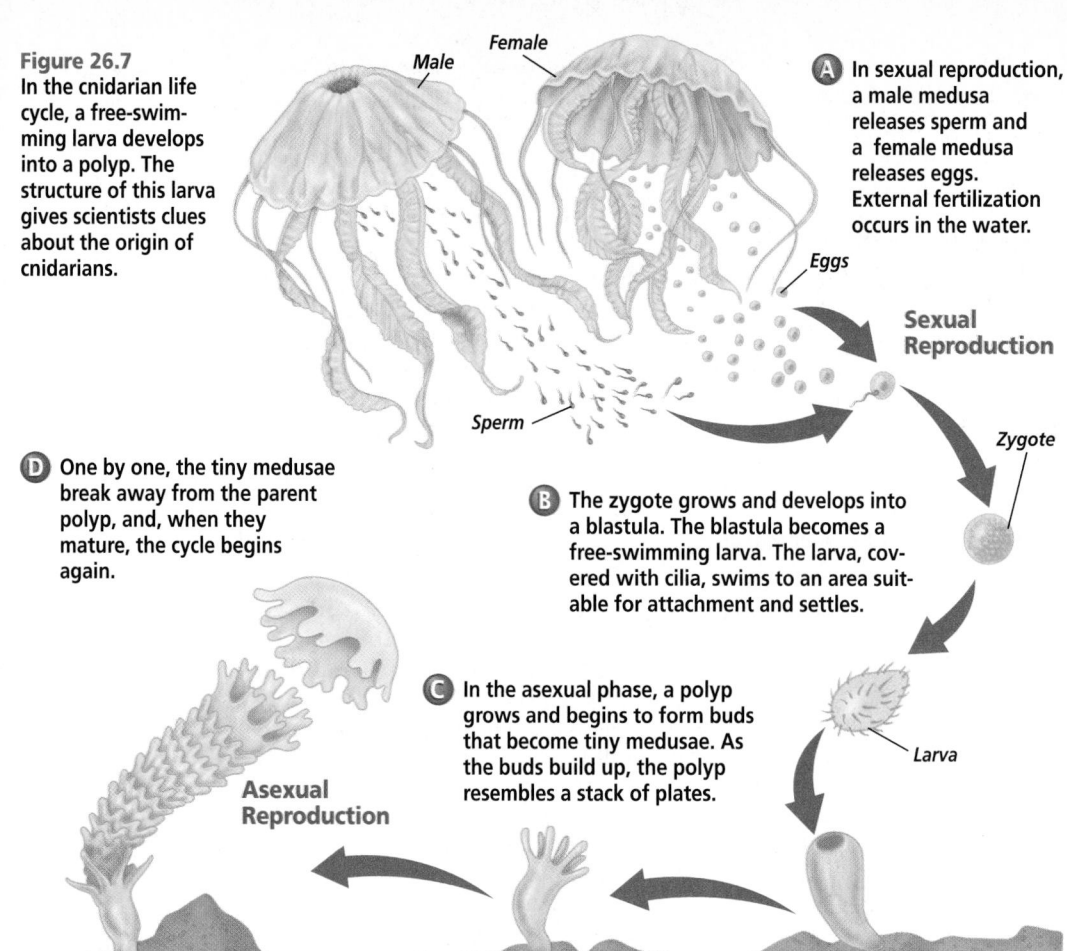

Figure 26.7
In the cnidarian life cycle, a free-swimming larva develops into a polyp. The structure of this larva gives scientists clues about the origin of cnidarians.

A In sexual reproduction, a male medusa releases sperm and a female medusa releases eggs. External fertilization occurs in the water.

Sexual Reproduction

B The zygote grows and develops into a blastula. The blastula becomes a free-swimming larva. The larva, covered with cilia, swims to an area suitable for attachment and settles.

C In the asexual phase, a polyp grows and begins to form buds that become tiny medusae. As the buds build up, the polyp resembles a stack of plates.

D One by one, the tiny medusae break away from the parent polyp, and, when they mature, the cycle begins again.

Asexual Reproduction

Male — *Female* — *Eggs* — *Sperm* — *Zygote* — *Larva*

Diversity of Cnidarians

Most of the 9000 described species of cnidarians belong to one of three classes: Hydrozoa, Scyphozoa, and Anthozoa.

Most hydrozoans form colonies

The class Hydrozoa includes two groups—the hydroids, such as hydra, and the siphonophores, including the Portuguese man-of-war. Most hydroids consist of branching polyp colonies that have formed by budding. The siphonophores include floating colonies that drift about on the ocean's surface as well as colonies that form swimming medusae. Hydrozoans have open gastrovascular cavities with no internal divisions.

It's difficult to understand how the organism shown in *Figure 26.8* could be a closely associated group of individual animals. The Portuguese man-of-war, *Physalia*, is an example of a siphonophore hydrozoan colony.

Each individual in a *Physalia* colony has a different function that helps the entire organism survive. For example, just one individual forms a large, blue, gas-filled float. Regulation of the gas in the float

Portfolio

Treating Jellyfish Stings

Interpersonal Ask a group of students to contact first aid stations on public beaches where jellyfishes are common. Have them prepare a report on the treatment given to victims of the stings of these animals.

Ask students to demonstrate the first aid procedures and explain why each procedure is performed. Students should write summaries for their portfolios. **L3** **ELL** **P**
COOP LEARN

allows the colony to dive to lower depths or rise to the surface. Other polyps hanging from the float have different functions, such as reproduction and feeding. The polyps all function together for the survival of the colony.

Scyphozoans are the jellyfishes

Have you ever seen a jellyfish like the one shown in *Figure 26.9?* Some jellyfishes are transparent, but others are pink, blue, or orange. The medusa stage is dominant in this class. Like other cnidarians, scyphozoans have musclelike cells in their outer cell layer that can contract. When these cells contract together, the medusa contracts, which propels the animal through the water. The fragile and sometimes luminescent bodies of jellyfishes can be beautiful, but most people know about jellyfishes more by their painful stings. Jellyfishes can be found everywhere in the oceans, from arctic to tropical waters, and have been seen as deep as 1000 m. The gastrovascular cavity of scyphozoans has four internal divisions.

Most anthozoans build coral reefs

Anthozoans are cnidarians that exhibit only the polyp form. All anthozoans have many divisions in their gastrovascular cavities.

Sea anemones are anthozoans that live as individual animals. Sea anemones are thought to live for centuries. Some tropical sea anemones may have a diameter of more than a meter. Sea anemones can be found in tropical, temperate, and arctic seas.

Corals are anthozoans that live in colonies of polyps in warm ocean waters around the world. They secrete cuplike calcium carbonate (limestone) shelters around their soft bodies for protection. Colonies of many coral species build the beautiful

Figure 26.8
Physalia colonies are found primarily in tropical waters, but they sometimes drift into temperate waters where they may be washed up on shore.

coral reefs that provide food and shelter for many other marine species. When a coral polyp dies, the limestone skeleton it leaves behind adds a tiny piece to the structure of the reef. The living portion of a coral reef is a thin, fragile layer that grows on top of the skeletons left behind by

Figure 26.9
The jellyfish *Chrysaora hysoscella* has the common name compass jellyfish due to the radiating brown lines on its medusa.

723

Purpose

Students will study two graphs that show how certain abiotic factors influence the location of coral species.

Process Skills

think critically, interpret data, compare and contrast, make and use graphs

Teaching Strategies

■ Review the meaning of symbiosis and mutualism, abiotic and biotic factors.

■ Suggest to students that they analyze each graph separately. However, point out that abiotic factors are interrelated.

Thinking Critically

1. abiotic—ocean depth, temperature; biotic—coral

2. The number of coral species decreases rapidly until about 20 m, when it begins to level off.

3. Temperatures between 22°C and 30°C seem to be more favorable to coral diversity. Fewer species survive at temperatures below 22°C.

4. Zooxanthellae require sunlight in order to carry out photosynthesis. The corals live near the surface of the ocean. Temperature decreases as depth increases. In tropical regions, corals that live close to the surface are more likely to be exposed to high temperatures and, therefore, are more at risk for bleaching.

What ocean conditions limit the number of coral species? All corals that build reefs have a mutualistic symbiotic relationship with zooxanthellae. Zooxanthellae within the coral carry on photosynthesis and provide some nutrients to the coral. Animals caught by the coral provide some nutrients to these protists.

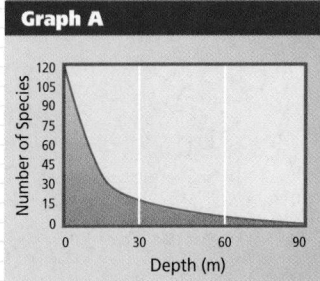
Graph A

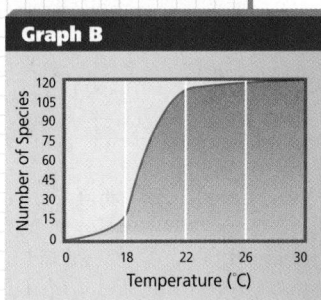
Graph B

Analysis

Graph A shows the number of species present in coral reefs at certain depths. Graph B shows the number of species present at different temperatures. All reef-building coral species have zooxanthellae. The effects of abiotic factors on organisms are usually related. For example, temperature and levels of illumination in an ocean vary with depth.

Thinking Critically

1. Identify the abiotic and biotic factors being studied in this ocean environment.

2. In Graph A, what seems to be the correlation between number of coral species present and depth? Use actual numbers from the graph in your answers.

3. In Graph B, what seems to be the correlation between number of species present and the temperature? Use actual numbers from the graph in your answer.

4. Bleaching occurs when coral polyps expel their zooxanthellae. Bleaching has been observed when water temperature exceeds 35°C. How might depth of coral be related to bleaching?

---

previous generations. Coral reefs grow very slowly, about 2 mm per year. It took centuries to form the reefs found today in tropical and subtropical oceans. Find out more about the fragility of coral reefs by reading the *Biology & Society* feature at the end of this chapter. Corals that form reefs are known as hard corals. Other corals are known as soft corals because they do not build such structures.

A coral polyp extends its tentacles to feed, as shown in *Figure 26.10*. Although coral reefs are often found in relatively shallow, nutrient-poor waters, the corals thrive because they form a symbiotic relationship with tiny organisms called zooxanthellae, photosynthetic protists. The zooxanthellae (zoh oh zan THEH lee) supply oxygen and food to the corals while using carbon dioxide and waste materials produced by the corals. These protists are primarily responsible for the bright colors found in coral reefs. Because the zooxanthellae are free-swimming, they sometimes leave the corals. Corals without these protists often die. You can find out how corals respond to changing environmental conditions in the *Problem-Solving Lab*.

Figure 26.10
Corals feed by extending their tentacles outside their limestone cups (a), but if they are threatened, they can retreat back into the cups (b) until danger has passed.

a

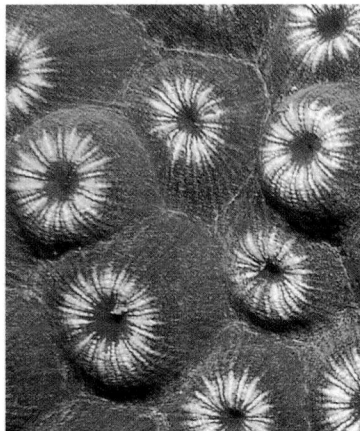
b

724

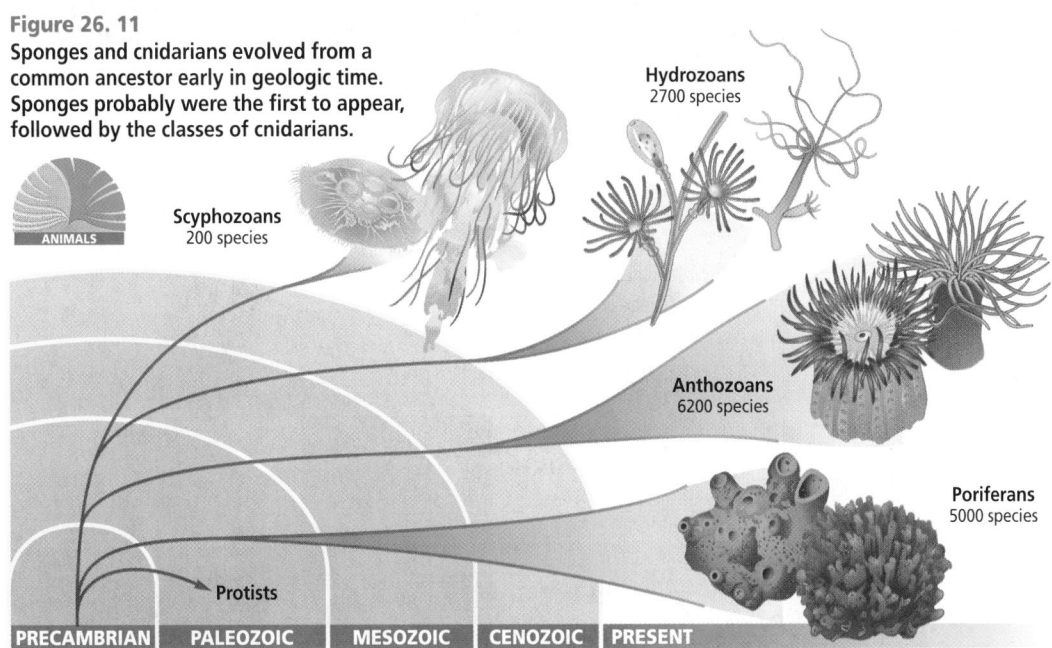

Figure 26. 11
Sponges and cnidarians evolved from a common ancestor early in geologic time. Sponges probably were the first to appear, followed by the classes of cnidarians.

ANIMALS

Scyphozoans
200 species

Hydrozoans
2700 species

Anthozoans
6200 species

Poriferans
5000 species

Protists

PRECAMBRIAN   PALEOZOIC   MESOZOIC   CENOZOIC   PRESENT

Origins of Sponges and Cnidarians

As shown in *Figure 26.11,* sponges represent an old animal phylum. The earliest fossil evidence for sponges dates this group to the Paleozoic Era, about 700 million years ago. Scientists infer that sponges may have evolved directly from a group of flagellated protists that today resemble the collar cells of sponges.

The earliest known cnidarians date to the Precambrian Era, about 630 million years ago. Because cnidarians are soft-bodied animals, they do not preserve well as fossils, and their origins are not well understood. The earliest coral species were not reef builders, so reefs cannot be used to date early cnidarians. The larval form of cnidarians resembles protists, and because of this, scientists consider cnidarians to have evolved from protists.

Section Assessment

Understanding Main Ideas
1. Compare the medusa and polyp forms of cnidarians.
2. Diagram the reproductive cycle of a jellyfish.
3. What are the advantages of a two-layered body in cnidarians?
4. How are corals different from other cnidarians?

Thinking Critically
5. Coral reefs are being destroyed at a rapid rate. What effect would you expect the

destruction of a large coral reef to have on other ocean life?

SKILL REVIEW

6. **Making and Using Tables** In a table, distinguish the three main groups of cnidarians, list their characteristics, and give examples of a member from each group. For more help, refer to *Organizing Information* in the **Skill Handbook.**

Section Assessment

1. Polyps, the sessile stage of cnidarians, have a mouth that points upward. The medusa, the free-swimming stage, has a mouth pointing downward.
2. Make sure students follow the steps shown in Figure 26.7.
3. The cell layers of cnidarians are organized into separate tissues, which enable an animal to be more efficient in carrying out life functions.
4. Corals secrete calcium carbonate shelters around their bodies.
5. Other ocean life will be destroyed because the marine life of the area is dependent upon coral reefs.
6. Make sure students have listed the characteristics and examples of the three cnidarian groups.

3 Assess

Check for Understanding
Have students draw a hydra cross section and add arrows to show the exchange of oxygen and carbon dioxide and how food reaches all cells. **L1 ELL**

Reteach
Interpersonal Draw a football field on the chalkboard. Divide the class into two teams. Ask questions about cnidarians. If a student answers correctly, advance the ball 10 yards towards that team's goalpost. If the answer is not correct, the question goes to the other team. The team that reaches its goalpost first wins. **L2 COOP LEARN**

Extension
Have students write about the different ways that corals obtain food. Ask them to describe how each of these food-getting strategies are used during a 24-hour cycle.

Assessment
Performance Ask students to make a travel brochure for tourists who wish to see cnidarians. They should include all cnidarian groups in their brochure. Use the Performance Task Assessment List for Booklet or Pamphlet in **PASC,** p. 57. **L2**

4 Close

Activity
Have students observe the movements of a live hydra in a deep-well 35 mm projector slide on a slide projector. Ask them to review the adaptations of the hydra to its environment. **L1**

Prepare

Key Concepts

■ In this section, students will study the adaptive structures of parasitic flatworms and planarians. They will learn about how these worms are adapted to their environments.

Planning

■ Purchase single-edged razors and spring water for the BioLab.

1 Focus

Bellringer

Before presenting the lesson, display **Section Focus Transparency 65** on the overhead projector and have students answer the accompanying questions. **L1** **ELL**

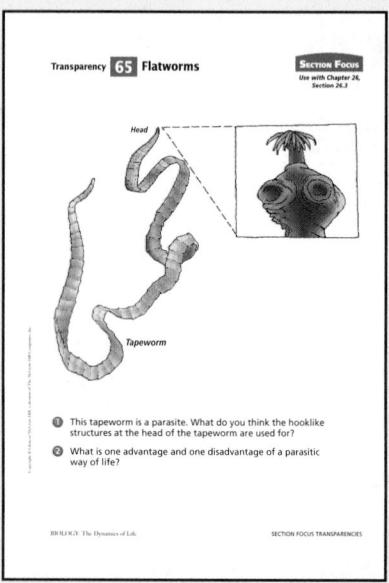

Resource Manager

Section Focus Transparency 65 and Master **L1** **ELL**
Laboratory Manual, pp. 189-190 **L2**

SECTION PREVIEW

Objectives
Distinguish the adaptive structures of parasitic flatworms and free-living planarians.
Explain how parasitic flatworms are adapted to their way of life.

Vocabulary
pharynx
regeneration
scolex
proglottid

Section
26.3 Flatworms

I*magine the ultimate couch potato among living organisms—a worm that never has to move using its own power, is always carried by another organism, is surrounded by food that is already digested, and never has to expend much energy. This describes a parasite called a tapeworm. The parasitic way of life has many advantages.*

Tapeworm scolex

Magnification: 11×

Magnification: 80×

What Is a Flatworm?

To most people, the word worm describes a long, spaghetti-shaped animal. Many animals have this general appearance, but now it is understood that many wormlike animals can be classified into different phyla.

The least complex worms belong to the phylum Platyhelminthes, *Figure 26.12.* These flatworms are acoelomates with thin, solid bodies. The most well-known members of this phylum are the parasitic tapeworms (class Cestoda) and flukes (class Trematoda), which cause

Figure 26.12
Tapeworms are parasites that invade and live in host organisms (a). Flukes usually require two hosts in a complex life cycle (b). Planarians are not parasitic, nor do they cause diseases (c).

a
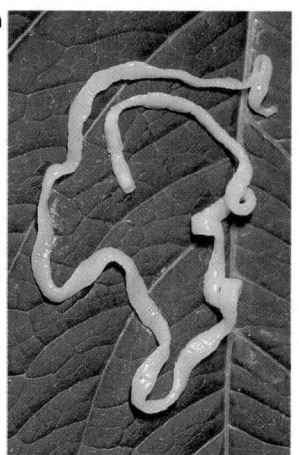

b
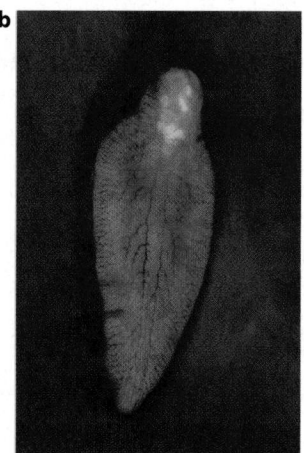

c
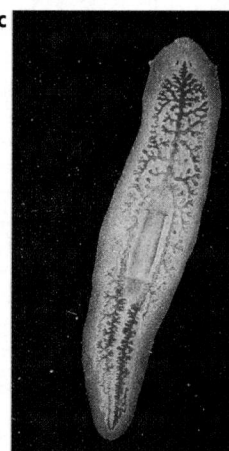

Magnification: 30×

Magnific

726

PROJECT

Planarian Behavior

Kinesthetic Have students design experiments that explore planarian behavior. They might investigate responses to food, touch, heat, cold, or other variables. Make sure they plan experiments in which there is only one variable, a control, and quantitative data to be collected. **L2** **ELL**

diseases in other animals, among them frogs and humans. The most commonly studied flatworms in biology classes are the free-living planarians (class Turbellaria). You can learn about the evolutionary relationships among these classes of flatworms in the *Problem-Solving Lab* on this page. Flatworms range in size from 1 mm up to several meters. There are approximately 14 500 species of flatworms found in marine and freshwater environments and in moist habitats on land.

Feeding and digestion in planarians

A planarian feeds on dead or slow-moving organisms. It extends a tube-like, muscular organ, called the **pharynx** (FAH rinx), out of its mouth. Enzymes released by the pharynx begin digesting food outside the animal's body. Then food is sucked into the gastrovascular cavity, where food particles are broken up. Cells lining the digestive tract obtain food by phagocytosis. Food is thus digested in individual cells.

Nervous control in planarians

Some flatworms have a nerve net, and others have the beginnings of a central nervous system. A planarian has a nervous system that includes two nerve cords that run the length of its body, as you can see in **Figure 26.13**, sensory pits that detect chemicals and movement in water, and eyespots that detect light and dark. At the anterior end of the nerve cord is a small swelling called a ganglion. Located in the head, the ganglion receives messages from the eyespots and sensory pits, then communicates with the rest of the body along the nerve cords. Messages from the nerve cords trigger responses in a planarian's muscle cells. The nervous

system enables a planarian to respond to the stimuli in its environment. Most of a planarian's nervous system is located in its head—a feature common to other bilaterally symmetrical animals.

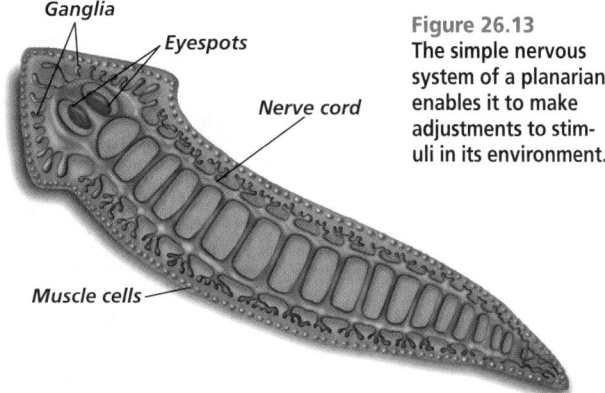

Ganglia
Eyespots
Nerve cord
Muscle cells

Figure 26.13
The simple nervous system of a planarian enables it to make adjustments to stimuli in its environment.

Problem-Solving Lab 26-3   Predicting

Which Came First? There are three classes of flatworms. Two classes are parasitic and the third is free-living. Free-living flatworms are grouped in the class Turbellaria. Trematoda and Cestoda are parasitic classes. These two classes often have humans or some other mammal as one of their hosts.

Analysis

Diagrams A, B, and C show a possible evolutionary relationship among the three classes. The class at the bottom of each diagram can be assumed to have evolved first.

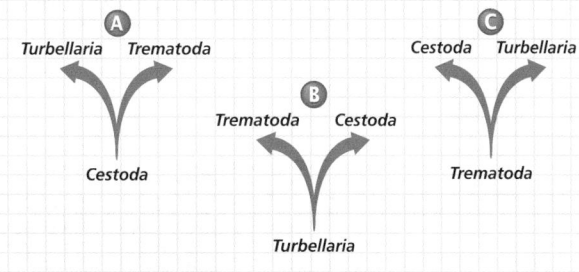

Thinking Critically

One of the three evolutionary patterns is correct. Pick the one that you consider to be correct. Defend your answer by explaining your reasoning. Include in your answer why the other two could not be correct.

---

2 Teach

Problem-Solving Lab 26-3

Purpose
Students are to determine which class of flatworms evolved first.

Process Skills
think critically, predict

Teaching Strategies
■ Students may need additional help in understanding the nature of the three diagrams.
■ Student groups of 2 or 3 may work well for this activity.

Thinking Critically
Diagram B is correct. Student explanations may vary but the main concept that should be explained is: Free-living classes of the flatworms evolved first. This class then evolved into the parasitic forms. It would be difficult to explain how a parasitic worm could evolve first with such a complex life cycle. If one assumes that flatworms evolved before mammals, then it would be impossible for the parasitic worms to have had any host available to them.

✔ Assessment

Knowledge Ask students to describe those traits that flatworms do and do not share with cnidarians. Use the Performance Task Assessment List for Making Observations and Inferences in **PASC**, p. 17. **L2**

✔ Portfolio

Marine Flatworms

 *Linguistic* Have students visit a marine aquarium, zoo, or pet store specializing in saltwater species to research marine flatworms. Have them write a summary of their findings to include in their portfolios. **L2** **ELL**
P

VIDEODISC
The Secret of Life
Flatworm Cross Section

Purpose

Students will study the structures and adaptations of planarians.

Teaching Strategies

■ Allow students to examine live planarians or a prepared slide of a planarian. Have them make drawings of their observations. **L1 ELL**

Visual Learning

■ Have students use the art in the feature to label the planarian they draw.

■ Point out various structures of the planarian. As you mention each structure, have students identify the function of the structure.

Critical Thinking

As a swimming animal moves forward, the head encounters new information first. As most sensory organs are located in the head, this information is relayed to the rest of the animal, enabling it to react appropriately.

Quick Demo

Place a live planarian in water in a 35mm deep-well slide that can be projected through a slide projector. Ask students to observe how the worm moves. Point out that the planarian has a head area. Remind them that sponges do not have a head and ask the survival advantage of having a head area on the body. **L1 ELL**

728

INSIDE STORY

A Planarian

f you've ever waded in a shallow stream and turned over some rocks, you may have found tiny, black organisms stuck to the bottom of the rocks. These organisms were most likely planarians. Planarians have many characteristics common to all species of flatworms. The bodies of planarians are flat, with both a dorsal and a ventral surface. All flatworms have bilateral symmetry.

Critical Thinking *Why is having a head an advantage to a swimming animal?*

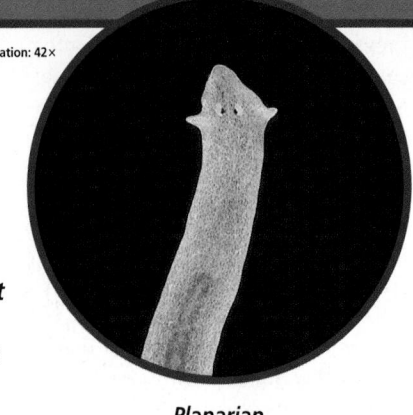

Magnification: 42×

Planarian

1 **Head** Flatworms have a clearly defined head. The head senses and responds to changes in the environment.

2 **Eyespots** Eyespots are sensitive to light and enable the animal to respond to the amount of light present.

3 **Sensory pits** Located on the sides of the head, sensory pits are used to detect food, chemicals, and movements in the environment.

6 **Pharynx** The pharynx is a muscular tube that can be extended outside the animal's body through its mouth. It is used to suck food into the planarian's gastrovascular cavity. Note that the mouth is a body opening located in the midsection of the planarian.

4 **Flame cells** Excess water is removed from the planarian's body by a system of flame cells. The water from flame cells collects in tubules and leaves the body through pores on the body surface. Flame cells are so named because the constant movement of the cilia inside flame cells resembles the flickering of a candle's flame.

Mouth

Extended pharynx

Digestive tract

Cilia

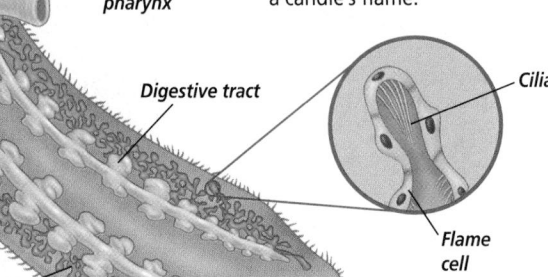

5 **Cilia** Hairlike cilia are located on the ventral surface of planarians. Cilia help the worm to pull itself along.

Excretory system

Flame cell

728 SPONGES, CNIDARIANS, FLATWORMS, AND ROUNDWORMS

Learning Disabled

 Naturalist Provide students with outlines of planarians. Ask students to use one outline to show the symmetry of the planarian as well as its anterior and posterior ends and dorsal and ventral sides. Have students use the Inside Story as a model to show individual body systems of planarians. **L1 ELL**

GLENCOE TECHNOLOGY

VIDEODISC
The Secret of Life
Six Kingdoms

Reproduction in planarians

Study the body of a planarian in the *Inside Story.* How does a planarian reproduce? Like many of the organisms studied in this chapter, most flatworms are hermaphrodites. During sexual reproduction, individual planarians exchange sperm, which travel along special tubes to reach the eggs. Fertilization occurs internally. The zygotes are released in capsules into the water, where they hatch into tiny planarians.

Planarians can also reproduce asexually. When a planarian is damaged, it has the ability to regenerate, or regrow, new body parts. **Regeneration** is the replacement or regrowth of missing body parts. Missing body parts are replaced through mitosis. If a planarian is cut horizontally, the section containing the head will grow a new tail, and the tail section will grow a new head. Thus, a planarian that is damaged or cut into two pieces may grow into two new organisms—a form of asexual reproduction. Go to the *BioLab* at the end of this chapter to observe regeneration in planarians.

Feeding and digestion in parasitic flatworms

Although the basic structure of a parasitic flatworm is similar to that of a planarian, it is adapted to obtaining nutrients from inside the bodies of one or two hosts. Recall that a parasite is an organism that lives on or in another organism and depends upon that host organism for its food. Parasitic flatworms have mouthparts with hooks that keep the worm firmly attached to the insides of its host. Parasitic flatworms do not have complex nervous or muscular tissue. Because they are surrounded by nutrients, they do not need to move to seek out or find food.

Diversity of Flatworms

Planarians are free-living flatworms. Tapeworms and flukes are parasitic flatworms. These parasites live in the bodies of many vertebrates including dogs, cats, cattle, monkeys, and people.

Tapeworm bodies have sections

Some adult tapeworms that live in animal intestines can grow to more than 10 m in length. The body of a tapeworm is made up of a head and individual repeating sections called proglottids, shown in *Figure 26.14.* The knob-shaped head of a tapeworm is called a **scolex** (SKOH leks). A **proglottid** (proh GLAH tihd) is a detachable section of a tapeworm that contains muscles, nerves, flame cells, and male and female reproductive organs. Each proglottid may contain up to 100 000 eggs, and some tapeworms consist of 2000 proglottids.

WORD *Origin*

scolex
From the Greek word *skolek*, meaning "worm." A scolex is the knob-shaped head of a tapeworm.

Figure 26.14
The scolex is covered with hooks and suckers that attach to the intestinal lining of the host (**a**). Mature proglottids full of fertilized eggs are shed (**b**). Eggs hatch when they are eaten by a secondary host.

Magnification: 29×

a

b

Magnification: 35×

Concept Development

Display several preserved flatworms. Ask students to observe the worms and list their similarities and differences. Have them speculate which ones are parasites and ask them to explain their choices. **L2** **ELL**

✔ Assessment

Knowledge Ask students to explain how regeneration of missing body parts in planarians can also be a form of asexual reproduction. *Students should be able to explain that when a planarian is cut into two halves, each half can regenerate the missing half through mitosis. This results in two new worms that are genetically identical to the original worm. In asexual reproduction, a parent produces offspring that are genetically identical to the parent.*

The BioLab at the end of the chapter can be used at this point in the lesson.

INVESTIGATE BioLab

GLENCOE TECHNOLOGY

CD-ROM
Biology: The Dynamics of Life
BioQuest: *Biodiversity Park*
Discs 3, 4

---

3 Assess

Check for Understanding

Naturalist Ask students to make and fill out a table about flatworms with the following headings: symmetry, habitat, food-getting, nervous control, digestion, reproduction, free-living or parasitic, examples. **L2**

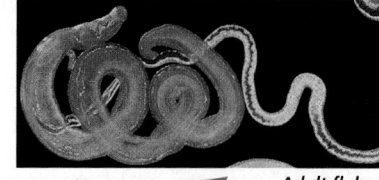

A Adult flukes are about 1 cm long and live in the blood vessels of the human intestine.

Human host

Adult flukes

B Fluke eggs pass out of the body with human wastes. If the eggs reach freshwater, they hatch.

Eggs hatch

C Eggs hatch into free-swimming larvae that enter their snail hosts.

Larva that infects intermediate host

D Larvae develop inside the snail and reproduce. New larvae leave the snail and enter the water.

E When humans walk through water with bare feet or legs, the fluke larvae bore through the skin, enter the bloodstream, and pass to the intestine, where they mature. Fertilized eggs pass out of the intestine and the cycle begins again.

Snail host

Larva that infects final host

Figure 26.15
The *Schistosoma* fluke requires two hosts to complete its life cycle.

The life cycle of a fluke

A fluke is a parasitic flatworm that invades the internal organs of a vertebrate such as a human or a sheep. It obtains its nutrition by embedding itself in organs where it feeds on cells, blood, and other fluids of the host organism. Flukes have a complex life cycle that may include one, two, or more hosts.

Blood flukes of the genus *Schistosoma*, shown in **Figure 26.15**, cause a disease in humans known as schistosomiasis. Schistosomiasis is common in countries where farmers grow rice. Farmers must work in standing water in rice fields during planting and harvesting. Blood flukes are common where the secondary host, snails, also are found.

Section Assessment

Understanding Main Ideas
1. Diagram and label the structures of a planarian.
2. Why don't tapeworms have a digestive system?
3. What is the adaptive advantage of a nervous system to a free-living flatworm?
4. How is the body of a tapeworm different from that of a planarian?

Thinking Critically
5. Examine the life cycle of a parasitic fluke, and suggest ways to prevent infection on a rice farm.

SKILL REVIEW
6. **Observing and Inferring** What can you infer about the way of life of an organism that has no mouth or digestive system, but is equipped with a sucker? For more help, refer to *Thinking Critically* in the **Skill Handbook**.

730 SPONGES, CNIDARIANS, FLATWORMS, AND ROUNDWORMS

26.4 Roundworms

H ave you ever been to the veterinarian to have your dog tested for heartworms? Perhaps you recall being warned not to eat uncooked pork products. Flatworms are not the only type of worms that can cause harm to humans and other vertebrates. It has been estimated that about one-third of the world's human population suffers from problems caused by roundworms.

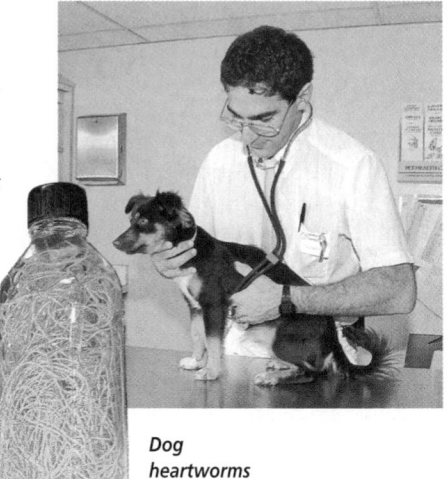
Dog heartworms

SECTION PREVIEW

Objectives

Compare the structural adaptations of roundworms and flatworms.

Identify the characteristics of four roundworm parasites.

Vocabulary
none

What Is a Roundworm?

Roundworms belong to the phylum Nematoda. They are widely distributed, living in soil, animals, and both freshwater and saltwater environments. More than 12 000 species of roundworms are known to scientists.

Most roundworm species are free-living, but many are parasitic, including those shown in *Figure 26.16*. In fact, virtually all plant and animal species are affected by parasitic roundworms.

Roundworms are tapered at both ends. They have a thick outer covering that protects them from being

Figure 26.16
Parasitic roundworms include *Ascaris* (a), *Trichinella* (b), hookworms (c), pinworms (d), and nematodes (e) that affect plants.

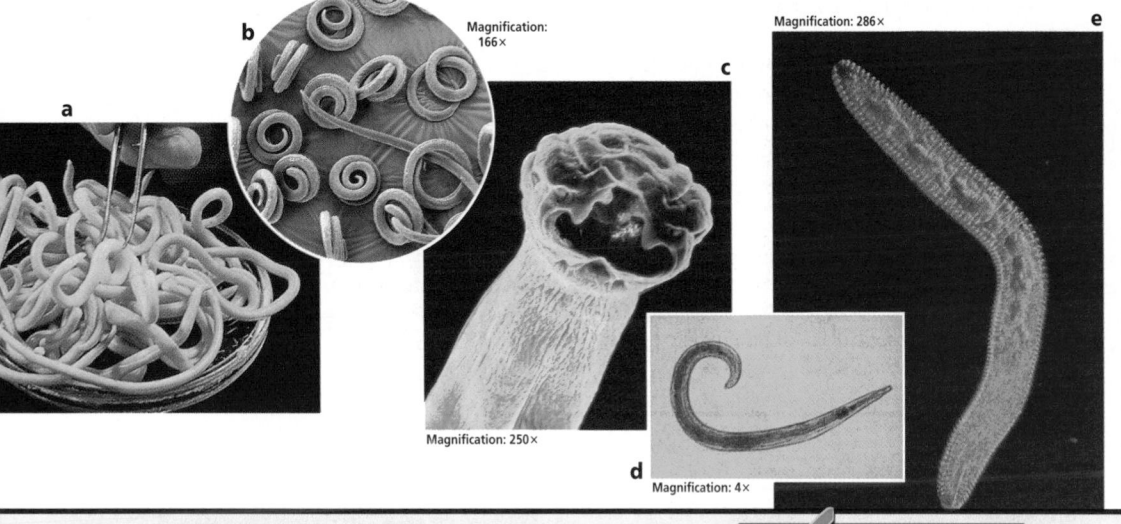

a

b   Magnification: 166×

c   Magnification: 250×

d   Magnification: 4×

e   Magnification: 286×

Section 26.4

Prepare

Key Concepts

Students will compare and contrast the structural adaptations of roundworms and learn about the characteristics of the roundworms *Ascaris*, *Trichinella*, hookworms, and pinworms.

Planning

■ Obtain a live vinegar eel culture for the Quick Demo.

1 Focus

Bellringer

Before presenting this lesson, display **Section Focus Transparency 66** on the overhead projector and have students answer the accompanying questions.
L1 ELL

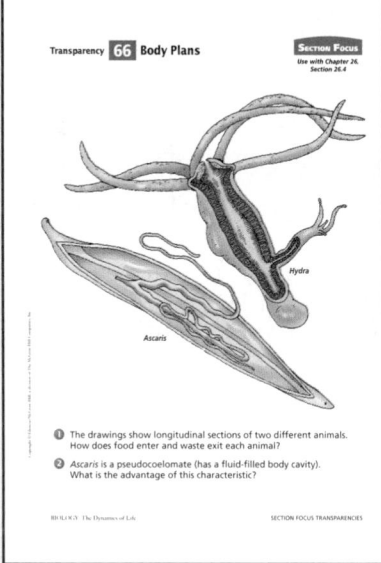

MEETING INDIVIDUAL NEEDS

Gifted/Visually Impaired/ Learning Disabled

Kinesthetic Have gifted students make clay models of the worms studied in this section. Provide these models to students who have visual problems or learning disabilities for use in studying these organisms. L1

✓ **Portfolio**

Roundworm Data Table

Naturalist Have students construct a table that lists the names of each roundworm discussed in the left column. Have students sketch the roundworm in the second column, describe its habitat in the third column, and explain how the worm affects humans in the fourth column. L2 P

2 Teach

Purpose

Students will observe the larval stage of the pork worm embedded in muscle tissue of its host.

Process Skills

observe and infer, measure in SI

Teaching Strategies

■ Prepared slides of pork worm are available from biological supply houses.

■ Certain larvae will appear as round segments due to the plane through which the slice was made in preparing the slide.

Expected Results

Students will not be able to see pork worm larvae without the aid of a microscope. Size of the larvae will be close to 100 mm.

Analysis

1. microscopic, round, like a hot dog, spiral
2. Pork worm larvae are microscopic and cannot be identified by a visual inspection.
3. Samples of muscle tissue can be taken and viewed under a microscope.

✓ Assessment

Portfolio Have students put their diagrams in their portfolios. Use the Performance Task Assessment List for Science Portfolio in **PASC**, p. 105. **L1** **P**

Problem-Solving Lab 26-4

Purpose

Students will study a diagram of the pork worm life cycle.

Process Skills

■ apply concepts, think critically, interpret scientific diagrams

Teaching Strategies

■ Show photographs of the pork worm adult to your students.

---

digested by their host organisms. On a flat surface, roundworms look like tiny, wriggling bits of sewing thread. They lack circular muscles but have pairs of lengthwise muscles. As one muscle of a pair contracts, the other muscle relaxes. This alternating contraction and relaxation of muscles causes roundworms to move in a thrashing fashion.

WORD *Origin*

Trichinella
From the Greek word *trichinos*, meaning "made of hair." *Trichinella* species are slender, hairlike, roundworms.

Roundworms have a pseudocoelom and are the simplest animals with a tubelike digestive system. Recall that a pseudocoelom is a body cavity partly lined with mesoderm. Unlike flatworms, roundworms have two body openings—a mouth and an anus. The free-living species have well-developed sense organs, such as eyespots, although these are reduced in parasitic forms.

MiniLab 26-2 Observing

Magnification: 195×

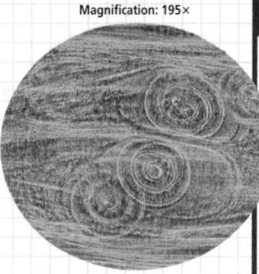

Pork Worm

Observing the Larval Stage of a Pork Worm You can observe the larval stage of a pork worm (*Trichinella spiralis*) embedded within the muscle tissue of its host. It will look like a curled up hot dog surrounded by muscle tissue.

Procedure

1. Examine a prepared slide of pork worm larvae under the low-power magnification of your microscope.
2. Locate several larvae by looking for "spiral worms enclosed in a sac." All other tissue is muscle.
3. Estimate the size of the larva in µm.
4. Diagram one larva. Indicate its size on the diagram.

Analysis

1. Describe the appearance of a pork worm larva.
2. Why might it be difficult to find larva embedded in muscle when meat inspectors use visual checking methods in packing houses to screen for pork worm contamination?
3. Suggest what inspectors might do to help detect pork worm larvae.

Diversity of Roundworms

Roundworms are found as parasites in most organisms on Earth. Approximately half of the described roundworm species are parasites, and about 50 species infect humans.

Roundworm parasites invade humans through a variety of methods

Ascaris mainly infects children who swallow eggs when they put their soiled hands into their mouths or eat vegetables that have not been washed. The eggs hatch in the intestines, move to the bloodstream, and then to the lungs, where they are coughed up and swallowed to begin the cycle again.

Hookworms commonly infect humans in warm climates where people walk on contaminated soil in bare feet. Hookworms cause people to feel weak and tired due to blood loss.

Pinworms are the most common parasites in children. Pinworms invade the intestinal tract when children eat something that has come in contact with contaminated soil. Female pinworms lay eggs near the anus, and reinfection is common because the worms cause itching.

Trichinella worms cause a disease called trichinosis. These worms enter the body in undercooked pork. Find out what these roundworms look like in the *MiniLab* on this page. Trichinosis is not as common in the United States as it once was because of stricter meat inspection standards. However, *Trichinella* worms are microscopic and may not be seen during a visual inspection. Trichinosis can be controlled by cooking pork long enough to kill any worms that may be present. You can learn more about controlling trichinosis in the *Problem-Solving Lab* on the next page.

Thinking Critically

1. impractical, unless there is a drug that can be given to the pig to destroy the adult worm; practical, make sure all scraps are cooked before feeding them to a pig; practical, advise consumers about proper cooking of pork; practical, the cycle ends here.

2. The cycle ends here in humans. Any larvae in human muscle tissue remains there.

✓ Assessment

Skill Ask students to describe any differences in the life cycle of the pork worm when in a pig host compared with its life cycle when in a human host. *They are essentially alike.* **L2**

Roundworm parasites of other organisms

Nematodes can infect and kill pine trees, cereal crops, and food plants such as potatoes. They are particularly attracted to plant roots. About 1200 species of nematodes cause diseases in plants. They also infect fungi and form symbiotic associations with bacteria. Soil nematodes invade roots of plants grown for food, as you can see in *Figure 26.17,* and cause a slow decline of the plant.

Figure 26.17
Roundworms that are plant parasites usually enter the roots, forming cysts that affect the plant's ability to absorb water.

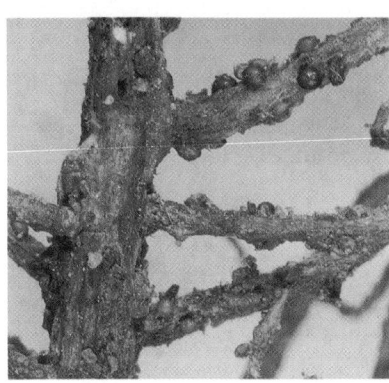

Problem-Solving Lab 26-4 — Interpreting Scientific Illustrations

How can the pork worm parasite be controlled? *Trichinella spiralis,* the pork worm, is contracted when humans eat raw or undercooked pork products.

Analysis
This diagram shows the life cycle of the pork worm. A cyst is a protective covering that encloses the dormant larval stage.

Life Cycle of a Pork Worm

Events in Pigs	① Pig is slaughtered	Events in Humans
② Uncooked pig scraps with larvae are fed to pigs.		⑥ Humans eat pork with cysts.
③ Larvae are released from cysts during digestion.		⑦ Larvae are released from cysts during digestion.
④ Larvae mature, mate, and produce thousands of new larvae.		⑧ Larvae mature, mate, and produce thousands of new larvae.
⑤ Larvae travel in blood stream to muscles and form cysts.		⑨ Larvae travel in blood stream to muscles and form cysts.

Thinking Critically
1. Would it be practical or impractical to disrupt the pork worm life cycle: At step 2? At step 4? At step 6? At step 9? Explain your answers.
2. Why doesn't the arrow return to the top of the diagram after step 9 as it does after step 5?

Section Assessment

Understanding Main Ideas
1. Compare the body structures of roundworms and flatworms.
2. Why do parents teach children to wash their hands before eating?
3. Describe the method of infection of one human roundworm parasite.
4. Compare how *Ascaris* and *Trichinella* are contracted.

Thinking Critically
5. An infection of pinworms is spreading to children who attend the same preschool.

Make a list of precautions that could be taken to help prevent its continued spread.

SKILL REVIEW

6. **Making and Using Tables** Make a table of the characteristics of four roundworm parasites, indicating the name of the worm, how it is contracted, the action of the parasite in the body, and means of prevention. For more help, refer to *Organizing Information* in the **Skill Handbook.**

Section Assessment

1. Roundworms have a pseudocoelom and two body openings. Flatworms are acoelomate with only one body opening.
2. It is important to wash hands before eating to prevent infection by parasitic worms and bacteria.
3. Roundworms can be contracted by eating improperly cooked pork and unwashed vegetables, walking barefoot, and by lack of good hygiene.
4. *Ascaris* is contracted by eating food contaminated by *Ascaris. Trichinella* is contracted by eating contaminated pork.
5. good personal hygiene such as washing hands, clothing, and bedding
6. Make sure students' tables include information from pages 731-733.

BIOLOGY JOURNAL

Roundworm Symmetry

Visual-Spatial Provide students with an outline drawing of a roundworm. Have students add a line to the diagram to show the bilateral symmetry of the worm. Ask students to label the worm's anterior, posterior, dorsal, and ventral areas. **L1 ELL**

3 Assess

Check for Understanding
Have students write true or false questions relating to roundworms. Review and explain the correct answers. **L2**

Reteach
Have students play bingo with a 16-square card with the name of a worm in each square. Read a list of statements and have students play bingo. **L2**

Extension
Ask students to do a videotape report on parasitic roundworms in humans. **L3**

✔ Assessment
Portfolio Ask students to write a paragraph to explain how the roundworm diseases they studied might be prevented. **L2 P**

4 Close

Activity
Set up stations around the classroom with photos or slides of the roundworms studied. Have students identify each worm. **L2 ELL**

Observing Planarian Regeneration

Time Allotment

30 minutes the first day, 10 minutes every day for 2 weeks

Process Skills

collect data, experiment, think critically, observe and infer, organize data

Safety Precautions

Make sure students work very carefully with razor blades. Remind students to use razor blades carefully to avoid injury. Caution them to always cut in a direction away from the body. Remind students they are working with live animals and to treat them gently. Have students wash their hands after working with planarians.

Preparation

■ Order planarians so that they arrive a day or two before they will be needed. Planarians are available from biological supply houses. Ask for the species, *Dugesia tigrina*.

■ DO NOT use tap water, distilled water, or spring water purchased from a grocery store. Purchase spring water from a biological supply house only.

Alternative Materials

■ Use a hand lens if binocular microscopes are not available.

■ A dropper or plastic pipette may be substituted for the camel hair brush.

Certain animals have the ability to replace lost body parts through regeneration. In regeneration, organisms regrow parts that were lost. This process occurs in a number of different phyla throughout the animal kingdom. Examples of animals that can regenerate include sponges, hydra, mudworms, sea stars, and reptiles. In this activity, you will observe regeneration in planarians. Planarians are able to form two new animals when one has been cut in half.

PREPARATION

Problem

How can you determine if the flatworm *Dugesia* is capable of regeneration?

Objectives

In this BioLab, you will:

■ **Observe** the flatworm, *Dugesia*.

■ **Conduct** an experiment to determine if planarians are capable of regeneration.

Materials

planarians
petri dish
springwater

camel hair brush
chilled glass slide
binocular microscope
marking pencil or labels
single-edged razor blade

Safety Precautions 🔥 🧤 🥽

Always wear goggles in the lab. Use extreme caution when cutting with a razor blade. Wash your hands both before and after working with planarians.

Skill Handbook

Use the **Skill Handbook** if you need additional help with this lab.

PROCEDURE

1. Obtain a planarian and place it in a petri dish containing a small amount of springwater. You can pick up a planarian easily with a small camel hair brush.

2. Use a binocular microscope to observe the planarian. Locate the animal's head and tail region and its "eyes." Use diagram **A** as a guide.

3. Place the animal on a chilled glass slide. This will cause it to stretch out.

4. Place the slide onto the microscope stage. While observing the worm through the microscope, use a single-edged razor to cut the animal in half across the midsection. Use diagram **B** as a guide.

734 SPONGES, CNIDARIANS, FLATWORMS, AND ROUNDWORMS

PROCEDURE

Teaching Strategies

■ Have students work in groups of 2 or 3.

■ Do not feed the planarians while they are undergoing regeneration. Replace any springwater lost through evaporation.

■ Store the regenerating pieces in a cool area (around 20°C) of the classroom and in subdued light.

Resource Manager

Section Focus Transparency 66 and Master L1 ELL

BioLab and MiniLab Worksheets pp. 118-119 L2

Reinforcement and Study Guide p. 118 L2

Content Mastery, pp. 129, 132 L1

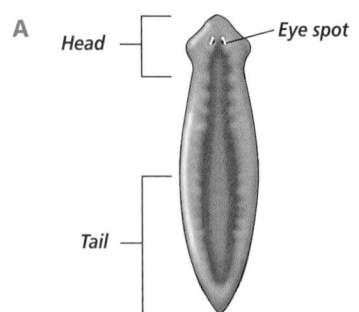

A

Head — Eye spot

Tail

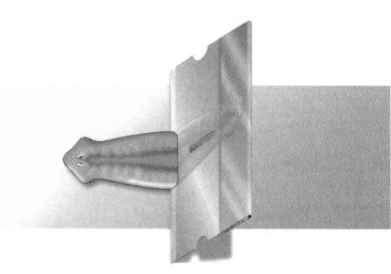

B

5. Remove the head end and place it in a petri dish filled with spring-water. Label the dish with the date, your name, and the word "head."

6. Add the tail section to a different petri dish and label it as in step 5, marking this dish "tail."

7. Repeat steps 3-6 with a second flatworm and add the correct pieces to the proper petri dishes.

8. Place the petri dishes in an area designated by your teacher.

9. Prepare a data table that will allow you to record the appearance of your flatworms every other day for two weeks. Include diagrams and the number of days since starting the experiment in your data table.

10. Observe your animals under a binocular microscope and record observations and diagrams in your data table.

ANALYZE AND CONCLUDE

1. **Knowledge** To what phylum do flatworms belong? Are planarians free living or parasitic? What is your evidence?

2. **Observing** What new part did each original head piece regenerate? What new part did each original tail piece regenerate?

3. **Observing** Which section, head or tail, regenerated new parts faster?

4. **Interpreting** Are planarians able to regenerate new parts? Would regeneration be by mitosis or meiosis? Explain.

5. **Thinking Critically** What might be the advantage for an animal that can grow new body parts through regeneration?

6. **Thinking Critically** Would the term "clone" be suitable in reference to the newly formed planarians? Explain your answer.

Going Further

Project Design and carry out an experiment that would test this hypothesis: If regenerating planarians are placed in a warmer environment, then the time needed for new parts to form would decrease.

BIOLOGY Online To find out more about regeneration, visit the Glencoe Science Web site.
science.glencoe.com

INVESTIGATE BioLab

ANALYZE AND CONCLUDE

1. Planarians belong to the phylum Playthelminthes. They are free-living and can be found in ponds and streams living on their own.

2. The head regenerated a new tail, and the tail regenerated a new head.

3. The head section regenerated a new tail faster.

4. Yes, planarians can regenerate new body parts. Regeneration occurs through mitosis; no sexual reproduction or formation of gametes was needed.

5. Answers may vary; there is no need to find a mate, able to replace lost body parts, faster than sexual reproduction, identical genetic makeup as original animal

6. Yes, clones may be formed asexually by mitosis. Regeneration in planarians is a type of cloning.

✓ Assessment

Portfolio Have students write a report on their experimental findings. Be sure to have them include their data table with recorded results. Use the Performance Task Assessment List for Lab Report in **PASC**, p. 47. **L2**

Going Further

Have students redesign the experiment by making different types of cuts and/or making cuts at different locations on the animals' bodies.

Data and Observations

Students will observe the flatworms and will be able to see those structures shown in Figure A. Data table design will vary from student to student. Encourage students to diagram their observations. At the end of two weeks, students will observe that a new head has formed on the original tail section, and a new tail on the original head section.

BIOLOGY Online Note Internet addresses that you find useful in the space below for quick reference.

_____

_____

_____

_____

_____

Biology & SOCIETY

Why are the corals dying?

Coral reefs are some of Earth's most spectacularly beautiful and productive ecosystems. A reef is composed of hundreds of corals that together create a structure of brightly colored shapes and patterns. In the reef's cracks and crevices live a dazzling array of fishes and invertebrates.

Purpose

Students learn about the importance of coral reefs, discover how they can be damaged, and explore methods for preventing damage.

Background

Corals contain zooxanthellae that have chlorophyll and carry out photosynthesis. During the day, most corals withdraw into the safety of their limestone cups and let the zooxanthellae make food for them. At night, corals use their tentacles and nematocysts to capture planktonic organisms that drift by.

Teaching Strategies

■ Point out to students that, since most coral reefs lie in shallow waters near the shoreline, they are easily affected by coastal activities. The clearing of coastal land for development may increase erosion. Siltation clouds the water and reduces the amount of sunlight available, limiting photosynthesis by zooxanthellae. Silt can also clog the tiny mouth opening of the coral polyps, affecting their ability to feed at night.

Investigating the Issue

Because coral reefs offer an enormous array of protected living spaces, the destruction of a reef would remove habitats needed by many species. Organisms that prey on any of those species would lose a food source. Coral reefs also absorb wave energy, so the destruction of a reef could leave the shore subject to erosion from wind and wave action.

Coral reefs protect nearby shore areas from erosion by breaking up the energy of incoming waves. But worldwide, coral reefs are increasingly being damaged and destroyed.

Physical Damage to Coral Reefs Hurricanes cause serious damage to coral reefs, as do large ships that run aground on reefs. Coral reefs lie close to the water's surface, and make attractive anchoring sites for boats. But when an anchor is pulled up, it may rip away a piece of the reef. In some parts of the world, explosives are used to mine coral limestone for building materials and fertilizers. Tropical aquarium fishes are sometimes collected by poisoning with cyanide, which stuns fishes and makes them easier to collect, but also kills corals. Collectors take pieces of coral for jewelry and souvenirs.

Damage from Disease In the 1970s, marine scientists began to realize that the world's coral reefs were being attacked by diseases no one had seen before.

Healthy coral reef (below), and diseased reef (right)

Black-band disease is caused by several species of cyanobacteria that combine to form a band of black filaments. This invading community slowly moves across the coral. White-band disease causes the living tissue of a coral to peel away from its skeleton; this may be caused by bacteria. Rapid-wasting disease, possibly caused by a fungus, forms white patches that consume not only the living tissue but also the top layers of the coral skeleton.

Many of the world's coral reefs are losing their beautiful colors in a process called bleaching. The corals become gray or white in color. Some scientists hypothesize that coral bleaching is the result of a loss of zooxanthellae, the symbiotic protists that live in coral and give it much of its color as well as nutrients.

Different Viewpoints

It is not easy to tell whether microorganisms are causing the diseases or are attacking already damaged or ailing corals that have lost their natural defenses. Most researchers hypothesize that coral diseases are on the increase because environmental changes, such as pollution in coastal runoff, higher water levels, or changes in ocean temperatures, make corals more vulnerable to opportunistic diseases.

INVESTIGATING THE ISSUE

Analyzing the Issue What effects might the death of a coral reef have on nearby ocean and coastal life?

BIOLOGY Online To find out more about coral reefs, visit the Glencoe Science Web site. **science.glencoe.com**

Going Further

Intrapersonal A great deal of research on methods for protecting and restoring coral reefs goes on in the Florida Keys. Invite students to use their Internet skills to find out about some of the problems and solutions scientists are exploring.

BIOLOGY Online Note Internet addresses that you find useful in the space below for quick reference.

_____

_____

_____

_____

_____

SUMMARY

Section 26.1

Sponges

Main Ideas
- A sponge is an aquatic, sessile, asymmetrical, filter-feeding invertebrate.
- Sponges are made of four types of cells. Each cell type contributes to the survival of the organism.
- Most sponges are hermaphroditic with free-swimming larvae.

Vocabulary
external fertilization (p. 716)
filter feeding (p. 713)
hermaphrodite (p. 716)
internal fertilization (p. 716)

Section 26.2

Cnidarians

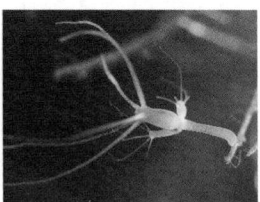

Main Ideas
- All cnidarians are radially symmetrical, aquatic invertebrates that display two basic forms: medusa and polyp.
- Cnidarians feed by stinging or entangling their prey with cells called nematocysts, usually located at the ends of their tentacles.
- The three primary classes of cnidarians include the hydrozoans, hydras; scyphozoans, jellyfishes; and anthozoans, corals and anemones.

Vocabulary
gastrovascular cavity (p. 719)
medusa (p. 719)
nematocyst (p. 719)
nerve net (p. 721)
polyp (p. 719)

Section 26.3

Flatworms

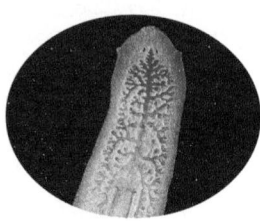

Main Ideas
- Flatworms are acoelomates with thin, solid bodies belonging to the phylum Platyhelminthes. They are grouped into three classes: free-living planarians, parasitic flukes, and tapeworms.
- Planarians have well-developed nervous and muscular systems. These systems are reduced in parasitic flatworms. Flukes and tapeworms have other structures adapted to their parasitic existence.

Vocabulary
pharynx (p. 727)
proglottid (p. 729)
regeneration (p. 729)
scolex (p. 729)

Section 26.4

Roundworms

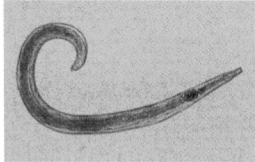

Main Ideas
- Roundworms are pseudocoelomate, cylindrical worms with lengthwise muscles, relatively complex digestive systems, and two body openings.
- Roundworm parasites include parasites of plants, fungi, and animals, including humans. *Ascaris*, hookworms, *Trichinella*, and pinworms are roundworm parasites of humans.

Vocabulary
none

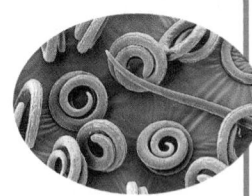

Main Ideas
Summary statements can be used by students to review the major concepts of the chapter.

Using the Vocabulary
To reinforce chapter vocabulary, use the Content Mastery Booklet and the activities in the Interactive Tutor for Biology: The Dynamics of Life on the Glencoe Science Web site.
science.glencoe.com

All Chapter Assessment questions and answers have been validated for accuracy and suitability by The Princeton Review.

GLENCOE TECHNOLOGY

VIDEOTAPE
MindJogger Videoquizzes
Chapter 26: *Sponges, Cnidarians, Flatworms and Roundworms*
Have students work in groups as they play the videoquiz game to review key chapter concepts.

Resource Manager

Chapter Assessment, pp. 151-156
MindJogger Videoquizzes
ExamView® Pro Software
BDOL Interactive CD-ROM, Chapter 26 quiz

UNDERSTANDING MAIN IDEAS

1. a
2. c
3. b
4. c
5. c
6. d
7. a
8. d
9. b
10. b
11. b
12. hermaphrodite
13. head
14. hooks
15. fluke
16. a
17. cnidarians
18. Eyespots, planarian
19. planarian, regenerate
20. internal fertilization

UNDERSTANDING MAIN IDEAS

1. Which of these is a type of cell found in sponges?
   a. epithelial cell
   c. stinging cell
   b. flame cell
   d. nematocyst

2. An individual sponge is a hermaphrodite because it _____.
   a. reproduces by budding
   b. can regenerate lost body parts
   c. can produce both eggs and sperm
   d. has both pore cells and stinging cells

3. Which of the following organisms obtain food by filter feeding?
   a. jellyfish
   c. pinworm
   b. sponge
   d. tapeworm

4. To what phylum do marine invertebrates with radial symmetry and two body layers belong?
   a. Porifera
   c. Cnidaria
   b. Platyhelminthes
   d. Cestoda

5. Both sponges and planarians have _____ larvae.
   a. sessile
   c. free-swimming
   b. polyp
   d. budding

6. In cnidarians, medusae reproduce sexually to produce polyps, which in turn reproduce asexually to form _____.
   a. buds
   c. hermaphrodites
   b. larvae
   d. new medusae

7. Sea anemones exhibit only the _____ type of body form.
   a. polyp
   c. bud
   b. medusa
   d. colony

THE PRINCETON REVIEW — **TEST-TAKING TIP**

Ignore Everyone
While you take a test, pay absolutely no attention to anyone else in the room. Don't worry if your friends finish a test before you do. If someone tries to talk with you during a test, don't answer. You run the risk of the teacher thinking you were cheating—even if you weren't.

8. Two basic body forms are found in _____.
   a. sponges
   c. roundworms
   b. flatworms
   d. cnidarians

9. Acoelomate worms called _____ have thin, solid bodies.
   a. roundworms
   c. nematodes
   b. flatworms
   d. hookworms

10. Of the following, which is a pseudocoelomate animal?
    a. fluke
    c. tapeworm
    b. roundworm
    d. planarian

11. Examine the diagram. The cell labeled _____ forms eggs and sperm.

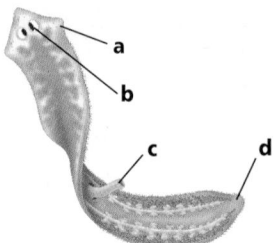

12. A _____ is an organism that has both male and female reproductive organs.

13. Unlike sponges and cnidarians, flatworms have a clearly defined _____.

14. Parasitic worms have mouthparts with _____.

15. A _____ is a parasitic worm that uses a snail as an intermediate host and has a larval stage that can bore through the skin of humans.

16. Examine the diagram. The structure labeled _____ is used to detect food, chemicals, and movement in the environment.

17. Hydrozoans and anthozoans are classes of _____.

18. _____, located on the head of a _____, are sensitive to light.

19. If a(n) _____ is cut accidentally, it can _____ lost tissue.

20. In _____, eggs and sperm meet inside an animal's body.

APPLYING MAIN IDEAS

21. In what ways are cnidarians more complex than sponges?

22. You are examining a wormlike animal found in the intestine of a sheep. It has a head with tiny hooks. What kind of worm is it?

23. Of what advantage is hermaphroditism to a sessile animal?

24. Describe the features that would be important for a predator of jellyfishes.

THINKING CRITICALLY

25. Observing and Inferring While examining soil from the bottom of a pond, you notice tiny red worms wriggling aimlessly in your petri dish. What kind of worms are they?

26. Recognizing Cause and Effect At what points could the life cycle of a blood fluke be interrupted so that disease would be prevented?

27. Observing and Inferring Why is the phylogeny of cnidarians so little understood?

28. Comparing and Contrasting Both sponges and hydras are sessile organisms that cannot pursue prey. Compare their methods of obtaining food.

29. Concept Mapping Complete the concept map by using the following vocabulary terms: medusa, nematocyst, polyp.

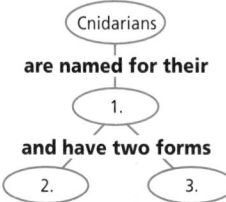

Cnidarians

are named for their

1.

and have two forms

2.   3.

CD-ROM

For additional review, use the assessment options for this chapter found on the ***Biology: The Dynamics of Life Interactive CD-ROM*** and on the Glencoe Science Web site.
science.glencoe.com

ASSESSING KNOWLEDGE & SKILLS

The diagram shows the life cycle for a beef tapeworm.

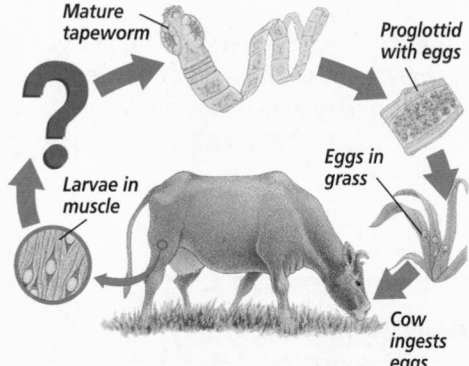

Mature tapeworm

Proglottid with eggs

Eggs in grass

Larvae in muscle

Cow ingests eggs

Interpreting Scientific Illustrations
Use the diagram to answer the following questions.

1. Which part of the life cycle for a beef tapeworm is missing?
 a. infection of the cow
 b. infection of the grass
 c. infection of the human host
 d. infection of the tapeworm

2. How do the tapeworm eggs get into the grass?
 a. from rainwater
 b. from feces of infected cattle
 c. from snails
 d. from dead cows

3. Beef tapeworm larvae get into human hosts when humans _____.
 a. eat beef   **c.** walk barefoot
 b. eat pork   **d.** go swimming

4. Completing a Diagram By making a diagram similar to the tapeworm life cycle on this page, trace the steps of a *Trichinella* infection.

THINKING CRITICALLY

25. They are roundworms. Roundworms move with a wriggling motion because they have no circular muscles as do earthworms.

26. Human wastes may be disposed of in such as way as to prevent fluke eggs from coming in contact with snails, or humans can avoid contact with water in which snails live.

27. Cnidarians are soft-bodied animals that do not preserve well as fossils. In corals, their calcium carbonate homes become fossils, showing where corals once lived, but the animals themselves are not fossilized.

28. Sponges are filter feeders. They pull water in through pores and filter out small particles of food. Hydras are cnidarians. They use nematocysts on long tentacles. When a waving tentacle touches a prey organism, the nematocyst discharges, and the tentacles bring the captured food to the hydra's mouth for digestion.

29. 1. Nematocysts; 2. Medusa; 3. Polyp

ASSESSING KNOWLEDGE & SKILLS

1. c
2. b
3. a
4. Life Cycle of *Trichinella*

Feeds contaminated pork scraps to pig

Human eats improperly cooked pork

Pig eats scraps containing eggs

Worm develops in pig muscle tissue

APPLYING MAIN IDEAS

21. The two cell layers of cnidarians are organized into tissues with specific functions. They have simple nervous systems, cells that can contract like muscles, nematocysts that are used to capture prey, and a gastrovascular cavity in which digestion occurs.

22. It is a parasitic flatworm. A head with hooks is characteristic of parasitic flatworms.

23. Animals that are sessile are unable to move and search for mates. If they are hermaphroditic, any individual can be a mate for any other individual of the same species.

24. Predators of jellyfishes must be immune to the toxins in jellyfishes, or have a protective covering such as mucus on the body that could absorb the nematocysts' toxins.

Chapter 27 Organizer

Refer to pages 4T-5T of the Teacher Guide for an explanation of the National Science Education Standards correlations.

Section	Objectives	Activities/Features
Section 27.1 **Mollusks** National Science Education Standards UCP.1, UCP.2, UCP.3, UCP.5; C.4, C.5, C.6 (1 session, $\frac{1}{2}$ block)	1. **Identify** the characteristics of mollusks. 2. **Compare** the adaptations of gastropod, bivalve, and cephalopod mollusks.	**Inside Story:** A Clam, p. 743 **Problem-Solving Lab 27-1**, p. 744 **MiniLab 27-1:** Identifying Mollusks, p. 746
Section 27.2 **Segmented Worms** National Science Education Standards UCP.1, UCP.2, UCP.4, UCP.5; A.1, A.2; C.3, C.5, C.6; E.1, E.2; F.6; G.1, G.3 (2 sessions, 1 block)	3. **Describe** the characteristics of segmented worms and their importance to the survival of these organisms. 4. **Compare and contrast** the classes of segmented worms.	**Problem-Solving Lab 27-2**, p. 749 **MiniLab 27-2:** A Different View of an Earthworm, p. 750 **Inside Story:** An Earthworm, p. 751 **Careers in Biology:** Microsurgeon, p. 752 **Design Your Own BioLab:** How do earthworms respond to their environment? p. 754 **Earth Science Connection:** Mollusks as Indicators, p. 756

Need Materials? Contact Carolina Biological Supply Company at 1-800-334-5551 or at **http://www.carolina.com**

MATERIALS LIST

BioLab
p. 754 live earthworms, paper towels, glass pan, sandpaper, culture dishes, thermometer, hand lens or stereomicroscope, dropper, penlight, ice, metric ruler, black paper, cotton swabs

MiniLabs
p. 746 dichotomous key transparency, overhead projector, marine shells
p. 750 cross-section diagrams of earthworm, longitudinal diagrams of earthworm

Alternative Lab
p. 744 land snails, clear plastic deli trays, wax marking pencil, lamp with 60-watt bulb, crushed ice, ring stand, black paper, sandpaper, metric ruler

Quick Demos
p. 742 land snail, petri dish, pencil, lettuce
p. 742 whole squid, knife

Key to Teaching Strategies

L1 Level 1 activities should be appropriate for students with learning difficulties.

L2 Level 2 activities should be within the ability range of all students.

L3 Level 3 activities are designed for above-average students.

ELL ELL activities should be within the ability range of English Language Learners.

COOP LEARN Cooperative Learning activities are designed for small group work.

P These strategies represent student products that can be placed into a best-work portfolio.

These strategies are useful in a block scheduling format.

Mollusks and Segmented Worms

Teacher Classroom Resources

Section	Reproducible Masters	Transparencies
Section 27.1 **Mollusks**	Reinforcement and Study Guide, pp. 119-121 `L2` BioLab and MiniLab Worksheets, p. 121 `L2` Laboratory Manual, pp. 191-198 `L2` Content Mastery, pp. 133-134, 136 `L1`	Section Focus Transparency 67 `L1` `ELL` Basic Concepts Transparency 48 `L2` `ELL` Reteaching Skills Transparency 40 `L1` `ELL`
Section 27.2 **Segmented Worms**	Reinforcement and Study Guide, p. 122 `L2` Concept Mapping, p. 27 `L3` `ELL` Critical Thinking/Problem Solving, p. 27 `L3` BioLab and MiniLab Worksheets, pp. 122-124 `L2` Content Mastery, pp. 133, 135-136 `L1`	Section Focus Transparency 68 `L1` `ELL`

Assessment Resources

Chapter Assessment, pp. 157-162
MindJogger Videoquizzes
Performance Assessment in the Biology Classroom
Alternate Assessment in the Science Classroom
ExamView® Pro Software 💾
BDOL Interactive CD-ROM, Chapter 27 quiz

Additional Resources

Spanish Resources `ELL`
English/Spanish Audiocassettes `ELL`
Cooperative Learning in the Science Classroom `COOP LEARN`
Lesson Plans/Block Scheduling

NATIONAL GEOGRAPHIC — Teacher's Corner

Index to National Geographic Magazine
The following articles may be used for research relating to this chapter:
"Money From the Sea," by Phil Nuytten, January 1993.
"The Pearl," by Fred Ward, August 1985.
"My Chesapeake—Queen of Bays," by Allan C. Fisher, Jr., October 1980.

GLENCOE TECHNOLOGY

The following multimedia resources are available from Glencoe.

Biology: The Dynamics of Life
CD-ROM `ELL`

　Exploration: *Mollusks*
　BioQuest: *Biodiversity Park*
　Exploration: *The Six Kingdoms*

Videodisc Program 📼

　Scallop Escape

The Infinite Voyage

　To the Edge of the Earth

The Secret of Life Series

　Molluscan Body Plan
　Earthworm
　Earthworm Segment

27 Mollusks and Segmented Worms

Chapter 27

GETTING STARTED DEMO

Display several seashells and have students speculate about what kinds of animals might have lived in them. Explain that the animals that lived in the shells are classified as mollusks.

Theme Development

The theme of **unity within diversity** is evident throughout the chapter. When comparing and contrasting these animal groups, similarities are pointed out while the unique characteristics of classes and species are emphasized. The theme of **evolution** is stressed through discussions of the origins of mollusks and the increasing complexity of the specialization of the body plans of mollusks and segmented worms.

0:00 OUT OF TIME?

If time does not permit teaching the entire chapter, use the BioDigest at the end of the unit as an overview.

READING BIOLOGY

Glencoe's *Biology: The Dynamics of Life* contains many resources to assist a student's reading skills. Each chapter contains figures with expanded captions that expand on written material. Word Origins, located along the side of text, expand knowledge of biology vocabulary. Glencoe's Content Mastery Booklet helps develop reading skills while reinforcing content. In addition, use the Interactive Tutor for *Biology: The Dynamics of Life* on the Glencoe Web site to reinforce vocabulary. **science.glencoe.com**

What You'll Learn

- You will distinguish among the classes of mollusks and segmented worms.
- You will compare and contrast the adaptations of mollusks and segmented worms.

Why It's Important

Mollusks are an important food source for many animals, including humans. Some mollusks are filter feeders that clean impurities out of their watery environment. Earthworms turn, aerate, and fertilize the soil in which they live.

READING BIOLOGY

Draw a line down the middle of a sheet of paper. Label one side "Mollusks" and the other "Segmented Worms." As you read the chapter, make a list of the different characteristics for each organism. Compare and contrast the two sets of characteristics.

To find out more about mollusks and segmented worms, visit the Glencoe Science Web site. **science.glencoe.com**

Mucus is an important adaptation for earthworms, as well as for slugs and snails. In addition to allowing earthworms to move through soil, mucus holds two earthworms together as they mate.

740

Multiple Learning Styles

Look for the following logos for strategies that emphasize different learning modalities.

 Kinesthetic Portfolio, pp. 745, 749; Meeting Individual Needs, p. 751

 Visual-Spatial Quick Demo, p. 742; Display, p. 745; Reteach, pp. 747, 753; Visual Learning, p. 752

 Intrapersonal Biology Journal, p. 742

Linguistic Biology Journal, p. 748

Logical-Mathematical Meeting Individual Needs, p. 746

Naturalist Project, p. 746

27.1 Mollusks

*I*f you are a shell collector, a walk on the beach as high tide begins to recede can reveal bountiful treasures. The shell sizes, shapes, and colors are clues to the many different kinds of animals that once inhabited these structures. How could the marine animal that lived in the fan-shaped shell be related to the common garden slug?

Shells (above) and a garden slug (inset)

SECTION PREVIEW

Objectives

Identify the characteristics of mollusks.

Compare the adaptations of gastropod, bivalve, and cephalopod mollusks.

Vocabulary

mantle
radula
open circulatory system
closed circulatory system
nephridia

What Is a Mollusk?

Slugs, snails, and animals that once lived in shells in the ocean or on the beach are all mollusks. These organisms belong to the phylum Mollusca. Members of this phylum range from the slow moving slug to the jet-propelled squid. Although most species live in the ocean, others live in freshwater and moist terrestrial habitats. Some aquatic mollusks, such as oysters and mussels, live firmly attached to the ocean floor or to the bases of docks or wooden boats. Others, such as the octopus, swim freely in the ocean. Land-dwelling slugs and snails can be found crawling slowly over leaves on the forest floor.

Examples of three classes of mollusks are shown in *Figure 27.1*.

WORD *Origin*

mollusk
From the Latin word *molluscus*, meaning "soft." Mollusks are animals with two body openings, a muscular foot, and a mantle.

Figure 27.1
With 100 000 described species, phylum Mollusca is second in size only to insects and their relatives.

B Oysters, clams, and scallops such as this one have two hinged shells.

A Snails, slugs, their shell-less relatives, and other one-shelled animals such as this limpet make up the largest class of mollusks.

C Predatory squids and octopuses are mollusks that do not have an external shell.

Assessment Planner

Portfolio Assessment
 Portfolio, TWE, pp. 745, 749
 Assessment, TWE, pp. 747
Performance Assessment
 Alternative Lab, TWE, p. 744
 MiniLabs, SE, pp. 746, 750
 BioLab, SE, pp. 754-755
 BioLab, TWE, pp. 754-755
 MiniLabs, TWE, pp. 746, 750

Knowledge Assessment
 Assessments, TWE, pp. 744, 747
 Problem-Solving Labs, TWE, pp. 744, 749
 Section Assessments, SE, pp. 747, 753
 Chapter Assessment, SE, pp. 757-759
Skill Assessment
 MiniLabs, TWE, pp. 746, 750
 Assessments, TWE, pp. 752, 753

Section 27.1

Prepare

Key Concepts

Students will study the general characteristics of mollusks and the traits that distinguish organisms in the three mollusk classes.

Planning

■ Gather assorted sea shells for the Getting Started Demo.
■ Order live land snails and purchase squid and live clams for the Quick Demos, Portfolio, and the Alternative Lab.
■ Purchase surgical gloves for the Building a Model.

---

1 Focus

Bellringer

Before presenting the lesson, display **Section Focus Transparency 67** on the overhead projector and have students answer the accompanying questions. **L1** **ELL**

Resource Manager

Section Focus Transparency 67 and Master **L1** **ELL**

Visual-Spatial Divide the class into groups. Give each group a live land snail on one half of a petri dish. Ask students to record their observations of the snail. Instruct students to observe the snail through the underside of the dish. Have them gently touch the antenna of the snail with the eraser end of a pencil and observe and describe its reaction. Finally, have them place the snail on a piece of lettuce to see if they can observe the snail feeding. Discuss all observations as a class.
L2 **ELL** **COOP LEARN**

Obtain a whole squid from a fish market. Point out the head and tentacles of the squid. Cut the squid open to reveal its transparent cuttlebone. Explain that many scientists consider the cuttlebone to be a remnant of a shell. **L1**

GLENCOE TECHNOLOGY

CD-ROM
Biology: The Dynamics of Life
Exploration: *Mollusks*
Disc 4

VIDEODISC
The Secret of Life
Molluscan Body Plan

Figure 27.2
A mollusk has a soft body composed of a foot, a mantle, and a visceral mass that contains internal organs. Some mollusks also have a shell. Compare the structures of a snail and a squid.

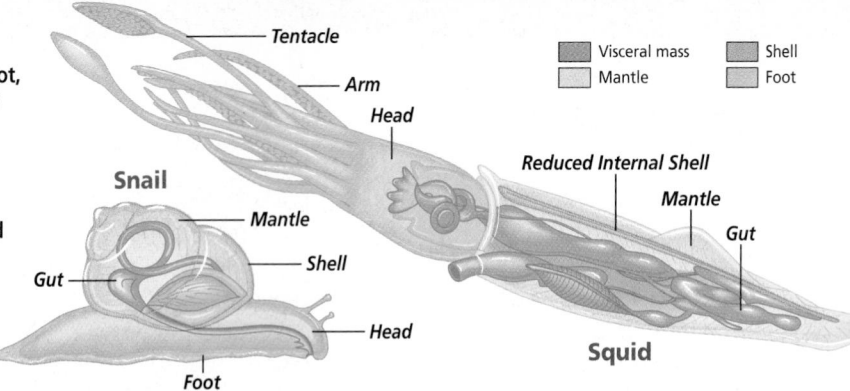

A Snails have a well-defined and developed head area in addition to a large foot.

B The foot area of the squid appears to have been modified into arms and tentacles that are used for capturing and holding prey.

Some mollusks have shells, and others, including slugs and squids, are adapted to life without a hard covering. All mollusks have bilateral symmetry, a coelom, two body openings, a muscular foot for movement, and a mantle. The **mantle** (MANT uhl) is a thin membrane that surrounds the internal organs of the mollusk. In shelled mollusks, the mantle secretes the shell.

Although mollusks look different from one another on the outside, they share many internal similarities. You can see the similarities and the differences in these body areas in *Figure 27.2* as you compare a snail and a squid. How does a clam buried in sand obtain its food? Find out in the *Inside Story* on the next page.

How mollusks obtain food

Have you ever watched a snail clean algae from the sides of an aquarium? Snails, like many mollusks, use a rasping structure called a radula to obtain food. A **radula** (RAJ uh luh), located within the mouth of a mollusk, is a tonguelike organ with rows of teeth. The radula is used to drill, scrape, grate, or cut food. *Figure 27.3* shows the results of the use of a radula. Octopuses and squids use their radulas to pull food they have captured into their mouths. Other mollusks are grazers, some are predators, and some are filter feeders. Bivalves do not have radulas; they obtain food by filtering it out of the water.

Reproduction in mollusks

Most mollusks have separate sexes and reproduce sexually. Eggs and sperm are released at the same time into the surrounding water, where external fertilization takes place. Many gastropods that live on land are hermaphrodites. The ability to produce both eggs and sperm is an adaptation commonly found in slow-moving animals because it increases the likelihood of fertilization.

Figure 27.3
Look at the clam shell in this photo and locate a small hole on its edge. This tiny hole was made by the radula of a mollusk that ate the clam, leaving its shell behind to tell the tale of the clam's fate.

BIOLOGY JOURNAL

Locating Mollusks

Intrapersonal Provide students with a blank outline map of the world. Have them conduct research to find out where five species of mollusks are commonly found. For example, the Atlantic bay scallop is commonly found from North Carolina to the West Indies and Brazil. Ask students to develop a key to indicate these locations on their maps. Have them locate both freshwater and saltwater species. Encourage students to combine their findings with those of two others in the class. If possible, provide students with nature and wildlife atlases to aid in their research. **L3** **ELL**

A Clam

Clams are bivalve mollusks. Bivalves include mussels, scallops, oysters, and other mollusks with two hinged shells. Clams, like oysters, can cover a foreign object, such as a grain of sand or a parasite, that has become lodged between its shell and its mantle with layers of shell that eventually form a pearl.

Critical Thinking *What function do gills have in digestion in a clam?*

Coquina clams

1 Shell The body of a bivalve is held between two hinged shells.

2 Heart A bivalve has a three-chambered heart and an open circulatory system. Its heart beats slowly, from 2 to 30 times per minute. Some mollusks, such as squids, have closed circulatory systems.

3 Nephridia Tube-like structures called nephridia form a clam's excretory system.

8 Mouth and Stomach Food particles pass from the gills, where they have been filtered from the water, to the mouth, and then to the stomach, where digestion begins.

Mantle

Excurrent Siphon

Incurrent Siphon

7 Muscle The two shells are drawn together by two muscles on opposite sides of the clam's body. The shells are kept closed when the animal requires protection. Bivalves can move by rapidly opening and shutting their hinged shells.

6 Foot A clam begins movement by extending a muscular foot from between its shells. When the foot is anchored in the mud or sand, the muscles in the foot shorten, thereby pulling the animal forward.

4 Siphons In a clam, the edges of the mantle form two siphons that let water in and out of the clam. In some clams, the siphons can be extended to the water above the sand where the clam is buried.

5 Gills Gas exchange occurs through the mantle and the gills. The gills also filter food particles from the water.

27.1 MOLLUSKS **743**

Purpose
Students study the functions of the organs of a bivalve mollusk.

Teaching Strategies

■ Ask students to explain the functions of the clam's incurrent and excurrent siphons. **L2**

■ Explain that the clam moves by extending its foot from the opening between the two shells.

Visual Learning

■ Have students examine the captions and make a simple drawing in their journals that shows the path that food takes in a clam. **L1 ELL**

Critical Thinking

Gills filter food particles from the water.

Building a Model

Fill a surgical glove with water. Squeeze the water in one of the fingers and have students observe how the water moves freely into the other parts of the glove. Explain that the glove roughly models an open circulatory system in which blood moves freely into open spaces surrounding organs. **L1**

GLENCOE TECHNOLOGY

VIDEODISC
Biology: The Dynamics of Life
Scallop Escape (Ch. 32)
Disc 1, Side 2, 11 sec.

Cultural Diversity

Pearl Cultivation

In your lessons on mollusk biology, discuss with students how pearls are formed and describe the pearl cultivation industry in Japan. Since 1893, pearl farming has been one of Japan's most famous industries.

To add interest, introduce students to the AMA women of Japan. The AMA is a group of diving women who collect pearls and valuable mollusks from the ocean. Diving women have operated in Japan for more than 2000 years. The divers take their name from the word *ama,* which in the ancient Japanese language meant "ocean" or "sky." The AMA of Japan have been known to dive to depths greater than 50 meters hundreds of times daily without the use of snorkels or air tanks.

Problem-Solving Lab 27-1

Purpose

Students will study the life cycle of larval development in a fresh-water mussel.

Process Skills

think critically, analyze data, interpret scientific drawings

Teaching Strategies

■ Remind students that all mollusks do not follow the pattern of reproduction and development illustrated here.

Thinking Critically

1. sperm
2. fertilization
3. They die.
4. They mature into adult mussels.
5. Although the animal produces many glochidia, most do not find a suitable host and thus do not survive to adult stage.

✔ Assessment

Knowledge Provide students with a sample of glochidia but do not tell them what they are looking at. Have them examine glochidia under the microscope and identify what the organism is. Preserved glochidia are available from biological supply houses. Use the Performance Task Assessment List for Making Observations and Inferences in **PASC**, p. 17. **L2** **ELL**

Problem-Solving Lab 27-1 — Observing and Inferring

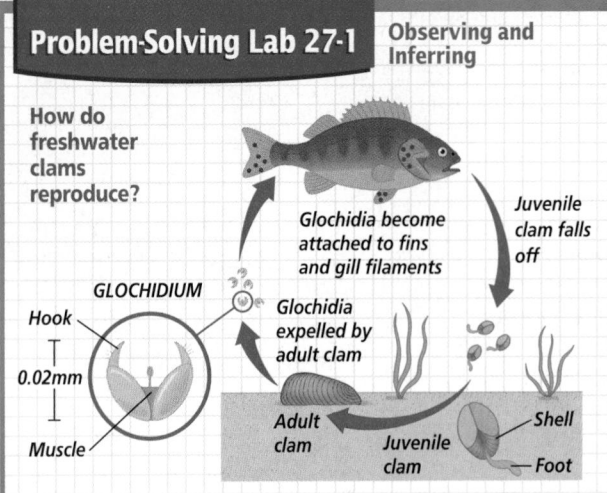

How do freshwater clams reproduce?

GLOCHIDIUM
Hook
0.02mm
Muscle

Glochidia become attached to fins and gill filaments

Juvenile clam falls off

Glochidia expelled by adult clam

Adult clam

Juvenile clam

Shell

Foot

Analysis

Examine the life cycle of the freshwater clam *Anodonta*. Freshwater clams are either male or female. Immature larvae, called glochidia, are formed within female clams' reproductive systems, then released in the surrounding water.

Thinking Critically

1. What cell type must enter a female clam's body in order for glochidia to form?
2. What reproductive process must occur prior to the formation of glochidia?
3. Glochidia attach to and feed off of a specific fish host. Predict what happens to glochidia if no host is available.
4. How do glochidia change while attached to their host?
5. It is estimated that a single clam can release over 1 000 000 glochidia. How might this be an adaptation to a life cycle that includes a parasitic stage?

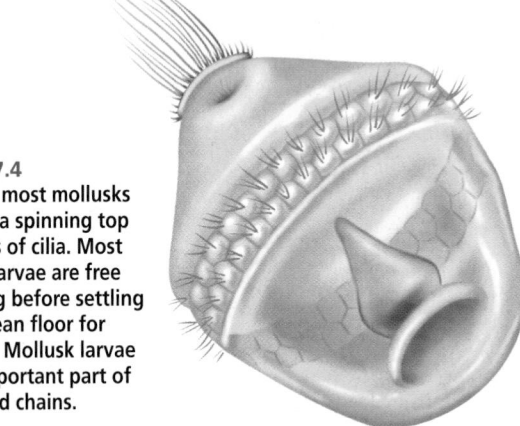

Figure 27.4
Larvae of most mollusks resemble a spinning top with tufts of cilia. Most of these larvae are free swimming before settling to the ocean floor for adult life. Mollusk larvae are an important part of many food chains.

Find out more about reproduction in mollusks by reading the *Problem-Solving Lab* on this page.

Although members of the phylum Mollusca have different appearances as adults, they all share similar developmental patterns. The larval stages of all mollusks are similar, as you can see in *Figure 27.4*.

Some marine mollusks have free-swimming larvae that propel themselves by cilia. In addition to larvae, most marine snails and bivalves have another developmental stage called a veliger in which the beginnings of a foot, shell, and mantle can be seen.

Nervous control in mollusks

Mollusks have simple nervous systems that include a brain and associated nerves that coordinate their movement and behavior. Most mollusks have paired eyes that range from simple cups that detect light to the complex eyes of octopuses that have irises, pupils, and retinas that function as well as those of humans.

Circulation in mollusks

Mollusks have a well-developed circulatory system that includes a three-chambered heart. In most mollusks, the heart pumps blood through an open circulatory system. In an **open circulatory system,** the blood moves through vessels and into open spaces around the body organs. This adaptation exposes body organs directly to blood that contains nutrients and oxygen, and removes metabolic wastes. Some mollusks, such as octopuses, move nutrients and oxygen through a closed circulatory system. In a **closed circulatory system,** blood moves through the body enclosed entirely in a series of blood vessels. A closed system provides an efficient means of gas exchange within the body.

Alternative Lab

Comparing Snail Speeds

Purpose

Students will compare the speed at which snails move under various environmental conditions.

Materials

land snails, clear plastic deli trays or large

deli containers, wax marking pencil, lamp with 60-watt bulb, crushed ice, ring stand, black construction paper, sandpaper, metric ruler

Procedure

Give students the following directions.

1. Make a table for distances traveled by the snail on a smooth surface, a rough surface, in cold conditions, and in warm conditions.
2. Make a hypothesis about the conditions

under which the snail will move the fastest.

3. With the wax marking pencil, mark an X in the middle of your tray. Place the snail on this X and measure how far it travels in 3 minutes.
4. Place a piece of black construction paper over the tray so that the snail is in the dark. Measure distance traveled in 3 minutes.
5. Cover the bottom of the tray with

Figure 27.5
Shelled gastropods vary from petite, thin-shelled species to large animals with thick shells.

A The pink conch is a large gastropod with a thick shell.

B The smooth dove shell is a small, delicate gastropod. These organisms can be found in the Florida Keys and West Indies.

Respiration in mollusks

Most mollusks have respiratory structures called gills. Gills are specialized parts of the mantle that consist of a system of filamentous projections that contain a rich supply of blood for the transport of gases. Gills increase the surface area through which gases can diffuse. In land snails and slugs, the mantle cavity appears to have evolved into a primitive lung.

Excretion in mollusks

Mollusks are the oldest known animals to have evolved excretory structures called nephridia. **Nephridia** (nih FRIHD ee uh) are organs that remove metabolic wastes from an animal's body. Mollusks have one or two nephridia that collect wastes from the coelom. Wastes are discharged into the mantle cavity, and expelled from the body by the pumping of the gills.

Diversity of Mollusks

Within the large phylum of mollusks, there are seven classes. The three classes that include the most common and well-known species are Gastropoda, Bivalvia, and Cephalopoda.

Gastropods: One-shelled mollusks

The largest class of mollusks is Gastropoda, or the stomach-footed mollusks. The name comes from the way the animal's large foot is positioned under the rest of its body. Most species of gastropods have a single shell. Other gastropod species, such as slugs, have no shell.

Shelled gastropods include snails, abalones, conches, periwinkles, whelks, limpets, cowries, and cones. They can be found in freshwater, saltwater, or moist terrestrial habitats. Shelled gastropods may be plant eaters, predators, or parasites. *Figure 27.5* shows two examples of shelled gastropods.

Instead of being protected by a shell, the body of a slug is protected by a thick layer of mucus. Colorful sea slugs, also called nudibranchs, are protected in another way. When certain species of sea slugs feed on jellyfishes, they incorporate the poisonous nematocysts of the jellyfish into their own tissues without causing these cells to discharge. Any fishes trying to eat the sea slugs are repelled when the nematocysts discharge into the unlucky predator. The bright colors of these gastropods warn predators of the potential danger, as shown in *Figure 27.6*.

Figure 27.6
Sea slugs such as this *Chromodoris* species live in the ocean. They eat hydras, sea anemones, and sea squirts.

Purpose

Students will use a dichotomous key to identify mollusks based on their shells.

Process Skills

observe and infer, compare and contrast, classify, use a dichotomous key

Teaching Strategies

■ Make an overhead transparency of a dichotomous key. Use the transparency to demonstrate how a dichotomous key is used.

■ Ask students who collect shells to bring in their shell collections and identify the shells for their classmates.

Expected Results

Students will classify the pictured shells using the dichotomous key provided.

Analysis

1. A dichotomous key divides a group into smaller and smaller groups until each organism is identified.
2. easy: 1, 2, 3, 4; more difficult, 5, because it requires more interpretation and closer comparison
3. one or two shells

✔ Assessment

Skill Give students a simple dichotomous key for several shells. Ask them to use the key to identify the shells. Use the Performance Task Assessment List for Making and Using a Classification System in **PASC,** p. 49. **L1**

Resource Manager

Reteaching Skills Transparency 40 and Master **L1** **ELL**

Biolab and MiniLab Worksheets, p. 122 **L2**

Reinforcement and Study Guide, pp. 119-121 **L2**

Content Mastery, p. 134 **L1**

Laboratory Manual, pp. 191-198 **L2**

MiniLab 27-1   Comparing and Contrasting

Identifying Mollusks Have you ever taken a walk on the beach and filled your pockets with shells, and as you examined them later, wondered what they were? Use the following dichotomous key to determine the names of the shells.

Procedure

1 To use a dichotomous key, begin with a choice from the first pair of descriptions.

2 Follow the instructions for the next choice. Notice that either a scientific name can be found at the end of each description, or directions will tell you to go on to another numbered set of choices.

1A One shell	....	Gastropods see 2
1B Two shells	....	Bivalves see 3
2A Flat coil	.... Sundial shell:	*Architectonica nobilis*
2B Thick coil	....	see 4
3A Shelf inside shell	.... Common Atlantic slipper:	*Crepidula fornicata*
3B No shelf inside shell	....	see 5
4A Spotted surface	.... Junonia shell:	*Scaphella junonia*
4B Lined surface	... Banded tulip shell:	*Fasciolaria hunteria*
5A Polished surface	... Sunray shell:	*Macrocallista mimbosa*
5B Rough surface	..... Lion's paw shell:	*Lyropecten nodosus*

Analysis

1. Why is a dichotomous key used for a variety of organisms?
2. What shell features were easy to pick out using the key? What features were more difficult?
3. What general feature was used to identify shells?

Bivalves: Two-shelled mollusks

Two-shelled mollusks such as clams, oysters, and scallops belong to the class Bivalvia, illustrated in *Figure 27.7*. Most bivalves are marine, but a few species live in freshwater habitats. Bivalves occur in a range of sizes. Some are less than 1 mm in length and others, such as the tropical giant clam, may be 1.5 m long. Bivalves have no distinct head or radula. Most use their large, muscular foot for burrowing in the mud or sand at the bottom of the ocean or a lake. A ligament, like a hinge, connects their shells; strong muscles allow the shell to open and close over the soft body. See if you can identify the shells pictured in the *MiniLab* by using the dichotomous key given.

One of the main differences between gastropods and bivalves is that bivalves are filter feeders that obtain food by filtering small particles from the surrounding water. Bivalve mollusks have several adaptations for filter feeding, including cilia that beat to draw water in through an incurrent siphon. As water moves over the gills, food and sediments become trapped in mucus. Cilia that line the gills push food particles to the stomach. Cilia also act as a sorting device. Large particles, sediment, and anything else that is rejected is transported to the mantle where it is expelled through the excurrent siphon, or to the foot, where it is eliminated from the animal's body.

Figure 27.7 In bivalves the mantle forms two siphons, one for incoming water and one for water that is excreted.

746 MOLLUSKS AND SEGMENTED WORMS

PROJECT

Classifying Mollusks Used as Food

Naturalist Have students photocopy a menu from a local seafood restaurant. Ask students to construct a key to identify the gastropods, bivalves, and cephalopods listed on the menu. Have students summarize the importance of mollusks as a food source. **L2** **ELL**

MEETING INDIVIDUAL NEEDS

Gifted

Logical-Mathematical Ask students to design an experiment to compare the strength of muscles in bivalves such as clams and scallops. Remind them to plan to collect quantitative data. **L3**

Figure 27.8
The class cephalopoda includes squids (**a**) and octopuses (**b**). The genus *Nautilus* is the only remaining living example of a cephalopod with an external shell (**c**). All other members of this class are extinct.

Cephalopods: Head-footed mollusks

The head-footed mollusks are in the class Cephalopoda. All cephalopods are marine organisms. This class includes the octopus, squid, cuttlefish, and chambered nautilus, as shown in *Figure 27.8*. The only cephalopod with a shell is the chambered nautilus, but some species, such as the cuttlefish, have a reduced internal shell. Scientists consider the cephalopods to have the most complex structures and to be the most recently evolved of all mollusks.

In cephalopods, the foot has evolved into tentacles with suckers, hooks, or adhesive structures. Cephalopods swim or walk over the ocean floor in pursuit of their prey, capturing it with their tentacles. Once tentacles have captured prey, it is brought to the mouth and bitten with the beaklike jaws. Then the food is pulled into the mouth by the radula.

Like bivalves, cephalopods have siphons that expel water. These mollusks can expel water forcefully in any direction, and move quickly by jet propulsion. Squids can attain speeds of 20 m per second using this system of movement. You may be aware that cephalopods use jet propulsion to escape from danger. They also can release a dark fluid to cloud the water. This "ink" helps to confuse their predators so they can make a quick escape.

Section Assessment

Understanding Main Ideas
1. Describe how mucus is important to some mollusks.
2. What adaptations make cephalopods effective predators?
3. Compare filter feeding with obtaining food by using a radula.
4. Compare how squids and sea slugs protect themselves.

Thinking Critically
5. How are the methods of movement for the snail, clam, and squid related to the structure of each one's foot?

SKILL REVIEW
6. **Classifying** Construct a key to identify the three classes of mollusks discussed. For more help, refer to *Organizing Information* in the **Skill Handbook**.

27.1 MOLLUSKS **747**

Section Assessment

1. Mucus enables mollusks to stick to surfaces and slide easily through or on materials in their habitats. Some mollusk mucus contains poisons.
2. jet propulsion-type swimming, tentacles with suckers, large eyes with a well-developed nervous system, radula for tearing apart prey
3. A filter feeder takes in water and filters out food. The radula is a tonguelike organ that scrapes food from surfaces.
4. Squids protect themselves by their ability to move quickly away from danger. Sea slugs are protected by their toxic mucous covering.
5. The muscular foot of the snail secretes mucus on which the snail glides slowly. The clam can burrow into sand with its muscular foot. The squid's foot is modified into tentacles that help obtain food.
6. Students' keys should include information found in the chapter under headings dealing with gastropods, bivalves, and cephalopods.

Prepare

Key Concepts

Students will learn the characteristics of segmented worms that enable them to survive in their environments. The classes of segmented worms will be compared and the traits of animals that are more complex than those studied in previous chapters will be emphasized.

Planning

- Set up an earthworm farm for the Quick Demo.
- Gather large jars and large earthworms for the Portfolio.
- Gather live earthworms, glass pans, sandpaper, penlights, culture dishes, droppers, ice, warm tap water, thermometers, and hand lenses or stereomicroscopes for the BioLab.

1 Focus

Bellringer

Before presenting the lesson, display **Section Focus Transparency 68** on the overhead projector and have students answer the accompanying questions.
L1 **ELL**

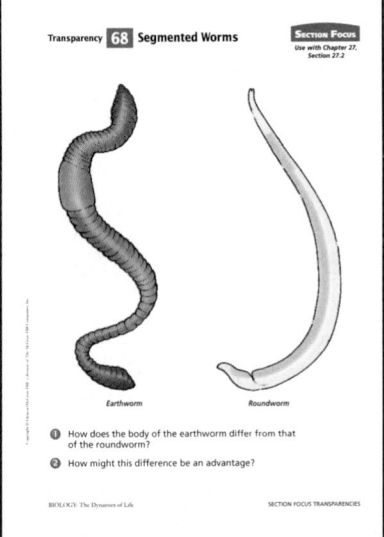

Transparency **68** Segmented Worms

SECTION FOCUS
Use with Chapter 27, Section 27.2

Earthworm   Roundworm

❶ How does the body of the earthworm differ from that of the roundworm?
❷ How might this difference be an advantage?

BIOLOGY: The Dynamics of Life   SECTION FOCUS TRANSPARENCIES

SECTION PREVIEW
Objectives
Describe the characteristics of segmented worms and their importance to the survival of these organisms.
Compare and contrast the classes of segmented worms.
Vocabulary
setae
gizzard

Section
27.2 Segmented Worms

Do earthworms have a front and a back end? Yes, they do. In fact, if you have ever watched one move, you know that it crawls by first stretching the front of its body forward, and then pulling the back of its body up to the front. A worm in motion looks a little like an accordion playing.

Earthworm

WORD *Origin*

annelid
From the Latin word *anellus*, meaning "tiny ring." The bodies of annelids, the segmented worms, look like stacks of tiny rings.

Figure 27.9
The phylum Annelida contains about 12 000 species, which are placed in three classes.

What Is a Segmented Worm?

Segmented worms are classified in the phylum Annelida. They include the earthworms, leeches, and bristleworms, shown in *Figure 27.9*. Segmented worms are bilaterally symmetrical and have a coelom and two body openings. Some have a larval stage that is similar to the larval stages of certain mollusks, suggesting a common ancestor.

The basic body plan of segmented worms is a tube within a tube. The internal tube, suspended within the coelom, is the digestive tract. Food is taken in by the mouth, an opening in the anterior end of the worm, and wastes are released through the anus, an opening at the posterior end.

Most segmented worms have tiny bristles called **setae** (SEE tee) on each segment. The setae help segmented worms move by providing a way to anchor their bodies in the soil so

Ⓐ Earthworms have only a few setae on each segment. An earthworm does not have a distinct head.

Ⓑ Leeches live in marine, freshwater, or terrestrial habitats. All leeches have 32 segments.

Ⓒ Bristleworms have distinct heads, eyes, and tentacles. They are mostly marine animals.

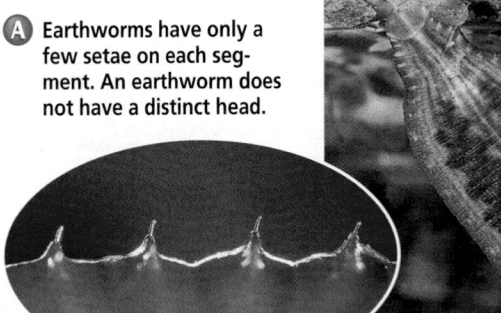

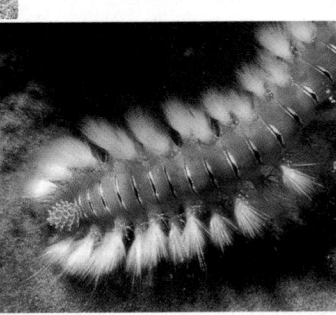

748

BIOLOGY JOURNAL

Earthworm Importance

TECH PREP

Linguistic Ask a group of students to interview a farmer or an agriculture professor about the importance of earthworms in agriculture. Have them present their findings to the class in an illustrated report. Ask other class members to take notes on the presentation. **L2**

GLENCOE TECHNOLOGY

VIDEODISC
The Secret of Life
Earthworm

each segment can move the animal along.

Segmented worms can be found in most environments, except in the frozen soil of the polar regions and the dry sand and soil of the deserts. You may be familiar with the earthworms in your garden, but these are just one of about 12 000 species of segmented worms that live in soil, freshwater, and the sea. Can you identify a segmented worm? Find out by reading the *Problem-Solving Lab* on this page.

Segmentation supports diversified functions

The most distinguishing characteristic of segmented worms is their cylindrical bodies that are divided into a series of ringed segments, as seen in the worms in *Figure 27.10*. This segmentation continues internally as each segment is separated from the others by a body partition. Segmentation is an important adaptation for movement because each segment has its own muscles, allowing shortening and lengthening of the body.

If you examine each segment of most annelids, you find that the body is made up of identical segments. Segmentation, however, also allows for specialization of body tissues. Groups of segments may be adapted for a particular function. Certain segments have modifications for functions such as sensing and reproduction.

Nervous system in segmented worms

Segmented worms have simple nervous systems in which organs in anterior segments have become modified for sensing the environment. Some sensory organs are sensitive to light, and eyes with lenses and retinas have evolved in certain species. In some

species there is a brain located in an anterior segment. Nerve cords connect the brain to nerve centers called ganglia, located in each segment. You can find out how earthworms respond to their environment in the *BioLab* at the end of this chapter.

Figure 27.10
Segmentation is easily seen in earthworms (a). The giant earthworm of Australia can be more than 3 m long (b).

2 Teach

Problem-Solving Lab 27-2

Purpose
Students will have to judge if certain traits are or are not associated with the phylum Annelida.

Process Skills
apply concepts, think critically, compare and contrast, draw a conclusion

Teaching Strategies
■ Be sure that students have read the entire section on annelids before attempting to do this lab.
■ Review any terms that may not be familiar to students.
■ Review the concept of a body cavity called a coelom.

Thinking Critically
A— no, all annelids have internal segmentation
B— no, all annelids have a coelom
C— undecided; traits may apply to other phyla as well as Annelida
D— no, annelids are not vertebrates
E— undecided; traits may apply to other phyla as well as annelids
F— yes, true only of annelids

✔ Assessment
Knowledge Ask students to list traits that are specific to Annelida and no other phyla. Ask students to list traits that are common to other phyla as well as Annelida. Use the Performance Task Assessment List for Making and Using a Classification System in **PASC,** p. 49. **L2**

Purpose

Students will use a diagram of an earthworm's internal anatomy to draw cross-section views.

Process Skills

analyze data, interpret scientific drawings, observe and infer

Teaching Strategies

■ Review the meaning of a cross-section view. Use a cucumber to show a longitudinal view. Then make cuts through the cucumber to illustrate cross-sectional slices.

■ Point out to students that the segments are numbered.

■ Make copies of an outline diagram of incomplete worm cross-sectional views for student use.

Expected Results

Student diagrams will reflect their ability to translate information from a longitudinal view to a cross-sectional view.

Analysis

Segment 8 will show: muscles, esophagus, heart, dorsal and ventral blood vessels, and nerve cord. Segment 12 will show all the parts from Segment 8 plus the seminal vesicle and calciferous gland.

Assessment

Skill Have students make cross-section diagrams of segments 3, 16, and 21. Use the Performance Task Assessment List for Scientific Drawing in **PASC**, p. 55. L2 ELL

Resource Manager

BioLab and MiniLab Worksheets, p. 122 L2

MiniLab 27-2 **Interpreting Scientific Diagrams**

A Different View of an Earthworm What does an earthworm look like internally? You could look at it many different ways—from the dorsal or ventral side, along the length of the animal (a longitudinal view), or in cross section through a segment.

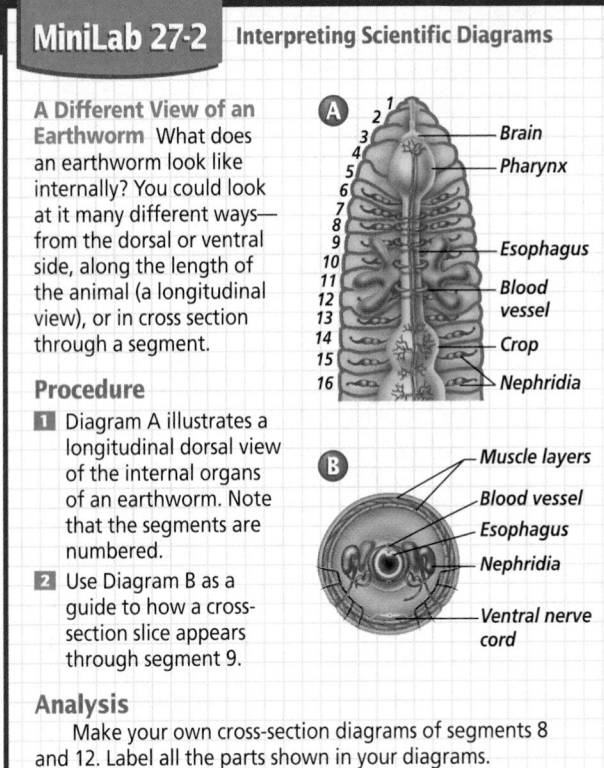

Brain
Pharynx
Esophagus
Blood vessel
Crop
Nephridia

Muscle layers
Blood vessel
Esophagus
Nephridia
Ventral nerve cord

Procedure

1 Diagram A illustrates a longitudinal dorsal view of the internal organs of an earthworm. Note that the segments are numbered.

2 Use Diagram B as a guide to how a cross-section slice appears through segment 9.

Analysis

Make your own cross-section diagrams of segments 8 and 12. Label all the parts shown in your diagrams.

Circulation and respiration

Segmented worms have a closed circulatory system. Blood carrying oxygen to and carbon dioxide from body cells flows through vessels to reach all parts of the body. Segmented worms must live in water or in wet areas on land because they also exchange gases directly through their moist skin.

Digestion and excretion

Segmented worms have a complete internal digestive tract that runs the length of the body. Food taken in by the mouth passes to the **gizzard,** a sac with muscular walls and hard particles that grind soil before the soil passes into the intestine. Undigested material and solid wastes pass out the

---

worm's body through the anus. Segmented worms also have two nephridia in each segment that collect waste products and transport them through the coelom and out of the body. Find out what an earthworm eats by reading the *Inside Story.*

Reproduction in segmented worms

Most segmented worms are hermaphrodites. During mating, two worms exchange sperm. Each worm forms a capsule for the eggs and sperm. The eggs are fertilized in each worm's capsule, then the capsule slips off the worm and is left behind in the soil. In two to three weeks, young worms emerge from the eggs. Earthworms and leeches both reproduce in this way.

However, bristleworms and their relatives have separate sexes and reproduce sexually, although mating occurs in only a few species. Usually eggs and sperm are released into the seawater, where fertilization takes place. Young bristleworms hatch in the sea.

Diversity of Segmented Worms

The phylum Annelida is divided into three classes: class Oligochaeta, earthworms; class Polychaeta, bristleworms and their relatives; and class Hirudinea, leeches.

Earthworms

Earthworms are the most well-known annelids because they can be seen easily by most people. Although earthworms have a definite anterior and posterior section, they do not have a distinct head. Earthworms have only a few setae on each segment. What does an earthworm look like internally? You can find out in the *MiniLab* on this page.

BIOLOGY Online Note Internet addresses that you find useful in the space below for quick reference.

_____

_____

_____

_____

_____

GLENCOE TECHNOLOGY

VIDEODISC
The Infinite Voyage: *To the Edge of the Earth*
Exploring the Galapagos Islands (Ch. 4)

 8 min.

INSIDE STORY

An Earthworm

As an earthworm burrows through soil, it loosens, aerates, and fertilizes the soil. Burrows provide passageways for plant roots and improve drainage of the soil.

Critical Thinking *In what way is segmentation an important advantage in earthworm movement?*

Earthworm

1 **Mouth** An earthworm takes soil into its mouth, the beginning of the digestive tract.

2 **Crop** The crop is a sac that holds soil temporarily before it is passed into the gizzard.

7 **Setae** An earthworm alternately contracts sets of longitudinal and circular muscles to move. First it contracts its longitudinal muscles on several segments, which bunch up. This causes tiny setae to protrude, anchoring the worm in the soil. Then the earthworm's circular muscles contract, the setae are withdrawn, and the worm moves forward.

6 **Nephridia** Nephridia are excretory structures that eliminate metabolic wastes from each segment.

3 **Gizzard** The gizzard grinds the organic matter, or food, into small pieces so that the nutrients in the food can be absorbed as it passes through the intestine. Undigested food and any remaining soil are eliminated through the anus.

4 **Circulatory system** The closed circulatory system consists of enlarged blood vessels that are heavily muscled. When these muscles contract, they help pump blood through the system, much as a heart does in other animals.

5 **Nervous system** An earthworm has a system of nerve fibers in each segment. The nerve fibers are coordinated by a simple brain that lies above the mouth. An earthworm also has a ventral nerve cord.

751

INSIDE STORY

Purpose

Students will examine the internal structures of the earthworm and their functions.

Teaching Strategies

■ Ask students to write a paragraph that explains how earthworm bodies show more complexity than the bodies of flatworms and roundworms. **L2**

■ Obtain a plastic model of an earthworm. Point out each structure discussed in the Inside Story on the model.

■ Challenge your advanced students to make a table that compares earthworms with the free-living flatworms and roundworms. Students should include the following in their tables: digestion, locomotion, circulation, excretion, and sensory functions. **L3**

Visual Learning

■ Make photocopies of the Inside Story diagram without the labels and captions. Have students label and describe the functions of each structure on the photocopy as it is discussed. **L1 ELL**

Critical Thinking

Earthworms have setae and muscles in each segment. They move by anchoring the setae in the ground, then contracting their muscles. The earthworm moves by alternately contracting and relaxing the muscles in each segment.

Resource Manager

Critical Thinking/Problem Solving, p. 27 **L3**
Concept Mapping, p. 27 **L1**
ELL

MEETING INDIVIDUAL NEEDS

Visually Impaired/Learning Disabled

Kinesthetic For students who are visually impaired, provide an earthworm for them to hold while you point out the main features of its structure and behavior. Ask the students to explain how the earthworm's shape and texture make it adapted for life in the soil. **L1 ELL**

GLENCOE TECHNOLOGY

VIDEODISC
The Secret of Life
Earthworm Segment

751

Courses in high school: advanced science and mathematics courses

College: bachelor's degree, medical degree, hospital residency, training in microsurgery

Career Issue

Skill in microsurgery allows surgeons to perform amazing repairs and corrections. Discuss with students whether all surgeons should be required to complete training in microsurgery.

For More Information

For more information about becoming a surgeon or microsurgeon, students might write to:

American College of Surgeons
55 East Erie Street
Chicago, IL 60611

Visual Learning

Visual-Spatial Ask students to compare the fan worm in Figure 27.11 with the earthworm in Figure 27.10. Use the photo of the fan worm to emphasize that not all segmented worms live in terrestrial habitats. **L1**

✓ Assessment

Skill Have students examine Figure 27.12 and create a circle graph that includes all the species of segmented worms and mollusks within the 360° circle. Ask them to indicate the portion of the total number of species each group contains by drawing lines inside the circle. When they have completed the graph, ask students to identify what percentage of the total is represented by gastropods and by cephalopods. Ask them to make a hypothesis about the reason for the relative sizes of these two groups. **L2**

Microsurgeon

Would you like to be able to reattach an accident victim's hand? Then you might consider a career as a microsurgeon.

Skills for the Job

Microsurgeons use high-powered microscopes and three-dimensional computer technology to see and repair tiny nerves and blood vessels. A microsurgeon in ophthalmology might repair a retina, while other microsurgeons remove tumors deep within a brain, or transplant organs. Microsurgeons who reattach hands, feet, and ears often use leeches after surgery to improve blood flow through the reattached body part. Microsurgeons must complete four years of college, four years of medical school, three to five years of a residency program, and special training in microsurgery. They must also pass an examination to become certified.

BIOLOGY Online

For more careers in related fields, be sure to check the Glencoe Science Web site.
science.glencoe.com

Word Origin

parapodia
From the Greek words *para*, meaning "before," and *podion*, meaning "foot." Polychaete worms move using fleshy, paddlelike flaps called parapodia.

Earthworms eat their way through soil. As they eat, they turn the soil and provide spaces for air to flow through soil. As soil passes through the organs of their digestive tract, nutrients are extracted from food and undigested materials pass out of the

Figure 27.11
The fan worm traps food in the mucus on its "fans." Disturbances in the water, such as a change in the direction of the current or the passing by of an organism, cause these worms to quickly withdraw into their tubes.

worm. The wastes of an earthworm are called castings. Castings help fertilize soil.

Bristleworms and their relatives

Bristleworms and their relatives are members of the phylum Polychaeta. Polychaetes are primarily marine organisms. Each segment of a polychaete has many setae, hence the name (polychaete means "many bristles"). This class includes bristleworms, lug worms, plumed worms, sea mice, and fan worms, shown in *Figure 27.11*. Each body segment of a polychaete also has a pair of appendages called parapodia, which can be used for swimming or crawling over corals and the bottom of the sea. Parapodia also function in gas exchange. A polychaete has a head with well-developed sense organs, including eyes. Eyes range from simple eyespots to larger eyes carried on stalks.

Leeches

Leeches are segmented worms with flattened bodies and no setae. Although these animals can be found in many different habitats, most leeches live in freshwater streams or rivers. Unlike earthworms, most species are parasites that suck blood or other body fluids from the bodies of their hosts, which include ducks, turtles, fishes, and people. Front and rear suckers enable leeches to attach themselves to their hosts.

You may cringe at the thought of being bitten by a leech, but the bite is not painful. This is because the saliva of the leech contains chemicals that act as an anesthetic. Other chemicals prevent the blood from clotting. A leech can ingest two to five times its own weight in one meal. Once fed, a leech will drop off its host. It may not eat again for a year.

TECHPREP

Ask students to visit a bait shop and find out where and how the shop gets bristle worms and earthworms, how they care for them, and what kinds of fishes are caught using these worms as bait. Have students go fishing with these worms as bait and report to the class. **L1**

GLENCOE TECHNOLOGY

CD-ROM
Biology: The Dynamics of Life
BioQuest: *Biodiversity Park*
Disc 3, 4

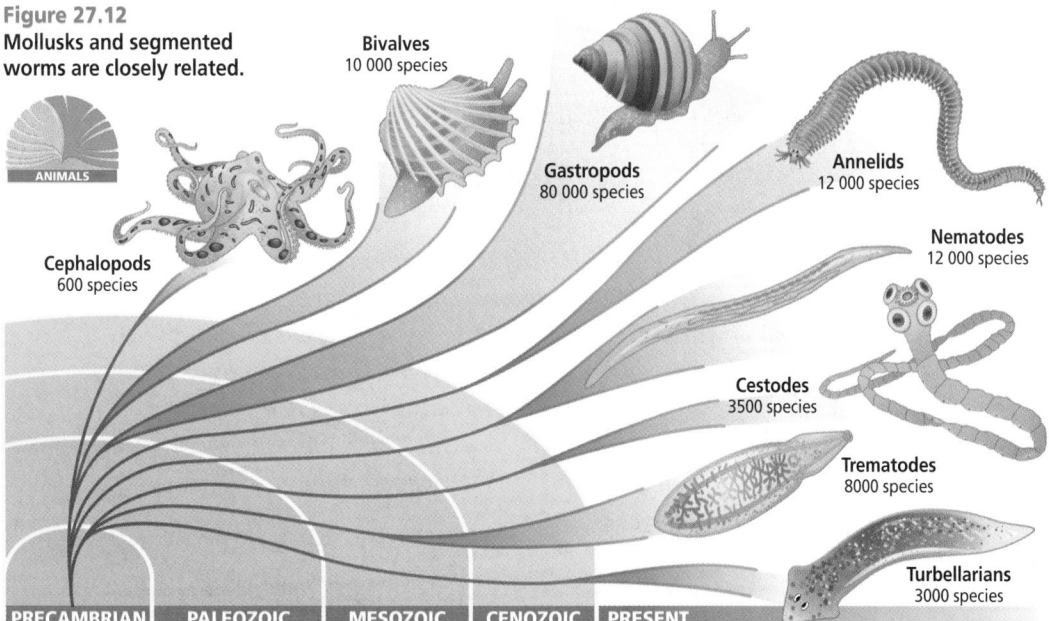

Figure 27.12
Mollusks and segmented worms are closely related.

Bivalves
10 000 species

Gastropods
80 000 species

Annelids
12 000 species

Cephalopods
600 species

Nematodes
12 000 species

Cestodes
3500 species

Trematodes
8000 species

Turbellarians
3000 species

PRECAMBRIAN | PALEOZOIC | MESOZOIC | CENOZOIC | PRESENT

Origins of Mollusks and Segmented Worms

Fossil records show that mollusks lived in great numbers as long as 500 million years ago. Gastropod, bivalve, and cephalopod fossils have been found in early Paleozoic deposits. Some species, such as the chambered nautilus, appear to have changed very little from related species that lived long ago. Find out how fossil mollusks are used to date rocks in the *Earth Science Connection* at the end of this chapter.

Annelids probably evolved in the sea, perhaps from larvae of ancestral flatworms. The fossil record for segmented worms is limited because segmented worms have almost no hard body parts. Tubes constructed by polychaetes are the most common fossils of this phylum. Some of these tubes appear in the fossil record as early as 620 million years ago, as you can see in *Figure 27.12*.

Section Assessment

Understanding Main Ideas
1. What is the most distinguishing characteristic of members of the phylum Annelida? Why is it important?
2. Describe how bristleworms reproduce.
3. How do earthworms improve soil fertility?
4. Why are leeches classified in phylum Annelida?

Thinking Critically
5. Polychaetes actively swim, burrow, and crawl.

How do parapodia support the active life that most polychaetes pursue?

SKILL REVIEW

6. **Interpreting Scientific Illustrations** Using the *Inside Story,* interpret how the two types of muscles in the earthworm are used to move the animal through the soil. For more help, refer to *Thinking Critically* in the **Skill Handbook.**

Section Assessment

1. segmentation; each segment has its own muscles that lengthen and shorten for efficient movement; groups of segments may take on specific functions
2. Bristleworms and their relatives have separate sexes and reproduce sexually. Eggs and sperm are released into the water, where fertilization takes place.
3. As the earthworm burrows through the soil, it loosens, aerates, and fertilizes the soil.
4. Leeches have segmented bodies just like other annelids.
5. Parapodia can be used for swimming or crawling and in gas exchange.
6. Circular muscles contract to move the worm forward. Longitudinal muscles contract to pull the worm's body along.

---

3 Assess

Check for Understanding
Show students cross-section slides of planarians, earthworms, nematodes, tapeworms, and leeches. Ask them to distinguish the segmented worms from the other worms. Have them explain their choices. **L2**

Reteach
Visual-Spatial Ask students to draw a large diagram that shows an earthworm's nervous, circulatory, muscular, digestive, and execretory systems. Have them label each structure and identify the system or systems to which it belongs. **L2** **ELL**

Extension
Ask students to interview a microsurgeon who uses leeches to increase the flow of blood to reattached body parts such as ears, fingers, and toes. Have them write about their interview as if it were going to appear in a magazine. They should ask the microsurgeon for information about the chemicals in leech saliva that dilate blood vessels to increase blood flow. **L3**

✔ Assessment
Skill Have students create a table that compares the characteristics of the three classes of annelids. **L1**

4 Close

Discussion
Discuss with students the characteristics that make annelids more evolutionarily advanced than flatworms or mollusks.

Resource Manager

Basic Concepts Transparency 48 and Master **L1** **ELL**
Reinforcement and Study Guide, p. 122 **L2**
Content Mastery, pp. 133, 135-136 **L1**

DESIGN YOUR OWN
BioLab

How do earthworms respond to their environment?

Time Allotment

One or two class periods

Process Skills

observe and infer, compare and contrast, recognize cause and effect, form a hypothesis, interpret data, design an experiment, separate controls and variables

Safety Precautions

■ Remind students to treat the earthworms in a humane manner at all times.

■ Make sure that students wash their hands both before and after the experiment.

PREPARATION

■ Keep earthworms in the refrigerator overnight, but remove them two hours prior to the lab.

Possible Hypotheses

Students may hypothesize that the worms will move toward the dark, move faster on a rough surface, choose a moist surface over a dry surface, and prefer cool versus warm conditions.

Resource Manager

BioLab and MiniLab Worksheets, p. 123 L2

An earthworm spends its time eating its way through soil, digesting organic matter, and passing inorganic matter through the digestive system and out of its body. Because earthworms are dependent on soil for food and shelter, they respond to stimuli in a way that will ensure a continuous supply of food and a safe place in which to live. These responses are genetically controlled. In this BioLab, you will design an experiment to determine the responses of earthworms to various stimuli.

PREPARATION

Problem

How do earthworms respond to light, different surfaces, moist and dry environments, and warm and cold environments?

Hypotheses

Place your worm in a tray with some moist soil. Watch your worm for about 5 minutes, and record what you observe. Make a hypothesis based on your observations about what the worm might do under conditions of light and dark, rough and smooth surfaces, moist and dry surfaces, and warm and cold conditions. Limit your investigation as time requires.

Objectives

In this BioLab, you will:

■ **Measure** the sensitivity of earthworms to different stimuli, including light, water, and temperature.

■ **Interpret** earthworm responses according to terms of adaptations that promote their survival.

Possible Materials

live earthworms	paper towels
glass pan	sandpaper
culture dishes	warm tap water
thermometer	water
dropper	penlight
ice	ruler
black paper	cotton swabs
hand lens or stereomicroscope	

754

PLAN THE EXPERIMENT

Teaching Strategies

■ To save time, have groups test only one or two variables and share their data.

■ Ask students to gently rub their fingers up and down the length of the ventral surface of the worms to feel their setae.

■ Review the terms *anterior, posterior, dorsal,* and *ventral.* Ask students to use these terms when recording their observations.

Possible Procedures

■ To test which surface enables a worm to move fastest, students may decide to measure how far the worm moves in a given period on surfaces such as sandpaper, the bottom of the dry glass pan, the bottom of a wet glass pan, and on wet and dry paper towels.

■ To test the worm's reaction to light, the

Safety Precautions

Be sure to treat the earthworm in a humane manner at all times. Wet your hands before handling earthworms. Always wear goggles in the lab.

Skill Handbook

Use the **Skill Handbook** if you need additional help with this lab.

PLAN THE EXPERIMENT

1. As a group, make a list of possible ways you might test your hypothesis. Keep the available materials in mind as you plan your procedure.
2. Be sure to design an experiment that will test one variable at a time. Plan to collect quantitative data. Make sure to incorporate a control.
3. Record your procedure and list materials and amounts you will need. Design and construct a data table for recording your findings.

Check the Plan

Discuss the following points with other group members.

1. What data will you collect, and how will they be recorded?
2. Does each test have one variable and a control? What are they?
3. Each test should include measurements of some kind. What are you measuring in each test?
4. How many trials will you run for each test?
5. Assign roles for this investigation.
6. *Make sure your teacher has approved your experimental plan before you proceed further.*
7. Carry out your experiment. **CAUTION:** *Return earthworms to the container the teacher has provided.*

ANALYZE AND CONCLUDE

1. **Checking Your Hypothesis** Which surface did the worm prefer? Explain.
2. **Interpreting Observations** In which temperature was the worm most active? Explain.
3. **Observing and Inferring** How did the earthworm respond to light? Of what survival value is this behavior?
4. **Observing and Inferring** How did the earthworm respond to dry and moist environments? Of what survival value is this behavior?
5. **Drawing Conclusions** Were your hypotheses supported by your data? Why or why not?

Going Further

Project Based on your experiment, design another experiment that would help to answer a question that arose from your work. You might want to try other variables similar to the ones you used, or you might choose to investigate a completely different variable.

BIOLOGY *Online* To find out more about segmented worms, visit the Glencoe Science Web site.

science.glencoe.com

ANALYZE AND CONCLUDE

Student answers may vary.

1. a rough surface; the worm moves more easily on a rough surface
2. an intermediate temperature; an earthworm is ectothermic so its level of activity depends upon the surrounding temperature
3. moved away from light; earthworms are safer from predators in the soil where it is dark
4. preferred a moist environment; earthworm's skin must remain moist or the animal will dry out and die
5. Students who made hypotheses that the worms would prefer moist environments, intermediate temperatures, darkness, and rough surfaces most likely would have their hypotheses supported by data.

✔ Assessment

Performance Ask students to design and then carry out an experiment to determine how earthworms respond to gravity. Use the Performance Task Assessment List for Designing an Experiment in **PASC**, p. 23. **L2**

Going Further

Have students design similar experiments for other invertebrates and compare their results with those obtained for the earthworm. **L3**

---

bottom of a pan may be covered with soil. Part of the pan may be covered with a piece of black construction paper while a penlight is shone on the other side. The amount of time a worm spends in the light and dark sides of the container may then be measured.

■ To determine the worm's preference for heat or cold, the glass pan could be placed on top of two culture dishes—one containing warm tap water and the other containing ice.

Data and Observations

Most likely, earthworms will avoid light and extremes of temperature, move quickly on a rough surface, and prefer a moist surface.

Purpose 📖

Students will learn how mollusks can be used to determine ancient climates and environments as well as radiometric ages.

Teaching Strategies

■ Provide students with an assortment of fossil mollusk shells as well as living examples of these organisms. Allow students to use hand lenses to observe the diversity in these organisms, especially the shells. If actual specimens are not readily available, provide students with color photographs of mollusks. Challenge students to classify the examples based on the relationship between the organisms' shells and soft body parts. Students should be able to identify organisms as belonging to the gastropod, bivalve, or cephalopod groups. **L2** **ELL**

■ Explain, if necessary, the method of absolute dating. Certain radioactive elements decay at a constant rate called a half-life. By measuring the amount of the original element left and the amount of its decay product, the age of a specimen can be determined.

■ Have a volunteer explain the observation made by Darwin. Students should recount that the shells were deposited in a body of water, probably an ocean. Earth processes, including uplift, caused this area of land, which was once below sea level, to be raised thousands of feet above the ocean's surface.

Connection to Biology

Most students should be able to deduce that ammonites are, in fact, excellent index fossils because they are readily preserved as fossils and lived for a geologically relatively short period of time.

Earth Science Connection

Mollusks as Indicators

"Finally, the shells in the Peuquenes or oldest ridge, prove, as before remarked, that it has been upraised 14 000 feet since a Secondary period…"
—Charles Darwin, in *The Voyage of the Beagle*

Although a few species of mollusks live on land, most mollusks are marine or freshwater organisms. How is it, then, that on one of his journeys to South America, Charles Darwin found aquatic mollusk shells thousands of feet above sea level? This observation by the famous naturalist helped to support Darwin's hypothesis that Earth has changed over time.

Mollusks once ruled earth Mollusks first appear in Earth's fossil record more than 500 million years ago. By 30 million years later, these shelled creatures had become the dominant life form on Earth. Thousands of species of mollusks evolved to fill available niches. Yet, numerous species of mollusks became extinct at the close of the Mesozoic era 66 million years ago. Today, the estimated number of mollusk species ranges between 50 000 and 130 000.

The present is the key to the past Because mollusks are generally well preserved in the fossil record, abundant, easy to recognize, and widely distributed geographically, they are excellent index fossils. Index fossils, together with their modern relatives, can be used to hypothesize about ancient climates and environments.

Mollusk shells can also provide information about the biotic, physical, and chemical changes that occur in an ecosystem. Modern mollusks, for example, have been used to determine the source and distribution of various aquatic pollutants.

Mollusks as timekeepers Mollusks can also be thought of as marine timekeepers. A mollusk shell grows only along one edge. The pigmented patterns produced by the animal along this growing edge rarely change. Thus, the pattern produced is not only specific to the species but also is a space and time record of the shell-producing process of that particular organism.

Mollusk shells can also be used to determine an exact age because these structures contain the radioactive element strontium. By measuring the amounts of different isotopes of strontium in the shell, scientists are able to compute the exact age of the organism, and, by extension, the exact age of the rocks containing the shell.

CONNECTION TO BIOLOGY

The fossil record shows that various species of ammonites lived from about 230 million years ago to about 66 million years ago. Ammonites are now extinct. Do you think these mollusks are good index fossils? Explain your answer.

BIOLOGY Online To find out more about mollusks and other index fossils, visit the Glencoe Science Web site. **science.glencoe.com**

Fossilized mollusk shells

BIOLOGY Online Note Internet addresses that you find useful in the space below for quick reference.

_____

_____

_____

_____

GLENCOE TECHNOLOGY

CD-ROM
Biology: The Dynamics of Life
Exploration: *The Six Kingdoms*
Disc 3

Chapter 27 Assessment

SUMMARY

Section 27.1

Mollusks

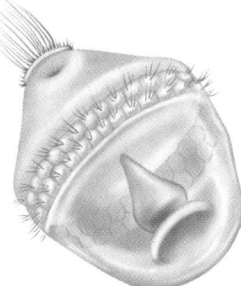

Main Ideas

- Mollusks have bilateral symmetry, a coelom, and two body openings. Many also have shells and similar larvae.

- Most gastropods have a shell, mantle, radula, open circulatory system, gills, and nephridia. Gastropods without shells are protected by a covering of mucus.

- Bivalve mollusks have two shells and are filter feeders. They have no radula.

- Cephalopods have tentacles with suckers, a beaklike mouth with a radula, and a closed circulatory system. They include the octopus, squid, and chambered nautilus.

Vocabulary

closed circulatory system (p. 744)
mantle (p. 742)
nephridia (p. 745)
open circulatory system (p. 744)
radula (p. 742)

Section 27.2

Segmented Worms

Main Ideas

- The phylum Annelida includes the earthworms, bristleworms and their relatives, and leeches. They are bilaterally symmetrical and have a coelom and two body openings; some have larvae that look like the larvae of mollusks. Their bodies are cylindrical and segmented.

- Earthworms have complex digestive, excretory, muscular, and circulatory systems.

- Bristleworms and their relatives are mostly marine species. They have many setae and parapodia that are used for crawling along.

- Leeches are flattened, segmented worms. Most are aquatic parasites.

- Fossil remains of mollusks show that they first lived 500 million years ago. Fossil records show that segmented worms first appeared 620 million years ago.

Vocabulary

gizzard (p. 750)
setae (p. 748)

Main Ideas

Summary statements can be used by students to review the major concepts of the chapter.

Using the Vocabulary

To reinforce chapter vocabulary, use the Content Mastery Booklet and the activities in the Interactive Tutor for Biology: The Dynamics of Life on the Glencoe Science Web site:
science.glencoe.com

THE PRINCETON REVIEW

All Chapter Assessment questions and answers have been validated for accuracy and suitability by The Princeton Review.

UNDERSTANDING MAIN IDEAS

1. b
2. c
3. c

UNDERSTANDING MAIN IDEAS

1. When an earthworm passes soil through its digestive tract, the soil does NOT go through the _____.
   a. stomach
   b. nephridia
   c. gizzard
   d. crop

2. Which of the following does NOT use a radula for feeding?
   a. snail
   b. slug
   c. oyster
   d. squid

3. Which of the following animals have setae?
   a. snails
   b. clams
   c. earthworms
   d. squids

GLENCOE TECHNOLOGY

 **VIDEOTAPE**
MindJogger Videoquizzes
Chapter 27: *Mollusks and Segmented Worms*
Have students work in groups as they play the videoquiz game to review key chapter concepts.

 ## Resource Manager

Chapter Assessment, pp. 157-162
MindJogger Videoquizzes
ExamView® Pro Software
BDOL Interactive CD-ROM, Chapter 27 quiz

4. c
5. b
6. a
7. d
8. c
9. c
10. c
11. open spaces
12. closed
13. external shell
14. larvae
15. segmented worms
16. Nephridia
17. mollusks
18. sperm
19. leeches
20. anesthetic

APPLYING MAIN IDEAS

21. Cnidarian tentacles contain stinging cells that immobilize prey, and octopus tentacles have suckers for capturing prey.
22. Sponges and bivalves are filter feeders that strain food from water currents.
23. Snails scrape algae from surfaces with their radulas.
24. Gastropods can withdraw into their shells. Bivalves can close up their shells. Cephalopods, such as squids, can swim very fast, and the octopus can use its tentacles for defense.

4. The _____ is a thin membrane that surrounds the internal organs of a mollusk.
   a. foot
   b. shell
   c. mantle
   d. siphon

5. Oysters, clams, and scallops are _____.
   a. gastropods
   b. bivalves
   c. cephalopods
   d. nematodes

6. Snails, slugs, and limpets are _____.
   a. gastropods
   b. cephalopods
   c. bivalves
   d. cestodes

7. A _____ is the tonguelike organ that assists gastropods to obtain food.
   a. foot
   b. shell
   c. siphon
   d. radula

8. Which of the following word pairs are most closely related?
   a. filter feeding—radula
   b. scraping algae—siphon
   c. predation—tentacle
   d. nephridia—gizzard

9. Which of the following is a gastropod?

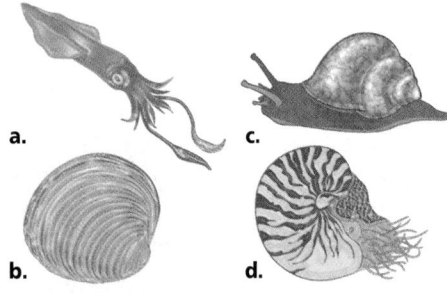

10. Animals with bilateral symmetry, a coelom, two body openings, a muscular foot, and a mantle are _____.
    a. segmented worms
    b. flatworms
    c. mollusks
    d. roundworms

THE PRINCETON REVIEW   **TEST-TAKING TIP**

What does the test expect of me?
Find out what concepts, objectives, or standards are being tested beforehand and keep those concepts in mind as you solve the questions. Stick to what the test is trying to test.

11. In an open circulatory system, the blood moves through vessels and into _____ around the body organs, whereas in a closed circulatory system, the blood remains in vessels.

12. The cephalopods circulate blood in a(n) _____ circulatory system.

13. This example of a cephalopod is the only animal of that group with a(n) _____.

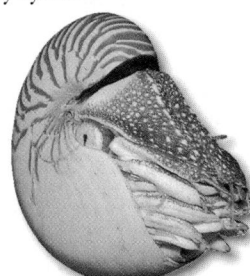

14. Segmented worms and mollusks both have bilateral symmetry, a coelom, and similar _____.

15. Animals distinguished by cylindrical bodies and ringed segments are _____.

16. _____ are excretory structures that remove wastes from an earthworm's body.

17. Annelids are probably most closely related to _____ because they have similar larvae.

18. During mating, two earthworms exchange _____.

19. Segmented worms with flattened bodies, suckers, and no bristles are called _____.

20. The saliva of leeches contains chemicals that act as a(n) _____.

APPLYING MAIN IDEAS

21. Compare how a cnidarian and an octopus use their tentacles to capture food.

22. Describe how sponges and bivalves have a similar way of obtaining food.

23. Why is it a good idea to keep a snail in an aquarium?

24. Compare the protective adaptations of gastropods and cephalopods.

25. Compare nephridia in mollusks and segmented worms.

THINKING CRITICALLY

26. **Observing and Inferring** Explain why the phylogeny of worms is not as well understood as the phylogeny of mollusks.

27. **Recognizing Cause and Effect** Explain how bivalves in salt marshes are important for the health of the other species that live there.

28. **Observing and Inferring** Suppose there are so many *Anodonta* clams in a stream that the fish population is reduced. How could you control the clam population without harming the fish?

29. **Recognizing Cause and Effect** Bivalves called scallops can escape from predators by clapping their shells together and forcibly expelling water. What structures in a bivalve allow scallops to behave in this manner?

30. **Concept Mapping** Complete the concept map by using the following vocabulary terms: mantle, closed circulatory system, open circulatory system.

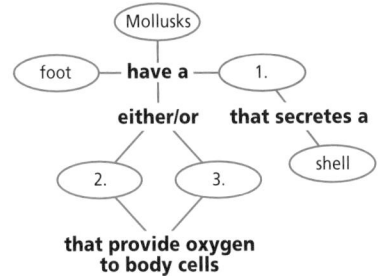

CD-ROM

For additional review, use the assessment options for this chapter found on the *Biology: The Dynamics of Life Interactive CD-ROM* and on the Glencoe Science Web site. science.glencoe.com

ASSESSING KNOWLEDGE & SKILLS

The graph below shows how a number of different animals respond to light.

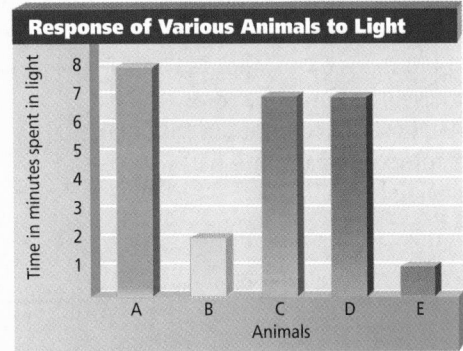

Response of Various Animals to Light

Using a Graph Study the graph and answer the following questions.

1. Which animals spend more time in the light?
   a. A, C, D   c. A, B, C, D, E
   b. B, E   d. C, D

2. Which animals do not spend as much time in the light?
   a. A, C, D   c. A, B, C, D, E
   b. B, E   d. C, D

3. Which animals might be nocturnal?
   a. A, C, D   c. A, B, C, D, E
   b. B, E   d. C, D

4. Which animals might live under a rock?
   a. A, C, D   c. A, B, C, D, E
   b. B, E   d. C, D

5. To which animal group might an earthworm belong?
   a. A   c. C
   b. B   d. D

6. **Making a Graph** Make a graph of the following data. Animal A spends 20 minutes in the dark. Animal B spends 15 minutes in the dark. Animal C spends two minutes in the dark. Animal D spends seven minutes in the dark.

25. Nephridia remove metabolic wastes from an animal's body. In segmented worms there are nephridia in each segment. Mollusks usually have one or two nephridia.

THINKING CRITICALLY

26. Soft-bodied worms do not fossilize as readily as mollusks, which have hard shells.

27. Bivalves filter organic matter from the water and break it down to smaller units that can be used by marsh grasses and algae.

28. Answers will vary. Students may suggest introducing a clam predator or a chemical to kill glochidia.

29. abductor muscles, incurrent and excurrent siphons

30. 1. Mantle, 2. Open circulatory system, 3. Closed circulatory system

ASSESSING KNOWLEDGE & SKILLS

1. a
2. b
3. b
4. b
5. b
6. **Response of Various Animals to Dark**

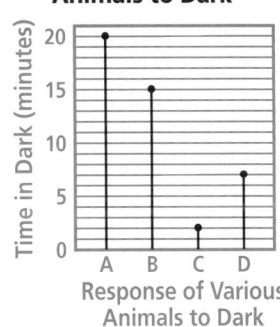

Response of Various Animals to Dark

Chapter 28 Organizer

Refer to pages 4T-5T of the Teacher Guide for an explanation of the National Science Education Standards correlations.

Section	Objectives	Activities/Features
Section 28.1 **Characteristics of Arthropods** National Science Education Standards UCP.1-5; A.1, A.2; C.3, C.5, C.6 (1 session, $\frac{1}{2}$ block)	1. **Relate** the structural and behavioral adaptations of arthropods to their ability to live in different habitats. 2. **Analyze** the adaptations that make arthropods an evolutionarily successful phylum.	**MiniLab 28-1:** Crayfish Characteristics, p. 763 **Problem-Solving Lab 28-1,** p. 766
Section 28.2 **Diversity of Arthropods** National Science Education Standards UCP.1-5; A.1, A.2; C.3, C.4, C.5, C.6; E.1, E.2; F.1, F.4, F.5, F.6; G.1 (2 sessions, 1 block)	3. **Compare and contrast** the similarities and differences among the major groups of arthropods. 4. **Explain** the adaptations of insects that contribute to their success.	**Inside Story:** A Spider, p. 769 **Inside Story:** A Grasshopper, p. 772 **MiniLab 28-2:** Comparing Patterns of Metamorphosis, p. 774 **Focus On** Insects, p. 776 **Design Your Own BioLab:** Will salt concentration affect brine shrimp hatching? p. 780 **Health Connection:** Terrible Ticks, p. 782

Need Materials? Contact Carolina Biological Supply Company at 1-800-334-5551 or at **http://www.carolina.com**

MATERIALS LIST

BioLab
p. 780 clear plastic trays, brine shrimp eggs, uniodized salt, balance, water, graduated cylinder, beakers, labels

MiniLabs
p. 763 preserved crayfish, forceps, pencil, paper
p. 774 life stage specimens of grasshopper, life stage specimens of moth, forceps, pencil, paper

Alternative Lab
p. 762 bess beatles, cloth towel, transparent tape, heavy thread, balance, pennies, plastic petri dish

Quick Demos
p. 762 arthropod specimens
p. 762 arthropod specimens, hand lenses
p. 768 crayfish, lobster, crab, spider
p. 773 raw meat, fly eggs, 2-L soda bottle
p. 773 butterfly chrysalis, terrarium

Key to Teaching Strategies

L1 Level 1 activities should be appropriate for students with learning difficulties.

L2 Level 2 activities should be within the ability range of all students.

L3 Level 3 activities are designed for above-average students.

ELL ELL activities should be within the ability range of English Language Learners.

COOP LEARN Cooperative Learning activities are designed for small group work.

P These strategies represent student products that can be placed into a best-work portfolio.

These strategies are useful in a block scheduling format.

Arthropods

Teacher Classroom Resources

Section	Reproducible Masters	Transparencies
Section 28.1 **Characteristics of Arthropods**	Reinforcement and Study Guide, pp. 123-124 L2 BioLab and MiniLab Worksheets, p. 125 L2 Content Mastery, pp. 137-138, 140 L1	Section Focus Transparency 69 L1 ELL Basic Concepts Transparency 49 L2 ELL Reteaching Skills Transparency 41 L1 ELL
Section 28.2 **Diversity of Arthropods**	Reinforcement and Study Guide, pp. 125-126 L2 Concept Mapping, p. 28 L3 ELL Critical Thinking/Problem Solving, p. 28 L3 BioLab and MiniLab Worksheets, pp. 126-128 L2 Laboratory Manual, pp. 199-204 L2 Content Mastery, pp. 137, 139-140 L1 Inside Story Poster ELL Tech Prep Applications, pp. 33-36 L2	Section Focus Transparency 70 L1 ELL Reteaching Skills Transparency 41 L1 ELL Reteaching Skills Transparency 42 L1 ELL

Assessment Resources

Chapter Assessment, pp. 163-168
MindJogger Videoquizzes
Performance Assessment in the Biology Classroom
Alternate Assessment in the Science Classroom
ExamView® Pro Software
BDOL Interactive CD-ROM, Chapter 28 quiz

Additional Resources

Spanish Resources L1 ELL
English/Spanish Audiocassettes L1 ELL
Cooperative Learning in the Science Classroom COOP LEARN
Lesson Plans/Block Scheduling

NATIONAL GEOGRAPHIC — Teacher's Corner

Products Available From Glencoe
To order the following products, call Glencoe at 1-800-334-7344:
CD-ROM
NGS PictureShow: Structure of Invertebrates
Transparency Set
NGS PicturePack: Structure of Invertebrates

Index to National Geographic Magazine
The following articles may be used for research relating to this chapter:
"The Changeless Horsheshoe Crab," by Anne Rudloe, Jr., April 1981.

GLENCOE TECHNOLOGY

The following multimedia resources are available from Glencoe.

Biology: The Dynamics of Life
CD-ROM ELL

Video: *Molting Crab*
Exploration: *Arthropods*
BioQuest: *Biodiversity Park*
Video: *Arthropods*
Exploration: *Classifying Beetles*
Video: *Web-Spinning Spider*
Video: *Gradual Metamorphosis*
Video: *Complete Metamorphosis*

Videodisc Program

Arthropods
Web-Spinning Spider
Gradual Metamorphosis
Complete Metamorphosis

Theme Development

The theme of **evolution** is stressed as the huge diversity of adaptations that arthropods have evolved is discussed. The theme of **homeostasis** is brought out through discussions of the organs that enable arthropods to maintain homeostasis with their environment.

0:00 OUT OF TIME?

If time does not permit teaching the entire chapter, use the BioDigest at the end of the unit as an overview.

READING BIOLOGY

Glencoe's *Biology: The Dynamics of Life* contains many resources to assist a student's reading skills. Each chapter contains figures with expanded captions that expand on written material. Word Origins, located along the side of text, expand knowledge of biology vocabulary. Glencoe's Content Mastery Booklet helps develop reading skills while reinforcing content. In addition, use the Interactive Tutor for *Biology: The Dynamics of Life* on the Glencoe Web site to reinforce vocabulary.
science.glencoe.com

Chapter
28 Arthropods

What You'll Learn

- You will distinguish among the adaptations that have made arthropods the most abundant and diverse animal phylum on Earth.
- You will compare and contrast different classes of arthropods.

Why It's Important

Arthropods are adapted to fill many important niches in every ecosystem in the world. Because arthropods occupy so many niches, they have an impact on all living things, including humans.

READING BIOLOGY

Skim through the new vocabulary terms in the section previews, noting any unfamiliar words. As you read, make an outline of the chapter. Try to include the definition of each new vocabulary word in the outline.

BIOLOGY Online

To find out more about arthropods, visit the Glencoe Science Web site.
science.glencoe.com

There are about 1 million known species of arthropods. How can we explain the enormous diversity of arthropods—a group that includes both spiders and lobsters?

760 ARTHROPODS

Multiple Learning Styles

Look for the following logos for strategies that emphasize different learning modalities.

Kinesthetic Meeting Individual Needs, p. 765; Tech Prep, p. 771; Extension, p. 775; Enrichment, p. 777

Visual-Spatial Quick Demo, pp. 762, 773; Reteach, p. 765; Display, p. 770; Portfolio, p. 773; Project, p. 773

Interpersonal Activity, p. 775; Project, p. 776

Intrapersonal Meeting Individual Needs, p. 770

Linguistic Portfolio, p. 764; Enrichment, pp. 768, 773; Biology Journal, pp. 768, 769, 777, 778

Naturalist Biology Journal, p. 764; Check for Understanding, p. 775

28.1 Characteristics of Arthropods

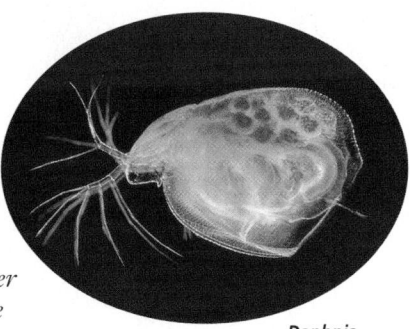

Daphnia

Two out of every three animals living on Earth today are arthropods. You can find arthropods deep in the ocean and on high mountaintops. They live in polar regions and in the tropics. Arthropods are adapted to living in air, on land, and in freshwater and saltwater environments. Arthropods range in size from the 0.3-mm-long spider mite to the giant Japanese spider crab, which measures 4 m across. This water flea, Daphnia, *lives in freshwater lakes and filters microscopic food from the water with its bristly legs.*

SECTION PREVIEW

Objectives

Relate the structural and behavioral adaptations of arthropods to their ability to live in different habitats.

Analyze the adaptations that make arthropods an evolutionarily successful phylum.

Vocabulary

appendage
molting
cephalothorax
tracheal tube
spiracle
book lung
pheromone
simple eye
compound eye
mandible
Malpighian tubule
parthenogenesis

What Is an Arthropod?

Arthropods pollinate many of the flowering plants on Earth. Some arthropods spread plant and animal diseases. Despite the enormous diversity of arthropods, they all share some common characteristics.

A typical arthropod is a segmented, coelomate invertebrate animal with bilateral symmetry, an exoskeleton, and jointed structures called appendages. An **appendage** (uh PEN dihj) is any structure, such as a leg or an antenna, that grows out of the body of an animal. In arthropods, appendages are adapted for a variety of purposes including sensing, walking, feeding, and mating. *Figure 28.1* shows some of these adaptations.

Figure 28.1
The development of jointed appendages was a major evolutionary step that led to the success of the arthropods.

A The powerful jointed legs of this crab are adapted for walking.

B Spiders hold their prey with jointed mouthparts while feeding.

C The antennae of a moth are adapted for the senses of touch and smell.

761

Section 28.1

Prepare

Key Concepts

Characteristics common to all arthropods are presented along with their specific adaptations to land, air, and water.

Planning

■ Obtain pennies, heavy thread, and plastic petri dishes for the Alternative Lab.

■ Purchase plastic arthropods for the Meeting Individual Needs.

---

1 Focus

Bellringer

Before presenting the lesson, display **Section Focus Transparency 69** on the overhead projector and have students answer the accompanying questions.
L1 ELL

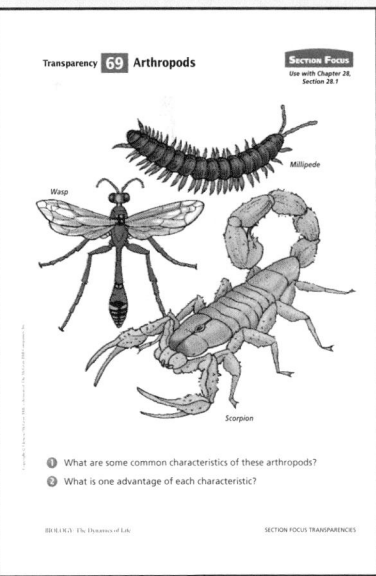

Resource Manager

Section Focus Transparency 69 and Master L1 ELL

2 Teach

Figure 28.2
Arthropods molt several times during their development. The old exoskeleton is discarded after a new one is formed underneath.

Arthropods are the earliest known invertebrates to exhibit jointed appendages. Joints are advantageous because they allow more flexibility in animals that have hard, rigid exoskeletons. Joints also allow powerful movements of appendages, and enable an appendage to be used in many different ways. For example, the second pair of appendages in spiders is used for sensing and for mating. In scorpions, this pair of appendages is used for seizing prey.

Arthropod exoskeletons provide protection

The success of arthropods as a group can be attributed in part to the presence of an exoskeleton. The exoskeleton is a hard, thick, outer covering made of protein and chitin (KITE un). Chitin is also found in the cell walls of fungi and in many other animals. In some species, the exoskeleton is a continuous covering over most of the body. In other species, the exoskeleton is made of separate plates held together by hinges. The exoskeleton protects and supports internal tissues and provides places for attachment of muscles. In many species that live on land, the exoskeleton is covered by a waxy layer that provides additional protection against water loss. In many aquatic species, the exoskeletons also contain calcium.

Why arthropods must molt

Exoskeletons are an important adaptation for arthropods, but they also have their disadvantages. First, they are relatively heavy structures. Many terrestrial and flying arthropods are adapted to their habitats by having a thinner, lighter-weight exoskeleton, which offers less protection but allows the animal more freedom to fly and jump.

More importantly, though, exoskeletons cannot grow, so arthropods must shed them periodically. Shedding the old exoskeleton is called **molting.** Before an arthropod molts, a new, soft exoskeleton is formed from chitin-secreting cells beneath the old one. When molting occurs, the animal contracts muscles in the rear part of its body, forcing blood forward. The forward part of the body swells, causing the old exoskeleton to split open, as *Figure 28.2* shows. The animal then climbs out of its old exoskeleton. Before the new exoskeleton hardens, the animal swallows air or water to puff itself up in size. Thus, the new exoskeleton hardens in a larger size, allowing some room for the animal to continue to grow.

Most arthropods molt four to seven times in their lives, and during these periods, they are particularly vulnerable to predators. When the new exoskeleton is soft, arthropods cannot protect themselves or escape from danger because they move by bracing muscles against the rigid exoskeleton. Therefore, many species hide or remain motionless for a few hours or days until the new exoskeleton hardens.

Alternative Lab

Beetle Strength

Purpose
Students will observe and compare the pulling power of a beetle and a human.

Materials
bess beetles, cloth toweling (30 cm²), clear tape, heavy thread (30 cm long), balance, pennies, smooth tabletop, plastic petri dish

Procedure
Give students the following directions.
1. Obtain the mass of the petri dish and the mass of the penny.
2. Place a beetle on its back in the petri dish and obtain the mass of the beetle and the dish. Calculate the mass of the beetle.
3. Make a slipknot loop on one end of the thread and put the loop over the head and body of the beetle so that it acts as a harness. Tape the ends of the thread inside the rim of the petri dish. Make a hypothesis about how many pennies the beetle will be able to pull on the petri dish sled.
4. Secure the cloth to the tabletop with tape. When the beetle begins to pull or move the sled by walking, slowly add pennies to the petri dish, one at a time, until you find the maximum mass the

Segmentation in arthropods

Most arthropods are segmented, but they do not have as many segments as you have seen in segmented worms. In most groups of arthropods, segments have become fused into three body sections—head, thorax, and abdomen. In other groups even these segments may be fused. Some arthropods show a head and a fused thorax and abdomen. In other groups, there is an abdomen and a fused head and thorax called a **cephalothorax** (sef uh luh THOR aks), as shown in *Figure 28.3*.

Fusion of the body segments is related to movement and protection. Species such as beetles that have separate head and thorax regions are more flexible than those with fused regions. Many species such as shrimps and lobsters have a cephalothorax, which protects the animal but which limits movement. Take a closer look at the fused body segments of an arthropod called a crayfish in the *MiniLab* on this page.

Figure 28.3
You can see the different body segments in these arthropods.

A A stag beetle shows fusion of body segments into a distinct head, thorax, and abdomen.

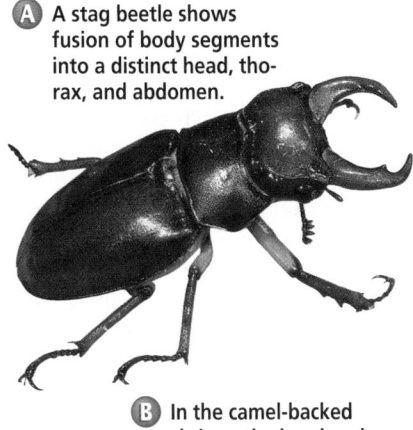

B In the camel-backed shrimp, the head and thorax are fused into a cephalothorax. The animal also has an abdomen.

MiniLab 28-1 Comparing and Contrasting

Crayfish Characteristics
There are more species of arthropods than all of the other animal species combined. This phylum includes a variety of adaptations that are not found in other animal phyla.

Blue Crayfish

Procedure
1. Examine a preserved crayfish. **CAUTION:** *Wear disposable latex gloves and use a forceps when handling preserved material.*
2. Prepare a data table with the following arthropod traits listed: body segmentation, jointed appendages, exoskeleton, sense organs, jaws.
3. Observe the crayfish. Fill in your data table, indicating which of the arthropod traits you observed.
4. Gently lift the edge of the body covering where the legs attach to the body. Look for feathery structures. These are gills and are part of the animal's respiratory system. **CAUTION:** *Wash hands with soap and water after handling preserved materials.*

Analysis
1. Do crayfish have all of the traits listed above?
2. Make a hypothesis as to how crayfish locate food.

Figure 28.4
Arthropods have a wide variety of respiratory structures.

B Tracheal tubes are inside the body, thereby reducing water loss through the respiratory surface while carrying air close to each cell.

A Gills, with their large surface area, enable a large amount of blood-rich tissue to be exposed to water containing oxygen.

C Book lungs are folded membranes that increase the surface area of blood-rich tissue exposed to air.

Arthropods have efficient gas exchange

Arthropods are generally quick, active animals. They crawl, run, climb, dig, swim, and fly. In fact, some flies beat their wings 1000 times per second. As you would expect, arthropods have efficient respiratory structures that ensure rapid oxygen delivery to cells. This large oxygen demand is needed to sustain the high levels of metabolism required for rapid movements.

Three types of respiratory structures for taking oxygen into their bodies have evolved in arthropods: gills, tracheal tubes, and book lungs. In some arthropods, air diffuses right through the body wall. Aquatic arthropods exchange gases through gills, which extract oxygen from water and release carbon dioxide into the water. Land arthropods have either a system of tracheal tubes or book lungs. Most insects have **tracheal tubes** (TRAY kee ul), branching networks of hollow air passages that carry air throughout the body. Muscle activity helps pump the air through the

tracheal tubes. Air enters and leaves the tracheal tubes through openings on the thorax and abdomen called **spiracles** (SPIHR ih kulz).

Most spiders and their relatives have **book lungs,** air-filled chambers that contain leaflike plates. The stacked plates of a book lung are arranged like pages of a book. All three types of respiration in arthropods are illustrated in *Figure 28.4*.

Arthropods have acute senses

Quick movements that are the result of strong muscular contractions enable arthropods to respond to a variety of stimuli. Movement, sound, and chemicals can be detected with great sensitivity by antennae, stalklike structures that detect changes in the environment.

Antennae are also used for communication among animals. Have you ever watched as a group of ants carried home a small piece of food? The ants were able to work together as a group because they were communicating with each other by **pheromones** (FER uh mohnz), chemical odor signals given off by animals. Antennae sense the odors of pheromones, which signal animals to engage in a variety of behaviors. Some pheromones are used as scent trails, such as in the group-feeding behavior of ants, and many are important in the mating behavior of arthropods.

Accurate vision is also important to the active lives of arthropods. Most arthropods have one pair of large compound eyes and from three to eight simple eyes. A **simple eye** is a visual structure with only one lens that is used for detecting light. A **compound eye** is a visual structure with many lenses. Each lens registers light from a tiny portion of the field of view. The total image that is formed is made up of thousands of parts. The multiple lenses of a flying arthropod, such as the dragonfly shown in *Figure 28.5,* enable it to analyze a fast-changing landscape during flight. Compound eyes can detect the movements of prey, mates, or predators, and can also detect colors.

Arthropod nervous systems are well developed

Arthropods have well-developed nervous systems that process infor-mation coming in from the sense organs. The nervous system consists of a double ventral nerve cord, an anterior brain, and several ganglia. Arthropods have ganglia that have become fused. These ganglia act as control centers for the body section in which they are located.

Arthropods have other complex body systems

Arthropod blood is pumped by one or more hearts in an open circulatory system with vessels that carry blood away from the heart. The blood flows out of the vessels, bathes the tissues of the body, and returns to the heart through open body spaces.

Arthropods have a complete diges-tive system with a mouth, stomach, intestine, and anus, together with various glands that produce digestive enzymes. The mouthparts of most arthropod groups include a variety of jaws called **mandibles** (MAND uh bulz). Mouthparts are adapted for holding, chewing, sucking, or biting the various foods eaten by arthro-pods, illustrated in *Figure 28.6.*

Most terrestrial arthropods excrete wastes through **Malpighian tubules** (mal PIGH ee un). In arthropods, the

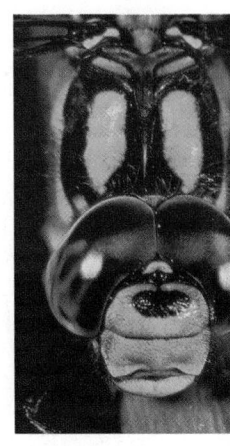

Figure 28.5
The compound eyes of this dragonfly cover most of its head and consist of about 30 000 lenses. However, the images formed by compound eyes are unclear.

Figure 28.6
Mouthparts of arthropods exhibit tremendous variation among species.

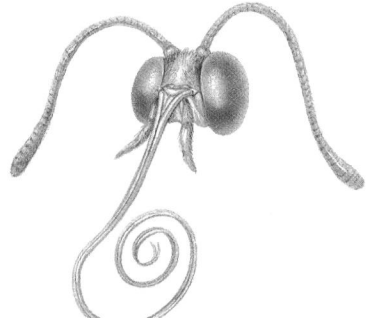

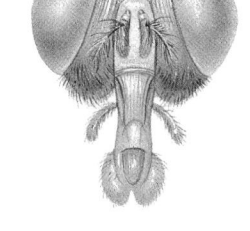

(A) Sand flies and other insects that feed by drawing blood have piercing blades or needle-like mouthparts.

(B) The rolled-up sucking tube of moths and butterflies can reach nectar at the bases of long, tubular flowers.

(C) The sponging tongue of the housefly has an opening between its two lobes through which food is lapped.

28.1   CHARACTERISTICS OF ARTHROPODS   **765**

Problem-Solving Lab 28-1 Using Numbers

How many are there? There are a lot of arthropod species on Earth. How do arthropods compare with other animals?

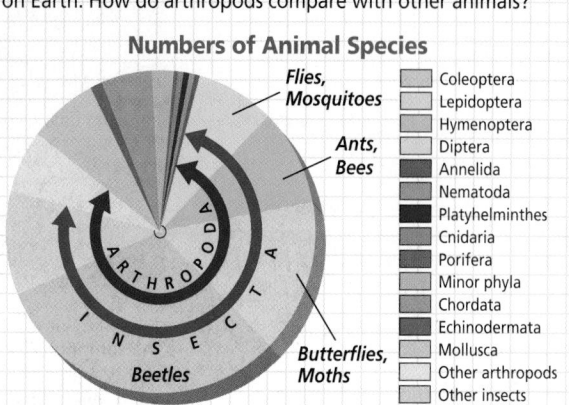

Numbers of Animal Species

	Coleoptera
	Lepidoptera
	Hymenoptera
	Diptera
	Annelida
	Nematoda
	Platyhelminthes
	Cnidaria
	Porifera
	Minor phyla
	Chordata
	Echinodermata
	Mollusca
	Other arthropods
	Other insects

Analysis

Look over the circle graph. Determine the number of species in each phylum or class by noting that each degree on the circle represents about 3000 species. Note: You will need a protractor.

Thinking Critically

1. About how many species of arthropods are known? What percentage of all animal species are arthropods?
2. Which class of arthropods makes up the larger category? How many species are in this class? What percentage of all arthropods is in this class? What percentage of all animal species is in this class?
3. Which order (Diptera, Hymenoptera) makes up the largest category? How many species are in this order? What percentage of all arthropods is in this order?
4. Formulate a hypothesis that explains why there are so many arthropod species.

---

tubules are all located in the abdomen rather than in each segment, as you have seen in segmented worms. Malpighian tubules are attached to and empty into the intestine.

Another well-developed system in arthropods is the muscular system. In a human limb, muscles are attached to the outer surfaces of internal bones. In an arthropod limb, the muscles are attached to the inner surface of the exoskeleton. An arthropod muscle is attached to the exoskeleton on both sides of the joint.

Arthropods reproduce sexually

Most arthropod species have separate males and females and reproduce sexually. Fertilization is usually internal in land species but is often external in aquatic species. A few species, such as barnacles, are hermaphrodites, animals with both male and female reproductive organs. Some species, including bees, ants, and wasps, exhibit **parthenogenesis** (par thuh noh JEN uh sus), a form of asexual reproduction in which a new individual develops from an unfertilized egg.

There are more arthropod species than all other animal species combined. Find out how many species of arthropods there are by reading the *Problem-Solving Lab* on this page.

Section Assessment

Understanding Main Ideas

1. Describe the pathway taken by the blood as it circulates through an arthropod's body.
2. Describe two features that are unique to arthropods.
3. What are the advantages and disadvantages of an exoskeleton?
4. Compare how spiders and crabs protect themselves.

Thinking Critically

5. What characteristics of arthropods might explain why they are the most successful animals in terms of population sizes and numbers of species?

SKILL REVIEW

6. **Comparing and Contrasting** Compare the adaptations for gas exchange in aquatic and land arthropods. For more help, refer to *Thinking Critically* in the **Skill Handbook.**

Section Assessment

28.2 Diversity of Arthropods

Female mosquitoes drink an average of 2.5 times their body weight in blood every day. Other arthropods feed on nectar, dead organic matter, oil, and just about every other substance you can imagine. The varied eating habits of arthropods reflect their huge diversity. The phylum Arthropoda includes these classes: Arachnida, spiders and their relatives; Crustacea, crabs and their relatives; Chilopoda, centipedes; Diplopoda, millipedes; Merostomata, horseshoe crabs; and Insecta, insects.

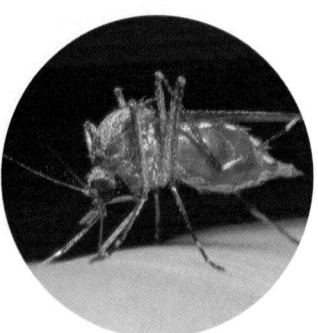

Female mosquito,
Aedes stimulans

SECTION PREVIEW

Objectives
Compare and contrast the similarities and differences among the major groups of arthropods.

Explain the adaptations of insects that contribute to their success.

Vocabulary
chelicerae
pedipalp
spinneret
metamorphosis
larva
pupa
nymph

Prepare

Key Concepts
Students will study the diversity and the structural and behavioral adaptations of arthropods. Spiders, ticks, mites, scorpions, crustaceans, centipedes, millipedes, and insects will all be examined.

Planning
- For Quick Demos and Meeting Individual Needs, obtain a lobster, raw meat, and 2L soft drink bottles.
- Purchase mealworms for an Assessment.
- For the BioLab, obtain clear plastic trays, brine shrimp eggs, noniodized salt, and plastic bottles.

1 Focus

Bellringer
Before presenting the lesson, display **Section Focus Transparency 70** on the overhead projector and have students answer the accompanying questions.
L1 **ELL**

Arachnids

Do you remember the last time you saw a spider? Did you draw back with a quick, fearful breath, or did you move a little closer, curious to see what it would do next? Of the 30 000 species of spiders, only about a dozen are dangerous to humans. In North America, you need to watch out for only the two species illustrated in *Figure 28.7*—the black widow and the brown recluse.

What is an arachnid?

Spiders, scorpions, mites, and ticks belong to the class Arachnida (uh RAK nu duh). Spiders are the largest group of arachnids. Spiders and other arachnids have only two

WORD *Origin*

arachnid
From the Greek word *arachne*, meaning "spider." Spiders, and their relatives the scorpions, ticks, and mites, are arachnids.

a

b

Figure 28.7
The black widow spider is shiny black with a red, hourglass-shaped spot on the underside of the abdomen (a). The brown recluse is brown to yellow and has a violin-shaped mark on its body (b). A bite from either spider will require prompt medical treatment.

28.2 DIVERSITY OF ARTHROPODS **767**

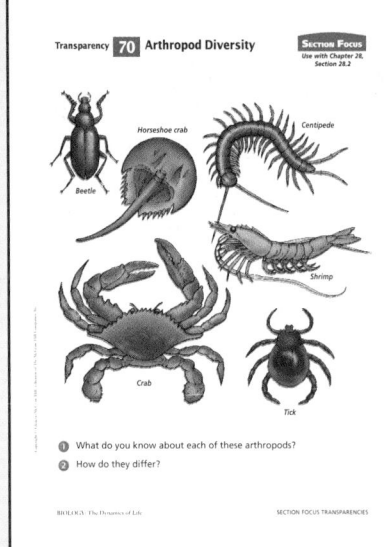

Transparency **70** Arthropod Diversity

GLENCOE TECHNOLOGY

 VIDEODISC
Biology: The Dynamics of Life
Arthropods (Ch. 35)
Disc 1, Side 2, 52 sec.

Resource Manager

Critical Thinking/Problem Solving, p. 28 **L3**
Section Focus Transparency 65 and Master **L1** **ELL**

2 Teach

Discussion

Ask students to relate their experiences with spiders. Remind them that spiders are predators and are able to inject poison that can cause itching and swelling.

Enrichment

🧠 *Linguistic* Ask groups of students to report on the current medical research on venom from spiders. **L3**

Misconception

Many students think that pill bugs, also known as wood lice, are insects. Point out that pill bugs have all the characteristics of crustaceans even though they live on land. Students may also think that pill bugs and sow bugs are the same organism. Although they have a similar appearance and habitat, pill bugs and sow bugs are different species.

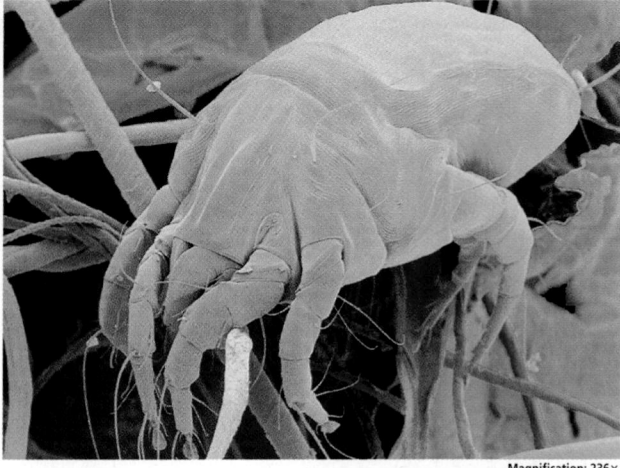
Magnification: 236×

Figure 28.8
Mites are distributed throughout the world and found in just about every habitat. House-dust mites feed on discarded skin cells that collect in dust on floors, in bedding, and on clothing. Some people are allergic to mite waste products.

body regions—the cephalothorax and the abdomen. Arachnids have six pairs of jointed appendages. The first pair of appendages, called **chelicerae** (kih LIHS uh ree), is located near the mouth. Chelicerae are often modified into pincers or fangs. Pincers are used to hold food, and fangs inject prey with poison. Spiders have no mandibles for chewing. Using a process of extracellular digestion, digestive enzymes from the spider's mouth liquefy the internal organs of the captured prey. The spider then sucks up the liquefied food.

The second pair of appendages, called the **pedipalps** (PED ih palpz), are adapted for handling food and for sensing. In male spiders, pedipalps are further modified to carry sperm during reproduction. The four remaining appendages in arachnids are modified as legs for locomotion. Arachnids have no antennae.

Most people know spiders for their ability to make elaborate webs. Although all spiders spin silk, not all make webs. Spider silk is secreted by silk glands in the abdomen. As silk is secreted, it is spun into thread by structures called **spinnerets,** located at the rear of the spider. How well

can a spider see? Find out by reading the *Inside Story* on the next page.

Ticks, mites, and scorpions: Spider relatives

Spiders are not the only arthropods classified as arachnids. Ticks, mites, and scorpions are arachnids, too. Ticks and mites differ from spiders in that they have only one body section, as shown in *Figure 28.8*. The head, thorax, and abdomen are completely fused. Ticks feed on blood from reptiles, birds, and mammals. They are small but capable of expanding up to 1 cm or more after a blood meal. Ticks also can spread diseases. You can find out more about ticks and disease in the *Health Connection* at the end of this chapter.

Mites are so small that they often are not visible to the naked human eye. However, you can certainly feel the bite of mites called chiggers if they get under your clothing while you are camping.

Scorpions are easily recognized by their many abdominal body segments and enlarged pincers. They have a long tail with a venomous stinger at the tip. Scorpions live in warm, dry climates and eat insects and spiders. They use the poison in their stingers to paralyze large prey organisms.

Crustaceans

Most crustaceans (krus TAY shuns) are aquatic and exchange gases as water flows over feathery gills. Crustaceans are the only arthropods that have two pairs of antennae for sensing. All crustaceans have mandibles for crushing food, and two compound eyes, which are usually located on movable stalks. Unlike the up-and-down movement of your jaws, crustacean mandibles open and close from side to side.

A Spider

The garden spider weaves an intricate and beautiful web, dribbles sticky glue on the spiraling silk threads, and waits for insects to crash into them. Spiders are predatory animals, feeding almost exclusively on other arthropods. Many spiders build unique webs, which are effective in trapping flying insects.

Critical Thinking *Explain how certain structures in spiders enable them to be effective predators.*

Wandering spider with cocoon

1 Simple eyes Spiders have six or eight simple eyes that, in most species, detect light but do not form images. Spiders have no compound eyes.

2 Book lungs Gas exchange in spiders takes place in book lungs.

3 Cocoon Female spiders wrap their eggs in a silken sac or cocoon, where the eggs remain until they hatch. Some spiders lay their eggs and never see their young. Others carry the sac around with them until the eggs hatch.

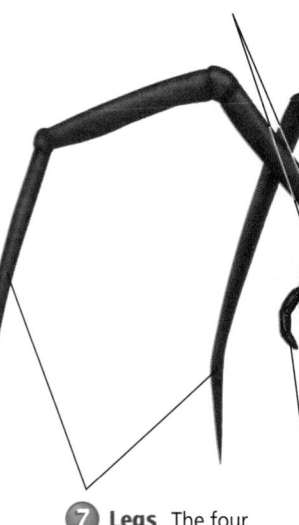

7 Legs The four pairs of walking legs are located on the cephalothorax of the spider.

6 Pedipalps A pair of pedipalps is used to hold and move food and also to function as sense organs. In males, pedipalps are bulbous and are used to carry sperm.

5 Chelicerae Chelicerae are the two biting appendages of arachnids. In spiders, they are modified into fangs. Poison glands are located near the tips of the fangs.

4 Silk glands Spiders have between two and six silk glands, which first release silk as a liquid. The silk then passes through as many as 100 small tubes before being spun into thread by the spinnerets.

BIOLOGY JOURNAL

Charlotte's Web

Linguistic Many students have read the book *Charlotte's Web* by E. B. White. Ask students to recall this story and list in their journals the scientific and factual information in the book. Have them make a second list of information that is not factual. Ask students to explain how this book portrays spiders. **L2**

Resource Manager

Reteaching Skills Transparency 41 and Master **L1** **ELL**

Purpose

Students learn about the structural and behavioral adaptations of spiders.

Teaching Strategies

■ Explain to students that not all spiders spin webs. The tarantula is a fierce predator that stalks its prey and does not build webs. Such predators usually have thicker legs than web-building spiders.

■ Have students create a table with these column heads: Head and Cephalothorax. Beneath each head, have them list the structures of a spider that connect to that body part. **L2**

Visual Learning

■ As you point out each structure on the spider, have students use the caption to identify its function.

Critical Thinking

Fangs with poison glands enable spiders to paralyze and subdue their prey. Spiders build webs that snare prey.

GLENCOE TECHNOLOGY

CD-ROM
Biology: The Dynamics of Life
Video: *Web-Spinning Spider*
Disc 4

VIDEODISC
Biology: The Dynamics of Life
Web-Spinning Spider (Ch. 36)
Disc 1, Side 2, 36 sec.

Figure 28.9
Adult barnacles are somewhat distinct in structure from other arthropods, but they have jointed limbs. Barnacles are filter feeders that trap food by extending feathery legs out of their shells.

Many crustaceans have five pairs of walking legs that are used for walking, for seizing prey, and for cleaning other appendages. The first pair of walking legs are often modified into strong claws for defense.

Crabs, lobsters, shrimps, crayfishes, barnacles, water fleas, and pill bugs are members of the class Crustacea, *Figure 28.9*. Some crustaceans have three body sections, and others have only two. Sow bugs and pill bugs, the only land crustaceans, must live where there is moisture, which aids in gas exchange. They are frequently found in damp areas around building foundations. You can observe crustaceans in the *BioLab* at the end of this chapter.

Centipedes and Millipedes

Centipedes, which belong to the class Chilopoda, and millipedes, members of the class Diplopoda, are shown in *Figure 28.10*. If you have ever turned over a rock on a damp forest floor, you may have seen the flattened bodies of centipedes wriggling along on their many tiny, jointed legs. Centipedes are carnivorous and eat soil arthropods, snails, slugs, and worms. The bite of a centipede is painful to humans. Like spiders, millipedes and centipedes have Malpighian tubules for excreting wastes. In contrast to spiders, centipedes and millipedes have tracheal tubes rather than book lungs for gas exchange.

A millipede eats mostly plants and dead material on damp forest floors. Millipedes do not bite, but they can spray obnoxious-smelling fluids from their defensive stink glands. You may have seen their cylindrical bodies walking with a slow, graceful motion.

Figure 28.10
A centipede may have from 15 to 181 body segments—always an odd number (**a**). A millipede may have more than 100 segments in its long abdomen, each with two spiracles and two pairs of legs (**b**).

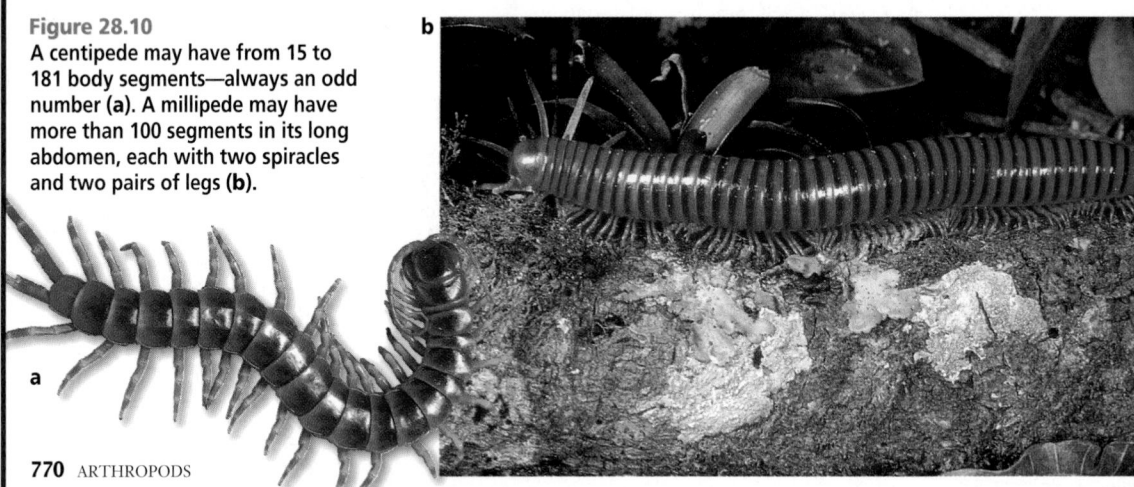

Horseshoe Crabs: Living Fossils

Horseshoe crabs, members of the class Merostomata, are considered to be living fossils because they have remained relatively unchanged since the Cambrian period, about 500 million years ago. They are similar to trilobites in that they are heavily protected by an extensive exoskeleton. Shown in *Figure 28.11*, horseshoe crabs forage on sandy or muddy ocean bottoms for seaweed, worms, and mollusks. These arthropods migrate to shallow water during mating season, and the females lay their eggs on land, buried in sand above the high water mark. Newly hatched horseshoe crabs look like trilobites.

Insects

Have you ever launched an ambush on a fly with your rolled-up newspaper? You swat with great accuracy and speed, yet your prey is now firmly attached upside down on the kitchen ceiling. How does a fly do this?

The fly approaches the ceiling right-side up at a steep angle. Just before impact, it reaches up with its front legs. The forelegs grip the ceiling with tiny claws and sticky hairs, while the other legs swing up into position. The flight mechanism shuts off, and the fly is safely out of swatting distance. Adaptations that enable flies to land on ceilings are among the many that make insects the most successful arthropod group. How is the ability to fly an adaptive advantage to insects? Find out by reading the *Inside Story* on the next page.

Flies, grasshoppers, lice, butterflies, bees, and beetles are just a few members of the class Insecta, by far the largest group of arthropods. There are more species of insects than all other classes of animals combined. You can find out more about insects in the *Focus On Insects* at the end of this section.

Insect reproduction

Insects mate once, or at most only a few times, during their lifetimes. The eggs are fertilized internally, and, in some species, shells form around them. Most insects lay a large number of eggs, which increases the chances that some offspring will survive long enough to reproduce. Many female insects are equipped with an appendage that is modified for piercing through the surface of the ground or into wood. The female lays eggs in the hole.

Metamorphosis: Change in body shape and form

After eggs are laid, the insect embryo develops and the eggs hatch. In some wingless insects, such as springtails and silverfish, development is direct; the eggs hatch into miniature forms that look just like tiny adults. These insects go through

Figure 28.11
Horseshoe crabs have a semicircular exoskeleton and a long, pointed tail. They have four pairs of walking legs, and five or six pairs of appendages that move water over their gills.

Cultural Diversity

Charles Henry Turner

Have students report on the important contributions of African-American biologist Charles Henry Turner (1867–1923) to our modern understanding of insect behavior. Turner's research included many species of insects such as ants, bees, and cockroaches, and he often developed unique and interesting experimental techniques to study them. Turner was a very prolific scientist; between 1892 and 1923, he published 49 articles in the leading scientific journals of his time. **L2**

Purpose

Students will learn about the structural adaptations of grass-hoppers.

Teaching Strategies

■ Provide students with a live grasshopper in a large, clear container. Ask students to observe and describe its behavior. Have them examine the grasshopper under a stereoscopic microscope and make a labeled sketch of its external body parts. Students should relate each structure to how the grasshopper is adapted to its environment. **L2**

■ Caution students that they are working with live animals and to treat them gently.

Visual Learning

■ Make an overhead trans-parency of the grasshopper with the labels covered. Ask students to label the lines pointing to grasshopper or-gans. Ask them to identify how each organ contributes to sur-vival of the grasshopper. **L1** **ELL**

Critical Thinking

Grasshoppers have a complex nervous system consisting of a brain, a double ventral nerve cord, and several ganglia that act as control centers for other body sections.

Resource Manager

Reteaching Skills Trans-parency 41 and Master **L1** **ELL**

772

A Grasshopper

Grasshoppers make rasping sounds either by rubbing their wings together or by rubbing small projections on their legs across a scraper on their wings. Most calls are made by males. Some aggressive calls are made when other males are close. Other calls attract females, and still others serve as an alarm to warn nearby grasshoppers of a predator in the area.

Green grasshopper nymph

Critical Thinking *Do grasshoppers have a well-developed nervous system?*

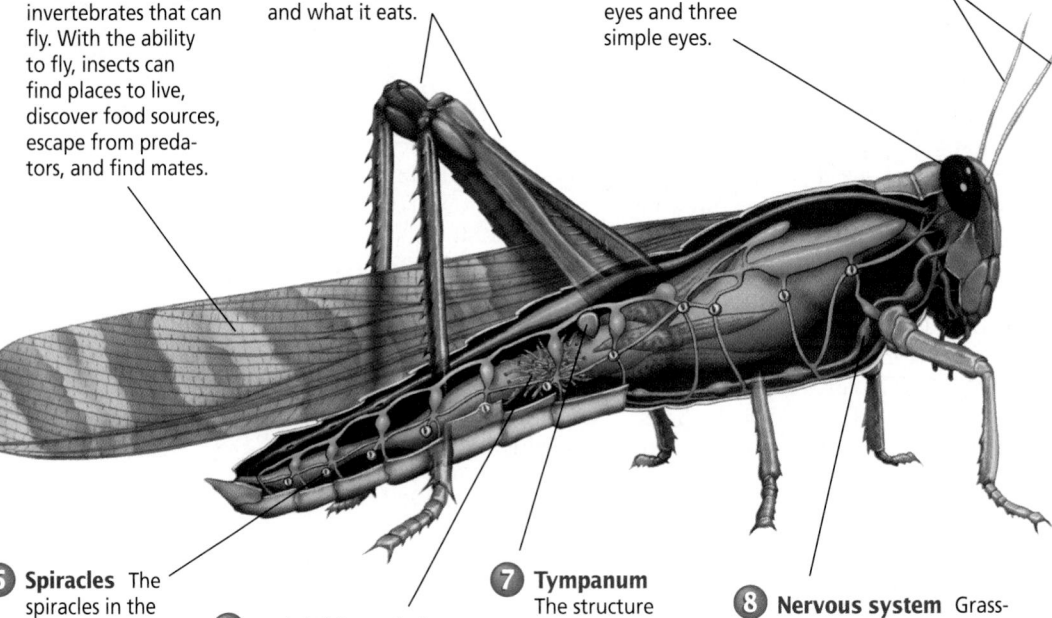

1 Ability to fly Insects are the only invertebrates that can fly. With the ability to fly, insects can find places to live, discover food sources, escape from preda-tors, and find mates.

2 Legs Insects have six legs. By looking at an insect's legs, you can sometimes tell how it moves about and what it eats.

3 Eyes Grass-hoppers have two compound eyes and three simple eyes.

4 Antennae Insects have one pair of antennae, which is used to sense vibrations, food, and pheromones in the environment.

5 Spiracles The spiracles in the abdomen open into a series of tracheal tubes.

6 Malpighian tubules Excretion takes place in Malpighian tubules. In the grasshopper, wastes are in the form of dry crystals of uric acid. Producing dry waste helps insects conserve water.

7 Tympanum The structure insects use for hearing is a flat membrane called a tympanum.

8 Nervous system Grass-hoppers, like other insects, have a complex nervous system that includes sev-eral ganglia that act as nerve control centers for the body sections in which they are located.

772 ARTHROPODS

Note Internet addresses that you find useful in the space below for quick reference.

_____

_____

_____

_____

_____

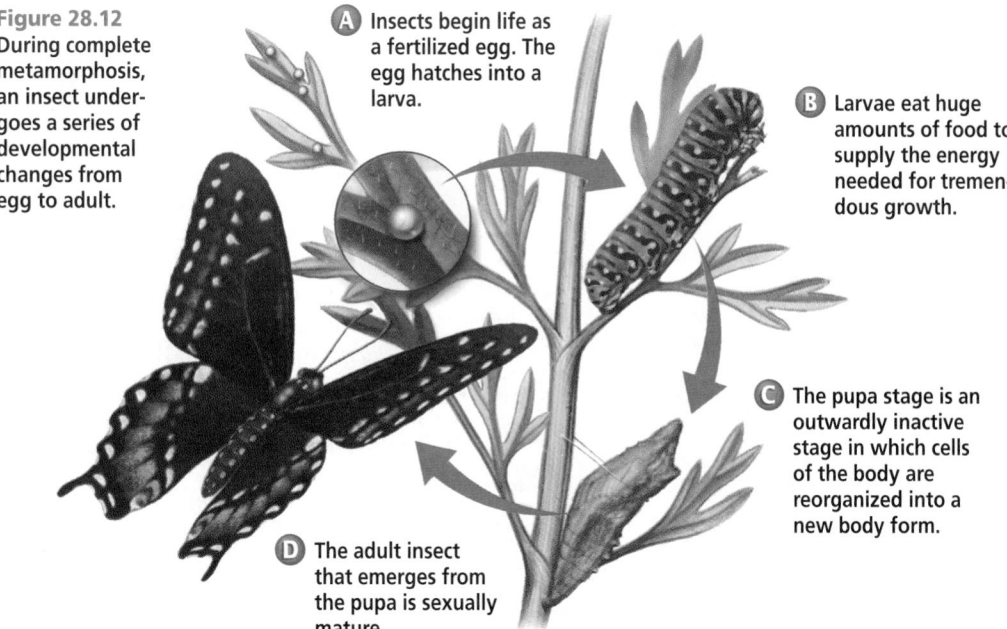

Figure 28.12 During complete metamorphosis, an insect undergoes a series of developmental changes from egg to adult.

A Insects begin life as a fertilized egg. The egg hatches into a larva.

B Larvae eat huge amounts of food to supply the energy needed for tremendous growth.

C The pupa stage is an outwardly inactive stage in which cells of the body are reorganized into a new body form.

D The adult insect that emerges from the pupa is sexually mature.

successive molts until the adult size is reached. Many other species of insects undergo a series of major changes in body structure as they develop. In some cases, the adult insect bears little resemblance to its juvenile stage. This series of changes, controlled by chemical substances in the animal, is called **metamorphosis** (met uh MOR fuh sus).

Insects that undergo metamorphosis usually go through four stages on their way to adulthood: egg, larva, pupa, and adult. The **larva** is the free-living, wormlike stage of an insect, often called a caterpillar. As the larva eats and grows, it molts several times. The **pupa** (PYEW puh) stage of insects is a period of reorganization in which the tissues and organs of the larva are broken down and replaced by adult tissues. Usually the insect does not move or feed during the pupa stage. After a period of time, a fully formed adult emerges from the pupa.

The series of changes that occur as an insect goes through the egg, larva,

pupa, and adult stages is known as complete metamorphosis. In winged insects that undergo complete metamorphosis, the wings do not appear until the adult stage. More than 90 percent of insects undergo complete metamorphosis. The complete metamorphosis of a butterfly is illustrated in *Figure 28.12.* Other insects that undergo complete metamorphosis include ants, beetles, flies, and wasps.

Complete metamorphosis is an advantage for arthropods because larvae do not compete with adults for the same food. For example, butterfly larvae (caterpillars) feed on leaves, but adult butterflies feed on nectar from flowers.

Incomplete metamorphosis has three stages

Many insect species, as well as other arthropods, undergo a gradual or incomplete metamorphosis, in which the insect goes through only three stages of development: egg, nymph, and adult, as shown in

Purpose

Students will compare and contrast patterns of metamorphosis.

Process Skills

compare and contrast, acquire information, collect data, hypothesize, observe and infer

Safety Precautions

Remind students to use tongs, forceps, and/or latex gloves when handling specimens.

Teaching Strategies

■ You will have to mark the different stages of each life cycle so students can identify them.

■ Legs may be a good indicator of the stage being capable of movement. Mouthparts may provide evidence of feeding.

Expected Results

See data table below.

Analysis

1. A grasshopper has incomplete metamorphosis. The moth has complete metamorphosis.
2. The stages that were able to move are the only stages in which feeding occurs.
3. The nymph stage looks like a small adult, but it lacks wings and is sexually immature.

✔ Assessment

Knowledge Show students photos of different insects undergoing metamorphosis. Ask them to determine if the animal shown exhibits complete or incomplete metamorphosis. Use the Performance Task Assessment List for Making Observations and Inferences in **PASC**, p. 17. **L2**

MiniLab 28-2 Comparing and Contrasting

Comparing Patterns of Metamorphosis Insects undergo a series of developmental changes called metamorphosis. But not all insects follow the same pattern of metamorphosis.

Procedure

1. Copy the data table.
2. Examine the three life stages of a grasshopper. Complete the information called for in your data table. **CAUTION:** *Wear disposable latex gloves and use forceps to handle preserved insects.*
3. Examine the four life stages of a moth. Complete the information called for in your data table.

Data Table							
Insect	Grasshopper			Moth			
Stage	egg	nymph	adult	egg	larva	pupa	adult
Locomotion Method							
Feeding Method							
Able to Reproduce							

Analysis

1. What are the differences between the stages of metamorphosis of a grasshopper and those of a moth?
2. Correlate the ability to move with ability to feed.
3. Compare a nymph stage with an adult stage.

Figure 28.13. A **nymph,** which hatches from an egg, has the same general appearance as the adult but is smaller. Nymphs may lack certain appendages, or have appendages not seen in adults, and they cannot reproduce. As the nymph eats and grows, it molts several times. With each molt, it comes to resemble the adult more and more. Wings begin to form, and an internal reproductive system develops. Gradually, the nymph becomes an adult. Grasshoppers and cockroaches are insects that undergo incomplete metamorphosis. You can compare the two types of metamorphosis in the *MiniLab* on this page.

Origins of Arthropods

Arthropods have been enormously successful in establishing themselves over the entire surface of Earth. Their ability to exploit just about every habitat is unequaled in the animal kingdom. The success of arthropods can be attributed in part to their varied life cycles, high reproductive output, and structural adaptations such as small size, a hard exoskeleton, and jointed appendages.

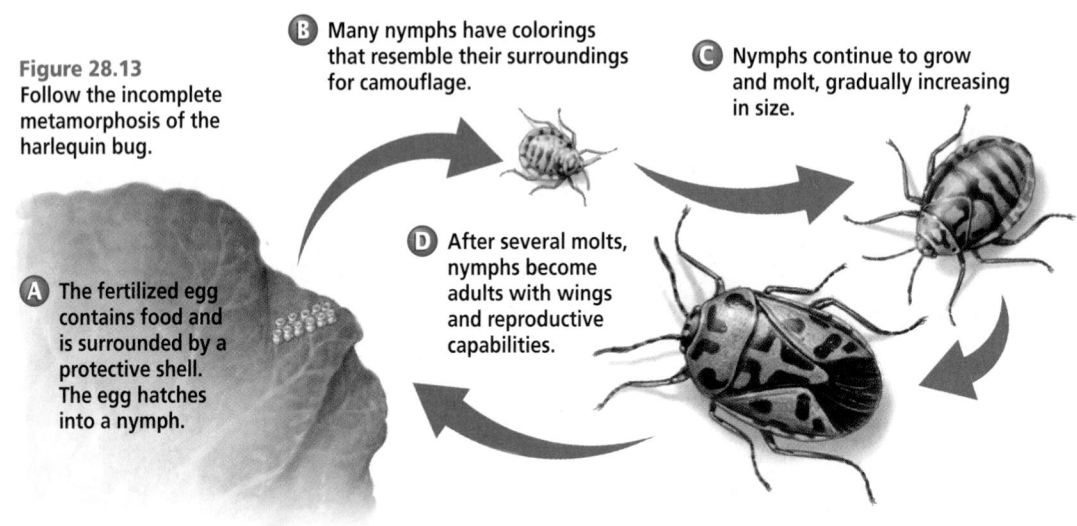

Figure 28.13
Follow the incomplete metamorphosis of the harlequin bug.

A The fertilized egg contains food and is surrounded by a protective shell. The egg hatches into a nymph.

B Many nymphs have colorings that resemble their surroundings for camouflage.

C Nymphs continue to grow and molt, gradually increasing in size.

D After several molts, nymphs become adults with wings and reproductive capabilities.

Data Table							
Insect	Grasshopper			Moth			
Stage	egg	nymph	adult	egg	larva	pupa	adult
Locomotion Method	no	yes	yes	no	yes	no	yes
Feeding Method	no	yes	yes	no	yes	no	yes
Able to Reproduce	no	no	yes	no	no	no	yes

Resource Manager

Content Mastery, pp. 137, 139-140 **L1**
Reinforcement and Study Guide, pp. 125-126 **L2**
Reteaching Skills Transparency 42 and Master **L1** **ELL**
Basic Concepts Transparency 49 and Master **L2** **ELL**
BioLab and MiniLab Worksheets, p. 126 **L2**

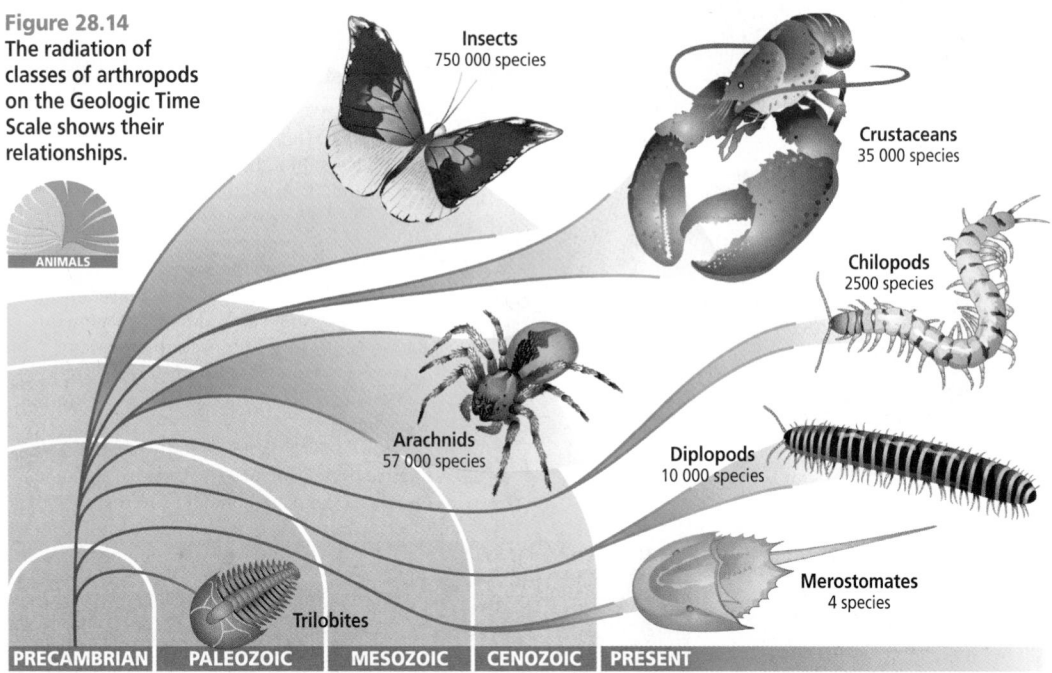

Figure 28.14
The radiation of classes of arthropods on the Geologic Time Scale shows their relationships.

Insects
750 000 species

Crustaceans
35 000 species

Chilopods
2500 species

Arachnids
57 000 species

Diplopods
10 000 species

Merostomates
4 species

Trilobites

ANIMALS

PRECAMBRIAN | PALEOZOIC | MESOZOIC | CENOZOIC | PRESENT

Arthropods most likely evolved from an ancestor of the annelids. As arthropods evolved, body segments fused and became adapted for certain functions such as locomotion, feeding, and sensing the environment. Segments in arthropods are more complex than in annelids, and arthropods have more developed nerve tissue and sensory organs such as eyes.

The exoskeleton of arthropods provides protection for their soft bodies. Muscles in arthropods are arranged in bands associated with particular segments and portions of appendages. The circular muscles of annelids do not exist in arthropods. Because arthropods have many hard parts, much is known about their evolutionary history. The trilobites shown in *Figure 28.14* were once an important group of ancient arthropods, but they have been extinct for 250 million years.

3 Assess

Check for Understanding
Naturalist Give students a list of traits for an arthropod. Have them determine which arthropod class the traits describe. **L2**

Reteach
Have students make a chart on arthropod classes. Have them include an illustration in each column that shows the number of body sections, antennae positions, and numbers of legs. **L1**

Extension
Kinesthetic Have students use modeling clay and pipe cleaners to make models of centipedes and millipedes. **L1 ELL**

✔ Assessment
Performance Have students develop a plan for constructing an insect exhibit for a zoo. They should choose insects, make a diagram of their exhibit, and explain how the insects must be cared for. **L2**

4 Close

Activity
Interpersonal Divide the class into two groups. Play the "Who am I?" game. A member of one group will call out clues, while a member of the second group guesses the class of the arthropod. A correct answer will earn the team one point. **L2** **COOP LEARN**

Section Assessment

Understanding Main Ideas
1. How are centipedes different from millipedes?
2. How are insects different from spiders?
3. Describe three sensory adaptations of insects.
4. Compare the stages of complete and incomplete metamorphosis.

Thinking Critically
5. Why might complete metamorphosis have greater adaptive value for an insect than incomplete metamorphosis?

SKILL REVIEW
6. **Recognizing Cause and Effect** Some plants produce substances that prevent insect larvae from forming pupae. How might this chemical production be a disadvantage to the plant? For more help, refer to *Thinking Critically* in the **Skill Handbook**.

Section Assessment

1. Centipedes have only one pair of legs per body segment, eat meat, and bite. Millipedes have two pairs of legs per body segment, eat plants and dead material, and do not bite.
2. Spiders have two body regions, six pairs of appendages, book lungs, simple eyes, and spin silk. Insects generally have three body regions, three pairs of legs, one pair of antennae, spiracles, and compound eyes.
3. Insects have compound eyes, antennae, tympanums, and sensitive hairs over parts of the body.
4. Complete metamorphosis has four stages: egg, larva, pupa, and adult. Incomplete metamorphosis has three stages: egg, nymph, and adult.
5. Complete metamorphosis is an advantage because the larvae do not compete with adults for food.
6. Although the larval stage is most destructive to plants, many plants require the adult insects for pollination.

NATIONAL GEOGRAPHIC

Focus On

Insects

Purpose 📖
Students will learn characteristics of insects and features of major groups of insects.

Background
Insects are essential to life, yet many are harmful. Their diversity in form makes insects not only successful, but also the largest animal group.

Teaching Strategies
■ Obtain a praying mantis egg case and have students make observations of hatching and development. Emphasize the adaptations that make the mantis an effective predator. **L1** **ELL**
■ Ask students to examine the mouthparts of preserved ants. Then have them design and conduct an experiment to determine food preferences of live ants. **L2**

Visual Learning
Ask students to discuss their most positive and most negative experiences with insects. Ask if they have seen any insects similar to ones in the photos in this Focus On. Have them explain how the insects were similar to and different from the photos. **L1**

COW KILLER ANT

FOCUS ON
Insects

Without insects, life as we know it would be impossible. Two-thirds of all flowering plants depend on insects to pollinate them. Insects also digest and degrade carrion, animal wastes, and plant matter. Their actions help fungi, bacteria, and other decomposers recycle nutrients and enrich the soil on which plants and all terrestrial organisms depend.

FLAME SKIMMER DRAGONFLY

CHARACTERISTICS
Insects have three body divisions—head, thorax, abdomen—and six legs attached to the thorax. The abdomen has multiple segments, the last ones often possessing external reproductive organs.

Most adult insects have wings, usually one or two pairs. An insect's skin, or integument, is hard yet flexible, and waterproof. Many insects must molt in order to grow larger before metamorphosing into adult forms.

Because of their ability to fly, a rapid reproductive cycle, and a tough, external skeleton, insects are both resilient and successful.

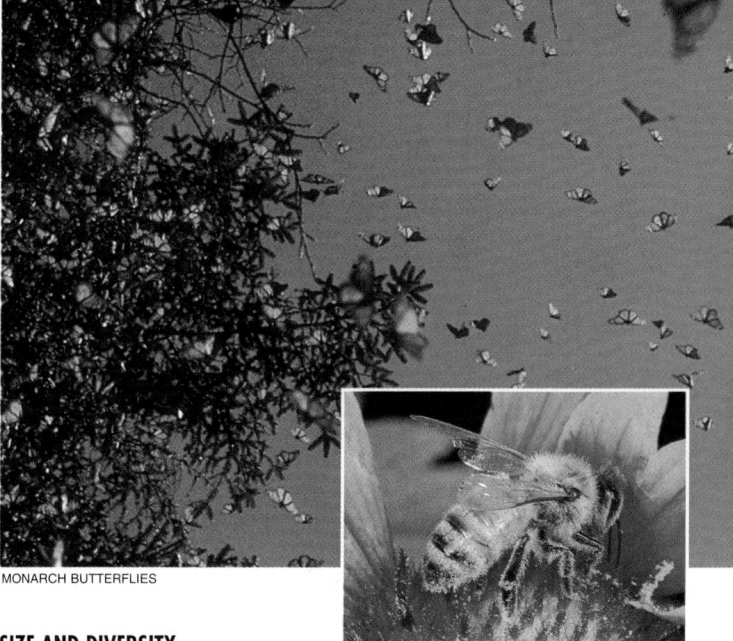

MONARCH BUTTERFLIES

HONEYBEE

SIZE AND DIVERSITY
Insects are members of the phylum Arthropoda and the class Insecta.
The most diverse class in the animal kingdom, Insecta is also the largest—it contains more species than all other animal groups combined.

PROJECT

Insect Scavenger Hunt
Interpersonal Ask a group of students to prepare a class scavenger hunt for insects and signs of insects. The scavenger hunt should require that classmates observe various insect behaviors such as movement and sound production, plus physical features of the insect and its habitat. Make sure the questions require that students actually observe the insect. For example: Describe two insect sounds that you hear outside. What insects are making these sounds? **L2** **COOP LEARN**

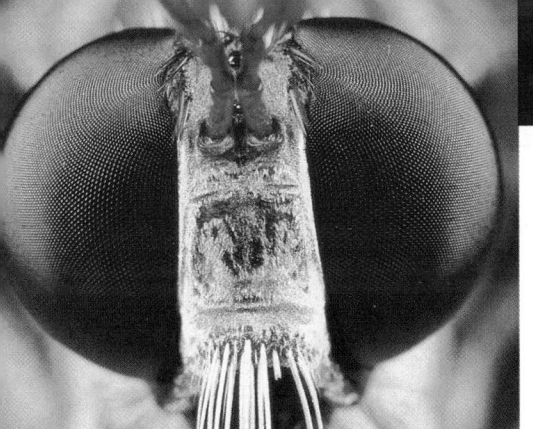
COMPOUND EYES OF GIANT RUDDER FLY

SENSE ORGANS

Insects gather information about their environment using a variety of sense organs that detect light, odors, sound, vibrations, temperature, and even humidity. Most adult insects have compound eyes, as well as two or three simple eyes on top of their heads. The compound eye of a large dragonfly contains a honeycomb of 28 000 lenses. The image from each lens is sent to the brain and somehow combined into a composite image, but we don't know exactly what such insects see. Some insects navigate by using sound waves or following odor trails. Katydids and crickets have "ears" on their front legs; houseflies have taste receptors on their feet.

ARCTIC WOOLLY BEAR CATERPILLAR

KATYDID

VERSATILITY

Some insects, such as the Arctic woolly bear caterpillar, can survive 10 months a year in subzero temperatures. Others, such as the monarch butterfly, migrate thousands of miles to warmer regions. Honeybees conserve heat in freezing temperatures by clumping into a ball that hums and churns all winter. Although some insects are plant pests, many others prey on their plant-munching relatives, and in so doing aid humans in the fight to control crop damage.

MOUTHPARTS

Insects get food by biting, lapping, and sucking. Some insects, such as grasshoppers and ants, have mouthparts for biting and chewing, with large mandibles for tearing into plant tissue or seizing prey. The powerful mandibles of bulldog ants, for example, are hinged at the sides of the head and bite inward—with great force—from side to side. Butterflies and honeybees have mouthparts shaped for lapping up nectar. Aphids and cicadas can pierce plant stems and then plant juices can be sucked like soda through a straw.

BLUE MORPHO BUTTERFLY

BULLDOG ANT

Enrichment

Kinesthetic Obtain preserved specimens of a variety of moths and butterflies and have students use a dissecting needle to unroll and measure the length of each proboscis. Have them hypothesize how proboscis length correlates with the type of flower each insect visits to sip nectar. **L2**

NATIONAL GEOGRAPHIC

 **VIDEODISC**
STV: Rain Forest
Forest Floor
Leaf-Cutting Ants
Unit 2, Side 1, 2 min. 30 sec.

Insects by Night and Day
Unit 2, Side 1, 1 min. 2 sec.

BIOLOGY JOURNAL

Flea Collar Safety

Linguistic Ask students to contact a local veterinarian and report in their journals about the safety of flea collars for pets. Students should ask about the insecticides used, how the flea collar works, why the flea collars are not dangerous to the health of the pet, and alternatives to flea collars. **L2**

BIOLOGY *Online* Note Internet addresses that you find useful in the space below for quick reference.

_____

_____

_____

_____

_____

Teaching Strategies

■ Ask if any students ever caught and kept an insect in a jar. Ask why they did this and what they learned.

■ Ask students to research the discovery of DDT and the factors that led to the banning of its use in the United States. **L3**

■ Have students work in groups and imagine that they are a team of inventors with backgrounds in entomology. They have been given unlimited funding to design the ultimate device for insect control. Students should use their creative imaginations as well as scientific knowledge to make a model of their bug catcher and explain its function to the class. **L2** **COOP LEARN**

GLENCOE
TECHNOLOGY

VIDEODISC

The Infinite Voyage:
*Insects: The Ruling Class
The Rothschild Legacy: Study of
Fleas and the Bubonic Plague*
(Ch. 2), 6 min. 30 sec.

Classifying Social Insects: Ants
(Ch. 3), 6 min. 30 sec.

Classifying Social Insects: Bees
(Ch. 4), 5 min. 30 sec.

*American Burying Beetles: An
Endangered Insect* (Ch. 7)
7 min. 30 sec.

SOLDIER BEETLE

HARMFUL VERSUS HELPFUL

Some beetles damage crops and spread disease. Spotted cucumber beetles, for example, devour leaves and flowers of cucumbers, melons, and squashes. They can also spread bacterial diseases to the plants they attack.

Many other beetles, such as ladybugs (also known as ladybirds), should be welcome visitors anywhere. Gardeners, farmers, and fruit-growers release thousands of ladybugs into gardens, fields, and orchards as a first line of defense against insect pests, especially aphids. The bright redorange of ladybug beetles is an unmistakable warning to potential predators that the beetles are extremely distasteful.

LEAF BEETLE

A SUPERLATIVE CRITTER

Some beetles can chew through lead or zinc or timber—not to mention whole fields of cotton. A leaf beetle in the Kalahari Desert produces a toxin powerful enough to fell an antelope. The American burying beetle can lift 200 times its weight. Among Earth's most recognizable beetles, fireflies light up summer evenings, and ladybugs control garden pests.

LADYBUG BEETLES

SPOTTED CUCUMBER BEETLE

778

BIOLOGY JOURNAL

Linguistic Have students contact their local agricultural extension service to ask about integrated pest management efforts in their area. Then have students write an advertisement for a gardening magazine to sell insects used for natural pest control. **L2**

WEEVIL BEETLE

COLORADO POTATO BEETLE

A SPECIAL NICHE

The Mesozoic era is often identified as the age of dinosaurs. But the truly colossal event during this period in Earth's history was the proliferation of flowering plants. Primary pollinators of the era, beetles most likely fueled this explosion of color and fragrance. Beetles fill critical ecological niches as scavengers and as harvesters of caterpillars and other pests, which, left untended, would devour thousands of acres of crops and forest trees each year. When a beetle species faces extinction—as nine species in the United States currently do—scientists see it as an early warning system alerting us to significant environmental change.

BODY ARMOR

Many scientists consider beetles to be evolution's biggest success story and think that thousands of additional species remain undiscovered. Beetles—all 350 000 described species—presently account for approximately 1 in 4 known animal species. Beetles thrive in deserts, under tropical forest canopies, and in water. One key to beetles' adaptability is their "shell"—actually a pair of hardened wings called elytra. Elytra permit some beetles to live in deserts by sealing in moisture and other species to breathe underwater by trapping air. Many beetles are remarkably resistant to pesticides.

LONG-HORNED BEETLE

EXPANDING Your View

1 **THINKING CRITICALLY** What are the advantages and disadvantages of an exoskeleton?

2 **JOURNAL WRITING** Research social behavior in insects and write a short essay to present to the class.

779

Enrichment

Ask students to design and conduct an experiment to determine the ideal temperature for hatching and developing flour beetle eggs. **L2**

Answers to Expanding Your View

1. Advantages of exoskeletons: protection, waterproofing. Disadvantages: heavy, can't grow very large, must molt to grow
2. Students might present information on the social structure of bees in a hive or an ant colony.

GLENCOE TECHNOLOGY

CD-ROM
Biology: The Dynamics of Life
Exploration: *Classifying Beetles*
Disc 3

✓ Portfolio

Insect Population

Ask students to do an insect population study by sampling one species of insect in a small area and then estimating the total number over a larger area. They should measure an area 1 m² and collect one type of insect, such as beetles. Then, knowing the size of the area, they can calculate how many insects of that type there are. Have students release the insects as soon as their calculations are complete. Challenge students to relate the number of insects counted to the insect's niche. Have students predict how a sudden change in insect population number would impact the ecology of the area. **L3** **P** **COOP LEARN**

Will salt concentration affect brine shrimp hatching?

Brine shrimp (Artemia salina) *belong to the class Crustacea. They are excellent experimental animals because their eggs hatch into visible swimming larvae within a very short time. Using the name as a clue, where might these animals normally be found?*

Time Allotment

One class period for set up, then 10 minutes per day for 3 days

Process Skills

collect data, identify and control variables, design an experiment, draw a conclusion, experiment, hypothesize, interpret data, observe and infer, organize data

Safety Precautions

Remind students to wear goggles to protect their eyes.

PREPARATION

- Purchase brine shrimp eggs from a pet store or biological supply house.
- Plan the experiment so that it starts on a Monday.
- Plastic trays are available from the meat or produce department in most grocery stores.
- Review the procedure for preparation of 1, 2, and 4% salt solutions.
- Use noniodized table salt (kosher or sea salt). Check to make sure that iodine has not been added.

Possible Hypotheses

Students may hypothesize that hatching will not occur in any salt solution. Students may hypothesize that hatching will occur only in salt concentrations higher than 10% but lower than 20%.

PREPARATION

Problem

How can you determine the optimum salt concentration for the hatching of brine shrimp eggs?

Hypothesis

Decide on one hypothesis that you will test. Your hypothesis might be that increased salt concentrations result in an increase in the number of eggs hatched.

Objectives

In this BioLab, you will:
- **Analyze** how salt concentration may affect brine shrimp hatching.
- **Interpret** your experimental findings.

Possible Materials

beakers or plastic bottles
labels or marking pencil
graduated cylinder
brine shrimp eggs
clear plastic trays
salt (noniodized)
balance
water

Safety Precautions

Wear protective eye goggles when preparing solutions.

Skill Handbook

Use the **Skill Handbook** if you need additional help with this lab.

PLAN THE EXPERIMENT

Teaching Strategies

- Have students work in small groups of 3 or 4 students.
- At the conclusion of the experiment, pool student data so students can see the results when a range of different salt concentrations were tested.
- Judging the amount or degree of hatching may be a problem. Brine shrimp hatchlings

cannot be seen by the naked eye. Observing trays under a binocular microscope will allow students to make a qualitative assessment of the amount of hatching.

Possible Procedures

- Students may prepare only 2 or 3 different salt concentrations. Make sure that the control dish contains only water and

PLAN THE EXPERIMENT

1. Decide on a way to test your group's hypothesis. Keep the available materials in mind as you plan your procedure. Be sure to include a control. For example, you might place brine shrimp eggs in two trays —one with the salt concentration of the water brine shrimp normally inhabit, and one with a different salt concentration.

2. Decide how long you will make observations and how you will judge the extent of egg hatching.

3. Decide on the number of different salt water concentrations to use and what these concentrations will be. Review the steps needed to prepare solutions of different concentrations.

Check the Plan

Discuss the following points

with other group members to decide on the final procedure for your experiment.

1. What is your one independent variable? Your dependent variable?

2. What will be your control?

3. How much water will you add to each tray and how will you measure the same number of eggs to be used in each tray?

4. Will it be necessary to control variables such as light and temperature?

5. What data will you collect and how will it be recorded?

6. *Make sure your teacher has approved your experimental plan before you proceed further.*

7. Carry out your experiment.

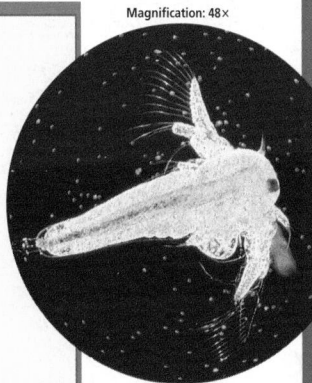

Magnification: 48×

Brine shrimp hatchling

ANALYZE AND CONCLUDE

1. **Interpreting Data** Using specific numbers from your data, explain how salt concentration affects brine shrimp hatching.

2. **Drawing a Conclusion** Was your hypothesis supported? Explain.

3. **Identifying and Controlling Variables** What were the independent and dependent variables? What were some of the variables that had to be controlled?

4. **Hypothesizing** Formulate a hypothesis that explains why high salt concentrations may be harmful to brine shrimp hatching.

5. **Classifying** Classify brine shrimp. Identify their kingdom, phylum, class, order, family, genus, and species.

Going Further

Project Design an experiment that you could perform to investigate the role that temperature plays in brine shrimp hatching. If you have all of the materials you will need, you may want to carry out the experiment.

BIOLOGY *Online*  To find out more about brine shrimp and other crustaceans, visit the Glencoe Science Web site.

science.glencoe.com

ANALYZE AND CONCLUDE

1. Students answers may vary; maximum hatching will occur at 1-4% salt concentration.

2. Student answers will vary depending upon their original hypothesis.

3. The independent variable was the salt concentration. The dependent variable was the number of eggs hatched. Other variables that had to be controlled were the temperature of the water, amount of water used, and number of eggs used.

4. Too high a salt concentration may result in water loss from the egg or larva, leading to dessication.

5. animal, Arthropoda, Crustacea, Anostraca, *Artemia salina*

✓ Assessment

Skill Have students design an experiment that would determine how much time brine shrimp larvae spend in this stage. Use the Performance Task Assessment List for Designing an Experiment in **PASC**, p. 23.

Going Further

Have students carry out an experiment to determine if brine shrimp larvae respond in a similar manner to different colors of light. **L2**

Resource Manager

BioLab and MiniLab Worksheets, pp. 127-128 **L2**

---

no salt.

■ Students will have to determine some method for controlling the same number of eggs used in each experimental tray. Counting eggs is difficult because of their small size. One suggestion is to touch the flat end of a toothpick to the eggs and transfer that same amount to each dish.

■ Binocular microscopes may be used to

observe the presence of hatched larvae.

■ Keep all trays at a temperature close to 21°C.

Data and Observations

Ideal salt concentrations for hatching will be in the 1-4% range. No salt present or concentrations higher than 4% will result in no or few larvae hatching.

Purpose

Students will learn about Lyme disease, the bacterium that causes it, and the ticks that transmit it.

Teaching Strategies

■ Be sure students understand that ticks do not cause Lyme disease; they carry the bacterium that causes the disease from animal host to human and animal hosts.

■ After students have read the feature, ask if they know how a tick should be removed. Placing several drops of vegetable oil on the tick will cause it to withdraw its head because it can no longer obtain enough oxygen. Then the tick can be removed easily.

■ Ask students to report about two other diseases transmitted by ticks: Rocky Mountain spotted fever and Colorado tick fever.
L3

Connection to Biology

Because deer are primary carriers of *Borrelia burgdorferi*, an increase in the deer population probably means an increase in infected ticks. This would likely translate into an increase in the incidence of Lyme disease.

Terrible Ticks

Magnification: 7×

Every American city, it seems, has its claim to fame. Chicago is recognized for its outstanding architecture. New York has long been thought of as the cultural center of the United States. Los Angeles is home to television production and to the nation's legendary movie industry. One American city, however, would probably just as soon forget its claim to fame. Lyme, Connecticut, will forever be associated with Lyme disease, a crippling bacterial malaise that was first identified in this town in 1975.

Lyme disease manifests itself in humans in three distinct stages. First, a circular, bull's-eye rash appears. The rash is generally accompanied by chills, fever, and aching joints. These symptoms may resemble symptoms that result from an infection with an influenza virus. This is the mildest form of the disease. If left untreated, Lyme disease progresses to a second stage. The joint pains become more severe and may be joined by neurological symptoms, such as memory disturbances and vision impairment. Stage three is the most severe form of the disease. Crippling arthritis, facial paralysis, heart abnormalities, and memory loss may result.

Tick transmission

The cause of this debilitating disease is *Borrelia burgdorferi*, a corkscrew-shaped bacterium that is transmitted to humans through the bite of ticks. The bacterium infects mostly deer and white-footed mice. Ticks pick up the bacteria by sucking the blood from these animals. When the same ticks bite humans, the bacteria are passed on, and the result is Lyme disease.

Where Lyme disease strikes

Most cases of Lyme disease are reported in the Northeast, Mid-Atlantic, and North Central regions, as seen on the map. Lyme disease is also on the rise in many other areas.

Numbers of cases of Lyme disease per 100 000 population
- More than 30 cases
- 15 to 30 cases
- 5 to 15 cases
- Fewer than 5 cases
- Area of heavy concentration

Prevention and treatment

Ticks live in weedy areas, low shrubs, and tall grasses. If you are entering this type of habitat, it is advisable to wear light colored clothing to easily detect darker ticks, and to tuck pants legs into socks. In addition, insect repellents containing the chemical DEET can be applied to clothing (but not to skin). Careful examination of the body for ticks is also important.

Like most bacteria, *Borrelia burgdorferi* responds to antibiotics. Early treatment with antibiotics usually prevents the disease from progressing to its second or third stages. A new vaccine has been developed that is effective in 90 percent of adults exposed to Lyme disease, but this vaccine has not yet been proven to be safe for those under age 18.

CONNECTION TO BIOLOGY

Since the turn of the century, the deer population in the United States has been increasing steadily. How might this increase affect the incidence of Lyme disease? Why?

BIOLOGY Online To find out more about ticks and Lyme disease, visit the Glencoe Science Web site.
science.glencoe.com

BIOLOGY Online Note Internet addresses that you find useful in the space below for quick reference.

_____

_____

_____

_____

Section 28.1

Characteristics of Arthropods

Main Ideas
- Arthropods have jointed appendages, exoskeletons, varied life cycles, and body systems adapted to life on land, water, or air.
- Arthropods are members of the most successful animal phylum in terms of diversity. This can be attributed in part to their structural and behavioral adaptations.

Vocabulary
appendage (p. 761)
book lung (p. 764)
cephalothorax (p. 763)
compound eye (p. 765)
mandible (p. 765)
Malpighian tubule (p. 765)
molting (p. 762)
parthenogenesis (p. 766)
pheromone (p. 764)
simple eye (p. 765)
spiracle (p. 764)
tracheal tube (p. 764)

Section 28.2

The Diversity of Arthropods

Main Ideas
- Spiders have two body regions with four pairs of walking legs. They spin silk. Ticks and mites have one body section. Scorpions have many abdominal segments, enlarged pincers, and a stinger at the end of the tail.
- Most crustaceans are aquatic and exchange gases in their gills. They include crabs, lobsters, shrimps, crayfishes, barnacles, and water fleas.
- Centipedes are carnivores with flattened, worm-like bodies. Millipedes are herbivores with cylindrical, wormlike bodies.
- Insects are the most successful arthropod class in terms of diversity. They have many structural and behavioral adaptations that allow them to exploit all habitats.

Vocabulary
chelicerae (p. 768)
larva (p.773)
metamorphosis (p. 773)
nymph (p. 774)
pedipalp (p. 768)
pupa (p. 773)
spinneret (p. 768)

Main Ideas

Summary statements can be used by students to review the major concepts of the chapter.

Using the Vocabulary

To reinforce chapter vocabulary, use the Content Mastery Booklet and the activities in the Interactive Tutor for Biology: The Dynamics of Life on the Glencoe Science Web site.
science.glencoe.com

 **THE PRINCETON REVIEW**

All Chapter Assessment questions and answers have been validated for accuracy and suitability by The Princeton Review.

UNDERSTANDING MAIN IDEAS

1. b
2. d
3. c
4. b
5. b

UNDERSTANDING MAIN IDEAS

1. Crustaceans are different from other arthropods because they have two _____ used for sensing.
   a. jointed appendages
   b. pairs of antennae
   c. pedipalps
   d. walking legs

2. Of the following, which are NOT appendages used by arthropods to obtain and eat food?
   a. chelicerae
   b. pedipalps
   c. mandibles
   d. spiracles

3. Jointed appendages allow for greater _____ and more powerful movements.
   a. mobility
   b. molting
   c. flexibility
   d. camouflage

4. _____ are arthropods with only one body section.
   a. Spiders
   b. Ticks and mites
   c. Scorpions
   d. Crustaceans

5. _____ are used by arthropods for gas exchange.
   a. Pedipalps
   b. Spiracles
   c. Chelicerae
   d. Spinnerets

GLENCOE TECHNOLOGY

 **VIDEOTAPE**
MindJogger Videoquizzes
Chapter 28: *Arthropods*
Have students work in groups as they play the videoquiz game to review key chapter concepts.

Resource Manager

Chapter Assessment, pp. 163-168
MindJogger Videoquizzes
ExamView® Pro Software
BDOL Interactive CD-ROM, Chapter 28 quiz

6. c
7. d
8. b
9. d
10. a
11. insect
12. exoskeleton
13. blood circulation
14. cephalothorax
15. Tracheal tubes
16. crustaceans
17. segmented bodies
18. Malpighian tubules
19. egg, nymph
20. oxygen, carbon dioxide

APPLYING MAIN IDEAS

21. An insect larva may eat crop plants, but an adult may pollinate the flowers.
22. They cannot move around to find mates. Having both male and female sex organs in the same animal means that every individual is a potential mate.
23. Exoskeletons in arthropods that swim can be heavier as the water will help support their weight. Arthropods that fly have lightweight exoskeletons. Arthropods that move on land have exoskeletons that are medium in weight.
24. Wings enable some arthropods to easily escape predators, find food sources inaccessible to terrestrial arthropods, and move easily to other areas to find mates and nesting areas.
25. The eyes would give the rigid crustacean a greater field of view.

6. What arthropod has many abdominal segments, enlarged pincers, and a stinger at the end of its tail?
   a. a spider   c. a scorpion
   b. a tick   d. a crab
7. Arthropods are so successful because of their _____.
   a. larvae   c. antennae
   b. book lungs   d. adaptations
8. Name an arthropod that is a carnivore and has a flattened, wormlike body.
   a. millipede   c. crustacean
   b. centipede   d. insect
9. Of the following, which is NOT an appendage of an arthropod?
   a. chelicerae   c. pedipalps
   b. antennae   d. cephalothorax
10. Arthropods with two body regions and four pairs of walking legs are called _____.
    a. spiders   c. scorpions
    b. ticks and mites d. crustaceans
11. The most diverse group of arthropods is the _____ class.
12. Molting occurs when an arthropod sheds its old _____ and grows a new one.
13. The structure labeled b in the diagram below is used for what purpose by the grasshopper?

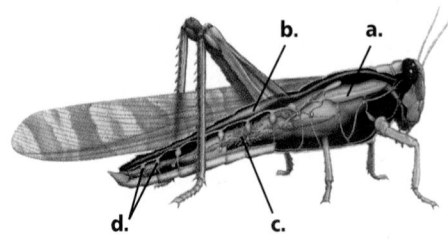

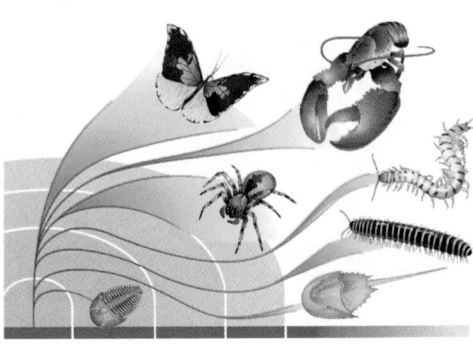

THE PRINCETON REVIEW **TEST-TAKING TIP**

Stock Up on Supplies
Be sure to supply yourself with the test-taking essentials: number two pencils, pens, erasers, a ruler, and a pencil sharpener. If the room doesn't have a pencil sharpener, a broken pencil can be upsetting.

14. Spiders have a fused head and thorax region called a _____.
15. _____ are the hollow passages that carry air through the body of an arthropod.
16. Study the diagram below. The group most closely related to insects is the _____.
17. Evolutionary biologists have hypothesized that arthropods may have evolved from annelids because both have _____.
18. A butterfly excretes wastes through _____.
19. List in the correct order the stages of incomplete metamorphosis: _____, _____, adult.
20. When water passes over gills, _____ and _____ are exchanged.

APPLYING MAIN IDEAS

21. Many insects are pests to humans when they are larvae but are beneficial when they are adults. Explain.
22. Why is it an adaptive advantage for barnacles to be hermaphrodites?
23. Relate differences in exoskeleton structure to the various modes of arthropod locomotion.
24. In what ways have wings been an adaptive advantage to the success of insects?
25. Of what advantage might movable, stalked eyes be to a crustacean that has a cephalothorax?

26. Interpreting Scientific Illustrations Identify each of the arthropods at right as an arachnid, crustacean, or insect. What are their distinguishing features?

27. Recognizing Cause and Effect What is the advantage to a plant of producing a chemical that is an effective insect repellent?

28. Recognizing Cause and Effect What might be the effect on plant and animal life if all insects were suddenly to die?

29. Observing and Inferring Evidence shows that deer, mice, and even household pets may harbor the bacteria that cause Lyme disease. How could pets become infected with these bacteria?

30. Concept Mapping Complete the concept map by using the following vocabulary terms: appendages, mandibles, chelicerae, pedipalps.

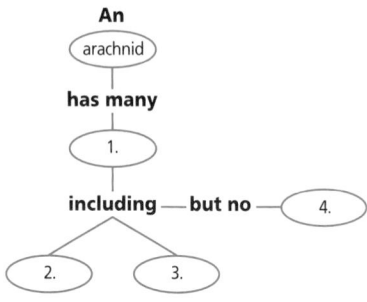

An
arachnid
has many
1.
including — but no — 4.
2.   3.

CD-ROM
For additional review, use the assessment options for this chapter found on the *Biology: The Dynamics of Life Interactive CD-ROM* and on the Glencoe Science Web site.
science.glencoe.com

The melting points of the waxy layers on certain insect exoskeletons are shown in the graph below. These melting points reflect the environments in which the insects were raised. Insects raised in warmer environments have wax that melts at higher temperatures than insects raised in cooler environments.

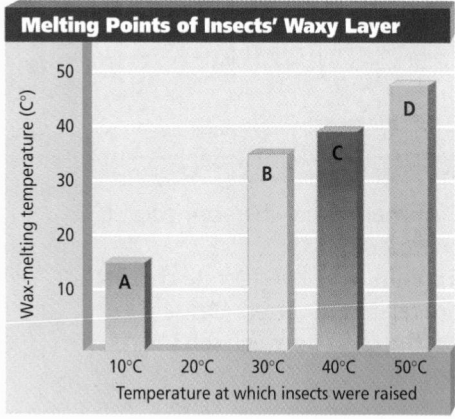

Melting Points of Insects' Waxy Layer

Interpreting Data
Study the graph and answer the following questions.

1. What is the melting point of the wax on insects in group B?
   a. 15°C   **c.** 35°C
   b. 50°C   **d.** 40°C

2. What is the melting point of the wax on insects in group C?
   a. 15°C   **c.** 35°C
   b. 50°C   **d.** 40°C

3. Which insects were raised at the lowest temperature?
   a. A   **b.** B   **c.** C   **d.** D

4. Making a Graph Make a graph of these data: insect exoskeletons found by a stream melt at 15°C; in a forested area at 20°C; in a grassy meadow at 40°C; and on roadside soil at 50°C.

26. An insect has six legs and a pair of antennae. An arachnid has eight legs and no antennae. A crustacean has five pairs of walking legs and stalked eyes.

27. The plant will not be eaten by insects.

28. Many plants that depend upon insects for pollination would be unable to reproduce, so there would be fewer plants. Plant- and insect-eating organisms would have to find new food sources or die.

29. Pets can pick up ticks when they are outdoors in an area inhabited by ticks.

30. 1. Appendages; 2. Pedipalps; 3. Chelicerae; 4. Mandibles

ASSESSING KNOWLEDGE & SKILLS

1. c
2. d
3. a
4. Melting Temperatures of Insects' Waxy Layers

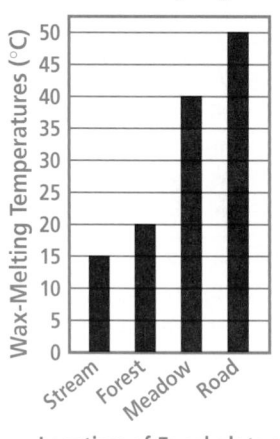

Location of Exoskeletons

Chapter 29 Organizer

Refer to pages 4T-5T of the Teacher Guide for an explanation of the National Science Education Standards correlations.

Section	Objectives	Activities/Features
Section 29.1 **Echinoderms** National Science Education Standards UCP.1-5; A.1, A.2; C.3, C.5, C.6 (1 session, $^1/_2$ block)	1. **Compare** similarities and differences among the classes of echinoderms. 2. **Interpret** the evidence biologists have for determining that echinoderms are close relatives of chordates.	**MiniLab 29-1:** Examining Pedicellariae, p. 788 **Inside Story:** A Sea Star, p. 790 **Problem-Solving Lab 29-1,** p. 792 **Investigate BioLab:** Observing Sea Urchin Gametes and Egg Development, p. 800 **Physics Connection:** Hydraulics of Sea Stars, p. 802
Section 29.2 **Invertebrate Chordates** National Science Education Standards UCP.1, UCP.2, UCP. 4, UCP.5; A.1, A.2; C.3, C.5, C.6; G.1-3 (2 sessions, $^1/_2$ block)	3. **Summarize** the characteristics of chordates. 4. **Explain** how invertebrate chordates are related to vertebrates. 5. **Distinguish** between sea squirts and lancelets.	**MiniLab 29-2:** Examining a Lancelet, p. 797 **Inside Story:** A Tunicate, p. 798 **Problem-Solving Lab 29-2,** p. 799

Need Materials? Contact Carolina Biological Supply Company at 1-800-334-5551 or at **http://www.carolina.com**

MATERIALS LIST

BioLab

p. 800 microscope, microscope slides (5), coverslips (5), petri dish, dropper (2), test tube, beaker (2), sea water, live sea urchins (male and female), syringe with needle, potassium chloride solution

MiniLabs

p. 788 microscope, prepared slide of sea star pedicellariae, paper, pencil
p. 797 stereomicroscope, microscope slide, forceps, metric ruler, preserved specimen of *Branchiostoma californiense (Amphioxus)*

Alternative Lab

p. 796 microscope, prepared slides of sea urchin development and lancelets

Quick Demos

p. 789 stereomicroscope, live sea urchin, toothpick
p. 789 dropper
p. 791 heavy book
p. 795 preserved specimens of tunicates and lancelets
p. 796 microscope, prepared slide of lancelet cross section

Key to Teaching Strategies

L1 Level 1 activities should be appropriate for students with learning difficulties.
L2 Level 2 activities should be within the ability range of all students.
L3 Level 3 activities are designed for above-average students.
ELL ELL activities should be within the ability range of English Language Learners.
COOP LEARN Cooperative Learning activities are designed for small group work.
P These strategies represent student products that can be placed into a best-work portfolio.
These strategies are useful in a block scheduling format.

Teacher Classroom Resources

Section	Reproducible Masters	Transparencies
Section 29.1 **Echinoderms**	Reinforcement and Study Guide, pp. 127-129 `L2` Critical Thinking/Problem Solving, p. 29 `L3` BioLab and MiniLab Worksheets, p. 129 `L2` Laboratory Manual, pp. 205-210 `L2` Content Mastery, pp. 141-142, 144 `L1`	Section Focus Transparency 71 `L1` `ELL` Basic Concepts Transparency 50 `L2` `ELL` Basic Concepts Transparency 51 `L2` `ELL`
Section 29.2 **Invertebrate Chordates**	Reinforcement and Study Guide, p. 130 `L2` Concept Mapping, p. 29 `L3` `ELL` Critical Thinking/Problem Solving, p. 29 `L3` BioLab and MiniLab Worksheets, pp. 130-132 `L2` Content Mastery, pp. 141, 143-144 `L1`	Section Focus Transparency 72 `L1` `ELL` Reteaching Skills Transparency 43 `L1` `ELL`

Assessment Resources

Chapter Assessment, pp. 169-174
MindJogger Videoquizzes
Performance Assessment in the Biology Classroom
Alternate Assessment in the Science Classroom
ExamView® Pro Software
BDOL Interactive CD-ROM, Chapter 29 quiz

Additional Resources

Spanish Resources `ELL`
English/Spanish Audiocassettes `ELL`
Cooperative Learning in the Science Classroom `COOP LEARN`
Lesson Plans/Block Scheduling

NATIONAL GEOGRAPHIC
Teacher's Corner

Products Available From Glencoe
To order the following products, call Glencoe at 1-800-334-7344:
CD-ROM
NGS PictureShow: Structure of Invertebrates
Transparency Set
NGS PicturePack: Structure of Invertebrates

Index to National Geographic Magazine
The following articles may be used for research relating to this chapter:
"Pillar of Life," by George Grall, July 1992.

GLENCOE TECHNOLOGY

The following multimedia resources are available from Glencoe.

Biology: The Dynamics of Life
CD-ROM `ELL`
 Exploration: *The Six Kingdoms*
Exploration: *Echinoderms*
Exploration: *Symmetry*
BioQuest: *Biodiversity Park*

Videodisc Program
 Starfishes

The Secret of Life Series
Sea Stars
Chordate Body Plan
Action of Tube Feet

Theme Development

Evolution is a major theme in this chapter. Students study how echinoderm larvae and invertebrate chordates show similarities to certain vertebrates. The theme of **systems and interactions** is obvious as students learn how the echinoderms and invertebrate chordates are adapted to and interact with their environments.

READING BIOLOGY

Glencoe's *Biology: The Dynamics of Life* contains many resources to assist a student's reading skills. Each chapter contains figures with expanded captions that expand on written material. Word Origins, located along the side of text, expand knowledge of biology vocabulary. Glencoe's Content Mastery Booklet helps develop reading skills while reinforcing content. In addition, use the Interactive Tutor for *Biology: The Dynamics of Life* on the Glencoe Web site to reinforce vocabulary. **science.glencoe.com**

Chapter

29 Echinoderms and Invertebrate Chordates

What You'll Learn

- You will compare and contrast the adaptations of echinoderms.
- You will distinguish the features of chordates by examining invertebrate chordates.

Why It's Important

By studying how echinoderms and invertebrate chordates function, you will enhance your understanding of evolutionary relationships between these two groups.

READING BIOLOGY

Scan the chapter, examining the illustrations and reading the captions. As you read, write down the key idea illustrated in each figure.

To find out more about echinoderms and invertebrate chordates, visit the Glencoe Science Web site.
science.glencoe.com

A sea star extends its stomach from its mouth and engulfs a sea urchin. Hours later, the sea star draws its stomach back in and moves away. All that's left of the urchin is the bumpy globe you see here. Even its spines are gone.

29.1 Echinoderms

T*hink about what the best defense might be for a small animal that moves slowly in tide pools on the seashore. Did you think of armor, spines, or perhaps poison as methods of protection? Sea urchins are masters of defense—some use all three methods. The sea urchin looks different from the feather star and from the sea star on the facing page, yet all three belong to the same phylum. What characteristics do they have in common? What features determine whether an animal is an echinoderm?*

Feather star (above) and sea urchin (inset)

SECTION PREVIEW

Objectives
Compare similarities and differences among the classes of echinoderms.
Interpret the evidence biologists have for determining that echinoderms are close relatives of chordates.

Vocabulary
ray
pedicellaria
tube feet
ampulla
water vascular system
madreporite

What Is an Echinoderm?

Members of the phylum Echinodermata have a number of unusual characteristics that easily distinguish them from members of any other animal phylum. Echinoderms move by means of hundreds of hydraulic, suction cup-tipped appendages and have skin covered with tiny, jawlike pincers. Echinoderms (ih KI nuh durmz) are found in all the oceans of the world.

Echinoderms have endoskeletons

If you were to examine the skin of several different echinoderms, you would find that they all have a hard, spiny, or bumpy endoskeleton covered by a thin epidermis. The long, pointed spines on a sea urchin are obvious. Sea stars, sometimes called

starfishes, may not appear spiny at first glance, but a close look reveals that their long, tapering arms, called **rays,** are covered with short, rounded spines. The spiny skin of a sea cucumber consists of soft tissue embedded with small, platelike structures that barely resemble spines. The endoskeleton of all echinoderms is made primarily of calcium carbonate, the compound that makes up limestone.

Some of the spines found on sea stars and sea urchins have become modified into pincerlike appendages called **pedicellariae** (PED ih sihl AHR ee ay). An echinoderm uses its jawlike pedicellariae for protection and for cleaning the surface of its body. You can examine these structures in the *MiniLab* on the following page.

WORD *Origin*
echinoderm
From the Greek words *echinos*, meaning "spiny," and *derma*, meaning "skin." Echinoderms are spiny-skinned animals.

pedicellariae
From the Latin word *pediculus*, meaning "little foot." Pedicellariae resemble little feet.

29.1 ECHINODERMS **787**

Section 29.1

Prepare

Key Concepts
The characteristics common to echinoderms are presented. The diversity of echinoderms is considered. Deuterostome development is examined in terms of its evolutionary significance.

Planning
- Obtain droppers for the Quick Demos.
- Make photocopies of the inner, shaded area of Figure 29.6 for the Reteach.

---

1 Focus

Bellringer
Before presenting the lesson, display **Section Focus Transparency 71** on the overhead projector and have students answer the accompanying questions.
L1 **ELL**

Resource Manager

Section Focus Transparency 71 and Master **L1** **ELL**

✓ Assessment Planner

Portfolio Assessment
  MiniLab, TWE, pp. 788, 797
  Assessment, TWE, p. 789
  Problem-Solving Lab, TWE, p. 792
  Portfolio, TWE, p. 792
Performance Assessment
  BioLab, SE, pp. 800-801
  MiniLab, SE, pp. 788-797
  Alternative Lab, TWE, pp. 796-797

  Assessment, TWE, pp. 793, 799
  Problem-Solving Lab, TWE, p. 798
Knowledge Assessment
  Section Assessment, SE, pp. 793, 799
  Chapter Assessment, SE, pp. 803-805
  Assessment, TWE, p. 796
  BioLab, TWE, pp. 800-801
Skill Assessment
  Alternative Lab, TWE, pp. 796-797

2 Teach

MiniLab 29-1

Purpose

Students will observe the pincherlike structure of pedicellariae.

Process Skills

observe and infer, compare and contrast, draw a conclusion, measure in SI

Teaching Strategies

■ Review the procedure for measuring objects under the microscope.

■ Prepared slides are available from biological supply houses.

Expected Results

Students will observe that pedicellariae have a pincherlike appearance and measure close to 250μm in length.

Analysis

1. Student answers will vary—pincherlike, plierslike, forcepslike
2. allows animal to grasp objects, clean itself of debris, protection
3. These pincherlike organs allow the sea star to pinch potential predators when they touch the animal. Their shape also allows the sea star to clean itself by picking off materials that become stuck to its body.

Assessment

Portfolio Have students write a lab report that summarizes their results and place it in their portfolios. Use the Performance Task Assessment List for Lab Report in **PASC**, p. 47. [L2] [P]

Visual Learning

Have students examine the animals shown in Figure 29.1 closely. Challenge them to identify the lines of symmetry for each organism that has radial symmetry. [L1]

788

MiniLab 29-1 · Observing and Inferring

Magnification: 22×

Examining Pedicellariae Echinoderms move by tube feet. They also have tiny pincers on their skin called pedicellariae.

Pedicellariae

Procedure

1. Observe a slide of sea star pedicellariae under low-power magnification. **CAUTION:** *Use caution when working with a microscope and slides.*
2. Record the general appearance of one pedicellaria. What does it look like?
3. Make a diagram of one pedicellaria under low-power magnification.
4. Record the size of one pedicellaria in micrometers.

Analysis

1. Describe the general appearance of one pedicellaria.
2. What is the function of this structure?
3. Explain how the structure of pedicellariae assists in their function.

Echinoderms have radial symmetry

You may remember that radial symmetry is an advantage to animals that are stationary or move slowly. Radial symmetry enables these animals to sense potential food, predators, and other aspects of their environment from all directions. Observe the radial symmetry, as well as the various sizes and shapes of spines, of each echinoderm pictured in *Figure 29.1.*

The water vascular system

Another characteristic unique to echinoderms is the water vascular system that enables them to move, exchange gases, capture food, and excrete wastes. Look at the close-up of the underside of a sea star in *Figure 29.2.* You can see that grooves filled with tube feet run from the area of the sea star's mouth to the tip of each ray. **Tube feet** are hollow, thin-walled tubes that end in a suction cup. Tube feet look somewhat like miniature droppers. The round, muscular structure called the **ampulla** (AM puh lah) works something like the bulb of a dropper. The end of a tube foot works like a tiny suction cup. Each tube foot works independently of the others, and the animal moves along slowly by alternately pushing out and pulling in its tube feet. You can learn more about the operation of the water vascular system in the *Physics*

Figure 29.1
All echinoderms have radial symmetry as adults and an endoskeleton composed primarily of calcium carbonate.

A. A brittle star's long, snakelike rays are composed of overlapping, calcified plates covered with a thin layer of skin cells.

B. A living sand dollar has a solid, immovable skeleton composed of flattened plates that are fused together.

C. A sea lily's feathery rays are composed of calcified skeletal plates covered with an epidermis.

788

MEETING INDIVIDUAL NEEDS

Visually Impaired

Kinesthetic Provide visually impaired students with dried specimens of various echinoderms. Allow students to handle the specimens so they can feel the shapes, sizes, and characteristic spiny skins of these animals. [L1] [ELL]

GLENCOE TECHNOLOGY

CD-ROM
Biology: The Dynamics of Life
Exploration: *The Six Kingdoms*
Disc 3

Connection at the end of this chapter.

The **water vascular system** is a hydraulic system that operates under water pressure. Water enters and leaves the water vascular system of a sea star through the **madreporite** (MAH dray pohr ite), a sievelike, disk-shaped opening on the upper surface of the echinoderm's body. You can think of this disk as being like the little strainer that fits into the drain in a sink and keeps large particles out of the pipes. You can find out how a sea star eliminates wastes by reading the *Inside Story* on the next page.

Finally, tube feet function in gas exchange and excretion. Gases are exchanged and wastes are eliminated by diffusion through the thin walls of the tube feet.

Echinoderms have varied nutrition

All echinoderms have a mouth, stomach, and intestines, but their methods of obtaining food vary. Sea stars are carnivorous and prey on worms or on mollusks such as clams. Most sea urchins are herbivores and graze on algae. Brittle stars, sea lilies, and sea cucumbers feed on dead and decaying matter that drifts down to the ocean floor. Sea lilies capture this suspended organic matter with their tentaclelike tube feet and move it to their mouths.

Echinoderms have a simple nervous system

Echinoderms have no head or brain, but they do have a central nerve ring that surrounds the mouth. Nerves extend from the nerve ring down each ray. Each radial nerve then branches into a nerve net that provides sensory information to the animal. Echinoderms have cells that detect light and touch, but most do not have sensory organs. Sea stars are an exception. A sea star's body con-

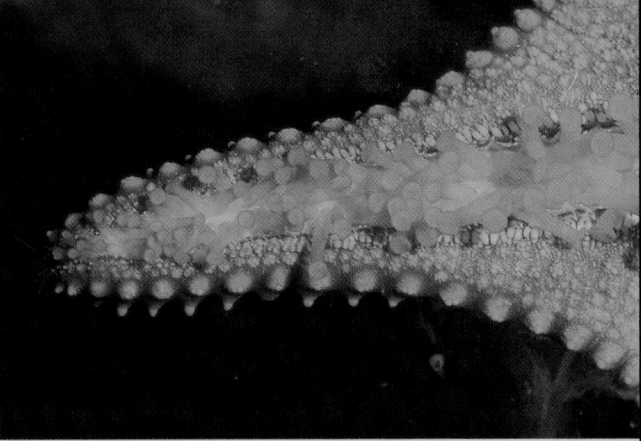

sists of long, tapering rays that extend from the animal's central disk. At the tip of each ray, on the underside, is an eyespot, a sensory organ consisting of a cluster of light-detecting cells. When walking, sea stars curve up the tips of their rays so that the eyespots are turned up and outward. This enables a sea star to detect the intensity of light coming from every direction.

Echinoderms have bilaterally symmetrical larvae

If you examine the larval stages of echinoderms, you will find that they have bilateral symmetry, a feature more common to chordates. The ciliated larva that develops from the fertilized egg of an echinoderm is shown in *Figure 29.3*. Through metamorphosis, the free-swimming larvae make dramatic changes in both body parts and in symmetry. The bilateral symmetry of echinoderm larvae indicates that echinoderm ancestors also may have had bilateral symmetry, suggesting a close relationship to the chordates. You can observe sea urchin development in the *BioLab* at the end of this chapter.

Figure 29.2
Tube feet enable sea stars and other echinoderms to creep along the ocean bottom or to pry open the shells of bivalves.

Figure 29.3
These sea urchin larvae are only 1 mm in size. The larval stage of echinoderms is bilateral, even though the adult stage has radial symmetry.

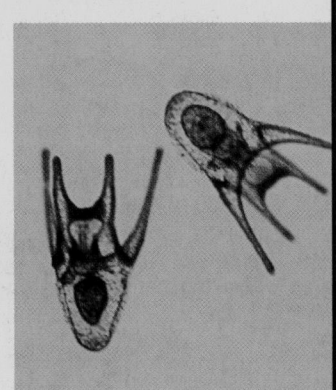

29.1 ECHINODERMS **789**

Quick Demo

Visual-Spatial Have students observe the pedicellariae of a live sea urchin under a stereomicroscope and make sketches of their observations. Have them touch one pedicellaria with a toothpick to observe the structure's response. **L2** **ELL**

Quick Demo

Kinesthetic Provide each student with a dropper. Have them squeeze the air from the dropper, and then, while still applying pressure to the rubber end of the dropper, touch the dropper tip to their finger. They should release the pressure on the dropper and observe how the dropper holds to their finger. Explain that this is similar to the suction action of the tube feet of a sea star. **L2** **ELL**

✔ Assessment

Portfolio Ask students to make a table showing the different groups of echinoderms, the type of food they eat, and whether or not they move around to find their food. Ask them to relate the type of food to the animals' methods of locomotion. Have them include their table and explanation in their portfolios. **L2** **P**

Resource Manager

BioLab and MiniLab Worksheets, p. 129 **L2**

TECHPREP

Raising Sea Stars

Visual-Spatial Ask students to visit a saltwater aquarium at a local zoo or pet shop. Have them visit when the sea stars and sea urchins are being fed. Ask students to observe and record sea star feeding behavior when live clams or oysters are added to the tank. Ask them to interview the person in charge of taking care of the aquarium to find out what daily tasks the caretaker must perform to maintain the sea stars and sea urchins. Have students report their findings to the class. **L2**

COOP LEARN

INSIDE STORY

Purpose

Students will learn about the structural and behavioral adaptations of a sea star.

Teaching Strategies

■ Ask students to observe a live sea star in action. They can use a hand lens to observe the tube feet as the sea star "walks" or "climbs" a surface of a marine aquarium. **L1** **ELL**

Visual Learning

■ Make photocopies of the Inside Story diagram without the labels of the structures. Number the parts and explain to students how each part functions and how the parts enable the sea star to survive in its environment. Have students use this information to label the diagram. **L1**

Critical Thinking

Radial symmetry enables animals that move slowly to sense and obtain potential food, escape predators, and sense other aspects of their environment from all directions.

GLENCOE TECHNOLOGY

CD-ROM
Biology: The Dynamics of Life

Exploration: *Echinoderms*
Disc 4

Exploration: *Symmetry*
Disc 4

INSIDE STORY

A Sea Star

If you ever tried to pull a sea star from a rock where it is attached, you would be impressed by how unyielding and rigid the animal seems to be. Yet at other times, the animal shows great flexibility, such as when it rights itself after being turned upside down.

Critical Thinking *How is radial symmetry useful to a sea star?*

Blood sea star

1 Endoskeleton A sea star can maintain a rigid structure or be flexible because it has an endoskeleton in the form of calcium carbonate plates just under its epidermis. The plates are connected by bands of soft tissue and muscle. When the muscles are contracted, the body becomes firm and rigid. When the muscles are relaxed, the body becomes flexible.

2 Madreporite Water flows in and out of the water vascular system through the madreporite.

3 Tube feet The suction action of tube feet, caused by the contraction and relaxation of the ampulla, is so strong that the sea star's muscles can open a clam or oyster shell.

4 Eyespots When moving, a sea star curves up the tips of its rays so that the eyespots are turned up and outward. Echinoderm eyespots distinguish between light and dark but do not form images.

8 Pedicellariae The pincerlike pedicellariae on the rays of the sea star will pinch any animal that tries to crawl over it.

7 Anus Waste products of digestion are eliminated through the anus.

6 Stomach To eat, a sea star pushes its stomach out of its mouth and spreads the stomach over the food. Powerful enzymes secreted by the digestive gland turn solid food into a soupy liquid that the stomach can easily absorb. Then the sea star pulls the stomach back into its body.

5 Digestive gland The digestive gland gives off chemicals for digestion.

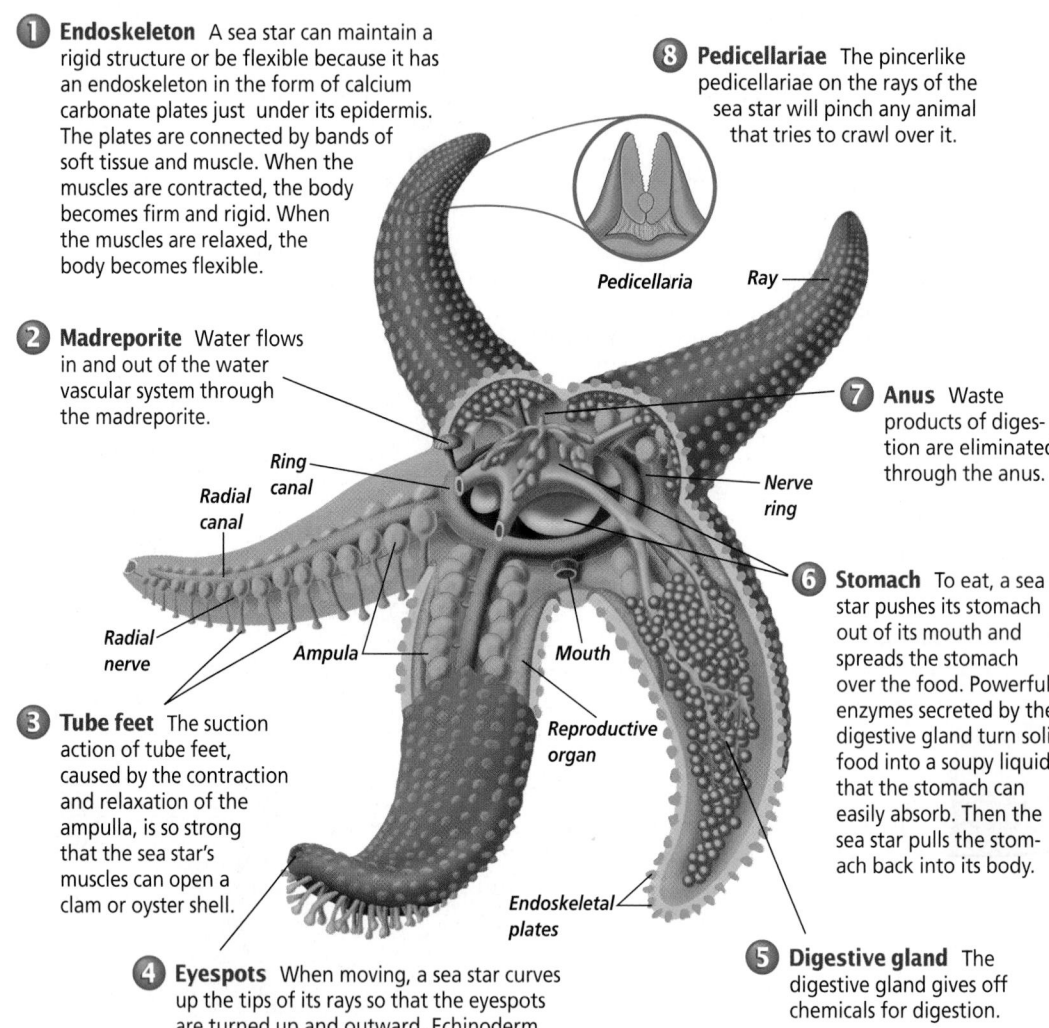

Pedicellaria   *Ray*

Ring canal

Radial canal

Radial nerve

Ampula

Mouth

Reproductive organ

Nerve ring

Endoskeletal plates

790

PROJECT

Echinoderm Display

Visual-Spatial Ask a group of students to create a bulletin board display of echinoderms. Have them highlight the features of echinoderms. Ask them to find examples of everyday objects that show similar traits. Students may use sandpaper to model the texture of echinoderm skin. **L2** **ELL** **COOP LEARN**

BIOLOGY JOURNAL

Sea Star Life

Linguistic Have students write an essay in which they describe one day in the life of a sea star. Allow a few students to read their essays. Have other students discuss which parts of the essays were scientific and which were creative fiction. **L3**

Diversity of Echinoderms

Approximately 6000 species of echinoderms exist today. More than one-third of these species are in the class Asteroidea (AS tuh royd ee uh), to which the sea stars belong. The four other classes of living echinoderms are Ophiuroidea (OH fee uh royd ee uh), the brittle stars; Echinoidea (eh kihn OYD ee uh), the sea urchins and sand dollars; Holothuroidea (HOH loh thuh royd ee uh), the sea cucumbers; and Crinoidea (cry NOYD ee uh), the sea lilies and feather stars.

Sea stars

Most species of sea stars have five rays, but some have more. Some species may have more than 40 rays. The rays are tapered and extend from the central disk. You have already read about the characteristics of sea stars that make them a typical example of echinoderms.

Brittle stars

As their name implies, brittle stars are extremely fragile, *Figure 29.4*. If you try to pick up a brittle star, parts of its rays will break off in your hand. This is an adaptation that helps the brittle star survive an attack by a predator. While the predator is busy with the broken-off ray, the brittle star can escape. A new ray will regenerate within weeks.

Brittle stars do not use their tube feet for locomotion. Instead, they propel themselves with the snakelike, slithering motion of their flexible rays. They use their tube feet to pass particles of food along the rays and into the mouth in the central disk.

Sea urchins and sand dollars

Sea urchins and sand dollars are globe- or disk-shaped animals covered with spines, as *Figure 29.4* shows. They do not have rays. The circular, flat skeletons of sand dollars have a five-petaled flower pattern on the surface. A living sand dollar is covered with minute, hairlike spines that are lost when the animal dies. A sand dollar has tube feet that protrude from the petal-like markings on its upper surface. These tube feet are modified into gills. Tube feet on the animal's bottom surface aid in bringing food particles to the mouth.

Sea urchins look like living pincushions, bristling with long, usually

Figure 29.4
Echinoderms are adapted to life in a variety of habitats.

Ⓐ Basket stars, a kind of brittle star, live on the soft substrate found below deep ocean waters.

Ⓑ Sea urchins often burrow into rocks to protect themselves from predators and rough water.

Ⓒ Sand dollars burrow into the sandy ocean bottom. They feed on tiny organic particles found in the sand.

29.1 ECHINODERMS **791**

Purpose

Students will design an experiment to determine the stimulus for release of sea cucumber gametes.

Process Skills

design an experiment, identify and control variables

Teaching Strategies

■ Ask students to think of strategies that animals may use to ensure aquatic fertilization. *Likely answers will include releasing large numbers of eggs and sperm, and releasing eggs and sperm at the same time.*

Thinking Critically

To design an experiment, students may decide to keep a group of female sea cucumbers in aquariums, control all environmental variables, and then release sperm into the water to see if this causes the release of eggs.

✓ Assessment

Portfolio Explain to students that sea stars have been found to spawn on the same day as sea cucumbers. Ask if this indicates that the stimulus for spawning is environmental or is in response to one male first releasing sperm. Ask students to write their responses in their portfolios. Use the Performance Task Assessment List for Making Observations and Inferences in **PASC**, p. 17. **L2** **P**

📁 Resource Manager

Content Mastery, p. 142 **L1**
Reinforcement and Study
Guide, pp. 127-129 **L2**
Basic Concepts Transparency
51 and Master **L2** **ELL**

Problem-Solving Lab 29-1 — Designing an Experiment

What makes sea cucumbers release gametes? The orange sea cucumber lives in groups of 100 or more per square meter. In the spring, these sea cucumbers produce large numbers of gametes (eggs and sperm), which they shed in the water all at the same time. The adaptive value of

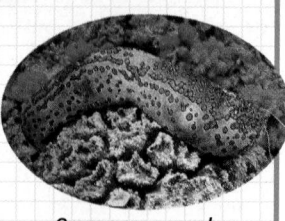

Orange sea cucumber

such behavior is that fertilization of many eggs is assured. When one male releases sperm, the other sea cucumbers in the population, both male and female, also release their gametes. Biologists do not know whether the sea cucumbers release their gametes in response to a seasonal cue, such as increasing day length or increasing water temperature, or whether they do this in response to the release of sperm by one sea cucumber.

Analysis

Design an experiment that will help to determine whether sea cucumbers release eggs and sperm in response to the release of sperm from one individual or in response to a seasonal cue.

Thinking Critically

If you find that female sea cucumbers release 200 eggs in the presence of male sperm and ten eggs in the presence of water that is warmer than the surrounding water, what would you do in your next experiment?

Figure 29.5
Sea lilies and feather stars use their feathery rays to capture downward-drifting organic particles (a). Sea cucumbers trap organic particles by sweeping their mucous-covered tentacles over the ocean bottom (b).

a

b

✓ Portfolio

Echinoderm Phylogeny

Naturalist Provide students with echinoderm fossils. Include as many different kinds as possible and use photographs of fossils you don't have. Ask students to arrange the fossils and photographs to illustrate the phylogeny of this group. **L3** **P**

---

pointed spines. They have long, slender tube feet that, along with the spines, aid the animal in locomotion.

The sea urchin's spines protect it from predators. In some species, sacs located near the tips of the spines contain a poisonous fluid that is injected into an attacker, further protecting the urchin. The spines also aid in locomotion and in burrowing. Burrowing species move their spines in a circular motion that grinds away the rock beneath them. This action, which is aided by a chewing action of the mouth, forms a depression in the rock that helps protect the urchin from predators and from wave action that could wash it out to sea.

Sea cucumbers

Sea cucumbers are so called because of their vegetablelike appearance, shown in *Figure 29.5.* Their leathery covering allows them to be more flexible than other echinoderms; they pull themselves along the ocean floor using tentacles and tube feet. When sea cucumbers are threatened, they exhibit a curious behavior. They may expel a tangled, sticky mass of tubes through the anus, or they may rupture, releasing some internal organs that are regenerated in a few weeks. These actions confuse their predators, giving the sea cucumber an opportunity to move away. Sea cucumbers reproduce by shedding eggs and sperm into the water, where fertilization occurs. You can find out more about sea cucumber reproduction in the *Problem-Solving Lab* on this page.

Sea lilies and feather stars

Sea lilies and feather stars resemble plants in some ways. Sea lilies are the only sessile echinoderms. Feather stars are sessile only in larval form. The adult feather star uses its feathery arms to swim from place to place.

GLENCOE TECHNOLOGY

CD-ROM
Biology: The Dynamics of Life
BioQuest: *Biodiversity Park*
Disc 3, 4

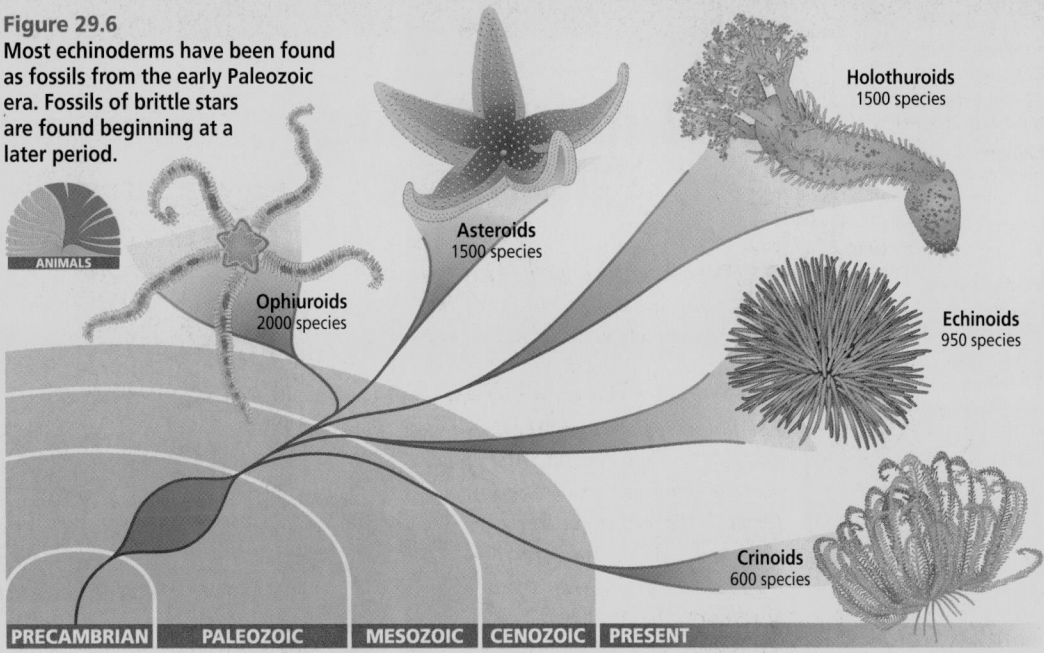

Figure 29.6
Most echinoderms have been found as fossils from the early Paleozoic era. Fossils of brittle stars are found beginning at a later period.

ANIMALS

Ophiuroids
2000 species

Asteroids
1500 species

Holothuroids
1500 species

Echinoids
950 species

Crinoids
600 species

PRECAMBRIAN   PALEOZOIC   MESOZOIC   CENOZOIC   PRESENT

Origins of Echinoderms

The earliest echinoderms may have been bilaterally symmetrical as adults, and probably were attached to the ocean floor by stalks. Another view of the earliest echinoderms is that they were bilateral and free-swimming. The development of bilateral larvae is one piece of evidence biologists have for placing echinoderms as the closest invertebrate relatives of the chordates.

Recall that most invertebrates show protostome development, whereas deuterostome development appears mainly in chordates. The echinoderms represent the only major group of deuterostome invertebrates.

Because the endoskeletons of echinoderms easily fossilize, there is a good record of this phylum. Echinoderms, as a group, date from the Paleozoic era, as shown in *Figure 29.6*. More than 13 000 fossil species have been identified.

Section Assessment

Understanding Main Ideas
1. How does a sea star move? Explain in terms of the water vascular system of echinoderms.
2. Describe the differences in symmetry between larval echinoderms and adult echinoderms.
3. How are sea cucumbers different from other echinoderms?
4. Compare how sea urchins and sea cucumbers obtain food.

Thinking Critically
5. How do the various defense mechanisms among the echinoderm classes help deter predators?

SKILL REVIEW
6. **Classifying** Prepare a key that distinguishes among classes of echinoderms. Include information on features you may find significant. For more help, refer to *Organizing Information* in the **Skill Handbook.**

Section Assessment

1. A sea star moves by regulation of its water vascular system. Tube feet attach to a surface, the sea star moves itself forward, and the suction is released.
2. Larval echinoderms are bilaterally symmetrical, whereas adult echinoderms are radially symmetrical.
3. Sea cucumbers are tubular and have a leathery outer covering instead of hard plates.
4. Sea urchins graze on algae; sea cucumbers feed on dead matter that drops to the ocean floor.
5. The rigid endoskeleton helps protect echinoderms from their enemies. Spines and poison glands also protect echinoderms. Adult echinoderms move by walking, whereas larval forms are free swimming. If an echinoderm such as a sea star loses part of a ray, it can be regenerated. Sea cucumbers can expel their digestive tracts and grow new ones.
6. Students' keys will vary considerably, but all should utilize the branching nature of keys described in the Skill Handbook.

793

Prepare

Key Concepts

Students will learn about invertebrate chordates. They will distinguish between sea squirts and lancelets and study the relationships of these animals to the vertebrates.

Planning

- Prepare syringes with potassium chloride and gather beakers, glass slides, coverslips, and test tubes for the BioLab.

1 Focus

Bellringer

Before presenting the lesson, display **Section Focus Transparency 72** on the overhead projector and have students answer the accompanying questions. **L1** **ELL**

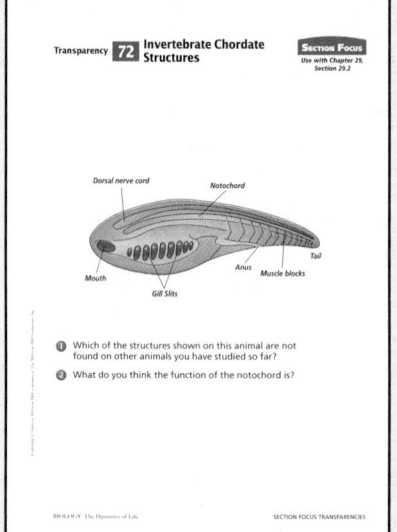

SECTION PREVIEW

Objectives
Summarize the characteristics of chordates.
Explain how invertebrate chordates are related to vertebrates.
Distinguish between sea squirts and lancelets.

Vocabulary
notochord
dorsal hollow nerve cord
gill slit

Section
29.2 Invertebrate Chordates

The brightly colored object pictured here is a sea squirt. As one of your closest invertebrate relatives, it is placed, along with humans, in the phylum Chordata. At first glance, this sea squirt may seem to resemble a sponge more than its fellow chordates. It is sessile, and it filters food particles from water it takes in through the opening at the top of its body. What characteristics could a human—or a fish or a lizard, for that matter—share with this colorful, ocean-dwelling organism?

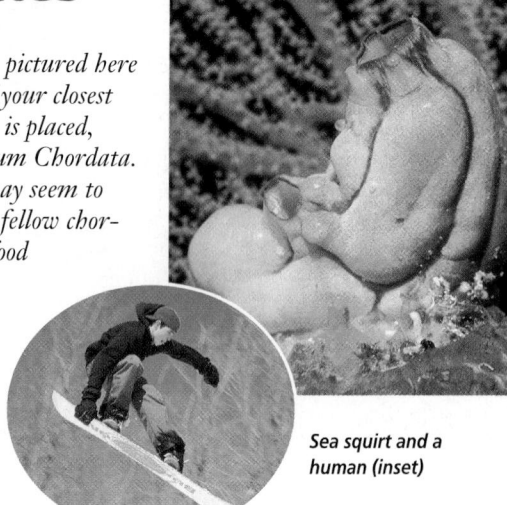

Sea squirt and a human (inset)

What Is an Invertebrate Chordate?

The chordates most familiar to you are the vertebrate chordates—chordates that have backbones, such as birds, fishes, and mammals, including humans. But the phylum Chordata (kor DAHT uh) includes three subphyla: Urochordata, the tunicates (sea squirts); Cephalochordata, the lancelets; and Vertebrata, the vertebrates. In this section you will examine the tunicates and lancelets—invertebrate chordates that have no backbones. You will study the vertebrate chordates in the next unit.

Invertebrate chordates may not look much like fishes, reptiles, or humans, but like all other chordates, they have a notochord, a dorsal hollow nerve cord, gill slits, and muscle blocks at some time during their development. In addition, all chordates have bilateral symmetry, a well-developed coelom, and segmentation. The features shared by invertebrate and vertebrate chordates are illustrated in *Figure 29.7.* You can observe these features in invertebrate chordates in the *Problem-Solving Lab* later in this section.

Figure 29.7
Chordate characteristics—the notochord, dorsal hollow nerve cord, gill slits, and muscle blocks—are shared by invertebrate as well as vertebrate chordates.

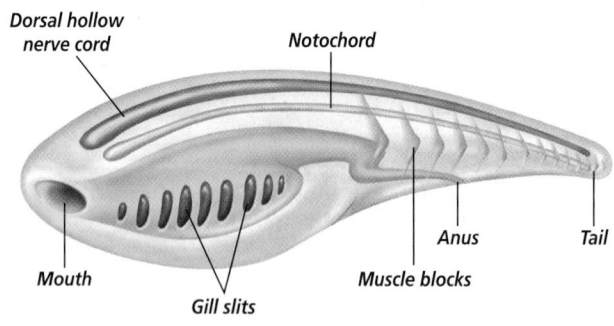

Dorsal hollow nerve cord   *Notochord*

Mouth   *Gill slits*   *Muscle blocks*   *Anus*   *Tail*

794 ECHINODERMS AND INVERTEBRATE CHORDATES

Resource Manager

Section Focus Transparency 72 and Master **L1** **ELL**
Concept Mapping, p. 29 **L3** **ELL**
Reteaching Skills Transparency 43 and Master **L1** **ELL**

BIOLOGY JOURNAL

Tunicate Research

Ask students to use the Internet and their local library to do research on tunicates. Ask them to find out where tunicates live and if any tunicate species are threatened or endangered. Have students place their findings in their journals. **L2**

All chordates have a notochord

All chordate embryos have a **noto-chord** (NOHT uh kord)—a long, semirigid, rodlike structure located between the digestive system and the dorsal hollow nerve cord. The noto-chord is made up of large, fluid-filled cells held within stiff, fibrous tissues. In invertebrate chordates, the noto-chord is retained into adulthood. But in vertebrate chordates, this structure is replaced by a backbone. Inverte-brate chordates do not develop a backbone.

The notochord develops just after the formation of a gastrula from mesoderm on what will be the dorsal side of the embryo. The physical sup-port provided by a notochord enables invertebrate chordates to make pow-erful side-to-side movements of the body. These movements propel the animal through the water at a great speed.

All chordates have a dorsal hollow nerve cord

The **dorsal hollow nerve cord** in chordates develops from a plate of ectoderm that rolls into a hollow tube. This occurs at the same time as the development of the notochord. The sequence of development of the dorsal hollow nerve cord is illustrated in *Figure 29.8*. This tube is composed of cells surrounding a fluid-filled canal that lies above the notochord. In most adult chordates, the cells in the posterior portion of the dorsal hollow nerve cord develop into the spinal cord. The cells in the anterior portion develop into a brain. A pair

Figure 29.8
After gastrulation, organs begin to form in a chordate embryo.

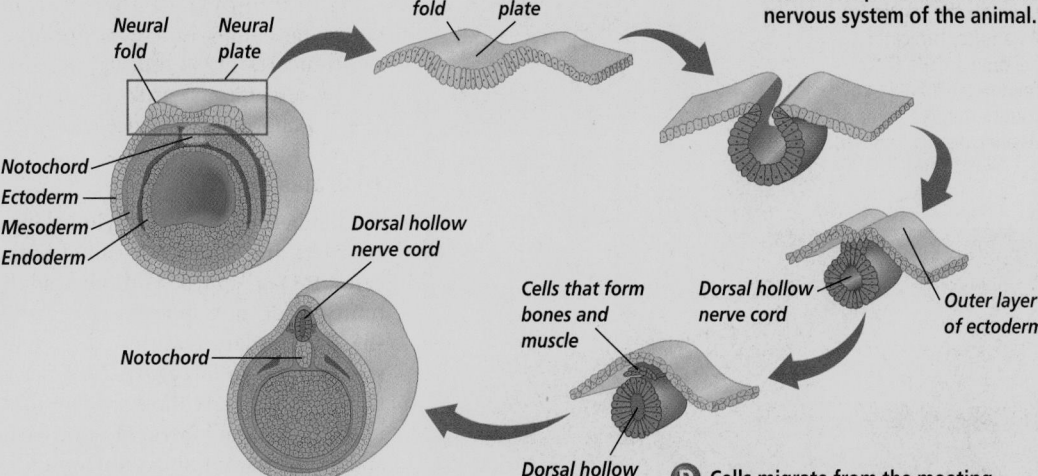

A The notochord forms from mesoderm on the dorsal side of a developing embryo.

Neural fold   Neural plate

Notochord
Ectoderm
Mesoderm
Endoderm

Dorsal hollow nerve cord

Notochord

Cells that form bones and muscle

Dorsal hollow nerve cord

B The dorsal hollow nerve cord originates as a plate of dorsal ectoderm just above the developing notochord.

Neural fold   Neural plate

C The edges of this plate of ectoderm fold inward, eventu-ally meeting to form a hollow tube surrounded by cells. The dorsal hollow nerve cord pinches off from the ectoderm and develops into the central nervous system of the animal.

Dorsal hollow nerve cord

Outer layer of ectoderm

D Cells migrate from the meeting margins of the neural tube and eventually form other organs, including bones and muscles.

of nerves connects the nerve cord to each block of muscles.

All chordates have gill slits

The **gill slits** of a chordate are paired openings located in the pharynx, behind the mouth. Many chordates have several pairs of gill slits only during embryonic development. Invertebrate chordates that have gill slits as adults use these structures to strain food from the water. In some vertebrates, especially the fishes, the gill slits develop into internal gills that are adapted to exchange gases during respiration.

All chordates have muscle blocks

Muscle blocks are modified body segments that consist of stacked muscle layers. You have probably seen muscle blocks when you ate a cooked fish. The blocks of muscle cause the

Figure 29.9
Tunicate larvae are about 1 cm long and are able to swim freely through the water (**a**). As adults, tunicates become sessile filter feeders enclosed in a tough, baglike layer of tissue called a tunic (**b**).

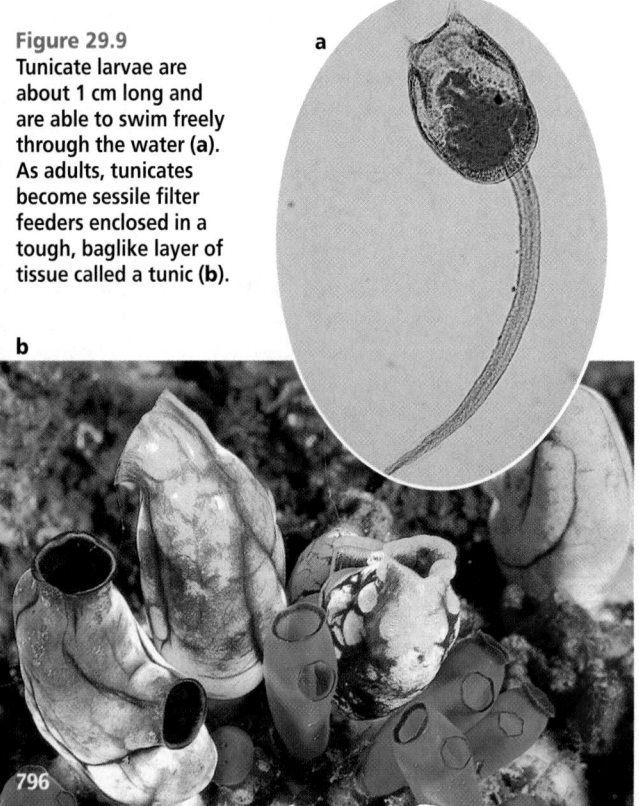

meat to separate easily into flakes. Muscle blocks are anchored by the notochord, which gives the muscles a firm structure to pull against. As a result, chordates tend to be more muscular than members of other phyla.

Muscle blocks also aid in movement of the tail. At some point in development, all chordates have a muscular tail. As you know, humans are chordates, and during the early development of the human embryo, there is a muscular tail that disappears as development continues. In most animals that have tails, the digestive system extends to the tip of the tail, where the anus is located. Chordates, however, usually have a tail that extends beyond the anus. You can observe many of the chordate traits in a lancelet in the *MiniLab* on the next page.

Diversity of Invertebrate Chordates

The invertebrate chordates belong to two subphyla of the phylum chordata: subphylum Urochordata, the tunicates (TEW nuh kaytz), also called sea squirts, and subphylum Cephalochordata, the lancelets.

Tunicates are sea squirts

Members of the subphylum Urochordata are commonly called tunicates, or sea squirts. Although adult tunicates do not appear to have any shared chordate features, the larval stage, as shown in *Figure 29.9*, has a tail that makes it look similar to a tadpole. Tunicate larvae do not feed, and are free swimming only for a few days after hatching. Then they settle and attach themselves with a sucker to boats, rocks, and the ocean bottom. Many adult tunicates secrete a tunic, a tough sac made of cellulose, around their bodies. Colonies of tunicates sometimes secrete just one big tunic that has a common opening to the outside. You can find out how tunicates eat in the *Inside Story* on the next page.

Only the gill slits in adult tunicates indicate their chordate relationship. Adult tunicates are small, tubular animals that range in size from microscopic to several centimeters long, about as big as a large potato. If you remove a tunicate from its sea home, it might squirt out a jet of water for protection—hence the name *sea squirt.*

Lancelets are similar to fishes

Lancelets belong to the subphylum Cephalochordata. They are small, streamlined, and common marine animals, usually about 5 cm long, as *Figure 29.10* shows. They spend most of their time buried in the sand with only their heads sticking out. Like tunicates, lancelets are filter feeders. Unlike tunicates, however, lancelets retain all their chordate features throughout life.

MiniLab 29-2 Observing

Examining a Lancelet *Branchiostoma californiense* is a small, sea-dwelling lancelet. At first glance, it appears to be a fish. However, its structural parts and appearance are quite different.

Procedure

1. Place the lancelet onto a glass slide. **CAUTION:** *Wear disposable latex gloves and handle preserved material with forceps.*
2. Use a dissecting microscope to examine the animal. **CAUTION:** *Use care when working with a microscope and slides.*
3. Prepare a data table that will allow you to record the following: General body shape, Length in mm, Head region present, Fins and tail present, Nature of body covering, Sense organs such as eyes present, Habitat, Segmented body.
4. Indicate on your data table if the following can easily be observed: gill slits, notochord, dorsal hollow nerve cord.

Analysis

1. How does *Branchiostoma* differ structurally from a fish? How are its general appearance and habitat similar to those of a fish?
2. Explain why you were not able to see gills, notochord, and a dorsal hollow nerve cord.
3. Using its scientific name as a guide, where might the habitat of this species be located?

Figure 29.10
Lancelets usually spend most of their time buried in the sand with only their heads sticking out so they can filter tiny morsels of food from the water (**a**). The lancelet's body looks very much like a typical chordate embryo (**b**).

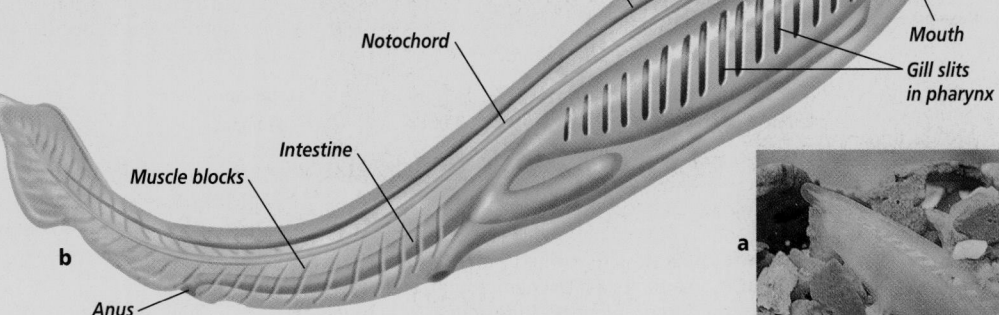

Oral hood with tentacles

Dorsal hollow nerve cord

Notochord

Intestine

Muscle blocks

Mouth

Gill slits in pharynx

b

Anus

a

MiniLab 29-2

Purpose
Students will observe the external appearance of lancelets.

Process Skills
compare and contrast, interpret data, make and use tables, measure in SI, observe and infer

Teaching Strategies
■ Specimens are available from biological supply houses.
■ A hand lens may be substituted for a binocular microscope.

Expected Results
Lancelets have a long, tubular body shape; length close to 50 mm, head region, tail-like posterior; smooth body; no sense organs. Students will not be able to see a notochord, gill slits, or a dorsal hollow nerve cord.

Analysis
1. It has no sense organs, gills, or fins. But it does have a distinct head area. Fishes are classified in the same phylum as lancelets—Chordata—but in subphylum Vertebrata, whereas lancelets are in subphylum Cephalochordata.
2. These structures are all internal organs.
3. in the ocean along the coast of California

✓ Assessment

Portfolio Have students write a paragraph explaining why lancelets are important in evolution. Use the Performance Task Assessment List for Writing in Science in **PASC**, p. 87. **L2** **P**

---

notochord, gill slits, muscle blocks, tail.
6. Note the fishlike form of the lancelet.

Expected Results
Students should see all of the features listed above.

Analysis
1. How does the size of the sea urchin blastula compare with that of the zygote? *The blastula and the zygote are the same size. The cells become smaller as development proceeds.*

2. What does the symmetry of the larva of a sea urchin imply about the evolutionary relationship between echinoderms and chordates? *Echinoderms are closely related to chordates.*

3. What features of the lancelet place it in the phylum Chordata? *dorsal hollow nerve cord, notochord, gill slits, muscle blocks, tail*

✓ Assessment

Skill Set up microscopes with slides showing features of lancelets and various stages of sea urchin development. Ask students to identify the part or sequence of development and state the importance of the part for chordates. Use the Performance Task Assessment List for Making Observations and Inferences in **PASC**, p. 17. **L2**

Purpose 📖

Students will learn about the structural and behavioral characteristics of tunicates.

Teaching Strategies

■ Ask students to compare the filter feeding of tunicates with the filter feeding of sponges and bivalves.

Visual Learning

■ Review with students the characteristics of tunicates that make them invertebrate chordates.

■ Show students live specimens of tunicates and have them compare the specimens with the diagram.

Critical Thinking

Sponges and tunciates are both filter feeders.

Problem-Solving Lab 29-2

Purpose 📖

Students will interpret a cross-section slice of a lancelet.

Process Skills

interpret scientific illustrations, observe and infer, think critically

Teaching Strategies

■ Review the procedure for making a cross-section slice.

■ Review the orientation of the cross-section slice in relation to the whole animal. Use a cucumber to illustrate the view and orientation by making a cross-section slice through the fruit.

Thinking Critically

1. Gill slits, muscle blocks, notochord, dorsal hollow nerve cord; student answers should agree that these are present.

A Tunicate

Tunicates, or sea squirts, are a group of about 1250 species that live in the ocean. They may live near the shore or at great depths. They may live individually, or several animals may share a tunic to form a colony.

Critical Thinking *In what ways are sponges and tunicates alike?*

Purple bell tunicate

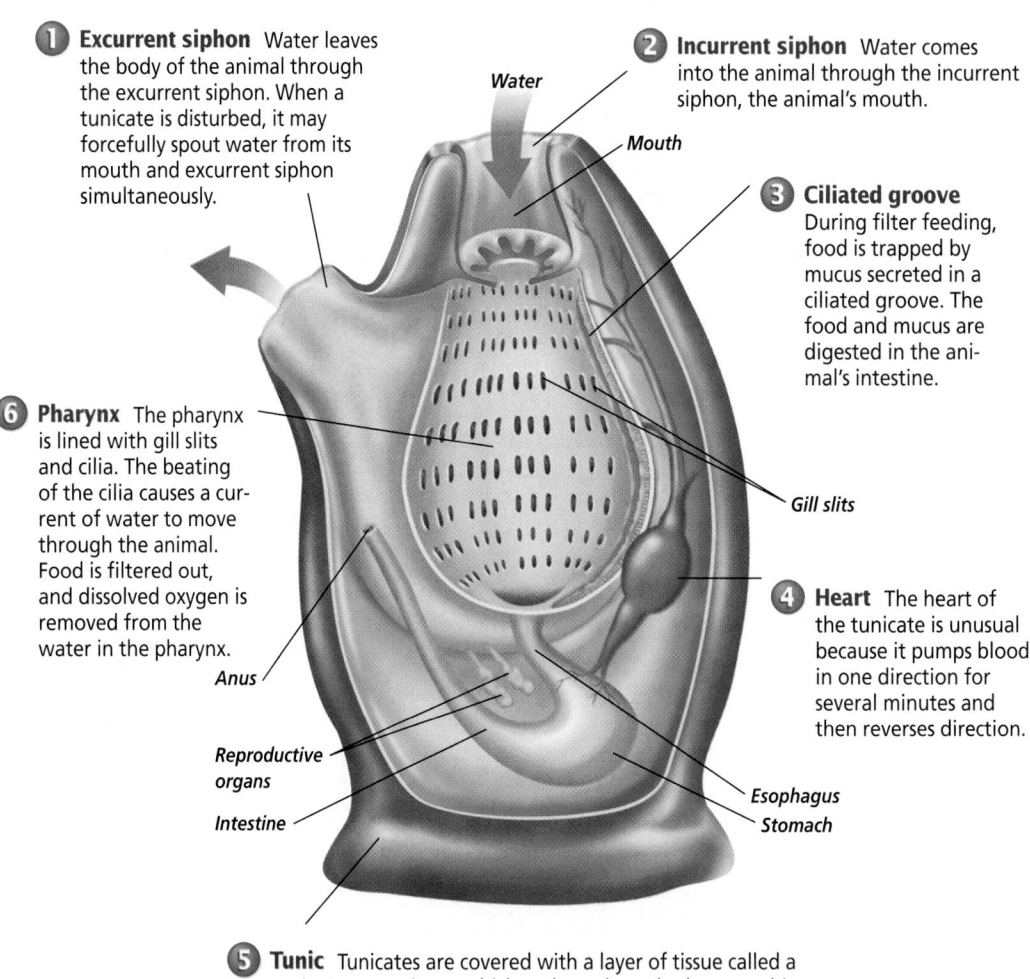

① Excurrent siphon Water leaves the body of the animal through the excurrent siphon. When a tunicate is disturbed, it may forcefully spout water from its mouth and excurrent siphon simultaneously.

② Incurrent siphon Water comes into the animal through the incurrent siphon, the animal's mouth.

③ Ciliated groove During filter feeding, food is trapped by mucus secreted in a ciliated groove. The food and mucus are digested in the animal's intestine.

④ Heart The heart of the tunicate is unusual because it pumps blood in one direction for several minutes and then reverses direction.

⑥ Pharynx The pharynx is lined with gill slits and cilia. The beating of the cilia causes a current of water to move through the animal. Food is filtered out, and dissolved oxygen is removed from the water in the pharynx.

Water

Mouth

Gill slits

Anus

Reproductive organs

Intestine

Esophagus

Stomach

⑤ Tunic Tunicates are covered with a layer of tissue called a tunic. Some tunics are thick and tough, and others are thin and translucent. All protect the animal from predators.

798 ECHINODERMS AND INVERTEBRATE CHORDATES

2. Echinoderms would have tube feet and a water vascular system, but no notochord, dorsal hollow nerve cord, gill slits, or muscle blocks.

3. An adult tunicate would have gill slits but no notochord or dorsal hollow nerve cord. A young developing tunicate has all of these structures.

✓ Assessment

Performance Ask students to draw an events chain that sequences the steps they would need to take to determine that a specimen is a lancelet rather than an immature fish. Use the Performance Task Assessment List for Events Chain in **PASC**, p. 91. **L2**

Although lancelets look somewhat similar to fishes, they have only one layer of skin, with no pigment and no scales. Lancelets do not have a distinct head, but they do have light sensitive cells on the anterior end. They also have a hood that covers the mouth and the sensory tentacles surrounding it. The tentacles direct the water current and food particles toward the animal's mouth.

Origins of Invertebrate Chordates

Because sea squirts and lancelets have no bones, shells, or other hard parts, their fossil record is incomplete. Biologists are not sure where sea squirts and lancelets fit in the phylogeny of chordates. According to one hypothesis, echinoderms, invertebrate chordates, and vertebrates all arose from ancestral sessile animals that fed by capturing food in tentacles. Modern vertebrates probably arose from the free-swimming larval stages of ancestral invertebrate chordates. Recent discoveries of fossil forms of organisms that are similar to living lancelets in rocks 550 million years old show that invertebrate chordates probably existed before vertebrate chordates.

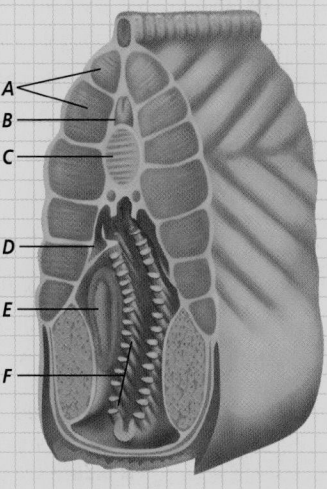
Section Assessment

Understanding Main Ideas
1. Describe the four features of chordates.
2. How are invertebrate chordates different from vertebrates?
3. Compare the physical features of sea squirts and lancelets.
4. How do sea squirts and lancelets protect themselves?

Thinking Critically
5. What features of chordates suggest that you are more closely related to invertebrate chordates than to echinoderms?

SKILL REVIEW

6. **Designing an Experiment** Assume that you have found some tadpolelike animals in the water near the seashore and that you can raise them in a laboratory. Design an experiment in which you will determine whether the animals are larvae or adults. For more help, refer to *Practicing Scientific Methods* in the **Skill Handbook**.

3 Assess

Check for Understanding
Ask students to list invertebrate chordate characteristics. **L1**

Reteach
Visual-Spatial Give student groups a large piece of paper. Have them diagram a lancelet and label its typical invertebrate chordate parts and their functions. **L2 ELL**

Extension
Ask students to research the phyla Hemichordata and Chaetognatha. Ask them to report about the structural and behavioral adaptations of these animals and their evolutionary relationships with echinoderms and chordates. **L3**

✓ Assessment
Performance Have students plan an exhibit for tunicates at a public aquarium. Ask them to include in their plan suggestions about capture, transport, habitat, feeding, and general maintenance. **L2 COOP LEARN**

4 Close

Activity
Develop a table that compares lancelets and tunicates, discussing which features are common to vertebrates and invertebrates.

Resource Manager
Content Mastery, pp. 141, 143-144 **L1**
Reinforcement and Study Guide, p. 130 **L2**

Section Assessment

1. notochord, dorsal hollow nerve cord, gill slits, muscle blocks
2. In invertebrate chordates, the notochord is not replaced by a backbone.
3. Sea squirts are small, tubular, stationary filter feeders. Lancelets are shaped like fishes and can swim freely, but they spend most of their time buried in the sand.
4. Sea squirts are covered with a thick tunic, and lancelets bury themselves in sand.
5. notochord, gill slits, muscle blocks, and dorsal hollow nerve cord
6. Watch the animals for several weeks to see if they change their body shape and symmetry. Watch for reproductive behavior or release of gametes.

Time Allotment

One class period the first day; 15 minutes the second day

Process Skills

make and use tables, collect data, communicate, compare and contrast, measure in SI, observe and infer, organize data, use numbers

Safety Precautions

Caution students to use care when working with syringes and chemicals. Make sure they return syringes to the designated place.

Alternative Materials

■ If time and cost prohibit the purchase of live sea urchins, the experiment can be done with prepared slides available from biological supply houses.

Resource Manager

BioLab and MiniLab Worksheets, pp. 131-132 L2

Observing Sea Urchin Gametes and Egg Development

Sea urchins are typical of most echinoderms in that their sexes are separate, fertilization is external, and development of a fertilized egg is quite rapid. Thus, these animals are excellent choices for studying gametes, watching fertilization, and observing changes occurring in a fertilized egg.

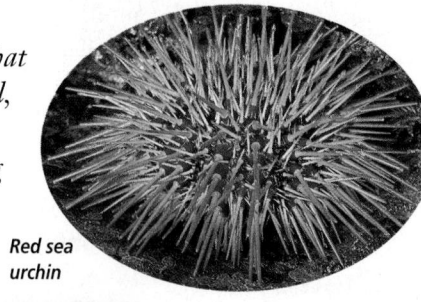

Red sea urchin

Problem

How can you induce a sea urchin to release its gametes?

Objectives

In this BioLab, you will:

■ **Induce** sea urchins to release their gamete cells.
■ **Observe** living sperm and egg cells under the microscope.
■ **Observe** developmental changes in a fertilized sea urchin egg.

Materials

live sea urchins
sea water
glass slides and
cover slips
syringe filled with
potassium chloride

beakers
petri dish
dropper
microscope
test tube

Safety Precautions

Always wear goggles in the lab.

Skill Handbook

Use the **Skill Handbook** if you need additional help with this lab.

1. Fill a small beaker (250 mL) with sea water.
2. Obtain a live sea urchin from your teacher and locate an area of soft tissue next to its mouth.
3. Using a syringe, your teacher will insert the needle into this soft tissue and inject the syringe contents into the sea urchin.
4. Turn your animal so that its mouth is facing up and place it in a petri dish. **CAUTION: *Use care in handling live animals.***
5. Wait a minute or two, then check the petri dish. If the sea

urchin is male, a milky white mass of sperm will be present in the dish. If it is female, a yellow orange mass of eggs will be seen.

6. If you have a female sea urchin, hold her upside down directly over the seawater-filled beaker and allow the eggs to fall directly into the water.
7. If your urchin is male, use a dropper to add several drops of sperm from the petri dish to your beaker of sea water.
8. Check with your classmates to see who has a male and who has a

---

Teaching Strategies

■ Student groups of at least four are recommended. Each student should be assigned a specific duty or role.
■ Sea urchins will typically arrive in a large foam container and can remain there for at least 24-48 hours before

using. Ordering so that animals arrive early in the week is recommended. Animals are available from biological supply houses.
■ Use 2 mL of a 0.5*M* potassium chloride (KCl) solution for each injection.
■ The container in which the urchins arrive will contain enough seawater for classroom use.

■ To reduce the number of sea urchins needed, have the first morning class do the injecting and capturing of gametes. Store sperm and eggs as follows for later classes: Cover petri dish with sperm, store in a refrigerator at 10°C; store eggs in beaker of seawater at same temperature as sperm. Try to use stored gametes the same day.

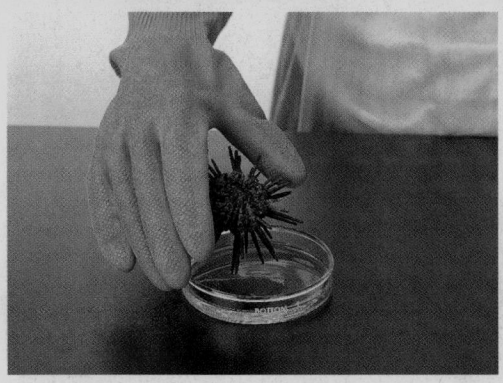

female sea urchin. Share gamete cells.

9. Use a clean dropper to transfer a drop of sperm from the beaker to a microscope slide. Observe under low power without a cover slip.

10. Add a cover slip and observe under high power. Note the movement of sperm. Draw several sperm cells and indicate their size in μm. Note the approximate number of sperm cells present.

11. Repeat steps 9 and 10 for egg cells. In step 10, use only low power to observe egg cells.

12. For this step, work with a partner. While one partner transfers some

sperm to the slide with egg cells using a clean dropper, the other partner should observe under low power.

13. Observe the process of fertilization and note any changes that occur to the egg. Record your observations in a data table.

14. When fertilization has been accomplished, place the fertilized eggs in a test tube filled with 10 mL of seawater. Label your tube and observe the eggs 24 hours later under low power. Record any changes that you see. **CAUTION: *Wash your hands immediately after working with animals.***

ANALYZE AND CONCLUDE

1. **Compare and Contrast** Compare eggs and sperm, noting numbers released, numbers observed under low power, size, and ability to move.

2. **Predicting** Based on the pattern of fertilization, predict the reason for the large number of gametes released in nature.

3. **Observing** Describe the behavior of sperm when they first come in contact with an egg.

4. **Observing** How does an unfertilized egg differ in appearance

from a fertilized egg? Draw both eggs in your data table.

Going Further

Project Continue to observe fertilized eggs and note the stages of development. Keep a record of time after fertilization and corresponding changes in development.

BIOLOGY *Online* To find out more about echinoderm development, visit the Glencoe Science Web site. **science.glencoe.com**

ANALYZE AND CONCLUDE

1. Sperm are motile, thousands are present on a slide, the size is close to 10 μm; eggs are not motile, fewer than 100 on a slide, size is close to 150 μm.

2. Student answers may vary; fertilization is external, so chances of a sperm meeting an egg are less; a large number of gametes released improves the chances of fertilization.

3. Sperm tend to cluster around the outer edge of the egg.

4. A fertilized egg forms a clear membrane around the outside edge of the egg shortly after fertilization.

✓ Assessment

Knowledge Have students correlate their observations of sea urchin gametes with the pattern seen throughout the animal kingdom. Is there that much variation between sperm and eggs of different species? Use the Performance Task Assessment List for Making Observations and Inferences in **PASC**, p. 17. **L2**

Going Further

Have students determine how long sperm cells will remain alive (as judged by their motility) after being released from the male sea urchin.

Troubleshooting

■ Some sea urchins may not release gametes with the initial injection. Repeat if necessary.

■ Make sure that you collect the same number of syringes as handed out.

Data and Observation

Students will observe the differences between egg and sperm. They will determine that developmental changes occur within minutes after fertilization.

Fertilization membrane appears	2-5 minutes
First cleavage	50-70 minutes
Second cleavage	78-107 minutes
Third cleavage	103-205 minutes
Blastula	6 hours
Gastrula	12-20 hours
Larva	24-48 hours

Physics
Connection

Purpose
Students learn the application of physical principles to biological systems.

Teaching Strategies

■ Have students refer to the Inside Story of a sea star earlier in this chapter to see how the water vascular system is connected to the ampullas and tube feet.

■ Explain that hydraulics is the practical application of liquids in motion and that in living systems the liquid is usually water. The term *hydraulic* means "operated, moved, or effected by water."

■ Have students discuss how flow of fluids in the body is essential to life. Topics might include the delivery of oral medication, distribution of hormones secreted by glands, or the removal of toxic substances in urine. **L1**

Connection to Biology
Scallops take in water and then clap their shells together, forcing water out quickly to move the animal in the opposite direction from the water flow. Earthworms have a hydrostatic skeleton (rigid, water-filled coelom) that their muscles brace against for movement through the soil.

Physics
Connection

Hydraulics of Sea Stars

Many organisms use hydraulic systems to supply food and oxygen to, and remove wastes from, cells lying deep within the body. Hydraulics is a branch of science that is concerned with the practical applications of liquids in motion. In living systems, hydraulics is usually concerned with the use of water to operate systems that help organisms find food and move from place to place.

The sea star uses a unique hydraulic mechanism called the water vascular system for movement and for obtaining food. The water vascular system provides the water pressure that operates the tube feet of sea stars and other echinoderms.

The water vascular system On the upper surface of a sea star is a sievelike disk, the madreporite, which opens into a fluid-filled ring. Extending from the ring are long radial canals running along a groove on the underside of each of the sea star's rays. Many small lateral canals branch off from the sides of the radial canals. Each lateral canal ends in a hollow tube foot. The tube foot has a small muscular bulb at one end, the ampulla, and a short, thin-walled tube at the other end that is usually flattened into a sucker. Each ray of the sea star has many tube feet arranged in two or four rows on the bottom side of the ray. The tube feet are extended or retracted by hydraulic pressure in the water vascular system.

Mechanics of the water vascular system The entire water vascular system is filled with water and acts as a hydraulic system, allowing the sea star to move. The muscular ampulla contracts and relaxes with an action similar to the squeezing of a dropper bulb. When the muscles in the wall of the ampulla contract, a valve between the lateral canal and the ampulla closes so that water does not flow backwards into the radial canal. The pressure from the walls of the ampulla acts

Sea star opening a mollusk to feed

on the water, forcing it into the tube foot's sucker end, causing it to extend.

When the extended tube foot touches a rock or a mollusk shell, the center of the foot is retracted slightly. This creates a vacuum, enabling the tube foot to adhere to the rock or shell. The tip of the tube foot also secretes a sticky substance that helps it adhere. To move forward, muscles in the ampulla relax, and muscles in the tube foot wall contract. These actions shorten the tube foot and pull the sea star forward. Water is forced back into the relaxed ampulla. When the muscles in the ampulla contract, the tube foot extends again. This pattern of extension and retraction of tube feet results in continuous movement. It is the coordinated movement of many tube feet that enable the sea star to move slowly along the ocean floor.

CONNECTION TO BIOLOGY

In what way do scallops and earthworms also use hydraulic pressure for locomotion?

BIOLOGY Online To find out more about hydraulic pressure systems, visit the Glencoe Science Web site. **science.glencoe.com**

GLENCOE TECHNOLOGY

VIDEODISC
The Secret of Life
Action of Tube Feet

BIOLOGY Online  Note Internet addresses that you find useful in the space below for quick reference.

_____

_____

_____

_____

_____

SUMMARY

Section 29.1

Echinoderms

Main Ideas

■ Echinoderms have spines or bumps on their endoskeletons, radial symmetry, and water vascular systems. Most move by means of the suction action of tube feet.

■ Echinoderms include sea stars, sea urchins, sand dollars, sea cucumbers, sea lilies, and feather stars.

■ Deuterostome development, an internal skeleton, and bilaterally symmetrical larvae are indicators of the close phylogenetic relationship between echinoderms and chordates.

Vocabulary

ampulla (p. 788)
madreporite (p. 789)
pedicellaria (p. 787)
ray (p. 787)
tube feet (p. 788)
water vascular system (p. 789)

Section 29.2

Invertebrate Chordates

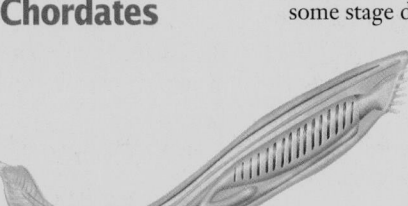

Main Ideas

■ Chordates have a dorsal hollow nerve cord, a notochord, muscle blocks, gill slits, and a tail at some stage during development.

■ Sea squirts and lancelets are invertebrate chordates.

■ Vertebrate chordates may have evolved from larval stages of ancestral invertebrate chordates.

Vocabulary

dorsal hollow nerve cord (p. 795)
gill slit (p. 796)
notochord (p. 795)

UNDERSTANDING MAIN IDEAS

1. Sea stars, sea urchins, sand dollars, sea cucumbers, sea lilies, and feather stars are examples of echinoderms that all have _____.
   a. exoskeletons
   b. jointed appendages
   c. tube feet
   d. larvae with radial symmetry

2. Of the following, which is NOT a characteristic of chordates?
   a. dorsal hollow nerve cord
   b. notochord
   c. pedicellariae
   d. muscle blocks

3. When a sea star loses a ray, it is replaced by the process of _____.
   a. regeneration   c. metamorphosis
   b. reproduction   d. parthenogenesis

4. Animals that have spines or bumps on their endoskeletons, radial symmetry, and water vascular systems are _____.
   a. invertebrate chordates
   b. chordates
   c. vertebrates
   d. echinoderms

5. A close phylogenetic relationship between echinoderms and some chordates is indicated by the fact that both have similar _____.
   a. habitats   c. sizes
   b. larvae   d. gills

Main Ideas

Summary statements can be used by students to review the major concepts of the chapter.

Using the Vocabulary

To reinforce chapter vocabulary, use the Content Mastery Booklet and the activities in the Interactive Tutor for Biology: The Dynamics of Life on the Glencoe Science Web site. science.glencoe.com

 **THE PRINCETON REVIEW** *All Chapter Assessment questions and answers have been validated for accuracy and suitability by The Princeton Review.*

UNDERSTANDING MAIN IDEAS

1. c
2. c
3. a
4. d
5. b

GLENCOE TECHNOLOGY

 **VIDEOTAPE**
MindJogger Videoquizzes
Chapter 29: *Echinoderms and Invertebrate Chordates*
Have students work in groups as they play the videoquiz game to review key chapter concepts.

Resource Manager

Chapter Assessment, pp. 169-174
MindJogger Videoquizzes
ExamView® Pro Software 💾
BDOL Interactive CD-ROM, Chapter 29 quiz

803

6. b
7. a
8. c
9. d
10. d
11. release
12. notochord
13. light or dark
14. strong and flexible
15. asteroids
16. filter feeding
17. evolved later
18. sand dollar
19. invertebrate chordates
20. bilateral symmetry

APPLYING MAIN IDEAS

21. Sea stars regenerate and the cut-up parts may become new sea stars.
22. Sea squirts are sessile, filter feeders that can shoot jets of water to protect themselves.

6. Spines on sea stars and sea urchins are modified into pedicellariae used for _____.
   a. feeding
   c. breathing
   b. protection
   d. reproduction

7. The water vascular system operates the tube feet of sea stars and other echinoderms by means of _____.
   a. water pressure
   c. water pumps
   b. water exchange
   d. water filtering

8. Tube feet, in addition to functioning in locomotion, also function in _____.
   a. gas exchange and digestion
   b. digestion and circulation
   c. gas exchange and excretion
   d. excretion and digestion

9. Water enters and leaves the water vascular system of a sea star through the _____.
   a. radial canal
   c. tube feet
   b. ampulla
   d. madreporite

10. Sea squirts and lancelets are invertebrate chordates that have _____.
    a. pedicellariae
    b. exoskeletons
    c. tube feet
    d. larvae with bilateral symmetry

11. When a sea cucumber is threatened, it can _____ its internal organs.

12. The _____ is a semirigid, rodlike structure common to all members of the phylum Chordata.

13. When a sea star lifts up the tips of its rays, it is detecting _____.

14. Muscle blocks attached to the notochord enable chordates to be more _____.

THE PRINCETON REVIEW — TEST-TAKING TIP

All or None
When filling in answer ovals, remember to fill in the whole oval. A computer will be scoring your answers. Don't give the right answer for a problem only to lose points on it because the computer couldn't read your oval.

15. Examine the diagram below. From which group did brittle stars most likely evolve?

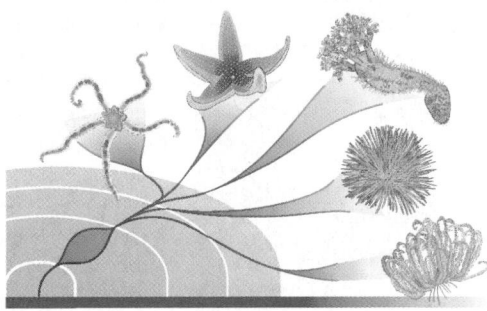

16. Tunicates and lancelets get food by _____.

17. Most echinoderms flourished in the Paleozoic era. Brittle stars require habitat similar to other echinoderms, but they did not flourish during the Paleozoic because they most likely _____.

18. A _____ is a flat, disc-shaped echinoderm without rays, and only minute hairlike spines.

19. Sea stars are more likely to leave a fossil record than _____ such as tunicates and lancelets.

20. The _____ of this larva shows its close relationship to chordates.

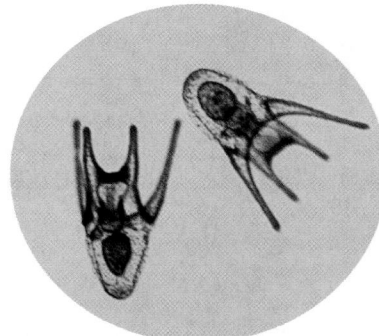

APPLYING MAIN IDEAS

21. If you were an oyster farmer, why would you be advised not to break apart and throw back any sea stars that were destroying the oyster beds?

22. How does a sessile animal such as a sea squirt protect itself?

23. Relate the various functions of the water vascular system to the environment in which echinoderms live.

24. How is the ability of echinoderms to regenerate an adaptive advantage to these animals?

25. Explain how a sea squirt maintains homeostasis.

THINKING CRITICALLY

26. Observing and Inferring Explain why the tube feet of a sand dollar are located on its upper surface as well as on its bottom surface.

27. Comparing and Contrasting Compare the pedicellariae of echinoderms with the nematocysts of cnidarians.

28. Concept Mapping Complete the concept map by using the following vocabulary terms: ampulla, madreporite, tube feet, water vascular system.

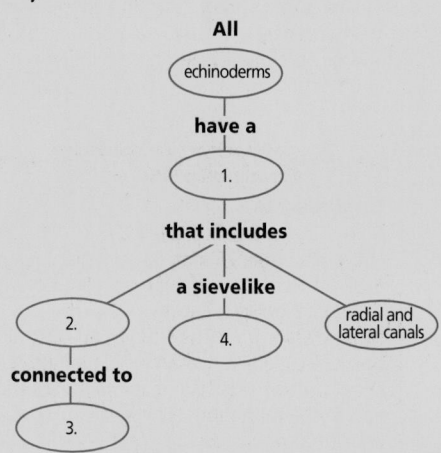

All

echinoderms

have a

1.

that includes

a sievelike

2.   4.   radial and lateral canals

connected to

3.

CD-ROM

For additional review, use the assessment options for this chapter found on the *Biology: The Dynamics of Life Interactive CD-ROM* and on the Glencoe Science Web site.
science.glencoe.com

ASSESSING KNOWLEDGE & SKILLS

The diagrams below represent cross sections of larvae. The intestines are shown in red and the nerve cords are shown in blue.

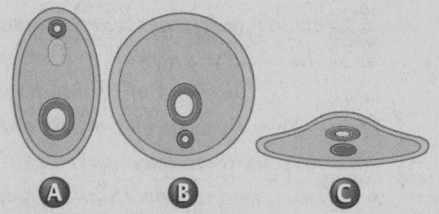

Ⓐ   Ⓑ   Ⓒ

Interpreting Scientific Illustrations Use the diagram to answer the following questions.

1. Which of the diagrams shows a cross section of a lancelet?
   a. A
   b. B
   c. C
   d. none of the diagrams

2. Which of the diagrams would represent segmented worms and echinoderms?
   a. A
   b. B
   c. C
   d. none of the diagrams

3. What does the yellow, solid area represent?
   a. nerve cord   **c.** notochord
   b. intestines   **d.** spinal cord

4. What is wrong with diagram C if it represents an invertebrate chordate?
   a. The notochord is ventral.
   b. The nerve cord is ventral and there is no notochord.
   c. It is too flat.
   d. The intestine should be round.

5. Interpreting Scientific Illustrations Using the same color code and the same three organs, draw a diagram of a cross section of a larval sea squirt, sea star, and earthworm.

23. Tube feet enable sea stars to move and open bivalve mollusks for food. Tube feet are also used for gas exchange and excretion. This water vascular system works well in a water environment.

24. Regeneration enables a sea star to survive and escape an attack by a predator. If a predator bites off a ray, the sea star can escape while the predator is feeding on that ray. A new ray will grow within weeks.

25. Sea squirts filter feed and take in oxygen from the water. They maintain a balance with their watery environment by filtering out what they require and giving off wastes into the water.

THINKING CRITICALLY

26. The tube feet on the upper surface are modified to function as gills.

27. Pedicellariae of echinoderms pinch potential predators but are not used in capturing prey and do not have immobilizing toxins as do nematocysts.

28. 1. Water vascular system; 2. Ampulla; 3. Tube feet; 4. Madreporite

ASSESSING KNOWLEDGE & SKILLS

1. a
2. b
3. c
4. b
5. The sea squirt will look like A, the sea star like B, and the earthworm like B.

BIODIGEST

For a **preview** of the invertebrate unit, study this BioDigest before you read the chapters. After you have studied the invertebrate chapters, you can use the BioDigest to **review** the unit.

National Science Education Standards: UCP.1, UCP.2, UCP.5, C.1, C.4, C.5, C.6, F.1

Prepare

Purpose

This BioDigest can be used as an overview of the invertebrate animal phyla. You may wish to use this unit summary to teach about invertebrates in place of the chapters in the Invertebrate unit.

Key Concepts

Students learn about the phyla of invertebrates, including sponges, cnidarians, flatworms, roundworms, arthropods, and echinoderms, as well as the invertebrate chordates. They learn about cell organization in animals and observe differences in physical forms as organisms become more complex, with cells organized into tissues, and tissues organized into organs and organ systems.

1 Focus

Bellringer

Bring an assortment of live invertebrates into the classroom and have students examine them. Ask students to identify traits that these animals share. They should discover that invertebrates are animals without backbones. **L1**
ELL

Invertebrates

How are jellyfishes, earthworms, sea stars, and butterflies alike? All of these animals are invertebrates—animals without backbones. The ancestors of all modern invertebrates had simple body plans. They lived in water and obtained food, oxygen, and other materials directly from their surroundings, just like present-day sponges, jellyfishes, and worms. Some invertebrates have external coverings such as shells and exoskeletons that provide protection and support.

Sponges

Sponges, phylum Porifera, are invertebrates made up of two cell layers. Most sponges are asymmetrical. They have no tissues, organs, or organ systems. Most adult sponges do not move from place to place.

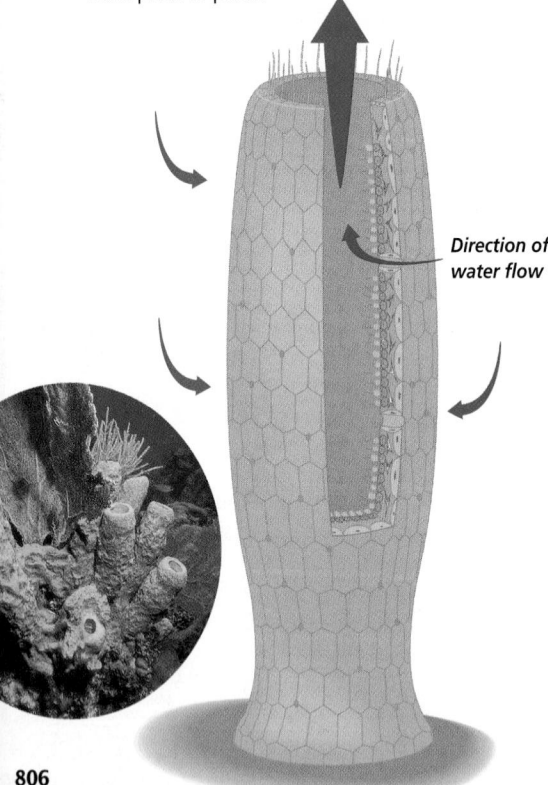

Direction of water flow

Cnidarians

Like sponges, cnidarians are made up of two cell layers and have only one body opening. The cell layers of a cnidarian, however, are organized into tissues with different functions. Cnidarians are named for stinging cells that contain nematocysts that are used to capture food. Jellyfishes, corals, sea anemones, and hydras belong to phylum Cnidaria.

VITAL STATISTICS

Cnidarians
Size ranges: Smallest: *Haliclystus salpinx,* jellyfish, diameter, 25 mm; largest: giant jellyfish medusa, diameter, 2 m; largest coral colony: Great Barrier Reef, length, 2027 km
Most poisonous: The sting of an Australian box jelly can kill a human within minutes.
Distribution: Worldwide in marine, brackish, and freshwater habitats.
Numbers of species:
Phylum Cnidaria
   Class Hydrozoa—hydroids, 2700 species
   Class Scyphozoa—jellyfishes, 200 species
   Class Anthozoa—sea anemones and corals, 6200 species

Sponges are filter feeders. A sponge takes in water through pores in the sides of its body, filters out food, and releases the water through the opening at the top.

806

Multiple Learning Styles

Look for the following logos for strategies that emphasize different learning modalities.

 Kinesthetic Activity, p. 808; Meeting Individual Needs, pp. 809, 810

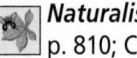 *Visual-Spatial* Activity, pp. 807, 808, 809, 810, 811; Quick Demo, pp. 810, 811

Interpersonal Visual Learning, p. 807

Linguistic Biology Journal, p. 808

Naturalist Visual Learning, p. 810; Check for Understanding, p. 812

Flatworms

Flatworms, phylum Platyhelminthes, include free-living planarians, parasitic tapeworms, and parasitic flukes. Flatworms are bilaterally symmetrical animals with flattened solid bodies and no body cavities. Flatworms have one body opening through which food enters and wastes leave.

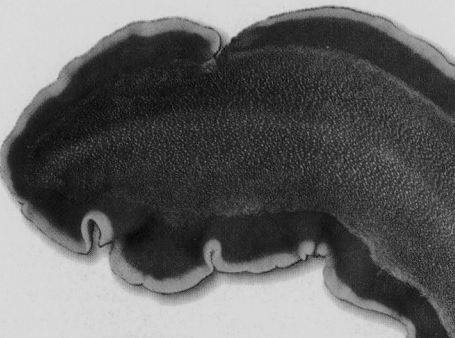

Free-living flatworms have a head end with organs that sense the environment. Flatworms can detect light, chemicals, food, and movements in their surroundings.

Prey

Nematocyst

Jellyfishes and other cnidarians have nematocysts on their tentacles.

Roundworms

Roundworms, phylum Nematoda, have a pseudocoelom and a tubelike digestive system with two body openings. Most roundworms are free-living, but many plants and animals are affected by parasitic roundworms.

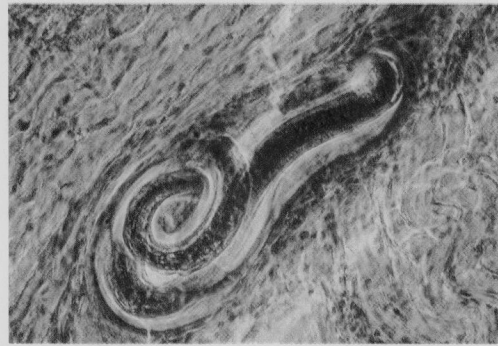

Parasitic roundworms such as this *Trichinella* are contracted by eating undercooked pork. Other roundworms can be contracted by walking barefoot on contaminated soil.

VITAL STATISTICS

Flatworms
Size ranges: Largest, beef tapeworm, length, 30 m
Distribution: Worldwide in soil, marine, brackish, and freshwater habitats
Numbers of species:
Phylum Platyhelminthes:
 Class Turbellaria—free-living planarians, 3000 species
 Class Cestoda—parasitic tapeworms, 3500 species
 Class Trematoda—parasitic flukes, 8000 species

807

GLENCOE TECHNOLOGY

VIDEODISC
The Secret of Life
Flatworm Cross Section

---

2 Teach

Activity

Visual-Spatial Many of the invertebrate phyla are best observed in nature films. Select several nature films that highlight invertebrates. Ask students to list all the invertebrates they see in the films, and group the animals into phyla. Ask them to identify the characteristics they used to classify the animals. **L2**

Visual Learning

Interpersonal Ask student groups to make a labeled diagram of a planarian worm on the chalkboard. Ask them how the shape of a planarian reflects its living habits. *Planarians are flattened, thereby enabling them to slip easily under rocks and debris in streams.* **L2** **ELL** **COOP LEARN**

Activity

Visual-Spatial Provide students with binocular microscopes, watch glasses, toothpicks, dropping pipettes, and a planarian worm culture. Ask them to place a few drops of water in the watch glass, then, using a toothpick, gently move a planarian to the watch glass. Have students observe the worm under the microscope and describe its movement. **L1** **ELL**

GLENCOE TECHNOLOGY

CD-ROM
Biology: The Dynamics of Life
Exploration: *The Six Kingdoms*
Disc 3
BioQuest: *Biodiversity Park*
Disc 3, 4
Video: *Ocean Cnidarians*
Disc 4

BioDigest

Activity

 *Visual-Spatial* Ask students to observe the movement of a live land snail. As the snail moves, ask students to describe its behavior when the antennae are touched with a cotton swab. **L1** **ELL**

Microscope Activity

Ask students to observe the development of snail eggs with hand lenses or binocular microscopes.

Activity

Kinesthetic Ask students to set up a saltwater aquarium and add marine mollusks that can be obtained live from the supermarket. Clams, mussels, and oysters are generally available. Have students care for and make observations of their aquarium. **L2** **ELL**

GLENCOE TECHNOLOGY

CD-ROM
Biology: The Dynamics of Life
Exploration: *Mollusks*
Disc 4

BioDigest

Mollusks

Slugs, snails, clams, squids, and octopuses are members of phylum Mollusca. All mollusks are bilaterally symmetrical and have a coelom, two body openings, a muscular foot for movement, and a mantle, which is a thin membrane that surrounds the internal organs. In shelled mollusks, the mantle secretes the shell.

Classes of Mollusks

The three major classes of mollusks are gastropods with one shell or no shell; bivalves with two hinged shells; and cephalopods. Cephalopods include octopuses, squids, and shelled nautiluses that all have muscular tentacles and are capable of swimming by jet propulsion. All mollusks, except bivalves, have a rough, tongue-like organ called a radula used for obtaining food.

Gastropods, such as snails, use their radulas to scrape algae from rock surfaces.

Bivalves, such as clams, strain food from water by filtering it through their gills.

VITAL STATISTICS

Mollusks
Size ranges: Largest: tropical giant clam, length, 1.5 m; North Atlantic giant squid, length, 18 m; Pacific giant octopus, length, 10 m; smallest: seed clam, length, less than 1 mm
Distribution: Worldwide in salt-, fresh-, and brackish water, and on land in moist temperate and tropical habitats.
Numbers of species:
Phylum Mollusca
 Class Gastropoda—snails and slugs, 80 000 species
 Class Bivalvia—bivalves, 10 000 species
 Class Cephalopoda—octopuses, squids, and nautiluses, 600 species

Cephalopods, such as octopuses, are predators. They capture prey using the suckers on their long tentacles.

FOCUS ON ADAPTATIONS

Body Cavities

Marine flatworm

The type of body cavity an animal has determines how large it can grow and how it takes in food and eliminates wastes. Acoelomate animals, such as planarians, have no body cavity. Water and digested food particles travel through a solid body by the process of diffusion.

Animals such as roundworms have a fluid-filled body cavity called a pseudocoelom that is partly lined with mesoderm. Mesoderm is a layer of cells between the ectoderm and endoderm that differentiates into muscles, circulatory vessels, and reproductive organs. The pseudocoelom provides support for the

808

BIOLOGY JOURNAL

Invertebrate Advertisements

Linguistic Ask students to prepare a classified newspaper advertisement for a homesite for three animals, one in each of the classes of mollusks. Provide students with a page of the real estate section of the newspaper to review how such advertisements for homes are written. Students should be creative with their language but keep the precise habitat requirements of each animal in mind. **L3**

Segmented Worms

Bristleworms, earthworms, and leeches are members of phylum Annelida, the segmented worms. Segmented worms are bilaterally symmetrical, coelomate animals that have segmented, cylindrical bodies with two body openings. Most annelids have setae, bristlelike hairs that extend from body segments, that help the worms move.

Segmentation is an adaptation that provides these animals with great flexibility. Each segment has its own muscles. Groups of segments have different functions, such as digestion or reproduction.

Classes of Segmented Worms

Phylum Annelida has three classes: Hirudinae, the leeches; Oligochaeta, the earthworms; and Polychaeta, the bristleworms.

Most bristleworms have a distinct head and a body with many setae.

Leeches have flattened bodies with no setae. Most species are parasites that suck blood and body fluids from ducks, turtles, fishes, and mammals.

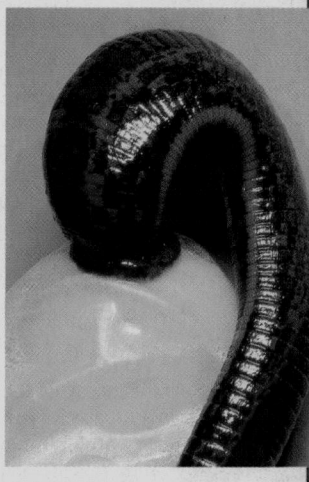

VITAL STATISTICS

Segmented Worms

Size ranges: Largest: giant tropical earthworm, length, 4 m; smallest: freshwater worm, *Aeolosoma*, length, 0.5 mm

Distribution: Terrestrial and marine, brackish, and freshwater habitats worldwide, except polar regions and deserts.

Numbers of species:

Phylum Annelida
- Class Hirudinea—leeches, 500 species
- Class Oligochaeta—earthworms, 3100 species
- Class Polychaeta—bristleworms, 8000 species

attachment of muscles, making movement more efficient. Earthworms have a coelom, a body cavity surrounded by mesoderm in which internal organs are suspended. The coelom acts as a watery skeleton against which muscles can work.

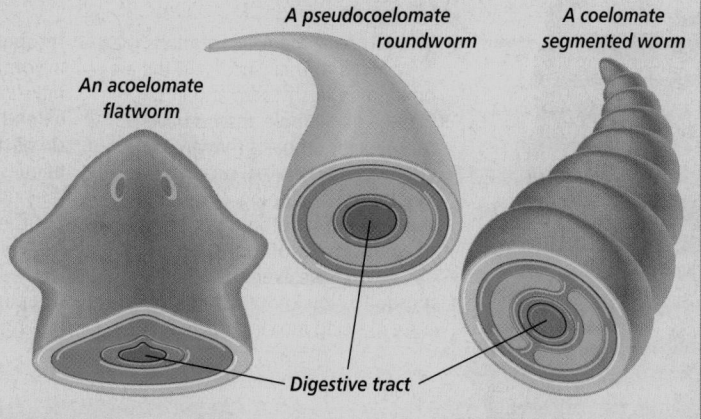

An acoelomate flatworm

A pseudocoelomate roundworm

A coelomate segmented worm

Digestive tract

809

BioDigest

Activity

Visual-Spatial Ask students to bring in jars of many types of arthropods. Number the jars and pass them from group to group, asking students to write the name of the arthropod and obvious arthropod features beside the name. For each arthropod, ask them to infer what features helped to make that particular arthropod successful. **L2**

Quick Demo

Visual-Spatial Show students a live water strider on the surface of pond water in a container with a large surface area. Ask students to hypothesize how the water strider stays on top of the water. Ask them to identify adaptations the animal has that enables it to skate on the water surface. **L2**

Visual Learning

Naturalist Ask students to examine the diagrams of arthropods on this page and identify adaptations that enable each animal to survive in its habitat. **L1**

Activity

Have students design and conduct an experiment to determine the effect of temperature on the jumping ability of crickets. **L2**

Arthropods

Arthropods are bilaterally symmetrical, coelomate invertebrates with tough outer coverings called exoskeletons and jointed appendages that are used for walking, sensing, feeding, and mating. Exoskeletons protect and support their soft internal tissues and organs. Jointed appendages allow for powerful and efficient movements.

Arthropod Diversity

Two out of three animals on Earth today are arthropods. The success of arthropods can be attributed to adaptations that provide efficient gas exchange, acute senses, and varied types of mouthparts for feeding. Arthropods include organisms such as spiders, crabs, lobsters, shrimps, crayfishes, centipedes, millipedes, and the enormously diverse group of insects.

The evolution of jointed appendages with many different functions probably led to the success of the arthropods as a group.

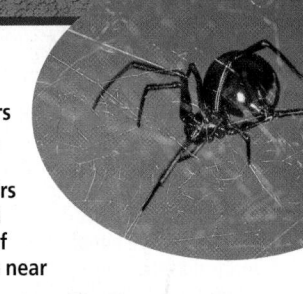

Like other members of class Arachnida, the black widow spider has four pairs of jointed legs and chelicerae, a pair of biting appendages near the mouth.

Lobsters, class Crustacea, have antennae and two compound eyes on movable stalks. Their mandibles move from side to side to seize prey.

FOCUS ON ADAPTATIONS

Insects

Insects have many adaptations that have led to their success in the air, on land, in freshwater, and in salt water. For example, insects have complex mouthparts that are well adapted for chewing, sucking, piercing, biting, or lapping. Different species have mouthparts adapted to eating a variety of foods.

If you have ever been bitten by a mosquito, you know that mosquitoes have piercing mouthparts that cut through your skin to suck up blood. In contrast, butterflies and moths have long, coiled tongues that they extend deep into tubular flowers to sip nectar. Grasshoppers and many beetles have hard, sharp mandibles they use to cut off and chew leaves. But the heavy mandibles of staghorn beetles no longer function as jaws; instead, they have become defensive weapons used for competition and mating purposes.

Staghorn beetle

MEETING INDIVIDUAL NEEDS

English Language Learners

Kinesthetic For students who are English language learners or are kinesthetic learners, purchase plastic arthropods. Ask them to work in groups and point out features of the arthropods as they pass them around their group. Features that they might be able to find include head, thorax, abdomen, spiracles, claws, flippers, jointed legs, cephalothorax, antennae, mouthparts, pedipalps, chelicerae, compound and simple eyes, wings, tympanum, and mandibles. **L2** **ELL** **COOP LEARN**

BIODIGEST

Arthropod Origins

Arthropods most likely evolved from segmented worms; they both show segmentation. However, an arthropod's segments are fused and have a greater complexity of structure than those of segmented worms. Because arthropods have exoskeletons, fossil arthropods are frequently found, and consequently more is known about their origins than about the phylogeny of worms.

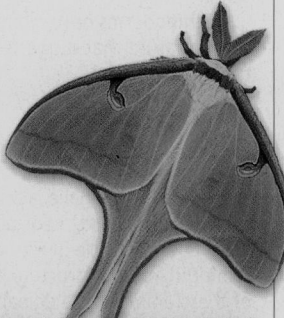

Members of class Insecta, the insects, such as this luna moth, have three pairs of jointed legs and one pair of antennae for sensing their environments.

VITAL STATISTICS

Arthropods

Size ranges: Largest insects: tropical stick insect, length, 33 cm; Goliath beetle, mass, 100 g; smallest insect: fairyfly wasp, length, 0.21 mm

Distribution: All habitats worldwide.

Numbers of species:

Phylum Arthropoda

    Class Arachnida—spiders and their relatives, 57 000 species

    Class Crustacea—crabs, shrimps, lobsters, crayfishes, 35 000 species

    Class Merostomata—horseshoe crabs, 4 species

    Class Chilopoda—centipedes, 2500 species

    Class Diplopoda—millipedes, 10 000 species

    Class Insecta—insects, 750 000 species

Millipedes, class Diplopoda, are herbivores. Millipedes have up to 100 body segments, and each segment has two pairs of legs.

Different Foods for Different Stages

Because insects undergo metamorphosis, they often utilize different food sources at different times of the year. For example, monarch butterfly larvae feed on milkweed leaves, whereas the adults feed on milkweed flower nectar. Apple blossom weevil larvae feed on the stamens and pistils of unopened flower buds, but the adult weevils eat apple leaves. Some adult insects, such as mayflies, do not eat at all! Instead, they rely on food stored in the larval stage for energy to mate and lay eggs.

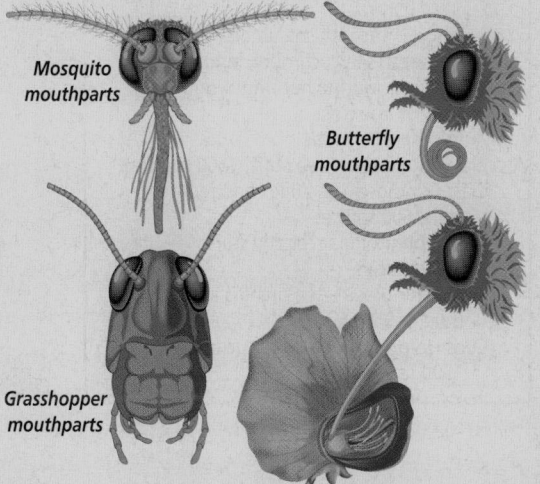

Mosquito mouthparts

Butterfly mouthparts

Grasshopper mouthparts

811

BioDigest

BioDigest

Assessment

Portfolio Have students visit a local zoo or pet shop that maintains a saltwater aquarium. Have them time their visit so they are present when it is feeding time for echinoderms such as sea stars, brittle stars, and feather stars. Have students watch one echinoderm as it feeds, then describe the behavior of the animal in their portfolios.

GLENCOE TECHNOLOGY

CD-ROM
Biology: The Dynamics of Life
Exploration: *Echinoderms*
Disc 4

3 Assess

Check for Understanding

 Naturalist Set up a lab practical with stations with preserved or live invertebrates. Ask students to identify the phylum and group within the phylum to which each animal belongs, and explain the visible features that helped them classify the invertebrates. **L2**

Echinoderms

Echinoderms, phylum Echinodermata, are radially symmetrical, coelomate animals with hard, bumpy, spiny endoskeletons covered by a thin epidermis. The endoskeleton is comprised of calcium carbonate. Echinoderms move using a unique water vascular system with tiny, suction-cuplike tube feet. Some echinoderms have long spines also used in locomotion.

The tube feet of a sea star operate by means of a hydraulic water vascular system. Sea stars move along slowly by alternately pushing out and pulling in their tube feet.

VITAL STATISTICS

Echinoderms
Size ranges: Largest: sea urchin, diameter, 19 cm; longest: sea cucumber, length, 60 cm
Distribution: Marine habitats worldwide.
Numbers of species:
Phylum Echinodermata
  Class Asteroidea—sea stars, 1500 species
  Class Crinoidea—sea lilies and feather stars, 600 species
  Class Ophiuroidea—brittle stars, 2000 species
  Class Echinoidea—sea urchins and sand dollars, 950 species
  Class Holothuroidea—sea cucumbers, 1500 species

Echinoderm Diversity

There are five major classes of echinoderms. They include sea stars, brittle stars, sea urchins, sea cucumbers, sand dollars, sea lilies, and feather stars.

Echinoderms have bilaterally symmetrical larvae, a feature that suggests a close relationship to the chordates.

Invertebrate Chordates

All chordates have, at one stage of their life cycles, a notochord, a dorsal hollow nerve cord, gill slits, and muscle blocks. A notochord is a long, semirigid, rodlike structure along the dorsal side of these animals. The dorsal hollow nerve cord is

Sea cucumbers have a leathery skin and are flexible. Like most echinoderms, they move using tube feet.

The long, thin arms of brittle stars are fragile and break easily, but they grow back. Brittle stars use their arms to walk along the ocean bottom.

 Note Internet addresses that you find useful in the space below for quick reference.

_____

_____

_____

_____

_____

a fluid-filled canal lying above the notochord. Gill slits are paired openings in the pharynx that, in some invertebrate chordates, are used to strain food from the water. In other chordates, gill slits develop into internal gills used for gas exchange. Muscle blocks are modified body segments consisting of stacked muscle layers. Muscle blocks are anchored by the notochord.

Invertebrate chordates have all of these features at some point in their life cycles. The invertebrate chordates include the lancelets and the tunicates, also known as sea squirts.

The lancelet is an example of an invertebrate chordate. Notice that the lancelet's body is shaped like that of a fish even though it is a burrowing filter feeder.

BioDigest Assessment

Understanding Main Ideas

1. An animal that is a filter feeder, takes in water through pores in the sides of its body, and releases water from the top is a _____.
   a. roundworm   c. sponge
   b. gastropod   d. lancelet

2. Nematocysts are unique to _____.
   a. sponges   c. annelids
   b. mollusks   d. cnidarians

3. An example of a free-living flatworm is a _____.
   a. planarian   c. nematode
   b. tapeworm   d. vinegar-eel

4. Which of the following is used by segmented worms for movement?
   a. chelicerae
   b. nematocysts
   c. setae
   d. water vascular system

5. Which of the following are invertebrate chordates?
   a. sea anemones   c. bivalves
   b. lancelets   d. squid

6. Parasitism is a way of life for most _____.
   a. flukes   c. cnidarians
   b. sponges   d. annelids

7. An example of an animal with no body cavity is a(n) _____.
   a. sea star   c. earthworm
   b. flatworm   d. clam

8. An octopus belongs to phylum Mollusca because it has a mantle, bilateral symmetry, two body openings, and _____.
   a. an external shell
   b. a muscular foot
   c. a pseudocoelom
   d. segmentation

9. Leeches feed by _____.
   a. grazing on aquatic plants
   b. stinging prey
   c. filter feeding
   d. sucking blood

10. Which of the following characteristics is unique to arthropods?
    a. nematocycts   c. filter feeding
    b. jointed appendages   d. tube feet

Thinking Critically

1. A radula is to a snail as a(n) _____ is to a jellyfish. Explain your answer.

2. Why is more known about animals with hard parts than is known about animals with only soft parts?

3. In what ways are echinoderms more similar to vertebrates than to other invertebrates?

4. You are examining a free-living animal that had a thin, solid body with two surfaces. Into what phylum is this organism classified? Explain.

5. In what two ways are spiders different from insects?

813

BioDigest Assessment

Understanding Main Ideas

1. c	4. c	7. b	10. b
2. d	5. b	8. b	
3. a	6. a	9. d	

Thinking Critically

1. nematocyst; a snail obtains food with its radula; a jellyfish obtains its food with nematocysts

2. Hard parts leave fossil evidence while soft parts leave little fossil evidence.

3. Echinoderms have bilaterally symmetrical larvae, as do chordates.

4. It belongs to the phylum Platyhelminthes because it is a free-living flatworm.

5. Spiders have four pairs of legs; insects have three. Spiders have chelicerae; insects have many different types of mouthparts. Insects have antennae; spiders do not.

813

Vertebrates

Unit Overview

Students learn about the diversity of animals with backbones in Unit 9. In Chapter 30, students begin to appreciate the evolutionary movement of animals from water to land environments as they explore the structure, adaptations, ecology, and phylogeny of fishes and amphibians. The movement of animals to land is completed with the reptiles, studied in Chapter 31. In this chapter, students also examine adaptations that enable birds to fly.

Included in Chapter 31 is Focus On Dinosaurs, a detailed examination of these remarkable, extinct animals.

In Chapter 32, students discover why mammals are able to occupy nearly all environments on Earth. The unit concludes in Chapter 33 with an exploration of animal behavior.

Introducing the Unit

Have examples of live vertebrates from different classes available for observation. Examples may include: mouse, frog, lizard, and goldfish. Ask students to work in groups to make a table comparing how the animals are alike and how they are different. Vertebrate skeletons may be purchased from biological supply houses. Ask students to compare and contrast the skeletons.

Vertebrates

Bears, salmon, gulls—and humans—belong to the group of animals known as vertebrates, animals with backbones. Their ability to survive the rigors of changing environments, such as the Alaskan tundra, and engage in complex behaviors, such as migration, are just a few of the characteristics that distinguish vertebrates from other members of the animal kingdom.

UNIT CONTENTS

30 Fishes and Amphibians

31 Reptiles and Birds

32 Mammals

33 Animal Behavior

BIODIGEST Vertebrates

UNIT PROJECT

BIOLOGY *Online* Use the Glencoe Science Web site for more project activities that are connected to this unit.
science.glencoe.com

814

Unit Projects

Animal Census/Habitat Analysis

Have students do one of the projects for this unit as described on the Glencoe Science Web site. As an alternative, they can do one of the projects described on these two pages.

814

Display

Visual-Spatial Have student groups do a census of animals on the school grounds or areas near the school. They should look for animals and their signs. Ask students to make a map of the area to scale, indicating where they found animals or their signs. **L2 ELL COOP LEARN**

Using the Library

Intrapersonal Ask students to find out the habitat requirements of the animals found in their study area. **L2**

Advance Planning

Chapter 30
- Order *Xenopus* eggs, preserved adult frogs, Ringer's solution, and tadpoles for the BioLab and MiniLab 30-2.
- Purchase plastic or clay models of the three orders of amphibians for Meeting Individual Needs.

Chapter 31
- none

Chapter 32
- Order prepared slides of longitudinal sections of human canine teeth for MiniLab 32-1.
- Order owl pellets from a biological supply house for MiniLab 32-2.
- Borrow a collection of mammal skulls from a local college or museum for the Focus On Placental Mammals.

Chapter 33
- none

Unit Projects

Modeling

Kinesthetic Have students make another map of the area showing how the area could be improved for wildlife living there. The map can be three-dimensional if materials are available. **L2**

Demonstration

Kinesthetic If student groups found animal tracks in their study area, have them make plaster of Paris molds of the tracks. Students can prepare molds and bring them into class. Have field guides available so students can compare their molds to tracks in these books. Make sure students can identify the tracks and name the animal that made their tracks. **L1 ELL**

Final Report

Ask students to present their findings to the owners of the property. **L3**

Chapter 30 Organizer

Refer to pages 4T-5T of the Teacher Guide for an explanation of the National Science Education Standards correlations.

Section	Objectives	Activities/Features
Section 30.1 **Fishes** National Science Education Standards UCP.1-5; A.1, A.2; C.3, C.5, C.6; F.4, F.5; G.3 ($1\frac{1}{2}$ sessions, $\frac{1}{2}$ block)	1. **Relate** the structural adaptations of fishes to their environments. 2. **Compare** and **contrast** the adaptations of the different groups of fishes. 3. **Interpret** the phylogeny of fishes.	**MiniLab 30-1:** Measuring Breathing Rate in Fishes, p. 818 **Problem-Solving Lab 30-1,** p. 820 **Inside Story:** A Bony Fish, p. 825
Section 30.2 **Amphibians** National Science Education Standards UCP.1-5; A.1, A.2; C.3, C.5, C.6; E.1, E.2; F.4-6; G.1-3 (2 sessions, 1 block)	4. **Relate** the demands of a terrestrial environment to the adaptations of amphibians. 5. **Relate** the evolution of the three-chambered heart to the amphibian lifestyle.	**Inside Story:** A Frog, p. 828 **MiniLab 30-2:** Looking at Frog and Tadpole Adaptations, p. 830 **Investigate BioLab:** Development of Frog Eggs, p. 834 **Chemistry Connection:** Killer Frogs, p. 836

Need Materials? Contact Carolina Biological Supply Company at 1-800-334-5551 or at **http://www.carolina.com**

MATERIALS LIST

BioLab

p. 834 Ringer's solution for frogs, culture dishes (4), light bulbs, thermometer, stereomicroscope, fertilized frog eggs (*Xenopus laevis*), flashlight

MiniLabs

p. 818 beaker, thermometer, goldfish, ice cube (2), pencil, paper
p. 830 preserved adult frog, preserved tadpole, pencil, paper

Alternative Lab

p. 822 aquarium, black gravel, white gravel, tropical fishes (zebra, catfish, gourami), timer or stopwatch, black construction paper, masking tape

Quick Demos

p. 821 stereomicroscope, microscope slides, coverslips, fish scales
p. 822 2-L soda bottle, goldfish, water, plastic ketchup packet
p. 829 live frog, clear plastic container
p. 831 tape recording of frog calls

Key to Teaching Strategies

L1 Level 1 activities should be appropriate for students with learning difficulties.

L2 Level 2 activities should be within the ability range of all students.

L3 Level 3 activities are designed for above-average students.

ELL ELL activities should be within the ability range of English Language Learners.

COOP LEARN Cooperative Learning activities are designed for small group work.

P These strategies represent student products that can be placed into a best-work portfolio.

These strategies are useful in a block scheduling format.

Fishes and Amphibians

Teacher Classroom Resources

Section	Reproducible Masters	Transparencies
Section 30.1 **Fishes**	Reinforcement and Study Guide, pp. 133-134 L2 Concept Mapping, p. 30 L3 ELL BioLab and MiniLab Worksheets, pp. 133-134 L2 Laboratory Manual, pp. 211-212 L2 Tech Prep Applications, pp. 37-38 L2 Content Mastery, pp. 149-150, 152 L1	Section Focus Transparency 73 L1 ELL Basic Concepts Transparency 52 L2 ELL Basic Concepts Transparency 53 L2 ELL
Section 30.2 **Amphibians**	Reinforcement and Study Guide, pp. 135-136 L2 Concept Mapping, p. 30 L3 ELL Critical Thinking/Problem Solving, p. 30 L3 BioLab and MiniLab Worksheets, pp. 135-138 L2 Laboratory Manual, pp. 213-218 L2 Content Mastery, pp. 149, 151-152 L1 Inside Story Poster ELL	Section Focus Transparency 74 L1 ELL Basic Concepts Transparency 53 L2 ELL Reteaching Skills Transparency 44 L1 ELL

Assessment Resources

Chapter Assessment, pp. 175-180

MindJogger Videoquizzes

Performance Assessment in the Biology Classroom

Alternate Assessment in the Science Classroom

ExamView® Pro Software 💾

BDOL Interactive CD-ROM, Chapter 30 quiz

Additional Resources

Spanish Resources ELL

English/Spanish Audiocassettes ELL

Cooperative Learning in the Science Classroom COOP LEARN

Lesson Plans/Block Scheduling

NATIONAL GEOGRAPHIC — Teacher's Corner

Products Available From Glencoe
To order the following products, call Glencoe at 1-800-334-7344:
CD-ROM
NGS PictureShow: Structure of Vertebrates 1
Curriculum Kit
GeoKit: Fish, Reptiles, and Amphibians
Transparency Set
NGS PicturePack: Structure of Vertebrates 1

Products Available From National Geographic Society
To order the following products,
call National Geographic Society at 1-800-368-2728:
Video
Reptiles and Amphibians

Index to National Geographic Magazine
The following articles may be used for research relating to this chapter:
"Coelacanths, The Fish That Time Forgot," by Hans Fricke, June 1988.
"The Preposterous Puffer," by Noel D. Vietmeyer, August 1984.

GLENCOE TECHNOLOGY

The following multimedia resources are available from Glencoe.

Biology: The Dynamics of Life
CD-ROM ELL

BioQuest: *Biodiversity Park*
Exploration: *The Six Kingdoms*
Video: *Salmon Migration*
Video: *Schooling*
Exploration: *Amphibians*
Video: *Feeding Frog*
Video: *Frog Behavior*

Videodisc Program 📼

Fish Schooling
Frog Behavior
Feeding Frog

30 Fishes and Amphibians

GETTING STARTED DEMO

Remove the hooks from a variety of fish-shaped lures all made of the same material. Have students work in small groups and drop each lure into a 1000 mL graduated cylinder filled with water. Ask them to measure the amount of time it takes for each of the lures to touch the bottom of the cylinder. Ask students to correlate their results to the shape of the lures. Ask them what other variables might be important in their observations.

Theme Development

Evolution is a major theme of this chapter. Emphasis is placed on how animals evolved adaptations to life on land. The theme of **unity within diversity** is apparent through the discussion of how varied the features of animals within a classification group can be even though all members of the group share many characteristics.

Resource Manager

Section Focus Transparency 73 and Master **L1** **ELL**

READING BIOLOGY

Glencoe's *Biology: The Dynamics of Life* contains many resources to assist a student's reading skills. Each chapter contains figures with expanded captions that expand on written material. Word Origins, located along the side of text, expand knowledge of biology vocabulary. Glencoe's Content Mastery Booklet helps develop reading skills while reinforcing content. In addition, use the Interactive Tutor for *Biology: The Dynamics of Life* on the Glencoe Web site to reinforce vocabulary. **science.glencoe.com**

What You'll Learn

- You will compare and contrast the adaptations of the different groups of fishes and amphibians.
- You will relate the move to land to the evolution of fishes and amphibians.

Why It's Important

Fishes are the most diverse and successful vertebrate group. Amphibians are adapted to live both in water and on land. The development of a bony endoskeleton in fishes and lungs in amphibians were major steps in animal evolution—steps that eventually led to human evolution.

READING BIOLOGY

Look through the vocabulary words, noting the familiar words. As you read the chapter, note how the words might be used differently in the study of biology. Write down the definitions to any new terms.

To find out more about fishes and amphibians, visit the Glencoe Science Web site. **science.glencoe.com**

Leopard frog eggs hatch into wiggling tadpoles. A few months later, they become four-legged, jumping, croaking animals.

Multiple Learning Styles

Look for the following logos for strategies that emphasize different learning modalities.

Kinesthetic Meeting Individual Needs, p. 827; Tech Prep, p. 832

Visual-Spatial Tech Prep, p. 819; Quick Demo, p. 821; Meeting Individual Needs, p. 821; Portfolio, pp. 824, 829; Biology Journal, p. 829

Linguistic Project, pp. 820, 831; Enrichment, p. 823; Biology Journal, pp. 824, 832;

Meeting Individual Needs, p. 825; Check for Understanding, p. 825

Auditory-Musical Quick Demo, p. 831

Naturalist Quick Demo, p. 829

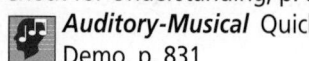

30.1 Fishes

H ave you ever visited
an aquarium to see
the amazing diversity
of fishes? As you pass tank
after tank, you can see fishes
of all shapes, sizes, and colors.
What's interesting is that
even though fishes share a
common environment, they
have evolved a variety of
different adaptations.
Although fishes may show
considerable variety in
structure and behavior,
they all share common
characteristics.

Saltwater aquarium
(above) and a blue
tang (inset)

SECTION PREVIEW

Objectives
Relate the structural
adaptations of fishes to
their environments.
Compare and contrast
the adaptations of the
different groups of
fishes.
Interpret the phylog-
eny of fishes.

Vocabulary
spawning
fin
lateral line system
scale
swim bladder
cartilage

What Is a Fish?

Fishes, like all vertebrates, are classified in the phylum Chordata. This
phylum includes three subphyla:
Urochordata, the tunicates; Cephalochordata, the lancelets; and Vertebrata, the vertebrates. Fishes belong
to the subphylum Vertebrata. In
addition to fishes, subphylum
Vertebrata includes amphibians, reptiles, birds, and mammals. Recall
from the previous chapter that all
chordates have four traits in common—a notochord, gill slits, muscle
blocks, and a dorsal hollow nerve
cord. In vertebrates, the notochord
of the embryo becomes a backbone

in adult animals. All vertebrates are
bilaterally symmetrical, coelomate
animals that have endoskeletons,
closed circulatory systems, nervous
systems with complex brains and
sense organs, and efficient respiratory systems.

Three classes of fishes

Fishes comprise three classes of
the subphylum Vertebrata: Class
Agnatha (AG nuh thuh), with jawless
fishes, lampreys and hagfishes; Class
Chondrichthyes (kahn DRIHK theez),
with cartilaginous fishes, sharks and
rays; and Class Osteichthyes (ahs tee
IHK theez), with bony fishes. An
example of each class is shown in

30.1 FISHES **817**

Assessment Planner

Portfolio Assessment
BioLab, TWE, pp. 834-835
Portfolio, TWE, pp. 824, 829
Assessment, TWE, 820, 830
Performance Assessment
MiniLab, SE, pp. 818, 830
BioLab, SE, pp. 834-835
Alternative Lab, TWE, pp. 822-823
Assessment, TWE, p. 832

Knowledge Assessment
Section Assessment, SE, pp. 826, 833
Chapter Assessment, SE, pp. 837-839
Assessment, TWE, pp. 824, 826
Skill Assessment
Assessment, TWE, pp. 818, 833

Section 30.1

Prepare

Key Concepts

Students learn about the characteristics all fishes have in common
while developing an understanding of the characteristics that distinguish the three classes of
fishes: jawless fishes, cartilaginous
fishes, and bony fishes.

Planning

■ Purchase small goldfishes for
MiniLab 30-1.
■ Purchase mounts of three
kinds of fish scales, fishing
lures, a model of a fish, and a
2 L plastic soda bottle for the
Demos.
■ Purchase tropical fishes such
as zebra fish, catfish, and
gourami for the Alternative
Lab.

1 Focus

Bellringer

Before presenting the lesson,
display **Section Focus Transparency 73** on the overhead projector and have students answer
the accompanying questions.
L1 **ELL**

Transparency **73** Fishes in the Sea

SECTION FOCUS
Use with Chapter 30,
Section 30.1

① Which of the animals in this picture are called fish, but really
aren't fish? Which are not called fish, but really are fish?

② How does the body plan of a fish differ from that of the
other animals pictured?

BIOLOGY: The Dynamics of Life

SECTION FOCUS TRANSPARENCIES

817

2 Teach

Purpose

Students will experimentally determine that fish breathing rate slows as water temperature decreases and oxygen content of the water increases.

Process Skills

experiment, hypothesize, sequence, acquire information, collect data, draw a conclusion, interpret data, measure in SI

Teaching Strategies

■ Inexpensive goldfish are available locally from most pet or aquarium shops.

■ 500 mL beakers are ideal in size, but smaller beakers may be used.

■ Review the location of the gill covering (operculum) with students.

■ Caution students about not "shaking down" thermometers.

■ Water temperature may be adjusted by adding ice cubes or cold water to the beakers.

Expected Results

Breathing rate will decrease as water temperature decreases.

Analysis

1. The breathing rate, as well as all body functions, decreases as the water temperature decreases.
2. Student answers will vary. Many students will have hypothesized that a decrease in water temperature results in an increase in breathing rate.
3. slower in cold water
4. Generally, there is more dissolved oxygen in cold water as compared with warm water.
5. An oxygen molecule in water passes over the gills, diffuses into gill capillaries, passes into the blood stream, and is pumped to body cells by heart.

MiniLab 30-1 Experimenting

Measuring Breathing Rate in Fishes Fishes are able to extract oxygen from water as it flows over their gills. Their rate of breathing is related to the availability of oxygen in the water. More oxygen results in a slower breathing rate. The breathing rate of a fish can be estimated by counting the number of times per minute its gill covers open to allow water to flow across its gills.

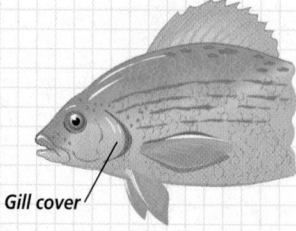

Gill cover

Procedure

1. Make a hypothesis about how the breathing rate of a fish may be influenced by a change in temperature of the water in which it is swimming. Record your hypothesis.
2. Fill a beaker with non-chlorinated water. Let the beaker sit until the water reaches room temperature (about 20°C). Measure the temperature with a thermometer and record it in a data table.
3. Add a small goldfish. **CAUTION:** *Handle animals with care.* Wait 5 minutes for the fish to acclimatize to the water temperature.
4. Count the number of times the goldfish's gill covers open in one minute. This is a measure of the rate at which the fish is breathing. Record your result in a data table.
5. Repeat step 4 four more times. Find an average for your trials and record these data in your table.
6. Remove the goldfish from the beaker. Add one ice cube made from non-chlorinated water to the beaker.
7. When the ice cube has melted, record the water's temperature.
8. Repeat steps 3-5. Remove goldfish from the beaker, and add another ice cube to the beaker. Repeat step 7.
9. Repeat steps 3-5. **CAUTION:** *Wash your hands after working with animals.*

Analysis

1. How does water temperature influence the rate at which breathing occurs in a goldfish?
2. Do your data support your hypothesis? Explain.
3. Was the breathing rate faster or slower in colder water compared with warmer water?
4. How might the amount of oxygen in cold water compare with that in warm water? Explain your answer.
5. Sequence the events associated with a fish obtaining oxygen. Start with a molecule of oxygen in water and be sure to include the capillaries located in the fish's gills.

Figure 30.1. More than 20 000 species of fishes exist. In fact, there are more fish species than all other kinds of vertebrates added together.

Fishes inhabit nearly every type of aquatic environment on Earth. They are adapted to living in shallow, warm water and deeper cold and sunless water. They are found in freshwater and salt water, and some fishes even survive in heavily polluted water.

Fishes breathe using gills

Agnathans, like all fishes, have gills made up of feathery gill filaments that contain tiny blood vessels. Gills are an important adaptation for fishes and other vertebrates that live in water. As a fish takes water in through its mouth, water passes over the gills and then out through slits at the side of the fish. Oxygen and carbon dioxide are exchanged through the capillaries in the gill filaments. You can find out more about respiration in fishes in the *MiniLab* on this page.

Fishes have two-chambered hearts

All fishes have two-chambered hearts, as shown in *Figure 30.2.* One chamber receives deoxygenated blood from the body tissues, and the second chamber pumps blood directly to the capillaries of the gills, where oxygen is picked up and carbon dioxide released. Oxygenated blood is carried from the gills to body tissues. Blood flow through the body of a fish is relatively slow because most of the heart's pumping action is used to push blood through the gills.

Fishes reproduce sexually

All fishes have separate sexes. Fertilization is external in most fishes, with eggs and sperm deposited in protected areas, such as on floating aquatic plants or in shallow nests of

---

Assessment

Skill Have students prepare a line graph of their data. Have them identify the independent and dependent variables and select and label the axes accordingly. Use the Performance Task Assessment List for Graph from Data in **PASC**, p. 39. **L2**

Resource Manager

BioLab and MiniLab Worksheets, pp. 133-134 **L2**
Concept Mapping, p. 30 **L3 ELL**
Laboratory Manual, pp. 211-212 **L2**

Figure 30.1
The three classes of fishes include jawless fishes, cartilaginous fishes, and bony fishes.

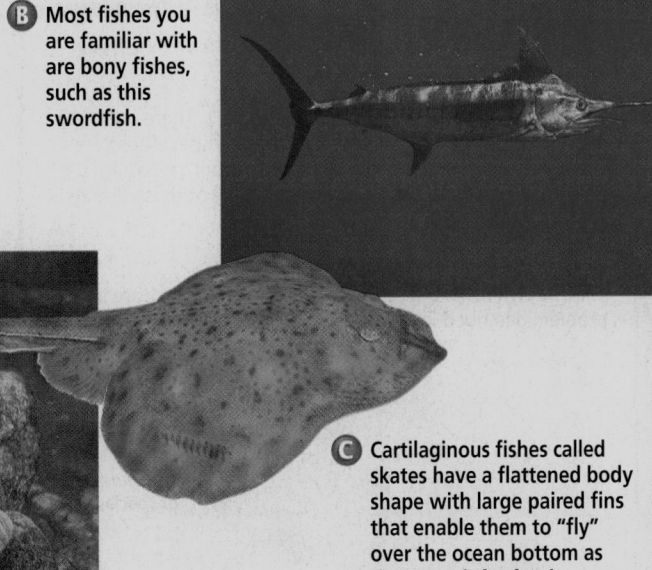

A Jawless fishes called lampreys have long, tubular bodies without paired fins, and no scales.

B Most fishes you are familiar with are bony fishes, such as this swordfish.

C Cartilaginous fishes called skates have a flattened body shape with large paired fins that enable them to "fly" over the ocean bottom as they search for food.

gravel on stream bottoms. Although most fishes produce large numbers of eggs at one time, agnathans called hagfishes produce small numbers of large eggs. Some cartilaginous fishes have internal fertilization. For example, some female sharks and rays produce as few as 20 eggs and keep them inside their bodies until they have hatched and developed to about 40 cm in length. These young, when released, behave like miniature adults, and many survive.

Most bony fishes have external fertilization. Reproduction in fishes and some other animals is called **spawning.** During spawning, some female bony fishes, such as cod, produce as many as 9 million eggs, of which only a small percentage survive. In some bony fishes, such as guppies, mollies, and swordtails, fertilization is internal and young fishes develop within the mother's body. These species are known as live-bearers because their offspring are born fully developed. Most fishes that produce millions of eggs provide no care for their offspring after spawning; in these species, only a few of the young survive to adulthood. But some, such as the mouth-brooding cichlids, stay with their young after hatching. When their young are threatened by predators, the parent fish scoop them into their mouths for protection.

Figure 30.2
In a fish's heart, deoxygenated blood flows from the first chamber to the second chamber, then on to the gills where it picks up oxygen. Blood in a fish flows in a one-way circuit throughout the body.

Circulation in a Fish

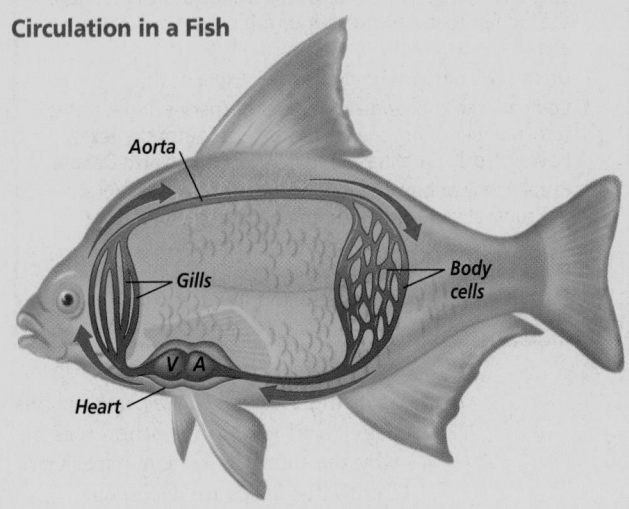

Aorta

Gills

Body cells

V A

Heart

Cultural Diversity

Samuel Milton Nabrit
African-American zoologist Samuel M. Nabrit conducted many experiments on tail regeneration in fishes. Nabrit earned a Ph.D. in Zoology in 1932 and became a researcher at Woods Hole Marine Biology Laboratory. Papers describing his work may be found in the *Journal of Experimental Zoology,* *Anatomical Record,* and *Biological Bulletin.* Nabrit is perhaps best known for his work toward promoting African Americans in science. He helped to organize the National Institute of Science in 1943, an organization designed to explore the teaching and research problems of African-American scientists.

819

Purpose

Students correlate the impact of human activities on the declining numbers of fish species in the U.S.

Process Skills

acquire information, analyze information, think critically, draw a conclusion, interpret data, interpret scientific illustrations

Teaching Strategies

■ Make sure that students are able to locate the various states referred to in the activity.

■ Remind students of terminology that can be used to describe general areas of the country, such as southwest, midwest, etc.

Thinking Critically

1. southeast, midwest, northeast, south-central, far west; southwest, northwest, southeast, midwest, northeast

2. no; one compares total species while the other compares species at risk

3. California: between 30 and 60; Texas: between 10 and 60; Hawaii: 30 or more; N. Carolina: between 20 to 60; Florida: between 10 and 60; Illinois: fewer than 10 to 60; N. Dakota: fewer than 10.

4. Fertilizers, pesticides, and herbicides wash into waterways and affect the fish. Dams and water diversions decrease the water available for reproduction and feeding and prevent fish from reaching breeding sites. Non-native fish species compete for food and breeding sites with native fish species and may prey upon native fishes.

5. Humans apply agricultural chemicals that run off into waterways. Humans build the dams and divert water for irrigation and drinking. Humans also introduce non-native species, sometimes for sport fishing.

820

Problem-Solving Lab 30-1   Cause and Effect

Why are fish species numbers declining? There are three major threats to the survival of freshwater fishes. These are: agricultural runoff; dams and water diversion; and competition from introduced non-native species. All of these threats exist in the southern United States.

Analysis

The two maps that appear here (A and B) illustrate the problem described in the introduction.

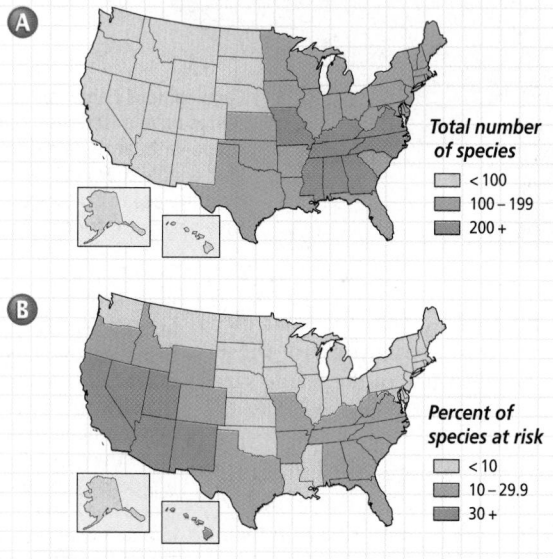

A

Total number of species
☐ < 100
☐ 100 – 199
☐ 200 +

B

Percent of species at risk
☐ < 10
☐ 10 – 29.9
☐ 30 +

Thinking Critically

1. Sequence the general areas of the United States that have the highest to lowest number of fish species and percentage of species at risk.

2. Do the two patterns tend to agree? Explain why.

3. Compare the approximate number of species at risk to the total number of fish species present in: California, Texas, Hawaii, North Carolina, Florida, Illinois, and North Dakota.

4. Explain why or how the following may be a problem to fish survival: agricultural runoff; dams and water diversion; introduction of non-native fish species.

5. Explain how the three problems associated with declining fish populations might be caused by humans.

In the *Problem-Solving Lab* on this page, find out some other reasons why the numbers of some species of freshwater fishes are declining.

820 FISHES AND AMPHIBIANS

Dorsal fin

Caudal fin

Pelvic fins

Pectoral fins

Figure 30.3
The paired fins of a fish include the pectoral fins and the pelvic fins. Fins found on the dorsal and ventral surfaces can include the dorsal fins and anal fin.

Most fishes have paired fins

Fishes in the classes Chondrichthyes and Osteichthyes have paired fins. **Fins** are fan-shaped membranes that are used for balance, swimming, and steering. Fins are attached to and supported by the endoskeleton and are important in locomotion. The paired fins of fishes, illustrated in *Figure 30.3*, foreshadowed the development of limbs for movement on land and ultimately of wings for flying.

Fishes have developed sensory systems

All fishes have highly developed sensory systems. Cartilaginous and bony fishes have an adaptation called the lateral line system that enables them to sense their environment. The **lateral line system** is a line of fluid-filled canals running along the sides of a fish that enable it to detect movement and vibrations in the water.

Some fishes also have an extremely sensitive sense of smell and can detect small amounts of chemicals in the water. Sharks can follow a trail of

---

✓ Assessment

Portfolio Have students examine their own state and describe the species of fish at risk versus total existing number of species. Have them write a paragraph explaining what steps might be taken to slow the decline of fish species in their area. Use the Performance Task Assessment List for Writing in Science in **PASC**, p. 87. **L2**

PROJECT

The Aquarium

Linguistic Ask students to prepare a written plan that identifies all materials needed to set up and maintain an aquarium, along with an explanation of why each material is needed. Allow students to set up and maintain their aquariums according to their plans. **L2**

TECH PREP

blood through the water for several kilometers. This ability helps them locate their prey.

Most fishes have scales

Cartilaginous and bony fishes have skin covered by intermittent or overlapping rows of scales. **Scales** are thin bony plates formed from the skin. Scales, shown in *Figure 30.4*, can be toothlike, diamond-shaped, cone-shaped, or round. Shark scales are similar to teeth found in other vertebrates. The age of some species of fishes can be estimated by counting annual growth rings in their scales.

Jaws evolved in fishes

Perhaps one of the most important events in vertebrate evolution was the evolution of jaws in ancestral fishes. The advantage of a jaw is that it enables an animal to grasp and crush its prey with great force. Sharks are able to eat large chunks of food. This, among other factors, explains why some early fishes were able to reach such great size. *Figure 30.5* shows the evolution of jaws in fishes.

When you think of a shark, do you imagine gaping jaws and rows of razor-sharp teeth? Sharks have six to 20 rows of teeth that are continually replaced. The teeth point backwards, preventing prey from escaping once

Figure 30.4
Fishes can be classified by the type of scales present. Diamond-shaped scales are common to bony fishes, such as gars (a). Bony fishes such as chinook salmon have either cone-shaped or round scales (b). Tooth-shaped scales are characteristic of the sharks (c).

caught. Sharks are among the most streamlined of all fishes and are well adapted for life as predators.

Most fishes have bony skeletons

The majority of the world's fishes belong to the class Osteichthyes, the bony fishes. Bony fishes, a successful and widely distributed class, differ greatly in habitat, size, feeding behavior, and shape, as *Figure 30.6* shows. All bony fishes have skeletons made of bone rather than the cartilaginous skeletons found in other

WORD *Origin*

scale
From the Old English word *sceala*, meaning "shell" or "husk." A scale is a thin, bony plate on the skin of a fish.

Figure 30.5
You can see how jaws evolved from the cartilaginous gill arches of early jawless fishes in this series of illustrations. Teeth evolved from skin.

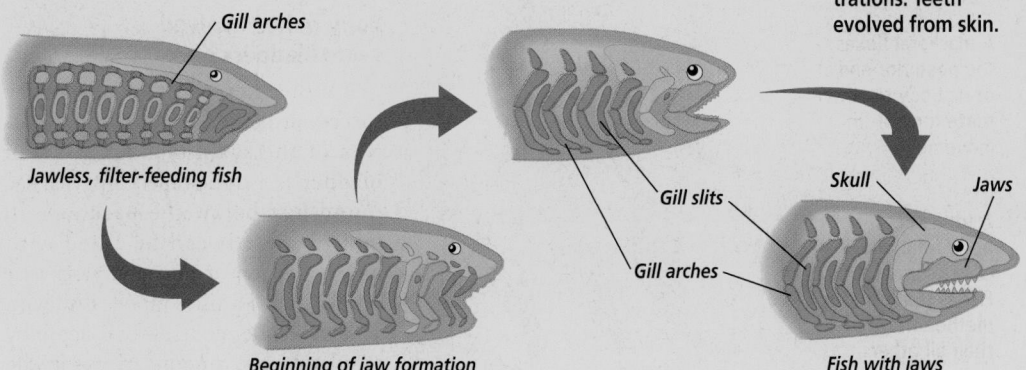

Gill arches

Jawless, filter-feeding fish

Beginning of jaw formation

Gill slits

Gill arches

Skull   Jaws

Fish with jaws

Figure 30.6
Bony fishes vary in appearance, behavior, and way of life.

A Eels are long and snakelike and can wriggle through mud and crevices in search of food.

B Sea horses move slowly through the underwater forests of seaweed where they live. They are unusual in that the males brood their young in stomach pouches.

C Predatory bony fishes, such as this pike, have sleek bodies with powerful muscles and tail fins for fast swimming.

classes of fishes. Bone is the hard, mineralized, living tissue that makes up the endoskeleton of most vertebrates. The appearance of bone was important for the evolution of fishes and vertebrates in general because it allowed fishes to adapt to a variety of aquatic environments, and finally even to land.

Figure 30.7
Most bony fishes swim in one of three possible ways.

A An eel moves its entire body in an S-shaped pattern.

B A mackerel flexes the posterior end of its body to accentuate the tail-fin movement.

C A tuna keeps its body rigid, moving only its powerful tail. Fishes that use this method move faster than all others.

Bony fishes have separate vertebrae that provide flexibility

The evolution of a backbone composed of separate, hard segments called vertebrae was significant in providing the major support structure of the vertebrate skeleton. Separate vertebrae provide great flexibility. This is especially important for fish locomotion, which involves continuous flexing of the backbone. You can see how modern bony fishes propel themselves in water in *Figure 30.7*. Some fishes are effective predators, in part because of the fast speeds they can attain as a result of having a flexible skeleton.

Bony fishes evolved swim bladders

Another key to the evolutionary success of bony fishes was the evolution of the swim bladder. A **swim bladder** is a thin-walled, internal sac found just below the backbone in bony fishes; it can be filled with mostly oxygen or nitrogen gases that diffuse out of a fish's blood. Fish with a swim bladder control their depth by regulating the amount of gas in the bladder. The gas works like the gas in a blimp that adjusts the height of the blimp above the ground.

Fishes that live in oxygen-poor water or in ponds or rivers that dry up in the hot season often have other ways to get oxygen. The African lungfish, for example, has a structure that allows it to obtain oxygen by gulping air. This structure is a modified swim bladder. The modified swim bladder is connected to the fish's mouth by a tube.

Diversity of Fishes

Fishes range in size from the tiny dwarf goby that is less than 1 cm long, to the huge whale shark that can reach a length of 15 m—the length of two school buses.

Agnathans are jawless fishes

Lampreys and hagfishes belong to the class Agnatha. Though they do not have jaws, they are voracious feeders. Hagfishes, *Figure 30.8*, have a slitlike, toothed mouth and feed on dead or dying fish by drilling a hole and sucking the blood and insides from the animal. Parasitic lampreys attack other fishes and attach them-

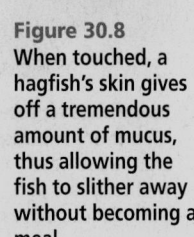

selves by their suckerlike mouths. They use their sharp teeth to scrape away the flesh and then suck out the prey's blood. The skeletons of agnathans, as well as of sharks and their relatives, are made of a tough, flexible material called **cartilage.**

Sharks and rays are cartilaginous fishes

Sharks, skates, and rays, like agnathans, possess skeletons composed entirely of cartilage. Sharks, skates, and rays belong to the class Chondrichthyes, illustrated in *Figure 30.9*. Because living sharks, skates, and rays are classified in the same genera as species that swam the seas more than 100 000 years ago, they are considered living fossils. Sharks are perhaps the most well-known

Figure 30.8
When touched, a hagfish's skin gives off a tremendous amount of mucus, thus allowing the fish to slither away without becoming a meal.

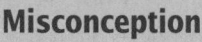

Figure 30.9
Cartilaginous fishes include sharks, skates, and rays.

Ⓐ The hammerhead shark is a large shark found in warm ocean water. It has eyes at the ends of its flattened, extended skull.

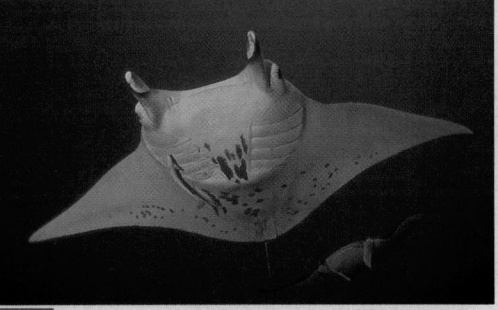

Ⓑ Most rays are ocean bottom dwellers, but the Atlantic manta ray prefers to glide along just under the water's surface.

aquarium that contains black gravel with black construction paper.
6. Develop a hypothesis about where the fish will spend its time, in the light or the dark. Then, observe the fish for two minutes and time in which part of the tank the fish spends its time. Record your data.
7. Compare your data with that of your classmates.

Expected Results
Catfish prefer the dark, but others prefer the light.
Analysis
1. Which fishes preferred dark gravel and the dark half of the tank? *catfish*
2. Do your data for each habitat type support your hypotheses? Explain. *Some students will find that their data supported their hypotheses.*

✔ **Assessment**
Performance Give students a fish they have not seen and ask them to determine the fish's preferred habitat. Ask them to explain how this preference may have adaptive value. Use the Performance Task Assessment List for Making Observations and Inferences in **PASC**, p. 17. **L2**

Figure 30.10
Lungfishes, lobe-finned fishes, and ray-finned fishes are the subclasses of the bony fishes.

A Lungfishes represent an ancient subclass that arose nearly 400 million years ago. Lungfishes such as this African lungfish have both gills and lungs.

B Lobe-finned fishes, such as this coelacanth, appeared in the fossil record about 395 million years ago. Long thought to be extinct, living examples of coelacanths were caught off the coast of Africa beginning in 1938.

C You can easily see the rays that support the pectoral fins of this flying fish, an example of a ray-finned fish.

predators of the oceans. Like sharks, most rays are predators and feed on or near the ocean floor. Rays have flat bodies and broad pectoral fins on their sides. By slowly flapping their fins up and down, rays can glide, searching for mollusks and crustaceans, along the ocean floor. Some species of rays have sharp spines with poison glands for defense on their long tails. Other species have organs that generate electricity to kill both prey and predators.

Three subclasses of bony fishes

Scientists recognize three subclasses of bony fishes—the lungfishes, the lobe-finned fishes, and the ray-finned fishes. *Figure 30.10* shows examples of each subclass. Lungfishes have both gills and lungs. The lobe-finned fishes are represented by only one living species. In the ray-finned fishes, fins are fan-shaped membranes supported by stiff spines called rays. Ray-finned fishes—such as catfish, perch, salmon, and cod—are more familiar to humans because most of the fishes we consume belong to this subclass. You can see the parts of a ray-finned fish in the *Inside Story.*

Origins of Fishes

Scientists have identified fossils of fishes that existed during the early Ordovician period, 500 million years ago. For 50 million years, ostracoderms (oh STRAHK oh durmz), early jawless fishes, were the only vertebrates on Earth. Although most ostracoderms became extinct at the end of the Devonian period, 400 million years ago, present-day agnathans appear to be their direct descendants.

Weighed down by heavy, bony external armor, ostracoderms, shown

824 FISHES AND AMPHIBIANS

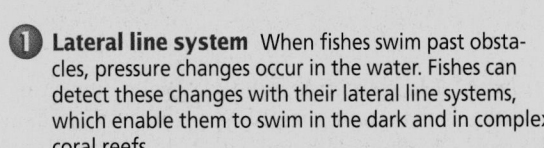
A Bony Fish

The bony fishes, class Osteichthyes, include some of the world's most familiar fishes, such as the bluegill, trout, minnow, bass, swordfish, and tuna. Though diverse in general appearance and behavior, bony fishes share some common adaptations with other fish classes.

Critical Thinking *Compare the ideal shapes for fishes that swim in rocky crevices and those that swim in open water.*

Rainbow trout,
Salmo gairdneri

1 Lateral line system When fishes swim past obstacles, pressure changes occur in the water. Fishes can detect these changes with their lateral line systems, which enable them to swim in the dark and in complex coral reefs.

2 Swim bladder The mass of a fish's tissue is greater than that of water; therefore, without a swim bladder, a fish would not be able to float. Gas pressure in the swim bladder alters the specific gravity of the whole body, enabling the fish to float at any depth.

3 Scales Scales are covered with slippery mucus, allowing a fish to move through water with minimal friction.

Kidney
Urinary bladder
Reproductive organ

5 Fins The structure and arrangement of fins are related to a particular type of locomotion. Tropical fishes that live among coral reefs tend to have small fins capable of maneuvering in this complex, three-dimensional environment, whereas a tuna has large, broad fins for moving quickly through open water.

Stomach
Intestine
Liver
Heart

4 Gills Gills are thin, blood vessel-rich tissues where gases are exchanged.

30.1 FISHES **825**

Resource Manager

Basic Concepts Transparency 52 and Master **L2** **ELL**

Purpose
Students study features of bony fishes and examine the adaptive advantages of these features.

Teaching Strategies
■ On the chalkboard, make a table that compares the features of bony fishes with those of jawless and cartilaginous fishes.

Visual Learning
■ Ask students to study the features shown. Have them identify the external features discussed on a live goldfish.
■ Purchase a fish from a fish market that has not been cleaned. Carefully cut open the fish to show students the swim bladder. The swim bladder is very fragile, so be careful as you cut into the fish with a razor blade or scalpel.

Critical Thinking
Fishes that swim in crevices may be laterally flattened, whereas fish that swim in the open may be wider but streamlined.

3 Assess

Check for Understanding
Linguistic Ask students to summarize in writing the characteristics of jawless fishes, cartilaginous fishes, and bony fishes. **L2**

Reteach
Have students work in groups to make a table that compares the characteristics of the three fish classes. Make sure they include features such as unique body structures, scale types, methods of getting food, and skeletal features. Have other groups construct a table that compares the three subclasses of bony fishes. Review the tables as a class.

825

Extension

Have interested students interview a fly-fishing expert to find out about the sport of fly-fishing. Encourage students to videotape the interview or take photographs of the techniques used in the sport. Have students report their findings to the class and include a demonstration of fly tying. **L2**

✓ Assessment

Knowledge Ask students to list adaptations fishes might have for each of the following niches: live on a muddy pond bottom; feed on other fishes; feed on clams and other mollusks; live in crevices in a coral reef; live in deep water where there is no light. **L2**

4 Close

Discussion

Ask students to speculate about why there are more fishes than any other type of vertebrate. *Responses may state that all life began in the water and this environment has undergone fewer changes than have terrestrial habitats, thus reducing the risk of extinction because of the inability to adapt to new conditions.* Have students predict how the extinction of fishes would affect world food supplies. *Because people in many parts of the world rely on fishes as a major food source, starvation would occur in these areas.*

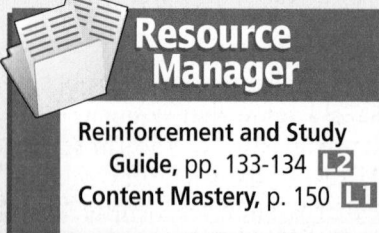

Resource Manager

Reinforcement and Study Guide, pp. 133-134 **L2**
Content Mastery, p. 150 **L1**

826

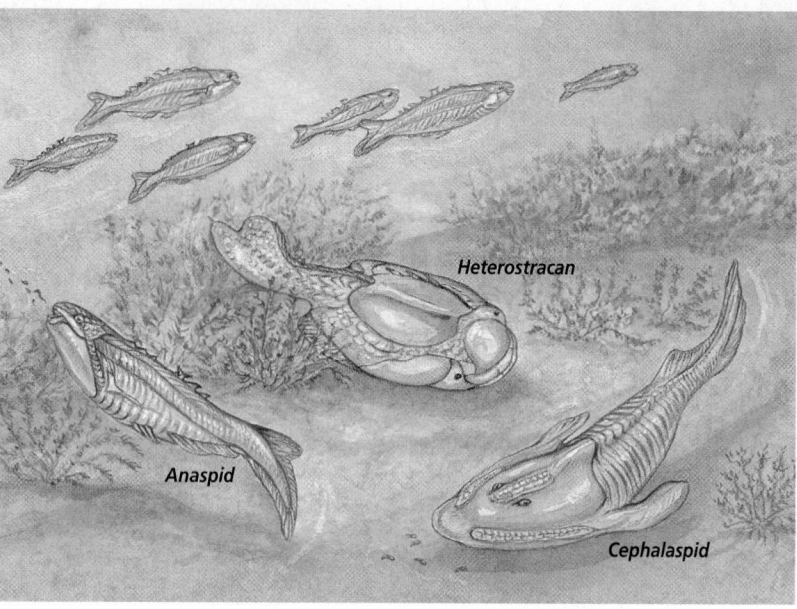

Figure 30.11
Ostracoderms, the earliest vertebrate fossils found, were characterized by bony, external plates covering the body and a jawless mouth. Lacking jaws, ostracoderms obtained food by sucking up bottom sediments and sorting out the nutrients.

Heterostracan

Anaspid

Cephalaspid

in *Figure 30.11*, were fearsome-looking animals that swam sluggishly over the murky seafloor. Although all ostracoderms had cartilaginous skeletons, they also had shields of bone covering their heads and necks. The development of bone in these animals was an important evolutionary step because bone provides a place for muscle attachment, which improves locomotion. In ancestral fishes, bone that formed into plates provided protection as well.

Lobe-finned fishes, the coelacanths (SEE luh kanthz), are another ancient group, appearing in the fossil record about 395 million years ago. They are characterized by lobelike, fleshy fins, and live at great depths where they are difficult to find. The limblike skeletal structure of fleshy fins is thought to be an ancestral condition of all tetrapods (animals with four limbs). The earliest tetrapods discovered also had gills and therefore were still aquatic.

Scientists hypothesize that the jawless ostracoderms were the common ancestors of all fishes. Modern cartilaginous and bony fishes evolved during the mid-Devonian period.

Section Assessment

Understanding Main Ideas
1. List three characteristics of fishes.
2. Compare how jawless fishes and cartilaginous fishes feed.
3. Why was the evolution of a swim bladder important to fishes?
4. How does a flexible skeleton aid swimming in fishes?

Thinking Critically
5. Why was the development of jaws an important step in the evolution of fishes?

SKILL REVIEW

6. **Making and Using Tables** Construct a table to compare the characteristics of the jawless, cartilaginous, and bony fishes. For more help, refer to *Organizing Information* in the **Skill Handbook**.

Section Assessment

1. They have scales, gills, and fins.
2. Lampreys are parasites. Hagfishes are scavengers. Cartilaginous fishes are predators.
3. It enables fishes to control their depth by regulating the amount of gas in the bladder.
4. Continuous flexing of the backbone enables fishes to propel themselves with muscles attached to the backbone.
5. Jaws enable fishes to grasp and crush prey with great force and to eat large chunks of food.
6. Make sure that students compare in their tables structural features such as skeleton, lateral line, scales, and swim bladder, as well as behavioral features such as ways of getting food and ways of swimming.

Section

30.2 Amphibians

If an alien visitor to our planet were to watch our television programs and read our children's literature, it might return home with wondrous stories of how frogs on Earth can talk and change by the touch of a kiss into princes. Frogs and toads don't talk, but they do change—from fishlike tadpoles to four-legged animals with bulging eyes, long tongues, loud songs, and remarkable jumping ability.

Pickerel frog (above) and tadpoles (inset)

What Is an Amphibian?

The striking transition from a completely aquatic larva to an air-breathing, semiterrestrial adult gives the class Amphibia (am FIHB ee uh) its name, which means "double life." The class Amphibia includes three orders: Caudata (kaw DAHT uh), with salamanders and newts; Anura (uh NUHR uh), with frogs and toads; and Apoda (uh POH duh), with legless caecilians, as shown in *Figure 30.12.* Amphibians have thin, moist skin and four legs. They have no claws on their toes. Although most adult amphibians are capable of a terrestrial existence, nearly all of them rely on water for reproduction. Fertilization in most amphibians is external, and water is needed as a medium for transporting sperm. Amphibian eggs

lack protective membranes and shells and must be laid in water to stay moist. How do frogs capture their food? Find out by reading the *Inside Story* on the next page.

Amphibians are ectotherms

Amphibians are more common in regions that have warm temperatures all year because they are ectotherms.

Figure 30.12
Caecilians, order Apoda, are long, limbless amphibians. They look like worms but have eyes that are covered by skin.

827

MEETING INDIVIDUAL NEEDS

Visually Impaired

Kinesthetic Provide visually impaired students with plastic or clay models of animals from the three orders of amphibians. Have students compare and contrast the features of each and describe their adaptations to moist habitats. **L1**

GLENCOE TECHNOLOGY

CD-ROM
Biology: The Dynamics of Life
Exploration: *Amphibians*
Disc 4

Section 30.2

Prepare

Key Concepts

Students will relate the demands of a terrestrial environment to the adaptations of amphibians. They will also study the diversity of amphibians.

Planning

■ Purchase a recording of frog calls and live frogs that have not been collected from the wild for the Quick Demos.

■ Gather culture dishes, thermometers, and binocular microscopes for the BioLab.

---

1 Focus

Bellringer

Before presenting the lesson, display **Section Focus Transparency 74** on the overhead projector and have students answer the accompanying questions. **L1 ELL**

Resource Manager

Section Focus Transparency 74 and Master **L1 ELL**

SECTION PREVIEW

Objectives
Relate the demands of a terrestrial environment to the adaptations of amphibians.
Relate the evolution of the three-chambered heart to the amphibian lifestyle.

Vocabulary
ectotherm
vocal cords

827

2 Teach

Purpose

Students learn about the important features of frogs.

Teaching Strategies

■ Have a live frog available as well as wall charts showing the internal and external anatomy of a frog. As each feature is discussed in the Inside Story, point out the corresponding feature on the live frog and wall charts.

■ Divide students into teams. Provide each team with birthday blowers that have a small piece of Velcro attached to the tip of the blowers. Attach the complementary piece of Velcro to a small piece of ribbon. Have one student on each team move the ribbon "fly" while another student uses the blower "tongue" to try to catch the "fly." Ask students how this model illustrates how a frog obtains food. Ask them to point out how the model is not like a real frog's tongue. **L1**
ELL **COOP LEARN**

Visual Learning

■ Provide each student with an acetate transparency and markers. Have them lay the transparency on the frog diagram shown and trace the outline of the frog on the acetate. Encourage students to label each of the structures shown in the diagram. **L1**

■ Ask students to use the transparencies they made to prepare a presentation about frogs for an elementary school science class. Have students outline the features to be covered in their presentations. **L2**

Critical Thinking

The frog's three-chambered heart can pump more oxygenated blood and thereby make more energy available to it.

828

A Frog

Many species of frogs look similar. As adults, they have short, bulbous bodies with no tails. This adaptation allows them to jump more easily.

Critical Thinking *How does a three-chambered heart benefit a frog?*

Green frog, *Rana clamitans*

1 Eyes Some frogs' eyes protrude from the tops of their heads—an adaptation that enables them to stay submerged in the water with only their eyes above the surface.

3 Tongue A frog's tongue is long, sticky, and fastened to the front of the mouth. These adaptations allow frogs to snare their prey, such as flies, with amazing accuracy.

4 Lungs Lungs enable adult amphibians to breathe air.

5 Calls Male frogs use sound to attract females. Females have distinct calls to indicate whether or not they are willing to mate.

2 Tympanic membrane Vibrations from water or air are picked up by the tympanic membrane and transmitted to the inner ear and then to the brain. The tympanic membrane also amplifies the sounds frogs make.

6 Legs The hind legs of a frog are muscular. If you have ever tried to catch a frog, you can appreciate the power in these leg muscles.

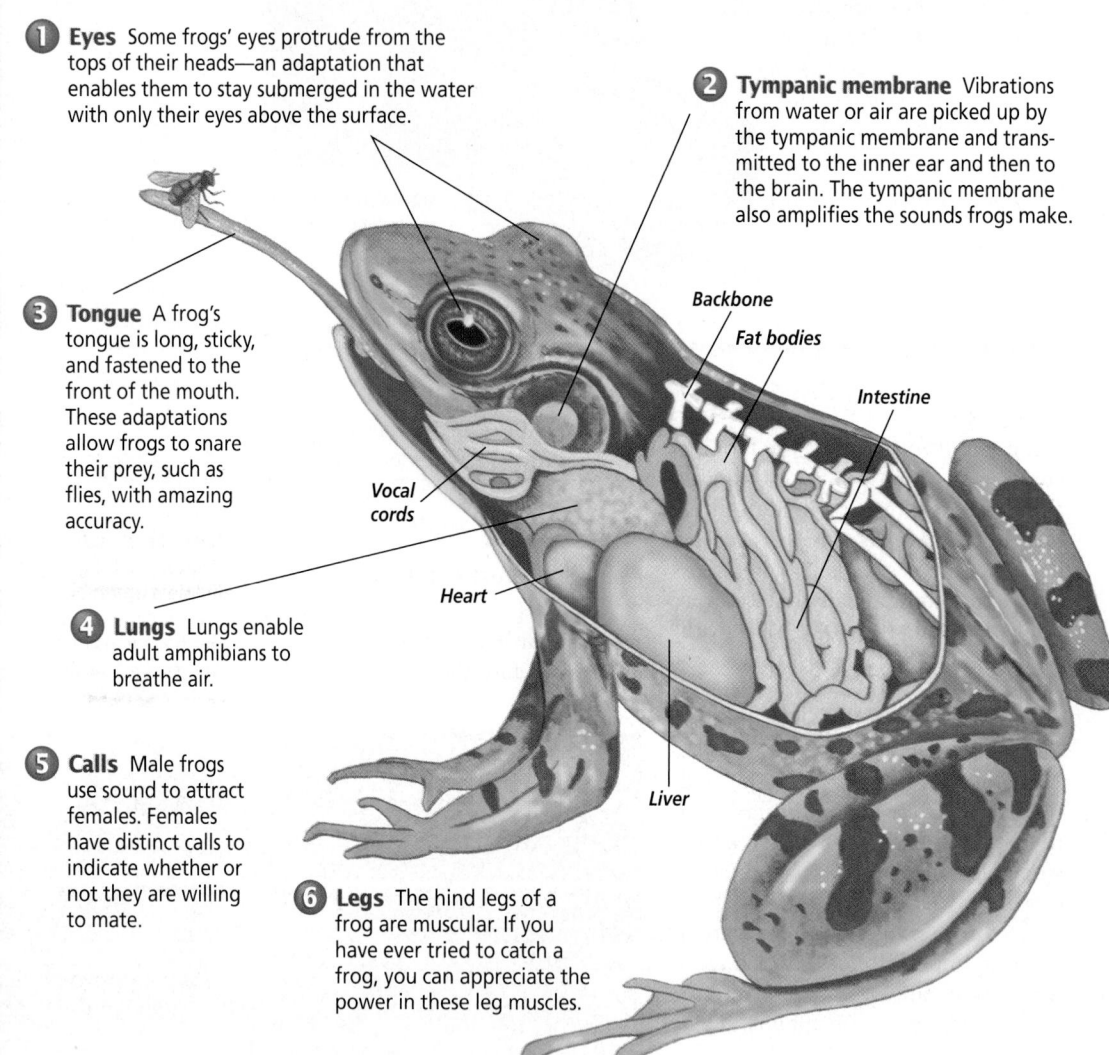

Backbone
Fat bodies
Intestine
Vocal cords
Heart
Liver

828 FISHES AND AMPHIBIANS

GLENCOE TECHNOLOGY

CD-ROM
Biology: The Dynamics of Life
Video: *Feeding Frog*
Video: *Frog Behavior*
Disc 4

Resource Manager

Reteaching Skills Transparency 44 and Master **L1** **ELL**
Critical Thinking/Problem Solving, p.30 **L3**
Concept Mapping, p. 30 **L3** **ELL**
Laboratory Manual, pp. 213-218 **L2**

An **ectotherm** (EK tuh thurm) is an animal in which the body temperature changes with the temperature of its surroundings. Because many biological processes require particular temperature ranges in order to function, amphibians become dormant in regions that are too hot or cold for part of the year. During such times, many amphibians burrow into the mud and stay there until suitable conditions return.

Amphibians undergo metamorphosis

Unlike fishes, most amphibians go through the process of metamorphosis. Fertilized eggs hatch into tadpoles, the aquatic stage of most amphibians. You can compare tadpoles with adult frogs in the *MiniLab* on the next page. Tadpoles possess fins, gills, and a two-chambered heart as seen in fishes. As tadpoles grow into adult

frogs and toads, they develop legs, lungs, and a three-chambered heart. *Figure 30.13* shows this life cycle. You can observe the development of frog eggs in the *Investigate BioLab* at the end of this chapter.

Young salamanders resemble adults, but they have gills and usually have a tail fin. Most adult salamanders lack gills and fins. Instead, they breathe through their moist skin or with lungs. Up to one-fourth of all salamanders have no lungs and breathe only through their skin. Most salamanders have four legs for moving about, but a few have only two front legs.

Walking requires more energy

The laborious walking of early amphibians required a great deal of energy from food and large amounts of oxygen for aerobic respiration. The evolution of the three-chambered heart in amphibians ensured

Figure 30.13
The amphibian life cycle includes an aquatic tadpole stage and a terrestrial adult stage.

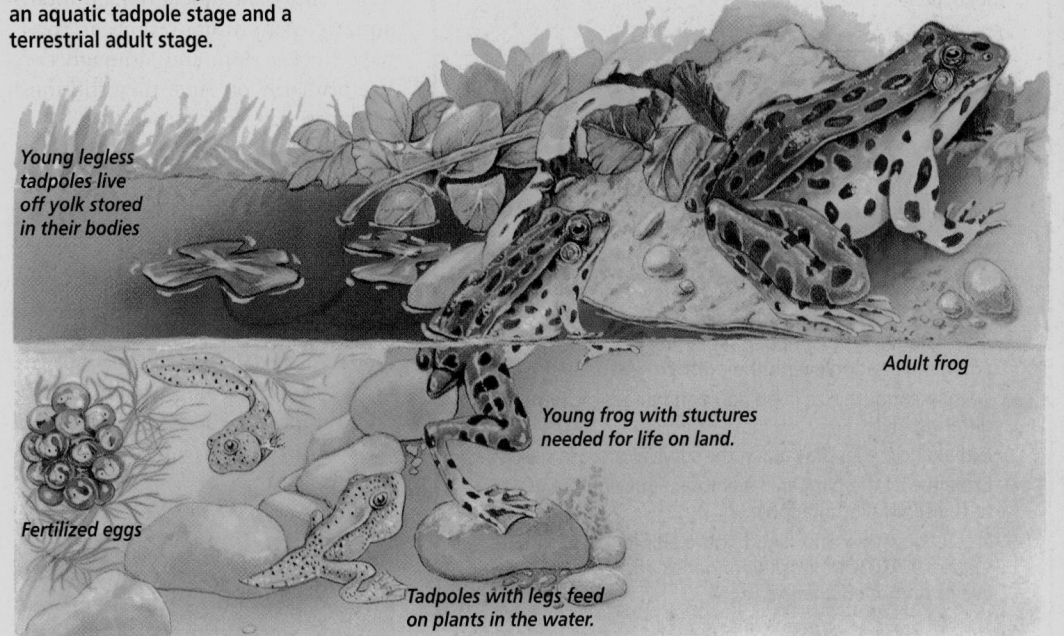

Young legless tadpoles live off yolk stored in their bodies

Fertilized eggs

Tadpoles with legs feed on plants in the water.

Young frog with stuctures needed for life on land.

Adult frog

 *Naturalist* Divide the class into groups. Give each group a live frog inside a large, clear container. Ask students to describe any behaviors they observe in the frog. *Students are likely to describe the frog jumping, moving about, or breathing.* Elicit from students how these activities may be suited to life on land. Make sure the frogs you use have not been collected from the wild. **L1**

The BioLab at the end of the chapter can be used at this point in the lesson. **INVESTIGATE BioLab**

Visual Learning

Figure 30.13 Ask students how having two different stages in a life cycle could have survival value for a species. *Responses may suggest that the two stages help the species survive during adverse or changing climate conditions on land or in water.*

GLENCOE TECHNOLOGY

 VIDEODISC
The Secret of Life
Fish/Amphibian—Air Exchange

Fish/Amphibian Heart

BIOLOGY JOURNAL

Mapping Amphibian Habitats
Visual-Spatial Give students a blank world map and ask them to use an atlas of world wildlife to find out where various amphibians live. Have them color code their maps for various types of amphibians and place their maps in their journals. **L2** **ELL**

Portfolio

Frog Life Cycle
Visual-Spatial Have students make a flowchart that traces the stages in the frog life cycle. Ask students to describe the features of each stage of the life cycle and how each feature benefits the organism at that stage. **L2** **P**

Purpose

Students will compare and contrast the appearance and traits of an adult frog and a tadpole.

Process Skills

compare and contrast, acquire information, recognize cause and effect, think critically

Teaching Strategies

■ Preserved materials are available from biological supply houses. Retain preserved specimens for the following year.
■ Place preserved materials on paper toweling or on trays.
■ Binocular microscopes may be used to observe the skin.
■ Provide references for the information regarding diet; larvae feed on algae, adults feed on insects, small fishes, and worms.

Expected Results

Tadpoles have a tail, gills, no tympanic membrane, and no limbs at early stages. Adults have lungs. All other traits are shared. Coloration of stages may vary with species being observed. Tadpoles feed on algae, adults on insects, worms, and small fishes. Tadpoles are confined to aquatic habitats; adults live both on land and in water.

Analysis

1. allows frog to escape predators, aids in swimming
2. Diffusion of oxygen occurs between lungs and air, and between gills and water.
3. A tadpole has a tail for swimming in water. An adult frog has legs for jumping away from predators and catching prey.
4. A tadpole has a small mouth for eating algae. An adult frog has a large mouth for eating insects, worms, and fishes.
5. seeing food and predators
6. An adult can hear mating calls to enable reproduction. Tadpoles do not reproduce.
7. protective coloration from predators, thin skin aids in gas exchange

830

MiniLab 30-2 — Comparing and Contrasting

Looking at Frog and Tadpole Adaptations
An adult frog and its larval stage are adapted to different habitats. How are the structures of a frog and a tadpole adapted to their environments?

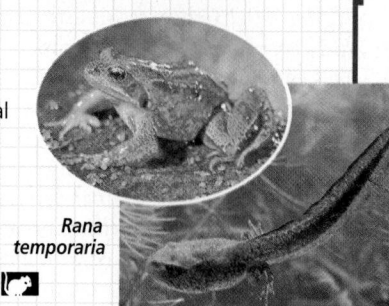
Rana temporaria

Procedure

1. Copy the data table.
2. Examine a living or preserved adult frog and larval (tadpole) stage. **CAUTION: *Wear disposable latex gloves and use a forceps when handling preserved specimens.***
3. Observe the first seven traits listed. Complete your data table for these observations.
4. Use references to fill in the information for the last three traits listed.

Data Table		
Trait or information	Tadpole	Adult
Limbs present?		
Eyes present?		
Tympanic membrane present?		
Tail present?		
Mouth present?		
Nature of skin (color and texture)		
General size		
Respiratory organ type		
Diet		
Habitat		

Analysis

1. Explain how hind leg musculature aids in adult frog survival.
2. Correlate the type of respiratory organ in an adult and a tadpole with their differing habitats.
3. Correlate the type of appendages (arm, leg, tail) in an adult and a tadpole with their differing habitats.
4. Correlate mouth size in an adult and a tadpole with their differing diet.
5. Explain how eyes may aid in the survival of both stages.
6. Explain why the tympanic membrane may not be essential to the survival of a tadpole.
7. Predict how skin color and texture aids in adult frog survival. **CAUTION: *Wash your hands after working with live or preserved animals.***

---

that cells received the proper amount of oxygen. This heart was an important evolutionary transition from the simple circulatory system of fishes.

In the three-chambered heart of amphibians, one chamber receives oxygen-rich blood from the lungs and skin, and another chamber receives oxygen-poor blood from the body tissues. Blood from both chambers then moves to the third chamber, which pumps oxygen-rich blood to body tissues and oxygen-poor blood back to the lungs and skin so it can pick up more oxygen. This results in some mixing of oxygen-rich and oxygen-poor blood in the amphibian heart and in blood vessels leading away from the heart. Thus, in amphibians, the skin is much more important than the lungs as an organ for gas exchange.

Because the skin of an amphibian must stay moist to exchange gases, most amphibians are limited to life on the water's edge. However, some newts and salamanders remain totally aquatic. Amphibians such as toads have thicker skin, and although they live primarily on land, they still must return to water to reproduce.

Diversity of Amphibians

Because amphibians still complete part of their life cycle in water, they are limited to the edges of ponds, lakes, streams, and rivers or to areas that remain damp during part of the year. Although they are not easily seen, amphibian species are numerous worldwide.

Frogs and toads belong to the order Anura

Frogs and toads are amphibians with no tails. Frogs have long hind legs and smooth, moist skin. Toads

---

✔ Assessment

Portfolio Have students locate references that describe time needed for metamorphosis to occur. Ask them to prepare a poster that compares length of metamorphosis for different frog species. Use the Performance Task Assessment List for Poster in **PASC**, p. 73. **L2**

GLENCOE TECHNOLOGY

VIDEODISC
Biology: The Dynamics of Life
Frog Behavior (Ch. 3)

Disc 2, Side 1
32 sec.

Feeding Frog (Ch. 4)

Disc 2, Side 1
13 sec.

have short legs and bumpy, dry skin. Like fishes, frogs and toads have jaws and teeth. All adult frogs and toads eat insects. Many species of frogs and toads secrete chemicals through their skin as a defense against predators. Some frogs and toads produce toxins that can kill predators, such as dogs, quickly. You can find out more about poisonous frogs in the *Chemistry Connection* at the end of this chapter.

Frogs and toads also have vocal cords that are capable of producing a wide range of sounds. **Vocal cords** are sound-producing bands of tissue in the throat. As air moves over the vocal cords, they vibrate and cause molecules in the air to vibrate. In many male frogs, air passes over the vocal cords, then passes into a pair of vocal sacs lying underneath the throat, illustrated in *Figure 30.14*.

Frogs and toads, like all amphibians, spend part of their life cycle in water and part on land. They breathe through lungs or through their thin skins. As a result, frogs and toads often are among the first organisms to be exposed to pollutants in the air, on land, or in the water. Declining numbers of frog species, or deformities in local frogs, sometimes indicate that the environment is no longer healthy.

Salamanders belong to the order Caudata

Unlike a frog or toad, a salamander has a long, slender body with a neck and tail. Salamanders resemble lizards, but have smooth, moist skin and lack claws. Some salamanders are totally aquatic, and others live in damp places on land. They range in size from a few centimeters in length up to 1.5 m. Newts are salamanders that live entirely in water. The young that hatch from salamander eggs look like small adults and are carnivorous.

Caecilians are limbless amphibians

Caecilians are amphibians that have no limbs, with a short, or no, tail. Caecilians are primarily tropical animals with small eyes that are often blind. They eat earthworms and other invertebrates found in the soil.

Origins of Amphibians

Imagine a time 350 million years ago when the inland, freshwater seas were filled with carnivorous fishes. One type of tetrapod had evolved that retained gills for breathing and a finned tail for swimming. In later fossils, the four limbs are found further below the body to lift it off the ground. Any animal that could move over land from the mud of a drying stream to another water source might survive. Most likely, amphibians arose as their ability to breathe air through well-developed lungs evolved. The success of inhabiting the land depended on adaptations that would provide support, protect membranes involved in respiration, and provide efficient circulation.

Figure 30.14
Most male frogs have throat pouches that, along with the tympanic membrane, increase the volume of their calls.

Figure 30.15
Adaptation to life on land involves the positioning of limbs. The evolution of tetrapods led to the diversification of land vertebrates.

A The salamander, an amphibian, has legs that extend at right angles to its body.

B Reptiles such as this crocodile have legs on the sides of their bodies, like amphibians, but the limbs have joints that enable them to bend and hold the body up off the ground.

C Like reptiles, mammal bodies are raised above the ground, but the limbs are positioned underneath the body. This position allows greater speed of locomotion, making mammals such as this cheetah the fastest-moving land animals.

Challenges of life on land

Life on land held many advantages for early amphibians. There was a large food supply, shelter, and no predators. In addition, there was much more oxygen in air than in water. However, land life also held many dangers. Unlike the temperature of water, which remains fairly constant, air temperatures vary a great deal. In addition, without the support of water, the body was clumsy and heavy. Some of the efforts to move on land by early amphibians probably were like movements of modern-day salamanders. The legs of salamanders are set at right angles to the body. You can see in *Figure 30.15* why the bellies of these animals may have dragged on the ground.

Amphibians first appeared about 360 million years ago. Amphibians probably evolved from an aquatic tetrapod, as shown in *Figure 30.16,* around the middle of the Paleozoic Era. At that time, the climate on Earth is known to have become warm and wet, ideally suited for an adaptive radiation of amphibians. Able to breathe through their lungs, gills, or skin, amphibians became, for a time, the dominant vertebrates on land.

Figure 30.16
Transitional fossils with four limbs from the Devonian period show that they had amphibian characteristics, but they also retained some fishlike features.

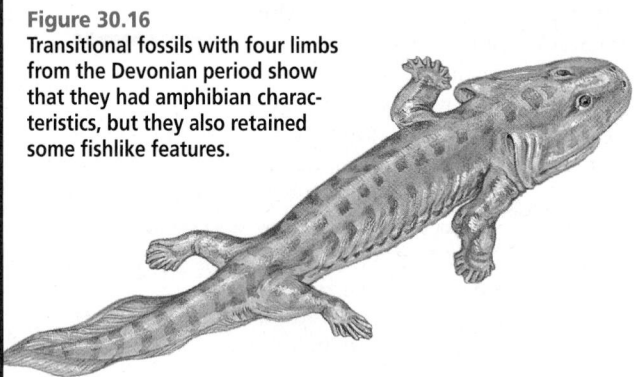

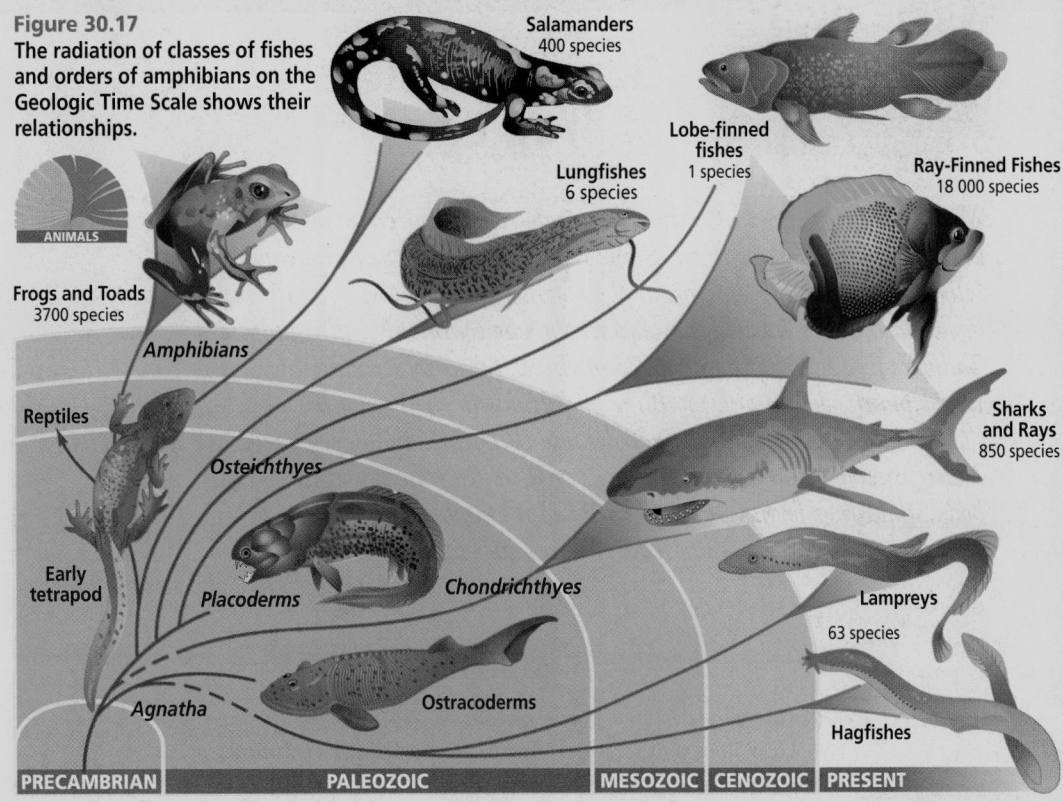

Figure 30.17
The radiation of classes of fishes and orders of amphibians on the Geologic Time Scale shows their relationships.

ANIMALS

Salamanders
400 species

Lobe-finned fishes
1 species

Lungfishes
6 species

Ray-Finned Fishes
18 000 species

Frogs and Toads
3700 species

Amphibians

Reptiles

Osteichthyes

Sharks and Rays
850 species

Early tetrapod

Placoderms

Chondrichthyes

Lampreys
63 species

Agnatha

Ostracoderms

Hagfishes

PRECAMBRIAN   PALEOZOIC   MESOZOIC   CENOZOIC   PRESENT

Recall that early vertebrates evolved from mud-sucking, swimming fishes to aquatic tetrapods. Scientists have found fossil evidence that supports the hypothesis that limbs first evolved in aquatic animals. Some of these aquatic vertebrates evolved into air-gulping animals that

crawled from one pond to another, and finally to fully developed amphibians that lived mainly on land. Although the fossil record for fishes and amphibians is incomplete, most scientists agree that the relationships shown in *Figure 30.17* represent the best fit for the available evidence.

Section Assessment

Understanding Main Ideas
1. Describe the events that may have led early animals to move to land.
2. List three characteristics of amphibians.
3. Name two ways that amphibians depend on water.
4. How does metamorphosis through two different forms benefit amphibians?

Thinking Critically
5. How does a three-chambered heart enable amphibians to obtain the energy needed for movement on land?

SKILL REVIEW
6. **Sequencing** Trace the evolutionary development of amphibians from lungfishes. For more help, refer to *Organizing Information* in the **Skill Handbook**.

Section Assessment

1. Shallow inland seas were affected by periodic droughts. Early tetrapods with gills and limbs were able to move briefly across land to another water source for survival.
2. three-chambered hearts, eggs without shells laid in water, smooth, moist skin
3. external fertilization, growth and development of eggs, development of larval stage
4. Tadpoles and adult frogs use different food sources and, therefore, do not compete with one another for food.
5. Cells obtain oxygen quickly in a three-chambered heart. The heart is a more efficient pump, enabling oxygen to reach cells quickly, thereby enabling the animal to move quickly.
6. Amphibians first appeared about 350 million years ago, probably evolving from the lungfishes. Over time, the lungfishes evolved legs and feet suited to a land environment.

Development of Frog Eggs

Most frogs breed in water. The male releases sperm over the female's eggs as she lays them. Some frogs lay up to a thousand eggs. A jellylike casing protects the eggs as they grow into embryos. When the embryos hatch, they develop into aquatic larvae commonly called tadpoles. Tadpoles feed by scooping algae from the water or by scraping algae from water plants with small, toothlike projections in their mouths.

During metamorphosis, tadpoles lose their tails and develop legs. Many internal changes take place as well. Gills are reabsorbed by the body, and lungs form. Development and metamorphosis to the adult frog take from three weeks up to several years, depending upon the species.

Time Allotment
First day—one class period. Every day for 2 weeks—10 minutes.

Process Skills
observe and infer, compare and contrast, form a hypothesis, interpret data, recognize cause and effect

Safety Precautions
■ Encourage students to treat living organisms in a humane manner at all times.
■ Students should wash their hands after each observation.
■ Caution students to avoid skin and eye contact with Ringer's solution.
■ Make sure students' hands and work area are dry when using electrical materials.

PREPARATION

■ Use sterile Ringer's solution to help prevent contamination by bacteria and fungi. Ask students to avoid touching the solution to prevent contamination.

Alternative Materials
■ Make sure to use frog eggs that have a short development time. It is best not to collect eggs from the wild as amphibians are declining worldwide. Order eggs from a biological supply house.

PREPARATION

Problem
How does temperature affect the development of frog eggs?

Objectives
In this BioLab, you will:
■ Compare development of frog eggs at varying temperatures.
■ Distinguish among various stages of development.

Materials
Ringer's solution
4 culture dishes
light bulbs with source of electricity
thermometer
binocular microscope
frog eggs, *Xenopus laevis*
flashlight

Safety Precautions
Always wear goggles in the lab. Wash hands before and after each observation.

Skill Handbook
Use the **Skill Handbook** if you need additional help with this lab.

PROCEDURE

1. Obtain four culture dishes of Ringer's solution and fertilized frog eggs.
2. Make a data table similar to the one shown for sketching of stages of development of the eggs.

3. Observe your eggs and determine their stage of development. At room temperature, you should see the two-cell stage about 1.5 hours after fertilization, the eight-cell stage at 2.25 hours, the

834 FISHES AND AMPHIBIANS

PROCEDURE

Teaching Strategies
■ Review frog development before students carry out the lab.
■ Have students work in cooperative groups to reduce the materials needed. Each person in the group should have a specific job during the data collection period.
■ Explain that the white part of the egg is the yolk and the black part is the developing tadpole. The yolk is used for food by the developing tadpole. The jellylike substance surrounds and protects the egg.
■ Students should be able to observe movement of the tadpole just before it hatches.

32-cell stage at 3 hours, the late gastrula stage at 9 hours, and a visible head area between 18 and 20 hours. Hatching will occur at about 50 hours (about two days).

4. Set up the appropriate numbers of light bulbs of different wattages over the water to keep the temperatures at 20°, 25°, and 30°C.

5. Place a dish of eggs in the refrigerator. Measure the temperature of your refrigerator. Keep a flashlight on in the refrigerator at all times so that you have only one variable, the temperature.

6. Make a hypothesis about how temperature will affect development of the eggs.

7. Set up a time schedule for observations and making sketches based on what stage you are observing. Make observations until the eggs hatch. Record your observations in a journal.

8. Observe your eggs under the microscope according to the schedule you have made. Draw sketches of your eggs in the data table.

Data Table

Temperature	Day 1	Day 2	Day 3
30° C			
25° C			
20° C			
refrigerator			

ANALYZE AND CONCLUDE

1. **Interpreting Observations** Which eggs develop the fastest? The slowest? Explain.

2. **Interpreting Observations** Did your data support your hypothesis? Explain.

3. **Drawing Conclusions** What advantage is it for frogs to have eggs that develop at different rates that correspond to different temperatures?

4. **Thinking Critically** What would happen if frog eggs developed

when the weather was still cold in the spring?

Going Further

Make a Hypothesis Make a hypothesis about what other environmental factors would affect the development of frog eggs. Explain your reasoning.

To find out more about metamorphosis, visit the Glencoe Science Web site.
science.glencoe.com

INVESTIGATE BioLab

ANALYZE AND CONCLUDE

1. Egg development and cell division occur more quickly at warmer temperatures. Cell division is slower at lower temperatures.

2. Check students' data to see how they supported their hypotheses.

3. The tadpoles will hatch when temperature is most conducive for survival. Perhaps more food is available at warmer temperatures.

4. The tadpoles might not survive or adequate food may not be available when the water is cold.

Error Analysis

Light bulbs and their distances from the dishes should not be disturbed during the experiment. Ask students why this is important. Ask students why it is important for them to wash their hands both before and after making observations.

✓ Assessment

Portfolio Ask students to summarize in their portfolios what they have learned in this lab. Use the Performance Task Assessment List for Lab Report in **PASC,** p. 47. **L2**

Going Further

Students' hypotheses may include that UV light or chemical pollution of water may destroy developing egg cells; or lack of oxygen in water due to decomposition of organic matter may affect egg development.

Data and Observations

Eggs develop more quickly at warmer temperatures and more slowly at cooler temperatures.

Resource Manager

BioLab and MiniLab Worksheets, pp. 137-138 **L2**

Chemistry
Connection

Killer Frogs

Purpose

Students will study the nature of poisonous frogs and how their secretions affect cell chemistry of the prey.

Teaching Strategies

■ Review details of ion channels in cell membranes.

■ Have students diagram the two different actions frog toxins may have on cells.

■ Poisons from all species of poison arrow frogs are extremely toxic. Barely 2 micrograms of these toxins can kill an adult human. Each frog can produce 200 micrograms of poison. Ask students to speculate about why these frogs are brightly colored. *The color advertises their deadly condition to potential predators.*

■ Have students research species of poison arrow frogs to find out which species are the most deadly.

■ Ask students whether scientists should lobby for tropical rain forest protection in view of the medical potential of some of these species.

Connection to Biology

Responses will vary depending upon the human diseases researched.

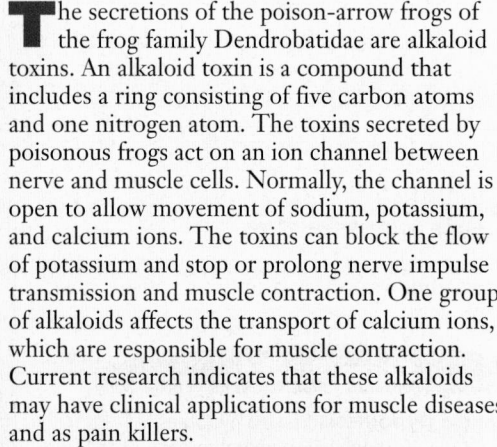

The most colorful frogs in the world are found in South and Central America. These poisonous frogs, including 130 species of the Dendrobatidae family, range in size from 1 to 5 cm. Although all frogs have glands that produce secretions, these frogs secrete toxic chemicals through their skin. A predator will usually drop the foul-tasting frog when it feels the numbing or burning effects of the poison in its mouth. The frogs advertise their poisonous personalities by bright coloration; they may be red or blue, solid colored, marked with stripes or spots, or have a mottled appearance. The poison secreted by these frogs is used by native peoples to coat the tips of the darts they use in their blow guns for hunting. Thus, these frogs are known as poison-arrow frogs.

Poison-arrow frog, *Epipedobates tricolor*

The secretions of the poison-arrow frogs of the frog family Dendrobatidae are alkaloid toxins. An alkaloid toxin is a compound that includes a ring consisting of five carbon atoms and one nitrogen atom. The toxins secreted by poisonous frogs act on an ion channel between nerve and muscle cells. Normally, the channel is open to allow movement of sodium, potassium, and calcium ions. The toxins can block the flow of potassium and stop or prolong nerve impulse transmission and muscle contraction. One group of alkaloids affects the transport of calcium ions, which are responsible for muscle contraction. Current research indicates that these alkaloids may have clinical applications for muscle diseases and as pain killers.

Frog poison eases pain Recent research shows that a drug derived from the extract from the poison-arrow frog, *Epipedobates tricolor*, works as a powerful pain killer. The drug ABT-594 may have the same benefits as morphine, but not the side effects. Morphine is the primary drug used to treat the severe and unrelenting pain caused by

cancer and serious injuries. Side effects of morphine include suppressed breathing and addiction. The "frog drug" does not interfere with breathing and does not appear to be addictive in initial testing. Another benefit of ABT-594 is that as it blocks pain, it does not block other sensations, such as touch or mild heat. One day pain you experience might be eased by a frog!

CONNECTION TO BIOLOGY

Research on newly discovered organisms such as poisonous frogs may result in drugs to treat specific disorders in human patients. Find out what human diseases are caused by problems in the transmission of nerve impulses and write an essay identifying the disorders that might be treated by toxins from poisonous frogs.

 To find out more about poisonous frogs, visit the Glencoe Science Web site. **science.glencoe.com**

VIDEODISC
STV: Rain Forest
Poison-Arrow Frog
Side 1

 Note Internet addresses that you find useful in the space below for quick reference.

_____

_____

_____

_____

_____

SUMMARY

Section 30.1

Fishes

Main Ideas

- Fishes are vertebrates with backbones and nerve cords that have expanded into brains.

- Fishes belong to three groups: the jawless lampreys and hagfishes, the cartilaginous sharks and rays, and the bony fishes. Bony fishes are made up of three groups: the lobe-finned fishes, the lung-fishes, and the ray-finned fishes.

- Jawless, cartilaginous, and bony fishes may have evolved from ancient ostracoderms.

Vocabulary

cartilage (p. 823)
fin (p. 820)
lateral line system (p. 820)
scale (p. 821)
spawning (p. 819)
swim bladder (p. 822)

Section 30.2

Amphibians

Main Ideas

- Adult amphibians have three-chambered hearts that provide oxygen to body tissues, but most gas exchange takes place through the skin.

- Land animals face problems of dehydration, gas exchange in the air, and support for heavy bodies. Amphibians possess adaptations well-suited for life on land.

- Amphibians probably evolved from ancient aquatic tetrapods.

Vocabulary

ectotherm (p. 829)
vocal cords (p. 831)

Main Ideas

Summary statements can be used by students to review the major concepts of the chapter.

Using the Vocabulary

To reinforce chapter vocabulary, use the Content Mastery Booklet and the activities in the Interactive Tutor for Biology: The Dynamics of Life on the Glencoe Science Web site.
science.glencoe.com

 **All Chapter Assessment questions and answers have been validated for accuracy and suitability by The Princeton Review.**

UNDERSTANDING MAIN IDEAS

1. c
2. a
3. c
4. c
5. d
6. b

UNDERSTANDING MAIN IDEAS

1. An animal with gill slits, a dorsal hollow nerve cord, and a backbone is a(n) _____.
   a. invertebrate
   b. invertebrate chordate
   c. vertebrate
   d. echinoderm

2. In addition to fishes, the subphylum Vertebrata includes _____.
   a. amphibians, reptiles, birds, and mammals
   b. echinoderms, reptiles, birds, and mammals
   c. tunicates, lancelets, reptiles, and birds
   d. tunicates, reptiles, birds, and mammals

3. Scientists hypothesize that the common ancestors to all fishes are the _____.
   a. amphibians   c. ostracoderms
   b. echinoderms   d. lancelets

4. How do jawless fishes obtain food?
   a. by injecting prey with poison from their hooklike fangs
   b. by using their round mouths like vacuum cleaners to suck up organic matter
   c. by drilling a hole and sucking out blood and insides of a prey animal
   d. by using their sharp teeth to grab and swallow smaller fishes

5. Which of the following is NOT a characteristic of most fishes?
   a. have scales
   b. have a two-chambered heart
   c. breathe using gills
   d. exchange gases through thin, moist skin

6. _____ use fins like the stabilizers of boats.
   a. Frogs   c. Sea stars
   b. Fishes   d. Lancelets

GLENCOE TECHNOLOGY

 **VIDEOTAPE**
MindJogger Videoquizzes
Chapter 30: *Fishes and Amphibians*
Have students work in groups as they play the videoquiz game to review key chapter concepts.

 **Resource Manager**

Chapter Assessment, pp. 175-180
MindJogger Videoquizzes
ExamView® Pro Software
BDOL Interactive CD-ROM, Chapter 30 quiz

7. d
8. a
9. b
10. a
11. cartilage
12. gills
13. two; fish
14. lamprey
15. gases; swim bladder
16. lateral line system
17. gills; limbs
18. cartilaginous
19. ectotherms
20. salamander

APPLYING MAIN IDEAS

21. They occupy different niches.
22. Advantages: egg and sperm are assured of coming together; male and female do not have to produce as many gametes. Disadvantages: male and female must find each other and be ready for reproduction at the same time.
23. Bone provides the support land vertebrates need for locomotion on land.

---

7. The lateral line system enables a _____ to detect movement and vibrations in the water.
   a. frog   c. toad
   b. salamander   d. fish

8. The class Amphibia is well-named because amphibians _____.
   a. spend part of their lives on land and part in water
   b. lay shelled eggs on land but develop in water
   c. spend part of their lives in the air and part on land
   d. use swim bladders for breathing in water but use lungs on land

9. Reptiles have legs on the sides of their bodies, but the limbs are flexed. Salamanders have legs that _____.
   a. extend straight out from the body and do not bend
   b. extend at right angles from the sides of the body
   c. are missing some of the key bones of reptiles
   d. evolved from shark's fins

10. The evolution of a three-chambered heart in amphibians ensured that cells receive the proper amount of _____.
    a. oxygen   c. blood
    b. carbon dioxide d. heat

11. The skeletons of lampreys, hagfishes, sharks, and their relatives are made of _____.

12. In all fishes, gas exchange takes place in the _____.

THE PRINCETON REVIEW — TEST-TAKING TIP

Skip Around, if You Can
Just because the questions are in order doesn't mean that you have to answer them that way. You may want to skip over hard questions and come back to them later, after you've answered all the easier questions that will guarantee you more points toward your score.

---

13. The _____-chambered heart pictured to the right belongs to a(n) _____.

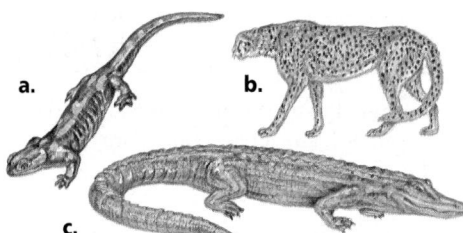

14. A fish that has a niche similar to a leech is a(n) _____.

15. Fishes control their depth in the water by altering the amount of _____ in their _____.

16. The _____ enables a fish to navigate in the dark.

17. Fossil evidence shows that early tetrapods had _____ like fishes and _____ like amphibians.

18. Both hagfishes and sharks have _____ skeletons.

19. Fishes and amphibians are _____ because their body temperature reflects the temperature of their surroundings.

20. The _____ in the diagram below has legs that show an earlier evolutionary origin than the other animals pictured.

a.

b.

c.

APPLYING MAIN IDEAS

21. What accounts for the different body shapes of fishes?
22. What are the advantages and disadvantages of internal fertilization?
23. Describe the importance of the evolution of bone in fishes to the evolution of vertebrates.

24. A male sea horse incubates fertilized eggs in a brood pouch. A codfish lays its eggs in the open sea. Which of these two types of fishes would need to lay more eggs? Why?

THINKING CRITICALLY

25. Using a Graph The following graph shows the number of leopard frogs in a wetland on a farm. One year, there was a prolonged drought in the farmer's area. What year did the drought occur? Explain.

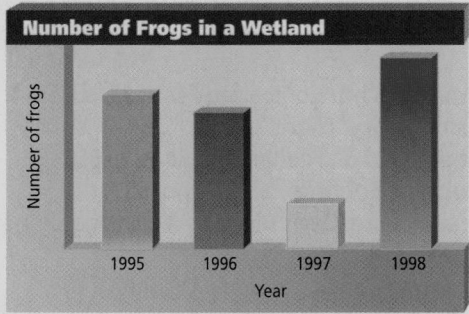

Number of Frogs in a Wetland

26. Concept Mapping Complete the concept map by using the following vocabulary terms: lateral line system, scales, swim bladder.

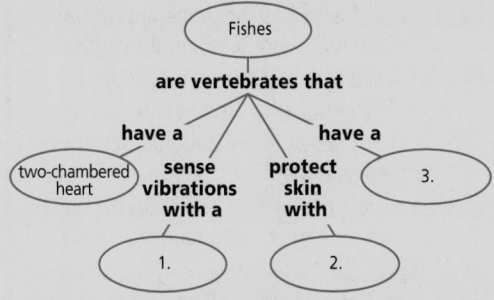

CD-ROM

For additional review, use the assessment options for this chapter found on the **Biology: The Dynamics of Life Interactive CD-ROM** and on the Glencoe Science Web site. **science.glencoe.com**

ASSESSING KNOWLEDGE & SKILLS

Amphibian populations are declining in many parts of the world. To reintroduce native amphibians into your state, assume you have a grant to breed frogs on a farm. To find out what temperature you need for optimum hatching of frog eggs, you test eggs at four different temperatures.

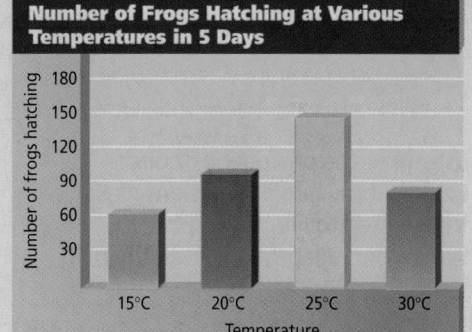

Number of Frogs Hatching at Various Temperatures in 5 Days

Interpreting Data Answer the following questions based on the graph.

1. How many eggs hatched at 25°C?
- **a.** 150
- **b.** 70
- **c.** 110
- **d.** 15

2. At which temperature did the fewest eggs hatch?
- **a.** 15°C
- **b.** 20°C
- **c.** 25°C
- **d.** 30°C

3. Based on the results of your experiment, what temperature will you use for optimum hatching of eggs?
- **a.** 15°C
- **b.** 20°C
- **c.** 25°C
- **d.** 30°C

4. Designing an Experiment You are trying to find out the optimum pH for hatching frog eggs. Design a controlled experiment that would give you quantitative data.

24. Cod are a type of fish that lay greater numbers of eggs. The eggs are unprotected in the ocean and most hatchlings die before reaching adult size.

THINKING CRITICALLY

25. Probably in 1997 because there was a decrease in the size of the frog population that year.

26. 1. Lateral line system; 2. Scales; 3. Swim bladder

ASSESSING KNOWLEDGE & SKILLS

1. a

2. a

3. c

4. Grow frog eggs in separate solutions with various pH levels. Count how many eggs hatch in each solution after a specific number of days.

Chapter 31 Organizer

Refer to pages 4T-5T of the Teacher Guide for an explanation of the National Science Education Standards correlations.

Section	Objectives	Activities/Features
Section 31.1 **Reptiles** National Science Education Standards UCP.1-5; C.3, C.5, C.6; G.3 (1½ sessions, ½ block)	1. **Compare** the characteristics of different groups of reptiles. 2. **Explain** how reptile adaptations make them suited to life on land.	**Inside Story:** An Amniotic Egg, p. 844 **Careers in Biology:** Wildlife Artist/ Photographer, p. 846 **Focus On** Dinosaurs, p. 850
Section 31.2 **Birds** National Science Education Standards UCP.1-5; A.1, A.2; C.3, C.5; F.4, F.5; G.1-3 (2½ sessions, 1 block)	3. **Interpret** the phylogeny of birds. 4. **Explain** how bird adaptations make them suited to life on land. 5. **Relate** bird adaptations to their ability to fly.	**MiniLab 31-1:** Comparing Feathers, p. 853 **Inside Story:** Flight, p. 855 **MiniLab 31-2:** Feeding the Birds, p. 856 **Problem-Solving Lab 31-1,** p. 857 **Design Your Own BioLab:** Which egg shape is best? p. 860 **Biology & Society:** Illegal Wildlife Trade, p. 862

Need Materials? Contact Carolina Biological Supply Company at 1-800-334-5551 or at **http://www.carolina.com**

MATERIALS LIST

BioLab
p. 860 modeling clay, cardboard, metric ruler, string, hard-boiled egg, Ping-Pong ball, golf ball, balance, protractor

MiniLabs
p. 853 contour feather, down feather, hand lens, paper, pencil
p. 856 1-gallon plastic milk bottles, wire, bird seed (assorted varieties)

Alternative Lab
p. 842 thermometer, black paper, white paper, transparent tape, lamp, metric ruler, modeling clay, small metal cans (2)

Quick Demos
p. 842 live or mounted reptile specimens
p. 848 plastic dinosaur models
p. 854 recording of bird songs and calls

Key to Teaching Strategies

L1 Level 1 activities should be appropriate for students with learning difficulties.

L2 Level 2 activities should be within the ability range of all students.

L3 Level 3 activities are designed for above-average students.

ELL ELL activities should be within the ability range of English Language Learners.

COOP LEARN Cooperative Learning activities are designed for small group work.

P These strategies represent student products that can be placed into a best-work portfolio.

These strategies are useful in a block scheduling format.

Reptiles and Birds

Teacher Classroom Resources

Section	Reproducible Masters	Transparencies
Section 31.1 **Reptiles**	Reinforcement and Study Guide, pp. 137-138 L2 Content Mastery, pp. 153-154, 156 L1	Section Focus Transparency 75 L1 ELL Basic Concepts Transparency 54 L2 ELL Basic Concepts Transparency 55 L2 ELL Reteaching Skills Transparency 45 L1 ELL
Section 31.2 **Birds**	Reinforcement and Study Guide, pp. 139-140 L2 Concept Mapping, p. 31 L3 ELL Critical Thinking/Problem Solving, p. 31 L3 BioLab and MiniLab Worksheets, pp. 139-142 L2 Laboratory Manual, pp. 219-228 L2 Content Mastery, pp. 153, 155-156 L1 Tech Prep Applications, pp. 39-40 L2	Section Focus Transparency 76 L1 ELL Basic Concepts Transparency 56 L2 ELL Basic Concepts Transparency 57 L2 ELL Reteaching Skills Transparency 46 L1 ELL

Assessment Resources

Chapter Assessment, pp. 181-186
MindJogger Videoquizzes
Performance Assessment in the Biology Classroom
Alternate Assessment in the Science Classroom
ExamView® Pro Software
BDOL Interactive CD-ROM, Chapter 31 quiz

Additional Resources

Spanish Resources ELL
English/Spanish Audiocassettes ELL
Cooperative Learning in the Science Classroom COOP LEARN
Lesson Plans/Block Scheduling

NATIONAL GEOGRAPHIC — Teacher's Corner

Products Available From Glencoe
To order the following products, call Glencoe at 1-800-334-7344:
CD-ROMs
NGS PictureShow: Structure of Vertebrates 1
NGS PictureShow: Structure of Vertebrates 2
Curriculum Kit
GeoKit: Fish, Reptiles, and Amphibians
Transparency Set
NGS PicturePack: Structure of Vertebrates 1
NGS PicturePack: Structure of Vertebrates 2

Products Available From National Geographic Society
To order the following products, call National Geographic Society at 1-800-368-2728:
Video
Reptiles and Amphibians

Index to National Geographic Magazine
The following articles may be used for research relating to this chapter:
"Ravens: Legendary Bird Brains," by Douglas H. Chadwick, January 1999.
"Dinosaurs Take Wing," by Jennifer Ackerman, July 1998.

GLENCOE TECHNOLOGY

The following multimedia resources are available from Glencoe.
Biology: The Dynamics of Life
CD-ROM ELL

Video: *Feeding Snake*
Exploration: *The Six Kingdoms*
BioQuest: *Biodiversity Park*
Video: *Sea Turtle*
Exploration: *Bird Adaptations*
Video: *Eagles*
Video: *Shorebirds*
Video: *Bird Courtship*

Videodisc Program

Snake Feeding
Sea Turtle

The Infinite Voyage

The Great Dinosaur Hunt

31 Reptiles and Birds

Theme Development

Evolution is one theme woven throughout the chapter and is apparent in the discussions of the movement of animals to land and how dinosaurs may have evolved into birds. **Unity within diversity** is exemplified by the features reptiles have in common and by a discussion of adaptations that led to the classification of reptiles into different orders. Similarly, birds have features in common but also have adaptations that make them suited for particular habitats.

Resource Manager

Section Focus Transparency 75 and Master **L1** **ELL**

READING BIOLOGY

Glencoe's *Biology: The Dynamics of Life* contains many resources to assist a student's reading skills. Each chapter contains figures with expanded captions that expand on written material. Word Origins, located along the side of text, expand knowledge of biology vocabulary. Glencoe's Content Mastery Booklet helps develop reading skills while reinforcing content. In addition, use the Interactive Tutor for *Biology: The Dynamics of Life* on the Glencoe Web site to reinforce vocabulary.
science.glencoe.com

840

What You'll Learn

- You will compare and contrast various reptiles and birds.
- You will identify reptile and bird adaptations that make these groups successful.

Why It's Important

Studying reptiles, the first animals to become independent of water, can help you understand the adaptations required for life on land. Birds are endotherms and have feathers and wings, adaptations that enable them to live anywhere on Earth.

READING BIOLOGY

Draw a line down the middle of a piece of paper. On one side, write "reptiles;" on the other, write "birds." Scan through the chapter, and write down different vocabulary that go with each category. As you read, list the characteristics for reptiles and birds on each side.

BIOLOGY Online

To find out more about reptiles and birds, visit the Glencoe Science Web site.
science.glencoe.com

What does a legless, cold-blooded viper with poisonous fangs have in common with a bird? Think about scales and shelled eggs. Scientists hypothesize that reptiles are the ancestors of birds.

840 REPTILES AND BIRDS

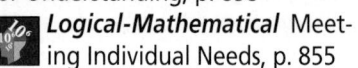

31.1 Reptiles

SECTION PREVIEW

Objectives

Compare the characteristics of different groups of reptiles.

Explain how reptile adaptations make them suited to life on land.

Vocabulary
amniotic egg
Jacobson's organ

You may remember seeing an adventure movie in which a ferocious crocodile devours a villain who is trying to swim across a jungle river. Moviemakers often use crocodiles, alligators, lizards, and snakes in their films to convey a sense of fear to the audience. However, only a few reptile species are capable of killing humans. Of the approximately 120 species of snakes found in the United States, the poisonous ones include the rattlesnakes, moccasins, copperheads, and coral snakes.

WORD *Origin*

reptile
From the Latin word *repere*, meaning "to creep." A reptile is an animal with dry skin, legs under the body, and amniotic eggs.

Nile crocodile and coral snake (inset)

What Is a Reptile?

At first glance, it may be difficult to determine how a legless snake is related to a tortoise. Snakes, turtles, alligators, and lizards are an extremely diverse group of animals, yet all share certain traits that place them in the class Reptilia.

Early reptiles, such as the cotylosaur (kaht ul oh SOR) shown in **Figure 31.1,** were the first animals to become adapted to life on land. All reptiles have adaptations that enable them to complete their life cycles entirely on land. These adaptations released the cotylosaurs and other

reptiles from the need to return to swamps, lakes, rivers, ponds, or oceans for reproduction.

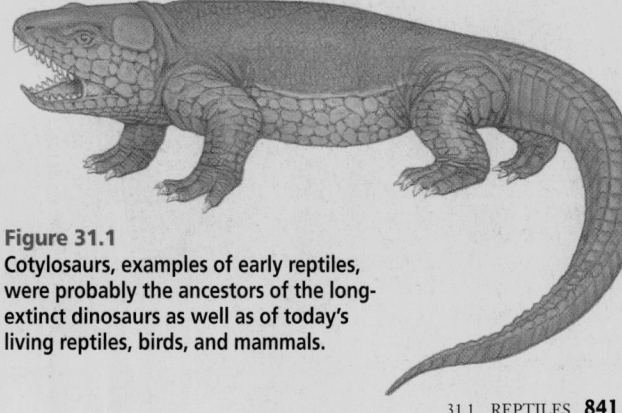

Figure 31.1
Cotylosaurs, examples of early reptiles, were probably the ancestors of the long-extinct dinosaurs as well as of today's living reptiles, birds, and mammals.

31.1 REPTILES **841**

Section 31.1

Prepare

Key Concepts

Students will study the features reptiles have in common and learn about the adaptations of crocodiles, alligators, lizards, snakes, and turtles. Origins of reptiles and their amniotic eggs are considered and discussed in terms of the movement of animals to land.

Planning

■ For the Quick Demos, borrow live or mounted reptiles and purchase small plastic dinosaurs.

■ For the Alternative Lab, gather thermometers, black and white paper, tape, lamps, rulers, empty frozen juice containers, and clay.

1 Focus

Bellringer

Before presenting the lesson, display **Section Focus Transparency 75** on the overhead projector and have students answer the accompanying questions.
L1 **ELL**

Transparency **75** Living on Land

SECTION FOCUS
Use with Chapter 31, Section 31.1

❶ The alligator lives both in and out of water. How does it differ from the frog?

❷ What adaptations do the snake and tortoise have for life in the desert?

BIOLOGY: The Dynamics of Life   SECTION FOCUS TRANSPARENCIES

✓ Assessment Planner

Portfolio Assessment
Alternative Lab, TWE, pp. 842-843
Assessment, TWE, pp. 847, 849, 856
Portfolio, TWE, pp. 847, 848, 850, 853, 856
MiniLab, TWE, p. 856

Performance Assessment
BioLab, SE, pp. 860-861
MiniLab, SE, pp. 853, 856
Alternative Lab, TWE, pp. 842-843
Problem-Solving Lab, TWE, p. 857

Assessment, TWE, p. 859
Knowledge Assessment
Section Assessment, SE, pp. 849, 859
Chapter Assessment, SE, pp. 863-865
Assessment, TWE, p. 858
BioLab, TWE, pp. 860-861

Skill Assessment
Assessment, TWE, p. 853
MiniLab, TWE, p. 853

2 Teach

Discussion

Ask students to share their experiences and feelings about reptiles, especially snakes. If they express fear, ask why they think that they have this fear. Point out that most primates show an instinctive fear of snakes. Ask students why they think this is so.

Reptiles have scaly skin

Unlike the moist, thin skin of amphibians, reptiles have a dry, thick skin covered with scales. Scaly skin, as shown in *Figure 31.2*, prevents the loss of body moisture and provides additional protection from predators. Because gas exchange cannot occur through scaly skin, reptiles are entirely dependent on lungs as their primary organ of gas exchange.

Skeletal changes in reptiles

Look again at the illustration of the cotylosaur. This reptile had legs that were placed more directly under the body rather than at right angles to the body as in early amphibians. This positioning of the legs provides greater body support and makes walking and running on land easier for reptiles. They have a good chance of catching prey or avoiding predators. Reptiles also have claws that help them obtain food and protect themselves. Additional evolutionary changes in the structure of the jaws and teeth of early reptiles allowed them to exploit other resources and niches on land.

Some reptiles have four-chambered hearts

Most reptiles, like amphibians, have three-chambered hearts. Some reptiles, notably the crocodilians, have a four-chambered heart that completely separates the supply of blood with oxygen from blood without oxygen. The separation enables more oxygen to reach body tissues. This separation is an adaptation that supports the higher level of energy use required by land animals.

Reptiles reproduce on land

Reptiles reproduce by laying eggs on land. Unlike amphibians, reptiles have no aquatic larval stage, and thus are not as vulnerable to water-dwelling predators as young amphibians are. Reptile hatchlings look just like adults, only smaller.

Although all of the adaptations discussed so far enabled reptiles to live successfully on land, the evolution of the amniotic egg was the adaptation that liberated reptiles from a dependence on water for reproduction. An **amniotic egg** (am nee OHT ihk) provides nourishment to the embryo and contains membranes that protect it while it develops in a terrestrial envi-

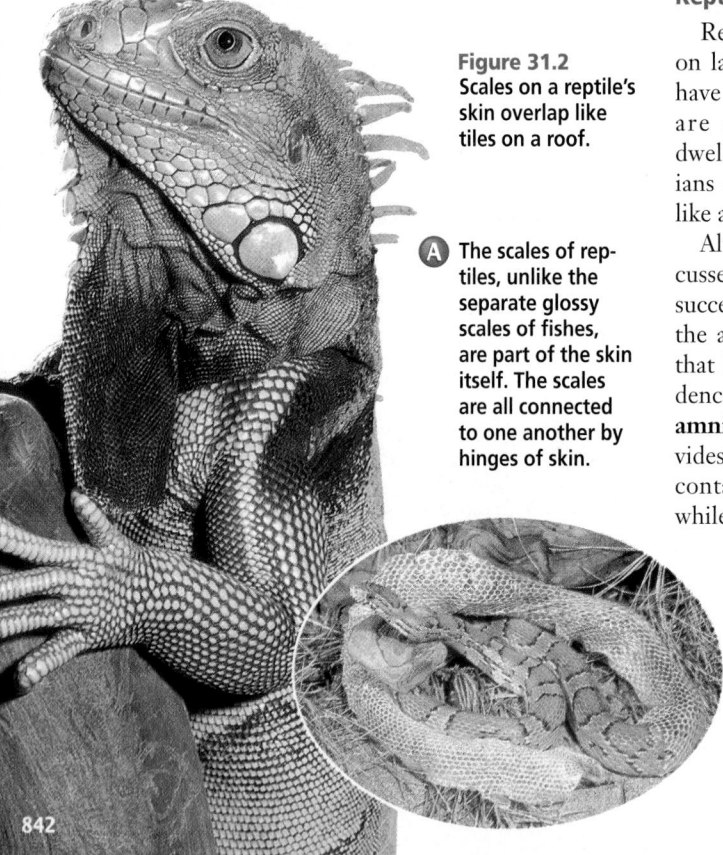

Figure 31.2
Scales on a reptile's skin overlap like tiles on a roof.

Ⓐ The scales of reptiles, unlike the separate glossy scales of fishes, are part of the skin itself. The scales are all connected to one another by hinges of skin.

Ⓑ To grow, young reptiles molt. Old scaly skin is replaced by new skin. Even the lenses of a reptile's eyes are replaced during a molt.

842

Alternative Lab

Body Color Adaptations

Purpose

Students observe how color affects the rate at which organisms absorb heat.

Materials

thermometer, black paper, white paper, tape, lamp, metric ruler, empty frozen juice containers with holes in one end, clay

Procedure

Give students the following directions.
1. Make a data table for starting temperature, final temperature, and total temperature change for a black can and a white can.
2. Cover one can with black paper and the other with white paper. Tape the paper in place.
3. Place a thermometer into the hole in each can top and secure it with clay. Make sure the hole is tightly sealed.
4. Place both cans about 5 cm away from the bulb of a lamp. Do not turn on the lamp.
5. Record the starting temperature of both cans in your data table.
6. Develop a hypothesis as to which can will show a higher temperature at the end of 10 minutes.

ronment. The egg functions as the embryo's total life-support system. Read the *Inside Story* on the next page to find out what the food supply is for a reptile embryo.

All reptiles have internal fertilization. The eggs are laid after fertilization, and embryos develop after eggs are laid. Most reptiles lay their eggs under rocks, bark, grasses, or other surface materials, but a few dig holes or collect materials for a nest. Most reptiles provide no care for hatchlings, but female crocodiles have been observed guarding their nests from predators.

Reptiles are ectotherms

Even though reptiles are different from amphibians in many ways, they are similar in one way. Both amphibians and reptiles are ectotherms. Their body temperatures depend on the temperatures of their environments. In the cool morning, a turtle might pull itself out of the pond or swamp and bask on a log in the sunlight until noon. Then, when the temperature gets a little too warm, the turtle may slip back into the cool water. This example shows that even though reptiles cannot control their body temperatures internally, they can use behavioral adaptations to compensate for changes in environmental temperature. *Figure 31.3* shows other examples of behavioral adjustment of body temperature in reptiles.

Because reptiles are dependent on the temperatures of their surroundings, they do not inhabit extremely cold regions. Reptiles are common in temperate and tropical regions, where climates are warm, and in hot desert climates. Many species of reptiles become dormant during cold periods in moderately cold environments such as in the northern United States.

Figure 31.3
Different reptiles regulate their body temperatures by a variety of behaviors.

A A chuckwalla keeps its temperature constant by lying in the shade of a large rock.

B A bearded lizard suns itself to get warm in the morning.

How reptiles obtain food

Like other animals, reptiles have adaptations that enable them to find food and to sense the world around them. Most turtles and tortoises are too slow to be effective predators, but that doesn't mean they go hungry. Most are herbivores, and those that are predators prey on worms and mollusks. Snapping turtles, however, are extremely aggressive, attacking fishes and amphibians, and even pulling ducklings under water.

Lizards primarily eat insects. The marine iguana of the Galapagos Islands is one of the few herbivorous lizards, feeding on marine algae. The Komodo dragon, the largest lizard, is found on several islands in Indonesia, north of Australia. It is an efficient predator, sometimes even of humans. Although lizards such as the Komodo dragon may look slow, they are capable of bursts of speed, which they use to catch their prey.

31.1 REPTILES **843**

7. Turn on the lamp and allow it to shine on both cans for 10 minutes.
8. Record the final temperature of both cans. Calculate the total change in temperature for each can and record it in your data table.

Expected Results
The black can will absorb more heat.

Analysis
1. Which color, black or white, had the greater temperature change after being heated for 10 minutes? *black*
2. Which color absorbs light better and warms up faster? *black*
3. Did your data support your hypothesis? Explain. *Yes, if students said that the black can would absorb more heat.*
4. Which should absorb more light from the sun, a dark animal or a light animal? Why? *Dark; dark colors absorb more light and heat up more quickly.*

INSIDE STORY

Purpose

Students will study the amniotic egg and learn how this adaptation enabled reptiles to live out of water.

Teaching Strategies

■ Ask students to carefully open a raw chicken egg and note the shell, albumen, yolk, and the tough membrane inside the shell. Ask what parts of the egg shown in the Inside Story are missing from their egg. **L1** **ELL**

Visual Learning

■ Challenge students to describe the differences between frog eggs and turtle eggs. *The frog eggs are surrounded by a jellylike substance that requires moisture. The reptile eggs are covered by a shell.* Ask how these differences might reflect the environment in which each of these animals live. *Frogs must live near water for reproduction; reptiles do not depend upon water for reproduction.*

Critical Thinking

Reptile eggs have a leathery, flexible shell. Bird eggs have a hard, but breakable, shell. Reptile eggs can withstand burial because they are flexible.

Resource Manager

Reteaching Skills Transparency 45 and Master **L1** **ELL**

INSIDE STORY

An Amniotic Egg

The evolution of the amniotic egg was a major step in reptilian adaptations to land environments. Amniotic eggs enclose the embryo in amniotic fluid, provide a source of food in the yolk, and surround both embryo and food with membranes and a tough, leathery shell. These structures in the egg help prevent injury and dehydration of the embryo as it develops on land.

Critical Thinking *How is the leathery covering of a reptile egg more suited to being laid deep in the sand than a hard-shelled bird egg would be?*

Hatchling turtle with egg tooth

1 Amnion The amnion (AM nee ahn) is a membrane filled with fluid that surrounds the developing embryo. The fluid-filled amnion cushions the embryo and prevents dehydration.

2 Shell The reptile egg is encased in a leathery shell. Most reptiles lay their eggs in protected places beneath sand, earth, gravel, or bark.

3 Yolk The main food supply for the embryo is the yolk, which is enclosed in a sac that is also attached to the embryo. The clear part of the egg is albumen (al BYEW mun), a source of additional food and water for the developing embryo.

6 Egg tooth A reptile hatches by breaking its shell with the horny tooth on its snout. This egg tooth drops off shortly after hatching.

5 Chorion The chorion (KOR ee ahn) is a membrane that forms around the yolk, allantois, amnion, and embryo. It allows gas exchange during respiration.

Embryo

4 Allantois The embryo's nitrogenous wastes are excreted into the allantois (uh LANT uh wus), a membranous sac that is associated with the embryo's gut. When a reptile hatches, it leaves behind the allantois with its collected wastes.

844 REPTILES AND BIRDS

BIOLOGY JOURNAL

Protecting Local Reptiles

Visual-Spatial Ask students to contact their state or local division of wildlife to find out about programs underway to protect locally endangered reptiles. Ask them to prepare an illustrated report of their findings that identifies the endangered reptiles in their area. **L2**

MEETING INDIVIDUAL NEEDS

Learning Disabled

Kinesthetic Have students make a cross-sectional model of an amniotic egg. They can use materials such as clay, plastic food wrap, and aluminum foil to represent membranes. **L1** **ELL**

Figure 31.4
Many reptiles are skillful predators that obtain prey in a variety of ways.

A The Komodo dragon is a predator that can kill animals as large as a deer or even a water buffalo. Adult Komodo dragons can reach a length of more than 3 m.

B The snapping turtle is common in North America. It has strong claws and a hooked beak that is sharp enough to bite through a person's fingers.

Snakes are also effective predators. Some, like the rattlesnake, have poison fangs that they use to subdue or kill their prey. A constrictor wraps its body around its prey, tightening its grip each time the prey animal exhales. Two of these reptiles are shown in *Figure 31.4.*

How reptiles use their sense organs

Reptiles have a variety of sense organs that help them detect danger or potential prey. How does a rattlesnake know you are nearby? The heads of some snakes, as shown in *Figure 31.5,* have heat-sensitive organs or pits that enable them to detect tiny variations in air temperature brought about by the presence of warm-blooded animals. Snakes and lizards are equipped with a keen sense of smell. Have you ever seen a snake flick out its tongue? The tongue is picking up chemical molecules in the air. The snake draws its tongue back into its mouth and inserts it into a structure called **Jacobson's organ,** described in *Figure 31.5.*

Figure 31.5
Snakes have sense organs that enable them to detect prey or identify chemicals in their environment.

A Rattlesnakes have a pair of heat-sensitive pits below their eyes that enable them to detect prey in total darkness.

B The long, flexible tongue of snakes and lizards picks up molecules in the air and transfers them to the Jacobson's organ in the roof of the animal's mouth for chemical analysis.

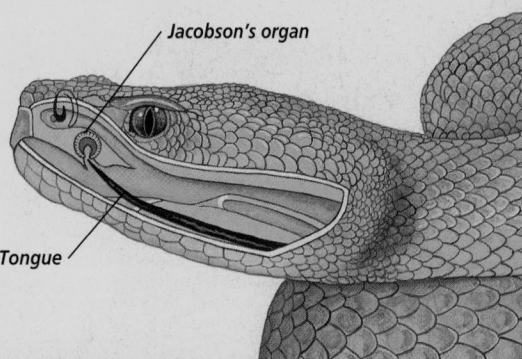
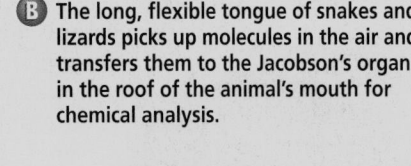

Jacobson's organ

Tongue

Visual Learning
Figure 31.4 Ask students why the snapping turtle is such an effective predator. *It has a flexible neck and a strong, horny beak.*

Misconception
Many people believe that the age of a rattlesnake can be determined by counting the number of rattles on its tail. However, because a new rattle forms each time the snake molts, and molting can occur many times each year, there is no correlation between the number of rattles and the age of the snake.

GLENCOE TECHNOLOGY

 CD-ROM
Biology: The Dynamics of Life
Video: *Feeding Snake*
Disc 4

 VIDEODISC
Biology: The Dynamics of Life
Snake Feeding (Ch. 6)
Disc 2, Side 1, 30 sec.

TECHPREP

Reptile Exhibits

Visual-Spatial Tell students that they are herpetologists at a large zoo. They must prepare plans and sketches of an exhibit for an unusual reptile. Have students select the reptile and then research its needs for space, food, and other care. They should show a care schedule on the sketch they make of the exhibit. **L2** **COOP LEARN**

Career Path

TECH PREP **Courses in high school:** art, photography, and biology.

College: courses in art, art history, photography, and zoology.

Other education sources: community art classes

Career Issue

Ask students if they think wildlife benefits or suffers when artists or photographers invade their habitats. Have the students provide examples to support their opinions.

For More Information

For more information on becoming a wildlife artist or photographer, students can write to:

Wildlife Artists Association
c/o Larry Waggoner
P.O. Box 33757
Granada Hills, CA 91394

Display

Visual-Spatial Assign groups of students orders of reptiles to research. Have them find out population distribution and prepare a map showing this distribution. Display these maps on the bulletin board. **L2 COOP LEARN**

Wildlife Artist/Photographer

If you are determined and patient, you can combine your love of nature and your artistic skills into a career as a wildlife artist or photographer.

Skills for the Job

Some wildlife artists/photographers spend weeks in the wilderness to find subjects for their art. Others draw, paint, or photograph animals in zoos or nature preserves. Becoming an artist or photographer depends more on your natural abilities than training, but art or photography courses can help strengthen your skills. Many wildlife artists also study biology or zoology so they can better understand their subjects. It can take years before artists are able to support themselves by selling their work, so many have another job, such as teaching art in a high school or college or giving private lessons.

BIOLOGY Online

For more careers in related fields, be sure to check the Glencoe Science Web site. science.glencoe.com

Diversity of Reptiles

Gracefully gliding snakes and quickly darting lizards are grouped together in the order Squamata. Turtles, slowly plodding and carrying heavy shells, belong to the order Chelonia. Basking crocodiles and alligators, classified in the order Crocodilia, may look clumsy but are surprisingly quick hunting machines that snap up fishes and lunge out at antelopes and other large animals that come to the river to drink.

Figure 31.6
In the past, sailors killed Galapagos tortoises for food. As a result, their numbers declined rapidly.

846

Turtles have shells

Turtles are the only reptiles protected by a shell made up of two parts. The dorsal part of the shell is the carapace, and the ventral part of the shell is the plastron. The vertebrae and ribs of turtles are fused to the inside of the carapace. A turtle's muscles are attached to the shell. Most turtles have a shell made out of hard, bony plates. In a few species, the shell is a covering made of tough, leathery skin.

Some turtles are aquatic, and some live on land. Turtles that live on land are called tortoises. Most turtles can draw their limbs, tail, and head into their shells for protection against predators. Although turtles have no teeth, they do have powerful jaws with a beaklike structure that is used to crush food.

Tortoises live on land, foraging for fruit, berries, and insects. The largest tortoises in the world, shown in *Figure 31.6*, are found on the Galapagos Islands off the coast of Ecuador.

Some adult marine turtles swim enormously long distances to lay their eggs. Like salmon, these turtles return from their feeding grounds to the place where they hatched. For example, green turtles travel from the coast of Brazil to Ascension Island in the Atlantic, a distance of more than 4000 km.

Crocodiles include the largest living reptiles

In contrast to marine turtles, crocodiles don't migrate. They may spend their days alternately basking in the sun on a riverbank and floating like motionless logs. Only their eyes and nostrils remain above water. Crocodiles can be identified by their long, slender snouts, whereas alligators have short, broad snouts. Both animals have powerful jaws with

PROJECT

Protecting the Environment

TECH PREP *Interpersonal* Ask students to role play a citizen group debating the protection of a beach where endangered green sea turtles nest. Have them work in groups of 3 to 4 and assume the role indicated on a card you have prepared. Have them do some research on their assigned roles. Cards should identify the following roles: teenage surfer, mayor, condominium developer, beach home owner, turtle biologist, president of the Turtle Protection Society, beach concession owner, state wildlife protection officer. Prior to role playing, students should prepare an outline of points that will be presented by the character he or she assumes. **L2 COOP LEARN**

Figure 31.7
Lizards have many adaptations that enable them to live in a variety of different habitats.

 Geckos are small, nocturnal lizards that live in warm climates, such as those of the southern United States, West Africa, and Asia. The toe pads of some geckos enable them to walk across walls and ceilings.

 Only the Gila monster of the southwestern United States and Mexico (shown here) and the beaded lizard of Mexico are poisonous lizards.

sharp teeth that can drag prey underwater and hold it there until it drowns. The American alligator is found throughout many of the freshwater habitats of the southeastern United States. The American crocodile can be found only in saltwater and estuarine habitats in southern Florida. The American alligator can reach a length of 5 m. Other crocodilians, such as the Nile crocodile of Africa, can grow even longer.

Both alligators and crocodiles lay eggs in nests on the ground. Unlike other reptiles, these animals stay close to their nests and guard them from predators. Several crocodilian species have been observed holding their newly-hatched offspring gently in their mouths as they carried them to the safety of the water.

Snakes and lizards are found in many environments

Lizards, shown in *Figure 31.7,* are found in many types of habitats in all but the polar regions of the world. Some live on the ground; some

burrow; some live in trees; and some are aquatic. Many are adapted to hot, dry climates.

Snakes, in contrast to most vertebrates, have no limbs and lack the bones to support limbs. Exceptions are pythons and boas, which retain bones of the pelvis. The many vertebrae of snakes permit fast undulations through grass and over rough terrain. Some snakes even swim and climb trees.

Snakes usually kill their prey in one of three ways. Remember that constrictors wrap themselves around their prey. If you ever watch someone handle a constrictor, you will notice that the handler never lets the snake start to wrap around his or her body. The snake is always held carefully so that its tail does not cross over its own head to begin a coil. Common constrictors include boas, pythons, and anacondas.

Venomous snakes use poison to kill their prey. These include rattlesnakes, cobras, and vipers, which inject poison from venom glands,

31.1 REPTILES **847**

Interpersonal Purchase a variety of plastic dinosaurs from a toy store. Have students form cooperative groups and give each group a different dinosaur. Ask them to select one of the features of their dinosaur and speculate about how it evolved. Ask them to present their explanations to the class. Follow this exercise with a review of the basic ideas about evolution. Challenge students to revise their explanations by including the terms variation and natural selection in their explanations. Have students conclude their explanations with the idea that only the dinosaurs best suited to their environments survived to produce offspring like themselves. **L2 COOP LEARN**

3 Assess

Check for Understanding

Naturalist Have students brainstorm a list of concepts they learned about reptiles. Write the list on the chalkboard and ask students to classify each concept as being related to adaptations, characteristics, origins, or the evolution of reptiles. **L3**

Reteach

Visual-Spatial For each reptile group, have students develop a table with the following heads: Representative organisms, Habitat, Food getting, Reproduction, Locomotion, Respiration, and Protection. Have students complete their tables and then review them as a class. **L2**

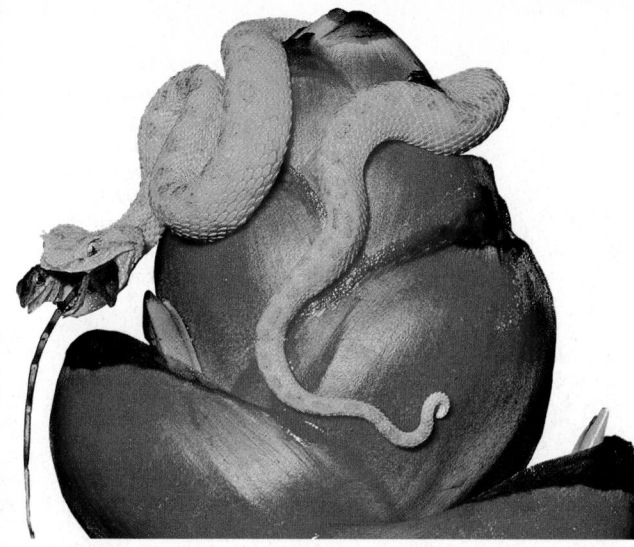

Figure 31.8
Many poisonous snakes have hollow fangs for injecting venom supplied by a venom gland. Venom may either paralyze the prey so it cannot run away or kill the prey immediately.

shown in *Figure 31.8*. Most snakes are neither constrictors nor poisonous. They get food by grabbing it with their mouths and swallowing it whole. Snakes eat rodents, amphibians, insects, fishes, eggs, and other reptiles.

The fourth order of reptiles, Rhynchocephalia, is represented by one living species, the tuatara, *Figure 31.9*. The tuatara is the only survivor of a primitive group of reptiles, most of which died out 100 million years ago.

Figure 31.9
The tuatara, *Sphenodon punctatus*, is found only in New Zealand. It has ancestral features, including teeth fused to the edge of the jaws, and a skull structure similar to that of early Permian reptiles.

Origins of Reptiles

You may have marveled at dinosaurs ever since you were very young. These animals were the most numerous land vertebrates during the Mesozoic Era. Some were the size of chickens, and others were the largest land dwellers that ever lived. Read the *Focus On* at the end of this section to learn more about dinosaurs.

The ancestors of snakes and lizards are traced to a group of early reptiles, called scaly reptiles, that branched off from the ancient cotylosaurs. The name "scaly reptiles" may be misleading because it implies that other reptiles lacked scales—which is not true. Although the evolutionary history of turtles is incomplete, scientists have suggested that they may also be descendants of cotylosaurs. Dinosaurs and crocodiles are the third group to descend from cotylosaurs, as you can see in *Figure 31.10*.

Although scientists used to think that birds arose as a separate group from this third branch, there is now much fossil evidence that leads biologists to suggest that birds are the living descendants of the dinosaurs.

Portfolio

Linguistic There are many misconceptions about reptiles. Ask students to look up and place in their portfolios information regarding misconceptions about reptiles. **L1 P**

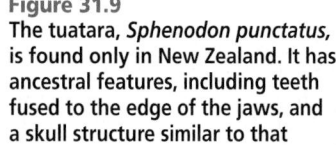

Resource Manager

Content Mastery, p. 154 **L1**
Reinforcement and Study Guide, pp. 137-138 **L2**

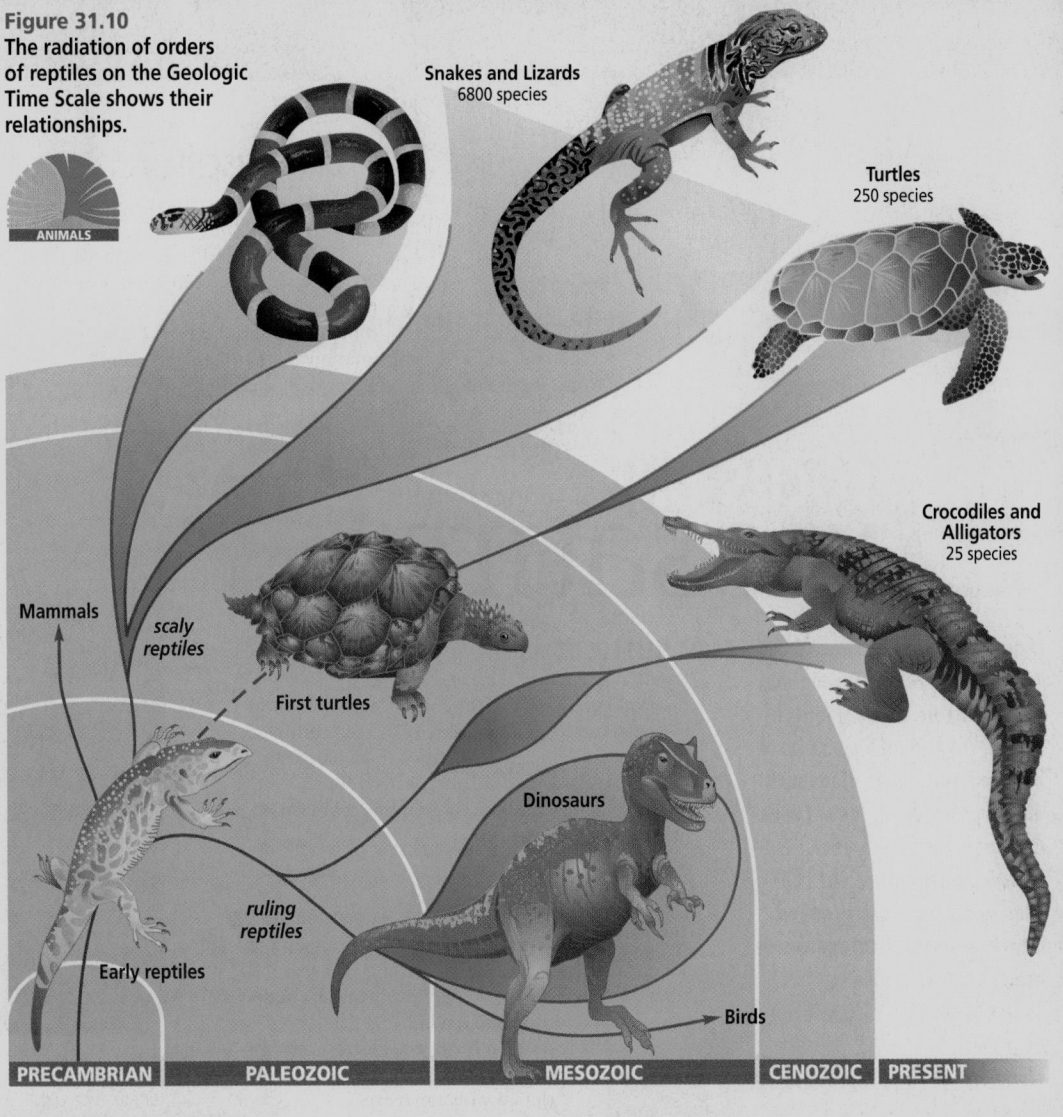

Figure 31.10
The radiation of orders of reptiles on the Geologic Time Scale shows their relationships.

ANIMALS

Snakes and Lizards
6800 species

Turtles
250 species

Crocodiles and Alligators
25 species

Mammals

scaly reptiles

First turtles

Dinosaurs

ruling reptiles

Early reptiles

Birds

PRECAMBRIAN | PALEOZOIC | MESOZOIC | CENOZOIC | PRESENT

Section Assessment

Understanding Main Ideas

1. Choose one adaptation of early reptiles and explain how it enabled these animals to live on land.
2. Describe two ways in which turtles protect themselves.
3. How do snakes use the Jacobson's organ for finding food?
4. How are modern reptiles like dinosaurs?

Thinking Critically

5. Explain how the development of a dry, thick skin was an adaptive advantage for reptiles.

SKILL REVIEW

6. **Classifying** Set up a classification key that allows you to identify a reptile as a snake, lizard, turtle, or crocodile. For more help, refer to *Organizing Information* in the **Skill Handbook**.

Linguistic Ask a group of students to find a children's story, folktale, or nursery rhyme that depicts reptiles in a negative way. Ask them to rewrite the story so the reptile is depicted in a positive way. Ask them to read the before and after versions of the story to the class. **L2**

Assessment

Portfolio Ask students to write a letter to the editor of their local newspaper to inform people about endangered or threatened reptiles that live in their area. Make sure students have researched the topic and include important facts in their letters. Have students include their letters in their portfolios.

4 Close

Activity

Naturalist Ask students in their groups to develop a television commercial in which a nonpoisonous reptile speaks about how it wants to be treated by people, what its habitat and food requirements are, and why it should not be feared. **L1**

Section Assessment

1. Reptile legs are located under the body rather than out to the sides, and they have clawed toes. The body structure and claws enhanced their movement on land.
2. Turtles can draw limbs, heads, and tails into their hard shells. Turtles may also use their powerful jaws to crush other animals.
3. The tongue picks up chemicals in the air. The snake then draws its tongue back into its mouth and inserts it into the Jacobson's organ for chemical analysis.
4. They have the same reptilian features as dinosaurs—dry, thick scaly skin, clawed toes, and the amniotic egg.
5. Dry, thick skin prevented reptiles from drying out on land.
6. Make sure that students start with general features of reptiles, such as scaly dry skin, claws, amniotic eggs, and being ectothermic. Then, they should use the main features for each group.

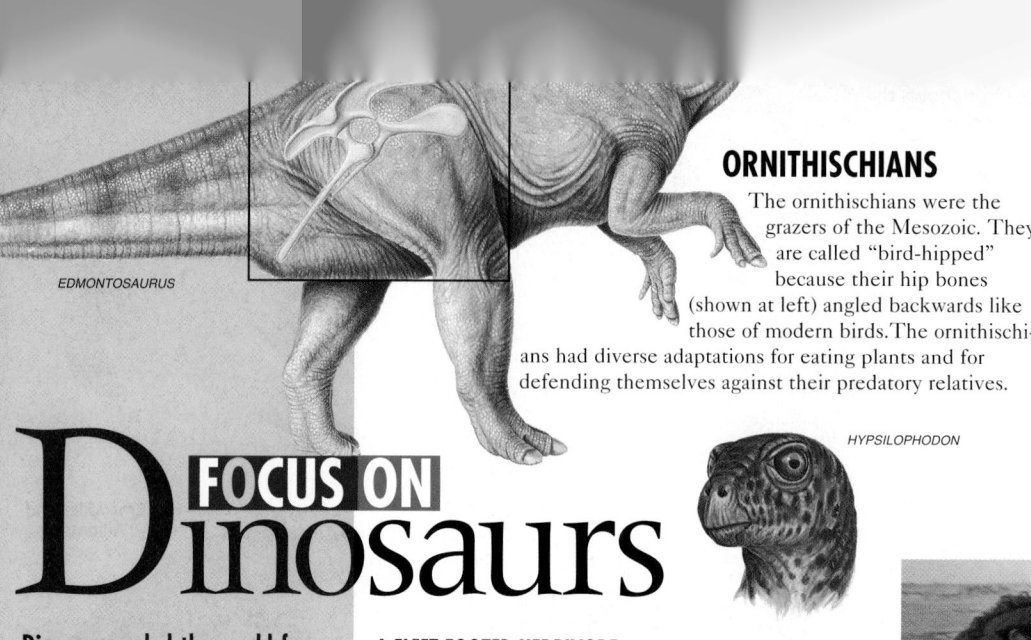

EDMONTOSAURUS

Focus On

Dinosaurs

Purpose 🗂

Students learn about dinosaurs, including how they are classified. They explore theories about how dinosaurs became extinct.

Background

Most students will be familiar with dinosaurs such as *Stegosaurus*, *Apatosaurus*, *Tyrannosaurus*, and *Diplodocus*. However, they may not be aware of the different time periods during which each of the more familiar dinosaurs lived. Explain that some dinosaur species evolved, lived, and often went extinct as other dinosaur species became the dominant land animals. When scientists discuss the extinction of the dinosaurs, they are talking about the species that existed about 66 million years ago, not all the dinosaurs that lived and became extinct before that time.

Teaching Strategies

■ Arrange a class trip to a museum that has a dinosaur exhibit. Ask students to examine the skulls and feet of the dinosaurs on display and hypothesize what their habitats and diets might have been. **L1**

■ Have interested students research and report on animals that were not dinosaurs that lived at the same time as the dinosaurs. Suggestions include: the flying pterosaurs, fishlike plesiosaurs, and sailback reptiles such as *Dimetrodon*. **L2**

■ Have students develop a time line that traces the development of animals from the earliest reptiles to the beginning of the Age of Mammals. Instruct students to include both plants and animals on their time lines. **L2**

FOCUS ON Dinosaurs

ORNITHISCHIANS

The ornithischians were the grazers of the Mesozoic. They are called "bird-hipped" because their hip bones (shown at left) angled backwards like those of modern birds. The ornithischians had diverse adaptations for eating plants and for defending themselves against their predatory relatives.

HYPSILOPHODON

Dinosaurs ruled the world for 130 million years, throughout the Mesozoic era. Paleontologists have identified several hundred species of dinosaurs, and about a dozen new types are unearthed each year. Descended from ancient reptiles, dinosaurs are grouped into two general categories—ornithischians and saurischians—based on the structure of their hip bones.

A FLEET-FOOTED HERBIVORE

Slender, graceful *Hypsilophodon* (above) was one of the fastest-moving ornithischians. With long hind legs adapted for running, this 1.5-meter herbivore probably was able to outdistance most predators with ease. *Hysilophodon* had a sharp beak and small overlapping teeth suited for grinding leaves and other plant material.

BODY ARMOR

Many slow-moving ornithischians had elaborate body armor. Seven meters long and built like an armored tank, *Euoplocephalus* was a peaceful grazer that must have frustrated many a hungry carnivore. Its body was completely encased in bony plates—even the eyelids were bone-reinforced. It had a tail tipped with a massive bone "club." When threatened, *Euoplocephalus* could have hugged the ground and swung its club-studded tail from side to side to protect itself.

EUOPLOCEPHALUS

PARASAUROLOPHUS

CORYTHOSAURUS

THE DUCKBILLS

There were many species of duck-billed dinosaurs. All had long tails, oddly shaped "bills," webbed fingers, and hooflike, three-toed hind feet. Despite the duckbills' webbed fingers, most paleontologists now think that duckbills lived on land. Some species, such as *Parasaurolophus* (left), had large, hollow crests on their heads that may have amplified whatever sounds these dinosaurs made. Fossil evidence indicates that duckbills were social animals that moved in herds and cared for their young.

 Portfolio

Modeling Dinosaur Bones

Kinesthetic Have students obtain information on the sizes and shapes of dinosaur bones. Ask them to work in small groups to use the information obtained to build a life-sized model of a dinosaur bone using pâpier maché and chicken wire. **L2** **P** **COOP LEARN**
🗂

GLENCOE TECHNOLOGY

VIDEODISC
The Infinite Voyage *The Great Dinosaur Hunt, Newborns: Examining Dinosaur Eggs* (Ch. 7)
8 min. 30 sec.

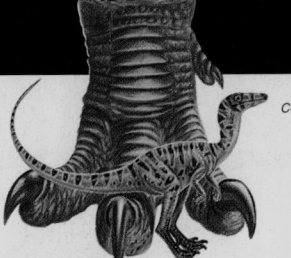

COMPSOGNATHUS

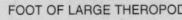

FOOT OF LARGE THEROPOD

SAURISCHIANS

The "lizard-hipped" dinosaurs, or saurischians, had hip bones (shown below) like those of modern-day lizards, with the pubic bone projecting forward. Two major groups of saurischians are the theropods, or three-toed carnivores, and the sauropods, or long-necked herbivores.

THE SMALLEST DINOSAUR

Less than a meter long, *Compsognathus* (above), which means "pretty-jawed," was the smallest of all dinosaurs. A delicate predator, this diminutive theropod probably hunted lizards and small mammals. A double-hinged jaw made it easy for *Compsognathus* to swallow its prey whole.

ALLOSAURUS

PALEONTOLOGIST SERENO AND TEAM IN NIGER

THE GIANT MEAT-EATERS

Big predatory theropods came in every imaginable shape and size. Fearsome *Allosaurus* and infamous *Tyrannosaurus* belong to this group, as does a new 12-meter-long dinosaur from the Sahara—*Suchomimus tenerensis*. Discovered in Niger in 1997 by a team of paleontologists led by Paul Sereno (shown in lower right of large photo), *Suchomimus* was a fish-eating predator with huge but narrow croc-odilelike jaws and powerful forelimbs with long thumb claws. Although *Suchomimus*, which means "crocodile mimic," was adapted to eating large fish, it probably stalked ter-restrial prey as well.

SERENO AT THE NATIONAL GEOGRAPHIC SOCIETY WITH CAST SKULL OF *SUCHOMIMUS TENERENSIS*

BIG BROWSERS

The largest dinosaurs ever to roam Earth's surface were the long-tailed, long-necked, barrel-bodied sauropods such as *Seismosaurus*, *Diplodocus*, and *Apatosaurus*. These enormous plant-eaters—*Seismosaurus* was 36 meters long and weighed between 80 and 100 tons—could have browsed on leaves high in the treetops. Sauropods had small jaws and teeth. Their leafy meals were ground up in their stomachs with the help of sharp-edged pebbles, called gastroliths, that were probably swallowed along with their food.

APATOSAURUS

EXPANDING Your View

1 THINKING CRITICALLY

Compare and contrast the feeding adaptations of a plant-eating ornithischian and a meat-eating saurischian.

2 JOURNAL WRITING

Research the hypothesis that the dinosaurs disappeared as the result of a mass extinction caused by a giant meteor or asteroid colliding with Earth. What evidence exists to support this hypothesis? Share your findings with the class in an oral report.

851

Display

Interpersonal Prepare a large outline map of the world. Post the map on a class-room wall or bulletin board. Provide students with the resources necessary to identify where in the world different types of dinosaur fossils have been found. Have students work together to create a key that identifies each location according to dinosaur type on the map. **L2**
COOP LEARN

Misconception

Explain to students that movies often lead people to believe that humans and dinosaurs existed at the same time. Emphasize that dinosaurs became extinct at the end of the Cretaceous period, 66 million years ago. Explain that humans have existed on Earth for fewer than 2 million years. Use these facts to demonstrate that humans and dinosaurs did not occupy Earth at the same time.

Answers to Expanding Your View

1. The ornithischians had adap-tations, such as duckbills, that enabled them to eat plants. The saurischians were carni-vores with heavy jaws and sharp teeth.
2. Dinosaurs became extinct at the end of the Cretaceous period, 66 million years ago. This corresponds to the age of a meteorite that hit Earth near the Yucatan Peninsula in Mexico.

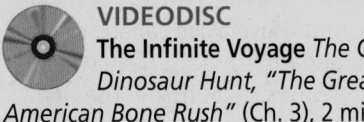

Prepare

Key Concepts

Students will study the adaptations that make birds suited to life on land and enable them to fly. Bird classification will be presented and the origins of birds will be considered.

Planning

■ Purchase recordings of bird calls for the Quick Demo.
■ Purchase down and contour feathers and gather bird field guides for the MiniLabs.
■ Gather clay, cardboard, rulers, string, hard-boiled eggs, Ping-pong balls, golf balls, balances, and protractors for the Bio-Lab.

1 Focus

Bellringer

Before presenting the lesson, display **Section Focus Transparency 76** on the overhead projector and have students answer the accompanying questions.
`L1` `ELL`

SECTION PREVIEW

Objectives

Interpret the phylogeny of birds.

Explain how bird adaptations make them suited to life on land.

Relate bird adaptations to their ability to fly.

Vocabulary
feather
sternum
endotherm
incubate

Section

31.2 Birds

Robin and chickadee (inset)

Have you ever seen a robin tug on a worm that is struggling to stay in the soil? Maybe you have had a chance to see the amusing antics of a chickadee hopping on a snow-covered branch in the forest. Almost everyone admires birds. The brilliant flash of a bluebird's wings, the uplifting sound of a bird's song that fills the woods on a spring morning, and the effortless soaring of a redtail hawk have always fascinated and delighted people.

What Is a Bird?

After conquering the sea and land, vertebrates took to the air, where there was a huge source of insect food and a refuge from land-dwelling predators. The existence of more than 8600 species of modern birds, class Aves, shows that flight was a successful adaptation for survival. Except for domestic animals and humans, the most common vertebrates you see in your daily life are birds. Biologists sometimes refer to birds as feathered dinosaurs. Fossil evidence seems to indicate that birds have evolved from small, two-legged dinosaurs called theropods, illustrated in *Figure 31.11.* Like reptiles, birds have clawed toes and scales on their feet. Fertilization is internal and shelled amniotic eggs are produced in both groups.

Birds have feathers

Birds can be defined simply as the only living organisms with feathers. A **feather** is a lightweight, modified scale that provides insulation and enables flight, illustrated in *Figure 31.12.* You may have seen a

Figure 31.11
Most scientists agree that birds evolved from a group of reptiles called theropod dinosaurs, as shown in this artist's rendition. The skeletons of birds and theropods are similar.

BIOLOGY JOURNAL

Observing Birds

 **Linguistic** Have a live canary, parakeet, or parrot in class. Have students observe behavior such as how it uses its beak in feeding and drinking. Have them also observe perching and reactions to objects in its cage, such as mirrors and bells. Ask students to write their observations in their journals. `L1` `ELL`

GLENCOE TECHNOLOGY

CD-ROM
Biology: The Dynamics of Life
Exploration: *Bird Adaptations*
Disc 4

bird running its bill through its feathers while sitting on a tree branch or on the shore of a pond. This process, called preening, keeps the feathers in good condition for flight. The bird also uses its beak to rub oil from a gland near the tail onto the feathers. This process is especially important for water birds as a way to waterproof the feathers. You can compare types of bird feathers in the *MiniLab* on this page.

Even with good care, feathers wear out and must be replaced. The shedding of old feathers with the growth of new ones is called molting. Most birds molt in late summer. However, most do not lose their feathers all at once and are able to fly while they are molting. Wing and tail feathers are usually lost in pairs so that the bird can maintain its balance in flight.

Birds have wings

A second adaptation for flight in birds is the modification of the front limbs into wings. Powerful flight muscles are attached to a large breastbone called the **sternum** and to the upper bone of each wing. The sternum looks like the keel of a sailing boat and is important because it supports the enormous thrust and power produced by the muscles as they move to generate the lift needed for flight.

Flight requires energy

Flight requires high levels of energy. Several factors are involved in maintaining these high energy levels. First, a bird's four-chambered, rapidly beating heart moves oxygenated blood quickly throughout the body. A chickadee's heart, for example, beats 1000 times a minute. Compare this to a human heart, which beats 70 times a minute. This efficient circulation supplies cells with the oxygen needed to produce energy.

MiniLab 31-1 — Comparing and Contrasting

Comparing Feathers Birds have two kinds of feathers. Contour feathers used for flight are found on a bird's body, wings, and tail. Down feathers lie under the contour feathers and insulate the body.

Magnification: 120×

Procedure

1. Examine a contour feather with a hand lens, and make a sketch of how the feather filaments are hooked together.
2. Examine a down feather with a hand lens. Draw a diagram of the filaments of the down feather.
3. Fan your face with each feather separately. Note how much air is moved past your face by each type of feather. **CAUTION: *Wash your hands after handling animal material.***

Analysis

1. How does the structure of a contour feather help a bird fly?
2. How does the structure of a down feather keep a bird warm?
3. How can you explain the differences you felt when fanning with each feather?

Figure 31.12
Feathers streamline a bird's body, making it possible for the bird to fly.

A Fluffy down feathers have no hooks to hold the filaments together. Down feathers act as insulators to keep a bird warm.

B A large bird can have 25 000 or more contour feathers with a million tiny hooks that interlock and make the feathers hold together, making a "fabric" suited for flight.

31.2 BIRDS **853**

Second, a bird's respiratory system supplies oxygenated air to the lungs when it inhales as well as when it exhales. A bird's respiratory system consists of lungs and anterior and posterior air sacs. During inhalation, oxygenated air passes through the trachea and into the lungs, where gas exchange occurs. Most of the air, however, moves through the lungs and passes directly into the posterior air sacs. When a bird exhales deoxygenated air from the lungs, oxygenated air returns to the lungs from the posterior air sacs. At the next inhalation, deoxygenated air in the lungs passes into the anterior air sacs. Finally, at the next exhalation, air passes from the anterior air sacs out of the trachea. Thus, air follows a one-way path in a bird. You can see the path air follows in a bird's respiratory system in *Figure 31.13*. How much of the air that a bird inhales is passed into the posterior air sacs? Find out in the *Inside Story*.

Birds are endotherms

Birds are able to maintain the high energy levels needed for flight because they are endotherms. An **endotherm** is an animal that maintains a constant body temperature that is not dependent on the environmental temperature.

Birds have a variety of ways to save or give off their body heat in order to maintain a constant body temperature. Feathers reduce heat loss in cold temperatures. The feathers fluff up to trap a layer of air that limits the amount of heat lost. Responses to high temperatures include flattening the feathers and holding the wings away from the body. Birds also pant to increase respiratory heat loss.

A major advantage of being endothermic is that birds can live in all environments, from the hot tropics to the frigid Antarctic. However, birds and other endotherms must eat large amounts of food to sustain these higher levels of energy. Find out what kinds of food local birds prefer by doing the *MiniLab* that follows the *Inside Story*.

Reproduction in birds

Birds, like reptiles, reproduce by internal fertilization and lay amniotic eggs inside a nest. Bird eggs are encased in a hard shell, unlike the leathery shell of a reptile. Bird nests may be made out of bits of straw and twigs, or they may consist of just a depression scratched into the sand. Some nests are elaborate structures that are added to yearly. Whatever the type of nest, birds do not leave the eggs to hatch on their own. Instead, birds **incubate** or sit on their eggs to keep them warm, turning the eggs

Figure 31.13
Birds require a great deal of oxygen because their large flight muscles expend huge amounts of ATP. Follow the arrows to see how air passes through a bird's respiratory system. Notice that when a bird inhales, inhalation cycles 1 and 2 occur simultaneously.

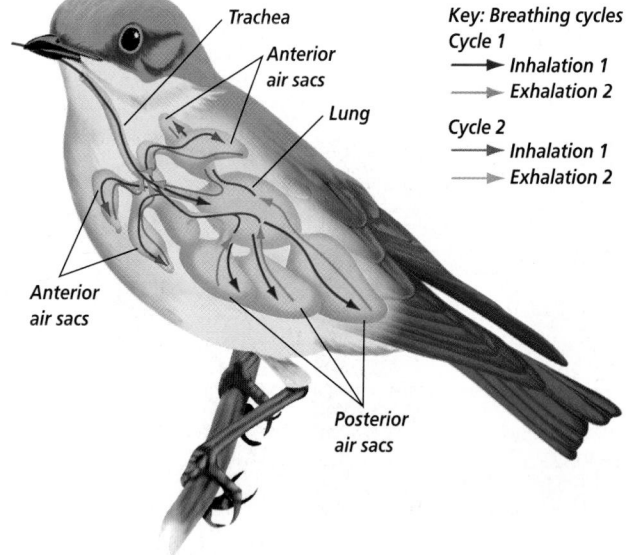

Trachea
Anterior air sacs
Lung
Anterior air sacs
Posterior air sacs

Key: Breathing cycles
Cycle 1
→ Inhalation 1
→ Exhalation 2
Cycle 2
→ Inhalation 1
→ Exhalation 2

854 REPTILES AND BIRDS

Male cardinal

Flight

Humans have always dreamed of being able to fly. The popularity of hang gliding and parachute jumping may reflect these dreams. For birds, the ability to fly is the result of complex selective pressures that led to the evolution of many adaptations.

Critical Thinking *Wing shapes reflect how birds fly. Describe as many ways as you can remember seeing birds fly.*

1 Wings Birds have a variety of wing shapes and sizes. Some birds have longer, narrower wings adapted for soaring on updrafts, whereas others have shorter, broader wings adapted for quick, short flights among forest trees.

2 Hollow bones Birds' bones are thin and hollow, thereby maintaining low weight and making flight easier. The hollow bones of birds are strengthened by bony crosspieces. The sternum is the large breastbone to which powerful flight muscles are attached.

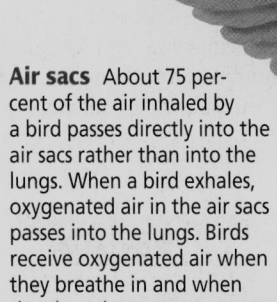

6 Air sacs About 75 percent of the air inhaled by a bird passes directly into the air sacs rather than into the lungs. When a bird exhales, oxygenated air in the air sacs passes into the lungs. Birds receive oxygenated air when they breathe in and when they breathe out.

3 Beaks Birds have beaks made out of a protein called keratin, but they do not have teeth. The lack of teeth or a heavy bony jaw reduces a bird's weight even further.

Crop
Lung
Sternum

Intestine   Gizzard

5 Digestion The digestive system of a bird is adapted for dealing with large quantities of food that must be eaten to maintain the level of energy necessary for flight. Because birds have no teeth, many swallow small stones that help to grind up food in the gizzard.

4 Legs The legs of birds are made up of mostly skin, bone, and tendons. The feet are adapted to perching, swimming, walking, or catching prey.

31.2 BIRDS **855**

Purpose
Students will gain an understanding of the adaptations that enable birds to fly.

Teaching Strategies
■ Ask students to bring beef and chicken bones to class. Have them compare the density of the two types of bones by finding the mass of each bone and dividing it by the volume of each bone as determined by water displacement in a graduated cylinder. *Students will find that the density of chicken bones is less than the density of beef bones.* **L2**

■ Obtain a turkey gizzard from a meat market. Use the gizzard to point out its thick muscles. Elicit reasons why the gizzard has such thick muscles.

Visual Learning
■ Obtain or borrow a mounted skeleton of a bird. Point out how the bones are thin and lightweight. Point out the size of the sternum and ask students to compare the sternum of a bird with their own sternum.

■ Obtain slides of birds in flight. Ask students to examine the wing shapes and determine the type of flight unique to each bird. Explain that elliptical wings are adapted for quick movements. Wings that sweep back and taper to a slender tip promote high speed. Soaring birds have broad wings.

Critical Thinking
soaring, diving, flapping, hovering, quick turning

MEETING INDIVIDUAL NEEDS

Gifted
Logical-Mathematical Ask students who have an interest in mathematics to make up a series of math problems dealing with wing beats for other students to solve. Provide students with the following data: number of wing beats per 10 seconds—crow 20, robin 23, pigeon 30, starling 45, chickadee 270, hummingbird 700. An example of a problem may be: If a crow, robin, pigeon, and starling each flew in the same direction at a starting speed of 48 km per hour, how many times would each flap its wings if they flew 24 km? **L3**

MiniLab 31-2

Purpose
Students will construct bird feeders and learn what types of foods birds prefer.

Process Skills
compare and contrast, recognize cause and effect, interpret data, classify

Teaching Strategies
■ Show photos or slides of common local birds.
■ Have students use binoculars to observe the birds that visit their feeders.

Expected Results
Depending on region, students may find cardinals, jays, woodpeckers, nuthatches, juncos, chickadees, sparrows, finches, tufted titmouse, mourning doves, and many other local and migrating bird species.

Analysis
1. Sunflower seeds will attract a large variety of birds.
2. It is likely that birds will visit the same feeder over and over again, unless the feeder runs out of food.
3. Ideal bird food would have a mixture of a variety of seeds that would appeal to many different birds.

✔ Assessment
Portfolio Ask students to summarize data collected by watching birds. They should record not only the types of birds seen and the foods preferred, but other interactions of birds, such as aggressive behavior and feeding methods. Use the Performance Task Assessment List for Making Observations and Inferences in **PASC**, p. 17. **L2** **P**

856

MiniLab 31-2 — Comparing and Contrasting

Feeding the Birds In the winter, it may be difficult for some birds to find food, especially if you live in an environment often blanketed with snow. Making a bird feeder and watching birds feed can be an enjoyable activity for you that may save some birds from starvation. If you do begin feeding birds in the winter, continue to feed them until natural food again becomes available in the spring.

Procedure
1. Obtain several large, plastic milk bottles. Cut two holes 5 cm from the base on opposite sides of each bottle, each about 8 cm². These are the openings birds will use to find the food inside.
2. Place small drainage holes in the bottom of each bottle. Hang the bottles from wires strung through small holes in the neck of each one.
3. Place a different kind of seed (sunflower seeds, hulled oats, cracked corn, wheat, thistle, millet) in different bottles. Add new seed when needed.
4. Using a bird guide, make a list of numbers and kinds of birds that frequent each feeder, noting the type of food offered.

Analysis
1. What type of seed attracted the largest variety of bird types?
2. Did any birds visit more than one feeder?
3. What do you think an ideal bird food would be?

---

periodically so that they develop properly. In some species of birds, both parents take turns incubating eggs; in others, only one parent does so. Bird eggs are distinctive, and often the species of bird can be identified just by the color, size, and shape of the eggs the bird lays. You can find out more about the adaptive value of bird egg shape in the *BioLab* at the end of this chapter.

Diversity of Birds

Unlike reptiles, which take on a wide variety of forms from legless snakes to shelled turtles, birds are all very much alike in their basic form and structure. You have no difficulty recognizing a bird.

In spite of the basic uniformity of birds, they do exhibit specific adaptations, depending on the environment in which they live and the food they eat. As shown in *Figure 31.14*, ptarmigans have feathered legs and feet that serve as snowshoes in the winter, making it easier for the birds to walk in the snow. Penguins are flightless birds with wings and feet modified for swimming and a body surrounded with a thick layer of insulating fat.

Figure 31.14
Examine these birds and infer where they live and how they are adapted to their environments.

Ptarmigan

Screech owls

Adélie penguins

✔ Portfolio

Making a Flyway Map

Visual-Spatial Provide students with a blank map of the world. Ask them to use a field guide to find the bird migration routes for the migrating birds of your area. Ask students to plot these data on their maps using a different color of pencil for each route. **L1** **ELL** **P**

GLENCOE TECHNOLOGY

CD-ROM
Biology: The Dynamics of Life
Video: *Penguins,* Disc 4
Video: *Eagles,* Disc 4
Video: *Shorebirds,* Disc 4

Large eyes, an acute sense of hearing, and sharp claws make owls well-adapted, nocturnal predators able to swoop with absolute precision onto their prey. However, many bird species are now threatened with extinction due to changes in their habitats. Read the *Problem-Solving Lab* to see in which countries birds are endangered. Then read the *Biology & Society* feature at the end of this chapter to learn how illegal trade in wildlife threatens birds and other animals.

The shape of a bird's beak gives clues to the kind of food the bird eats, as you can see in *Figure 31.15*. Hummingbirds have long beaks that are used for dipping into flowers to obtain nectar. Hawks have large, curved beaks that are adapted for tearing apart their prey. Pelicans have huge beaks with pouches that they use as nets for capturing fish. The short, stout beak of a goldfinch is adapted to cracking seeds.

Figure 31.15
The beaks of different species of birds are adapted to eating different kinds of food.

Hummingbird
Pelican
Hawk
Goldfinch

Problem-Solving Lab 31-1

Analyzing Information

Where are the most endangered bird species? More than 100 bird species have become extinct in the last 400 years.

Analysis

Examine the world map. The key at the bottom right shows the number of bird species that are currently threatened with extinction. The numbers appearing on the map indicate the actual number of threatened bird species in specific countries.

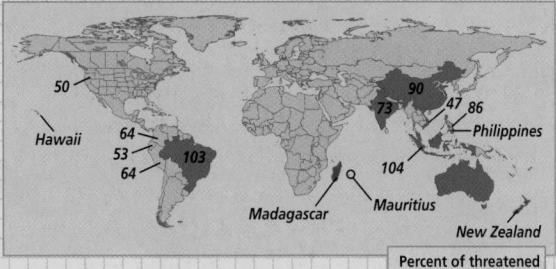

50
Hawaii
64
53
64
103
73
90
47 86
Philippines
104
Madagascar   *Mauritius*
New Zealand

Percent of threatened bird species
☐ Fewer than 5%
▨ 5 to 9.9%
■ More than 10%

Thinking Critically

1. If 50 species are threatened, what is the approximate number of bird species in the United States? (Hint: 2.5 percent of the bird species in the U.S. are threatened.)

2. It is estimated that about 11 percent or 1107 of the world's bird species are threatened. About how many bird species are there in the world?

3. Hawaii, the Philippines, New Zealand, and Madagascar all show the highest percent of threatened species. What common geographical feature do these four areas share?

4. Use the map to support the fact that many areas have a lower number of threatened species and offer an explanation as to why this is so.

857

Resource Manager

BioLab and MiniLab Worksheets, p. 140 **L2**

Problem-Solving Lab 31-1

Purpose
Students will determine that bird species decline is most apparent on islands because of habitat destruction.

Process Skills
think critically, recognize cause and effect, analyze information, apply concepts, use numbers, define operationally, predict

Teaching Strategies
■ Review the mathematics needed to calculate the total number of species if students know the number threatened and that this number is close to 2.5%. Give them a sample problem to work, such as: 50 species: 2.5% = total U.S. species: 100%.
■ Review the mathematics needed to calculate the total number of bird species. Give them a sample problem to work, such as: 1107 species: 11% = total world species: 100%.
■ Provide students with references or dictionaries to look up the definitions asked for.

Thinking Critically
1. close to 2000
2. close to 10 000
3. all are islands
4. Europe and Africa have very low numbers of threatened species. Both parts of the world involve large areas that still provide enough new space for species to move to when their natural habitat is destroyed.

✔ **Assessment**
Performance Ask students to research the success of capture, breeding, and release programs for certain bird species in the United States, such as the whooping crane, and write a newspaper article describing these programs. Use the Performance Task Assessment List for Newspaper Article in **PASC**, p. 69. **L2**

TECHPREP

Photographing Birds

Visual-Spatial Provide students with equipment to photograph or videotape local birds in their habitats. Have students display their work and discuss in groups the habitat of each bird, the type of food it eats, and adaptations it has for its way of life. L2

COOP LEARN

Assessment

Knowledge Have each student write two questions about birds. Divide the class into pairs of students, and have them quiz each other. L1

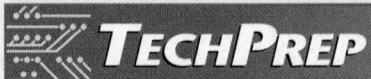

3 Assess

Check for Understanding

Linguistic Ask students to write a letter that begins, "This is everything I know about birds...." Have them summarize what they have learned about birds in this letter. Have students then write a second paragraph that begins, "What I still don't understand about birds is...." Have them exchange letters with another student. Each student should write a response to the other in which they try to explain what their partner doesn't understand. If they both have the same area of weaknesses, they should exchange information with someone else. L2

Reteach

Visual-Spatial Give students field guides to birds of your area. Ask them to observe birds for several days, identify them, and explain how each is adapted to its way of life. L1

Origins of Birds

Current thoughts about bird evolution are illustrated in ***Figure 31.16.*** Scientists hypothesize that today's birds are derived from an evolutionary line of dinosaurs that did not become extinct. ***Figure 31.17*** shows the earliest known bird in the fossil record, *Archaeopteryx*. At first, scientists thought that *Archaeopteryx* was a direct ancestor of modern birds; however,

Figure 31.16
The radiation of orders of birds on the Geologic Time Scale shows their relationships.

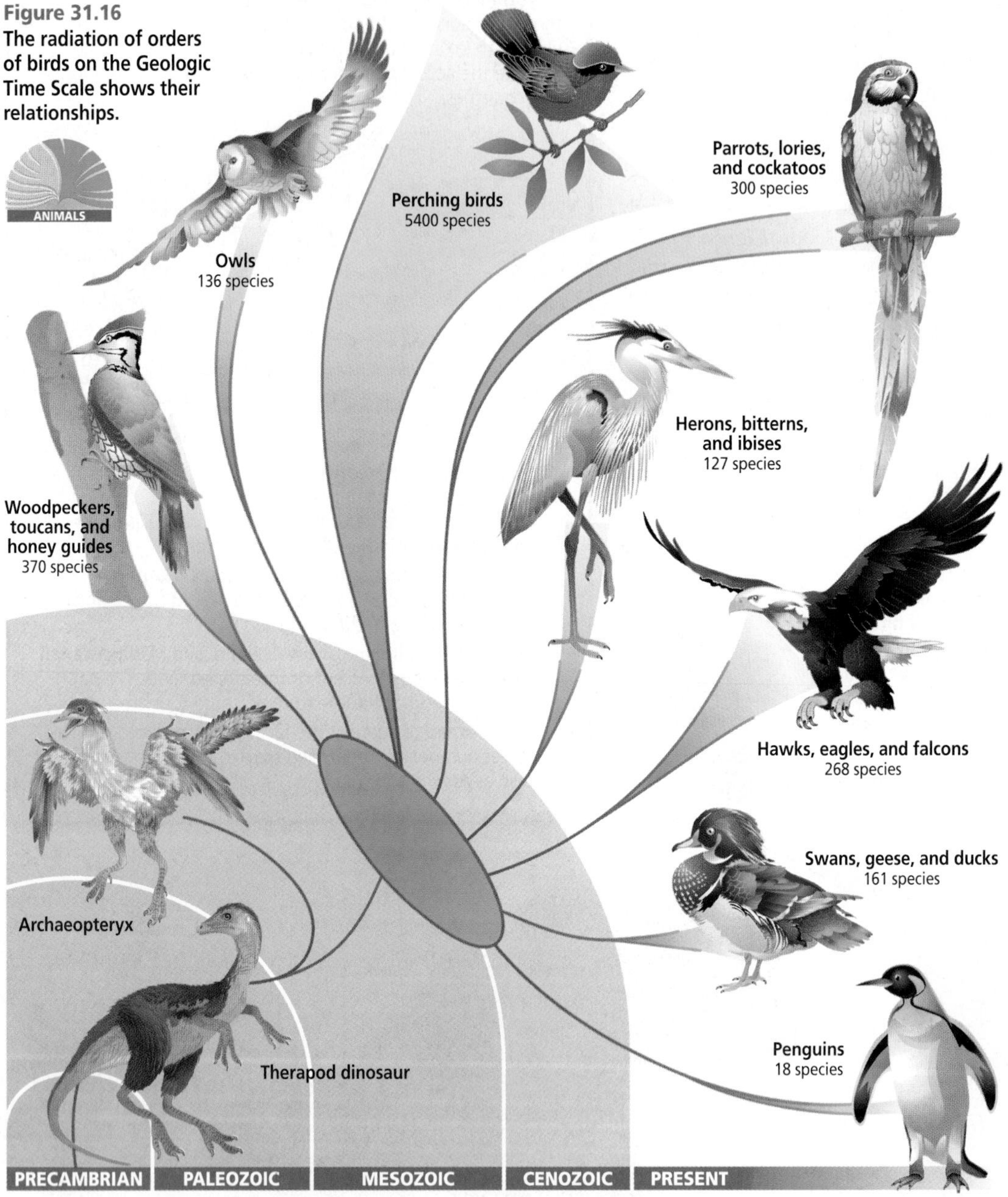

ANIMALS
Perching birds
5400 species
Parrots, lories, and cockatoos
300 species
Owls
136 species
Herons, bitterns, and ibises
127 species
Woodpeckers, toucans, and honey guides
370 species
Hawks, eagles, and falcons
268 species
Swans, geese, and ducks
161 species
Archaeopteryx
Penguins
18 species
Therapod dinosaur
PRECAMBRIAN
PALEOZOIC
MESOZOIC
CENOZOIC
PRESENT

Cultural Diversity

Sankar Chatterjee and Bird Evolution

The evolutionary history of birds is currently under debate by scientists. In your discussions of bird evolution, introduce students to the research and hypotheses of Indian-American paleontologist, Sankar Chatterjee.

Chatterjee is best known for his 1986 discovery of *Protoavis*, a 225 million-year-old fossil that may turn out to be the earliest known bird. Have students research the various hypotheses of bird evolution and initiate a discussion about the evidence used in each hypothesis. L3

Figure 31.17
The fossil bones of *Archaeopteryx* show that it was definitely a bird, whereas those of *Caudipteryx zoui* indicate that it was a feathered theropod dinosaur.

Caudipteryx zoui

Archaeopteryx

some paleontologists now think that it most likely did not give rise to any other bird groups. *Archaeopteryx* was about the size of a crow and had feathers and wings like a modern bird. But it also had teeth, a long tail, and clawed front toes, much like a reptile.

Fossil finds in China support the idea that birds evolved from dinosaurs. The fossil theropod shown in *Figure 31.17* was a two-legged, meat-eating, running dinosaur. It had feathers similar to those of modern

birds. Scientists hypothesize that these early feathers helped insulate the animal, or perhaps were adapted for camouflage or courtship behavior. Most scientists studying the origins of birds hypothesize that feathers came first, and flight evolved later.

But feathers aren't the only features shared by modern birds and some theropod dinosaurs. Both also have a sternum, a wishbone, shoulder blades, flexible wrists, and three fingers on each hand.

Section Assessment

Understanding Main Ideas
1. Why is the lack of teeth in birds an adaptation for flight?
2. Explain how air sacs improve a bird's ability to obtain the energy necessary for flight.
3. How does being an endotherm have adaptive value for birds that live in polar regions?
4. What features of birds enable them to live on land?

Thinking Critically
5. Large, flightless birds once were common

in areas that did not have large, carnivorous predators. Many of these birds are now extinct. What hypothesis can you suggest for the evolution and extinction of large, flightless birds?

SKILL REVIEW
6. **Making and Using Tables** Make a table that summarizes the adaptations birds have that enable them to fly. For more help, refer to *Organizing Information* in the **Skill Handbook**.

Extension

Kinesthetic Have students obtain small birds such as Cornish game hens, chickens, or quail from a meat market or grocery store. The birds should be boiled and all meat removed from the bones. Have students assemble the skeleton using sturdy glue and thin wire. L2 ELL

✔ Assessment

Performance Have students sketch a scale map of the school grounds or a spot near the school on a sheet of butcher paper. Have students indicate how the area could be made a more suitable bird habitat. Have them work in groups to add features to their maps that would make their areas more attractive to birds. Books about attracting birds in all types of environments from the inner city to the suburbs are available at libraries. L2 ELL COOP LEARN

4 Close

Activity
Ask students to develop a hypothesis and experimental plan to determine why flamingos stand on one leg. L2

Resource Manager

Content Mastery, pp. 153, 155-156 L1
Reinforcement and Study Guide, pp. 139-140 L2
Basic Concepts Transparency 57 and Master L2 ELL

Section Assessment

1. Without teeth, birds are lighter for flight.
2. When a bird exhales, oxygenated air in the air sacs passes into the lungs, thereby increasing the amount of oxygen available to the bird to generate energy needed for flight.
3. Endotherms can regulate their body temperature regardless of what the environmental temperature is, and therefore,

can keep warm even in cold areas.
4. lungs, wings, legs, beaks, internal fertilization, shelled, amniotic eggs
5. They filled a niche not otherwise occupied (large ground feeder). Wings became unimportant to survival, becoming vestigial over time.
6. Make sure students' tables include all the features listed on pages 852 through 854.

Time Allotment
One class period

Process Skills
collect data, compare and contrast, identify and control variables, design an experiment, draw a conclusion, experiment, formulate models, hypothesize, interpret data, measure in SI, organize data

PREPARATION

Possible Hypotheses
If shape controls rolling, then different shapes will roll different distances and different shapes will form different patterns.

GLENCOE
TECHNOLOGY

CD-ROM
Biology: The Dynamics of Life
Video: *Bird Courtship*
Disc 4

Resource Manager

BioLab and MiniLab Worksheets, pp. 141-142 L2

DESIGN YOUR OWN
BioLab

Which egg shape is best?

Not all bird eggs have the same shape. An ostrich egg is almost totally round. Chicken eggs are almost a perfect oval on one end. Cliff-dwelling birds such as the common guillemot (Uria aalge) have eggs that come almost to a point on one end. Why the variety of shapes? Is there any adaptive benefit to this variety of shapes? Could egg shape be related to where the bird nests?

PREPARATION

Problem
What shape would be best for an egg to reduce the distance it could roll if pushed from a nest?

Hypotheses
There are several hypotheses that you can test. Your hypothesis might be that egg shape influences the distance an egg rolls, or that shape determines the tightness of circular rolling patterns.

Objectives
In this BioLab you will:
■ **Design** an experiment to test your hypotheses.
■ **Model** different egg shapes and egg masses.
■ **Experiment** to test your hypotheses.
■ **Draw conclusions** based on your experimental data.

Possible Materials
clay
cardboard ramp
ruler
string
hard-boiled egg
Ping-Pong ball
golf ball
balance
protractor

Safety Precautions
Always wear goggles in the lab.

Skill Handbook
Use the **Skill Handbook** if you need additional help with this lab.

860 REPTILES AND BIRDS

PLAN THE EXPERIMENT

Teaching Strategies
■ Students should work in groups of three or four with specific duties being assigned to each student within the group.
■ Balances may be shared among groups.
■ Class discussion at the conclusion would be meaningful to see the variety of hypotheses tested and conclusions reached.

Possible Procedures
■ To make egg models, students could mold clay around golf or ping pong balls or real eggs.
■ The mass of the models can be kept fairly constant by adding or removing clay.

Data and Observations
Round eggs will roll the farthest. Eggs of

PLAN THE EXPERIMENT

1. Decide on a way to test your group's hypothesis. Keep the list of available materials in mind as you plan your procedure.

2. There are a number of questions to be asked before starting. Here are some suggestions. How will you incorporate a control? How many egg shapes will you test? How will you model your egg shapes? How many trials will you perform? How might you keep egg models identical in mass? How will you measure the angle of the cardboard ramp? Where will you start to measure distance rolled?

Check the Plan

Discuss the following points with other group members to decide the final procedure for each of your experiments.

1. What is your independent and dependent variable?
2. How will you eliminate all other variables?
3. What data will you collect? How many trials will you run?
4. Will you need a data table and how might it be organized?
5. *Make sure your teacher has approved your experimental plan before you proceed further.*
6. Carry out your experiments.

ANALYZE AND CONCLUDE

1. **Hypothesizing** Record your hypothesis.
2. **Interpreting Data** Describe your results after testing your hypothesis.
3. **Concluding** Do your data support your hypothesis? Explain using both quantitative and qualitative observations.
4. **Identifying Variables** What were your independent and dependent variables?
5. **Concluding** In general, how does mass influence the distance an egg will roll? How does egg shape influence the distance an egg will roll or the pattern taken

when it rolls?
6. **Predicting** Predict why egg shape or mass may be helpful adaptations when considering the variety of habitats where birds live.

Going Further

Knowledge Find out the chemical and physical nature of bird shells. Find out how and where birds produce a shell.

BIOLOGY Online To find out more about birds and bird eggs, visit the Glencoe Science Web site.
science.glencoe.com

ANALYZE AND CONCLUDE

1. Student answers will vary. Example: Egg shape will not influence path that the egg follows.
2. Student answers will vary. Example: Round eggs roll in straight lines whereas oval or pointed eggs roll in a circular pattern.
3. Student answers will vary. Example: Heavy round eggs rolled farther (26 cm) than lighter round eggs (22 cm).
4. Student answers will vary. Example: independent variable, egg shape; dependent variable, distance rolled.
5. A greater mass results in a longer distance rolled. Oval or pointed eggs roll a shorter distance and form a circular path; the more pointed the end, the tighter the circular path.
6. Round shapes aid birds that nest on flat ground. Oval or pointed eggs aid birds that nest on slanted ground or on cliffs.

✓ Assessment

Knowledge Ask students to write a short report on their experimental findings and to emphasize how egg shape has adaptive value for birds. Use the Performance Task Assessment List for Lab Report in **PASC**, p. 47. **L2**

higher mass will roll farther than lighter eggs. Pointed eggs or oval (normal egg shape) eggs will roll in a circular pattern and total distance will be less than total distance rolled for round eggs.

Going Further

Have students research the variety in egg shell coloration. Have them correlate this variety in coloration with natural selection and species survival. **L3**

Purpose

Encourage students to become aware of the illegal wildlife trade, and how consumer demand for wildlife and wildlife products drives the market for products made from threatened and endangered species.

Background

It may come as a surprise to students that the United States is probably the largest single consumer of wildlife in the world. Wildlife trade in the U.S. represents roughly one-fifth of the global wildlife market. Many animals and wildlife products are bought and sold legally in the U.S., but many others are not. In a recent survey conducted by TRAFFIC, the wildlife trade monitoring organization supported by the World Wildlife Fund and the World Conservation Union (IUCN), traditional packaged medicines containing (or purporting to contain) tiger or rhino body parts were easy to find for sale in many large cities, even though the trade of such products is illegal. Fewer than 6000 tigers and 12 000 rhinos remain in the wild.

Teaching Strategies

■ As a class project, contact the U.S. Fish and Wildlife Service for information about the illegal wildlife trade. Have students create posters showing ways in which people can help reduce the worldwide demand for products made from threatened and endangered animals and plants.

■ Visit a zoo that participates in a nationally accredited captive breeding program for threatened or endangered species.

Biology & SOCIETY

Illegal Wildlife Trade

In May 1998, the U.S. Fish and Wildlife Service broke up an international smuggling ring. Their three-year investigation—code-named Operation Jungle Trade—ended in the arrest of smugglers operating in a dozen countries. In what illegal products were these criminals trafficking? Not diamonds or drugs, but rare birds.

An international treaty called CITES (the Convention on International Trade in Endangered Species of Wild Fauna and Flora) makes it illegal to buy or sell many of the world's threatened and endangered animals and plants, or products made from them. Despite this global agreement, the illegal wildlife trade is a multi-billion-dollar-a-year business.

Species for sale Some people pay large sums of money for parrots, tropical fish, monkeys, snakes, and lizards to add to animal collections or keep as exotic pets. Worldwide, millions of illegal wildlife products—from jewelry made from sea turtle shells to snow leopard coats and lizard skin belts—are bought and sold annually on the wildlife black market. Many traditional remedies manufactured in certain countries are made with body parts from rare species, including endangered tigers and rhinos.

Confiscated products from endangered species and a jaguar skin (inset)

Different Viewpoints

For some people, owning an unusual pet or wearing clothing or jewelry made from endangered species is a status symbol. They feel if they have the money, they should be able to buy whatever they want. Far more people support the illegal wildlife trade without realizing it by buying sea shells, coral jewelry, ivory trinkets, and animal skin accessories sold as souvenirs in many countries. Some users of traditional remedies believe strongly in the power of parts of certain endangered animals or plants to enhance physical attractiveness or treat physical conditions and don't think about protecting endangered wildlife.

The ethics of wildlife trade The illegal wildlife trade is pushing many species to the brink of extinction. With every extinction, Earth's already dwindling biodiversity shrinks a little more. Once a given species is gone, wildlife traders will turn to a different one to try to meet the demands of consumers. Education, stricter laws regulating wildlife trade, and better law enforcement are needed to bring the booming illegal wildlife trade under control.

INVESTIGATING THE ISSUE

Analyzing the Issue In the United States alone, hundreds of different kinds of plants are collected and sold for use in the medicinal plant trade. In small groups, discuss how the increasing popularity of herbal medicines—many of which are made from plants that grow in the wild—could endanger numerous plant species.

BIOLOGY Online To find out more about the illegal wildlife trade, visit the Glencoe Science Web site. **science.glencoe.com**

862 REPTILES AND BIRDS

Investigating the Issue

Students should understand that using plant and animal resources wisely and sustainably will help ensure a continuing supply of medicines.

Going Further

Linguistic Have students write an essay on the following question: Would you buy a product if you were aware that it was made from the body parts of an endangered species? **L2**

SUMMARY

Section 31.1
Reptiles

Main Ideas

- Reptiles are ectotherms that have dry, scaly skin; legs under the body; internal fertilization; and amniotic eggs. Most reptiles have three-chambered hearts. Some reptiles have four-chambered hearts.
- Present-day reptiles belong to one of four groups. Turtles have shells and no teeth. Crocodiles and alligators have streamlined bodies and powerful, toothed jaws. Lizards have a variety of adaptations, including long bodies, tails, and short limbs. Snakes have no limbs.
- The ancestors of present-day reptiles arose from ancient cotylosaurs, which were also the ancestors of the dinosaurs.

Vocabulary
amniotic egg (p.842)
Jacobson's organ (p.845)

Section 31.2
Birds

Main Ideas

- Birds have adaptations for flight including feathers; a keel-shaped sternum; a four-chambered heart; endothermy; thin, hollow bones; a beak; and air sacs.
- Birds may be derived from a line of dinosaurs that did not become extinct.

Vocabulary
endotherm (p.854)
feather (p.852)
incubate (p.854)
sternum (p.853)

Main Ideas
Summary statements can be used by students to review the major concepts of the chapter.

Using the Vocabulary
To reinforce chapter vocabulary, use the Content Mastery Booklet and the activities in the Interactive Tutor for Biology: The Dynamics of Life on the Glencoe Science Web site.
science.glencoe.com

 THE PRINCETON REVIEW *All Chapter Assessment questions and answers have been validated for accuracy and suitability by The Princeton Review.*

UNDERSTANDING MAIN IDEAS

1. c
2. c
3. d
4. d
5. b

UNDERSTANDING MAIN IDEAS

1. Scientists hypothesize that _____ were the ancestors of birds.
  a. fishes
  b. amphibians
  c. reptiles
  d. mammals

2. Of the following, which is NOT an example of a reptile?
  a. snake
  b. turtle
  c. salamander
  d. lizard

3. For gas exchange, reptiles are dependent on _____.
  a. gills
  b. skin
  c. skin and lungs
  d. lungs

4. Which of the following is NOT a characteristic of reptiles?
  a. has three- or four-chambered heart
  b. lays amniotic eggs
  c. has legs flexed under the body
  d. has external fertilization

5. Why don't reptiles inhabit extremely cold regions on Earth?
  a. They have moist skin that would freeze in the cold.
  b. They are ectotherms.
  c. They lay eggs in water and water would freeze in the cold.
  d. They are endotherms.

GLENCOE TECHNOLOGY

 **VIDEOTAPE**
MindJogger Videoquizzes
Chapter 30: *Reptiles and Birds*
Have students work in groups as they play the videoquiz game to review key chapter concepts.

Resource Manager

Chapter Assessment, pp. 181-186
MindJogger Videoquizzes
ExamView® Pro Software 💾
BDOL Interactive CD-ROM, Chapter 31 quiz

6. a
7. b
8. b
9. a
10. c
11. endotherm
12. heat-sensitive
13. nesting in trees
14. Crocodilians
15. flight
16. removes waste (allantois)
17. scales, claws, amniotic eggs
18. membranes
19. feathers
20. Jacobson's

APPLYING MAIN IDEAS

21. Birds are endotherms and maintain a constant body temperature regardless of the temperature of their environment.

22. Sea turtles might not recognize their nesting beaches and they may not lay eggs. Sea turtles migrate back to the beaches where they hatched to lay eggs. The release of chemicals into the water may alter the water chemistry in such a way that turtles may not be able to recognize the area.

23. The allantois collects nitrogenous wastes. The embryo inside an amniotic egg gets food from the yolk. The chorion permits gas exchange. The shell, fluids, and membranes cushion and protect the developing embryo.

24. *Caudipteryx* is a fossil dinosaur with feathers similar to those of modern birds. *Archaeopteryx* is the earliest fossil bird with feathers and wings.

---

6. Eggs that cushion the embryo in fluid and protect it with membranes and a shell developed first in _____.
  a. reptiles   **c.** amphibians
  b. birds   **d.** mammals

7. When a snake flicks out its tongue, it is using its sense of _____.
  a. vision   **c.** hearing
  b. smell   **d.** touch

8. From what group of reptiles do most scientists agree birds evolved?
  a. cotylosaurs   **c.** therapsids
  b. theropods   **d.** turtles

9. The function of down feathers is _____.
  a. insulation   **c.** preening
  b. flight   **d.** molting

10. The earliest known fossil bird with feathers is _____.
  a. *Caudipteryx*   **c.** *Archaeopteryx*
  b. an ostrich   **d.** a pigeon

11. A penguin can live in Antarctica because it is a(n) _____.

12. Some snakes can find their prey even in the dark because they have _____ pits along their upper jaws.

13. In the cladogram below, you can see that a shorter beak and _____ evolved after skeletal fusion.

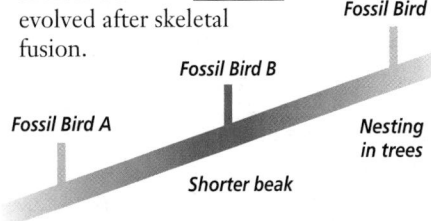

Fossil Bird C
Fossil Bird B
Fossil Bird A
Nesting in trees
Shorter beak
More skeletal fusion

THE PRINCETON REVIEW — TEST-TAKING TIP

Your Answers Are Better Than the Test's
When you know the answer, answer the question in your own words before looking at the answer choices. Often, more than one answer choice will look good, so arm yourself with yours before looking.

---

14. _____ move faster than all other reptiles because their cells get more oxygen through their hearts.

15. The large breastbone of a bird is an adaptation that aids in _____.

16. What function does the structure labeled A in the diagram below perform in the amniotic egg?

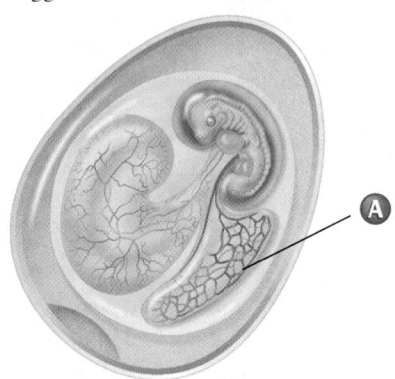

17. What are three features that modern-day reptiles and birds share?

18. In the amniotic egg, the chorion and the amnion are _____ that permit gas exchange and protect the developing embryo.

19. Birds can be distinguished from all other living animals because they have _____.

20. A snake's sense of smell is located in its _____ organ.

APPLYING MAIN IDEAS

21. Why are birds able to inhabit more diverse environments than reptiles?

22. Newly developed industries that locate on shorelines often release chemicals into the ocean. How might this development affect sea turtles that use these areas to breed? Explain.

23. Explain how the amniotic egg maintains homeostasis.

24. Discuss why the fossils of *Archaeopteryx* and *Caudipteryx* are significant in explaining the evolutionary history of birds.

25. Interpreting Data A biologist counts the feathers on the bodies of two different species of birds. The data are represented in the graph below. What might be inferred about the type of environments in which the birds live?

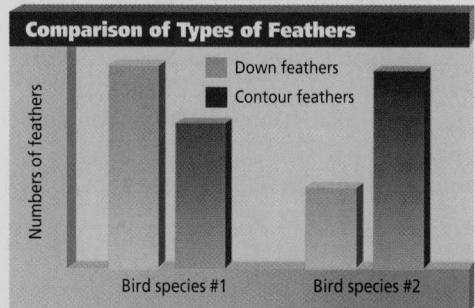

Comparison of Types of Feathers

- Down feathers
- Contour feathers

Numbers of feathers

Bird species #1   Bird species #2

26. Comparing and Contrasting Most reptiles lay between one and 200 eggs at a time. Amphibians lay thousands of eggs at a time. Is there an adaptive advantage to laying fewer eggs on land? Explain.

27. Concept Mapping Complete the concept map by using the following vocabulary terms: sternum, feathers, endotherms.

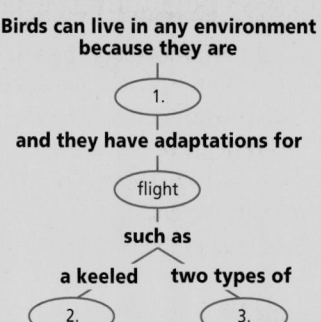

Birds can live in any environment because they are

1.

and they have adaptations for

flight

such as

a keeled   two types of

2.   3.

CD-ROM

For additional review, use the assessment options for this chapter found on the *Biology: The Dynamics of Life Interactive CD-ROM* and on the Glencoe Science Web site.
science.glencoe.com

A biologist is comparing yearly censuses of owls and mice found in one area. The data obtained are represented in the graph below.

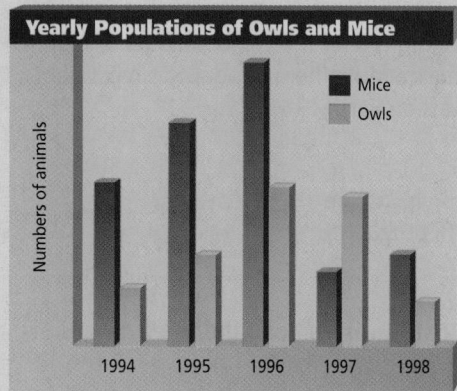

Yearly Populations of Owls and Mice

- Mice
- Owls

Numbers of animals

1994   1995   1996   1997   1998

Interpreting Data Study the graph and answer the following questions.

1. In most years, there are _____.
   - **a.** more mice than owls
   - **b.** more owls than mice
   - **c.** the same number of owls as mice
   - **d.** twice as many owls as mice

2. The best explanation for the fluctuations in owl populations is that the owl population increases and decreases in response to _____.
   - **a.** the size of the mouse population
   - **b.** the decrease in the mouse population
   - **c.** the increase in the mouse population
   - **d.** the 4-year cycle of the mouse population

3. **Formulating Hypotheses** Examine the graph again and assume that the key is changed to the following: purple represents a certain type of eagle and yellow represents rabbits. Make a hypothesis about the relationship between eagles and rabbits.

25. Bird species 1 probably lives in a colder environment than does bird species 2.

26. The shelled eggs of the reptile are more protected than the amphibian eggs, so they have a greater chance of producing live young. Also, energy is conserved when fewer eggs are produced.

27. 1. Endotherms; 2. Sternum; 3. Feathers

1. a
2. a
3. There are more eagles than rabbits in any one year, except 1997. Eagles exert pressure on the rabbit population, but because eagles also eat many other animals, they do not completely decimate the rabbit population in any one year. In years with fewer eagles, more rabbits survive.

Chapter 32 Organizer

Refer to pages 4T-5T of the Teacher Guide for an explanation of the National Science Education Standards correlations.

Section	Objectives	Activities/Features
Section 32.1 **Mammal Characteristics** National Science Education Standards UCP.1-5; A.1, A.2; B.2; C.3, C.5, C.6 (1 session, 1 block)	1. **Distinguish** mammalian characteristics. 2. **Explain** how the characteristics of mammals enable them to adapt to most habitats on Earth.	**MiniLab 32-1:** Anatomy of a Tooth, p. 869 **Problem-Solving Lab 32-1,** p. 870 **MiniLab 32-2:** Mammal Skeletons, p. 871 **Inside Story:** A Mammal, p. 872 **Careers in Biology:** Animal Trainer, p. 873 **Internet BioLab:** Domestic Dogs Wanted, p. 882 **Biology & Society:** Do we need zoos? p. 884
Section 32.2 **Diversity of Mammals** National Science Education Standards UCP.1-5; C.3, C.5, C.6; D.3; F.3-6; G.1-3 (3 sessions, ½ block)	3. **Distinguish** among the three groups of living mammals. 4. **Compare** reproduction in egg-laying, pouched, and placental mammals.	**Focus On** Placental Mammals, p. 876

Need Materials? Contact Carolina Biological Supply Company at 1-800-334-5551 or at **http://www.carolina.com**

MATERIALS LIST

BioLab

p. 882 Internet access

MiniLabs

p. 869 microscope, prepared slide of human tooth cross section, paper, pencil
p. 871 owl pellet, paper towel, forceps, microscope, microscope slide, coverslip, dropper, water

Alternative Lab

p. 870 mouse, terrarium, sawdust or sand, assorted objects, timer or clock with second hand

Quick Demos

p. 868 microscope, microscope slides, coverslips, dog hair, cat hair, glycerol
p. 875 animal field guides

Key to Teaching Strategies

L1 Level 1 activities should be appropriate for students with learning difficulties.

L2 Level 2 activities should be within the ability range of all students.

L3 Level 3 activities are designed for above-average students.

ELL ELL activities should be within the ability range of English Language Learners.

COOP LEARN Cooperative Learning activities are designed for small group work.

P These strategies represent student products that can be placed into a best-work portfolio.

These strategies are useful in a block scheduling format.

Teacher Classroom Resources

Section	Reproducible Masters	Transparencies
Section 32.1 **Mammal Characteristics**	Reinforcement and Study Guide, pp. 141-142 L2 Concept Mapping, p. 32 L3 ELL Critical Thinking/Problem Solving, p. 32 L3 BioLab and MiniLab Worksheets, pp. 143-144 L2 Laboratory Manual, pp. 229-238 L2 Content Mastery, pp. 157-158, 160 L1 Tech Prep Applications, pp. 41-42 L2	Section Focus Transparency 77 L1 ELL Basic Concepts Transparency 58 L2 ELL Basic Concepts Transparency 59 L2 ELL Reteaching Skills Transparency 47 L1 ELL
Section 32.2 **Diversity of Mammals**	Reinforcement and Study Guide, pp. 143-144 L2 BioLab and MiniLab Worksheets, pp. 145-146 L2 Content Mastery, pp. 157, 159-160 L1	Section Focus Transparency 78 L1 ELL

Assessment Resources

Chapter Assessment, pp. 187-192
MindJogger Videoquizzes
Performance Assessment in the Biology Classroom
Alternate Assessment in the Science Classroom
ExamView® Pro Software
BDOL Interactive CD-ROM, Chapter 32 quiz

Additional Resources

Spanish Resources ELL
English/Spanish Audiocassettes ELL
Cooperative Learning in the Science Classroom COOP LEARN
Lesson Plans/Block Scheduling

NATIONAL GEOGRAPHIC — Teacher's Corner

Products Available From Glencoe
To order the following products, call Glencoe at 1-800-334-7344:
CD-ROM
Mammals: A Multimedia Encyclopedia
Videodisc
STV: Animals

Products Available From National Geographic Society
To order the following products, call National Geographic Society at 1-800-368-2728:

Book
National Geographic Book of Mammals
Video
Wild Survivors: Camouflage and Mimicry

Index to National Geographic Magazine
The following articles may be used for research relating to this chapter:
"Between Monterey Tides," by Rick Gore, February 1990.

GLENCOE TECHNOLOGY

The following multimedia resources are available from Glencoe.

Biology: The Dynamics of Life
CD-ROM ELL
BioQuest: *Biodiversity Park*
Video: *Primate Charactersistics*
Video: *Beaver*
Video: *Dolphin*
Video: *Shrew*
Video: *Bat*
Video: *Elephant*
Video: *Elephant Behavior*
Video: *Duck-Billed Platypus*

Videodisc Program
Duck-Billed Platypus

The Infinite Voyage
The Keepers of Eden

GETTING STARTED DEMO

 Visual-Spatial Show students a mouse, rat, gerbil, or hamster. Offer the animal food, and ask students to observe in groups its behavior for five minutes. Ask students to list the adaptations of the animal to its environment.

Theme Development

The theme of **unity within diversity** is obvious in the discussions of the traits all mammals share. The theme of **homeostasis** is woven throughout the chapter via a discussion of mammal endothermy. This mammalian trait is one of the characteristics that enables mammals to inhabit a wide variety of habitats.

Resource Manager

Section Focus Transparency 77 and Master **L1** **ELL**

READING BIOLOGY

Glencoe's *Biology: The Dynamics of Life* contains many resources to assist a student's reading skills. Each chapter contains figures with expanded captions that expand on written material. Word Origins, located along the side of text, expand knowledge of biology vocabulary. Glencoe's Content Mastery Booklet helps develop reading skills while reinforcing content. In addition, use the Interactive Tutor for *Biology: The Dynamics of Life* on the Glencoe Web site to reinforce vocabulary. **science.glencoe.com**

Chapter

32 Mammals

What You'll Learn

- You will distinguish the characteristics of mammals.
- You will compare and contrast three groups of living mammals and examine their relationships to their ancient ancestors.

Why It's Important

Mammals play a major role in most ecosystems on Earth because they are at the top of many food chains. Humans are mammals, so studying mammal characteristics provides information about humans as well.

READING BIOLOGY

Look through the chapter and write down different section heads and subheads. As you read, make an outline, using the heads and new vocabulary as guides.

To find out more about mammals, visit the Glencoe Science Web site.
science.glencoe.com

African elephants live in social groups and have strong family ties. A cheetah and her cubs rest after eating the prey she has killed. Elephants and cheetahs share other characteristics that classify both as mammals.

Multiple Learning Styles

Look for the following logos for strategies that emphasize different learning modalities.

 *Kinesthetic* Reteach, p. 881

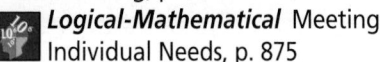 *Visual-Spatial* Quick Demo, pp. 868, 875; Portfolio, p. 868; Reteach, p. 873; Project, p. 878

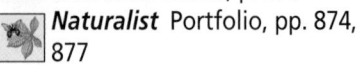 *Intrapersonal* Project, p. 869

Linguistic Biology Journal, pp. 868, 875, 876; Check for Understanding, p. 873

Logical-Mathematical Meeting Individual Needs, p. 875

Naturalist Portfolio, pp. 874, 877

32.1 Mammal Characteristics

Objectives

Distinguish mammalian characteristics.

Explain how the characteristics of mammals enable them to adapt to most habitats on Earth.

Vocabulary

gland
diaphragm
mammary gland

Prepare

Key Concepts

Mammalian characteristics such as hair, endothermy, a muscular diaphragm, mammary glands, and intelligence are presented. Adaptations for obtaining and consuming food are discussed for specific mammals.

Planning

■ Borrow or buy a small mammal such as a mouse for the Getting Started Demo.

■ Obtain hair from a variety of mammals, and gather slides, coverslips, and glycerol for the Quick Demo.

■ Obtain a mouse, a terrarium with sand, and small objects for burying for the Alternative Lab.

Have you ever read a children's story in which the children feared a big, bad wolf? The wolf characters in stories like Little Red Riding Hood *or* The Three Little Pigs *are usually portrayed as evil, cunning, and vicious killers. But real wolves kill prey in order to eat, as all predators must. Wolves prey mostly on small animals such as rabbits and pikas, rodents such as mice and voles, and occasionally deer or elk. Both wolves and their prey are mammals.*

Wolf eating prey (above) and a pika (inset)

What Is a Mammal?

Mammals, like birds, are endotherms. The ability to maintain a fairly constant body temperature enables mammals to live in almost every possible environment on Earth. Polar bears of the Arctic, tigers in tropical jungles, and dolphins that roam the Atlantic Ocean are able to live in these varied environments because they are endotherms. Mammals also share other important characteristics.

Mammals have hair

Have you ever heard someone complain about a pet that is shedding its hair? There's no doubt that such a pet is a mammal because only mammals have hair. You have read that

birds' feathers probably evolved from scalelike structures such as those of reptiles. Mammalian hair, made out of keratin, is also thought to have evolved from scales. The structure of hair provides insulation and waterproofing and thereby conserves body heat. If you have ever worn a wool sweater made from the hair of a sheep, you know how warm wool can be on a cold day. As shown in *Figure 32.1,* hair also serves other functions.

Although hair helps retain body heat, mammals also have adaptations that aid in cooling off the body when it gets too warm. Mammals cool off by panting and through the action of sweat glands. Panting releases water from the lungs, which results in a loss of body heat. Sweat glands help

WORD Origin

mammal
From the Latin word *mamma,* meaning "breast." A mammal is an animal with hair that feeds its young with milk from mammary glands.

1 Focus

Bellringer

Before presenting the lesson, display **Section Focus Transparency 77** on the overhead projector and have students answer the accompanying questions.
L1 ELL

Transparency **77** Scales, Feathers, Hair — SECTION FOCUS — Use with Chapter 32, Section 32.1

❶ Identify the mammal in this picture.
❷ In what major ways do mammals differ from the other animals shown?

BIOLOGY: The Dynamics of Life — SECTION FOCUS TRANSPARENCIES

32.1 MAMMAL CHARACTERISTICS **867**

✔ Assessment Planner

Portfolio Assessment
 Portfolio, TWE, pp. 868, 874, 877
 MiniLab, TWE, pp. 869, 871
 Assessment, TWE, pp. 873, 881
 Problem-Solving Lab, TWE, p. 870
Performance Assessment
 BioLab, SE, pp. 882-883
 MiniLab, SE, pp. 869, 871

Alternative Lab, TWE, pp. 870-871
 Assessment, TWE, p. 868
Knowledge Assessment
 Section Assessment, SE, pp. 873, 881
 Chapter Assessment, SE, pp. 885-887
 BioLab, TWE, pp. 882-883
Skill Assessment
 Assessment, TWE, p. 880

Figure 32.1
Hair helps in maintaining a constant body temperature. It also serves a variety of other purposes.

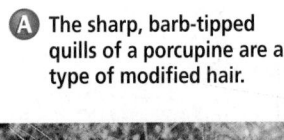

A The sharp, barb-tipped quills of a porcupine are a type of modified hair.

B The black stripes of a tiger's fur aid in camouflaging this beautiful cat as it hunts for prey.

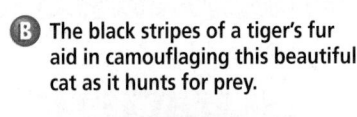

C The white patch of hair on the rump of the fleeing pronghorn signals danger to other members of the herd.

regulate body temperature by secreting water onto the surface of the skin. As the water evaporates, it transfers heat from the body to the surrounding air.

Mammals nurse their young

Mammals have organs called glands that secrete various substances needed by the animal. A **gland** is an organ that secretes substances inside or outside the body. Mammals have several kinds of glands, including glands that produce saliva, sweat, digestive enzymes, and hormones. You have already learned how sweat glands help keep a mammal cool.

Mammals also feed their young from mammary glands. In female mammals, the **mammary glands** (MAM uh ree) secrete milk, a liquid that is rich in fats, sugars, proteins, minerals, and vitamins. Mammals

Figure 32.2
Large mammals usually have few young. Mammals that are prey for many predators tend to have larger litters.

A An Indian rhinoceros usually has one calf at a time. Calves begin to graze at two months of age.

B Mice have four to nine offspring in each litter, and up to 17 litters a year. The young nurse for just a few weeks.

nurse their young until they are mature enough to find food for themselves. *Figure 32.2* shows that the number of young each mother has and the length of time she nurses her young vary among species.

Respiration and circulation in mammals

Mammals need a high level of energy for heating and cooling their bodies, as well as for locomotion. This energy level is sustained when large amounts of oxygen enter the body and reach all the cells. One way mammals accomplish this is by using a diaphragm that expands and contracts the chest cavity. A **diaphragm** (DI uh fram) is the sheet of muscle located beneath the lungs. The diaphragm separates the chest cavity from the abdominal cavity, where other organs are located.

Mammals, like birds, have four-chambered hearts in which oxygenated blood is kept entirely separate from deoxygenated blood. This ensures that mammals receive a good supply of oxygen to support their endothermic metabolism.

Mammals have different types of teeth

Teeth are a distinguishing feature of most mammals. Although fishes and reptiles have teeth, their teeth are relatively uniform and are used primarily for tearing, grasping, and holding prey.

Mammals have different kinds of teeth adapted to the type of food they eat. Think of the different tools you might use to build a piece of furniture, such as a chisel for scraping or a saw for cutting. Like a cabinetmaker's tools, teeth are shaped to match the types of jobs they do. The pointed incisors of moles grasp and hold small prey. The chisel-like incisors of

beavers are modified for gnawing. A lion's canines puncture and tear the flesh of its prey. Premolars and molars are used for slicing or shearing, crushing, and grinding. You can get a closer look at mammalian teeth in the *MiniLab* on this page. By examining the teeth of a mammal, a scientist can determine what kind of food it eats.

Many hoofed mammals have an adaptation called cud chewing that enables them to break down the cellulose of plant cell walls into nutrients

---

MiniLab 32-1 Observing

Magnification: 8×

Anatomy of a Tooth Most mammals have teeth. Teeth typically have evolved into a variety of shapes, depending on the diet of the animal.

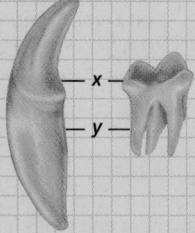
Longitudinal section of a canine tooth

Procedure

1. Examine a prepared slide of a human tooth under low-power magnification. The slide is a longitudinal section view of the tooth. **CAUTION: *Use caution when working with a microscope and slides.***
2. Locate the different areas that form the human tooth. From outside to inside they are: enamel, dentine, and pulp cavity. Each area varies in thickness and texture. You will have to move the slide from left to right or up and down to see all areas.
3. Diagram the appearance of the entire longitudinal section as it appears under low power, labeling the three areas.

Analysis

1. Which area is the thickest? Thinnest?
2. Describe the nature of the three areas. Use such terms as compact, loose, bonelike, vascular (with blood vessels).
3. Which area of the three would you expect to be the toughest? Explain why.
4. Draw cross sections of these two tooth shapes at the points marked *x* and *y* on each diagram. Include the three areas just studied and label each area.

MiniLab 32-1

Purpose
Students will identify the three major areas of a typical mammalian tooth.

Process Skills
interpret scientific diagrams, observe and infer, predict

Teaching Strategies
■ Prepared slides of tooth longitudinal sections are available from biological supply houses. You could use 35 mm slides instead of prepared slides.
■ Have students observe slides using dissecting microscopes, if available.
■ Reduce the cost of purchasing class sets of slides by using a microprojector or microvideo camera. Have the class view the slide material as a group rather than individually.

Expected Results
Enamel is the outermost, most compact and bonelike thinnest layer. Dentine is the thickest layer, also compact and bonelike. Pulp is the center layer and appears vascular and the least dense of the three layers.

Analysis
1. Dentine is thickest; enamel or pulp is thinnest.
2. Enamel is compact and bonelike; dentine is also compact and bonelike; pulp is vascular and looser.
3. Enamel; the outermost layer receives the most wear.
4. Examine student drawings to make sure they have located each layer appropriately.

✔ **Assessment**

Portfolio Have students draw longitudinal sections of all four types of human teeth—incisors, canines, premolars, and molars. Use the Performance Task Assessment List for Scientific Drawing in **PASC**, p. 55. **L2** **ELL** **P**

PROJECT

Zookeeper

Intrapersonal Have students imagine they are the director of a zoo that will be adding an unusual mammal exhibit. Ask them to select a mammal, design the exhibit, and determine the animal's feeding schedule and other special needs. Finally, they should prepare an exhibit sign for visitors. **L2** **COOP LEARN**

GLENCOE TECHNOLOGY

VIDEODISC
The Secret of Life
Circulatory System

Human Respiratory System

Purpose 🖐

Students compare the length of herbivore and carnivore digestive systems.

Process Skills

analyze information, compare and contrast, hypothesize, interpret data, predict, think critically

Teaching Strategies

■ You may wish to mention the need for several stomachs in ruminants to help with the digestion of cellulose.

Thinking Critically

1. Herbivores have a longer digestive system.

2. Answers may vary; hypotheses will be supported if students predicted herbivores have a longer digestive system.

3. The more difficult a food is to digest, the longer the digestive system will be.

4. Yes; for the comparison to be valid, one must compare animals of equal size. An animal of larger size, such as a lion, would have a much longer digestive system than a rabbit due to its larger size.

✔ Assessment

Knowledge Have students predict the relative lengths of the digestive system in these pairs: deer and lion, fox and dog, cat and rabbit, mouse and vole, horse and bear. Use the Performance Task Assessment List for Analyzing the Data in **PASC**, p. 27. **L2**

Problem-Solving Lab 32-1   Analyzing Information

Which animal has the longer digestive system? A mammal may be an herbivore, carnivore, or omnivore. Is there a relationship between length of a mammal's digestive system and its diet? Make a hypothesis as to what that correlation might be.

Analysis

The following data table provides general information on digestive systems for several mammals.

Data Table			
Animal	Length of digestive system	Diet category	Animal weight
Koala	305 cm	herbivore	10 kg
Dog	135 cm	carnivore	11 kg
Rabbit	272 cm	herbivore	9 kg
Bobcat	145 cm	carnivore	12 kg

Thinking Critically

1. What does the relationship between diet and digestive system length appear to be?

2. Do the data support your hypothesis? Explain your answer.

3. Formulate a hypothesis that explains the relationship between digestive system length and difficulty in digesting food type. (Hint: Does cellulose take longer to digest?)

4. Was the mass of all animals relatively close when compared? Explain why mass is important.

Figure 32.3
Mammal groups are distinguished by number and types of teeth.

(A) Premolars and molars are the predominant teeth in horses and other herbivores. These crushing and grinding teeth are covered with hard enamel.

(B) Carnivores, such as this tiger, have canine teeth that stab and pierce food, and premolars and molars adapted for chewing.

---

that they can use and absorb. In cud chewing, plant material that has been swallowed is brought back up to the mouth and chewed again. Have you ever seen cows slowly chewing and chewing while lying in a pasture? When grass is swallowed, cellulose in the cell walls is broken down by bacteria in one of several pouches in the stomach. The food, called cud, is then brought up into the mouth. After more chewing, the cud is swallowed again and passed to three other stomach areas, where digestion continues. You can learn more about digestion in mammals in the *Problem-Solving Lab* on this page. The various types of teeth in different mammals can be seen in *Figure 32.3*.

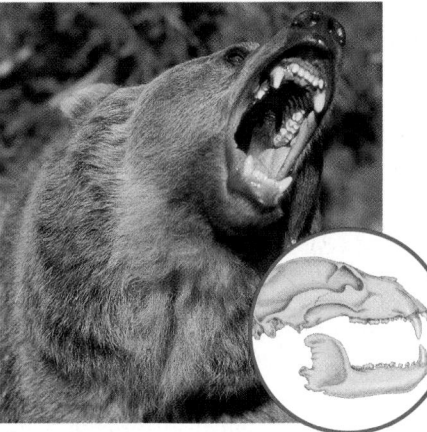

(C) Bears, like humans, have incisors, canines, premolars, and molars. Bears and humans are omnivores.

Alternative Lab

Burying Behavior

Purpose 🖐

Students will observe defensive burying behavior in mice and determine the types of objects that cause this behavior.

870

Materials 🔲 🔲 🔲

mouse, terrarium with 3-5 cm of sawdust or sand, objects for burying, timer

Preparation

When mice are confronted by potential predators such as snakes, they sometimes spray sand or soil at the predator. This behavior is called defensive burying. Defensive burying can be elicited by objects that resemble predators, as well as by predators themselves.

Procedure

Give students the following directions.

1. Make a data table that lists objects, the time it takes the mouse to start burying, and the time spent burying.

2. Develop a hypothesis about which objects a mouse will spend the most time burying.

3. Place one of the objects on the floor of the terrarium at the end farthest from the animal. **CAUTION:** *Always use care*

Figure 32.4
In moles, the front limbs are powerful and short, with large claws. Bats have elongated finger bones that support the flight membranes.

Mammals have modified limbs

Mammals have several adaptations that help them meet their energy needs. For example, mammal limbs are adapted for a variety of methods of food gathering. Recall that primates use their opposable thumb to grasp objects—including fruits and other foods. *Figure 32.4* illustrates other limb modifications in mammals that help them to capture food or avoid becoming food for other animals. You can see the bones of mammal limbs and other parts of mammalian skeletons in the *MiniLab* on this page. Refer to the *Inside Story* on the next page for a summary of mammalian characteristics.

Mammals can learn

One reason mammals are successful is that they guard their young fiercely and teach them survival skills. Mammals can accomplish complex behaviors, such as learning and remembering what they have learned. Have you ever attended an aquarium show or watched a movie about performing dolphins and whales? Dolphins exhibit a wide variety of learned behaviors, including the behaviors performed for films or in aquarium shows.

MiniLab 32-2 Observing

Mammal Skeletons Owls are predators. They feed on small mammals as well as on birds. After eating a meal, the tough indigestible parts of prey, such as bones, are regurgitated as pellets. The skeletons of a variety of small mammals can be studied by examining owl pellets.

Owl pellet

Procedure

1. Place an owl pellet onto a sheet of paper toweling.
2. Use a forceps to remove a very small amount of the outer covering.
3. Prepare a wet mount of this material and observe under low-, then high-power magnification. Diagram what you see. Use forceps to open the pellet and remove all bones that are present. Look especially for skulls. (Note: Skulls may be small and certain parts, such as lower jaws, will be separated.)
4. Identify each mammalian skull using Diagrams A and B as a guide.
5. Attempt to reconstruct the skeleton for an entire animal. You may wish to glue the skeletal pieces onto a piece of cardboard. **CAUTION:** *Wash your hands after handling animal materials.*

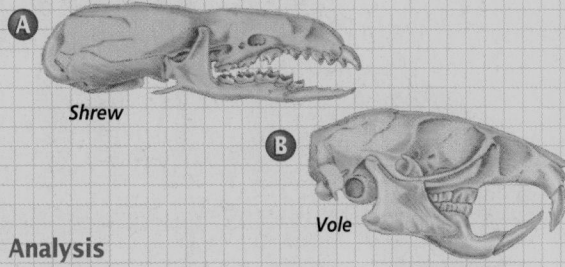

Ⓐ *Shrew*

Ⓑ *Vole*

Analysis

1. What was the outer covering on the pellet? What does this tell you about the contents of the pellet? Explain.
2. How many vole and shrew skulls were present in your pellet?
3. How were you able to differentiate the skulls of these two mammals?
4. Predict if voles and shrews are herbivores or carnivores based on appearance of their teeth. Explain.

MiniLab 32-2

Purpose 🔖
Students examine the disarticulated skeletons of small mammals in an owl pellet.

Process Skills
experiment, draw a conclusion, observe and infer, predict

Teaching Strategies
■ Owl pellets are available from biological supply houses.
■ Typically, there may be several disarticulated skeletons within each pellet. Small bird skeletons may also be present.
■ You may wish to have students tally and correlate the average number of vole and shrew skeletons per pellet.

Expected Results
Hair comprises the outer covering of the pellet. Students will be able to differentiate between vole and shrew skulls.

Analysis
1. Hair; the prey must be a mammal if hair is present.
2. Answers may vary; there may be 1-4 skeletons per pellet.
3. by size and dentition patterns
4. Both are herbivores because they lack canines for capturing or tearing prey.

✔ Assessment
Portfolio Ask students to research the dental formula for adult humans and make a poster. The dental formula indicates the number of incisors, canines, premolars, and molars. Use the Performance Task Assessment List for Poster in **PASC,** p. 73. **L2**

---

when handling live animals.

4. Note the time it takes for the mouse to begin burying behavior and the time it spends burying.
5. Test one object at a time.
6. After filling out the data table, make a bar graph to illustrate the total amount of time the mouse spent burying each object.

Analysis
1. Which objects did your mouse spend the most time burying? The least time? *objects that looked more like predators; least like predators*
2. How do your results compare with those of the class as a whole? *Answers should be similar.*
3. Do your data support your hypothesis? *Data will support hypotheses that stated "predatorlike" objects would cause more burying behavior.*

✔ Assessment
Performance Ask students to design and conduct an experiment that will determine the types of objects that cause more burying behavior. Use the Performance Task Assessment List for Assessing a Whole Experiment and Planning the Next Experiment in **PASC,** p. 33. **L2**

Purpose

Students will review the main characteristics of mammals.

Teaching Strategies

■ Ask groups of students to prepare detailed reports on the life histories of various types of foxes in the United States. **L1**
■ Ask students to speculate about the adaptive advantage of the fox's bushy tail. **L2**

Visual Learning

■ Obtain models that show the teeth of mammals that eat different foods. Have students examine the models. Next, ask them to identify the model that most closely represents the structure of the teeth of a fox.

Critical Thinking

sharp canines and premolars and molars for chewing

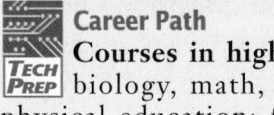

CAREERS IN BIOLOGY

Career Path

Courses in high school: biology, math, English, physical education; for some positions, SCUBA certification
College: bachelor's degree in biological sciences or psychology; business management degree to manage a theme park; public speaking or drama experience

Career Issue

Some people feel that holding ocean-dwelling animals in small enclosures and teaching them "tricks" to perform for the public is wrong. Other people feel such close encounters with people serve as environmental education—that is, the more people who know about these mammals, the more likely it is they will want to protect them in the wild.

INSIDE STORY

A Mammal

A red fox, a member of the dog family, can be found in open country and forests throughout the United States. Red foxes are active at night, and feed on insects, birds, rodents, rabbits, berries, and fruit.

Critical Thinking *Based on its diet, what kinds of teeth would you expect a fox to have?*

Red fox

1 Glands Most mammals have sweat, oil, and scent glands. Sweat glands help mammals cool off. Oil glands lubricate the hair and skin. Foxes use their scent glands to mark new territories.

2 Diaphragm The diaphragm is a muscle that helps the chest cavity expand to take in large amounts of oxygen used to maintain the high metabolism of all mammals.

3 Heart The four-chambered heart of mammals enables them to have the high rate of metabolism necessary for regulation of their body temperature.

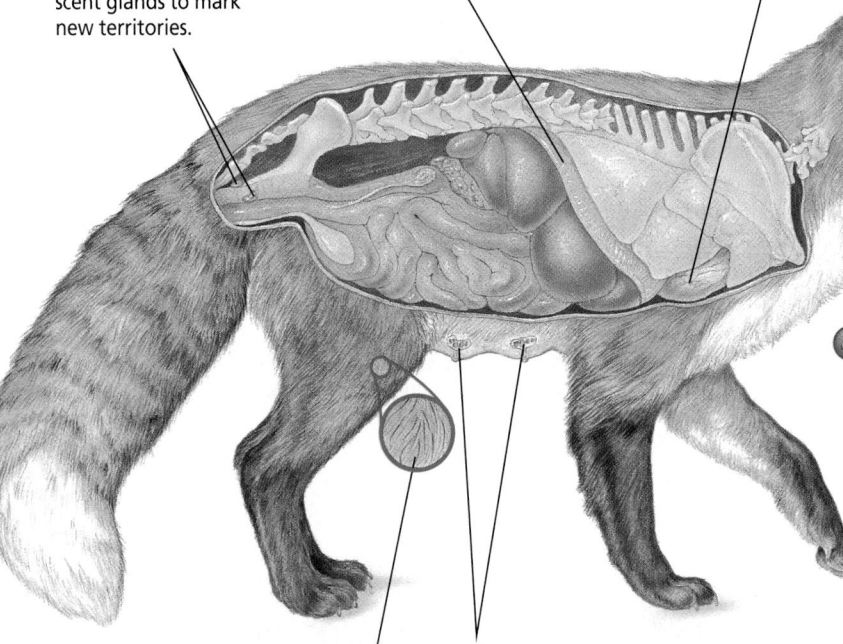

4 Teeth A fox's teeth indicate what it eats and how it gets food. Because a fox preys on anything it can catch, it is found in unlikely environments, such as suburban areas where it can easily catch mice, shrews, and even small birds.

6 Hair Dense, soft underhair insulates the fox by trapping warm air next to its body. The coarse, long guard hairs protect against wear and may be colored for camouflage. The fox sheds its coat little by little during the summer.

5 Mammary glands Like all female mammals, a female fox nourishes her young with milk from her mammary glands.

For More Information

For more information on becoming an animal trainer, students can write to:

IMATA (International Marine Animal Trainers Association)
1720 South Shores Road
San Diego, CA 92109-7995

Resource Manager

Basic Concepts Transparency and 58 Master **L2** **ELL**
Reinforcement and Study Guide, pp. 141-142 **L2**
Content Mastery, p. 158 **L1**
Concept Mapping, p. 32 **L3** **ELL**
BioLab and MiniLab Worksheets, p. 144 **L2**

Primates, including humans, are perhaps the most intelligent animals. Chimpanzees, for example, can use tools, illustrated in *Figure 32.5*, work machines, and use sign language to communicate with humans. Mammalian intelligence is a result of complex nervous systems and highly developed brains. The outer layer of a mammalian brain often is folded, forming ridges and grooves. These ridges and grooves increase by three times the brain's active surface area.

Figure 32.5
A chimpanzee using a stick to get insects out of a tree trunk demonstrates that mammals other than humans are also intelligent enough to make and use tools.

CAREERS IN BIOLOGY

Animal Trainer

If you seem to get along well with animals, consider a career as an animal trainer. A job as a trainer can be hard to find, but fun and rewarding.

Skills for the Job

Some animal trainers work in zoos or aquariums, teaching monkeys, parrots, seals, and other animals to perform specific behaviors. Some trainers work with race horses, and others teach police dogs to sniff out explosives or guide dogs to help people with physical limitations. Many trainers begin as animal keepers. A position at a dog obedience school may require only on-the-job experience, but if you want to train guide dogs, you must complete a three-year course of study. To train dolphins at a large aquarium, you must have a two- or four-year degree in psychology or biology. If you want to narrate the shows, you must also be a good public speaker.

For more careers in related fields, be sure to check the Glencoe Science Web site.
science.glencoe.com

Section Assessment

Understanding Main Ideas
1. Name four characteristics of all mammals.
2. Describe three mammal adaptations for obtaining and consuming food.
3. Describe how endothermy has contributed to the success of mammals.
4. How does intelligence benefit mammals?

Thinking Critically
5. Suppose you are a mammal that feeds on pine seeds and lives in a forest in a cold region. Describe the adaptations that would help you survive.

SKILL REVIEW
6. **Observing and Inferring** On an archaeological dig, you find a skull about 5 cm long with two chisel-shaped front teeth and several flattened back teeth. Is this a skull from a mammal? Explain your answer. For more help, refer to *Thinking Critically* in the **Skill Handbook**.

Check for Understanding
Linguistic Ask students to summarize in two paragraphs why mammals live everywhere. Have them include a discussion of mammals that live below the surface of Earth and in lakes, oceans, and deserts. **L2**

Reteach
Visual-Spatial Bring in a variety of pictures of the limbs of mammals. Ask students working in groups to examine the pictures and tell as much as they can about the life of that mammal based on the structure of its limbs. **L1** **COOP LEARN**

Extension
Dolphins, sea lions, and beluga whales are being trained by the U.S. Navy to watch over Navy submarines while they are in harbor. Have students research the work of the U.S. Navy in training aquatic mammals for various tasks, such as spying and delivering messages. **L3**

✔ Assessment
Portfolio Ask students to prepare an illustrated essay on an ideal mammal pet created by genetic engineering in the year 3000. **L2** **P**

4 Close

Discussion
Have students propose a hypothesis to explain why deer eat more in winter than in summer. **L2**

Section Assessment

1. hair, mammary glands, diaphragm, and a uterus in the female
2. specialized teeth, modified limbs, endothermy
3. Endothermy enables mammals to live in all environments on Earth.
4. They can learn, remember, communicate, and use tools.
5. forefeet with sharp claws that can hold a pinecone; sharp front teeth to get seeds inside; thick fur; short extremities including ears, legs, and nose; a long, bushy tail that can help with balance as the mammal moves in the trees and protects the face from cold as the animal sleeps.
6. Yes. One mammal may have varied types of teeth in its skull.

Prepare

Key Concepts

The three groups of living mammals are introduced and discussed. In addition, reproduction in egg-laying, pouched, and placental mammals is compared.

Planning

- Obtain animal track guide books for the Quick Demo.
- Borrow a collection of mammal skulls for the Focus On.

1 Focus

Bellringer

Before presenting the lesson, display **Section Focus Transparency 78** on the overhead projector and have students answer the accompanying questions.
L1 ELL

Transparency **78** Mammals: Land, Sea, and Air

SECTION FOCUS
Use with Chapter 32, Section 32.2

① Identify the mammals pictured. What adaptations does each have for the environment in which it lives?

② How does each animal obtain food?

BIOLOGY: The Dynamics of Life   SECTION FOCUS TRANSPARENCIES

Resource Manager

Section Focus Transparency 78 and Master **L1 ELL**

Objectives

Distinguish among the three groups of living mammals.

Compare reproduction in egg-laying, pouched, and placental mammals.

Vocabulary

uterus
placental mammal
placenta
gestation
marsupial
monotreme
therapsid

Section

32.2 Diversity of Mammals

All the animals in these photographs are mammals, yet they don't look much alike. What characteristics do bison and antelopes share? You know that both of these mammals have hair, mammary glands, and various kinds of teeth. A major characteristic that separates mammals from all other animals is how they nourish their young. Among mammals there are differences in methods of reproduction. These differences aid scientists in tracing the phylogeny of mammal groups.

Plains of America with bison (above) and antelopes (inset)

Mammal Classification

Living in the United States, you are probably familiar with only one of the three groups, or subclasses, of the class Mammalia. Scientists place mammals into one of three subclasses based on their method of reproduction.

Figure 32.6
The length of gestation varies from species to species in placental mammals. These raccoon kits were born after nine weeks of gestation. Gestation in mice is 21 days, whereas gestation for a rhinoceros is 19 months.

874 MAMMALS

Placental mammals: A great success

The kits shown in *Figure 32.6* were born after a period of development within the uterus of their mother. The **uterus** (YEWT uh rus) is a hollow, muscular organ in which offspring develop. Development inside the mother's body is an adaptation that played a major role in the success of mammals as they spread throughout the world. It ensures that the offspring are protected from predators and the environment during the early stages of growth.

A mammal that carries its young inside the uterus until development is almost complete is known as a **placental mammal.** Nourishment of the young inside the uterus occurs through an organ called the **placenta** (pluh SENT uh), which develops

---

Portfolio

Care of Young

Naturalist Ask students to visit or contact a zoo nursery and compare the care provided for various young zoo animals. Ask students to compare these methods with the care mammals in the wild provide to their young. Students should each prepare an illustrated report for the class. **L2 P**

GLENCOE *TECHNOLOGY*

CD-ROM
Biology: The Dynamics of Life
BioQuest: *Biodiversity Park*
Disc 3, 4

Figure 32.7

In Australia and Tasmania, many marsupials fill niches that are occupied by placental mammals on other continents.

A The giant anteater of Mexico has a long, sticky tongue that it uses to collect ants and termites from their nests.

B The numbat, a marsupial, lives in Australia. It has a long, sticky tongue that it uses to eat termites and ants.

C The spotted cuscus of Australia, a marsupial, lives in trees. It is a solitary, nocturnal animal that eats fruit, leaves, bark, insects, small mammals, reptiles, and birds.

D The ring tailed lemur, a placental mammal, lives in trees on the island of Madagascar. It is active by day, and eats fruits, leaves, and insects.

during pregnancy. The placenta is also instrumental in passing oxygen to and removing wastes from the developing embryo. The time during which placental mammals develop inside the uterus is called **gestation** (jeh STAY shun). About 95 percent of all mammals are placentals. You can learn more about placental mammals in the *Focus On Placental Mammals*.

Pouched mammals: The marsupials

Marsupials make up the second subclass of mammals. A **marsupial** (mar SEW pee uhl) is a mammal in which the young have a short period of development within the mother's body, followed by a second period of development inside a pouch made of skin and hair found on the outside of the mother's body. You may have seen the only North American marsupial, the opossum. Most marsupials are found in Australia and surrounding

islands. The theory of plate tectonics explains why most marsupials are found in Australia today. Scientists have found fossil marsupials on the continents that once made up Gondwana. These fossils support the idea that marsupial mammals originated in South America, moved across Antarctica, and populated Australia before Gondwana broke up.

Ancestors of today's marsupials were able to populate the landmass that became Australia without having to share the area with the competitive placental mammals that evolved in other places. They successfully spread out and filled niches similar to those that placental mammals filled in all other parts of the world, as you can see in *Figure 32.7*. In fact, since humans introduced sheep, rabbits, and other placental mammals to Australia, many of the native marsupial species have become threatened, endangered, or even extinct.

WORD *Origin*

gestation
From the Latin word *gestare*, meaning "to bear." Gestation is time during which a placental mammal develops in a uterus.

---

2 Teach

Quick Demo

Visual-Spatial Provide students with guides to animal tracks. Ask them to pick an animal that lives in your area, trace the track pattern, and later see if they can find these tracks outdoors in wet and/or muddy or snowy areas. Point out that the shape of the track, size, numbers of toes, and claw marks will be significant in their identification of a track. **L2**

TECHPREP

Agricultural Mammals

Ask students to visit a dairy farm or sheep ranch and report on the latest techniques and strategies used in caring for and enhancing production of farm animals. If there are no farms nearby, provide library resources and information from a university cooperative extension unit. **L2**

NATIONAL GEOGRAPHIC

VIDEODISC
GTV: Planetary Manager
Animal, Side 2

BIOLOGY JOURNAL

A Marsupial Adventure

Linguistic Have students make a travel brochure of a photo safari to Australia. They should describe the animals they promise to show tourists and include a map of their proposed travels. **L2**

MEETING INDIVIDUAL NEEDS

Gifted

Logical-Mathematical Ask students to look up survey methods for mammals in a wildlife management-techniques manual. Ask them to make a management plan for a particular mammal in your area that would ensure that their mammal has adequate habitat over the next 25 years. **L3**

Focus On

Placental Mammals

Purpose

Students will examine the main orders of placental mammals.

Background

Once in their own order, the pinnipeds (seals, sea lions, and walruses) have been subsumed by order Carnivora. Scientists continue to refine their classification as more information becomes available. Besides the 12 orders shown here, the placental mammals also include the following orders: Dermoptera, the gliding lemurs; Pholidota, the scaly anteaters (also called pangolins); Tubulidentata, the aardvarks; Hyracoidea, the hyraxes; Scandentia, the elephant shrews; and Macroscelidea, the tree shrews.

Whales are divided into two subgroups based upon how they feed. One group of whales has teeth; this group includes the killer whales and dolphins shown here. These toothed whales feed on fishes, mollusks, and other aquatic mammals. The other subgroup of whales is made up of whales that have baleen rather than teeth. Baleen acts like a giant strainer that captures tiny organisms, such as krill, present in the water. As a baleen whale swims, it gulps water containing small organisms. As the water passes through the baleen, the organisms become trapped, becoming the whale's next meal.

JACKRABBIT

FOCUS ON
Placental Mammals

ORDER LAGOMORPHA

From a standstill, a jackrabbit can leap straight into the air. Rabbits, pikas, and hares belong to the lagomorph order. Most lagomorphs have hind legs suited for leaping. They also have two pairs of chisellike front teeth that grow throughout their lives.

Most of the more than 4300 species of mammals are placental mammals—whose young are nourished by a placenta while they complete development within the mother's uterus. At birth, however, newborn placental mammals vary widely. Newborn gazelles can run fast enough to keep up with the herd within days of birth. Young kittens are blind and helpless. A human baby spends many years dependent on its parents before it can take care of itself. Many mammalogists recognize 18 orders of placental mammals, of which 12 are shown here.

CHIMPANZEE

ORDER PRIMATES

Chimpanzees communicate, walk upright, and make and use tools. The outstanding characteristic of these mammals is their keen intelligence. Most primates have complex social lives. Chimps—like orangutans, gorillas, and gibbons—are apes. Along with apes, the primate order includes lemurs, Old and New World monkeys, and humans.

JAGUAR

ORDER CARNIVORA

Powerful and golden-eyed, the jaguar is the largest cat in the Western Hemisphere. It prowls many habitats in Central and South America. Like all carnivores, a jaguar has long, pointed canines and incisors and strong jaws suited for cutting and tearing flesh. Some carnivores have claws that help them seize their prey. Most of these mammals are meat eaters; some, such as bears, consume plant material as well.

876

BIOLOGY JOURNAL

Endangered Mammals

Linguistic Provide students with a current list of endangered mammal species in the United States. Identify the state(s) in which each mammal is found. Have students write the names of the mammals in their correct state(s) on an outline map. Next, explain that the state in which a particular mammal lives wants to remove the animal from the endangered species list in order to develop the area for housing, malls, and businesses. Ask students to assume they are going to testify as expert mammalogists in favor of preserving the endangered animal's habitat. Have them write their testimony in their journals. L2

AFRICAN CRESTED PORCUPINE

ORDER RODENTIA

Needle-sharp quills protect porcupines from enemies. This African crested porcupine—shown here eating a desert melon—is larger and heavier than its distant relatives in North and South America and has much longer quills. Porcupines, beavers, and chipmunks are rodents, along with rats and mice. Rodents, the largest order of mammals, live in all environments. Rodents have continuously growing, razor-sharp teeth, which they use to gnaw on hard seeds, bark, twigs, and roots.

BOTTLENOSE DOLPHINS

MANATEE

ORDER CETACEA

The bottlenose dolphin, a kind of toothed whale, uses squeaks, growls, whistles, and other sounds to communicate. Animal trainers and scientists report that dolphins are very intelligent and learn quickly. Dolphins, porpoises, and whales—all members of the order Cetacea—have large complex brains. Little or no hair, and the ability to breathe through blowholes on the tops of their heads are other characteristics they share.

ORDER SIRENIA

Slow-moving manatees cruise warm, tropical waters—near the surface of bays, rivers, and coastal areas—and are often injured by speedboats. They can nap underwater for up to 15 minutes, but they must come to the surface to breathe. They have tails and front flippers like whales, distinct heads with a snout that points downward, and short necks. Manatees, nicknamed sea cows, and dugongs belong to the order Sirenia, which includes only four species.

877

Teaching Strategies

■ Ask students to solve the following problem: A little brown bat's diet consists of 20% mosquitoes. It eats 4 grams of food per night. How many mosquitoes does it eat in one night if a mosquito weighs 2.2 milligrams? *354 per night* How many mosquitoes will it eat in one summer during June, July, and August? *33 488 per summer*

■ Borrow a collection of mammal skulls from a local college or museum. Ask students to examine the teeth in each skull. In their groups, have them hypothesize about the type of food each mammal eats.

■ Have students working in groups create a photo essay that shows examples of mammals classified in each order discussed in this feature. Each group should focus on a different mammal order. Have students combine their materials into a class bulletin board display. **L2** **ELL**

Visual Learning

Ask students to look at the jackrabbit and the porcupine on these two pages. Remind them that rodents include mice, rats, gerbils, squirrels, chipmunks, and so on. Ask students why they think rabbits are not included with the rodents in one order. *There are differences in bone structure and teeth.*

✔ Portfolio

Lemur Classification

 Naturalist The Dermoptera, or gliding lemurs, at times have been placed into several different orders. Have students research gliding lemurs and find out and report on why they are not classified along with other lemurs into the order Primates. **L3** **P** 📋

Background

Armadillos are the only members of the order Edentata that live in the United States. Although the armadillo has hair, its most distinguishing feature is its hardened skin, which is arranged in a pattern of either six or nine bands around the armadillo's body. This tough, banded skin acts like a suit of armor to protect the armadillo from predators. When frightened, the animal rolls itself up into a tight ball, with its exterior armor protecting its head, limbs, and soft underbelly.

ORDER INSECTIVORA

HAIRY-TAILED MOLE

The hairy-tailed mole of North America seldom comes out of its tunnel. Designed for digging, the mole's front feet are powerful earth movers. Its eyes are nearly covered by thick, soft fur, and its eyesight is poor. Underground, moles use their senses of smell and touch to find food. Although they all eat insects and most have pointed snouts and sharp claws for digging, insectivores are a mix of mammals with few other shared characteristics.

ORDER PERISSODACTYLA

The wild Przewalski (Pruz WOL skee) horse from Mongolia looks similar to the ancestor of all modern horses. These hairy horses have thick legs and sturdy bodies. Zoos throughout the world have breeding programs to save this species. Several other species of mammals had toenails that evolved into hooves. Most hoofed mammals are herbivores with molars for grinding. Hooved mammals with an odd number of toes belong to the order Perissodactyla.

PRZEWALSKI HORSES

ARMADILLO

ORDER EDENTATA

Don't let its armor fool you—this is a mammal. Plates of skin-covered bone protect most of the armadillo's body. But like anteaters and sloths—the other members of this order—armadillos are well equipped for digging. Edentates are found in Central and South America and in southern regions of North America.

878

PROJECT

Lead and Mammal Development

Visual-Spatial Mammalian development is adversely affected by high levels of lead. Mammal bones and teeth absorb lead because it is similar to calcium. When calcium is replaced by lead, toxic effects result. Have students use the Internet to research a study of California sea otters in which lead was found in their teeth. Have students create a poster that shows the pathway that lead takes from the environment into the bodies of the otters. Have them infer from the results of this otter study how lead could affect their own health. **L3**

HIPPO

ORDER ARTIODACTYLA

Found along African rivers, the hippo is one of the largest land mammals. Hippos eat enormous amounts of grass and water vegetation. They spend warm days in the water so their skins don't dry out or get sunburned. An artiodactyl is a hoofed animal with an even number of toes on each foot. There are about 200 species of artiodactyls.

SHORT-TAILED LEAFNOSED FRUIT BAT

ORDER CHIROPTERA

The short-tailed, leafnosed fruit bat—the nocturnal equivalent of the hummingbird—feeds on fruit, pollen, and nectar. Many plants depend on bats for pollination and would become extinct without them. Fruit bats usually use visual navigation rather than echolocation, a technique involving high-frequency sounds and their echoes. Most insect-eating bats, however, use echolocation to navigate. In flight, these bats emit short, high-pitched cries. When the sounds hit an object, echoes bounce back to the bat, allowing it to locate the object. A bat has skin that stretches from body, legs, and tail to arms and fingers to form thin, membranous wings. Chiropterans are the only mammals that fly.

ELEPHANTS

ORDER PROBOSCIDEA

Proboscideans use their flexible trunks mostly to gather plants for eating and to suck in water for drinking. One pair of incisors is modified into large tusks for digging up roots and stripping bark from trees. The largest land animals, elephants spend most of their time eating. African elephants are distinguished from Asian elephants by their larger ears.

EXPANDING Your View

1 **UNDERSTANDING CONCEPTS** Explain, by using examples, how mammals have become so successful.

2 **WRITING ABOUT BIOLOGY** Go to a toy store and inventory the kinds of mammals in the stuffed animal section. Write a report detailing what the data might reflect about the view of society towards mammals.

Answers to Expanding Your View

1. Mammals such as rabbits and jaguars have young that complete development internally and then the mother provides nourishment and protection. Mammals such as humans and chimpanzees are intelligent and have communication abilities. Responses should also indicate that mammals have adapted to a large variety of environments. For example, some carnivores, such as seals, have evolved modified flippers that permit life in water as well as on land. Other carnivores have developed teeth that are well-adapted to the foods they eat. Mammals, like insects, have become so successful because they can survive in almost every environment on Earth. They have varied teeth, modified limbs, modified digestive systems, and intelligence that enables them to respond to changes in their environment.

2. Student answers will vary, but most will find that more stuffed animals will be furry bears and other furry mammals than almost any other kind of animal.

BIOLOGY Online Note Internet addresses that you find useful in the space below for quick reference.

Figure 32.8
Present-day monotremes include one species of platypus and two species of echidnas. This echidna species is found only in Australia (a). The duck-billed platypus has several physical features that seem to belong to a variety of other animals (b).

Monotremes: The egg layers

Do you think the animal shown in *Figure 32.8b* is a mammal? It has hair and mammary glands, yet it lays eggs. The duck-billed platypus is a monotreme. A **monotreme** (MAHN uh treem) is a mammal that reproduces by laying eggs. Monotremes are found only in Australia, Tasmania, and New Guinea. Spiny anteaters, also called echidnas, belong to this subclass as well. One of the two species of spiny anteaters can be found only in New Guinea. Only three species of monotremes are alive today.

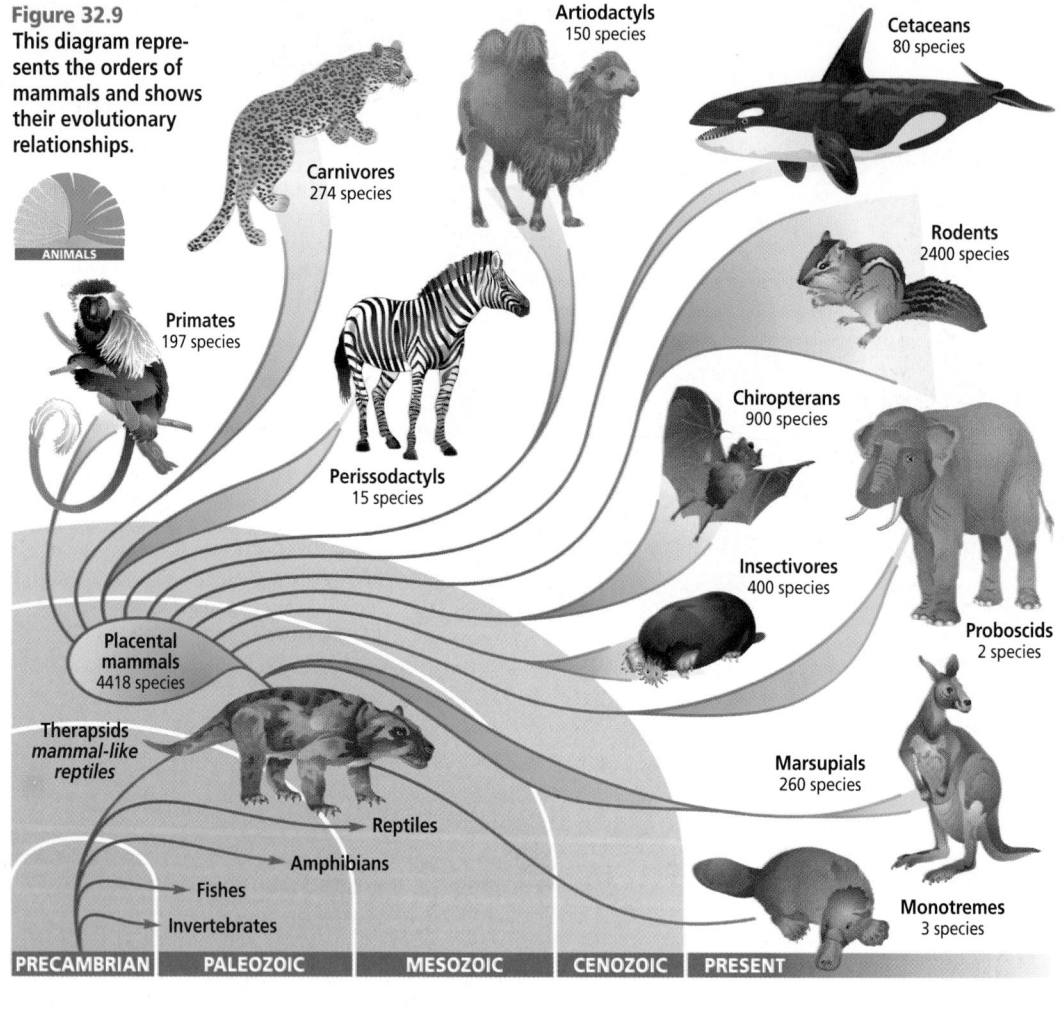

Figure 32.9
This diagram represents the orders of mammals and shows their evolutionary relationships.

Artiodactyls
150 species

Cetaceans
80 species

Carnivores
274 species

Rodents
2400 species

Primates
197 species

Chiropterans
900 species

Perissodactyls
15 species

Insectivores
400 species

Proboscids
2 species

Placental mammals
4418 species

Therapsids
mammal-like reptiles

Marsupials
260 species

Reptiles

Amphibians

Fishes

Invertebrates

Monotremes
3 species

PRECAMBRIAN | PALEOZOIC | MESOZOIC | CENOZOIC | PRESENT

880 MAMMALS

The platypus, a mostly aquatic animal, has a broad, flat tail, much like that of a beaver. Its rubbery snout resembles the bill of a duck. The platypus has webbed front feet for swimming through water, but it also has sharp claws on its front and hind feet for digging and burrowing into the soil. Much of its body is covered with thick, brown fur.

The spiny anteater has coarse, brown hair, and its back and sides are covered with sharp spines that it can erect for defensive purposes when threatened by enemies. From its mouth, the anteater extends its long, sticky tongue to catch insects.

Origins of Mammals

Present-day relationships of mammal orders are shown in *Figure 32.9*. The first true mammals appeared in the fossil record about 200 million years ago. Scientists can trace the origins of mammals from an insect-eating animal to a group of reptilian ancestors called therapsids. **Therapsids** (ther AP sidz), represented in *Figure 32.10*, were fierce-looking, heavy-set animals that had features of both reptiles and mammals. They existed between 270 and 180 million years ago.

Figure 32.10
Therapsids were the ancestors of mammals. The lower jaw and middle ear bones of therapsids were like those of reptiles. However, they had straighter legs than reptiles and held them closer to the body.

The mass extinction of the dinosaurs at the end of the Mesozoic era, along with the breaking apart of Pangaea and changes in climate, opened up new niches for early mammals to fill. The appearance of flowering plants at the end of this era supplied new living areas, food sources, and shelter. Some mammals that moved into the drier grasslands became fast-running grazers, browsers, and predators. The Cenozoic era is sometimes called the golden age of mammals because of the dramatic increase in their numbers and diversity.

Section Assessment

Understanding Main Ideas
1. Describe the characteristics of placental mammals.
2. Compare monotremes and marsupials.
3. Monotremes reproduce by laying eggs. Why are they classified as mammals?
4. What are therapsids and what is their relationship to mammals?

Thinking Critically
5. There are several marsupial species in South America, but only one species is native to North America. Make a hypothesis about the presence or absence of marsupial species in Europe. How could you test your hypothesis?

SKILL REVIEW
6. **Observing and Inferring** You find a mammal fossil and observe the following traits: hooves, flattened teeth, skeleton the size of a large dog. What can you infer about its way of life? For more help, refer to *Thinking Critically* in the **Skill Handbook**.

Check for Understanding
Ask students to list the major orders of placental mammals and explain how each is adapted to its niche. L2

Reteach
Kinesthetic Have students draw a bingo card with 16 squares and the name of a mammal written in each square. Call out different mammal characteristics such as "lays eggs." As you call out each characteristic, have students cover the squares of the mammals that have that trait. The first person to cover four squares in a row wins. L1 ELL

Extension
Ask students to investigate and report on two mammals in the same order that are adapted to very different habitats, such as the jackrabbit and the Arctic hare. Have students report on the adaptations each animal has for survival in its habitat. L2

✔ **Assessment**
Portfolio Ask students to write a paragraph speculating why there are fewer big mammals than small mammals. L2 P

4 Close

Discussion
Ask students to decide which mammal they would most like to have as a pet. Have them explain their choices.

Section Assessment

1. A placental mammal carries its young inside the uterus until development is nearly complete.
2. Monotremes lay eggs. Marsupials give birth to immature young that continue development in the mother's external pouch.
3. They have other mammalian characteristics, such as hair and mammary glands.
4. Therapsids are the reptilian ancestors of mammals.
5. Marsupials developed in South America and Australia, continents in Gondwana; Europe was part of Laurasia. Students should infer that this suggests that there are no marsupials in Europe.
6. It is an herbivore that can run fast. It may have multiple stomachs and chew cud.

INTERNET

BioLab

Domestic Dogs Wanted

Time Allotment

One class period

Process Skills

make and use tables, observe and infer, make hypotheses, make predictions, observe and record data, use the Internet

PROCEDURE

Teaching Strategies

■ Remind students that the research they conduct is only an estimate and there will be some variations in their findings.
■ Be sure to carefully monitor students' use of the Internet. Follow the Acceptable Use Policy of your school.
■ If possible, have students research the animals in a shelter in their area.
■ When students come across a mutt or a mixed breed on an Internet site, it may list the dominant breed only. Have them make observations based on the dominant breed.

Resource Manager

BioLab and MiniLab Worksheets, pp. 145-146 **L2**

*D*ogs make great companions. They provide their owners with an opportunity to love and nurture another living thing. In return they are loyal, offer protection to their owners, and are fun to have around. Dogs come in all different shapes, sizes, and colors. However, all these different breeds have the same basic dog characteristics. Unfortunately, sometimes dogs are abandoned or run away from home. Many of these unwanted and stray dogs end up in animal shelters where they can be adopted by new owners.

PREPARATION

Problem

What are the most popular breeds of dog?

Objectives

In this BioLab, you will:
■ **Observe** the characteristics of dogs.
■ **Record** the popularity of different dog breeds.
■ **Use the Internet** to collect and compare data from other students.

■ **Compare** the most popular dog breeds with the most common breeds found in animal shelters.

Materials

computer access to the Internet

Skill Handbook

Use the **Skill Handbook** if you need additional help with this lab.

PROCEDURE

1. Make a copy of both data tables.
2. Visit the Glencoe Science Web Site at the address shown below to find Internet links to sites that rank the most popular breeds of dog. Also find links to animal shelters and pet adoption agencies.
3. From your data, determine the five most popular dog breeds. Record your findings in Data Table I.

4. Find pictures of these dogs and record the physical characteristics unique to their breeds in Data Table I.
5. At the Glencoe Science Web Site find links to five web sites of animal shelters and pet adoption agencies across the country. Many of these sites post pictures and other information about the dogs that are up for adoption.
6. From your data, determine which breeds are most commonly found

882

Data and Observations

Make sure students are sampling data from different animal shelters around the county. Many shelters will list only the breed of dog and possibly a picture. Ask students to infer the different dog characteristics according to the breed.

Data Table I

Most Popular Dog Breeds		
Rank	**Breed**	**General Characteristics**

Data Table II

Most Common Dog Breeds Found in Animal Shelters		
Rank	**Breed**	**General Characteristics**

A domestic dog, *Canus familiaris*

in animal shelters. Record the five most common breeds in Data Table II.

7. Find pictures of these dog breeds and record their unique characteristics in Data Table II.

ANALYZE AND CONCLUDE

1. **Using the Internet** Compare the data in Tables I and II. Are there similarities? How might you explain any similarities or differences? Consider the time of year, the age of the dog, and any breeds of dog that may be popularized by television or other advertising while formulating your answer.

2. **Thinking Critically** Are there any breeds that are popular as pets that are commonly found in animal shelters? Propose an explanation for this.

3. **Comparing and Contrasting** What characteristics do domestic dogs have that are the same as all other carnivores? As all other mammals?

4. **Problem Solving** Propose a local program to reduce the number of stray and abandoned dogs.

Sharing Your Data

BIOLOGY *Online* Find this BioLab on the Glencoe Science Web site at **science.glencoe.com**. Post your data in the data tables provided for this activity. Briefly describe your plan to reduce the numbers of stray animals in your community. Compare plans with those posted by other students.

ANALYZE AND CONCLUDE

1. Student answers will vary depending on the information they find on the Internet.

2. Some large dogs, such as German shepherds and golden retrievers, are often bought as puppies but abandoned or given up when they grow too large.

3. carnivores: canine teeth, claws, strong jaws; mammals: hair, warm-blooded, teeth

4. Many programs of this sort can be run through a local humane society or "no-kill" animal shelter. Often these programs involve educating the public on spaying or neutering their animals to reduce unwanted litters.

✓ Assessment

Knowledge Certain breeds of dogs are made popular in movies and television. Some recent examples include Dalmatians, Chihuahuas, and Jack Russell terriers. Ask students what effect this has on the number of stray and abandoned dogs. *Students should infer that people often adopt or purchase dogs because they have been featured on TV shows or in movies. Unfortunately, a featured dog breed may not be the correct breed for a family pet. As a result, many of these popular breeds are abandoned when the animals become adults.* Use the Performance Task Assessment List for Analyzing the Data in **PASC**, p. 27.

Sharing Your Data

BIOLOGY *Online* To navigate to the Internet BioLabs, choose the *Biology: The Dynamics of Life* icon at the Glencoe Science Web site. Click on the student icon, then the BioLabs icon. Ask students to compare their results with those of students around the county. Do different cities have different results? Do different regions of the country have different results? Do animal shelters have the same general programs around the country, or do they differ? Explain.

Purpose

Students will explore problems faced by modern zoos and analyze what critics say about zoos.

Teaching Strategies

■ Ask students to recount why they think zoos should or should not be preserved. Have them consider a zoo's educational value and the opportunity it affords for the community.

■ Explain that most zoos are financially unable to reform their programs to meet their new mission as wildlife conservationists. Ask what kind of fund-raising efforts are appropriate for zoos. *They might solicit donations from people who visit the zoo and from people in the community. They might also raise money through the sale of books, magazines, or objects created to increase awareness about animals and about zoo functions.*

Investigating the Issue

One side should defend the idea that the zoo is a wildlife preserve helping to preserve endangered species by captive breeding. The other side should defend the idea that captive species promote species favoritism and that the money spent on zoos should be used to preserve natural wildlife habitats.

GLENCOE TECHNOLOGY

VIDEODISC
The Infinite Voyage:
The Keepers of Eden
The Cheetah: Using DNA Profiles to Research Production (Ch. 5), 9 min.

Biology & SOCIETY

Do we need zoos?

Faced with rising costs and criticism, modern zoos are struggling to survive financially. Many metropolitan zoos are questioning their existence and redefining their missions. The most dedicated zoos have started natural habitat programs and are experimenting with captive breeding.

Originally created for public entertainment, zoos began educational programs and species conservation in the 1940s. Zoo reform began in the 1980s in response to public outrage over poor conditions.

Modern zoos The modern zoo is a place for scientists and citizens to study zoology, paleontology, and animal behavior. Zoos also provide a place where interested people can hear lectures and take classes on topics as varied as classification, biodiversity, and habitat conservation. Today's most modern zoos—such as the San Diego and Cincinnati Zoos—are using the latest reproductive techniques, including artificial insemination and test-tube fertilization, to help save endangered species and improve the gene pools of others. Animals such as the California condor, black-footed ferret, and the red wolf probably could not survive without captive breeding. To minimize inbreeding, zoos have a computerized mating sys-

Captive breeding in a zoo and a black-footed ferret (inset)

tem that keeps track of the genetic background of their captive animals. Zoos borrow suitable mates to pass on genes in "breeding loans."

Reduced need to capture wild animals
Through captive breeding programs, zoos are able to lessen their reliance on wild populations for their exhibits. The more sophisticated breeding laboratories are able to freeze and save genetic materials from endangered species for future breeding purposes.

Different Viewpoints
Captive breeding of endangered species leads to other problems. The very success of captive breeding results in the production of surplus animals. Surplus animals often have no place to go. They cannot be given to other zoos due to limited budgets for the care and feeding of these species. They also cannot be returned to the wild in most cases because there is no longer natural habitat available for many of these species and because most captive-raised animals do not have the necessary survival skills.

Those against zoo policy say that biodiversity is more likely to be preserved through public education than through captive breeding programs. They accuse zoo managers of species favoritism, saying that because the managers are unable to save all endangered species, they tend to select only the most attractive or popular species.

INVESTIGATING THE ISSUE

Debating the Issue What should be the role of modern zoos?

 BIOLOGY *Online* To find out more about modern zoos, visit the Glencoe Science Web site. **science.glencoe.com**

884

Going Further

If you have a local zoo, ask students to find out more about their zoo's captive-breeding programs or other ongoing research. **L2**

GLENCOE TECHNOLOGY

VIDEODISC **The Infinite Voyage**
The Keepers of Eden: The Transformation of Zoos (Ch. 3), 3 min.

Cincinnati Zoo's Cat House: Innovative Breeding Techniques (Ch. 6), 6 min.

SUMMARY

Section 32.1

Mammal Characteristics

Main Ideas
- The first mammal-like animals were therapsids.
- A mammal is an endotherm with hair, a diaphragm, modified limbs and teeth, a highly developed nervous system and senses, and mammary glands.

Vocabulary
diaphragm (p. 869)
gland (p. 868)
mammary gland (p. 868)

Section 32.2

Diversity of Mammals

Main Ideas
- Mammals are classified into three subclasses—monotremes, marsupials, and placentals—based on the way they reproduce.
- Placental mammals carry young inside the uterus until development is nearly complete. The young are nourished through an organ called the placenta.
- Monotremes are egg-laying mammals found only in Australia, Tasmania, and New Guinea.
- Marsupials carry partially developed young in pouches on the outside of the mother's body.

Vocabulary
gestation (p. 875)
marsupial (p. 875)
monotreme (p. 880)
placenta (p. 874)
placental mammal (p. 874)
therapsid (p. 881)
uterus (p. 874)

Main Ideas

Summary statements can be used by students to review the major concepts of the chapter.

Using the Vocabulary

To reinforce chapter vocabulary, use the Content Mastery Booklet and the activities in the Interactive Tutor for Biology: The Dynamics of Life on the Glencoe Science Web site.
science.glencoe.com

 *All Chapter Assessment questions and answers have been validated for accuracy and suitability by The Princeton Review.*

UNDERSTANDING MAIN IDEAS

1. b
2. a
3. d
4. c
5. a
6. a

UNDERSTANDING MAIN IDEAS

1. Which of the following is NOT a characteristic of mammals?
   a. endothermic
   b. three-chambered heart
   c. hair
   d. mammary glands
2. Which of these is NOT an endothermic animal?
   a. rattlesnake   c. cat
   b. penguin   d. gorilla
3. Hair helps mammals by providing camouflage and helping them to maintain _____.
   a. evolution   c. reproduction
   b. running speed   d. body temperature

4. _____ are examples of egg-laying mammals.
   a. Numbats and lemurs
   b. Anteaters and shrews
   c. Platypuses and spiny anteaters
   d. Seals and whales
5. Which pair of terms is most closely related?
   a. gland—secretion
   b. diaphragm—heart
   c. placenta—Golgi bodies
   d. gestation—molars
6. Which pair of terms is most closely related?
   a. cud chewing—cellulose
   b. canines—diaphragm
   c. incisors—sipping nectar
   d. monotreme—placenta

GLENCOE TECHNOLOGY

 **VIDEOTAPE**
MindJogger Videoquizzes
Chapter 32: *Mammals*
Have students work in groups as they play the videoquiz game to review key chapter concepts.

Resource Manager

Chapter Assessment, pp. 187-196
MindJogger Videoquizzes
ExamView® Pro Software 💾
BDOL Interactive CD-ROM, Chapter 31 quiz

7. a
8. b
9. b
10. c
11. lays eggs
12. diaphragm
13. placenta
14. niches
15. uterus
16. Marsupials; pouch
17. internally (in the uterus)
18. glands
19. teeth
20. intelligence

APPLYING MAIN IDEAS

21. Developing inside the uterus, the young of mammals are protected from environmental conditions, whereas the eggs of reptiles, developing externally, are subject to environmental conditions.

22. They do this to cool themselves. The evaporating water cools the skin and the mud protects the skin, which has little hair, from the sun.

23. Lower jaws, middle ear bones, and teeth of mammals are different from those of reptiles. Mammals also have straighter legs that are placed underneath the body.

24. Animals that are endotherms can maintain their body temperature at a constant level no matter what the environmental temperature.

25. First, some early fishes probably evolved with appendages suited to propelling themselves in shallow water and muddy wetlands. These fishes probably also had primitive lungs. Amphibians have lungs that enable them to breathe air, and legs that enable them to move on land, but they are still restricted to life in the water because their eggs must be laid in water. Reptiles do not need water because they have internal fertilization and their eggs

7. Which type of teeth pictured below would be most suited for feeding on grasses?

a. 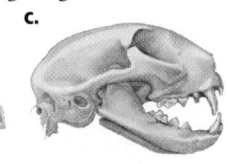   c.

b. 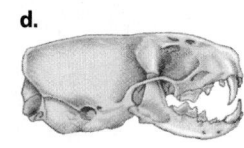   d.

8. Like bird feathers, mammalian hair probably evolved from _____.
   a. teeth   c. claws
   b. scales   d. setae

9. _____ is a behavioral adaptation for cooling off a mammal's body.
   a. Running   c. Jogging
   b. Panting   d. Gnawing

10. Scientists hypothesize that mammals descended from a group of reptiles called _____.
    a. theropods   c. therapsids
    b. cotylosaurs d. dinosaurs

11. A duck-billed platypus is classified as a monotreme because it _____.

12. Mammals are able to maintain high energy levels because they breathe using a(n) _____.

13. Dogs and cats nourish their unborn young through the _____.

THE PRINCETON REVIEW   **TEST-TAKING TIP**

If It Looks Too Good to Be True. . .
Beware of answer choices that seem obvious. Remember that only one answer choice of the several that you're offered for each question is correct. The others are made up by the test-makers to challenge you. Check each answer choice carefully before finally selecting it.

14. Mammals increased in numbers and varieties during the Cenozoic era because new _____ opened.

15. Protection of developing young occurs in many mammals because the females have an internal _____ in which the young develop.

16. _____ are able to give birth early in development because the young continue to develop in a(n) _____.

17. Gestation is different in mammals, reptiles, and birds because in mammals it occurs _____.

18. Mammals use fluids from their _____ to mark territory.

19. The kind of food a mammal eats can be determined by examining its _____.

20. What characteristic of mammals is illustrated in the photo to the right?

APPLYING MAIN IDEAS

21. How does development of young inside a uterus enable mammals to adapt to environments that reptiles cannot?

22. Most African elephants live in hot grasslands of Africa. They can often be seen using their trunks to slap watery globs of mud onto their heads and backs. Make a hypothesis that explains this behavior.

23. You find the skeleton of an animal. What features would indicate that it is a mammal rather than a reptile?

24. How does endothermy enable mammals and birds to survive in the Arctic and in the Sahara Desert of Africa?

25. Trace the main paths of evolution from fishes to mammals to explain how mammals may have become adapted to life on land.

are encased in leathery shells. Legs under the body make locomotion on land more efficient. The first four-chambered hearts, enabling more oxygen to reach cells to produce more energy and quicker movements, appeared in reptiles. Mammals are endotherms and so can maintain body temperature in a wide variety of environments, especially when protected from heat loss by fur or hair. Young develop inside the body,

thereby giving them protection against predation and harsh environmental conditions.

THINKING CRITICALLY

26. Interpreting Scientific Illustrations Examine the teeth in the diagram of a skull below and infer whether the mammal was an herbivore, carnivore, or omnivore.

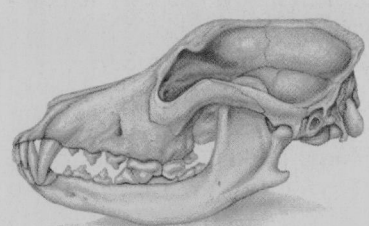

27. Comparing and Contrasting Mammals and insects are both considered to be extraordinarily successful animals. Explain the criteria used in both cases that give them this distinction.

28. Seal, walruses, and sea lions once were classified in their own order, Pinnepedia. Scientists now place them with dogs, cats, and bears in order Carnivora. What features of the pinnepeds caused scientists to classify them with the carnivores?

29. Concept Mapping Complete the concept map by using the following vocabulary terms: marsupials, monotremes, placental mammals, placenta.

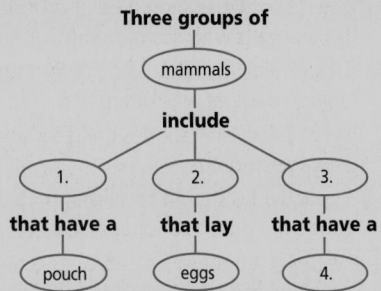

CD-ROM

For additional review, use the assessment options for this chapter found on the *Biology: The Dynamics of Life Interactive CD-ROM* and on the Glencoe Science Web site. **science.glencoe.com**

ASSESSING KNOWLEDGE & SKILLS

Interpreting Scientific Illustrations
Examine *Figure 32.9* in the text to answer the following questions.

1. Egg-laying mammals, placental mammals, and marsupials developed from _____.
- **a.** therapsids
- **b.** insectivores
- **c.** edentates
- **d.** proboscids

2. During which geological time period did placental mammals evolve?
- **a.** Precambrian
- **b.** Paleozoic
- **c.** Mesozoic
- **d.** Cenozoic

3. To which group are perissodactyls more closely related?
- **a.** chiropterans
- **b.** edentates
- **c.** artiodactyls
- **d.** lagomorphs

4. Which group of animals was not common in the Paleozoic era?
- **a.** fishes
- **b.** amphibians
- **c.** invertebrates
- **d.** mammals

5. Which group of placental mammals had more diversity during an earlier time?
- **a.** primates
- **b.** edentates
- **c.** perissodactyls
- **d.** sirenians

6. Making a Graph Using the data in *Table 32.1,* make a graph that shows the numbers of species in each mammal order.

Table 32.1 Number of mammal species	
Order	**Number of species**
Insectivora	428
Chiroptera	925
Primates	233
Edentata	29
Lagomorpha	80
Rodentia	2021
Cetacea	78

THINKING CRITICALLY

26. carnivore
27. They both have enormous diversity of species and live in all types of environments.
28. Seals, walruses, and sea lions are predators that eat meat and have sharp canine teeth used to pierce and stab their prey.
29. 1. Marsupials; 2. Monotremes; 3. Placental mammals; 4. Placenta

ASSESSING KNOWLEDGE & SKILLS

1. a
2. c
3. c
4. d
5. c
6. Make sure students' graphs reflect the data and relationships shown in the table.

Note Internet addresses that you find useful in the space below for quick reference.

Chapter 33 Organizer

Refer to pages 4T-5T of the Teacher Guide for an explanation of the National Science Education Standards correlations.

Section	Objectives	Activities/Features
Section 33.1 **Innate Behavior** National Science Education Standards UCP.2-4; A.1, A.2; C.3, C.6; F.4; G.1, G.2 (1 session)	1. **Distinguish** among the types of innate behavior. 2. **Demonstrate,** by example, the adaptive value of innate behavior.	**MiniLab 33-1:** Testing an Isopod's Response to Light, p. 890 **Problem-Solving Lab 33-1,** p. 897 **Investigate BioLab:** Behavior of a Snail, p. 904 **BioTechnology:** Tracking Sea Turtles, p. 906
Section 33.2 **Learned Behavior** National Science Education Standards UCP.2, UCP.3; A.1, A.2; C.6; E.1, E.2; F.4, F.6; G.1-3 (2 sessions)	3. **Distinguish** among types of learned behavior. 4. **Demonstrate,** by example, types of learned behavior.	**MiniLab 33-2:** Solving a Puzzle, p. 900 **Problem-Solving Lab 33-2,** p. 902

Need Materials? Contact Carolina Biological Supply Company at 1-800-334-5551 or at **http://www.carolina.com**

MATERIALS LIST

BioLab

p. 904 snails, dropper, spring water, plastic petri dish, scissors, stereomicroscope, pencil, rubber band, masking tape

MiniLabs

p. 890 isopods (5), plastic petri dish, black paper, paper towel, water, transparent tape, paper, pencil
p. 900 paper puzzle, clock with second hand, paper, pencil

Alternative Lab

p. 894 terrarium with cover, sand, crickets (1 female, 5 male), dry oatmeal, apple slices, cellulose sponge, matchbox (4), jar with lid (5), nail polish (4 colors)

Quick Demos

p. 891 assorted animals
p. 891 plexiglass sheet, paper
p. 901 none

Key to Teaching Strategies

L1 Level 1 activities should be appropriate for students with learning difficulties.

L2 Level 2 activities should be within the ability range of all students.

L3 Level 3 activities are designed for above-average students.

ELL ELL activities should be within the ability range of English Language Learners.

COOP LEARN Cooperative Learning activities are designed for small group work.

P These strategies represent student products that can be placed into a best-work portfolio.

These strategies are useful in a block scheduling format.

Teacher Classroom Resources

Section	Reproducible Masters	Transparencies
Section 33.1 **Innate Behavior**	Reinforcement and Study Guide, pp. 145-146 `L2` Concept Mapping, p. 33 `L3` `ELL` Critical Thinking/Problem Solving, p. 33 `L3` BioLab and MiniLab Worksheets, p. 147 `L2` Laboratory Manual, pp. 239-242 `L2` Content Mastery, pp. 161-162, 164 `L1`	Section Focus Transparency 79 `L1` `ELL`
Section 33.2 **Learned Behavior**	Reinforcement and Study Guide, pp. 147-148 `L2` Critical Thinking/Problem Solving, p. 33 `L3` BioLab and MiniLab Worksheets, pp. 148-150 `L2` Laboratory Manual, pp. 243-246 `L2` Content Mastery, pp. 161, 163-164 `L1`	Section Focus Transparency 80 `L1` `ELL` Basic Concepts Transparency 60 `L2` `ELL` Reteaching Skills Transparency 48 `L1` `ELL`

Assessment Resources

Chapter Assessment, pp. 193-198
MindJogger Videoquizzes
Performance Assessment in the Biology Classroom
Alternate Assessment in the Science Classroom
ExamView® Pro Software 💾
BDOL Interactive CD-ROM, Chapter 33 quiz

Additional Resources

Spanish Resources `ELL`
English/Spanish Audiocassettes `ELL`
Cooperative Learning in the Science Classroom `COOP LEARN`
Lesson Plans/Block Scheduling

■ NATIONAL GEOGRAPHIC — Teacher's Corner

Products Available From Glencoe
To order the following products, call Glencoe at 1-800-334-7344:
CD-ROM
Mammals: A Multimedia Encyclopedia
Videodisc
STV: Animals

Products Available From National Geographic Society
To order the following products, call National Geographic Society at 1-800-368-2728:
Book
National Geographic Book of Mammals

Videos
Predators of North America
Strange Creatures of the Night

Index to National Geographic Magazine
The following articles may be used for research relating to this chapter:
"Animals at Play," by Stuart L. Brown, December 1994.
"Secrets of Animal Navigation," by Michael E. Long, June 1991.

GLENCOE TECHNOLOGY

The following multimedia resources are available from Glencoe.

Biology: The Dynamics of Life
CD-ROM `ELL`

Video: *Bird Courtship*
Video: *Territorial Behavior*
Video: *Salmon Migration*
Exploration: *Learned Behavior*
Video: *Elephant Behavior*

Videodisc Program

Bird Courtship
Territorial Behavior
Salmon Migration

Theme Development

Students will examine the theme of **unity within diversity** as they consider the kinds of behaviors animals have in common and behaviors that are unique to a species. The theme of **evolution** is important to the study of behavior because of the adaptive value of behavior and the fact that behavior, just like physical features of animals, evolves.

READING BIOLOGY

Glencoe's **Biology: The Dynamics of Life** contains many resources to assist a student's reading skills. Each chapter contains figures with expanded captions that expand on written material. Word Origins, located along the side of text, expand knowledge of biology vocabulary. Glencoe's Content Mastery Booklet helps develop reading skills while reinforcing content. In addition, use the Interactive Tutor for **Biology: The Dynamics of Life** on the Glencoe Web site to reinforce vocabulary.
science.glencoe.com

33 Animal Behavior

What You'll Learn

- You will distinguish between innate and learned behavior.
- You will identify the adaptive value of specific types of behavior.

Why It's Important

Animals have patterns of behavior that help them survive and reproduce. Some of these behavior patterns are inherited and some are learned. You will recognize that humans, like other animals, have both types of behavior, and that these behavior patterns enable you to survive as well.

READING BIOLOGY

Read through the new vocabulary in the section previews. Try to figure out the word meanings by determining possible word origins. Compare the possible meanings with the definitions given in the text.

To find out more about animal behavior, visit the Glencoe Science Web site.
science.glencoe.com

Prairie dogs bark to warn others of approaching predators. Vultures soar high above the prairie. Some animal behavior is inherited and is performed correctly right away. Some behavior is learned through a lifetime of practice.

888 ANIMAL BEHAVIOR

33.1 Innate Behavior

SECTION PREVIEW

Objectives
Distinguish among the types of innate behavior.
Demonstrate, by example, the adaptive value of innate behavior.

Vocabulary
behavior
innate behavior
reflex
fight-or-flight response
instinct
courtship behavior
territory
aggressive behavior
dominance hierarchy
circadian rhythm
migration
hibernation
estivation

Have you ever watched a bird feed its young? Nestlings greet a parent returning to the nest with cries and open beaks. Parent birds practically stuff the food down their offsprings' throats, then fly off to find more food. Why do baby birds open their beaks wide? Why do parent birds respond to open beaks by feeding their offspring? These actions are examples of behavior that appears in birds without being taught or learned. Animals exhibit many kinds of behavior in nature, both inherited and learned.

Cedar waxwings and their nestlings (inset)

What Is Behavior?

A peacock displaying his colorful tail, a whale spending the winter months in the ocean off the coast of southern California, and a lizard seeking shade from the hot desert sun are all examples of animal behavior. **Behavior** is anything an animal does in response to a stimulus in its environment. The presence of a peahen stimulates a peacock to open its tail feathers and strut. Environmental cues, such as a change in daylength, might be the stimulus that causes the whale to leave its summertime arctic habitat. Heat stimulates the lizard to seek shade. The illustrations in *Figure 33.1* show two examples of stimuli that affect animal behavior.

Figure 33.1
Animals exhibit a variety of behavioral responses.

A This butterfly exposes eyespots on its wings, and a predatory bird stops its pursuit of the insect. The eyespots look like the eyes of an owl.

B The onset of short days and cold weather stimulates squirrels to collect acorns and walnuts and store them. What is the adaptive value of the squirrel's behavior?

2 Teach

Purpose

Students will observe the innate response of an isopod to light.

Process Skills

experiment, collect data, analyze information, draw a conclusion

Teaching Strategies

■ Isopods may be collected locally or purchased from biological supply houses.

■ Use plastic or glass petri dishes for chambers or collect empty plastic food cartons. Cardboard boxes such as shoe boxes will also work well. Use black construction paper to block out light.

■ Tape the toweling edges prior to moistening.

■ If there is a problem of getting all isopods aligned at the center of the dish when first starting, use the species *Armadillidium* (available from supply houses). This species curls up into a ball when touched. Thus, curled up isopods can all be placed at the center at the same time.

Expected Results

Isopods will move toward low light.

Analysis

1. dark areas; majority of isopods were found under dark paper side
2. Explanations may vary, but the behavior is innate. This type of behavior would not have been learned during the experiment because isopods have this behavior as soon as they hatch and it may not be under their conscious control.
3. Isopods must remain moist to survive. Dark areas in nature are more likely to be moist than light areas. Through natural selection, those isopods that innately moved toward dark areas survived and passed this genetic trait on to their offspring.
4. Students' bar graphs should depict the data they collected.

890

MiniLab 33-1 Experimenting

Testing an Isopod's Response to Light Isopods, the pill bugs and sow bugs, are common arthropods on sidewalks or patios. They are actually land crustaceans and respire through gill-like organs that must be kept moist at all times.

Procedure

1. Copy the data table.
2. Prepare a plastic dish using the diagram as a guide. Moisten the paper toweling.

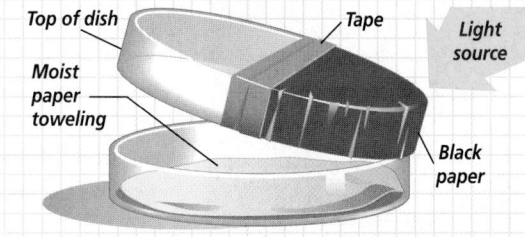

Top of dish / Tape / Light source / Moist paper toweling / Black paper

3. Place six isopods in the center of the dish and add the cover. Place the dish near a lamp or next to a classroom window with light. Have the light strike the dish as shown in the diagram. **CAUTION: *Treat isopods gently.***
4. Wait five minutes and observe the dish. Count and record in your data table the number of isopods on the dark or light side. This is your "five minute observation."
5. Repeat step 4 three more times, waiting five minutes before each observation.

Data Table		
Observation in minutes	**Number of isopods present**	
	Light side	**Dark side**
5		
10		
15		
20		
25		

Analysis

1. Do isopods tend to move toward light or dark areas? Support your answer with specific numbers from your data.
2. Might the behavior of isopods toward light or darkness be innate or learned? Explain your answer.
3. What might be the adaptive advantage for the observed isopod behavior and their response to light? Explain how natural selection may have influenced this isopod behavior.
4. Prepare a bar graph that depicts your data.

890 ANIMAL BEHAVIOR

Animals carry on many activities—such as getting food, avoiding predators, caring for young, finding shelter, and attracting mates—that enable them to survive. These behavior patterns, therefore, have adaptive value. For example, a parent gull that is not incubating eggs or caring for chicks joins a noisy flock of gulls to dive for fishes. If the parent cannot catch a lot of fishes, not only will it die, but its chicks will not survive either. Therefore, this feeding behavior has adaptive value for the gull.

Inherited Behavior

Inheritance plays an important role in the ways animals behave. You don't expect a duck to tunnel underground or a mouse to fly. Yet, why does a mouse run away when a cat appears? Why does a mallard duck fly south for the winter? These behavior patterns are genetically programmed. An animal's genetic makeup determines how that animal reacts to certain stimuli.

Natural selection favors certain behaviors

Often, a behavior exhibited by an animal species is the result of natural selection. The variability of behavior among individuals affects their ability to survive and reproduce. Individuals with behavior that makes them more successful at surviving and reproducing will produce more offspring. These offspring will inherit the genetic basis for the successful behavior. Individuals without the behavior will die or fail to reproduce. You can observe the behavior of isopods in the *MiniLab* on this page.

Inherited behavior of animals is called **innate behavior** (ihn AYT). A toad captures prey by flipping out its sticky tongue. To capture prey, a toad

---

must first be able to detect and follow its movement. Toads have "insect detector" cells in the retinas of their eyes. As an insect moves across a toad's line of sight, the "insect detector" cells signal the brain of the prey's changing position, thus releasing an innate response; the toad's tongue flips out. Toads capture prey through an innate behavior known as a fixed action pattern, *Figure 33.2*.

Genes form the basis of behavior

Through experiments, scientists have found that an animal's hormonal balance and its nervous system—especially the sense organs responsible for sight, touch, sound, or odor identification—affect how sensitive the individual is to certain stimuli. Because genes control the production of an animal's hormones and development of its nervous system, it's logical to conclude that genes indirectly control behavior. Innate behavior includes both automatic responses and instinctive behaviors. You can observe the response of animals to certain stimuli in the *BioLab* at the end of this chapter.

Automatic Responses

What happens if something quickly passes in front of your eyes or if something is thrown at your face? Your first reaction is to blink and jerk back your head. Even if a protective clear shield is placed in front of you, you can't stop yourself from behaving this way when the object is thrown. This reaction is an example of the simplest form of innate behavior, called a reflex. A **reflex** (REE fleks) is a simple, automatic response that involves no conscious control. *Figure 33.3* shows an example of a reflex.

The adaptive value of another automatic response is obvious. Think

about a time when you were suddenly scared. Immediately, your heart began to beat faster. Your skin got cold and clammy, your respiration increased, and maybe you trembled. You were having a **fight-or-flight response**. A fight-or-flight response mobilizes the body for greater activity. Your body is being prepared to either fight or run from the danger. A fight-or-flight response is automatic and controlled by hormones.

Figure 33.2
A toad can starve even though it is surrounded by dead insects because it cannot recognize non-moving animals as prey.

Figure 33.3
Reflexes have survival value for animals. When you accidentally touch a hot stove, you jerk your hand away from the hot surface. The movement saves your body from serious injury.

Instinctive Behavior

Compare the fixed action pattern of a toad capturing prey with a fight-or-flight response. Both are quick, automatic responses to stimuli. But some behaviors take a longer time because they involve more complex actions. An **instinct** (IHN stingt) is a complex pattern of innate behavior. Instinctive behavior patterns may have several parts and may take weeks to complete. Instinctive behavior begins when the animal recognizes a stimulus and continues until all parts of the behavior have been performed.

As shown in *Figure 33.4,* greylag geese instinctively retrieve eggs that have rolled from the nest and will go through the motions of egg retrieval even when the eggs are taken away. You can see that survival of the young may be dependent on this behavior.

WORD *Origin*

instinct
From the Latin word *instinctus,* meaning "impulse." An instinct is a complex pattern of innate behavior.

Figure 33.4
The female greylag goose instinctively retrieves an egg that has rolled out of the nest by arching her neck around the stray egg and moving it like a hockey player advancing a puck. The female goose will retrieve many objects outside the nest, including baseballs and tin cans.

Courtship behavior ensures reproduction

Much of an animal's courtship behavior is instinctive. **Courtship behavior** is the behavior that males and females of a species carry out before mating. Like other instinctive behaviors, courtship has evolved through natural selection. Imagine what would happen to the survival of a species if members were unable to recognize other members of that same species. Individuals often can recognize one another by the behavior patterns each performs. In courtship, behavior ensures that members of the same species find each other and mate. Obviously, such behavior has an adaptive value for the species. Different species of fireflies, for example, can be seen at dusk flashing distinct light patterns. However, female fireflies of one species respond only to those males exhibiting the species-correct flashing pattern.

Some courtship behaviors help prevent females from killing males before they have had the opportunity to mate. For example, in some spiders, the male is smaller than the female and risks the chance of being eaten if he approaches her. Before mating, the male in some species presents the female with a nuptial gift, an insect wrapped in a silk web. While the female is unwrapping and eating the insect, the male is able to mate with her without being attacked. After mating, however, the male may be eaten by the female anyway.

In some species, nuptial gifts play an important role in allowing the female to exercise a choice as to which male to choose for a mating partner. The hanging fly, shown in *Figure 33.5*, is such a species.

Territoriality reduces competition

You may have seen a chipmunk chase another chipmunk away from seeds on the ground under a bird feeder. The chipmunk was defending its territory. **A territory** is a physical space an animal defends against other members of its species. It may contain the animal's breeding area, feeding area, and potential mates, or all three.

Animals that have territories will defend their space by driving away other individuals of the same species. For example, a male sea lion patrols the area of beach where his harem of female sea lions rests. He does not bother a neighboring male that has a harem of his own because both have marked their territories, and each respects the common boundaries. But if an unattached, young male tries to enter the sea lion's territory, the owner of the territory will attack and drive the intruder away from his harem.

Although it may not appear so, setting up territories actually reduces conflicts, controls population growth, and provides for efficient use of environmental resources. When animals space themselves out, they don't compete for the same resources within a limited space. This behavior improves the chances of survival of the young, and, therefore, survival of the species. If the male has selected an appropriate site and the young survive, they may inherit his ability to select an appropriate territory. Therefore, territorial behavior has survival value, not only for individuals, but also for the species. The male stickleback shown in *Figure 33.6* is another animal that exhibits territoriality, especially during breeding season.

Recall that pheromones are chemicals that communicate information

among individuals of the same species. Many animals produce pheromones to mark territorial boundaries. For example, wolf urine contains pheromones that warn other wolves to stay away. The male pronghorn antelope uses a pheromone secreted from facial glands. One advantage of using pheromones is that they work both day and night,

Figure 33.5
Female hanging flies instinctively favor the male that supplies the largest nuptial gift—in this case, a moth. The amount of sperm the female will accept from the male is determined by the size of the gift.

Figure 33.6
The male three-spined stickleback displays a red belly to other breeding males near his territory. The male instinctively responds to other red-bellied males by attacking and driving them away.

Visual Learning

Figure 33.7 Have students examine Figure 33.7 and explain how symbolic behavior contributes to survival.

Display

Interpersonal Ask a group of students to make a photo collage of aggressive behaviors of pets and other local animals. Have them post their display on the classroom bulletin board.
L2 **ELL** **COOP LEARN**

Figure 33.7
In many species, such as bighorn sheep, individuals have innate inhibitions that make them fight in relatively harmless ways among themselves.

Figure 33.8
A dominance hierarchy often prevents continuous fighting because submissive birds give way peacefully in confrontations. The hierarchy also may provide a way for females to choose the best males.

and whether or not the animal that made the mark is present.

Aggressive behavior threatens other animals

Animals occasionally engage in aggression. **Aggressive behavior** is used to intimidate another animal of the same species. Animals fight or threaten one another in order to defend their young, their territory, or a resource such as food. Aggressive behaviors, such as bird calling, teeth baring, or growling, deliver the message to others of the same species to keep away.

When a male gorilla is threatened by another male moving into his territory, for example, he does not kill

the invader. Animals of the same species rarely fight to the death. The fights are usually symbolic, as shown in *Figure 33.7*. Male gorillas do not usually even injure one another. Why does aggressive behavior rarely result in serious injury? One answer is that the defeated individual shows signs of submission to the victor. These signs inhibit further aggressive actions by the victor. Continued fighting might result in serious injury for the victor; thus, its best interests are served by stopping the fight.

Submission leads to dominance hierarchies

Do you have an older or younger sibling? Who wins when you argue? In animals, it is usually the oldest or strongest that wins the argument. But what happens when several individuals are involved in the argument? Sometimes, aggressive behavior among several individuals results in a grouping in which there are different levels of dominant and submissive animals. A **dominance hierarchy** (DAHM uh nunts • HI rar kee) is a form of social ranking within a group in which some individuals are more subordinate than others. Usually, one animal is the top-ranking, dominant individual. This animal might lead

894 ANIMAL BEHAVIOR

Alternative Lab

Cricket Hierarchies

Purpose
Students will observe crickets setting up a dominance hierarchy.

Materials
glass terrarium or aquarium tank with cover, sand, 1 female and 5 male field crickets, dry oatmeal in a jar lid, apple slices, clean sponge pieces soaked in water, 4 matchboxes, 5 jars with lids, 4 different colors of nail polish

Procedure
Give students the following directions.
1. Set up a terrarium with about 2 cm of sand in the bottom and numbered matchboxes in all 4 corners.
2. Place different colored spots of nail polish on the thoraxes of four male crickets. One male will not need polish. The female can be identified by her long ovipositor at the end of her abdomen.
3. Keep the five crickets in separate jars for at least one day prior to beginning the experiment.
4. Place four males in the terrarium and

Figure 33.9
A variety of animals respond to the urge to migrate.

A Canadian and Alaskan caribou migrate from their winter homes in the taiga forests to the tundra for the summer.

B Both the freshwater eel and all species of salmon migrate to their spawning grounds.

C Adult monarch butterflies fly southward where they roost. In the spring, their young fly back north.

others to food, water, and shelter. A dominant male often sires most or all of the offspring. There might be several levels in the hierarchy, with individuals in each level subordinate to the one above. The ability to form a dominance hierarchy is innate, but the position each animal assumes may be learned.

The term *pecking order* comes from a dominance hierarchy that is formed by chickens, illustrated in *Figure 33.8*. The top-ranking chicken can peck any other chicken. The chicken lowest in the hierarchy is pecked at by all the other chickens in the group.

Behavior resulting from internal and external cues

Some instinctive behavior is exhibited in animals in response to internal, biological rhythms. Behavior based on a 24-hour day/night cycle is one example. Many animals, humans included, sleep at night and are awake during the day. Other animals, such as owls, reverse this pattern and are awake at night. A 24-hour cycle of behavior is called a **circadian rhythm** (sur KAYD ee uhn). Most animals come close to this 24-hour cycle of sleeping and wakefulness. Experiments have shown that in laboratory settings with no windows to show night and day, animals continue to behave on a 24-hour cycle.

Rhythms also can occur on a yearly or seasonal cycle. **Migration,** for example, occurs on a seasonal cycle. Migration is the instinctive, seasonal movement of animals, shown in *Figure 33.9*. In the United States, about two-thirds of bird species fly south in the fall to areas such as South America where food is available during the winter. The birds fly north in the spring to areas where they breed during the summer. Whales migrate seasonally, as well. Change in day length is thought to

Enrichment
Ask a group of students to make a trip to a local zoo to observe and photograph pecking orders of animals, and report back to the class. **L3** **ELL** **COOP LEARN**

Visual Learning
Figure 33.9 After studying Figure 33.9, have students explain how migration can contribute to survival.

GLENCOE
TECHNOLOGY

CD-ROM
Biology: The Dynamics of Life
Video: *Salmon Migration*
Disc 4

VIDEODISC
Biology: The Dynamics of Life
Salmon Migration (Ch. 26)
Disc 2, Side 1, 28 sec.

The Infinite Voyage *The Living Clock Changing Circadian Rhythms Through Altering DNA* (Ch. 5), 6 min.

How Living Things Convert Light into Time (Ch. 6), 5 min. 30 sec.

observe and record their behavior for 15 minutes.

5. Examine the terrarium for about 10 minutes each day for 5 days. Note which cricket becomes dominant and the behavior it exhibits.

Expected Results
One cricket will become dominant and the others will avoid him.

Analysis
1. How could you tell that your crickets set up a dominance hierarchy? *One cricket chirped more, was initially more aggressive, and later others avoided him.*
2. Describe the differences in behavior of the crickets before and after the hierarchy was established. *Before: much aggression and chirping; after: other crickets avoided the dominant male.*

✓ Assessment
Performance Ask students to hypothesize how the crickets will respond to the female and then introduce her to the terrarium. *The dominant male may mate with the female.* Use the Performance Task Assessment List for Formulating a Hypothesis in **PASC,** p. 21. **L2**

Purpose 🖐

Students will determine that hibernation in squirrels is controlled by an internal annual biological clock.

Process Skills

think critically, analyze information, identify and control variables, design an experiment, interpret scientific illustrations

Teaching Strategies

■ Point out that not all months of the year are listed.
■ Remind students that the dark bands correspond to periods of hibernation.
■ Discuss human circadian patterns as an introduction to periods of hibernation.
■ Help students determine the approximately annual cycle of behavior by noting the peaks of body weight and the time lapse between these peaks.

Thinking Critically

1. Squirrels spend about 120 days in hibernation and close to 200 days out of hibernation.
2. They were controlling variables by eliminating them from the experiment. Outside stimuli could provide clues as to when it might be time to hibernate.
3. to show that temperature was not influencing start of hibernation; changing length of light time to more or less than 12 hours per day, altering food selection
4. Place newborn squirrels into identical experimental chambers as those used in the other experiments and observe if squirrels hibernate at the proper time of the year.

stimulate the onset of migration in the same way that it controls the flowering of plants. You can find out how migrating turtles are tracked in the *BioTechnology* at the end of this chapter.

Migration calls for remarkable strength and endurance. The Arctic tern migrates between the Arctic Circle and the Antarctic, a one-way flight of almost 18 000 km.

Animals navigate in a variety of ways. Some use the positions of the sun and stars to navigate. They may use geographic clues, such as mountain ranges. Some bird species seem to be guided by Earth's magnetic field. You might think of this as being guided by an internal compass.

Biological rhythms are clearly governed by a combination of internal and external cues—that is, by both innate and learned behavior. Animals that migrate might be responding to colder temperatures and shorter days, as well as to hormones. Young animals may learn when and where to migrate by following their parents. You can easily see why animals migrate from a cold place to a warmer place, yet most animals do not migrate. How animals cope with

winter is another example of instinctive behavior.

You know that many animals store food in burrows and nests. But other animals survive the winter by undergoing physiological changes that reduce their need for energy. Many mammals, some birds, and a few other types of animals go into a deep sleep during the cold winter months. Other animals experience hibernation. **Hibernation** (hi bur NAY shun) is a state in which the body temperature drops substantially, oxygen consumption decreases, and breathing rates decline to a few breaths per minute. Hibernation conserves energy. Animals that hibernate typically eat vast amounts of food to build up body fat before entering hibernation. This fat fuels the animal's body while it is in this state. The golden-mantled ground squirrel shown in *Figure 33.10* is an example of an animal that hibernates. You can find out more about hibernation in the *Problem-Solving Lab* on the next page.

What happens to animals that live year-round in hot environments? Some of these animals respond in a way that is similar to hibernation. **Estivation** (es tuh VAY shun) is a state

Figure 33.10
The golden-mantled ground squirrel has a normal body temperature of around 37°C. When the day length shortens in the fall, the ground squirrel's temperature drops to 2°C, and it goes into hibernation.

896 ANIMAL BEHAVIOR

Figure 33.11
Australian long-necked turtles are among the reptiles and amphibians that respond to hot and dry summer conditions by estivating.

of reduced metabolism that occurs in animals living in conditions of intense heat. Desert animals appear to estivate sometimes in response to lack of food or periods of drought. However, Australian long-necked turtles, shown in *Figure 33.11*, will estivate even when they are kept in a laboratory with constant food and water. Clearly, estivation is an innate behavior that depends on both internal and external cues.

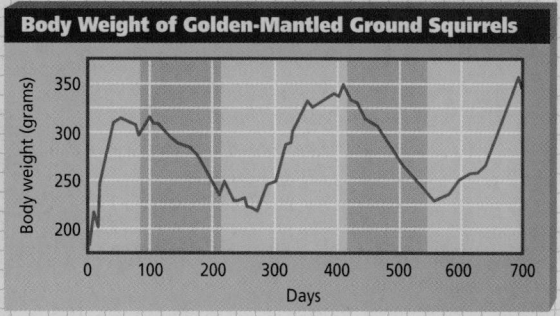

---

3 Assess

Check for Understanding
Ask students to prepare a concept map using all the vocabulary words in this section. **L2**

Reteach
Have students make a table that lists types of innate behaviors down the left side. Across the top, have them write the following heads: Definition, Example, Outcome of behavior, Survival value. Have them fill in the table. **L1**

Extension
Visual-Spatial Ask students to go to a nearby zoo. As they observe animals, have them note innate behaviors and explain their survival value. **L3**

✔ Assessment
Portfolio Have students list five major groups (phylum or class) of animals they have studied. For each phylum or class, ask them to identify one innate behavior and explain its adaptive value. **L2 P**

---

4 Close

Discussion
Male katydids sing to attract females. In Panamanian forests where bats are common, male katydids on plants shake their bodies vigorously to attract females. The females detect the shaking of the plant and respond to the male. The bats cannot detect the shaking. Ask students to explain the behavior of the male katydids.

Prepare

Key Concepts

Students will study various types of learned behavior. They will learn about the adaptive value of habituation, imprinting, trial-and-error learning, conditioning, insight, and communication.

Planning

- Prepare puzzle pieces for MiniLab 33-2.
- Gather tie shoes for the Quick Demo.
- Purchase snails and gather binocular microscopes, spring water, droppers, small dishes, and scissors for the BioLab. Prepare probes from tape, rubber bands, and pencils.

1 Focus

Bellringer 🖌

Before presenting the lesson, display **Section Focus Transparency 80** on the overhead projector and have students answer the accompanying questions. **L1** **ELL**

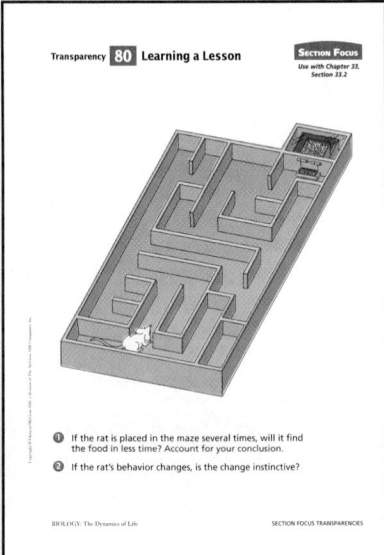

Transparency **80** Learning a Lesson

SECTION FOCUS
Use with Chapter 33,
Section 33.2

❶ If the rat is placed in the maze several times, will it find the food in less time? Account for your conclusion.

❷ If the rat's behavior changes, is the change instinctive?

BIOLOGY: The Dynamics of Life   SECTION FOCUS TRANSPARENCIES

Objectives

Distinguish among types of learned behavior.

Demonstrate, by example, types of learned behavior.

Vocabulary

habituation
imprinting
trial-and-error learning
motivation
conditioning
insight
communication
language

Section
33.2 Learned Behavior

You were born knowing how to cry, but were you born knowing how to tie your shoes or read? Behavior controlled by instinct, as you now know, occurs automatically. However, some behavior is the direct result of the previous experiences of an animal. A dog that has picked a poisonous frog up in its mouth will never do so again. You may know how to play a musical instrument after a few months or it may take several years. These behaviors are a result of learning.

Both the teen and the dog are exhibiting learned behavior.

Figure 33.12
By examining the graph, you can see that humans demonstrate the most learned behavior. Insects, fishes, amphibians, and reptiles demonstrate the most innate behavior.

Comparison of Animal Behaviors

Types of behavior:
Reasoning
Learning
Instinct
Reflex

Invertebrates   Vertebrates

Protists Worms Insects Fishes, reptiles, amphibians Birds Mammals Primates Humans

898 ANIMAL BEHAVIOR

What Is Learned Behavior?

Learning, or learned behavior, takes place when behavior changes through practice or experience. The more complex an animal's brain, the more elaborate the patterns of its learned behavior. As you can see in *Figure 33.12,* innate types of behavior are more common in invertebrates, and learned types of behavior are more common in vertebrates. In humans, many behaviors are learned. Reading, writing, and playing a sport are all learned.

Learning has survival value for all animals in changing environments because it permits behavior to change

in response to varied conditions. Learning allows an animal to adapt to change, an ability that is especially important for animals with long life spans. The longer that an animal lives, the greater the chance that its environment will change and that it will encounter unfamiliar situations.

Kinds of Learned Behavior

Just as there are several types of innate behavior, there are several types of learned behavior. Some learned behavior is simple and some is complex. Which group of animals do you think carries out the most-complex type of learned behavior?

Habituation: A simple form of learning

Horses normally shy away from an object that suddenly appears from the trees or bushes, yet after a while they disregard noisy cars that speed by the pasture honking their horns. This lack of response is called habituation. **Habituation** (huh BIT yew ay shun) illustrated in *Figure 33.13*, occurs when an animal is repeatedly given a stimulus that is not associated with any punishment or reward. An animal has become habituated to a stimulus when it finally ceases to respond to the stimulus.

Imprinting: A permanent attachment

Have you ever seen young ducklings following their mother? This behavior is the result of imprinting. **Imprinting** is a form of learning in which an animal, at a specific critical time of its life, forms a social attachment to another object. Many kinds of birds and mammals do not innately know how to recognize members of their own species. Instead, they

learn to make this distinction early in life. Imprinting takes place only during a specific period of time in the animal's life and is usually irreversible. For example, birds that leave the nest immediately after hatching, such as geese, imprint on their mother. They learn to recognize and follow her within a day of hatching.

In birds such as ducks, imprinting takes place during the first day or two after hatching. A duckling rapidly learns to recognize and follow the first conspicuous moving object it sees. Normally, that object is the duckling's mother. Learning to recognize their mother and follow her ensures that food and protection will always be nearby.

Learning by trial and error

Do you remember when you first learned how to ride a bicycle? You probably tried many times before being able to successfully complete the task. Nest building, like riding a bicycle, may be a learning experience. The first time a jackdaw builds

Figure 33.13
Habituation is a loss of sensitivity to certain stimuli. Young horses often are afraid of cars and noisy streets. Gradually, they become habituated to the city and ignore normal sights and sounds.

Purpose

Students will conduct an experiment to test the nature of trial-and-error learning.

Process Skills

interpret data, think critically, collect data, draw a conclusion

Teaching Strategies

■ Have students work in teams of two. One can be the time keeper while the other does the puzzle. The time keeper should not watch the student doing the assembly but should be told when the assembler begins and completes the puzzle.

■ Prepare puzzle pieces in advance and place on card stock. Save puzzle pieces in plastic bags for students in later classes.

■ Enlarge puzzle pieces shown here to approximately twice their size for your students.

Expected Results

Student times needed to complete the puzzle will decline with each trial.

Analysis

1. times decreased
2. No, the ability to work the puzzle was an example of learned behavior.
3. Yes, because the time it took to do the puzzle decreased as learning took place.
4. Student answers may vary. Imprinting and conditioning were not involved with this type of learning. Trial-and-error learning was operating with the initial solving of the puzzle, but later trials may have relied upon insight.

MiniLab 33-2 Experimenting

Solving a Puzzle You are given a bunch of keys and asked to open a door. How do you go about finding the right key? Several attempts are needed and then finally the door opens. The next time you are asked to perform the same task, can you go directly to the correct key? Chances are, you can. You have learned how to solve this problem.

Procedure

1. Copy the data table below.
2. Obtain a paper puzzle from your teacher.
3. Time how long it takes you to assemble the puzzle pieces into a perfect square.
4. Record the time it took and call this trial 1.
5. Disassemble the square and mix the pieces.
6. Repeat step 3 for four more trials.

Data Table	
Trial	Time needed to complete square puzzle
1	
2	
3	
4	
5	

Analysis

1. Using your data, explain how the time needed to complete the puzzle changed from trial 1 to trial 5.
2. Was the final completion of the puzzle an example of innate behavior? Explain your answer.
3. Was the final completion of the puzzle an example of learned behavior? Explain your answer.
4. Analyze your behavior when solving the puzzle as to the role that imprinting, trial and error, conditioning, and insight may have played in improving your trial times.

Figure 33.14
Mice soon learn where grain is stored in a barn and are motivated by hunger to chew through the storage containers.

---

a nest, it uses grass, bits of glass, stones, empty cans, old light bulbs, and anything else it can find. With experience, the bird finds that grasses and twigs make a better nest than do light bulbs. The jackdaw has used **trial-and-error learning** in which an animal receives a reward for making a particular response. When an animal tries one solution and then another in the course of obtaining a reward, in this case a suitable nest, it is learning by trial and error. Find out for yourself how trial and error learning works in the *MiniLab* on this page.

Learning happens more quickly if there is a reason to learn or be successful. **Motivation,** an internal need that causes an animal to act, is necessary for learning to take place. In most animals, motivation often involves satisfying a physical need, such as hunger or thirst. If an animal isn't motivated, it won't learn. Animals that aren't hungry won't respond to a food reward. Mice living in a barn, shown in *Figure 33.14,* discover that they can eat all the grain they like if they first chew through the container in which the grain is stored.

Figure 33.15
In 1900, Ivan Pavlov, a Russian biologist, first demonstrated conditioning in dogs.

(A) Pavlov noted that dogs salivate when they smell food. Responding to the smell of food is a reflex, an example of innate behavior.

(B) By ringing a bell each time he presented food to a dog, Pavlov established an association between the food and the ringing bell.

(C) Eventually, the dog salivated at the sound of the bell alone. The dog had been conditioned to respond to a stimulus that it did not normally associate with food.

Conditioning: Learning by association

When you first got a new kitten, it would meow as soon as it smelled the aroma of cat food in the can you were opening. After a few weeks, the sound of the can opener alone attracted your kitten, causing it to meow. Your kitten had become conditioned to respond to a stimulus other than the smell of food. **Conditioning** is learning by association. A well-known example of an early experiment in conditioning is illustrated in *Figure 33.15*.

Insight: The most complex type of learning

In a classic study of animal behavior, a chimpanzee was given two bamboo poles, neither of which was long enough to reach some fruit placed outside its cage. By connecting the two tapering short pieces to make one longer pole, the chimpanzee learned to solve the problem of how to reach the fruit. This type of learning is called insight. **Insight** is learning in which an animal uses previous experience to respond to a new situation.

Much of human learning is based on insight. When you were a baby, you learned a great deal by trial and error. As you grew older, you relied more on insight. Solving math problems is a daily instance of using insight. Probably your first experience with mathematics was when you learned to count. Based on your concept of numbers, you then learned to add, subtract, multiply, and divide. Years later, you continue to solve problems in mathematics based on your past experiences. When you encounter a problem you have never experienced before, you solve the problem through insight.

33.2 LEARNED BEHAVIOR **901**

901

Purpose

Students will determine that the song of certain bird species is partially innate but mostly learned behavior.

Process Skills

analyze information, apply concepts, think critically, interpret scientific illustrations

Teaching Strategies

■ Advise students that the term "wild" refers to the sparrow raised in its natural surroundings.
■ Units along the left axis are kilocycles per second. Units along the bottom axis are in seconds of time.

Expected Results

Students will recognize that a small portion of the sparrow's song is innate, but the major portion of the song must be learned.

Analysis

1. Segment A is similar in both songs. Segments B and C are quite different.

2. Segment A appears to be innate; it would be impossible for the bird raised in isolation to have learned the first part of the song. Segments B and C must be learned, because the bird raised in isolation cannot match these segments of the wild bird's song.

3. The majority of the bird's song appears to be learned because segments B and C are longer than segment A.

4. The bird will learn the correct sparrow song.

✓ Assessment

Performance Have students design an experiment to test the hypothesis that birds raised in isolation will learn their proper song from a recording only immediately after hatching. Use the Performance Task Assessment List for Designing an Experiment in **PASC**, p. 23. **L2**

Problem-Solving Lab 33-2   Interpreting Data

Do birds learn how to sing? Do birds learn how to sing, or is this innate behavior? Most experimental evidence points to the fact that singing may be a combination of the two types of behavior, but in certain species, learning is critical in order to sing the species song correctly.

Analysis

Bird sound spectrograms allow scientists to record and visually study the song patterns of birds. Using this tool, they recorded spectrograms for white-crowned sparrows. The top spectrogram is that of a wild white-crowned sparrow. The bottom spectrogram is that of a white-crowned sparrow hatched and raised in total isolation from all other birds. Segments of the song have been identified with the letters A-C.

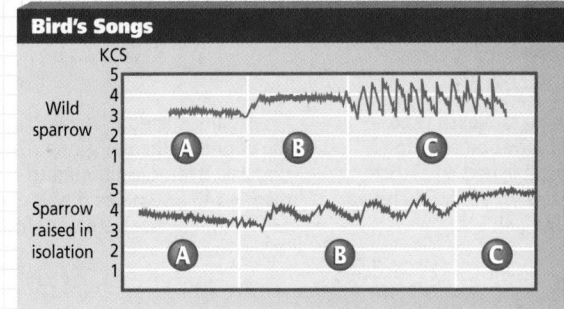

Bird's Songs

Thinking Critically

1. In general, how do the two spectrograms compare?
2. Which segment of the sparrow's song may be innate? Learned? Explain your answers.
3. Does it appear that the majority of the sparrow's song is learned or innate? Explain your answer.
4. In a different experiment, a recording of white-crowned sparrow song was repeatedly played for a young bird raised in isolation. If bird song is mainly learned, predict the outcome of the experiment.

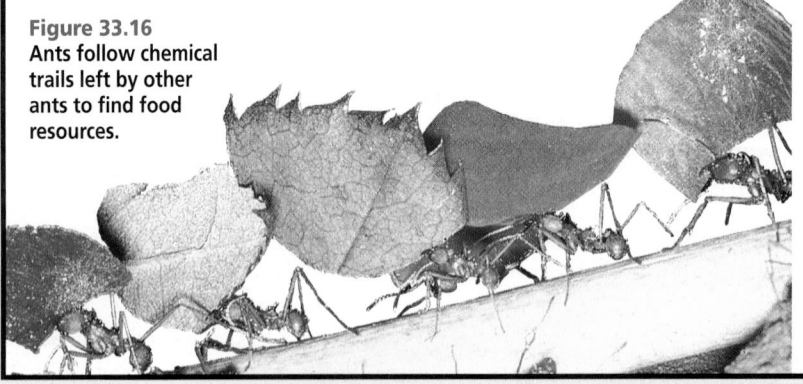

Figure 33.16
Ants follow chemical trails left by other ants to find food resources.

The Role of Communication

When you think about the interactions that happen among animals as a result of their behavior, you realize that some sort of communication has taken place. **Communication** is an exchange of information that results in a change of behavior. Black-headed gulls visually communicate their availability for mating with instinctive courtship behavior. The pat on the head from a dog's owner after the dog retrieves a stick signals a job well done.

Most animals communicate

Animals have several channels of communication open to them. They signal each other by sounds, sights, touches, or smells. Sounds radiate in all directions and can be heard a long way off. The sounds of the humpback whale can be heard 1200 km away. Sounds such as songs, roars, and calls communicate a lot of information quickly. For example, the song of a male cricket tells his sex, his location, his social status, and, because communication by sound is usually species specific, his species.

Signals that involve odors may be broadcast widely and carry a general message. Ants, shown in **Figure 33.16**, leave odor trails that are followed by other members of their nest. Some odors may also be species specific. As you know, pheromones, such as those of moths, may be used to attract mates. Because only small amounts of pheromones are needed, other animals, especially predators, may be unable to detect the odor.

Cultural Diversity

Bertram Fraser-Reid

Discuss with students the important contributions of African American organic chemist, Bertram O. Fraser-Reid. His most important work involved synthesizing artificial insect pheromones as substitutes for dangerous insecticides. In Canada, the western pine beetle causes billions of dollars of damage to trees each year. Fraser-Reid reasoned that if he released artificial pheromones of female pine beetles in a part of the forest that contained no females, it might attract male pine beetles to the spot, thus preventing them from mating. Fraser-Reid's initial research laid the groundwork for future studies.

Using both innate and learned behavior

Some communication is a combination of both innate and learned behavior. In some species of songbirds, as shown in *Figure 33.17*, males automatically sing when they reach sexual maturity. Their songs are specific to their species, and singing is innate behavior. Yet members of the same species that live in different regions learn different variations of the song. They learn to sing with a regional dialect. In other species, birds raised in isolation never learn to sing their species song. Find out more about the songs birds sing in the *Problem-Solving Lab* on the facing page.

Figure 33.17
The Indigo bunting sings a high-pitched series of notes that descend the scale, then ascend again at the end of the song.

Some animals use language

Language, the use of symbols to represent ideas, is present primarily in animals with complex nervous systems, memory, and insight. Humans, with the help of spoken and written language, can benefit from what other people and cultures have learned and don't have to experience everything for themselves. We can use accumulated knowledge as a basis on which to build new knowledge, illustrated in *Figure 33.18*.

Figure 33.18
English and other languages are made up of words that have specific meanings. An amazing number of meanings can be communicated using words of any human language.

Section Assessment

Understanding Main Ideas
1. What is the difference between imprinting and other types of learned behavior?
2. How does learning have survival value in a changing environment? Explain your answer by using an example from your daily life.
3. Explain by example the difference between trial-and-error learning and conditioning.
4. What is the difference between communication and language?

Thinking Critically
5. How would a cat respond if the mice in a barn no longer entered at the usual places?

SKILL REVIEW
6. **Observing and Inferring** Two dog trainers teach dogs to do tricks. One trainer gives her dog a food treat whenever the dog correctly performs the trick. The other trainer does not use treats. Which trainer will be more successful at dog training? Why? For more help, refer to *Thinking Critically* in the **Skill Handbook.**

Section Assessment

1. Imprinting can take place only during a specific time during an animal's life. Other learned behaviors may develop later.
2. The behavior can be modified to deal with a changing environment.
3. You learn to ride a bicycle by trial-and-error learning; an animal may be conditioned to the sound of a can opener.
4. Communication is an exchange of information that results in a change of behavior, whereas language is the use of symbols to represent ideas.
5. The cat might wait at the usual places for a while; after not having any success finding mice, the cat would look for new places mice might enter.
6. The dog will be more easily trained if it is rewarded when it responds correctly.

Time Allotment

One class period

Process Skills

collect data, experiment, hypothesize, define operationally, observe and infer, conclude, think critically, use numbers

PREPARATION

■ No special housing or food are needed if snails are used almost immediately upon arrival from a supply house.

PROCEDURE

Teaching Strategies

■ Small disposable plastic petri dishes (60 mm diameter × 15 mm deep) or Syracuse watch glasses are ideal.

■ Allow students to prepare the probe. Substitute wooden match or Q-tip for the pencil portion of the probe. Simply snip a rubber band in half to obtain the needed section.

■ Have students work in small groups of three or four.

Troubleshooting

■ Students must be able to tell front and rear ends. They are easy to tell apart macroscopically. Review the differences with students.

■ Students must be cautioned about not striking the snail too hard with the rubber band probe.

■ Some stimulations will result in no response. Students should record exactly what they observe. Habituation will be difficult to achieve, if at all. Students will tire before the snail does.

Behavior of a Snail

*L*and snails are members of the mollusk class Gastropoda. Land snails live on or near the ground, feed on decaying organic matter, and breathe with gills or, in some cases, with a simple lung. Land snails sense their environment with a pair of antennae and eyes. Snails are excellent organisms for behavioral studies because they show a variety of consistent responses to certain stimuli.

PREPARATION

Problem

How can you test the behavior of snails to touch stimuli?

Objectives

In this BioLab you will:
■ **Test** the response of snails to touch.
■ **Measure** the time needed for habituation to occur after repeated touch stimuli.

Materials

snails   spring water
dropper   small dish

scissors
dissecting microscope
probe constructed from tape, rubber band, and pencil

Safety Precautions

Always wear goggles in the lab. Wash your hands both before and after handling any animals. Use caution when working with live animals.

Skill Handbook

Use the **Skill Handbook** if you need additional help with this lab.

A land snail

PROCEDURE

904

1. Copy the data table.
2. Prepare a stimulator probe by taping a small piece of a cut rubber band to the tip of a pencil.
3. Cover the bottom of a small dish with spring water.
4. Obtain a snail from your teacher and place it in the dish.
5. Use a dissecting microscope to examine and locate its head. Its head has two antennae that it can extend and retract.

Data and Observations

Both ends are sensitive to touch. Students may detect no response when performing a specific trial. Habituation will usually not occur.

6. Place the dish on your desk.

7. Lightly touch the snail's anterior end using the end of the rubber band probe. Note if it responds (yes or no), and record the direction the snail moves. Consider this trial 1.

8. Repeat step 7 for four more trials.

9. Lightly touch the snail's posterior end using the rubber band probe. Record your observations and conduct a total of five trials.

10. Repeat step 9, touching the middle of the snail's body.

11. Test the snail's ability to become habituated from stimulation to its anterior end.

   a. Continue to touch the snail's anterior end with the probe every 10 seconds until habituation occurs. Continue testing for a reasonable length of time if habituation does not occur.

   b. Count and record the number of stimulations needed for habituation.

Data Table

Body area	Response to touch		
Trial	Anterior	Posterior	Middle
1			
2			
3			
4			
5			
Habituation studies			
Rate of stimulation	Number of stimulations needed to reach habituation		

ANALYZE AND CONCLUDE

1. **Hypothesizing** Are the responses shown by snails to touch learned or innate? Explain your answer.

2. **Observing** Describe the direction that a snail moves when its anterior and posterior ends are stimulated. Does one end appear to be more sensitive than the other? Is the middle sensitive to touch? Is the speed of response slow or rapid?

3. **Hypothesizing** Explain how the behavior of responding to touch may be an adaptation for survival.

4. **Experimenting** Why did you perform several trials for each experiment involving stimulation of the anterior, posterior, and middle of the snail?

5. **Defining Operationally** Define the term habituation.

6. **Concluding** Explain how your data may be used to support the observation that snails are not easily habituated to touch. Use actual data to support your answer.

7. **Predicting** How might this lack of habituation serve as an adaptation for survival?

Going Further

Experimentation Form a hypothesis regarding snail behavior when given a choice between light and dark conditions. Design and carry out an experiment to test your hypothesis.

 **BIOLOGY** *Online* To find out more about animal behavior, visit the Glencoe Science Web site.
science.glencoe.com

33.2 LEARNED BEHAVIOR **905**

ANALYZE AND CONCLUDE

1. Student answers may vary. These responses are innate because the animals lack a large enough brain to learn behavior; they do not have the opportunity to learn correct behavior because the first time they do not respond correctly, they may be eaten by a predator.

2. The snail moves backward when the anterior end is touched. The snail moves forward when the posterior end is touched. The anterior appears to be more sensitive to touch. The middle is less sensitive to touch.

3. The snail will move away from predators rapidly.

4. Student answers will vary; good experimental technique, failure to make contact with probe, etc.

5. Habituation is the lack of a response after repeated stimulation.

6. Snails continued to respond after several minutes of stimulation to their anterior ends.

7. Continued stimulation from a predator would bring about continued response of moving away from that predator.

Resource Manager

BioLab and MiniLab Worksheets, pp. 149-150 L2

✔ Assessment

Performance Ask students to plan an experiment that could be used to test a snail's response to rapid changes in light intensity. Use the Performance Task Assessment List for Designing an Experiment in **PASC**, p. 23. L2

Going Further

Snails can also respond by withdrawing into their shells. Have students determine under what conditions snails withdraw rather than move away. L1 ELL

BIO Technology

Purpose

Students become familiar with tagging and with satellite telemetry as methods of tracking animal migrations.

Background

Four species of sea turtles found along the U.S. coast are on the endangered species list: Kemp's ridley, hawksbill, Florida green, and leatherback. A fifth species, the loggerhead, is threatened.

The Florida green turtle nests from June to October, and a female will nest every two to four years. During a nesting year, the female will lay clutches of about 115 eggs at 12-day intervals.

Teaching Strategies

■ Have students locate the Caribbean Sea on a map. Point out that satellite tracking is especially useful for following the movements of animals that travel large distances.
■ Go over with students the definition of the word telemetry. Telemetry is the science of transmitting data over a distance via radio waves. It comes from the Greek words *tele*, meaning "distance," and *metron*, meaning "measure."

Investigating the Technology

Students are likely to hypothesize that these areas are green turtle feeding grounds. The hypothesis could be tested by going to the area to see if turtles are feeding there, or by sampling the area for the presence of organisms on which the turtles are known to feed.

BIO Technology

Tracking Sea Turtles

The Florida green turtle (Chelonia mydas mydas) is an endangered species that nests on sandy beaches. It is found in temperate and tropical waters, including the southeastern coast of the United States. Like other sea turtles, the Florida green turtle spends virtually all of its life at sea; however, adult females visit beaches several times a year to lay their eggs.

Studying sea turtles presents a challenge because they spend so little time on land. Research is most easily conducted on the beach, where the nesting behavior of the females can be directly observed. These observations have provided important information about how to protect the nesting sites from human disturbance or predation. But more information about the Florida green turtle is needed because protecting an endangered species requires knowing what environmental factors are crucial to its survival.

Tagging To study these animals, researchers affix a small metal tag onto the flipper of a captured turtle. The tag is etched with an identification number. If the animal is captured again, the date and location are shared with other turtle researchers. But even when a tagged turtle is recaptured, the route the animal took to move from one location to another remains unknown.

Satellite tracking Recent improvements in satellite telemetry are making it possible for researchers to keep much better track of individual turtles as they swim from place to place. A transmitter the size of a small, portable cassette player is attached to the shell behind the turtle's neck. The battery-powered transmitter will work for six to ten months before it falls off. When the turtle comes to the surface to breathe, the transmitter broadcasts data in the form of a digital signal to an orbiting communications satellite. The satellite transmits the data to a receiving station on Earth.

The digital signal contains information about the turtle's latitude and longitude, the number of dives it made in the past 24 hours, the length of its most recent dive, and the temperature of the water. By plotting the location of data transmissions on a map, researchers can track the direction and speed in which the animal is moving.

Problems with satellite tracking Sometimes a transmitter stops working after just a few weeks, and there are problems with the data itself. Increasingly accurate information will become available as the technology improves and as more turtles are included in satellite tracking efforts.

INVESTIGATING THE TECHNOLOGY

Thinking Critically Telemetry data from a Florida green turtle indicate the animal has spent the past several days in an offshore location characterized by coral reefs and seagrass meadows. Past telemetry data from other green turtles indicate that these animals periodically interrupt their travels to stop at this location and at other coral reefs and seagrass meadows. Form a hypothesis that could explain this behavior. How could you test your hypothesis?

 BIOLOGY Online To find out more about sea turtle migration, visit the Glencoe Science Web site. **science.glencoe.com**

906

Florida green turtle

Going Further

Invite students to learn more about the life cycle of the Florida green turtle, or to find out what other species of animals are currently being tracked via satellite telemetry.

GLENCOE TECHNOLOGY

 **VIDEODISC**
The Secret of Life
Tracking Equipment

SUMMARY

Section 33.1

Innate Behavior

Main Ideas

- Behavior is anything an animal does in response to a stimulus. Many behaviors have adaptive value and are shaped by natural selection.

- Innate behavior is inherited. Innate behaviors include automatic responses and instincts. Automatic responses include reflexes and fight-or-flight responses.

- An instinct is a complex pattern of innate behaviors.

- Behaviors such as courtship rituals, displays of aggressive behavior, territoriality, dominance hierarchies, hibernation, and migration are all forms of instinctive behavior.

Vocabulary

aggressive behavior (p. 894)
behavior (p. 889)
circadian rhythm (p. 895)
courtship behavior (p. 892)
dominance hierarchy (p. 894)
estivation (p. 896)
fight-or-flight response (p. 891)
hibernation (p. 896)
innate behavior (p. 890)
instinct (p. 892)
migration (p. 895)
reflex (p. 891)
territory (p. 893)

Section 33.2

Learned Behavior

Main Ideas

- Learning takes place when behavior changes through practice or experience. Learned behavior has adaptive value.

- Learning includes habituation, imprinting, trial and error, and conditioning. The most complex type of learning is learning by insight.

- Some animals use language, whereas most communicate by either visual, auditory, or chemical signals.

Vocabulary

communication (p. 902)
conditioning (p. 901)
habituation (p. 899)
imprinting (p. 899)
insight (p. 901)
language (p. 903)
motivation (p. 900)
trial-and-error learning (p. 900)

Main Ideas

Summary statements can be used by students to review the major concepts of the chapter.

Using the Vocabulary

To reinforce chapter vocabulary, use the Content Mastery Booklet and the activities in the Interactive Tutor for Biology: The Dynamics of Life on the Glencoe Science Web site. **science.glencoe.com**

 **THE PRINCETON REVIEW** *All Chapter Assessment questions and answers have been validated for accuracy and suitability by The Princeton Review.*

UNDERSTANDING MAIN IDEAS

1. b
2. c
3. a
4. d
5. d

UNDERSTANDING MAIN IDEAS

1. Your adult dog is chewing on a bone when a puppy approaches. Your dog growls at the puppy. What type of behavior is your dog exhibiting?
 a. conditioning **c.** habituation
 b. aggressive behavior **d.** fighting

2. A change in temperature or the presence of a female may be the _____ that results in a change in an animal's behavior.
 a. response **c.** stimulus
 b. reflex **d.** rhythm

3. Animals with behavior that makes them more successful at surviving and reproducing will produce more _____.
 a. offspring **c.** territory
 b. aggression **d.** eggs

4. All inherited behavior of animals is _____ behavior.
 a. instinctive **c.** conditioned
 b. learned **d.** innate

5. When a toad flips out its sticky tongue to catch an insect flying past, it is exhibiting _____.
 a. learned behavior **c.** territoriality
 b. courtship behavior **d.** innate behavior

GLENCOE TECHNOLOGY

VIDEOTAPE
MindJogger Videoquizzes
Chapter 33: *Animal Behavior*
Have students work in groups as they play the videoquiz game to review key chapter concepts.

Resource Manager

Chapter Assessment, pp. 193-198
MindJogger Videoquizzes
ExamView® Pro Software
BDOL Interactive CD-ROM, Chapter 33 quiz

6. c
7. c
8. d
9. c
10. b
11. insight
12. attract
13. visual
14. instinctive
15. motivation
16. Hibernation
17. aggressive
18. language
19. courtship
20. territorial

APPLYING MAIN IDEAS

21. A dominance hierarchy would reduce aggression at common feeding sites.

22. No, most likely they would already have imprinted on their own mother by the time they are five days old.

23. Ivan Pavlov observed dogs' natural behavior when food was presented. He hypothesized that the dogs could be conditioned to respond to the sound of a bell as if it were food. He designed an experiment to test his hypothesis. He rang a bell each time he fed the dog, then again when the dog smelled food. Finally, when he rang the bell with no food stimulus, the dog salivated. He concluded that the dog had been conditioned to respond to a stimulus that it did not normally associate with food.

6. What type of behavior is shown in this diagram?

   a. conditioning
   b. imprinting
   c. instinctive
   d. habituation

7. Caribou are _____ when they move from their winter homes in the forests to the tundra for the summer.
   a. hibernating   c. migrating
   b. imprinting   d. learning

8. Of these, which is NOT an example of instinctive behavior resulting from internal or external cues?
   a. circadian rhythm   c. hibernation
   b. migration   d. habituation

9. Establishing _____ reduces the need for aggressive behavior among members of the same species.
   a. reflexes   c. territories
   b. conditioning   d. habituation

10. Your cat exhibits _____ when it runs for its food dish upon hearing the sound of the can opener.
    a. insight   c. habituation
    b. conditioning   d. imprinting

11. You use _____ when you solve a math problem that you have never seen before.

12. Male moths use pheromones to _____ female moths.

13. Black-headed sea gulls use _____ cues as a means of communicating their availability for mating.

THE PRINCETON REVIEW — TEST-TAKING TIP

Use Roots to Learn
The root of a word can help you group words together as you learn them. If you learn that *trans-* means *across*, as in *transfer*, you might then remember the meaning of words like *transgenic*, *translocation*, and *transpiration*.

14. The ability to form a dominance hierarchy is an example of a(n) _____ behavior pattern because it is inherited.

15. When a mouse is learning to go through a maze, the food reward at the end of the maze is _____ because it is based on a need the animal has.

16. _____ is different from deep sleep because an animal exhibiting this behavior has a lowered body temperature.

17. The bighorn sheep below are demonstrating _____ behavior.

18. When a chimpanzee learns to use a computer, it is learning _____ because it is using symbols that represent ideas.

19. Certain ground-dwelling birds stamp their feet and fluff their feathers repeatedly at a certain time of the year. This may be _____ behavior if potential mates are attracted to this display.

20. If your dog continually chases other dogs away from your front yard, your dog may be exhibiting _____ behavior because it has also marked certain areas of the yard with urine.

APPLYING MAIN IDEAS

21. What would be the advantage of a dominance hierarchy in members of a species that are not defending a territory?

22. If you found a nest of five-day-old goslings, would they imprint on you and follow you home? Explain.

23. Explain how Ivan Pavlov used the methods of science to study conditioning behavior.

24. By accident, a gull drops a snail on the road. The snail's shell breaks, and the gull eats the snail. The gull continues to drop mollusks on the road. What type of behavior is this?

THINKING CRITICALLY

25. Comparing and Contrasting Ducklings display an alarm reaction when a model of a hawk is flown over their heads, and no alarm reaction when a model of a goose is flown over their heads. After several days, neither model causes any reaction. Compare the effects of the two models during the first two days with the effects of the same models two weeks later.

26. Recognizing Cause and Effect When Charles Darwin visited the Galapagos Islands in 1835, he was amazed that the animals would allow him to touch them. Why were they not afraid?

27. Concept Mapping Complete the concept map by using the following vocabulary terms: innate behavior, imprinting, habituation, conditioning, insight.

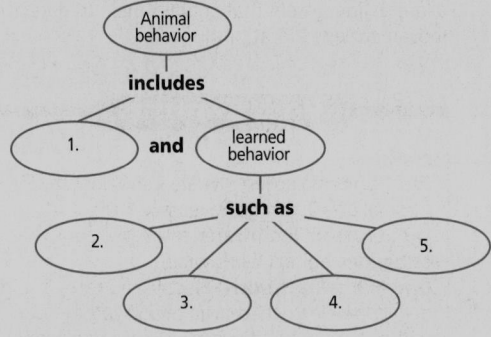

CD-ROM

For additional review, use the assessment options for this chapter found on the **Biology: The Dynamics of Life Interactive CD-ROM** and on the Glencoe Science Web site.
science.glencoe.com

ASSESSING KNOWLEDGE & SKILLS

Chickens that show submission find themselves at lower levels in the barnyard pecking order.

Interpreting Scientific Illustrations Use the diagram above to answer the following questions.

1. Which chicken is dominant?
 a. 1   **c.** 3
 b. 2   **d.** 4

2. Which chicken or chickens would chicken number 4 peck?
 a. only 3   **c.** 1, 2, and 3
 b. only 2 and 3   **d.** none

3. Which chickens could be described as submissive?
 a. only 2, 3, and 4   **c.** only 4
 b. only 3 and 4   **d.** only 1

4. What type of behavior is illustrated in the diagram?
 a. learned behavior
 b. courtship behavior
 c. instinctive behavior
 d. reflex behavior

5. Creating a Model Under crowded conditions, mice form dominance hierarchies. Create a model showing six mice that have set up a dominance hierarchy in a small cage. Write a caption that explains what is happening.

24. learned behavior

THINKING CRITICALLY

25. They show alarm with the hawk model, but not with the goose because a hawk is a potential predator of the duckling, and the goose is not. This is instinctive behavior. When there are no dangerous results, the ducklings become habituated to the hawk model.

26. They had learned not to fear large animals because there were no large predators on the islands.

27. 1. Innate behavior; 2. Imprinting; 3. Insight; 4. Habituation; 5. Conditioning

ASSESSING KNOWLEDGE & SKILLS

1. a
2. d
3. a
4. c
5. Make sure that the mice are exhibiting threatening or aggressive posture from one to another and that the order of the dominance hierarchy is clear.

Note Internet addresses that you find useful in the space below for quick reference.

_____
_____
_____
_____

National Science Education Standards:
UCP.1, UCP.2, UCP.3, UPC.4,
UCP.5, C.3, C.4, C.5, C.6

Prepare

Purpose

This BioDigest can be used as an introduction to or as an overview of the vertebrate classes. You may wish to use this unit summary to teach about vertebrates in place of the chapters in the Vertebrates unit.

Key Concepts

Students will study vertebrates by comparing and contrasting the features of fishes, amphibians, reptiles, birds, and mammals. Students learn that there are three classes of fishes as compared with one class each of amphibians, reptiles, birds, and mammals. Students will trace the evolution of vertebrates and identify the similar traits that are found within all seven vertebrate classes. Students will see differences in form that correspond to adaptations to different habitats as vertebrate ancestors moved from life in the water to life on land.

1 Focus

Bellringer

Collect pictures of many different kinds of vertebrates and place them on the bulletin board. Number each picture. As students enter the classroom, have them take a sheet of notebook paper and list the names of the numbered animals. Ask students what all these animals have in common. They should be able to list some vertebrate characteristics, such as having a backbone. **L1**

For a **preview** of the vertebrate unit, study this BioDigest before you read the chapters. After you have studied the vertebrate chapters, you can use the BioDigest to **review** the unit.

Vertebrates

Like all chordates, vertebrates have a notochord, gill slits, a dorsal hollow nerve cord, and muscle blocks. However, in vertebrates the notochord is replaced during development by a backbone. All vertebrates are bilaterally symmetrical, coelomate animals that have an endoskeleton, a closed circulatory system, an efficient respiratory system, and a complex brain and nervous system.

Fishes

All fishes are ectotherms, animals with body temperatures dependent upon the temperature of their surroundings. Fishes have two-chambered hearts and breathe through gills. Fishes are grouped into three different classes.

Jawless Fishes

Lampreys and hagfishes are jawless fishes. Jawless fishes have endoskeletons made of cartilage, like sharks and rays, but they do not have jaws.

Cartilaginous Fishes

Sharks, skates, and rays are cartilaginous fishes. Fossil evidence shows that jaws first

Cartilaginous fishes such as this whitetip reef shark are more dense than water. Sharks will sink if they stop swimming.

evolved in these fishes. Cartilaginous fishes have endoskeletons made of cartilage, paired fins, and a lateral line system that enables them to detect movement and vibrations in water.

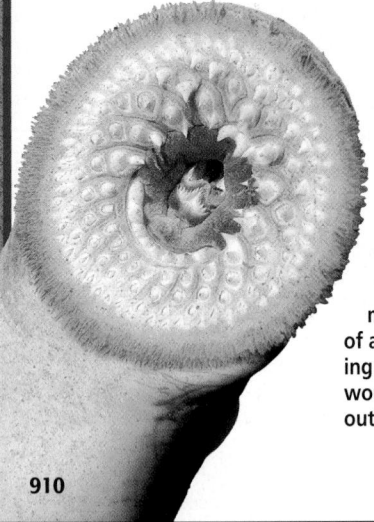

Lacking jaws, this sea lamprey obtains food by clamping its round mouth onto the side of a fish, using its rasping tongue to make a wound, then sucking out the prey's blood.

VITAL STATISTICS

Fishes
Size ranges: Largest: Whale shark, length, 15 m; smallest: Dwarf goby, length, 1 cm.
Distribution: Freshwater, saltwater, and estuarine habitats worldwide.
Unusual adaptations: Electric eels can deliver an electrical charge of 650 volts, which stuns or kills their prey. Some deep-sea fishes have their own bioluminescent lures to catch prey.
Longest-lived: Lake sturgeon, 80 years.
Numbers of species:
Class Agnatha—jawless fishes, 63 species
Class Chondrichthyes—cartilaginous fishes, 850 species
Class Osteichthyes—bony fishes, 18 000 species

Multiple Learning Styles

Look for the following logos for strategies that emphasize different learning modalities.

Kinesthetic Meeting Individual Needs, p. 912; Building a Model, p. 913; Activity, p. 915; Extension, p. 918

Visual-Spatial Visual Learning, pp. 911, 915; Display, p. 911, 912; Quick Demo, p. 912; Biology Journal, p. 915; Portfolio, p. 918

Interpersonal Tech Prep, p. 913; Project, p. 916
Linguistic Tech Prep, p. 914; Biology Journal, p. 918
Auditory-Musical Meeting Individual Needs, p. 914
Naturalist Field Trip, p. 911; Biology Journal, p. 916; Meeting Individual Needs, p. 917

BioDigest

Most fishes fertilize their eggs externally and leave their survival to chance. Female sea horses deposit their eggs directly into brood pouches found underneath the tails of the males. The eggs develop inside the brood pouch.

Bony Fishes

Most fish species belong to the bony fishes. All bony fishes have a bony skeleton, gills, paired fins, flattened bony scales, and a lateral line system. Bony fishes breathe by drawing water into their mouths, then passing it over gills where gas exchange occurs. They adjust their depth in the water by regulating the amount of gas that diffuses out of their blood into a swim bladder.

Amphibians

Amphibians are ectothermic vertebrates with three-chambered hearts, lungs, and thin, moist skin. Although they have lungs, most gas exchange in amphibians is carried out through the skin. As adults, amphibians live on land but rely on water for reproduction. Almost all amphibians go through metamorphosis, in which the young hatch into tadpoles, which gradually lose their tails and gills as they develop legs, lungs, and other adult structures.

Although salamanders resemble lizards, they have smooth, moist skin and lack claws on their toes, features used to classify salamanders as amphibians.

Amphibian Classification

Amphibians are classified into three orders: Anura, the frogs and toads; Caudata, the salamanders and newts; and Apoda, the legless caecilians. Frogs and toads have vocal cords that can produce a wide range of sounds. Frogs have thin, smooth, moist skin and toads have thick, bumpy skin with poison glands. Salamanders have long, slender bodies with a neck and tail. Caecilians are amphibians with long, wormlike bodies and no legs.

VITAL STATISTICS

Amphibians
Size ranges: Largest: Goliath frog, length, 30 cm; Chinese giant salamander, length, 1.8 m; Smallest frog: *Psyllophryne didactyla*, length, 9.8 mm.
Distribution: Tropical and temperate regions worldwide.
Numbers of species:
Class Amphibia
   Order Anura—frogs and toads, 3700 species
   Order Caudata—salamanders and newts, 400 species
   Order Apoda—legless caecilians, 168 species

Vibrations from air or water are picked up by the frog's tympanic membrane and transmitted to the inner ear. The tympanic membrane is located behind and below the frog's eye.

911

BioDigest

2 Teach

Field Trip

Naturalist Arrange a class trip to a local zoological park, nature center, or other park or natural area where students will be able to see representatives of the classes of vertebrates discussed in this unit. Ask students to list as many vertebrates as they see, then divide the animals into classes. Ask them to identify the characteristics they used to classify the animals. **L2**

Display

Visual-Spatial Ask students to prepare a bulletin board display of bony fishes by cutting out pictures of bony fishes from nature magazines. Fishes should be identified and each fish should have one part labeled. Have students use the following labels: scales, lateral line, gills, swim bladder, and fins. **L1** **ELL**

Assessment Planner

Portfolio Assessment
   Portfolio, TWE, p. 918
Performance Assessment
   Assessment, TWE, pp. 917, 918
   BioDigest Assessment, TWE, p. 919
Knowledge Assessment
   BioDigest Assessment, SE, p. 919

BIODIGEST

Visual Learning

Direct students' attention to the photos of the snake and the crocodile. Ask them to explain how the snake and the crocodile are alike and how they are different. *They are both ectotherms with skin that is dry, thick, and covered with scales. They both have lungs and lay eggs with leathery shells. The snake does not have legs, whereas the crocodile has four legs. The snake has a three-chambered heart and the crocodile has a four-chambered heart.*

Quick Demo

 Visual-Spatial Show students a live snake, lizard, or turtle from the local environment or borrowed from a pet shop. Have students point out the reptile features of the animal.

Display

 Visual-Spatial Have students prepare a bulletin board display of local reptiles. Have them do research to determine if there are any local endangered reptiles.
L1

GLENCOE TECHNOLOGY

CD-ROM
Biology: The Dynamics of Life
Exploration: *The Six Kingdoms*
Disc 3
BioQuest: *Biodiversity Park*
Disc 3, 4

BIODIGEST

Reptiles

Reptiles are ectotherms with dry, scaly skin and clawed toes. They include snakes, lizards, turtles, crocodiles, and alligators. With the exception of snakes, all reptiles have four legs that are positioned somewhat underneath their bodies. Most reptiles have a three-chambered heart, but crocodilians have a four-chambered heart in which oxygenated blood is kept entirely separate from blood without oxygen. The scaly skin of reptiles reduces the loss of body moisture on land, but scales also prevent the skin from absorbing or releasing gases to the air. Reptiles are entirely dependent upon lungs for this essential gas exchange.

Constrictors, such as this emerald tree boa, hold prey with their mouths, then wrap coils around the prey's body. The snake tightens its coils, preventing inhalation, and the prey suffocates.

All crocodilians have nostrils and eyes that extend above the rest of their faces. This enables the animal to breathe and see while most of its body is under water.

FOCUS ON ADAPTATIONS

The Amniotic Egg

Nile crocodile hatchling

Reptiles were the first group of vertebrates to live entirely on land. They evolved a thick, scaly skin that prevented water loss from body tissues. They evolved strong skeletons, with limbs positioned somewhat underneath their bodies. These limbs enabled them to move quickly on land, avoiding or seeking the sun as their body temperatures demanded. But perhaps their most important adaptation to life on land was the development of the amniotic egg.

Protecting the Embryo
An amniotic egg encloses the embryo in amniotic fluid; provides the yolk, a source of food for the embryo; and surrounds both embryo and food with membranes and a tough, leathery shell. These structures in the egg

MEETING INDIVIDUAL NEEDS

English Language Learners/ Visually Impaired

 Kinesthetic Give each group of students several scientifically accurate reptile models that are available in toy stores. As they handle the models, ask them to make a list of the reptilian features of each model. **L1** **ELL**

GLENCOE TECHNOLOGY

CD-ROM
Biology: The Dynamics of Life
Exploration: *Amphibians*
Disc 4
Video: *Sea Turtle*
Disc 4

Marine turtles, such as this olive ridley, come ashore only to lay eggs in nests they dig on sandy beaches. Once the eggs are laid and covered with sand, mother turtles head back to sea.

The shell of a turtle is unique among vertebrates. Consisting of bony plates covered with horny shields, the turtle's shell provides protection from most predators, but it also prevents turtles from making rapid movements.

Internal Fertilization

All reptiles have internal fertilization and lay eggs. The development of the amniotic egg enabled reptiles to move away from a dependence upon water for reproduction. The amniotic egg provides nourishment to the embryo and protects it from drying out as it develops.

VITAL STATISTICS

Reptiles

Size ranges: Largest: Anaconda snake, length, 9 m; Leatherback turtle, weight, 680 kg; smallest: Thread snake, length, 1.3 cm.

Distribution: Temperate and tropical forests, deserts, and grasslands, and freshwater, saltwater, and estuarine habitats worldwide.

Reptile that causes most human death: King cobra, 7500 deaths per year.

Numbers of species:
Class Reptilia
 Order Squamata—snakes and lizards, 6800 species
 Order Chelonia—turtles, 250 species
 Order Crocodilia—crocodiles and alligators, 25 species
 Order Rhynchocephalia—tuataras, 1 species

---

help prevent dehydration and injury to the embryo as it develops on land. Most reptiles lay their eggs in protected places beneath sand, soil, gravel, or bark.

Membranes Inside the Egg

Membranes found inside the amniotic egg include the amnion, the chorion, and the allantois. The amnion is a membrane filled with fluid that surrounds the developing embryo. The embryo's nitrogenous wastes are excreted into a membranous sac called the allantois. The chorion surrounds the yolk, allantois, amnion, and embryo. With this egg, reptiles do not need water for reproduction. The evolution of the amniotic egg completed the move of tetrapods from water to land.

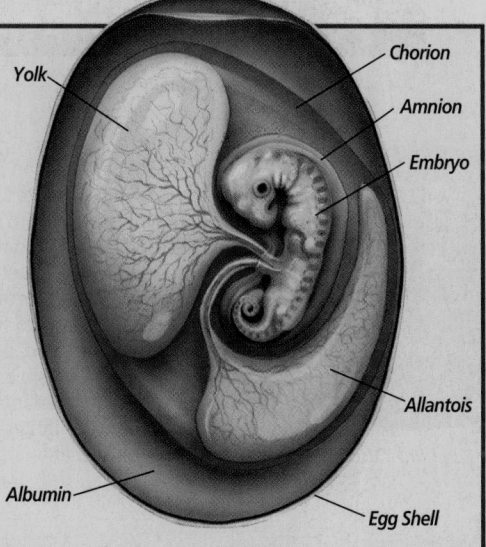

Yolk · Chorion · Amnion · Embryo · Allantois · Albumin · Egg Shell

913

Building a Model

Kinesthetic Ask students in groups to make a model of an amniotic egg using the following materials: a small lump of clay, two small zipper-type plastic bags, two pill vials or plastic film containers, and a larger lunch-size paper bag.

Do not tell them how to make their model, but give hints if they have trouble. Ask each group to explain their model. A model might be constructed as follows. The lump of clay represents the embryo. The pill vials stick into the lump of clay, one representing the yolk, and the other the allantois. The clay with attached vials can be placed into one plastic bag that represents the amnion. The amniotic sac and its contents can be placed inside the other plastic bag, which represents the chorion. The chorion and its contents can be placed inside the paper bag, which is folded closed to represent the shell. **L1** **ELL**
COOP LEARN

TECHPREP

First Aid for Snake Bite

Interpersonal Have a group of students investigate and demonstrate first aid for snake bites to the class.
L2 **ELL**

PROJECT

Reptile Misconceptions

Ask students to research and report on misconceptions about reptiles such as: that snakes sting with their tongues and you can tell the age of a rattlesnake by counting its rattles.

BIODIGEST

TechPrep

Endangered State Birds

Linguistic Have students do research to find out which endangered birds live in your state. Have them prepare a videotaped public service announcement about the birds' status with information about what can be done to enhance their chances of survival. **L3** **COOP LEARN**

GLENCOE
TECHNOLOGY

CD-ROM
Biology: The Dynamics of Life
Exploration: *Bird Adaptations*
Disc 4

BIODIGEST

Birds

Birds are the only class of animals with feathers. Feathers, which are lightweight, modified scales, help insulate birds and enable them to fly. Birds have forelimbs that are modified into wings. Like reptiles, birds have scales on their feet and clawed toes; unlike reptiles, they are endotherms, animals that maintain a constant body temperature. Endotherms must eat frequently to provide the energy needed for producing body heat.

Feathers keep birds warm and streamline them for flight. Feather colors are often important in courtship or camouflage. The peacock attracts the peahen with its display of tail feathers.

Penguins are flightless birds with wings and feet modified for swimming and a body surrounded by a thick layer of insulating fat. This young emperor penguin may reach a height of 1 m and weigh nearly 34 kg.

Bird Flight

Birds have thin, hollow bones with cross braces that provide support for strong flight muscles while reducing their body weight. Birds also have a four-chambered heart and a unique respiratory system in which oxygen is available during both inhalation and exhalation.

FOCUS ON ADAPTATIONS

Bird Flight

Peregrine falcon

What selection pressure may have resulted in bird flight? Maybe an early bird's need to escape from a predator caused it to run so fast its feet left the ground. Whatever caused birds to evolve an ability to fly, there must first have been adaptations that made flight possible. What are some of these adaptations? A bird's body is lighter in weight than any other animals' of the same size because it has hollow bones and air sacs throughout its body. It also has a beak instead of a heavy jaw with teeth, and its legs are made mostly of skin, bone, and tendons.

Efficient Respiration

Birds receive oxygenated air when they breathe in as well as when they breathe out. Air sacs enable birds to get more oxygen because

914

MEETING INDIVIDUAL NEEDS

Visually Impaired

Auditory-Musical Play a recording of various types of bird songs and calls. Ask students to distinguish among distress calls, calls made to the young, courtship songs, and territorial calls. **L2** **ELL**

BioDigest

Nest Builders

Like reptiles, birds lay amniotic eggs. Unlike reptiles, birds incubate their eggs in nests, keeping eggs warm until the young birds hatch.

The cedar waxwing is found in open woodlands, orchards, and backyards across the United States. They spend most of the year in flocks, descending upon orchards and eating until the fruit is gone.

VITAL STATISTICS

Birds

Size ranges: Largest: Ostrich, height, 2.4 m, mass, 156 kg; smallest: Bee hummingbird, length, 57 mm, mass, 1.5 g.
Distribution: Worldwide in all habitats.
Widest wingspan: Wandering albatross, 3.7 m.
Fastest flyer: White-throated spinetail swift, 171 kph.
Largest egg: Ostrich, length, 13.5 cm, mass, 1.5 kg.
Longest yearly migration: Arctic tern, 40 000 km.
Numbers of species:
Class Aves—8600 species in 27 present-day orders
    Order Passeriformes—perching song birds, 5400 species
    Order Ciconiiformes—herons, bitterns, ibises, 127 species
    Order Anseriformes—swans, ducks, geese, 161 species
    Order Falconiformes—eagles, hawks, falcons, 298 species

The largest bird's nest is built by the bald eagle. Every year, eagles add another layer of sticks to the nest until some nests are 2 m across and 2 tons in mass.

---

75 percent of the air inhaled by a bird passes directly into posterior air sacs rather than into its lungs. When a bird exhales, oxygenated air in the sacs passes into its lungs, then into anterior air sacs and out through the trachea. This one-way flow of air provides the large amounts of oxygen that birds need to power flight muscles.

Wings Adapted for Flight

Flight is also supported by feathers that streamline a bird's body and shape the wings. Wing shape and size determine the type of flight a bird is capable of. Birds that fly through the branches of trees in a forest, such as finches, have elliptically shaped wings adapted to quick changes of direction. Wings of swallows and terns have shapes that sweep back and taper to a slender tip, promoting high speed in open areas. The broad wings of hawks, eagles, and owls provide strong lift and slow speeds. These birds are predators that carry prey while in flight.

Arctic tern

915

BIOLOGY JOURNAL

Observing Feathers

Visual-Spatial Provide students with contour and down feathers and a magnifying glass. Feathers may be purchased from craft shops. Ask students to examine both feathers with a magnifying glass and draw diagrams of them in their journals. Ask them to hypothesize how the function of each feather differs. *Down feathers keep birds warm by trapping warm air near their bodies. Contour feathers are used for flight.* **L1** **ELL**

Visual Learning

Visual-Spatial Show students slides or pictures of birds in flight and ask them to hypothesize about what types of flight each bird could undertake: quick movements, high speeds, or soaring. *Elliptical wings are adapted to quick movements. Wings that sweep back and taper to a slender tip promote high speed. Soaring birds have broad wings.* **L1**

Activity

Kinesthetic Have students determine what kinds of birds live on school grounds. Look up specifications for houses for these birds and build bird houses to place on school grounds. **L1**

BioDigest

Concept Development

In one year, 45 bears were killed on the roads of Florida. Accidents between large mammals and vehicles are common. Ask students to find out which large mammals are killed each year in their state. Ask them to find out what measures are being, or could be, taken to prevent these accidents.

GLENCOE
TECHNOLOGY

VIDEODISC
The Infinite Voyage
The Keepers of Eden
Preserves of Endangered
Species: San Diego and Kenya
(Ch. 8), 8 min. 30 sec.

BioDigest

Mammals

Mammals are endotherms. Mammals are named for their mammary glands, which produce milk to feed their young. Most mammals have hair that helps insulate their bodies and sweat glands that help keep them cool. Mammals need a high level of energy for maintaining body temperature and high speeds of locomotion. An efficient four-chambered heart and the muscular diaphragm beneath the lungs help to deliver the necessary oxygen for these activities.

Mammal Diversity

All mammals have internal fertilization, and the young begin development inside the mother's uterus. But from that point, developmental patterns in mammals diverge. Mammals are classified into three groups. Monotremes are mammals that lay eggs. Marsupials are mammals in which the young complete a second stage of development after birth in a pouch made of skin and hair on the outside of the mother's body. Placental mammals carry their young inside the uterus until development is nearly complete.

Female mammals, such as this moose, feed their young milk secreted from mammary glands. Often, the young are cared for until they become adults.

FOCUS ON ADAPTATIONS

Endothermy

Both birds and mammals are endotherms. Endotherms have internal processes that maintain a constant body temperature. Just as a thermostat controls the temperature of your home, internal processes cool endotherms if they are too warm, and warm them if they are too cool, thus maintaining homeostasis.

A variety of adaptations enables mammals to maintain body temperature. Hair helps many mammals conserve heat. The thick coat of a polar bear is an adaptation to living in a cold climate. Small ears and an accumulation of body fat under the skin also help prevent heat loss. Small ears have less surface area than large ears from which body heat can escape.

Polar bear

PROJECT

Campaign to Save Mammals

Interpersonal Ask students to develop a campaign to save a local mammal from extinction. They should prepare a list of recommendations, publicity in the form of commercials, bumper stickers, and a poster and suggest an original idea to raise money to create a sanctuary. **L2** **COOP LEARN**

BIOLOGY JOURNAL

Mammal Range Maps

Naturalist Give students a blank map of the United States and field guides. Ask them to make a list of local mammals. Have them look up each mammal on their list in field guides and indicate the range of each animal on their maps. They should also make a color-coded key for their maps. **L2** **ELL**

BioDigest

The duck-billed platypus is a monotreme with webbed front feet adapted for swimming, and sharp claws for digging and burrowing into the soil.

Most mammals are placental mammals. They have extraordinary ranges in sizes and body structures. Many hoofed mammals, such as this deer, have an adaptation known as cud chewing that enables them to break down the cellulose of plants to make nutrients available to the animal.

This young wallaby is a marsupial that is old enough to survive outside its mother's pouch, but it still seeks protection there when danger threatens.

Hibernating bat

Hibernation Many rodents hibernate during periods of extreme cold. During hibernation, the body temperature lowers. For example, when the surrounding temperature drops to about 0°C, a ground squirrel's temperature drops to 2°C, and it goes into hibernation, which conserves the animal's energy.

Estivation In hot desert environments, where water is limited, some small rodents survive without drinking. They obtain enough water from the foods they eat. Other desert mammals, such as the fennec fox, have large ears that aid in heat loss. During periods of intense heat, some desert mammals go into a state of reduced metabolism called estivation. As a result, the animal's body temperature is lowered and energy is conserved.

Fennec fox

917

BioDigest

BioDigest

✓ **Assessment**

Performance Have students sketch, on a sheet of butcher paper, a scale map of the school grounds or a spot near the school. Have them indicate how the area could be made more suitable for local wildlife. Have students work in groups to add features to their maps that would make the area more attractive to wildlife. Books about wildlife habitat restoration are available in libraries. **L2** **ELL** 📋 **COOP LEARN**

Visual Learning

Show students slides of local mammals and ask them to point out how they are adapted to their habitats. *Make sure they point out features that help them move, feed, and give them protection from predators and environmental elements.*

MEETING INDIVIDUAL NEEDS

Gifted

Naturalist Ask students to develop a wildlife-management plan for a particular mammal in your area. They should develop and carry out a census. Then, based on the census, they should develop a management plan that will protect and maintain appropriate habitat for 25 years.

L3 📋

BioDigest

3 Assess

Check for Understanding

Show students a variety of mammal skulls or transparencies of mammal skulls you have made from field guides to mammals. Direct their attention to how teeth reflect what the animal eats. Ask them to make comparisons of teeth in carnivores and herbivores. *Carnivores have sharp, pointed canines that enable them to tear apart prey. Herbivores have premolars and molars adapted to grinding the plant materials they eat.*

Reteach

Show students slides or photos of the three groups of mammals and ask them to identify them by their features.

Extension

Kinesthetic Mammals must use their teeth exclusively to eat as most mammals do not have hands or opposable thumbs to assist. Ask students to shell a peanut with their fingers taped together so that only the top joint of the thumb can move. Ask them why diversified teeth are important to mammals. **L1 ELL**

✔ Assessment

Performance Ask students to design an experiment that would test the suitability of fat as insulation. Lard or cooking fat can be used in their experiments. **L2**

4 Close

Audiovisual

Rent a feature film in which a wild mammal or mammals play an important role. Ask students to identify important features of the mammals in the film.

BioDigest

Mammal Teeth

Mammals can be classified by the number and type of teeth. All mammals have diversified teeth used for different purposes. Incisors are used to cut food. Canines—long, pointed teeth—are used to stab or hold food. Molars and premolars have flat surfaces with ridges and are used to grind and chew food.

The chisel-like incisors of beavers and other rodents never stop growing.

Carnivores, such as this coyote, have canine teeth that stab and pierce food.

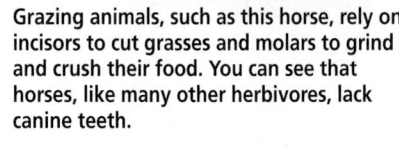

Grazing animals, such as this horse, rely on incisors to cut grasses and molars to grind and crush their food. You can see that horses, like many other herbivores, lack canine teeth.

VITAL STATISTICS

Mammals
Size ranges: Largest: Blue whale, length, 30 m, mass, 190 metric tons; smallest: Etruscan shrew, length, 6 cm, mass, 1.5 g.
Distribution: Worldwide in all habitats.
Fastest: Cheetah, 110 kph.
Longest-lived: Asiatic elephant, 80 years; humans, up to 120 years.
Numbers of species:
Class Mammalia
   Order Monotremata—egg-laying mammals, 3 species
   Order Marsupialia—pouched mammals, 260 species
   Orders of Placental Mammals—4418 species

✔ Portfolio

Photo Essay

Visual-Spatial Ask a group of students to go to a nearby zoo and take photographs of the animals. Have them present a photo essay to the class. Ask them to discuss whether each animal is endangered or threatened and any programs undertaken for these animals. **L2**
ELL **P** **COOP LEARN**

Biology Journal

Create a Pet

Linguistic Ask students to use their creativity and knowledge of vertebrates to create their ideal vertebrate pet. They should write a description and draw a diagram of this pet in their journals. Have them label their diagrams with all the features unique to the vertebrate group to which their pet belongs. **L2 ELL**

BioDigest

BioDigest Assessment

Understanding Main Ideas

1. Which of the following animals are ectotherms?
   a. fishes, amphibians, reptiles, and birds
   b. birds and mammals
   c. fishes, amphibians, and reptiles
   d. fishes, amphibians, reptiles, birds, and mammals

2. Which of the following fishes have jaws?
   a. lampreys, sharks, and bony fishes
   b. lampreys only
   c. sharks, skates, and rays only
   d. cartilaginous and bony fishes only

3. Which of the following animals have eggs without shells?
   a. lizards, snakes, and turtles
   b. lizards, frogs, and toads
   c. frogs, toads, and salamanders
   d. frogs, snakes, and alligators

4. Which amphibian has thick, bumpy skin with poison glands?
   a. frogs   c. lizards
   b. toads   d. salamanders

5. The first animals to lay amniotic eggs were _____.
   a. fishes   c. reptiles
   b. amphibians   d. birds

6. Which reptile has a four-chambered heart?
   a. duck-billed platypus
   b. lizard
   c. snake
   d. alligator

7. The air sacs of birds enable them to _____.
   a. eat more food
   b. receive more oxygen
   c. hide from predators
   d. build large nests

8. Both birds and reptiles lay shelled, amniotic eggs; however, only birds sit on their eggs to _____ them, keeping them warm until they hatch.
   a. guard   c. protect
   b. incubate   d. nurse

9. Because the hearts of all mammals have four chambers, more _____ is delivered to their cells.
   a. energy   c. heat
   b. oxygen   d. sweat

10. Mammals are classified into subclasses based on their method of _____.
    a. locomotion   c. breathing
    b. feeding   d. reproduction

Thinking Critically

1. Why are endothermic animals able to live in areas of extreme temperatures such as the Arctic and the tropics? Explain.

2. Explain how the development of the amniotic egg was important for the transition of animals from life in the water to life on land.

3. How are reptiles and amphibians alike? How are they different?

4. If you found a mammal skull with chisel-like incisors, what type of feeding habits might this animal have?

5. Describe three structures of fishes not found in mollusks.

Three-toed sloth

919

BioDigest

BioDigest Assessment

Understanding Main Ideas

1. c	5. c	8. b
2. d	6. d	9. b
3. c	7. b	10. d
4. b		

Thinking Critically

1. When an animal can control its body temperature, it is able to live in habitats with temperature extremes without upsetting homeostasis of the body.

2. A food source, protective membranes and fluids, and a tough outer shell on the egg help prevent injury and dehydration of the embryo as it develops on land.

3. They are both ectotherms. All amphibians have a three-chambered heart, as do reptiles, with the exception of crocodilians. Reptiles have claws on their toes, whereas amphibians do not. Reptiles have thick, dry skin covered with scales, whereas amphibians have smooth, moist skin.

4. It might be an animal that gnaws on woody branches and bark.

5. Fishes have internal skeletons, fins, and lateral line systems.

Resource Manager

Reinforcement and Study Guide,
    pp. 149-150 **L2**
Content Mastery, pp. 165-168 **L1**

Unit 10

The Human Body

Unit Overview

Unit 10 describes the organs and systems of the human body and how they interact with one another. Chapter 34 describes the skin, skeletal, and muscular systems. The digestive and endocrine systems are covered in Chapter 35.

Chapter 36 looks at the nervous system's control of the body and the effects of drugs on body systems. Included is a Focus On The Brain, which describes functions of the brain in more detail. Respiration, Circulation, and Excretion follow in Chapter 37.

Chapter 38 provides a discussion of human reproduction and development. Chapter 39 outlines the function of the immune system and ends with a discussion of how AIDS affects the immune system.

Introducing the Unit

Ask students which systems are important for a runner to succeed at a track meet. Then discuss the following questions with the class. Which systems are involved in the excitement and tension the runner feels before the race? Which systems are involved when the runner is racing?

Unit 10

The Human Body

As the sprinter crosses the finish line, the crowd cheers. Winning a race requires the coordination of many different body systems—systems that don't work independently, but interact in hundreds of complex ways.

UNIT CONTENTS

UNIT PROJECT

BIOLOGY *Online*   Use the Glencoe Science Web site for more project activities that are connected to this unit.
science.glencoe.com

920

Unit Projects

Interactions of Body Systems

Have students do one of the projects for this unit as described on the Glencoe Science Web site. As an alternative, students can do one of the projects described on these two pages.

Display

Visual-Spatial Students can collect pictures of buildings from magazines and make a display. For each picture, encourage students to draw analogies between parts of the building and parts of the body that have similar functions. **L1 ELL**

Modeling

Kinesthetic Have student groups design and make a model of one body system, such as the nervous or endocrine system. They may use any materials they wish. **L2 ELL**

Advance Planning

Chapter 34
■ Order slides of human skin for the Alternative Lab.

Chapter 35
■ Borrow a human skull or a large animal skull for the first Quick Demo.
■ Order slides of the pancreas for the second Quick Demo.
■ Purchase whole kidneys that still have the adrenal glands attached for use in the third Quick Demo.
■ Order thyroid/parathyroid slides for MiniLab 35-2.

Chapter 36
■ Acquire models of ears, eyes, skin, and nose for the Display.
■ Order *Daphnia* for the BioLab.

Chapter 37
■ Order *Daphnia* for the Microscope Activity.
■ Order a whole kidney for the Quick Demo.

Chapter 38
■ Purchase a microscope slide showing a cross-section of testis for the Quick Demo and the Inside Story.
■ Purchase a microscope slide showing a cross-section of ovaries for the Inside Story.
■ Order sea urchin sperm and eggs for MiniLab 38-1.
■ Obtain slides of sea star development for the Quick Demo.

Chapter 39
■ Purchase bacterial culture and antibiotic disks for the Alternative Lab.
■ Purchase blood smear slides for MiniLab 39-2.

Unit Projects

Using the Library
Linguistic Have student groups prepare a report on how an injury to one system structure—such as torn cartilage or a ruptured spleen—can disrupt the whole system. **L3**

Interview
Interpersonal Have student groups interview a nurse or physician about a disease or disorder that affects mainly one body system. This could be an inherited disorder, such as sickle cell anemia, or a disease that disrupts the system, such as pneumonia. Have students present their findings in a report. **L2**

Final Report
Have student groups present their findings about body systems to the other students in the class. **L2**

921

Chapter 34 Organizer

Section	Objectives	Activities/Features
Section 34.1 **Skin: The Body's Protection** National Science Education Standards UCP.1, UCP.2, UCP.5; A.1, A.2; C.5; F.1, F.5; G.1 (1 session, ½ block)	1. **Compare** the makeup and functions of the dermis and epidermis. 2. **Recognize** the role of the skin in responding to external stimuli. 3. **Outline** the healing process that takes place when the skin is injured.	**Inside Story:** The Skin, p. 924 **MiniLab 34-1:** Examine Your Fingerprints, p. 925 **Problem-Solving Lab 34-1,** p. 926
Section 34.2 **Bones: The Body's Support** National Science Education Standards UCP.1, UCP.2, UCP.5; B.2, B.6; C.5; D.1; E.1; F.1, F.6 (2 sessions, 1 block)	4. **Identify** the structure and functions of the skeleton. 5. **Compare** the different types of movable joints. 6. **Recognize** how bone is formed.	**Problem-Solving Lab 34-2,** p. 933 **Physics Connection:** X rays—The Painless Probe, p. 942
Section 34.3 **Muscles for Locomotion** National Science Education Standards UCP.1-3, UCP.5; A.1, A.2; C.5; E.1, E.2; F.1, F.6; G.1 (2 sessions, 1½ blocks)	7. **Classify** the three types of muscles. 8. **Analyze** the structure of a myofibril. 9. **Interpret** the sliding filament theory.	**Problem-Solving Lab 34-3,** p. 936 **MiniLab 34-2:** Look at Muscle Contraction, p. 937 **Inside Story:** A Muscle, p. 938 **Design Your Own BioLab:** Does fatigue affect the ability to perform an exercise? p. 940

Need Materials? Contact Carolina Biological Supply Company at 1-800-334-5551 or at **http://www.carolina.com**

MATERIALS LIST

BioLab

p. 940 stopwatch or clock with second hand, graph paper, small weights, wooden box or step stool

MiniLabs

p. 925 ink pad, index cards, magnifying glass, acetone or nail polish remover

p. 937 metric ruler

Alternative Lab

p. 924 microscope, prepared slide of human skin

Quick Demos

p. 927 none

p. 930 X ray of human body

p. 930 skeleton

p. 931 chicken bone, vinegar or 10% hydrochloric acid

p. 932 fresh beef bone(2)

p. 936 fresh chicken feet, fresh chicken wings

Key to Teaching Strategies

L1 Level 1 activities should be appropriate for students with learning difficulties.

L2 Level 2 activities should be within the ability range of all students.

L3 Level 3 activities are designed for above-average students.

ELL ELL activities should be within the ability range of English Language Learners.

COOP LEARN Cooperative Learning activities are designed for small group work.

P These strategies represent student products that can be placed into a best-work portfolio.

These strategies are useful in a block scheduling format.

Teacher Classroom Resources

Section	Reproducible Masters	Transparencies
Section 34.1 **Skin: The Body's Protection**	Reinforcement and Study Guide, p. 151 **L2** BioLab and MiniLab Worksheets, p. 151 **L2** Content Mastery, pp. 169-170, 172 **L1**	Section Focus Transparency 81 **L1** **ELL** Basic Concepts Transparency 61 **L2** **ELL**
Section 34.2 **Bones: The Body's Support**	Reinforcement and Study Guide, pp. 152-153 **L2** Concept Mapping, p. 34 **L3** **ELL** Critical Thinking/Problem Solving, p. 34 **L3** Laboratory Manual, pp. 247-250 **L2** Tech Prep Applications, pp. 43-44 **L2** Content Mastery, pp. 169, 171-172 **L1**	Section Focus Transparency 82 **L1** **ELL** Basic Concepts Transparency 62 **L2** **ELL** Basic Concepts Transparency 63 **L2** **ELL**
Section 34.3 **Muscles for Locomotion**	Reinforcement and Study Guide, p. 154 **L2** BioLab and MiniLab Worksheets, pp. 152-154 **L2** Laboratory Manual, pp. 251-254 **L2** Content Mastery, pp. 169, 171-172 **L1** Inside Story Poster **ELL** Tech Prep Applications, pp. 43-44 **L2**	Section Focus Transparency 83 **L1** **ELL** Basic Concepts Transparency 64 **L2** **ELL** Reteaching Skills Transparency 49 **L1** **ELL**

Assessment Resources

Chapter Assessment, pp. 199-204
MindJogger Videoquizzes
Performance Assessment in the Biology Classroom
Alternate Assessment in the Science Classroom
ExamView® Pro Software 💾
BDOL Interactive CD-ROM, Chapter 34 quiz

Additional Resources

Spanish Resources **ELL**
English/Spanish Audiocassettes **ELL**
Cooperative Learning in the Science Classroom **COOP LEARN**
Lesson Plans/Block Scheduling

NATIONAL GEOGRAPHIC — Teacher's Corner

Products Available From Glencoe
To order the following products, call Glencoe at 1-800-334-7344:
CD-ROM
NGS PictureShow: Human Body 1
Curriculum Kit
GeoKit: Human Body 2
Transparency Set
NGS PicturePack: Human Body 1
Videodisc
STV: Human Body

Products Available From National Geographic Society
To order the following products, call National Geographic Society at 1-800-368-2728:
Videos
Incredible Human Machine
Muscular and Skeletal Systems (Human Body Series)

GLENCOE TECHNOLOGY

The following multimedia resources are available from Glencoe.

Biology: The Dynamics of Life
CD-ROM **ELL**

BioQuest: *Body Systems*
Exploration: *Bones: The Body's Support*
Animation: *Paired Skeletal Muscles*
Animation: *Sliding Filament Theory*

Videodisc Program 📦

Paired Skeletal Muscles
Sliding Filament Theory

Chapter

34 Protection, Support, and Locomotion

GETTING STARTED DEMO

Call students' attention to the photos of athletes on this page and in the magazines they bring to class. Discuss how muscle and skeletal shapes/sizes differ in the various types of athletes.

Theme Development

The major themes of the chapter are **systems** and **interactions.** Various body structures are discussed in relation to the systems they form. Interactions among and within systems are also emphasized.

0:00 OUT OF TIME?

If time does not permit teaching the entire chapter, use the BioDigest at the end of the unit as an overview.

READING BIOLOGY

Glencoe's *Biology: The Dynamics of Life* contains many resources to assist a student's reading skills. Each chapter contains figures with expanded captions that expand on written material. Word Origins, located along the side of text, expand knowledge of biology vocabulary. Glencoe's Content Mastery Booklet helps develop reading skills while reinforcing content. In addition, use the Interactive Tutor for *Biology: The Dynamics of Life* on the Glencoe Web site to reinforce vocabulary. **science.glencoe.com**

What You'll Learn

- You will recognize the skin's role in protecting your body.
- You will identify the many functions of your skeleton.
- You will classify the different types of muscles in your body.

Why It's Important

Your skin, skeleton, and muscles work together to protect, support, and move your body. A knowledge of each system helps you understand how your body is able to accomplish such a variety of activities.

READING BIOLOGY

As you look through the chapter, carefully look at some of the diagrams that appear. Do the detailed drawings of human skin, bones and muscle differ from what you expected? Write down any new vocabulary terms or questions that arise. Answer the questions as you read.

BIOLOGY Online

To find out more about skin, bones, and muscles, visit the Glencoe Science Web site. **science.glencoe.com**

Strong muscles and bones allow a basketball player to leap above a defender, or a tennis player to serve a ball at breakneck speed.

922

Multiple Learning Styles

Look for the following logos for strategies that emphasize different learning modalities.

Kinesthetic Portfolio, pp. 926, 931; Quick Demo, p. 927

Visual-Spatial Building a Model, p. 930; Quick Demo, p. 931; Project, p. 938

Interpersonal Meeting Individual Needs, p. 932; Reteach, p. 934; Activity, p. 934

Intrapersonal Tech Prep, p. 926; Meeting Individual Needs, p. 927

Linguistic Biology Journal, p. 927; Cultural Diversity, p. 930; Enrichment, p. 931; Portfolio, p. 936

Logical-Mathematical Reteach, p. 928; Tech Prep, p. 933; Biology Journal, p. 937

34.1 Skin: The Body's Protection

L ike all land animals, you live in a harsh, dry world. One of the many challenges you face is maintaining a moist environment inside your body. In addition, you must regulate your internal body temperature, keeping it within a certain range. Your skin helps you accomplish both of these tasks. Although it may seem like just a wrapping on the surface of your body, you'll see that skin is actually a complex organ that performs a variety of functions.

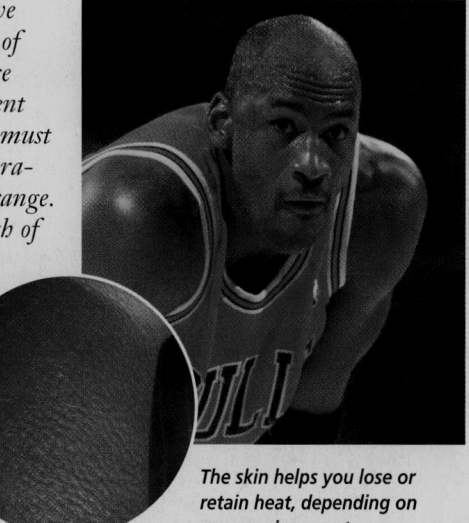

The skin helps you lose or retain heat, depending on your environment.

SECTION PREVIEW

Objectives

Compare the makeup and functions of the dermis and epidermis.

Recognize the role of the skin in responding to external stimuli.

Outline the healing process that takes place when the skin is injured.

Vocabulary
epidermis
keratin
melanin
dermis
hair follicle

Structure and Function of the Skin

Skin is composed of layers of the four types of body tissues: epithelial, connective, muscle, and nervous. Recall that epithelial tissue is derived from the ectodermal layer of the embryo and functions to cover surfaces of the body. Connective tissue, which consists of both tough and flexible protein fibers, serves as a sort of organic glue, holding your body together. Muscle tissues interact with hairs on the skin to respond to stimuli such as cold and fright. Nervous tissue helps us detect external stimuli, such as pain or pressure. As you can see, skin is a flexible and responsive organ.

Skin is composed of two principal layers—the epidermis and dermis. Each layer performs a different function in the body.

Epidermis: The outer layer of skin

The layer of skin that you see covering your body is called the epidermis. The **epidermis** is the outermost layer of the skin, and is made up of two parts—an exterior and interior portion. The exterior layer of the epidermis consists of 25 to 30 layers of dead, flattened cells that are continually being shed. Although dead, these cells still serve an important function as they contain a protein called **keratin** (KER uh tun).

WORD *Origin*

epidermis
From the Greek words *epi*, meaning "on," and *derma*, meaning "skin." The epidermis covers other layers of skin.

34.1 SKIN: THE BODY'S PROTECTION **923**

Section 34.1

Prepare

Key Concepts
Students look at how the skin functions to protect the body.

Planning
■ Gather ink pads, fingernail polish remover, cotton balls, index cards, and hand lenses for MiniLab 34-1.

1 Focus

Bellringer
Before presenting the lesson, display **Section Focus Transparency 81** on the overhead projector and have students answer the accompanying questions.
L1 **ELL**

Resource Manager

Section Focus Transparency 81 and Master **L1** **ELL**

✓ Assessment Planner

Portfolio Assessment
  Portfolio, TWE, pp. 926, 931, 936
  BioLab, TWE, pp. 940-941
Performance Assessment
  MiniLab, SE, pp. 925, 937
  BioLab, SE, pp. 940-941
  MiniLab, TWE, p. 925
  Assessment, TWE, pp. 932, 939
  Alternative Lab, TWE, pp. 924-925

Knowledge Assessment
  Assessment, TWE, pp. 934, 938
  Section Assessment, SE, pp. 928, 934, 939
  Chapter Assessment, SE, pp. 943-945
  Problem-Solving Lab, TWE, pp. 926, 933, 936
  MiniLab, TWE, p. 937
Skill Assessment
  Alternative Lab, TWE, pp. 924-925
  Assessment, TWE, pp. 927, 928

2 Teach

Purpose

Students gain a further understanding of how structures of the skin perform their functions.

Teaching Strategies

■ Ask students to identify the primary function of hair. **L2**

■ Use the diagram to point out to students where pimples may form. *in blocked hair follicles*

Visual Learning

■ If you set up the bulletin board display showing skin problems, challenge students to identify which skin structure is involved in each problem shown. **L2**

Critical Thinking

The skin gets sweaty in order to control its internal temperature as evaporated sweat lowers body temperature. The skin secretes oil to keep from drying out.

Resource Manager

BioLab and MiniLab Worksheets, p. 151 **L2**

Basic Concepts Transparency 61 and Master **L2** **ELL**

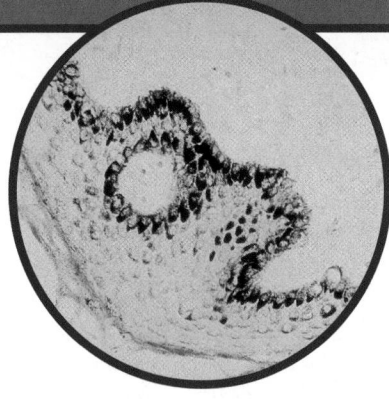

INSIDE STORY

The Skin

The skin is an organ because it consists of tissues joined together to perform specific activities. It is the largest organ of the body; the average adult's skin covers one to two square meters.

Critical Thinking *Why does your skin get sweaty and oily?*

Melanin in skin cells.

1 Melanin Differences in skin color are due to the amount of the pigment melanin produced by the cells. Exposure to sunlight causes an increase in melanin production, and the skin becomes darker.

2 Oil glands Most oil glands are connected to hair follicles. Oil prevents hair from drying out and keeps the skin soft and pliable. It also inhibits the growth of certain bacteria.

3 Hair Hair's primary function is protection of the skin from injury and damaging solar rays. It also provides an insulating layer of air just above the surface of the skin.

4 Elasticity The connective tissue of the dermis contains many elastic fibers that allow the skin to return to its original shape after being stretched.

5 Sweat glands Sweat glands are located deep in the dermis and open up through pores onto the surface of the skin. On average, a person loses about 900 mL of sweat each day.

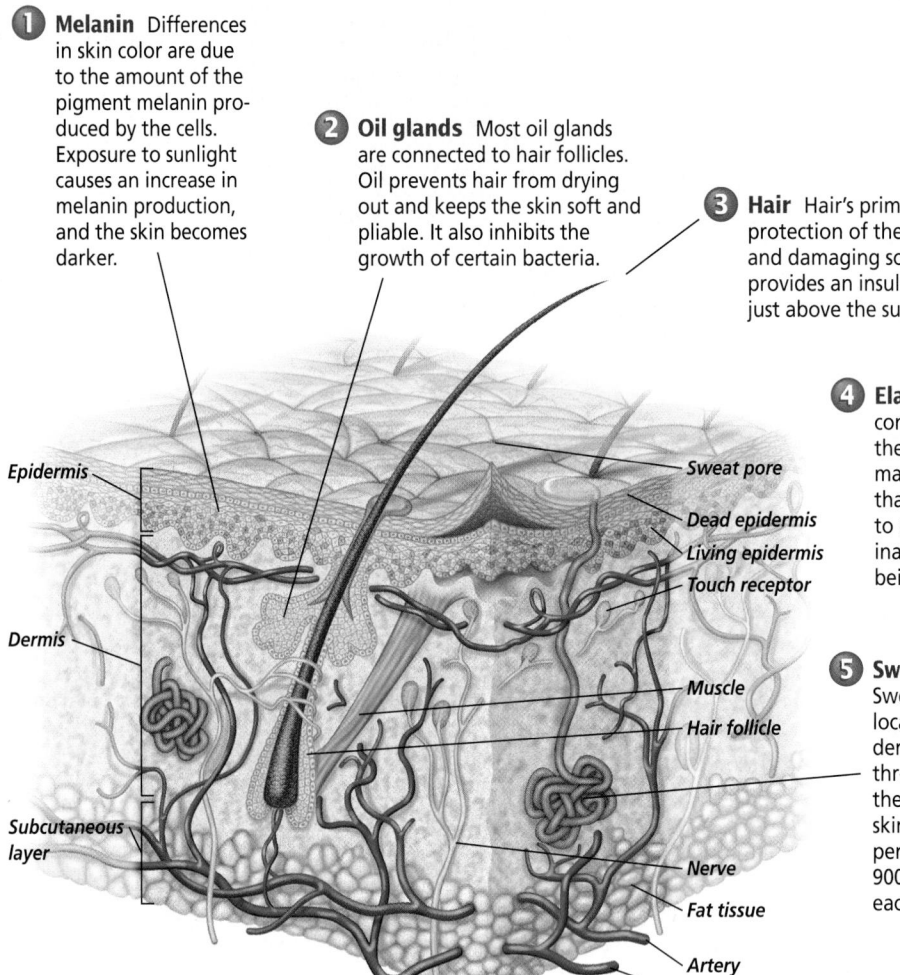

Epidermis

Dermis

Subcutaneous layer

Sweat pore
Dead epidermis
Living epidermis
Touch receptor
Muscle
Hair follicle
Nerve
Fat tissue
Artery
Vein

Alternative Lab

The Structure of Skin

Purpose

In this lab, students will observe the structure of human skin.

Preparation

You may choose to view the slides yourself before giving them to students to view. Look for the structures named in the steps below. Remind students to handle prepared slides with extreme caution.

Materials

microscope slide of human skin, microscope

Procedure

Give students the following directions.

1. View the human skin slide on low power. Focus first on the epidermis and switch to high power.

2. Then switch back to low power and locate the dermis of the skin. Switch to high power.

3. Find the following structures: hair follicle, hair shaft, sweat gland, blood vessels, fat deposits, oil gland, and elastic fibers.

4. Notice that the outer portion of the dermis has numerous projections, similar to hills and valleys. These projections, or papillae, form no pattern

Keratin helps protect the living cell layers underneath and contributes to skin's elasticity.

The interior layer of the epidermis contains living cells that continually divide to replace the dead cells. Some of these cells contain **melanin,** a pigment that colors the skin and protects the cells from damage by solar radiation. As the newly formed cells are pushed toward the skin's surface, the nuclei degenerate and the cells die. Once they reach the outermost epidermal layer, the cells are shed. This entire process takes about 28 days. Therefore, every four weeks, all cells of the epidermis are replaced by new cells.

Look at your fingertips. The epidermis on the fingers and palms of your hands, and on the toes and soles of your feet, contains ridges and grooves that are formed before birth. These epidermal ridges are important for gripping as they increase friction. As shown in *Figure 34.1,* footprints, as well as fingerprints, are often used to identify individuals as each person's pattern is unique. Make a set of your own fingerprints while doing the *MiniLab* on this page.

Dermis: The inner layer of skin

The second principal layer of the skin is the dermis. The **dermis** is the inner, thicker portion of the skin. The thickness of the dermis varies in different parts of the body, depending on the function of that part.

The dermis contains structures such as blood vessels, nerves, nerve endings, hair follicles, sweat glands, and oil glands. Why do some people have dark skin while others are pale? Find out by reading the *Inside Story.* Some areas of your skin may be looser and more flexible than others. This is because different amounts of fat lie underneath the dermis in dif-

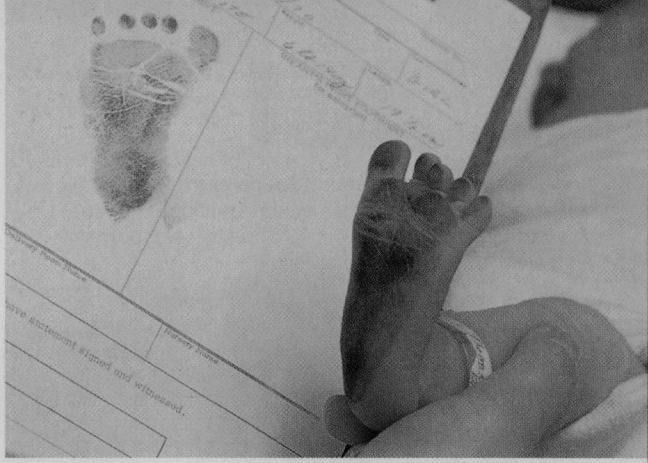

Figure 34.1
Babies' footprints are recorded at birth to establish an identification record for them in the future.

MiniLab 34-1 Comparing

Examine Your Fingerprints Fingerprints play a major role in any police investigation. Because a fingerprint is an individual characteristic, extensive FBI fingerprint files are used for identification in criminal cases.

Procedure
1. Press your thumb lightly on the surface of an ink pad.
2. Roll your thumb from left to right across the corner of an index card, then immediately lift your thumb straight up from the paper.
3. Repeat the steps above for your other four fingers, placing the prints in order across the card.
4. Examine your fingerprints with a magnifying lens, identifying the patterns in each by comparing them with the diagrams below.
5. Compare your fingerprints with those of your classmates.

| Arch | Whorl | Loop | Combination |

Analysis
1. Are the fingerprint patterns on your five fingers all the same?
2. Do any of your fingerprints show the same patterns as those of a classmate?
3. Why is a fingerprint a good way to identify a person?

Purpose
Students will observe, describe, and compare similarities and differences among fingerprint patterns to determine the uniqueness of such patterns.

Process Skills
observe and infer

Safety Precautions
Be sure students do not breathe the fingernail polish remover fumes. When this substance is in use, the room should be well ventilated. Keep chemicals away from open flames.

Teaching Strategies
■ Students should use nail polish remover to clean the ink off their fingers before washing them with soap and water.

Expected Results
Each finger has a unique fingerprint and each individual has a unique set of fingerprints.

Analysis
1. Although each finger has a unique fingerprint, the general patterns may be the same on different fingers.
2. It is possible that students may share patterns, but not identical fingerprints.
3. It is unique for each person.

✓ Assessment
Performance Have students design an experiment to determine if the print patterns of toes are unique to each individual. Use the Performance Task Assessment List for Designing an Experiment in **PASC,** p. 23. **L3**

---

except on the epidermis of the fingertips, palms, and soles of the feet, where they create parallel ridges that facilitate gripping.

Analysis
1. What structures form the fingerprints seen on fingers? *The outer portion of the dermis has projections, or papillae, that form the epidermal ridges seen in fingerprints.*
2. What is the function of the oil glands of the skin? *Oil secreted by the glands keeps hair from drying out and keeps the skin soft and pliable.*
3. How does sunlight affect the amount of melanin in skin? *Sunlight causes an increased production of melanin in skin cells.*

✓ Assessment
Skill Have students draw, label, and color human skin as seen under the microscope. Students should place their drawings in their journals. Use the Performance Task Assessment List for Scientific Drawing in **PASC,** p. 55. **L1**
ELL

Purpose

Students will analyze the series of events that occur in the body as internal body temperature increases.

Process Skills

think critically, recognize cause and effect, concept map, apply concepts, predict, sequence

Teaching Strategies

■ Make sure that students are familiar with all the terms used in the lab, including *dilate* and *capillaries*.

■ Verify that the concept of perspiring as a way of lowering body temperature is understood by students.

■ This lab would be suitable for small cooperative groups.

Thinking Critically

1. skin (dermis and epidermis) and circulatory (blood capillaries)

2. The brain is the structure that gives the signal to the sweat glands (to produce sweat) and to the capillaries (to dilate).

3. No, perspiring is a mechanism used for lowering body temperature.

4. Cold causes body temperature to drop ⇒ brain detects drop in blood temperature ⇒ message sent to blood capillaries ⇒ capillaries constrict ⇒ heat not lost from skin ⇒ body warms.

✔ Assessment

Knowledge Ask students to explain why a person becomes flushed when he or she has a fever. What evidence do they have that body temperature is controlled automatically and not consciously? Use the Performance Task Assessment List for Formulating a Hypothesis in **PASC**, p. 21. **L2**

926

Problem-Solving Lab 34-1

Recognizing Cause and Effect

How does your body respond to too much heat? As you exercise vigorously, you notice two things happening. One, you start to perspire. Two, your face becomes red. Both reactions are your body's way of cooling off.

Analysis

At right is a diagram of the events that take place in your body without your having to tell it to respond to a rise in internal temperature.

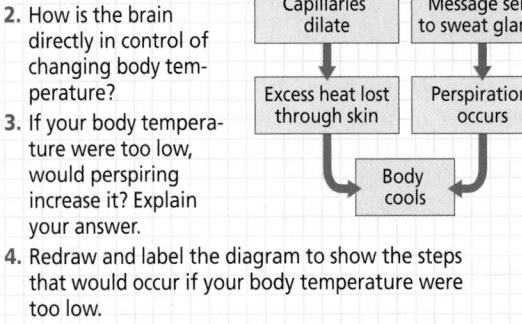

Thinking Critically

1. What two systems (not including the nervous system) work together to cool the body?

2. How is the brain directly in control of changing body temperature?

3. If your body temperature were too low, would perspiring increase it? Explain your answer.

4. Redraw and label the diagram to show the steps that would occur if your body temperature were too low.

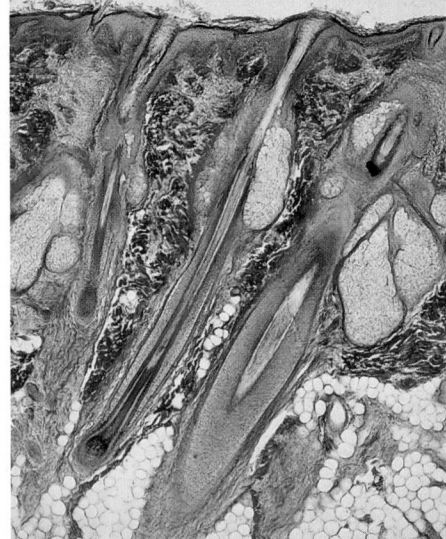

Figure 34.2
Photomicrograph of a cross section of human skin with hair follicle and hair.

926

---

ferent parts of your body. These fat deposits help the body absorb impact, retain heat, and store food.

Hair grows out of narrow cavities in the dermis called **hair follicles,** as shown in *Figure 34.2*. As hair follicles develop, they are supplied with blood vessels and nerves and become attached to muscle tissue. Most hair follicles have an oil gland associated with them. When oil and dead cells block the opening of the hair follicle, blackheads or pimples may form.

The skin's vital functions

One major function of skin is to regulate your internal body temperature. Think about how your body warms up as you exercise. When your body temperature rises, the many small blood vessels in the dermis dilate, blood flow increases, and body heat is lost by radiation. This mechanism also works in reverse. When you are cold, the blood vessels in the skin constrict and heat is conserved.

Another noticeable thing that happens to your skin as your body heats up is that it becomes wet. Glands in the dermis produce sweat in response to an increase in body temperature. As sweat evaporates, the body cools. This is because as water changes state from liquid to vapor, heat is lost. Investigate further the role of skin in cooling the body by carrying out the *Problem-Solving Lab* on this page.

Of course, anyone who has ever stepped on a sharp object or been burned by a hot pot handle knows that skin also functions as a sense organ. Nerve cells in the dermis receive stimuli from the external environment and relay information about pressure, pain, and temperature to the brain.

✔ Portfolio

Modeling Skin Structure

Kinesthetic Have students use a variety of materials to create three-dimensional models of the skin. Encourage students to use the illustrations found in the Inside Story as a guide. Students should include labels that identify each structure in their models as well as its function. **L2**

ELL **P** **COOP LEARN**

Acne Treatments

Intrapersonal Ask students to search the Internet and teen magazines to find various acne treatments. Have them record in their journals how these treatments prevent or treat acne. **L2** **ELL**

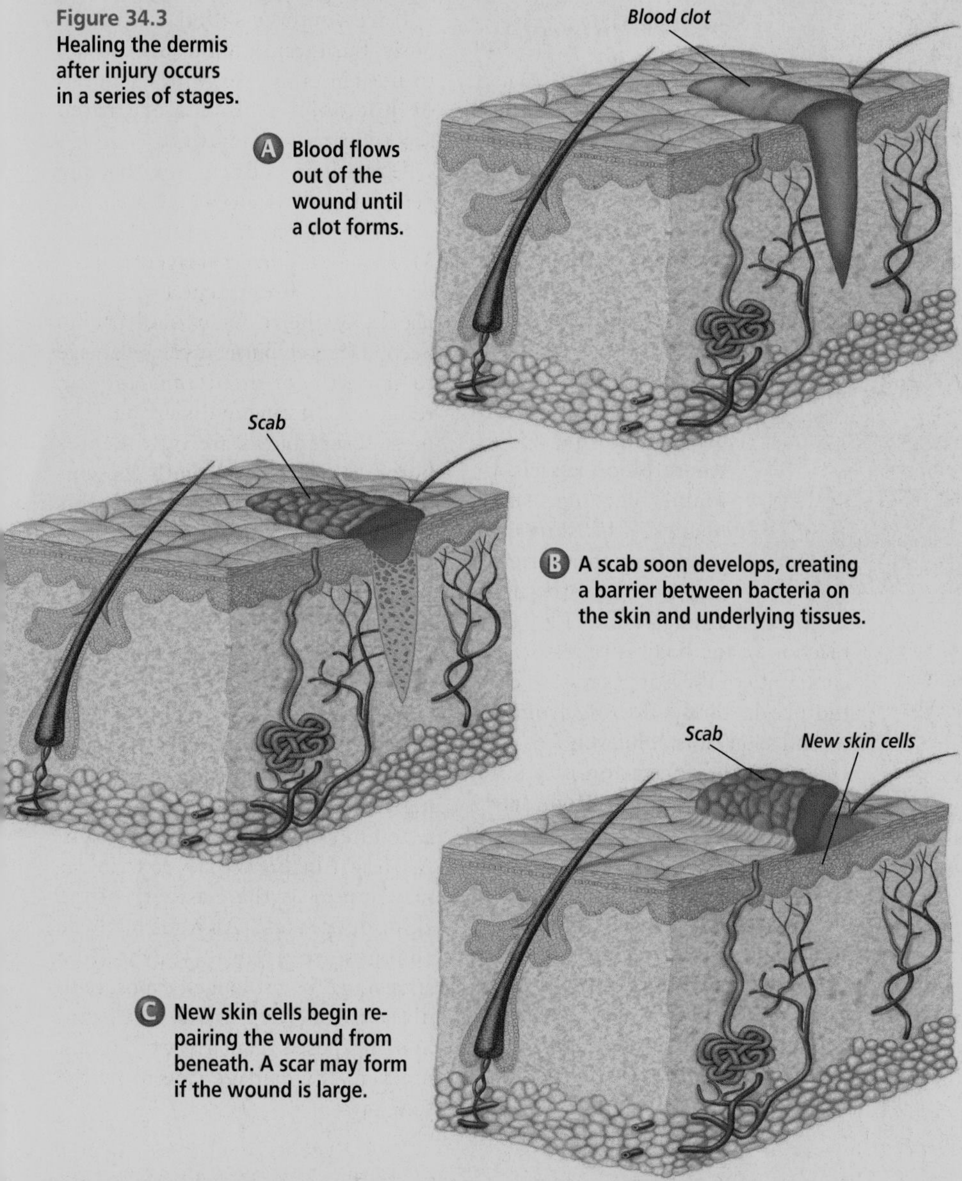
Blood clot
A Blood flows out of the wound until a clot forms.
Scab
B A scab soon develops, creating a barrier between bacteria on the skin and underlying tissues.
Scab
New skin cells
C New skin cells begin repairing the wound from beneath. A scar may form if the wound is large.

Figure 34.3
Healing the dermis after injury occurs in a series of stages.

Skin also plays a role in producing essential vitamins. When exposed to ultraviolet light, skin cells produce vitamin D, a nutrient that aids the absorption of calcium into the bloodstream. Because an individual's exposure to sunlight varies, daily intake of vitamin D supplements is sometimes needed to meet requirements.

Skin also serves as a protective layer to underlying tissues. It shields the body from physical and chemical damage and from invasion by microbes. Cuts or other openings in the skin surface allow bacteria to enter the body and so must be repaired quickly. ***Figure 34.3*** shows the stages involved in skin repair.

Quick Demo

Kinesthetic Ask students to examine and compare the skin that covers various parts of their bodies and describe whether the skin is the same in all places. *Likely responses will state that, in different parts of the body, skin has different thicknesses, varies in moisture, is tight or loose, and is either covered with hair or hairless.* L1 ELL

Assessment

Skill Have students make a diagram outlining the process whereby a skin injury is healed. L2 ELL

Enrichment

Have students conduct research on artificial skin. Students should report how artificial skin is made and how its structure compares with that of natural skin. L3

GLENCOE TECHNOLOGY

VIDEODISC
The Infinite Voyage:
Miracles by Design, Burn Patients and Artificial Skin (Ch.5)
8 min.

3 Assess

Check for Understanding

On the chalkboard, list the structures of the skin and ask students to copy the list. Have students identify the function of each structure. L1

Meeting Individual Needs

Gifted

Intrapersonal Have gifted students research the current treatments being used by regional burn centers for people with life-threatening third-degree burns. This research can be done in the library or on the Internet. L3

Biology Journal

Writing a Story

Linguistic Have students write a story about skin injury and healing from the viewpoint of the skin. Students should include the steps outlined in Figure 34.3 in their stories. L2

928

Reteach

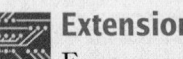 **Logical-Mathematical** Have students organize the following information about skin thicknesses in a table. The thickness of the epidermis on the back is 0.25 mm, on the face and scalp 0.12 mm, on the palm 0.5 mm. The thickness of the dermis on the back is 3.75 mm, on the face and scalp 1.6 mm, and on the palm 1.0 mm. Discuss why skin thickness varies in different parts of the body by relating this phenomenon to the function of the skin in each area. **L2**

Extension

TECH PREP Encourage interested students to research what sort of special training dermatologists receive in medical school. What kinds of skin problems do they treat? *skin cancer, burns, rashes, acne, among others* **L3**

✔ Assessment

Skill Ask students to write a paragraph listing the functions of the skin as they relate to skin structures. **L2**

4 Close

Discussion

Discuss the relationship between the structures of the skin and the skin's functions. Relate the presence of sweat glands, hair, oil glands, nerves, and blood vessels to the functions of skin. Ask students what problems a burn victim faces. *infection, fluid loss*

Resource Manager

Reinforcement and Study Guide, p. 151 **L2**
Content Mastery, p. 170 **L1**

Figure 34.4
As people age, their skin loses its elasticity and begins to wrinkle.

Skin Injury and Healing

If you've ever had a mild scrape, you know that it doesn't take long for the wound to heal. When the epidermis sustains a mild injury, such as a scrape, the deepest layer of epidermal cells divide to help fill in the gap left by the abrasion. If, however, the injury extends into the dermis, where blood vessels are found, bleeding usually occurs. The skin then goes through a series of stages to heal the damaged tissue. The first reaction of the body is to restore the continuity of the skin, that is, to close the break. Blood flowing from the wound soon clots. The wound is then closed by the formation of a scab, which prevents bacteria from entering the body. Dilated blood vessels then allow infection-fighting white blood cells to migrate to the wound site. Soon after, skin cells beneath the scab begin to multiply and fill in the gap. Eventually, the scab falls off to expose newly formed skin. If a wound is large, high amounts of dense connective tissue fibers used to close the wound may leave a scar.

Have you ever suffered a painful burn? Burns can result from exposure to the sun or contact with chemicals or hot objects. Burns are rated according to their severity.

First-degree burns involve the death of epidermal cells and are characterized by redness and mild pain. Most people have received a first-degree burn at one time or another, usually as the result of sunbathing. Second-degree burns involve damage to skin cells of the dermis and can result in blistering and scarring. The most severe burns are third-degree burns, which destroy both the epidermis and the dermis. With this type of burn, skin function is lost, and skin grafts may be required to replace lost skin. In some cases, healthy skin can be removed from another area of the patient's body and transplanted to a burned area.

As people get older, their skin changes. It becomes drier as glands decrease their production of lubricating skin oils—a mixture of fats, cholesterol, proteins and inorganic salts. As shown in *Figure 34.4*, wrinkles may appear as the elasticity of the skin decreases. Although these changes are natural, they can be accelerated by prolonged exposure to ultraviolet rays from the sun. Sunblock, used regularly, can prevent much of the damage caused by the sun's rays.

Section Assessment

Understanding Main Ideas
1. Which skin structures are damaged from a third-degree burn?
2. How does skin help control body temperature?
3. How do pimples form?
4. What happens to the dermal blood vessels when you are cold? Hot?

Thinking Critically
5. Why do some areas of your body contain more fat deposits than others?

SKILL REVIEW
6. **Sequencing** Outline the steps that occur when a cut in the skin heals. For more help, refer to *Organizing Information* in the **Skill Handbook**.

Section Assessment

1. All of the structures found in both the epidermis and the dermis are destroyed in a third-degree burn.
2. by producing sweat and by constriction and dilation of skin capillaries
3. Pimples result from bacteria growing in the hair follicle when oil and dead cells block the follicle's opening to the skin's surface.
4. The blood vessels constrict when you are cold and dilate when you are hot.
5. Fat deposits can serve to protect organs in particular areas of the body. These deposits are also used to store energy.
6. (1) formation of blood clot, (2) blood vessels dilate and infection-fighting blood cells rush to the wound, (3) epidermal cells divide to fill in the wound

34.2 Bones: The Body's Support

Would you believe that you had more bones when you were born than you have now? That's because some of your bones have grown together since then. Your head, for example, had soft spots when you were an infant. It feels solid now because the soft membranes were gradually replaced by bone. Remodeling of the skeleton occurs throughout life. In fact, your skeleton hasn't completely formed yet. You will not have a solid, fused skeleton until about age 25.

As a baby develops in the uterus, its bones fuse together.

SECTION PREVIEW

Objectives

Identify the structure and functions of the skeleton.

Compare the different types of movable joints.

Recognize how bone is formed.

Vocabulary

axial skeleton
appendicular skeleton
joint
ligament
bursa
tendon
compact bone
spongy bone
osteoblast
osteocyte
red marrow
yellow marrow

Skeletal System Structure

The adult human skeleton contains about 206 bones. Its two main parts are shown in *Figure 34.5* on the next page. The **axial skeleton** includes the skull and the bones that support it, such as the vertebral column, the ribs, and the sternum. The **appendicular skeleton** (a pen DIHK yuh lur) includes the bones of the arms and legs and structures associated with them, such as the shoulder and hip bones, wrists, ankles, fingers, toes and so on.

Joints: Where bones meet

Next time you open a door, notice how it is connected to the door frame. A metal joint positioned where the door and frame meet allows the door to move easily back and forth. In vertebrates, **joints** are found where two or more bones meet. Most joints facilitate the movement of bones in relation to one another. The joints of the skull, on the other hand, are fixed, as the bones of the skull don't move. These immovable joints are actually held together by the intergrowth of bone, or by fibrous cartilage.

Section 34.2

Prepare

Key Concepts

Students become acquainted with the structure of bones and their functions.

Planning

- Obtain X rays for the Quick Demo.
- Borrow a human skeleton for the Quick Demo and Meeting Individual Needs.
- Save two toilet paper tubes for Building a Model.
- Soak a chicken bone in vinegar or HCl for the Quick Demo.
- Have beef bones cut by a butcher for the Quick Demo.

---

1 Focus

Bellringer

Before presenting the lesson, display **Section Focus Transparency 82** on the overhead projector and have students answer the accompanying questions.
L1 ELL

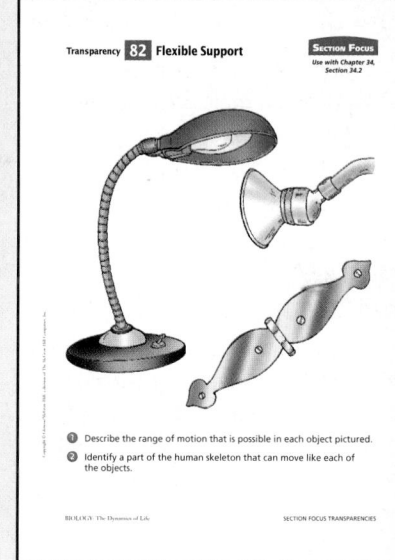

Transparency **82** Flexible Support   **SECTION FOCUS**
Use with Chapter 34, Section 34.2

① Describe the range of motion that is possible in each object pictured.
② Identify a part of the human skeleton that can move like each of the objects.

BIOLOGY: The Dynamics of Life   SECTION FOCUS TRANSPARENCIES

Resource Manager

Section Focus Transparency 82 and Master L1 ELL
Laboratory Manual, pp. 247-250 L2

GLENCOE TECHNOLOGY

CD-ROM
Biology: The Dynamics of Life
BioQuest: *Body Systems*
Disc 1-5
Exploration: *Bones: The Body's Support*
Disc 5

Project X rays or hang them in windows on the day that you begin discussing the skeleton. Identify the bones shown in each X ray and the types of information X rays can reveal about bones.

Visual Learning

Have students examine Figure 34.5. Ask: To which skeletal group do the phalanges belong? *appendicular*

Building a Model

Visual-Spatial To demonstrate the strength of hollow bones, have students balance a heavy book on the open ends of two toilet paper tubes placed side by side. **L1** **ELL**

Revealing Misconceptions

Many students believe that males have fewer ribs than females. Explain to students that males and females have an equal number of ribs.

Point out the differences between a male and female pelvis using skeletal diagrams.

Tying to Previous Knowledge

Compare the anatomy of the front limbs of a frog, bird, cat, whale, and bat with that of the human arm. Compare the number, shapes, and arrangements of bones.

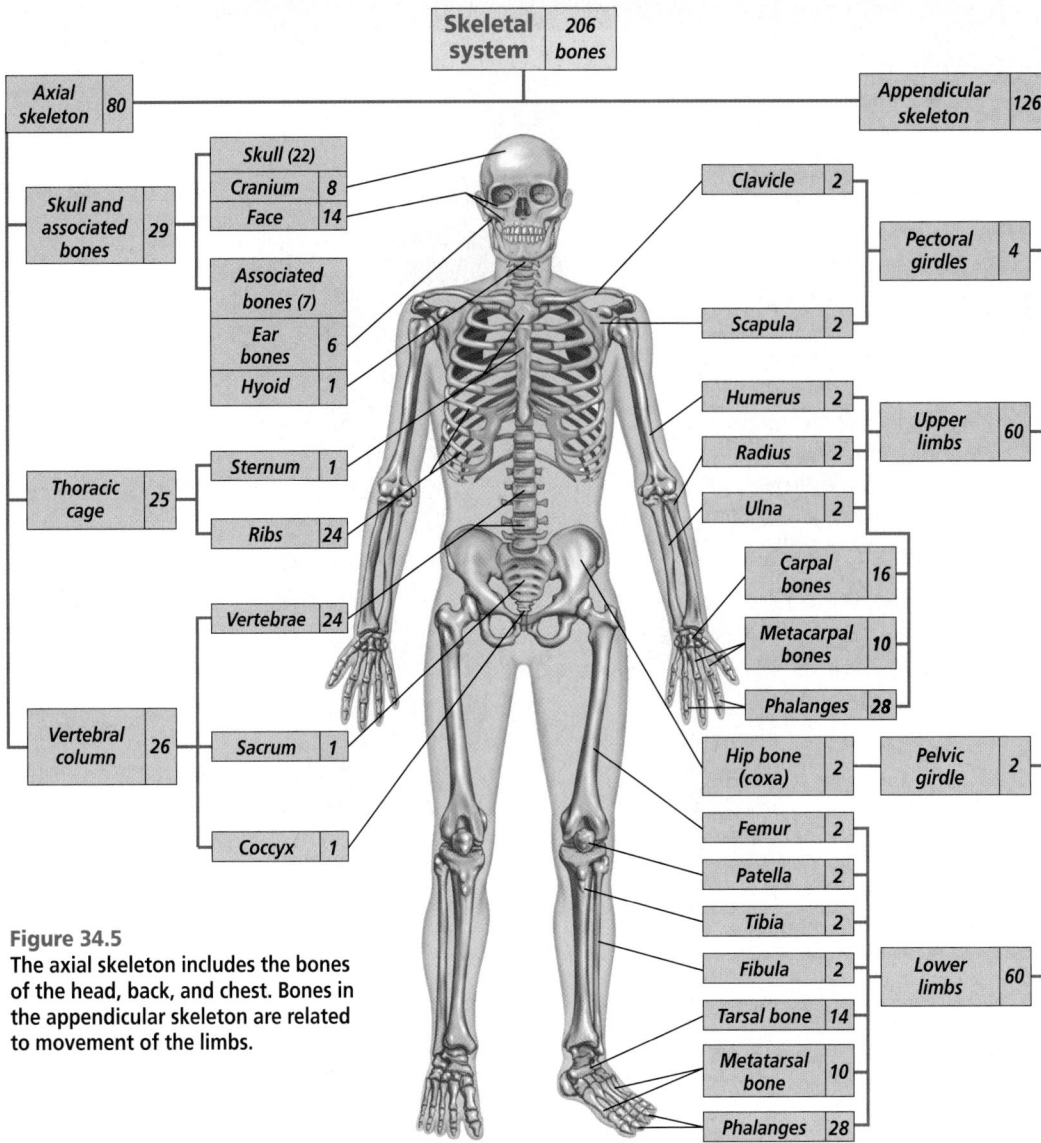

Figure 34.5
The axial skeleton includes the bones of the head, back, and chest. Bones in the appendicular skeleton are related to movement of the limbs.

WORD *Origin*

arthritis
From the Greek words *arthron*, meaning "joint," and *itis*, meaning "swelling disease." Arthritis is a swelling disease of the joints.

Joints are often held together and enclosed by ligaments. A **ligament** is a tough band of connective tissue that attaches one bone to another. Joints with large ranges of motion, such as the knee, typically have more ligaments surrounding them. In movable joints, the ends of bones are covered with a layer of cartilage, which allows for smooth movement between the bones. In addition, joints such as those of the shoulder and knee have fluid-filled sacs called **bursae** located between the bones. The bursae act to absorb shock and keep bones from rubbing against each other. **Tendons,** which are thick bands of connective tissue, attach muscles to bones. *Figure 34.6* shows the different movable joints in the skeleton.

Cultural Diversity

Human Skeletal Variation

Linguistic Variation exists in human body form, especially within the skeletal system. Differences in skeletal morphology are related to the geographical origins of populations. For example, a leaner body form is often observed in people who live in arid regions where greater skin-surface area in proportion to body weight facilitates heat loss. A stockier build is more adaptive for inhabitants of cold areas. Differences in body form and skin color have been used as a justification for racism. Have students write short essays about how their attitudes on racism have been affected by an understanding of human variation. **L2**

Figure 34.6
Body movements are made possible by joints that allow bones to move in several different directions.

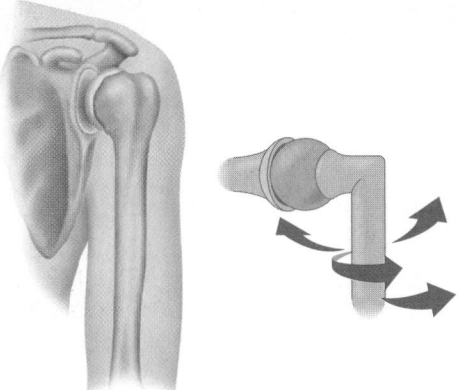

A Ball-and-socket joints allow rotational movement. The joints of the hips and shoulders are ball-and-socket joints; they allow you to swing your arms and legs around in a circular motion.

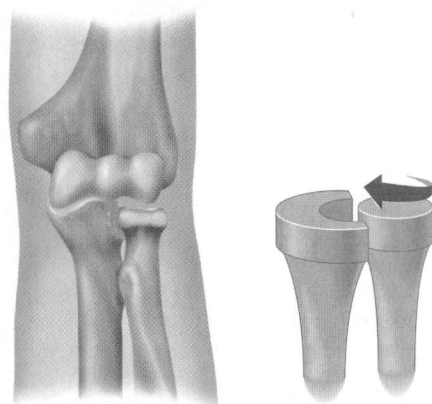

B Pivot joints allow bones to twist around each other. One of the joints in your elbow is a pivot joint. It allows you to twist your lower arm around.

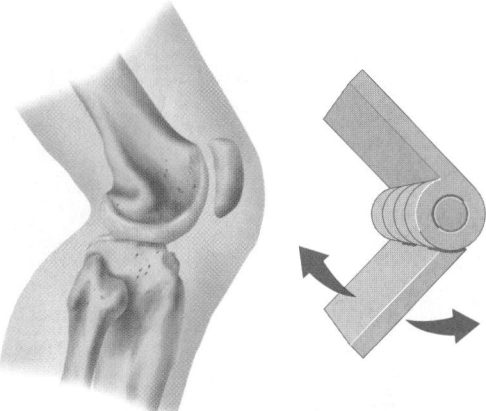

C Hinge joints are found in the elbows, knees, fingers, and toes. They allow back-and-forth movement like that of a door hinge.

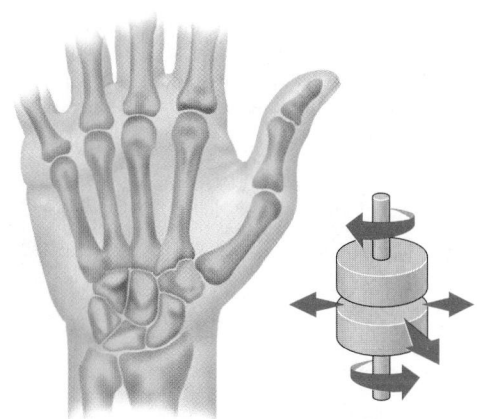

D Gliding joints, found in the wrists and ankles, allow bones to slide past each other.

Forcible twisting of a joint, called a sprain, can result in injury to the bursae, ligaments, or tendons. A sprain most often occurs at joints with large ranges of motion such as the wrist, ankle, and knee.

Besides injury, joints are also subject to disease. One common joint disease is arthritis, an inflammation of the joints. It can be caused by infections, aging, or injury. One kind of arthritis results in bone spurs, or splinters of bone, inside the joints. Such arthritis is especially painful, and often limits the patient's ability to move his or her joints.

34.2 BONES: THE BODY'S SUPPORT **931**

Compact and spongy bone

Although bones may appear uniform, they are actually composed of two different types of bone tissue, as shown in *Figure 34*.7. Surrounding every bone is a layer of hard bone, or **compact bone.** Running the length of compact bone are tubular structures known as osteon systems, as shown in *Figure 34.7.* Compact bone is made up of repeating units of osteon systems. Living bone cells, or **osteocytes,** receive oxygen and nutrients from small blood vessels running within the osteon systems. Nerves in the canals conduct impulses to and from each bone cell. Compact bone surrounds less dense bone known as **spongy bone,** so called because it is filled with many holes and spaces, like those in a sponge.

Formation of Bone

The skeleton of a vertebrate embryo is made of cartilage. By the ninth week of human development, bone begins to replace cartilage. Blood vessels penetrate the membrane covering the cartilage and stimulate its cells to become potential bone cells called **osteoblasts** (AHS tee oh blastz). These potential bone cells secrete a protein called collagen in which minerals in the bloodstream begin to be deposited. The deposition of calcium salts and other ions hardens the newly formed bone cells, now called osteocytes. The adult skeleton is almost all bone, with cartilage found only in places where flexibility is needed—regions such as the nose tip, external ears, discs between vertebrae, and movable joint linings.

Figure 34.7
A bone has several components, including compact bone, spongy bone, and osteon systems. The osteon systems contain blood vessels that nourish bone cells and nerve tissues that conduct impulses to and from each cell.

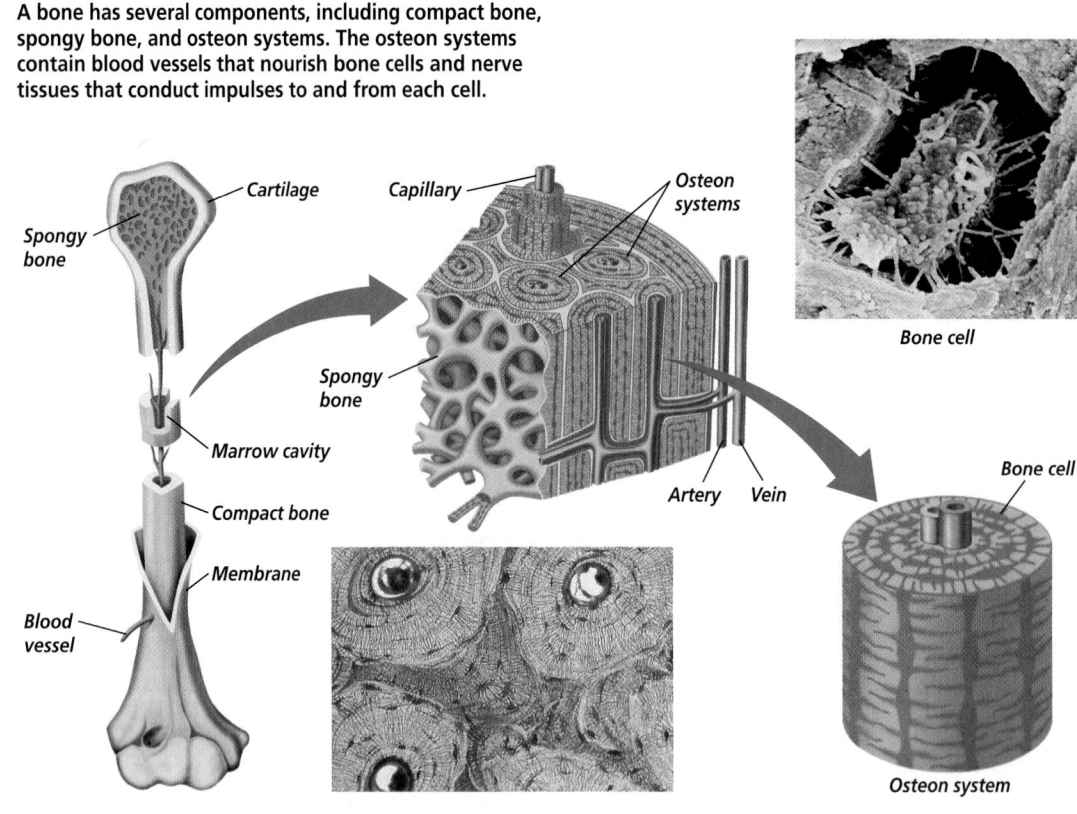

Cartilage
Capillary
Osteon systems
Spongy bone
Spongy bone
Marrow cavity
Artery Vein
Compact bone
Bone cell
Membrane
Blood vessel
Bone cell
Osteon system

Bone growth

Your bones grow in both length and diameter. Growth in length occurs at the ends of bones in cartilage plates. Growth in diameter occurs on the outer surface of the bone. The increased production of sex hormones during your teen years causes the osteoblasts to divide more rapidly, resulting in a growth spurt. By age 20, about 98 percent of your skeleton growth will be completed. However, these same hormones will also cause the growth centers at the ends of your bones to degenerate. As these cells die, your growth will slow. After growth stops, bone-forming cells are involved mainly in repair and maintenance of bone. Learn more about how bones age by performing the *Problem-Solving Lab* on this page.

Skeletal System Functions

The primary function of your skeleton is to provide a framework for the tissues of your body. The skeleton also protects your internal organs, including your heart, lungs, and brain.

The arrangement of the human skeleton allows for efficient body movement. Muscles that move the body need firm points of attachment to pull against so they can work effectively. The skeleton provides these attachment points.

Bones also produce blood cells. **Red marrow**—found in the humerus, femur, sternum, ribs, vertebrae, and pelvis—is the production site for red blood cells, white blood cells, and cell fragments involved in blood clotting. **Yellow marrow,** found in many other bones, consists of stored fat as shown in *Figure 34.8.*

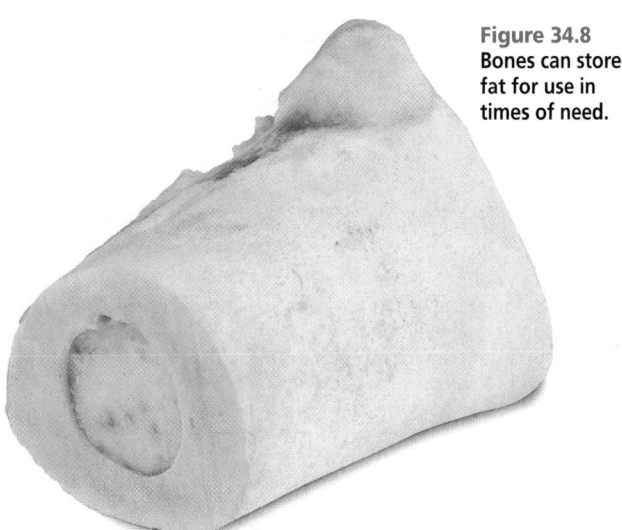

Figure 34.8 Bones can store fat for use in times of need.

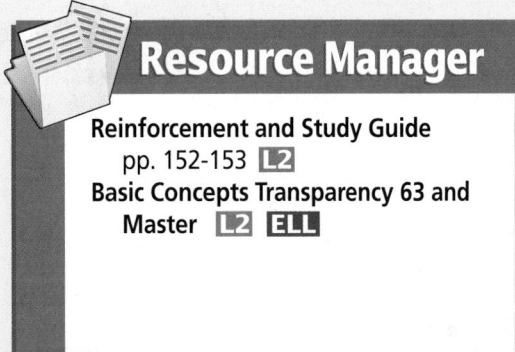

3 Assess

Check for Understanding

Have students fill in the names of the bones and label an example of each of the major types of joints on a diagram of the skeleton. **L2**

Reteach

Interpersonal Have students make flash cards containing the names of bones. On the opposite side of each card, have students describe where each bone is located and identify its common name. Have students work in pairs to review the information on the flash cards. **L1** **COOP LEARN**

Extension

Have students conduct research on the various types of arthritis and their causes and treatments. Have students construct a table of their findings. **L3**

✓ **Assessment**

Knowledge Provide students with an outline showing the bones of the human skeleton. Have students create a color key to show which bones make up the appendicular skeleton and which make up the axial skeleton **L1** **ELL**

4 Close

Activity

Interpersonal Play bone and bone anatomy Password. Divide the class into groups of two. Have one partner try to get the other to say a certain word by giving other one-word clues. For example, if the password were "skeleton," the clues might include "bones," "support," and so on. After they guess the word, switch players. **L2** **COOP LEARN**

934

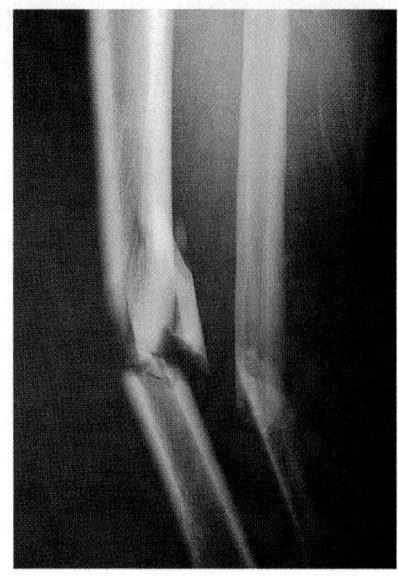

Figure 34.9
The X ray on the left shows a leg bone that has completely fractured. The X ray on the right shows the bone (with a supporting rod) after it has healed. The arrow indicates the area where the break healed.

Bones store minerals

Finally, your bones serve as storehouses for minerals, including calcium and phosphates. Calcium is needed to form strong, healthy bones and is therefore an important part of your diet. Sources of calcium include milk, yogurt, cheese, lettuce, spinach, and other assorted leafy vegetables.

Bone injury and disease

Bones tend to become more brittle as their composition changes with age. For example, a disease called osteoporosis (ahs tee oh puh ROH sus) involves a loss of bone volume and mineral content, causing the bones to become more porous and brittle. Osteoporosis is most common in older women because they produce lesser amounts of estrogen—a hormone that aids in bone formation.

When bones are broken, as shown by the X ray images in *Figure 34.9*, a doctor moves them back into position and immobilizes them with a cast or splint until the bone tissue regrows. Read more about the use of X rays in the diagnosis of broken bones in the *Physics Connection* at the end of this chapter.

Section Assessment

Understanding Main Ideas
1. Distinguish between the appendicular skeleton and the axial skeleton.
2. List the four main kinds of movable joints and provide an example of each.
3. In what way do bones change as a person ages?
4. How is compact bone structurally different from spongy bone?

Thinking Critically
5. Why would it be impossible for bones to grow from within?

SKILL REVIEW
6. **Sequencing** Outline the steps involved in bone formation—from cartilage to bone. For more help, refer to *Organizing Information* in the **Skill Handbook**.

934 PROTECTION, SUPPORT, AND LOCOMOTION

Section Assessment

1. The axial skeleton includes the skull and the bones that support it. The appendicular skeleton includes the bones associated with the appendages.
2. ball-and-socket, shoulder or hip; pivot joint, elbow; hinge joint, elbow, knees, fingers and toes; gliding, wrists and ankles
3. Bones become more brittle with age.
4. Compact bone is made up of osteon systems. Spongy bone is filled with holes.
5. The inflexible structure of compact bone would not allow growth from the inside.
6. (1) embryo skeleton is cartilage; (2) bone begins to replace cartilage—osteoblasts secrete a material in which calcium salts and other ions are deposited and harden to form bone.

SECTION PREVIEW

Objectives
Classify the three types of muscles.
Analyze the structure of a myofibril.
Interpret the sliding filament theory.

Vocabulary
smooth muscle
involuntary muscle
cardiac muscle
skeletal muscle
voluntary muscle
myofibril
myosin
actin
sarcomere
sliding filament theory

Have you seen the Olympic games on television? Think about the different athletes involved in the games. You may be able to tell what sports some of them participate in just by looking at their body shapes. For example, swimmers and ice skaters have different shapes because ice skaters develop strong leg muscles over many months of training, whereas swimmers develop larger shoulder muscles.

A swimmer has well-developed shoulder muscles

Three Types of Muscles

Nearly half of your body mass is muscle. A muscle consists of groups of fibers bound together. Almost all of the muscle fibers you will ever have were present at birth.

Figure 34.10 shows the three main kinds of muscles in your body. One type of tissue, **smooth muscle,** is found in the walls of your

Figure 34.10
Muscles are under either voluntary or involuntary control and differ in their structure and appearance.

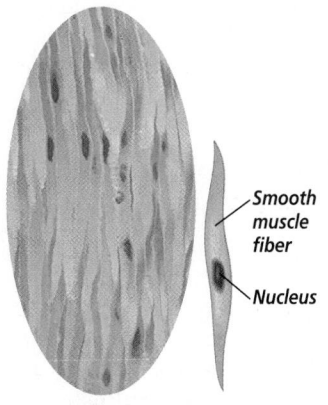

Smooth muscle fiber

Nucleus

Magnification: 4500×

Ⓐ Smooth muscle fibers are under involuntary control and appear spindle-shaped.

Cardiac muscle fiber

Striation

Nucleus

Magnification: 27 000×

Ⓑ Cardiac muscle fibers, which are also under involuntary control, appear striated or striped when magnified.

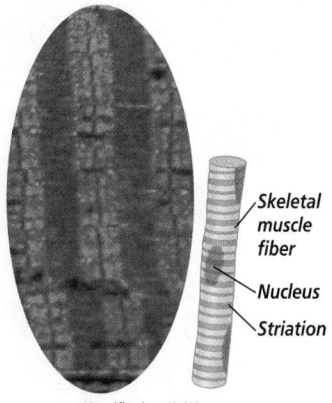

Skeletal muscle fiber

Nucleus

Striation

Magnification: 12 600×

Ⓒ Skeletal muscle fibers, while also striated, are under voluntary control.

34.3 MUSCLES FOR LOCOMOTION **935**

Section 34.3

Prepare

Key Concepts

Students learn to distinguish among the three types of muscles. Emphasis is placed on the structure and function of skeletal muscle and the current sliding filament theory of muscle contraction.

Planning

■ Obtain chicken feet and wings for the Quick Demo.

1 Focus

Bellringer 👆

Before presenting the lesson, display **Section Focus Transparency 83** on the overhead projector and have students answer the accompanying questions.
L1 **ELL**

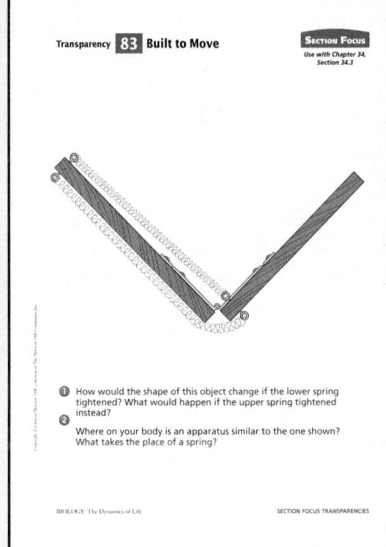

Quick Demo

📺 Obtain some chicken feet from a butcher. Cut the upper end so that the tendons are free. Pull on each tendon to demonstrate the movement of the toes. Use a chicken wing to demonstrate muscle pairing in the same manner. 📦

Problem-Solving Lab 34-3

Purpose 📦

Students will conclude how rigor mortis is used to estimate time of death.

Process Skills

analyze, draw a conclusion

Background

When a person dies, the respiratory and circulatory systems that obtain and deliver oxygen to the muscles stop functioning. As a result, the muscles do not receive ATP and they stiffen. Scientists believe that stiffening rates are related to muscle fiber length, shorter fibers stiffening before longer ones. Full rigor mortis takes about 10-12 hours to set in.

Teaching Strategies

■ Draw an outline of a human body on the chalkboard. On the diagram, trace the path of the stiffening of muscles in the body that indicates the progression of rigor mortis.

Thinking Critically

Accept all answers in the range of 12-16 hours.

✔ Assessment

Knowledge Have students list the order and time frame in which rigor mortis subsides in the body. Use the Performance Task Assessment List for Events Chain in **PASC**, p. 91. **L2**

936

Problem-Solving Lab 34-3   Drawing a Conclusion

How is rigor mortis used to estimate time of death? Rigor mortis is the stiffening of both voluntary and involuntary muscles due to lack of ATP after death. Stiffening develops about an hour or two after death, and usually proceeds from the upper body to the lower body. Rigor mortis affects the face area first, the neck, chest, and arms next, and the legs last. Stiffening gradually subsides in the same order it develops, disappearing completely after 24 to 36 hours.

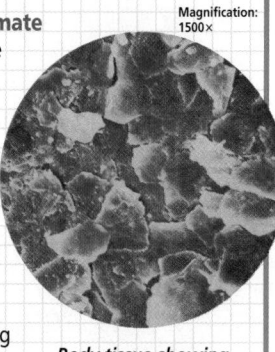

Magnification: 1500×

Body tissue showing rigor mortis

Analysis

Coroners determine a victim's time of death by examining body tissues under a microscope and observing the degree of rigor mortis.

Thinking Critically

A murder victim is found with relaxed face muscles, but a stiff upper and lower body. How long might it have been since the crime was committed?

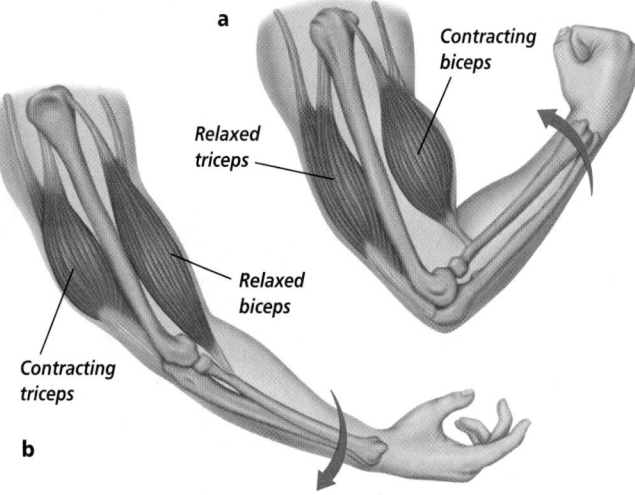

a
Contracting biceps
Relaxed triceps
Relaxed biceps
Contracting triceps
b

Figure 34.11
When the biceps muscle contracts, the lower arm is moved upward (a). When the triceps muscle on the back of the upper arm contracts, the lower arm moves downward (b).

💿 CD-ROM

View an animation of muscle contraction in the Presentation Builder of the Interactive CD-ROM.

---

internal organs and blood vessels. Smooth muscle is made up of sheets of cells that are ideally shaped to form a lining for organs such as the digestive tract and the reproductive tract. The most common function of smooth muscle is to squeeze, exerting pressure on the space inside the tube or organ it surrounds in order to move material through it. Because contractions of smooth muscle are not under conscious control, smooth muscle is considered an **involuntary muscle.**

Another type of involuntary muscle is the **cardiac muscle,** which makes up your heart. Cardiac muscle fibers are interconnected and form a network that helps the heart muscle contract efficiently. Cardiac muscle is found only in the heart and is adapted to generate and conduct electrical impulses necessary for its rhythmic contraction. The third type of muscle tissue, **skeletal muscle,** is the type that is attached to and moves your bones. The majority of the muscles in your body are skeletal muscles, and, as you know, you can control their contractions. A muscle that contracts under conscious control is called a **voluntary muscle.** Find out how coroners use muscle tissue samples to determine a victim's time of death in the *Problem-Solving Lab* on this page.

Skeletal Muscle Contraction

Whether you are playing tennis, pushing a lawn mower, or writing, your muscles are contracting as they perform the action. *Figure 34.11* shows the movement of the lower arm as controlled by opposing muscles in the upper arm. The majority of skeletal muscles work in opposing pairs.

Muscle tissue is made up of muscle fibers, which are actually just very

---

✔ Portfolio

Sports Medicine

🧠 *Linguistic* Ask students to prepare and conduct an interview with a person who works in the field of sports medicine or in physical education. Students should include questions on the type of training the specialists have and how their programs relate to muscle fitness. **L2** **P** 📦 **COOP LEARN**

GLENCOE TECHNOLOGY

💿 **CD-ROM**
Biology: The Dynamics of Life
Animation: *Paired Skeletal Muscles*
Animation: *Sliding Filament Theory*
Disc 5

long fused muscle cells. Each fiber is made up of smaller units called **myofibrils** (mi oh FIBE rulz). Myofibrils are themselves composed of even smaller protein filaments that can be either thick or thin. The thicker filaments are made of the protein **myosin,** and the thinner filaments are made of the protein **actin.** The arrangement of myosin and actin gives skeletal muscle its striated, or striped, appearance. Each myofibril can be divided into sections called **sarcomeres** (SAR kuh meerz), the functional units of muscle. How do nerves signal muscles to contract? Find out in the *Inside Story* on the next page.

The sliding filament theory currently offers the best explanation for how muscle contraction occurs. The **sliding filament theory** states that, when signaled, the actin filaments within the sarcomere slide toward one another, shortening the sarcomere and causing the muscle to contract. The myosin filaments, on the other hand, do not move. Learn more about the sliding filament theory in the *MiniLab* on this page.

Muscle Strength and Exercise

How can you increase the strength of your muscles? Muscle strength does not depend on the number of fibers in a muscle. It has been shown that this number is basically fixed before you are born. Rather, muscle strength depends on the thickness of the fibers and on how many of them contract at one time. Regular exercise stresses muscle fibers slightly; to compensate for this added workload, the fibers increase in diameter.

Recall that ATP is produced during cellular respiration. Muscle cells

are continually supplied with ATP from both aerobic and anaerobic processes. However, the aerobic respiration process dominates when adequate oxygen is delivered to muscle cells, as when a muscle is at rest or during moderate activity. When an adequate supply of oxygen is unavailable, such as during vigorous activity, anaerobic respiration—specifically the process of lactic acid fermentation—becomes the primary source of ATP production.

MiniLab 34-2   Interpreting

Look at Muscle Contraction Muscle fibers are composed of a number of small functional units called sarcomeres. Sarcomeres, in turn, are composed of protein filaments called actin and myosin. The sliding action of these filaments in relation to one another results in muscle contraction.

Procedure
1. Look at Diagrams A and B. Diagram A shows a sarcomere in a relaxed muscle. Diagram B shows a sarcomere in a flexed muscle.
2. Using a centimeter ruler, measure and record the length of a: sarcomere, a myosin filament, and an actin filament in diagram A. Record your data in a table.
3. Repeat step 2 for Diagram B.

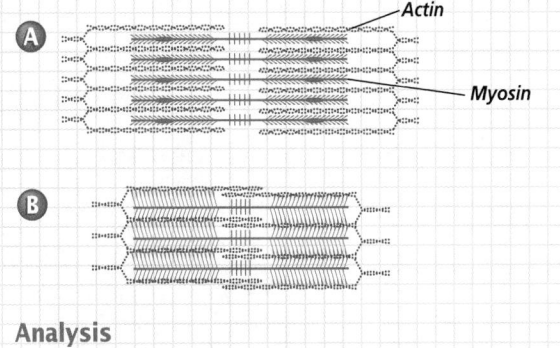

Analysis
1. When a muscle contracts, do actin or myosin filaments shorten? Use specific data from your model to support your answer.
2. How does the sarcomere shorten when the parts that make it up don't shorten?

WORD *Origin*

myofibril
From the Greek words *mys*, meaning "muscle," and *fibrilla*, meaning "small fiber." A myofibril is a small part of a muscle fiber.

MiniLab 34-2

Purpose
Students will measure the length of a sarcomere during muscle relaxation and contraction.

Process Skills
acquire information, apply concepts, collect data, compare and contrast, formulate models, interpret data, measure in SI, predict

Teaching Strategies
■ Provide students with rulers.
■ Review vocabulary terms prior to the start of this activity.

Expected Results
Actin filaments slide over myosin filaments resulting in shortening of sarcomeres during contraction.

Analysis
1. Students will learn that actin and myosin do not change in length.
2. Actin filaments slide over myosin filaments, shortening the sarcomere.

✓ Assessment

Knowledge Using the diagram of relaxed muscle, have students explain why skeletal muscle has a striated appearance when viewed under the microscope. Provide students with a slide of striated muscle tissue to verify their explanations. Use the Performance Task Assessment List for Making Observations and Inferences in **PASC**, p. 17. **L2**

GLENCOE TECHNOLOGY

VIDEODISC
Biology: The Dynamics of Life
Sliding Filament Theory (Ch. 30)
Disc 2, Side 1, 67 sec.

Paired Skeletal Muscles (Ch. 29)
Disc 2, Side 1, 22 sec.

MEETING INDIVIDUAL NEEDS

Learning Disabled/English Language Learners

Linguistic Have students who are having difficulty make flash cards that show an anatomical term on one side of the card and its meaning and location on the other side. Have students work in pairs to review the cards. **L1 ELL**

BIOLOGY JOURNAL

Evaluating Exercise Programs

Logical-Mathematical Have students evaluate their own exercise program in their journal. The American Heart Association recommends a minimum of 30 minutes of moderate activity per day. **L1**

Purpose

Students gain further understanding of the way muscle cells contract.

Teaching Strategies

■ Have students describe the structure of skeletal muscle. **L2**

■ Ask students to describe the differences between actin and myosin filaments. **L2**

Visual Learning

■ Ask students what makes up a single muscle fiber. *many myofibril units* Elicit what chemical stimulates the formation of attachments between myosin and actin filaments. *calcium*

Critical Thinking

The nerve signal causes calcium to be released in the muscle. The calcium causes actin and myosin filaments to bind together. The actin filaments are then pulled inward, resulting in a muscle contraction.

✓ Assessment

Knowledge Have students write a short summary of why lactic acid builds up in muscle cells during intense exercise.

The BioLab at the end of the chapter can be used at this point in the lesson.

Tying to Previous Knowledge

Point out that myofibrils of the muscles are a special type of microfilament, as discussed previously. This is a good time to review ATP, along with aerobic and anaerobic metabolism.

A Muscle

Locomotion is made possible by the contraction and relaxation of muscles. The sliding filament theory of how muscles contract can be better understood by examining the detailed structure of a skeletal muscle.

Critical Thinking *How does a nerve signal cause a skeletal muscle to contract?*

Magnification: 1500×

Skeletal muscle is responsible for moving your bones.

1 **Muscle structure** When you tease apart a typical skeletal muscle and view it under a microscope, you can see that it consists of bundles of fibers. A single fiber is made up of myofibrils which, in turn, are made up of actin or myosin filaments. Each myofibril can be broken up into functional units called sarcomeres.

2 **Nerve signal** When a skeletal muscle receives a signal from a nerve, calcium is released inside the muscle fibers, causing them to contract.

3 **Contraction** The presence of calcium causes attachments to form between the thick myosin and thin actin filaments. The actin filaments are then pulled inward toward the center of each sarcomere, shortening the sarco-mere and producing a muscle contraction. When the muscle relaxes, the filaments slide back into their original positions.

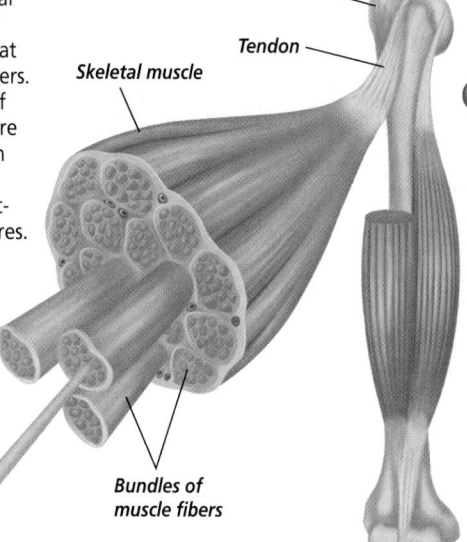

Bone

Tendon

Skeletal muscle

Bundles of muscle fibers

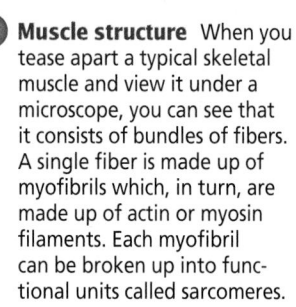

Myofibril

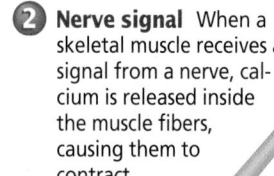

Filaments   Actin

Myosin

Sarcomere

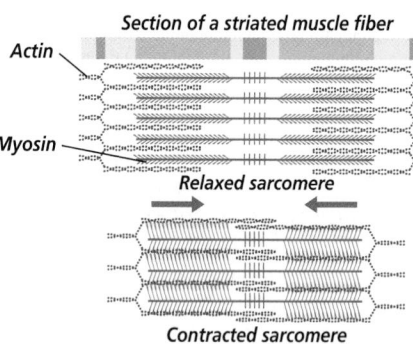

Section of a striated muscle fiber

Actin

Myosin

Relaxed sarcomere

Contracted sarcomere

PROJECT

Modeling Body Movement

Visual-Spatial Have students working in groups construct models to show how bones and muscles work together to move appendages. More advanced students may wish to work together to prepare a model that demonstrates the sliding filament theory. **L2**

ELL **COOP LEARN**

MEETING INDIVIDUAL NEEDS

Physically Challenged

Locate a physically challenged spokesperson who has a musculoskeletal disease and who is willing to speak to the class about the disorder. The focus of the discussion should be on the nature of the disease, restrictions the disease places on the person, and treatments for the disease.

Think about what happens when you are running in gym class or around the track at school. *Figure 34.12A* illustrates how an athlete's need for oxygen changes as the intensity of his or her workout increases. At some point, your muscles are not able to get oxygen fast enough to sustain aerobic respiration and produce adequate ATP. Thus, the amount of available ATP becomes limited. For your muscle cells to get the energy they need, they must rely on anaerobic respiration as well. *Figure 34.12B* indicates how, at a certain intensity, the body shifts from aerobic respiration to the anaerobic process of lactic acid fermentation for its energy needs.

During exercise, lactic acid builds up in muscle cells. As the excess lactic acid is passed into the bloodstream, the blood becomes more acidic, rapid breathing is stimulated, and cramping can occur. As you catch your breath following exercise, adequate amounts of oxygen are supplied to your muscles and lactic acid is broken down. Regular exercise can result in improved performance of muscles. Do the *BioLab* at the end of the chapter to find out how muscle fatigue affects the amount of exercise your muscles can accomplish.

Figure 34.12
Athletic trainers use information about muscle functioning during exercise to establish appropriate levels of intensity for training.

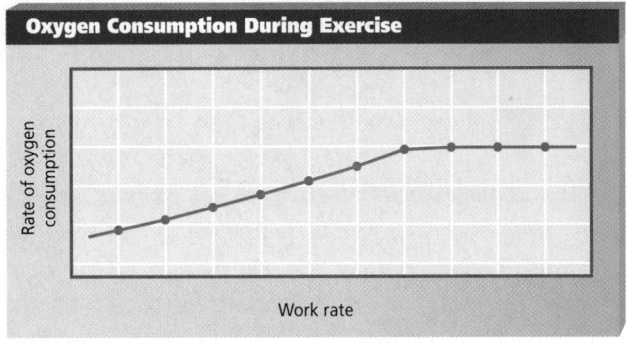

Oxygen Consumption During Exercise

Rate of oxygen consumption

Work rate

A As an individual increases the intensity of his or her workout, the need for oxygen goes up in predictable increments.

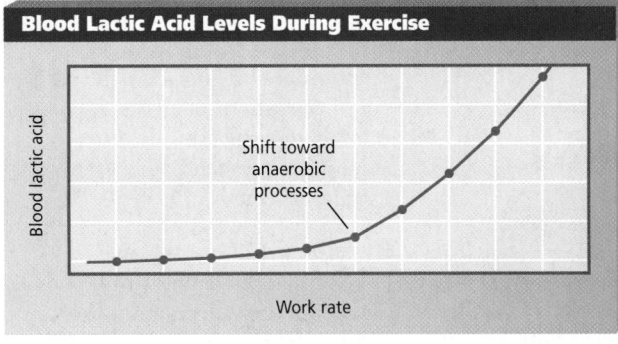

Blood Lactic Acid Levels During Exercise

Blood lactic acid

Shift toward anaerobic processes

Work rate

B An upswing in the presence of lactic acid in the bloodstream can be used to indicate the point at which anaerobic respiration becomes the dominant means of ATP production.

3 Assess

Check for Understanding
Ask students to diagram and explain the sliding filament theory of muscle contraction. **L2**

Reteach
Review the three types of muscles and where in the body each is found.

Extension
Have interested students research diseases and disorders that affect the muscular system. Have them prepare a report on one of the disorders for their portfolios. **L3**

✔ Assessment
Performance Provide students with a diagram of muscles of the human body. Have students make a key to identify each type of muscle. **L1**

4 Close

Discussion
Discuss the function of each type of muscle. Help students recognize the relationship between the skeleton and skeletal muscles. **L1**

Resource Manager
Reinforcement and Study Guide, p. 154 **L2**
Content Mastery, pp. 169, 171-172 **L1**
Reteaching Skills Transparency 49 and Master L1 ELL
Laboratory Manual, pp. 251-254 **L2**

Section Assessment

Understanding Main Ideas
1. Compare the structure and functions of the three main types of muscles.
2. Summarize the sliding filament theory of muscle contraction.
3. How can exercise change muscle strength? How can it change muscle function?
4. What, for the most part, determines muscle strength?

Thinking Critically
5. Why would a disease that causes paralysis of smooth muscles be life threatening?

SKILL REVIEW

6. **Interpreting Scientific Illustrations** Diagram the composition of muscle fibers as shown in the *Inside Story*. For more help, refer to *Thinking Critically* in the **Skill Handbook**.

Section Assessment

1. Smooth muscle cells are spindle shaped and form the linings of organs. They apply pressure and squeeze. Cardiac muscle cells are striated and form a contraction network in the heart. Skeletal muscle cells are striated. They move the body.
2. Calcium is released into the muscle fiber, actin and myosin filaments form attachments, and the actin filaments are pulled inward toward the center of each sarcomere.
3. Exercise stresses muscles and causes fibers to increase in size and strength. Exercise cannot change muscle function.
4. Muscle strength depends on the thickness of the muscle fibers and how many of the fibers contract at one time.
5. Internal organs utilize smooth muscle to perform their functions, such as breathing and digestion.
6. The drawing should reflect that muscle fibers are made of tiny cylinders called myofibrils. Each myofibril is made of myosin and actin filaments arranged in a pattern within the sarcomere.

Time Allotment

One class period

Process Skills

observe and infer, communicate, measure in SI, predict, form a hypothesis, design an experiment

PREPARATION

Alternative Materials

Stopwatches are not necessary if a clock with a second hand is visible to all students.

Possible Hypotheses

If a muscle becomes fatigued, then the muscle will not be able to do as much work.

If a muscle becomes fatigued, its capacity to do work will not be diminished.

Resource Manager

BioLab and MiniLab Worksheets, pp. 153-154 L2

DESIGN YOUR OWN BioLab

Does fatigue affect the ability to perform an exercise?

The movement of body parts results from the contraction and relaxation of muscles. In this process, muscles use energy from aerobic and anaerobic respiration. When exercise is continued for a long period of time, the waste products of fermentation accumulate and muscle fibers are stressed, causing fatigue. Fatigue affects the various muscles differently. It also affects individuals differently, even when they are performing the same tasks. Muscular strength, muscular endurance, and the amount of effort required to perform a task are all variables to consider.

PREPARATION

Problem

How does fatigue affect the number of repetitions of an exercise you can accomplish? How do different amounts of resistance affect rate of fatigue?

Hypotheses

Hypothesize whether or not muscle fatigue has any effect on the amount of exercise muscles can accomplish. Consider whether fatigue occurs within minutes or hours.

Objectives

In this BioLab, you will:
- **Hypothesize** whether or not muscle fatigue affects the amount of exercise muscles can accomplish.
- **Measure** the amount of exercise done by a group of muscles.
- **Make a graph** to show the amount of exercise done by a group of muscles.

Possible Materials

stopwatch or clock with second hand
graph paper
small weights
wooden box or step stool

Skill Handbook

Use the **Skill Handbook** if you need additional help with this lab.

940 PROTECTION, SUPPORT, AND LOCOMOTION

PLAN THE EXPERIMENT

Teaching Strategies

■ One student should act as timekeeper. The exerciser should count the number of times he or she can carry out the exercise in a given time period, such as 3 minutes.
■ Trials should be performed as closely together as possible, so that little muscle rest occurs between trials.
■ Place a sample graph on the chalkboard

to help students prepare their graphs.

Possible Procedures

■ Students should choose an exercise such as how many times they can lift a weight in 3 minutes. They should repeat the experiment 4 or 5 times so that the muscles become fatigued.

PLAN THE EXPERIMENT

1. Design a repetitive exercise for a particular group of muscles. Make sure you can count single repetitions of the exercise (for example, one sit-up or one jumping jack) over time.
2. Work in pairs, with one member of the team being a timekeeper and the other member performing the exercise.
3. Consider setting up your experiment so that the amount of resistance is the independent variable. Compare your design with those of other groups.

Check the Plan

1. Be sure that the exercises are ones that can be done rapidly and cause a minimum of disruption to other groups in the classroom.

Sample Data Table

Time interval	Number of repetitions
First minute	
Second minute	
Third minute	
Fourth minute	
Fifth minute	

2. Consider how long you will do the activity and how often you will record measurements.
3. ***Make sure your teacher has approved your experimental plan before you proceed further.***
4. Make a table in which you can record the number of exercise repetitions per time interval.
5. Carry out the experiment.
6. On a piece of graph paper, plot the number of repetitions on the vertical axis and the time intervals on the horizontal axis.

ANALYZE AND CONCLUDE

1. **Making Inferences** What effect did repeating the exercise over time have on the muscle group?
2. **Comparing and Contrasting** As you repeated the exercise over time, how did your muscles feel?
3. **Recognizing Cause and Effect** What physiological factors are responsible for fatigue?
4. **Thinking Critically** How well do you think your fatigued muscles

would work after 30 minutes of rest? Explain your answer.

Going Further

Project Design an experiment that will enable you to measure the strength of muscle contractions.

BIOLOGY *Online* To find out more about muscles, visit the Glencoe Science Web site.

science.glencoe.com

ANALYZE AND CONCLUDE

1. The muscle groups became fatigued and the number of repetitions of the activity went down as more trials were run.
2. The muscles felt tired and may even have begun to hurt toward the end. It became harder to do the activity and the strength of each contraction was reduced.
3. Cells are running out of oxygen and accumulating toxic products such as lactic acid and carbon dioxide as the muscles change to anaerobic processes.
4. Fatigued muscles should work as well after a rest of 30 minutes as they did before fatigue became a factor. The accumulated lactic acid has been broken down as the oxygen supply in the muscle cells has been replenished during the rest.

Error Analysis

Students must put real effort into their activity in order to see the results of fatigue.

✓ Assessment

Portfolio Have students place their laboratory reports, including their tables and graphs, in their journals. Use the Performance Task Assessment List for Science Journal in **PASC**, p. 103. **L2**

Going Further

Ask students how athletes such as marathon runners continue to exercise at a high level for 3 or more hours. Why don't their muscles fatigue after a few minutes?

Data and Observations

Student graphs should show that the number of exercise repetitions goes down over time as the muscles become tired.

Physics Connection

Purpose

Students will learn about some uses of X rays in medicine.

Teaching Strategies

■ Ask students to share experiences about when they have had X rays taken. Ask students to explain why the X rays were needed.

■ Hang up X rays on the windows or show them on the overhead as you discuss the Physics Connection. Point out how soft tissues do not show up on the X rays.

Connection to Biology

An X ray will show the image of a broken bone because X rays don't pass through dense bone. X rays will not, however, show if a ligament is sprained because the X rays pass through soft tissue and thus do not create an image.

Physics Connection

X Rays— The Painless Probe

X rays are a form of radiation emitted by X-ray tubes and by some astronomical objects such as stars. Machines that use X rays to view concealed objects are so common that you have probably had contact with one recently. Dentists use them to examine teeth, doctors to inspect bones and organs, and airports to look inside your carry-on baggage.

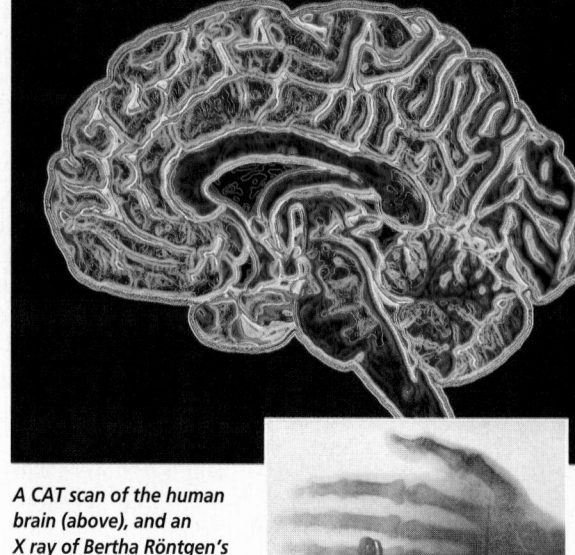

A CAT scan of the human brain (above), and an X ray of Bertha Röntgen's hand (right).

Wilhelm Röntgen, a German physics professor, accidentally discovered the X ray in 1895. As he was studying cathode rays in a high-voltage vacuum tube, he noticed that a screen lying nearby was giving off fluorescent light. He eventually determined that special rays given off by the tube were able to penetrate the black box that enclosed it and strike the screen, causing it to glow. Because he did not know what these rays were, he called them X rays, "X" standing for "unknown." He made a film of his wife's hand (shown here), exposing the bones—the first permanent X ray of a human. Two months later, he published a short paper. Within a month of its publication, doctors in Europe and the U.S. were using X rays in their work.

Noninvasive diagnosis In medicine, X rays are passed through the body to photographic film. Bones and other dense objects show up as white areas on the film. As a result, the position and nature of a break is clearly visible. The contours of organs such as the stomach can be seen when a patient ingests a high-contrast liquid; other organs can be marked with special dyes.

X rays generate a two-dimensional photograph with no depth. On the other hand, a computerized scan system, such as a CAT scanner, can make a 3-D image by scanning an area many times at different angles. These 3-D images are useful for viewing tissues shadowed by overlying organs.

Radiation treatments As X rays bombard atoms of tissues, electrons are knocked from their orbits, resulting in damage to the exposed tissue cells. To protect healthy tissues, absorptive metals are used as shields. You've probably had a dental X ray where the dental assistant spread a heavy lead apron across your chest. The destructive nature of X rays has proven useful in the treatment of cancers, where cancerous cells are targeted and destroyed.

CONNECTION TO BIOLOGY

Why would an X ray be useful in diagnosing a fractured bone but not a sprained ligament?

BIOLOGY *Online*  To find out more about X rays, visit the Glencoe Science Web site.
science.glencoe.com

BIOLOGY *Online* Note Internet addresses that you find useful in the space below for quick reference.

_____

_____

_____

_____

_____

SUMMARY

Section 34.1
Skin: The Body's Protection

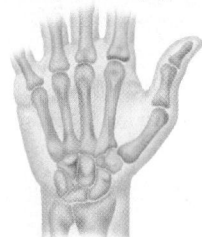

Main Ideas
- Skin regulates body temperature, protects the body, and functions as a sense organ.
- Skin cells are constantly being shed and replaced.
- Skin responds to injury by producing new cells and signaling a response to fight infection.

Vocabulary
dermis (p.925)
epidermis (p.923)
hair follicle (p.926)
keratin (p.923)
melanin (p.925)

Section 34.2
Bones: The Body's Support

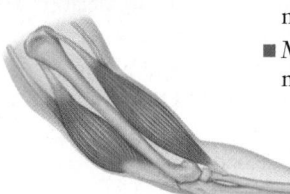

Main Ideas
- The skeleton is made up of the axial and appendicular skeletons.
- The skeleton supports the body, provides a place for muscle attachment, protects vital organs, manufactures blood cells, and serves as a storehouse for calcium and phosphorus.
- Joints allow movement between two or more bones where they meet.

Vocabulary
appendicular skeleton (p.929)
axial skeleton (p.929)
bursa (p.930)
compact bone (p.932)
joint (p.929)
ligament (p.930)
osteoblast (p.932)
osteocyte (p.932)
red marrow (p.933)
spongy bone (p.932)
tendon (p.930)
yellow marrow (p.933)

Section 34.3
Muscles for Locomotion

Main Ideas
- There are three types of tissue: smooth, cardiac, and skeletal.
- Muscles contract as filaments within the myofibrils slide toward one another.
- Muscle strength depends on muscle fiber thickness and the number of fibers contracting.

Vocabulary
actin (p.937)
cardiac muscle (p.936)
involuntary muscle (p.936)
myofibril (p.937)
myosin (p.937)
sarcomere (p.937)
skeletal muscle (p.936)
sliding filament theory (p.937)
smooth muscle (p.935)
voluntary muscle (p.936)

Main Ideas
Summary statements can be used by students to review the major concepts of the chapter.

Using the Vocabulary
To reinforce chapter vocabulary, use the Content Mastery Booklet and the activities in the Interactive Tutor for Biology: The Dynamics of Life on the Glencoe Science Web site. **science.glencoe.com**

All Chapter Assessment questions and answers have been validated for accuracy and suitability by The Princeton Review.

UNDERSTANDING MAIN IDEAS

1. d
2. b

UNDERSTANDING MAIN IDEAS

1. All of the following are tissues found in the skin except:
   a. connective   **c.** muscle
   b. epithelial   **d.** cardiac

2. Which of the following is NOT a function of the skeletal system?
   a. provide a framework for body tissues
   b. regulate temperature
   c. produce blood cells
   d. act as a storehouse for minerals

CHAPTER 34 ASSESSMENT **943**

GLENCOE TECHNOLOGY

 **VIDEOTAPE**
MindJogger Videoquizzes
Chapter 34: *Protection, Support, and Locomotion*
Have students work in groups as they play the videoquiz game to review key chapter concepts.

Resource Manager

Chapter Assessment, pp. 199-204
MindJogger Videoquizzes
ExamView® Pro Software
BDOL Interactive CD-ROM, Chapter 34 quiz

3. c
4. c
5. b
6. d
7. a
8. a
9. b
10. b
11. epidermis
12. Osteoblasts
13. Third-degree, dermis
14. cartilage
15. axial
16. Red marrow
17. ectodermal
18. bursa
19. tendons
20. voluntary, involuntary

APPLYING MAIN IDEAS

21. Ankles, wrists and knees are highly movable joints that are especially vulnerable to excessive twisting. The ankles and knees are also weight-bearing joints, which makes them even more vulnerable.

22. If red bone marrow were destroyed, the production of red blood cells, white blood cells, and blood-clotting cell fragments would be impaired.

---

3. Which of the following is a skin pigment that protects cells from solar radiation damage?
   a. keratin
   c. melanin
   b. epidermis
   d. dermis

4. Which of the following is NOT found in the dermis?
   a. blood vessels
   c. keratin
   b. nerves
   d. oil glands

5. Which of the following are nourished by blood vessels in the central canal of this structure?
   a. dermis
   b. bone cells
   c. epidermis
   d. sarcomeres

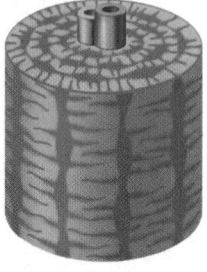

6. When exposed to sunlight, skin is able to produce which vitamin?
   a. Vitamin C
   c. Vitamin B
   b. Vitamin A
   d. Vitamin D

7. Which type of joint allows rotational movement?

a.
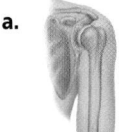

c.

b.

d.
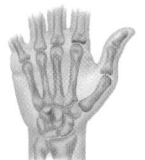

THE PRINCETON REVIEW   **TEST-TAKING TIP**

Warm Up Before the Race
On the day of your exam, arrive at the site early enough to relax, get settled, and go over your notes. It will give you time to relax and prepare your mind for the test.

---

8. All of the following are types of muscle except:
   a. epidermal
   c. smooth
   b. cardiac
   d. skeletal

9. As a muscle relaxes after contraction, actin filaments:
   a. slide in toward each other
   b. slide away from each other
   c. are digested into their component proteins
   d. lengthen and become invisible under the microscope

10. Skin plays a role in:
    a. storing calcium
    b. regulating body temperature
    c. manufacturing blood cells
    d. supporting the body

11. Ridges in the _____ of your skin give you a unique fingerprint.

12. _____ grow to become mature bone cells.

13. _____ burns are the most severe, resulting in destroyed epidermis and _____.

14. The fetal appendicular skeleton is mostly made of _____.

15. The _____ skeleton includes the skull, ribs, vertebral column, and sternum.

16. _____ fills the cavities of bones and is responsible for the production of red and white blood cells.

17. Epidermal cells in the skin are derived from the _____ layer of the embryo.

18. In movable joints, a fluid-filled sac called a _____ prevents bones from rubbing against each other.

19. Muscles are attached to bones by _____.

20. Skeletal muscle is under _____ control, while cardiac muscle is under _____ control.

APPLYING MAIN IDEAS

21. Why do sprains usually occur in the ankles, wrists, and knees?

22. How would the destruction of red marrow affect other systems of the body?

23. How could you use a person's skeleton to determine his or her age?

24. Outline how the skin helps the body maintain body temperature.

25. Interpreting Data The graph below shows the time relationship between muscle force and calcium levels inside a contracting muscle fiber. Describe the cause-and-effect relationship shown between the development of muscle force and calcium levels in the fiber.

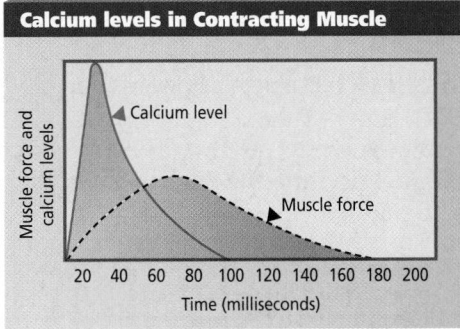

Calcium levels in Contracting Muscle

Muscle force and calcium levels (y-axis)

Calcium level

Muscle force

Time (milliseconds): 20 40 60 80 100 120 140 160 180 200

26. Concept Mapping Complete the concept map by using the following vocabulary terms: actin, myofibrils, skeletal muscles, myosin.

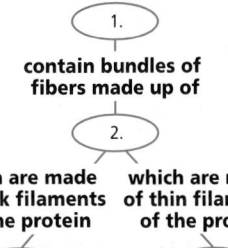

1.

contain bundles of
fibers made up of

2.

which are made
of thick filaments
of the protein

which are made
of thin filaments
of the protein

3.

4.

CD-ROM

For additional review, use the assessment options for this chapter found on the *Biology: The Dynamics of Life Interactive CD-ROM* and on the Glencoe Science Web site.
science.glencoe.com

ASSESSING KNOWLEDGE & SKILLS

Bone is living tissue that includes different kinds of cells and structures.

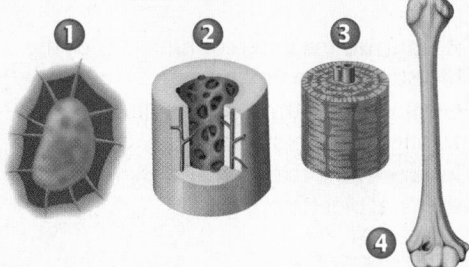

Interpreting Scientific Illustrations
Use the diagram above to answer the following questions.

1. Place the structures of the human body shown in the diagram in order from least to most complex.
  a. 1, 2, 3, 4   **c.** 4, 3, 2, 1
  b. 1, 3, 2, 4   **d.** 4, 2, 3, 1

2. Which diagram represents one osteon system?
  a. 1   **c.** 3
  b. 2   **d.** 4

3. Which of the following demonstrates that bones are alive?
  a. Bones grow in both length and diameter.
  b. Bones are able to repair themselves.
  c. Bones contain living cells called osteocytes.
  d. All of the above.

4. Interpreting Data The center of most long bones is hollow. What advantage might having bones with hollow centers confer to humans and other mammals?

23. Degree of bone fusion can be used to identify the age of a skeleton. The more fusion, the older the skeleton. The amount of cartilage present and brittleness of bone also may be used to determine age.

24. Sweat evaporates from the body, cooling it in the process. Blood vessels in the skin can also constrict to conserve heat or dilate to increase heat loss.

THINKING CRITICALLY

25. A large increase in muscle cell calcium precedes (and initiates) muscle contraction.

26. 1. Skeletal muscles; 2. Myofibrils; 3. Myosin; 4. Actin

ASSESSING KNOWLEDGE & SKILLS

1. b
2. c
3. d
4. Hollow bones allow humans and other mammals to have strength with reduced weight.

Chapter 35 Organizer

Refer to pages 4T-5T of the Teacher Guide for an explanation of the National Science Education Standards correlations.

Section	Objectives	Activities/Features
Section 35.1 **Following Digestion of a Meal** National Science Education Standards UCP.1-3, UCP.5; B.3; C.5; F.1; G.1 (1 session, $^1/_2$ block)	1. **Recognize** the different functions of the digestive system organs. 2. **Outline** the pathway food follows through the digestive tract. 3. **Interpret** the role of enzymes in chemical digestion.	**Inside Story:** Your Mouth, p. 949 **Problem-Solving Lab 35-1**, p. 952
Section 35.2 **Nutrition** National Science Education Standards UCP.2, UCP.3; A.1, A.2; B.2, B.3, B.6; C.5, C.6; F.1 ($^1/_2$ session)	4. **Summarize** the contribution of the six classes of nutrients to body nutrition. 5. **Identify** the role of the liver in food storage. 6. **Relate** caloric intake to weight loss or gain.	**MiniLab 35-1:** Evaluate a Bowl of Soup, p. 957 **Problem-Solving Lab 35-2**, p. 958 **Biology & Society:** The Promise of Weight Loss, p. 968
Section 35.3 **The Endocrine System** National Science Education Standards UCP.1-3, UCP.5; A.1, A.2; B.2, B.3, B.6; C.1, C.5, C.6; F.1; G.1 ($2^1/_2$ sessions, 1 block)	7. **Identify** the functions of some of the hormones secreted by endocrine glands. 8. **Summarize** the negative feedback mechanism controlling hormone levels in the body. 9. **Contrast** the actions of steroid and amino acid hormones.	**Problem-Solving Lab 35-3**, p. 962 **MiniLab 35-2:** Compare Thyroid and Parathyroid Tissue, p. 964 **Investigate BioLab:** Average Growth Rate in Humans, p. 966

Need Materials? Contact Carolina Biological Supply Company at 1-800-334-5551 or at **http://www.carolina.com**

MATERIALS LIST

BioLab
p. 966 blue pencil, red pencil, graph paper, ruler

MiniLabs
p. 957 paper, pencil
p. 964 microscope, prepared slide of thyroid and parathyroid tissue, paper, pencil

Alternative Lab
p. 950 Lactaid solution, glucose test paper, glucose solution, milk, graduated cylinder (2), test tubes (4), dropper

Quick Demos
p. 948 barium X ray of digestive tract
p. 950 dialysis bag (2), starch solution, pancreatic enzyme solution, beaker, distilled water, Benedict's solution
p. 955 paper grocery bags, assorted food samples, water
p. 960 human skull
p. 961 microprojector, prepared slide of pancreas
p. 963 preserved whole kidney

Key to Teaching Strategies

L1 Level 1 activities should be appropriate for students with learning difficulties.

L2 Level 2 activities should be within the ability range of all students.

L3 Level 3 activities are designed for above-average students.

ELL ELL activities should be within the ability range of English Language Learners.

COOP LEARN Cooperative Learning activities are designed for small group work.

P These strategies represent student products that can be placed into a best-work portfolio.

These strategies are useful in a block scheduling format.

The Digestive and Endocrine Systems

Teacher Classroom Resources

Section	Reproducible Masters	Transparencies
Section 35.1 **Following Digestion of a Meal**	Reinforcement and Study Guide, pp. 155-156 **L2** Content Mastery, pp. 173-174, 176 **L1** Tech Prep Applications, pp. 45-46 **L2**	Section Focus Transparency 84 **L1** **ELL** Reteaching Skills Transparency 50 **L1** **ELL**
Section 35.2 **Nutrition**	Reinforcement and Study Guide, p. 157 **L2** Concept Mapping, p. 35 **L3** **ELL** BioLab and MiniLab Worksheets, p. 155 **L2** Laboratory Manual, pp. 255-260 **L2** Content Mastery, pp. 173, 175-176 **L1**	Section Focus Transparency 85 **L1** **ELL** Reteaching Skills Transparency 51 **L1** **ELL**
Section 35.3 **The Endocrine System**	Reinforcement and Study Guide, p. 158 **L2** Critical Thinking/Problem Solving, p. 35 **L3** BioLab and MiniLab Worksheets, pp. 156-158 **L2** Content Mastery, pp. 173, 175-176 **L1**	Section Focus Transparency 86 **L1** **ELL** Basic Concepts Transparency 65 **L2** **ELL**

Assessment Resources

Chapter Assessment, pp. 205-210
MindJogger Videoquizzes
Performance Assessment in the Biology Classroom
Alternate Assessment in the Science Classroom
ExamView® Pro Software 💾
BDOL Interactive CD-ROM, Chapter 35 quiz

Additional Resources

Spanish Resources **ELL**
English/Spanish Audiocassettes **ELL**
Cooperative Learning in the Science Classroom **COOP LEARN**
Lesson Plans/Block Scheduling

 **NATIONAL GEOGRAPHIC** **Teacher's Corner**

Products Available From Glencoe
To order the following products, call Glencoe at 1-800-334-7344:
CD-ROMs
NGS PictureShow: Human Body 1
NGS PictureShow: Human Body 2
Curriculum Kit
GeoKit: Human Body 1
Transparency Set
NGS PicturePack: Human Body 1
NGS PicturePack: Human Body 2
Videodisc
STV: Human Body

Products Available From National Geographic Society
To order the following products, call National Geographic Society at 1-800-368-2728:
Videos
Digestive System (Human Body Series)
Our Immune System

GLENCOE TECHNOLOGY

The following multimedia resources are available from Glencoe.

Biology: The Dynamics of Life
CD-ROM **ELL**
 Video: *X-Ray of Swallowing*
BioQuest: *Body Systems*
Exploration: *Nutrition*

Videodisc Program
 X-Ray of Swallowing

The Infinite Voyage
 A Taste of Health

Visual-Spatial Test a cracker for starch by adding a few drops of iodine to the ground crumbs. A dark brown/purple indicates the presence of starch. To demonstrate the effects of saliva on starch, add about 1 g of ground crackers to 1-2 ml of water. Add a small amount of amylase (available from biological supply companies) and let sit for a few minutes. You can test for the breakdown of starch by periodically using the iodine test.

Theme Development

The theme of **systems and interactions** is developed by looking at the role of the digestive and endocrine systems in regulating body functions.

0:00 OUT OF TIME?

If time does not permit teaching the entire chapter, use the BioDigest at the end of the unit as an overview.

READING BIOLOGY

Glencoe's *Biology: The Dynamics of Life* contains many resources to assist a student's reading skills. Each chapter contains figures with expanded captions that expand on written material. Word Origins, located along the side of text, expand knowledge of biology vocabulary. Glencoe's Content Mastery Booklet helps develop reading skills while reinforcing content. In addition, use the Interactive Tutor for *Biology: The Dynamics of Life* on the Glencoe Web site to reinforce vocabulary.
science.glencoe.com

Chapter

35 The Digestive and Endocrine Systems

What You'll Learn
- You will trace the journey of a meal through the digestive system.
- You will recognize different nutrients and their uses in the body.
- You will outline how endocrine hormones control internal body processes.

Why It's Important
By examining the functions of your digestive and endocrine systems, you will understand how your body obtains energy from food and how it controls your behavior and development.

READING BIOLOGY

Scan the chapter sections and make an outline using the headings in each. Make notes on this outline as you read the chapter.

To find out more about digestion and the endocrine system, visit the Glencoe Science Web site.
science.glencoe.com

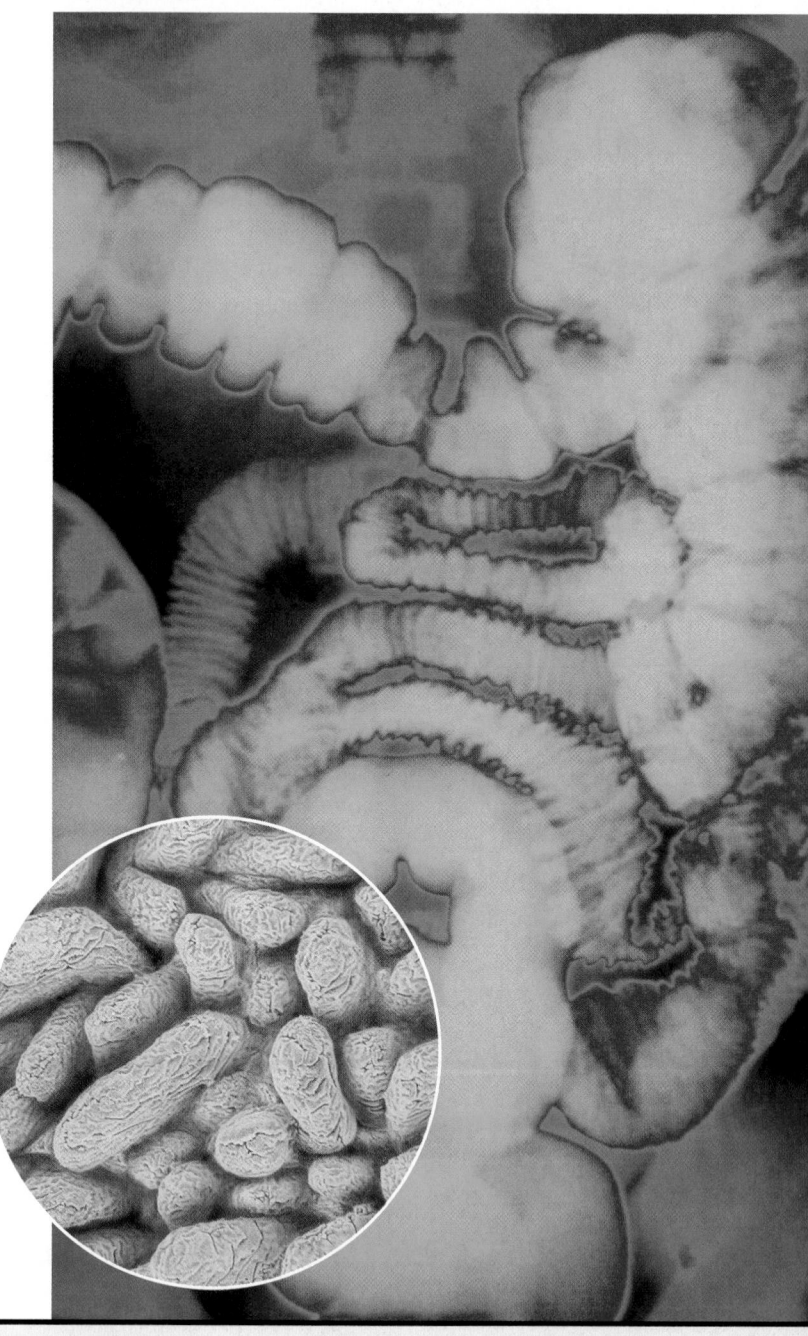

Magnification: 70×

These projections, called villi, line the walls of your small intestine, absorbing food particles as they pass by.

946

Multiple Learning Styles

Look for the following logos for strategies that emphasize different learning modalities.

Kinesthetic Meeting Individual Needs, p. 948; Quick Demo, pp. 950, 955

Visual-Spatial Getting Started Demo, p. 946; Quick Demo, pp. 948, 960, 961; Portfolio, p. 949; Meeting Individual Needs, pp. 954, 960; Reteach, p. 957

Interpersonal Check for Understanding, p. 953;

Enrichment, p. 955; Biology Journal, p. 956

Intrapersonal Project, p. 952; Cultural Diversity, p. 955; Tech Prep, p. 962; Portfolio, p. 964

Linguistic Biology Journal, pp. 955, 961, 963; Tech Prep, p. 956; Extension, p. 957

Logical-Mathematical Project, p. 957

35.1 Following Digestion of a Meal

E ating is something you probably spend a lot of time thinking about—what you will eat, when you will eat, and who you will eat with. Your digestive system helps turn food into energy for your body. As in many animals you have studied, the human digestive system is essentially a specialized tube that has evolved over millions of years to form digestive organs, each of which performs a unique function.

These gastric pits in the stomach secrete acid needed for digestion.

Magnification: 55×

SECTION PREVIEW

Objectives

Recognize the different functions of the digestive system organs.
Outline the pathway food follows through the digestive tract.
Interpret the role of enzymes in chemical digestion.

Vocabulary

amylase
esophagus
peristalsis
epiglottis
stomach
pepsin
small intestine
pancreas
liver
bile
gallbladder
villus
large intestine
rectum

Section 35.1

Prepare

Key Concepts

The structures and functions of the organs of the digestive system are presented as students follow a meal through the digestive tract.

Planning

■ Acquire old barium X rays for the first Quick Demo.
■ Prepare materials for the second Quick Demo.
■ Purchase milk and Lactaid for the Alternative Lab.

1 Focus

Bellringer

Before presenting the lesson, display **Section Focus Transparency 84** on the overhead projector and have students answer the accompanying questions.
[L1] [ELL]

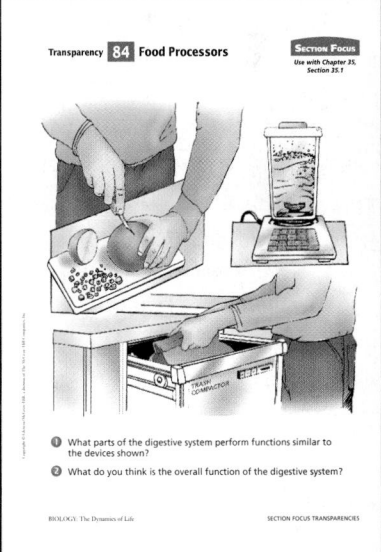

Functions of the Digestive System

The main function of the digestive system is to disassemble the food you eat into its component molecules so that it can be used as energy for your body. In this sense, your digestive system can be thought of as a sort of disassembly line.

Digestion is accomplished through a number of steps. First, the system takes ingested food and begins moving it through the digestive tract. As it does so, it digests—or breaks down mechanically and chemically—the complex food molecules. Then, the system absorbs the digested food and distributes it to your cells. Finally, it eliminates undigested materials from your body. As you read about each digestive organ, use *Figure 35.1* to locate its position within the system.

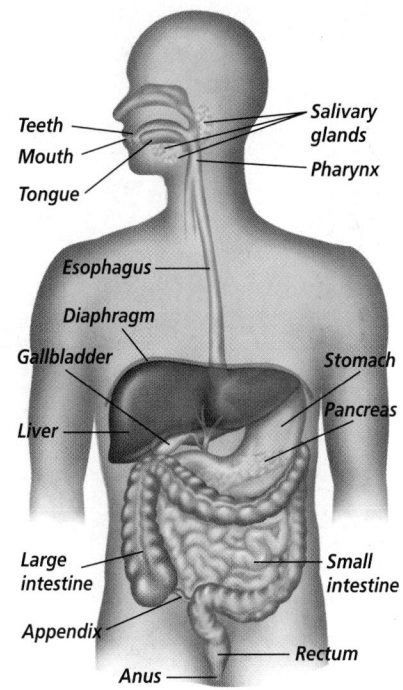

Teeth
Mouth
Tongue
Salivary glands
Pharynx
Esophagus
Diaphragm
Gallbladder
Liver
Stomach
Pancreas
Large intestine
Small intestine
Appendix
Rectum
Anus

Figure 35.1
All the digestive organs work together to break down food into simpler compounds that can be absorbed by the body.

Assessment Planner

Portfolio Assessment
  Portfolio, TWE, pp. 949, 957, 964
  Alternative Lab, TWE, pp. 950-951
  Problem-Solving Lab, TWE, p. 958
  Assessment, TWE, p. 957
  BioLab, TWE, pp. 966-967
Performance Assessment
  MiniLab, SE, p. 964
  Alternative Lab, TWE, pp. 950-951

Problem-Solving Lab, TWE, p. 952
MiniLab, TWE, pp. 956, 964
Knowledge Assessment
  Section Assessment, SE, pp. 953, 958, 965
  Chapter Assessment, SE, pp. 969-971
  Assessment, TWE, pp. 956, 965
  Problem-Solving Lab, TWE, p. 962
Skill Assessment
  Assessment, TWE, pp. 951, 953, 961

Visual Learning

Explain that the pancreas is 15 cm long and produces 1 L of pancreatic juice each day. Have students use Table 35.1 to list the chemicals that compose pancreatic juice. *pancreatic amylase, trypsin, pancreatic lipase, nucleases* **L1**

Reinforcement

Have two students hold a 9-m long piece of string. Explain that the string represents the length of the human digestive tract. Briefly discuss how food moves along the length of the digestive tract and the changes that occur within the tract.

The Mouth

The first stop along the digestive disassembly line is your mouth. Suppose it's lunchtime and you have just prepared a bacon, lettuce, and tomato sandwich. The first thing you do is bite off a piece and chew it.

What happens as you chew?

In your mouth, your tongue moves the food around and helps position it between your teeth so that it can be chewed. Chewing is a form of mechanical digestion, the physical process of breaking food into smaller pieces. Mechanical digestion prepares food particles for chemical digestion. Chemical digestion is the process of changing food on a

WORD *Origin*

peristalsis
From the Greek word *peri*, meaning "around," and *stellein*, meaning "to draw in." Peristalsis propels food in one direction.

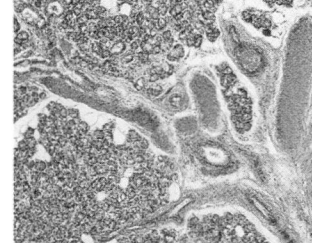

Figure 35.2
Salivary glands secrete saliva, a watery substance that contains the enzyme amylase.

Magnification: 90×

molecular level through the action of enzymes. What purpose do the different structures inside your mouth serve? Find out by reading the *Inside Story*.

Chemical digestion begins in the mouth

Some of the nutrients in your sandwich are starches, large molecules known as polysaccharides. As you chew your bite of sandwich, salivary glands in your mouth secrete saliva, as shown in **Figure 35.2**. Saliva contains a digestive enzyme, called **amylase**, which breaks down starch into smaller molecules such as di- or monosaccharides. In the stomach, amylase continues to digest starch in the swallowed food for about 30 minutes. **Table 35.1** lists some digestive enzymes that act to break food molecules apart.

Swallowing your food

Once you've thoroughly chewed your bite of sandwich, your tongue shapes it into a ball and moves it to the back of your mouth to be swallowed. Swallowing forces food from your mouth into your **esophagus,** a muscular tube that connects your mouth to your stomach. Food moves down the esophagus by way of

Table 35.1 Digestive enzymes			
Organ	**Enzyme**	**Molecules digested**	**Product**
Salivary glands	Salivary amylase	Starch	Disaccharide
Stomach	Pepsin	Proteins	Peptides
Pancreas	Pancreatic amylase Trypsin Pancreatic lipase Nucleases	Starch Proteins Fats Nucleic acids	Disaccharide Peptides Fatty acids and glycerol Sugar and nitrogen bases
Small intestine	Maltase Sucrase Lactase Peptidase Nuclease	Disaccharide Disaccharide Disaccharide Peptides Nucleic acids	Monosaccharide Monosaccharide Monosaccharide Amino acids Sugar and nitrogen bases

Your Mouth

Your mouth houses many structures involved in other functions besides digestion. Some of these structures protect against foreign materials invading your body; others help you taste the food you eat.

Critical Thinking *Why do your teeth come in various shapes?*

Magnification: 725×

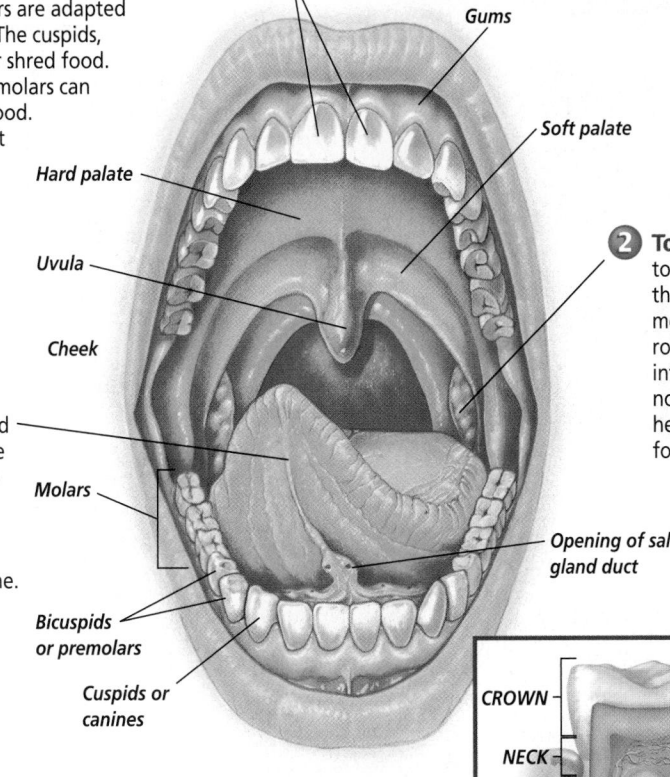

The tongue is covered by projections that contain numerous taste receptor cells like the one shown here.

1 Teeth The incisors are adapted for cutting food. The cuspids, or canines, tear or shred food. The three sets of molars can crush and grind food. Often, there is not enough room for the third set of molars, called wisdom teeth, which then must be removed.

3 Tongue The tongue is attached to the floor of the mouth. It is made of numerous skeletal muscles covered with a mucous membrane.

Incisors
Gums
Soft palate
Hard palate
Uvula
Cheek
Molars
Bicuspids or premolars
Cuspids or canines
Tonsils
Opening of salivary gland duct

2 Tonsils A pair of tonsils is located at the back of the mouth. They play a role in preventing infections in the nose and mouth by helping to eliminate foreign bacteria.

4 Structure of a tooth Teeth are made mainly of dentin, a bonelike substance that gives a tooth its shape and strength. The dentin encloses a space filled with pulp, a tissue that contains blood vessels and nerves. The dentin of the crown is covered with an enamel that consists mostly of calcium salts. Tooth enamel is the hardest substance in the body.

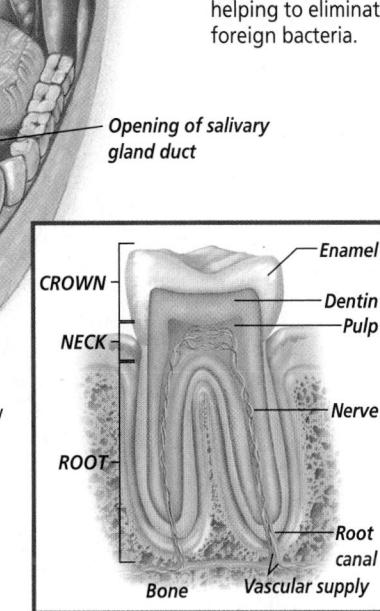

CROWN
NECK
ROOT
Enamel
Dentin
Pulp
Nerve
Root canal
Bone
Vascular supply

Purpose
Students examine teeth and other structures of the mouth.

Teaching Strategies
■ Bring in models of animal skulls, including a human skull, to compare teeth. Ask students whether other mammals have baby or milk teeth. *Yes. Some students may recall their puppies or kittens losing baby teeth.*

Misconception
Students may believe that cavities are the greatest threat to teeth. Explain that while cavities do destroy teeth, periodontal disease of the gums is much more serious and far more prevalent today.

Visual Learning
■ Point out the structures of the mouth shown in the diagram. As each structure is mentioned, have a volunteer read the description of the structure.
■ Ask students to name the three types of teeth shown. *incisors, cuspids, and molars*
■ Elicit from students the function of the tonsils and their location. *The tonsils are located in the back of the mouth and function to remove bacteria that enter the mouth or nose.*

Critical Thinking
Different teeth perform different functions—incisors are shaped for cutting, while molars are shaped for crushing and grinding.

Portfolio

Primary Teeth

Visual-Spatial Have students investigate how the number and location of primary teeth in humans differ from the secondary teeth. Ask students to diagram a mouth that shows only primary teeth. Have students identify the types of teeth shown in the diagram.

L2 P

Figure 35.3
Smooth muscle contractions are responsible for moving food through the digestive system.

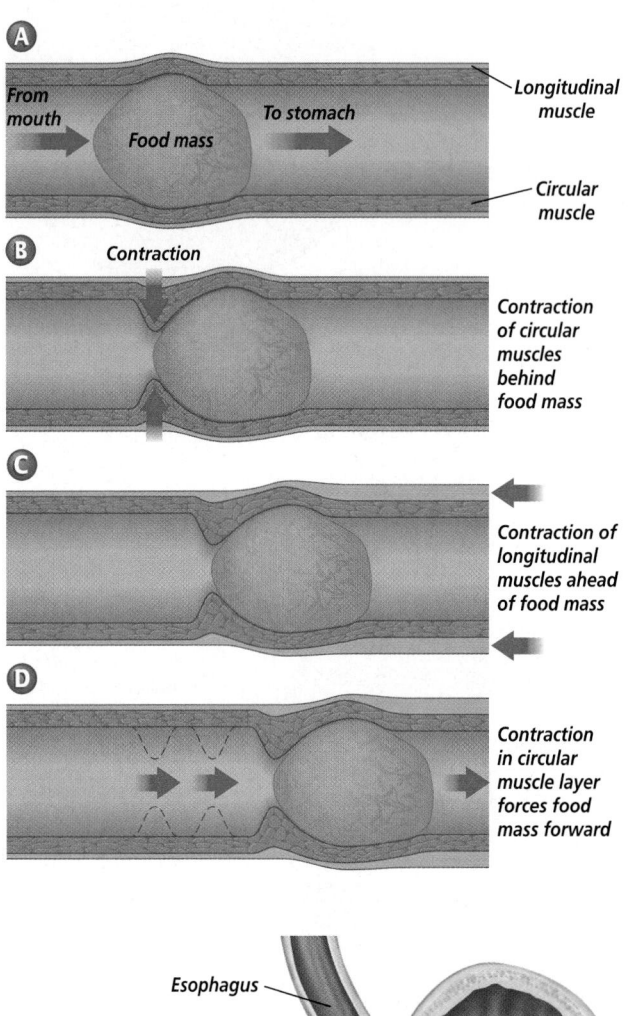

Ⓐ
From mouth
Food mass
To stomach
Longitudinal muscle
Circular muscle

Ⓑ Contraction
Contraction of circular muscles behind food mass

Ⓒ
Contraction of longitudinal muscles ahead of food mass

Ⓓ
Contraction in circular muscle layer forces food mass forward

Figure 35.4
Smooth muscle contractions churn the food in the stomach until it becomes a thin liquid.

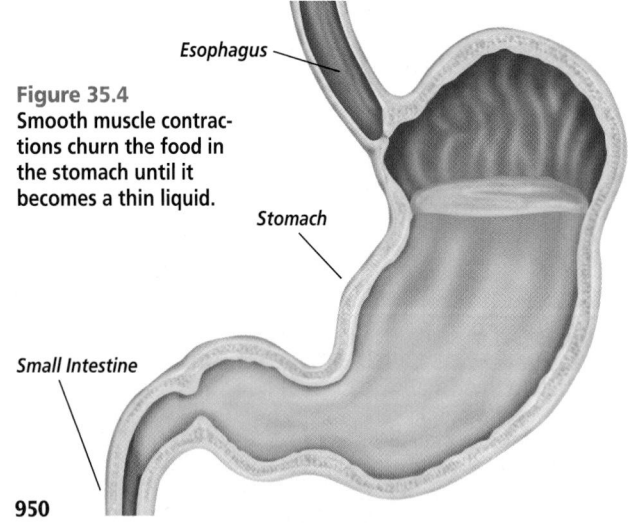

Esophagus
Stomach
Small Intestine

950

peristalsis (per uh STAHL sus), a series of involuntary smooth muscle contractions along the walls of the digestive tract. *Figure 35.3* shows how the food is moved along from the mouth to the stomach. The contractions occur in waves: first, circular muscles relax and longitudinal muscles contract; then circular muscles contract and longitudinal muscles relax.

Have you ever had food go down the wrong way? When you swallow, the food enters the esophagus. Usually, a flap of cartilage called the epiglottis (ep uh GLAHT us) closes over the opening to the respiratory tract as you swallow, preventing food from entering. After the food passes into your esophagus, the epiglottis opens again. But if you talk or laugh as you swallow, the epiglottis may open, allowing food to enter the upper portion of the respiratory tract. Your response, a reflex, is to choke and cough, forcing the food out of the respiratory tube.

The Stomach

When the chewed food reaches the end of your esophagus, it enters the stomach. The **stomach** is a muscular, pouchlike enlargement of the digestive tract. Both physical and chemical digestion take place in the stomach.

Muscular churning

Three layers of involuntary muscles, lying across one another, are located within the wall of the stomach. When these muscles contract, as shown in *Figure 35.4*, they work to physically break down the swallowed food, creating smaller pieces. As the muscles continue to work the food pieces, they mix them with digestive juices produced by the stomach.

Chemical digestion in the stomach

The inner lining of the stomach contains millions of glands that secrete a mixture of chemicals called gastric juice. Gastric juice contains pepsin and hydrochloric acid. **Pepsin** is an enzyme that begins the chemical digestion of proteins in food. Pepsin works best in the acidic environment provided by hydrochloric acid, which increases the acidity of the stomach contents to pH 2.

Knowing that the stomach secretes acids and enzymes, you may be wondering why the stomach doesn't digest itself. The stomach lining is protected by mucus that forms a layer between it and the acidic environment of the stomach. The mucus is secreted by the stomach lining itself.

Food remains in your stomach for approximately two to four hours. By the time the food is ready to leave the stomach, it is about the consistency of tomato soup. At that time, the peristaltic waves gradually become more vigorous and begin to force small amounts of liquid out of the lower end of the stomach and into the small intestine.

The Small Intestine

From your stomach, the liquid food moves into your **small intestine,** a muscular tube about 6 m long. This section of the intestine is called *small* not because of its length, but because of its narrow diameter—only 2.5 cm. Digestion of your meal is completed within the small intestine. Muscle contractions contribute to further mechanical breakdown of the food; at the same time, carbohydrates and proteins undergo further chemical digestion with the help of enzymes produced and secreted by the pancreas and liver.

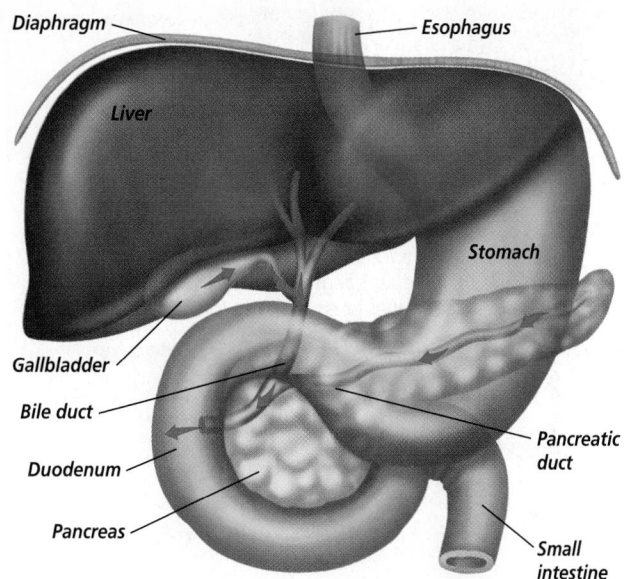

Diaphragm — Esophagus
Liver
Stomach
Gallbladder
Bile duct
Duodenum
Pancreas
Pancreatic duct
Small intestine

Figure 35.5
Both the pancreas and the liver produce chemicals needed for digestion in the small intestine.

Chemical action

The first 25 cm of the small intestine is called the duodenum (doo AHD un um). Most of the enzymes and chemicals that function in the duodenum enter it through ducts that collect juices from the pancreas, liver, and gallbladder. These organs, shown in *Figure 35.5,* play important roles in digestion, even though food does not pass directly through them.

Secretions of the pancreas

The **pancreas** is a soft, flattened gland that secretes both digestive enzymes and hormones, which you will learn more about in the last section of this chapter. The mixture of enzymes it secretes breaks down carbohydrates, proteins, and fats. Alkaline pancreatic juices also help to neutralize the acidity of the liquid food, stopping any further action of pepsin.

Secretions of the liver

The **liver** is a large, complex organ that, among its many functions, produces bile. **Bile** is a chemical

Analysis

1. What did Lactaid do to the lactose in milk? *It broke down the lactose.*
2. What is the function of test tubes *1, 2,* and *4*? *They are controls.*
3. How will Lactaid help people who cannot digest lactose. *Lactaid will help break down the lactose present in dairy products.*

Purpose

Students compare the path of bile from the liver to the duodenum with and without the gallbladder.

Process Skills

analyze information, compare and contrast, draw a conclusion, interpret scientific illustrations, sequence, think critically

Teaching Strategies

■ Demonstrate the role of bile by adding a small amount of oil to a test tube of water. Shake it and have students note what happens to the oil after waiting a few minutes. (Oil reforms on top of the water.) Repeat the demonstration using liquid detergent to simulate bile. (Oil will remain as small droplets.)

■ Point out to students that the gallbladder is under involuntary control—it releases bile when stimulated by a hormone.

Thinking Critically

1. liver; gallbladder
2. Bile physically changes fat into smaller droplets.
3. liver, hepatic duct, gallbladder, bile duct, duodenum
4. liver, hepatic duct, bile duct, duodenum
5. Only a small, but continuous, amount of bile reaches the duodenum. Large amounts needed to break up fats efficiently are not available.

952

Problem-Solving Lab 35-1 — Sequencing

Is it possible to live without a gallbladder? Apparently yes, as many people have had this organ surgically removed and are still alive.

Analysis

The following diagrams show the appearance of a normal liver and gallbladder (diagram A) and the appearance when the gallbladder has been removed (diagram B).

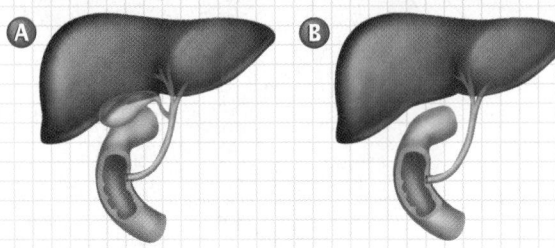

Thinking Critically

1. Where is bile produced? Where is bile stored?
2. Does bile bring about a chemical or physical change to fat? Explain.
3. Sequence the pathway for bile from the liver to the duodenum in the person with a gallbladder.
4. Sequence the pathway for bile from the liver to the duodenum in the person with no gallbladder.
5. The gallbladder is a muscular sac. It squeezes and discharges a large quantity of bile when fats are present in the duodenum. Explain why a person without a gallbladder is unable to digest fats as efficiently as someone who has a gallbladder.

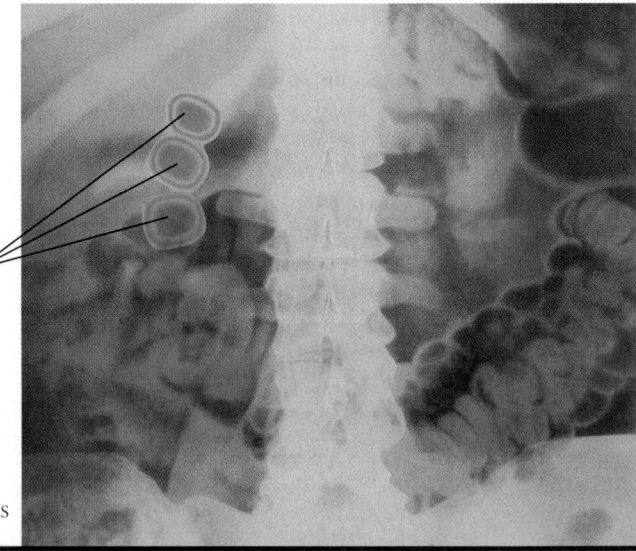

Figure 35.6
Gallstones can form in the gallbladder or bile duct. They consist mainly of precipitated bile salts.

Gallstones

952 THE DIGESTIVE AND ENDOCRINE SYSTEMS

---

substance that helps break down fats. Once made in the liver, bile is stored in a small organ called the **gallbladder,** from which it passes into the duodenum. Bile causes further mechanical digestion by breaking apart large drops of fat into smaller droplets. If bile becomes too concentrated due to high levels of cholesterol in the diet, or if the gallbladder becomes inflamed, gallstones can form, as seen in *Figure 35.6.* Can a person live without a gallbladder? Find out in the *Problem-Solving Lab* on this page.

Absorption of food

Liquid food stays in your small intestine for three to five hours and is slowly moved along its length by peristalsis. As digested food moves through the intestine, it passes over thousands of tiny fingerlike structures called villi. A **villus** is a single projection on the lining of the small intestine that functions in the absorption of digested food. The villi greatly increase the surface area of the small intestine, allowing for a greater absorption rate. Because the digested food is now in the form of small molecules, it can be absorbed

---

PROJECT

Meat Tenderizers

Intrapersonal Have students conduct library research to find out how meat tenderizer works. Have them use the information they gather to design a demonstration that can be used to explain the process to others. Encourage students to write out the procedural steps for their demonstration and explain the purpose for each step. Finally, have students complete their demonstration with information that relates the function of meat tenderizer to its complementary organ of the digestive system. **L3**
COOP LEARN

directly into the cells of the villi, as shown in *Figure 35.7*. The food molecules then diffuse into the blood vessels of the villus and enter the bloodstream. As you can see, the villi are the link between the digestive system and the circulatory system.

As your lunch comes to the end of its passage through the small intestine, only the indigestible materials remain in the digestive tract.

The Large Intestine

The indigestible material from your meal now passes into your **large intestine,** a muscular tube that is also called the colon. Even though the large intestine is only about 1.5 m long, it is much wider than the small intestine—about 6.5 cm in diameter. The appendix, a small, tubelike extension off the large intestine, is thought to be an evolutionary remnant from our herbivorous ancestors as it seems to serve no function in human digestion.

Water absorption

As the indigestible mixture passes through the large intestine, water is absorbed by the intestine walls, leaving behind a more solid material. In this way, the water is not wasted. A secondary function of the large

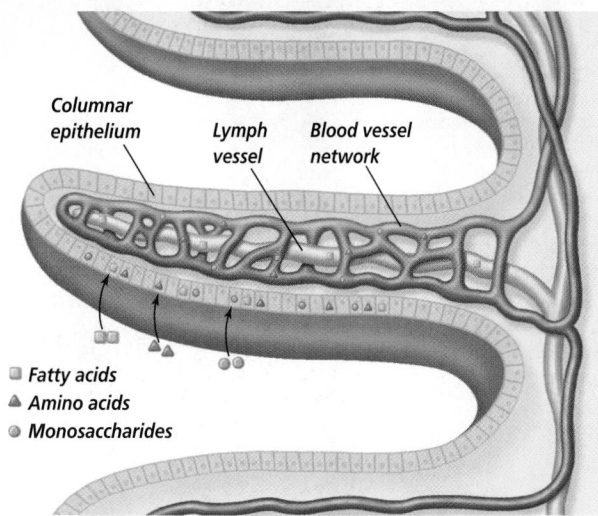

Columnar epithelium

Lymph vessel

Blood vessel network

■ Fatty acids
▲ Amino acids
● Monosaccharides

intestine is vitamin synthesis. Anaerobic bacteria in the large intestine synthesize some B vitamins and vitamin K, which are absorbed as needed by the body.

Elimination of wastes

After 18 to 24 hours in the large intestine, the remaining indigestible material, now called feces, reaches the rectum. The **rectum** is the last part of the digestive system. Feces are eliminated from the rectum through the anus. Your meal's entire journey from the beginning of the digestive tract to the end has taken between 24 and 33 hours.

Figure 35.7
Once food has been fully digested in the small intestine, it is in the form of molecules small enough to enter the body's bloodstream.

Section Assessment

Understanding Main Ideas
1. Sequence the organs of your digestive system according to the order in which food passes through them.
2. In which sections of the digestive system are starches digested? Which enzymes break down starches?
3. How do villi of the small intestine increase the rate of nutrient absorption?
4. What role does the pancreas play in digestion?

Thinking Critically
5. How would chronic diarrhea affect the balance of fluids in your body?

SKILL REVIEW
6. **Making and Using Graphs** Prepare a circle graph representing the time food remains in each part of the digestive tract. For more help, refer to *Organizing Information* in the **Skill Handbook.**

Prepare

Key Concepts

Students become familiar with the six classes of nutrients. They also relate these nutrients to Calories and metabolism.

Planning

- Purchase a school lunch for Meeting Individual Needs.
- Acquire foods and brown paper for the Quick Demo.
- Gather Calorie charts for the Project.

1 Focus

Bellringer 🔥

Before presenting the lesson, display **Section Focus Transparency 85** on the overhead projector and have students answer the accompanying questions. **L1** **ELL**

Transparency **85** All the Things You Eat — **Section Focus** Use with Chapter 35, Section 35.2

1 What two or three essential nutrients are most abundant in each of the foods shown?

2 What essential nutrient is available in all the foods except the grapefruit?

BIOLOGY: The Dynamics of Life — SECTION FOCUS TRANSPARENCIES

Objectives

Summarize the contribution of the six classes of nutrients to body nutrition.

Identify the role of the liver in food storage.

Relate caloric intake to weight loss or gain.

Vocabulary

mineral
vitamin
Calorie

Section

35.2 Nutrition

What's your favorite food? How often do you eat it? Of what nutritional value is it? The food pyramid is a diagram that indicates the number of servings a person should have daily from each of the food groups. How do your meals fit into this pyramid?

USE SPARINGLY

☐ Fat (naturally occurring and added during cooking)
▽ Sugars (added to foods)

2-3 SERVINGS 2-3 SERVINGS

3-5 SERVINGS 2-4 SERVINGS

6-11 SERVINGS

The food pyramid

Figure 35.8
Select foods from the five food groups and you'll have a healthful diet that supplies the six essential nutrients your body needs.

The Vital Nutrients

Six basic kinds of nutrients can be found in foods: carbohydrates, fats, proteins, minerals, vitamins, and water. These substances are essential to proper body function. You supply your body with these nutrients when you eat foods from the five main food groups shown in *Figure 35.8*.

Carbohydrates

Perhaps your favorite food is pasta, fresh-baked bread, or corn on the cob. If so, your favorite food contains carbohydrates, important sources of energy for your body cells. Recall that carbohydrates are starches and sugars. Starches are complex carbohydrates found in bread, cereal, potatoes, rice, corn, beans, and pasta. Sugars are simple carbohydrates found mainly in fruits, such as plums, strawberries, and oranges.

During digestion, complex carbohydrates are broken down into simple sugars such as glucose, fructose, and galactose. Absorbed into the bloodstream through the villi of the small intestine, these sugar molecules circulate to

---

MEETING INDIVIDUAL NEEDS

Hearing Impaired/English Language Learners

Visual-Spatial Bring in a sample school lunch. Ask students to list, on a sheet of paper, the nutrients found in each of the foods provided. Next, have them point out on a diagram of the digestive system where each type of nutrient is digested. **L1** **ELL**

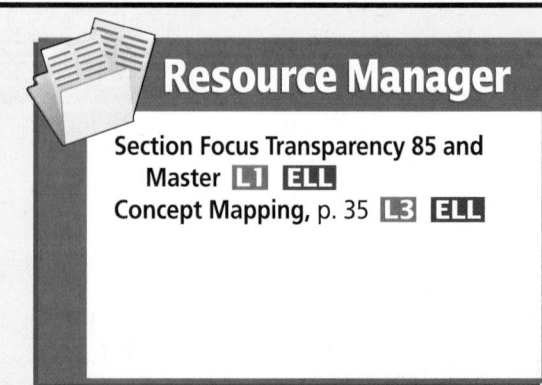

Resource Manager

Section Focus Transparency 85 and Master **L1** **ELL**
Concept Mapping, p. 35 **L3** **ELL**

fuel body functions. Some sugar is carried to the liver where it is stored as glycogen.

Cellulose, another complex carbohydrate, is found in all plant cell walls and is not digestible by humans. However, cellulose (also known as fiber) is still an important item to include in the diet as it helps in the elimination of wastes. Sources of fiber include bran and spinach.

Fats

Many people think that eating fat means getting fat, yet fats are an essential nutrient. They provide energy for your body and are also used as building materials. Recall that fats are essential building blocks of the cell membrane. They are also needed to synthesize hormones, protect body organs against injury, and insulate the body from cold.

Sources of fat in the diet include meats, nuts, and dairy products, as well as cooking oils. In the digestive system, fats are broken down into fatty acids and glycerol and absorbed by the villi of the small intestine. Eventually, some of these fatty acids end up in the liver. The liver converts them to glycogen or stores them as fat throughout your body.

Proteins

Your body has many uses for proteins. Enzymes, antibodies, many hormones, and substances that help the blood to clot, are all proteins. Proteins form part of muscles and many cell structures, including the cell membrane.

During digestion, proteins are broken down into amino acids. After the amino acids have been absorbed by the small intestine, they enter the bloodstream and are carried to the liver. The liver can convert amino acids to fats or glucose, both of which can be used

Fluorine (F)
Dental cavity reduction
Fluoridated water

Iodine (I)
Formation of thyroid hormone
Seafood, eggs, iodized salt, milk group

Iron (Fe)
Formation of hemoglobin (carries oxygen to body cells) and cytochromes (ATP formation)
Liver, egg yolk, grain and meat group, leafy vegetables

Sodium (Na)
Nerve activity, body pH regulation
Bacon, butter, table salt, vegetable group

Magnesium (Mg)
Muscle and nerve activity, bone formation, enzyme function
Fruit, vegetable and grain groups

Calcium (Ca)
Teeth and bone formation, muscle and nerve activity, blood clotting
Milk and grain group

Phosphorus (P)
Teeth and bone formation, blood pH, muscle and nerve activity, part of enzymes and nucleic acids
Milk, grain and vegetable group

Copper (Cu)
Development of red blood cells, formation of some respiratory enzymes
Grain group, liver

Potassium (K)
Nerve and muscle activity
Vegetable group, bananas

Sulfur (S)
Builds hair, nails and skin, component of insulin
Grain and fruit group, eggs, cheese

for energy. However, your body uses amino acids for energy only if other energy sources are depleted. Most amino acids are absorbed by cells and used for protein synthesis. The human body needs 20 different amino acids to carry out protein synthesis, but it can make only 12 of them. The rest must be consumed in the diet and so are called essential amino acids. Sources of essential amino acids include meats, dried beans, whole grains, eggs and dairy products.

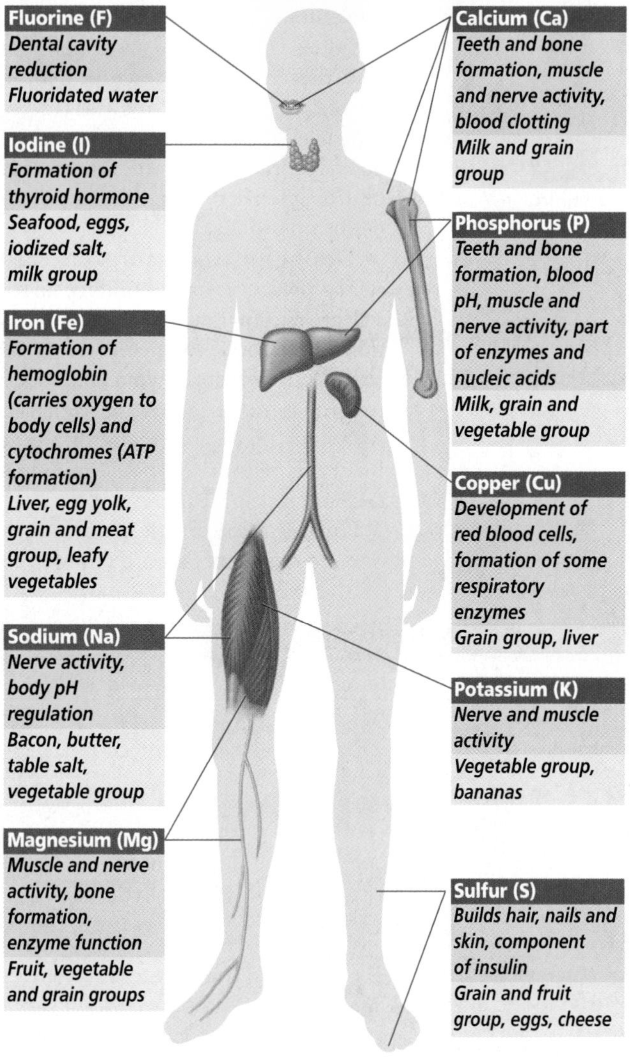

Figure 35.9
Minerals serve many vital functions.

Quick Demo

Kinesthetic Test various foods for the presence of fat using brown paper (from grocery sacks). Smear the food on the brown paper and allow to dry. If the food is dry, grind a small portion with some water. After the paper has had sufficient time to dry, hold it up to the light. If the paper is translucent, the food tested contains fat.

Visual Learning

Refer to the periodic table on page A9 of Appendix D as you discuss Figure 35.9. Challenge students to locate the minerals on the periodic table.

Enrichment

Interpersonal Have students work in groups to investigate and prepare a report on one vitamin deficiency, such as scurvy, beriberi, rickets, night blindness, polyneuritis, or pellagra. **L2**
COOP LEARN

GLENCOE
TECHNOLOGY

CD-ROM
Biology: The Dynamics of Life
Exploration: *Nutrition*
Disc 5

Cultural Diversity

Food Preferences

Intrapersonal Populations develop food preferences as they try to meet their nutritional needs using foods that are locally available. Have students find out how different cultures choose combinations of foods and flavors. Have them bring in samples of foods from various cultures. **L2**

BIOLOGY JOURNAL

Cancer and Nutrients

Linguistic Have students research how antioxidants destroy free radicals that can damage cells and lead to cancer. Ask them to find out how vitamins C and E and beta-carotene destroy these substances. **L3**

Knowledge Have students list three good sources of each mineral (or vitamin) discussed in the text. **L2**

MiniLab 35-1

Purpose
Students evaluate the nutritional value of a food based on its label.

Process Skills
observe and infer, analyze, interpret data

Teaching Strategies
■ Explain the concept of daily value (DV) to students.
■ You may wish to have students work in pairs.

Expected Results
Students should discover that the soup represented is basically a healthy meal.

Analysis
1. Yes, sodium.
2. A serving of soup contains about 39% of its Calories from saturated fat. (6 g/serving x 9 Cal/g = 54 Calories of saturated fat per serving. 54 ÷ 140 = 39%)
3. Basically yes, although the sodium content is high.

✓ **Assessment**

Performance Have students repeat the MiniLab for three foods found in their homes. Use the Performance Task Assessment List for Carrying Out a Strategy and Collecting Data in **PASC**, p. 25. **L2**

Resource Manager

Laboratory Manual, pp. 255-260 **L2**

Minerals and vitamins

When you think of minerals, you may picture substances that people mine, or extract from Earth. As shown in *Figure 35.9* on the previous page, the same minerals can also be extracted from foods and put to use by your body.

A **mineral** is an inorganic substance that serves as a building material or takes part in a chemical reaction in the body. Minerals make up about four percent of your total body weight, most of it in your skeleton. Although they serve many different functions within the body, minerals are not used as an energy source.

Unlike minerals, **vitamins** are organic nutrients that are required in small amounts to maintain growth and metabolism. The two main groups of vitamins are fat-soluble and water-soluble, as shown in *Table 35.2*. Although fat-soluble vitamins can be stored in the liver, the accumulation of excess amounts can prove toxic. Water-soluble vitamins cannot be stored in the body and so must be included regularly in the diet. *Table 35.2* lists foods that contain fat-soluble and water-soluble vitamins.

Vitamin D, a fat-soluble vitamin, is synthesized in your skin. Vitamin K and some B vitamins are made by bacteria in your large intestine. The rest of the vitamins must be consumed in your diet.

WORD Origin

vitamin
From the Latin word *vita*, meaning "life." Vitamins are necessary for life.

Table 35.2 Vitamins		
Vitamin	**Function**	**Source**
Fat-soluble		
A	maintain health of epithelial cells; formation of light-absorbing pigment; growth of bones and teeth	liver, broccoli, green and yellow vegetables, tomatoes, butter, egg yolk
D	absorption of calcium and phosphorus in digestive tract	egg yolk, shrimp, yeast, liver, fortified milk; produced in the skin upon exposure to ultraviolet rays in sunlight
E	formation of DNA, RNA, and red blood cells	leafy vegetables, milk, butter
K	blood clotting	green vegetables, tomatoes; produced by intestinal bacteria
Water-soluble		
B_1	sugar metabolism; synthesis of neurotransmitters	ham, eggs, green vegetables, chicken, raisins, seafood, soybeans, milk
B_2 (riboflavin)	sugar and protein metabolism in cells of eyes, skin, intestines, blood	green vegetables, meats, yeast, eggs
Niacin	energy-releasing reactions; fat metabolism	yeast, meats, liver, fish, whole-grain cereals, nuts
B_6	fat metabolism	salmon, yeast, tomatoes, corn, spinach, liver, yogurt, wheat bran, whole-grain cereals and breads
B_{12}	red blood cell formation; metabolism of amino acids	liver, milk, cheese, eggs, meats
Pantothenic acid	aerobic respiration; synthesis of hormones	milk, liver, yeast, green vegetables, whole-grain cereals and breads
Folic acid	synthesis of DNA and RNA; production of red and white blood cells	liver, leafy green vegetables, nuts, orange juice
Biotin	aerobic respiration; fat metabolism	yeast, liver, egg yolk
C	protein metabolism; wound healing	citrus fruits, tomatoes, leafy green vegetables, broccoli, potatoes, peppers

956 THE DIGESTIVE AND ENDOCRINE SYSTEMS

TECHPREP

Professional Nutritionist

Linguistic Students interested in nutrition can interview a nutritionist working for a food company. They may ask what tests are carried out to determine the nutritional content of food produced by the company. **L2**

BIOLOGY JOURNAL

Nutritional Issues

Interpersonal Divide the class into groups. Have each group prepare a presentation on a nutrition issue such as cholesterol, HDLs versus LDLs, anorexia, bulimia, weight gain, junk foods, vegetarian diets, or food additives. Have students record their notes for the group report in their journals. **L2** **COOP LEARN**

Water

Water is the most abundant substance in your body, making up 60 percent of red blood cells and 75 percent of muscle cells. Water facilitates the chemical reactions in your body and is necessary for the breakdown of foods during digestion. Water is also an excellent solvent; oxygen and nutrients from food could not enter your cells if they did not first dissolve in water.

Recall that water absorbs and releases heat slowly. It is this characteristic that helps water maintain your body's internal temperature. A large amount of heat is needed to raise the temperature of water. Because the body contains so much water, it takes a lot of added energy to raise its internal temperature. Your body loses about 2.5 L of water per day through exhalation, sweat, and urine. As a result, water must be replaced constantly.

Calories and Metabolism

The energy content of food is measured in units of heat called **Calories,** each of which represents a kilocalorie, or 1000 calories (written with a small c). A calorie is the amount of heat required to raise the temperature of 1 mL of water by 1°C. Some foods, especially those with fats, contain more Calories than others. In general, 1 g of fat contains nine Calories, while 1 g of carbohydrate or protein contains four Calories. To learn more about Calories in meals, complete the *MiniLab* on this page.

The number of Calories needed each day varies from person to person, depending on the person's metabolism, or the rate at which they burn energy. Metabolic rate, in turn, is determined by a person's body mass, age, sex, and level of physical activity. In general, males need more Calories per day than females, teenagers need more than adults, and active people need more than inactive people. Physicians have determined that many Americans are overweight. Calculate your Body Mass Index by doing the *Problem-Solving Lab* on the next page.

MiniLab 35-1 Interpreting the Data

Evaluate a Bowl of Soup As a consumer, you are bombarded by advertising that promotes the nutritional benefits of specific food products. Choosing a food to eat on the basis of such ads may not make nutritional sense. By examining the ingredients of processed foods, you can learn about their actual nutritional content.

Table 35.3

Percentage of Daily Value (DV)

Carbohydrates	60%
Fat	30%
Saturated Fats	10%
Cholesterol	1.5%
Protein	10%
Total Calories	2000

NUTRITION FACTS
Serving Size: 2 cups (452g)
Servings Per Container: 1

Amount Per Serving

Calories 140	Calories from Fat 54

	% Daily Value*
Total Fat 8g	12%
Saturated Fat 6g	30%
Cholesterol 20mg	7%
Sodium 1640 mg	68%
Total Carbohydrate 22g	7%
Dietary Fiber 5g	20%
Sugars 5g	
Protein 6g	

Vitamin A	50%	Vitamin C	4%
Calcium	2%	Iron	2%

* Percent Daily Values are based on a 2,000 calorie diet. Your daily values may be higher or lower depending on your calorie needs:

		Calories	2,000	2,500
Total Fat	Less than		65g	80g
Sat Fat	Less than		20g	25g
Cholesterol	Less than		300mg	300mg
Sodium	Less than		2,400mg	2,400mg
Total Carbohydrate			300g	375g
Fiber			25g	30g

Calories per gram:
Fat 9 * Carbohydrates 4 * Protein 4

Procedure

1. Examine the information in the table listing the daily value (DV) of various nutrients. DV expresses what percent of Calories should come from certain nutrients. For instance, in the proposed diet of 2000 Calories, 60 percent of the Calories should come from carbohydrates.

2. Examine the nutritional information on the soup can label and compare it with the DV table.

Analysis

1. Does your bowl of soup provide more than 30 percent of any of the daily nutrients? Which ones?

2. Evaluate the percentage of calories in soup that are provided by saturated fat.

3. Is soup a nutritious meal? Explain your answer.

Check for Understanding

Give the students a sample lunch that has nutritional deficiencies. Ask them to evaluate the lunch. Ask what nutrients are in excess and what nutrients should be added. Elicit what foods could provide these nutrients. **L1**

Reteach

Visual-Spatial Have students prepare a chart of the six nutrients. Have them identify the functions of these nutrients and identify foods that contain each. **L1**

 Extension

TECH PREP *Linguistic* Have students interview a nutritionist to find out what criteria the nutritionist uses when planning meals that meet nutritional needs. **L2**

✓ **Assessment**

Portfolio Ask students to write a summary of which substances need to be included in a daily balanced diet and what each nutrient does. **L2** **P**

GLENCOE TECHNOLOGY

VIDEODISC
The Infinite Voyage
*A Taste of Health,
Consequences of a Fatty Diet*
(Ch. 4) 3 min.

Ancel Keys: Pioneer in Nutrition
(Ch. 5) 6 min.

Assessing Dietary Intake Through Analysis of Garbage (Ch. 8) 4 min.

Understanding How to Eat and Live (Ch. 9) 6 min.

PROJECT

Calories

Logical-Mathematical Ask students to list all the foods they eat during a 24-hour period. Distribute Calorie charts. Have students use the charts to figure out how many Calories they ate, as well as calculate the grams of carbohydrates, fats, and proteins they consumed. **L2**

Resource Manager

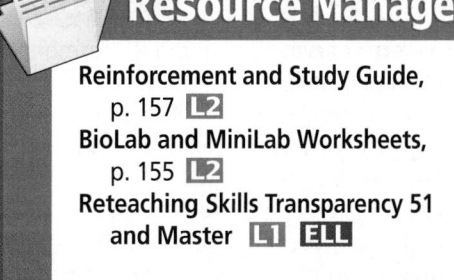

Reinforcement and Study Guide, p. 157 **L2**

BioLab and MiniLab Worksheets, p. 155 **L2**

Reteaching Skills Transparency 51 and Master **L1** **ELL**

Problem-Solving Lab 35-2

Purpose

Students learn how to calculate their Body Mass Index and determine whether or not their weight is within accepted values.

Process Skills

think critically, apply concepts, collect data, interpret data, recognize cause and effect, use numbers

Teaching Strategies

■ Allow students to use calculators.
■ Review the technique for squaring a number.
■ You may wish to illustrate the process of calculating BMI by working an example on the board or overhead projector.

Thinking Critically

1. Student answers will vary.
2. The person could reduce his or her Calorie intake, and/or increase activity level.
3. His Calorie intake is in balance with his expenditure.
4. Student answers may include the idea that fatty food intake has increased even as daily exercise has decreased.

✔ Assessment

Portfolio Have student groups plan a general long-range program that will either maintain their BMI if below 25 or reduce their BMI if over 25. Use the Performance Task Assessment List for Group Work in **PASC**, p. 97. **L3**

---

4 Close

Discussion

Ask students to evaluate an average American's diet compared with that of someone living in a developing nation. Have students consider differences in protein, mineral, and vitamin content.

Problem-Solving Lab 35-2 — Using Numbers

What is your BMI? Fifty-five percent of adults in the United States are considered overweight. How can you tell if you fall into this category? Use the following equation to find out where you rank in relation to the rest of the population.

Analysis

Compute your BMI, or Body Mass Index, using the following formula:

$$\frac{\text{weight (in pounds)}}{\text{height (in inches)}^2} \times 704.5 = BMI$$

The federal guidelines are as follows:

A BMI
• 25 or below = normal weight
• from 25 to 29.9 = overweight
• 30 or over = obese

Thinking Critically

1. According to federal guidelines, are you normal weight, overweight, or obese?
2. How might a person with a BMI of 27 reduce his or her BMI? Consider both nutritional intake and physical activity.
3. Fred has a BMI of 22. How do you suppose his Calorie intake compares to his Calorie expenditure?
4. Since 1960, the population of obese individuals in the United States has risen from 13 to 22 percent. Formulate a hypothesis that may explain this rise.

Figure 35.10
When the energy taken in is greater than the energy expended, a person gains weight.

What happens if you eat more Calories than your body can metabolize? As **Figure 35.10** shows, you store the extra energy as body fat and gain weight. On the other hand, if you eat fewer Calories than your body can metabolize, you use some of the energy stored in your body as fat and lose weight.

Millions of people put themselves on diets every year in hopes of losing weight. While many diets are nutritionally sound, others prescribe eating habits that are not sensible and usually fail to produce the desired result. Read more about fad diets in the *Biology & Society* section at the end of this chapter.

Section Assessment

Understanding Main Ideas
1. In what ways are proteins used in the body?
2. Why is it important to eat cellulose even though it has no nutritional value?
3. What happens when a person takes in more food energy than his or her body needs?
4. Why are fats needed in the diet?

Thinking Critically
5. A person can live several weeks without food, but can live only days without water. Why is the constant intake of water necessary for the body?

SKILL REVIEW

6. **Classifying** Prepare a chart of food groups, each group showing foods rich in one of the six nutrients discussed in this section. For more help, refer to *Organizing Information* in the **Skill Handbook**.

Section Assessment

1. to form enzymes, antibodies, hormones, clotting chemicals, and cell structures
2. Cellulose provides bulk in the diet, helping with the elimination of wastes.
3. That person gains weight.
4. Fats provide energy for the body, act as building materials, protect body organs against injury, and insulate the body from cold.
5. Water is needed for oxygen and nutrients to enter cells and to maintain body temperature.
6. *carbohydrates:* bread, pasta, cereal, fruits and vegetables; *fats:* dairy, meat, oil, butter; *protein:* meat, dairy, beans; *minerals:* fruits and vegetables, meat, dairy; *vitamins:* fruits and vegetables, grains, dairy, meat; and so on

35.3 The Endocrine System

SECTION PREVIEW

Objectives

Identify the functions of some of the hormones secreted by endocrine glands.

Summarize the negative feedback mechanism controlling hormone levels in the body.

Contrast the actions of steroid and amino acid hormones.

Vocabulary

endocrine glands
pituitary gland
hypothalamus
target tissues
receptor
negative feedback system
adrenal glands
thyroid gland
parathyroid glands

I*magine yourself as a quarterback behind these linemen. Could you see over everyone's head? Do you wonder how tall you will end up being? Because you continue growing until about age 25, you may still be a few inches shy of your adult height. What controls your growth? In part, chemical messages within your body. They can affect how long and at what rate you grow.*

Pituitary gland

The pituitary gland releases chemicals important in controlling the growth of these linemen.

Control of the Body

Internal control of the body is directed by two systems: the nervous system, which you will learn more about later, and the endocrine system. The endocrine system is made up of a series of glands, called **endocrine glands,** that release chemicals directly into the bloodstream. These chemicals act as messengers, relaying information to other parts of the body. Whereas the nervous system produces an immediate response, the endocrine system induces gradual change. Let's take one of the football players as an example: While his nervous system is directing his legs to run in order to catch a forward pass, his endocrine system is controlling the rate at which he grows. The first response is instant; the second takes years.

Interaction of the nervous system and endocrine system

Although endocrine glands are found throughout the body, most of them are controlled by the action of the **pituitary gland** (puh TEW uh ter ee), the master endocrine gland. Because there are two control systems within the body—nervous and endocrine—coordination is needed.

WORD *Origin*

endocrine
From the Greek words *endo*, meaning "within," and *krinein*, meaning "to separate." The endocrine glands secrete hormones into the blood.

BIOLOGY *Online*

Note Internet addresses that you find useful in the space below for quick reference.

Section 35.3

Prepare

Key Concepts
Students study how the nervous system and hormones control metabolic processes in the body.

Planning
- Borrow a human skull or a large animal skull for the first Quick Demo.
- Order slides of the pancreas for the second Quick Demo.
- Purchase a whole kidney that still has the adrenal gland attached for use in the third Quick Demo.
- Order slides of thyroid/parathyroid tissue for the MiniLab.

1 Focus

Bellringer
Before presenting the lesson, display **Section Focus Transparency 86** on the overhead projector and have students answer the accompanying questions.
L1 **ELL**

Transparency **86** Negative Feedback   Section Focus

❶ What is the internal result of each of these actions?
❷ Describe the feedback loop involved in each situation.

2 Teach

Using Science Terms

Explain the meaning of the following terms: *Endo* is Greek for "within" and *crine* (krinein) is Greek for "to separate." Insulin gets its name from the Latin *insula*, meaning "island." Insulin is made in the small islands or islets of beta cells in the pancreas. *Hypo* is Greek for "under" and *thalamus* is Greek for "the inner room." The thalamus was an inner room in a Greek ship.

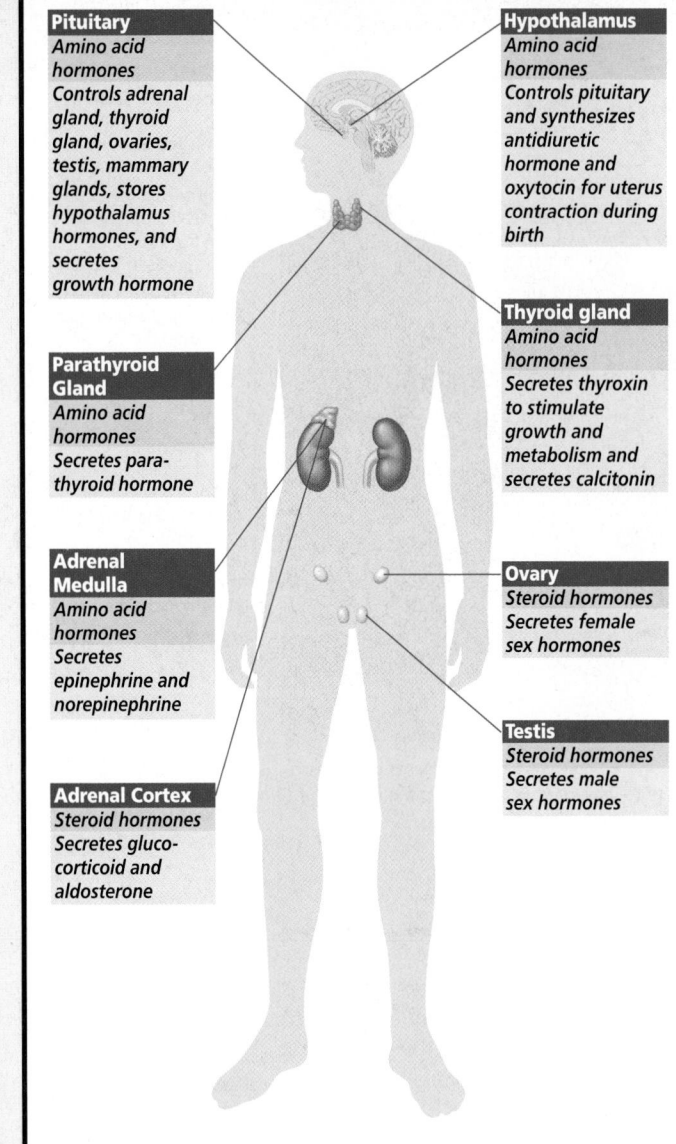

Pituitary
Amino acid hormones
Controls adrenal gland, thyroid gland, ovaries, testis, mammary glands, stores hypothalamus hormones, and secretes growth hormone

Parathyroid Gland
Amino acid hormones
Secretes parathyroid hormone

Adrenal Medulla
Amino acid hormones
Secretes epinephrine and norepinephrine

Adrenal Cortex
Steroid hormones
Secretes glucocorticoid and aldosterone

Hypothalamus
Amino acid hormones
Controls pituitary and synthesizes antidiuretic hormone and oxytocin for uterus contraction during birth

Thyroid gland
Amino acid hormones
Secretes thyroxin to stimulate growth and metabolism and secretes calcitonin

Ovary
Steroid hormones
Secretes female sex hormones

Testis
Steroid hormones
Secretes male sex hormones

Figure 35.11
This diagram shows the principal human endocrine glands. The top label indicates the name of the gland, the middle label indicates the type of hormone(s) secreted, and the bottom label tells the action of the gland/hormone.

The **hypothalamus** is the portion of the brain that controls the pituitary gland. The pituitary gland is located in the skull just beneath the hypothalamus, and the two are connected by nerves and blood vessels. The hypothalamus sends messages to the pituitary, which then releases its own chemicals, or stimulates other glands to release theirs. Other endocrine glands under control of the pituitary include the thyroid gland, the adrenal glands, and glands associated with reproduction.

Endocrine control of the body

The chemicals secreted by endocrine glands into the bloodstream are called hormones. Recall that a hormone is a chemical released in one part of an organism that affects another part. Hormones convey information to other cells in your body, giving them instructions regarding your growth, development, and behavior. Once released by the glands, the hormones travel in the bloodstream and then attach to specific binding sites found on the plasma membranes, or in the nuclei, of **target tissue** cells. These binding sites on cells are called **receptors**. *Figure 35.11* summarizes the action of endocrine hormones.

Example of endocrine control

Human growth hormone (hGH) is a good example of an endocrine system hormone. When your body is actively growing, blood glucose levels are slightly lowered as the growing cells use up the sugar. This low blood glucose level is detected by the hypothalamus, which stimulates the production and release of hGH from the pituitary into the bloodstream. hGH binds to receptors on the plasma membranes of liver cells, stimulating the liver cells to release glucose into your blood. Your cells need the glucose in order to continue growing. *Figure 35.12* summarizes the control of hGH by the pituitary gland. You can further

investigate growth rate in humans by doing the *BioLab* at the end of this chapter.

Negative Feedback Control

The amount of hormone released by an endocrine gland is determined by your body's demand for that hormone at a given time. In this way, the endocrine system ensures that the appropriate amounts of hormone are present in the system at all times.

How do your endocrine glands know when you need a certain hormone? The endocrine system is controlled by a self-regulating system called the **negative feedback system.** The negative feedback system is a system in which the hormones, or their effects, are fed back to inhibit the original signal. The thermostat in your home is controlled by a similar negative feedback system. It maintains the room at a set temperature. When the temperature drops, the thermostat senses the lack of thermal energy and signals the heater to increase its output. When the thermal energy of the room rises again to a certain point, the thermostat no longer stimulates the heater, which shuts off. When the temperature drops again, the process repeats itself. In this negative feedback system, the increase in temperature "feeds back" to signal the thermostat to stop stimulating thermal energy production.

Feedback control of hormones

The majority of endocrine glands operate under negative feedback systems. A gland synthesizes and secretes its hormone, which travels in the blood to the target tissue where the appropriate response occurs. Information regarding the hormone

① *Low blood sugar is detected*

② *Hypothalamus stimulates pituitary to release hGH*

Pituitary

④ *Increased sugar in blood signals back to hypothalamus to slow stimulation of pituitary*

③ *hGH stimulates liver to convert glycogen into glucose and release glucose into blood*

level or its effect on the target tissue is fed back, usually to the hypothalamus or pituitary gland, to regulate the gland's production of the hormone.

Figure 35.12
The hypothalamus and pituitary gland control the amount of human growth hormone (hGH) in your blood.

Control of blood water levels

Let's take a look at an example of a hormone that is controlled by a negative feedback system. After working out in the gym and building up a sweat, you are thirsty. This is because the water content of your blood has been reduced. The hypothalamus, which is able to sense the concentration of water in your blood, determines that your body is dehydrated. In response, it stimulates the pituitary gland to release antidiuretic (ANT ih di uh reht ihk) hormone (ADH). ADH reduces the amount of

Concept Development
Insulin-dependent diabetes is also called Type I, or juvenile, diabetes. One major complication of Type I diabetes is loss of vision due to cataracts. The excessive blood glucose chemically attaches to the lens proteins, clouding the lens. Type I diabetes often causes kidney damage, also. Noninsulin-dependent diabetes is called Type II diabetes. Because this type of diabetes is most common in elderly people, it is sometimes called late-onset diabetes.

Quick Demo

Visual-Spatial Using a projection microscope viewer, show a section of the pancreas. Point out the islets (small islands) containing the hormone-producing cells. These cells are surrounded by other cells that produce digestive enzymes.

✔ Assessment
Skill Have students make a graph plotting blood glucose levels against production of hGH by the pituitary. **L2**

Resource Manager
Critical Thinking/Problem Solving, p. 35 **L3**

BIOLOGY JOURNAL

Hormone Actions

Linguistic Have students write a paragraph describing what happens to insulin and glucagon levels in their bodies while they sleep. **L2**

Purpose

Students relate changes in blood insulin and glucagon levels during prolonged exercise to the body's need to get glucose to its cells.

Process Skills

recognize cause and effect, interpret data, analyze

Teaching Strategies

■ Ask students to list on the chalkboard the changes that occur in the body during exercise. Identify the changes that require increased glucose inside cells. **L1**

Thinking Critically

Glucagon causes blood glucose levels to rise by increasing the conversion of glycogen into glucose. The body needs more glucose during exercise. Insulin acts to lower blood glucose levels by converting glucose to glycogen. Consequently, its levels are reduced during exercise.

962

Problem-Solving Lab 35-3   Interpreting Data

What are the effects of glucagon and insulin during exercise? Exercise represents a special example of rapid fuel mobilization in the body. The body must gear up to supply great amounts of glucose and oxygen for muscle metabolism. The glucose use in a resting muscle is generally low but changes dramatically with exercise. Within ten minutes of beginning exercise, glucose uptake from the blood may increase by fifteenfold; within 60 minutes, it may increase by thirtyfold.

Analysis

The graph here shows the effects of prolonged exercise on blood insulin and glucagon levels in humans.

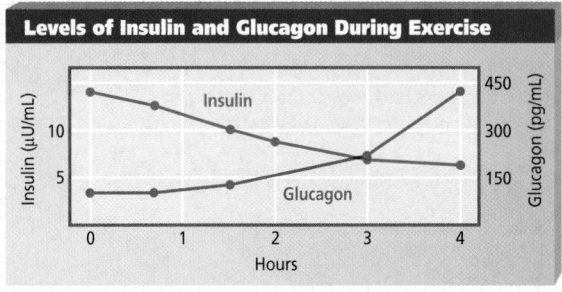

Levels of Insulin and Glucagon During Exercise

Thinking Critically

Explain why glucagon concentration goes up and insulin concentration goes down during exercise, and how these actions help get glucose to the body cells.

---

water in your urine. It does so by binding to receptors in kidney cells, promoting their absorption of water and reducing the amount of water excreted in urine. Information regarding blood water levels is constantly fed back to the hypothalamus so it can regulate the pituitary's release of ADH. If the body becomes overhydrated, the hypothalamus stops stimulating release of ADH.

Control of blood glucose levels

Another example of a negative feedback system involves the regulation of blood glucose levels. Unlike most other endocrine glands, the

---

pancreas is not controlled by the pituitary gland. When you have just eaten and your blood glucose levels are high, your pancreas releases the hormone insulin. Insulin signals liver and muscle cells to take in glucose, thus lowering blood glucose levels. When blood glucose levels become too low, another pancreatic hormone, glucagon, is released. Glucagon binds to liver cells, signaling them to release stored glycogen as glucose. Learn more about glucose storage and release by doing the *Problem-Solving Lab* on this page.

Hormone Action

Once hormones are released by an endocrine gland, they travel to the target tissue and cause a change. Hormones can be grouped into two basic types according to how they act on their target cells: steroid hormones and amino acid hormones.

Action of steroid hormones

Hormones that are made from lipids are called steroid hormones. Steroid hormones are lipid-soluble and therefore diffuse freely into cells through their plasma membranes, as shown in *Figure 35.13.* There they bind to a hormone receptor inside the cell. The hormone-receptor complex then travels to the nucleus where it activates the synthesis of specific messenger RNA molecules. The mRNA molecules move out to the cytoplasm where they activate the synthesis of the required proteins.

Action of amino acid hormones

The second group of hormones is made from amino acids. Recall that amino acids can be strung together in chains and that proteins are made from long chains of amino acids. Some hormones are short chains

---

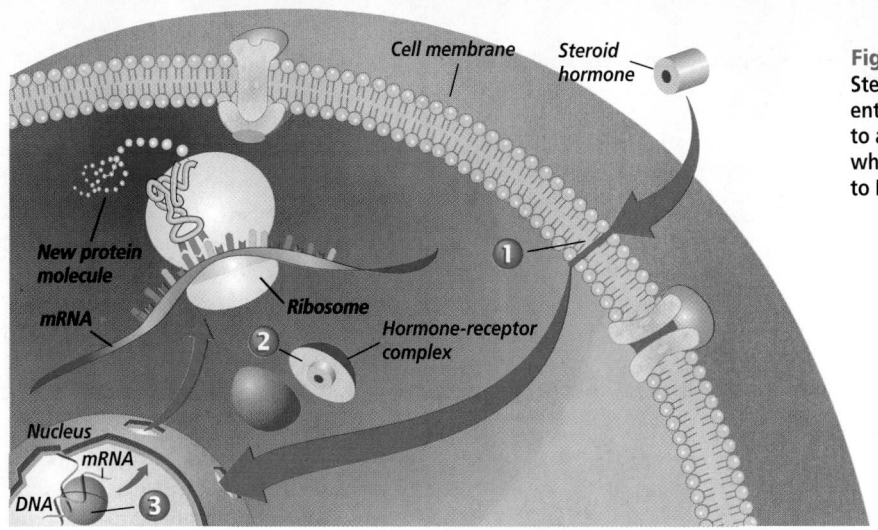

Figure 35.13
Steroid hormones enter a cell (1), bind to a receptor (2), which in turn binds to DNA (3).

Cell membrane

Steroid hormone

New protein molecule

mRNA

Ribosome

Hormone-receptor complex

Nucleus

mRNA

DNA

of amino acids and others are large chains. These amino acid hormones, once secreted into the bloodstream, bind to receptors embedded in the plasma membrane of the target cell, as shown in *Figure 35.14*. From there, they open ion channels in the membrane, or route signals down from the surface of the membrane to activate enzymes inside the cell. The enzymes, in turn, alter the behavior of other molecules inside the cell. In both of these ways, the hormone is able to control what goes on inside the target cell.

Adrenal Hormones and Stress

You are sitting in math class and the teacher is about to hand out the semester test. Because this test is an important one, you have spent many hours studying for it. Like most of your classmates, you are a little nervous as the test is being passed down the row. Your heart is beating fast and your hands are a little sweaty. As you review the first problem, however, you begin to calm down because you know how to solve it.

The **adrenal glands** play an important role in preparing your body for stressful situations. The adrenal glands, located on top of the kidneys, consist of two parts—an inner portion and an outer portion. The outer portion secretes steroid hormones, including glucocorticoids (glew ko KOR tuh koydz) and aldosterone (ahl duh STEER ohn). These steroid

Figure 35.14
When an amino acid hormone binds to the receptor on the cell membrane (1), it can open ion channels (2), or activate enzymes (3).

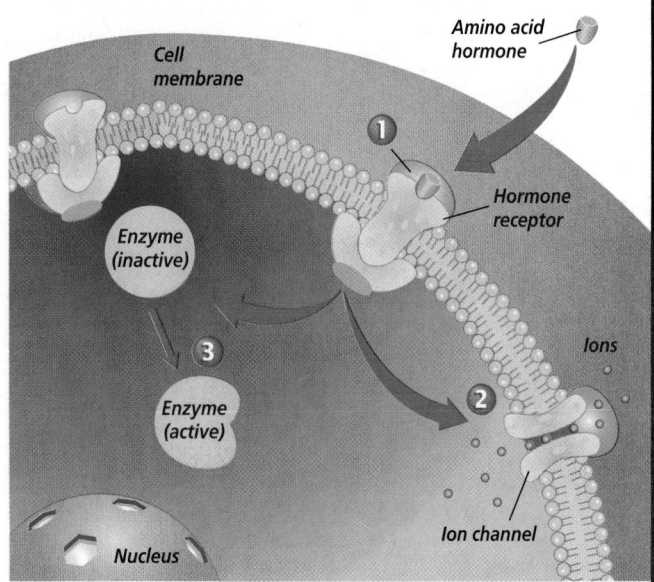

Cell membrane

Amino acid hormone

Enzyme (inactive)

Hormone receptor

Ions

Enzyme (active)

Ion channel

Nucleus

BIOLOGY JOURNAL

Hormone Release

Linguistic Have students write a story that describes how and when stress hormones are released in their bodies and the effects of these hormones. **L2**

Purpose

Students analyze a prepared slide of thyroid and parathyroid tissue.

Process Skills

compare and contrast, observe, apply concepts, hypothesize, interpret scientific illustrations

Teaching Strategies

■ Prepared slides of thyroid and parathyroid tissue are available from biological supply houses. It is cheaper to purchase the combined slide of both tissues rather than separate slides of each.

■ To further reduce costs, project one prepared slide onto a TV screen using a video camera.

Expected Results

Students will be able to differentiate between thyroid and parathyroid tissue.

Analysis

1. Student may notice that thyroid tissue contains many large spaces surrounded by a thin band while parathyroid tissue contains no large spaces or follicles.
2. **a.** thyroid
   b. stored hormones (thyroxine or calcitonin)
   c. No. No, cells, not storage areas, would be needed to produce the hormone.
   d. It makes the hormones.
3. Both glands are located in the same general area of the neck. The parathyroids lie on the thyroid gland itself.

✔ Assessment

Performance Ask students to research the cause and appearance of thyroid goiter. Provide students with prepared slides of normal thyroid tissue and thyroid tissue exhibiting a goiter. Have students compare and contrast the two tissues. Ask them to relate their microscopic observations to the macroscopic appearance of a goiter. Use the Performance Task Assessment List for Making Observations and Inferences in **PASC**, p. 17.

964

MiniLab 35-2 Observing

Compare Thyroid and Parathyroid Tissue Although their names seem somewhat similar, the thyroid and parathyroid glands perform rather different functions within the body.

Procedure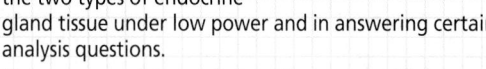

1. Copy the data table.
2. Use low-power magnification to examine a prepared slide of thyroid and parathyroid endocrine gland tissue. Note: Both tissues appear on the same slide. **CAUTION: Use caution when working with a microscope and prepared slides.**
3. The image on the right is a photograph of thyroid and parathyroid tissue. Use it as a guide in locating the two types of endocrine gland tissue under low power and in answering certain analysis questions.
4. Now locate each type of gland tissue under high-power magnification. Draw what you see in the data table. Then use what you learned in the chapter to identify the names of the hormones produced by each gland.

Magnification: 33×

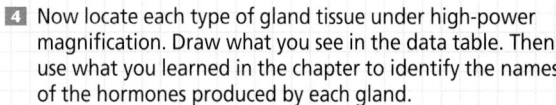

Parathyroid (A) and thyroid (B) tissue

Data Table

Tissue	Drawing	Name of hormone(s) produced
Thyroid		
Parathyroid		

Analysis

1. Compare the microscopic appearance of parathyroid tissue to that of thyroid tissue.
2. **a.** Which tissue type contains follicles (large liquid storage areas)?
   b. What may be present within the follicles?
   c. Are follicles composed of cells? Could they produce the hormone associated with this gland tissue? Explain your answer.
   d. Hypothesize what the function may be for the thin layer of tissue that surrounds each follicle.
3. How might you explain the fact that both thyroid and parathyroid tissue can be seen on the same slide?

---

hormones cause an increase in available glucose and raise blood pressure. In this way, they help the body combat stresses such as fright, temperature extremes, bleeding, infection, disease, and even test anxiety.

The inner portion of the adrenal gland secretes two amino acid hormones: epinephrine (ep uh NEF run)—often called adrenaline—and norepinephrine. Recall the fight-or-flight response discussed in the animal behavior chapter. During such a response, the hypothalamus relays impulses to the nervous system, which in turn stimulates the adrenal glands to increase their output of epinephrine and norepinephrine. These hormones increase heart rate, blood pressure, and rate of respiration; increase efficiency of muscle contractions; and increase blood sugar levels. If you have ever had to perform in front of a large audience, you may have experienced these symptoms, often referred to collectively as an "adrenaline rush." This is how the body prepares itself to face or flee a stressful situation.

Thyroid and Parathyroid Hormones

The **thyroid gland,** located in the neck, regulates metabolism, growth, and development. The main metabolic and growth hormone of the thyroid is thyroxine. This hormone affects the rate at which the body uses energy and determines your food intake requirements.

The thyroid gland also secretes calcitonin (kal suh TONE un)—a hormone that regulates calcium levels in the blood. Calcium is a mineral the body needs for blood clotting, formation of bones and teeth, and normal nerve and muscle function. Calcitonin binds to the membranes

---

Resource Manager

Reinforcement and Study Guide,
p. 158 **L2**
Content Mastery, pp. 173, 175-176 **L1**
BioLab and MiniLab Worksheets,
p. 156 **L2**

✔ Portfolio

Parathyroid Hormone

Intrapersonal Have students sequence the pathway of parathyroid hormone from the parathyroid glands to its target tissues. Have students caption their flowcharts with a summary of the effects the hormone has on its target tissues. **L2** **P**

of kidney cells and causes an increase in calcium excretion. Calcitonin also binds to bone-forming cells, causing them to increase calcium absorption and synthesize new bone.

Another hormone involved in mineral regulation, the parathyroid hormone (PTH), is produced by the **parathyroid glands,** which are closely associated with the thyroid gland. It increases the rate of calcium, phosphate, and magnesium absorption in the intestines and causes the release of calcium and phosphate from bone tissue. It also increases the rate at which the kidneys remove calcium and magnesium from urine and return them to the blood.

The overall effect of parathyroid hormone and calcitonin hormone interaction in the body is shown in *Figure 35.15.* Take a closer look at thyroid and parathyroid tissue by completing the *MiniLab* on the previous page.

As you can see, hormones associated with the endocrine system are responsible for controlling many different functions in your body. Different hormones may play more important roles during some periods in your life than others. In any case, they remain the principal biological influence on your behavior and development.

Figure 35.15
Calcitonin and parathyroid hormone (PTH) have opposite effects on blood calcium levels.

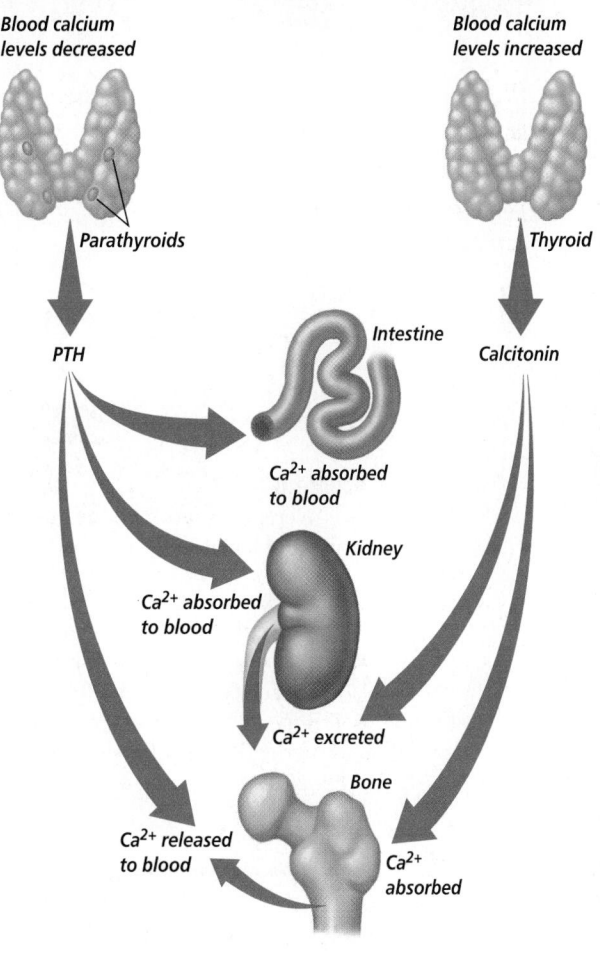

Blood calcium levels decreased

Blood calcium levels increased

Parathyroids

Thyroid

PTH

Intestine

Calcitonin

Ca²⁺ absorbed to blood

Kidney

Ca²⁺ absorbed to blood

Ca²⁺ excreted

Bone

Ca²⁺ released to blood

Ca²⁺ absorbed

Section Assessment

Understanding Main Ideas

1. How does a steroid hormone affect its target cell? How does this action differ from how an amino acid hormone affects its target cell?
2. Explain how the nervous system helps to control the endocrine system.
3. How does the negative feedback system work to control hormone levels in the blood?
4. What glands and hormones are involved in stress reactions?

Thinking Critically

5. Hormones continually make adjustments in blood glucose levels. Why must blood glucose levels be kept fairly constant?

SKILL REVIEW

6. **Comparing and Contrasting** What effects do calcitonin and parathyroid hormone have on blood calcium levels? For more information, refer to *Thinking Critically* in the **Skill Handbook.**

3 Assess

Check for Understanding
Have students make a diagram that summarizes the control of calcium levels in the body. **L1** **ELL**

Reteach
Have students go around the room, with the first student naming a gland, the second naming a hormone, and the third naming the function of the hormone. **L2**

Extension
Have students look up information on scientists who have discovered or synthesized endocrine hormones. F. G. Banting and C. H. Best discovered insulin, while E. C. Kendall isolated thyroxine and cortisone, and P. S. Hench discovered that cortisone had a beneficial effect on inflamed tissues. **L3**

✔ **Assessment**
Knowledge Ask students to summarize hormonal control of blood sugar levels. **L2**

4 Close

Discussion
Discuss with students what might happen if their thyroid gland became over- or underactive.

Section Assessment

1. A steroid hormone passes through the target cell membrane and activates protein synthesis. An amino acid hormone activates ion channels or enzyme pathways in the cell from its position on the membrane.
2. The hypothalamus, part of the brain (central nervous system), controls the pituitary, or master, endocrine gland.
3. In a negative feedback system, when a hormone reaches an appropriate level, it or its effects feed back to inhibit the release of more hormone.
4. The adrenal glands secrete glucocorticoids and epinephrine, hormones involved in stress reactions.
5. Glucose is the fuel for body cells and a constant level needs to be maintained for normal body functions.
6. Parathyroid hormone raises blood calcium levels by increasing the rate of absorption in the intestines, while calcitonin lowers blood calcium levels by increasing its excretion rate.

INVESTIGATE BioLab

Time Allotment
One class period

Process Skills
make and use graphs, interpret data, analyze

PREPARATION

- Collect two sheets of graph paper and two different colored pencils for each student.

Resource Manager

BioLab and MiniLab Worksheets, pp. 157-158 L2

INVESTIGATE BioLab

Average Growth Rate in Humans

Human growth results from more than one hormone. Human growth hormone, thyroid hormones, and the reproductive hormones that are produced during puberty are all important in human growth at various ages. Together, these hormones stimulate the growth of bone and cartilage, protein synthesis, and the addition of muscle mass. Because the reproductive hormones are involved in human growth, perhaps there is a difference in the growth rate between males and females.

PREPARATION

Problem
Is average growth rate the same in males and females?

Objectives
In this BioLab, you will:
- **Graph** the average growth rates in males and females.
- **Identify** any differences in the average growth rates of males and females.

Materials
blue pencils
graph paper
red pencils
ruler

Skill Handbook
Use the **Skill Handbook** if you need additional help with this lab.

PROCEDURE

1. Construct a graph for the growth rate data that shows mass on the vertical axis and age on the horizontal axis.
2. On the graph, plot the data shown in the table for the average female growth in mass from ages 8 to 18. Use a ruler to connect the data points with a straight red line.
3. On the same graph, plot the data for the average male growth in mass from ages 8 to 18. Connect these data points with a straight blue line.

PROCEDURE

Teaching Strategies
- Have students refer to the Making and Using Graphs section of the **Skill Handbook** for help.

Troubleshooting
- As students connect their data points, they should draw a line closest to the given set of points. Remind them that not all of the points will be on the line.

Data and Observations
Students will observe that females have an earlier growth spurt than males, but on the average, males grow taller and heavier than females.

4. Construct a second graph that shows height on the vertical axis and age on the horizontal axis.
5. Plot the data for the average female growth in height from ages 8 to 18. Connect the data points with a straight red line.
6. Plot the data for the average male growth in height from ages 8 to 18. Connect these data points with a straight blue line.

Data Table: Averages for growth in humans

| | Mass (kg) | | Height (cm) | |
Age	Female	Male	Female	Male
8	25	25	123	124
9	28	28	129	130
10	31	31	135	135
11	35	37	140	140
12	40	38	147	145
13	47	43	155	152
14	50	50	159	161
15	54	57	160	167
16	57	62	163	172
17	58	65	163	174
18	58	68	163	178

ANALYZE AND CONCLUDE

1. **Analyzing Data** During what ages do females and males increase the most in mass? In height?
2. **Analyzing Data** Interpret the data to find if the average growth rate is the same in males and females.
3. **Thinking Critically** How can you explain the differences in growth rates between males and females?
4. **Relating Concepts** Why do you think male and female growth rates increase during the teen years?

Going Further

Application Determine the height of all the students in your biology class. Compare the range of heights in your class to the statistical average.

BIOLOGY *Online* To find out more about human growth, visit the Glencoe Science Web site.
science.glencoe.com

35.3 THE ENDOCRINE SYSTEM **967**

ANALYZE AND CONCLUDE

1. Mass: females, ages 11-13; males, ages 12-14. Height: females, ages 9-14; males, ages 12-15
2. Average growth is the same until puberty. At puberty, females have an earlier growth spurt than males, but on the average, males grow taller and have greater mass than females.
3. Timing of puberty and the production of sex hormones differ between males and females. For example, males produce more of the hormone testosterone than females. Testosterone increases growth in muscle and bone mass in males.
4. The reproductive hormones released during the teen years increase the growth rate.

✓ Assessment

Portfolio Have students write summaries of the BioLab and include them, their graphs, and their answers to the questions posed in the Analyze and Conclude section in their portfolios. Use the Performance Task Assessment List for Lab Report in **PASC,** p. 47. **L2**

Going Further

Ask students to research which factors are known to affect growth rate in humans, including hormones, genetics, and diet. **L2**

Purpose

Students evaluate diet programs in light of what they know about good nutrition.

Background

Several years ago, nutritionists belonging to the Food and Nutrition Science Alliance (FANSA) put together a list of "warning signs" to help consumers evaluate various diet programs. Claims to watch out for include the following:

- Recommendations that promise a quick fix
- Dire warnings of danger from a single product or regimen
- Claims that sound too good to be true
- Simplistic conclusions drawn from a complex study
- Recommendations based on a single study
- Dramatic statements that are refuted by reputable scientific organizations
- Lists of "good" and "bad" foods
- Recommendations made to help sell a product
- Recommendations based on studies published without peer review
- Recommendations from studies that ignore differences among individuals or groups

Teaching Strategies

■ Invite a nutritionist to your classroom to talk about the risks of fad diets and to suggest strategies for designing healthy eating plans.

■ Divide the class into three groups. Have each group research a popular fad diet and then present their findings regarding the likelihood that the diet will result in safe, long-term weight loss.

Biology & SOCIETY

The Promise of Weight Loss

"Lose ten pounds in one week!"
"Shed weight without going hungry!"
"Burn fat while you sleep!"

You've probably come across statements like these in magazine and television advertisements. Take a pill, sip a shake, or follow a certain eating plan and those extra pounds will just slip away—or so the headlines claim.

The appeal of fad diets Many people who are overweight (or who simply think they are) are often willing to do almost anything to lose unwanted pounds. Most fad diets look like a fast and easy solution to a weight-loss problem. But do fad diets work as advertisements claim they do? And are they safe?

Types of fad diets Some fad diets involve fasting—going without food for a period of time. Some require taking diet pills that depress the appetite, or that cause the body to lose water. Other fad diets revolve around eating only one food, or a certain kind of food. Then there are liquid diets, in which a special drink replaces breakfast and lunch, and a dieter eats only one meal of solid food each day.

A temporary solution Many people who start a fad diet shed weight quickly in the first week or two. After that, however, weight loss usually slows dramatically. This is because the initial weight loss is mostly due to loss of water, not fat. When people quit a fad diet, they usually return to their old eating habits and rapidly regain the lost weight.

What the advertisements don't say Nearly all fad diets are based on unhealthy nutritional principles. People on fad diets usually are not eating a balanced diet, and, therefore, not getting proper amounts of vitamins, minerals, and other important compounds their bodies need to grow and function properly.

Some fad diets also can cause serious health problems. High-protein diets, for example, are very high in fat and cholesterol, substances that promote heart disease and circulatory problems.

Fad diets may help some people lose a few pounds temporarily. But for safe, long-term weight loss, nutritionists recommend a diet based on healthy eating habits: balanced, regular meals rich in fruits, vegetables, whole grains, sufficient protein, and small amounts of fat.

INVESTIGATING THE ISSUE

Analyzing Information Collect advertisements for three different fad diets that promise "miracle" results. Based on what you know about good nutrition, would you recommend any of these diets to a friend who is trying to lose weight? Why or why not?

BIOLOGY Online To find out more about fad diets, visit the Glencoe Science Web site.
science.glencoe.com

■ Lead the class in a discussion about societal pressures to be thin, and why Americans spend millions of dollars every year trying to lose weight.

Investigating the Issue

In most cases, students will be able to identify serious flaws in the nutritional soundness of fad diets, as well as potential health risks if the diet is followed for an extended period of time.

Going Further

Have students check out Internet sites that look at the topic of weight loss or other health issues. **L2**

SUMMARY

Section 35.1

Following Digestion of a Meal

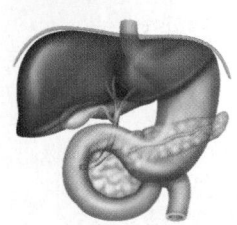

Main Ideas
- Digestion begins in the mouth with both mechanical and chemical action. The esophagus transports food from the mouth to the stomach.
- Chemical and mechanical digestion continue in the acidic environment of the stomach.
- In the small intestine, digestion is completed and food is absorbed. The liver and pancreas play key roles in digestion.
- The large intestine absorbs water before indigestible materials are eliminated.

Vocabulary
amylase (p. 948)
bile (p. 951)
epiglottis (p. 950)
esophagus (p. 948)
gallbladder (p. 952)
large intestine (p. 953)
liver (p. 951)
pancreas (p. 951)
pepsin (p. 951)
peristalsis (p. 950)
rectum (p. 953)
small intestine (p. 951)
stomach (p. 950)
villus (p. 952)

Section 35.2

Nutrition

Main Ideas
- Carbohydrates are the body's main source of energy. Fats are used to store energy. Proteins are used as building materials.
- Minerals serve as structural materials or take part in chemical reactions. Vitamins are needed for growth and metabolism.
- Metabolic rate determines how quickly energy is burned.

Vocabulary
Calorie (p. 957)
mineral (p. 956)
vitamin (p. 956)

Section 35.3

The Endocrine System

Main Ideas
- The endocrine glands work with the nervous system to regulate body functions.
- Blood hormone levels are controlled by a negative feedback system.
- Steroid hormones bind to receptors inside the target cells and amino acid hormones bind to plasma membrane receptors.

Vocabulary
adrenal glands (p. 963)
endocrine glands (p. 959)
hypothalamus (p. 960)
negative feedback system (p. 961)
parathyroid glands (p. 965)
pituitary gland (p. 959)
receptor (p. 960)
target tissues (p. 960)
thyroid gland (p. 964)

Main Ideas

Summary statements can be used by students to review the major concepts of the chapter.

Using the Vocabulary

To reinforce chapter vocabulary, use the Content Mastery Booklet and the activities in the Interactive Tutor for Biology: The Dynamics of Life on the Glencoe Science Web site. **science.glencoe.com**

 *All Chapter Assessment questions and answers have been validated for accuracy and suitability by The Princeton Review.*

UNDERSTANDING MAIN IDEAS

1. c
2. d

UNDERSTANDING MAIN IDEAS

1. Which of these is an example of mechanical digestion?
   a. peristalsis   **c.** chewing
   b. coughing   **d.** epiglottis

2. Which of these is NOT a function of the digestive system?
   a. eliminating wastes
   b. absorbing nutrients
   c. digesting food
   d. regulating metabolism

GLENCOE TECHNOLOGY

 **VIDEOTAPE**
MindJogger Videoquizzes
Chapter 35: *The Digestive and Endocrine Systems*
Have students work in groups as they play the videoquiz game to review key chapter concepts.

Resource Manager

Chapter Assessment, pp. 205-210
MindJogger Videoquizzes
ExamView® Pro Software 💾
BDOL Interactive CD-ROM, Chapter 35 quiz

3. a
4. (lemon juice)
5. c
6. d
7. a
8. d
9. c
10. c
11. endocrine, target tissues
12. amylase, starches
13. epiglottis, respiratory tract
14. glycogen, glucose
15. insulin, glucose
16. mineral
17. Steroid, amino acids
18. hypothalamus, pituitary gland
19. pituitary, hypothalamus
20. esophagus, stomach

3. Even if you were standing on your head, this process would still move food along your digestive tract.
   a. peristalsis   **c.** secretion
   b. swallowing   **d.** absorption

4. Which of the substances listed in *Table 35.4* have a pH near that of your stomach during digestion?

Table 35.4

Substance	pH
black coffee	5
bleach	12
lemon juice	2
baking soda	9

5. Which of these enzymes functions best in the acidic pH of the stomach?
   a. lipase   **c.** pepsin
   b. lactase   **d.** amylase

6. What is the primary function of the large intestine?
   a. food absorption   **c.** vitamin synthesis
   b. food digestion   **d.** water absorption

7. Which of the following is under the control of the hypothalamus?
   a. pituitary gland   **c.** pancreas
   b. taste buds   **d.** liver

8. Which of these is NOT a function of the thyroid gland?
   a. controls growth and development
   b. regulates metabolism
   c. regulates blood calcium levels
   d. responds to stressful situations

9. What unit is used to measure the energy content of food?
   a. temperature   **c.** Calorie
   b. grams   **d.** mass

THE PRINCETON REVIEW   **TEST-TAKING TIP**

Where's the fire?
Slow down! Go back over reading passages and double-check your math. Remember that doing most of the questions and getting them right is always preferable to doing all the questions and getting lots of them wrong.

10. What is the most abundant substance in the human body?
    a. carbohydrates   **c.** water
    b. vitamins   **d.** proteins

11. Hormones released by _____ glands affect specific areas known as _____.

12. Salivary glands in your mouth produce _____, an enzyme that breaks down _____.

13. The _____ prevents swallowed food from entering the _____.

14. When your body needs energy, it breaks down _____ in the liver and releases _____ into the bloodstream.

15. The pancreas releases the hormone _____, which removes _____ from the blood.

16. A(n) _____ is an inorganic substance that serves as a building material or takes part in a chemical reaction in the body.

17. _____ hormones are made from lipids; the other group of hormones is made from _____.

18. A negative feedback system controls the level of hormones by feeding back information to the _____ or the _____.

19. The _____ gland is controlled by the _____, tying the endocrine and nervous systems together.

20. The _____ is a muscular tube of the digestive system that connects the mouth to the _____.

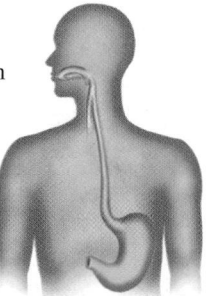

APPLYING MAIN IDEAS

21. Achlorhydria is a condition in which the stomach fails to secrete hydrochloric acid. How would this condition affect digestion?

22. How could removal of the parathyroid glands affect muscle contraction?

THINKING CRITICALLY

23. **Recognizing Cause and Effect** How is the role of pancreatic hormones in glucose regulation important for homeostasis?

24. **Interpreting Data** The relationship between parathyroid hormone secretion and blood calcium levels is shown in the graph below. To what level does the blood calcium level have to fall in order to get maximum parathyroid hormone secretion?

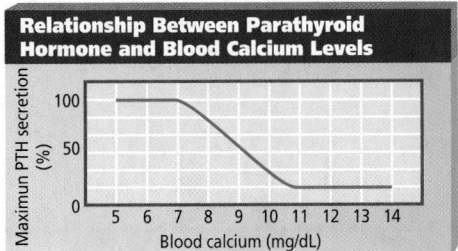

Relationship Between Parathyroid Hormone and Blood Calcium Levels

25. **Concept Mapping** Complete the concept map by using the following vocabulary terms: liver, bile, small intestine, stomach, esophagus, gallbladder.

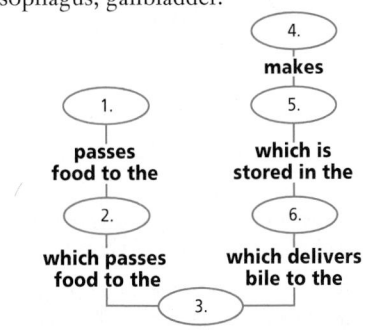

CD-ROM

For additional review, use the assessment options for this chapter found on the *Biology: The Dynamics of Life Interactive CD-ROM* and on the Glencoe Science Web site. **science.glencoe.com**

ASSESSING KNOWLEDGE & SKILLS

The following table contains nutritional information for a meal of macaroni and cheese.

Table 35.5 Macaroni and cheese nutrition
Serving size: 8 ounces
Calories per serving: 280

Nutrient	Grams per serving	Calories per gram
Protein	7	4
Carbohydrate	35	4
Fat	12	9
Sodium	1.540	0

Interpreting Data Use the data in *Table 35.5* to answer the following questions.

1. How many Calories are there in one serving of macaroni and cheese?
   a. 7   c. 280
   b. 35   d. 1540

2. Being fond of macaroni and cheese, Juan eats five servings each day. Assuming this is all he eats, how many Calories does Juan eat each day?
   a. 8   c. 300
   b. 35   d. 1400

3. What percent of Juan's Calories are derived from fat?
   a. less than 1 percent
   b. approximately 10 percent
   c. approximately 38 percent
   d. more than 50 percent

4. Juan should eat only 1800 Calories a day. What proportion of his daily diet is derived from his five servings of macaroni and cheese?
   a. 10 percent   c. 50 percent
   b. 38 percent   d. 78 percent

5. **Interpreting Data** The recommended daily allowance of sodium is approximately 2.4 g. Make a statement that describes Juan's sodium intake.

APPLYING MAIN IDEAS

21. The enzyme pepsin would not have the proper environment in which to function. As a result, proteins would not be digested in the stomach.

22. The parathyroid glands secrete PTH, which increases the calcium level in the blood. Calcium is required for muscle contraction. Removal of the gland could greatly reduce blood calcium concentration and interfere with muscle contraction.

THINKING CRITICALLY

23. The glucose concentration in the blood is maintained in a very narrow range. The pancreas releases insulin to remove glucose from the blood and glucagon to cause the release of glucose into the blood.

24. Blood calcium levels must drop below 8 mg/dL to get maximum parathyroid hormone secretion.

25. 1. Esophagus; 2. Stomach; 3. Small intestine; 4. Liver; 5. Bile; 6. Gallbladder

ASSESSING KNOWLEDGE & SKILLS

1. c
2. d
3. c
4. d
5. Juan's diet contains too much sodium.

Chapter 36 Organizer

Refer to pages 4T-5T of the Teacher Guide for an explanation of the National Science Education Standards correlations.

Section	Objectives	Activities/Features
Section 36.1 **The Nervous System** National Science Education Standards UCP.1-3, UCP.5; A.1; B.1, B.3; C.1, C.5, C.6 (2 sessions, 1 block)	1. **Analyze** how nerve impulses travel within the nervous system. 2. **Recognize** the functions of the major parts of the nervous system. 3. **Compare** voluntary responses and involuntary responses.	**Focus On** Evolution of the Brain, p. 978 **MiniLab 36-1:** Distractions and Reaction Time, p. 980 **BioTechnology:** Scanning the Mind, p. 998
Section 36.2 **The Senses** National Science Education Standards UCP.1-3, UCP.5; C.1, C.5, C.6 (1 session)	4. **Define** the role of the senses in the human nervous system. 5. **Recognize** how senses detect chemical, light, and mechanical stimulation. 6. **Identify** ways in which the senses work together to gather information.	**Inside Story:** The Eye, p. 985 **Problem-Solving Lab 36-1,** p. 986
Section 36.3 **The Effects of Drugs** National Science Education Standards UCP.1, UCP.2; A.1, A.2; C.6; E.1, E.2; F.1, F.5, F.6; G.1, G.2 (2 sessions, 1/2 block)	7. **Recognize** the medicinal uses of drugs. 8. **Identify** the different classes of drugs. 9. **Interpret** the effect of drug misuse and abuse on the body.	**Problem-Solving Lab 36-2,** p. 989 **Careers in Biology:** Pharmacist, p. 990 **MiniLab 36-2:** Interpret a Drug Label, p. 991 **Design Your Own BioLab:** What drugs affect the heart rate of *Daphnia?* p. 996

Need Materials? Contact Carolina Biological Supply Company at 1-800-334-5551 or at **http://www.carolina.com**

MATERIALS LIST

BioLab

p. 996 microscope, microscope slides, dropper, aged tap water, *Daphnia* culture, dilute solutions of coffee, tea, cola, ethyl alcohol, tobacco, and cough medicine

MiniLabs

p. 980 meterstick, paper, pencil
p. 991 paper, pencil

Alternative Lab

p. 980 paper clips, metric ruler, paper, pencil

Quick Demos

p. 975 telephone cable
p. 981 rubber hammer
p. 984 tape recording of common sounds
p. 990 samples of non prescription drugs

Key to Teaching Strategies

L1 Level 1 activities should be appropriate for students with learning difficulties.

L2 Level 2 activities should be within the ability range of all students.

L3 Level 3 activities are designed for above-average students.

ELL ELL activities should be within the ability range of English Language Learners.

COOP LEARN Cooperative Learning activities are designed for small group work.

P These strategies represent student products that can be placed into a best-work portfolio.

These strategies are useful in a block scheduling format.

The Nervous System

Teacher Classroom Resources

Section	Reproducible Masters	Transparencies
Section 36.1 **The Nervous System**	Reinforcement and Study Guide, pp. 159-160 **L2** Critical Thinking/Problem Solving, p. 36 **L3** BioLab and MiniLab Worksheets, p. 159 **L2** Laboratory Manual, pp. 261-264 **L2** Content Mastery, pp. 177-178, 180 **L1**	Section Focus Transparency 87 **L1** **ELL** Basic Concepts Transparency 67 **L2** **ELL**
Section 36.2 **The Senses**	Reinforcement and Study Guide, p. 161 **L2** Concept Mapping, p. 36 **L3** **ELL** Laboratory Manual, pp. 265-268 **L2** Content Mastery, pp. 177, 179-180 **L1** Tech Prep Applications, pp. 47-48 **L2**	Section Focus Transparency 88 **L1** **ELL** Basic Concepts Transparency 68 **L2** **ELL** Basic Concepts Transparency 69 **L2** **ELL** Reteaching Skills Transparency 52 **L1** **ELL** Reteaching Skills Transparency 53 **L1** **ELL**
Section 36.3 **The Effects of Drugs**	Reinforcement and Study Guide, p. 162 **L2** BioLab and MiniLab Worksheets, pp. 160-162 **L2** Content Mastery, pp. 177, 180 **L1** Tech Prep Applications, pp. 49-50 **L2**	Section Focus Transparency 89 **L1** **ELL**

Assessment Resources

Chapter Assessment, pp. 211-216
MindJogger Videoquizzes
Performance Assessment in the Biology Classroom
Alternate Assessment in the Science Classroom
ExamView® Pro Software 💾
BDOL Interactive CD-ROM, Chapter 36 quiz

Additional Resources

Spanish Resources **ELL**
English/Spanish Audiocassettes **ELL**
Cooperative Learning in the Science Classroom **COOP LEARN**
Lesson Plans/Block Scheduling

NATIONAL GEOGRAPHIC
Teacher's Corner

Products Available From Glencoe
To order the following products, call Glencoe at 1-800-334-7344:
CD-ROM
NGS PictureShow: Human Body 1
Curriculum Kit
GeoKit: Human Body 2
Transparency Set
NGS PicturePack: Human Body 1
Videodisc
STV: Human Body

Products Available From National Geographic Society
To order the following products, call National Geographic Society at 1-800-368-2728:
Videos
Incredible Human Machine
Nervous System (Human Body Series)

GLENCOE TECHNOLOGY

The following multimedia resources are available from Glencoe.

Biology: The Dynamics of Life
CD-ROM **ELL**

BioQuest: *Body Systems*
Animation: *Impulse Transmission in a Motor Neuron*
Animation: *Impulse Transmission Across a Synapse*
Animation: *The Sense of Sight*
Animation: *The Sense of Hearing*

Videodisc Program
Impulse Transmission
Impulse Transmission: *Synapse*
Sense of Sight

Chapter

36 The Nervous System

Create a loud noise by slamming a book on your desk. Use student reaction to begin a discussion of how the nervous system reacts to environmental stimuli, including loud noises.

Theme Development

The major theme in this chapter is **systems and interactions**. Stress that the nervous system receives information from all body systems and responds in order to maintain homeostasis in the body.

`0:00` OUT OF TIME?

If time does not permit teaching the entire chapter, use the BioDigest at the end of the unit as an overview.

READING BIOLOGY

Glencoe's *Biology: The Dynamics of Life* contains many resources to assist a student's reading skills. Each chapter contains figures with expanded captions that expand on written material. Word Origins, located along the side of text, expand knowledge of biology vocabulary. Glencoe's Content Mastery Booklet helps develop reading skills while reinforcing content. In addition, use the Interactive Tutor for *Biology: The Dynamics of Life* on the Glencoe Web site to reinforce vocabulary. **science.glencoe.com**

What You'll Learn

■ You will relate the structure of a nerve cell to the transmission of a nerve signal.
■ You will identify the senses and their signal pathways.
■ You will compare and contrast various types of drugs and their effects on the nervous system.

Why It's Important

The nervous system helps you perceive and react to the world around you. By understanding how drugs—both legal and illegal—affect the function of the nervous system, you will discover their role in treating medical disorders, and the danger they pose if misused.

READING BIOLOGY

Look at several of the diagrams throughout the chapter. Write down any new terms that are used. As you read, jot down the definitions for the new vocabulary. How do the diagrams illustrate the words' meanings?

BIOLOGY Online

To find out more about the nervous system, visit the Glencoe Science Web site.
science.glencoe.com

Magnification: 5550×

When nerve cells (inset) in your eyes detect something scary, they relay a message to your brain, which, in turn, tells your muscles to tense up and your heart to beat faster.

972

Multiple Learning Styles

Look for the following logos for strategies that emphasize different learning modalities.

Kinesthetic Portfolio, p. 975; Quick Demo, p. 981; Project, p. 983

Visual-Spatial Meeting Individual Needs, pp. 974, 988; Reinforcement, p. 975; Display, p. 984; Portfolio, p. 985; Check for Understandning, p. 987

Interpersonal Biology Journal, p. 994

Intrapersonal Enrichment, p. 994; Concept Development, p. 994

Linguistic Biology Journal, pp. 974, 976, 977, 984, 990, 992; Enrichment, pp. 976, 977; Tech Prep, p. 976; Project, p. 977; Portfolio, p. 993; Check for Understanding, p. 994; Reteach, p. 994

Auditory-Musical Quick Demo, p. 984

36.1 The Nervous System

SECTION PREVIEW

Objectives

Analyze how nerve impulses travel within the nervous system.

Recognize the functions of the major parts of the nervous system.

Compare voluntary responses and involuntary responses.

Vocabulary

neuron
dendrite
axon
synapse
neurotransmitter
central nervous system
peripheral nervous
  system
cerebrum
cerebellum
medulla oblongata
somatic nervous system
reflex
autonomic nervous
  system
sympathetic nervous
  system
parasympathetic
  nervous system

What do you use the telephone for? To communicate with a friend in another location. You may know that your message is transmitted as an electrical impulse across the telephone wires. Would it surprise you to know that a similar electrical impulse travels through your body, helping some parts to communicate with others?

Like telephone wires between homes, nerve cells relay messages within the human body.

Neurons: Basic Units of the Nervous System

The basic unit of structure and function in the nervous system is the neuron, or nerve cell. **Neurons** (NYU ronz) conduct impulses throughout the nervous system. As shown in *Figure 36.1,* a neuron is a long cell that consists of three regions: a cell body, dendrites, and an axon.

Dendrites (DEN drites) are branchlike extensions of the neuron that receive impulses and carry them toward the cell body. The **axon** is a single extension of the neuron that carries impulses away from the cell body and toward other neurons, muscles, or glands.

Neurons fall into three categories: sensory neurons, motor neurons, and interneurons. Sensory neurons carry impulses from the body to the spinal cord and brain. Interneurons are found within the brain and spinal cord. They process incoming impulses and pass response impulses on to motor neurons. Motor neurons carry the response impulses away from the brain and spinal cord to a muscle or gland.

Figure 36.1
Dendrites and axons are extensions that branch out from the cell body of a neuron.

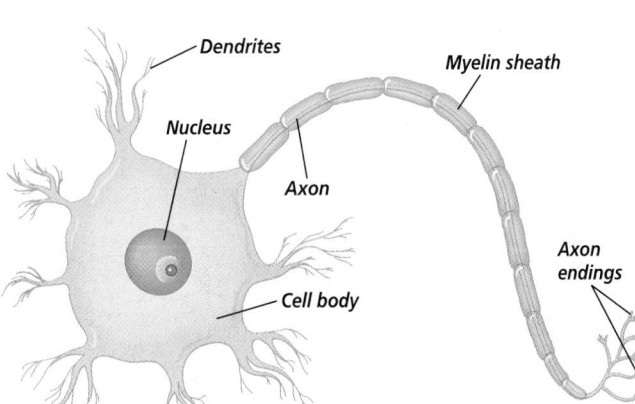

Dendrites

Nucleus

Myelin sheath

Axon

Axon endings

Cell body

36.1 THE NERVOUS SYSTEM **973**

Prepare

Key Concepts

The method by which impulses travel in the nervous system—including electrical transmission along the neuron and chemical transmission at the synapse—are covered in this section. The organization and function of the major parts of the nervous system are discussed.

Planning

■ Obtain a piece of cable for the first Quick Demo.

■ Collect enough metersticks for student use in MiniLab 36-1.

■ Acquire a rubber hammer for the second Quick Demo.

1 Focus

Bellringer

Before presenting the lesson, display **Section Focus Transparency 87** on the overhead projector and have students answer the accompanying questions.
L1 **ELL**

---

2 Teach

Visual Learning

Figure 36.2 Spend a few minutes reviewing the sequence of events involved in the nervous system's response to a stimulus. Ask students what path the nervous system would take to turn the head in response to a sound such as a honking car horn. *The path would be the same except for the initial receptors, which would be located in the ear rather than in the skin.*

Visual Learning

Figure 36.3 Ask: What effect does the Na$^+$/K$^+$ pump have on the charge of the normal resting neuron? *It maintains a positive charge on the outside of the membrane and a negative charge within the membrane of a resting neuron.*

Resource Manager

Section Focus Transparency 87 and Master **L1** **ELL**
Critical Thinking/Problem Solving, p. 36 **L3**

Figure 36.2
The nervous system sorts and interprets incoming information before directing a response.

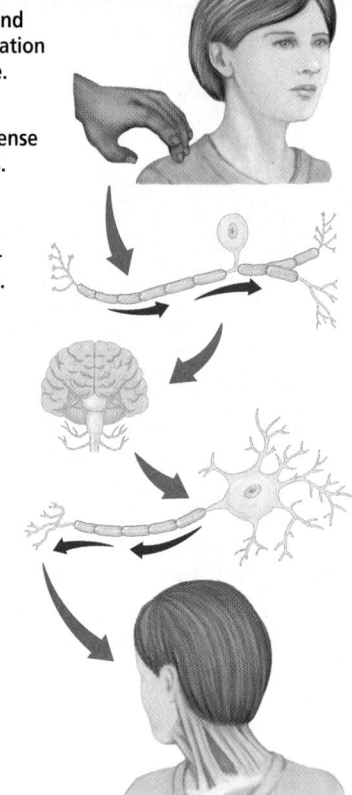

A Receptors in the skin sense a tap or other stimulus.

B Sensory neurons transmit the touch message.

C The message is interpreted. A response is sent to the motor neurons.

D Motor neurons transmit a response message to the shoulder muscles.

E The neck muscles are activated, causing the head to turn.

Figure 36.3
The membrane of a neuron contains open as well as gated channels that allow movement of sodium (Na$^+$) and potassium (K$^+$) ions into and out of the cell.

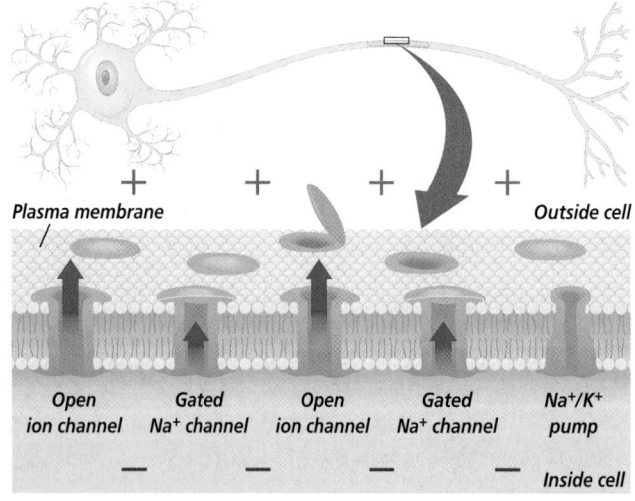

Plasma membrane
Outside cell

Open ion channel | Gated Na$^+$ channel | Open ion channel | Gated Na$^+$ channel | Na$^+$/K$^+$ pump

Inside cell

Relaying an impulse

Suppose you're in a crowded, noisy store and you feel a tap on your shoulder. Turning your head, you see the smiling face of a good friend. How did the shoulder tap get your attention? The touch stimulated sensory receptors located in the skin of your shoulder to produce an impulse. The sensory impulse was carried to the spinal cord and then up to your brain. From your brain, an impulse was sent out to your motor neurons, which then transmitted the impulse to muscles in your neck. Your neck muscles then turned your head. *Figure 36.2* shows how a stimulus, such as a tap on the shoulder, is transmitted through your nervous system.

A neuron at rest

First, let's look at a resting neuron—one that is not transmitting an impulse. You have learned that the plasma membrane controls the concentration of ions inside a cell. Because the plasma membrane of a neuron is more permeable to potassium ions (K$^+$) than it is to sodium ions (Na$^+$), more potassium ions exist within the cell membrane than outside it. Similarly, more sodium ions exist outside the cell membrane than within it.

The neuron membrane also contains an active transport system, called the sodium/potassium (Na$^+$/K$^+$) pump, which uses ATP to pump three sodium ions out of the cell for every two potassium ions it pumps in. As you can see in *Figure 36.3,* the action of the pump increases the concentration of positive charges on the outside of the membrane. In addition, the presence of many negatively charged proteins and organic phosphates means that the inside of the membrane is more negatively charged than the outside. Under these conditions, which exist when the cell is at rest,

---

the plasma membrane is said to be polarized. A polarized membrane has the potential to transmit an impulse.

How an impulse is transmitted

When a stimulus excites a neuron, gated sodium channels in the membrane open up and sodium ions rush into the cell. As the positive sodium ions build up inside the membrane, the inside of the cell becomes more positively charged than the outside. This change in charge, called depolarization, moves like a wave down the length of the axon, as seen in *Figure 36.4*. As the wave passes, gated channels and the Na+/K+ pump act to return the neuron to its resting state, with the inside of the cell negatively charged and the outside positively charged.

An impulse can move down the complete length of an axon only when stimulation of the neuron is strong enough. If the threshold level—the level at which depolarization occurs—is not reached, the impulse quickly dies out.

White matter and gray matter

Most axons are surrounded by a white covering of cells called the myelin sheath. Like the plastic coating on an electric wire, the myelin sheath tightly insulates the axon, hindering the movement of ions across its plasma membrane. The ions move quickly down the axon until they reach a gap in the sheath. At this point, the ions pass through the plasma membrane of the nerve cell and depolarization occurs. As a result, the impulse jumps from gap to gap, greatly increasing the speed at which it travels.

CD-ROM
View an animation of nerve depolarization in the Presentation Builder of the Interactive CD-ROM.

Figure 36.4
A wave of depolarization moves down the axon of a neuron.

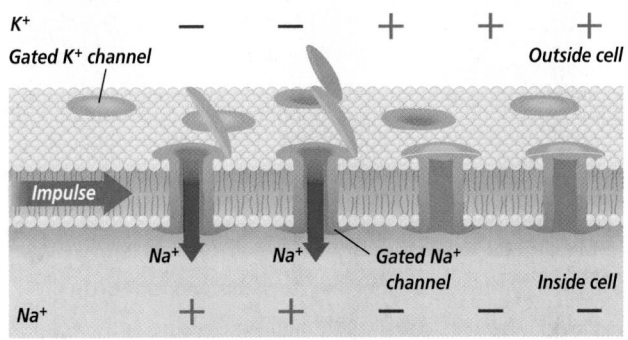

A Gated sodium channels open, allowing sodium ions to enter and make the inside of the cell positively charged and the outside negatively charged.

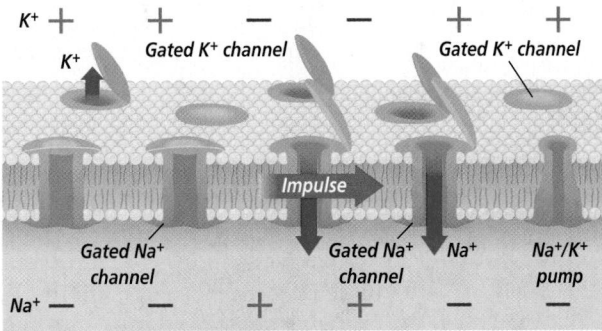

B As the impulse passes, gated sodium channels close, stopping the influx of sodium ions. Gated potassium channels open, letting potassium ions out of the cell. This action repolarizes the cell.

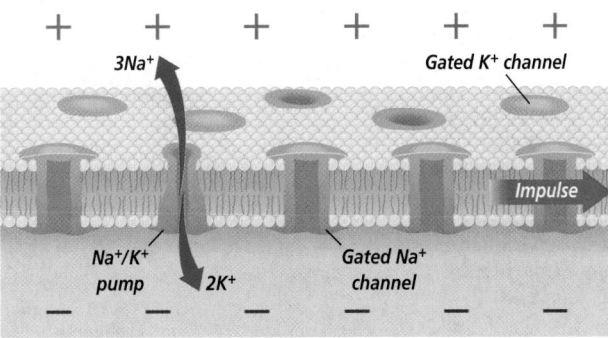

C As gated potassium channels close, the Na+/K+ pump restores the ion distribution.

36.1 THE NERVOUS SYSTEM **975**

The myelin sheath gives axons a white appearance. In the brain and spinal cord, masses of myelinated axons make up what is called "white matter." Has anyone ever told you to "use your gray matter"? They were actually referring to a specific part of your brain. The absence of myelin in masses of neurons accounts for the grayish color of "gray matter" in the brain.

Connections between neurons

Although neurons lie end to end—axons to dendrites—they don't actually touch. A tiny space lies between one neuron's axon and another neuron's dendrites. This junction between neurons is called a **synapse.** Impulses traveling to and from the brain must move across the synaptic space that separates the axon and dendrites. How do they make this leap?

As an impulse reaches the end of an axon, calcium channels open, allowing calcium to enter the end of the axon. As shown in *Figure 36.5*, the calcium causes vesicles in the axon to fuse with the plasma membrane, releasing their chemicals into the synaptic space by exocytosis. These chemicals, called **neurotransmitters,** diffuse across the space to the dendrites of the next neuron. As the neurotransmitters reach the dendrites, they signal receptor sites to open the ion channels. These open channels change the polarity in the neuron, initiating a new impulse. Enzymes in the synapse typically break down the neurotransmitters shortly after transmission, preventing the continual firing of impulses.

The Central Nervous System

When you make a call to a friend, your call travels through wires to a control center where it is switched over to wires that connect with your friend's telephone. In the same manner, an impulse traveling through neurons in your body usually reaches the control center of the nervous system—your brain—before being rerouted. The brain and the spinal cord together make up the **central nervous system,** which coordinates all your body's activities.

Figure 36.5 Neurotransmitters are released into the synaptic space by the axon. When the neurotransmitters bind to receptors on the dendrites of the next neuron, the ion channels open, changing the polarity of the dendrites. In this way, nerve impulses move from neuron to neuron.

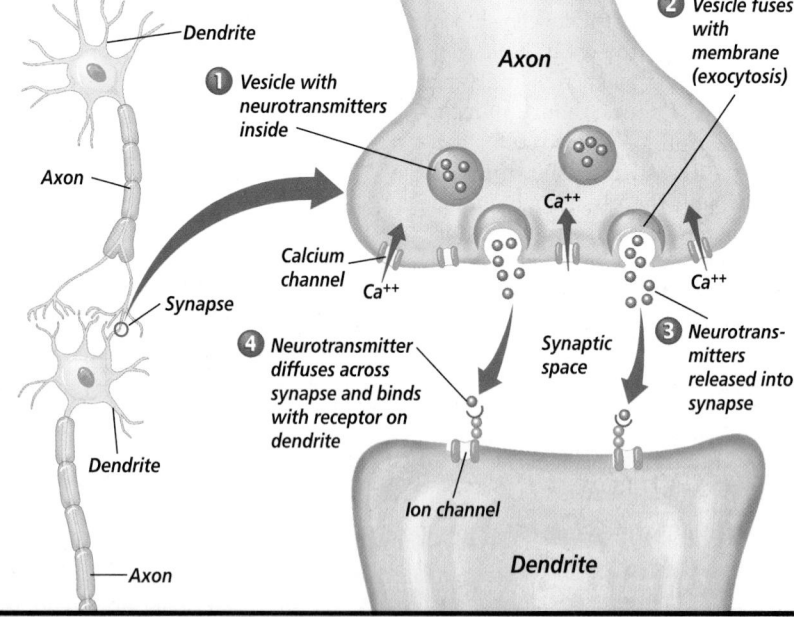

Dendrite
Axon
Axon
Synapse
Dendrite
Axon

① Vesicle with neurotransmitters inside
② Vesicle fuses with membrane (exocytosis)
Ca++
Calcium channel
Ca++
Ca++
③ Neurotransmitters released into synapse
④ Neurotransmitter diffuses across synapse and binds with receptor on dendrite
Synaptic space
Ion channel
Dendrite

Two systems work together

Another division of your nervous system, called the **peripheral nervous system** (puh RIHF rul), is made up of all the nerves that carry messages to and from the central nervous system. It is similar to the telephone wires that run between a phone system's control center and the phones in individual homes. Together, the central nervous system (CNS) and the peripheral nervous system (PNS), shown in *Figure 36.6,* respond to stimuli from the external environment.

Anatomy of the brain

The brain is the control center of the entire nervous system. For descriptive purposes, it is useful to divide the brain into three main sections: the cerebrum, the cerebellum, and the brain stem.

The **cerebrum** (suh REE brum) is divided into two halves, called hemispheres, that are connected by bundles of nerves. Your conscious activities, intelligence, memory, language, skeletal muscle movements, and senses are all controlled by the cerebrum. The outer surface of the cerebrum, called the cerebral cortex, is made up of gray matter. The cerebral cortex contains countless folds and grooves that increase its total surface area. This increase in surface area played an important role in the evolution of human intelligence as greater surface area allowed more and more complex thought processes.

The **cerebellum** (ser uh BEL um), located at the back of your brain, controls your balance, posture, and coordination. If the cerebellum is injured, your movements become jerky.

The brain stem is made up of the medulla oblongata, the pons, and the midbrain. The **medulla oblongata** (muh DUL uh • ahb long GAHT uh) is the part of the brain that controls

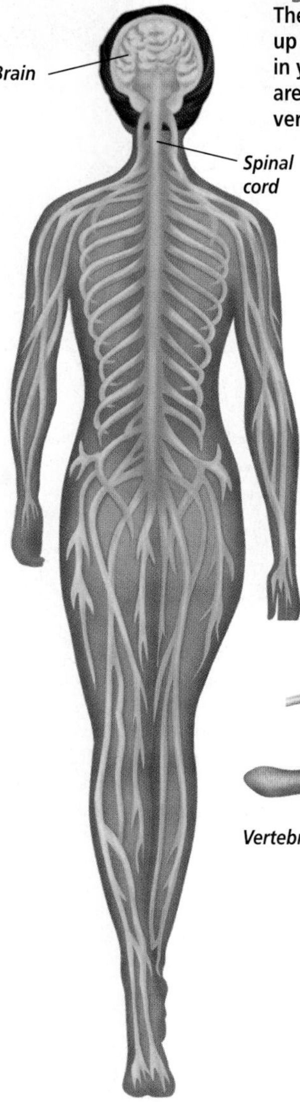

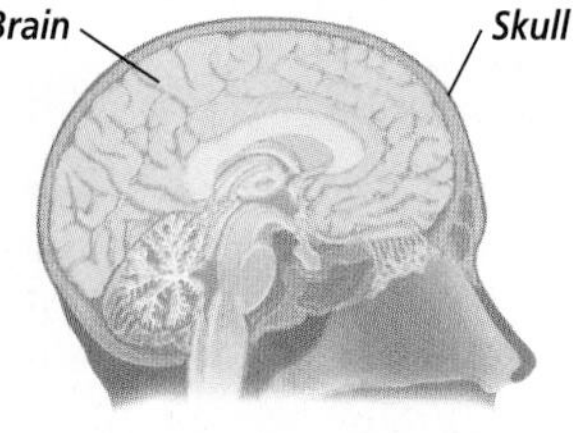

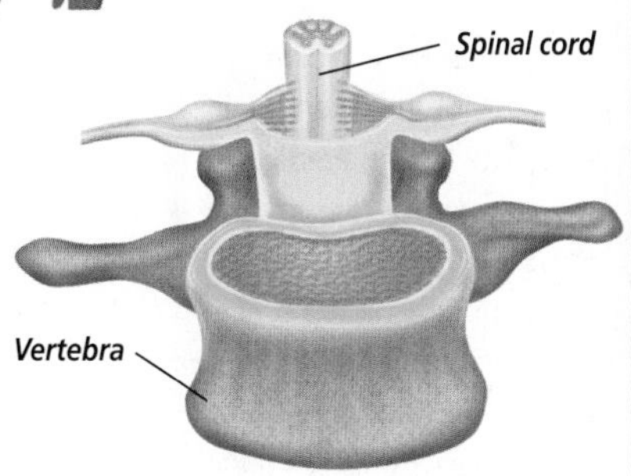

Figure 36.6
The human nervous system is made up of the CNS, in pink, and the PNS, in yellow. The brain and spinal cord are protected by the skull and the vertebrae, respectively.

Brain

Spinal cord

Brain

Skull

Spinal cord

Vertebra

involuntary activities such as breathing and heart rate. The pons and midbrain act as pathways connecting various parts of the brain with each other. Find out more about how the brain evolved by reading *Focus On Evolution of the Brain.* For the latest on technological advances in brain imaging, check out the *BioTechnology* section at the end of the chapter.

Enrichment

 Linguistic The Chinese have used acupuncture as a complete system of medicine for thousands of years. The procedure involves inserting long, thin needles into specific areas of the patient's body. Electrical impulses are then sent down the length of the needles. Acupuncture is used to relieve pain, cure cancer, and everything in between. It is also currently being used in China as a form of anesthesia for patients undergoing surgery.

Have students research and report on ancient methods of acupuncture. How did the methods used then differ from those used today? *The needles used to be turned to achieve the desired effect. Now electrical impulses are used instead.* **L3**

✓ **Assessment**

Performance Assessment in the Biology Classroom, p. 49, *Preparing and Teaching a Lesson About the Nervous System.* Have students carry out this activity to demonstrate their knowledge of the structure and function of the nervous system. **L2**

NATIONAL GEOGRAPHIC

 **VIDEODISC**
STV: Human Body Vol. 2
Nervous System
Unit 2, Side 2, 1 min. 59 sec.
The Brain and Its Parts

 **Resource Manager**

Basic Concepts Transparency 66 and Master **L2** **ELL**

BIOLOGY JOURNAL

Involuntary Actions

Linguistic Ask students to list activities their bodies do without conscious thought. To get them started, have them consider what their bodies do while they are asleep. **L1**

PROJECT

Neuroscience Research

Linguistic Have students research and prepare a report, a visual device, or an audiovisual presentation on one of the following topics: sleep, split brain experiments, dementia, Alzheimer's disease, retrograde amnesia, or Penfield maps of the brain. Have students present their findings to the class. **L2**

Focus On
Evolution of the Brain

Purpose
Students will explore the evolution of the brain by comparing the brains of different animals. They will also learn about brain structure and function.

Background
Functionally, the human brain can be divided into the lobes of the cerebrum (including the cerebral cortex), the cerebellum, and the brain stem—pons, midbrain, and medulla oblongata.

Over the course of human evolution, the human brain has tripled in size. Scientists surmise that the increase in brain size accompanied the developed use of tools and other skills that enabled our ancestors to live in a greater variety of habitats.

Teaching Strategies
■ Have students in groups draw a large outline of the brain on a piece of poster paper. Have them cut out magazine pictures that relate to the specializations of the brain and glue them on their outline. For instance, they could cut out eyes and place them in the occipital lobe to indicate that is where vision is interpreted.

■ Have students make scale models of the cerebrums in the different animals shown. Then have them compare the size of the cerebrum with the size of the animal. Which animal has the largest cerebrum for its body size? Which has the smallest?

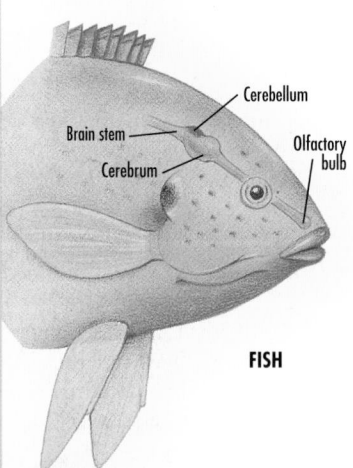

PLANARIAN

Nerve cords

Ganglion

FOCUS ON

Evolution
of the Brain

As animals have evolved over hundreds of millions of years, there has been a tendency toward ever-increasing complexity in the nervous system, and especially, in the brain. Brains had their beginnings as relatively simple bundles of nerve cells. But over time, the brains of vertebrate animals have become more complex and specialized. Humans possess the most complex brain in the animal kingdom, a remarkable organ that enables us to reason, wonder, and dream.

THE EVOLVING BRAIN
Jumping ahead millions of years to when the vertebrates emerged, the five brains shown here illustrate how evolution has transformed a simple ganglion to a complex brain. As the brain evolved, areas that control senses, behavior, and coordination became predominant.

Notice that in humans the brain is proportionally much larger than it is in many other vertebrates and that the area dedicated to thinking, the cerebrum, covers and dominates everything else.

THE SIMPLEST BRAIN
Flatworms are the simplest animals that have an identifiable brain. A planarian, for example, has a mass of nerve tissue called a ganglion that lies beneath each eyespot. Extending back from these ganglia are long nerve cords that run the length of the body. Between the cords are cross connections that make the planarian nervous system look like a ladder.

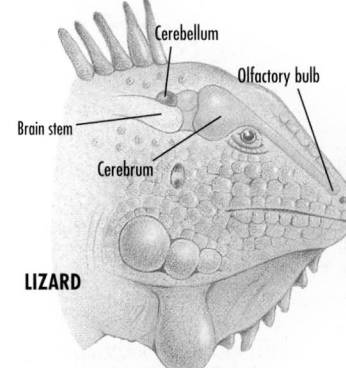

Cerebellum

Olfactory bulb

Brain stem

Cerebrum

LIZARD

OLYMPIC GYMNAST

Cerebellum

Brain stem

Olfactory bulb

Cerebrum

FISH

978

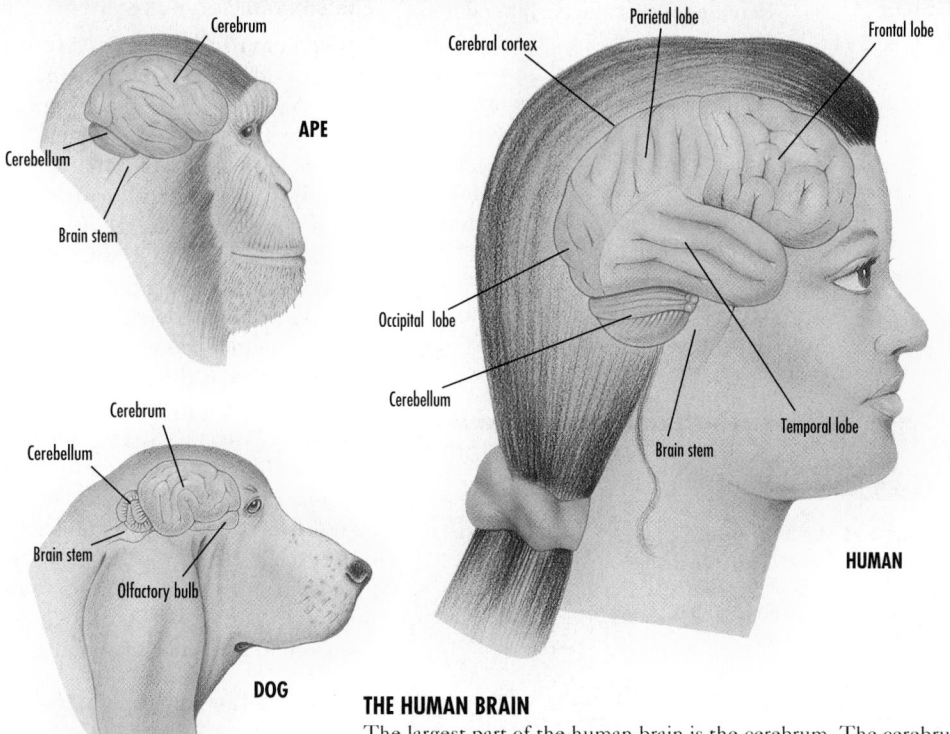

APE

Cerebrum
Cerebellum
Brain stem

DOG

Cerebrum
Cerebellum
Brain stem
Olfactory bulb

HUMAN

Parietal lobe
Cerebral cortex
Frontal lobe
Occipital lobe
Cerebellum
Brain stem
Temporal lobe

PET SCAN

HEARING WORDS
SEEING WORDS
SPEAKING WORDS
GENERATING WORDS

GIRL PLAYING A VIOLIN

THE HUMAN BRAIN

The largest part of the human brain is the cerebrum. The cerebrum is divided into two hemispheres, left and right. But the feature that makes the cerebrum unique is an outer, folded layer less than 5 mm thick—the cerebral cortex. Because of the cortex, you can remember, reason, organize, communicate, understand, and create.

When you watch an Olympic gymnast perform on the balance beam, you are witnessing the work of a well-trained cerebellum. It is here that muscles are coordinated and the memories of physical skills are stored.

The brain stem consists of the medulla, pons, and midbrain. The brain stem regulates breathing, heart rate, circulation, and other vital body processes.

COMPLEX COORDINATION

The cerebral cortex may look like a uniform mass of nerve tissue. But different areas of the cortex receive and process different types of sensory, motor, and integrative nerve impulses. Using a technological tool known as a PET (positron emission tomography) scan (above left), scientists can pinpoint areas of increased metabolic activity in the brain. In so doing they can identify specific regions of the cortex that are involved in complex behaviors such as playing—from memory—a musical composition on the violin.

EXPANDING Your View

1. **THINKING CRITICALLY** Fish, reptiles, and dogs rely far more on their sense of smell to monitor their environments than humans do. How do the brain structures of these different animals reflect this fact?

2. **JOURNAL WRITING** In your journal, record your predictions about how a person's behavior might be affected by an injury to the cerebellum.

GLENCOE TECHNOLOGY

VIDEODISC

The Infinite Voyage *Unseen Worlds*
Magnetic Resonance Imaging:
MRI a Medical Breakthrough (Ch. 6)
2 min. 30 sec.

Unseen Worlds
Brain Tumor Surgery: Made Possible by MRI
(Ch. 7)
2 min. 30 sec.

The Peripheral Nervous System

Remember that the peripheral nervous system carries impulses between the body and the central nervous system. For example, when a stimulus is picked up by receptors in your skin, it initiates an impulse in the sensory neurons. The impulse is carried to the CNS. There, the impulse transfers to motor neurons that carry the impulse to a muscle.

The peripheral nervous system can be separated into two divisions—the somatic nervous system and the autonomic nervous system.

The somatic nervous system

The **somatic nervous system** is made up of 12 pairs of cranial nerves from the brain, 31 pairs of spinal nerves from the spinal cord, and all of their branches. These nerves are actually bundles of neuron axons bound together by connective tissue. The cell bodies of the neurons are found in clusters along the spinal column. Most nerves contain both sensory and motor axons.

The nerves of the somatic system relay information mainly between your skin, the CNS, and skeletal muscles. This pathway is voluntary, meaning that you can decide whether or not to move body parts under the control of this system. Try the *MiniLab* on this page to find out how distractions can affect the time it takes you to respond to a stimulus.

Reflexes in the somatic system

Sometimes a stimulus results in an automatic, unconscious response within the somatic system. When you touch something hot, you automatically jerk your hand away. Such an action is a **reflex**, an automatic response to a stimulus. Rather than proceeding to the brain for interpretation, a reflex impulse travels to the spinal column where it is sent directly back out to a muscle. The brain becomes aware of the reflex only after it occurs. *Figure 36.7* on the next page shows the shortened route of a reflex impulse.

The autonomic nervous system

Imagine that you are spending the night alone in a creepy old house. Suddenly, a creak comes from the attic and you think you hear footsteps. Your heart begins

---

MiniLab 36-1 Experimenting

Distractions and Reaction Time
Have you ever tried to read while someone is talking to you? What effect does such a distracting stimulus have on your reaction time?

Procedure
1. Work with a partner. Sit facing your partner as he or she stands.
2. Have your partner hold the top of a meterstick above your hand. Hold your thumb and index finger about 2.5 cm away from either side of the lower end of the meterstick without touching it.
3. Tell your partner to drop the meterstick straight down between your fingers.
4. Catch the meterstick between your thumb and finger as soon as it begins to fall. Measure how far it falls before you catch it. Practice several times.
5. Run ten trials, recording the number of centimeters the meterstick drops each time. Average the results.
6. Repeat the experiment, this time counting backwards from 100 by fives (100, 95, 90, . . .) as you wait for your partner to release the meterstick.

Analysis
1. Did your reaction time improve with practice? Explain.
2. How was your reaction time affected by the distraction (counting backward)?
3. What other factors, besides distractions, would increase reaction time?

Alternative Lab

Skin Sensitivity

Materials
paper clips unfolded in U-shapes, metric ruler

Background
Certain areas of the skin contain sensory neurons that are packed closely together, whereas other areas have neurons scattered up to centimeters apart. When two different stimuli depolarize parts of the same neuron, the brain interprets them as if they were one stimulus.

Procedure
Give students the following directions.
1. Working with a partner, plan ten areas of the skin to test for the distance between sensory neurons.
2. Have your partner close his or her eyes. With the paper clip ends close together, test your partner by gently touching his or her skin with the opened paper clip.
3. Continually spread the two ends of the

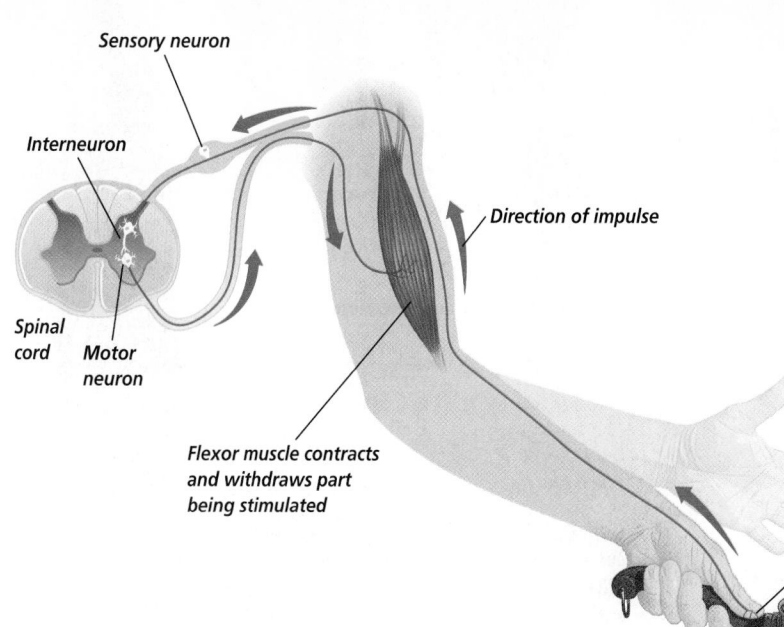

Sensory neuron

Interneuron

Spinal cord

Motor neuron

Flexor muscle contracts and withdraws part being stimulated

Direction of impulse

Pain receptors in skin

to pound. Your breathing becomes rapid. Your thoughts race wildly as you try to figure out what to do—stay and confront the unknown, or run out of the house!

Your internal reactions to this scary situation are being controlled by your autonomic nervous system. The **autonomic nervous system** carries impulses from the CNS to internal organs. These impulses produce responses that are involuntary, or not under conscious control.

There are two divisions of the autonomic nervous system—the **sympathetic nervous system** and the parasympathetic nervous system. The sympathetic nervous system controls many internal functions during times of stress. When you are frightened, the sympathetic nervous system causes the release of a hormone that results in the fight-or-flight response you learned about earlier, as shown in **Figure 36.8**.

The **parasympathetic nervous system** on the other hand, controls many of the body's internal functions

when it is at rest. It is in control when you are relaxing on a warm summer day after a picnic or reading quietly in your room. Both the sympathetic and parasympathetic systems send signals to the same internal organs. The resulting activity of the organ depends on the intensities of the opposing signals.

Figure 36.8
A fight-or-flight response to a rattlesnake will increase heart and breathing rates.

Concept Development
Have students name some of the reflexes an infant has from birth. List these reflexes on the chalkboard. Ask students how these behaviors are different from conscious behaviors such as walking and talking. *Reflexes are not learned behaviors.*

Quick Demo

Kinesthetic Demonstrate the knee-jerk reflex using a rubber hammer. Point out that students did not choose to move their legs—they did so automatically.

3 Assess

Check for Understanding
Make sure students understand that some parts of the nervous system are not under conscious control. Test their understanding by asking students to list body functions that they can and cannot control. Have them identify the part of the brain that controls each function. **L2**

Resource Manager

Laboratory Manual, pp. 261-264 **L2**

paper clip farther apart until your partner can feel two points rather than one.

4. Determine the distance between sensory neurons in a certain body area by measuring the distance between the two ends of the paper clip.

5. In a data table, record the distance between sensory neurons in different parts of the body.

Analysis

1. Which areas of the skin did you find most sensitive? Least sensitive? *Responses will vary depending upon the areas tested. Regions of the back and inner arms will be less sensitive than areas of the palms and fingers.*

2. Explain why two stimuli are felt as two points when the ends of the paper clip are moved farther apart. *The two signals are detected by different sensory neurons.*

Assessment
Portfolio Have students include a summary of the lab, the data table, and answers to Analysis questions in their portfolios. Have them also write a paragraph explaining possible advantages of some areas of the skin being more sensitive than others. Use the Performance Task Assessment List for Lab Report in **PASC**, p. 47. **L2 P**

Reteach

Have students make a table that lists the three major parts of the brain and the functions of each. **L2**

Extension

Have interested students find out about the function of various neurotransmitters such as dopamine, serotonin, norepinephrine, adenosine, or GABA. **L3**

✔ Assessment

Skill Prepare a handout showing a resting neuron, a stimulated neuron, and a synapse. Have students label the steps of neurotransmission. **L1**

4 Close

Discussion

Ask students what would happen if their cerebellum were damaged. For example, could they play a video game? *probably not, because the cerebellum controls coordination*

Resource Manager

Reinforcement and Study Guide, pp. 159-160 **L2**
Content Mastery, p. 178 **L1**
Basic Concepts Transparency 67 and Master **L2** **ELL**

Figure 36.9
Understanding the organization of your nervous system can be made easier by studying the different divisions of the nervous system.

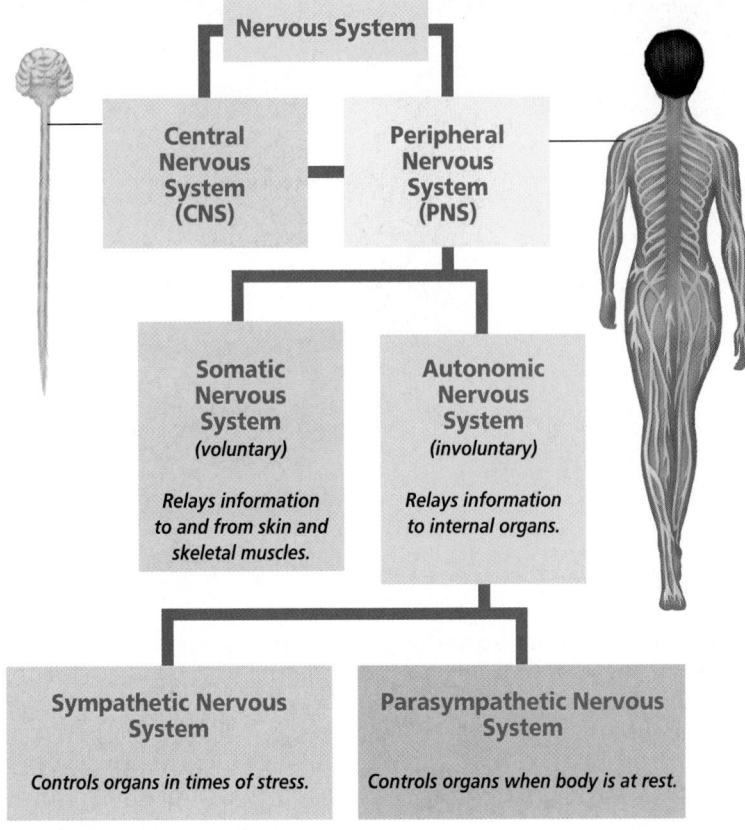

Nervous System

Central Nervous System (CNS)

Peripheral Nervous System (PNS)

Somatic Nervous System *(voluntary)*
Relays information to and from skin and skeletal muscles.

Autonomic Nervous System *(involuntary)*
Relays information to internal organs.

Sympathetic Nervous System
Controls organs in times of stress.

Parasympathetic Nervous System
Controls organs when body is at rest.

The different divisions and subsystems of your nervous system are summarized in *Figure 36.9.* Each division plays a key role in communication and control within your body.

Note that the sympathetic and parasympathetic systems are part of the autonomic nervous system. The autonomic and somatic systems are part of the peripheral nervous system. The peripheral nervous system feeds information to the central nervous system.

Section Assessment

Understanding Main Ideas

1. Summarize the charge distribution that exists inside and outside a resting neuron.
2. Outline the functions of the three major parts of the brain.
3. Contrast the functions of the two divisions of the autonomic nervous system.
4. How does the Na$^+$/K$^+$ pump affect ion distribution in a neuron?

Thinking Critically

5. Why is it nearly impossible to stop a reflex from taking place?

SKILL REVIEW

6. **Sequencing** Sequence the events as a nerve impulse moves from one neuron to another. For more help, refer to *Organizing Information* in the **Skill Handbook.**

Section Assessment

1. The inside of the neuron is more negatively charged than the outside.
2. The cerebrum controls conscious activities, intelligence, memory, language, movement, and the senses. The cerebellum controls balance, posture, and coordination. The medulla oblongata mainly controls involuntary activities.

3. The sympathetic nervous system controls functions in times of stress, while the parasympathetic controls functions while the body is at rest.
4. The pump moves three sodium ions out of the cell for every two potassium ions it pumps into the cell.
5. A reflex is an involuntary action that does not involve conscious control by

the brain. You cannot stop it.
6. An impulse reaches the end of an axon, opening calcium channels. This causes vesicles to fuse with the membrane and release their chemicals. These chemicals diffuse across the synaptic space to the dendrites of the next neuron.

P icture yourself in a park on a beautiful summer day. Stretching out on the grass, you look up and see white clouds float by against a background of blue. You feel grass tickling your toes and hear a breeze rustling the leaves of the trees. Sipping on a glass of lemonade, you note its tangy odor and sweet flavor. All these sensations are made possible by your senses: sight, touch, hearing, smell, and taste. Senses enable you to interpret your environment.

Magnification: 1040×

Taste buds (inset) respond to different chemical stimuli.

SECTION PREVIEW

Objectives

Define the role of the senses in the human nervous system.

Recognize how senses detect chemical, light, and mechanical stimulation.

Identify ways in which the senses work together to gather information.

Vocabulary
taste bud
retina
rods
cones
cochlea
semicircular canals

Sensing Chemicals

How are you able to smell and taste the lemonade? Chemical molecules of lemonade contact receptors in your nose and mouth as you sniff and drink the beverage. The receptors for smell are hairlike nerve endings located in the upper portion of your nose, as shown in *Figure 36.10.* Chemicals acting on these nerve endings initiate impulses in the olfactory nerve, which is connected to your brain. In the brain, this signal is interpreted as a particular odor.

The senses of taste and smell are closely linked. Think about what your sense of taste is like when your nose is stuffed up and you can smell

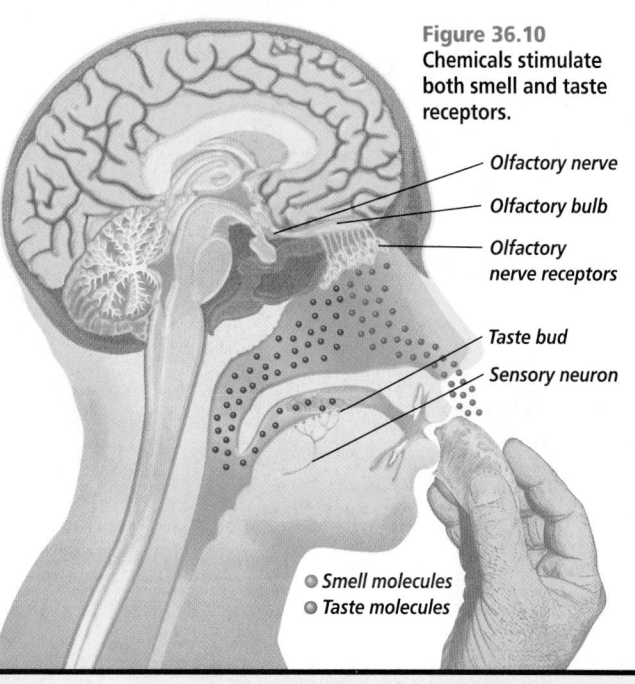

Figure 36.10
Chemicals stimulate both smell and taste receptors.

Olfactory nerve
Olfactory bulb
Olfactory nerve receptors
Taste bud
Sensory neuron

● Smell molecules
● Taste molecules

PROJECT

Modeling the Senses

 Kinesthetic Have student groups select and make a model of one of the senses. The model may demonstrate the anatomy or the function of the sense. Have students explain and demonstrate their models to the class. **L2** **ELL**
COOP LEARN

GLENCOE TECHNOLOGY

VIDEODISC
The Infinite Voyage *A Taste of Health, Developing Taste* (Ch. 3)
7 min. 30 sec.

Key Concepts

The anatomy and physiology of the major senses are presented. The senses that detect chemicals (taste, smell), the sense that detects light (vision), and the senses that detect mechanical stimulation (hearing, touch, and balance) are examined.

Planning

■ Make or purchase a tape of common sounds for the first Quick Demo.

■ Acquire models of ears, eyes, skin, and nose for the Display.

■ Order cow or sheep eyes for the *Inside Story.*

1 Focus

Bellringer

Before presenting the lesson, display **Section Focus Transparency 88** on the overhead projector and have students answer the accompanying questions. **L1** **ELL**

Transparency **88** Common Senses

SECTION FOCUS
Use with Chapter 36, Section 36.2

❶ List the senses that you think the person in this scene is using.

❷ Identify the stimulus for each of the senses you listed in question 1.

BIOLOGY: The Dynamics of Life   SECTION FOCUS TRANSPARENCIES

2 Teach

Concept Development

Ask students what kind of information the sense organs keep the body informed of. *They inform the body of changes that occur in the surroundings.* What is the reason for keeping the body informed? *So that it can respond to changes in the environment.*

Enrichment

TECH PREP Invite an audiologist to speak to the class about sound levels, how they are measured, and damage that may result from high levels. Have the audiologist demonstrate how hearing is tested. **L1**

Display

Visual-Spatial Obtain display models of the ear, eye, skin, or nose. Have students examine the models as they read about each sense. **L1 ELL**

✔ Assessment

Knowledge Have students write two questions about the sense of sight. Students should ask these questions of a classmate and then exchange questions with him or her. **L2**

984

little, if anything. Because much of what you taste depends on your sense of smell, your sense of taste may also be dulled. You taste something when chemicals dissolved in saliva contact sensory receptors on your tongue called **taste buds.** Signals from your taste buds travel to the cerebrum. There, the signal is interpreted, and you notice a particular taste.

Tastes that you experience can be divided into four basic categories: sour, salty, bitter, and sweet. Certain regions of your tongue react more strongly to particular categories. Bitter flavors are most likely to be sensed at the back of the tongue, sour on the sides, and sweet and salty on the tip.

Sensing Light

How are you able to see? Your sense of sight depends on receptors in your eyes that respond to light energy. The **retina,** found at the back of the eye, is a thin layer of tissue made up of light receptors and sensory neurons.

Figure 36.11
A cross section through the human eye shows the path light takes as it enters through the pupil.

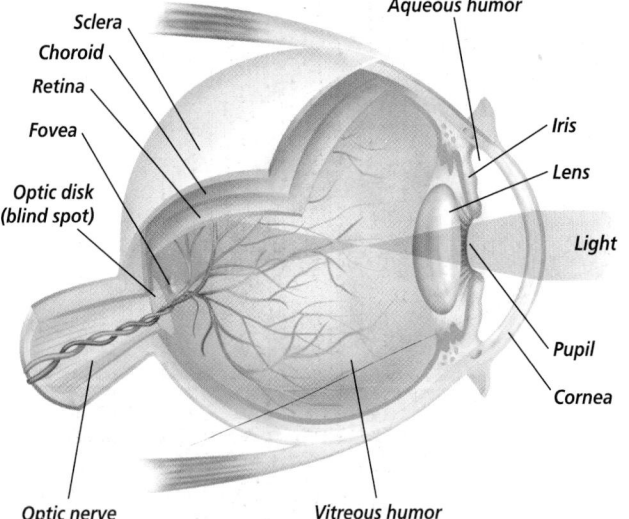

Sclera
Choroid
Retina
Fovea
Optic disk (blind spot)
Optic nerve
Aqueous humor
Iris
Lens
Light
Pupil
Cornea
Vitreous humor

Light enters the eye through the pupil and is focused by the lens onto the back of the eye, where it strikes the retina. Follow the pathway of light to the retina in *Figure 36.11.*

The retina contains two types of light receptor cells—rods and cones. **Rods** are receptor cells adapted for vision in dim light. They help you detect shape and movement. **Cones** are receptor cells adapted for sharp vision in bright light. They also help you detect color.

At the back of the eye, retinal tissue comes together to form the optic nerve, which leads to the brain, where images are interpreted. Can you see as well with one eye as with two? To find out more about how the brain forms a visual image, read the *Inside Story.*

Sensing Mechanical Stimulation

How are you able to hear the leaves rustle and feel the grass as you relax in the park? These senses—hearing and touch—depend on receptors that respond to mechanical stimulation.

Your sense of hearing

Every sound causes the air around it to vibrate. These vibrations travel outward from the source in waves, called sound waves. Sound waves enter your outer ear and travel down to the end of the ear canal, where they strike a membrane called the eardrum and cause it to vibrate. The vibrations then pass to three small bones in the middle ear—the malleus, the incus, and the stapes. As the stapes vibrates, it causes fluid in the **cochlea,** a snail-shaped structure in the inner ear, to move like a wave against the hair cells that line the cochlea's circular walls. Pressed by the fluid, the hairs bend.

The Eye

*T*he light energy that reaches your retina is converted into nerve impulses, which are interpreted by your brain, allowing you to see the world around you.

Critical Thinking *How would a person's vision be affected if his or her rod cells didn't function?*

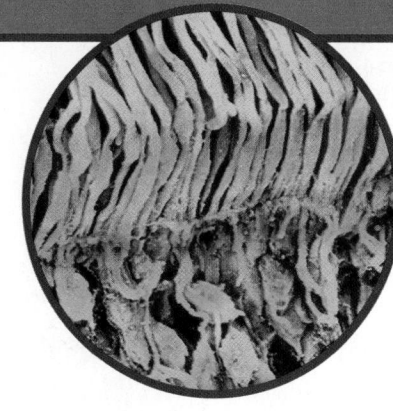

Rod and cone cells

① Rod and cone cells Rod cells in the retina are excited by low levels of light. These cells convert light signals into nerve impulses and relay them to the brain. Your brain interprets the information as a black and white picture. Your cone cells respond to bright light. They provide the brain with information about color.

② Visual field Close one eye. Everything you can see with one eye open is the visual field of that eye. The visual field of each eye can be divided into two parts: a lateral and a medial part. As shown, the lateral half of the visual field projects onto the medial portion of the retina, and the medial half of the visual field projects onto the lateral portion of the retina.

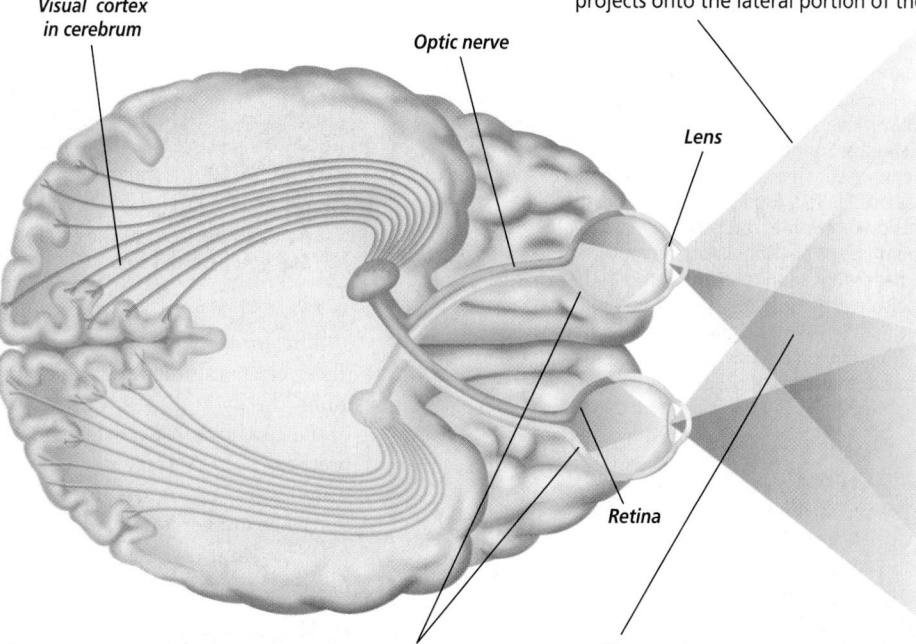

Visual cortex in cerebrum

Optic nerve

Lens

Retina

④ Brain image projections The right half of the retina in each eye is connected to the right side of the visual cortex in the cerebrum. The left half of the retina is similarly connected to the left side of the visual cortex. Thus images entering the eye from the right half of each visual field project to the left half of the brain, and vice versa.

③ Depth perception The visual fields of the eyes partially overlap, each eye seeing about two-thirds of the total field. This overlap allows your brain to judge the depth of your visual field.

36.2 THE SENSES **985**

Purpose

Students examine the function and structure of rods and cones and the divided projection of the visual field.

Teaching Strategies

■ Ask students which cells are most active when they are reading color comics. *cone cells*

■ Have students close one eye at a time and note the visual field they see out of each.

■ Ask students to explain how they are able to judge depths using both eyes together. **L2**

Visual Learning

■ Dissect a cow or sheep eye. Cow and sheep eyes have a layer on the inner choroid coat that is not present in humans. Explain that this iridescent layer enhances night vision by reflecting some light back into the retina.

Critical Thinking

Because rod cells function in low levels of light, the person would be virtually blind in dimly lit areas.

Resource Manager

Section Focus Transparency 88 and Master **L1** **ELL**
Basic Concepts Transparency 68 and Master **L2** **ELL**
Tech Prep Applications, p. 47 **L2**

Portfolio

Creating a Flowchart

Visual-Spatial Have students use Figures 36.11 and 36.12 to create flowcharts that show the series of events involved in seeing an object or hearing a sound. Encourage students to include the names of the eye, ear, and brain parts involved in these processes. **L2** **P**

GLENCOE TECHNOLOGY

VIDEODISC
Biology: The Dynamics of Life
Sense of Sight (Ch. 39)
Disc 2, Side 1, 30 sec.

Purpose

Students gain practice in determining whether statements are observations or inferences.

Process Skills

acquire information, define operationally, observe and infer, recognize cause and effect, think critically

Teaching Strategies

■ Refer students to the "Observing and Inferring" section in the Skill Handbook at the back of this text. The entire report can also be accessed on the Internet.

Thinking Critically

1. Student choices must include a specific number. A definition of the word *quantitative* should also be included in their answer.

2. Student choices should not include a specific number. A definition of the word *qualitative* should also be included.

3. Student may cite sentences that include "may be" or "can." A definition of the word *inference* should also be included in their answer.

4. Student answers may include the wearing of protective hearing devices and reducing volume or exposure time.

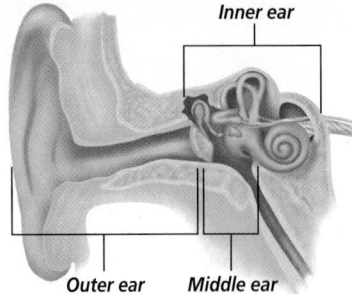

Figure 36.12 The internal structure of the human ear is divided into three areas: the outer ear, middle ear, and inner ear. Follow the pathway sound waves take as they move through your ear.

Inner ear

Outer ear   Middle ear

A Sounds waves strike a membrane called the eardrum and cause it to vibrate.

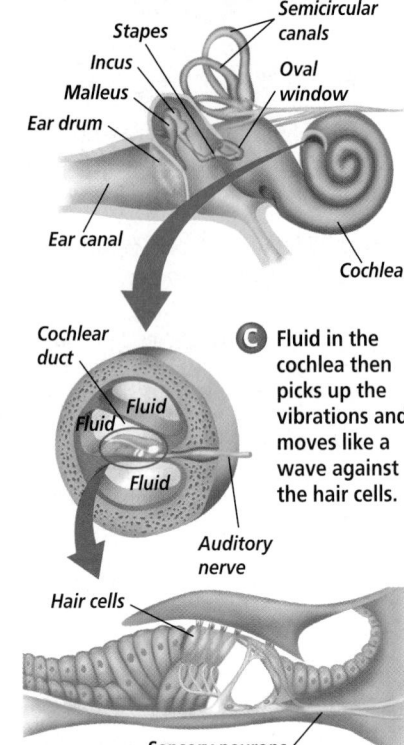

B The vibrations are passed on to three bones in the middle ear–the malleus, the incus, and the stapes.

Stapes
Semicircular canals
Incus
Oval window
Malleus
Ear drum
Ear canal
Cochlea

Cochlear duct
Fluid
Fluid
Fluid
Fluid
Auditory nerve
Hair cells
Sensory neurons

C Fluid in the cochlea then picks up the vibrations and moves like a wave against the hair cells.

D Impulses travel along the auditory nerve to the sides of the cerebrum, and you hear the sound.

Problem-Solving Lab 36-1
Interpreting and Analyzing

When are loud sounds dangerous to our hearing? Observations may be described as either qualitative or quantitative. A qualitative observation about a woman's height might be that she is tall. A quantitative observation about the same person might be that she is 1.92 m tall.

Analysis

"Hearing loss afflicts approximately 28 million people in the United States. Approximately 10 million of these impairments may be partially attributable to damage from exposure to loud sounds. Sounds that are sufficiently loud to damage sensitive inner ear structures can produce hearing loss that is not reversible. Very loud sounds of short duration, such as an explosion or gunfire, can produce immediate, severe, and permanent loss of hearing. Longer exposure to less intense but still hazardous sounds encountered in the workplace or during leisure activities, exacts a gradual toll on hearing sensitivity, initially without the victim's awareness. Live or recorded high-volume music, lawn-care equipment, and airplanes are examples of potentially hazardous noise."
"Noise and Hearing Loss," NIH Consensus Statement, January 22-24, 1990.

Thinking Critically

1. Choose and record two sentences or phrases from the passage above that provide examples of quantitative observations. Explain your selections.
2. Choose and record two sentences or phrases that provide examples of qualitative observations. Explain your selections.
3. Choose and record one sentence or phrase that provides an example of an inference. Explain your selection.
4. Suggest ways to minimize the type of noise exposure discussed in the last sentence.

The movement of the hairs produces electrical impulses, which travel along the auditory nerve to the sides of the cerebrum, where they are interpreted as sound. Trace the pathway of sound waves in *Figure 36.12*. To find out what impact loud sounds have on your hearing, do the *Problem-Solving Lab* on this page.

Your sense of balance

The inner ear also converts information about the position of your head into nerve impulses which travel to your brain, informing it about your body's equilibrium.

Maintaining balance is the function of your **semicircular canals.** Like the cochlea, the semicircular canals are also filled with a thick fluid and lined with hair cells. When you tilt your head, the fluid moves, causing the hairs to bend. This movement stimulates the hair cells to produce impulses. Neurons carry the impulses to the brain, which sends an impulse to stimulate your neck muscles and readjust the position of your head.

Your sense of touch

Like the ear, your skin also responds to mechanical stimulation with receptors that convert the stimulus into a nerve impulse. Receptors in the dermis of the skin respond to changes in temperature, pressure, and pain. It is with the help of these receptors, shown in *Figure 36.13,* that your body is able to respond to its external environment.

Although some receptors are found all over your body, those responsible for responding to particular stimuli are usually concentrated within certain areas of your body. For example, receptors that respond to light pressure are numerous in the dermis of your fingertips, eyelids, lips, the tip of your tongue, and the palms of your hands. When these receptors are stimulated, you perceive sensations of light touch.

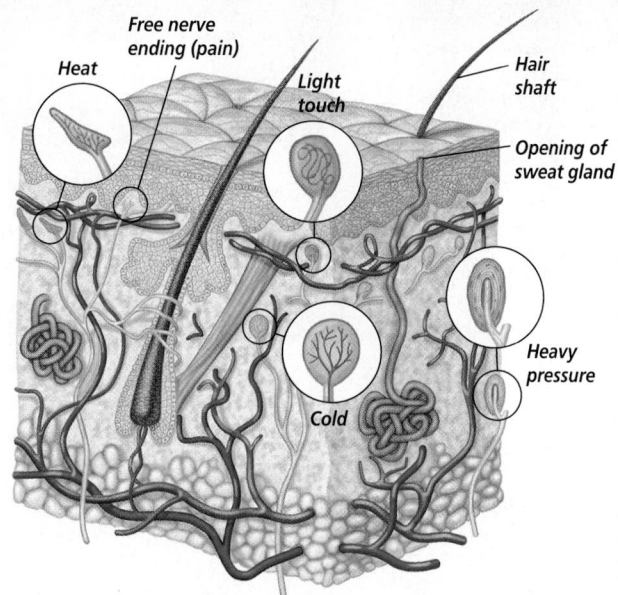

Receptors that respond to heavier pressure are found inside your joints, in muscle tissue, and in certain organs. They are also abundant on the skin of your palms and fingers and on the soles of your feet. When these receptors are stimulated, you perceive heavy pressure.

Free nerve endings extend into the lower layers of the epidermis. These nerve endings act as pain receptors. You have two other kinds of receptors that respond to temperature. Heat receptors are found deep in the dermis, while cold receptors are found closer to the surface of your skin.

Figure 36.13
Many kinds of receptors are located throughout the dermis of your skin. Some receptors detect gentle touches; others respond to heavy pressure. The sensations of pain, heat, and cold are sensed by other kinds of receptors.

Section Assessment

Understanding Main Ideas
1. Summarize the types of messages the senses receive.
2. When you have a cold, why is it hard to taste food?
3. Explain how your eyes detect light and images.
4. What types of receptors are found in the skin?

Thinking Critically
5. Why might an ear infection lead to problems with balance?

SKILL REVIEW

6. **Sequencing** List the sequence of structures through which sound waves pass to reach the auditory nerve. For more help, refer to *Organizing Information* in the **Skill Handbook.**

36.2 THE SENSES **987**

Section Assessment

1. The eyes respond to light. The ears respond to sound. Touch receptors respond to mechanical stimulation. The tongue and nose respond to chemicals.
2. The tasting of food involves both the sense of smell and the sense of taste.
3. Light stimulates the rod or cone cells in the retina, which transmit a signal to the brain by way of the optic nerve.
4. touch, temperature, pressure, and pain
5. Swelling associated with an ear infection could cause fluid in the ear to put pressure on the semicircular canals and cause the hairs in the canals to signal a false sense of balance in the brain.
6. outer ear, eardrum, malleus, incus, and stapes, fluid of cochlea, hairs of cochlea, auditory nerve to the brain

---

3 Assess

Check for Understanding
☑ *Visual-Spatial* Have students label diagrams of the eye, ear, and nose. **L2** **ELL**

Extension
Have students research and report on the causes and treatments of cataracts, glaucoma, or vertigo. **L3**

✔ Assessment
Knowledge Provide students with a list of sensations (such as pain, pressure, cold, odor, and so on) that can be detected by the body. Have them match each sensation with the sense organ that detects it. **L1**

4 Close

Discussion
Ask students to identify which of the following professions could be undertaken by someone who has lost the sense of sight: architect, violinist, mathematician, public speaker, professional athlete, accountant, physicist. What special tools (e.g., talking computers) would someone without sight have to use for each of the professions listed?

Resource Manager

Reinforcement and Study Guide, p. 161 **L2**
Content Mastery, p. 179 **L1**
Concept Mapping, p. 36 **L3** **ELL**
Basic Concepts Transparency 69 and Master **L2 **ELL**
Reteaching Skills Transparencies 52, 53 and Masters **L1 **ELL**
Laboratory Manual, pp. 265-268 **L2**

987

Prepare

Key Concepts

This section summarizes the medicinal uses of drugs and explains how addictive drugs affect the body. The major classes of misused and abused drugs are discussed.

Planning

- Bring in various nonprescription drugs for students to compare in the first Quick Demo.
- Have students bring in magazines for the third Biology Journal activity.
- Order *Daphnia* and purchase coffee, tea, cola, unfiltered cigarettes, and cough syrup with dextromethorphen hydrobromide for the BioLab.

1 Focus

Bellringer

Before presenting the lesson, display **Section Focus Transparency 89** on the overhead projector and have students answer the accompanying questions.
L1 ELL

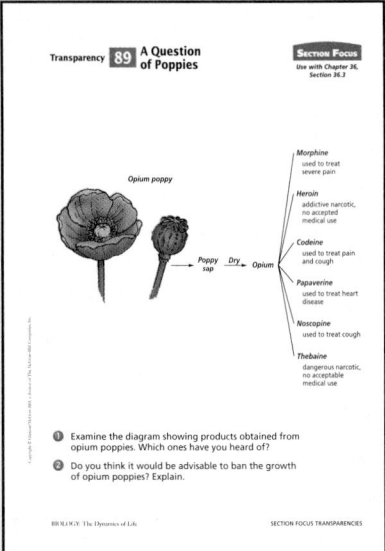

SECTION PREVIEW

Objectives

Recognize the medicinal uses of drugs.
Identify the different classes of drugs.
Interpret the effect of drug misuse and abuse on the body.

Vocabulary

drug
narcotic
stimulant
depressant
addiction
tolerance
withdrawal
hallucinogen

Section

36.3 The Effects of Drugs

Nerves frazzled? Feeling tense? For a person trying to quit a smoking habit, the desire for a cigarette is almost overpowering. Without the person realizing it, an occasional cigarette has grown into a pack-a-day habit. The body now cries out for nicotine. Why do people start smoking even though nicotine is so addictive?

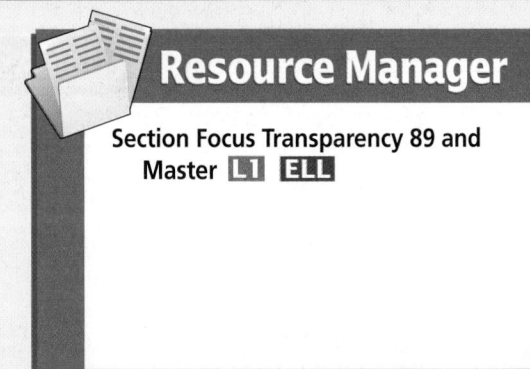

Cigarettes, made from tobacco leaves (top), contain the addictive drug nicotine.

Drugs Act on the Body

You probably hear the word *drug* used often, maybe even every day. A **drug** is a chemical that affects the body's functions. Most drugs interact with receptor sites on cells, probably the same ones used by neurotransmitters of the nervous system or hormones of the endocrine system. Some drugs increase the rate at which neurotransmitters are synthesized and released, or slow the rate at which they are broken down, as illustrated in *Figure 36.14.* Other drugs interfere with a neurotransmitter's ability to interact with its receptor. Explore how these different drugs work on neurotransmitters by doing the *Problem-Solving Lab* on the next page.

Figure 36.14
Drugs can increase neurotransmitter levels in the synapse by stimulating their synthesis, increasing their release, or by slowing their breakdown by enzymes.

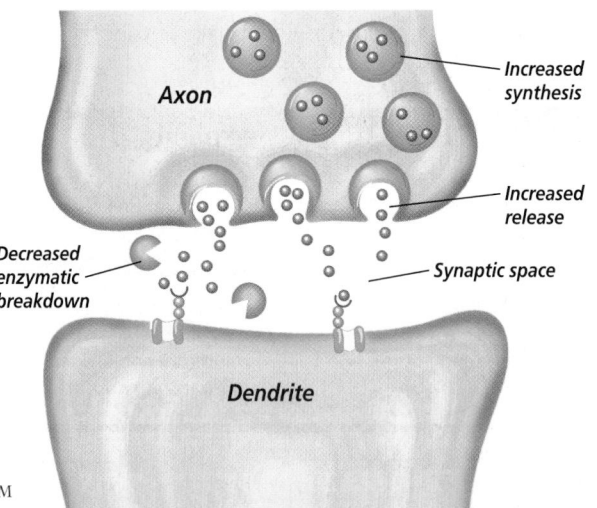

Axon

Increased synthesis

Increased release

Synaptic space

Decreased enzymatic breakdown

Dendrite

988 THE NERVOUS SYSTEM

MEETING INDIVIDUAL NEEDS

English Language Learners

Visual-Spatial Have students with limited English proficiency make a poster that summarizes the effects on different body systems of one drug that is misused or abused. Have students combine their posters to create a bulletin board display. **L1 ELL**

Resource Manager

Section Focus Transparency 89 and Master **L1 ELL**

Medicinal Uses of Drugs

A medicine is a drug that, when taken into the body, helps prevent, cure, or relieve a medical problem. Some of the many kinds of medicines used to relieve medical conditions are discussed below.

Relieving pain

Headache, muscle ache, cramps—all are common pain sensations. You just studied how pain receptors in your body send signals to your brain. Medicines that relieve pain manipulate either the receptors that initiate the impulses or the central nervous system that receives them.

Pain relievers that do not cause a loss of consciousness are called analgesics. Some analgesics, like aspirin, work by inhibiting receptors at the site of pain from producing nerve impulses. Analgesics that work on the central nervous system are called **narcotics.** Many narcotics are made from the opium poppy flower, shown in *Figure 36.15*. Opiates, as they are called, can be useful in controlled medical therapy because only these drugs are able to relieve severe pain.

Figure 36.15
Sticky sap from the fruit of an opium poppy is used to make drugs called opiates.

How do different drugs affect the levels of neurotransmitters in synapses? Drugs can act on neurotransmitters in a number of different ways. For example, they may speed up the reabsorption of a neurotransmitter back into the dendrite end of a neuron. Or, they may block the release of the neurotransmitter from the axon end of a neuron. They may also prevent the breakdown of the neurotransmitter by blocking the enzyme responsible for this action.

Analysis

Examine the diagram shown here, which illustrates how neurotransmitters work.

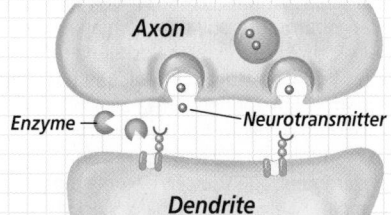

Axon

Enzyme — Neurotransmitter

Dendrite

Thinking Critically

1. Design three different drugs:
   a. One that will block the enzyme from breaking down the neurotransmitter.
   b. One that will prevent the neurotransmitter from reaching the receptor site on the dendrite.
   c. One that will block the release of the neurotransmitter from the axon.
2. Draw three separate figures to show how each of your drugs works.
3. Predict the effects of each drug on the body. Explain your answer.

Purpose

Students design drugs that will interfere with the action of neurotransmitters.

Process Skills

apply concepts, formulate models, hypothesize, interpret scientific illustrations, predict, recognize cause and effect, think critically

Teaching Strategies

■ Review the nature of the synapse and how neurotransmitters work normally.
■ Encourage students to consider the shapes of the neurotransmitter molecules, enzymes, and binding sites when designing their blockers.
■ Allow students to work in small groups.

Thinking Critically

1. Student models will vary. Make sure that each one produces the effect described.
2. Student figures should indicate how each of their drugs works.
3. (a) Message will be transmitted for a longer period of time than normal; (b) Message will not be received by dendrite; (c) Message will not be delivered to dendrite.

✔ Assessment

Knowledge Ask students to make a diagram showing some type of neurotransmitter blockage. Ask them to interpret the drawing. Use the Performance Task Assessment List for Scientific Drawing in **PASC**, p. 55. **L2**

BIOLOGY *Online* Note Internet addresses that you find useful in the space below for quick reference.

_____
_____
_____
_____

GLENCOE TECHNOLOGY

VIDEODISC
The Infinite Voyage *Prisoners of the Brain: Neurotransmitters and the Chemical Basis of Mental Illness* (Ch. 3), 3 min.

Dopamine and Antipsychotic Drugs (Ch. 4), 7 min.

Using PCP to Study Schizophrenia (Ch. 5), 7 min. 30 sec.

CAREERS IN BIOLOGY

Pharmacist

Would you like to help people get well, but can't stand the sight of blood? Then consider a career as a pharmacist.

Skills for the Job

Pharmacists read prescriptions written by doctors and other health professionals and carefully prepare containers with the correct medicine. (Few pharmacists still mix the medicine themselves; that is done by the drug manufacturer.) Pharmacists must know how drugs interact and guide people in avoiding harmful combinations. They also help customers select over-the-counter medicines. Besides drugstores, pharmacists work in hospitals and nursing homes, and for drug companies and government agencies. To become a pharmacist, you must complete a five-year bachelor's degree in pharmacy. You must also pass a state examination.

BIOLOGY Online

For more careers in related fields, be sure to check the Glencoe Science Web site. **science.glencoe.com**

 **Origin**

cardiovascular
From the Greek words *kardia*, meaning "the heart," and *vasculum*, meaning "small vessel." Cardiovascular drugs treat problems associated with blood vessels of the heart.

Treating circulatory problems

Many drugs have been developed to treat heart and circulatory problems such as high blood pressure. These medicines are called cardiovascular drugs. In addition to treating high blood pressure, cardiovascular drugs may be used to normalize an irregular heartbeat, increase the heart's pumping capacity, or enlarge small blood vessels. Discover how various types of drugs can affect heart rate by doing the *BioLab* at the end of this chapter.

Treating nervous disorders

Several kinds of medicines are used to help relieve symptoms of nervous system problems. Among these medicines are stimulants and depressants.

Drugs that increase the activity of the central and sympathetic nervous systems are called **stimulants.** Amphetamines (am FET uh meenz)

are synthetic stimulants that increase the output of CNS neurotransmitters. Amphetamines are seldom prescribed because they can lead to dependence, which you'll read more about later in this chapter. However, because they increase wakefulness and alertness, amphetamines are sometimes used to treat patients with sleep disorders.

Drugs that lower, or depress, the activity of the nervous system are called **depressants,** or sedatives. The primary medicinal uses of depressants are to encourage calmness and produce sleep. For some people, the symptoms of anxiety are so extreme that they interfere with the person's ability to function effectively. By slowing down the activities of the CNS, a depressant can temporarily relieve some of this anxiety.

The Misuse and Abuse of Drugs

The misuse or abuse of drugs can cause serious health problems—even death. Drug misuse occurs when a medicine is taken for an unintended use. For example, giving your prescription medicine to someone else, not following the prescribed dosage by taking too much or too little, and mixing medicines, are all instances of drug misuse. You must pay careful attention to the specific instructions given on the label of a drug you are taking. The *MiniLab* on the next page shows you how to analyze such a label.

Drug abuse is the inappropriate self-administration of a drug for non-medical purposes. Drug abuse may involve use of an illegal drug, such as cocaine; use of an illegally obtained medicine, such as someone else's prescribed drugs; or excessive use of a legal drug, such as alcohol or nicotine. Drugs abused in this way can

have powerful effects on the nervous system and other systems of the body, as described in *Figure 36.16*.

Addiction to drugs

When a person believes he or she needs a drug in order to feel good or function normally, that person is psychologically dependent on the drug. When a person's body develops a chemical need for the drug in order to function normally, the person is physiologically dependent. Psychological and physiological dependence are both forms of **addiction.**

Tolerance and withdrawal

When a drug user experiences tolerance or withdrawal to a frequently used drug, that person is addicted to the drug. **Tolerance** occurs when a person needs larger or more frequent doses of a drug to achieve the same effect. The dosage increases are necessary because the body becomes less responsive to the drug. **Withdrawal** occurs when the person stops taking the drug and actually becomes ill.

Figure 36.16
The use of anabolic steroids without careful guidance from a physician is illegal. Some dangerous side effects of steroid abuse include cardiovascular disease, kidney damage, and cancer.

MiniLab 36-2 Analyzing Information

Interpret a Drug Label One common misuse of drugs is not following the instructions that accompany them. Over-the-counter medicines can be harmful—even fatal—if they are not used as directed. The Food and Drug Administration requires that certain information about a drug be provided on its label to help the consumer use the medicine properly and safely.

Procedure

1. The photograph below shows a label from an over-the-counter drug. Read it carefully.
2. Make a data table like the one shown. Then fill in the table using information on the label.

Information from a drug label				
People with these conditions should avoid this drug	Possible side effects	This drug should not be taken with these medicines	Symptoms this drug will relieve	Correct dosage

Analysis

1. What is a side effect? What side effects are caused by this drug?
2. Why should a person never take more than the recommended dosage?
3. How are over-the-counter drugs different from prescription drugs?

Purpose
Students study information found on the labels of over-the-counter drugs.

Process Skills
observe and infer, analyze

Teaching Strategies
■ Have students look at the package labels of other over-the-counter drugs.

Expected Results
Student tables should show the following information: among others, children under 12 years, people with high blood pressure, heart disease, diabetes, thyroid disease, asthma, or emphysema should avoid this drug; drowsiness is a possible side effect; should not be taken with antihypertensive or antidepressant drugs; will relieve nasal congestion associated with the common cold, hay fever, or sinusitis; the correct dosage is one tablet every 12 hours.

Analysis

1. A side effect is any effect of the drug other than the one it is designed to produce. Side effects of this drug include drowsiness.
2. The recommended dosage is the one that has been tested and proven to be safe. Over-the-counter medicines are potentially harmful if not used as directed.
3. Over-the-counter drugs are available without a doctor's prescription. They are not as strong as prescription medicines.

✔ **Assessment**
Performance Ask students to repeat the MiniLab for other drug labels and include the tables in their portfolios. Use the Performance Task Assessment List for Carrying Out a Strategy and Collecting Data in **PASC,** p. 25. **L2** **ELL**

Cultural Diversity

Solomon Carter Fuller

Students may obtain a more thorough understanding of brain function by learning about common neuropathologies, such as Alzheimer's disease. During your discussions, emphasize the work of African American psychiatrist and researcher, Solomon Carter Fuller (1872–1953).

Fuller taught pathology, neurology, and psychiatry at Boston University for nearly 40 years. He was best known for expanding our medical knowledge in the fields of neuropathology and psychiatry. His research on degenerative diseases of the brain—including Alzheimer's disease—was considered pioneering work. In 1913, Fuller became the editor of the *Westborough State Hospital Papers,* an influential publication specializing in mental diseases.

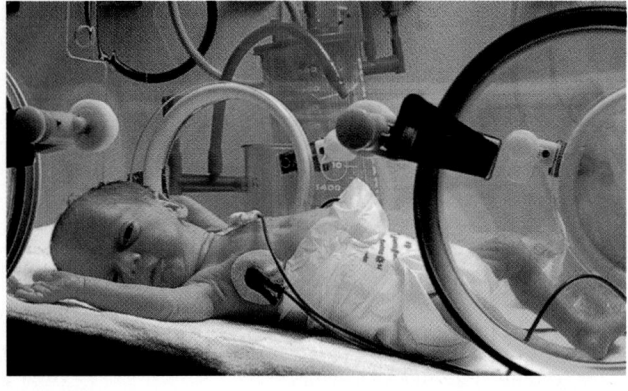

Figure 36.17
Babies born addicted to crack cocaine are usually low in birth weight, continually irritable, and may shake constantly.

Figure 36.18
Caffeine can trigger a condition called tachycardia, when the heart beats more than 100 times per minute.

Classes of Commonly Abused Drugs

Each class of drug produces its own special effect on the body, and its own particular symptoms of withdrawal.

Stimulants: Cocaine, amphetamines, caffeine, and nicotine

You already know that stimulants increase the activity of the central nervous system and the sympathetic nervous system. Increased CNS stimulation can result in mild elevation of alertness, increased nervousness, anxiety, or even convulsions.

Cocaine stimulates the CNS by working on the part of the inner brain that governs emotions and basic drives, such as hunger and thirst. When these needs are met under normal circumstances, neurotransmitters—such as dopamine—are released to reward centers and the person experiences pleasure. Cocaine artificially increases levels of these neurotransmitters in the brain. As a result, false messages are sent to reward centers indicating that a basic drive has been satisfied. The user quickly feels a euphoric high called a rush. This sense of intense pleasure and satisfaction cannot be maintained, however, and soon the effects of the drug change. Physical hyperactivity follows, and the user is unable to sit still. Often, anxiety and depression set in.

Cocaine also disrupts the body's circulatory system by interfering with the sympathetic nervous system. Although initially causing a slowing of the heart rate, it soon produces a great increase in heart rate and a narrowing of blood vessels, known as vasoconstriction. The result is high blood pressure. Heavy use of this drug compromises the immune system and often leads to heart abnormalities. Cocaine affects more than just the people who use it. As *Figure 36.17* shows, babies of addicted mothers are sometimes born already dependent on this drug.

As you've already learned, amphetamines are stimulants that increase levels of CNS neurotransmitters. Like cocaine, amphetamines also cause vasoconstriction, a racing heart, and increased blood pressure. Other adverse side effects of amphetamine abuse include irregular heartbeat, chest pain, paranoia, hallucinations, and convulsions.

Not all stimulants are illegal. As shown in *Figure 36.18,* one stimulant in particular is as close as the nearest coffee maker or candy machine. Caffeine—a substance found in coffee, cola-flavored drinks,

992 THE NERVOUS SYSTEM

cocoa, and tea—is a CNS stimulant. Its effects include increased alertness and some mood elevation. Caffeine also causes an increase in heart rate and urine production, which can lead to dehydration.

Nicotine, a substance found in tobacco, also is a stimulant. By increasing the release of the hormone epinephrine (adrenaline), nicotine increases heart rate, blood pressure, breathing rate, and stomach acid secretion. Although nicotine is the addictive substance in tobacco, it is only one of about 3000 known chemicals found in cigarettes, many of which are also harmful. Smoking cigarettes is legal for adults. But our ever-increasing knowledge of the effects of smoke not only on the bodies of smokers but on nonsmokers as well has made cigarette smoking an increasingly unaccepted social habit.

Depressants: Alcohol and barbiturates

As you already know, depressants slow down the activities of the CNS. All CNS depressants relieve anxiety, but most also produce noticeable sedation.

One of the most widely abused drugs in the world today is alcohol. Easily produced from various grains and fruits, such as those shown in *Figure 36.19,* this depressant is distributed throughout a person's body via the bloodstream.

Unlike other drugs that act on specific receptors, alcohol probably acts on the brain by dissolving through the membranes of neurons. Once inside a neuron, alcohol disrupts important cellular functions. For instance, it appears to block the movement of sodium and calcium ions, which are important in the transmission of impulses and the release of neurotransmitters.

Figure 36.19
Most spirits are made from fermented grains such as wheat, barley, and rye. Wine is made from grapes, and cider from apples.

Tolerance to the effects of alcohol develops as a result of heavy consumption. Addiction to alcohol—alcoholism—can cause the destruction of nerve cells and brain damage. A number of organ diseases are directly attributable to chronic alcohol use. For example, cirrhosis, a hardening of the tissues of the liver, is a common affliction of alcoholics. *Figure 36.20* shows another tragic outcome of alcoholism.

Figure 36.20
Alcohol depresses the nervous system, impairing sensory perception and delaying reaction time.

Bioethics

Part of the procedure in getting a drug approved by the Food and Drug Administration (FDA) involves human drug tests. The deaths of five out of 15 participants in National Institute of Health (NIH) drug trial in the summer of 1993 has raised questions as to the safety of these tests. Even though participants sign a consent form, desperately ill patients may be willing to place themselves in inordinately dangerous situations in hopes of helping to find a cure. As a result, researchers question whether these patients can be truly "informed" about the risks involved.

Ask students to research what consenting to drug testing by the FDA involves. Have students present their findings in a written report. **L3**

Portfolio

Alcohol in the Body

Linguistic Ask students to write a skit about the travel route of alcohol in the body—from the mouth, into the bloodstream, and to the brain and liver. Have them place the written lines and directions for the skits in their portfolios.

L2 **P**

994

Enrichment

Intrapersonal Have interested students research how the following toxins affect the nervous system and what they are used for: saxitoxin (from red tide), physostigmine, alpha-bungarotoxin, tetrodotoxin, and diisopropyl fluorophosphate. **L3**

Concept Development

Intrapersonal Have students find out how computers are allowing pharmacologists to design new drugs. **L3**

3 Assess

Check for Understanding

Linguistic Have students make a list of stimulants and depressants and write a paragraph about how these substances affect the body. **L2**

Reteach

Linguistic Have students write a paragraph explaining how nicotine replacement therapy helps smokers to stop smoking. **L2**

Extension

Have students research information about genetic susceptibility to alcoholism. **L2**

Barbiturates (bar BIHCH uh ruts) are sedatives and anti-anxiety drugs. When barbiturates are used in excess, the user's respiratory and circulatory systems become depressed. Chronic use results in both tolerance and addiction.

Narcotics: Opiates

Most narcotics are opiates, that is, they are derived from the opium poppy. They act directly on the brain. The most abused narcotic in the United States is heroin. It depresses the CNS, slows breathing, and lowers heart rate. Tolerance develops quickly, and withdrawal from heroin is painful.

Hallucinogens: Natural and synthetic

Natural hallucinogens have been known and used for thousands of years, but the abuse of hallucinogenic drugs did not become widespread in the United States until the 1960s, when new synthetic versions became widely available.

Hallucinogens (huh LEWS un uh junz) stimulate the CNS—altering moods, thoughts, and sensory perceptions. Quite simply, the user sees, hears, feels, tastes, or smells things that are not actually there. This disorientation can impair the user's judgement and place him or her in some potentially dangerous situations. Hallucinogens also increase heart rate, blood pressure, respiratory rate, and body temperature, and sometimes cause sweating, salivation, nausea, and vomiting. After large enough doses, convulsions of the body may even occur.

Unlike the hallucinogens shown in **Figure 36.21,** LSD—or acid—is a synthetic drug. The mechanism by which LSD produces hallucinations is still debated, but it may involve the blocking of a CNS neurotransmitter.

Figure 36.21
Some hallucinogens are found in nature.

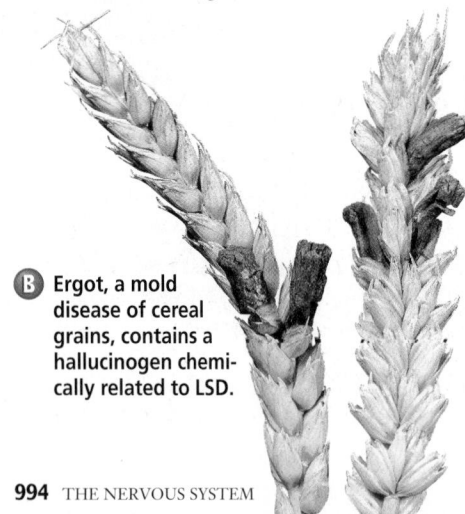

A Mushrooms of the genus *Psilocybe* contain the CNS hallucinogen psilocybin. These mushrooms are considered sacred by certain Native American tribes, who use them in traditional religious rites.

B Ergot, a mold disease of cereal grains, contains a hallucinogen chemically related to LSD.

BIOLOGY JOURNAL

Evaluating Advertising

Interpersonal Ask students to cut out cigarette ads from magazines. Post the ads on a bulletin board so all students can view them. In groups, have students discuss who the ads are likely to influence (the targeted audience). Ask students to select one ad and discuss its effectiveness. **L2**

Breaking the Habit

Once a person has become addicted to a drug, breaking the habit can be very difficult. Recall that an addiction can involve both physiological and psychological dependencies. Besides the desire to break the addiction, studies have shown that people usually need both medical and psychological therapy—such as counseling—to be successful in their treatment. Support groups such as Alcoholics Anonymous allow addicts to share their experiences in an effort to maintain sobriety. Often, people going through the same recovery are able to offer the best support.

Nicotine replacement therapy

Nicotine replacement therapy is one example of a relatively successful drug treatment approach. People who are trying to break their addiction to tobacco often go through stressful withdrawals when they stop smoking cigarettes. To ease the intensity of the withdrawal symptoms, patients wear adhesive patches that slowly release small amounts of nicotine into their bloodstream, as shown in *Figure 36.22.* Alternatively, pieces of nicotine-containing gum are chewed periodically to temporarily relieve cravings.

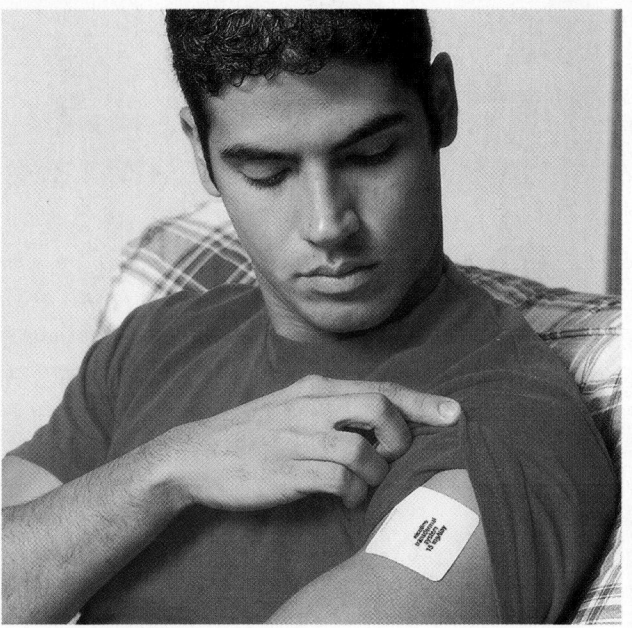

Figure 36.22
To help break an addiction to tobacco, this patient is wearing a patch on his arm that releases small amounts of nicotine directly into his bloodstream.

Nicotine inhalers—similar to asthma inhalers—provide immediate relief. By gradually decreasing the amount of nicotine released by the patches—or the number of gum pieces chewed—cigarette smokers are able to minimize the withdrawal symptoms that often result in a failure to quit.

Section Assessment

Understanding Main Ideas
1. In what ways can drugs be used to treat a cardiovascular problem?
2. What is the difference between aspirin and a narcotic?
3. How does nicotine affect the body?
4. How can drugs affect levels of neurotransmitters between neurons?

Thinking Critically
5. Suggest why a physician but not a pharmacist is legally permitted to write prescriptions.

SKILL REVIEW
6. **Comparing and Contrasting** Distinguish between stimulants and depressants, comparing their effects on the body. For more help, refer to *Thinking Critically* in the **Skill Handbook.**

Section Assessment

1. Drugs are used to normalize an irregular heartbeat, increase the heart's pumping capacity, or enlarge small blood vessels.
2. Aspirin inhibits the production of impulses at the site of pain, while narcotics work on the central nervous system to relieve pain.
3. Nicotine will stimulate the central nervous system, causing an increase in heart rate, blood pressure, and breathing rate.
4. Drugs can increase or decrease the amount of neurotransmitters found between neurons.
5. A physician knows the medical history of the patient and can treat problems that might occur as side effects of the drug.
6. Stimulants increase the activity of the central and sympathetic nervous systems. Depressants decrease the activity of the central nervous system and increase activity of the parasympathetic nervous system. Stimulants can increase alertness, nervousness, anxiety, heart rate, and breathing rate. Depressants do the opposite. **995**

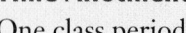

Time Allotment

One class period

Process Skills

observe and infer, form a hypothesis, communicate, predict, interpret data, experiment, and analyze

Safety Precautions

Caution students not to drink any of the solutions tested and advise them to wash their hands at the conclusion of the lab.

PREPARATION

Solutions

- Ethyl alcohol—add 2 mL of ethyl alcohol to 98 mL of distilled water.
- Nicotine—soak an unfiltered cigarette in 100 mL of warm distilled water for one hour; then filter the solution.
- Prepare weak solutions of coffee and tea.
- Dilute cola l part water to 1 part cola.
- Cough medicine—add 2 mL of cough medicine to 98 mL of distilled water.

Possible Hypotheses

- Students' hypotheses should categorize each drug according to whether it will increase, decrease, or not affect heart rate. Coffee, tea, cola, tobacco, and possibly cough medicine are stimulants and will increase heart rate. Ethyl alcohol and possibly cough medicine (with dextromethorphen hydrobromide) are depressants and will decrease heart rate.

What drugs affect the heart rate of *Daphnia?*

Depending on their chemical composition, drugs affect different parts of your body. Stimulants and depressants are drugs that affect the central nervous system and the autonomic nervous system. Stimulants increase the activity of the sympathetic nervous system, which is responsible for the fight-or-flight response. They cause an increase in your breathing rate and in your heart rate. Depressants, on the other hand, decrease the activity of the sympathetic nervous system, reducing your breathing and heart rates.

PREPARATION

Problem

What legally available drugs are stimulants to the heart? What legal drugs are depressants? Because these drugs are legally available, are they less dangerous?

Hypotheses

Based on what you learned in this chapter, which of the drugs listed under Possible Materials do you think are stimulants? Which are depressants? How will they affect the heart rate in *Daphnia?* Make a hypothesis concerning how each of the drugs listed will affect heart rate.

Objectives

In this BioLab, you will:
- **Measure** the resting heart rate in *Daphnia.*
- **Compare** the resting heart rate with the heart rate when a drug is applied.

Possible Materials

aged tap water
Daphnia culture
dilute solutions of coffee, tea, cola, ethyl alcohol, tobacco, and cough medicine (containing dextromethorphan)
dropper
microscope
microscope slide

Safety Precautions

Do not drink any of the solutions used in this lab. Always wear goggles in the lab. Use caution when working with a microscope, microscope slides, and glassware.

Skill Handbook

Use the **Skill Handbook** if you need additional help with this lab.

996 THE NERVOUS SYSTEM

PLAN THE EXPERIMENT

Teaching Strategies

- Age tap water by leaving it in a beaker overnight.
- Set up a distribution station for all the solutions being tested.
- *Daphnia* that are placed in the aged tap water after being tested with a drug can be reused again later.

Possible Procedures

- Students should measure the resting heart rate of each *Daphnia* before adding several drops of one of the solutions and measuring the heart rate again.

PLAN THE EXPERIMENT

1. Using a dropper, place a single *Daphnia* crustacean on a slide.
2. Observe the animal on low power and find its heart.
3. Design an experiment to measure the effect on heart rate of four of the drug-containing substances in the Possible Materials list.
4. Design and construct a data table for recording your data.

Check the Plan

1. Be sure to consider what you will use as a control.
2. Plan to add two drops of a drug-containing substance directly to the slide.
3. When you are finished testing one drug, you will need to flush the used *Daphnia* with the solution into a beaker of aged tap water provided by your teacher. Plan to use a new *Daphnia* for each substance tested.

4. *Make sure your teacher has approved your experimental plan before you proceed further.*
5. Carry out your experiment. **CAUTION:** *Wash your hands with soap and water immediately after making observations.*

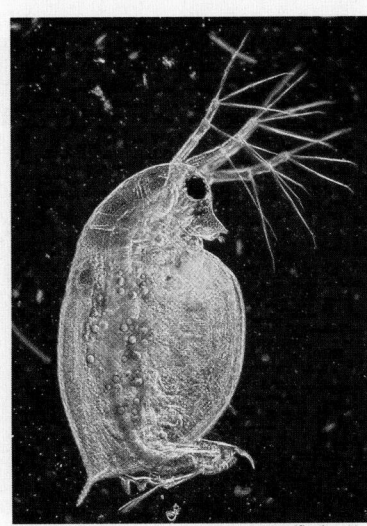

Magnification: 30×

ANALYZE AND CONCLUDE

1. **Making Inferences** Which drugs are stimulants? Which are depressants?
2. **Checking Your Hypotheses** Compare your predicted results with the experimental data. Explain whether or not your data support your hypotheses regarding the drugs' effects.
3. **Drawing Conclusions** How do the drugs affect the heart rate of this animal?
4. **Analyzing the Procedure** How would you alter your experiment if you did it again?

Going Further

Changing Variables Many other over-the-counter drugs are available. You may wish to test their effect on the heart rate of *Daphnia*.

BIOLOGY *Online* To find out more about drug effects, visit the Glencoe Science Web site.
science.glencoe.com

ANALYZE AND CONCLUDE

1. Stimulants are coffee, tea, cola, and tobacco. Cough medicine may also be listed. Depressants are ethyl alcohol and cough medicine if it contains dextromethorphen hydrobromide.
2. Some students' hypotheses will be confirmed by their data; others' will be rejected.
3. Stimulants speed up the animal's heart rate. Depressants slow the heart rate.
4. Answers will vary.

Error Analysis

Advise students they must have only a small amount of water on each slide with their *Daphnia*. If they have too much water, it will dilute the solutions tested.

✔ Assessment

Performance Have students each prepare a laboratory report that includes the experimental plan, the data table, and the answers to Analyze and Conclude, to be placed in their journals. Use the Performance Task Assessment List for Lab Report in **PASC**, p. 47. **L2**

Going Further

Have students compare the effects of four over-the-counter cough medicines on the heart rate of *Daphnia*. They should carefully make equal dilutions of each medicine after determining the concentration of the active drug in each (listed on the package). Tablet cough medications list milligrams of medication in each tablet. **L2**

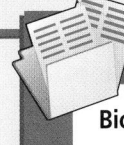

Resource Manager

BioLab and MiniLab Worksheets,
p. 161-162 **L2**

Data and Observations

Drug	Heart rate/min
No drug	240
Coffee	270
Cola	270
Tea	260
Ethyl alcohol	215
Tobacco	300
Cough medicine	heart rate varies with brand

Purpose

Students learn how radioactive isotopes are used in positron emission tomography (PET).

Background

■ To make use of PET scanners, patients are injected with short-lived radioisotopes such as C-11, N-13, or O-15. As the radioisotope circulates through the body, it emits positively charged particles called *positrons*. The positrons collide with electrons in body tissues, causing the release of gamma rays that are detected by PET receptors.

Teaching Strategies

■ PET scanners can be used to measure activities that involve circulation of chemicals or chemical reactions in the body. Ask students to list some body activities that doctors might be able to monitor with PET scanners.

Investigating the Technology

Thinking Critically Any place where blood flows and metabolic activities occur. Students may mention different organ systems.

Going Further

Students can find out how PET scans are used to distinguish between the two types of breast tumor: those that have estrogen receptors and those that do not have estrogen receptors.

 **Scanning the Mind**

Advancements in medical technology have led to instruments—such as X-ray and magnetic resonance imaging (MRI) machines—that can examine the human body in a noninvasive way. Recently, another technology has been added to the medical toolbox— positron emission tomography (PET). This instrument is unique in that it allows a physician to view internal body tissues while they carry out their normal daily functions.

PET scanners are excellent tools for studying the human brain. By monitoring either the blood flow to an area or the amount of glucose being metabolized there, doctors are able to pinpoint active sections of the brain.

Here's how it works: The patient is injected with a compound containing radioactive isotopes. Because these isotopes emit detectable radiation, they can be tracked by the sensitive PET scanner. Computers create a picture of brain activity by converting the energy emitted from the radioisotopes into a colorful map. The image indicates the location of an activity, such as glucose utilization, and its relative intensity in various regions.

Applications for the Future

PET scanners are important in brain research, including the detection and diagnosis of brain tumors, the evaluation of damage due to stroke, and the mapping of brain functions.

PET scans can also be used to see how learning takes place in the brain. The images on this page show activity in the left and right brains of two different people. Each person was given a list of nouns and asked to visualize them. The unpracticed brain (top) had no previous experience with this exercise and thus was forced to engage in a high level of brain activity to perform the task. The practiced brain (bottom), by comparison, was able to picture the words with much less brain activity. Biologists can discover functions of different parts of the brain and their roles in learning.

PET scans are also proving useful in the study of drug and alcohol addiction. Addicts can be given the addictive drug and then asked questions

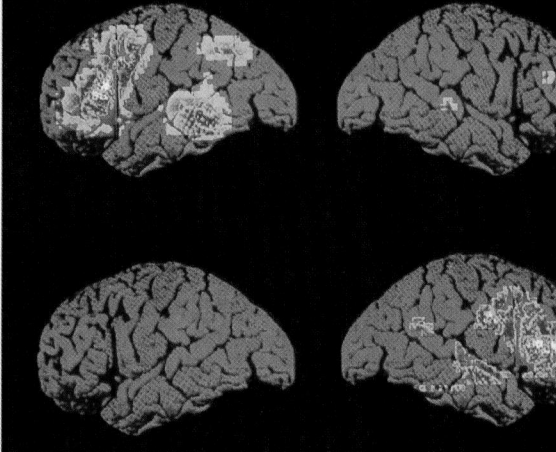

PET scans

about their physical and emotional status while the scanner records metabolic activity in the brain. Researchers hope that information gained from the study of drug addiction will provide help in diagnosing and treating manic-depressive psychosis and schizophrenia.

INVESTIGATING THE TECHNOLOGY

Thinking Critically What other body systems might be examined with the aid of a PET scanner?

 To find out more about PET scans, visit the Glencoe Science Web site.
science.glencoe.com

GLENCOE TECHNOLOGY

 VIDEODISC
The Infinite Voyage
Fires of the Mind: Positron Emission Tomography (PET) (Ch. 4)
3 min.

SUMMARY

Section 36.1

The Nervous System

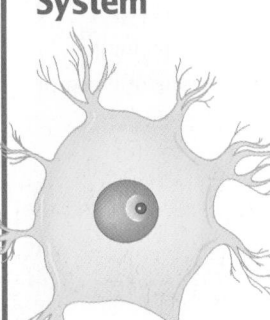

Main Ideas

- The neuron is the basic structural unit of the nervous system. Impulses move along a neuron in a wave of changing charges.
- The central nervous system consists of the brain and spinal cord.
- The peripheral nervous system relays messages to and from the central nervous system. It consists of the somatic and autonomic nervous systems.

Vocabulary

autonomic nervous system (p. 981)
axon (p. 973)
central nervous system (p. 976)
cerebellum (p. 977)
cerebrum (p. 977)
dendrite (p. 973)
medulla oblongata (p. 977)
neuron (p. 973)
neurotransmitter (p. 976)
parasympathetic nervous system (p. 981)
peripheral nervous system (p. 977)
reflex (p. 980)
somatic nervous system (p. 980)
sympathetic nervous system (p. 981)
synapse (p. 976)

Section 36.2

The Senses

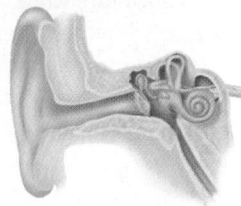

Main Ideas

- Taste and smell are senses that respond to chemical stimulation.
- Sight is a sense that responds to light stimulation.
- Hearing, balance, and touch are senses that respond to mechanical stimulation.

Vocabulary

cochlea (p. 984)
cones (p. 984)
retina (p. 984)
rods (p. 984)
semicircular canals (p. 987)
taste bud (p. 984)

Section 36.3

The Effects of Drugs

Main Ideas

- Drugs act on the body's nervous system.
- Some medicinal uses of drugs include relieving pain and treating cardiovascular problems and nervous disorders.
- The misuse of drugs involves taking a medicine for an unintended use. Drug abuse involves using a drug for a nonmedical purpose.
- Addictions can be broken with the aid of counseling and drug therapy.

Vocabulary

addiction (p. 991)
depressant (p. 990)
drug (p. 988)
hallucinogen (p. 994)
narcotic (p. 989)
stimulant (p. 990)
tolerance (p. 991)
withdrawal (p. 991)

Main Ideas

Summary statements can be used by students to review the major concepts of the chapter.

Using the Vocabulary

To reinforce chapter vocabulary, use the Content Mastery Booklet and the activities in the Interactive Tutor for *Biology: The Dynamics of Life* on the Glencoe Science Web site. **science.glencoe.com**

THE PRINCETON REVIEW

All Chapter Assessment questions and answers have been validated for accuracy and suitability by The Princeton Review.

GLENCOE TECHNOLOGY

VIDEOTAPE
MindJogger Videoquizzes
Chapter 36: *The Nervous System*
Have students work in groups as they play the videoquiz game to review key chapter concepts.

Resource Manager

Chapter Assessment, pp. 211-216
MindJogger Videoquizzes
ExamView® Pro Software 💾
BDOL Interactive CD-ROM, Chapter 36 quiz

UNDERSTANDING MAIN IDEAS

1. b
2. d
3. cerebellum
4. c
5. semicircular canals
6. c
7. a
8. b
9. c
10. a
11. neuron
12. axons, myelin
13. sodium
14. central, brain
15. axon, away from
16. synaptic space
17. neurotransmitters
18. addiction, drug
19. taste buds, brain
20. depressant, central nervous system

UNDERSTANDING MAIN IDEAS

1. Which of the following is NOT part of the brain?
   a. cerebrum
   b. cochlea
   c. cerebellum
   d. pons

2. Which of the following is NOT a type of neuron?
   a. interneuron
   b. sensory neuron
   c. motor neuron
   d. stimulus neuron

3. Which portion of the brain controls balance, posture, and coordination?

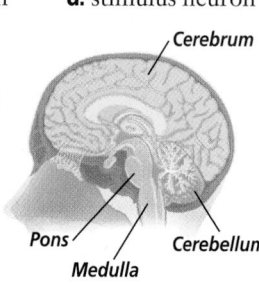

Cerebrum
Pons
Cerebellum
Medulla oblongata

4. Which vision cells allow humans to see color?
   a. thalamic cells
   b. rod cells
   c. cone cells
   d. cortex cells

5. Which part of the ear is involved in maintaining balance?

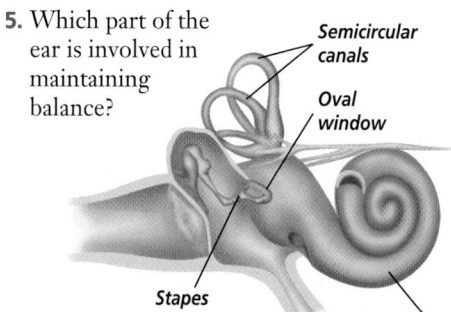

Semicircular canals
Oval window
Stapes
Cochlea

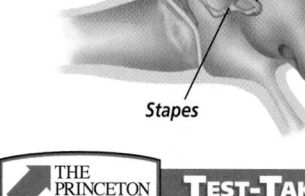

6. Which drug type relieves pain by inhibiting receptors at the site of pain?
   a. stimulants
   b. depressants
   c. aspirin
   d. narcotics

7. What type of drug is nicotine?
   a. stimulant
   b. depressant
   c. analgesic
   d. hallucinogen

8. Which of these is NOT a type of receptor found in the dermis of the skin?
   a. pain
   b. light
   c. pressure
   d. temperature

9. Which of the following drugs depresses the activities of the CNS?
   a. cocaine
   b. aspirin
   c. alcohol
   d. opiate

10. Which type of neuron carries impulses toward the brain?
    a. sensory
    b. motor
    c. association
    d. none of the above

11. The basic unit of structure and function in the nervous system is the _____, or nerve cell.

12. Most _____ are surrounded by a white covering called the _____ sheath.

13. When a stimulus excites a neuron, _____ ions rush into the cell.

14. The _____ nervous system is made up of the spinal cord and the _____.

15. A(n) _____ is a single extension of a neuron that carries messages _____ the cell body.

16. A _____ is the region between one neuron's axon and another neuron's dendrites.

17. Chemicals called _____ diffuse across synapses and stimulate neurons.

18. A(n) _____ is a psychological or physiological dependence on a _____.

19. The _____ on the tongue contain sensory receptors that send taste signals to the _____.

20. Alcohol is a _____ that slows the activities of the _____.

APPLYING MAIN IDEAS

21. You are making a ferry crossing during rough weather and the horizon seems to be moving up and down as you hold on to the railing. You begin to feel seasick. Explain what is going on inside your body that might be causing this sensation.

22. Tetrodotoxin, a chemical produced by the puffer fish, blocks sodium channels. How does this toxin help the fish capture its prey?

23. Explain how alcohol disrupts the normal functions of a neuron.

THINKING CRITICALLY

24. Observing and Inferring A medicine has this precaution on its label: "Avoid driving a motor vehicle while taking this medicine as it may cause drowsiness." What type of drug does this medicine contain? Explain.

25. Concept Mapping Complete the concept map by using the following vocabulary terms: neurons, neurotransmitters, axons, dendrites, synapses.

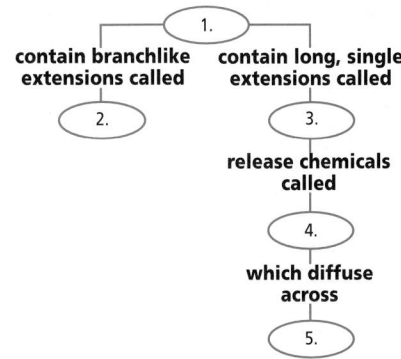

CD-ROM

For additional review, use the assessment options for this chapter found on the *Biology: The Dynamics of Life Interactive CD-ROM* and on the Glencoe Science Web site.
science.glencoe.com

ASSESSING KNOWLEDGE & SKILLS

As a part of his job in building a new highway, a construction worker is planning to light a fuse hooked to some TNT to blast out a portion of rock.

Comparing and Contrasting Use the illustration above to answer the following questions.

1. How does the steady burning of the fuse resemble the depolarization of a neuron?
 a. Both involve sodium channels.
 b. Both are under voluntary control.
 c. Both are self-propagating.
 d. Both involve the combustion of oxygen.

2. How do a neuron and the fuse compare in terms of repeated use?
 a. Neither the fuse nor the neuron can be used repeatedly.
 b. The fuse can be used over and over, but the neuron must regrow before being reused.
 c. The neuron can be used after recovery, but the fuse is consumed and cannot be reused.
 d. Both the neuron and the fuse can be used repeatedly.

3. Applying Concepts In what ways is the movement of an impulse down an axon similar to the movement of an electric current in a wire?

APPLYING MAIN IDEAS

21. Your body is constantly moving up, down, and sideways. As a result, your semicircular canals are over-stimulated and you feel seasick.

22. Because the prey cannot transport sodium across its membranes in order to relay an impulse to move, it is effectively paralyzed.

23. Alcohol blocks the movement of sodium and calcium ions, which are responsible for transmitting impulses.

THINKING CRITICALLY

24. The medicine may contain a depressant or mild narcotic.

25. 1. Neurons; 2. Dendrites; 3. Axons; 4. Neurotransmitters; 5. Synapses

ASSESSING KNOWLEDGE & SKILLS

1. c

2. c

3. Both are due to the flow of charged particles. The electrical current is due to the flow of electrons, whereas the impulse is due to the flow of charged ions.

BIOLOGY Online

Note Internet addresses that you find useful in the space below for quick reference.

Chapter 37 Organizer

Refer to pages 4T-5T of the Teacher Guide for an explanation of the National Science Education Standards correlations.

Section	Objectives	Activities/Features
Section 37.1 **The Respiratory System** National Science Education Standards UCP.1, UCP.2, UCP.5; B.3; C.1, C.5; F.1, F.4, F.5 (1½ sessions, 1 block)	1. **List** the structures involved in external respiration. 2. **Explain** the mechanics of breathing. 3. **Contrast** external and cellular respiration.	**Problem-Solving Lab 37-1,** p. 1005 **Careers in Biology:** Registered Nurse, p. 1006 **Investigate BioLab:** Measuring Respiration, p. 1020
Section 37.2 **The Circulatory System** National Science Education Standards UCP.1, UCP.2, UCP.5; A.1, A.2; B.3; C.1, C.5; E.1, E.2; F.1, F.5; G.1, G.2 (1½ sessions, 1 block)	4. **Distinguish** among the various components of blood and among blood groups. 5. **Trace** the route blood takes through the body and heart. 6. **Explain** how heart rate is controlled.	**MiniLab 37-1:** Checking Your Pulse, p. 1013 **Inside Story:** Your Heart, p. 1014 **Problem-Solving Lab 37-2,** p. 1015
Section 37.3 **The Urinary System** National Science Education Standards UCP.1, UCP.2, UCP.3, UCP.5; A.1, A.2; C.5; E.1, E.2; F.1, F.6; G.1 (2 sessions, 1 block)	7. **Describe** the structures and functions of the urinary system. 8. **Explain** the kidneys' role in maintaining homeostasis.	**MiniLab 37-2:** Testing Urine for Glucose, p. 1019 **Biology & Society:** Finding Transplant Donors, p. 1022

Need Materials? Contact Carolina Biological Supply Company at 1-800-334-5551 or at **http://www.carolina.com**

MATERIALS LIST

BioLab
p. 1020 round balloon, string, metric ruler, watch with second hand

MiniLabs
p. 1013 pencil, paper, watch with second hand
p. 1019 microscope slide, grease pencil, droppers (2), normal "urine," abnormal "urine," glucose test paper, pencil, paper

Alternative Lab
p. 1014 distilled water, 250 mL flask, plastic straw, droppers (2), phenolphthalein indicator, 0.04% NaOH solution

Quick Demos
p. 1004 respiratory system model
p. 1010 watch with second hand, drinking glasses, water
p. 1019 preserved sheep kidney

Key to Teaching Strategies

L1 Level 1 activities should be appropriate for students with learning difficulties.

L2 Level 2 activities should be within the ability range of all students.

L3 Level 3 activities are designed for above-average students.

ELL ELL activities should be within the ability range of English Language Learners.

COOP LEARN Cooperative Learning activities are designed for small group work.

P These strategies represent student products that can be placed into a best-work portfolio.

These strategies are useful in a block scheduling format.

Teacher Classroom Resources

Section	Reproducible Masters	Transparencies
Section 37.1 **The Respiratory System**	Reinforcement and Study Guide, p. 163 **L2** Critical Thinking/Problem Solving, p. 37 **L3** Content Mastery, pp. 181-182, 184 **L1** Tech Prep Applications, pp. 51-52 **L2**	Section Focus Transparency 90 **L1** **ELL**
Section 37.2 **The Circulatory System**	Reinforcement and Study Guide, pp. 164-165 **L2** Concept Mapping, p. 37 **L3** **ELL** Critical Thinking/Problem Solving, p. 37 **L3** BioLab and MiniLab Worksheets, pp. 163-164 **L2** Laboratory Manual, pp. 269-276 **L2** Content Mastery, pp. 181, 183-184 **L1** Inside Story Poster **ELL**	Section Focus Transparency 91 **L1** **ELL** Basic Concepts Transparency 70 **L2** **ELL** Basic Concepts Transparency 71 **L2** **ELL** Basic Concepts Transparency 72 **L2** **ELL** Reteaching Skills Transparency 54 **L1** **ELL**
Section 37.3 **The Urinary System**	Reinforcement and Study Guide, p. 166 **L2** Critical Thinking/Problem Solving, p. 37 **L3** BioLab and MiniLab Worksheets, pp. 165-168 **L2** Content Mastery, pp. 181-182, 184 **L1**	Section Focus Transparency 92 **L1** **ELL** Basic Concepts Transparency 73 **L2** **ELL**

Assessment Resources

Chapter Assessment, pp. 217-222
MindJogger Videoquizzes
Performance Assessment in the Biology Classroom
Alternate Assessment in the Science Classroom
ExamView® Pro Software 💾
BDOL Interactive CD-ROM, Chapter 37 quiz

Additional Resources

Spanish Resources **ELL**
English/Spanish Audiocassettes **ELL**
Cooperative Learning in the Science Classroom **COOP LEARN**
Lesson Plans/Block Scheduling

NATIONAL GEOGRAPHIC — Teacher's Corner

Products Available From Glencoe
To order the following products, call Glencoe at 1-800-334-7344:
CD-ROM
NGS PictureShow: Human Body 2
Curriculum Kit
GeoKit: Human Body 1
Transparency Set
NGS PicturePack: Human Body 2
Videodisc
STV: Human Body

Products Available From National Geographic Society
To order the following products, call National Geographic Society at 1-800-368-2728:
Videos
Incredible Human Machine
Circulatory and Respiratory Systems (Human Body Series)

GLENCOE TECHNOLOGY

The following multimedia resources are available from Glencoe.

Biology: The Dynamics of Life
CD-ROM **ELL**

Animation: *The Mechanics of Breathing*
BioQuest: *Body Systems*
BioQuest: *Triathalon*
Video: *Lymphocytes*
Exploration: *Blood Types*
Video: *Capillaries*
Animation: *One Way Valves*

Videodisc Program

The Mechanics of Breathing
Capillaries
One Way Valves

Chapter

37 Respiration, Circulation, and Excretion

Theme Development

Homeostasis is the major theme of the chapter. Emphasis is placed on how the respiratory, circulatory, and urinary systems maintain balance in the body. **Systems and interactions** is shown by the interactions among the three systems discussed and how these systems work in conjunction with the nervous system.

0:00 OUT OF TIME?

If time does not permit teaching the entire chapter, use the BioDigest at the end of the unit as an overview.

READING BIOLOGY

Glencoe's *Biology: The Dynamics of Life* contains many resources to assist a student's reading skills. Each chapter contains figures with expanded captions that expand on written material. Word Origins, located along the side of text, expand knowledge of biology vocabulary. Glencoe's Content Mastery Booklet helps develop reading skills while reinforcing content. In addition, use the Interactive Tutor for *Biology: The Dynamics of Life* on the Glencoe Web site to reinforce vocabulary. **science.glencoe.com**

What You'll Learn

- You will explain the mechanics of breathing.
- You will distinguish the types of blood cells and trace the pathway of blood circulation through the body.
- You will describe the structure and function of the urinary system.

Why It's Important

With a knowledge of how your circulatory, respiratory, and urinary systems function, you will understand how your cells receive, deliver, and remove materials to maintain your body's homeostasis.

READING BIOLOGY

As you read, make an outline of the chapter. For each section, include the key ideas, and new vocabulary terms that appear. Jot down any questions that arise while reading.

To find out more about respiration, circulation, and excretion, visit the Glencoe Science Web site. **science.glencoe.com**

Magnification: 40 000×

These biconcave blood cells carry vital oxygen to the rock climber's muscles.

1002

Multiple Learning Styles

Look for the following logos for strategies that emphasize different learning modalities.

Kinesthetic Quick Demo, p. 1010

Visual-Spatial Portfolio, pp. 1005, 1009, 1012, 1018; Meeting Individual Needs, pp. 1005, 1009; Microscope Activity, p. 1011

Interpersonal Meeting Individual Needs, p. 1017

Intrapersonal Project, p. 1004

Linguistic Biology Journal, pp. 1004, 1013, 1017; Tech Prep, p. 1010; Project, p. 1011

Logical-Mathematical Portfolio, p. 1010; Going Further, p. 1021

SECTION PREVIEW

Objectives

List the structures involved in external respiration.

Explain the mechanics of breathing.

Contrast external and cellular respiration.

Vocabulary

trachea
alveoli

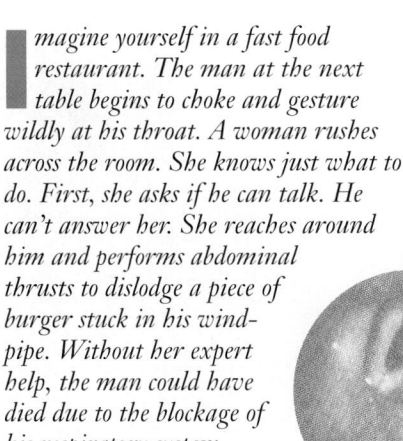

I*magine yourself in a fast food restaurant. The man at the next table begins to choke and gesture wildly at his throat. A woman rushes across the room. She knows just what to do. First, she asks if he can talk. He can't answer her. She reaches around him and performs abdominal thrusts to dislodge a piece of burger stuck in his wind-pipe. Without her expert help, the man could have died due to the blockage of his respiratory system.*

Abdominal thrusts (above) are used to clear obstructions from the trachea (inset).

Passageways and Lungs

Your respiratory system is made up of a pair of lungs, a series of passageways into your body, and a thin sheet of skeletal muscle called the diaphragm. When you hear the word respiration, you probably think of breathing. But breathing is just part of the process of respiration that an oxygen-dependent organism carries out. Respiration includes all the mechanisms involved in getting oxygen to the cells of your body and getting rid of carbon dioxide. Recall that respiration also involves the formation of ATP within cells.

The path air takes

The first step in the process of respiration involves taking air into your body through your nose or mouth. Air flows into the pharynx, passes the epiglottis, and moves through the larynx. It then travels down the windpipe, or **trachea** (TRAY kee uh), a tubelike passageway that leads to two bronchi (BRAHN ki) tubes, which lead into the lungs. When you swallow food, the epiglottis covers the entrance to the trachea. It prevents food and other large materials from getting into the air passages.

Cleaning dirty air

The air you breathe is far from clean. It is estimated that an individual living in an urban area breathes in 20 million particles of foreign matter each day. To prevent most of this material from reaching your lungs, the nasal cavity, trachea, and bronchi

37.1 THE RESPIRATORY SYSTEM **1003**

Assessment Planner

Portfolio Assessment
Portfolio, TWE, pp. 1005, 1010, 1012, 1018, 1019
Performance Assessment
Assessment, TWE, pp. 1004, 1008
Alternative Lab, TWE, pp. 1014-1015
BioLab, SE, pp. 1020-1021
BioLab, TWE, p. 1021
MiniLab, SE, pp. 1013, 1019
MiniLab, TWE, pp. 1013, 1018

Knowledge Assessment
Assessment, TWE, p. 1016
Section Assessment, SE, pp. 1006, 1016, 1019
Chapter Assessment, SE, pp. 1023-1025
Skill Assessment
Assessment, TWE, p. 1006
Problem-Solving Lab, SE, pp. 1005, 1015

Prepare

Key Concepts

Students will learn about the mechanics of breathing and the exchange of gases. The control of the respiratory system by the nervous system is also discussed.

Planning

■ Obtain a model of the respiratory system for the Quick Demo and the Meeting Individual Needs.

■ Buy balloons, straws, and string for the BioLab.

---

1 Focus

Bellringer

Before presenting the lesson, display **Section Focus Transparency 90** on the overhead projector and have students answer the accompanying questions. **L1** **ELL**

Resource Manager

Section Focus Transparency 90 and Master **L1** **ELL**

2 Teach

Quick Demo

Use an anatomical model of the respiratory organs to point out the parts of the respiratory system as they are mentioned in the chapter.

GLENCOE TECHNOLOGY

CD-ROM
Biology: The Dynamics of Life
Animation: *The Mechanics of Breathing*, Disc 5

VIDEODISC
Biology: The Dynamics of Life
The Mechanics of Breathing (Ch. 32), Disc 2, Side 1, 1 min.

✓ Assessment

Performance Assessment in the Biology Classroom, p. 43, *Making a Model of Inhalation and Exhalation*. Have students carry out this activity to model the respiratory system. **L2**

Resource Manager

Reinforcement and Study Guide, p. 163 **L2**
Tech Prep Applications, p. 51 **L2**
Content Mastery, pp. 181-182 **L1**
BioLab and MiniLab Worksheets, pp. 163-164 **L2**

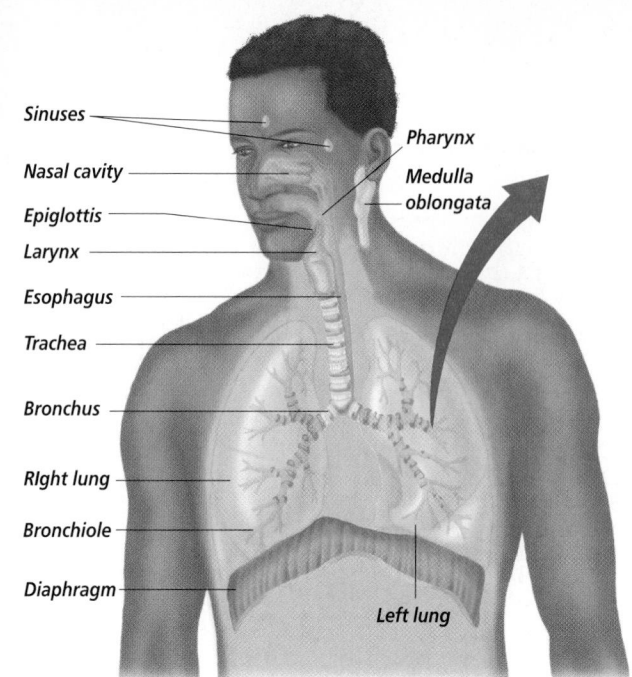

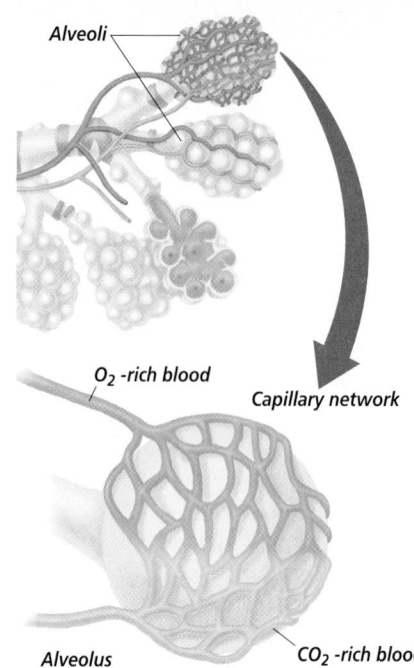

Figure 37.1
As air passes through the respiratory system, it travels through narrower and narrower passageways until it reaches the alveoli.

WORD Origin

alveoli
From the Latin word *alveolus*, meaning "a small hollow sac." The alveoli are small air sacs of the lung.

are lined with ciliated cells that secrete mucus. The cilia constantly beat upward in the direction of your throat, where foreign material can be swallowed or expelled by coughing or sneezing. Follow the passage of air through the lungs in *Figure 37.1*.

Alveoli: The place of gas exchange

Like the branches of a tree, each bronchus branches into bronchioles, which in turn branch into numerous microscopic tubules that eventually expand into thousands of thin-walled sacs called alveoli. **Alveoli** (al VEE uh li) are the sacs of the lungs where oxygen and carbon dioxide are exchanged by diffusion between the air and blood. The clusters of alveoli are surrounded by networks of tiny blood vessels. Blood in these vessels has come from the cells of the body and contains wastes from cellular respiration. Diffusion of gases takes place easily because the wall of each

alveolus, and the walls of each capillary, are only one cell thick. External respiration involves the exchange of oxygen or carbon dioxide between the air in the alveoli and the blood that circulates through the walls of the alveoli, and is shown in the inset portion of *Figure 37.1*.

Blood transport of gases

Once oxygen from the air diffuses into the blood vessels surrounding the alveoli, it is pumped by the heart to the body cells, where it is used for cellular respiration. In an earlier chapter, you learned that cellular respiration is the process by which cells use oxygen to break down glucose and release energy in the form of ATP. Carbon dioxide and water are waste products of this process. The water stays in the cell or diffuses into the blood. The carbon dioxide diffuses into the blood, which carries it back to the lungs.

1004 RESPIRATION, CIRCULATION, AND EXCRETION

PROJECT

Smoking and Lung Capacity

Intrapersonal Ask students to form a hypothesis about the lung capacity of smokers versus nonsmokers. Have students design an experiment to test their hypothesis. **L2**

BIOLOGY JOURNAL

Respiratory Diseases

Linguistic Have students compose letters to the American Lung Association and American Cancer Society asking for posters and information on the respiratory system and diseases that affect it. Have students include the letters in their journals. **L2**

As a result, the blood that comes to the alveoli from the body's cells is high in carbon dioxide and low in oxygen. Carbon dioxide from the body diffuses from the blood into the air spaces in the alveoli. During exhalation, this carbon dioxide is removed from your body. At the same time, oxygen diffuses from the air in the alveoli into the blood, making the blood rich in oxygen. Use the *Problem-Solving Lab* on this page to find out more about how the composition of air changes as it passes through the lungs.

The Mechanics of Breathing

The action of your diaphragm and the muscles between your ribs enable you to breathe in and breathe out. *Figure 37.2* shows how air is drawn in or forced out of the lungs as a result of the diaphragm's position.

When you inhale, the muscles between your ribs contract and your

Problem-Solving Lab 37-1   Interpreting Data

How do inhaled and exhaled air compare? Air is composed of a number of different gases. During respiration, the lungs absorb some of these gases, but not others.

Analysis

Study *Table 37.1* below. It compares the relative percentages of gases in inhaled and exhaled air.

Table 37.1 Comparison of gases in inhaled and exhaled air

Gas	Inhaled air	Exhaled air
Nitrogen	78.00 %	78.00 %
Oxygen	21.00 %	16.54 %
Carbon dioxide	0.03 %	4.49 %
Other gases	0.97 %	0.97 %

Thinking Critically

1. What information about respiration is conveyed by the data in the table?
2. Trace the pathway that a molecule of nitrogen follows when entering and leaving the lungs. Start with air in the bronchus. (Note: Normally, nitrogen does enter the blood stream.)
3. Explain why carbon dioxide levels are higher in exhaled air.

Figure 37.2
Air pressure in the lungs is varied by changes in the volume of the chest cavity.

CD-ROM
View an animation of the mechanics of breathing in the Presentation Builder of the Interactive CD-ROM.

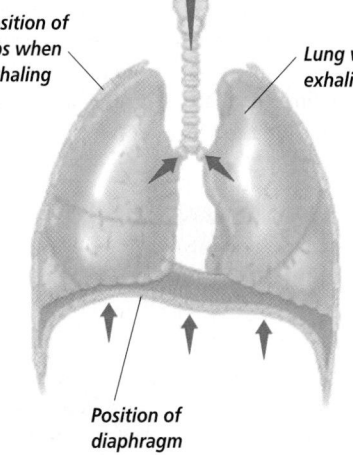

Position of ribs when exhaling

Lung when exhaling

Position of diaphragm when exhaling

A When relaxed, your diaphragm is positioned in a dome shape beneath your lungs.

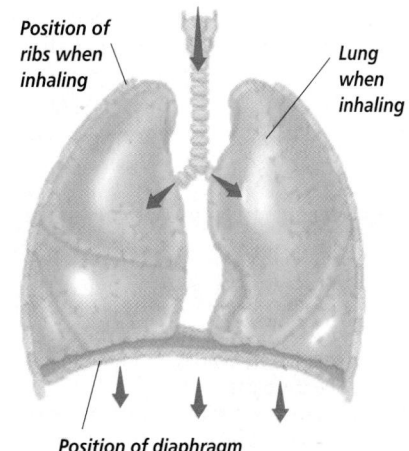

Position of ribs when inhaling

Lung when inhaling

Position of diaphragm when inhaling

B When contracting, the diaphragm flattens, enlarging the chest cavity and drawing air into the lungs.

37.1   THE RESPIRATORY SYSTEM **1005**

Problem-Solving Lab 37-1

Purpose
Students use data to support the conclusion that the body uses oxygen and produces carbon dioxide.

Process Skills
analyze information, think critically, compare and contrast, draw a conclusion, sequence

Teaching Strategies
■ Tell students that "other gases" include inert gases such as helium, neon, and argon.
■ Make certain students understand that the respiratory system delivers oxygen to the circulatory system for distribution throughout the body.

Thinking Critically
1. The same volume of air is inhaled and exhaled, but its composition changes; the percentage of oxygen is higher in inhaled air, and the percentage of carbon dioxide is higher in exhaled air.
2. bronchus, bronchiole, alveoli, capillary network in alveoli, heart, circulatory system, heart, lung, capillary network, alveoli, bronchiole, bronchus, exhaled to air
3. Carbon dioxide is a waste produced by body cells that exits the body via the lungs.

✔ Assessment
Skill Ask students to trace the pathway oxygen follows inside the body. Use the Performance Task Assessment List for Events Chain in **PASC**, p. 91. **L2**

3 Assess

Check for Understanding
Have students trace the pathway of a carbon dioxide molecule from a body cell to the nose.

MEETING INDIVIDUAL NEEDS

Visually/Hearing Impaired
Visual-Spatial Pair visually impaired or hearing impaired students with other students. Have them use a model of the respiratory system to explain the mechanics of breathing. Have students point to or touch the various anatomical parts as they are discussed. **L1 ELL**

✔ Portfolio

Tracing the Path of Air
Visual-Spatial Ask students to make a circular flowchart describing the pathway and processes involved in respiration. One-half of the circle should show events that take place as the diaphragm contracts, the other half should show events that take place as the diaphragm relaxes. Have them include the charts in their portfolios. **L2 P**

1005

4 Close

CAREERS IN BIOLOGY

Registered Nurse

If you want a fast-paced, hands-on career that puts your "people skills" to work, consider becoming a registered nurse.

Skills for the Job

Nurses give care, support, and advice as they help their patients get well or stay well. They may work in medical or psychiatric hospitals, doctors' offices, schools, hospice programs, nursing homes, rehabilitation centers, public health agencies, and other settings. Registered nurses (RNs) must complete a two-year associate degree program, a two- or three-year diploma program, or a four-year bachelor's degree at a college or a hospital-based nursing school. To become licensed, they must also pass a national test. After earning a master's degree, RNs can become nurse practitioners or specialize in areas such as anesthesia or midwifery.

For more careers in related fields, be sure to check the Glencoe Science Web site.
science.glencoe.com

rib cage rises. At the same time, the diaphragm muscle contracts, becomes flattened, and moves lower in the chest cavity. These actions increase the space in the chest cavity, which creates a slight vacuum. Air rushes into your lungs because the air pressure outside your body is greater than the air pressure inside your lungs.

When you exhale, the muscles associated with the ribs relax, and your ribs drop down in the chest cavity. Your diaphragm relaxes, returning to its resting position. The relaxation of these muscles decreases the volume of the chest cavity and forces most of the air out of the alveoli.

The alveoli in healthy lungs are elastic, like balloons. They stretch as you inhale and return to their original size as you exhale. A balloon that has had the air let out of it does not go totally flat. Similarly, the alveoli still contain a small amount of air after you exhale. Measure your breathing rate in the *BioLab* at the end of this chapter.

Control of Respiration

Breathing is usually an involuntary process. It is controlled by the chemistry of your blood as it interacts with a part of your brain called the medulla oblongata. The medulla oblongata helps maintain homeostasis. It responds to higher levels of carbon dioxide in your blood by sending nerve signals to the rib muscles and diaphragm. The nerve signals cause these muscles to contract, and you inhale. When breathing becomes more rapid, as during exercise, a more rapid exchange of gases between air and blood occurs.

Section Assessment

Understanding Main Ideas
1. Describe the path an oxygen molecule takes as it travels from your nose to a body cell. List each structure of the respiratory system through which it passes.
2. Compare and contrast external respiration and cellular respiration.
3. Explain the process by which gases are exchanged in the lungs.
4. Describe how air in the respiratory tract is cleaned before it reaches the lungs.

Thinking Critically
5. During a temper tantrum, four-year-old Jamal tries to hold his breath. His parents are afraid that he will be harmed by this behavior. How will Jamal be affected by holding his breath?

SKILL REVIEW

6. **Sequencing** What is the sequence of muscle actions that takes place during inhalation and exhalation? For more help, refer to *Organizing Information* in the **Skill Handbook**.

Section Assessment

1. nose, pharynx, epiglottis, larynx, trachea, bronchi, alveoli, blood, cells
2. External respiration is the exchange of oxygen and carbon dioxide in the alveoli. Internal respiration uses oxygen to break down glucose inside the cells.
3. In the alveoli, oxygen from the air diffuses into, and carbon dioxide diffuses out of, the blood.
4. Particles are trapped in mucus. Cilia beat the mucus up the throat where it can be expelled or swallowed.
5. As carbon dioxide builds up in the blood, the child's medulla will stimulate his muscles to contract so that he inhales.
6. During inhalation, muscles between the ribs and the diaphragm contract. During exhalation, these muscles relax.

Section

37.2 The Circulatory System

Blood flowed freely from this injury until direct pressure was applied. Pressure limits bleeding until the blood's adaptive ability to clot takes over. Your blood has other life-supporting qualities. As it travels through your body, it carries oxygen from your lungs and nutrients from your digestive system to your cells, then hauls away cell wastes. Together, your blood, your heart, and a network of blood vessels make up your circulatory system.

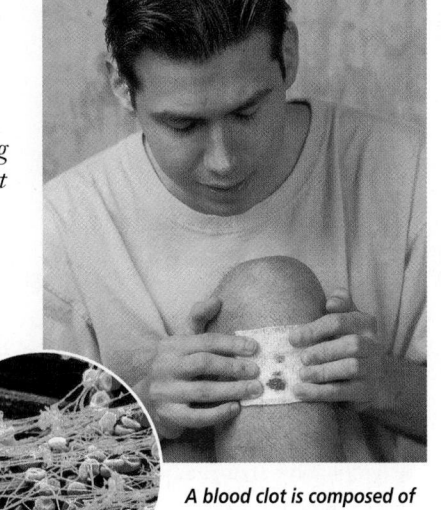

A blood clot is composed of a network of fibers in which blood cells are trapped.

Magnification: 5500×

SECTION PREVIEW

Objectives
Distinguish among the various components of blood and among blood groups.
Trace the route blood takes through the body and heart.
Explain how heart rate is controlled.

Vocabulary
plasma
red blood cell
hemoglobin
white blood cell
platelet
antigen
antibody
artery
capillary
vein
atrium
ventricle
vena cava
aorta
pulse
blood pressure

Your Blood: Fluid Transport

Your blood is a tissue composed of fluid, cells, and fragments of cells. *Table 37.2* summarizes information about the components of human blood. The fluid portion of blood is called **plasma.** Plasma is straw colored and makes up about 55 percent of the total volume of blood. Blood cells—both red and white—and cell fragments are suspended in plasma.

Red blood cells: Oxygen carriers

The round, disk-shaped cells in blood are red blood cells. **Red blood cells** carry oxygen to body cells. They make up 44 percent of the total volume of your blood, and are produced

in the red bone marrow of your ribs, humerus, femur, sternum, and other long bones.

Red blood cells in humans have nuclei only during an early stage in each cell's development. The nucleus is lost before the cell enters the bloodstream. Red blood cells remain active

Table 37.2 Blood components

Components	Characteristics
Red blood cells	Transport oxygen and some carbon dioxide; lack a nucleus; contain hemoglobin
White blood cells	Large; several different types; all contain nuclei; defend the body against disease
Platelets	Cell fragments needed for blood clotting
Plasma	Liquid; contains proteins; transports red and white blood cells, platelets, nutrients, enzymes, hormones, gases, and inorganic salts

37.2 THE CIRCULATORY SYSTEM **1007**

Section 37.2

Prepare

Key Concepts
Students will examine the three major components of the circulatory system: the blood, the vessels, and the heart. They will examine the composition and functions of blood and learn the importance of blood groups. Students will trace the path of blood through the body and heart. They will examine how heart rate is controlled.

Planning
- Borrow a stethoscope and a heart model for the Inside Story.
- Obtain live *Daphnia* for the Microscope Activity.
- Purchase soda straws and gather other materials for the Alternative Lab.

---

1 Focus

Bellringer
Before presenting the lesson, display **Section Focus Transparency 91** on the overhead projector and have students answer the accompanying questions.
L1 **ELL**

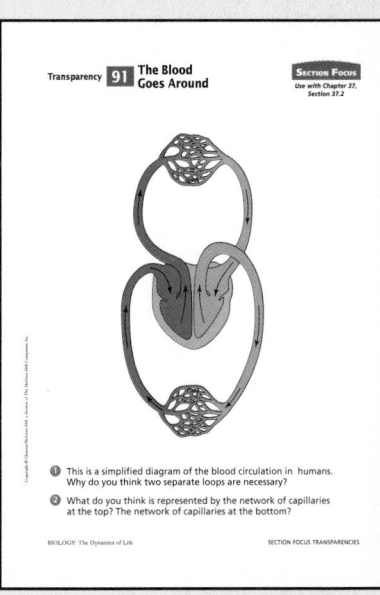

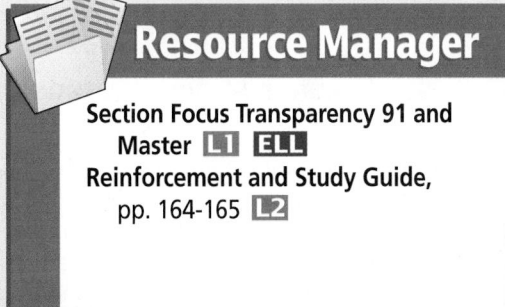

BIOLOGY Online Note Internet addresses that you find useful in the space below for quick reference.

_____

_____

_____

_____

Resource Manager

Section Focus Transparency 91 and Master **L1** **ELL**
Reinforcement and Study Guide, pp. 164-165 **L2**

Tying to Previous Knowledge

Review diffusion from Chapter 6. Relate this process to how oxygen and carbon dioxide get into the blood.

Assessment

Performance Have each student write two questions about blood transport. Arrange students in pairs and have them quiz each other. **L2**

GLENCOE TECHNOLOGY

CD-ROM
Biology: The Dynamics of Life
Bioquest: *Body Systems*
Disc 1-5
BioQuest: *Triathalon*
Disc 5
Video: *Lymphocytes*
Disc 5

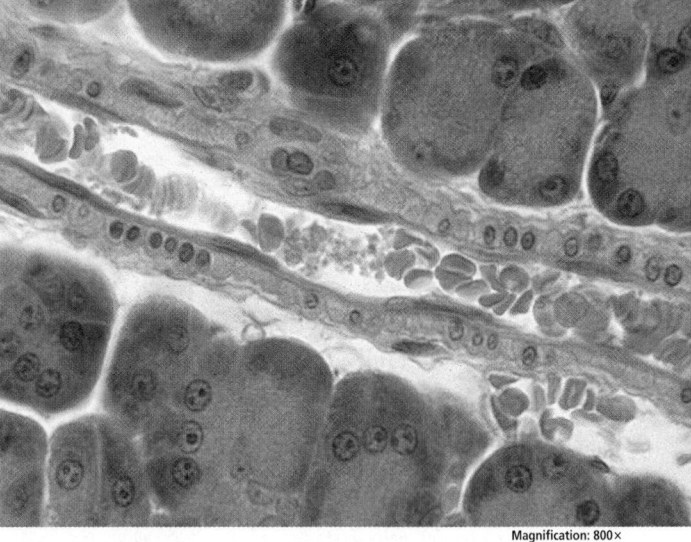

Figure 37.3
Red blood cells are donut-shaped cells that carry oxygen to tissue cells through thin-walled capillaries.

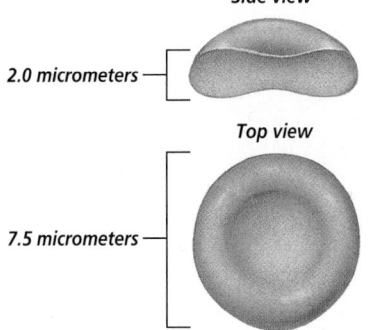

Side view

2.0 micrometers

Top view

7.5 micrometers

Magnification: 800×

Figure 37.4
Compared with red blood cells, white blood cells are larger in size and far fewer in number. White blood cells have a nucleus; mature red blood cells do not have a nucleus.

Magnification: 900×

in the bloodstream for about 120 days, then they break down and are removed as waste. Old red blood cells are destroyed in your spleen, an organ of the lymphatic system, and in your liver.

Oxygen in the blood

How is oxygen carried by the blood? Red blood cells like those shown in *Figure 37.3* are equipped with an iron-containing protein molecule called **hemoglobin** (HEE muh gloh bun). Oxygen becomes loosely bound to the hemoglobin in blood cells that have entered the lung. These oxygenated blood cells carry oxygen from the lungs to the body's cells. As blood passes through body tissues with low oxygen concentrations, oxygen is released from the hemoglobin and diffuses into the tissues.

Carbon dioxide in the blood

Hemoglobin carries some carbon dioxide as well as oxygen. You have already learned that, once biological work has been done in a cell, wastes in the form of carbon dioxide diffuse into the blood and are carried in the bloodstream to the lungs. About 70 percent of this carbon dioxide combines with water in the blood plasma to form bicarbonate. The remaining 30 percent travels back to the lungs dissolved in the plasma or attached to hemoglobin molecules that have already released their oxygen into the tissues.

White blood cells: Infection fighters

White blood cells shown in *Figure 37.4,* play a major role in protecting your body from foreign substances and from microscopic organisms that cause disease. They make up only one percent of the total volume of your blood.

1008 RESPIRATION, CIRCULATION, AND EXCRETION

Cultural Diversity

Carlos Monge and High-Altitude Physiology

Much of what is known about high-altitude or environmental physiology is the result of research by Peruvian scientist Carlos Monge (1884–1970), the first to describe high-altitude sickness, or hypoxia, and the founder of the Institute of Andean Biology and

Pathology. Introduce students to Monge's research and major findings. Discuss with students the physiological adaptations of people who live at high altitudes, such as the Aymara and Quechua Indians of South America. Have students relate these adaptations to the proper functioning of the respiratory system.

Blood clotting

Think about what happens when you cut yourself. If the cut is slight, you usually bleed for a short while, until the blood clots. That's because, in addition to red and white blood cells, your blood also contains small cell fragments called **platelets,** which help blood clot after an injury, as shown in *Figure 37.5.* Platelets help link together a sticky network of protein fibers called fibrin, which forms a web over the wound that traps escaping blood cells. Eventually, a dry, leathery scab forms. Platelets are produced from cells in bone marrow. They have a short life span, remaining in the blood for only about one week.

ABO Blood Groups

If a person is injured so severely that a massive amount of blood is lost, a transfusion of blood from a second person may be necessary. Whenever blood is transfused from one person to another, it is important to know to which blood group each person belongs. You have already learned about the four human blood groups—A, B, AB, and O. You inherited the characteristics of one of these blood groups from your parents. Sometimes, the term *blood type* is used to describe the blood group to which a person belongs. If your blood falls into group O, for example, you are said to have type O blood.

Blood surface antigens determine blood group

Differences in blood groups are due to the presence or absence of proteins, called antigens, on the membranes of red blood cells. **Antigens** are substances that stimulate an immune response in the body. As you'll learn in a later chapter, an immune response defends the body against foreign

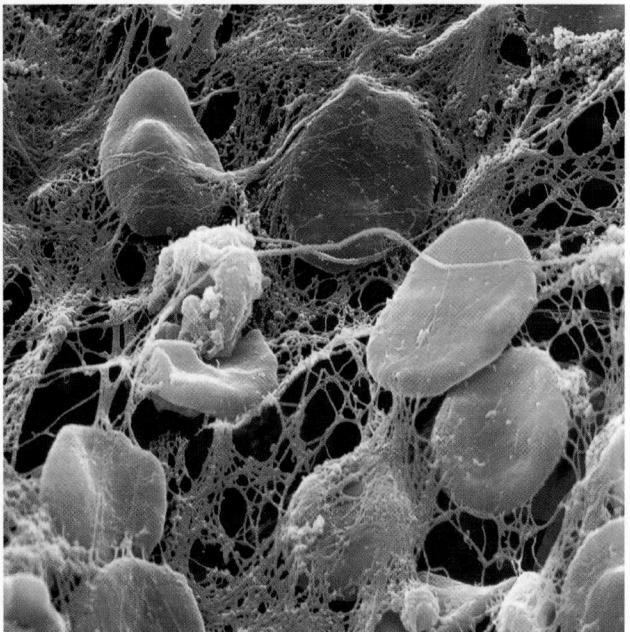

Magnification: 9800×

Figure 37.5
Blood platelets—the whitish, globular structures shown adhering to red blood cells—help clot the blood by linking together a threadlike network of fibrin that traps escaping blood cells.

proteins. The letters A and B stand for the types of blood surface antigens found on human red blood cells.

Blood plasma contains proteins, called **antibodies** (ANT ih bahd eez), that are shaped to correspond with the different blood surface antigens. The antibody in the blood plasma reacts with its matching antigen on red blood cells if they are brought into contact with one another. This reaction results in clumped blood cells that can no longer function. Each blood group contains antibodies for the blood surface antigens found only in the other blood groups—not for antigens found on its own red blood cells.

For example, if you have type A blood, you have the A antigen on your red blood cells and the anti-B antibody in your plasma. If you had anti-A antibodies, they would react with your own type A red blood cells. However, if you have type A blood and anti-A is added to it by way of a

Reinforcement

A for *artery* and *A* for *away* is a mnemonic device that helps students remember that arteries carry blood away from the heart.

TECHPREP

American Red Cross

Linguistic Have students interested in a health career interview an American Red Cross representative and find out about the work of the American Red Cross in the United States as well as internationally. Have them place a copy of their interview questions and answers in their portfolios. **L2**

GLENCOE TECHNOLOGY

VIDEODISC
The Secret of Life
Pregnancy and Rh Disease

transfusion of type B blood, an antigen-antibody reaction will occur, resulting in clumped blood cells like those shown in **Figure 37.6A.** Clumped blood cells cannot carry oxygen or nutrients to body cells. Similarly, if you have type B blood, you have the B antigen on your red blood cells and the anti-A antibody in your blood plasma. **Figure 37.6** illustrates the antigens and antibodies present in each blood group.

Rh factor

Another characteristic of red blood cells involves the presence or absence of an antigen called Rh, or Rhesus factor. Rh factor is an inherited characteristic. People are Rh positive (Rh^+) if they have the Rh antigen factor on their red blood cells. They are Rh negative (Rh^-) if they don't.

Rh factor can cause complications in some pregnancies. Problems occur only if an Rh^- mother becomes pregnant with an Rh^+ baby. At birth, the Rh^+ baby's blood mixes with the Rh^- blood of the mother, as **Figure 37.7A** illustrates. Upon exposure to the baby's Rh^+ antigen factor, the mother will make anti-Rh^+ antibodies like those shown in **Figure 37.7B.** Should the mother become pregnant again, these antibodies will cross the

Figure 37.6
Blood contains both antigens and antibodies.

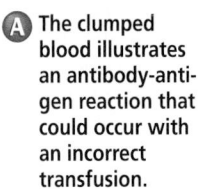

A The clumped blood illustrates an antibody-antigen reaction that could occur with an incorrect transfusion.

C The four types of blood groups have different antigens and antibodies.

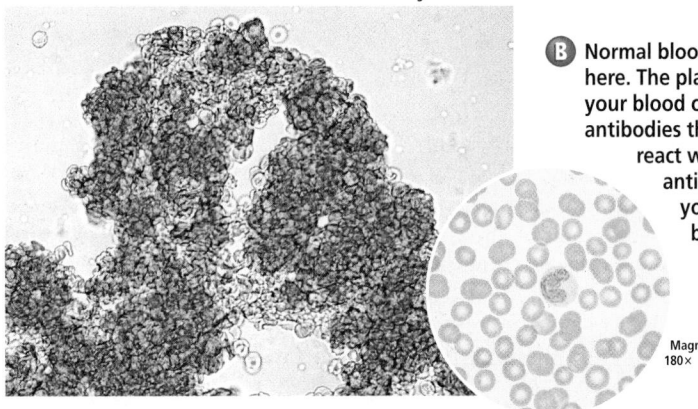

Magnification: 200×

B Normal blood is shown here. The plasma in your blood contains antibodies that do not react with the antigens on your own red blood cells.

Magnification: 180×

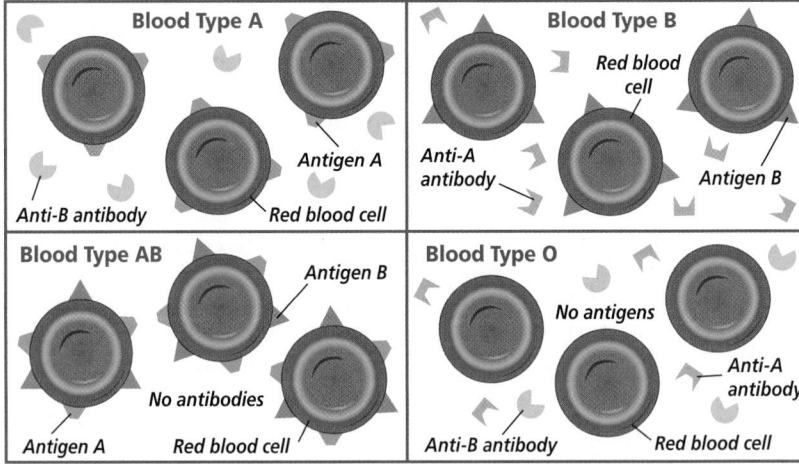

Blood Type A
Antigen A
Anti-B antibody
Red blood cell

Blood Type B
Red blood cell
Anti-A antibody
Antigen B

Blood Type AB
Antigen B
No antibodies
Antigen A
Red blood cell

Blood Type O
No antigens
Anti-A antibody
Anti-B antibody
Red blood cell

Portfolio

Daily Heart Output

Logical-Mathematical Have students calculate the daily heart output of a person whose pulse rate is 72 beats per minute and blood volume is 70 mL per beat. They should show and label the steps in their calculations. **L3**

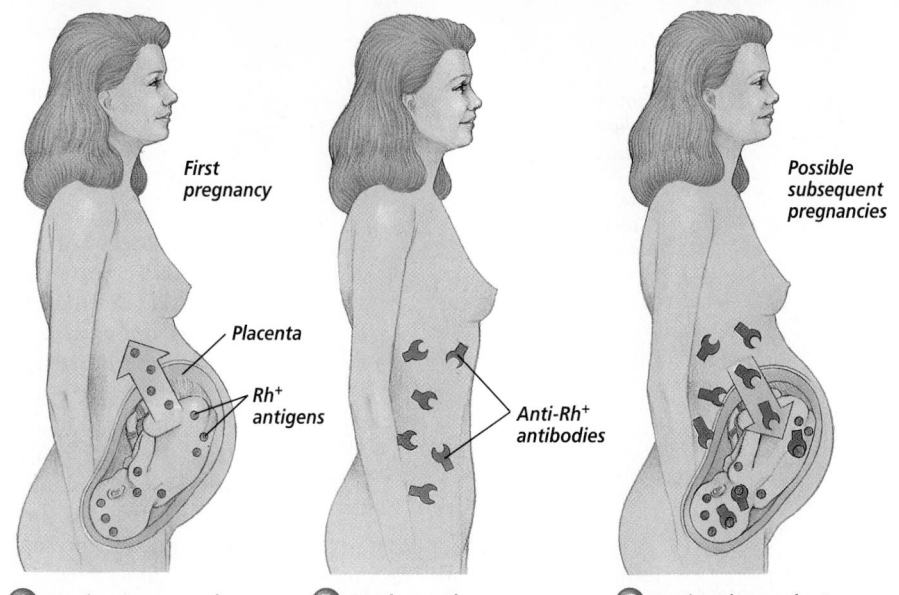

Figure 37.7 If a baby inherits Rh⁺ blood from the father and the mother is Rh⁻, problems can develop if the blood cells of mother and baby mix during birth.

First pregnancy

Placenta

Rh⁺ antigens

A Mother is exposed to Rh antigens at the birth of her Rh⁺ baby.

Anti-Rh⁺ antibodies

B Mother makes anti-Rh⁺ antibodies.

Possible subsequent pregnancies

C During the mother's next pregnancy, Rh antibodies can cross the placenta and endanger the fetus.

placenta. If the new fetus is Rh⁺, the anti-Rh⁺ antibodies from the mother will destroy red blood cells in the fetus, as shown in *Figure 37.7 C*.

Treatment for this problem is available. When the Rh⁺ fetus is 28 weeks old, and again shortly after the Rh+ baby is born, the Rh⁻ mother is given a substance that removes the Rh antibodies from her blood. As a result, the next fetus will not be in danger.

Your Blood Vessels: Pathways of Circulation

Because blood is fluid, it must be channeled through blood vessels like those shown in *Figure 37.8*. The three main types of blood vessels are arteries, capillaries, and veins. Each is different in structure and function.

Arteries are large, thick-walled, muscular, elastic vessels that carry oxygenated blood away from the heart. The blood that they carry is

Figure 37.8 Arteries carry blood away from the heart, whereas veins carry blood toward the heart. Capillaries form an extensive web in the tissues.

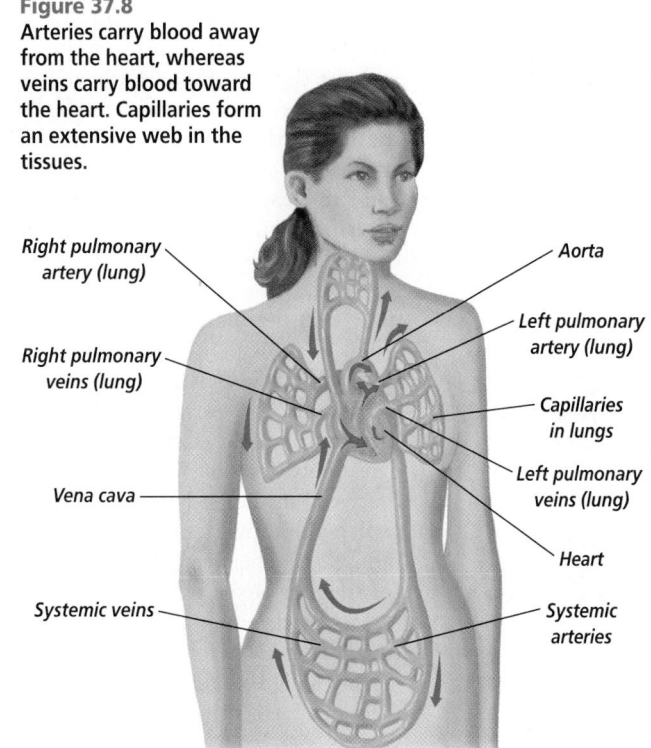

Right pulmonary artery (lung)

Right pulmonary veins (lung)

Vena cava

Systemic veins

Aorta

Left pulmonary artery (lung)

Capillaries in lungs

Left pulmonary veins (lung)

Heart

Systemic arteries

1011

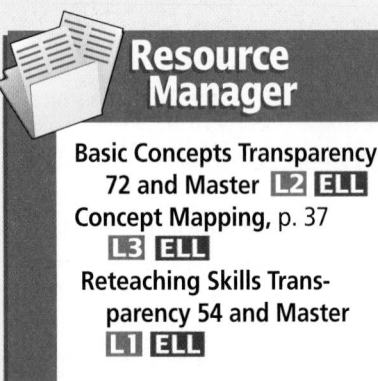
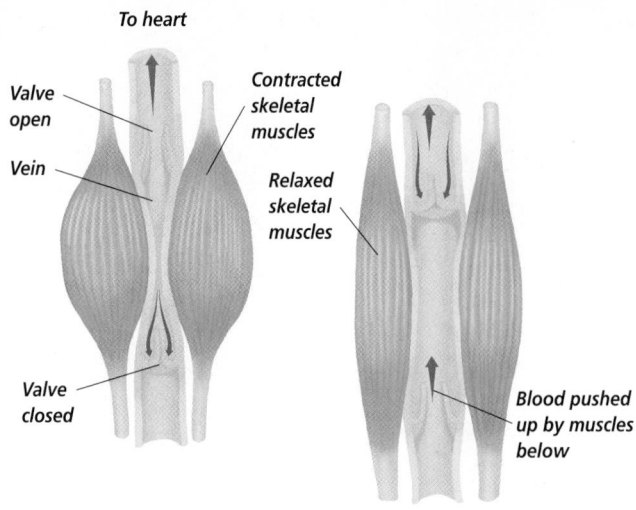

Figure 37.9
Veins contain one-way valves that work in conjunction with skeletal muscles.

CD-ROM
View an animation of the veins' one-way valves in the Presentation Builder of the Interactive CD-ROM.

WORD Origin

vena cava
From the Latin words *vena*, meaning "vein," and *cava*, meaning "empty." Each vena cava empties blood into the heart.

under great pressure. As the heart contracts, it pushes blood through the arteries. Each artery's elastic walls expand slightly. As the heart relaxes, the artery shrinks a bit, which also helps push the blood forward. As a result, blood surges through the arteries in pulses that correspond with the rhythm of the heartbeat.

After the arteries branch off from the heart, they divide into smaller arteries that, in turn, divide into even smaller vessels called arterioles. Arterioles enter tissues, where they branch into the smallest blood vessels, the capillaries. **Capillaries** (KAP ul ler eez) are microscopic blood vessels with walls that are only one cell thick. These vessels are so tiny that red blood cells must move through them in single file. Capillaries form a dense network that reaches virtually every cell in the body. Thin capillary walls enable nutrients and gases to diffuse easily between blood cells and surrounding tissue cells.

As blood leaves the tissues, the capillaries join to form slightly larger vessels called venules. The venules merge to form **veins,** the large blood vessels that carry blood from the tissues back toward the heart. Blood in veins is not under pressure as great as that in the arteries. In some veins, especially those in your arms and legs, blood travels uphill against gravity. These veins are equipped with valves that prevent blood from flowing backward. *Figure 37.9* shows how these valves function. When your skeletal muscles contract, the valves open, and blood is forced toward the heart. When the skeletal muscles relax, the valves close to prevent blood from flowing backward, away from the heart. If you remain inactive for too long, blood may pool in your feet and lower legs, which can sometimes result in swollen feet or ankles.

Your Heart: The Vital Pump

The thousands of blood vessels in your body would be of little use if there were not a way to move blood through them. The main function of the heart is to keep blood moving constantly throughout the body. Well adapted for its job, the heart is a large organ made of cardiac muscle cells that are rich in energy-producing mitochondria.

All mammalian hearts, including yours, have four chambers. The two upper chambers of the heart are the **atria.** The two lower chambers are the **ventricles.** The walls of each atrium are thinner and less muscular than those of each ventricle. As you will see, the ventricles perform more work than the atria, a factor that helps explain the thickness of their muscles. The left ventricle pumps blood to the entire body, so its muscles are thicker than those of the right ventricle, which pumps blood to the lungs. As a result, your heart is somewhat lopsided.

Blood's path through the heart

Blood enters the heart through the atria and leaves it through the ventricles. Both atria fill up with blood at the same time. The right atrium receives oxygen-poor blood from the head and body through two large veins called the **venae cavae** (vee nee KAY vee). The left atrium receives oxygen-rich blood from the lungs through four pulmonary veins. These veins are the only veins that carry blood rich in oxygen. After they have filled with blood, the two atria then contract, pushing the blood down into the two ventricles.

After the ventricles have filled with blood, they contract simultaneously. When the right ventricle contracts, it pushes the oxygen-poor blood from the right ventricle against gravity out of the heart and toward the lungs through the pulmonary arteries. These arteries are the only arteries that carry blood poor in oxygen. At the same time, the left ventricle forcefully pushes oxygen-rich blood from the left ventricle out of the heart through the **aorta** to the arteries. The aorta is the largest blood vessel in the body. How does a drop of blood move through the heart? To find out, read the *Inside Story* on the next page.

Heartbeat regulation

Each time the heart beats, a surge of blood flows from the left ventricle into the aorta and then into the arteries. Because the radial artery in the arm and carotid arteries near the jaw are fairly close to the surface of the body, the surge of blood can be felt as it moves through them. This surge of blood through an artery is called a **pulse**. Find out more about how the pulse is used to measure heart rate by conducting the *MiniLab* on this page.

MiniLab 37-1   Experimenting

Checking Your Pulse The heart speeds up when the blood volume reaching your right atrium increases. It also speeds up when the level of carbon dioxide in the blood rises. The number of heartbeats per minute is your heart rate, which can be measured by taking your pulse.

Procedure

1. Copy the data table.
2. Have a classmate take your resting pulse for 60 seconds while you are sitting at your lab table or desk. Use the photo above as a guide to finding your radial pulse.
3. Record your pulse in the table.
4. Repeat steps 2 and 3 four more times, then calculate your average resting pulse rate. Switch roles and take your classmate's resting pulse.
5. Exercise by doing "jumping jacks" for one minute.
6. Have your classmate take your pulse for 60 seconds immediately after exercising and record the value in the data table.
7. Repeat steps 5 and 6 four more times. Switch roles again with your classmate.

Data Table

Heart rate (beats per minute)

Trial	Resting	After exercise
1		
2		
3		
4		
5		
Total		
Average		

Analysis

1. Explain why your pulse is a means of indirectly measuring heart rate.
2. Use actual values from your data table to describe the changes that occur to your heart rate when exercising.
3. Suppose the amount of blood pumped by your left ventricle each time it contracts is 70 mL. Calculate your cardiac output (70 mL × heart rate per minute) while at rest and just after exercise.

MiniLab 37-1

Purpose
Students will measure their resting pulse and compare it with their pulse after exercise.

Process Skills
compare and contrast, hypothesize, experiment, analyze information, apply concepts, collect data, use numbers, recognize cause and effect

Teaching Strategies
■ Excuse students with health or physical impairments from the exercise portion of this activity.
■ Help students find their radial pulse. If difficulty continues, have them locate their carotid or neck pulse. Advise students not to use their thumb when taking a pulse.
■ Pulse recording should begin immediately after exercising is finished. Each exercise trial must be preceded by a full minute of exercise.

Expected Results
Resting pulses will be around 80-90 beats per minute. After exercise, beats will be close to 120 per minute.

Analysis
1. A surge of blood through arteries corresponds with each ejection of blood from the heart.
2. Answers will vary. Average pulse after exercise is higher than resting pulse.
3. 70 mL × average resting pulse per minute = resting cardiac output in mL per minute
70 mL × average exercising pulse per minute = exercising cardiac output in mL per minute

---

Resource Manager

BioLab and MiniLab Worksheets, pp. 165-166 **L2**
Content Mastery, pp. 183-184 **L1**
Laboratory Manual, pp. 269-276 **L2**

✓ Assessment

Performance Have students design and carry out an experiment that determines the time needed after exercise for heart rate to return to its resting rate. This period of time is called the heart's recovery time. Use the Performance Task Assessment List for Designing an Experiment in **PASC**, p. 23. **L2**

Purpose

Students will examine heart structure and function.

Teaching Strategies

■ Heart valves close due to blood pressure changes in the chambers, creating the characteristic lubb-dupp sounds that can be heard with a stethoscope. As the atria contract, they force blood against the valves, causing them to open. As the ventricles contract, blood flows back against the heart valves, causing them to close.

■ Allow students to listen to their heartbeats with stethoscopes. The earpiece should be cleaned with alcohol before each use.

Visual Learning

■ Use a heart model to trace the pathway of blood as it flows through the heart.

■ A fetal heart has an opening between the two atria. Contrast how fetal heart circulation is different from adult heart circulation. Have students research blue babies. **L2**

Critical Thinking

Evaluate students' concept maps. The maps should show an understanding of the pathway of blood through the body.

INSIDE STORY

Your Heart

Your heart is about 12 cm by 8 cm—roughly the size of your fist. It lies in your chest cavity, just behind the breastbone and between the lungs, and is essentially a large muscle completely under involuntary control.

Critical Thinking *Construct a concept map for the pathway of blood through the body described on this page.*

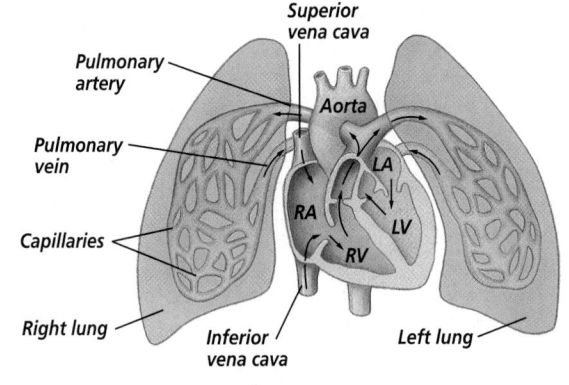

Superior vena cava
Pulmonary artery
Pulmonary vein
Capillaries
Right lung
Aorta
LA
RA
LV
RV
Inferior vena cava
Left lung

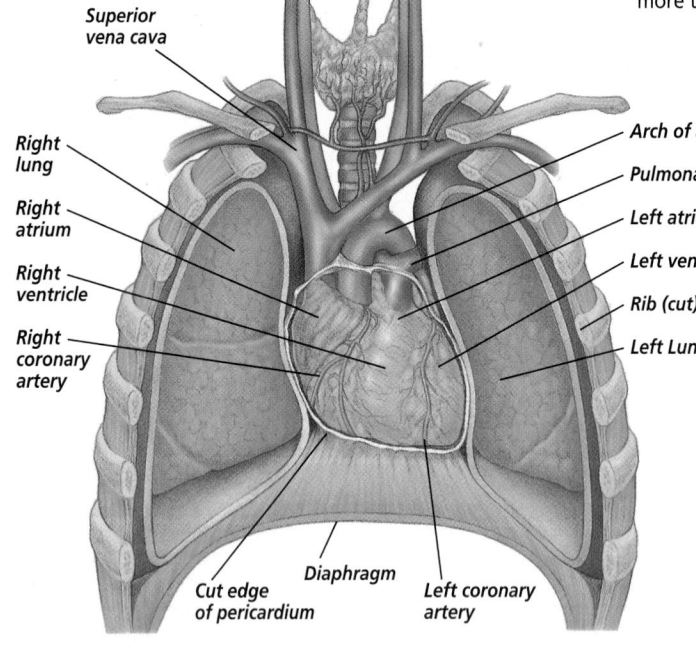

Superior vena cava
Right lung
Right atrium
Right ventricle
Right coronary artery
Arch of aorta
Pulmonary trunk
Left atrium
Left ventricle
Rib (cut)
Left Lung
Cut edge of pericardium
Diaphragm
Left coronary artery

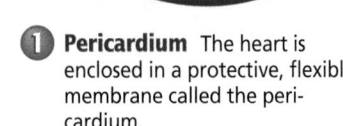

① **Pericardium** The heart is enclosed in a protective, flexible membrane called the pericardium.

② **The passage of blood** If you were to trace the path of a drop of blood through the heart, you could begin with blood coming back from the body through a vena cava. The drop travels first into the right atrium, then into the right ventricle, and then through a pulmonary artery to one of the lungs. In the lungs, the blood drops off its carbon dioxide and picks up oxygen. Then it moves through a pulmonary vein to the left atrium, into the left ventricle, and finally out to the body through the aorta, eventually returning once more to the heart.

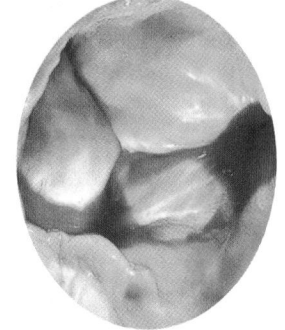

③ **Heart valves** Between the atria and ventricles are one-way valves that keep blood from flowing back into the atria. Sets of valves also lie between the ventricles and the arteries leaving the heart.

1014 RESPIRATION, CIRCULATION, AND EXCRETION

Alternative Lab

Exercise and CO₂

Purpose

Students will determine how exercise affects the amount of carbon dioxide exhaled.

Materials

distilled water, 250 mL flask, plastic straw, dropper bottle of phenolphthalein indicator, dropper bottle of 0.04% NaOH. For preparation instructions for NaOH solution, see page 40T of the Teacher Guide.

Procedure

Give students the following directions.

CAUTION: *Put on goggles. NaOH can cause burns. Always place the dropper back in the bottle. Flush all spills with excess water. If NaOH spills on skin or clothing, rinse off immediately and notify your teacher.*

1. Place 200 mL of distilled water in a 250 mL flask and add 5 drops of phenolphthalein indicator. This indicator is clear in the presence of an acid and pink in the presence of a base.

2. Add NaOH drop by drop to the flask while gently swirling until a faint pink color remains in the solution for one minute.

3. Exhale through the straw into the flask for 5 seconds, then remove your mouth from the straw and inhale for 5 seconds. Repeat

The heart rate is set by the pacemaker, a bundle of nerve cells located at the top of the right atrium. This pacemaker generates an electrical impulse that spreads over both atria. The impulse signals the two atria to contract at almost the same time. The impulse also triggers a second set of cells at the base of the right atrium to send the same electrical impulse over the ventricles, causing them to contract. These electrical signals can be measured and recorded by a machine called an electrocardiograph. This recording, shown in **Figure 37.10,** is called an electrocardiogram (ECG).

The ECG is an important tool used in diagnosing abnormal heart rhythms or patterns. Each peak or valley in the ECG tracing represents a particular electrical activity that takes place during a heartbeat. You can learn how ECG tracings are analyzed by carrying out the *Problem-Solving Lab* on this page.

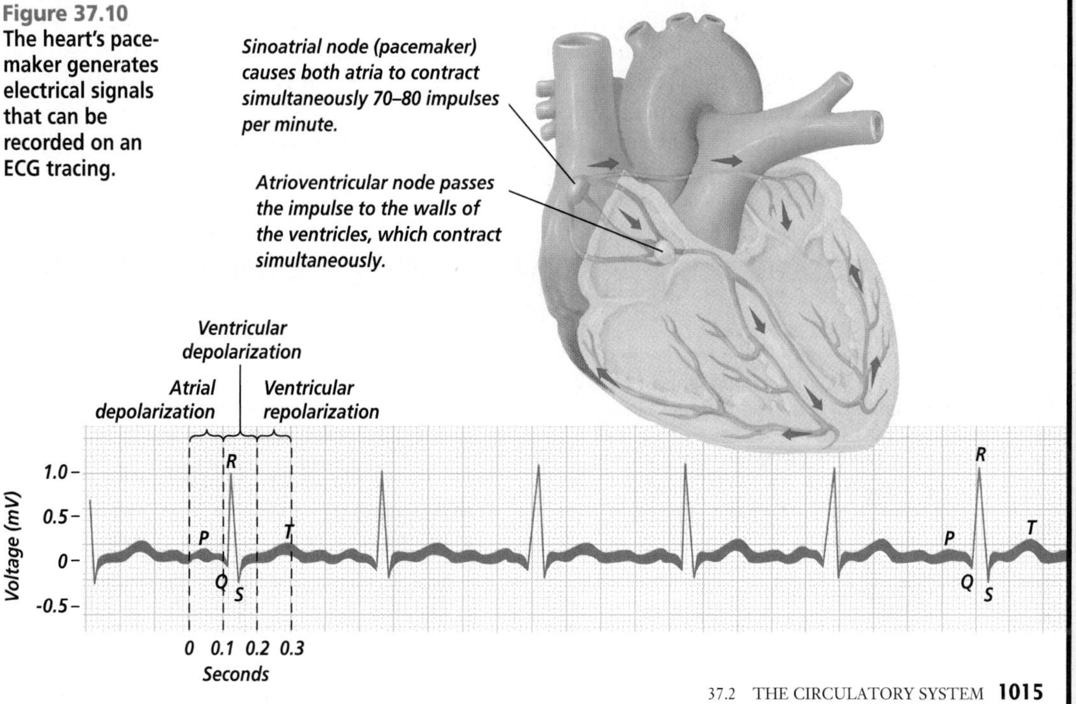

Figure 37.10
The heart's pacemaker generates electrical signals that can be recorded on an ECG tracing.

Sinoatrial node (pacemaker) causes both atria to contract simultaneously 70–80 impulses per minute.

Atrioventricular node passes the impulse to the walls of the ventricles, which contract simultaneously.

37.2 THE CIRCULATORY SYSTEM **1015**

until one minute has elapsed. **CAUTION: Do not inhale through the straw.**

4. Again add NaOH drop by drop, while swirling the flask, until the solution again remains pink for one minute.
5. Record the number of drops of NaOH needed to produce the pink color.
6. Rinse the flask. Repeat steps 2 through 5 after jogging in place for 1 minute. Record the number of drops of NaOH needed after jogging. **CAUTION:** *Wash hands with soap*

and water at the end of the lab.

Analysis
1. Did the number of drops of NaOH needed increase or decrease after exercise? *increase*
2. NaOH neutralizes the carbonic acid formed by carbon dioxide and water. Why is more carbon dioxide produced during exercise? *As more energy is used, more carbon dioxide is produced.*

✔ Assessment
Performance Have students write a summary of the lab for their journal, including answers to the Analysis questions. Use the Performance Task Assessment List for Lab Report in **PASC,** p. 47. **L2**

3 Assess

Check for Understanding

Have students trace the pathway of blood on a diagram of the heart. **L1**

Reteach

Play "Wheel of Circulation" using a cardboard wheel. You can be Pat Pacemaker or Vanna Valve. Have students make up questions. **L2**

Extension

Have students find out more about heart transplants, high blood pressure, and the use of blood typing in criminal investigations. **L3**

✔ Assessment

Knowledge Ask students to summarize the role of the three components of the circulatory system and explain how each structure is related to its function. **L2**

4 Close

Debate

Divide the class into two groups. Have each group take a stand on the question: Does death occur when the heart stops beating? **L2**

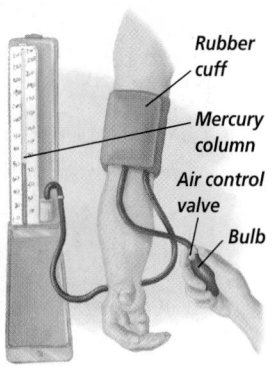

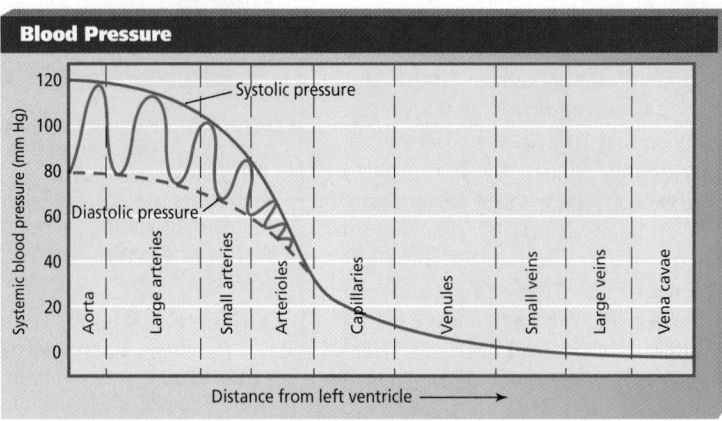

Blood Pressure

Figure 37.11
Blood pressure rises and falls with each heart beat. Pressure is exerted on all vessels throughout the body. However, blood vessels near the left ventricle are subjected to higher pressure than vessels that are farther away. Blood pressure is measured in the artery of the upper arm.

Blood pressure

A pulse beat represents the pressure that blood exerts as it pushes against the walls of an artery. **Blood pressure** is the force that the blood exerts on the blood vessels. As *Figure 37.11* shows, blood pressure rises and falls as the heart contracts and then relaxes.

Blood pressure rises sharply when the ventricles contract, pushing blood through the arteries. The high pressure is called systolic pressure. Blood pressure then drops dramatically as the ventricles relax. The lowest pressure occurs just before the ventricles contract again and is called diastolic pressure.

Control of the heart

Whereas the pacemaker controls the heartbeat, a portion of the brain called the medulla oblongata regulates the rate of the pacemaker, speeding or slowing its nerve impulses. If the heart beats too fast, sensory cells in arteries near the heart become stretched. Via the nervous system, these cells send a signal to the medulla oblongata, which in turn sends signals that slow the pacemaker. If the heart slows down too much, blood pressure in the arteries drops, signalling the medulla oblongata to speed up the pacemaker and increase the heart rate.

Section Assessment

Understanding Main Ideas
1. Summarize the distinguishing features and role of each of the four components of blood: plasma, platelets, and red and white blood cells.
2. Distinguish between an artery and a vein.
3. Outline the path taken by a red blood cell as it passes from the left atrium to the right ventricle of the heart.
4. Describe the location and function of the two pacemakers in the heart.

Thinking Critically
5. The level of carbon dioxide in the blood affects breathing rate. It also affects the heart rate. How would you expect high levels of carbon dioxide to affect the heart rate?

SKILL REVIEW
6. **Making and Using Graphs** Make a circle graph showing the relative proportions of the components of blood. For more help, refer to *Organizing Information* in the **Skill Handbook**.

1016 RESPIRATION, CIRCULATION, AND EXCRETION

Section Assessment

1. Plasma is a fluid that contains proteins and carries blood cells, platelets, enzymes, hormones, gases, and inorganic salts. Red blood cells lack a nucleus, contain hemoglobin, and transport oxygen. White blood cells contain nuclei and defend against disease. Platelets are cell fragments needed for blood clotting.

2. An artery is a vessel that carries blood away from the heart. A vein is a vessel that carries blood toward the heart.

3. left atrium to left ventricle to arteries to arterioles to capillaries to venules to veins to right atrium to right ventricle

4. The main pacemaker is located at the top of the right atrium. A second pacemaker is located at the base of the right atrium. These cells send impulses to signal the chambers of the heart to contract.

5. High levels of carbon dioxide increase the heart rate.

6. The circle graph should show the following proportions: red blood cells—44%; white blood cells—1%; plasma—55%.

37.3 The Urinary System

W ater consumption helps speed the filtering process of the kidneys and maintain their efficiency. Any disruption in kidney function is potentially serious because these organs are essential to maintaining the balance of fluids in the body. The kidneys are the most important organs of the human urinary system. They perform a major cleanup job for the body.

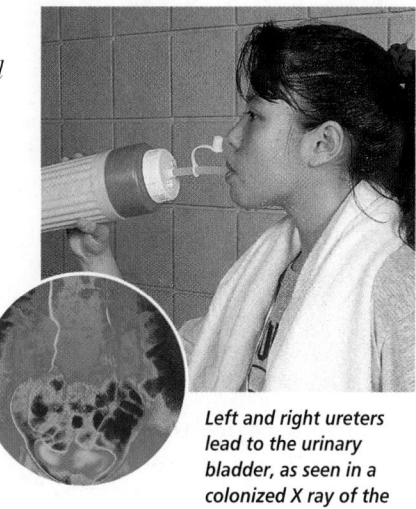

Left and right ureters lead to the urinary bladder, as seen in a colonized X ray of the human urinary system.

SECTION PREVIEW

Objectives
Describe the structures and functions of the urinary system.
Explain the kidneys' role in maintaining homeostasis.

Vocabulary
kidney
ureter
urinary bladder
nephron
urine
urethra

Kidneys: The Body's Janitors

The urinary system is made up of two kidneys, a pair of ureters, the urinary bladder, and the urethra, which you can see in *Figure 37.12*. The **kidneys** filter the blood to remove wastes from it, thus maintaining the homeostasis of body fluids. Your kidneys are located just above the waist, behind the stomach. One kidney lies on each side of the spine, partially surrounded by ribs. Each kidney is connected to a tube called a **ureter,** which leads to the urinary bladder. The **urinary bladder** is a smooth muscle bag that stores a solution of wastes called urine.

Nephron: The unit of the kidney

Have you ever seen an air filter on a car or a water filter in an aquarium?

Figure 37.12
The paired kidneys are reddish-colored organs that resemble kidney beans in shape.

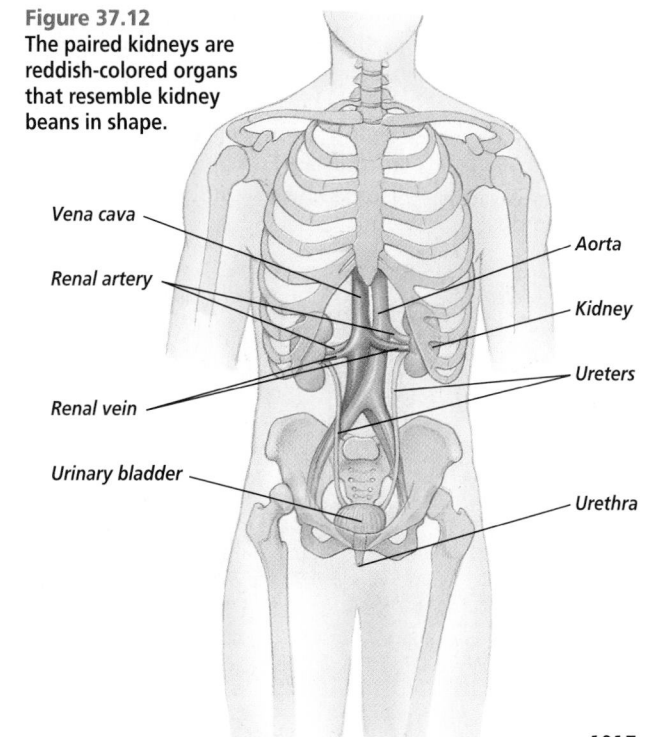

Vena cava

Renal artery

Renal vein

Urinary bladder

Aorta

Kidney

Ureters

Urethra

MEETING INDIVIDUAL NEEDS

English Language Learners

Interpersonal Have students in teams play Hangman or Password with the terms found in this section. Once a team gets the word, they must properly define the term to get the point. This will also help students who are having difficulty with the terms. **L1** **ELL**

BIOLOGY JOURNAL

Using an Analogy

Linguistic Ask students to write an essay in which they compare a kidney to a recycling center. Have them consider substances that can be reused and those that cannot be recycled. **L2**

Section 37.3

Prepare

Key Concepts

Students will survey the structure and function of the urinary system and learn about the kidneys' role in maintaining homeostasis.

Planning

■ Obtain whole kidney for Quick Demo.
■ Acquire materials for MiniLab 37-2.

---

1 Focus

Bellringer

Before presenting the lesson, display **Section Focus Transparency 92** on the overhead projector and have students answer the accompanying questions. **L1** **ELL**

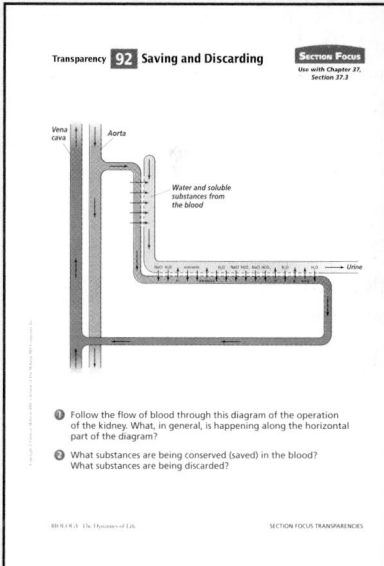

Resource Manager

Section Focus Transparency 92 and Master **L1** **ELL**
Reinforcement and Study Guide, p. 166 **L2**
Basic Concepts Transparency 73 and Master **L2** **ELL**

2 Teach

MiniLab 37-2

Purpose 🔲

Students will use commercial glucose testing paper to detect glucose in simulated urine.

Process Skills

collect data, compare and contrast, draw a conclusion, experiment, interpret data, observe and infer, predict

Safety Precautions

Remind students to use care when handling glass slides and to wash their hands at the end of the lab.

Teaching Strategies

■ Glucose test paper is available from drug stores or biological supply houses.

■ To prepare abnormal urine, add 2 tsp. glucose and 2-3 drops of yellow food coloring to 500 mL water. To prepare normal urine, add 2-3 drops yellow food coloring to 500 mL water.

■ Prepare unknowns—some containing glucose and some without—and place in Barnes dropping bottles. Use a letter or number code to keep track of the contents of each bottle.

Expected Results

Test paper turns green in the presence of glucose. Water causes no color change.

Analysis

1. Any unknown that tests positive for glucose could indicate diabetes. High sugar intake may also result in glucose elimination in urine.

2. testing of normal and abnormal urine before testing of unknowns

3. Possible answers: error in labeling, contamination of samples, normal urine was mixed with abnormal urine by accident, not using clean droppers

1018

Figure 37.13
Each kidney receives a large blood supply through the renal artery for filtering. The blood leaves the kidney by the renal vein.

CD-ROM
View an animation of kidney function in the Presentation Builder of the Interactive CD-ROM.

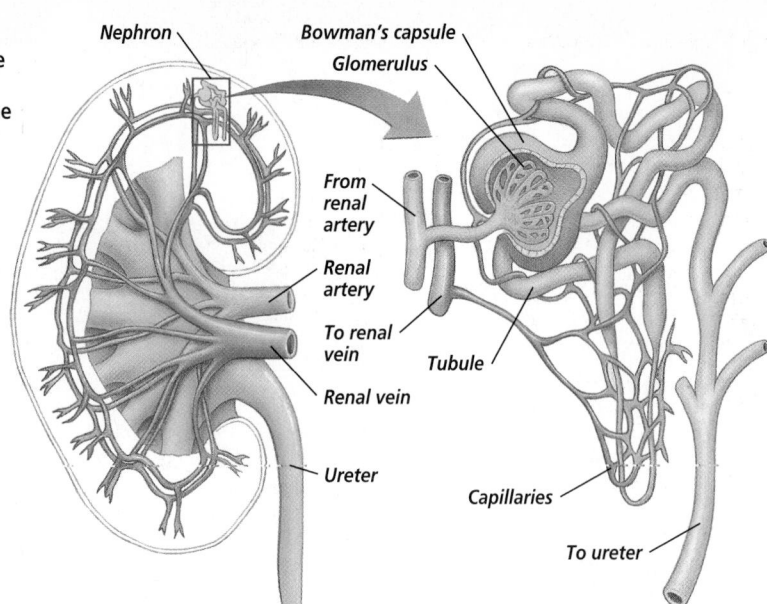

WORD *Origin*
nephron
From the Greek word *nephros*, meaning "kidney." A nephron is a functional unit of a kidney.

Figure 37.14
Filtration and reabsorption take place in the nephron.

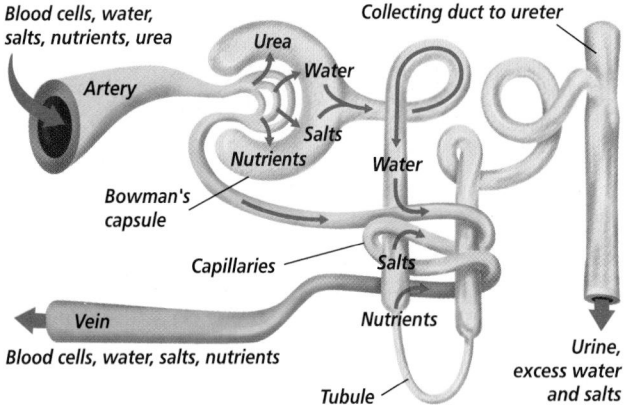

A filter is a device that removes impurities from a solution. Within your body, each kidney is made up of about 1 million tiny filters. Each filtering unit in the kidney is called a **nephron**. *Figure 37.13* shows the parts of a typical nephron.

Blood entering a nephron carries wastes produced by body cells. As blood enters the nephron, it is under high pressure and immediately flows into a bed of capillaries called the glomerulus. Because of the pressure, water, glucose, vitamins, amino acids, protein waste products, salts, and ions from the blood pass out of the capillaries into a part of the nephron called the Bowman's capsule. Blood cells and most proteins are too large to pass through the walls of a capillary, so these components stay within the blood vessels.

The liquid forced into the Bowman's capsule passes through a narrow, U-shaped tubule. As the liquid moves along the tubule, most of the ions and water, and all of the glucose and amino acids, are reabsorbed into the bloodstream. This reabsorption of substances is the process by which the body's water is conserved and homeostasis is maintained. Small molecules, including water, move back into the capillaries by diffusion. Other molecules and ions move back into the capillaries by active transport.

The formation of urine

The liquid that remains in the tubules—composed of excess water, waste molecules, and excess ions—is **urine**. The production of urine is shown in *Figure 37.14*. You produce

1018 RESPIRATION, CIRCULATION, AND EXCRETION

✔ **Assessment**

Performance Have students design an experiment to test the lowest glucose concentration that can be detected using glucose test paper. Use the Performance Task Assessment List for Designing an Experiment in **PASC**, p. 23. **L3**

✔ **Portfolio**

Diagramming the Urinary System

Visual-Spatial Have students make a labeled drawing of the urinary system. Have them place arrows on the diagram to show the flow of fluids through the nephron to the bladder and then out of the body. **L1** **P** 🔲

about 2 L of urine a day. This waste fluid flows out of the kidneys, through the ureter, and into the urinary bladder, where it may be stored. Urine passes from the urinary bladder out of the body through a tube called the **urethra** (yoo REE thruh).

The Urinary System and Homeostasis

The major waste products of cells are nitrogenous wastes, which come from the breakdown of proteins. These wastes include ammonia and urea. Both compounds are toxic to your body and, therefore, must be removed from the blood regularly.

In addition to removing these wastes, the kidneys control the level of sodium in blood by removing and reabsorbing sodium ions. This helps control the osmotic pressure of the blood. The kidneys also regulate the pH of blood by filtering out hydrogen ions and allowing bicarbonate to be reabsorbed back into the blood. Glucose is a sugar that is not usually filtered out of the blood by the kidneys. Individuals who have the disease known as diabetes have excess levels of glucose in their blood. The *MiniLab* on this page shows how urine is used to test for diabetes.

MiniLab 37-2   Experimenting

Testing Urine for Glucose Glucose is a sugar that is needed by the body and is normally not present in the urine. When the concentration of glucose becomes too high in the blood, as happens with diabetes, glucose is filtered out by the kidneys.

Procedure

Data Table

Urine sample	Color of test paper	Glucose present?
Normal (N)		
Abnormal (A)		
Unknown X		
Unknown Y		
Unknown Z		

1. Copy the data table.
2. Using a grease pencil, draw two circles on a glass slide. Mark one circle N, the other A.
3. Use a clean dropper to add two drops of "normal urine" to the circle marked N.
4. Use a clean dropper to add two drops of "abnormal urine" to the circle marked A.
5. Hold a small strip of glucose test paper in a forceps and touch it to the liquid in the drop labeled N. Remove it, wait 30 seconds, and record the color. A green color means glucose is present.
6. Use a new strip of glucose test paper to test drop A and record the color.
7. Test several unknown "urine" samples for the presence of glucose. Use a clean slide for each test.

Analysis

1. Which of the "unknown" samples could be from a person who has diabetes?
2. Which part of the test procedure could be considered your control? Explain your answer.
3. How could you explain your results if a test of normal urine indicated the presence of glucose?

Section Assessment

Understanding Main Ideas

1. Identify the organs that make up the urinary system and describe the function of each.
2. What is the function of a nephron in the kidney? Describe what happens in the glomerulus, Bowman's capsule, and U-shaped tubule.
3. Identify the major components of urine, and explain why it is considered a waste fluid.
4. What is the kidney's role in maintaining homeostasis in the body?

Thinking Critically

5. During a routine physical, a urine test indicates the presence of proteins in the patient's urine. Explain what this could indicate about the patient's health.

SKILL REVIEW

6. **Sequencing** Trace the sequence of urinary waste from a cell to the outside of the body. For more help, refer to *Organizing Information* in the **Skill Handbook**.

---

3 Assess

Check for Understanding

Have students trace the pathway of glucose, water, etc. through the kidney. **L1**

Reteach

Have students discuss the effect on the amount and concentration of urine of a meal rich in proteins, drinking large amounts of water, or high blood pressure. **L2** **COOP LEARN**

Extension

Have interested students research how renin and angiotensin affect the kidneys. **L3**

✔ Assessment

Portfolio Have students make a chart comparing the composition of blood plasma, the fluid inside the Bowman's capsule, and urine. Ask them to write a paragraph explaining the differences. **L2** **P**

4 Close

Discussion

Have students orally summarize the role of the kidneys in maintaining homeostasis. **L2**

Section Assessment

1. two kidneys, two ureters, urinary bladder, urethra. Kidneys filter blood to remove waste. Waste moves into the ureters and is collected in the bladder. When the bladder contracts, urine flows out of the body through the urethra.
2. to filter wastes from the blood plasma
3. urea, water, nitrogenous wastes, excess salt; contains toxic products from the metabolic breakdown of proteins
4. removes urinary waste; maintains blood pH, sodium levels, and osmotic balance
5. High blood pressure could force proteins into Bowman's capsule.
6. cell to blood to nephron to ureter to urinary bladder to urethra

---

Measuring Respiration

*T*he exchange of oxygen and carbon dioxide between the body and the atmosphere is external respiration. It should not be confused with the processes of cellular respiration, which are the chemical reactions that take place within cells to provide energy. The amount of air exchanged between the atmosphere and the blood during external respiration can be measured using a clinical machine called a spirometer. It also can be measured, although less accurately, using a balloon.

Time Allotment
One class period

Process Skills
communicate, measure in SI, use numbers, interpret data, experiment, formulate a model

Safety Precautions
Provide plenty of balloons; students should not share balloons.

PREPARATION

■ Ask students to bring calculators to lab for use in making calculations.
■ Use large-capacity balloons.

Alternative Materials
■ To decrease the number of balloons needed, cut a straw in half and use a rubber band to attach the balloon to the straw. Blow up the balloon through the straw, then replace only the straw for the next student.

Resource Manager

BioLab and MiniLab Worksheets, pp. 167-168 L2

PREPARATION

Problem
How can you measure respiratory rate and estimate tidal volume?

Objectives
In this BioLab, you will:
■ **Measure** resting breathing rate.
■ **Estimate** tidal volume by exhaling into a balloon.
■ **Calculate** the amount of air inhaled per minute.

Materials
round balloon
string (1 m)
metric ruler
clock or watch with
  second hand

Skill Handbook
Use the **Skill Handbook** if you need additional help with this lab.

PROCEDURE

Data Table 1

Resting breathing rate	
Trial	Inhalations in 30 s
1	
2	
3	
Average number breaths	
Breaths per minute	

number of times you inhale in 30s.
3. Repeat step 2 two more times.
4. Calculate the average number of breaths.
5. Multiply the average number of breaths by two to get the average resting breathing rate in breaths per minute.

Part A: Breathing Rate at Rest
1. Copy Data Table 1.
2. Have your partner count the

Part B: Estimating Tidal Volume
1. Copy Data Table 2.
2. Take a regular breath and exhale

PROCEDURE

Teaching Strategies
■ Explain that there are about 300 million alveoli in a human lung with a surface area of more than 50 m². Have students measure the classroom, calculate the number of square meters, and compare this figure with that of a human lung.
■ Have students work in pairs and, if time allows, reverse roles.

Data and Observations
Answers among students could vary greatly. Average breathing rate is 11 to 12 breaths per minute. Tidal volume should be approximately 280 mL. The amount of air inhaled should be 5 to 6 L per minute.

normally into the balloon. Pinch the balloon closed.

3. Have a partner fit the string around the balloon at the widest part.

4. Measure the length of the string, in centimeters, around the circumference of the balloon. Record this measurement.

5. Repeat steps 2-4 four more times.

6. Calculate the average circumference of the five measurements.

7. Calculate the average radius of the balloon by dividing the average circumference by 6.28 (which is approximately equal to 2 π).

8. Tidal volume is the amount of air expelled during a normal breath. Tidal volume can be determined using the balloon radius and the formula for determining the volume of a sphere.

$$\text{Volume} = \frac{4\pi r^3}{3}$$

where r = radius and π = 3.14. Calculate the average tidal volume using the average balloon radius.

9. Your calculated volume will be in cubic centimeters: 1 cm³ = 1 mL.

Data Table 2

Tidal volume	
Trial	**String measurement**
1	
2	
3	
4	
5	
Average circumference	
Average radius	
Average tidal volume	

Part C: Amount of Air Inhaled

1. Copy Data Table 3.

2. Multiply the average tidal volume by the average number of breaths per minute to calculate the amount of air you inhale per minute.

3. Divide the number of milliliters of air by 1000 to get the number of liters of air you inhale per minute.

Data Table 3

Amount of air inhaled	
mL/min	
L/min	

ANALYZE AND CONCLUDE

1. **Making Comparisons** Compare your average number of breaths per minute and tidal volume per minute with those of other students.

2. **Thinking Critically** An average adult inhales 6000 mL of air per minute. Compare your estimated average volume of air with this figure. What factors could account for any differences?

3. **Making Predictions** Predict what would happen to your resting breathing rate after exercise.

Going Further

Applying Concepts The largest amount of air that can be exhaled is called vital capacity. Determine your estimated average vital capacity by following a procedure similar to the one you used to determine average tidal volume.

BIOLOGY *Online* To find out more about respiratory volumes, visit the Glencoe Science Web site.
science.glencoe.com

INVESTIGATE BioLab

ANALYZE AND CONCLUDE

1. Average breaths per minute and tidal volume per minute will differ among students.

2. Answers may vary from the average due to age, sex, size, and athletic condition.

3. After exercising, the breathing rate will be higher.

✔ **Assessment**

Performance Have students write a summary of the lab, including the three tables and the answers to the Analyze and Conclude. Use the Performance Task Assessments List for Lab Report in **PASC**, p. 47. **L2**

Going Further

Logical-Mathematical Students can calculate their tidal volume after exercising and compare this with their tidal volume during rest. **L3**

Biology & SOCIETY

Purpose

Students explore the connection between medical and social issues and learn about the need for organ donors

Background

There are two types of donors. Cadaveric donors are patients who have suffered brain death and will not survive. Living donors are healthy individuals willing to donate a kidney or a portion of their bone marrow. Humans can thrive with only a single kidney. Bone marrow is easily replaced by a healthy body.

Teaching Strategies

■ Explain to students that most donor organs come from individuals who have suffered massive injuries from which they can never recover. The body may continue to function for a time, but only if it is kept on artificial life support. A patient is never taken off life support if there is any hope of recovery, even if it means that his or her organs would become useless for transplanting.

Investigating the Issue

Students may encounter a variety of opinions and beliefs. People who have received a donated organ, or know someone who has, are likely to encourage others to become donors. Some people may be uncomfortable talking about the subject. Students may discover that organ donation is one of the important decisions a family must make when one of its members has suffered brain death and is being kept alive by life support.

Biology & SOCIETY

Finding Transplant Donors

The ability to replace a diseased heart, liver, kidney, pancreas, or other organ with a healthy organ from a human donor is one of the most important of medical science's recent advances. Organs suitable for transplanting are scarce, and there are thousands of transplant patients waiting for them.

Transplant recipients include children born with malformed hearts or digestive systems, patients suffering from severe liver or heart disease, burn victims in desperate need of skin grafts, or people whose kidneys have stopped functioning.

The waiting list A patient who is a good candidate for a transplant is placed on a waiting list. A national computer database keeps track of each patient's blood and tissue type, organ size, and other medical factors. When an organ becomes available, the computer produces a list of all the patients for whom the organ is medically suitable. Patients are ranked according to the severity of their illness, the length of time they have been waiting, and the distance between the donor and the transplant hospital.

Recovering transplant patient

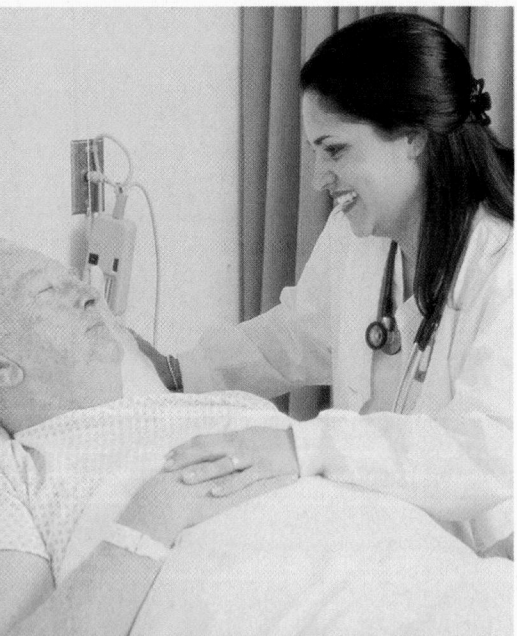

Different Viewpoints

A donor organ must be transplanted within hours. Deciding who should receive an organ involves questions of medical urgency and logistics. In most cases, organs are offered first to patients who are located closest to the donor. This sometimes means that patients who have been waiting longer or are more seriously ill are passed over because they live farther away.

Should the most seriously ill be first? One of the problems with selecting organ recipients by geography is that more donor organs become available in the most densely populated regions of the country. A seriously ill patient who lives far from a large population center might have to wait longer than a patient who lives in a big city. In 1998, the Department of Health and Human Services asked the national organization that coordinates the matching of organs with patients to revise their policies and make organs available to patients regardless of location. If the most seriously ill patient is so far away that the transplant cannot be completed within the time limit, then another patient must be selected. Final decisions are always left to medical professionals who are experts in transplant surgery.

INVESTIGATING THE ISSUE

Analyzing the Issue Interview friends and family about the reasons why they would or would not consider becoming an organ donor. Organize a class discussion to share what you learned.

 To find out more about organ transplants, visit the Glencoe Science Web site.
science.glencoe.com

Going Further

Invite interested students to develop an education program to increase awareness in their community about the need for organ and tissue donors. **L3**

Main Ideas
Summary statements can be used by students to review the major concepts of the chapter.

Using the Vocabulary
To reinforce chapter vocabulary, use the Content Mastery Booklet and the activities in the Interactive Tutor for Biology: The Dynamics of Life on the Glencoe Science Web site.
science.glencoe.com

SUMMARY

Section 37.1

The Respiratory System

Main Ideas

- External respiration involves taking in air through the passageways of the respiratory system and exchanging gases in the alveoli of the lungs.
- Breathing involves contraction of the diaphragm, the rush of air into the lungs, relaxation of the diaphragm, and air being pushed out of the lungs.
- Breathing is controlled by the chemistry of the blood.

Vocabulary
alveoli (p. 1004)
trachea (p. 1003)

 **All Chapter Assessment** questions and answers have been validated for accuracy and suitability by the Princeton Review.

Section 37.2

The Circulatory System

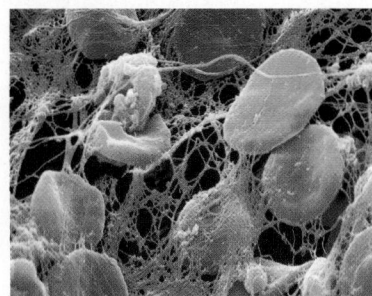

Main Ideas

- Blood is composed of red and white blood cells, platelets, and plasma. Blood carries oxygen, carbon dioxide, and other substances through the body.
  - Blood cell antigens determine blood group and are important in blood transfusions.
  - Blood is carried by arteries, veins, and capillaries.
  - Blood is pushed through the vessels by the heart.

Vocabulary
antibody (p. 1009)
antigen (p. 1009)
aorta (p. 1013)
artery (p. 1011)
atrium (p. 1012)
blood pressure (p. 1016)
capillary (p. 1012)
hemoglobin (p. 1008)
plasma (p. 1007)
platelet (p. 1009)
pulse (p. 1013)
red blood cell (p. 1007)
vein (p. 1012)
vena cava (p. 1013)
ventricle (p. 1012)
white blood cell (p. 1008)

- UNDERSTANDING MAIN IDEAS -

1. a
2. c

Section 37.3

The Urinary System

Main Ideas

- The nephrons of the kidneys filter wastes from the blood.
- The urinary system helps maintain the homeostasis of body fluids.

Vocabulary
kidney (p. 1017)
nephron (p. 1018)
ureter (p. 1017)
urethra (p. 1019)
urinary bladder (p. 1017)
urine (p. 1018)

- UNDERSTANDING MAIN IDEAS -

1. If you have type O blood, what blood group could be used if you needed a transfusion?
   a. type O   **c.** type B
   b. type A   **d.** type AB

2. The primary function of the kidneys is to rid the body of _____.
   a. carbon dioxide wastes
   b. undigested food
   c. wastes in the blood
   d. excess enzymes

CHAPTER 37 ASSESSMENT **1023**

GLENCOE *TECHNOLOGY*

 **VIDEOTAPE**
MindJogger Videoquizzes
Chapter 37: *Respiration, Circulation, and Excretion*
Have students work in groups as they play the videoquiz game to review key chapter concepts.

 **Resource Manager**

Chapter Assessment, pp. 217-222
MindJogger Videoquizzes
ExamView® Pro Software
BDOL Interactive CD-ROM, Chapter 37 quiz

3. nuclei
4. c
5. d
6. b
7. b
8. medulla oblongata
9. a
10. b
11. trachea
12. alveoli, blood
13. carbon dioxide
14. lack a nucleus
15. antigens
16. arteries, blood
17. aorta
18. alveoli, lungs
19. blood, plasma
20. arterioles, capillaries

APPLYING MAIN IDEAS

21. Pulmonary arteries carry oxygen-poor blood from the heart to the lungs, where excess carbon dioxide is released.

22. High blood pressure can damage the capsule by forcing fluids too rapidly through the capsule tissue.

3. White blood cells differ from red blood cells in that white blood cells have _____.

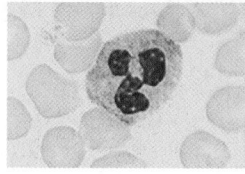

4. Blood cells receive oxygen in the _____.
   a. bronchi   c. alveoli
   b. trachea   d. pharynx

5. Which cell parts are involved in blood clotting?
   a. ribosomes   c. nucleus
   b. mitochondria d. platelets

6. Oxygen travels in the blood attached to _____.
   a. white blood cells   c. platelets
   b. hemoglobin   d. wastes

7. The basic filtering unit of the kidney is the _____.
   a. Bowman's capsule   c. bladder
   b. nephron   d. glomerulus

8. Breathing is an involuntary process that is controlled by the _____.

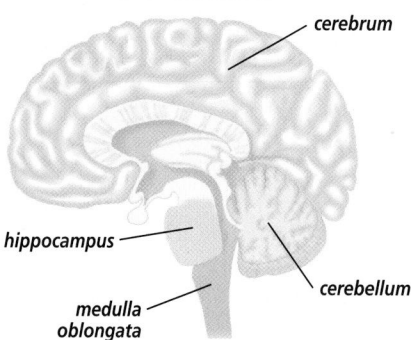

cerebrum

hippocampus

medulla oblongata

cerebellum

 **TEST-TAKING TIP**

Don't Use Outside Knowledge
When answering questions for a reading passage, do not use anything you know about the subject of the passage, or any opinions you have about it. Always return to the passage to reread and get the details from there.

9. Which of the following molecules would NOT be found in normal urine?
   a. glucose   c. salt
   b. water   d. urea

10. The renal artery branches and sends a knot of capillaries into the _____.
    a. glomerulus   c. nephron
    b. Bowman's capsule d. tubules

11. The _____ divides into narrower tubes called bronchi.

12. External respiration involves an exchange of gases between air in the _____ and the _____.

13. Blood that comes to the alveoli from body cells is high in _____.

14. Red blood cells are different from other body cells because they _____.

15. Differences in blood groups are due to the presence or absence of proteins called _____ on red blood cell membranes.

16. _____ are large, elastic blood vessels that carry _____ away from the heart.

17. The _____ is the largest blood vessel in the body.

18. The _____ are small sacs in the _____ where oxygen and carbon dioxide are exchanged.

19. The liquid portion of the _____ is called _____ and it carries blood cells around the body.

20. Arteries branch into _____, which in turn branch into _____.

APPLYING MAIN IDEAS

21. Explain why all blood in all arteries is not oxygen-rich. Where in the circulatory system do arteries carry oxygen-poor blood?

22. Think about the structure of a nephron, and suggest a reason why high blood pressure could damage the kidneys.

23. A diet low in saturated fats and cholesterol helps maintain elasticity and prevent clogging

of the blood vessels. Why is this type of diet considered healthy for your heart?

24. In some large cities, citizens are advised to avoid heavy outdoor exercise when air pollution levels are high. Explain why it would be wise to heed such a warning.

THINKING CRITICALLY

25. **Recognizing Cause and Effect** What activities besides exercise could affect breathing rate?

26. **Designing an Experiment** Design an experiment that would test the effect of various types of music on heartbeat rate.

27. **Formulating Hypotheses** The evolution of the mammalian kidney has allowed mammals to live efficiently on land and lose little water. Explain how the mammalian kidney conserves water.

28. **Concept Mapping** Complete the concept map using the following vocabulary terms: aorta, arteries, left atrium, left ventricle.

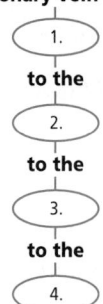

Blood moves from the pulmonary vein to the

1.

to the

2.

to the

3.

to the

4.

CD-ROM

For additional review, use the assessment options for this chapter found on the *Biology: The Dynamics of Life Interactive CD-ROM* and on the Glencoe Science Web site.
science.glencoe.com

ASSESSING KNOWLEDGE & SKILLS

The following graph shows blood pressure fluctuation in blood vessels.

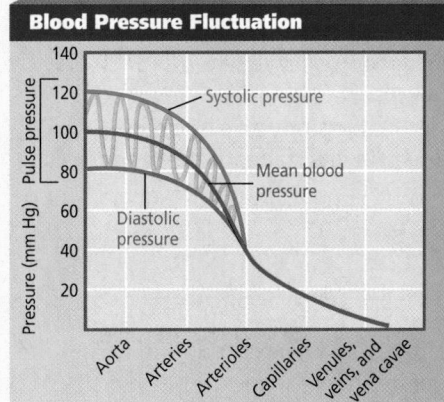

Blood Pressure Fluctuation

Interpreting Data Use the graph to answer the following questions.

1. Which of the variables shown on the graph is the dependent variable?
   a. type of blood vessel
   b. systolic pressure
   c. diastolic pressure
   d. blood pressure

2. In which blood vessel is blood pressure greatest?
   a. the aorta   c. the capillaries
   b. the arteries   d. the veins

3. What is the mean blood pressure in the arteries?
   a. 120 mm Hg   c. 80 mm Hg
   b. 90 mm Hg   d. 70 mm Hg

4. **Observing and Inferring** Explain why the blood pressure changes periodically in the aorta and the small arteries, but not in the capillaries and veins.

23. Diets high in saturated fats and cholesterol are associated with increased risk of atherosclerosis.

24. Most air pollution contains high levels of carbon monoxide and other toxic substances. Exercise would increase the amount of these toxic gases drawn into the lungs.

THINKING CRITICALLY

25. sleeping, eating, watching an exciting movie, use of stimulants or depressants

26. Experiments might include playing various types of music while taking pulse rate and comparing with pulse rate without music.

27. In the tubules of the nephron, much of the water present in the filtrate is reabsorbed.

28. 1. Left atrium; 2. Left ventricle; 3. Aorta; 4. Arteries

ASSESSING KNOWLEDGE & SKILLS

1. d
2. a
3. b
4. The muscular aorta and arteries expand when the left ventricle forces blood into them. The muscles of these vessels recoil in between heartbeats, helping to push the blood along.

Chapter 38 Organizer

Refer to pages 4T-5T of the Teacher Guide for an explanation of the National Science Education Standards correlations.

Section	Objectives	Activities/Features
Section 38.1 **Human Reproductive Systems** National Science Education Standards UCP.1-3, UCP.5; C.1, C.5; F.1 (2 sessions, 1½ blocks)	1. **Identify** the parts of the male and female reproductive systems. 2. **Summarize** the negative feedback control of reproductive hormones. 3. **Sequence** the stages of the menstrual cycle.	**Inside Story:** Sex Cell Production, p. 1033 **Problem-Solving Lab 38-1,** p. 1035
Section 38.2 **Development Before Birth** National Science Education Standards UCP.1-3, UCP.5; A.1, A.2; C.1, C.5, C.6; E.1, E.2; F.1, F.6; G.1 (2 sessions, 1 block)	4. **Summarize** the events during each trimester of pregnancy.	**MiniLab 38-1:** Examining Sperm and Egg Attraction, p. 1038 **MiniLab 38-2:** Making a Graph of Fetal Size, p. 1042 **Problem-Solving Lab 38-2,** p. 1043 **Investigate BioLab:** What hormone is produced by an embryo? p. 1048 **BioTechnology:** Frozen Embryos, p. 1050
Section 38.3 **Birth, Growth, and Aging** National Science Education Standards UCP.1, UCP.3, UCP.5; A.1, A.2; C.6; E.1, E.2; F.1, F.6; G.1-3 (2 sessions, ½ block)	5. **Describe** the three stages of birth. 6. **Summarize** the developmental stages of humans after they are born.	**Careers in Biology:** Midwife, p. 1046

Need Materials? Contact Carolina Biological Supply Company at 1-800-334-5551 or at **http://www.carolina.com**

MATERIALS LIST

BioLab
p. 1048 scissors, heavy paper, tracing paper

MiniLabs
p. 1038 microscope, microscope slide, droppers (2), live sea urchin eggs, live sea urchin sperm
p. 1042 graph paper, pencil

Alternative Lab
p. 1030 graph paper, colored pencils (4 colors), data sheet

Quick Demos
p. 1029 microprojector, prepared slide of testis cross section
p. 1039 microprojector, prepared slides of sea star embryos
p. 1045 overhead projector, photos of infants and elderly persons

Key to Teaching Strategies

L1 Level 1 activities should be appropriate for students with learning difficulties.

L2 Level 2 activities should be within the ability range of all students.

L3 Level 3 activities are designed for above-average students.

ELL ELL activities should be within the ability range of English Language Learners.

COOP LEARN Cooperative Learning activities are designed for small group work.

P These strategies represent student products that can be placed into a best-work portfolio.

These strategies are useful in a block scheduling format.

Teacher Classroom Resources

Section	Reproducible Masters	Transparencies
Section 38.1 **Human Reproductive Systems**	Reinforcement and Study Guide, pp. 167-168 L2 Concept Mapping, p. 38 L3 ELL Critical Thinking/Problem Solving, p. 38 L3 Content Mastery, pp. 185-186, 188 L1	Section Focus Transparency 93 L1 ELL Basic Concepts Transparency 74 L2 ELL Basic Concepts Transparency 75 L2 ELL Reteaching Skills Transparency 55 L1 ELL
Section 38.2 **Development Before Birth**	Reinforcement and Study Guide, p. 169 L2 BioLab and MiniLab Worksheets, pp. 169-170 L2 Laboratory Manual, pp. 277-284 L2 Content Mastery, pp. 185, 187-188 L1	Section Focus Transparency 94 L1 ELL Reteaching Skills Transparency 56a, 56b L1 ELL
Section 38.3 **Birth, Growth, and Aging**	Reinforcement and Study Guide, p. 170 L2 BioLab and MiniLab Worksheets, pp. 171-172 L2 Content Mastery, pp. 185, 187-188 L1 Tech Prep Applications, pp. 53-54 L2	Section Focus Transparency 95 L1 ELL

Assessment Resources

Chapter Assessment, pp. 223-228
MindJogger Videoquizzes
Performance Assessment in the Biology Classroom
Alternate Assessment in the Science Classroom
ExamView® Pro Software 💾
BDOL Interactive CD-ROM, Chapter 38 quiz

Additional Resources

Spanish Resources ELL
English/Spanish Audiocassettes ELL
Cooperative Learning in the Science Classroom COOP LEARN
Lesson Plans/Block Scheduling

NATIONAL GEOGRAPHIC — Teacher's Corner

Products Available From Glencoe
To order the following products, call Glencoe at 1-800-334-7344:
Curriculum Kit
GeoKit: Human Body 2
Videodisc
STV: Human Body

Products Available From National Geographic Society
To order the following products, call National Geographic Society at 1-800-368-2728:
Videos
Incredible Human Machine
Reproductive Systems (Human Body Series)

GLENCOE TECHNOLOGY

The following multimedia resources are available from Glencoe.

Biology: The Dynamics of Life
CD-ROM ELL
Video: *Fetal Development*

Videodisc Program
Human Fertilization
Fetal Development

The Secret of Life Series
Testis
 Cross Section of Ovary

38 Reproduction and Development

Theme Development

The themes of **homeostasis** and **systems and interactions** are evident in the study of the hormone regulation of the male and female reproductive systems and in the examination of embryonic membranes, fetal development, growth, and aging.

What You'll Learn

- You will compare and contrast the anatomy, control, and function of the male and female reproductive systems.
- You will distinguish the stages of development before birth.
- You will summarize the processes of birth, growth, and aging.

Why It's Important

As you grow and develop, your reproductive system is maturing. The human reproductive system prepares sex cells—sperm or eggs—which, when combined, ensure the continuation of our species.

READING BIOLOGY

Skim the chapter, writing down some new terms that are used. Next, look at the list and separate the words into categories for male and female. How do the characteristics of the male and female reproductive systems differ? How are they similar?

BIOLOGY Online

To find out more about reproduction and development, visit the Glencoe Science Web site.
science.glencoe.com

Like a NASA astronaut in a space suit, a human fetus is protected inside a controlled environment.

1026

38.1 Human Reproductive Systems

As the small sperm approach the large egg, their size difference becomes very apparent. Yet, a sperm and an egg each carry half of the chromosomes needed for the growth and development of a complete individual. As a sperm merges with an egg, a new life is launched.

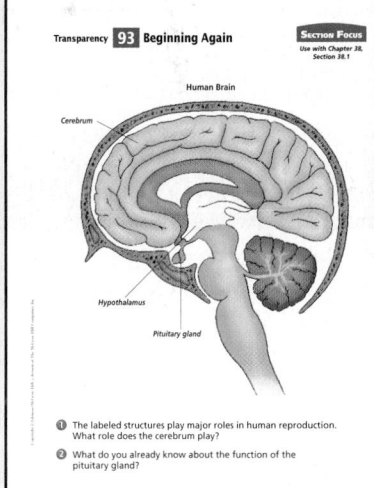

Egg surrounded by sperm and sperm in the testes (inset).

Magnification: 165×

Magnification: 9750×

SECTION PREVIEW

Objectives

Identify the parts of the male and female reproductive systems.

Summarize the negative feedback control of reproductive hormones.

Sequence the stages of the menstrual cycle.

Vocabulary

scrotum
epididymis
vas deferens
seminal vesicle
prostate gland
bulbourethral gland
semen
puberty
oviduct
cervix
follicle
ovulation
menstrual cycle
corpus luteum

Human Male Anatomy

The ultimate goal of the reproductive process is the formation and union of egg and sperm, development of the fetus, and birth of the infant. The organs, glands, and hormones of the male reproductive system are instrumental in meeting this goal. Their main functions are the production of sperm—the male sex cells—and their delivery to the female.

Where sperm form

Sperm production takes place in the testes, which are located in the scrotum. The **scrotum** is a sac that contains the testes and is suspended directly behind the base of the penis.

Before birth, the testes form in the embryo's abdomen and then descend into the scrotum. Because sperm can develop only in an environment with a temperature about 3°C lower than normal body temperature, the scrotum is positioned outside the abdomen. Muscles in the walls of the scrotum help maintain the proper temperature. The muscles contract in response to cold temperatures, pulling the scrotum closer to the body for warmth. The muscles relax in response to warm temperatures, lowering the scrotum to allow air to circulate and cool both testes and sperm.

Figure 38.1 shows the organs and glands of the male reproductive system.

38.1 HUMAN REPRODUCTIVE SYSTEMS **1027**

Section 38.1

Prepare

Key Concepts

This section focuses on the anatomy and physiology of the male and female reproductive systems. It describes the negative-feedback control of sexual hormones, and hormonal control of the menstrual cycle.

Planning

■ Make photocopies of a SEM of a sperm cell for the Getting Started Demo.

---

1 Focus

Bellringer 👆

Before presenting the lesson, display **Section Focus Transparency 93** on the overhead projector and have students answer the accompanying questions.
L1 **ELL**

Transparency **93** Beginning Again

SECTION FOCUS
Use with Chapter 38, Section 38.1

Human Brain

Cerebrum

Hypothalamus

Pituitary gland

❶ The labeled structures play major roles in human reproduction. What role does the cerebrum play?

❷ What do you already know about the function of the pituitary gland?

BIOLOGY: The Dynamics of Life

SECTION FOCUS TRANSPARENCIES

Resource Manager

Section Focus Transparency 93 and Master **L1** **ELL**

✓ Assessment Planner

Portfolio Assessment
  Portfolio, TWE, pp. 1032, 1034, 1039, 1045
Performance Assessment
  Assessment, TWE, pp. 1036, 1042
  Alternative Lab, TWE pp. 1030-1031
  BioLab, TWE, pp. 1048-1049
  BioLab, SE, pp. 1048-1049
  MiniLab, TWE, pp. 1038, 1042
  MiniLab, SE, pp. 1038, 1042
  Problem-Solving Lab, TWE, p. 1043

Knowledge Assessment
  Assessment, TWE, pp. 1041, 1047
  Section Assessment, SE, pp. 1036, 1043, 1047
  Chapter Assessment, SE, pp. 1051-1053
Skill Assessment
  Assessment, TWE, pp. 1028, 1034, 1035, 1038, 1039, 1045
  Problem-Solving Lab, TWE, p. 1035
  Problem-Solving Lab, SE, pp. 1035, 1043

1027

Visual Learning

Figure 38.1 Have students use the illustration of the male reproductive system to trace the path of sperm from the testes to the outside of the body. *testes, epididymus, vas deferens, urethra* **L1**

Using Science Terms

The Latin word *testis* means "witness," while *testicle* means "little witness." Explain that these terms may have developed as a result of the Romans allowing only men to testify in court.

Tying to Previous Knowledge

Ask students to recall how sex cells are produced by meiosis. If necessary, remind them that meiosis produces haploid cells.

Skill Have students make a table with the following headings: Organ, Reproductive Function. In the first column, have students list the organs of the male reproductive system. In the second column, they should describe each organ's function. **L2**

NATIONAL GEOGRAPHIC

VIDEODISC
STV: Human Body Vol. 3
Reproductive Systems
Unit 2, Side 2, 5 min. 41 sec.
The Male System

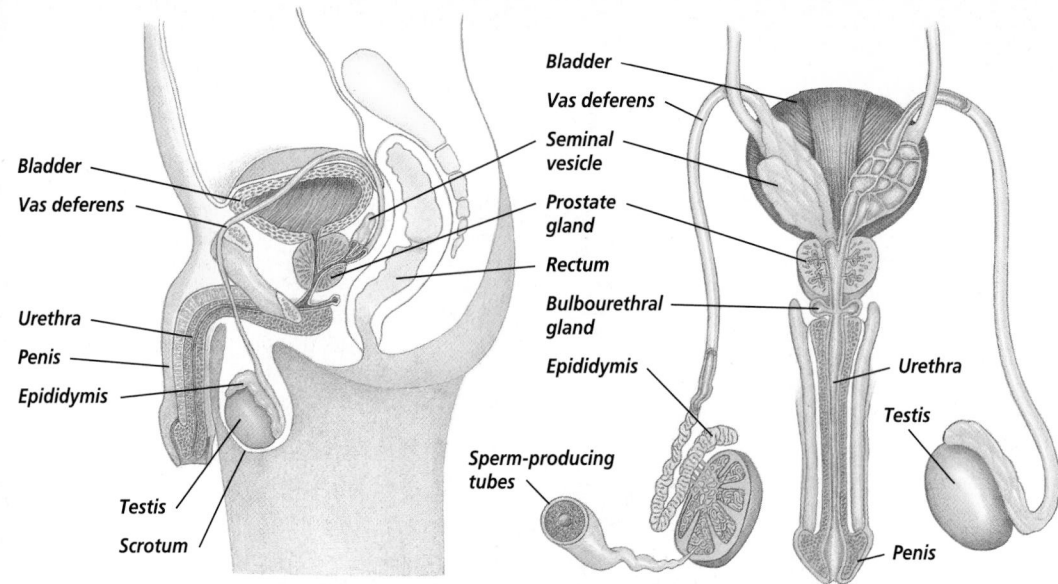

Figure 38.1
The organs and glands of the male reproductive system are shown in front and side views.

Within each testis is a fine network of highly coiled tubes. Sperm are produced by meiosis of the cells that line these tubes. Recall that meiosis produces haploid cells. When a single cell in the testis divides by meiosis, it produces four haploid cells. All four of these cells develop into mature sperm over a period of about 74 days. A sexually mature human male can produce about 300 million mature sperm per day, each day of his life.

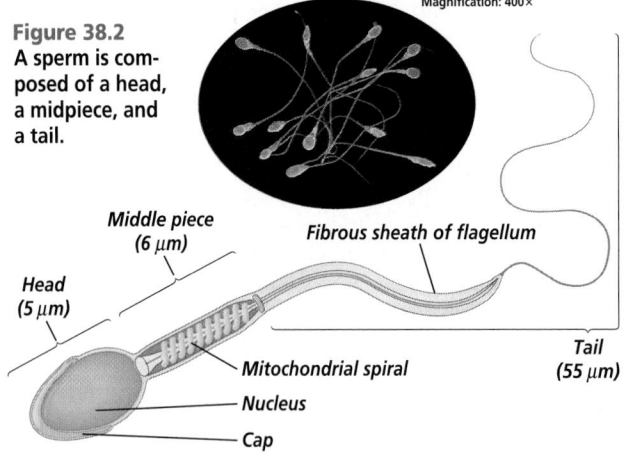

Figure 38.2
A sperm is composed of a head, a midpiece, and a tail.

Magnification: 400×

Middle piece (6 μm)

Head (5 μm)

Fibrous sheath of flagellum

Mitochondrial spiral

Nucleus

Cap

Tail (55 μm)

As you can see in **Figure 38.2**, a sperm is highly adapted for reaching and entering the female egg. The head portion of a sperm contains the nucleus and is covered by a cap containing enzymes that help penetrate the egg. A number of mitochondria are found in the midpiece of the sperm; they provide energy for locomotion. The tail is a typical flagellum that propels the sperm along its way. Sperm can live for about 48 hours inside the female reproductive tract.

How sperm leave the testes

Before the sperm mature, they move out of the testes through a series of coiled ducts that empty into a single tube called the epididymis. The **epididymis** (ep uh DIHD uh mus) is a coiled tube within the scrotum in which the sperm complete their maturation.

When sperm are released from the epididymis, they enter the vas deferens, where they are stored until they are released from the body. The **vas deferens** (vas DEF uh runz) is a

1028 REPRODUCTION AND DEVELOPMENT

MEETING INDIVIDUAL NEEDS

Learning Disabled

Visual-Spatial Have students use tracing paper to trace the structures of the male reproductive system shown in Figure 38.1. Ask them to label each structure and write a description of its function beside the label. Have them include the diagrams they have created in their journals and use them as study tools. **L1**

Gifted

Linguistic Have students conduct research on prostate cancer. Ask them to find out what factors place a man at risk for this type of cancer, the possible treatments, and what factors determine the likelihood of survival with treatment. Have students prepare written reports of their findings for their portfolios. **L3**

duct that transports sperm from the epididymis toward the ejaculatory ducts and the urethra. Peristaltic contractions of the vas deferens force the sperm along. The urethra is a tube in the penis that transports sperm out of the male's body. Notice in *Figure 38.1* that the urethra also transports urine from the urinary bladder. A muscle located at the base of the bladder prevents urine and sperm from mixing.

Fluids that help transport sperm

As sperm travel from the testes, they mix with fluids that are secreted by several different glands. The **seminal vesicles** are a pair of glands located at the base of the urinary bladder. They secrete a mucouslike fluid into the vas deferens. The fluid is rich in the sugar fructose, which provides energy for the sperm cells.

The **prostate gland** is a single, doughnut-shaped structure that lies below the urinary bladder and surrounds the top portion of the urethra. The prostate secretes a thinner, alkaline fluid that helps sperm move and survive. Two tiny **bulbourethral** (bul boh yoo REE thrul) **glands** are located beneath the prostate. These

glands secrete a clear, sticky, alkaline fluid that protects sperm by neutralizing the acidic environment of the vagina. The combination of sperm and all of these fluids is called **semen.**

Hormonal Control

In an earlier chapter, you learned that the glands of the endocrine system release hormones, which play a key role in the regulation of body functions, metabolism, and homeostasis. Hormones also control the development and activity of the male reproductive system.

Hormones and male puberty

It's obvious from the physical appearance of young children that they are not sexually mature. In the early teen years, as shown in *Figure 38.3*, changes to a child's body begin to occur. Puberty begins. **Puberty** refers to the time when secondary sex characteristics begin to develop so that sexual maturity—the potential for sexual reproduction—is reached. The changes associated with puberty are controlled by sex hormones secreted by the endocrine system.

WORD *Origin*

epididymis
From the Greek words *epi*, meaning "upon," and *didymos*, meaning "testis." The epididymis tube is on top of the testis.

Figure 38.3
Puberty results in many physical and emotional changes. Generally, males undergo puberty sometime between the ages of 13 and 16.

1029

Reinforcement

Visual-Spatial Have students use the diagram in Figure 38.4B and the text description of negative feedback systems to develop a flowchart that shows the sequence of events involved in the production and release of the hormones FSH and LH. **L2**

Revealing Misconceptions

Many people believe that only the sperm is an active player in the fusion of egg and sperm. Explain that a team of researchers at Johns Hopkins University have determined that sperm that reach an egg are held to the egg by receptor molecules on the egg's surface. These molecules hook together with counterparts on the surface of the sperm. The receptor molecules fasten the sperm tightly until it can be absorbed by the egg.

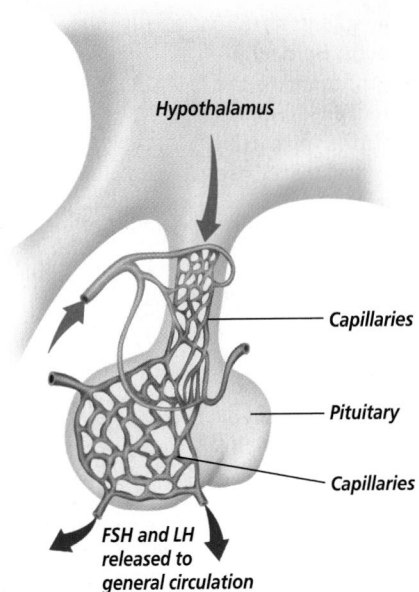

Hypothalamus

Capillaries

Pituitary

Capillaries

FSH and LH released to general circulation

A In the hypothalamus, neurons secrete releasing and inhibiting hormones. These hormones travel to the pituitary gland, where they increase or decrease the secretion of FSH and LH.

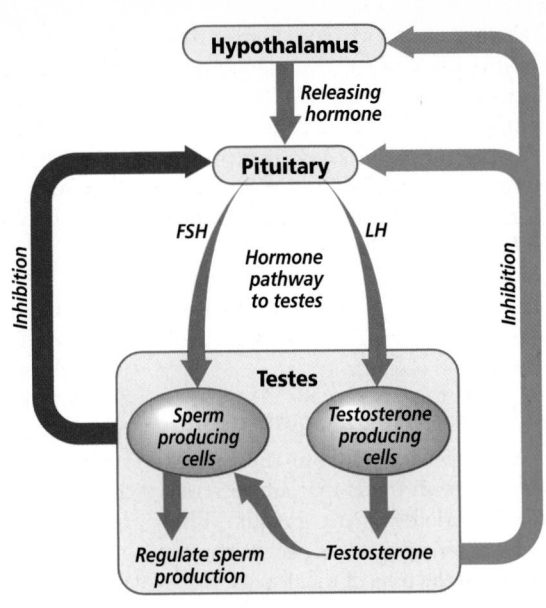

B Release of FSH and LH from the pituitary gland stimulates production of sperm and of testosterone. As testosterone levels increase, production of FSH and LH slows. When sperm and testosterone levels drop, production of FSH and LH increases again.

Figure 38.4
The activity of the male reproductive system is controlled by the hypothalamus and the pituitary gland in the brain.

Hormones and the male reproductive system

In males, the onset of puberty causes the hypothalamus to produce several kinds of hormones that interact with the pituitary gland, which influences many physiological processes of the body. As shown in *Figure 38.4A*, the hypothalamus secretes a hormone that causes the pituitary to release two other hormones: follicle-stimulating hormone (FSH) and luteinizing (LEW teen i zing) hormone (LH). When released into the bloodstream, FSH and LH are transported to the testes. In the testes, FSH causes the production of sperm cells. LH causes endocrine cells in the testes to produce the male hormone, testosterone (tehs TAHS tuh rohn), which in turn influences sperm cell production.

The levels of these hormones in the body are regulated by a negative-feedback system. As the testosterone levels in the blood increase, the production of FSH and LH is inhibited, or decreased. Increased production of sperm in the testis also feeds back into the system to inhibit production of FSH and LH, as *Figure 38.4B* illustrates. When testosterone levels in the blood drop, production of FSH and LH increases.

Testosterone is the steroid hormone responsible for the growth and development of secondary sex characteristics in a male. These characteristics include growth and maintenance of male sex organs; the production of sperm; an increase in body hair, especially on the face, under the arms, and in the pubic area; an increase in muscle mass; increased growth of the long bones of the arms and legs; and deepening of the voice.

Alternative Lab

Tracking Hormone Levels

Purpose
Students will graph and analyze patterns of change that take place in female hormones during the menstrual cycle.

Materials
graph paper, colored pencils (red, yellow, blue, green), data sheet

Preparation
Prepare data sheets for students that include the information given here.

Data Table														
Day	2	4	6	8	10	12	14	16	18	20	22	24	26	28
LH	17	17	17	17	17	46	35	20	19	18	17	16	14	13
FSH	14	14	14	13	10	8	15	8	7	7	6	6	6	7
Estrogen	4	4	5	6	10	13	13	10	9	10	11	11	11	8
Progesterone	1	1	1	1	1	1	2	4	7	12	14	14	9	3

Human Female Anatomy

The main functions of the female reproductive system are to produce eggs, which are the female sex cells, and to provide an environment in which a fertilized egg can develop. Egg production takes place in the two ovaries. Each ovary is about the size and shape of an almond. One ovary is located on each side of the lower part of the abdomen.

As you can see in *Figure 38.6,* the open end of an oviduct is located close to each ovary. The **oviduct** is a tube that transports eggs from the ovary to the uterus. Peristaltic contractions of the muscles in the wall of the oviduct combine with beating cilia to move the egg through the tube.

You learned earlier that female mammals have a uterus in which the fetus develops during pregnancy. The human uterus is situated between the urinary bladder and the rectum and is shaped like an inverted pear. The uterine wall is composed of three layers: an outer layer of connective tissue; a thick, muscular middle layer; and a thin, inner lining called the endometrium (en doh MEE tree um). The lower end of the uterus, called the **cervix,** tapers to a narrow opening into the vagina, which is a passageway to the outside of the female's body.

Puberty in Females

As in males, puberty in females begins when the hypothalamus signals the pituitary to produce and release the hormones FSH and LH. These are the same hormones that are produced in males; however, in females, FSH stimulates the development of follicles in the ovary. A **follicle** is a group of epithelial cells that surround a developing egg cell. FSH also causes the release of the hormone estrogen from the ovary. Estrogen is the steroid hormone responsible for the secondary sex characteristics of females. These characteristics include the growth and maintenance of female sex

Figure 38.6
The female reproductive system includes two ovaries, two oviducts—sometimes called fallopian tubes—the uterus, and the vagina.

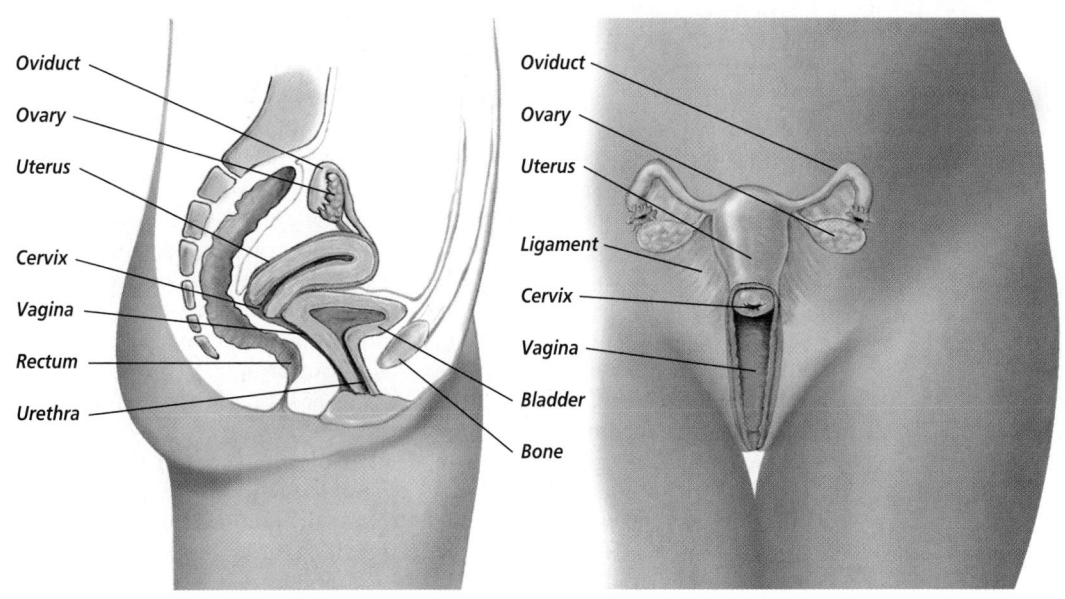

Oviduct
Ovary
Uterus
Cervix
Vagina
Rectum
Urethra

Oviduct
Ovary
Uterus
Ligament
Cervix
Vagina
Bladder
Bone

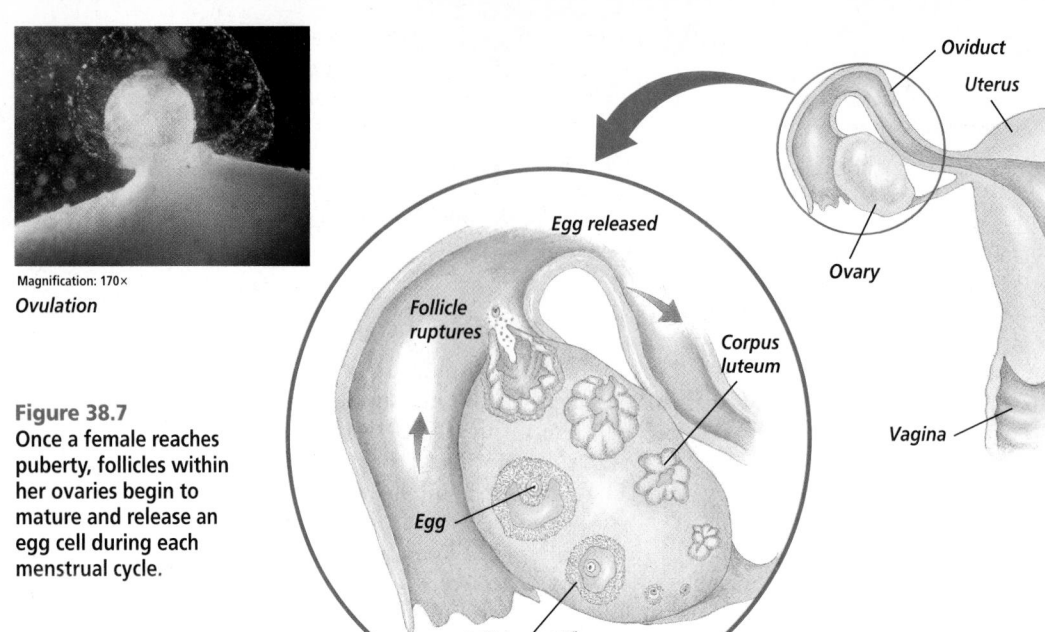

Magnification: 170×
Ovulation

Figure 38.7
Once a female reaches puberty, follicles within her ovaries begin to mature and release an egg cell during each menstrual cycle.

organs; an increase in body hair, especially under the arms and in the pubic area; an increase in the growth rates of the long bones of the arms and legs; a broadening of the hips; an increase in fat deposits in the breasts, buttocks, and thighs; and the onset of the menstrual cycle.

Production of eggs

Recall that sperm production does not begin in males until they reach puberty, after which time it continues for the rest of their lives. Egg production is different. Even before a female is born, her body begins to develop eggs. During this prenatal period, cells in her ovaries divide until the first stage of meiosis, prophase I, is reached. At this point, the cells go into a resting stage. At birth, a female's ovaries contain about two million of these potential eggs, which are called primary oocytes. Many of these break down, or degenerate. At

puberty, a female's ovaries contain about 40 000 primary oocytes. How does the production of sperm differ from the production of egg cells? To find out, read the *Inside Story.*

How eggs are released

About once a month, beginning at puberty, the process of meiosis starts up again in several of the prophase I cells. Each cell completes meiosis I and begins meiosis II. During meiosis II, one of the egg cells ruptures from the ovary and passes into the oviduct. The process of the egg rupturing through the ovary wall and moving into the oviduct is called **ovulation.** A total of about 400 eggs are ovulated during the reproductive life of a female. Fertilization, if it takes place, occurs in the oviduct. *Figure 38.*7 shows the process leading to ovulation. Usually, only one follicle matures and releases an egg each month.

INSIDE STORY

Sex Cell Production

As with many other animals, human sex cells are produced by meiosis. A mature male produces millions of swimming sperm cells each day. A mature female usually releases only one mature egg each month.

Critical Thinking *Compare the number of sex cells produced by each meiotic division in the testes and the ovaries.*

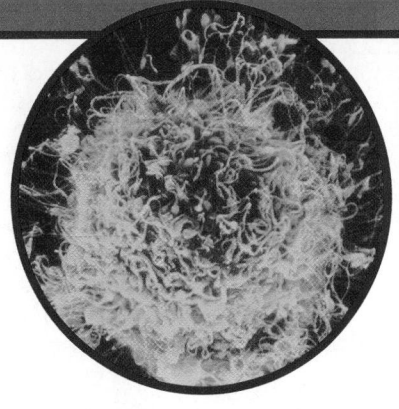

Human egg and sperm

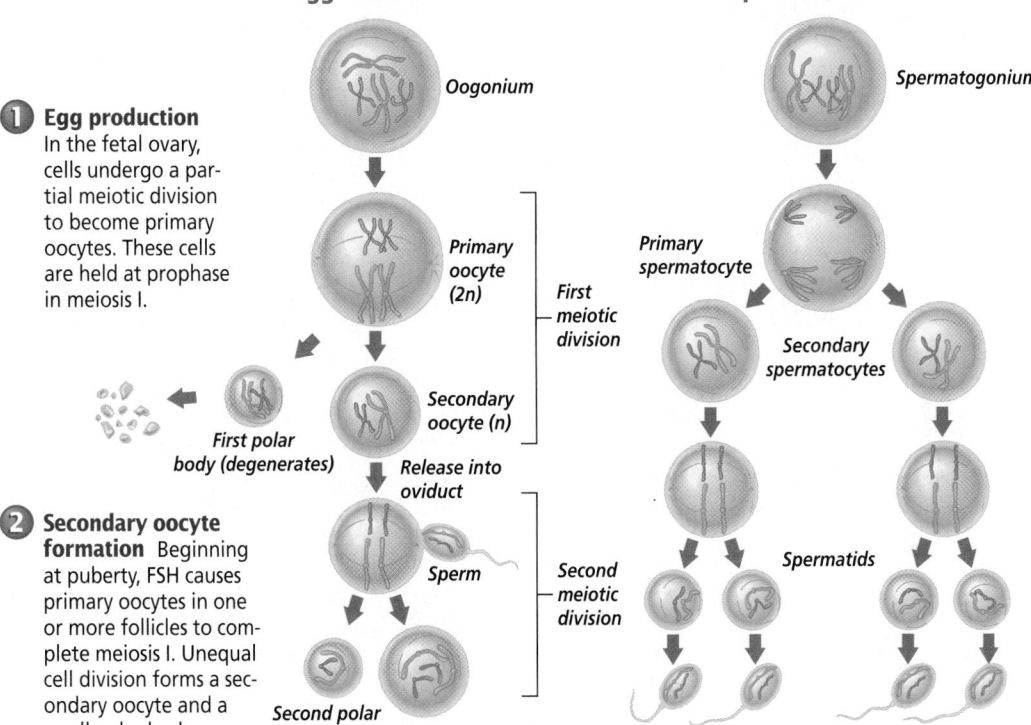

Egg Formation

Oogonium

1 Egg production In the fetal ovary, cells undergo a partial meiotic division to become primary oocytes. These cells are held at prophase in meiosis I.

Primary oocyte (2n)

First meiotic division

First polar body (degenerates)

Secondary oocyte (n)

Release into oviduct

2 Secondary oocyte formation Beginning at puberty, FSH causes primary oocytes in one or more follicles to complete meiosis I. Unequal cell division forms a secondary oocyte and a small polar body.

Sperm

Second meiotic division

Second polar body (degenerates)

Zygote

3 Ovulation Usually, only one secondary oocyte is released from the ovary at ovulation. That oocyte will complete meiosis II only if fertilization takes place. Meiosis II produces another polar body.

Sperm Formation

Spermatogonium

Primary spermatocyte

Secondary spermatocytes

Spermatids

Mature sperm cells

4 Sperm production Once puberty begins, cells within the testes undergo meiosis daily to begin the formation of sperm. Meiosis I and II take place in the tubules of the testes. Maturation takes place in the epididymis. Mature sperm are stored in the vas deferens.

38.1 HUMAN REPRODUCTIVE SYSTEMS **1033**

INSIDE STORY

Purpose
Students will compare and contrast human egg production with sperm production.

Teaching Strategies
■ Have students identify similarities and differences in egg and sperm production. *Egg production begins in the female fetus; sperm production begins in the male at puberty. Mature sperm are produced daily; mature eggs are produced only once a month.*

Visual Learning
■ Using a microscope projector, show students a cross-section slide of an ovary. Point out developing follicles that contain a cell ready to complete meiosis (future egg cell). Project a cross-section of a testis, showing the sperm cells that result from meiosis.

Critical Thinking
For every cell starting meiosis in the male, there are four sperm cells produced; in the female, only one egg is produced. The other three cells in the female are polar bodies, which disintegrate.

The Menstrual Cycle

The series of changes in the female reproductive system that includes producing an egg and preparing the uterus for receiving it is known as the **menstrual cycle.** The entire menstrual cycle repeats about once a month. Once an egg has been released during ovulation, the part of the follicle that remains in the ovary develops into a structure called the **corpus luteum.** The corpus luteum secretes the hormones estrogen and progesterone. Progesterone causes changes to occur in the lining of the uterus that prepare it for receiving a fertilized egg. The menstrual cycle begins during puberty and continues for 30 to 40 years, until menopause. At menopause, the female stops releasing eggs and the secretion of female hormones decreases.

The length of each menstrual cycle varies from female to female, but the average is 28 days. If the egg released at ovulation is not fertilized, the lining of the uterus is shed, causing some bleeding for a few days. The entire menstrual cycle can be divided into three phases: the flow phase, the follicular phase, and the luteal phase, illustrated in *Figure 38.8*. The timing of each phase of the menstrual cycle correlates with hormone output from the pituitary gland, changes in the ovary, and changes in the uterus. *Figure 38.9* shows how the cycle is altered when fertilization occurs. Carry out the *Problem-Solving Lab* to find out how the phases of the menstrual cycle can vary in length.

Flow phase

Day 1 of the menstrual cycle is the day menstrual flow begins. Menstrual flow is the shedding of blood, tissue fluid, mucus, and epithelial cells that made up the lining of the uterus, the endometrium. This flow passes from the uterus through the cervix and the vagina to the outside of the body. Contractions of the uterine muscle help expel the uterine lining and can cause discomfort in some females. Generally, menstrual flow ends by day

Figure 38.8
Changes in the uterine lining, follicles, and hormone levels take place during each phase of the menstrual cycle.

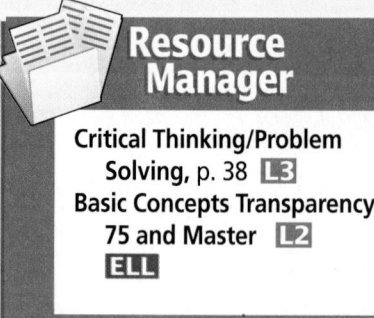

5 of the cycle. During the flow phase, the level of FSH in the blood begins to rise, and another follicle in one of the ovaries begins to mature as meiosis of the prophase I cell proceeds.

Follicular phase

The second phase of the menstrual cycle is more varied in length than the other phases. In a 28-day cycle, it lasts from about day 6 to day 14. As the follicle containing a primary oocyte continues to develop, it secretes estrogen, which stimulates the repair of the endometrial lining of the uterus. The endometrial cells undergo mitosis, and the lining thickens. The steady increase in estrogen also feeds back to the hypothalamus and pituitary gland, which slows the production of FSH and LH. Just before ovulation, estrogen levels peak, stimulating a sudden, sharp increase in the release of LH.

Ovulation occurs at about day 14, when the sharp increase in LH causes the follicle to rupture and release the egg into the oviduct. At this time, the female's body temperature increases

Problem-Solving Lab 38-1 — Applying Concepts

What happens when the menstrual cycle is not exactly 28 days? How does the number of days spent in each phase differ?

Analysis

The graph compares menstrual cycles of different lengths. Study the graph and then answer the questions that follow.

Variation in Phase Lengths for the Menstrual Cycle

24-day cycle: M F L — Back to M
28-day cycle: M F L — Back to M
35-day cycle: M F L — Back to M

14 days
Ovulation

M = flow phase   F = follicular phase   L = luteal phase

Thinking Critically

1. Which phase does not vary in length, regardless of the total time for a cycle? Which hormones are associated with this phase?
2. Offer a possible explanation for why the length of the follicular phase may vary.
3. How would these events differ for the cycle during which a female becomes pregnant?

Figure 38.9
Negative feedback controls the levels of hormones in the female during her menstrual cycle. If fertilization occurs, this cycle is broken and hormone levels change in preparation for the developing embryo and fetus.

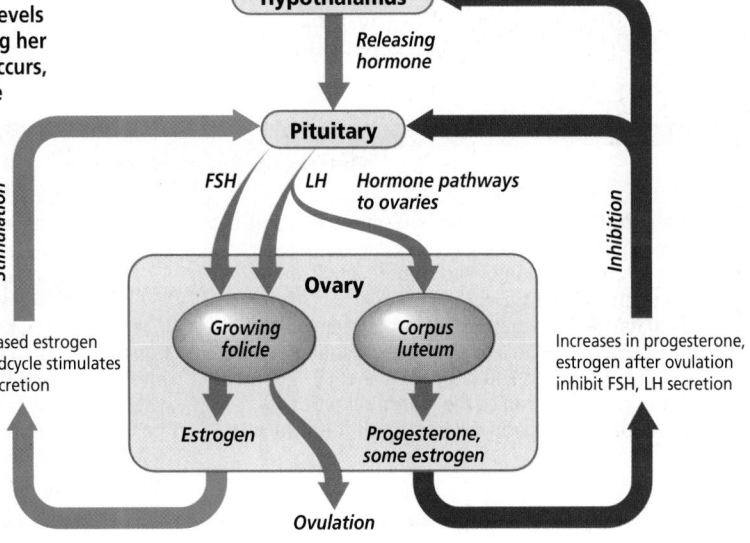

Hypothalamus
Releasing hormone
Pituitary
FSH   LH   Hormone pathways to ovaries
Stimulation
Inhibition
Ovary
Growing folicle
Corpus luteum
Increased estrogen at midcycle stimulates LH secretion
Increases in progesterone, estrogen after ovulation inhibit FSH, LH secretion
Estrogen
Progesterone, some estrogen
Ovulation

Problem-Solving Lab 38-1

Purpose
Students will study similarities and differences among menstrual cycles of differing lengths of time.

Process Skills
analyze information, apply concepts, compare and contrast, think critically, draw a conclusion, interpret data, make and use graphs

Teaching Strategies
■ Make sure students are familiar with the three phases of the menstrual cycle.
■ Suggest that students refer to the graph and diagrams in Figures 38.7 and 38.8 for further information, if needed.

Thinking Critically
1. luteal phase; progesterone and estrogen
2. Student answers may vary. The three hormones interact, and may influence one another so as to alter the exact timing of the follicular phase
3. The corpus luteum does not degenerate and continues to secrete hormones. The lining of the uterus is not shed, so there is no flow phase.

✓ Assessment
Skill Have students prepare a graph that depicts the sequence of events in a menstrual cycle during which a female becomes pregnant. Use the Performance Task Assessment List for Graph from Data in **PASC,** p. 39. **L2**

BIOLOGY JOURNAL

Diagramming the Menstrual Cycle

Visual-Spatial Have students create flowcharts or concept maps showing the sequence of events that occurs during each phase of the menstrual cycle. Ask students to include information about all hormones involved in the regulation of the cycle. **L2**

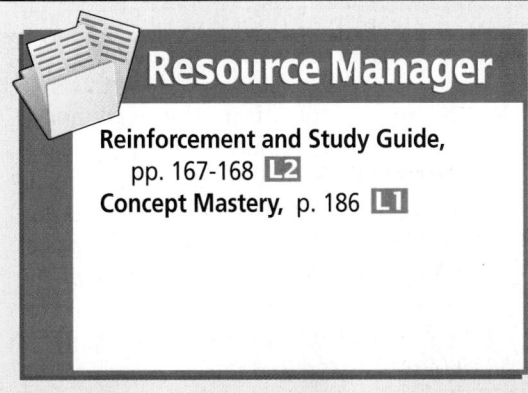

Resource Manager

Reinforcement and Study Guide, pp. 167-168 **L2**
Concept Mastery, p. 186 **L1**

Check for Understanding

Visual-Spatial Display a diagram of the female reproductive system. Ask students to identify where an egg would be located within the system at various stages of the menstrual cycle. **L1**

Reteach

Visual-Spatial Provide students with diagrams of the male and female reproductive systems. Have them draw a line to show how sperm and an unfertilized egg travel through their respective reproductive systems. **L1**

Extension

Have students research the development and specialization of sperm cells within the testis and epididymis. Ask them to prepare a report of their findings. **L3**

✔ Assessment

Performance Ask students to label diagrams of the female and male reproductive systems. Have them indicate, using colored pencils and arrows, the pathways of sperm and egg through these systems. **L2**

4 Close

Discussion

Have each student anonymously submit three questions written on index cards. Read the questions aloud and have volunteers suggest possible answers. Clear up any misconceptions students may have. As an alternative, invite the school nurse or a physician to visit to address the questions.

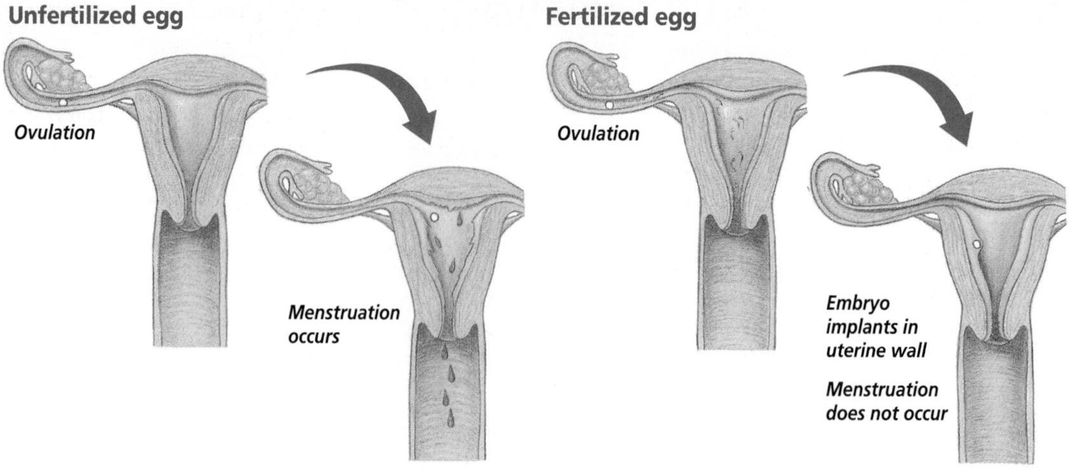

Unfertilized egg — Ovulation — Menstruation occurs

Fertilized egg — Ovulation — Embryo implants in uterine wall — Menstruation does not occur

Figure 38.10
Events that take place in the uterine wall after ovulation depend on whether or not fertilization has occurred.

about 0.5°C. In addition, the cells of the cervix produce large amounts of mucus. Some females also experience discomfort in the area of one or both ovaries around the time of ovulation.

Luteal phase

The last stage of the menstrual cycle, from days 15 to 28, is named the luteal phase, for the corpus luteum. During the luteal phase, LH stimulates the corpus luteum to develop from the ruptured follicle. The corpus luteum produces progesterone and some estrogen. Progesterone increases the blood supply of the endometrium, causing it to accumulate lipids and tissue fluid. These changes correspond to the arrival of a fertilized egg. Through negative feedback, progesterone prevents the production of LH.

If the egg is not fertilized, the rising levels of progesterone and estrogen from the corpus luteum cause the hypothalamus to inhibit the release of FSH and LH. The corpus luteum degenerates and stops secreting progesterone or estrogen. As hormone levels drop, the thick lining of the uterus begins to shed. If fertilization occurs, as shown in **Figure 38.10,** the endometrium begins secreting a fluid rich in nutrients for the zygote.

Section Assessment

Understanding Main Ideas

1. Describe the pathway of an unfertilized egg as it travels through the female reproductive system. Explain the function of each reproductive organ through which it travels.
2. Summarize the negative-feedback system of hormone regulation in a male, including the roles of the hypothalamus and pituitary gland.
3. What is the function of the menstrual cycle?
4. Describe how a sperm cell is adapted for its function.

Thinking Critically

5. What might happen to sperm production if a male has a high fever?

SKILL REVIEW

6. **Interpreting Scientific Illustrations** Study **Figure 38.1.** Using the terms dorsal, ventral, anterior, posterior, superior, and inferior, describe where the epididymis is located in relation to the vas deferens. Describe where the prostate is located in relation to the testes. For more help, refer to *Thinking Critically* in the **Skill Handbook.**

1036 REPRODUCTION AND DEVELOPMENT

Section Assessment

1. The egg ruptures from the ovary and is swept into the oviduct. Beating cilia move the egg toward the uterus, where a fertilized egg implants and develops. An unfertilized egg or fully developed fetus passes through the vagina to the outside of the body.
2. Increased testosterone levels cause a decrease in FSH and LH; decreased testosterone results in an increase in FSH and LH.
3. produce an egg; prepare for pregnancy
4. streamlined, with mitochondria to provide ATP and a flagellum for movement
5. Sperm may be killed by the high fever.
6. The epididymis is inferior (or ventral) to the vas deferens. The prostate is superior and dorsal to the testes.

38.2 Development Before Birth

What do you have in common with a period at the end of a sentence? You were once about the same size. You started out life as a single, microscopic fertilized egg. That one cell went through numerous mitotic divisions to produce the trillions of cells that make up your body today. It all began when an egg from your mother was fertilized by a sperm from your father.

Magnification: 225×

A human blastocyst (inset) emerges from its protective membrane to travel to the uterus.

Fertilization and Implantation

After an egg ruptures from a follicle, it is able to stay alive for about 24 hours. For fertilization to occur, sperm must be present in the oviduct at some point during those first hours after ovulation. Sperm enter the vagina of the female's reproductive system when strong, muscular contractions ejaculate semen from the male's penis. As many as 350 million sperm are forced out of the male's penis and into the female's vagina during intercourse. Because sperm can live for 48 hours after ejaculation, fertilization can occur if intercourse occurs anywhere from a few days before to a day after ovulation.

One sperm plus one egg

How is it possible that, of the millions of sperm released into the vagina during ejaculation, only one fertilizes the mature egg? One reason is that the fluids secreted by the vagina are acidic and destroy most of the delicate sperm. Yet, some sperm survive because of the buffering effect of semen. The surviving sperm swim up the vagina into the uterus. Of the sperm that reach the uterus, only a few hundred pass into the two oviducts. The egg is present in one of them. To examine the attraction between sperm and egg, carry out the *MiniLab* on the next page.

Recall that the head of the sperm contains enzymes that help the sperm penetrate the egg. As the sperm

---

BIOLOGY JOURNAL

Life Before Birth

Linguistic Ask students to write an imaginary story about what they think it is like to be an embryo or fetus in the uterus. The article "Sensing in the Womb," by Jacqueline S. Palmer, *The American Biology Teacher,* vol. 49, no. 7 (October 1987), may be a helpful resource for students. **L2**

GLENCOE TECHNOLOGY

CD-ROM
Biology: The Dynamics of Life
Animation: *Human Fertilization*
Video: *Human Fertilization*
Disc 5

Section 38.2

Prepare

Key Concepts

This section summarizes fertilization and implantation of the egg and development of the human fetus, and discusses the importance of genetic counseling.

Planning

■ Save articles about unusual births for the Discussion.
■ Gather images for the Display and reteach.
■ Obtain modeling clay for the Project.
■ Gather materials for the BioLab.

---

1 Focus

Bellringer

Before presenting the lesson, display **Section Focus Transparency 94** on the overhead projector and have students answer the accompanying questions. **L1 ELL**

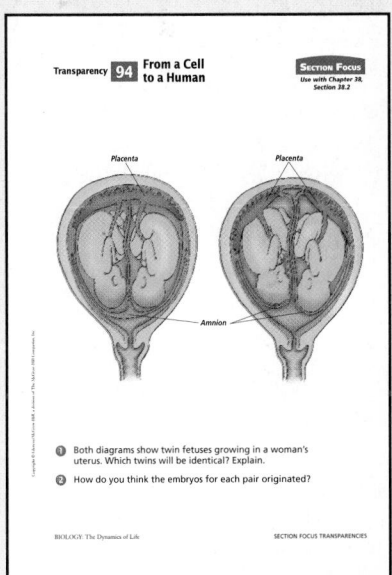

Resource Manager

Section Focus Transparency 94 and Master L1 ELL

2 Teach

MiniLab 38-1

Purpose
Students will determine whether sperm are attracted to eggs.

Process Skills
observe and infer, recognize cause and effect, experiment, analyze

Safety Precautions
Caution students to be careful with microscopes and microscope slides. Have students wear a lab apron and safety goggles, and wash their hands at the end of the lab.

Teaching Strategies
■ Fertilize some of the eggs a few hours before class so students can observe various stages of cleavage. If the eggs are kept at room temperature, the first cleavage takes about 50-60 minutes; second cleavage, 1 1/2 hours; third cleavage, 1 3/4 hours. The blastula forms after about 6 hours.

Expected Results
Sperm will collect around the egg, indicating that they are attracted to the egg.

Analysis
1. The sperm swim like tadpoles with tails whipping back and forth.
2. mitochondria
3. Yes, they swim toward and gather around the egg.

✔ Assessment
Performance Have students write a summary of the MiniLab and place it in their journals along with their answers to the Analysis questions. Use the Performance Task Assessment List for Lab Report in **PASC**, p. 47. **L1**

MiniLab 38-1 Observing and Inferring

Examining Sperm and Egg Attraction Most animals that reproduce by external fertilization live in water and release their sperm and eggs into the water. Somehow, a sperm and egg must meet. One adaptation of these animals that helps ensure fertilization is the release of thousands of sex cells at one time. The large number of sex cells increases the odds that at least some eggs will be fertilized. Chemical attraction could be another adaptation that encourages fertilization. If eggs give off a chemical that attracts sperm, each sperm would have help "finding" an egg.

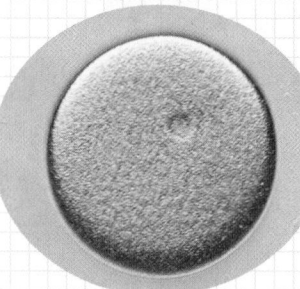

Magnification: 230×

Sea urchin egg

Procedure
1. Place a dropperful of sea urchin eggs on a microscope slide. **CAUTION: *Use care when working with a microscope and microscope slides.***
2. While observing the eggs under the microscope, add a drop of sea urchin sperm to the eggs.

Analysis
1. Describe the motion of a single sperm.
2. What cell structures are involved in providing energy for the sperm motion?
3. Are the sperm attracted to the eggs? How do you know?

---

crosses the cell membrane of the egg, it loses its midpiece and tail. Once one sperm has entered the egg, the electrical charge of the egg's membrane changes, thus preventing other sperm from entering. The sperm's nucleus then combines with the egg's nucleus to form a zygote.

The fertilized egg travels to the uterus

As the zygote passes down the oviduct, it begins to divide by repeated mitotic division. During its journey, pictured in *Figure 38.11*, the zygote obtains nutrients from flu-

1038 REPRODUCTION AND DEVELOPMENT

---

ids secreted by the mother. By the sixth day, the zygote passes into the uterus. Continuous cell divisions result in the formation of a hollow ball of cells called a blastocyst. Blastocyst is the term used when discussing human embryonic development. Recall that the term blastula is used for the embryonic development of other animals.

The blastocyst attaches to the uterine lining seven to eight days after fertilization. Attachment of the blastocyst to the lining of the uterus is called **implantation.** A small, inner mass of cells within the blastocyst will soon become a human embryo.

Embryonic Membranes and the Placenta

You have already learned about the importance of the amniotic egg to the evolutionary advancement of animals. Membranes that are similar to those of the amniotic egg form around the human embryo, protecting and nourishing it. The amnion is a thin, inner membrane filled with a clear, watery amniotic fluid. Amniotic fluid serves as a shock absorber and helps regulate the body temperature of the developing embryo.

The allantois membrane is an outgrowth of the digestive tract of the embryo. Blood vessels of the allantois form the **umbilical cord,** a ropelike structure that attaches the embryo to the wall of the uterus. The chorion is the outer membrane that surrounds the amniotic sac and the embryo within it. About 14 days after fertilization, fingerlike projections of the chorion, called chorionic villi, begin to grow into the uterine wall, as shown in *Figure 38.12*. The chorionic villi combine with part of the uterine lining to form the placenta.

GLENCOE TECHNOLOGY

VIDEODISC
Biology: The Dynamics of Life
Human Fertilization (Ch. 41)
Disc 2, Side 1, 40 sec.

VIDEODISC
Biology: The Dynamics of Life
Human Fertilization (Ch. 42)
Disc 2, Side 1, 20 sec.

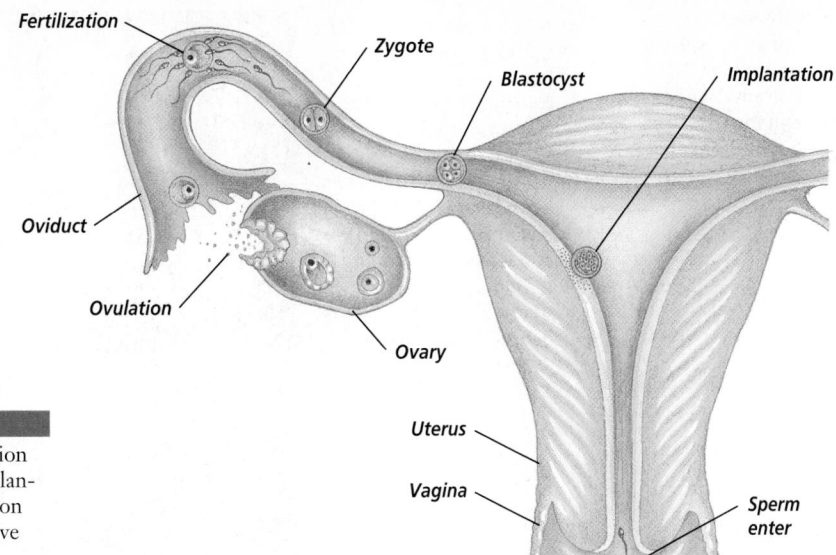

Figure 38.11 To reach the egg, sperm travel through the female's reproductive system, greatly decreasing in number along the way.

Fertilization

Zygote

Blastocyst

Implantation

Oviduct

Ovulation

Ovary

Uterus

Vagina

Sperm enter

CD-ROM

View an animation of fertilization and implantation in the Presentation Builder of the Interactive CD-ROM.

Exchange between embryo and mother

To survive and grow, the embryo must obtain the proper nutrients and eliminate the wastes its cells produce. The placenta delivers nutrients to the embryo and carries wastes away.

In the placenta, blood vessels from the mother's uterine wall lie close to the blood vessels of the embryo's chorionic villi. Although they are close together, they are not directly connected to one another. Instead, oxygen and nutrients transported by the mother's blood diffuse into the blood vessels of the chorionic villi in the placenta. These vital substances are then carried by the blood in the umbilical cord to the embryo. In turn, waste products from the embryo travel in the umbilical blood vessels to the placenta. Here they diffuse out of the vessels in the chorionic villi into the blood of the mother. These waste products are then removed by the mother's excretory system.

Figure 38.12 The placenta contains tissue from both mother and fetus (a). This embryo is shown at approximately seven weeks of development (b).

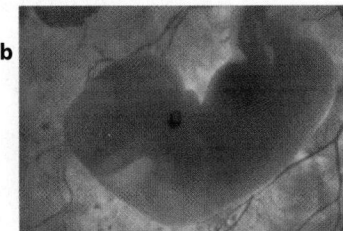

b

Embryo with Chorion

Fetus with Placenta

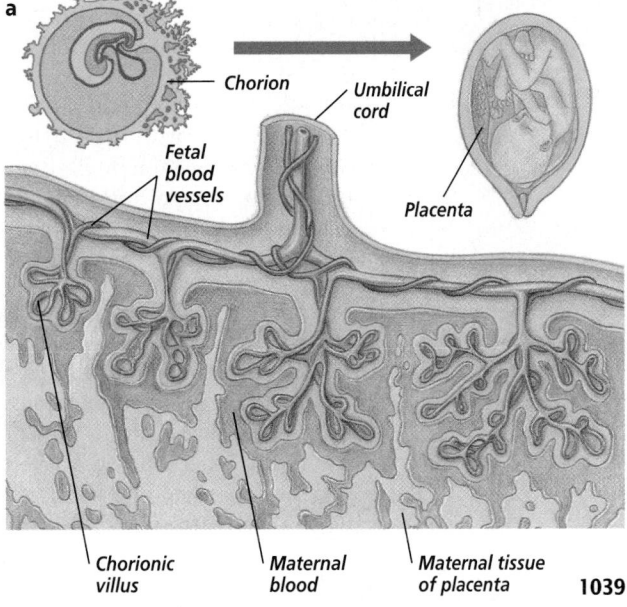

a

Chorion

Umbilical cord

Fetal blood vessels

Placenta

Chorionic villus

Maternal blood

Maternal tissue of placenta

1039

Quick Demo

Visual-Spatial Using a microscope projector, show students images of sea star embryos at various stages of development, such as the two-cell stage and four-cell stage.

Display

Prepare a bulletin board display that shows the development of the human fetus. Use medical journals or photographs from biological supply houses. Refer to the display as various stages of development are discussed.

Discussion

Using recent newspaper and magazine articles, initiate a class discussion of how a woman's body changes during pregnancy and the importance of prenatal care. Elicit what factors might prevent women from receiving proper prenatal care and what the results of such neglect might be.

Assessment

Skill Ask students to prepare a time line that plots the changes that take place in a developing embryo and fetus. Have them use different-colored pencils to indicate clearly where each trimester of pregnancy begins and ends. **L2** **ELL**

GLENCOE TECHNOLOGY

VIDEODISC
The Secret of Life
Placenta Portion

Portfolio

Body System Formation

Visual-Spatial Assign each student a body system. Have students research the fetal development of the assigned body system and prepare a visual display that traces the development of the system. Have them include labels and captioned summaries of the changes that occur at different stages. **L2** **ELL** **P**

Resource Manager

BioLab and MiniLab Worksheets, p. 169 **L2**
Reteaching Skills Transparencies 56a, 56b and Masters **L1** **ELL**

The BioLab at the end of the chapter can be used at this point in the lesson.

1040

A A five-week-old embryo is about 7 mm long. The heart—the large, red, circular structure protruding out of the embryo—begins as two muscular tubes. It starts to beat on about the 21st day of development. The arms and legs are beginning to bud, and the tissue that will form the eyes is beginning to darken.

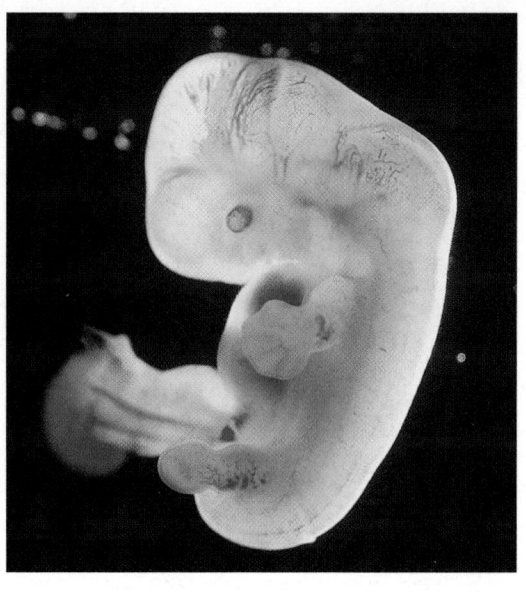

Hormonal maintenance of pregnancy

Remember that estrogen, and especially progesterone, cause the uterine lining to thicken in preparation for implantation. Once the blastocyst implants, the chorion membrane of the embryo starts to secrete the hormone human chorionic gonadotropin (hCG). This hormone keeps the corpus luteum alive so that it continues to secrete progesterone. Learn how this hormone is an indicator of pregnancy in the *BioLab* at the end of this chapter. By the third month, the placenta takes over for the corpus luteum, secreting enough estrogen and progesterone to maintain the pregnancy.

Fetal Development

When you think of an embryo growing within the mother's body, you may not realize that its development involves three different processes: growth, development, and cellular differentiation. Growth refers to the actual increase in the number of cells. As the cells develop, they move within the embryo's body and arrange themselves into specific organs. In addition, each cell becomes specialized to perform specific tasks and functions. All three processes begin with fertilization.

Pregnancy in humans usually lasts about 280 days, calculated from the first day of the mother's last menstrual period. The baby actually develops for about 266 days, calculated from the time of fertilization to birth. This time span is divided into three trimesters, each about three months in length. Each trimester brings significant advancement in the development of the embryo and fetus.

First trimester: Organ systems form

During the first trimester, all the organ systems of the embryo begin to form. A five-week embryo is shown in *Figure 38.13A*. During this time of development, the woman may not even realize she is pregnant. Yet, the first seven weeks following

---

PROJECT

Modeling Fetal Development

Kinesthetic Have students use modeling clay to show the changes a human egg undergoes fertilization to implantation. Students should label the zygote and embryo stages, and summarize on index cards the changes and approximate time period involved between changes. Have students work in groups to carry out the project by having gifted students work with less able students. If necessary, have students refer to discussions of embryonic development elsewhere in this textbook. Have students wear a lab apron and safety goggles when working with modeling materials. Have them wash their hands after working with the modeling materials. **L2** **ELL**
COOP LEARN

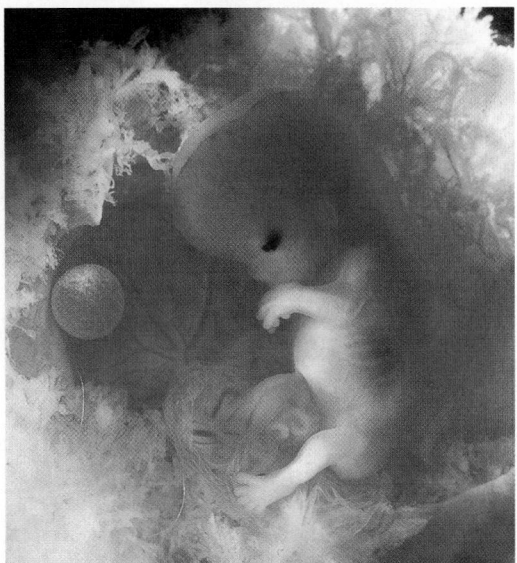

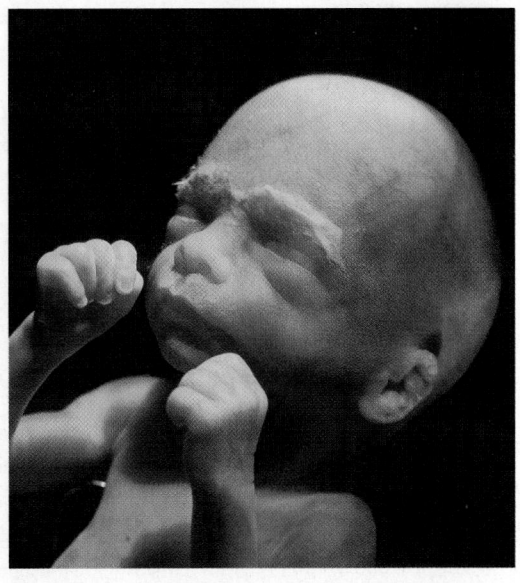

B A two-month-old fetus is 4 cm long. The heart is fully formed, bones are beginning to harden, and nearly all muscles have appeared. As a result, the fetus can move spontaneously.

C A second-trimester fetus is 15 to 30 cm long. Its skin is covered by a white, fatty substance that protects it from the amniotic fluid. Movements are commonly felt by the mother as the fetus exercises its muscles.

fertilization are critical because during this time, the embryo is more sensitive to outside influences—such as alcohol, tobacco, and other drugs that cause malformations—than at any other time.

By the eighth week, all the organ systems have been formed, and the embryo is now referred to as a fetus. You can see this stage of fetal development in *Figure 38.13B.* At the end of the first trimester, the fetus weighs about 28 g and is about 7.5 cm long from the top of its head to its buttocks. The sex of the fetus can be determined by the appearance of the external sex organs when viewed by ultrasound.

Second trimester:
A time of growth

For the most part, fetal development during the next three months is limited to body growth. Growth is rapid during the fourth month, but then slows by the beginning of the fifth month. At this point, it is possible for the fetus to survive outside the uterus, but it would require a great amount of medical assistance, and the mortality rate is high. The fetus's metabolism cannot yet maintain a constant body temperature, and its lungs have not matured enough to provide a regular respiratory rate. *Figure 38.13C* shows a fetus during the second trimester. By the end of the second trimester, the fetus weighs about 650 g and is about 34 cm long.

Third trimester:
Continued growth

During the last trimester, the mass of the fetus more than triples. By the beginning of the seventh month, the fetus kicks, stretches, and moves freely within the amniotic cavity, somewhat like the astronaut in the

Check for Understanding

Linguistic Ask students to describe orally how a fertilized egg changes after it is implanted in the uterus. **L2**

Reteach

Visual-Spatial Photocopy and distribute sketches showing fetal development in the later stages of pregnancy. Have students label the embryonic membranes and the placenta as you discuss them. **L1**

Extension

Linguistic Ask student groups to interview one of their mothers to find out what types of prenatal care and tests she received during her pregnancy. Ask students to record this information in their journals. **L2** **COOP LEARN**

✔ Assessment

Knowledge Ask students to write a summary of the role of the embryonic membranes and hormones during pregnancy. **L2**

Resource Manager

Laboratory Manual, pp. 277–284 **L2**

✔ Portfolio

The Story of Life

Linguistic Have students write a story or poem using any topic in this section. Encourage them to include some of the biology of this section in their writings. **L2** **P**

MEETING INDIVIDUAL NEEDS

Gifted

Intrapersonal Have students research the medical techniques used to examine fetal development and genetics. Techniques may include ultrasound, amniocentesis, and chorionic villi sampling. **L3**

Purpose

Students will graph and evaluate the growth of a human embryo.

Process Skills

make and use graphs, interpret data, analyze

Teaching Strategies

■ Be sure students recognize the change in units between the 6th and 7th weeks of development.

Expected Results

Graphs will show that the embryo grows fastest during the early periods of development.

Analysis

1. The embryo doubles in size during two 1-week periods: 3 to 4 weeks and 7 to 8 weeks.
2. All body systems are beginning to form.
3. 5th month

Assessment

Performance Have students write a summary of the MiniLab and place it in their journals along with the graph and the answers to the Analysis questions. Use the Performance Task Assessment List for Lab Report in **PASC**, p. 47. **L2**

Resource Manager

BioLab and MiniLab Worksheets, p. 170 **L2**
Reinforcement and Study Guide, p. 169 **L2**

MiniLab 38-2 Making and Using Graphs

Making a Graph of Fetal Size You started out as a single cell. That cell divided by the process of mitosis to produce organ systems capable of maintaining an independent existence outside your mother's uterus. During the time you were in your mother's uterus, major changes took place. One of these changes involved your growth in length.

Table 38.1 Growth of a Fetus		
Source of sample	Time after fertilization	Size
First trimester	3 weeks	3 mm
	4 weeks	6 mm
	6 weeks	12 mm
	7 weeks	2 cm
	8 weeks	4 cm
	9 weeks	5 cm
	3 months	7.5 cm
Second trimester	4 months	15 cm
	5 months	25 cm
	6 months	30 cm
Third trimester	7 months	35 cm
	8 months	40 cm
	9 months	51 cm

Procedure

1. Prepare a graph that plots time on the horizontal axis and length in centimeters on the vertical axis. Equally divide the horizontal axis into nine months. Then equally divide each of the first three months into four weeks.
2. Plot the data in **Table 38.1** on your graph.

Analysis

1. When is the fastest period of growth?
2. What structures are developing during this period of growth?
3. At what point does growth begin to slow down?

Figure 38.14
Genetic counselors use a variety of medical tests to provide couples with information about the risks of hereditary disorders.

---

space suit at the beginning of the chapter. During the eighth month, its eyes open. To examine the growth of a fetus, graph the data in the *MiniLab*.

During the final weeks of pregnancy, the fetus has grown large enough to fill the space within the embryonic membranes. Sometime during the ninth month, the fetus rotates its position so that its head is down, partly as a result of the shape of the uterus, but also because the head is the heaviest part of the body. By the end of the third trimester, the fetus weighs about 3300 g and is about 51 cm long. All of its body systems have developed, and it can now survive independently outside the uterus.

Genetic counseling

Most expectant parents desperately want just one thing—a healthy, normal baby. With our increasing knowledge of human heredity and advancing technology, determining that a newborn will be healthy and normal is much more possible today than it was in the past.

Genetic disorders can be predicted

Generally, people in the industrialized nations are aware of the possible genetic disorders that can affect a child. For many, this awareness has made them eager to know whether a potential child will be healthy. As advances have been made in the detection and treatment of genetic disorders, including those you studied earlier, the demand for genetic services, especially prenatal testing, has increased.

Most people do not even think about genetics until they are consider-

---

TECHPREP

Genetic Counseling

Linguistic Have students interested in a career in genetic counseling visit with a local genetic counselor or school counselor to find out what a specialty in genetic counseling involves. **L2**

GLENCOE TECHNOLOGY

VIDEODISC
VIDEOTAPE
The Secret of Life
Tinkering With Our Genes: Genetic Medicine

ing having children or are already expecting a child. If there is no history of genetic disorders in the family of either prospective parent, there may be no need for genetic services. However, if one or both prospective parents have a family history of some genetic disorder, they will likely want to get additional information. To find out how an expectant mother can help prevent one type of birth defect, try the *Problem-Solving Lab* shown here.

The job of a genetic counselor

Couples who seek information from trained professionals about the probabilities of hereditary disorders and what can be done if they occur are receiving **genetic counseling.** Genetic counselors like the one shown in *Figure 38.14* have a medical background with additional training in genetics. Sometimes, a team of professionals works with prospective parents. The team may include geneticists, clinical psychologists, social workers, and other consultants.

How do genetic counselors go about their work? First, they develop medical histories of both families. These histories may include pedigrees, biochemical analyses of blood, and karyotypes. Once the counselor has collected and analyzed all the available information, he or she explains to the couple their risk factors for giving birth to children with genetic disorders. If the probabilities of having a severely affected child are high, a couple must decide whether or not to have biological children together.

Problem-Solving Lab 38-2 — Interpreting Data

How can pregnant women reduce certain birth defects?
Ten of every 10,000 American babies are born with neural-tube defects. One of the defects included in this group is known as spina bifida. This condition occurs if, during early embryonic development, the bones of the spine fail to form properly. As a result, the spinal cord forms outside of the spinal column rather than inside it. How can pregnant women decrease the occurrence of neural-tube defects?

Analysis

Research findings about how neural-tube defects can be almost completely eliminated are provided in *Table 38.2.*

Table 38.2 Effect of folic acid on birth defects	
Folic acid used before or during pregnancy	**Neural-tube defects per 1000 births**
Did use	0.9
Did not use	3.5

Thinking Critically

1. Folic acid is a vitamin. Can a mother's diet during pregnancy influence her fetus? Explain your answer.
2. Does the use of folic acid totally prevent neural-tube defects? Explain your answer, using data to support it.
3. What additional questions might scientists want to ask regarding folic acid's role in fetal development?

Section Assessment

Understanding Main Ideas
1. What changes occur in the zygote as it passes along the oviduct and into the uterus?
2. What is the function of the placenta?
3. Why is an embryo most vulnerable to drugs and other harmful substances taken by its mother when it is between two and seven weeks old?
4. What is the function of human chorionic gonadotropin?

Thinking Critically
5. Compare the functions of human embryonic membranes with those inside a bird's egg.

SKILL REVIEW
6. **Sequencing** Prepare a table listing the events in the three trimesters of pregnancy. For more help, refer to *Organizing Information* in the **Skill Handbook.**

Problem-Solving Lab 38-2

Purpose
Students will learn that folic acid can reduce the incidence of certain birth defects.

Process Skills
analyze information, compare and contrast, draw a conclusion, think critically

Teaching Strategies
■ Relate the body's need for folic acid to a healthy diet. Foods rich in folic acid include spinach, beans, whole grains, oranges.
■ Explain that, during the third week of development, a groove forms in the embryo. Cells grow across the groove opening to form the neural tube, which will become the spine. Tube formation begins at the middle of the embryo and moves up toward the head. Incomplete closure results in neural-tube birth defects.

Thinking Critically
1. Increasing folic acid helps prevent neural tube defects.
2. No, the incidence of defects with folic acid is still 0.9.
3. Possible question: How does folic acid decrease incidence of neural-tube defects?

✔ Assessment
Performance Have students look up the recommended daily allowance (RDA) of folic acid in the normal diet and in the diet of pregnant females. Ask them to write an article informing people why pregnant women should receive plenty of folic acid. Use the Performance Task Assessment List for Newspaper Article in **PASC,** p. 69. **L3**

Section Assessment

1. The zygote divides as it passes down the oviduct; cells organize into a small hollow ball.
2. Nutrients and wastes are exchanged between the mother's and embryo's blood.
3. This is when all body systems are forming.
4. It keeps the corpus luteum alive to maintain pregnancy.
5. both protect the developing embryo and allow for the exchange of materials
6. First trimester: organ systems start forming; partially formed heart begins beating. External sex organs become apparent. Second trimester: heart is fully formed; size increases; bones harden and fetus begins to move. Third trimester: Growth continues; fetus moves freely; eyes open.

---

4 Close

Discussion
Have students summarize the development of the human fetus. Discuss how changes in this orderly development may result in birth defects. **L2**

Prepare

Key Concepts

This section describes the three stages of birth—dilation, expulsion, and the placental stage—and summarizes the developmental stages of humans after birth.

Planning

- Gather additional pictures to add to the Display begun in Section 38.2.
- Gather photographs of newborn infants and elderly individuals for the Quick Demo.

1 Focus

Bellringer

Before presenting the lesson, display **Section Focus Transparency 95** on the overhead projector and have students answer the accompanying questions.
L1 **ELL**

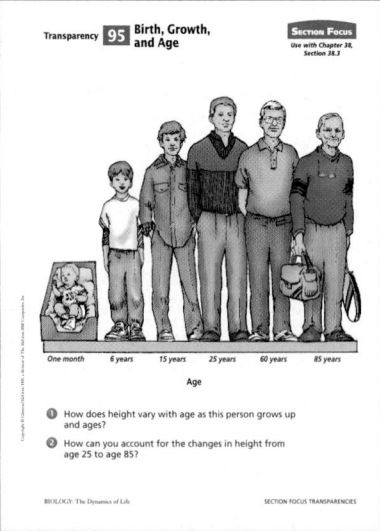

Resource Manager

Section Focus Transparency 95 and Master **L1** **ELL**

SECTION PREVIEW

Objectives
Describe the three stages of birth.
Summarize the developmental stages of humans after they are born.

Vocabulary
labor

Section

38.3 Birth, Growth, and Aging

How often have you heard the comment, "My, you've grown since I last saw you"? It may seem that you have grown a lot in the last few years. Yet the most rapid stage of growth in the life cycle of a human takes place within the uterus. From fertilization to birth, mass increases about 3000 times. Even so, although growth slows after birth, changes certainly do not stop.

The human body changes throughout life.

Figure 38.15
A newborn infant continues growth and development outside the mother's body.

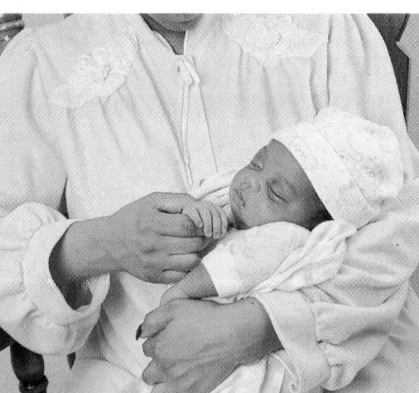

Birth

Birth is the process by which a fetus is pushed out of the uterus and the mother's body and into the outside world, like the newborn in *Figure 38.15*. What triggers the onset of birth is not fully understood. Different hormones released from the pituitary gland, uterus, and placenta may all be involved in stimulating the uterus. Birth occurs in three recognizable stages: dilation, expulsion, and the placental stage.

Dilation of the cervix

The physiological and physical changes a female goes through to give birth are called **labor.** Labor begins with a series of mild contractions of the uterine muscles. These contractions are stimulated by oxytocin, a peptide hormone released by the pituitary. The contractions open, or dilate, the cervix to allow for passage of the baby, as shown in *Figure 38.16A*. As labor progresses, the contractions begin to occur at regular intervals and intensify as the time between them shortens. When the opening of the cervix is about 10 cm, it is fully dilated. Usually, the amniotic sac ruptures and the amniotic fluid is released through the vagina, which is also referred to as the birth canal. This first stage of labor is usually the longest, sometimes lasting up to 24 hours.

BIOLOGY JOURNAL

Aging Assessment

Visual-Spatial Have students write an essay expressing their personal views on aging. Ask them to explain why they feel as they do. **L2**

MEETING INDIVIDUAL NEEDS

English Language Learners/ Learning Disabled

Visual-Spatial Have students prepare flash cards of the vocabulary terms from this chapter by writing the term on one side of the card and its definition on the other side. Have student groups review the terms and their meanings. **L1** **ELL**
COOP LEARN

Expulsion of the baby

Expulsion occurs when the involuntary uterine contractions become so forceful that they push the baby through the cervix into the birth canal. The mother assists with expelling the baby by contracting her abdominal muscles in time with the uterine contractions. As shown in *Figure 38.16B,* the baby moves from the uterus, through the birth canal, and out of the mother's body. The expulsion stage usually lasts from 20 minutes to an hour.

Placental stage

As shown in *Figure 38.16C,* within ten to 15 minutes after the birth of the baby, the placenta separates from the uterine wall and is expelled with the remains of the embryonic membranes. Collectively, these materials are known as the afterbirth. The uterine muscles continue to contract forcefully, constricting uterine blood vessels to prevent the mother from hemorrhaging. After the baby is born, the umbilical cord is clamped and cut near the baby's abdomen. The bit of cord that is left eventually dries up and falls off, leaving an abdominal scar called the navel.

Growth and Aging

Once a baby is born, growth continues and learning begins. Human growth varies with age and is somewhat sex dependent.

A hormone controls growth

Human growth is regulated by human growth hormone (hGH), a protein secreted by the pituitary gland. Although hGH causes all body cells to grow, it acts principally on the skeleton and skeletal muscles. The hormone works by increasing the rate of protein synthesis and the metabolism of fat molecules. Other hormones that influence growth include thyroxin, estrogen, and testosterone.

Figure 38.16
The stages of birth are dilation, expulsion, and the placental stage.

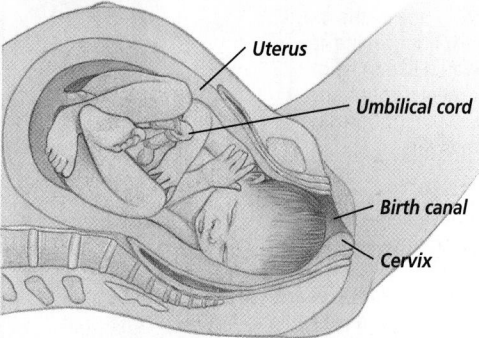

Uterus
Umbilical cord
Birth canal
Cervix

A **Dilation** Labor contractions open the cervix.

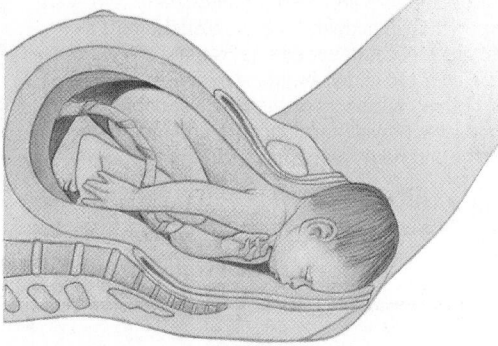

B **Expulsion** The baby rotates as it moves through the birth canal, making expulsion easier.

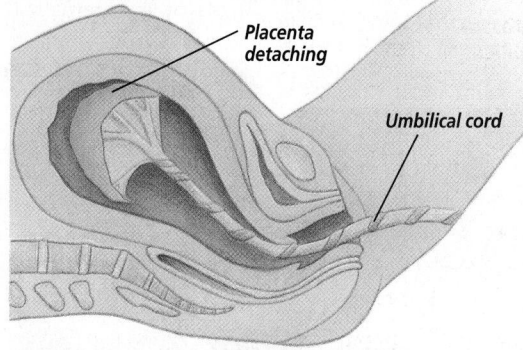

Placenta detaching
Umbilical cord

C **Placental stage** During the placental stage, the placenta and umbilical cord are expelled.

38.3 BIRTH, GROWTH, AND AGING **1045**

Career Path

TECH PREP **Courses in high school:** sciences, mathematics, psychology

College: bachelor's degree to become a registered nurse; up to two years of training in midwifery

Career Issue

Tell students that some people think midwives should not be allowed to deliver babies without a doctor present. Invite the class to explore both sides of this issue.

For More Information

To learn more about becoming a midwife, students can contact:

American College of Nurse Midwives
818 Connecticut Avenue, NW
Suite 900
Washington, DC 20006

Resource Manager

Tech Prep Applications, p. 53 **L2**

Reinforcement and Study Guide, p. 170 **L2**

Content Mastery, pp. 185, 187-188 **L1**

Midwife

Giving birth is one of the most exciting experiences life can offer. If you would like to help expectant mothers through the birth process, you might consider becoming a midwife.

Skills for the Job

Midwives have helped women give birth since ancient times, sometimes with little training. However, today's midwives are professionally trained and well able to guide women with low-risk pregnancies safely through the birth process. Midwives first become registered nurses and then complete up to two years of clinical instruction in midwifery. They must also pass a national test and meet state requirements before they can become certified nurse-midwives. Many midwives work in hospitals or birthing centers; some help deliver babies at home. All midwives provide care, support, and monitoring throughout the pregnancy and afterward.

BIOLOGY Online

For more careers in related fields, be sure to check the Glencoe Science Web site.
science.glencoe.com

The first stage of growth: Infancy

The first two years of life are known as infancy. During infancy, a child shows tremendous growth as well as an increase in physical coordination and mental development. Generally, an infant will double its birth weight by the time it is five months of age, and triple its weight in a year. By two years of age, most infants weigh approximately four times their birth weight. During this time, the infant learns to control its limbs, roll over, sit, crawl, and walk. By the end of infancy, the child also utters his or her first words.

From child to adult

Childhood is the period of growth and development that extends from infancy to adolescence, when puberty begins. Physically, the childhood years are a period of relatively steady growth. Mentally, a child develops the ability to reason and to solve problems.

Figure 38.17
Changes in the size and shape of the body are associated with growth. These photos show the changes that Shirley Temple Black has undergone as she has aged.

Cultural Diversity

Rites of Passage

In many cultures and religions, traditional celebrations mark the transition from childhood to adulthood. For example, in Mexican tradition, a girl celebrates her transition from childhood to adulthood on her fifteenth birthday in a celebration known as quinceanera. Discuss with students examples of initiation rites in various cultures. You may wish to have students research topics by asking them to identify a culture and then determine whether that culture has a traditional rite of passage.

Figure 38.18
In 1962, astronaut John Glenn made history as the first American to orbit the Earth. He enthusiastically returned to space in 1998, at the age of 77, when he joined the crew of space shuttle *Discovery*.

Check for Understanding
Ask students to describe the aging process. **L2**

Reteach
Have students choose a system of the body and describe how it changes from birth through old age. **L2**

Extension
Ask students to research and report on current theories about infant learning. Encourage them to include illustrations in their reports. **L3**

✓ Assessment
Knowledge Ask students to choose a hormone discussed in this chapter and diagram a negative-feedback loop for the hormone. **L2**

4 Close

Discussion
Ask students to state several misconceptions about aging. List their statements on the chalkboard. As a class, discuss evidence that supports why each statement is a misconception.

Adolescence follows childhood. At puberty, the onset of adolescence, there is often a growth spurt, sometimes quite a dramatic one. Increases of 5 to 8 cm of height in one year are not uncommon in teenage boys. During the teen years, adolescents reach their maximum physical stature, which is determined by heredity, nutrition, and their environment. By the time a young person reaches adulthood, his or her organs have reached their maximum mass, and physical growth is complete. You can see in *Figure 38.17* how the physical appearance of a person changes from birth to adulthood.

An adult ages

As an adult ages, his or her body undergoes many distinct changes. Metabolism and digestion become slower. The skin loses some of its elasticity, and less pigment is produced in the hair follicles; that is, the hair turns white. Bones often become thinner and more brittle, resulting in an increased risk of fracture. Stature may shorten because the disks between the vertebrae become compressed. Vision and hearing might diminish, but, as *Figure 38.18* shows, many people continue to be both intellectually and physically active as they grow older.

WORD Origin
infancy
From the Latin word *infantia*, meaning "inability to speak." Infants are in the growth stage that involves learning to speak.

Section Assessment

Understanding Main Ideas
1. What events occur during the dilation stage of the birth process?
2. How does the human growth hormone produce growth?
3. How does the human body change during childhood?
4. What changes to the body are usually associated with aging?

Thinking Critically
5. Compare the birth of a human baby with that of a marsupial mammal.

SKILL REVIEW
6. **Recognizing Cause and Effect** Someone tells you that as people age, their personalities normally change. Do you think this statement is valid? Why or why not? For more help, refer to *Thinking Critically* in the **Skill Handbook**.

Section Assessment

1. Muscle contractions expand the cervix to about 10 cm.
2. Human growth hormone increases the rate of protein synthesis inside cells and increases the metabolism of fat molecules.
3. Muscle coordination, reasoning, and problem-solving abilities increase; the body grows; the child develops the ability to reason and solve problems.
4. Metabolism and digestion slow. Skin loses elasticity and hair follicles lose pigment. Bones may become thin and more brittle. Vision and hearing may diminish.
5. When compared with humans, marsupials are born earlier in development. After birth, marsupials continue to develop within the mother's pouch. Both humans and marsupials are dependent upon the mother at birth.
6. Students are likely to agree that most people function effectively throughout life without experiencing changes in personality.

Time Allotment

One class period

Process Skills

acquire information, apply concepts, compare and contrast, draw a conclusion, formulate models, think critically

Safety Precautions

Remind students to use caution when working with scissors.

PREPARATION

To save class time, trace models A, B, and C onto tracing paper prior to class. Provide each student group with one set of tracing paper shapes to use in making their copies on heavy paper.

Resource Manager

BioLab and MiniLab Worksheets, pp. 171-172 **L2**

What hormone is produced by an embryo?

The chorion of an eight-day-old embryo produces a hormone called human chorionic gonadotropin (hCG). This hormone stimulates the corpus luteum to continue its production of progesterone, which in turn maintains the attachment of the embryo to the uterine lining. There is such a high concentration of hCG present in the blood of the mother that the kidneys excrete it in urine.

PREPARATION

Problem

How can you test for the presence of hCG?

Objectives

In this BioLab, you will:
■ **Model** the chemicals used to test for the presence of hCG.
■ **Interpret** the results of chemical reactions involving hCG in a pregnant and nonpregnant female.

Materials

scissors
heavy paper
tracing paper

Safety Precautions

Handle scissors with caution.

Skill Handbook

Use the **Skill Handbook** if you need additional help with this lab.

PROCEDURE

1. Copy the data table.
2. Copy models **A**, **B**, and **C** onto tracing paper.

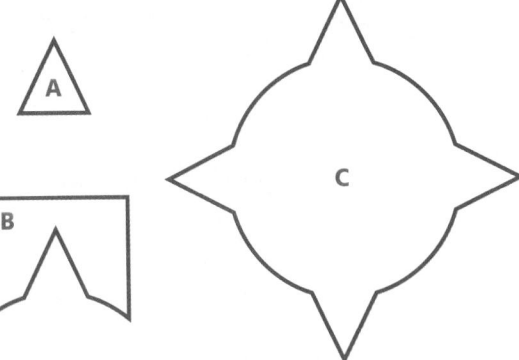

3. Copy the tracings onto heavy paper and cut them out. You will need 4 models of **A**, 4 of **B**, and 1 of **C**.
4. Model **A** represents a molecule of the hCG hormone. Model **B** represents a chemical called anti-hCG hormone. Model **C** represents a chemical that has four hCG molecules attached to it.
5. Note that the shapes of hCG and anti-hCG join together like puzzle pieces. These two chemicals react, or join together, when both are present in a solution. The shapes of anti-hCG and

---

PROCEDURE

Teaching Strategies

■ This lab simulates a home pregnancy test. Each manufacturer of home pregnancy tests has a variation on the final color and its interpretation. Students may be familiar with TV commercials for products in which the final color may not be green or colorless.
■ Students do not have to diagram all four molecules of chemical A and B on the data

chart. However, they must diagram at least one to represent the concept of what is occurring.

Troubleshooting

■ Remind students that each model represents a molecule. When two models fit together, this represents a chemical reaction.

Data Table

Condition	hCG in urine?	+ Anti-hCG	= Joined hCG and anti-hCG?	+ Chemical C with anti-hCG?	Color
Not pregnant					
Pregnant					

Chemical C also join, indicating that they chemically react when both are present. The combination of Chemical C and anti-hCG is green. Chemical C without anti-hCG attached is colorless.

6. Model the following events for the "Not pregnant" condition. Record them in the data table using drawings of the models.
   a. The hormone hCG is not present in the urine.
   b. Anti-hCG is added to a urine sample, then chemical C is added.
   c. Draw the resulting chemical in the data table and indicate the color that appears.
7. Model the following events for the "Pregnant" condition. Record

them in the data table using drawings of the models.
   a. The hormone hCG is present.
   b. Anti-hCG is added to urine, then Chemical C is added.
   c. Draw the resulting chemical in the data table and indicate the color that appears.

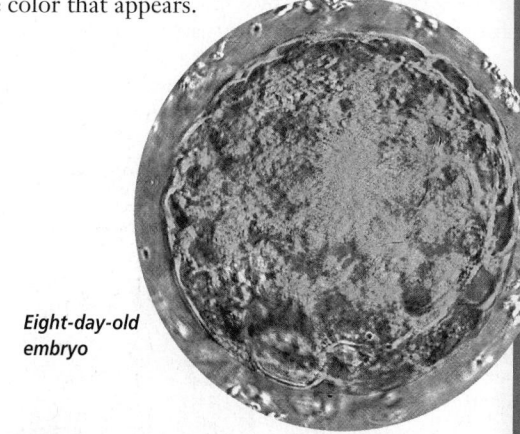

Eight-day-old embryo

ANALYZE AND CONCLUDE

1. **Analyzing** Explain the origin of hCG in a pregnant female.
2. **Analyzing** Explain why hCG is absent in a nonpregnant female.
3. **Concluding** Describe the roles of anti-hCG and Chemical C in both tests.
4. **Observing and Inferring** Explain why anti-hCG is added to the sample before Chemical C is added.

Going Further

Analyzing Information Using references, look up the meaning of the words "chorionic" and "gonadotropin." Explain why the name hCG suits this hormone.

BIOLOGY *Online* To find out more about embryonic hormones, visit the Glencoe Science Web site.
science.glencoe.com

ANALYZE AND CONCLUDE

1. The chorion begins to release hCG as early as eight days after fertilization.
2. There is no chorion to release hCG.
3. If pregnant, hCG joins with anti-hCG, preventing attachment to chemical C's hCG. No color appears. If not pregnant, anti-hCG attaches to chemical C's hCG and color appears.
4. To get an accurate result, hCG and anti-hCG must have a chance to bind before chemical C is added.

✓ Assessment

Performance Have students prepare a puzzle-piece model to show how the presence of a hormone such as FSH or estrogen might be detected. Use the Performance Task Assessment List for Model in **PASC**, p. 51. **L2**

Going Further

Ask students who are interested to check on the meaning of "antigen-antibody reaction" and determine if the completed activity fits this description. **L3**

■ Students must realize that there is no hCG in the urine of a non-pregnant female.
■ When joined together, hCG and anti-hCG cannot attach to chemical C.

Data Table

Condition	hCG in urine?	+ Anti-hCG	= Joined hCG and anti-hCG?	+ Chemical C with anti-hCG?	Color
Not pregnant	—	⟁⟁ ⟁⟁	—	◇	Green
Pregnant	△ ▷ ◁ △	⟁⟁ ⟁⟁	⟁⟁ ⟁⟁	◇	Colorless

BIO Technology

Purpose ⬚

Students learn how the process of cryopreservation is used to preserve embryos that are implanted during *in vitro* fertilization.

Background

Cryopreservation could help save critically endangered animals. In zoos around the world, eggs and sperm of endangered species are being frozen and stored. One day, it may be possible to implant embryos produced from these rare cells into other species that will act as surrogate mothers. This approach may be the only option for the continuation of extremely rare species.

Teaching Strategies

■ Have students research the use of cryopreservation in the breeding of racehorses and cattle.
■ Have students research the properties of liquid nitrogen to learn why it is used for cryopreservation.

Investigating the Technology

Possible answers: The protective chemicals that are mixed with the embryos cause the concentration of water outside the cells to be lower than the concentration of water inside the cells. Because water flows from high to low concentration in osmosis, it flows out of the cells into the surrounding environment.

BIO Technology — **Frozen Embryos**

In February 1998, an 8-pound, 15-ounce baby boy was born in Los Angeles, California. What's so special about that? This particular baby developed from an embryo that had been frozen for more than seven years. Freezing human embryos that can be implanted in a mother's uterus at some later date has become a fairly routine part of in vitro fertilization.

In vitro fertilization is a process in which a man's sperm and a woman's eggs are combined in the laboratory. Eggs are removed from the woman's ovaries, fertilized with sperm, and allowed to divide in culture dishes. Up to four of the resulting embryos are then carefully transferred into the woman's uterus. If all goes well, at least one of the embryos will implant in the uterine wall and grow into a healthy baby.

Cryopreservation Many embryos may be produced during the laboratory phase of *in vitro* fertilization. The "extra" embryos are frozen and stored by cryopreservation, a process in which living tissue is frozen at an extremely low temperature so that the tissue can be revived and restored to the same condition as before it was stored. Embryos are mixed with protective chemicals and submerged in liquid nitrogen,

which has a temperature of –196° C. During the freezing process, water moves from inside the cells to outside by osmosis, dehydrating and shrinking the cells. Metabolism stops. As the temperature decreases, the water around the outside of the cells freezes. The rate of cooling must be carefully controlled. If the rate of cooling is too slow, too much dehydration occurs. On the other hand, too rapid cooling causes the formation of ice crystals inside the cells, which might tear the cell membranes. A cooling rate of –1° C per minute has been found to be optimal. When the frozen tissue is needed, it is removed from liquid nitrogen storage and warmed rapidly to room temperature. Cells begin to grow within one or two days.

As many as 100 000 human embryos are currently "on ice" in the United States, and about 10 000 are added every year. For couples who are trying to get pregnant through *in vitro* fertilization, being able to keep some of their embryos frozen has advantages. If the first attempt at implantation is unsuccessful (and it often is), some of the frozen embryos can be thawed and the transfer process repeated fairly easily—hopefully with better results.

INVESTIGATING THE TECHNOLOGY

Thinking Critically Use your knowledge of osmosis to explain why water leaves the cells of an embryo during cryopreservation.

BIOLOGY Online To find out more about cryopreservation, visit the Glencoe Science Web site. **science.glencoe.com**

Frozen embryos are stored in liquid nitrogen

Going Further

Have interested students make a poster explaining the osmotic relationships between the embryo and the surrounding environment during cryopreservation. **L2**

BIOLOGY Online Note Internet addresses that you find useful in the space below for quick reference.

_____

_____

_____

_____

_____

SUMMARY

Section 38.1

Human Reproductive Systems

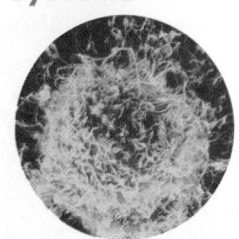

Main Ideas

■ The male reproductive system produces sperm and the female reproductive system produces eggs.

■ Through the control of the hypothalamus and pituitary, hormones act on the reproductive system as well as on other body systems. Hormone levels are regulated by negative feedback.

■ Changes in males and females at puberty are the result of the production of FSH, LH, and other sex hormones.

■ Under the control of hormones, the menstrual cycle produces a mature egg and prepares the uterus for receiving a fertilized egg.

Vocabulary

bulbourethral gland (p.1029)
cervix (p.1031)
corpus luteum (p.1034)
epididymis (p.1028)
follicle (p.1031)
menstrual cycle (p.1034)
oviduct (p.1031)
ovulation (p.1032)
prostate gland (p.1029)
puberty (p.1029)
scrotum (p.1027)
semen (p.1029)
seminal vesicle (p.1029)
vas deferens (p.1028)

Section 38.2

Development Before Birth

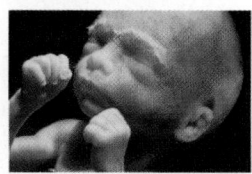

Main Ideas

■ Fertilization occurs in the oviduct. The ball of cells that develops from the fertilized egg implants in the uterine wall.

■ The embryo changes from a small ball of cells to a well-developed fetus over the course of nine months.

■ Genetic counseling offers people information about their chances of having a child with a genetic disorder.

Vocabulary

genetic counseling (p.1043)
implantation (p.1038)
umbilical cord (p.1038)

Section 38.3

Birth, Growth, and Aging

Main Ideas

■ Birth involves dilation of the cervix, expulsion of the baby, and expulsion of the placenta.

■ Infancy, childhood, adolescence, and adulthood are the stages of human development. Human growth hormone (hGH) produces growth in all body cells, especially in cells of the skeleton and muscles.

Vocabulary

labor (p.1044)

Main Ideas
Summary statements can be used by students to review the major concepts of the chapter.

Using the Vocabulary
To reinforce chapter vocabulary, use the Content Mastery Booklet and the activities in the Interactive Tutor for Biology: The Dynamics of Life on the Glencoe Science Web site. **science.glencoe.com**

 *All Chapter Assessment questions and answers have been validated for accuracy and suitability by The Princeton Review.*

UNDERSTANDING MAIN IDEAS

1. d
2. b
3. d

UNDERSTANDING MAIN IDEAS

1. Which of the following is NOT a part of the male reproductive system?
   a. scrotum   c. testis
   b. vas deferens   d. cervix

2. Which of these is NOT found in a sperm?
   a. head   c. tail
   b. chloroplasts   d. mitochondria

3. What tubule transports both urine and semen?
   a. testes   c. epididymis
   b. vas deferens   d. urethra

 VIDEOTAPE
GLENCOE TECHNOLOGY
MindJogger Videoquizzes
Chapter 38: *Reproduction and Development*
Have students work in groups as they play the videoquiz game to review key chapter concepts.

 **Resource Manager**

Chapter Assessment, pp. 223-228
MindJogger Videoquizzes
ExamView® Pro Software
BDOL Interactive CD-ROM, Chapter 38 quiz

4. a
5. d
6. c
7. b
8. b
9. a
10. testes
11. vas deferens
12. puberty
13. testosterone
14. ovary, oviduct
15. follicle
16. oviduct, uterus
17. implantation, uterus
18. ovulation, oviduct
19. infancy
20. luteinizing, testosterone

APPLYING MAIN IDEAS

21. This hormone is secreted by the embryo/fetus, so it indicates pregnancy.

4. Fertilization of an egg to form a zygote takes place in _____.
   a. oviduct
   b. uterus
   c. vagina
   d. ovary

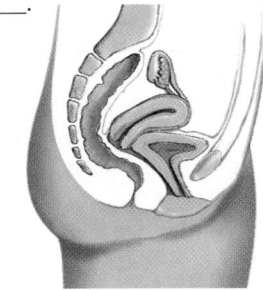

5. Which of these does NOT produce a fluid that surrounds sperm as they travel from the testes?
   a. seminal vesicles
   b. bulbourethral glands
   c. prostate gland
   d. pituitary gland

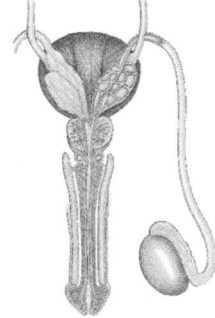

6. When there is a surge of LH during the menstrual cycle, what event occurs?
   a. luteinization   c. ovulation
   b. fertilization   d. menstruation

7. What happens to a follicle after it releases an egg?
   a. It degenerates.
   b. It changes into the corpus luteum.
   c. It is released into the oviduct.
   d. It turns into the placenta after fertilization of the egg.

THE PRINCETON REVIEW — **TEST-TAKING TIP**

The "Best" Answer Is Often the One "Left Over"
If none of the answer choices look right, use the process of elimination to eliminate the worst ones. The one you've got left is the best choice.

8. Once one sperm enters an egg, no other sperm can enter because the egg's membrane _____.
   a. changes its shape
   b. changes its electrical charge
   c. hardens
   d. closes off its pores

9. During which trimester does the embryo's heart start beating?
   a. first   c. third
   b. second   d. fourth

10. Sperm production takes place in the _____.

11. The _____ transports sperm from the testes to the urethra.

12. Secondary sex characteristics begin to develop during _____.

13. When luteinizing hormone is released by the pituitary, it causes cells in the testes to produce _____.

14. In females, eggs are produced in the _____, and when they mature, the eggs rupture into the _____.

15. An undeveloped egg is surrounded by a group of epithelial cells known as a(n) _____.

16. The _____ is a tube that transports the egg from the ovary to the _____ for implantation.

17. _____ occurs when the blastocyst attaches to the lining of the _____.

18. _____ occurs when an egg ruptures through the ovary wall and moves into the _____.

19. Human development occurs most rapidly during the _____ stage.

20. _____ hormone stimulates the testis to produce the hormone _____.

APPLYING MAIN IDEAS

21. Why does the presence of human chorionic gonadotropin in the urine indicate pregnancy?

22. A pregnant woman tells her physician that she is 50 days past the first day of her last menstrual period. How many days has the embryo been developing?

23. Interpreting Data This graph indicates changes in cardiac output and heartbeat rate in a woman over the course of her pregnancy. Explain the changes.

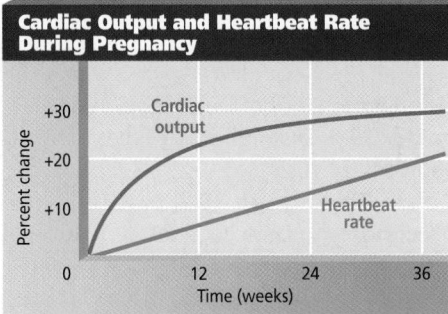

24. Concept Mapping Complete the concept map by using the following terms: follicle, ovulation, oviduct.

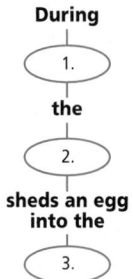

CD-ROM

For additional review, use the assessment options for this chapter found on the *Biology: The Dynamics of Life Interactive CD-ROM* and on the Glencoe Science Web site.
science.glencoe.com

The following graph represents the average blood concentration of four circulating hormones collected from 50 healthy adult women who were not pregnant.

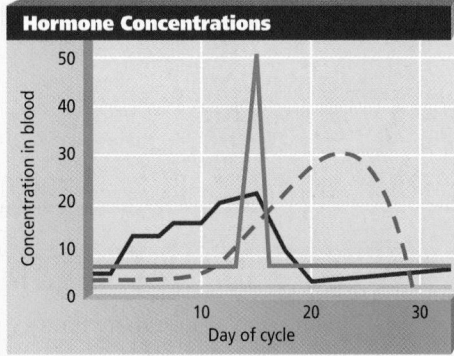

Interpreting Data Study the graph and answer the following questions.

1. Which line represents luteinizing hormone?
   a. red line   **c.** yellow line
   b. blue line   **d.** green line

2. Which hormone increases during the last half of the menstrual cycle?
   a. estrogen   **c.** LH
   b. progesterone   **d.** FSH

3. Which hormone is responsible for stimulating the egg development each month?
   a. estrogen   **c.** LH
   b. progesterone   **d.** FSH

4. Interpreting Scientific Illustrations
   The yellow line represents the hormone human chorionic gonadotropin. Explain why this hormone remained at an extremely low level during the women's menstrual cycles.

23. During pregnancy, the heartbeat rate and cardiac output increase. This increase compensates for blood delivery to the developing fetus.

24. 1. Ovulation; 2. Follicle; 3. Oviduct

   1. d
   2. b
   3. d
   4. This hormone is secreted only when women are pregnant.

Chapter 39 Organizer

Refer to pages 4T-5T of the Teacher Guide for an explanation of the National Science Education Standards correlations.

Section	Objectives	Activities/Features
Section 39.1 **The Nature of Disease** National Science Education Standards UCP.1, UCP.2, UCP.5; A.1, A.2; C.1, C.4, C.6; F.1, F.5; G.1-3 (1 session)	1. **Outline** the steps of Koch's postulates. 2. **Describe** how infections are transmitted. 3. **Explain** what causes the symptoms of a disease.	**Problem-Solving Lab 39-1,** p. 1059 **MiniLab 39-1:** Testing How Diseases are Spread, p. 1060 **Internet BioLab:** Getting On-line for Information on Diseases, p. 1074
Section 39.2 **Defense Against Infectious Diseases** National Science Education Standards UCP.1-3, UCP.5; A.1, A.2; C.1, C.4, C.5, C.6; E.1, E.2; F.1, F.5, F.6; G.1-3 (3 sessions)	4. **Compare** nonspecific and specific immune responses. 5. **Compare** innate and acquired immune responses. 6. **Distinguish** between antibody and cellular immunity.	**Inside Story:** Lines of Defense, p. 1066 **MiniLab 39-2:** Distinguishing Types of White Blood Cells, p. 1067 **Problem-Solving Lab 39-2,** p. 1072 **BioTechnology:** New Vaccines, p. 1076

Need Materials? Contact Carolina Biological Supply Company at 1-800-334-5551 or at **http://www.carolina.com**

MATERIALS LIST

BioLab

p. 1074 Internet access, paper, pencil

MiniLabs

p. 1060 fresh apples (4), rotting apple, zipper-lock plastic bags (4), cotton ball, ethanol

p. 1067 microscope, prepared slide of blood cells, paper, pencil

Alternative Lab

p. 1064 *E. coli* culture, antibiotic disks (3 types), sterile untreated disks, petri dish with sterile nutrient agar, cotton swabs, forceps, ethanol, transparent tape, marking pen, incubator

Quick Demos

p. 1056 petri dish with sterile nutrient agar, incubator

p. 1064 microprojector, prepared slides of red and white blood cells

Key to Teaching Strategies

L1	Level 1 activities should be appropriate for students with learning difficulties.
L2	Level 2 activities should be within the ability range of all students.
L3	Level 3 activities are designed for above-average students.
ELL	ELL activities should be within the ability range of English Language Learners.
COOP LEARN	Cooperative Learning activities are designed for small group work.
P	These strategies represent student products that can be placed into a best-work portfolio.
	These strategies are useful in a block scheduling format.

Teacher Classroom Resources

Section	Reproducible Masters	Transparencies
Section 39.1 The Nature of Disease	Reinforcement and Study Guide, p. 171 L2 BioLab and MiniLab Worksheets, p. 173 L2 Laboratory Manual, pp. 285-292 L2 Content Mastery, pp. 189-190, 192 L1	Section Focus Transparency 96 L1 ELL Basic Concepts Transparency 76 L2 ELL
Section 39.2 Defense Against Infectious Diseases	Reinforcement and Study Guide, pp. 172-174 L2 Concept Mapping, p. 39 L3 ELL Critical Thinking/Problem Solving, p. 39 L3 BioLab and MiniLab Worksheets, pp. 174-176 L2 Content Mastery, pp. 189, 191-192 L1	Section Focus Transparency 97 L1 ELL Basic Concepts Transparency 77 L2 ELL Basic Concepts Transparency 78a, 78b L2 ELL Basic Concepts Transparency 79 L2 ELL Reteaching Skills Transparency 57 L1 ELL Reteaching Skills Transparency 58 L1 ELL

Assessment Resources

Chapter Assessment, pp. 229-234
MindJogger Videoquizzes
Performance Assessment in the Biology Classroom
Alternate Assessment in the Science Classroom
ExamView® Pro Software 💾
BDOL Interactive CD-ROM, Chapter 39 quiz

Additional Resources

Spanish Resources ELL
English/Spanish Audiocassettes ELL
Cooperative Learning in the Science Classroom COOP LEARN
Lesson Plans/Block Scheduling

NATIONAL GEOGRAPHIC — Teacher's Corner

Products Available From Glencoe
To order the following products, call Glencoe at 1-800-334-7344:
CD-ROM
NGS PictureShow: Human Body 3
Curriculum Kit
GeoKit: Human Body 1
Transparency Set
NGS PicturePack: Human Body 3
Videodisc
STV: Human Body

Products Available From National Geographic Society
To order the following products, call National Geographic Society at 1-800-368-2728:
Videos
Incredible Human Machine
Our Immune System

Index to National Geographic Magazine
The following articles may be used for research relating to this chapter:
"Our Immune System: The Wars Within," by Peter Jaret, June 1986.

GLENCOE TECHNOLOGY

The following multimedia resources are available from Glencoe.

Biology: The Dynamics of Life
CD-ROM ELL

Video: *Lymphocyctes*
Animation: *Antibody Immunity*
Animation: *Cellular Immunity*

Videodisc Program

Lymphocyctes
Antibody Immunity
Cellular Immunity

The Secret of Life Series

Tinkering With Our Genes: Genetic Medicine
Nothing to Sneeze At: Viruses

Theme Development

The mechanisms by which parts of various body systems work together to fight disease are discussed, emphasizing the theme of **systems and interactions. Homeostasis** is emphasized as the immune system's response to disease is discussed.

READING BIOLOGY

Glencoe's *Biology: The Dynamics of Life* contains many resources to assist a student's reading skills. Each chapter contains figures with expanded captions that expand on written material. Word Origins, located along the side of text, expand knowledge of biology vocabulary. Glencoe's Content Mastery Booklet helps develop reading skills while reinforcing content. In addition, use the Interactive Tutor for *Biology: The Dynamics of Life* on the Glencoe Web site to reinforce vocabulary. **science.glencoe.com**

Chapter
39 Immunity from Disease

What You'll Learn

- You will describe how infections are transmitted and what causes the symptoms of diseases
- You will explain the various types of innate and acquired immune responses
- You will compare antibody and cellular immunity

Why It's Important

Your body constantly faces attack from disease-causing organisms. A knowledge of your immune system will help you understand how your body defends itself.

READING BIOLOGY

Scan the chapter, and write down some of the infectious diseases mentioned. Note where you may have heard these terms before. As you read, note which diseases come from bacteria and which come from viruses.

BIOLOGY *Online*

To find out more about the immune system, visit the Glencoe Science Web site. **science.glencoe.com**

Magnification: 17 350×

The immune system, like a castle, protects the body against invasion. Like the knight protecting his castle, the neutrophil shown here is on constant surveillance against foreign invaders.

1054 IMMUNITY FROM DISEASE

E *veryone occasionally gets a cold. Cold viruses enter your body by way of your nose and are swept to the back of your throat by hairlike cilia. Some are washed down your esophagus and destroyed by your digestive system. Others become lodged in your nasal passages. These viruses enter cells that line your nasal cavity. Once inside, the viruses unleash their genes, taking over the DNA of your cells. Soon you have a sore throat, stuffy nose, headache, and mild fever. How did the infection cause these symptoms?*

Cold viruses (inset) make a person feel sick.

<blockquote>**SECTION PREVIEW**

Objectives

Outline the steps of Koch's postulates.

Describe how infections are transmitted.

Explain what causes the symptoms of a disease.

Vocabulary

pathogen
infectious disease
Koch's postulates
endemic disease
epidemic
antibiotic</blockquote>

What Is an Infectious Disease?

The cold virus causes a disease—a change that disrupts the homeostasis in the body. Disease-producing agents such as bacteria, protozoans, fungi, viruses, and other parasites are called **pathogens.** The main sources of pathogens are soil, contaminated water, and infected animals, including other people.

Not all microorganisms are pathogenic. In fact, the presence of some microorganisms in your body is beneficial. Before birth, your body is free of pathogens. At birth, microorganisms establish themselves on your skin and in your upper respiratory system, lower urinary and reproductive tracts, and lower intestinal tract. *Figure 39.1* shows some common bacteria that live on your skin. They have a symbiotic relationship with

Figure 39.1
These bacteria establish a more-or-less permanent residence in or on your skin, but do not cause disease under normal conditions.

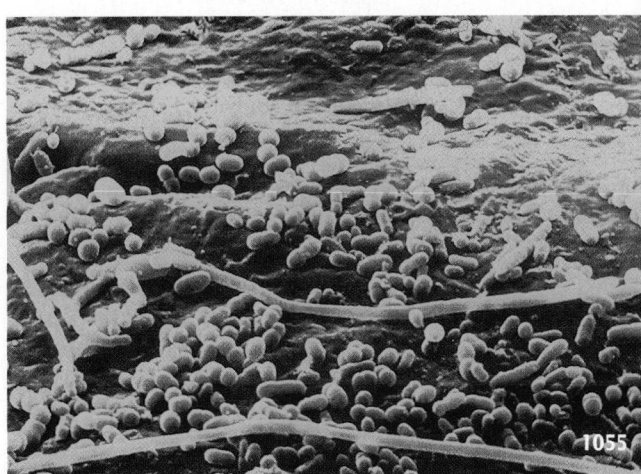

1055

Section 39.1

Prepare

Key Concepts

The steps of Koch's postulates are presented. The spread of disease is covered with emphasis on how infections are transmitted.

Planning

- Obtain printed materials from the local health department, doctors' offices, hospitals, or pharmacies for the Display.
- Prepare nutrient agar plates for the Quick Demo and the Project.
- Purchase apples and plastic bags for MiniLab 39-1.

1 Focus

Bellringer

Before presenting the lesson, display **Section Focus Transparency 96** on the overhead projector and have students answer the accompanying questions.
L1 **ELL**

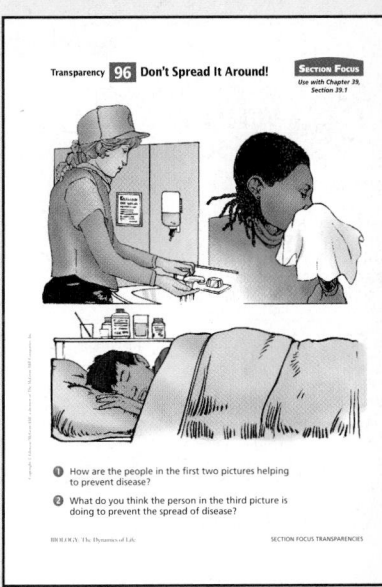

Discussion

Explain the term plague and ask why some people survived plagues while many others perished. *Students may suggest that people in better health may have been more resistant to infection or possessed some unusual trait that prevented them from succumbing to infection. Accept all logical responses.* Initiate a general discussion about the body's defenses to disease.

Quick Demo

Touch the surface of a plate of nutrient agar with the palm of your hand or your fingertips, or have a student do so. Cover and incubate for 24-48 hours. Show students the colonies of bacteria that grow on the agar.

 The BioLab at the end of the chapter can be used at this point in the lesson.

GLENCOE TECHNOLOGY

VIDEODISC
VIDEOTAPE
The Secret of Life
Tinkering With Our Genes: Genetic Medicine

Resource Manager

Section Focus Transparency 96 and Master **L1** **ELL**
Reinforcement and Study Guide, pp. 171-172 **L2**

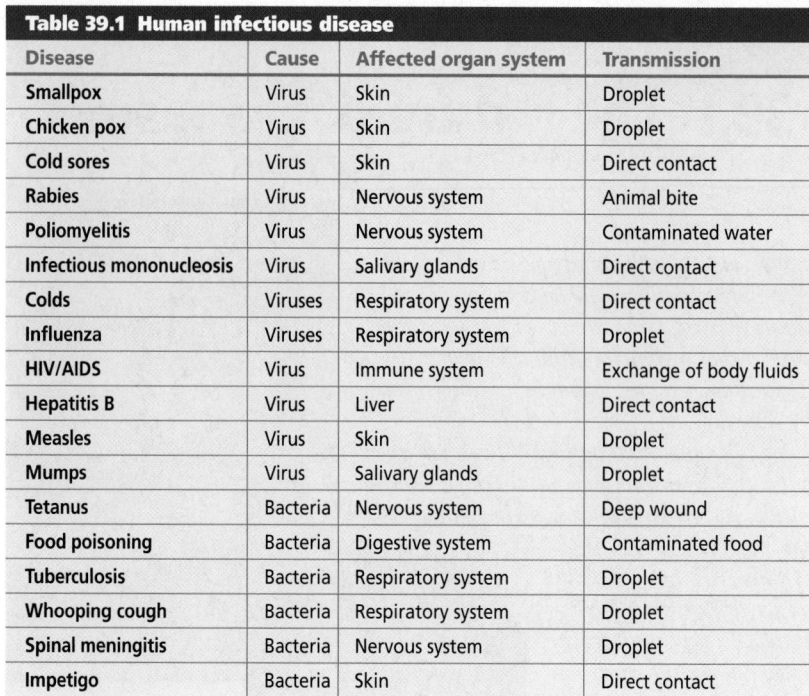

Table 39.1 Human infectious disease

Disease	Cause	Affected organ system	Transmission
Smallpox	Virus	Skin	Droplet
Chicken pox	Virus	Skin	Droplet
Cold sores	Virus	Skin	Direct contact
Rabies	Virus	Nervous system	Animal bite
Poliomyelitis	Virus	Nervous system	Contaminated water
Infectious mononucleosis	Virus	Salivary glands	Direct contact
Colds	Viruses	Respiratory system	Direct contact
Influenza	Viruses	Respiratory system	Droplet
HIV/AIDS	Virus	Immune system	Exchange of body fluids
Hepatitis B	Virus	Liver	Direct contact
Measles	Virus	Skin	Droplet
Mumps	Virus	Salivary glands	Droplet
Tetanus	Bacteria	Nervous system	Deep wound
Food poisoning	Bacteria	Digestive system	Contaminated food
Tuberculosis	Bacteria	Respiratory system	Droplet
Whooping cough	Bacteria	Respiratory system	Droplet
Spinal meningitis	Bacteria	Nervous system	Droplet
Impetigo	Bacteria	Skin	Direct contact

WORD Origin

pathogen
From the Greek words *pathos*, meaning "to suffer," and *geneia*, meaning "producing." Pathogens cause diseases.

your body. However, if you become weakened or injured, these same organisms can become potential pathogens.

Any disease caused by the presence of pathogens in the body is called an **infectious disease** (ihn FEK shus). *Table 39.1* lists some of the infectious diseases that occur in humans.

Determining What Causes a Disease

One of the first problems scientists face when studying a disease is finding out what causes the disease. Not all diseases are caused by pathogens. Disorders such as hemophilia (hee muh FIHL ee uh), which is caused by a recessive allele on the X chromosome, and sickle cell anemia are inherited. Others, such as osteoarthritis (ahs tee oh ar THRITE us), may be caused by wear and tear on the body as it ages.

Pathogens cause infectious diseases and some cancers. In fact, about half of all human diseases are infectious. In order to determine which pathogen causes a specific disease, scientists follow a standard set of procedures.

First pathogen identified

The first proof that pathogens actually cause disease came from the work of Robert Koch in 1876. Koch, a German physician, was looking for the cause of anthrax, a deadly disease that affects mainly cattle and sheep but can also occur in humans. Koch discovered a rod-shaped bacterium in the blood of cattle that had died of anthrax. He cultured the bacteria on nutrients and then injected samples of the culture into healthy animals. When these animals became sick and died, Koch isolated the bacteria in their blood and compared them with

1056 IMMUNITY FROM DISEASE

BIOLOGY JOURNAL

Diseases and Their Causes

Visual-Spatial Have students make a table that lists all the infectious diseases discussed in this section. As students read, have them fill in their tables with the cause of each disease, its vector (if applicable), and the body systems it infects. Have students use their tables as study tools. **L1**

the bacteria he had originally isolated from anthrax victims. He found that the two sets of blood cultures contained the same bacteria.

A procedure to establish the cause of a disease

Koch established experimental steps, shown in *Figure 39.2,* for directly relating a specific pathogen to a specific disease. These steps, first published in 1884, are known today as **Koch's postulates:**

1. The pathogen must be found in the host in every case of the disease.
2. The pathogen must be isolated from the host and grown in a pure culture—that is, a culture containing no other organisms.
3. When the pathogen from the pure culture is placed in a healthy host, it must cause the disease.
4. The pathogen must be isolated from the new host and be shown to be the original pathogen.

Exceptions to Koch's postulates

Although Koch's postulates are useful in determining the cause of most diseases, some exceptions exist. Some organisms, such as the pathogenic bacterium that causes the sexually transmitted disease syphilis (SIHF uh lus), have never been grown on an artificial medium. Viral pathogens also cannot be cultured this way because they multiply only within cells. As a result, living tissue must be used as a culture medium for viruses.

The Spread of Infectious Diseases

For a disease to continue and spread, there must be a continual source of the disease organisms. This source can be either a living organism or an inanimate object on which the pathogen can survive.

Figure 39.2
Koch's postulates are steps used to identify an infectious pathogen.

Display
Prepare a bulletin board with printed information about specific diseases. Brochures on diseases may be available from pharmaceutical companies, doctors' offices, or hospitals. Refer to the display as you discuss pathogens and diseases.

Visual Learning
Figure 39.2 Have students describe each of the four steps of Koch's postulates shown in the diagram.

Brainstorming
Have students discuss the importance of each step of Koch's postulates and how skipping any of them would affect the results of a search for the pathogen that causes a particular disease.

Resource Manager
Laboratory Manual,
pp. 285-292 L2
Basic Concepts Transparency
76 and Master L2 ELL

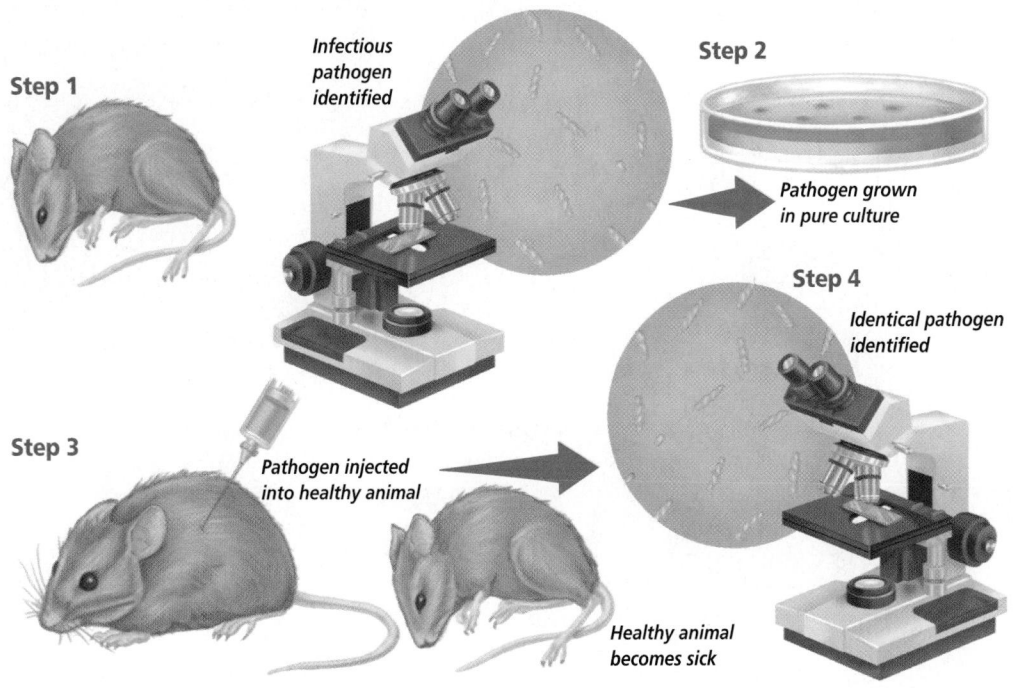

Step 1

Infectious pathogen identified

Step 2

Pathogen grown in pure culture

Step 4

Identical pathogen identified

Step 3

Pathogen injected into healthy animal

Healthy animal becomes sick

MEETING INDIVIDUAL NEEDS

Gifted

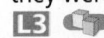 *Linguistic* Have gifted students state their views on using humans to test unproved drugs and other therapies for life-threatening diseases, such as AIDS. Have them consider whether or not they would agree to participate in such tests if they were suffering from a fatal disease.
L3

Discussion

Elicit from students why Koch's postulates are not useful for identifying viral pathogens. *Viruses multiply only within living cells.*

Enrichment

 *Linguistic* Have students research and report on the recent increase in tuberculosis (TB) infections in the United States. Since 1986, TB infections have increased 20%, creating an urban epidemic. Tuberculosis is caused by an airborne bacteria. Infection can be latent or active. Once active, it infects the lungs and other vital organs and can cause death if not cured. Among all the world's infectious diseases, TB remains the number-one cause of death. The majority of new cases in the United States occur among foreign-born groups, especially Asian-born refugees. The disease has also become common in the homeless populations of many urban centers. **L2**

NATIONAL GEOGRAPHIC

VIDEODISC
STV: Human Body Vol. 3
AIDS Virus (tinted blue) 1

Reservoirs of pathogens

The main source of human disease pathogens is the human body itself. In fact, the body can be a reservoir of disease-causing organisms. People may transmit pathogens directly or indirectly to other people. Sometimes, people can harbor pathogens without exhibiting any signs of the illness and unknowingly transmit the pathogens to others. These people are called carriers and are a significant reservoir of infectious diseases.

Other people may unknowingly pass on a disease during its first stage, before they begin to experience symptoms. This symptom-free period, while the pathogens are multiplying within the body, is called an incubation period. Humans can unknowingly pass on the pathogens that cause colds, streptococcal throat infections, and sexually transmitted diseases (STDs) such as gonorrhea and AIDS during the incubation periods of these diseases.

Animals are other living reservoirs of microorganisms that cause disease in humans. For example, some types of influenza, commonly known as the flu, and rabies are often transmitted to humans from animals. The major nonliving reservoirs of infectious diseases are soil and water. Soil harbors pathogens such as fungi and the bacterium that causes botulism, a type of food poisoning. Water contaminated by feces of humans and other animals is a reservoir for several pathogens, especially those responsible for intestinal diseases.

Transmission of disease

How are pathogens transmitted from a reservoir to a human host? Pathogens can be transmitted from reservoirs in four main ways: by direct contact, by an object, through

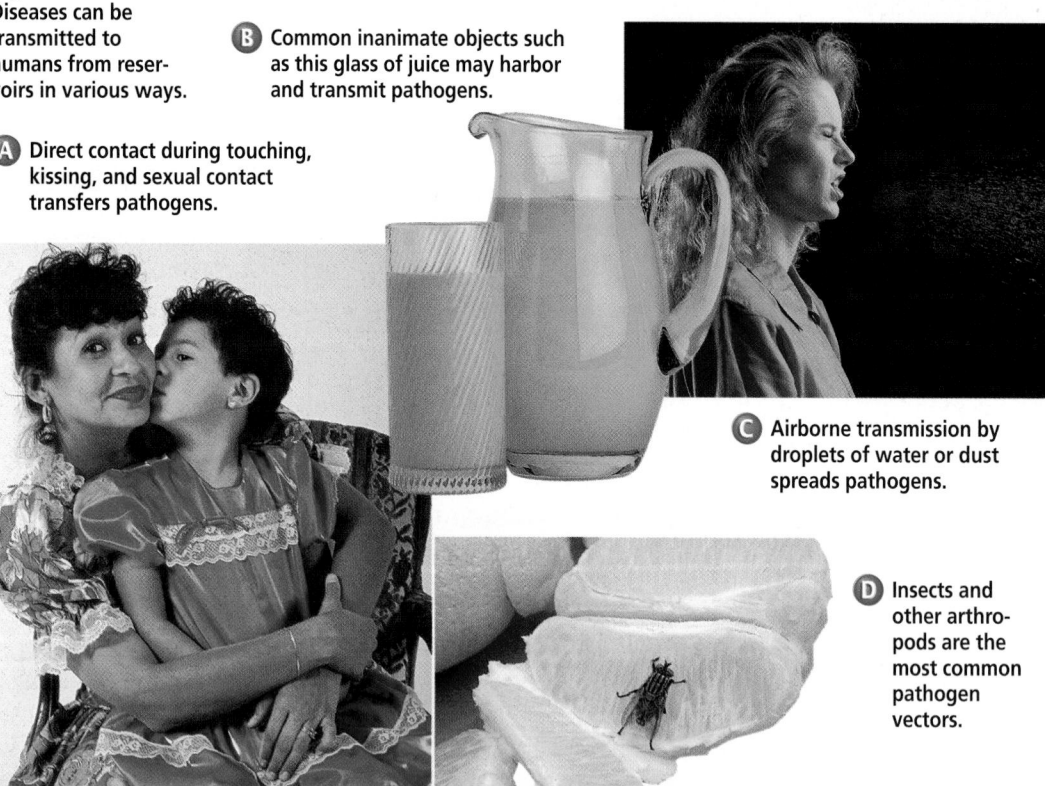

Figure 39.3 Diseases can be transmitted to humans from reservoirs in various ways.

A Direct contact during touching, kissing, and sexual contact transfers pathogens.

B Common inanimate objects such as this glass of juice may harbor and transmit pathogens.

C Airborne transmission by droplets of water or dust spreads pathogens.

D Insects and other arthropods are the most common pathogen vectors.

TECHPREP

Disease Prevention

Interpersonal Students who have part-time jobs or are looking for a career that includes disease prevention can find out what measures are used to prevent the spread of disease. Examples of work places in which disease prevention is practiced include plant nurseries, veterinarian's offices, and the food industry. **L2**

the air, or by an intermediate organism called a *vector*. *Figure 39.3* illustrates each of these.

The common cold, influenza, and STDs are spread by direct contact. STDs, such as genital herpes and the virus that causes AIDS, are usually transmitted by the exchange of body fluids, especially during sexual intercourse.

Food poisoning is a common example of a disease transmitted by an object. This disease is often transmitted when food is contaminated by a food handler. To help prevent the spread of these types of diseases, restaurants have equipment that cleans and disinfects dishes and utensils. Today, laws require food handlers to wash their hands thoroughly before preparing food, and frequent inspections of restaurants help prevent unsanitary conditions.

Diseases transmitted by vectors are most commonly spread by insects and arthropods. You have read about malaria, which is transmitted by mosquitoes, and about Lyme disease, which is transmitted by ticks. The bubonic plague—a disease that swept through Europe in the 1600s, killing up to one-third of the population—was transmitted from infected rats to humans by fleas. Flies also are significant vectors of disease. They transmit pathogens when they land on infected materials, such as animal wastes, and then land on fresh food that is eaten by humans. To learn more about how diseases are spread, refer to the *Problem-Solving Lab* here and the *MiniLab* on the next page.

What Causes the Symptoms of a Disease?

When a pathogen invades your body, it encounters your immune system. If the pathogen overcomes the defenses of your immune system, it can metabolize and multiply, causing damage to the tissues it has invaded, and even killing host cells.

Damage to the host by viruses and bacteria

You already know that viruses cause damage by taking over a host cell's genetic and metabolic machinery. Many viruses also cause the eventual death of the cells they invade.

Problem-Solving Lab 39-1 — Designing an Experiment

How does the herpes simplex virus spread? Herpes simplex virus, which causes cold sores, infects a person for life, occasionally reproducing and then spreading to other cells in the body of its host. Scientists have been interested for a long time in how the herpes virus actually enters a cell.

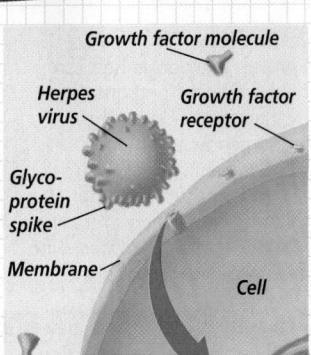

Analysis
Scientists have found that the herpes virus infects a cell in one of two possible ways. It may latch onto a cell receptor with its own glycoprotein spike, or it may use this spike to grab a growth factor molecule that latches onto the receptor, as shown in the diagram.

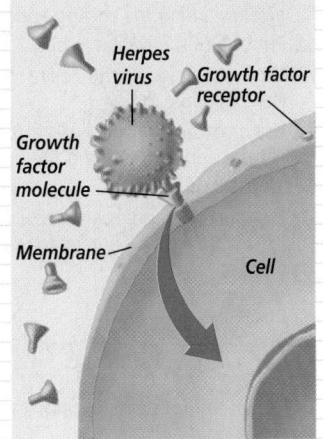

Thinking Critically
Design an experiment to determine which method the herpes virus uses to get into a cell.

Purpose
Students will design an experiment to determine how a virus enters a cell.

Process Skills
observe and infer, formulate models

Teaching Strategies
■ Ask students to include a control setup in their procedures.

Thinking Critically
Student answers will vary. One experiment might be to remove the growth factor from the cell to see if the virus can still get in. Another possibility would be to remove the receptor sites from the cell or to remove the glycoprotein spikes from the virus to see if the virus can still get in.

✓ Assessment
Portfolio Ask students to write a summary of the lab. Have them include the summaries of their experimental designs in their Portfolios. Use the Performance Task Assessment List for Lab Report in **PASC**, p. 47. **L2** **P**

NATIONAL GEOGRAPHIC

VIDEODISC
STV: Human Body Vol. 3
Cell Bursting, Releasing Cold Virus

BIOLOGY JOURNAL

Disease Transmission
Linguistic Ask students to list the ways in which diseases are transmitted and give an example of a disease transmitted in each way. Students may need help giving the examples. **L2**

TECHPREP

Role Playing
Linguistic Have students prepare a skit demonstrating the ways diseases might spread in a restaurant from the viewpoint of a health inspector, a restaurant owner, a restaurant cook, or a disease organism. Methods of prevention should also be included. **L2** **ELL**

MiniLab 39-1 — Experimenting

Testing How Diseases are Spread Microorganisms cannot travel over long distances by themselves. Unless they are somehow transferred from one animal or plant to another, infections will not spread. One method of transmission is by direct contact with an infected animal or plant.

Rotten apples

Procedure
1. Label four plastic bags 1 to 4.
2. Put a fresh apple in bag 1 and seal the bag.
3. Rub a rotting apple over the entire surface of the remaining three apples. The rotting apple is your source of pathogens. **CAUTION:** *Make sure to wash your hands after handling the rotting apple.*
4. Put one of the apples in bag 2.
5. Drop one apple to the floor from a height of about 2 m. Put this apple in bag 3.
6. Use a cotton ball to spread alcohol over the last apple. Let the apple air-dry and then place it in bag 4.
7. Store all of the bags in a dark place for one week.
8. Compare the apples and record your observations. **CAUTION:** *Give all apples to your teacher for proper disposal.*

Analysis
1. What was the purpose of the fresh apple in bag 1?
2. Explain what happened to the rest of the apples.
3. Why is it important to clean a wound with disinfectant?

Most of the damage done to host cells by bacteria is inflicted by toxins. Toxins are poisonous substances that are sometimes produced by microorganisms. These poisons are transported by the blood and can cause serious and sometimes fatal effects. Some toxins produce fever and cardiovascular disturbances. Toxins can also inhibit protein synthesis in the host cell, destroy blood cells and blood vessels, or cause spasms by disrupting the nervous system.

For example, the toxin produced by tetanus bacteria affects nerve cells and produces uncontrollable muscle contractions. If the condition is left untreated, paralysis and death occur. Tetanus bacteria are normally present in soil, as *Figure 39.4* illustrates. If dirt transfers the bacteria into a deep wound on your body, the bacteria begin to produce the toxin in the wounded area. A small amount of this toxin, about the same amount as the ink used to make a period on this page, could kill 30 people. That is why you should be vaccinated for tetanus.

Patterns of Diseases
In today's highly mobile world, diseases can spread rapidly. Contaminated water, for example, can affect

Figure 39.4
Conditions of a battlefield are ideal for tetanus bacteria. Before the days of modern medicine, wounded soldiers faced the additional, deadly danger of becoming infected with these bacteria.

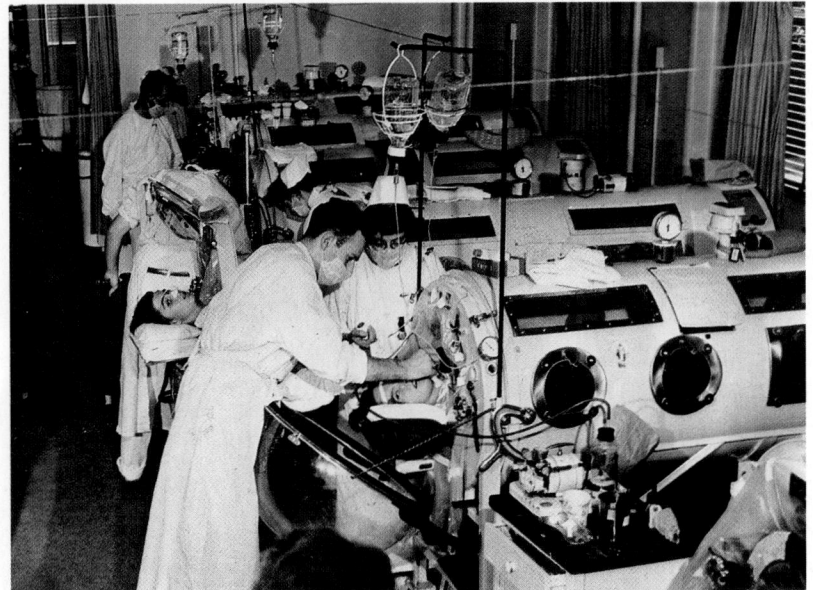

many thousands of people quickly. Therefore, identifying a pathogen, its method of transmission, and the geographic distribution of the disease it causes are major concerns of government health departments. The Centers for Disease Control and Prevention, the central source of disease information in the United States, publishes a weekly report about the incidence of specific diseases.

Some diseases, such as typhoid fever, occur only occasionally in the United States. These periodic outbreaks often occur because someone traveling in a foreign country has brought the disease back home. On the other hand, many diseases are constantly present in the population. Such a disease is called an endemic disease. The common cold is an **endemic disease.**

Sometimes, an **epidemic** breaks out. An epidemic occurs when many people in a given area are afflicted with the same disease at about the same time. Influenza is a disease that

often achieves epidemic status, sometimes spreading to many parts of the world. During the 1950s, a polio epidemic spread across the United States. Victims of this disease were paralyzed or died when the polio virus attacked the nerve cells of the brain and spinal cord. Many survived only after being placed in an iron lung—a machine that allowed the patient to continue to breathe, as shown in *Figure 39.5.*

Treating Diseases

A person who becomes sick often can be treated with medicinal drugs, such as antibiotics. An **antibiotic** is a substance produced by a microorganism that, in small amounts, will kill or inhibit the growth and reproduction of other microorganisms, especially bacteria. Antibiotics are produced naturally by various species of bacteria and fungi. Although antibiotics can be used to cure some bacterial infections, antibiotics do not affect viruses.

WORD *Origin*
epidemic
From the Greek words *epi*, meaning "upon," and *demos*, meaning "people." An epidemic is a disease found among many people in an area.

Figure 39.6
This graph shows the occurrence of penicillin-resistant gonorrhea. Notice the increase in the number of reported cases of gonorrhea in the United States.

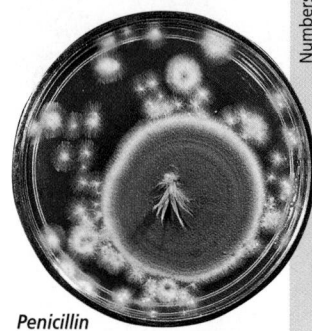
Penicillin

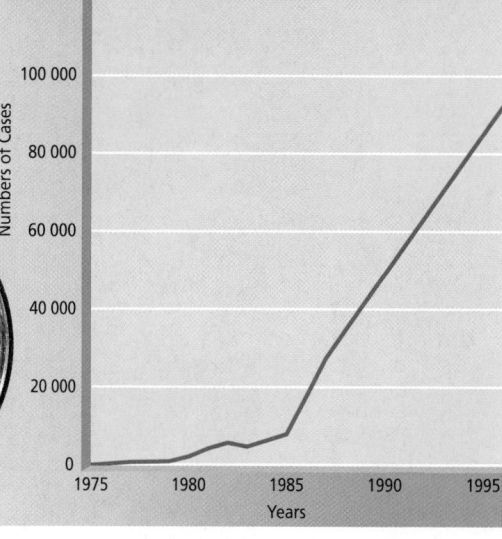
Cases of Penicillin-Resistant Gonorrhea

Numbers of Cases — 100 000, 80 000, 60 000, 40 000, 20 000, 0
Years — 1975, 1980, 1985, 1990, 1995

WORD *Origin*
antibiotic
From the Greek words *anti*, meaning "against," and *bios*, meaning "life." An antibiotic is given to control a bacterial infection.

A problem that sometimes occurs with the continued use of antibiotics is that the bacteria become resistant to the drugs. That means the drugs become ineffective. Penicillin, an antibiotic produced by a fungus, was used for the first time in the 1940s and is still one of the most effective antibiotics known. However, penicillin has now been in use for more than 50 years, and more and more types of bacteria have evolved that are resistant to it. Bacteria that are resistant to penicillin produce an enzyme that breaks down this antibiotic. In certain infections, such as the STD gonorrhea, this resistance is a problem because, until now, penicillin has been the most successful drug in treating the infection. The increase in penicillin-resistant gonorrhea is graphed in *Figure 39.6*.

The use of antibiotics is only one way to fight infections. Your body also has its own built-in defense system—the immune system—that is continually working to keep you healthy.

Section Assessment

Understanding Main Ideas
1. What are the major reservoirs of pathogens?
2. In what way can a family member who is in the incubation period for strep throat be a threat to your good health?
3. When does a disease become an epidemic?
4. In what ways are diseases transmitted?

Thinking Critically
5. Many patients enter the hospital with one medical problem but contract an infection while in the hospital. What are possible ways in which a disease might be transmitted to a hospital patient?

SKILL REVIEW
6. **Designing an Experiment** Design an experiment that could be conducted to determine whether a recently identified bacterium causes a type of pneumonia. For more help, refer to *Practicing Scientific Methods* in the **Skill Handbook**.

1062 IMMUNITY FROM DISEASE

Section Assessment

1. infected animals, soil, and contaminated water
2. They can spread the disease to you.
3. when many people in a given area are infected with the same disease in a relatively short period of time
4. direct contact, by an object, through the air, or by a vector
5. by direct contact with employees or patients, through contaminated objects, or by airborne transmission
6. Take the suspected organisms from an infected animal, grow them in a pure culture, then inject them into another animal. If that animal develops the same disease, the organisms probably caused it. To be sure, isolate the organisms and compare with the original organisms.

39.2 Defense Against Infectious Diseases

Y ou can't see it, but a war is going on around these teenagers. In fact, the same sort of war is occurring around you. Hordes of unseen enemies are present everywhere—in the air, on the ground, and even on your clothes. Defenders ready to protect you from the onset of attack are inside your body. How does your body save you from the microscopic foes that cause infectious diseases? How do the body's defenses protect you from these unseen enemies?

Magnification: 3700×

Lymphocytes (inset) help protect you from diseases.

SECTION PREVIEW

Objectives

Compare nonspecific and specific immune responses.

Compare innate and acquired immune responses.

Distinguish between antibody and cellular immunity.

Vocabulary

innate immunity
phagocyte
macrophage
pus
interferons
acquired immunity
tissue fluid
lymph
lymph node
lymphocyte
T cell
B cell
vaccine

Innate Immunity

Your body produces a variety of white blood cells that defend it against invasion by pathogens that are constantly bombarding you. No matter what pathogens are present, your body is always ready. The body's earliest lines of defense against any and all pathogens make up your nonspecific, **innate immunity.**

Skin and body secretions

When a potential pathogen contacts your body, often the first barrier it must penetrate is your skin. Like the walls of a castle, intact skin is a formidable physical barrier to the entrance of microorganisms.

In addition to the skin, pathogens also encounter your body's secretions of mucus, sweat, tears, and saliva. The main function of mucus is to prevent various areas of the body from drying out. Because mucus is slightly viscous (thick), it also traps many microorganisms and other foreign substances that enter the respiratory and digestive tracts. Mucus is continually swallowed and passed to the stomach, where acidic gastric juice (made of hydrochloric acid and other fluids) destroys most bacteria and their toxins. Sweat, tears, and saliva all contain the enzyme lysozyme, which is capable of breaking down the cell walls of some bacteria.

BIOLOGY Online Note Internet addresses that you find useful in the space below for quick reference.

_____

_____

_____

_____

Resource Manager

Section Focus Transparency 97 and Master **L1** **ELL**

Basic Concepts Transparency 77 and Master **L2** **ELL**

Reinforcement and Study Guide, pp. 173-174 **L2**

Section 39.2

Prepare

Key Concepts

The anatomy and physiology of the lymphatic system are discussed in this section. Various types of nonspecific defense mechanisms are described along with the specific defenses of antibody and cellular immunity. The last part of the section discusses AIDS and its effect on the immune system.

Planning

■ Collect articles about AIDS for the Display and Biology and Society.

■ Prepare nutrient agar plates for the Alternative Lab.

---

1 Focus

Bellringer

Before presenting the lesson, display **Section Focus Transparency 97** on the overhead projector and have students answer the accompanying questions. **L1** **ELL**

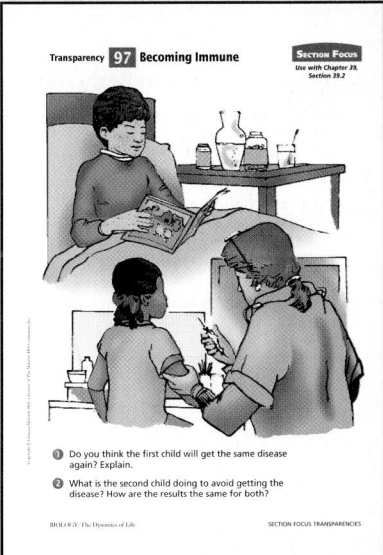

Transparency **97** Becoming Immune

1 Do you think the first child will get the same disease again? Explain.

2 What is the second child doing to avoid getting the disease? How are the results the same for both?

Display

Assemble a bulletin board display of newspaper and magazine articles about AIDS. Refer to the articles when discussing AIDS and other viral infections.

Tying to Previous Knowledge

Ask students why, when a patient complains of a sore throat, the physician feels under the patient's chin and behind his or her ears and looks down the patient's throat. *The doctor is looking for swollen lymph glands and for redness and blotchy white spots in the throat that would indicate a bacterial infection.*

Quick Demo

Visual-Spatial Using a microprojector, demonstrate the differences between white blood cells (nucleated) and red blood cells. You can also use 35-mm slides or have students examine prepared slides.

NATIONAL GEOGRAPHIC

VIDEODISC
STV: Human Body Vol. 3
Immune System
Unit 1, Side 1, 1 min. 56 sec.
Phagocytes

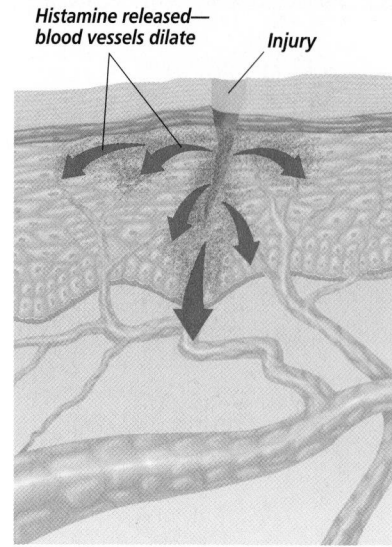
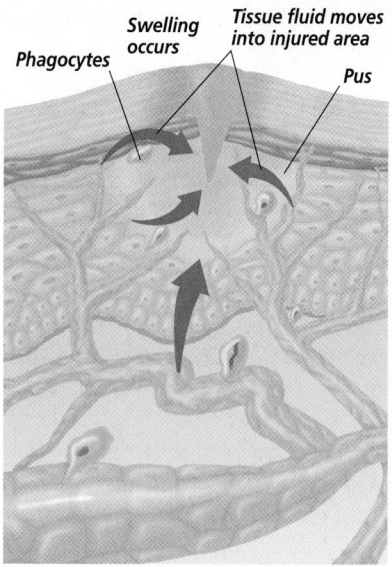

Figure 39.8
When tissues become inflamed, histamine release causes blood vessels to dilate. Tissue fluid leaks out of the vessels into the injured area, causing swelling.

Histamine released—blood vessels dilate

Injury

Phagocytes

Swelling occurs

Tissue fluid moves into injured area

Pus

Inflammation of body tissues

If a pathogen manages to get past the skin and body secretions, your body has several other nonspecific defense mechanisms that can destroy the invader and restore homeostasis. Think about what happens when you get a splinter. When bacteria or other pathogens damage body tissues, inflammation (ihn fluh MAY shun) may result. Inflammation is characterized by four symptoms—redness, swelling, pain, and heat. As *Figure 39.8* shows, inflammation begins when damaged tissue cells, and white blood cells called basophils, release histamine (HIHS tuh meen). Histamine causes blood vessels in the injured area to dilate, which makes them more permeable to tissue fluid. These dilated blood vessels cause the redness of an inflamed area. Fluid that leaks from the vessels into the injured tissue helps the body destroy toxic agents and restore homeostasis. This increase in tissue fluid causes swelling and pain, and may also cause a local temperature increase. Inflammation can occur as a reaction to other types of injury as well as infections. Physical force, chemical substances, extreme temperatures, and radiation may also inflame body tissues.

Phagocytosis of pathogens

Pathogens that enter your body may encounter cells that carry on phagocytosis. Recall that phagocytosis occurs when a cell engulfs a particle. **Phagocytes** (FAG uh sites) are white blood cells that destroy pathogens by surrounding and engulfing them. Phagocytes include macrophages, neutrophils, monocytes, and eosinophils. Macrophages are present in body tissues. The other types of phagocytes circulate in the blood.

Macrophages are white blood cells that provide the first line of defense against pathogens that have managed to enter the tissues. Macrophages, shown in *Figure 39.9*, are found in the tissues of the body. They are sometimes called giant scavengers, or

Alternative Lab

Antibiotics and Bacteria

Purpose
Students will determine which antibiotics are most effective in inhibiting bacterial growth.

Materials
E. coli bacterial culture, 3 types of antibiotic disks, sterile untreated disks, cotton swabs, forceps, ethanol, transparent tape, plates of nutrient agar, marking pen, incubator

Procedure
Give students the following directions:
CAUTION: *Wear lab aprons, safety goggles, and disposable latex gloves. Do not open petri dishes after they have been sealed. Wash hands after inoculating petri dishes*

and after making observations. Used petri dishes and toothpicks should be autoclaved before disposal.

1. Use the marking pen to make a cross on the bottom of an agar plate, dividing it into four sections.
2. Using a cotton swab, gently transfer some of the bacterial culture onto the agar plate.
3. Spread the culture evenly over the surface of the agar plate using the cotton swab.

big eaters, because of the manner in which they engulf pathogens or damaged cells. Lysosomal enzymes inside the macrophage digest the particles it has engulfed.

If the infection is not stopped by the tissue macrophages, another type of phagocyte, called a neutrophil, is attracted to the site. Neutrophils constitute the second line of defense. They also destroy pathogens by engulfing and digesting them.

If the infection is not stopped by tissue macrophages and neutrophils, there is a third line of defense. A different type of phagocyte begins to arrive on the scene. Monocytes are small, immature macrophages that circulate in the bloodstream. These cells squeeze through blood vessel walls to move into the infected area. Once they reach the site of the infection, they mature, becoming as large as tissue macrophages. They then begin consuming pathogens and dead neutrophils by phagocytosis. Once the infection is over, some monocytes mature into tissue macrophages that

remain in the area, prepared to fend off a new infection.

After a macrophage has destroyed large numbers of pathogens, dead neutrophils, and damaged tissue cells, it dies. After a few days, infected tissue harbors a collection of dead macrophages and body fluids called **pus**. Pus formation usually continues until the infection subsides. Eventually, the pus is cleared away by macrophages.

Which white blood cells are involved in each of the body's lines of defense against pathogens? Find out by reading the *Inside Story* on the following page and carry out the *MiniLab* that follows to observe the different types of white blood cells.

Protective Proteins

When an infection is caused by a virus, your body faces a problem. Phagocytes cannot destroy viruses. Recall that a virus multiplies within a host cell. A phagocyte that engulfs a virus will itself be destroyed if the virus multiplies within it. One way

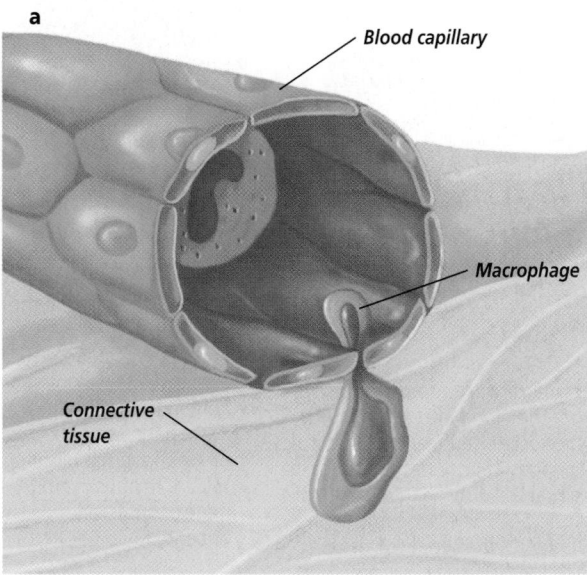

a

Blood capillary

Macrophage

Connective tissue

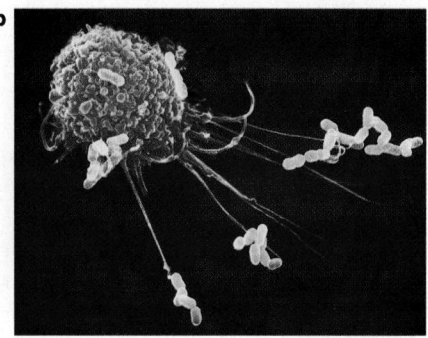

Magnification: 6000✓

Figure 39.9
Macrophages move out of blood vessels by squeezing between the cells of the vessel walls (**a**). Macrophages will attack anything they recognize as foreign, including microorganisms and dust particles that are breathed into the lungs (**b**).

4. Dip the forceps into alcohol and allow them to air dry without touching any surface.
5. Use the forceps to transfer two antibiotic disks into each of three sections of the plate. Label the bottom with the type of antibiotic used in each section.
6. Use the forceps to transfer two untreated disks to the fourth section of the agar plate.

7. Tape the plates closed and incubate for 48 hours at 37°C.
8. Look for regions of inhibition near the disks.

Analysis
1. Why were untreated disks placed in one section? *This section was the control.*
2. Did any of the antibiotics inhibit the growth of the bacteria? *Answers will vary.*

Purpose

Students visualize the various types of white blood cells and their role in the immune system.

Teaching Strategies

■ Review this page with students before and after completing MiniLab 39-2

■ Have students prepare a table of the white blood cells that compares their characteristics, functions, and where they are found.

■ Have students discuss what would happen to the numbers of white blood cells during an infection. *The number of white blood cells per milliliter of blood increases during an infection.*

Visual Learning

■ Using a microscope projector, point out the various types of white blood cells on a blood smear slide.

Critical Thinking

Mature tissue macrophages are not found in the blood. Immature macrophages, called monocytes, are found in the bloodstream.

BIOLOGY *Online* Note Internet addresses that you find useful in the space below for quick reference.

_____

_____

_____

_____

_____

INSIDE STORY

Lines of Defense

White blood cells play a major role in protecting your body against disease. Many of these cells leave the bloodstream to fight disease organisms in the tissues.

Critical Thinking *Why would you not expect to see tissue macrophages in a sample of blood cells?*

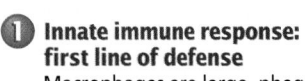

A graze breaks the protective barrier of the skin

1 **Innate immune response: first line of defense**
Macrophages are large, phagocytic white blood cells found in the tissues. They are the first to arrive at the site of an infection. Basophils, found in the blood, are not phagocytic. They are filled with granules that release histamine at an infection site. Eosinophils are also granular and also play a role in inflammation.

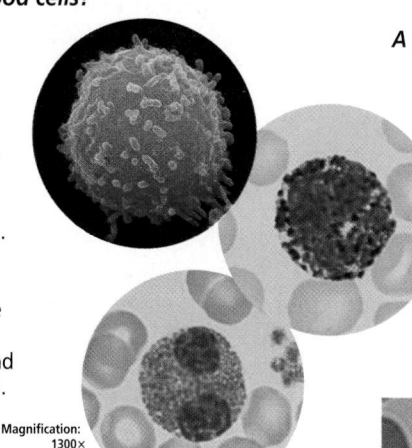

Magnification: 1400×

Tissue macrophage (top), basophil (middle), and eosinophil (bottom)

Magnification: 1300×

Magnification: 7000×

Magnification: 2300×

Magnification: 6900×

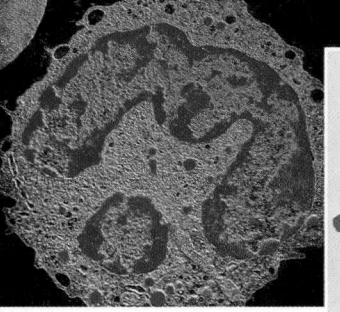

2 **Second line of defense** A neutrophil (above) is a phagocytic white blood cell with a nucleus that has several lobes.

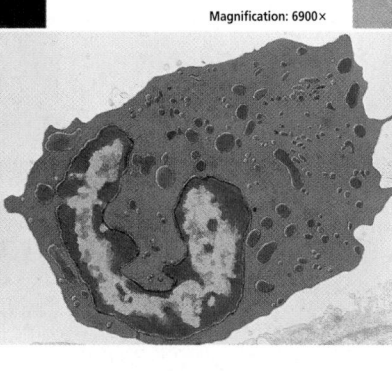

3 **Third line of defense** After moving from the blood into an infected area, monocytes (above) mature into macrophages. Monocytes are two to three times larger than other blood cells and have large nuclei. They replenish the supply of tissue macrophages following an infection.

4 **Acquired immune response** Lymphocytes (above) are cells with nuclei that nearly fill the cell. They include B cells and T cells and are involved in developing immunity to specific pathogens. Lymphocytes are found in the blood, spleen, thymus, lymph nodes, tonsils, and appendix.

Cultural Diversity

Susumu Tonagawa and Immunity

Introduce students to the important contributions Japanese immunologist Susumu Tonagawa has made toward our understanding of the immune system. Tonagawa's major contribution ended the debate on whether individuals inherit separate genes for each of the millions of antibody molecules, or whether individuals inherit a small number of genes that diversify in specialized somatic cells. Tonagawa was able to show that somatic cells use a multi-step process to rearrange fragments from different genes, in different ways, to produce many different antibodies. For this work, Tonagawa received a Nobel prize in 1987.

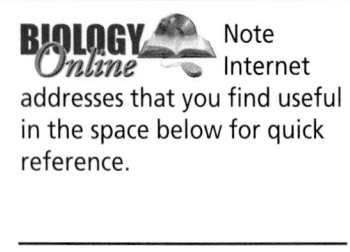

your body can counteract viral infections is with interferons. **Interferons** are proteins that protect cells from viruses. They are host-cell specific; that is, human interferons will protect human cells from viruses but will do little to protect cells of other species from the same virus.

Interferon is produced by a body cell that has been infected by the virus. The interferon diffuses into uninfected neighboring cells, which then produce antiviral proteins that can prevent the virus from multiplying.

Acquired Immunity

The cells of your innate immune system continually survey your body for foreign invaders. When a pathogen is detected, these cells begin defending your body right away. Meanwhile, as the infection continues, another type of immune response that counteracts the invading pathogen is also mobilized. Certain white blood cells gradually develop the ability to recognize a specific foreign substance. This acquired immune response enables these white blood cells to inactivate or destroy the pathogen. Defending against a specific pathogen by gradually building up a resistance to it is called **acquired immunity.**

Normally, the immune system recognizes components of the body as self, and foreign substances, called antigens, as nonself. Antigens are usually proteins present on the surfaces of whole organisms, such as bacteria, or on parts of organisms, such as the pollen grains of plants. An acquired immune response occurs when the immune system recognizes an antigen and responds to it by producing antibodies against it. Antigens are foreign substances that stimulate

an immune response, and antibodies are proteins in the blood that correspond specifically to each antigen. The development of acquired immunity is the job of the lymphatic system. The process of acquiring immunity to a specific disease can take days or weeks.

39.2 DEFENSE AGAINST INFECTIOUS DISEASES **1067**

MiniLab 39-2 Observing and Inferring

Distinguishing Types of White Blood Cells The human immune system includes five types of white blood cells found in the bloodstream: basophils, neutrophils, monocytes, eosinophils, and lymphocytes.

A blood smear

Procedure

1. Copy the data table below.
2. Mount a prepared slide of blood cells on the microscope and focus on low power. Turn to high power and look for white blood cells. **CAUTION:** *Use care when working with microscope slides.*
3. Find a neutrophil, monocyte, eosinophil, and lymphocyte. You may see a basophil, although they are rare. Refer to the *Inside Story* for photos of these cells.
4. Count a total of 50 white blood cells, and record how many of each type you see.
5. Calculate the percentage by multiplying the number of each cell type by two. Record the percentages. Diagram each cell type.

Data Table

Type of white blood cell	Number counted	Percent	Diagram
Neutrophil			
Monocyte			
Basophil			
Lymphocyte			
Eosinophil			

Analysis

1. Which type of white blood cell was most common? Second most common?
2. How do red and white blood cells differ?

MiniLab 39-2

Purpose
Students will determine the percentages of different kinds of white blood cells.

Process Skills
classify, observe and infer, compare and contrast, collect and organize data

Safety Precautions
Caution students to use care when working with prepared slides and microscopes. Special care should be taken when viewing slides under high power so the objective does not break the slide.

Teaching Strategies
■ It would be helpful for student identification to set up a prepared microscope slide showing both an eosinophil and a basophil.

Expected Results
The common percentages of white blood cells are: neutrophils, 60-70%; lymphocytes, 20-25%; monocytes, 3-8%; eosinophils, 2-4%; and basophils, 0.5-1%.

Analysis
1. neutrophils, lymphocytes
2. White blood cells have a nucleus; red blood cells don't.

✓ Assessment

Performance Have students include summaries of the lab, their data tables, and their answers to the Analysis questions in their Biology Journals. Use the Performance Task Assessment List for Lab Report in **PASC**, p. 47. L2

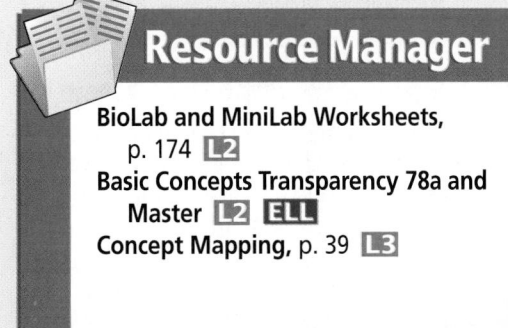

✓ Portfolio

Scientific Illustration

Visual-Spatial Have students sketch the different types of blood cells they observe when viewing a blood smear in MiniLab 39-2. Ask them to add these illustrations to their portfolios. L1 ELL
P

Resource Manager

BioLab and MiniLab Worksheets, p. 174 L2
Basic Concepts Transparency 78a and Master L2 ELL
Concept Mapping, p. 39 L3

Discussion

Ask students to compare and contrast the functions of the circulatory system and the lymphatic system. Have students as a class discuss the two systems.

Enrichment

Have students research the life of David, the "bubble boy" of Houston, Texas, who lived in a sterile plastic bubble for all but 15 days of his life. He suffered from severe combined immune deficiency (SCID), a rare genetic disorder related to a dysfunction of the lymphocyte-producing cells of the bone marrow. At birth, David was placed in a sterile plastic bubble. As he grew, he was placed in increasingly larger bubbles. At the age of 12, he left his bubble and underwent a bone marrow transplant using tissue from his sister. David died four months later, in February, 1984, of blood cancer, which was unrelated to his transplant. Encourage students to find out about current treatments for SCID. **L3**

Figure 39.10
The lymphatic system is spread throughout the body. As you read about the various vessels, tissues, and organs of the lymphatic system, locate them on this diagram.

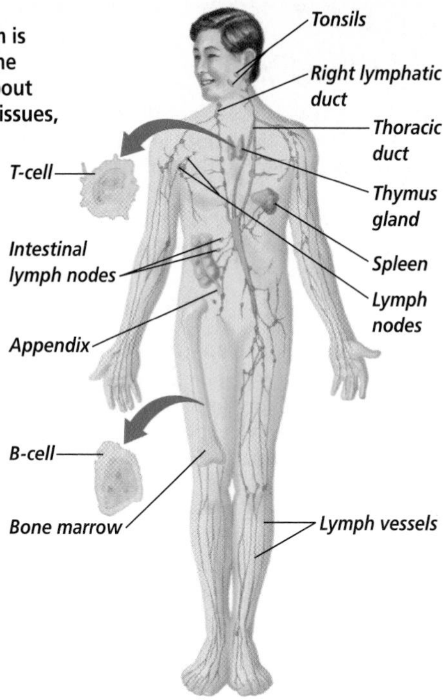

Tonsils
Right lymphatic duct
Thoracic duct
Thymus gland
Spleen
Lymph nodes
T-cell
Intestinal lymph nodes
Appendix
B-cell
Bone marrow
Lymph vessels

The lymphatic system

Your lymphatic (lihm FAT ihk) system not only helps the body defend itself against disease, but also maintains homeostasis by keeping body fluids at a constant level. *Figure 39.10* shows the major glands and vessels

Figure 39.11
As lymph filters through a lymph node, the lymphocytes in the node trap and kill pathogens.

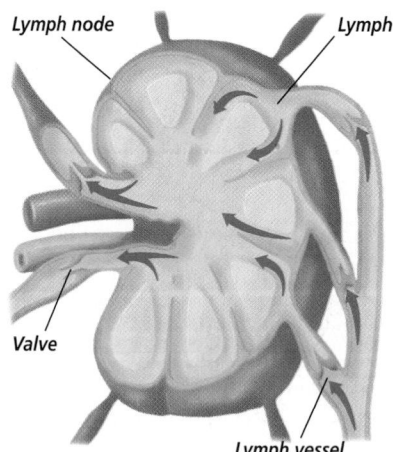

Lymph node
Lymph
Valve
Lymph vessel

that make up the lymphatic system.

Your body's cells are constantly bathed with fluid. This **tissue fluid** forms when water and dissolved substances diffuse from the blood into the spaces between the cells that make up the surrounding tissues. This tissue fluid collects in open-ended lymph capillaries. Once the tissue fluid enters the lymph vessels, it is called **lymph**. *Figure 39.10* shows the major glands and vessels that make up the lymphatic system.

Lymph capillaries meet to form larger vessels called lymph veins. The flow of lymph is only toward the heart, so there are no lymph arteries. The lymph veins converge to form two major lymph ducts. These ducts return the lymph to the bloodstream in the shoulder area, after it has been filtered through various lymph glands.

Glands of the lymphatic system

At locations along the lymphatic pathways, the lymph vessels pass through lymph nodes. A **lymph node** is a small mass of tissue that contains lymphocytes and filters pathogens from the lymph, as shown in *Figure 39.11.* Lymph nodes are made of an interlaced network of connective tissue fibers that holds lymphocytes. A **lymphocyte** (LIHM fuh site) is a type of white blood cell that defends the body against foreign substances.

Have you ever had a sore throat caused by infected tonsils? The tonsils are large clusters of lymph tissue located at the back of the mouth cavity and at the back of the throat. They form a protective ring around the openings of the nasal and oral cavities. Your tonsils provide protection against bacteria and other pathogens that enter your nose and mouth.

The spleen is an organ in which certain types of lymphocytes are

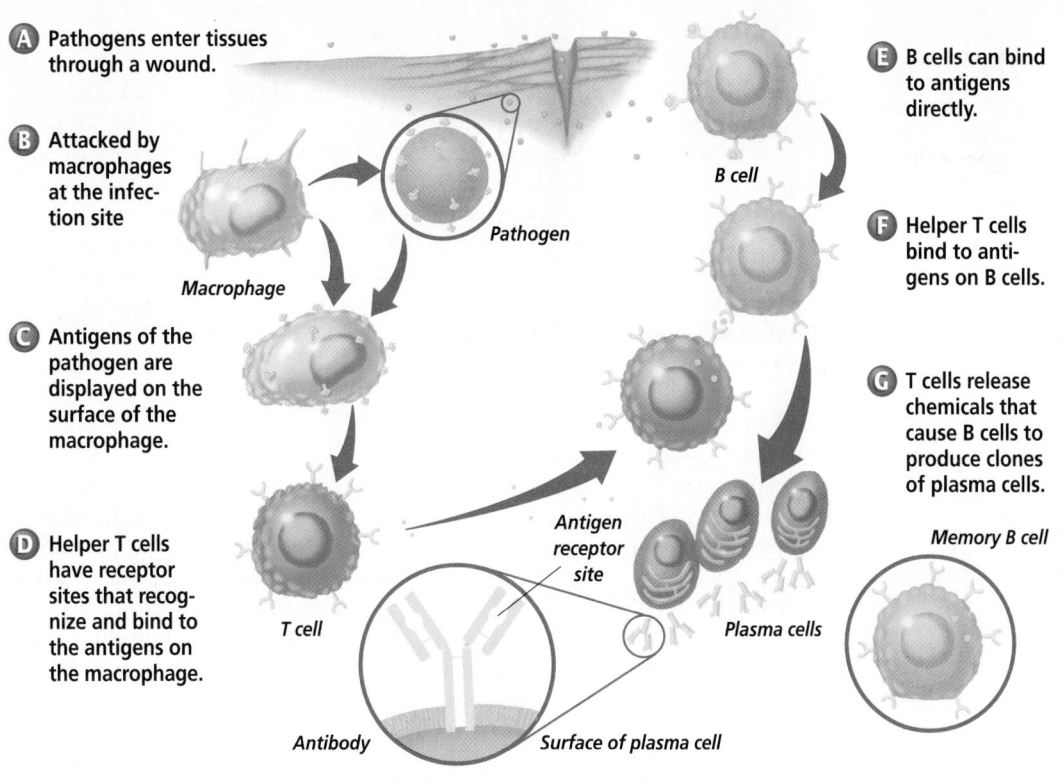

A Pathogens enter tissues through a wound.

B Attacked by macrophages at the infection site

C Antigens of the pathogen are displayed on the surface of the macrophage.

D Helper T cells have receptor sites that recognize and bind to the antigens on the macrophage.

Macrophage

Pathogen

T cell

Antigen receptor site

Antibody

Surface of plasma cell

E B cells can bind to antigens directly.

B cell

F Helper T cells bind to antigens on B cells.

G T cells release chemicals that cause B cells to produce clones of plasma cells.

Plasma cells

Memory B cell

H Most plasma cells secrete antibodies that bind to antigens on infected cells.

I Each plasma cell secretes more than 2000 antibodies per second in the blood. Memory B cells and antibodies remain in the blood and respond to future invasions by the same pathogen.

stored. The spleen also filters out and destroys bacteria and worn-out red blood cells and acts as a blood reservoir. Unlike lymph nodes, the spleen does not filter lymph.

Another important component of the lymphatic system is the thymus gland, which is located above the heart. The thymus gland stores immature lymphocytes until they mature and are released into the body's defense system.

Antibody Immunity

Acquired immunity involves the production of two kinds of immune responses: antibody immunity and cellular immunity. Antibody immunity

is a type of chemical warfare within your body that involves several types of cells. Follow the steps of antibody immunity illustrated in *Figure 39.12.*

When a pathogen invades your body, it is first attacked by the cells of your innate immune system, as shown in *Figure 39.12A, B,* and *C.* If the infection is not controlled, then your body builds up acquired immunity to the antigen by producing antibodies to it. A type of lymphocyte called a T cell becomes involved. A **T cell** is a lymphocyte that is produced in bone marrow and processed in the thymus gland. Two kinds of T cells play different roles in immunity.

Figure 39.12
Antibody immunity utilizes B cells and antibodies in defending your body against invading pathogens.

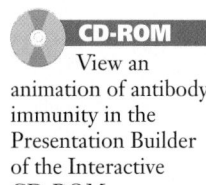

CD-ROM
View an animation of antibody immunity in the Presentation Builder of the Interactive CD-ROM.

Visual Learning
Figure 39.12 Ask students to write a summary of what is happening in this figure.

Reinforcement
Compare the antigen-antibody reaction to the lock-and-key fit of enzymes and substrates.

✓ Assessment
Performance Assessment in the Biology Classroom, p. 45, *Vaccine Models and the Common Cold.* Have students carry out this activity to find out why a single vaccine will not combat the common cold. **L2**

Concept Development
Explain to students that normally the body's immune response helps maintain homeostasis. However, sometimes the body loses its ability to discriminate between self and nonself, which leads to autoimmunity. Autoimmunity is a response by antibodies or sensitized T cells against a person's own tissue antigens. It is involved in multiple sclerosis, Graves' disease of the thyroid, rheumatic fever (in which antibodies are formed against the heart), and juvenile diabetes (in which antibodies are formed against the pancreas).

GLENCOE
TECHNOLOGY

VIDEODISC
Biology: The Dynamics of Life
Antibody Immunity (Ch. 45)
Disc 2, Side 1, 52 sec.

MEETING INDIVIDUAL NEEDS

Visually Impaired
Visual-Spatial Photocopy and enlarge Figures 39.12 and 39.13. Use a felt-tip pen to outline the edges of the cells and the arrows in the illustrations for use by visually impaired students. **L1 ELL**

Physically Challenged
Linguistic Ask students who are physically challenged by allergies or asthma to speak to the class about how the disorder affects their lives.

One kind of T cell, called a helper T cell, interacts with B cells, shown in *Figure 39.12D, E,* and *F.* A **B cell** is a lymphocyte that, when activated by a T cell, becomes a plasma cell and produces antibodies. B cells are produced in the bone marrow. Plasma cells, shown in *Figure 39.12G* and *H,* release antibodies into the bloodstream and tissue spaces. Some activated B cells do not become plasma cells but remain in the bloodstream as memory B cells. Memory B cells are ready and armed to respond rapidly if the same pathogen invades the body at a later time. The response to a second invasion is immediate and rapid, usually without any symptoms.

Cellular Immunity

Like antibody immunity, cellular immunity also involves T cells with antigens on their surfaces. The T cells involved in cellular immunity are cytotoxic, or killer, T cells. T cells stored in the lymph nodes, spleen, and tonsils transform into cytotoxic T cells that are specific for a single antigen. However, unlike B cells, they do not form antibodies. Cytotoxic T cells differentiate and produce identical clones. They travel to the infection site and release enzymes directly into the pathogens, causing them to lyse and die. The steps in cellular immunity are illustrated in *Figure 39.13.*

Passive and Active Immunity

Perhaps you had chicken pox as a child. Many children have had chicken pox by the time they enter school. Most people don't have chicken pox a second time because they have become immune to the chicken pox virus. Acquired immunity to a disease may be either passive or active. Passive acquired immunity develops as a result of acquiring antibodies generated in another host. For

Figure 39.13
Cellular immunity involves T cells that transform into cytotoxic T cells. These T cells release perforin, which pokes holes in cells invaded by pathogens.

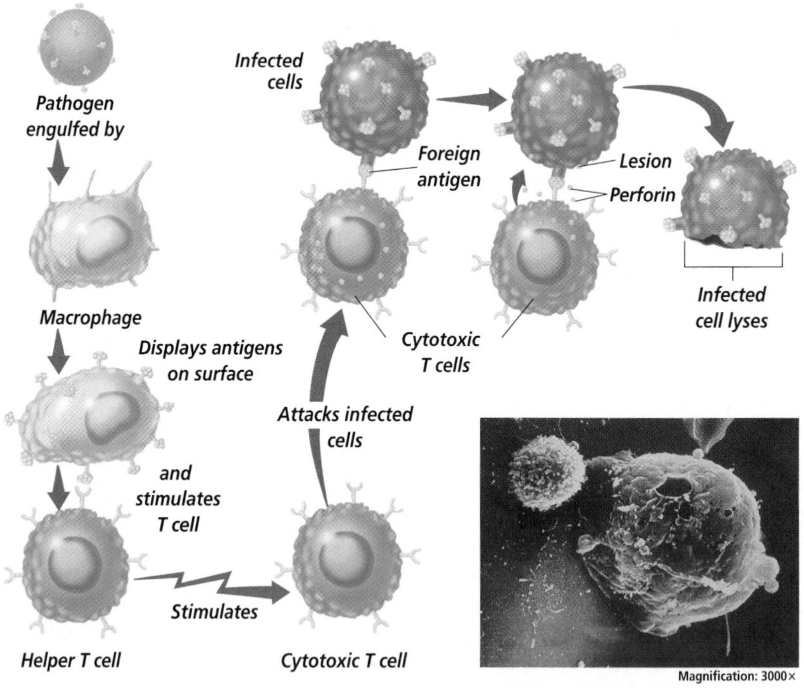

CD-ROM
View an animation of cellular immunity in the Presentation Builder of the Interactive CD-ROM.

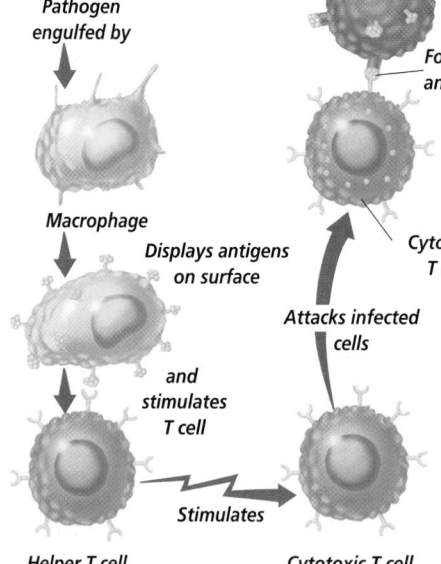

Pathogen engulfed by

Macrophage

Displays antigens on surface

and stimulates T cell

Helper T cell

Stimulates

Cytotoxic T cell

Infected cells

Foreign antigen

Attacks infected cells

Cytotoxic T cells

Lesion
Perforin

Infected cell lyses

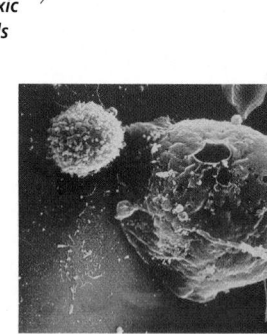

Magnification: 3000×

BIOLOGY JOURNAL

Immune System Roles

Linguistic Have students write a story about the various immune system "characters" involved in the immune response to a viral disease, such as chicken pox or AIDS. **L2** **COOP LEARN**

Portfolio

Demonstrating Immune Response

Interpersonal Divide students into groups to write and perform a short play that demonstrates aspects of the immune system and its responses. They should each include a copy of the play in their portfolios. **L2** **P** **COOP LEARN**

Table 39.2 Recommended childhood immunizations		
Immunization	**Agent**	**Protection against**
Acellular DPT or Tetrammune	Bacteria	Diphtheria, pertussis (whooping cough), tetanus (lockjaw)
MMR	Virus	Measles, mumps, rubella
OPV	Virus	Poliomyelitis (polio)
HBV	Virus	Hepatitis B
HIB or Tetrammune	Bacteria	*Haemophilis infuenzae B* (spinal meningitis)

example, at birth, a human infant acquires passive immunity to disease from its mother. Active acquired immunity develops when your body is directly exposed to antigens and produces antibodies in response to those antigens.

Passive immunity

Passive immunity may develop in two ways. Natural passive immunity develops when antibodies are transferred from a mother to her unborn baby through the placenta or to a newborn infant through the mother's milk. Artificial passive immunity involves injecting into the body antibodies that come from an animal or a human who is already immune to the disease. For example, a person who is bitten by a snake might be injected with antibodies from a horse that is immune to the snake venom.

Active immunity

Active immunity is obtained naturally when a person is exposed to antigens. The body produces antibodies that correspond specifically to these antigens. Once the person recovers from the infection, he or she will be immune if exposed to the pathogen again.

Active immunity can be induced artificially by vaccines. A **vaccine** is a substance consisting of weakened, dead, or incomplete portions of pathogens or antigens that, when

injected into the body, cause an immune response. Vaccines produce immunity because they prompt the body to react as if it were naturally infected. *Table 39.2* lists some common vaccines.

In 1798, Edward Jenner, an English country doctor, demonstrated the first safe vaccination procedure. Jenner knew that dairy workers who acquired cowpox from infected cows were resistant to catching smallpox during epidemics. Cowpox is a disease similar to, but milder than, smallpox. To test whether immunity to cowpox also caused immunity to smallpox, Jenner infected a young boy with cowpox. The boy developed a mild cowpox infection. Six weeks later, Jenner scratched the skin of the boy with viruses from a smallpox victim, as depicted in *Figure 39.14.*

Figure 39.14
This portrait shows Jenner vaccinating a young boy with smallpox. A worldwide attack on the disease through vaccinations brought an end to it. Because of the efforts of the World Health Organization, smallpox has been eliminated.

Purpose

Students will determine that a person's immune system responds to a disease faster and more vigorously if he or she has been vaccinated against that disease.

Process Skills

analyze information, apply concepts, define operationally, compare and contrast, draw a conclusion, make and use graphs, recognize cause and effect, think critically

Teaching Strategies

■ Remind students of the difference between a toxin and toxoid.
■ Review the difference between antigen and antibody.

Thinking Critically

1. They illustrate acquired immunity. Innate immunity consists of skin barriers or body secretions. Acquired immunity is the process by which the body adapts to a specific antigen by forming antibodies against it.

2. A toxin is a poison released by a pathogen that may cause disease symptoms. A toxoid is a treated form of the toxin that does not harm the body but stimulates the immune system to form antibodies against the toxin.

3. Line A-B involves macrophages, T cells, and B cells. Line B-C probably involves memory B cells and plasma cells.

1072

Problem-Solving Lab 39-2
Analyzing Information

Get a Shot or Get the Disease? You have had personal experience with injections for immunization. You know that these shots prevent you from catching a particular disease. But how do they actually work? Why are you protected against a specific disease when you receive a vaccination for that disease?

Analysis

Diphtheria toxoid is a modified form of the toxin produced by the bacterium *Cornyebacterium diphtheriae*. The toxin causes the symptoms associated with the disease diphtheria. When injected into your body, the toxoid prompts the immune system to respond as if it were being attacked by the diphtheria toxin. The graph below shows the human body's response to receiving an immunization shot of the diphtheria toxoid and then to being infected later on by the diphtheria bacterium.

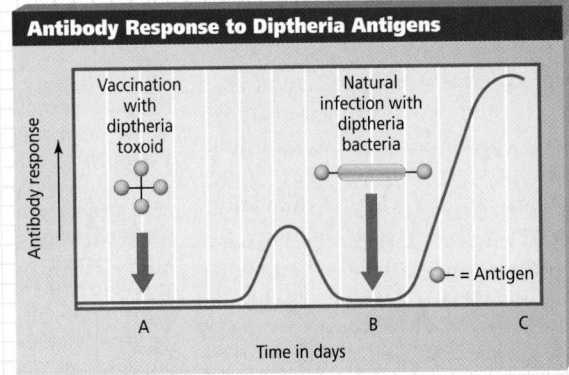

Antibody Response to Diptheria Antigens

Thinking Critically

1. Are the events in the graph illustrating innate or acquired immunity? Explain your answer.
2. What is the difference between a toxin and a toxoid?
3. Which cells associated with the immune system are most likely involved with line A-B? With line B-C?

The viruses for cowpox and smallpox are so similar that the immune system cannot tell them apart. The boy, therefore, did not get sick because he had artificially acquired active immunity to the disease. To learn more about how vaccines work, try the *Problem-Solving Lab* here.

AIDS and the Immune System

In 1981, an unusual cluster of cases of a rare pneumonia caused by a protozoan appeared in the San Francisco area. Medical investigators soon related the appearance of this disease with the incidence of a rare form of skin cancer called Kaposi's sarcoma. Both diseases seemed associated with a general lack of function of the body's immune system. By 1983, the pathogen causing this immune system disease had been identified as a retrovirus, now known as Human Immunodeficiency (ihm yew noh dih FIHSH un see) Virus, or HIV. HIV kills helper T cells and leads to the disorder known as Acquired Immune Deficiency Syndrome, or AIDS.

HIV is transmitted when blood or body fluids from an infected person are passed to another person through direct contact, or through contact with objects that have been contaminated by infected blood or body fluids. Methods of transmission include intimate sexual contact, contaminated intravenous needles, and blood-to-blood contact, such as through transfusions of contaminated blood. Since 1985, careful screening measures have been instituted by blood banks in the United States to help keep HIV-infected blood from being given to people who need transfusions. A pregnant woman infected with the virus can transmit it to her fetus. The virus can also be transmitted through breast milk.

Abstinence from intimate sexual contact provides protection from HIV and other sexually transmitted diseases. Among illegal drug users, HIV transmission can be prevented by not sharing needles.

The HIV virus in *Figure 39.15* is basically two copies of RNA wrapped

---

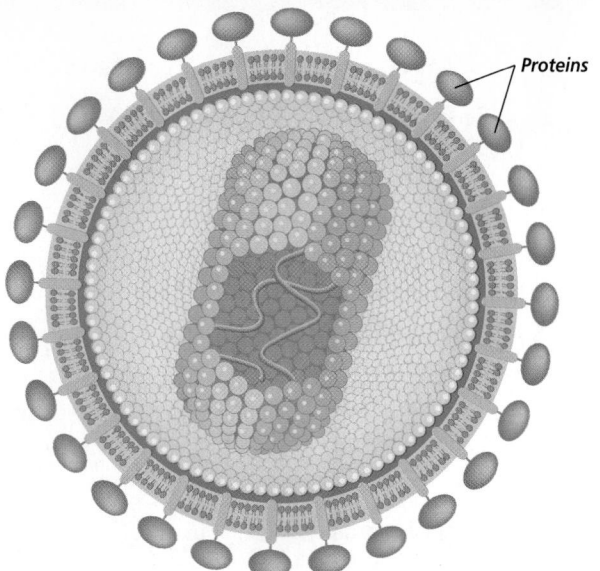

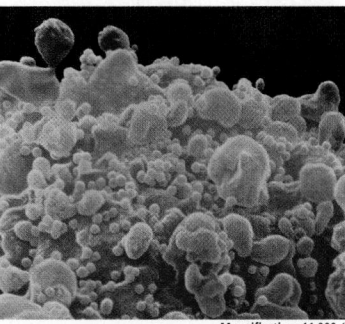

Figure 39.15
HIV is a retrovirus with an outer envelope covered with knoblike attachment
proteins. Researchers are studying these
proteins to find ways to stop the spread
of the virus in humans. The photo shows
a T-lymphocyte with yellow-green HIV
particles on its surface.

Proteins

Magnification: 11 000✓

in proteins, then further wrapped in a lipid coat. The knoblike outer proteins of the virus attach to a receptor on a helper T cell. The virus can then penetrate the cell, where it may remain inactive for months. HIV contains the enzyme reverse transcriptase, which allows the virus to use its RNA to synthesize viral DNA in the host cell.

The first symptoms of AIDS may not appear for eight to ten years after initial HIV infection. During this time, the virus reproduces and infects an increasing number of T cells.

Infected persons may eventually develop AIDS. During the early stages of the disease, symptoms may include swollen lymph nodes, a loss of appetite and weight, fever, rashes, night sweats, and fatigue.

It is not known what percentage of persons infected with HIV will develop AIDS, but present indications are that the majority will. Almost all who develop AIDS die, usually because of infectious diseases or certain forms of cancer that take advantage of the body's weakened immune system.

3 Assess

Check for Understanding
Visual-Spatial Ask students to create a flowchart of B and T cell immune reactions. **L1**

Reteach
Have students prepare a table of the components of the lymphatic system and their functions. **L1**

Extension
Ask students to find out the current status of an AIDS vaccine. Have them look in the latest issues of medical and scientific journals. **L3**

✔ Assessment
Skill Have students explain how the AIDS virus overcomes both nonspecific and specific immunity. **L2**

4 Close

Discussion
Ask students to summarize the role of B and T cells in immune reactions. **L2**

Section Assessment

Understanding Main Ideas
1. What role do phagocytes play in defending the body against disease?
2. What role does a lymph node play in defending your body against microorganisms?
3. What is the difference between naturally acquired passive immunity and naturally acquired active immunity?
4. How does histamine release lead to inflammation of a wound?

Thinking Critically
5. Why is it adaptive for memory cells to remain in the immune system after an invasion by pathogens?

SKILL REVIEW
6. **Sequencing** Sequence the events that occur in the formation of antibody immunity. For more help, refer to *Organizing Information* in the **Skill Handbook**.

Section Assessment

1. Phagocytes are white blood cells that ingest and destroy pathogens by surrounding and engulfing them.
2. A lymph node is a small mass of tissue that filters lymph and traps and destroys microorganisms.
3. Naturally acquired passive immunity involves a mother passing antibodies to her baby through the placenta or breast milk. Naturally acquired active immunity involves having the disease and forming your own antibodies.
4. Histamine causes blood vessels in the injured area to dilate, making them more permeable to tissue fluid, which leaks out, causing swelling.
5. Memory cells remain in case the body encounters the same antigen again.

The second response will be rapid, before the antigen is able to cause disease.

6. Helper T cells interact with macrophages that have "self" and pathogen antigens on their surfaces. The helper T cells activate B cells which produce antibodies and cloned memory B cells.

INTERNET
BioLab

Getting On-Line for Information on Diseases

There are two main categories of disease, infectious and noninfectious. Infectious diseases are caused by pathogens. These diseases, like the common cold or AIDS, are said to be communicable because they can be passed from one person to another. Noninfectious diseases are not caused by a pathogen. These diseases, like cancer or arthritis, are said to be noncommunicable because they are not passed from one person to another.

Time Allotment
One class period

Process Skills
acquire information, classify, collect data, communicate, compare and contrast, make and use tables, think critically, write about biology

PREPARATION

Alternative Materials
■ Textbooks and materials from the school or public library may be substituted for Internet information if computers are not available to all students.

Resource Manager

BioLab and MiniLab Work-sheets, pp. 175-176 **L2**

PREPARATION

Problem
How can you use the Internet to obtain current research information on different diseases?

Objectives
In this BioLab, you will:
■ **Choose** five communicable and five noncommunicable diseases for study.
■ **Use the Internet** to gather information.

■ **Collect data** on the ten diseases and record in a table.

Materials
access to the Internet

Skill Handbook
Use the **Skill Handbook** if you need additional help with this lab.

PROCEDURE

1. Copy the two data tables on the next page.
2. Choose five communicable and five noncommunicable diseases you wish to investigate.

3. List the diseases in your data tables. Try to make your disease choices as specific as possible. For example, cancer as a topic is too broad. Instead, limit your choice to a specific type of cancer, such as breast cancer, prostate cancer, or Hodgkin's disease.
4. Go to the Glencoe Science Web site **to find links** that will provide you with information for this BioLab.
5. Be sure to complete the last two rows asking for current research findings and your sources of information.

PROCEDURE

Teaching Strategies
■ You may wish to have students complete all or part of this activity at home if they have access to computers, or in the school library or computer center.
■ Point out to students that their data tables will have to be expanded in size to accommodate all the information they obtain.

■ Have students share their findings with classmates at the conclusion of the activity.

Data and Observations
Student data and observations will vary, depending on the diseases selected and the information resources used.

Data Table 1

Communicable diseases	1	2	3	4	5
Disease name					
Organism responsible					
Classification of organism					
Mode of transmission					
Symptoms					
Treatment					
Current research					
Source of information					

Data Table 2

Noncommunicable diseases	1	2	3	4	5
Disease name					
Symptoms					
Organs affected					
Age group affected					
Treatment					
Current research					
Source of information					

Children with chicken pox

Elderly woman with osteoporosis

ANALYZE AND CONCLUDE

1. **Defining** What is a pathogen? Provide several examples.
2. **Comparing** Describe the difference between a communicable and a noncommunicable disease. Provide several examples of each.
3. **Thinking Critically** What are vectors? Are they associated with communicable or noncommunicable diseases? Explain your answer.
4. **Applying Concepts** Explain why the table for noncommunicable diseases does not have a column for organism responsible or method of transmission.
5. **Using the Internet** What is one advantage of getting information on disease research by way of the Internet rather than from textbooks or an encyclopedia?

Sharing Your Data

BIOLOGY Online Find this BioLab on the Glencoe Science Web site at **science.glencoe.com**. Post your findings in the data table provided for this activity. Add additional data from other students to your data table. Analyze the data to help you answer the problem posed for this BioLab.

INTERNET BioLab

ANALYZE AND CONCLUDE

1. A pathogen is a disease-producing agent. Pathogens include bacteria, viruses, parasites, and fungi.
2. Communicable diseases, (such as flu, TB, AIDS) can be passed from one person to another while noncommunicable diseases (such as cancer, arthritis) cannot.
3. Vectors are carriers of a pathogen. Since noncommunicable diseases are not caused by pathogens, there are no vectors involved in their transmission.
4. Noncommunicable diseases are not caused by an organism or pathogen.
5. Internet information can be updated more frequently than information printed in textbooks and encyclopedias.

Assessment

Skill Have students combine their lists of communicable diseases so that a total of at least 20 diseases are included. Have students use the list to compare characteristics of viral versus bacterial diseases. Use the Performance Task Assessment List for Analyzing the Data in **PASC**, p. 27. **L2**

Sharing Your Data

BIOLOGY Online To navigate to the Internet BioLabs, choose the *Biology: The Dynamics of Life* icon at the Glencoe Science Web Site. Click on the student site icon, then the BioLabs icon. Ask students to research current efforts to develop an AIDS vaccine or a cure for cancer or other disease.

BIOLOGY Online Note Internet addresses that you find useful in the space below for quick reference.

_____

_____

_____

_____

_____

Purpose

Students learn that genetic engineering technology can be used to develop new vaccines.

Background

Traditional vaccines against rabies, measles, mumps, and many other diseases are made from live, attenuated (weakened) viruses. The virus is grown in laboratory cultures of nonhuman cells and allowed to mutate over several generations. Then a strain is selected that does not produce the disease but does induce an immune response. This strain is used to make the vaccine. There is a danger that the virus could mutate back into the form that produces disease. The advantage of using recombinant DNA to make attenuated vaccines is that mutations in the viral genome can be engineered to ensure that reverse mutation back to a pathogenic form is virtually impossible.

Teaching Strategies

■ Have students research their own vaccination histories and make a chart that lists each vaccine they have received since birth and approximately when they received it. **L1**

Investigating the Technology

A person could be vaccinated against a number of diseases with a single innoculation.

 New Vaccines

Greater understanding of how the immune system works and rapid advances in gene technology have paved the way for the development of new types of vaccines that offer hope in the fight against some of the world's most deadly and widespread diseases.

Traditionally, most vaccines have been made from weakened or killed forms of a disease-causing virus or bacterium, or from some of its cellular components or toxins. Although these types of vaccines have helped to prevent disease, they sometimes cause severe side effects. Furthermore, it hasn't been possible to create vaccines for diseases such as malaria and AIDS using traditional methods. With the help of genetic engineering technology, researchers can now manipulate microbial genes to create entirely new kinds of vaccines.

Recombinant vaccines One revolutionary approach to developing vaccines uses recombinant DNA technology, a process in which genes from one organism are inserted into another organism. The hepatitis B virus vaccine was the first genetically engineered vaccine to be produced in this way. Researchers isolated the gene in the hepatitis virus that codes for the production of the antigen protein that stimulates an immune response. Then they inserted that gene into yeast cells. Like tiny microbial machines, the genetically altered yeast cells produce great quantities of pure hepatitis B antigen, which is then used to make a vaccine.

Applications for the Future

An antigen-coding gene from a disease-causing virus such as HIV can be inserted into a non-disease-causing virus such as cowpox virus. When a vaccine made from a carrier virus is injected into a host, the virus replicates and in the process produces the antigen protein, which causes an immune response. This type of vaccine, called a live vector vaccine, shows promise against AIDS.

DNA Vaccines DNA vaccines differ from other vaccines in that only the cloned segment of DNA that codes for a disease-causing antigen is injected into a host—the DNA itself is the vaccine. The DNA can be injected through a hypodermic needle into muscle tissue, or tiny DNA-coated metal beads can be fired into muscle cells using a "gene gun." Once in the cells, the foreign DNA is expressed as antigen protein that induces an immune response. Researchers currently are working on DNA vaccines for cancer and tuberculosis.

INVESTIGATING THE TECHNOLOGY

Thinking Critically It is possible to insert antigen-coding genes for several different diseases into one virus carrier that can be used to make a vaccine. What would be an advantage of such a vaccine?

 To find out more about vaccines, visit the Glencoe Science Web site.
science.glencoe.com

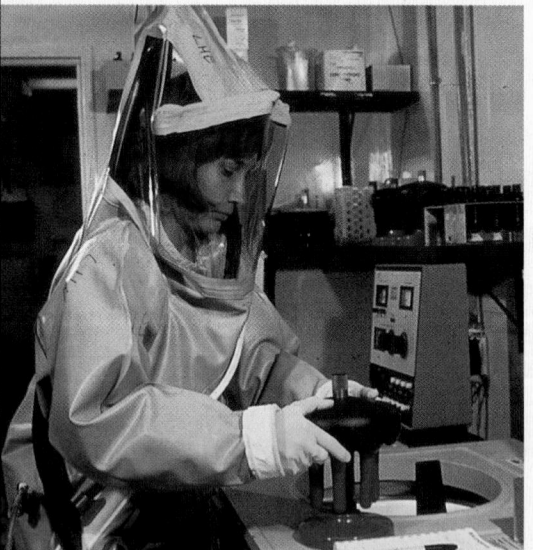

Researchers who work with viruses must wear protective clothing.

Going Further

Encourage students to use the library or the Internet to find out about new vaccines that have been or are being developed using genetic engineering technologies.
L2

SUMMARY

Section 39.1
The Nature of Disease

Main Ideas
- Infectious diseases are caused by the presence of pathogens in the body.
- The cause of an infection can be established by following Koch's postulates.
- Animals, including humans, and nonliving objects can serve as reservoirs of pathogens. Pathogens can be transmitted by direct contact, by a contaminated object, through the air, or by a vector.
- Symptoms of a disease are caused by direct damage to cells or by toxins produced by the pathogen.
- Some diseases occur periodically, whereas others are endemic. Occasionally, a disease reaches epidemic proportions.
- Some infectious diseases can be treated with antibiotics, but pathogens may become resistant to these drugs.

Vocabulary
antibiotic (p. 1061)
endemic disease (p. 1061)
epidemic (p. 1061)
infectious disease (p. 1056)
Koch's postulates (p. 1057)
pathogen (p. 1055)

Section 39.2
Defense Against Infectious Diseases

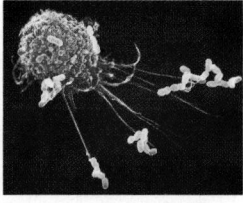

Main Ideas
- The lymphatic system consists of the lymphatic vessels and the lymphatic organs: lymph nodes, tonsils, spleen, and thymus.
- Innate immunity provides general protection against various pathogens.
- Acquired immunity provides a way of fighting specific pathogens by recognizing invaders as nonself. It includes the production of antibodies and cellular immunity.
- AIDS is caused by HIV, which damages the immune system and allows other infections to invade the body.

Vocabulary
acquired immunity (p. 1067)
B cell (p. 1070)
innate immunity (p. 1063)
interferons (p. 1067)
lymph (p. 1068)
lymph node (p. 1068)
lymphocyte (p. 1068)
macrophage (p. 1064)
phagocyte (p. 1064)
pus (p. 1065)
T cell (p. 1069)
tissue fluid (p. 1068)
vaccine (p. 1071)

Main Ideas
Summary statements can be used by students to review the major concepts of the chapter.

Using the Vocabulary
To reinforce chapter vocabulary, use the Content Mastery Booklet and the activities in the Interactive Tutor for Biology: The Dynamics of Life on the Glencoe Science Web site.
science.glencoe.com

THE PRINCETON REVIEW *All Chapter Assessment questions and answers have been validated for accuracy and suitability by The Princeton Review.*

UNDERSTANDING MAIN IDEAS

1. d
2. b
3. a

UNDERSTANDING MAIN IDEAS

1. Koch's postulates are a series of steps a scientist takes to relate a specific _____ to a specific disease.
   a. host
   b. medium
   c. epidemic
   d. pathogen

2. Bacteria, viruses, and other disease-producing agents are called _____.
   a. parasites
   b. pathogens
   c. antibodies
   d. lymph

3. Bacteria damage host cells by producing _____.
   a. toxins
   b. antibodies
   c. hormones
   d. tRNA

GLENCOE TECHNOLOGY

 **VIDEOTAPE**
MindJogger Videoquizzes
Chapter 39: *Immunity from Disease*
Have students work in groups as they play the videoquiz game to review key chapter concepts.

 **Resource Manager**

Chapter Assessment, pp. 229-234
MindJogger Videoquizzes
ExamView® Pro Software 💾
BDOL Interactive CD-ROM, Chapter 39 quiz

4. c
5. b
6. b
7. a
8. d
9. a
10. b
11. infectious
12. endemic
13. antibiotic
14. tissue, lymph
15. basophil
16. lysozyme, cell walls
17. inflammation
18. acquired immunity
19. macrophage
20. T helper cell

▶ APPLYING MAIN IDEAS ◀

21. The disease tetanus is due to a toxin produced by the bacteria. Killing the bacteria does not affect the toxin that has already been produced.
22. A burn patient loses protective layers of skin, exposing the body to the possibility of massive infection.

4. Which of these diseases is caused by a pathogen?
   a. osteoarthritis   **c.** smallpox
   b. hemophilia   **d.** cystic fibrosis
5. Which of these diseases is spread only through sexual intercourse?
   a. food poisoning   **c.** tetanus
   b. genital herpes   **d.** mumps
6. Of these, which is NOT a component of the innate immune system?
   a. phagocytosis   **c.** skin
   b. antibodies   **d.** mucus
7. What is produced by a body cell that has been infected with a virus?
   a. interferon   **c.** lysozyme
   b. toxins   **d.** histamine
8. What scientist demonstrated the first safe vaccination procedure?
   a. Koch   **c.** Mendel
   b. Pasteur   **d.** Jenner
9. Of the following, which is NOT a reservoir for pathogens?

   a.    **c.**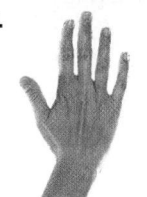

   b.    **d.**

THE PRINCETON REVIEW   **TEST-TAKING TIP**

Read the Instructions
No matter how many times you've taken a particular test or practiced for an exam, it's always a good idea to skim through the instructions provided at the beginning of each section. It only takes a moment.

10. How is malaria transmitted?

   a.    **c.**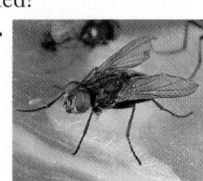

   b.    **d.**

11. Any disease caused by microorganisms in the body is known as a(n) _____ disease.
12. Diseases that are constantly present in the population are _____ diseases.
13. A(n) _____ is a substance produced by a microorganism that inhibits the growth of other microorganisms.
14. When _____ fluid collects in open-ended vessels, it is called _____.
15. A(n) _____ is a type of white blood cell that secretes histamine.
16. _____, an enzyme produced in sweat, tears, and saliva, can break down the _____ of some bacteria.
17. _____ is a body response to an injury characterized by redness, swelling, pain, and heat.
18. Building up a resistance to a specific pathogen is called _____ _____.
19. A tissue _____ combats invading pathogens by engulfing them.
20. A _____ is a type of lymphocyte destroyed by HIV.

▶ APPLYING MAIN IDEAS ◀

21. If the bacteria that cause tetanus are easily killed by penicillin, why doesn't penicillin cure the disease tetanus?
22. Why must severe burn victims be kept in pathogen-free isolation?

23. While building a tree house, you get a tiny splinter in your finger. Two days later, the area is swollen and pus leaks out. Why is there pus around the splinter?

24. A month after buying a new pet parakeet, Susan experienced pains in her legs, followed by chills, fever, diarrhea, and a headache. She recovered after two weeks of antibiotics. When she next visited the pet store, many of the parakeets were ill. How could researchers find out if Susan had the same disease as the birds?

THINKING CRITICALLY

25. **Observing and Inferring** A new mother had chicken pox as a child. Why doesn't her newborn infant get the disease, even after being exposed to the virus that causes it?

26. **Observing and Inferring** How does AIDS upset homeostasis in the body?

27. **Concept Mapping** Complete the concept map by using the following vocabulary terms: infectious disease, pathogen, endemic, Koch's postulates, epidemic.

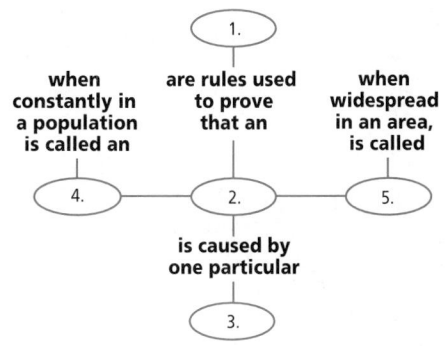

CD-ROM

For additional review, use the assessment options for this chapter found on the *Biology: The Dynamics of Life Interactive CD-ROM* and on the Glencoe Science Web site.
science.glencoe.com

ASSESSING KNOWLEDGE & SKILLS

The graph below shows the progress of a typical HIV infection with signs of the various stages of the AIDS disease.

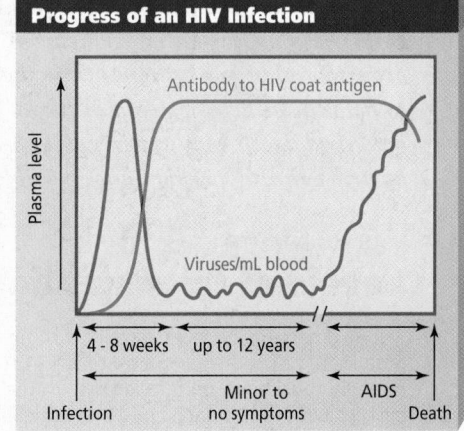

Progress of an HIV Infection

Interpreting Data Use the graph to answer these questions.

1. Which variable is the dependent variable?
   a. time
   c. level in blood
   b. symptoms
   d. antibody

2. When does the antibody level begin to rise?
   a. at about 4 weeks
   b. at about 8 weeks
   c. during the AIDS symptom stage
   d. at death

3. The HIV virus attacks _____.
   a. red blood cells
   c. T cells
   b. B cells
   d. epithelial cells

4. What type of molecule are antibodies?
   a. carbohydrates
   c. proteins
   b. fats
   d. nucleic acids

5. **Observing and Inferring** Explain why AIDS is considered a syndrome rather than a single disease.

23. Macrophages that came to the injured area engulfed microbes and damaged tissue, then died. Dead macrophages and body fluids collectively are called pus.

24. Researchers would have to isolate the organism from the parakeets and run Koch's postulates on them. The organism can also be exposed to human cell cultures to determine susceptibility.

THINKING CRITICALLY

25. The mother's antibodies to chicken pox crossed the placenta and are in the baby's blood, protecting the baby against chicken pox.

26. AIDS destroys the immune system, which constantly monitors the body for invasion by pathogens. If the immune system is not functioning, the body is extremely susceptible to disease.

27. 1. Koch's postulates; 2. Infectious disease; 3. Pathogen; 4. Endemic; 5. Epidemic

ASSESSING KNOWLEDGE & SKILLS

1. c
2. a
3. c
4. c
5. The HIV infection's destruction of the immune system is not directly life threatening; the secondary infections are.

Prepare

Purpose

This BioDigest can be used as an overview of the structures and functions of the human body systems. You may wish to use this unit summary to teach human biology in place of the chapters in the Human Biology unit.

Key Concepts

Students learn about the level of organization in body systems. They are then introduced to the structures and functions of 11 major body systems. Various vital statistics provide students with interesting facts about their body systems.

1 Focus

Bellringer

Using a large picture of a human body or a plastic model of a torso, ask students to name as many body systems as they can. Have students list organs under each body system. **L1** **ELL**

GLENCOE
TECHNOLOGY

CD-ROM
Biology: The Dynamics of Life
BioQuest: *Body Systems*
Disc 1-5

For a **preview** of the human body unit, study this BioDigest before you read the chapters. After you have studied the human biology chapters, you can use the BioDigest to **review** the unit.

The Human Body

How do the human body systems function together? When an Olympic ice-skater performs on the ice, the cells, tissues, organs, and organ systems of the skater's body function together to help the athlete perform at his or her best and perhaps win a gold medal. All body systems must work together to make an award-winning performance possible.

Levels of Organization

All organisms are made of cells. In complex organisms, such as humans, most cells are organized into functional units called tissues. The four basic tissues of the human body are epithelium, muscle, connective, and nervous tissues. Epithelium covers the body and lines organs, vessels, and body cavities. Muscle tissue is contractile and is found attached to bones and in the walls of organs, such as the heart. Connective tissue is widely distributed throughout the body. It produces blood and provides support, binding, and storage. Nervous tissue transmits impulses that coordinate, regulate, and integrate body systems.

Tissues to Systems

Groups of tissues that perform specialized functions are called organs. Your stomach and eyes are examples of organs. Most organs contain all four basic tissue types. Each of the body's organs is part of an organ system. An organ system contains a group of organs that work together to carry out a major life function. The eleven major organ systems of the human body are described in this BioDigest.

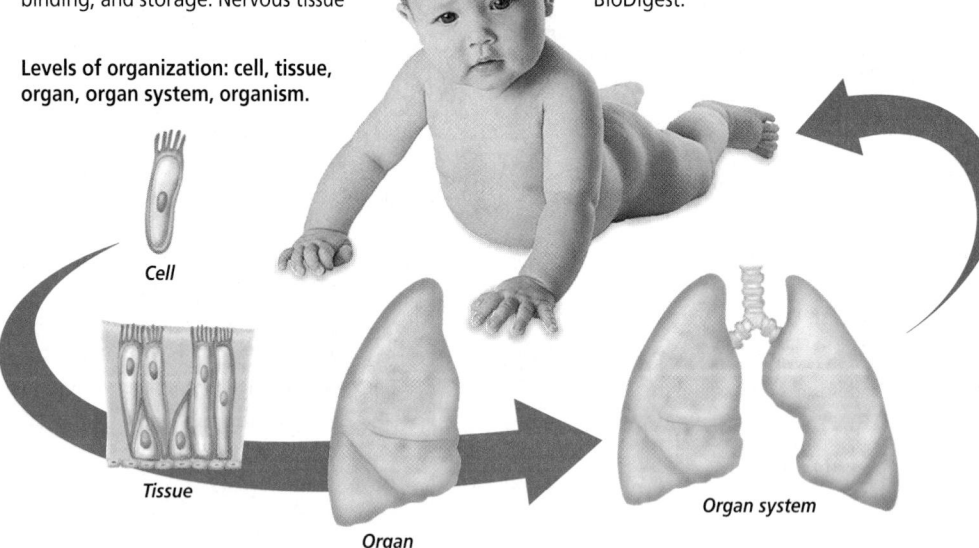

Levels of organization: cell, tissue, organ, organ system, organism.

Cell

Tissue

Organ

Organ system

Multiple Learning Styles

Look for the following logos for strategies that emphasize different learning modalities.

Kinesthetic Meeting Individual Needs, p. 1084; Quick Demo, p. 1085

Visual-Spatial Project, p. 1082; Meeting Individual Needs, p. 1082; Reteach, p. 1088

Interpersonal Reinforcement, p. 1086; Tech Prep, p. 1086

Linguistic Biology Journal, pp. 1084, 1088; Project, p. 1085; Meeting Individual Needs, p. 1087; Portfolio, p. 1088

Logical-Mathematical Project, p. 1083

Skin

The skin and its associated structures, including hair, nails, sweat glands, and oil glands, are important in maintaining homeostasis in the body. The skin protects tissues and organs, helps regulate body temperature, produces vitamin D, and contains sensory receptors.

Skeletal System

The skeletal system consists of the axial skeleton and appendicular skeleton. The axial skeleton supports the head and includes the skull and the bones of the back and chest. The appendicular skeleton contains the bones associated with the limbs. The entire skeleton, which is made up of 206 bones, has many functions. It provides support for the softer, underlying tissues; provides a place for muscle attachment; protects vital organs; manufactures blood cells; and serves as a storehouse for calcium and phosphorus.

Joints: Where Bones Meet

The place where two bones meet is called a joint. Joints can be immovable, such as the joints in the skull, or movable, such as the shoulder joints. The shoulder joint is called a ball-and-socket joint; the elbow joint is a hinge joint. The wrists have gliding joints, and the neck has pivot joints.

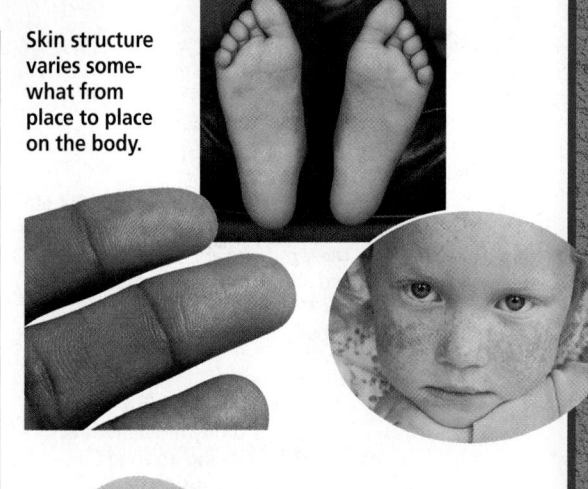

Skin structure varies somewhat from place to place on the body.

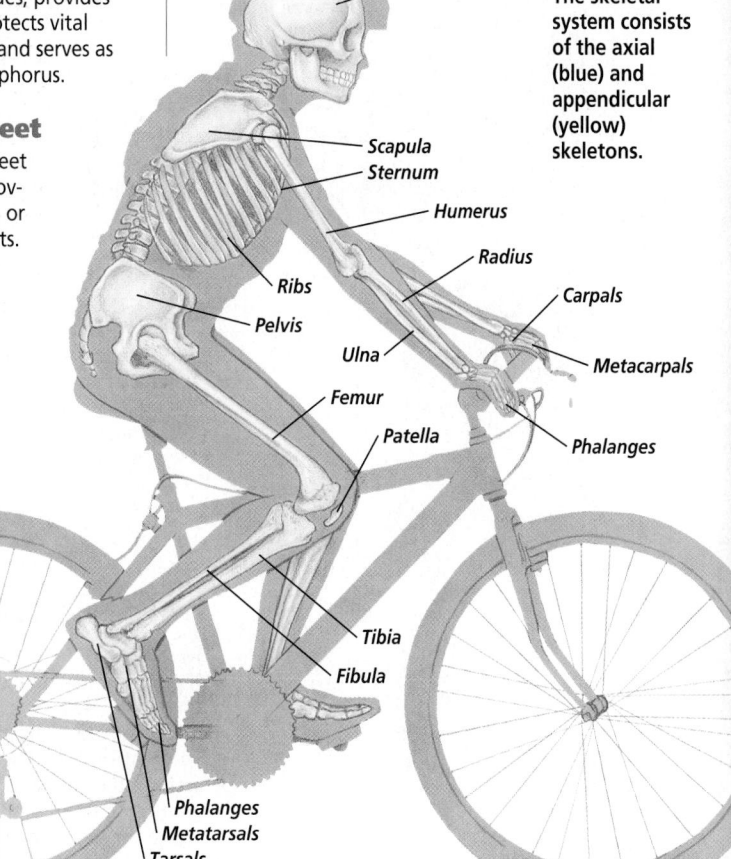

The skeletal system consists of the axial (blue) and appendicular (yellow) skeletons.

Skull
Scapula
Sternum
Humerus
Radius
Ribs
Carpals
Pelvis
Ulna
Metacarpals
Femur
Phalanges
Patella

Tibia
Fibula
Phalanges
Metatarsals
Tarsals

1081

BioDigest

BioDigest

Muscular System

The muscular system includes three types of muscles: smooth, cardiac, and skeletal.

Smooth Muscle

Smooth muscles are found in the walls of hollow internal organs, such as inside the stomach or blood vessels. These muscles are not under conscious control and are called involuntary muscles. Smooth muscle cells are spindle shaped and contain a single nucleus.

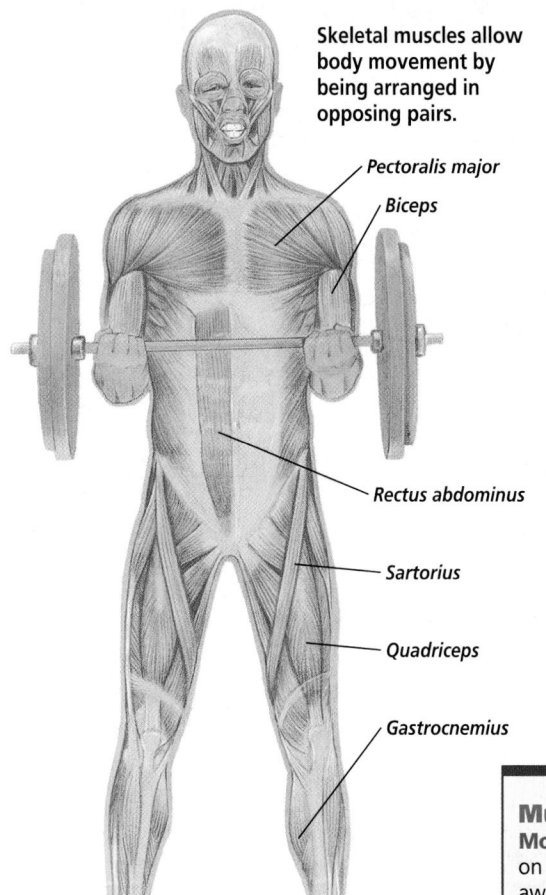

Skeletal muscles allow body movement by being arranged in opposing pairs.

- Pectoralis major
- Biceps
- Rectus abdominus
- Sartorius
- Quadriceps
- Gastrocnemius

During physical activity, almost every muscle can be involved, either voluntarily or by reflex actions.

Skeletal Muscle

Skeletal muscles are usually attached to bones. They can be controlled by conscious effort so they are called voluntary muscles. Skeletal muscle tissue is made up of long, threadlike cells, called fibers, which have alternating dark and light striations. Each fiber has many nuclei.

Heart Muscle

Cardiac muscle tissue is found only in the heart. These cells contain a single nucleus and striations made up of organized protein filaments that are involved in contraction of the muscle. Like smooth muscle, cardiac muscle is involuntary muscle. Cardiac muscle has the unique ability to contract without first being stimulated by nervous tissue.

VITAL STATISTICS

Muscles
Most powerful skeletal muscle: The muscle you sit on is the gluteus maximus; it moves the thighbone away from the body and straightens the hip joint.
Longest muscle: The sartorius muscle runs from the waist to the knee and flexes the hip and knee.
A broad smile: A smile uses 17 facial muscles; a frown uses more than 40.

1082

Digestive System

The digestive system receives food and breaks it down so it can be absorbed by the body's cells. The digestive system also eliminates food materials that are not digested or absorbed. Foods are broken down into simpler molecules that can move through cell membranes and be transported to all parts of the body by the bloodstream or the lymphatic vessels. The digestive system includes the mouth, tongue, teeth, salivary glands, pharynx, esophagus, stomach, liver, gallbladder, pancreas, and small and large intestines.

The digestive system breaks down food particles so that they can enter the body's cells.

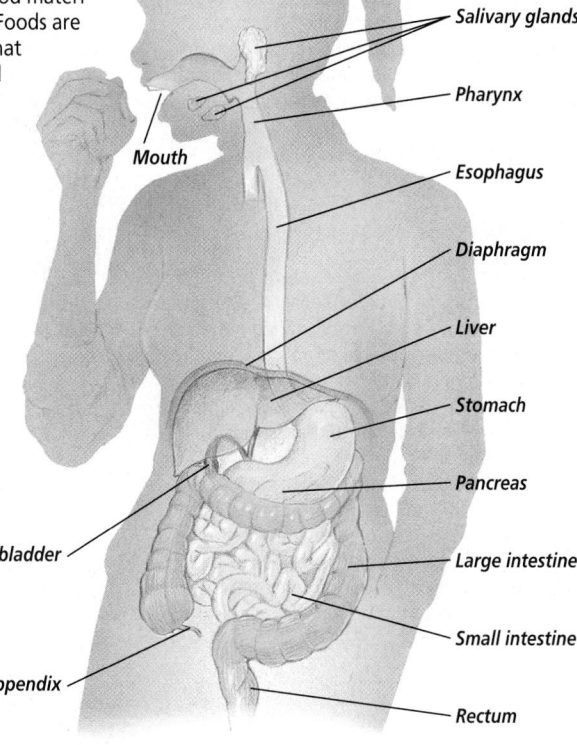

Salivary glands
Pharynx
Mouth
Esophagus
Diaphragm
Liver
Stomach
Pancreas
Gallbladder
Large intestine
Small intestine
Appendix
Rectum

FOCUS ON HEALTH

Blood Glucose Levels

A healthy breakfast can be an important supply of carbohydrates.

Levels of glucose in the blood are maintained all day long by hormones secreted by the pancreas. After a meal, the sugars from the food are transported into the blood, raising the blood glucose level. The sugars are either used immediately for activity or stored in the liver for later use. The pancreas secretes insulin, which helps the body's cells take up the sugar or convert it to glycogen in the liver for storage.

Between meals, when blood glucose levels go down, the pancreas secretes glucagon. Glucagon causes the glycogen in the liver to be broken down into glucose, which is then released into the bloodstream and made available to the body's cells. The control of blood sugar levels in the body is an example of a feedback mechanism that is vital for maintaining homeostasis.

1083

PROJECT

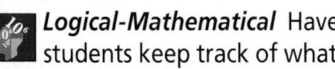

Daily Calorie Intake

Logical-Mathematical Have students keep track of what they eat and of all of their activities (including resting, sleeping, watching TV, etc.) for 48 hours. Students can use a calorie guide to estimate their daily calorie intake and an activity guide to calculate their calorie usage. **L1**

Quick Demo

X rays of the skeletal system and barium X rays of the digestive system can be viewed by hanging them on classroom windows or projecting them with an overhead projector. Use the X rays to discuss the parts and functions of each of these systems.

Enrichment

TECH PREP

Have a doctor from a sports medicine clinic speak to the class concerning fitness and the effects of sports injuries on the muscular system and other body systems.

Chalkboard Activity

Make a list of the items students have eaten in the last two days that would increase blood glucose levels. Discuss what happens in terms of insulin after the body takes in sugar.

NATIONAL GEOGRAPHIC

VIDEODISC
STV: Human Body Vol. 1
Digestive System
Unit 2, Side 2, 15 min. 55 sec.
Digestive System
(In its entirety)

GLENCOE TECHNOLOGY

CD-ROM
Biology: The Dynamics of Life
Exploration: *Nutrition*
Disc 5

BIODIGEST

Chalkboard Example

Draw a neuron on the chalkboard and label the parts. Using arrows, indicate the direction an impulse travels along a neuron. As you discuss transmission across the synapse, bring in the action of various drugs on the synapse. You can find more information in the chapter on the nervous system.

Quick Demo

Kinesthetic To demonstrate a protective reflex, have a student hold up a piece of Plexiglas in front of his or her face. Have another student throw a soft ball, such as Nerf ball, at the Plexiglas. The student behind the glass will blink automatically, even if trying not to blink.

NATIONAL GEOGRAPHIC

VIDEODISC
STV: Human Body Vol. 2
Nervous System
Unit 2, Side 2, 16 min. 19 sec.
Nervous System
(In its entirety)

BIODIGEST

Endocrine System

The endocrine system controls all of the metabolic activities of body structures. This system includes all of the glands in the body that secrete chemical messengers called hormones. Hormones travel in the bloodstream to target tissues, where they alter the metabolism of the target tissue. Some of the major endocrine glands include the pituitary, thyroid, parathyroids, adrenals, pancreas, ovaries, and testes.

Nervous System

The organs of the nervous system include the brain, spinal cord, nerves, and sensory receptors. These organs contain nerve cells, called neurons, that conduct impulses. Nerve impulses allow the neurons to communicate with each other and with the cells of muscles and glands. Each impulse consists of an electrical charge that travels the length of a neuron's cell membrane.

Between two neurons there is a small gap called a synapse. When one neuron is stimulated, it releases chemicals called neurotransmitters into the synapse, which stimulates a change in electrical charge in the next neuron. Nerve impulses travel through the body this way, from neuron to neuron.

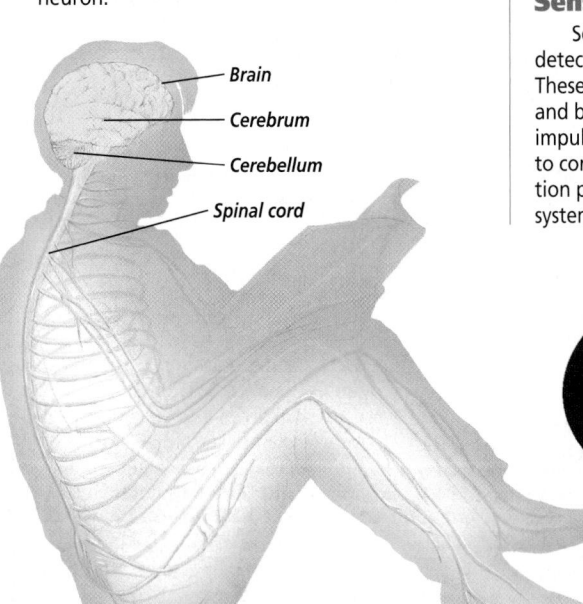

Brain
Cerebrum
Cerebellum
Spinal cord

1084

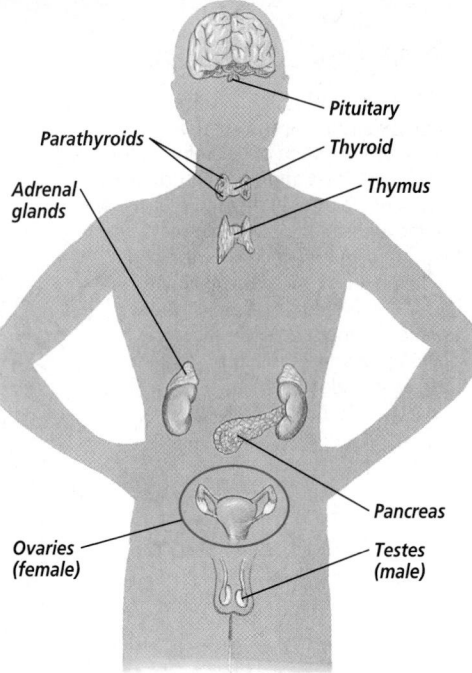

Pituitary
Parathyroids
Thyroid
Adrenal glands
Thymus
Pancreas
Ovaries (female)
Testes (male)

The major glands of the endocrine system secrete hormones that regulate body functions.

Sensory Receptors

Some nerve cells act as sensory receptors that detect changes inside and outside of the body. These neurons carry impulses to the spinal cord and brain. The brain and spinal cord then send impulses to muscles or glands, stimulating them to contract or secrete hormones. This interconnection provides coordination between the nervous system and the endocrine system.

Interpreting and acting on information sent to the central nervous system (brain and spinal cord) is the major job of the nervous system.

MEETING INDIVIDUAL NEEDS

Visually Impaired

Kinesthetic Pair a visually impaired student with a peer to go over the organs of the respiratory and nervous systems using a model of a human torso. Ask the students to consider the functions of each organ. **L1** **ELL**

BIOLOGY JOURNAL

The Brain

Linguistic Have students write a skit in their journal with a major player being the brain and how it interacts with and controls the body systems. **L2**
COOP LEARN

Respiratory System

The organs of the respiratory system exchange gases between blood and the air. During inhalation, oxygen in the air passes into the blood from small air sacs called alveoli in the lungs. Body cells use oxygen to break down glucose to make ATP needed for metabolism.

Carbon dioxide (CO_2) is produced by the breakdown of glucose and is transported to the lungs by the blood. In the lungs, carbon dioxide diffuses out of the blood and into the alveoli. It is forced out of the lungs during exhalation. The major organs of the respiratory system are the nasal cavity in the nose, the pharynx, larynx, trachea, bronchi, and lungs.

Nasal cavity
Epiglottis
Larynx
Trachea
Right lung
Bronchus
Left lung
Diaphragm

The respiratory system filters the air as it passes into the nose, down the air passages, and into the lungs.

CO_2 rich blood
Bronchioles
O_2 rich blood
Alveoli

The lungs contain many small sacs called alveoli, where gas exchange with the blood occurs.

VITAL STATISTICS

Respiration

Breathing: At rest, humans inhale and exhale about 12 to 20 times per minute, moving about 15 L of air per minute, and inhaling 21.6 cubic meters of air each day.

Lungs: Lungs weigh about 2.2 kg each. The right lung has three lobes and the left lung has two lobes. There are 300 million alveoli in the lungs. Flattened out, they would cover 360 square meters.

Sneezes: A sneeze ejects particles at 165.76 km/hr.

A swimmer comes up for air between strokes.

1085

Visual Learning

Review the process of diffusion using the diffusion of oxygen out of the alveoli into the blood and carbon dioxide out of the blood into the alveoli.

Quick Demo

 *Kinesthetic* Have students work in pairs. Have one student count the breathing rate of the other student, and vice versa. Compare individual (anonymous) breathing rates to the average breathing rate of the class. Discuss factors that could affect breathing rates, such as size, activity level when the breath rate is measured, and congestion from a cold or allergies. **L2** **ELL**

Enrichment

Have students choose a disease (infectious or noninfectious) of the respiratory system to research in the library or on the Internet.

NATIONAL GEOGRAPHIC

VIDEODISC
STV: Human Body Vol. 1

Circulatory and Respiratory Systems
Unit 1, Side 1, 16 min. 29 sec.
Circulatory and Respiratory Systems (In its entirety)

PROJECT

Tobacco and Cancer

 *Linguistic* Many representatives of the tobacco industry claim there is no proof that use of tobacco causes cancer. After researching the issue, have students write an essay in their journals on how they would respond to the assertion that tobacco usage does not cause cancer.
L2

BioDigest

BioDigest

Reinforcement

Interpersonal Play a body systems password game. Divide students into teams of two. One person on each team gives one-word clues to his or her partner, whose job is to guess the password with as few clues as possible. Each team plays against one other team. Each team works on a separate term and alternates giving clues to their partners. The first team to guess their word gets a point, then both teams go onto another password. Tally scores at the end of the game. **L2** **COOP LEARN**

Enrichment

Have students research the various treatments for kidney disease, including portable kidney dialysis and kidney transplants.

GLENCOE
TECHNOLOGY

CD-ROM
Biology: The Dynamics of Life
BioQuest: *Triathalon*
Disc 5

NATIONAL GEOGRAPHIC

VIDEODISC
STV: Human Body Vol. 3
Reproductive Systems
Unit 2, Side 2, 19 min. 57 sec.
Reproductive Systems
(In its entirety)

Urinary System

Metabolic waste products are created during the breakdown of amino acids. The urinary system removes these metabolic wastes from the blood, maintains the balance of water and salts in the blood, stores wastes in the form of urine, and transports urine out of the body.

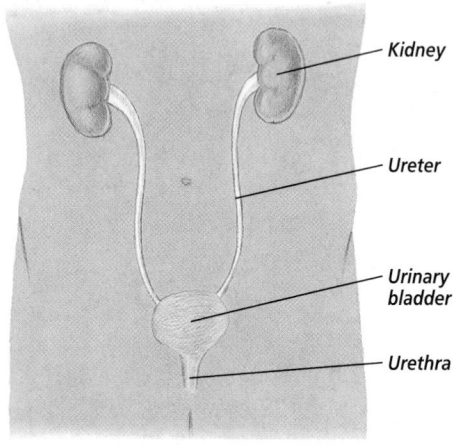

- Kidney
- Ureter
- Urinary bladder
- Urethra

The urinary system filters the blood, collects urine, and excretes urine from the body.

Reproductive System

The reproductive system is involved in the production of gametes. The male reproductive system produces and maintains sperm cells and transfers them into the female reproductive tract. The female reproductive system produces and maintains egg cells, receives and transports sperm cells, and supports the development of the fetus.

The male reproductive system.

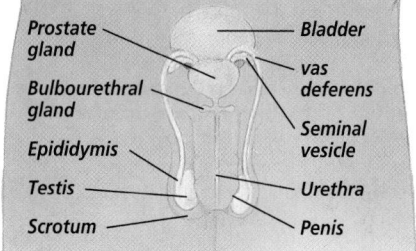

- Prostate gland
- Bulbourethral gland
- Epididymis
- Testis
- Scrotum
- Bladder
- vas deferens
- Seminal vesicle
- Urethra
- Penis

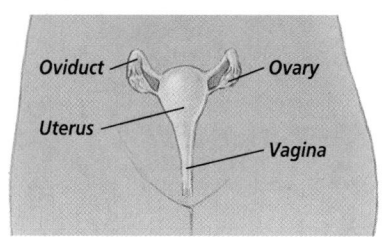

- Oviduct
- Uterus
- Ovary
- Vagina

The female reproductive system.

FOCUS ON HEALTH

Your Blood Pressure

Blood pressure measurements give an indication of the health of the heart and blood vessels.

Systolic Pressure When the cuff of the blood pressure machine squeezes the arm, it blocks the blood flow in an artery. As the pressure in the cuff is released, a gauge attached to the cuff measures the pressure in the artery as blood flows back into the artery. This is the systolic pressure, which is a measure of the pressure when the right and left ventricles contract.

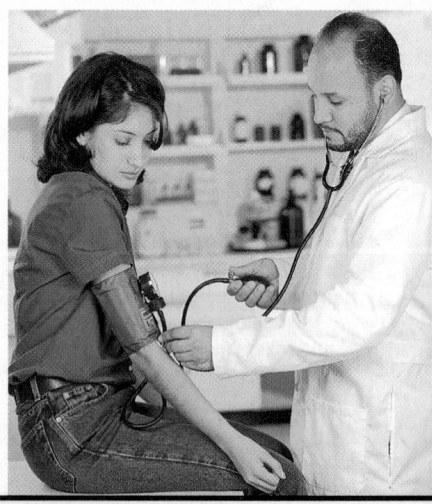

1086

TechPrep

Health Professionals

Interpersonal Have students prepare for an interview with a health professional such as a nurse, doctor, emergency room personnel, ambulance driver, EMT, physician's assistant, or phlebotomist. They should prepare questions about what the person does and what education is needed for the job. The information can be shared with the whole class. **L2**

BioDigest

BioDigest

Reproduction

Testes: The testes contain 244 m of tubules in which sperm cells are continually produced by meiosis.

Ovaries: At birth, a female already has about 2 million eggs. About 300 000 survive to puberty, but only 450 or so mature and are expelled from the ovary during her lifetime.

Circulatory System

The circulatory system includes the heart, blood vessels (arteries, veins, and capillaries), and blood. The muscular heart pumps blood through the blood vessels. The blood carries oxygen from the lungs and nutrients from the digestive tract to all the body cells. Blood also carries hormones to their target tissues, carbon dioxide back to the lungs, and other waste products to the excretory system.

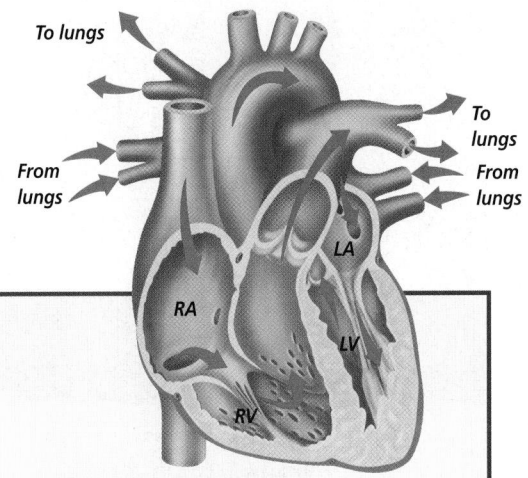

To lungs

To lungs

From lungs

From lungs

LA

RA

LV

RV

Diastolic Pressure

When the first rush of blood through the arteries slows, the gauge measures a pressure called the diastolic pressure. This is the lowest pressure in the vessels, just before the two ventricles contract again. Blood pressure readings give both the systolic and the diastolic pressure of the arteries. Blood pressure is used to evaluate artery condition.

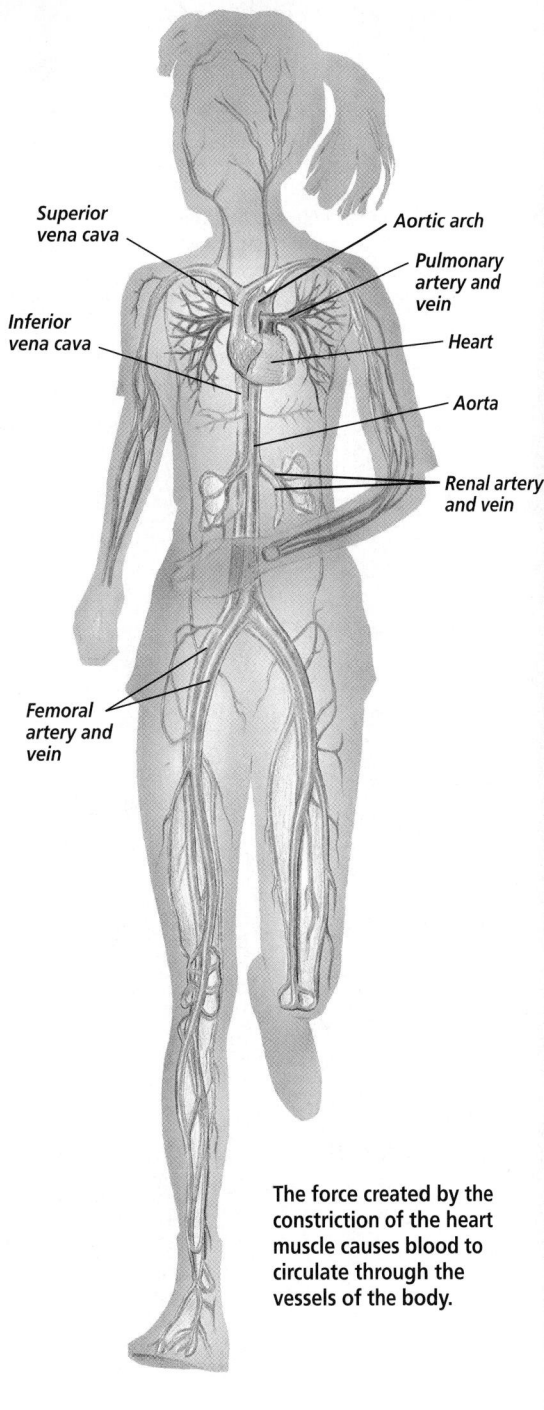

Superior vena cava

Aortic arch

Pulmonary artery and vein

Inferior vena cava

Heart

Aorta

Renal artery and vein

Femoral artery and vein

The force created by the constriction of the heart muscle causes blood to circulate through the vessels of the body.

1087

TECH PREP Ask the school nurse to demonstrate measuring blood pressure on a few students in the class.

Enrichment

Have students use the Internet to research the factors involved in the development of arteriosclerosis and its treatment.

✔ Assessment

Knowledge Assign each student one component of a body system. Have students write a description of their assigned component and how it interacts with its body system and the entire organism. **L2**

Extension

Have students interested in a career in sports medicine visit with a trainer from their high school or local college team to find out what type of education is needed to become an athletic trainer. **L2**

Resource Manager

Reinforcement and Study Guide, pp. 175-176 **L2**
Content Mastery, pp. 193-196 **L1**

Gifted

Linguistic Have gifted students research premature infant care costs. They can write a position paper on whether Americans should spend more or less on dramatic lifesaving measures and why. **L3**

BioDigest

3 Assess

Check for Understanding

Provide groups of students with diagrams of the different human body systems. Have them use colored pencils to shade various organs of particular systems. For example, the components of the circulatory system could be colored red (oxygenated blood) and blue (nonoxygenated blood).

Reteach

Students can be broken into groups that are assigned to summarize the components of each of the body's systems and present the summaries to the class.

Extension

Ask the school nurse to speak to the class on prevention of disease transmission and vaccination.

✔ Assessment

Performance If necessary, explain to students that AIDS is caused by human immunodeficiency virus (HIV). Ask the students to explain why, since an infection with HIV is itself not fatal, a person who develops AIDS could die. *AIDS suppresses the immune system. A person with AIDS usually dies of an infectious disease or cancer.* **L2**

4 Close

Reteach

Visual-Spatial Have students make a table based on this BioDigest with three columns: System, Major Parts, and Function. **L1**

Lymphatic System

Fluids leak out of capillaries and bathe body tissues. The lymphatic system, also known as the immune system, transports this tissue fluid back into the bloodstream. As tissue fluids pass through lymphatic vessels and lymph nodes, disease-causing pathogens and other foreign substances are filtered out and destroyed.

Innate immunity involves the action of several types of white blood cells that protect the body against any type of pathogen. Macrophages and neutrophils engulf foreign substances that enter the body. If the infection persists, the lymphatic system becomes involved. The body develops an acquired immune response that defends against the specific pathogen.

Acquired immunity involves helper T cells that pass on chemical information about the pathogen to B cells. B cells produce antibodies that disarm or destroy the invaders. Some B cells remain in the body as memory B cells that recognize the antigens if they ever invade the body again. This process provides the body with acquired natural immunity against disease.

The lymphatic system includes lymph nodes, tonsils, the thymus gland, and spleen. T cells mature in the thymus. The spleen stores both T cells and B cells.

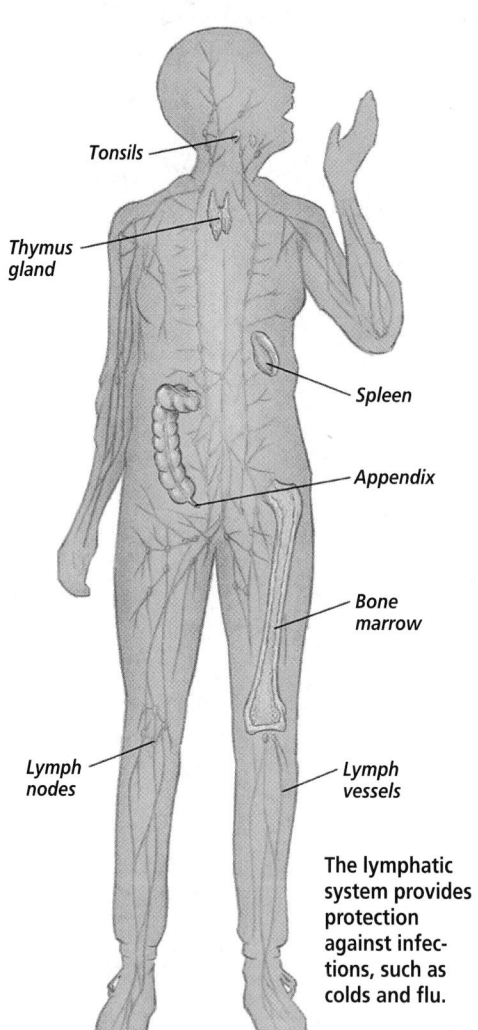

Tonsils

Thymus gland

Spleen

Appendix

Bone marrow

Lymph nodes

Lymph vessels

The lymphatic system provides protection against infections, such as colds and flu.

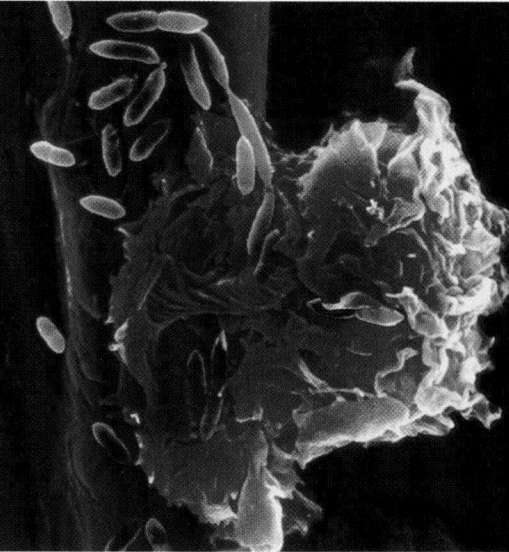

Magnification: 4200×

A lymphocyte attacks path

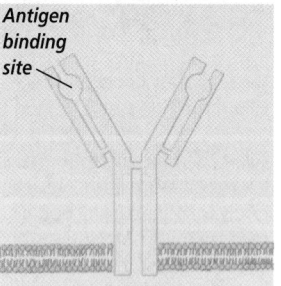

An antibody has an antigen binding site that varies from one antibody to the next.

Antigen binding site

✔ Portfolio

Health Myths

Linguistic Have students look up health myths from the 1700-1800s. Books on the history of medicine, surgery, or disease are good sources of this information. Have the class discuss modern treatments and how our understanding of the human body has changed over time. **L2** **P**

Biology Journal

Summarizing Body Systems

Linguistic Ask students to write a paragraph summarizing the function of each of the body systems. **L2**

BioDigest

BioDigest Assessment

BioDigest Assessment

Understanding Main Ideas

1. Which of the following is NOT one of the levels of organization of cells in the human body?
   a. tissue   c. organ system
   b. organ   d. receptor

2. Which of the following systems manufactures blood cells?
   a. skin   c. circulatory system
   b. skeletal system   d. respiratory system

3. Which type of muscle lines hollow internal organs?
   a. smooth   c. cardiac
   b. skeletal d. voluntary

4. Which of the following organs is NOT a part of the digestive system?
   a. tongue   c. spleen
   b. saliva glands d. pancreas

5. Oxygen is needed by your body cells to _____.
   a. produce carbon dioxide in the cells
   b. break down glucose to make ATP
   c. exchange gas in the alveoli of the lungs
   d. provide muscles with energy to contract

6. What type of event occurs at the synapse between two neurons?
   a. Calcium passes from one cell to another cell.
   b. A neurotransmitter passes from one neuron to the next neuron.
   c. A wave of electrical charges passes from one cell to the next cell.
   d. Sensory receptors detect changes inside the body.

7. Which system secretes hormones to control the metabolic activities of the body structures?
   a. endocrine system
   b. nervous system
   c. circulatory system
   d. excretory system

8. Which type of immune cell creates antibodies against foreign invaders?
   a. red blood cells   c. spleen cells
   b. T cells   d. B cells

9. Urine contains the metabolic waste products from the digestion of _____.
   a. glucose   c. amino acids
   b. fats   d. water

10. The highest blood pressure, systolic pressure, is the force created by _____.
    a. the lungs   c. the two ventricles
    b. the two atria   d. the arteries

Thinking Critically

1. Describe how both the nervous system and endocrine system are involved in controlling all other body systems.

2. AIDS is a viral disease that attacks and kills T cells. Why does a person with AIDS usually die from the inability to fight off infection?

3. Which systems are involved in excretion of waste materials?

4. How does the respiratory system work with the circulatory system?

5. How might an injury to the skeletal system affect the circulatory system?

Understanding Main Ideas

1. d   5. b   8. d
2. b   6. b   9. c
3. a   7. a   10. c
4. c

Thinking Critically

1. The nervous system receives information from the inside and outside of the body, interprets it, and acts on the information by stimulating muscles or glands. The endocrine system secretes hormones that regulate metabolic activities of body structures.

2. The disease destroys T cells, thereby removing the very immune system cells capable of killing agents that cause infections.

3. digestive, respiratory, and urinary

4. The respiratory system delivers oxygen, which is transported by the circulatory system to the body cells. The circulatory system delivers carbon dioxide from the body cells to the lungs for elimination from the body.

5. Breaking a bone will disrupt capillaries, causing bleeding. Blood clots must be formed until the capillaries can be healed. Breaking a bone could also disrupt blood cell production from that area until healing restores the bone.

APPENDIX

Contents

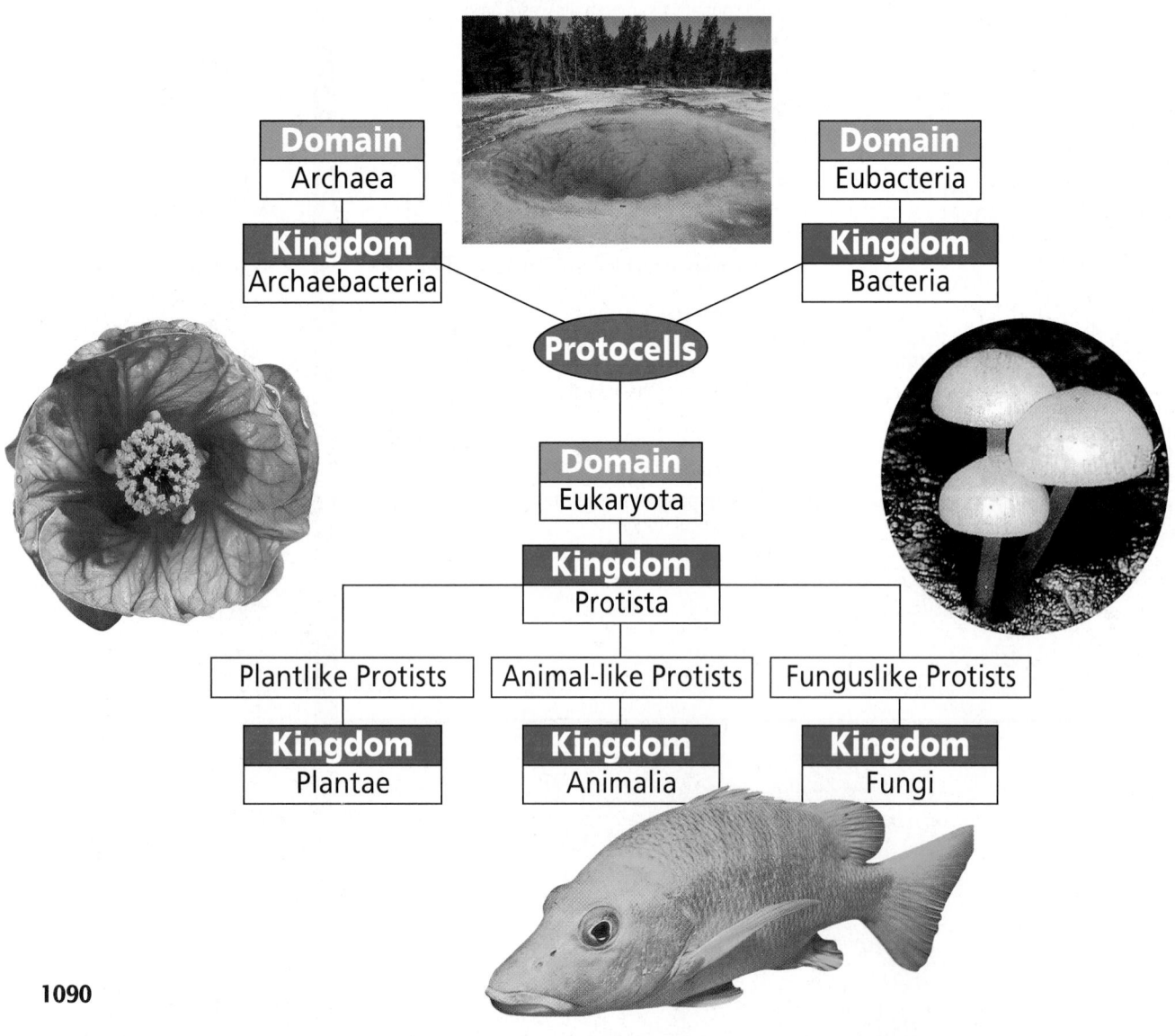

Domain
Archaea

Kingdom
Archaebacteria

Domain
Eubacteria

Kingdom
Bacteria

Protocells

Domain
Eukaryota

Kingdom
Protista

Plantlike Protists

Animal-like Protists

Funguslike Protists

Kingdom
Plantae

Kingdom
Animalia

Kingdom
Fungi

The Six-Kingdom Classification

The classification used in this text combines information gathered from the systems of many different fields of biology. For example, phycologists, biologists who study algae, have developed their own system of classification, as have mycologists, biologists who study fungi. The naming of animals and plants is controlled by two completely different sets of rules. The six-kingdom system, although not ideal for reflecting the phylogeny of all life, is useful for showing relationships. Taxonomy is an area of biology that evolves just like the species it studies. In this Appendix, only the major phyla are listed, and at least one genus is named as an example. For more information about each taxon, refer to the chapter in the text in which the group is described.

Kingdom Eubacteria
True Bacteria
Phylum Actinobacteria
Example: *Mycobacterium*
Phylum Omnibacteria
Example: *Salmonella*
Phylum Spirochaetae (Spirochaetes)
Example: *Treponema*
Phylum Chloroxybacteria (Grass-green Bacteria)
Example: *Prochloron*
Phylum Cyanobacteria (Blue-green Algae)
Example: *Nostoc*

Kingdom Archaebacteria
Ancient Prokaryotes
Phylum Aphragmabacteria (Thermoacidophiles)
Example: *Mycoplasma*
Phylum Halobacteria (Halophiles)
Example: *Halobacterium*
Phylum Methanocreatrices (Methanogens)
Example: *Methanobacillus*

Kingdom Protista
Animal-like Protists
Phylum Rhizopoda (Amoebas)
Example: *Amoeba*
Phylum Ciliophora (Ciliates)
Example: *Paramecium*
Phylum Sporozoa (Sporozoans)
Example: *Plasmodium*
Phylum Zoomastigina (Flagellates)
Example: *Trypanosoma*

Plantlike Protists
Phylum Euglenophyta (Euglenoids)
Example: *Euglena*
Phylum Bacillariophyta (Diatoms)
Example: *Navicula*
Phylum Dinoflagellata (Dinoflagellates)
Example: *Gonyaulax*
Phylum Rhodophyta (Red Algae)
Example: *Chondrus*
Phylum Phaeophyta (Brown Algae)
Example: *Laminaria*
Phylum Chlorophyta (Green Algae)
Example: *Ulva*

Funguslike Protists
Phylum Acrasiomycota (Cellular Slime Molds)
Example: *Dictyostelium*
Phylum Myxomycota (Plasmodial Slime Molds)
Example: *Physarum*
Phylum Oomycota (Water Molds, Mildews, Rusts)
Example: *Phytophthora*

Kingdom Fungi

Division Zygomycota (Sporangium Fungi)
Example: *Rhizopus*
Division Ascomycota (Cup Fungi and Yeasts)
Example: *Saccharomyces*
Division Basidiomycota (Club Fungi)
Example: *Amanita*
Division Deuteromycota (Imperfect Fungi)
Example: *Penicillium*
Division Mycophycota (Lichens)
Example: *Cladonia*

Kingdom Plantae

Nonseed Plants

Division Hepatophyta (Liverworts)
Example: *Pellia*
Division Anthocerophyta
Example: *Anthoceros*
Division Bryophyta (Mosses)
Example: *Polytrichum*
Division Psilophyta (Whisk Ferns)
Example: *Psilotum*
Division Lycophyta
(Club Mosses)
Example: *Lycopodium*
Division Sphenophyta
(Horsetails)
Example: *Equisetum*
Division Pterophyta
(Ferns)
Example: *Polypodium*

Seed Plants

Division Ginkgophyta
(Ginkgoes)
Example: *Ginkgo*
Division Cycadophyta (Cycads)
Example: *Cycas*
Division Coniferophyta
(Conifers)
Example: *Pinus*
Division Gnetophyta
Example: *Welwitschia*

Division Anthophyta (Flowering Plants)
Class Dicotyledones (Dicots)
Family Magnoliaceae (Magnolias)
Example: *Magnolia*
Family Fagaceae (Beeches)
Example: *Quercus*
Family Cactaceae (Cacti)
Example: *Opuntia*
Family Malvaceae (Mallows)
Example: *Gossypium*
Family Brassicaceae (Mustards)
Example: *Brassica*
Family Rosaceae (Roses)
Example: *Rosa*
Family Fabaceae (Peas)
Example: *Arachis*
Family Aceracea (Maples)
Example: *Acer*
Family Lamiaceae (Mints)
Example: *Thymus*
Family Asteraceae (Daisies)
Example: Helianthus
Class Monocotyledones (Monocots)
Family Poaceae (Grasses)
Example: *Triticum*
Family Palmae (Palms)
Example: *Phoenix*
Family Liliaceae (Lilies)
Example: *Asparagus*
Family Orchidaceae (Orchids)
Example: *Cypripedium*

Kingdom Animalia

Invertebrates

Phylum Porifera (Sponges)
    Example: *Spongilla*

Phylum Cnidaria (Corals, Jellyfishes, Hydras)
    Class Hydrozoa (Hydroids)
      Example: *Hydra*
    Class Scyphozoa (Jellyfishes)
      Example: *Aurelia*
    Class Anthozoa (Sea Anemones, Corals)
      Example: *Corallium*

Phylum Platyhelminthes (Flatworms)
    Class Turbellaria (Free-living Flatworms)
      Example: *Dugesia*
    Class Trematoda (Flukes)
      Example: *Fasciola*
    Class Cestoda (Tapeworms)
      Example: *Taenia*

Phylum Nematoda (Roundworms)
    Example: *Trichinella*

Phylum Mollusca (Mollusks)
    Class Gastropoda (Snails and Slugs)
      Example: *Helix*
    Class Bivalvia (Bivalves)
      Example: *Arca*
    Class Cephalopoda (Octopuses, Squid)
      Example: *Nautilus*

Phylum Annelida (Annelids)
    Class Polychaeta (Polychaetes)
      Example: *Nereis*
    Class Oligochaete (Earthworms)
      Example: *Lumbricus*
    Class Hirudinea (Leeches)
      Example: *Hirudo*

Phylum Arthropoda (Arthropods)
    Class Arachnida (Spiders, Mites, Scorpions)
      Example: *Latrodectus*
    Class Merostomata (Horseshoe Crabs)
      Example: *Limulus*
    Class Crustacea (Lobsters, Crayfishes, Crabs)
      Example: *Homarus*
    Class Chilopoda (Centipedes)
      Example: *Scutigerella*

    Class Diplopoda (Millipedes)
      Example: *Julus*
    Class Insecta (Insects)
      Example: *Bombus*

Phylum Echinodermata (Echinoderms)
    Class Crinoidea (Sea Lilies, Feather Stars)
      Example: *Ptilocrinus*
    Class Asteroidea (Sea Stars)
      Example: *Asterias*
    Class Ophiuroidea (Brittle Stars)
      Example: *Ophiura*
    Class Echinoidea (Sea Urchins and Sand Dollars)
      Example: *Arbacia*
    Class Holothuroidea (Sea Cucumbers)
      Example: *Cucumaria*

Vertebrates

Phylum Chordata (Chordates)
Subphylum Urochordata (Tunicates)
  Example: *Polycarpa*
Subphylum Cephalochordata (Lancelets)
  Example: *Branchiostoma*
Subphylum Vertebrata (Vertebrates)
  Class Agnatha (Lampreys and Hagfishes)
    Example: *Petromyzon*
  Class Chondrichthyes (Sharks, Rays)
    Example: *Squalus*
  Class Osteichthyes (Bony Fishes)
    Example: *Hippocampas*
  **Subclass Crossopterygii
    (Lobe-finned Fishes)**
    Example: *Latimeria*
  Subclass Dipneusti (Lungfishes)
    Example: *Neoceratodus*
  **Subclass Actinopterygii
    (Ray-finned Fishes)**
    Example: *Acipenser*
  Class Amphibia (Newts, Frogs, Toads)
    Example: *Rana*
  **Class Reptilia (Turtles, Snakes, Lizards,
    Crocodiles, Alligators)**
    Example: *Anolis*

Class Aves (Birds)
**Order Anseriformes
  (Ducks, Geese, Swans)**
  Example: *Olor*
Order Falconiformes (Hawks, Eagles)
  Example: *Falco*
Order Galliformes (Ground Birds)
  Example: *Perdix*
Order Passeriformes (Perching Birds)
  Example: *Spizella*
Class Mammalia (Mammals)
Order Monotremata (Monotremes)
  Example: *Ornithorhynchus*
Order Marsupialia (Marsupials)
  Example: *Didelphis*
Order Insectivora (Insect Eaters)
  Example: *Scapanus*
Order Chiroptera (Bats)
  Example: *Desmodus*
Order Carnivora (Carnivores)
  Example: *Ursus*
Order Rodentia (Rodents)
  Example: *Castor*
Order Cetacea (Whales, Dolphins)
  Example: *Delphinus*
Order Primates (Primates)
  Example: *Gorilla*

Origins of Scientific Terms

This list of Greek and Latin roots will help you interpret the meaning of biological terms. The column headed Root gives many of the actual Greek (GK) or Latin (L) root words used in science. The letter groups that follow are forms in which the root word is most often found combined in science words. In the second column is the meaning of the root as it is used in science. The third column shows a typical science word containing the root from the first column. These root words are used throughout your textbook.

ROOT	MEANING	EXAMPLE
A		
ad (L)	next to	adaxial
aeros (GK)	air	aerobic
an (GK)	without	anaerobic
ana (L)	away	anaphase
andro (GK)	male	androecium
angio (GK)	vessel	angiosperm
anthos (GK)	flower	anthophyte
anti (GK)	against	antibody
aqua (L)	water	aquatic
archae (GK)	ancient	archaebacteria
arthron (GK)	jointed	arthropod
artios (GK)	even	artiodactyl
askos (GK)	bag	ascospore
aster (GK)	star	Asteroidea
autos (GK)	self	autoimmune
B		
bi (L)	two	bipedal
bios (GK)	life	biology
C		
carn (L)	flesh	carnivore
ceph (GK)	head	cephalopoda
chloros (GK)	pale green	chlorophyll
chroma, (GK)	colored	chromosome
cide (L)	kill	insecticide
circa (L)	about	circadian
con (L)	together	convergent
cyte (GK)	cell	cytoplasm
D		
de (L)	remove	decompose
dendron (GK)	tree	dendrite
dens (L)	tooth	edentate
derma (GK)	skin	epidermis
di (GK)	two	disaccharide
dia (GK)	apart	diastolic
dormire (L)	sleep	dormancy

ROOT	MEANING	EXAMPLE
E		
echinos (GK)	spine	echinoderm
eco (GK)	house	ecosystem
ella (GK)	small	organelle
endo (GK)	within	endosperm
epi (GK)	upon	epidermis
eu (GK)	true	eukaryote
exo (GK)	outside	exoskeleton
F		
ferre (L)	carry	porifera
G		
gastro (GK)	stomach	gastropoda
genesis (L)	origin	parthenogenesis
genos (GK)	make	genotype
gons (GK)	reproductive	archegonium
gravis (L)	heavy	gravitropism
gymnos (GK)	naked	gymnosperm
gyne (GK)	female	gynoecium
H		
halo (GK)	salt	halophile
haplo (GK)	single	haploid
hemi (GK)	half	hemisphere
hemo (GK)	blood	hemoglobin
herba (L)	plant	herbivore
heteros (GK)	mixed	heterotrophic
homeo (GK, L)	same	homeostasis
homo (L)	human	hominid
homos (GK)	alike	homozygous
hydro (GK)	water	hydrolysis
I		
inter (L)	between	internode
intra (L)	within	intracellular
isos (GK)	equal	isotonic

ROOT	MEANING	EXAMPLE
J		
jugare *(L)*	join	conjugate
K		
karyon *(GK)*	seed	prokaryote
keras *(GK)*	horn	chelicerae
kokkus *(GK)*	berry-shape	streptococcus
L		
leukos *(GK)*	white	leukocyte
logy *(GK)*	study	biology
lympha *(L)*	water	lymphocyte
lysis *(GK)*	loosen	lysosome
M		
makros *(GK)*	large	macromolecule
megas *(GK)*	large	megaspore
mesos *(GK)*	middle	mesophyll
meta *(GK)*	after	metaphase
micro *(GK)*	small	microscope
monos *(GK)*	single	monocotyledon
morphe *(GK)*	form	morphology
N		
nema *(GK)*	thread	nematode
neuro *(GK)*	nerve	neuron
nodus *(L)*	knot	internode
nomia *(GK)*	law	taxonomy
O		
oligos *(GK)*	few	oligochaete
omnis *(L)*	all	omnivore
ornis *(GK)*	bird	ornithology
osteon *(GK)*	bone	osteocyte
ovum *(L)*	egg	oviduct
P		
paleo *(GK)*	ancient	paleontology
para *(GK)*	beside	parasite
pathos *(GK)*	disease	pathogen
pedis *(L)*	foot	pseudopodia
per *(L)*	through	permeable
peri *(GK)*	around	peristalsis
phago *(GK)*	eat	phagocyte
photos *(GK)*	light	phototropism
phulon *(GK)*	related group	phylogeny
phyllon *(GK)*	leaf	chlorophyll

ROOT	MEANING	EXAMPLE
P (CONT'D)		
phyton *(GK)*	plant	epiphyte
pilus *(L)*	hair	pili
pinna *(L)*	feather	pinnate
plasma *(GK)*	mold	plasmodium
pod *(GK)*	foot	gastropoda
polys *(GK)*	many	polymer
post *(L)*	after	posterior
pro *(GK, L)*	before	prokaryote
protos *(GK)*	first	protocells
pseudes *(GK)*	false	pseudopodium
R		
re *(L)*	again	reproduce
rhiza *(GK)*	root	mycorrhiza
S		
scop *(GK)*	look	microscope
soma *(GK)*	body	lysosome
sperma *(GK)*	seed	gymnosperm
stasis *(GK)*	staying	homeostasis
stoma *(GK)*	mouth	stomata
syn *(GK)*	together	synapse
T		
telos *(GK)*	end	telophase
terra *(L)*	Earth	terrestrial
therme *(GK)*	heat	endotherm
thulakos *(GK)*	pouch	thylakoid
trans *(L)*	across	transpiration
trich *(GK)*	hair	trichome
trope *(GK)*	turn	gravitropism
trophe *(GK)*	nourish	heterotrophic
U		
uni *(L)*	one	unicellular
V		
vacca *(L)*	cow	vaccine
vorare *(L)*	devour	omnivore
X		
xeros *(GK)*	dry	xerophyte
Z		
zoon *(GK)*	animal	zoology
zygous *(GK)*	joined	homozygous

Safety in the Laboratory

These safety symbols are used in the lab activities to indicate possible hazards. Learn the meaning of each symbol.

SAFETY SYMBOLS	HAZARD	EXAMPLES	PRECAUTION	REMEDY
DISPOSAL	Special disposal considerations required	chemicals, broken glass, living organisms such as bacterial cultures, protists, etc.	Plan to dispose of wastes as directed by your teacher.	Ask your teacher how to dispose of laboratory materials.
BIOLOGICAL	Organisms or organic materials that can harm humans	bacteria, fungus, blood, raw organs, plant material	Avoid skin contact with organisms or material. Wear dust mask or gloves. Wash hands thoroughly.	Notify your teacher if you suspect contact.
EXTREME TEMPERATURE	Objects that can burn skin by being too cold or too hot	boiling liquids, hot plates, liquid nitrogen, dry ice, all burners	Use proper protection when handling. Remove flammables from area around open flames or spark sources.	Go to your teacher for first aid.
SHARP OBJECT	Use of tools or glassware that can easily puncture or slice skin	razor blade, scalpel, awl, nails, push pins	Practice common sense behavior and follow guidelines for use of the tool.	Go to your teacher for first aid.
FUME	Potential danger to olfactory tract from fumes	ammonia, heating sulfur, moth balls, nail polish remover, acetone	Make sure there is good ventilation and never smell fumes directly.	Leave foul area and notify your teacher immediately.
ELECTRICAL	Possible danger from electrical shock or burn	improper grounding, liquid spills, short circuits	Double-check setup with instructor. Check condition of wires and apparatus.	Do not attempt to fix electrical problems. Notify your teacher immediately.
IRRITANT	Substances that can irritate the skin or mucus membranes	pollen, mothballs, steel wool, potassium permanganate	Dust mask or gloves are advisable. Practice extra care when handling these materials.	Go to your teacher for first aid.
CORROSIVE	Substances (acids and bases) that can react with and destroy tissue and other materials	acids such as vinegar, hydrochloric acid, hydrogen peroxide; bases such as bleach, soap, sodium hydroxide	Wear goggles and an apron.	Immediately begin to flush with water and notify your teacher.
TOXIC	Poisonous substance that can be acquired through skin absorption, inhalation, or ingestion	mercury, many metal compounds, iodine, Poinsettia leaves	Follow your teacher's instructions. Always wash hands thoroughly after use.	Go to your teacher for first aid.
FLAMMABLE	Flammable and combustible materials that may ignite if exposed to an open flame or spark	alcohol, powders, kerosene, potassium permanganate	Avoid flames and heat sources. Be aware of locations of fire safety equipment.	Notify your teacher immediately. Use fire safety equipment if applicable.

 **Eye Safety**
This symbol appears when a danger to eyes exists.

 **Clothing Protection**
This symbol appears when substances could stain or burn clothing.

 **Animal Safety**
This symbol appears when safety of live animals and students must be ensured.

Safety in the Laboratory

The biology laboratory is a safe place to work if you are aware of important safety rules and if you are careful. You must be responsible for your own safety and for the safety of others. The safety rules given here will protect you and others from harm in the lab. While carrying out procedures in any of the BioLabs, notice the safety symbols and caution statements. The safety symbols are explained on the previous page.

The Ten Rules of Safety

1. Always obtain your teacher's permission to begin a lab.
2. Study the procedure. If you have questions, ask your teacher. Be sure you understand all safety symbols shown.
3. Use the safety equipment provided for you. Goggles and a safety apron should be worn when any lab calls for using chemicals.
4. When you are heating a test tube, always slant it so the mouth points away from you and others.
5. Never eat or drink in the lab. Never inhale chemicals. Do not taste any substance or draw any material into your mouth.
6. If you spill any chemical, wash it off immediately with water. Report the spill immediately to your teacher.
7. Know the location and proper use of the fire extinguisher, safety shower, fire blanket, first aid kit, and fire alarm.
8. Keep all materials away from open flames. Tie back long hair.
9. If a fire should break out in the lab, or if your clothing should catch fire, smother it with the fire blanket or a coat, or get under a safety shower. **NEVER RUN.**
10. Report any accident or injury, no matter how small, to your teacher.

Procedures for Clean-Up Time

1. Turn off the water and gas. Disconnect electrical devices.
2. Return materials to their places.
3. Dispose of chemicals and other materials as directed by your teacher. Place broken glass and solid substances in the proper containers. Never discard materials in the sink.
4. Clean your work area.
5. Wash your hands thoroughly after working in the laboratory.

Table C–1 First aid in the laboratory	
Injury	**Safe response**
Burns	Apply cold water. Call your teacher immediately.
Cuts and bruises	Stop any bleeding by applying direct pressure. Cover cuts with a clean dressing. Apply cold compresses to bruises. Call your teacher immediately.
Fainting	Leave the person lying down. Loosen any tight clothing and keep crowds away. Call your teacher immediately.
Foreign matter in eye	Flush with plenty of water. Use an eyewash bottle or fountain.
Poisoning	Note the suspected poisoning agent and call your teacher immediately.
Any spills on skin	Flush with large amounts of water or use safety shower. Call your teacher immediately.

Periodic Table of the Elements

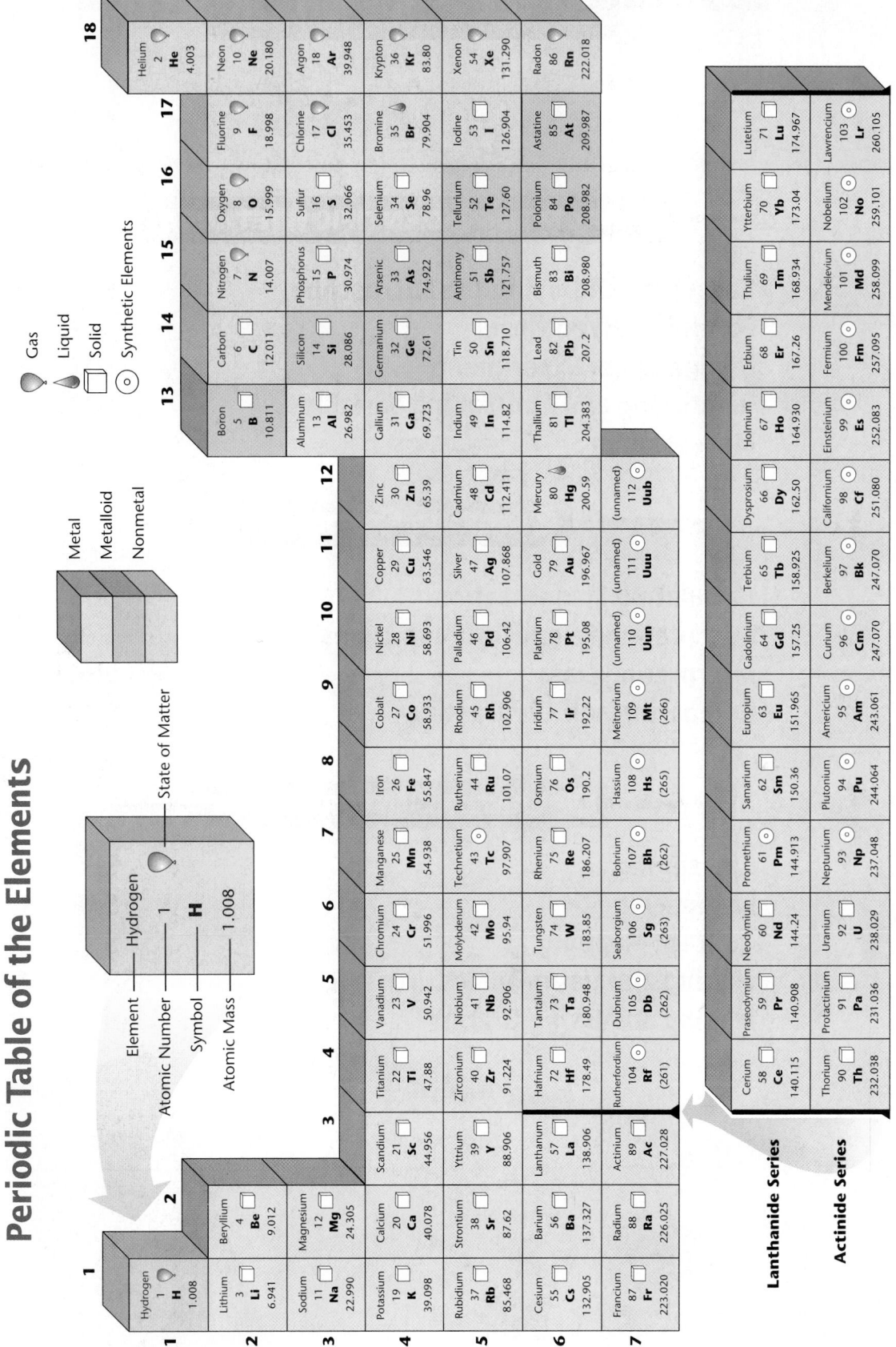

SKILL HANDBOOK

Contents

PRACTICING SCIENTIFIC METHODS

THINKING CRITICALLY

ORGANIZING INFORMATION

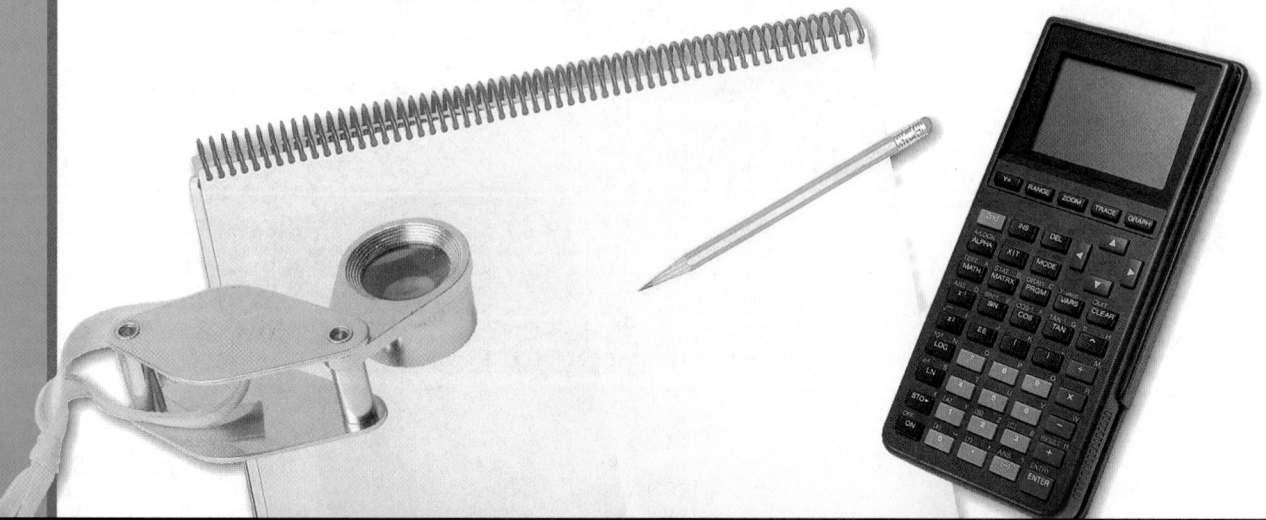

PRACTICING SCIENTIFIC METHODS

Caring for and Using the Microscope

1. Always carry the microscope by holding the arm with one hand and supporting the base with the other hand.

2. Place the microscope on a flat surface. The arm should be toward you.

3. Look through the eyepiece. Adjust the diaphragm so that light comes through the opening in the stage.

4. Place a slide on the stage so that the specimen is in the field of view. Hold it firmly in place by using the stage clips.

5. Always focus first with the coarse adjustment and the low-power objective lens. Once the object is in focus on low power, turn the nosepiece until the high-power objective is in place. Then use ONLY the fine adjustment to focus with this lens.

6. Store the microscope covered.

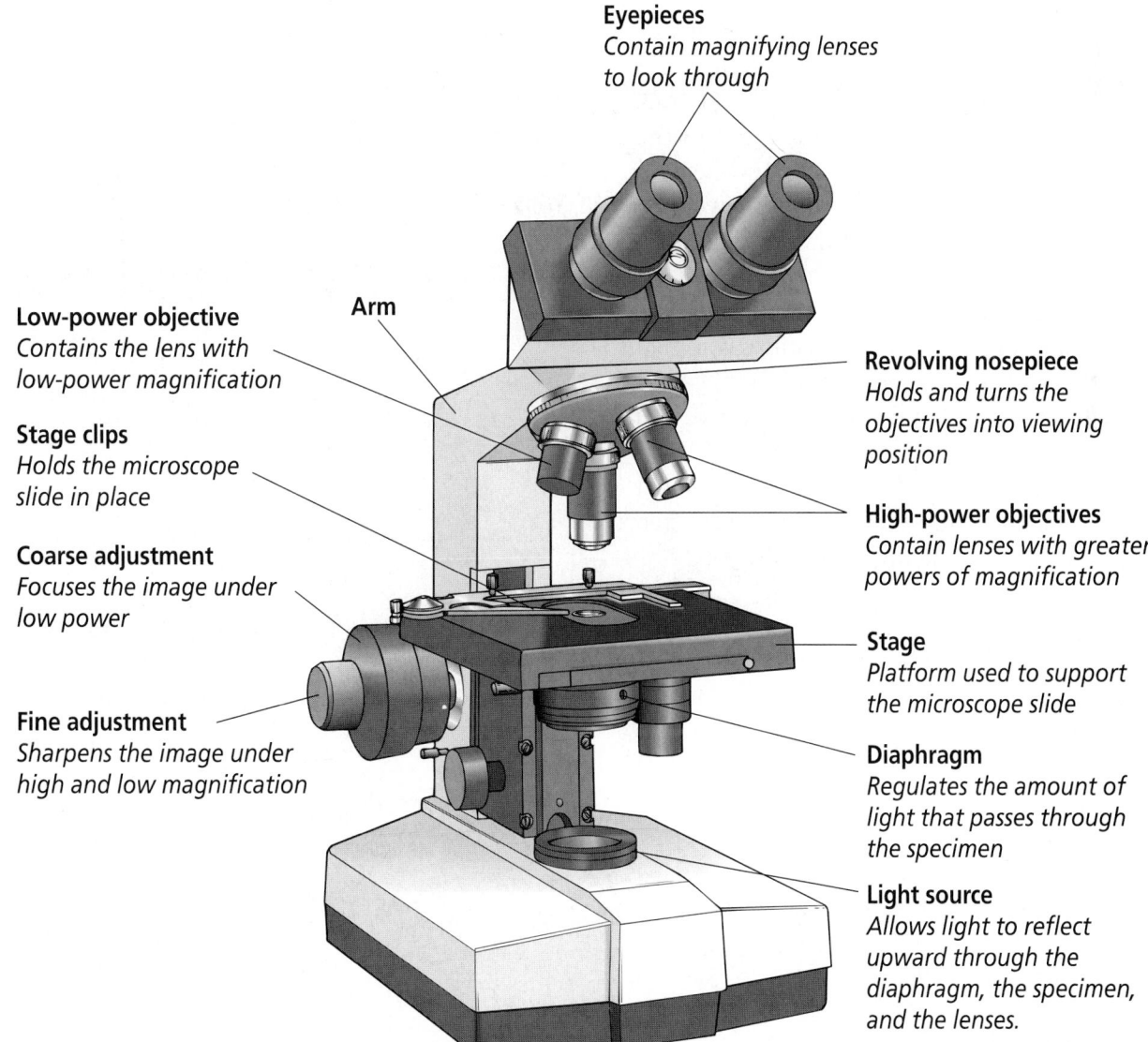

Eyepieces
Contain magnifying lenses to look through

Low-power objective
Contains the lens with low-power magnification

Arm

Stage clips
Holds the microscope slide in place

Coarse adjustment
Focuses the image under low power

Fine adjustment
Sharpens the image under high and low magnification

Revolving nosepiece
Holds and turns the objectives into viewing position

High-power objectives
Contain lenses with greater powers of magnification

Stage
Platform used to support the microscope slide

Diaphragm
Regulates the amount of light that passes through the specimen

Light source
Allows light to reflect upward through the diaphragm, the specimen, and the lenses.

PRACTICING SCIENTIFIC METHODS

Forming a Hypothesis

Suppose you wanted to earn a perfect score on a biology exam. You think of several ways to accomplish a perfect score. You base these choices on past experiences and observations of your friends' results. All of the following are hypotheses you might consider that could explain how it would be possible to score 100 percent on your exam:

If the exam is easy, then I will get a good grade.
If I am intelligent, then I will get a good grade.
If I study hard, then I will get a good grade.

How could you test if any of these hypotheses are valid? How would you judge if the exam were easy? How do you measure intelligence? What does it mean to study hard? There is one definite way to judge the success of any of these hypotheses—your final score. To test these hypotheses scientifically, you would have to take the exam many times and under many different conditions!

Let's look at something a little more scientific. If you made an observation in nature, you might make a hypothesis to explain it. Perhaps you observed that your pet fish were less active than they usually seem after you changed the water in their tank. You might form a hypothesis that says: If fishes are exposed to a different temperature of their water, their activity will change.

Designing an Experiment

Once you have stated a hypothesis, you will want to find out whether or not it explains your observation. This requires a test. To be valid, a hypothesis must be testable by experimentation. Let's consider how you would

conduct an experiment to test the hypothesis about the effects of water temperature on fishes that you have in an aquarium.

First, obtain several identical, clear glass containers, and fill each of them with the same amount of tap water. Leave the containers for a day to allow the water to come to room temperature. On the day of your experiment, measure and record the temperature of the water in the aquarium. Heat and cool the other containers, adjusting the water temperatures in the test containers so that two have higher temperatures and two have lower temperatures than the aquarium water temperature.

You place a fish from your aquarium in each container. You count the number of horizontal and vertical movements each fish makes during five minutes and record your data in a table. Your data table might look like this.

Table SH.1 Number of fish movements

Container	Temperature (°C)	Number of movements
Aquarium	20	56
A	22	61
B	24	70
C	18	46
D	16	42

From the data you recorded, you will draw a conclusion and make a statement about your results. If your conclusion supports your

PRACTICING SCIENTIFIC METHODS

number of movements during the same amount of time. Scientists can know only that the independent variable caused the change in the dependent variable if they keep all other factors the same in an experiment.

Dependent Variable The dependent variable is any change that results from manipulating the independent variables. In the case of the fish experiment, the dependent variable is the level of activity in the fishes.

Control Scientists also use a control to be certain that the observed changes resulted only from the manipulation of one variable. A control is a sample that is treated exactly like the experimental group except that the independent variable is not applied to the control. The control in the fish experiment is the activity of the fishes in the aquarium. After the experiment, if there has been any change in the dependent variable of the control sample, this change is considered when evaluating the results in the experimental group. Controls allow scientists to see the effect of the independent variable. For example, if the fishes in the aquarium had increased their level of activity during the experiment, the cause of this change could not be associated with temperature change.

hypothesis, then you can say that your hypothesis is reliable. If it did not support your hypothesis, then you would have to make new observations and state a new hypothesis, one that you could also test. Do the data in the table support the hypothesis that different water temperatures affect fish activity?

Identifying and Controlling Variables

When scientists perform experiments, they must be careful to manipulate or change only one condition and keep all other conditions in the experiment the same.

Independent Variable The condition that is manipulated is called the independent variable. In the case of the fish experiment, the independent variable is the temperature. All other conditions that are kept the same during an experiment are called constants. The constants in the fish experiment are using the same size and shape containers, the same kind of fish, using equal amounts of water, and counting the

Why is it important to know the best temperature for aquarium fishes? If you have an aquarium at home, you have probably learned how different fishes need different conditions to survive. They probably came from many different parts of the world and are adapted to living in very different environments. As a responsible pet owner, it is important to provide your pets with the environment that best suits their needs for a healthy life.

PRACTICING SCIENTIFIC METHODS

Measuring in SI

The International System of Measurement (commonly abbreviated to SI) is accepted as the standard for measurement throughout most of the world. Four of the base units in SI are the meter, liter, kilogram, and second. The size of a unit can be determined from the prefix used with the base unit name. For example: kilo means one thousand; milli means one-thousandth; micro means one-millionth; and centi means one-hundredth. The table below gives the standard symbols for these SI units and some of their equivalents.

Larger and smaller units of measurement in SI are obtained by multiplying or dividing the base unit by some multiple of ten. Multiply to change from larger units to smaller units. Divide to change from smaller units to larger units. For example, to change 1 km to meters, you would multiply 1 km by 1000 to obtain 1000 m. To change 10 g to kilograms, you would divide 10 g by 1000 to obtain 0.01 kg.

Table SH.2 Common SI units

Measurement	Unit	Symbol	Equivalents
Length	1 millimeter	mm	1000 micrometers (μm)
	1 centimeter	cm	10 millimeters (mm)
	1 meter	m	100 centimeters (cm)
	1 kilometer	km	1000 meters (m)
Volume	1 milliliter	mL	1 cubic centimeter (cm^3 or cc)
	1 liter	L	1000 milliliters (mL)
Mass	1 gram	g	1000 milligrams (mg)
	1 kilogram	kg	1000 grams (g)
	1 metric ton	t	1000 kilograms (kg)
Time	1 second	s	
Area	1 square meter	m^2	10 000 square centimeters (cm^2)
	1 square kilometer	km^2	1 000 000 square meters (m^2)
	1 hectare	ha	10 000 square meters (m^2)
Temperature	1 Kelvin	K	1 degree Celsius (°C)

The top of the thermometer is marked off in degrees Fahrenheit (°F). To read the corresponding temperature in degrees Celsius (°C), look at the bottom side of the thermometer.

For example, 50°F is the same temperature as 10°C. You may also use the formulas shown here for conversions.

Conversion of Fahrenheit to Celsius

$$°C = \frac{5}{9}(°F - 32)$$

Conversion of Celsius to Fahrenheit

$$°F = \left(\frac{9}{5}°C\right) + 32$$

°F   0 20 40 60 80 100 120 140 160 180 200 210

°C −20 −10 0 10 20 30 40 50 60 70 80 90 100

Freezing point of water | *Normal human body temperature*
Room temperature

THINKING CRITICALLY

Calculating Magnification

Objects viewed under the microscope appear larger than normal because they are magnified. Total magnification describes how much larger an object appears when viewed through the microscope.

Look for a number marked with an × on the eyepiece, the low-power objective, and the high-power objective. The × stands for how many times the lens of each microscope part magnifies an object.

To calculate total magnification of any object viewed under your microscope, multiply the number on the eyepiece by the number on the objective. For example, if the eyepiece magnification is 4×, the low-power objective magnification is 10×, and the high-power objective magnification is 40×, then:

a. The magnification under low power is 4× for the eyepiece times 10× for the low-power objective = 40. (4 × 10 = 40)

b. The magnification under high power is 4 × 40 = 160.

To measure the field of view of a microscope, you must use a unit called a micrometer. A micrometer equals 0.001 mm; in other words, there are 1000 micrometers (µm) in a millimeter. Place a millimeter section of a plastic ruler over the central opening of your microscope stage. Using low power, locate the measured lines of the ruler in the center of the field of view. Move the ruler so that one of the lines representing a millimeter is visible at one edge of the field of view.

Remember that the distance between two lines is one millimeter, and estimate

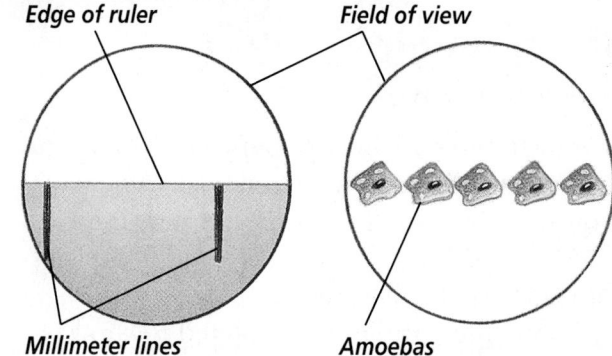

Edge of ruler   Field of view

Millimeter lines   Amoebas

the diameter in millimeters of the field of view on low power. Calculate the diameter in micrometers. For example, if the distance is 1.5 mm, then the diameter of the field of view at low power is 1500 µm. [1.5 × 1000]

To calculate the diameter of the high-power field, divide the magnification of your high power (40×) by the magnification of the low power (10×); 40 ÷ 10 = 4. Then, divide the diameter of the low-power field in micrometers (1500 µm) by this quotient (4). The answer is the diameter of the high-power field in micrometers. In this example, the diameter of the high-power field is 1500 ÷ 4 = 375 µm.

You can calculate the diameters of microscopic specimens, such as pollen grains or amoebas, viewed under low and high power by estimating how many of them could fit end to end across the field of view. Divide the diameter of the field of view by the number of specimens.

If you want to know the actual size of any specimen shown in an electron micrograph in this textbook, first measure the diameter of the structure in millimeters, multiply this number by 1000 to convert the measurement to micrometers, and then divide this number by the magnification given next to the photograph.

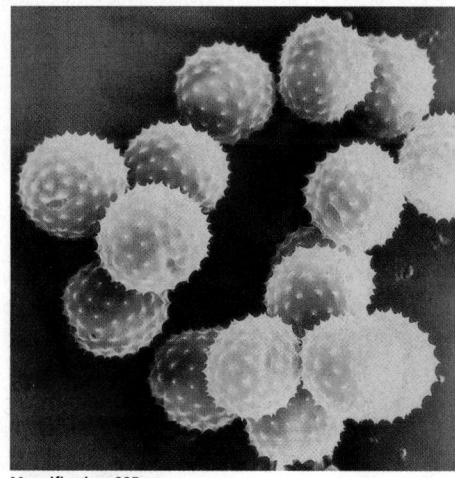

Magnification: 285×

THINKING CRITICALLY

Interpreting Scientific Illustrations

Illustrations are included in your textbook to help you understand what you read. Whenever you encounter an illustration, examine it carefully and read the caption. The caption explains or identifies the illustration.

Some illustrations are designed to show you how the internal parts of a structure are arranged. Look below at the illustrations of a squash. The squash has been cut lengthwise so that it shows a section that runs along the length of the squash. This type of illustration is called a longitudinal section. Cutting the squash crosswise at right angles to the length produces a cross section.

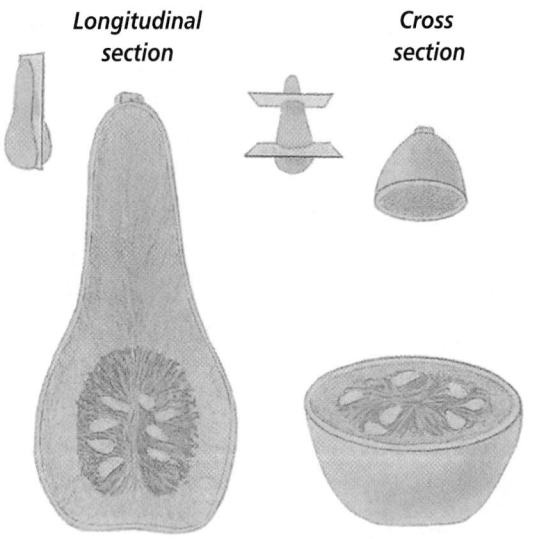

Longitudinal section   Cross section

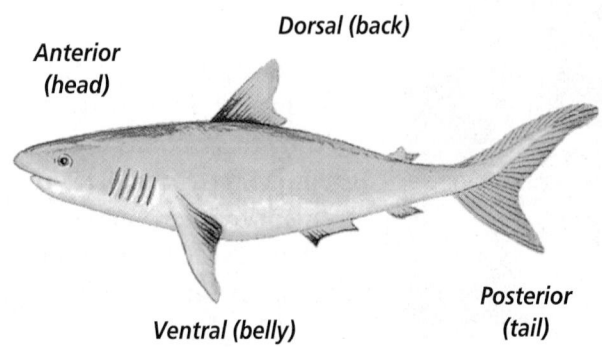

Dorsal (back)
Anterior (head)
Ventral (belly)
Posterior (tail)

You will sometimes see terms that refer to the orientation of an organism. For example, the word dorsal refers to the upper side or back of an animal. Ventral refers to the lower side or belly of the animal. The illustration of the shark shows both dorsal and ventral sides.

Interpreting Data

Observing and Inferring Scientists try to make careful and accurate observations. When possible, they use instruments, such as microscopes, binoculars, and tape recorders, to extend their senses. Other instruments, such as a thermometer or a pan balance, are used to measure observations. Measurements provide numerical data, a concrete means of comparing collected data that can be checked and repeated.

Scientists often use their observations to make inferences. An inference is an attempt to explain or interpret observations. For example, if you observed a bird feeding at a bird feeder, you might infer that the bird's nest is close by. But, the bird may just be passing through the area. The only way to be sure your inference is correct is to investigate further.

When investigating, be certain to make accurate observations and to record them carefully. Collect all the information you can. Then, based on everything you know, try to explain or interpret what you observed. If possible, investigate further to determine whether your inference is correct.

THINKING CRITICALLY

Comparing and Contrasting

Observations can be analyzed and then organized by noting the similarities and differences between two or more objects or situations. When you examine objects or situations to determine similarities, you are comparing. When you look at similar objects or situations for differences, you are contrasting.

Compare and contrast a grasshopper and a dragonfly. First make your observations. Divide a piece of paper into two columns and list the ways the two insects are similar in one column and ways they are different in the other. After completing your lists, you might report your findings in a table or in a graph.

Recognizing Cause and Effect Have you ever observed something happen and then tried to determine why or how it came about? If so, you have observed an event and inferred a reason for the event. The event or result of an action is an effect, and the reason for the event is the cause.

Suppose that every time your teacher fed fish in a classroom aquarium, she tapped the food container on the edge. Then, one day she tapped the edge of the aquarium to make a point about an ecology lesson. You observe the fish swim to the surface of the aquarium to feed. You might infer that the tapping on the aquarium by the teacher caused the fish to swim to the surface, as if to find food. You would have made a logical inference based on careful observations.

Is it possible that the fish swam to the surface for another reason? Was there another cause for this effect that you may not have noticed? When scientists are unsure of the cause for a certain event, they often design controlled experiments. Although you have made a sound judgment, you would have to perform an experiment to be certain that it was the tapping that caused the effect you observed.

ORGANIZING INFORMATION

Classifying

You may not realize it, but you impose order on the world around you. When you stack your favorite CDs in groups according to recording artist, or when you separate your socks from your shirts, you have used the skill of classifying.

Classifying is grouping objects or events based on common features. For example, how would you classify a collection of CDs? When classifying, first make careful observations of the group of items to be classified. Select one feature that is shared by some items in the group but not others. For example, you might classify good dance CDs in one subgroup and great CDs for relaxation in another. You now have made two subgroups. Ideally, the items in each subgroup will have some features in common. Now examine the CDs for other features and form further subgroups. For example, the CDs you like to dance to could be subdivided into rap or rock subgroups. Continue to identify subgroups until the items can no longer be distinguished enough to identify them as distinct.

Remember, when you classify, you are grouping objects or events for a purpose. The purpose could be general, such as for ease of finding an item. The classification of books in a library is a general-purpose classification. The classification may have a special purpose. For example, plants may be classified as poisonous or harmless to humans. All species have been classified in many different ways depending on the purpose of the classification. They are classified for identification, or to show their evolutionary relationships. One is a general purpose classification, the other is a phylogenetic classification. Both systems have been used in this textbook.

Sequencing

A sequence is an arrangement of things or events in a particular order. A common sequence with which you may be familiar is the sequence of the seasons in a north temperate climate—spring, summer, autumn, winter.

You will be following sequences of steps whenever you carry out a MiniLab or BioLab in this textbook. You will also learn about sequences of events in nature. When you are asked to follow a sequence of events, you must identify what comes first. You then decide what should come second. Continue to choose things or events until they are all in order. Then, go back over the sequence to make sure each thing or event logically leads to the next.

Concept Mapping

If you were taking an automobile trip, you would probably take along a road map. The road map shows your location, your destination, and other places along the way. By examining the map, you can understand where you are in relation to other locations.

A concept map is similar to a road map. But, a concept map shows the relationship among ideas (or concepts) rather than places. A concept map is a diagram that shows visually how concepts are related. By developing your own concept maps as you read this textbook, you will better understand the relationships among

ORGANIZING INFORMATION

biology concepts. There are three styles of concept map that you might find useful— an events chain, a network tree, and a cycle concept map.

An Events Chain An events chain map is used to describe a sequence of events. These events could include the steps in a procedure or the stages of a process. When making an events chain map, you first must find the one event that starts the chain. This event is called the initiating event. You then find the next event in the chain and continue until you reach an outcome. Think of the process of observing a butterfly. An events chain map might look like the one shown here. Connecting words may not always be necessary.

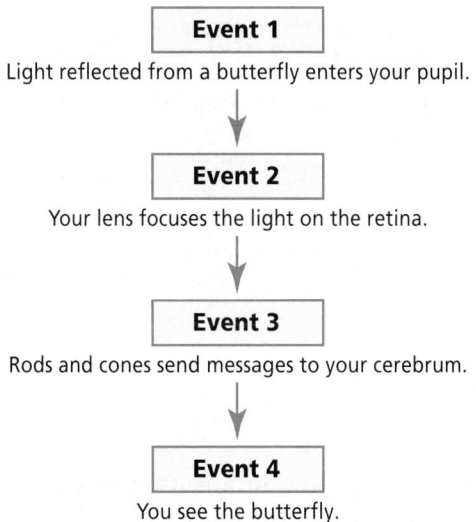

A Network Tree If you want to diagram the relationships among major concepts in a chapter, you might use the network tree. Notice how some words in the concept map below are circled. The circled words are science concepts. The lines in the map show related concepts, and the words written on the lines describe relationships between the concepts.

To construct this kind of concept map, you would first select the topic and choose the major concepts. Find related concepts and put them in order from general to specific. Branch the related concepts from the major concept, and describe the relationship on the connecting lines. Continue to place the concepts in order of how they relate until all concepts are mapped.

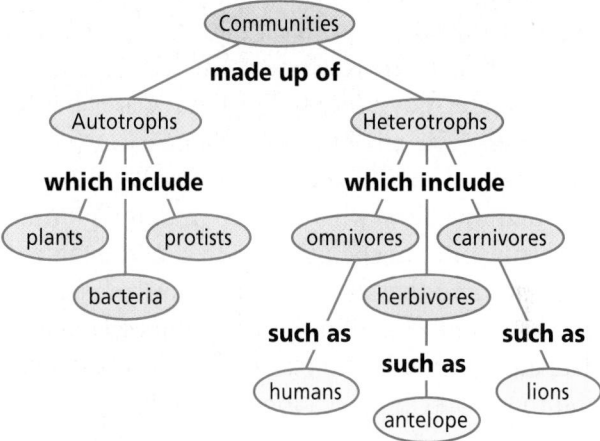

Cycle Concept Map In a cycle concept map, the series of events do not produce a final outcome. The last event in the chain relates back to the initiating event. Because there is no outcome and the last event relates back to the initiating event, the cycle repeats itself. Follow the stages shown in the cycle map of insect metamorphosis.

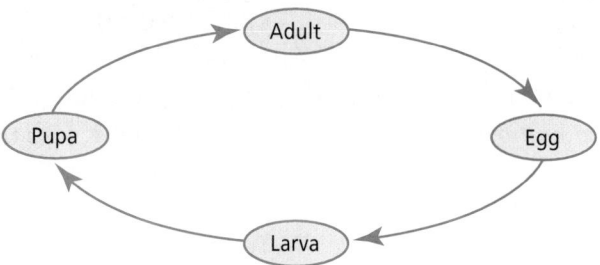

Concept maps are useful in understanding the ideas you have read about. As you construct a map, you are organizing knowledge. Once concept maps are constructed, you can use them to review and study and to test your knowledge. The construction of concept maps is a useful learning tool.

ORGANIZING INFORMATION

Making and Using Tables

Browse through your textbook, and you will notice many tables both in the text and in the labs. The tables in the text arrange data in a way that makes it easier for you to compare all the information. Also, many labs in your text have data tables to complete. Lab tables will help you organize the data you collect during the lab so that it can be interpreted more easily.

Each table has a title telling you what is being presented. The table itself is divided into columns and rows. Each column usually has a title that describes what is listed in that column. Each row of items being compared in the table has a heading that describes specific characteristics. Within the grid of the table, the collected data are recorded. Look at the following table, and then study the questions that follow it.

Table SH.3 Effect of exercise on heart rate		
Pulse taken	Individual heart rate	Class average
At rest	73	72
After exercise	110	112
1 minute after exercise	94	90
5 minutes after exercise	76	75

What is the title of this table? The title is "Effect of exercise on heart rate." What items are being compared? The heart rate for an individual and the class average are being compared at rest and for several durations after exercise.

What is the average heart rate of the class one minute after exercise? To find the answer, you must locate the column labeled "class average" and the row "1 minute after exercise." The number contained in the box where the column and row intersect is the answer. Whose heart rate was recorded as 110 after exercise? If you answered "the experimenter's," you have an understanding of how to use a table.

Making and Using Graphs

After scientists organize data in tables, they often manipulate and organize and then display the data in graphs. A graph is a diagram that shows a comparison among variables. Because graphs show a picture of collected data, they make interpretation and analysis of the data easier. The three basic types of graphs used in science are the line graph, bar graph, and circle graph.

Line Graphs A line graph is used to show the relationship between two variables. The variables being compared go on two axes of the graph. The independent variable always goes on the horizontal axis, called the x-axis. The independent variable, such as temperature, is the condition that is manipulated. The dependent variable always goes on the vertical axis, the y-axis. The dependent variable, such as growth, is any change that results from manipulating the independent variable.

Suppose a school started a peer-study program with a class of students to see how the

ORGANIZING INFORMATION

program affected their science grades. You could make a graph of the grades of students in the program over a period of time. The grading period is the independent variable and should be placed on the *x*-axis of your graph. Instead of four grading periods, we could look at average grades for the week or month or year. In this way, we would be manipulating the independent variable. The average grade of the students in the program is the dependent variable and would go on the *y*-axis.

Plain or graph paper can be used to construct graphs. After drawing your axes, you would label each axis with a scale. The *x*-axis simply lists the grading periods. To make a scale of grades on the *y*-axis, you must look at the data values provided in the data table.

Table SH.4 Average grades of students in study program	
Grading period	Average science grade
First	81
Second	85
Third	86
Fourth	89

Because the lowest grade was 81 and the highest was 89, you know that you will have to start numbering at least at 81 and go through 89. You decide to start numbering at 80 and number by twos spaced at equal distances through 90.

Average grades of students in study program

You next must plot the data points. The first pair of data you want to plot is the first grading period and 81. Locate "first" on the *x*-axis and 81 on the *y*-axis. Where an imaginary vertical line from the *x*-axis and an imaginary horizontal line from the *y*-axis would meet, place the first data point. Place the other data points the same way. After all the points are plotted, connect them with a smooth line.

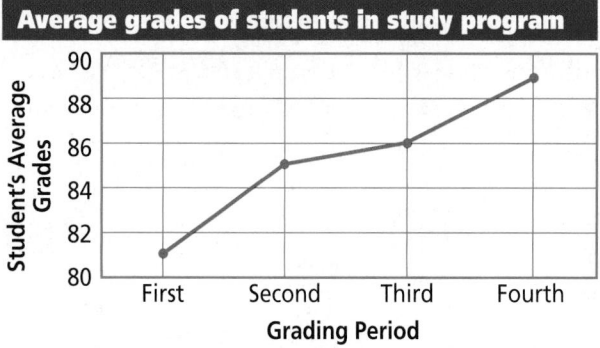

Average grades of students in study program

What if you wanted to compare the average grades of the class in the study group with the grades of another class? The data of the other class can be plotted on the same graph to make the comparison. You must include a key with two different lines, each indicating a different set of data.

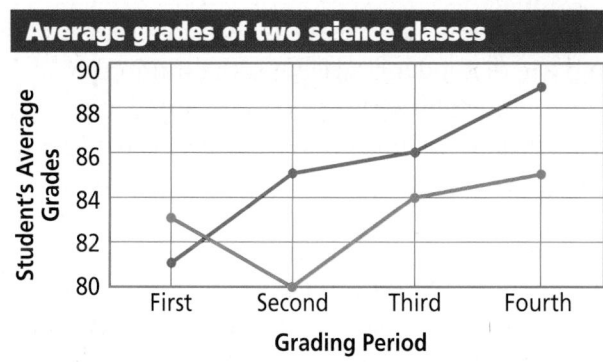

Average grades of two science classes

KEY: Class of study students ——— Regular class ———

Bar Graphs A bar graph is similar to a line graph, except it is used to show comparisons among data or to display data that do not continuously change. In a bar graph, thick bars

ORGANIZING INFORMATION

rather than data points show the relationships among data. To make a bar graph, set up the *x*-axis and *y*-axis as you did for the line graph. The data are plotted by drawing thick bars from the *x*-axis up to an imaginary point where the *y*-axis would intersect the bar if it were extended.

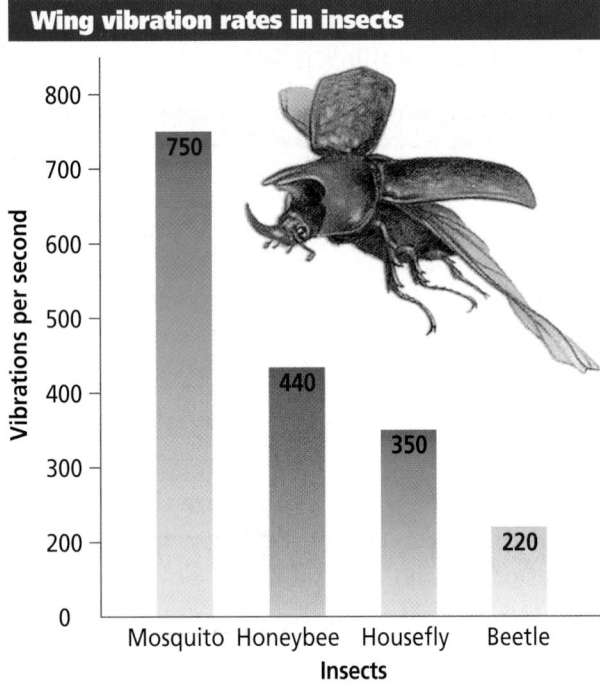

Wing vibration rates in insects

Look at the bar graph above, comparing the wing vibration rates for different insects. The independent variable is the type of insect, and the dependent variable is the number of wing vibrations per second. The number of wing vibrations for different insects is being compared.

Circle Graphs A circle graph uses a circle divided into sections to display data. Each section represents a part of the whole. When all the sections are placed together, they equal 100 percent of the whole.

Suppose you wanted to make a circle graph to show the number of seeds that germinate in a package. You would have to determine the total number of seeds and the number of seeds that germinate out of the total. You count the seeds and find that the package contains 143 seeds. Therefore, the whole circle will represent this amount. You plant the seeds and determine that 129 seeds germinate. The group of seeds that germinated will make up one section of the circle graph, and the group of seeds that did not germinate will make up another section.

To find out how much of the circle each section should take, you must divide the number of seeds germinated by the total number of seeds. You then multiply your answer by 360, the number of degrees in a circle. Round your answer to the nearest whole number. The segment of the circle for germinated seeds is then determined as follows:

$$129 \div 143 \times 360 = 324.75 \text{ or } 325°$$

To draw your circle graph, you need a compass and a protractor. Use the compass to draw a circle. Then, draw a straight line from the center to the edge of the circle. Place your protractor on this line, and use it to mark a point on the edge of the circle at 325°. Connect this point with a straight line to the center of the circle. This is the section for the group of seeds that germinated. The other section represents the group of seeds that did not germinate. Complete the graph by labeling the sections of your graph and giving the graph a title.

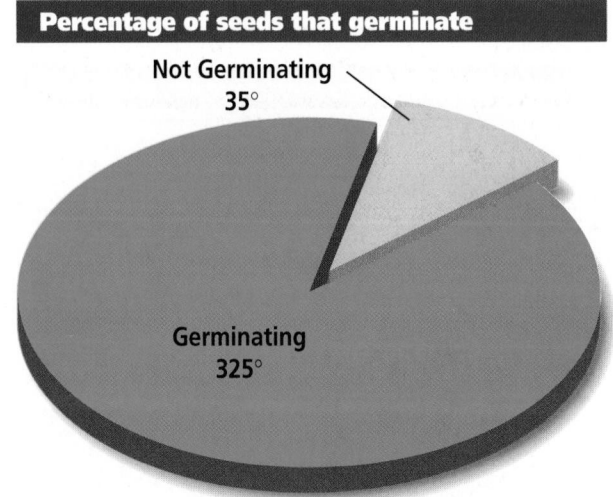

Percentage of seeds that germinate

Not Germinating
35°

Germinating
325°

Pronunciation Key

Use the following key to help you sound out words in the glossary.

a	back (bak)	**ew**	food (fewd)
ay	day (day)	**yoo**	pure (pyoor)
ah	father (fahth ur)	**yew**	few (fyew)
ow	flower (flow ur)	**uh**	comma (cahm uh)
ar	car (car)	**u** (+con)	flower (flow ur)
e	less (les)	**sh**	shelf (shelf)
ee	leaf (leef)	**ch**	nature (nay chur)
ih	trip (trihp)	**g**	gift (gihft)
i (i+con+e)	idea, life (i dee uh, life)	**j**	gem (jem)
oh	go (goh)	**ing**	sing (sing)
aw	soft (sawft)	**zh**	vision (vihzh un)
or	orbit (or but)	**k**	cake (kayk)
oy	coin (coyn)	**s**	seed, cent (seed, sent)
oo	foot (foot)	**z**	zone, raise (zohn, rayz)

abiotic factors (ay bi AHT ihk): nonliving parts of an organism's environment; temperature, moisture, light, and soil are examples. (Section 2.1)

acid: any substance that forms hydrogen ions (H⁺) in water and has a pH below 7. (Section 6.1)

acid precipitation: rain, snow, sleet, or fog with a pH below 7; causes the deterioration of forests, lakes, statues, and buildings. (Section 5.1)

acoelomate (ay SEE lum ate): animals with three cell layers—ectoderm, endoderm, and mesoderm, but no body cavities; organs are embedded in solid tissues; nutrients, water, and oxygen move by diffusion. (Section 25.2)

acquired immunity: gradual build-up of resistance to a specific pathogen over time. (Section 39.2)

actin: structural protein in muscle cells that makes up the thin filaments of myofibrils; functions in muscle contraction. (Section 34.3)

active transport: energy-expending process by which cells transport materials across the cell membrane against a concentration gradient. (Section 8.1)

adaptation (ad ap TAY shun): evolution of a structure, behavior, or internal process that enables an organism to respond to stimuli and better survive in an environment. (Section 1.1)

adaptive radiation: divergent evolution in which ancestral species evolve into an array of species to fit a number of diverse habitats. (Section 15.2)

addiction: psychological and/or physiological drug dependence. (Section 36.3)

ADP (adenosine diphosphate): molecule formed from the breaking off of a phosphate group for ATP; results in a large release of energy that is used for biological reactions. (Section 9.1)

adrenal glands: pair of glands located on top of the kidneys that secrete hormones, such as adrenaline, that prepare the body for stressful situations. (Section 35.3)

aerobic: chemical reactions that require the presence of oxygen. (Section 9.3)

age structure: proportions of a population that are at different age levels. (Section 4.2)

aggressive behavior: innate behavior used to intimidate another animal of the same species in order to defend young, territory, or resources. (Section 33.1)

alcoholic fermentation: anaerobic process where cells convert pyruvic acid into carbon dioxide and ethyl alcohol; carried out by many bacteria and fungi such as yeasts. (Section 9.3)

algae (AL jee): photosynthetic, plantlike, autotrophic protists. (Section 19.1)

allele (uh LEEL): alternative forms of a gene for each variation of a trait of an organism. (Section 10.1)

GLOSSARY

allelic frequency: percentage of any allele in a population's gene pool. (Section 15.2)

alternation of generations: type of life cycle found in some algae, fungi, and all plants where an organism alternates between a haploid (*n*) gametophyte generation and a diploid (2*n*) sporophyte generation. (Section 19.2)

alveoli (al vee OH li): sacs in the lungs where oxygen diffuses into the blood and carbon dioxide diffuses into the air. (Section 37.1)

amino acids: basic building blocks of protein molecules. (Section 6.3)

amniotic egg (am nee OHT ihk): major adaptation in land animals; amniotic sac encloses an embryo and provides nutrition and protection from the outside environment. (Section 31.1)

ampulla (am PUH lah): in echinoderms, the round, muscular structure on a tube foot that aids in locomotion. (Section 29.1)

amylase: digestive enzyme found in saliva and stomach fluids; breaks starches into smaller molecules such as disaccharides and monosaccharides. (Section 35.1)

anaerobic: chemical reactions that do not require the presence of oxygen. (Section 9.3)

analogous structures: structures that do not have a common evolutionary origin but are similar in function (Section 15.1)

anaphase: phase of mitosis where the centromeres split and the chromatid pairs of each chromosome are pulled apart by microtubules. (Section 8.2)

annual: anthophyte that lives for one year or less. (Section 22.3)

anterior: head end of bilateral animals where sensory organs are often located. (Section 25.2)

anther: pollen-producing structure located at the tip of a flower's stamen. (Section 24.2)

antheridium (an thuh RIHD ee um): male reproductive structure where sperm develop in the male gametophyte. (Section 22.1)

anthropoids (AN thruh poydz): humanlike primates that include New World monkeys, Old World monkeys, and hominids. (Section 16.1)

antibiotics: substances produced by a microorganism that, in small amounts, will kill or inhibit growth and reproduction of other microorganisms. (Section 39.1)

antibodies (ANT ih bahd eez): proteins in the blood plasma produced in reaction to antigens that react with and disable antigens. (Section 37.2)

antigens: foreign substances that stimulate an immune response in the body. (Section 37.2)

aorta: largest blood vessel in the body; transports oxygen-rich blood from the left ventricle of the heart to the arteries. (Section 37.2)

aphotic zone: portion of the marine biome that is too deep for sunlight to penetrate. (Section 3.2)

apical meristem: regions of actively dividing cells near the tips of roots and stems; allows roots and stems to increase in length. (Section 23.1)

appendage (uh PEN dihj): any structure, such as a leg or an antenna, that grows out of an animal's body. (Section 28.1)

appendicular skeleton (a pen DIHK yuh lur): one of two main parts of the human skeleton, includes the bones of the arms and legs and associated structures, such as the shoulders and hip bones. (Section 34.2)

archaebacteria (ar kee bac TIHR ee uh): chemosynthetic prokaryotes that live in harsh environments, such as deep-sea vents and hot springs. (Section 14.2)

archegonium (ar kee GOH nee um): female reproductive structure where eggs develop in the female gametophyte. (Section 22.1)

artery: large, thick-walled muscular vessels that carry oxygenated blood away from the heart. (Section 37.2)

artificial selection: process of breeding organisms with specific traits in order to produce offspring with those same traits. (Section 15.1)

ascospores: sexual spores of ascomycete fungi that develop within an ascus. (Section 20.2)

ascus: tiny, saclike structures in ascomycetes in which ascospores develop. (Section 20.2)

asexual reproduction: type of reproduction where one parent produces one or more identical offspring without the fusion of gametes. (Section 19.1)

atoms: smallest particle of an element that has the characteristics of that element; basic building block of all matter. (Section 6.1)

ATP (adenosine triphosphate): energy-storing molecule in cells composed of an adenosine molecule, a ribose sugar and three phosphate groups; energy is stored in the molecule's chemical bonds and can be used quickly and easily by cells. (Section 9.1)

atria: two upper chambers of the mammalian heart through which blood enters. (Section 37.2)

australopithecine (ah stra loh PIH thuh sine): early African hominid, genus *Australopithecus*, that had both apelike and humanlike qualities. (Section 16.2)

automatic nervous system (ANS): in humans, portion of the peripheral nervous system that carries impulses from the central nervous system to internal organs; produces involuntary responses. (Section 36.1)

autosomes: pairs of matching homologous chromosomes in somatic cells. (Section 12.2)

autotrophs (AWT uh trohfs): organisms that use energy from the sun or energy stored in chemical compounds to manufacture their own nutrients. (Section 2.2)

auxins (AWK sunz): group of plant hormones that promote stem elongation. (Section 23.3)

axial skeleton: one of two main parts of the human skeleton, includes the skull and the bones that support it, such as the vertebral column, ribs, and sternum. (Section 34.2)

axon: a single cytoplasmic extension of a neuron; carries impulses away from a nerve cell. (Section 36.1)

B

bacteriophage (bak TIHR ee uh fayj): also called phages, viruses that infect and destroy bacteria. (Section 18.1)

base: any substance that forms hydroxide ions (OH⁻) in water and has a pH above 7. (Section 6.1)

basidia (buh SIHD ee uh): club-shaped hyphae of basidiomycete fungi that produce spores. (Section 20.2)

basidiospores: spores produced in the basidia of basidiomycetes during sexual reproduction. (Section 20.2)

behavior: anything an animal does in response to a stimulus in its environment. (Section 33.1)

biennial: anthophyte that flowers only after two years of growth. (Section 22.3)

bilateral symmetry (bi LAT uh rul): animals with a body plan that can be divided down its length into two similar right and left halves that form mirror images of each other. (Section 25.2)

bile: chemical substance produced by the liver and stored in the gallbladder that helps break down fats during digestion. (Section 35.1)

binary fission: asexual reproductive process in which one cell divides into two separate genetically identical cells. (Section 18.2)

binomial nomenclature: two-word system developed by Carolus Linnaeus to name species; first word identifies the genus of the organism, the second word is often a descriptive word that describes a characteristic of the organism. (Section 17.1)

biodiversity: variety of life in an area; usually measured as the number of species that live in an area. (Section 5.1)

biogenesis (bi oh JEN uh sus): idea that living organisms come only from other living organisms. (Section 14.2)

biology: the study of life that seeks to provide an understanding of the natural world. (Section 1.1)

biome: group of ecosystems with the same climax communities; biomes on land are called terrestrial biomes, those in water are called aquatic biomes. (Section 3.2)

biosphere (BI uh sfihr): portion of Earth that supports life; extends from the atmosphere to the bottom of the oceans. (Section 2.1)

biotic factors (by AHT ihk): all the living organisms that inhabit an environment. (Section 2.1)

bipedal: ability to walk on two legs; leaves arms and hands free for other activities such as hunting, protecting young, and using tools. (Section 16.2)

blastula (BLAS chuh luh): hollow ball of cells in a layer surrounding a fluid-filled space; an animal embryo after cleavage but before the formation of the gastrula. (Section 25.1)

blood pressure: force that blood exerts on blood vessels; rises and falls as the heart contracts and relaxes. (Section 37.2)

book lungs: gas exchange system found in some arthropods where air-filled chambers have plates of folded membranes that increase the surface area of tissue exposed to the air. (Section 28.1)

budding: type of asexual reproduction in unicellular yeasts and some other organisms in which a cell or group of cells pinch off from the parent to form a new individual. (Section 20.1)

bulbourethral glands (bul boh yoo REE thrul): glands located beneath the prostrate that secrete a clear, sticky, alkaline fluid that protects sperm by neutralizing the acidic environment of the vagina. (Section 38.1)

bursa: fluid-filled sac located between the bones that absorb shock and keep bones from rubbing against each other. (Section 34.2)

C

Calorie: unit of heat used to measure the energy content of food, each Calorie represents a kilocalorie, or 1000 calories; the amount of heat required to raise the temperature of 1 mL of water by 1°C. (Section 35.2)

Calvin cycle: series of reactions during the light-independent phase of photosynthesis in which simple sugars are formed from carbon dioxide using ATP

GLOSSARY

and hydrogen from the light-dependent reactions. (Section 9.2)

camouflage (KAM uh flahj): structural adaptation that enables species to blend with their surroundings; allows a species to avoid detection by predators. (Section 15.1)

cancer: uncontrolled cell division that may be caused by environmental factors and/or changes in enzyme production in the cell cycle. (Section 8.3)

capillaries (KAP ul ler eez): microscopic blood vessels with walls only one cell thick that allow diffusion of gases and nutrients between the blood and surrounding tissues. (Section 37.2)

capsid: outer coat of proteins that surrounds a virus's inner core of nucleic acid; arrangement of capsid proteins determines the virus's shape. (Section 18.1)

captivity: when members of a species are held by people in zoos or other conservation facilities. (Section 5.2)

carbohydrate (car boh HI drayt): organic compound used by cells to store and release energy; composed of carbon, hydrogen, and oxygen. (Section 6.3)

cardiac muscle: type of involuntary muscle found only in the heart; composed of interconnected cardiac muscle fibers; adapted to generate and conduct electrical impulses for muscle contraction. (Section 34.3)

carrier: an individual heterozygous for a specific trait. (Section 12.1)

carrying capacity: number of organisms of one species that an environment can support; populations below carrying capacity tend to increase; those above carrying capacity tend to decrease. (Section 4.1)

cartilage: tough flexible material making up the skeletons of agnathans, sharks, and their relatives, as well as portions of bony-animal skeletons. (Section 30.1)

cell: basic unit of all organisms; all living things are composed of cells. (Section 7.1)

cell cycle: continuous sequence of growth (interphase) and division (mitosis) in a cell. (Section 8.2)

cell theory: the theory that (1) all organisms are composed of cells, (2) the cell is the basic unit of organization of organisms, (3) all cells come from preexisting cells. (Section 7.1)

cell wall: firm, fairly rigid structure located outside the plasma membrane of plants, fungi, most bacteria, and some protists; provides support and protection. (Section 7.3)

cellular respiration: chemical process where mitochondria break down food molecules to produce ATP; the three stages of cellular respiration are glycolysis, the citric acid cycle, and the electron transport chain. (Section 9.3)

central nervous system (CNS): in humans, the central control center of the nervous system made up of the brain and spinal cord. (Section 36.1)

centrioles (SEN tree ohlz): in animal cells, a pair of small cylindrical structures composed of microtubules that duplicate during interphase and move to opposite ends of the cell during prophase. (Section 8.2)

centromere (SEN truh meer): cell structure that joins two sister chromatids of a chromosome. (Section 8.2)

cephalothorax (sef uh luh THOR aks): structure in some arthropods formed by the fusion of the head and thorax. (Section 28.1)

cerebellum (ser uh BEL um): rear portion of the brain; controls balance, posture and coordination. (Section 36.1)

cerebrum (suh REE brum): largest part of the brain, composed of two hemispheres connected by bundles of nerves; controls conscious activities, intelligence, memory, language, skeletal muscle movements, and the senses. (Section 36.1)

cervix: lower end of the uterus that tapers to a narrow opening into the vagina. (Section 38.1)

chelicerae (kih LIHS uh ree): first pair of an arachnid's six pairs of appendages; located near the mouth, they are often modified into pincers or fangs. (Section 28.2)

chemosynthesis (kee moh SIHN thuh sus): autotrophic process where organisms obtain energy from the breakdown of inorganic compounds containing sulfur and nitrogen. (Section 18.2)

chitin (KITE un): complex carbohydrate that makes up the cell walls of fungi. (Section 20.1)

chlorophyll: light-absorbing pigment in plants and some protists that is required for photosynthesis; absorbs most wavelengths of light except for green. (Section 9.2)

chloroplasts: chlorophyll-containing cell organelles found in the cells of green plants and some protists; capture light energy from the sun, which is converted to chemical energy in food molecules. (Section 7.3)

chromatin (KROH muh tihn): long, tangled strands of DNA found in the eukaryotic cell nucleus during interphase. (Section 7.3)

chromosomal mutations: mutation that occurs at the chromosome level resulting in changes in the gene distribution to gametes during meiosis; caused when parts of chromosomes break off or rejoin incorrectly. (Section 11.3)

chromosomes (KROH muh sohmz): cell structures that carry the genetic material that is copied and passed from generation to generation of cells. (Section 8.2)

cilia (SIH lee uh): short, numerous, hairlike projections composed of pairs of microtubles; frequently move in a wavelike motion; aid in feeding and locomotion. (Section 7.3)

ciliates: group of protozoans of the phylum Ciliophora that have a covering of cilia that aids in locomotion. (Section 19.1)

circadian rhythm (sur KAYD ee uhn): innate behavior based on the 24-hour cycle of the day; may determine when an animal eats, sleeps, and wakes. (Section 33.1)

citric acid cycle: in cellular respiration, series of chemical reactions that break down glucose and produce ATP; energizes electron carriers that pass the energized electrons on to the electron transport chain. (Section 9.3)

cladistics (kla DIHS tiks): biological classification system based on phylogeny; assumes that as groups of organisms diverge and evolve from a common ancestral group, they retain derived traits. (Section 17.2)

cladogram (KLAY duh gram): branching diagram that models the phylogeny of a species based on the derived traits of a group of organisms. (Section 17.2)

class: taxonomic grouping of similar orders. (Section 17.1)

classification: grouping of objects or information based on similarities. (Section 17.1)

climax community: a stable, mature community that undergoes little or no change in species over time. (Section 3.1)

clones: genetically identical copies of an organism. (Section 13.2)

closed circulatory system: system in which blood moves through the body enclosed entirely in a series of blood vessels; provides an efficient means of gas exchange within the body. (Section 27.1)

cochlea (KAWK lee uh): snail-shaped, structure in the inner ear containing fluid and hairs; produces electrical impulses that the brain interprets as sound. (Section 36.2)

codominant alleles: pattern where the phenotypes of both homozygote parents are produced in heterozygous offspring so that both alleles are equally expressed. (Section 12.2)

codon: set of three nitrogen bases that represents an amino acid; order of nitrogen bases in mRNA determines the type and order of amino acids in a protein. (Section 11.2)

coelom (SEE lohm): fluid-filled body cavity completely surrounded by mesoderm; provides space for the development and suspension of organs and organ systems. (Section 25.2)

collenchyma (coh LENG kuh muh): long, flexible plant cells with unevenly thickened cell walls; most common in actively growing tissues. (Section 23.1)

colony: group of unicellular or multicellular organisms that live together in a close association. (Section 19.2)

commensalism (kuh MEN suh lihz um): symbiotic relationship in which one species benefits and the other species is neither harmed nor benefited. (Section 2.1)

communication: exchange of information that results in a change of behavior. (Section 33.2)

community: collection of several interacting populations that inhabit a common environment. (Section 2.1)

compact bone: layer of protective hard bone tissue surrounding every bone; composed of repeating units of osteon systems. (Section 34.2)

companion cells: nucleated cells that help manage the transport of sugars and other organic compounds through the sieve cells of the phloem. (Section 23.1)

compound: substance composed of atoms of two or more different elements that are chemically combined. (Section 6.1)

compound eye: in arthropods, a visual system composed of multiple lenses; each lens registers light from a small portion of the field of view, creating an image composed of thousands of parts. (Section 28.1)

compound light microscope: instrument that uses visible light and a series of lenses to magnify objects in steps; can magnify an object up to 1500 times its original size. (Section 7.1)

conditioning: response to a stimulus learned by association with a specific action. (Section 33.2)

cones: scaly structures produced by some seed plants that support male or female reproductive structures and are the sites of seed production. (Section 21.2); receptor cells in the retina adapted for sharp vision in bright light and color detection. (Section 36.2)

conidia (kuh NIHD ee uh): chains or clusters of asexual ascomycete spores that develop on the tips of conidiophores. (Section 20.2)

conidiophores (kuh NIHD ee uh forz): in ascomycetes, elongated, upright hyphae that produce conidia at their tips. (Section 20.2)

conjugation (kahn juh GAY shun): form of sexual reproduction in some bacteria where one bacterium transfers all or part of its genetic material to another through a bridgelike structure called a pilus. (Section 18.2)

conservation biology: field of biology that studies methods and implements plans to protect biodiversity. (Section 5.2)

control: in an experiment, the standard in which all of the conditions are kept the same. (Section 1.2)

convergent evolution: evolution in which distantly related organisms evolve similar traits; occurs when unrelated species occupy similar environments. (Section 15.2)

cork cambium: lateral meristem that produces a tough protective covering for the surface of stems and roots. (Section 23.1)

corpus luteum: part of an ovarian follicle that remains in the ovary after ovulation; produces estrogen and progesterone. (Section 38.1)

cortex: layer of ground tissue in the root that is involved in the transport of water and ions into the vascular core at the center of the root. (Section 23.2)

cotyledons (kah tuh LEE dunz): structure of seed plant embryo that stores or absorbs food for the developing embryo; may become the plant' first leaves when the plant emerges from the soil. (Section 22.3)

courtship behavior: an instinctive behavior that males and females of a species carry out before mating. (Section 33.1)

covalent bond (koh VAY lunt): chemical bond formed when two atoms combine by sharing electrons. (Section 6.1)

Cro-Magnon: modern form of *Homo sapiens* that spread throughout Europe between 35 000 to 40 000 years ago; were identical to modern humans in height, skull and tooth structure, and brain size. (Section 16.2)

crossing over: exchange of genetic material between nonsister chromatids from homologous chromosomes during prophase I of meiosis; results in new allele combinations. (Section 10.2)

cuticle (KYEWT ih kul): protective, waxy coating on the outer surface of the epidermis of most stems, leaves, and fruits; important adaptation in reducing water loss. (Section 21.1)

cytokinesis (site uh kih NEE sus): cell process following meiosis or mitosis in which the cell's cytoplasm divides and separates into new cells. (Section 8.2)

cytokinins: group of hormones that stimulate cell division. (Section 23.3)

cytoplasm: clear, gelatinous fluid in eukaryotic cells that suspends the cell's organelles and is the site of numerous chemical reactions. (Section 7.3)

cytoskeleton: cellular framework found within the cytoplasm composed of microtubules and microfilaments. (Section 7.3)

data: information obtained from experiments, sometimes called experimental results. (Section 1.2)

day-neutral plants: plants in which temperature, moisture, or factors other than day length induce flowering. (Section 24.2)

deciduous plants: plants that lose all of their leaves at the same time; an adaptation for reducing water loss when water is unavailable. (Section 22.3)

decomposers: organisms, such as fungi, that break down and absorb nutrients from dead organisms. (Section 2.2)

demography (dem AW graf ee): study of population characteristics such as growth rate, age structure, and geographic distribution. (Section 4.2)

dendrite (DEN drite): branchlike cytoplasmic extension of a neuron; transports impulses toward the cell body. (Section 36.1)

density-dependent factor: factor such as disease, parasites, or food supply that has an increasing effect as populations increase. (Section 4.1)

density-independent factor: factor such as temperature, storms, floods, drought, or habitat disruption that affects all populations, regardless of their density. (Section 4.1)

dependent variable: in an experiment, the condition that results from changes in the independent variable. (Section 1.2)

depressant: type of drug that lowers or depresses the activity of the nervous system. (Section 36.3)

dermis (DUR mus): inner, thicker portion of the skin that contains structures such as blood vessels, nerves, nerve endings, hair follicles, sweat glands, and oil glands. (Section 34.1)

desert: arid region with sparse to almost nonexistent plant life; the driest biome, it gets less than 25 cm of precipitation annually. (Section 3.2)

deuterostome (DEW tihr uh stohm): animal whose mouth develops from cells other than the those at the opening of the gastrula. (Section 25.1)

development: all of the changes that take place during the life of an organism; a characteristic of all living things. (Section 1.1)

diaphragm (DI uh fram): in mammals, the sheet of muscles located beneath the lungs that separates the chest cavity from the abdominal cavity; expands and contracts the chest cavity, which increases the amount of oxygen entering the body. (Section 32.1)

dicotyledon: class of anthophytes that have two cotyledons, reticulate leaf venation, and flower parts in multiples of four or five. (Section 22.3)

diffusion: net, random movement of particles from an area of higher concentration to an area of lower concentration, eventually resulting in even distribution. (Section 6.2)

diploid: cell with two of each kind of chromosome; is said to contain a diploid, or $2n$, number of chromosomes. (Section 10.2)

directional selection: natural selection that favors extreme variations of a trait; can lead to rapid evolution in a population. (Section 15.2)

disruptive selection: natural selection that favors individuals with either extreme of a trait; tends to eliminate intermediate phenotypes. (Section 15.2)

divergent evolution: evolution in which species that once were similar to an ancestral species diverge; occurs when populations adapt to different environmental conditions; may result in the formation of a new species. (Section 15.2)

division: taxonomic grouping of similar classes; term used instead of phyla by plant taxonomists. (Section 17.1)

DNA replication: process in which chromosomal DNA is copied before mitosis or meiosis. (Section 11.1)

dominance hierarchy (DAHM uh nunts • HI rar kee): innate behavior by which animals form a social ranking within a group in which some individuals are more subordinate than others; usually has one top-ranking individual. (Section 33.1)

dominant: observed trait of an organism that masks the recessive form of a trait. (Section 10.1)

dormancy: period of inactivity in a mature seed prior to germination; seed remains dormant until conditions are favorable for growth and development of the new plant. (Section 24.3)

dorsal (DOR sul): back surface of bilaterally symmetric animals. (Section 25.2)

dorsal hollow nerve cord: nerve cord found in all chordates that forms the spinal cord and brain. (Section 29.2)

double fertilization: fertilization process unique to anthophytes in which one sperm fertilizes the haploid egg and the other sperm joins with the diploid central cell; results in the formation of a diploid ($2n$) zygote and a triploid ($3n$) endosperm. (Section 24.3)

double helix: shape of a DNA molecule formed when two twisted DNA strands are coiled into a springlike structure and held together by hydrogen bonds between the bases. (Section 11.1)

drug: chemical substance that affects body functions. (Section 36.3)

dynamic equilibrium: result of diffusion where there is continuous movement of particles but no overall change in concentration. (Section 6.2)

ecology: scientific study of interactions among organisms and their environments. (Section 2.1)

ecosystem: interactions among populations in a community; the community's physical surroundings, or abiotic factors. (Section 2.1)

ectoderm: layer of cells on the outer surface of the gastrula; eventually develops into the skin and nervous tissue of an animal. (Section 25.1)

ectotherm (EK tuh thurm): animal whose body temperature changes with the temperature of its surroundings. (Section 30.2)

edge effect: different environmental conditions that occur along the boundaries of an ecosystem. (Section 5.1)

egg: haploid female sex cell produced by meiosis. (Section 10.2)

electron microscope: instrument that uses a beam of electrons instead of natural light to magnify structures up to 500 000 times actual size; allows scientists to view structures within a cell. (Section 7.1)

electron transport chain: series of proteins embedded in a membrane along which energized electrons are transported; as electrons are passed from molecule to molecule, energy is released. (Section 9.2)

element: substance that that can't be broken down into simpler chemical substances. (Section 6.1)

embryo: earliest stage of growth and development of both plants and animals; differences and similarities among embryos can provide evidence of evolution. (Section 15.1); the young diploid sporophyte of a plant. (Section 22.3)

emigration: movement of individuals from a population. (Section 4.2)

endangered species: a species in which the number of individuals falls so low that extinction is possible. (Section 5.1)

endemic disease: diseases that are constantly present in the population. (Section 39.1)

endocrine glands: series of ductless glands that make up the endocrine system; release chemicals directly into the bloodstream where they relay messages to other parts of the body. (Section 35.3)

endocytosis (en doh si TOH sus): active transport process where a cell engulfs materials with a portion of the cell's plasma membrane and releases the contents inside of the cell. (Section 8.1)

endoderm: layer of cells on the inner surface of the gastrula; will eventually develop into the lining of the animal's digestive tract and organs associated with digestion. (Section 25.1)

endodermis: single layer of cells that forms a waterproof seal around a root's vascular tissue; controls the flow of water and dissolved ions into the root. (Section 23.2)

endoplasmic reticulum (ER): organelle in eukaryotic cells with a series of highly folded membranes surrounded in cytoplasm; site of cellular chemical reactions; can either be rough (with ribosomes) or smooth (without ribosomes). (Section 7.3)

endoskeleton: internal skeleton of vertebrates and some invertebrates; provides support, protects internal organs, and acts as an internal brace for muscles to pull against. (Section 25.2)

endosperm: triploid ($3n$) nucleus formed during double fertilization in anthophytes; forms a food storage tissue that supports development of the growing embryo. (Section 24.3)

endospore: structure formed by bacteria during unfavorable conditions that contains DNA and a small amount of cytoplasm encased by a protective outer covering; germinates during favorable conditions. (Section 18.2)

endotherm: animal that maintains a constant body temperature and is not dependent on the environmental temperature. (Section 31.2)

energy: the ability to do work; organisms use energy to perform biological functions. (Section 1.1)

environment: biotic and abiotic surroundings to which an organism must constantly adjust; includes air, water, weather, temperature, other organisms, and many other factors. (Section 1.1)

enzymes: type of protein found in all living things that increases the rate of chemical reactions. (Section 6.3)

epidemic: occurs when many people in a given area are afflicted with the same disease at about the same time. (Section 39.1)

epidermis: in plants, the outmost layer of cells that covers and protects all parts of the plant. (Section 23.1); in humans and some other animals, the outermost protective layer composed of an outer layer of dead cells and an inner layer of living cells. (Section 34.1)

epididymis (ep uh DIHD uh mus): in human males, the coiled tube within the scrotum in which the sperm complete maturation. (Section 38.1)

epiglottis (ep uh GLAHT us): flap of cartilage that closes over the opening of the respiratory tract during swallowing; prevents food from entering the respiratory tract. (Section 35.1)

esophagus: muscular tube that connects the mouth to the stomach; moves food by peristalsis. (Section 35.1)

estivation (es tuh VAY shun): state of reduced metabolism that occurs in animals living in conditions of intense heat. (Section 33.1)

estuary (ES chuh wer ee): coastal body of water, partially surrounded by land, in which freshwater and salt water mix. (Section 3.2)

ethics: the moral principles and values held by humans. (Section 1.3)

ethylene (ETH uh leen): plant hormone that speeds the ripening of fruits. (Section 23.3)

eubacteria (yew bak TEER ee uh): group of prokaryotes with strong cell walls and a variety of structures, may be autotrophs (chemosynthetic or photosynthetic) or heterotrophs. (Section 17.2)

eukaryotes: unicellular or multicellular organisms, such as yeast, plants, and animals, composed of eukaryotic cells, which contain a true nucleus and membrane-bound organelles. (Section 7.1)

evolution (ev uh LEW shun): gradual accumulation of adaptations over time. (Section 1.1)

exocytosis: active transport process by which materials are secreted or expelled from a cell. (Section 8.1)

exoskeleton: hard, waxy coating on the outside of some animals, including spiders and mollusks; provides a framework for support, protects soft body tissues, and provides a place for muscle attachment. (Section 25.2)

exotic species: nonnative species in an area; may take over niches of native species in an area and eventually replace them. (Section 5.1)

experiment: procedure that tests a hypothesis by collecting information under controlled conditions. (Section 1.2)

exponential growth: population growth pattern where a population grows faster as it increases in size; graph of a growing population resembles a J-shaped curve. (Section 4.1)

external fertilization: type of fertilization in which eggs and sperm are both released into water; fertilization occurs outside the animal's body. (Section 26.1)

extinction (ek STINGK shun): when the last members of a species die. (Section 5.1)

facilitated diffusion: passive transport of materials across a plasma membrane by transport proteins embedded in the plasma membrane. (Section 8.1)

feather: lightweight, modified scale found only on birds; provide insulation and enable flight. (Section 31.2)

fertilization: fusion of male and female gametes. (Section 10.1)

fetus: a developing mammal from nine weeks to birth. (Section 12.1)

fight-or-flight response: automatic response controlled by hormones that prepares the body to either fight or run from danger. (Section 33.1)

filter feeding: method in which food particles are filtered from water as it passes by or through some part of the organism. (Section 26.1)

fins: in fishes, fan-shaped membranes used for balance, swimming, and steering. (Section 30.1)

flagella (fluh JEL uh): long, hairlike projections composed of pairs of microtubules; found on some cell surfaces; they help propel cells and organisms by a whiplike motion. (Section 7.3)

fluid mosaic model: structural model of the plasma membrane where molecules are free to move sideways within a lipid bilayer. (Section 7.2)

follicle: in human females, group of epithelial cells that surround a developing egg cell. (Section 38.1)

food chain: simple model that shows how matter and energy move through an ecosystem; can consist of three steps, but must have no more than five steps. (Section 2.2)

food web: model that expresses all the possible feeding relationships at each trophic level in a community. (Section 2.2)

fossil: physical evidence of an organism that lived long ago that scientists use to study the past; evidence may appear in rocks, amber, or ice. (Section 14.1)

fragmentation: type of asexual reproduction in algae where an individual breaks into pieces and each piece grows into a new individual. (Section 19.2)

frameshift mutation: mutation that occurs when a single base is added or deleted from DNA; causes a shift in the reading of codons by one base. (Section 11.3)

fronds: in ferns, large, complex leaves that grow upward from the rhizome; often divided into pinnae that are attached to a central stipe. (Section 21.2)

fruit: seed-containing ripened ovary of an anthophyte flower; may be fleshy or dry. (Section 22.3)

fungus: group of unicellular or multicellular heterotrophic eukaryotes that do not move from place to place; absorb nutrients from organic materials in the environment. (Section 17.2)

gallbladder: small organ that stores bile before the bile passes into the duodenum of the small intestine. (Section 35.1)

gametangium (gam ee TAN ghee uhm): structure that contains a haploid nucleus; formed by the fusion of haploid hyphae. (Section 20.2)

gametes: male and female sex cells; sperm and eggs. (Section 10.1)

gametophyte: haploid form of an organism in alternation of generations that produces gametes. (Section 19.2)

gastrovascular cavity (gas troh VAS kyuh lur): in cnidarians, a large cavity in which digestion takes place. (Section 26.2)

gastrula (GAS truh luh): animal embryo development stage where cells on one side of the blastula fold inward forming a cavity of two or three layers of cells with an opening at one end. (Section 25.1)

gene: segment of DNA that controls the protein production and the cell cycle. (Section 8.3)

gene pool: all of the alleles in a population's genes. (Section 15.2)

gene splicing: in recombinant DNA technology, the rejoining of DNA fragments by vectors and other enzymes. (Section 13.2)

genetic counseling: counseling by trained professionals about the probabilities of hereditary disorders and what can be done if they occur. (Section 38.2)

genetic drift: alteration of allelic frequencies in a population by chance events; results in disruption of genetic equilibrium. (Section 15.2)

genetic engineering: method of cutting DNA from one organism and inserting the DNA fragment into a host organism of the same or a different species. (Section 13.2)

genetic equilibrium: condition in which the frequency of alleles in a population remains the same over generations. (Section 15.2)

genetic recombination: major source of genetic variation among organisms caused by reassortment or crossing over during meiosis. (Section 10.2)

gene therapy: insertion of normal genes into human cells to correct genetic disorders. (Section 13.3)

genetics: branch of biology that studies heredity. (Section 10.1)

genotype (JEE nuh tipe): combination of genes in an organism. (Section 10.1)

genus (JEE nus): first word of a two-part scientific name used to identify a group of similar species. (Section 17.1)

geographic isolation: occurs whenever a physical barrier divides a population, which results in individuals no longer being able to mate; can lead to the formation of a new species. (Section 15.2)

germination: beginning of the development of an embryo into a new plant. (Section 24.3)

gestation (jeh STAY shun): time during which placental mammals develop inside the uterus. (Section 32.2)

gibberellins (jihb uh REL uns): group of plant hormones that cause plants to grow taller by stimulating cell elongation. (Section 23.3)

gill slits: paired openings found behind the mouth of all chordates; in invertebrate chordates, they are used to strain food and water; in some vertebrates they are used for gas exchange and respiration. (Section 29.2)

gizzard: sac with muscular walls and hard particles that grind soil before it passes into the intestine; common in birds and annelids such as earthworms. (Section 27.2)

gland: in mammals, a cell or group of cells that secretes fluids. (Section 32.1)

glycolysis (GLI kol ih sis): in cellular respiration, series of anaerobic chemical reactions in the cytoplasm that break down glucose into pyruvic acid; forms a net profit of two ATP molecules. (Section 9.3)

Golgi apparatus (GALW jee): organelle in eukaryotic cells with a system of flattened tubular membranes; modifies proteins and sends them to their appropriate destinations. (Section 7.3)

gradualism: idea that species originate through a gradual accumulation of adaptations. (Section 15.2)

grasslands: biome composed of large communities covered with grasses and similar small plants; receives 25–75 cm of precipitation annually. (Section 3.2)

growth: increase in the amount of living material and formation of new structures in an organism; a characteristic of all living things. (Section 1.1)

guard cells: cells that control the opening and closing of the stomata; regulate the flow of water vapor from leaf tissue. (Section 23.1)

habitat (HAB uh tat): place where an organism lives out its life. (Section 2.1)

habitat corridors: natural strips of land that allow the migration of organisms from one wilderness area to another. (Section 5.2)

habitat degradation: damage to a habitat by air, water, and land pollution. (Section 5.1)

habitat fragmentation: separation of wilderness areas from each other; may cause problems for organisms that need large areas for food or mating. (Section 5.1)

habituation (huh BIT yew ay shun): learned behavior that occurs when an animal is repeatedly given a stimulus not associated with any punishment or reward. (Section 33.2)

hair follicle: narrow cavities in the dermis from which hair grows. (Section 34.1)

hallucinogen (huh LEWS un uh junz): drug that stimulates the central nervous system so that the user becomes disoriented and sees, hears, feels, tastes, or smells things that are not there. (Section 36.3)

haploid: cell with one of each kind of chromosome; is said to contain a haploid or *n*, number of chromosomes. (Section 10.2)

haustoria (huh STOR ee uh): in parasitic fungi, hyphae that grow into host cells and absorb nutrients and minerals from the host. (Section 20.1)

hemoglobin (hee muh GLOH bun): iron-containing protein molecule in red blood cells that binds to oxygen and carries it from the lungs to the body's cells. (Section 37.2)

heredity: passing on of characteristics from parents to offspring. (Section 10.1)

hermaphrodite (hur MAF ruh dite): individual animal that can produce both eggs and sperm; increases the likelihood of fertilization in sessile or slow-moving animals. (Section 26.1)

heterotrophs (HET uh ruh trohfs): organisms that cannot make their own food and must feed on other organisms for energy and nutrients. (Section 2.2)

heterozygous (het uh roh ZI gus): when there are two different alleles for a trait. (Section 10.1)

hibernation (hi bur NAY shun): state of reduced metabolism occurring in animals that sleep through cold winter conditions; an animal's temperature drops, oxygen consumption decreases, and breathing rate declines. (Section 33.1)

homeostasis (hoh mee oh STAY sus): organism's regulation of its internal environment to maintain conditions suitable for survival; a characteristic of all living things. (Section 1.1); process of maintaining equilibrium in an organism's internal environment. (Section 7.2)

hominids (HOH mihn udz): group of primates that can walk upright on two legs; includes gorillas, chimpanzees, and humans. (Section 16.2)

homologous chromosomes (huh MAHL uh gus): paired chromosomes with genes for the same traits arranged in the same order. (Section 10.2)

homologous structures: structures with common evolutionary origins; can be similar in arrangement, in function, or both; provides evidence of evolution from a common ancestor; forelimbs of crocodiles, whales, and birds are examples. (Section 15.1)

homozygous (hoh muh ZI gus): when there are two identical alleles for a trait. (Section 10.1)

hormone: chemical produced in one part of an organism and transported to another part, where it causes a physiological change. (Section 23.3)

host cell: living cell in which a virus replicates. (Section 18.1)

human genome: map of the approximately 80 000 genes on 46 human chromosomes that when mapped and sequenced, may provide information on the treatment or cure of genetic disorders. (Section 13.3)

hybrid: offspring formed by parents having different forms of a trait. (Section 10.1)

hydrogen bond: weak chemical bond formed by the attraction of positively charged hydrogen atoms to other negatively charged atoms. (Section 6.2)

hypertonic solution: in cells, solution in which the concentration of dissolved substances outside the cell is higher than the concentration inside the cell; causes a cell to shrink as water leaves the cell. (Section 8.1)

hyphae (HI fee): threadlike filaments that are the basic structural units of multicellular fungi. (Section 20.1)

hypocotyl (HI poh kaht ul): portion of the stem near the seed in young plants; in some plants, it is the first part of the stem to push above the ground. (Section 24.3)

hypothalamus: part of the brain that controls the pituitary gland by sending messages to the pituitary, which then releases its own chemicals or stimulates other glands to release chemicals. (Section 35.3)

hypothesis (hi PAHTH us sus): explanation for a question or a problem that can be formally tested. (Section 1.2)

hypotonic solution: in cells, solution in which the concentration of dissolved substances is lower in the solution outside the cell than the concentration inside the cell; causes a cell to swell and possibly burst as water enters the cell. (Section 8.1)

immigration: movement of individuals into a population. (Section 4.2)

implantation: in females, the attachment of a blastocyst to the lining of the uterus. (Section 38.2)

imprinting: learned behavior in which an animal, at a specific critical time of its life, forms a social attachment to another object; usually occurs early in life and allows an animal to recognize its mother and others of its species. (Section 33.2)

inbreeding: mating between closely related individuals; insures that offspring are homozygous for most traits, but also brings out harmful, recessive traits. (Section 13.1)

incomplete dominance: inheritance pattern where the phenotype of a heterozygote is intermediate between those of the two homozygotes; neither allele of the pair is dominant but combine and display a new trait. (Section 12.2)

incubate: process of keeping eggs laid outside of the body warm; also involves periodic turning of the eggs to ensure proper development. (Section 31.1 in reptiles; 31.2 in birds)

independent variable: in an experiment, the condition that is changed because it affects the outcome of the experiment. (Section 1.2)

infectious disease (ihn FEK shus): any disease caused by pathogens in the body. (Section 39.1)

innate behavior: an inherited, genetically based behavior in animals; includes automatic responses and instinctive behaviors. (Section 33.1)

innate immunity: body's earliest lines of defense against any and all pathogens; includes skin and body secretions, inflammation of body tissues, and phagocytosis of pathogens. (Section 39.2)

insight: type of learning in which an animal uses previous experiences to respond to a new situation. (Section 33.2)

instinct (IHN stingt): complex innate behavior pattern that begins when an animal recognizes a stimulus and performs an action until all parts of the behavior have been formed. (Section 33.1)

interferons: host-cell specific proteins that protect cells from viruses. (Section 39.2)

internal fertilization: type of fertilization in which eggs remain inside the female's body and sperm are carried to the eggs by water or other fluid. (Section 26.1)

interphase: cell growth phase where a cell increases in size, carries on metabolism, and duplicates chromosomes prior to division. (Section 8.2)

intertidal zone: portion of the shoreline that lies between high tide and low tide lines; size of zone depends on slope of the land and height of the tide. (Section 3.2)

invertebrate: animal that does not have a backbone. (Section 25.2)

involuntary muscle: muscle whose contractions are not under conscious control. (Section 34.3)

ion: atom or group of atoms that gain or lose electrons; has a positive or negative electrical charge. (Section 6.1)

ionic bond: chemical bond formed by the attractive forces between two ions of opposite charge. (Section 6.1)

isomers (I suh murz): compounds with the same simple formula but different three-dimensional structures resulting in different physical and chemical properties. (Section 6.3)

isotonic solution: in cells, solution in which the concentration of dissolved substance in the solution is the same as the concentration of dissolved substances inside a cell. (Section 8.1)

isotopes (I suh tohps): atoms of the same element that have different numbers of neutrons in the nucleus. (Section 6.1)

Jacobson's organ: in snakes, a pitlike sense organ on the roof of the mouth that picks up and analyzes airborne chemicals. (Section 31.1)

joints: point where two or more bones meet; can be fixed or facilitate movement of bones in relation to one another. (Section 34.2)

karyotype (KAYR ee uh tipe): chart of metaphase chromosome pairs arranged according to length and location of the centromere; used to pinpoint unusual chromosome numbers in cells. (Section 12.3)

keratin (KER uh tun): protein found in the exterior portion of the epidermis that helps protect living cells in the interior epidermis and contributes to the skin's elasticity. (Section 34.1)

kidneys: organs of the vertebrate urinary system; remove wastes, control sodium levels of the blood, and regulate blood pH levels. (Section 37.3)

kingdom: taxonomic grouping of similar phyla or divisions. (Section 17.1)

Koch's postulates: experimental steps relating a specific pathogen to a specific disease. (Section 39.1)

labor: physiological and physical changes a female goes through during the birthing process. (Section 38.3)

lactic acid fermentation: series of anaerobic chemical reactions in which pyruvic acid uses NADH to form lactic acid and NAD$^+$, which is then used in glycolysis; supplies energy when oxygen for aerobic respiration is scarce. (Section 9.3)

language: use of symbols to represent ideas; usually present in animals with complex nervous systems, memory, and insight. (Section 33.2)

large intestine: muscular tube through which indigestible materials are passed to the rectum for excretion. (Section 35.1)

larva: in insects, the freeliving, wormlike stage of metamorphosis, often called a caterpillar. (Section 28.2)

lateral line system: line of fluid-filled canals running along the sides of a fish that enable the fish to detect movement and vibrations in the water. (Section 30.1)

law of independent assortment: Mendelian principle stating that genes for different traits are inherited independently of each other. (Section 10.1)

law of segregation: Mendelian principle explaining that because each plant has two different alleles, it can produce two different types of gametes. During fertilization, male and female gametes randomly pair to produce four combinations of alleles. (Section 10.1)

leaf: broad, flat plant organ supported by the stem that grows upward toward sunlight and traps light energy for photosynthesis. (Section 21.1)

lichen (LI kun): organism formed from a symbiotic association between a fungus, usually an ascomycete, and a photosynthetic green alga or cyanobacteria. (Section 20.2)

ligament: tough band of connective tissue that attaches one bone to another; joints are often held together and enclosed by ligaments. (Section 34.2)

light-dependent reactions: phase of photosynthesis where light energy is converted to chemical energy in the form of ATP; results in the splitting of water and the release of oxygen. (Section 9.2)

light-independent reactions: phase of photosynthesis where energy from light-dependent reactions is used to produce glucose and additional ATP molecules. (Section 9.2)

limiting factor: any biotic or abiotic factor that restricts the existence, numbers, reproduction, or distribution of organisms. (Section 3.1)

linkage map: genetic map that shows the location of genes on a chromosome. (Section 13.3)

lipids: organic compounds commonly called fats and oils; are insoluble in water and used by cells for long-term energy storage, insulation, and protective coatings, such as in membranes. (Section 6.3)

liver: large, complex organ of the digestive system that produces many chemicals for digestion, including bile. (Section 35.1)

long-day plants: plants that are induced to flower by exposure to short nights. (Section 24.2)

lymph: tissue fluids composed of water and dissolved substances from the blood that have collected and entered the lymph vessels. (Section 39.2)

lymph node: small mass of tissue that contains lymphocytes and filters pathogens from the lymph; made of a network of connective tissue fibers that contain lymphocytes. (Section 39.2)

lymphocyte (LIHM fuh site): type of white blood cell stored in lymph nodes that defends the body against foreign agents. (Section 39.2)

lysogenic cycle: viral replication cycle in which the virus's nucleic acid is integrated into the host cell's chromosome; a provirus is formed and replicated each time the host cell reproduces; the host cell is not killed until the lytic cycle is activated. (Section 18.1)

lysosomes: organelles that contain digestive enzymes; digest excess or worn out organelles, food particles, and engulfed viruses or bacteria. (Section 7.3)

lytic cycle (LIH tik): viral replication cycle in which a virus takes over a host cell's genetic material and uses the host cell's structures and energy to replicate until the host cell bursts, killing it. (Section 18.1)

macrophages: type of phagocyte that engulfs damaged cells or pathogens that have entered the body's tissues. (Section 39.2)

madreporite (mah DREH pohr ite): in echinoderms, the sievelike, disk-shaped opening through which water flows in and out of the water vascular system; helps filter out large particles from entering the body. (Section 29.1)

Malpighian tubules (mal PIGH ee un): in arthropods, tubules located in the abdomen that are attached to and empty waste into the intestine. (Section 28.1)

mammary gland: milk-producing glands in mammals; mammals nurse their young until they are mature enough to find food for themselves. (Section 32.1)

mandibles (MAND uh bulz): in most arthropods, mouthparts adapted for holding, chewing, sucking, or biting various foods. (Section 28.1)

mantle (MANT uhl): thin membrane that surrounds the internal organs of mollusks; in mollusks with shells, it secretes the shell. (Section 27.1)

marsupial (mar SEW pee uhl): subclass of mammals in which young develop for a short time in the uterus and complete their development outside of the mother's body inside a pouch made of skin and hair. (Section 32.2)

medulla oblongata (muh DUL uh • ahb long GAHT uh): part of the brain stem that controls involuntary activities such as breathing and heart rate. (Section 36.1)

megaspore: large haploid spore formed by some plants that develops into a female gametophyte. (Section 24.1)

meiosis (mi OH sus): type of cell division where one body cell produces four gametes, each containing half the number of chromosomes as a parent's body cell. (Section 10.2)

melanin: pigment found in cells of the interior layer of the epidermis; protects cells from solar-radiation damage. (Section 34.1)

menstrual cycle: in human females, the monthly cycle that includes the production of an egg, the preparation of the uterus to receive an egg, and the shedding of an egg if it remains unfertilized. (Section 38.1)

meristems: regions of actively dividing cells in plants. (Section 23.1)

mesoderm (MEZ uh durm): middle cell layer in the gastrula, between the ectoderm and the endoderm; develops into the muscles, circulatory system, excretory system, and in some animals, the respiratory system. (Section 25.1)

mesophyll (MEZ uh fihl): chlorophyll-rich photosynthetic tissue of a leaf; may have a palisade or spongy form. (Section 23.2)

messenger RNA (mRNA): RNA that transports information from DNA in the nucleus to the cell's cytoplasm. (Section 11.2)

metabolism: all of the chemical reactions that occur within an organism. (Section 6.1)

metamorphosis (met uh MOR fuh sus): in insects, series of chemically-controlled changes in body structure from juvenile to adult; may be complete or incomplete. (Section 28.2)

metaphase: short second phase of mitosis where doubled chromosomes move to the equator of the spindle and chromatids are attached by centromeres to a separate spindle fiber. (Section 8.2)

microfilaments: thin, solid protein fibers that provide structural support for eukaryotic cells. (Section 7.3)

micropyle (MI kruh pile): in many seed plants, the opening in the ovule of a female gametophyte through which the pollen grain enters. (Section 24.1)

microspore: small haploid spore formed by some plants that develops into a male gametophyte. (Section 24.1)

microtubules: thin, hollow cylinders made of protein that provide structural support for eukaryotic cells. (Section 7.3)

migration: instinctive seasonal movements of animals from place to place. (Section 33.1)

mimicry: structural adaptation evolved in some species where one species resembles another; may provide protection from predators or other advantages. (Section 15.1)

minerals: inorganic substances that are important for chemical reactions or as building materials in the body. (Section 35.2)

mitochondria: eukaryotic membrane-bound organelles that transform energy stored in food molecules into ATP; has a highly folded inner membrane that produces energy-storing molecules. (Section 7.3)

mitosis (mi TOH sus): period of nuclear cell division in which two daughter cells are formed, each containing a complete set of chromosomes. (Section 8.2)

mixture: combination of substances in which individual components do not combine chemically and retain their own properties. (Section 6.1)

molecule: group of atoms held together by covalent bonds; has no overall charge. (Section 6.1)

molting: in arthropods, the periodic shedding of an old exoskeleton. (Section 28.1)

monocotyledons: class of anthophytes that have one cotyledon, parallel leaf venation, and flower parts in multiples of three. (Section 22.3)

monotreme (MAHN uh treem): subclass of mammals that have hair and mammary glands but reproduce by laying eggs. (Section 32.2)

motivation: internal need that causes an animal to act and that is necessary for learning to take place; often involves hunger or thirst. (Section 33.2)

multiple alleles: presence of more that two alleles for a genetic trait. (Section 12.2)

mutagen (MYEWT uh jun): any agent that can cause a change in DNA; includes high-energy radiation, chemicals, or high temperatures. (Section 11.3)

mutation: any change or random error in a DNA sequence. (Section 11.3)

mutualism (MYEW chuh lihz um): a symbiotic relationship in which both species benefit. (Section 2.1)

mycelium (mi SEE lee um): in fungi, a complex network of branching hyphae; may serve to anchor the fungus, invade food sources, or form reproductive structures. (Section 20.1)

mycorrhiza (my kuh RHY zuh): mutualistic relationship in which a fungus lives symbiotically with a plant. (Section 20.2)

myofibril (mi oh FIBE rul): unit of muscle fibers composed of thick myosin protein filaments and thin actin protein filaments. (Section 34.3)

myosin: structural protein that makes up the thick filaments of myofibrils; functions in muscle contraction. (Section 34.3)

NADP⁺ (nicotinamide adenine dinucleotide phosphate): electron carrier molecule; when carrying excited electrons it becomes NADPH. (Section 9.2)

narcotic: type of pain-relief drug that affects the central nervous system. (Section 36.3)

nastic movement: temporary, responsive movement of a plant not dependent on the direction of the stimulus. (Section 23.3)

natural selection: mechanism for change in populations; occurs when organisms with certain variations survive, reproduce, and pass their variations to the next generation; can be directional, disruptive, or stabilizing. (Section 15.1)

Neanderthal: archaic *Homo sapiens* that lived from 35 000 to 100 000 years ago in Europe, Asia, and the Middle East; had thick bones and large faces with prominent noses and brains at least as large as those of modern humans. (Section 16.2)

negative feedback system: system in which a substance is fed back to inhibit the original signal and reduce production of a substance; examples include the endocrine system and a thermostat. (Section 35.3)

nematocyst (nih MAT uh sihst): in cnidarians, a capsule that contains a coiled, threadlike tube that may be sticky, barbed, or contain poisons; used in capturing and digesting prey. (Section 26.2)

nephridia (nih FRIHD ee uh): organs that remove metabolic wastes from an animal's body. (Section 27.1)

nephron: individual filtering unit of the kidneys. (Section 37.3)

nerve net: simple netlike nervous system in cnidarians that conducts nerve impulses from all parts of the cnidarian's body; impulses are not controlled by a central nervous center. (Section 26.2)

neurons (NYU ronz): basic unit of structure and function in the nervous system; conducts impulses throughout the nervous system; composed of dendrites, a cell body, and an axon. (Section 36.1)

neurotransmitters: chemicals released from an axon that diffuse across a synapse to the next neuron's dendrites to initiate a new impulse. (Section 36.1)

niche (nich): role and position a species has in its environment; includes all biotic and abiotic interactions as an animal meets its needs for survival and reproduction. (Section 2.1)

nitrogen base: carbon ring structure found in DNA or RNA that contains one or more atoms of nitrogen; includes adenine, guanine, cytosine, thymine, and uracil. (Section 11.1)

nitrogen fixation: metabolic process in which bacteria use enzymes to convert atmospheric nitrogen (N_2) into ammonia (NH_3). (Section 18.2)

nondisjunction: failure of homologous chromosomes to separate properly during meiosis; results in gametes with too many or too few chromosomes. (Section 10.2)

nonvascular plants: plants that do not have vascular tissues but rely on the relatively slow processes of osmosis and diffusion to transport water, food, and other materials throughout the plant. (Section 21.1)

notochord (NOHT uh kord): long, semirigid, rodlike structure found in all chordate embryos that is located between the digestive system and the dorsal hollow nerve cord. (Section 29.2)

nucleic acid (noo KLAY ihk): complex macromolecules, such as RNA and DNA, that store genetic information in cells in the form of a code (Section 6.3)

nucleolus (noo klee OH lus): organelle in eukaryotic cell nucleus that produces ribosomes. (Section 7.3)

nucleotides: subunits of nucleic acid formed from a simple sugar, a phosphate group, and a nitrogen base. (Section 6.3)

nucleus (NEW klee us): positively charged center of an atom composed of neutrons and positively charged protons, and surrounded by a cloud of negatively charged electrons. (Section 6.1); in eukaryotic cells, the central membrane-bound organelle that manages cellular functions and contains DNA. (Section 7.1)

nymph: stage of incomplete metamorphosis where an insect hatching from an egg has the same general appearance as the adult insect but is smaller and sexually immature. (Section 28.2)

GLOSSARY

O

obligate aerobes: bacteria that require oxygen for cellular respiration. (Section 18.2)

obligate anaerobes: bacteria that are killed by oxygen and can survive only in oxygen-free environments. (Section 18.2)

open circulatory system: system where blood moves through vessels into open spaces around the body organs. (Section 27.1)

opposable thumb: primate characteristic of having a thumb that can meet the other fingertips; enables animal to grasp and cling to objects. (Section 16.1)

order: taxonomic grouping of similar families. (Section 17.1)

organ system: multiple organs that work together to perform a specific life function. (Section 8.2)

organ: group of two or more tissues organized to perform complex activities within an organism. (Section 8.2)

organelles: membrane-bound structures within eukaryotic cells. (Section 7.1)

organism: anything that possesses all the characteristics of life; all organisms have an orderly structure, produce offspring, grow, develop, and adjust to changes in the environment. (Section 1.1)

organization: orderly structure of cells in an organism; a characteristic of all living things. (Section 1.1)

osmosis (ahs MOH sus): diffusion of water across a selectively permeable membrane depending on the concentration of solutes on either side of the membrane. (Section 8.1)

osteoblasts (AHS tee oh blastz): potential bone-forming cells that secrete collagen in which minerals in the bloodstream can be deposited. (Section 34.2)

osteocytes: newly formed bone cells. (Section 34.2)

ovary: in plants, the bottom portion of the pistil of a flower that enlarges into one or more ovules each containing one egg. (Section 24.2)

oviduct: in females, the tube that transports eggs from the ovary to the uterus. (Section 38.1)

ovulation: in females, the process of an egg rupturing through the ovary wall and moving into the oviduct. (Section 38.1)

ovule: in seed plants, the sporophyte structure surrounding the developing female gametophyte; forms the seed after fertilization. (Section 22.3)

ozone layer: layer of the atmosphere that helps to protect living organisms on Earth's surface from damaging doses of ultraviolet radiation from the sun. (Section 5.1)

P

pancreas: soft, flattened gland that secretes digestive enzymes and hormones; products help break down carbohydrates, proteins, and fats. (Section 35.1)

parasitism (PAH ruh suh tihz um): symbiotic relationship in which one organism benefits at the expense of the other species. (Section 2.1)

parasympathetic nervous system (PNS): division of the automatic nervous system that controls many of the body's internal functions when the body is at rest. (Section 36.1)

parathyroid glands: produce parathyroid hormone (PTH), which is involved the regulation of minerals in the body. (Section 35.3)

parenchyma (puh RENG kuh muh): most abundant type of plant cell; spherical cells with thin, cell walls and a large central vacuole; important for storage and food production. (Section 23.1)

parthenogenesis (par thuh noh JEN uh sus): type of asexual reproduction in which a new individual develops from an unfertilized egg. (Section 28.1)

passive transport: movement of particles across cell membranes by diffusion or osmosis; the cell uses no energy to move particles across the membrane. (Section 8.1)

pathogens: disease-producing agents such as bacteria, protozoans, fungi, viruses, and other parasites. (Section 39.1)

pedicellariae (ped ih sihl AHR ee ee): pincerlike appendages on echinoderms used for protection and cleaning. (Section 29.1)

pedigree: graphic representation of genetic inheritance used by geneticists to map genetic traits. (Section 12.1)

pedipalps (PED ih palpz): second pair of an arachnids six pairs of appendages that are often adapted for handling food and sensing. (Section 28.2)

pepsin: enzyme found in gastric juices; begins the chemical digestion of proteins in food; most effective in acidic environments. (Section 35.1)

peptide bond: covalent bond formed between amino acids (Section 6.3)

perennial: anthophyte that lives for several years. (Section 22.3)

pericycle: in plants, the layer of cells just within the endodermis that gives rise to lateral roots. (Section 23.2)

peripheral nervous system (puh RIHF rul) (PNS): division of the central nervous system made up of all the nerves that carry messages to and from the central nervous system. (Section 36.1)

peristalsis (per uh STAHL sus): series of involuntary smooth muscle contractions along the walls of the digestive tract that move food from the mouth to the stomach. (Section 35.1)

permafrost: layer of permanently frozen ground that lies underneath the topsoil of the tundra. (Section 3.2)

petals: leaflike flower organs, usually brightly colored structures arranged in a circle around the top of a flower stem. (Section 24.2)

petiole (PET ee ohl): in plants, the stalk that joins the leaf blade to the stem. (Section 23.2)

pH: measure of how acidic or basic a solution is; the scale ranges from 0 to 14; solution with pH of 7 is neutral; above 7, basic, and below 7, acidic. (Section 6.1)

phagocytes (FAG uh sites): white blood cells that destroy pathogens by surrounding and engulfing them; include macrophages, neutrophils, monocytes, and eosinophils. (Section 39.2)

pharynx (FAH rinx): in planarians, the tubelike, muscular organ that extends from the mouth; aids in feeding and digestion. (Section 26.3)

phenotype (FEE nuh tipe): outward appearance of an organism, regardless of its genes. (Section 10.1)

pheromones (FER uh mohnz): chemical signals given off by animals that signal animals to engage in specific behaviors. (Section 28.1)

phloem: vascular plant tissue composed of tubular cells joined end to end; transports sugars from the leaves to all parts of the plant. (Section 23.1)

phospholipids: lipids with an attached phosphate group; plasma membranes are composed of phospholipid bilayer with embedded proteins. (Section 7.2)

photic zone: portion of the marine biome that is shallow enough for sunlight to penetrate. (Section 3.2)

photolysis (FO tohl ih sis): reaction taking place in the thylakoid membranes of a chloroplast during light-dependent reactions where two molecules of water are split to form oxygen, hydrogen ions, and electrons. (Section 9.2)

photoperiodism: flowering plant response to differences in the length of day and night. (Section 24.2)

photosynthesis: process by which autotrophs, such as algae and plants, trap energy from sunlight with chlorophyll and use this energy to convert carbon dioxide and water into simple sugars. (Section 9.2)

phylogeny (fy LOH juh nee): evolutionary history of a species based on comparative relationships of structures and comparisons of modern life forms with fossils. (Section 17.2)

phylum (FI lum): taxonomic grouping of similar classes. (Section 17.1)

pigments: molecules that absorb specific wavelengths of sunlight. (Section 9.2)

pistil: female reproduction structure of a flower; bottom portion forms the ovary. (Section 24.2)

pituitary gland (puh TEW uh ter ee): main gland of the endocrine system that controls many other endocrine glands. (Section 35.3)

placenta (pluh SENT uh): organ that provides food and oxygen to and removes waste from young inside the uterus of placental mammals. (Section 32.2)

placental mammal: mammal that carries its young inside the uterus until development is almost complete. (Section 32.2)

plankton: small organisms that live in the waters of the photic zone; includes both autotrophic and heterotrophic organisms. (Section 3.2)

plasma: fluid portion of the blood that makes up about 55 percent of the total volume of the blood; contains fragments of red and white blood cells. (Section 37.2)

plasma membrane: serves as boundary between the cell and its environment; allows materials such as water and nutrients to enter and waste products to leave. (Section 7.2)

plasmid: small ring of DNA found in a bacterial cell that is used as a biological vector. (Section 13.2)

plasmodium (plaz MOHD ee um): in plasmodial slime molds, the mass of cytoplasm that contains many diploid nuclei but no cell walls or membranes. (Section 19.3)

plastids: group of plant organelles that are used for storage of starches, lipids, or pigments. (Section 7.3)

plate tectonics (tek TAHN ihks): geological explanation for the movement of continents over Earth's thick, liquid interior. (Section 14.1)

platelets: small cell fragments in the blood that help blood clot after an injury. (Section 37.2)

point mutation: mutation in a DNA sequence; occurs from a change in a single base pair. (Section 11.3)

GLOSSARY

polar molecule: molecule with an unequal distribution of charge, resulting in the molecule having a positive end and a negative end. (Section 6.2)

polar nuclei: two of the six nuclei in the center of the egg sac of a flowering plant that becomes the triploid ($3n$) endosperm when joined with a sperm during double fertilization. (Section 24.3)

pollen grain: in seed plants, structure in which the male gametophyte develops; consists of sperm cells, nutrients, and a protective outer covering. (Section 22.3)

pollination: transfer of male pollen grains to the pistil of a flower. (Section 10.1)

polygenic inheritance: inheritance pattern of a trait controlled by two or more genes; genes may be on the same or different chromosomes. (Section 12.2)

polymer: large molecule formed when many smaller molecules bond together. (Section 6.3)

polyp (PAHL up): in the life cycle of cnidarians, the sessile stage where the body is shaped like an umbrella with tentacles hanging downward. (Section 26.2)

polyploid: any species with multiple sets of the normal set of chromosomes; results from errors during mitosis or meiosis. (Section 15.2)

population: group of organisms of one species that interbreed and live in the same place at the same time. (Section 2.1)

posterior: tail end of bilaterally symmetric animals. (Section 25.2)

prehensile tail (pree HEN sul): long muscular tail used as a fifth limb for grasping and wrapping around objects; characteristic of many New World monkeys. (Section 16.1)

primary succession: colonization of new land that is exposed by avalanches, volcanoes, or glaciers by pioneer organisms. (Section 3.1)

primate: group of mammals including lemurs, monkeys, apes, and humans that evolved from a common ancestor; shared characteristics include a rounded head, a flattened face, fingernails, flexible shoulder joints, opposable thumbs or big toes, and a large, complex brain. (Section 16.1)

proglottid (proh GLAH tihd): detachable section of a tapeworm that contains muscles, nerves, flame cells, and reproductive organs. (Section 26.3)

prokaryotes: unicellular organisms, such as bacteria, composed of prokaryotic cells. Prokaryotic cells lack internal membrane-bound structures. (Section 7.1)

prophase: first and longest phase of mitosis where chromatin coils into visible chromosomes. (Section 8.2)

prostate gland: in human males, single gland that lies below the bladder and surrounds the top portion of the urethra; secretes a thin, alkaline fluid that helps sperm move and survive. (Section 38.1)

protein: large, complex polymer essential to all life composed of carbon, hydrogen, oxygen, nitrogen, and usually sulfur; provides structure for tissues and organs and helps carry out cell metabolism. (Section 6.3)

prothallus (proh THAL us): gametophyte structure in non-seed vascular plants that produces antheridia and/or archegonia. (Section 22.2)

protist: diverse group of multicullular or unicellular eukaryotes that lack complex organ systems and live in moist environments; may be autotrophic or heterotrophic. (Section 17.2)

protocell: large, ordered structure, enclosed by a membrane, that carries out some life activities, such as growth and division. (Section 14.2)

protonema (proht uh NEE muh): in mosses, a small, green filament of haploid cells that develops from a spore; develops into a male or female gametophyte. (Section 24.1)

protostome (PROHT uh stohm): animal with a mouth that develops from the opening in the gastrula. (Section 25.1)

protozoan (proht uh ZOH uhn): unicellular, heterotrophic, animal-like protist. (Section 19.1)

provirus: viral DNA that is integrated into a host cell's chromosome and replicated each time the host cell replicates. (Section 18.1)

pseudocoelom (SEWD uh see lohm): fluid-filled body cavity partly lined with mesoderm; enables organism to move quickly. (Section 25.2)

pseudopodia (sewd uh POHD ee uh): in protozoans, cytoplasm-containing extensions of the plasma membrane; aid in locomotion and feeding. (Section 19.1)

puberty: in humans, the period when secondary sex characteristics begin to appear; changes are controlled by sex hormones secreted by the endocrine system. (Section 38.1)

pulse: surge of blood through an artery that can be felt on the surface of the body. (Section 37.2)

punctuated equilibrium: idea that periods of speciation occur relatively quickly with long periods of genetic equilibrium in between. (Section 15.2)

pupa (PYEW puh): stage of insect metamorphosis where tissues and organs are broken down and replaced by adult tissues; larvae emerges from pupa as a mature adult. (Section 28.2)

pus: collection of dead macrophages and body fluids that forms in infected tissues. (Section 39.2)

R

radial symmetry (RAYD ee uhl): an animal's body plan that can be divided along any plane, through a central axis, into roughly equal halves; an adaptation that enables an animal to detect and capture prey coming toward it from any direction. (Section 25.2)

radicle (RAD ih kul): embryonic root of an anthophyte embryo; the first part of the young sporophyte to emerge during germination. (Section 24.3)

radula (RAJ uh luh): in some snails and mollusks, the rasping, tonguelike organ used to drill, scrape, grate, or cut food. (Section 27.1)

rays: long tapered arms of some echinoderms that are covered with short, rounded spines. (Section 29.1)

receptors: binding sites on target tissue cells that bind with specific hormones. (Section 35.3)

recessive: trait of an organism that can be masked by the dominant form of a trait. (Section 10.1)

recombinant DNA (ree KAHM buh nunt): DNA made by recombining fragments of DNA from different sources. (Section 13.2)

rectum: last part of the digestive system through which feces passes before it exits the body through the anus. (Section 35.1)

red blood cells: round, disk-shaped cells in the blood that carry oxygen to body cells; make up 44 percent of the total volume of the blood. (Section 37.2)

red marrow: marrow found in the humerus, femur, sternum, ribs, vertebrae, and pelvis that produces red blood cells, white blood cells, and cell fragments involved in blood clotting. (Section 34.2)

reflex (REE fleks): simple, automatic response in an animal that involves no conscious control; usually acts to protect an animal from serious injury. (Section 33.1); automatic response to a stimulus; reflex stimulus travels to the spinal column and sent directly back to the muscle. (Section 36.1)

regeneration: replacement or regrowth of missing body parts by mitosis. (Section 26.3)

reintroduction programs: programs that release organisms into an area where their species once lived in hopes of reestablishing naturally reproducing populations. (Section 5.2)

reproduction: production of offspring by an organism; a characteristic of all living things. (Section 1.1)

reproductive isolation: occurs when formerly inter-breeding organisms can no longer produce fertile offspring due to an incompatibility of their genetic material or by differences in mating behavior. (Section 15.2)

response: an organism's reaction to a change in its internal or external environment. (Section 1.1)

restriction enzymes: DNA-cutting enzymes that can cut both strands of a DNA molecule at a specific nucleotide sequence. (Section 13.2)

retina: thin layer of tissue found at the back of the eye made up of light receptors and sensory neurons. (Section 36.2)

retrovirus (reh tro VY rus): type of viral replication where a virus uses reverse transcriptase to make DNA from viral RNA; the retroviral DNA is then integrated into the host cell's chromosome. (Section 18.1)

reverse transcriptase (trans KRIHP tayz): enzyme carried in the capsid of a retrovirus that helps produce viral DNA from viral RNA. (Section 18.1)

rhizoids (RI zoyds): fungal hyphae that penetrate food and anchor a mycelium; secrete enzymes for extracellular digestion and absorb nutrients. (Section 20.2)

rhizome: thick, underground stem produced by ferns and other vascular plants; often functions as an organ for food storage. (Section 22.2)

ribosomal RNA (rRNA): RNA that makes up the ribosomes; clamps onto mRNA and uses its information to assemble amino acids in the correct order. (Section 11.2)

ribosomes: nonmembrane-bound organelles in the nucleus where enzymes and other proteins are assembled. (Section 7.3)

rods: receptor cells in the retina that are adapted for vision in dim light; also help detect shape and movement. (Section 36.2)

root: plant organ that absorbs water and minerals from the soil, transports those nutrients to the stem, and anchors the plant in the ground; may also serve as food storage organs. (Section 21.1)

root cap: tough, protective layer of parenchyma cells that covers the tip of a root. (Section 23.2)

safety symbol: symbol that warns you about a danger that may exist, from chemicals, electricity, heat, or experimental procedures. (Section 1.2)

sarcomere (SAR kuh meer): each section of a myofibril in muscle. (Section 34.3)

scales: thin bony plates that come in a variety of shapes and sizes formed from the skin of many fishes and reptiles. (Section 30.1)

scavengers: animals that feed on animals that have already died. (Section 2.2)

scientific methods: series of steps that biologists and other scientists use to gather information and answer questions; include observing and hypothesizing, experimenting, and gathering and interpreting results. (Section 1.2)

sclerenchyma (skler ENG kuh muh): plant cells with thick, rigid cell walls; provide support for the plant and are a major component of vascular tissue; includes fibers and stone cells. (Section 23.1)

scolex (SKOH leks): knob-shaped head of a tapeworm. (Section 26.3)

scrotum: in males, the sac suspended directly behind the base of the penis that contains the testes. (Section 38.1)

secondary succession: sequence of community changes that take place after a community is disrupted by natural disasters or human actions. (Section 3.1)

seed: adaptive reproductive structure of seed plants consisting of an embryo, a food supply, and a protective coat; protects zygote or embryo from drying out and may also aid in dispersal. (Section 21.1)

selective permeability: feature of the plasma membrane that maintains homeostasis within a cell by allowing some molecules into the cell while keeping others out. (Section 7.2)

semen: combination of sperm and fluids from the seminal vesicles, prostate gland, and bulbourethral glands. (Section 38.1)

semicircular canals: structures in the inner ear containing fluid and hairs that help the body maintain balance. (Section 36.2)

seminal vesicles: in males, pair of glands located at the base of the urinary bladder that secrete a mucouslike fluid into the vas deferens. (Section 38.1)

sepals: leaflike, usually green structures arranged in a circle around the top of a flower stem but below the petals. (Section 24.2)

sessile (SES ul): organism that doesn't move from place to place but is permanently attached to a surface. (Section 25.1)

setae (SEE tee): tiny bristles that help segmented worms move by anchoring their bodies in the soil so each segment can move the animal along. (Section 27.2)

sex chromosomes: in humans, the 23rd pair of chromosomes; determine the sex of an individual and carry sex-linked characteristics. (Section 12.2)

sex-linked traits: traits controlled by genes located on sex chromosomes. (Section 12.2)

sexual reproduction: pattern of reproduction that involves the production and subsequent fusion of haploid sex cells. (Section 10.2)

short-day plants: plants that are induced to flower by exposure to long nights. (Section 24.2)

sieve tube members: tubular cells in phloem that conduct sugars and other organic compounds. (Section 23.1)

simple eye: visual structure in arthropods that uses one lens to detect light and focus. (Section 28.1)

sink: any part of a plant that stores sugars produced during photosynthesis. (Section 23.2)

sister chromatids: identical halves of a duplicated parent chromosome formed during the prophase stage of mitosis; the halves are held together by a centromere. (Section 8.2)

skeletal muscle: a type of voluntary muscle that is attached to and moves the bones of the skeleton. (Section 34.3)

sliding filament theory: theory that actin filaments slide toward each other during muscle contraction while the myosin filaments are still. (Section 34.3)

small intestine: muscular tube about 6 m long where digestion is completed; connects the stomach and the large intestine. (Section 35.1)

smooth muscle: type of involuntary muscle found in the walls of internal organs and blood vessels; most common function is to squeeze, exerting pressure inside the tube or organ it surrounds. (Section 34.3)

solution: mixture in which one or more substances (solutes) are distributed evenly in another substance (solvent). (Section 6.1)

somatic nervous system: portion of the nervous system composed of cranial nerves, spinal nerves, and all of their branches; voluntary pathway that relays information mainly between the skin, the CNS, and skeletal muscles. (Section 36.1)

sorus: clusters of sporangia usually found on the surface of fern fronds. (Section 22.2)

spawning: method of reproduction in fishes and some other animals where a large number of eggs are fertilized outside of the body. (Section 30.1)

speciation (spee shee AY shun): process of evolution of new species that occurs when members of similar populations no longer interbreed to produce fertile offspring. (Section 15.2)

species (SPEE sheez): group of organisms that can interbreed and produce fertile offspring in nature. (Section 1.1)

sperm: haploid male sex cells produced by meiosis. (Section 10.2)

spindle: cell structures composed of microtubule fibers; form between the centrioles during prophase and shorten during anaphase, pulling apart sister chromatids. (Section 8.2)

spinnerets: silk-producing glands located at the rear of a spider. (Section 28.2)

spiracles (SPIHR ih kulz): in arthropods, openings on the thorax and abdomen through which air enters and leaves the tracheal tubes. (Section 28.1)

spongy bone: soft bone filled with many holes and spaces surrounded by a layer of more dense compact bone. (Section 34.2)

spontaneous generation: mistaken idea that life can arise from nonliving materials. (Section 14.2)

sporangium (spuh RAN jee uhm): in fungi, a sac or case of hyphae in which spores are produced. (Section 20.1)

spore: type of haploid (*n*) reproductive cell with a hard outer coat that forms a new organism without the fusion of gametes. (Section 19.1)

sporophyte: in algae and plants, the diploid (*2n*) form of an organism in alternation of generation that produces spores. (Section 19.2)

sporozoans: group of parasitic protozoans of the phylum Sporozoa that reproduce by spore production. (Section 19.1)

stabilizing selection: natural selection that favors average individuals in a population; results in a decline in population variation. (Section 15.2)

stamen: male reproductive structure of a flower consisting of an anther and a filament. (Section 24.2)

stem: plant organ that provides structural support for upright growth and contains tissues for transporting food, water, and other materials from one part of the plant to another; may also serve as a food storage organ. (Section 21.1)

sternum: large breastbone that provides a site for muscle attachment; provides support for the thrust and power produced by birds as they generate motion for flight. (Section 31.2)

stimulant: drug that increases the activity of the central and sympathetic nervous systems. (Section 36.3)

stimulus: any condition in an environment that requires an organism to adjust. (Section 1.1)

stolons (STOH lunz): fungal hyphae that grow horizontally along a surface and rapidly produce a mycelium. (Section 20.2)

stomach: muscular, pouchlike enlargement of the digestive tract where chemical and physical digestion take place. (Section 35.1)

stomata (STOH mut uh): openings in the cuticle of a leaf epidermis that control gas exchange for respiration and photosynthesis. (Section 23.1)

strobilus (STROH bih lus): compact cluster of spore-bearing leaves produced by some non-seed vascular plants. (Section 22.2)

succession (suk SESH un): orderly, natural changes, and species replacements that take place in ecosystem communities over time. (Section 3.1)

sustainable use: philosophy that promotes letting people use resources in wilderness areas in ways that will not damage the ecosystem. (Section 5.2)

swim bladder: thin-walled, internal sac found just below the backbone in bony fishes; helps fishes control their swimming depth. (Section 30.1)

symbiosis (sihm bee OH sus): permanent, close association between two or more organisms of different species. (Section 2.1)

symmetry (SIH muh tree): balance of proportions of an object or organism. (Section 25.2)

sympathetic nervous system: division of the automatic nervous system that controls many of the body's internal functions during times of stress. (Section 36.1)

synapse: tiny space between one neuron's axon and another neuron's dendrites over which a nerve impulse must pass. (Section 36.1)

T cell: lymphocyte produced in bone marrow and processed in the thymus that plays a role in immunity; includes helper T cells and killer T cells. (Section 39.2)

taiga (TI guh): biome just south of the tundra; characterized by a northern coniferous forest composed of pine, fir, hemlock, and spruce trees and acidic, mineral-poor topsoils. (Section 3.2)

target tissues: tissues with target cells that have receptors on their plasma membranes or in their nuclei for specific endocrine hormones. (Section 35.3)

taste buds: sensory receptors located on the tongue that result in taste perception. (Section 36.2)

taxonomy (tak SAHN uh mee): branch of biology that groups and names organisms based on studies of their shared characteristics; biologists who study taxonomy are called taxonomists. (Section 17.1)

technology (tek NAHL uh jee): application of scientific research to society's needs and problems. (Section 1.3)

telophase: final phase of mitosis during which new cells prepare for their own independent existence. (Section 8.2)

temperate forests: biome composed of forests of broad-leaved hardwood trees that lose their foliage annually; receives 70–150 cm of precipitation annually. (Section 3.2)

tendons: thick bands of connective tissue that attach muscles to bones. (Section 34.2)

territory: physical space an animal defends against other members of its species; may contain an animal's breeding area, feeding area, potential mates, or all three. (Section 33.1)

test cross: mating of an individual of unknown genotypes with an individual of known genotype; can help determine the unknown genotype of the parent. (Section 13.1)

thallus: body structure produced by some plants and some other organisms that lacks roots, stems, and leaves. (Section 19.2)

theory: explanation of natural phenomenon supported by a large body of scientific evidence obtained from many different investigations and observations. (Section 1.2)

therapsids (ther AP sidz): reptilian ancestors of mammals that had features of both reptiles and mammals. (Section 32.2)

threatened species: species that have rapidly decreasing numbers of individuals. (Section 5.1)

thyroid gland: gland located in the neck; regulates metabolism, growth, and development. (Section 35.3)

tissue: groups of cells that work together to perform a specific function. (Section 8.2)

tissue fluid: fluid that bathes the cells of the body; formed when water and dissolved substances diffuse from the blood into the spaces between the cells that make up the surrounding tissues. (Section 39.2)

tolerance: as the body becomes less responsive to drug and an individual needs larger or more frequent doses of the drug to achieve the same effect. (Section 36.3)

toxin: poison produced by a bacterium. (Section 18.2)

trachea (TRAY kee uh): tubelike passageway for air flow that connects with two bronchi tubes that lead into the lungs. (Section 37.1)

tracheal tube (TRAY kee ul): hollow passages in some arthropods that transport air throughout the body. (Section 28.1)

tracheids: tubular, water-conducting sclerenchyma cells in the xylem of conifers and ferns; have tapered ends and are dead at maturity. (Section 23.1)

trait: characteristic that is inherited; can be either dominant or recessive. (Section 10.1)

transcription (trans KRIHP shun): process in the cell nucleus where enzymes make an RNA copy of a DNA strand. (Section 11.2)

transfer RNA (tRNA): RNA that transports amino acids to the ribosomes to be assembled into proteins. (Section 11.2)

transgenic organisms: organisms that contain functional recombinant DNA from a different organism. (Section 13.2)

translation: process of converting information in mRNA into a sequence of amino acids in a protein. (Section 11.2)

translocation (trans loh KAY shun): movement of sugars from photosynthesis from the leaves through the phloem of a plant. (Section 23.2)

transpiration: in plants, the loss of water through leaf stomata by evaporation. (Section 23.2)

transport proteins: proteins that span the plasma membrane creating a selectively permeable membrane that regulates which molecules enter and leave a cell. (Section 7.2)

trial-and-error: type of learning in which an animal receives a reward for making a particular response. (Section 33.2)

trichomes: hairlike projections that extend from a plant's epidermis; help reduce water evaporation and may provide protection from herbivores. (Section 23.1)

trophic level (TROHF ihk): organism in a food chain that represents a feeding step in the passage of energy and materials through an ecosystem. (Section 2.2)

tropical rain forests: biome near the equator with warm temperatures, wet weather, and lush plant growth; receives at least 200 cm of rain annually;

contains more species of organisms than any other place on Earth. (Section 3.2)

tropism: growth response of a plant to an external stimulus that comes from a particular direction; examples include phototropism and gravitropism. (Section 23.3)

tube feet: in echinoderms, hollow, thin-walled tubes that end in a suction cup; part of the water vascular system, they also aid in locomotion, gas exchange, and excretion. (Section 29.1)

tundra (TUN druh): biome that surrounds the north and south poles; treeless land with long summer days and short periods of winter sunlight, beneath the topsoil is a layer of permafrost. (Section 3.2)

umbilical cord: ropelike structure that attaches the embryo to the wall of the uterus; supplies a developing embryo with oxygen and nutrients and removes waste products. (Section 38.2)

ureter: tube that transports urine from each kidney to the urinary bladder. (Section 37.3)

urethra (yoo REE thruh): tube through which urine is passed from the urinary bladder to the outside of the body. (Section 37.3)

urinary bladder: smooth muscle bag that stores urine until it is expelled from the body. (Section 37.3)

urine: liquid composed of wastes that is filtered from the blood by the kidneys, stored in the urinary bladder, and eliminated through the urethra. (Section 37.3)

uterus (YEWT uh rus): in females, the hollow, muscular organ in which the offspring of placental mammals develop. (Section 32.2)

vaccine: substance consisting of weakened, dead, or incomplete portions of pathogens or antigens that produce an immune response when injected into the body. (Section 39.2)

vacuole: membrane-bound fluid-filled space in the cytoplasm of plant cells used for the temporary storage of materials. (Section 7.3)

vas deferens (vas • DEF uh runz): in males, duct that transports sperm from the epididymis towards the ejaculatory ducts of the urethra. (Section 38.1)

vascular cambium: lateral meristem that produces new xylem and phloem cells in the stem and roots. (Section 23.1)

vascular plants: plants that have vascular tissues for the transport of water, food, and other materials throughout the plant; enables taller growth and survival on land. (Section 21.1)

vascular tissues (VAS kyuh lur): tissues found in vascular plants composed of tubelike, elongated cells through which water, food, and other materials are transported throughout the plant; include xylem and phloem. (Section 21.1)

vector: means by which DNA from another species can be carried into the host cell; may be biological or mechanical. (Section 13.2)

vegetative reproduction: type of asexual reproduction in plants where a new plant is produced from existing vegetative structures. (Section 24.1)

veins: large blood vessels that carry blood from the tissues back toward the heart. (Section 37.2)

venae cavae (vee nee • KAY vee): two large veins that fill the right atrium of the mammalian heart with oxygen-poor blood from the head and body. (Section 37.2)

ventral (VEN trul): belly surface of bilaterally symmetric animals. (Section 25.2)

ventricles: two lower chambers of the mammalian heart; receive blood from the atria and send it to the lungs and arteries. (Section 37.2)

vertebrate (VURT uh brayt): an animal with a backbone. (Section 25.2)

vessel elements: hollow, tubular cells in the xylem of anthophytes; conduct water and minerals from the roots to the stem; have open ends through which water passes freely from cell to cell. (Section 23.1)

vestigial structure (veh STIHJ ee ul): body structure that has no function in a present-day organism but was probably useful to an ancestor; provides evidence of evolution. (Section 15.1)

villus: single projection on the lining of the small intestine that functions in the absorption of digested food; they increase the surface area of the small intestine and increase the absorption rate. (Section 35.1)

viruses: disease-causing, nonliving particles composed of an inner core of nucleic acids surrounded by a capsid; replicate inside living cells called host cells. (Section 18.1)

vitamins: organic nutrients required in small amounts to maintain growth and metabolism; are either fat-soluble or fat-insoluble vitamins. (Section 35.2)

GLOSSARY

vocal cords: sound-producing bands of tissue in the throat that produce sound as air passes over them. (Section 30.2)

voluntary muscle: muscle that contracts under conscious control. (Section 34.3)

water vascular system: in echinoderms, the hydraulic system that operates under water pressure; aids in locomotion, gas exchange, and excretion. (Section 29.1)

white blood cells: large, nucleated blood cells that play a major role in protecting the body from foreign substances and microscopic organisms; make up only one percent of the total volume of the blood. (Section 37.2)

withdrawal: psychological response or physiological illness that occurs when a person stops taking a drug. (Section 36.3)

xylem: vascular plant tissue composed of tubular cells that transport water and minerals from the roots to the rest of the plant; water conducting cells include tracheids and vessel elements. (Section 23.1)

yellow marrow: marrow composed of stored fats found in many bones. (Section 34.2)

Z

zygospores (ZI guh sporz): thick-walled spores of zygomycetes that can withstand unfavorable conditions. (Section 20.2)

zygote (ZI goht): diploid cell formed when a sperm fertilizes an egg. (Section 10.2)

674; dormancy, 675, *illus.* 675; formation, 672; germination, 674–677, *illus.* 676, *lab* 677

Seed plant, 587–589, 686–688; characteristics, 587, 608–609, *illus.* 609; divisions, A2, *illus.* 587, 588–589, 610–616; fertilization and reproduction, 608–609; phylogeny, 617, *illus.* 617

Segmented worm, 748–753, 809, *illus.* 809; characteristics, 748–751, *illus.* 749, *lab* 749, *lab* 750, *illus.* 751, *lab* 754–755; classes, 748, *illus.* 748, 750, 752; phylogeny, 753, *illus.* 753

Selective breeding, 343–346, *illus.* 344–345, *illus.* 346, 680

Selective permeability, 181–182, *illus.* 182, *lab* 182

Self-pollination, 260

Semen, 1029

Semicircular canal, 987

Seminal vesicle, *illus.* 1028, 1029

Senses, 983–987

Sensory neuron, 973, *illus.* 974, 1084

Sensory pit, 727, *illus.* 728

Sepal, 661, *illus.* 665, 687, *illus.* 687

Septa, 547, *illus.* 547

Sequencing, DNA, 353, *illus.* 354

Sequencing skills, SH8

Sessile organism, 694

Setae, 748–749, *illus.* 748, 751

Sex cell, 259, 271

Sex chromosome, 324–325, *illus.* 324, 335

Sex-linked inheritance, 325–326, *illus.* 325, 332–333, *illus.* 332, *lab* 332, 338, 371

Sexual reproduction, 272, *illus.* 272

Sexually transmitted disease (STD), 1058, 1059, *lab* 1059, 1062, *illus.* 1062

Short-day plant, 666, *illus.* 666

SI measurements, SH4, 22–23

Sickle-cell anemia, 329–330, *illus.* 330, 370

Sieve plate, 630, *illus.* 630

Sieve tube member, 630, *illus.* 630

Sight, 984, *illus.* 984, *illus.* 985

Silk gland, 768, 769, *illus.* 769

Silviculture, *illus.* 582

Simple eye, 765

Sink, 637, *illus.* 637

Siphon, 743, *illus.* 743, 746, *illus.* 746, 747

Siphonophore, 721

Sister chromatid, 212, *illus.* 212

Six kingdom system, A1–A4, 485; development of, *illus.* 460–463; phylogenetic diagram, *illus.* 468–469, *illus.* 485; survey, 470–473

Skeletal muscle, 936–937, *lab* 936, *lab* 937, 938, 1082

Skeletal system, 929–934, *illus.* 930, *illus.* 931, 1081, *illus.* 1081

Skin, 923–928, *illus.* 924, *illus.* 925, *lab* 925, *illus.* 926, *lab* 926, 1063, 1081, *illus.* 1081

Skin cancer, 222, 254, *illus.* 254

Skull, primate, 439–440, *illus.* 440, *lab* 446–447, 484

Sliding filament theory, 937, *lab* 937, 938

Slime mold, 533–535, *illus.* 534, *lab* 534, *illus.* 535, 569, *illus.* 569

Small intestine, *table* 948, 951–953, *illus.* 951, *illus.* 953

Smell, 983–984, *illus.* 983

Smoking, 993, 995, *illus.* 995

Smooth muscle, *illus.* 935, 936, 1082

Snake, 845, *illus.* 845, 847–848, *illus.* 848, 913, *illus.* 913

Society and technology, 24–25

Solute, 153, *illus.* 153

Solution, 153, *illus.* 153

Solvent, 153, *illus.* 153

Somatic nervous system, 980, *lab* 980, *illus.* 981, *illus.* 982

Sorus, 606, *illus.* 606, 656, *illus.* 656

Spawning, 819

Speciation, 417–419

Species, 7; classification, 457, *lab* 457, 464–465, 484; diversity, 115–117, *lab* 116; evolution, 417–419; extinction, 119, *lab* 119; naming, 454–455; number in each kingdom, *lab* 471

Sperm, 271

Sphenophyta (horsetails), *illus.* 585, 586–587, *lab* 586, 603, *illus.* 603, 685, *illus.* 685

Spicule, 714, 717, *illus.* 717

Spider, 767–768, *illus.* 767, 769, *illus.* 769

Spinal cord, 1084, *illus.* 1084

Spindle, 214

Spinneret, 768

Spiracle, 764, *illus.* 772

Spirometer, *lab* 1020–1021

Spleen, 1068–1069, 1088

Sponge: characteristics, 713–717, *illus.* 713, *lab* 715, 806, *illus.* 806; phylogeny, 715, 725, *illus.* 725

Spongy bone, 932, *illus.* 932

Spontaneous generation, 388–389, *illus.* 388, 481

Sporangium: fern, 606, *lab* 606; fungi, 549–550, 570

Spore, 584; algae, 532, *illus.* 532; fungus, 546, *illus.* 546, 549–550, *illus.* 550, *lab* 550; plant, 578, 579, 584, 608, 653, 658, 667, 658, 669; sporozoan, 524

Sporophyte, 532, *illus.* 532, 579, *lab* 579, *lab* 588; non-seed vascular plant, 601–602, *illus.* 601; nonvascular plant, 597, *illus.* 597, *lab* 598; seed plant, 608

Sporozoan, 524–525, 568

Stabilizing selection, 416, *illus.* 416, 483

Stamen, 662, *illus.* 665, 687

Starch, 163, *illus.* 163, 251, 954

Starfish. *See* Sea star

Stem, 577–578, *illus.* 577, 635–637, *illus.* 635, *illus.* 636, *illus.* 637

Steppe, 83

Sternum, 853

Steroid hormone, 962, *illus.* 963

Stigma, *illus.* 665

Stimulant, 990, 992–993, *illus.* 992

Stimulus, 9, *illus.* 9, *lab* 26–27, 32

Stipe: brown algae, 531; fern, 605, *illus.* 605; mushroom, *illus.* 555

Stolon, 551

Stomach, 950–951; digestive enzymes, *table* 948, 951; muscular churning, 950, *illus.* 950

Stomata, 628, *illus.* 628, *lab*

Design and Production: DECODE, Inc.

Art Credits

Morgan Cain, 242, 714,716, 719, 720, 722, 727(b), 728, 730, 735(tl,tr), 737, 738(tr,br), A9

Ortelius Design, 402, 434, 448, 782

John Edwards, 55(b), 57, 58, 59, 77(b), 86, 187(t), 195, 272, 403, 409, 426(tl,tr), 685, 687(t,b), 688, 689, 701(tl,tcr,tr), 764, 769, 772, 773, 774, 784(bl), 869, 870, 871, 886, 887, 913

Tom Gagliano, 98, 419, 425(b), 468-469, 480(b), 485, 537, 559, 600, 607, 617, 664, 725, 753, 775, 784(tr), 793, 804, 825, 828, 833, 849, 852, 858, 859, 863, 880, 881, A1(tl,tr,bcr), A3, A4, SH6(bl)

Precision Graphics, xx, xxi, 62, 177, 180, 182(t,b), 184, 185, 186(cr,b), 187(b), 188, 189(t,b), 190, 191, 192, 198, 197(t,cr,b), 199, 202(b), 203(c), 206, 207, 211, 212(t), 213, 214, 216, 221(tl,tr,crr), 222, 230, 260(t,b), 261, 262, 263, 264, 265, 266, 267, 268(tl,crr), 270, 271, 273, 274, 275, 276, 277, 278, 283(t,cr), 284(tl,crr), 360, 368, 369, 373, 416, 418, 433, 443, 445, 483(crl), 491(tl,tr,bl,br), 492, 493, 495, 503, 516, 517, 521, 523, 524, 525, 527, 528, 532, 534, 535, 542(tl,tr), 546, 548, 552, 555, 558, 564, 566, 570, 578(t), 586, 598, 601, 602, 605, 613, 622, 625, 628, 629, 630(t,b), 631, 634, 636, 637, 638, 639(b), 641, 642, 644(t,b), 650(l,r), 653, 654, 655, 657, 660, 663(t,b), 668, 669(t,b), 671, 672, 681, 682, 695(tr), 696, 698, 703, 711, 742, 743, 744(t,b), 750, 751, 757, 758, 790, 794, 795, 797, 798, 799, 803, 805, 806, 807, 809, 811, 818, 819, 820, 821, 822, 838(t), 844, 854, 855, 857, 864(tr), 901, 924, 927(t,cr,b), 930, 931(tl,tr,bl,br), 932, 935(bl,bcr,br), 936(br,bl), 937, 938, 943(cr,b), 944(t,b), 945, 949(br), 950(t,b), 951, 952, 953, 955, 957, 958, 959, 960, 961, 965, 969(t,cr), 970, 973, 974(t,b), 975(t,cr,b), 976, 977, 981, 982, 983, 984, 985, 986, 987, 988, 989, 999(t,cr), 1000(t,b), 1001, 1028(b), 1030(tl), 1033, 1073, 1080, 1087(l), A2

Rolin Graphics, 70, 212(b), 223(cr), 398(tr), 410, 446, 635, 1004, 1005, 1011(t,b), 1012, 1014, 1015(t), 1016(tl), 1017, 1018(t), 1023, 1024, 1028(t), 1031, 1032, 1036, 1039 (t,b), 1045, 1052(t,cr), 1057, 1059, 1064, 1065, 1068(t,b), 1069, 1070, 1077, 1082, 1083, 1084(t,b), 1085, 1086, 1087(r), 1088(l,br), SH1

Nancy Heim/158 Street Design Group, 1008

Tom Kennedy/Romark Illustrations, vii, 22(t,b), 31, 39, 65, 69, 70(t), 74, 77(t), 96, 97, 99, 102, 104, 107, 112(t,b), 113, 116, 134, 135, 147, 149, 150, 151, 152, 153, 157, 159, 160, 161, 162, 163(t,b), 164, 165, 166(t,b), 167, 171(t,cr,b), 172, 173, 183, 202(t,bl), 203(tl,bl), 204, 205, 208, 209, 210, 218, 223(t), 227, 229(t,b), 232, 233(t,b), 234, 235, 238(t,b), 239, 240, 247(t,cr), 248, 249, 251(t,cr), 252, 253(cr), 288, 290, 291, 292-293, 295, 296, 297, 299, 300-301, 304, 306(t,b), 308, 311, 312(tr,bl), 313(crl,tr), 315, 316(t,b), 318, 320, 324, 326, 331, 332(t.b), 334, 339(br), 340(tr,bl), 341, 350, 351, 352, 354, 355, 361, 363, 365, 366, 367, 370, 371(bl,br), 372, 388, 389, 390, 391, 392, 393, 397, 399, 407, 411, 427(tr,br), 434(t), 450(t), 459, 467, 470, 475(br), 484, 543, 561, 588, 595, 623(t), 639(t), 651, 683, 724, 727(t), 759, 766, 781(r), 836, 839(r,l), 865(r,l), 890, 897, 898, 900, 902, 939(t,b), 945, 970, 962, 963(t,b), 971, 1010, 1015(b), 1016(tr), 1018(b),1025, 1034, 1035(t,b), 1053(r,l), 1062, SH12(tl,br)

Susan Moore/Lisa Freeman Inc., 322, 665, 676

Laurie O'Keefe, 53, 54(t,b), 55(t), 138, 337, 345, 380, 381, 404-405, 413, 417, 420, 425(t), 431, 450(b), 466, 475(t), 478, 480(t), 483(b), 702, 710(l,r), 765, 826, 832, 838(b), 872, 892, 894, 907

Felipe Passalacqua, 50-51, 561, 695(tl), 739, 829, 949(cr), SH5(tr), SH6(crl)

James Shough & Associates, 538, 539, SH2, SH4

Jackie Urbanovic, 24

Unknown, 140, 576

Photo Credits

List of Abbreviations:
AA=Animals Animals; CBS=Carolina Biological Supply; CB=Corbis-Bettmann; CP=Color-Pic; CMSP=Custom Medical Stock Photo; DRK=DRK Photo; ES=Earth Scenes; FPG=FPG International; GH=Grant Heilman Photography; LI=Liaison International; MP=Minden Pictures; OSF=Oxford Scientific Films; PA=Peter Arnold, Inc.; PR=Photo Researchers; PT=Phototake, NYC; SPL=Science Photo Library; SS=Science Source; TSM=The Stock Market; TSA=Tom Stack & Associates; TSI=Tony Stone Images; VU=Visuals Unlimited

Cover "Salmon Hunter" by Randall Scott courtesy Wild Wings, Inc.; **i** "Salmon Hunter" by Randall Scott courtesy Wild Wings, Inc.; **iv** (tl) Jan Hinsch/SPL,(tr) Stephen Dalton/AA, (bl) Wolfgang Kaehler, (bc) Biophoto Assoc./PR, (br) Rod Planck/TSA; **v** (t) Mark M. Moffett, (c) Brian Kenney, (bl) Runk Schoenberger from GH, (bc) VU/Doug Sokell, (br) Barbara Cushman Rowell/DRK; **vi** (l) Pat O'Hara/DRK, (r) Adam Jones/PR; **vii** (t) Biophoto Assoc./PR, (b) VU/Fred Hossler; **viii** (t) John Readers/SPL/PR, (b) Daniel J. Cox/Natural Exposures; **ix** Jeremy Burgess/PR; **x** William Leonard/DRK; **xi** (t) Oliver Meckes/Ottawa/PR, (b) Alfred Pasieka/SPL/PR; **xii** Michio Hoshino/MP; **xiii** Matt Meadows; **xiv** S. Nelson/DRK; **xv** VU/V. Ahmadijan **xvi** (t) Ed Reschke/PA (b) Jan Hinsch/SPL; **xvii** (t) Dr. E.R. Degginger/AA, (b) Oliver Meckes/Ottawa/PR; **xix** (l) Mark Newman/TSA, (c) Victoria McCormick/AA, (r) Joe Donald/TSA; **xxii** SSEC/UW/Madison; **xxii** (t) Michelle Del Guerico/CMSP, (b) Science Museum/Science & Society Picture Library; **xxiv** (t) M. Wendler/Okapia/PR, (b) W. Gregory Brown/AA; **xxv** John Gerlach/DRK; **xxvi** (t) Patti Murray/ES, (b) Marcia Griffen/AA; **xxvi-1** John Eastcott & Yva Momatiuk/DRK; **2** Carl R. Sams II/PA, (inset)) Fred Unverhau/ES, **3** (t) Kenneth M. Highfill/PR, (b) Steve E. Ross/PR; **4** (tl) Steve Kaufman/DRK, (tc) John Gerlach/DRK, (tr) Johnny Johnson/AA, (bl) Norbert Wu/Mo Yung Productions, (br) Jeff Lepore/PR; **5** (t) Bob Daemmrich, (b) VU/William J. Weber; **6** (l) VU/John Sohlden, (r) Kjell B. Dandved/PR; **7** (t) OSF/AA, (b) Tom McHugh/PR; **8** (t) Tom Bean/DRK, (b) John Gerlach/DRK; **9** (l) John Gerlach/DRK, (r) Tom Brakefield/DRK; **10** (l) VU/T.E. Adams, (c) Renee Lynn/PR, (r) S. Fried/PR; **11** M.C. Chamberlain/DRK; **12** Gerry Ellis/ENP Images; **13** (t) OSU/OARDC, (b) Tom Brakefield/DRK; **14** KS Studio; **15** (t) Phillippe Plailly/SPL/PR, (c) Phil Degginger/CP, (b) Tom Pantages; **16** Salamander Picture Library; **16-17** (t) Vito Palmisano/TSI, (b) Francois Gohier; **17** (tl) John Cancalosi/TSA, (tr) Francois Gohier, (c) Allen B. Smith/TSA, (b) Rich Frishman/TSI; **18** KS Studio; **19** (t) VU/Science VU, (b) KS Studio; **20** Gregory G. Dimijian/PR; **21** (t) Will & Deni McIntyre, (b) Belinda Wright/DRK; **22** M. Abbey/PR; **23** (l) Art Wolfe, (r) Luiz C. Marigo/PA; **25** Mark E. Gibson; **26** J.H. Robinson/PR; **28** (t) KS Studio, (b) VU/Jeff Greenberg; **29** (t) Steve E. Ross/PR, (b) Phil Degginger/CP; **30** (t) Dr. E.R. Degginger/PR, (b) Carolyn A. McKeone/PR; **32** (t) Arthur C. Twomey/PR, (b) Jeff Lepore/PR; **33** Telegraph Colour Library/FPG; **34-35** Frans Lanting/MP; **36** Michael Durham/ENP Images, (inset) Hans Pfletschinger/PA; **37** (tl) Michael Gallagher/LI, (tr) Joe McDonald/DRK, (b) British Library/Bridgeman Art Library, London/NY; **38** (l) VU/David Sieren, (r) Marc Epstein/DRK; **39** Jim Brandenburg/MP; **40** (t) Aaron Haupt, (b) Michael Habicht/ES; **41** (tl, tr) Barbara Gerlach/DRK, (c) IFA/PA, (bl) Earth Satellite Corporation/SPL/PR, (br) Frans Lanting/MP; **42** (t) Michael Gadomski/ES, (bl) Breck P. Kent/AA, (br) Runk/Schoenberger from GH; **43** (t) Grant Heilman from GH, (bl) Michael P. Gadomski/PR, (bc) Rod Planck/PR, (br) Pat O'Hara/DRK; **44** (t) Breck P. Kent/AA, (b) VU/John D. Cunningham; **44-45** Matt Meadows; **45** (t) Stanley Breeden/DRK, (b) Zig Leszcznski/AA; **46** (tl) Margaret Conte/AA, (tr) Dr. E.R. Degginger/CP, (b) John Eastcott & Yva Momatiuk/ES; **47** (t) Michael Fogden/DRK, (b) Donald Specker/AA; **48** (l) Patti Murray/AA, (r) Richard & Susan Day/AA; **49** (t) M.P. Kahl/DRK, (b) Brad Fowle; **56** Matt Meadows; **57** Michael P. Gadomski/PR; **60** (t) Dr. E.R. Degginger/CP, (b) VU/M. Abbey; **61** Biophoto Assoc./PR; **62** James R. Fisher/DRK; **63** (t) Zig Leszcznski/AA, (b) Richard & Susan Day/AA; **64** Adam Jones/PR; **66** Breck P. Kent/ES; **67** (tl) Emil Muench/PR, (tr) Phil Degginger/CP, (b) Johnny Johnson/DRK; **68** Jeff Lepore/PR; **69** (t) William H. Mullins/PR, (b) VU/V. Ahmadjian; **71** VU/Richard Thom; **72** (l) D. Northcott/DRK, (r) Tom Bean/DRK; **73** Altitude (Y. Arthus B.)/PA; **74** (l) Norbert Wu/PA, (r) Robert P. Comport/ES; **75** VU/Science VU; **76** Doug Sokell/TSA; **78** (tl) Wolfgang Kaehler, (tr) Jean-Claude Carton/Bruce Coleman, Inc./PNI, (c) Joseph Schuyler/Stock Boston/PNI, (b) Raymond Gehman; **78-79** Chris Newbert; **79** (t) Douglas Faulkner/SS/PR, (c) George Herben/Alaska Stock, (b) Gary Braasch/TSI; **80** (t) Johnny Johnson, (bl) Dr. E.R. Degginger/AA, (br) Tom McHugh/PR; **81** (t) Dr. E.R. Degginger/CP, (bl) Tom & Pat Leeson/DRK, (bc) Alan D. Carey/PR, (br) Johnny Johnson; **82** (t) Zig Leszczynski/ES, (bl) Joe McDonald/DRK, (br) Kennan Ward/DRK; **83** (t) Steve Kaufam/PA, (b) Tom Bean/DRK; **84** (t) Barbara Cushman Rowell/DRK, (bl) Jeff Lepore/PR, (bc) M H Sharp/PR, (br) VU/Joe McDonald; **85** Gregory G. Dimijian/PR; **87** (l) Kevin Schafer/PA, (c r) Dr. E.R. Degginger/CP; **88** Bob Daemmrich; **90** (t) Porterfield-Chickering/PR, (b) VU/William J. Weber; **91** (t) VU/Richard Thom, (b) Barbara Cushman Rowell/DRK; **92** (l) Simon Fraser/SPL/PR, (r) John Kieffer/PA; **94** Gary Gray/DRK; **95** (t) Dr. E.R. Degginger/ES, (c) Stephen J. Krasemann/DRK, (b) Biophoto Assoc./PR; **96** KS Studio; **98** Francois Gohier/PR; **99** John Gerlach/DRK; **100** (tl) Thomas Kitchin/TSA, (tr) Blair Seitz/PR, (b) David Cavagnaro/DRK; **101** (l) Vanessa Serra/LI, (r) Jesse R. Lee/PR; **102** (l) Robert Wilson/AA, (r) Stephen J. Krasemann/PA; **103** (t) KS Studio, (b) Marcie Griffen/AA; **104** Tom Van Sant/Geosphere Project, Santa Monica CA/SPL/PR; **106** X. Zimbardo/LI; **108 110** Aaron Haupt; **111** Thomas Kitchin/TSA; **114** John Cancalosi/DRK, (inset) George Bernard/SPL/PR; **115** (t) Gary Gray/DRK, (b) Envision/B.W. Hoffman; **116** Envision/George Mattei; **117** (l) A. & F. Michler/PA, (r) Douglas Faulkner/PR; **118** (l) Tom & Pat Leeson/PR, (c) Patti Murray/ES, (r) Don & Pat Valenti/DRK; **119** (l) Michael Sewell/PA (r) Doug Perrine/DRK; **120** (l) Tom McHugh/PR, (r) Fred Bavendam/PA; **121** (t) Steve Wolper/DRK, (b) Yann Arthus-Bertrand/Photo Researchers; **122** Will & Deni McIntyre/PR; **123** (t) Matt Meadows, (b) Doug Sokell/TSA; **124** (l) Scott Camazine/PR, (r) G. Carleton Ray/PR; **125** (t) Victoria McCormick/AA, (b) Jeff Lepore/PR; **126** Matt Meadows; **127** (t) Shin Yoshino/MP, (bl) Tony Morrison/South American Pictures, (br) KS Studio; **128** (l) M.C. Chamberlain/DRK, (r) Alan D. Carey/PR; **129** Sid Greenberg/PR; **130** (t) Aldo Brando/PA, (c) George Calef/DRK, (b) Renee Stockdale/AA; **131** (t) Renee Stockdale/AA, (b) Fred Bruemmer/DRK; **132** Art Wolfe; **133** (t) Michael Sewell/PA, (b) Jeff Lepore/PR; **134** Bates Littlehales/AA; **136** (t) Fred Bavendam/MP, (c) Nigel J. Dennis, (bl) D. R. Specker/AA, (br) W. Gregory Brown/AA; **137** (t) Dana White/ImageQuest, (b) VU/Doug Sokell; **138** (t) J & B Photographers/AA, (b) VU/Doug Sokell; **139** (t) Rod Planck/PR, (c) Arthur Gloor/AA, (b) B.W. Hoffman/Envision; **140** (t) Tom McHugh/PR, (c) Stephen J. Krasemann/PR, (b) CBS/PT; **141** Michael P. Gadomski/PR; **142-143** Dan McCoy/Rainbow; **144** Michael Sewell/PA, (inset) VU Science/M. Obtsuki; **145** (t) CBS/PT, (b) Laguna Design/SPL/PR; **146** (l) A. B. Joyce/PR; (c) F. Stuart Westmoreland/PR, (r) Dwight Kuhn; **148** (tl) Laure Communications, (tr) Aaron Haupt, (bl, bc) VU/SIU, (br) Matt Meadows/PA; **149** Chip Clark; **150** William J. Weber; **151** Doug Martin; **153** (t) Matt Meadows, (b) Aaron Haupt; **154** (top, left to right) Amanita Pictures, Mark C. Burnett, Studiohio, Glencoe photo, Mark C. Burnett, Aaron Haupt, (b) Aaron Haupt; **155** (l) Elaine Shay, (c, r) Aaron Haupt; **156** Jim Steinberg/PR, (inset) Aaron Haupt; **158, 159** Aaron Haupt; **161** (t) Aaron Haupt, (b) Elaine Shay; **163** (t) Joyce Photographics/PR, (tc, bc) Aaron Haupt, (b) C. Allan Morgan/PA; **164** (tr) Elaine Shay, (others) Aaron Haupt; **165** (l) Aaron Haupt, (r) William S. Lea; **168 169** Matt Meadows; **170** Doug Martin; **174** Biophoto Assoc./PR, (inset) VU/R. Kessel-G. Shih; **175** Matt Meadows; **176** (t) VU/Cabisco, (b) SPL/PR; **177** VU/George Musil; **178** (tl) Culver Pictures, Inc., (tr) Science Museum/Science & Society Picture Library, (b) VU/K. Talaro; **179** (clockwise) VU/M. Bessir-D. Fawcett, VU/Fred Hossler, Oliver Meckes/PR, VU/David M. Phillips, Caltech/SPL/SS/PR, Takeshi Takahara/PR; **181** Aaron Haupt; **182** C. C. Lockwood/DRK; **185** (t) Aaron Haupt, (b) VU/Don W. Fawcett; **186** VU/D. E. Akin; **187** VU/Don W. Fawcett; **188** VU/R. Bolander-Don W. Fawcett; **189** (t) VU/Don W.Fawcett, (bl) Biophoto Assoc./PR, (br) VU/David M. Phillips; **190** Jeremy Burgess/SPL/PR; **191** Keith R. Porter/PR; **192** Johnny Johnson; **193** (l) VU/David M. Phillips, (c) Professor P. Motta/Department of Anatomy/University of La Sapienza, Rome/SPL/PR, (r) VU/M. Abbey; **195** Breck P. Kent/AA; **196** (l) SSEC/UW-Madison, (r) Alfred Pasieka/SS/PR; **198** VU/R. Bolender-Don W. Fawcett; **200** Jean Claude Revy/PT, (inset) Dr. D. Spector/PA; **201** (t) Betty Derig/PR, (b) VU/Jeff J. Daly; **202 203** Joseph Kurantsin-Mills, M.Sc.Ph.D; **204** KS Studio; **208** VU/Michael Abbey; **211** (t) Ed Reschke/PA, (b) J. L. Carson/CMSP; **212** VU/Science VU; **213** (br) CBS/PT, (others) Ed Reschke/PA; **214** Barry King, U. of CA School of Medicine/BPS; **215** (t) Ed Reschke/PA, (b) VU/David M. Phillips; **217** (t) OSF/AA, (b) Luis M. De La Maza, PhD.MD./PT; **219** KS Studio/Bob Mullenix; **222** Dixie Knight/The Stock Shop/Medichrome; **222-223** Mitsuaki Iwago/MP; **223** OSF/AA; **224** (t) Ed Reschke/PA, (b) Barry King, U. of CA School of Medicine/BPS; **226** Grant Heilman from GH, (inset) William Allen; **227** Aaron Haupt; **228** Mel Victor/The Stock Shop/Medichrome; **230** (t) Dr. K. S. Kim/PA, (b) Noah Poritz/Macro World/PR; (r) M. Doolittle/Rainbow **231** (t) Geoff Butler, (b) James Westwater; **232** James Westwater; **234** CB-UPI; **235** Dr. Jeremy Burgess/SPL/PR; **236** (t) Matt Meadows, (b) CB-UPI; **237** (t) Matt Meadows, (b) VU/T. Kanariki-D.W.Fawcett; **239** Profs. P. Motta & T. Naguro/SPL/CMSP; **241 242** Matt Meadows; **243** Glencoe photo; **244** Matt Meadows;

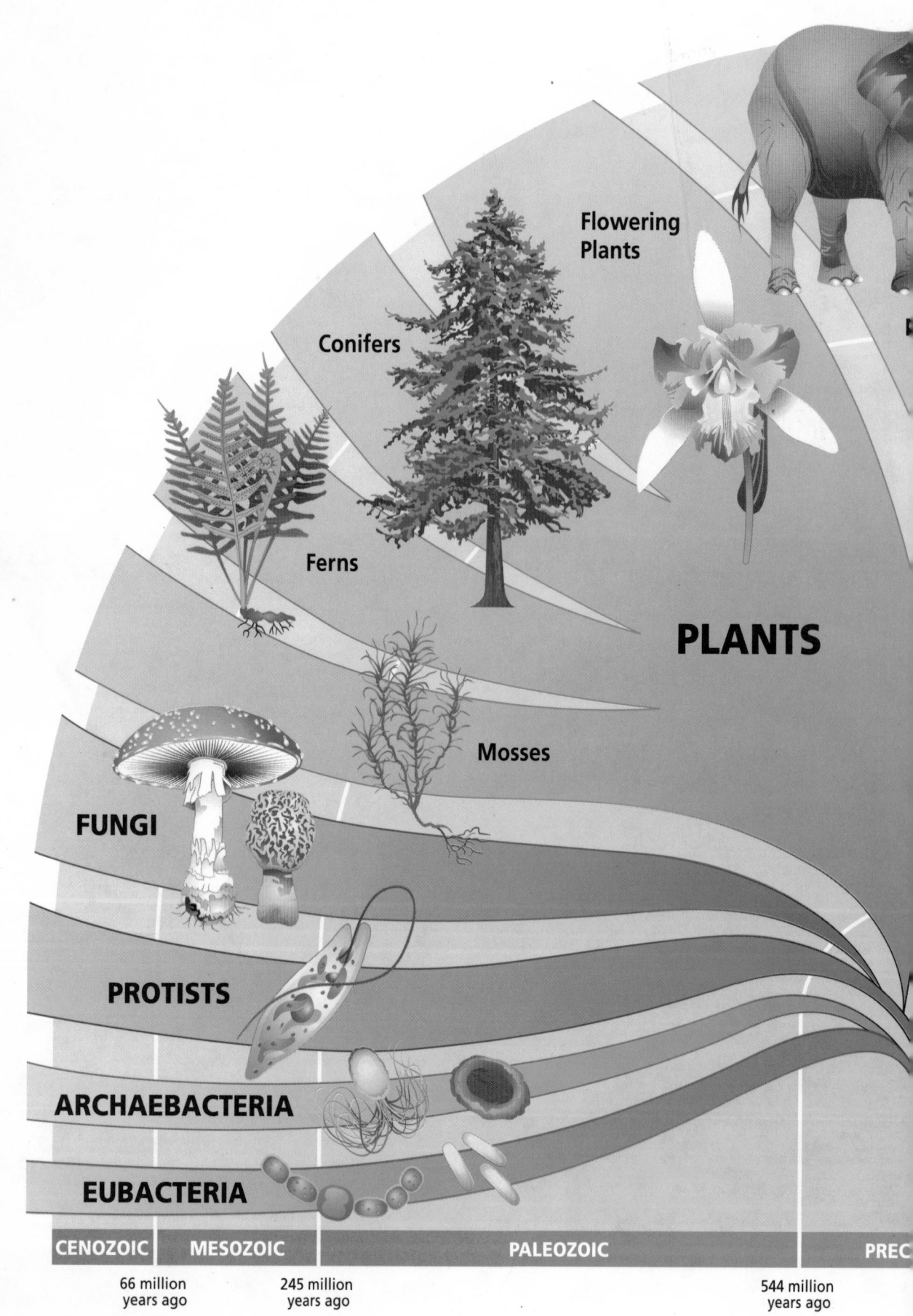

Flowering
Plants

Conifers

Ferns

PLANTS

Mosses

FUNGI

PROTISTS

ARCHAEBACTERIA

EUBACTERIA

CENOZOIC	MESOZOIC	PALEOZOIC	PREC

66 million
years ago

245 million
years ago

544 million
years ago